FORMULAS FROM GEOMETRY

area A perimeter P circumference C volume V curved surface area S altitude h radius r

RIGHT TRIANGLE

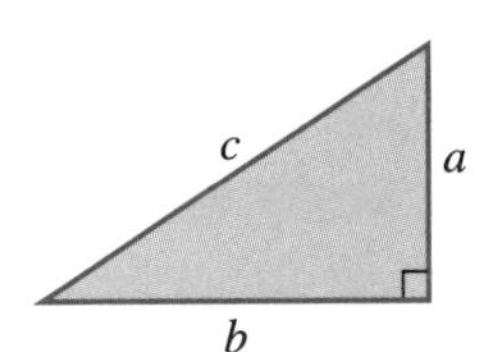

Pythagorean Theorem: $c^2 = a^2 + b^2$

TRIANGLE

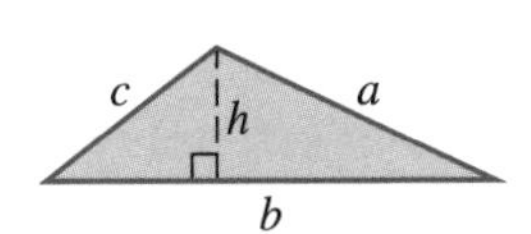

$A = \frac{1}{2}bh \qquad P = a + b + c$

EQUILATERAL TRIANGLE

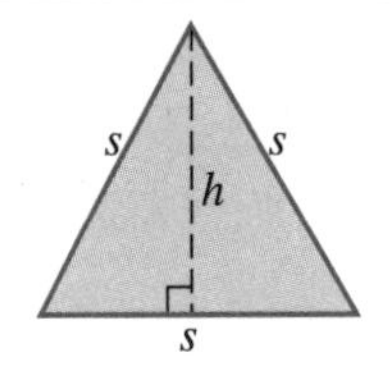

$h = \frac{\sqrt{3}}{2}s \qquad A = \frac{\sqrt{3}}{4}s^2$

RECTANGLE

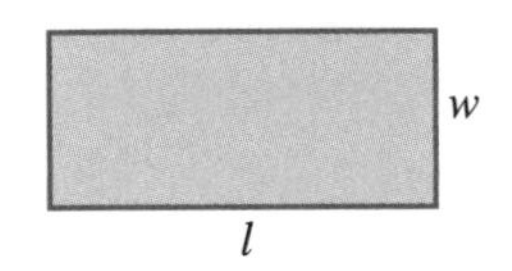

$A = lw \qquad P = 2l + 2w$

PARALLELOGRAM

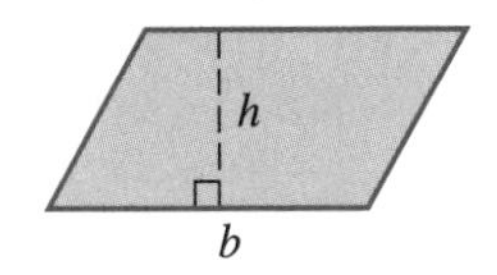

$A = bh$

TRAPEZOID

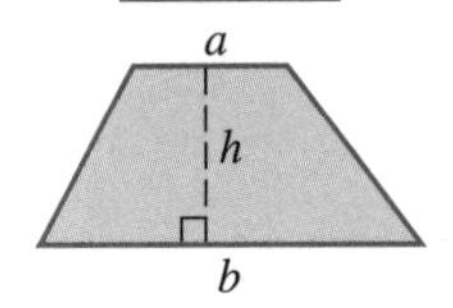

$A = \frac{1}{2}(a + b)h$

CIRCLE

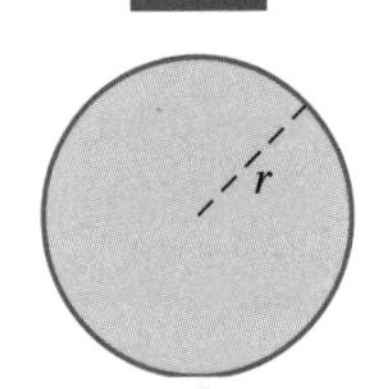

$A = \pi r^2 \qquad C = 2\pi r$

CIRCULAR SECTOR

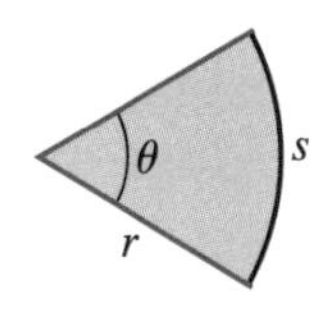

$A = \frac{1}{2}r^2\theta \qquad s = r\theta$

CIRCULAR RING

$A = \pi(R^2 - r^2)$

RECTANGULAR BOX

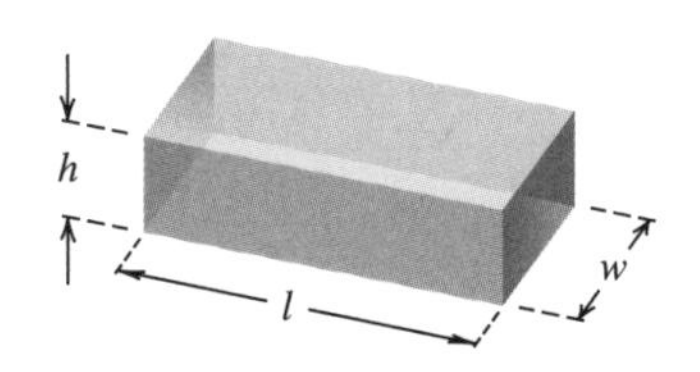

$V = lwh \qquad S = 2(hl + lw + hw)$

SPHERE

$V = \frac{4}{3}\pi r^3 \qquad S = 4\pi r^2$

RIGHT CIRCULAR CYLINDER

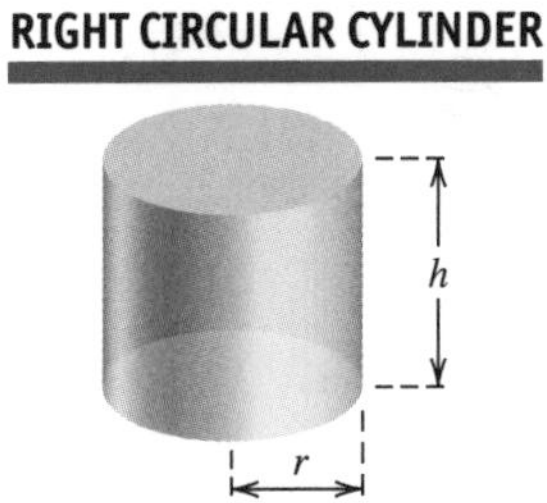

$V = \pi r^2 h \qquad S = 2\pi rh$

RIGHT CIRCULAR CONE

$V = \frac{1}{3}\pi r^2 h \qquad S = \pi r\sqrt{r^2 + h^2}$

FRUSTUM OF A CONE

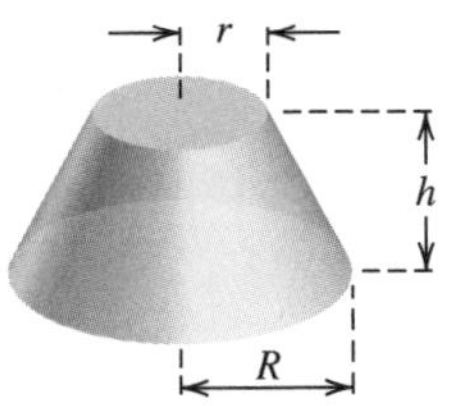
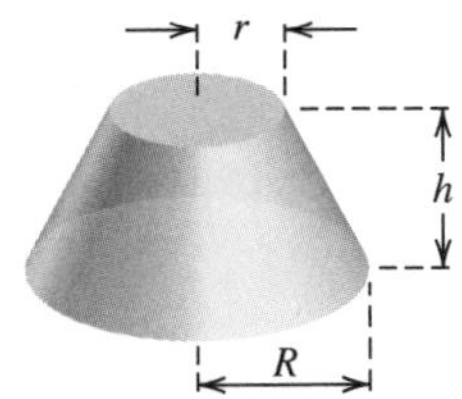

$V = \frac{1}{3}\pi h(r^2 + rR + R^2)$

PRISM

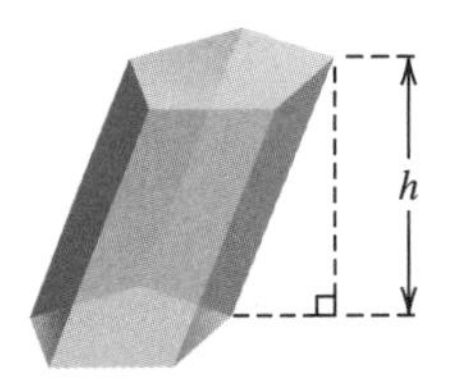

$V = Bh$ with B the area of the base

DISTANCE FORMULA

$$d(P_1, P_2) = \sqrt{(x_2 - x_1)^2 + (y_2 - y_1)^2}$$

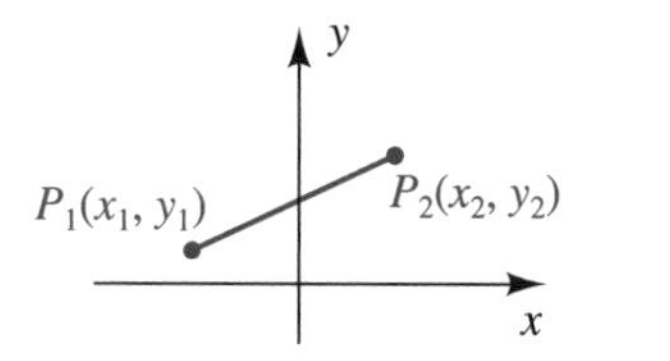

EQUATION OF A CIRCLE

$$(x - h)^2 + (y - k)^2 = r^2$$

SLOPE m OF A LINE

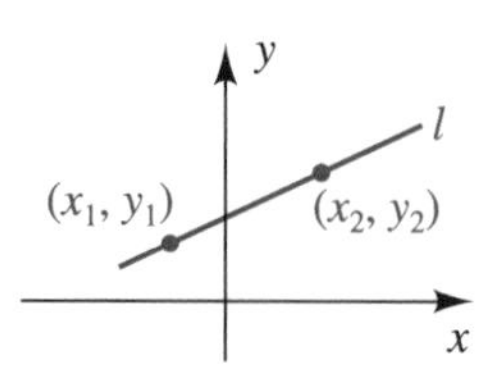

$$m = \frac{y_2 - y_1}{x_2 - x_1}$$

GRAPH OF A QUADRATIC FUNCTION

$y = ax^2,\ a > 0$

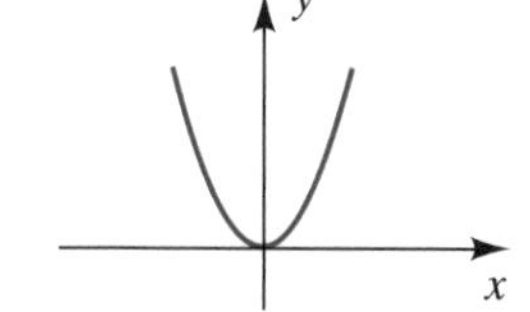

$y = ax^2 + bx + c,\ a > 0$

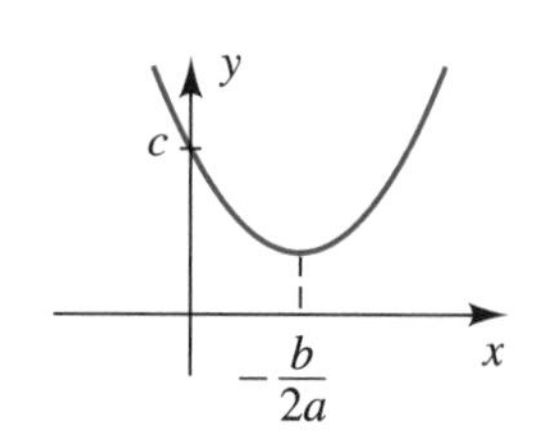

POINT-SLOPE FORM OF A LINE

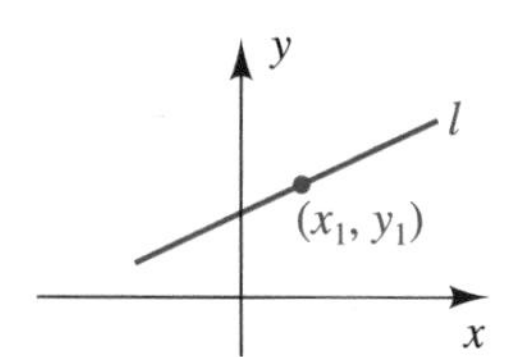

$$y - y_1 = m(x - x_1)$$

SLOPE-INTERCEPT FORM OF A LINE

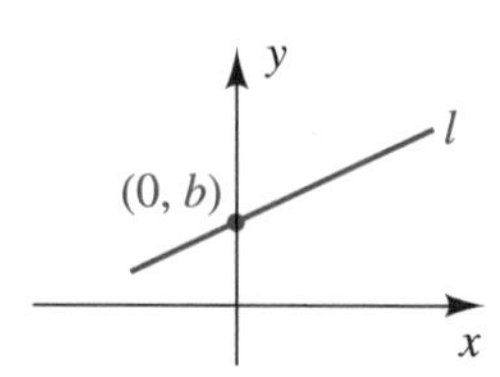

$$y = mx + b$$

INTERCEPT FORM OF A LINE

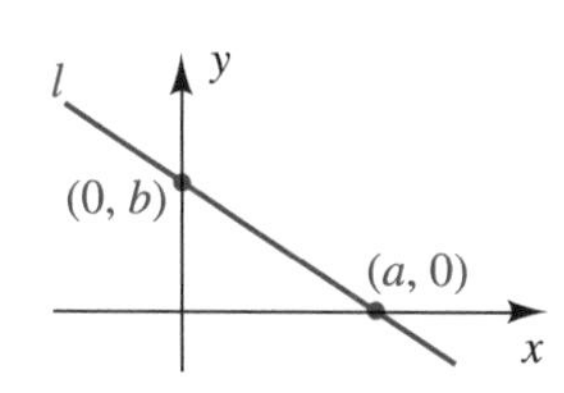

$$\frac{x}{a} + \frac{y}{b} = 1 \quad (a \neq 0, b \neq 0)$$

CONSTANTS

$\pi \approx 3.14159$

$e \approx 2.71828$

CONVERSIONS

1 centimeter $\approx$ 0.3937 inch

1 meter $\approx$ 3.2808 feet

1 kilometer $\approx$ 0.6214 mile

1 gram $\approx$ 0.0353 ounce

1 kilogram $\approx$ 2.2046 pounds

1 liter $\approx$ 0.2642 gallon

1 milliliter $\approx$ 0.0381 fluid ounce

1 joule $\approx$ 0.7376 foot-pound

1 newton $\approx$ 0.2248 pound

1 lumen $\approx$ 0.0015 watt

1 acre = 43,560 square feet

PRECALCULUS

FUNCTIONS AND GRAPHS

ELEVENTH EDITION

PRECALCULUS

FUNCTIONS AND GRAPHS

EARL W. SWOKOWSKI

JEFFERY A. COLE
Anoka-Ramsey Community College

THOMSON
BROOKS/COLE

Australia • Canada • Mexico • Singapore • Spain
United Kingdom • United States

Precalculus: Functions and Graphs, 11e

Earl W. Swokowski and Jeffery A. Cole

Acquisitions Editor: *Gary Whalen*
Development Editors: *Kari Hopperstead/Leslie Lahr*
Assistant Editor: *Dianna Muhammad*
Editorial Assistant: *Lynh Pham*
Technology Project Manager: *Donna Kelley*
Marketing Manager: *Joe Rogove*
Marketing Assistant: *Jennifer Liang*
Marketing Communications Manager: *Darlene Amidon-Brent*
Content Project Manager: *Janet Hill*
Creative Director: *Vernon Boes*
Print Buyer: *Judy Inouye*
Production Service: *Lifland et al., Bookmakers*
Text Designers: *Andrew Ogus/Rokusek*
Cover Designer: *Bill Stanton*
Cover Image: *Scott Tysick/Masterfile*
Compositor: *Better Graphics, Inc.*
Cover and Interior Printer: *R R Donnelley—Willard*

Printed in the United States of America

1 2 3 4 5 6 7 11 10 09 08 07

Library of Congress Control Number: 2007922096

Student Edition:
ISBN-13: 978-0-495-10837-5
ISBN-10: 0-495-10837-5

Thomson Higher Education
10 Davis Drive
Belmont, CA 94002-3098
USA

For more information about our products, contact us at:
Thomson Learning Academic Resource Center
1-800-423-0563

For permission to use material from this text or product, submit a request online at **http://www.thomsonrights.com**. Any additional questions about permissions can be submitted by email to **thomsonrights@thomson.com**.

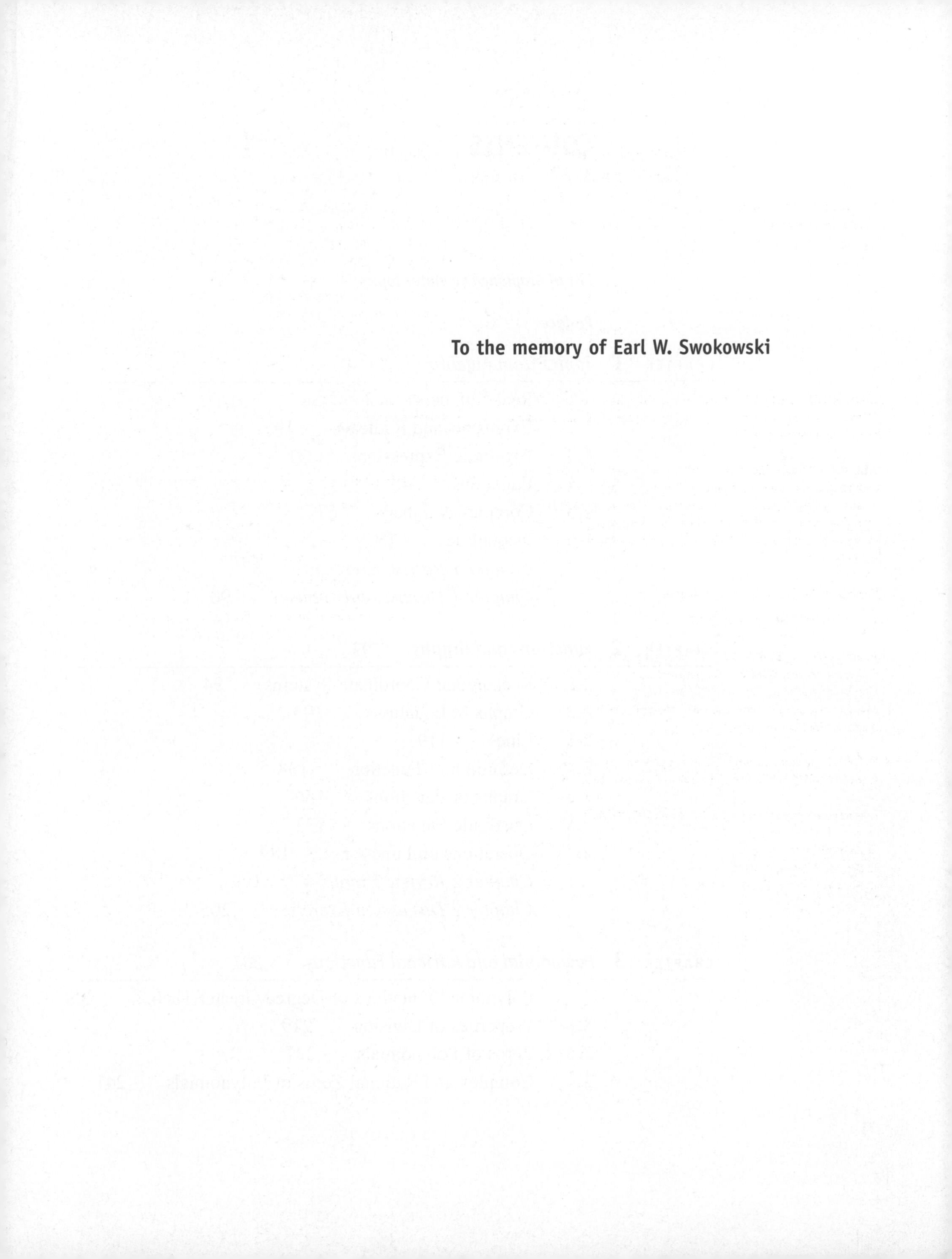

To the memory of Earl W. Swokowski

CONTENTS

LIST OF GRAPHING CALCULATOR TOPICS

There are many other places where a graphing calculator is used—these are the ones that include specific keystrokes.

PREFACE

The eleventh edition of *Precalculus: Functions and Graphs* includes over 100 new or revised examples and exercises, many of these resulting from suggestions of users and reviewers of the tenth edition. All have been incorporated without sacrificing the mathematical soundness that has been paramount to the success of this text.

The inclusion of graphing calculator examples and inserts, which feature specific color-coded keystrokes and screens for the TI-83/4 Plus and the TI-86, has proven to give added value to the text for students—especially those who are working with a graphing calculator for the first time. It also gives professors more flexibility in terms of the way they approach a solution. The design of the text makes the technology inserts easily identifiable, and they are listed in a special technology table of contents to make looking them up easier.

Below is a brief overview of the chapters, followed by a short description of the College Algebra course that I teach at Anoka-Ramsey Community College, and then a list of the general features of the text.

Overview

Chapter 1 This chapter contains a summary of some basic algebra topics. Students should be familiar with much of this material, but also challenged by some of the exercises that prepare them for calculus. Graphing calculator operations are introduced and used to verify algebraic operations. Equations and inequalities are solved algebraically and numerically in this chapter with technology support; they will be solved graphically in subsequent chapters. Students will extend their knowledge of these topics; for example, they have worked with the quadratic formula, but will be asked to relate it to factoring and work with coefficients that are not real numbers (see Examples 6 and 7 in Section 1.4).

Chapter 2 Two-dimensional graphs and functions are introduced in this chapter. Specific graphing calculator directions are given for most of the basic graphing features, such as finding zeros and points of intersection, as well as some of the more difficult topics, such as finding a regression model and graphing a piecewise-defined function. See the updated Example 10 in Section 2.5 for a topical application (taxes) that relates tables, formulas, and graphs.

Chapter 3 This chapter begins with a discussion of polynomial functions and some polynomial theory. A thorough treatment of rational functions is given in Section 3.5. This is followed by a section on variation, which includes graphs of simple polynomial and rational functions.

Chapter 4 Inverse functions is the first topic of discussion, followed by several sections that deal with exponential and logarithmic functions. Modeling an exponential function is given additional attention in this chapter (see Example 8 in Section 4.2), as well as in Chapter 8.

Chapter 5 Angles are the first topic in this chapter. Next, the trigonometric functions are introduced using a right triangle approach and then defined in terms of a unit circle. Basic trigonometric identities appear throughout the chapter. The chapter concludes with sections on trigonometric graphs and applied problems.

Chapter 6 This chapter consists mostly of trigonometric identities, formulas, and equations. The last section contains definitions, properties, and applications of the inverse trigonometric functions.

Chapter 7 The law of sines and the law of cosines are used to solve oblique triangles. Vectors are then introduced and used in applications. The last two sections relate the trigonometric functions and complex numbers.

Chapter 8 Systems of inequalities and linear programming immediately follow solving systems by substitution and elimination. Next, matrices are introduced and used to solve systems. This chapter concludes with a discussion of determinants and partial fractions.

Chapter 9 This chapter begins with a discussion of sequences, and substantial technology support has been included. Mathematical induction and the binomial theorem are next, followed by counting topics (see Example 3 in Section 9.7 for an example involving both combinations and permutations). The last section is about probability and includes topics such as odds and expected value.

Chapter 10 Sections on the parabola, ellipse, and hyperbola begin this chapter. Two different ways of representing functions are given in the next sections on parametric equations and polar coordinates.

My Course

At Anoka-Ramsey Community College in Coon Rapids, Minnesota, College Algebra I is a one-semester 3-credit course. For students intending to take Calculus, this course is followed by a one-semester 4-credit course, College Algebra II and Trigonometry. This course also serves as a terminal math course for many students.

The sections covered in College Algebra I are

2.1–2.7, 3.1, 3.5 (part), 3.6, 4.1–4.6, 8.1–8.4, 9.1–9.3, and 9.5–9.8.

Chapter 1 is used as review material in some classes, and the remaining sections are taught in the following course. We also offer a one-semester 5-credit course, Precalculus, that covers much of the same material and is an alternative track to Calculus. A graphing calculator is required in some sections and optional in others.

Features

A Separate List of Graphing Calculator Topics On pages x and xi, there is a list of graphing calculator topics for quick reference.

Illustrations Brief demonstrations of the use of definitions, laws, and theorems are provided in the form of illustrations.

Charts Charts give students easy access to summaries of properties, laws, graphs, relationships, and definitions. These charts often contain simple illustrations of the concepts that are being introduced.

Examples Titled for easy reference, all examples provide detailed solutions of problems similar to those that appear in exercise sets. Many examples include graphs, charts, or tables to help the student understand procedures and solutions. Most examples have online tutorials associated with them. These examples are identified in the text with icons, making it simple for students to get extra practice as needed. The tutorials are also assignable through WebAssign and/or ThomsonNOW.

Step-by-Step Explanations In order to help students follow them more easily, many of the solutions in examples contain step-by-step explanations.

Discussion Exercises Each chapter ends with several exercises that are suitable for small-group discussions. These exercises range from easy to difficult and from theoretical to application-oriented.

Checks The solutions to some examples are explicitly checked, to remind students to verify that their solutions satisfy the conditions of the problems.

Graphing Calculator Examples Wherever appropriate, examples requiring the use of a graphing utility are included in the text. These are designated by a calculator icon (shown to the left) and illustrated with a figure reproduced from a graphing calculator screen.

Graphing Calculator Inserts In addition to the graphing calculator examples, these inserts are included to highlight some of the capabilities of graphing calculators and/or illustrate their use in performing the operations under discussion. See, for example, "Using the TI-86 POLY Feature" in Section 3.1 and "Using the TI-83/4 Plus Sequence Mode" in Section 9.1.

Graphing Calculator Exercises Exercises specifically designed to be solved with a graphing utility are included in appropriate sections. These exercises are also designated by a calculator icon (shown to the left).

Applications To arouse student interest and to help students relate the exercises to current real-life situations, applied exercises have been titled. One look at the Index of Applications in the back of the book reveals the wide array of topics. Many professors have indicated that the applications constitute one of the strongest features of the text.

Exercises Hundreds of exercises have been updated with new data and new applications to increase relevance. Exercise sets begin with routine drill problems and gradually progress to more difficult problems. An ample number of exercises contain graphs and tabular data; others require the student to find a mathematical model for the given data. Many exercises have online tutorials associated with them; these are identified in the text with icons. These tutorials are also assignable through WebAssign and/or ThomsonNOW.

Applied problems generally appear near the end of an exercise set, to allow students to gain confidence in working with the new ideas that have been presented before they attempt problems that require greater analysis and synthesis of these ideas. Review exercises at the end of each chapter may be used to prepare for examinations.

Guidelines Boxed guidelines enumerate the steps in a procedure or technique to help students solve problems in a systematic fashion.

Warnings Interspersed throughout the text are warnings to alert students to common mistakes.

Text Art Forming a total art package that is second to none, figures and graphs have been computer-generated for accuracy, using the latest technology. Colors are employed to distinguish between different parts of figures. For example, the graph of one function may be shown in blue and that of a second function in red. Labels are the same color as the parts of the figure they identify.

Text Design The text has been designed to ensure that discussions are easy to follow and important concepts are highlighted. Color is used pedagogically to clarify complex graphs and to help students visualize applied problems. Previous adopters of the text have confirmed that the text strikes a very appealing balance in terms of color use.

Endpapers The endpapers in the front and back of the text provide useful summaries from algebra, geometry, and trigonometry.

Appendixes Appendix I, "Common Graphs and Their Equations," is a pictorial summary of graphs and equations that students commonly encounter in precalculus mathematics. Appendix II, "A Summary of Graph Transformations," is an illustrative synopsis of the basic graph transformations discussed in the text: shifting, stretching, compressing, and reflecting. Appendix III, "Graphs of Trigonometric Functions and Their Inverses," contains graphs, domains, and ranges of the six trigonometric functions and their inverses. Appendix IV, "Values of the Trigonometric Functions of Special Angles on a Unit Circle," is a full-page reference for the most common angles on a unit circle—valuable for students who are trying to learn the basic trigonometric functions values.

Answer Section The answer section at the end of the text provides answers for most of the odd-numbered exercises, as well as answers for all chapter review exercises. Considerable thought and effort were devoted to making this section a learning device for the student instead of merely a place to check answers. For instance, proofs are given for mathematical induction problems. Numerical answers for many exercises are stated in both an exact and an approximate form. Graphs, proofs, and hints are included whenever appropriate. Author-prepared solutions and answers ensure a high degree of consistency among the text, the solutions manuals, and the answers.

Teaching Tools for the Instructor

Annotated Instructor's Edition (ISBN 0-495-38288-4) This special version of the complete student text contains the answers printed in blue next to the respective exercises. Graphs, tables, and other long answers appear in a special answer section in the back of the text. In addition, all of the exercises available for online testing and homework through WebAssign and ThomsonNOW are indicated by a blue shaded box behind the question number.

Instructor's Solutions Manual by Jeffery A. Cole (ISBN 0-495-38290-6) This author-prepared manual includes answers to all text exercises and detailed solutions to most exercises. The manual has been thoroughly reviewed for accuracy.

Test Bank (ISBN 0-495-38291-4) The *Test Bank* includes six tests per chapter as well as three final exams, all consisting of a combination of multiple-choice, free response, true/false, and fill-in-the-blank questions.

ExamView® (ISBN 0-495-38293-0) Create, deliver, and customize tests and study guides (both in print and online) in minutes with this easy-to-use assessment and tutorial system, which contains all questions from the *Test Bank* in electronic format.

Enhanced WebAssign (ISBN 0-495-38324-4) WebAssign, the most widely used homework system in higher education, allows you to assign, collect, grade, and record homework assignments via the web. Through a partnership between WebAssign and Thomson Brooks/Cole, this proven homework system has been enhanced to include links to textbook sections, video examples, and problem-specific tutorials.

ThomsonNOW™ (ISBN 0-495-39280-4) ThomsonNOW saves you time and provides students with an efficient way to learn. Based on answers to chapter pre-tests, personalized study plans direct students to interactive tutorials and videos that they need to review. When students get stuck on a problem or concept, they can access vMentor™ for live online tutoring by an experienced mathematics instructor. An ideal self-study resource, ThomsonNOW requires no instructor setup or involvement.

Text-Specific DVD (ISBN 0-495-38289-2) This DVD supports your teaching and saves you time by offering assistance to your students outside of class. It presents the material in each text chapter, broken down into 10- to 20-minute problem-solving lessons that cover each section.

JoinIn™ on TurningPoint® (ISBN 0-495-38295-7) JoinIn™ content for student classroom response systems tailored to this text allows you to transform your classroom and assess students' progress with instant in-class quizzes and polls. Pose book-specific questions and display students' answers seamlessly within the Microsoft® PowerPoint® slides of your own lecture, in conjunction with the "clicker" hardware of your choice.

Website The Book Companion Website includes study hints, review material, instructions for using various graphing calculators, a tutorial quiz for each chapter of the text, and other materials for students and instructors.

Learning Tools for the Student

Student Solutions Manual (ISBN 0-495-38287-6) This author-prepared manual provides solutions for all of the odd-numbered exercises in the text, strategies for solving additional exercises, and many helpful hints and warnings.

ThomsonNOW™ (ISBN 0-495-39280-4) This online suite of resources (including an integrated eBook and a personalized study plan) gives students the choices and tools they need to study smarter and get the grade. When students get stuck on a problem or concept, they can access vMentor™ for live online tutoring by an experienced mathematics instructor. An ideal self-study resource, ThomsonNOW requires no instructor setup or involvement.

Website The Book Companion Website contains study hints, review material, instructions for using various graphing calculators, a tutorial quiz for each chapter of the text, and other materials for students and instructors.

Acknowledgments

Many thanks go to the reviewers of this edition: Brenda Burns-Williams, North Carolina State University; Gregory Cripe, Spokane Falls Community College; George DeRise, Thomas Nelson Community College; Ronald Dotzel, University of Missouri, St. Louis; Hamidullah Farhat, Hampton University; Sherry Gale, University of Cincinnati; Carole Krueger, University of Texas, Arlington; Sheila Ledford, Coastal Georgia Community College; Christopher Reisch, Jamestown Community College; Beverly Shryock, University of North Carolina, Chapel Hill; Hanson Umoh, Delaware State University; Beverly Vredevelt, Spokane Falls Community College; and Limin Zhang, Columbia Basin Community College.

Thanks are also due to reviewers of past editions, who have helped increase the usefulness of the text for the students over the years: Jean H. Bevis, Georgia State University; David Boliver, University of Central Oklahoma; Randall Dorman, Cochise College; Karen Hinz, Anoka-Ramsey Community College; Sudhir Goel, Valdosta State University; John W. Horton, Sr., St. Petersburg College; Robert Jajcay, Indiana State University; Conrad D. Krueger, San Antonio College; Susan McLoughlin, Union County College; Lakshmi Nigam, Quinnipiac University; Wesley J. Orser, Clark College; Don E. Soash, Hillsborough Community College; Thomas A. Tredon, Lord Fairfax Community College; and Fred Worth, Henderson State University. In addition, I thank Marv Riedesel and Mary Johnson for their precise accuracy checking of new and revised examples and exercises.

I am thankful for the excellent cooperation of the staff of Brooks/Cole, especially the editorial group of Charlie Van Wagner, Gary Whalen, and Kari Hopperstead. Donna Kelley and Dianna Muhammad managed the excellent ancillary package that accompanies the text. Special thanks go to Leslie Lahr for the time and energy she put into research and other contributions to the project. Sally Lifland, Gail Magin, and Denise Throckmorton, of Lifland et al., Bookmakers, saw the book through all the stages of production, took exceptional care in seeing that no inconsistencies occurred, and offered many helpful suggestions. The late George Morris, of Scientific Illustrators, created the mathematically precise art package and updated all the art through several editions. This tradition of excellence is carried on by his son Brian.

In addition to all the persons named here, I would like to express my sincere gratitude to the many students and teachers who have helped shape my views on mathematics education. Please feel free to write to me about any aspect of this text—I value your opinion.

Jeffery A. Cole

PRECALCULUS

FUNCTIONS AND GRAPHS

1

Topics from Algebra

The word *algebra* comes from *ilm al-jabr w'al muqabala,* the title of a book written in the ninth century by the Arabian mathematician al-Khworizimi. The title has been translated as the science of restoration and reduction, which means transposing and combining similar terms (of an equation). The Latin transliteration of *al-jabr* led to the name of the branch of mathematics we now call algebra.

We begin this chapter with a review of real numbers and their properties, which will be used throughout this text and further mathematics courses. A discussion of some fundamental algebraic techniques follows before we turn our attention to solving basic equations and inequalities—those whose solutions are subsets of the one-dimensional real number line.

1.1
Real Numbers

Real numbers are used throughout mathematics, and you should be acquainted with symbols that represent them, such as

$$1, \quad 73, \quad -5, \quad \tfrac{49}{12}, \quad \sqrt{2}, \quad 0, \quad \sqrt[3]{-85}, \quad 0.33333\ldots, \quad 596.25,$$

and so on. The **positive integers, or natural numbers,** are

$$1, \quad 2, \quad 3, \quad 4, \quad \ldots.$$

The **whole numbers** (or *nonnegative integers*) are the natural numbers combined with the number 0. The **integers** are often listed as follows:

$$\ldots, \quad -4, \quad -3, \quad -2, \quad -1, \quad 0, \quad 1, \quad 2, \quad 3, \quad 4, \quad \ldots$$

Throughout this text lowercase letters $a, b, c, x, y, \ldots$ represent arbitrary real numbers (also called *variables*). If a and b denote the same real number, we write $a = b$, which is read "a **is equal to** b" and is called an **equality**. The notation $a \neq b$ is read "a **is *not* equal to** b."

If a, b, and c are integers and $c = ab$, then a and b are **factors, or divisors,** of c. For example, since

$$6 = 2 \cdot 3 = (-2)(-3) = 1 \cdot 6 = (-1)(-6),$$

we know that $1, -1, 2, -2, 3, -3, 6$, and -6 are factors of 6.

A positive integer p different from 1 is **prime** if its only positive factors are 1 and p. The first few primes are 2, 3, 5, 7, 11, 13, 17, and 19. The **Fundamental Theorem of Arithmetic** states that every positive integer different from 1 can be expressed as a product of primes in one and only one way (except for order of factors). Some examples are

$$12 = 2 \cdot 2 \cdot 3, \quad 126 = 2 \cdot 3 \cdot 3 \cdot 7, \quad 540 = 2 \cdot 2 \cdot 3 \cdot 3 \cdot 3 \cdot 5.$$

A **rational number** is a real number that can be expressed in the form a/b, where a and b are integers and $b \neq 0$. Note that every integer a is a rational number, since it can be expressed in the form $a/1$. Every real number can be expressed as a decimal, and the decimal representations for rational numbers are either *terminating* or *nonterminating and repeating*. For example, we can show by using the arithmetic process of division that

$$\tfrac{5}{4} = 1.25 \quad \text{and} \quad \tfrac{177}{55} = 3.2181818\ldots,$$

where the digits 1 and 8 in the representation of $\frac{177}{55}$ repeat indefinitely (sometimes written $3.2\overline{18}$).

Real numbers that are not rational are **irrational numbers.** Decimal representations for irrational numbers are always *nonterminating and nonrepeating.* One common irrational number, denoted by π, is the ratio of the circumference of a circle to its diameter. We sometimes use the notation $\pi \approx 3.1416$ to indicate that π **is approximately equal to** 3.1416.

In technical writing, the use of the symbol $\doteq$ *for* **is approximately equal to** *is convenient.*

There is no *rational* number b such that $b^2 = 2$, where b^2 denotes $b \cdot b$. However, there is an *irrational* number, denoted by $\sqrt{2}$ (the **square root** of 2), such that $(\sqrt{2})^2 = 2$.

The system of **real numbers** consists of all rational and irrational numbers. Relationships among the types of numbers used in algebra are illustrated in the diagram in Figure 1, where a line connecting two rectangles means that the numbers named in the higher rectangle include those in the lower rectangle. The complex numbers, discussed in Section 1.5, contain all real numbers.

Figure 1 Types of numbers used in algebra

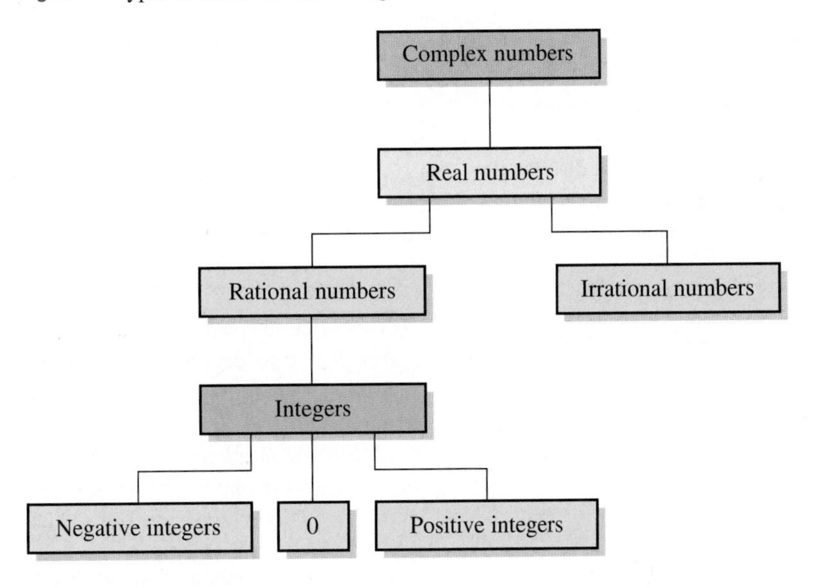

The real numbers are **closed relative to the operation of addition** (denoted by $+$); that is, to every pair a, b of real numbers there corresponds exactly one real number $a + b$ called the **sum** of a and b. The real numbers are also **closed relative to multiplication** (denoted by $\cdot$); that is, to every pair a, b of real numbers there corresponds exactly one real number $a \cdot b$ (also denoted by ab) called the **product** of a and b.

Important properties of addition and multiplication of real numbers are listed in the following chart.

Properties of Real Numbers

Terminology	General case	Meaning
(1) Addition is **commutative.**	$a + b = b + a$	Order is immaterial when adding two numbers.
(2) Addition is **associative.**	$a + (b + c) = (a + b) + c$	Grouping is immaterial when adding three numbers.
(3) 0 is the **additive identity.**	$a + 0 = a$	Adding 0 to any number yields the same number.
(4) $-a$ is the **additive inverse,** or **negative,** of a.	$a + (-a) = 0$	Adding a number and its negative yields 0.
(5) Multiplication is **commutative.**	$ab = ba$	Order is immaterial when multiplying two numbers.
(6) Multiplication is **associative.**	$a(bc) = (ab)c$	Grouping is immaterial when multiplying three numbers.
(7) 1 is the **multiplicative identity.**	$a \cdot 1 = a$	Multiplying any number by 1 yields the same number.
(8) If $a \neq 0$, $\frac{1}{a}$ is the **multiplicative inverse,** or **reciprocal,** of a.	$a\left(\frac{1}{a}\right) = 1$	Multiplying a nonzero number by its reciprocal yields 1.
(9) Multiplication is **distributive** over addition.	$a(b + c) = ab + ac$ and $(a + b)c = ac + bc$	Multiplying a number and a sum of two numbers is equivalent to multiplying each of the two numbers by the number and then adding the products.

Since $a + (b + c)$ and $(a + b) + c$ are always equal, we may use $a + b + c$ to denote this real number. We use abc for either $a(bc)$ or $(ab)c$. Similarly, if four or more real numbers a, b, c, d are added or multiplied, we may write $a + b + c + d$ for their sum and $abcd$ for their product, regardless of how the numbers are grouped or interchanged.

The distributive properties are useful for finding products of many types of expressions involving sums. The next example provides one illustration.

EXAMPLE 1 Using distributive properties

If p, q, r, and s denote real numbers, show that

$$(p + q)(r + s) = pr + ps + qr + qs.$$

▶ SOLUTION We use both of the distributive properties listed in (9) of the preceding chart:

$$\begin{aligned}
&(p + q)(r + s) \\
&\quad = p(r + s) + q(r + s) && \text{second distributive property, with } c = r + s \\
&\quad = (pr + ps) + (qr + qs) && \text{first distributive property} \\
&\quad = pr + ps + qr + qs && \text{remove parentheses}
\end{aligned}$$

▶ Tutorial available at www.thomsonedu.com/login

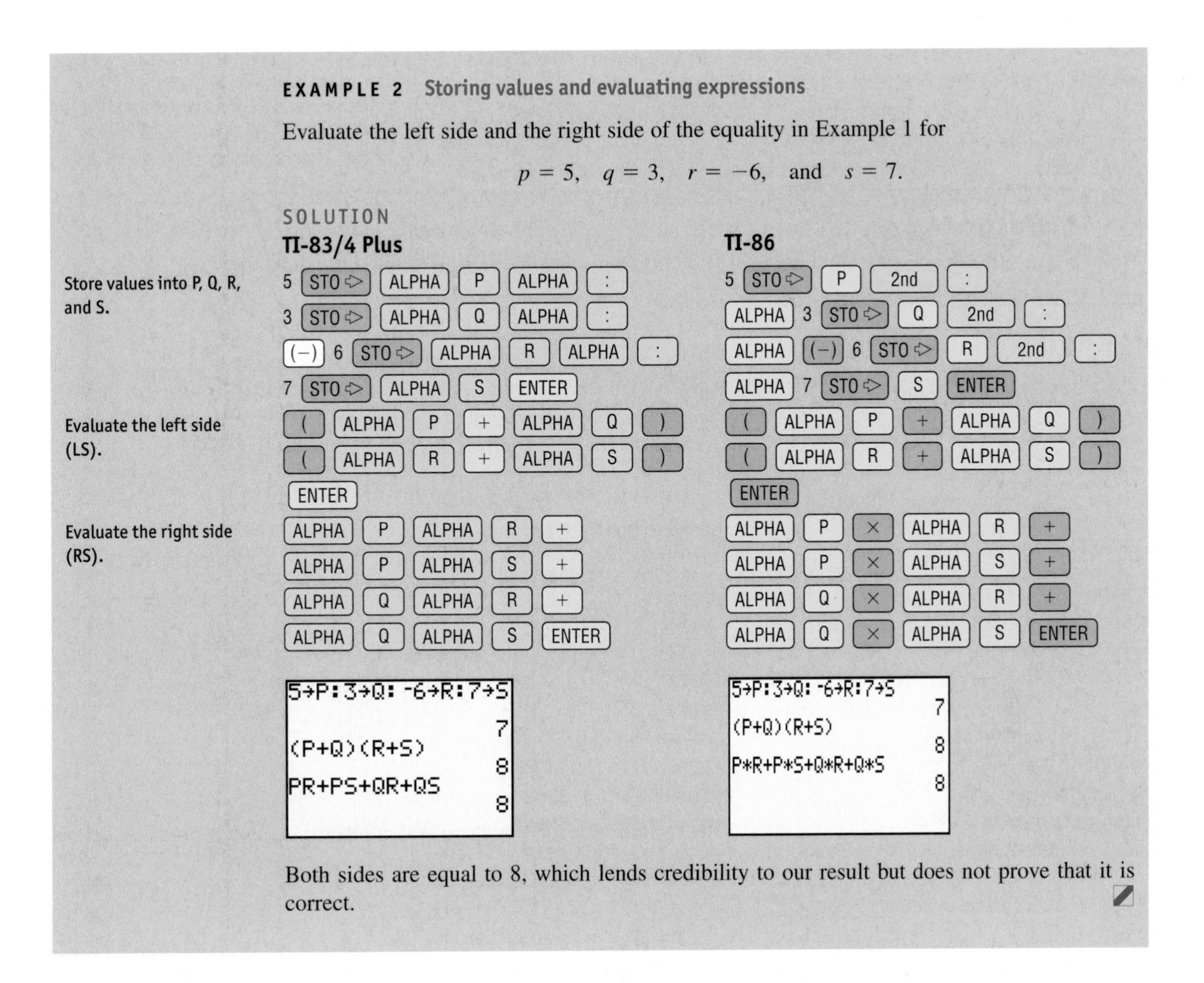

EXAMPLE 2 **Storing values and evaluating expressions**

Evaluate the left side and the right side of the equality in Example 1 for

$$p = 5, \quad q = 3, \quad r = -6, \quad \text{and} \quad s = 7.$$

SOLUTION

Store values into P, Q, R, and S.

Evaluate the left side (LS).

Evaluate the right side (RS).

Both sides are equal to 8, which lends credibility to our result but does not prove that it is correct.

The following are basic properties of equality.

Properties of Equality	If $a = b$ and c is any real number, then **(1)** $a + c = b + c$ **(2)** $ac = bc$

Properties 1 and 2 state that the same number may be added to both sides of an equality, and both sides of an equality may be multiplied by the same number. We will use these properties extensively throughout the text to help find solutions of equations.

The next result can be proved.

Products Involving Zero	
	(1) $a \cdot 0 = 0$ for every real number a.
	(2) If $ab = 0$, then either $a = 0$ or $b = 0$.

When we use the word *or* as we do in (2), we mean that *at least* one of the factors a and b is 0. We will refer to (2) as the *zero factor theorem* in future work.

Some properties of negatives are listed in the following chart.

Properties of Negatives

Property	Illustration
(1) $-(-a) = a$	$-(-3) = 3$
(2) $(-a)b = -(ab) = a(-b)$	$(-2)3 = -(2 \cdot 3) = 2(-3)$
(3) $(-a)(-b) = ab$	$(-2)(-3) = 2 \cdot 3$
(4) $(-1)a = -a$	$(-1)3 = -3$

The reciprocal $\dfrac{1}{a}$ of a nonzero real number a is often denoted by a^{-1}, as in the next chart.

Notation for Reciprocals

Definition	Illustrations
If $a \neq 0$, then $a^{-1} = \dfrac{1}{a}$.	$2^{-1} = \dfrac{1}{2}$
	$\left(\dfrac{3}{4}\right)^{-1} = \dfrac{1}{3/4} = \dfrac{4}{3}$

Note that if $a \neq 0$, then

$$a \cdot a^{-1} = a\left(\frac{1}{a}\right) = 1.$$

Reciprocals

TI-83/4 Plus

2 [STO▷] [ALPHA] [A] [ENTER]
[x^{-1}] [ENTER]
[ALPHA] [A] [x^{-1}] [ENTER]

```
2→A
                 2
Ans⁻¹
                .5
A⁻¹
                .5
```

TI-86

2 [STO▷] [A] [ENTER]
[2nd] [x^{-1}] [ENTER]
[ALPHA] [A] [2nd] [x^{-1}] [ENTER]

```
2→A
                 2
Ans⁻¹
                .5
A⁻¹
                .5
```

From either figure, we see two ways to calculate the reciprocal: (1) By merely pressing [x^{-1}], we obtain the reciprocal of the last answer, which is stored in [ANS]. (2) We can enter a variable (or just a number) and then find its reciprocal.

The operations of **subtraction** ($-$) and **division** ($\div$) are defined as follows.

Subtraction and Division

Definition	Meaning	Illustration
$a - b = a + (-b)$	To subtract one number from another, add the negative.	$3 - 7 = 3 + (-7)$
$a \div b = a \cdot \left(\frac{1}{b}\right)$ $= a \cdot b^{-1}; b \neq 0$	To divide one number by a nonzero number, multiply by the reciprocal.	$3 \div 7 = 3 \cdot \left(\frac{1}{7}\right)$ $= 3 \cdot 7^{-1}$

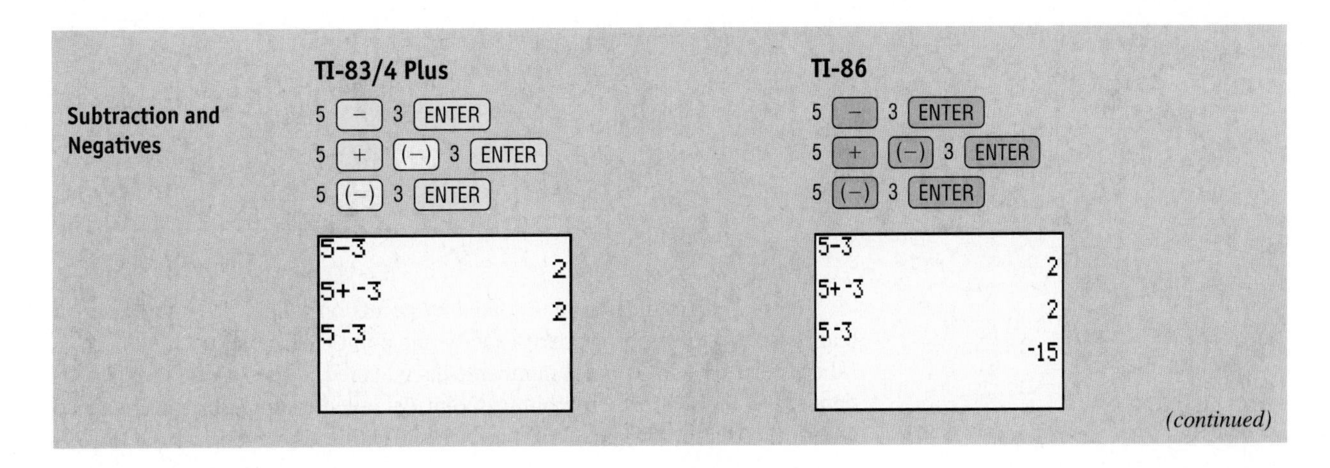

Execution of the last statement produces a SYNTAX error on the TI-83/4 Plus and a product on the TI-86. Use the [−] (minus) key for the operation of subtraction and the [(−)] (negation) key for negative numbers. We will often omit the negation key from here on and simply write -3.

We use either a/b or $\frac{a}{b}$ for $a \div b$ and refer to a/b as the **quotient of *a* and *b*** or the **fraction *a* over *b*.** The numbers a and b are the **numerator** and **denominator,** respectively, of a/b. Since 0 has no multiplicative inverse, a/b is not defined if $b = 0$; that is, *division by zero is not defined.* It is for this reason that the real numbers are not closed relative to division. Note that

$$1 \div b = \frac{1}{b} = b^{-1} \quad \text{if} \quad b \neq 0.$$

The following properties of quotients are true, provided all denominators are nonzero real numbers.

Properties of Quotients

Property	Illustration
(1) $\frac{a}{b} = \frac{c}{d}$ if $ad = bc$	$\frac{2}{5} = \frac{6}{15}$ because $2 \cdot 15 = 5 \cdot 6$
(2) $\frac{ad}{bd} = \frac{a}{b}$	$\frac{2 \cdot 3}{5 \cdot 3} = \frac{2}{5}$
(3) $\frac{a}{-b} = \frac{-a}{b} = -\frac{a}{b}$	$\frac{2}{-5} = \frac{-2}{5} = -\frac{2}{5}$
(4) $\frac{a}{b} + \frac{c}{b} = \frac{a + c}{b}$	$\frac{2}{5} + \frac{9}{5} = \frac{2 + 9}{5} = \frac{11}{5}$
(5) $\frac{a}{b} + \frac{c}{d} = \frac{ad + bc}{bd}$	$\frac{2}{5} + \frac{4}{3} = \frac{2 \cdot 3 + 5 \cdot 4}{5 \cdot 3} = \frac{26}{15}$
(6) $\frac{a}{b} \cdot \frac{c}{d} = \frac{ac}{bd}$	$\frac{2}{5} \cdot \frac{7}{3} = \frac{2 \cdot 7}{5 \cdot 3} = \frac{14}{15}$
(7) $\frac{a}{b} \div \frac{c}{d} = \frac{a}{b} \cdot \frac{d}{c} = \frac{ad}{bc}$	$\frac{2}{5} \div \frac{7}{3} = \frac{2}{5} \cdot \frac{3}{7} = \frac{6}{35}$

Real numbers may be represented by points on a line l such that to each real number a there corresponds exactly one point on l and to each point P on l there corresponds one real number. This is called a **one-to-one correspondence.** We first choose an arbitrary point O, called the **origin,** and associate with it the real number 0. Points associated with the integers are then determined by laying off successive line segments of equal length on either side of O, as illustrated in Figure 2. The point corresponding to a rational number, such as $\frac{23}{5}$, is obtained by subdividing these line segments. Points associated with certain irrational numbers, such as $\sqrt{2}$, can be found by construction (see Exercise 45).

Figure 2

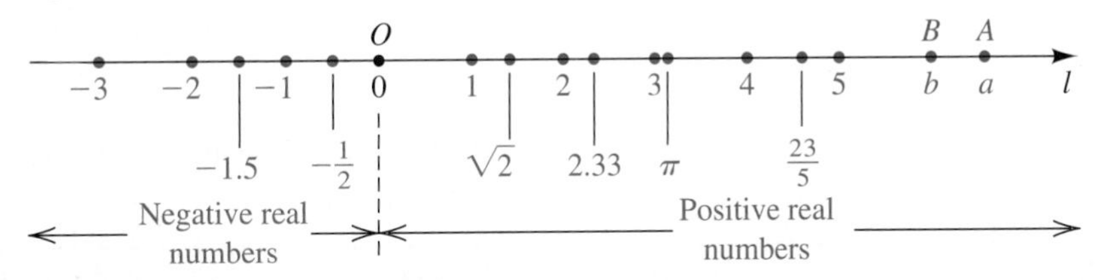

The number a that is associated with a point A on l is the **coordinate** of A. We refer to these coordinates as a **coordinate system** and call l a **coordinate line** or a **real line.** A direction can be assigned to l by taking the **positive direction** to the right and the **negative direction** to the left. The positive direction is noted by placing an arrowhead on l, as shown in Figure 2.

The numbers that correspond to points to the right of O in Figure 2 are **positive real numbers.** Numbers that correspond to points to the left of O are **negative real numbers.** *The real number* 0 *is neither positive nor negative.*

Note the difference between a negative real number and the *negative of* a real number. In particular, the negative of a real number a can be positive. For example, if a is negative, say $a = -3$, then the negative of a is $-a = -(-3) = 3$, which is positive. In general, we have the following relationships.

Relationships Between a and $-a$	**(1)** If a is positive, then $-a$ is negative. **(2)** If a is negative, then $-a$ is positive.

In the following chart we define the notions of **greater than** and **less than** for real numbers a and b. The symbols $>$ and $<$ are **inequality signs,** and the expressions $a > b$ and $a < b$ are called **(strict) inequalities.**

Greater Than or Less Than

Notation	Definition	Terminology
$a > b$	$a - b$ is positive	a is greater than b
$a < b$	$a - b$ is negative	a is less than b

If points A and B on a coordinate line have coordinates a and b, respectively, then $a > b$ is equivalent to the statement "A is to the *right* of B," whereas $a < b$ is equivalent to "A is to the *left* of B."

ILLUSTRATION **Greater Than (>) and Less Than (<)**

- $5 > 3$, since $5 - 3 = 2$ is positive.
- $-6 < -2$, since $-6 - (-2) = -6 + 2 = -4$ is negative.
- $\frac{1}{3} > 0.33$, since $\frac{1}{3} - 0.33 = \frac{1}{3} - \frac{33}{100} = \frac{1}{300}$ is positive.
- $7 > 0$, since $7 - 0 = 7$ is positive.
- $-4 < 0$, since $-4 - 0 = -4$ is negative.

The next law enables us to compare, or *order*, any two real numbers.

Trichotomy Law	If a and b are real numbers, then exactly one of the following is true: $a = b$, $a > b$, or $a < b$

Testing Inequalities and the Trichotomy Law

TI-83/4 Plus

5 [2nd] [TEST] [3] 3 [ENTER]
5 [2nd] [TEST] [5] 3 [ENTER]
5 [2nd] [TEST] [1] 3 [ENTER]

```
5>3
                1
5<3
                0
5=3
                0
```

TI-86

5 [2nd] [TEST] [> (F3)] 3 [ENTER]
5 [2nd] [TEST] [< (F2)] 3 [ENTER]
5 [2nd] [TEST] [== (F1)] 3 [ENTER]

```
5>3
                1
5<3
                0
5==3
                0
 ==  <  >  ≤  ≥
```

The results indicate that "1" represents *true* and "0" represents *false*. Only one of the above statements can be true by the Law of Trichotomy. As illustrated above, we will use the notation [n] for menu choices on the TI-83/4 Plus and [symbol (Fn)] on the TI-86. Note that the TI-86 uses == for a relational operator (testing equality) since it uses = for an assignment operator (storing values).

We refer to the **sign** of a real number as positive if the number is positive, or negative if the number is negative. Two real numbers have *the same sign* if both are positive or both are negative. The numbers have *opposite signs* if one is positive and the other is negative. The following results about the signs of products and quotients of two real numbers a and b can be proved using properties of negatives and quotients.

Laws of Signs

(1) If a and b have the same sign, then ab and $\frac{a}{b}$ are positive.

(2) If a and b have opposite signs, then ab and $\frac{a}{b}$ are negative.

The **converses*** of the laws of signs are also true. For example, if a quotient is negative, then the numerator and denominator have opposite signs.

The notation $a \geq b$, read "a **is greater than or equal to** b," means that either $a > b$ or $a = b$ (but not both). For example, $a^2 \geq 0$ for every real number a. The symbol $a \leq b$, which is read "a **is less than or equal to** b," means that either $a < b$ or $a = b$. Expressions of the form $a \geq b$ and $a \leq b$ are called **nonstrict inequalities,** since a may be equal to b. As with the equality symbol, we may negate any inequality symbol by putting a slash through it—that is, $\not>$ means not greater than.

An expression of the form $a < b < c$ is called a **continued inequality** and means that both $a < b$ *and* $b < c$; we say "b **is between** a and c." Similarly, the expression $c > b > a$ means that both $c > b$ and $b > a$.

ILLUSTRATION **Ordering Three Real Numbers**

- $1 < 5 < \frac{11}{2}$
- $-4 < \frac{2}{3} < \sqrt{2}$
- $3 > -6 > -10$

There are other types of inequalities. For example, $a < b \leq c$ means both $a < b$ and $b \leq c$. Similarly, $a \leq b < c$ means both $a \leq b$ and $b < c$. Finally, $a \leq b \leq c$ means both $a \leq b$ and $b \leq c$.

EXAMPLE 3 **Determining the sign of a real number**

If $x > 0$ and $y < 0$, determine the sign of $\frac{x}{y} + \frac{y}{x}$.

▶ **SOLUTION** Since x is a positive number and y is a negative number, x and y have opposite signs. Thus, both x/y and y/x are negative. The sum of two negative numbers is a negative number, so

$$\text{the sign of } \frac{x}{y} + \frac{y}{x} \text{ is negative.}$$

Figure 3

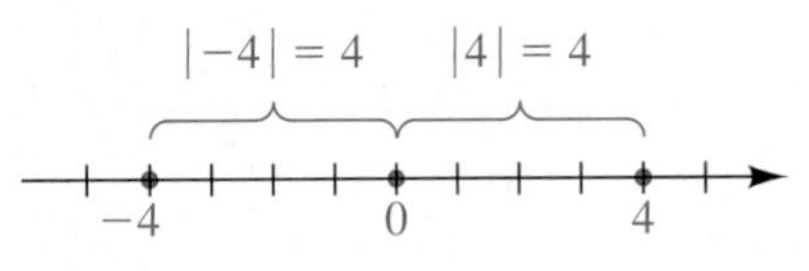

If a is an integer, then it is the coordinate of some point A on a coordinate line, and the symbol $|a|$ denotes the number of units between A and the origin, without regard to direction. The nonnegative number $|a|$ is called the *absolute value of a*. Referring to Figure 3, we see that for the point with coordinate -4 we have $|-4| = 4$. Similarly, $|4| = 4$. In general, *if a is negative, we change its sign to find* $|a|$; *if a is nonnegative, then* $|a| = a$. The next definition extends this concept to every real number.

*If a theorem is written in the form "if P, then Q," where P and Q are mathematical statements called the *hypothesis* and *conclusion*, respectively, then the *converse* of the theorem has the form "if Q, then P." If both the theorem and its converse are true, we often write "P if and only if Q" (denoted P iff Q).

Definition of Absolute Value	The **absolute value** of a real number a, denoted by $\lvert a \rvert$, is defined as follows. **(1)** If $a \geq 0$, then $\lvert a \rvert = a$. **(2)** If $a < 0$, then $\lvert a \rvert = -a$.

Since a is negative in part (2) of the definition, $-a$ represents a *positive* real number. Some special cases of this definition are given in the following illustration.

ILLUSTRATION **The Absolute Value Notation $\lvert a \rvert$**

- $\lvert 3 \rvert = 3$, since $3 > 0$.
- $\lvert -3 \rvert = -(-3)$, since $-3 < 0$. Thus, $\lvert -3 \rvert = 3$.
- $\lvert 2 - \sqrt{2} \rvert = 2 - \sqrt{2}$, since $2 - \sqrt{2} > 0$.
- $\lvert \sqrt{2} - 2 \rvert = -(\sqrt{2} - 2)$, since $\sqrt{2} - 2 < 0$.
  Thus, $\lvert \sqrt{2} - 2 \rvert = 2 - \sqrt{2}$.

In the preceding illustration, $\lvert 3 \rvert = \lvert -3 \rvert$ and $\lvert 2 - \sqrt{2} \rvert = \lvert \sqrt{2} - 2 \rvert$. In general, we have the following:

$$\lvert a \rvert = \lvert -a \rvert, \text{ for every real number } a$$

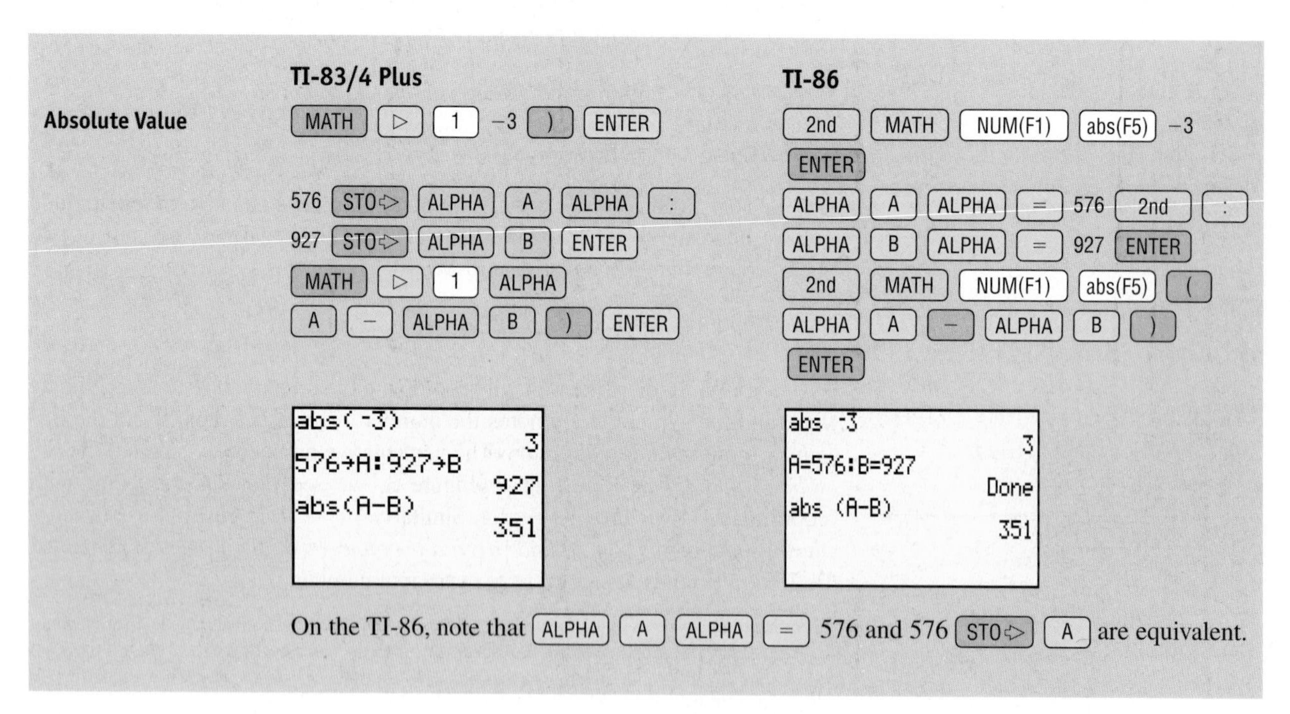

EXAMPLE 4 Removing an absolute value symbol

If $x < 1$, rewrite $|x - 1|$ without using the absolute value symbol.

▶ SOLUTION If $x < 1$, then $x - 1 < 0$; that is, $x - 1$ is negative. Hence, by part (2) of the definition of absolute value,

$$|x - 1| = -(x - 1) = -x + 1 = 1 - x.$$

Figure 4

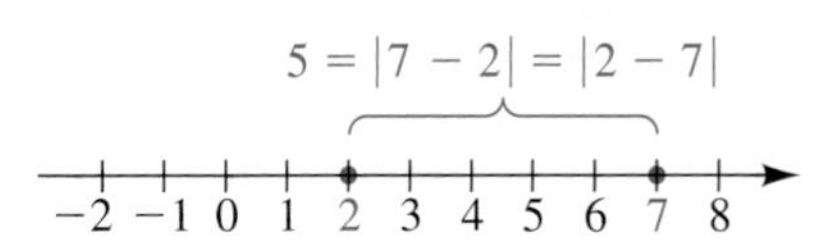

We shall use the concept of absolute value to define the distance between any two points on a coordinate line. First note that the distance between the points with coordinates 2 and 7, shown in Figure 4, equals 5 units. This distance is the difference obtained by subtracting the smaller (leftmost) coordinate from the larger (rightmost) coordinate ($7 - 2 = 5$). If we use absolute values, then, since $|7 - 2| = |2 - 7|$, it is unnecessary to be concerned about the order of subtraction. This fact motivates the next definition.

Definition of the Distance Between Points on a Coordinate Line	Let a and b be the coordinates of two points A and B, respectively, on a coordinate line. The **distance between A and B,** denoted by $d(A, B)$, is defined by $$d(A, B) = \|b - a\|.$$

The number $d(A, B)$ is the length of the line segment AB.

Since $d(B, A) = |a - b|$ and $|b - a| = |a - b|$, we see that

$$d(A, B) = d(B, A).$$

Note that the distance between the origin O and the point A is

$$d(O, A) = |a - 0| = |a|,$$

which agrees with the geometric interpretation of absolute value illustrated in Figure 4. The formula $d(A, B) = |b - a|$ is true regardless of the signs of a and b, as illustrated in the next example.

EXAMPLE 5 Finding distances between points

Figure 5

A B O C D
−5 −3 0 1 6

Let A, B, C, and D have coordinates -5, -3, 1, and 6, respectively, on a coordinate line, as shown in Figure 5. Find $d(A, B)$, $d(C, B)$, $d(O, A)$, and $d(C, D)$.

▶ SOLUTION Using the definition of the distance between points on a coordinate line, we obtain the distances:

$$\begin{aligned} d(A, B) &= |-3 - (-5)| = |-3 + 5| = |2| = 2 \\ d(C, B) &= |-3 - 1| = |-4| = 4 \\ d(O, A) &= |-5 - 0| = |-5| = 5 \\ d(C, D) &= |6 - 1| = |5| = 5 \end{aligned}$$

▶ **Tutorial available at www.thomsonedu.com/login**

The concept of absolute value has uses other than finding distances between points; it is employed whenever we are interested in the magnitude or numerical value of a real number without regard to its sign.

In the next section we shall discuss the *exponential notation* a^n, where a is a real number (called the *base*) and n is an integer (called an *exponent*). In particular, for base 10 we have

$$10^0 = 1, \quad 10^1 = 10, \quad 10^2 = 10 \cdot 10 = 100, \quad 10^3 = 10 \cdot 10 \cdot 10 = 1000,$$

and so on. For negative exponents we use the reciprocal of the corresponding positive exponent, as follows:

$$10^{-1} = \frac{1}{10^1} = \frac{1}{10}, \quad 10^{-2} = \frac{1}{10^2} = \frac{1}{100}, \quad 10^{-3} = \frac{1}{10^3} = \frac{1}{1000}$$

We can use this notation to write any finite decimal representation of a real number as a sum of the following type:

$$\begin{aligned} 437.56 &= 4(100) + 3(10) + 7(1) + 5\left(\tfrac{1}{10}\right) + 6\left(\tfrac{1}{100}\right) \\ &= 4(10^2) + 3(10^1) + 7(10^0) + 5(10^{-1}) + 6(10^{-2}) \end{aligned}$$

In the sciences it is often necessary to work with very large or very small numbers and to compare the relative magnitudes of very large or very small quantities. We usually represent a large or small positive number a in *scientific form,* using the symbol $\times$ to denote multiplication.

Scientific Form	$a = c \times 10^n$, where $1 \le c < 10$ and n is an integer

The distance a ray of light travels in one year is approximately 5,900,000,000,000 miles. This number may be written in scientific form as 5.9×10^{12}. The positive exponent 12 indicates that the decimal point should be moved 12 places to the *right.* The notation works equally well for small numbers. The weight of an oxygen molecule is estimated to be

$$0.000\,000\,000\,000\,000\,000\,000\,053 \text{ gram,}$$

or, in scientific form, 5.3×10^{-23} gram. The negative exponent indicates that the decimal point should be moved 23 places to the *left.*

ILLUSTRATION **Scientific Form**

- $513 = 5.13 \times 10^2$
- $7.3 = 7.3 \times 10^0$
- $93{,}000{,}000 = 9.3 \times 10^7$
- $20{,}700 = 2.07 \times 10^4$
- $0.000\,000\,000\,43 = 4.3 \times 10^{-10}$
- $0.000\,648 = 6.48 \times 10^{-4}$

Figure 6

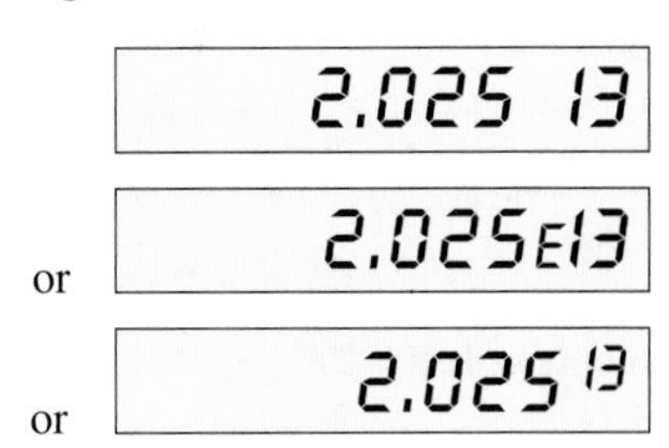

Many calculators use scientific form in their display panels. For the number $c \times 10^n$, the 10 is suppressed and the exponent is often shown preceded by the letter E. For example, to find $(4{,}500{,}000)^2$ on a scientific calculator, we could enter the integer 4,500,000 and press the x^2 (or squaring) key, obtaining a display similar to one of those in Figure 6. We would translate this as 2.025×10^{13}. Thus,

$$(4{,}500{,}000)^2 = 20{,}250{,}000{,}000{,}000.$$

Calculators may also use scientific form in the entry of numbers. The user's manual for your calculator should give specific details.

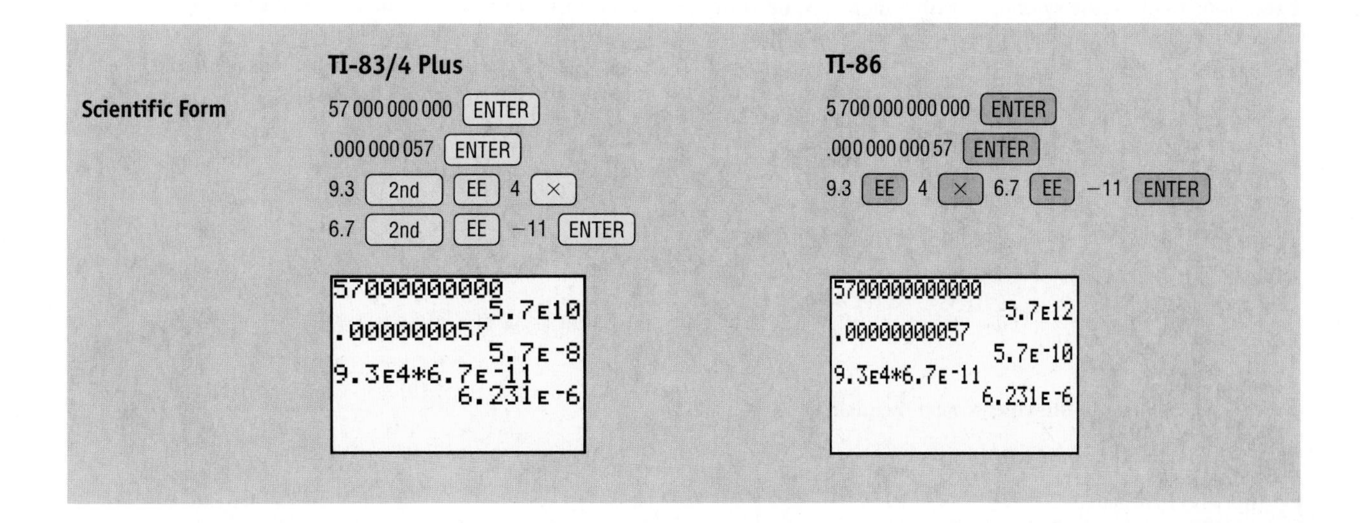

Before we conclude this section, we should briefly consider the issue of rounding off results. Applied problems often include numbers that are obtained by various types of measurements and, hence, are *approximations* to exact values. Such answers should be rounded off, since the final result of a calculation cannot be more accurate than the data that have been used. For example, if the length and width of a rectangle are measured to two-decimal-place accuracy, we cannot expect more than two-decimal-place accuracy in the calculated value of the area of the rectangle. For purely *mathematical* work, if values of the length and width of a rectangle are given, we assume that the dimensions are *exact,* and no rounding off is required.

If a number a is written in scientific form as $a = c \times 10^n$ for $1 \leq c < 10$ and if c is rounded off to k decimal places, then we say that a is accurate (or has been rounded off) to $k + 1$ **significant figures,** or **digits.** For example, 37.2638 rounded to 5 significant figures is 3.7264×10^1, or 37.264; to 3 significant figures, 3.73×10^1, or 37.3; and to 1 significant figure, 4×10^1, or 40.

1.1 Exercises

Exer. 1–2: If $x < 0$ and $y > 0$, determine the sign of the real number.

1 (a) xy (b) x^2y (c) $\frac{x}{y} + x$ (d) $y - x$

2 (a) $\frac{x}{y}$ ▶ (b) xy^2 (c) $\frac{x - y}{xy}$ (d) $y(y - x)$

Exer. 3–6: Replace the symbol □ with either <, >, or = to make the resulting statement true.

3 (a) $-7 \square -4$ (b) $\frac{\pi}{2} \square 1.57$ (c) $\sqrt{225} \square 15$

4 (a) $-3 \square -5$ (b) $\frac{\pi}{4} \square 0.8$ (c) $\sqrt{289} \square 17$

5 (a) $\frac{1}{11} \square 0.09$ (b) $\frac{2}{3} \square 0.6666$ (c) $\frac{22}{7} \square \pi$

6 (a) $\frac{1}{7} \square 0.143$ (b) $\frac{5}{6} \square 0.833$ (c) $\sqrt{2} \square 1.4$

Exer. 7–8: Express the statement as an inequality.

7 (a) x is negative.

(b) y is nonnegative.

(c) q is less than or equal to π.

▶ (d) d is between 4 and 2.

(e) t is not less than 5.

(f) The negative of z is not greater than 3.

(g) The quotient of p and q is at most 7.

(h) The reciprocal of w is at least 9.

(i) The absolute value of x is greater than 7.

8 (a) b is positive.

(b) s is nonpositive.

(c) w is greater than or equal to -4.

(d) c is between $\frac{1}{5}$ and $\frac{1}{3}$.

(e) p is not greater than -2.

(f) The negative of m is not less than -2.

(g) The quotient of r and s is at least $\frac{1}{5}$.

(h) The reciprocal of f is at most 14.

(i) The absolute value of x is less than 4.

Exer. 9–14: Rewrite the number without using the absolute value symbol, and simplify the result.

9 (a) $|-3 - 2|$ (b) $|-5| - |2|$ (c) $|7| + |-4|$

10 (a) $|-11 + 1|$ (b) $|6| - |-3|$ (c) $|8| + |-9|$

▶ 11 (a) $(-5)|3 - 6|$ (b) $|-6|/(-2)$ (c) $|-7| + |4|$

12 (a) $(4)|6 - 7|$ (b) $5/|-2|$ (c) $|-1| + |-9|$

13 (a) $|4 - \pi|$ (b) $|\pi - 4|$ (c) $|\sqrt{2} - 1.5|$

14 (a) $|\sqrt{3} - 1.7|$ (b) $|1.7 - \sqrt{3}|$ (c) $|\frac{1}{5} - \frac{1}{3}|$

Exer. 15–18: The given numbers are coordinates of points A, B, and C, respectively, on a coordinate line. Find the distance.

(a) $d(A, B)$ **(b) $d(B, C)$**

(c) $d(C, B)$ **(d) $d(A, C)$**

15 3, 7, -5

16 -6, -2, 4

17 -9, 1, 10

18 8, -4, -1

Exer. 19–24: The two given numbers are coordinates of points A and B, respectively, on a coordinate line. Express the indicated statement as an inequality involving the absolute value symbol.

19 x, 7; $d(A, B)$ is less than 5

20 x, $-\sqrt{2}$; $d(A, B)$ is greater than 1

21 x, -3; $d(A, B)$ is at least 8

22 x, 4; $d(A, B)$ is at most 2

23 4, x; $d(A, B)$ is not greater than 3

24 -2, x; $d(A, B)$ is not less than 2

Exer. 25–32: Rewrite the expression without using the absolute value symbol, and simplify the result.

▶ 25 $|3 + x|$ if $x < -3$

26 $|5 - x|$ if $x > 5$

27 $|2 - x|$ if $x < 2$

28 $|7 + x|$ if $x \geq -7$

29 $|a - b|$ if $a < b$

30 $|a - b|$ if $a > b$

31 $|x^2 + 4|$

▶ 32 $|-x^2 - 1|$

Exer. 33–40: Replace the symbol □ with either = or ≠ to make the resulting statement true for all real numbers *a*, *b*, *c*, and *d*, whenever the expressions are defined.

▶ 33 $\dfrac{ab + ac}{a} \;\square\; b + ac$

34 $\dfrac{ab + ac}{a} \;\square\; b + c$

35 $\dfrac{b + c}{a} \;\square\; \dfrac{b}{a} + \dfrac{c}{a}$

36 $\dfrac{a + c}{b + d} \;\square\; \dfrac{a}{b} + \dfrac{c}{d}$

37 $(a \div b) \div c \;\square\; a \div (b \div c)$

38 $(a - b) - c \;\square\; a - (b - c)$

39 $\dfrac{a - b}{b - a} \;\square\; -1$

40 $-(a + b) \;\square\; -a + b$

Exer. 41–42: Approximate the real-number expression to four decimal places.

▶ 41 (a) $|3.2^2 - \sqrt{3.15}|$

(b) $\sqrt{(15.6 - 1.5)^2 + (4.3 - 5.4)^2}$

42 (a) $\dfrac{3.42 - 1.29}{5.83 + 2.64}$

(b) π^3

Exer. 43–44: Approximate the real-number expression. Express the answer in scientific notation accurate to four significant figures.

▶ 43 (a) $\dfrac{1.2 \times 10^3}{3.1 \times 10^2 + 1.52 \times 10^3}$

(b) $(1.23 \times 10^{-4}) + \sqrt{4.5 \times 10^3}$

44 (a) $\sqrt{|3.45 - 1.2 \times 10^4| + 10^5}$

(b) $(1.791 \times 10^2) \times (9.84 \times 10^3)$

45 The point on a coordinate line corresponding to $\sqrt{2}$ may be determined by constructing a right triangle with sides of length 1, as shown in the figure. Determine the points that correspond to $\sqrt{3}$ and $\sqrt{5}$, respectively. (*Hint:* Use the Pythagorean theorem.)

Exercise 45

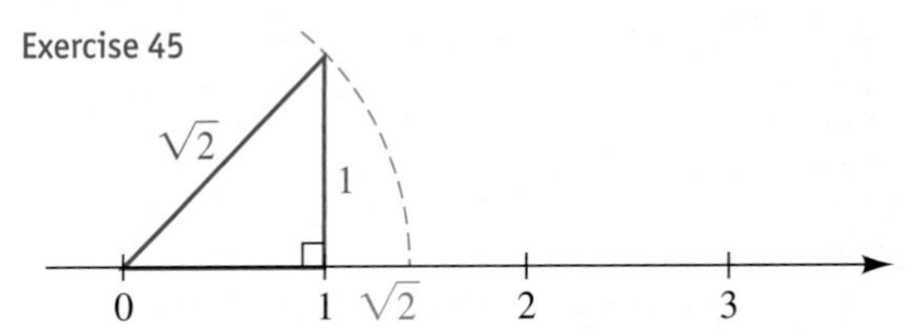

46 A circle of radius 1 rolls along a coordinate line in the positive direction, as shown in the figure. If point P is initially at the origin, find the coordinate of P after one, two, and ten complete revolutions.

Exercise 46

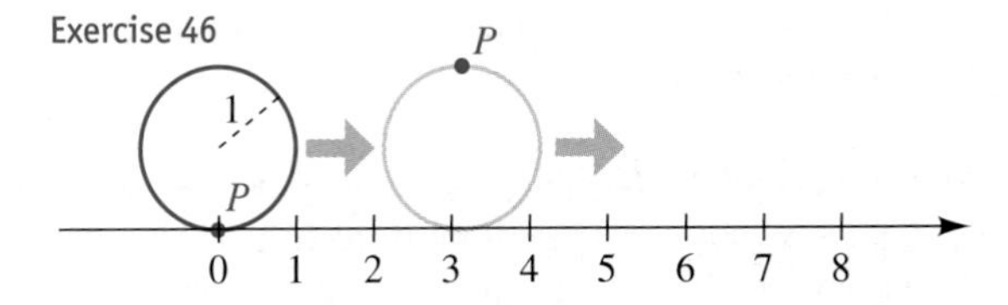

47 Geometric proofs of properties of real numbers were first given by the ancient Greeks. In order to establish the distributive property $a(b + c) = ab + ac$ for positive real numbers a, b, and c, find the area of the rectangle shown in the figure in two ways.

Exercise 47

a
b
c

48 Rational approximations to square roots can be found using a formula discovered by the ancient Babylonians. Let x_1 be the first rational approximation for $\sqrt{n}$. If we let

$$x_2 = \frac{1}{2}\left(x_1 + \frac{n}{x_1}\right),$$

then x_2 will be a better approximation for $\sqrt{n}$, and we can repeat the computation with x_2 replacing x_1. Starting with $x_1 = \frac{3}{2}$, find the next two rational approximations for $\sqrt{2}$.

Exer. 49–50: Express the number in scientific form.

49 (a) 427,000 (b) 0.000 000 098 (c) 810,000,000

50 (a) 85,200 (b) 0.000 005 5 (c) 24,900,000

Exer. 51–52: Express the number in decimal form.

51 (a) 8.3×10^5 (b) 2.9×10^{-12} (c) 5.63×10^8

▶ 52 (a) 2.3×10^7 (b) 7.01×10^{-9} (c) 1.23×10^{10}

53 Mass of a hydrogen atom The mass of a hydrogen atom is approximately

0.000 000 000 000 000 000 000 001 7 gram.

Express this number in scientific form.

54 Mass of an electron The mass of an electron is approximately 9.1×10^{-31} kilogram. Express this number in decimal form.

55 Light year In astronomy, distances to stars are measured in light years. One light year is the distance a ray of light travels in one year. If the speed of light is approximately 186,000 miles per second, estimate the number of miles in one light year.

56 Milky Way galaxy

(a) Astronomers have estimated that the Milky Way galaxy contains 100 billion stars. Express this number in scientific form.

(b) The diameter d of the Milky Way galaxy is estimated as 100,000 light years. Express d in miles. (Refer to Exercise 55.)

57 Avogadro's number The number of hydrogen atoms in a mole is Avogadro's number, 6.02×10^{23}. If one mole of the gas has a mass of 1.01 grams, estimate the mass of a hydrogen atom.

58 Fish population The population dynamics of many fish are characterized by extremely high fertility rates among adults and very low survival rates among the young. A mature halibut may lay as many as 2.5 million eggs, but only 0.00035% of the offspring survive to the age of 3 years. Use scientific form to approximate the number of offspring that live to age 3.

59 Frames in a movie film One of the longest movies ever made is a 1970 British film that runs for 48 hours. Assuming that the film speed is 24 frames per second, approximate the total number of frames in this film. Express your answer in scientific form.

60 Large prime numbers The number $2^{44,497} - 1$ is prime. At the time that this number was determined to be prime, it took one of the world's fastest computers about 60 days to verify that it was prime. This computer was capable of performing 2×10^{11} calculations per second. Use scientific form to estimate the number of calculations needed to perform this computation. (More recently, in 2005, $2^{30,402,457} - 1$, a number containing 9,152,052 digits, was shown to be prime.)

61 Tornado pressure When a tornado passes near a building, there is a rapid drop in the outdoor pressure and the indoor pressure does not have time to change. The resulting difference is capable of causing an outward pressure of 1.4 lb/in^2 on the walls and ceiling of the building.

(a) Calculate the force in pounds exerted on 1 square foot of a wall.

(b) Estimate the tons of force exerted on a wall that is 8 feet high and 40 feet wide.

62 Cattle population A rancher has 750 head of cattle consisting of 400 adults (aged 2 or more years), 150 yearlings, and 200 calves. The following information is known about this particular species. Each spring an adult female gives birth to a single calf, and 75% of these calves will survive the first year. The yearly survival percentages for yearlings and adults are 80% and 90%, respectively. The male-female ratio is one in all age classes. Estimate the population of each age class

(a) next spring (b) last spring

1.2 Exponents and Radicals

If n is a positive integer, the exponential notation a^n, defined in the following chart, represents the product of the real number a with itself n times. We refer to a^n as ***a* to the *n*th power** or, simply, *a to the n*. The positive integer n is called the **exponent,** and the real number a is called the **base.**

Exponential Notation

General case (n is any positive integer)	Special cases
$a^n = \underbrace{a \cdot a \cdot a \cdot \cdots \cdot a}_{n \text{ factors of } a}$	$a^1 = a$ $a^2 = a \cdot a$ $a^3 = a \cdot a \cdot a$ $a^6 = a \cdot a \cdot a \cdot a \cdot a \cdot a$

The next illustration contains several numerical examples of exponential notation.

ILLUSTRATION The Exponential Notation a^n

- $5^4 = 5 \cdot 5 \cdot 5 \cdot 5 = 625$
- $\left(\frac{1}{2}\right)^5 = \frac{1}{2} \cdot \frac{1}{2} \cdot \frac{1}{2} \cdot \frac{1}{2} \cdot \frac{1}{2} = \frac{1}{32}$
- $(-3)^3 = (-3)(-3)(-3) = -27$
- $\left(-\frac{1}{3}\right)^4 = \left(-\frac{1}{3}\right)\left(-\frac{1}{3}\right)\left(-\frac{1}{3}\right)\left(-\frac{1}{3}\right) = \left(\frac{1}{9}\right)\left(\frac{1}{9}\right) = \frac{1}{81}$

It is important to note that if n is a positive integer, then an expression such as $3a^n$ means $3(a^n)$, *not* $(3a)^n$. The real number 3 is the **coefficient** of a^n in the expression $3a^n$. Similarly, $-3a^n$ means $(-3)a^n$, *not* $(-3a)^n$.

ILLUSTRATION The Notation ca^n

- $5 \cdot 2^3 = 5 \cdot 8 = 40$
- $-5 \cdot 2^3 = -5 \cdot 8 = -40$
- $-2^4 = -(2^4) = -16$
- $3(-2)^3 = 3(-2)(-2)(-2) = 3(-8) = -24$

TI-83/4 Plus and TI-86

Exponential Notation

$\left(\frac{1}{2}\right)^5 \rightarrow$

```
(-3)²
                9
-3²
               -9
(1/2)^5
           .03125
```

Note that the expression on the second line, -3^2, is equivalent to $-1 \cdot 3^2$.

We next extend the definition of a^n to nonpositive exponents.

Zero and Negative (Nonpositive) Exponents

Definition ($a \neq 0$)	Illustrations
$a^0 = 1$	$3^0 = 1, \quad (-\sqrt{2})^0 = 1$
$a^{-n} = \dfrac{1}{a^n}$	$5^{-3} = \dfrac{1}{5^3}, \quad (-3)^{-5} = \dfrac{1}{(-3)^5}$

If m and n are positive integers, then

$$a^m a^n = \underbrace{a \cdot a \cdot a \cdot \cdots \cdot a}_{m \text{ factors of } a} \cdot \underbrace{a \cdot a \cdot a \cdot \cdots \cdot a}_{n \text{ factors of } a}.$$

Since the total number of factors of a on the right is $m + n$, this expression is equal to a^{m+n}; that is,

$$a^m a^n = a^{m+n}.$$

We can extend this formula to $m \leq 0$ or $n \leq 0$ by using the definitions of the zero exponent and negative exponents. This gives us law 1, stated in the next chart.

To prove law 2, we may write, for m and n positive,

$$(a^m)^n = \underbrace{a^m \cdot a^m \cdot a^m \cdot \cdots \cdot a^m}_{n \text{ factors of } a^m}$$

and count the number of times a appears as a factor on the right-hand side. Since $a^m = a \cdot a \cdot a \cdot \cdots \cdot a$, with a occurring as a factor m times, and since the number of such groups of m factors is n, the total number of factors of a is $m \cdot n$. Thus,

$$(a^m)^n = a^{mn}.$$

The cases $m \leq 0$ and $n \leq 0$ can be proved using the definition of nonpositive exponents. The remaining three laws can be established in similar fashion by counting factors. In laws 4 and 5 we assume that denominators are not 0.

Laws of Exponents for Real Numbers *a* and *b* and Integers *m* and *n*

Law	Illustration
(1) $a^m a^n = a^{m+n}$	$2^3 \cdot 2^4 = 2^{3+4} = 2^7 = 128$
(2) $(a^m)^n = a^{mn}$	$(2^3)^4 = 2^{3 \cdot 4} = 2^{12} = 4096$
(3) $(ab)^n = a^n b^n$	$(20)^3 = (2 \cdot 10)^3 = 2^3 \cdot 10^3 = 8 \cdot 1000 = 8000$
(4) $\left(\dfrac{a}{b}\right)^n = \dfrac{a^n}{b^n}$	$\left(\dfrac{2}{5}\right)^3 = \dfrac{2^3}{5^3} = \dfrac{8}{125}$
(5) (a) $\dfrac{a^m}{a^n} = a^{m-n}$	$\dfrac{2^5}{2^3} = 2^{5-3} = 2^2 = 4$
(b) $\dfrac{a^m}{a^n} = \dfrac{1}{a^{n-m}}$	$\dfrac{2^3}{2^5} = \dfrac{1}{2^{5-3}} = \dfrac{1}{2^2} = \dfrac{1}{4}$

We usually use 5(a) if $m > n$ and 5(b) if $m < n$.

We can extend laws of exponents to obtain rules such as $(abc)^n = a^n b^n c^n$ and $a^m a^n a^p = a^{m+n+p}$. Some other examples of the laws of exponents are given in the next illustration.

ILLUSTRATION **Laws of Exponents**

- $x^5 x^6 x^2 = x^{5+6+2} = x^{13}$
- $(y^5)^7 = y^{5 \cdot 7} = y^{35}$
- $(3st)^4 = 3^4 s^4 t^4 = 81 s^4 t^4$
- $\left(\dfrac{p}{2}\right)^5 = \dfrac{p^5}{2^5} = \dfrac{p^5}{32}$
- $\dfrac{c^8}{c^3} = c^{8-3} = c^5$
- $\dfrac{u^3}{u^8} = \dfrac{1}{u^{8-3}} = \dfrac{1}{u^5}$

To **simplify** an expression involving powers of real numbers means to change it to an expression in which each real number appears only once and all exponents are positive. *We shall assume that denominators always represent nonzero real numbers.*

EXAMPLE 1 **Simplifying expressions containing exponents**

Use laws of exponents to simplify each expression:

(a) $(3x^3y^4)(4xy^5)$ **(b)** $(2a^2b^3c)^4$ **(c)** $\left(\dfrac{2r^3}{s}\right)^2\left(\dfrac{s}{r^3}\right)^3$ **(d)** $(u^{-2}v^3)^{-3}$

▶ SOLUTION

(a)

$$\begin{aligned} (3x^3y^4)(4xy^5) &= (3)(4)x^3xy^4y^5 && \text{rearrange factors} \\ &= 12x^4y^9 && \text{law 1} \end{aligned}$$

(b)

$$\begin{aligned} (2a^2b^3c)^4 &= 2^4(a^2)^4(b^3)^4c^4 && \text{law 3} \\ &= 16a^8b^{12}c^4 && \text{law 2} \end{aligned}$$

(c)

$$\begin{aligned} \left(\frac{2r^3}{s}\right)^2\left(\frac{s}{r^3}\right)^3 &= \frac{(2r^3)^2}{s^2} \cdot \frac{s^3}{(r^3)^3} && \text{law 4} \\ &= \frac{2^2(r^3)^2}{s^2} \cdot \frac{s^3}{(r^3)^3} && \text{law 3} \\ &= \left(\frac{4r^6}{s^2}\right)\left(\frac{s^3}{r^9}\right) && \text{law 2} \\ &= 4\left(\frac{r^6}{r^9}\right)\left(\frac{s^3}{s^2}\right) && \text{rearrange factors} \\ &= 4\left(\frac{1}{r^3}\right)(s) && \text{laws 5(b) and 5(a)} \\ &= \frac{4s}{r^3} && \text{rearrange factors} \end{aligned}$$

(continued)

(d) $(u^{-2}v^{3})^{-3} = (u^{-2})^{-3}(v^{3})^{-3}$ law 3

$= u^{6}v^{-9}$ law 2

$= \dfrac{u^{6}}{v^{9}}$ definition of a^{-n}

The following theorem is useful for problems that involve negative exponents.

Theorem on Negative Exponents

$$\text{(1)}\ \frac{a^{-m}}{b^{-n}} = \frac{b^{n}}{a^{m}} \qquad \text{(2)}\ \left(\frac{a}{b}\right)^{-n} = \left(\frac{b}{a}\right)^{n}$$

PROOFS Using properties of negative exponents and quotients, we obtain

$$\text{(1)}\ \frac{a^{-m}}{b^{-n}} = \frac{1/a^{m}}{1/b^{n}} = \frac{1}{a^{m}} \cdot \frac{b^{n}}{1} = \frac{b^{n}}{a^{m}}$$

$$\text{(2)}\ \left(\frac{a}{b}\right)^{-n} = \frac{a^{-n}}{b^{-n}} = \frac{b^{n}}{a^{n}} = \left(\frac{b}{a}\right)^{n}$$

EXAMPLE 2 Simplifying expressions containing negative exponents

Simplify:

(a) $\dfrac{8x^{3}y^{-5}}{4x^{-1}y^{2}}$ (b) $\left(\dfrac{u^{2}}{2v}\right)^{-3}$

▶ SOLUTION We apply the theorem on negative exponents and the laws of exponents.

(a) $\dfrac{8x^{3}y^{-5}}{4x^{-1}y^{2}} = \dfrac{8x^{3}}{4y^{2}} \cdot \dfrac{y^{-5}}{x^{-1}}$ rearrange quotients so that negative exponents are in one fraction

$= \dfrac{8x^{3}}{4y^{2}} \cdot \dfrac{x^{1}}{y^{5}}$ theorem on negative exponents (1)

$= \dfrac{2x^{4}}{y^{7}}$ law 1 of exponents

(b) $\left(\dfrac{u^{2}}{2v}\right)^{-3} = \left(\dfrac{2v}{u^{2}}\right)^{3}$ theorem on negative exponents (2)

$= \dfrac{2^{3}v^{3}}{(u^{2})^{3}}$ laws 4 and 3 of exponents

$= \dfrac{8v^{3}}{u^{6}}$ law 2 of exponents

▶ Tutorial available at www.thomsonedu.com/login

We next define the **principal *n*th root** $\sqrt[n]{a}$ of a real number a.

Definition of $\sqrt[n]{a}$	Let n be a positive integer greater than 1, and let a be a real number. **(1)** If $a = 0$, then $\sqrt[n]{a} = 0$. **(2)** If $a > 0$, then $\sqrt[n]{a}$ is the *positive* real number b such that $b^n = a$. **(3)** **(a)** If $a < 0$ and n is odd, then $\sqrt[n]{a}$ is the *negative* real number b such that $b^n = a$. **(b)** If $a < 0$ and n is even, then $\sqrt[n]{a}$ is not a real number.

Complex numbers, discussed in Section 1.5, are needed to define $\sqrt[n]{a}$ if $a < 0$ and n is an *even* positive integer, because for all real numbers b, $b^n \geq 0$ whenever n is even.

If $n = 2$, we write $\sqrt{a}$ instead of $\sqrt[2]{a}$ and call $\sqrt{a}$ the **principal square root** of a or, simply, the **square root** of a. The number $\sqrt[3]{a}$ is the (principal) **cube root** of a.

ILLUSTRATION **The Principal *n*th Root $\sqrt[n]{a}$**

- $\sqrt{16} = 4$, since $4^2 = 16$.
- $\sqrt[5]{\frac{1}{32}} = \frac{1}{2}$, since $\left(\frac{1}{2}\right)^5 = \frac{1}{32}$.
- $\sqrt[3]{-8} = -2$, since $(-2)^3 = -8$.
- $\sqrt[4]{-16}$ is not a real number.

Note that $\sqrt{16} \neq \pm 4$, since, by definition, roots of positive real numbers are positive. The symbol $\pm$ is read "plus or minus."

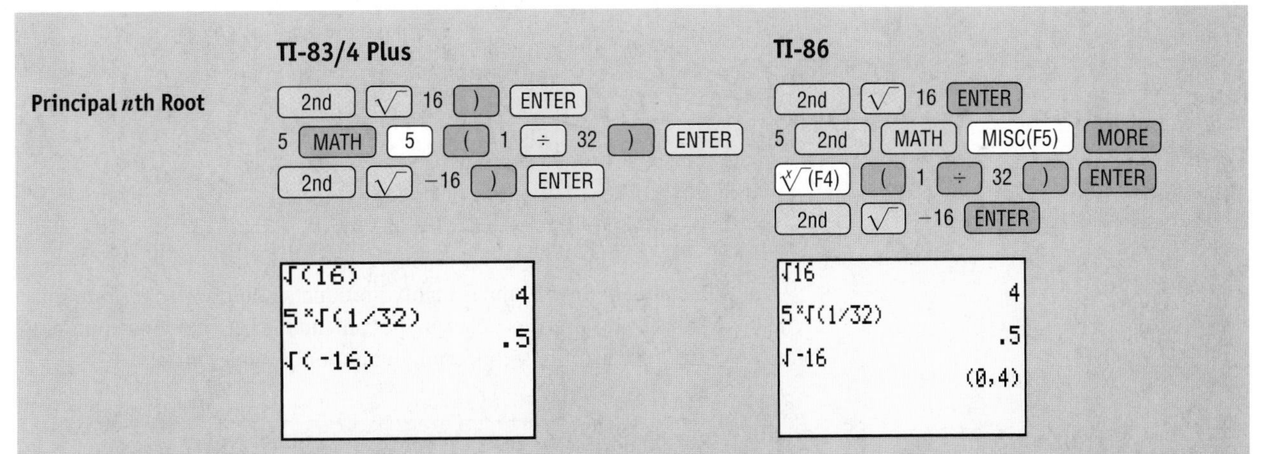

When the last line is executed on the TI-83/4 Plus, the error message NONREAL ANS is given since this expression represents a complex number, not a real number (covered in Section 1.5). The answer on the TI-86, (0, 4), represents $0 + 4i$.

To complete our terminology, the expression $\sqrt[n]{a}$ is a **radical,** the number a is the **radicand,** and n is the **index** of the radical. The symbol $\sqrt{\ }$ is called a **radical sign.**

If $\sqrt{a} = b$, then $b^2 = a$; that is, $(\sqrt{a})^2 = a$. If $\sqrt[3]{a} = b$, then $b^3 = a$, or $(\sqrt[3]{a})^3 = a$. Generalizing this pattern gives us property 1 in the next chart.

Properties of $\sqrt[n]{a}$ (n is a positive integer)

Property	Illustrations	
(1) $(\sqrt[n]{a})^n = a$ if $\sqrt[n]{a}$ is a real number	$(\sqrt{5})^2 = 5$,	$(\sqrt[3]{-8})^3 = -8$
(2) $\sqrt[n]{a^n} = a$ if $a \geq 0$	$\sqrt{5^2} = 5$,	$\sqrt[3]{2^3} = 2$
(3) $\sqrt[n]{a^n} = a$ if $a < 0$ and n is odd	$\sqrt[3]{(-2)^3} = -2$,	$\sqrt[5]{(-2)^5} = -2$
(4) $\sqrt[n]{a^n} = \lvert a \rvert$ if $a < 0$ and n is even	$\sqrt{(-3)^2} = \lvert -3 \rvert = 3$,	$\sqrt[4]{(-2)^4} = \lvert -2 \rvert = 2$

If $a \geq 0$, then property 4 reduces to property 2. We also see from property 4 that

$$\sqrt{x^2} = \lvert x \rvert$$

for every real number x. In particular, if $x \geq 0$, then $\sqrt{x^2} = x$; however, if $x < 0$, then $\sqrt{x^2} = -x$, which is positive.

The three laws listed in the next chart are true for positive integers m and n, *provided the indicated roots* ***exist***—that is, provided the roots are real numbers.

Laws of Radicals

Law	Illustrations
(1) $\sqrt[n]{ab} = \sqrt[n]{a}\,\sqrt[n]{b}$	$\sqrt{50} = \sqrt{25 \cdot 2} = \sqrt{25}\,\sqrt{2} = 5\sqrt{2}$ $\sqrt[3]{-108} = \sqrt[3]{(-27)(4)} = \sqrt[3]{-27}\,\sqrt[3]{4} = -3\sqrt[3]{4}$
(2) $\sqrt[n]{\dfrac{a}{b}} = \dfrac{\sqrt[n]{a}}{\sqrt[n]{b}}$	$\sqrt[3]{\dfrac{5}{8}} = \dfrac{\sqrt[3]{5}}{\sqrt[3]{8}} = \dfrac{\sqrt[3]{5}}{2}$
(3) $\sqrt[m]{\sqrt[n]{a}} = \sqrt[mn]{a}$	$\sqrt{\sqrt[3]{64}} = \sqrt[2(3)]{64} = \sqrt[6]{2^6} = 2$

The radicands in laws 1 and 2 involve products and quotients. Care must be taken if sums or differences occur in the radicand. The following chart contains two particular warnings concerning commonly made mistakes.

Warning!

If $a \neq 0$ and $b \neq 0$	Illustration
(1) $\sqrt{a^2 + b^2} \neq a + b$	$\sqrt{3^2 + 4^2} = \sqrt{25} = 5 \neq 3 + 4 = 7$
(2) $\sqrt{a + b} \neq \sqrt{a} + \sqrt{b}$	$\sqrt{4 + 9} = \sqrt{13} \neq \sqrt{4} + \sqrt{9} = 5$

If c is a real number and c^n occurs as a factor in a radical of index n, then we can remove c from the radicand if the sign of c is taken into account. For example, if $c > 0$ or if $c < 0$ and n is *odd,* then

$$\sqrt[n]{c^n d} = \sqrt[n]{c^n}\,\sqrt[n]{d} = c\sqrt[n]{d},$$

provided $\sqrt[n]{d}$ exists. If $c < 0$ and n is *even,* then

$$\sqrt[n]{c^n d} = \sqrt[n]{c^n}\,\sqrt[n]{d} = |c|\sqrt[n]{d},$$

provided $\sqrt[n]{d}$ exists.

ILLUSTRATION **Removing nth Powers from $\sqrt[n]{}$**

- $\sqrt[5]{x^7} = \sqrt[5]{x^5 \cdot x^2} = \sqrt[5]{x^5}\,\sqrt[5]{x^2} = x\sqrt[5]{x^2}$
- $\sqrt[3]{x^7} = \sqrt[3]{x^6 \cdot x} = \sqrt[3]{(x^2)^3 x} = \sqrt[3]{(x^2)^3}\,\sqrt[3]{x} = x^2\sqrt[3]{x}$
- $\sqrt{x^2 y} = \sqrt{x^2}\,\sqrt{y} = |x|\sqrt{y}$
- $\sqrt{x^6} = \sqrt{(x^3)^2} = |x^3|$
- $\sqrt[4]{x^6 y^3} = \sqrt[4]{x^4 \cdot x^2 y^3} = \sqrt[4]{x^4}\,\sqrt[4]{x^2 y^3} = |x|\sqrt[4]{x^2 y^3}$

Note: To avoid considering absolute values, *in examples and exercises involving radicals in this chapter, we shall assume that all letters—a, b, c, d, x, y, and so on—that appear in radicands represent positive real numbers, unless otherwise specified.*

As shown in the preceding illustration and in the following examples, if the index of a radical is n, then we rearrange the radicand, isolating a factor of the form p^n, where p may consist of several letters. We then remove $\sqrt[n]{p^n} = p$ from the radical, as previously indicated. Thus, in Example 3(b) the index of the radical is 3 and we rearrange the radicand into *cubes,* obtaining a factor p^3, with $p = 2xy^2z$. In part (c) the index of the radical is 2 and we rearrange the radicand into squares, obtaining a factor p^2, with $p = 3a^3b^2$.

To *simplify a radical* means to remove factors from the radical until no factor in the radicand has an exponent greater than or equal to the index of the radical and the index is as low as possible.

EXAMPLE 3 **Removing factors from radicals**

Simplify each radical (all letters denote positive real numbers):

(a) $\sqrt[3]{320}$ (b) $\sqrt[3]{16x^3y^8z^4}$ (c) $\sqrt{3a^2b^3}\,\sqrt{6a^5b}$

SOLUTION

▶ (a)

$\sqrt[3]{320} = \sqrt[3]{64 \cdot 5}$	factor out the largest cube in 320
$= \sqrt[3]{4^3}\,\sqrt[3]{5}$	law 1 of radicals
$= 4\sqrt[3]{5}$	property 2 of $\sqrt[n]{}$

(continued)

▶ **Tutorial available at www.thomsonedu.com/login**

(b) $\sqrt[3]{16x^3y^8z^4} = \sqrt[3]{(2^3x^3y^6z^3)(2y^2z)}$ rearrange radicand into cubes

$= \sqrt[3]{(2xy^2z)^3(2y^2z)}$ laws 2 and 3 of exponents

$= \sqrt[3]{(2xy^2z)^3}\sqrt[3]{2y^2z}$ law 1 of radicals

$= 2xy^2z\sqrt[3]{2y^2z}$ property 2 of $\sqrt[n]{\ }$

(c) $\sqrt{3a^2b^3}\sqrt{6a^5b} = \sqrt{3a^2b^3 \cdot 2 \cdot 3a^5b}$ law 1 of radicals

$= \sqrt{(3^2a^6b^4)(2a)}$ rearrange radicand into squares

$= \sqrt{(3a^3b^2)^2(2a)}$ laws 2 and 3 of exponents

$= \sqrt{(3a^3b^2)^2}\sqrt{2a}$ law 1 of radicals

$= 3a^3b^2\sqrt{2a}$ property 2 of $\sqrt[n]{\ }$

If the denominator of a quotient contains a factor of the form $\sqrt[n]{a^k}$, with $k < n$ and $a > 0$, then multiplying the numerator and denominator by $\sqrt[n]{a^{n-k}}$ will eliminate the radical from the denominator, since

$$\sqrt[n]{a^k}\sqrt[n]{a^{n-k}} = \sqrt[n]{a^{k+n-k}} = \sqrt[n]{a^n} = a.$$

This process is called **rationalizing a denominator.** Some special cases are listed in the following chart.

Rationalizing Denominators of Quotients ($a > 0$)

Factor in denominator	Multiply numerator and denominator by	Resulting factor
$\sqrt{a}$	$\sqrt{a}$	$\sqrt{a}\sqrt{a} = \sqrt{a^2} = a$
$\sqrt[3]{a}$	$\sqrt[3]{a^2}$	$\sqrt[3]{a}\sqrt[3]{a^2} = \sqrt[3]{a^3} = a$
$\sqrt[7]{a^3}$	$\sqrt[7]{a^4}$	$\sqrt[7]{a^3}\sqrt[7]{a^4} = \sqrt[7]{a^7} = a$

The next example illustrates this technique.

EXAMPLE 4 Rationalizing denominators

Rationalize each denominator:

(a) $\dfrac{1}{\sqrt{5}}$ (b) $\dfrac{1}{\sqrt[3]{x}}$ (c) $\sqrt{\dfrac{2}{3}}$ (d) $\sqrt[5]{\dfrac{x}{y^2}}$

SOLUTION

(a) $\dfrac{1}{\sqrt{5}} = \dfrac{1}{\sqrt{5}}\dfrac{\sqrt{5}}{\sqrt{5}} = \dfrac{\sqrt{5}}{\sqrt{5^2}} = \dfrac{\sqrt{5}}{5}$

(b) $\dfrac{1}{\sqrt[3]{x}} = \dfrac{1}{\sqrt[3]{x}}\dfrac{\sqrt[3]{x^2}}{\sqrt[3]{x^2}} = \dfrac{\sqrt[3]{x^2}}{\sqrt[3]{x^3}} = \dfrac{\sqrt[3]{x^2}}{x}$

(c) $\sqrt{\dfrac{2}{3}} = \dfrac{\sqrt{2}}{\sqrt{3}} = \dfrac{\sqrt{2}\sqrt{3}}{\sqrt{3}\sqrt{3}} = \dfrac{\sqrt{2 \cdot 3}}{\sqrt{3^2}} = \dfrac{\sqrt{6}}{3}$

(d) $\sqrt[5]{\dfrac{x}{y^2}} = \dfrac{\sqrt[5]{x}}{\sqrt[5]{y^2}} = \dfrac{\sqrt[5]{x}\sqrt[5]{y^3}}{\sqrt[5]{y^2}\sqrt[5]{y^3}} = \dfrac{\sqrt[5]{xy^3}}{\sqrt[5]{y^5}} = \dfrac{\sqrt[5]{xy^3}}{y}$

If we use a calculator to find decimal approximations of radicals, there is no advantage in rationalizing denominators, such as $1/\sqrt{5} = \sqrt{5}/5$ or $\sqrt{2/3} = \sqrt{6}/3$, as we did in Example 4(a) and (c). However, for *algebraic* simplifications, changing expressions to such forms is sometimes desirable. Similarly, in advanced mathematics courses such as calculus, changing $1/\sqrt[3]{x}$ to $\sqrt[3]{x^2}/x$, as in Example 4(b), could make a problem *more* complicated. In such courses it is simpler to work with the expression $1/\sqrt[3]{x}$ than with its rationalized form.

We next use radicals to define *rational exponents.*

Definition of Rational Exponents

Let m/n be a rational number, where n is a positive integer greater than 1. If a is a real number such that $\sqrt[n]{a}$ exists, then

(1) $a^{1/n} = \sqrt[n]{a}$

(2) $a^{m/n} = \left(\sqrt[n]{a}\right)^m = \sqrt[n]{a^m}$

(3) $a^{m/n} = (a^{1/n})^m = (a^m)^{1/n}$

When evaluating $a^{m/n}$ in (2), we usually use $\left(\sqrt[n]{a}\right)^m$; that is, we take the nth root of a first and then raise that result to the mth power, as shown in the following illustration.

ILLUSTRATION **The Exponential Notation $a^{m/n}$**

- $x^{1/3} = \sqrt[3]{x}$ ■ $x^{3/5} = \left(\sqrt[5]{x}\right)^3 = \sqrt[5]{x^3}$
- $125^{2/3} = \left(\sqrt[3]{125}\right)^2 = \left(\sqrt[3]{5^3}\right)^2 = 5^2 = 25$
- $\left(\frac{32}{243}\right)^{3/5} = \left(\sqrt[5]{\frac{32}{243}}\right)^3 = \left(\sqrt[5]{\left(\frac{2}{3}\right)^5}\right)^3 = \left(\frac{2}{3}\right)^3 = \frac{8}{27}$

Rational Exponents

TI-83/4 Plus

8 [^] [(] 1 [÷] 3 [)] [ENTER]
−8 [^] [(] 1 [÷] 3 [)] [ENTER]
[(] 32 [÷] 243 [)] [^] [(] 3 [÷] 5 [)]
[MATH] [1] [ENTER]

```
8^(1/3)
                2
-8^(1/3)
               -2
(32/243)^(3/5)▸F
rac
             8/27
```

TI-86

8 [^] [(] 1 [÷] 3 [)] [ENTER]
−8 [^] [(] 1 [÷] 3 [)] [ENTER]
[(] 32 [÷] 243 [)] [^] [(] 3 [÷] 5 [)]
[2nd] [MATH] [MISC(F5)]
[MORE] [Frac(F1)] [ENTER]

```
8^(1/3)
                2
-8^(1/3)
               -2
(32/243)^(3/5)▸Frac
             8/27
```

The Frac command changes a decimal representation to a fractional representation.

The laws of exponents are true for rational exponents and also for *irrational* exponents, such as $3^{\sqrt{2}}$ or 5^{π}, considered in Chapter 4.

To simplify an expression involving rational powers of letters that represent real numbers, we change it to an expression in which each letter appears only once and all exponents are positive. As we did with radicals, we shall assume that all letters represent positive real numbers unless otherwise specified.

EXAMPLE 5 Simplifying rational powers

Simplify:

(a) $(-27)^{2/3}(4)^{-5/2}$ **(b)** $(r^2s^6)^{1/3}$ **(c)** $\left(\dfrac{2x^{2/3}}{y^{1/2}}\right)^2\left(\dfrac{3x^{-5/6}}{y^{1/3}}\right)$

▶ **SOLUTION**

(a)

$$\begin{aligned}(-27)^{2/3}(4)^{-5/2} &= \left(\sqrt[3]{-27}\right)^2\left(\sqrt{4}\right)^{-5} && \text{definition of rational exponents}\\ &= (-3)^2(2)^{-5} && \text{take roots}\\ &= \frac{(-3)^2}{2^5} && \text{definition of negative exponents}\\ &= \frac{9}{32} && \text{take powers}\end{aligned}$$

(b)

$$\begin{aligned}(r^2s^6)^{1/3} &= (r^2)^{1/3}(s^6)^{1/3} && \text{law 3 of exponents}\\ &= r^{2/3}s^2 && \text{law 2 of exponents}\end{aligned}$$

(c)

$$\begin{aligned}\left(\frac{2x^{2/3}}{y^{1/2}}\right)^2\left(\frac{3x^{-5/6}}{y^{1/3}}\right) &= \left(\frac{4x^{4/3}}{y}\right)\left(\frac{3x^{-5/6}}{y^{1/3}}\right) && \text{laws of exponents}\\ &= \frac{(4\cdot 3)x^{4/3-5/6}}{y^{1+(1/3)}} && \text{law 1 of exponents}\\ &= \frac{12x^{8/6-5/6}}{y^{4/3}} && \text{common denominator}\\ &= \frac{12x^{1/2}}{y^{4/3}} && \text{simplify}\end{aligned}$$

Rational exponents are useful for problems involving radicals that do not have the same index, as illustrated in the next example.

EXAMPLE 6 Combining radicals

Change to an expression containing one radical of the form $\sqrt[n]{a^m}$:

(a) $\sqrt[3]{a}\sqrt{a}$ **(b)** $\dfrac{\sqrt[4]{a}}{\sqrt[3]{a^2}}$

▶ **SOLUTION** Introducing rational exponents, we obtain

(a) $\sqrt[3]{a}\sqrt{a} = a^{1/3}a^{1/2} = a^{(1/3)+(1/2)} = a^{5/6} = \sqrt[6]{a^5}$

(b) $\dfrac{\sqrt[4]{a}}{\sqrt[3]{a^2}} = \dfrac{a^{1/4}}{a^{2/3}} = a^{(1/4)-(2/3)} = a^{-5/12} = \dfrac{1}{a^{5/12}} = \dfrac{1}{\sqrt[12]{a^5}}$

▶ **Tutorial available at www.thomsonedu.com/login**

In Exercises 1.2, whenever an index of a radical is even (or a rational exponent m/n with n even is employed), assume that the letters that appear in the radicand denote positive real numbers unless otherwise specified.

1.2 *Exercises*

Exer. 1–10: Express the number in the form a/b, where a and b are integers.

1 $\left(-\frac{2}{3}\right)^4$

2 $(-3)^3$

3 $\dfrac{2^{-3}}{3^{-2}}$

4 $\dfrac{2^0 + 0^2}{2 + 0}$

5 $-2^4 + 3^{-1}$

6 $\left(-\frac{3}{2}\right)^4 - 2^{-4}$

7 $16^{-3/4}$

8 $9^{5/2}$

9 $(-0.008)^{2/3}$

10 $(0.008)^{-2/3}$

Exer. 11–46: Simplify.

11 $\left(\frac{1}{2}x^4\right)(16x^5)$

12 $(-3x^{-2})(4x^4)$

13 $\dfrac{(2x^3)(3x^2)}{(x^2)^3}$

14 $\dfrac{(2x^2)^3}{4x^4}$

15 $\left(\frac{1}{6}a^5\right)(-3a^2)(4a^7)$

16 $(-4b^3)\left(\frac{1}{6}b^2\right)(-9b^4)$

17 $\dfrac{(6x^3)^2}{(2x^2)^3} \cdot (3x^2)^0$

18 $\dfrac{(3y^3)(2y^2)^2}{(y^4)^3} \cdot (y^3)^0$

19 $(3u^7v^3)(4u^4v^{-5})$

20 $(x^2yz^3)(-2xz^2)(x^3y^{-2})$

21 $(8x^4y^{-3})\left(\frac{1}{2}x^{-5}y^2\right)$

22 $\left(\dfrac{4a^2b}{a^3b^2}\right)\left(\dfrac{5a^2b}{2b^4}\right)$

23 $\left(\frac{1}{3}x^4y^{-3}\right)^{-2}$

24 $(-2xy^2)^5\left(\dfrac{x^7}{8y^3}\right)$

25 $(3y^3)^4(4y^2)^{-3}$

26 $(-3a^2b^{-5})^3$

27 $(-2r^4s^{-3})^{-2}$

28 $(2x^2y^{-5})(6x^{-3}y)\left(\frac{1}{3}x^{-1}y^3\right)$

29 $(5x^2y^{-3})(4x^{-5}y^4)$

30 $(-2r^2s)^5(3r^{-1}s^3)^2$

31 $\left(\dfrac{3x^5y^4}{x^0y^{-3}}\right)^2$

32 $(4a^2b)^4\left(\dfrac{-a^3}{2b}\right)^2$

33 $(4a^{3/2})(2a^{1/2})$

34 $(-6x^{7/5})(2x^{8/5})$

35 $(3x^{5/6})(8x^{2/3})$

36 $(8r)^{1/3}(2r^{1/2})$

37 $(27a^6)^{-2/3}$

38 $(25z^4)^{-3/2}$

39 $(8x^{-2/3})x^{1/6}$

40 $(3x^{1/2})(-2x^{5/2})$

41 $\left(\dfrac{-8x^3}{y^{-6}}\right)^{2/3}$

42 $\left(\dfrac{-y^{3/2}}{y^{-1/3}}\right)^3$

43 $\left(\dfrac{x^6}{9y^{-4}}\right)^{-1/2}$

44 $\left(\dfrac{c^{-4}}{16d^8}\right)^{3/4}$

45 $\dfrac{(x^6y^3)^{-1/3}}{(x^4y^2)^{-1/2}}$

46 $a^{4/3}a^{-3/2}a^{1/6}$

Exer. 47–52: Rewrite the expression using rational exponents.

47 $\sqrt[4]{x^3}$

48 $\sqrt[3]{x^5}$

49 $\sqrt[3]{(a+b)^2}$

50 $\sqrt{a + \sqrt{b}}$

51 $\sqrt{x^2 + y^2}$

52 $\sqrt[3]{r^3 - s^3}$

Exer. 53–56: Rewrite the expression using a radical.

53 (a) $4x^{3/2}$ (b) $(4x)^{3/2}$

54 (a) $4 + x^{3/2}$ (b) $(4 + x)^{3/2}$

55 (a) $8 - y^{1/3}$ (b) $(8 - y)^{1/3}$

56 (a) $8y^{1/3}$ (b) $(8y)^{1/3}$

Exer. 57–80: Simplify the expression, and rationalize the denominator when appropriate.

57 $\sqrt{81}$

58 $\sqrt[3]{-125}$

59 $\sqrt[5]{-64}$

60 $\sqrt[4]{256}$

61 $\dfrac{1}{\sqrt[3]{2}}$

62 $\sqrt{\dfrac{1}{7}}$

63 $\sqrt{9x^{-4}y^6}$

64 $\sqrt{16a^8b^{-2}}$

65 $\sqrt[3]{8a^6b^{-3}}$

66 $\sqrt[4]{81r^5s^8}$

67 $\sqrt{\dfrac{3x}{2y^3}}$

68 $\sqrt{\dfrac{1}{3x^3y}}$

69 $\sqrt[3]{\dfrac{2x^4y^4}{9x}}$

70 $\sqrt[3]{\dfrac{3x^2y^5}{4x}}$

71 $\sqrt[4]{\dfrac{5x^8y^3}{27x^2}}$

72 $\sqrt[4]{\dfrac{x^7y^{12}}{125x}}$

73 $\sqrt[5]{\dfrac{5x^7y^2}{8x^3}}$

74 $\sqrt[5]{\dfrac{3x^{11}y^3}{9x^2}}$

75 $\sqrt[4]{(3x^5y^{-2})^4}$

76 $\sqrt[6]{(2u^{-3}v^4)^6}$

77 $\sqrt[5]{\dfrac{8x^3}{y^4}}\sqrt[5]{\dfrac{4x^4}{y^2}}$

78 $\sqrt{5xy^7}\sqrt{10x^3y^3}$

79 $\sqrt[3]{3t^4v^2}\sqrt[3]{-9t^{-1}v^4}$

80 $\sqrt[3]{(2r-s)^3}$

Exer. 81–84: Simplify the expression, assuming x and y may be negative.

81 $\sqrt{x^6y^4}$

82 $\sqrt{x^4y^{10}}$

83 $\sqrt[4]{x^8(y-1)^{12}}$

84 $\sqrt[4]{(x+2)^{12}y^4}$

Exer. 85–90: Replace the symbol □ with either = or ≠ to make the resulting statement true, whenever the expression has meaning. Give a reason for your answer.

85 $(a^r)^2 \;\square\; a^{(r^2)}$

86 $(a^2+1)^{1/2} \;\square\; a+1$

87 $a^xb^y \;\square\; (ab)^{xy}$

88 $\sqrt{a^r} \;\square\; \left(\sqrt{a}\right)^r$

89 $\sqrt[n]{\dfrac{1}{c}} \;\square\; \dfrac{1}{\sqrt[n]{c}}$

90 $a^{1/k} \;\square\; \dfrac{1}{a^k}$

Exer. 91–92: In evaluating negative numbers raised to fractional powers, it may be necessary to evaluate the root and integer power separately. For example, $(-3)^{2/5}$ can be evaluated successfully as $[(-3)^{1/5}]^2$ or $[(-3)^2]^{1/5}$, whereas an error message might otherwise appear. Approximate the real-number expression to four decimal places.

91 (a) $(-3)^{2/5}$ (b) $(-5)^{4/3}$

92 (a) $(-1.2)^{3/7}$ (b) $(-5.08)^{7/3}$

Exer. 93–94: Approximate the real-number expression to four decimal places.

93 (a) $\sqrt{\pi+1}$ (b) $\sqrt[3]{15.1}+5^{1/4}$

94 (a) $(2.6-1.9)^{-2}$ (b) $5^{\sqrt{7}}$

95 Savings account One of the oldest banks in the United States is the Bank of America, founded in 1812. If \$200 had been deposited at that time into an account that paid 4% annual interest, then 180 years later the amount would have grown to $200(1.04)^{180}$ dollars. Approximate this amount to the nearest cent.

96 Viewing distance On a clear day, the distance d (in miles) that can be seen from the top of a tall building of height h (in feet) can be approximated by $d = 1.2\sqrt{h}$. Approximate the distance that can be seen from the top of the Chicago Sears Tower, which is 1454 feet tall.

97 Length of a halibut The length-weight relationship for Pacific halibut can be approximated by the formula $L = 0.46\sqrt[3]{W}$, where W is in kilograms and L is in meters. The largest documented halibut weighed 230 kilograms. Estimate its length.

98 Weight of a whale The length-weight relationship for the sei whale can be approximated by $W = 0.0016L^{2.43}$, where W is in tons and L is in feet. Estimate the weight of a whale that is 25 feet long.

99 Weight lifters' handicaps O'Carroll's formula is used to handicap weight lifters. If a lifter who weighs b kilograms lifts w kilograms of weight, then the handicapped weight W is given by

$$W = \frac{w}{\sqrt[3]{b-35}}.$$

Suppose two lifters weighing 75 kilograms and 120 kilograms lift weights of 180 kilograms and 250 kilograms, respectively. Use O'Carroll's formula to determine the superior weight lifter.

100 Body surface area A person's body surface area S (in square feet) can be approximated by

$$S = (0.1091)w^{0.425}h^{0.725},$$

where height h is in inches and weight w is in pounds.

(a) Estimate S for a person 6 feet tall weighing 175 pounds.

(b) If a person is 5 feet 6 inches tall, what effect does a 10% increase in weight have on S?

101 Men's weight The average weight W (in pounds) for men with height h between 64 and 79 inches can be approximated using the formula $W = 0.1166h^{1.7}$. Construct a table for W by letting $h = 64, 65, \ldots, 79$. Round all weights to the nearest pound.

Height	Weight	Height	Weight
64		72	
65		73	
66		74	
67		75	
68		76	
69		77	
70		78	
71		79	

102 Women's weight The average weight W (in pounds) for women with height h between 60 and 75 inches can be approximated using the formula $W = 0.1049h^{1.7}$. Construct a table for W by letting $h = 60, 61, \ldots, 75$. Round all weights to the nearest pound.

Height	Weight	Height	Weight
60		68	
61		69	
62		70	
63		71	
64		72	
65		73	
66		74	
67		75	

1.3 Algebraic Expressions

We sometimes use the notation and terminology of sets to describe mathematical relationships. A **set** is a collection of objects of some type, and the objects are called **elements** of the set. Capital letters $R, S, T, \ldots$ are often used to denote sets, and lowercase letters $a, b, x, y, \ldots$ usually represent elements of sets. Throughout this book, $\mathbb{R}$ denotes the set of real numbers and $\mathbb{Z}$ denotes the set of integers.

Two sets S and T are **equal,** denoted by $S = T$, if S and T contain exactly the same elements. We write $S \neq T$ if S and T are not equal. Additional notation and terminology are listed in the following chart.

Notation or terminology	Meaning	Illustrations
$a \in S$	a is an element of S	$3 \in \mathbb{Z}$
$a \notin S$	a is not an element of S	$\frac{3}{5} \notin \mathbb{Z}$
S is a **subset** of T	Every element of S is an element of T	$\mathbb{Z}$ is a subset of $\mathbb{R}$
Constant	A letter or symbol that represents a *specific* element of a set	$5, -\sqrt{2}, \pi$
Variable	A letter or symbol that represents *any* element of a set	Let x denote any real number

We usually use letters near the end of the alphabet, such as x, y, and z, for variables, and letters near the beginning of the alphabet, such as a, b, and c, for constants. Throughout this text, unless otherwise specified, variables represent real numbers.

If the elements of a set S have a certain property, we sometimes write $S = \{x: \}$ and state the property describing the variable x in the space after the colon. The expression involving the braces and colon is read "the set of all x such that . . . ," where we complete the phrase by stating the desired property. For example, $\{x: x > 3\}$ is read "the set of all x such that x is greater than 3."

$\{x \mid x > 3\}$ is an equivalent notation.

For finite sets, we sometimes list all the elements of the set within braces. Thus, if the set T consists of the first five positive integers, we may write $T = \{1, 2, 3, 4, 5\}$. When we describe sets in this way, the order used in listing the elements is irrelevant, so we could also write $T = \{1, 3, 2, 4, 5\}$, $T = \{4, 3, 2, 5, 1\}$, and so on.

If we begin with any collection of variables and real numbers, then an **algebraic expression** is the result obtained by applying additions, subtractions, multiplications, divisions, powers, or the taking of roots to this collection. If specific numbers are substituted for the variables in an algebraic expression, the resulting number is called the **value** of the expression for these numbers. The **domain** of an algebraic expression consists of all real numbers that may represent the variables. Thus, unless otherwise specified, *we assume that the domain consists of the real numbers that, when substituted for the variables, do not make the expression meaningless, in the sense that denominators cannot equal zero and roots always exist.* Two illustrations are given in the following chart.

Algebraic Expressions

Illustration	Domain	Typical value
$x^3 - 5x + \dfrac{6}{\sqrt{x}}$	all $x > 0$	At $x = 4$: $4^3 - 5(4) + \dfrac{6}{\sqrt{4}} = 64 - 20 + 3 = 47$
$\dfrac{2xy + (3/x^2)}{\sqrt[3]{y} - 1}$	all $x \neq 0$ and all $y \neq 1$	At $x = 1$ and $y = 9$: $\dfrac{2(1)(9) + (3/1^2)}{\sqrt[3]{9 - 1}} = \dfrac{18 + 3}{\sqrt[3]{8}} = \dfrac{21}{2}$

If x is a variable, then a **monomial** in x is an expression of the form ax^n, where a is a real number and n is a nonnegative integer. A **binomial** is a sum of two monomials, and a **trinomial** is a sum of three monomials. A *polynomial in x* is a sum of any number of monomials in x. Another way of stating this is as follows.

Definition of Polynomial

A **polynomial in x** is a sum of the form

$$a_nx^n + a_{n-1}x^{n-1} + \cdots + a_1x + a_0,$$

where n is a nonnegative integer and each coefficient a_k is a real number. If $a_n \neq 0$, then the polynomial is said to have **degree n.**

Each expression a_kx^k in the sum is a **term** of the polynomial. If a coefficient a_k is zero, we usually delete the term a_kx^k. The coefficient a_k of the highest power of x is called the **leading coefficient** of the polynomial.

The following chart contains specific illustrations of polynomials.

Polynomials

Example	Leading coefficient	Degree
$3x^4 + 5x^3 + (-7)x + 4$	3	4
$x^8 + 9x^2 + (-2)x$	1	8
$-5x^2 + 1$	-5	2
$7x + 2$	7	1
8	8	0

By definition, two polynomials are **equal** if and only if they have the same degree and the coefficients of like powers of x are equal. If all the coefficients of a polynomial are zero, it is called the **zero polynomial** and is denoted by 0. However, by convention, the degree of the zero polynomial is *not* zero but, instead, is undefined. If c is a *nonzero real number,* then c is a polynomial of degree 0. Such polynomials (together with the zero polynomial) are **constant polynomials.**

If a coefficient of a polynomial is negative, we usually use a minus sign between appropriate terms. To illustrate,

$$3x^2 + (-5)x + (-7) = 3x^2 - 5x - 7.$$

We may also consider polynomials in variables other than x. For example, $\frac{2}{5}z^2 - 3z^7 + 8 - \sqrt{5}z^4$ is a polynomial in z of degree 7. We often arrange the terms of a polynomial in order of decreasing powers of the variable; thus, we write

$$\tfrac{2}{5}z^2 - 3z^7 + 8 - \sqrt{5}z^4 = -3z^7 - \sqrt{5}z^4 + \tfrac{2}{5}z^2 + 8.$$

We may regard a polynomial in x as an algebraic expression obtained by employing a finite number of additions, subtractions, and multiplications involving x. If an algebraic expression contains divisions or roots involving a variable x, then it is not a polynomial in x.

ILLUSTRATION **Nonpolynomials**

- $\dfrac{1}{x} + 3x$
- $\dfrac{x - 5}{x^2 + 2}$
- $3x^2 + \sqrt{x} - 2$

Since polynomials represent real numbers, we may use the properties described in Section 1.1. In particular, if additions, subtractions, and multiplications are carried out with polynomials, we may simplify the results by using properties of real numbers, as demonstrated in the following example.

EXAMPLE 1 **Multiplying polynomials**

Find the product: $(x^2 + 5x - 4)(2x^3 + 3x - 1)$

▶ SOLUTION

We begin by using a distributive property, treating the polynomial $2x^3 + 3x - 1$ as a single real number:

$$\begin{aligned}&(x^2 + 5x - 4)(2x^3 + 3x - 1)\\ &\quad = x^2(2x^3 + 3x - 1) + 5x(2x^3 + 3x - 1) - 4(2x^3 + 3x - 1)\end{aligned}$$

We next use another distributive property three times and simplify the result, obtaining

$$\begin{aligned}&(x^2 + 5x - 4)(2x^3 + 3x - 1)\\ &\quad = 2x^5 + 3x^3 - x^2 + 10x^4 + 15x^2 - 5x - 8x^3 - 12x + 4\\ &\quad = 2x^5 + 10x^4 - 5x^3 + 14x^2 - 17x + 4.\end{aligned}$$

Calculator check for Example 1: Store 7 in X and show that the original expression and the final expression both equal 56,480.

Note that the three monomials in the first polynomial were multiplied by each of the three monomials in the second polynomial, giving us a total of nine terms.

We may consider polynomials in more than one variable. For example, a polynomial in *two* variables, x and y, is a finite sum of terms, each of the form ax^my^k for some real number a and nonnegative integers m and k. An example is

$$3x^4y + 2x^3y^5 + 7x^2 - 4xy + 8y - 5.$$

Other polynomials may involve three variables—such as x, y, z—or, for that matter, *any* number of variables. Addition, subtraction, and multiplication are performed using properties of real numbers, just as for polynomials in one variable.

The products listed in the next chart occur so frequently that they deserve special attention. You can check the validity of each formula by multiplication. In (2) and (3), we use either the top sign on both sides or the bottom sign on both sides. Thus, (2) is actually *two* formulas:

$$(x + y)^2 = x^2 + 2xy + y^2 \qquad \text{and} \qquad (x - y)^2 = x^2 - 2xy + y^2$$

Similarly, (3) represents two formulas.

▶ **Tutorial available at www.thomsonedu.com/login**

Product Formulas

Formula	Illustration
(1) $(x + y)(x - y) = x^2 - y^2$	$(2a + 3)(2a - 3) = (2a)^2 - 3^2 = 4a^2 - 9$
(2) $(x \pm y)^2 = x^2 \pm 2xy + y^2$	$(2a - 3)^2 = (2a)^2 - 2(2a)(3) + (3)^2$ $= 4a^2 - 12a + 9$
(3) $(x \pm y)^3 = x^3 \pm 3x^2y + 3xy^2 \pm y^3$	$(2a + 3)^3 = (2a)^3 + 3(2a)^2(3) + 3(2a)(3)^2 + (3)^3$ $= 8a^3 + 36a^2 + 54a + 27$

If a polynomial is a product of other polynomials, then each polynomial in the product is a **factor** of the original polynomial. **Factoring** is the process of expressing a sum of terms as a product. For example, since $x^2 - 9 = (x + 3)(x - 3)$, the polynomials $x + 3$ and $x - 3$ are factors of $x^2 - 9$.

Factoring is an important process in mathematics, since it may be used to reduce the study of a complicated expression to the study of several simpler expressions. For example, properties of the polynomial $x^2 - 9$ can be determined by examining the factors $x + 3$ and $x - 3$. As we shall see later in this chapter, another important use for factoring is in finding solutions of equations.

We shall be interested primarily in **nontrivial factors** of polynomials—that is, factors that contain polynomials of positive degree. However, if the coefficients are restricted to *integers,* then we usually remove a common integral factor from each term of the polynomial. For example,

$$4x^2y + 8z^3 = 4(x^2y + 2z^3).$$

A polynomial with coefficients in some set S of numbers is **prime,** or **irreducible** over S, if it cannot be written as a product of two polynomials of positive degree with coefficients in S. A polynomial may be irreducible over one set S but not over another. For example, $x^2 - 2$ is irreducible over the rational numbers, since it cannot be expressed as a product of two polynomials of positive degree that have *rational* coefficients. However, $x^2 - 2$ is *not* irreducible over the real numbers, since we can write

$$x^2 - 2 = \left(x + \sqrt{2}\right)\left(x - \sqrt{2}\right).$$

Similarly, $x^2 + 1$ is irreducible over the real numbers, but, as we shall see in Section 1.5, not over the complex numbers.

Every polynomial $ax + b$ of degree 1 is irreducible.

Before we factor a polynomial, we must specify the number system (or set) from which the coefficients of the factors are to be chosen. In this chapter we shall use the rule that *if a polynomial has integral coefficients, then the factors should be polynomials with integral coefficients.* To **factor a polynomial** means to express it as a product of irreducible polynomials.

The **greatest common factor (gcf)** of an expression is the product of the factors that appear in each term, with each of these factors raised to the smallest nonzero exponent appearing in any term. In factoring polynomials, it is advisable to first factor out the gcf, as shown in the last two parts of the following illustration. In the first two parts, we have factored trinomials by the **method of trial and error**.

ILLUSTRATION Factored Polynomials

- $x^2 + x - 6 = (x + 3)(x - 2)$
- $6x^2 - 7x - 3 = (2x - 3)(3x + 1)$
- $12x^3 - 36x^2y + 27xy^2 = 3x(4x^2 - 12xy + 9y^2)$
  $= 3x(2x - 3y)(2x - 3y) = 3x(2x - 3y)^2$
- $4x^4y - 11x^3y^2 + 6x^2y^3 = x^2y(4x^2 - 11xy + 6y^2)$
  $= x^2y(4x - 3y)(x - 2y)$

It is usually difficult to factor polynomials of degree greater than 2. In simple cases, the following factoring formulas may be useful. It can be shown that the factors $x^2 + xy + y^2$ and $x^2 - xy + y^2$ in the difference and sum of two cubes, respectively, are irreducible over the real numbers.

Factoring Formulas

Formula	Illustration
(1) Difference of two squares: $x^2 - y^2 = (x + y)(x - y)$	$9a^2 - 16 = (3a)^2 - (4)^2 = (3a + 4)(3a - 4)$
(2) Difference of two cubes: $x^3 - y^3 = (x - y)(x^2 + xy + y^2)$	$8a^3 - 27 = (2a)^3 - (3)^3$ $= (2a - 3)[(2a)^2 + (2a)(3) + (3)^2]$ $= (2a - 3)(4a^2 + 6a + 9)$
(3) Sum of two cubes: $x^3 + y^3 = (x + y)(x^2 - xy + y^2)$	$a^3 + 64b^3 = a^3 + (4b)^3$ $= (a + 4b)[a^2 - a(4b) + (4b)^2]$ $= (a + 4b)(a^2 - 4ab + 16b^2)$

Checking a Factoring Result

TI-83/4 Plus | **TI-86**

We can check a factoring result by multiplying the proposed answer and comparing it to the original expression. Here we will substitute values for the variables and evaluate the original expression and the proposed answer.

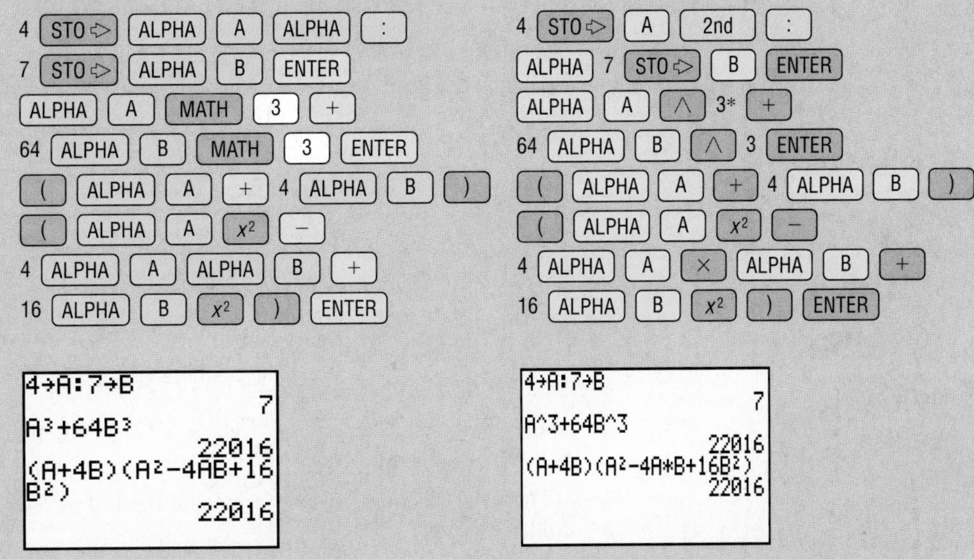

*There is no special cube feature for the TI-86.

Don't pick values such as 0, 1, or 2 for A and B—it's too easy to get the same value for the original expression and the proposed answer. For example, if we substituted 1 for A and 0 for B and incorrectly factored $A^3 + 64B^3$ as $(A + 4B)(A^2 + 16B^2)$, both expressions would equal 1 and we would be misled in thinking that we had correctly factored $A^3 + 64B^3$.

If a sum contains four or more terms, it may be possible to group the terms in a suitable manner and then find a factorization by using distributive properties. This technique, called **factoring by grouping,** is illustrated in the next example.

EXAMPLE 2 **Factoring by grouping**

Factor:

(a) $4ac + 2bc - 2ad - bd$ (b) $3x^3 + 2x^2 - 12x - 8$

(c) $x^2 - 16y^2 + 10x + 25$

SOLUTION

(a) We group the first two terms and the last two terms and then proceed as follows:

$$\begin{aligned} 4ac + 2bc - 2ad - bd &= (4ac + 2bc) - (2ad + bd) \\ &= 2c(2a + b) - d(2a + b) \end{aligned}$$

At this stage we have not factored the given expression because the right-hand side has the form

$$2ck - dk \quad \text{with } k = 2a + b.$$

However, if we factor out k, then

$$2ck - dk = (2c - d)k = (2c - d)(2a + b).$$

Hence,

$$\begin{aligned} 4ac + 2bc - 2ad - bd &= 2c(2a + b) - d(2a + b) \\ &= (2c - d)(2a + b). \end{aligned}$$

Note that if we factor $2ck - dk$ as $k(2c - d)$, then the last expression is $(2a + b)(2c - d)$.

(b) We group the first two terms and the last two terms and then proceed as follows:

$$\begin{aligned} 3x^3 + 2x^2 - 12x - 8 &= (3x^3 + 2x^2) - (12x + 8) \\ &= x^2(3x + 2) - 4(3x + 2) \\ &= (x^2 - 4)(3x + 2) \end{aligned}$$

Finally, using the difference of two squares formula for $x^2 - 4$, we obtain the factorization:

$$3x^3 + 2x^2 - 12x - 8 = (x + 2)(x - 2)(3x + 2)$$

(c) First we rearrange and group terms, and then we apply the difference of two squares formula, as follows:

$$\begin{aligned} x^2 - 16y^2 + 10x + 25 &= (x^2 + 10x + 25) - 16y^2 \\ &= (x + 5)^2 - (4y)^2 \\ &= [(x + 5) + 4y][(x + 5) - 4y] \\ &= (x + 4y + 5)(x - 4y + 5) \end{aligned}$$

A **fractional expression** is a quotient of two algebraic expressions. As a special case, a **rational expression** is a quotient p/q of two *polynomials* p and q. Since division by zero is not allowed, the domain of p/q consists of all real numbers except those that make the denominator zero. Two illustrations are given in the chart.

Rational Expressions

Quotient	Denominator is zero if	Domain
$\dfrac{6x^2 - 5x + 4}{x^2 - 9}$	$x = \pm 3$	All $x \neq \pm 3$
$\dfrac{x^3 - 3x^2y + 4y^2}{y - x^3}$	$y = x^3$	All x and y such that $y \neq x^3$

In most of our work we will be concerned with rational expressions in which both numerator and denominator are polynomials in only one variable.

Since the variables in a rational expression represent real numbers, we may use the properties of quotients in Section 1.1, replacing the letters a, b, c, and d with polynomials. The following property is of particular importance, where $bd \neq 0$:

$$\frac{ad}{bd} = \frac{a}{b} \cdot \frac{d}{d} = \frac{a}{b} \cdot 1 = \frac{a}{b}$$

We sometimes describe this simplification process by saying that *a common nonzero factor in the numerator and denominator of a quotient may be canceled.*

A rational expression is *simplified,* or *reduced to lowest terms,* if the numerator and denominator have no common polynomial factors of positive degree and no common integral factors greater than 1. To simplify a rational expression, we factor both the numerator and the denominator into prime factors and then, assuming the factors in the denominator are not zero, cancel common factors, as in the following example.

EXAMPLE 3 Simplifying products and quotients of rational expressions

Perform the indicated operation and simplify:

(a) $\dfrac{x^2 - 6x + 9}{x^2 - 1} \cdot \dfrac{2x - 2}{x - 3}$ (b) $\dfrac{x + 2}{2x - 3} \div \dfrac{x^2 - 4}{2x^2 - 3x}$

SOLUTION

(a) $\dfrac{x^2 - 6x + 9}{x^2 - 1} \cdot \dfrac{2x - 2}{x - 3} = \dfrac{(x^2 - 6x + 9)(2x - 2)}{(x^2 - 1)(x - 3)}$ property of quotients

$= \dfrac{(x - 3)^{2} \cdot 2(x - 1)}{(x + 1)(x - 1)(x - 3)}$ factor all polynomials

if $x \neq 3, x \neq 1$

$= \dfrac{2(x - 3)}{x + 1}$ cancel common factors

(continued)

Tutorial available at www.thomsonedu.com/login

(b) $\dfrac{x+2}{2x-3} \div \dfrac{x^2-4}{2x^2-3x} = \dfrac{x+2}{2x-3} \cdot \dfrac{2x^2-3x}{x^2-4}$ property of quotients

$= \dfrac{(x+2)x(2x-3)}{(2x-3)(x+2)(x-2)}$ property of quotients; factor all polynomials

if $x \neq -2, x \neq 3/2$

$\downarrow$
$= \dfrac{x}{x-2}$ cancel common factors

To add or subtract two rational expressions, we usually find a *common denominator* and use the following properties of quotients:

$$\frac{a}{d} + \frac{c}{d} = \frac{a+c}{d} \quad \text{and} \quad \frac{a}{d} - \frac{c}{d} = \frac{a-c}{d}$$

If the denominators of the expressions are not the same, we may obtain a common denominator by multiplying the numerator and denominator of each fraction by a suitable expression. We usually use the ***least* common denominator (lcd)** of the two quotients. To find the lcd, we factor each denominator into primes and then form the product of the different prime factors, using the *largest* exponent that appears with each prime factor.

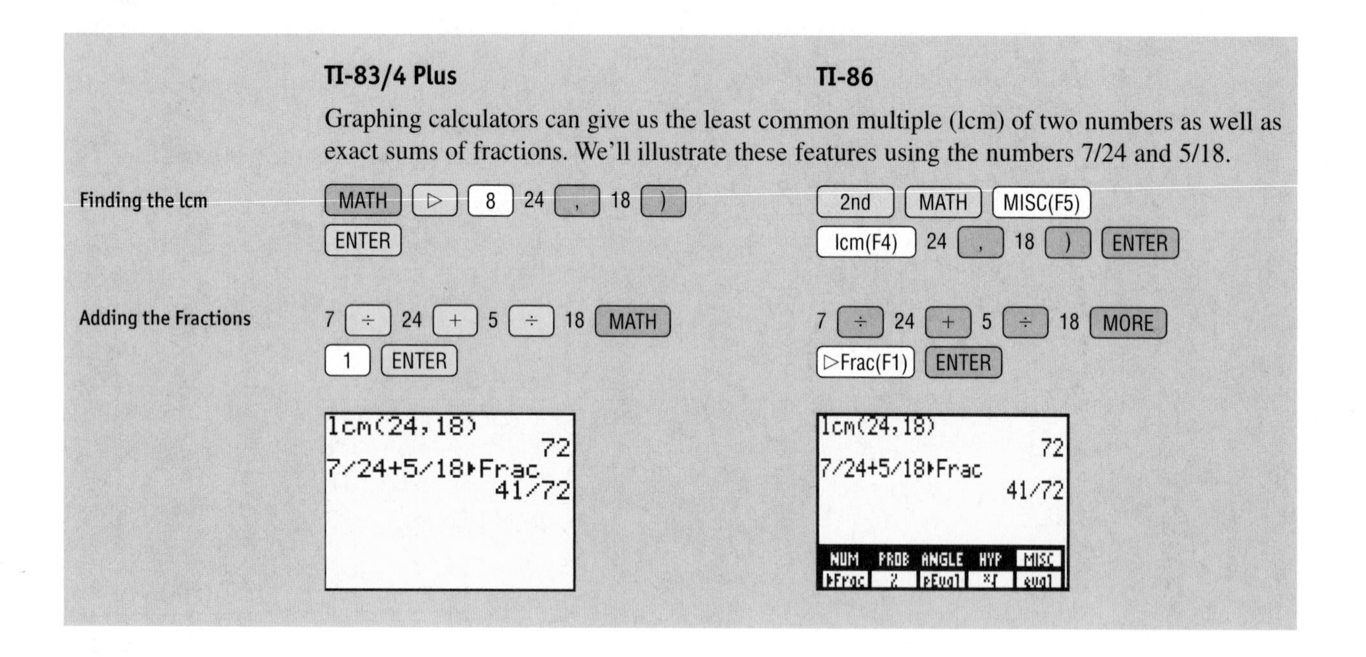

EXAMPLE 4 **Simplifying sums and differences of rational expressions**

Perform the operations and simplify:

$$\frac{6}{x(3x-2)}+\frac{5}{3x-2}-\frac{2}{x^2}$$

▶ **SOLUTION** The denominators are already in factored form. The lcd is $x^2(3x-2)$. To obtain three quotients having the denominator $x^2(3x-2)$, we multiply the numerator and denominator of the first quotient by x, those of the second by x^2, and those of the third by $3x-2$, which gives us

$$\begin{aligned}\frac{6}{x(3x-2)}+\frac{5}{3x-2}-\frac{2}{x^2} &= \frac{6}{x(3x-2)}\cdot\frac{x}{x}+\frac{5}{3x-2}\cdot\frac{x^2}{x^2}-\frac{2}{x^2}\cdot\frac{3x-2}{3x-2}\\ &= \frac{6x}{x^2(3x-2)}+\frac{5x^2}{x^2(3x-2)}-\frac{2(3x-2)}{x^2(3x-2)}\\ &= \frac{6x+5x^2-2(3x-2)}{x^2(3x-2)}\\ &= \frac{5x^2+4}{x^2(3x-2)}.\end{aligned}$$

TI-83/4 Plus **TI-86**

Creating a Table

Let's check the simplification of Example 4 by creating and comparing tables of values for the original expression and the final expression. We'll assign these expressions to Y_1 and Y_2 (later called *functions*) and compare their values for $x = 1, 2, 3, \ldots$.

Make Y assignments.

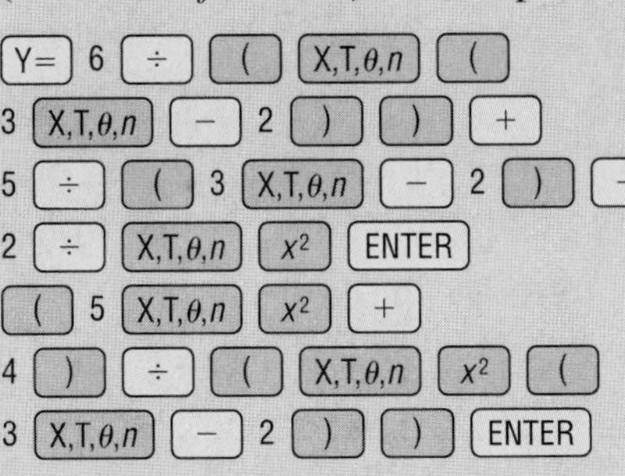

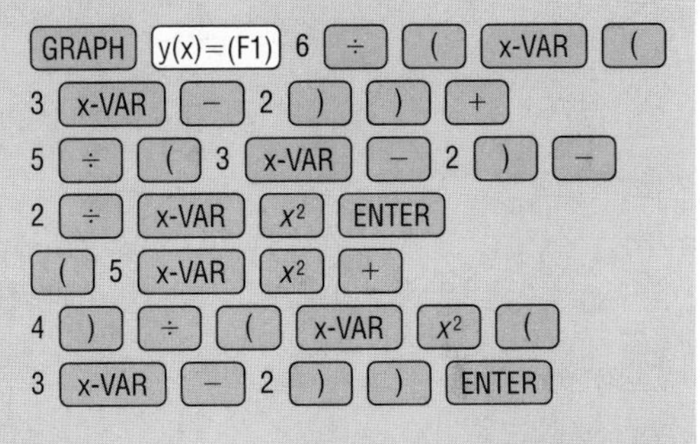

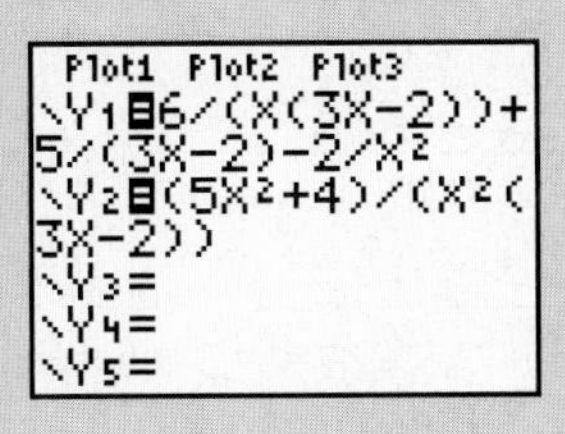

(continued)

▶ **Tutorial available at www.thomsonedu.com/login**

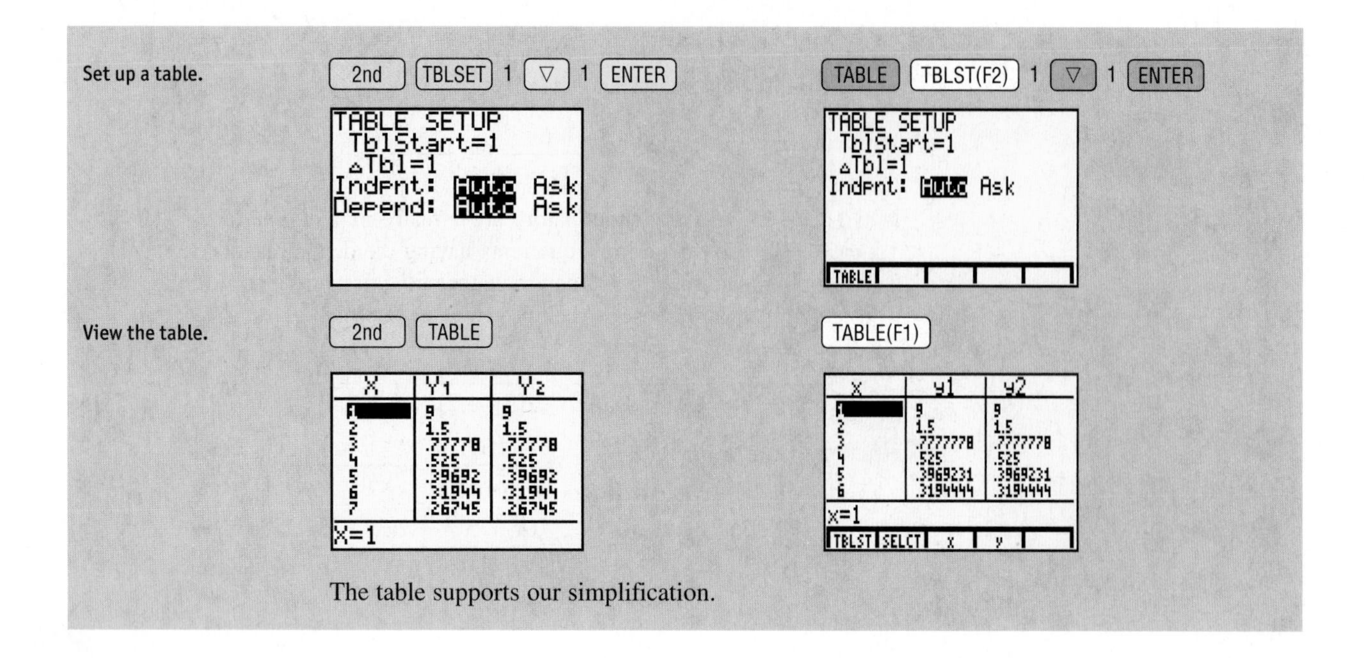

The table supports our simplification.

A **complex fraction** is a quotient in which the numerator and/or the denominator is a fractional expression. Certain problems in calculus require simplifying complex fractions of the type given in the next example.

EXAMPLE 5 Simplifying a complex fraction

Simplify the complex fraction:

$$\frac{\dfrac{2}{x+3} - \dfrac{2}{a+3}}{x-a}$$

▶ **SOLUTION** We change the numerator of the given expression into a single quotient and then use a property for simplifying quotients:

$$\frac{\dfrac{2}{x+3} - \dfrac{2}{a+3}}{x-a} = \frac{\dfrac{2(a+3)-2(x+3)}{(x+3)(a+3)}}{x-a} \quad \text{combine fractions in the numerator}$$

$$= \frac{2a-2x}{(x+3)(a+3)} \cdot \frac{1}{x-a} \quad \text{simplify; property of quotients}$$

$$= \frac{2(a-x)}{(x+3)(a+3)(x-a)} \quad \text{factor } 2a-2x\text{; property of quotients}$$

$$\overset{\text{if } x \neq a}{=} -\frac{2}{(x+3)(a+3)} \quad \text{replace } \frac{a-x}{x-a} \text{ with } -1$$

▶ **Tutorial available at www.thomsonedu.com/login**

An alternative method is to multiply the numerator and denominator of the given expression by $(x + 3)(a + 3)$, the lcd of the numerator and denominator, and then simplify the result.

Some quotients that are not rational expressions contain denominators of the form $a + \sqrt{b}$ or $\sqrt{a} + \sqrt{b}$; as in the next example, these quotients can be simplified by multiplying the numerator and denominator by the **conjugate** $a - \sqrt{b}$ or $\sqrt{a} - \sqrt{b}$, respectively. Of course, if $a - \sqrt{b}$ appears, multiply by $a + \sqrt{b}$ instead.

In Example 4 of Section 1.2, we rationalized denominators. In calculus it is sometimes necessary to rationalize the *numerator* of a quotient, as shown in the following example.

EXAMPLE 6 Rationalizing a numerator

If $h \neq 0$, rationalize the numerator of

$$\frac{\sqrt{x + h} - \sqrt{x}}{h}.$$

SOLUTION

$$\frac{\sqrt{x + h} - \sqrt{x}}{h} = \frac{\sqrt{x + h} - \sqrt{x}}{h} \cdot \frac{\sqrt{x + h} + \sqrt{x}}{\sqrt{x + h} + \sqrt{x}}$$ multiply numerator and denominator by the conjugate of $\sqrt{x + h} - \sqrt{x}$

$$= \frac{(\sqrt{x + h})^2 - (\sqrt{x})^2}{h(\sqrt{x + h} + \sqrt{x})}$$ property of quotients and difference of squares

$$= \frac{(x + h) - x}{h(\sqrt{x + h} + \sqrt{x})}$$ law of radicals

$$= \frac{\cancel{h}}{\cancel{h}(\sqrt{x + h} + \sqrt{x})}$$ simplify

$$= \frac{1}{\sqrt{x + h} + \sqrt{x}}$$ cancel $h \neq 0$

It may seem as though we have accomplished very little, since radicals occur in the denominator. In calculus, however, it is of interest to determine what is true if h is very close to zero. Note that if we use the *given* expression we obtain the following:

$$\text{If} \quad h \approx 0, \quad \text{then} \quad \frac{\sqrt{x + h} - \sqrt{x}}{h} \approx \frac{\sqrt{x + 0} - \sqrt{x}}{0} = \frac{0}{0},$$

(continued)

Tutorial available at www.thomsonedu.com/login

a meaningless expression. If we use the *rationalized* form, however, we obtain the following information:

$$\text{If} \quad h \approx 0, \quad \text{then} \quad \frac{\sqrt{x+h}-\sqrt{x}}{h} = \frac{1}{\sqrt{x+h}+\sqrt{x}}$$

$$\approx \frac{1}{\sqrt{x}+\sqrt{x}} = \frac{1}{2\sqrt{x}}.$$

Problems of the type given in the next example also occur in calculus.

EXAMPLE 7 Simplifying a fractional expression

Simplify:

$$\frac{3x^2(2x+5)^{1/2} - x^3\left(\frac{1}{2}\right)(2x+5)^{-1/2}(2)}{[(2x+5)^{1/2}]^2}$$

SOLUTION One way to simplify the expression is as follows:

$$\frac{3x^2(2x+5)^{1/2} - x^3\left(\frac{1}{2}\right)(2x+5)^{-1/2}(2)}{[(2x+5)^{1/2}]^2}$$

$$= \frac{3x^2(2x+5)^{1/2} - \dfrac{x^3}{(2x+5)^{1/2}}}{2x+5} \qquad \text{definition of negative exponents}$$

$$= \frac{\dfrac{3x^2(2x+5) - x^3}{(2x+5)^{1/2}}}{2x+5} \qquad \text{combine terms in numerator}$$

$$= \frac{6x^3+15x^2-x^3}{(2x+5)^{1/2}} \cdot \frac{1}{2x+5} \qquad \text{property of quotients}$$

$$= \frac{5x^3+15x^2}{(2x+5)^{3/2}} \qquad \text{simplify}$$

$$= \frac{5x^2(x+3)}{(2x+5)^{3/2}} \qquad \text{factor numerator}$$

An alternative simplification is to eliminate the negative power, $-\frac{1}{2}$, in the given expression, as follows:

$$\frac{3x^2(2x+5)^{1/2} - x^3\left(\frac{1}{2}\right)(2x+5)^{-1/2}(2)}{[(2x+5)^{1/2}]^2} \cdot \frac{(2x+5)^{1/2}}{(2x+5)^{1/2}} \qquad \text{multiply numerator and denominator by } (2x+5)^{1/2}$$

$$= \frac{3x^2(2x+5) - x^3}{(2x+5)(2x+5)^{1/2}} \qquad \text{property of quotients and law of exponents}$$

The remainder of the simplification is similar.

A third method of simplification is to first factor out the gcf. In this case, the common factors are x and $(2x + 5)$, and the smallest exponents are 2 and

$-\frac{1}{2}$, respectively. Thus, the gcf is $x^2(2x+5)^{-1/2}$, and we factor the numerator and simplify as follows:

$$\frac{x^2(2x+5)^{-1/2}[3(2x+5)^1 - x]}{(2x+5)^1} = \frac{x^2(5x+15)}{(2x+5)^{3/2}} = \frac{5x^2(x+3)}{(2x+5)^{3/2}}$$

One of the problems in calculus is determining the values of x that make the numerator equal to zero. The simplified form helps us answer this question with relative ease—the values are 0 and -3.

1.3 Exercises

Exer. 1–12: Express as a polynomial.

1 $(2u+3)(u-4)+4u(u-2)$

2 $(3u-1)(u+2)+7u(u+1)$

3 $\dfrac{8x^2y^3 - 10x^3y}{2x^2y}$

4 $\dfrac{6x^2yz^3 - xy^2z}{xyz}$

5 $(2x+3y)(2x-3y)$

6 $(5x+4y)(5x-4y)$

7 $(3x+2y)^2$

8 $(5x-4y)^2$

9 $(\sqrt{x}+\sqrt{y})(\sqrt{x}-\sqrt{y})$

10 $(\sqrt{x}+\sqrt{y})^2(\sqrt{x}-\sqrt{y})^2$

11 $(x-2y)^3$

12 $(x+3y)^3$

Exer. 13–30: Factor the polynomial.

13 $8x^2-53x-21$

14 $7x^2+10x-8$

15 x^2+3x+4

16 $3x^2-4x+2$

17 $4x^2-20x+25$

18 $9x^2+24x+16$

19 x^4-4x^2

20 x^3-25x

21 $64x^3-y^6$

22 x^6-27y^3

23 $64x^3+27$

24 $343x^3+y^9$

25 $3x^3+3x^2-27x-27$

26 $5x^3+10x^2-20x-40$

27 a^6-b^6

28 x^8-16

29 $x^2+4x+4-9y^2$

30 x^2-4y^2-6x+9

Exer. 31–60: Simplify the expression.

▶ 31 $\dfrac{y^2-25}{y^3-125}$

32 $\dfrac{12+r-r^2}{r^3+3r^2}$

33 $\dfrac{9x^2-4}{3x^2-5x+2}\cdot\dfrac{9x^4-6x^3+4x^2}{27x^4+8x}$

34 $\dfrac{5a^2+12a+4}{a^4-16}\div\dfrac{25a^2+20a+4}{a^2-2a}$

35 $\dfrac{2}{3s+1}-\dfrac{9}{(3s+1)^2}$

36 $\dfrac{4}{(5s-2)^2}+\dfrac{s}{5s-2}$

37 $\dfrac{2}{x}+\dfrac{3x+1}{x^2}-\dfrac{x-2}{x^3}$

38 $\dfrac{5}{x}-\dfrac{2x-1}{x^2}+\dfrac{x+5}{x^3}$

39 $\dfrac{3t}{t+2}+\dfrac{5t}{t-2}-\dfrac{40}{t^2-4}$

40 $\dfrac{t}{t+3}+\dfrac{4t}{t-3}-\dfrac{18}{t^2-9}$

41 $\dfrac{4x}{3x-4}+\dfrac{8}{3x^2-4x}+\dfrac{2}{x}$

42 $\dfrac{12x}{2x+1}-\dfrac{3}{2x^2+x}+\dfrac{5}{x}$

43 $\dfrac{2x}{x+2}-\dfrac{8}{x^2+2x}+\dfrac{3}{x}$

44 $\dfrac{5x}{2x+3}-\dfrac{6}{2x^2+3x}+\dfrac{2}{x}$

45 $3+\dfrac{5}{u}+\dfrac{2u}{3u+1}$

46 $4+\dfrac{2}{u}-\dfrac{3u}{u+5}$

47 $\dfrac{2x+1}{x^2+4x+4}-\dfrac{6x}{x^2-4}+\dfrac{3}{x-2}$

48 $\dfrac{2x+6}{x^2+6x+9}+\dfrac{5x}{x^2-9}+\dfrac{7}{x-3}$

▶ **Tutorial available at www.thomsonedu.com/login**

49 $\dfrac{\dfrac{b}{a}-\dfrac{a}{b}}{\dfrac{1}{a}-\dfrac{1}{b}}$

50 $\dfrac{\dfrac{x}{y^2}-\dfrac{y}{x^2}}{\dfrac{1}{y^2}-\dfrac{1}{x^2}}$

51 $\dfrac{y^{-1}+x^{-1}}{(xy)^{-1}}$

52 $\dfrac{y^{-2}-x^{-2}}{y^{-2}+x^{-2}}$

53 $\dfrac{\dfrac{r}{s}+\dfrac{s}{r}}{\dfrac{r^2}{s^2}-\dfrac{s^2}{r^2}}$

54 $\dfrac{\dfrac{3}{w}-\dfrac{6}{2w+1}}{\dfrac{5}{w}+\dfrac{8}{2w+1}}$

55 $\dfrac{(x+h)^2-3(x+h)-(x^2-3x)}{h}$

56 $\dfrac{(x+h)^3+5(x+h)-(x^3+5x)}{h}$

57 $\dfrac{\dfrac{3}{x-1}-\dfrac{3}{a-1}}{x-a}$

58 $\dfrac{\dfrac{x+2}{x}-\dfrac{a+2}{a}}{x-a}$

59 $\dfrac{\dfrac{1}{(x+h)^3}-\dfrac{1}{x^3}}{h}$

60 $\dfrac{\dfrac{1}{x+h}-\dfrac{1}{x}}{h}$

Exer. 61–64: Rationalize the denominator.

61 $\dfrac{\sqrt{t}+5}{\sqrt{t}-5}$

62 $\dfrac{16x^2-y^2}{2\sqrt{x}-\sqrt{y}}$

63 $\dfrac{1}{\sqrt[3]{a}-\sqrt[3]{b}}$ (*Hint:* Multiply numerator and denominator by $\sqrt[3]{a^2}+\sqrt[3]{ab}+\sqrt[3]{b^2}$.)

64 $\dfrac{1}{\sqrt[3]{x}+\sqrt[3]{y}}$

Exer. 65–68: Rationalize the numerator.

65 $\dfrac{\sqrt{a}-\sqrt{b}}{a^2-b^2}$

66 $\dfrac{\sqrt{b}+\sqrt{c}}{b^2-c^2}$

67 $\dfrac{\sqrt{2(x+h)+1}-\sqrt{2x+1}}{h}$

68 $\dfrac{\sqrt{x}-\sqrt{x+h}}{h\sqrt{x}\,\sqrt{x+h}}$

Exer. 69–72: Express as a sum of terms of the form ax^r, where r is a rational number.

69 $\dfrac{4x^2-x+5}{x^{2/3}}$

70 $\dfrac{x^2+4x-6}{\sqrt{x}}$

71 $\dfrac{(x^2+2)^2}{x^5}$

72 $\dfrac{(\sqrt{x}-3)^2}{x^3}$

Exer. 73–76: Express as a quotient.

73 $x^{-3}+x^2$

74 $x^{-4}-x$

75 $x^{-1/2}-x^{3/2}$

76 $x^{-2/3}+x^{7/3}$

Exer. 77–90: Simplify the expression.

77 $(2x^2-3x+1)(4)(3x+2)^3(3)+(3x+2)^4(4x-3)$

78 $(6x-5)^3(2)(x^2+4)(2x)+(x^2+4)^2(3)(6x-5)^2(6)$

79 $(x^2-4)^{1/2}(3)(2x+1)^2(2)+(2x+1)^3\left(\frac{1}{2}\right)(x^2-4)^{-1/2}(2x)$

80 $(3x+2)^{1/3}(2)(4x-5)(4)+(4x-5)^2\left(\frac{1}{3}\right)(3x+2)^{-2/3}(3)$

81 $(3x+1)^6\left(\frac{1}{2}\right)(2x-5)^{-1/2}(2)+(2x-5)^{1/2}(6)(3x+1)^5(3)$

82 $(x^2+9)^4\left(-\frac{1}{3}\right)(x+6)^{-4/3}+(x+6)^{-1/3}(4)(x^2+9)^3(2x)$

83 $\dfrac{(6x+1)^3(27x^2+2)-(9x^3+2x)(3)(6x+1)^2(6)}{(6x+1)^6}$

84 $\dfrac{(x^2-1)^4(2x)-x^2(4)(x^2-1)^3(2x)}{(x^2-1)^8}$

85 $\dfrac{(x^2+2)^3(2x)-x^2(3)(x^2+2)^2(2x)}{[(x^2+2)^3]^2}$

86 $\dfrac{(x^2 - 5)^4(3x^2) - x^3(4)(x^2 - 5)^3(2x)}{[(x^2 - 5)^4]^2}$

87 $\dfrac{(x^2 + 4)^{1/3}(3) - (3x)(\frac{1}{3})(x^2 + 4)^{-2/3}(2x)}{[(x^2 + 4)^{1/3}]^2}$

88 $\dfrac{(1 - x^2)^{1/2}(2x) - x^2(\frac{1}{2})(1 - x^2)^{-1/2}(-2x)}{[(1 - x^2)^{1/2}]^2}$

89 $\dfrac{(4x^2 + 9)^{1/2}(2) - (2x + 3)(\frac{1}{2})(4x^2 + 9)^{-1/2}(8x)}{[(4x^2 + 9)^{1/2}]^2}$

90
$$\frac{(3x + 2)^{1/2}(\frac{1}{3})(2x + 3)^{-2/3}(2) - (2x + 3)^{1/3}(\frac{1}{2})(3x + 2)^{-1/2}(3)}{[(3x + 2)^{1/2}]^2}$$

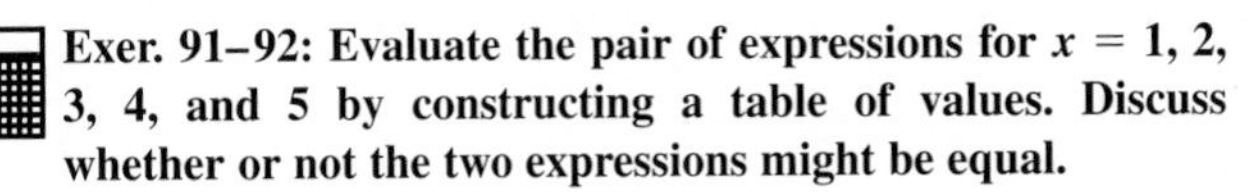

Exer. 91–92: Evaluate the pair of expressions for $x = 1, 2, 3, 4,$ and 5 by constructing a table of values. Discuss whether or not the two expressions might be equal.

91 $\dfrac{113x^3 + 280x^2 - 150x}{22x^3 + 77x^2 - 100x - 350}$, $\quad \dfrac{3x}{2x + 7} + \dfrac{4x^2}{1.1x^2 - 5}$

92 $\dfrac{20x^2 + 41x + 31}{10x^3 + 10x^2}$, $\quad \dfrac{1}{x} + \dfrac{1}{x + 1} + \dfrac{3.2}{x^2}$

Exer. 93–94: The ancient Greeks gave geometric proofs of the factoring formulas for the difference of two squares and the difference of two cubes. Establish the formula for the special case described.

93 Find the areas of regions I and II in the figure to establish the difference of two squares formula for the special case $x > y$.

Exercise 93

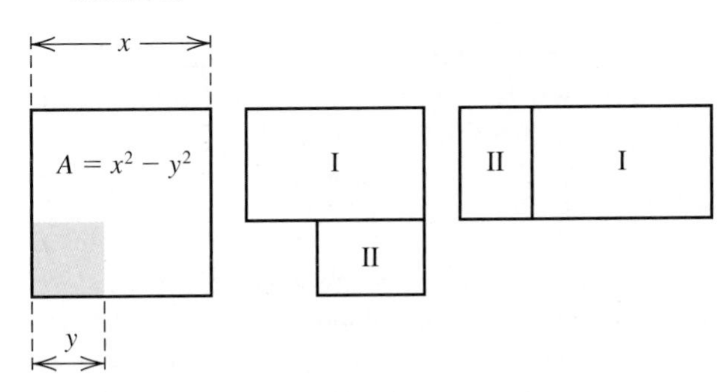

94 Find the volumes of boxes I, II, and III in the figure to establish the difference of two cubes formula for the special case $x > y$.

Exercise 94

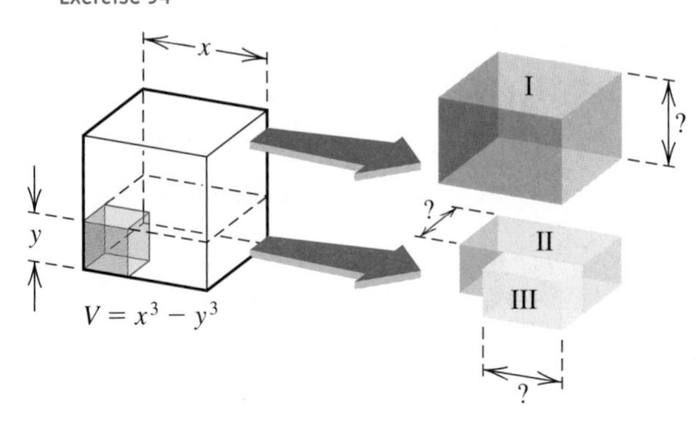

95 Calorie requirements The basal energy requirement for an individual indicates the minimum number of calories necessary to maintain essential life-sustaining processes such as circulation, regulation of body temperature, and respiration. Given a person's sex, weight w (in kilograms), height h (in centimeters), and age y (in years), we can estimate the basal energy requirement in calories using the following formulas, where C_f and C_m are the calories necessary for females and males, respectively:

$$C_f = 66.5 + 13.8w + 5h - 6.8y$$
$$C_m = 655 + 9.6w + 1.9h - 4.7y$$

(a) Determine the basal energy requirements first for a 25-year-old female weighing 59 kilograms who is 163 centimeters tall and then for a 55-year-old male weighing 75 kilograms who is 178 centimeters tall.

(b) Discuss why, in both formulas, the coefficient for y is negative but the other coefficients are positive.

1.4 Equations

An **equation** (or **equality**) is a statement that two quantities or expressions are equal. Equations are employed in every field that uses real numbers.

The following chart applies to a variable x, but any other variable may be considered. The abbreviations LS and RS in the second illustration stand for the equation's left side and right side, respectively.

Terminology	Definition	Illustration
Equation in x	A statement of equality involving one variable, x	$x^2 - 5 = 4x$
Solution, or **root,** of an equation in x	A number b that yields a true statement when substituted for x	5 is a solution of $x^2 - 5 = 4x$, since substitution gives us LS: $5^2 - 5 = 25 - 5 = 20$ and RS: $4 \cdot 5 = 20$, and $20 = 20$ is a true statement.
A number b **satisfies** an equation in x	b is a solution of the equation	5 satisfies $x^2 - 5 = 4x$.
Equivalent equations	Equations that have exactly the same solutions	$2x + 1 = 7$ $2x = 7 - 1$ $2x = 6$ $x = 3$
Solve an equation in x	Find all solutions of the equation	To solve $(x + 3)(x - 5) = 0$, set each factor equal to 0: $x + 3 = 0, x - 5 = 0$, obtaining the solutions -3 and 5.

An **algebraic equation** in x contains only algebraic expressions such as polynomials, rational expressions, radicals, and so on. An equation of this type is called a **conditional equation** if there are numbers in the domains of the expressions that are not solutions. For example, the equation $x^2 = 9$ is conditional, since the number $x = 4$ (and others) is not a solution. If *every* number in the domains of the expressions in an algebraic equation is a solution, the equation is called an **identity.**

We shall assume some experience in finding solutions of equations in one variable. To solve a **linear equation** $ax + b = 0$, where a and b are real numbers and $a \neq 0$, we subtract b from both sides and divide by a as follows:

$$ax + b = 0, \quad ax = -b, \quad x = -\frac{b}{a}$$

Thus, the linear equation $ax + b = 0$ has exactly one solution, $-b/a$.

In the following examples, the phrases in red indicate how an equivalent equation was obtained from the preceding equation. To shorten these phrases we have, as in Example 1, used "divide by 2" instead of the more accurate but lengthy *divide both sides by* 2.

If an equation contains rational expressions, we often eliminate denominators by multiplying both sides by the lcd of these expressions. If we multiply both sides by an expression that equals zero for some value of x, then the resulting equation may *not* be equivalent to the original equation, as illustrated in the following example.

EXAMPLE 1 **An equation with no solutions**

Solve the equation $\dfrac{3x}{x-2} = 1 + \dfrac{6}{x-2}$.

▶ **SOLUTION**

$$\frac{3x}{x-2} = 1 + \frac{6}{x-2} \qquad \text{given}$$

$$\left(\frac{3x}{x-2}\right)(x-2) = (1)(x-2) + \left(\frac{6}{x-2}\right)(x-2) \qquad \text{multiply by } x-2$$

$$3x = (x-2) + 6 \qquad \text{simplify}$$

$$3x = x + 4 \qquad \text{simplify}$$

$$2x = 4 \qquad \text{subtract } x$$

$$x = 2 \qquad \text{divide by 2}$$

✔ Check $x = 2$ LS: $\dfrac{3(2)}{(2)-2} = \dfrac{6}{0}$

Since division by 0 is not permissible, $x = 2$ is not a solution ($x = 2$ is called an **extraneous solution**, or **extraneous root**, of the given equation). Hence, *the given equation has no solutions.*

Formulas involving several variables occur in many applications of mathematics. Sometimes it is necessary to solve for a specific variable in terms of the remaining variables that appear in the formula, as the next example illustrates.

EXAMPLE 2 **Resistors connected in parallel**

Figure 1

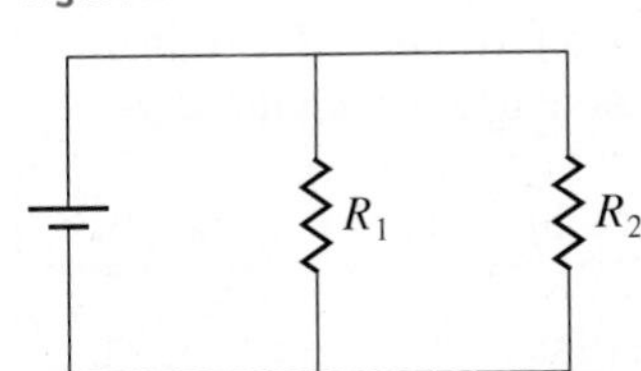

In electrical theory, the formula

$$\frac{1}{R} = \frac{1}{R_1} + \frac{1}{R_2}$$

is used to find the total resistance R when two resistors R_1 and R_2 are connected in parallel, as illustrated in Figure 1. Solve for R_1.

(continued)

▶ **Tutorial available at www.thomsonedu.com/login**

▶ SOLUTION

$$\frac{1}{R} \cdot RR_1R_2 = \frac{1}{R_1} \cdot RR_1R_2 + \frac{1}{R_2} \cdot RR_1R_2 \quad \text{multiply by the lcd, } RR_1R_2$$

$$R_1R_2 = RR_2 + RR_1 \quad \text{cancel common factors}$$

$$R_1R_2 - RR_1 = RR_2 \quad \text{collect terms with } R_1 \text{ on one side}$$

$$R_1(R_2 - R) = RR_2 \quad \text{factor out } R_1$$

$$R_1 = \frac{RR_2}{R_2 - R} \quad \text{divide by } R_2 - R$$

An alternative method of solution is to first solve the given formula for $1/R_1$, combine the fractions $1/R$ and $-1/R_2$, and then apply the property that if two nonzero numbers are equal, so are their reciprocals. The same result is obtained.

Equations are often used to solve *applied problems*—that is, problems that involve applications of mathematics to other fields. Because of the unlimited variety of applied problems, it is difficult to state specific rules for finding solutions. The following guidelines may be helpful, provided the problem can be formulated in terms of an equation in one variable. As you read the solution in Example 3, identify the specific guidelines used.

Guidelines for Solving Applied Problems

1. If the problem is stated in writing, read it carefully several times and think about the given facts, together with the unknown quantity that is to be found.
2. Introduce a letter to denote the unknown quantity. This is one of the most crucial steps in the solution. Phrases containing words such as *what, find, how much, how far,* or *when* should alert you to the unknown quantity.
3. If appropriate, draw a picture and label it.
4. List the known facts, together with any relationships that involve the unknown quantity. A relationship may be described by an equation in which written statements, instead of letters or numbers, appear on one or both sides of the equals sign.
5. After analyzing the list in guideline 4, formulate an equation that describes precisely what is stated in words.
6. Solve the equation formulated in guideline 5.
7. Check the solutions obtained in guideline 6 by referring to the original statement of the problem. Verify that the solution agrees with the stated conditions.

Banks and other financial institutions pay interest on investments. Usually this interest is *compounded* (as described in Section 4.2); however, if money is

▶ Tutorial available at www.thomsonedu.com/login

invested or loaned for a short period of time, *simple interest* may be paid, using the following formula.

Simple Interest Formula	If a sum of money P (the **principal**) is invested at a simple interest rate r (expressed as a decimal), then the **simple interest** I at the end of t years is $$I = Prt.$$

EXAMPLE 3 Investing money in two stocks

An investment firm has \$100,000 to invest for a client and decides to invest it in two stocks, A and B. The expected annual rate of return, or simple interest, for stock A is 15%, but there is some risk involved, and the client does not wish to invest more than \$50,000 in this stock. The annual rate of return on the more stable stock B is anticipated to be 10%. Determine whether there is a way of investing the money so that the annual interest is

(a) \$12,000 **(b)** \$13,000

SOLUTION The annual interest is given by $I = Pr$, which comes from the simple interest formula $I = Prt$ with $t = 1$. If we let x denote the amount invested in stock A, then $100{,}000 - x$ will be invested in stock B. This leads to the following equalities:

$$\begin{aligned} x &= \text{amount invested in stock A at } 15\% \\ 100{,}000 - x &= \text{amount invested in stock B at } 10\% \\ 0.15x &= \text{annual interest from stock A} \\ 0.10(100{,}000 - x) &= \text{annual interest from stock B} \end{aligned}$$

Adding the interest from both stocks, we obtain

$$\text{total annual interest} = 0.15x + 0.10(100{,}000 - x).$$

Simplifying the right-hand side gives us

$$\text{total annual interest} = 10{,}000 + 0.05x. \qquad (*)$$

▶ **(a)** The total annual interest is \$12,000 if

$$\begin{aligned} 10{,}000 + 0.05x &= 12{,}000 && \text{from } (*) \\ 0.05x &= 2000 && \text{subtract } 10{,}000 \\ x &= \frac{2000}{0.05} = 40{,}000. && \text{divide by } 0.05 \end{aligned}$$

Thus, \$40,000 should be invested in stock A, and the remaining \$60,000 should be invested in stock B. Since the amount invested in stock A is not more than \$50,000, this manner of investing the money meets the requirement of the client.

(continued)

▶ **Tutorial available at www.thomsonedu.com/login**

Check If \$40,000 is invested in stock A and \$60,000 in stock B, then the total annual interest is

$$40{,}000(0.15) + 60{,}000(0.10) = 6000 + 6000 = 12{,}000.$$

(b) The total annual interest is \$13,000 if

$$
\begin{aligned}
10{,}000 + 0.05x &= 13{,}000 && \text{from } (*) \\
0.05x &= 3000 && \text{subtract } 10{,}000 \\
x &= \frac{3000}{0.05} = 60{,}000. && \text{divide by } 0.05
\end{aligned}
$$

Thus, \$60,000 should be invested in stock A and the remaining \$40,000 in stock B. This plan does *not* meet the client's requirement that no more than \$50,000 be invested in stock A. Hence, the firm cannot invest the client's money in stocks A and B such that the total annual interest is \$13,000.

Figure 2

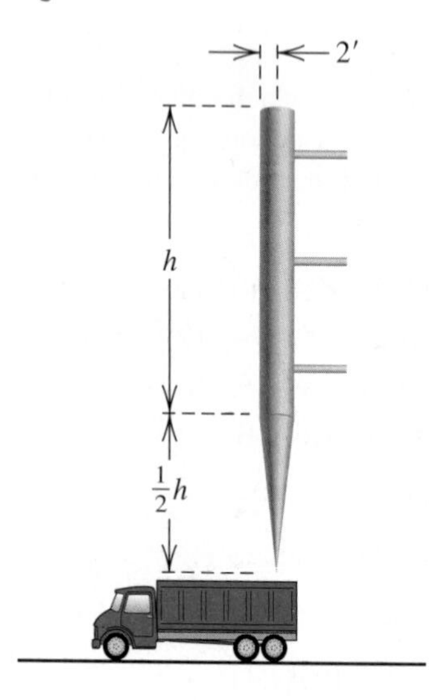

EXAMPLE 4 Constructing a grain-elevator hopper

A grain-elevator hopper is to be constructed as shown in Figure 2, with a right circular cylinder of radius 2 feet and altitude h feet on top of a right circular cone whose altitude is one-half that of the cylinder. What value of h will make the total volume V of the hopper 500 ft^3?

SOLUTION If V_{cylinder} and V_{cone} denote the volumes (in ft^3) and h_{cylinder} and h_{cone} denote the heights (in feet) of the cylinder and cone, respectively, then, using the formulas for volume stated on the endpapers at the front of the text, we obtain the following:

$$
\begin{aligned}
V_{\text{cylinder}} &= \pi r^2 h_{\text{cylinder}} = \pi(2)^2 h = 4\pi h \\
V_{\text{cone}} &= \tfrac{1}{3}\pi r^2 h_{\text{cone}} = \tfrac{1}{3}\pi(2)^2\left(\tfrac{1}{2}h\right) = \tfrac{2}{3}\pi h
\end{aligned}
$$

Since the total volume V of the hopper is to be 500 ft^3, we must have

$$
\begin{aligned}
4\pi h + \tfrac{2}{3}\pi h &= 500 && V_{\text{cylinder}} + V_{\text{cone}} = V_{\text{total}} \\
12\pi h + 2\pi h &= 1500 && \text{multiply by } 3 \\
14\pi h &= 1500 && \text{combine terms} \\
h &= \frac{1500}{14\pi} \approx 34.1 \text{ ft.} && \text{divide by } 14\pi
\end{aligned}
$$

A **quadratic equation** in x is an equation that can be written in the form $ax^2 + bx + c = 0$, where $a \neq 0$. This form is called the **standard form** of a quadratic equation in x.

To enable us to solve quadratic and other types of equations, we will make use of the next theorem.

Tutorial available at www.thomsonedu.com/login

Zero Factor Theorem

If p and q are algebraic expressions, then

$$pq = 0 \quad \text{if and only if} \quad p = 0 \quad \text{or} \quad q = 0.$$

The zero factor theorem can be extended to any number of algebraic expressions—that is,

$$pqr = 0 \quad \text{if and only if} \quad p = 0 \quad \text{or} \quad q = 0 \quad \text{or} \quad r = 0,$$

and so on. It follows that if $ax^2 + bx + c$ can be written as a product of two first-degree polynomials, then solutions can be found by setting each factor equal to 0, as illustrated below. This technique is called the **method of factoring.** To use the method of factoring, *it is essential that only the number* 0 *appear on one side of the equation.*

ILLUSTRATION **Solving Quadratic Equations by Factoring**

■
$$\begin{aligned} 3x^2 &= 10 - x \\ 3x^2 + x - 10 &= 0 \\ (3x - 5)(x + 2) &= 0 \\ x &= \tfrac{5}{3} \quad \text{or} \quad x = -2 \end{aligned}$$

■
$$\begin{aligned} x^2 + 16 &= 8x \\ x^2 - 8x + 16 &= 0 \\ (x - 4)(x - 4) &= 0 \\ x &= 4 \end{aligned}$$

Since $x - 4$ appears as a factor twice in the second part of the previous illustration, we call 4 a **double root** or **root of multiplicity 2** of the equation $x^2 + 16 = 8x$.

If a quadratic equation has the form $x^2 = d$ for some number $d > 0$, then $x^2 - d = 0$ or, equivalently,

$$\left(x + \sqrt{d}\right)\left(x - \sqrt{d}\right) = 0.$$

Setting each factor equal to zero gives us the solutions $-\sqrt{d}$ and $\sqrt{d}$. We frequently use the symbol $\pm\sqrt{d}$ (*plus or minus* $\sqrt{d}$) to represent both $\sqrt{d}$ and $-\sqrt{d}$. Thus, for $d > 0$, we have proved the following result. (The case $d < 0$ requires the system of complex numbers discussed in Section 1.5.)

A Special Quadratic Equation

If $x^2 = d$, then $x = \pm\sqrt{d}$.

Note on Notation: It is common practice to allow one variable to represent more than one value, as in $x = \pm 3$. A more descriptive notation is $x_{1,2} = \pm 3$, implying that $x_1 = 3$ and $x_2 = -3$.

The process of solving $x^2 = d$ as indicated in the preceding box is referred to as *taking the square root of both sides of the equation.* Note that if

$d > 0$ we obtain both a positive square root and a negative square root, not just the principal square root defined in Section 1.2.

ILLUSTRATION **Solving Equations of the Form $x^2 = d$**

■ $x^2 = 169$
$x = \pm\sqrt{169}$
$= \pm 13$

■ $(x + 3)^2 = 5$
$x + 3 = \pm\sqrt{5}$
$x = -3 \pm \sqrt{5}$

The solutions of a **quadratic equation** $ax^2 + bx + c = 0$, for $a \neq 0$, may be obtained by means of the following formula.

Quadratic Formula

If $a \neq 0$, the roots of $ax^2 + bx + c = 0$ are given by

$$x = \frac{-b \pm \sqrt{b^2 - 4ac}}{2a}.$$

The quadratic formula gives us two solutions of the equation

$$ax^2 + bx + c = 0.$$

They are $x = x_1, x_2$, where

$$x_1 = \frac{-b + \sqrt{b^2 - 4ac}}{2a}$$

and

$$x_2 = \frac{-b - \sqrt{b^2 - 4ac}}{2a}.$$

PROOF We shall assume that $b^2 - 4ac \geq 0$ so that $\sqrt{b^2 - 4ac}$ is a real number. (The case in which $b^2 - 4ac < 0$ will be discussed in the next section.) Let us proceed as follows:

$ax^2 + bx + c = 0$	given
$ax^2 + bx = -c$	subtract c
$x^2 + \frac{b}{a}x = -\frac{c}{a}$	divide by a
$x^2 + \frac{b}{a}x + \left(\frac{b}{2a}\right)^2 = \left(\frac{b}{2a}\right)^2 - \frac{c}{a}$	complete the square*
$\left(x + \frac{b}{2a}\right)^2 = \frac{b^2 - 4ac}{4a^2}$	equivalent equation
$x + \frac{b}{2a} = \pm\sqrt{\frac{b^2 - 4ac}{4a^2}}$	take the square root
$x = -\frac{b}{2a} \pm \sqrt{\frac{b^2 - 4ac}{4a^2}}$	subtract $\frac{b}{2a}$

We may write the radical in the last equation as

$$\pm\sqrt{\frac{b^2 - 4ac}{4a^2}} = \pm\frac{\sqrt{b^2 - 4ac}}{\sqrt{(2a)^2}} = \pm\frac{\sqrt{b^2 - 4ac}}{|2a|}.$$

*Recall, from a previous mathematics class, that to **complete the square** means to add the square of half the coefficient of x.

Since $|2a| = 2a$ if $a > 0$ or $|2a| = -2a$ if $a < 0$, we see that in all cases

$$x = -\frac{b}{2a} \pm \frac{\sqrt{b^2 - 4ac}}{2a} = \frac{-b \pm \sqrt{b^2 - 4ac}}{2a}.$$

Note that if the quadratic formula is executed properly, it is unnecessary to check the solutions because no extraneous solutions are introduced. The number $b^2 - 4ac$ under the radical sign in the quadratic formula is called the **discriminant** of the quadratic equation. If the discriminant is positive, there are two real and unequal roots of the quadratic equation. If it is zero, there is one root of multiplicity 2; and if it is negative, there are no real roots of the quadratic equation.

EXAMPLE 5 Using the quadratic formula

Solve the equation $2x(3 - x) = 3$.

Note that

$$\frac{3 \pm \sqrt{3}}{2} \neq \frac{3}{2} \pm \sqrt{3}.$$

The 2 in the denominator must be divided into ***both*** *terms of the numerator, so*

$$\frac{3 \pm \sqrt{3}}{2} = \frac{3}{2} \pm \frac{1}{2}\sqrt{3}.$$

▶ **SOLUTION** To use the quadratic formula, we must write the equation in the standard form $ax^2 + bx + c = 0$. Doing so gives us the equation $-2x^2 + 6x = 3$ or, equivalently, $2x^2 - 6x + 3 = 0$. We now let $a = 2$, $b = -6$, and $c = 3$ in the quadratic formula, obtaining

$$x = \frac{-(-6) \pm \sqrt{(-6)^2 - 4(2)(3)}}{2(2)} = \frac{6 \pm \sqrt{12}}{4} = \frac{6 \pm 2\sqrt{3}}{4} = \frac{3 \pm \sqrt{3}}{2}.$$

Hence, the solutions are $\frac{3 + \sqrt{3}}{2} \approx 2.37$ and $\frac{3 - \sqrt{3}}{2} \approx 0.63$.

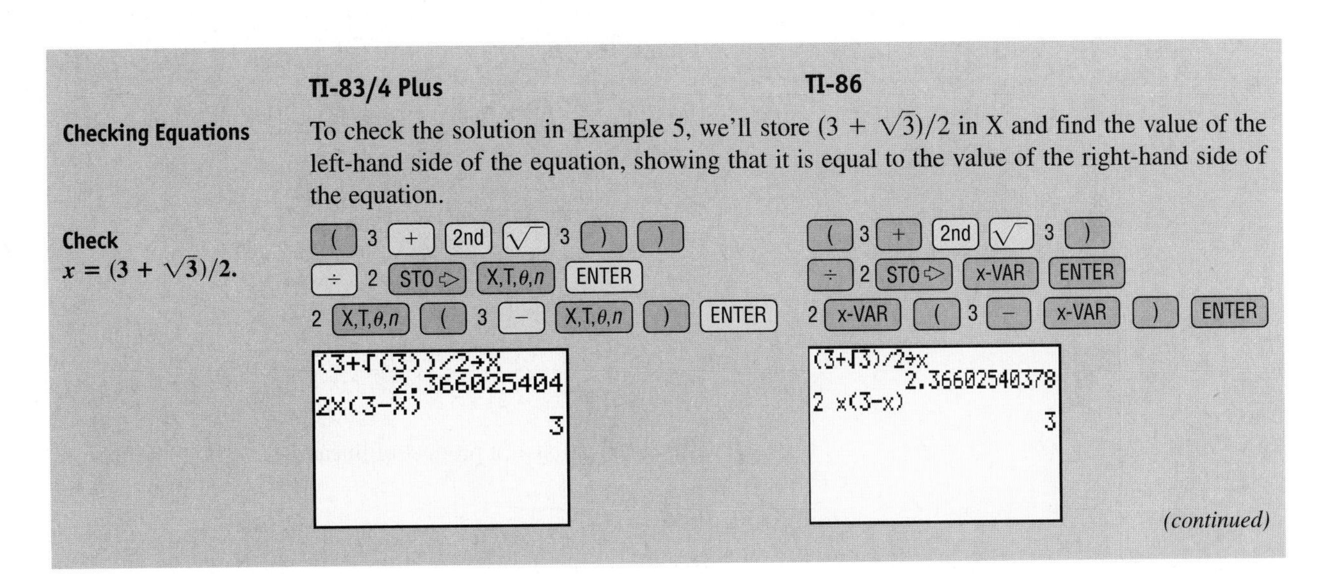

▶ Tutorial available at www.thomsonedu.com/login

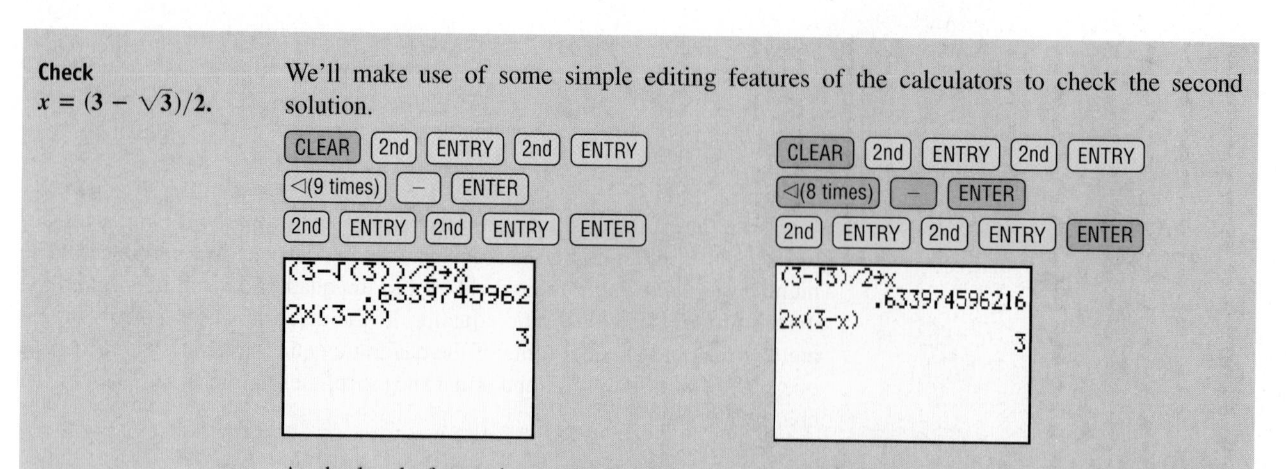

The next example shows how the quadratic formula can be used to help factor trinomials.

EXAMPLE 6 **Factoring with the quadratic formula**

Factor the polynomial $21x^2 - 13x - 20$.

▶ **SOLUTION** We solve the associated quadratic equation,

$$21x^2 - 13x - 20 = 0,$$

by using the quadratic formula:

$$\begin{aligned} x &= \frac{-(-13) \pm \sqrt{(-13)^2 - 4(21)(-20)}}{2(21)} \\ &= \frac{13 \pm \sqrt{169 + 1680}}{42} \\ &= \frac{13 \pm \sqrt{1849}}{42} \\ &= \frac{13 \pm 43}{42} = \frac{56}{42}, -\frac{30}{42} = \frac{4}{3}, -\frac{5}{7} \end{aligned}$$

We now write the equation as a product of linear factors, both of the form $(x - \text{solution})$:

$$\left(x - \tfrac{4}{3}\right)\left(x - \left(-\tfrac{5}{7}\right)\right) = 0$$

▶ **Tutorial available at www.thomsonedu.com/login**

Eliminate the denominators by multiplying both sides by $3 \cdot 7$:

$$3 \cdot 7\left(x - \tfrac{4}{3}\right)\left(x + \tfrac{5}{7}\right) = 0 \cdot 3 \cdot 7$$

$$3\left(x - \tfrac{4}{3}\right) \cdot 7\left(x + \tfrac{5}{7}\right) = 0$$

$$(3x - 4)(7x + 5) = 0$$

The left side is the desired factoring—that is,

$$21x^2 - 13x - 20 = (3x - 4)(7x + 5).$$

In the next example, we use the quadratic formula to solve an equation that contains more than one variable.

EXAMPLE 7 **Using the quadratic formula**

Solve $y = x^2 - 6x + 5$ for x, where $x \leq 3$.

▶ **SOLUTION** The equation can be written in the form

$$x^2 - 6x + 5 - y = 0,$$

so it is a quadratic equation in x with coefficients $a = 1$, $b = -6$, and $c = 5 - y$. Notice that y is considered to be a constant since we are solving for the variable x. Now we use the quadratic formula:

$$x = \frac{-(-6) \pm \sqrt{(-6)^2 - 4(1)(5 - y)}}{2(1)} \qquad x = \frac{-b \pm \sqrt{b^2 - 4ac}}{2a}$$

$$= \frac{6 \pm \sqrt{16 + 4y}}{2} \qquad \text{simplify } b^2 - 4ac$$

$$= \frac{6 \pm \sqrt{4}\sqrt{4 + y}}{2} \qquad \text{factor out } \sqrt{4}$$

$$= \frac{6 \pm 2\sqrt{4 + y}}{2} \qquad \sqrt{4} = 2$$

$$= 3 \pm \sqrt{4 + y} \qquad \text{divide 2 into } \textit{both} \text{ terms}$$

Since $\sqrt{4 + y}$ is nonnegative, $3 + \sqrt{4 + y}$ is greater than or equal to 3 and $3 - \sqrt{4 + y}$ is less than or equal to 3. Because the given restriction is $x \leq 3$, we have

$$x = 3 - \sqrt{4 + y}.$$

Many applied problems lead to quadratic equations. One is illustrated in the following example.

▶ **Tutorial available at www.thomsonedu.com/login**

Figure 3

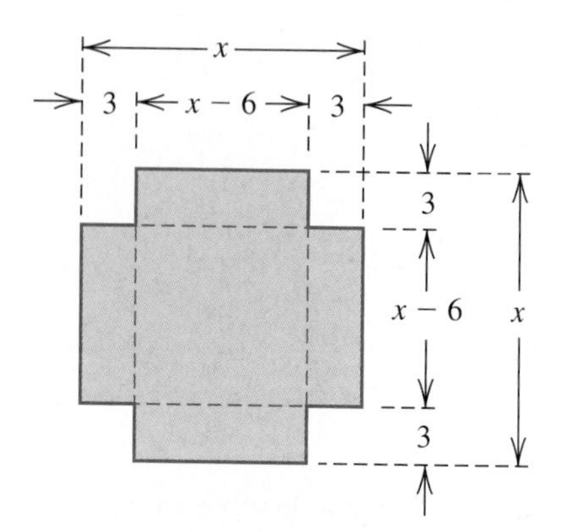

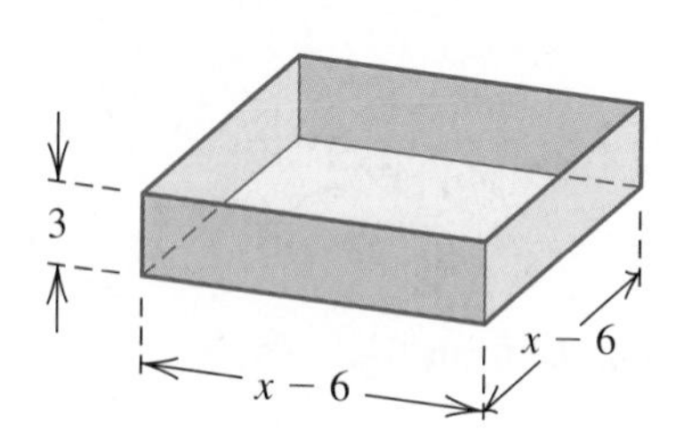

EXAMPLE 8 **Constructing a rectangular box**

A box with a square base and no top is to be made from a square piece of tin by cutting out a 3-inch square from each corner and folding up the sides. If the box is to hold 48 in^3, what size piece of tin should be used?

SOLUTION We begin by drawing the picture in Figure 3, letting x denote the unknown length of the side of the piece of tin. Subsequently, each side of the base of the box will have length $x - 3 - 3 = x - 6$.

Since the area of the base of the box is $(x - 6)^2$ and the height is 3, we obtain

$$\text{volume of box} = 3(x - 6)^2.$$

Since the box is to hold 48 in^3,

$$3(x - 6)^2 = 48.$$

We now solve for x:

$$(x - 6)^2 = 16 \quad \text{divide by 3}$$
$$x - 6 = \pm 4 \quad \text{take the square root}$$
$$x = 6 \pm 4 \quad \text{add 6}$$

Consequently,

$$x = 10 \quad \text{or} \quad x = 2.$$

Check Referring to Figure 3, we see that $x = 2$ is unacceptable, since no box is possible in this case. However, if we begin with a 10-inch square of tin, cut out 3-inch corners, and fold, we obtain a box having dimensions 4 inches, 4 inches, and 3 inches. The box has the desired volume of 48 in^3. Thus, a 10-inch square is the answer to the problem.

The equations considered so far are inadequate for many problems. For example, in applications it is often necessary to consider powers x^k with $k > 2$. Some equations involve absolute values or radicals. We conclude this section by giving examples of equations of various types that can be solved using elementary methods.

EXAMPLE 9 **Solving an equation containing an absolute value**

Solve the equation $2|x - 5| + 3 = 9$.

▶ SOLUTION We first isolate the absolute value expression by subtracting 3 and dividing by 2, to obtain

$$|x - 5| = \frac{9 - 3}{2} = 3.$$

▶ Tutorial available at www.thomsonedu.com/login

If a and b are real numbers with $b > 0$, then $|a| = b$ if and only if $a = b$ or $a = -b$. Hence, if $|x - 5| = 3$, then either

$$x - 5 = 3 \quad \text{or} \quad x - 5 = -3.$$

Solving for x gives us

$$x = 5 + 3 = 8 \quad \text{or} \quad x = 5 - 3 = 2.$$

Thus, the given equation has two solutions, 8 and 2.

EXAMPLE 10 **Solving an equation using grouping**

Solve the equation $x^3 + 2x^2 - x - 2 = 0$.

▶ **SOLUTION**

$$\begin{aligned} x^3 + 2x^2 - x - 2 &= 0 && \text{given} \\ x^2(x + 2) - 1(x + 2) &= 0 && \text{group terms} \\ (x^2 - 1)(x + 2) &= 0 && \text{factor out } x + 2 \\ (x + 1)(x - 1)(x + 2) &= 0 && \text{factor } x^2 - 1 \\ x + 1 = 0, \quad x - 1 = 0, \quad x + 2 &= 0 && \text{zero factor theorem} \\ x = -1, \quad x = 1, \quad x &= -2 && \text{solve for } x \end{aligned}$$

EXAMPLE 11 **Solving an equation containing rational exponents**

Solve the equation $x^{3/2} = x^{1/2}$.

▶ **SOLUTION**

$$\begin{aligned} x^{3/2} &= x^{1/2} && \text{given} \\ x^{3/2} - x^{1/2} &= 0 && \text{subtract } x^{1/2} \\ x^{1/2}(x - 1) &= 0 && \text{factor out } x^{1/2} \\ x^{1/2} = 0, \quad x - 1 &= 0 && \text{zero factor theorem} \\ x = 0, \quad x &= 1 && \text{solve for } x \end{aligned}$$

In Example 11 it would have been incorrect to divide both sides of the equation $x^{3/2} = x^{1/2}$ by $x^{1/2}$, obtaining $x = 1$, since the solution $x = 0$ would be lost. In general, *avoid dividing both sides of an equation by an expression that contains variables*—always *factor* instead.

If an equation involves radicals or fractional exponents, we often raise both sides to a positive power. The solutions of the new equation always contain the solutions of the given equation. For example, the solutions of

$$2x - 3 = \sqrt{x + 6}$$

are also solutions of

$$(2x - 3)^2 = \left(\sqrt{x + 6}\right)^2.$$

In some cases the new equation has *more* solutions than the given equation. To illustrate, if we are given the equation $x = 3$ and we square both sides, we obtain $x^2 = 9$. Note that the given equation $x = 3$ has only one solution, 3, but the new equation $x^2 = 9$ has two solutions, 3 and -3. Any solution of the new equation that is not a solution of the given equation is an extraneous solution.

Raising both sides of an equation to an ***odd*** *power can introduce imaginary solutions. For example, cubing both sides of $x = 1$ gives us $x^3 = 1$, which is equivalent to $x^3 - 1 = 0$. This equation has three solutions, of which two are imaginary (see Example 7 in Section 1.5).*

▶ **Tutorial available at www.thomsonedu.com/login**

Since extraneous solutions may occur, *it is absolutely essential to check all solutions obtained after raising both sides of an equation to an even power.* Such checks are unnecessary if both sides are raised to an *odd* power, because in this case extraneous (real number) solutions are not introduced.

In general, for the equation $x^{m/n} = a$, where x is a real number, we raise both sides to the power n/m (the reciprocal of m/n) to solve for x. If m is odd, we obtain $x = a^{n/m}$, but if m is even, we have $x = \pm a^{n/m}$. If n is even, extraneous solutions may occur—for example, if $x^{3/2} = -8$, then $x = (-8)^{2/3} = (\sqrt[3]{-8})^2 = (-2)^2 = 4$. However, 4 is not a solution of $x^{3/2} = -8$ since $4^{3/2} = 8$, not -8.

ILLUSTRATION **Solving $x^{m/n} = a$, m odd, x real**

Equation	Solution
■ $x^{3/1} = 64$	$x = 64^{1/3} = \sqrt[3]{64} = 4$
■ $x^{3/2} = 64$	$x = 64^{2/3} = (\sqrt[3]{64})^2 = 4^2 = 16$

ILLUSTRATION **Solving $x^{m/n} = a$, m even, x real**

Equation	Solution
■ $x^{4/1} = 16$	$x = \pm 16^{1/4} = \pm\sqrt[4]{16} = \pm 2$
■ $x^{2/3} = 16$	$x = \pm 16^{3/2} = \pm(\sqrt{16})^3 = \pm 4^3 = \pm 64$

In the next example, before we raise both sides of the equation to a power, we *isolate a radical*—that is, we consider an equivalent equation in which only the radical appears on one side.

EXAMPLE 12 **Solving an equation containing a radical**

Solve the equation $3 + \sqrt{3x + 1} = x$.

▶ **SOLUTION**

$$\sqrt{3x+1} = x - 3 \quad \text{isolate the radical}$$
$$\left(\sqrt{3x+1}\right)^2 = (x-3)^2 \quad \text{square both sides}$$
$$3x + 1 = x^2 - 6x + 9 \quad \text{simplify}$$
$$x^2 - 9x + 8 = 0 \quad \text{subtract } 3x + 1$$
$$(x-1)(x-8) = 0 \quad \text{factor}$$
$$x - 1 = 0, \quad x - 8 = 0 \quad \text{zero factor theorem}$$
$$x = 1, \quad x = 8 \quad \text{solve for } x$$

We raised both sides to an even power, so checks are required.

Check $x = 1$ LS: $3 + \sqrt{3(1) + 1} = 3 + \sqrt{4} = 3 + 2 = 5$
RS: 1

Since $5 \neq 1$, $x = 1$ is not a solution.

Check $x = 8$ LS: $3 + \sqrt{3(8) + 1} = 3 + \sqrt{25} = 3 + 5 = 8$
RS: 8

Since $8 = 8$ is a true statement, $x = 8$ is a solution.

Hence, the given equation has one solution, $x = 8$.

▶ Tutorial available at www.thomsonedu.com/login

An equation is of **quadratic type** if it can be written in the form

$$au^2 + bu + c = 0,$$

where $a \neq 0$ and u is an expression in some variable. If we find the solutions in terms of u, then the solutions of the given equation can be obtained by referring to the specific form of u, as demonstrated in the next example.

EXAMPLE 13 **Solving an equation of quadratic type**

Solve the equation $x^6 + 7x^3 = 8$.

SOLUTION Since $x^6 = (x^3)^2$, the form of the equation suggests that we let $u = x^3$, as in the second line below:

$$
\begin{array}{rll}
x^6 + 7x^3 - 8 = 0 & & \text{make one side 0} \\
u^2 + 7u - 8 = 0 & & \text{let } u = x^3 \\
(u + 8)(u - 1) = 0 & & \text{factor} \\
u = -8, & u = 1 & \text{solve for } u \\
x^3 = -8, & x^3 = 1 & u = x^3 \\
x = -2, & x = 1 & \text{take the cube root}
\end{array}
$$

1.4 *Exercises*

Exer. 1–10: Solve the equation.

1 $4x - 3 = -5x + 6$

▶ 2 $5x - 4 = 2(x - 2)$

3 $(3x - 2)^2 = (x - 5)(9x + 4)$

4 $(x + 5)^2 + 3 = (x - 2)^2$

5 $\dfrac{3x + 1}{6x - 2} = \dfrac{2x + 5}{4x - 13}$

6 $\dfrac{5x + 2}{10x - 3} = \dfrac{x - 8}{2x + 3}$

7 $\dfrac{4}{x + 2} + \dfrac{1}{x - 2} = \dfrac{5x - 6}{x^2 - 4}$

8 $\dfrac{2}{2x + 5} + \dfrac{3}{2x - 5} = \dfrac{10x + 5}{4x^2 - 25}$

9 $\dfrac{5}{2x + 3} + \dfrac{4}{2x - 3} = \dfrac{14x + 3}{4x^2 - 9}$

10 $\dfrac{-3}{x + 4} + \dfrac{7}{x - 4} = \dfrac{-5x + 4}{x^2 - 16}$

Exer. 11–14: Solve the equation by factoring.

11 $75x^2 + 35x - 10 = 0$

12 $48x^2 + 12x - 90 = 0$

13 $\dfrac{2x}{x + 3} + \dfrac{5}{x} - 4 = \dfrac{18}{x^2 + 3x}$

▶ 14 $\dfrac{3x}{x - 2} + \dfrac{1}{x + 2} = \dfrac{-4}{x^2 - 4}$

Exer. 15–18: Solve the equation by using the special quadratic equation on page 53.

15 $25x^2 = 9$

16 $16x^2 = 49$

17 $(x - 3)^2 = 17$

18 $(x + 4)^2 = 31$

Exer. 19–20: Solve by using the quadratic formula.

19 $x^2 + 4x + 2 = 0$

20 $x^2 - 6x - 3 = 0$

▶ **Tutorial available at www.thomsonedu.com/login**

Exer. 21–24: Use the quadratic formula to factor the expressions.

21 $x^2 + x - 30$

22 $x^2 + 7x$

23 $12x^2 - 16x - 3$

24 $15x^2 + 34x - 16$

Exer. 25–42: Solve the equation.

25 $|3x - 2| + 3 = 7$

▶ 26 $2|5x + 2| - 1 = 5$

27 $3|x + 1| - 2 = -11$

28 $|x - 2| + 5 = 5$

29 $9x^3 - 18x^2 - 4x + 8 = 0$

30 $4x^4 + 10x^3 = 6x^2 + 15x$

31 $y^{3/2} = 5y$

32 $y^{4/3} = -3y$

33 $\sqrt{7 - 5x} = 8$

34 $\sqrt{3 - x} - x = 3$

35 $x = 3 + \sqrt{5x - 9}$

36 $x + \sqrt{5x + 19} = -1$

37 $5y^4 - 7y^2 + 1 = 0$

38 $3y^4 - 5y^2 + 1 = 0$

39 $36x^{-4} - 13x^{-2} + 1 = 0$

40 $x^{-2} - 2x^{-1} - 35 = 0$

41 $3x^{2/3} + 4x^{1/3} - 4 = 0$

42 $2y^{1/3} - 3y^{1/6} + 1 = 0$

Exer. 43–44: Find the real solutions of the equation.

43 (a) $x^{5/3} = 32$ (b) $x^{4/3} = 16$

(c) $x^{2/3} = -36$ (d) $x^{3/4} = 125$

(e) $x^{3/2} = -27$

44 (a) $x^{3/5} = -27$ (b) $x^{2/3} = 25$

(c) $x^{4/3} = -49$ (d) $x^{3/2} = 27$

(e) $x^{3/4} = -8$

Exer. 45–46: Use the quadratic formula to solve the equation for (a) x in terms of y and (b) y in terms of x.

45 $4x^2 - 4xy + 1 - y^2 = 0$

46 $2x^2 - xy = 3y^2 + 1$

Exer. 47–48: When computations are carried out on a calculator, the quadratic formula will not always give accurate results if b^2 is large in comparison to ac, because one of the roots will be close to zero and difficult to approximate.

(a) Use the quadratic formula to approximate the roots of the given equation.

(b) To obtain a better approximation for the root near zero, rationalize the numerator to change

$$x = \frac{-b \pm \sqrt{b^2 - 4ac}}{2a} \quad \text{to} \quad x = \frac{2c}{-b \mp \sqrt{b^2 - 4ac}},$$

and use the second formula.

47 $x^2 + 4{,}500{,}000x - 0.96 = 0$

48 $x^2 - 73{,}000{,}000x + 2.01 = 0$

Exer. 49–52: Solve the formula for the specified variable.

49 $EK + L = D - TK$ for K

50 $CD + C = PC + N$ for C

51 $M = \dfrac{Q + 1}{Q}$ for Q

52 $\beta = \dfrac{\alpha}{1 - \alpha}$ for α

Exer. 53–64: The formula occurs in the indicated application. Solve for the specified variable.

▶ 53 $A = P + Prt$ for r (principal plus interest)

54 $s = \frac{1}{2}gt^2 + v_0t$ for v_0 (distance an object falls)

55 $S = \dfrac{p}{q + p(1 - q)}$ for q (Amdahl's law for supercomputers)

▶ 56 $S = 2(lw + hw + hl)$ for h (surface area of a rectangular box)

57 $\dfrac{1}{f} = \dfrac{1}{p} + \dfrac{1}{q}$ for q (lens equation)

58 $\dfrac{1}{R} = \dfrac{1}{R_1} + \dfrac{1}{R_2} + \dfrac{1}{R_3}$ for R_2 (three resistors connected in parallel)

59 $K = \frac{1}{2}mv^2$ for v (kinetic energy)

60 $F = g\dfrac{mM}{d^2}$ for d (Newton's law of gravitation)

▶ Tutorial available at www.thomsonedu.com/login

61 $A = 2\pi r(r + h)$ for r (surface area of a closed cylinder)

62 $s = \frac{1}{2}gt^2 + v_0t$ for t (distance an object falls)

63 $d = \frac{1}{2}\sqrt{4R^2 - C^2}$ for C (segments of circles)

64 $S = \pi r\sqrt{r^2 + h^2}$ for h (surface area of a cone)

65 **Cost of insulation** The cost of installing insulation in a particular two-bedroom home is \$2400. Present monthly heating costs average \$200, but the insulation is expected to reduce heating costs by 10%. How many months will it take to recover the cost of the insulation?

▶ 66 **Municipal funding** A city government has approved the construction of an \$800 million sports arena. Up to \$480 million will be raised by selling bonds that pay simple interest at a rate of 6% annually. The remaining amount (up to \$640 million) will be obtained by borrowing money from an insurance company at a simple interest rate of 5%. Determine whether the arena can be financed so that the annual interest is \$42 million.

67 **Walking rates** Two children, who are 224 meters apart, start walking toward each other at the same instant at rates of 1.5 m/sec and 2 m/sec, respectively (see the figure).

(a) When will they meet?

(b) How far will each have walked?

Exercise 67

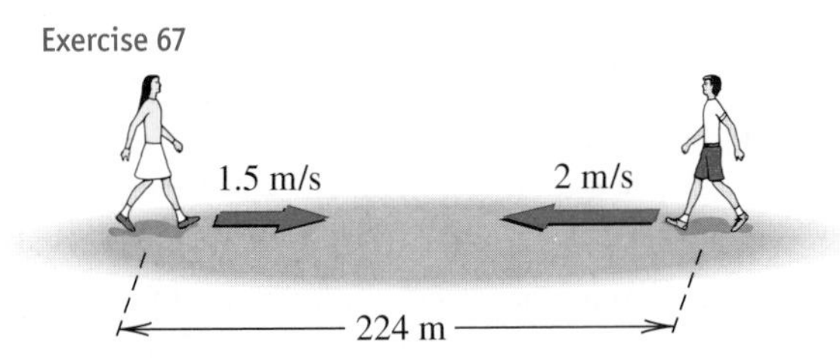

68 **Fencing a region** A farmer plans to use 180 feet of fencing to enclose a rectangular region, using part of a straight river bank instead of fencing as one side of the rectangle, as shown in the figure. Find the area of the region if the length of the side parallel to the river bank is

(a) twice the length of an adjacent side.

(b) one-half the length of an adjacent side.

(c) the same as the length of an adjacent side.

Exercise 68

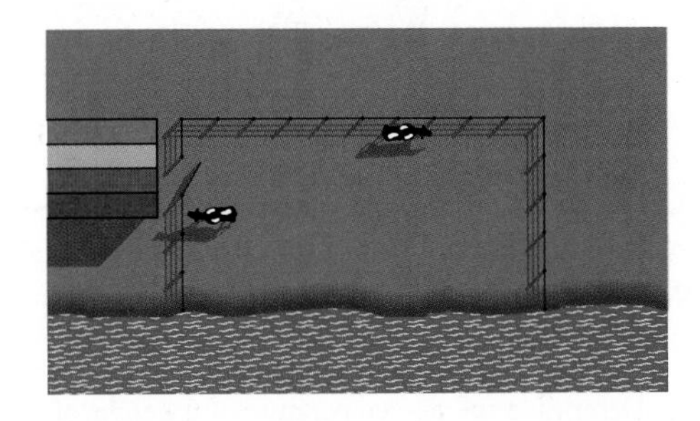

69 **Distance to a target** A bullet is fired horizontally at a target, and the sound of its impact is heard 1.5 seconds later. If the speed of the bullet is 3300 ft/sec and the speed of sound is 1100 ft/sec, how far away is the target?

70 **Jogging rates** A woman begins jogging at 3:00 P.M., running due north at a 6-minute-mile pace. Later, she reverses direction and runs due south at a 7-minute-mile pace. If she returns to her starting point at 3:45 P.M., find the total number of miles run.

71 **Drainage ditch dimensions** Every cross section of a drainage ditch is an isosceles trapezoid with a small base of 3 feet and a height of 1 foot, as shown in the figure. Determine the width of the larger base that would give the ditch a cross-sectional area of 5 ft^2.

Exercise 71

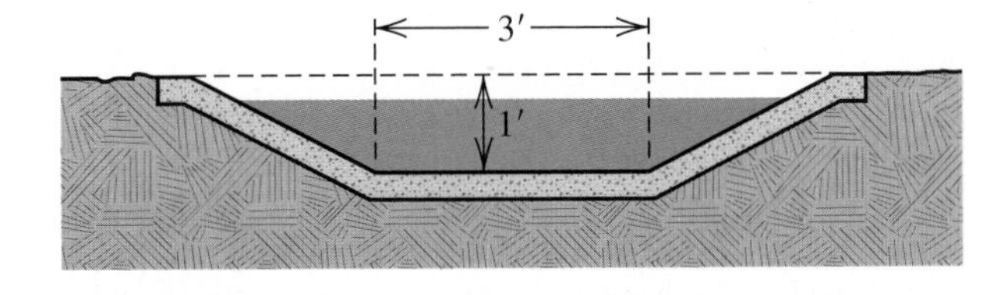

▶ 72 **Constructing a silo** A large grain silo is to be constructed in the shape of a circular cylinder with a hemisphere attached to the top (see the figure). The diameter of the silo is to be 30 feet, but the height is yet to be determined. Find the height h of the silo that will result in a capacity of $11{,}250\pi$ ft^3.

▶ **Tutorial available at www.thomsonedu.com/login**

Exercise 72

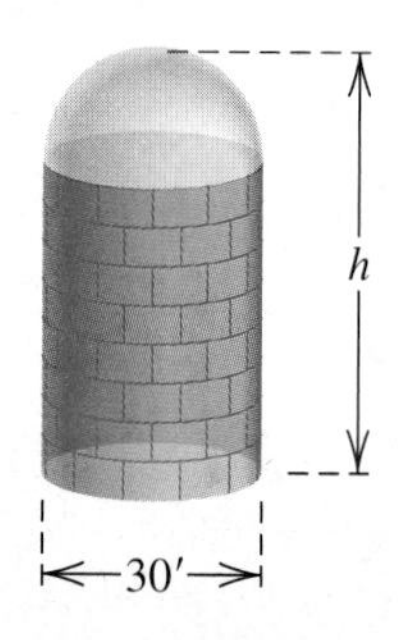

73 Air temperature Below the cloud base, the air temperature T (in °F) at height h (in feet) can be approximated by the equation $T = T_0 - \left(\frac{5.5}{1000}\right)h$, where T_0 is the temperature at ground level.

(a) Determine the air temperature at a height of 1 mile if the ground temperature is 70°F.

(b) At what altitude is the temperature freezing?

74 Height of a cloud The height h (in feet) of the cloud base can be estimated using $h = 227(T - D)$, where T is the ground temperature and D is the dew point.

(a) If the temperature is 70°F and the dew point is 55°F, find the height of the cloud base.

(b) If the dew point is 65°F and the cloud base is 3500 feet, estimate the ground temperature.

75 A cloud's temperature The temperature T within a cloud at height h (in feet) above the cloud base can be approximated using the equation $T = B - \left(\frac{3}{1000}\right)h$, where B is the temperature of the cloud at its base. Determine the temperature at 10,000 feet in a cloud with a base temperature of 55°F and a base height of 4000 feet. **Note:** For an interesting application involving the three preceding exercises, see Exercise 14 in the Discussion Exercises at the end of the chapter.

76 Bone-height relationship Archeologists can determine the height of a human without having a complete skeleton. If an archeologist finds only a humerus, then the height of the individual can be determined by using a simple linear relationship. (The humerus is the bone between the shoulder and the elbow.) For a female, if x is the length of the humerus (in centimeters), then her height h (in centimeters) can be determined using the formula $h = 65 + 3.14x$. For a male, $h = 73.6 + 3.0x$ should be used.

(a) A female skeleton having a 30-centimeter humerus is found. Find the woman's height at death.

(b) A person's height will typically decrease by 0.06 centimeter each year after age 30. A complete male skeleton is found. The humerus is 34 centimeters, and the man's height was 174 centimeters. Determine his approximate age at death.

77 Braking distance The distance that a car travels between the time the driver makes the decision to hit the brakes and the time the car actually stops is called the braking distance. For a certain car traveling v mi/hr, the braking distance d (in feet) is given by $d = v + (v^2/20)$.

(a) Find the braking distance when v is 55 mi/hr.

(b) If a driver decides to brake 120 feet from a stop sign, how fast can the car be going and still stop by the time it reaches the sign?

78 Temperature of boiling water The temperature T (in °C) at which water boils is related to the elevation h (in meters above sea level) by the formula

$$h = 1000(100 - T) + 580(100 - T)^2$$

for $95 \le T \le 100$.

(a) At what elevation does water boil at a temperature of 98°C?

(b) The elevation of Mt. Everest is approximately 8840 meters. Estimate the temperature at which water boils at the top of this mountain. (*Hint:* Use the quadratic formula with $x = 100 - T$.)

79 Distance between airplanes An airplane flying north at 200 mi/hr passed over a point on the ground at 2:00 P.M. Another airplane at the same altitude passed over the point at 2:30 P.M., flying east at 400 mi/hr (see the figure).

(a) If t denotes the time in hours after 2:30 P.M., express the distance d between the airplanes in terms of t.

(b) At what time after 2:30 P.M. were the airplanes 500 miles apart?

Exercise 79

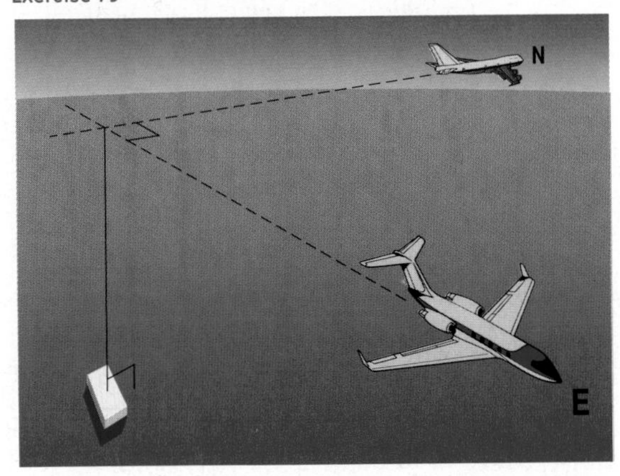

80 Height of a cliff When a rock is dropped from a cliff into an ocean, it travels approximately $16t^2$ feet in t seconds. If the splash is heard 4 seconds later and the speed of sound is 1100 ft/sec, approximate the height of the cliff.

81 Price of a CD player When a popular brand of CD player is priced at \$300 per unit, a store sells 15 units per week. Each time the price is reduced by \$10, however, the sales increase by 2 per week. What selling price will result in weekly revenues of \$7000?

82 Dimensions of an oil drum A closed right circular cylindrical oil drum of height 4 feet is to be constructed so that the total surface area is 10π ft^2. Find the diameter of the drum.

83 Dimensions of a vitamin tablet The rate at which a tablet of vitamin C begins to dissolve depends on the surface area of the tablet. One brand of tablet is 2 centimeters long and is in the shape of a cylinder with hemispheres of diameter 0.5 centimeter attached to both ends, as shown in the figure. A second brand of tablet is to be manufactured in the shape of a right circular cylinder of altitude 0.5 centimeter.

(a) Find the diameter of the second tablet so that its surface area is equal to that of the first tablet.

(b) Find the volume of each tablet.

Exercise 83

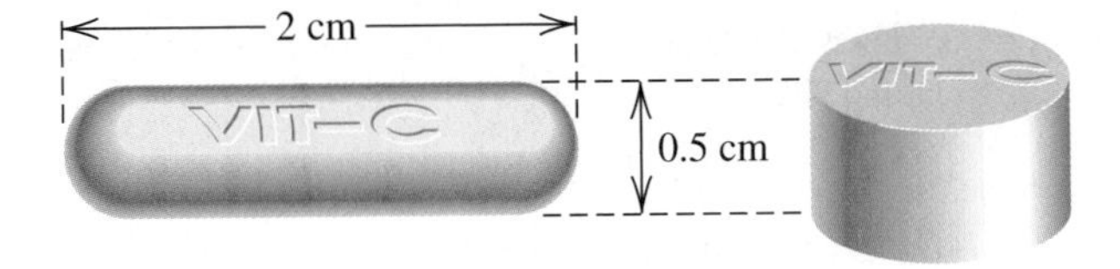

84 Withdrawal resistance of nails The *withdrawal resistance* of a nail indicates its holding strength in wood. A formula that is used for bright common nails is $P = 15{,}700S^{5/2}RD$, where P is the maximum withdrawal resistance (in pounds), S is the specific gravity of the wood at 12% moisture content, R is the radius of the nail (in inches), and D is the depth (in inches) that the nail has penetrated the wood. A 6d (sixpenny) bright, common nail of length 2 inches and diameter 0.113 inch is driven completely into a piece of Douglas fir. If it requires a maximum force of 380 pounds to remove the nail, approximate the specific gravity of Douglas fir.

85 Ladder height The recommended distance d that a ladder should be placed away from a vertical wall is 25% of its length L. Approximate the height h that can be reached by relating h as a percentage of L.

Exercise 85

86 The urban heat island Urban areas have higher average air temperatures than rural areas, as a result of the presence of buildings, asphalt, and concrete. This phenomenon has become known as the *urban heat island.* The temperature difference T (in °C) between urban and rural areas near Montreal, with a population P between 1000 and 1,000,000, can be described by the formula $T = 0.25P^{1/4}/\sqrt{v}$, where v is the average wind speed (in mi/hr) and $v \geq 1$. If $T = 3$ and $v = 5$, find P.

87 Installing a power line A power line is to be installed across a river that is 1 mile wide to a town that is 5 miles downstream (see the figure). It costs $7500 per mile to lay the cable underwater and $6000 per mile to lay it overland. Determine how the cable should be installed if $35,000 has been allocated for this project.

Exercise 87

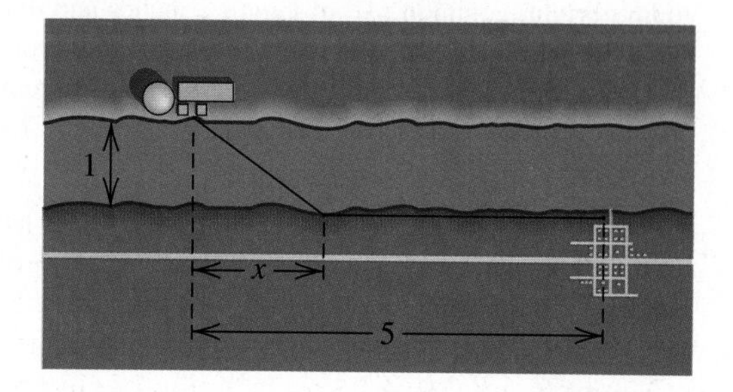

Exer. 88–89: Choose the equation that best describes the table of data. (*Hint:* Make assignments to Y_1–Y_4 and examine a table of their values.)

88

x	y
1	0.8
2	−0.4
3	−1.6
4	−2.8
5	−4.0

(1) $y = -1.2x + 2$

(2) $y = -1.2x^2 + 2$

(3) $y = 0.8\sqrt{x}$

(4) $y = x^{3/4} - 0.2$

89

x	y
1	−9
2	−4
3	11
4	42
5	95

(1) $y = 13x - 22$

(2) $y = x^2 - 2x - 8$

(3) $y = 4\sqrt{x} - 13$

(4) $y = x^3 - x^2 + x - 10$

90 **Temperature-latitude relationships** The table contains average annual temperatures for the northern and southern hemispheres at various latitudes.

Latitude	N. hem.	S. hem.
85°	−8°F	−5°F
75°	13°F	10°F
65°	30°F	27°F
55°	41°F	42°F
45°	57°F	53°F
35°	68°F	65°F
25°	78°F	73°F
15°	80°F	78°F
5°	79°F	79°F

(a) Which of the following equations more accurately predicts the average annual temperature in the southern hemisphere at latitude L?

(1) $T_1 = -1.09L + 96.01$

(2) $T_2 = -0.011L^2 - 0.126L + 81.45$

(b) Approximate the average annual temperature in the southern hemisphere at latitude 50°.

91 **Daylight-latitude relationships** The table gives the numbers of minutes of daylight occurring at various latitudes in the northern hemisphere at the summer and winter solstices.

Latitude	Summer	Winter
0°	720	720
10°	755	685
20°	792	648
30°	836	604
40°	892	548
50°	978	462
60°	1107	333

(a) Which of the following equations more accurately predicts the length of day at the summer solstice at latitude L?

(1) $D_1 = 6.096L + 685.7$

(2) $D_2 = 0.00178L^3 - 0.072L^2 + 4.37L + 719$

(b) Approximate the length of daylight at 35° at the summer solstice.

92 **Volume of a box** From a rectangular piece of metal having dimensions 24 inches by 36 inches, an open box is to be made by cutting out an identical square of area x^2 from each corner and turning up the sides.

(a) Determine an equation for the volume V of the box in terms of x.

(b) Use a table utility to approximate the value of x within ± 0.1 inch that will produce the maximum volume.

93 **Constructing a box** A cardboard box with an open top and a square bottom is to have a volume of 25 ft³. Use a table utility to determine the dimensions of the box to the nearest 0.1 foot that will minimize the amount of cardboard used to construct the box.

1.5 Complex Numbers

Complex numbers are needed to find solutions of equations that cannot be solved using only the set $\mathbb{R}$ of real numbers. The following chart illustrates several simple quadratic equations and the types of numbers required for solutions.

Equation	Solutions	Type of numbers required
$x^2 = 9$	$3, -3$	Integers
$x^2 = \frac{9}{4}$	$\frac{3}{2}, -\frac{3}{2}$	Rational numbers
$x^2 = 5$	$\sqrt{5}, -\sqrt{5}$	Irrational numbers
$x^2 = -9$	?	Complex numbers

The solutions of the first three equations in the chart are in $\mathbb{R}$; however, since squares of real numbers are never negative, $\mathbb{R}$ does not contain the solutions of $x^2 = -9$. To solve this equation, we need the **complex number system** $\mathbb{C}$, which contains both $\mathbb{R}$ and numbers whose squares are negative.

We begin by introducing the **imaginary unit,** denoted by i, which has the following properties.

Properties of i	$i = \sqrt{-1}, \quad i^2 = -1$

Because its square is negative, the letter i does not represent a real number. It is a new mathematical entity that will enable us to obtain $\mathbb{C}$. Since i, together with $\mathbb{R}$, is to be contained in $\mathbb{C}$, we must consider products of the form bi for a real number b and also expressions of the form $a + bi$ for real numbers a and b. The next chart provides definitions we shall use.

Terminology	Definition	Example(s)
Complex number	$a + bi$, where a and b are real numbers and $i^2 = -1$	$3,\ 2 + i,\ 2i$
Imaginary number	$a + bi$ with $b \neq 0$	$3 + 2i,\ -4i$
Pure imaginary number	bi with $b \neq 0$	$-4i,\ \sqrt{3}\,i,\ i$
Equality	$a + bi = c + di$ if and only if $a = c$ and $b = d$	$x + yi = 3 + 4i$ iff $x = 3$ and $y = 4$
Sum	$(a + bi) + (c + di) = (a + c) + (b + d)i$	see Example 1(a)
Product	$(a + bi)(c + di) = (ac - bd) + (ad + bc)i$	see Example 1(b)

Note that the pure imaginary numbers are a subset of the imaginary numbers and the imaginary numbers are a subset of the complex numbers. We use the phrase *nonreal complex number* interchangeably with *imaginary number.*

It is not necessary to memorize the definitions of addition and multiplication of complex numbers given in the preceding chart. Instead, *we may treat all symbols as having properties of real numbers, with exactly one exception: We replace i^2 by -1*. Thus, for the product $(a + bi)(c + di)$ we simply use the distributive laws and the fact that

$$(bi)(di) = bdi^2 = bd(-1) = -bd.$$

EXAMPLE 1 Addition and multiplication of complex numbers

Express in the form $a + bi$, where a and b are real numbers:

(a) $(3 + 4i) + (2 + 5i)$ **(b)** $(3 + 4i)(2 + 5i)$

▶ **SOLUTION**

(a) $(3 + 4i) + (2 + 5i) = (3 + 2) + (4 + 5)i = 5 + 9i$

(b)
$$\begin{aligned}(3 + 4i)(2 + 5i) &= (3 + 4i)(2) + (3 + 4i)(5i)\\ &= 6 + 8i + 15i + 20i^2\\ &= 6 + 23i + 20(-1)\\ &= -14 + 23i\end{aligned}$$

The set $\mathbb{R}$ of real numbers may be identified with the set of complex numbers of the form $a + 0i$. It is also convenient to denote the complex number $0 + bi$ by bi. Thus,

$$(a + 0i) + (0 + bi) = (a + 0) + (0 + b)i = a + bi.$$

Hence, we may regard $a + bi$ as the sum of two complex numbers a and bi (that is, $a + 0i$ and $0 + bi$). For the complex number $a + bi$, we call a the **real part** and b the **imaginary part.**

EXAMPLE 2 Equality of complex numbers

Find the values of x and y, where x and y are real numbers:

$$(2x - 4) + 9i = 8 + 3yi$$

▶ **Tutorial available at www.thomsonedu.com/login**

▶ SOLUTION We begin by equating the real parts and the imaginary parts of each side of the equation:

$$2x - 4 = 8 \quad \text{and} \quad 9 = 3y$$

Since $2x - 4 = 8$, $2x = 12$ and $x = 6$. Since $9 = 3y$, $y = 3$. The values of x and y that make the complex numbers equal are

$$x = 6 \quad \text{and} \quad y = 3.$$

With complex numbers, we are now able to solve an equation such as $x^2 = -9$. Specifically, since

$$(3i)(3i) = 3^2i^2 = 9(-1) = -9,$$

we see that one solution is $3i$ and another is $-3i$.

In the next chart we define the difference of complex numbers and multiplication of a complex number by a real number.

Terminology	Definition
Difference	$(a + bi) - (c + di) = (a - c) + (b - d)i$
Multiplication by a real number k	$k(a + bi) = ka + (kb)i$

If we are asked to write an expression in the form $a + bi$, the form $a - di$ is acceptable, since $a - di = a + (-d)i$.

EXAMPLE 3 Operations with complex numbers

Express in the form $a + bi$, where a and b are real numbers:

(a) $4(2 + 5i) - (3 - 4i)$ **(b)** $(4 - 3i)(2 + i)$ **(c)** $i(3 - 2i)^2$

(d) i^{51} **(e)** i^{-13}

SOLUTION

▶ **(a)** $4(2 + 5i) - (3 - 4i) = 8 + 20i - 3 + 4i = 5 + 24i$

▶ **(b)** $(4 - 3i)(2 + i) = 8 - 6i + 4i - 3i^2 = 11 - 2i$

▶ **(c)** $i(3 - 2i)^2 = i(9 - 12i + 4i^2) = i(5 - 12i) = 5i - 12i^2 = 12 + 5i$

(d) Taking successive powers of i, we obtain

$$i^1 = i, \quad i^2 = -1, \quad i^3 = -i, \quad i^4 = 1,$$

and then the cycle starts over:

$$i^5 = i, \quad i^6 = i^2 = -1, \quad \text{and so on.}$$

In particular,

$$i^{51} = i^{48}i^3 = (i^4)^{12}i^3 = (1)^{12}i^3 = (1)(-i) = -i.$$

▶ **Tutorial available at www.thomsonedu.com/login**

(e) In general, multiply i^{-a} by i^b, where $a \le b \le a + 3$ and b is a multiple of 4 (so that $i^b = 1$). For i^{-13}, choose $b = 16$.

$$i^{-13} \cdot i^{16} = i^3 = -i$$

The following concept has important uses in working with complex numbers.

Definition of the Conjugate of a Complex Number

If $z = a + bi$ is a complex number, then its **conjugate,** denoted by $\bar{z}$, is $a - bi$.

Since $a - bi = a + (-bi)$, it follows that the conjugate of $a - bi$ is

$$a - (-bi) = a + bi.$$

Therefore, *$a + bi$ and $a - bi$ are conjugates of each other.* Some properties of conjugates are given in Exercises 57–62.

ILLUSTRATION Conjugates

Complex number	Conjugate
$5 + 7i$	$5 - 7i$
$5 - 7i$	$5 + 7i$
$4i$	$-4i$
3	3

Complex Number Operations

TI-83/4 Plus

First, change to the complex mode.

[MODE] [▽ (6 times)] [▷] [ENTER]

```
Normal Sci Eng
Float 0123456789
Radian Degree
Func Par Pol Seq
Connected Dot
Sequential Simul
Real a+bi re^θi
Full Horiz G-T
```

The i is on the decimal point key.

4 [(] 2 [+] 5 [2nd] [i] [)] [−]
[(] 3 [−] 4 [2nd] [i] [)] [ENTER]
[2nd] [i] [∧] 51 [ENTER]
[MATH] [▷] [▷] [1]
5 [−] 7 [2nd] [i] [)] [ENTER]

TI-86

Complex numbers are entered in the form (real, imaginary).

4 [(] 2 [,] 5 [)] [−] [(] 3 [,] −4
[)] [ENTER]
[(] 0 [,] 1 [)] [∧] 51 [ENTER]
[2nd] [CPLX] [conj(F1)]
[(] 5 [,] −7 [)] [ENTER]

```
4(2+5i)-(3-4i)
             5+24i
i^51
          1E-13-i   ← 0 − i
conj(5-7i)
              5+7i
```

```
4(2,5)-(3,-4)
              (5,24)
(0,1)^51
              (0,-1)
conj (5,-7)
               (5,7)
conj  real  imag  abs  angle
```

On the TI-83/4 Plus, note that the second answer is equivalent to $0 - i$. We know this from Example 3(d), where we saw that the real part of a power of i must be 0, 1, or -1. Be on the lookout for such small inconsistencies.

The following two properties are consequences of the definitions of the sum and the product of complex numbers.

Properties of conjugates	Illustration
$(a + bi) + (a - bi) = 2a$	$(4 + 3i) + (4 - 3i) = 4 + 4 = 2 \cdot 4$
$(a + bi)(a - bi) = a^2 + b^2$	$(4 + 3i)(4 - 3i) = 4^2 - (3i)^2 = 4^2 - 3^2i^2 = 4^2 + 3^2$

Note that *the sum and the product of a complex number and its conjugate are real numbers.* Conjugates are useful for finding the **multiplicative inverse** of $a + bi$, $1/(a + bi)$, or for simplifying the quotient of two complex numbers. As illustrated in the next example, we may think of these types of simplifications as merely *rationalizing the denominator,* since we are multiplying the quotient by the conjugate of the denominator divided by itself.

EXAMPLE 4 Quotients of complex numbers

Express in the form $a + bi$, where a and b are real numbers:

(a) $\dfrac{1}{9 + 2i}$ (b) $\dfrac{7 - i}{3 - 5i}$

SOLUTION

▶ (a) $$\frac{1}{9 + 2i} = \frac{1}{9 + 2i} \cdot \frac{9 - 2i}{9 - 2i} = \frac{9 - 2i}{81 + 4} = \frac{9}{85} - \frac{2}{85}i$$

(b) $$\frac{7 - i}{3 - 5i} = \frac{7 - i}{3 - 5i} \cdot \frac{3 + 5i}{3 + 5i} = \frac{21 + 35i - 3i - 5i^2}{9 + 25}$$

$$= \frac{26 + 32i}{34} = \frac{13}{17} + \frac{16}{17}i$$

▶ **Tutorial available at www.thomsonedu.com/login**

If p is a positive real number, then the equation $x^2 = -p$ has solutions in $\mathbb{C}$. One solution is $\sqrt{p}\,i$, since

$$\left(\sqrt{p}\,i\right)^2 = \left(\sqrt{p}\right)^2 i^2 = p(-1) = -p.$$

Similarly, $-\sqrt{p}\,i$ is also a solution.

The definition of $\sqrt{-r}$ in the next chart is motivated by $\left(\sqrt{r}\,i\right)^2 = -r$ for $r > 0$. When using this definition, take care *not* to write $\sqrt{ri}$ when $\sqrt{r}\,i$ is intended.

Terminology	Definition	Illustrations
Principal square root $\sqrt{-r}$ for $r > 0$	$\sqrt{-r} = \sqrt{r}\,i$	$\sqrt{-9} = \sqrt{9}\,i = 3i$ $\sqrt{-5} = \sqrt{5}\,i$ $\sqrt{-1} = \sqrt{1}\,i = i$

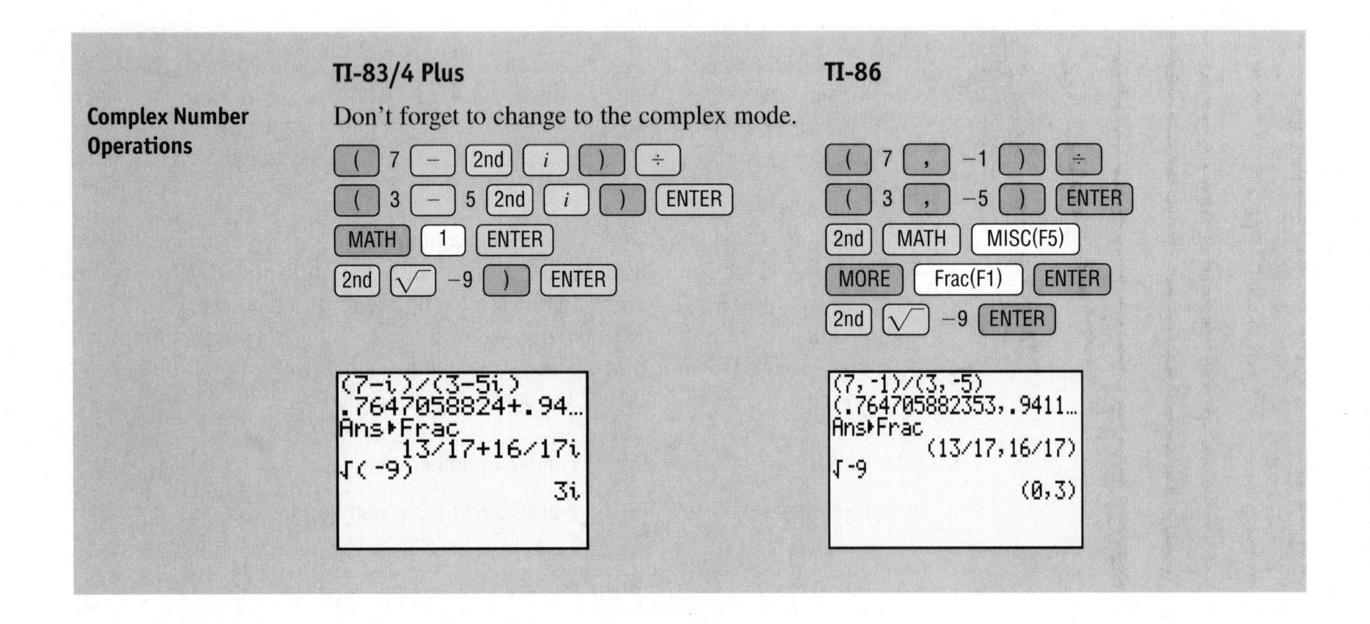

The radical sign must be used with caution when the radicand is negative. For example, the formula $\sqrt{a}\,\sqrt{b} = \sqrt{ab}$, which holds for positive real numbers, is *not* true when a and b are both negative, as shown below:

$$\sqrt{-3}\,\sqrt{-3} = \left(\sqrt{3}\,i\right)\left(\sqrt{3}\,i\right) = \left(\sqrt{3}\right)^2 i^2 = 3(-1) = -3$$

But $\qquad \sqrt{(-3)(-3)} = \sqrt{9} = 3.$

Hence, $\qquad \sqrt{-3}\,\sqrt{-3} \neq \sqrt{(-3)(-3)}.$

If only *one* of a or b is negative, then $\sqrt{a}\sqrt{b} = \sqrt{ab}$. In general, we shall not apply laws of radicals if radicands are negative. Instead, we shall change the form of radicals before performing any operations, as illustrated in the next example.

EXAMPLE 5 Working with square roots of negative numbers

Express in the form $a + bi$, where a and b are real numbers:

$$\left(5 - \sqrt{-9}\right)\left(-1 + \sqrt{-4}\right)$$

▶ SOLUTION First we use the definition $\sqrt{-r} = \sqrt{r}\,i$, and then we simplify:

$$\begin{aligned}\left(5 - \sqrt{-9}\right)\left(-1 + \sqrt{-4}\right) &= \left(5 - \sqrt{9}\,i\right)\left(-1 + \sqrt{4}\,i\right)\\ &= (5 - 3i)(-1 + 2i)\\ &= -5 + 10i + 3i - 6i^2\\ &= -5 + 13i + 6 = 1 + 13i\end{aligned}$$

In Section 1.4 we stated that if the discriminant $b^2 - 4ac$ of the quadratic equation $ax^2 + bx + c = 0$ is negative, then there are no real roots of the equation. In fact, the solutions of the equation are two *imaginary* numbers. Moreover, the solutions are conjugates of each other, as shown in the next example.

EXAMPLE 6 A quadratic equation with complex solutions

Solve the equation $5x^2 + 2x + 1 = 0$.

SOLUTION Applying the quadratic formula with $a = 5$, $b = 2$, and $c = 1$, we see that

$$\begin{aligned}x &= \frac{-2 \pm \sqrt{2^2 - 4(5)(1)}}{2(5)}\\ &= \frac{-2 \pm \sqrt{-16}}{10} = \frac{-2 \pm 4i}{10} = \frac{-1 \pm 2i}{5} = -\frac{1}{5} \pm \frac{2}{5}i.\end{aligned}$$

Thus, the solutions of the equation are $-\frac{1}{5} + \frac{2}{5}i$ and $-\frac{1}{5} - \frac{2}{5}i$.

EXAMPLE 7 An equation with complex solutions

Solve the equation $x^3 - 1 = 0$.

Difference of two cubes:

$a^3 - b^3 = (a - b)(a^2 + ab + b^2)$

▶ SOLUTION Using the difference of two cubes factoring formula with $a = x$ and $b = 1$, we write $x^3 - 1 = 0$ as

$$(x - 1)(x^2 + x + 1) = 0.$$

(continued)

▶ **Tutorial available at www.thomsonedu.com/login**

Setting each factor equal to zero and solving the resulting equations, we obtain the solutions

$$1, \quad \frac{-1 \pm \sqrt{1-4}}{2} = \frac{-1 \pm \sqrt{3}\,i}{2}$$

or, equivalently,

$$1, \quad -\frac{1}{2} + \frac{\sqrt{3}}{2}i, \quad -\frac{1}{2} - \frac{\sqrt{3}}{2}i.$$

Since the number 1 is called the **unit real number** and the given equation may be written as $x^3 = 1$, we call these three solutions the **cube roots of unity.**

In Section 1.3 we mentioned that $x^2 + 1$ is irreducible over the *real* numbers. However, if we factor over the *complex* numbers, then $x^2 + 1$ may be factored as follows:

$$x^2 + 1 = (x + i)(x - i)$$

1.5 *Exercises*

Exer. 1–34: Write the expression in the form $a + bi$, where a and b are real numbers.

1 $(5 - 2i) + (-3 + 6i)$

2 $(-5 + 7i) + (4 + 9i)$

3 $(7 - 6i) - (-11 - 3i)$

4 $(-3 + 8i) - (2 + 3i)$

5 $(3 + 5i)(2 - 7i)$

6 $(-2 + 6i)(8 - i)$

7 $(1 - 3i)(2 + 5i)$

8 $(8 + 2i)(7 - 3i)$

9 $(5 - 2i)^2$

10 $(6 + 7i)^2$

11 $i(3 + 4i)^2$

12 $i(2 - 7i)^2$

13 $(3 + 4i)(3 - 4i)$

14 $(4 + 9i)(4 - 9i)$

15 (a) i^{43} (b) i^{-20}

16 (a) i^{92} (b) i^{-33}

17 (a) i^{73} (b) i^{-46}

18 (a) i^{66} (b) i^{-55}

19 $\dfrac{3}{2 + 4i}$

20 $\dfrac{5}{2 - 7i}$

21 $\dfrac{1 - 7i}{6 - 2i}$

22 $\dfrac{2 + 9i}{-3 - i}$

23 $\dfrac{-4 + 6i}{2 + 7i}$

24 $\dfrac{-3 - 2i}{5 + 2i}$

25 $\dfrac{4 - 2i}{-5i}$

26 $\dfrac{-2 + 6i}{3i}$

27 $(2 + 5i)^3$

28 $(3 - 2i)^3$

29 $\left(2 - \sqrt{-4}\right)\left(3 - \sqrt{-16}\right)$

30 $\left(-3 + \sqrt{-25}\right)\left(8 - \sqrt{-36}\right)$

31 $\dfrac{4 + \sqrt{-81}}{7 - \sqrt{-64}}$

32 $\dfrac{5 - \sqrt{-121}}{1 + \sqrt{-25}}$

33 $\dfrac{\sqrt{-36}\,\sqrt{-49}}{\sqrt{-16}}$

34 $\dfrac{\sqrt{-25}}{\sqrt{-16}\,\sqrt{-81}}$

Exer. 35–38: Find the values of x and y, where x and y are real numbers.

35 $4 + (x + 2y)i = x + 2i$

36 $(x - y) + 3i = 7 + yi$

37 $(2x - y) - 16i = 10 + 4yi$

38 $8 + (3x + y)i = 2x - 4i$

Exer. 39–56: Find the solutions of the equation.

39 $x^2 - 6x + 13 = 0$

▶ 40 $x^2 - 2x + 26 = 0$

41 $x^2 + 4x + 13 = 0$

42 $x^2 + 8x + 17 = 0$

43 $x^2 - 5x + 20 = 0$

44 $x^2 + 3x + 6 = 0$

45 $4x^2 + x + 3 = 0$

46 $-3x^2 + x - 5 = 0$

47 $x^3 + 125 = 0$

48 $x^3 - 27 = 0$

49 $27x^3 = (x + 5)^3$

50 $16x^4 = (x - 4)^4$

51 $x^4 = 256$

52 $x^4 = 81$

53 $4x^4 + 25x^2 + 36 = 0$

54 $27x^4 + 21x^2 + 4 = 0$

55 $x^3 + 3x^2 + 4x = 0$

56 $8x^3 - 12x^2 + 2x - 3 = 0$

Exer. 57–62: Verify the property.

57 $\overline{z + w} = \overline{z} + \overline{w}$

58 $\overline{z - w} = \overline{z} - \overline{w}$

59 $\overline{z \cdot w} = \overline{z} \cdot \overline{w}$

60 $\overline{z/w} = \overline{z}/\overline{w}$

61 $\overline{z} = z$ if and only if z is real.

62 $\overline{z^2} = (\overline{z})^2$

1.6 *Inequalities*

An **inequality** is a statement that two quantities or expressions are not equal. It may be the case that one quantity is less than ($<$), less than or equal to ($\leq$), greater than ($>$), or greater than or equal to ($\geq$) another quantity. Consider the inequality

$$2x + 3 > 11,$$

where x is a variable. If a true statement is obtained when a number b is substituted for x, then b is a **solution** of the inequality. Thus, $x = 5$ is a solution of $2x + 3 > 11$ since $13 > 11$ is true, but $x = 3$ is not a solution since $9 > 11$ is false. To **solve** an inequality means to find *all* solutions. Two inequalities are **equivalent** if they have exactly the same solutions.

Most inequalities have an infinite number of solutions. To illustrate, the solutions of the continued inequality

$$2 < x < 5$$

consist of *every* real number x between 2 and 5. We call this set of numbers an **open interval** and denote it by $(2, 5)$. The **graph** of the open interval $(2, 5)$ is the set of all points on a coordinate line that lie between—but do not include—the points corresponding to $x = 2$ and $x = 5$. The graph is represented by shading an appropriate part of the axis, as shown in Figure 1. We refer to this process as **sketching the graph** of the interval. The numbers 2 and 5 are called the **endpoints** of the interval $(2, 5)$. The parentheses in the notation $(2, 5)$ and in Figure 1 are used to indicate that the endpoints of the interval are not included.

Figure 1

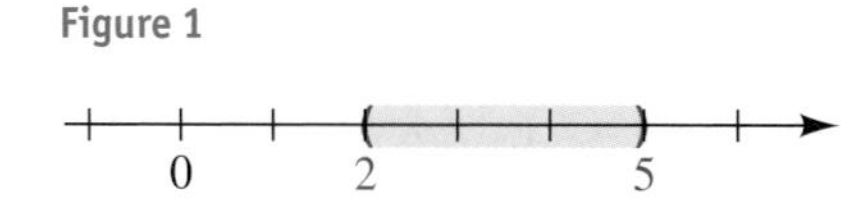

If we wish to include an endpoint, we use a bracket instead of a parenthesis. For example, the solutions of the inequality $2 \leq x \leq 5$ are denoted by $[2, 5]$ and are referred to as a **closed interval.** The graph of $[2, 5]$ is sketched

▶ **Tutorial available at www.thomsonedu.com/login**

Figure 2

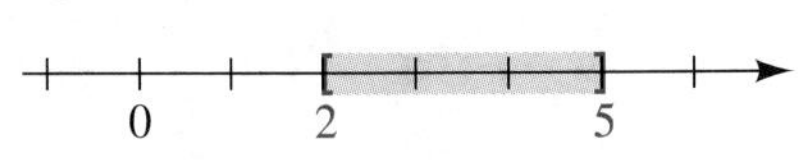

in Figure 2, where brackets indicate that endpoints are included. We shall also consider **half-open intervals** $[a, b)$ and $(a, b]$ and **infinite intervals,** as described in the following chart. The symbol ∞ (read "infinity") used for infinite intervals is merely a notational device and does *not* represent a real number.

Intervals

Notation	Inequality	Graph
(1) (a, b)	$a < x < b$	a b
(2) $[a, b]$	$a \le x \le b$	a b
(3) $[a, b)$	$a \le x < b$	a b
(4) $(a, b]$	$a < x \le b$	a b
(5) (a, ∞)	$x > a$	a
(6) $[a, \infty)$	$x \ge a$	a
(7) $(-\infty, b)$	$x < b$	b
(8) $(-\infty, b]$	$x \le b$	b
(9) $(-\infty, \infty)$	$-\infty < x < \infty$	

Methods for solving inequalities in x are similar to those used for solving equations. In particular, we often use properties of inequalities to replace a given inequality with a list of equivalent inequalities, ending with an inequality from which solutions are easily obtained. The properties in the following chart can be proved for real numbers a, b, c, and d.

Properties of Inequalities

Property	Illustration
(1) If $a < b$ and $b < c$, then $a < c$.	$2 < 5$ and $5 < 9$, so $2 < 9$.
(2) If $a < b$, then $a + c < b + c$ and $a - c < b - c$.	$2 < 7$, so $2 + 3 < 7 + 3$ and $2 - 3 < 7 - 3$.
(3) If $a < b$ and $c > 0$, then $ac < bc$ and $\frac{a}{c} < \frac{b}{c}$.	$2 < 5$ and $3 > 0$, so $2 \cdot 3 < 5 \cdot 3$ and $\frac{2}{3} < \frac{5}{3}$.
(4) If $a < b$ and $c < 0$, then $ac > bc$ and $\frac{a}{c} > \frac{b}{c}$.	$2 < 5$ and $-3 < 0$, so $2(-3) > 5(-3)$ and $\frac{2}{-3} > \frac{5}{-3}$.

Reverse the inequality when multiplying or dividing by a negative number.

It is important to remember that multiplying or dividing both sides of an inequality by a negative real number *reverses* the inequality sign (see property 4). Properties similar to those above are true for other inequalities and for $\leq$ and $\geq$. Thus, if $a > b$, then $a + c > b + c$; if $a \geq b$ and $c < 0$, then $ac \leq bc$; and so on.

If x represents a real number, then, by property 2, adding or subtracting the same expression containing x on both sides of an inequality yields an equivalent inequality. By property 3, we may multiply or divide both sides of an inequality by an expression containing x if we are certain that the expression is positive for all values of x under consideration. To illustrate, multiplication or division by $x^4 + 3x^2 + 5$ would be permissible, since this expression is always positive. If we multiply or divide both sides of an inequality by an expression that is always negative, such as $-7 - x^2$, then, by property 4, the inequality is reversed.

In examples we shall describe solutions of inequalities by means of intervals and also represent them graphically.

EXAMPLE 1 Solving an inequality

Solve the inequality $-4x - 3 > -2x - 11$.

SOLUTION

$-4x - 3 > -2x - 11$	given
$-2x > -8$	add $2x + 3$
$x < 4$	divide by -2; reverse the inequality sign

Figure 3

Hence, the solutions of the given inequality consist of all real numbers x such that $x < 4$. This is the interval $(-\infty, 4)$ sketched in Figure 3.

EXAMPLE 2 Solving a continued inequality

Solve the continued inequality $-5 \leq \dfrac{4 - 3x}{2} < 1$.

SOLUTION A number x is a solution of the given inequality if and only if

$$-5 \leq \frac{4 - 3x}{2} \quad \text{and} \quad \frac{4 - 3x}{2} < 1.$$

We can either work with each inequality separately or solve both inequalities simultaneously, as follows (keep in mind that our goal is to isolate x):

$-5 \leq \dfrac{4 - 3x}{2} < 1$	given
$-10 \leq 4 - 3x < 2$	multiply by 2
$-14 \leq -3x < -2$	subtract 4
$\frac{14}{3} \geq x > \frac{2}{3}$	divide by -3; reverse the inequality signs
$\frac{2}{3} < x \leq \frac{14}{3}$	equivalent inequality

(continued)

Tutorial available at www.thomsonedu.com/login

Figure 4

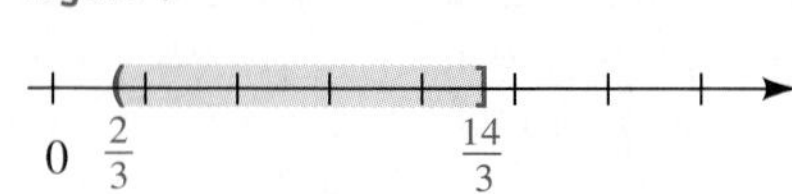

Thus, the solutions of the inequality are all numbers in the half-open interval $\left(\frac{2}{3}, \frac{14}{3}\right]$ sketched in Figure 4.

EXAMPLE 3 Solving a rational inequality

Solve the inequality $\dfrac{1}{x-2} > 0$.

▶ SOLUTION Since the numerator is positive, the fraction is positive if and only if the denominator, $x - 2$, is also positive. Thus, $x - 2 > 0$ or, equivalently, $x > 2$, and the solutions are all numbers in the infinite interval $(2, \infty)$ sketched in Figure 5.

Figure 5

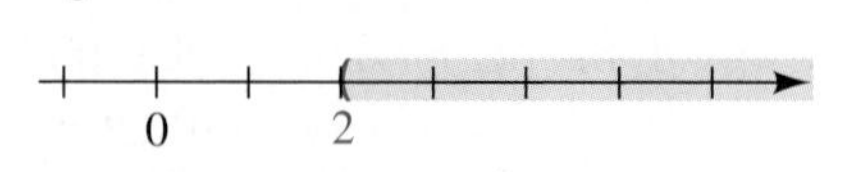

EXAMPLE 4 Using a lens formula

Figure 6

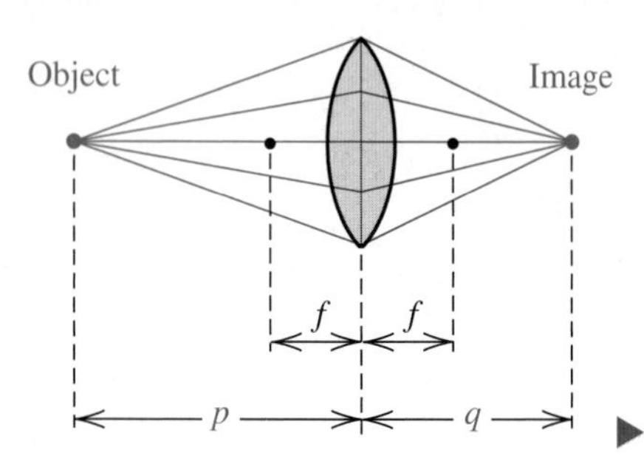

As illustrated in Figure 6, if a convex lens has focal length f centimeters and if an object is placed a distance p centimeters from the lens with $p > f$, then the distance q from the lens to the image is related to p and f by the formula

$$\frac{1}{p} + \frac{1}{q} = \frac{1}{f}.$$

If $f = 5$ cm, how close must the object be to the lens for the image to be more than 12 centimeters from the lens?

▶ SOLUTION Since $f = 5$, the given formula may be written as

$$\frac{1}{p} + \frac{1}{q} = \frac{1}{5}.$$

We wish to determine the values of q such that $q > 12$. Let us first solve the equation for q:

$$5q + 5p = pq \qquad \text{multiply by the lcd, } 5pq$$

$$q(5 - p) = -5p \qquad \text{collect } q \text{ terms on one side and factor}$$

$$q = -\frac{5p}{5-p} = \frac{5p}{p-5} \qquad \text{divide by } 5 - p$$

To solve the inequality $q > 12$, we proceed as follows:

$$\frac{5p}{p-5} > 12 \qquad q = \frac{5p}{p-5}$$

$$5p > 12(p - 5) \qquad \text{allowable, since } p > f \text{ implies } p - 5 > 0$$

$$-7p > -60 \qquad \text{multiply factors and collect } p \text{ terms on one side}$$

$$p < \tfrac{60}{7} \qquad \text{divide by } -7\text{; reverse the inequality}$$

Combining the last inequality with the fact that p is greater than 5, we obtain the solution

$$5 < p < \tfrac{60}{7}.$$

▶ Tutorial available at www.thomsonedu.com/login

Figure 7

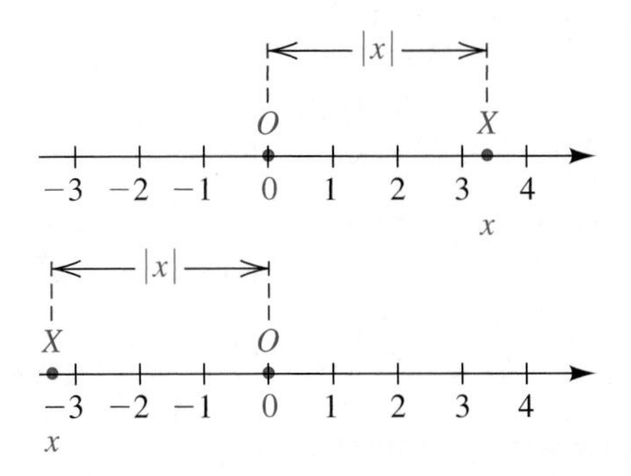

If a point X on a coordinate line has coordinate x, as shown in Figure 7, then X is to the right of the origin O if $x > 0$ and to the left of O if $x < 0$. From Section 1.1, the distance $d(O, X)$ between O and X is the *nonnegative* real number given by

$$d(O, X) = |x - 0| = |x|.$$

It follows that the solutions of an inequality such as $|x| < 3$ consist of the coordinates of all points whose distance from O is less than 3. This is the open interval $(-3, 3)$ sketched in Figure 8. Thus,

$$|x| < 3 \quad \text{is equivalent to} \quad -3 < x < 3.$$

Figure 8

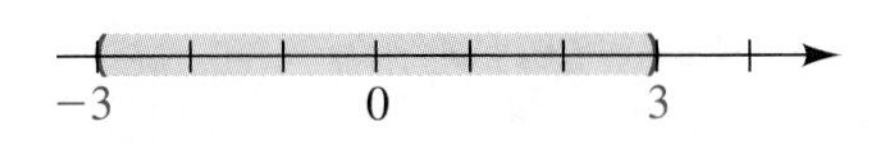

Similarly, for $|x| > 3$, the distance between O and a point with coordinate x is greater than 3; that is,

$$|x| > 3 \quad \text{is equivalent to} \quad x < -3 \text{ or } x > 3.$$

Figure 9

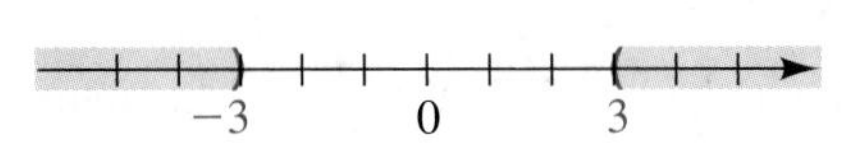

The graph of the solutions to $|x| > 3$ is sketched in Figure 9. We often use the **union symbol** $\cup$ and write

$$(-\infty, -3) \cup (3, \infty)$$

to denote all real numbers that are in either $(-\infty, -3)$ or $(3, \infty)$.

The notation

$$(-\infty, 2) \cup (2, \infty)$$

represents the set of all real numbers except 2.

The **intersection symbol** $\cap$ is used to denote the elements that are *common* to two sets. For example,

$$(-\infty, 3) \cap (-3, \infty) = (-3, 3),$$

since the intersection of $(-\infty, 3)$ and $(-3, \infty)$ consists of all real numbers x such that both $x < 3$ *and* $x > -3$.

The preceding discussion may be generalized to obtain the following properties of absolute values.

Properties of Absolute Values ($b > 0$)

(1) $|a| < b$ is equivalent to $-b < a < b$.

(2) $|a| > b$ is equivalent to $a < -b$ or $a > b$.

In the next example we use property 1 with $a = x - 3$ and $b = 0.5$.

EXAMPLE 5 Solving an inequality containing an absolute value

Solve the inequality $|x - 3| - 0.1 < 0.4$.

▶ **SOLUTION**

$$|x - 3| < 0.5 \qquad \text{isolate } |x - 3|$$
$$-0.5 < x - 3 < 0.5 \qquad \text{property 1}$$
$$2.5 < x < 3.5 \qquad \text{isolate } x \text{ by adding 3}$$

Figure 10

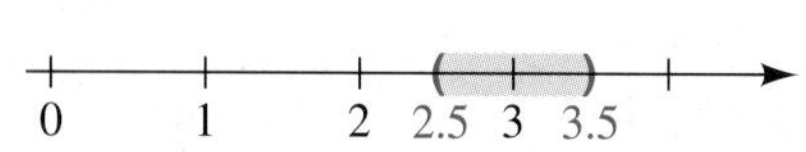

Thus, the solutions are the real numbers in the open interval (2.5, 3.5). The graph is sketched in Figure 10.

In the next example we use property 2 with $a = 2x + 3$ and $b = 9$.

EXAMPLE 6 Solving an inequality containing an absolute value

Solve the inequality $|2x + 3| > 9$.

▶ **SOLUTION**

$$|2x + 3| > 9 \qquad \text{given}$$
$$2x + 3 < -9 \quad \text{or} \quad 2x + 3 > 9 \qquad \text{property 2}$$
$$2x < -12 \quad \text{or} \quad 2x > 6 \qquad \text{subtract 3}$$
$$x < -6 \quad \text{or} \quad x > 3 \qquad \text{divide by 2}$$

Figure 11

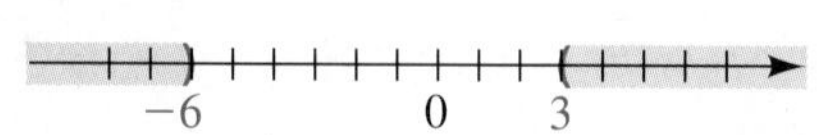

Consequently, the solutions of the inequality $|2x + 3| > 9$ consist of the numbers in $(-\infty, -6) \cup (3, \infty)$. The graph is sketched in Figure 11.

The trichotomy law in Section 1.1 states that for any real numbers a and b exactly one of the following is true:

$$a > b, \qquad a < b, \qquad \text{or} \qquad a = b$$

Thus, after solving $|2x + 3| > 9$ in Example 6, we readily obtain the solutions for $|2x + 3| < 9$ and $|2x + 3| = 9$—namely, $(-6, 3)$ and $\{-6, 3\}$, respectively. Note that the union of these three sets of solutions is necessarily the set $\mathbb{R}$ of real numbers.

When using the notation $a < x < b$, we must have $a < b$. Thus, *it is incorrect to write the solution* $x < -6$ *or* $x > 3$ (in Example 6) *as* $3 < x < -6$. Another misuse of inequality notation is to write $a < x > b$, since when several inequality symbols are used in one expression, *they must point in the same direction.*

To solve an inequality involving polynomials of degree greater than 1, we shall express each polynomial as a product of linear factors $ax + b$ and/or irreducible quadratic factors $ax^2 + bx + c$. If any such factor is not zero in an interval, then it is either positive throughout the interval or negative throughout the interval. Hence, if we choose any k in the interval and if the factor is positive (or negative) for $x = k$, then it is positive (or negative) throughout the interval. The value of the factor at $x = k$ is called a **test value** of the factor at the test number k. This concept is exhibited in the following example.

EXAMPLE 7 Solving a quadratic inequality

Solve the inequality $2x^2 - x < 3$.

▶ **SOLUTION** To use test values, *it is essential to have* 0 *on one side of the inequality sign.* Thus, we proceed as follows:

$$2x^2 - x - 3 < 0 \quad \text{make one side 0}$$
$$(x + 1)(2x - 3) < 0 \quad \text{factor}$$

Figure 12

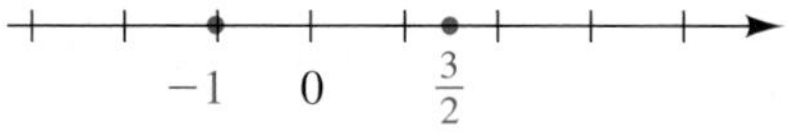

$-1 \quad 0 \quad \frac{3}{2}$

The factors $x + 1$ and $2x - 3$ are zero at -1 and $\frac{3}{2}$, respectively. The corresponding points on a coordinate line (see Figure 12) determine the nonintersecting intervals

$$(-\infty, -1), \qquad \left(-1, \tfrac{3}{2}\right), \quad \text{and} \quad \left(\tfrac{3}{2}, \infty\right).$$

We may find the signs of $x + 1$ and $2x - 3$ in each interval by using a test value taken from each interval. To illustrate, if we choose $k = -10$ in $(-\infty, -1)$, the values of both $x + 1$ and $2x - 3$ are negative, and hence they are negative throughout $(-\infty, -1)$. A similar procedure for the remaining two intervals gives us the following *sign chart,* where the term *resulting sign* in the last row refers to the sign obtained by applying laws of signs to the product of the factors. Note that the resulting sign is positive or negative according to whether the number of negative signs of factors is even or odd, respectively.

Interval	$(-\infty, -1)$	$\left(-1, \frac{3}{2}\right)$	$\left(\frac{3}{2}, \infty\right)$
Sign of $x + 1$	$-$	$+$	$+$
Sign of $2x - 3$	$-$	$-$	$+$
Resulting sign	$+$	$-$	$+$

Sometimes it is convenient to represent the signs of $x + 1$ and $2x - 3$ by using a coordinate line and a *sign diagram,* of the type illustrated in Figure 13. The vertical lines indicate where the factors are zero, and signs of factors are shown above the coordinate line. The resulting signs are shown in red.

Figure 13

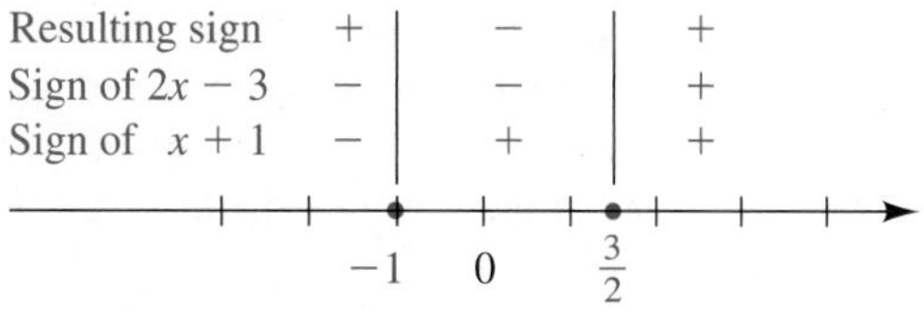

Resulting sign $+$ | $-$ | $+$
Sign of $2x - 3$ $-$ | $-$ | $+$
Sign of $x + 1$ $-$ | $+$ | $+$

$-1 \quad 0 \quad \frac{3}{2}$

The solutions of $(x + 1)(2x - 3) < 0$ are the values of x for which the product of the factors is *negative*—that is, where the resulting sign is negative. This corresponds to the open interval $\left(-1, \frac{3}{2}\right)$.

Back on page 53, we discussed the zero factor theorem, which dealt with *equalities.* It is a common mistake to extend this theorem to *inequalities.* The following warning shows this incorrect extension applied to the inequality in Example 7.

▶ **Tutorial available at www.thomsonedu.com/login**

Warning!

$(x + 1)(2x - 3) < 0$ is ***not*** equivalent to $x + 1 < 0$ or $2x - 3 < 0$

In future examples (here and in Chapter 3) we will use either a sign chart or a sign diagram, but not both. When working exercises, you should choose the method of solution with which you feel most comfortable.

EXAMPLE 8 **Using a sign diagram to solve an inequality**

Solve the inequality $\dfrac{(x + 2)(3 - x)}{(x + 1)(x^2 + 1)} \le 0.$

▶ **SOLUTION** Since 0 is already on the right side of the inequality and the left side is factored, we may proceed directly to the sign diagram in Figure 14, where the vertical lines indicate the zeros (-2, -1, and 3) of the factors.

Figure 14

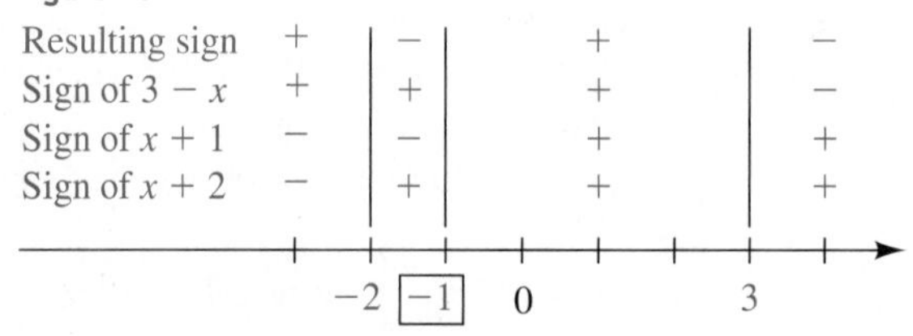

The frame around the -1 indicates that -1 makes a factor in the denominator of the original inequality equal to 0. Since the quadratic factor $x^2 + 1$ is always positive, it has no effect on the sign of the quotient and hence may be omitted from the diagram.

The various signs of the factors can be found using test values. Alternatively, we need only remember that as x increases, the sign of a linear factor $ax + b$ changes from negative to positive if the coefficient a of x is positive, and the sign changes from positive to negative if a is negative.

To determine where the quotient is less than or equal to 0, we first note from the sign diagram that it is *negative* for numbers in $(-2, -1) \cup (3, \infty)$. Since the quotient is 0 at $x = -2$ and $x = 3$, the numbers -2 and 3 are also solutions and must be *included* in our solution. Lastly, the quotient is *undefined* at $x = -1$, so -1 must be *excluded* from our solution. Thus, the solutions of the given inequality are given by

$$[-2, -1) \cup [3, \infty).$$

EXAMPLE 9 **Using a sign diagram to solve an inequality**

Solve the inequality $\dfrac{(2x + 1)^2(x - 1)}{x(x^2 - 1)} \ge 0.$

▶ **SOLUTION** Rewriting the inequality as

$$\frac{(2x + 1)^2(x - 1)}{x(x + 1)(x - 1)} \ge 0,$$

▶ **Tutorial available at www.thomsonedu.com/login**

we see that $x - 1$ is a factor of both the numerator and the denominator. Thus, *assuming that* $x - 1 \neq 0$ (that is, $x \neq 1$), we may cancel this factor and reduce our search for solutions to the case

$$\frac{(2x+1)^2}{x(x+1)} \geq 0 \qquad \text{and} \qquad x \neq 1.$$

We next observe that this quotient is 0 if $2x + 1 = 0$ $\left(\text{that is, if } x = -\frac{1}{2}\right)$. Hence, $-\frac{1}{2}$ is a solution. To find the remaining solutions, we construct the sign diagram in Figure 15. We do not include $(2x + 1)^2$ in the sign diagram, since this expression is always positive if $x \neq -\frac{1}{2}$ and so has no effect on the sign of the quotient. Referring to the resulting sign and remembering that $-\frac{1}{2}$ is a solution but 1 is *not* a solution, we see that the solutions of the given inequality are given by

$$(-\infty, -1) \cup \left\{-\tfrac{1}{2}\right\} \cup (0, 1) \cup (1, \infty).$$

Figure 15

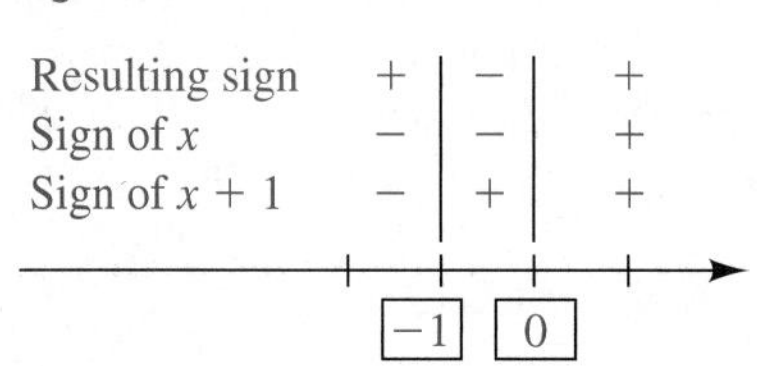

EXAMPLE 10 **Using a sign diagram to solve an inequality**

Solve the inequality $\dfrac{x+1}{x+3} \leq 2$.

SOLUTION A common mistake in solving such an inequality is to first multiply both sides by $x + 3$. If we did so, we would have to consider two cases, since $x + 3$ may be positive or negative (assuming $x + 3 \neq 0$), and we might have to reverse the inequality. A simpler method is to first obtain an equivalent inequality that has 0 on the right side and proceed from there:

$$\frac{x+1}{x+3} \leq 2 \quad \text{given}$$

$$\frac{x+1}{x+3} - 2 \leq 0 \quad \text{make one side 0}$$

$$\frac{x+1-2(x+3)}{x+3} \leq 0 \quad \text{combine into one fraction}$$

$$\frac{-x-5}{x+3} \leq 0 \quad \text{simplify}$$

$$\frac{x+5}{x+3} \geq 0 \quad \text{multiply by } -1$$

Note that the direction of the inequality is changed in the last step, since we multiplied by a negative number. This multiplication was performed for convenience, so that all factors would have positive coefficients of x.

The factors $x + 5$ and $x + 3$ are 0 at $x = -5$ and $x = -3$, respectively. This leads to the sign diagram in Figure 16, where the signs are determined as in previous examples. We see from the diagram that the resulting sign, and hence the sign of the quotient, is positive in $(-\infty, -5) \cup (-3, \infty)$. The quotient is 0 at $x = -5$ (include -5) and undefined at $x = -3$ (exclude -3). Hence, the solution of $(x + 5)/(x + 3) \geq 0$ is $(-\infty, -5] \cup (-3, \infty)$.

(continued)

Tutorial available at www.thomsonedu.com/login

Figure 16

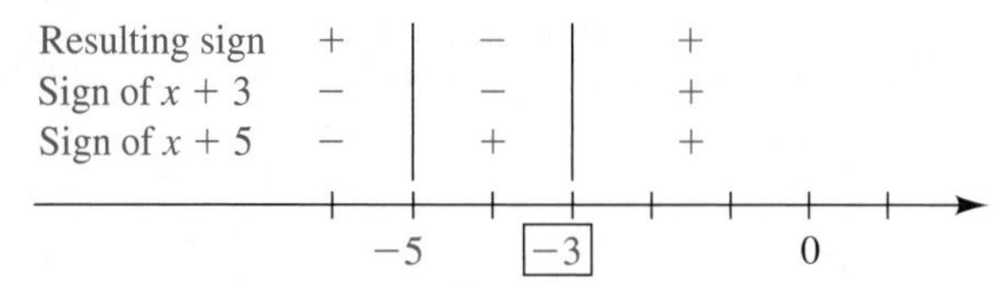

An alternative method of solution is to begin by multiplying both sides of the given inequality by $(x + 3)^2$, *assuming that* $x \neq -3$. In this case, $(x + 3)^2 > 0$ and the multiplication is permissible; however, after the resulting inequality is solved, the value $x = -3$ must be excluded.

EXAMPLE 11 Determining minimum therapeutic levels

For a drug to have a beneficial effect, its concentration in the bloodstream must exceed a certain value, which is called the *minimum therapeutic level.* Suppose that the concentration c (in mg/L) of a particular drug t hours after it is taken orally is given by

$$c = \frac{20t}{t^2 + 4}.$$

If the minimum therapeutic level is 4 mg/L, determine when this level is exceeded.

▶ SOLUTION The minimum therapeutic level, 4 mg/L, is exceeded if $c > 4$. Thus, we must solve the inequality

$$\frac{20t}{t^2 + 4} > 4.$$

Since $t^2 + 4 > 0$ for every t, we may multiply both sides by $t^2 + 4$ and proceed as follows:

$$20t > 4t^2 + 16 \qquad \text{allowable, since } t^2 + 4 > 0$$

$$-4t^2 + 20t - 16 > 0 \qquad \text{make one side 0}$$

$$t^2 - 5t + 4 < 0 \qquad \text{divide by the common factor } -4$$

$$(t - 1)(t - 4) < 0 \qquad \text{factor}$$

The factors in the last inequality are 0 when $t = 1$ and $t = 4$. These are the times at which c is *equal to* 4. As in previous examples, we may use a sign chart or sign diagram (with $t \geq 0$) to show that $(t - 1)(t - 4) < 0$ for every t in the interval $(1, 4)$. Hence, the minimum therapeutic level is exceeded if $1 < t < 4$.

Because graphs in a coordinate plane are introduced in the next chapter, it would be premature to demonstrate here the use of a graphing calculator or

▶ Tutorial available at www.thomsonedu.com/login

computer software to solve inequalities in x. Such methods will be considered later in the text.

Some basic properties of inequalities were stated at the beginning of this section. The following additional properties are helpful for solving certain inequalities. Proofs of the properties are given after the chart.

Additional Properties of Inequalities

Property	Illustration
(1) If $0 < a < b$, then $\frac{1}{a} > \frac{1}{b}$.	If $0 < \frac{1}{x} < 4$, then $\frac{1}{1/x} > \frac{1}{4}$, or $x > \frac{1}{4}$.
(2) If $0 < a < b$, then $0 < a^2 < b^2$.	If $0 < \sqrt{x} < 4$, then $0 < (\sqrt{x})^2 < 4^2$, or $0 < x < 16$.
(3) If $0 < a < b$, then $0 < \sqrt{a} < \sqrt{b}$.	If $0 < x^2 < 4$, then $0 < \sqrt{x^2} < \sqrt{4}$, or $0 < \lvert x \rvert < 2$.

PROOFS

(1) If $0 < a < b$, then multiplying by $1/(ab)$ yields

$$a \cdot \frac{1}{ab} < b \cdot \frac{1}{ab}, \quad \text{or} \quad \frac{1}{b} < \frac{1}{a}; \qquad \text{that is,} \quad \frac{1}{a} > \frac{1}{b}.$$

(2) If $0 < a < b$, then multiplying by a yields $a \cdot a < a \cdot b$ and multiplying by b yields $b \cdot a < b \cdot b$, so $a^2 < ab < b^2$ and hence $a^2 < b^2$.

(3) If $0 < a < b$, then $b - a > 0$ or, equivalently,

$$(\sqrt{b} + \sqrt{a})(\sqrt{b} - \sqrt{a}) > 0.$$

Dividing both sides of the last inequality by $\sqrt{b} + \sqrt{a}$, we obtain $\sqrt{b} - \sqrt{a} > 0$; that is, $\sqrt{b} > \sqrt{a}$.

1.6 *Exercises*

1–4: Express the inequality as an interval, and sketch its graph.

1 $x < -2$

2 $x \geq 4$

3 $-3 \geq x > -5$

4 $3 \leq x \leq 7$

Exer. 5–6: Express the interval as an inequality in the variable x.

5 $(-5, 8]$

6 $(-3, \infty)$

Exer. 7–50: Solve the inequality, and express the solutions in terms of intervals whenever possible.

7 $2x + 5 < 3x - 7$

8 $x - 8 > 5x + 3$

9 $3 \leq \frac{2x - 3}{5} < 7$

10 $-2 < \frac{4x + 1}{3} \leq 0$

▶ 11 $\frac{4}{3x + 2} \geq 0$

12 $\frac{3}{2x + 5} \leq 0$

▶ **Tutorial available at www.thomsonedu.com/login**

13 $\dfrac{-2}{4-3x} > 0$

14 $\dfrac{-3}{2-x} < 0$

15 $\dfrac{2}{(1-x)^2} > 0$

16 $\dfrac{4}{x^2+4} < 0$

17 $|x+3| < 0.01$

18 $|x-4| \le 0.03$

19 $|3x-7| \ge 5$

20 $2|-11-7x| - 2 > 10$

21 $|7x+2| > -2$

22 $|6x-5| \le -2$

23 $|3x-9| > 0$

24 $|5x+2| \le 0$

25 $-2 < |x| < 4$

26 $1 < |x| < 5$

27 $(3x+1)(5-10x) > 0$

28 $(x+2)(x-1)(4-x) \le 0$

29 $x^2 - x - 6 < 0$

30 $x^2 + 4x + 3 \ge 0$

31 $x(2x+3) \ge 5$

32 $6x - 8 > x^2$

33 $25x^2 - 9 < 0$

34 $25x^2 - 9x < 0$

35 $\dfrac{x^2(x+2)}{(x+2)(x+1)} \le 0$

36 $\dfrac{(x^2+1)(x-3)}{x^2-9} \ge 0$

37 $\dfrac{x^2-x}{x^2+2x} \le 0$

38 $\dfrac{(x+3)^2(2-x)}{(x+4)(x^2-4)} \le 0$

39 $\dfrac{x-2}{x^2-3x-10} \ge 0$

40 $\dfrac{x+5}{x^2-7x+12} \le 0$

41 $\dfrac{-3x}{x^2-9} > 0$

42 $\dfrac{2x}{16-x^2} < 0$

▶ 43 $\dfrac{x+1}{2x-3} > 2$

44 $\dfrac{x-2}{3x+5} \le 4$

45 $\dfrac{1}{x-2} \ge \dfrac{3}{x+1}$

46 $\dfrac{2}{2x+3} \le \dfrac{2}{x-5}$

47 $\dfrac{x}{3x-5} \le \dfrac{2}{x-1}$

48 $\dfrac{x}{2x-1} \ge \dfrac{3}{x+2}$

49 $x^3 > x$

50 $x^4 \ge x^2$

Exer. 51–52: Solve part (a) and use that answer to determine the answers to parts (b) and (c).

51 (a) $|x+5| = 3$ (b) $|x+5| < 3$
(c) $|x+5| > 3$

52 (a) $|x-3| < 2$ (b) $|x-3| = 2$
(c) $|x-3| > 2$

Exer. 53–54: Express the statement in terms of an inequality involving an absolute value.

53 The weight w of a wrestler must be within 2 pounds of 148 pounds.

▶ 54 The radius r of a ball bearing must be within 0.01 centimeter of 1 centimeter.

55 **Linear magnification** Shown in the figure is a simple magnifier consisting of a convex lens. The object to be magnified is positioned so that the distance p from the lens is less than the focal length f. The linear magnification M is the ratio of the image size to the object size. It is shown in physics that $M = f/(f-p)$. If $f = 6$ cm, how far should the object be placed from the lens so that its image appears at least three times as large? (Compare with Example 4.)

Exercise 55

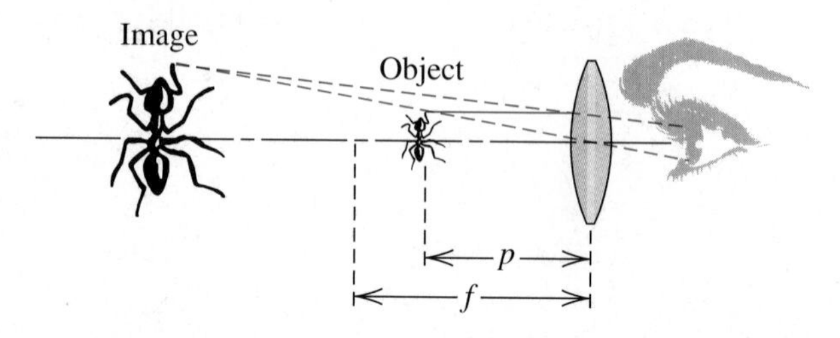

56 **Drug concentration** To treat arrhythmia (irregular heartbeat), a drug is fed intravenously into the bloodstream. Suppose that the concentration c of the drug after t hours is given by $c = 3.5t/(t+1)$ mg/L. If the minimum therapeutic level is 1.5 mg/L, determine when this level is exceeded.

57 **Business expenditure** A construction firm is trying to decide which of two models of a crane to purchase. Model A costs \$100,000 and requires \$8000 per year to maintain. Model B has an initial cost of \$80,000 and a maintenance cost of \$11,000 per year. For how many years must model A be used before it becomes more economical than B?

▶ 58 **Buying a car** A consumer is trying to decide whether to purchase car A or car B. Car A costs \$20,000 and has an mpg rating of 30, and insurance is \$1000 per year. Car B costs \$24,000 and has an mpg rating of 50, and insurance is \$1200 per year. Assume that the consumer drives 15,000 miles per year and that the price of gas remains constant at \$3 per gallon. Based only on these facts, determine how long it will take for the total cost of car B to become less than that of car A.

59 **Vertical leap record** *Guinness Book of World Records* reports that German shepherds can make vertical leaps of over 10 feet when scaling walls. If the distance s (in feet) off the ground after t seconds is given by the equation $s = -16t^2 + 24t + 1$, for how many seconds is the dog more than 9 feet off the ground?

60 **Height of a projected object** If an object is projected vertically upward from ground level with an initial velocity of 320 ft/sec, then its distance s above the ground after t seconds is given by $s = -16t^2 + 320t$. For what values of t will the object be more than 1536 feet above the ground?

61 **Braking distance** The braking distance d (in feet) of a certain car traveling v mi/hr is given by the equation $d = v + (v^2/20)$. Determine the velocities that result in braking distances of less than 75 feet.

62 **Gas mileage** The number of miles M that a certain compact car can travel on 1 gallon of gasoline is related to its speed v (in mi/hr) by

$$M = -\tfrac{1}{30}v^2 + \tfrac{5}{2}v \quad \text{for } 0 < v < 70.$$

For what speeds will M be at least 45?

63 **Decreasing height** A person's height will typically decrease by 0.024 inch each year after age 30.

(a) If a woman was 5 feet 9 inches tall at age 30, predict her height at age 70.

(b) A 50-year-old man is 5 feet 6 inches tall. Determine an inequality for the range of heights (in inches) that this man will experience between the ages of 30 and 70.

64 **Aircraft's landing speed** In the design of certain small turboprop aircraft, the landing speed V (in ft/sec) is determined by the formula $W = 0.00334V^2S$, where W is the gross weight (in pounds) of the aircraft and S is the surface area (in ft^2) of the wings. If the gross weight of the aircraft is between 7500 pounds and 10,000 pounds and $S = 210\ \text{ft}^2$, determine the range of the landing speeds in miles per hour.

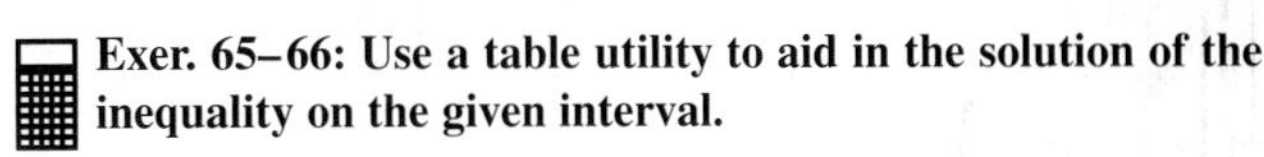

Exer. 65–66: Use a table utility to aid in the solution of the inequality on the given interval.

65 $\dfrac{(2 - x)(3x - 9)}{(1 - x)(x + 1)} > 0, \quad [-2, 3.5]$

66 $x^4 - x^3 - 16x^2 + 4x + 48 < 0, \quad [-3.5, 5]$

CHAPTER 1 REVIEW EXERCISES

▶ **Online support materials can be found at www.thomsonedu.com/login**

Exer. 1–2: Rewrite the expression without using the absolute value symbol, and simplify the result.

1 $|x + 3|$ if $x \leq -3$

2 $|(x - 2)(x - 3)|$ if $2 < x < 3$

Exer. 3–14: Simplify the expression, and rationalize the denominator when appropriate.

3 $\left(\dfrac{a^{2/3}b^{3/2}}{a^2b}\right)^6$

4 $(-2p^2q)^3\left(\dfrac{p}{4q^2}\right)^2$

5 $\left(\dfrac{xy^{-1}}{\sqrt{z}}\right)^4 \div \left(\dfrac{x^{1/3}y^2}{z}\right)^3$

6 $\left(\dfrac{-64x^3}{z^6y^9}\right)^{2/3}$

7 $[(a^{2/3}b^{-2})^3]^{-1}$

8 $x^{-2} - y^{-1}$

9 $\sqrt[3]{8x^5y^3z^4}$

10 $\sqrt[4]{(-4a^3b^2c)^2}$

11 $\dfrac{1}{\sqrt{t}}\left(\dfrac{1}{\sqrt{t}} - 1\right)$

12 $\sqrt{\sqrt[3]{(c^3d^6)^4}}$

▶ **Tutorial available at www.thomsonedu.com/login**

13 $\dfrac{\sqrt{12x^4y}}{\sqrt{3x^2y^5}}$

14 $\dfrac{3+\sqrt{x}}{3-\sqrt{x}}$

Exer. 15–22: Express as a polynomial.

15 $(3x^3 - 4x^2 + x - 7) + (x^4 - 2x^3 + 3x^2 + 5)$

16 $(x + 4)(x + 3) - (2x - 1)(x - 5)$

17 $(3a - 5b)(2a + 7b)$

18 $(4r^2 - 3s)^2$

19 $(13a^2 + 4b)(13a^2 - 4b)$

20 $(2a + b)^3$

21 $(3x + 2y)^2(3x - 2y)^2$

22 $(a + b + c + d)^2$

Exer. 23–30: Factor the polynomial.

23 $60xw + 70w$

24 $16a^4 + 24a^2b^2 + 9b^4$

25 $8x^3 + 64y^3$

26 $u^3v^4 - u^6v$

27 $p^8 - q^8$

28 $x^4 - 8x^3 + 16x^2$

29 $x^2 - 49y^2 - 14x + 49$

30 $x^5 - 4x^3 + 8x^2 - 32$

Exer. 31–36: Simplify the expression.

31 $\dfrac{2}{4x-5} - \dfrac{5}{10x+1}$

32 $\dfrac{7}{x+2} + \dfrac{3x}{(x+2)^2} - \dfrac{5}{x}$

33 $\dfrac{x + x^{-2}}{1 + x^{-2}}$

34 $(a^{-1} + b^{-1})^{-1}$

35 $\dfrac{\dfrac{x}{x+2} - \dfrac{4}{x+2}}{x - 3 - \dfrac{6}{x+2}}$

36 $\dfrac{(4 - x^2)\left(\frac{1}{3}\right)(6x + 1)^{-2/3}(6) - (6x + 1)^{1/3}(-2x)}{(4 - x^2)^2}$

Exer. 37–60: Solve the equation or inequality. Express the solutions in terms of intervals whenever possible.

37 $\dfrac{3x+1}{5x+7} = \dfrac{6x+11}{10x-3}$

38 $2x^2 + 5x - 12 = 0$

39 $x(3x + 4) = 5$

40 $4x^4 - 33x^2 + 50 = 0$

41 $20x^3 + 8x^2 - 35x - 14 = 0$

42 $|4x - 1| = 7$

43 $2|2x + 1| + 1 = 19$

44 $\dfrac{1}{x} + 6 = \dfrac{5}{\sqrt{x}}$

45 $\sqrt{7x + 2} + x = 6$

46 $\sqrt{3x + 1} - \sqrt{x + 4} = 1$

47 $10 - 7x < 4 + 2x$

48 $-\dfrac{1}{2} < \dfrac{2x+3}{5} < \dfrac{3}{2}$

49 $\dfrac{6}{10x+3} < 0$

50 $|4x + 7| < 21$

51 $2|3 - x| + 1 > 5$

52 $|16 - 3x| \geq 5$

53 $10x^2 + 11x > 6$

54 $x(x - 3) \leq 10$

55 $\dfrac{x^2(3-x)}{x+2} \leq 0$

56 $\dfrac{x^2 - x - 2}{x^2 + 4x + 3} \leq 0$

57 $\dfrac{3}{2x+3} < \dfrac{1}{x-2}$

58 $\dfrac{x+1}{x^2 - 25} \leq 0$

59 $x^3 > x^2$

60 $(x^2 - x)(x^2 - 5x + 6) < 0$

Exer. 61–64: Solve for the specified variable.

61 $P + N = \dfrac{C+2}{C}$ for C

62 $A = B\sqrt[3]{\dfrac{C}{D}} - E$ for D

63 $F = \dfrac{\pi P R^4}{8VL}$ for R (Poiseuille's law for fluids)

64 $V = \frac{1}{3}\pi h(r^2 + R^2 + rR)$ for r (volume of a frustum of a cone)

Exer. 65–68: Express in the form $a + bi$, where a and b are real numbers.

65 $(3 + 8i)^2$

66 $\dfrac{1}{9 - \sqrt{-4}}$

67 $\dfrac{6 - 3i}{2 + 7i}$

68 $\dfrac{20 - 8i}{4i}$

69 Investment income An investor has a choice of two investments: a bond fund and a stock fund. The bond fund yields 7.186% interest annually, which is nontaxable at both the federal and state levels. Suppose the investor pays federal income tax at a rate of 28% and state income tax at a rate of 7%. Determine what the annual yield must be on the taxable stock fund so that the two funds pay the same amount of net interest income to the investor.

70 Gold and silver mixture A ring that weighs 80 grams is made of gold and silver. By measuring the displacement of the ring in water, it has been determined that the ring has a volume of 5 cm^3. Gold weighs 19.3 g/cm^3, and silver weighs 10.5 g/cm^3. How many grams of gold does the ring contain?

71 Preparing hospital food A hospital dietitian wishes to prepare a 10-ounce meat-vegetable dish that will provide 7 grams of protein. If an ounce of the vegetable portion supplies $\frac{1}{2}$ gram of protein and an ounce of meat supplies 1 gram of protein, how much of each should be used?

72 Solar heating A large solar heating panel requires 120 gallons of a fluid that is 30% antifreeze. The fluid comes in either a 50% solution or a 20% solution. How many gallons of each should be used to prepare the 120-gallon solution?

73 Passing speed An automobile 20 feet long overtakes a truck 40 feet long that is traveling at 50 mi/hr (see the figure). At what constant speed must the automobile travel in order to pass the truck in 5 seconds?

Exercise 73

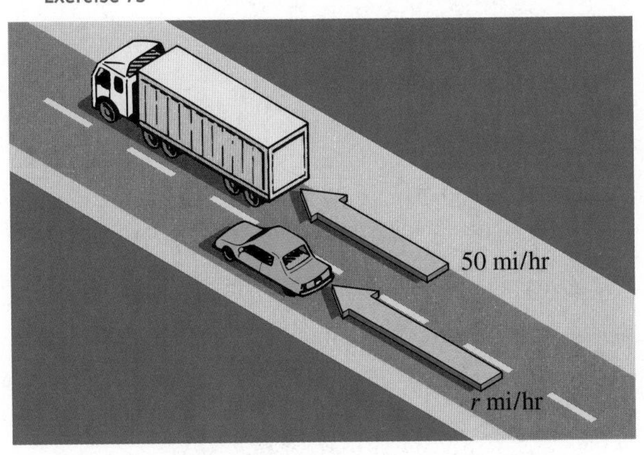

74 Filling a bin An extruder can fill an empty bin in 2 hours, and a packaging crew can empty a full bin in 5 hours. If a bin is half full when an extruder begins to fill it and a crew begins to empty it, how long will it take to fill the bin?

75 Highway travel A north-south highway intersects an east-west highway at a point P. An automobile crosses P at 10 A.M., traveling east at a constant rate of 20 mi/hr. At the same instant another automobile is 2 miles north of P, traveling south at 50 mi/hr.

(a) Find a formula for the distance d between the automobiles t hours after 10:00 A.M.

(b) At approximately what time will the automobiles be 104 miles apart?

76 Fencing a kennel A kennel owner has 270 feet of fencing material to be used to divide a rectangular area into 10 equal pens, as shown in the figure. Find dimensions that would allow 100 ft^2 for each pen.

Exercise 76

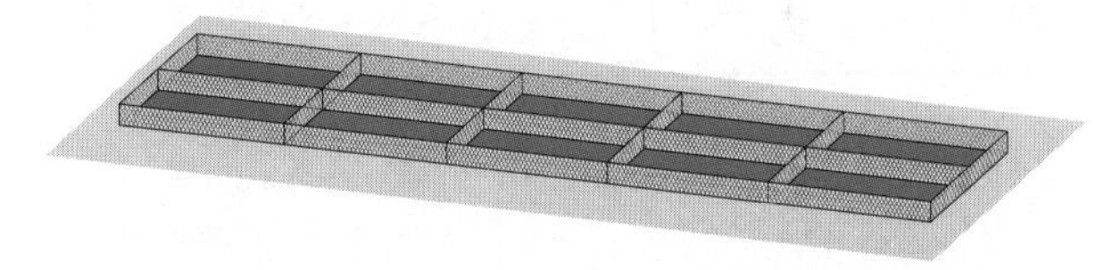

77 Dimensions of an aquarium An open-topped aquarium is to be constructed with 6-foot-long sides and square ends, as shown in the figure.

(a) Find the height of the aquarium if the volume is to be 48 ft^3.

(b) Find the height if 44 ft^2 of glass is to be used.

Exercise 77

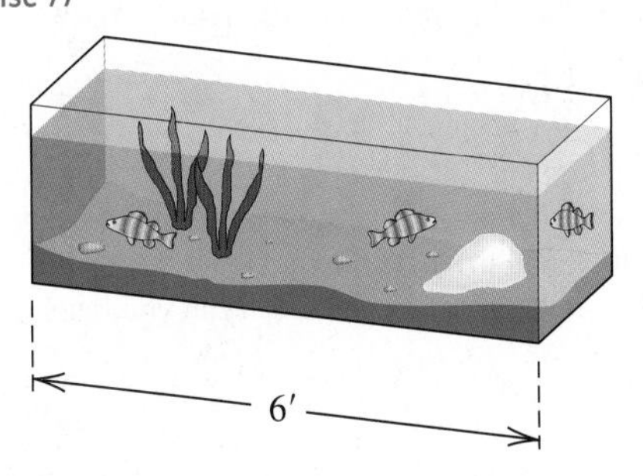

78 Boyle's law Boyle's law for a certain gas states that if the temperature is constant, then $pv = 200$, where p is the pressure (in lb/in^2) and v is the volume (in in^3). If $25 \le v \le 50$, what is the corresponding range for p?

79 **Sales commission** A recent college graduate has job offers for a sales position in two computer firms. Job A pays \$50,000 per year plus 10% commission. Job B pays only \$40,000 per year, but the commission rate is 20%. How much yearly business must the salesman do for the second job to be more lucrative?

80 **Speed of sound** The speed of sound in air at 0°C (or 273 K) is 1087 ft/sec, but this speed increases as the temperature rises. The speed v of sound at temperature T in K is given by $v = 1087\sqrt{T/273}$. At what temperatures does the speed of sound exceed 1100 ft/sec?

81 **Planting an apple orchard** The owner of an apple orchard estimates that if 24 trees are planted per acre, then each mature tree will yield 600 apples per year. For each additional tree planted per acre, the number of apples produced by each tree decreases by 12 per year. How many trees should be planted per acre to obtain at least 16,416 apples per year?

82 **Apartment rentals** A real estate company owns 218 efficiency apartments, which are fully occupied when the rent is \$940 per month. The company estimates that for each \$25 increase in rent, 5 apartments will become unoccupied. What rent should be charged in order to pay the monthly bills, which total \$205,920?

83 Choose the equation that best describes the table of data.

x	y
1	2.1213
2	3.6742
3	4.7434
4	5.6125
5	6.3640

(1) $y = 1.5529x + 0.5684$

(2) $y = \dfrac{3}{x} + x^2 - 1$

(3) $y = 3\sqrt{x - 0.5}$

(4) $y = 3x^{1/3} + 1.1213$

CHAPTER 1 DISCUSSION EXERCISES

1 **Surface area of a tank** You know that a spherical tank holds 10,000 gallons of water. What do you need to know to determine the surface area of the tank? Estimate the surface area of the tank.

2 Determine the conditions under which $\sqrt{a^2 + b^2} = a + b$.

3 Show that the sum of squares $x^2 + 25$ can be factored by adding and subtracting a particular term and following the method demonstrated in Example 2(c) of Section 1.3.

4 What is the difference between the expressions $\dfrac{1}{x + 1}$ and $\dfrac{x - 1}{x^2 - 1}$?

5 Write the quotient of two arbitrary second-degree polynomials in x, and evaluate the quotient with several large values of x. What general conclusion can you reach about such quotients?

6 Simplify the expression $\dfrac{3x^2 - 5x - 2}{x^2 - 4}$. Now evaluate both expressions with a value of x ($x \neq \pm 2$). Discuss what this evaluation proves (or doesn't) and what your simplification proves (or doesn't).

7 **Party trick** To guess your partner's age and height, have him/her do the following:

1 Write down his/her age.

2 Multiply it by 2.

3 Add 5.

4 Multiply this sum by 50.

5 Subtract 365.

6 Add his/her height (in inches).

7 Add 115.

The first two digits of the result equal his/her age, and the last two digits equal his/her height. Explain why this is true.

8 **Circuits problem** In a particular circuits problem, the output voltage is defined by

$$V_{\text{out}} = I_{\text{in}}\left(-\frac{RXi}{R - Xi}\right),$$

where $I_{\text{in}} = \dfrac{V_{\text{in}}}{Z_{\text{in}}}$ and $Z_{\text{in}} = \dfrac{R^2 - X^2 - 3RXi}{R - Xi}$. Find a formula for V_{out} in terms of V_{in} when R is equal to X.

9 When we factor the sum or difference of cubes, $x^3 \pm y^3$, is the factor $(x^2 \mp xy + y^2)$ ever factorable over the real numbers?

10 What is the average of the two solutions of the arbitrary quadratic equation $ax^2 + bx + c = 0$? Discuss how this knowledge can help you easily check the solutions to a quadratic equation.

11 (a) Find an expression of the form $p + qi$ for the multiplicative inverse of $\dfrac{a + bi}{c + di}$, where a, b, c, and d are real numbers.

(b) Does the expression you found apply to real numbers of the form a/c?

(c) Are there any restrictions on your answer for part (a)?

12 In solving the inequality $\dfrac{x - 1}{x - 2} \geq 3$, what is wrong with employing $x - 1 \geq 3(x - 2)$ as a first step?

13 Consider the inequality $ax^2 + bx + c \geq 0$, where a, b, and c are real numbers with $a \neq 0$. Suppose the associated equality $ax^2 + bx + c = 0$ has discriminant D. Categorize the solutions of the inequality according to the signs of a and D.

14 Freezing level in a cloud Refer to Exercises 73–75 in Section 1.4.

(a) Approximate the height of the freezing level in a cloud if the ground temperature is 80°F and the dew point is 68°F.

(b) Find a formula for the height h of the freezing level in a cloud for ground temperature G and dew point D.

15 Explain why you should not try to solve one of these equations.

$$\sqrt{2x - 3} + \sqrt{x + 5} = 0$$

$$\sqrt[3]{2x - 3} + \sqrt[3]{x + 5} = 0$$

16 Solve the equation

$$\sqrt{x} = cx - 2/c$$

for x, where $c = 2 \times 10^{500}$. Discuss why one of your positive solutions is extraneous.

17 Relating baseball records Based on the number of runs scored (S) and runs allowed (A), the Pythagorean winning percentage estimates what a baseball team's winning percentage should be. This formula, developed by baseball statistician Bill James, has the form

$$\frac{S^x}{S^x + A^x}.$$

James determined that $x = 1.83$ yields the most accurate results.

The 1927 New York Yankees are generally regarded as one of the best teams in baseball history. Their record was 110 wins and 44 losses. They scored 975 runs while allowing only 599.

(a) Find their Pythagorean win–loss record.

(b) Estimate the value of x (to the nearest 0.01) that best predicts the 1927 Yankees' actual win–loss record.

2

Functions and Graphs

The mathematical term *function* (or its Latin equivalent) dates back to the late seventeenth century, when calculus was in the early stages of development. This important concept is now the backbone of advanced courses in mathematics and is indispensable in every field of science.

In this chapter we study properties of functions using algebraic and graphical methods that include plotting points, determining symmetries, and making horizontal and vertical shifts. These techniques are adequate for obtaining rough sketches of graphs that help us understand properties of functions; modern-day methods, however, employ sophisticated computer software and advanced mathematics to generate extremely accurate graphical representations of functions.

2.1 Rectangular Coordinate Systems

In Section 1.1 we discussed how to assign a real number (coordinate) to each point on a line. We shall now show how to assign an **ordered pair** (a, b) of real numbers to each point in a plane. Although we have also used the notation (a, b) to denote an open interval, there is little chance for confusion, since it should always be clear from our discussion whether (a, b) represents a point or an interval.

We introduce a **rectangular,** or **Cartesian,*** **coordinate system** in a plane by means of two perpendicular coordinate lines, called **coordinate axes,** that intersect at the **origin** O, as shown in Figure 1. We often refer to the horizontal line as the ***x*-axis** and the vertical line as the ***y*-axis** and label them x and y, respectively. The plane is then a **coordinate plane,** or an ***xy*-plane.** The coordinate axes divide the plane into four parts called the **first, second, third,** and **fourth quadrants,** labeled I, II, III, and IV, respectively (see Figure 1). Points on the axes do not belong to any quadrant.

Each point P in an xy-plane may be assigned an ordered pair (a, b), as shown in Figure 1. We call a the ***x*-coordinate** (or **abscissa**) of P, and b the ***y*-coordinate** (or **ordinate**). We say that P *has coordinates* (a, b) and refer to the *point* (a, b) or the *point* $P(a, b)$. Conversely, every ordered pair (a, b) determines a point P with coordinates a and b. We **plot a point** by using a dot, as illustrated in Figure 2.

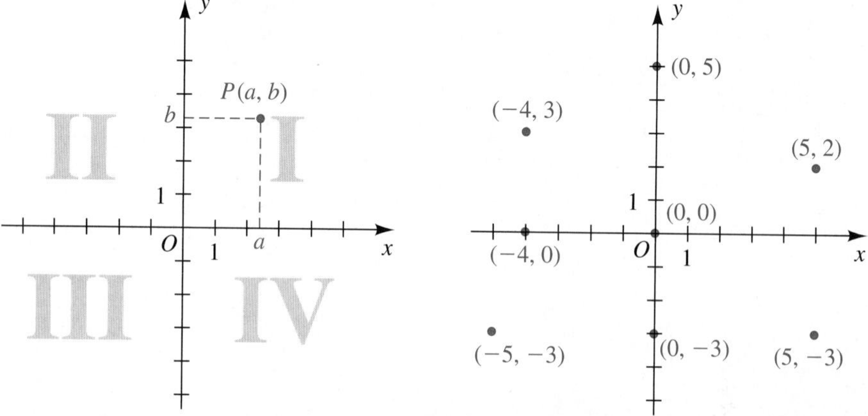

We may use the following formula to find the distance between two points in a coordinate plane.

Distance Formula

The distance $d(P_1, P_2)$ between any two points $P_1(x_1, y_1)$ and $P_2(x_2, y_2)$ in a coordinate plane is

$$d(P_1, P_2) = \sqrt{(x_2 - x_1)^2 + (y_2 - y_1)^2}.$$

*The term *Cartesian* is used in honor of the French mathematician and philosopher René Descartes (1596–1650), who was one of the first to employ such coordinate systems.

Figure 3

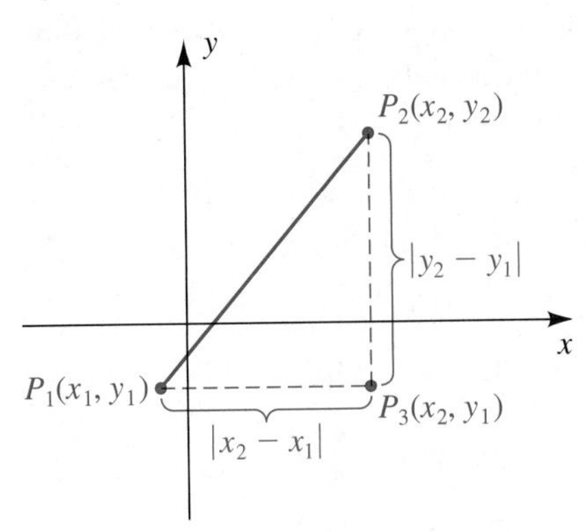

PROOF If $x_1 \neq x_2$ and $y_1 \neq y_2$, then, as illustrated in Figure 3, the points P_1, P_2, and $P_3(x_2, y_1)$ are vertices of a right triangle. By the Pythagorean theorem,

$$[d(P_1, P_2)]^2 = [d(P_1, P_3)]^2 + [d(P_3, P_2)]^2.$$

From the figure we see that

$$d(P_1, P_3) = |x_2 - x_1| \quad \text{and} \quad d(P_3, P_2) = |y_2 - y_1|.$$

Since $|a|^2 = a^2$ for every real number a, we may write

$$[d(P_1, P_2)]^2 = (x_2 - x_1)^2 + (y_2 - y_1)^2.$$

Taking the square root of each side of the last equation and using the fact that $d(P_1, P_2) \geq 0$ gives us the distance formula.

If $y_1 = y_2$, the points P_1 and P_2 lie on the same horizontal line, and

$$d(P_1, P_2) = |x_2 - x_1| = \sqrt{(x_2 - x_1)^2}.$$

Similarly, if $x_1 = x_2$, the points are on the same vertical line, and

$$d(P_1, P_2) = |y_2 - y_1| = \sqrt{(y_2 - y_1)^2}.$$

These are special cases of the distance formula.

Although we referred to the points shown in Figure 3, our proof is independent of the positions of P_1 and P_2. ■

When applying the distance formula, note that $d(P_1, P_2) = d(P_2, P_1)$ and, hence, the order in which we subtract the x-coordinates and the y-coordinates of the points is immaterial. We may think of the distance between two points as the length of the hypotenuse of a right triangle.

Figure 4

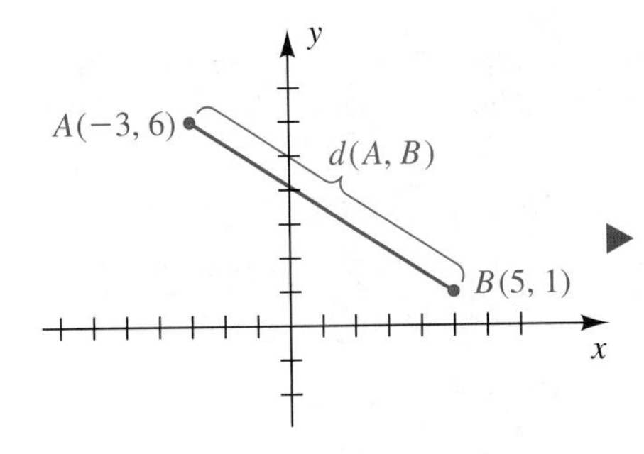

EXAMPLE 1 Finding the distance between points

Plot the points $A(-3, 6)$ and $B(5, 1)$, and find the distance $d(A, B)$.

▶ SOLUTION The points are plotted in Figure 4. By the distance formula,

$$\begin{aligned} d(A, B) &= \sqrt{[5 - (-3)]^2 + (1 - 6)^2} \\ &= \sqrt{8^2 + (-5)^2} \\ &= \sqrt{64 + 25} = \sqrt{89} \approx 9.43. \end{aligned}$$

■

EXAMPLE 2 Showing that a triangle is a right triangle

(a) Plot $A(-1, -3)$, $B(6, 1)$, and $C(2, -5)$, and show that triangle ABC is a right triangle.

(b) Find the area of triangle ABC.

▶ Tutorial available at www.thomsonedu.com/login

Figure 5

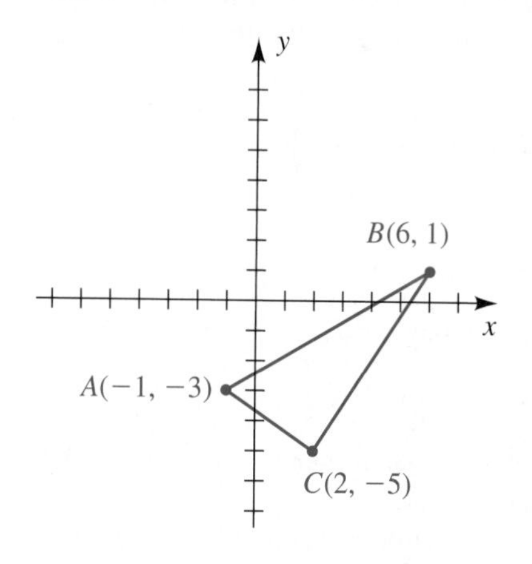

▶ **SOLUTION**

(a) The points are plotted in Figure 5. From geometry, triangle ABC is a right triangle if the sum of the squares of two of its sides is equal to the square of the remaining side. By the distance formula,

$$d(A, B) = \sqrt{(6 + 1)^2 + (1 + 3)^2} = \sqrt{49 + 16} = \sqrt{65}$$
$$d(B, C) = \sqrt{(2 - 6)^2 + (-5 - 1)^2} = \sqrt{16 + 36} = \sqrt{52}$$
$$d(A, C) = \sqrt{(2 + 1)^2 + (-5 + 3)^2} = \sqrt{9 + 4} = \sqrt{13}.$$

Since $d(A, B) = \sqrt{65}$ is the largest of the three values, the condition to be satisfied is

$$[d(A, B)]^2 = [d(B, C)]^2 + [d(A, C)]^2.$$

Substituting the values found using the distance formula, we obtain

$$[d(A, B)]^2 = \left(\sqrt{65}\right)^2 = 65$$

and $[d(B, C)]^2 + [d(A, C)]^2 = \left(\sqrt{52}\right)^2 + \left(\sqrt{13}\right)^2 = 52 + 13 = 65.$

Thus, the triangle is a right triangle with hypotenuse AB.

Area of a triangle:

$$A = \tfrac{1}{2}bh$$

(b) The area of a triangle with base b and altitude h is $\frac{1}{2}bh$. Referring to Figure 5, we let

$$b = d(B, C) = \sqrt{52} \quad \text{and} \quad h = d(A, C) = \sqrt{13}.$$

Hence, the area of triangle ABC is

$$\tfrac{1}{2}bh = \tfrac{1}{2}\sqrt{52}\,\sqrt{13} = \tfrac{1}{2} \cdot 2\sqrt{13}\,\sqrt{13} = 13.$$

EXAMPLE 3 Applying the distance formula

Given $A(1, 7)$, $B(-3, 2)$, and $C\left(4, \frac{1}{2}\right)$, prove that C is on the perpendicular bisector of segment AB.

Figure 6

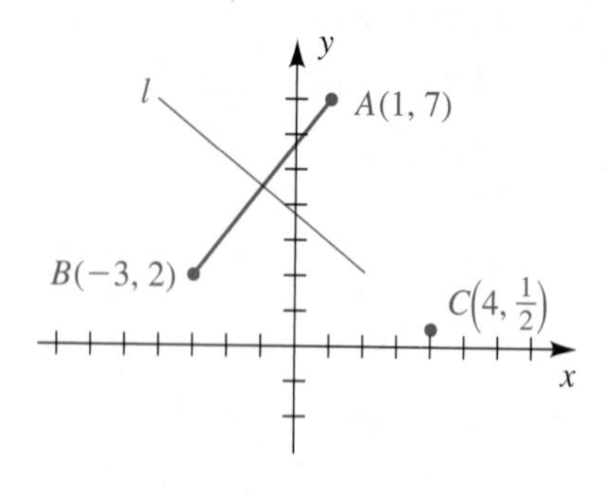

▶ **SOLUTION** The points A, B, C and the *perpendicular bisector* l are illustrated in Figure 6. From plane geometry, l can be characterized by either of the following conditions:

(1) l is the line perpendicular to segment AB at its midpoint.

(2) l is the set of all points equidistant from the endpoints of segment AB.

We shall use condition 2 to show that C is on l by verifying that

$$d(A, C) = d(B, C).$$

We apply the distance formula:

$$d(A, C) = \sqrt{(4 - 1)^2 + \left(\tfrac{1}{2} - 7\right)^2} = \sqrt{3^2 + \left(-\tfrac{13}{2}\right)^2} = \sqrt{9 + \tfrac{169}{4}} = \sqrt{\tfrac{205}{4}}$$
$$d(B, C) = \sqrt{[4 - (-3)]^2 + \left(\tfrac{1}{2} - 2\right)^2} = \sqrt{7^2 + \left(-\tfrac{3}{2}\right)^2} = \sqrt{49 + \tfrac{9}{4}} = \sqrt{\tfrac{205}{4}}$$

Thus, C is equidistant from A and B, and the verification is complete.

▶ **Tutorial available at www.thomsonedu.com/login**

EXAMPLE 4 **Finding a formula that describes a perpendicular bisector**

Given $A(1, 7)$ and $B(-3, 2)$, find a formula that expresses the fact that an arbitrary point $P(x, y)$ is on the perpendicular bisector l of segment AB.

▶ **SOLUTION** By condition 2 of Example 3, $P(x, y)$ is on l if and only if $d(A, P) = d(B, P)$; that is,

$$\sqrt{(x-1)^2 + (y-7)^2} = \sqrt{[x-(-3)]^2 + (y-2)^2}.$$

To obtain a simpler formula, let us square both sides and simplify terms of the resulting equation, as follows:

$$\begin{aligned}(x-1)^2 + (y-7)^2 &= [x-(-3)]^2 + (y-2)^2 \\ x^2 - 2x + 1 + y^2 - 14y + 49 &= x^2 + 6x + 9 + y^2 - 4y + 4 \\ -2x + 1 - 14y + 49 &= 6x + 9 - 4y + 4 \\ -8x - 10y &= -37 \\ 8x + 10y &= 37\end{aligned}$$

Note that, in particular, the last formula is true for the coordinates of the point $C\left(4, \frac{1}{2}\right)$ in Example 3, since if $x = 4$ and $y = \frac{1}{2}$, substitution in $8x + 10y$ gives us

$$8 \cdot 4 + 10 \cdot \tfrac{1}{2} = 37.$$

In Example 9 of Section 2.3, we will find a formula for the perpendicular bisector of a segment using condition 1 of Example 3.

We can find the midpoint of a line segment by using the following formula.

Midpoint Formula

The midpoint M of the line segment from $P_1(x_1, y_1)$ to $P_2(x_2, y_2)$ is

$$\left(\frac{x_1 + x_2}{2}, \frac{y_1 + y_2}{2}\right).$$

PROOF The lines through P_1 and P_2 parallel to the y-axis intersect the x-axis at $A_1(x_1, 0)$ and $A_2(x_2, 0)$. From plane geometry, the line through the midpoint M parallel to the y-axis bisects the segment A_1A_2 at point M_1 (see Figure 7). If $x_1 < x_2$, then $x_2 - x_1 > 0$, and hence $d(A_1, A_2) = x_2 - x_1$. Since M_1 is halfway from A_1 to A_2, the x-coordinate of M_1 is equal to the x-coordinate of A_1 plus one-half the distance from A_1 to A_2; that is,

$$x\text{-coordinate of } M_1 = x_1 + \tfrac{1}{2}(x_2 - x_1).$$

(continued)

▶ **Tutorial available at www.thomsonedu.com/login**

Figure 7

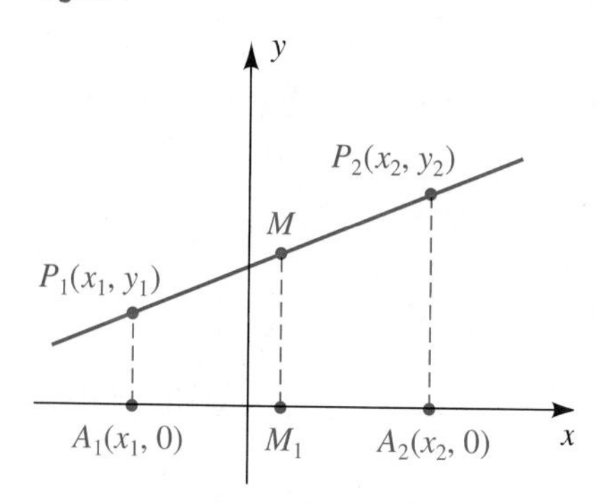

The expression on the right side of the last equation simplifies to

$$\frac{x_1 + x_2}{2}.$$

This quotient is the *average* of the numbers x_1 and x_2. It follows that the x-coordinate of M is also $(x_1 + x_2)/2$. Similarly, the y-coordinate of M is $(y_1 + y_2)/2$. These formulas hold for all positions of P_1 and P_2. ■

To apply the midpoint formula, it may suffice to remember that

the x-coordinate of the midpoint = the *average* of the x-coordinates,

and that

the y-coordinate of the midpoint = the *average* of the y-coordinates.

EXAMPLE 5 Finding a midpoint

Find the midpoint M of the line segment from $P_1(-2, 3)$ to $P_2(4, -2)$, and verify that $d(P_1, M) = d(P_2, M)$.

Figure 8

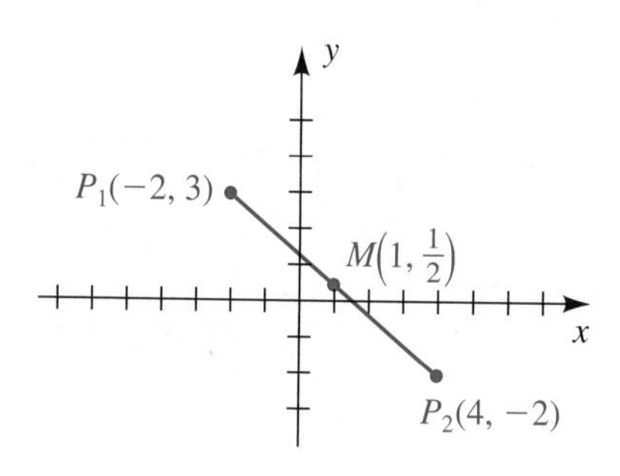

▶ SOLUTION By the midpoint formula, the coordinates of M are

$$\left(\frac{-2+4}{2}, \frac{3+(-2)}{2}\right), \quad \text{or} \quad \left(1, \frac{1}{2}\right).$$

The three points P_1, P_2, and M are plotted in Figure 8. By the distance formula,

$$d(P_1, M) = \sqrt{(1+2)^2 + \left(\tfrac{1}{2} - 3\right)^2} = \sqrt{9 + \tfrac{25}{4}}$$

$$d(P_2, M) = \sqrt{(1-4)^2 + \left(\tfrac{1}{2} + 2\right)^2} = \sqrt{9 + \tfrac{25}{4}}.$$

Hence, $d(P_1, M) = d(P_2, M)$. ■

The term **graphing utility** refers to either a graphing calculator or a computer equipped with appropriate software packages. The **viewing rectangle** of a graphing utility is the portion of the xy-plane shown on the screen. The boundaries (sides) of the viewing rectangle can be manually set by assigning a minimum x value (Xmin), a maximum x value (Xmax), the difference between the tick marks on the x-axis (Xscl), a minimum y value (Ymin), a maximum y value (Ymax), and the difference between the tick marks on the y-axis (Yscl). In examples, we often use the standard (or default) values for the viewing rectangle. These values depend on the dimensions (measured in pixels) of the graphing utility screen. If we want a different view of the graph, we use the phrase

"using [Xmin, Xmax, Xscl] by [Ymin, Ymax, Yscl]"

to indicate the change in the viewing rectangle. If Xscl and/or Yscl are omitted, the default value is 1.

▶ **Tutorial available at www.thomsonedu.com/login**

EXAMPLE 6 **Plotting points on a graphing calculator**

The United States population estimates for July 1 of several years are listed in the table.

(a) Plot the data.

(b) Use the midpoint formula to estimate the population in 2003.

(c) Find the percentage increase in population from 2004 to 2005.

Year	Population
2001	285,107,923
2002	287,984,799
2004	293,656,842
2005	296,410,404

SOLUTION

Enter the data.

TI-83/4 Plus

(a) Put years in L1 (list 1), populations in L2.

[STAT] [1] 2001 [ENTER] 2002 [ENTER] 2004 [ENTER] 2005 [ENTER] [△ (4 times)] [▷] 285 107 923 [ENTER] 287 984 799 [ENTER] 293 656 842 [ENTER] 296 410 404 [ENTER]

L1	L2	L3 2
2001	2.85E8	------
2002	2.88E8	
2004	2.94E8	
2005	2.96E8	

L2(5) =

TI-86

Put years in xStat, populations in yStat.

[2nd] [STAT] [EDIT(F2)] 2001 [ENTER] 2002 [ENTER] 2004 [ENTER] 2005 [ENTER] [△ (4 times)] [▷] 285 107 923 [ENTER] 287 984 799 [ENTER] 293 656 842 [ENTER] 296 410 404 [ENTER]

xStat	yStat	fStat 2
2001	2.8511E8	1
2002	2.8798E8	1
2004	2.9366E8	1
2005	2.9641E8	1
--------		--------

yStat(5) =

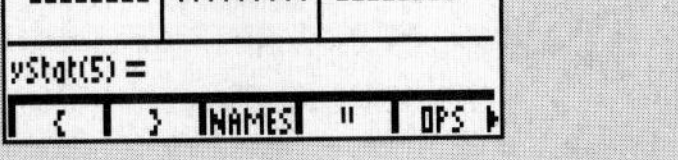

Turn STAT PLOT 1 on.

[2nd] [STAT PLOT] [1] [ENTER]

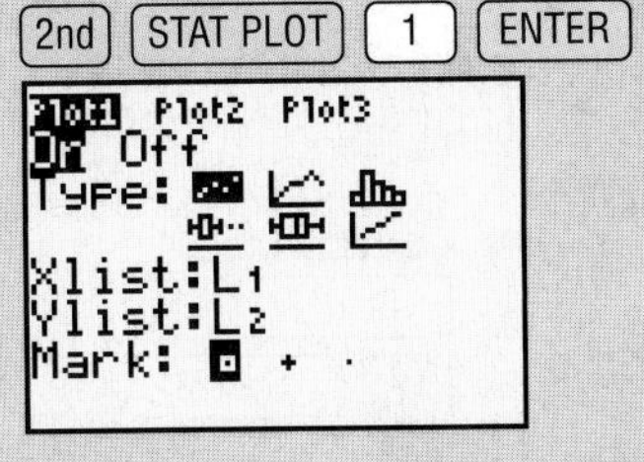

[2nd] [STAT] [PLOT(F3)] [PLOT1(F1)] [ENTER]

Plot the data.

Be sure to turn off or delete all Y assignments. If you use ZOOM STAT or ZDATA, the calculator will automatically select the viewing rectangle so that all the data are displayed.

[ZOOM] [9]

[GRAPH] [ZOOM(F3)] [MORE] [ZDATA(F5)] [CLEAR]

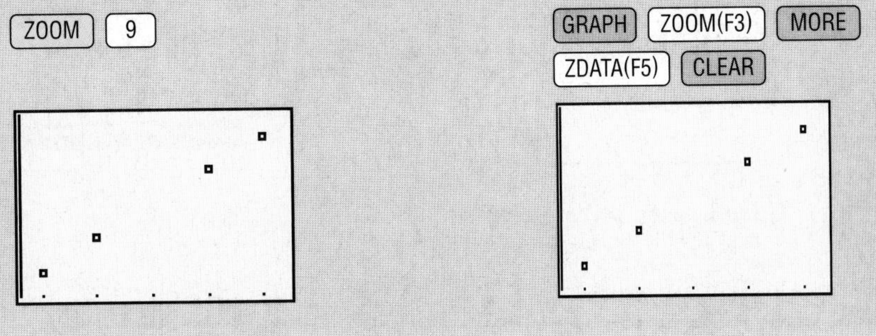

(continued)

Check the window values.

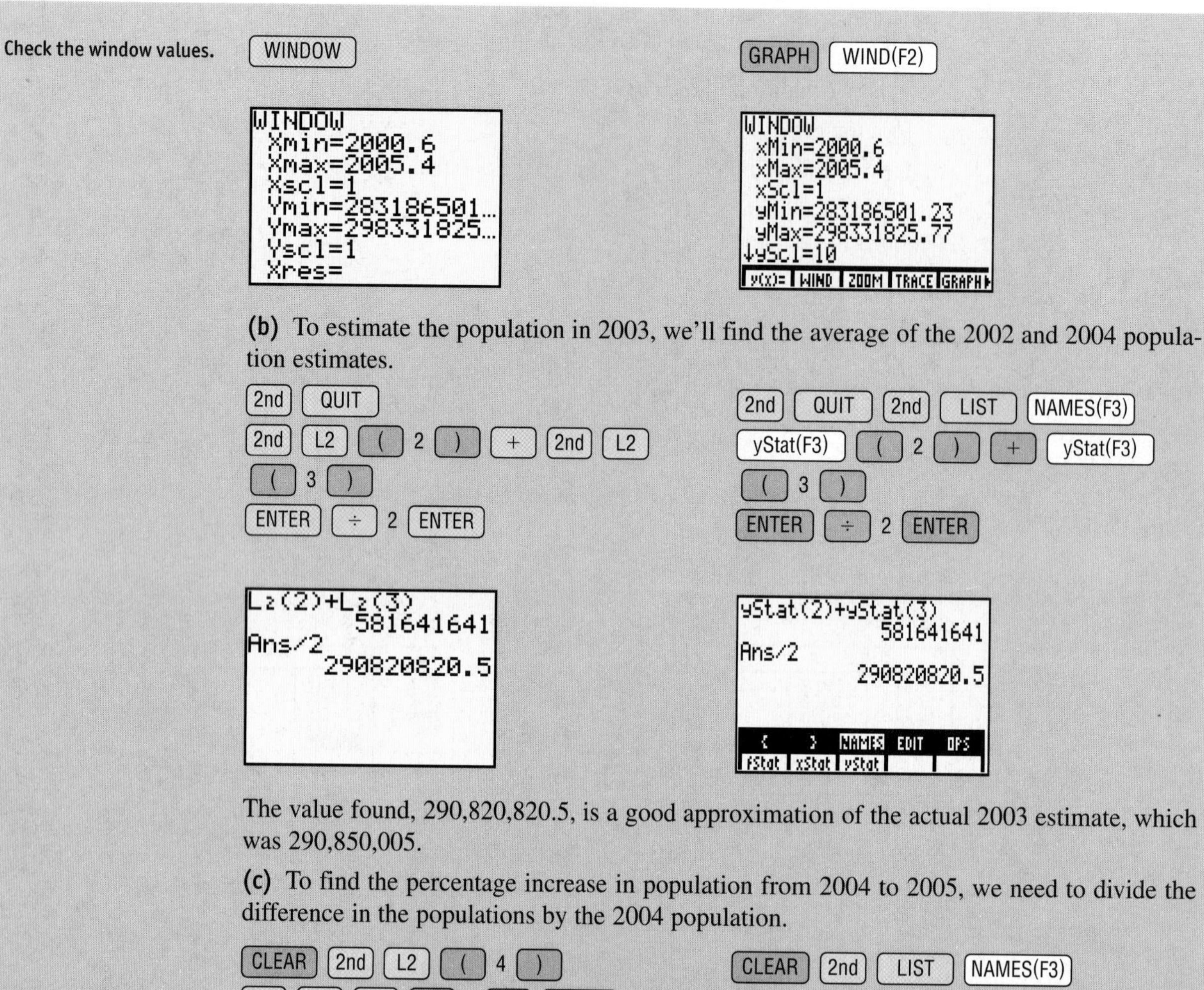

(b) To estimate the population in 2003, we'll find the average of the 2002 and 2004 population estimates.

The value found, 290,820,820.5, is a good approximation of the actual 2003 estimate, which was 290,850,005.

(c) To find the percentage increase in population from 2004 to 2005, we need to divide the difference in the populations by the 2004 population.

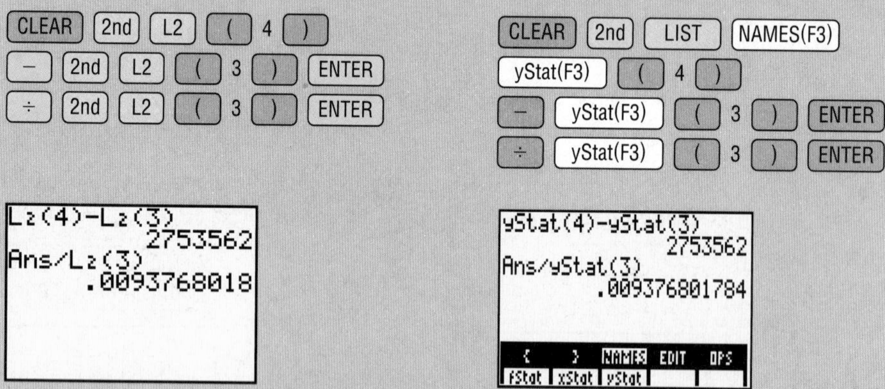

There was an increase of about 0.94% from 2004 to 2005.

2.1 Exercises

1 Plot the points $A(5, -2)$, $B(-5, -2)$, $C(5, 2)$, $D(-5, 2)$, $E(3, 0)$, and $F(0, 3)$ on a coordinate plane.

2 Plot the points $A(-3, 1)$, $B(3, 1)$, $C(-2, -3)$, $D(0, 3)$, and $E(2, -3)$ on a coordinate plane. Draw the line segments AB, BC, CD, DE, and EA.

3 Plot the points $A(0, 0)$, $B(1, 1)$, $C(3, 3)$, $D(-1, -1)$, and $E(-2, -2)$. Describe the set of all points of the form (a, a), where a is a real number.

4 Plot the points $A(0, 0)$, $B(1, -1)$, $C(3, -3)$, $D(-1, 1)$, and $E(-3, 3)$. Describe the set of all points of the form $(a, -a)$, where a is a real number.

Exer. 5–6: Find the coordinates of the points A–F.

5

6

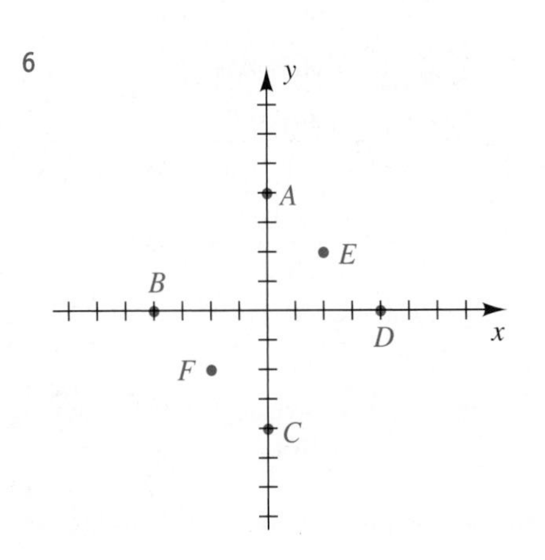

Exer. 7–8: Describe the set of all points $P(x, y)$ in a coordinate plane that satisfy the given condition.

7 (a) $x = -2$ (b) $y = 3$ (c) $x \geq 0$

(d) $xy > 0$ (e) $y < 0$ (f) $x = 0$

8 (a) $y = -2$ (b) $x = -4$ (c) $x/y < 0$

(d) $xy = 0$ (e) $y > 1$ (f) $y = 0$

Exer. 9–14: (a) Find the distance $d(A, B)$ between A and B. (b) Find the midpoint of the segment AB.

9 $A(4, -3)$, $B(6, 2)$

10 $A(-2, -5)$, $B(4, 6)$

11 $A(-5, 0)$, $B(-2, -2)$

12 $A(6, 2)$, $B(6, -2)$

13 $A(7, -3)$, $B(3, -3)$

14 $A(-4, 7)$, $B(0, -8)$

Exer. 15–16: Show that the triangle with vertices A, B, and C is a right triangle, and find its area.

15

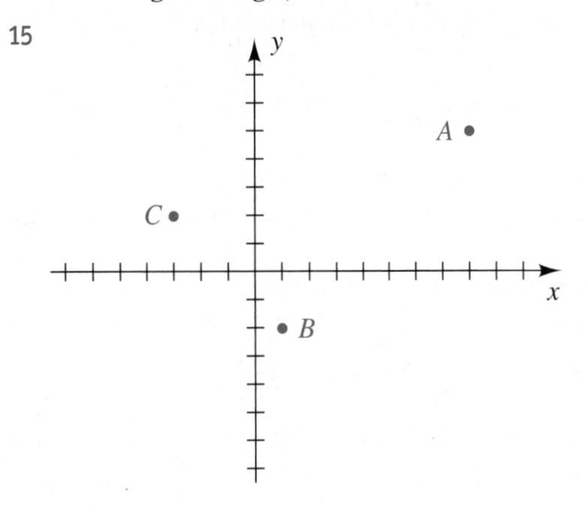

16

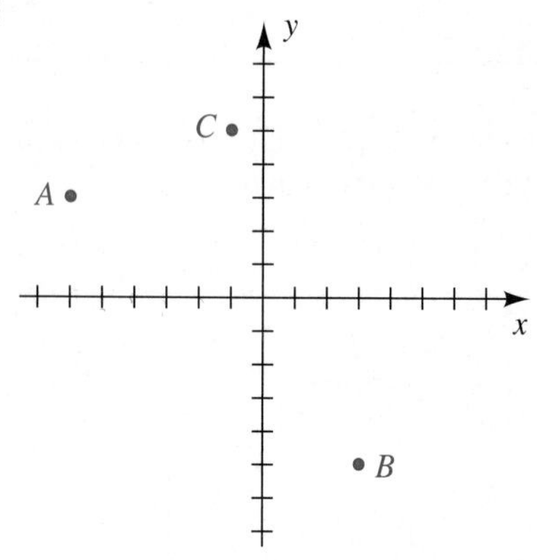

17 Show that $A(-4, 2)$, $B(1, 4)$, $C(3, -1)$, and $D(-2, -3)$ are vertices of a square.

18 Show that $A(-4, -1)$, $B(0, -2)$, $C(6, 1)$, and $D(2, 2)$ are vertices of a parallelogram.

▶ 19 Given $A(-3, 8)$, find the coordinates of the point B such that $C(5, -10)$ is the midpoint of segment AB.

20 Given $A(5, -8)$ and $B(-6, 2)$, find the point on segment AB that is three-fourths of the way from A to B.

Exer. 21–22: Prove that C is on the perpendicular bisector of segment AB.

21 $A(-4, -3)$, $B(6, 1)$, $C(5, -11)$

22 $A(-3, 2)$, $B(5, -4)$, $C(7, 7)$

Exer. 23–24: Find a formula that expresses the fact that an arbitrary point $P(x, y)$ is on the perpendicular bisector l of segment AB.

▶ 23 $A(-4, -3)$, $B(6, 1)$

24 $A(-3, 2)$, $B(5, -4)$

▶ 25 Find a formula that expresses the fact that $P(x, y)$ is a distance 5 from the origin. Describe the set of all such points.

26 Find a formula that states that $P(x, y)$ is a distance $r > 0$ from a fixed point $C(h, k)$. Describe the set of all such points.

27 Find all points on the y-axis that are a distance 6 from $P(5, 3)$.

28 Find all points on the x-axis that are a distance 5 from $P(-2, 4)$.

▶ 29 Find the point with coordinates of the form $(2a, a)$ that is in the third quadrant and is a distance 5 from $P(1, 3)$.

30 Find all points with coordinates of the form (a, a) that are a distance 3 from $P(-2, 1)$.

31 For what values of a is the distance between $P(a, 3)$ and $Q(5, 2a)$ greater than $\sqrt{26}$?

32 Given $A(-2, 0)$ and $B(2, 0)$, find a formula not containing radicals that expresses the fact that the sum of the distances from $P(x, y)$ to A and to B, respectively, is 5.

33 Prove that the midpoint of the hypotenuse of any right triangle is equidistant from the vertices. (*Hint:* Label the vertices of the triangle $O(0, 0)$, $A(a, 0)$, and $B(0, b)$.)

34 Prove that the diagonals of any parallelogram bisect each other. (*Hint:* Label three of the vertices of the parallelogram $O(0, 0)$, $A(a, b)$, and $C(0, c)$.)

Exer. 35–36: Plot the points in the given viewing rectangle.

35 $A(-5, -3.5)$, $B(-2, 2)$, $C(1, 0.5)$, $D(4, 1)$, and $E(7, 2.5)$ in $[-10, 10]$ by $[-10, 10]$

36 $A(-10, 4)$, $B(-7, -1.1)$, $C(0, -6)$, $D(3, -5.1)$, and $E(9, 2.1)$ in $[-12, 12]$ by $[-8, 8]$

37 Households with a computer The table lists the number of U.S. households with a computer for selected years.

Year	Households
1997	36,600,000
1998	42,100,000
2000	51,000,000
2001	56,300,000
2003	61,800,000

(a) Plot the data in the viewing rectangle $[1996, 2004]$ by $[35 \times 10^6, 63 \times 10^6, 1 \times 10^6]$.

(b) Discuss how the number of households is changing.

▶ **Tutorial available at www.thomsonedu.com/login**

38 **Published newspapers** The table lists the number of daily newspapers published in the United States for various years.

(a) Plot the data in the viewing rectangle [1895, 2005, 10] by [0, 3000, 1000].

(b) Use the midpoint formula to estimate the number of newspapers in 1930. Compare your answer with the true value, which is 1942.

Year	Newspapers
1900	2226
1920	2042
1940	1878
1960	1763
1980	1745
2000	1480

2.2 Graphs of Equations

Graphs are often used to illustrate changes in quantities. A graph in the business section of a newspaper may show the fluctuation of the Dow-Jones average during a given month; a meteorologist might use a graph to indicate how the air temperature varied throughout a day; a cardiologist employs graphs (electrocardiograms) to analyze heart irregularities; an engineer or physicist may turn to a graph to illustrate the manner in which the pressure of a confined gas increases as the gas is heated. Such visual aids usually reveal the behavior of quantities more readily than a long table of numerical values.

Two quantities are sometimes related by means of an equation or formula that involves two variables. In this section we discuss how to represent such an equation geometrically, by a graph in a coordinate plane. The graph may then be used to discover properties of the quantities that are not evident from the equation alone. The following chart introduces the basic concept of the graph of an equation in two variables x and y. Of course, other letters can also be used for the variables.

Terminology	Definition	Illustration
Solution of an equation in x and y	An ordered pair (a, b) that yields a true statement if $x = a$ and $y = b$	$(2, 3)$ is a solution of $y^2 = 5x - 1$, since substituting $x = 2$ and $y = 3$ gives us LS: $3^2 = 9$ RS: $5(2) - 1 = 10 - 1 = 9$.

For each solution (a, b) of an equation in x and y there is a point $P(a, b)$ in a coordinate plane. The set of all such points is called the **graph** of the equation. To *sketch the graph of an equation,* we illustrate the significant features of the graph in a coordinate plane. In simple cases, a graph can be sketched by plotting few, if any, points. For a complicated equation, plotting points may give very little information about the graph. In such cases, methods of calculus or computer graphics are often employed. Let us begin with a simple example.

EXAMPLE 1 Sketching a simple graph by plotting points

Sketch the graph of the equation $y = 2x - 1$.

Figure 1

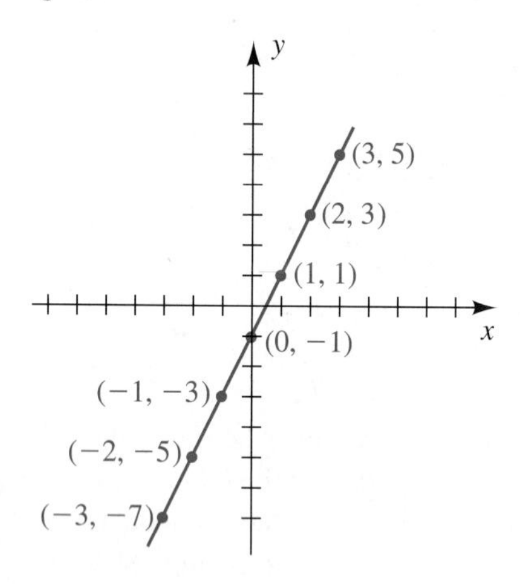

SOLUTION We wish to find the points (x, y) in a coordinate plane that correspond to the solutions of the equation. It is convenient to list coordinates of several such points in a table, where for each x we obtain the value for y from $y = 2x - 1$:

x	−3	−2	−1	0	1	2	3
y	−7	−5	−3	−1	1	3	5

The points with these coordinates appear to lie on a line, and we can sketch the graph in Figure 1. Ordinarily, the few points we have plotted would not be enough to illustrate the graph of an equation; however, in this elementary case we can be reasonably sure that the graph is a line. In the next section we will establish this fact.

It is impossible to sketch the entire graph in Example 1, because we can assign values to x that are numerically as large as desired. Nevertheless, we call the drawing in Figure 1 *the graph of the equation* or *a sketch of the graph.* In general, the sketch of a graph should illustrate its essential features so that the remaining (unsketched) parts are self-evident. For instance, in Figure 1, the **end behavior**—the pattern of the graph as x assumes large positive and negative values (that is, the shape of the right and left ends)—is apparent to the reader.

If a graph terminates at some point (as would be the case for a half-line or line segment), we place a dot at the appropriate *endpoint* of the graph. As a final general remark, *if ticks on the coordinate axes are not labeled* (as in Figure 1), *then each tick represents one unit.* We shall label ticks only when different units are used on the axes. For *arbitrary* graphs, where units of measurement are irrelevant, we omit ticks completely (see, for example, Figures 5 and 6).

Figure 2

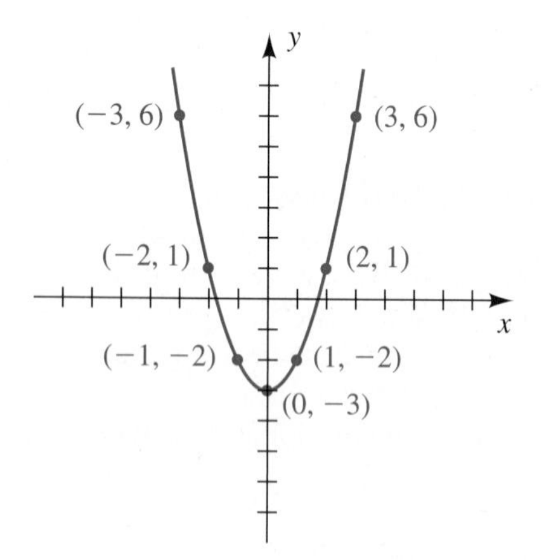

EXAMPLE 2 Sketching the graph of an equation

Sketch the graph of the equation $y = x^2 - 3$.

SOLUTION Substituting values for x and finding the corresponding values of y using $y = x^2 - 3$, we obtain a table of coordinates for several points on the graph:

x	−3	−2	−1	0	1	2	3
y	6	1	−2	−3	−2	1	6

Larger values of $|x|$ produce larger values of y. For example, the points (4, 13), (5, 22), and (6, 33) are on the graph, as are (−4, 13), (−5, 22), and (−6, 33). Plotting the points given by the table and drawing a smooth curve through these points (in the order of increasing values of x) gives us the sketch in Figure 2.

The graph in Figure 2 is a **parabola,** and the y-axis is the **axis of the parabola.** The lowest point (0, −3) is the **vertex** of the parabola, and we say that the parabola *opens upward.* If we invert the graph, then the parabola *opens downward* and the vertex is the highest point on the graph. In general, the graph of *any* equation of the form $y = ax^2 + c$ with $a \neq 0$ is a parabola with vertex $(0, c)$, opening upward if $a > 0$ or downward if $a < 0$. If $c = 0$, the equation reduces to $y = ax^2$ and the vertex is at the origin $(0, 0)$. Parabolas may also open to the right or to the left (see Example 5) or in other directions.

We shall use the following terminology to describe where the graph of an equation in x and y intersects the x-axis or the y-axis.

Intercepts of the Graph of an Equation in x and y

Terminology	Definition	Graphical interpretation	How to find
x-intercepts	The x-coordinates of points where the graph intersects the x-axis		Let $y = 0$ and solve for x. Here, a and c are x-intercepts.
y-intercepts	The y-coordinates of points where the graph intersects the y-axis		Let $x = 0$ and solve for y. Here, b is the y-intercept.

An x-intercept is sometimes referred to as a *zero* of the graph of an equation or as a *root* of an equation. When using a graphing utility to find an x-intercept, we will say that we are using a *root feature.*

EXAMPLE 3 Finding x-intercepts and y-intercepts

Find the x- and y-intercepts of the graph of $y = x^2 - 3$.

▶ **SOLUTION** The graph is sketched in Figure 2 (Example 2). We find the intercepts as stated in the preceding chart.

(1) x-intercepts:

$$
\begin{aligned}
y &= x^2 - 3 && \text{given} \\
0 &= x^2 - 3 && \text{let } y = 0 \\
x^2 &= 3 && \text{equivalent equation} \\
x &= \pm\sqrt{3} \approx \pm 1.73 && \text{take the square root}
\end{aligned}
$$

(continued)

▶ **Tutorial available at www.thomsonedu.com/login**

Thus, the x-intercepts are $-\sqrt{3}$ and $\sqrt{3}$. The points at which the graph crosses the x-axis are $(-\sqrt{3}, 0)$ and $(\sqrt{3}, 0)$.

(2) y-intercepts:

$$y = x^2 - 3 \qquad \text{given}$$

$$y = 0 - 3 = -3 \qquad \text{let } x = 0$$

Thus, the y-intercept is -3, and the point at which the graph crosses the y-axis is $(0, -3)$.

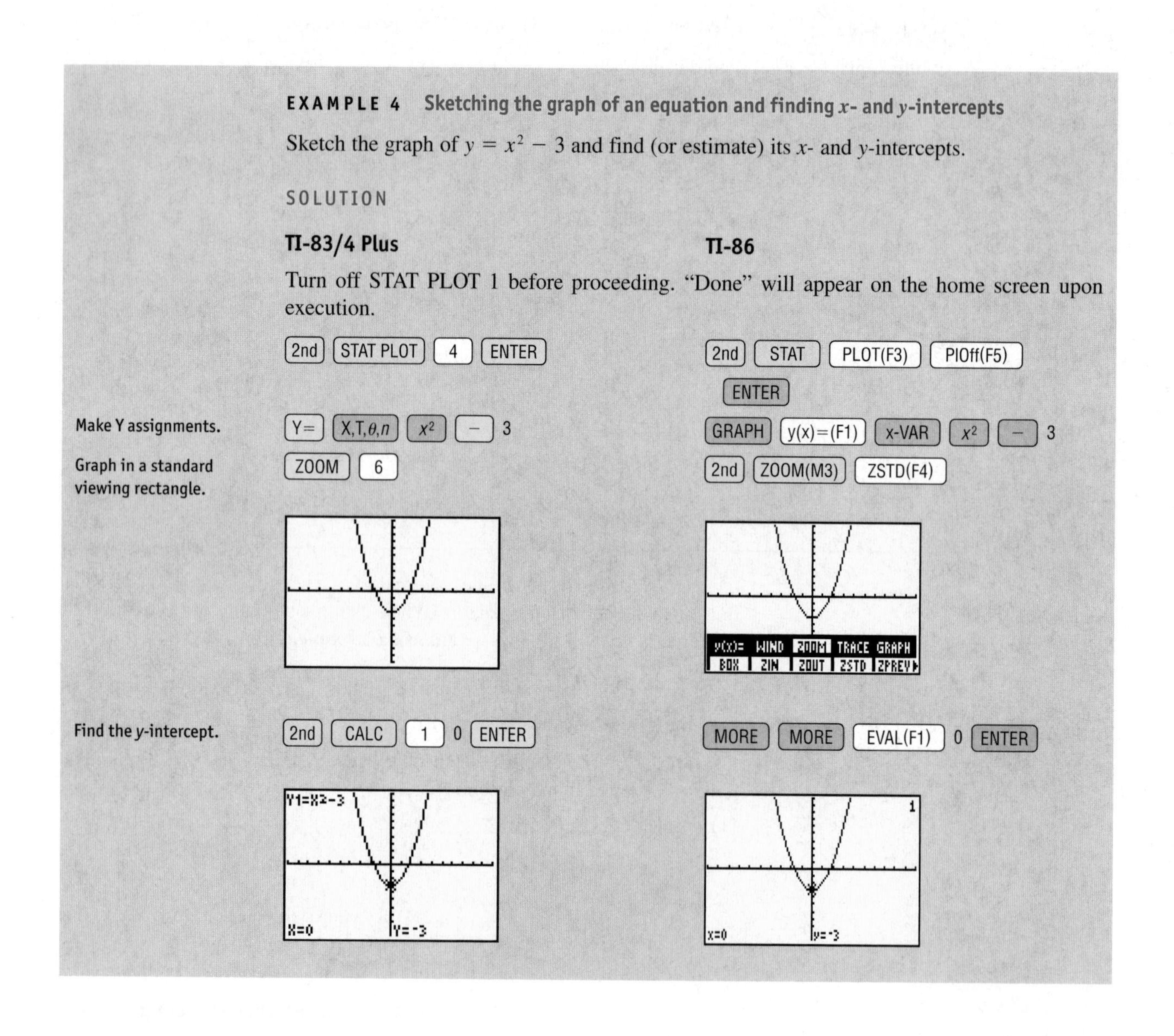

EXAMPLE 4 Sketching the graph of an equation and finding x- and y-intercepts

Sketch the graph of $y = x^2 - 3$ and find (or estimate) its x- and y-intercepts.

SOLUTION

TI-83/4 Plus | **TI-86**

Turn off STAT PLOT 1 before proceeding. "Done" will appear on the home screen upon execution.

TI-83/4 Plus: 2nd STAT PLOT 4 ENTER

TI-86: 2nd STAT PLOT(F3) PlOff(F5) ENTER

Make Y assignments.

TI-83/4 Plus: Y= X,T,θ,n x^2 − 3

TI-86: GRAPH y(x)=(F1) x-VAR x^2 − 3

Graph in a standard viewing rectangle.

TI-83/4 Plus: ZOOM 6

TI-86: 2nd ZOOM(M3) ZSTD(F4)

Find the y-intercept.

TI-83/4 Plus: 2nd CALC 1 0 ENTER

TI-86: MORE MORE EVAL(F1) 0 ENTER

Estimate the x-intercepts.

2nd | CALC | 2   GRAPH | MORE | MATH(F1) | ROOT(F1)

We'll find the positive x-intercept. In response to "Left Bound?" move the cursor to the right so that the y-coordinate is a small negative number and then press ENTER.

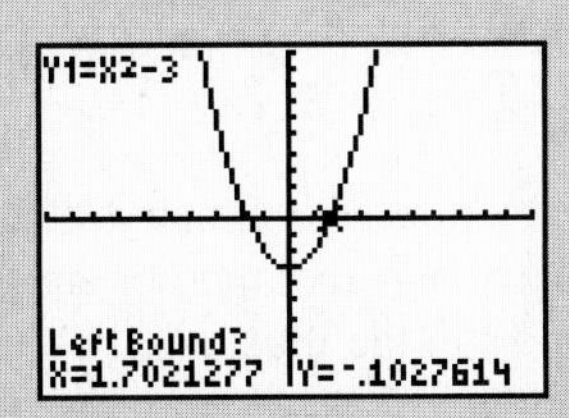

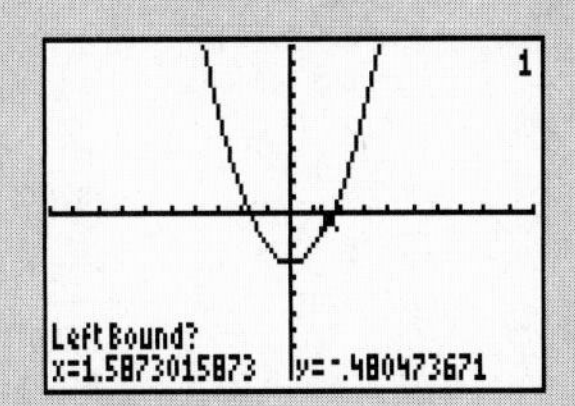

In response to "Right Bound?" move the cursor to the right so that the y-coordinate is a small positive number and then press ENTER.

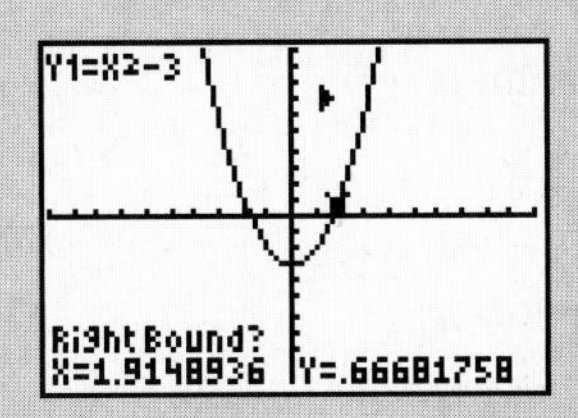

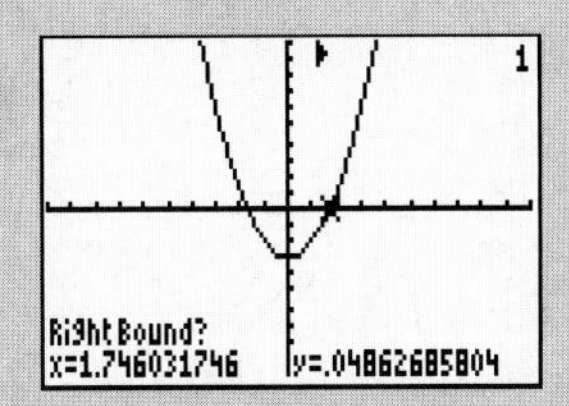

In response to "Guess?" just press ENTER, since we are very close to the x-intercept.

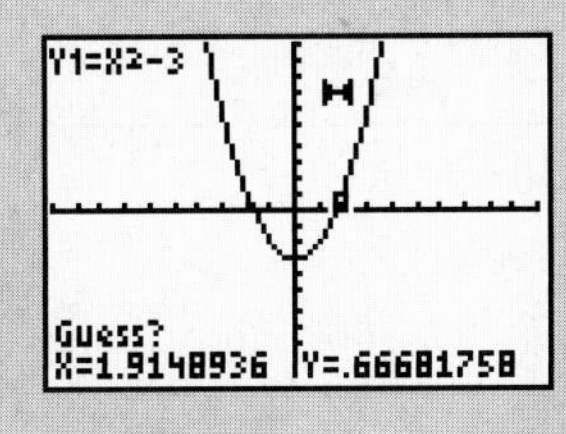

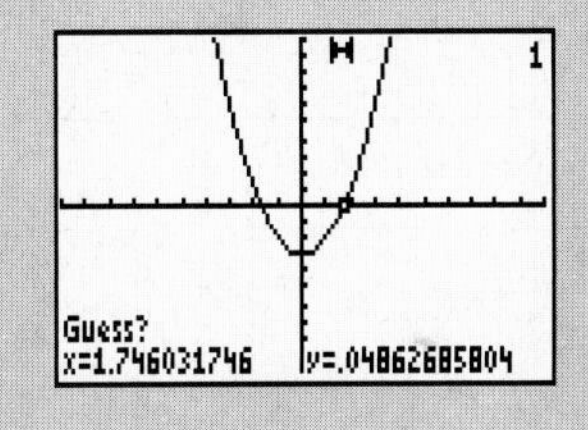

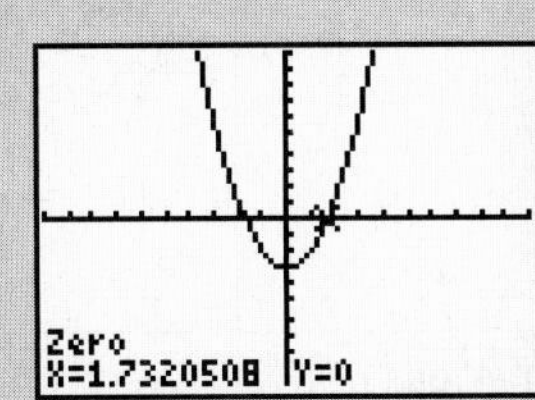

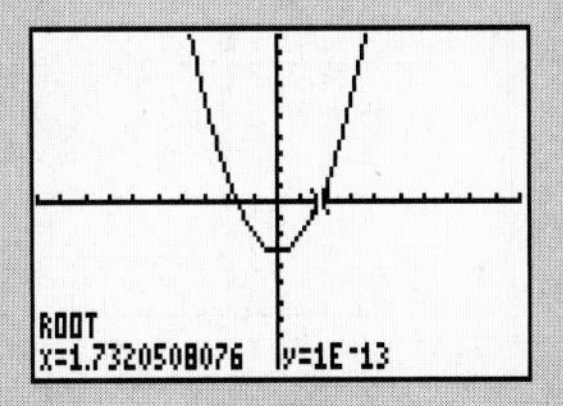

From the previous example, we know that the x-intercepts are approximately ± 1.73.

Calculator Note: If you know an approximation of the x-intercept, then you can enter x-values for your responses. The following responses produce the same result as above.

Left bound?	1 ENTER
Right bound?	2 ENTER
Guess?	1.5 ENTER

If the coordinate plane in Figure 2 is folded along the y-axis, the graph that lies in the left half of the plane coincides with that in the right half, and we say that *the graph is symmetric with respect to the y-axis.* A graph is symmetric with respect to the y-axis provided that the point $(-x, y)$ is on the graph whenever (x, y) is on the graph. The graph of $y = x^2 - 3$ in Example 2 has this property, since substitution of $-x$ for x yields the same equation:

$$y = (-x)^2 - 3 = x^2 - 3$$

This substitution is an application of symmetry test 1 in the following chart. Two other types of symmetry and the appropriate tests are also listed. The graphs of $x = y^2$ and $4y = x^3$ in the illustration column are discussed in Examples 5 and 6, respectively.

Symmetries of Graphs of Equations in x and y

Terminology	Graphical interpretation	Test for symmetry	Illustration
The graph is symmetric with respect to the y-axis.	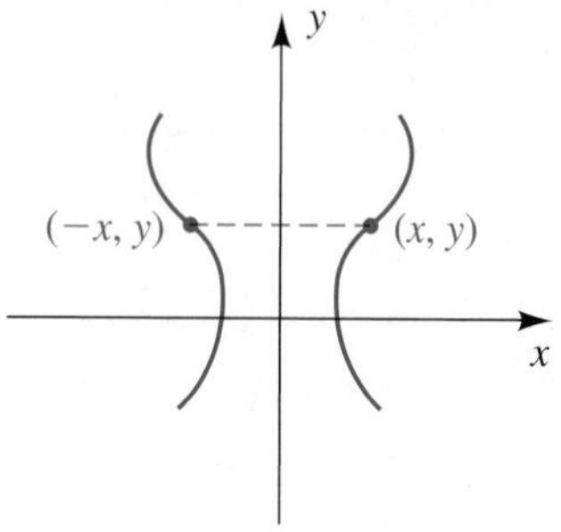 y; $(-x, y)$; (x, y); x	**(1)** Substitution of $-x$ for x leads to the same equation.	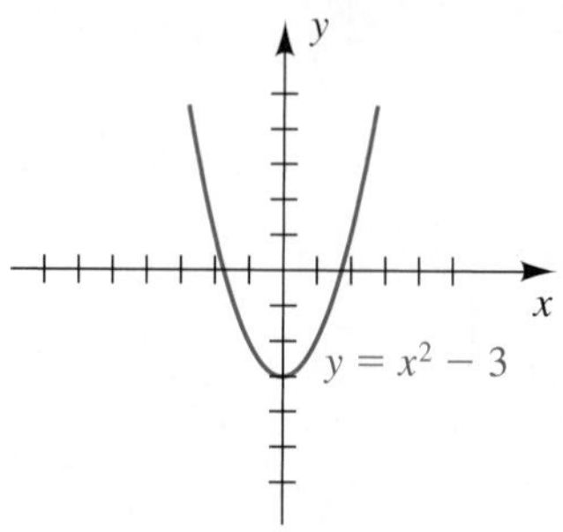 y; x; $y = x^2 - 3$
The graph is symmetric with respect to the x-axis.	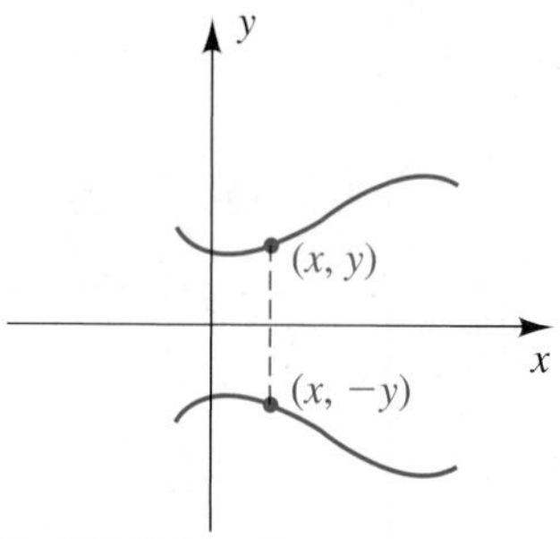 y; (x, y); x; $(x, -y)$	**(2)** Substitution of $-y$ for y leads to the same equation.	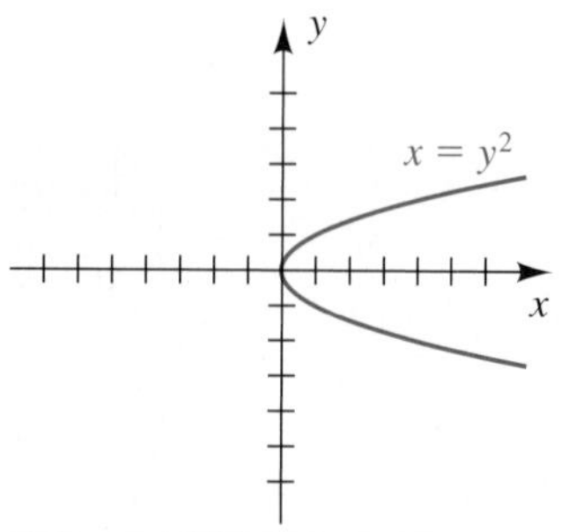 y; $x = y^2$; x
The graph is symmetric with respect to the origin.	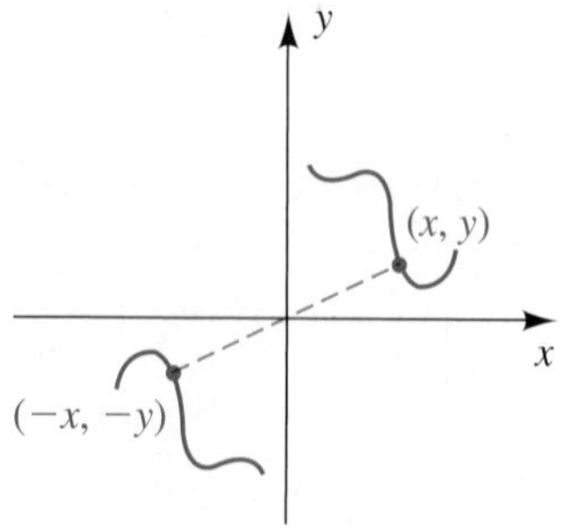 y; (x, y); x; $(-x, -y)$	**(3)** Simultaneous substitution of $-x$ for x and $-y$ for y leads to the same equation.	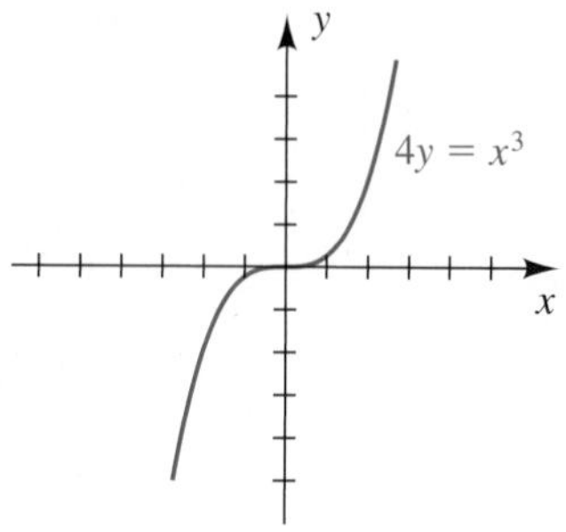 y; $4y = x^3$; x

If a graph is symmetric with respect to an axis, it is sufficient to determine the graph in half of the coordinate plane, since we can sketch the remainder of the graph by taking a *mirror image,* or *reflection,* through the appropriate axis.

EXAMPLE 5 **A graph that is symmetric with respect to the x-axis**

Sketch the graph of the equation $y^2 = x$.

SOLUTION Since substitution of $-y$ for y does not change the equation, the graph is symmetric with respect to the x-axis (see symmetry test 2). Hence, if the point (x, y) is on the graph, then the point $(x, -y)$ is on the graph. Thus, it is sufficient to find points with nonnegative y-coordinates and then reflect through the x-axis. The equation $y^2 = x$ is equivalent to $y = \pm\sqrt{x}$. The y-coordinates of points *above* the x-axis (y is *positive*) are given by $y = \sqrt{x}$, whereas the y-coordinates of points *below* the x-axis (y is *negative*) are given by $y = -\sqrt{x}$. Coordinates of some points on the graph are listed below. The graph is sketched in Figure 3.

Figure 3

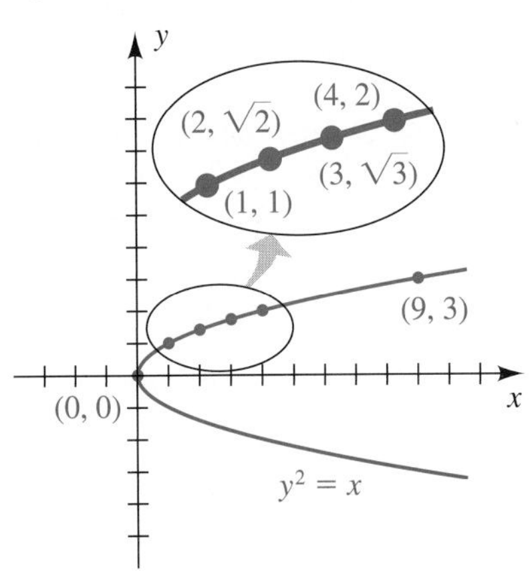

x	0	1	2	3	4	9
y	0	1	$\sqrt{2} \approx 1.4$	$\sqrt{3} \approx 1.7$	2	3

The graph is a parabola that opens to the right, with its vertex at the origin. In this case, the x-axis is the axis of the parabola.

EXAMPLE 6 **A graph that is symmetric with respect to the origin**

Sketch the graph of the equation $4y = x^3$.

SOLUTION If we simultaneously substitute $-x$ for x and $-y$ for y, then

$$4(-y) = (-x)^3 \qquad \text{or, equivalently,} \qquad -4y = -x^3.$$

Multiplying both sides by -1, we see that the last equation has the same solutions as the equation $4y = x^3$. Hence, from symmetry test 3, the graph is symmetric with respect to the origin—and if the point (x, y) is on the graph, then the point $(-x, -y)$ is on the graph. The following table lists coordinates of some points on the graph.

Figure 4

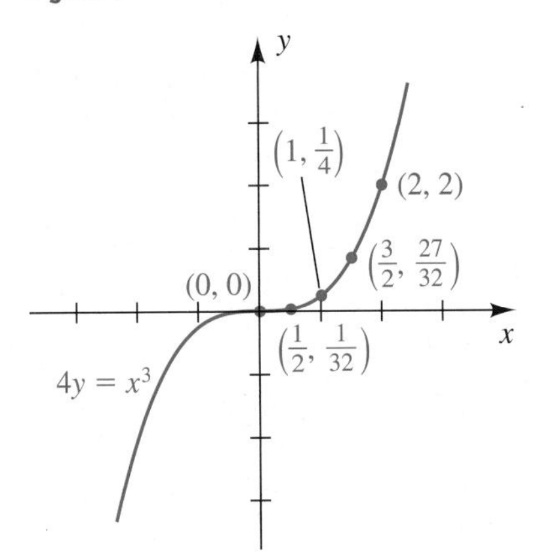

x	0	$\frac{1}{2}$	1	$\frac{3}{2}$	2	$\frac{5}{2}$
y	0	$\frac{1}{32}$	$\frac{1}{4}$	$\frac{27}{32}$	2	$\frac{125}{32}$

Because of the symmetry, we can see that the points $\left(-1, -\frac{1}{4}\right)$, $(-2, -2)$, and so on, are also on the graph. The graph is sketched in Figure 4.

Tutorial available at www.thomsonedu.com/login

Figure 5

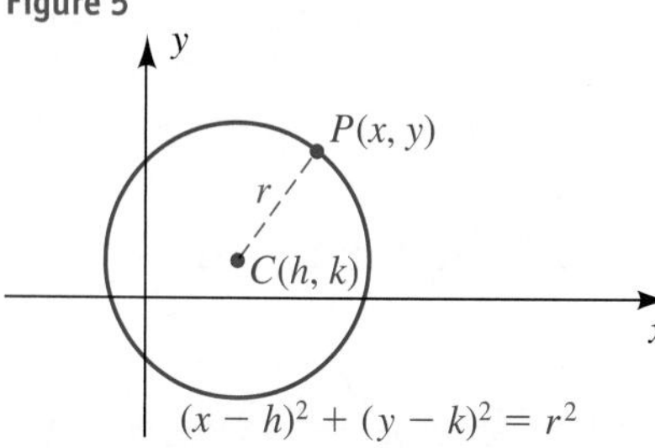

If $C(h, k)$ is a point in a coordinate plane, then a circle with center C and radius $r > 0$ consists of all points in the plane that are r units from C. As shown in Figure 5, a point $P(x, y)$ is on the circle provided $d(C, P) = r$ or, by the distance formula,

$$\sqrt{(x - h)^2 + (y - k)^2} = r.$$

The above equation is equivalent to the following equation, which we will refer to as the **standard equation of a circle.**

Standard Equation of a Circle with Center (h, k) and Radius r	$(x - h)^2 + (y - k)^2 = r^2$

Figure 6

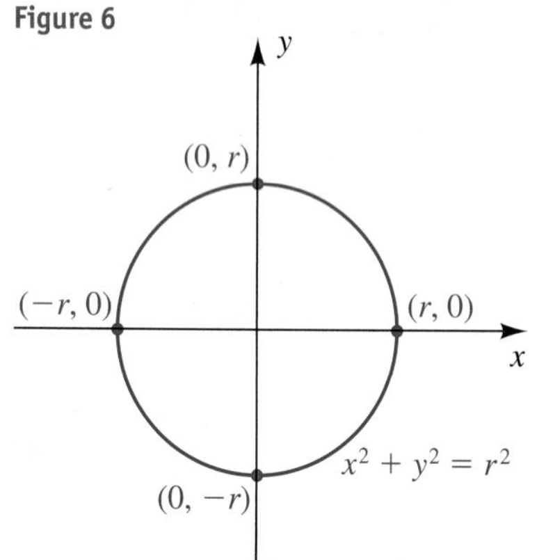

If $h = 0$ and $k = 0$, this equation reduces to $x^2 + y^2 = r^2$, which is an equation of a circle of radius r with center at the origin (see Figure 6). If $r = 1$, we call the graph a **unit circle.**

EXAMPLE 7 Finding an equation of a circle

Find an equation of the circle that has center $C(-2, 3)$ and contains the point $D(4, 5)$.

▶ SOLUTION The circle is shown in Figure 7. Since D is on the circle, the radius r is $d(C, D)$. By the distance formula,

$$r = \sqrt{(4 + 2)^2 + (5 - 3)^2} = \sqrt{36 + 4} = \sqrt{40}.$$

Using the standard equation of a circle with $h = -2$, $k = 3$, and $r = \sqrt{40}$, we obtain

$$(x + 2)^2 + (y - 3)^2 = 40.$$

By squaring terms and simplifying the last equation, we may write it as

$$x^2 + y^2 + 4x - 6y - 27 = 0.$$

Figure 7

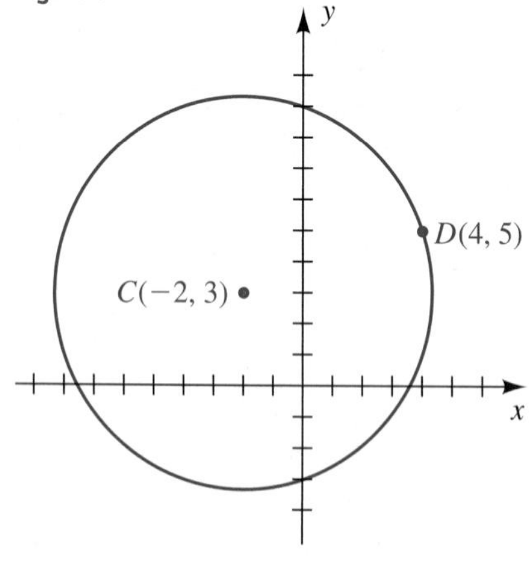

As in the solution to Example 7, squaring terms of an equation of the form $(x - h)^2 + (y - k)^2 = r^2$ and simplifying leads to an equation of the form

$$x^2 + y^2 + ax + by + c = 0,$$

where a, b, and c are real numbers. Conversely, if we begin with this equation, it is always possible, by *completing squares,* to obtain an equation of the form

$$(x - h)^2 + (y - k)^2 = d.$$

This method will be illustrated in Example 8. If $d > 0$, the graph is a circle with center (h, k) and radius $r = \sqrt{d}$. If $d = 0$, the graph consists of only the point (h, k). Finally, if $d < 0$, the equation has no real solutions, and hence there is no graph.

EXAMPLE 8 Finding the center and radius of a circle

Find the center and radius of the circle with equation

$$3x^2 + 3y^2 - 12x + 18y = 9.$$

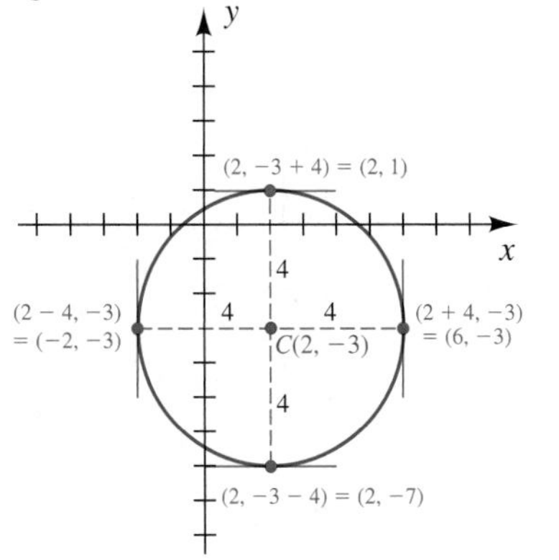

▶ SOLUTION Since it is easier to complete the square if the coefficients of x^2 and y^2 are 1, we begin by dividing the given equation by 3, obtaining

$$x^2 + y^2 - 4x + 6y = 3.$$

Next, we rewrite the equation as follows, where the underscored spaces represent numbers to be determined:

$$(x^2 - 4x + __) + (y^2 + 6y + __) = 3 + __ + __$$

We then complete the squares for the expressions within parentheses, taking care to add the appropriate numbers to *both* sides of the equation. To complete the square for an expression of the form $x^2 + ax$, we add the square of half the coefficient of x (that is, $(a/2)^2$) to both sides of the equation. Similarly, for $y^2 + by$, we add $(b/2)^2$ to both sides. In this example, $a = -4$, $b = 6$, $(a/2)^2 = (-2)^2 = 4$, and $(b/2)^2 = 3^2 = 9$. These additions lead to

$$(x^2 - 4x + \underline{4}) + (y^2 + 6y + \underline{9}) = 3 + \underline{4} + \underline{9} \quad \text{completing the squares}$$

$$(x - 2)^2 + (y + 3)^2 = 16. \quad \text{equivalent equation}$$

Comparing the last equation with the standard equation of a circle, we see that $h = 2$ and $k = -3$ and conclude that the circle has center $(2, -3)$ and radius $\sqrt{16} = 4$. A sketch of this circle is shown in Figure 8.

Figure 8

$(2, -3 + 4) = (2, 1)$
$(2 - 4, -3) = (-2, -3)$
$C(2, -3)$
$(2 + 4, -3) = (6, -3)$
$(2, -3 - 4) = (2, -7)$

*Recall that a **tangent line** to a circle is a line that contains exactly one point of the circle. Every circle has four points of tangency associated with horizontal and vertical lines. It is helpful to plot these points when sketching the graph of a circle.*

In some applications it is necessary to work with only one-half of a circle—that is, a **semicircle.** The next example indicates how to find equations of semicircles for circles with centers at the origin.

EXAMPLE 9 Finding equations of semicircles

Find equations for the upper half, lower half, right half, and left half of the circle $x^2 + y^2 = 81$.

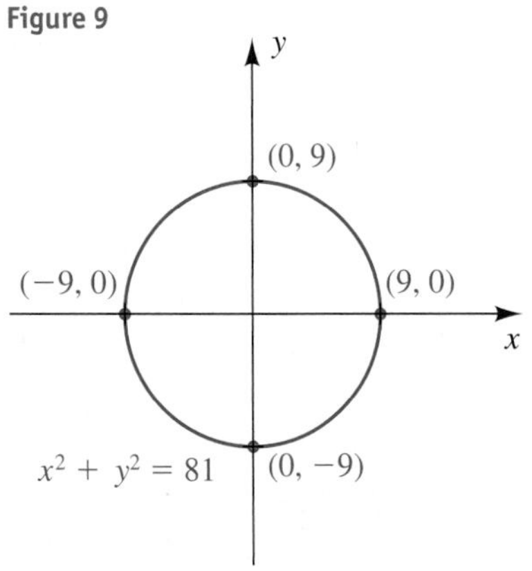

▶ SOLUTION The graph of $x^2 + y^2 = 81$ is a circle of radius 9 with center at the origin (see Figure 9). To find equations for the upper and lower halves, we solve for y in terms of x:

$$x^2 + y^2 = 81 \quad \text{given}$$

$$y^2 = 81 - x^2 \quad \text{subtract } x^2$$

$$y = \pm\sqrt{81 - x^2} \quad \text{take the square root}$$

Since $\sqrt{81 - x^2} \geq 0$, it follows that the upper half of the circle has the equation $y = \sqrt{81 - x^2}$ (y is positive) and the lower half is given by $y = -\sqrt{81 - x^2}$ (y is negative), as illustrated in Figure 10(a) and (b).

(continued)

Figure 9

$(0, 9)$ $(-9, 0)$ $(9, 0)$ $(0, -9)$
$x^2 + y^2 = 81$

▶ **Tutorial available at www.thomsonedu.com/login**

Figure 10

(a) $y = \sqrt{81 - x^2}$

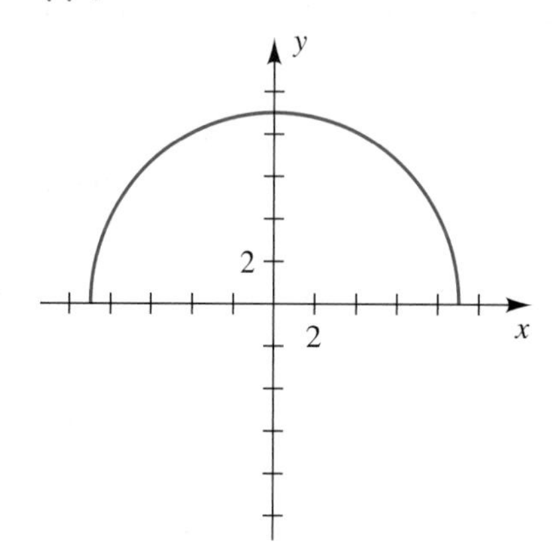

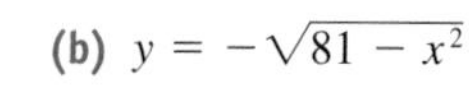

(b) $y = -\sqrt{81 - x^2}$

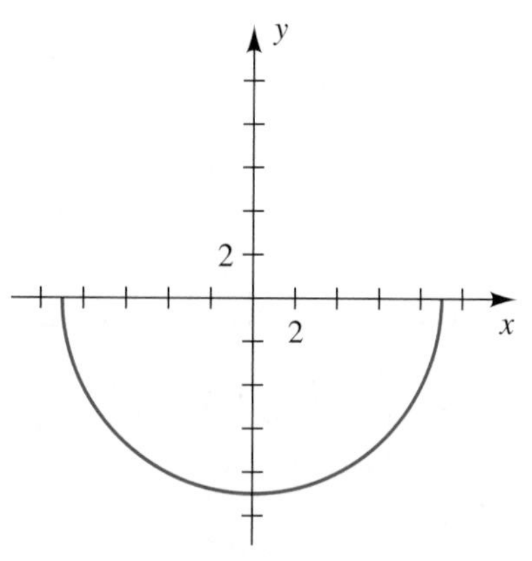

(c) $x = \sqrt{81 - y^2}$

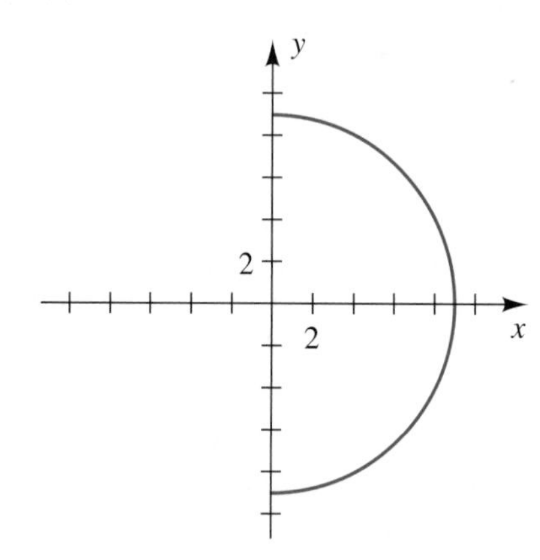

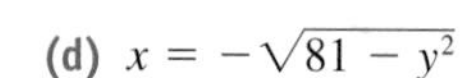

(d) $x = -\sqrt{81 - y^2}$

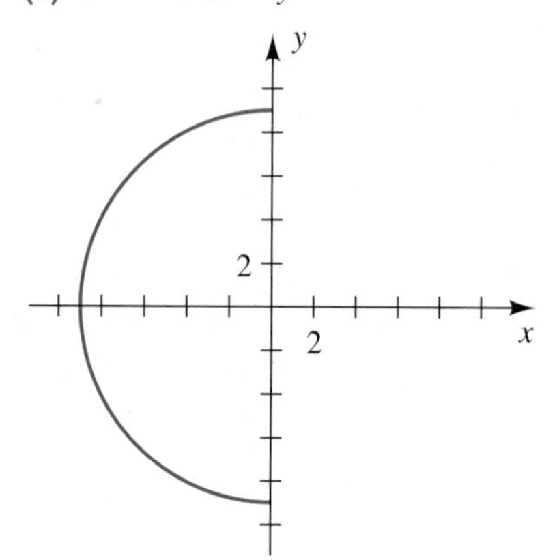

Similarly, to find equations for the right and left halves, we solve $x^2 + y^2 = 81$ for x in terms of y, obtaining

$$x = \pm\sqrt{81 - y^2}.$$

Since $\sqrt{81 - y^2} \geq 0$, it follows that the right half of the circle has the equation $x = \sqrt{81 - y^2}$ (x is positive) and the left half is given by the equation $x = -\sqrt{81 - y^2}$ (x is negative), as illustrated in Figure 10(c) and (d). ■

In many applications it is essential to find the points at which the graphs of two equations in x and y intersect. To approximate such points of intersection with a graphing utility, it is often necessary to solve each equation for y in terms of x. For example, suppose one equation is

$$4x^2 - 3x + 2y + 6 = 0.$$

Solving for y gives us

$$y = \frac{-4x^2 + 3x - 6}{2} = -2x^2 + \frac{3}{2}x - 3.$$

The graph of the equation is then found by making the assignment

$$Y_1 = -2x^2 + \tfrac{3}{2}x - 3$$

in the graphing utility. (The symbol Y_1 indicates the *first* equation, or the first y value.) We also solve the second equation for y in terms of x and make the assignment

$$Y_2 = \text{an expression in } x.$$

Pressing appropriate keys gives us sketches of the graphs, which we refer to as the graphs of Y_1 and Y_2. We then use a graphing utility feature such as *intersect* to estimate the coordinates of the points of intersection.

In the next example we demonstrate this technique for the graphs discussed in Examples 1 and 2.

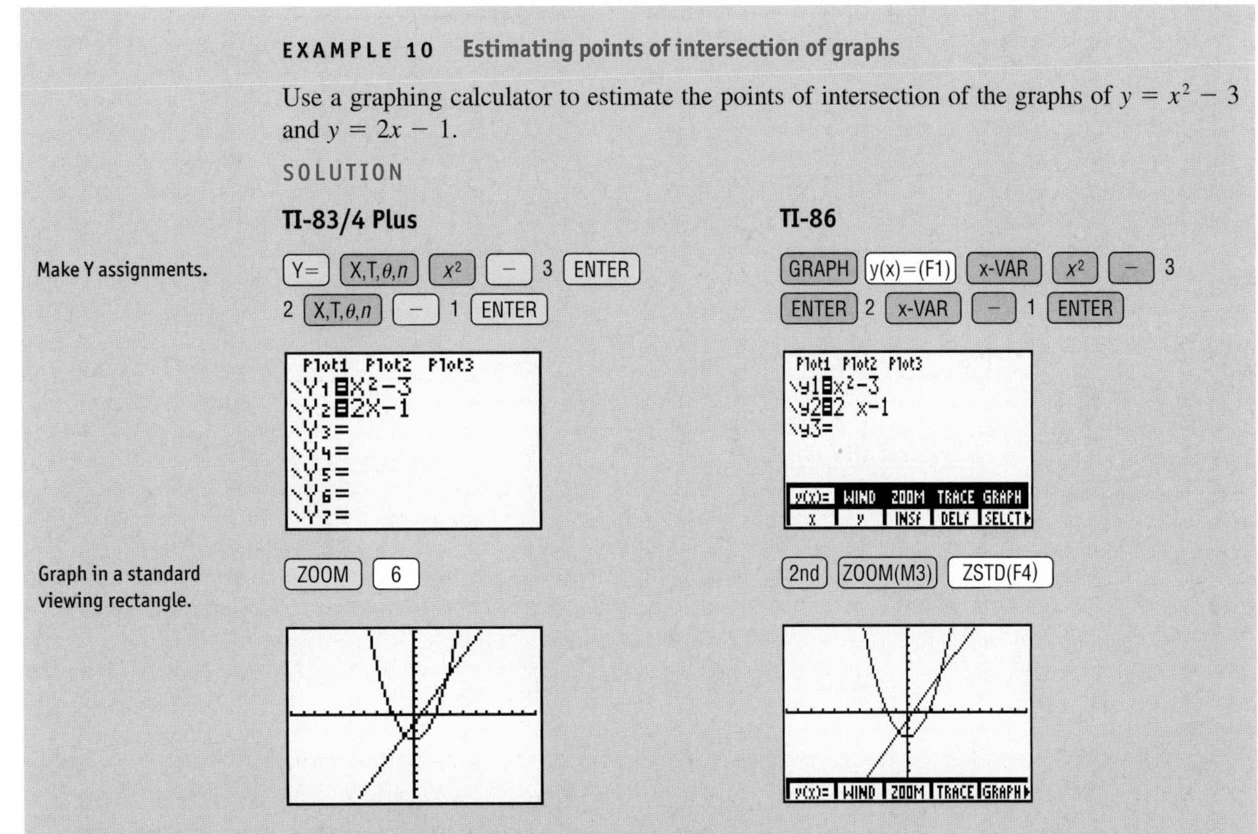

EXAMPLE 10 Estimating points of intersection of graphs

Use a graphing calculator to estimate the points of intersection of the graphs of $y = x^2 - 3$ and $y = 2x - 1$.

SOLUTION

	TI-83/4 Plus	TI-86
Make Y assignments.	Y= X,T,θ,n x² − 3 ENTER 2 X,T,θ,n − 1 ENTER	GRAPH y(x)=(F1) x-VAR x² − 3 ENTER 2 x-VAR − 1 ENTER
Graph in a standard viewing rectangle.	ZOOM 6	2nd ZOOM(M3) ZSTD(F4)

We see from the graphs of Y_1 and Y_2 that there are two points of intersection: P_1 in quadrant I and P_2 in quadrant III. We'll find P_1.

(continued)

Find one point of intersection.

[2nd] [CALC] [5] [MORE] [MATH(F1)] [MORE] [ISECT(F3)]

In response to "First curve?" just press [ENTER] to indicate that Y_1 is the first curve.

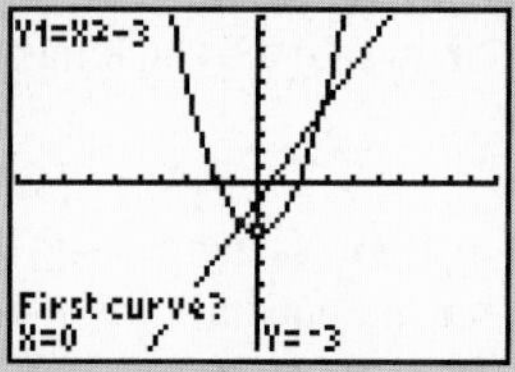

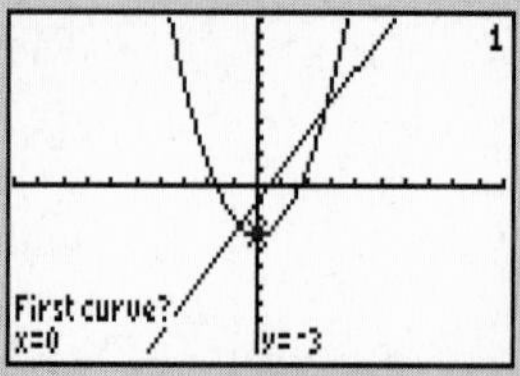

In response to "Second curve?" just press [ENTER] to indicate that Y_2 is the second curve.

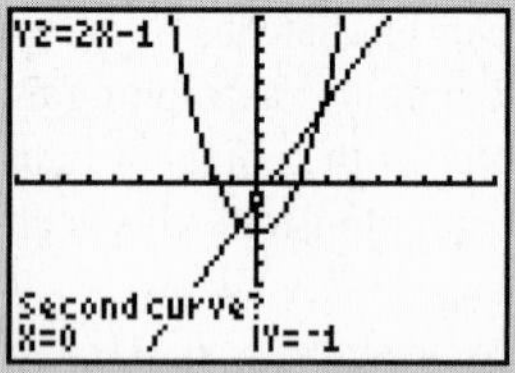

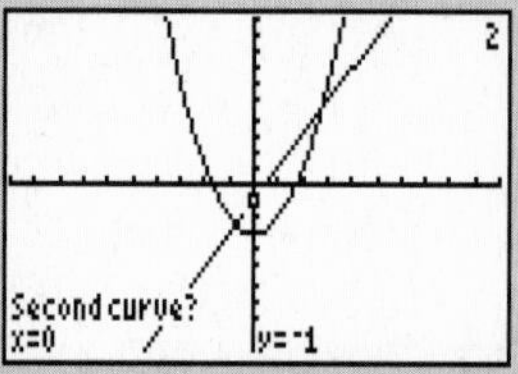

In response to "Guess?" move the cursor close to P_1 and then press [ENTER].

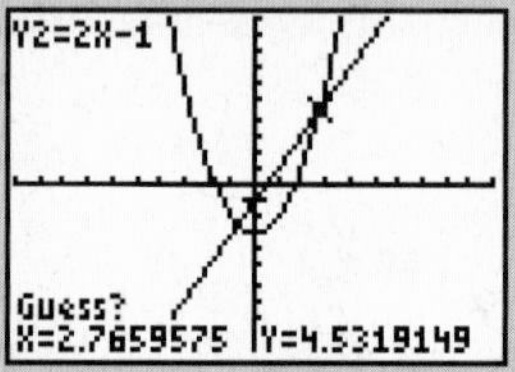

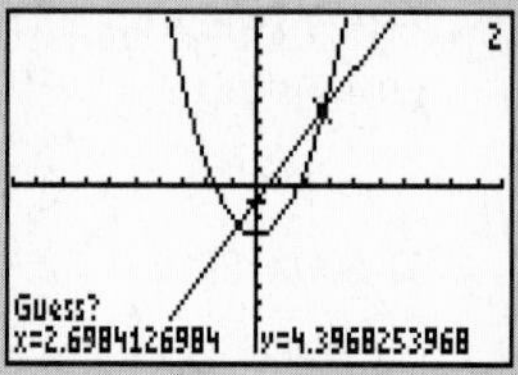

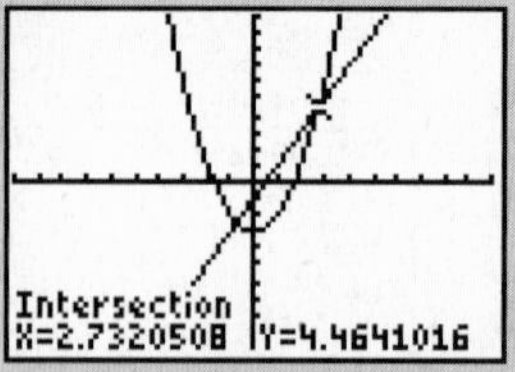

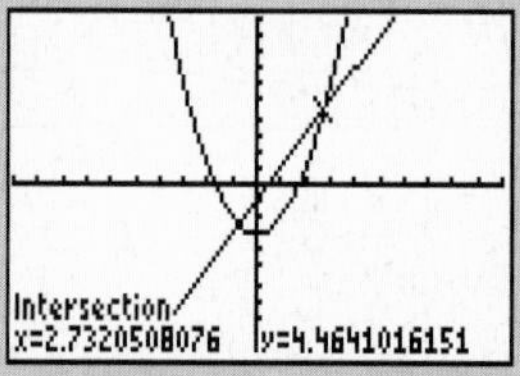

We estimate the coordinates of P_1 as (2.73, 4.46). Then we use the intersect feature again to obtain $(-0.73, -2.46)$ as approximate coordinates of P_2.

Calculator Note: An alternative response to "Guess?" is to enter an estimate of the x-value of the point of intersection. The following response produces the same result as above:

Guess? 3 [ENTER]

EXAMPLE 11 Estimating points of intersection of graphs

Use a graphing calculator to estimate the points of intersection of the circles $x^2 + y^2 = 25$ and $x^2 + y^2 - 4y = 12$.

SOLUTION As in Example 9, we solve $x^2 + y^2 = 25$ for y in terms of x, obtaining

$$y = \pm\sqrt{25 - x^2},$$

and make the following assignments:

$$Y_1 = \sqrt{25 - x^2} \quad \text{and} \quad Y_2 = -Y_1$$

(We often assign Y_2 in terms of Y_1 to avoid repetitive keystroking.)

We may regard the equation of the second circle as a quadratic equation of the form $ay^2 + by + c = 0$ in y by rearranging terms as follows:

$$y^2 - 4y + (x^2 - 12) = 0$$

Applying the quadratic formula with $a = 1$, $b = -4$, and $c = x^2 - 12$ ($x^2 - 12$ is considered to be the constant term, since it does not contain the variable y) gives us

$$\begin{aligned} y &= \frac{-(-4) \pm \sqrt{(-4)^2 - 4(1)(x^2 - 12)}}{2(1)} \\ &= \frac{4 \pm \sqrt{16 - 4(x^2 - 12)}}{2} = \frac{4 \pm 2\sqrt{4 - (x^2 - 12)}}{2} = 2 \pm \sqrt{16 - x^2}. \end{aligned}$$

(It is unnecessary to simplify the equation as much as we have, but the simplified form is easier to enter in a graphing calculator.)

We now make the assignments

$$Y_3 = \sqrt{16 - x^2}, \quad Y_4 = 2 + Y_3, \quad \text{and} \quad Y_5 = 2 - Y_3.$$

TI-83/4 Plus

Make Y assignments.

[Y=] [2nd] [√] 25 [−] [X,T,θ,n] [x²]
[)] [▽] [(−)] [VARS] [▷] [1] [1] [▽]
[2nd] [√] 16 [−] [X,T,θ,n] [x²] [)] [▽]
2 [+] [VARS] [▷] [1] [3] [▽]
2 [−] [VARS] [▷] [1] [3]

Turn off Y_3.

[△ (3 times)] [◁] [ENTER]

```
Plot1 Plot2 Plot3
\Y1=√(25-X²)
\Y2=-Y1
\Y3=√(16-X²)
\Y4=2+Y3
\Y5=2-Y3
\Y6=
\Y7=
```

TI-86

[GRAPH] [y(x)=(F1)] [2nd] [√] [(] 25 [−]
[x-VAR] [x²] [)] [▽] [(−)] [y(F2)] 1 [▽]
[2nd] [√] [(] 16 [−] [x-VAR] [x²] [)]
[▽] 2 [+] [y(F2)] 3 [▽]
2 [−] [y(F2)] 3
[△ (2 times)] [SELCT(F5)]

```
Plot1 Plot2 Plot3
\y1=√(25-x²)
\y2=-y1
\y3=√(16-x²)
\y4=2+y3
\y5=2-y3
y(x)= WIND ZOOM TRACE GRAPH
x y INSf DELf SELCT▸
```

We will use a square viewing rectangle so that the circles look like circles instead of ovals.

(continued)

Graph in a square viewing rectangle.

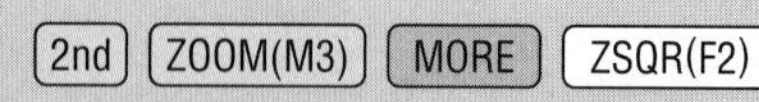

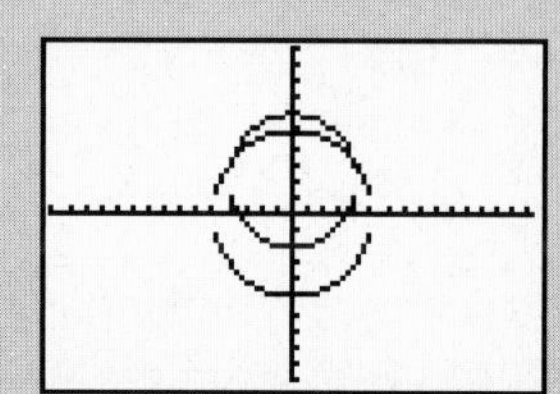

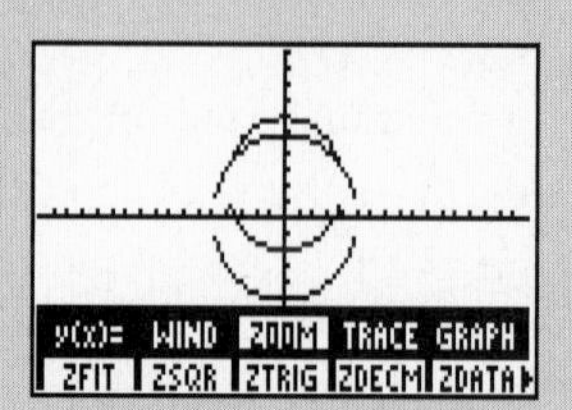

We see from the graphs of the circles that there are two points of intersection: P_1 in quadrant I and P_2 in quadrant II. Again, we'll find P_1.

Find one point of intersection.

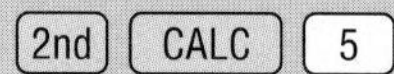

GRAPH MORE MATH(F1) MORE ISECT(F3)

In response to "First curve?" just press ENTER to indicate that Y_1 is the first curve. In response to "Second curve?" press ▽ to skip Y_2 as the selection for the second curve, since it does not intersect Y_1. Now press ENTER to select Y_4 as the second curve. In response to "Guess?" move the cursor close to P_1 and then press ENTER or just type 3.5 for a guess and press ENTER.

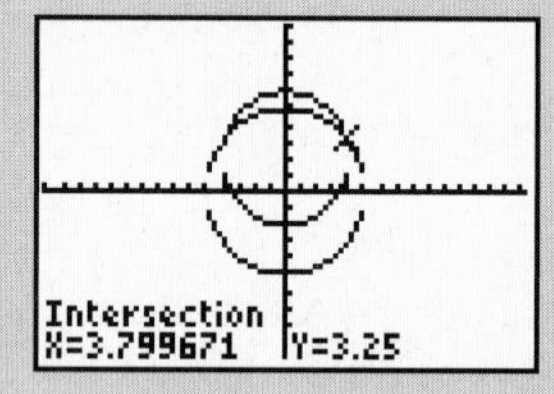

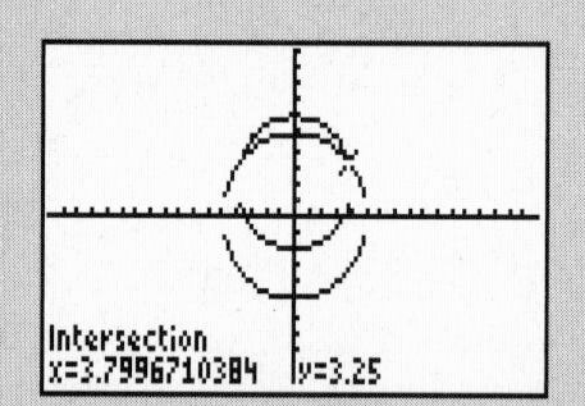

Thus, we estimate the coordinates of P_1 as (3.8, 3.25). Since both circles are symmetric with respect to the y-axis, P_2 is approximately (−3.8, 3.25).

It should be noted that the approximate solutions found in Examples 10 and 11 do not satisfy the given equations because of the inaccuracy of the estimates made from the graph. In a later chapter we will discuss how to find the *exact* values for the points of intersection.

2.2 *Exercises*

Exer. 1–20: Sketch the graph of the equation, and label the x- and y-intercepts.

1 $y = 2x - 3$

2 $y = 3x + 2$

3 $y = -x + 1$

4 $y = -2x - 3$

5 $y = -4x^2$

6 $y = \frac{1}{3}x^2$

7 $y = 2x^2 - 1$

8 $y = -x^2 + 2$

9 $x = \frac{1}{4}y^2$

10 $x = -2y^2$

11 $x = -y^2 + 3$

12 $x = 2y^2 - 4$

13 $y = -\frac{1}{2}x^3$

14 $y = \frac{1}{2}x^3$

15 $y = x^3 - 8$

16 $y = -x^3 + 1$

17 $y = \sqrt{x}$

18 $y = \sqrt{-x}$

19 $y = \sqrt{x} - 4$

20 $y = \sqrt{x - 4}$

Exer. 21–22: Use tests for symmetry to determine which graphs in the indicated exercises are symmetric with respect to (a) the *y*-axis, (b) the *x*-axis, and (c) the origin.

21 The odd-numbered exercises in 1–20

22 The even-numbered exercises in 1–20

Exer. 23–34: Sketch the graph of the circle or semicircle.

23 $x^2 + y^2 = 11$

24 $x^2 + y^2 = 7$

25 $(x + 3)^2 + (y - 2)^2 = 9$

26 $(x - 4)^2 + (y + 2)^2 = 4$

27 $(x + 3)^2 + y^2 = 16$

28 $x^2 + (y - 2)^2 = 25$

29 $4x^2 + 4y^2 = 25$

30 $9x^2 + 9y^2 = 1$

31 $y = -\sqrt{16 - x^2}$

32 $y = \sqrt{4 - x^2}$

33 $x = \sqrt{9 - y^2}$

34 $x = -\sqrt{25 - y^2}$

Exer. 35–46: Find an equation of the circle that satisfies the stated conditions.

35 Center $C(2, -3)$, radius 5

36 Center $C(-4, 1)$, radius 3

▶ 37 Center $C\left(\frac{1}{4}, 0\right)$, radius $\sqrt{5}$

38 Center $C\left(\frac{3}{4}, -\frac{2}{3}\right)$, radius $3\sqrt{2}$

39 Center $C(-4, 6)$, passing through $P(1, 2)$

40 Center at the origin, passing through $P(4, -7)$

▶ 41 Center $C(-3, 6)$, tangent to the *y*-axis

42 Center $C(4, -1)$, tangent to the *x*-axis

43 Tangent to both axes, center in the second quadrant, radius 4

44 Tangent to both axes, center in the fourth quadrant, radius 3

45 Endpoints of a diameter $A(4, -3)$ and $B(-2, 7)$

46 Endpoints of a diameter $A(-5, 2)$ and $B(3, 6)$

Exer. 47–56: Find the center and radius of the circle with the given equation.

47 $x^2 + y^2 - 4x + 6y - 36 = 0$

48 $x^2 + y^2 + 8x - 10y + 37 = 0$

49 $x^2 + y^2 + 4y - 117 = 0$

50 $x^2 + y^2 - 10x + 18 = 0$

51 $2x^2 + 2y^2 - 12x + 4y - 15 = 0$

52 $9x^2 + 9y^2 + 12x - 6y + 4 = 0$

53 $x^2 + y^2 + 4x - 2y + 5 = 0$

54 $x^2 + y^2 - 6x + 4y + 13 = 0$

55 $x^2 + y^2 - 2x - 8y + 19 = 0$

56 $x^2 + y^2 + 4x + 6y + 16 = 0$

Exer. 57–60: Find equations for the upper half, lower half, right half, and left half of the circle.

57 $x^2 + y^2 = 36$

▶ 58 $(x + 3)^2 + y^2 = 64$

59 $(x - 2)^2 + (y + 1)^2 = 49$

60 $(x - 3)^2 + (y - 5)^2 = 4$

Exer. 61–64: Find an equation for the circle or semicircle.

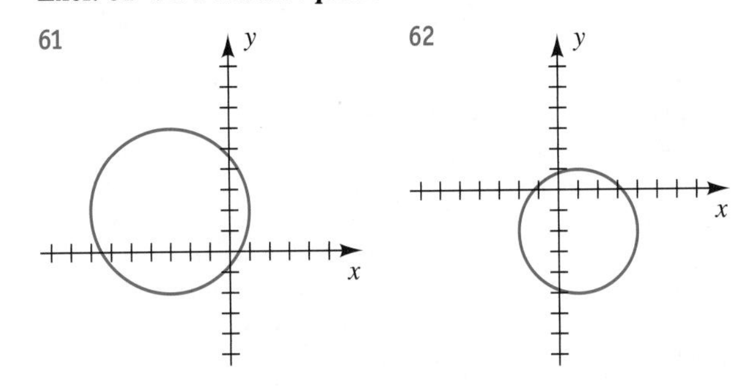

▶ **Tutorial available at www.thomsonedu.com/login**

63 64

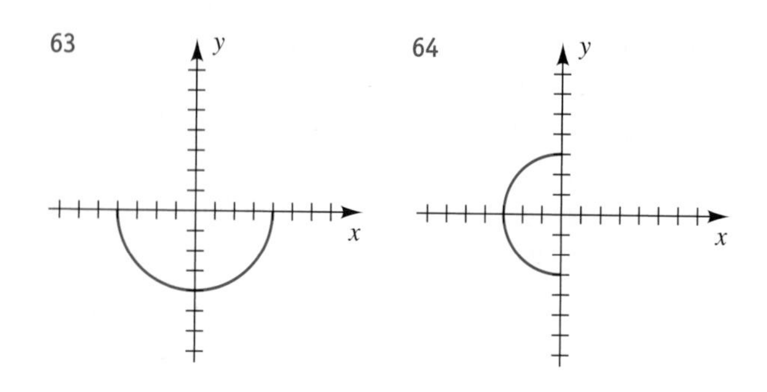

Exer. 65–66: Determine whether the point P is inside, outside, or on the circle with center C and radius r.

▶ 65 (a) $P(2, 3)$, $C(4, 6)$, $r = 4$

(b) $P(4, 2)$, $C(1, -2)$, $r = 5$

(c) $P(-3, 5)$, $C(2, 1)$, $r = 6$

66 (a) $P(3, 8)$, $C(-2, -4)$, $r = 13$

(b) $P(-2, 5)$, $C(3, 7)$, $r = 6$

(c) $P(1, -2)$, $C(6, -7)$, $r = 7$

Exer. 67–68: For the given circle, find (a) the x-intercepts and (b) the y-intercepts.

▶ 67 $x^2 + y^2 - 4x - 6y + 4 = 0$

68 $x^2 + y^2 - 10x + 4y + 13 = 0$

69 Find an equation of the circle that is concentric (has the same center) with $x^2 + y^2 + 4x - 6y + 4 = 0$ and passes through $P(2, 6)$.

70 Radio broadcasting ranges The signal from a radio station has a circular range of 50 miles. A second radio station, located 100 miles east and 80 miles north of the first station, has a range of 80 miles. Are there locations where signals can be received from both radio stations? Explain your answer.

71 A circle C_1 of radius 5 has its center at the origin. Inside this circle there is a first-quadrant circle C_2 of radius 2 that is tangent to C_1. The y-coordinate of the center of C_2 is 2. Find the x-coordinate of the center of C_2.

72 A circle C_1 of radius 5 has its center at the origin. Outside this circle is a first-quadrant circle C_2 of radius 2 that is tangent to C_1. The y-coordinate of the center of C_2 is 3. Find the x-coordinate of the center of C_2.

Exer. 73–76: Express, in interval form, the x-values such that $y_1 < y_2$. Assume all points of intersection are shown on the interval $(-\infty, \infty)$.

73

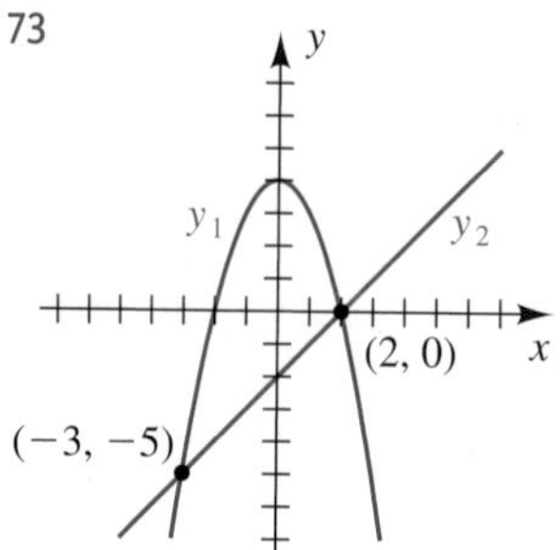

74

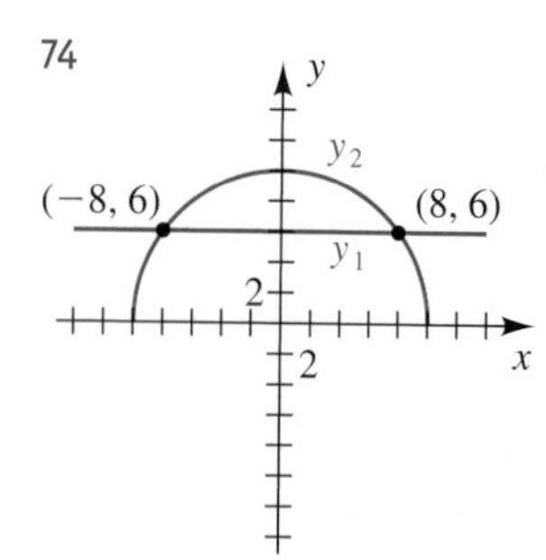

75

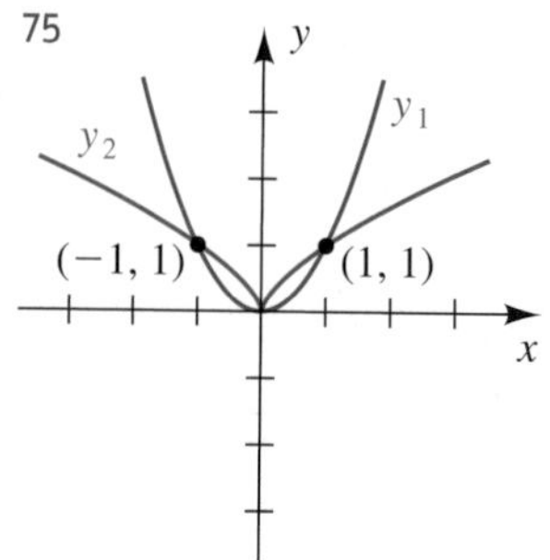

76

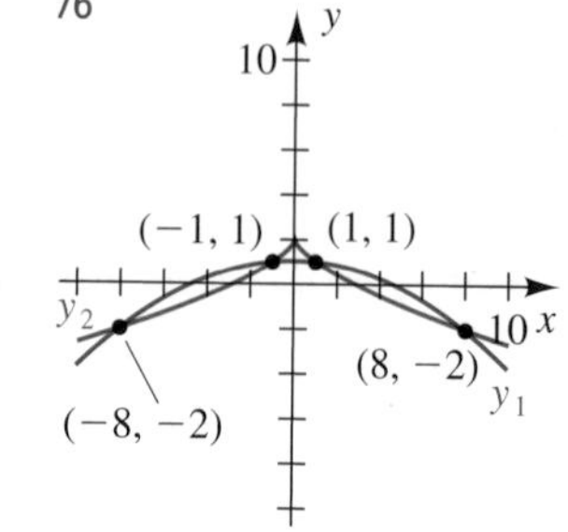

77 Graph the unit circle $x^2 + y^2 = 1$ using the equations $Y_1 = \sqrt{1 - x^2}$ and $Y_2 = -Y_1$ in the given viewing rectangle. Then discuss how the viewing rectangle affects the graph, and determine the viewing rectangle that results in a graph that most looks like a circle.

(1) $[-2, 2]$ by $[-2, 2]$ (2) $[-3, 3]$ by $[-2, 2]$

(3) $[-2, 2]$ by $[-5, 5]$ (4) $[-5, 5]$ by $[-2, 2]$

78 Graph the equation $|x| + |y| = 5$, using the equations $Y_1 = 5 - |x|$ and $Y_2 = -Y_1$ in the viewing rectangle $[-5, 5]$ by $[-5, 5]$.

(a) Find the number of x- and y-intercepts.

(b) Use the graph to determine the region where $|x| + |y| < 5$.

Exer. 79–80: Graph the equation, and estimate the x-intercepts.

79 $y = x^3 - \frac{9}{10}x^2 - \frac{43}{25}x + \frac{24}{25}$

80 $y = x^4 + 0.85x^3 - 2.46x^2 - 1.07x + 0.51$

▶ Tutorial available at www.thomsonedu.com/login

Exer. 81–84: Graph the two equations on the same coordinate plane, and estimate the coordinates of their points of intersection.

81 $y = x^3 + x$; $\quad x^2 + y^2 = 1$

82 $y = 3x^4 - \frac{3}{2}$; $\quad x^2 + y^2 = 1$

83 $x^2 + (y - 1)^2 = 1$; $\quad \left(x - \frac{5}{4}\right)^2 + y^2 = 1$

84 $(x + 1)^2 + (y - 1)^2 = \frac{1}{4}$; $\quad \left(x + \frac{1}{2}\right)^2 + \left(y - \frac{1}{2}\right)^2 = 1$

85 **Distance between cars** The distance D (in miles) between two cars meeting on the same highway at time t (in minutes) is described by the equation $D = |2t - 4|$ on the interval $[0, 4]$. Graph D, and describe the motion of the cars.

86 **Water in a pool** The amount of water A in a swimming pool on day x is given by $A = 12{,}000x - 2000x^2$, where A is in gallons and $x = 0$ corresponds to noon on Sunday. Graph A on the interval $[0, 6]$, and describe the amount of water in the pool.

87 **Speed of sound** The speed of sound v in air varies with temperature. It can be calculated in ft/sec using the equation $v = 1087\sqrt{\dfrac{T + 273}{273}}$, where T is temperature (in °C).

(a) Approximate v when $T = 20°\text{C}$.

(b) Determine the temperature to the nearest degree, both algebraically and graphically, when the speed of sound is 1000 ft/sec.

88 The area A of an equilateral triangle with a side of length s is $A = \dfrac{\sqrt{3}}{4}s^2$. Suppose that A must be equal to 100 ft^2 with an error of at most ± 1 ft^2. Determine graphically how accurately s must be measured in order to satisfy this error requirement. (*Hint:* Graph $y = A$, $y = 99$, and $y = 101$.)

2.3 Lines

One of the basic concepts in geometry is that of a *line*. In this section we will restrict our discussion to lines that lie in a coordinate plane. This will allow us to use algebraic methods to study their properties. Two of our principal objectives may be stated as follows:

(1) Given a line l in a coordinate plane, find an equation whose graph corresponds to l.

(2) Given an equation of a line l in a coordinate plane, sketch the graph of the equation.

The following concept is fundamental to the study of lines.

Definition of Slope of a Line

Let l be a line that is not parallel to the y-axis, and let $P_1(x_1, y_1)$ and $P_2(x_2, y_2)$ be distinct points on l. The **slope m** of l is

$$m = \frac{y_2 - y_1}{x_2 - x_1}.$$

If l is parallel to the y-axis, then the slope of l is not defined.

The Greek letter Δ (delta) is used in mathematics to denote "change in." Thus, we can think of the slope m as

$$m = \frac{\Delta y}{\Delta x} = \frac{\text{change in } y}{\text{change in } x}.$$

Figure 1
(a) Positive slope (line rises)

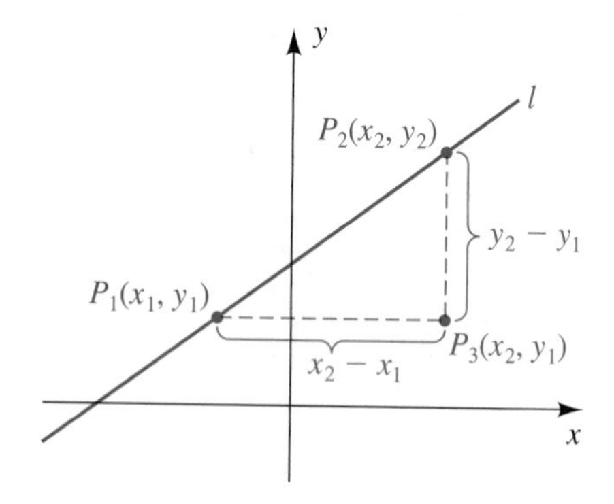

(b) Negative slope (line falls)

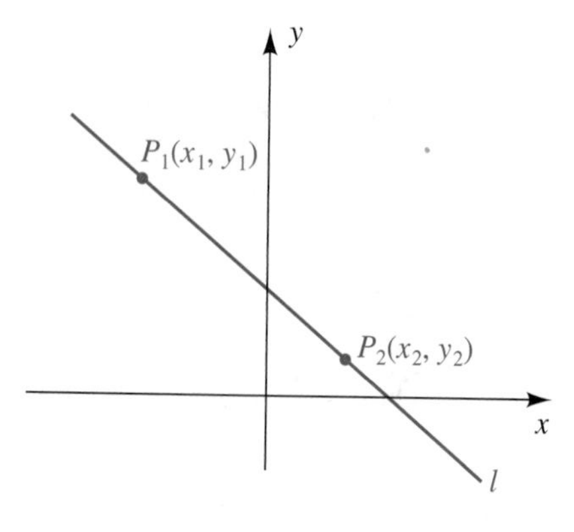

Typical points P_1 and P_2 on a line l are shown in Figure 1. The numerator $y_2 - y_1$ in the formula for m is the vertical change in direction from P_1 to P_2 and may be positive, negative, or zero. The denominator $x_2 - x_1$ is the horizontal change from P_1 to P_2, and it may be positive or negative, but never zero, because l is not parallel to the y-axis if a slope exists. In Figure 1(a) the slope is positive, and we say that the line *rises*. In Figure 1(b) the slope is negative, and the line *falls*.

In finding the slope of a line it is immaterial which point we label as P_1 and which as P_2, since

$$\frac{y_2 - y_1}{x_2 - x_1} = \frac{y_2 - y_1}{x_2 - x_1} \cdot \frac{(-1)}{(-1)} = \frac{y_1 - y_2}{x_1 - x_2}.$$

If the points are labeled so that $x_1 < x_2$, as in Figure 1, then $x_2 - x_1 > 0$, and hence the slope is positive, negative, or zero, depending on whether $y_2 > y_1$, $y_2 < y_1$, or $y_2 = y_1$, respectively.

The definition of slope is independent of the two points that are chosen on l. If other points $P_1'(x_1', y_1')$ and $P_2'(x_2', y_2')$ are used, then, as in Figure 2, the triangle with vertices P_1', P_2', and $P_3'(x_2', y_1')$ is similar to the triangle with vertices P_1, P_2, and $P_3(x_2, y_1)$. Since the ratios of corresponding sides of similar triangles are equal,

$$\frac{y_2 - y_1}{x_2 - x_1} = \frac{y_2' - y_1'}{x_2' - x_1'}.$$

Figure 2

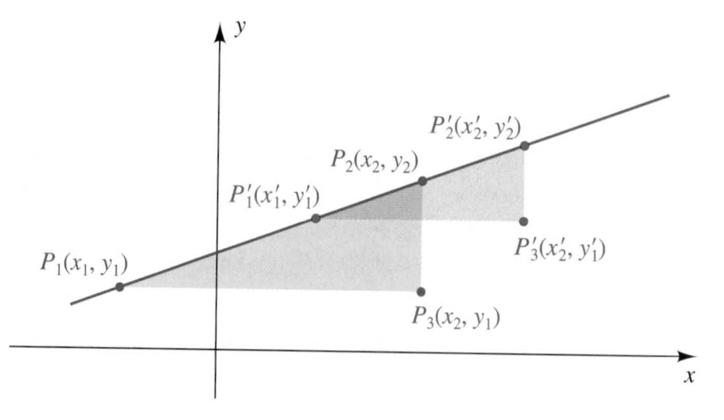

EXAMPLE 1 Finding slopes

Sketch the line through each pair of points, and find its slope m:

(a) $A(-1, 4)$ and $B(3, 2)$ **(b)** $A(2, 5)$ and $B(-2, -1)$

(c) $A(4, 3)$ and $B(-2, 3)$ **(d)** $A(4, -1)$ and $B(4, 4)$

SOLUTION The lines are sketched in Figure 3. We use the definition of slope to find the slope of each line.

Figure 3

(a) $m = -\frac{1}{2}$

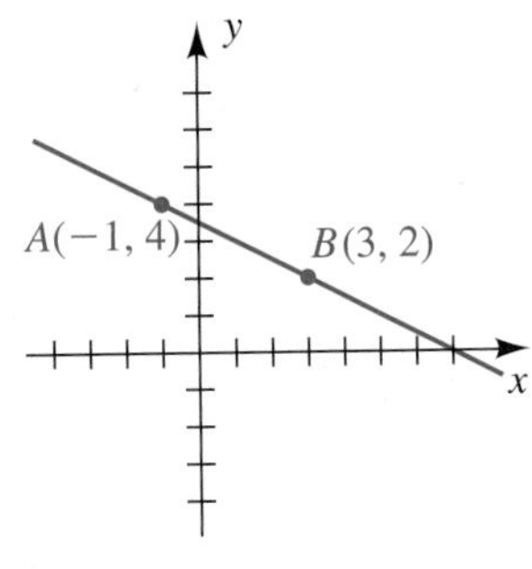

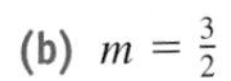

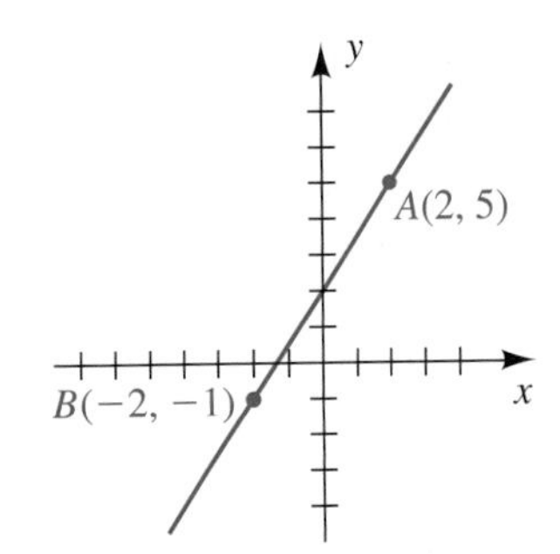

(c) $m = 0$

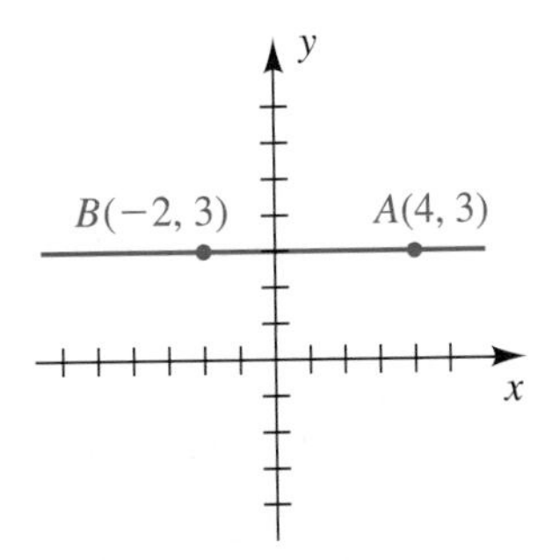

(d) m undefined

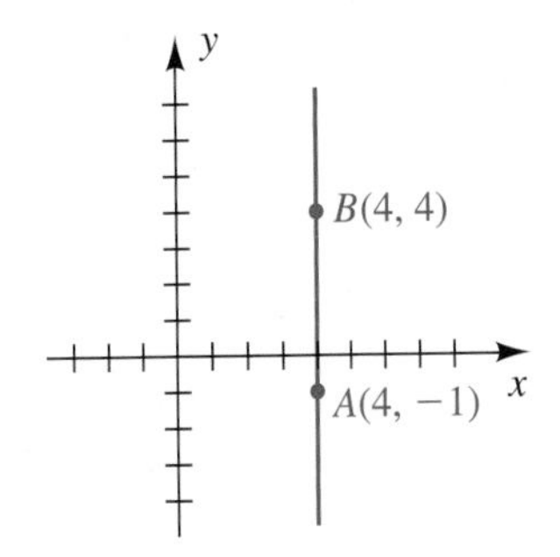

▶ **(a)** $m = \dfrac{2 - 4}{3 - (-1)} = \dfrac{-2}{4} = -\dfrac{1}{2}$

(b) $m = \dfrac{5 - (-1)}{2 - (-2)} = \dfrac{6}{4} = \dfrac{3}{2}$

▶ **(c)** $m = \dfrac{3 - 3}{-2 - 4} = \dfrac{0}{-6} = 0$

(d) The slope is undefined because the line is parallel to the y-axis. Note that if the formula for m is used, the denominator is zero.

EXAMPLE 2 Sketching a line with a given slope

Sketch a line through $P(2, 1)$ that has

(a) slope $\frac{5}{3}$ **(b)** slope $-\frac{5}{3}$

SOLUTION If the slope of a line is a/b and b is positive, then for every change of b units in the horizontal direction, the line rises or falls $|a|$ units, depending on whether a is positive or negative, respectively.

(a) If $P(2, 1)$ is on the line and $m = \frac{5}{3}$, we can obtain another point on the line by starting at P and moving 3 units to the right and 5 units *upward*. This gives us the point $Q(5, 6)$, and the line is determined as in Figure 4(a).

(continued)

▶ **Tutorial available at www.thomsonedu.com/login**

(b) If $P(2, 1)$ is on the line and $m = -\frac{5}{3}$, we move 3 units to the right and 5 units *downward,* obtaining the line through $Q(5, -4)$, as in Figure 4(b).

Figure 4
(a) $m = \frac{5}{3}$ **(b)** $m = -\frac{5}{3}$

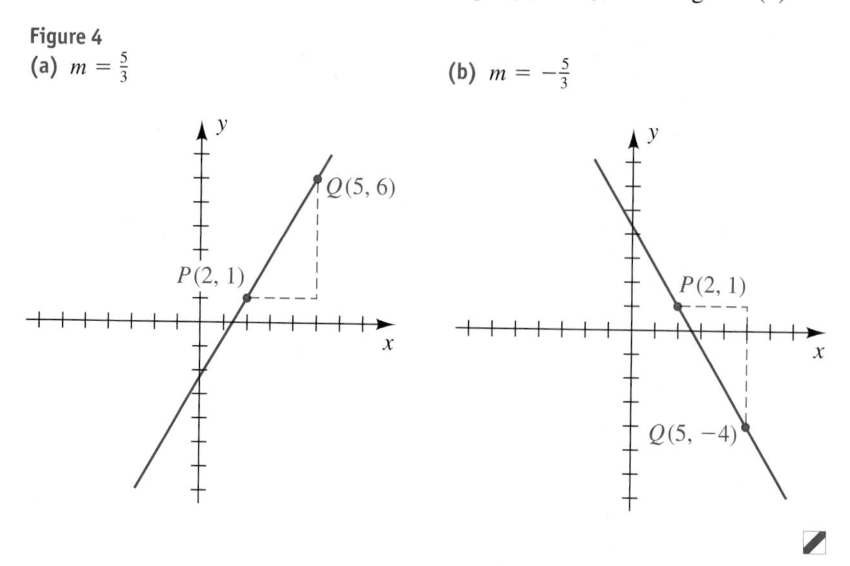

The diagram in Figure 5 indicates the slopes of several lines through the origin. The line that lies on the x-axis has slope $m = 0$. If this line is rotated about O in the *counterclockwise* direction (as indicated by the blue arrow), the slope is positive and increases, reaching the value 1 when the line bisects the first quadrant and continuing to increase as the line gets closer to the y-axis. If we rotate the line of slope $m = 0$ in the *clockwise* direction (as indicated by the red arrow), the slope is negative, reaching the value -1 when the line bisects the second quadrant and becoming large and negative as the line gets closer to the y-axis.

Figure 5

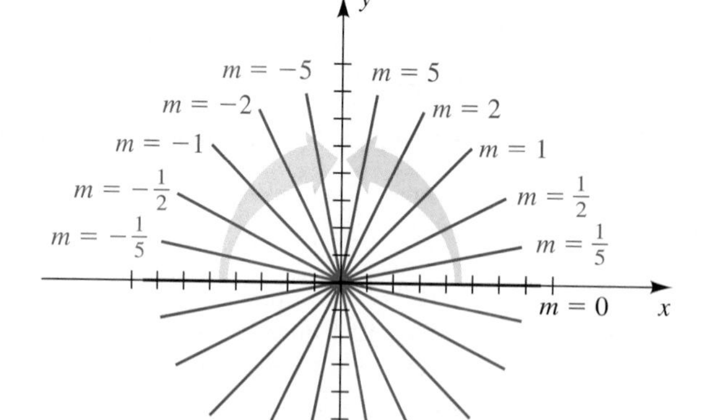

Lines that are horizontal or vertical have simple equations, as indicated in the following chart.

Terminology	Definition	Graph	Equation	Slope
Horizontal line	A line parallel to the x-axis		$y = b$ y-intercept is b	Slope is 0
Vertical line	A line parallel to the y-axis		$x = a$ x-intercept is a	Slope is undefined

Figure 6

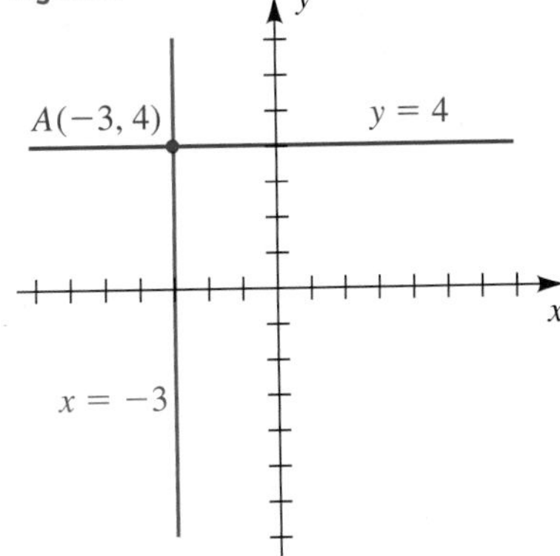

A common error is to regard the graph of $y = b$ as consisting of only the one point $(0, b)$. If we express the equation in the form $0 \cdot x + y = b$, we see that the value of x is immaterial; thus, the graph of $y = b$ consists of the points (x, b) for *every* x and hence is a horizontal line. Similarly, the graph of $x = a$ is the vertical line consisting of all points (a, y), where y is a real number.

EXAMPLE 3 Finding equations of horizontal and vertical lines

Find an equation of the line through $A(-3, 4)$ that is parallel to

(a) the x-axis **(b)** the y-axis

▶ **SOLUTION** The two lines are sketched in Figure 6. As indicated in the preceding chart, the equations are $y = 4$ for part (a) and $x = -3$ for part (b).

Figure 7

Let us next find an equation of a line l through a point $P_1(x_1, y_1)$ with slope m. If $P(x, y)$ is any point with $x \neq x_1$ (see Figure 7), then P is on l if and only if the slope of the line through P_1 and P is m—that is, if

$$\frac{y - y_1}{x - x_1} = m.$$

This equation may be written in the form

$$y - y_1 = m(x - x_1).$$

▶ **Tutorial available at www.thomsonedu.com/login**

Note that (x_1, y_1) is a solution of the last equation, and hence the points on l are precisely the points that correspond to the solutions. This equation for l is referred to as the **point-slope form.**

Point-Slope Form for the Equation of a Line	An equation for the line through the point (x_1, y_1) with slope m is $$y - y_1 = m(x - x_1).$$

The point-slope form is only one possibility for an equation of a line. There are many equivalent equations. We sometimes simplify the equation obtained using the point-slope form to either

$$ax + by = c \qquad \text{or} \qquad ax + by + d = 0,$$

where a, b, and c are integers with no common factor, $a > 0$, and $d = -c$.

EXAMPLE 4 Finding an equation of a line through two points

Find an equation of the line through $A(1, 7)$ and $B(-3, 2)$.

Figure 8

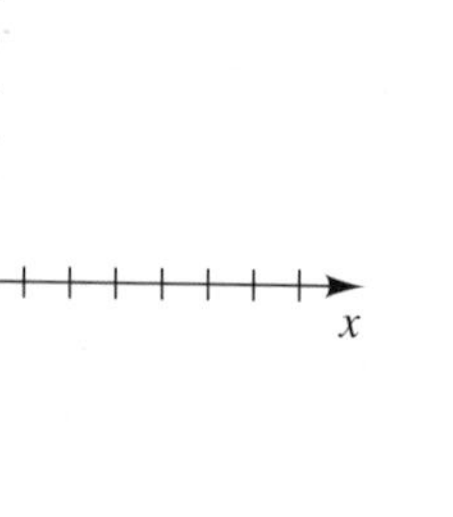

▶ **SOLUTION** The line is sketched in Figure 8. The formula for the slope m gives us

$$m = \frac{7-2}{1-(-3)} = \frac{5}{4}.$$

We may use the coordinates of either A or B for (x_1, y_1) in the point-slope form. Using $A(1, 7)$ gives us

$$\begin{aligned} y - 7 &= \tfrac{5}{4}(x - 1) && \text{point-slope form} \\ 4(y - 7) &= 5(x - 1) && \text{multiply by 4} \\ 4y - 28 &= 5x - 5 && \text{multiply factors} \\ -5x + 4y &= 23 && \text{subtract } 5x \text{ and add 28} \\ 5x - 4y &= -23 && \text{multiply by } -1 \end{aligned}$$

The last equation is one of the desired forms for an equation of a line. Another is $5x - 4y + 23 = 0$.

Figure 9

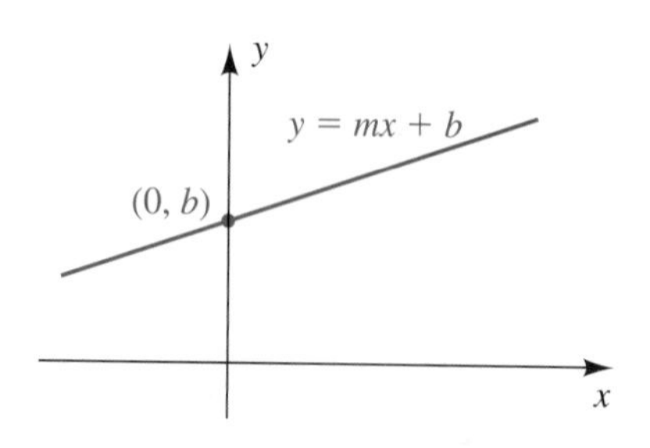

The point-slope form for the equation of a line may be rewritten as $y = mx - mx_1 + y_1$, which is of the form

$$y = mx + b$$

with $b = -mx_1 + y_1$. The real number b is the y-intercept of the graph, as indicated in Figure 9. Since the equation $y = mx + b$ displays the slope m and

▶ **Tutorial available at www.thomsonedu.com/login**

y-intercept b of l, it is called the **slope-intercept form** for the equation of a line. Conversely, if we start with $y = mx + b$, we may write

$$y - b = m(x - 0).$$

Comparing this equation with the point-slope form, we see that the graph is a line with slope m and passing through the point $(0, b)$. We have proved the following result.

Slope-Intercept Form for the Equation of a Line	The graph of $y = mx + b$ is a line having slope m and y-intercept b.

EXAMPLE 5 Expressing an equation in slope-intercept form

Express the equation $2x - 5y = 8$ in slope-intercept form.

▶ SOLUTION Our goal is to solve the given equation for y to obtain the form $y = mx + b$. We may proceed as follows:

$$2x - 5y = 8 \qquad \text{given}$$

$$-5y = -2x + 8 \qquad \text{subtract } 2x$$

$$y = \left(\frac{-2}{-5}\right)x + \left(\frac{8}{-5}\right) \qquad \text{divide by } -5$$

$$y = \tfrac{2}{5}x + \left(-\tfrac{8}{5}\right) \qquad \text{equivalent equation}$$

The last equation is the slope-intercept form $y = mx + b$ with slope $m = \frac{2}{5}$ and y-intercept $b = -\frac{8}{5}$.

It follows from the point-slope form that every line is a graph of an equation

$$ax + by = c,$$

where a, b, and c are real numbers and a and b are not both zero. We call such an equation a **linear equation** in x and y. Let us show, conversely, that the graph of $ax + by = c$, with a and b not both zero, is always a line. If $b \neq 0$, we may solve for y, obtaining

$$y = \left(-\frac{a}{b}\right)x + \frac{c}{b},$$

which, by the slope-intercept form, is an equation of a line with slope $-a/b$ and y-intercept c/b. If $b = 0$ but $a \neq 0$, we may solve for x, obtaining $x = c/a$, which is the equation of a vertical line with x-intercept c/a. This discussion establishes the following result.

▶ **Tutorial available at www.thomsonedu.com/login**

General Form for the Equation of a Line	The graph of a linear equation $ax + by = c$ is a line, and conversely, every line is the graph of a linear equation.

For simplicity, we use the terminology *the line* $ax + by = c$ rather than *the line with equation* $ax + by = c$.

EXAMPLE 6 Sketching the graph of a linear equation

Sketch the graph of $2x - 5y = 8$.

Figure 10

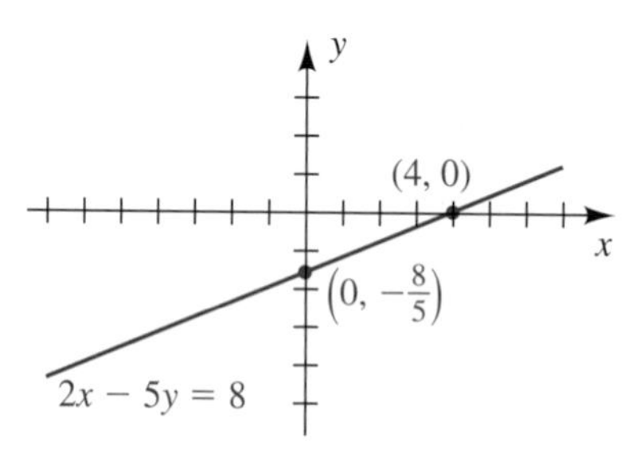

SOLUTION We know from the preceding discussion that the graph is a line, so it is sufficient to find two points on the graph. Let us find the x- and y-intercepts by substituting $y = 0$ and $x = 0$, respectively, in the given equation, $2x - 5y = 8$.

$$x\text{-intercept:}\quad \text{If } y = 0, \text{ then } 2x = 8, \text{ or } x = 4.$$
$$y\text{-intercept:}\quad \text{If } x = 0, \text{ then } -5y = 8, \text{ or } y = -\tfrac{8}{5}.$$

Plotting the points $(4, 0)$ and $\left(0, -\frac{8}{5}\right)$ and drawing a line through them gives us the graph in Figure 10.

The following theorem specifies the relationship between **parallel lines** (lines in a plane that do not intersect) and slope.

Theorem on Slopes of Parallel Lines	Two nonvertical lines are parallel if and only if they have the same slope.

Figure 11

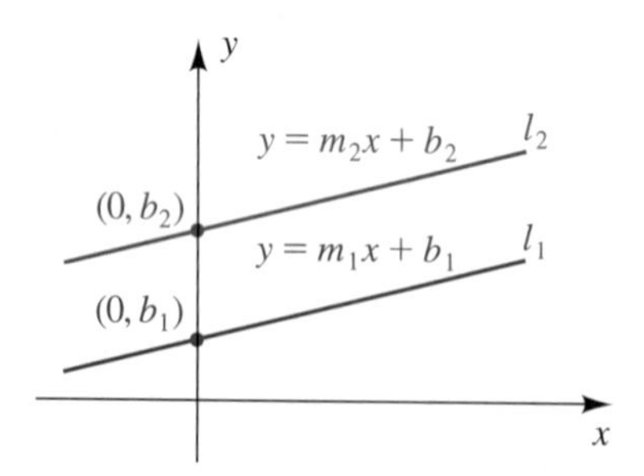

PROOF Let l_1 and l_2 be distinct lines of slopes m_1 and m_2, respectively. If the y-intercepts are b_1 and b_2 (see Figure 11), then, by the slope-intercept form, the lines have equations

$$y = m_1x + b_1 \qquad \text{and} \qquad y = m_2x + b_2.$$

The lines intersect at some point (x, y) if and only if the values of y are equal for some x—that is, if

$$m_1x + b_1 = m_2x + b_2,$$

or

$$(m_1 - m_2)x = b_2 - b_1.$$

The last equation can be solved for x if and only if $m_1 - m_2 \neq 0$. We have shown that the lines l_1 and l_2 intersect if and only if $m_1 \neq m_2$. Hence, they do *not* intersect (are parallel) if and only if $m_1 = m_2$.

EXAMPLE 7 **Finding an equation of a line parallel to a given line**

Find an equation of the line through $P(5, -7)$ that is parallel to the line $6x + 3y = 4$.

▶ **SOLUTION** We first express the given equation in slope-intercept form:

$$\begin{aligned} 6x + 3y &= 4 && \text{given} \\ 3y &= -6x + 4 && \text{subtract } 6x \\ y &= -2x + \tfrac{4}{3} && \text{divide by 3} \end{aligned}$$

Figure 12

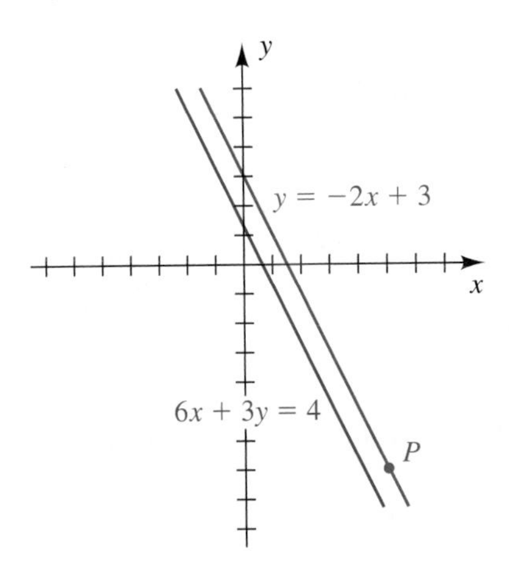

The last equation is in slope-intercept form, $y = mx + b$, with slope $m = -2$ and y-intercept $\frac{4}{3}$. Since parallel lines have the same slope, the required line also has slope -2. Using the point $P(5, -7)$ gives us the following:

$$\begin{aligned} y - (-7) &= -2(x - 5) && \text{point-slope form} \\ y + 7 &= -2x + 10 && \text{simplify} \\ y &= -2x + 3 && \text{subtract 7} \end{aligned}$$

The last equation is in slope-intercept form and shows that the parallel line we have found has y-intercept 3. This line and the given line are sketched in Figure 12.

As an alternative solution, we might use the fact that lines of the form $6x + 3y = k$ have the same slope as the given line and hence are parallel to it. Substituting $x = 5$ and $y = -7$ into the equation $6x + 3y = k$ gives us $6(5) + 3(-7) = k$ or, equivalently, $k = 9$. The equation $6x + 3y = 9$ is equivalent to $y = -2x + 3$.

If the slopes of two nonvertical lines are not the same, then the lines are not parallel and intersect at exactly one point.

The next theorem gives us information about **perpendicular lines** (lines that intersect at a right angle).

Theorem on Slopes of Perpendicular Lines	Two lines with slope m_1 and m_2 are perpendicular if and only if $$m_1m_2 = -1.$$

Figure 13

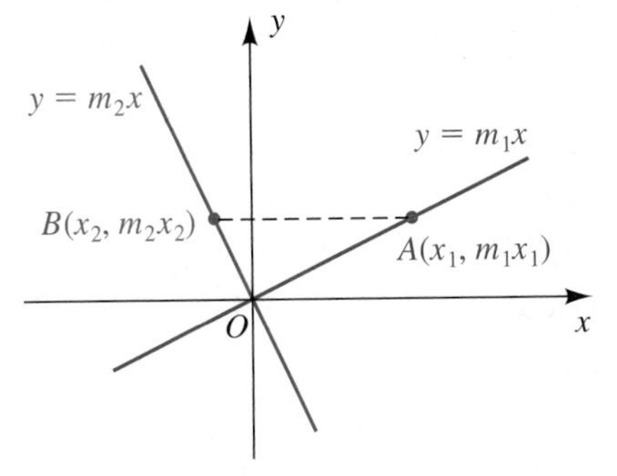

PROOF For simplicity, let us consider the special case of two lines that intersect at the origin O, as illustrated in Figure 13. Equations of these lines are $y = m_1x$ and $y = m_2x$. If, as in the figure, we choose points $A(x_1, m_1x_1)$ and $B(x_2, m_2x_2)$ different from O on the lines, then the lines are perpendicular if and only if angle AOB is a right angle. Applying the Pythagorean theorem, we know that angle AOB is a right angle if and only if

$$[d(A, B)]^2 = [d(O, B)]^2 + [d(O, A)]^2$$

or, by the distance formula,

$$(x_2 - x_1)^2 + (m_2x_2 - m_1x_1)^2 = x_2^2 + (m_2x_2)^2 + x_1^2 + (m_1x_1)^2.$$

(continued)

▶ **Tutorial available at www.thomsonedu.com/login**

Figure 14

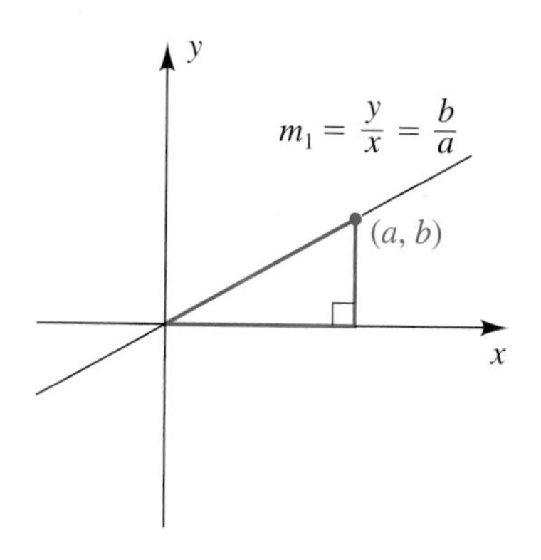

Squaring terms, simplifying, and factoring gives us

$$-2m_1m_2x_1x_2 - 2x_1x_2 = 0$$
$$-2x_1x_2(m_1m_2 + 1) = 0.$$

Since both x_1 and x_2 are not zero, we may divide both sides by $-2x_1x_2$, obtaining $m_1m_2 + 1 = 0$. Thus, the lines are perpendicular if and only if $m_1m_2 = -1$.

The same type of proof may be given if the lines intersect at *any* point (a, b).

Figure 15

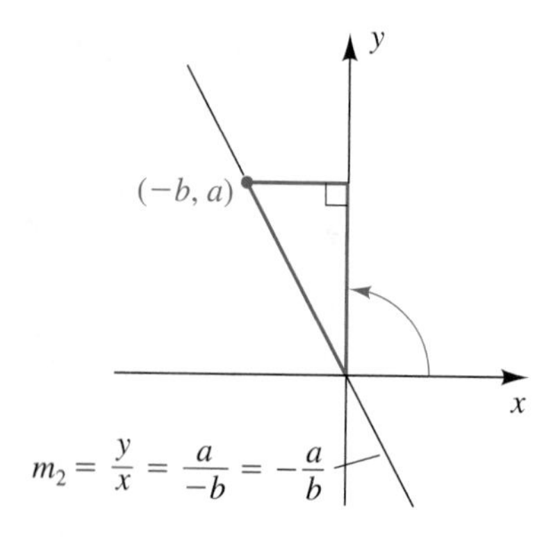

A convenient way to remember the conditions on slopes of perpendicular lines is to note that m_1 and m_2 must be *negative reciprocals* of each other—that is, $m_1 = -1/m_2$ and $m_2 = -1/m_1$.

We can visualize the result of the last theorem as follows. Draw a triangle as in Figure 14; the line containing its hypotenuse has slope $m_1 = b/a$. Now rotate the triangle 90° as in Figure 15. The line now has slope $m_2 = a/(-b)$, the negative reciprocal of m_1.

EXAMPLE 8 Finding an equation of a line perpendicular to a given line

Find the slope-intercept form for the line through $P(5, -7)$ that is perpendicular to the line $6x + 3y = 4$.

▶ **SOLUTION** We considered the line $6x + 3y = 4$ in Example 7 and found that its slope is -2. Hence, the slope of the required line is the negative reciprocal $-[1/(-2)]$, or $\frac{1}{2}$. Using $P(5, -7)$ gives us the following:

Figure 16

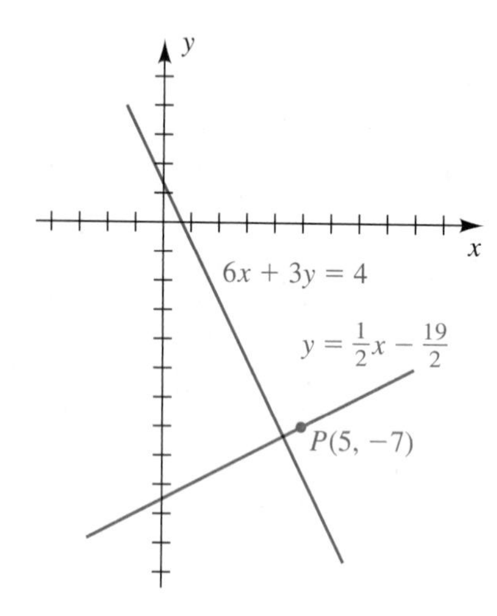

$$y - (-7) = \tfrac{1}{2}(x - 5) \quad \text{point-slope form}$$
$$y + 7 = \tfrac{1}{2}x - \tfrac{5}{2} \quad \text{simplify}$$
$$y = \tfrac{1}{2}x - \tfrac{19}{2} \quad \text{put in slope-intercept form}$$

The last equation is in slope-intercept form and shows that the perpendicular line has y-intercept $-\frac{19}{2}$. This line and the given line are sketched in Figure 16.

EXAMPLE 9 Finding an equation of a perpendicular bisector

Given $A(-3, 1)$ and $B(5, 4)$, find the general form of the perpendicular bisector l of the line segment AB.

Figure 17

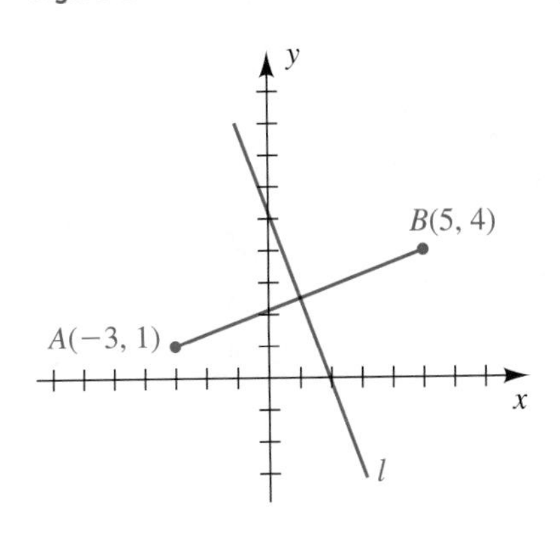

▶ **SOLUTION** The line segment AB and its perpendicular bisector l are shown in Figure 17. We calculate the following, where M is the midpoint of AB:

Coordinates of M:	$\left(\dfrac{-3+5}{2}, \dfrac{1+4}{2}\right) = \left(1, \dfrac{5}{2}\right)$	midpoint formula
Slope of AB:	$\dfrac{4-1}{5-(-3)} = \dfrac{3}{8}$	slope formula
Slope of l:	$-\dfrac{1}{\frac{3}{8}} = -\dfrac{8}{3}$	negative reciprocal of $\frac{3}{8}$

Using the point $M\left(1, \frac{5}{2}\right)$ and slope $-\frac{8}{3}$ gives us the following equivalent equations for l:

$$y - \tfrac{5}{2} = -\tfrac{8}{3}(x-1) \quad \text{point-slope form}$$
$$6y - 15 = -16(x-1) \quad \text{multiply by the lcd, 6}$$
$$6y - 15 = -16x + 16 \quad \text{multiply}$$
$$16x + 6y = 31 \quad \text{put in general form}$$

Two variables x and y are **linearly related** if $y = ax + b$, where a and b are real numbers and $a \neq 0$. Linear relationships between variables occur frequently in applied problems. The following example gives one illustration.

EXAMPLE 10 **Relating air temperature to altitude**

The relationship between the air temperature T (in °F) and the altitude h (in feet above sea level) is approximately linear for $0 \le h \le 20{,}000$. If the temperature at sea level is 60°, an increase of 5000 feet in altitude lowers the air temperature about 18°.

(a) Express T in terms of h, and sketch the graph on an hT-coordinate system.

(b) Approximate the air temperature at an altitude of 15,000 feet.

(c) Approximate the altitude at which the temperature is 0°.

SOLUTION

(a) If T is linearly related to h, then

$$T = ah + b$$

for some constants a and b (a represents the slope and b the T-intercept). Since $T = 60°$ when $h = 0$ ft (sea level), the T-intercept is 60, and the temperature T for $0 \le h \le 20{,}000$ is given by

$$T = ah + 60.$$

From the given data, we note that when the altitude $h = 5000$ ft, the temperature $T = 60° - 18° = 42°$. Hence, we may find a as follows:

$$42 = a(5000) + 60 \quad \text{let } T = 42 \text{ and } h = 5000$$
$$a = \frac{42-60}{5000} = -\frac{9}{2500} \quad \text{solve for } a$$

(continued)

▶ **Tutorial available at www.thomsonedu.com/login**

Figure 18

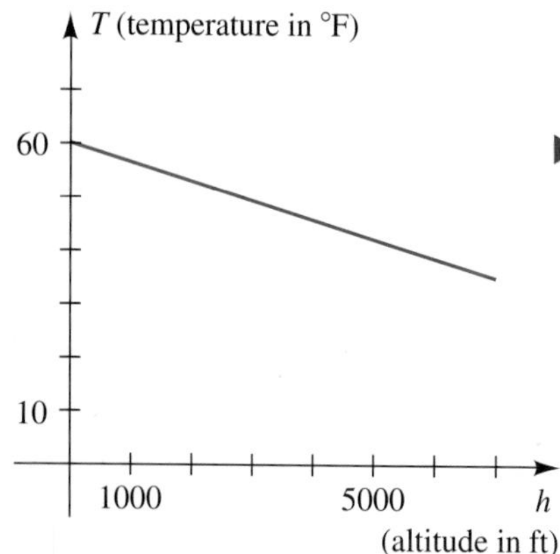

Substituting for a in $T = ah + 60$ gives us the following formula for T:

$$T = -\tfrac{9}{2500}h + 60$$

The graph is sketched in Figure 18, with different scales on the axes.

(b) Using the last formula for T obtained in part (a), we find that the temperature (in °F) when $h = 15{,}000$ is

$$T = -\tfrac{9}{2500}(15{,}000) + 60 = -54 + 60 = 6.$$

(c) To find the altitude h that corresponds to $T = 0°$, we proceed as follows:

$$T = -\tfrac{9}{2500}h + 60 \qquad \text{from part (a)}$$

$$0 = -\tfrac{9}{2500}h + 60 \qquad \text{let } T = 0$$

$$\tfrac{9}{2500}h = 60 \qquad \text{add } \tfrac{9}{2500}h$$

$$h = 60 \cdot \tfrac{2500}{9} \qquad \text{multiply by } \tfrac{2500}{9}$$

$$h = \frac{50{,}000}{3} \approx 16{,}667 \text{ ft} \qquad \text{simplify and approximate}$$

A **mathematical model** is a mathematical description of a problem. For our purposes, these descriptions will be graphs and equations. In the last example, the equation $T = -\frac{9}{2500}h + 60$ *models* the relationship between air temperature and altitude.

In the next example, we find a model of the form $y = mx + b$, called the *linear regression line.* We can think of this line as *the line of best fit*—that is, the unique line that best describes the behavior of the data.

EXAMPLE 11 Finding a line of best fit

(a) Find the line of best fit that approximates the following data on world record times for the women's 100-meter dash.

Year (x)	Runner	Time in seconds (y)
1952	Marjorie Jackson	11.4
1960	Wilma Rudolph	11.3
1972	Renate Stecher	11.07
1984	Evelyn Ashford	10.76

(b) Graph the data and the regression line.

(c) Wyomia Tyus held the record in 1968 at 11.08 seconds. What time does the model predict for 1968? This question calls for **interpolation,** since we must estimate a value between known values. What time does the model predict for 1988? This question calls for **extrapolation,** since we must estimate a value outside known values.

(d) Interpret the slope of the line.

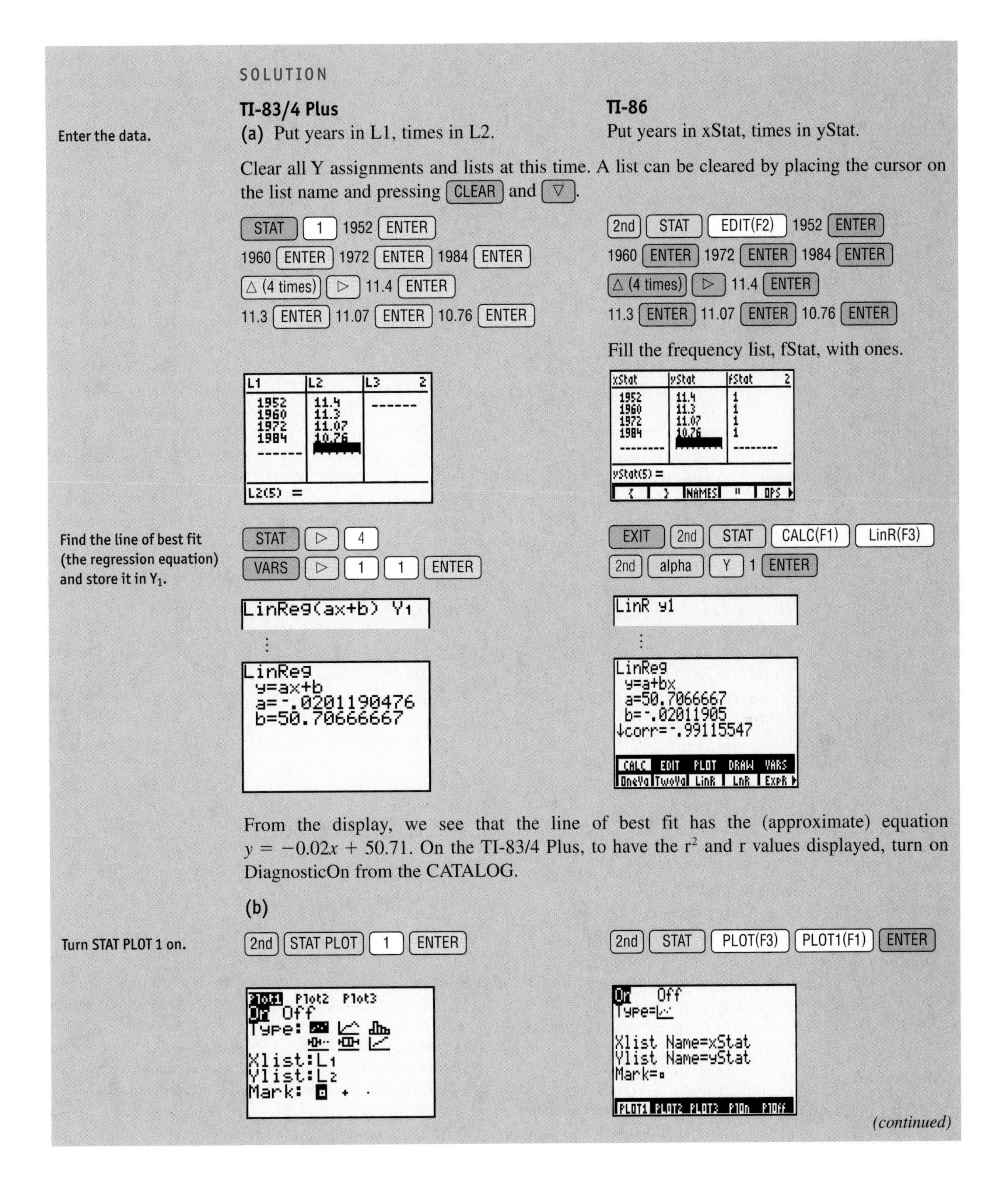

SOLUTION

TI-83/4 Plus

Enter the data.

(a) Put years in L1, times in L2.

TI-86

Put years in xStat, times in yStat.

Clear all Y assignments and lists at this time. A list can be cleared by placing the cursor on the list name and pressing CLEAR and ▽.

TI-83/4 Plus:

STAT 1 1952 ENTER

1960 ENTER 1972 ENTER 1984 ENTER

△ (4 times) ▷ 11.4 ENTER

11.3 ENTER 11.07 ENTER 10.76 ENTER

TI-86:

2nd STAT EDIT(F2) 1952 ENTER

1960 ENTER 1972 ENTER 1984 ENTER

△ (4 times) ▷ 11.4 ENTER

11.3 ENTER 11.07 ENTER 10.76 ENTER

Fill the frequency list, fStat, with ones.

Find the line of best fit (the regression equation) and store it in Y_1.

TI-83/4 Plus:

STAT ▷ 4

VARS ▷ 1 1 ENTER

TI-86:

EXIT 2nd STAT CALC(F1) LinR(F3)

2nd alpha Y 1 ENTER

From the display, we see that the line of best fit has the (approximate) equation $y = -0.02x + 50.71$. On the TI-83/4 Plus, to have the r^2 and r values displayed, turn on DiagnosticOn from the CATALOG.

(b)

Turn STAT PLOT 1 on.

TI-83/4 Plus:

2nd STAT PLOT 1 ENTER

TI-86:

2nd STAT PLOT(F3) PLOT1(F1) ENTER

(continued)

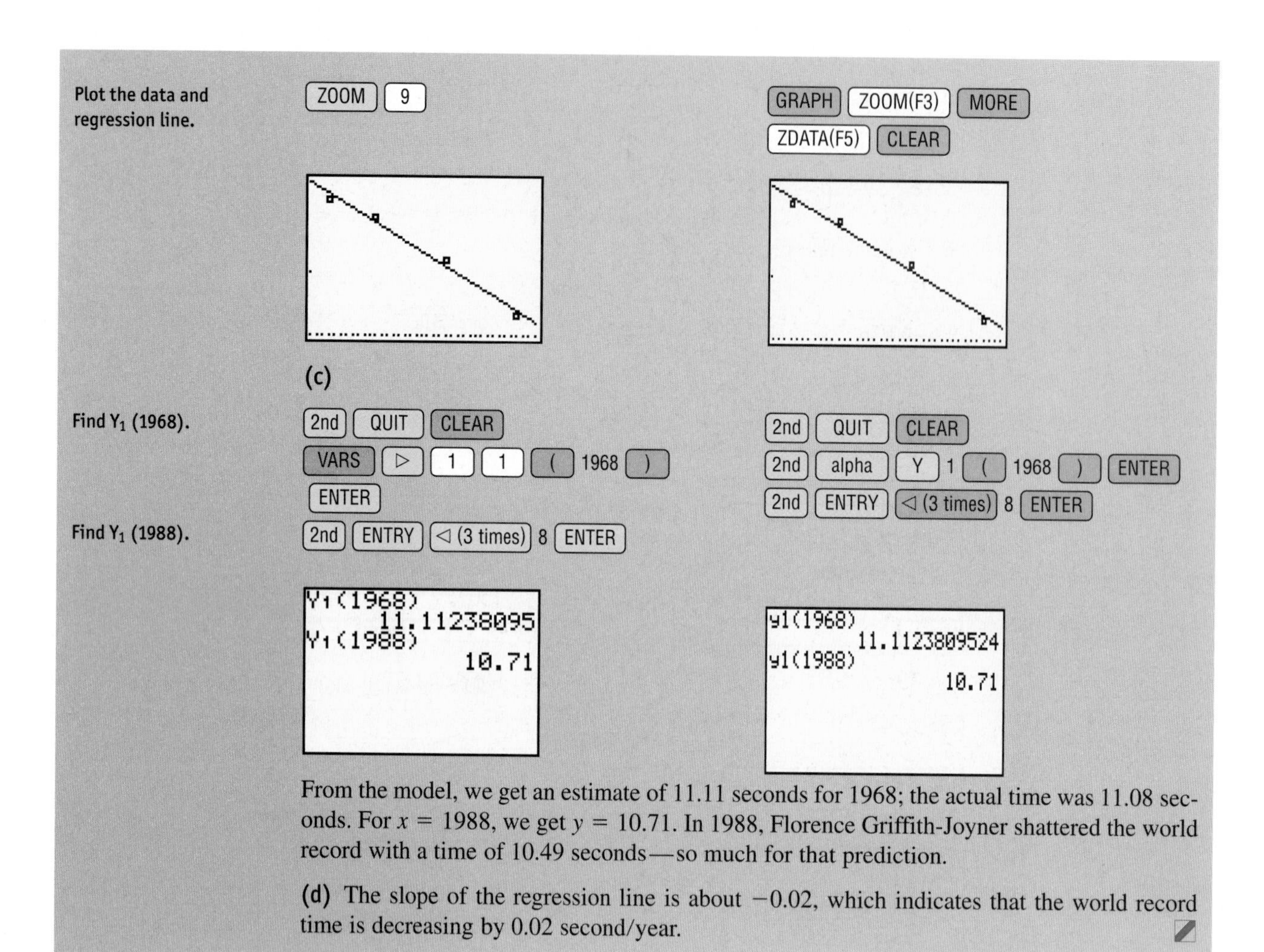

From the model, we get an estimate of 11.11 seconds for 1968; the actual time was 11.08 seconds. For $x = 1988$, we get $y = 10.71$. In 1988, Florence Griffith-Joyner shattered the world record with a time of 10.49 seconds—so much for that prediction.

(d) The slope of the regression line is about -0.02, which indicates that the world record time is decreasing by 0.02 second/year.

2.3 *Exercises*

Exer. 1–6: Sketch the line through A and B, and find its slope m.

1 $A(-3, 2)$, $B(5, -4)$

2 $A(4, -1)$, $B(-6, -3)$

3 $A(2, 5)$, $B(-7, 5)$

4 $A(5, -1)$, $B(5, 6)$

5 $A(-3, 2)$, $B(-3, 5)$

6 $A(4, -2)$, $B(-3, -2)$

Exer. 7–10: Use slopes to show that the points are vertices of the specified polygon.

7 $A(-3, 1)$, $B(5, 3)$, $C(3, 0)$, $D(-5, -2)$; parallelogram

8 $A(2, 3)$, $B(5, -1)$, $C(0, -6)$, $D(-6, 2)$; trapezoid

9 $A(6, 15)$, $B(11, 12)$, $C(-1, -8)$, $D(-6, -5)$; rectangle

10 $A(1, 4)$, $B(6, -4)$, $C(-15, -6)$; right triangle

11 If three consecutive vertices of a parallelogram are $A(-1, -3)$, $B(4, 2)$, and $C(-7, 5)$, find the fourth vertex.

12 Let $A(x_1, y_1)$, $B(x_2, y_2)$, $C(x_3, y_3)$, and $D(x_4, y_4)$ denote the vertices of an arbitrary quadrilateral. Show that the line segments joining midpoints of adjacent sides form a parallelogram.

Exer. 13–14: Sketch the graph of $y = mx$ for the given values of m.

13 $m = 3, -2, \frac{2}{3}, -\frac{1}{4}$

14 $m = 5, -3, \frac{1}{2}, -\frac{1}{3}$

Exer. 15–16: Sketch the graph of the line through P for each value of m.

15 $P(3, 1)$; $m = \frac{1}{2}, -1, -\frac{1}{5}$

16 $P(-2, 4)$; $m = 1, -2, -\frac{1}{2}$

Exer. 17–18: Write equations of the lines.

17

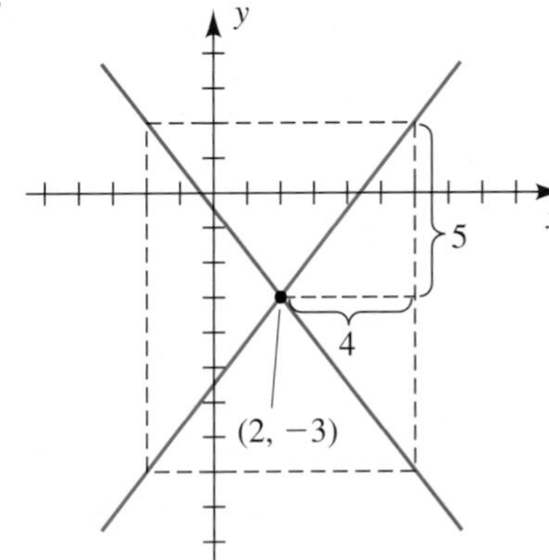

18

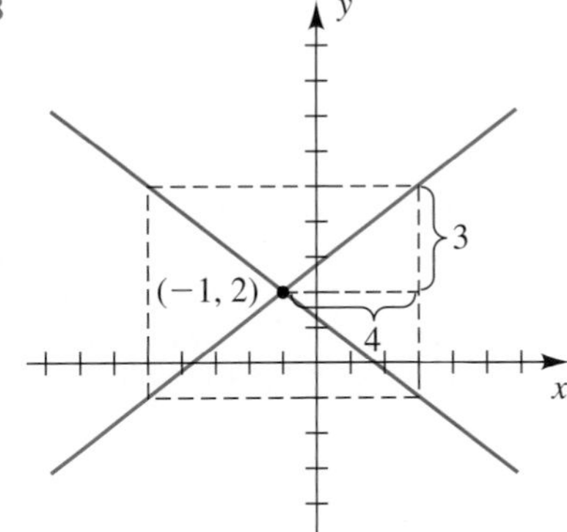

Exer. 19–20: Sketch the graphs of the lines on the same coordinate plane.

19 $y = x + 3$, $y = x + 1$, $y = -x + 1$

20 $y = -2x - 1$, $y = -2x + 3$, $y = \frac{1}{2}x + 3$

Exer. 21–32: Find a general form of an equation of the line through the point A that satisfies the given condition.

21 $A(5, -2)$

(a) parallel to the y-axis

(b) perpendicular to the y-axis

22 $A(-4, 2)$

(a) parallel to the x-axis

(b) perpendicular to the x-axis

23 $A(5, -3)$; slope -4

24 $A(-1, 4)$; slope $\frac{2}{3}$

25 $A(4, 0)$; slope -3

26 $A(0, -2)$; slope 5

27 $A(4, -5)$; through $B(-3, 6)$

28 $A(-1, 6)$; x-intercept 5

29 $A(2, -4)$; parallel to the line $5x - 2y = 4$

30 $A(-3, 5)$; parallel to the line $x + 3y = 1$

31 $A(7, -3)$; perpendicular to the line $2x - 5y = 8$

32 $A(4, 5)$; perpendicular to the line $3x + 2y = 7$

Exer. 33–36: Find the slope-intercept form of the line that satisfies the given conditions.

33 x-intercept 4, y-intercept -3

34 x-intercept -5, y-intercept -1

35 Through $A(5, 2)$ and $B(-1, 4)$

36 Through $A(-2, 1)$ and $B(3, 7)$

Exer. 37–38: Find a general form of an equation for the perpendicular bisector of the segment AB.

37 $A(3, -1)$, $B(-2, 6)$

38 $A(4, 2)$, $B(-2, 10)$

Exer. 39–40: Find an equation for the line that bisects the given quadrants.

39 II and IV

40 I and III

Exer. 41–44: Use the slope-intercept form to find the slope and *y*-intercept of the given line, and sketch its graph.

41 $2x = 15 - 3y$

42 $7x = -4y - 8$

43 $4x - 3y = 9$

44 $x - 5y = -15$

Exer. 45–46: Find an equation of the line shown in the figure.

45 (a) (b)

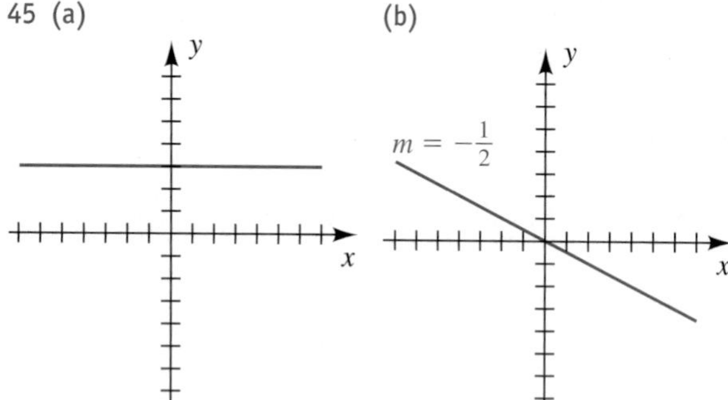

(c) (d)

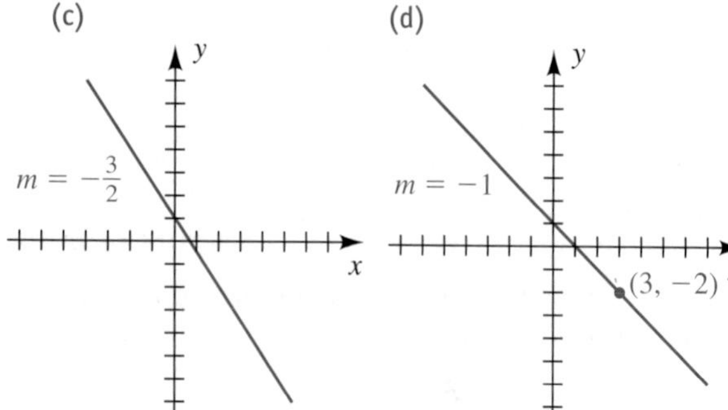

46 (a) (b)

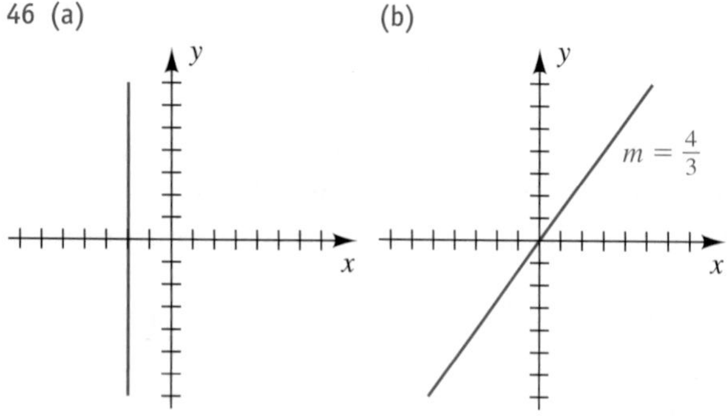

(c) (d)

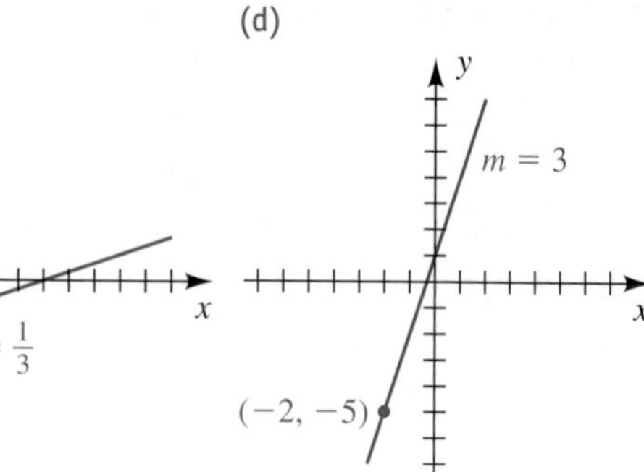

Exer. 47–48: If a line *l* has nonzero *x*- and *y*-intercepts *a* and *b*, respectively, then its *intercept form* is

$$\frac{x}{a} + \frac{y}{b} = 1.$$

Find the intercept form for the given line.

47 $4x - 2y = 6$

48 $x - 3y = -2$

49 Find an equation of the circle that has center $C(3, -2)$ and is tangent to the line $y = 5$.

50 Find an equation of the line that is tangent to the circle $x^2 + y^2 = 25$ at the point $P(3, 4)$.

51 Fetal growth The growth of a fetus more than 12 weeks old can be approximated by the formula $L = 1.53t - 6.7$, where L is the length (in centimeters) and t is the age (in weeks). Prenatal length can be determined by ultrasound. Approximate the age of a fetus whose length is 28 centimeters.

52 Estimating salinity Salinity of the ocean refers to the amount of dissolved material found in a sample of seawater. Salinity S can be estimated from the amount C of chlorine in seawater using $S = 0.03 + 1.805C$, where S and C are measured by weight in parts per thousand. Approximate C if S is 0.35.

53 Weight of a humpback whale The expected weight W (in tons) of a humpback whale can be approximated from its length L (in feet) by using $W = 1.70L - 42.8$ for $30 \leq L \leq 50$.

(a) Estimate the weight of a 40-foot humpback whale.

(b) If the error in estimating the length could be as large as 2 feet, what is the corresponding error for the weight estimate?

54 **Growth of a blue whale** Newborn blue whales are approximately 24 feet long and weigh 3 tons. Young whales are nursed for 7 months, and by the time of weaning they often are 53 feet long and weigh 23 tons. Let L and W denote the length (in feet) and the weight (in tons), respectively, of a whale that is t months of age.

(a) If L and t are linearly related, express L in terms of t.

(b) What is the daily increase in the length of a young whale? (Use 1 month = 30 days.)

(c) If W and t are linearly related, express W in terms of t.

(d) What is the daily increase in the weight of a young whale?

55 **Baseball stats** Suppose a major league baseball player has hit 5 home runs in the first 14 games, and he keeps up this pace throughout the 162-game season.

(a) Express the number y of home runs in terms of the number x of games played.

(b) How many home runs will the player hit for the season?

56 **Cheese production** A cheese manufacturer produces 18,000 pounds of cheese from January 1 through March 24. Suppose that this rate of production continues for the remainder of the year.

(a) Express the number y of pounds of cheese produced in terms of the number x of the day in a 365-day year.

(b) Predict, to the nearest pound, the number of pounds produced for the year.

57 **Childhood weight** A baby weighs 10 pounds at birth, and three years later the child's weight is 30 pounds. Assume that childhood weight W (in pounds) is linearly related to age t (in years).

(a) Express W in terms of t.

(b) What is W on the child's sixth birthday?

(c) At what age will the child weigh 70 pounds?

(d) Sketch, on a tW-plane, a graph that shows the relationship between W and t for $0 \leq t \leq 12$.

58 **Loan repayment** A college student receives an interest-free loan of \$8250 from a relative. The student will repay \$125 per month until the loan is paid off.

(a) Express the amount P (in dollars) remaining to be paid in terms of time t (in months).

(b) After how many months will the student owe \$5000?

(c) Sketch, on a tP-plane, a graph that shows the relationship between P and t for the duration of the loan.

59 **Vaporizing water** The amount of heat H (in joules) required to convert one gram of water into vapor is linearly related to the temperature T (in °C) of the atmosphere. At 10°C this conversion requires 2480 joules, and each increase in temperature of 15°C lowers the amount of heat needed by 40 joules. Express H in terms of T.

60 **Aerobic power** In exercise physiology, aerobic power P is defined in terms of maximum oxygen intake. For altitudes up to 1800 meters, aerobic power is optimal—that is, 100%. Beyond 1800 meters, P decreases linearly from the maximum of 100% to a value near 40% at 5000 meters.

(a) Express aerobic power P in terms of altitude h (in meters) for $1800 \leq h \leq 5000$.

(b) Estimate aerobic power in Mexico City (altitude: 2400 meters), the site of the 1968 Summer Olympic Games.

61 **Urban heat island** The urban heat island phenomenon has been observed in Tokyo. The average temperature was 13.5°C in 1915, and since then has risen 0.032°C per year.

(a) Assuming that temperature T (in °C) is linearly related to time t (in years) and that $t = 0$ corresponds to 1915, express T in terms of t.

(b) Predict the average temperature in the year 2010.

62 **Rising ground temperature** In 1870 the average ground temperature in Paris was 11.8°C. Since then it has risen at a nearly constant rate, reaching 13.5°C in 1969.

(a) Express the temperature T (in °C) in terms of time t (in years), where $t = 0$ corresponds to the year 1870 and $0 \leq t \leq 99$.

(b) During what year was the average ground temperature 12.5°C?

63 **Business expenses** The owner of an ice cream franchise must pay the parent company \$1000 per month plus 5% of the monthly revenue R. Operating cost of the franchise includes a fixed cost of \$2600 per month for items such as

utilities and labor. The cost of ice cream and supplies is 50% of the revenue.

(a) Express the owner's monthly expense E in terms of R.

(b) Express the monthly profit P in terms of R.

(c) Determine the monthly revenue needed to break even.

64 **Drug dosage** Pharmacological products must specify recommended dosages for adults and children. Two formulas for modification of adult dosage levels for young children are

$$\text{Cowling's rule:} \quad y = \tfrac{1}{24}(t + 1)a$$

and

$$\text{Friend's rule:} \quad y = \tfrac{2}{25}ta,$$

where a denotes adult dose (in milligrams) and t denotes the age of the child (in years).

(a) If $a = 100$, graph the two linear equations on the same coordinate plane for $0 \le t \le 12$.

(b) For what age do the two formulas specify the same dosage?

65 **Video game** In the video game shown in the figure, an airplane flies from left to right along the path given by $y = 1 + (1/x)$ and shoots bullets in the tangent direction at creatures placed along the x-axis at $x = 1, 2, 3, 4$.

Exercise 65

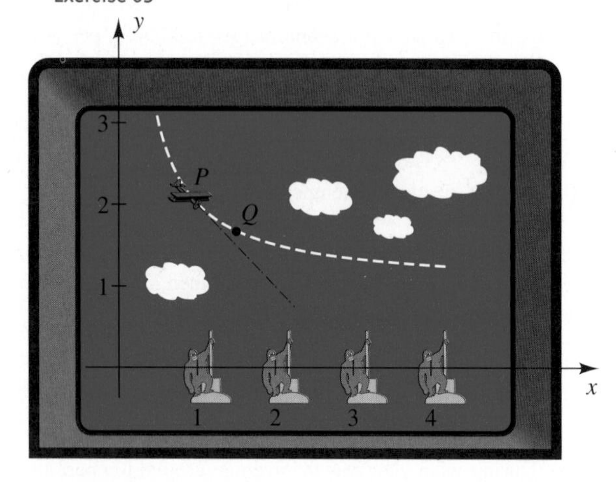

From calculus, the slope of the tangent line to the path at $P(1, 2)$ is $m = -1$ and at $Q\left(\frac{3}{2}, \frac{5}{3}\right)$ is $m = -\frac{4}{9}$. Determine whether a creature will be hit if bullets are shot when the airplane is at

(a) P (b) Q

66 **Temperature scales** The relationship between the temperature reading F on the Fahrenheit scale and the temperature reading C on the Celsius scale is given by $C = \frac{5}{9}(F - 32)$.

(a) Find the temperature at which the reading is the same on both scales.

(b) When is the Fahrenheit reading twice the Celsius reading?

67 **Vertical wind shear** Vertical wind shear occurs when wind speed varies at different heights above the ground. Wind shear is of great importance to pilots during takeoffs and landings. If the wind speed is v_1 at height h_1 and v_2 at height h_2, then the average wind shear s is given by the slope formula

$$s = \frac{v_2 - v_1}{h_2 - h_1}.$$

If the wind speed at ground level is 22 mi/hr and s has been determined to be 0.07, find the wind speed 185 feet above the ground.

68 **Vertical wind shear** In the study of vertical wind shear, the formula

$$\frac{v_1}{v_2} = \left(\frac{h_1}{h_2}\right)^P$$

is sometimes used, where P is a variable that depends on the terrain and structures near ground level. In Montreal, the average daytime value for P with north winds over 29 mi/hr was determined to be 0.13. If a 32 mi/hr north wind is measured 20 feet above the ground, approximate the average wind shear (see Exercise 67) between 20 feet and 200 feet.

Exer. 69–70: The given points were found using empirical methods. Determine whether they lie on the same line $y = ax + b$, and if so, find the values of a and b.

69 $A(-1.3, -1.3598)$, $B(-0.55, -1.11905)$, $C(1.2, -0.5573)$, $D(3.25, 0.10075)$

70 $A(-0.22, 1.6968)$, $B(-0.12, 1.6528)$, $C(1.3, 1.028)$, $D(1.45, 0.862)$

Exer. 71–72: Graph the lines on the same coordinate plane, and find the coordinates of the points of intersection (the coordinates are integers).

71 $x - 3y = -58$; $3x - y = -70$

72 $x + 10y = 123$; $2x - y = -6$

Exer. 73–74: Graph the lines on the same coordinate plane, and estimate the coordinates of the points of intersection. Identify the polygon determined by the lines.

73 $2x - y = -1$; $x + 2y = -2$; $3x + y = 11$

74 $10x - 42y = -7.14$; $8.4x + 2y = -3.8$;
$0.5x - 2.1y = 2.73$; $16.8x + 4y = 14$

Exer. 75–76: For the data table, determine a line in the form $y = ax + b$ that approximately models the data. Plot the line together with the data on the same coordinate axes. *Note:* For exercises requiring an approximate model, answers may vary depending on the data points selected.

75

x	y
−7	−25
−5.8	−21
−5	−18.5
−4	−15.4
0.6	−0.58
1.8	3.26
3	7.1
4.6	12.2

76

x	y
0.4	2.88
1.2	2.45
2.2	1.88
3.6	1.12
4.4	0.68
6.2	−0.30

77 **Super Bowl TV costs** The following table gives the cost (in thousands of dollars) for a 30-second television advertisement during the Super Bowl for various years.

Year	Cost
1986	550
1996	1085
2001	2100
2005	2400

(a) Plot the data on the xy-plane.

(b) Determine a line in the form $y = ax + b$, where x is the year and y is the cost that models the data. Graph this line together with the data on the same coordinate axes. Answers may vary.

(c) Use this line to predict the cost of a 30-second commercial in 2002 and 2003. Compare your answers to the actual values of \$2,200,000 and \$2,150,000, respectively.

78 **Record times in the mile** World record times (in seconds) for the mile run are listed in the table.

Year	Time
1913	254.4
1934	246.8
1954	238.0
1975	229.4
1999	223.1

(a) Plot the data.

(b) Find a line of the form $T = aY + b$ that approximates these data, where T is the time and Y is the year. Graph this line together with the data on the same coordinate axes.

(c) Use the line to predict the record time in 1985, and compare it with the actual record of 226.3 seconds.

(d) Interpret the slope of this line.

2.4 Definition of Function

The notion of **correspondence** occurs frequently in everyday life. Some examples are given in the following illustration.

ILLUSTRATION Correspondence

- To each book in a library there corresponds the number of pages in the book.
- To each human being there corresponds a birth date.
- If the temperature of the air is recorded throughout the day, then to each instant of time there corresponds a temperature.

Each correspondence in the previous illustration involves two sets, D and E. In the first illustration, D denotes the set of books in a library and E the set of positive integers. To each book x in D there corresponds a positive integer y in E—namely, the number of pages in the book.

Figure 1

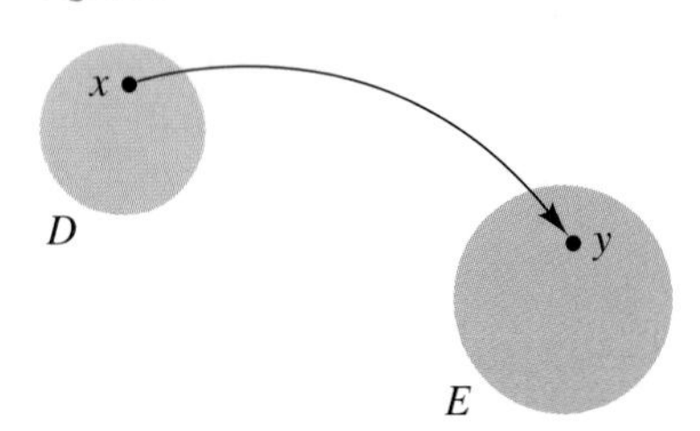

We sometimes depict correspondences by diagrams of the type shown in Figure 1, where the sets D and E are represented by points within regions in a plane. The curved arrow indicates that the element y of E corresponds to the element x of D. The two sets may have elements in common. As a matter of fact, we often have $D = E$. It is important to note that *to each x in D there corresponds exactly one y in E*. However, the same element of E may correspond to different elements of D. For example, two books may have the same number of pages, two people may have the same birthday, and the temperature may be the same at different times.

In most of our work, D and E will be sets of numbers. To illustrate, let both D and E denote the set $\mathbb{R}$ of real numbers, and to each real number x let us assign its square x^2. This gives us a correspondence from $\mathbb{R}$ to $\mathbb{R}$.

Each of our illustrations of a correspondence is a *function,* which we define as follows.

Definition of Function	A **function** f from a set D to a set E is a correspondence that assigns to each element x of D exactly one element y of E.

*For many cases, we can simply remember that the **domain** is the set of x-values and the **range** is the set of y-values.*

The element x of D is the **argument** of f. The set D is the **domain** of the function. The element y of E is the **value** of f at x (or the **image** of x under f) and is denoted by $f(x)$, read "f of x." The **range** of f is the subset R of E consisting of all possible values $f(x)$ for x in D. Note that there may be elements in the set E that are not in the range R of f.

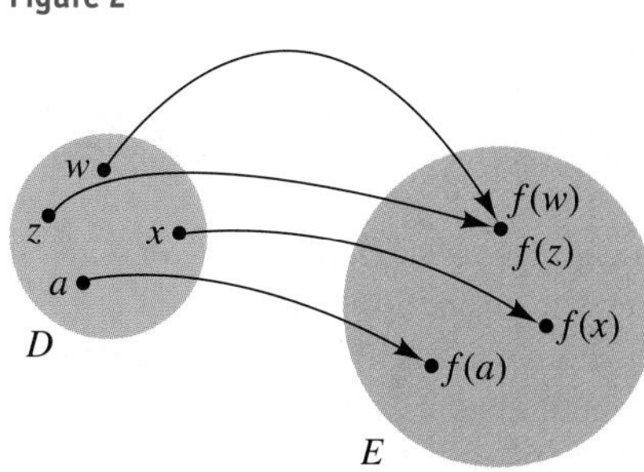

Figure 2

Consider the diagram in Figure 2. The curved arrows indicate that the elements $f(w)$, $f(z)$, $f(x)$, and $f(a)$ of E correspond to the elements w, z, x, and a of D. *To each element in D there is assigned exactly one function value in E;* however, different elements of D, such as w and z in Figure 2, may have the same value in E.

The symbols

$$D \xrightarrow{f} E, \qquad f\colon D \to E, \qquad \text{and}$$

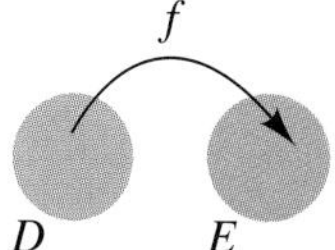

signify that f is a function from D to E, and we say that f **maps** *D into E.* Initially, the notations f and $f(x)$ may be confusing. Remember that f is used to represent the function. It is neither in D nor in E. However, $f(x)$ is an element of the range R—the element that the function f assigns to the element x, which is in the domain D.

Two functions f and g from D to E are **equal,** and we write

$$f = g \text{ provided } \quad f(x) = g(x) \quad \text{for every } x \text{ in } D.$$

For example, if $g(x) = \frac{1}{2}(2x^2 - 6) + 3$ and $f(x) = x^2$ for every x in $\mathbb{R}$, then $g = f$.

EXAMPLE 1 Finding function values

Let f be the function with domain $\mathbb{R}$ such that $f(x) = x^2$ for every x in $\mathbb{R}$.

(a) Find $f(-6)$, $f(\sqrt{3})$, $f(a + b)$, and $f(a) + f(b)$, where a and b are real numbers.

(b) What is the range of f?

SOLUTION

▶ **(a)** We find values of f by substituting for x in the equation $f(x) = x^2$:

$$f(-6) = (-6)^2 = 36$$
$$f(\sqrt{3}) = (\sqrt{3})^2 = 3$$
$$f(a + b) = (a + b)^2 = a^2 + 2ab + b^2$$
$$f(a) + f(b) = a^2 + b^2$$

Note that, in general,

$$f(a + b) \neq f(a) + f(b).$$

(b) By definition, the range of f consists of all numbers of the form $f(x) = x^2$ for x in $\mathbb{R}$. Since the square of every real number is nonnegative, the range is contained in the set of all nonnegative real numbers. Moreover, every nonnegative real number c is a value of f, since $f(\sqrt{c}) = (\sqrt{c})^2 = c$. Hence, the range of f is the set of all nonnegative real numbers.

▶ Tutorial available at www.thomsonedu.com/login

If a function is defined as in Example 1, the symbols used for the function and variable are immaterial; that is, expressions such as $f(x) = x^2$, $f(s) = s^2$, $g(t) = t^2$, and $k(r) = r^2$ all define the same function. This is true because if a is any number in the domain, then the same value a^2 is obtained regardless of which expression is employed.

In the remainder of our work, the phrase *f is a function* will mean that the domain and range are sets of real numbers. If a function is defined by means of an expression, as in Example 1, and the domain D is not stated, then we will consider D to be the totality of real numbers x such that $f(x)$ is real. This is sometimes called the **implied domain** of f. To illustrate, if $f(x) = \sqrt{x-2}$, then the implied domain is the set of real numbers x such that $\sqrt{x-2}$ is real—that is, $x - 2 \geq 0$, or $x \geq 2$. Thus, the domain is the infinite interval $[2, \infty)$. If x is in the domain, we say that *f is defined at x* or that *f(x) exists.* If a set S is contained in the domain, *f is defined on S.* The terminology *f is undefined at x* means that x is not in the domain of f.

EXAMPLE 2 Finding function values

Let $g(x) = \dfrac{\sqrt{4+x}}{1-x}$.

(a) Find the domain of g.

(b) Find $g(5)$, $g(-2)$, $g(-a)$, and $-g(a)$.

SOLUTION

(a) The expression $\sqrt{4+x}/(1-x)$ is a real number if and only if the radicand $4 + x$ is nonnegative and the denominator $1 - x$ is not equal to 0. Thus, $g(x)$ exists if and only if

$$4 + x \geq 0 \quad \text{and} \quad 1 - x \neq 0$$

or, equivalently,

$$x \geq -4 \quad \text{and} \quad x \neq 1.$$

We may express the domain in terms of intervals as $[-4, 1) \cup (1, \infty)$.

(b) To find values of g, we substitute for x:

$$g(5) = \frac{\sqrt{4+5}}{1-5} = \frac{\sqrt{9}}{-4} = -\frac{3}{4}$$

$$g(-2) = \frac{\sqrt{4+(-2)}}{1-(-2)} = \frac{\sqrt{2}}{3}$$

$$g(-a) = \frac{\sqrt{4+(-a)}}{1-(-a)} = \frac{\sqrt{4-a}}{1+a}$$

$$-g(a) = -\frac{\sqrt{4+a}}{1-a} = \frac{\sqrt{4+a}}{a-1}$$

Tutorial available at www.thomsonedu.com/login

Figure 3

Functions are commonplace in everyday life and show up in a variety of forms. For instance, the menu in a restaurant (Figure 3) can be considered to be a function f from a set of items to a set of prices. Note that f is given in a tablet format. Here $f(\text{Hamburger}) = 1.69$, $f(\text{French fries}) = 0.99$, and $f(\text{Soda}) = 0.79$.

An example of a function given by a rule can be found in the federal tax tables (Figure 4). Specifically, in 2006, for a single person with a taxable income of \$120,000, the tax due was given by the rule

\$15,107.50 plus 28% of the amount over \$74,200.

Figure 4

2006 Federal Tax Rate Schedules

Schedule X –Use if your Filing status is **single**

If taxable income is over–	*But not over–*	*The tax is:*	*of the amount over–*
\$0	\$7,550	-------- **10%**	**\$0**
7,550	30,650	**\$755.00 + 15%**	**7,550**
30,650	74,200	**\$4,220.00 + 25%**	**30,650**
74,200	154,800	**15,107.50 + 28%**	**74,200**
154,800	336,550	**37,675.50 + 33%**	**154,800**
336,550	-------	**97,653.00 + 35%**	**336,550**

In this case, the tax would be

$$\$15,107.50 + 0.28(\$120,000 - \$74,200) = \$27,931.50.$$

Figure 5

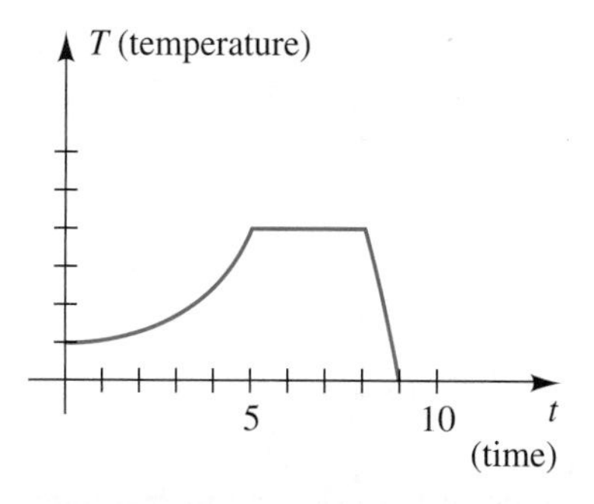

Graphs are often used to describe the variation of physical quantities. For example, a scientist may use the graph in Figure 5 to indicate the temperature T of a certain solution at various times t during an experiment. The sketch shows that the temperature increased gradually for time $t = 0$ to time $t = 5$, did not change between $t = 5$ and $t = 8$, and then decreased rapidly from $t = 8$ to $t = 9$.

Similarly, if f is a function, we may use a graph to indicate the change in $f(x)$ as x varies through the domain of f. Specifically, we have the following definition.

Definition of Graph of a Function

The **graph of a function** f is the graph of the equation $y = f(x)$ for x in the domain of f.

We often attach the label $y = f(x)$ to a sketch of the graph. If $P(a, b)$ is a point on the graph, then the y-coordinate b is the function value $f(a)$, as illustrated in Figure 6 on the next page. The figure displays the domain of f (the set of possible values of x) and the range of f (the corresponding values of y). Although we have pictured the domain and range as closed intervals, they may be infinite intervals or other sets of real numbers.

Since there is exactly one value $f(a)$ for each a in the domain of f, only *one* point on the graph of f has x-coordinate a. In general, we may use the following graphical test to determine whether a graph is the graph of a function.

Vertical Line Test	The graph of a set of points in a coordinate plane is the graph of a function if every vertical line intersects the graph in at most one point.

Figure 6

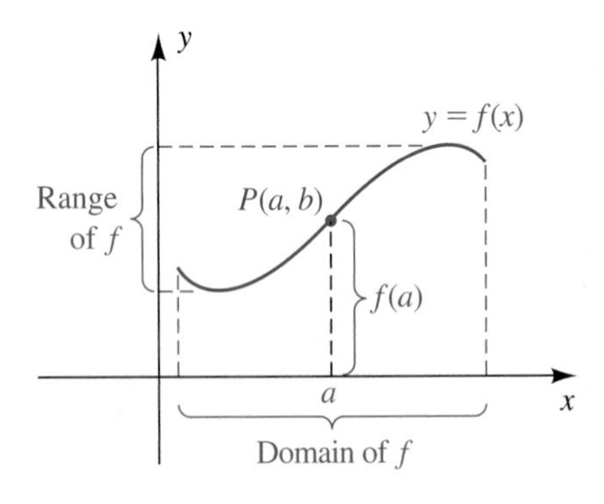

Thus, *every vertical line intersects the graph of a function in at most one point.* Consequently, the graph of a function cannot be a figure such as a circle, in which a vertical line may intersect the graph in more than one point.

The x-intercepts of the graph of a function f are the solutions of the equation $f(x) = 0$. These numbers are called the **zeros** of the function. The y-intercept of the graph is $f(0)$, if it exists.

EXAMPLE 3 Sketching the graph of a function

Let $f(x) = \sqrt{x - 1}$.

(a) Sketch the graph of f.

(b) Find the domain and range of f.

Figure 7

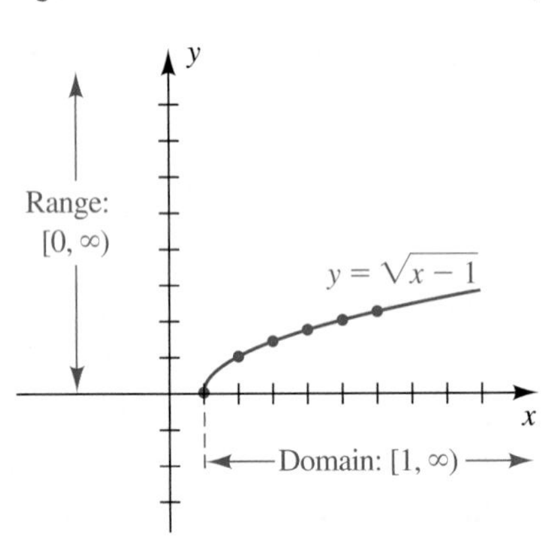

▶ SOLUTION

(a) By definition, the graph of f is the graph of the equation $y = \sqrt{x - 1}$. The following table lists coordinates of several points on the graph.

x	1	2	3	4	5	6
$y = f(x)$	0	1	$\sqrt{2} \approx 1.4$	$\sqrt{3} \approx 1.7$	2	$\sqrt{5} \approx 2.2$

Plotting points, we obtain the sketch shown in Figure 7. Note that the x-intercept is 1 and there is no y-intercept.

(b) Referring to Figure 7, note that the domain of f consists of all real numbers x such that $x \geq 1$ or, equivalently, the interval $[1, \infty)$. The range of f is the set of all real numbers y such that $y \geq 0$ or, equivalently, $[0, \infty)$.

The **square root function,** defined by $f(x) = \sqrt{x}$, has a graph similar to the one in Figure 7, but the endpoint is at $(0, 0)$. The y-value of a point on this graph is the number displayed on a calculator when a square root is requested. This graphical relationship may help you remember that $\sqrt{9}$ is 3 and that $\sqrt{9}$ is *not* ± 3. Similarly, $f(x) = x^2$, $f(x) = x^3$, and $f(x) = \sqrt[3]{x}$ are often referred to as the **squaring function,** the **cubing function,** and the **cube root function,** respectively.

▶ Tutorial available at www.thomsonedu.com/login

In Example 3, as x increases, the function value $f(x)$ also increases, and we say that the graph of f *rises* (see Figure 7). A function of this type is said to be *increasing*. For certain functions, $f(x)$ decreases as x increases. In this case the graph *falls,* and f is a *decreasing* function. In general, we shall consider functions that increase or decrease on an interval I, as described in the following chart, where x_1 and x_2 denote numbers in I.

Increasing, Decreasing, and Constant Functions

Terminology	Definition	Graphical interpretation
f is **increasing** on an interval I	$f(x_1) < f(x_2)$ whenever $x_1 < x_2$	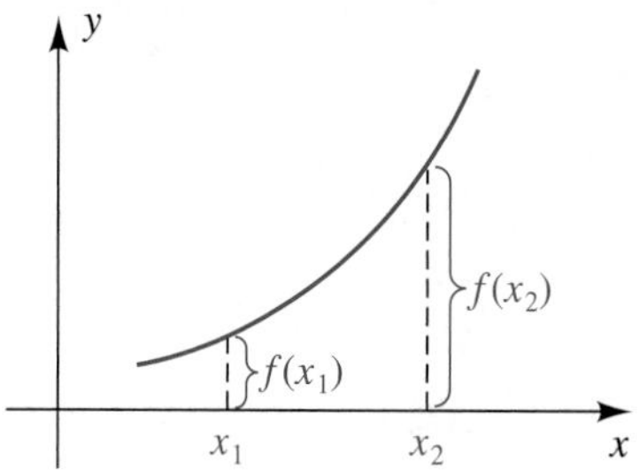 y, $f(x_1)$, $f(x_2)$, x_1, x_2, x
f is **decreasing** on an interval I	$f(x_1) > f(x_2)$ whenever $x_1 < x_2$	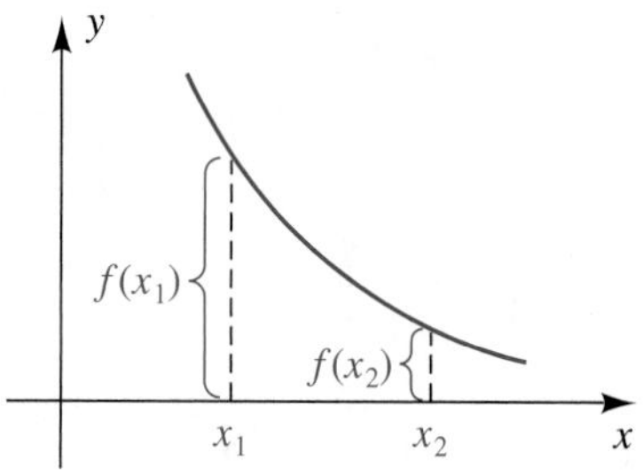 y, $f(x_1)$, $f(x_2)$, x_1, x_2, x
f is **constant** on an interval I	$f(x_1) = f(x_2)$ for every x_1 and x_2	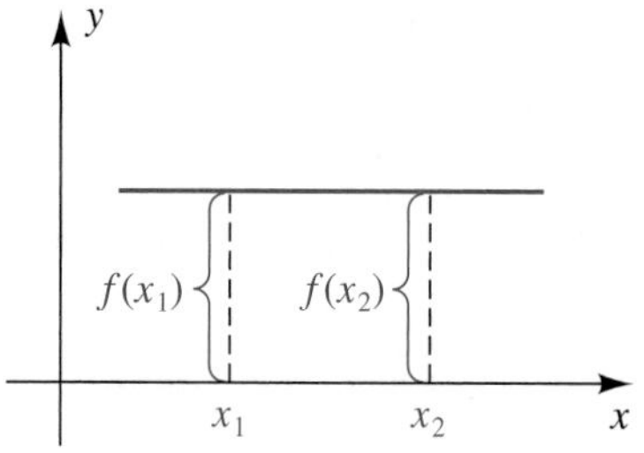 y, $f(x_1)$, $f(x_2)$, x_1, x_2, x

An example of an *increasing function* is the **identity function,** whose equation is $f(x) = x$ and whose graph is the line through the origin with slope 1. An example of a *decreasing function* is $f(x) = -x$, an equation of the line through the origin with slope -1. If $f(x) = c$ for every real number x, then f is called a *constant function.*

We shall use the phrases *f is increasing* and *f(x) is increasing* interchangeably. We shall do the same with the terms *decreasing* and *constant.*

EXAMPLE 4 Using a graph to find domain, range, and where a function increases or decreases

Let $f(x) = \sqrt{9 - x^2}$.

(a) Sketch the graph of f.

(b) Find the domain and range of f.

(c) Find the intervals on which f is increasing or is decreasing.

SOLUTION

(a) By definition, the graph of f is the graph of the equation $y = \sqrt{9 - x^2}$. We know from our work with circles in Section 2.2 that the graph of $x^2 + y^2 = 9$ is a circle of radius 3 with center at the origin. Solving the equation $x^2 + y^2 = 9$ for y gives us $y = \pm\sqrt{9 - x^2}$. It follows that the graph of f is the *upper half* of the circle, as illustrated in Figure 8.

Figure 8

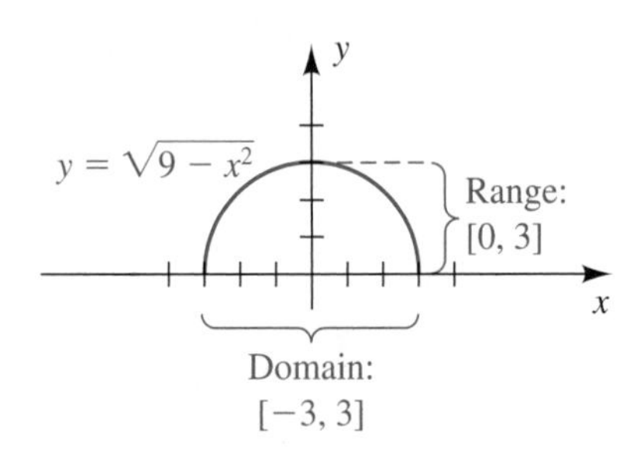

(b) Referring to Figure 8, we see that the domain of f is the closed interval $[-3, 3]$, and the range of f is the interval $[0, 3]$.

(c) The graph rises as x increases from -3 to 0, so f is increasing on the closed interval $[-3, 0]$. Thus, as shown in the preceding chart, if $x_1 < x_2$ in $[-3, 0]$, then $f(x_1) < f(x_2)$ (note that *possibly* $x_1 = -3$ or $x_2 = 0$).

The graph falls as x increases from 0 to 3, so f is decreasing on the closed interval $[0, 3]$. In this case, the chart indicates that if $x_1 < x_2$ in $[0, 3]$, then $f(x_1) > f(x_2)$ (note that *possibly* $x_1 = 0$ or $x_2 = 3$).

Of special interest in calculus is a problem of the following type.

Problem: Find the slope of the secant line through the points P and Q shown in Figure 9.

Figure 9

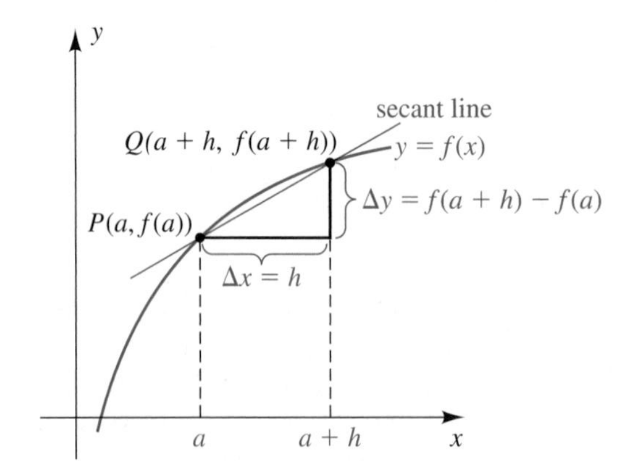

Tutorial available at www.thomsonedu.com/login

The slope m_{PQ} is given by

$$m_{PQ} = \frac{\Delta y}{\Delta x} = \frac{f(a+h) - f(a)}{h}.$$

The last expression (with $h \neq 0$) is commonly called a **difference quotient.** Let's take a look at the algebra involved in simplifying a difference quotient. (See Discussion Exercise 5 at the end of the chapter for a related problem.)

EXAMPLE 5 Simplifying a difference quotient

Simplify the difference quotient

$$\frac{f(x+h) - f(x)}{h}$$

using the function $f(x) = x^2 + 6x - 4$.

SOLUTION

$$\frac{f(x+h) - f(x)}{h} = \frac{[(x+h)^2 + 6(x+h) - 4] - [x^2 + 6x - 4]}{h}$$

definition of f

$$= \frac{(x^2 + 2xh + h^2 + 6x + 6h - 4) - (x^2 + 6x - 4)}{h}$$

expand numerator

$$= \frac{(\cancel{x^2} + 2xh + h^2 + \cancel{6x} + 6h - \cancel{4}) - (\cancel{x^2} + \cancel{6x} - \cancel{4})}{h}$$

subtract terms

$$= \frac{2xh + h^2 + 6h}{h} \quad \text{simplify}$$

$$= \frac{h(2x + h + 6)}{h} \quad \text{factor out } h$$

$$= 2x + h + 6 \quad \text{cancel } h \neq 0$$

The following type of function is one of the most basic in algebra.

Definition of Linear Function	A function f is a **linear function** if $$f(x) = ax + b,$$ where x is any real number and a and b are constants.

The graph of f in the preceding definition is the graph of $y = ax + b$, which, by the slope-intercept form, is a line with slope a and y-intercept b.

Tutorial available at www.thomsonedu.com/login

Thus, *the graph of a linear function is a line.* Since $f(x)$ exists for every x, the domain of f is $\mathbb{R}$. As illustrated in the next example, if $a \neq 0$, then the range of f is also $\mathbb{R}$.

EXAMPLE 6 Sketching the graph of a linear function

Let $f(x) = 2x + 3$.

(a) Sketch the graph of f.

(b) Find the domain and range of f.

(c) Determine where f is increasing or is decreasing.

Figure 10

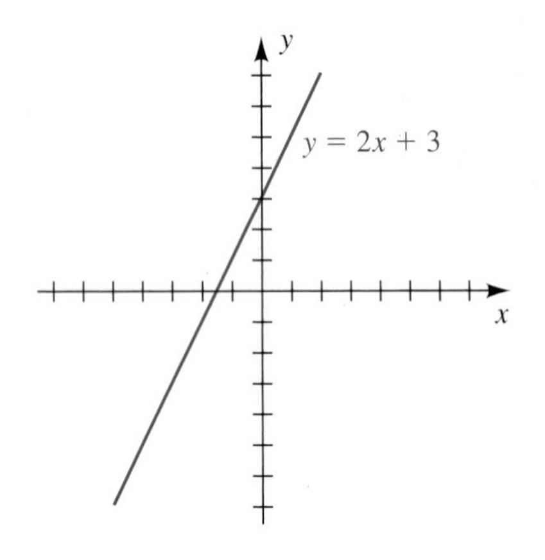

SOLUTION

(a) Since $f(x)$ has the form $ax + b$, with $a = 2$ and $b = 3$, f is a linear function. The graph of $y = 2x + 3$ is the line with slope 2 and y-intercept 3, illustrated in Figure 10.

(b) We see from the graph that x and y may be any real numbers, so both the domain and the range of f are $\mathbb{R}$.

(c) Since the slope a is positive, the graph of f rises as x increases; that is, $f(x_1) < f(x_2)$ whenever $x_1 < x_2$. Thus, f is increasing throughout its domain.

In applications it is sometimes necessary to determine a specific linear function from given data, as in the next example.

EXAMPLE 7 Finding a linear function

If f is a linear function such that $f(-2) = 5$ and $f(6) = 3$, find $f(x)$, where x is any real number.

▶ SOLUTION By the definition of linear function, $f(x) = ax + b$, where a and b are constants. Moreover, the given function values tell us that the points $(-2, 5)$ and $(6, 3)$ are on the graph of f—that is, on the line $y = ax + b$ illustrated in Figure 11. The slope a of this line is

Figure 11

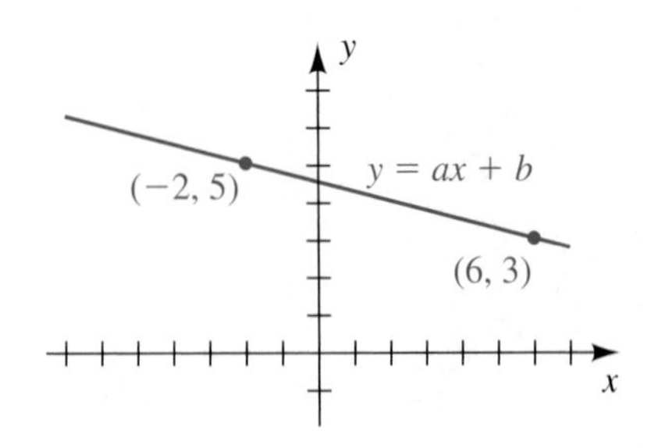

$$a = \frac{5 - 3}{-2 - 6} = \frac{2}{-8} = -\frac{1}{4},$$

and hence $f(x)$ has the form

$$f(x) = -\tfrac{1}{4}x + b.$$

To find the value of b, we may use the fact that $f(6) = 3$, as follows:

$$f(6) = -\tfrac{1}{4}(6) + b \quad \text{let } x = 6 \text{ in } f(x) = -\tfrac{1}{4}x + b$$

$$3 = -\tfrac{3}{2} + b \quad f(6) = 3$$

$$b = 3 + \tfrac{3}{2} = \tfrac{9}{2} \quad \text{solve for } b$$

▶ Tutorial available at www.thomsonedu.com/login

Thus, the linear function satisfying $f(-2) = 5$ and $f(6) = 3$ is

$$f(x) = -\tfrac{1}{4}x + \tfrac{9}{2}.$$

Many formulas that occur in mathematics and the sciences determine functions. For instance, the formula $A = \pi r^2$ for the area A of a circle of radius r assigns to each positive real number r exactly one value of A. This determines a function f such that $f(r) = \pi r^2$, and we may write $A = f(r)$. The letter r, which represents an arbitrary number from the domain of f, is called an **independent variable.** The letter A, which represents a number from the range of f, is a **dependent variable,** since its value depends on the number assigned to r. If two variables r and A are related in this manner, we say that A *is a function of* r. In applications, the independent variable and dependent variable are sometimes referred to as the **input variable** and **output variable,** respectively. As another example, if an automobile travels at a uniform rate of 50 mi/hr, then the distance d (miles) traveled in time t (hours) is given by $d = 50t$, and hence *the distance* d *is a function of time* t.

EXAMPLE 8 Expressing the volume of a tank as a function of its radius

A steel storage tank for propane gas is to be constructed in the shape of a right circular cylinder of altitude 10 feet with a hemisphere attached to each end. The radius r is yet to be determined. Express the volume V (in ft^3) of the tank as a function of r (in feet).

Figure 12

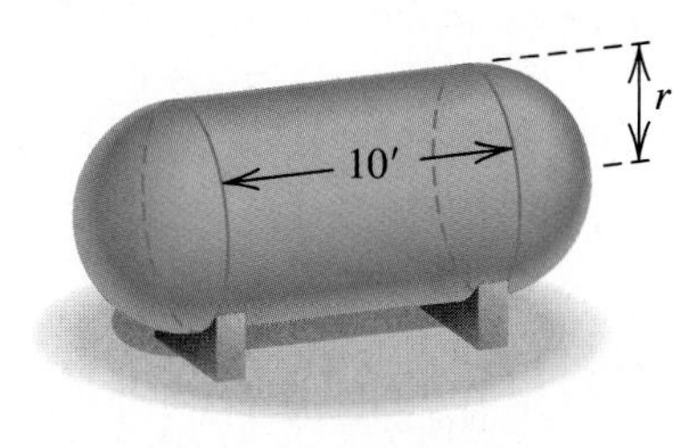

SOLUTION The tank is illustrated in Figure 12. We may find the volume of the cylindrical part of the tank by multiplying the altitude 10 by the area πr^2 of the base of the cylinder. This gives us

$$\text{volume of cylinder} = 10(\pi r^2) = 10\pi r^2.$$

The two hemispherical ends, taken together, form a sphere of radius r. Using the formula for the volume of a sphere, we obtain

$$\text{volume of the two ends} = \tfrac{4}{3}\pi r^3.$$

Thus, the volume V of the tank is

$$V = \tfrac{4}{3}\pi r^3 + 10\pi r^2.$$

This formula expresses V as a function of r. In factored form,

$$V(r) = \tfrac{1}{3}\pi r^2(4r + 30) = \tfrac{2}{3}\pi r^2(2r + 15).$$

EXAMPLE 9 Expressing a distance as a function of time

Two ships leave port at the same time, one sailing west at a rate of 17 mi/hr and the other sailing south at 12 mi/hr. If t is the time (in hours) after their departure, express the distance d between the ships as a function of t.

Tutorial available at www.thomsonedu.com/login

Figure 13

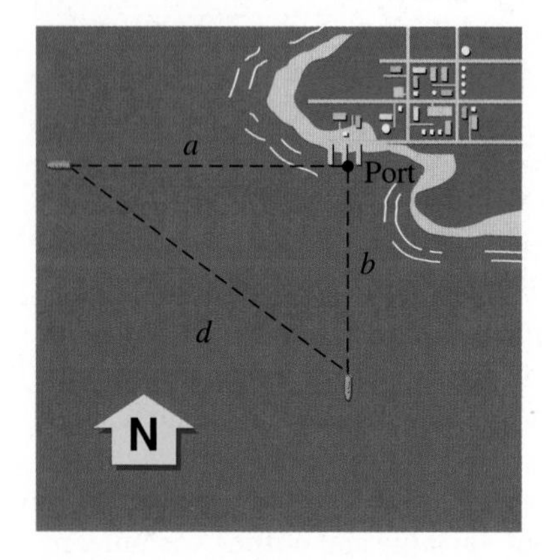

▶ SOLUTION To help visualize the problem, we begin by drawing a picture and labeling it, as in Figure 13. By the Pythagorean theorem,

$$d^2 = a^2 + b^2, \qquad \text{or} \qquad d = \sqrt{a^2 + b^2}.$$

Since distance = (rate)(time) and the rates are 17 and 12, respectively,

$$a = 17t \qquad \text{and} \qquad b = 12t.$$

Substitution in $d = \sqrt{a^2 + b^2}$ gives us

$$d = \sqrt{(17t)^2 + (12t)^2} = \sqrt{289t^2 + 144t^2} = \sqrt{433t^2} \approx (20.8)t.$$

Ordered pairs can be used to obtain an alternative approach to functions. We first observe that a function f from D to E determines the following set W of ordered pairs:

$$W = \{(x, f(x)) \colon x \text{ is in } D\}$$

Thus, W consists of all ordered pairs such that the first number x is in D and the second number is the function value $f(x)$. In Example 1, where $f(x) = x^2$, W is the set of all ordered pairs of the form (x, x^2). It is important to note that, *for each x, there is exactly one ordered pair (x, y) in W having x in the first position.*

Conversely, if we begin with a set W of ordered pairs such that each x in D appears exactly once in the first position of an ordered pair, then W determines a function. Specifically, for each x in D there is exactly one pair (x, y) in W, and by letting y correspond to x, we obtain a function with domain D. The range consists of all real numbers y that appear in the second position of the ordered pairs.

It follows from the preceding discussion that the next statement could also be used as a definition of function.

Alternative Definition of Function	A **function** with domain D is a set W of ordered pairs such that, for each x in D, there is exactly one ordered pair (x, y) in W having x in the first position.

In terms of the preceding definition, the ordered pairs $\left(x, \sqrt{x-1}\right)$ determine the function of Example 3 given by $f(x) = \sqrt{x-1}$. Note, however, that if

$$W = \{(x, y) \colon x^2 = y^2\},$$

then W is *not* a function, since for a given x there may be more than one pair in W with x in the first position. For example, if $x = 2$, then both $(2, 2)$ and $(2, -2)$ are in W.

▶ Tutorial available at www.thomsonedu.com/login

In the next example we illustrate how some of the concepts presented in this section may be studied with the aid of a graphing calculator. Hereafter, when making assignments on a graphing utility, we will frequently refer to variables such as Y_1 and Y_2 as the *functions* Y_1 and Y_2.

EXAMPLE 10 Analyzing the graph of a function

Let $f(x) = x^{2/3} - 3$.

(a) Find $f(-2)$.

(b) Sketch the graph of f.

(c) State the domain and range of f.

(d) State the intervals on which f is increasing or is decreasing.

(e) Estimate the x-intercepts of the graph to one-decimal-place accuracy.

SOLUTION

TI-83/4 Plus **TI-86**

(a) Shown below are four representations of f. All of these are valid on the TI-83/4 Plus and the TI-86. On some older graphing calculator models, you may get only the right-hand side of the graph in Figure 14 on the next page. If that happens, change your representation of f.

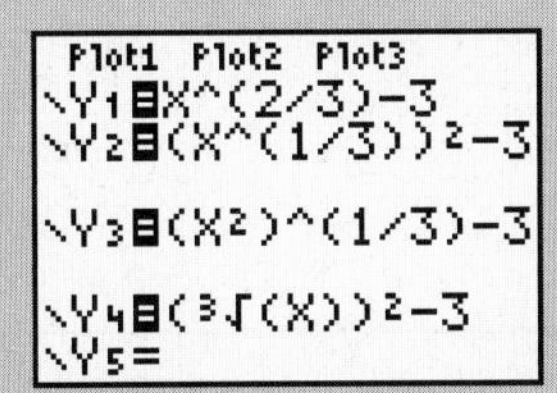

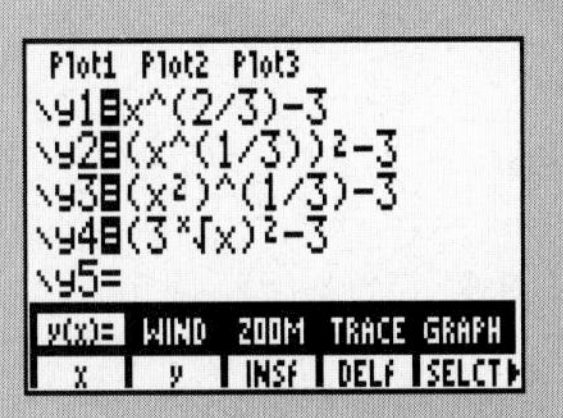

Shown below are two methods of finding a function value. In the first method, we simply find the value of $Y_1(-2)$. In the second method, we store -2 in X and then find the value of Y_1.

VARS ▷ 1 1 (−2) ENTER

2nd alpha Y 1 (−2) ENTER

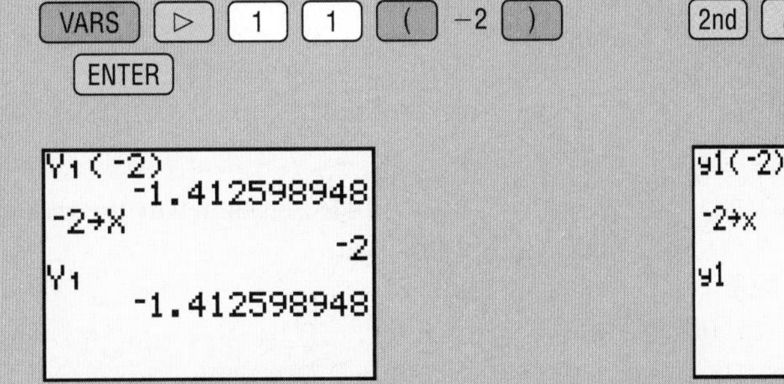

Y1(−2)
−1.412598948
−2→X
−2
Y1
−1.412598948

y1(−2)
−1.41259894803
−2→x
−2
y1
−1.41259894803

(b) Using the viewing rectangle $[-15, 15]$ by $[-10, 10]$ to graph Y_1 gives us a display similar to that of Figure 14. The v-shaped part of the graph of f at $x = 0$ is called a **cusp.**

(c) The domain of f is $\mathbb{R}$, since we may input any value for x. The figure indicates that $y \geq -3$, so we conclude that the range of f is $[-3, \infty)$. *(continued)*

(d) From the figure, we see that f is decreasing on $(-\infty, 0]$ and is increasing on $[0, \infty)$.

(e) Using the root feature, we find that the positive x-intercept in Figure 14 is approximately 5.2. Since f is symmetric with respect to the y-axis, the negative x-intercept is about -5.2.

Figure 14 $[-15, 15]$ by $[-10, 10]$

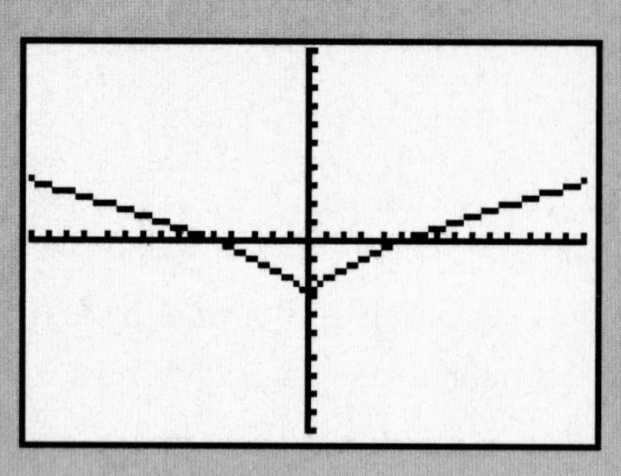

As a reference aid, some common graphs and their equations are listed in Appendix I. Many of these graphs are graphs of functions.

2.4 *Exercises*

1 If $f(x) = -x^2 - x - 4$, find $f(-2)$, $f(0)$, and $f(4)$.

2 If $f(x) = -x^3 - x^2 + 3$, find $f(-3)$, $f(0)$, and $f(2)$.

3 If $f(x) = \sqrt{x - 4} - 3x$, find $f(4)$, $f(8)$, and $f(13)$.

4 If $f(x) = \dfrac{x}{x - 3}$, find $f(-2)$, $f(0)$, and $f(3)$.

Exer. 5–10: If a and h are real numbers, find

(a) $f(a)$ **(b)** $f(-a)$ **(c)** $-f(a)$ **(d)** $f(a + h)$

(e) $f(a) + f(h)$ **(f)** $\dfrac{f(a + h) - f(a)}{h}$, **if** $h \neq 0$

5 $f(x) = 5x - 2$

6 $f(x) = 3 - 4x$

7 $f(x) = -x^2 + 4$

8 $f(x) = 3 - x^2$

9 $f(x) = x^2 - x + 3$

10 $f(x) = 2x^2 + 3x - 7$

Exer. 11–14: If a is a positive real number, find

(a) $g\left(\dfrac{1}{a}\right)$ **(b)** $\dfrac{1}{g(a)}$ **(c)** $g(\sqrt{a})$ **(d)** $\sqrt{g(a)}$

11 $g(x) = 4x^2$

12 $g(x) = 2x - 5$

13 $g(x) = \dfrac{2x}{x^2 + 1}$

14 $g(x) = \dfrac{x^2}{x + 1}$

Exer. 15–16: Explain why the graph is or is not the graph of a function.

15

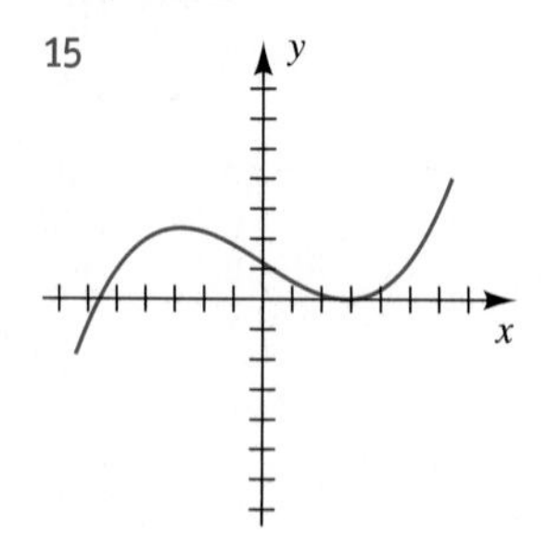

16

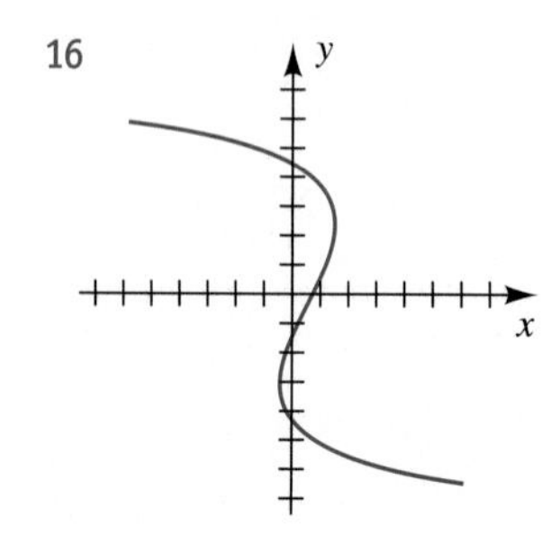

Exer. 17–18: Determine the domain D and range R of the function shown in the figure.

17 18

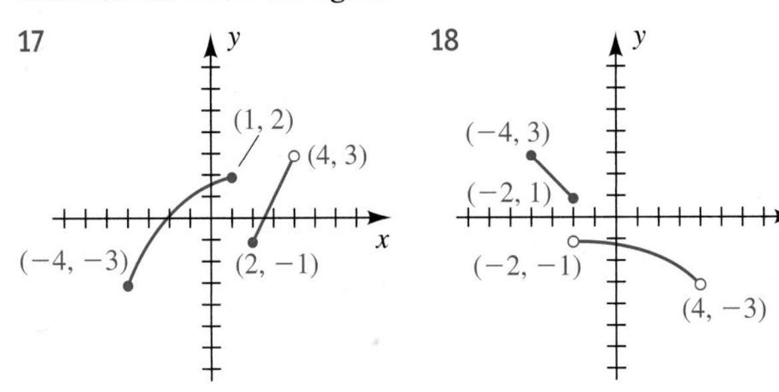

Exer. 19–20: For the graph of the function f sketched in the figure, determine

(a) the domain **(b) the range** **(c) $f(1)$**

(d) all x such that $f(x) = 1$

(e) all x such that $f(x) > 1$

19

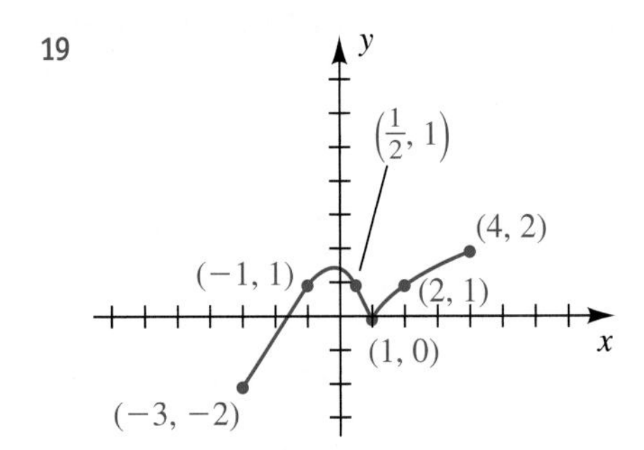

20

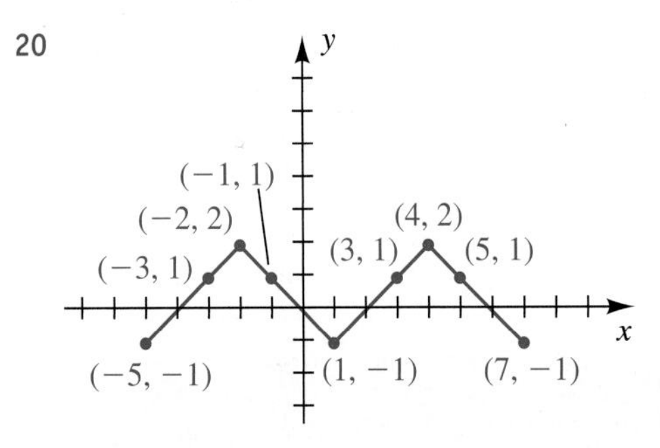

Exer. 21–32: Find the domain of f.

21 $f(x) = \sqrt{2x + 7}$

22 $f(x) = \sqrt{8 - 3x}$

23 $f(x) = \sqrt{9 - x^2}$

24 $f(x) = \sqrt{x^2 - 25}$

25 $f(x) = \dfrac{x + 1}{x^3 - 4x}$

26 $f(x) = \dfrac{4x}{6x^2 + 13x - 5}$

27 $f(x) = \dfrac{\sqrt{2x - 3}}{x^2 - 5x + 4}$

28 $f(x) = \dfrac{\sqrt{4x - 3}}{x^2 - 4}$

29 $f(x) = \dfrac{x - 4}{\sqrt{x - 2}}$

30 $f(x) = \dfrac{1}{(x - 3)\sqrt{x + 3}}$

31 $f(x) = \sqrt{x + 2} + \sqrt{2 - x}$

▶ 32 $f(x) = \sqrt{(x - 2)(x - 6)}$

Exer. 33–34: (a) Find the domain D and range R of f. (b) Find the intervals on which f is increasing, is decreasing, or is constant.

33

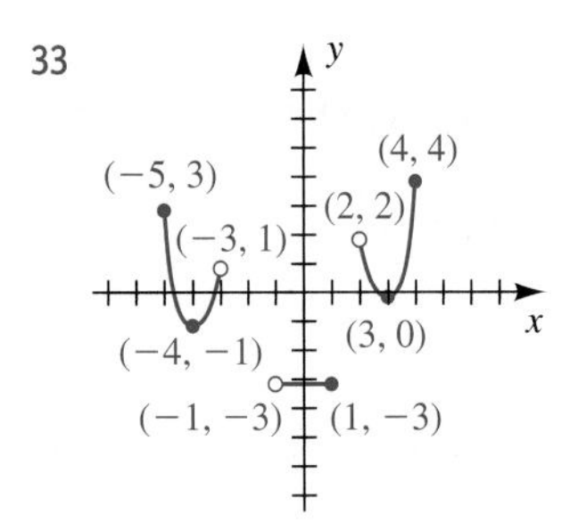

34

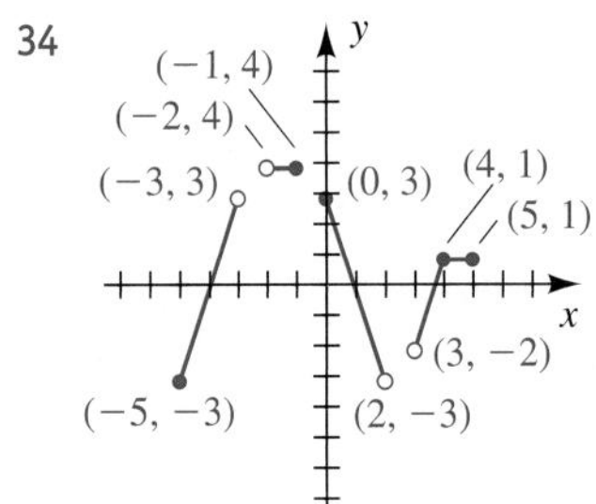

35 Sketch the graph of a function that is increasing on $(-\infty, -3]$ and $[2, \infty)$ and is decreasing on $[-3, 2]$.

36 Sketch the graph of a function that is decreasing on $(-\infty, -2]$ and $[1, 4]$ and is increasing on $[-2, 1]$ and $[4, \infty)$.

Exer. 37–46: (a) Sketch the graph of f. (b) Find the domain D and range R of f. (c) Find the intervals on which f is increasing, is decreasing, or is constant.

37 $f(x) = 3x - 2$

38 $f(x) = -2x + 3$

39 $f(x) = 4 - x^2$

40 $f(x) = x^2 - 1$

41 $f(x) = \sqrt{x + 4}$

42 $f(x) = \sqrt{4 - x}$

43 $f(x) = -2$

44 $f(x) = 3$

45 $f(x) = -\sqrt{36 - x^2}$

46 $f(x) = \sqrt{16 - x^2}$

Exer. 47–48: Simplify the difference quotient $\dfrac{f(2 + h) - f(2)}{h}$ if $h \neq 0$.

47 $f(x) = x^2 - 3x$

48 $f(x) = -2x^2 + 3$

Exer. 49–50: Simplify the difference quotient $\dfrac{f(x + h) - f(x)}{h}$ if $h \neq 0$.

49 $f(x) = x^2 + 5$

50 $f(x) = 1/x^2$

Exer. 51–52: Simplify the difference quotient $\dfrac{f(x) - f(a)}{x - a}$ if $x \neq a$.

51 $f(x) = \sqrt{x - 3}$ (*Hint:* Rationalize the numerator.)

52 $f(x) = x^3 - 2$

Exer. 53–54: If a linear function f satisfies the given conditions, find $f(x)$.

53 $f(-3) = 1$ and $f(3) = 2$

▶ 54 $f(-2) = 7$ and $f(4) = -2$

Exer. 55–64: Determine whether the set W of ordered pairs is a function in the sense of the alternative definition of function on page 148.

55 $W = \{(x, y): 2y = x^2 + 5\}$

56 $W = \{(x, y): x = 3y + 2\}$

57 $W = \{(x, y): x^2 + y^2 = 4\}$

58 $W = \{(x, y): y^2 - x^2 = 1\}$

59 $W = \{(x, y): y = 3\}$

60 $W = \{(x, y): x = 3\}$

61 $W = \{(x, y): xy = 0\}$

62 $W = \{(x, y): x + y = 0\}$

63 $W = \{(x, y): |y| = |x|\}$

64 $W = \{(x, y): y < x\}$

65 **Constructing a box** From a rectangular piece of cardboard having dimensions 20 inches × 30 inches, an open box is to be made by cutting out an identical square of area x^2 from each corner and turning up the sides (see the figure). Express the volume V of the box as a function of x.

Exercise 65

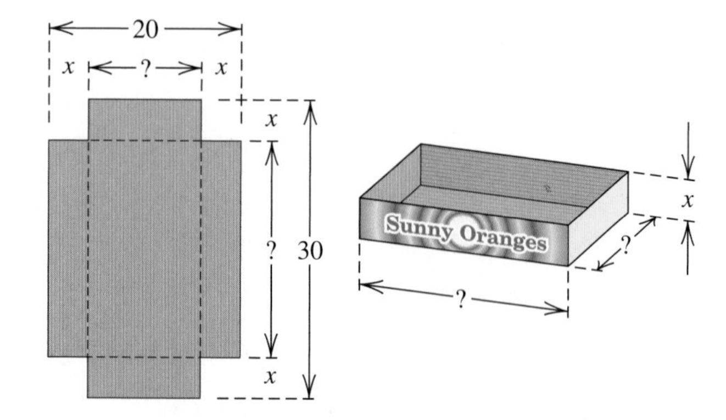

66 **Constructing a storage tank** Refer to Example 8. A steel storage tank for propane gas is to be constructed in the shape of a right circular cylinder of altitude 10 feet with a hemisphere attached to each end. The radius r is yet to be determined. Express the surface area S of the tank as a function of r.

67 **Dimensions of a building** A small office unit is to contain 500 ft^2 of floor space. A simplified model is shown in the figure.

(a) Express the length y of the building as a function of the width x.

(b) If the walls cost \$100 per running foot, express the cost C of the walls as a function of the width x. (Disregard the wall space above the doors and the thickness of the walls.)

Exercise 67

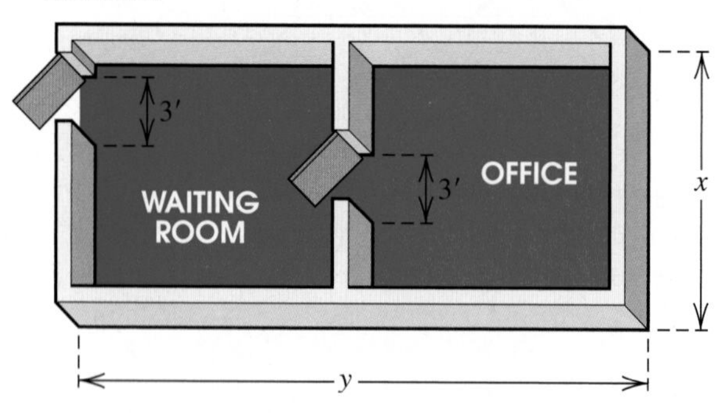

68 Dimensions of an aquarium An aquarium of height 1.5 feet is to have a volume of 6 ft^3. Let x denote the length of the base and y the width (see the figure).

(a) Express y as a function of x.

(b) Express the total number S of square feet of glass needed as a function of x.

Exercise 68

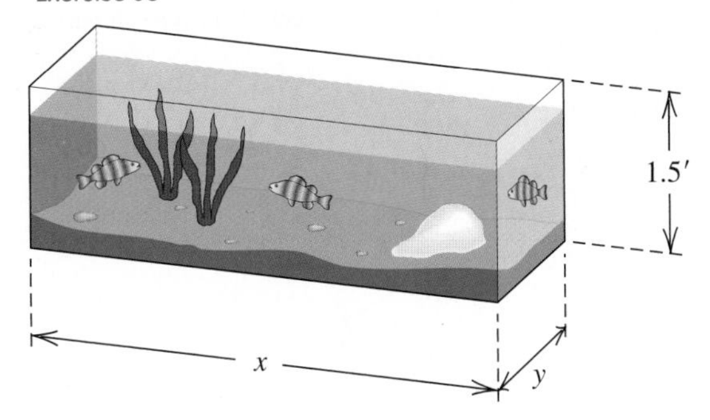

69 Skyline ordinance A city council is proposing a new skyline ordinance. It would require the setback S for any building from a residence to be a minimum of 100 feet, plus an additional 6 feet for each foot of height above 25 feet. Find a linear function for S in terms of h.

Exercise 69

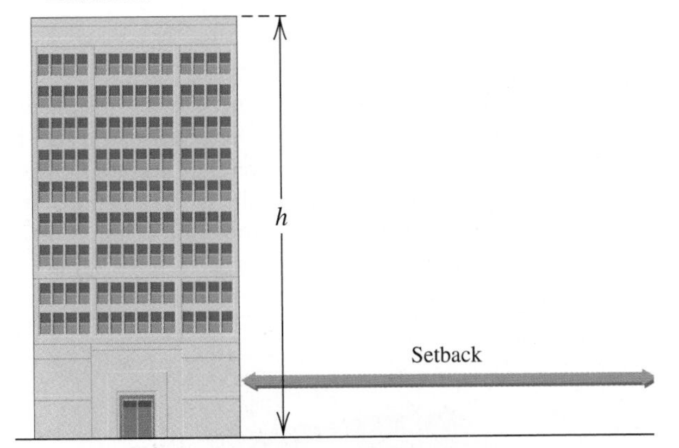

70 Energy tax A proposed energy tax T on gasoline, which would affect the cost of driving a vehicle, is to be computed by multiplying the number x of gallons of gasoline that you buy by 125,000 (the number of BTUs per gallon of gasoline) and then multiplying the total BTUs by the tax—34.2 cents per million BTUs. Find a linear function for T in terms of x.

71 Childhood growth For children between ages 6 and 10, height y (in inches) is frequently a linear function of age t (in years). The height of a certain child is 48 inches at age 6 and 50.5 inches at age 7.

(a) Express y as a function of t.

(b) Sketch the line in part (a), and interpret the slope.

(c) Predict the height of the child at age 10.

72 Radioactive contamination It has been estimated that 1000 curies of a radioactive substance introduced at a point on the surface of the open sea would spread over an area of 40,000 km^2 in 40 days. Assuming that the area covered by the radioactive substance is a linear function of time t and is always circular in shape, express the radius r of the contamination as a function of t.

73 Distance to a hot-air balloon A hot-air balloon is released at 1:00 P.M. and rises vertically at a rate of 2 m/sec. An observation point is situated 100 meters from a point on the ground directly below the balloon (see the figure). If t denotes the time (in seconds) after 1:00 P.M., express the distance d between the balloon and the observation point as a function of t.

Exercise 73

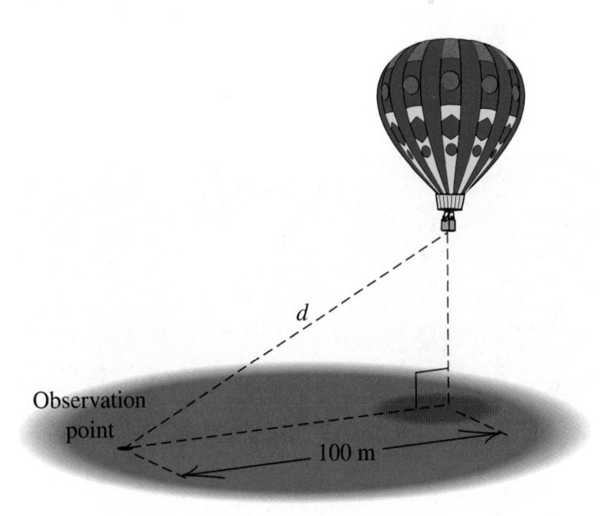

74 Triangle ABC is inscribed in a semicircle of diameter 15 (see the figure).

(a) If x denotes the length of side AC, express the length y of side BC as a function of x. (*Hint:* Angle ACB is a right angle.)

(b) Express the area $\mathscr{A}$ of triangle ABC as a function of x, and state the domain of this function.

Exercise 74

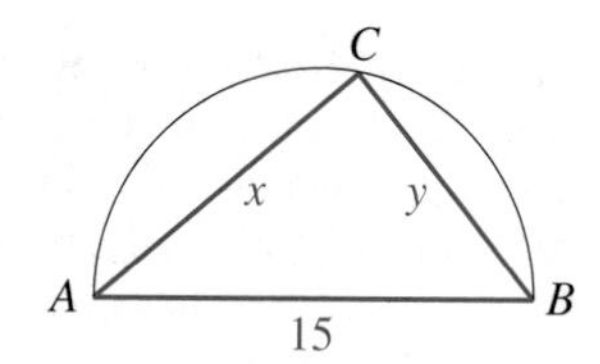

75 **Distance to the earth** From an exterior point P that is h units from a circle of radius r, a tangent line is drawn to the circle (see the figure). Let y denote the distance from the point P to the point of tangency T.

(a) Express y as a function of h. (*Hint:* If C is the center of the circle, then PT is perpendicular to CT.)

(b) If r is the radius of the earth and h is the altitude of a space shuttle, then y is the maximum distance to the earth that an astronaut can see from the shuttle. In particular, if $h = 200$ mi and $r \approx 4000$ mi, approximate y.

Exercise 75

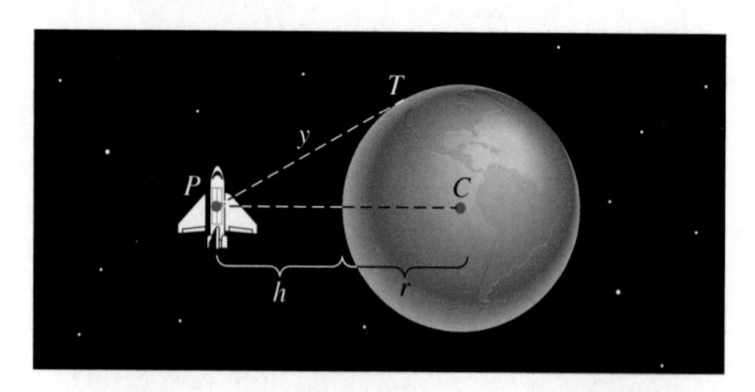

76 **Length of a tightrope** The figure illustrates the apparatus for a tightrope walker. Two poles are set 50 feet apart, but the point of attachment P for the rope is yet to be determined.

(a) Express the length L of the rope as a function of the distance x from P to the ground.

(b) If the total walk is to be 75 feet, determine the distance from P to the ground.

Exercise 76

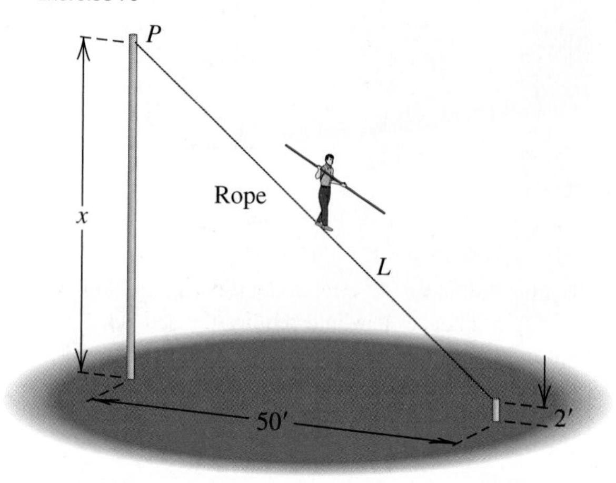

77 **Airport runway** The relative positions of an aircraft runway and a 20-foot-tall control tower are shown in the figure. The beginning of the runway is at a perpendicular distance of 300 feet from the base of the tower. If x denotes the distance an airplane has moved down the runway, express the distance d between the airplane and the top of the control tower as a function of x.

Exercise 77

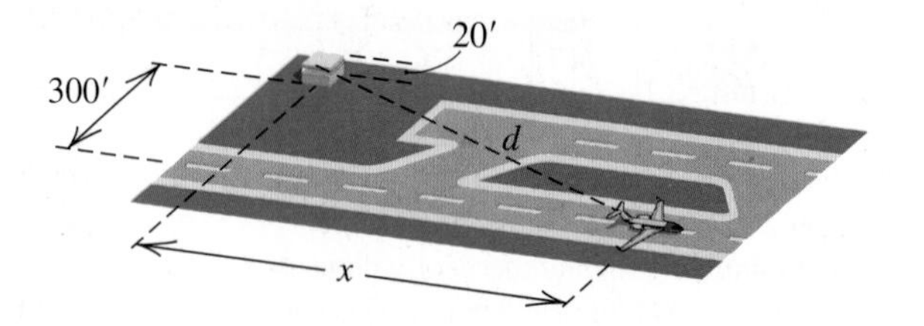

78 Destination time A man in a rowboat that is 2 miles from the nearest point A on a straight shoreline wishes to reach a house located at a point B that is 6 miles farther down the shoreline (see the figure). He plans to row to a point P that is between A and B and is x miles from the house, and then he will walk the remainder of the distance. Suppose he can row at a rate of 3 mi/hr and can walk at a rate of 5 mi/hr. If T is the total time required to reach the house, express T as a function of x.

Exercise 78

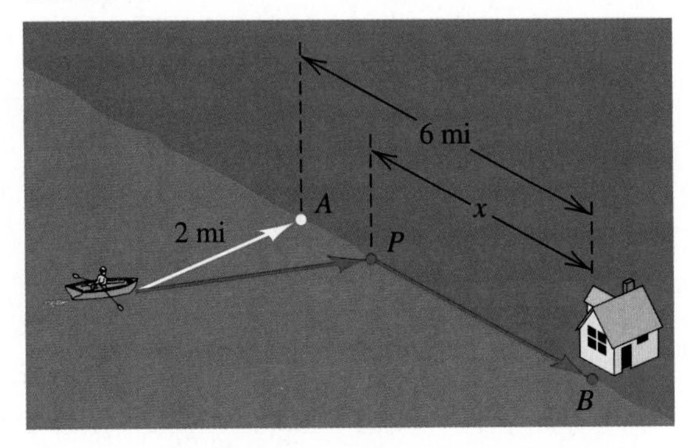

Exer. 79–82: (a) Sketch the graph of f on the given interval $[a, b]$. (b) Estimate the range of f on $[a, b]$. (c) Estimate the intervals on which f is increasing or is decreasing.

79 $f(x) = \dfrac{x^{1/3}}{1 + x^4}$; $[-2, 2]$

80 $f(x) = x^4 - 0.4x^3 - 0.8x^2 + 0.2x + 0.1$; $[-1, 1]$

81 $f(x) = x^5 - 3x^2 + 1$; $[-0.7, 1.4]$

82 $f(x) = \dfrac{1 - x^3}{1 + x^4}$; $[-4, 4]$

Exer. 83–84: In Exercises 43–44 of Section 1.4, algebraic methods were used to find solutions to each of the following equations. Now solve the equation graphically by assigning the expression on the left side to Y_1 and the number on the right side to Y_2 and then finding the x-coordinates of all points of intersection of the two graphs.

83 (a) $x^{5/3} = 32$ (b) $x^{4/3} = 16$ (c) $x^{2/3} = -36$

(d) $x^{3/4} = 125$ (e) $x^{3/2} = -27$

84 (a) $x^{3/5} = -27$ (b) $x^{2/3} = 25$ (c) $x^{4/3} = -49$

(d) $x^{3/2} = 27$ (e) $x^{3/4} = -8$

85 Calculator screen A particular graphing calculator screen is 95 pixels wide and 63 pixels high.

(a) Find the total number of pixels in the screen.

(b) If a function is graphed in dot mode, determine the maximum number of pixels that would typically be darkened on the calculator screen to show the function.

86 Stopping distances The table lists the practical stopping distances D (in feet) for cars at speeds S (in miles per hour) on level surfaces, as used by the American Association of State Highway and Transportation Officials.

S	20	30	40	50	60	70
D	33	86	167	278	414	593

(a) Plot the data.

(b) Determine whether stopping distance is a linear function of speed.

(c) Discuss the practical implications of these data for safely driving a car.

87 New car prices In 1993 and 2000, the average prices paid for a new car were \$16,871 and \$20,356, respectively. Assume the average price increased linearly.

(a) Find a function f that models the average price paid for a new car. Graph f together with the two data points.

(b) Interpret the slope of the graph of f.

(c) Graphically approximate the year when the average price paid would be \$25,000.

2.5 Graphs of Functions

In this section we discuss aids for sketching graphs of certain types of functions. In particular, a function f is called **even** if $f(-x) = f(x)$ for every x in its domain. In this case, the equation $y = f(x)$ is not changed if $-x$ is substituted for x, and hence, from symmetry test 1 of Section 2.2, the graph of an even function is symmetric with respect to the y-axis.

A function f is called **odd** if $f(-x) = -f(x)$ for every x in its domain. If we apply symmetry test 3 of Section 2.2 to the equation $y = f(x)$, we see that the graph of an odd function is symmetric with respect to the origin.

These facts are summarized in the first two columns of the next chart.

Even and Odd Functions

Terminology	Definition	Illustration	Type of symmetry of graph
f is an ***even* function.**	$f(-x) = f(x)$ for every x in the domain.	$y = f(x) = x^2$	with respect to the y-axis
f is an ***odd* function.**	$f(-x) = -f(x)$ for every x in the domain.	$y = f(x) = x^3$	with respect to the origin

EXAMPLE 1 Determining whether a function is even or odd

Determine whether f is even, odd, or neither even nor odd.

(a) $f(x) = 3x^4 - 2x^2 + 5$ (b) $f(x) = 2x^5 - 7x^3 + 4x$

(c) $f(x) = x^3 + x^2$

SOLUTION In each case the domain of f is $\mathbb{R}$. To determine whether f is even or odd, we begin by examining $f(-x)$, where x is any real number.

▶ (a)

$$\begin{aligned} f(-x) &= 3(-x)^4 - 2(-x)^2 + 5 && \text{substitute } -x \text{ for } x \text{ in } f(x) \\ &= 3x^4 - 2x^2 + 5 && \text{simplify} \\ &= f(x) && \text{definition of } f \end{aligned}$$

Since $f(-x) = f(x)$, f is an even function.

(b)

$$\begin{aligned} f(-x) &= 2(-x)^5 - 7(-x)^3 + 4(-x) && \text{substitute } -x \text{ for } x \text{ in } f(x) \\ &= -2x^5 + 7x^3 - 4x && \text{simplify} \\ &= -(2x^5 - 7x^3 + 4x) && \text{factor out } -1 \\ &= -f(x) && \text{definition of } f \end{aligned}$$

Since $f(-x) = -f(x)$, f is an odd function.

▶ **(c)** $f(-x) = (-x)^3 + (-x)^2$ substitute $-x$ for x in $f(x)$

$= -x^3 + x^2$ simplify

Since $f(-x) \neq f(x)$, and $f(-x) \neq -f(x)$ (note that $-f(x) = -x^3 - x^2$), the function f is neither even nor odd.

In the next example we consider the **absolute value function** f, defined by $f(x) = |x|$.

EXAMPLE 2 **Sketching the graph of the absolute value function**

Let $f(x) = |x|$.

(a) Determine whether f is even or odd.

(b) Sketch the graph of f.

(c) Find the intervals on which f is increasing or is decreasing.

SOLUTION

(a) The domain of f is $\mathbb{R}$, because the absolute value of x exists for every real number x. If x is in $\mathbb{R}$, then

$$f(-x) = |-x| = |x| = f(x).$$

Thus, f is an even function, since $f(-x) = f(x)$.

(b) Since f is even, its graph is symmetric with respect to the y-axis. If $x \geq 0$, then $|x| = x$, and therefore the first quadrant part of the graph coincides with the line $y = x$. Sketching this half-line and using symmetry gives us Figure 1.

Figure 1

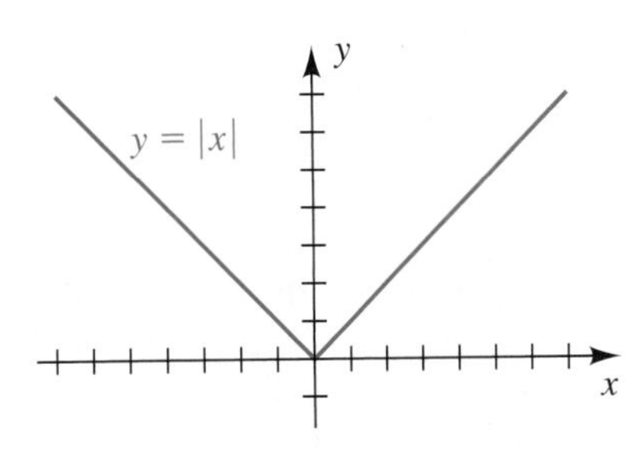

(c) Referring to the graph, we see that f is decreasing on $(-\infty, 0]$ and is increasing on $[0, \infty)$.

If we know the graph of $y = f(x)$, it is easy to sketch the graphs of

$$y = f(x) + c \qquad \text{and} \qquad y = f(x) - c$$

for any positive real number c. As in the next chart, for $y = f(x) + c$, we add c to the y-coordinate of each point on the graph of $y = f(x)$. This *shifts* the graph of f *upward* a distance c. For $y = f(x) - c$ with $c > 0$, we subtract c from each y-coordinate, thereby shifting the graph of f a distance c *downward*. These are called **vertical shifts** of graphs.

▶ **Tutorial available at www.thomsonedu.com/login**

Vertically Shifting the Graph of $y = f(x)$

Equation	$y = f(x) + c$ with $c > 0$	$y = f(x) - c$ with $c > 0$
Effect on graph	The graph of f is shifted vertically upward a distance c.	The graph of f is shifted vertically downward a distance c.
Graphical interpretation	y $y = f(x) + c$ $(a, b + c)$ $c > 0$ (a, b) $y = f(x)$ x	y (a, b) $c > 0$ $y = f(x)$ $(a, b - c)$ x $y = f(x) - c$

Figure 2

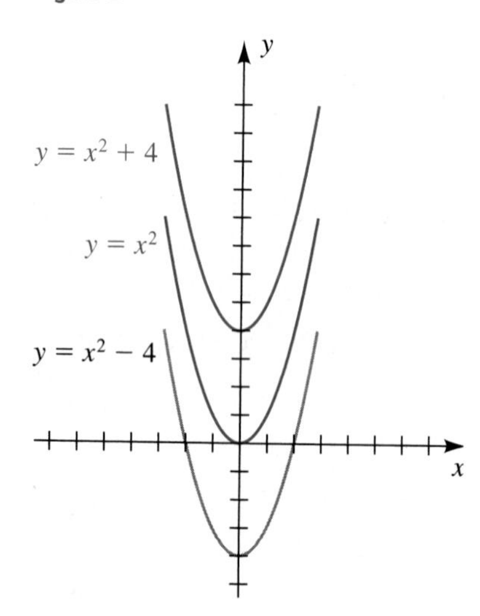

EXAMPLE 3 **Vertically shifting a graph**

Sketch the graph of f:

(a) $f(x) = x^2$ **(b)** $f(x) = x^2 + 4$ **(c)** $f(x) = x^2 - 4$

SOLUTION We shall sketch all graphs on the same coordinate plane.

(a) Since

$$f(-x) = (-x)^2 = x^2 = f(x),$$

the function f is even, and hence its graph is symmetric with respect to the y-axis. Several points on the graph of $y = x^2$ are $(0, 0)$, $(1, 1)$, $(2, 4)$, and $(3, 9)$. Drawing a smooth curve through these points and reflecting through the y-axis gives us the sketch in Figure 2. The graph is a parabola with vertex at the origin and opening upward.

(b) To sketch the graph of $y = x^2 + 4$, we add 4 to the y-coordinate of each point on the graph of $y = x^2$; that is, we shift the graph in part (a) upward 4 units, as shown in the figure.

(c) To sketch the graph of $y = x^2 - 4$, we decrease the y-coordinates of $y = x^2$ by 4; that is, we shift the graph in part (a) downward 4 units.

We can also consider **horizontal shifts** of graphs. Specifically, if $c > 0$, consider the graphs of $y = f(x)$ and $y = g(x) = f(x - c)$ sketched on the same coordinate plane, as illustrated in the next chart. Since

$$g(a + c) = f([a + c] - c) = f(a),$$

we see that the point with x-coordinate a on the graph of $y = f(x)$ has the same y-coordinate as the point with x-coordinate $a + c$ on the graph of

$y = g(x) = f(x - c)$. This implies that the graph of $y = g(x) = f(x - c)$ can be obtained by shifting the graph of $y = f(x)$ *to the right* a distance c. Similarly, the graph of $y = h(x) = f(x + c)$ can be obtained by shifting the graph of f *to the left* a distance c, as shown in the chart.

Horizontally Shifting the Graph of $y = f(x)$

Equation	Effect on graph	Graphical interpretation
$y = g(x)$ $= f(x - c)$ with $c > 0$	The graph of f is shifted horizontally to the *right* a distance c.	y; $y = f(x)$; $y = g(x) = f(x - c)$; (a, b); $(a + c, b)$; $f(a)$; $g(a + c)$; a; $a + c$; x; $c > 0$
$y = h(x)$ $= f(x + c)$ with $c > 0$	The graph of f is shifted horizontally to the *left* a distance c.	y; $y = h(x) = f(x + c)$; $y = f(x)$; $(a - c, b)$; (a, b); $h(a - c)$; $f(a)$; $a - c$; a; x; $c > 0$

Horizontal and vertical shifts are also referred to as *translations*.

Figure 3

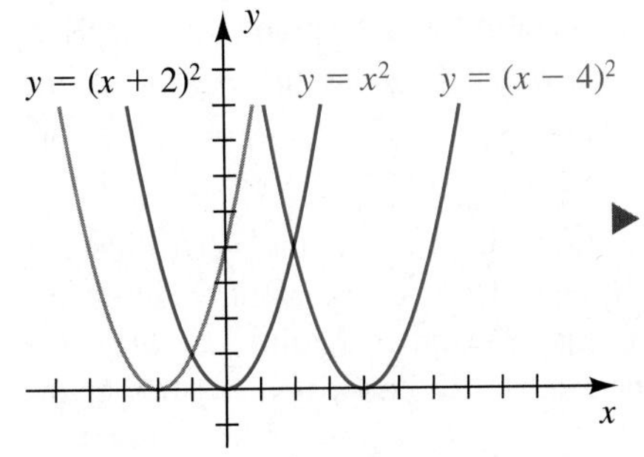

EXAMPLE 4 Horizontally shifting a graph

Sketch the graph of f:

(a) $f(x) = (x - 4)^2$ **(b)** $f(x) = (x + 2)^2$

▶ **SOLUTION** The graph of $y = x^2$ is sketched in Figure 3.

(a) Shifting the graph of $y = x^2$ to the right 4 units gives us the graph of $y = (x - 4)^2$, shown in the figure.

(b) Shifting the graph of $y = x^2$ to the left 2 units leads to the graph of $y = (x + 2)^2$, shown in the figure.

▶ **Tutorial available at www.thomsonedu.com/login**

To obtain the graph of $y = cf(x)$ for some real number c, we may *multiply* the y-coordinates of points on the graph of $y = f(x)$ by c. For example, if $y = 2f(x)$, we double the y-coordinates; or if $y = \frac{1}{2}f(x)$, we multiply each y-coordinate by $\frac{1}{2}$. This procedure is referred to as **vertically stretching** the graph of f (if $c > 1$) or **vertically compressing** the graph (if $0 < c < 1$) and is summarized in the following chart.

Vertically Stretching or Compressing the Graph of $y = f(x)$

Equation	$y = cf(x)$ with $c > 1$	$y = cf(x)$ with $0 < c < 1$
Effect on graph	The graph of f is stretched vertically by a factor c.	The graph of f is compressed vertically by a factor $1/c$.
Graphical interpretation	(a, cb), (a, b), $y = cf(x)$ with $c > 1$, $y = f(x)$, x, y	(a, b), (a, cb), $y = cf(x)$ with $0 < c < 1$, $y = f(x)$, x, y

EXAMPLE 5 Vertically stretching or compressing a graph

Sketch the graph of the equation:

(a) $y = 4x^2$ **(b)** $y = \frac{1}{4}x^2$

Figure 4

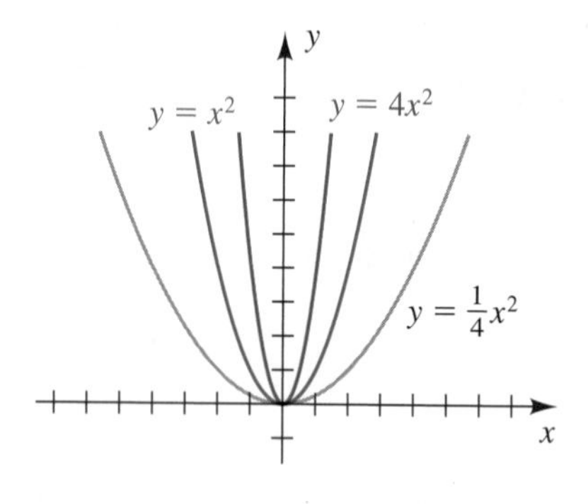

SOLUTION

(a) To sketch the graph of $y = 4x^2$, we may refer to the graph of $y = x^2$ in Figure 4 and multiply the y-coordinate of each point by 4. This stretches the graph of $y = x^2$ vertically by a factor 4 and gives us a narrower parabola that is sharper at the vertex, as illustrated in the figure.

(b) The graph of $y = \frac{1}{4}x^2$ may be sketched by multiplying the y-coordinates of points on the graph of $y = x^2$ by $\frac{1}{4}$. This compresses the graph of $y = x^2$ vertically by a factor $1/\frac{1}{4} = 4$ and gives us a wider parabola that is flatter at the vertex, as shown in Figure 4. ◢

Replacing y with $-y$ reflects the graph of $y = f(x)$ through the x-axis.

We may obtain the graph of $y = -f(x)$ by multiplying the y-coordinate of each point on the graph of $y = f(x)$ by -1. Thus, every point (a, b) on the graph of $y = f(x)$ that lies above the x-axis determines a point $(a, -b)$ on the graph of $y = -f(x)$ that lies below the x-axis. Similarly, if (c, d) lies below the x-axis (that is, $d < 0$), then $(c, -d)$ lies above the x-axis. The graph of $y = -f(x)$ is a **reflection** of the graph of $y = f(x)$ through the x-axis.

Figure 5

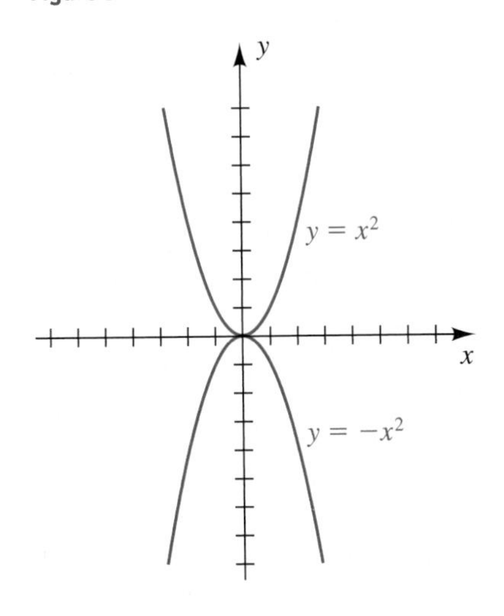

EXAMPLE 6 **Reflecting a graph through the x-axis**

Sketch the graph of $y = -x^2$.

SOLUTION The graph may be found by plotting points; however, since the graph of $y = x^2$ is familiar to us, we sketch it as in Figure 5 and then multiply the y-coordinates of points by -1. This procedure gives us the reflection through the x-axis indicated in the figure.

Sometimes it is useful to compare the graphs of $y = f(x)$ and $y = f(cx)$ if $c \neq 0$. In this case the function values $f(x)$ for

$$a \leq x \leq b$$

are the same as the function values $f(cx)$ for

$$a \leq cx \leq b \qquad \text{or, equivalently,} \qquad \frac{a}{c} \leq x \leq \frac{b}{c}.$$

This implies that the graph of f is **horizontally compressed** (if $c > 1$) or **horizontally stretched** (if $0 < c < 1$), as summarized in the following chart.

Horizontally Compressing or Stretching the Graph of $y = f(x)$

Equation	Effect on graph	Graphical interpretation
$y = f(cx)$ with $c > 1$	The graph of f is compressed horizontally by a factor c.	y; x; $y = f(x)$; $y = f(cx)$ with $c > 1$; $\left(\frac{a}{c}, b\right)$; (a, b)
$y = f(cx)$ with $0 < c < 1$	The graph of f is stretched horizontally by a factor $1/c$.	y; x; $y = f(x)$; $y = f(cx)$ with $0 < c < 1$; (a, b); $\left(\frac{a}{c}, b\right)$

Replacing x with $-x$ reflects the graph of $y = f(x)$ through the y-axis.

If $c < 0$, then the graph of $y = f(cx)$ may be obtained by reflecting the graph of $y = f(|c|x)$ through the y-axis. For example, to sketch the graph of $y = f(-2x)$, we reflect the graph of $y = f(2x)$ through the y-axis. As a special case, the graph of $y = f(-x)$ is a **reflection** of the graph of $y = f(x)$ through the y-axis.

EXAMPLE 7 Horizontally stretching or compressing a graph

If $f(x) = x^3 - 4x^2$, sketch the graphs of $y = f(x)$, $y = f(2x)$, and $y = f\left(\frac{1}{2}x\right)$.

SOLUTION We have the following:

$$y = f(x) = x^3 - 4x^2 = x^2(x - 4)$$
$$y = f(2x) = (2x)^3 - 4(2x)^2 = 8x^3 - 16x^2 = 8x^2(x - 2)$$
$$y = f\left(\tfrac{1}{2}x\right) = \left(\tfrac{1}{2}x\right)^3 - 4\left(\tfrac{1}{2}x\right)^2 = \tfrac{1}{8}x^3 - x^2 = \tfrac{1}{8}x^2(x - 8)$$

Figure 6
$[-6, 15]$ by $[-10, 4]$

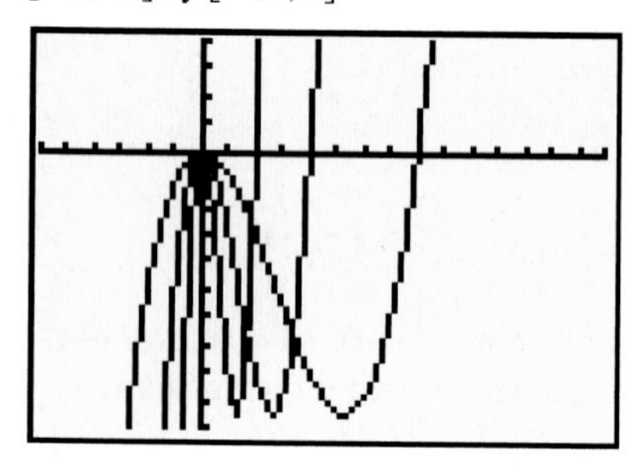

Note that the x-intercepts of the graph of $y = f(2x)$ are 0 and 2, which are $\frac{1}{2}$ the x-intercepts of 0 and 4 for $y = f(x)$. This indicates a horizontal compression by a factor 2.

The x-intercepts of the graph of $y = f\left(\frac{1}{2}x\right)$ are 0 and 8, which are 2 times the x-intercepts for $y = f(x)$. This indicates a horizontal stretching by a factor $1/\frac{1}{2} = 2$.

The graphs, obtained by using a graphing calculator with viewing rectangle $[-6, 15]$ by $[-10, 4]$, are shown in Figure 6.

Functions are sometimes described by more than one expression, as in the next examples. We call such functions **piecewise-defined functions.**

EXAMPLE 8 Sketching the graph of a piecewise-defined function

Sketch the graph of the function f if

$$f(x) = \begin{cases} 2x + 5 & \text{if } x \le -1 \\ x^2 & \text{if } |x| < 1 \\ 2 & \text{if } x \ge 1 \end{cases}$$

Figure 7

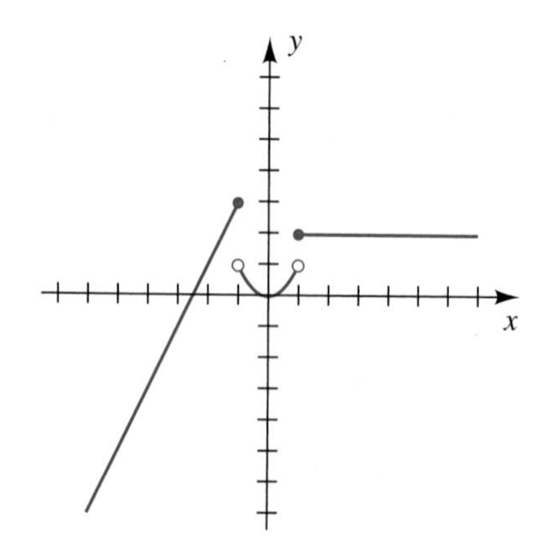

SOLUTION If $x \le -1$, then $f(x) = 2x + 5$ and the graph of f coincides with the line $y = 2x + 5$ and is represented by the portion of the graph to the left of the line $x = -1$ in Figure 7. The small dot indicates that the point $(-1, 3)$ is on the graph.

If $|x| < 1$ (or, equivalently, $-1 < x < 1$), we use x^2 to find values of f, and therefore this part of the graph of f coincides with the parabola $y = x^2$, as indicated in the figure. Note that the points $(-1, 1)$ and $(1, 1)$ are *not* on the graph.

Finally, if $x \ge 1$, the values of f are always 2. Thus, the graph of f for $x \ge 1$ is the horizontal half-line in Figure 7.

Note: When you finish sketching the graph of a piecewise-defined function, check that it passes the vertical line test.

The next example shows how we can graph the piecewise-defined function in the last example on a graphing calculator.

EXAMPLE 9 Sketching the graph of a piecewise-defined function

Sketch the graph of the function f if

$$f(x) = \begin{cases} 2x + 5 & \text{if } x \le -1 \\ x^2 & \text{if } |x| < 1 \\ 2 & \text{if } x \ge 1 \end{cases}$$

SOLUTION We begin by making the assignment

$$Y_1 = \underbrace{(2x + 5)(x \le -1)}_{\text{first piece}} + \underbrace{x^2(\text{abs}(x) < 1)}_{\text{second piece}} + \underbrace{2(x \ge 1)}_{\text{third piece}}.$$

Make Y assignment.

TI-83/4 Plus

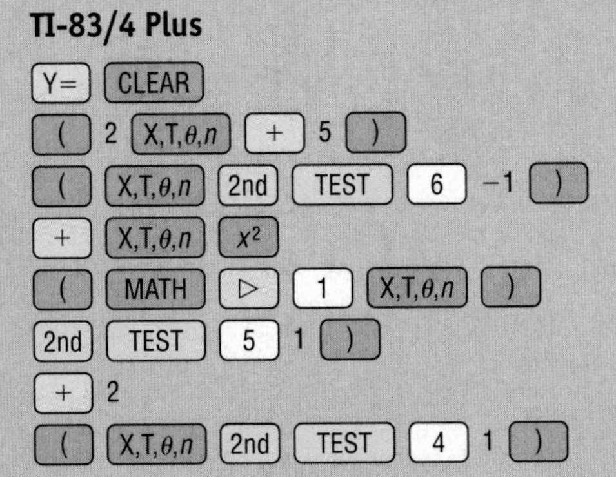

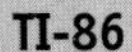

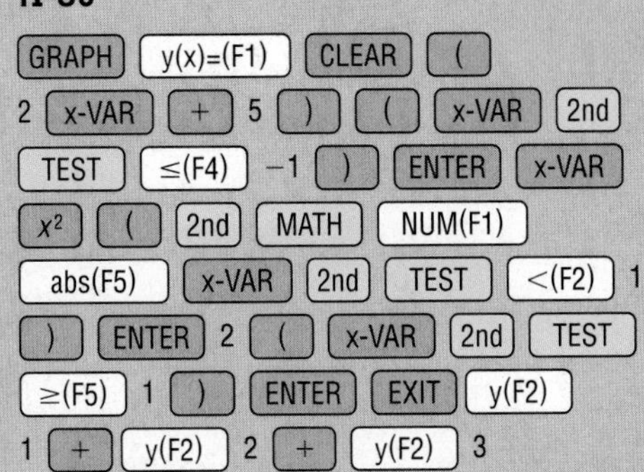

See the Note on page 165 concerning turning off y1, y2, and y3. We could type the whole function in Y_1, as shown in the TI-83/4 Plus figure to the left.

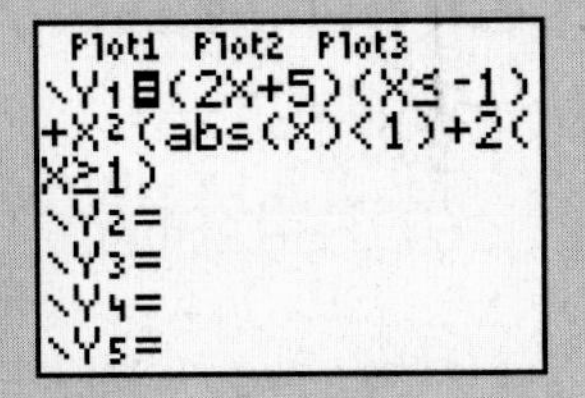

As the variable x takes on values from Xmin to Xmax, the inequality $x \le -1$ in the first piece will have a value of either 1 (if $x \le -1$) or 0 (if $x > -1$). This value is multiplied by the value of $2x + 5$ and assigned to Y_1. In the second piece, note that *both* $-1 < x$ and $x < 1$ (equivalent to $|x| < 1$) must be true for the value of x^2 to be assigned to Y_1 (y2 for the TI-86). The general idea is that each piece is "on" only when x takes on the associated domain values.

(continued)

Set the viewing rectangle.

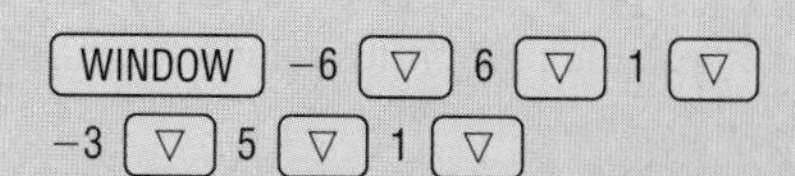

−3 ▽ 5 ▽ 1 ▽

2nd WIND(M2) −6 ▽ 6 ▽ 1 ▽
−3 ▽ 5 ▽ 1

WINDOW
Xmin=-6
Xmax=6
Xscl=1
Ymin=-3
Ymax=5
Yscl=1
Xres=1

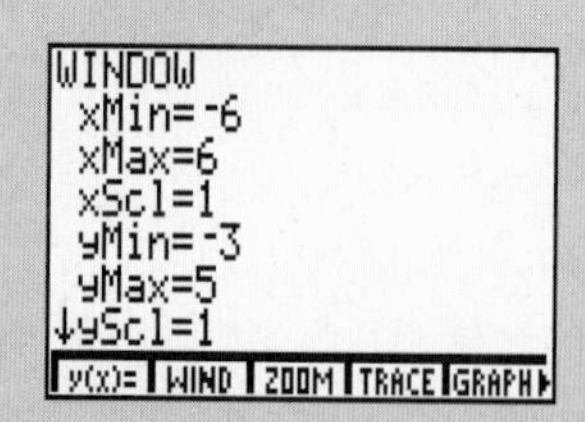

Graphing the function in the standard *connected mode* allows us to see the most important features of the graph. In connected mode, the calculator includes lines between the endpoints of the pieces. Press GRAPH or GRAPH(F5).

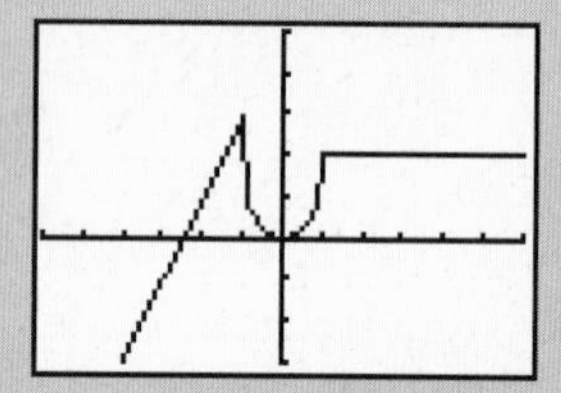

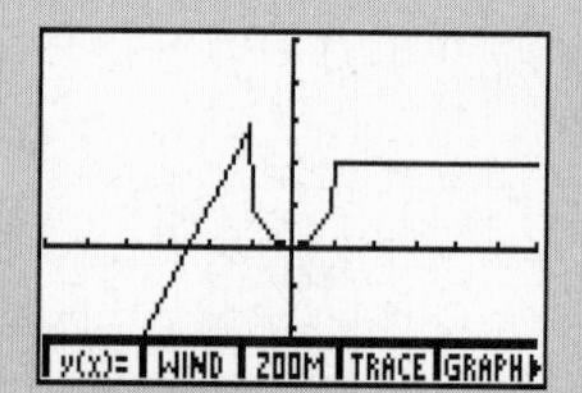

To eliminate these lines, we can change to *dot mode* and re-graph. Note that the graphing calculator makes no distinction between including and excluding an endpoint (some software packages do).

Change to dot mode.

MODE ▽ (4 times) ▷ ENTER
GRAPH

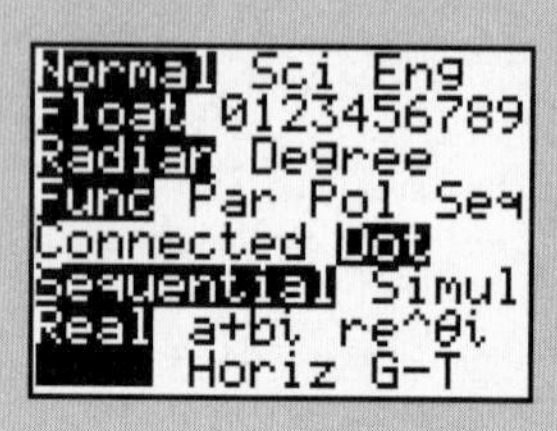

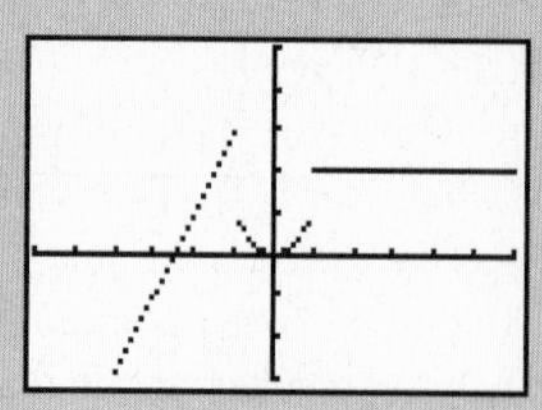

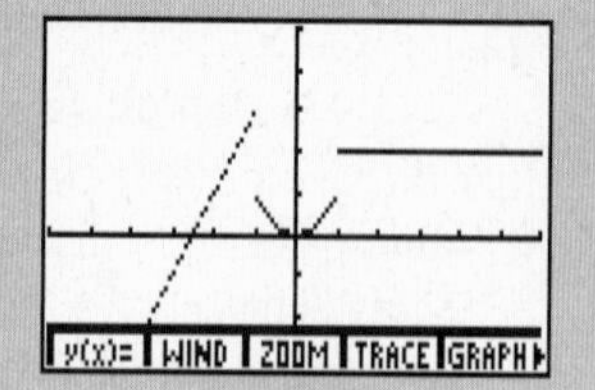

Note: As shown for the TI-86, an alternative method for representing the function f is to assign each piece to a Y-value as follows:

$$Y_1 = (2x + 5)(x \leq -1), Y_2 = x^2(\text{abs}(x) < 1), Y_3 = 2(x \geq 1)$$

The graphing of the three screens, however, is a rather slow process. Speed can be improved by graphing $Y_4 = Y_1 + Y_2 + Y_3$ to obtain the graph of f (be sure to turn off Y_1, Y_2, and Y_3). To turn off Y_1 on the TI-83/4 Plus, place the cursor on the = sign to the right of Y_1 and press [ENTER]. On the TI-86, place the cursor anywhere on the line for y1 and press [SELCT(F5)].

Yet another method for representing the function f is to assign each piece to a Y-value using division, as follows:

$$Y_1 = (2x + 5)/(x \leq -1), \quad Y_2 = x^2/(\text{abs}(x) < 1), \quad Y_3 = 2/(x \geq 1)$$

Graphing the three Y-values gives us the graph of f once more. The advantage of this method is apparent when you use the connected mode—try it!

Calculator Note: Recall that $|x| < 1$ or, equivalently, $-1 < x < 1$ can also be written as "$-1 < x$ and $x < 1$." The operators "and" and "or" are found under the TEST LOGIC menu on the TI-83/4 Plus and under the BASE BOOL menu on the TI-86. We can use "and" to make an alternative assignment for the function in this example, as shown in the figure.

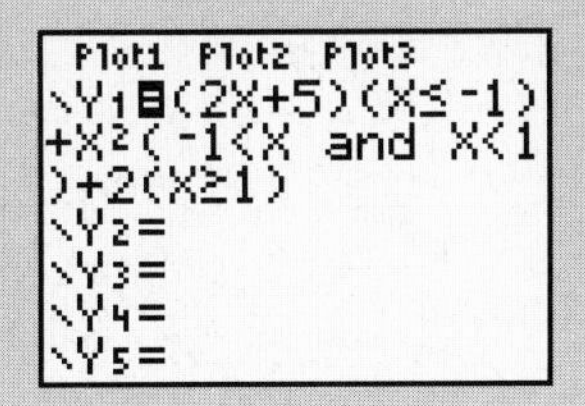

It is a common misconception to think that if you move up to a higher tax bracket, *all* your income is taxed at the higher rate. The following example of a graph of a piecewise-defined function helps dispell that notion.

EXAMPLE 10 Application using a piecewise-defined function

Sketch a graph of the 2006 Federal Tax Rate Schedule X, shown in Figure 8. Let x represent the taxable income and T represent the amount of tax. (Assume the domain is the set of nonnegative real numbers.)

Figure 8

2006 Federal Tax Rate Schedules

Schedule X –Use if your Filing status is **single**

If taxable income is over–	*But not over–*	*The tax is:*	*of the amount over–*
$0	$7,550	-------- **10%**	**$0**
7,550	30,650	**$755.00 + 15%**	**7,550**
30,650	74,200	**$4,220.00 + 25%**	**30,650**
74,200	154,800	**15,107.50 + 28%**	**74,200**
154,800	336,550	**37,675.50 + 33%**	**154,800**
336,550	-------	**97,653.00 + 35%**	**336,550**

SOLUTION The tax table can be represented by a piecewise-defined function as follows:

$$T(x) = \begin{cases} 0 & \text{if } x \leq 0 \\ 0.10x & \text{if } 0 < x \leq 7550 \\ 755.00 + 0.15(x - 7550) & \text{if } 7550 < x \leq 30{,}650 \\ 4220.00 + 0.25(x - 30{,}650) & \text{if } 30{,}650 < x \leq 74{,}200 \\ 15{,}107.50 + 0.28(x - 74{,}200) & \text{if } 74{,}200 < x \leq 154{,}800 \\ 37{,}675.50 + 0.33(x - 154{,}800) & \text{if } 154{,}800 < x \leq 336{,}550 \\ 97{,}653.00 + 0.35(x - 336{,}550) & \text{if } x > 336{,}550 \end{cases}$$

Note that the assignment for the 15% tax bracket is *not* $0.15x$, but 10% of the first \$7550 in taxable income plus 15% of the amount *over* \$7550; that is,

$$0.10(7550) + 0.15(x - 7550) = 755.00 + 0.15(x - 7550).$$

The other pieces can be established in a similar fashion. The graph of T is shown in Figure 9; note that the slope of each piece represents the tax rate.

Figure 9

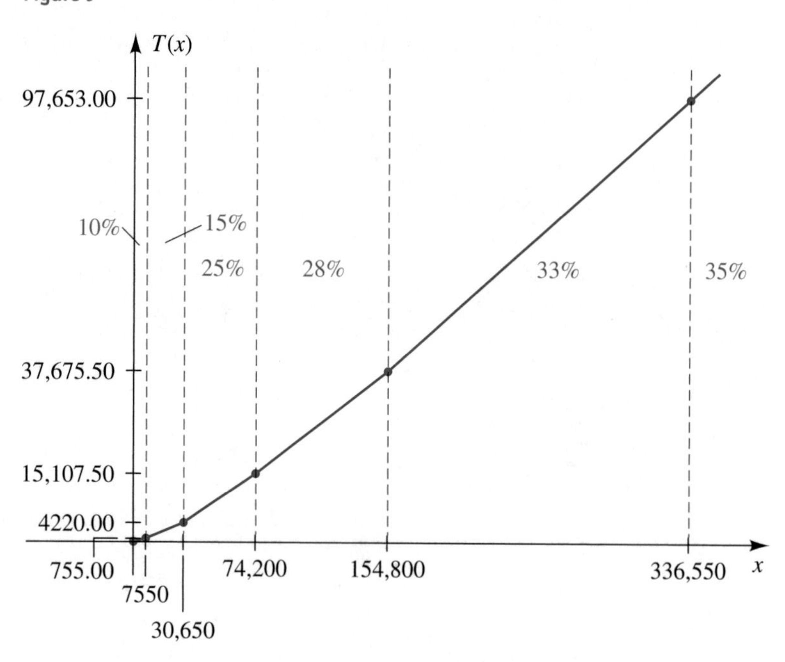

If x is a real number, we define the symbol $[\![x]\!]$ as follows:

$$[\![x]\!] = n, \qquad \text{where } n \text{ is the greatest integer such that } n \leq x$$

If we identify $\mathbb{R}$ with points on a coordinate line, then n is the first integer to the *left* of (or *equal* to) x.

ILLUSTRATION The Symbol $[\![x]\!]$

To graph $y = [\![x]\!]$, graph $Y_1 = \text{int}(X)$ in dot mode. On the TI-83/4 Plus and the TI-86, int *is under* MATH, NUM.

- $[\![0.5]\!] = 0$
- $[\![1.8]\!] = 1$
- $[\![\sqrt{5}]\!] = 2$
- $[\![3]\!] = 3$
- $[\![-3]\!] = -3$
- $[\![-2.7]\!] = -3$
- $[\![-\sqrt{3}]\!] = -2$
- $[\![-0.5]\!] = -1$

The **greatest integer function** f is defined by $f(x) = [\![x]\!]$.

Figure 10

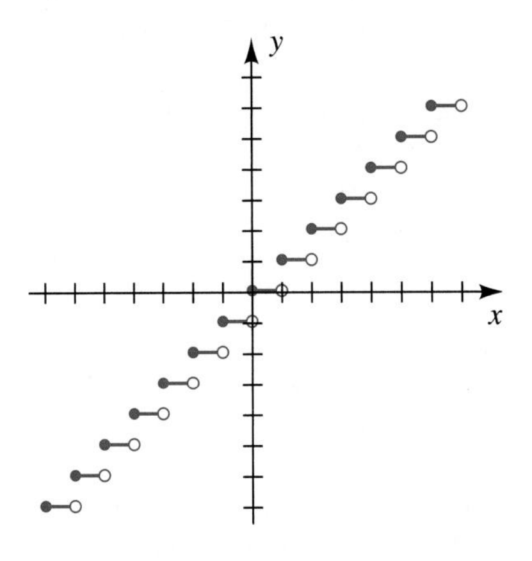

EXAMPLE 11 Sketching the graph of the greatest integer function

Sketch the graph of the greatest integer function.

SOLUTION The x- and y-coordinates of some points on the graph may be listed as follows:

Values of x	$f(x) = [\![x]\!]$
⋮	⋮
$-2 \le x < -1$	-2
$-1 \le x < 0$	-1
$0 \le x < 1$	0
$1 \le x < 2$	1
$2 \le x < 3$	2
⋮	⋮

Whenever x is between successive integers, the corresponding part of the graph is a segment of a horizontal line. Part of the graph is sketched in Figure 10. The graph continues indefinitely to the right and to the left.

The next example involves absolute values.

Figure 11

(a)

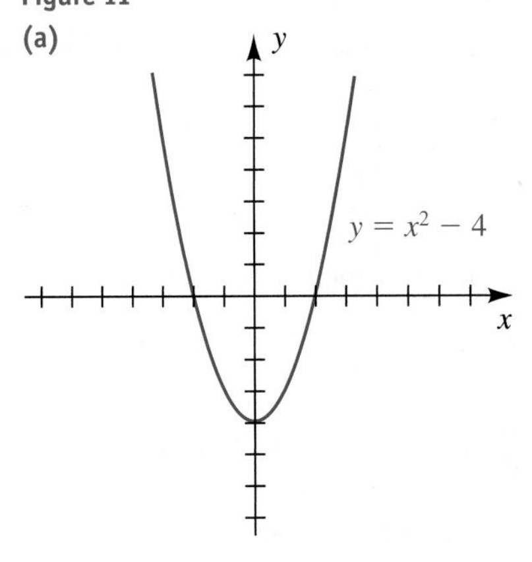

(b)

EXAMPLE 12 Sketching the graph of an equation containing an absolute value

Sketch the graph of $y = |x^2 - 4|$.

▶ SOLUTION The graph of $y = x^2 - 4$ was sketched in Figure 2 and is re-sketched in Figure 11(a). We note the following facts:

(1) If $x \le -2$ or $x \ge 2$, then $x^2 - 4 \ge 0$, and hence $|x^2 - 4| = x^2 - 4$.

(2) If $-2 < x < 2$, then $x^2 - 4 < 0$, and hence $|x^2 - 4| = -(x^2 - 4)$.

It follows from (1) that the graphs of $y = |x^2 - 4|$ and $y = x^2 - 4$ coincide for $|x| \ge 2$. We see from (2) that if $|x| < 2$, then the graph of $y = |x^2 - 4|$ is the reflection of the graph of $y = x^2 - 4$ through the x-axis. This gives us the sketch in Figure 11(b).

▶ **Tutorial available at www.thomsonedu.com/login**

In general, if the graph of $y = f(x)$ contains a point $P(c, -d)$ with d positive, then the graph of $y = |f(x)|$ contains the point $Q(c, d)$—that is, Q is the reflection of P through the x-axis. Points with nonnegative y-values are the same for the graphs of $y = f(x)$ and $y = |f(x)|$.

In Chapter 1 we used algebraic methods to solve inequalities involving absolute values of polynomials of degree 1, such as

$$|2x - 5| < 7 \qquad \text{and} \qquad |5x + 2| \geq 3.$$

Much more complicated inequalities can be investigated using a graphing utility, as illustrated in the next example.

EXAMPLE 13 Solving an absolute value inequality graphically

Estimate the solutions of

$$|0.14x^2 - 13.72| > |0.58x| + 11.$$

SOLUTION To solve the inequality, we make the assignments

$$Y_1 = \text{ABS}(0.14x^2 - 13.72) \qquad \text{and} \qquad Y_2 = \text{ABS}(0.58x) + 11$$

Figure 12
$[-30, 30, 5]$ by $[0, 40, 5]$

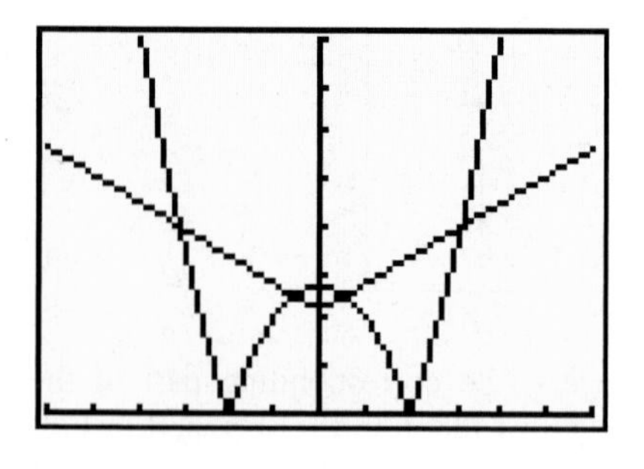

and estimate the values of x for which the graph of Y_1 is *above* the graph of Y_2 (since we want Y_1 *greater* than Y_2). After perhaps several trials, we choose the viewing rectangle $[-30, 30, 5]$ by $[0, 40, 5]$, obtaining graphs similar to those in Figure 12. Since there is symmetry with respect to the y-axis, it is sufficient to find the x-coordinates of the points of intersection of the graphs for $x > 0$. Using an intersect feature, we obtain $x \approx 2.80$ and $x \approx 15.52$. Referring to Figure 12, we obtain the (approximate) solution

$$(-\infty, -15.52) \cup (-2.80, 2.80) \cup (15.52, \infty).$$

Graphing $y = f(|x|)$

Later in this text and in calculus, you will encounter functions such as

$$g(x) = \ln |x| \qquad \text{and} \qquad h(x) = \sin |x|.$$

Both functions are of the form $y = f(|x|)$. The effect of substituting $|x|$ for x can be described as follows: If the graph of $y = f(x)$ contains a point $P(c, d)$ with c positive, then the graph of $y = f(|x|)$ contains the point $Q(-c, d)$—that is, Q is the reflection of P through the y-axis. Points on the y-axis ($x = 0$) are the same for the graphs of $y = f(x)$ and $y = f(|x|)$. Points with negative x-values on the graph of $y = f(x)$ are not on the graph of $y = f(|x|)$, since the result of the absolute value is always nonnegative.

The processes of shifting, stretching, compressing, and reflecting a graph may be collectively termed *transforming* a graph, and the resulting graph is called a **transformation** of the original graph. A graphical summary of the types of transformations encountered in this section appears in Appendix II.

2.5 Exercises

Exer. 1–2: Suppose f is an even function and g is an odd function. Complete the table, if possible.

1

x	-2	2
$f(x)$		7
$g(x)$		-6

2

x	-3	3
$f(x)$		-5
$g(x)$		15

Exer. 3–12: Determine whether f is even, odd, or neither even nor odd.

3 $f(x) = 5x^3 + 2x$

▶ 4 $f(x) = |x| - 3$

5 $f(x) = 3x^4 + 2x^2 - 5$

6 $f(x) = 7x^5 - 4x^3$

7 $f(x) = 8x^3 - 3x^2$

8 $f(x) = 12$

9 $f(x) = \sqrt{x^2 + 4}$

10 $f(x) = 3x^2 - 5x + 1$

11 $f(x) = \sqrt[3]{x^3 - x}$

12 $f(x) = x^3 - \dfrac{1}{x}$

Exer. 13–26: Sketch, on the same coordinate plane, the graphs of f for the given values of c. (Make use of symmetry, shifting, stretching, compressing, or reflecting.)

13 $f(x) = |x| + c;\quad c = -3, 1, 3$

14 $f(x) = |x - c|;\quad c = -3, 1, 3$

15 $f(x) = -x^2 + c;\quad c = -4, 2, 4$

16 $f(x) = 2x^2 - c;\quad c = -4, 2, 4$

17 $f(x) = 2\sqrt{x} + c;\quad c = -3, 0, 2$

18 $f(x) = \sqrt{9 - x^2} + c;\quad c = -3, 0, 2$

19 $f(x) = \frac{1}{2}\sqrt{x - c};\quad c = -2, 0, 3$

20 $f(x) = -\frac{1}{2}(x - c)^2;\quad c = -2, 0, 3$

21 $f(x) = c\sqrt{4 - x^2};\quad c = -2, 1, 3$

22 $f(x) = (x + c)^3;\quad c = -2, 1, 2$

23 $f(x) = cx^3;\quad c = -\frac{1}{3}, 1, 2$

24 $f(x) = (cx)^3 + 1;\quad c = -1, 1, 4$

25 $f(x) = \sqrt{cx} - 1;\quad c = -1, \frac{1}{9}, 4$

26 $f(x) = -\sqrt{16 - (cx)^2};\quad c = 1, \frac{1}{2}, 4$

Exer. 27–32: If the point P is on the graph of a function f, find the corresponding point on the graph of the given function.

▶ 27 $P(0, 5);\quad y = f(x + 2) - 1$

▶ 28 $P(3, -1);\quad y = 2f(x) + 4$

29 $P(3, -2);\quad y = 2f(x - 4) + 1$

30 $P(-2, 4);\quad y = \frac{1}{2}f(x - 3) + 3$

31 $P(3, 9);\quad y = \frac{1}{3}f\left(\frac{1}{2}x\right) - 1$

▶ 32 $P(-2, 1);\quad y = -3f(2x) - 5$

Exer. 33–40: Explain how the graph of the function compares to the graph of $y = f(x)$. For example, for the equation $y = 2f(x + 3)$, the graph of f is shifted 3 units to the left and stretched vertically by a factor of 2.

33 $y = f(x - 2) + 3$

34 $y = 3f(x - 1)$

35 $y = f(-x) - 2$

36 $y = -f(x + 4)$

37 $y = -\frac{1}{2}f(x)$

38 $y = f\left(\frac{1}{2}x\right) - 3$

39 $y = -2f\left(\frac{1}{3}x\right)$

40 $y = \frac{1}{3}|f(x)|$

▶ Tutorial available at www.thomsonedu.com/login

Exer. 41–42: The graph of a function f with domain [0, 4] is shown in the figure. Sketch the graph of the given equation.

41

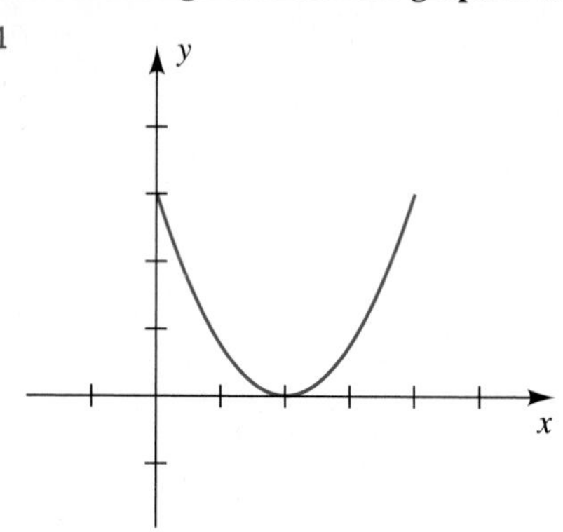

(a) $y = f(x + 3)$ (b) $y = f(x - 3)$

(c) $y = f(x) + 3$ (d) $y = f(x) - 3$

(e) $y = -3f(x)$ (f) $y = -\frac{1}{3}f(x)$

(g) $y = f\left(-\frac{1}{2}x\right)$ (h) $y = f(2x)$

(i) $y = -f(x + 2) - 3$ (j) $y = f(x - 2) + 3$

(k) $y = |f(x)|$ (l) $y = f(|x|)$

42

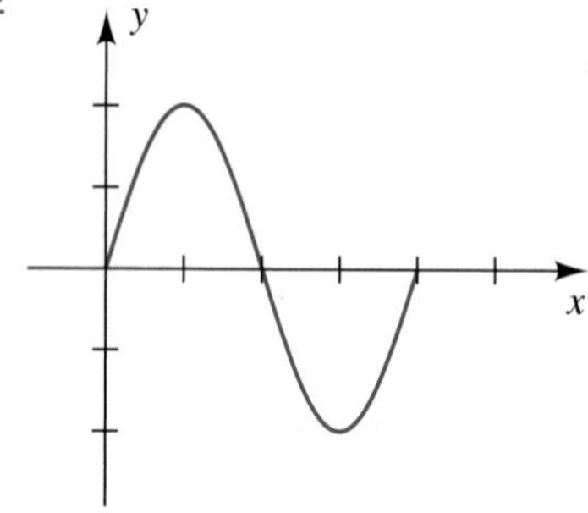

(a) $y = f(x - 2)$ (b) $y = f(x + 2)$

(c) $y = f(x) - 2$ (d) $y = f(x) + 2$

(e) $y = -2f(x)$ (f) $y = -\frac{1}{2}f(x)$

(g) $y = f(-2x)$ (h) $y = f\left(\frac{1}{2}x\right)$

(i) $y = -f(x + 4) - 2$ (j) $y = f(x - 4) + 2$

(k) $y = |f(x)|$ (l) $y = f(|x|)$

Exer. 43–46: The graph of a function f is shown, together with graphs of three other functions (a), (b), and (c). Use properties of symmetry, shifts, and reflecting to find equations for graphs (a), (b), and (c) in terms of f.

▶ 43

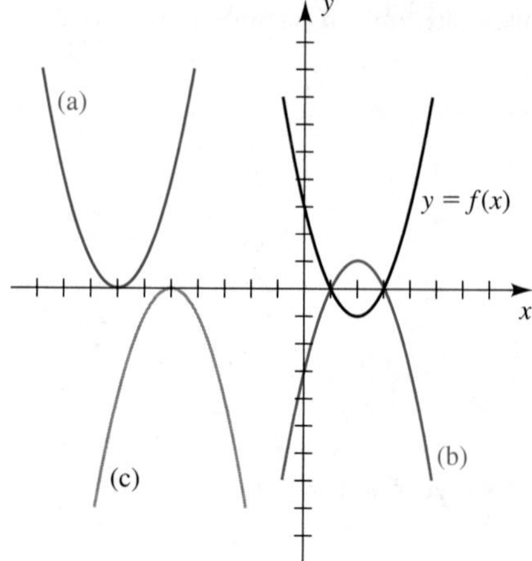

44

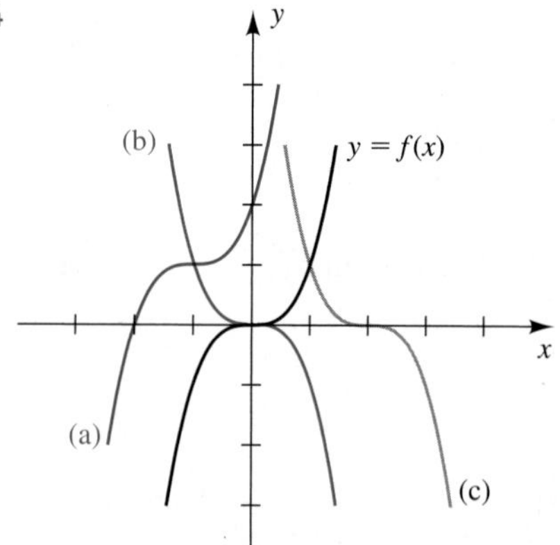

45

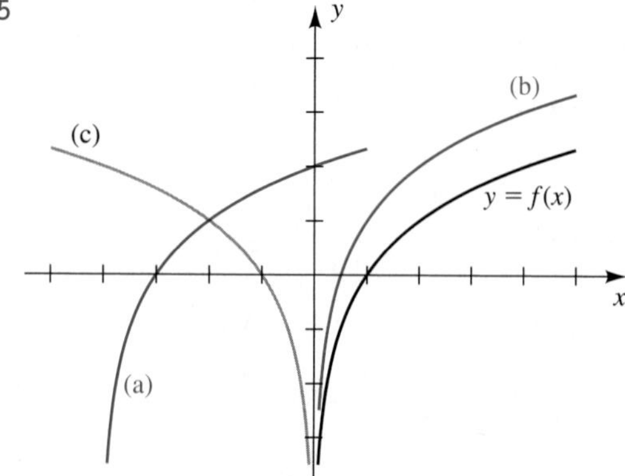

46

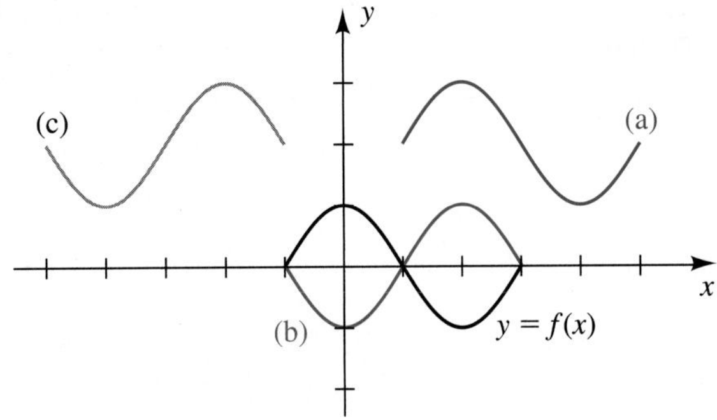

Exer. 47–52: Sketch the graph of *f*.

47 $f(x) = \begin{cases} 3 & \text{if } x \leq -1 \\ -2 & \text{if } x > -1 \end{cases}$

48 $f(x) = \begin{cases} -1 & \text{if } x \text{ is an integer} \\ -2 & \text{if } x \text{ is not an integer} \end{cases}$

49 $f(x) = \begin{cases} 3 & \text{if } x < -2 \\ -x + 1 & \text{if } |x| \leq 2 \\ -3 & \text{if } x > 2 \end{cases}$

50 $f(x) = \begin{cases} -2x & \text{if } x < -1 \\ x^2 & \text{if } -1 \leq x < 1 \\ -2 & \text{if } x \geq 1 \end{cases}$

51 $f(x) = \begin{cases} x + 2 & \text{if } x \leq -1 \\ x^3 & \text{if } |x| < 1 \\ -x + 3 & \text{if } x \geq 1 \end{cases}$

52 $f(x) = \begin{cases} x - 3 & \text{if } x \leq -2 \\ -x^2 & \text{if } -2 < x < 1 \\ -x + 4 & \text{if } x \geq 1 \end{cases}$

Exer. 53–54: The symbol $[\![x]\!]$ denotes values of the greatest integer function. Sketch the graph of *f*.

53 (a) $f(x) = [\![x - 3]\!]$ (b) $f(x) = [\![x]\!] - 3$
(c) $f(x) = 2[\![x]\!]$ (d) $f(x) = [\![2x]\!]$
(e) $f(x) = [\![-x]\!]$

54 (a) $f(x) = [\![x + 2]\!]$ (b) $f(x) = [\![x]\!] + 2$
(c) $f(x) = \frac{1}{2}[\![x]\!]$ (d) $f(x) = [\![\frac{1}{2}x]\!]$
(e) $f(x) = -[\![-x]\!]$

Exer. 55–56: Explain why the graph of the equation is not the graph of a function.

55 $x = y^2$

56 $x = -|y|$

Exer. 57–58: For the graph of $y = f(x)$ shown in the figure, sketch the graph of $y = |f(x)|$.

57

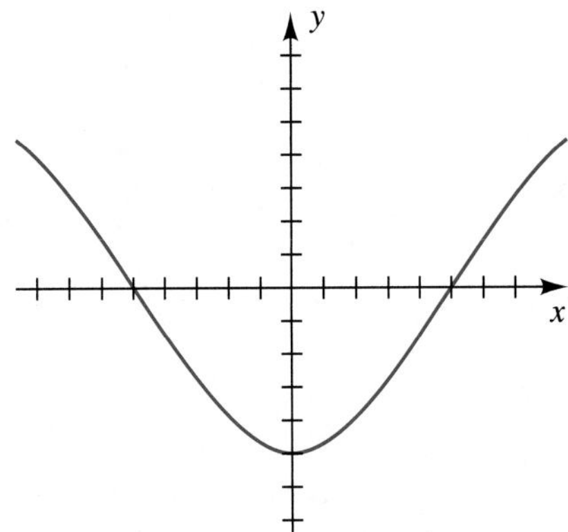

58

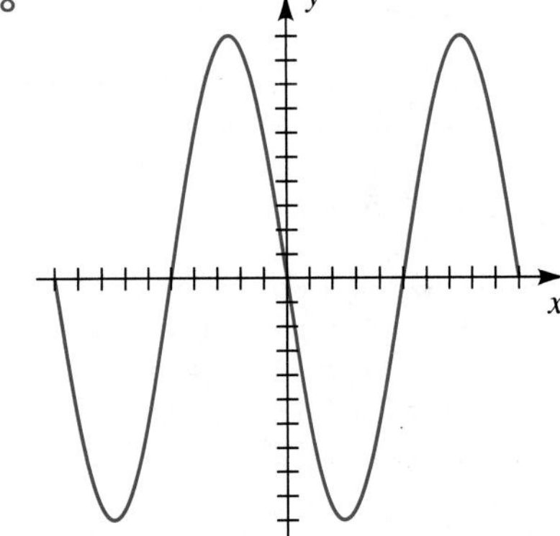

Exer. 59–62: Sketch the graph of the equation.

59 $y = |9 - x^2|$

60 $y = |x^3 - 1|$

61 $y = |\sqrt{x} - 1|$

62 $y = ||x| - 1|$

63 Let $y = f(x)$ be a function with domain $D = [-2, 6]$ and range $R = [-4, 8]$. Find the domain D and range R for each function. Assume $f(2) = 8$ and $f(6) = -4$.

(a) $y = -2f(x)$ (b) $y = f(\frac{1}{2}x)$

(c) $y = f(x - 3) + 1$ (d) $y = f(x + 2) - 3$

(e) $y = f(-x)$ (f) $y = -f(x)$

(g) $y = f(|x|)$ (h) $y = |f(x)|$

64 Let $y = f(x)$ be a function with domain $D = [-6, -2]$ and range $R = [-10, -4]$. Find the domain D and range R for each function.

▶ (a) $y = \frac{1}{2}f(x)$ ▶ (b) $y = f(2x)$

▶ (c) $y = f(x - 2) + 5$ (d) $y = f(x + 4) - 1$

(e) $y = f(-x)$ (f) $y = -f(x)$

(g) $y = f(|x|)$ (h) $y = |f(x)|$

65 Income tax rates A certain country taxes the first \$20,000 of an individual's income at a rate of 15%, and all income over \$20,000 is taxed at 20%. Find a piecewise-defined function T that specifies the total tax on an income of x dollars.

66 Property tax rates A certain state taxes the first \$500,000 in property value at a rate of 1%; all value over \$500,000 is taxed at 1.25%. Find a piecewise-defined function T that specifies the total tax on a property valued at x dollars.

67 Royalty rates A certain paperback sells for \$12. The author is paid royalties of 10% on the first 10,000 copies sold, 12.5% on the next 5000 copies, and 15% on any additional copies. Find a piecewise-defined function R that specifies the total royalties if x copies are sold.

68 Electricity rates An electric company charges its customers \$0.0577 per kilowatt-hour (kWh) for the first 1000 kWh used, \$0.0532 for the next 4000 kWh, and \$0.0511 for any kWh over 5000. Find a piecewise-defined function C for a customer's bill of x kWh.

Exer. 69–72: Estimate the solutions of the inequality.

69 $|1.3x + 2.8| < 1.2x + 5$

70 $|0.3x| - 2 > 2.2 - 0.63x^2$

71 $|1.2x^2 - 10.8| > 1.36x + 4.08$

72 $|\sqrt{16 - x^2} - 3| < 0.12x^2 - 0.3$

Exer. 73–78: Graph f in the viewing rectangle $[-12, 12]$ by $[-8, 8]$. Use the graph of f to predict the graph of g. Verify your prediction by graphing g in the same viewing rectangle.

73 $f(x) = 0.5x^3 - 4x - 5;$ $g(x) = 0.5x^3 - 4x - 1$

74 $f(x) = 0.25x^3 - 2x + 1;$ $g(x) = -0.25x^3 + 2x - 1$

75 $f(x) = x^2 - 5;$ $g(x) = \frac{1}{4}x^2 - 5$

76 $f(x) = |x + 2|;$ $g(x) = |x - 3| - 3$

77 $f(x) = x^3 - 5x;$ $g(x) = |x^3 - 5x|$

78 $f(x) = 0.5x^2 - 2x - 5;$ $g(x) = 0.5x^2 + 2x - 5$

79 Car rental charges There are two car rental options available for a four-day trip. Option I is \$45 per day, with 200 free miles and \$0.40 per mile for each additional mile. Option II is \$58.75 per day, with a charge of \$0.25 per mile.

(a) Determine the cost of a 500-mile trip for both options.

(b) Model the data with a cost function for each four-day option.

(c) Make a table that lists the mileage and the charge for each option for trips between 100 and 1200 miles, using increments of 100 miles.

(d) Use the table to determine the mileages at which each option is preferable.

80 Traffic flow Cars are crossing a bridge that is 1 mile long. Each car is 12 feet long and is required to stay a distance of at least d feet from the car in front of it (see figure).

(a) Show that the largest number of cars that can be on the bridge at one time is $[\![5280/(12 + d)]\!]$, where $[\![\ ]\!]$ denotes the greatest integer function.

(b) If the velocity of each car is v mi/hr, show that the maximum traffic flow rate F (in cars/hr) is given by $F = [\![5280v/(12 + d)]\!]$.

Exercise 80

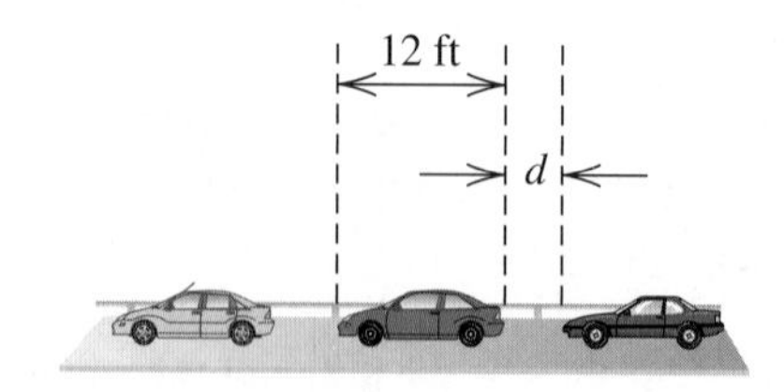

2.6 Quadratic Functions

If $a \neq 0$, then the graph of $y = ax^2$ is a parabola with vertex at the origin (0, 0), a vertical axis, opening upward if $a > 0$ or downward if $a < 0$ (see, for example, Figures 4 and 5 in Section 2.5). In this section we show that the graph of an equation of the form

$$y = ax^2 + bx + c$$

can be obtained by vertical and/or horizontal shifts of the graph of $y = ax^2$ and hence is also a parabola. An important application of such equations is to describe the trajectory, or path, of an object near the surface of the earth when the only force acting on the object is gravitational attraction. To illustrate, if an outfielder on a baseball team throws a ball into the infield, as illustrated in Figure 1, and if air resistance and other outside forces are negligible, then the path of the ball is a parabola. If suitable coordinate axes are introduced, then the path coincides with the graph of the equation $y = ax^2 + bx + c$ for some a, b, and c. We call the function determined by this equation a *quadratic function.*

Figure 1

Definition of Quadratic Function	A function f is a **quadratic function** if $$f(x) = ax^2 + bx + c,$$ where a, b, and c are real numbers with $a \neq 0$.

If $b = c = 0$ in the preceding definition, then $f(x) = ax^2$, and the graph is a parabola with vertex at the origin. If $b = 0$ and $c \neq 0$, then

$$f(x) = ax^2 + c,$$

and, from our discussion of vertical shifts in Section 2.5, the graph is a parabola with vertex at the point $(0, c)$ on the y-axis. The following example contains specific illustrations.

Figure 2

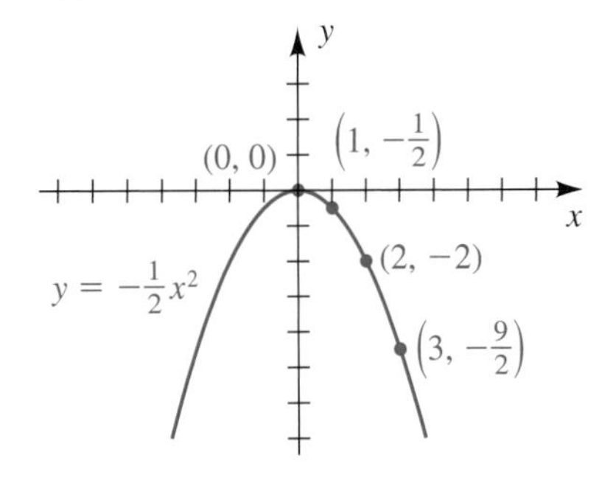

Figure 3

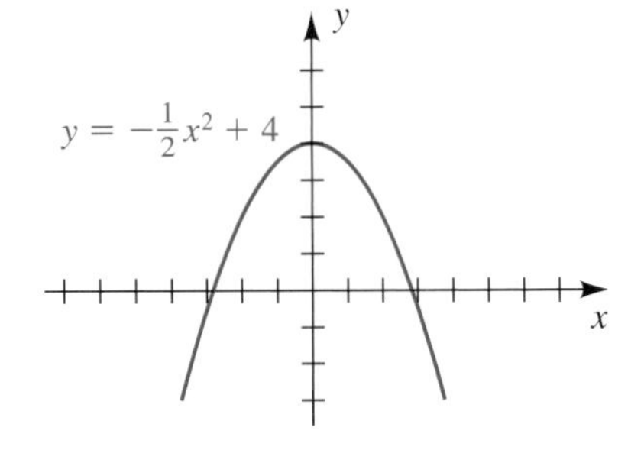

EXAMPLE 1 Sketching the graph of a quadratic function

Sketch the graph of f if

(a) $f(x) = -\frac{1}{2}x^2$ **(b)** $f(x) = -\frac{1}{2}x^2 + 4$

SOLUTION

(a) Since f is even, the graph of f $\left(\text{that is, of } y = -\frac{1}{2}x^2\right)$ is symmetric with respect to the y-axis. It is similar in shape to but wider than the parabola $y = -x^2$, sketched in Figure 5 of Section 2.5. Several points on the graph are $(0, 0)$, $\left(1, -\frac{1}{2}\right)$, $(2, -2)$, and $\left(3, -\frac{9}{2}\right)$. Plotting and using symmetry, we obtain the sketch in Figure 2.

(b) To find the graph of $y = -\frac{1}{2}x^2 + 4$, we shift the graph of $y = -\frac{1}{2}x^2$ upward a distance 4, obtaining the sketch in Figure 3.

If $f(x) = ax^2 + bx + c$ and $b \neq 0$, then, by completing the square, we can change the form to

$$f(x) = a(x - h)^2 + k$$

for some real numbers h and k. This technique is illustrated in the next example.

EXAMPLE 2 **Expressing a quadratic function as $f(x) = a(x - h)^2 + k$**

If $f(x) = 3x^2 + 24x + 50$, express $f(x)$ in the form $a(x - h)^2 + k$.

▶ **SOLUTION 1** Before completing the square, *it is essential that we factor out the coefficient of x^2 from the first two terms of $f(x)$,* as follows:

$$\begin{aligned} f(x) &= 3x^2 + 24x + 50 && \text{given} \\ &= 3(x^2 + 8x + \quad) + 50 && \text{factor out 3 from } 3x^2 + 24x \end{aligned}$$

We now complete the square for the expression $x^2 + 8x$ within the parentheses by adding the square of half the coefficient of x—that is, $\left(\frac{8}{2}\right)^2$, or 16. However, if we add 16 to the expression within parentheses, then, because of the factor 3, we are actually adding 48 to $f(x)$. Hence, we must compensate by subtracting 48:

$$\begin{aligned} f(x) &= 3(x^2 + 8x + \quad) + 50 && \text{given} \\ &= 3(x^2 + 8x + 16) + (50 - 48) && \text{complete the square for } x^2 + 8x \\ &= 3(x + 4)^2 + 2 && \text{equivalent equation} \end{aligned}$$

The last expression has the form $a(x - h)^2 + k$ with $a = 3$, $h = -4$, and $k = 2$.

SOLUTION 2 We begin by dividing both sides by the coefficient of x^2.

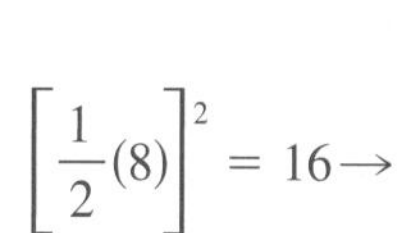

$$\begin{aligned} f(x) &= 3x^2 + 24x + 50 && \text{given} \\ \frac{f(x)}{3} &= x^2 + 8x + \frac{50}{3} && \text{divide by 3} \\ &= x^2 + 8x + \underline{16} + \frac{50}{3} \underline{-16} && \text{add and subtract 16, the number that completes the square for } x^2 + 8x \\ &= (x + 4)^2 + \frac{2}{3} && \text{equivalent equation} \\ f(x) &= 3(x + 4)^2 + 2 && \text{multiply by 3} \end{aligned}$$

If $f(x) = ax^2 + bx + c$, then, by completing the square as in Example 2, we see that the graph of f is the same as the graph of an equation of the form

$$y = a(x - h)^2 + k.$$

The graph of this equation can be obtained from the graph of $y = ax^2$ shown in Figure 4(a) by means of a horizontal and a vertical shift, as follows. First,

▶ **Tutorial available at www.thomsonedu.com/login**

as in Figure 4(b), we obtain the graph of $y = a(x - h)^2$ by shifting the graph of $y = ax^2$ either to the left or to the right, depending on the sign of h (the figure illustrates the case with $h > 0$). Next, as in Figure 4(c), we shift the graph in (b) vertically a distance $|k|$ (the figure illustrates the case with $k > 0$). It follows that *the graph of a quadratic function is a parabola with a vertical axis.*

Figure 4

(a)

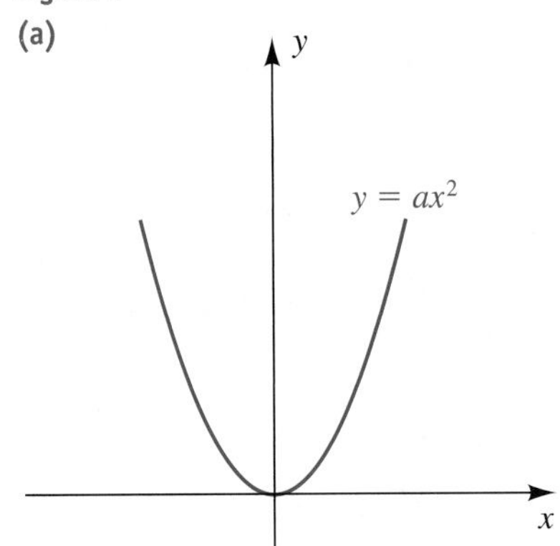

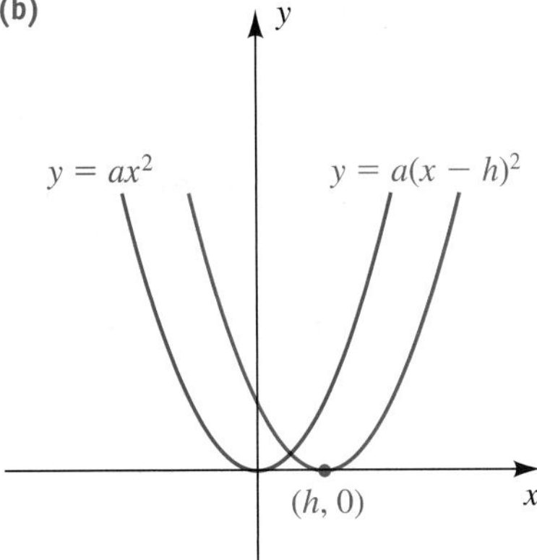

(c)

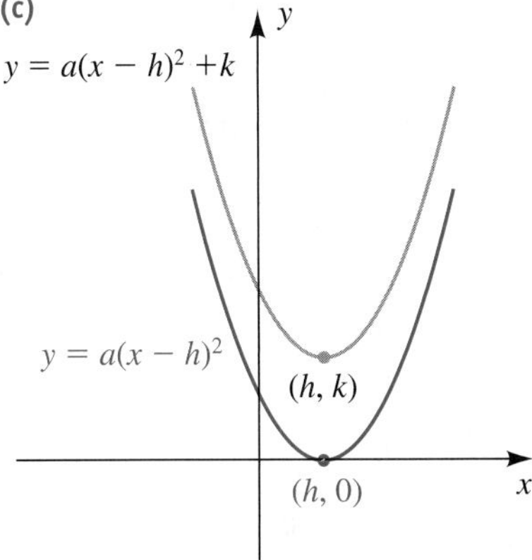

The sketch in Figure 4(c) illustrates one possible graph of the equation $y = ax^2 + bx + c$. If $a > 0$, the point (h, k) is the lowest point on the parabola, and the function f has a **minimum value** $f(h) = k$. If $a < 0$, the parabola opens downward, and the point (h, k) is the highest point on the parabola. In this case, the function f has a **maximum value** $f(h) = k$.

We have obtained the following result.

Standard Equation of a Parabola with Vertical Axis

The graph of the equation

$$y = a(x - h)^2 + k$$

for $a \neq 0$ is a parabola that has vertex $V(h, k)$ and a vertical axis. The parabola opens upward if $a > 0$ or downward if $a < 0$.

For convenience, we often refer to the *parabola* $y = ax^2 + bx + c$ when considering the graph of this equation.

EXAMPLE 3 **Finding a standard equation of a parabola**

Express $y = 2x^2 - 6x + 4$ as a standard equation of a parabola with a vertical axis. Find the vertex and sketch the graph.

Figure 5

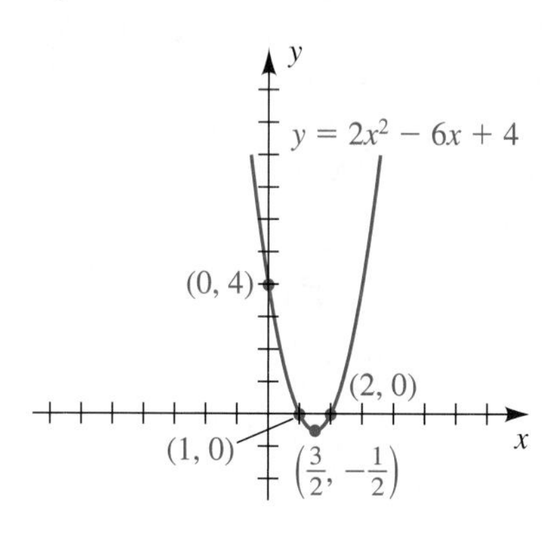

Figure 6

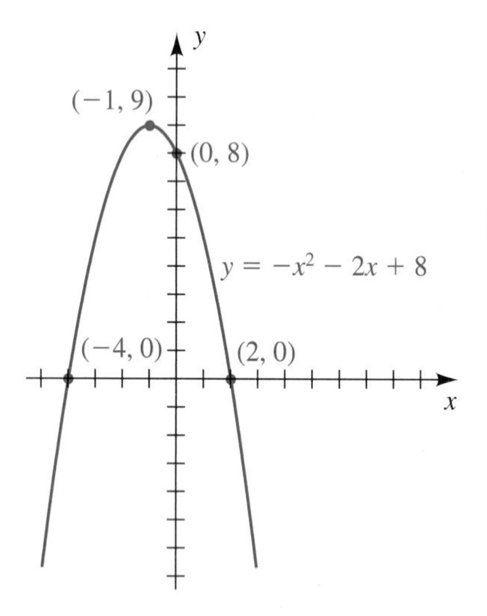

Figure 7

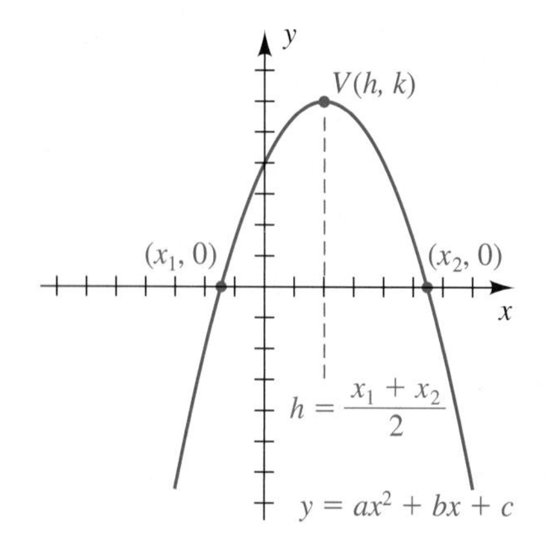

▶ **SOLUTION**

$$
\begin{aligned}
y &= 2x^2 - 6x + 4 && \text{given} \\
&= 2(x^2 - 3x + \quad) + 4 && \text{factor out 2 from } 2x^2 - 6x \\
&= 2\left(x^2 - 3x + \tfrac{9}{4}\right) + \left(4 - \tfrac{9}{2}\right) && \text{complete the square for } x^2 - 3x \\
&= 2\left(x - \tfrac{3}{2}\right)^2 - \tfrac{1}{2} && \text{equivalent equation}
\end{aligned}
$$

The last equation has the form of the standard equation of a parabola with $a = 2$, $h = \frac{3}{2}$, and $k = -\frac{1}{2}$. Hence, the vertex $V(h, k)$ of the parabola is $V\left(\frac{3}{2}, -\frac{1}{2}\right)$. Since $a = 2 > 0$, the parabola opens upward.

To find the y-intercept of the graph of $y = 2x^2 - 6x + 4$, we let $x = 0$, obtaining $y = 4$. To find the x-intercepts, we let $y = 0$ and solve the equation $2x^2 - 6x + 4 = 0$ or the equivalent equation $2(x - 1)(x - 2) = 0$, obtaining $x = 1$ and $x = 2$. Plotting the vertex and using the x- and y-intercepts provides enough points for a reasonably accurate sketch (see Figure 5).

EXAMPLE 4 Finding a standard equation of a parabola

Express $y = -x^2 - 2x + 8$ as a standard equation of a parabola with a vertical axis. Find the vertex and sketch the graph.

▶ **SOLUTION**

$$
\begin{aligned}
y &= -x^2 - 2x + 8 && \text{given} \\
&= -(x^2 + 2x + \quad) + 8 && \text{factor out } -1 \text{ from } -x^2 - 2x \\
&= -(x^2 + 2x + 1) + (8 + 1) && \text{complete the square for } x^2 + 2x \\
&= -(x + 1)^2 + 9 && \text{equivalent equation}
\end{aligned}
$$

This is the standard equation of a parabola with $h = -1$, $k = 9$, and hence the vertex is $(-1, 9)$. Since $a = -1 < 0$, the parabola opens downward.

The y-intercept of the graph of $y = -x^2 - 2x + 8$ is the constant term, 8. To find the x-intercepts, we solve $-x^2 - 2x + 8 = 0$ or, equivalently, $x^2 + 2x - 8 = 0$. Factoring gives us $(x + 4)(x - 2) = 0$, and hence the intercepts are $x = -4$ and $x = 2$. Using this information gives us the sketch in Figure 6.

If a parabola $y = ax^2 + bx + c$ has x-intercepts x_1 and x_2, as illustrated in Figure 7 for the case $a < 0$, then the axis of the parabola is the vertical line $x = (x_1 + x_2)/2$ through the midpoint of $(x_1, 0)$ and $(x_2, 0)$. Therefore, the x-coordinate h of the vertex (h, k) is $h = (x_1 + x_2)/2$. Some special cases are illustrated in Figures 5 and 6.

In the following example we find an equation of a parabola from given data.

EXAMPLE 5 Finding an equation of a parabola with a given vertex

Find an equation of a parabola that has vertex $V(2, 3)$ and a vertical axis and passes through the point $(5, 1)$.

Figure 8

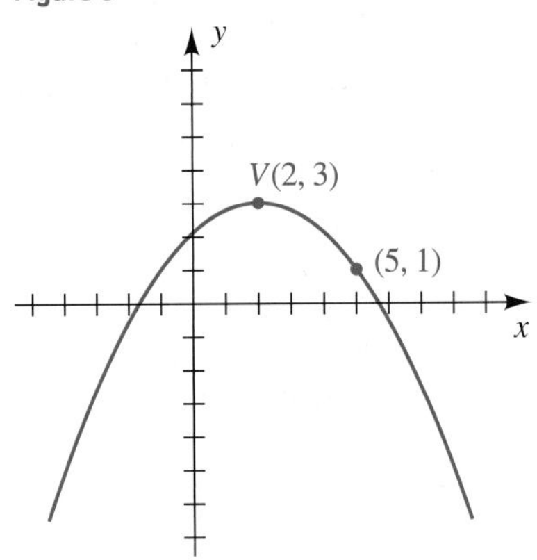

▶ SOLUTION Figure 8 shows the vertex V, the point (5, 1), and a possible position of the parabola. Using the standard equation

$$y = a(x - h)^2 + k$$

with $h = 2$ and $k = 3$ gives us

$$y = a(x - 2)^2 + 3.$$

To find a, we use the fact that (5, 1) is on the parabola and so is a solution of the last equation. Thus,

$$1 = a(5 - 2)^2 + 3, \qquad \text{or} \qquad a = -\tfrac{2}{9}.$$

Hence, an equation for the parabola is

$$y = -\tfrac{2}{9}(x - 2)^2 + 3.$$

The next theorem gives us a simple formula for locating the vertex of a parabola.

Theorem for Locating the Vertex of a Parabola

The vertex of the parabola $y = ax^2 + bx + c$ has x-coordinate

$$-\frac{b}{2a}.$$

PROOF Let us begin by writing $y = ax^2 + bx + c$ as

$$y = a\left(x^2 + \frac{b}{a}x + \quad\right) + c.$$

Next we complete the square by adding $\left(\frac{1}{2}\frac{b}{a}\right)^2$ to the expression within parentheses:

$$y = a\left(x^2 + \frac{b}{a}x + \frac{b^2}{4a^2}\right) + \left(c - \frac{b^2}{4a}\right)$$

Note that if $b^2/(4a^2)$ is added *inside* the parentheses, then, because of the factor a on the *outside*, we have actually added $b^2/(4a)$ to y. Therefore, we must compensate by subtracting $b^2/(4a)$. The last equation may be written

$$y = a\left(x + \frac{b}{2a}\right)^2 + \left(c - \frac{b^2}{4a}\right).$$

This is the equation of a parabola that has vertex (h, k) with $h = -b/(2a)$ and $k = c - b^2/(4a)$.

▶ **Tutorial available at www.thomsonedu.com/login**

It is unnecessary to remember the formula for the y-coordinate of the vertex of the parabola in the preceding result. Once the x-coordinate has been found, we can calculate the y-coordinate by substituting $-b/(2a)$ for x in the equation of the parabola.

EXAMPLE 6 Finding the vertex of a parabola

Find the vertex of the parabola $y = 2x^2 - 6x + 4$.

▶ SOLUTION We considered this parabola in Example 3 and found the vertex by completing the square. We shall use the vertex formula with $a = 2$ and $b = -6$, obtaining the x-coordinate

$$\frac{-b}{2a} = \frac{-(-6)}{2(2)} = \frac{6}{4} = \frac{3}{2}.$$

We next find the y-coordinate by substituting $\frac{3}{2}$ for x in the given equation:

$$y = 2\left(\tfrac{3}{2}\right)^2 - 6\left(\tfrac{3}{2}\right) + 4 = -\tfrac{1}{2}$$

Thus, the vertex is $\left(\frac{3}{2}, -\frac{1}{2}\right)$ (see Figure 5).

Since the graph of $f(x) = ax^2 + bx + c$ for $a \neq 0$ is a parabola, we can use the vertex formula to help find the maximum or minimum value of a quadratic function. Specifically, since the x-coordinate of the vertex V is $-b/(2a)$, the y-coordinate of V is the function value $f(-b/(2a))$. Moreover, since the parabola opens downward if $a < 0$ and upward if $a > 0$, this function value is the maximum or minimum value, respectively, of f. We may summarize these facts as follows.

Theorem on the Maximum or Minimum Value of a Quadratic Function

If $f(x) = ax^2 + bx + c$, where $a \neq 0$, then $f\left(-\dfrac{b}{2a}\right)$ is

(1) the maximum value of f if $a < 0$

(2) the minimum value of f if $a > 0$

We shall use this theorem in the next two examples.

EXAMPLE 7 Finding a maximum (or minimum) value

Find the vertex of the parabola $y = f(x) = -2x^2 - 12x - 13$.

SOLUTION Since the coefficient of x^2 is -2 and $-2 < 0$, the parabola opens downward and the y-value of the vertex is a *maximum* value. We assign $-2x^2 - 12x - 13$ to Y_1 and graph Y_1 in a standard viewing rectangle.

▶ Tutorial available at www.thomsonedu.com/login

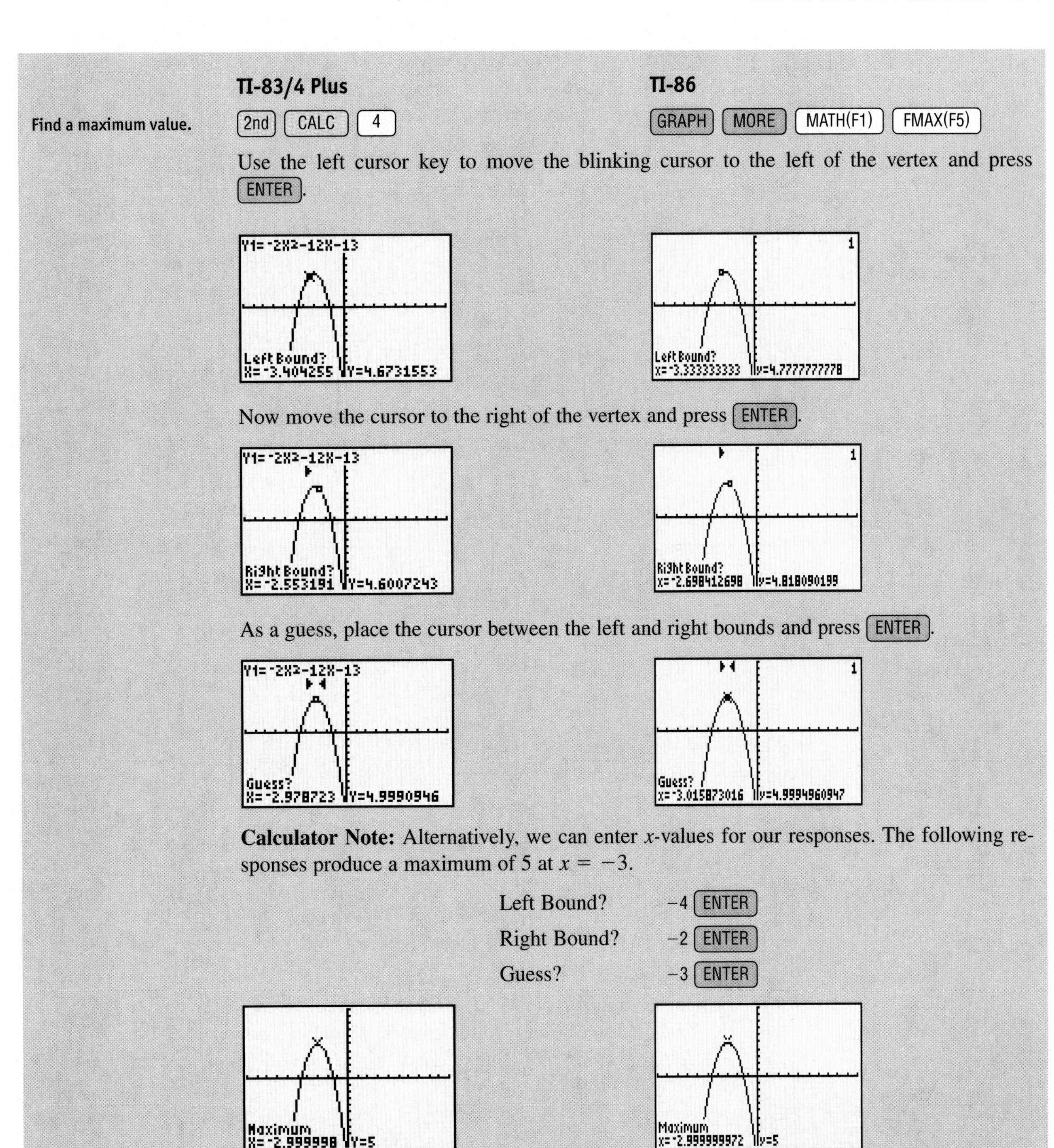

	TI-83/4 Plus	**TI-86**
Find a maximum value.	2nd CALC 4	GRAPH MORE MATH(F1) FMAX(F5)

Use the left cursor key to move the blinking cursor to the left of the vertex and press ENTER.

Now move the cursor to the right of the vertex and press ENTER.

As a guess, place the cursor between the left and right bounds and press ENTER.

Calculator Note: Alternatively, we can enter x-values for our responses. The following responses produce a maximum of 5 at $x = -3$.

Left Bound?	-4 ENTER
Right Bound?	-2 ENTER
Guess?	-3 ENTER

The calculator indicates that the vertex is about $(-3, 5)$. (You may get different results depending on your cursor placements.)

(continued)

We can also find a maximum value from the home screen as follows. (Assume we have looked at the graph and estimated that the x-coordinate of the vertex lies between -3.5 and -2.5.) First we find the x-value of the vertex.

Use the function maximum feature.

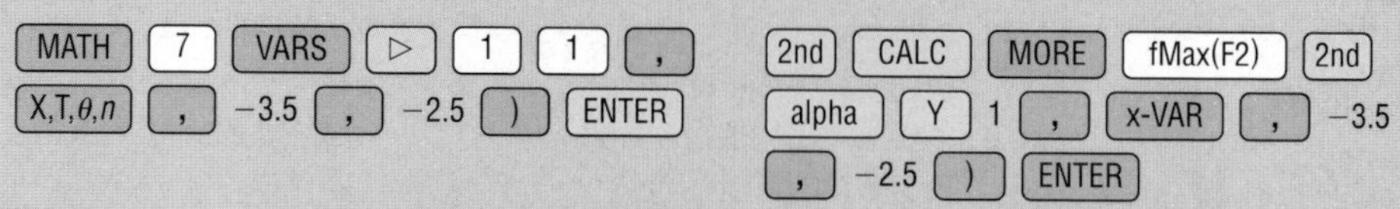

Next we find the y-value of the vertex using the result from fMax (it's stored in ANS).

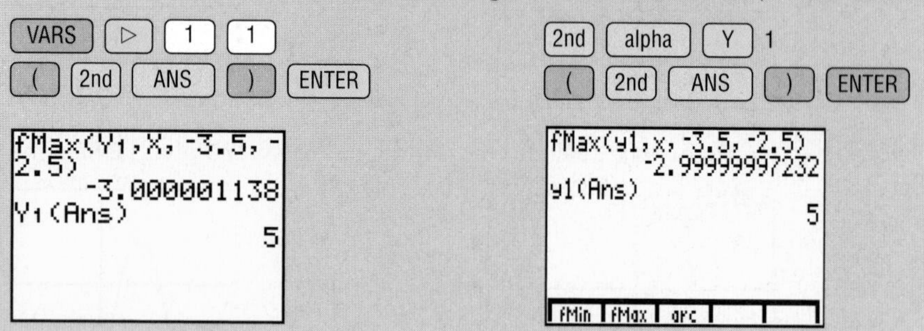

Notice the "strange" results given for fMax. (Your professor will not be too impressed if you say that the vertex is $(-3.000001138, 5)$.) In this case a calculator is helpful, but it is easy to calculate that

$$-\frac{b}{2a} = -\frac{-12}{2(-2)} = -3 \quad \text{and} \quad f(-3) = 5,$$

which gives us a vertex of $(-3, 5)$ (and an answer that will please your professor).

EXAMPLE 8 Finding the maximum value of a quadratic function

A long rectangular sheet of metal, 12 inches wide, is to be made into a rain gutter by turning up two sides so that they are perpendicular to the sheet. How many inches should be turned up to give the gutter its greatest capacity?

Figure 9

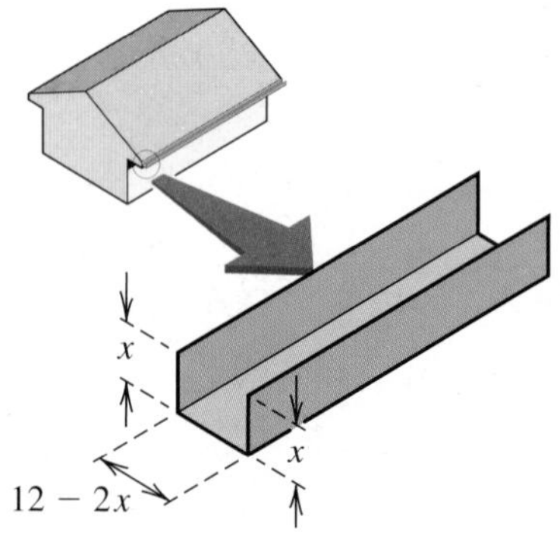

▶ **SOLUTION** The gutter is illustrated in Figure 9. If x denotes the number of inches turned up on each side, the width of the base of the gutter is $12 - 2x$ inches. The capacity will be greatest when the cross-sectional area of the rectangle with sides of lengths x and $12 - 2x$ has its greatest value. Letting $f(x)$ denote this area, we have

$$\begin{aligned} f(x) &= x(12 - 2x) \\ &= 12x - 2x^2 \\ &= -2x^2 + 12x, \end{aligned}$$

which has the form $f(x) = ax^2 + bx + c$ with $a = -2$, $b = 12$, and $c = 0$. Since f is a quadratic function and $a = -2 < 0$, it follows from the preceding

▶ Tutorial available at www.thomsonedu.com/login

theorem that the maximum value of f occurs at

$$x = -\frac{b}{2a} = -\frac{12}{2(-2)} = 3.$$

Thus, 3 inches should be turned up on each side to achieve maximum capacity.

As an alternative solution, we may note that the graph of the function $f(x) = x(12 - 2x)$ has x-intercepts at $x = 0$ and $x = 6$. Hence, the average of the intercepts,

$$x = \frac{0 + 6}{2} = 3,$$

is the x-coordinate of the vertex of the parabola and the value that yields the maximum capacity.

In Chapter 1 we solved quadratic equations and inequalities algebraically. The next example indicates how they can be solved with the aid of a graphing calculator.

EXAMPLE 9 Analyzing the flight of a projectile

A projectile is fired vertically upward from a height of 600 feet above the ground. Its height $h(t)$ in feet above the ground after t seconds is given by

$$h(t) = -16t^2 + 803t + 600.$$

(a) Determine a reasonable viewing rectangle that includes all pertinent features of the graph of h.

(b) Estimate when the height of the projectile is 5000 feet above the ground.

(c) Determine when the projectile will be more than 5000 feet above the ground.

(d) How long will the projectile be in flight?

SOLUTION

(a) The graph of h is a parabola that opens downward. To estimate Ymax (note that we use x and y interchangeably with t and h), let us approximate the maximum value of h. Using

$$t = -\frac{b}{2a} = -\frac{803}{2(-16)} \approx 25.1,$$

we see that the maximum height is approximately $h(25) = 10{,}675$.

The projectile rises for approximately the first 25 seconds, and because its height at $t = 0$, 600 feet, is small in comparison to 10,675, it will take only slightly more than an additional 25 seconds to fall to the ground. Since h and t are positive, a reasonable viewing rectangle is

$$[0, 60, 5] \quad \text{by} \quad [0, 11{,}000, 1000].$$

(continued)

Calculator Note: Once we determine the Xmin and Xmax values, we can use the ZoomFit feature to graph a function over the interval [Xmin, Xmax]. In this example, assign 0 to Xmin and 51 to Xmax and then select ZoomFit under the ZOOM menu.

(b) We wish to estimate where the graph of h intersects the horizontal line $h(t) = 5000$, so we make the assignments

$$Y_1 = -16x^2 + 803x + 600 \quad \text{and} \quad Y_2 = 5000$$

Figure 10
[0, 60, 5] by [0, 11,000, 1000]

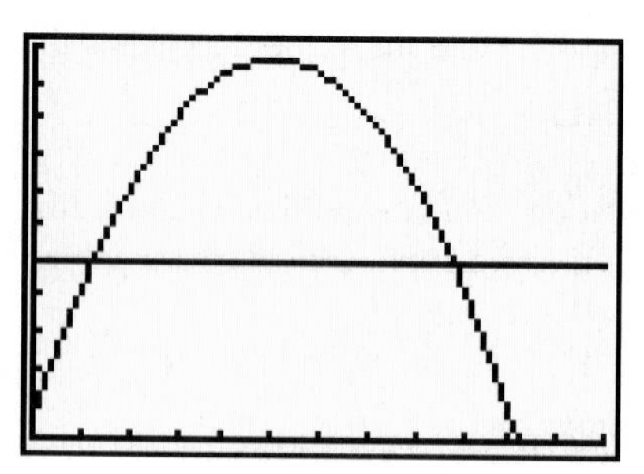

and obtain a display similar to Figure 10. It is important to remember that the graph of Y_1 shows only the height at time t—it is *not* the path of the projectile, which is vertical. Using an intersect feature, we find that the smallest value of t for which $h(t) = 5000$ is about 6.3 seconds.

Since the vertex is on the axis of the parabola, the other time at which $h(t)$ is 5000 is approximately $25.1 - 6.3$, or 18.8, seconds *after* $t = 25.1$—that is, at $t \approx 25.1 + 18.8 = 43.9$ sec.

(c) The projectile is more than 5000 feet above the ground when the graph of the parabola in Figure 10 is above the horizontal line—that is, when

$$6.3 < t < 43.9.$$

(d) The projectile will be in flight until $h(t) = 0$. This corresponds to the x-intercept in Figure 10. Using a root or zero feature, we obtain $t \approx 50.9$ sec. (Note that since the y-intercept is not zero, it is incorrect to merely double the t value of the vertex to find the total time of the flight; however, this *would* be acceptable for problems in which $h(0) = 0$.)

When working with quadratic functions, we are often most interested in finding the vertex and the x-intercepts. Typically, a given quadratic function closely resembles one of the three forms listed in the following chart.

Relationship Between Quadratic Function Forms and Their Vertex and x-intercepts

Form	Vertex (h, k)	x-intercepts (if there are any)
(1) $y = f(x) = a(x - h)^2 + k$	h and k as in the form	$x = h \pm \sqrt{-k/a}$ (see below)
(2) $y = f(x) = a(x - x_1)(x - x_2)$	$h = \dfrac{x_1 + x_2}{2}$, $k = f(h)$	$x = x_1, x_2$
(3) $y = f(x) = ax^2 + bx + c$	$h = -\dfrac{b}{2a}$, $k = f(h)$	$x = -\dfrac{b}{2a} \pm \dfrac{\sqrt{b^2 - 4ac}}{2a}$ (see below)

If the radicands in (1) or (3) are negative, then there are no x-intercepts. To find the x-intercepts with form (1), use the special quadratic equation on page 53. If you have a quadratic function in form (3) and want to find the vertex and the x-intercepts, it may be best to first find the x-intercepts by using the quadratic formula. Then you can easily obtain the x-coordinate of the vertex, h, since

$$-\frac{b}{2a} \pm \frac{\sqrt{b^2 - 4ac}}{2a} = h \pm \frac{\sqrt{b^2 - 4ac}}{2a}.$$

Of course, if the function in form (3) is easily factorable, it is not necessary to use the quadratic formula.

We will discuss parabolas further in a later chapter.

2.6 *Exercises*

Exer. 1–4: Find the standard equation of any parabola that has vertex V.

▶ 1 $V(-3, 1)$ 2 $V(4, -2)$

3 $V(0, -3)$ 4 $V(-2, 0)$

Exer. 5–12: Express $f(x)$ in the form $a(x - h)^2 + k$.

5 $f(x) = -x^2 - 4x - 8$ 6 $f(x) = x^2 - 6x + 11$

7 $f(x) = 2x^2 - 12x + 22$ 8 $f(x) = 5x^2 + 20x + 17$

9 $f(x) = -3x^2 - 6x - 5$

▶ 10 $f(x) = -4x^2 + 16x - 13$

11 $f(x) = -\frac{3}{4}x^2 + 9x - 34$ 12 $f(x) = \frac{2}{5}x^2 - \frac{12}{5}x + \frac{23}{5}$

Exer. 13–22: (a) Use the quadratic formula to find the zeros of f. (b) Find the maximum or minimum value of $f(x)$. (c) Sketch the graph of f.

13 $f(x) = x^2 - 4x$ 14 $f(x) = -x^2 - 6x$

15 $f(x) = -12x^2 + 11x + 15$

16 $f(x) = 6x^2 + 7x - 24$

17 $f(x) = 9x^2 + 24x + 16$ 18 $f(x) = -4x^2 + 4x - 1$

19 $f(x) = x^2 + 4x + 9$ 20 $f(x) = -3x^2 - 6x - 6$

21 $f(x) = -2x^2 + 20x - 43$

22 $f(x) = 2x^2 - 4x - 11$

Exer. 23–26: Find the standard equation of the parabola shown in the figure.

 23

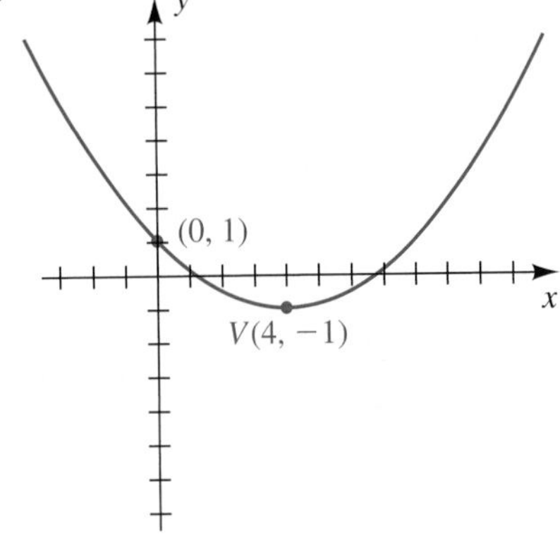

▶ 24

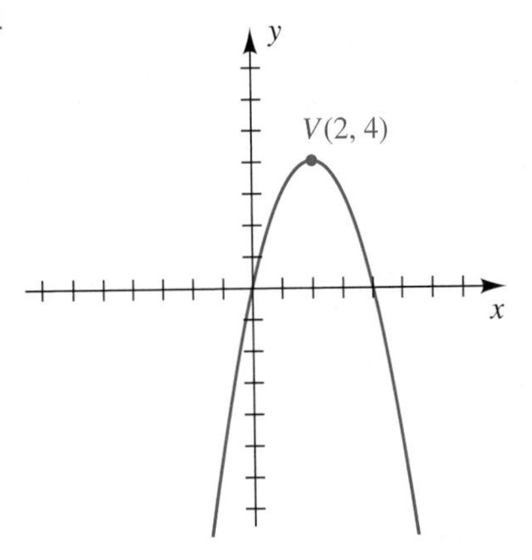

▶ Tutorial available at www.thomsonedu.com/login

25

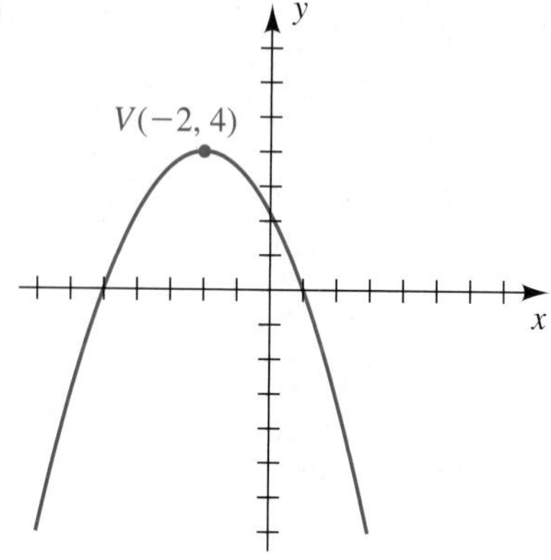

26

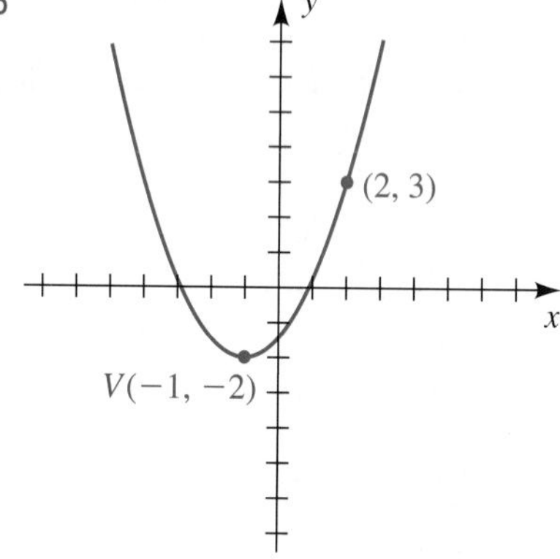

Exer. 27–28: Find an equation of the form

$$y = a(x - x_1)(x - x_2)$$

of the parabola shown in the figure. See the chart on page 182.

27

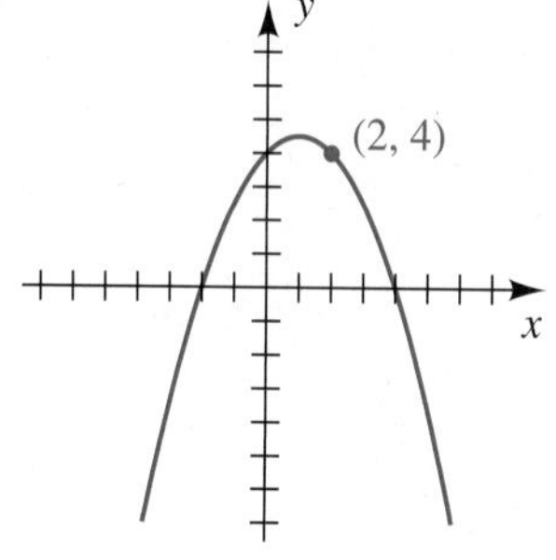

28

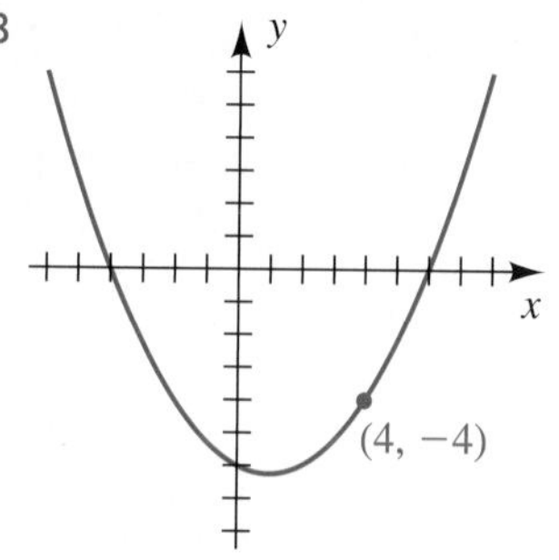

Exer. 29–34: Find the standard equation of a parabola that has a vertical axis and satisfies the given conditions.

29 Vertex $(0, -2)$, passing through $(3, 25)$

30 Vertex $(0, 5)$, passing through $(2, -3)$

31 Vertex $(3, 5)$, x-intercept 0

32 Vertex $(4, -7)$, x-intercept -4

33 x-intercepts -3 and 5, highest point has y-coordinate 4

34 x-intercepts 8 and 0, lowest point has y-coordinate -48

Exer. 35–36: Find the maximum vertical distance *d* between the parabola and the line for the green region.

35

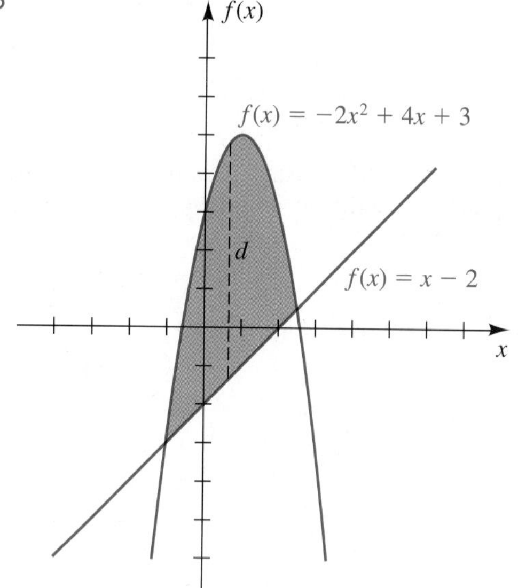

36

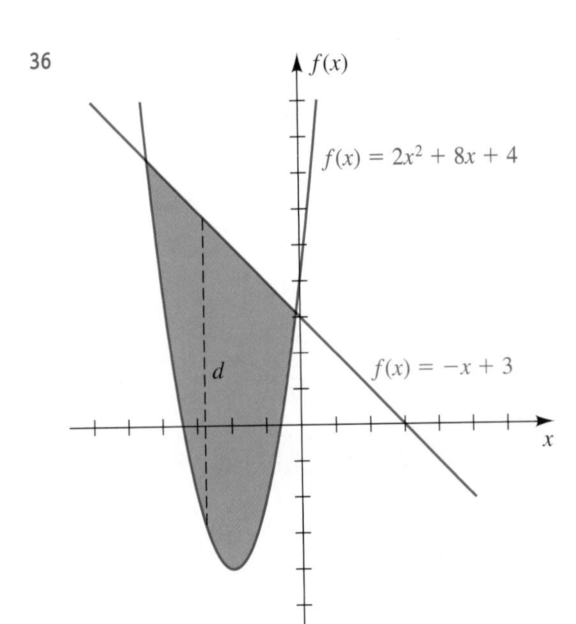

Exer. 37–38: Ozone occurs at all levels of the earth's atmosphere. The density of ozone varies both seasonally and latitudinally. At Edmonton, Canada, the density $D(h)$ of ozone (in 10^{-3} cm/km) for altitudes h between 20 kilometers and 35 kilometers was determined experimentally. For each $D(h)$ and season, approximate the altitude at which the density of ozone is greatest.

37 $D(h) = -0.058h^2 + 2.867h - 24.239$ (autumn)

38 $D(h) = -0.078h^2 + 3.811h - 32.433$ (spring)

39 **Infant growth rate** The growth rate y (in pounds per month) of an infant is related to present weight x (in pounds) by the formula $y = cx(21 - x)$, where c is a positive constant and $0 < x < 21$. At what weight does the maximum growth rate occur?

40 **Gasoline mileage** The number of miles M that a certain automobile can travel on one gallon of gasoline at a speed of v mi/hr is given by

$$M = -\tfrac{1}{30}v^2 + \tfrac{5}{2}v \qquad \text{for } 0 < v < 70.$$

(a) Find the most economical speed for a trip.

(b) Find the largest value of M.

41 **Height of a projectile** An object is projected vertically upward from the top of a building with an initial velocity of 144 ft/sec. Its distance $s(t)$ in feet above the ground after t seconds is given by the equation

$$s(t) = -16t^2 + 144t + 100.$$

(a) Find its maximum distance above the ground.

(b) Find the height of the building.

42 **Flight of a projectile** An object is projected vertically upward with an initial velocity of v_0 ft/sec, and its distance $s(t)$ in feet above the ground after t seconds is given by the formula $s(t) = -16t^2 + v_0t$.

(a) If the object hits the ground after 12 seconds, find its initial velocity v_0.

(b) Find its maximum distance above the ground.

43 Find two positive real numbers whose sum is 40 and whose product is a maximum.

44 Find two real numbers whose difference is 40 and whose product is a minimum.

45 **Constructing cages** One thousand feet of chain-link fence is to be used to construct six animal cages, as shown in the figure.

(a) Express the width y as a function of the length x.

(b) Express the total enclosed area A of the cages as a function of x.

(c) Find the dimensions that maximize the enclosed area.

Exercise 45

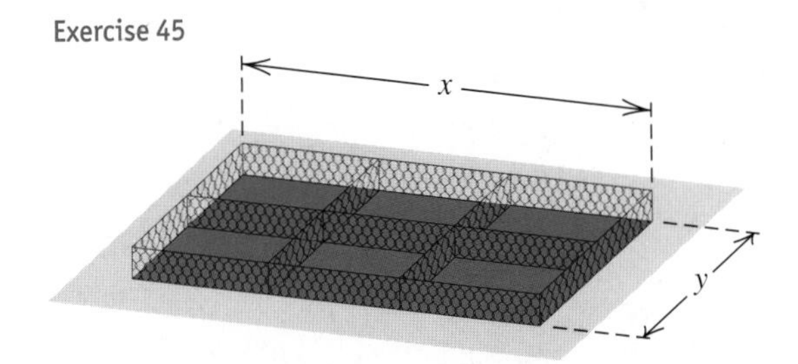

46 **Fencing a field** A farmer wishes to put a fence around a rectangular field and then divide the field into three rectangular plots by placing two fences parallel to one of the sides. If the farmer can afford only 1000 yards of fencing, what dimensions will give the maximum rectangular area?

47 **Leaping animals** Flights of leaping animals typically have parabolic paths. The figure on the next page illustrates a frog jump superimposed on a coordinate plane. The length of the leap is 9 feet, and the maximum height off the ground is 3 feet. Find a standard equation for the path of the frog.

Exercise 47

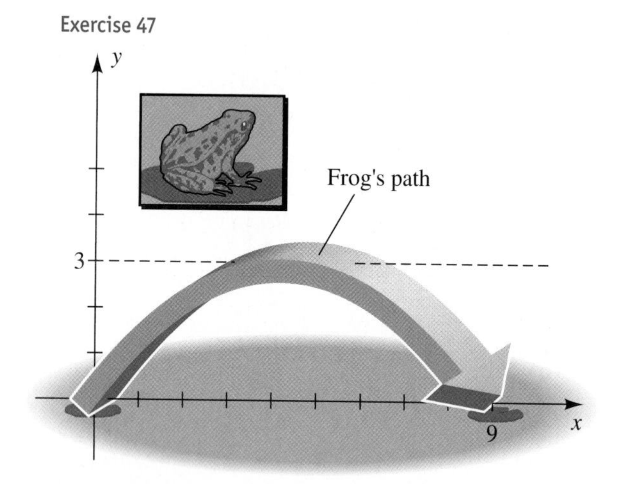

48 The human cannonball In the 1940s, the human cannonball stunt was performed regularly by Emmanuel Zacchini for The Ringling Brothers and Barnum & Bailey Circus. The tip of the cannon rose 15 feet off the ground, and the total horizontal distance traveled was 175 feet. When the cannon is aimed at an angle of 45°, an equation of the parabolic flight (see the figure) has the form $y = ax^2 + x + c$.

(a) Use the given information to find an equation of the flight.

(b) Find the maximum height attained by the human cannonball.

Exercise 48

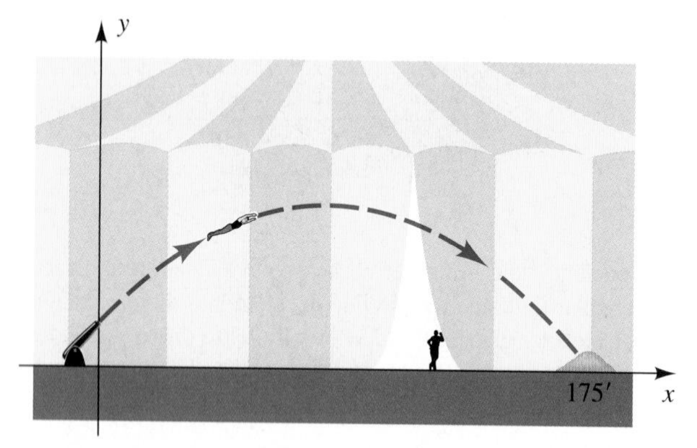

49 Shape of a suspension bridge One section of a suspension bridge has its weight uniformly distributed between twin towers that are 400 feet apart and rise 90 feet above the horizontal roadway (see the figure). A cable strung between the tops of the towers has the shape of a parabola, and its center point is 10 feet above the roadway. Suppose coordinate axes are introduced, as shown in the figure.

Exercise 49

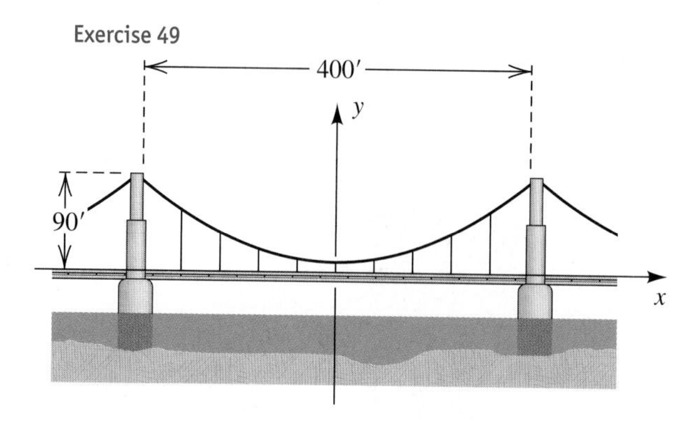

(a) Find an equation for the parabola.

(b) Nine equally spaced vertical cables are used to support the bridge (see the figure). Find the total length of these supports.

50 Designing a highway Traffic engineers are designing a stretch of highway that will connect a horizontal highway with one having a 20% grade (that is, slope $\frac{1}{5}$), as illustrated in the figure. The smooth transition is to take place over a horizontal distance of 800 feet, with a parabolic piece of highway used to connect points A and B. If the equation of the parabolic segment is of the form $y = ax^2 + bx + c$, it can be shown that the slope of the tangent line at the point $P(x, y)$ on the parabola is given by $m = 2ax + b$.

(a) Find an equation of the parabola that has a tangent line of slope 0 at A and $\frac{1}{5}$ at B.

(b) Find the coordinates of B.

Exercise 50

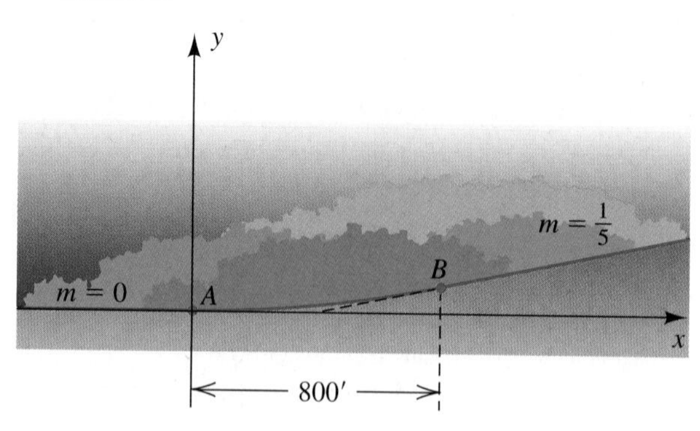

51 **Parabolic doorway** A doorway has the shape of a parabolic arch and is 9 feet high at the center and 6 feet wide at the base. If a rectangular box 8 feet high must fit through the doorway, what is the maximum width the box can have?

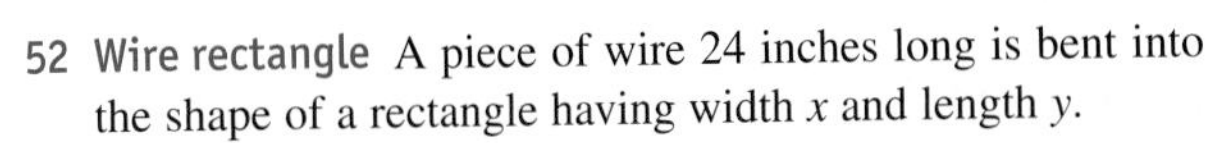

52 **Wire rectangle** A piece of wire 24 inches long is bent into the shape of a rectangle having width x and length y.

(a) Express y as a function of x.

(b) Express the area A of the rectangle as a function of x.

(c) Show that the area A is greatest if the rectangle is a square.

53 **Quantity discount** A company sells running shoes to dealers at a rate of \$40 per pair if fewer than 50 pairs are ordered. If a dealer orders 50 or more pairs (up to 600), the price per pair is reduced at a rate of 4 cents times the number ordered. What size order will produce the maximum amount of money for the company?

54 **Group discount** A travel agency offers group tours at a rate of \$60 per person for the first 30 participants. For larger groups—up to 90—each person receives a \$0.50 discount for every participant in excess of 30. For example, if 31 people participate, then the cost per person is \$59.50. Determine the size of the group that will produce the maximum amount of money for the agency.

55 **Cable TV fee** A cable television firm presently serves 8000 households and charges \$50 per month. A marketing survey indicates that each decrease of \$5 in the monthly charge will result in 1000 new customers. Let $R(x)$ denote the total monthly revenue when the monthly charge is x dollars.

(a) Determine the revenue function R.

(b) Sketch the graph of R and find the value of x that results in maximum monthly revenue.

56 **Apartment rentals** A real estate company owns 218 efficiency apartments, which are fully occupied when the rent is \$940 per month. The company estimates that for each \$25 increase in rent, 5 apartments will become unoccupied. What rent should be charged so that the company will receive the maximum monthly income?

Exer. 57–58: Graph $y = x^3 - x^{1/3}$ and f on the same coordinate plane, and estimate the points of intersection.

57 $f(x) = x^2 - x - \frac{1}{4}$

58 $f(x) = -x^2 + 0.5x + 0.4$

59 Graph, on the same coordinate plane, $y = ax^2 + x + 1$ for $a = \frac{1}{4}, \frac{1}{2}$, 1, 2, and 4, and describe how the value of a affects the graph.

60 Graph, on the same coordinate plane, $y = x^2 + bx + 1$ for $b = 0, \pm 1, \pm 2$, and ± 3, and describe how the value of b affects the graph.

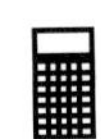

61 **Precipitation in Seattle** The average monthly precipitation (in inches) for Seattle is listed in the following table. (*Note:* April average is not given.)

(a) Plot the average monthly precipitation.

(b) Model the data with a quadratic function of the form $f(x) = a(x - h)^2 + k$. Graph f and the data on the same coordinate axes.

(c) Use f to predict the average rainfall in April. Compare your prediction with the actual value of 2.55 inches.

Month	Precipitation
Jan.	5.79
Feb.	4.02
Mar.	3.71
April	
May	1.70
June	1.46
July	0.77
Aug.	1.10
Sept.	1.72
Oct.	3.50
Nov.	5.97
Dec.	5.81

62 Handgun homicides The annual numbers of handgun homicides (in thousands) from 1982 to 1993 are listed in the table. (After this period, the number of handgun homicides decreased and leveled off to values similar to those in the mid-1980s.)

Year	Homicides
1982	8.3
1983	8.0
1984	7.6
1985	7.9
1986	8.3
1987	8.0
1988	8.3
1989	9.2
1990	10.0
1991	11.6
1992	12.5
1993	13.3

(a) Plot the data. Discuss any overall trends in the data.

(b) Model these data with a quadratic function of the form $f(x) = a(x - h)^2 + k$.

(c) Graph f together with the data.

63 Crest vertical curves When engineers plan highways, they must design hills so as to ensure proper vision for drivers. Hills are referred to as *crest vertical curves.* Crest vertical curves change the slope of a highway. Engineers use a parabolic shape for a highway hill, with the vertex located at the top of the crest. Two roadways with different slopes are to be connected with a parabolic crest curve. The highway passes through the points $A(-800, -48)$, $B(-500, 0)$, $C(0, 40)$, $D(500, 0)$, and $E(800, -48)$, as shown in the figure. The roadway is linear between A and B, parabolic between B and D, and then linear between D and E.

Exercise 63

(a) Find a piecewise-defined function f that models the roadway between the points A and E.

(b) Graph f in the viewing rectangle $[-800, 800, 100]$ by $[-100, 200, 100]$.

64 Sag vertical curves Refer to Exercise 63. Valleys or dips in highways are referred to as *sag vertical curves.* Sag vertical curves are also modeled using parabolas. Two roadways with different grades meeting at a sag curve need to be connected. The highway passes through the points $A\left(-500, 243\frac{1}{3}\right)$, $B(0, 110)$, $C(750, 10)$, $D(1500, 110)$, and $E\left(2000, 243\frac{1}{3}\right)$, as shown in the figure. The roadway is linear between A and B, parabolic between B and D, and linear between D and E.

Exercise 64

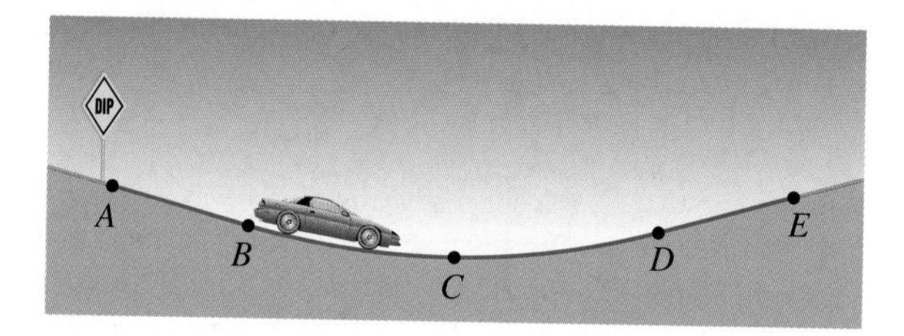

(a) Find a piecewise-defined function f that models the roadway between the points A and E.

(b) Graph f in the viewing rectangle $[-500, 2000, 500]$ by $[0, 800, 100]$.

65 Parabolic path Under ideal conditions an object thrown from level ground will follow a parabolic path of the form $f(x) = ax^2 + bx$, where a and b are constants and x represents the horizontal distance traveled by the object.

(a) Determine a and b so that the object reaches a maximum height of 100 feet and travels a horizontal distance of 150 feet before striking the ground.

(b) Graph $f(x) = ax^2 + bx$ in the viewing rectangle $[0, 180, 50]$ by $[0, 120, 50]$.

(c) Graph $y = kax^2 + bx$, where $k = \frac{1}{4}, \frac{1}{2}, 1, 2, 4$, in the same viewing rectangle of $[0, 600, 50]$ by $[0, 400, 50]$. How does the constant k affect the path of the object?

2.7 Operations on Functions

Functions are often defined using sums, differences, products, and quotients of various expressions. For example, if

$$h(x) = x^2 + \sqrt{5x + 1},$$

we may regard $h(x)$ as a sum of values of the functions f and g given by

$$f(x) = x^2 \qquad \text{and} \qquad g(x) = \sqrt{5x + 1}.$$

We call h the *sum* of f and g and denote it by $f + g$. Thus,

$$h(x) = (f + g)(x) = x^2 + \sqrt{5x + 1}.$$

In general, if f and g are *any* functions, we use the terminology and notation given in the following chart.

While it is true that

$$(f + g)(x) = f(x) + g(x),$$

remember that, in general,

$$f(a + b) \neq f(a) + f(b).$$

Sum, Difference, Product, and Quotient of Functions

Terminology	Function value
sum $f + g$	$(f + g)(x) = f(x) + g(x)$
difference $f - g$	$(f - g)(x) = f(x) - g(x)$
product fg	$(fg)(x) = f(x)g(x)$
quotient $\dfrac{f}{g}$	$\left(\dfrac{f}{g}\right)(x) = \dfrac{f(x)}{g(x)}, \ g(x) \neq 0$

The domains of $f + g$, $f - g$, and fg are the intersection I of the domains of f and g—that is, the numbers that are *common* to both domains. The domain of f/g is the subset of I consisting of all x in I such that $g(x) \neq 0$.

EXAMPLE 1 **Finding function values of $f + g$, $f - g$, fg, and f/g**

If $f(x) = 3x - 2$ and $g(x) = x^3$, find $(f + g)(2)$, $(f - g)(2)$, $(fg)(2)$, and $(f/g)(2)$.

SOLUTION Since $f(2) = 3(2) - 2 = 4$ and $g(2) = 2^3 = 8$, we have

$$(f + g)(2) = f(2) + g(2) = 4 + 8 = 12$$

$$(f - g)(2) = f(2) - g(2) = 4 - 8 = -4$$

$$(fg)(2) = f(2)g(2) = (4)(8) = 32$$

$$\left(\frac{f}{g}\right)(2) = \frac{f(2)}{g(2)} = \frac{4}{8} = \frac{1}{2}.$$

EXAMPLE 2 **Finding $(f + g)(x)$, $(f - g)(x)$, $(fg)(x)$, and $(f/g)(x)$**

If $f(x) = \sqrt{4 - x^2}$ and $g(x) = 3x + 1$, find $(f + g)(x)$, $(f - g)(x)$, $(fg)(x)$, and $(f/g)(x)$, and state the domains of the respective functions.

Tutorial available at www.thomsonedu.com/login

▶ SOLUTION The domain of f is the closed interval $[-2, 2]$, and the domain of g is $\mathbb{R}$. The intersection of these domains is $[-2, 2]$, which is the domain of $f + g, f - g$, and fg. For the domain of f/g, we exclude each number x in $[-2, 2]$ such that $g(x) = 3x + 1 = 0$ (namely, $x = -\frac{1}{3}$). Thus, we have the following:

$$(f + g)(x) = \sqrt{4 - x^2} + (3x + 1), \qquad -2 \le x \le 2$$

$$(f - g)(x) = \sqrt{4 - x^2} - (3x + 1), \qquad -2 \le x \le 2$$

$$(fg)(x) = \sqrt{4 - x^2}(3x + 1), \qquad -2 \le x \le 2$$

$$\left(\frac{f}{g}\right)(x) = \frac{\sqrt{4 - x^2}}{3x + 1}, \qquad -2 \le x \le 2 \text{ and } x \ne -\frac{1}{3}$$

A function f is a **polynomial function** if $f(x)$ is a polynomial—that is, if

$$f(x) = a_n x^n + a_{n-1} x^{n-1} + \cdots + a_1 x + a_0,$$

where the coefficients $a_0, a_1, \ldots, a_n$ are real numbers and the exponents are nonnegative integers. A polynomial function may be regarded as a sum of functions whose values are of the form cx^k, where c is a real number and k is a nonnegative integer. Note that the quadratic functions considered in the previous section are polynomial functions.

An **algebraic function** is a function that can be expressed in terms of finite sums, differences, products, quotients, or roots of polynomial functions.

ILLUSTRATION **Algebraic Function**

■ $f(x) = 5x^4 - 2\sqrt[3]{x} + \dfrac{x(x^2 + 5)}{\sqrt{x^3 + \sqrt{x}}}$

Functions that are not algebraic are **transcendental.** The exponential and logarithmic functions considered in Chapter 4 are examples of transcendental functions.

In the remainder of this section we shall discuss how two functions f and g may be used to obtain the *composite functions* $f \circ g$ and $g \circ f$ (read "f circle g" and "g circle f," respectively). Functions of this type are very important in calculus. The function $f \circ g$ is defined as follows.

Definition of Composite Function

The **composite function** $f \circ g$ of two functions f and g is defined by

$$(f \circ g)(x) = f(g(x)).$$

The domain of $f \circ g$ is the set of all x in the domain of g such that $g(x)$ is in the domain of f.

▶ **Tutorial available at www.thomsonedu.com/login**

*A number x is in the domain of $(f \circ g)(x)$ if and only if **both** $g(x)$ **and** $f(g(x))$ are defined.*

Figure 1

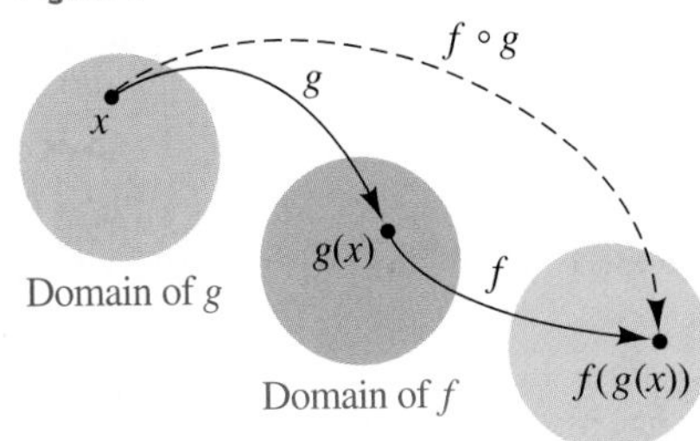

Figure 1 is a schematic diagram that illustrates relationships among f, g, and $f \circ g$. Note that for x in the domain of g, *first we find* $g(x)$ (which must be in the domain of f) and then, *second, we find* $f(g(x))$.

For the composite function $g \circ f$, we reverse this order, first finding $f(x)$ and second finding $g(f(x))$. The domain of $g \circ f$ is the set of all x in the domain of f such that $f(x)$ is in the domain of g.

Since the notation $g(x)$ is read "g of x," we sometimes say that *g is a function of x*. For the composite function $f \circ g$, the notation $f(g(x))$ is read "f of g of x," and we could regard f as a function of $g(x)$. In this sense, *a composite function is a function of a function* or, more precisely, a function of another function's values.

EXAMPLE 3 Finding composite functions

Let $f(x) = x^2 - 1$ and $g(x) = 3x + 5$.

(a) Find $(f \circ g)(x)$ and the domain of $f \circ g$.

(b) Find $(g \circ f)(x)$ and the domain of $g \circ f$.

(c) Find $f(g(2))$ in two different ways: first using the functions f and g separately and second using the composite function $f \circ g$.

SOLUTION

▶ **(a)**

$$\begin{aligned} (f \circ g)(x) &= f(g(x)) && \text{definition of } f \circ g \\ &= f(3x + 5) && \text{definition of } g \\ &= (3x + 5)^2 - 1 && \text{definition of } f \\ &= 9x^2 + 30x + 24 && \text{simplify} \end{aligned}$$

The domain of both f and g is $\mathbb{R}$. Since for each x in $\mathbb{R}$ (the domain of g), the function value $g(x)$ is in $\mathbb{R}$ (the domain of f), the domain of $f \circ g$ is also $\mathbb{R}$. Note that *both* $g(x)$ *and* $f(g(x))$ are defined for all real numbers.

(b)

$$\begin{aligned} (g \circ f)(x) &= g(f(x)) && \text{definition of } g \circ f \\ &= g(x^2 - 1) && \text{definition of } f \\ &= 3(x^2 - 1) + 5 && \text{definition of } g \\ &= 3x^2 + 2 && \text{simplify} \end{aligned}$$

Since for each x in $\mathbb{R}$ (the domain of f), the function value $f(x)$ is in $\mathbb{R}$ (the domain of g), the domain of $g \circ f$ is $\mathbb{R}$. Note that *both* $f(x)$ *and* $g(f(x))$ are defined for all real numbers.

▶ **(c)** To find $f(g(2))$ using $f(x) = x^2 - 1$ and $g(x) = 3x + 5$ separately, we may proceed as follows:

$$g(2) = 3(2) + 5 = 11$$
$$f(g(2)) = f(11) = 11^2 - 1 = 120$$

To find $f(g(2))$ using $f \circ g$, we refer to part (a), where we found

$$(f \circ g)(x) = f(g(x)) = 9x^2 + 30x + 24.$$

(continued)

▶ **Tutorial available at www.thomsonedu.com/login**

Hence,

$$f(g(2)) = 9(2)^2 + 30(2) + 24$$
$$= 36 + 60 + 24 = 120.$$

Note that in Example 3, $f(g(x))$ and $g(f(x))$ are not always the same; that is, $f \circ g \neq g \circ f$.

If two functions f and g both have domain $\mathbb{R}$, then the domain of $f \circ g$ and $g \circ f$ is also $\mathbb{R}$. This was illustrated in Example 3. The next example shows that the domain of a composite function may differ from those of the two given functions.

EXAMPLE 4 Finding composite functions

Let $f(x) = x^2 - 16$ and $g(x) = \sqrt{x}$.

(a) Find $(f \circ g)(x)$ and the domain of $f \circ g$.

(b) Find $(g \circ f)(x)$ and the domain of $g \circ f$.

SOLUTION We first note that the domain of f is $\mathbb{R}$ and the domain of g is the set of all nonnegative real numbers—that is, the interval $[0, \infty)$. We may proceed as follows.

(a)

$$\begin{aligned} (f \circ g)(x) &= f(g(x)) && \text{definition of } f \circ g \\ &= f(\sqrt{x}) && \text{definition of } g \\ &= (\sqrt{x})^2 - 16 && \text{definition of } f \\ &= x - 16 && \text{simplify} \end{aligned}$$

If we consider only the final expression, $x - 16$, we might be led to believe that the domain of $f \circ g$ is $\mathbb{R}$, since $x - 16$ is defined for every real number x. However, this is not the case. By definition, the domain of $f \circ g$ is the set of all x in $[0, \infty)$ (the domain of g) such that $g(x)$ is in $\mathbb{R}$ (the domain of f). Since $g(x) = \sqrt{x}$ is in $\mathbb{R}$ for every x in $[0, \infty)$, it follows that the domain of $f \circ g$ is $[0, \infty)$. Note that *both* $g(x)$ *and* $f(g(x))$ are defined for x in $[0, \infty)$.

(b)

$$\begin{aligned} (g \circ f)(x) &= g(f(x)) && \text{definition of } g \circ f \\ &= g(x^2 - 16) && \text{definition of } f \\ &= \sqrt{x^2 - 16} && \text{definition of } g \end{aligned}$$

By definition, the domain of $g \circ f$ is the set of all x in $\mathbb{R}$ (the domain of f) such that $f(x) = x^2 - 16$ is in $[0, \infty)$ (the domain of g). The statement "$x^2 - 16$ is in $[0, \infty)$" is equivalent to each of the inequalities

$$x^2 - 16 \geq 0, \qquad x^2 \geq 16, \qquad |x| \geq 4.$$

Thus, the domain of $g \circ f$ is the union $(-\infty, -4] \cup [4, \infty)$. Note that *both* $f(x)$ *and* $g(f(x))$ are defined for x in $(-\infty, -4] \cup [4, \infty)$. Also note that this domain is different from the domains of both f and g.

The next example illustrates how special values of composite functions may sometimes be obtained from tables.

Tutorial available at www.thomsonedu.com/login

EXAMPLE 5 Finding composite function values from tables

Several values of two functions f and g are listed in the following tables.

x	1	2	3	4
$f(x)$	3	4	2	1

x	1	2	3	4
$g(x)$	4	1	3	2

Find $(f \circ g)(2)$, $(g \circ f)(2)$, $(f \circ f)(2)$, and $(g \circ g)(2)$.

▶ **SOLUTION** Using the definition of composite function and referring to the tables above, we obtain

$$(f \circ g)(2) = f(g(2)) = f(1) = 3$$
$$(g \circ f)(2) = g(f(2)) = g(4) = 2$$
$$(f \circ f)(2) = f(f(2)) = f(4) = 1$$
$$(g \circ g)(2) = g(g(2)) = g(1) = 4.$$

In some applied problems it is necessary to express a quantity y as a function of time t. The following example illustrates that it is often easier to introduce a third variable x, express x as a function of t (that is, $x = g(t)$), express y as a function of x (that is, $y = f(x)$), and finally form the composite function given by $y = f(x) = f(g(t))$.

EXAMPLE 6 Using a composite function to find the volume of a balloon

A meteorologist is inflating a spherical balloon with helium gas. If the radius of the balloon is changing at a rate of 1.5 cm/sec, express the volume V of the balloon as a function of time t (in seconds).

▶ **SOLUTION** Let x denote the radius of the balloon. If we assume that the radius is 0 initially, then after t seconds

$$x = 1.5t. \quad \text{radius of balloon after } t \text{ seconds}$$

To illustrate, after 1 second, the radius is 1.5 centimeters; after 2 seconds, it is 3.0 centimeters; after 3 seconds, it is 4.5 centimeters; and so on.

Next we write

$$V = \tfrac{4}{3}\pi x^3. \quad \text{volume of a sphere of radius } x$$

This gives us a composite function relationship in which V is a function of x, and x is a function of t. By substitution, we obtain

$$V = \tfrac{4}{3}\pi x^3 = \tfrac{4}{3}\pi(1.5t)^3 = \tfrac{4}{3}\pi\left(\tfrac{3}{2}t\right)^3 = \tfrac{4}{3}\pi\left(\tfrac{27}{8}t^3\right).$$

Simplifying, we obtain the following formula for V as a function of t:

$$V(t) = \tfrac{9}{2}\pi t^3$$

▶ **Tutorial available at www.thomsonedu.com/login**

If f and g are functions such that

$$y = f(u) \qquad \text{and} \qquad u = g(x),$$

then substituting for u in $y = f(u)$ yields

$$y = f(g(x)).$$

For certain problems in calculus we *reverse* this procedure; that is, given $y = h(x)$ for some function h, we find a *composite function form* $y = f(u)$ and $u = g(x)$ such that $h(x) = f(g(x))$.

EXAMPLE 7 Finding a composite function form

Express $y = (2x + 5)^8$ as a composite function form.

SOLUTION Suppose, for a real number x, we wanted to evaluate the expression $(2x + 5)^8$ by using a calculator. We would first calculate the value of $2x + 5$ and then raise the result to the eighth power. This suggests that we let

$$u = 2x + 5 \qquad \text{and} \qquad y = u^8,$$

which is a composite function form for $y = (2x + 5)^8$.

The method used in the preceding example can be extended to other functions. In general, suppose we are given $y = h(x)$. To choose the *inside* expression $u = g(x)$ in a composite function form, ask the following question: If a calculator were being used, which part of the expression $h(x)$ would be evaluated first? This often leads to a suitable choice for $u = g(x)$. After choosing u, refer to $h(x)$ to determine $y = f(u)$. The following illustration contains typical problems.

ILLUSTRATION Composite Function Forms

Function value	Choice for $u = g(x)$	Choice for $y = f(u)$
■ $y = (x^3 - 5x + 1)^4$	$u = x^3 - 5x + 1$	$y = u^4$
■ $y = \sqrt{x^2 - 4}$	$u = x^2 - 4$	$y = \sqrt{u}$
■ $y = \dfrac{2}{3x + 7}$	$u = 3x + 7$	$y = \dfrac{2}{u}$

The composite function form is never unique. For example, consider the first expression in the preceding illustration:

$$y = (x^3 - 5x + 1)^4$$

If n is any nonzero integer, we could choose

$$u = (x^3 - 5x + 1)^n \qquad \text{and} \qquad y = u^{4/n}.$$

Thus, there are an *unlimited* number of composite function forms. Generally, our goal is to choose a form such that the expression for y is simple, as we did in the illustration.

The next example illustrates how a graphing utility can help determine the domain of a composite function. We use the same functions that appeared in Example 4.

EXAMPLE 8 **Graphically analyzing a composite function**

Let $f(x) = x^2 - 16$ and $g(x) = \sqrt{x}$.

(a) Find $f(g(3))$.

(b) Sketch $y = (f \circ g)(x)$, and use the graph to find the domain of $f \circ g$.

SOLUTION

(a) We begin by making the assignments

$$Y_1 = \sqrt{x} \qquad \text{and} \qquad Y_2 = (Y_1)^2 - 16.$$

Note that we have substituted Y_1 for x in $f(x)$ and assigned this expression to Y_2, much the same way as we substituted $g(x)$ for x in Example 4.

Next we store the value 3 in the memory location for x and then query the value of Y_2. We see that the value of Y_2 at 3 is -13; that is, $f(g(3)) = -13$.

Figure 2
$[-10, 50, 5]$ by $[-20, 20, 5]$

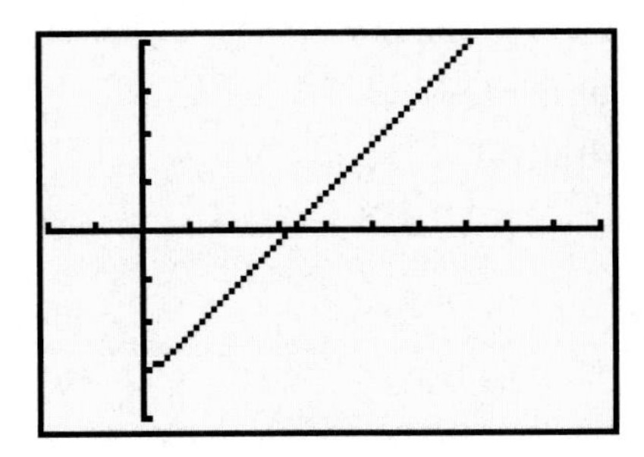

(b) To determine a viewing rectangle for the graph of $f \circ g$, we first note that $f(x) \geq -16$ for all x and therefore choose Ymin *less* than -16; say, Ymin $= -20$. If we want the rectangle to have a vertical dimension of 40, we must choose Ymax $= 20$.

If your screen is in 1:1 proportion (horizontal:vertical), then a reasonable choice for [Xmin, Xmax] would be $[-10, 30]$, a horizontal dimension of 40. If your screen is in 3:2 proportion, choose [Xmin, Xmax] to be $[-10, 50]$, a horizontal dimension of 60.

Selecting Y_2 and then displaying the graph of Y_2 using the viewing rectangle $[-10, 50, 5]$ by $[-20, 20, 5]$ gives us a graph similar to Figure 2. We see that the graph is a half-line with endpoint $(0, -16)$. Thus, the domain of Y_2 is all $x \geq 0$.

The next example demonstrates how to use a graphing utility to graph composite functions of the form $af(bx)$. We will use the function from Example 7 of Section 2.5.

EXAMPLE 9 **Graphing composite functions**

If $f(x) = x^3 - 4x^2$, sketch the graph of $y = -\frac{1}{2}f\left(\frac{1}{3}x\right)$.

Figure 3
$[-7, 14]$ by $[-3, 11]$

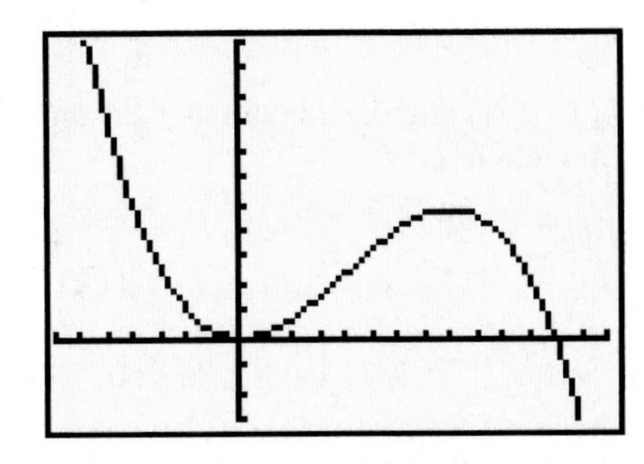

SOLUTION From our discussion on compressing and stretching graphs in Section 2.5, we recognize that the graph of f will be vertically compressed by a factor 2 and horizontally stretched by a factor 3. To relate this problem to composite functions, we may think of

$$y = -\tfrac{1}{2}f\left(\tfrac{1}{3}x\right) \qquad \text{as} \qquad y = -\tfrac{1}{2}f(g(x)), \qquad \text{where } g(x) = \tfrac{1}{3}x.$$

The last equation for y suggests the assignments

$$Y_1 = \tfrac{1}{3}x, \quad Y_2 = (Y_1)^3 - 4(Y_1)^2, \quad \text{and} \quad Y_3 = -\tfrac{1}{2}Y_2.$$

Note that $Y_2 = f(Y_1) = f(g(x))$. We select only Y_3 to be graphed and choose the viewing rectangle $[-7, 14]$ by $[-3, 11]$, to obtain Figure 3.

(continued)

Figure 4
$[-1, 3]$ by $[-5, 1]$

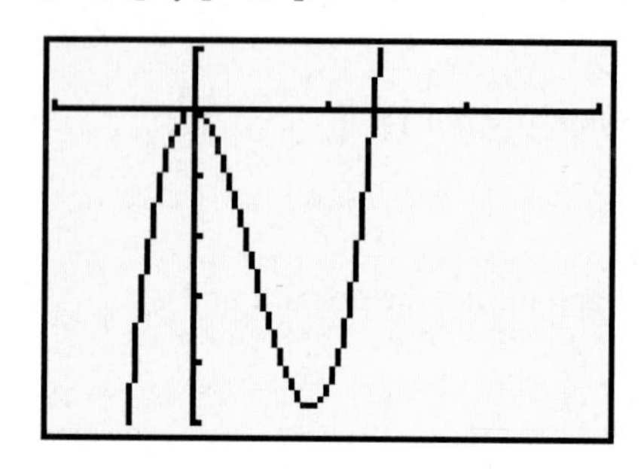

There are two advantages of assigning the functions in the fashion above:

(1) We did not actually have to compute the polynomial function to be graphed, as we did in Example 7 of Section 2.5.

(2) By merely changing the coefficients in Y_1 and Y_3, we can easily examine their effect on the graph of Y_3.

As an illustration of item (2), you should try graphing $y = \frac{1}{2}f(3x)$ by changing Y_1 to $3x$, Y_3 to $\frac{1}{2}Y_2$, and the viewing rectangle to $[-1, 3]$ by $[-5, 1]$ and then graphing Y_3, to obtain Figure 4.

2.7 Exercises

Exer. 1–2: Find
(a) $(f + g)(3)$ **(b)** $(f - g)(3)$
(c) $(fg)(3)$ **(d)** $(f/g)(3)$

1 $f(x) = x + 3, \quad g(x) = x^2$

2 $f(x) = -x^2, \quad g(x) = 2x - 1$

Exer. 3–8: Find
(a) $(f + g)(x)$, $(f - g)(x)$, $(fg)(x)$, and $(f/g)(x)$
(b) the domain of $f + g$, $f - g$, and fg
(c) the domain of f/g

3 $f(x) = x^2 + 2, \quad g(x) = 2x^2 - 1$

4 $f(x) = x^2 + x, \quad g(x) = x^2 - 3$

5 $f(x) = \sqrt{x + 5}, \quad g(x) = \sqrt{x + 5}$

6 $f(x) = \sqrt{3 - 2x}, \quad g(x) = \sqrt{x + 4}$

▶ 7 $f(x) = \dfrac{2x}{x - 4}, \quad g(x) = \dfrac{x}{x + 5}$

8 $f(x) = \dfrac{x}{x - 2}, \quad g(x) = \dfrac{3x}{x + 4}$

Exer. 9–10: Find
(a) $(f \circ g)(x)$ **(b)** $(g \circ f)(x)$
(c) $(f \circ f)(x)$ **(d)** $(g \circ g)(x)$

9 $f(x) = 2x - 1, \quad g(x) = -x^2$

10 $f(x) = 3x^2, \quad g(x) = x - 1$

Exer. 11–20: Find
(a) $(f \circ g)(x)$ **(b)** $(g \circ f)(x)$
(c) $f(g(-2))$ **(d)** $g(f(3))$

11 $f(x) = 2x - 5, \quad g(x) = 3x + 7$

12 $f(x) = 5x + 2, \quad g(x) = 6x - 1$

13 $f(x) = 3x^2 + 4, \quad g(x) = 5x$

14 $f(x) = 3x - 1, \quad g(x) = 4x^2$

15 $f(x) = 2x^2 + 3x - 4, \quad g(x) = 2x - 1$

16 $f(x) = 5x - 7, \quad g(x) = 3x^2 - x + 2$

17 $f(x) = 4x, \quad g(x) = 2x^3 - 5x$

▶ 18 $f(x) = x^3 + 2x^2, \quad g(x) = 3x$

19 $f(x) = |x|, \quad g(x) = -7$

20 $f(x) = 5, \quad g(x) = x^2$

Exer. 21–34: Find (a) $(f \circ g)(x)$ and the domain of $f \circ g$ and (b) $(g \circ f)(x)$ and the domain of $g \circ f$.

21 $f(x) = x^2 - 3x, \quad g(x) = \sqrt{x + 2}$

22 $f(x) = \sqrt{x - 15}, \quad g(x) = x^2 + 2x$

▶ **Tutorial available at www.thomsonedu.com/login**

23 $f(x) = x^2 - 4$, $g(x) = \sqrt{3x}$

24 $f(x) = -x^2 + 1$, $g(x) = \sqrt{x}$

25 $f(x) = \sqrt{x - 2}$, $g(x) = \sqrt{x + 5}$

26 $f(x) = \sqrt{3 - x}$, $g(x) = \sqrt{x + 2}$

27 $f(x) = \sqrt{3 - x}$, $g(x) = \sqrt{x^2 - 16}$

28 $f(x) = x^3 + 5$, $g(x) = \sqrt[3]{x - 5}$

29 $f(x) = \dfrac{3x + 5}{2}$, $g(x) = \dfrac{2x - 5}{3}$

30 $f(x) = \dfrac{1}{x - 1}$, $g(x) = x - 1$

31 $f(x) = x^2$, $g(x) = \dfrac{1}{x^3}$

32 $f(x) = \dfrac{x}{x - 2}$, $g(x) = \dfrac{3}{x}$

33 $f(x) = \dfrac{x - 1}{x - 2}$, $g(x) = \dfrac{x - 3}{x - 4}$

34 $f(x) = \dfrac{x + 2}{x - 1}$, $g(x) = \dfrac{x - 5}{x + 4}$

Exer. 35–36: Solve the equation $(f \circ g)(x) = 0$.

35 $f(x) = x^2 - 2$, $g(x) = x + 3$

36 $f(x) = x^2 - x - 2$, $g(x) = 2x - 1$

37 Several values of two functions f and g are listed in the following tables:

x	5	6	7	8	9
$f(x)$	8	7	6	5	4

x	5	6	7	8	9
$g(x)$	7	8	6	5	4

If possible, find

(a) $(f \circ g)(6)$ (b) $(g \circ f)(6)$ (c) $(f \circ f)(6)$

(d) $(g \circ g)(6)$ (e) $(f \circ g)(9)$

38 Several values of two functions T and S are listed in the following tables:

t	0	1	2	3	4
$T(t)$	2	3	1	0	5

x	0	1	2	3	4
$S(x)$	1	0	3	2	5

If possible, find

(a) $(T \circ S)(1)$ (b) $(S \circ T)(1)$ (c) $(T \circ T)(1)$

(d) $(S \circ S)(1)$ (e) $(T \circ S)(4)$

▶ 39 If $D(t) = \sqrt{400 + t^2}$ and $R(x) = 20x$, find $(D \circ R)(x)$.

40 If $S(r) = 4\pi r^2$ and $D(t) = 2t + 5$, find $(S \circ D)(t)$.

41 If f is an odd function and g is an even function, is fg even, odd, or neither even nor odd?

42 There is one function with domain $\mathbb{R}$ that is both even and odd. Find that function.

43 **Payroll functions** Let the social security tax function SSTAX be defined as $\text{SSTAX}(x) = 0.0765x$, where $x \geq 0$ is the weekly income. Let ROUND2 be the function that rounds a number to two decimal places. Find the value of $(\text{ROUND2} \circ \text{SSTAX})(525)$.

44 **Computer science functions** Let the function CHR be defined by CHR(65) = "A", CHR(66) = "B", . . . , CHR(90) = "Z". Then let the function ORD be defined by ORD("A") = 65, ORD("B") = 66, . . . , ORD("Z") = 90. Find

(a) (CHR ∘ ORD)("C") (b) CHR(ORD("A") + 3)

45 **Spreading fire** A fire has started in a dry open field and is spreading in the form of a circle. If the radius of this circle increases at the rate of 6 ft/min, express the total fire area A as a function of time t (in minutes).

46 **Dimensions of a balloon** A spherical balloon is being inflated at a rate of $\frac{9}{2}\pi$ ft^3/min. Express its radius r as a function of time t (in minutes), assuming that $r = 0$ when $t = 0$.

47 **Dimensions of a sand pile** The volume of a conical pile of sand is increasing at a rate of 243π ft^3/min, and the height of the pile always equals the radius r of the base. Express r as a function of time t (in minutes), assuming that $r = 0$ when $t = 0$.

48 **Diagonal of a cube** The diagonal d of a cube is the distance between two opposite vertices. Express d as a function of the edge x of the cube. (*Hint:* First express the diagonal y of a face as a function of x.)

49 **Altitude of a balloon** A hot-air balloon rises vertically from ground level as a rope attached to the base of the balloon is released at the rate of 5 ft/sec (see the figure). The pulley that releases the rope is 20 feet from a platform where passengers board the balloon. Express the altitude h of the balloon as a function of time t.

Exercise 49

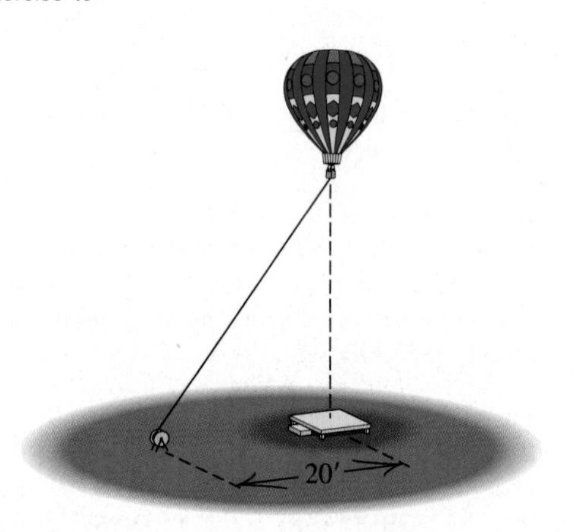

50 **Tightrope walker** Refer to Exercise 76 of Section 2.4. Starting at the lowest point, the tightrope walker moves up the rope at a steady rate of 2 ft/sec. If the rope is attached 30 feet up the pole, express the height h of the walker above the ground as a function of time t. (*Hint:* Let d denote the total distance traveled along the wire. First express d as a function of t, and then h as a function of d.)

51 **Airplane take-off** Refer to Exercise 77 of Section 2.4. When the airplane is 500 feet down the runway, it has reached a speed of 150 ft/sec (or about 102 mi/hr), which it will maintain until take-off. Express the distance d of the plane from the control tower as a function of time t (in seconds). (*Hint:* In the figure, first write x as a function of t.)

52 **Cable corrosion** A 100-foot-long cable of diameter 4 inches is submerged in seawater. Because of corrosion, the surface area of the cable decreases at the rate of 750 in^2 per year. Express the diameter d of the cable as a function of time t (in years). (Disregard corrosion at the ends of the cable.)

Exer. 53–60: Find a composite function form for y.

53 $y = (x^2 + 3x)^{1/3}$

54 $y = \sqrt[4]{x^4 - 16}$

55 $y = \dfrac{1}{(x - 3)^4}$

56 $y = 4 + \sqrt{x^2 + 1}$

57 $y = (x^4 - 2x^2 + 5)^5$

58 $y = \dfrac{1}{(x^2 + 3x - 5)^3}$

59 $y = \dfrac{\sqrt{x + 4} - 2}{\sqrt{x + 4} + 2}$

60 $y = \dfrac{\sqrt[3]{x}}{1 + \sqrt[3]{x}}$

61 If $f(x) = \sqrt{x} - 1$ and $g(x) = x^3 + 1$, approximate $(f \circ g)(0.0001)$. In order to avoid calculating a zero value for $(f \circ g)(0.0001)$, rewrite the formula for $f \circ g$ as

$$\frac{x^3}{\sqrt{x^3 + 1} + 1}.$$

62 If $f(x) = \dfrac{x^3}{x^2 + x + 2}$ and $g(x) = \left(\sqrt{3x} - x^3\right)^{3/2}$, approximate

$$\frac{(f + g)(1.12) - (f/g)(1.12)}{[(f \circ f)(5.2)]^2}.$$

63 Refer to Exercise 63 of Section 2.5. Make the assignments $Y_1 = x$ and $Y_2 = 3\sqrt{(Y_1 + 2)(6 - Y_1)} - 4$. Determine assignments for Y_1 (and Y_3 if necessary) that will enable you to graph each function in (a)–(h), and then graph the function. (Check the domain and range with the previously listed answer.)

(a) $y = -2f(x)$
(b) $y = f(\frac{1}{2}x)$
(c) $y = f(x - 3) + 1$
(d) $y = f(x + 2) - 3$
(e) $y = f(-x)$
(f) $y = -f(x)$
(g) $y = f(|x|)$
(h) $y = |f(x)|$

64 Refer to Exercise 64 of Section 2.5. Make the assignments $Y_1 = x$ and $Y_2 = 3\sqrt{(-Y_1 - 6)(Y_1 + 2)} - 10$. Determine assignments for Y_1 and Y_3 that will enable you to graph each function, and then graph the function.

(a) $y = \frac{1}{2}f(x)$
(b) $y = f(2x)$
(c) $y = f(x - 2) + 5$
(d) $y = f(x + 4) - 1$
(e) $y = f(-x)$
(f) $y = -f(x)$
(g) $y = f(|x|)$
(h) $y = |f(x)|$

CHAPTER 2 REVIEW EXERCISES

▶ **Online support materials can be found at www.thomsonedu.com/login**

1 Describe the set of all points (x, y) in a coordinate plane such that $y/x < 0$.

2 Show that the triangle with vertices $A(3, 1)$, $B(-5, -3)$, and $C(4, -1)$ is a right triangle, and find its area.

3 Given $P(-5, 9)$ and $Q(-8, -7)$, find

(a) the distance $d(P, Q)$

(b) the midpoint of the segment PQ

(c) a point R such that Q is the midpoint of PR

4 Find all points on the y-axis that are a distance 13 from $P(12, 6)$.

5 For what values of a is the distance between $P(a, 1)$ and $Q(-2, a)$ less than 3?

6 Find an equation of the circle that has center $C(7, -4)$ and passes through $P(-3, 3)$.

7 Find an equation of the circle that has endpoints of a diameter $A(8, 10)$ and $B(-2, -14)$.

8 Find an equation for the left half of the circle given by $(x + 2)^2 + y^2 = 9$.

9 Find the slope of the line through $C(11, -5)$ and $D(-8, 6)$.

10 Show that $A(-3, 1)$, $B(1, -1)$, $C(4, 1)$, and $D(3, 5)$ are vertices of a trapezoid.

11 Find an equation of the line through $A(\frac{1}{2}, -\frac{1}{3})$ that is

(a) parallel to the line $6x + 2y + 5 = 0$

(b) perpendicular to the line $6x + 2y + 5 = 0$

12 Express $8x + 3y - 24 = 0$ in slope-intercept form.

13 Find an equation of the circle that has center $C(-5, -1)$ and is tangent to the line $x = 4$.

14 Find an equation of the line that has x-intercept -3 and passes through the center of the circle that has equation $x^2 + y^2 - 4x + 10y + 26 = 0$.

15 Find a general form of an equation of the line through $P(4, -3)$ with slope 5.

16 Given $A(-1, 2)$ and $B(3, -4)$, find a general form of an equation for the perpendicular bisector of segment AB.

Exer. 17–18: Find the center and radius of the circle with the given equation.

17 $x^2 + y^2 - 12y + 31 = 0$

18 $4x^2 + 4y^2 + 24x - 16y + 39 = 0$

19 If $f(x) = \dfrac{x}{\sqrt{x+3}}$, find

(a) $f(1)$ (b) $f(-1)$ (c) $f(0)$ (d) $f(-x)$

(e) $-f(x)$ (f) $f(x^2)$ (g) $[f(x)]^2$

Exer. 20–21: Find *the sign* of $f(4)$ without actually finding $f(4)$.

20 $f(x) = \dfrac{-32(x^2-4)}{(9-x^2)^{5/3}}$

21 $f(x) = \dfrac{-2(x^2-20)(5-x)}{(6-x^2)^{4/3}}$

22 Find the domain and range of f if

(a) $f(x) = \sqrt{3x-4}$ (b) $f(x) = \dfrac{1}{(x+3)^2}$

Exer. 23–24: Find $\dfrac{f(a+h)-f(a)}{h}$ if $h \neq 0$.

23 $f(x) = -x^2 + x + 5$

24 $f(x) = \dfrac{1}{x+2}$

25 Find a linear function f such that $f(1) = 2$ and $f(3) = 7$.

26 Determine whether f is even, odd, or neither even nor odd.

(a) $f(x) = \sqrt[3]{x^3 + 4x}$ (b) $f(x) = \sqrt[3]{3x^2 - x^3}$

(c) $f(x) = \sqrt[3]{x^4 + 3x^2 + 5}$

Exer. 27–40: Sketch the graph of the equation, and label the x- and y-intercepts.

27 $x + 5 = 0$ 28 $2y - 7 = 0$

29 $2y + 5x - 8 = 0$ 30 $x = 3y + 4$

31 $9y + 2x^2 = 0$ 32 $3x - 7y^2 = 0$

33 $y = \sqrt{1-x}$ 34 $y = (x-1)^3$

35 $y^2 = 16 - x^2$

36 $x^2 + y^2 + 4x - 16y + 64 = 0$

37 $x^2 + y^2 - 8x = 0$ 38 $x = -\sqrt{9 - y^2}$

39 $y = (x-3)^2 - 2$ 40 $y = -x^2 - 2x + 3$

41 Find the center of the small circle.

Exercise 41

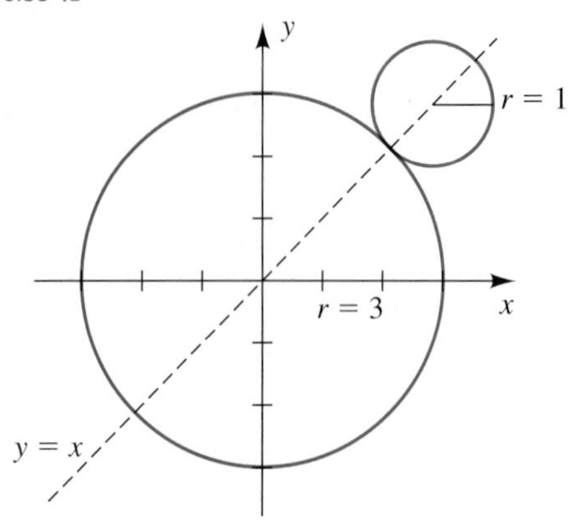

42 Explain how the graph of $y = -f(x-2)$ compares to the graph of $y = f(x)$.

Exer. 43–52: (a) Sketch the graph of f. (b) Find the domain D and range R of f. (c) Find the intervals on which f is increasing, is decreasing, or is constant.

43 $f(x) = \dfrac{1-3x}{2}$ 44 $f(x) = 1000$

45 $f(x) = |x+3|$ 46 $f(x) = -\sqrt{10 - x^2}$

47 $f(x) = 1 - \sqrt{x+1}$ 48 $f(x) = \sqrt{2-x}$

49 $f(x) = 9 - x^2$ 50 $f(x) = x^2 + 6x + 16$

51 $f(x) = \begin{cases} x^2 & \text{if } x < 0 \\ 3x & \text{if } 0 \le x < 2 \\ 6 & \text{if } x \ge 2 \end{cases}$ 52 $f(x) = 1 + 2[\![x]\!]$

53 Sketch the graphs of the following equations, making use of shifting, stretching, or reflecting:

(a) $y = \sqrt{x}$ (b) $y = \sqrt{x+4}$

(c) $y = \sqrt{x} + 4$ (d) $y = 4\sqrt{x}$

(e) $y = \frac{1}{4}\sqrt{x}$ (f) $y = -\sqrt{x}$

54 The graph of a function f with domain $[-3, 3]$ is shown in the figure. Sketch the graph of the given equation.

(a) $y = f(x - 2)$ (b) $y = f(x) - 2$

(c) $y = f(-x)$ (d) $y = f(2x)$

(e) $y = f\left(\frac{1}{2}x\right)$ (f) $y = |f(x)|$

(g) $y = f(|x|)$

Exercise 54

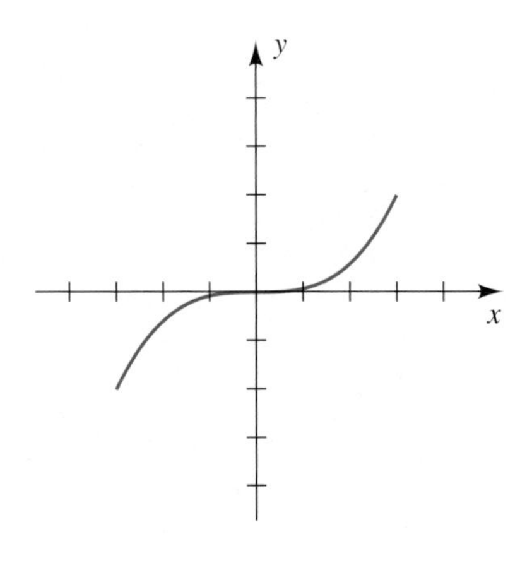

Exer. 55–58: Find an equation for the graph shown in the figure.

55

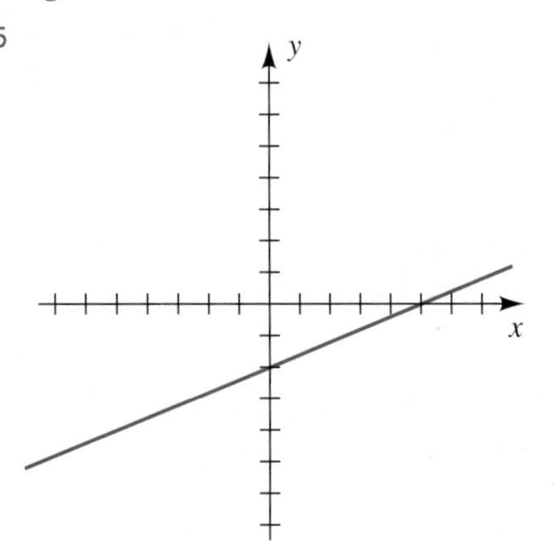

56

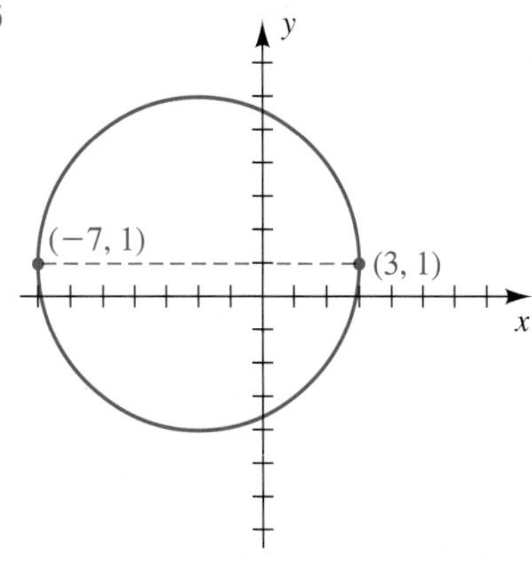

57

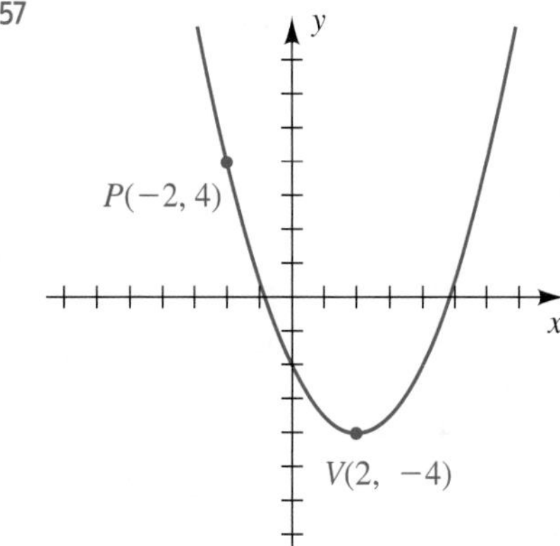

58

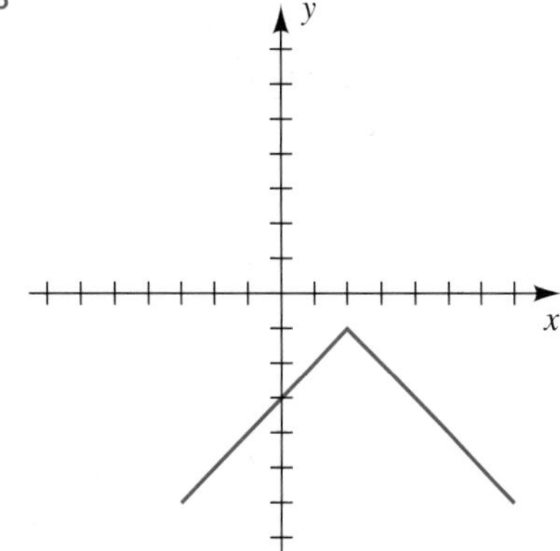

Exer. 59–62: Find the maximum or minimum value of $f(x)$.

59 $f(x) = 5x^2 + 30x + 49$

60 $f(x) = -3x^2 + 30x - 82$

61 $f(x) = -12(x + 1)^2 - 37$

62 $f(x) = 3(x + 2)(x - 10)$

63 Express the function $f(x) = -2x^2 + 12x - 14$ in the form $a(x - h)^2 + k$.

64 Find the standard equation of a parabola with a vertical axis that has vertex $V(3, -2)$ and passes through $(5, 4)$.

65 If $f(x) = \sqrt{4 - x^2}$ and $g(x) = \sqrt{x}$, find the domain of

(a) fg (b) f/g

66 If $f(x) = 8x - 1$ and $g(x) = \sqrt{x - 2}$, find

(a) $(f \circ g)(2)$ (b) $(g \circ f)(2)$

Exer. 67–68: Find (a) $(f \circ g)(x)$ and (b) $(g \circ f)(x)$.

67 $f(x) = 2x^2 - 5x + 1, \quad g(x) = 3x + 2$

68 $f(x) = \sqrt{3x + 2}, \quad g(x) = 1/x^2$

Exer. 69–70: Find (a) $(f \circ g)(x)$ and the domain of $f \circ g$ and (b) $(g \circ f)(x)$ and the domain of $g \circ f$.

69 $f(x) = \sqrt{25 - x^2}, \quad g(x) = \sqrt{x - 3}$

70 $f(x) = \dfrac{x}{3x + 2}, \quad g(x) = \dfrac{2}{x}$

71 Find a composite function form for $y = \sqrt[3]{x^2 - 5x}$.

72 Wheelchair ramp The Americans with Disabilities Act of 1990 guarantees all persons the right of accessibility of public accommodations. Providing access to a building often involves building a wheelchair ramp. Ramps should have approximately 1 inch of vertical rise for every 12–20 inches of horizontal run. If the base of an exterior door is located 3 feet above a sidewalk, determine the range of appropriate lengths for a wheelchair ramp.

73 Discus throw Based on Olympic records, the winning distance for the discus throw can be approximated by the equation $d = 181 + 1.065t$, where d is in feet and $t = 0$ corresponds to the year 1948.

(a) Predict the winning distance for the Summer Olympics in the year 2008.

(b) Estimate the Olympic year in which the winning distance will be 265 feet.

74 House appreciation Six years ago a house was purchased for \$179,000. This year it was appraised at \$215,000. Assume that the value V of the house after its purchase is a linear function of time t (in years).

(a) Express V in terms of t.

(b) How many years after the purchase date was the house worth \$193,000?

75 Temperature scales The freezing point of water is 0°C, or 32°F, and the boiling point is 100°C, or 212°F.

(a) Express the Fahrenheit temperature F as a linear function of the Celsius temperature C.

(b) What temperature increase in °F corresponds to an increase in temperature of 1°C?

76 Gasoline mileage Suppose the cost of driving an automobile is a linear function of the number x of miles driven and that gasoline costs \$3 per gallon. A certain automobile presently gets 20 mi/gal, and a tune-up that will improve gasoline mileage by 10% costs \$120.

(a) Express the cost C_1 of driving without a tune-up in terms of x.

(b) Express the cost C_2 of driving with a tune-up in terms of x.

(c) How many miles must the automobile be driven after a tune-up to make the cost of the tune-up worthwhile?

77 Constructing a storage shelter An open rectangular storage shelter, consisting of two 4-foot-wide vertical sides and a flat roof, is to be attached to an existing structure, as illustrated in the figure. The flat roof is made of tin and costs \$5 per square foot, and the two sides are made of plywood costing \$2 per square foot.

(a) If \$400 is available for construction, express the length y as a function of the height x.

(b) Express the volume V inside the shelter as a function of x.

Exercise 77

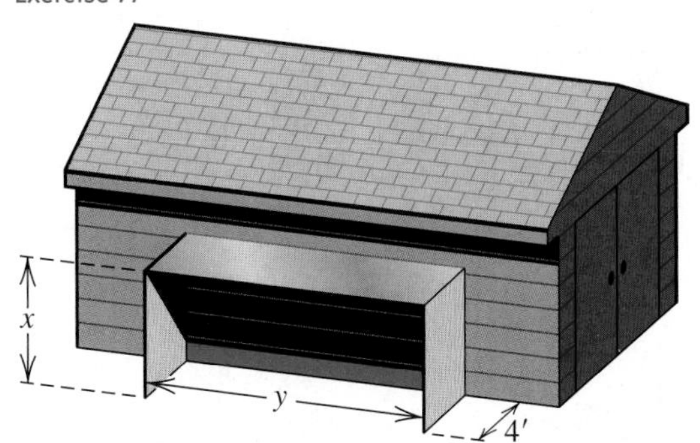

78 Constructing a cylindrical container A company plans to manufacture a container having the shape of a right circular cylinder, open at the top, and having a capacity of 24π in^3. If the cost of the material for the bottom is \$0.30/in^2 and that for the curved sides is \$0.10/in^2, express the total cost C of the material as a function of the radius r of the base of the container.

79 Filling a pool A cross section of a rectangular pool of dimensions 80 feet by 40 feet is shown in the figure. The pool is being filled with water at a rate of 10 ft^3/min.

Exercise 79

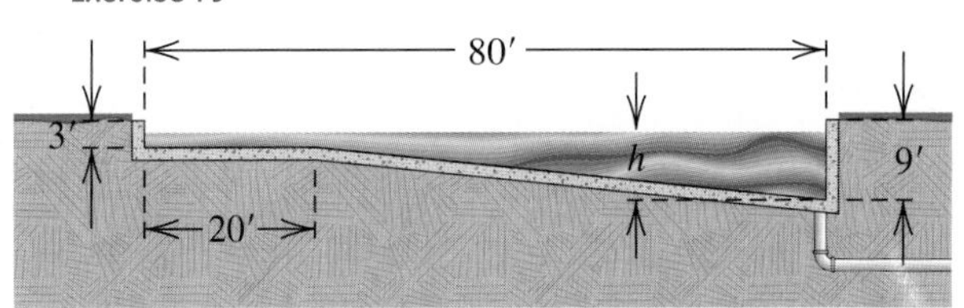

(a) Express the volume V of the water in the pool as a function of time t.

(b) Express V as a function of the depth h at the deep end for $0 \leq h \leq 6$ and then for $6 < h \leq 9$.

(c) Express h as a function of t for $0 \leq h \leq 6$ and then for $6 < h \leq 9$.

80 Filtering water Suppose 5 in^3 of water is poured into a conical filter and subsequently drips into a cup, as shown in the figure. Let x denote the height of the water in the filter, and let y denote the height of the water in the cup.

(a) Express the radius r shown in the figure as a function of x. (*Hint:* Use similar triangles.)

(b) Express the height y of the water in the cup as a function of x. (*Hint:* What is the sum of the two volumes shown in the figure?)

Exercise 80

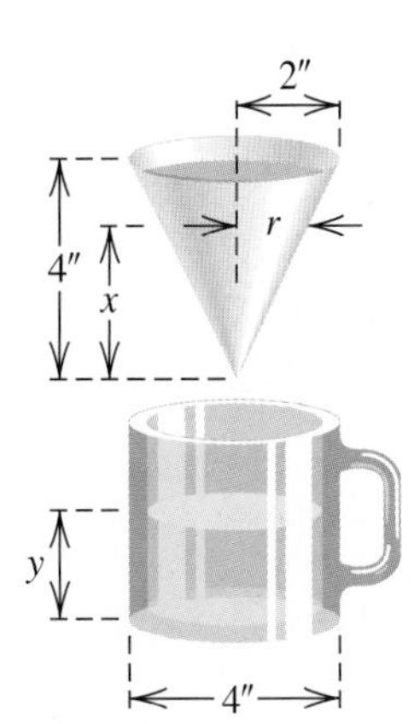

81 Frustum of a cone The shape of the first spacecraft in the Apollo program was a frustum of a right circular cone—a solid formed by truncating a cone by a plane parallel to its base. For the frustum shown in the figure, the radii a and b have already been determined.

Exercise 81

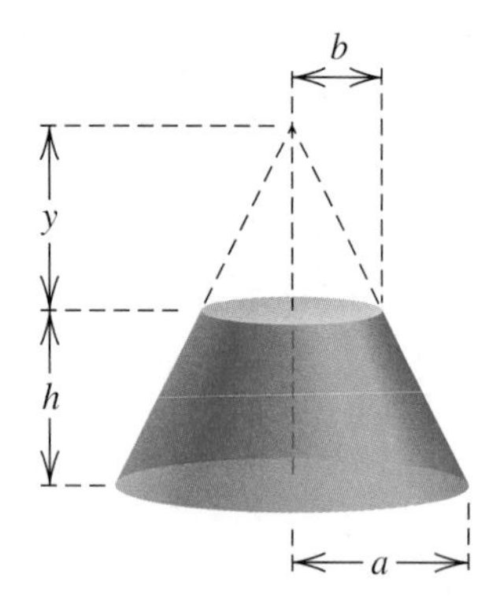

(a) Use similar triangles to express y as a function of h.

(b) Derive a formula for the volume of the frustum as a function of h.

(c) If $a = 6$ ft and $b = 3$ ft, for what value of h is the volume of the frustum 600 ft^3?

82 Distance between ships At 1:00 P.M. ship A is 30 miles due south of ship B and is sailing north at a rate of 15 mi/hr. If ship B is sailing west at a rate of 10 mi/hr, find the time at which the distance d between the ships is minimal (see the figure).

Exercise 82

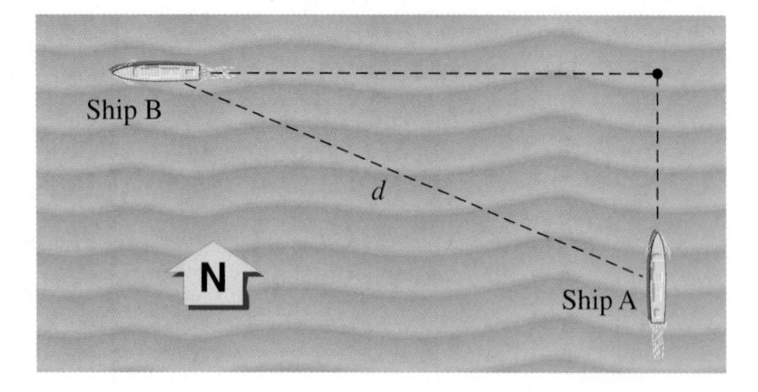

83 Dimensions of a race track The interior of a half-mile race track consists of a rectangle with semicircles at two opposite ends. Find the dimensions that will maximize the area of the rectangle.

84 Vertical leaps When a particular basketball player leaps straight up for a dunk, the player's distance $f(t)$ (in feet) off the floor after t seconds is given by the formula $f(t) = -\frac{1}{2}gt^2 + 16t$, where g is a gravitational constant.

(a) If $g = 32$, find the player's hang time—that is, the total number of seconds that the player is in the air.

(b) Find the player's vertical leap—that is, the maximum distance of the player's feet from the floor.

(c) On the moon, $g = \frac{32}{6}$. Rework parts (a) and (b) for the player on the moon.

85 Trajectory of a rocket A rocket is fired up a hillside, following a path given by $y = -0.016x^2 + 1.6x$. The hillside has slope $\frac{1}{5}$, as illustrated in the figure.

(a) Where does the rocket land?

(b) Find the maximum height of the rocket *above the ground.*

Exercise 85

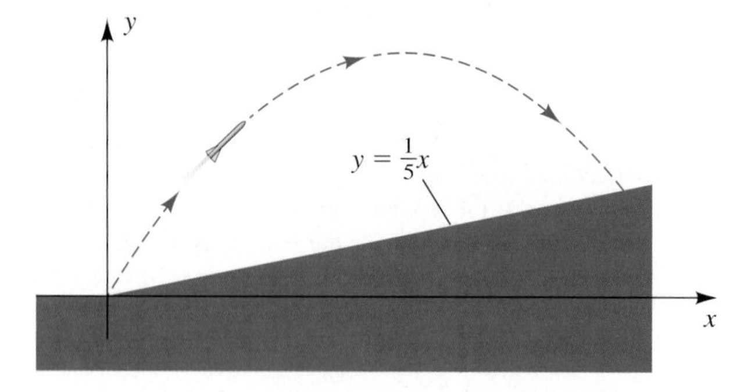

CHAPTER 2 DISCUSSION EXERCISES

1 Compare the graphs of $y = \sqrt[3]{x}$, $y = \sqrt{x}$, $y = x$, $y = x^2$, and $y = x^3$ on the interval $0 \leq x \leq 2$. Write a generalization based on what you find out about graphs of equations of the form $y = x^{p/q}$, where $x \geq 0$ and p and q are positive integers.

2 Write an expression for $g(x)$ if the graph of g is obtained from the graph of $f(x) = \frac{1}{2}x - 3$ by reflecting f about the

(a) x-axis (b) y-axis

(c) line $y = 2$ (d) line $x = 3$

3 Consider the graph of $g(x) = \sqrt{f(x)}$, where f is given by $f(x) = ax^2 + bx + c$. Discuss the general shape of g, including its domain and range. Discuss the advantages and disadvantages of graphing g as a composition of the functions $h(x) = \sqrt{x}$ and $f(x)$. (*Hint:* You may want to use the following expressions for f: $x^2 - 2x - 8$, $-x^2 + 2x + 8$, $x^2 - 2x + 2$, $-x^2 + 2x - 2$.)

4 Simplify the difference quotient in Exercises 49 and 50 of Section 2.4 for an arbitrary quadratic function of the form $f(x) = ax^2 + bx + c$.

5 Refer to Example 5 in Section 2.4. Geometrically, what does the expression $2x + h + 6$ represent on the graph of f? What do you think it represents if $h = 0$?

6 The midpoint formula could be considered to be the "halfway" formula since it gives us the point that is $\frac{1}{2}$ of the distance from the point $P(x_1, y_1)$ to the point $Q(x_2, y_2)$. Develop an "m-nth way" formula that gives the point $R(x_3, y_3)$ that is m/n of the distance from P to Q (assume m and n are positive integers with $m < n$).

7 Consider the graphs of equations of the quadratic form $y = ax^2 + bx + c$ that have two x-intercepts. Let d denote the distance from the axis of the parabola to either of the x-intercepts, and let h denote the value of the y-coordinate of the vertex. Explore the relationship between d and h for several specific equations, and then develop a formula for this relationship.

8 **Billing for service** A common method of billing for service calls is to charge a flat fee plus an additional fee for each quarter-hour spent on the call. Create a function for a washer repair company that charges \$40 plus \$20 for each quarter-hour or portion thereof—for example, a 30-minute repair call would cost \$80, while a 31-minute repair call would cost \$100. The input to your function is any positive integer. (*Hint:* See Exercise 54(e) of Section 2.5.)

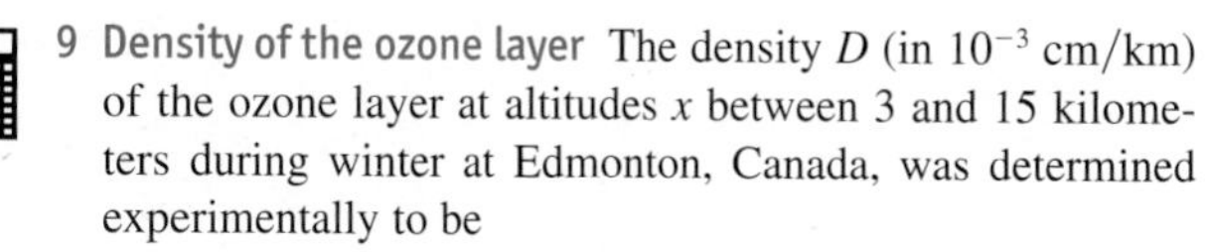

9 **Density of the ozone layer** The density D (in 10^{-3} cm/km) of the ozone layer at altitudes x between 3 and 15 kilometers during winter at Edmonton, Canada, was determined experimentally to be

$$D = 0.0833x^2 - 0.4996x + 3.5491.$$

Express x as a function of D.

10 **Precipitation in Minneapolis** The average monthly precipitation in inches in Minneapolis is listed in the table.

Month	Precipitation
Jan.	0.7
Feb.	0.8
Mar.	1.5
Apr.	1.9
May	3.2
June	4.0
July	3.3
Aug.	3.2
Sept.	2.4
Oct.	1.6
Nov.	1.4
Dec.	0.9

(a) Plot the average monthly precipitation.

(b) Model these data with a piecewise function f that is first quadratic and then linear.

(c) Graph f together with the data.

3

Polynomial and Rational Functions

Polynomial functions are the most basic functions in mathematics, because they are defined only in terms of addition, subtraction, and multiplication. In applications it is often necessary to sketch their graphs and to find (or approximate) their zeros. In the first part of this chapter we discuss results that are useful in obtaining this information. We then turn our attention to quotients of polynomial functions—that is, rational functions. The last section, on variation, contains applications of simple polynomial and rational functions.

3.1

Polynomial Functions of Degree Greater Than 2

If f is a polynomial function with real coefficients of degree n, then

$$f(x) = a_n x^n + a_{n-1}x^{n-1} + \cdots + a_1 x + a_0,$$

with $a_n \neq 0$. The special cases listed in the following chart were previously discussed.

Degree of f	Form of $f(x)$	Graph of f (with y-intercept a_0)
0	$f(x) = a_0$	A horizontal line
1	$f(x) = a_1 x + a_0$	A line with slope a_1
2	$f(x) = a_2 x^2 + a_1 x + a_0$	A parabola with a vertical axis

Figure 1

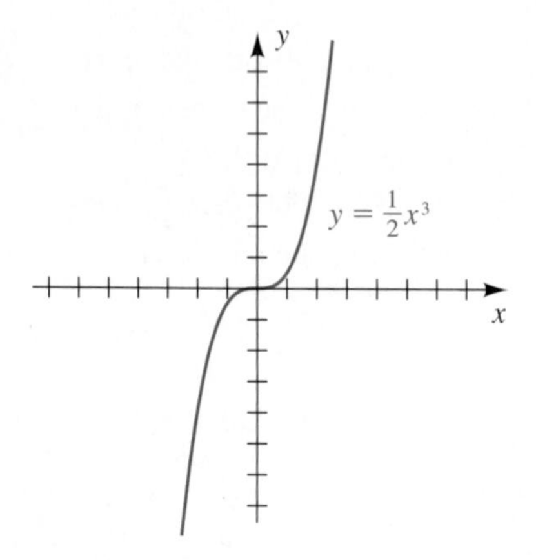

In this section we shall discuss graphs of polynomial functions of degree greater than 2. All polynomial functions are **continuous functions**—that is, their graphs can be drawn without any breaks.

If f has degree n *and all the coefficients except* a_n *are zero,* then

$$f(x) = ax^n \quad \text{for some} \quad a = a_n \neq 0.$$

In this case, if $n = 1$, the graph of f is a line through the origin. If $n = 2$, the graph is a parabola with vertex at the origin. Two illustrations with $n = 3$ (**cubic polynomials**) are given in the next example.

EXAMPLE 1 Sketching graphs of $y = ax^3$

Sketch the graph of f if

(a) $f(x) = \frac{1}{2}x^3$ **(b)** $f(x) = -\frac{1}{2}x^3$

SOLUTION

Figure 2

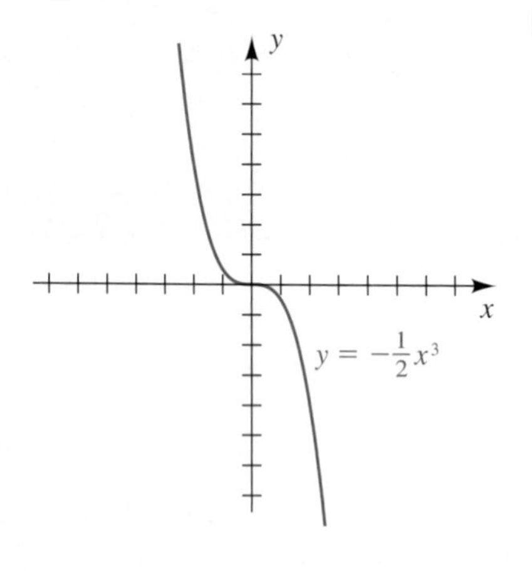

(a) The following table lists several points on the graph of $y = \frac{1}{2}x^3$.

x	0	$\frac{1}{2}$	1	$\frac{3}{2}$	2	$\frac{5}{2}$
y	0	$\frac{1}{16} \approx 0.06$	$\frac{1}{2}$	$\frac{27}{16} \approx 1.7$	4	$\frac{125}{16} \approx 7.8$

Since f is an odd function, the graph of f is symmetric with respect to the origin, and hence points such as $\left(-\frac{1}{2}, -\frac{1}{16}\right)$ and $\left(-1, -\frac{1}{2}\right)$ are also on the graph. The graph is sketched in Figure 1.

(b) If $y = -\frac{1}{2}x^3$, the graph can be obtained from that in part (a) by multiplying all y-coordinates by -1 (that is, by reflecting the graph in part (a) through the x-axis). This gives us the sketch in Figure 2.

If $f(x) = ax^n$ and n is an *odd* positive integer, then f is an odd function and the graph of f is symmetric with respect to the origin, as illustrated in Figures 1

Tutorial available at www.thomsonedu.com/login

and 2. For $a > 0$, the graph is similar in shape to that in Figure 1; however, as either n or a increases, the graph rises more rapidly for $x > 1$. If $a < 0$, we reflect the graph through the x-axis, as in Figure 2.

If $f(x) = ax^n$ and n is an *even* positive integer, then f is an even function and the graph of f is symmetric with respect to the y-axis, as illustrated in Figure 3 for the case $a = 1$ and $n = 4$. Note that as the exponent increases, the graph becomes flatter at the origin. It also rises more rapidly for $x > 1$. If $a < 0$, we reflect the graph through the x-axis. Also note that the graph *intersects* the x-axis at the origin, but it does not *cross* the x-axis (change sign).

Figure 3

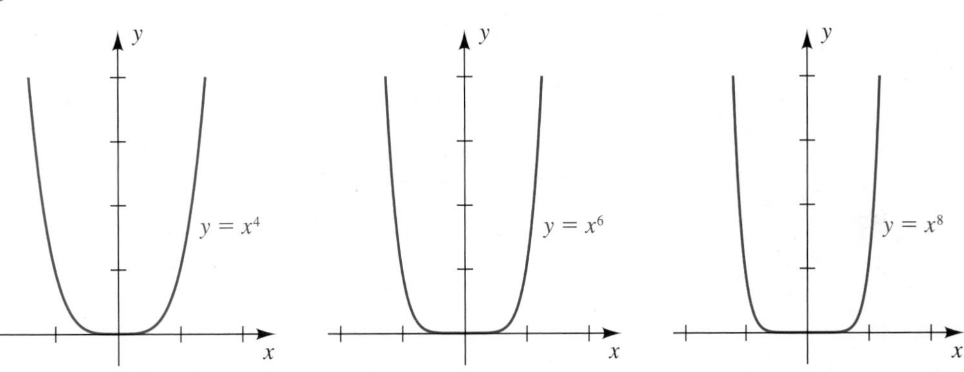

Figure 4

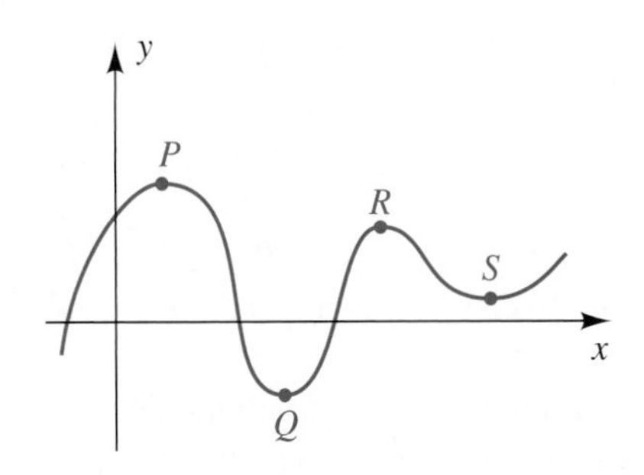

A complete analysis of graphs of polynomial functions of degree greater than 2 requires methods that are used in calculus. As the degree increases, the graphs usually become more complicated. They always have a smooth appearance, however, with a number of high points and low points, such as P, Q, R, and S in Figure 4. Such points are sometimes called **turning points** for the graph. It should be noted that an n-degree polynomial has at most $n - 1$ turning points. Each function value (y-coordinate) corresponding to a high or low point is called an **extremum** of the function f. At an extremum, f changes from an increasing function to a decreasing function, or vice versa.

The intermediate value theorem specifies another important property of polynomial functions.

Intermediate Value Theorem for Polynomial Functions	If f is a polynomial function and $f(a) \neq f(b)$ for $a < b$, then f takes on every value between $f(a)$ and $f(b)$ in the interval $[a, b]$.

The intermediate value theorem for polynomial functions states that if w is any number between $f(a)$ and $f(b)$, there is at least one number c between a and b such that $f(c) = w$. If we regard the graph of f as extending continuously

Figure 5

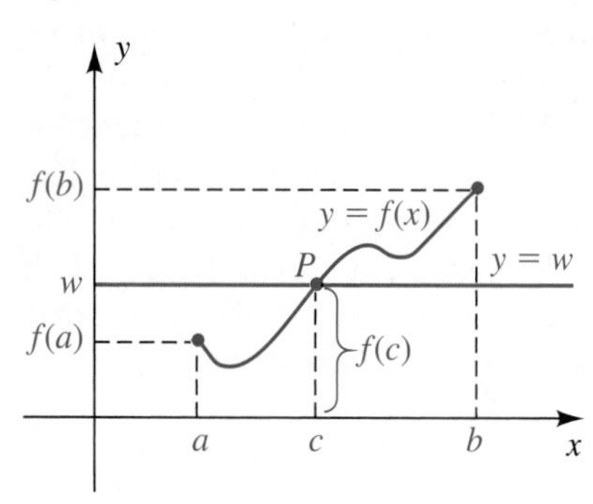

from the point $(a, f(a))$ to the point $(b, f(b))$, as illustrated in Figure 5, then for any number w between $f(a)$ and $f(b)$, the horizontal line $y = w$ intersects the graph in at least one point P. The x-coordinate c of P is a number such that $f(c) = w$.

A consequence of the intermediate value theorem is that if $f(a)$ and $f(b)$ have opposite signs (one positive and one negative), there is at least one number c between a and b such that $f(c) = 0$; that is, *f has a* **zero** *at c*. Thus, if the point $(a, f(a))$ lies below the x-axis and the point $(b, f(b))$ lies above the x-axis, or vice versa, the graph crosses the x-axis at least once between $x = a$ and $x = b$, as illustrated in Figure 6.

Figure 6

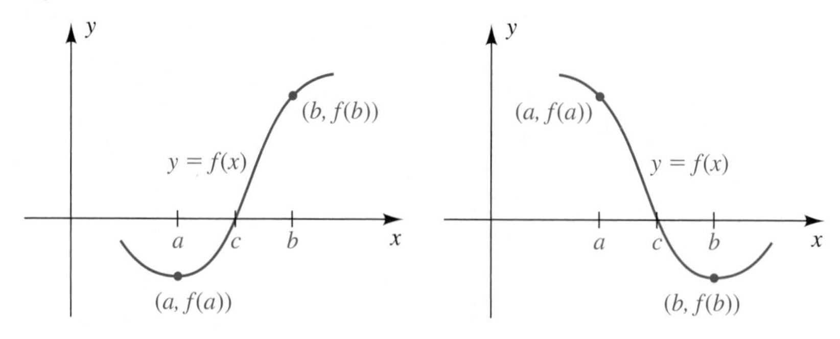

EXAMPLE 2 **Using the intermediate value theorem**

Show that $f(x) = x^5 + 2x^4 - 6x^3 + 2x - 3$ has a zero between 1 and 2.

▶ SOLUTION Substituting 1 and 2 for x gives us the following function values:

$$f(1) = 1 + 2 - 6 + 2 - 3 = -4$$
$$f(2) = 32 + 32 - 48 + 4 - 3 = 17$$

Since $f(1)$ and $f(2)$ have opposite signs ($f(1) = -4 < 0$ and $f(2) = 17 > 0$), we see that $f(c) = 0$ for at least one real number c between 1 and 2.

Example 2 illustrates a method for locating real zeros of polynomials. By using *successive approximations,* we can approximate each zero at any degree of accuracy by locating it in smaller and smaller intervals.

If c and d are *successive* at real zeros of $f(x)$—that is, there are no other zeros between c and d—then *$f(x)$ does not change sign on the interval (c, d)*. Thus, if we choose any number k such that $c < k < d$ and if $f(k)$ is positive, then $f(x)$ is positive throughout (c, d). Similarly, if $f(k)$ is negative, then $f(x)$ is negative throughout (c, d). We shall call $f(k)$ a **test value** for $f(x)$ on the interval (c, d). Test values may also be used on infinite intervals of the form $(-\infty, a)$ or (a, ∞), provided that $f(x)$ has no zeros on these intervals. The use of test values in graphing is similar to the technique used for inequalities in Section 1.6.

▶ **Tutorial available at www.thomsonedu.com/login**

EXAMPLE 3 **Sketching the graph of a polynomial function of degree 3**

Let $f(x) = x^3 + x^2 - 4x - 4$. Find all values of x such that $f(x) > 0$ and all x such that $f(x) < 0$, and then sketch the graph of f.

SOLUTION We may factor $f(x)$ as follows:

$$\begin{aligned} f(x) &= x^3 + x^2 - 4x - 4 && \text{given} \\ &= (x^3 + x^2) + (-4x - 4) && \text{group terms} \\ &= x^2(x + 1) - 4(x + 1) && \text{factor out } x^2 \text{ and } -4 \\ &= (x^2 - 4)(x + 1) && \text{factor out } (x + 1) \\ &= (x + 2)(x - 2)(x + 1) && \text{difference of squares} \end{aligned}$$

Figure 7

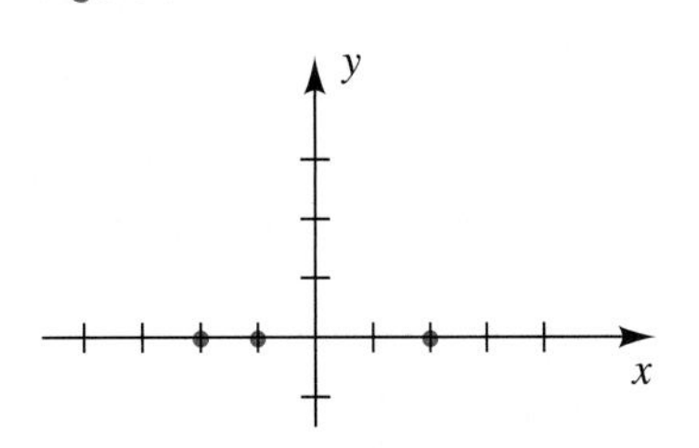

We see from the last equation that the zeros of $f(x)$ (the x-intercepts of the graph) are -2, -1, and 2. The corresponding points on the graph (see Figure 7) divide the x-axis into four parts, and we consider the open intervals

$$(-\infty, -2), \quad (-2, -1), \quad (-1, 2), \quad (2, \infty).$$

As in our work with inequalities in Section 1.6, the sign of $f(x)$ in each of these intervals can be determined by using a sign chart. The graph of f lies above the x-axis for values of x such that $f(x) > 0$, and it lies below the x-axis for all x such that $f(x) < 0$.

Interval	$(-\infty, -2)$	$(-2, -1)$	$(-1, 2)$	$(2, \infty)$
Sign of $x + 2$	$-$	$+$	$+$	$+$
Sign of $x + 1$	$-$	$-$	$+$	$+$
Sign of $x - 2$	$-$	$-$	$-$	$+$
Sign of $f(x)$	$-$	$+$	$-$	$+$
Position of graph	Below x-axis	Above x-axis	Below x-axis	Above x-axis

Figure 8

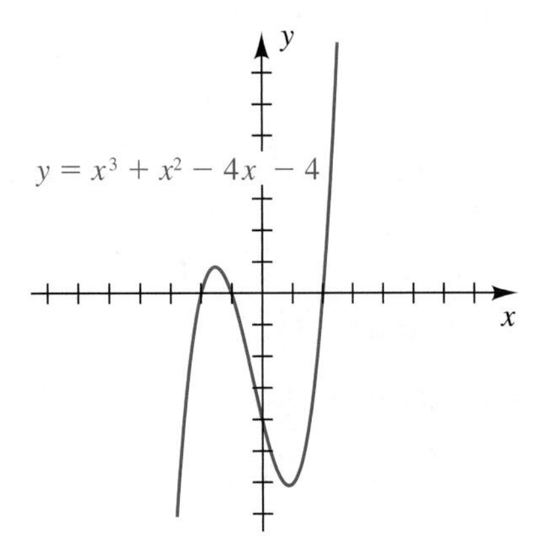

Referring to the sign of $f(x)$ in the chart, we conclude that

$$f(x) > 0 \quad \text{if } x \text{ is in } (-2, -1) \cup (2, \infty)$$

and

$$f(x) < 0 \quad \text{if } x \text{ is in } (-\infty, -2) \cup (-1, 2).$$

Using this information leads to the sketch in Figure 8. To find the turning points on the graph, it would be necessary to use a computational device (as we will do in Example 6) or methods developed in calculus.

The graph of every polynomial function of degree 3 has an appearance similar to that of Figure 8, or it has an inverted version of that graph if the coefficient of x^3 is negative. Sometimes, however, the graph may have only one x-intercept or the shape may be elongated, as in Figures 1 and 2.

Tutorial available at www.thomsonedu.com/login

EXAMPLE 4 Sketching the graph of a polynomial function of degree 4

Let $f(x) = x^4 - 4x^3 + 3x^2$. Find all values of x such that $f(x) > 0$ and all x such that $f(x) < 0$, and then sketch the graph of f.

▶ **SOLUTION** We begin by factoring $f(x)$:

$$\begin{aligned} f(x) &= x^4 - 4x^3 + 3x^2 && \text{given} \\ &= x^2(x^2 - 4x + 3) && \text{factor out } x^2 \\ &= x^2(x-1)(x-3) && \text{factor } x^2 - 4x + 3 \end{aligned}$$

Next, we construct the sign diagram in Figure 9, where the vertical lines indicate the zeros 0, 1, and 3 of the factors. Since the factor x^2 is always positive if $x \neq 0$, it has no effect on the sign of the product and hence may be omitted from the diagram.

Figure 9

		0		1		3	
Sign of $f(x)$	+		+		−		+
Sign of $x - 3$	−		−		−		+
Sign of $x - 1$	−		−		+		+

Figure 10

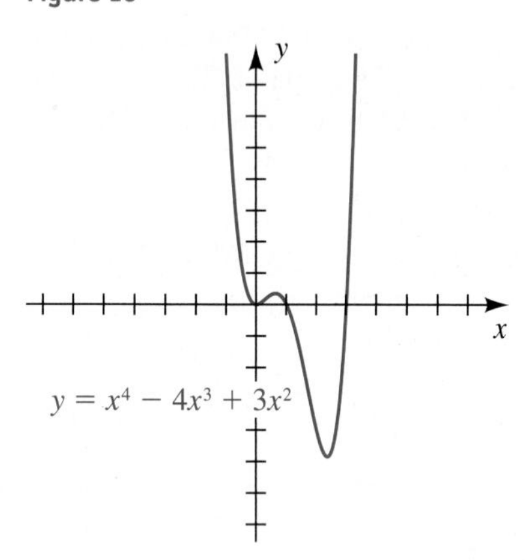

Referring to the sign of $f(x)$ in the diagram, we see that

$$f(x) > 0 \quad \text{if } x \text{ is in } (-\infty, 0) \cup (0, 1) \cup (3, \infty)$$

and

$$f(x) < 0 \quad \text{if } x \text{ is in } (1, 3).$$

Note that the sign of $f(x)$ does not change at $x = 0$. Making use of these facts leads to the sketch in Figure 10.

In the next example we construct a graph of a polynomial knowing only its sign.

EXAMPLE 5 Sketch the graph of a polynomial knowing its sign

Given the sign diagram in Figure 11, sketch a possible graph of the polynomial f.

Figure 11

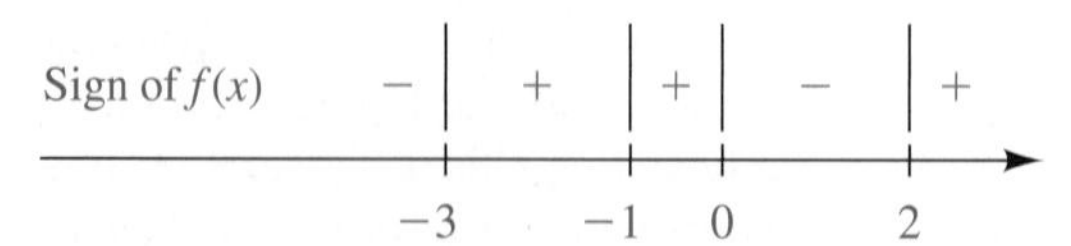

▶ **Tutorial available at www.thomsonedu.com/login**

Figure 12

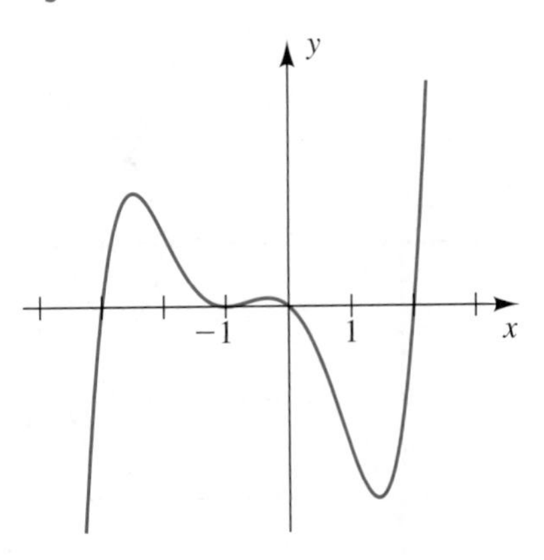

▶ SOLUTION Since the sign of $f(x)$ is *negative* in the interval $(-\infty, -3)$, the graph of f must be *below* the x-axis, as shown in Figure 12. In the interval $(-3, -1)$, the sign of $f(x)$ is *positive*, so the graph of f is *above* the x-axis.

The sign of $f(x)$ is also *positive* in the next interval, $(-1, 0)$. Thus, the graph of f must touch the x-axis at the x-intercept -1 and then remain *above* the x-axis. (The graph of f is *tangent* to the x-axis at $x = -1$.)

In the interval $(0, 2)$, the sign of $f(x)$ is *negative*, so the graph of f is *below* the x-axis. Lastly, the sign of $f(x)$ is *positive* in the interval $(2, \infty)$, and the graph of f is *above* the x-axis.

In the last example we used the function

$$f(x) = (x + 3)(x + 1)^2(x)(x - 2).$$

Note how the graph of f relates to the solutions of the following inequalities.

Inequality	Solution	Position of graph in relation to the x-axis
(1) $f(x) > 0$	$(-3, -1) \cup (-1, 0) \cup (2, \infty)$	Above
(2) $f(x) \geq 0$	$[-3, 0] \cup [2, \infty)$	Above or on
(3) $f(x) < 0$	$(-\infty, -3) \cup (0, 2)$	Below
(4) $f(x) \leq 0$	$(-\infty, -3] \cup \{-1\} \cup [0, 2]$	Below or on

Notice that every real number must be in the solution to either inequality (1) or inequality (4)—the same can be said for inequalities (2) and (3).

In the following example we use a graphing utility to estimate coordinates of important points on a graph.

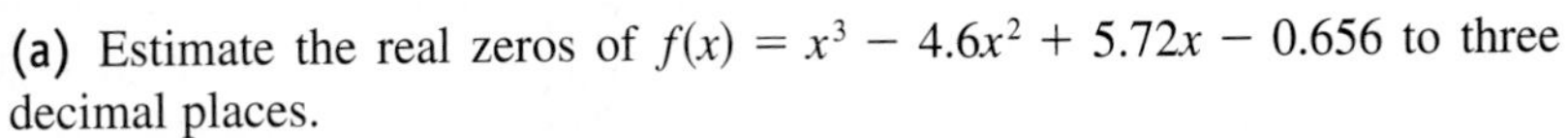

EXAMPLE 6 Estimating zeros and turning points

(a) Estimate the real zeros of $f(x) = x^3 - 4.6x^2 + 5.72x - 0.656$ to three decimal places.

(b) Estimate the coordinates of the turning points on the graph.

SOLUTION

(a) We assign $f(x)$ to Y_1 and use a standard viewing rectangle to obtain a display similar to Figure 13(a). Since all the real roots appear to lie between 0 and 3, let us regraph, using the viewing rectangle $[-1, 3]$ by $[-1, 3]$. This gives us a display similar to Figure 13(b), which shows that there is only one x-intercept

(continued)

▶ **Tutorial available at www.thomsonedu.com/login**

and hence one real root. Using a zero or root feature, we estimate the real zero as 0.127.

Figure 13

(a) $[-15, 15]$ by $[-10, 10]$

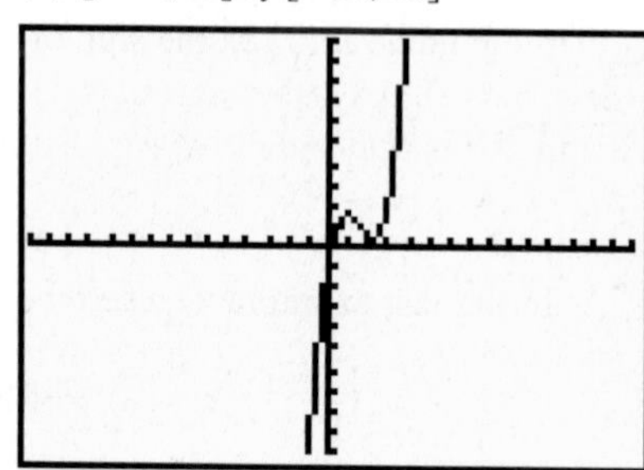

(b) $[-1, 3]$ by $[-1, 3]$

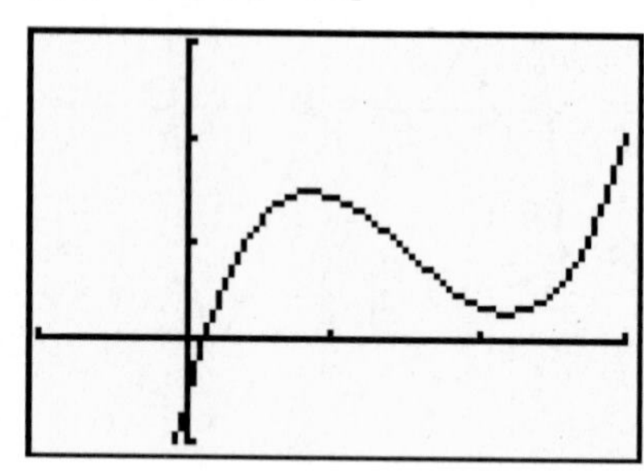

(b) Using a maximum feature, we estimate the high point to be (0.867, 1.497), and using a minimum feature, we estimate the low point to be (2.200, 0.312).

In Section 1.6 we solved inequalities similar to the one in the next example, but we relied heavily on the fact that we could somehow factor the expression. We now use a graphing utility to solve an inequality involving an expression (a cubic polynomial) that is not easily factored.

EXAMPLE 7 **Solving an inequality graphically**

Estimate the solutions of the inequality

$$6x^2 - 3x^3 < 2.$$

SOLUTION Let us subtract 2 from both sides and consider the equivalent inequality

$$6x^2 - 3x^3 - 2 < 0.$$

Figure 14

$[-2, 3]$ by $[-3, 3]$

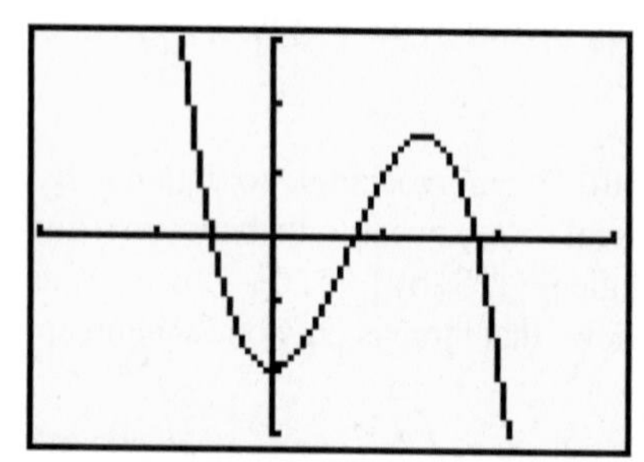

We assign $6x^2 - 3x^3 - 2$ to Y_1 and use the viewing rectangle $[-2, 3]$ by $[-3, 3]$ to obtain a display similar to Figure 14. We see that there are three x-intercepts. If we denote them by x_1, x_2, and x_3 (with $x_1 < x_2 < x_3$), then the solutions to the inequality are given by

$$(x_1, x_2) \cup (x_3, \infty),$$

since these are the intervals on which Y_1 is less than 0 (the graph is below the x-axis). Using a zero or root feature for each x-intercept, we find that

$$x_1 \approx -0.515, \qquad x_2 \approx 0.722, \qquad x_3 \approx 1.793.$$

Using the TI-86 POLY Feature

The TI-86 has a special POLY feature that calculates the zeros of a polynomial. Let's apply this feature to the polynomial $6x^2 - 3x^3 - 2$ from Example 7, which can be written as $-3x^3 + 6x^2 + 0x - 2$.

Enter the degree of the polynomial.

2nd POLY 3 ENTER

```
POLY
order=3
```

Enter the coefficients of the polynomial.

−3 ▽ 6 ▽ 0 ▽ −2

```
a3x^3+…+a1x+a0=0
 a3=-3
 a2=6
 a1=0
 a0=-2
CLRa        SOLVE
```

Calculate the zeros of the polynomial.

SOLVE(F5)

```
a3x^3+…+a1x+a0=0
 x1=1.79251721397
 x2=.722351724464
 x3=-.514868938439
COEFS STOa
```

3.1 Exercises

Exer. 1–4: Sketch the graph of f for the indicated value of c or a.

▶ 1 $f(x) = 2x^3 + c$

(a) $c = 3$ (b) $c = -3$

2 $f(x) = -2x^3 + c$

(a) $c = -2$ (b) $c = 2$

3 $f(x) = ax^3 + 2$

(a) $a = 2$ (b) $a = -\frac{1}{3}$

4 $f(x) = ax^3 - 3$

(a) $a = -2$ (b) $a = \frac{1}{4}$

Exer. 5–10: Use the intermediate value theorem to show that f has a zero between a and b.

▶ 5 $f(x) = x^3 - 4x^2 + 3x - 2$; $a = 3$, $b = 4$

6 $f(x) = 2x^3 + 5x^2 - 3$; $a = -3$, $b = -2$

▶ 7 $f(x) = -x^4 + 3x^3 - 2x + 1$; $a = 2$, $b = 3$

8 $f(x) = 2x^4 + 3x - 2$; $a = \frac{1}{2}$, $b = \frac{3}{4}$

9 $f(x) = x^5 + x^3 + x^2 + x + 1$; $a = -\frac{1}{2}$, $b = -1$

10 $f(x) = x^5 - 3x^4 - 2x^3 + 3x^2 - 9x - 6$; $a = 3$, $b = 4$

▶ Tutorial available at www.thomsonedu.com/login

Exer. 11–12: Match each graph with an equation.

11

(a)

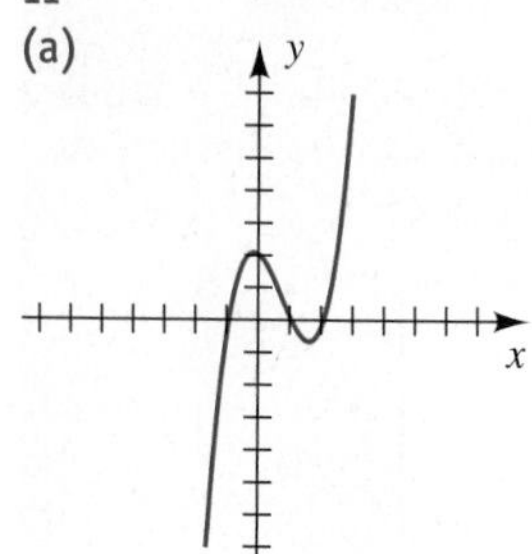

(b)

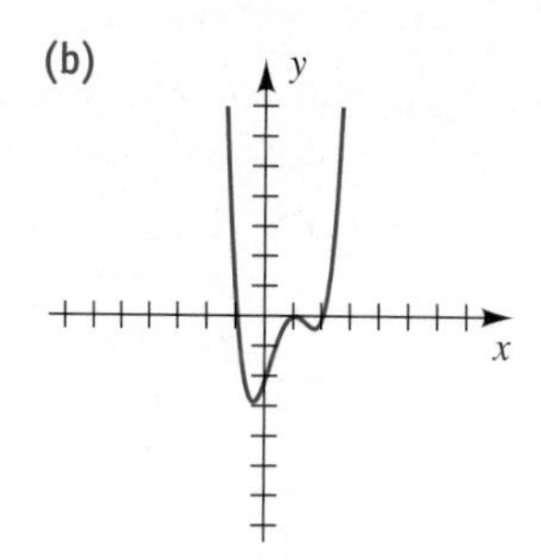

(c)

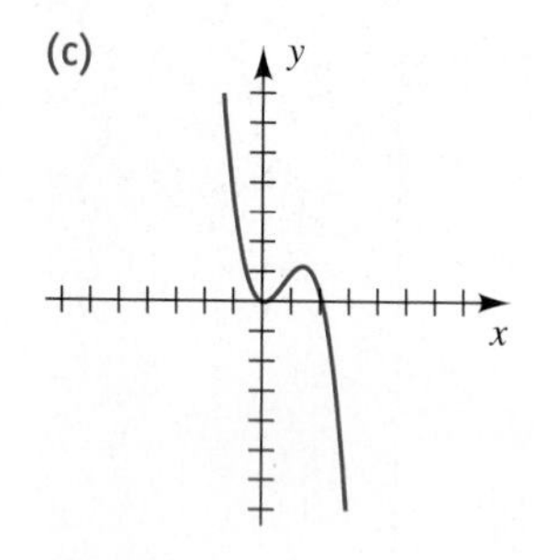

(d)

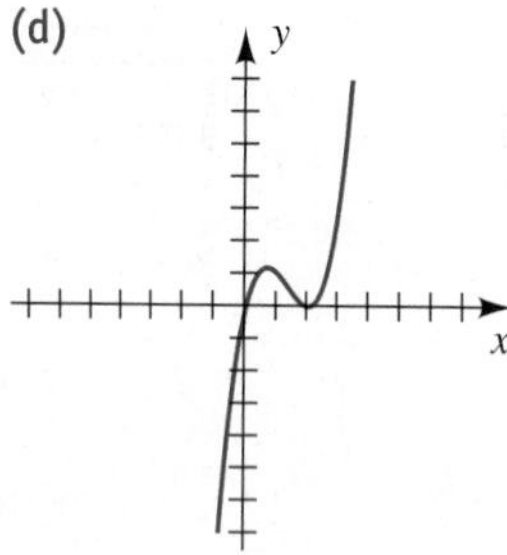

(A) $f(x) = x(x - 2)^2$

(B) $f(x) = -x^2(x - 2)$

(C) $f(x) = (x + 1)(x - 1)(x - 2)$

(D) $f(x) = (x + 1)(x - 1)^2(x - 2)$

12

(a)

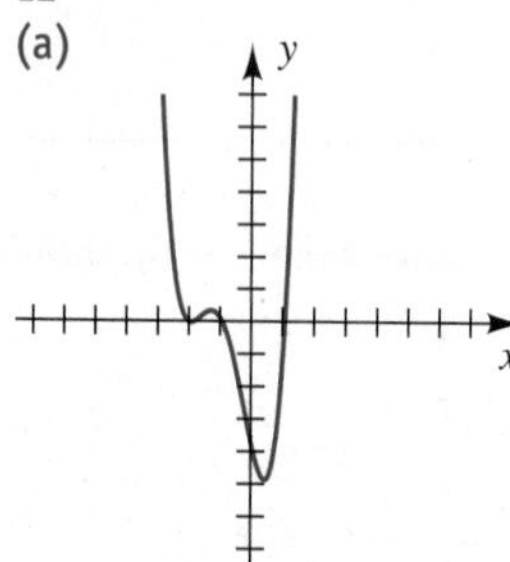

(b)

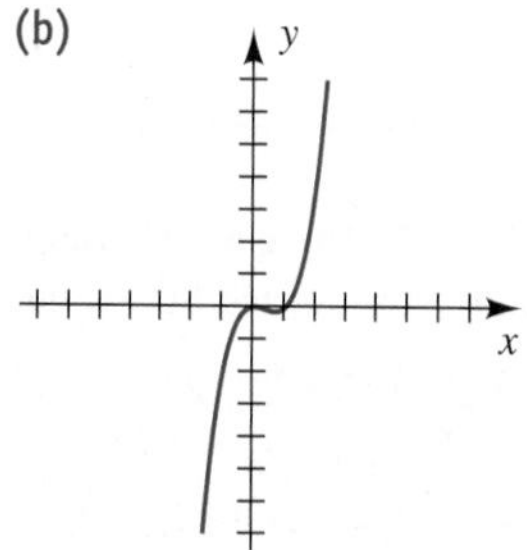

(c)

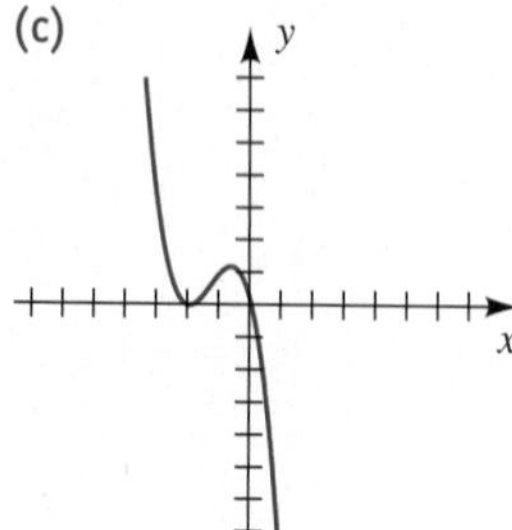

(d)

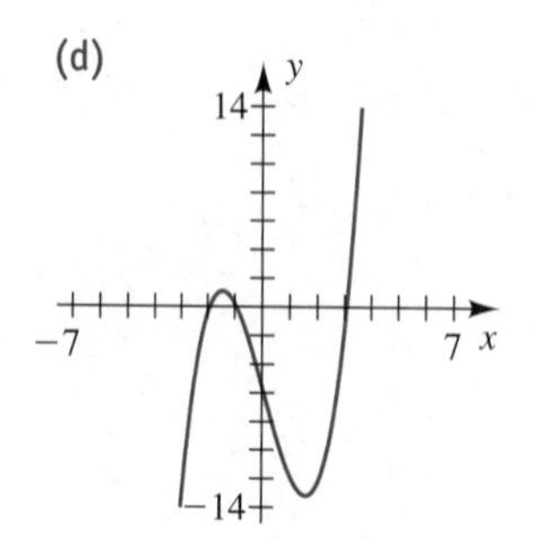

(A) $f(x) = x^2(x - 1)$

(B) $f(x) = -x(x + 2)^2$

(C) $f(x) = (x + 2)(x + 1)(x - 3)$

(D) $f(x) = (x + 2)^2(x + 1)(x - 1)$

Exer. 13–28: Find all values of x such that $f(x) > 0$ and all x such that $f(x) < 0$, and sketch the graph of f.

13 $f(x) = \frac{1}{4}x^3 - 2$

14 $f(x) = -\frac{1}{9}x^3 - 3$

15 $f(x) = -\frac{1}{16}x^4 + 1$

▶ 16 $f(x) = x^5 + 1$

▶ 17 $f(x) = x^4 - 4x^2$

18 $f(x) = 9x - x^3$

19 $f(x) = -x^3 + 3x^2 + 10x$

20 $f(x) = x^4 + 3x^3 - 4x^2$

21 $f(x) = \frac{1}{6}(x + 2)(x - 3)(x - 4)$

22 $f(x) = -\frac{1}{8}(x + 4)(x - 2)(x - 6)$

23 $f(x) = x^3 + 2x^2 - 4x - 8$

24 $f(x) = x^3 - 3x^2 - 9x + 27$

25 $f(x) = x^4 - 6x^2 + 8$

26 $f(x) = -x^4 + 12x^2 - 27$

27 $f(x) = x^2(x + 2)(x - 1)^2(x - 2)$

28 $f(x) = x^3(x + 1)^2(x - 2)(x - 4)$

Exer. 29–30: Sketch the graph of a polynomial given the sign diagram.

29

Sign of $f(x)$	+	−4	−	0	−	1	+	3	−

30

Sign of $f(x)$	+	−3	+	−2	−	0	+	2	−

31 (a) Sketch a graph of

$$f(x) = (x - a)(x - b)(x - c),$$

where $a < 0 < b < c$.

(b) What is the y-intercept?

▶ Tutorial available at www.thomsonedu.com/login

(c) What is the solution to $f(x) < 0$?

(d) What is the solution to $f(x) \geq 0$?

32 (a) Sketch a graph of

$$f(x) = (x - a)^2(x - b)(x - c),$$

where $a < b < 0 < c$.

(b) What is the y-intercept?

(c) What is the solution to $f(x) > 0$?

(d) What is the solution to $f(x) \leq 0$?

33 Let $f(x)$ be a polynomial such that the coefficient of every odd power of x is 0. Show that f is an even function.

34 Let $f(x)$ be a polynomial such that the coefficient of every even power of x is 0. Show that f is an odd function.

35 If $f(x) = 3x^3 - kx^2 + x - 5k$, find a number k such that the graph of f contains the point $(-1, 4)$.

▶ 36 If $f(x) = kx^3 + x^2 - kx + 2$, find a number k such that the graph of f contains the point $(2, 12)$.

37 If one zero of $f(x) = x^3 - 2x^2 - 16x + 16k$ is 2, find two other zeros.

38 If one zero of $f(x) = x^3 - 3x^2 - kx + 12$ is -2, find two other zeros.

39 **A Legendre polynomial** The third-degree Legendre polynomial $P(x) = \frac{1}{2}(5x^3 - 3x)$ occurs in the solution of heat transfer problems in physics and engineering. Find all values of x such that $P(x) > 0$ and all x such that $P(x) < 0$, and sketch the graph of P.

40 **A Chebyshev polynomial** The fourth-degree Chebyshev polynomial $f(x) = 8x^4 - 8x^2 + 1$ occurs in statistical studies. Find all values of x such that $f(x) > 0$. (*Hint:* Let $z = x^2$, and use the quadratic formula.)

41 **Constructing a box** From a rectangular piece of cardboard having dimensions 20 inches $\times$ 30 inches, an open box is to be made by cutting out identical squares of area x^2 from each corner and turning up the sides (see Exercise 65 of Section 2.4).

(a) Show that the volume of the box is given by the function $V(x) = x(20 - 2x)(30 - 2x)$.

(b) Find all positive values of x such that $V(x) > 0$, and sketch the graph of V for $x > 0$.

42 **Constructing a crate** The frame for a shipping crate is to be constructed from 24 feet of 2 $\times$ 2 lumber (see the figure).

(a) If the crate is to have square ends of side x feet, express the outer volume V of the crate as a function of x (disregard the thickness of the lumber).

(b) Sketch the graph of V for $x > 0$.

Exercise 42

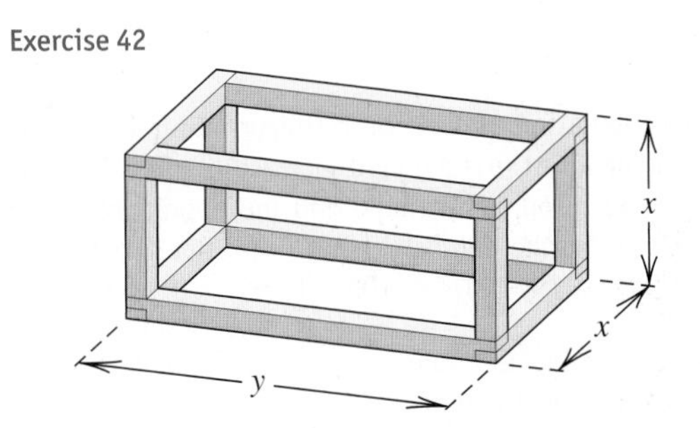

43 **Determining temperatures** A meteorologist determines that the temperature T (in °F) for a certain 24-hour period in winter was given by the formula $T = \frac{1}{20}t(t - 12)(t - 24)$ for $0 \leq t \leq 24$, where t is time in hours and $t = 0$ corresponds to 6 A.M.

(a) When was $T > 0$, and when was $T < 0$?

(b) Sketch the graph of T.

(c) Show that the temperature was 32°F sometime between 12 noon and 1 P.M. (*Hint:* Use the intermediate value theorem.)

44 **Deflections of diving boards** A diver stands at the very end of a diving board before beginning a dive. The deflection d

Exercise 44

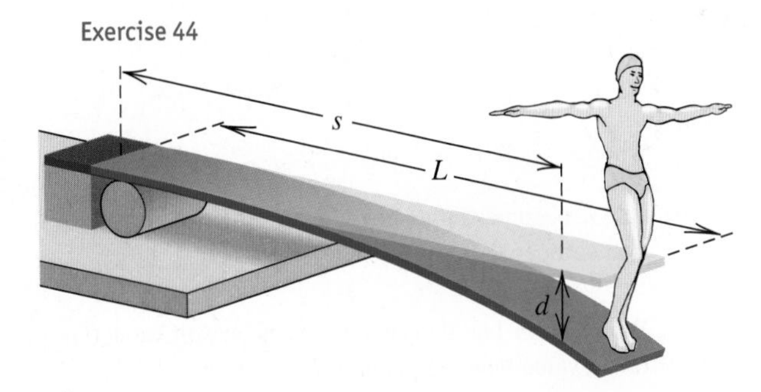

of the board at a position s feet from the stationary end is given by $d = cs^2(3L - s)$ for $0 \leq s \leq L$, where L is the length of the board and c is a positive constant that depends on the weight of the diver and on the physical properties of the board (see the figure). Suppose the board is 10 feet long.

(a) If the deflection at the end of the board is 1 foot, find c.

(b) Show that the deflection is $\frac{1}{2}$ foot somewhere between $s = 6.5$ and $s = 6.6$.

45 **Deer population** A herd of 100 deer is introduced onto a small island. At first the herd increases rapidly, but eventually food resources dwindle and the population declines. Suppose that the number $N(t)$ of deer after t years is given by $N(t) = -t^4 + 21t^2 + 100$, where $t > 0$.

(a) Determine the values of t for which $N(t) > 0$, and sketch the graph of N.

(b) Does the population become extinct? If so, when?

46 **Deer population** Refer to Exercise 45. It can be shown by means of calculus that the rate R (in deer per year) at which the deer population changes at time t is given by $R = -4t^3 + 42t$.

(a) When does the population cease to grow?

(b) Determine the positive values of t for which $R > 0$.

47 (a) Construct a table containing the values of the fourth-degree polynomials

$$f(x) = 2x^4,$$
$$g(x) = 2x^4 - 5x^2 + 1,$$
$$h(x) = 2x^4 + 5x^2 - 1,$$

and

$$k(x) = 2x^4 - x^3 + 2x,$$

when $x = \pm 20, \pm 40$, and ± 60.

(b) As $|x|$ becomes large, how do the values for each function compare?

(c) Which term has the greatest influence on each function's value when $|x|$ is large?

48 (a) Graph the cubic polynomials

$$f(x) = -3x^3,$$
$$g(x) = -3x^3 - x^2 + 1,$$
$$h(x) = -3x^3 + x^2 - 1,$$

and

$$k(x) = -3x^3 - 2x^2 + 2x$$

in the same coordinate plane, using each of the following viewing rectangles:

(1) $[-2, 2]$ by $[-2, 2]$
(2) $[-10, 10]$ by $[-10, 10]$
(3) $[-50, 50, 10]$ by $[-5000, 5000, 1000]$
(4) $[-100, 100, 10]$ by $[-5 \times 10^5, 5 \times 10^5, 10^5]$

(b) As the viewing rectangle increases in size, how do the graphs of the four functions compare?

(c) Which term has the greatest influence on each function's value when $|x|$ is large?

49 (a) Graph each of the following cubic polynomials f in the viewing rectangle $[-9, 9]$ by $[-6, 6]$.

(1) $f(x) = x^3 - x + 1$
(2) $f(x) = -x^3 + 4x^2 - 3x - 1$
(3) $f(x) = 0.1x^3 - 1$
(4) $f(x) = -x^3 + 4x + 2$

(b) Discuss the shape of the graph of f as $|x|$ becomes large.

(c) Make a generalization about the end behavior of the function $f(x) = ax^3 + bx^2 + cx + d$.

50 (a) Graph each of the following fourth-degree polynomials f in the viewing rectangle $[-9, 9]$ by $[-6, 6]$.

(1) $f(x) = -x^4 - 2x^3 + 5x^2 + 6x - 3$
(2) $f(x) = x^4 - 2x^3 + 1$
(3) $f(x) = -\frac{1}{2}x^4 + 2x^2 - x + 1$
(4) $f(x) = \frac{1}{5}x^4 - \frac{1}{2}x^3 - \frac{7}{3}x^2 + \frac{7}{2}x + 3$

(b) Discuss the shape of the graph of f as $|x|$ becomes large.

(c) Make a generalization about the end behavior of the function $f(x) = ax^4 + bx^3 + cx^2 + dx + e$.

Exer. 51–54: Graph f, and estimate its zeros.

51 $f(x) = x^3 + 0.2x^2 - 2.6x + 1.1$

52 $f(x) = -x^4 + 0.1x^3 + 4x^2 - 0.5x - 3$

53 $f(x) = x^3 - 3x + 1$

54 $f(x) = 2x^3 - 4x^2 - 3x + 1$

Exer. 55–58: Graph f, and estimate all values of x such that $f(x) > k$.

55 $f(x) = x^3 + 5x - 2;$ $k = 1$

56 $f(x) = x^4 - 4x^3 + 3x^2 - 8x + 5;$ $k = 3$

57 $f(x) = x^5 - 2x^2 + 2;$ $k = -2$

58 $f(x) = x^4 - 2x^3 + 10x - 26;$ $k = -1$

Exer. 59–60: Graph f and g on the same coordinate plane, and estimate the points of intersection.

59 $f(x) = x^3 - 2x^2 - 1.5x + 2.8;$
$g(x) = -x^3 - 1.7x^2 + 2x + 2.5$

60 $f(x) = x^4 - 5x^2 + 4;$
$g(x) = x^4 - 3x^3 - 0.25x^2 + 3.75x$

61 **Medicare recipients** The function f given by

$$f(x) = -0.000\,015z^3 - 0.005z^2 + 0.75z + 23.5,$$

where $z = x - 1973$, approximates the total number of Medicare recipients in millions, from $x = 1973$ to $x = 2005$. There were 23,545,363 Medicare recipients in 1973 and 42,394,926 in 2005.

(a) Graph f, and discuss how the number of Medicare recipients has changed over this time period.

(b) Create a linear model similar to f that approximates the number of Medicare recipients. Which model is more realistic?

62 **Head Start participants** The function f given by

$$f(x) = -0.11x^4 - 46x^3 + 4000x^2 - 76{,}000x + 760{,}000$$

approximates the total number of preschool children participating in the government program Head Start between 1966 and 2005, where $x = 0$ corresponds to the year 1966.

(a) Graph f on the interval [0, 40]. Discuss how the number of participants has changed between 1966 and 2005.

(b) Approximate the number of children enrolled in 1986.

(c) Estimate graphically the years in which there were 500,000 children enrolled in Head Start.

3.2 Properties of Division

In this section we use $f(x)$, $g(x)$, and so on, to denote polynomials in x. If $g(x)$ is a factor of $f(x)$, then $f(x)$ is **divisible** by $g(x)$. For example, $x^4 - 16$ is divisible by $x^2 - 4$, by $x^2 + 4$, by $x + 2$, and by $x - 2$.

The polynomial $x^4 - 16$ is not divisible by $x^2 + 3x + 1$; however, we can use the process called **long division** to find a *quotient* and a *remainder*, as in the following illustration, where we have inserted terms with zero coefficients.

ILLUSTRATION Long Division of Polynomials

$$
\begin{array}{r|l}
 & \overbrace{x^2 - 3x + 8}^{\text{quotient}} \\
x^2 + 3x + 1 & x^4 + 0x^3 + 0x^2 + 0x - 16 \\
 & \underline{x^4 + 3x^3 + x^2} \qquad x^2(x^2 + 3x + 1) \\
 & -3x^3 - x^2 \qquad \text{subtract} \\
 & \underline{-3x^3 - 9x^2 - 3x} \qquad -3x(x^2 + 3x + 1) \\
 & 8x^2 + 3x - 16 \qquad \text{subtract} \\
 & \underline{8x^2 + 24x + 8} \qquad 8(x^2 + 3x + 1) \\
 & \underbrace{-21x - 24}_{\text{remainder}} \qquad \text{subtract}
\end{array}
$$

■

The long division process ends when we arrive at a polynomial (the remainder) that either is 0 or has smaller degree than the divisor. The result of the long division in the preceding illustration can be written

$$\frac{x^4 - 16}{x^2 + 3x + 1} = (x^2 - 3x + 8) + \left(\frac{-21x - 24}{x^2 + 3x + 1}\right).$$

Multiplying both sides of this equation by $x^2 + 3x + 1$, we obtain

$$x^4 - 16 = (x^2 + 3x + 1)(x^2 - 3x + 8) + (-21x - 24).$$

This example illustrates the following theorem.

Division Algorithm for Polynomials

If $f(x)$ and $p(x)$ are polynomials and if $p(x) \neq 0$, then there exist unique polynomials $q(x)$ and $r(x)$ such that

$$f(x) = p(x) \cdot q(x) + r(x),$$

where either $r(x) = 0$ or the degree of $r(x)$ is less than the degree of $p(x)$. The polynomial $q(x)$ is the **quotient,** and $r(x)$ is the **remainder** in the division of $f(x)$ by $p(x)$.

A useful special case of the division algorithm for polynomials occurs if $f(x)$ is divided by $x - c$, where c is a real number. If $x - c$ is a factor of $f(x)$, then

$$f(x) = (x - c)q(x)$$

for some quotient $q(x)$, and the remainder $r(x)$ is 0. If $x - c$ is not a factor of $f(x)$, then the degree of the remainder $r(x)$ is less than the degree of $x - c$, and hence $r(x)$ must have degree 0. This means that the remainder is a nonzero number. Consequently, for every $x - c$ we have

$$f(x) = (x - c)q(x) + d,$$

where the remainder d is a real number (possibly $d = 0$). If we substitute c for x, we obtain

$$\begin{aligned} f(c) &= (c - c)q(c) + d \\ &= 0 \cdot q(c) + d \\ &= 0 + d = d. \end{aligned}$$

This proves the following theorem.

Remainder Theorem

If a polynomial $f(x)$ is divided by $x - c$, then the remainder is $f(c)$.

EXAMPLE 1 Using the remainder theorem

If $f(x) = x^3 - 3x^2 + x + 5$, use the remainder theorem to find $f(2)$.

▶ SOLUTION According to the remainder theorem, $f(2)$ is the remainder when $f(x)$ is divided by $x - 2$. By long division,

$$
\begin{array}{r|l}
\multicolumn{1}{r}{x^2 - x - 1} & \\
x - 2 \overline{) x^3 - 3x^2 + x + 5} & \\
\underline{x^3 - 2x^2} \qquad\qquad & x^2(x-2) \\
-x^2 + x \qquad & \text{subtract} \\
\underline{-x^2 + 2x} \qquad & -x(x-2) \\
-x + 5 & \text{subtract} \\
\underline{-x + 2} & (-1)(x-2) \\
3 & \text{subtract}
\end{array}
$$

Hence, $f(2) = 3$. We may check this fact by direct substitution:

$$f(2) = 2^3 - 3(2)^2 + 2 + 5 = 3$$

We shall use the remainder theorem to prove the following important result.

Factor Theorem	A polynomial $f(x)$ has a factor $x - c$ if and only if $f(c) = 0$.

PROOF By the remainder theorem,

$$f(x) = (x - c)q(x) + f(c)$$

for some quotient $q(x)$.

If $f(c) = 0$, then $f(x) = (x - c)q(x)$; that is, $x - c$ is a factor of $f(x)$. Conversely, if $x - c$ is a factor of $f(x)$, then the remainder upon division of $f(x)$ by $x - c$ must be 0, and hence, by the remainder theorem, $f(c) = 0$.

The factor theorem is useful for finding factors of polynomials, as illustrated in the next example.

EXAMPLE 2 Using the factor theorem

Show that $x - 2$ is a factor of $f(x) = x^3 - 4x^2 + 3x + 2$.

▶ SOLUTION Since $f(2) = 8 - 16 + 6 + 2 = 0$, we see from the factor theorem that $x - 2$ is a factor of $f(x)$. Another method of solution would be to divide $f(x)$ by $x - 2$ and show that the remainder is 0. The quotient in the division would be another factor of $f(x)$.

▶ Tutorial available at www.thomsonedu.com/login

EXAMPLE 3 Finding a polynomial with prescribed zeros

Find a polynomial $f(x)$ of degree 3 that has zeros 2, -1, and 3.

▶ SOLUTION By the factor theorem, $f(x)$ has factors $x - 2$, $x + 1$, and $x - 3$. Thus,

$$f(x) = a(x - 2)(x + 1)(x - 3),$$

where any nonzero value may be assigned to a. If we let $a = 1$ and multiply, we obtain

$$f(x) = x^3 - 4x^2 + x + 6.$$

To apply the remainder theorem it is necessary to divide a polynomial $f(x)$ by $x - c$. The method of **synthetic division** may be used to simplify this work. The following guidelines state how to proceed. The method can be justified by a careful (and lengthy) comparison with the method of long division.

Guidelines for Synthetic Division of $a_nx^n + a_{n-1}x^{n-1} + \cdots + a_1x + a_0$ by $x - c$

1 Begin with the following display, supplying zeros for any missing coefficients in the given polynomial.

$$\begin{array}{l|llllll} c & a_n & a_{n-1} & a_{n-2} & \cdots & a_1 & a_0 \\ & & & & & & \\ \hline & a_n & & & & & \end{array}$$

2 Multiply a_n by c, and place the product ca_n underneath a_{n-1}, as indicated by the arrow in the following display. (This arrow, and others, is used only to clarify these guidelines and will not appear in *specific* synthetic divisions.) Next find the sum $b_1 = a_{n-1} + ca_n$, and place it below the line as shown.

$$\begin{array}{l|lllllll} c & a_n & a_{n-1} & a_{n-2} & \cdots & & a_1 & a_0 \\ & & ca_n & cb_1 & cb_2 & \cdots & cb_{n-2} & cb_{n-1} \\ \hline & a_n & b_1 & b_2 & \cdots & b_{n-2} & b_{n-1} & r \end{array}$$

3 Multiply b_1 by c, and place the product cb_1 underneath a_{n-2}, as indicated by the second arrow. Proceeding, we next find the sum $b_2 = a_{n-2} + cb_1$ and place it below the line as shown.

4 Continue this process, as indicated by the arrows, until the final sum $r = a_0 + cb_{n-1}$ is obtained. The numbers

$$a_n, \quad b_1, \quad b_2, \quad \ldots, \quad b_{n-2}, \quad b_{n-1}$$

are the coefficients of the quotient $q(x)$; that is,

$$q(x) = a_nx^{n-1} + b_1x^{n-2} + \cdots + b_{n-2}x + b_{n-1},$$

and r is the remainder.

▶ Tutorial available at www.thomsonedu.com/login

Synthetic division does not replace long division; it is merely a faster method and is applicable only when the divisor is of the form $x - c$.

The following examples illustrate synthetic division for some special cases.

EXAMPLE 4 Using synthetic division to find a quotient and remainder

Use synthetic division to find the quotient $q(x)$ and remainder r if the polynomial $2x^4 + 5x^3 - 2x - 8$ is divided by $x + 3$.

▶ **SOLUTION** Since the divisor is $x + 3 = x - (-3)$, the value of c in the expression $x - c$ is -3. Hence, the synthetic division takes this form:

$$\begin{array}{r|rrrrr} -3 & 2 & 5 & 0 & -2 & -8 \\ & & -6 & 3 & -9 & 33 \\ \hline & 2 & -1 & 3 & -11 & 25 \end{array}$$

coefficients of quotient (2, −1, 3, −11); remainder (25)

As we have indicated, the first four numbers in the third row are the coefficients of the quotient $q(x)$, and the last number is the remainder r. Thus,

$$q(x) = 2x^3 - x^2 + 3x - 11 \quad \text{and} \quad r = 25.$$

Synthetic division can be used to find values of polynomial functions, as illustrated in the next example.

EXAMPLE 5 Using synthetic division to find values of a polynomial

If $f(x) = 3x^5 - 38x^3 + 5x^2 - 1$, use synthetic division to find $f(4)$.

▶ **SOLUTION** By the remainder theorem, $f(4)$ is the remainder when $f(x)$ is divided by $x - 4$. Dividing synthetically, we obtain

$$\begin{array}{r|rrrrrr} 4 & 3 & 0 & -38 & 5 & 0 & -1 \\ & & 12 & 48 & 40 & 180 & 720 \\ \hline & 3 & 12 & 10 & 45 & 180 & 719 \end{array}$$

coefficients of quotient (3, 12, 10, 45, 180); remainder (719)

Consequently, $f(4) = 719$.

Synthetic division may be used to help find zeros of polynomials. By the method illustrated in the preceding example, $f(c) = 0$ if and only if the remainder in the synthetic division by $x - c$ is 0.

EXAMPLE 6 Using synthetic division to find zeros of a polynomial

Show that -11 is a zero of the polynomial

$$f(x) = x^3 + 8x^2 - 29x + 44.$$

▶ **Tutorial available at www.thomsonedu.com/login**

▶ **SOLUTION** Dividing synthetically by $x - (-11) = x + 11$ gives us

$$\begin{array}{r|rrrr} -11 & 1 & 8 & -29 & 44 \\ & & -11 & 33 & -44 \\ \hline & 1 & -3 & 4 & 0 \end{array}$$

coefficients of quotient (1, −3, 4); remainder (0)

Thus, $f(-11) = 0$, and -11 is a zero of f.

Example 6 shows that the number -11 is a solution of the equation $x^3 + 8x^2 - 29x + 44 = 0$. In Section 3.4 we shall use synthetic division to find rational solutions of equations.

At this stage you should recognize that the following three statements are equivalent for a polynomial function f whose graph is the graph of the equation $y = f(x)$.

equivalent statements for $f(a) = b$

(1) The point (a, b) is on the graph of f.
(2) The value of f at $x = a$ equals b; that is, $f(a) = b$.
(3) If $f(x)$ is divided by $x - a$, then the remainder is b.

Furthermore, if b is equal to 0, then the next four statements are also equivalent.

additional equivalent statements for $f(a) = 0$

(1) The number a is a zero of the function f.
(2) The point $(a, 0)$ is on the graph of f; that is, a is an x-intercept.
(3) The number a is a solution of the equation $f(x) = 0$.
(4) The binomial $x - a$ is a factor of the polynomial $f(x)$.

You should become familiar with these statements—so familiar that if you know one of them is true, you can easily recall and apply any appropriate equivalent statement.

EXAMPLE 7 Relating a graph to division

Use the graph of

$$f(x) = 0.5x^5 + 3.5x^4 - 5.5x^3 - 7.5x^2 + 2x + 2$$

to approximate (to two decimal places) the remainder if $f(x)$ is divided by $x + 1.37$.

Figure 1
$[-10, 10]$ by $[-10, 10]$

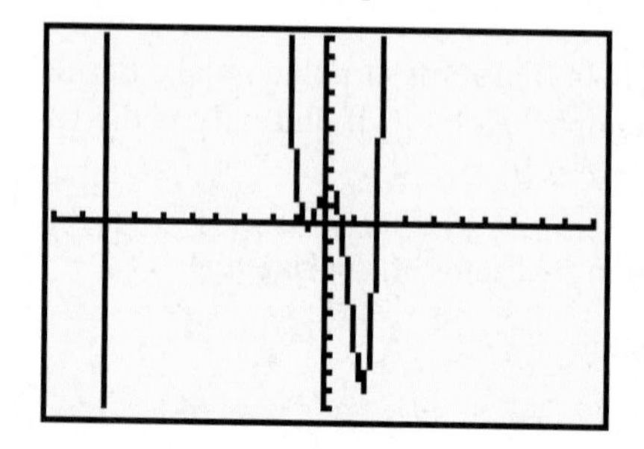

SOLUTION We assign $f(x)$ to Y_1 and graph f with a standard viewing rectangle, as shown in Figure 1. From the preceding discussion, we know that to find a remainder b by utilizing a graph, we should find the point (a, b) that corresponds to dividing $f(x)$ by $x - a$. In this case $a = -1.37$, and the point on the graph with x-coordinate -1.37 is approximately $(-1.37, 9.24)$. Hence, the remainder b is approximately 9.24.

▶ **Tutorial available at www.thomsonedu.com/login**

The easiest way to find the remainder using a graphing utility is to simply find the function value Y_1 when $x = -1.37$. However, the purpose of this example was to point out the graphical relationship to the division process.

3.2 Exercises

Exer. 1–8: Find the quotient and remainder if $f(x)$ is divided by $p(x)$.

1 $f(x) = 2x^4 - x^3 - 3x^2 + 7x - 12;\quad p(x) = x^2 - 3$

2 $f(x) = 3x^4 + 2x^3 - x^2 - x - 6;\quad p(x) = x^2 + 1$

▶ 3 $f(x) = 3x^3 + 2x - 4;\quad p(x) = 2x^2 + 1$

4 $f(x) = 3x^3 - 5x^2 - 4x - 8;\quad p(x) = 2x^2 + x$

5 $f(x) = 7x + 2;\quad p(x) = 2x^2 - x - 4$

6 $f(x) = -5x^2 + 3;\quad p(x) = x^3 - 3x + 9$

7 $f(x) = 9x + 4;\quad p(x) = 2x - 5$

8 $f(x) = 7x^2 + 3x - 10;\quad p(x) = x^2 - x + 10$

Exer. 9–12: Use the remainder theorem to find $f(c)$.

9 $f(x) = 3x^3 - x^2 + 5x - 4;\quad c = 2$

10 $f(x) = 2x^3 + 4x^2 - 3x - 1;\quad c = 3$

11 $f(x) = x^4 - 6x^2 + 4x - 8;\quad c = -3$

12 $f(x) = x^4 + 3x^2 - 12;\quad c = -2$

Exer. 13–16: Use the factor theorem to show that $x - c$ is a factor of $f(x)$.

13 $f(x) = x^3 + x^2 - 2x + 12;\quad c = -3$

14 $f(x) = x^3 + x^2 - 11x + 10;\quad c = 2$

▶ 15 $f(x) = x^{12} - 4096;\quad c = -2$

16 $f(x) = x^4 - 3x^3 - 2x^2 + 5x + 6;\quad c = 2$

Exer. 17–20: Find a polynomial $f(x)$ with leading coefficient 1 and having the given degree and zeros.

17 degree 3; zeros $-2, 0, 5$

18 degree 3; zeros $\pm 2, 3$

19 degree 4; zeros $-2, \pm 1, 4$

▶ 20 degree 4; zeros $-3, 0, 1, 5$

Exer. 21–28: Use synthetic division to find the quotient and remainder if the first polynomial is divided by the second.

21 $2x^3 - 3x^2 + 4x - 5;\quad x - 2$

22 $3x^3 - 4x^2 - x + 8;\quad x + 4$

23 $x^3 - 8x - 5;\quad x + 3$

24 $5x^3 - 6x^2 + 15;\quad x - 4$

25 $3x^5 + 6x^2 + 7;\quad x + 2$

26 $-2x^4 + 10x - 3;\quad x - 3$

27 $4x^4 - 5x^2 + 1;\quad x - \frac{1}{2}$

28 $9x^3 - 6x^2 + 3x - 4;\quad x - \frac{1}{3}$

Exer. 29–34: Use synthetic division to find $f(c)$.

29 $f(x) = 2x^3 + 3x^2 - 4x + 4;\quad c = 3$

30 $f(x) = -x^3 + 4x^2 + x;\quad c = -2$

31 $f(x) = 0.3x^3 + 0.04x - 0.034;\quad c = -0.2$

32 $f(x) = 8x^5 - 3x^2 + 7;\quad c = \frac{1}{2}$

33 $f(x) = x^2 + 3x - 5;\quad c = 2 + \sqrt{3}$

34 $f(x) = x^3 - 3x^2 - 8;\quad c = 1 + \sqrt{2}$

Exer. 35–38: Use synthetic division to show that c is a zero of $f(x)$.

35 $f(x) = 3x^4 + 8x^3 - 2x^2 - 10x + 4;\quad c = -2$

36 $f(x) = 4x^3 - 9x^2 - 8x - 3;\quad c = 3$

▶ Tutorial available at www.thomsonedu.com/login

37 $f(x) = 4x^3 - 6x^2 + 8x - 3;$ $\quad c = \frac{1}{2}$

38 $f(x) = 27x^4 - 9x^3 + 3x^2 + 6x + 1;$ $\quad c = -\frac{1}{3}$

Exer. 39–40: Find all values of k such that $f(x)$ is divisible by the given linear polynomial.

39 $f(x) = kx^3 + x^2 + k^2x + 3k^2 + 11;$ $\quad x + 2$

40 $f(x) = k^2x^3 - 4kx + 3;$ $\quad x - 1$

Exer. 41–42: Show that $x - c$ is not a factor of $f(x)$ for any real number c.

41 $f(x) = 3x^4 + x^2 + 5$ $\qquad$ 42 $f(x) = -x^4 - 3x^2 - 2$

43 Find the remainder if the polynomial

$$3x^{100} + 5x^{85} - 4x^{38} + 2x^{17} - 6$$

is divided by $x + 1$.

Exer. 44–46: Use the factor theorem to verify the statement.

44 $x - y$ is a factor of $x^n - y^n$ for every positive integer n.

45 $x + y$ is a factor of $x^n - y^n$ for every positive even integer n.

46 $x + y$ is a factor of $x^n + y^n$ for every positive odd integer n.

47 Let $P(x, y)$ be a first-quadrant point on $y = 6 - x$, and consider the vertical line segment PQ shown in the figure.

(a) If PQ is rotated about the y-axis, determine the volume V of the resulting cylinder.

▶ (b) For what point $P(x, y)$ with $x \neq 1$ is the volume V in part (a) the same as the volume of the cylinder of radius 1 and altitude 5 shown in the figure?

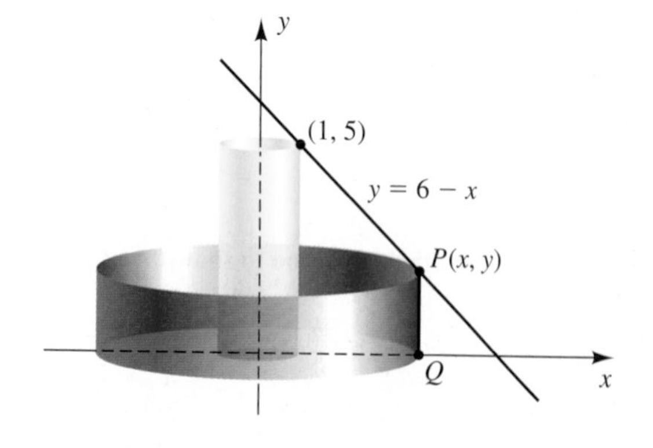

Exercise 47

48 Strength of a beam The strength of a rectangular beam is directly proportional to the product of its width and the square of the depth of a cross section (see the figure). A beam of width 1.5 feet has been cut from a cylindrical log of radius 1 foot. Find the width of a second rectangular beam of equal strength that could have been cut from the log.

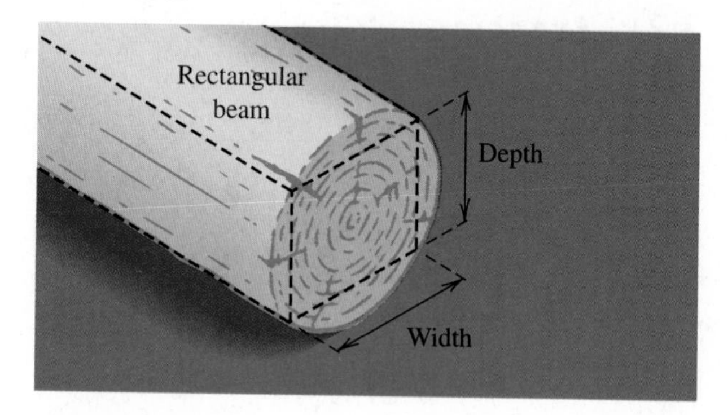

Exercise 48

49 Parabolic arch An arch has the shape of the parabola $y = 4 - x^2$. A rectangle is fit under the arch by selecting a point (x, y) on the parabola (see the figure).

(a) Express the area A of the rectangle in terms of x.

(b) If $x = 1$, the rectangle has base 2 and height 3. Find the base of a second rectangle that has the same area.

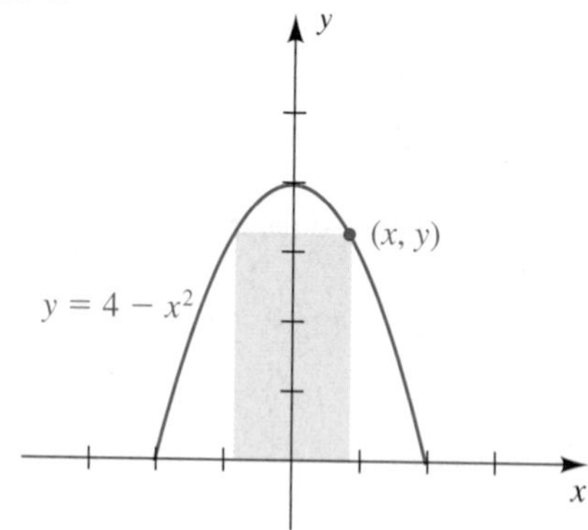

Exercise 49

50 Dimensions of a capsule An aspirin tablet in the shape of a right circular cylinder has height $\frac{1}{3}$ centimeter and radius $\frac{1}{2}$ centimeter. The manufacturer also wishes to market the

▶ **Tutorial available at www.thomsonedu.com/login**

aspirin in capsule form. The capsule is to be $\frac{3}{2}$ centimeters long, in the shape of a right circular cylinder with hemispheres attached at both ends (see the figure).

(a) If r denotes the radius of a hemisphere, find a formula for the volume of the capsule.

(b) Find the radius of the capsule so that its volume is equal to that of the tablet.

Exercise 50

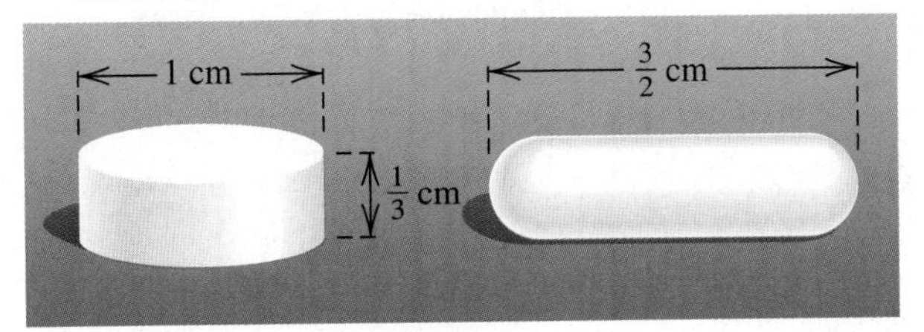

Exer. 51–52: Use the graph of f to approximate the remainder if f is divided by $x - 0.21$.

51 $f(x) = x^8 - 7.9x^5 - 0.8x^4 + x^3 + 1.2x - 9.81$

52 $f(x) = 3.33x^6 - 2.5x^5 + 6.9x^3 - 4.1x^2 + 1.22x - 6.78$

Exer. 53–54: Use the graph of f to approximate all values of k such that $f(x)$ is divisible by the given linear polynomial.

53 $f(x) = x^3 + k^3x^2 + 2kx - 2k^4$; $\quad x - 1.6$

54 $f(x) = k^5x^3 - 2.1x^2 + k^3x - 1.2k^2$; $\quad x + 0.4$

3.3 *Zeros of Polynomials*

The **zeros of a polynomial** $f(x)$ are the solutions of the equation $f(x) = 0$. Each real zero is an x-intercept of the graph of f. In applied fields, calculators and computers are usually used to find or approximate zeros. Before using a calculator, however, it is worth knowing what type of zeros to expect. Some questions we could ask are

(1) How many zeros of $f(x)$ are real? imaginary?

(2) How many real zeros of $f(x)$ are positive? negative?

(3) How many real zeros of $f(x)$ are rational? irrational?

(4) Are the real zeros of $f(x)$ large or small in value?

In this and the following section we shall discuss results that help answer some of these questions. These results form the basis of the *theory of equations*.

The factor and remainder theorems can be extended to the system of complex numbers. Thus, a complex number $c = a + bi$ is a zero of a polynomial $f(x)$ if and only if $x - c$ is a factor of $f(x)$. Except in special cases, zeros of polynomials are very difficult to find. For example, there are no obvious zeros of $f(x) = x^5 - 3x^4 + 4x^3 - 4x - 10$. Although we have no formula that can be used to find the zeros, the next theorem states that there is at *least* one zero c, and hence, by the factor theorem, $f(x)$ has a factor of the form $x - c$.

Fundamental Theorem of Algebra	If a polynomial $f(x)$ has positive degree and complex coefficients, then $f(x)$ has at least one complex zero.

The standard proof of this theorem requires results from an advanced field of mathematics called *functions of a complex variable*. A prerequisite for studying this field is a strong background in calculus. The first proof of the fundamental theorem of algebra was given by the German mathematician Carl Friedrich Gauss (1777–1855), who is considered by many to be the greatest mathematician of all time.

As a special case of the fundamental theorem of algebra, if all the coefficients of $f(x)$ are real, then $f(x)$ has at least one complex zero. If $a + bi$ is a complex zero, it may happen that $b = 0$, in which case the number a is a real zero.

The fundamental theorem of algebra enables us, at least in theory, to express every polynomial $f(x)$ of positive degree as a product of polynomials of degree 1, as in the next theorem.

Complete Factorization Theorem for Polynomials

If $f(x)$ is a polynomial of degree $n > 0$, then there exist n complex numbers $c_1, c_2, \ldots, c_n$ such that

$$f(x) = a(x - c_1)(x - c_2) \cdots (x - c_n),$$

where a is the leading coefficient of $f(x)$. Each number c_k is a zero of $f(x)$.

PROOF If $f(x)$ has degree $n > 0$, then, by the fundamental theorem of algebra, $f(x)$ has a complex zero c_1. Hence, by the factor theorem, $f(x)$ has a factor $x - c_1$; that is,

$$f(x) = (x - c_1)f_1(x),$$

where $f_1(x)$ is a polynomial of degree $n - 1$. If $n - 1 > 0$, then, by the same argument, $f_1(x)$ has a complex zero c_2 and therefore a factor $x - c_2$. Thus,

$$f_1(x) = (x - c_2)f_2(x),$$

where $f_2(x)$ is a polynomial of degree $n - 2$. Hence,

$$f(x) = (x - c_1)(x - c_2)f_2(x).$$

Continuing this process, after n steps we arrive at a polynomial $f_n(x)$ of degree 0. Thus, $f_n(x) = a$ for some nonzero number a, and we may write

$$f(x) = a(x - c_1)(x - c_2) \cdots (x - c_n),$$

where each complex number c_k is a zero of $f(x)$. The leading coefficient of the polynomial on the right-hand side in the last equation is a, and therefore a is the leading coefficient of $f(x)$.

ILLUSTRATION **Complete Factorization Theorem for Polynomials**

A Polynomial $f(x)$	A Factored Form of $f(x)$	Zeros of $f(x)$
$3x^2 - (12 + 6i)x + 24i$	$3(x - 4)(x - 2i)$	$4, 2i$
$-6x^3 - 2x^2 - 6x - 2$	$-6\left(x + \frac{1}{3}\right)(x + i)(x - i)$	$-\frac{1}{3}, \pm i$
$5x^3 - 30x^2 + 65x$	$5(x - 0)[x - (3 + 2i)][x - (3 - 2i)]$	$0, 3 \pm 2i$
$\frac{2}{3}x^3 + 8x^2 - \frac{2}{3}x - 8$	$\frac{2}{3}(x + 12)(x + 1)(x - 1)$	$-12, \pm 1$

We may now prove the following.

Theorem on the Maximum Number of Zeros of a Polynomial

A polynomial of degree $n > 0$ has at most n different complex zeros.

PROOF We will give an indirect proof; that is, we will suppose $f(x)$ has *more* than n different complex zeros and show that this supposition leads to a contradiction. Let us choose $n + 1$ of the zeros and label them $c_1, c_2, \ldots, c_n$, and c. We may use the c_k to obtain the factorization indicated in the statement of the complete factorization theorem for polynomials. Substituting c for x and using the fact that $f(c) = 0$, we obtain

$$0 = a(c - c_1)(c - c_2) \cdots (c - c_n).$$

However, each factor on the right-hand side is different from zero because $c \neq c_k$ for every k. Since the product of nonzero numbers cannot equal zero, we have a contradiction.

EXAMPLE 1 **Finding a polynomial with prescribed zeros**

Find a polynomial $f(x)$ in factored form that has degree 3; has zeros 2, -1, and 3; and satisfies $f(1) = 5$.

SOLUTION By the factor theorem, $f(x)$ has factors $x - 2$, $x + 1$, and $x - 3$. No other factors of degree 1 exist, since, by the factor theorem, another linear factor $x - c$ would produce a fourth zero of $f(x)$, contrary to the preceding theorem. Hence, $f(x)$ has the form

$$f(x) = a(x - 2)(x + 1)(x - 3)$$

for some number a. Since $f(1) = 5$, we see that

$$5 = a(1 - 2)(1 + 1)(1 - 3) \quad \text{let } x = 1 \text{ in } f(x)$$
$$5 = a(-1)(2)(-2) \quad \text{simplify}$$
$$a = \tfrac{5}{4} \quad \text{solve for } a$$

(continued)

Tutorial available at www.thomsonedu.com/login

Consequently,

$$f(x) = \tfrac{5}{4}(x - 2)(x + 1)(x - 3).$$

If we multiply the factors, we obtain the polynomial

$$f(x) = \tfrac{5}{4}x^3 - 5x^2 + \tfrac{5}{4}x + \tfrac{15}{2}.$$

The numbers $c_1, c_2, \ldots, c_n$ in the complete factorization theorem are not necessarily all different. To illustrate, $f(x) = x^3 + x^2 - 5x + 3$ has the factorization

$$f(x) = (x + 3)(x - 1)(x - 1).$$

If a factor $x - c$ occurs m times in the factorization, then c is a **zero of multiplicity m** of the polynomial $f(x)$, or a **root of multiplicity m** of the equation $f(x) = 0$. In the preceding display, 1 is a zero of multiplicity 2, and -3 is a zero of multiplicity 1.

If c is a real zero of $f(x)$ of multiplicity m, then $f(x)$ has the factor $(x - c)^m$ and the graph of f has an x-intercept c. The general shape of the graph at $(c, 0)$ depends on whether m is an odd integer or an even integer. If m is odd, then $(x - c)^m$ changes sign as x increases through c, and hence the graph of f crosses the x-axis at $(c, 0)$, as indicated in the first row of the following chart. The figures in the chart do not show the complete graph of f, but only its general shape near $(c, 0)$. If m is even, then $(x - c)^m$ does not change sign at c and the graph of f near $(c, 0)$ has the appearance of one of the two figures in the second row.

Factor of $f(x)$	**General shape of the graph of f near $(c, 0)$**
$(x - c)^m$, with m odd and $m \neq 1$	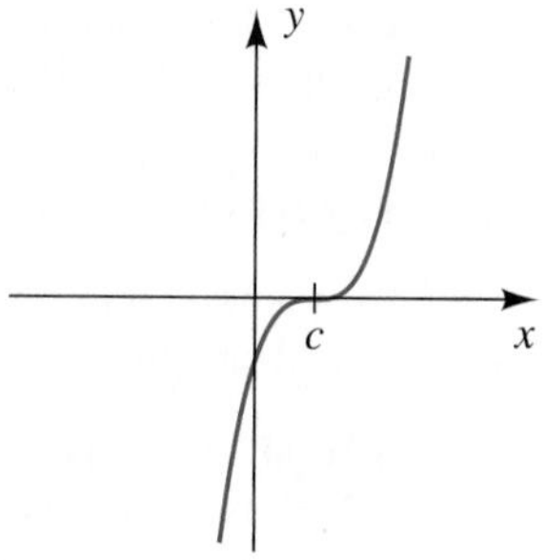 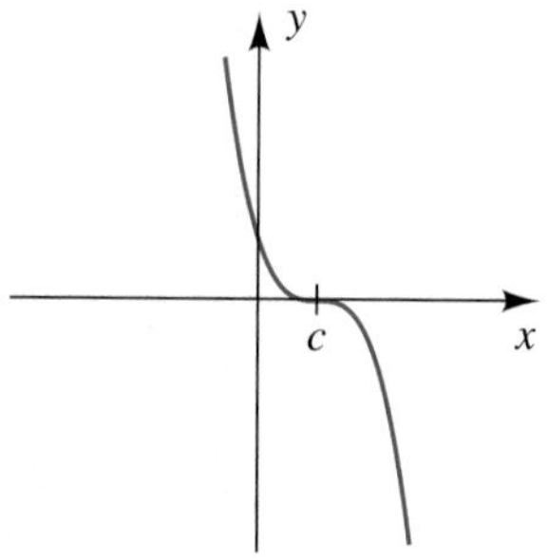
$(x - c)^m$, with m even	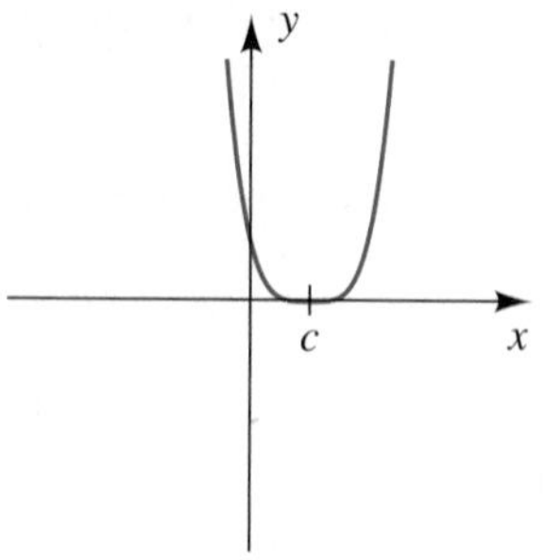 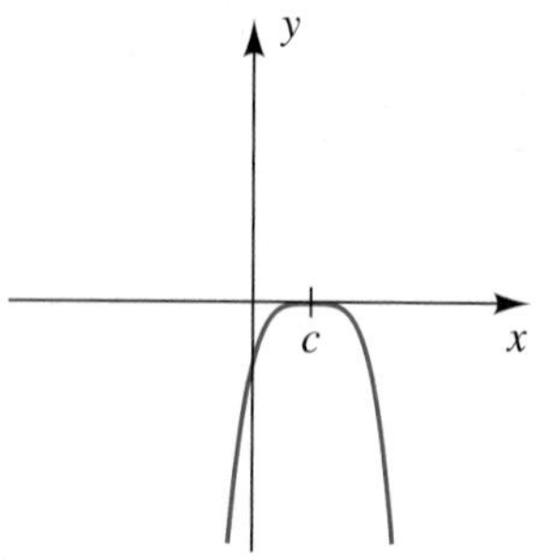

EXAMPLE 2 **Finding multiplicities of zeros**

Find the zeros of the polynomial $f(x) = \frac{1}{16}(x - 2)(x - 4)^3(x + 1)^2$, state the multiplicity of each, and then sketch the graph of f.

Figure 1

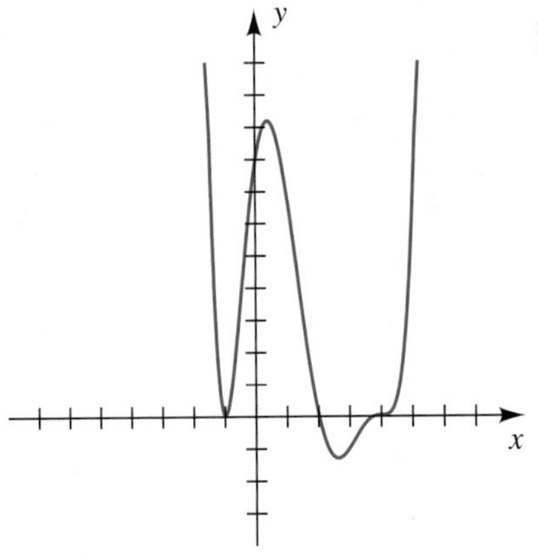

▶ SOLUTION We see from the factored form that $f(x)$ has three distinct zeros, 2, 4, and -1. The zero 2 has multiplicity 1, the zero 4 has multiplicity 3, and the zero -1 has multiplicity 2. Note that $f(x)$ has degree 6.

The x-intercepts of the graph of f are the real zeros -1, 2, and 4. Since the multiplicity of -1 is an even integer, the graph intersects, but does not cross, the x-axis at $(-1, 0)$. Since the multiplicities of 2 and 4 are odd, the graph crosses the x-axis at $(2, 0)$ and $(4, 0)$. (Note that the graph is "flatter" at 4 than at 2.) The y-intercept is $f(0) = \frac{1}{16}(-2)(-4)^3(1)^2 = 8$. The graph is shown in Figure 1.

If $f(x) = a(x - c_1)(x - c_2) \cdots (x - c_n)$ is a polynomial of degree n, then the n complex numbers $c_1, c_2, \ldots, c_n$ are zeros of $f(x)$. Counting a zero of multiplicity m as m zeros tells us that $f(x)$ has at least n zeros (not necessarily all different). Combining this fact with the fact that $f(x)$ has at most n zeros gives us the next result.

Theorem on the Exact Number of Zeros of a Polynomial

If $f(x)$ is a polynomial of degree $n > 0$ and if a zero of multiplicity m is counted m times, then $f(x)$ has precisely n zeros.

Notice how the polynomial of degree 6 in Example 2 relates to the last theorem. The multiplicities are 1, 3, and 2, so f has precisely $1 + 3 + 2 = 6$ zeros.

EXAMPLE 3 **Finding the zeros of a polynomial**

Express $f(x) = x^5 - 4x^4 + 13x^3$ as a product of linear factors, and find the five zeros of $f(x)$.

▶ SOLUTION We begin by factoring out x^3:

$$f(x) = x^3(x^2 - 4x + 13)$$

By the quadratic formula, the zeros of the polynomial $x^2 - 4x + 13$ are

$$\frac{-(-4) \pm \sqrt{(-4)^2 - 4(1)(13)}}{2(1)} = \frac{4 \pm \sqrt{-36}}{2} = \frac{4 \pm 6i}{2} = 2 \pm 3i.$$

Hence, by the factor theorem, $x^2 - 4x + 13$ has factors $x - (2 + 3i)$ and $x - (2 - 3i)$, and we obtain the factorization

$$f(x) = x \cdot x \cdot x \cdot (x - 2 - 3i)(x - 2 + 3i).$$

Since $x - 0$ occurs as a factor three times, the number 0 is a zero of multiplicity 3, and the five zeros of $f(x)$ are 0, 0, 0, $2 + 3i$, and $2 - 3i$.

▶ Tutorial available at www.thomsonedu.com/login

TI-86 Note: The screen below shows the output given by the POLY feature for the polynomial in Example 3. The notation for the fifth zero, $(2, -3)$, represents the zero $2 - 3i$. (For more information on the POLY feature, see the TI-86 note after Example 7 in Section 3.1.)

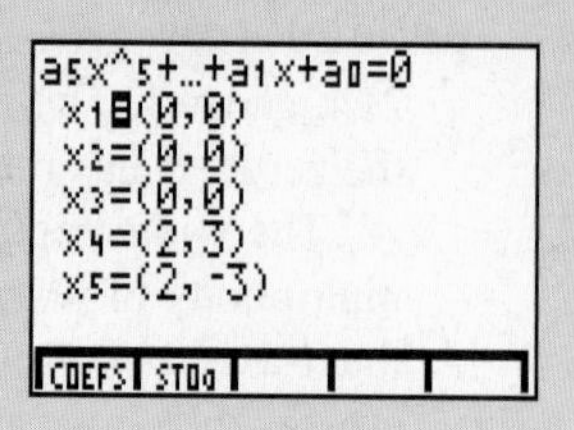

We next show how to use *Descartes' rule of signs* to obtain information about the zeros of a polynomial $f(x)$ with real coefficients. In the statement of the rule we assume that the terms of $f(x)$ are arranged in order of decreasing powers of x and that terms with zero coefficients are deleted. We also assume that the **constant term**—that is, the term that does not contain x—is different from 0. We say there is a **variation of sign** in $f(x)$ if two consecutive coefficients have opposite signs. To illustrate, the polynomial $f(x)$ in the following illustration has *three* variations of sign, as indicated by the braces—one variation from $2x^5$ to $-7x^4$, a second from $-7x^4$ to $3x^2$, and a third from $6x$ to -5.

ILLUSTRATION **Variations of Sign in $f(x) = 2x^5 - 7x^4 + 3x^2 + 6x - 5$**

$$f(x) = \overbrace{2x^5 \quad - 7x^4}^{+\text{ to }-} \quad \overbrace{\quad + 3x^2}^{-\text{ to }+} \quad \overbrace{\quad + 6x}^{\text{no variation}} \quad \overbrace{\quad - 5}^{+\text{ to }-}$$

■

Descartes' rule also refers to the variations of sign in $f(-x)$. Using the previous illustration, note that

$$\begin{aligned} f(-x) &= 2(-x)^5 - 7(-x)^4 + 3(-x)^2 + 6(-x) - 5 \\ &= -2x^5 - 7x^4 + 3x^2 - 6x - 5. \end{aligned}$$

Hence, as indicated in the next illustration, there are *two* variations of sign in $f(-x)$—one from $-7x^4$ to $3x^2$ and a second from $3x^2$ to $-6x$.

ILLUSTRATION **Variations of Sign in $f(-x)$ if $f(x) = 2x^5 - 7x^4 + 3x^2 + 6x - 5$**

$$f(-x) = \overbrace{-2x^5 \quad - 7x^4}^{\text{no variation}} \quad \overbrace{\quad + 3x^2}^{-\text{ to }+} \quad \overbrace{\quad - 6x}^{+\text{ to }-} \quad \overbrace{\quad - 5}^{\text{no variation}}$$

■

We may state Descartes' rule as follows.

Descartes' Rule of Signs

Let $f(x)$ be a polynomial with real coefficients and a nonzero constant term.

(1) The number of *positive* real zeros of $f(x)$ either is equal to the number of variations of sign in $f(x)$ or is less than that number by an even integer.

(2) The number of *negative* real zeros of $f(x)$ either is equal to the number of variations of sign in $f(-x)$ or is less than that number by an even integer.

A proof of Descartes' rule will not be given.

EXAMPLE 4 Using Descartes' rule of signs

Discuss the number of possible positive and negative real solutions and imaginary solutions of the equation $f(x) = 0$, where

$$f(x) = 2x^5 - 7x^4 + 3x^2 + 6x - 5.$$

SOLUTION The polynomial $f(x)$ is the one given in the two previous illustrations. Since there are three variations of sign in $f(x)$, the equation has either three positive real solutions or one positive real solution.

Since $f(-x)$ has two variations of sign, the equation has either two negative solutions or no negative solution. Because $f(x)$ has degree 5, there are a total of 5 solutions. The solutions that are not positive or negative real numbers are imaginary numbers. The following table summarizes the various possibilities that can occur for solutions of the equation.

Number of positive real solutions	3	3	1	1
Number of negative real solutions	2	0	2	0
Number of imaginary solutions	0	2	2	4
Total number of solutions	5	5	5	5

Descartes' rule stipulates that the constant term of the polynomial $f(x)$ is different from 0. If the constant term is 0, as in the equation

$$x^4 - 3x^3 + 2x^2 - 5x = 0,$$

we factor out the lowest power of x, obtaining

$$x(x^3 - 3x^2 + 2x - 5) = 0.$$

Thus, one solution is $x = 0$, and we apply Descartes' rule to the polynomial $x^3 - 3x^2 + 2x - 5$ to determine the nature of the remaining three solutions.

Tutorial available at www.thomsonedu.com/login

When applying Descartes' rule, we count roots of multiplicity k as k roots. For example, given $x^2 - 2x + 1 = 0$, the polynomial $x^2 - 2x + 1$ has two variations of sign, and hence the equation has either two positive real roots or none. The factored form of the equation is $(x - 1)^2 = 0$, and hence 1 is a root of multiplicity 2.

We next discuss the *bounds* for the real zeros of a polynomial $f(x)$ that has real coefficients. By definition, a real number b is an **upper bound** for the zeros if no zero is greater than b. A real number a is a **lower bound** for the zeros if no zero is less than a. Thus, if r is any real zero of $f(x)$, then $a \leq r \leq b$; that is, r is in the closed interval $[a, b]$, as illustrated in Figure 2. Note that upper and lower bounds are not unique, since any number greater than b is also an upper bound and any number less than a is also a lower bound.

Figure 2

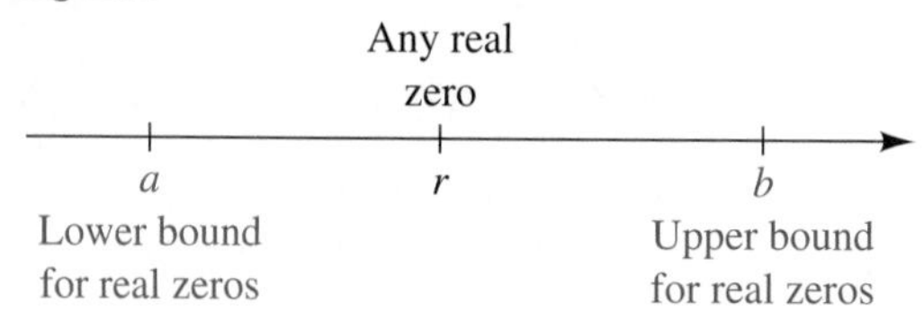

We may use synthetic division to find upper and lower bounds for the zeros of $f(x)$. Recall that if we divide $f(x)$ synthetically by $x - c$, the third row in the division process contains the coefficients of the quotient $q(x)$ together with the remainder $f(c)$. The following theorem indicates how this third row may be used to find upper and lower bounds for the real solutions.

First Theorem on Bounds for Real Zeros of Polynomials

Suppose that $f(x)$ is a polynomial with real coefficients and a positive leading coefficient and that $f(x)$ is divided synthetically by $x - c$.

(1) If $c > 0$ and if all numbers in the third row of the division process are either positive or zero, then c is an upper bound for the real zeros of $f(x)$.

(2) If $c < 0$ and if the numbers in the third row of the division process are alternately positive and negative (and a 0 in the third row is considered to be either positive or negative), then c is a lower bound for the real zeros of $f(x)$.

EXAMPLE 5 **Finding bounds for the solutions of an equation**

Find upper and lower bounds for the real solutions of the equation $f(x) = 0$, where $f(x) = 2x^3 + 5x^2 - 8x - 7$.

▶ **SOLUTION** We divide $f(x)$ synthetically by $x - 1$ and $x - 2$.

$$\begin{array}{r|rrrr} 1 & 2 & 5 & -8 & -7 \\ & & 2 & 7 & -1 \\ \hline & 2 & 7 & -1 & -8 \end{array} \qquad \begin{array}{r|rrrr} 2 & 2 & 5 & -8 & -7 \\ & & 4 & 18 & 20 \\ \hline & 2 & 9 & 10 & 13 \end{array}$$

▶ Tutorial available at www.thomsonedu.com/login

The third row of the synthetic division by $x - 1$ contains negative numbers, and hence part (1) of the theorem on bounds for real zeros of polynomials does not apply. However, since all numbers in the third row of the synthetic division by $x - 2$ are positive, it follows from part (1) that 2 is an upper bound for the real solutions of the equation. This fact is also evident if we express the division by $x - 2$ in the division algorithm form

$$2x^3 + 5x^2 - 8x - 7 = (x - 2)(2x^2 + 9x + 10) + 13,$$

for if $x > 2$, then the right-hand side of the equation is positive (why?), and hence $f(x)$ is not zero.

We now find a lower bound. After some trial-and-error attempts using $x - (-1)$, $x - (-2)$, and $x - (-3)$, we see that synthetic division of f by $x - (-4)$ gives us

$$\begin{array}{r|rrrr} -4 & 2 & 5 & -8 & -7 \\ & & -8 & 12 & -16 \\ \hline & 2 & -3 & 4 & -23 \end{array}$$

Figure 3

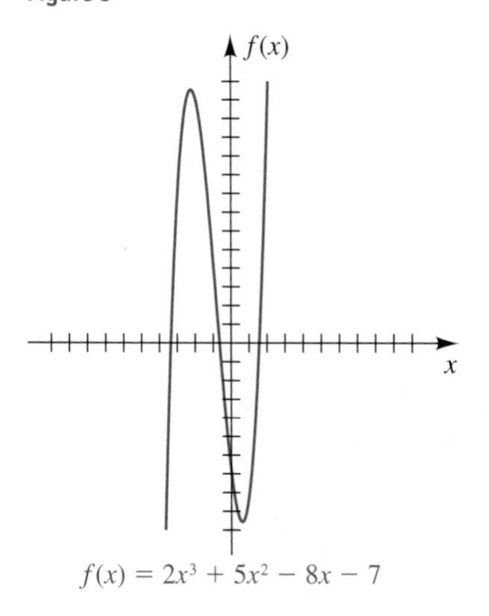

$f(x) = 2x^3 + 5x^2 - 8x - 7$

Since the numbers in the third row are alternately positive and negative, it follows from part (2) of the preceding theorem that -4 is a lower bound for the real solutions. This can also be proved by expressing the division by $x + 4$ in the form

$$2x^3 + 5x^2 - 8x - 7 = (x + 4)(2x^2 - 3x + 4) - 23,$$

for if $x < -4$, then the right-hand side of this equation is negative (why?), and hence $f(x)$ is not zero.

Since lower and upper bounds for the real solutions are -4 and 2, respectively, it follows that all real solutions are in the closed interval $[-4, 2]$.

The graph of f in Figure 3 shows that the three zeros of f are in the intervals $[-4, -3]$, $[-1, 0]$, and $[1, 2]$, respectively.

When a graphing utility is used, the following theorem is helpful in finding a viewing rectangle that shows all the zeros of a polynomial.

Second Theorem on Bounds for Real Zeros of Polynomials

Suppose $f(x) = a_nx^n + a_{n-1}x^{n-1} + \cdots + a_1x + a_0$ is a polynomial with real coefficients. All of the real zeros of $f(x)$ are in the interval

$$(-M, M),$$

where $M = \dfrac{\max(|a_n|, |a_{n-1}|, \ldots, |a_1|, |a_0|)}{|a_n|} + 1.$

In words, the value of M is equal to the ratio of the largest coefficient (in magnitude) to the absolute value of the leading coefficient, plus 1. For example, using the polynomial $f(x) = 2x^3 + 5x^2 - 8x - 7$ in Example 5, we have

$$M = \frac{|-8|}{|2|} + 1 = 4 + 1 = 5.$$

When a graphing utility is used *only* to find the zeros of a polynomial $f(x)$, it is not necessary to see the turning points of the polynomial. Hence, you might begin looking for the zeros of $f(x)$ by using the viewing rectangle dimensions

$$[-M, M] \text{ by } [-1, 1].$$

By graphing $Y_1 = f(x) = 2x^3 + 5x^2 - 8x - 7$ (from Example 5) in the viewing rectangle $[-5, 5]$ by $[-1, 1, 0.5]$ in Figure 4, you can almost "eyeball" the approximate solutions -3.4, -0.7, and 1.5.

Figure 4
$[-5, 5]$ by $[-1, 1, 0.5]$

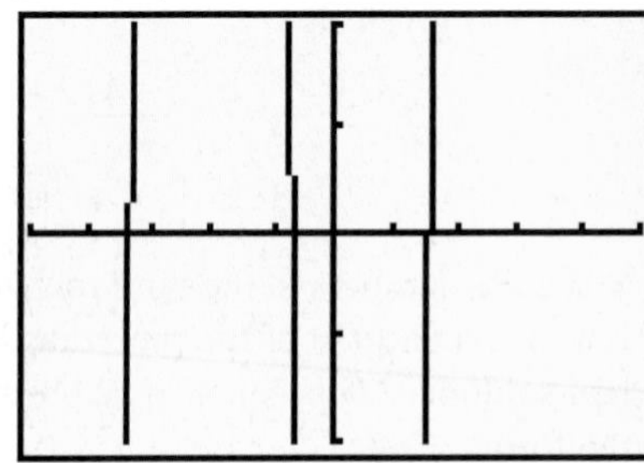

EXAMPLE 6 Finding a polynomial from a graph

Shown in Figure 5 are all the zeros of a polynomial function f.

(a) Find a factored form for f that has minimal degree.

(b) Assuming the leading coefficient of f is 1, find the y-intercept.

Figure 5
$[-4, 4]$ by $[-35, 35, 5]$

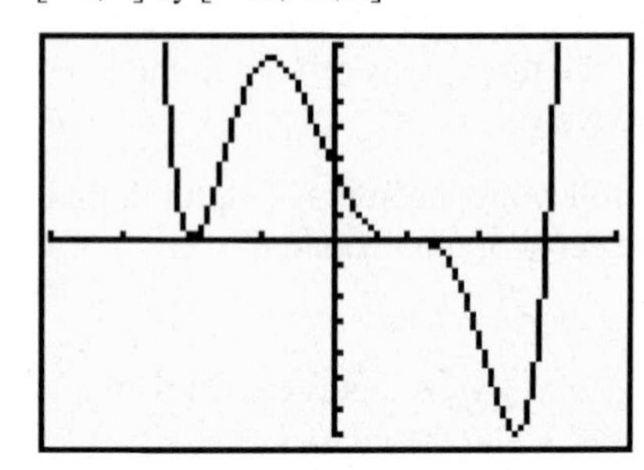

▶ **SOLUTION**

(a) The zero at $x = -2$ must have a multiplicity that is an even number, since f does not change sign at $x = -2$. The zero at $x = 1$ must have an odd multiplicity of 3 or greater, since f changes sign at $x = 1$ and levels off. The zero at $x = 3$ is of multiplicity 1, since f changes sign and does not level off. Thus, a factored form of f is

$$f(x) = a(x + 2)^m(x - 1)^n(x - 3)^1.$$

Because we desire the function having minimal degree, we let $m = 2$ and $n = 3$, obtaining

$$f(x) = a(x + 2)^2(x - 1)^3(x - 3),$$

which is a sixth-degree polynomial.

(b) If the leading coefficient of f is to be 1, then, from the complete factorization theorem for polynomials, we know that the value of a is 1. To find the y-intercept, we let $x = 0$ and compute $f(0)$:

$$f(0) = 1(0 + 2)^2(0 - 1)^3(0 - 3) = 1(4)(-1)(-3) = 12$$

Hence, the y-intercept is 12.

▶ **Tutorial available at www.thomsonedu.com/login**

EXAMPLE 7 **Exploring the graph of a polynomial**

Find the zeros of $f(x) = x^3 - 1000x^2 - x + 1000$.

Figure 6
$[-15, 15]$ by $[-10, 10]$

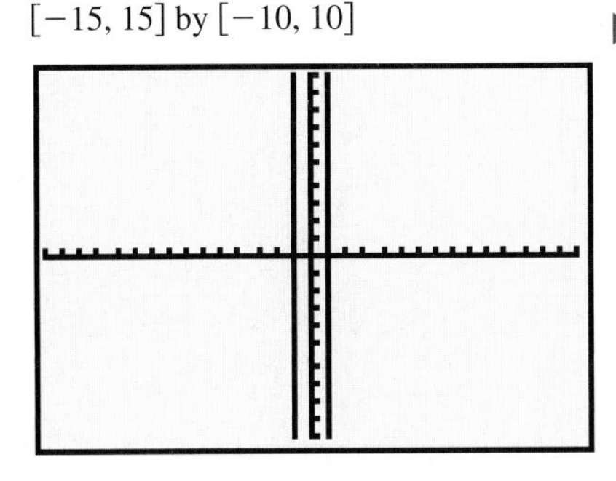

SOLUTION We assign $f(x)$ to Y_1 and use a standard viewing rectangle to obtain Figure 6. It appears that 1 is a root of f, and we can prove this fact by using synthetic division:

$$\begin{array}{r|rrrr} 1 & 1 & -1000 & -1 & 1000 \\ & & 1 & -999 & -1000 \\ \hline & 1 & -999 & -1000 & 0 \end{array}$$

Using the **depressed equation,** $x^2 - 999x - 1000 = 0$, we can also show that -1 is a root of f:

$$\begin{array}{r|rrr} -1 & 1 & -999 & -1000 \\ & & -1 & 1000 \\ \hline & 1 & -1000 & 0 \end{array}$$

From the last synthetic division, we see that $x - 1000$ is a factor of f, and hence the third root is 1000.

Because of the relative sizes of the roots 1 and 1000, it is very difficult to obtain a viewing rectangle that shows all three zeros. However, by setting Xmin to -50, Xmax to 1050, and Xscl to 100 and using ZoomFit (choice 0 on the TI-83/4 Plus or F1 under the second submenu of the ZOOM menu on the TI-86), we obtain the sketch of f in Figure 7, showing its zeros and turning points.

Now check the values of Ymin and Ymax to see the necessary viewing rectangle.

Figure 7 Using ZoomFit
$[-50, 1050, 100]$ by $[?, ?, ?]$

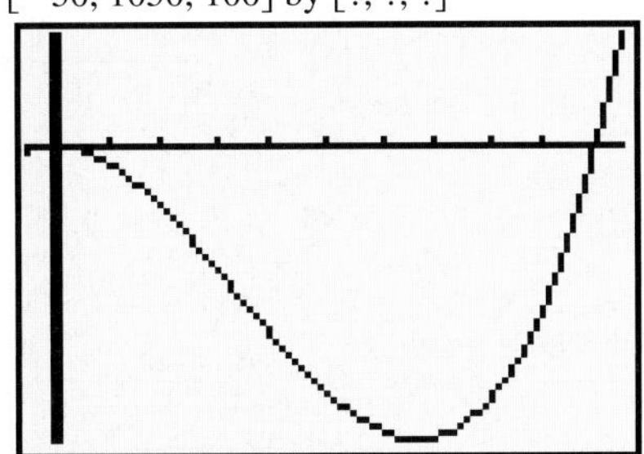

3.3 *Exercises*

Exer. 1–6: Find a polynomial $f(x)$ of degree 3 that has the indicated zeros and satisfies the given condition.

1 $-1, 2, 3$; $f(-2) = 80$

2 $-5, 2, 4$; $f(3) = -24$

3 $-4, 3, 0$; $f(2) = -36$

4 $-3, -2, 0$; $f(-4) = 16$

5 $-2i, 2i, 3$; $f(1) = 20$

Tutorial available at www.thomsonedu.com/login

6 $-3i, 3i, 4;\quad f(-1) = 50$

7 Find a polynomial $f(x)$ of degree 4 with leading coefficient 1 such that both -4 and 3 are zeros of multiplicity 2, and sketch the graph of f.

8 Find a polynomial $f(x)$ of degree 4 with leading coefficient 1 such that both -5 and 2 are zeros of multiplicity 2, and sketch the graph of f.

9 Find a polynomial $f(x)$ of degree 6 such that 0 and 3 are both zeros of multiplicity 3 and $f(2) = -24$. Sketch the graph of f.

10 Find a polynomial $f(x)$ of degree 7 such that -2 and 2 are both zeros of multiplicity 2, 0 is a zero of multiplicity 3, and $f(-1) = 27$. Sketch the graph of f.

▶ 11 Find the third-degree polynomial function whose graph is shown in the figure.

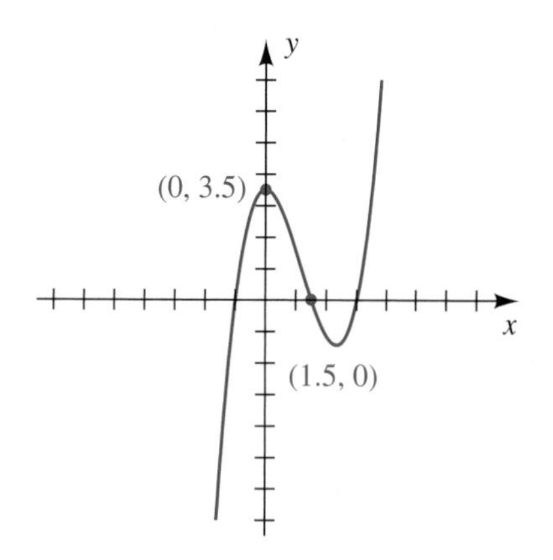

12 Find the fourth-degree polynomial function whose graph is shown in the figure.

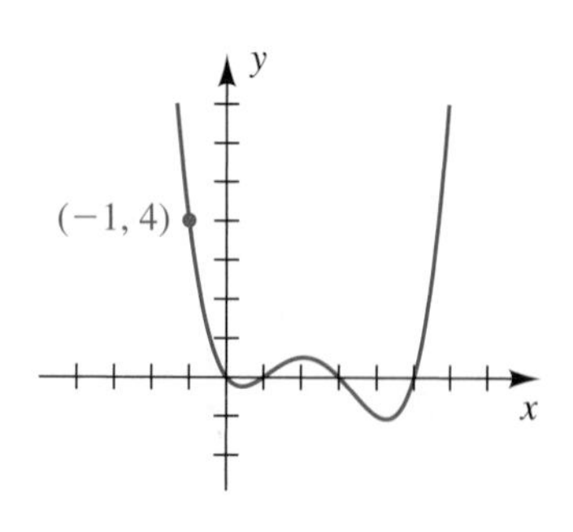

Exer. 13–14: Find the polynomial function of degree 3 whose graph is shown in the figure.

13

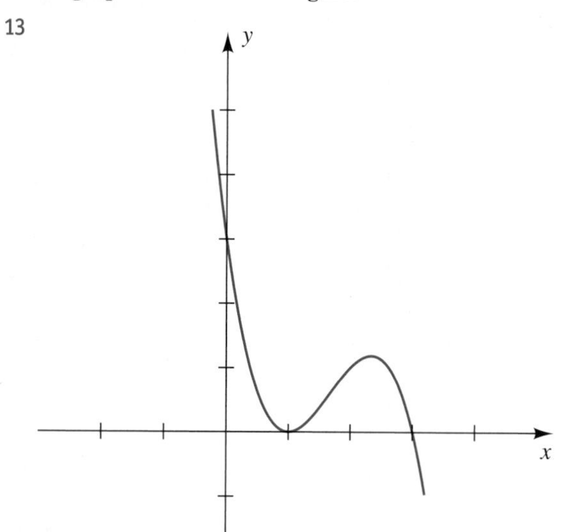

14

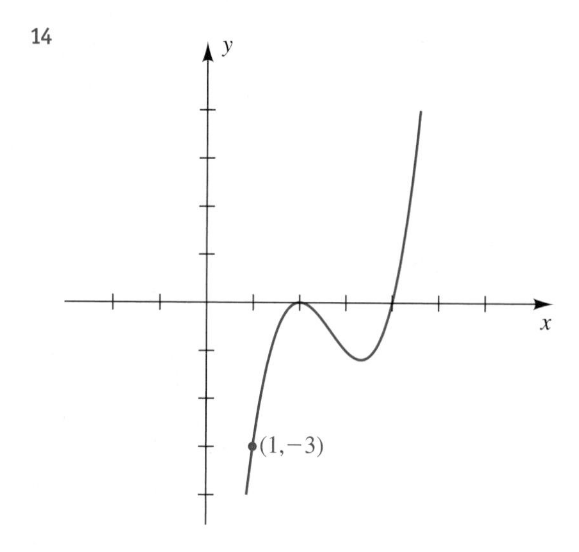

Exer. 15–22: Find the zeros of $f(x)$, and state the multiplicity of each zero.

15 $f(x) = x^2(3x + 2)(2x - 5)^3$

16 $f(x) = x(x + 1)^4(3x - 7)^2$

17 $f(x) = 4x^5 + 12x^4 + 9x^3$

18 $f(x) = (4x^2 - 5)^2$

▶ **Tutorial available at www.thomsonedu.com/login**

19 $f(x) = (x^2 + x - 12)^3(x^2 - 9)^2$

20 $f(x) = (6x^2 + 7x - 5)^4(4x^2 - 1)^2$

21 $f(x) = x^4 + 7x^2 - 144$

22 $f(x) = x^4 + 21x^2 - 100$

Exer. 23–26: Show that the number is a zero of $f(x)$ of the given multiplicity, and express $f(x)$ as a product of linear factors.

▶ 23 $f(x) = x^4 + 7x^3 + 13x^2 - 3x - 18$; -3 (multiplicity 2)

24 $f(x) = x^4 - 9x^3 + 22x^2 - 32$; 4 (multiplicity 2)

25 $f(x) = x^6 - 4x^5 + 5x^4 - 5x^2 + 4x - 1$; 1 (multiplicity 5)

26 $f(x) = x^5 + x^4 - 6x^3 - 14x^2 - 11x - 3$; -1 (multiplicity 4)

Exer. 27–34: Use Descartes' rule of signs to determine the number of possible positive, negative, and nonreal complex solutions of the equation.

27 $4x^3 - 6x^2 + x - 3 = 0$

28 $5x^3 - 6x - 4 = 0$

29 $4x^3 + 2x^2 + 1 = 0$

30 $3x^3 - 4x^2 + 3x + 7 = 0$

31 $3x^4 + 2x^3 - 4x + 2 = 0$

32 $2x^4 - x^3 + x^2 - 3x + 4 = 0$

33 $x^5 + 4x^4 + 3x^3 - 4x + 2 = 0$

34 $2x^6 + 5x^5 + 2x^2 - 3x + 4 = 0$

Exer. 35–40: Applying the first theorem on bounds for real zeros of polynomials, determine the smallest and largest integers that are upper and lower bounds, respectively, for the real solutions of the equation. With the aid of a graphing utility, discuss the validity of the bounds.

35 $x^3 - 4x^2 - 5x + 7 = 0$

36 $2x^3 - 5x^2 + 4x - 8 = 0$

37 $x^4 - x^3 - 2x^2 + 3x + 6 = 0$

38 $2x^4 - 9x^3 - 8x - 10 = 0$

39 $2x^5 - 13x^3 + 2x - 5 = 0$

40 $3x^5 + 2x^4 - x^3 - 8x^2 - 7 = 0$

Exer. 41–42: Find a factored form for a polynomial function f that has a minimal degree. Assume that the intercept values are integers and that Xscl = Yscl = 1.

41

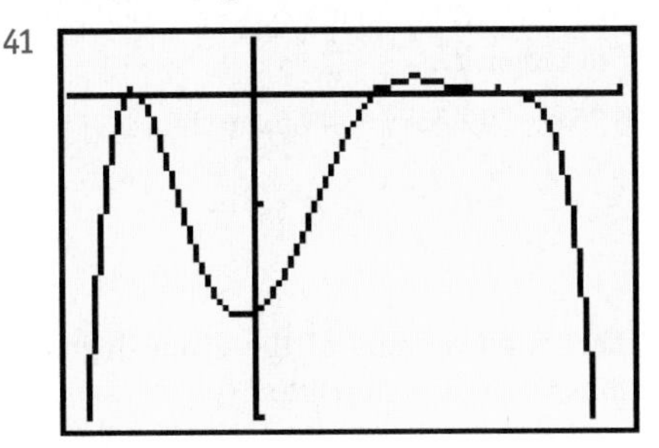

42

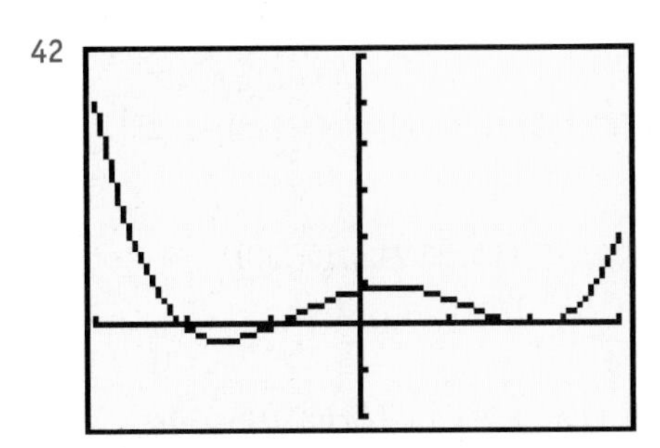

Exer. 43–44: (a) Find a factored form for a polynomial function f that has minimal degree. Assume that the intercept values are integers, Xscl = 1, and Yscl = 5. (b) If the leading coefficient of f is a, find the y-intercept.

43 $a = 1$

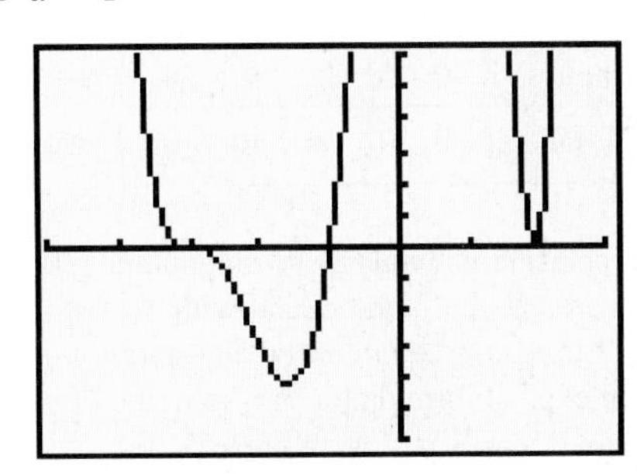

▶ Tutorial available at www.thomsonedu.com/login

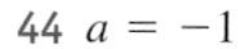

44 $a = -1$

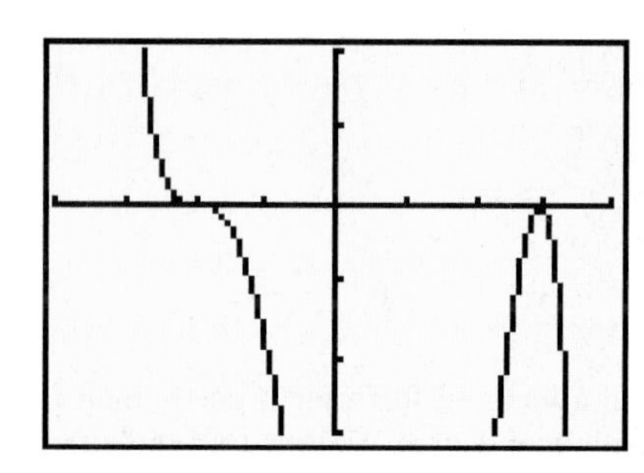

Exer. 45–46: The polynomial function f has only real zeros. Use the graph of f to factor it.

45 $f(x) = x^5 - 16.75x^3 + 12.75x^2 + 49.5x - 54$

46 $f(x) = x^5 - 2.5x^4 - 12.75x^3 + 19.625x^2 + 27.625x + 7.5$

Exer. 47–50: Is there a polynomial of the given degree n whose graph contains the indicated points?

47 $n = 4$;
$(-2, 0), (0, -24), (1, 0), (3, 0), (2, 0), (-1, -52)$

48 $n = 5$;
$(0, 0), (-3, 0), (-1, 0), (2, 0), (3, 0), (-2, 5), (1, 2)$

49 $n = 3$;
$(1.1, -49.815), (2, 0), (3.5, 25.245), (5.2, 0), (6.4, -29.304), (10.1, 0)$

50 $n = 4$;
$(1.25, 0), (2, 0), (2.5, 56.25), (3, 128.625), (6.5, 0), (9, -307.75), (10, 0)$

51 **Using limited data** A scientist has limited data on the temperature T (in °C) during a 24-hour period. If t denotes time in hours and $t = 0$ corresponds to midnight, find the fourth-degree polynomial that fits the information in the following table.

t (hours)	0	5	12	19	24
T (°C)	0	0	10	0	0

52 **Lagrange interpolation polynomial** A polynomial $f(x)$ of degree 3 with zeros at c_1, c_2, and c_3 and with $f(c) = 1$ for $c_2 < c < c_3$ is a third-degree *Lagrange interpolation polynomial*. Find an explicit formula for $f(x)$ in terms of c_1, c_2, c_3, and c.

Exer. 53–54: Graph f for each value of n on the same coordinate plane, and describe how the multiplicity of a zero affects the graph of f.

53 $f(x) = (x - 0.5)^n(x^2 + 1); \quad n = 1, 2, 3, 4$

54 $f(x) = (x - 1)^n(x + 1)^n; \quad n = 1, 2, 3, 4$

Exer. 55–56: Graph f, estimate all real zeros, and determine the multiplicity of each zero.

55 $f(x) = x^3 + 1.3x^2 - 1.2x - 1.584$

56 $f(x) = x^5 - \frac{1}{4}x^4 - \frac{19}{8}x^3 - \frac{9}{32}x^2 + \frac{405}{256}x + \frac{675}{1024}$

57 **Greenhouse effect** Because of the combustion of fossil fuels, the concentration of carbon dioxide in the atmosphere is increasing. Research indicates that this will result in a *greenhouse effect* that will change the average global surface temperature. Assuming a vigorous expansion of coal use, the future amount $A(t)$ of atmospheric carbon dioxide concentration can be approximated (in parts per million) by

$$A(t) = -\frac{1}{2400}t^3 + \frac{1}{20}t^2 + \frac{7}{6}t + 340,$$

where t is in years, $t = 0$ corresponds to 1980, and $0 \le t \le 60$. Use the graph of A to estimate the year when the carbon dioxide concentration will be 400.

58 **Greenhouse effect** The average increase in global surface temperature due to the greenhouse effect can be approximated by

$$T(t) = \frac{21}{5{,}000{,}000}t^3 - \frac{127}{1{,}000{,}000}t^2 + \frac{1293}{50{,}000}t,$$

where $0 \le t \le 60$ and $t = 0$ corresponds to 1980. Use the graph of T to estimate the year when the average temperature will have risen by 1°C.

Exer. 59–60: The average monthly temperatures in °F for two Canadian locations are listed in the following tables.

Month	Jan.	Feb.	Mar.	Apr.
Arctic Bay	−22	−26	−18	−4
Trout Lake	−11	−6	7	25

Month	May	June	July	Aug.
Arctic Bay	19	36	43	41
Trout Lake	39	52	61	59

Month	Sept.	Oct.	Nov.	Dec.
Arctic Bay	28	12	−8	−17
Trout Lake	48	34	16	−4

(a) **If January 15 corresponds to $x = 1$, February 15 to $x = 2, \ldots,$ and December 15 to $x = 12$, determine graphically which of the three polynomials given best models the data.**

(b) **Use the intermediate value theorem for polynomial functions to approximate an interval for x when an average temperature of 0°F occurs.**

(c) **Use your choice from part (a) to estimate x when the average temperature is 0°F.**

59 Arctic Bay temperatures

(1) $f(x) = -1.97x^2 + 28x - 67.95$

(2) $g(x) = -0.23x^3 + 2.53x^2 + 3.6x - 36.28$

(3) $h(x) = 0.089x^4 - 2.55x^3 + 22.48x^2 - 59.68x + 19$

60 Trout Lake temperatures

(1) $f(x) = -2.14x^2 + 28.01x - 55$

(2) $g(x) = -0.22x^3 + 1.84x^2 + 11.70x - 29.90$

(3) $h(x) = 0.046x^4 - 1.39x^3 + 11.81x^2 - 22.2x + 1.03$

Exer. 61–62: A solid wood sphere whose density is less than that of water will float. The depth d that the sphere will sink into the water is determined by the equation

$$\frac{4k}{3}\pi r^3 - \pi d^2 r + \frac{1}{3}\pi d^3 = 0,$$

where r is the radius of the sphere and k is a positive constant less than or equal to 1. If $r = 6$ cm, graphically estimate d for each constant k.

61 Pine sphere in water $k = 0.7$

62 Oak sphere in water $k = 0.85$

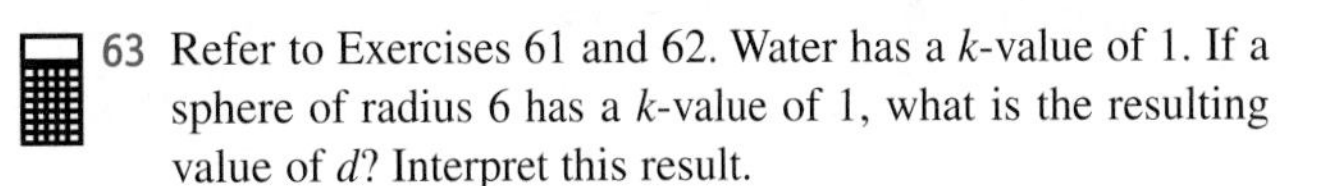

63 Refer to Exercises 61 and 62. Water has a k-value of 1. If a sphere of radius 6 has a k-value of 1, what is the resulting value of d? Interpret this result.

3.4 Complex and Rational Zeros of Polynomials

Example 3 of the preceding section illustrates an important fact about polynomials with real coefficients: The two complex zeros $2 + 3i$ and $2 - 3i$ of $x^5 - 4x^4 + 13x^3$ are conjugates of each other. The relationship is not accidental, since the following general result is true.

Theorem on Conjugate Pair Zeros of a Polynomial	If a polynomial $f(x)$ of degree $n > 1$ has real coefficients and if $z = a + bi$ with $b \neq 0$ is a complex zero of $f(x)$, then the conjugate $\bar{z} = a - bi$ is also a zero of $f(x)$.

A proof is left as a discussion exercise at the end of the chapter.

EXAMPLE 1 Finding a polynomial with prescribed zeros

Find a polynomial $f(x)$ of degree 4 that has real coefficients and zeros $2 + i$ and $-3i$.

▶ SOLUTION By the theorem on conjugate pair zeros of a polynomial, $f(x)$ must also have zeros $2 - i$ and $3i$. Applying the factor theorem, we find that $f(x)$ has the following factors:

$$x - (2 + i), \quad x - (2 - i), \quad x - (-3i), \quad x - (3i)$$

(continued)

▶ Tutorial available at www.thomsonedu.com/login

Multiplying these four factors gives us

$$\begin{aligned} f(x) &= [x - (2 + i)][x - (2 - i)](x + 3i)(x - 3i) \\ &= (x^2 - 4x + 5)(x^2 + 9) \qquad (*) \\ &= x^4 - 4x^3 + 14x^2 - 36x + 45. \end{aligned}$$

Note that in $(*)$ the symbol i does not appear. This is not a coincidence, since if $a + bi$ is a zero of a polynomial with real coefficients, then $a - bi$ is also a zero and we can multiply the associated factors as follows:

$$[x - (a + bi)][x - (a - bi)] = x^2 - 2ax + a^2 + b^2$$

In Example 1 we have $a = 2$ and $b = 1$, so $-2a = -4$ and $a^2 + b^2 = 5$ and the associated quadratic factor is $x^2 - 4x + 5$. This resulting quadratic factor will always have real coefficients, as stated in the next theorem.

Theorem on Expressing a Polynomial as a Product of Linear and Quadratic Factors

Every polynomial with real coefficients and positive degree n can be expressed as a product of linear and quadratic polynomials with real coefficients such that the quadratic factors are irreducible over $\mathbb{R}$.

PROOF Since $f(x)$ has precisely n complex zeros $c_1, c_2, \ldots, c_n$, we may write

$$f(x) = a(x - c_1)(x - c_2) \cdots (x - c_n),$$

where a is the leading coefficient of $f(x)$. Of course, some of the zeros may be real. In such cases we obtain the linear factors referred to in the statement of the theorem.

If a zero c_k is not real, then, by the theorem on conjugate pair zeros of a polynomial, the conjugate $\overline{c_k}$ is also a zero of $f(x)$ and hence must be one of the numbers $c_1, c_2, \ldots, c_n$. This implies that both $x - c_k$ and $x - \overline{c_k}$ appear in the factorization of $f(x)$. If those factors are multiplied, we obtain

$$(x - c_k)(x - \overline{c_k}) = x^2 - (c_k + \overline{c_k})x + c_k\overline{c_k},$$

which has *real* coefficients, since $c_k + \overline{c_k}$ and $c_k\overline{c_k}$ are real numbers. Thus, if c_k is a complex zero, then the product $(x - c_k)(x - \overline{c_k})$ is a quadratic polynomial that is irreducible over $\mathbb{R}$. This completes the proof.

EXAMPLE 2 Expressing a polynomial as a product of linear and quadratic factors

Express $x^5 - 4x^3 + x^2 - 4$ as a product of

(a) linear and quadratic polynomials with real coefficients that are irreducible over $\mathbb{R}$

(b) linear polynomials

▶ SOLUTION

(a) $x^5 - 4x^3 + x^2 - 4$

$= (x^5 - 4x^3) + (x^2 - 4)$	group terms
$= x^3(x^2 - 4) + 1(x^2 - 4)$	factor out x^3
$= (x^3 + 1)(x^2 - 4)$	factor out $(x^2 - 4)$
$= (x + 1)(x^2 - x + 1)(x + 2)(x - 2)$	factor as the sum of cubes and the difference of squares

Using the quadratic formula, we see that the polynomial $x^2 - x + 1$ has the complex zeros

$$\frac{-(-1) \pm \sqrt{(-1)^2 - 4(1)(1)}}{2(1)} = \frac{1 \pm \sqrt{3}i}{2} = \frac{1}{2} \pm \frac{\sqrt{3}}{2}i$$

and hence is irreducible over $\mathbb{R}$. Thus, the desired factorization is

$$(x + 1)(x^2 - x + 1)(x + 2)(x - 2).$$

(b) Since the polynomial $x^2 - x + 1$ in part (a) has zeros $\frac{1}{2} \pm \left(\sqrt{3}/2\right)i$, it follows from the factor theorem that the polynomial has factors

$$x - \left(\frac{1}{2} + \frac{\sqrt{3}}{2}i\right) \quad \text{and} \quad x - \left(\frac{1}{2} - \frac{\sqrt{3}}{2}i\right).$$

Substituting in the factorization found in part (a), we obtain the following complete factorization into linear polynomials:

$$(x + 1)\left(x - \frac{1}{2} - \frac{\sqrt{3}}{2}i\right)\left(x - \frac{1}{2} + \frac{\sqrt{3}}{2}i\right)(x + 2)(x - 2)$$

We previously pointed out that it is generally very difficult to find the zeros of a polynomial of high degree. If all the coefficients are integers, however, there is a method for finding the *rational* zeros, if they exist. The method is a consequence of the following result.

Theorem on Rational Zeros of a Polynomial

If the polynomial

$$f(x) = a_n x^n + a_{n-1}x^{n-1} + \cdots + a_1 x + a_0$$

has *integer* coefficients and if c/d is a rational zero of $f(x)$ such that c and d have no common prime factor, then

(1) the numerator c of the zero is a factor of the constant term a_0

(2) the denominator d of the zero is a factor of the leading coefficient a_n

PROOF Assume that $c > 0$. (The proof for $c < 0$ is similar.) Let us show that c is a factor of a_0. The case $c = 1$ is trivial, since 1 is a factor of *any*

(continued)

▶ **Tutorial available at www.thomsonedu.com/login**

number. Thus, suppose $c \neq 1$. In this case $c/d \neq 1$, for if $c/d = 1$, we obtain $c = d$, and since c and d have no prime factor in common, this implies that $c = d = 1$, a contradiction. Hence, in the following discussion we have $c \neq 1$ and $c \neq d$.

Since $f(c/d) = 0$,

$$a_n\frac{c^n}{d^n} + a_{n-1}\frac{c^{n-1}}{d^{n-1}} + \cdots + a_1\frac{c}{d} + a_0 = 0.$$

We multiply by d^n and then add $-a_0d^n$ to both sides:

$$a_nc^n + a_{n-1}c^{n-1}d + \cdots + a_1cd^{n-1} = -a_0d^n$$
$$c(a_nc^{n-1} + a_{n-1}c^{n-2}d + \cdots + a_1d^{n-1}) = -a_0d^n$$

The last equation shows that c is a factor of the integer a_0d^n. Since c and d have no common factor, c is a factor of a_0. A similar argument may be used to prove that d is a factor of a_n. ✎

As an aid in listing the possible rational zeros, remember the following quotient:

$$\text{Possible rational zeros} = \frac{\text{factors of the constant term } a_0}{\text{factors of the leading coefficient } a_n}$$

The theorem on rational zeros of a polynomial may be applied to equations with rational coefficients by merely multiplying both sides of the equation by the lcd of all the coefficients to obtain an equation with integral coefficients.

EXAMPLE 3 Showing a polynomial has no rational zeros

Show that $f(x) = x^3 - 4x - 2$ has no rational zeros.

▶ **SOLUTION** If $f(x)$ has a rational zero c/d such that c and d have no common prime factor, then, by the theorem on rational zeros of a polynomial, c is a factor of the constant term -2 and hence is either 2 or -2 (which we write as ± 2) or ± 1. The denominator d is a factor of the leading coefficient 1 and hence is ± 1. Thus, the only possibilities for c/d are

$$\frac{\pm 1}{\pm 1} \quad \text{and} \quad \frac{\pm 2}{\pm 1} \qquad \text{or, equivalently,} \qquad \pm 1 \quad \text{and} \quad \pm 2.$$

Substituting each of these numbers for x, we obtain

$$f(1) = -5, \quad f(-1) = 1, \quad f(2) = -2, \quad \text{and} \quad f(-2) = -2.$$

Since $f(\pm 1) \neq 0$ and $f(\pm 2) \neq 0$, it follows that $f(x)$ has no rational zeros. ✎

▶ **Tutorial available at www.thomsonedu.com/login**

In the solution of the following example we will assume that a graphing utility is not available. In Example 5 we will rework the problem to demonstrate the advantage of using a graphing utility.

EXAMPLE 4 **Finding the rational solutions of an equation**

Find all rational solutions of the equation

$$3x^4 + 14x^3 + 14x^2 - 8x - 8 = 0.$$

▶ SOLUTION The problem is equivalent to finding the rational zeros of the polynomial on the left-hand side of the equation. If c/d is a rational zero and c and d have no common factor, then c is a factor of the constant term -8 and d is a factor of the leading coefficient 3. All possible choices are listed in the following table.

Choices for the numerator c	$\pm1, \pm2, \pm4, \pm8$
Choices for the denominator d	$\pm1, \pm3$
Choices for c/d	$\pm1, \pm2, \pm4, \pm8, \pm\frac{1}{3}, \pm\frac{2}{3}, \pm\frac{4}{3}, \pm\frac{8}{3}$

We can reduce the number of choices by finding upper and lower bounds for the real solutions; however, we shall not do so here. It is necessary to determine which of the choices for c/d, if any, are zeros. We see by substitution that neither 1 nor -1 is a solution. If we divide synthetically by $x + 2$, we obtain

$$\begin{array}{r|rrrrr} -2 & 3 & 14 & 14 & -8 & -8 \\ & & -6 & -16 & 4 & 8 \\ \hline & 3 & 8 & -2 & -4 & 0 \end{array}$$

This result shows that -2 is a zero. Moreover, the synthetic division provides the coefficients of the quotient in the division of the polynomial by $x + 2$. Hence, we have the following factorization of the given polynomial:

$$(x + 2)(3x^3 + 8x^2 - 2x - 4)$$

The remaining solutions of the equation must be zeros of the second factor, so we use that polynomial to check for solutions. *Do not* use the polynomial in the original equation. (Note that $\pm\frac{8}{3}$ are no longer candidates, since the numerator must be a factor of 4.) Again proceeding by trial and error, we ultimately find that synthetic division by $x + \frac{2}{3}$ gives us the following result:

$$\begin{array}{r|rrrr} -\frac{2}{3} & 3 & 8 & -2 & -4 \\ & & -2 & -4 & 4 \\ \hline & 3 & 6 & -6 & 0 \end{array}$$

Therefore, $-\frac{2}{3}$ is also a zero.

(continued)

▶ **Tutorial available at www.thomsonedu.com/login**

Using the coefficients of the quotient, we know that the remaining zeros are solutions of the equation $3x^2 + 6x - 6 = 0$. Dividing both sides by 3 gives us the equivalent equation $x^2 + 2x - 2 = 0$. By the quadratic formula, this equation has solutions

$$\frac{-2 \pm \sqrt{2^2 - 4(1)(-2)}}{2(1)} = \frac{-2 \pm \sqrt{12}}{2} = \frac{-2 \pm 2\sqrt{3}}{2} = -1 \pm \sqrt{3}.$$

Hence, the given polynomial has two rational roots, -2 and $-\frac{2}{3}$, and two irrational roots, $-1 + \sqrt{3} \approx 0.732$ and $-1 - \sqrt{3} \approx -2.732$.

EXAMPLE 5 Finding the rational solutions of an equation

Find all rational solutions of the equation

$$3x^4 + 14x^3 + 14x^2 - 8x - 8 = 0.$$

Figure 1
$[-7.5, 7.5]$ by $[-5, 5]$

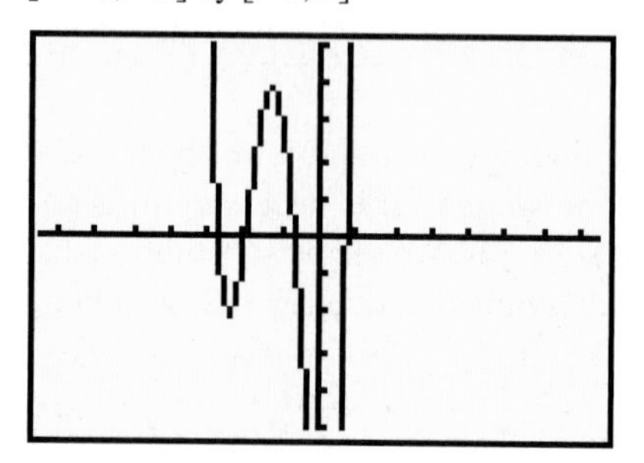

SOLUTION Assigning the indicated polynomial to Y_1 and choosing the viewing rectangle $[-7.5, 7.5]$ by $[-5, 5]$, we obtain a sketch similar to Figure 1. The graph indicates that -2 is a solution and that there is one solution in each of the intervals $(-3, -2)$, $(-1, 0)$, and $(0, 1)$. From Example 4 we know that the possible rational zeros are

$$\pm 1, \quad \pm 2, \quad \pm 4, \quad \pm 8, \quad \pm\tfrac{1}{3}, \quad \pm\tfrac{2}{3}, \quad \pm\tfrac{4}{3}, \quad \pm\tfrac{8}{3}.$$

We conclude that the only possibilities are $-\frac{8}{3}$ in $(-3, -2)$, $-\frac{2}{3}$ in $(-1, 0)$, and $\frac{2}{3}$ in $(0, 1)$. Thus, by referring to the graph, we have reduced the number of choices for zeros from sixteen to three. Synthetic division can now be used to determine that the only rational solutions are -2 and $-\frac{2}{3}$.

EXAMPLE 6 Finding the radius of a grain silo

Figure 2

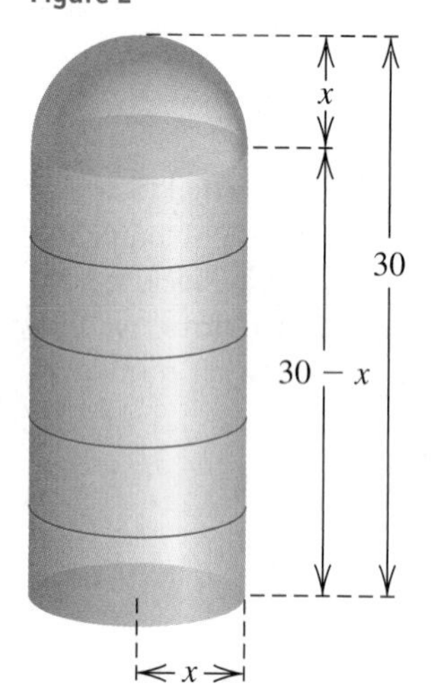

A grain silo has the shape of a right circular cylinder with a hemisphere attached to the top. If the total height of the structure is 30 feet, find the radius of the cylinder that results in a total volume of 1008π ft^3.

SOLUTION Let x denote the radius of the cylinder as shown in Figure 2. The volume of the cylinder is $\pi r^2 h = \pi x^2(30 - x)$, and the volume of the hemisphere is $\frac{2}{3}\pi r^3 = \frac{2}{3}\pi x^3$, so we solve for x as follows:

$$\begin{aligned} \pi x^2(30 - x) + \tfrac{2}{3}\pi x^3 &= 1008\pi && \text{total volume is } 1008\pi \\ 3x^2(30 - x) + 2x^3 &= 3024 && \text{multiply by } \frac{3}{\pi} \\ 90x^2 - x^3 &= 3024 && \text{simplify} \\ x^3 - 90x^2 + 3024 &= 0 && \text{equivalent equation} \end{aligned}$$

Since the leading coefficient of the polynomial on the left-hand side of the last equation is 1, any rational root has the form $c/1 = c$, where c is a factor of

3024. If we factor 3024 into primes, we find that $3024 = 2^4 \cdot 3^3 \cdot 7$. It follows that some of the positive factors of 3024 are

$$1, \quad 2, \quad 3, \quad 4, \quad 6, \quad 7, \quad 8, \quad 9, \quad 12, \quad \ldots .$$

To help us decide which of these numbers to test first, let us make a rough estimate of the radius by assuming that the silo has the shape of a right circular cylinder of height 30 feet. In that case, the volume would be $\pi r^2 h = 30\pi r^2$. Since this volume should be close to 1008π, we see that

$$30r^2 = 1008, \qquad \text{or} \qquad r^2 = 1008/30 = 33.6.$$

This suggests that we use 6 in our first synthetic division, as follows:

$$\begin{array}{r|rrrr} 6 & 1 & -90 & 0 & 3024 \\ & & 6 & -504 & -3024 \\ \hline & 1 & -84 & -504 & 0 \end{array}$$

Figure 3
[0, 10] by [0, 4000, 500]

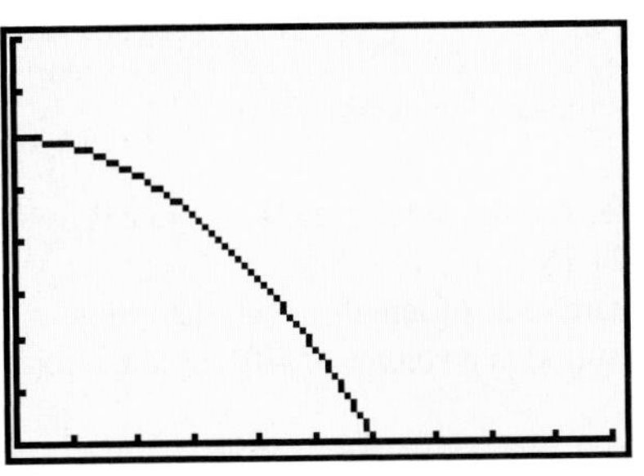

Thus, 6 is a solution of the equation $x^3 - 90x^2 + 3024 = 0$.

The remaining two solutions of the equation can be found by solving the depressed equation $x^2 - 84x - 504 = 0$. These zeros are approximately -5.62 and 89.62—neither of which satisfies the conditions of the problem. Hence, the desired radius is 6 feet.

The graph of $f(x) = x^3 - 90x^2 + 3024$ in Figure 3 shows the zero $x = 6$. An extended graph would also indicate the other two zeros.

3.4 Exercises

Exer. 1–10: A polynomial $f(x)$ with real coefficients and leading coefficient 1 has the given zero(s) and degree. Express $f(x)$ as a product of linear and quadratic polynomials with real coefficients that are irreducible over $\mathbb{R}$.

▶ 1 $3 + 2i$; degree 2

2 $-4 + 3i$; degree 2

▶ 3 $2, -2 - 5i$; degree 3

4 $-3, 1 - 7i$; degree 3

▶ 5 $-1, 0, 3 + i$; degree 4

6 $0, 2, -2 - i$; degree 4

7 $4 + 3i, -2 + i$; degree 4

8 $3 + 5i, -1 - i$; degree 4

9 $0, -2i, 1 - i$; degree 5

▶ 10 $0, 3i, 4 + i$; degree 5

Exer. 11–14: Show that the equation has no rational root.

▶ 11 $x^3 + 3x^2 - 4x + 6 = 0$

12 $3x^3 - 4x^2 + 7x + 5 = 0$

13 $x^5 - 3x^3 + 4x^2 + x - 2 = 0$

14 $2x^5 + 3x^3 + 7 = 0$

Exer. 15–24: Find all solutions of the equation.

▶ 15 $x^3 - x^2 - 10x - 8 = 0$

16 $x^3 + x^2 - 14x - 24 = 0$

17 $2x^3 - 3x^2 - 17x + 30 = 0$

18 $12x^3 + 8x^2 - 3x - 2 = 0$

19 $x^4 + 3x^3 - 30x^2 - 6x + 56 = 0$

▶ **Tutorial available at www.thomsonedu.com/login**

20 $3x^5 - 10x^4 - 6x^3 + 24x^2 + 11x - 6 = 0$

21 $6x^5 + 19x^4 + x^3 - 6x^2 = 0$

22 $6x^4 + 5x^3 - 17x^2 - 6x = 0$

23 $8x^3 + 18x^2 + 45x + 27 = 0$

24 $3x^3 - x^2 + 11x - 20 = 0$

Exer. 25–26: Find a factored form with integer coefficients of the polynomial f shown in the figure. Assume that Xscl = Yscl = 1.

25 $f(x) = 6x^5 - 23x^4 + 24x^3 + x^2 - 12x + 4$

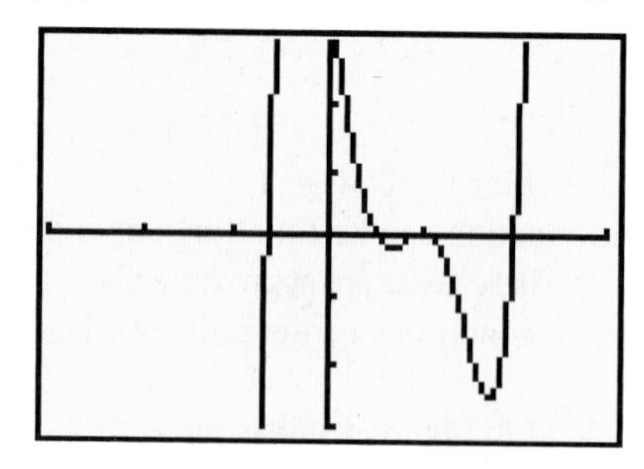

26 $f(x) = -6x^5 + 5x^4 + 14x^3 - 8x^2 - 8x + 3$

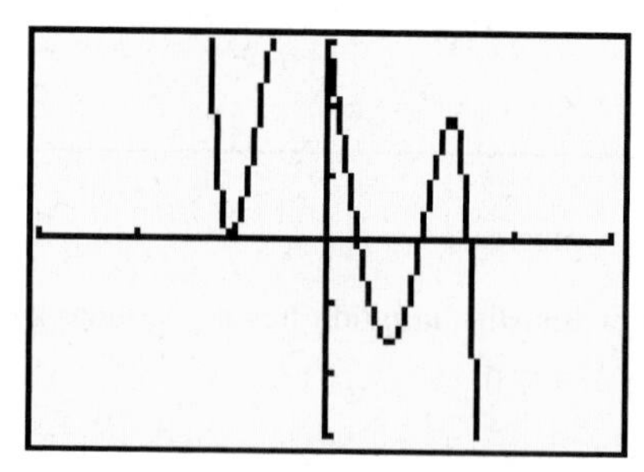

Exer. 27–28: The polynomial function f has only real zeros. Use the graph of f to factor it.

27 $f(x) = 2x^3 - 25.4x^2 + 3.02x + 24.75$

28 $f(x) = 0.5x^3 + 0.65x^2 - 5.365x + 1.5375$

29 Does there exist a polynomial of degree 3 with real coefficients that has zeros 1, -1, and i? Justify your answer.

30 The polynomial $f(x) = x^3 - ix^2 + 2ix + 2$ has the complex number i as a zero; however, the conjugate $-i$ of i is not a zero. Why doesn't this result contradict the theorem on conjugate pair zeros of a polynomial?

31 If n is an odd positive integer, prove that a polynomial of degree n with real coefficients has at least one real zero.

32 If a polynomial of the form

$$x^n + a_{n-1}x^{n-1} + \cdots + a_1x + a_0,$$

where each a_k is an integer, has a rational root r, show that r is an integer and is a factor of a_0.

33 **Constructing a box** From a rectangular piece of cardboard having dimensions 20 inches $\times$ 30 inches, an open box is to be made by removing squares of area x^2 from each corner and turning up the sides. (See Exercise 41 of Section 3.1.)

(a) Show that there are two boxes that have a volume of 1000 in^3.

(b) Which box has the smaller surface area?

34 **Constructing a crate** The frame for a shipping crate is to be constructed from 24 feet of 2×2 lumber. Assuming the crate is to have square ends of length x feet, determine the value(s) of x that result(s) in a volume of 4 ft^3. (See Exercise 42 of Section 3.1.)

35 A right triangle has area 30 ft^2 and a hypotenuse that is 1 foot longer than one of its sides.

(a) If x denotes the length of this side, then show that $2x^3 + x^2 - 3600 = 0$.

(b) Show that there is a positive root of the equation in part (a) and that this root is less than 13.

(c) Find the lengths of the sides of the triangle.

36 **Constructing a storage tank** A storage tank for propane gas is to be constructed in the shape of a right circular cylinder of altitude 10 feet with a hemisphere attached to each end. Determine the radius x so that the resulting volume is 27π ft^3. (See Example 8 of Section 2.4.)

37 **Constructing a storage shelter** A storage shelter is to be constructed in the shape of a cube with a triangular prism forming the roof (see the figure). The length x of a side of the cube is yet to be determined.

(a) If the total height of the structure is 6 feet, show that its volume V is given by $V = x^3 + \frac{1}{2}x^2(6 - x)$.

(b) Determine x so that the volume is 80 ft^3.

Exercise 37

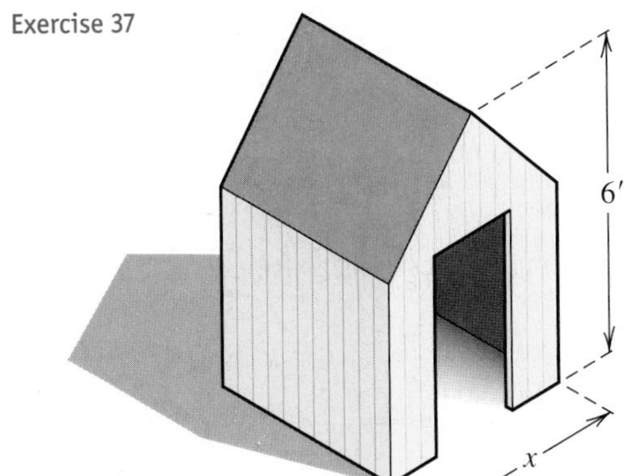
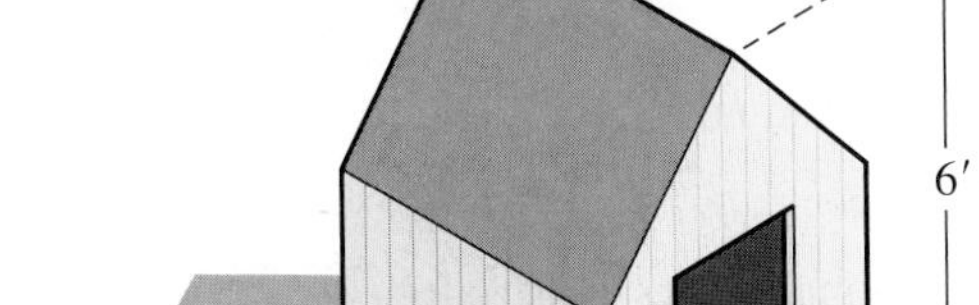

38 **Designing a tent** A canvas camping tent is to be constructed in the shape of a pyramid with a square base. An 8-foot pole will form the center support, as illustrated in the figure. Find the length x of a side of the base so that the total amount of canvas needed for the sides and bottom is 384 ft^2.

Exercise 38

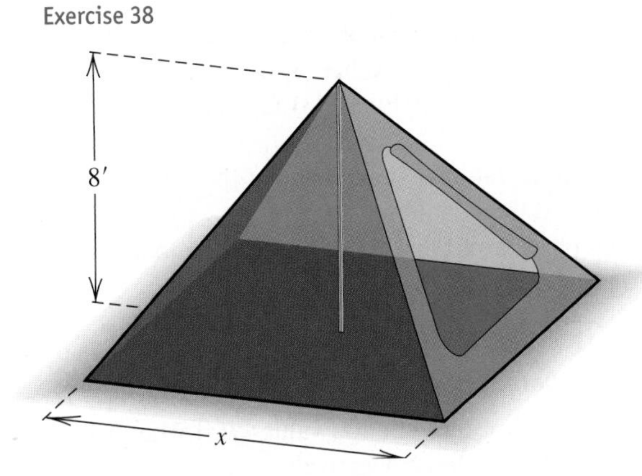

Exer. 39–40: Use a graph to determine the number of nonreal solutions of the equation.

39 $x^5 + 1.1x^4 - 3.21x^3 - 2.835x^2 + 2.7x + 0.62 = -1$

40 $x^4 - 0.4x^3 - 2.6x^2 + 1.1x + 3.5 = 2$

Exer. 41–42: Use a graph and synthetic division to find all solutions of the equation.

41 $x^4 + 1.4x^3 + 0.44x^2 - 0.56x - 0.96 = 0$

42 $x^5 + 1.1x^4 - 2.62x^3 - 4.72x^2 - 0.2x + 5.44 = 0$

43 **Atmospheric density** The density $D(h)$ (in kg/m^3) of the earth's atmosphere at an altitude of h meters can be approximated by

$$D(h) = 1.2 - ah + bh^2 - ch^3,$$

where

$$a = 1.096 \times 10^{-4},\ b = 3.42 \times 10^{-9},\ c = 3.6 \times 10^{-14},$$

and $0 \le h \le 30{,}000$. Use the graph of D to approximate the altitude h at which the density is 0.4.

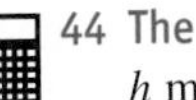

44 **The earth's density** The earth's density $D(h)$ (in g/cm^3) h meters underneath the surface can be approximated by

$$D(h) = 2.84 + ah + bh^2 - ch^3,$$

where

$$a = 1.4 \times 10^{-3},\ b = 2.49 \times 10^{-6},\ c = 2.19 \times 10^{-9},$$

and $0 \le h \le 1000$. Use the graph of D to approximate the depth h at which the density of the earth is 3.7.

3.5

Rational Functions

A function f is a **rational function** if

$$f(x) = \frac{g(x)}{h(x)},$$

where $g(x)$ and $h(x)$ are polynomials. The domain of f consists of all real numbers *except* the zeros of the denominator $h(x)$.

ILLUSTRATION **Rational Functions and Their Domains**

■ $f(x) = \dfrac{1}{x - 2}$; *domain:* all x *except* $x = 2$

(continued)

Figure 1

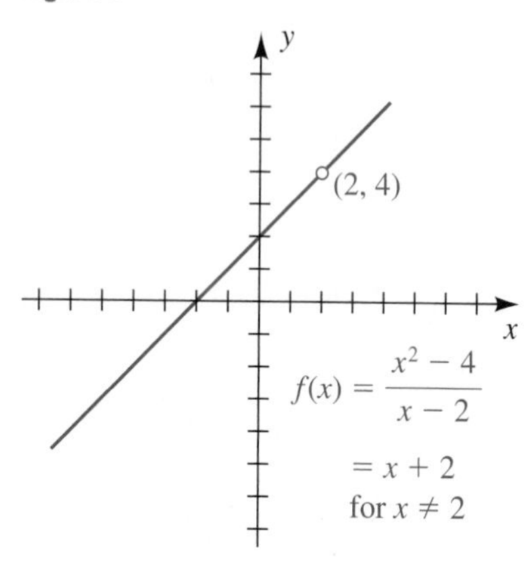

Figure 2

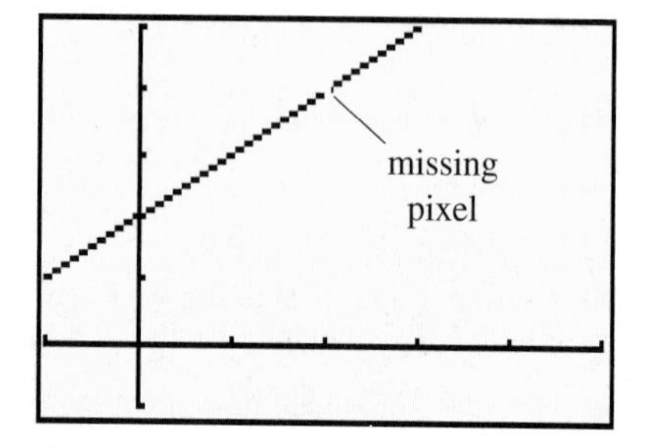

Figure 3

X	Y1	
-1	1	
0	2	
1	3	
2	ERROR	
3	5	
4	6	
5	7	

Y1 =ERROR

- $f(x) = \dfrac{5x}{x^2 - 9}$; *domain:* all x *except* $x = \pm 3$
- $f(x) = \dfrac{x^3 - 8}{x^2 + 4}$; *domain:* all real numbers x

Previously we simplified rational expressions as follows:

$$\frac{x^2 - 4}{x - 2} = \frac{(x + 2)(x - 2)}{x - 2} \overset{\text{if } x \neq 2}{=} \frac{x + 2}{1} = x + 2$$

If we let $f(x) = \dfrac{x^2 - 4}{x - 2}$ and $g(x) = x + 2$, then the domain of f is all x except $x = 2$ and the domain of g is all real numbers. These domains and the above simplification suggest that the graphs of f and g are the same except for $x = 2$. What happens to the graph of f at $x = 2$? There is a *hole* in the graph—that is, a single point is missing. To find the y-value of the hole, we can substitute 2 for x in the reduced function, which is simply $g(2) = 4$. A graph of f is shown in Figure 1.

To alert the user to the presence of a hole in the graph, some graphing utilities will actually draw a hole, as in Figure 1; others simply omit a pixel, as in Figure 2. Checking a table of values for f (Figure 3) indicates that f is undefined for $x = 2$.

We now turn our attention to rational functions that do not have a common factor in the numerator and the denominator.

When sketching the graph of a rational function f, it is important to answer the following two questions.

Question 1 What can be said of the function values $f(x)$ when x is close to (but not equal to) a zero of the denominator?

Question 2 What can be said of the function values $f(x)$ when x is large positive or when x is large negative?

As we shall see, if a is a zero of the denominator, one of several situations often occurs. These are shown in Figure 4, where we have used notations from the following chart.

Notation	Terminology
$x \to a^-$	x approaches a from the left (through values *less* than a).
$x \to a^+$	x approaches a from the right (through values *greater* than a).
$f(x) \to \infty$	$f(x)$ increases without bound (can be made as large positive as desired).
$f(x) \to -\infty$	$f(x)$ decreases without bound (can be made as large negative as desired).

Figure 4

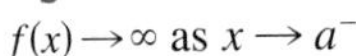

$f(x) \to \infty$ as $x \to a^-$

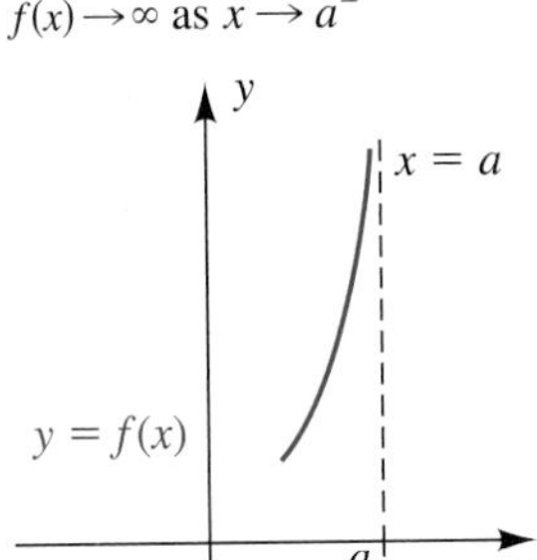

$f(x) \to \infty$ as $x \to a^+$

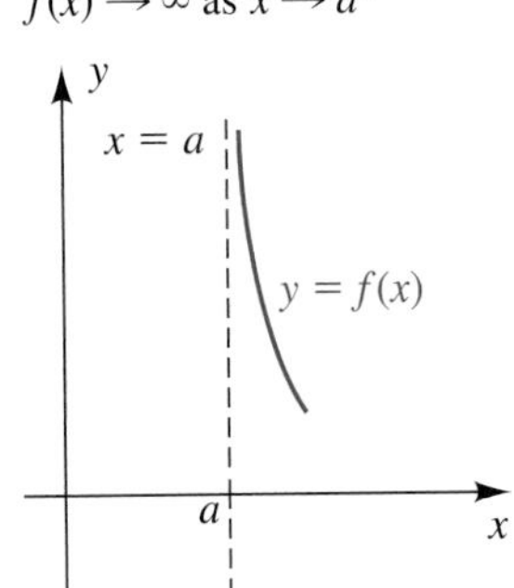

$f(x) \to -\infty$ as $x \to a^-$

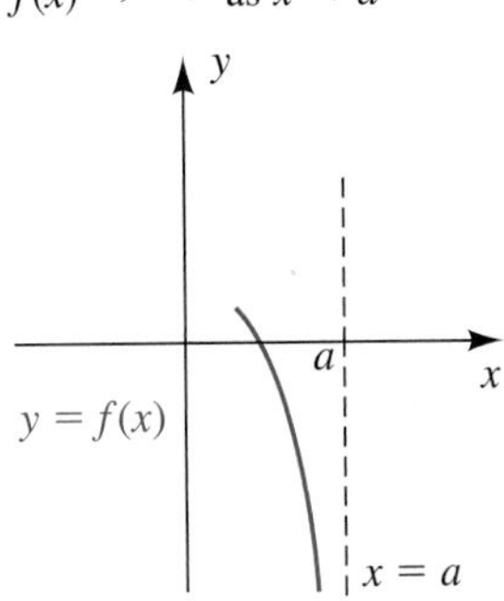

$f(x) \to -\infty$ as $x \to a^+$

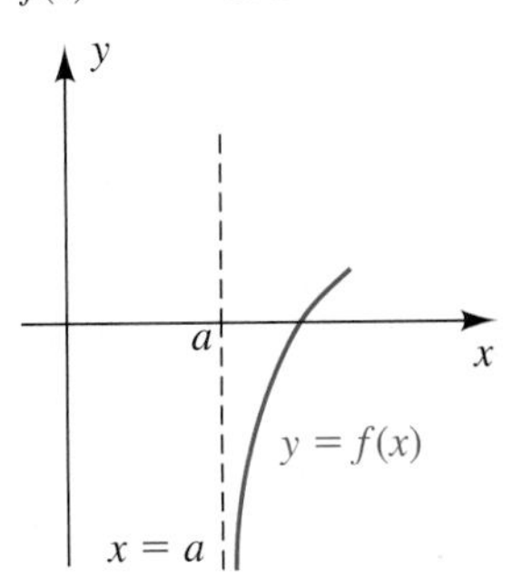

The symbols ∞ (read "infinity") and $-\infty$ (read "minus infinity") do not represent real numbers; they simply specify certain types of behavior of functions and variables.

The dashed line $x = a$ in Figure 4 is called a *vertical asymptote,* as in the following definition.

Definition of Vertical Asymptote	The line $x = a$ is a **vertical asymptote** for the graph of a function f if $$f(x) \to \infty \qquad \text{or} \qquad f(x) \to -\infty$$ as x approaches a from either the left or the right.

Thus, the answer to Question 1 is that if a is a zero of the denominator of a rational function f, then the graph of f *may* have a vertical asymptote $x = a$. There are rational functions where this is *not* the case (as in Figure 1 of this section). If the numerator and denominator have no common factor, then f *must* have a vertical asymptote $x = a$.

Let us next consider Question 2. For x *large positive* or *large negative,* the graph of a rational function may look like one of those in Figure 5, where the notation

$$f(x) \to c \quad \text{as} \quad x \to \infty$$

is read "$f(x)$ approaches c as x increases without bound" or "$f(x)$ approaches c as x approaches infinity," and the notation

$$f(x) \to c \quad \text{as} \quad x \to -\infty$$

is read "$f(x)$ approaches c as x decreases without bound."

Figure 5

$f(x) \to c$ as $x \to \infty$

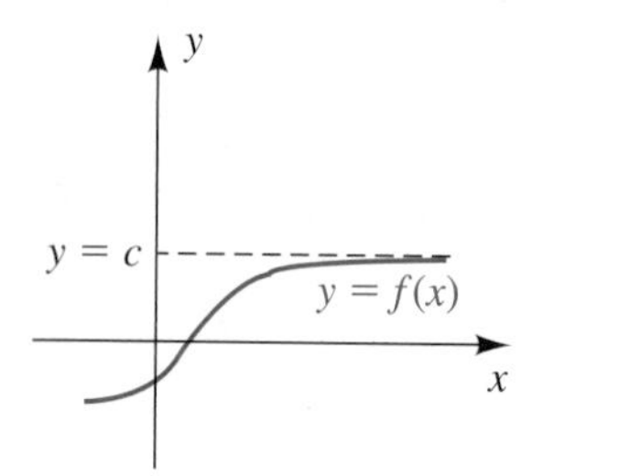

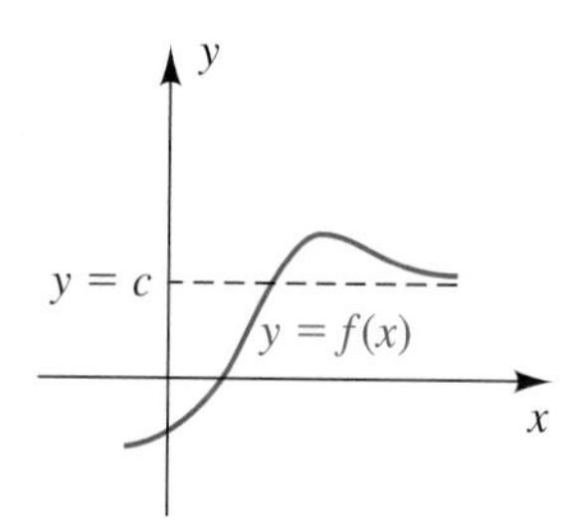

$f(x) \to c$ as $x \to -\infty$

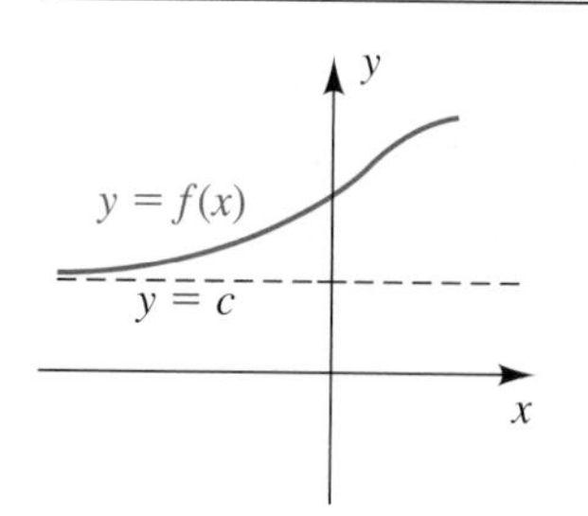

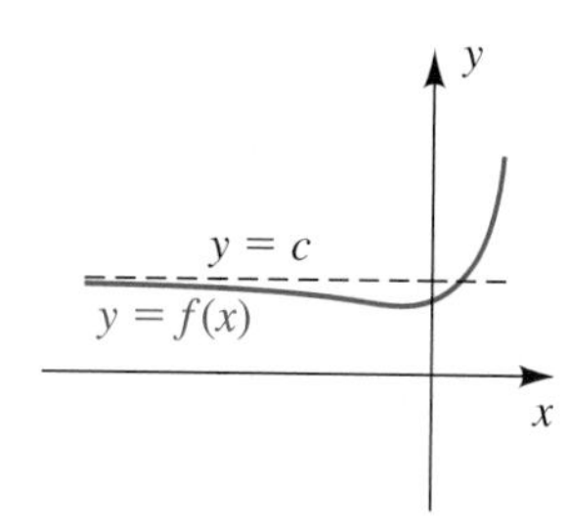

We call the dashed line in Figure 5 a *horizontal asymptote,* as in the next definition.

Definition of Horizontal Asymptote

The line $y = c$ is a **horizontal asymptote** for the graph of a function f if

$$f(x) \to c \quad \text{as} \quad x \to \infty \quad \text{or as} \quad x \to -\infty.$$

Thus, the answer to Question 2 is that $f(x)$ *may* be very close to some number c when x is large positive or large negative; that is, the graph of f may have a horizontal asymptote $y = c$. There are rational functions where this is *not* the case (as in Examples 2(c) and 9).

Note that, as in the second and fourth sketches in Figure 5, the graph of f may cross a horizontal asymptote.

In the next example we find the asymptotes for the graph of a simple rational function.

EXAMPLE 1 Sketching the graph of a rational function

Sketch the graph of f if

$$f(x) = \frac{1}{x - 2}.$$

▶ **SOLUTION** Let us begin by considering Question 1, stated at the beginning of this section. The denominator $x - 2$ is zero at $x = 2$. If x is close to 2 and $x > 2$, then $f(x)$ is large positive, as indicated in the following table.

x	2.1	2.01	2.001	2.0001	2.00001
$\dfrac{1}{x-2}$	10	100	1000	10,000	100,000

▶ Tutorial available at www.thomsonedu.com/login

Since we can make $1/(x - 2)$ as large as desired by taking x close to 2 (and $x > 2$), we see that

$$f(x) \to \infty \quad \text{as} \quad x \to 2^+.$$

If $f(x)$ is close to 2 and $x < 2$, then $f(x)$ is large negative; for example, $f(1.9999) = -10{,}000$ and $f(1.99999) = -100{,}000$. Thus,

$$f(x) \to -\infty \quad \text{as} \quad x \to 2^-.$$

The line $x = 2$ is a vertical asymptote for the graph of f, as illustrated in Figure 6.

We next consider Question 2. The following table lists some approximate values for $f(x)$ when x is large and positive.

Figure 6

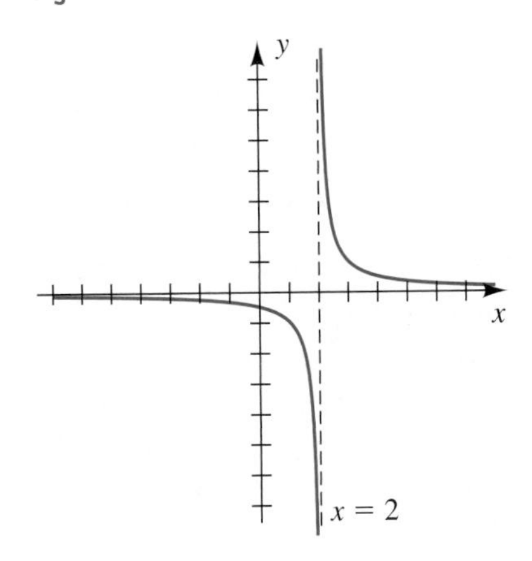

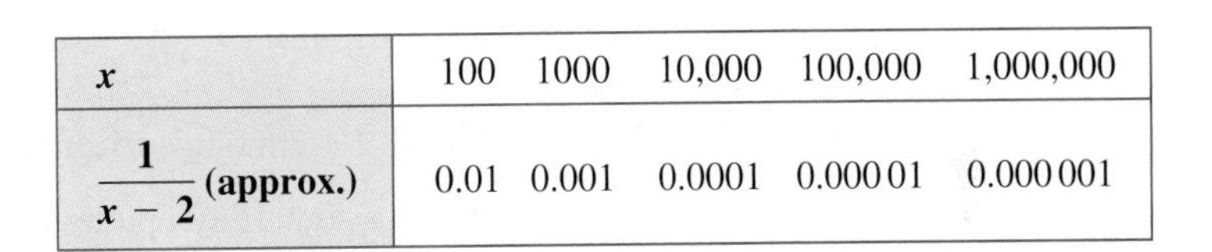

x	100	1000	10,000	100,000	1,000,000
$\dfrac{1}{x-2}$ (approx.)	0.01	0.001	0.0001	0.000 01	0.000 001

We may describe this behavior of $f(x)$ by writing

$$f(x) \to 0 \quad \text{as} \quad x \to \infty.$$

Similarly, $f(x)$ is close to 0 when x is large negative; for example, $f(-100{,}000) \approx -0.00001$. Thus,

$$f(x) \to 0 \quad \text{as} \quad x \to -\infty.$$

The line $y = 0$ (the x-axis) is a horizontal asymptote, as shown in Figure 6.

Plotting the points $(1, -1)$ and $(3, 1)$ helps give us a rough sketch of the graph.

The function considered in Example 1, $f(x) = 1/(x - 2)$, closely resembles one of the simplest rational functions, the **reciprocal function.** The reciprocal function has equation $f(x) = 1/x$, vertical asymptote $x = 0$ (the y-axis), and horizontal asymptote $y = 0$ (the x-axis). The graph of the reciprocal function (shown in Appendix I) is the graph of a *hyperbola* (discussed later in the text). Note that we can obtain the graph of $y = 1/(x - 2)$ by shifting the graph of $y = 1/x$ to the right 2 units.

The following theorem is useful for finding the horizontal asymptote for the graph of a rational function.

Theorem on Horizontal Asymptotes

Let $f(x) = \dfrac{a_nx^n + a_{n-1}x^{n-1} + \cdots + a_1x + a_0}{b_kx^k + b_{k-1}x^{k-1} + \cdots + b_1x + b_0}$, where $a_n \neq 0$ and $b_k \neq 0$.

(1) If $n < k$, then the x-axis (the line $y = 0$) is the horizontal asymptote for the graph of f.

(2) If $n = k$, then the line $y = a_n/b_k$ (the ratio of leading coefficients) is the horizontal asymptote for the graph of f.

(3) If $n > k$, the graph of f has no horizontal asymptote. Instead, either $f(x) \to \infty$ or $f(x) \to -\infty$ as $x \to \infty$ or as $x \to -\infty$.

Proofs for each part of this theorem may be patterned after the solutions in the next example. Concerning part (3), if $q(x)$ is the quotient obtained by dividing the numerator by the denominator, then $f(x) \to \infty$ if $q(x) \to \infty$ or $f(x) \to -\infty$ if $q(x) \to -\infty$.

EXAMPLE 2 Finding horizontal asymptotes

Find the horizontal asymptote for the graph of f, if it exists.

(a) $f(x) = \dfrac{3x - 1}{x^2 - x - 6}$ **(b)** $f(x) = \dfrac{5x^2 + 1}{3x^2 - 4}$

(c) $f(x) = \dfrac{2x^4 - 3x^2 + 5}{x^2 + 1}$

SOLUTION

▶ **(a)** The degree of the numerator, 1, is less than the degree of the denominator, 2, so, by part (1) of the theorem on horizontal asymptotes, the x-axis is a horizontal asymptote. To verify this directly, we divide the numerator and denominator of the quotient by x^2 (since 2 is the highest power on x in the denominator), obtaining

$$f(x) = \frac{\dfrac{3x - 1}{x^2}}{\dfrac{x^2 - x - 6}{x^2}} = \frac{\dfrac{3}{x} - \dfrac{1}{x^2}}{1 - \dfrac{1}{x} - \dfrac{6}{x^2}} \quad \text{for} \quad x \neq 0.$$

If x is large positive or large negative, then $3/x$, $1/x^2$, $1/x$, and $6/x^2$ are close to 0, and hence

$$f(x) \approx \frac{0 - 0}{1 - 0 - 0} = \frac{0}{1} = 0.$$

Thus,

$$f(x) \to 0 \quad \text{as} \quad x \to \infty \quad \text{or as} \quad x \to -\infty.$$

Since $f(x)$ is the y-coordinate of a point on the graph, the last statement means that the line $y = 0$ (that is, the x-axis) is a horizontal asymptote.

▶ **Tutorial available at www.thomsonedu.com/login**

▶ **(b)** If $f(x) = (5x^2 + 1)/(3x^2 - 4)$, then the numerator and denominator have the same degree, 2, and the leading coefficients are 5 and 3, respectively. Hence, by part (2) of the theorem on horizontal asymptotes, the line $y = \frac{5}{3}$ is the horizontal asymptote. We could also show that $y = \frac{5}{3}$ is the horizontal asymptote by dividing the numerator and denominator of $f(x)$ by x^2, as in part (a).

(c) The degree of the numerator, 4, is greater than the degree of the denominator, 2, so, by part (3) of the theorem on horizontal asymptotes, the graph has no horizontal asymptote. If we use long division, we obtain

$$f(x) = 2x^2 - 5 + \frac{10}{x^2 + 1}.$$

As either $x \to \infty$ or $x \to -\infty$, the quotient $2x^2 - 5$ increases without bound and $10/(x^2 + 1) \to 0$. Hence, $f(x) \to \infty$ as $x \to \infty$ or as $x \to -\infty$.

We next list some guidelines for sketching the graph of a rational function. Their use will be illustrated in Examples 3, 6, and 7.

Guidelines for Sketching the Graph of a Rational Function

Assume that $f(x) = \dfrac{g(x)}{h(x)}$, where $g(x)$ and $h(x)$ are polynomials that have no common factor.

1 Find the x-intercepts—that is, the real zeros of the numerator $g(x)$—and plot the corresponding points on the x-axis.

2 Find the real zeros of the denominator $h(x)$. For each real zero a, sketch the vertical asymptote $x = a$ with dashes.

3 Find the y-intercept $f(0)$, if it exists, and plot the point $(0, f(0))$ on the y-axis.

4 Apply the theorem on horizontal asymptotes. If there is a horizontal asymptote $y = c$, sketch it with dashes.

5 If there is a horizontal asymptote $y = c$, determine whether it intersects the graph. The x-coordinates of the points of intersection are the solutions of the equation $f(x) = c$. Plot these points, if they exist.

6 Sketch the graph of f in each of the regions in the xy-plane determined by the vertical asymptotes in guideline 2. If necessary, use the sign of specific function values to tell whether the graph is above or below the x-axis or the horizontal asymptote. Use guideline 5 to decide whether the graph approaches the horizontal asymptote from above or below.

In the following examples our main objective is to determine the general shape of the graph, paying particular attention to how the graph approaches the asymptotes. We will plot only a few points, such as those corresponding to the x-intercepts and y-intercept or the intersection of the graph with a horizontal asymptote.

EXAMPLE 3 Sketching the graph of a rational function

Sketch the graph of f if

$$f(x) = \frac{3x + 4}{2x - 5}.$$

SOLUTION We follow the guidelines.

Guideline 1 To find the x-intercepts we find the zeros of the numerator. Solving $3x + 4 = 0$ gives us $x = -\frac{4}{3}$, and we plot the point $\left(-\frac{4}{3}, 0\right)$ on the x-axis, as shown in Figure 7.

Figure 7

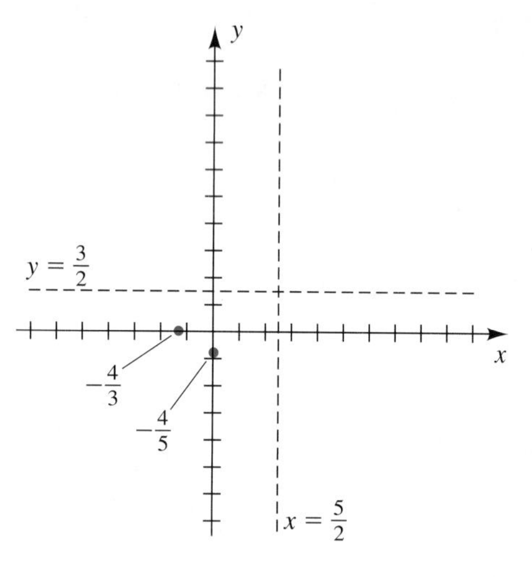

Guideline 2 The denominator has zero $\frac{5}{2}$, so the line $x = \frac{5}{2}$ is a vertical asymptote. We sketch this line with dashes, as in Figure 7.

Guideline 3 The y-intercept is $f(0) = -\frac{4}{5}$, and we plot the point $\left(0, -\frac{4}{5}\right)$ in Figure 7.

Guideline 4 The numerator and denominator of $f(x)$ have the same degree, 1. The leading coefficients are 3 and 2, so by part (2) of the theorem on horizontal asymptotes, the line $y = \frac{3}{2}$ is a horizontal asymptote. We sketch the line with dashes in Figure 7.

Guideline 5 The x-coordinates of the points where the graph intersects the horizontal asymptote $y = \frac{3}{2}$ are solutions of the equation $f(x) = \frac{3}{2}$. We solve this equation as follows:

$$\begin{aligned}
\frac{3x + 4}{2x - 5} &= \frac{3}{2} && \text{let } f(x) = \frac{3}{2} \\
2(3x + 4) &= 3(2x - 5) && \text{multiply by } 2(2x - 5) \\
6x + 8 &= 6x - 15 && \text{multiply} \\
8 &= -15 && \text{subtract } 6x
\end{aligned}$$

Since $8 \neq -15$ for any value of x, this result indicates that the graph of f does *not* intersect the horizontal asymptote. As an aid in sketching, we can now think of the horizontal asymptote as a boundary that cannot be crossed.

Figure 8

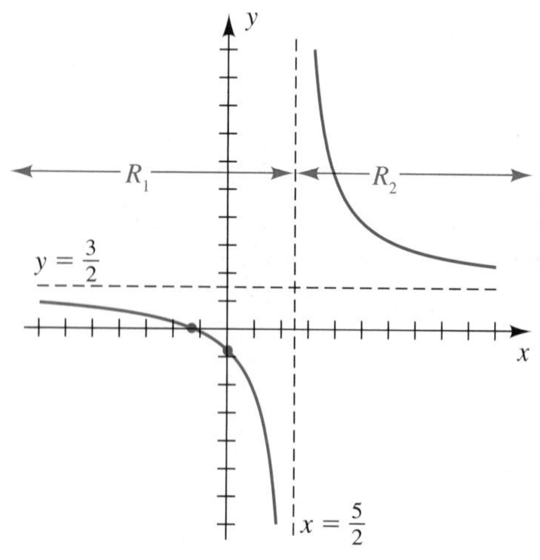

Guideline 6 The vertical asymptote in Figure 7 divides the xy-plane into two regions:

R_1: the region to the left of $x = \frac{5}{2}$

R_2: the region to the right of $x = \frac{5}{2}$

For R_1, we have the two points $\left(-\frac{4}{3}, 0\right)$ and $\left(0, -\frac{4}{5}\right)$ that the graph of f must pass through, as well as the two asymptotes that the graph must approach. This portion of f is shown in Figure 8.

Tutorial available at www.thomsonedu.com/login

Figure 9

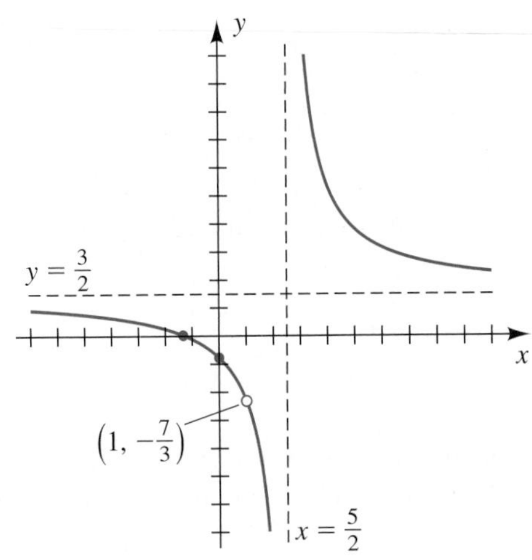

For R_2, the graph must again approach the two asymptotes. Since the graph cannot cross the x-axis (there is no x-intercept in R_2), it must be above the horizontal asymptote, as shown in Figure 8.

EXAMPLE 4 **Sketching a graph that has a hole**

Sketch the graph of g if

$$g(x) = \frac{(3x + 4)(x - 1)}{(2x - 5)(x - 1)}.$$

SOLUTION The domain of g is all real numbers except $\frac{5}{2}$ and 1. If g is reduced, we obtain the function f in the previous example. The only difference between the graphs of f and g is that g has a hole at $x = 1$. Since $f(1) = -\frac{7}{3}$, we need only make a hole on the graph in Figure 8 to obtain the graph of g in Figure 9.

EXAMPLE 5 **Finding an equation of a rational function satisfying prescribed conditions**

Find an equation of a rational function f that satisfies the following conditions:

$$x\text{-intercept: } 4, \quad \text{vertical asymptote: } x = -2,$$

$$\text{horizontal asymptote: } y = -\tfrac{3}{5}, \text{ and a hole at } x = 1$$

SOLUTION An x-intercept of 4 implies that $x - 4$ must be a factor in the numerator, and a vertical asymptote of $x = -2$ implies that $x + 2$ is a factor in the denominator. So we can start with the form

$$\frac{x - 4}{x + 2}.$$

The horizontal asymptote is $y = -\frac{3}{5}$. We can multiply the numerator by -3 and the denominator by 5 to get the form

$$\frac{-3(x - 4)}{5(x + 2)}.$$

(Do *not* write $(-3x - 4)/(5x + 2)$, since that would change the x-intercept and the vertical asymptote.) Lastly, since there is a hole at $x = 1$, we must have a factor of $x - 1$ in both the numerator and the denominator. Thus, an equation for f is

$$f(x) = \frac{-3(x - 4)(x - 1)}{5(x + 2)(x - 1)} \quad \text{or, equivalently,} \quad f(x) = \frac{-3x^2 + 15x - 12}{5x^2 + 5x - 10}.$$

Tutorial available at www.thomsonedu.com/login

EXAMPLE 6 Sketching the graph of a rational function

Sketch the graph of f if

$$f(x) = \frac{x - 1}{x^2 - x - 6}.$$

▶ SOLUTION It is useful to express both numerator and denominator in factored form. Thus, we begin by writing

$$f(x) = \frac{x - 1}{x^2 - x - 6} = \frac{x - 1}{(x + 2)(x - 3)}.$$

Figure 10

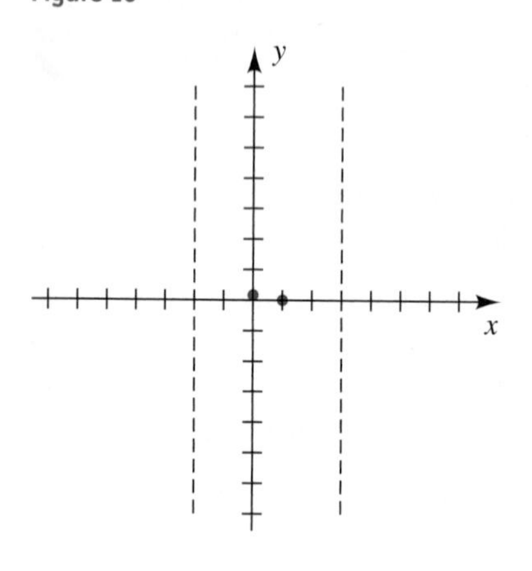

Guideline 1 To find the x-intercepts we find the zeros of the numerator. Solving $x - 1 = 0$ gives us $x = 1$, and we plot the point $(1, 0)$ on the x-axis, as shown in Figure 10.

Guideline 2 The denominator has zeros -2 and 3. Hence, the lines $x = -2$ and $x = 3$ are vertical asymptotes; we sketch them with dashes, as in Figure 10.

Guideline 3 The y-intercept is $f(0) = \frac{1}{6}$, and we plot the point $\left(0, \frac{1}{6}\right)$, shown in Figure 10.

Guideline 4 The degree of the numerator of $f(x)$ is less than the degree of the denominator, so, by part (1) of the theorem on horizontal asymptotes, the x-axis is the horizontal asymptote.

Guideline 5 The points where the graph intersects the horizontal asymptote (the x-axis) found in guideline 4 correspond to the x-intercepts. We already plotted the point $(1, 0)$ in guideline 1.

Guideline 6 The vertical asymptotes in Figure 10 divide the xy-plane into three regions:

R_1: the region to the left of $x = -2$

R_2: the region between $x = -2$ and $x = 3$

R_3: the region to the right of $x = 3$

For R_1, we have $x < -2$. There are only two choices for the shape of the graph of f in R_1: as $x \to -\infty$, the graph approaches the x-axis either from above or from below. To determine which choice is correct, we will examine the *sign* of a typical function value in R_1. Choosing -10 for x, we use the factored form of $f(x)$ to find the sign of $f(-10)$ (this process is similar to the one used in Section 1.6):

$$f(-10) = \frac{(-)}{(-)(-)} = -$$

The negative value of $f(-10)$ indicates that the graph approaches the horizontal asymptote from *below* as $x \to -\infty$. Moreover, as $x \to -2^-$, the graph

▶ **Tutorial available at www.thomsonedu.com/login**

extends *downward;* that is, $f(x) \to -\infty$. A sketch of f on R_1 is shown in Figure 11(a).

Figure 11

(a)

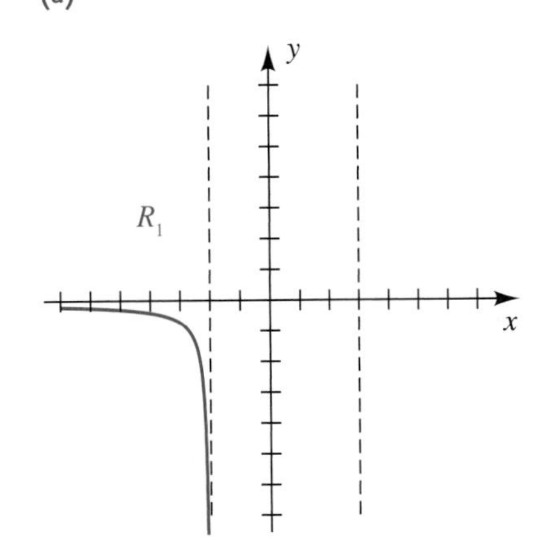

(b)

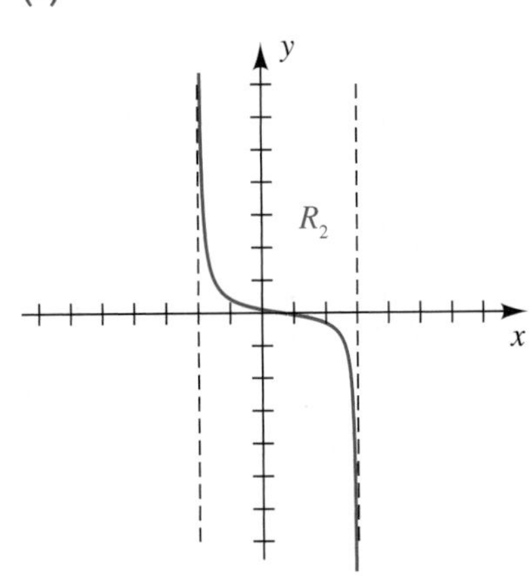

(c)

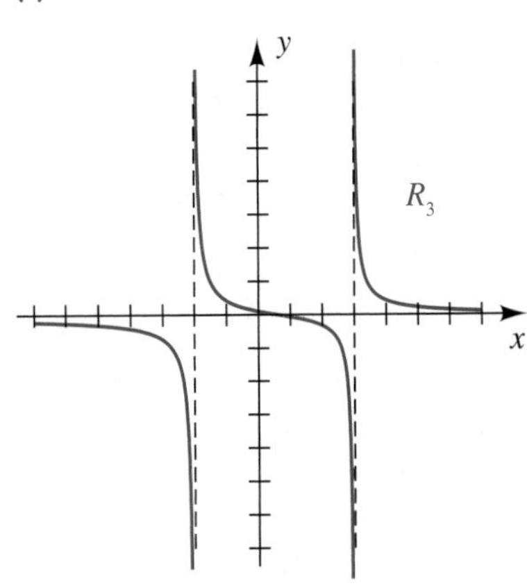

In R_2, we have $-2 < x < 3$, and the graph crosses the x-axis at $x = 1$. Since, for example, $f(0)$ is positive, it follows that the graph lies *above* the x-axis if $-2 < x < 1$. Thus, as $x \to -2^+$, the graph extends *upward;* that is, $f(x) \to \infty$. Since $f(2)$ can be shown to be negative, the graph lies *below* the x-axis if $1 < x < 3$. Hence, as $x \to 3^-$, the graph extends *downward;* that is, $f(x) \to -\infty$. A sketch of f on R_2 is shown in Figure 11(b).

Finally, in R_3, $x > 3$, and the graph does not cross the x-axis. Since, for example, $f(10)$ can be shown to be positive, the graph lies *above* the x-axis. It follows that $f(x) \to \infty$ as $x \to 3^+$ and that the graph approaches the horizontal asymptote from *above* as $x \to \infty$. The graph of f is sketched in Figure 11(c).

EXAMPLE 7 Sketching the graph of a rational function

Sketch the graph of f if

$$f(x) = \frac{x^2}{x^2 - x - 2}.$$

▶ SOLUTION Factoring the denominator gives us

$$f(x) = \frac{x^2}{x^2 - x - 2} = \frac{x^2}{(x + 1)(x - 2)}.$$

We again follow the guidelines.

(continued)

▶ **Tutorial available at www.thomsonedu.com/login**

Figure 12

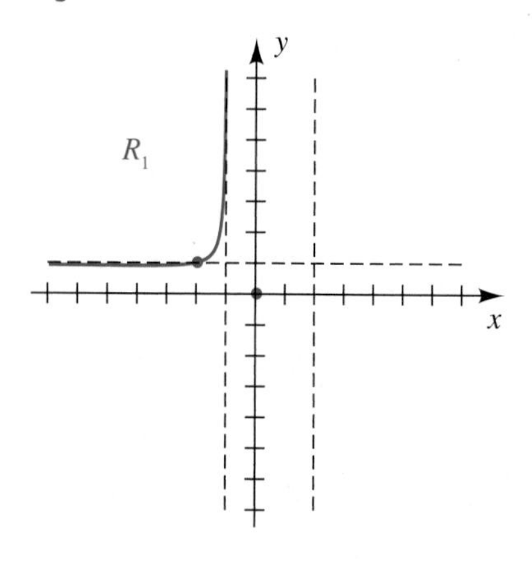

Guideline 1 To find the x-intercepts we find the zeros of the numerator. Solving $x^2 = 0$ gives us $x = 0$, and we plot the point $(0, 0)$ on the x-axis, as shown in Figure 12.

Guideline 2 The denominator has zeros -1 and 2. Hence, the lines $x = -1$ and $x = 2$ are vertical asymptotes, and we sketch them with dashes, as in Figure 12.

Guideline 3 The y-intercept is $f(0) = 0$. This gives us the same point $(0, 0)$ found in guideline 1.

Guideline 4 The numerator and denominator of $f(x)$ have the same degree, and the leading coefficients are both 1. Hence, by part (2) of the theorem on horizontal asymptotes, the line $y = \frac{1}{1} = 1$ is a horizontal asymptote. We sketch the line with dashes, as in Figure 12.

Guideline 5 The x-coordinates of the points where the graph intersects the horizontal asymptote $y = 1$ are solutions of the equation $f(x) = 1$. We solve this equation as follows:

$$\frac{x^2}{x^2 - x - 2} = 1 \qquad \text{let } f(x) = 1$$

$$x^2 = x^2 - x - 2 \qquad \text{multiply by } x^2 - x - 2$$

$$x = -2 \qquad \text{subtract } x^2 \text{ and add } x$$

This result indicates that the graph intersects the horizontal asymptote $y = 1$ *only* at $x = -2$; hence, we plot the point $(-2, 1)$ shown in Figure 12.

Guideline 6 The vertical asymptotes in Figure 12 divide the xy-plane into three regions:

R_1: the region to the left of $x = -1$

R_2: the region between $x = -1$ and $x = 2$

R_3: the region to the right of $x = 2$

For R_1, let us first consider the portion of the graph that corresponds to $-2 < x < -1$. From the point $(-2, 1)$ on the horizontal asymptote, the graph must extend *upward* as $x \to -1^-$ (it cannot extend downward, since there is no x-intercept between $x = -2$ and $x = -1$). As $x \to -\infty$, there will be a low point on the graph between $y = 0$ and $y = 1$, and then the graph will approach the horizontal asymptote $y = 1$ from *below.* It is difficult to see where the low point occurs in Figure 12 because the function values are very close to one another. Using calculus, it can be shown that the low point is $\left(-4, \frac{8}{9}\right)$.

In R_2, we have $-1 < x < 2$, and the graph intersects the x-axis at $x = 0$. Since the function does not cross the horizontal asymptote in this region, we know that the graph extends *downward* as $x \to -1^+$ and as $x \to 2^-$, as shown in Figure 13(a).

Figure 13

(a)

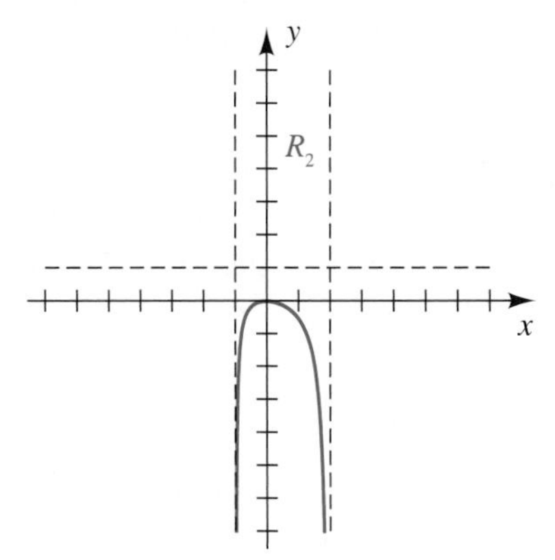

(b)

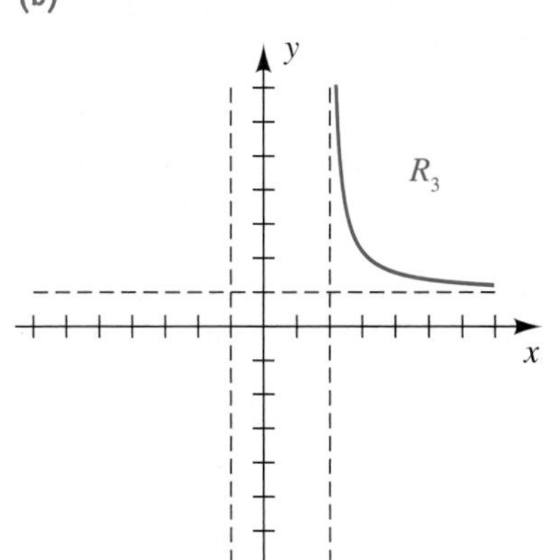

(c)

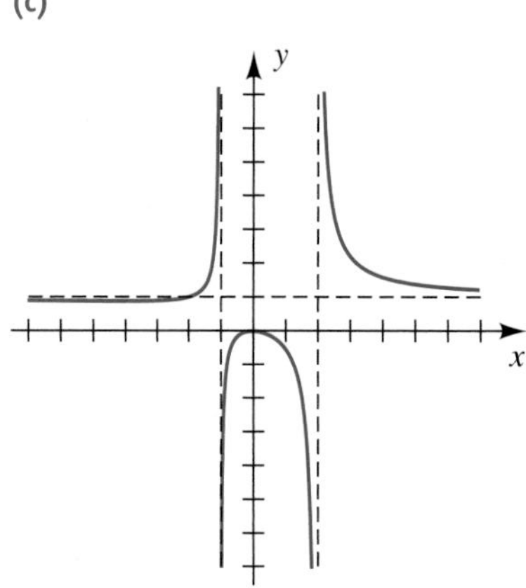

In R_3, the graph approaches the horizontal asymptote $y = 1$ (from either above or below) as $x \to \infty$. Furthermore, the graph must extend *upward* as $x \to 2^+$ because there are no x-intercepts in R_3. This implies that as $x \to \infty$, the graph approaches the horizontal asymptote from *above,* as in Figure 13(b).

The graph of f is sketched in Figure 13(c).

In the remaining solutions we will not formally write down each guideline.

EXAMPLE 8 Sketching the graph of a rational function

Sketch the graph of f if

$$f(x) = \frac{2x^4}{x^4 + 1}.$$

Figure 14

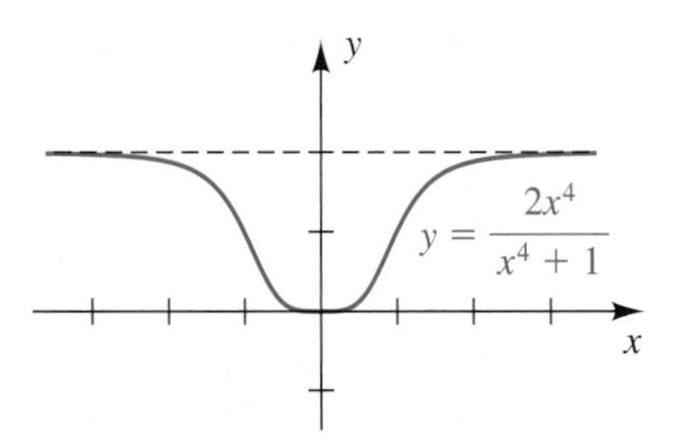

▶ **SOLUTION** Note that since $f(-x) = f(x)$, the function is even, and hence the graph is symmetric with respect to the y-axis.

The graph intersects the x-axis at $(0, 0)$. Since the denominator of $f(x)$ has no real zero, the graph has no vertical asymptote.

The numerator and denominator of $f(x)$ have the same degree. Since the leading coefficients are 2 and 1, respectively, the line $y = \frac{2}{1} = 2$ is the horizontal asymptote. The graph does not cross the horizontal asymptote $y = 2$, since the equation $f(x) = 2$ has no real solution.

Plotting the points $(1, 1)$ and $\left(2, \frac{32}{17}\right)$ and making use of symmetry leads to the sketch in Figure 14.

An **oblique asymptote** for a graph is a line $y = ax + b$, with $a \neq 0$, such that the graph approaches this line as $x \to \infty$ or as $x \to -\infty$. (If the graph is a line, we consider it to be its own asymptote.) If the rational function

▶ **Tutorial available at www.thomsonedu.com/login**

$f(x) = g(x)/h(x)$ for polynomials $g(x)$ and $h(x)$ and *if the degree of* $g(x)$ *is one greater than the degree of* $h(x)$, then the graph of f has an oblique asymptote. To find this oblique asymptote we may use long division to express $f(x)$ in the form

$$f(x) = \frac{g(x)}{h(x)} = (ax + b) + \frac{r(x)}{h(x)},$$

where either $r(x) = 0$ or the degree of $r(x)$ is less than the degree of $h(x)$. From part (1) of the theorem on horizontal asymptotes,

$$\frac{r(x)}{h(x)} \to 0 \quad \text{as} \quad x \to \infty \quad \text{or as} \quad x \to -\infty.$$

Consequently, $f(x)$ approaches the line $y = ax + b$ as x increases or decreases without bound; that is, $y = ax + b$ is an oblique asymptote.

EXAMPLE 9 Finding an oblique asymptote

Find all the asymptotes and sketch the graph of f if

$$f(x) = \frac{x^2 - 9}{2x - 4}.$$

Figure 15

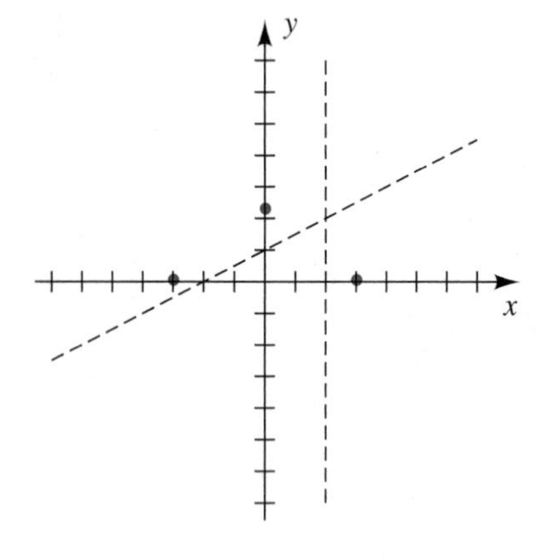

SOLUTION A vertical asymptote occurs if $2x - 4 = 0$ (that is, if $x = 2$).

The degree of the numerator of $f(x)$ is greater than the degree of the denominator. Hence, by part (3) of the theorem on horizontal asymptotes, there is no *horizontal* asymptote; but since the degree of the numerator, 2, is *one* greater than the degree of the denominator, 1, the graph has an *oblique* asymptote. By long division we obtain

$$\begin{array}{r|l} & \tfrac{1}{2}x + 1 \\ 2x - 4 & x^2 \qquad - 9 \\ & x^2 - 2x \qquad (\tfrac{1}{2}x)(2x - 4) \\ & \quad 2x - 9 \qquad \text{subtract} \\ & \quad 2x - 4 \qquad (1)(2x - 4) \\ & \quad\;\; -5 \qquad \text{subtract} \end{array}$$

Therefore,

$$\frac{x^2 - 9}{2x - 4} = \left(\frac{1}{2}x + 1\right) - \frac{5}{2x - 4}.$$

Figure 16

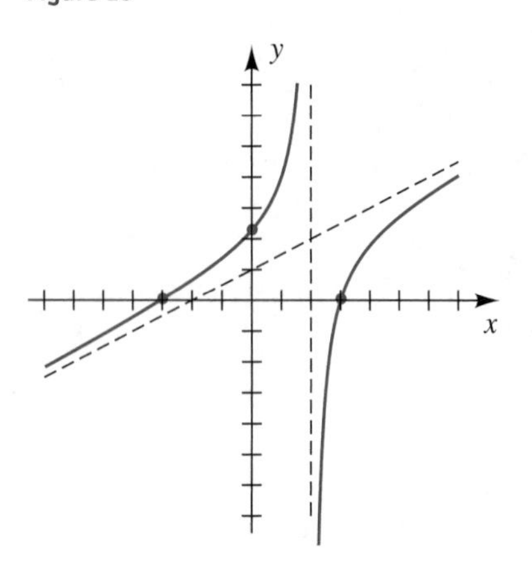

As we indicated in the discussion preceding this example, the line $y = \frac{1}{2}x + 1$ is an oblique asymptote. This line and the vertical asymptote $x = 2$ are sketched with dashes in Figure 15.

The x-intercepts of the graph are the solutions of $x^2 - 9 = 0$ and hence are 3 and -3. The y-intercept is $f(0) = \frac{9}{4}$. The corresponding points are plotted in Figure 15. We may now show that the graph has the shape indicated in Figure 16.

Tutorial available at www.thomsonedu.com/login

In Example 9, the graph of f approaches the line $y = \frac{1}{2}x + 1$ *asymptotically* as $x \to \infty$ or as $x \to -\infty$. Graphs of rational functions may approach different types of curves asymptotically. For example, if

$$f(x) = \frac{x^4 - x}{x^2} = x^2 - \frac{1}{x},$$

then for large values of $|x|$, $1/x \approx 0$ and hence $f(x) \approx x^2$. Thus, the graph of f approaches the parabola $y = x^2$ asymptotically as $x \to \infty$ or as $x \to -\infty$. In general, if $f(x) = g(x)/h(x)$ and if $q(x)$ is the quotient obtained by dividing $g(x)$ by $h(x)$, then the graph of f approaches the graph of $y = q(x)$ asymptotically as $x \to \infty$ or as $x \to -\infty$.

EXAMPLE 10 Sketching the graph of a rational function

Sketch the graph of f if

$$f(x) = \frac{x^2 - x}{9x^3 - 9x^2 - 22x + 8},$$

and find equations of the vertical asymptotes.

SOLUTION We begin by making the assignments

$$Y_1 = x^2 - x, \quad Y_2 = 9x^3 - 9x^2 - 22x + 8, \quad \text{and} \quad Y_3 = Y_1/Y_2.$$

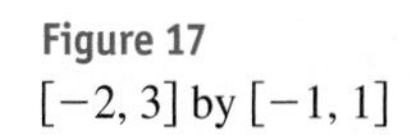

Figure 17
$[-2, 3]$ by $[-1, 1]$

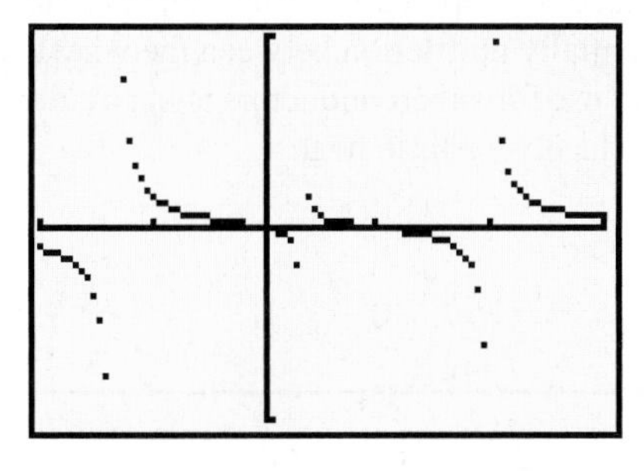

Selecting only Y_3 to be graphed (turn off Y_1 and Y_2) and using a standard viewing rectangle, we obtain a graph that gives us virtually no indication of the true shape of f. Changing to a viewing rectangle of $[-6, 6]$ by $[-4, 4]$ gives us a hint that the vertical asymptotes are confined to the interval $-2 < x < 3$.

Using a viewing rectangle of $[-2, 3]$ by $[-1, 1]$ and changing to *dot mode* (so as not to graph the function across the vertical asymptotes) leads us to the sketch in Figure 17. Since the degree of the numerator, 2, is less than the degree of the denominator, 3, we know that the horizontal asymptote is the x-axis. The zeros of the numerator, 0 and 1, are the only x-intercepts.

Figure 18
$[-2, 3]$ by $[-1, 1]$

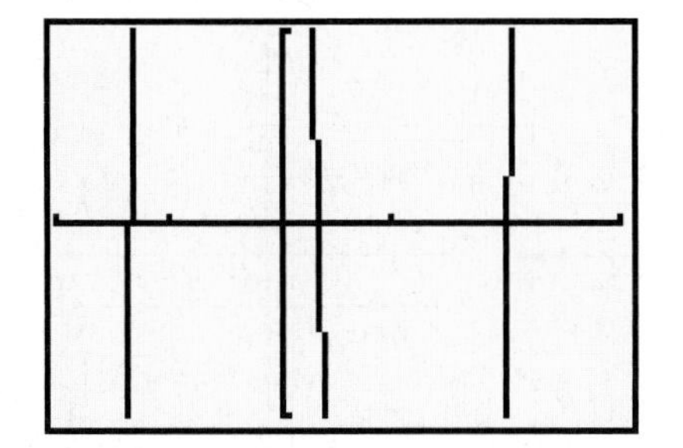

To determine the equations of the vertical asymptotes, we will abandon the graph of Y_3 and examine the graph of Y_2—looking for its zeros. Graphing Y_2 with the same viewing rectangle, but using *connected mode,* gives us Figure 18.

By the theorem on rational zeros of a polynomial, we know that the possible rational roots of $9x^3 - 9x^2 - 22x + 8 = 0$ are

$$\pm 1, \pm 2, \pm 4, \pm 8, \pm\tfrac{1}{3}, \pm\tfrac{2}{3}, \pm\tfrac{4}{3}, \pm\tfrac{8}{3}, \pm\tfrac{1}{9}, \pm\tfrac{2}{9}, \pm\tfrac{4}{9}, \pm\tfrac{8}{9}.$$

From the graph, we see that the only reasonable choice for the zero in the interval $(-2, -1)$ is $-\frac{4}{3}$. The number 2 appears to be a zero, and using a zero

(continued)

or root feature indicates that $\frac{1}{3}$ is also a good candidate for a zero. We can prove that $-\frac{4}{3}, \frac{1}{3}$, and 2 are zeros of Y_2 by using synthetic division. Thus, equations of the vertical asymptotes are

$$x = -\tfrac{4}{3}, \quad x = \tfrac{1}{3}, \quad \text{and} \quad x = 2.$$

Graphs of rational functions may become increasingly complicated as the degrees of the polynomials in the numerator and denominator increase. Techniques developed in calculus are very helpful in achieving a more thorough treatment of such graphs.

Formulas that represent physical quantities may determine rational functions. For example, consider Ohm's law in electrical theory, which states that $I = V/R$, where R is the resistance (in ohms) of a conductor, V is the potential difference (in volts) across the conductor, and I is the current (in amperes) that flows through the conductor. The resistance of certain alloys approaches zero as the temperature approaches absolute zero (approximately $-273°C$), and the alloy becomes a *superconductor* of electricity. If the voltage V is fixed, then, for such a superconductor,

$$I = \frac{V}{R} \to \infty \quad \text{as} \quad R \to 0^+;$$

that is, as R approaches 0, the current increases without bound. Superconductors allow very large currents to be used in generating plants and motors. They also have applications in experimental high-speed ground transportation, where the strong magnetic fields produced by superconducting magnets enable trains to levitate so that there is essentially no friction between the wheels and the track. Perhaps the most important use for superconductors is in circuits for computers, because such circuits produce very little heat.

3.5 *Exercises*

Exer. 1–2: (a) Sketch the graph of f. (b) Find the domain D and range R of f. (c) Find the intervals on which f is increasing or is decreasing.

1 $f(x) = \dfrac{4}{x}$

2 $f(x) = \dfrac{1}{x^2}$

Exer. 3–4: Identify any vertical asymptotes, horizontal asymptotes, and holes.

3 $f(x) = \dfrac{-2(x+5)(x-6)}{(x-3)(x-6)}$

4 $f(x) = \dfrac{2(x+4)(x+2)}{5(x+2)(x-1)}$

Exer. 5–6: All asymptotes, intercepts, and holes of a rational function f are labeled in the figure. Sketch a graph of f and find a formula for f.

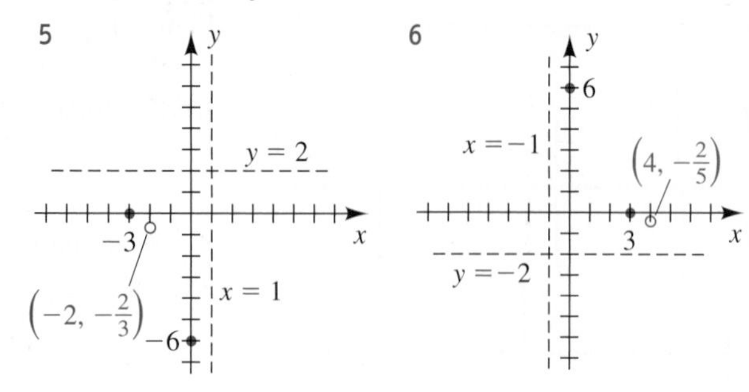

Exer. 7–32: Sketch the graph of f.

7 $f(x) = \dfrac{3}{x - 4}$

8 $f(x) = \dfrac{-3}{x + 3}$

9 $f(x) = \dfrac{-3x}{x + 2}$

10 $f(x) = \dfrac{4x}{2x - 5}$

11 $f(x) = \dfrac{4x - 1}{2x + 3}$

12 $f(x) = \dfrac{5x + 3}{3x - 7}$

13 $f(x) = \dfrac{(4x - 1)(x - 2)}{(2x + 3)(x - 2)}$

14 $f(x) = \dfrac{(5x + 3)(x + 1)}{(3x - 7)(x + 1)}$

15 $f(x) = \dfrac{x - 2}{x^2 - x - 6}$

16 $f(x) = \dfrac{x + 1}{x^2 + 2x - 3}$

17 $f(x) = \dfrac{-4}{(x - 2)^2}$

18 $f(x) = \dfrac{2}{(x + 1)^2}$

19 $f(x) = \dfrac{x - 3}{x^2 - 1}$

20 $f(x) = \dfrac{x + 4}{x^2 - 4}$

21 $f(x) = \dfrac{2x^2 - 2x - 4}{x^2 + x - 12}$

22 $f(x) = \dfrac{-3x^2 - 3x + 6}{x^2 - 9}$

23 $f(x) = \dfrac{-x^2 - x + 6}{x^2 + 3x - 4}$

24 $f(x) = \dfrac{x^2 - 3x - 4}{x^2 + x - 6}$

25 $f(x) = \dfrac{3x^2 - 3x - 36}{x^2 + x - 2}$

26 $f(x) = \dfrac{2x^2 + 4x - 48}{x^2 + 3x - 10}$

27 $f(x) = \dfrac{-2x^2 + 10x - 12}{x^2 + x}$

28 $f(x) = \dfrac{2x^2 + 8x + 6}{x^2 - 2x}$

29 $f(x) = \dfrac{x - 1}{x^3 - 4x}$

30 $f(x) = \dfrac{x^2 - 2x + 1}{x^3 - 9x}$

31 $f(x) = \dfrac{-3x^2}{x^2 + 1}$

32 $f(x) = \dfrac{x^2 - 4}{x^2 + 1}$

Exer. 33–36: Find the oblique asymptote, and sketch the graph of f.

33 $f(x) = \dfrac{x^2 - x - 6}{x + 1}$

34 $f(x) = \dfrac{2x^2 - x - 3}{x - 2}$

35 $f(x) = \dfrac{8 - x^3}{2x^2}$

36 $f(x) = \dfrac{x^3 + 1}{x^2 - 9}$

Exer. 37–44: Simplify $f(x)$, and sketch the graph of f.

37 $f(x) = \dfrac{2x^2 + x - 6}{x^2 + 3x + 2}$

38 $f(x) = \dfrac{x^2 - x - 6}{x^2 - 2x - 3}$

39 $f(x) = \dfrac{x - 1}{1 - x^2}$

40 $f(x) = \dfrac{x + 2}{x^2 - 4}$

41 $f(x) = \dfrac{x^2 + x - 2}{x + 2}$

42 $f(x) = \dfrac{x^3 - 2x^2 - 4x + 8}{x - 2}$

43 $f(x) = \dfrac{x^2 + 4x + 4}{x^2 + 3x + 2}$

44 $f(x) = \dfrac{(x^2 + x)(2x - 1)}{(x^2 - 3x + 2)(2x - 1)}$

Exer. 45–48: Find an equation of a rational function f that satisfies the given conditions.

45 vertical asymptote: $x = 4$
horizontal asymptote: $y = -1$
x-intercept: 3

46 vertical asymptotes: $x = -2$, $x = 0$
horizontal asymptote: $y = 0$
x-intercept: 2; $f(3) = 1$

47 vertical asymptotes: $x = -3$, $x = 1$
horizontal asymptote: $y = 0$
x-intercept: -1; $f(0) = -2$
hole at $x = 2$

48 vertical asymptotes: $x = -1$, $x = 3$
horizontal asymptote: $y = 2$
x-intercepts: -2, 1; hole at $x = 0$

49 **A container for radioactive waste** A cylindrical container for storing radioactive waste is to be constructed from lead. This container must be 6 inches thick. The volume of the outside cylinder shown in the figure is to be 16π ft^3.

(a) Express the height h of the inside cylinder as a function of the inside radius r.

(b) Show that the inside volume $V(r)$ is given by

$$V(r) = \pi r^2\left[\frac{16}{(r + 0.5)^2} - 1\right].$$

(c) What values of r must be excluded in part (b)?

Exercise 49

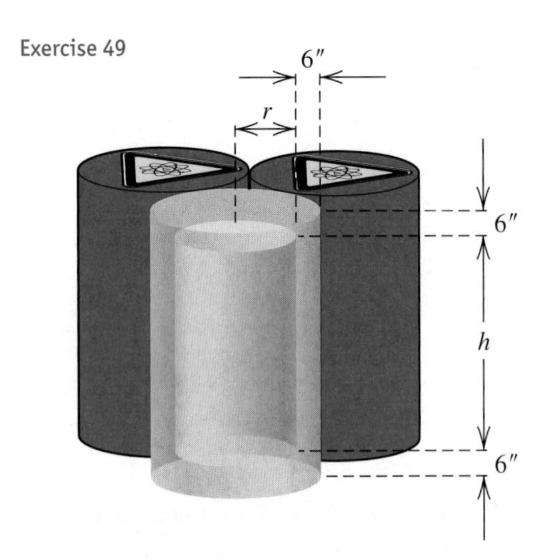

50 **Drug dosage** Young's rule is a formula that is used to modify adult drug dosage levels for young children. If a denotes the adult dosage (in milligrams) and if t is the age of the child (in years), then the child's dose y is given by the equation $y = ta/(t + 12)$. Sketch the graph of this equation for $t > 0$ and $a = 100$.

51 **Salt concentration** Salt water of concentration 0.1 pound of salt per gallon flows into a large tank that initially contains 50 gallons of pure water.

(a) If the flow rate of salt water into the tank is 5 gal/min, find the volume $V(t)$ of water and the amount $A(t)$ of salt in the tank after t minutes.

(b) Find a formula for the salt concentration $c(t)$ (in lb/gal) after t minutes.

(c) Discuss the variation of $c(t)$ as $t \to \infty$.

52 **Amount of rainfall** The total number of inches $R(t)$ of rain during a storm of length t hours can be approximated by

$$R(t) = \frac{at}{t + b},$$

where a and b are positive constants that depend on the geographical locale.

(a) Discuss the variation of $R(t)$ as $t \to \infty$.

(b) The intensity I of the rainfall (in in./hr) is defined by $I = R(t)/t$. If $a = 2$ and $b = 8$, sketch the graph of R and I on the same coordinate plane for $t > 0$.

53 **Salmon propagation** For a particular salmon population, the relationship between the number S of spawners and the number R of offspring that survive to maturity is given by the formula

$$R = \frac{4500S}{S + 500}.$$

(a) Under what conditions is $R > S$?

(b) Find the number of spawners that would yield 90% of the greatest possible number of offspring that survive to maturity.

(c) Work part (b) with 80% replacing 90%.

(d) Compare the results for S and R (in terms of percentage increases) from parts (b) and (c).

54 **Population density** The population density D (in people/mi^2) in a large city is related to the distance x (in miles) from the center of the city by

$$D = \frac{5000x}{x^2 + 36}.$$

(a) What happens to the density as the distance from the center of the city changes from 20 miles to 25 miles?

(b) What eventually happens to the density?

(c) In what areas of the city does the population density exceed 400 people/mi^2?

Exer. 55–58: Graph f, and find equations of the vertical asymptotes.

55 $f(x) = \dfrac{20x^2 + 80x + 72}{10x^2 + 40x + 41}$

56 $f(x) = \dfrac{15x^2 - 60x + 68}{3x^2 - 12x + 13}$

57 $f(x) = \dfrac{(x - 1)^2}{(x - 0.999)^2}$

58 $f(x) = \dfrac{x^2 - 9.01}{x - 3}$

59 Let $f(x)$ be the polynomial

$$(x + 3)(x + 2)(x + 1)(x)(x - 1)(x - 2)(x - 3).$$

(a) Describe the graph of $g(x) = f(x)/f(x)$.

(b) Describe the graph of $h(x) = g(x)p(x)$, where $p(x)$ is a polynomial function.

60 Refer to Exercise 59.

(a) Describe the graph of $y = f(x)$.

(b) Describe the graph of $k(x) = 1/f(x)$.

61 Grade point average (GPA)

(a) A student has finished 48 credit hours with a GPA of 2.75. How many additional credit hours y at 4.0 will raise the student's GPA to some desired value x? (Determine y as a function of x.)

(b) Create a table of values for x and y, starting with $x = 2.8$ and using increments of 0.2.

(c) Graph the function in part (a) in the viewing rectangle [2, 4] by [0, 1000, 100].

(d) What is the vertical asymptote of the graph in part (c)?

(e) Explain the practical significance of the value $x = 4$.

3.6 Variation

In some scientific investigations, the terminology of *variation* or *proportion* is used to describe relationships between variable quantities. In the following chart, k is a nonzero real number called a **constant of variation** or a **constant of proportionality.**

Terminology	General formula	Illustration
y **varies directly** as x, or y is **directly proportional** to x	$y = kx$	$C = 2\pi r$, where C is the circumference of a circle, r is the radius, and $k = 2\pi$
y **varies inversely** as x, or y is **inversely proportional** to x	$y = \dfrac{k}{x}$	$I = \dfrac{110}{R}$, where I is the current in an electrical circuit, R is the resistance, and $k = 110$ is the voltage

The variable x in the chart can also represent a power. For example, the formula $A = \pi r^2$ states that the area A of a circle varies directly as the *square* of the radius r, where π is the constant of variation. Similarly, the formula $V = \frac{4}{3}\pi r^3$ states that the volume V of a sphere is directly proportional to the *cube* of the radius. In this case the constant of proportionality is $\frac{4}{3}\pi$.

In general, graphs of variables related by *direct variation* resemble graphs of **power functions** of the form $y = x^n$ with $n > 0$ (such as $y = \sqrt{x}$ or $y = x^2$ for nonnegative x-values, as shown in Figure 1). With direct variation, as one variable increases, so does the other variable. An example of two quantities that are directly related is the number of miles run and the number of calories burned.

Figure 1
As x increases, y increases, *or* as x decreases, y decreases

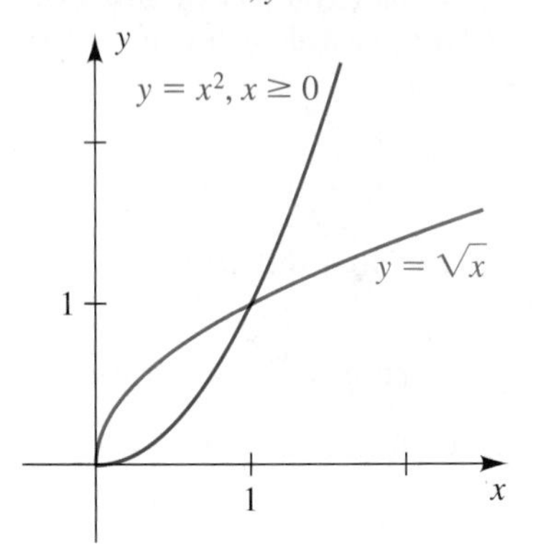

Figure 2
As x increases, y decreases, *or* as x decreases, y increases

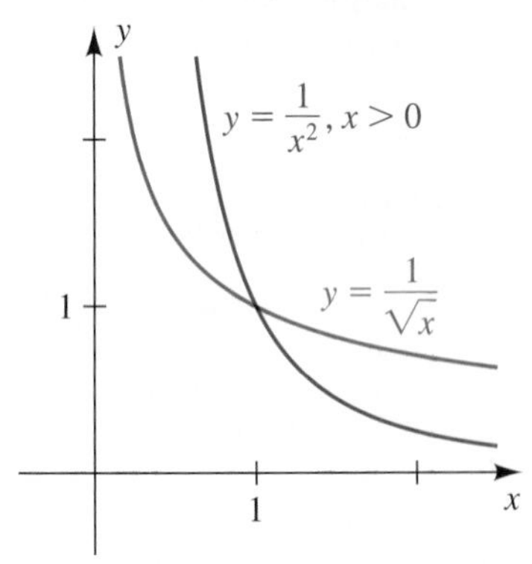

Graphs of variables related by *inverse variation* resemble graphs of power functions of the form $y = x^n$ with $n < 0$ (such as $y = 1/\sqrt{x}$ or $y = 1/x^2$ for positive x-values, as shown in Figure 2). In this case, as one variable increases, the other variable decreases. An example of two quantities that are inversely related is the number of inches of rainfall and the number of grass fires.

EXAMPLE 1 Directly proportional variables

Suppose a variable q is directly proportional to a variable z.

(a) If $q = 12$ when $z = 5$, determine the constant of proportionality.

(b) Find the value of q when $z = 7$ and sketch a graph of this relationship.

▶ **SOLUTION** Since q is directly proportional to z,

$$q = kz,$$

where k is a constant of proportionality.

(a) Substituting $q = 12$ and $z = 5$ gives us

$$12 = k \cdot 5, \qquad \text{or} \qquad k = \tfrac{12}{5}.$$

(b) Since $k = \frac{12}{5}$, the formula $q = kz$ has the specific form

$$q = \tfrac{12}{5}z.$$

Thus, when $z = 7$,

$$q = \tfrac{12}{5} \cdot 7 = \tfrac{84}{5} = 16.8.$$

Figure 3 illustrates the relationship of the variables q and z—a simple linear relationship.

Figure 3

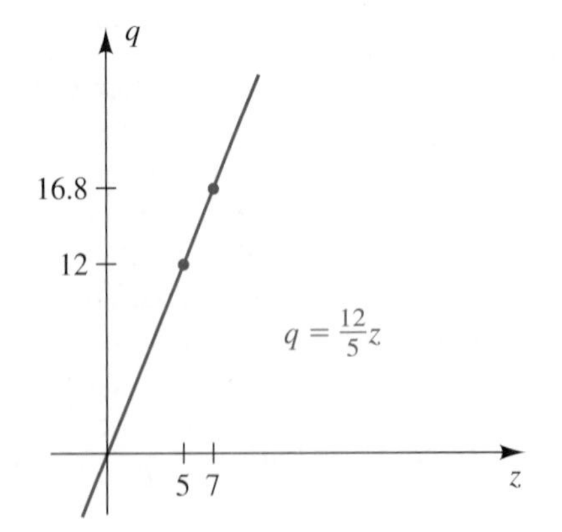

▶ **Tutorial available at www.thomsonedu.com/login**

The following guidelines may be used to solve applied problems that involve variation or proportion.

Guidelines for Solving Variation Problems

1. Write a *general* formula that involves the variables and a constant of variation (or proportion) k.
2. Find the value of k in guideline 1 by using the initial data given in the statement of the problem.
3. Substitute the value of k found in guideline 2 into the formula of guideline 1, obtaining a *specific* formula that involves the variables.
4. Use the new data to solve the problem.

We shall follow these guidelines in the solution of the next example.

EXAMPLE 2 Pressure and volume as inversely proportional quantities

If the temperature remains constant, the pressure of an enclosed gas is inversely proportional to the volume. The pressure of a certain gas within a spherical balloon of radius 9 inches is 20 lb/in^2. If the radius of the balloon increases to 12 inches, approximate the new pressure of the gas. Sketch a graph of the relationship between the pressure and the volume.

SOLUTION

Guideline 1 If we denote the pressure by P (in lb/in^2) and the volume by V (in in^3), then since P is inversely proportional to V,

$$P = \frac{k}{V}$$

for some constant of proportionality k.

Guideline 2 We find the constant of proportionality k in guideline 1. Since the volume V of a sphere of radius r is $V = \frac{4}{3}\pi r^3$, the initial volume of the balloon is $V = \frac{4}{3}\pi(9)^3 = 972\pi\ \text{in}^3$. This leads to the following:

$$20 = \frac{k}{972\pi} \qquad P = 20 \text{ when } V = 972\pi$$

$$k = 20(972\pi) = 19{,}440\pi \qquad \text{solve for } k$$

Guideline 3 Substituting $k = 19{,}440\pi$ into $P = k/V$, we find that the pressure corresponding to any volume V is given by

$$P = \frac{19{,}440\pi}{V}.$$

(continued)

Tutorial available at www.thomsonedu.com/login

Guideline 4 If the new radius of the balloon is 12 inches, then

$$V = \tfrac{4}{3}\pi(12)^3 = 2304\pi \text{ in}^3.$$

Substituting this number for V in the formula obtained in guideline 3 gives us

$$P = \frac{19{,}440\pi}{2304\pi} = \frac{135}{16} = 8.4375.$$

Thus, the pressure decreases to approximately 8.4 lb/in^2 when the radius increases to 12 inches.

Figure 4 illustrates the relationship of the variables P and V for $V > 0$. Since $P = 19{,}440\pi/V$ and $V = \frac{4}{3}\pi r^3$, we can show that $(P \circ V)(r) = 14{,}580/r^3$, so we could also say that P is inversely proportional to r^3. Note that this is a graph of a simple rational function.

Figure 4

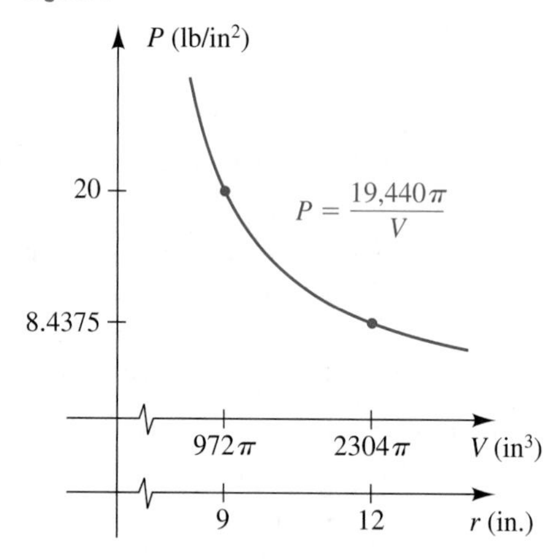

There are other types of variation. If x, y, and z are variables and $y = kxz$ for some real number k, we say that y *varies directly as the product of* x *and* z or y **varies jointly as** x **and** z. If $y = k(x/z)$, then y *varies directly as* x *and inversely as* z. As a final illustration, if a variable w varies directly as the product of x and the cube of y and inversely as the square of z, then

$$w = k\frac{xy^3}{z^2},$$

where k is a constant of proportionality. Graphs of equations for these types of variation will not be considered in this text.

EXAMPLE 3 Combining several types of variation

A variable w varies directly as the product of u and v and inversely as the square of s.

(a) If $w = 20$ when $u = 3$, $v = 5$, and $s = 2$, find the constant of variation.

(b) Find the value of w when $u = 7$, $v = 4$, and $s = 3$.

▶ **SOLUTION** A general formula for w is

$$w = k\frac{uv}{s^2},$$

where k is a constant of variation.

(a) Substituting $w = 20$, $u = 3$, $v = 5$, and $s = 2$ gives us

$$20 = k\frac{3 \cdot 5}{2^2}, \quad \text{or} \quad k = \frac{80}{15} = \frac{16}{3}.$$

(b) Since $k = \frac{16}{3}$, the specific formula for w is

$$w = \frac{16}{3}\frac{uv}{s^2}.$$

Thus, when $u = 7$, $v = 4$, and $s = 3$,

$$w = \frac{16}{3}\frac{7 \cdot 4}{3^2} = \frac{448}{27} \approx 16.6.$$

In the next example we again follow the guidelines stated in this section.

EXAMPLE 4 Finding the support load of a rectangular beam

The weight that can be safely supported by a beam with a rectangular cross section varies directly as the product of the width and square of the depth of the cross section and inversely as the length of the beam. If a 2-inch by 4-inch beam that is 8 feet long safely supports a load of 500 pounds, what weight can be safely supported by a 2-inch by 8-inch beam that is 10 feet long? (Assume that the width is the *shorter* dimension of the cross section.)

▶ **SOLUTION**

Guideline 1 If the width, depth, length, and weight are denoted by w, d, l, and W, respectively, then a general formula for W is

$$W = k\frac{wd^2}{l},$$

where k is a constant of variation.

Guideline 2 To find the value of k in guideline 1, we see from the given data that

$$500 = k\frac{2(4^2)}{8}, \quad \text{or} \quad k = 125.$$

Guideline 3 Substituting $k = 125$ into the formula of guideline 1 gives us the specific formula

$$W = 125\frac{wd^2}{l}.$$

Guideline 4 To answer the question, we substitute $w = 2$, $d = 8$, and $l = 10$ into the formula found in guideline 3, obtaining

$$W = 125 \cdot \frac{2 \cdot 8^2}{10} = 1600 \text{ lb.}$$

▶ **Tutorial available at www.thomsonedu.com/login**

3.6 Exercises

Exer. 1–12: Express the statement as a formula that involves the given variables and a constant of proportionality k, and then determine the value of k from the given conditions.

▶ 1 u is directly proportional to v. If $v = 30$, then $u = 12$.

2 s varies directly as t. If $t = 10$, then $s = 18$.

3 r varies directly as s and inversely as t. If $s = -2$ and $t = 4$, then $r = 7$.

4 w varies directly as z and inversely as the square root of u. If $z = 2$ and $u = 9$, then $w = 6$.

5 y is directly proportional to the square of x and inversely proportional to the cube of z. If $x = 5$ and $z = 3$, then $y = 25$.

▶ 6 q is inversely proportional to the sum of x and y. If $x = 0.5$ and $y = 0.7$, then $q = 1.4$.

7 z is directly proportional to the product of the square of x and the cube of y. If $x = 7$ and $y = -2$, then $z = 16$.

8 r is directly proportional to the product of s and v and inversely proportional to the cube of p. If $s = 2$, $v = 3$, and $p = 5$, then $r = 40$.

9 y is directly proportional to x and inversely proportional to the square of z. If $x = 4$ and $z = 3$, then $y = 16$.

10 y is directly proportional to x and inversely proportional to the sum of r and s. If $x = 3$, $r = 5$, and $s = 7$, then $y = 2$.

▶ 11 y is directly proportional to the square root of x and inversely proportional to the cube of z. If $x = 9$ and $z = 2$, then $y = 5$.

12 y is directly proportional to the square of x and inversely proportional to the square root of z. If $x = 5$ and $z = 16$, then $y = 10$.

13 **Liquid pressure** The pressure P acting at a point in a liquid is directly proportional to the distance d from the surface of the liquid to the point.

(a) Express P as a function of d by means of a formula that involves a constant of proportionality k.

(b) In a certain oil tank, the pressure at a depth of 2 feet is 118 lb/ft^2. Find the value of k in part (a).

(c) Find the pressure at a depth of 5 feet for the oil tank in part (b).

(d) Sketch a graph of the relationship between P and d for $d \geq 0$.

14 **Hooke's law** Hooke's law states that the force F required to stretch a spring x units beyond its natural length is directly proportional to x.

(a) Express F as a function of x by means of a formula that involves a constant of proportionality k.

(b) A weight of 4 pounds stretches a certain spring from its natural length of 10 inches to a length of 10.3 inches. Find the value of k in part (a).

(c) What weight will stretch the spring in part (b) to a length of 11.5 inches?

(d) Sketch a graph of the relationship between F and x for $x \geq 0$.

▶ 15 **Electrical resistance** The electrical resistance R of a wire varies directly as its length l and inversely as the square of its diameter d.

(a) Express R in terms of l, d, and a constant of variation k.

(b) A wire 100 feet long of diameter 0.01 inch has a resistance of 25 ohms. Find the value of k in part (a).

(c) Sketch a graph of the relationship between R and d for $l = 100$ and $d > 0$.

(d) Find the resistance of a wire made of the same material that has a diameter of 0.015 inch and is 50 feet long.

16 **Intensity of illumination** The intensity of illumination I from a source of light varies inversely as the square of the distance d from the source.

▶ **Tutorial available at www.thomsonedu.com/login**

(a) Express I in terms of d and a constant of variation k.

(b) A searchlight has an intensity of 1,000,000 candle-power at a distance of 50 feet. Find the value of k in part (a).

(c) Sketch a graph of the relationship between I and d for $d > 0$.

(d) Approximate the intensity of the searchlight in part (b) at a distance of 1 mile.

17 **Period of a pendulum** The period P of a simple pendulum—that is, the time required for one complete oscillation—is directly proportional to the square root of its length l.

(a) Express P in terms of l and a constant of proportionality k.

(b) If a pendulum 2 feet long has a period of 1.5 seconds, find the value of k in part (a).

(c) Find the period of a pendulum 6 feet long.

18 **Dimensions of a human limb** A circular cylinder is sometimes used in physiology as a simple representation of a human limb.

(a) Express the volume V of a cylinder in terms of its length L and the square of its circumference C.

(b) The formula obtained in part (a) can be used to approximate the volume of a limb from length and circumference measurements. Suppose the (average) circumference of a human forearm is 22 centimeters and the average length is 27 centimeters. Approximate the volume of the forearm to the nearest cm^3.

19 **Period of a planet** Kepler's third law states that the period T of a planet (the time needed to make one complete revolution about the sun) is directly proportional to the $\frac{3}{2}$ power of its average distance d from the sun.

(a) Express T as a function of d by means of a formula that involves a constant of proportionality k.

(b) For the planet Earth, $T = 365$ days and $d = 93$ million miles. Find the value of k in part (a).

(c) Estimate the period of Venus if its average distance from the sun is 67 million miles.

20 **Range of a projectile** It is known from physics that the range R of a projectile is directly proportional to the square of its velocity v.

(a) Express R as a function of v by means of a formula that involves a constant of proportionality k.

(b) A motorcycle daredevil has made a jump of 150 feet. If the speed coming off the ramp was 70 mi/hr, find the value of k in part (a).

(c) If the daredevil can reach a speed of 80 mi/hr coming off the ramp and maintain proper balance, estimate the possible length of the jump.

21 **Automobile skid marks** The speed V at which an automobile was traveling before the brakes were applied can sometimes be estimated from the length L of the skid marks. Assume that V is directly proportional to the square root of L.

(a) Express V as a function of L by means of a formula that involves a constant of proportionality k.

(b) For a certain automobile on a dry surface, $L = 50$ ft when $V = 35$ mi/hr. Find the value of k in part (a).

(c) Estimate the initial speed of the automobile in part (b) if the skid marks are 150 feet long.

22 **Coulomb's law** Coulomb's law in electrical theory states that the force F of attraction between two oppositely charged particles varies directly as the product of the magnitudes Q_1 and Q_2 of the charges and inversely as the square of the distance d between the particles.

(a) Find a formula for F in terms of Q_1, Q_2, d, and a constant of variation k.

(b) What is the effect of reducing the distance between the particles by a factor of one-fourth?

23 **Threshold weight** Threshold weight W is defined to be that weight beyond which risk of death increases significantly. For middle-aged males, W is directly proportional to the third power of the height h.

(a) Express W as a function of h by means of a formula that involves a constant of proportionality k.

(b) For a 6-foot male, W is about 200 pounds. Find the value of k in part (a).

(continued)

(c) Estimate, to the nearest pound, the threshold weight for an individual who is 5 feet 6 inches tall.

24 **The ideal gas law** The ideal gas law states that the volume V that a gas occupies is directly proportional to the product of the number n of moles of gas and the temperature T (in K) and is inversely proportional to the pressure P (in atmospheres).

(a) Express V in terms of n, T, P, and a constant of proportionality k.

(b) What is the effect on the volume if the number of moles is doubled and both the temperature and the pressure are reduced by a factor of one-half?

25 **Poiseuille's law** Poiseuille's law states that the blood flow rate F (in L/min) through a major artery is directly proportional to the product of the fourth power of the radius r of the artery and the blood pressure P.

(a) Express F in terms of P, r, and a constant of proportionality k.

(b) During heavy exercise, normal blood flow rates sometimes triple. If the radius of a major artery increases by 10%, approximately how much harder must the heart pump?

26 **Trout population** Suppose 200 trout are caught, tagged, and released in a lake's general population. Let T denote the number of tagged fish that are recaptured when a sample of n trout are caught at a later date. The validity of the mark-recapture method for estimating the lake's total trout population is based on the assumption that T is directly proportional to n. If 10 tagged trout are recovered from a sample of 300, estimate the total trout population of the lake.

27 **Radioactive decay of radon gas** When uranium disintegrates into lead, one step in the process is the radioactive decay of radium into radon gas. Radon enters through the soil into home basements, where it presents a health hazard if inhaled. In the simplest case of radon detection, a sample of air with volume V is taken. After equilibrium has been established, the radioactive decay D of the radon gas is counted with efficiency E over time t. The radon concentration C present in the sample of air varies directly as the product of D and E and inversely as the product of V and t. For a fixed radon concentration C and time t, find the change in the radioactive decay count D if V is doubled and E is reduced by 20%.

28 **Radon concentration** Refer to Exercise 27. Find the change in the radon concentration C if D increases by 30%, t increases by 60%, V decreases by 10%, and E remains constant.

29 **Density at a point** A thin flat plate is situated in an xy-plane such that the density d (in lb/ft^2) at the point $P(x, y)$ is inversely proportional to the square of the distance from the origin. What is the effect on the density at P if the x- and y-coordinates are each multiplied by $\frac{1}{3}$?

30 **Temperature at a point** A flat metal plate is positioned in an xy-plane such that the temperature T (in °C) at the point (x, y) is inversely proportional to the distance from the origin. If the temperature at the point $P(3, 4)$ is 20°C, find the temperature at the point $Q(24, 7)$.

Exer. 31–34: Examine the expression for the given set of data points of the form (x, y). Find the constant of variation and a formula that describes how y varies with respect to x.

31 y/x; $\{(0.6, 0.72), (1.2, 1.44), (4.2, 5.04), (7.1, 8.52), (9.3, 11.16)\}$

32 xy; $\{(0.2, -26.5), (0.4, -13.25), (0.8, -6.625), (1.6, -3.3125), (3.2, -1.65625)\}$

33 x^2y; $\{(0.16, -394.53125), (0.8, -15.78125), (1.6, -3.9453125), (3.2, -0.986328125)\}$

34 y/x^3; $\{(0.11, 0.00355377), (0.56, 0.46889472), (1.2, 4.61376), (2.4, 36.91008)\}$

35 **Stopping distances** Refer to Exercise 86 in Section 2.4. The distance D (in feet) required for a car to safely stop varies directly with its speed S (in mi/hr).

(a) Use the table to determine an approximate value for k in the variation formula $D = kS^{2.3}$.

S	20	30	40	50	60	70
D	33	86	167	278	414	593

(b) Check your approximation by graphing both the data and D on the same coordinate axes.

CHAPTER 3 REVIEW EXERCISES

▶ Online support materials can be found at www.thomsonedu.com/login

Exer. 1–6: Find all values of x such that $f(x) > 0$ and all x such that $f(x) < 0$, and sketch the graph of f.

1 $f(x) = (x + 2)^3$

2 $f(x) = x^6 - 32$

3 $f(x) = -\frac{1}{4}(x + 2)(x - 1)^2(x - 3)$

4 $f(x) = 2x^2 + x^3 - x^4$

5 $f(x) = x^3 + 2x^2 - 8x$

6 $f(x) = \frac{1}{15}(x^5 - 20x^3 + 64x)$

7 If $f(x) = x^3 - 5x^2 + 7x - 9$, use the intermediate value theorem for polynomial functions to prove that there is a real number a such that $f(a) = 100$.

8 Prove that the equation $x^5 - 3x^4 - 2x^3 - x + 1 = 0$ has a solution between 0 and 1.

Exer. 9–10: Find the quotient and remainder if $f(x)$ is divided by $p(x)$.

9 $f(x) = 3x^5 - 4x^3 + x + 5$; $p(x) = x^3 - 2x + 7$

10 $f(x) = 4x^3 - x^2 + 2x - 1$; $p(x) = x^2$

11 If $f(x) = -4x^4 + 3x^3 - 5x^2 + 7x - 10$, use the remainder theorem to find $f(-2)$.

12 Use the factor theorem to show that $x - 3$ is a factor of $f(x) = 2x^4 - 5x^3 - 4x^2 + 9$.

Exer. 13–14: Use synthetic division to find the quotient and remainder if $f(x)$ is divided by $p(x)$.

13 $f(x) = 6x^5 - 4x^2 + 8$; $p(x) = x + 2$

14 $f(x) = 2x^3 + 5x^2 - 2x + 1$; $p(x) = x - \sqrt{2}$

Exer. 15–16: A polynomial $f(x)$ with real coefficients has the indicated zero(s) and degree and satisfies the given condition. Express $f(x)$ as a product of linear and quadratic polynomials with real coefficients that are irreducible over $\mathbb{R}$.

15 $-3 + 5i, -1$; degree 3; $f(1) = 4$

16 $1 - i, 3, 0$; degree 4; $f(2) = -1$

17 Find a polynomial $f(x)$ of degree 7 with leading coefficient 1 such that -3 is a zero of multiplicity 2 and 0 is a zero of multiplicity 5, and sketch the graph of f.

18 Show that 2 is a zero of multiplicity 3 of the polynomial $f(x) = x^5 - 4x^4 - 3x^3 + 34x^2 - 52x + 24$, and express $f(x)$ as a product of linear factors.

Exer. 19–20: Find the zeros of $f(x)$, and state the multiplicity of each zero.

19 $f(x) = (x^2 - 2x + 1)^2(x^2 + 2x - 3)$

20 $f(x) = x^6 + 2x^4 + x^2$

Exer. 21–22: (a) Use Descartes' rule of signs to determine the number of possible positive, negative, and nonreal complex solutions of the equation. (b) Find the smallest and largest integers that are upper and lower bounds, respectively, for the real solutions of the equation.

21 $2x^4 - 4x^3 + 2x^2 - 5x - 7 = 0$

22 $x^5 - 4x^3 + 6x^2 + x + 4 = 0$

23 Show that $7x^6 + 2x^4 + 3x^2 + 10$ has no real zero.

Exer. 24–26: Find all solutions of the equation.

24 $x^4 + 9x^3 + 31x^2 + 49x + 30 = 0$

25 $16x^3 - 20x^2 - 8x + 3 = 0$

26 $x^4 - 7x^2 + 6 = 0$

Exer. 27–28: Find an equation for the sixth-degree polynomial f shown in the figure.

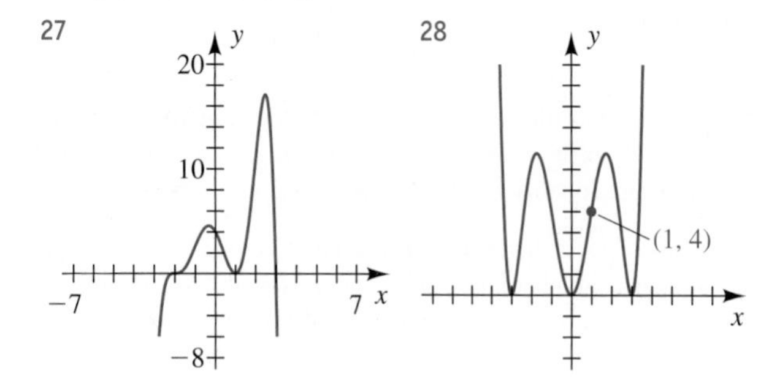

29 Identify any vertical asymptotes, horizontal asymptotes, intercepts, and holes for $f(x) = \dfrac{4(x+2)(x-1)}{3(x+2)(x-5)}$.

Exer. 30–39: Sketch the graph of *f*.

30 $f(x) = \dfrac{-2}{(x+1)^2}$

31 $f(x) = \dfrac{1}{(x-1)^3}$

32 $f(x) = \dfrac{3x^2}{16-x^2}$

33 $f(x) = \dfrac{x}{(x+5)(x^2-5x+4)}$

34 $f(x) = \dfrac{x^3-2x^2-8x}{-x^2+2x}$

35 $f(x) = \dfrac{x^2-2x+1}{x^3-x^2+x-1}$

36 $f(x) = \dfrac{3x^2+x-10}{x^2+2x}$

37 $f(x) = \dfrac{-2x^2-8x-6}{x^2-6x+8}$

38 $f(x) = \dfrac{x^2+2x-8}{x+3}$

39 $f(x) = \dfrac{x^4-16}{x^3}$

40 Find an equation of a rational function f that satisfies the given conditions.

vertical asymptote: $x = -3$
horizontal asymptote: $y = \frac{3}{2}$
x-intercept: 5
hole at $x = 2$

41 Suppose y is directly proportional to the cube root of x and inversely proportional to the square of z. Find the constant of proportionality if $y = 6$ when $x = 8$ and $z = 3$.

42 Suppose y is inversely proportional to the square of x. Sketch a graph of this relationship for $x > 0$, given that $y = 18$ when $x = 4$. Include a point for $x = 12$.

43 **Deflection of a beam** A horizontal beam l feet long is supported at one end and unsupported at the other end (see the figure). If the beam is subjected to a uniform load and if y denotes the deflection of the beam at a position x feet from the supported end, then it can be shown that

$$y = cx^2(x^2 - 4lx + 6l^2),$$

where c is a positive constant that depends on the weight of the load and the physical properties of the beam.

(a) If the beam is 10 feet long and the deflection at the unsupported end of the beam is 2 feet, find c.

(b) Show that the deflection is 1 foot somewhere between $x = 6.1$ and $x = 6.2$.

Exercise 43

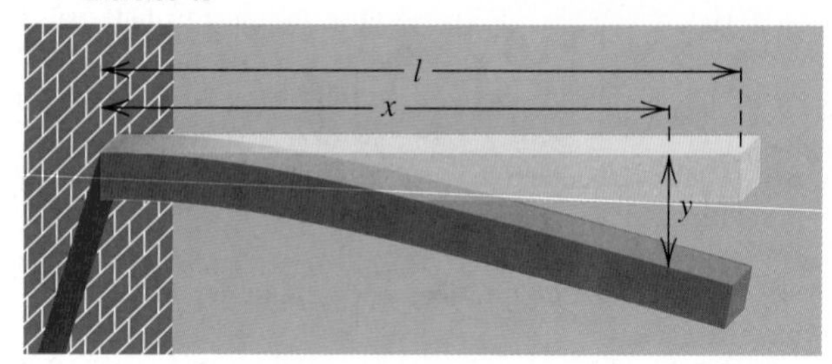

44 **Elastic cylinder** A rectangle made of elastic material is to be made into a cylinder by joining edge AD to edge BC, as shown in the figure. A wire of fixed length l is placed along the diagonal of the rectangle to support the structure. Let x denote the height of the cylinder.

(a) Express the volume V of the cylinder in terms of x.

(b) For what positive values of x is $V > 0$?

Exercise 44

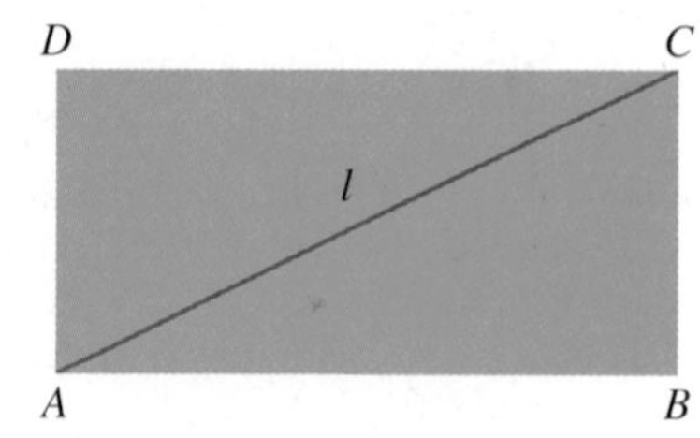

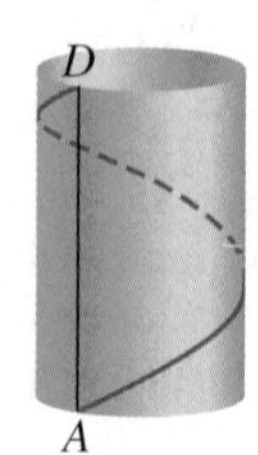

45 **Determining temperatures** A meteorologist determines that the temperature T (in °F) for a certain 24-hour period in winter was given by the formula $T = \frac{1}{20}t(t - 12)(t - 24)$ for $0 \le t \le 24$, where t is time in hours and $t = 0$ corresponds to 6 A.M. At what time(s) was the temperature 32°F?

46 **Deer propagation** A herd of 100 deer is introduced onto a small island. Assuming the number $N(t)$ of deer after t years is given by $N(t) = -t^4 + 21t^2 + 100$ (for $t > 0$), determine when the herd size exceeds 180.

47 **Threshold response curve** In biochemistry, the general threshold response curve is the graph of an equation

$$R = \frac{kS^n}{S^n + a^n},$$

where R is the chemical response when the level of the substance being acted on is S and a, k, and n are positive constants. An example is the removal rate R of alcohol from the bloodstream by the liver when the blood alcohol concentration is S.

(a) Find an equation of the horizontal asymptote for the graph.

(b) In the case of alcohol removal, $n = 1$ and a typical value of k is 0.22 gram per liter per minute. What is the interpretation of k in this setting?

48 **Oil spill clean-up** The cost $C(x)$ of cleaning up x percent of an oil spill that has washed ashore increases greatly as x approaches 100. Suppose that

$$C(x) = \frac{0.3x}{101 - x} \text{ (million dollars).}$$

(a) Compare $C(100)$ to $C(90)$.

(b) Sketch the graph of C for $0 < x < 100$.

49 **Telephone calls** In a certain county, the average number of telephone calls per day between any two cities is directly proportional to the product of their populations and inversely proportional to the square of the distance between them. Cities A and B are 25 miles apart and have populations of 10,000 and 5000, respectively. Telephone records indicate an average of 2000 calls per day between the two cities. Estimate the average number of calls per day between city A and another city of 15,000 people that is 100 miles from A.

50 **Power of a wind rotor** The power P generated by a wind rotor is directly proportional to the product of the square of the area A swept out by the blades and the third power of the wind velocity v. Suppose the diameter of the circular area swept out by the blades is 10 feet, and $P = 3000$ watts when $v = 20$ mi/hr. Find the power generated when the wind velocity is 30 mi/hr.

CHAPTER 3 DISCUSSION EXERCISES

1 Compare the domain, range, number of x-intercepts, and general shape of even-degreed polynomials and odd-degreed polynomials.

2 When using synthetic division, could you use a complex number c rather than a real number in $x - c$?

3 Discuss how synthetic division can be used to help find the quotient and remainder when $4x^3 - 8x^2 - 11x + 9$ is divided by $2x + 3$. Discuss how synthetic division can be used with any linear factor of the form $ax + b$.

4 Draw (by hand) a graph of a polynomial function of degree 3 that has x-intercepts 1, 2, and 3, has a y-intercept of 6, and passes through the point $(-1, 25)$. Can you actually have the graph you just drew?

5 How many different points do you need to specify a polynomial of degree n?

6 Prove the theorem on conjugate pair zeros of a polynomial. (*Hint:* For an arbitrary polynomial f, examine the conjugates of both sides of the equation $f(z) = 0$.)

7 Give an example of a rational function that has a common factor in the numerator and denominator, but does *not* have a hole in its graph. Discuss, in general, how this can happen.

8 (a) Can the graph of $f(x) = \dfrac{ax + b}{cx + d}$ (where $ax + b \neq cx + d$) cross its horizontal asymptote? If yes, then where?

(b) Can the graph of $f(x) = \dfrac{ax^2 + bx + c}{dx^2 + ex + f}$ (assume there are no like factors) cross its horizontal asymptote? If yes, then where?

9 **Gambling survival formula** An empirical formula for the bankroll B (in dollars) that is needed to survive a gambling session with confidence C (a percent expressed as a decimal) is given by the formula

$$B = \frac{GW}{29.3 + 53.1E - 22.7C},$$

where G is the number of games played in the session, W is the wager per game, and E is the player's edge on the game (expressed as a decimal).

(a) Approximate the bankroll needed for a player who plays 500 games per hour for 3 hours at \$5 per game with a -5% edge, provided the player wants a 95% chance of surviving the 3-hour session.

(b) Discuss the validity of the formula; a table and graph may help.

10 Multiply three consecutive integers together and then add the second integer to that product. Use synthetic division to help prove that the sum is the cube of an integer, and determine which integer.

11 **Personal tax rate** Assume the total amount of state tax paid consists of an amount P for personal property and S percent of income I.

(a) Find a function that calculates an individual's state tax rate R—that is, the percentage of the individual's income that is paid in taxes. (It is helpful to consider specific values to create the function.)

(b) What happens to R as I gets very large?

(c) Discuss the statement "Rich people pay a lower percentage of their income in state taxes than any other group."

12 **NFL passer rating** The National Football League ranks its passers by assigning a passer rating R based on the numbers of completions C, attempts A, yards Y, touchdowns T, and interceptions I. In a normal situation, it can be shown that the passer rating can be calculated using the formula

$$R = \frac{25(A + 40C + 2Y + 160T - 200I)}{12A}.$$

(a) In 1994, Steve Young completed 324 of 461 passes for 3969 yards and had 35 touchdown passes as well as 10 interceptions. Calculate his record-setting rating.

(b) How many more yards would he have needed to obtain a passer rating of at least 113?

(c) If he could make one more touchdown pass, how long would it have to be for him to obtain a passer rating of at least 114?

4

Inverse, Exponential, and Logarithmic Functions

Exponential and logarithmic functions are transcendental functions, since they cannot be defined in terms of only addition, subtraction, multiplication, division, and rational powers of a variable x, as is the case for the algebraic functions considered in previous chapters. Such functions are of major importance in mathematics and have applications in almost every field of human endeavor. They are especially useful in the fields of chemistry, biology, physics, and engineering, where they help describe the manner in which quantities in nature grow or decay. As we shall see in this chapter, there is a close relationship between specific exponential and logarithmic functions—they are inverse functions of each other.

4.1 Inverse Functions

A function f may have the same value for different numbers in its domain. For example, if $f(x) = x^2$, then $f(2) = 4$ and $f(-2) = 4$, but $2 \neq -2$. For *the inverse of a function* to be defined, it is essential that different numbers in the domain *always* give different values of f. Such functions are called *one-to-one functions.*

Definition of One-to-One Function	A function f with domain D and range R is a **one-to-one function** if either of the following equivalent conditions is satisfied: **(1)** Whenever $a \neq b$ in D, then $f(a) \neq f(b)$ in R. **(2)** Whenever $f(a) = f(b)$ in R, then $a = b$ in D.

Figure 1

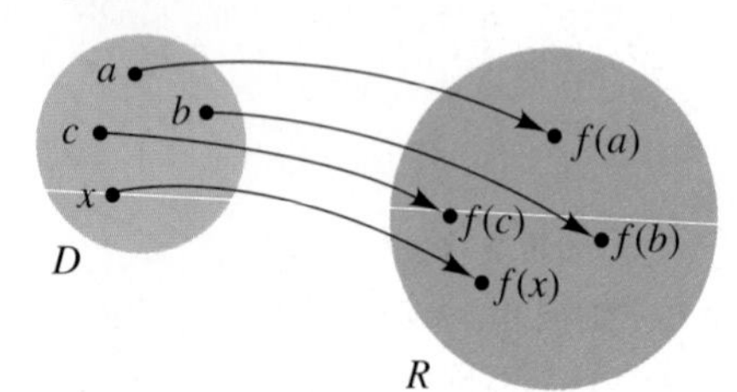

The arrow diagram in Figure 1 illustrates a one-to-one function. Note that each function value in the range R corresponds to *exactly one* element in the domain D. The function illustrated in Figure 2 of Section 2.4 is not one-to-one, since $f(w) = f(z)$, but $w \neq z$.

EXAMPLE 1 Determining whether a function is one-to-one

(a) If $f(x) = 3x + 2$, prove that f is one-to-one.

(b) If $g(x) = x^2 - 3$, prove that g is not one-to-one.

SOLUTION

(a) We shall use condition 2 of the preceding definition. Thus, suppose that $f(a) = f(b)$ for some numbers a and b in the domain of f. This gives us

$$\begin{aligned} 3a + 2 &= 3b + 2 && \text{definition of } f(x) \\ 3a &= 3b && \text{subtract 2} \\ a &= b && \text{divide by 3} \end{aligned}$$

Since we have concluded that a must equal b, f is one-to-one.

(b) Showing that a function *is* one-to-one requires a *general* proof, as in part (a). To show that g is *not* one-to-one we need only find two distinct real numbers in the domain that produce the same function value. For example, $-1 \neq 1$, but $g(-1) = g(1)$. In fact, since g is an even function, $g(-a) = g(a)$ for every real number a.

Figure 2

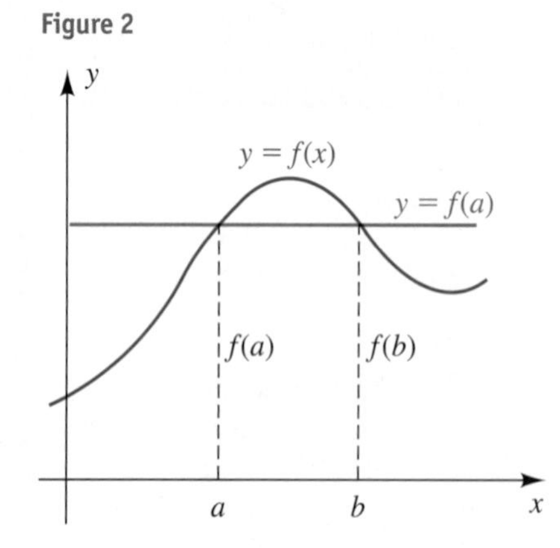

If we know the graph of a function f, it is easy to determine whether f is one-to-one. For example, the function whose graph is sketched in Figure 2 is not one-to-one, since $a \neq b$, but $f(a) = f(b)$. Note that the horizontal line $y = f(a)$ (or $y = f(b)$) intersects the graph in more than one point. In general, we may use the following graphical test to determine whether a function is one-to-one.

Horizontal Line Test	A function f is one-to-one if and only if every horizontal line intersects the graph of f in at most one point.

Let's apply the horizontal line test to the functions in Example 1.

EXAMPLE 2 Using the horizontal line test

Use the horizontal line test to determine if the function is one-to-one.

(a) $f(x) = 3x + 2$

(b) $g(x) = x^2 - 3$

SOLUTION

(a) The graph of $f(x) = 3x + 2$ is a line with y-intercept 2 and slope 3, as shown in Figure 3. We see that any horizontal line intersects the graph of f in at most one point. Thus, f is one-to-one.

Figure 3

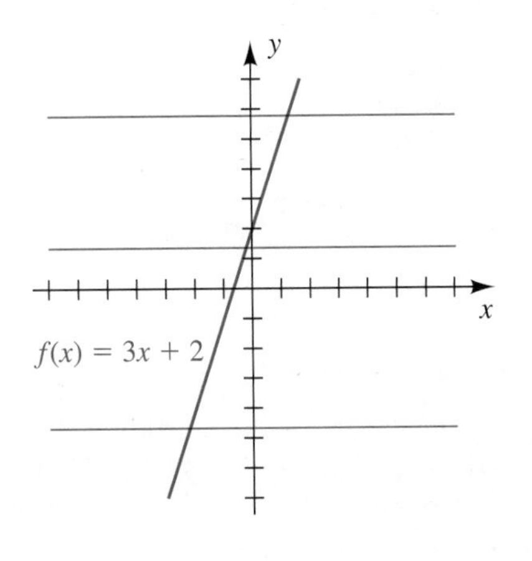

Figure 4

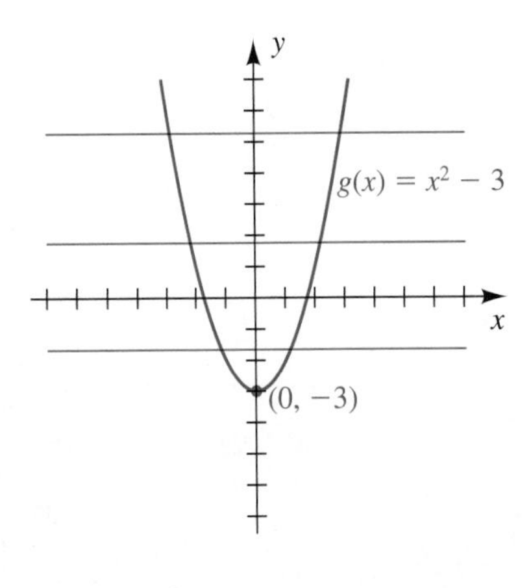

(b) The graph of $g(x) = x^2 - 3$ is a parabola opening upward with vertex $(0, -3)$, as shown in Figure 4. In this case, any horizontal line with equation $y = k$, where $k > -3$, will intersect the graph of g in two points. Thus, g is *not* one-to-one.

Tutorial available at www.thomsonedu.com/login

We may surmise from Example 2 that every increasing function or decreasing function passes the horizontal line test. Hence, we obtain the following result.

Theorem: Increasing or Decreasing Functions Are One-to-One

(1) A function that is increasing throughout its domain is one-to-one.

(2) A function that is decreasing throughout its domain is one-to-one.

Let f be a one-to-one function with domain D and range R. Thus, for each number y in R, there is *exactly one* number x in D such that $y = f(x)$, as illustrated by the arrow in Figure 5(a). We may, therefore, define a function g from R to D by means of the following rule:

$$x = g(y)$$

As in Figure 5(b), *g reverses the correspondence given by f.* We call g the *inverse function* of f, as in the next definition.

Figure 5

(a) $y = f(x)$

(b) $x = g(y)$

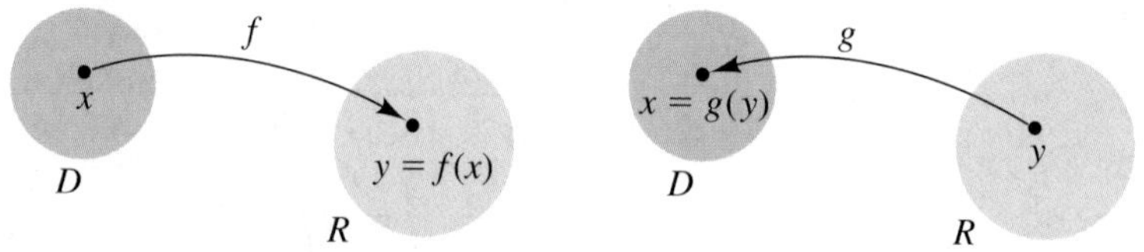

Definition of Inverse Function

Let f be a one-to-one function with domain D and range R. A function g with domain R and range D is the **inverse function** of f, provided the following condition is true for every x in D and every y in R:

$$y = f(x) \quad \text{if and only if} \quad x = g(y)$$

Remember that for the inverse of a function f to be defined, *it is absolutely essential that f be one-to-one*. The following theorem, stated without proof, is useful to verify that a function g is the inverse of f.

Theorem on Inverse Functions	Let f be a one-to-one function with domain D and range R. If g is a function with domain R and range D, then g is the inverse function of f if and only if both of the following conditions are true: **(1)** $g(f(x)) = x$ for every x in D **(2)** $f(g(y)) = y$ for every y in R

Conditions 1 and 2 of the preceding theorem are illustrated in Figure 6(a) and (b), respectively, where the blue arrow indicates that f is a function from D to R and the red arrow indicates that g is a function from R to D.

Figure 6

(a) First f, then g

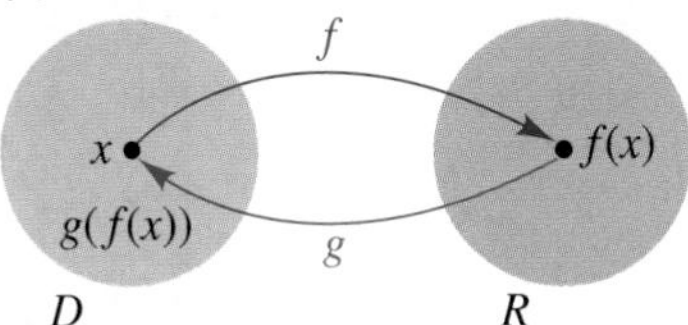

(b) First g, then f

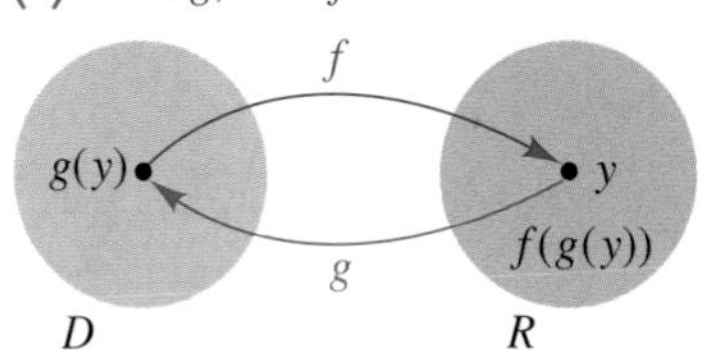

Note that in Figure 6(a) we first apply f to the number x in D, obtaining the function value $f(x)$ in R, and then apply g to $f(x)$, obtaining the number $g(f(x))$ in D. Condition 1 of the theorem states that $g(f(x)) = x$ for every x; that is, g *reverses* the correspondence given by f.

In Figure 6(b) we use the opposite order for the functions. We first apply g to the number y in R, obtaining the function value $g(y)$ in D, and then apply f to $g(y)$, obtaining the number $f(g(y))$ in R. Condition 2 of the theorem states that $f(g(y)) = y$ for every y; that is, f *reverses* the correspondence given by g.

If a function f has an inverse function g, we often denote g by f^{-1}. The -1 used in this notation should not be mistaken for an exponent; that is,

$$f^{-1}(y) \textit{ does not mean } 1/[f(y)].$$

The reciprocal $1/[f(y)]$ may be denoted by $[f(y)]^{-1}$. It is important to remember the following facts about the domain and range of f and f^{-1}.

Domain and Range of f and f^{-1}	domain of f^{-1} = range of f range of f^{-1} = domain of f

When we discuss functions, we often let x denote an arbitrary number in the domain. Thus, for the inverse function f^{-1}, we may wish to consider $f^{-1}(x)$, *where x is in the domain R of f^{-1}*. In this event, the two conditions in the theorem on inverse functions are written as follows:

(1) $f^{-1}(f(x)) = x$ for every x in the domain of f

(2) $f(f^{-1}(x)) = x$ for every x in the domain of f^{-1}

Figure 6 contains a hint for finding the inverse of a one-to-one function in certain cases: If possible, *we solve the equation $y = f(x)$ for x in terms of y*, obtaining an equation of the form $x = g(y)$. If the two conditions $g(f(x)) = x$ and $f(g(x)) = x$ are true for every x in the domains of f and g, respectively, then g is the required inverse function f^{-1}. The following guidelines summarize this procedure; in guideline 2, in anticipation of finding f^{-1}, we write $x = f^{-1}(y)$ instead of $x = g(y)$.

Guidelines for Finding f^{-1} in Simple Cases

1. Verify that f is a one-to-one function throughout its domain.
2. Solve the equation $y = f(x)$ for x in terms of y, obtaining an equation of the form $x = f^{-1}(y)$.
3. Verify the following two conditions:
   (a) $f^{-1}(f(x)) = x$ for every x in the domain of f
   (b) $f(f^{-1}(x)) = x$ for every x in the domain of f^{-1}

The success of this method depends on the nature of the equation $y = f(x)$, since we must be able to solve for x in terms of y. For this reason, we include the phrase *in simple cases* in the title of the guidelines. We shall follow these guidelines in the next three examples.

EXAMPLE 3 Finding the inverse of a function

Let $f(x) = 3x - 5$. Find the inverse function of f.

SOLUTION

Guideline 1 The graph of the linear function f is a line of slope 3, and hence f is increasing throughout $\mathbb{R}$. Thus, f is one-to-one and the inverse function f^{-1} exists. Moreover, since the domain and range of f are $\mathbb{R}$, the same is true for f^{-1}.

Guideline 2 Solve the equation $y = f(x)$ for x:

$$y = 3x - 5 \quad \text{let } y = f(x)$$

$$x = \frac{y+5}{3} \quad \text{solve for } x \text{ in terms of } y$$

We now formally let $x = f^{-1}(y)$; that is,

$$f^{-1}(y) = \frac{y+5}{3}.$$

Since the symbol used for the variable is immaterial, we may also write

$$f^{-1}(x) = \frac{x+5}{3},$$

where x is in the domain of f^{-1}.

Tutorial available at www.thomsonedu.com/login

Guideline 3 Since the domain and range of both f and f^{-1} are $\mathbb{R}$, we must verify conditions (a) and (b) for every real number x. We proceed as follows:

(a)
$$\begin{aligned} f^{-1}(f(x)) &= f^{-1}(3x - 5) && \text{definition of } f \\ &= \frac{(3x-5)+5}{3} && \text{definition of } f^{-1} \\ &= x && \text{simplify} \end{aligned}$$

(b)
$$\begin{aligned} f(f^{-1}(x)) &= f\left(\frac{x+5}{3}\right) && \text{definition of } f^{-1} \\ &= 3\left(\frac{x+5}{3}\right) - 5 && \text{definition of } f \\ &= x && \text{simplify} \end{aligned}$$

These verifications prove that the inverse function of f is given by

$$f^{-1}(x) = \frac{x+5}{3}.$$

EXAMPLE 4 Finding the inverse of a function

Let $f(x) = x^2 - 3$ for $x \geq 0$. Find the inverse function of f.

▶ **SOLUTION**

Figure 7

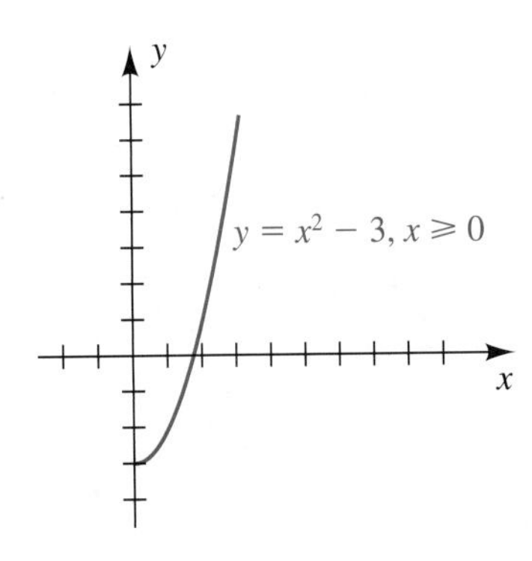

Guideline 1 The graph of f is sketched in Figure 7. The domain of f is $[0, \infty)$, and the range is $[-3, \infty)$. Since f is increasing, it is one-to-one and hence has an inverse function f^{-1} with domain $[-3, \infty)$ and range $[0, \infty)$.

Guideline 2 We consider the equation

$$y = x^2 - 3$$

and solve for x, obtaining

$$x = \pm\sqrt{y+3}.$$

Since x is nonnegative, we reject $x = -\sqrt{y+3}$ and let

$$f^{-1}(y) = \sqrt{y+3} \qquad \text{or, equivalently,} \qquad f^{-1}(x) = \sqrt{x+3}.$$

(Note that if the function f had domain $x \leq 0$, we would choose the function $f^{-1}(x) = -\sqrt{x+3}$.)

Guideline 3 We verify conditions (a) and (b) for x in the domains of f and f^{-1}, respectively.

(a) $f^{-1}(f(x)) = f^{-1}(x^2 - 3)$
$$= \sqrt{(x^2-3)+3} = \sqrt{x^2} = x \text{ for } x \geq 0$$

(b) $f(f^{-1}(x)) = f\left(\sqrt{x+3}\right)$
$$= \left(\sqrt{x+3}\right)^2 - 3 = (x+3) - 3 = x \text{ for } x \geq -3$$

Thus, the inverse function is given by

$$f^{-1}(x) = \sqrt{x+3} \quad \text{for } x \geq -3.$$

▶ Tutorial available at www.thomsonedu.com/login

Figure 8

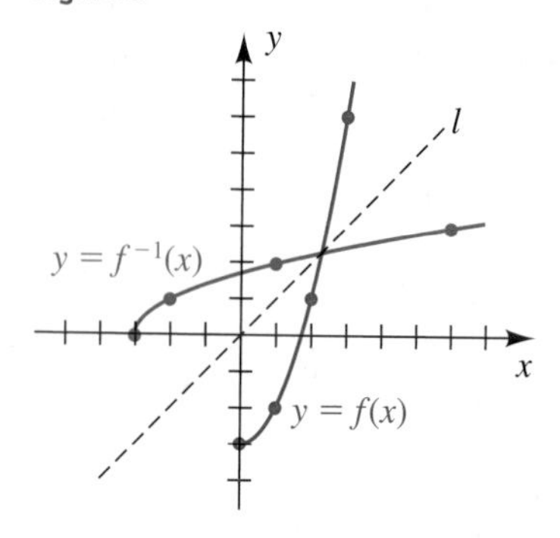

Note that the graphs of f and f^{-1} intersect on the line $y = x$.

There is an interesting relationship between the graph of a function f and the graph of its inverse function f^{-1}. We first note that $b = f(a)$ is equivalent to $a = f^{-1}(b)$. These equations imply that *the point (a, b) is on the graph of f if and only if the point (b, a) is on the graph of f^{-1}.*

As an illustration, in Example 4 we found that the functions f and f^{-1} given by

$$f(x) = x^2 - 3 \quad \text{and} \quad f^{-1}(x) = \sqrt{x + 3}$$

are inverse functions of each other, provided that x is suitably restricted. Some points on the graph of f are $(0, -3)$, $(1, -2)$, $(2, 1)$, and $(3, 6)$. Corresponding points on the graph of f^{-1} are $(-3, 0)$, $(-2, 1)$, $(1, 2)$, and $(6, 3)$. The graphs of f and f^{-1} are sketched on the same coordinate plane in Figure 8. If the page is folded along the line $y = x$ that bisects quadrants I and III (as indicated by the dashes in the figure), then the graphs of f and f^{-1} coincide. The two graphs are *reflections* of each other through the line $y = x$, or are *symmetric* with respect to this line. This is typical of the graph of every function f that has an inverse function f^{-1} (see Exercise 50).

Figure 9

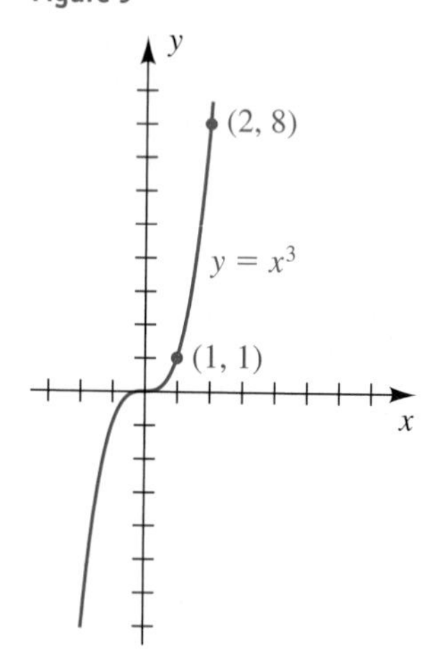

EXAMPLE 5 The relationship between the graphs of f and f^{-1}

Let $f(x) = x^3$. Find the inverse function f^{-1} of f, and sketch the graphs of f and f^{-1} on the same coordinate plane.

▶ **SOLUTION** The graph of f is sketched in Figure 9. Note that f is an odd function, and hence the graph is symmetric with respect to the origin.

Guideline 1 Since f is increasing throughout its domain $\mathbb{R}$, it is one-to-one and hence has an inverse function f^{-1}.

Guideline 2 We consider the equation

$$y = x^3$$

and solve for x by taking the cube root of each side, obtaining

$$x = y^{1/3} = \sqrt[3]{y}.$$

We now let

$$f^{-1}(y) = \sqrt[3]{y} \quad \text{or, equivalently,} \quad f^{-1}(x) = \sqrt[3]{x}.$$

Guideline 3 We verify conditions (a) and (b):

(a) $f^{-1}(f(x)) = f^{-1}(x^3) = \sqrt[3]{x^3} = x$ for every x in $\mathbb{R}$

(b) $f(f^{-1}(x)) = f(\sqrt[3]{x}) = (\sqrt[3]{x})^3 = x$ for every x in $\mathbb{R}$

The graph of f^{-1} (that is, the graph of the equation $y = \sqrt[3]{x}$) may be obtained by reflecting the graph in Figure 9 through the line $y = x$, as shown in Figure 10. Three points on the graph of f^{-1} are $(0, 0)$, $(1, 1)$, and $(8, 2)$.

Figure 10

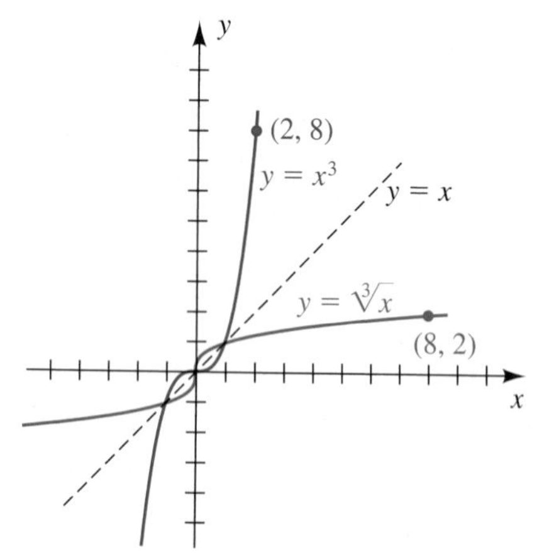

The next example shows how to graph the inverse of a function using a graphing calculator.

▶ **Tutorial available at www.thomsonedu.com/login**

EXAMPLE 6 **Graphing the inverse of a function**

(a) Sketch the graph of the inverse function of

$$f(x) = \frac{1}{35}(x^3 + 9x).$$

(b) Approximate the solutions of the equation $f(x) = f^{-1}(x)$.

SOLUTION

(a) We will assign $(x^3 + 9x)/35$ to Y_1, assign x to Y_2, set the viewing rectangle to $[-12, 12]$ by $[-8, 8]$, and graph the functions.

TI-83/4 Plus

TI-86

Make Y assignments.

```
Plot1 Plot2 Plot3
\Y1■(X³+9X)/35
\Y2■X
\Y3=
\Y4=
\Y5=
\Y6=
\Y7=
```

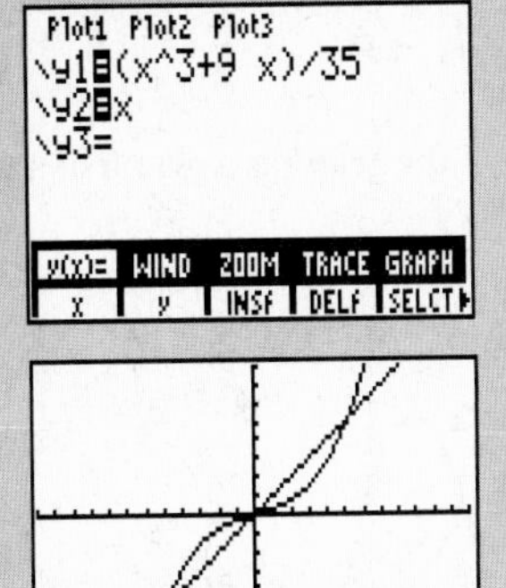

Graph the functions.

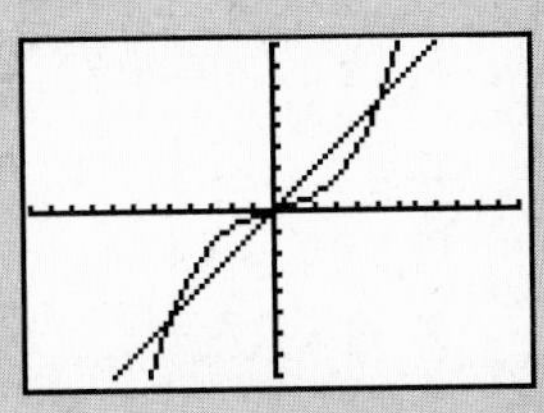

Graph the inverse.

Since f is increasing throughout its domain, it is one-to-one and has an inverse. If f were not one-to-one and we entered the following keystrokes, then the calculator would draw the inverse relation, but it would not be a function.

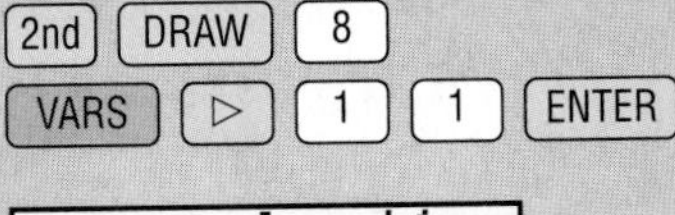

MORE DRAW(F2) MORE (3 times)
DrInv(F3) 2nd alpha Y 1 ENTER

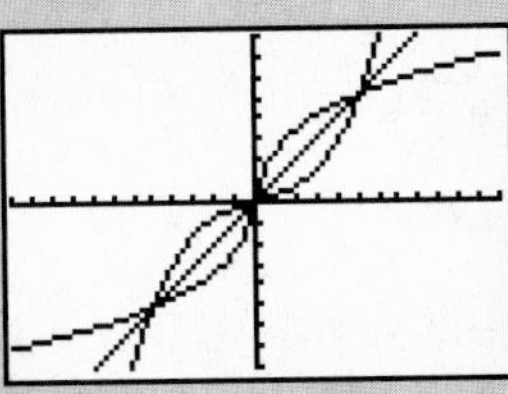

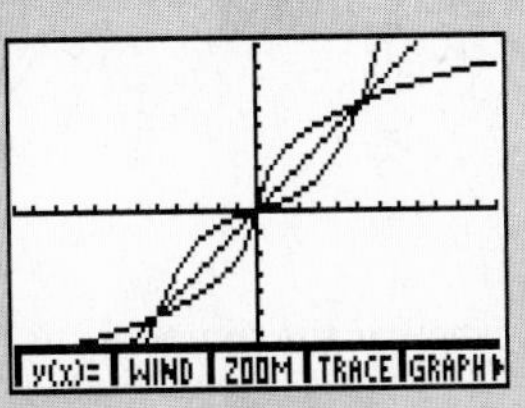

(b) $f(x) = f^{-1}(x)$ on the line $y = x$. Using the intersect feature with Y_1 and Y_2 gives the solution $x \approx 5.1$. By the symmetry of the graphs, we have the solutions $x = 0$ and $x \approx \pm 5.1$.

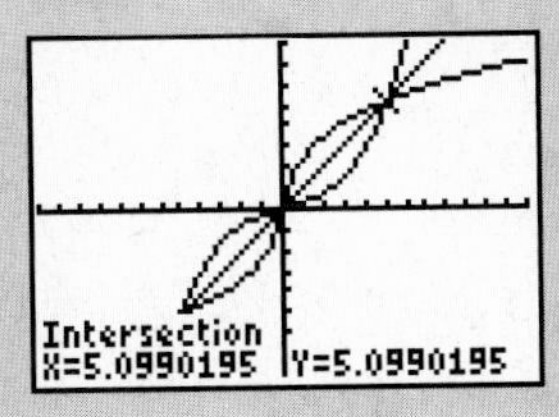

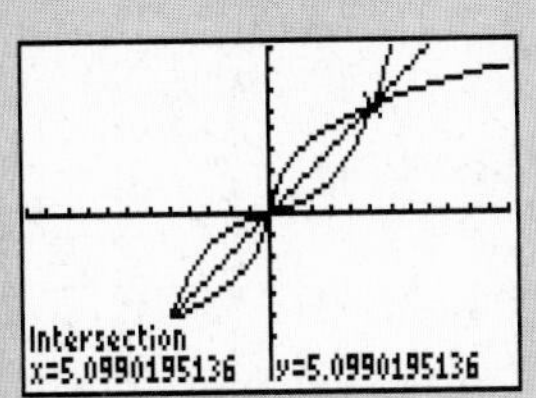

4.1 Exercises

Exer. 1–2: If possible, find
(a) $f^{-1}(5)$ (b) $g^{-1}(6)$

1

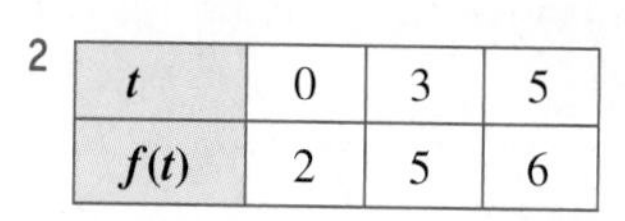

x	2	4	6
$f(x)$	3	5	9

x	1	3	5
$g(x)$	6	2	6

2

t	0	3	5
$f(t)$	2	5	6

t	1	2	4
$g(t)$	3	6	6

Exer. 3–4: Determine if the graph is a graph of a one-to-one function.

3 (a) (b) (c)

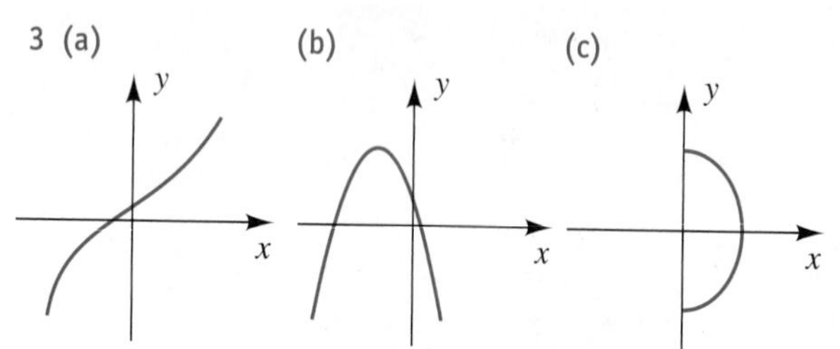

4 (a) (b) (c)

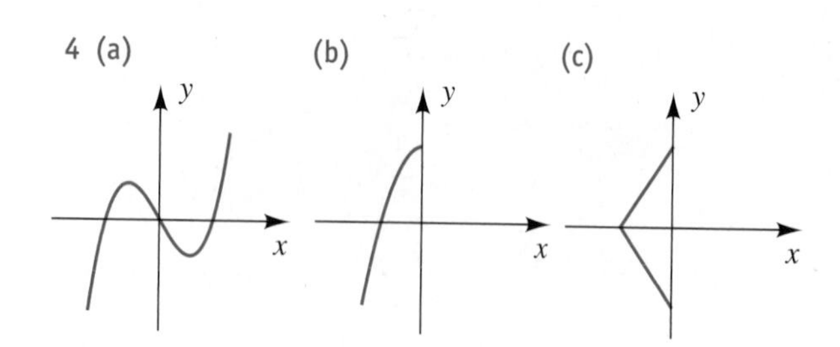

Exer. 5–16: Determine whether the function f is one-to-one.

5 $f(x) = 3x - 7$

6 $f(x) = \dfrac{1}{x-2}$

7 $f(x) = x^2 - 9$

8 $f(x) = x^2 + 4$

9 $f(x) = \sqrt{x}$

10 $f(x) = \sqrt[3]{x}$

▶ 11 $f(x) = |x|$

12 $f(x) = 3$

13 $f(x) = \sqrt{4 - x^2}$

14 $f(x) = 2x^3 - 4$

▶ 15 $f(x) = \dfrac{1}{x}$

16 $f(x) = \dfrac{1}{x^2}$

Exer. 17–20: Use the theorem on inverse functions to prove that f and g are inverse functions of each other, and sketch the graphs of f and g on the same coordinate plane.

17 $f(x) = 3x - 2$; $g(x) = \dfrac{x+2}{3}$

18 $f(x) = x^2 + 5,\ x \le 0$; $g(x) = -\sqrt{x-5},\ x \ge 5$

19 $f(x) = -x^2 + 3,\ x \ge 0$; $g(x) = \sqrt{3-x},\ x \le 3$

20 $f(x) = x^3 - 4$; $g(x) = \sqrt[3]{x+4}$

Exer. 21–24: Determine the domain and range of f^{-1} for the given function without actually finding f^{-1}. *Hint:* First find the domain and range of f.

21 $f(x) = -\dfrac{2}{x-1}$

22 $f(x) = \dfrac{5}{x+3}$

23 $f(x) = \dfrac{4x+5}{3x-8}$

24 $f(x) = \dfrac{2x-7}{9x+1}$

Exer. 25–42: Find the inverse function of f.

25 $f(x) = 3x + 5$

26 $f(x) = 7 - 2x$

27 $f(x) = \dfrac{1}{3x-2}$

28 $f(x) = \dfrac{1}{x+3}$

29 $f(x) = \dfrac{3x+2}{2x-5}$

30 $f(x) = \dfrac{4x}{x-2}$

31 $f(x) = 2 - 3x^2,\ x \le 0$

32 $f(x) = 5x^2 + 2,\ x \ge 0$

33 $f(x) = 2x^3 - 5$

34 $f(x) = -x^3 + 2$

35 $f(x) = \sqrt{3-x}$

36 $f(x) = \sqrt{4-x^2},\ 0 \le x \le 2$

37 $f(x) = \sqrt[3]{x} + 1$

38 $f(x) = (x^3 + 1)^5$

39 $f(x) = x$

40 $f(x) = -x$

▶ Tutorial available at www.thomsonedu.com/login

▶ 41 $f(x) = x^2 - 6x, x \geq 3$

42 $f(x) = x^2 - 4x + 3, x \leq 2$

Exer. 43–44: Let $h(x) = 4 - x$. Use h, the table, and the graph to evaluate the expression.

x	2	3	4	5	6
$f(x)$	-1	0	1	2	3

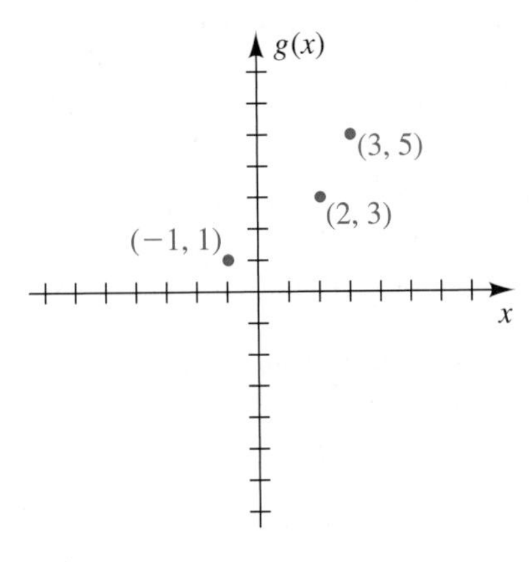

43 (a) $(g^{-1} \circ f^{-1})(2)$ (b) $(g^{-1} \circ h)(3)$

(c) $(h^{-1} \circ f \circ g^{-1})(3)$

44 (a) $(g \circ f^{-1})(-1)$ (b) $(f^{-1} \circ g^{-1})(3)$

(c) $(h^{-1} \circ g^{-1} \circ f)(6)$

Exer. 45–48: The graph of a one-to-one function f is shown. (a) Use the reflection property to sketch the graph of f^{-1}. (b) Find the domain D and range R of the function f. (c) Find the domain D_1 and range R_1 of the inverse function f^{-1}.

45

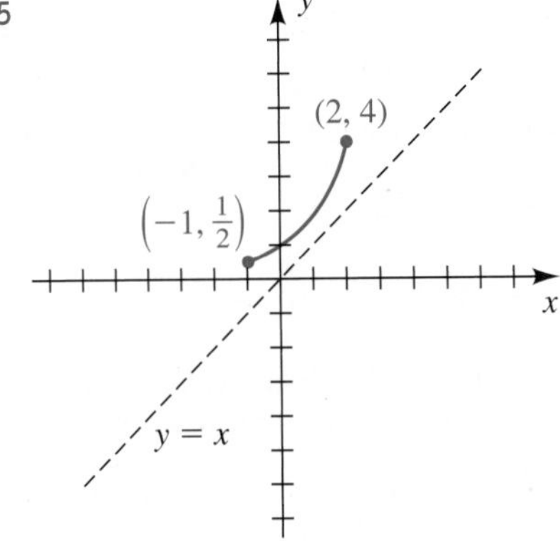

46

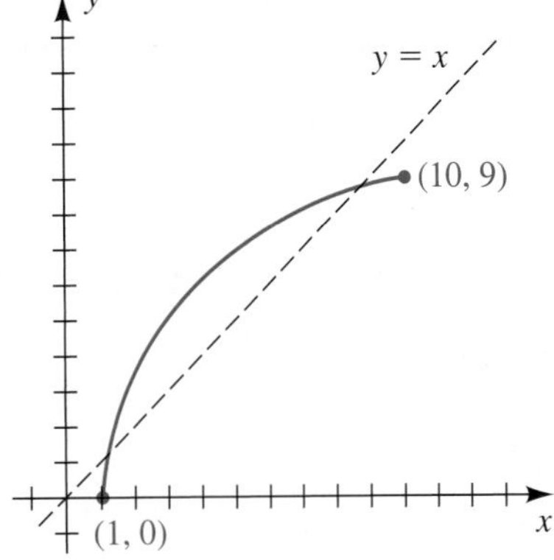

47

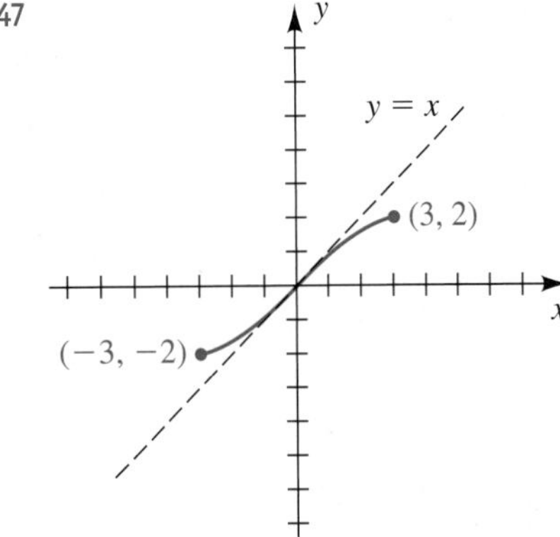

48

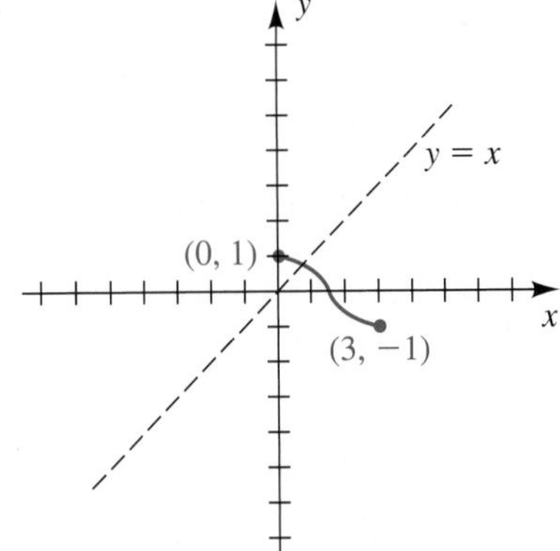

49 (a) Prove that the function defined by $f(x) = ax + b$ (a linear function) for $a \neq 0$ has an inverse function, and find $f^{-1}(x)$.

(b) Does a constant function have an inverse? Explain.

▶ **Tutorial available at www.thomsonedu.com/login**

50 Show that the graph of f^{-1} is the reflection of the graph of f through the line $y = x$ by verifying the following conditions:

(1) If $P(a, b)$ is on the graph of f, then $Q(b, a)$ is on the graph of f^{-1}.

(2) The midpoint of line segment PQ is on the line $y = x$.

(3) The line PQ is perpendicular to the line $y = x$.

51 Verify that $f(x) = f^{-1}(x)$ if

(a) $f(x) = -x + b$ (b) $f(x) = \dfrac{ax + b}{cx - a}$ for $c \neq 0$

(c) $f(x)$ has the following graph:

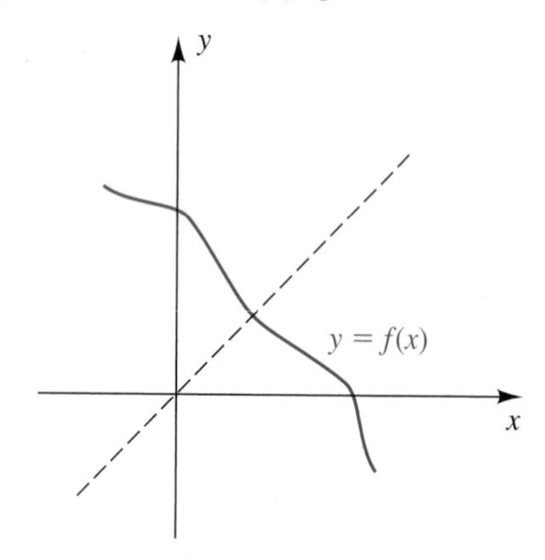

52 Let n be any positive integer. Find the inverse function of f if

(a) $f(x) = x^n$ for $x \geq 0$

(b) $f(x) = x^{m/n}$ for $x \geq 0$ and m any positive integer

Exer. 53–54: Use the graph of f to determine whether f is one-to-one.

53 $f(x) = 0.4x^5 - 0.4x^4 + 1.2x^3 - 1.2x^2 + 0.8x - 0.8$

54 $f(x) = \dfrac{x - 8}{x^{2/3} + 4}$

Exer. 55–56: Graph f on the given interval. (a) Estimate the largest interval $[a, b]$ with $a < 0 < b$ on which f is one-to-one. (b) If g is the function with domain $[a, b]$ such that $g(x) = f(x)$ for $a \leq x \leq b$, estimate the domain and range of g^{-1}.

55 $f(x) = 2.1x^3 - 2.98x^2 - 2.11x + 3$; $[-1, 2]$

56 $f(x) = 0.05x^4 - 0.24x^3 - 0.15x^2 + 1.18x + 0.24$; $[-2, 2]$

Exer. 57–58: Graph f in the given viewing rectangle. Use the graph of f to predict the shape of the graph of f^{-1}. Verify your prediction by graphing f^{-1} and the line $y = x$ in the same viewing rectangle.

57 $f(x) = \sqrt[3]{x - 1}$; $[-12, 12]$ by $[-8, 8]$

58 $f(x) = 2(x - 2)^2 + 3$, $x \geq 2$; $[0, 12]$ by $[0, 8]$

59 **Ventilation requirements** Ventilation is an effective way to improve indoor air quality. In nonsmoking restaurants, air circulation requirements (in ft^3/min) are given by the function $V(x) = 35x$, where x is the number of people in the dining area.

(a) Determine the ventilation requirements for 23 people.

▶ (b) Find $V^{-1}(x)$. Explain the significance of V^{-1}.

(c) Use V^{-1} to determine the maximum number of people that should be in a restaurant having a ventilation capability of 2350 ft^3/min.

60 **Radio stations** The table lists the total numbers of radio stations in the United States for certain years.

Year	Number
1950	2773
1960	4133
1970	6760
1980	8566
1990	10,770
2000	12,717

(a) Plot the data.

(b) Determine a linear function $f(x) = ax + b$ that models these data, where x is the year. Plot f and the data on the same coordinate axes.

(c) Find $f^{-1}(x)$. Explain the significance of f^{-1}.

(d) Use f^{-1} to predict the year in which there were 11,987 radio stations. Compare it with the true value, which is 1995.

▶ **Tutorial available at www.thomsonedu.com/login**

4.2

Exponential Functions

Previously, we considered functions having terms of the form

$$\text{variable base}^{\text{constant power}},$$

such as x^2, $0.2x^{1.3}$, and $8x^{2/3}$. We now turn our attention to functions having terms of the form

$$\text{constant base}^{\text{variable power}},$$

such as 2^x, $(1.04)^{4x}$, and 3^{-x}. Let us begin by considering the function f defined by

$$f(x) = 2^x,$$

where x is restricted to *rational* numbers. (Recall that if $x = m/n$ for integers m and n with $n > 0$, then $2^x = 2^{m/n} = \left(\sqrt[n]{2}\right)^m$.) Coordinates of several points on the graph of $y = 2^x$ are listed in the following table.

x	-10	-3	-2	-1	0	1	2	3	10
$y = 2^x$	$\frac{1}{1024}$	$\frac{1}{8}$	$\frac{1}{4}$	$\frac{1}{2}$	1	2	4	8	1024

Figure 1

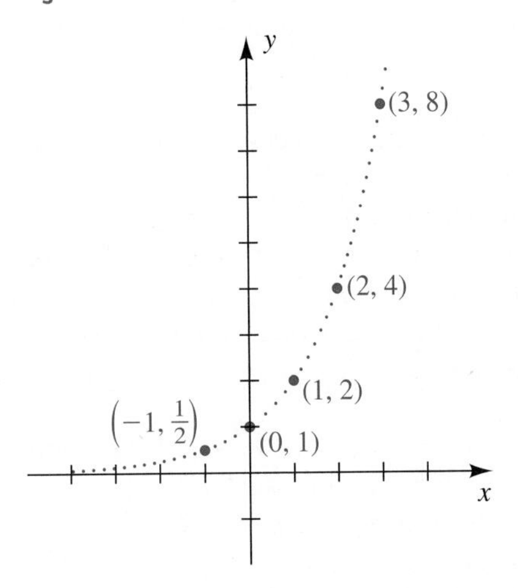

Other values of y for x rational, such as $2^{1/3}$, $2^{-9/7}$, and $2^{5.143}$, can be approximated with a calculator. We can show algebraically that if x_1 and x_2 are rational numbers such that $x_1 < x_2$, then $2^{x_1} < 2^{x_2}$. Thus, f is an increasing function, and its graph rises. Plotting points leads to the sketch in Figure 1, where the small dots indicate that only the points with *rational* x-coordinates are on the graph. There is a *hole* in the graph whenever the x-coordinate of a point is irrational.

To extend the domain of f to all real numbers, it is necessary to define 2^x for every *irrational* exponent x. To illustrate, if we wish to define 2^π, we could use the nonterminating decimal representing 3.1415926 . . . for π and consider the following *rational* powers of 2:

$$2^3,\quad 2^{3.1},\quad 2^{3.14},\quad 2^{3.141},\quad 2^{3.1415},\quad 2^{3.14159},\quad \ldots$$

Figure 2

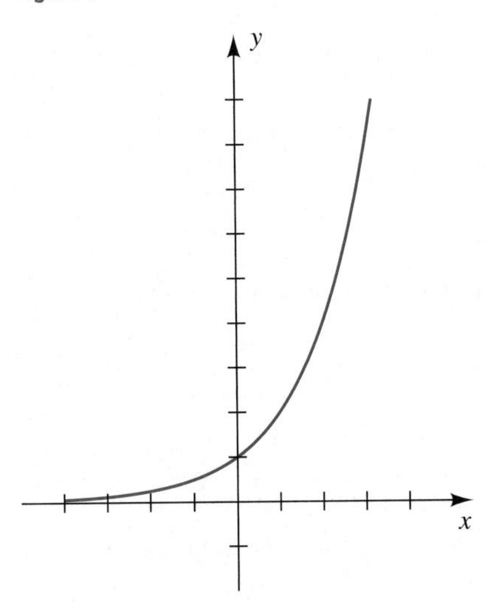

It can be shown, using calculus, that each successive power gets closer to a unique real number, denoted by 2^π. Thus,

$$2^x \to 2^\pi \quad \text{as} \quad x \to \pi, \quad \text{with } x \text{ rational.}$$

The same technique can be used for any other irrational power of 2. To sketch the graph of $y = 2^x$ with x *real,* we replace the holes in the graph in Figure 1 with points, and we obtain the graph in Figure 2. The function f defined by $f(x) = 2^x$ for every real number x is called the **exponential function** *with base* 2.

Let us next consider *any* base a, where a is a positive real number different from 1. As in the preceding discussion, to each real number x there corresponds exactly one positive number a^x such that the laws of exponents are true. Thus, as in the following chart, we may define a function f whose domain is $\mathbb{R}$ and range is the set of positive real numbers.

Terminology	Definition	Graph of f for $a > 1$	Graph of f for $0 < a < 1$
Exponential function f with base a	$f(x) = a^x$ for every x in $\mathbb{R}$, where $a > 0$ and $a \neq 1$		

The graphs in the chart show that if $a > 1$, then f is increasing on $\mathbb{R}$, and if $0 < a < 1$, then f is decreasing on $\mathbb{R}$. (These facts can be proved using calculus.) The graphs merely indicate the *general* appearance—the *exact* shape depends on the value of a. Note, however, that since $a^0 = 1$, the y-intercept is 1 for every a.

Note that if $a > 1$, then $a = 1 + d$ $(d > 0)$ and the base a in $y = a^x$ can be thought of as representing multiplication by more than 100% as x increases by 1, so the function is increasing. For example, if $a = 1.15$, then $y = (1.15)^x$ can be considered to be a 15% per year growth function. More details on this concept appear later.

If $a > 1$, then as x *decreases* through negative values, the graph of f approaches the x-axis (see the third column in the chart). Thus, the x-axis is a *horizontal asymptote*. As x increases through positive values, the graph rises rapidly. This type of variation is characteristic of the **exponential law of growth,** and f is sometimes called a **growth function.**

If $0 < a < 1$, then as x *increases,* the graph of f approaches the x-axis asymptotically (see the last column in the chart). This type of variation is known as **exponential decay.**

When considering a^x we exclude the cases $a \leq 0$ and $a = 1$. Note that if $a < 0$, then a^x is not a real number for many values of x such as $\frac{1}{2}$, $\frac{3}{4}$, and $\frac{11}{6}$. If $a = 0$, then $a^0 = 0^0$ is undefined. Finally, if $a = 1$, then $a^x = 1$ for every x, and the graph of $y = a^x$ is a horizontal line.

The graph of an exponential function f is either increasing throughout its domain or decreasing throughout its domain. Thus, f is one-to-one by the theorem on page 282. Combining this result with the definition of a one-to-one function (see page 280) gives us parts (1) and (2) of the following theorem.

Theorem: Exponential Functions Are One-to-One

The exponential function f given by

$$f(x) = a^x \quad \text{for} \quad 0 < a < 1 \quad \text{or} \quad a > 1$$

is one-to-one. Thus, the following equivalent conditions are satisfied for real numbers x_1 and x_2.

(1) If $x_1 \neq x_2$, then $a^{x_1} \neq a^{x_2}$.

(2) If $a^{x_1} = a^{x_2}$, then $x_1 = x_2$.

When using this theorem as a reason for a step in the solution to an example, we will state that *exponential functions are one-to-one*.

ILLUSTRATION **Exponential Functions Are One-to-One**

- If $7^{3x} = 7^{2x+5}$, then $3x = 2x + 5$, or $x = 5$.

In the following example we solve a simple *exponential equation*—that is, an equation in which the variable appears in an exponent.

EXAMPLE 1 **Solving an exponential equation**

Solve the equation $3^{5x-8} = 9^{x+2}$.

▶ **SOLUTION**

$3^{5x-8} = 9^{x+2}$	given
$3^{5x-8} = (3^2)^{x+2}$	express both sides with the same base
$3^{5x-8} = 3^{2x+4}$	law of exponents
$5x - 8 = 2x + 4$	exponential functions are one-to-one
$3x = 12$	subtract $2x$ and add 8
$x = 4$	divide by 3

Note that the solution in Example 1 depended on the fact that the base 9 could be written as 3 to some power. We will consider only exponential equations of this type for now, but we will solve more general exponential equations later in the chapter.

In the next two examples we sketch the graphs of several different exponential functions.

Figure 3

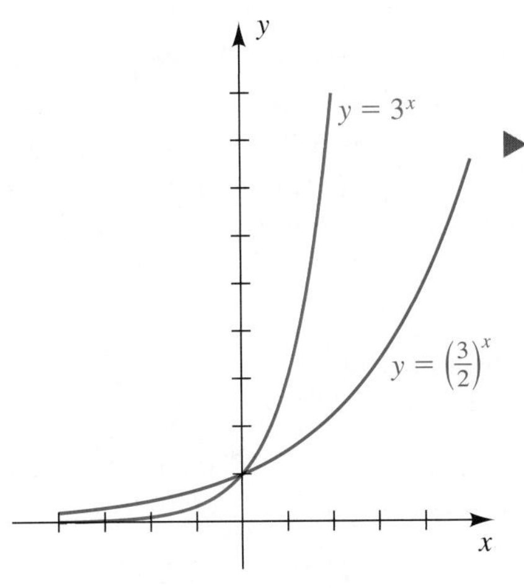

EXAMPLE 2 **Sketching graphs of exponential functions**

If $f(x) = \left(\frac{3}{2}\right)^x$ and $g(x) = 3^x$, sketch the graphs of f and g on the same coordinate plane.

▶ **SOLUTION** Since $\frac{3}{2} > 1$ and $3 > 1$, each graph *rises* as x increases. The following table displays coordinates for several points on the graphs.

x	-2	-1	0	1	2	3	4
$y = \left(\frac{3}{2}\right)^x$	$\frac{4}{9} \approx 0.4$	$\frac{2}{3} \approx 0.7$	1	$\frac{3}{2}$	$\frac{9}{4} \approx 2.3$	$\frac{27}{8} \approx 3.4$	$\frac{81}{16} \approx 5.1$
$y = 3^x$	$\frac{1}{9} \approx 0.1$	$\frac{1}{3} \approx 0.3$	1	3	9	27	81

Plotting points and being familiar with the general graph of $y = a^x$ leads to the graphs in Figure 3.

Example 2 illustrates the fact that if $1 < a < b$, then $a^x < b^x$ for positive values of x and $b^x < a^x$ for negative values of x. In particular, since $\frac{3}{2} < 2 < 3$, the graph of $y = 2^x$ in Figure 2 lies between the graphs of f and g in Figure 3.

▶ Tutorial available at www.thomsonedu.com/login

Figure 4

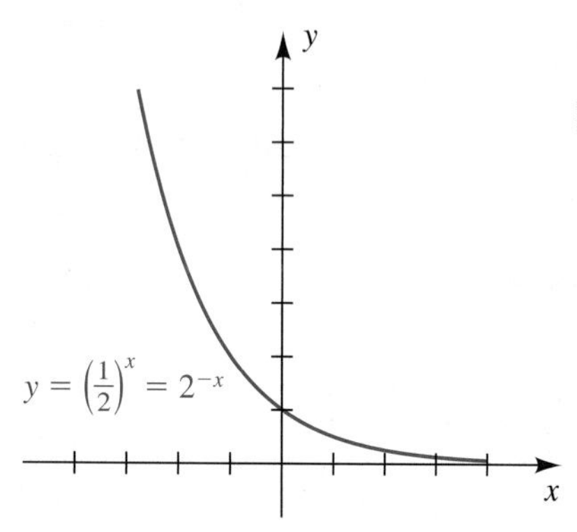

EXAMPLE 3 Sketching the graph of an exponential function

Sketch the graph of the equation $y = \left(\frac{1}{2}\right)^x$.

▶ **SOLUTION** Since $0 < \frac{1}{2} < 1$, the graph *falls* as x increases. Coordinates of some points on the graph are listed in the following table.

x	-3	-2	-1	0	1	2	3
$y = \left(\frac{1}{2}\right)^x$	8	4	2	1	$\frac{1}{2}$	$\frac{1}{4}$	$\frac{1}{8}$

The graph is sketched in Figure 4. Since $\left(\frac{1}{2}\right)^x = (2^{-1})^x = 2^{-x}$, the graph is the same as the graph of the equation $y = 2^{-x}$. Note that the graph is a reflection through the y-axis of the graph of $y = 2^x$ in Figure 2.

Figure 5

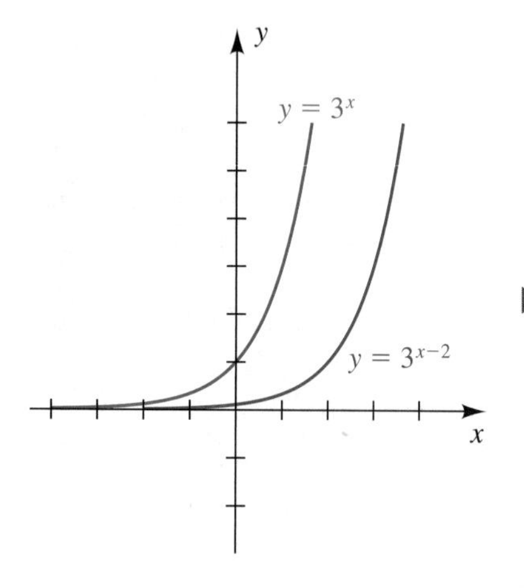

Equations of the form $y = a^u$, where u is some expression in x, occur in applications. The next two examples illustrate equations of this form.

EXAMPLE 4 Shifting graphs of exponential functions

Sketch the graph of the equation:

(a) $y = 3^{x-2}$ **(b)** $y = 3^x - 2$

▶ **SOLUTION**

(a) The graph of $y = 3^x$, sketched in Figure 3, is resketched in Figure 5. From the discussion of horizontal shifts in Section 2.5, we can obtain the graph of $y = 3^{x-2}$ by shifting the graph of $y = 3^x$ two units to the right, as shown in Figure 5.

The graph of $y = 3^{x-2}$ can also be obtained by plotting several points and using them as a guide to sketch an exponential-type curve.

(b) From the discussion of vertical shifts in Section 2.5, we can obtain the graph of $y = 3^x - 2$ by shifting the graph of $y = 3^x$ two units downward, as shown in Figure 6. Note that the y-intercept is -1 and the line $y = -2$ is a horizontal asymptote for the graph.

Figure 6

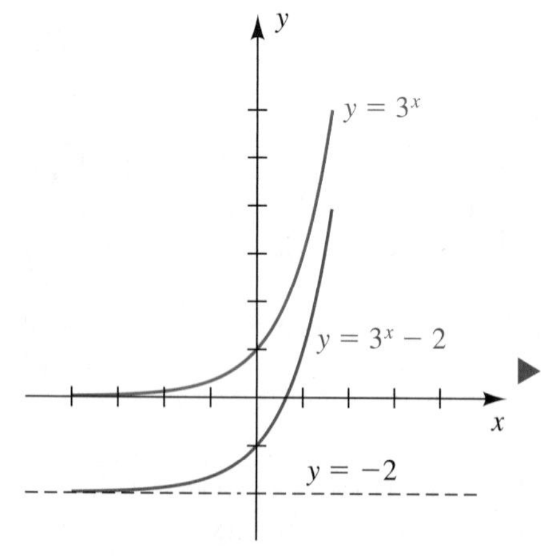

EXAMPLE 5 Finding an equation of an exponential function satisfying prescribed conditions

Find an exponential function of the form $f(x) = ba^{-x} + c$ that has horizontal asymptote $y = -2$, y-intercept 16, and x-intercept 2.

▶ **SOLUTION** The horizontal asymptote of the graph of an exponential function of the form $f(x) = ba^{-x}$ is the x-axis—that is, $y = 0$. Since the desired horizontal asymptote is $y = -2$, we must have $c = -2$, so $f(x) = ba^{-x} - 2$.

Because the y-intercept is 16, $f(0)$ must equal 16. But $f(0) = ba^{-0} - 2 = b - 2$, so $b - 2 = 16$ and $b = 18$. Thus, $f(x) = 18a^{-x} - 2$.

▶ **Tutorial available at www.thomsonedu.com/login**

Figure 7

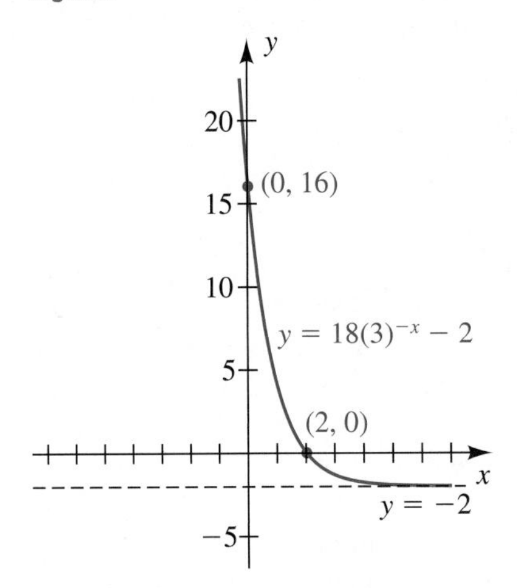

Lastly, we find the value of a:

$$f(x) = 18a^{-x} - 2 \quad \text{given form of } f$$

$$0 = 18(a)^{-2} - 2 \quad f(2) = 0 \text{ since 2 is the } x\text{-intercept}$$

$$2 = 18 \cdot \frac{1}{a^2} \quad \text{add 2; definition of negative exponent}$$

$$a^2 = 9 \quad \text{multiply by } a^2/2$$

$$a = \pm 3 \quad \text{take square root}$$

Since a must be positive, we have

$$f(x) = 18(3)^{-x} - 2.$$

Figure 7 shows a graph of f that satisfies all of the conditions in the problem statement. Note that $f(x)$ could be written in the equivalent form

$$f(x) = 18\left(\tfrac{1}{3}\right)^x - 2.$$

The bell-shaped graph of the function in the next example is similar to a *normal probability curve* used in statistical studies.

EXAMPLE 6 **Sketching a bell-shaped graph**

Figure 8

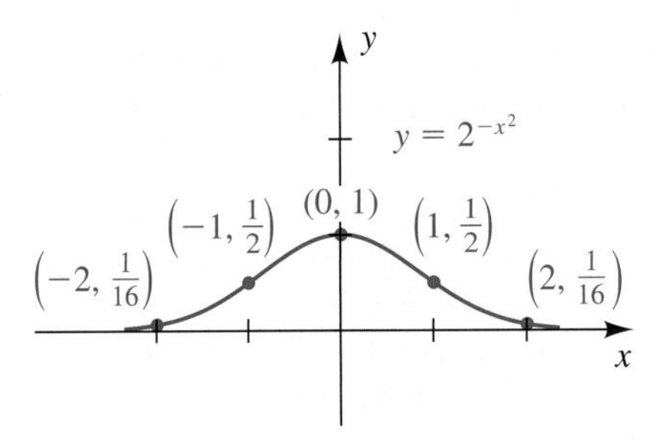

If $f(x) = 2^{-x^2}$, sketch the graph of f.

SOLUTION If we rewrite $f(x)$ as

$$f(x) = \frac{1}{2^{(x^2)}},$$

we see that as x increases through positive values, $f(x)$ decreases rapidly; hence the graph approaches the x-axis asymptotically. Since x^2 is smallest when $x = 0$, the maximum value of f is $f(0) = 1$. Since f is an even function, the graph is symmetric with respect to the y-axis. Some points on the graph are $(0, 1)$, $\left(1, \frac{1}{2}\right)$, and $\left(2, \frac{1}{16}\right)$. Plotting and using symmetry gives us the sketch in Figure 8.

APPLICATION **Bacterial Growth**

Exponential functions may be used to describe the growth of certain populations. As an illustration, suppose it is observed experimentally that the number of bacteria in a culture doubles every day. If 1000 bacteria are present at the start, then we obtain the following table, where t is the time in days and $f(t)$ is the bacteria count at time t.

t (time in days)	0	1	2	3	4
$f(t)$ (bacteria count)	1000	2000	4000	8000	16,000

Figure 9

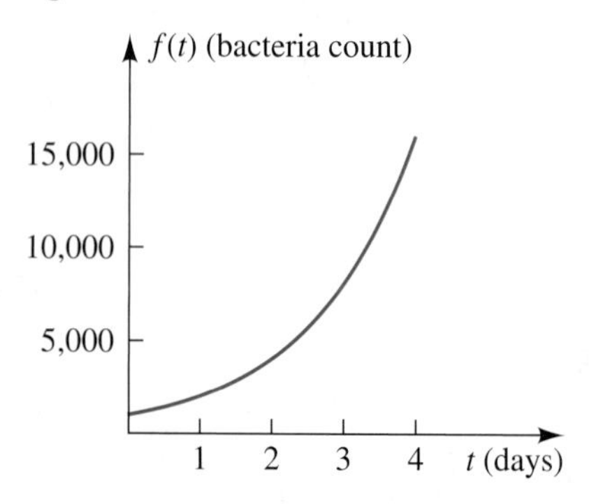

It appears that $f(t) = (1000)2^t$. With this formula we can predict the number of bacteria present at any time t. For example, at $t = 1.5 = \frac{3}{2}$,

$$f(t) = (1000)2^{3/2} \approx 2828.$$

The graph of f is sketched in Figure 9.

APPLICATION **Radioactive Decay**

Certain physical quantities *decrease* exponentially. In such cases, if a is the base of the exponential function, then $0 < a < 1$. One of the most common examples of exponential decrease is the decay of a radioactive substance, or isotope. The **half-life** of an isotope is the time it takes for one-half the original amount in a given sample to decay. The half-life is the principal characteristic used to distinguish one radioactive substance from another. The polonium isotope ^{210}Po has a half-life of approximately 140 days; that is, given any amount, one-half of it will disintegrate in 140 days. If 20 milligrams of ^{210}Po is present initially, then the following table indicates the amount remaining after various intervals of time.

Figure 10

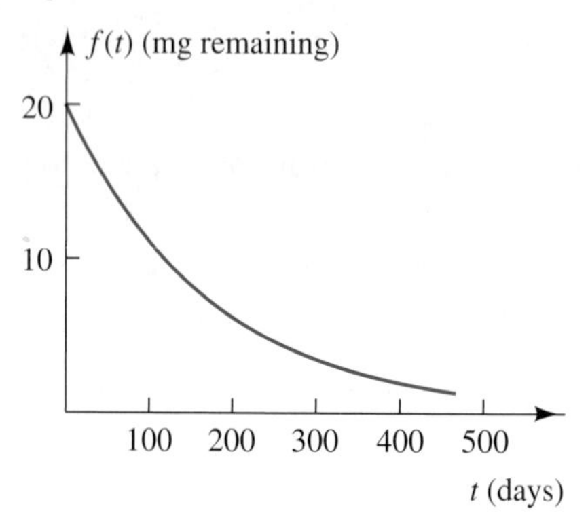

t (time in days)	0	140	280	420	560
$f(t)$ (mg remaining)	20	10	5	2.5	1.25

The sketch in Figure 10 illustrates the exponential nature of the disintegration.

Other radioactive substances have much longer half-lives. In particular, a by-product of nuclear reactors is the radioactive plutonium isotope ^{239}Pu, which has a half-life of approximately 24,000 years. It is for this reason that the disposal of radioactive waste is a major problem in modern society.

APPLICATION **Compound Interest**

Compound interest provides a good illustration of exponential growth. If a sum of money P, the *principal,* is invested at a *simple* interest rate r, then the interest at the end of one interest period is the product Pr when r is expressed as a decimal. For example, if $P = \$1000$ and the interest rate is 9% per year, then $r = 0.09$, and the interest at the end of one year is \$1000(0.09), or \$90.

If the interest is reinvested with the principal at the end of the interest period, then the new principal is

$$P + Pr \quad \text{or, equivalently,} \quad P(1 + r).$$

Note that to find the new principal we may multiply the original principal by $(1 + r)$. In the preceding example, the new principal is \$1000(1.09), or \$1090.

After another interest period has elapsed, the new principal may be found by multiplying $P(1 + r)$ by $(1 + r)$. Thus, the principal after two interest periods is $P(1 + r)^2$. If we continue to reinvest, the principal after three periods is $P(1 + r)^3$; after four it is $P(1 + r)^4$; and, in general, the amount A accumulated after k interest periods is

$$A = P(1 + r)^k.$$

Interest accumulated by means of this formula is **compound interest.** Note that A is expressed in terms of an exponential function with base $1 + r$. The interest period may be measured in years, months, weeks, days, or any other suitable unit of time. When applying the formula for A, remember that *r is the interest rate per interest period expressed as a decimal.* For example, if the rate is stated as 6% *per year compounded monthly,* then the rate per month is $\frac{6}{12}$% or, equivalently, 0.5%. Thus, $r = 0.005$ and k is the number of months. If \$100 is invested at this rate, then the formula for A is

$$A = 100(1 + 0.005)^k = 100(1.005)^k.$$

In general, we have the following formula.

Compound Interest Formula

$$A = P\left(1 + \frac{r}{n}\right)^{nt},$$

where P = principal
r = annual interest rate expressed as a decimal
n = number of interest periods per year
t = number of years P is invested
A = amount after t years.

The next example illustrates a special case of the compound interest formula.

EXAMPLE 7 Using the compound interest formula

Suppose that \$1000 is invested at an interest rate of 9% compounded monthly. Find the new amount of principal after 5 years, after 10 years, and after 15 years. Illustrate graphically the growth of the investment.

▶ **SOLUTION** Applying the compound interest formula with $r = 9\% = 0.09$, $n = 12$, and $P = \$1000$, we find that the amount after t years is

$$A = 1000\left(1 + \frac{0.09}{12}\right)^{12t} = 1000(1.0075)^{12t}.$$

Substituting $t = 5$, 10, and 15 and using a calculator, we obtain the following table.

Number of years	Amount
5	$A = \$1000(1.0075)^{60} = \$1565.68$
10	$A = \$1000(1.0075)^{120} = \$2451.36$
15	$A = \$1000(1.0075)^{180} = \$3838.04$

Note that when working with monetary values, we use = instead of ≈ and round to two decimal places.

(continued)

▶ **Tutorial available at www.thomsonedu.com/login**

Figure 11
Compound interest: $A = 1000(1.0075)^{12t}$

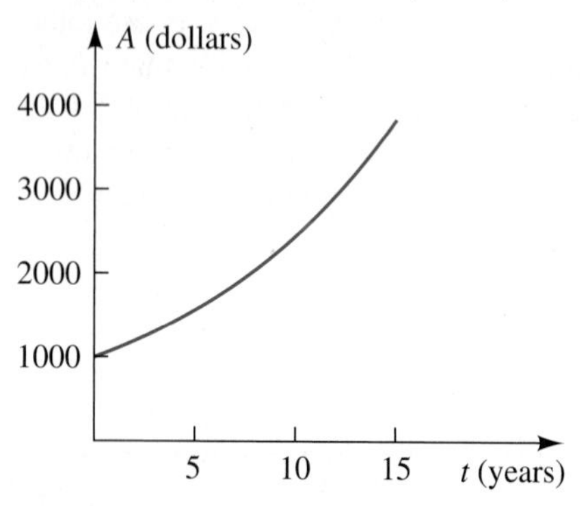

The exponential nature of the increase is indicated by the fact that during the first five years, the growth in the investment is \$565.68; during the second five-year period, the growth is \$885.68; and during the last five-year period, it is \$1386.68.

The sketch in Figure 11 illustrates the growth of \$1000 invested over a period of 15 years.

EXAMPLE 8 Finding an exponential model

In 1938, a federal law establishing a minimum wage was enacted, and the wage was set at \$0.25 per hour; the wage had risen to \$5.15 per hour by 1997. Find a simple exponential function of the form $y = ab^t$ that models the federal minimum wage for 1938–1997.

SOLUTION

$y = ab^t$	given
$0.25 = ab^0$	let $t = 0$ for 1938
$0.25 = a$	$b^0 = 1$
$y = 0.25b^t$	replace a with 0.25
$5.15 = 0.25b^{59}$	$t = 1997 - 1938 = 59$
$b^{59} = \dfrac{5.15}{0.25} = 20.6$	divide by 0.25
$b = \sqrt[59]{20.6}$	take 59th root
$b \approx 1.0526$	approximate

We obtain the model $y = 0.25(1.0526)^t$, which indicates that the federal minimum wage rose about 5.26% per year from 1938 to 1997. A graph of the model is shown in Figure 12. Do you think this model will hold true through the year 2010?

Figure 12

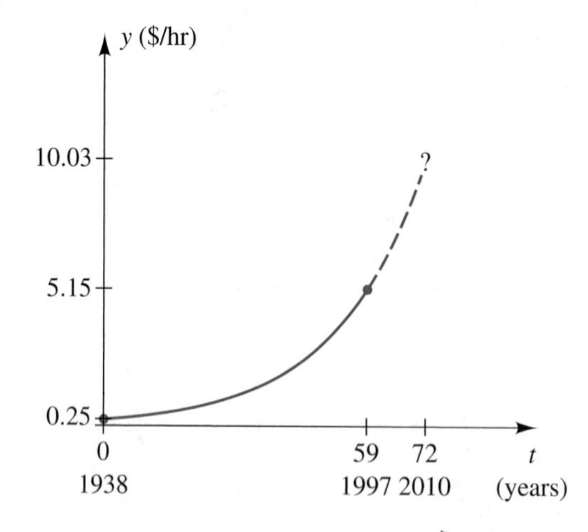

Tutorial available at www.thomsonedu.com/login

We conclude this section with an example involving a graphing utility.

EXAMPLE 9 **Estimating amounts of a drug in the bloodstream**

If an adult takes a 100-milligram tablet of a certain prescription drug orally, the rate R at which the drug enters the bloodstream t minutes later is predicted to be

$$R = 5(0.95)^t \text{ mg/min}.$$

It can be shown using calculus that the amount A of the drug in the bloodstream at time t can be approximated by

$$A = 97.4786[1 - (0.95)^t] \text{ mg}.$$

(a) Estimate how long it takes for 50 milligrams of the drug to enter the bloodstream.

(b) Estimate the number of milligrams of the drug in the bloodstream when the drug is entering at a rate of 3 mg/min.

Figure 13
[0, 100, 10] by [0, 100, 10]

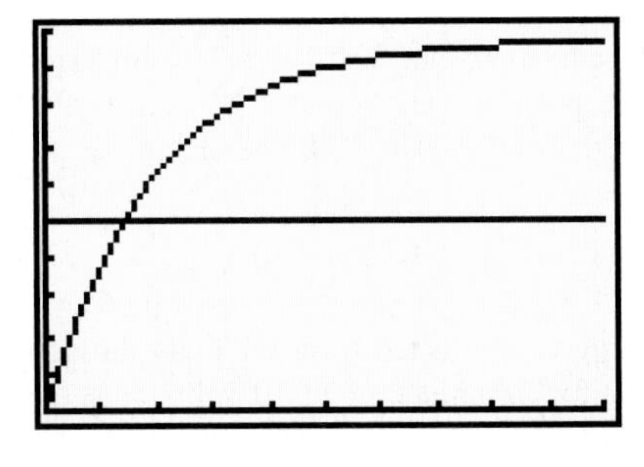

Figure 14
[0, 15] by [0, 5]

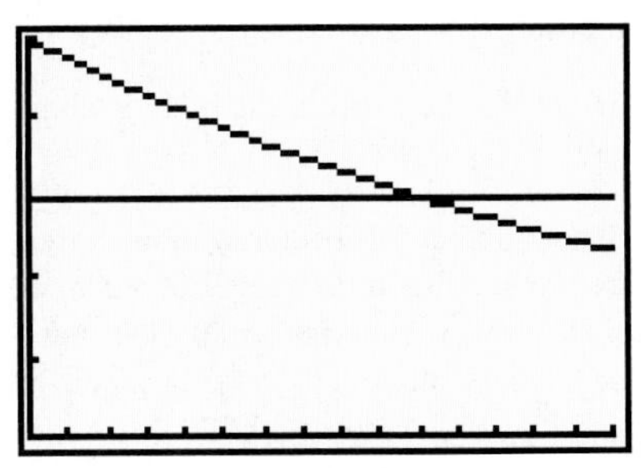

SOLUTION

(a) We wish to determine t when A is equal to 50. Since the value of A cannot exceed 97.4786, we choose the viewing rectangle to be [0, 100, 10] by [0, 100, 10].

We next assign $97.4786[1 - (0.95)^x]$ to Y_1, assign 50 to Y_2, and graph Y_1 and Y_2, obtaining a display similar to that in Figure 13 (note that $x = t$). Using the intersect feature, we estimate that $A = 50$ mg when $x \approx 14$ min.

(b) We wish to determine t when R is equal to 3. Let us first assign $5(0.95)^x$ to Y_3 and 3 to Y_4. Since the maximum value of Y_3 is 5 (at $t = 0$), we use a viewing rectangle of dimensions [0, 15] by [0, 5] and obtain a display similar to that in Figure 14. Using the intersect feature again, we find that $y = 3$ when $x \approx 9.96$. Thus, after almost 10 minutes, the drug will be entering the bloodstream at a rate of 3 mg/min. (Note that the initial rate, at $t = 0$, is 5 mg/min.) Finding the value of Y_1 at $x = 10$, we see that there is almost 39 milligrams of the drug in the bloodstream after 10 minutes.

4.2 *Exercises*

Exer. 1–10: Solve the equation.

1 $7^{x+6} = 7^{3x-4}$

2 $6^{7-x} = 6^{2x+1}$

3 $3^{2x+3} = 3^{(x^2)}$

4 $9^{(x^2)} = 3^{3x+2}$

▶ 5 $2^{-100x} = (0.5)^{x-4}$

6 $\left(\frac{1}{2}\right)^{6-x} = 2$

7 $4^{x-3} = 8^{4-x}$

8 $27^{x-1} = 9^{2x-3}$

9 $4^x \cdot \left(\frac{1}{2}\right)^{3-2x} = 8 \cdot (2^x)^2$

▶ 10 $9^{2x} \cdot \left(\frac{1}{3}\right)^{x+2} = 27 \cdot (3^x)^{-2}$

▶ **Tutorial available at www.thomsonedu.com/login**

11 Sketch the graph of f if $a = 2$.

(a) $f(x) = a^x$ (b) $f(x) = -a^x$

(c) $f(x) = 3a^x$ (d) $f(x) = a^{x+3}$

(e) $f(x) = a^x + 3$ (f) $f(x) = a^{x-3}$

(g) $f(x) = a^x - 3$ (h) $f(x) = a^{-x}$

(i) $f(x) = \left(\frac{1}{a}\right)^x$ (j) $f(x) = a^{3-x}$

12 Work Exercise 11 if $a = \frac{1}{2}$.

Exer. 13–24: Sketch the graph of f.

13 $f(x) = \left(\frac{2}{5}\right)^{-x}$ 14 $f(x) = \left(\frac{2}{5}\right)^x$

15 $f(x) = 5\left(\frac{1}{2}\right)^x + 3$ 16 $f(x) = 8(4)^{-x} - 2$

17 $f(x) = -\left(\frac{1}{2}\right)^x + 4$ 18 $f(x) = -3^x + 9$

19 $f(x) = 2^{|x|}$ 20 $f(x) = 2^{-|x|}$

21 $f(x) = 3^{1-x^2}$ 22 $f(x) = 2^{-(x+1)^2}$

23 $f(x) = 3^x + 3^{-x}$ 24 $f(x) = 3^x - 3^{-x}$

Exer. 25–28: Find an exponential function of the form $f(x) = ba^x$ or $f(x) = ba^x + c$ that has the given graph.

25

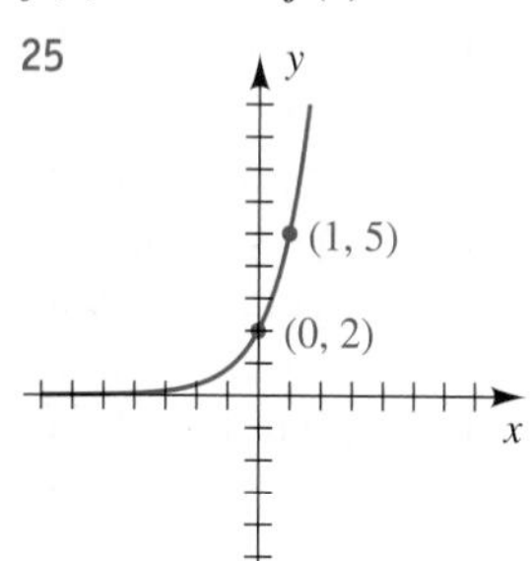

26

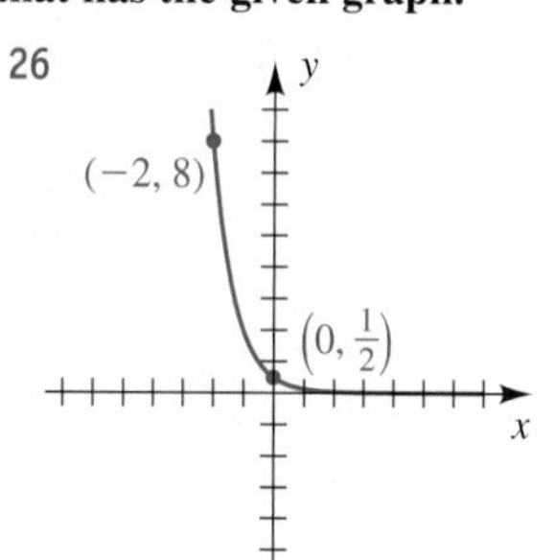

27

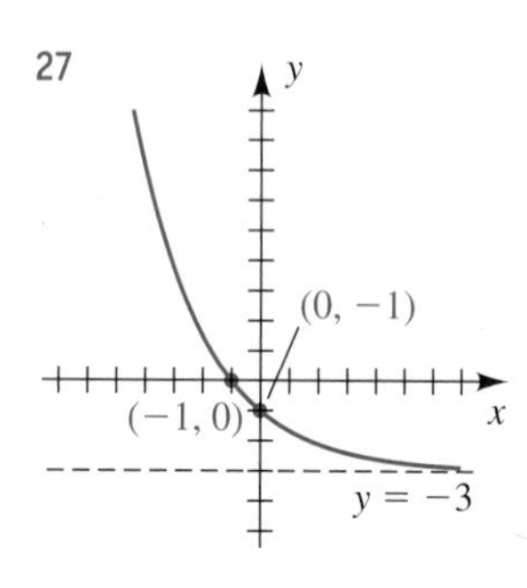

28

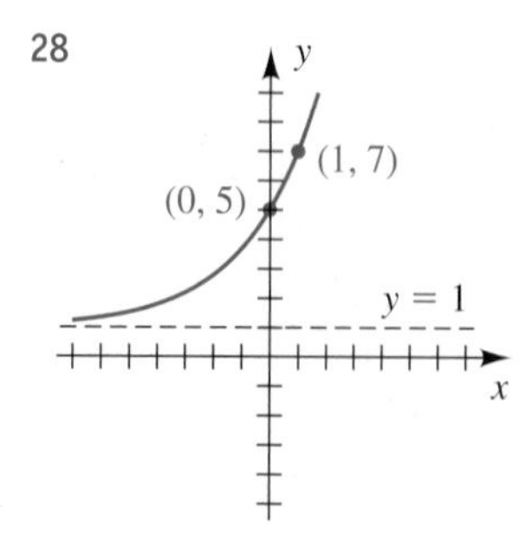

Exer. 29–30: Find an exponential function of the form $f(x) = ba^x$ that has the given y-intercept and passes through the point P.

29 y-intercept 8; $P(3, 1)$

30 y-intercept 6; $P\left(2, \frac{3}{32}\right)$

Exer. 31–32: Find an exponential function of the form $f(x) = ba^{-x} + c$ that has the given horizontal asymptote and y-intercept and passes through point P.

31 $y = 32$; y-intercept 212; $P(2, 112)$

32 $y = 72$; y-intercept 425; $P(1, 248.5)$

33 **Elk population** One hundred elk, each 1 year old, are introduced into a game preserve. The number $N(t)$ alive after t years is predicted to be $N(t) = 100(0.9)^t$. Estimate the number alive after

(a) 1 year (b) 5 years (c) 10 years

34 **Drug dosage** A drug is eliminated from the body through urine. Suppose that for an initial dose of 10 milligrams, the amount $A(t)$ in the body t hours later is given by $A(t) = 10(0.8)^t$.

(a) Estimate the amount of the drug in the body 8 hours after the initial dose.

(b) What percentage of the drug still in the body is eliminated each hour?

▶ 35 **Bacterial growth** The number of bacteria in a certain culture increased from 600 to 1800 between 7:00 A.M. and 9:00 A.M. Assuming growth is exponential, the number $f(t)$ of bacteria t hours after 7:00 A.M. is given by $f(t) = 600(3)^{t/2}$.

(a) Estimate the number of bacteria in the culture at 8:00 A.M., 10:00 A.M., and 11:00 A.M.

(b) Sketch the graph of f for $0 \le t \le 4$.

36 **Newton's law of cooling** According to Newton's law of cooling, the rate at which an object cools is directly proportional to the difference in temperature between the object and the surrounding medium. The face of a household iron cools from 125° to 100° in 30 minutes in a room that remains at a constant temperature of 75°. From calculus, the temperature $f(t)$ of the face after t hours of cooling is given by $f(t) = 50(2)^{-2t} + 75$.

▶ **Tutorial available at www.thomsonedu.com/login**

(a) Assuming $t = 0$ corresponds to 1:00 P.M., approximate to the nearest tenth of a degree the temperature of the face at 2:00 P.M., 3:30 P.M., and 4:00 P.M.

(b) Sketch the graph of f for $0 \leq t \leq 4$.

37 **Radioactive decay** The radioactive bismuth isotope ^{210}Bi has a half-life of 5 days. If there is 100 milligrams of ^{210}Bi present at $t = 0$, then the amount $f(t)$ remaining after t days is given by $f(t) = 100(2)^{-t/5}$.

(a) How much ^{210}Bi remains after 5 days? 10 days? 12.5 days?

(b) Sketch the graph of f for $0 \leq t \leq 30$.

38 **Light penetration in an ocean** An important problem in oceanography is to determine the amount of light that can penetrate to various ocean depths. The Beer-Lambert law asserts that the exponential function given by $I(x) = I_0c^x$ is a model for this phenomenon (see the figure). For a certain location, $I(x) = 10(0.4)^x$ is the amount of light (in calories/cm^2/sec) reaching a depth of x meters.

(a) Find the amount of light at a depth of 2 meters.

(b) Sketch the graph of I for $0 \leq x \leq 5$.

Exercise 38

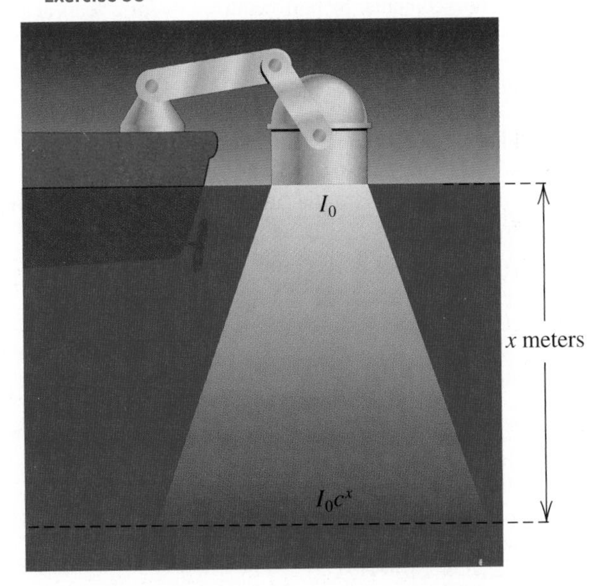

39 **Decay of radium** The half-life of radium is 1600 years. If the initial amount is q_0 milligrams, then the quantity $q(t)$ remaining after t years is given by $q(t) = q_0 2^{kt}$. Find k.

40 **Dissolving salt in water** If 10 grams of salt is added to a quantity of water, then the amount $q(t)$ that is undissolved after t minutes is given by $q(t) = 10\left(\frac{4}{5}\right)^t$. Sketch a graph that shows the value $q(t)$ at any time from $t = 0$ to $t = 10$.

41 **Compound interest** If \$1000 is invested at a rate of 7% per year compounded monthly, find the principal after

(a) 1 month (b) 6 months

(c) 1 year (d) 20 years

42 **Compound interest** If a savings fund pays interest at a rate of 6% per year compounded semiannually, how much money invested now will amount to \$5000 after 1 year?

43 **Automobile trade-in value** If a certain make of automobile is purchased for C dollars, its trade-in value $V(t)$ at the end of t years is given by $V(t) = 0.78C(0.85)^{t-1}$. If the original cost is \$25,000, calculate, to the nearest dollar, the value after

(a) 1 year (b) 4 years (c) 7 years

44 **Real estate appreciation** If the value of real estate increases at a rate of 5% per year, after t years the value V of a house purchased for P dollars is $V = P(1.05)^t$. A graph for the value of a house purchased for \$80,000 in 1986 is shown in the figure. Approximate the value of the house, to the nearest \$1000, in the year 2010.

Exercise 44

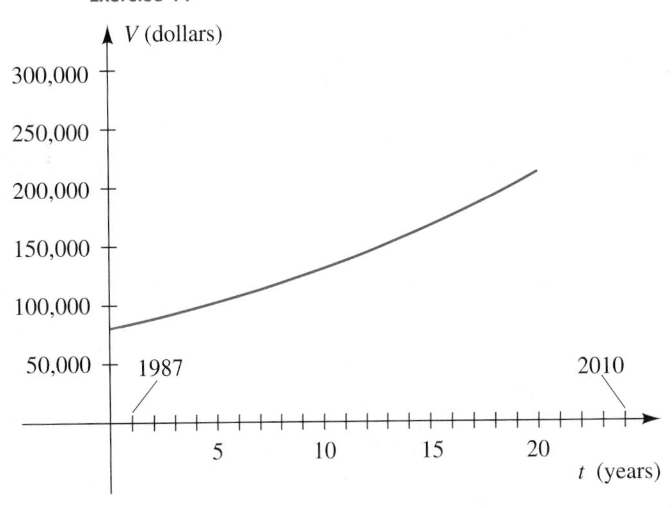

45 **Manhattan Island** The Island of Manhattan was sold for \$24 in 1626. How much would this amount have grown to by 2006 if it had been invested at 6% per year compounded quarterly?

46 **Credit-card interest** A certain department store requires its credit-card customers to pay interest on unpaid bills at the rate of 18% per year compounded monthly. If a customer buys a television set for \$500 on credit and makes no payments for one year, how much is owed at the end of the year?

47 **Depreciation** The declining balance method is an accounting method in which the amount of depreciation taken each year is a fixed percentage of the present value of the item. If y is the value of the item in a given year, the depreciation taken is ay for some depreciation rate a with $0 < a < 1$, and the new value is $(1 - a)y$.

(a) If the initial value of the item is y_0, show that the value after n years of depreciation is $(1 - a)^n y_0$.

(b) At the end of T years, the item has a salvage value of s dollars. The taxpayer wishes to choose a depreciation rate such that the value of the item after T years will equal the salvage value (see the figure). Show that $a = 1 - \sqrt[T]{s/y_0}$.

Exercise 47

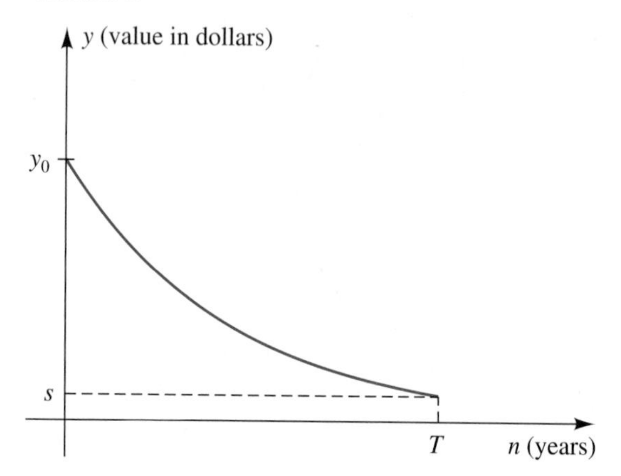

48 **Language dating** Glottochronology is a method of dating a language at a particular stage, based on the theory that over a long period of time linguistic changes take place at a fairly constant rate. Suppose that a language originally had N_0 basic words and that at time t, measured in millennia (1 millennium = 1000 years), the number $N(t)$ of basic words that remain in common use is given by $N(t) = N_0(0.805)^t$.

(a) Approximate the percentage of basic words lost every 100 years.

(b) If $N_0 = 200$, sketch the graph of N for $0 \le t \le 5$.

Exer. 49–52: Some lending institutions calculate the monthly payment M on a loan of L dollars at an interest rate r (expressed as a decimal) by using the formula

$$M = \frac{Lrk}{12(k - 1)},$$

where $k = [1 + (r/12)]^{12t}$ and t is the number of years that the loan is in effect.

49 **Home mortgage**

(a) Find the monthly payment on a 30-year \$250,000 home mortgage if the interest rate is 8%.

(b) Find the total interest paid on the loan in part (a).

50 **Home mortgage** Find the largest 25-year home mortgage that can be obtained at an interest rate of 7% if the monthly payment is to be \$1500.

51 **Car loan** An automobile dealer offers customers no-down-payment 3-year loans at an interest rate of 10%. If a customer can afford to pay \$500 per month, find the price of the most expensive car that can be purchased.

52 **Business loan** The owner of a small business decides to finance a new computer by borrowing \$3000 for 2 years at an interest rate of 7.5%.

(a) Find the monthly payment.

(b) Find the total interest paid on the loan.

Exer. 53–54: Approximate the function at the value of x to four decimal places.

53 (a) $f(x) = 13^{\sqrt{x+1.1}}$, $x = 3$

(b) $g(x) = \left(\frac{5}{42}\right)^{-x}$, $x = 1.43$

(c) $h(x) = (2^x + 2^{-x})^{2x}$, $x = 1.06$

54 (a) $f(x) = 2^{\sqrt[3]{1-x}}$, $x = 2.5$

(b) $g(x) = \left(\frac{2}{25} + x\right)^{-3x}$, $x = 2.1$

(c) $h(x) = \dfrac{3^{-x} + 5}{3^x - 16}$, $x = \sqrt{2}$

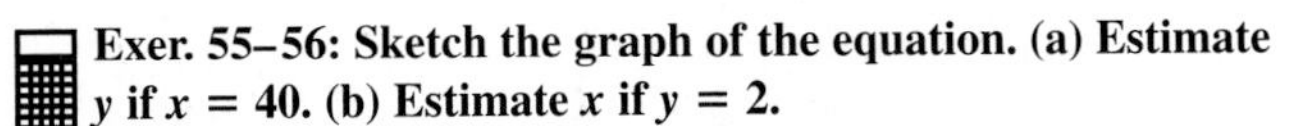

Exer. 55–56: Sketch the graph of the equation. (a) Estimate y if $x = 40$. (b) Estimate x if $y = 2$.

55 $y = (1.085)^x$

56 $y = (1.0525)^x$

Exer. 57–58: Use a graph to estimate the roots of the equation.

57 $1.4x^2 - 2.2^x = 1$

58 $1.21^{3x} + 1.4^{-1.1x} - 2x = 0.5$

Exer. 59–60: Graph f on the given interval. (a) Determine whether f is one-to-one. (b) Estimate the zeros of f.

59 $f(x) = \dfrac{3.1^x - 2.5^{-x}}{2.7^x + 4.5^{-x}}; \quad [-3, 3]$

60 $f(x) = \pi^{0.6x} - 1.3^{(x^{1.8})}; \quad [-4, 4]$

(*Hint:* Change $x^{1.8}$ to an equivalent form that is defined for $x < 0$.)

Exer. 61–62: Graph f on the given interval. (a) Estimate where f is increasing or is decreasing. (b) Estimate the range of f.

61 $f(x) = 0.7x^3 + 1.7^{(-1.8x)}, \quad [-4, 1]$

62 $f(x) = \dfrac{3.1^{-x} - 4.1^x}{4.4^{-x} + 5.3^x}; \quad [-3, 3]$

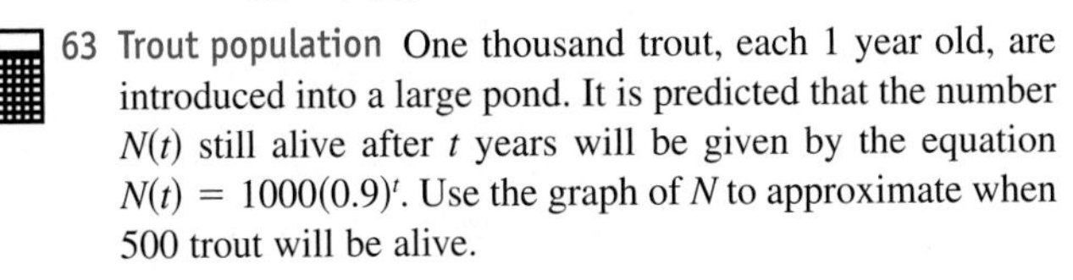

63 Trout population One thousand trout, each 1 year old, are introduced into a large pond. It is predicted that the number $N(t)$ still alive after t years will be given by the equation $N(t) = 1000(0.9)^t$. Use the graph of N to approximate when 500 trout will be alive.

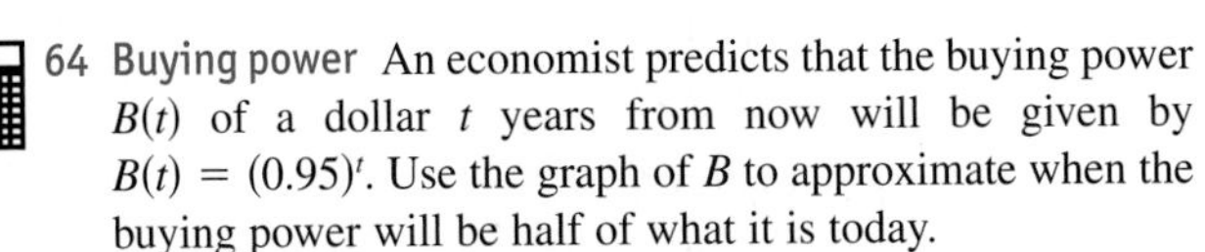

64 Buying power An economist predicts that the buying power $B(t)$ of a dollar t years from now will be given by $B(t) = (0.95)^t$. Use the graph of B to approximate when the buying power will be half of what it is today.

65 Gompertz function The **Gompertz function,**

$$y = ka^{(b^x)} \quad \text{with } k > 0,\ 0 < a < 1, \text{ and } 0 < b < 1,$$

is sometimes used to describe the sales of a new product whose sales are initially large but then level off toward a maximum saturation level. Graph, on the same coordinate plane, the line $y = k$ and the Gompertz function with $k = 4$, $a = \frac{1}{8}$, and $b = \frac{1}{4}$. What is the significance of the constant k?

66 Logistic function The **logistic function,**

$$y = \frac{1}{k + ab^x} \quad \text{with } k > 0,\ a > 0, \text{ and } 0 < b < 1,$$

is sometimes used to describe the sales of a new product that experiences slower sales initially, followed by growth toward a maximum saturation level. Graph, on the same coordinate plane, the line $y = 1/k$ and the logistic function with $k = \frac{1}{4}$, $a = \frac{1}{8}$, and $b = \frac{5}{8}$. What is the significance of the value $1/k$?

Exer. 67–68: If monthly payments p are deposited in a savings account paying an annual interest rate r, then the amount A in the account after n years is given by

$$A = \frac{p\left(1 + \frac{r}{12}\right)\left[\left(1 + \frac{r}{12}\right)^{12n} - 1\right]}{\frac{r}{12}}.$$

Graph A for each value of p and r, and estimate n for $A = \$100{,}000$.

67 $p = 100, \quad r = 0.05$

68 $p = 250, \quad r = 0.09$

69 Government receipts Federal government receipts (in billions of dollars) for selected years are listed in the table.

Year	1910	1930	1950	1970
Receipts	0.7	4.1	39.4	192.8

Year	1980	1990	2000
Receipts	517.1	1032.0	2025.2

(a) Let $x = 0$ correspond to the year 1910. Plot the data, together with the functions f and g:

(1) $f(x) = 0.786(1.094)^x$

(2) $g(x) = 0.503x^2 - 27.3x + 149.2$

(b) Determine whether the exponential or quadratic function better models the data.

(c) Use your choice in part (b) to graphically estimate the year in which the federal government first collected \$1 trillion.

70 Epidemics In 1840, Britain experienced a bovine (cattle and oxen) epidemic called epizooty. The estimated number of new cases every 28 days is listed in the table. At the time, the *London Daily* made a dire prediction that the number of new cases would continue to increase indefinitely. William Farr correctly predicted when the number of new cases would peak. Of the two functions

$$f(t) = 653(1.028)^t$$

and

$$g(t) = 54{,}700e^{-(t-200)^2/7500}$$

one models the newspaper's prediction and the other models Farr's prediction, where t is in days with $t = 0$ corresponding to August 12, 1840.

Date	New cases
Aug. 12	506
Sept. 9	1289
Oct. 7	3487
Nov. 4	9597
Dec. 2	18,817
Dec. 30	33,835
Jan. 27	47,191

(a) Graph each function, together with the data, in the viewing rectangle [0, 400, 100] by [0, 60,000, 10,000].

(b) Determine which function better models Farr's prediction.

(c) Determine the date on which the number of new cases peaked.

71 **Cost of a stamp** The price of a first-class stamp was 3¢ in 1958 and 39¢ in 2006 (it was 2¢ in 1885). Find a simple exponential function of the form $y = ab^t$ that models the cost of a first-class stamp for 1958–2006, and predict its value for 2010.

72 **Consumer Price Index** The CPI is the most widely used measure of inflation. In 1970, the CPI was 37.8, and in 2000, the CPI was 168.8. This means that an urban consumer who paid $37.80 for a market basket of consumer goods and services in 1970 would have needed $168.80 for similar goods and services in 2000. Find a simple exponential function of the form $y = ab^t$ that models the CPI for 1970–2000, and predict its value for 2010.

73 **Inflation comparisons** In 1974, Johnny Miller won 8 tournaments on the PGA tour and accumulated $353,022 in official season earnings. In 1999, Tiger Woods accumulated $6,616,585 with a similar record.

(a) Suppose the monthly inflation rate from 1974 to 1999 was 0.0025 (3%/yr). Use the compound interest formula to estimate the equivalent value of Miller's winnings in the year 1999. Compare your answer with that from an inflation calculation on the web (e.g., bls.gov/cpi/home.htm).

(b) Find the annual interest rate needed for Miller's winnings to be equivalent in value to Woods's winnings.

(c) What type of function did you use in part (a)? part (b)?

4.3 The Natural Exponential Function

The *compound interest formula* discussed in the preceding section is

$$A = P\left(1 + \frac{r}{n}\right)^{nt},$$

where P is the principal invested, r is the annual interest rate (expressed as a decimal), n is the number of interest periods per year, and t is the number of years that the principal is invested. The next example illustrates what happens if the rate and total time invested are fixed, but the *interest period* is varied.

EXAMPLE 1 Using the compound interest formula

Suppose $1000 is invested at a compound interest rate of 9%. Find the new amount of principal after one year if the interest is compounded quarterly, monthly, weekly, daily, hourly, and each minute.

▶ **SOLUTION** If we let $P = \$1000$, $t = 1$, and $r = 0.09$ in the compound interest formula, then

$$A = \$1000\left(1 + \frac{0.09}{n}\right)^{n}$$

▶ **Tutorial available at www.thomsonedu.com/login**

for n interest periods per year. The values of n we wish to consider are listed in the following table, where we have assumed that there are 365 days in a year and hence $(365)(24) = 8760$ hours and $(8760)(60) = 525{,}600$ minutes. (In many business transactions an investment year is considered to be only 360 days.)

Interest period	Quarter	Month	Week	Day	Hour	Minute
n	4	12	52	365	8760	525,600

Using the compound interest formula (and a calculator), we obtain the amounts given in the following table.

Interest period	**Amount after one year**
Quarter	$\$1000\left(1 + \frac{0.09}{4}\right)^{4} = \$1093.08$
Month	$\$1000\left(1 + \frac{0.09}{12}\right)^{12} = \$1093.81$
Week	$\$1000\left(1 + \frac{0.09}{52}\right)^{52} = \$1094.09$
Day	$\$1000\left(1 + \frac{0.09}{365}\right)^{365} = \$1094.16$
Hour	$\$1000\left(1 + \frac{0.09}{8760}\right)^{8760} = \$1094.17$
Minute	$\$1000\left(1 + \frac{0.09}{525{,}600}\right)^{525{,}600} = \$1094.17$

Note that, in the preceding example, after we reach an interest period of one hour, the number of interest periods per year has no effect on the final amount. If interest had been compounded each *second,* the result would still be \$1094.17. (Some decimal places *beyond* the first two *do* change.) Thus, the amount approaches a fixed value as n increases. Interest is said to be **compounded continuously** if the number n of time periods per year increases without bound.

If we let $P = 1$, $r = 1$, and $t = 1$ in the compound interest formula, we obtain

$$A = \left(1 + \frac{1}{n}\right)^{n}.$$

The expression on the right-hand side of the equation is important in calculus. In Example 1 we considered a similar situation: as n increased, A approached a limiting value. The same phenomenon occurs for this formula, as illustrated by the following table.

n	Approximation to $\left(1 + \dfrac{1}{n}\right)^n$
1	2.00000000
10	2.59374246
100	2.70481383
1000	2.71692393
10,000	2.71814593
100,000	2.71826824
1,000,000	2.71828047
10,000,000	2.71828169
100,000,000	2.71828181
1,000,000,000	2.71828183

In calculus it is shown that as n increases without bound, the value of the expression $[1 + (1/n)]^n$ approaches a certain irrational number, denoted by e. The number e arises in the investigation of many physical phenomena. An approximation is $e \approx 2.71828$. Using the notation we developed for rational functions in Section 3.5, we denote this fact as follows.

The Number e

If n is a positive integer, then

$$\left(1 + \frac{1}{n}\right)^n \to e \approx 2.71828 \quad \text{as} \quad n \to \infty.$$

In the following definition we use e as a base for an important exponential function.

Definition of the Natural Exponential Function

The **natural exponential function** f is defined by

$$f(x) = e^x$$

for every real number x.

The natural exponential function is one of the most useful functions in advanced mathematics and applications. Since $2 < e < 3$, the graph of $y = e^x$

Figure 1

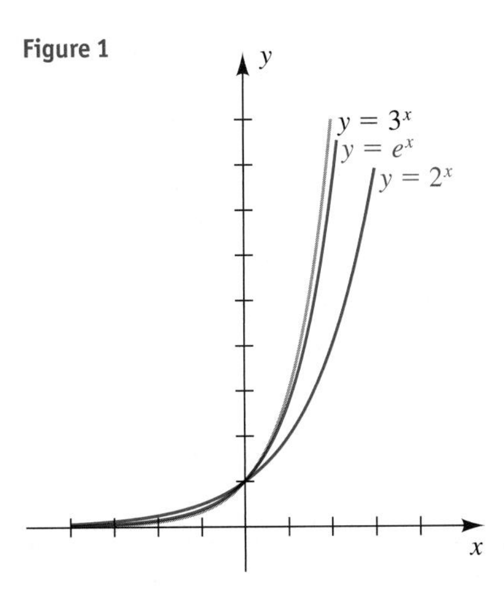

The $\boxed{e^x}$ *key can be accessed by pressing* $\boxed{\text{2nd}}$ $\boxed{\text{LN}}$.

lies between the graphs of $y = 2^x$ and $y = 3^x$, as shown in Figure 1. Scientific and graphing calculators have an $\boxed{e^x}$ key for approximating values of the natural exponential function.

APPLICATION **Continuously Compounded Interest**

The compound interest formula is

$$A = P\left(1 + \frac{r}{n}\right)^{nt}.$$

If we let $1/k = r/n$, then $k = n/r$, $n = kr$, and $nt = krt$, and we may rewrite the formula as

$$A = P\left(1 + \frac{1}{k}\right)^{krt} = P\left[\left(1 + \frac{1}{k}\right)^k\right]^{rt}.$$

For continuously compounded interest we let n (the number of interest periods per year) increase without bound, denoted by $n \to \infty$ or, equivalently, by $k \to \infty$. Using the fact that $[1 + (1/k)]^k \to e$ as $k \to \infty$, we see that

$$P\left[\left(1 + \frac{1}{k}\right)^k\right]^{rt} \to P[e]^{rt} = Pe^{rt} \quad \text{as} \quad k \to \infty.$$

This result gives us the following formula.

Continuously Compounded Interest Formula

$$A = Pe^{rt},$$

where P = principal
r = annual interest rate expressed as a decimal
t = number of years P is invested
A = amount after t years.

The next two examples illustrate the use of this formula.

EXAMPLE 2 Using the continuously compounded interest formula

Suppose \$20,000 is deposited in a money market account that pays interest at a rate of 6% per year compounded continuously. Determine the balance in the account after 5 years.

▶ SOLUTION Applying the formula for continuously compounded interest with $P = 20{,}000$, $r = 0.06$, and $t = 5$, we have

$$A = Pe^{rt} = 20{,}000e^{0.06(5)} = 20{,}000e^{0.3}.$$

Using a calculator, we find that $A = \$26{,}997.18$.

EXAMPLE 3 Using the continuously compounded interest formula

An investment of \$10,000 increased to \$28,576.51 in 15 years. If interest was compounded continuously, find the interest rate.

▶ SOLUTION We apply the formula for continuously compounded interest with $P = \$10{,}000$, $A = 28{,}576.51$, and $t = 15$:

$$A = Pe^{rt} \quad \text{formula}$$

$$28{,}576.51 = 10{,}000e^{r(15)} \quad \text{substitute for } A, P, t$$

Figure 2

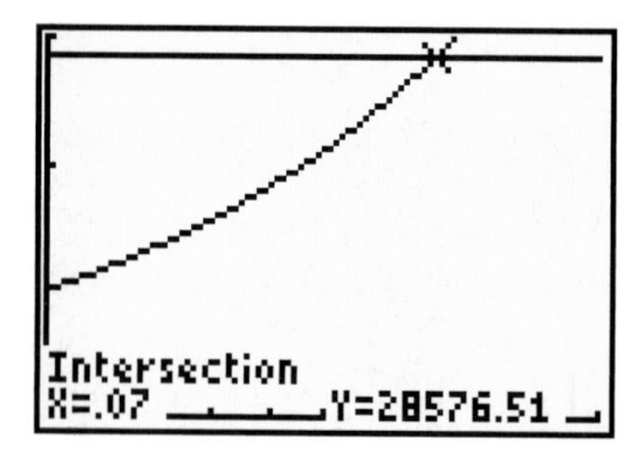

At this point, we could divide by 10,000, but that would leave us with an equation that we can't solve (yet). So we'll graph both $Y_1 = 28{,}576.51$ and $Y_2 = 10{,}000e$^$(15x)$ and find their point of intersection. As r is an interest rate, we'll start with a viewing rectangle of [0, 0.10, 0.01] by [0, 30,000, 10,000]. Using an intersect feature, we find that $Y_1 = Y_2$ for $x = 0.07$ in Figure 2. Thus, the interest rate is 7%.

The continuously compounded interest formula is just one specific case of the following law.

Law of Growth (or Decay) Formula	Let q_0 be the value of a quantity q at time $t = 0$ (that is, q_0 is the initial amount of q). If q changes instantaneously at a rate proportional to its current value, then $q = q(t) = q_0e^{rt}$, where $r > 0$ is the rate of growth (or $r < 0$ is the rate of decay) of q.

EXAMPLE 4 Predicting the population of a city

The population of a city in 1970 was 153,800. Assuming that the population increases continuously at a rate of 5% per year, predict the population of the city in the year 2010.

▶ **Tutorial available at www.thomsonedu.com/login**

▶ SOLUTION We apply the growth formula $q = q_0e^{rt}$ with initial population $q_0 = 153{,}800$, rate of growth $r = 0.05$, and time $t = 2010 - 1970 = 40$ years. Thus, a prediction for the population of the city in the year 2010 is

$$153{,}800e^{(0.05)(40)} = 153{,}800e^2 \approx 1{,}136{,}437.$$

The function f in the next example is important in advanced applications of mathematics.

EXAMPLE 5 **Sketching a graph involving two exponential functions**

Sketch the graph of f if

$$f(x) = \frac{e^x + e^{-x}}{2}.$$

▶ SOLUTION Note that f is an even function, because

$$f(-x) = \frac{e^{-x} + e^{-(-x)}}{2} = \frac{e^{-x} + e^x}{2} = f(x).$$

Thus, the graph is symmetric with respect to the y-axis. Using a calculator, we obtain the following approximations of $f(x)$.

x	0	0.5	1.0	1.5	2.0
$f(x)$ (approx.)	1	1.13	1.54	2.35	3.76

Figure 3

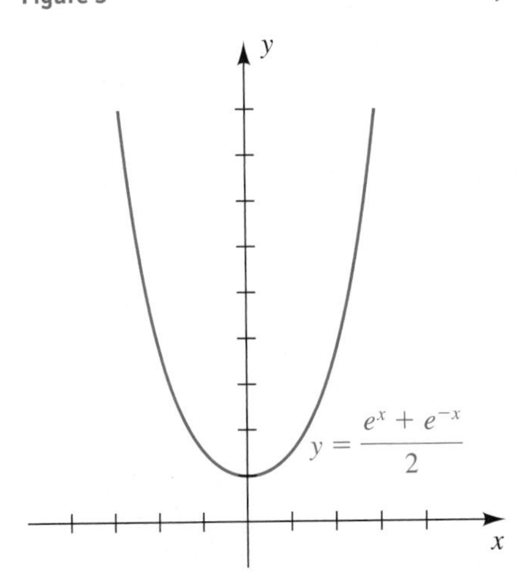

Plotting points and using symmetry with respect to the y-axis gives us the sketch in Figure 3. The graph *appears* to be a parabola; however, this is not actually the case.

APPLICATION **Flexible Cables**

The function f of Example 5 occurs in applied mathematics and engineering, where it is called the **hyperbolic cosine function.** This function can be used to describe the shape of a uniform flexible cable or chain whose ends are supported from the same height, such as a telephone or power line cable (see Figure 4). If we introduce a coordinate system, as indicated in the figure, then it can be shown that an equation that corresponds to the shape of the cable is

$$y = \frac{a}{2}(e^{x/a} + e^{-x/a}),$$

where a is a real number. The graph is called a **catenary,** after the Latin word for *chain.* The function in Example 5 is the special case in which $a = 1$. See Discussion Exercise 3 at the end of this chapter for an application involving a catenary.

Figure 4

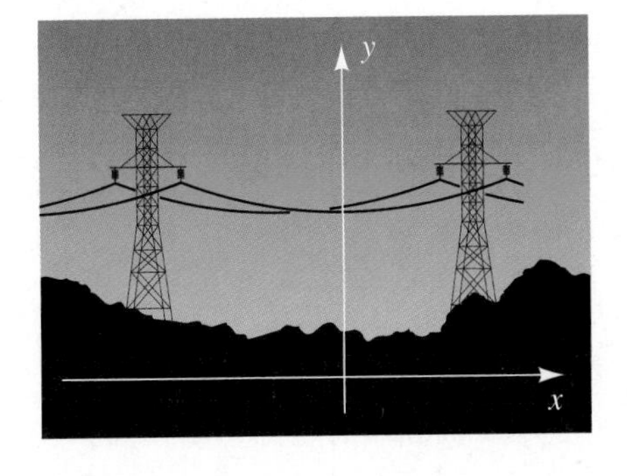

▶ Tutorial available at www.thomsonedu.com/login

APPLICATION **Radiotherapy**

Exponential functions play an important role in the field of *radiotherapy,* the treatment of tumors by radiation. The fraction of cells in a tumor that survive a treatment, called the *surviving fraction,* depends not only on the energy and nature of the radiation, but also on the depth, size, and characteristics of the tumor itself. The exposure to radiation may be thought of as a number of potentially damaging events, where at least one *hit* is required to kill a tumor cell. For instance, suppose that each cell has exactly one *target* that must be hit. If k denotes the average target size of a tumor cell and if x is the number of damaging events (the *dose*), then the surviving fraction $f(x)$ is given by

$$f(x) = e^{-kx}.$$

This is called the *one target–one hit surviving fraction.*

Suppose next that each cell has n targets and that each target must be hit once for the cell to die. In this case, the *n target–one hit surviving fraction* is given by

$$f(x) = 1 - (1 - e^{-kx})^n.$$

The graph of f may be analyzed to determine what effect increasing the dosage x will have on decreasing the surviving fraction of tumor cells. Note that $f(0) = 1$; that is, if there is no dose, then all cells survive. As an example, if $k = 1$ and $n = 2$, then

$$\begin{aligned} f(x) &= 1 - (1 - e^{-x})^2 \\ &= 1 - (1 - 2e^{-x} + e^{-2x}) \\ &= 2e^{-x} - e^{-2x}. \end{aligned}$$

Figure 5
Surviving fraction of tumor cells after a radiation treatment

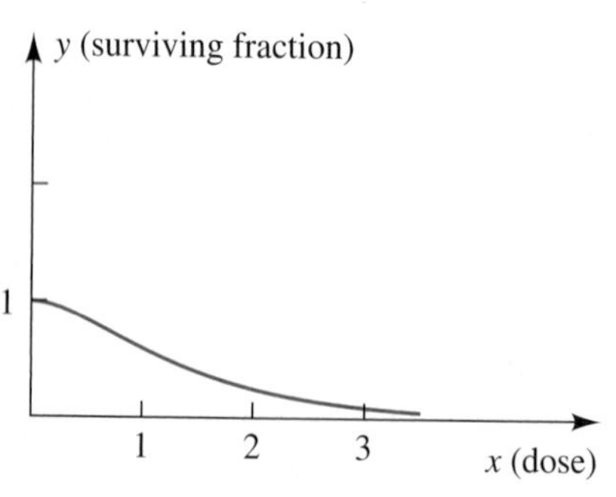

A complete analysis of the graph of f requires calculus. The graph is sketched in Figure 5. The *shoulder* on the curve near the point (0, 1) represents the threshold nature of the treatment—that is, a small dose results in very little tumor cell elimination. Note that for a large x, an increase in dosage has little effect on the surviving fraction. To determine the ideal dose to administer to a patient, specialists in radiation therapy must also take into account the number of healthy cells that are killed during a treatment.

Problems of the type illustrated in the next example occur in the study of calculus.

EXAMPLE 6 **Finding zeros of a function involving exponentials**

If $f(x) = x^2(-2e^{-2x}) + 2xe^{-2x}$, find the zeros of f.

▶ **SOLUTION** We may factor $f(x)$ as follows:

$$\begin{aligned} f(x) &= 2xe^{-2x} - 2x^2e^{-2x} && \text{given} \\ &= 2xe^{-2x}(1 - x) && \text{factor out } 2xe^{-2x} \end{aligned}$$

▶ **Tutorial available at www.thomsonedu.com/login**

To find the zeros of f, we solve the equation $f(x) = 0$. Since $e^{-2x} > 0$ for every x, we see that $f(x) = 0$ if and only if $x = 0$ or $1 - x = 0$. Thus, the zeros of f are 0 and 1.

EXAMPLE 7 **Sketching a Gompertz growth curve**

In biology, the **Gompertz growth function** G, given by

$$G(t) = ke^{(-Ae^{-Bt})}$$

where k, A, and B are positive constants, is used to estimate the size of certain quantities at time t. The graph of G is called a **Gompertz growth curve.** The function is always positive and increasing, and as t increases without bound, $G(t)$ levels off and approaches the value k. Graph G on the interval $[0, 5]$ for $k = 1.1$, $A = 3.2$, and $B = 1.1$, and estimate the time t at which $G(t) = 1$.

SOLUTION We begin by assigning

$$1.1e^{(-3.2e^{-1.1t})}$$

to Y_1. Since we wish to graph G on the interval $[0, 5]$, we choose Xmin = 0 and Xmax = 5. Because $G(t)$ is always positive and does not exceed the value $k = 1.1$, we choose Ymin = 0 and Ymax = 2. Hence, the viewing rectangle dimensions are $[0, 5]$ by $[0, 2]$. Graphing G gives us a display similar to Figure 6. The endpoint values of the graph are approximately $(0, 0.045)$ and $(5, 1.086)$.

To determine the time when $y = G(t) = 1$, we use an intersect feature, with $Y_2 = 1$, to obtain $x = t \approx 3.194$.

Figure 6
$[0, 5]$ by $[0, 2]$

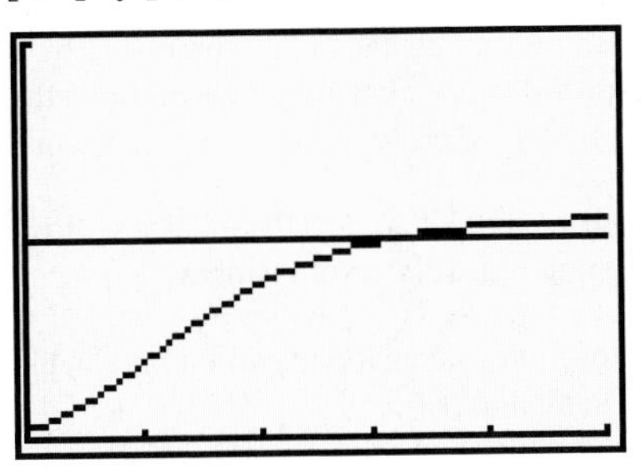

4.3 Exercises

Exer. 1–4: Use the graph of $y = e^x$ to help sketch the graph of f.

1 (a) $f(x) = e^{-x}$ (b) $f(x) = -e^x$

2 (a) $f(x) = e^{2x}$ (b) $f(x) = 2e^x$

3 (a) $f(x) = e^{x+4}$ (b) $f(x) = e^x + 4$

4 (a) $f(x) = e^{-2x}$ (b) $f(x) = -2e^x$

Exer. 5–6: If P dollars is deposited in a savings account that pays interest at a rate of r% per year compounded continuously, find the balance after t years.

5 $P = 1000$, $r = 8\frac{1}{4}$, $t = 5$

6 $P = 100$, $r = 6\frac{1}{2}$, $t = 10$

Exer. 7–8: How much money, invested at an interest rate of r% per year compounded continuously, will amount to A dollars after t years?

▶ 7 $A = 100{,}000$, $r = 6.4$, $t = 18$

8 $A = 15{,}000$, $r = 5.5$, $t = 4$

Exer. 9–10: An investment of P dollars increased to A dollars in t years. If interest was compounded continuously, find the interest rate.

9 $A = 13{,}464$, $P = 1000$, $t = 20$

10 $A = 890.20$, $P = 400$, $t = 16$

Exer. 11–12: Solve the equation.

▶ 11 $e^{(x^2)} = e^{7x-12}$

12 $e^{3x} = e^{2x-1}$

▶ **Tutorial available at www.thomsonedu.com/login**

Exer. 13–16: Find the zeros of *f*.

13 $f(x) = xe^x + e^x$

14 $f(x) = -x^2e^{-x} + 2xe^{-x}$

▶ 15 $f(x) = x^3(4e^{4x}) + 3x^2e^{4x}$

16 $f(x) = x^2(2e^{2x}) + 2xe^{2x} + e^{2x} + 2xe^{2x}$

Exer. 17–18: Simplify the expression.

17 $\dfrac{(e^x + e^{-x})(e^x + e^{-x}) - (e^x - e^{-x})(e^x - e^{-x})}{(e^x + e^{-x})^2}$

18 $\dfrac{(e^x - e^{-x})^2 - (e^x + e^{-x})^2}{(e^x + e^{-x})^2}$

19 **Crop growth** An exponential function W such that $W(t) = W_0e^{kt}$ for $k > 0$ describes the first month of growth for crops such as maize, cotton, and soybeans. The function value $W(t)$ is the total weight in milligrams, W_0 is the weight on the day of emergence, and t is the time in days. If, for a species of soybean, $k = 0.2$ and $W_0 = 68$ mg, predict the weight at the end of 30 days.

20 **Crop growth** Refer to Exercise 19. It is often difficult to measure the weight W_0 of a plant from when it first emerges from the soil. If, for a species of cotton, $k = 0.21$ and the weight after 10 days is 575 milligrams, estimate W_0.

21 **U.S. population growth** The 1980 population of the United States was approximately 231 million, and the population has been growing continuously at a rate of 1.03% per year. Predict the population $N(t)$ in the year 2020 if this growth trend continues.

22 **Population growth in India** The 1985 population estimate for India was 766 million, and the population has been growing continuously at a rate of about 1.82% per year. Assuming that this rapid growth rate continues, estimate the population $N(t)$ of India in the year 2015.

▶ 23 **Longevity of halibut** In fishery science, a cohort is the collection of fish that results from one annual reproduction. It is usually assumed that the number of fish $N(t)$ still alive after t years is given by an exponential function. For Pacific halibut, $N(t) = N_0e^{-0.2t}$, where N_0 is the initial size of the cohort. Approximate the percentage of the original number still alive after 10 years.

24 **Radioactive tracer** The radioactive tracer ^{51}Cr can be used to locate the position of the placenta in a pregnant woman. Often the tracer must be ordered from a medical laboratory. If A_0 units (microcuries) are shipped, then because of the radioactive decay, the number of units $A(t)$ present after t days is given by $A(t) = A_0e^{-0.0249t}$.

(a) If 35 units are shipped and it takes 2 days for the tracer to arrive, approximately how many units will be available for the test?

(b) If 35 units are needed for the test, approximately how many units should be shipped?

25 **Blue whale population growth** In 1980, the population of blue whales in the southern hemisphere was thought to number 4500. The population $N(t)$ has been decreasing according to the formula $N(t) = 4500e^{-0.1345t}$, where t is in years and $t = 0$ corresponds to 1980. Predict the population in the year 2015 if this trend continues.

26 **Halibut growth** The length (in centimeters) of many common commercial fish t years old can be approximated by a von Bertalanffy growth function having an equation of the form $f(t) = a(1 - be^{-kt})$, where a, b, and k are constants.

(a) For Pacific halibut, $a = 200$, $b = 0.956$, and $k = 0.18$. Estimate the length of a 10-year-old halibut.

(b) Use the graph of f to estimate the maximum attainable length of the Pacific halibut.

27 **Atmospheric pressure** Under certain conditions the atmospheric pressure p (in inches) at altitude h feet is given by $p = 29e^{-0.000034h}$. What is the pressure at an altitude of 40,000 feet?

28 **Polonium isotope decay** If we start with c milligrams of the polonium isotope ^{210}Po, the amount remaining after t days may be approximated by $A = ce^{-0.00495t}$. If the initial amount is 50 milligrams, approximate, to the nearest hundredth, the amount remaining after

(a) 30 days (b) 180 days (c) 365 days

29 **Growth of children** The Jenss model is generally regarded as the most accurate formula for predicting the height of preschool children. If y is height (in centimeters) and x is age (in years), then

$$y = 79.041 + 6.39x - e^{3.261-0.993x}$$

for $\frac{1}{4} \le x \le 6$. From calculus, the rate of growth R (in cm/year) is given by $R = 6.39 + 0.993e^{3.261-0.993x}$. Find the height and rate of growth of a typical 1-year-old child.

▶ **Tutorial available at www.thomsonedu.com/login**

30 **Particle velocity** A very small spherical particle (on the order of 5 microns in diameter) is projected into still air with an initial velocity of v_0 m/sec, but its velocity decreases because of drag forces. Its velocity t seconds later is given by $v(t) = v_0 e^{-at}$ for some $a > 0$, and the distance $s(t)$ the particle travels is given by

$$s(t) = \frac{v_0}{a}(1 - e^{-at}).$$

The stopping distance is the total distance traveled by the particle.

(a) Find a formula that approximates the stopping distance in terms of v_0 and a.

(b) Use the formula in part (a) to estimate the stopping distance if $v_0 = 10$ m/sec and $a = 8 \times 10^5$.

31 **Minimum wage** In 1971 the minimum wage in the United States was \$1.60 per hour. Assuming that the rate of inflation is 5% per year, find the equivalent minimum wage in the year 2010.

32 **Land value** In 1867 the United States purchased Alaska from Russia for \$7,200,000. There is 586,400 square miles of land in Alaska. Assuming that the value of the land increases continuously at 3% per year and that land can be purchased at an equivalent price, determine the price of 1 acre in the year 2010. (One square mile is equivalent to 640 acres.)

Exer. 33–34: The *effective yield* (or effective annual interest rate) for an investment is the simple interest rate that would yield at the end of one year the same amount as is yielded by the compounded rate that is actually applied. Approximate, to the nearest 0.01%, the effective yield corresponding to an interest rate of r% per year compounded (a) quarterly and (b) continuously.

33 $r = 7$ 34 $r = 12$

Exer. 35–36: Sketch the graph of the equation.

35 $y = e^{1000x}$ 36 $y = e^{-1000x}$

Exer. 37–38: Sketch the graph of the equation. (a) Estimate y if $x = 40$. (b) Estimate x if $y = 2$.

37 $y = e^{0.085x}$ 38 $y = e^{0.0525x}$

Exer. 39–41: (a) Graph f using a graphing utility. (b) Sketch the graph of g by taking the reciprocals of y-coordinates in (a), *without* using a graphing utility.

39 $f(x) = \dfrac{e^x - e^{-x}}{2}$; $g(x) = \dfrac{2}{e^x - e^{-x}}$

40 $f(x) = \dfrac{e^x + e^{-x}}{2}$; $g(x) = \dfrac{2}{e^x + e^{-x}}$

41 $f(x) = \dfrac{e^x - e^{-x}}{e^x + e^{-x}}$; $g(x) = \dfrac{e^x + e^{-x}}{e^x - e^{-x}}$

42 **Probability density function** In statistics, the probability density function for the normal distribution is defined by

$$f(x) = \frac{1}{\sigma\sqrt{2\pi}} e^{-z^2/2} \quad \text{with} \quad z = \frac{x - \mu}{\sigma},$$

where μ and σ are real numbers (μ is the *mean* and σ^2 is the *variance* of the distribution). Sketch the graph of f for the case $\sigma = 1$ and $\mu = 0$.

Exer. 43–44: Graph f and g on the same coordinate plane, and estimate the solutions of the equation $f(x) = g(x)$.

43 $f(x) = e^{0.5x} - e^{-0.4x}$; $g(x) = x^2 - 2$

44 $f(x) = 0.3e^x$; $g(x) = x^3 - x$

Exer. 45–46: The functions f and g can be used to approximate e^x on the interval $[0, 1]$. Graph f, g, and $y = e^x$ on the same coordinate plane, and compare the accuracy of $f(x)$ and $g(x)$ as an approximation to e^x.

45 $f(x) = x + 1$; $g(x) = 1.72x + 1$

46 $f(x) = \frac{1}{2}x^2 + x + 1$; $g(x) = 0.84x^2 + 0.878x + 1$

Exer. 47–48: Graph f, and estimate its zeros.

47 $f(x) = x^2e^x - xe^{(x^2)} + 0.1$

48 $f(x) = x^3e^x - x^2e^{2x} + 1$

Exer. 49–50: Graph f on the interval $(0, 200]$. Find an approximate equation for the horizontal asymptote.

49 $f(x) = \left(1 + \dfrac{1}{x}\right)^x$ 50 $f(x) = \left(1 + \dfrac{2}{x}\right)^x$

Exer. 51–52: Approximate the real root of the equation.

51 $e^{-x} = x$

52 $e^{3x} = 5 - 2x$

Exer. 53–54: Graph f, and determine where f is increasing or is decreasing.

53 $f(x) = xe^{x}$

54 $f(x) = x^2e^{-2x}$

55 **Pollution from a smokestack** The concentration C (in units/m^3) of pollution near a ground-level point that is downwind from a smokestack source of height h is sometimes given by

$$C = \frac{Q}{\pi vab} e^{-y^2/(2a^2)}\left[e^{-(z-h)^2/(2b^2)} + e^{-(z+h)^2/(2b^2)}\right],$$

where Q is the source strength (in units/sec), v is the average wind velocity (in m/sec), z is the height (in meters) above the downwind point, y is the distance from the downwind point in the direction that is perpendicular to the wind (the cross-wind direction), and a and b are constants that depend on the downwind distance (see the figure).

(a) How does the concentration of pollution change at the ground-level, downwind position ($y = 0$ and $z = 0$) if the height of the smokestack is increased?

(b) How does the concentration of pollution change at ground level ($z = 0$) for a smokestack of fixed height h if a person moves in the cross-wind direction, thereby increasing y?

Exercise 55

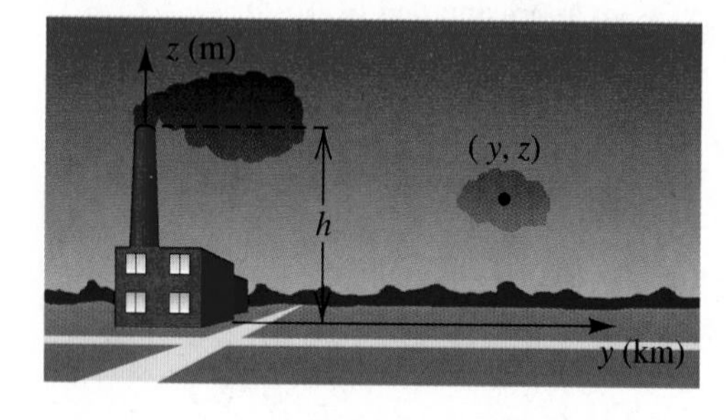

56 **Pollution concentration** Refer to Exercise 55. If the smokestack height is 100 meters and $b = 12$, use a graph to estimate the height z above the downwind point ($y = 0$) where the maximum pollution concentration occurs. (*Hint:* Let $h = 100$, $b = 12$, and graph the equation $C = e^{-(z-h)^2/(2b^2)} + e^{-(z+h)^2/(2b^2)}$.)

57 **Atmospheric density** The atmospheric density at altitude x is listed in the table.

Altitude (m)	0	2000	4000
Density (kg/m^3)	1.225	1.007	0.819

Altitude (m)	6000	8000	10,000
Density (kg/m^3)	0.660	0.526	0.414

(a) Find a function $f(x) = C_0e^{kx}$ that approximates the density at altitude x, where C_0 and k are constants. Plot the data and f on the same coordinate axes.

(b) Use f to predict the density at 3000 and 9000 meters. Compare the predictions to the actual values of 0.909 and 0.467, respectively.

58 **Government spending** Federal government expenditures (in billions of dollars) for selected years are listed in the table.

Year	1910	1930	1950	1970
Expenditures	0.7	3.3	42.6	195.6

Year	1980	1990	2000
Expenditures	590.9	1253.1	1789.1

(a) Let $x = 0$ correspond to the year 1910. Find a function $A(x) = A_0e^{kx}$ that approximates the data, where A_0 and k are constants. Plot the data and A on the same coordinate axes.

(b) Use A to predict graphically the year in which the federal government first spent \$1 trillion. (The actual year was 1987.)

4.4 Logarithmic Functions

In Section 4.2 we observed that the exponential function given by $f(x) = a^x$ for $0 < a < 1$ or $a > 1$ is one-to-one. Hence, f has an inverse function f^{-1} (see Section 4.1). This inverse of the exponential function with base a is called the **logarithmic function with base a** and is denoted by $\log_a$. Its values are written $\log_a (x)$ or $\log_a x$, read "the logarithm of x with base a." Since, by the definition of an inverse function f^{-1},

$$y = f^{-1}(x) \quad \text{if and only if} \quad x = f(y),$$

the definition of $\log_a$ may be expressed as follows.

Definition of $\log_a$

Let a be a positive real number different from 1. The **logarithm of x with base a** is defined by

$$y = \log_a x \quad \text{if and only if} \quad x = a^y$$

for every $x > 0$ and every real number y.

Note that the two equations in the definition are equivalent. We call the first equation the **logarithmic form** and the second the **exponential form.** You should strive to become an expert in changing each form into the other. The following diagram may help you achieve this goal.

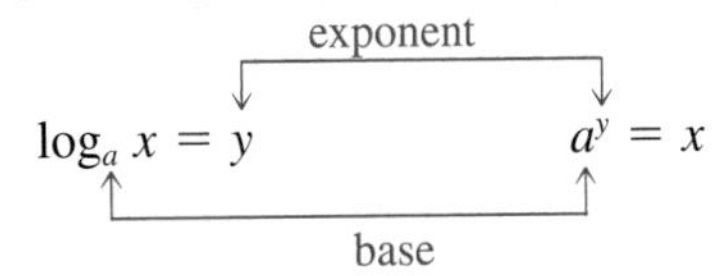

Observe that when forms are changed, *the bases of the logarithmic and exponential forms are the same.* The number y (that is, $\log_a x$) corresponds to the exponent in the exponential form. In words, $\log_a x$ is *the exponent to which the base a must be raised to obtain x.* This is what people are referring to when they say "Logarithms are exponents."

The following illustration contains examples of equivalent forms.

ILLUSTRATION **Equivalent Forms**

Logarithmic form	Exponential form
■ $\log_5 u = 2$	$5^2 = u$
■ $\log_b 8 = 3$	$b^3 = 8$
■ $r = \log_p q$	$p^r = q$
■ $w = \log_4 (2t + 3)$	$4^w = 2t + 3$
■ $\log_3 x = 5 + 2z$	$3^{5+2z} = x$

The next example contains an application that involves changing from an exponential form to a logarithmic form.

EXAMPLE 1 Changing exponential form to logarithmic form

The number N of bacteria in a certain culture after t hours is given by $N = (1000)2^t$. Express t as a logarithmic function of N with base 2.

▶ **SOLUTION**

$$N = (1000)2^t \quad \text{given}$$

$$\frac{N}{1000} = 2^t \quad \text{isolate the exponential expression}$$

$$t = \log_2 \frac{N}{1000} \quad \text{change to logarithmic form}$$

Some special cases of logarithms are given in the next example.

EXAMPLE 2 Finding logarithms

Find the number, if possible.

(a) $\log_{10} 100$ **(b)** $\log_2 \frac{1}{32}$ **(c)** $\log_9 3$ **(d)** $\log_7 1$ **(e)** $\log_3 (-2)$

SOLUTION In each case we are given $\log_a x$ and must find the exponent y such that $a^y = x$. We obtain the following.

▶ **(a)** $\log_{10} 100 = 2$ because $10^2 = 100$.

(b) $\log_2 \frac{1}{32} = -5$ because $2^{-5} = \frac{1}{32}$.

▶ **(c)** $\log_9 3 = \frac{1}{2}$ because $9^{1/2} = 3$.

(d) $\log_7 1 = 0$ because $7^0 = 1$.

(e) $\log_3 (-2)$ is not possible because $3^y \neq -2$ for any real number y.

The following general properties follow from the interpretation of $\log_a x$ as an exponent.

Property of $\log_a x$	Reason	Illustration
(1) $\log_a 1 = 0$	$a^0 = 1$	$\log_3 1 = 0$
(2) $\log_a a = 1$	$a^1 = a$	$\log_{10} 10 = 1$
(3) $\log_a a^x = x$	$a^x = a^x$	$\log_2 8 = \log_2 2^3 = 3$
(4) $a^{\log_a x} = x$	as follows	$5^{\log_5 7} = 7$

The reason for property 4 follows directly from the definition of $\log_a$, since

$$\text{if} \quad y = \log_a x, \quad \text{then} \quad x = a^y, \quad \text{or} \quad x = a^{\log_a x}.$$

The logarithmic function with base a is the inverse of the exponential function with base a, so the graph of $y = \log_a x$ can be obtained by reflecting the graph of $y = a^x$ through the line $y = x$ (see Section 4.1). This procedure is illustrated in Figure 1 for the case $a > 1$. Note that the x-intercept of the graph is 1, the domain is the set of positive real numbers, the range is $\mathbb{R}$, and the

Figure 1

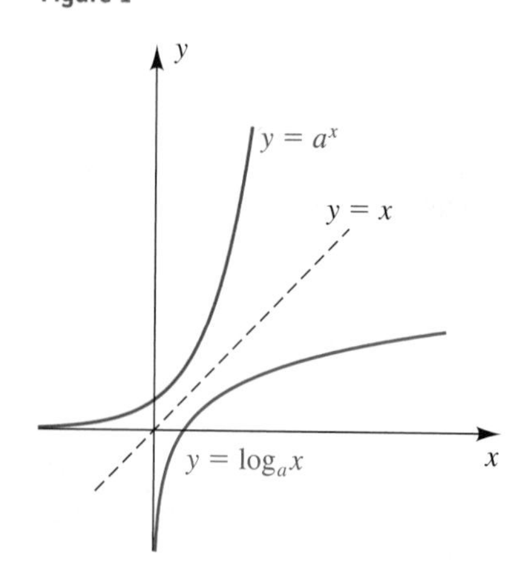

▶ Tutorial available at www.thomsonedu.com/login

y-axis is a vertical asymptote. Logarithms with base $0 < a < 1$ are seldom used, so we will not emphasize their graphs.

We see from Figure 1 that if $a > 1$, then $\log_a x$ is increasing on $(0, \infty)$ and hence is one-to-one by the theorem on page 282. Combining this result with parts (1) and (2) of the definition of one-to-one function on page 280 gives us the following theorem, which can also be proved if $0 < a < 1$.

Theorem: Logarithmic Functions Are One-to-One

The logarithmic function with base a is one-to-one. Thus, the following equivalent conditions are satisfied for positive real numbers x_1 and x_2.

(1) If $x_1 \neq x_2$, then $\log_a x_1 \neq \log_a x_2$.

(2) If $\log_a x_1 = \log_a x_2$, then $x_1 = x_2$.

When using this theorem as a reason for a step in the solution to an example, we will state that *logarithmic functions are one-to-one.*

In the following example we solve a simple *logarithmic equation*—that is, an equation involving a logarithm of an expression that contains a variable. Extraneous solutions may be introduced when logarithmic equations are solved. Hence, we must check solutions of logarithmic equations to make sure that we are taking logarithms of *only positive real numbers;* otherwise, a logarithmic function is not defined.

EXAMPLE 3 Solving a logarithmic equation

Solve the equation $\log_6 (4x - 5) = \log_6 (2x + 1)$.

▶ **SOLUTION**

$\log_6 (4x - 5) = \log_6 (2x + 1)$	given
$4x - 5 = 2x + 1$	logarithmic functions are one-to-one
$2x = 6$	subtract $2x$; add 5
$x = 3$	divide by 2

✔ **Check** $x = 3$ LS: $\log_6 (4 \cdot 3 - 5) = \log_6 7$

RS: $\log_6 (2 \cdot 3 + 1) = \log_6 7$

Since $\log_6 7 = \log_6 7$ is a true statement, $x = 3$ is a solution.

When we check the solution $x = 3$ in Example 3, it is not required that the solution be positive. But it is required that the two expressions, $4x - 5$ and $2x + 1$, be positive after we substitute 3 for x. If we extend our idea of *argument* from variables to expressions, then when checking solutions, we can simply remember that *arguments must be positive.*

In the next example we use the definition of logarithm to solve a logarithmic equation.

▶ **Tutorial available at www.thomsonedu.com/login**

EXAMPLE 4 Solving a logarithmic equation

Solve the equation $\log_4(5 + x) = 3$.

▶ SOLUTION

$$\log_4(5 + x) = 3 \quad \text{given}$$
$$5 + x = 4^3 \quad \text{change to exponential form}$$
$$x = 59 \quad \text{solve for } x$$

Check $x = 59$ LS: $\log_4(5 + 59) = \log_4 64 = \log_4 4^3 = 3$

RS: 3

Since $3 = 3$ is a true statement, $x = 59$ is a solution.

We next sketch the graph of a specific logarithmic function.

EXAMPLE 5 Sketching the graph of a logarithmic function

Sketch the graph of f if $f(x) = \log_3 x$.

▶ SOLUTION We will describe three methods for sketching the graph.

Method 1 Since the functions given by $\log_3 x$ and 3^x are inverses of each other, we proceed as we did for $y = \log_a x$ in Figure 1; that is, we first sketch the graph of $y = 3^x$ and then reflect it through the line $y = x$. This gives us the sketch in Figure 2. Note that the points $(-1, 3^{-1})$, $(0, 1)$, $(1, 3)$, and $(2, 9)$ on the graph of $y = 3^x$ reflect into the points $(3^{-1}, -1)$, $(1, 0)$, $(3, 1)$, and $(9, 2)$ on the graph of $y = \log_3 x$.

Figure 2

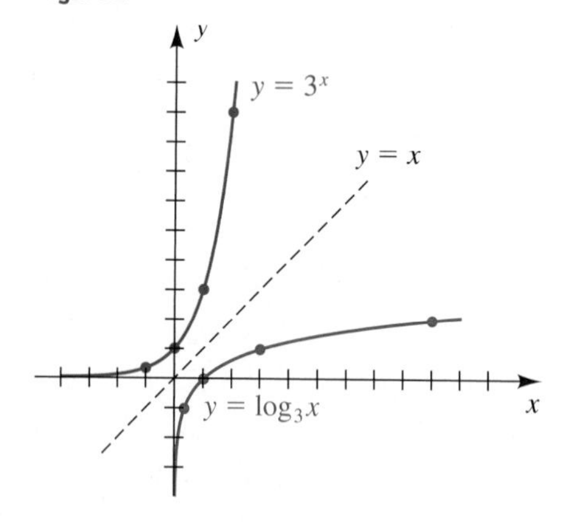

Method 2 We can find points on the graph of $y = \log_3 x$ by letting $x = 3^k$, where k is a real number, and then applying property 3 of logarithms on page 316, as follows:

$$y = \log_3 x = \log_3 3^k = k$$

▶ Tutorial available at www.thomsonedu.com/login

Using this formula, we obtain the points on the graph listed in the following table.

$x = 3^k$	3^{-3}	3^{-2}	3^{-1}	3^0	3^1	3^2	3^3
$y = \log_3 x = k$	-3	-2	-1	0	1	2	3

Figure 3

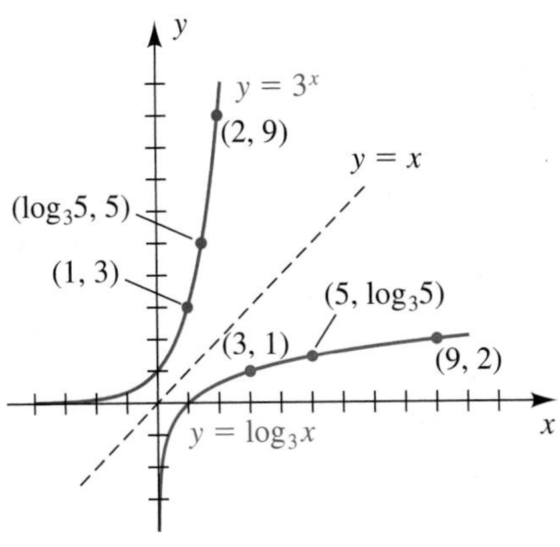

This gives us the same points obtained using the first method.

Method 3 We can sketch the graph of $y = \log_3 x$ by sketching the graph of the equivalent exponential form $x = 3^y$.

Before proceeding, let's plot one more point on $y = \log_3 x$ in Figure 2. If we let $x = 5$, then $y = \log_3 5$ (see Figure 3). (We see that $\log_3 5$ is a number between 1 and 2; we'll be able to better approximate $\log_3 5$ in Section 4.6.) Now on the graph of $y = 3^x$ we have the point $(x, y) = (\log_3 5, 5)$, so $5 = 3^{\log_3 5}$, which illustrates property 4 of logarithms on page 316 and reinforces the claim that *logarithms are exponents*.

As in the following examples, we often wish to sketch the graph of $f(x) = \log_a u$, where u is some expression involving x.

Figure 4

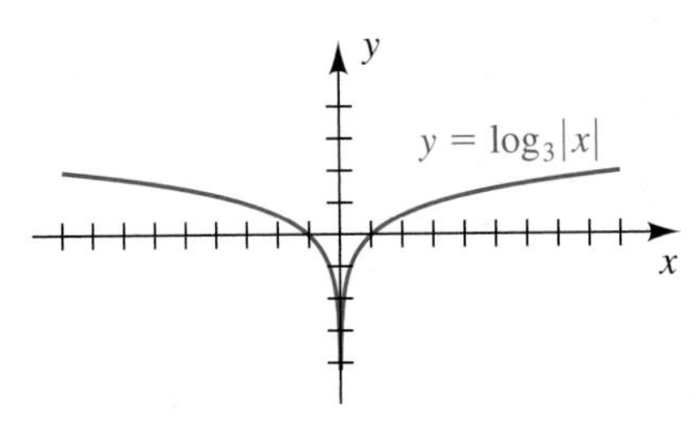

EXAMPLE 6 Sketching the graph of a logarithmic function

Sketch the graph of f if $f(x) = \log_3 |x|$ for $x \neq 0$.

SOLUTION The graph is symmetric with respect to the y-axis, since

$$f(-x) = \log_3 |-x| = \log_3 |x| = f(x).$$

If $x > 0$, then $|x| = x$ and the graph coincides with the graph of $y = \log_3 x$ sketched in Figure 2. Using symmetry, we reflect that part of the graph through the y-axis, obtaining the sketch in Figure 4.

Alternatively, we may think of this function as $g(x) = \log_3 x$ with $|x|$ substituted for x (refer to the discussion on page 168). Since all points on the graph of g have positive x-coordinates, we can obtain the graph of f by combining g with the reflection of g through the y-axis.

Figure 5

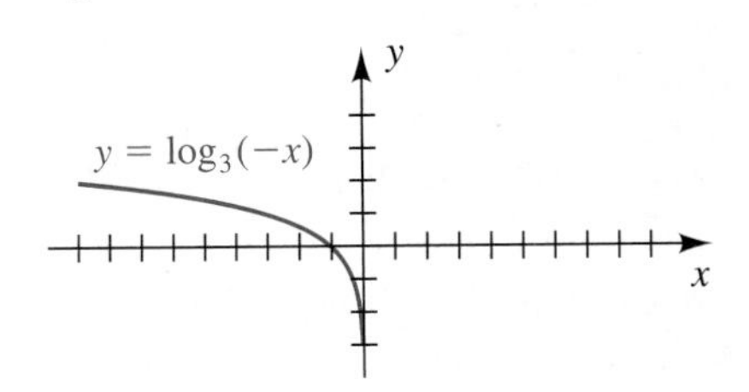

EXAMPLE 7 Reflecting the graph of a logarithmic function

Sketch the graph of f if $f(x) = \log_3 (-x)$.

SOLUTION The domain of f is the set of negative real numbers, since $\log_3 (-x)$ exists only if $-x > 0$ or, equivalently, $x < 0$. We can obtain the graph of f from the graph of $y = \log_3 x$ by replacing each point (x, y) in Figure 2 by $(-x, y)$. This is equivalent to reflecting the graph of $y = \log_3 x$ through the y-axis. The graph is sketched in Figure 5.

Another method is to change $y = \log_3 (-x)$ to the exponential form $3^y = -x$ and then sketch the graph of $x = -3^y$.

EXAMPLE 8 **Shifting graphs of logarithmic equations**

Sketch the graph of the equation:

(a) $y = \log_3 (x - 2)$ **(b)** $y = \log_3 x - 2$

SOLUTION

Figure 6

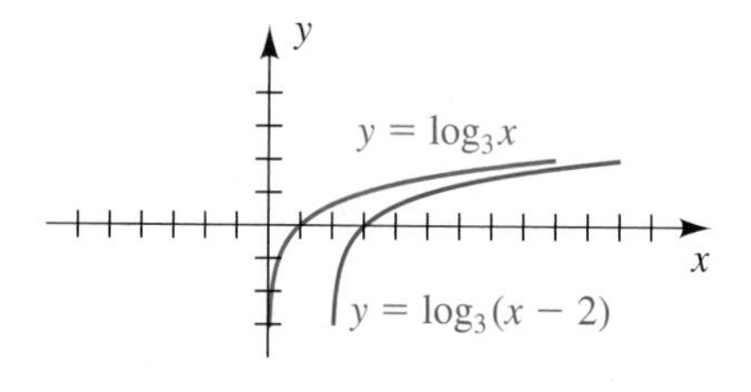

(a) The graph of $y = \log_3 x$ was sketched in Figure 2 and is resketched in Figure 6. From the discussion of horizontal shifts in Section 2.5, we can obtain the graph of $y = \log_3 (x - 2)$ by shifting the graph of $y = \log_3 x$ two units to the right, as shown in Figure 6.

(b) From the discussion of vertical shifts in Section 2.5, the graph of the equation $y = \log_3 x - 2$ can be obtained by shifting the graph of $y = \log_3 x$ two units downward, as shown in Figure 7. Note that the x-intercept is given by $\log_3 x = 2$, or $x = 3^2 = 9$.

Figure 7

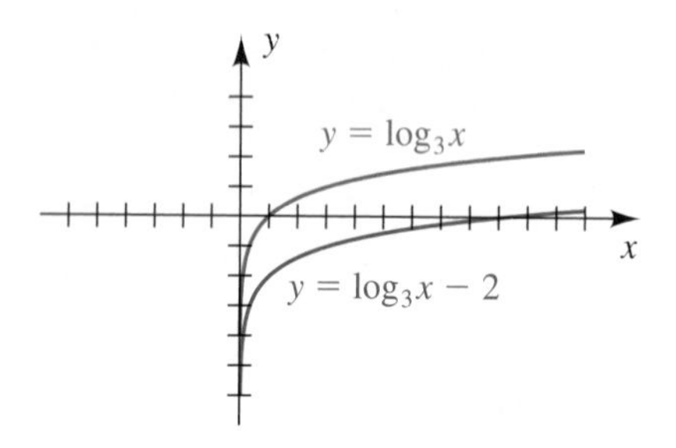

EXAMPLE 9 **Reflecting the graph of a logarithmic function**

Sketch the graph of f if $f(x) = \log_3 (2 - x)$.

SOLUTION If we write

$$f(x) = \log_3 (2 - x) = \log_3 [-(x - 2)],$$

then, by applying the same technique used to obtain the graph of the equation $y = \log_3 (-x)$ in Example 7 (with x replaced by $x - 2$), we see that the graph of f is the reflection of the graph of $y = \log_3 (x - 2)$ through the vertical line $x = 2$. This gives us the sketch in Figure 8.

Figure 8

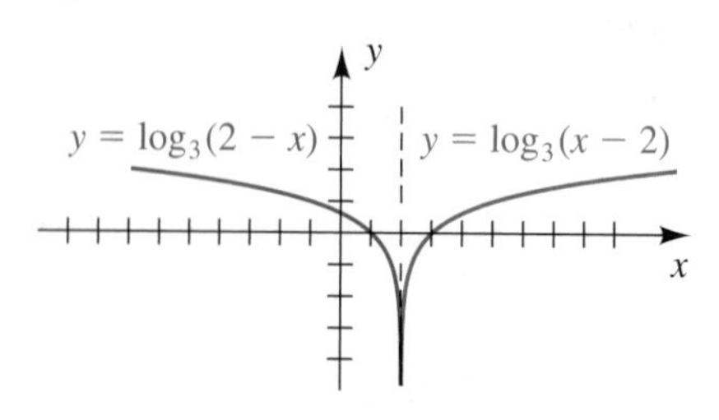

Another method is to change $y = \log_3 (2 - x)$ to the exponential form $3^y = 2 - x$ and then sketch the graph of $x = 2 - 3^y$.

Before electronic calculators were invented, logarithms with base 10 were used for complicated numerical computations involving products, quotients, and powers of real numbers. Base 10 was used because it is well suited for numbers that are expressed in scientific form. Logarithms with base 10 are called **common logarithms.** The symbol $\log x$ is used as an abbreviation for $\log_{10} x$, just as $\sqrt{\ }$ is used as an abbreviation for $\sqrt[2]{\ }$.

Definition of Common Logarithm	$\log x = \log_{10} x$ for every $x > 0$

Since inexpensive calculators are now available, there is no need for common logarithms as a tool for computational work. Base 10 does occur in applications, however, and hence many calculators have a LOG key, which can be used to approximate common logarithms.

The natural exponential function is given by $f(x) = e^x$. The logarithmic function with base e is called the **natural logarithmic function.** The symbol **ln x** (read "ell-en of x") is an abbreviation for $\log_e x$, and we refer to it as the **natural logarithm of x.** Thus, *the natural logarithmic function and the natural exponential function are inverse functions of each other.*

Definition of Natural Logarithm	$\ln x = \log_e x$ for every $x > 0$

Most calculators have a key labeled LN, which can be used to approximate natural logarithms. The next illustration gives several examples of equivalent forms involving common and natural logarithms.

ILLUSTRATION Equivalent Forms

Logarithmic form	Exponential form
$\log x = 2$	$10^2 = x$
$\log z = y + 3$	$10^{y+3} = z$
$\ln x = 2$	$e^2 = x$
$\ln z = y + 3$	$e^{y+3} = z$

To find x when given $\log x$ or $\ln x$, we may use the 10^x key or the e^x key, respectively, on a calculator, as in the next example. If your calculator has an INV key (for inverse), you may enter x and successively press INV LOG or INV LN.

EXAMPLE 10 Solving a simple logarithmic equation

Find x if

(a) $\log x = 1.7959$ (b) $\ln x = 4.7$

SOLUTION

(a) Changing $\log x = 1.7959$ to its equivalent exponential form gives us

$$x = 10^{1.7959}.$$

Evaluating the last expression to three-decimal-place accuracy yields

$$x \approx 62.503.$$

(b) Changing $\ln x = 4.7$ to its equivalent exponential form gives us

$$x = e^{4.7} \approx 109.95.$$

Tutorial available at www.thomsonedu.com/login

The following chart lists common and natural logarithmic forms for the properties on page 316.

Logarithms with base *a*	Common logarithms	Natural logarithms
(1) $\log_a 1 = 0$	$\log 1 = 0$	$\ln 1 = 0$
(2) $\log_a a = 1$	$\log 10 = 1$	$\ln e = 1$
(3) $\log_a a^x = x$	$\log 10^x = x$	$\ln e^x = x$
(4) $a^{\log_a x} = x$	$10^{\log x} = x$	$e^{\ln x} = x$

The last property for natural logarithms allows us to write the number a as $e^{\ln a}$, so the exponential function $f(x) = a^x$ can be written as $f(x) = (e^{\ln a})^x$ or as $f(x) = e^{x \ln a}$. Many calculators compute an exponential regression model of the form $y = ab^x$. If an exponential model with base e is desired, we can write the model

$$y = ab^x \quad \text{as} \quad y = ae^{x \ln b}.$$

ILLUSTRATION **Converting to Base e Expressions**

- 3^x is equivalent to $e^{x \ln 3}$
- x^3 is equivalent to $e^{3 \ln x}$
- $4 \cdot 2^x$ is equivalent to $4 \cdot e^{x \ln 2}$

Figure 9

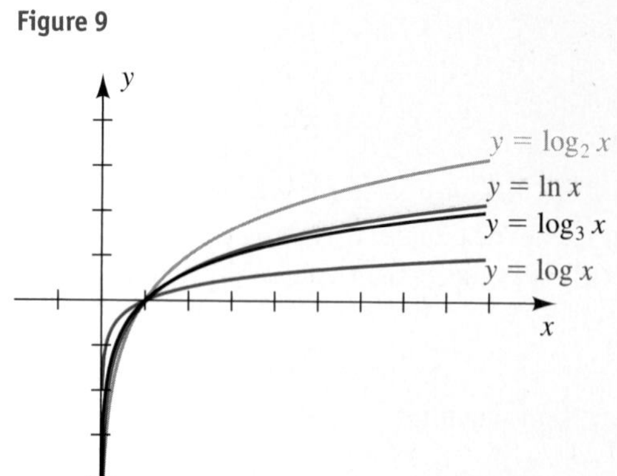

Figure 9 shows four logarithm graphs with base $a > 1$. Note that for $x > 1$, as the base of the logarithm increases, the graphs increase more slowly (they are more horizontal). This makes sense when we consider the graphs of the inverses of these functions: $y = 2^x$, $y = e^x$, $y = 3^x$, and $y = 10^x$. Here, for $x > 0$, as the base of the exponential increases, the graphs increase faster (they are more vertical).

The next four examples illustrate applications of common and natural logarithms.

EXAMPLE 11 **The Richter scale**

On the Richter scale, the magnitude R of an earthquake of intensity I is given by

$$R = \log \frac{I}{I_0},$$

where I_0 is a certain minimum intensity.

(a) If the intensity of an earthquake is $1000I_0$, find R.

(b) Express I in terms of R and I_0.

SOLUTION

▶ (a) $R = \log \frac{I}{I_0}$ given

$= \log \frac{1000I_0}{I_0}$ let $I = 1000I_0$

$= \log 1000$ cancel I_0

$= \log 10^3$ $1000 = 10^3$

$= 3$ $\log 10^x = x$ for every x

From this result we see that a tenfold increase in intensity results in an increase of 1 in magnitude (if 1000 were changed to 10,000, then 3 would change to 4).

(b) $R = \log \frac{I}{I_0}$ given

$\frac{I}{I_0} = 10^R$ change to exponential form

$I = I_0 \cdot 10^R$ multiply by I_0

EXAMPLE 12 Newton's law of cooling

Newton's law of cooling states that the rate at which an object cools is directly proportional to the difference in temperature between the object and its surrounding medium. Newton's law can be used to show that under certain conditions the temperature T (in °C) of an object at time t (in hours) is given by $T = 75e^{-2t}$. Express t as a function of T.

▶ SOLUTION $T = 75e^{-2t}$ given

$e^{-2t} = \frac{T}{75}$ isolate the exponential expression

$-2t = \ln \frac{T}{75}$ change to logarithmic form

$t = -\frac{1}{2} \ln \frac{T}{75}$ divide by -2

EXAMPLE 13 Approximating a doubling time

Assume that a population is growing continuously at a rate of 4% per year. Approximate the amount of time it takes for the population to double its size—that is, its **doubling time.**

▶ SOLUTION Note that an initial population size is not given. Not knowing the initial size does not present a problem, however, since we wish only to determine the time needed to obtain a population size *relative* to the initial population size. Using the growth formula $q = q_0e^{rt}$ with $r = 0.04$ gives us

$2q_0 = q_0e^{0.04t}$ let $q = 2q_0$

$2 = e^{0.04t}$ divide by q_0 ($q_0 \neq 0$)

(continued)

▶ **Tutorial available at www.thomsonedu.com/login**

$$0.04t = \ln 2 \qquad \text{change to logarithmic form}$$

$$t = 25 \ln 2 \approx 17.3 \text{ yr.} \qquad \text{multiply by } \frac{1}{0.04} = 25$$

The fact that q_0 did not have any effect on the answer indicates that the doubling time for a population of 1000 is the same as the doubling time for a population of 1,000,000 or any other reasonable initial population.

From the last example we may obtain a general formula for the doubling time of a population—namely,

$$rt = \ln 2 \qquad \text{or, equivalently,} \qquad t = \frac{\ln 2}{r}.$$

Since $\ln 2 \approx 0.69$, we see that the doubling time t for a growth of this type is approximately $0.69/r$. Because the numbers 70 and 72 are close to 69 but have more divisors, some resources refer to this doubling relationship as the **rule of 70** or the **rule of 72.** As an illustration of the rule of 72, if the growth rate of a population is 8%, then it takes about $72/8 = 9$ years for the population to double. More precisely, this value is

$$\frac{\ln 2}{8} \cdot 100 \approx 8.7 \text{ yr.}$$

EXAMPLE 14 Determining the half-life of a radioactive substance

A physicist finds that an unknown radioactive substance registers 2000 counts per minute on a Geiger counter. Ten days later the substance registers 1500 counts per minute. Using calculus, it can be shown that after t days the amount of radioactive material, and hence the number of counts per minute $N(t)$, is directly proportional to e^{ct} for some constant c. Determine the half-life of the substance.

SOLUTION Since $N(t)$ is directly proportional to e^{ct},

$$N(t) = ke^{ct},$$

where k is a constant. Letting $t = 0$ and using $N(0) = 2000$, we obtain

$$2000 = ke^{c0} = k \cdot 1 = k.$$

Hence, the formula for $N(t)$ may be written

$$N(t) = 2000e^{ct}.$$

Since $N(10) = 1500$, we may determine c as follows:

$$1500 = 2000e^{c \cdot 10} \qquad \text{let } t = 10 \text{ in } N(t)$$

$$\tfrac{3}{4} = e^{10c} \qquad \text{isolate the exponential expression}$$

$$10c = \ln \tfrac{3}{4} \qquad \text{change to logarithmic form}$$

$$c = \tfrac{1}{10} \ln \tfrac{3}{4} \qquad \text{divide by 10}$$

Finally, since the half-life corresponds to the time t at which $N(t)$ is equal to 1000, we have the following:

$$
\begin{aligned}
1000 &= 2000e^{ct} && \text{let } N(t) = 1000 \\
\tfrac{1}{2} &= e^{ct} && \text{isolate the exponential expression} \\
ct &= \ln \tfrac{1}{2} && \text{change to logarithmic form} \\
t &= \frac{1}{c} \ln \frac{1}{2} && \text{divide by } c \\
&= \frac{1}{\frac{1}{10} \ln \frac{3}{4}} \ln \frac{1}{2} && c = \tfrac{1}{10} \ln \tfrac{3}{4} \\
&\approx 24 \text{ days} && \text{approximate}
\end{aligned}
$$

The following example is a good illustration of the power of a graphing utility, since it is impossible to find the exact solution using only algebraic methods.

EXAMPLE 15 Approximating a solution to an inequality

Graph $f(x) = \log(x + 1)$ and $g(x) = \ln(3 - x)$, and estimate the solution of the inequality $f(x) \geq g(x)$.

Figure 10
$[-1, 3]$ by $[-2, 2]$

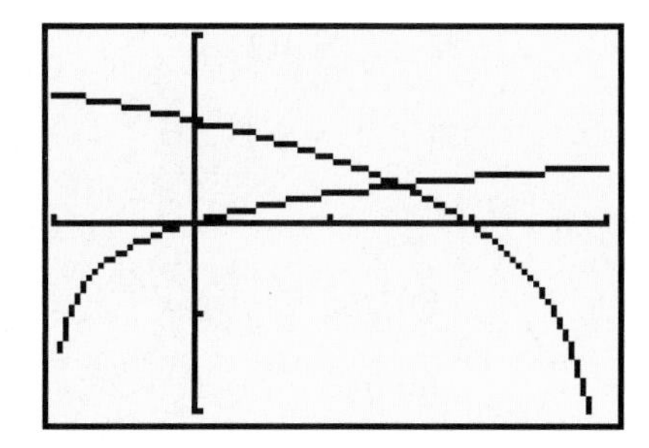

SOLUTION We begin by making the assignments

$$Y_1 = \log(x + 1) \quad \text{and} \quad Y_2 = \ln(3 - x).$$

Since the domain of f is $(-1, \infty)$ and the domain of g is $(-\infty, 3)$, we choose the viewing rectangle $[-1, 3]$ by $[-2, 2]$ and obtain the graph in Figure 10. Using an intersect feature, we find that the point of intersection is approximately $(1.51, 0.40)$. Thus, the approximate solution of $f(x) \geq g(x)$ is the interval

$$1.51 < x < 3.$$

4.4 *Exercises*

Exer. 1–2: Change to logarithmic form.

1 (a) $4^3 = 64$ (b) $4^{-3} = \frac{1}{64}$ (c) $t^r = s$

(d) $3^x = 4 - t$ (e) $5^{7t} = \dfrac{a + b}{a}$ (f) $(0.7)^t = 5.3$

2 (a) $3^5 = 243$ (b) $3^{-4} = \frac{1}{81}$ (c) $c^p = d$

(d) $7^x = 100p$ (e) $3^{-2x} = \dfrac{P}{F}$ (f) $(0.9)^t = \frac{1}{2}$

Exer. 3–4: Change to exponential form.

3 (a) $\log_2 32 = 5$ (b) $\log_3 \frac{1}{243} = -5$

(c) $\log_t r = p$ (d) $\log_3 (x + 2) = 5$

(e) $\log_2 m = 3x + 4$ (f) $\log_b 512 = \frac{3}{2}$

4 (a) $\log_3 81 = 4$ (b) $\log_4 \frac{1}{256} = -4$

(c) $\log_v w = q$ (d) $\log_6 (2x - 1) = 3$

(e) $\log_4 p = 5 - x$ (f) $\log_a 343 = \frac{3}{4}$

Exer. 5–10: Solve for t using logarithms with base a.

5 $2a^{t/3} = 5$

6 $3a^{4t} = 10$

7 $K = H - Ca^t$

8 $F = D + Ba^t$

9 $A = Ba^{Ct} + D$

10 $L = Ma^{t/N} - P$

Exer. 11–12: Change to logarithmic form.

11 (a) $10^5 = 100{,}000$ (b) $10^{-3} = 0.001$

(c) $10^x = y + 1$ (d) $e^7 = p$

(e) $e^{2t} = 3 - x$

12 (a) $10^4 = 10{,}000$ (b) $10^{-2} = 0.01$

(c) $10^x = 38z$ (d) $e^4 = D$

(e) $e^{0.1t} = x + 2$

Exer. 13–14: Change to exponential form.

13 (a) $\log x = 50$ (b) $\log x = 20t$

(c) $\ln x = 0.1$ (d) $\ln w = 4 + 3x$

(e) $\ln (z - 2) = \frac{1}{6}$

14 (a) $\log x = -8$ (b) $\log x = y - 2$

(c) $\ln x = \frac{1}{2}$ (d) $\ln z = 7 + x$

(e) $\ln (t - 5) = 1.2$

Exer. 15–16: Find the number, if possible.

15 (a) $\log_5 1$ (b) $\log_3 3$ (c) $\log_4 (-2)$

(d) $\log_7 7^2$ (e) $3^{\log_3 8}$ (f) $\log_5 125$

(g) $\log_4 \frac{1}{16}$

16 (a) $\log_8 1$ (b) $\log_9 9$ (c) $\log_5 0$

(d) $\log_6 6^7$ (e) $5^{\log_5 4}$ (f) $\log_3 243$

(g) $\log_2 128$

Exer. 17–18: Find the number.

17 (a) $10^{\log 3}$ (b) $\log 10^5$ (c) $\log 100$

(d) $\log 0.0001$ (e) $e^{\ln 2}$ (f) $\ln e^{-3}$

(g) $e^{2+\ln 3}$

18 (a) $10^{\log 7}$ (b) $\log 10^{-6}$ (c) $\log 100{,}000$

(d) $\log 0.001$ (e) $e^{\ln 8}$ (f) $\ln e^{2/3}$

(g) $e^{1+\ln 5}$

Exer. 19–34: Solve the equation.

19 $\log_4 x = \log_4 (8 - x)$

20 $\log_3 (x + 4) = \log_3 (1 - x)$

21 $\log_5 (x - 2) = \log_5 (3x + 7)$

22 $\log_7 (x - 5) = \log_7 (6x)$

23 $\log x^2 = \log (-3x - 2)$

24 $\ln x^2 = \ln (12 - x)$

25 $\log_3 (x - 4) = 2$

26 $\log_2 (x - 5) = 4$

27 $\log_9 x = \frac{3}{2}$

28 $\log_4 x = -\frac{3}{2}$

29 $\ln x^2 = -2$

30 $\log x^2 = -4$

31 $e^{2 \ln x} = 9$

32 $e^{-\ln x} = 0.2$

33 $e^{x \ln 3} = 27$

34 $e^{x \ln 2} = 0.25$

35 Sketch the graph of f if $a = 4$:

(a) $f(x) = \log_a x$ (b) $f(x) = -\log_a x$

(c) $f(x) = 2 \log_a x$ (d) $f(x) = \log_a (x + 2)$

(e) $f(x) = (\log_a x) + 2$ (f) $f(x) = \log_a (x - 2)$

(g) $f(x) = (\log_a x) - 2$ (h) $f(x) = \log_a |x|$

(i) $f(x) = \log_a (-x)$ (j) $f(x) = \log_a (3 - x)$

(k) $f(x) = |\log_a x|$ (l) $f(x) = \log_{1/a} x$

36 Work Exercise 35 if $a = 5$.

Exer. 37–42: Sketch the graph of f.

37 $f(x) = \log (x + 10)$

38 $f(x) = \log (x + 100)$

39 $f(x) = \ln |x|$

40 $f(x) = \ln |x - 1|$

41 $f(x) = \ln e + x$

42 $f(x) = \ln (e + x)$

Exer. 43–44: Find a logarithmic function of the form $f(x) = \log_a x$ for the given graph.

43

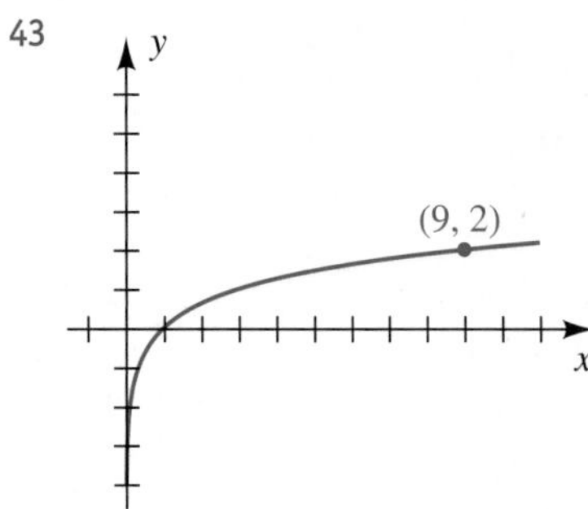

44

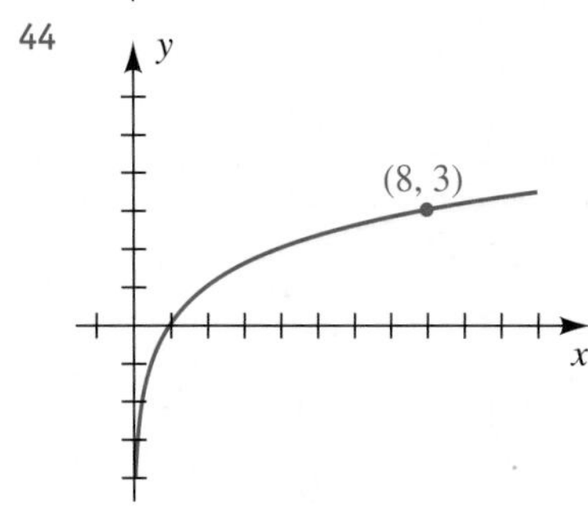

Exer. 45–50: Shown in the figure is the graph of a function f. Express $f(x)$ in terms of F.

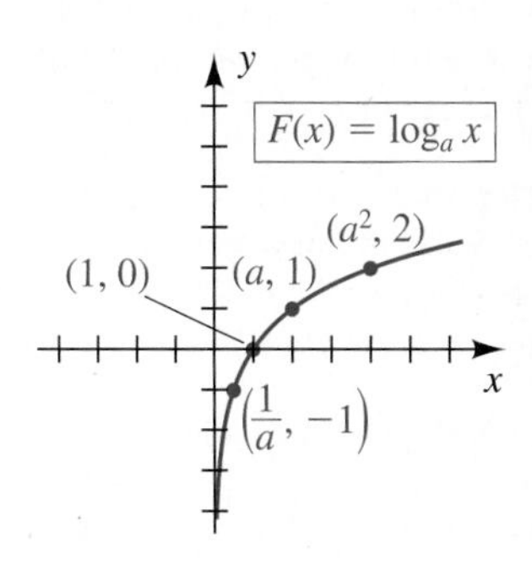

45

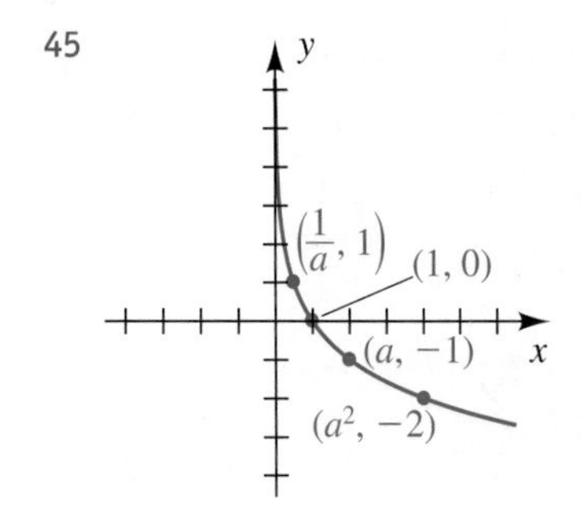

46

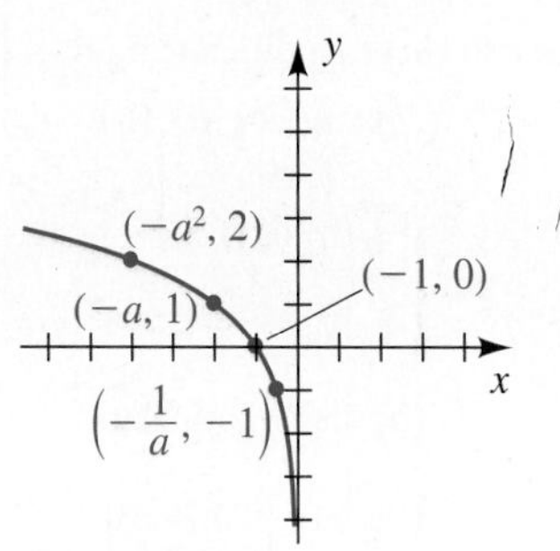

47

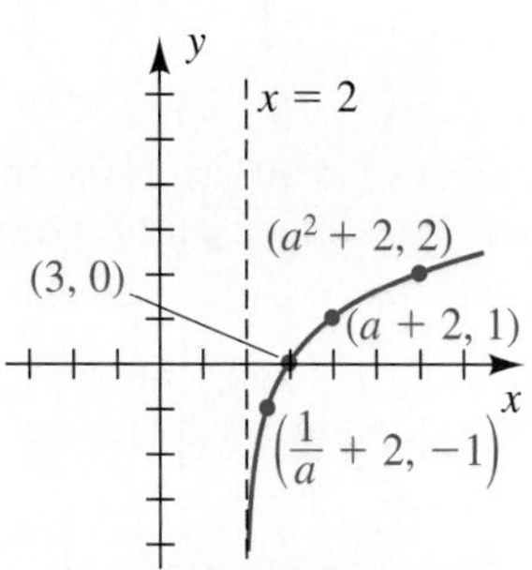

48

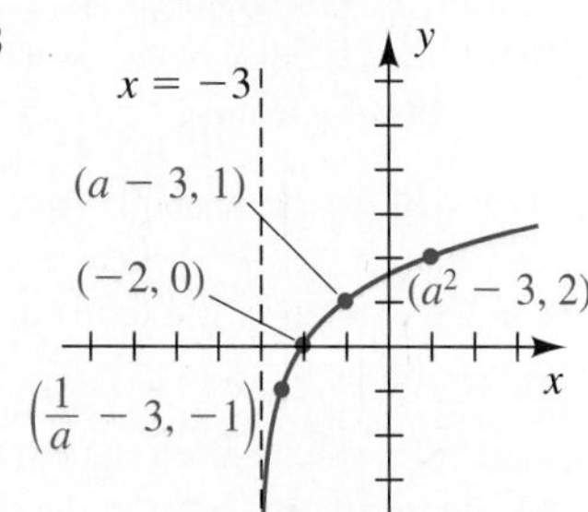

49

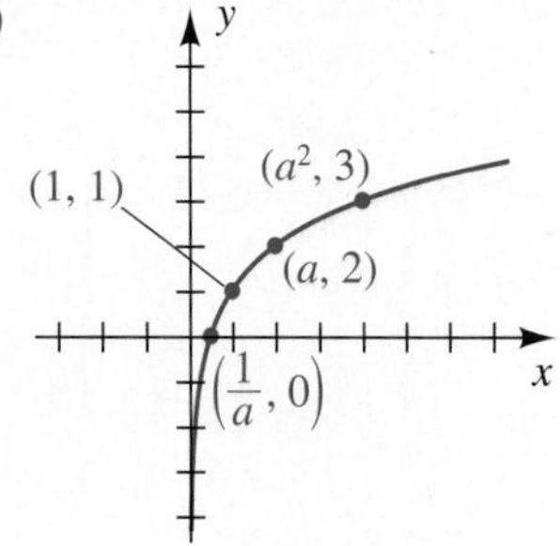

50

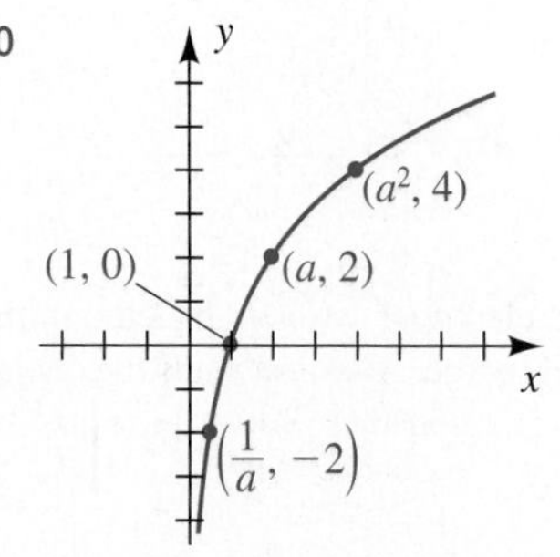

Exer. 51–52: Approximate x to three significant figures.

51 (a) $\log x = 3.6274$ (b) $\log x = 0.9469$

(c) $\log x = -1.6253$ (d) $\ln x = 2.3$

(e) $\ln x = 0.05$ (f) $\ln x = -1.6$

52 (a) $\log x = 1.8965$ (b) $\log x = 4.9680$

(c) $\log x = -2.2118$ (d) $\ln x = 3.7$

(e) $\ln x = 0.95$ (f) $\ln x = -5$

53 **Finding a growth rate** Change $f(x) = 1000(1.05)^x$ to an exponential function with base e and approximate the growth rate of f.

54 **Finding a decay rate** Change $f(x) = 100(\frac{1}{2})^x$ to an exponential function with base e and approximate the decay rate of f.

55 **Radium decay** If we start with q_0 milligrams of radium, the amount q remaining after t years is given by the formula $q = q_0(2)^{-t/1600}$. Express t in terms of q and q_0.

56 **Bismuth isotope decay** The radioactive bismuth isotope ^{210}Bi disintegrates according to $Q = k(2)^{-t/5}$, where k is a constant and t is the time in days. Express t in terms of Q and k.

57 **Electrical circuit** A schematic of a simple electrical circuit consisting of a resistor and an inductor is shown in the figure. The current I at time t is given by the formula $I = 20e^{-Rt/L}$, where R is the resistance and L is the inductance. Solve this equation for t.

Exercise 57

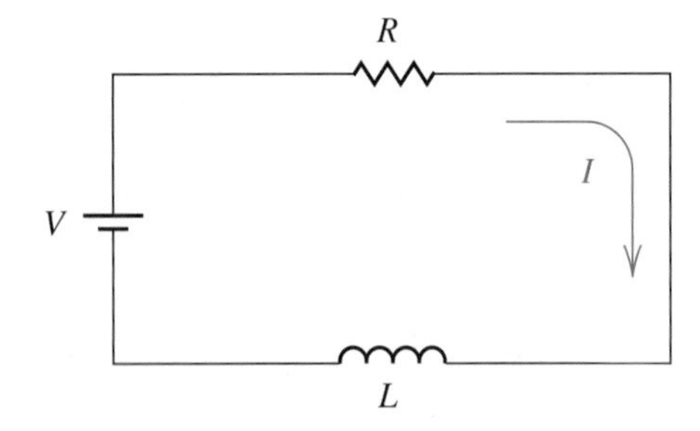

58 **Electrical condenser** An electrical condenser with initial charge Q_0 is allowed to discharge. After t seconds the charge Q is $Q = Q_0e^{kt}$, where k is a constant. Solve this equation for t.

59 **Richter scale** Use the Richter scale formula $R = \log(I/I_0)$ to find the magnitude of an earthquake that has an intensity

(a) 100 times that of I_0

(b) 10,000 times that of I_0

(c) 100,000 times that of I_0

60 **Richter scale** Refer to Exercise 59. The largest recorded magnitudes of earthquakes have been between 8 and 9 on the Richter scale. Find the corresponding intensities in terms of I_0.

61 **Sound intensity** The loudness of a sound, as experienced by the human ear, is based on its intensity level. A formula used for finding the intensity level α (in decibels) that corresponds to a sound intensity I is $\alpha = 10 \log(I/I_0)$, where I_0 is a special value of I agreed to be the weakest sound that can be detected by the ear under certain conditions. Find α if

(a) I is 10 times as great as I_0

(b) I is 1000 times as great as I_0

(c) I is 10,000 times as great as I_0 (This is the intensity level of the average voice.)

62 **Sound intensity** Refer to Exercise 61. A sound intensity level of 140 decibels produces pain in the average human ear. Approximately how many times greater than I_0 must I be in order for α to reach this level?

63 **U.S. population growth** The population $N(t)$ (in millions) of the United States t years after 1980 may be approximated by the formula $N(t) = 231e^{0.0103t}$. When will the population be twice what it was in 1980?

64 **Population growth in India** The population $N(t)$ (in millions) of India t years after 1985 may be approximated by the formula $N(t) = 766e^{0.0182t}$. When will the population reach 1.5 billion?

65 **Children's weight** The Ehrenberg relation

$$\ln W = \ln 2.4 + (1.84)h$$

is an empirically based formula relating the height h (in meters) to the average weight W (in kilograms) for children 5 through 13 years old.

(a) Express W as a function of h that does not contain ln.

(b) Estimate the average weight of an 8-year-old child who is 1.5 meters tall.

66 Continuously compounded interest If interest is compounded continuously at the rate of 6% per year, approximate the number of years it will take an initial deposit of \$6000 to grow to \$25,000.

67 Air pressure The air pressure $p(h)$ (in lb/in^2) at an altitude of h feet above sea level may be approximated by the formula $p(h) = 14.7e^{-0.0000385h}$. At approximately what altitude h is the air pressure

(a) 10 lb/in^2?

(b) one-half its value at sea level?

68 Vapor pressure A liquid's vapor pressure P (in lb/in^2), a measure of its volatility, is related to its temperature T (in °F) by the Antoine equation

$$\log P = a + \frac{b}{c + T},$$

where a, b, and c are constants. Vapor pressure increases rapidly with an increase in temperature. Express P as a function of T.

69 Elephant growth The weight W (in kilograms) of a female African elephant at age t (in years) may be approximated by

$$W = 2600(1 - 0.51e^{-0.075t})^3.$$

(a) Approximate the weight at birth.

(b) Estimate the age of a female African elephant weighing 1800 kilograms by using (1) the accompanying graph and (2) the formula for W.

Exercise 69

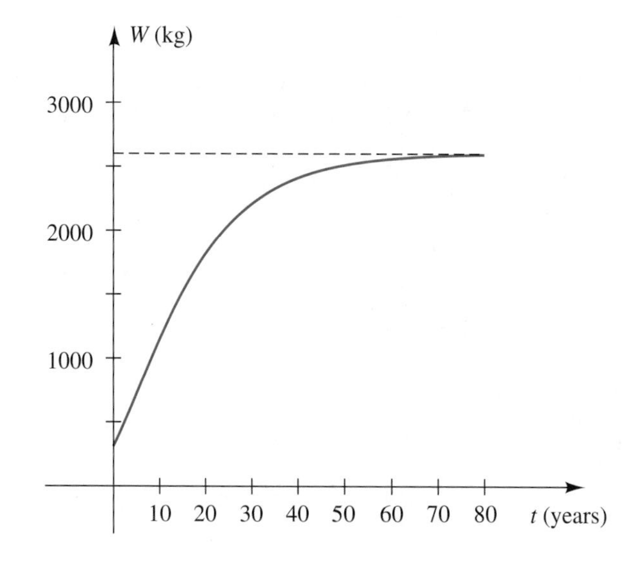

70 Coal consumption A country presently has coal reserves of 50 million tons. Last year 6.5 million tons of coal was consumed. Past years' data and population projections suggest that the rate of consumption R (in million tons/year) will increase according to the formula $R = 6.5e^{0.02t}$, and the total amount T (in million tons) of coal that will be used in t years is given by the formula $T = 325(e^{0.02t} - 1)$. If the country uses only its own resources, when will the coal reserves be depleted?

71 Urban population density An urban density model is a formula that relates the population density D (in thousands/mi^2) to the distance x (in miles) from the center of the city. The formula $D = ae^{-bx}$ for the central density a and coefficient of decay b has been found to be appropriate for many large U.S. cities. For the city of Atlanta in 1970, $a = 5.5$ and $b = 0.10$. At approximately what distance was the population density 2000 per square mile?

72 Brightness of stars Stars are classified into categories of brightness called magnitudes. The faintest stars, with light flux L_0, are assigned a magnitude of 6. Brighter stars of light flux L are assigned a magnitude m by means of the formula

$$m = 6 - 2.5 \log \frac{L}{L_0}.$$

(a) Find m if $L = 10^{0.4}L_0$.

(b) Solve the formula for L in terms of m and L_0.

73 Radioactive iodine decay Radioactive iodine ^{131}I is frequently used in tracer studies involving the thyroid gland. The substance decays according to the formula $A(t) = A_0a^{-t}$, where A_0 is the initial dose and t is the time in days. Find a, assuming the half-life of ^{131}I is 8 days.

74 Radioactive contamination Radioactive strontium ^{90}Sr has been deposited in a large field by acid rain. If sufficient amounts make their way through the food chain to humans, bone cancer can result. It has been determined that the radioactivity level in the field is 2.5 times the safe level S. ^{90}Sr decays according to the formula

$$A(t) = A_0e^{-0.0239t},$$

where A_0 is the amount currently in the field and t is the time in years. For how many years will the field be contaminated?

75 Walking speed In a survey of 15 cities ranging in population P from 300 to 3,000,000, it was found that the average walking speed S (in ft/sec) of a pedestrian could be approximated by $S = 0.05 + 0.86 \log P$.

(a) How does the population affect the average walking speed?

(b) For what population is the average walking speed 5 ft/sec?

76 Computer chips For manufacturers of computer chips, it is important to consider the fraction F of chips that will fail after t years of service. This fraction can sometimes be approximated by the formula $F = 1 - e^{-ct}$, where c is a positive constant.

(a) How does the value of c affect the reliability of a chip?

(b) If $c = 0.125$, after how many years will 35% of the chips have failed?

Exer. 77–78: Approximate the function at the value of x to four decimal places.

77 (a) $f(x) = \ln(x + 1) + e^x$, $x = 2$

(b) $g(x) = \dfrac{(\log x)^2 - \log x}{4}$, $x = 3.97$

78 (a) $f(x) = \log(2x^2 + 1) - 10^{-x}$, $x = 1.95$

(b) $g(x) = \dfrac{x - 3.4}{\ln x + 4}$, $x = 0.55$

Exer. 79–80: Approximate the real root of the equation.

79 $x \ln x = 1$

80 $\ln x + x = 0$

Exer. 81–82: Graph f and g on the same coordinate plane, and estimate the solution of the inequality $f(x) \geq g(x)$.

81 $f(x) = 2.2 \log(x + 2)$; $g(x) = \ln x$

82 $f(x) = x \ln |x|$; $g(x) = 0.15e^x$

83 Cholesterol level in women Studies relating serum cholesterol level to coronary heart disease suggest that a risk factor is the ratio x of the total amount C of cholesterol in the blood to the amount H of high-density lipoprotein cholesterol in the blood. For a female, the lifetime risk R of having a heart attack can be approximated by the formula

$$R = 2.07 \ln x - 2.04 \quad \text{provided} \quad 0 \leq R \leq 1.$$

For example, if $R = 0.65$, then there is a 65% chance that a woman will have a heart attack over an average lifetime.

(a) Calculate R for a female with $C = 242$ and $H = 78$.

(b) Graphically estimate x when the risk is 75%.

84 Cholesterol level in men Refer to Exercise 83. For a male, the risk can be approximated by the formula $R = 1.36 \ln x - 1.19$.

(a) Calculate R for a male with $C = 287$ and $H = 65$.

(b) Graphically estimate x when the risk is 75%.

4.5 Properties of Logarithms

In the preceding section we observed that $\log_a x$ can be interpreted as an exponent. Thus, it seems reasonable to expect that the laws of exponents can be used to obtain corresponding laws of logarithms. This is demonstrated in the proofs of the following laws, which are fundamental for all work with logarithms.

Laws of Logarithms

If u and w denote positive real numbers, then

(1) $\log_a (uw) = \log_a u + \log_a w$

(2) $\log_a \left(\dfrac{u}{w}\right) = \log_a u - \log_a w$

(3) $\log_a (u^c) = c \log_a u$ for every real number c

PROOFS For all three proofs, let

$$r = \log_a u \qquad \text{and} \qquad s = \log_a w.$$

The equivalent exponential forms are

$$u = a^r \qquad \text{and} \qquad w = a^s.$$

We now proceed as follows:

(1)

$uw = a^r a^s$	definition of u and w
$uw = a^{r+s}$	law 1 of exponents
$\log_a (uw) = r + s$	change to logarithmic form
$\log_a (uw) = \log_a u + \log_a w$	definition of r and s

(2)

$\dfrac{u}{w} = \dfrac{a^r}{a^s}$	definition of u and w
$\dfrac{u}{w} = a^{r-s}$	law 5(a) of exponents
$\log_a \left(\dfrac{u}{w}\right) = r - s$	change to logarithmic form
$\log_a \left(\dfrac{u}{w}\right) = \log_a u - \log_a w$	definition of r and s

(3)

$u^c = (a^r)^c$	definition of u
$u^c = a^{cr}$	law 2 of exponents
$\log_a (u^c) = cr$	change to logarithmic form
$\log_a (u^c) = c \log_a u$	definition of r

The laws of logarithms for the special cases $a = 10$ (common logs) and $a = e$ (natural logs) are written as shown in the following chart.

Common logarithms	Natural logarithms
(1) $\log (uw) = \log u + \log w$	**(1)** $\ln (uw) = \ln u + \ln w$
(2) $\log \left(\dfrac{u}{w}\right) = \log u - \log w$	**(2)** $\ln \left(\dfrac{u}{w}\right) = \ln u - \ln w$
(3) $\log (u^c) = c \log u$	**(3)** $\ln (u^c) = c \ln u$

As indicated by the following warning, there are no laws for expressing $\log_a (u + w)$ or $\log_a (u - w)$ in terms of simpler logarithms.

Warning!

$$\log_a (u + w) \neq \log_a u + \log_a w$$
$$\log_a (u - w) \neq \log_a u - \log_a w$$

The following examples illustrate uses of the laws of logarithms.

EXAMPLE 1 Using laws of logarithms

Express $\log_a \dfrac{x^3\sqrt{y}}{z^2}$ in terms of logarithms of x, y, and z.

▶ **SOLUTION** We write $\sqrt{y}$ as $y^{1/2}$ and use laws of logarithms:

$$
\begin{aligned}
\log_a \frac{x^3\sqrt{y}}{z^2} &= \log_a (x^3y^{1/2}) - \log_a z^2 && \text{law 2}\\
&= \log_a x^3 + \log_a y^{1/2} - \log_a z^2 && \text{law 1}\\
&= 3\log_a x + \tfrac{1}{2}\log_a y - 2\log_a z && \text{law 3}
\end{aligned}
$$

Note that if a term with a positive exponent (such as x^3) is in the numerator of the original expression, it will have a positive coefficient in the expanded form, and if it is in the denominator (such as z^2), it will have a negative coefficient in the expanded form.

EXAMPLE 2 Using laws of logarithms

Express as one logarithm:

$$\tfrac{1}{3}\log_a (x^2 - 1) - \log_a y - 4\log_a z$$

▶ **SOLUTION** We apply the laws of logarithms as follows:

$$
\begin{aligned}
&\tfrac{1}{3}\log_a (x^2 - 1) - \log_a y - 4\log_a z\\
&\quad = \log_a (x^2 - 1)^{1/3} - \log_a y - \log_a z^4 && \text{law 3}\\
&\quad = \log_a \sqrt[3]{x^2 - 1} - (\log_a y + \log_a z^4) && \text{algebra}\\
&\quad = \log_a \sqrt[3]{x^2 - 1} - \log_a (yz^4) && \text{law 1}\\
&\quad = \log_a \frac{\sqrt[3]{x^2 - 1}}{yz^4} && \text{law 2}
\end{aligned}
$$

In Figure 1 we perform a simple calculator check of Example 2 by assigning arbitrary values to X, Y, and Z and then evaluating the given expression and our answer. This doesn't prove that we are correct, but lends credibility to our result (not to mention peace of mind).

Figure 1

```
73→X:101→Y:7→Z
                7
(1/3)log(X²-1)-l
og(Y)-4log(Z)
     -4.142525462
log(³√((X²-1)/(Y*
Z^4))
     -4.142525462
```

▶ **Tutorial available at www.thomsonedu.com/login**

EXAMPLE 3 **Solving a logarithmic equation**

Solve the equation $\log_5 (2x + 3) = \log_5 11 + \log_5 3$.

▶ SOLUTION

$\log_5 (2x + 3) = \log_5 11 + \log_5 3$	given
$\log_5 (2x + 3) = \log_5 (11 \cdot 3)$	law 1 of logarithms
$2x + 3 = 33$	logarithmic functions are one-to-one
$x = 15$	solve for x

✓ Check $x = 15$ LS: $\log_5 (2 \cdot 15 + 3) = \log_5 33$
RS: $\log_5 11 + \log_5 3 = \log_5 (11 \cdot 3) = \log_5 33$

Since $\log_5 33 = \log_5 33$ is a true statement, $x = 15$ is a solution.

The laws of logarithms were proved for logarithms of *positive* real numbers u and w. If we apply these laws to equations in which u and w are expressions involving a variable, then extraneous solutions may occur. Answers should therefore be substituted for the variable in u and w to determine whether these expressions are defined.

EXAMPLE 4 **Solving a logarithmic equation**

Solve the equation $\log_2 x + \log_2 (x + 2) = 3$.

▶ SOLUTION

$\log_2 x + \log_2 (x + 2) = 3$	given
$\log_2 [x(x + 2)] = 3$	law 1 of logarithms
$x(x + 2) = 2^3$	change to exponential form
$x^2 + 2x - 8 = 0$	multiply and set equal to 0
$(x - 2)(x + 4) = 0$	factor
$x - 2 = 0, \quad x + 4 = 0$	zero factor theorem
$x = 2, \quad x = -4$	solve for x

✓ Check $x = 2$ LS: $\log_2 2 + \log_2 (2 + 2) = 1 + \log_2 4$
$= 1 + \log_2 2^2 = 1 + 2 = 3$

RS: 3

Since $3 = 3$ is a true statement, $x = 2$ is a solution.

✓ Check $x = -4$ LS: $\log_2 (-4) + \log_2 (-4 + 2)$

Since logarithms of negative numbers are undefined, $x = -4$ is not a solution.

EXAMPLE 5 **Solving a logarithmic equation**

Solve the equation $\ln (x + 6) - \ln 10 = \ln (x - 1) - \ln 2$.

▶ Tutorial available at www.thomsonedu.com/login

▶ **SOLUTION**

$$\ln(x+6) - \ln(x-1) = \ln 10 - \ln 2 \qquad \text{rearrange terms}$$

$$\ln\left(\frac{x+6}{x-1}\right) = \ln\frac{10}{2} \qquad \text{law 2 of logarithms}$$

$$\frac{x+6}{x-1} = 5 \qquad \text{ln is one-to-one}$$

$$x + 6 = 5x - 5 \qquad \text{multiply by } x - 1$$

$$x = \tfrac{11}{4} \qquad \text{solve for } x$$

✓ **Check** Since both $\ln(x+6)$ and $\ln(x-1)$ are defined at $x = \frac{11}{4}$ (they are logarithms of positive real numbers) and since our algebraic steps are correct, it follows that $\frac{11}{4}$ is a solution of the given equation. (Figure 2 shows a calculator check for Example 5.)

Figure 2

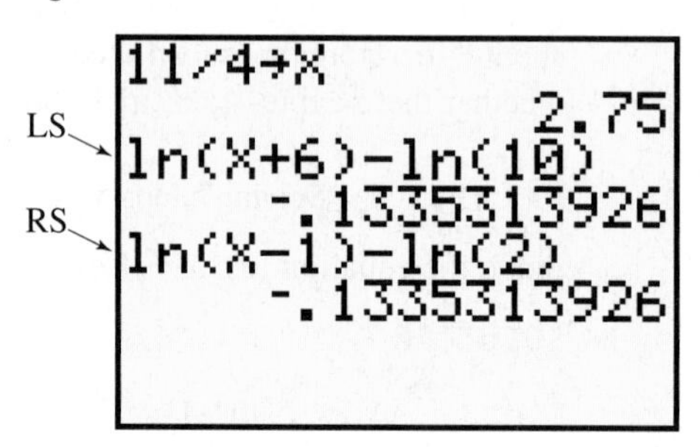

EXAMPLE 6 Shifting the graph of a logarithmic equation

Sketch the graph of $y = \log_3(81x)$.

▶ **SOLUTION** We may rewrite the equation as follows:

$$\begin{aligned} y &= \log_3(81x) && \text{given} \\ &= \log_3 81 + \log_3 x && \text{law 1 of logarithms} \\ &= \log_3 3^4 + \log_3 x && 81 = 3^4 \\ &= 4 + \log_3 x && \log_a a^x = x \end{aligned}$$

Thus, we can obtain the graph of $y = \log_3(81x)$ by vertically shifting the graph of $y = \log_3 x$ in Figure 2 in Section 4.4 upward four units. This gives us the sketch in Figure 3.

Figure 3

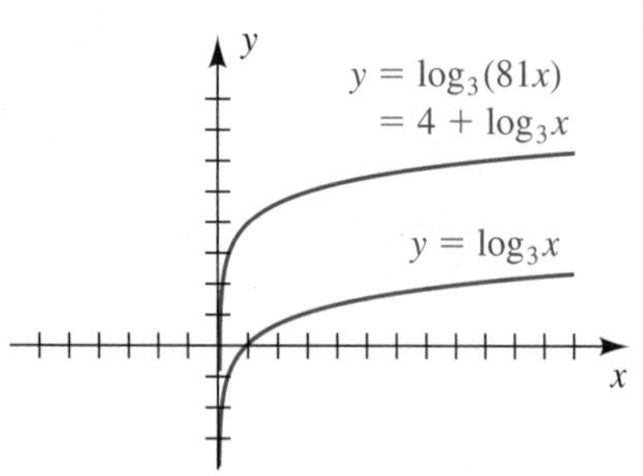

EXAMPLE 7 Sketching graphs of logarithmic equations

Sketch the graph of the equation:

(a) $y = \log_3(x^2)$ **(b)** $y = 2\log_3 x$

▶ **Tutorial available at www.thomsonedu.com/login**

▶ SOLUTION

(a) Since $x^2 = |x|^2$, we may rewrite the given equation as

$$y = \log_3 |x|^2.$$

Using law 3 of logarithms, we have

$$y = 2 \log_3 |x|.$$

We can obtain the graph of $y = 2 \log_3 |x|$ by multiplying the y-coordinates of points on the graph of $y = \log_3 |x|$ in Figure 4 of Section 4.4 by 2. This gives us the graph in Figure 4(a).

Figure 4

(a)

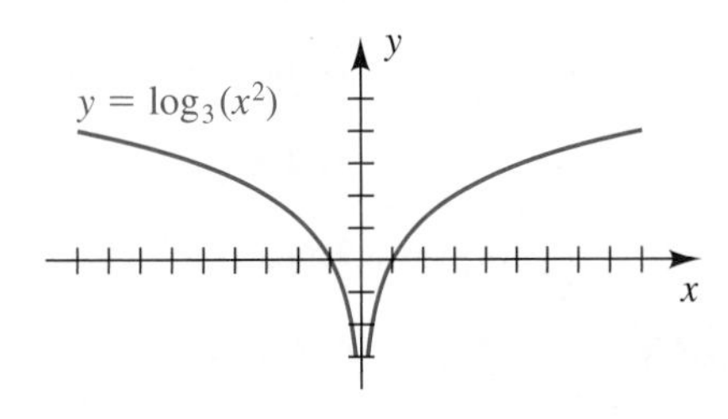

(b)

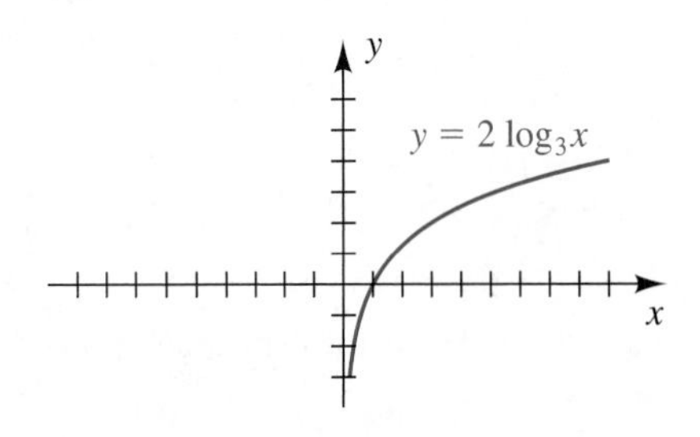

(b) If $y = 2 \log_3 x$, then x must be positive. Hence, the graph is identical to that part of the graph of $y = 2 \log_3 |x|$ in Figure 4(a) that lies to the right of the y-axis. This gives us Figure 4(b).

EXAMPLE 8 A relationship between selling price and demand

In the study of economics, the demand D for a product is often related to its selling price p by an equation of the form

$$\log_a D = \log_a c - k \log_a p,$$

where a, c, and k are positive constants.

(a) Solve the equation for D.

(b) How does increasing or decreasing the selling price affect the demand?

SOLUTION

▶ **(a)**

$\log_a D = \log_a c - k \log_a p$	given
$\log_a D = \log_a c - \log_a p^k$	law 3 of logarithms
$\log_a D = \log_a \dfrac{c}{p^k}$	law 2 of logarithms
$D = \dfrac{c}{p^k}$	$\log_a$ is one-to-one

(b) If the price p is increased, the denominator p^k in $D = c/p^k$ will also increase and hence the demand D for the product will decrease. If the price is decreased, then p^k will decrease and the demand D will increase.

▶ **Tutorial available at www.thomsonedu.com/login**

4.5 *Exercises*

Exer. 1–8: Express in terms of logarithms of x, y, z, or w.

1 (a) $\log_4 (xz)$ (b) $\log_4 (y/x)$ (c) $\log_4 \sqrt[3]{z}$

2 (a) $\log_3 (xyz)$ (b) $\log_3 (xz/y)$ (c) $\log_3 \sqrt[5]{y}$

3 $\log_a \dfrac{x^3 w}{y^2 z^4}$

4 $\log_a \dfrac{y^5 w^2}{x^4 z^3}$

5 $\log \dfrac{\sqrt[3]{z}}{x\sqrt{y}}$

6 $\log \dfrac{\sqrt{y}}{x^4 \sqrt[3]{z}}$

7 $\ln \sqrt[4]{\dfrac{x^7}{y^5 z}}$

8 $\ln x \sqrt[3]{\dfrac{y^4}{z^5}}$

Exer. 9–16: Write the expression as one logarithm.

9 (a) $\log_3 x + \log_3 (5y)$ (b) $\log_3 (2z) - \log_3 x$

(c) $5 \log_3 y$

10 (a) $\log_4 (3z) + \log_4 x$ (b) $\log_4 x - \log_4 (7y)$

(c) $\frac{1}{3} \log_4 w$

11 $2 \log_a x + \frac{1}{3} \log_a (x - 2) - 5 \log_a (2x + 3)$

12 $5 \log_a x - \frac{1}{2} \log_a (3x - 4) - 3 \log_a (5x + 1)$

▶ 13 $\log (x^3 y^2) - 2 \log x\sqrt[3]{y} - 3 \log \left(\dfrac{x}{y}\right)$

14 $2 \log \dfrac{y^3}{x} - 3 \log y + \dfrac{1}{2} \log x^4 y^2$

15 $\ln y^3 + \frac{1}{3} \ln (x^3 y^6) - 5 \ln y$

16 $2 \ln x - 4 \ln (1/y) - 3 \ln (xy)$

Exer. 17–34: Solve the equation.

17 $\log_6 (2x - 3) = \log_6 12 - \log_6 3$

18 $\log_4 (3x + 2) = \log_4 5 + \log_4 3$

19 $2 \log_3 x = 3 \log_3 5$

20 $3 \log_2 x = 2 \log_2 3$

21 $\log x - \log (x + 1) = 3 \log 4$

22 $\log (x + 2) - \log x = 2 \log 4$

23 $\ln (-4 - x) + \ln 3 = \ln (2 - x)$

24 $\ln x + \ln (x + 6) = \frac{1}{2} \ln 9$

25 $\log_2 (x + 7) + \log_2 x = 3$

26 $\log_6 (x + 5) + \log_6 x = 2$

27 $\log_3 (x + 3) + \log_3 (x + 5) = 1$

28 $\log_3 (x - 2) + \log_3 (x - 4) = 2$

29 $\log (x + 3) = 1 - \log (x - 2)$

30 $\log (57x) = 2 + \log (x - 2)$

31 $\ln x = 1 - \ln (x + 2)$

32 $\ln x = 1 + \ln (x + 1)$

33 $\log_3 (x - 2) = \log_3 27 - \log_3 (x - 4) - 5^{\log_5 1}$

34 $\log_2 (x + 3) = \log_2 (x - 3) + \log_3 9 + 4^{\log_4 3}$

Exer. 35–46: Sketch the graph of f.

35 $f(x) = \log_3 (3x)$

36 $f(x) = \log_4 (16x)$

37 $f(x) = 3 \log_3 x$

38 $f(x) = \frac{1}{3} \log_3 x$

39 $f(x) = \log_3 (x^2)$

40 $f(x) = \log_2 (x^2)$

41 $f(x) = \log_2 (x^3)$

42 $f(x) = \log_3 (x^3)$

43 $f(x) = \log_2 \sqrt{x}$

44 $f(x) = \log_2 \sqrt[3]{x}$

45 $f(x) = \log_3 \left(\dfrac{1}{x}\right)$

46 $f(x) = \log_2 \left(\dfrac{1}{x}\right)$

▶ **Tutorial available at www.thomsonedu.com/login**

Exer. 47–50: Shown in the figure is the graph of a function f. Express $f(x)$ as one logarithm with base 2.

47

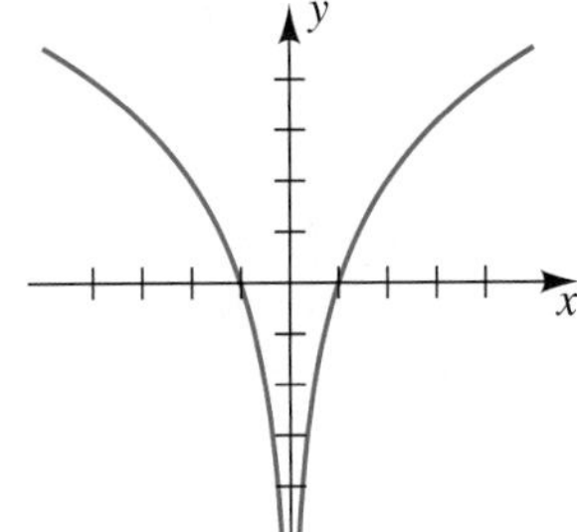

48

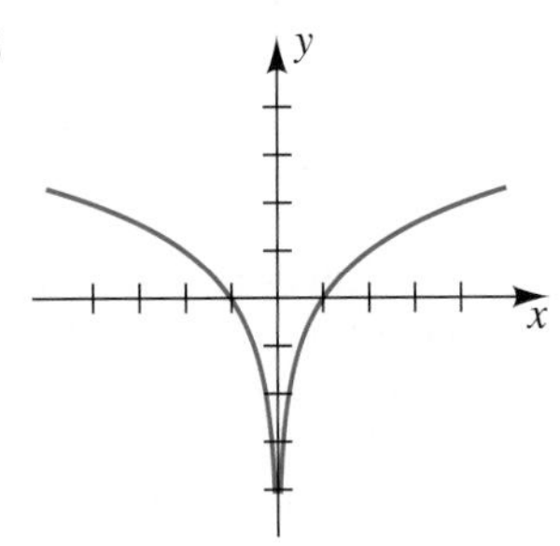

49

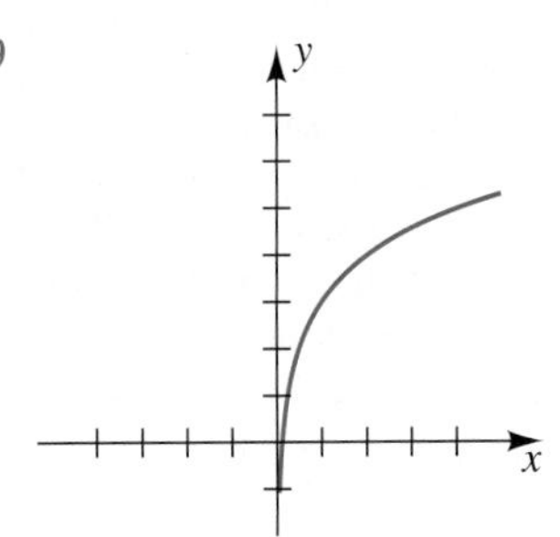

50

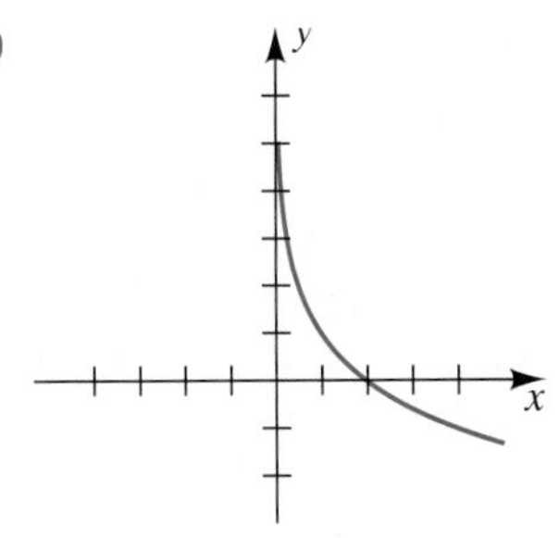

51 **Volume and decibels** When the volume control on a stereo system is increased, the voltage across a loudspeaker changes from V_1 to V_2, and the decibel increase in gain is given by

$$\text{db} = 20 \log \frac{V_2}{V_1}.$$

Find the decibel increase if the voltage changes from 2 volts to 4.5 volts.

52 **Volume and decibels** Refer to Exercise 51. What voltage ratio k is needed for a $+20$ decibel gain? for a $+40$ decibel gain?

53 **Pareto's law** Pareto's law for capitalist countries states that the relationship between annual income x and the number y of individuals whose income exceeds x is

$$\log y = \log b - k \log x,$$

where b and k are positive constants. Solve this equation for y.

54 **Price and demand** If p denotes the selling price (in dollars) of a commodity and x is the corresponding demand (in number sold per day), then the relationship between p and x is sometimes given by $p = p_0 e^{-ax}$, where p_0 and a are positive constants. Express x as a function of p.

55 **Wind velocity** If v denotes the wind velocity (in m/sec) at a height of z meters above the ground, then under certain conditions $v = c \ln (z/z_0)$, where c is a positive constant and z_0 is the height at which the velocity is zero. Sketch the graph of this equation on a zv-plane for $c = 0.5$ and $z_0 = 0.1$ m.

56 **Eliminating pollution** If the pollution of Lake Erie were stopped suddenly, it has been estimated that the level y of pollutants would decrease according to the formula $y = y_0 e^{-0.3821t}$, where t is the time in years and y_0 is the pollutant level at which further pollution ceased. How many years would it take to clear 50% of the pollutants?

57 **Reaction to a stimulus** Let R denote the reaction of a subject to a stimulus of strength x. There are many possibilities for R and x. If the stimulus x is saltiness (in grams of salt per liter), R may be the subject's estimate of how salty the solution tasted, based on a scale from 0 to 10. One relationship between R and x is given by the Weber-Fechner formula, $R(x) = a \log (x/x_0)$, where a is a positive constant and x_0 is called the threshold stimulus.

(a) Find $R(x_0)$.

(b) Find a relationship between $R(x)$ and $R(2x)$.

58 **Electron energy** The energy $E(x)$ of an electron after passing through material of thickness x is given by the equation $E(x) = E_0e^{-x/x_0}$, where E_0 is the initial energy and x_0 is the radiation length.

(a) Express, in terms of E_0, the energy of an electron after it passes through material of thickness x_0.

(b) Express, in terms of x_0, the thickness at which the electron loses 99% of its initial energy.

▶ 59 **Ozone layer** One method of estimating the thickness of the ozone layer is to use the formula

$$\ln I_0 - \ln I = kx,$$

where I_0 is the intensity of a particular wavelength of light from the sun before it reaches the atmosphere, I is the intensity of the same wavelength after passing through a layer of ozone x centimeters thick, and k is the absorption constant of ozone for that wavelength. Suppose for a wavelength of 3176×10^{-8} cm with $k \approx 0.39$, I_0/I is measured as 1.12. Approximate the thickness of the ozone layer to the nearest 0.01 centimeter.

60 **Ozone layer** Refer to Exercise 59. Approximate the percentage decrease in the intensity of light with a wavelength of 3176×10^{-8} centimeter if the ozone layer is 0.24 centimeter thick.

Exer. 61–62: Graph f and g on the same coordinate plane, and estimate the solution of the inequality $f(x) \geq g(x)$.

61 $f(x) = x^3 - 3.5x^2 + 3x;\quad g(x) = \log 3x$

62 $f(x) = 3^{-0.5x};\quad g(x) = \log x$

Exer. 63–64: Use a graph to estimate the roots of the equation on the given interval.

63 $e^{-x} - 2 \log (1 + x^2) + 0.5x = 0;\quad [0, 8]$

64 $0.3 \ln x + x^3 - 3.1x^2 + 1.3x + 0.8 = 0;\quad (0, 3)$

Exer. 65–66: Graph f on the interval [0.2, 16]. (a) Estimate the intervals where f is increasing or is decreasing. (b) Estimate the maximum and minimum values of f on [0.2, 16].

65 $f(x) = 2 \log 2x - 1.5x + 0.1x^2$

66 $f(x) = 1.1^{3x} + x - 1.35^x - \log x + 5$

Exer. 67–68: Solve the equation graphically.

67 $x \log x - \log x = 5$

68 $0.3e^x - \ln x = 4 \ln (x + 1)$

Exer. 69–70: Bird calls decrease in intensity (loudness) as they travel through the atmosphere. The farther a bird is from an observer, the softer the sound. This decrease in intensity can be used to estimate the distance between an observer and a bird. A formula that can be used to measure this distance is

$$I = I_0 - 20 \log d - kd \quad \text{provided} \quad 0 \leq I \leq I_0,$$

where I_0 represents the intensity (in decibels) of the bird at a distance of one meter (I_0 is often known and usually depends only on the type of bird), I is the observed intensity at a distance d meters from the bird, and k is a positive constant that depends on the atmospheric conditions such as temperature and humidity. Given I_0, I, and k, graphically estimate the distance d between the bird and the observer.

69 $I_0 = 70,\quad I = 20,\quad k = 0.076$

70 $I_0 = 60,\quad I = 15,\quad k = 0.11$

4.6 Exponential and Logarithmic Equations

In this section we shall consider various types of exponential and logarithmic equations and their applications. When solving an equation involving exponential expressions with constant bases and variables appearing in the exponent(s), we often *equate the logarithms of both sides* of the equation. When we do so, the variables in the exponent become multipliers, and the resulting equation is usually easier to solve. We will refer to this step as simply "take log of both sides."

EXAMPLE 1 Solving an exponential equation

Solve the equation $3^x = 21$.

▶ SOLUTION

$$\begin{aligned} 3^x &= 21 && \text{given} \\ \log(3^x) &= \log 21 && \text{take log of both sides} \\ x\log 3 &= \log 21 && \text{law 3 of logarithms} \\ x &= \frac{\log 21}{\log 3} && \text{divide by log 3} \end{aligned}$$

We could also have used natural logarithms to obtain

$$x = \frac{\ln 21}{\ln 3}.$$

Using a calculator gives us the approximate solution $x \approx 2.77$. A partial check is to note that since $3^2 = 9$ and $3^3 = 27$, the number x such that $3^x = 21$ must be between 2 and 3, somewhat closer to 3 than to 2.

We could also have solved the equation in Example 1 by changing the exponential form $3^x = 21$ to logarithmic form, as we did in Section 4.4, obtaining

$$x = \log_3 21.$$

This is, in fact, the solution of the equation; however, since calculators typically have keys only for log and ln, we cannot approximate $\log_3 21$ directly. The next theorem gives us a simple *change of base formula* for finding $\log_b u$ if $u > 0$ and b is *any* logarithmic base.

Theorem: Change of Base Formula

If $u > 0$ and if a and b are positive real numbers different from 1, then

$$\log_b u = \frac{\log_a u}{\log_a b}.$$

PROOF We begin with the equivalent equations

$$w = \log_b u \qquad \text{and} \qquad b^w = u$$

and proceed as follows:

$$\begin{aligned} b^w &= u && \text{given} \\ \log_a b^w &= \log_a u && \text{take } \log_a \text{ of both sides} \\ w\log_a b &= \log_a u && \text{law 3 of logarithms} \\ w &= \frac{\log_a u}{\log_a b} && \text{divide by } \log_a b \end{aligned}$$

Since $w = \log_b u$, we obtain the formula.

▶ **Tutorial available at www.thomsonedu.com/login**

The following special case of the change of base formula is obtained by letting $u = a$ and using the fact that $\log_a a = 1$:

$$\log_b a = \frac{1}{\log_a b}$$

The change of base formula is sometimes confused with law 2 of logarithms. The first of the following warnings could be remembered with the phrase "a quotient of logs is *not* the log of the quotient."

$$\frac{\log_a u}{\log_a b} \neq \log_a \frac{u}{b}; \qquad \frac{\log_a u}{\log_a b} \neq \log_a (u - b)$$

The most frequently used special cases of the change of base formula are those for $a = 10$ (common logarithms) and $a = e$ (natural logarithms), as stated in the next box.

Special Change of Base Formulas

$$(1)\quad \log_b u = \frac{\log_{10} u}{\log_{10} b} = \frac{\log u}{\log b} \qquad (2)\quad \log_b u = \frac{\log_e u}{\log_e b} = \frac{\ln u}{\ln b}$$

Next, we will rework Example 1 using a change of base formula.

EXAMPLE 2 Using a change of base formula

Solve the equation $3^x = 21$.

SOLUTION We proceed as follows:

$$3^x = 21 \qquad \text{given}$$

$$x = \log_3 21 \qquad \text{change to logarithmic form}$$

$$= \frac{\log 21}{\log 3} \qquad \text{special change of base formula 1}$$

Another method is to use special change of base formula 2, obtaining

$$x = \frac{\ln 21}{\ln 3}.$$

Logarithms with base 2 are used in computer science. The next example indicates how to approximate logarithms with base 2 using change of base formulas.

EXAMPLE 3 Approximating a logarithm with base 2

Approximate $\log_2 5$ using

(a) common logarithms **(b)** natural logarithms

Tutorial available at www.thomsonedu.com/login

SOLUTION Using special change of base formulas 1 and 2, we obtain the following:

(a) $\log_2 5 = \dfrac{\log 5}{\log 2} \approx 2.322$

▶ (b) $\log_2 5 = \dfrac{\ln 5}{\ln 2} \approx 2.322$

EXAMPLE 4 Solving an exponential equation

Solve the equation $5^{2x+1} = 6^{x-2}$.

▶ SOLUTION We can use either common or natural logarithms. Using common logarithms gives us the following:

$5^{2x+1} = 6^{x-2}$	given
$\log(5^{2x+1}) = \log(6^{x-2})$	take log of both sides
$(2x+1)\log 5 = (x-2)\log 6$	law 3 of logarithms
$2x\log 5 + \log 5 = x\log 6 - 2\log 6$	multiply
$2x\log 5 - x\log 6 = -\log 5 - 2\log 6$	get all terms with x on one side
$x(\log 5^2 - \log 6) = -(\log 5 + \log 6^2)$	factor, and use law 3 of logarithms
$x = -\dfrac{\log(5 \cdot 36)}{\log \frac{25}{6}}$	solve for x, and use laws of logarithms

Figure 1

```
-log(180)/log(25
/6)→X
      -3.638776075
5^(2X+1)
    4.094297034E-5
6^(X-2)
    4.094297034E-5
```

An approximation is $x \approx -3.64$. Figure 1 shows a calculator check for this example. We deduce from the check that the graphs of $y = 5^{2x+1}$ and $y = 6^{x-2}$ intersect at approximately $(-3.64, 0.00004)$.

EXAMPLE 5 Solving an exponential equation

Solve the equation $\dfrac{5^x - 5^{-x}}{2} = 3$.

▶ SOLUTION

$\dfrac{5^x - 5^{-x}}{2} = 3$	given
$5^x - 5^{-x} = 6$	multiply by 2
$5^x - \dfrac{1}{5^x} = 6$	definition of negative exponent
$5^x(5^x) - \dfrac{1}{5^x}(5^x) = 6(5^x)$	multiply by the lcd, 5^x
$(5^x)^2 - 6(5^x) - 1 = 0$	simplify and subtract $6(5^x)$

(continued)

▶ **Tutorial available at www.thomsonedu.com/login**

Note that $(5^x)^2$ can be written as 5^{2x}.

We recognize this form of the equation as a quadratic in 5^x and proceed as follows:

$$\begin{aligned}
(5^x)^2 - 6(5^x) - 1 &= 0 && \text{law of exponents}\\
5^x &= \frac{6 \pm \sqrt{36+4}}{2} && \text{quadratic formula}\\
5^x &= 3 \pm \sqrt{10} && \text{simplify}\\
5^x &= 3 + \sqrt{10} && 5^x > 0, \text{ but } 3 - \sqrt{10} < 0\\
\log 5^x &= \log\left(3 + \sqrt{10}\right) && \text{take log of both sides}\\
x \log 5 &= \log\left(3 + \sqrt{10}\right) && \text{law 3 of logarithms}\\
x &= \frac{\log\left(3 + \sqrt{10}\right)}{\log 5} && \text{divide by log 5}
\end{aligned}$$

An approximation is $x \approx 1.13$.

EXAMPLE 6 Solving an equation involving logarithms

Solve the equation $\log \sqrt[3]{x} = \sqrt{\log x}$ for x.

▶ SOLUTION

$$\begin{aligned}
\log x^{1/3} &= \sqrt{\log x} && \sqrt[n]{x} = x^{1/n}\\
\tfrac{1}{3} \log x &= \sqrt{\log x} && \log x^r = r \log x\\
\tfrac{1}{9}(\log x)^2 &= \log x && \text{square both sides}\\
(\log x)^2 &= 9 \log x && \text{multiply by 9}\\
(\log x)^2 - 9 \log x &= 0 && \text{make one side 0}\\
(\log x)(\log x - 9) &= 0 && \text{factor out } \log x\\
\log x = 0, \qquad \log x - 9 &= 0 && \text{set each factor equal to 0}\\
\log x &= 9 && \text{add 9}\\
x = 10^0 = 1 \quad \text{or} \quad x &= 10^9 && \log_{10} x = a \Longleftrightarrow x = 10^a
\end{aligned}$$

✓ Check $x = 1$ LS: $\log \sqrt[3]{1} = \log 1 = 0$
RS: $\sqrt{\log 1} = \sqrt{0} = 0$

✓ Check $x = 10^9$ LS: $\log \sqrt[3]{10^9} = \log 10^3 = 3$
RS: $\sqrt{\log 10^9} = \sqrt{9} = 3$

The equation has two solutions, 1 and 1 billion.

The function $y = 2/(e^x + e^{-x})$ is called the **hyperbolic secant function.** In the next example we solve this equation for x in terms of y. Under suitable restrictions, this gives us the inverse function.

▶ **Tutorial available at www.thomsonedu.com/login**

EXAMPLE 7 Finding an inverse hyperbolic function

Solve $y = 2/(e^x + e^{-x})$ for x in terms of y.

▶ **SOLUTION**

$$y = \frac{2}{e^x + e^{-x}} \quad \text{given}$$

$$ye^x + ye^{-x} = 2 \quad \text{multiply by } e^x + e^{-x}$$

$$ye^x + \frac{y}{e^x} = 2 \quad \text{definition of negative exponent}$$

$$ye^x(e^x) + \frac{y}{e^x}(e^x) = 2(e^x) \quad \text{multiply by the lcd, } e^x$$

$$y(e^x)^2 - 2e^x + y = 0 \quad \text{simplify and subtract } 2e^x$$

We recognize this form of the equation as a quadratic in e^x with coefficients $a = y$, $b = -2$, and $c = y$. Note that we are solving for e^x, not x.

$$e^x = \frac{-(-2) \pm \sqrt{(-2)^2 - 4(y)(y)}}{2(y)} \quad \text{quadratic formula}$$

$$= \frac{2 \pm \sqrt{4 - 4y^2}}{2y} \quad \text{simplify}$$

$$= \frac{2 \pm \sqrt{4}\sqrt{1 - y^2}}{2y} \quad \text{factor out } \sqrt{4}$$

$$e^x = \frac{1 \pm \sqrt{1 - y^2}}{y} \quad \text{cancel a factor of 2}$$

$$x = \ln \frac{1 \pm \sqrt{1 - y^2}}{y} \quad \text{take ln of both sides}$$

Figure 2

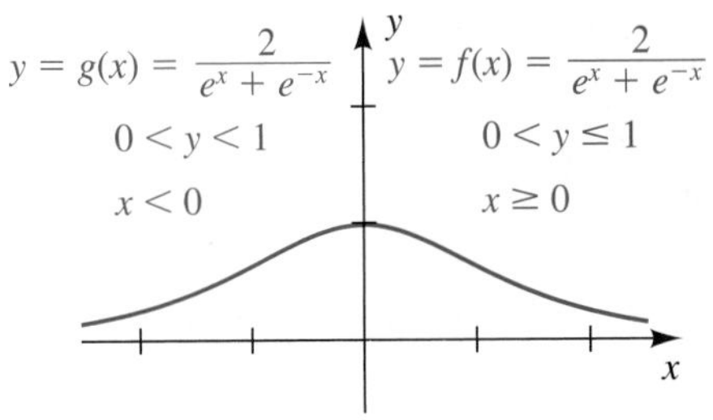

$y = g(x) = \frac{2}{e^x + e^{-x}}$, $0 < y < 1$, $x < 0$

$y = f(x) = \frac{2}{e^x + e^{-x}}$, $0 < y \le 1$, $x \ge 0$

Figure 3

$y = f^{-1}(x) = \ln \frac{1 + \sqrt{1 - x^2}}{x}$, $0 < x \le 1$, $y \ge 0$

$y = g^{-1}(x) = \ln \frac{1 - \sqrt{1 - x^2}}{x}$, $0 < x < 1$, $y < 0$

For the blue curve $y = f(x)$ in Figure 2, the inverse function is

$$y = f^{-1}(x) = \ln \frac{1 + \sqrt{1 - x^2}}{x},$$

shown in blue in Figure 3. Notice the domain and range relationships. For the red curve $y = g(x)$ in Figure 2, the inverse function is

$$y = g^{-1}(x) = \ln \frac{1 - \sqrt{1 - x^2}}{x},$$

shown in red in Figure 3. Since the hyperbolic secant is not one-to-one, it cannot have one simple equation for its inverse.

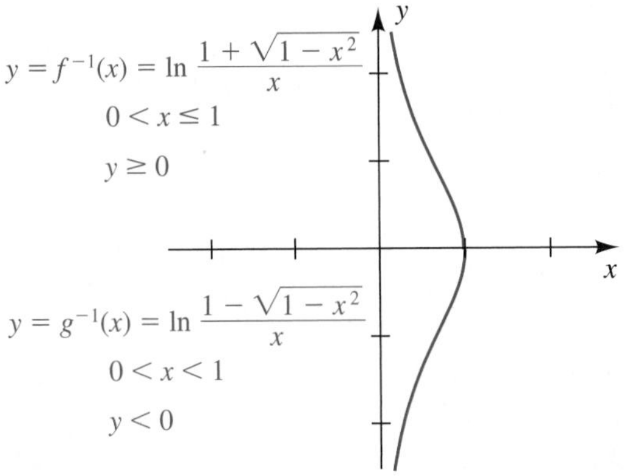

▶ **Tutorial available at www.thomsonedu.com/login**

The inverse hyperbolic secant is part of the equation of the curve called a **tractrix.** The curve is associated with Gottfried Wilhelm von Leibniz's (1646–1716) solution to the question "What is the path of an object dragged along a horizontal plane by a string of constant length when the end of the string not joined to the object moves along a straight line in the plane?"

EXAMPLE 8 Approximating light penetration in an ocean

The Beer-Lambert law states that the amount of light I that penetrates to a depth of x meters in an ocean is given by $I = I_0c^x$, where $0 < c < 1$ and I_0 is the amount of light at the surface.

(a) Solve for x in terms of common logarithms.

(b) If $c = \frac{1}{4}$, approximate the depth at which $I = 0.01I_0$. (This determines the photic zone where photosynthesis can take place.)

SOLUTION

(a)

$I = I_0c^x$	given
$\dfrac{I}{I_0} = c^x$	isolate the exponential expression
$x = \log_c \dfrac{I}{I_0}$	change to logarithmic form
$= \dfrac{\log (I/I_0)}{\log c}$	special change of base formula 1

▶ **(b)** Letting $I = 0.01I_0$ and $c = \frac{1}{4}$ in the formula for x obtained in part (a), we have

$x = \dfrac{\log (0.01I_0/I_0)}{\log \frac{1}{4}}$	substitute for I and c
$= \dfrac{\log (0.01)}{\log 1 - \log 4}$	cancel I_0; law 2 of logarithms
$= \dfrac{\log 10^{-2}}{0 - \log 4}$	property of logarithms
$= \dfrac{-2}{-\log 4}$	$\log 10^x = x$
$= \dfrac{2}{\log 4}.$	simplify

An approximation is $x \approx 3.32$ m.

▶ **Tutorial available at www.thomsonedu.com/login**

EXAMPLE 9 Comparing light intensities

If a beam of light that has intensity I_0 is projected vertically downward into water, then its intensity $I(x)$ at a depth of x meters is $I(x) = I_0 e^{-1.4x}$ (see Figure 4). At what depth is the intensity one-half its value at the surface?

SOLUTION At the surface, $x = 0$, and the intensity is

$$\begin{aligned} I(0) &= I_0 e^0 \\ &= I_0. \end{aligned}$$

Figure 4

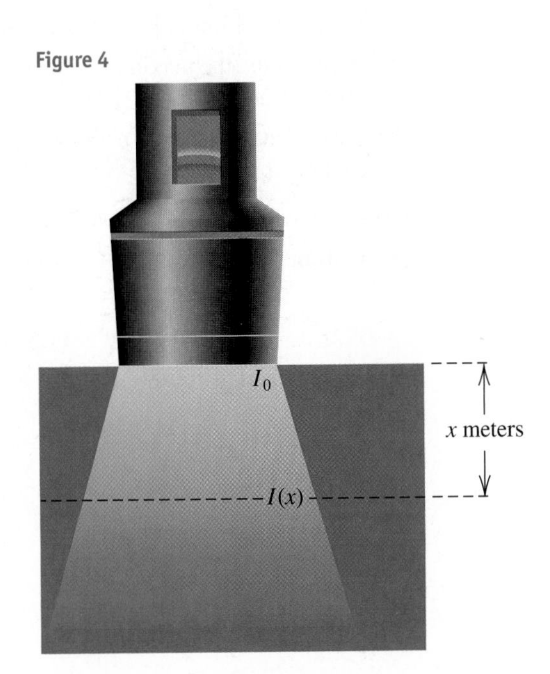

We wish to find the value of x such that $I(x) = \frac{1}{2}I_0$. This leads to the following:

$$\begin{aligned} I(x) &= \tfrac{1}{2}I_0 && \text{desired intensity} \\ I_0 e^{-1.4x} &= \tfrac{1}{2}I_0 && \text{formula for } I(x) \\ e^{-1.4x} &= \tfrac{1}{2} && \text{divide by } I_0 \ (I_0 \neq 0) \\ -1.4x &= \ln \tfrac{1}{2} && \text{change to logarithmic form} \\ x &= \frac{\ln \frac{1}{2}}{-1.4} && \text{divide by } -1.4 \end{aligned}$$

An approximation is $x \approx 0.495$ m.

Tutorial available at www.thomsonedu.com/login

EXAMPLE 10 A logistic curve

A **logistic curve** is the graph of an equation of the form

$$y = \frac{k}{1 + be^{-cx}},$$

where k, b, and c are positive constants. Such curves are useful for describing a population y that grows rapidly initially, but whose growth rate decreases after x reaches a certain value. In a famous study of the growth of protozoa by Gause, a population of *Paramecium caudata* was found to be described by a logistic equation with $c = 1.1244$, $k = 105$, and x the time in days.

(a) Find b if the initial population was 3 protozoa.

(b) In the study, the maximum growth rate took place at $y = 52$. At what time x did this occur?

(c) Show that after a long period of time, the population described by any logistic curve approaches the constant k.

SOLUTION

▶ **(a)** Letting $c = 1.1244$ and $k = 105$ in the logistic equation, we obtain

$$y = \frac{105}{1 + be^{-1.1244x}}.$$

We now proceed as follows:

$$3 = \frac{105}{1 + be^{0}} = \frac{105}{1 + b} \qquad y = 3 \text{ when } x = 0$$

$$1 + b = 35 \qquad \text{multiply by } \frac{1 + b}{3}$$

$$b = 34 \qquad \text{solve for } b$$

(b) Using the fact that $b = 34$ leads to the following:

$$52 = \frac{105}{1 + 34e^{-1.1244x}} \qquad \text{let } y = 52 \text{ in part (a)}$$

$$1 + 34e^{-1.1244x} = \frac{105}{52} \qquad \text{multiply by } \frac{1 + 34e^{-1.1244x}}{52}$$

$$e^{-1.1244x} = \left(\tfrac{105}{52} - 1\right) \cdot \tfrac{1}{34} = \tfrac{53}{1768} \qquad \text{isolate } e^{-1.1244x}$$

$$-1.1244x = \ln \tfrac{53}{1768} \qquad \text{change to logarithmic form}$$

$$x = \frac{\ln \frac{53}{1768}}{-1.1244} \approx 3.12 \text{ days} \qquad \text{divide by } -1.1244$$

(c) As $x \to \infty$, $e^{-cx} \to 0$. Hence,

$$y = \frac{k}{1 + be^{-cx}} \to \frac{k}{1 + b \cdot 0} = k.$$

▶ **Tutorial available at www.thomsonedu.com/login**

In the next example we graph the equation obtained in part (a) of the preceding example.

EXAMPLE 11 **Sketching the graph of a logistic curve**

Graph the logistic curve given by

$$y = \frac{105}{1 + 34e^{-1.1244x}},$$

and estimate the value of x for $y = 52$.

Figure 5
[0, 10] by [0, 105, 10]

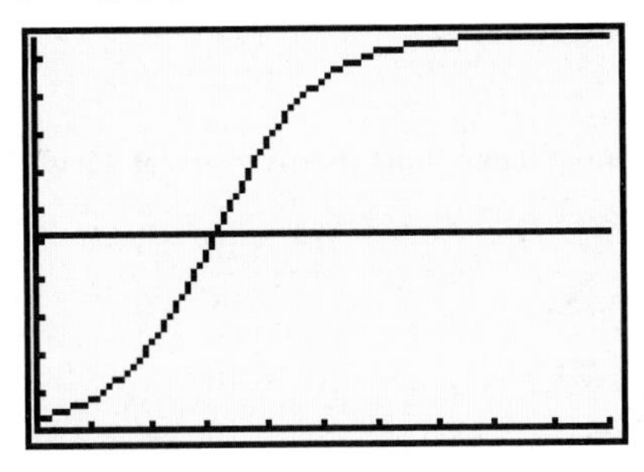

SOLUTION We begin by assigning

$$\frac{105}{1 + 34e^{-1.1244x}}$$

to Y_1 and 52 to Y_2. Since the time x is nonnegative, we choose Xmin = 0. We select Xmax = 10 in order to include the value of x found in part (b) of Example 10. By part (c), we know that the value of y cannot exceed 105. Thus, we choose Ymin = 0 and Ymax = 105 and obtain a display similar to Figure 5.

Using an intersect feature, we see that for $y = 52$, the value of x is approximately 3.12, which agrees with the approximation found in (b) of Example 10.

The following example shows how a change of base formula may be used to enable us to graph logarithmic functions with bases other than 10 and e on a graphing utility.

EXAMPLE 12 **Estimating points of intersection of logarithmic graphs**

Estimate the point of intersection of the graphs of

$$f(x) = \log_3 x \qquad \text{and} \qquad g(x) = \log_6 (x + 2).$$

SOLUTION Most graphing utilities are equipped to work with only common and natural logarithmic functions. Thus, we first use a change of base formula to rewrite f and g as

$$f(x) = \frac{\ln x}{\ln 3} \qquad \text{and} \qquad g(x) = \frac{\ln (x + 2)}{\ln 6}.$$

Figure 6
[−2, 4] by [−2, 2]

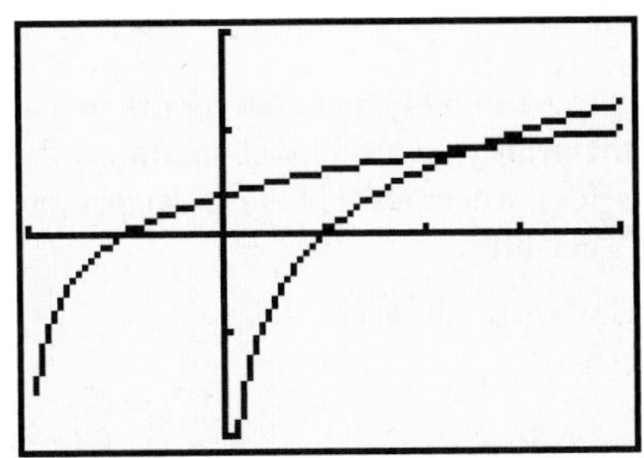

We next assign $(\ln x)/\ln 3$ and $(\ln (x + 2))/\ln 6$ to Y_1 and Y_2, respectively. After graphing Y_1 and Y_2 using a standard viewing rectangle, we see that there is a point of intersection in the first quadrant with $2 < x < 3$. Using an intersect feature, we find that the point of intersection is approximately (2.52, 0.84).

Figure 6 was obtained using viewing rectangle dimensions [−2, 4] by [−2, 2]. There are no other points of intersection, since f increases more rapidly than g for $x > 3$.

4.6 Exercises

Exer. 1–4: Find the exact solution and a two-decimal-place approximation for it by using (a) the method of Example 1 and (b) the method of Example 2.

1 $5^x = 8$

2 $4^x = 3$

3 $3^{4-x} = 5$

4 $\left(\frac{1}{3}\right)^x = 100$

Exer. 5–8: Estimate using the change of base formula.

5 $\log_5 6$

6 $\log_2 20$

7 $\log_9 0.2$

8 $\log_6 \frac{1}{2}$

Exer. 9–10: Evaluate using the change of base formula (without a calculator).

9 $\dfrac{\log_5 16}{\log_5 4}$

10 $\dfrac{\log_7 243}{\log_7 3}$

Exer. 11–24: Find the exact solution, using common logarithms, and a two-decimal-place approximation of each solution, when appropriate.

11 $3^{x+4} = 2^{1-3x}$

12 $4^{2x+3} = 5^{x-2}$

13 $2^{2x-3} = 5^{x-2}$

14 $3^{2-3x} = 4^{2x+1}$

15 $2^{-x} = 8$

16 $2^{-x^2} = 5$

17 $\log x = 1 - \log (x - 3)$

18 $\log (5x + 1) = 2 + \log (2x - 3)$

19 $\log (x^2 + 4) - \log (x + 2) = 2 + \log (x - 2)$

20 $\log (x - 4) - \log (3x - 10) = \log (1/x)$

21 $5^x + 125(5^{-x}) = 30$

22 $3(3^x) + 9(3^{-x}) = 28$

23 $4^x - 3(4^{-x}) = 8$

24 $2^x - 6(2^{-x}) = 6$

Exer. 25–32: Solve the equation without using a calculator.

25 $\log (x^2) = (\log x)^2$

26 $\log \sqrt{x} = \sqrt{\log x}$

27 $\log (\log x) = 2$

28 $\log \sqrt{x^3 - 9} = 2$

29 $x^{\sqrt{\log x}} = 10^8$

30 $\log (x^3) = (\log x)^3$

31 $e^{2x} + 2e^x - 15 = 0$

32 $e^x + 4e^{-x} = 5$

Exer. 33–34: Solve the equation.

33 $\log_3 x - \log_9 (x + 42) = 0$

34 $\log_4 x + \log_8 x = 1$

Exer. 35–38: Use common logarithms to solve for x in terms of y.

35 $y = \dfrac{10^x + 10^{-x}}{2}$

36 $y = \dfrac{10^x - 10^{-x}}{2}$

37 $y = \dfrac{10^x - 10^{-x}}{10^x + 10^{-x}}$

38 $y = \dfrac{10^x + 10^{-x}}{10^x - 10^{-x}}$

Exer. 39–42: Use natural logarithms to solve for x in terms of y.

39 $y = \dfrac{e^x - e^{-x}}{2}$

40 $y = \dfrac{e^x + e^{-x}}{2}$

41 $y = \dfrac{e^x + e^{-x}}{e^x - e^{-x}}$

42 $y = \dfrac{e^x - e^{-x}}{e^x + e^{-x}}$

Exer. 43–44: Sketch the graph of f, and use the change of base formula to approximate the y-intercept.

43 $f(x) = \log_2 (x + 3)$

44 $f(x) = \log_3 (x + 5)$

Exer. 45–46: Sketch the graph of f, and use the change of base formula to approximate the x-intercept.

45 $f(x) = 4^x - 3$

46 $f(x) = 3^x - 6$

Exer. 47–50: Chemists use a number denoted by pH to describe quantitatively the acidity or basicity of solutions. By definition, $\text{pH} = -\log [H^+]$, where $[H^+]$ is the hydrogen ion concentration in moles per liter.

47 Approximate the pH of each substance.

(a) vinegar: $[H^+] \approx 6.3 \times 10^{-3}$

(b) carrots: $[H^+] \approx 1.0 \times 10^{-5}$

(c) sea water: $[H^+] \approx 5.0 \times 10^{-9}$

48 Approximate the hydrogen ion concentration $[H^+]$ of each substance.

(a) apples: pH $\approx$ 3.0

(b) beer: pH $\approx$ 4.2

(c) milk: pH $\approx$ 6.6

49 A solution is considered basic if $[H^+] < 10^{-7}$ or acidic if $[H^+] > 10^{-7}$. Find the corresponding inequalities involving pH.

50 Many solutions have a pH between 1 and 14. Find the corresponding range of $[H^+]$.

51 **Compound interest** Use the compound interest formula to determine how long it will take for a sum of money to double if it is invested at a rate of 6% per year compounded monthly.

52 **Compound interest** Solve the compound interest formula

$$A = P\left(1 + \frac{r}{n}\right)^{nt}$$

for t by using natural logarithms.

53 **Photic zone** Refer to Example 8. The most important zone in the sea from the viewpoint of marine biology is the photic zone, in which photosynthesis takes place. The photic zone ends at the depth where about 1% of the surface light penetrates. In very clear waters in the Caribbean, 50% of the light at the surface reaches a depth of about 13 meters. Estimate the depth of the photic zone.

54 **Photic zone** In contrast to the situation described in the previous exercise, in parts of New York harbor, 50% of the surface light does not reach a depth of 10 centimeters. Estimate the depth of the photic zone.

55 **Drug absorption** If a 100-milligram tablet of an asthma drug is taken orally and if none of the drug is present in the body when the tablet is first taken, the total amount A in the bloodstream after t minutes is predicted to be

$$A = 100[1 - (0.9)^t] \quad \text{for} \quad 0 \leq t \leq 10.$$

(a) Sketch the graph of the equation.

(b) Determine the number of minutes needed for 50 milligrams of the drug to have entered the bloodstream.

56 **Drug dosage** A drug is eliminated from the body through urine. Suppose that for a dose of 10 milligrams, the amount $A(t)$ remaining in the body t hours later is given by $A(t) = 10(0.8)^t$ and that in order for the drug to be effective, at least 2 milligrams must be in the body.

(a) Determine when 2 milligrams is left in the body.

(b) What is the half-life of the drug?

57 **Genetic mutation** The basic source of genetic diversity is mutation, or changes in the chemical structure of genes. If a gene mutates at a constant rate m and if other evolutionary forces are negligible, then the frequency F of the original gene after t generations is given by $F = F_0(1 - m)^t$, where F_0 is the frequency at $t = 0$.

(a) Solve the equation for t using common logarithms.

(b) If $m = 5 \times 10^{-5}$, after how many generations does $F = \frac{1}{2}F_0$?

58 **Employee productivity** Certain learning processes may be illustrated by the graph of an equation of the form $f(x) = a + b(1 - e^{-cx})$, where a, b, and c are positive constants. Suppose a manufacturer estimates that a new employee can produce five items the first day on the job. As the employee becomes more proficient, the daily production increases until a certain maximum production is reached. Suppose that on the nth day on the job, the number $f(n)$ of items produced is approximated by

$$f(n) = 3 + 20(1 - e^{-0.1n}).$$

(a) Estimate the number of items produced on the fifth day, the ninth day, the twenty-fourth day, and the thirtieth day.

(b) Sketch the graph of f from $n = 0$ to $n = 30$. (Graphs of this type are called *learning curves* and are used frequently in education and psychology.)

(c) What happens as n increases without bound?

59 Height of trees The growth in height of trees is frequently described by a logistic equation. Suppose the height h (in feet) of a tree at age t (in years) is

$$h = \frac{120}{1 + 200e^{-0.2t}},$$

as illustrated by the graph in the figure.

(a) What is the height of the tree at age 10?

(b) At what age is the height 50 feet?

Exercise 59

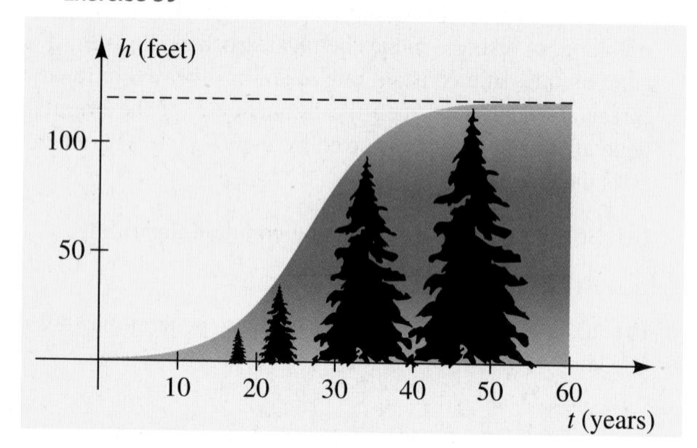

60 Employee productivity Manufacturers sometimes use empirically based formulas to predict the time required to produce the nth item on an assembly line for an integer n. If $T(n)$ denotes the time required to assemble the nth item and T_1 denotes the time required for the first, or prototype, item, then typically $T(n) = T_1 n^{-k}$ for some positive constant k.

(a) For many airplanes, the time required to assemble the second airplane, $T(2)$, is equal to $(0.80)T_1$. Find the value of k.

(b) Express, in terms of T_1, the time required to assemble the fourth airplane.

(c) Express, in terms of $T(n)$, the time $T(2n)$ required to assemble the $(2n)$th airplane.

61 Vertical wind shear Refer to Exercises 67–68 in Section 2.3. If v_0 is the wind speed at height h_0 and if v_1 is the wind speed at height h_1, then the vertical wind shear can be described by the equation

$$\frac{v_0}{v_1} = \left(\frac{h_0}{h_1}\right)^P,$$

where P is a constant. During a one-year period in Montreal, the maximum vertical wind shear occurred when the winds at the 200-foot level were 25 mi/hr while the winds at the 35-foot level were 6 mi/hr. Find P for these conditions.

62 Vertical wind shear Refer to Exercise 61. The average vertical wind shear is given by the equation

$$s = \frac{v_1 - v_0}{h_1 - h_0}.$$

Suppose that the velocity of the wind increases with increasing altitude and that all values for wind speeds taken at the 35-foot and 200-foot altitudes are greater than 1 mi/hr. Does increasing the value of P produce larger or smaller values of s?

Exer. 63–64: An economist suspects that the following data points lie on the graph of $y = c2^{kx}$, where c and k are constants. If the data points have three-decimal-place accuracy, is this suspicion correct?

63 (0, 4), (1, 3.249), (2, 2.639), (3, 2.144)

64 (0, −0.3), (0.5, −0.345), (1, −0.397), (1.5, −0.551), (2, −0.727)

Exer. 65–66: It is suspected that the following data points lie on the graph of $y = c \log (kx + 10)$, where c and k are constants. If the data points have three-decimal-place accuracy, is this suspicion correct?

65 (0, 1.5), (1, 1.619), (2, 1.720), (3, 1.997)

66 (0, 0.7), (1, 0.782), (2, 0.847), (3, 0.900), (4, 0.945)

Exer. 67–68: Approximate the function at the value of x to four decimal places.

67 $h(x) = \log_4 x - 2 \log_8 1.2x$; $x = 5.3$

68 $h(x) = 3 \log_3 (2x - 1) + 7 \log_2 (x + 0.2)$; $x = 52.6$

Exer. 69–70: Use a graph to estimate the roots of the equation on the given interval.

69 $x - \ln (0.3x) - 3 \log_3 x = 0$; $(0, 9)$

70 $2 \log 2x - \log_3 x^2 = 0$; $(0, 3)$

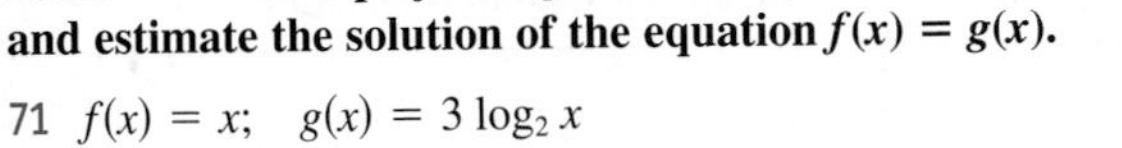

Exer. 71–72: Graph f and g on the same coordinate plane, and estimate the solution of the equation $f(x) = g(x)$.

71 $f(x) = x;\quad g(x) = 3 \log_2 x$

72 $f(x) = x;\quad g(x) = -x^2 - \log_5 x$

Exer. 73–74: Graph f and g on the same coordinate plane, and estimate the solution of the inequality $f(x) > g(x)$.

73 $f(x) = 3^{-x} - 4^{0.2x};\quad g(x) = \ln(1.2) - x$

74 $f(x) = 3 \log_4 x - \log x;\quad g(x) = e^x - 0.25x^4$

75 **Human memory** A group of elementary students were taught long division over a one-week period. Afterward, they were given a test. The average score was 85. Each week thereafter, they were given an equivalent test, without any review. Let $n(t)$ represent the average score after $t \geq 0$ weeks. Graph each $n(t)$, and determine which function best models the situation.

(1) $n(t) = 85e^{t/3}$

(2) $n(t) = 70 + 10 \ln(t + 1)$

(3) $n(t) = 86 - e^t$

(4) $n(t) = 85 - 15 \ln(t + 1)$

76 **Cooling** A jar of boiling water at 212°F is set on a table in a room with a temperature of 72°F. If $T(t)$ represents the temperature of the water after t hours, graph $T(t)$ and determine which function best models the situation.

(1) $T(t) = 212 - 50t$

(2) $T(t) = 140e^{-t} + 72$

(3) $T(t) = 212e^{-t}$

(4) $T(t) = 72 + 10 \ln(140t + 1)$

CHAPTER 4 REVIEW EXERCISES

▶ **Online support materials can be found at www.thomsonedu.com/login**

1 Is $f(x) = 2x^3 - 5$ a one-to-one function?

2 The graph of a function f with domain $[-3, 3]$ is shown in the figure. Sketch the graph of $y = f^{-1}(x)$.

Exercise 2

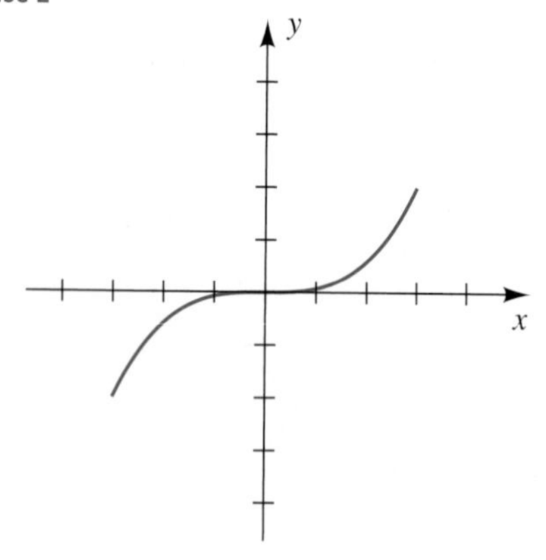

Exer. 3–4: (a) Find $f^{-1}(x)$. (b) Sketch the graphs of f and f^{-1} on the same coordinate plane.

3 $f(x) = 10 - 15x$ 4 $f(x) = 9 - 2x^2, x \leq 0$

5 Refer to the figure to determine each of the following:

(a) $f(1)$ (b) $(f \circ f)(1)$ (c) $f^{-1}(4)$

(d) all x such that $f(x) = 4$

(e) all x such that $f(x) > 4$

Exercise 5

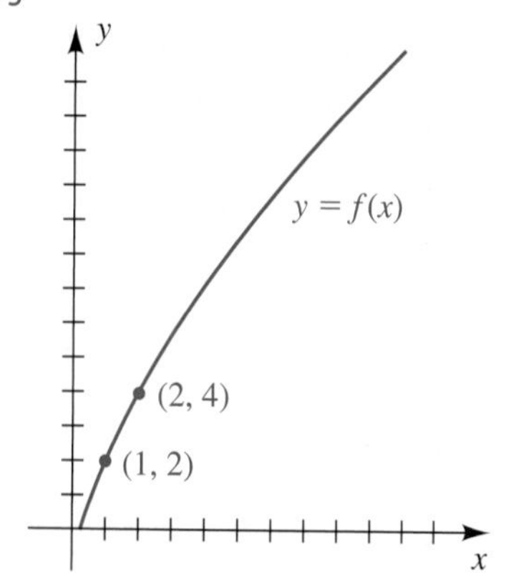

6 Suppose f and g are one-to-one functions such that $f(2) = 7$, $f(4) = 2$, and $g(2) = 5$. Find the value, if possible.

(a) $(g \circ f^{-1})(7)$ (b) $(f \circ g^{-1})(5)$

(c) $(f^{-1} \circ g^{-1})(5)$ (d) $(g^{-1} \circ f^{-1})(2)$

Exer. 7–22: Sketch the graph of f.

7 $f(x) = 3^{x+2}$ 8 $f(x) = \left(\frac{3}{5}\right)^x$

9 $f(x) = \left(\frac{3}{2}\right)^{-x}$ 10 $f(x) = 3^{-2x}$

11 $f(x) = 3^{-x^2}$ 12 $f(x) = 1 - 3^{-x}$

13 $f(x) = e^{x/2}$ 14 $f(x) = \frac{1}{2}e^x$

15 $f(x) = e^{x-2}$ 16 $f(x) = e^{2-x}$

17 $f(x) = \log_6 x$ 18 $f(x) = \log_6 (36x)$

19 $f(x) = \log_4 (x^2)$ 20 $f(x) = \log_4 \sqrt[3]{x}$

21 $f(x) = \log_2 (x + 4)$ 22 $f(x) = \log_2 (4 - x)$

Exer. 23–24: Evaluate without using a calculator.

23 (a) $\log_2 \frac{1}{16}$ (b) $\log_\pi 1$ (c) $\ln e$

(d) $6^{\log_6 4}$ (e) $\log 1{,}000{,}000$ (f) $10^{3 \log 2}$

(g) $\log_4 2$

24 (a) $\log_5 \sqrt[3]{5}$ (b) $\log_5 1$ (c) $\log 10$

(d) $e^{\ln 5}$ (e) $\log \log 10^{10}$ (f) $e^{2 \ln 5}$

(g) $\log_{27} 3$

Exer. 25–44: Solve the equation without using a calculator.

25 $2^{3x-1} = \frac{1}{2}$ 26 $8^{2x} \cdot \left(\frac{1}{4}\right)^{x-2} = 4^{-x} \cdot \left(\frac{1}{2}\right)^{2-x}$

27 $\log \sqrt{x} = \log (x - 6)$ 28 $\log_8 (x - 5) = \frac{2}{3}$

29 $\log_4 (x + 1) = 2 + \log_4 (3x - 2)$

30 $2 \ln (x + 3) - \ln (x + 1) = 3 \ln 2$

31 $\ln (x + 2) = \ln e^{\ln 2} - \ln x$ 32 $\log \sqrt[4]{x + 1} = \frac{1}{2}$

33 $2^{5-x} = 6$ 34 $3^{(x^2)} = 7$

35 $2^{5x+3} = 3^{2x+1}$

36 $\log_3 (3x) = \log_3 x + \log_3 (4 - x)$

37 $\log_4 x = \sqrt[3]{\log_4 x}$

38 $e^{x+\ln 4} = 3e^x$

39 $10^{2 \log x} = 5$

40 $e^{\ln (x+1)} = 3$

41 $x^2(-2xe^{-x^2}) + 2xe^{-x^2} = 0$

42 $e^x + 2 = 8e^{-x}$

43 (a) $\log x^2 = \log (6 - x)$ (b) $2 \log x = \log (6 - x)$

44 (a) $\ln (e^x)^2 = 16$ (b) $\ln e^{(x^2)} = 16$

45 Express $\log x^4 \sqrt[3]{y^2/z}$ in terms of logarithms of x, y, and z.

46 Express $\log (x^2/y^3) + 4 \log y - 6 \log \sqrt{xy}$ as one logarithm.

47 Find an exponential function that has y-intercept 6 and passes through the point (1, 8).

48 Sketch the graph of $f(x) = \log_3(x + 2)$.

Exer. 49–50: Use common logarithms to solve the equation for x in terms of y.

49 $y = \dfrac{1}{10^x + 10^{-x}}$

50 $y = \dfrac{1}{10^x - 10^{-x}}$

Exer. 51–52: Approximate x to three significant figures.

51 (a) $x = \ln 6.6$ (b) $\log x = 1.8938$

(c) $\ln x = -0.75$

52 (a) $x = \log 8.4$ (b) $\log x = -2.4260$

(c) $\ln x = 1.8$

Exer. 53–54: (a) Find the domain and range of the function. (b) Find the inverse of the function and its domain and range.

53 $y = \log_2 (x + 1)$

54 $y = 2^{3-x} - 2$

55 **Bacteria growth** The number of bacteria in a certain culture at time t (in hours) is given by $Q(t) = 2(3^t)$, where $Q(t)$ is measured in thousands.

(a) What is the number of bacteria at $t = 0$?

(b) Find the number of bacteria after 10 minutes, 30 minutes, and 1 hour.

56 **Compound interest** If \$1000 is invested at a rate of 8% per year compounded quarterly, what is the principal after one year?

57 **Radioactive iodine decay** Radioactive iodine ^{131}I, which is frequently used in tracer studies involving the thyroid gland, decays according to $N = N_0(0.5)^{t/8}$, where N_0 is the initial dose and t is the time in days.

(a) Sketch the graph of the equation if $N_0 = 64$.

(b) Find the half-life of ^{131}I.

58 **Trout population** A pond is stocked with 1000 trout. Three months later, it is estimated that 600 remain. Find a formula of the form $N = N_0a^{ct}$ that can be used to estimate the number of trout remaining after t months.

59 **Continuously compounded interest** Ten thousand dollars is invested in a savings fund in which interest is compounded continuously at the rate of 7% per year.

(a) When will the account contain \$35,000?

(b) How long does it take for money to double in the account?

60 **Ben Franklin's will** In 1790, Ben Franklin left \$4000 with instructions that it go to the city of Philadelphia in 200 years. It was worth about \$2 million at that time. Approximate the annual interest rate for the growth.

61 **Electrical current** The current $I(t)$ in a certain electrical circuit at time t is given by $I(t) = I_0e^{-Rt/L}$, where R is the resistance, L is the inductance, and I_0 is the initial current at $t = 0$. Find the value of t, in terms of L and R, for which $I(t)$ is 1% of I_0.

62 **Sound intensity** The sound intensity level formula is $\alpha = 10 \log (I/I_0)$.

(a) Solve for I in terms of α and I_0.

(b) Show that a one-decibel rise in the intensity level α corresponds to a 26% increase in the intensity I.

63 **Fish growth** The length L of a fish is related to its age by means of the von Bertalanffy growth formula

$$L = a(1 - be^{-kt}),$$

where a, b, and k are positive constants that depend on the type of fish. Solve this equation for t to obtain a formula that can be used to estimate the age of a fish from a length measurement.

64 **Earthquake area in the West** In the western United States, the area A (in mi^2) affected by an earthquake is related to the magnitude R of the quake by the formula

$$R = 2.3 \log (A + 3000) - 5.1.$$

Solve for A in terms of R.

65 **Earthquake area in the East** Refer to Exercise 64. For the eastern United States, the area-magnitude formula has the form

$$R = 2.3 \log (A + 34{,}000) - 7.5.$$

If A_1 is the area affected by an earthquake of magnitude R in the West and A_2 is the area affected by a similar quake in the East, find a formula for A_1/A_2 in terms of R.

66 **Earthquake area in the Central states** Refer to Exercise 64. For the Rocky Mountain and Central states, the area-magnitude formula has the form

$$R = 2.3 \log (A + 14{,}000) - 6.6.$$

If an earthquake has magnitude 4 on the Richter scale, estimate the area A of the region that will feel the quake.

67 **Atmospheric pressure** Under certain conditions, the atmospheric pressure p at altitude h is given by the formula $p = 29e^{-0.000034h}$. Express h as a function of p.

68 **Rocket velocity** A rocket of mass m_1 is filled with fuel of initial mass m_2. If frictional forces are disregarded, the total mass m of the rocket at time t after ignition is related to its upward velocity v by $v = -a \ln m + b$, where a and b are constants. At ignition time $t = 0$, $v = 0$ and $m = m_1 + m_2$. At burnout, $m = m_1$. Use this information to find a formula, in terms of one logarithm, for the velocity of the rocket at burnout.

69 **Earthquake frequency** Let n be the average number of earthquakes per year that have magnitudes between R and $R + 1$ on the Richter scale. A formula that approximates the relationship between n and R is

$$\log n = 7.7 - (0.9)R.$$

(a) Solve the equation for n in terms of R.

(b) Find n if $R = 4$, 5, and 6.

70 **Earthquake energy** The energy E (in ergs) released during an earthquake of magnitude R may be approximated by using the formula

$$\log E = 11.4 + (1.5)R.$$

(a) Solve for E in terms of R.

(b) Find the energy released during the earthquake off the coast of Sumatra in 2004, which measured 9.0 on the Richter scale.

71 **Radioactive decay** A certain radioactive substance decays according to the formula $q(t) = q_0e^{-0.0063t}$, where q_0 is the initial amount of the substance and t is the time in days. Approximate the half-life of the substance.

72 **Children's growth** The Count Model is a formula that can be used to predict the height of preschool children. If h is height (in centimeters) and t is age (in years), then

$$h = 70.228 + 5.104t + 9.222 \ln t$$

for $\frac{1}{4} \le t \le 6$. From calculus, the rate of growth R (in cm/year) is given by $R = 5.104 + (9.222/t)$. Predict the height and rate of growth of a typical 2-year-old.

73 **Electrical circuit** The current I in a certain electrical circuit at time t is given by

$$I = \frac{V}{R}(1 - e^{-Rt/L}),$$

where V is the electromotive force, R is the resistance, and L is the inductance. Solve the equation for t.

74 **Carbon 14 dating** The technique of carbon 14 (^{14}C) dating is used to determine the age of archaeological and geological specimens. The formula $T = -8310 \ln x$ is sometimes used to predict the age T (in years) of a bone fossil, where x is the percentage (expressed as a decimal) of ^{14}C still present in the fossil.

(a) Estimate the age of a bone fossil that contains 4% of the ^{14}C found in an equal amount of carbon in present-day bone.

(b) Approximate the percentage of ^{14}C present in a fossil that is 10,000 years old.

75 **Population of Kenya** Based on present birth and death rates, the population of Kenya is expected to increase according to the formula $N = 30.7e^{0.022t}$, with N in millions and $t = 0$ corresponding to 2000. How many years will it take for the population to double?

76 **Language history** Refer to Exercise 48 of Section 4.2. If a language originally had N_0 basic words of which $N(t)$ are still in use, then $N(t) = N_0(0.805)^t$, where time t is measured in millennia. After how many years are one-half the basic words still in use?

CHAPTER 4 DISCUSSION EXERCISES

1 (a) Sketch the graph of $f(x) = -(x-1)^3 + 1$ along with the graph of $y = f^{-1}(x)$.

(b) Discuss what happens to the graph of $y = f^{-1}(x)$ (in general) as the graph of $y = f(x)$ is increasing or is decreasing.

(c) What can you conclude about the intersection points of the graphs of a function and its inverse?

2 Graph $y = (-3)^x$ in $[-4.7, 4.7]$ by $[-3.1, 3.1]$. Trace the graph for $x = 0, 0.1, 0.2, \ldots, 0.9, 1$. Discuss how the graph relates to the graphs of $y = 3^x$ and $y = -3^x$. Also, discuss how these results relate to the restriction $a > 0$ for exponential functions of the form $f(x) = a^x$.

3 **Catenary** Refer to the catenary discussion on page 309 and Figure 4 in Section 4.3.

(a) Describe the graph of the displayed equation for increasing values of a.

(b) Find an equation of the cable in the figure such that the lowest point on the cable is 30 feet off the ground and the difference between the highest point on the cable (where it is connected to the tower) and the lowest point is less than 2 feet, provided the towers are 40 feet apart.

4 Refer to Exercise 70 of Section 4.4. Discuss how to solve this exercise *without* the use of the formula for the total amount T. Proceed with your solution, and compare your answer to the answer arrived at using the formula for T.

5 Shown in the figure is a graph of $f(x) = (\ln x)/x$ for $x > 0$. The maximum value of $f(x)$ occurs at $x = e$.

(a) The integers 2 and 4 have the unusual property that $2^4 = 4^2$. Show that if $x^y = y^x$ for positive real numbers x and y, then $(\ln x)/x = (\ln y)/y$.

(b) Use the graph of f (a table is helpful) to find another pair of real numbers x and y (to two decimal places) such that $x^y \approx y^x$.

(c) Explain why many pairs of real numbers satisfy the equation $x^y = y^x$.

Exercise 5

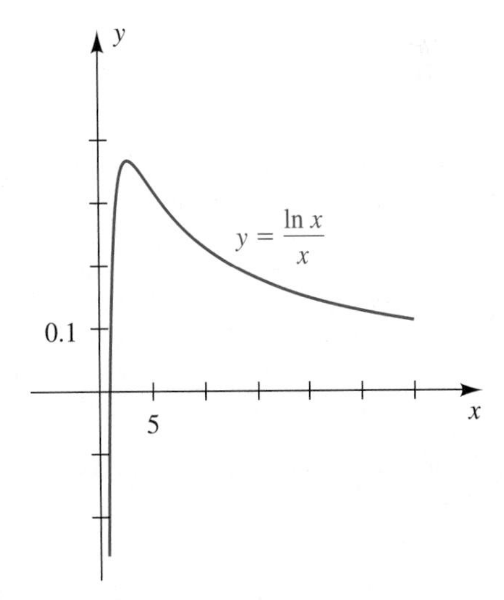

6 (a) Compare the results of Exercise 55 in Section 4.2 and Exercise 37 in Section 4.3. Explain the difference between the two functions.

(b) Now suppose you are investing money at 8.5% per year compounded monthly. How would a graph of this growth compare with the two graphs in part (a)?

(c) Using the function described in part (b), mentally estimate the answers to parts (a) and (b) of Exercise 37 in Section 4.3, and explain why you believe they are correct before actually calculating them.

7 Since $y = \log_3 (x^2)$ is equivalent to $y = 2 \log_3 x$ by law 3 of logarithms, why aren't the graphs in Figure 4(a) and (b) of Section 4.5 the same?

8 **Unpaid balance on a mortgage** When lending institutions loan money, they expect to receive a return equivalent to the amount given by the compound interest formula. The borrower accumulates money "against" the original amount by making a monthly payment M that accumulates according to

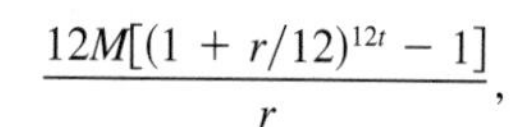

$$\frac{12M[(1 + r/12)^{12t} - 1]}{r},$$

where r is the annual rate of interest and t is the number of years of the loan.

(a) Create a formula for the unpaid balance U of a loan.

(b) Graph the unpaid balance for the home mortgage loan in Exercise 49(a) of Section 4.2.

(c) What is the unpaid balance after 10 years? Estimate the number of years it will take to pay off one-half of the loan.

(d) Discuss the conditions your graph must satisfy to be correct.

(e) Discuss the validity of your results obtained from the graph.

9 Discuss *how many times* the graphs of

$$y = 0.01(1.001)^x \quad \text{and} \quad y = x^3 - 99x^2 - 100x$$

intersect. Approximate the points of intersection. In general, compare the growth of polynomial functions and exponential functions.

10 Discuss *how many times* the graphs of

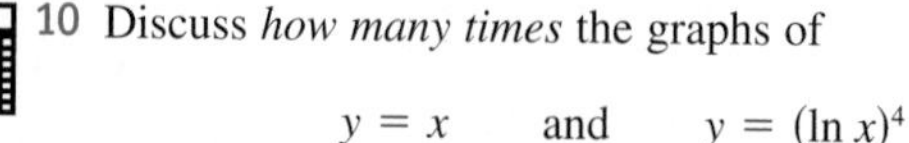

$$y = x \quad \text{and} \quad y = (\ln x)^4$$

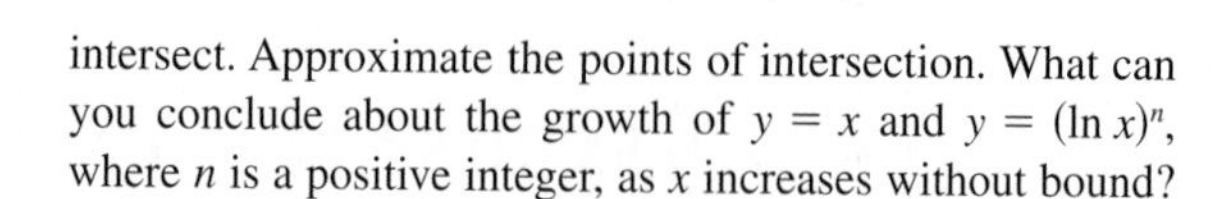

intersect. Approximate the points of intersection. What can you conclude about the growth of $y = x$ and $y = (\ln x)^n$, where n is a positive integer, as x increases without bound?

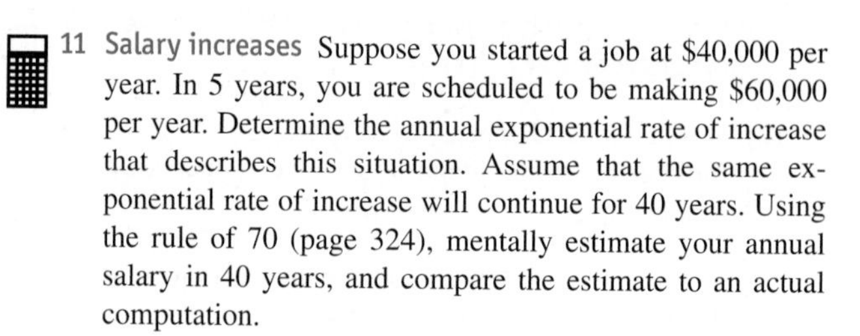

11 **Salary increases** Suppose you started a job at \$40,000 per year. In 5 years, you are scheduled to be making \$60,000 per year. Determine the annual exponential rate of increase that describes this situation. Assume that the same exponential rate of increase will continue for 40 years. Using the rule of 70 (page 324), mentally estimate your annual salary in 40 years, and compare the estimate to an actual computation.

12 **Energy release** Consider these three events:

(1) On May 18, 1980, the volcanic eruption of Mount St. Helens in Washington released approximately 1.7×10^{18} joules of energy.

(2) When a 1-megaton nuclear bomb detonates, it releases about 4×10^{15} joules of energy.

(3) The 1989 San Francisco earthquake registered 7.1 on the Richter scale.

(a) Make some comparisons (i.e., how many of one event is equivalent to another) in terms of energy released. (*Hint:* Refer to Exercise 70 in Chapter 4 Review Exercises.) *Note:* The atomic bombs dropped in World War II were 1-kiloton bombs (1000 1-kiloton bombs = 1 1-megaton bomb).

(b) What reading on the Richter scale would be equivalent to the Mount St. Helens eruption? Has there ever been a reading that high?

13 Dow-Jones average The Dow-Jones industrial average is an index of 30 of America's biggest corporations and is the most common measure of stock performance in the United States. The following table contains some 1000-point milestone dates for the Dow.

Dow-Jones average	Day first reached	Number of days from previous milestone
1003.16	11/14/72	—
2002.25	1/8/87	5168
3004.46	4/17/91	1560
4003.33	2/23/95	1408
5023.55	11/21/95	271
6010.00	10/14/96	328
7022.44	2/13/97	122
8038.88	7/16/97	153
9033.23	4/6/98	264
10,006.78	3/29/99	357
11,014.69	5/3/99	35
12,011.73	10/20/06	2727

Find an exponential model for these data, and use it to predict when the Dow will reach 20,000. Find the average yearly rate of return according to the Dow. Discuss some of the practical considerations pertaining to these calculations.

14 Nasdaq average The Nasdaq stock market's composite index experienced a period of phenomenal growth (shown in the last few lines of the table).

Nasdaq average	Day first reached	Number of days from previous milestone
100 (birth)	2/5/71	—
200.25	11/13/80	3569
501.62	4/12/91	3802
1005.89	7/17/95	1557
2000.56	7/16/98	1095
3028.51	11/3/99	475
4041.46	12/29/99	56
5046.86	3/9/00	71

The technology-driven index is considered by some to be the fastest growing indicator of the entire United States stock market.

Find an exponential regression model for the data. Discuss the fit of the model to the data and possible reasons for the quality of the fit.

15 Total world population The United States Bureau of the Census provided the following estimates and predictions for the total world population.

Year	Population
1950	2,556,518,868
1960	3,040,617,514
1970	3,707,921,742
1980	4,447,068,714
1990	5,274,320,491
2000	6,073,265,234
2010	6,838,220,183
2020	7,608,075,253
2030	8,295,925,812
2040	8,897,180,403
2050	9,404,296,384

(a) Let $t = 0$ correspond to 1950 and plot the data in the viewing rectangle $[-10, 110, 10]$ by $[0, 10^{10}, 10^9]$.

(b) Discuss whether an exponential or logistic model is more appropriate and why.

(c) Find a model of the type you selected in part (b), and graph it with the data.

(d) According to the model, what will the population approach after a long period of time?

16 Discuss how many solutions the equation

$$\log_5 x + \log_7 x = 11$$

has. Solve the equation using the change of base formula.

17 Find the inverse function of $f(x) = \dfrac{9x}{\sqrt{x^2 + 1}}$ and identify any asymptotes of the graph of f^{-1}. How do they relate to the asymptotes of the graph of f?

5

The Trigonometric Functions

Trigonometry was invented over 2000 years ago by the Greeks, who needed precise methods for measuring angles and sides of triangles. In fact, the word *trigonometry* was derived from the two Greek words *trigonon* (triangle) and *metria* (measurement). This chapter begins with a discussion of angles and how they are measured. We next introduce the trigonometric functions by using ratios of sides of a right triangle. After extending the domains of the trigonometric functions to arbitrary angles and real numbers, we consider their graphs and graphing techniques that make use of amplitudes, periods, and phase shifts. The chapter concludes with a section on applied problems.

5.1 Angles

In geometry an **angle** is defined as the set of points determined by two rays, or half-lines, l_1 and l_2, having the same endpoint O. If A and B are points on l_1 and l_2, as in Figure 1, we refer to **angle *AOB*** (denoted $\angle AOB$). An angle may also be considered as two finite line segments with a common endpoint.

Figure 1

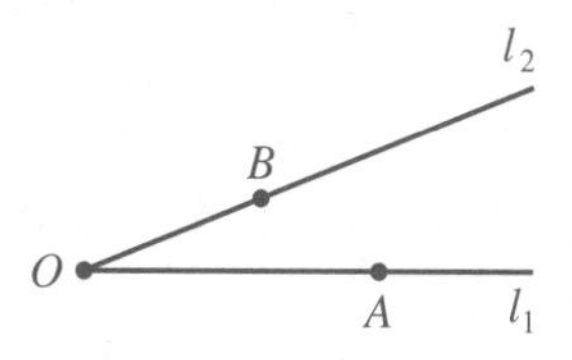

In trigonometry we often interpret angles as rotations of rays. Start with a fixed ray l_1, having endpoint O, and rotate it about O, in a plane, to a position specified by ray l_2. We call l_1 the **initial side,** l_2 the **terminal side,** and O the **vertex** of $\angle AOB$. The amount or direction of rotation is not restricted in any way. We might let l_1 make several revolutions in either direction about O before coming to position l_2, as illustrated by the curved arrows in Figure 2. Thus, many different angles have the same initial and terminal sides. Any two such angles are called **coterminal angles.** A **straight angle** is an angle whose sides lie on the same straight line but extend in opposite directions from its vertex.

Figure 2
Coterminal angles

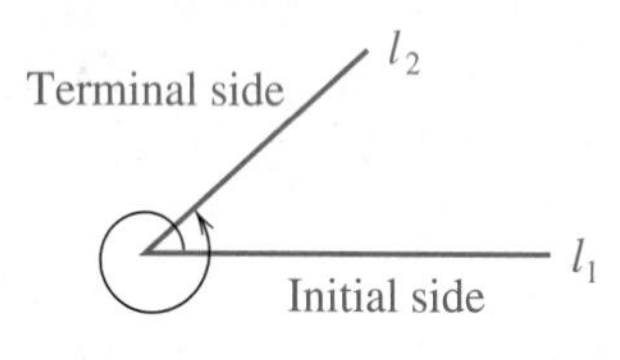

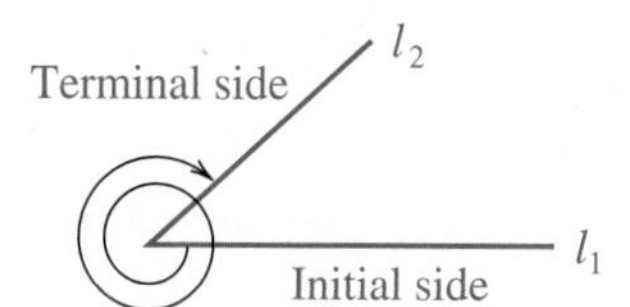

If we introduce a rectangular coordinate system, then the **standard position** of an angle is obtained by taking the vertex at the origin and letting the initial side l_1 coincide with the positive x-axis. If l_1 is rotated in a *counterclockwise* direction to the terminal position l_2, then the angle is considered **positive.** If l_1 is rotated in a *clockwise* direction, the angle is **negative.** We often denote angles by lowercase Greek letters such as α (*alpha*), β (*beta*), γ (*gamma*), θ (*theta*), ϕ (*phi*), and so on. Figure 3 contains sketches of two positive angles, α and β, and a negative angle, γ. If the terminal side of an angle in standard position is in a certain quadrant, we say that the *angle* is in that quadrant. In Figure 3, α is in quadrant III, β is in quadrant I, and γ is in quadrant II. An angle is called a **quadrantal angle** if its terminal side lies on a coordinate axis.

Figure 3 Standard position of an angle

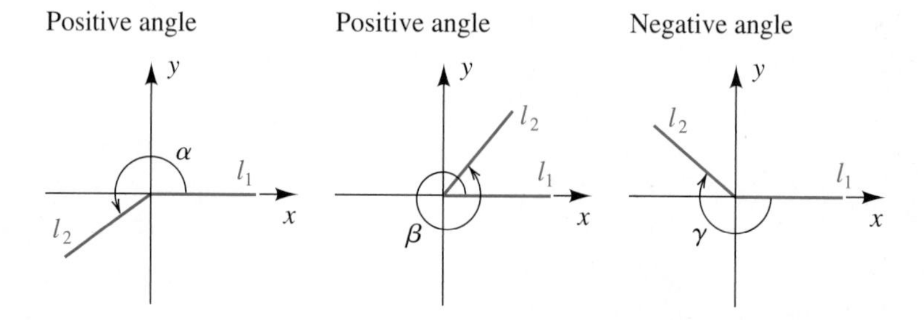

One unit of measurement for angles is the **degree.** The angle in standard position obtained by one complete revolution in the counterclockwise direction has measure 360 degrees, written 360°. Thus, an angle of measure 1 degree (1°) is obtained by $\frac{1}{360}$ of one complete counterclockwise revolution. In Figure 4, several angles measured in degrees are shown in standard position on rectangular coordinate systems. Note that the first three are quadrantal angles.

Figure 4

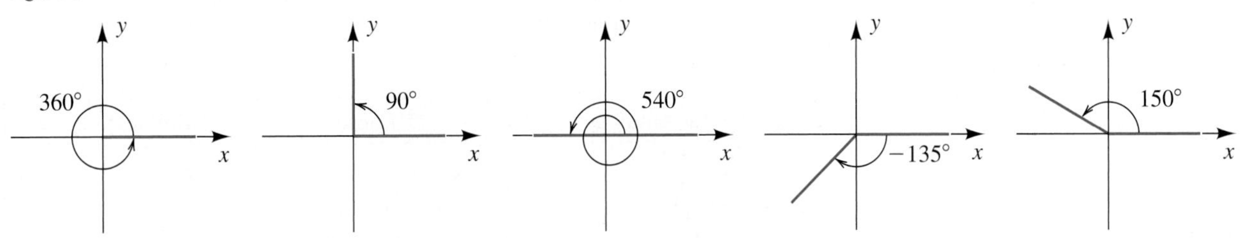

Throughout our work, a notation such as $\theta = 60°$ specifies an angle θ whose measure is 60°. We also refer to *an angle of* 60° or *a* 60° *angle,* instead of using the more precise (but cumbersome) phrase *an angle having measure* 60°.

EXAMPLE 1 **Finding coterminal angles**

If $\theta = 60°$ is in standard position, find two positive angles and two negative angles that are coterminal with θ.

▶ SOLUTION The angle θ is shown in standard position in the first sketch in Figure 5. To find positive coterminal angles, we may add 360° or 720° (or any other positive integer multiple of 360°) to θ, obtaining

$$60° + 360° = 420° \qquad \text{and} \qquad 60° + 720° = 780°.$$

These coterminal angles are also shown in Figure 5.

To find negative coterminal angles, we may add −360° or −720° (or any other negative integer multiple of 360°), obtaining

$$60° + (-360°) = -300° \qquad \text{and} \qquad 60° + (-720°) = -660°,$$

as shown in the last two sketches in Figure 5.

Figure 5

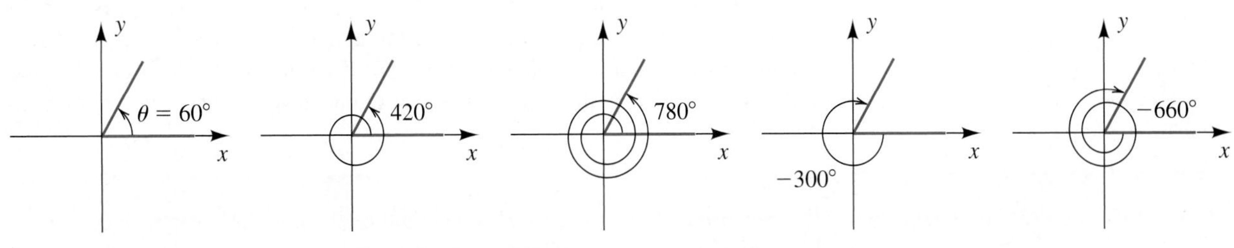

▶ **Tutorial available at www.thomsonedu.com/login**

A **right angle** is half of a straight angle and has measure 90°. The following chart contains definitions of other special types of angles.

Terminology	Definition	Illustration
acute angle θ	$0° < \theta < 90°$	12°; 37°
obtuse angle θ	$90° < \theta < 180°$	95°; 157°
complementary angles α, β	$\alpha + \beta = 90°$	20°, 70°; 7°, 83°
supplementary angles α, β	$\alpha + \beta = 180°$	115°, 65°; 18°, 162°

If smaller measurements than the degree are required, we can use tenths, hundredths, or thousandths of degrees. Alternatively, we can divide the degree into 60 equal parts, called **minutes** (denoted by $'$), and each minute into 60 equal parts, called **seconds** (denoted by $''$). Thus, $1° = 60'$, and $1' = 60''$. The notation $\theta = 73°56'18''$ refers to an angle θ that has measure 73 degrees, 56 minutes, 18 seconds.

EXAMPLE 2 **Finding complementary angles**

Find the angle that is complementary to θ:

(a) $\theta = 25°43'37''$ (b) $\theta = 73.26°$

SOLUTION We wish to find $90° - \theta$. It is convenient to write 90° as an equivalent measure, $89°59'60''$.

(a)
$$\begin{aligned} 90° &= 89°59'60'' \\ \underline{\theta\quad} &= \underline{25°43'37''} \\ 90° - \theta &= 64°16'23'' \end{aligned}$$

(b)
$$\begin{aligned} 90° &= 90.00° \\ \underline{\theta\quad} &= \underline{73.26°} \\ 90° - \theta &= 16.74° \end{aligned}$$

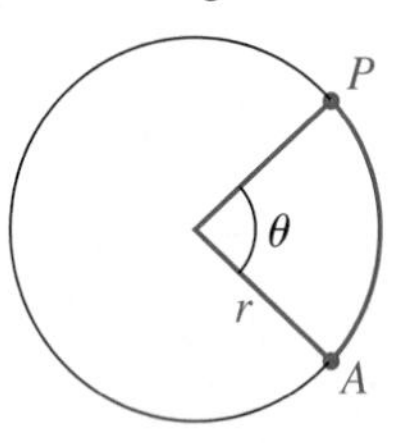

Figure 6
Central angle θ

Degree measure for angles is used in applied areas such as surveying, navigation, and the design of mechanical equipment. In scientific applications that require calculus, it is customary to employ *radian measure.* To define an angle of radian measure 1, we consider a circle of any radius r. A **central angle** of a circle is an angle whose vertex is at the center of the circle. If θ is the central angle shown in Figure 6, we say that the **arc AP** (denoted $\overparen{AP}$) of the circle **subtends** θ or that θ **is subtended by** $\overparen{AP}$. If the length of $\overparen{AP}$ is equal to the radius r of the circle, then θ has a measure of one radian, as in the next definition.

Definition of Radian Measure	**One radian** is the measure of the central angle of a circle subtended by an arc equal in length to the radius of the circle.

Tutorial available at www.thomsonedu.com/login

If we consider a circle of radius r, then an angle α whose measure is 1 radian intercepts an arc AP of length r, as illustrated in Figure 7(a). The angle β in Figure 7(b) has radian measure 2, since it is subtended by an arc of length $2r$. Similarly, γ in (c) of the figure has radian measure 3, since it is subtended by an arc of length $3r$.

Figure 7

(a) $\alpha = 1$ radian

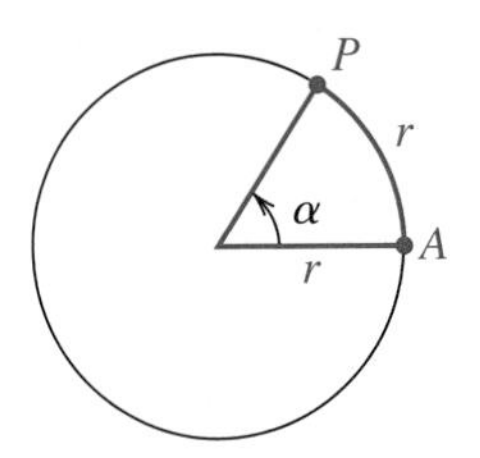

(b) $\beta = 2$ radians

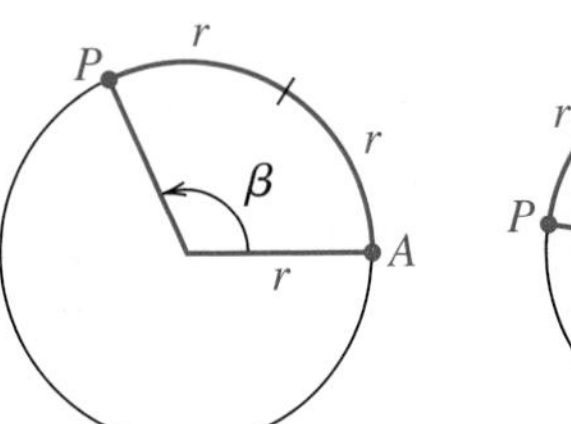

(c) $\gamma = 3$ radians

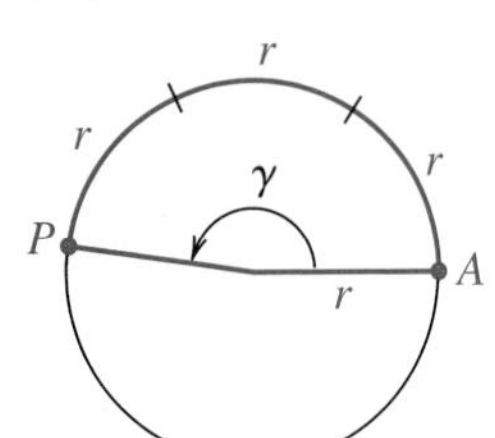

(d) $360° = 2\pi \approx 6.28$ radians

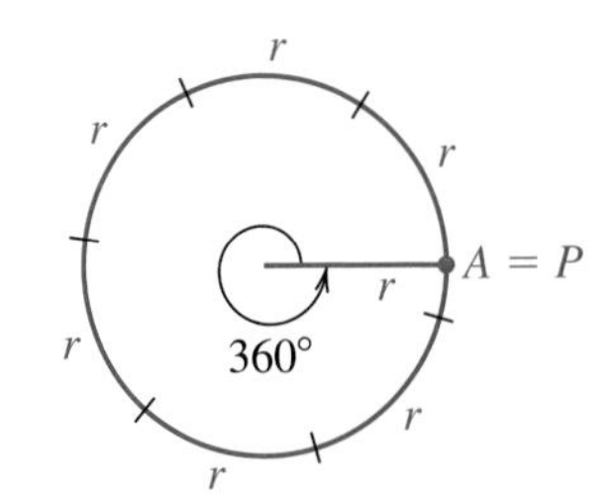

To find the radian measure corresponding to 360°, we must find the number of times that a circular arc of length r can be laid off along the circumference (see Figure 7(d)). This number is not an integer or even a rational number. Since the circumference of the circle is $2\pi r$, the number of times r units can be laid off is 2π. Thus, an angle of measure 2π radians corresponds to the degree measure 360°, and we write $360° = 2\pi$ radians. This result gives us the following relationships.

Relationships Between Degrees and Radians

(1) $180° = \pi$ radians

(2) $1° = \frac{\pi}{180}$ radian ≈ 0.0175 radian

(3) 1 radian $= \left(\frac{180°}{\pi}\right) \approx 57.2958°$

When radian measure of an angle is used, no units will be indicated. Thus, if an angle has radian measure 5, we write $\theta = 5$ instead of $\theta = 5$ *radians.* There should be no confusion as to whether radian or degree measure is being used, since if θ has *degree* measure 5°, we write $\theta = 5°$, and *not* $\theta = 5$.

The next chart illustrates how to change from one angular measure to another.

Changing Angular Measures

To change	Multiply by	Illustration
degrees to radians	$\frac{\pi}{180^\circ}$	$150^\circ = 150^\circ\left(\frac{\pi}{180^\circ}\right) = \frac{5\pi}{6}$ $225^\circ = 225^\circ\left(\frac{\pi}{180^\circ}\right) = \frac{5\pi}{4}$
radians to degrees	$\frac{180^\circ}{\pi}$	$\frac{7\pi}{4} = \frac{7\pi}{4}\left(\frac{180^\circ}{\pi}\right) = 315^\circ$ $\frac{\pi}{3} = \frac{\pi}{3}\left(\frac{180^\circ}{\pi}\right) = 60^\circ$

We may use the techniques illustrated in the preceding chart to obtain the following table, which displays the corresponding radian and degree measures of special angles.

Radians	0	$\frac{\pi}{6}$	$\frac{\pi}{4}$	$\frac{\pi}{3}$	$\frac{\pi}{2}$	$\frac{2\pi}{3}$	$\frac{3\pi}{4}$	$\frac{5\pi}{6}$	π	$\frac{7\pi}{6}$	$\frac{5\pi}{4}$	$\frac{4\pi}{3}$	$\frac{3\pi}{2}$	$\frac{5\pi}{3}$	$\frac{7\pi}{4}$	$\frac{11\pi}{6}$	2π
Degrees	0°	30°	45°	60°	90°	120°	135°	150°	180°	210°	225°	240°	270°	300°	315°	330°	360°

Several of these special angles, in radian measure, are shown in standard position in Figure 8.

Figure 8

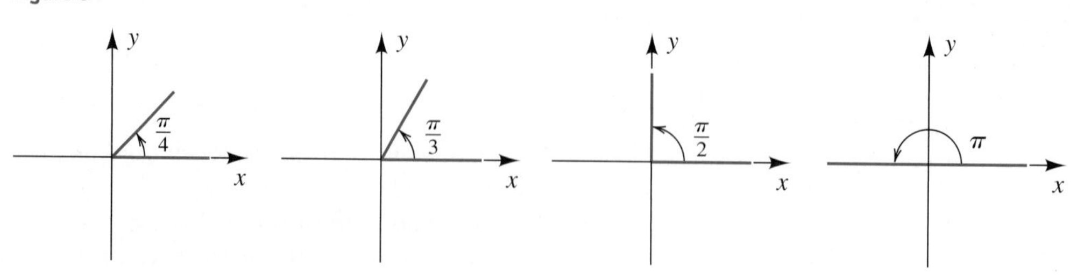

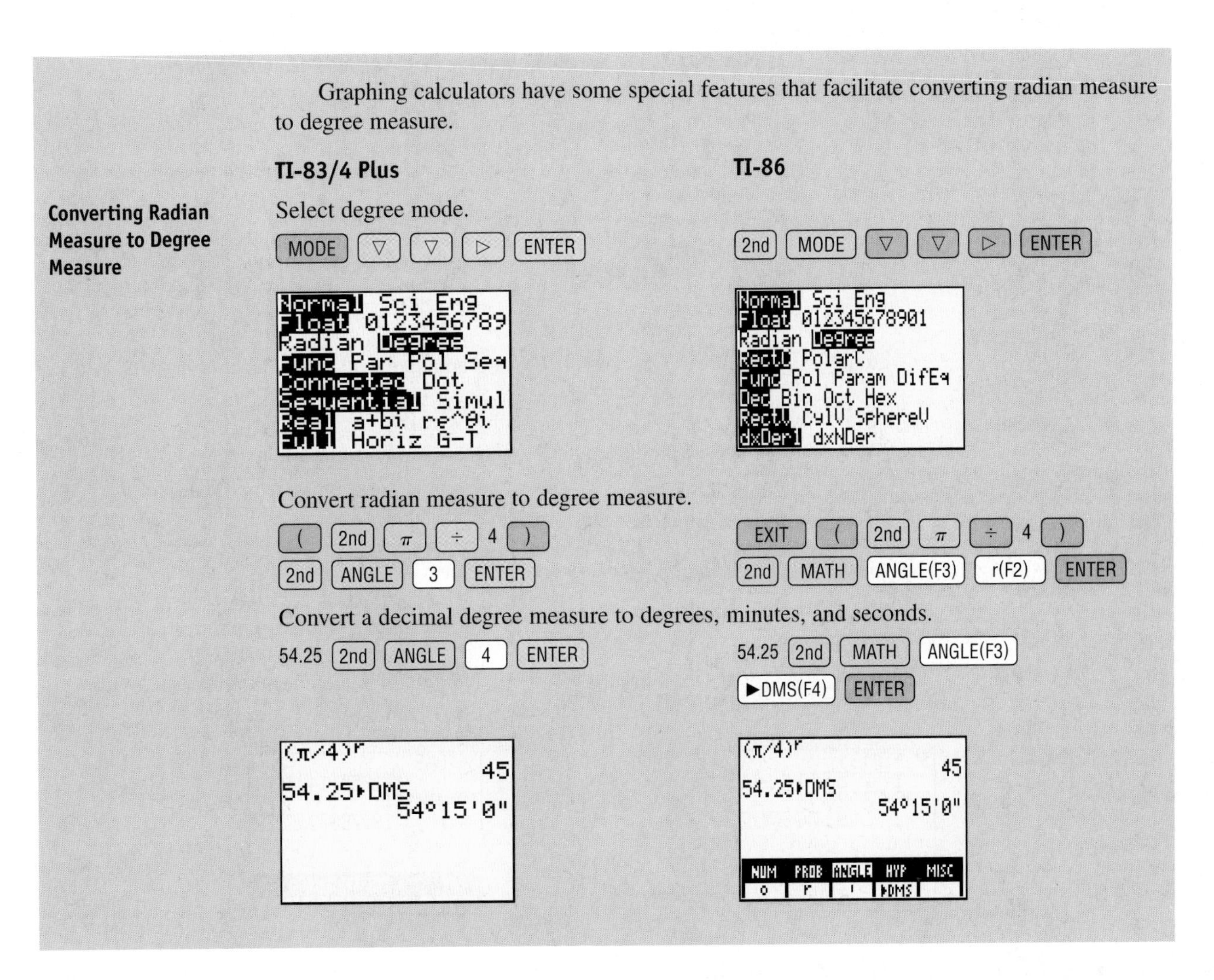

EXAMPLE 3 **Changing radians to degrees, minutes, and seconds**

If $\theta = 3$, approximate θ in terms of degrees, minutes, and seconds.

▶ **SOLUTION**

$$
\begin{aligned}
3 \text{ radians} &= 3\left(\frac{180^\circ}{\pi}\right) && \text{multiply by } \frac{180^\circ}{\pi} \\
&\approx 171.8873^\circ && \text{approximate} \\
&= 171^\circ + (0.8873)(60') && 1^\circ = 60' \\
&= 171^\circ + 53.238' && \text{multiply} \\
&= 171^\circ + 53' + (0.238)(60'') && 1' = 60'' \\
&= 171^\circ 53' + 14.28'' && \text{multiply} \\
&\approx 171^\circ 53' 14'' && \text{approximate}
\end{aligned}
$$

▶ **Tutorial available at www.thomsonedu.com/login**

EXAMPLE 4 **Expressing minutes and seconds as decimal degrees**

Express 19°47′23″ as a decimal, to the nearest ten-thousandth of a degree.

▶ SOLUTION Since $1' = \left(\frac{1}{60}\right)^\circ$ and $1'' = \left(\frac{1}{60}\right)' = \left(\frac{1}{3600}\right)^\circ$,

$$\begin{aligned} 19^\circ 47' 23'' &= 19^\circ + \left(\tfrac{47}{60}\right)^\circ + \left(\tfrac{23}{3600}\right)^\circ \\ &\approx 19^\circ + 0.7833^\circ + 0.0064^\circ \\ &= 19.7897^\circ. \end{aligned}$$

Examples 3 and 4 are easily handled by a graphing calculator (in degree mode).

TI-83/4 Plus	**TI-86**
Convert the radian measure in Example 3 to degrees, minutes, and seconds.	
3 [2nd] [ANGLE] [3] [2nd] [ANGLE] [4] [ENTER]	3 [2nd] [MATH] [ANGLE(F3)] [r(F2)] [▶DMS(F4)] [ENTER]
Express the angle in Example 4 as a decimal degree.	
19 [2nd] [ANGLE] [1] 47 [2nd] [ANGLE] [2] 23 [ALPHA] ["(on+key)] [ENTER]	19 [2nd] [MATH] [ANGLE(F3)] ['(F3)] 47 ['(F3)] 23 ['(F3)] [ENTER] Note that the angle is entered in degrees′minutes′seconds′ format.

```
3r▶DMS
       171°53'14.419"
19°47'23"
          19.78972222
```

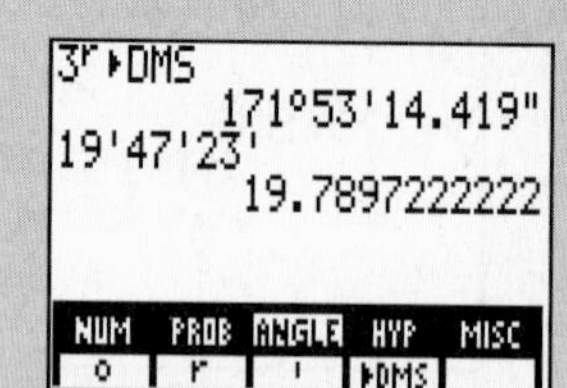

A mnemonic device for remembering $s = r\theta$ is SRO (Standing Room Only).

The next result specifies the relationship between the length of a circular arc and the central angle that it subtends.

Formula for the Length of a Circular Arc	If an arc of length s on a circle of radius r subtends a central angle of radian measure θ, then $$s = r\theta.$$

Figure 9

(a) θ, r, s (b) θ_1, r, s_1

PROOF A typical arc of length s and the corresponding central angle θ are shown in Figure 9(a). Figure 9(b) shows an arc of length s_1 and central angle θ_1. If radian measure is used, then, from plane geometry, the ratio of the lengths of the arcs is the same as the ratio of the angular measures; that is,

$$\frac{s}{s_1} = \frac{\theta}{\theta_1}, \quad \text{or} \quad s = \frac{\theta}{\theta_1} s_1.$$

▶ Tutorial available at www.thomsonedu.com/login

If we consider the special case in which θ_1 has radian measure 1, then, from the definition of radian, $s_1 = r$ and the last equation becomes

$$s = \frac{\theta}{1} \cdot r = r\theta.$$

Notice that if $\theta = 2\pi$, then the formula for the length of a circular arc becomes $s = r(2\pi)$, which is simply the formula for the circumference of a circle, $C = 2\pi r$.

The next formula is proved in a similar manner.

Formula for the Area of a Circular Sector

If θ is the radian measure of a central angle of a circle of radius r and if A is the area of the circular sector determined by θ, then

$$A = \tfrac{1}{2}r^2\theta.$$

Figure 10

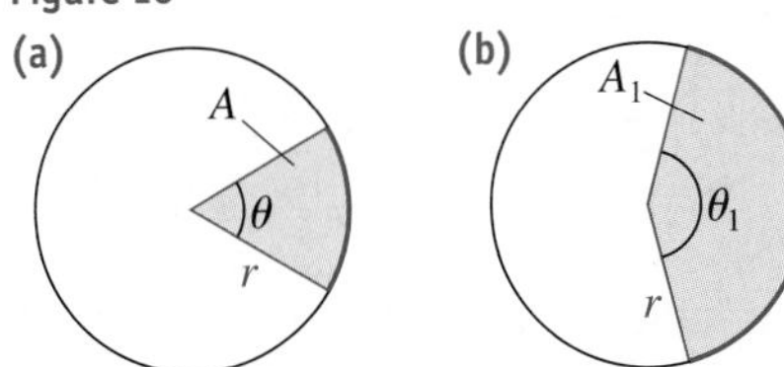

PROOF If A and A_1 are the areas of the sectors in Figures 10(a) and 10(b), respectively, then, from plane geometry,

$$\frac{A}{A_1} = \frac{\theta}{\theta_1}, \quad \text{or} \quad A = \frac{\theta}{\theta_1}A_1.$$

If we consider the special case $\theta_1 = 2\pi$, then $A_1 = \pi r^2$ and

$$A = \frac{\theta}{2\pi} \cdot \pi r^2 = \frac{1}{2}r^2\theta.$$

When using the preceding formulas, it is important to remember to use the radian measure of θ rather than the degree measure, as illustrated in the next example.

EXAMPLE 5 Using the circular arc and sector formulas

In Figure 11, a central angle θ is subtended by an arc 10 centimeters long on a circle of radius 4 centimeters.

(a) Approximate the measure of θ in degrees.

(b) Find the area of the circular sector determined by θ.

Figure 11

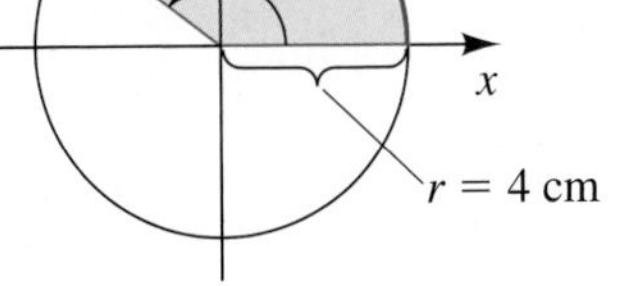

▶ SOLUTION We proceed as follows:

(a)

$s = r\theta$ length of a circular arc formula

$\theta = \dfrac{s}{r}$ solve for θ

$= \frac{10}{4} = 2.5$ let $s = 10$, $r = 4$

(continued)

▶ Tutorial available at www.thomsonedu.com/login

This is the *radian* measure of θ. Changing to degrees, we have

$$\theta = 2.5\left(\frac{180^\circ}{\pi}\right) = \frac{450^\circ}{\pi} \approx 143.24^\circ.$$

(b)

$$\begin{aligned} A &= \tfrac{1}{2}r^2\theta && \text{area of a circular sector formula} \\ &= \tfrac{1}{2}(4)^2(2.5) && \text{let } r = 4,\ \theta = 2.5 \text{ radians} \\ &= 20\ \text{cm}^2 && \text{multiply} \end{aligned}$$

The **angular speed** of a wheel that is rotating at a constant rate is the angle generated in one unit of time by a line segment from the center of the wheel to a point P on the circumference (see Figure 12). The **linear speed** of a point P on the circumference is the distance that P travels per unit of time. By dividing both sides of the formula for a circular arc by time t, we obtain a relationship for linear speed and angular speed; that is,

Figure 12

$$\frac{s}{t} = \frac{r\theta}{t}, \quad \text{or, equivalently,} \quad \underset{\text{linear speed}}{\frac{s}{t}} = r \cdot \underset{\text{angular speed}}{\frac{\theta}{t}}.$$

EXAMPLE 6 Finding angular and linear speeds

Suppose that the wheel in Figure 12 is rotating at a rate of 800 rpm (revolutions per minute).

(a) Find the angular speed of the wheel.

(b) Find the linear speed (in in./min and mi/hr) of a point P on the circumference of the wheel.

SOLUTION

(a) Let O denote the center of the wheel, and let P be a point on the circumference. Because the number of revolutions per minute is 800 and because each revolution generates an angle of 2π radians, the angle generated by the line segment OP in one minute has radian measure $(800)(2\pi)$; that is,

$$\text{angular speed} = \frac{800 \text{ revolutions}}{1 \text{ minute}} \cdot \frac{2\pi \text{ radians}}{1 \text{ revolution}} = 1600\pi \text{ radians per minute.}$$

Note that the diameter of the wheel is irrelevant in finding the angular speed.

(b)

$$\begin{aligned} \text{linear speed} &= \text{radius} \cdot \text{angular speed} \\ &= (12 \text{ in.})(1600\pi \text{ rad/min}) \\ &= 19{,}200\pi \text{ in./min} \end{aligned}$$

Converting in./min to mi/hr, we get

$$\frac{19{,}200\pi \text{ in.}}{1 \text{ min}} \cdot \frac{60 \text{ min}}{1 \text{ hr}} \cdot \frac{1 \text{ ft}}{12 \text{ in.}} \cdot \frac{1 \text{ mi}}{5280 \text{ ft}} \approx 57.1 \text{ mi/hr}.$$

Unlike the angular speed, the linear speed *is* dependent on the diameter of the wheel.

▶ Tutorial available at www.thomsonedu.com/login

5.1 *Exercises*

Exer. 1–4: If the given angle is in standard position, find two positive coterminal angles and two negative coterminal angles.

1 (a) $120°$ (b) $135°$ (c) $-30°$

2 (a) $240°$ (b) $315°$ (c) $-150°$

3 (a) $620°$ (b) $\frac{5\pi}{6}$ (c) $-\frac{\pi}{4}$

4 (a) $570°$ (b) $\frac{2\pi}{3}$ (c) $-\frac{5\pi}{4}$

Exer. 5–6: Find the angle that is complementary to θ.

5 (a) $\theta = 5°17'34''$ (b) $\theta = 32.5°$

6 (a) $\theta = 63°4'15''$ (b) $\theta = 82.73°$

Exer. 7–8: Find the angle that is supplementary to θ.

7 (a) $\theta = 48°51'37''$ (b) $\theta = 136.42°$

8 (a) $\theta = 152°12'4''$ (b) $\theta = 15.9°$

Exer. 9–12: Find the exact radian measure of the angle.

9 (a) $150°$ (b) $-60°$ (c) $225°$

10 (a) $120°$ (b) $-135°$ (c) $210°$

11 (a) $450°$ (b) $72°$ (c) $100°$

12 (a) $630°$ (b) $54°$ (c) $95°$

Exer. 13–16: Find the exact degree measure of the angle.

13 (a) $\frac{2\pi}{3}$ (b) $\frac{11\pi}{6}$ (c) $\frac{3\pi}{4}$

14 (a) $\frac{5\pi}{6}$ (b) $\frac{4\pi}{3}$ (c) $\frac{11\pi}{4}$

15 (a) $-\frac{7\pi}{2}$ (b) 7π (c) $\frac{\pi}{9}$

16 (a) $-\frac{5\pi}{2}$ (b) 9π (c) $\frac{\pi}{16}$

Exer. 17–20: Express θ in terms of degrees, minutes, and seconds, to the nearest second.

17 $\theta = 2$

18 $\theta = 1.5$

19 $\theta = 5$

20 $\theta = 4$

Exer. 21–24: Express the angle as a decimal, to the nearest ten-thousandth of a degree.

21 $37°41'$

22 $83°17'$

23 $115°26'27''$

24 $258°39'52''$

Exer. 25–28: Express the angle in terms of degrees, minutes, and seconds, to the nearest second.

25 $63.169°$

26 $12.864°$

27 $310.6215°$

28 $81.7238°$

Exer. 29–30: If a circular arc of the given length s subtends the central angle θ on a circle, find the radius of the circle.

29 $s = 10$ cm, $\theta = 4$

30 $s = 3$ km, $\theta = 20°$

Exer. 31–32: (a) Find the length of the arc of the colored sector in the figure. (b) Find the area of the sector.

31

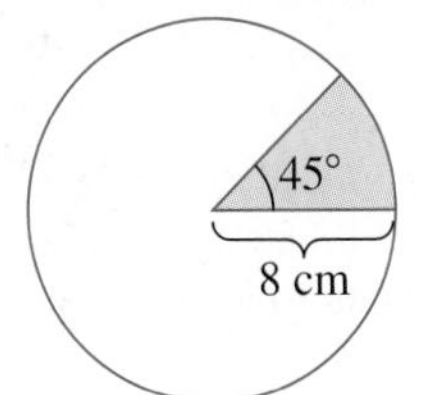

32

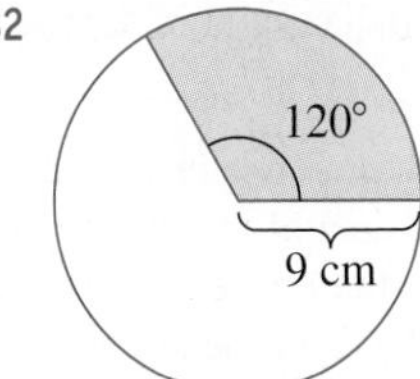

Exer. 33–34: (a) Find the radian and degree measures of the central angle θ subtended by the given arc of length s on a circle of radius r. (b) Find the area of the sector determined by θ.

33 $s = 7$ cm, $r = 4$ cm

34 $s = 3$ ft, $r = 20$ in.

Exer. 35–36: (a) Find the length of the arc that subtends the given central angle θ on a circle of diameter d. (b) Find the area of the sector determined by θ.

35 $\theta = 50°$, $d = 16$ m

36 $\theta = 2.2$, $d = 120$ cm

▶ 37 Measuring distances on Earth The distance between two points A and B on Earth is measured along a circle having center C at the center of Earth and radius equal to the distance from C to the surface (see the figure). If the diameter of Earth is approximately 8000 miles, approximate the distance between A and B if angle ACB has the indicated measure:

(a) 60° (b) 45° (c) 30° (d) 10° (e) 1°

Exercise 37

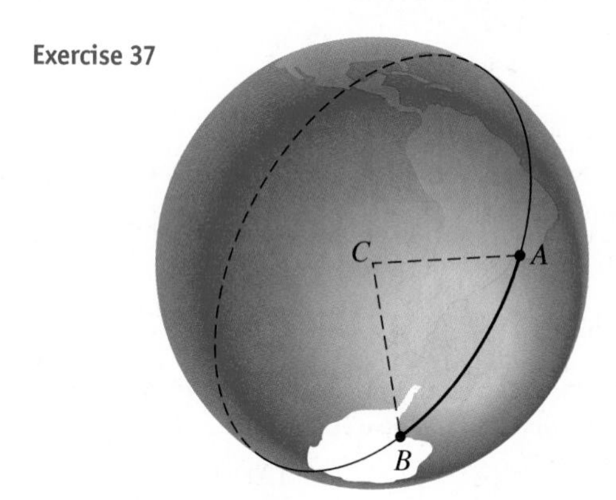

38 Nautical miles Refer to Exercise 37. If angle ACB has measure $1'$, then the distance between A and B is a nautical mile. Approximate the number of land (statute) miles in a nautical mile.

39 Measuring angles using distance Refer to Exercise 37. If two points A and B are 500 miles apart, express angle ACB in radians and in degrees.

40 A hexagon is inscribed in a circle. If the difference between the area of the circle and the area of the hexagon is 24 m^2, use the formula for the area of a sector to approximate the radius r of the circle.

41 Window area A rectangular window measures 54 inches by 24 inches. There is a 17-inch wiper blade attached by a 5-inch arm at the center of the base of the window, as shown in the figure. If the arm rotates 120°, approximate the percentage of the window's area that is wiped by the blade.

Exercise 41

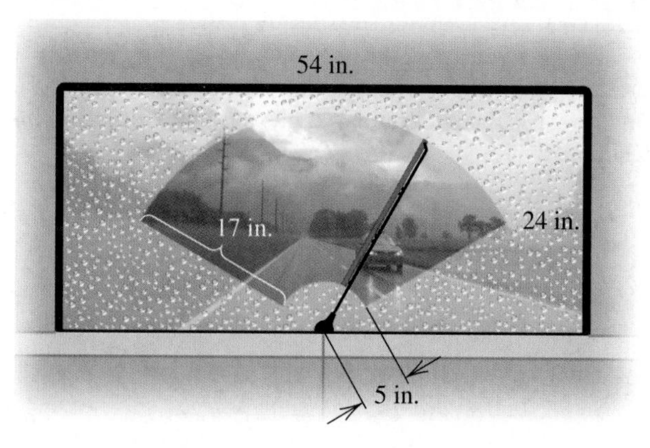

42 A tornado's core A simple model of the core of a tornado is a right circular cylinder that rotates about its axis. If a tornado has a core diameter of 200 feet and maximum wind speed of 180 mi/hr (or 264 ft/sec) at the perimeter of the core, approximate the number of revolutions the core makes each minute.

43 Earth's rotation Earth rotates about its axis once every 23 hours, 56 minutes, and 4 seconds. Approximate the number of radians Earth rotates in one second.

44 Earth's rotation Refer to Exercise 43. The equatorial radius of Earth is approximately 3963.3 miles. Find the linear speed of a point on the equator as a result of Earth's rotation.

Exer. 45–46: A wheel of the given radius is rotating at the indicated rate.
(a) Find the angular speed (in radians per minute).
(b) Find the linear speed of a point on the circumference (in ft/min).

45 radius 5 in., 40 rpm

46 radius 9 in., 2400 rpm

47 Rotation of compact discs (CDs) The drive motor of a particular CD player is controlled to rotate at a speed of 200 rpm when reading a track 5.7 centimeters from the center of the CD. The speed of the drive motor must vary so that the reading of the data occurs at a constant rate.

(a) Find the angular speed (in radians per minute) of the drive motor when it is reading a track 5.7 centimeters from the center of the CD.

▶ Tutorial available at www.thomsonedu.com/login

(b) Find the linear speed (in cm/sec) of a point on the CD that is 5.7 centimeters from the center of the CD.

(c) Find the angular speed (in rpm) of the drive motor when it is reading a track 3 centimeters from the center of the CD.

(d) Find a function S that gives the drive motor speed in rpm for any radius r in centimeters, where $2.3 \leq r \leq 5.9$. What type of variation exists between the drive motor speed and the radius of the track being read? Check your answer by graphing S and finding the speeds for $r = 3$ and $r = 5.7$.

48 **Tire revolutions** A typical tire for a compact car is 22 inches in diameter. If the car is traveling at a speed of 60 mi/hr, find the number of revolutions the tire makes per minute.

49 **Cargo winch** A large winch of diameter 3 feet is used to hoist cargo, as shown in the figure.

(a) Find the distance the cargo is lifted if the winch rotates through an angle of radian measure $7\pi/4$.

(b) Find the angle (in radians) through which the winch must rotate in order to lift the cargo d feet.

Exercise 49

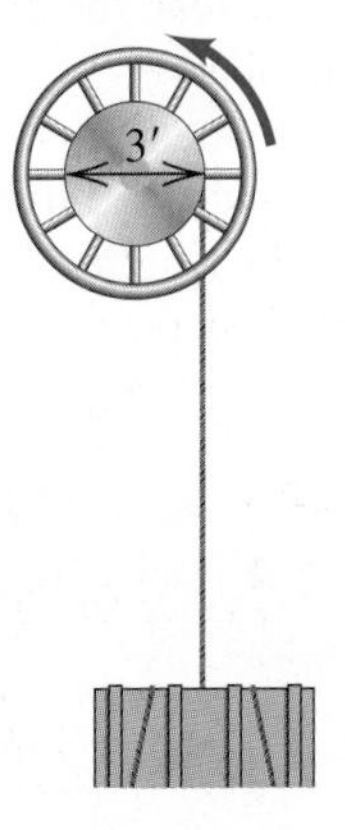

50 **Pendulum's swing** A pendulum in a grandfather clock is 4 feet long and swings back and forth along a 6-inch arc. Approximate the angle (in degrees) through which the pendulum passes during one swing.

51 **Pizza values** A vender sells two sizes of pizza by the slice. The *small* slice is $\frac{1}{6}$ of a circular 18-inch-diameter pizza, and it sells for \$2.00. The *large* slice is $\frac{1}{8}$ of a circular 26-inch-diameter pizza, and it sells for \$3.00. Which slice provides more pizza per dollar?

52 **Bicycle mechanics** The sprocket assembly for a bicycle is shown in the figure. If the sprocket of radius r_1 rotates through an angle of θ_1 radians, find the corresponding angle of rotation for the sprocket of radius r_2.

Exercise 52

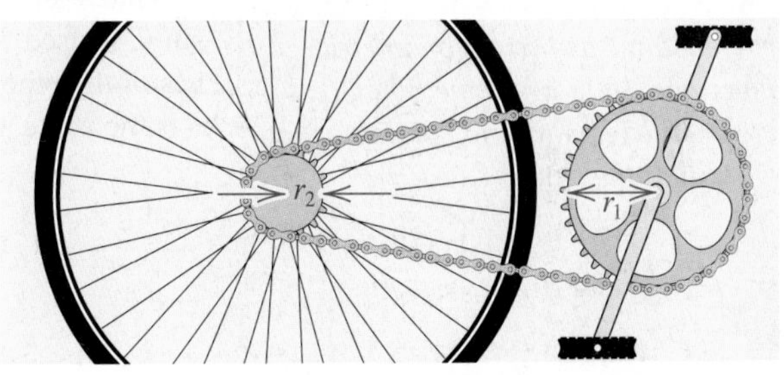

53 **Bicycle mechanics** Refer to Exercise 52. An expert cyclist can attain a speed of 40 mi/hr. If the sprocket assembly has $r_1 = 5$ in., $r_2 = 2$ in., and the wheel has a diameter of 28 inches, approximately how many revolutions per minute of the front sprocket wheel will produce a speed of 40 mi/hr? (*Hint:* First change 40 mi/hr to in./sec.)

54 **Magnetic pole drift** The geographic and magnetic north poles have different locations. Currently, the magnetic north pole is drifting westward through 0.0017 radian per year, where the angle of drift has its vertex at the center of Earth. If this movement continues, approximately how many years will it take for the magnetic north pole to drift a total of 5°?

5.2 *Trigonometric Functions of Angles*

We shall introduce the trigonometric functions in the manner in which they originated historically—as ratios of sides of a right triangle. A triangle is a **right triangle** if one of its angles is a right angle. If θ is any acute angle, we may consider a right triangle having θ as one of its angles, as in Figure 1,

Figure 1

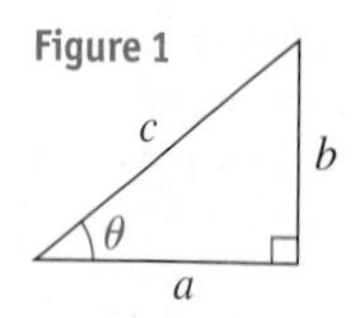

where the symbol ⌐ specifies the 90° angle. Six ratios can be obtained using the lengths a, b, and c of the sides of the triangle:

$$\frac{b}{c},\quad \frac{a}{c},\quad \frac{b}{a},\quad \frac{a}{b},\quad \frac{c}{a},\quad \frac{c}{b}$$

Figure 2

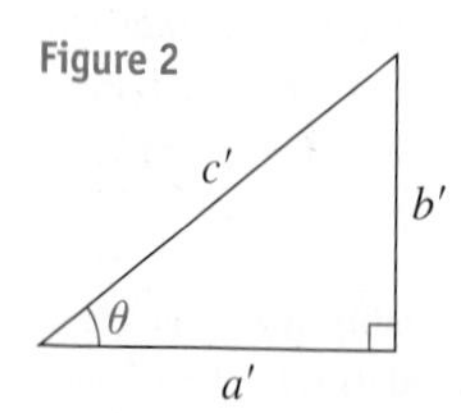

We can show that these ratios depend only on θ, and not on the size of the triangle, as indicated in Figure 2. Since the two triangles have equal angles, they are similar, and therefore ratios of corresponding sides are proportional. For example,

$$\frac{b}{c} = \frac{b'}{c'},\quad \frac{a}{c} = \frac{a'}{c'},\quad \frac{b}{a} = \frac{b'}{a'}.$$

Thus, for each θ, the six ratios are uniquely determined and hence are functions of θ. They are called the **trigonometric functions*** and are designated as the **sine, cosine, tangent, cotangent, secant,** and **cosecant** functions, abbreviated **sin, cos, tan, cot, sec,** and **csc,** respectively. The symbol sin (θ), or $\sin\theta$, is used for the ratio b/c, which the sine function associates with θ. Values of the other five functions are denoted in similar fashion. To summarize, if θ is the acute angle of the right triangle in Figure 1, then, by definition,

$$\sin\theta = \frac{b}{c} \qquad \cos\theta = \frac{a}{c} \qquad \tan\theta = \frac{b}{a}$$

$$\csc\theta = \frac{c}{b} \qquad \sec\theta = \frac{c}{a} \qquad \cot\theta = \frac{a}{b}.$$

We will refer to these six trigonometric functions as* **the *trigonometric functions. Here are some other, less common trigonometric functions that we will not use in this text:*

$$\text{vers}\ \theta = 1 - \cos\theta$$

$$\text{covers}\ \theta = 1 - \sin\theta$$

$$\text{exsec}\ \theta = \sec\theta - 1$$

$$\text{hav}\ \theta = \tfrac{1}{2}\,\text{vers}\ \theta$$

The domain of each of the six trigonometric functions is the set of all acute angles. Later in this section we will extend the domains to larger sets of angles, and in the next section, to real numbers.

If θ is the angle in Figure 1, we refer to the sides of the triangle of lengths a, b, and c as the **adjacent side, opposite side,** and **hypotenuse,** respectively. We shall use **adj, opp,** and **hyp** to denote the lengths of the sides. We may then represent the triangle as in Figure 3. With this notation, the trigonometric functions may be expressed as follows.

Figure 3

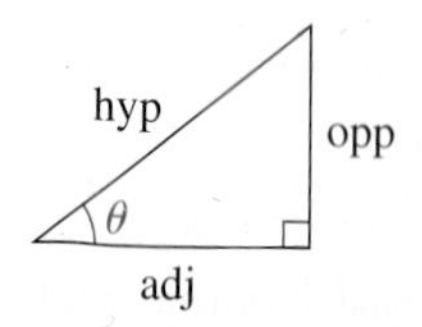

Definition of the Trigonometric Functions of an Acute Angle of a Right Triangle

$$\sin\theta = \frac{\text{opp}}{\text{hyp}} \qquad \cos\theta = \frac{\text{adj}}{\text{hyp}} \qquad \tan\theta = \frac{\text{opp}}{\text{adj}}$$

$$\csc\theta = \frac{\text{hyp}}{\text{opp}} \qquad \sec\theta = \frac{\text{hyp}}{\text{adj}} \qquad \cot\theta = \frac{\text{adj}}{\text{opp}}$$

A mnemonic device for remembering the top row in the definition is

SOH CAH TOA,

where SOH is an abbreviation for <u>S</u>in θ = <u>O</u>pp/<u>H</u>yp, *and so forth.*

The formulas in the preceding definition can be applied to any right triangle without attaching the labels a, b, c to the sides. Since the lengths of the sides of a triangle are positive real numbers, *the values of the six trigonometric functions are positive for every acute angle* θ. Moreover, the hypotenuse is always greater than the adjacent or opposite side, and hence $\sin\theta < 1$, $\cos\theta < 1$, $\csc\theta > 1$, and $\sec\theta > 1$ for every acute angle θ.

Note that since

$$\sin\theta = \frac{\text{opp}}{\text{hyp}} \quad \text{and} \quad \csc\theta = \frac{\text{hyp}}{\text{opp}},$$

$\sin\theta$ and $\csc\theta$ are reciprocals of each other, giving us the two identities in the left-hand column of the next box. Similarly, $\cos\theta$ and $\sec\theta$ are reciprocals of each other, as are $\tan\theta$ and $\cot\theta$.

Reciprocal Identities

$$\sin\theta = \frac{1}{\csc\theta} \qquad \cos\theta = \frac{1}{\sec\theta} \qquad \tan\theta = \frac{1}{\cot\theta}$$

$$\csc\theta = \frac{1}{\sin\theta} \qquad \sec\theta = \frac{1}{\cos\theta} \qquad \cot\theta = \frac{1}{\tan\theta}$$

Several other important identities involving the trigonometric functions will be discussed at the end of this section.

EXAMPLE 1 Finding trigonometric function values

If θ is an acute angle and $\cos\theta = \frac{3}{4}$, find the values of the trigonometric functions of θ.

Figure 4

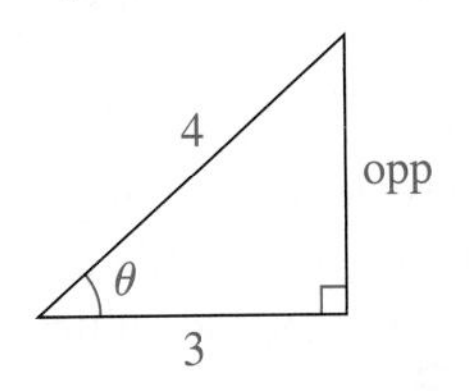

▶ **SOLUTION** We begin by sketching a right triangle having an acute angle θ with adj $= 3$ and hyp $= 4$, as shown in Figure 4, and proceed as follows:

$$\begin{aligned} 3^2 + (\text{opp})^2 &= 4^2 && \text{Pythagorean theorem} \\ (\text{opp})^2 &= 16 - 9 = 7 && \text{isolate } (\text{opp})^2 \\ \text{opp} &= \sqrt{7} && \text{take the square root} \end{aligned}$$

Applying the definition of the trigonometric functions of an acute angle of a right triangle, we obtain the following:

$$\sin\theta = \frac{\text{opp}}{\text{hyp}} = \frac{\sqrt{7}}{4} \qquad \cos\theta = \frac{\text{adj}}{\text{hyp}} = \frac{3}{4} \qquad \tan\theta = \frac{\text{opp}}{\text{adj}} = \frac{\sqrt{7}}{3}$$

$$\csc\theta = \frac{\text{hyp}}{\text{opp}} = \frac{4}{\sqrt{7}} \qquad \sec\theta = \frac{\text{hyp}}{\text{adj}} = \frac{4}{3} \qquad \cot\theta = \frac{\text{adj}}{\text{opp}} = \frac{3}{\sqrt{7}}$$

In Example 1 we could have rationalized the denominators for $\csc\theta$ and $\cot\theta$, writing

$$\csc\theta = \frac{4\sqrt{7}}{7} \quad \text{and} \quad \cot\theta = \frac{3\sqrt{7}}{7}.$$

However, in most examples and exercises we will leave expressions in unrationalized form. An exception to this practice is the special trigonometric function values corresponding to 60°, 30°, and 45°, which are obtained in the following example.

▶ **Tutorial available at www.thomsonedu.com/login**

EXAMPLE 2 Finding trigonometric function values of 60°, 30°, and 45°

Find the values of the trigonometric functions that correspond to θ:

(a) $\theta = 60°$ **(b)** $\theta = 30°$ **(c)** $\theta = 45°$

Figure 5

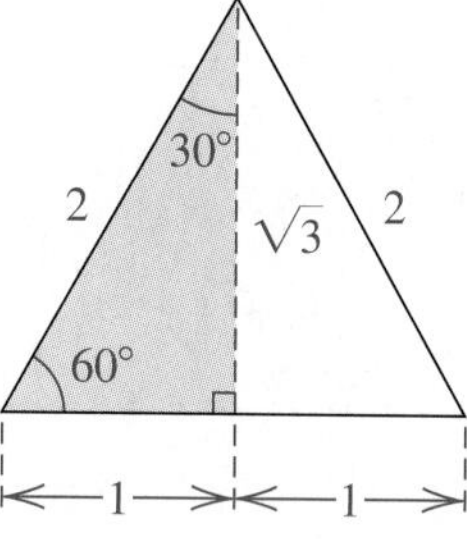

SOLUTION Consider an equilateral triangle with sides of length 2. The median from one vertex to the opposite side bisects the angle at that vertex, as illustrated by the dashes in Figure 5. By the Pythagorean theorem, the side opposite 60° in the shaded right triangle has length $\sqrt{3}$. Using the formulas for the trigonometric functions of an acute angle of a right triangle, we obtain the values corresponding to 60° and 30° as follows:

▶ **(a)** $\sin 60° = \dfrac{\sqrt{3}}{2}$ $\cos 60° = \dfrac{1}{2}$ $\tan 60° = \dfrac{\sqrt{3}}{1} = \sqrt{3}$

$\csc 60° = \dfrac{2}{\sqrt{3}} = \dfrac{2\sqrt{3}}{3}$ $\sec 60° = \dfrac{2}{1} = 2$ $\cot 60° = \dfrac{1}{\sqrt{3}} = \dfrac{\sqrt{3}}{3}$

(b) $\sin 30° = \dfrac{1}{2}$ $\cos 30° = \dfrac{\sqrt{3}}{2}$ $\tan 30° = \dfrac{1}{\sqrt{3}} = \dfrac{\sqrt{3}}{3}$

$\csc 30° = \dfrac{2}{1} = 2$ $\sec 30° = \dfrac{2}{\sqrt{3}} = \dfrac{2\sqrt{3}}{3}$ $\cot 30° = \dfrac{\sqrt{3}}{1} = \sqrt{3}$

Figure 6

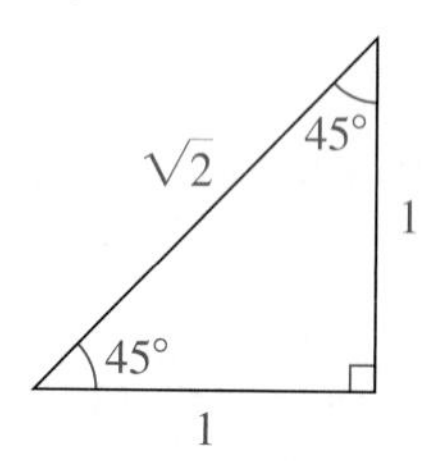

(c) To find the values for $\theta = 45°$, we may consider an isosceles right triangle whose two equal sides have length 1, as illustrated in Figure 6. By the Pythagorean theorem, the length of the hypotenuse is $\sqrt{2}$. Hence, the values corresponding to 45° are as follows:

$\sin 45° = \dfrac{1}{\sqrt{2}} = \dfrac{\sqrt{2}}{2} = \cos 45°$ $\tan 45° = \dfrac{1}{1} = 1$

$\csc 45° = \dfrac{\sqrt{2}}{1} = \sqrt{2} = \sec 45°$ $\cot 45° = \dfrac{1}{1} = 1$

For reference, we list the values found in Example 2, together with the radian measures of the angles, in the following table. Two reasons for stressing these values are that they are exact and that they occur frequently in work involving trigonometry. Because of the importance of these special values, it is a good idea either to memorize the table or to learn to find the values quickly by using triangles, as in Example 2.

▶ **Tutorial available at www.thomsonedu.com/login**

Special Values of the Trigonometric Functions

θ (radians)	θ (degrees)	$\sin\theta$	$\cos\theta$	$\tan\theta$	$\cot\theta$	$\sec\theta$	$\csc\theta$
$\frac{\pi}{6}$	$30°$	$\frac{1}{2}$	$\frac{\sqrt{3}}{2}$	$\frac{\sqrt{3}}{3}$	$\sqrt{3}$	$\frac{2\sqrt{3}}{3}$	2
$\frac{\pi}{4}$	$45°$	$\frac{\sqrt{2}}{2}$	$\frac{\sqrt{2}}{2}$	1	1	$\sqrt{2}$	$\sqrt{2}$
$\frac{\pi}{3}$	$60°$	$\frac{\sqrt{3}}{2}$	$\frac{1}{2}$	$\sqrt{3}$	$\frac{\sqrt{3}}{3}$	2	$\frac{2\sqrt{3}}{3}$

The next example illustrates a practical use for trigonometric functions of acute angles. Additional applications involving right triangles will be considered in Section 5.7.

EXAMPLE 3 Finding the height of a flagpole

A surveyor observes that at a point A, located on level ground a distance 25.0 feet from the base B of a flagpole, the angle between the ground and the top of the pole is 30°. Approximate the height h of the pole to the nearest tenth of a foot.

Figure 7

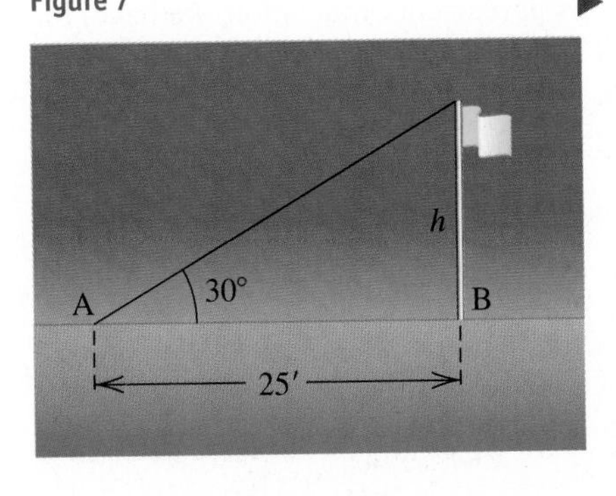

▶ **SOLUTION** Referring to Figure 7, we see that we want to relate the opposite side and the adjacent side, h and 25, respectively, to the 30° angle. This suggests that we use a trigonometric function involving those two sides—namely, tan or cot. It is usually easier to solve the problem if we select the function for which the variable is in the numerator. Hence, we have

$$\tan 30° = \frac{h}{25} \quad \text{or, equivalently,} \quad h = 25 \tan 30°.$$

We use the value of tan 30° from Example 2 to find h:

$$h = 25\left(\frac{\sqrt{3}}{3}\right) \approx 14.4 \text{ ft}$$

It is possible to approximate, to any degree of accuracy, the values of the trigonometric functions for any acute angle. Calculators have keys labeled SIN, COS, and TAN that can be used to approximate values of these functions. The values of csc, sec, and cot may then be found by means of the reciprocal key. *Before using a calculator to find function values that correspond to the radian measure of an acute angle, be sure that the calculator is in radian mode. For values corresponding to degree measure, select degree mode.*

Figure 8
In degree mode

```
sin(30)
                .5
sin(60)
      .8660254038
```

As an illustration (see Figure 8), to find sin 30° on a typical calculator, we place the calculator in degree mode and use the SIN key to obtain $\sin 30° = 0.5$, which is the exact value. Using the same procedure for 60°, we obtain a decimal approximation to $\sqrt{3}/2$, such as

$$\sin 60° \approx 0.8660.$$

▶ **Tutorial available at www.thomsonedu.com/login**

Figure 9
In radian mode

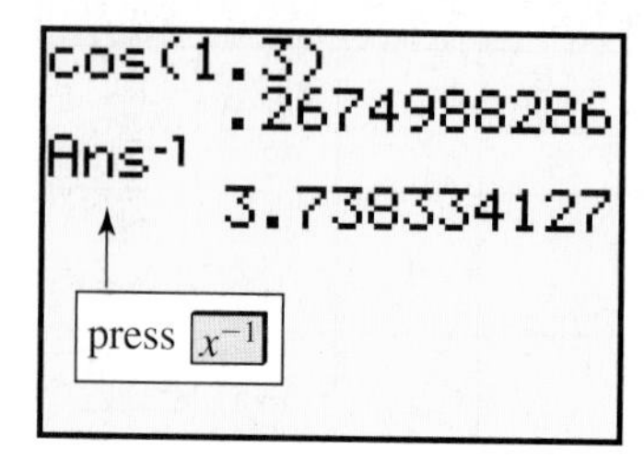

Most calculators give eight- to ten-decimal-place accuracy for such function values; throughout the text, however, we will usually round off values to four decimal places.

To find a value such as cos 1.3 (see Figure 9), where 1.3 is the radian measure of an acute angle, we place the calculator in radian mode and use the [COS] key, obtaining

$$\cos 1.3 \approx 0.2675.$$

For sec 1.3, we could find cos 1.3 and then use the reciprocal key, usually labeled [1/x] or [x^{-1}] (as shown in Figure 9), to obtain

$$\sec 1.3 = \frac{1}{\cos 1.3} \approx 3.7383.$$

The formulas listed in the box on the next page are, without doubt, the most important identities in trigonometry, because they can be used to simplify and unify many different aspects of the subject. Since the formulas are part of the foundation for work in trigonometry, they are called the *fundamental identities.*

Three of the fundamental identities involve squares, such as $(\sin \theta)^2$ and $(\cos \theta)^2$. In general, if n is an integer different from -1, then a power such as $(\cos \theta)^n$ is written $\cos^n \theta$. The symbols $\sin^{-1} \theta$ and $\cos^{-1} \theta$ are reserved for inverse trigonometric functions, which we will discuss in Section 5.4 and treat thoroughly in the next chapter. With this agreement on notation, we have, for example,

$$\cos^2 \theta = (\cos \theta)^2 = (\cos \theta)(\cos \theta)$$

$$\tan^3 \theta = (\tan \theta)^3 = (\tan \theta)(\tan \theta)(\tan \theta)$$

$$\sec^4 \theta = (\sec \theta)^4 = (\sec \theta)(\sec \theta)(\sec \theta)(\sec \theta).$$

Evaluating Powers of Trigonometric Functions (in degree mode)

Caution must be used when evaluating powers of trigonometric functions on calculators. For example, consider the expression $\sin^2 30°$. Since $\sin 30° = \frac{1}{2}$, we have

$$\sin^2 30° = \left(\tfrac{1}{2}\right)^2 = \tfrac{1}{4}.$$

The way the expression is written in the first entry on each screen below, we would expect the calculator to evaluate 30^2 and then take the sine of 900°, and that is what happens. However, we would expect the same in the second entry, where the TI-83/4 Plus gives us the value of $\sin^2 30°$. So in the future, to evaluate $\sin^2 30°$, we'll use the format shown in the third entry.

TI-83/4 Plus

```
sin((30)²)
                0
sin(30)²
              .25
(sin(30))²
              .25
```

TI-86

```
sin 30²
                0
sin (30)²
                0
(sin 30)²
              .25
```

Let us next list all the fundamental identities and then discuss the proofs. These identities are true for every acute angle θ, and θ may take on various forms. For example, using the first Pythagorean identity with $\theta = 4\alpha$, we know that

$$\sin^2 4\alpha + \cos^2 4\alpha = 1.$$

We shall see later that these identities are also true for other angles and for real numbers.

The Fundamental Identities

(1) The reciprocal identities:

$$\csc\theta = \frac{1}{\sin\theta} \qquad \sec\theta = \frac{1}{\cos\theta} \qquad \cot\theta = \frac{1}{\tan\theta}$$

(2) The tangent and cotangent identities:

$$\tan\theta = \frac{\sin\theta}{\cos\theta} \qquad \cot\theta = \frac{\cos\theta}{\sin\theta}$$

(3) The Pythagorean identities:

$$\sin^2\theta + \cos^2\theta = 1 \qquad 1 + \tan^2\theta = \sec^2\theta \qquad 1 + \cot^2\theta = \csc^2\theta$$

PROOFS

(1) The reciprocal identities were established earlier in this section.

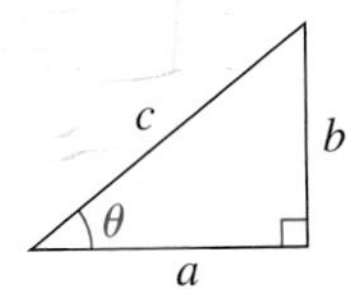

Figure 10

(2) To prove the tangent identity, we refer to the right triangle in Figure 10 and use definitions of trigonometric functions as follows:

$$\tan\theta = \frac{b}{a} = \frac{b/c}{a/c} = \frac{\sin\theta}{\cos\theta}$$

To verify the cotangent identity, we use a reciprocal identity and the tangent identity:

$$\cot\theta = \frac{1}{\tan\theta} = \frac{1}{\sin\theta/\cos\theta} = \frac{\cos\theta}{\sin\theta}$$

(3) The Pythagorean identities are so named because of the first step in the following proof. Referring to Figure 10, we obtain

$$b^2 + a^2 = c^2 \qquad \text{Pythagorean theorem}$$

$$\left(\frac{b}{c}\right)^2 + \left(\frac{a}{c}\right)^2 = \left(\frac{c}{c}\right)^2 \qquad \text{divide by } c^2$$

$$(\sin\theta)^2 + (\cos\theta)^2 = 1 \qquad \text{definitions of } \sin\theta \text{ and } \cos\theta$$

$$\sin^2\theta + \cos^2\theta = 1. \qquad \text{equivalent notation}$$

We may use this identity to verify the second Pythagorean identity as follows:

$$\frac{\sin^2\theta + \cos^2\theta}{\cos^2\theta} = \frac{1}{\cos^2\theta} \qquad \text{divide by } \cos^2\theta$$

$$\frac{\sin^2\theta}{\cos^2\theta} + \frac{\cos^2\theta}{\cos^2\theta} = \frac{1}{\cos^2\theta} \qquad \text{equivalent equation}$$

$$\left(\frac{\sin\theta}{\cos\theta}\right)^2 + \left(\frac{\cos\theta}{\cos\theta}\right)^2 = \left(\frac{1}{\cos\theta}\right)^2 \qquad \text{law of exponents}$$

$$\tan^2\theta + 1 = \sec^2\theta \qquad \text{tangent and reciprocal identities}$$

To prove the third Pythagorean identity, $1 + \cot^2\theta = \csc^2\theta$, we could divide both sides of the identity $\sin^2\theta + \cos^2\theta = 1$ by $\sin^2\theta$.

We can use the fundamental identities to express each trigonometric function in terms of any other trigonometric function. Two illustrations are given in the next example.

EXAMPLE 4 Using fundamental identities

Let θ be an acute angle.

(a) Express $\sin\theta$ in terms of $\cos\theta$.

(b) Express $\tan\theta$ in terms of $\sin\theta$.

▶ **SOLUTION**

(a) We may proceed as follows:

$$\sin^2\theta + \cos^2\theta = 1 \qquad \text{Pythagorean identity}$$

$$\sin^2\theta = 1 - \cos^2\theta \qquad \text{isolate } \sin^2\theta$$

$$\sin\theta = \pm\sqrt{1 - \cos^2\theta} \qquad \text{take the square root}$$

$$\sin\theta = \sqrt{1 - \cos^2\theta} \qquad \sin\theta > 0 \text{ for acute angles}$$

Later in this section (Example 12) we will consider a simplification involving a *non*-acute angle θ.

(b) If we begin with the fundamental identity

$$\tan\theta = \frac{\sin\theta}{\cos\theta},$$

then all that remains is to express $\cos\theta$ in terms of $\sin\theta$. We can do this by solving $\sin^2\theta + \cos^2\theta = 1$ for $\cos\theta$, obtaining

$$\cos\theta = \sqrt{1 - \sin^2\theta} \quad \text{for} \quad 0 < \theta < \frac{\pi}{2}.$$

▶ **Tutorial available at www.thomsonedu.com/login**

Hence,

$$\tan\theta = \frac{\sin\theta}{\cos\theta} = \frac{\sin\theta}{\sqrt{1-\sin^2\theta}} \quad \text{for} \quad 0 < \theta < \frac{\pi}{2}.$$

Just as we have with algebraic manipulations, we can lend numerical support to the results of our trigonometric manipulations by examining a table of values. The following screens show that the result from Example 4(a), that $\sin\theta = \sqrt{1-\cos^2\theta}$ for θ acute, is supported by the equality of Y_1 and Y_2 in the table of selected values. We will discuss graphical support later in the text.

Plot1 Plot2 Plot3
\Y1■sin(X)
\Y2■√(1−(cos(X))²)
\Y3=
\Y4=
\Y5=
\Y6=

X	Y1	Y2
0	0	0
15	.25882	.25882
30	.5	.5
45	.70711	.70711
60	.86603	.86603
75	.96593	.96593
90	1	1

Y1■sin(X)

Fundamental identities are often used to simplify expressions involving trigonometric functions, as illustrated in the next example.

EXAMPLE 5 **Showing that an equation is an identity**

Show that the following equation is an identity by transforming the left-hand side into the right-hand side:

$$(\sec\theta + \tan\theta)(1 - \sin\theta) = \cos\theta$$

▶ **SOLUTION** We begin with the left-hand side and proceed as follows:

$$\begin{aligned}
(\sec\theta + \tan\theta)(1-\sin\theta) &= \left(\frac{1}{\cos\theta} + \frac{\sin\theta}{\cos\theta}\right)(1-\sin\theta) && \text{reciprocal and tangent identities} \\
&= \left(\frac{1+\sin\theta}{\cos\theta}\right)(1-\sin\theta) && \text{add fractions} \\
&= \frac{1-\sin^2\theta}{\cos\theta} && \text{multiply} \\
&= \frac{\cos^2\theta}{\cos\theta} && \sin^2\theta + \cos^2\theta = 1 \\
&= \cos\theta && \text{cancel } \cos\theta
\end{aligned}$$

▶ **Tutorial available at www.thomsonedu.com/login**

Let's examine the result of Example 5 from a numerical point of view. We assign the left-hand side to Y_1 and the right-hand side to Y_2 and create a table of values for $\theta = 0°$ to $\theta = 90°$. Notice that the values of Y_1 and Y_2 in the third screen are equal except for $\theta = 90°$. The ERROR message occurs because sec 90° and tan 90° are undefined.

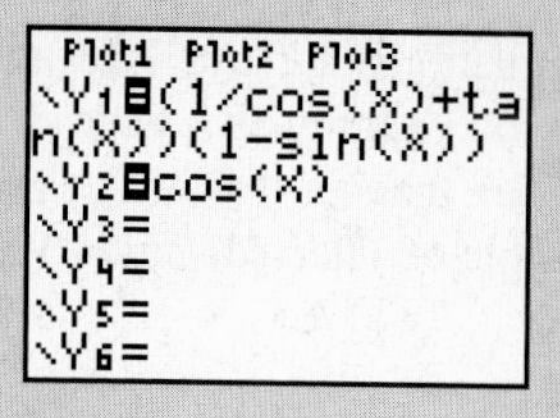

X	Y1	Y2
0	1	1
15	.96593	.96593
30	.86603	.86603
45	.70711	.70711
60	.5	.5
75	.25882	.25882
90	ERROR	0

Y1=ERROR

There are other ways to simplify the expression on the left-hand side in Example 5. We could first multiply the two factors and then simplify and combine terms. The method we employed—changing all expressions to expressions that involve only sines and cosines—is often useful. However, that technique does not always lead to the shortest possible simplification.

Hereafter, we shall use the phrase *verify an identity* instead of *show that an equation is an identity*. When verifying an identity, we often use fundamental identities and algebraic manipulations to simplify expressions, as we did in the preceding example. As with the fundamental identities, we understand that an identity that contains fractions is valid for all values of the variables such that no denominator is zero.

EXAMPLE 6 Verifying an identity

Verify the following identity by transforming the left-hand side into the right-hand side:

$$\frac{\tan\theta + \cos\theta}{\sin\theta} = \sec\theta + \cot\theta$$

SOLUTION We may transform the left-hand side into the right-hand side as follows:

$$\begin{aligned}
\frac{\tan\theta + \cos\theta}{\sin\theta} &= \frac{\tan\theta}{\sin\theta} + \frac{\cos\theta}{\sin\theta} && \text{divide numerator by } \sin\theta \\
&= \frac{\left(\dfrac{\sin\theta}{\cos\theta}\right)}{\sin\theta} + \cot\theta && \text{tangent and cotangent identities} \\
&= \frac{\sin\theta}{\cos\theta}\cdot\frac{1}{\sin\theta} + \cot\theta && \text{rule for quotients} \\
&= \frac{1}{\cos\theta} + \cot\theta && \text{cancel } \sin\theta \\
&= \sec\theta + \cot\theta && \text{reciprocal identity}
\end{aligned}$$

Figure 11

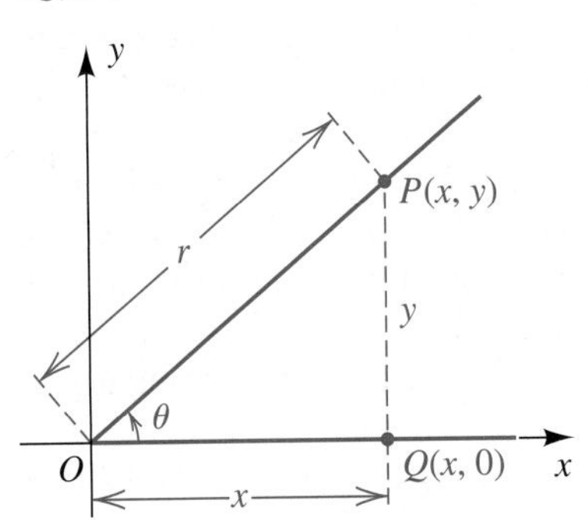

In Section 6.1 we will verify many other identities using methods similar to those used in Examples 5 and 6.

Since many applied problems involve angles that are not acute, it is necessary to extend the definition of the trigonometric functions. We make this extension by using the standard position of an angle θ on a rectangular coordinate system. If θ is acute, we have the situation illustrated in Figure 11, where we have chosen a point $P(x, y)$ on the terminal side of θ and where $d(O, P) = r = \sqrt{x^2 + y^2}$. Referring to triangle OQP, we have

$$\sin\theta = \frac{\text{opp}}{\text{hyp}} = \frac{y}{r}, \quad \cos\theta = \frac{\text{adj}}{\text{hyp}} = \frac{x}{r}, \quad \text{and} \quad \tan\theta = \frac{\text{opp}}{\text{adj}} = \frac{y}{x}.$$

We now wish to consider angles of the types illustrated in Figure 12 (or any *other* angle, either positive, negative, or zero). Note that in Figure 12 the value of x or y may be negative. In each case, side QP (opp in Figure 11) has length $|y|$, side OQ (adj in Figure 11) has length $|x|$, and the hypotenuse OP has length r. We shall define the six trigonometric functions so that their values agree with those given previously whenever the angle is acute. It is understood that if a zero denominator occurs, then the corresponding function value is undefined.

Figure 12

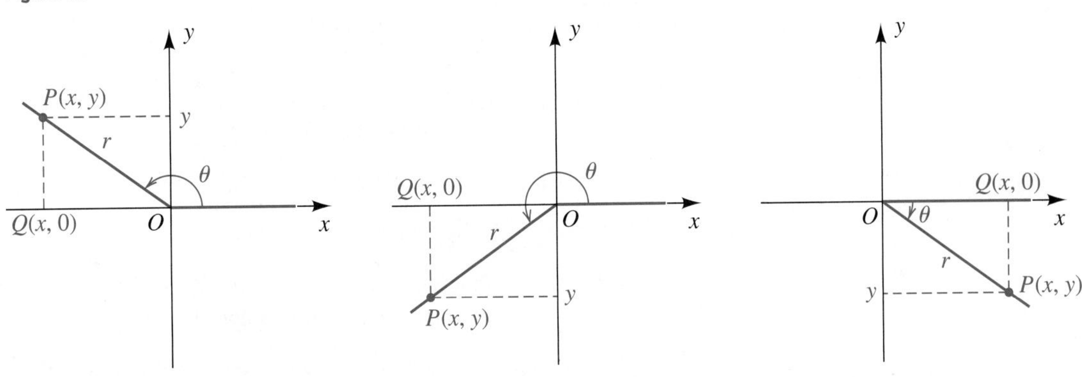

Definition of the Trigonometric Functions of Any Angle

Let θ be an angle in standard position on a rectangular coordinate system, and let $P(x, y)$ be any point other than the origin O on the terminal side of θ. If $d(O, P) = r = \sqrt{x^2 + y^2}$, then

$$\sin\theta = \frac{y}{r} \qquad \cos\theta = \frac{x}{r} \qquad \tan\theta = \frac{y}{x} \quad (\text{if } x \neq 0)$$

$$\csc\theta = \frac{r}{y} \quad (\text{if } y \neq 0) \qquad \sec\theta = \frac{r}{x} \quad (\text{if } x \neq 0) \qquad \cot\theta = \frac{x}{y} \quad (\text{if } y \neq 0).$$

We can show, using similar triangles, that the formulas in this definition do not depend on the point $P(x, y)$ that is chosen on the terminal side of θ. The fundamental identities, which were established for acute angles, are also true for trigonometric functions of any angle.

The domains of the sine and cosine functions consist of all angles θ. However, $\tan\theta$ and $\sec\theta$ are undefined if $x = 0$ (that is, if the terminal side of θ is on the y-axis). Thus, the domains of the tangent and the secant functions consist of all angles *except* those of radian measure $(\pi/2) + \pi n$ for any integer n. Some special cases are $\pm\pi/2$, $\pm 3\pi/2$, and $\pm 5\pi/2$. The corresponding degree measures are $\pm 90°$, $\pm 270°$, and $\pm 450°$.

The domains of the cotangent and cosecant functions consist of all angles except those that have $y = 0$ (that is, all angles except those having terminal sides on the x-axis). These are the angles of radian measure πn (or degree measure $180° \cdot n$) for any integer n.

Our discussion of domains is summarized in the following table, where n denotes any integer.

Function		Domain
sine,	cosine	every angle θ
tangent,	secant	every angle θ except $\theta = \dfrac{\pi}{2} + \pi n = 90° + 180° \cdot n$
cotangent,	cosecant	every angle θ except $\theta = \pi n = 180° \cdot n$

For any point $P(x, y)$ in the preceding definition, $|x| \le r$ and $|y| \le r$ or, equivalently, $|x/r| \le 1$ and $|y/r| \le 1$. Thus,

$$|\sin\theta| \le 1, \quad |\cos\theta| \le 1, \quad |\csc\theta| \ge 1, \quad \text{and} \quad |\sec\theta| \ge 1$$

for every θ in the domains of these functions.

EXAMPLE 7 Finding trigonometric function values of an angle in standard position

If θ is an angle in standard position on a rectangular coordinate system and if $P(-15, 8)$ is on the terminal side of θ, find the values of the six trigonometric functions of θ.

Figure 13

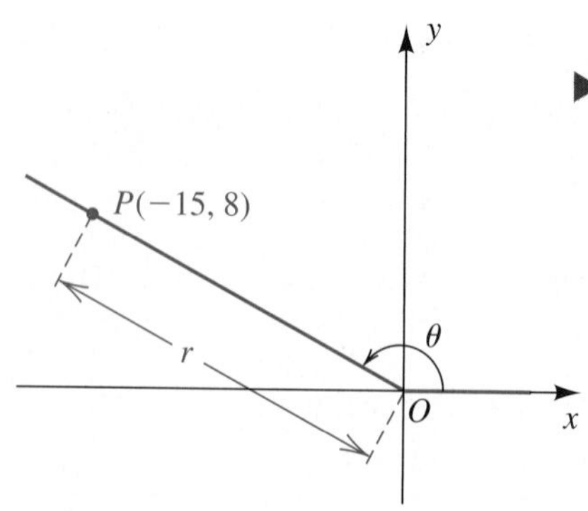

▶ **SOLUTION** The point $P(-15, 8)$ is shown in Figure 13. Applying the definition of the trigonometric functions of any angle with $x = -15$, $y = 8$, and

$$r = \sqrt{x^2 + y^2} = \sqrt{(-15)^2 + 8^2} = \sqrt{289} = 17,$$

we obtain the following:

$$\sin\theta = \frac{y}{r} = \frac{8}{17} \qquad \cos\theta = \frac{x}{r} = -\frac{15}{17} \qquad \tan\theta = \frac{y}{x} = -\frac{8}{15}$$

$$\csc\theta = \frac{r}{y} = \frac{17}{8} \qquad \sec\theta = \frac{r}{x} = -\frac{17}{15} \qquad \cot\theta = \frac{x}{y} = -\frac{15}{8}$$

▶ **Tutorial available at www.thomsonedu.com/login**

EXAMPLE 8 Finding trigonometric function values of an angle in standard position

An angle θ is in standard position, and its terminal side lies in quadrant III on the line $y = 3x$. Find the values of the trigonometric functions of θ.

Figure 14

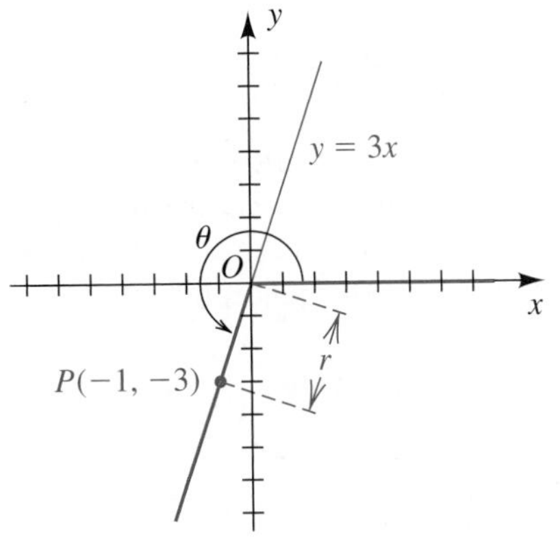

▶ SOLUTION The graph of $y = 3x$ is sketched in Figure 14, together with the initial and terminal sides of θ. Since the terminal side of θ is in quadrant III, we begin by choosing a convenient negative value of x, say $x = -1$. Substituting in $y = 3x$ gives us $y = 3(-1) = -3$, and hence $P(-1, -3)$ is on the terminal side. Applying the definition of the trigonometric functions of any angle with

$$x = -1, \quad y = -3, \quad \text{and} \quad r = \sqrt{x^2 + y^2} = \sqrt{(-1)^2 + (-3)^2} = \sqrt{10}$$

gives us

$$\sin\theta = -\frac{3}{\sqrt{10}} \qquad \cos\theta = -\frac{1}{\sqrt{10}} \qquad \tan\theta = \frac{-3}{-1} = 3$$

$$\csc\theta = -\frac{\sqrt{10}}{3} \qquad \sec\theta = -\frac{\sqrt{10}}{1} \qquad \cot\theta = \frac{-1}{-3} = \frac{1}{3}.$$

The definition of the trigonometric functions of any angle may be applied if θ is a quadrantal angle. The procedure is illustrated by the next example.

EXAMPLE 9 Finding trigonometric function values of a quadrantal angle

If $\theta = 3\pi/2$, find the values of the trigonometric functions of θ.

Figure 15

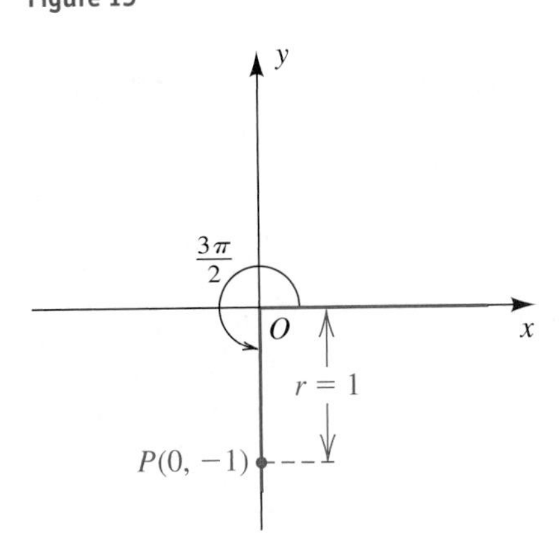

SOLUTION Note that $3\pi/2 = 270°$. If θ is placed in standard position, the terminal side of θ coincides with the negative y-axis, as shown in Figure 15. To apply the definition of the trigonometric functions of any angle, we may choose *any* point P on the terminal side of θ. For simplicity, we use $P(0, -1)$. In this case, $x = 0$, $y = -1$, $r = 1$, and hence

$$\sin\frac{3\pi}{2} = \frac{-1}{1} = -1 \qquad \cos\frac{3\pi}{2} = \frac{0}{1} = 0$$

$$\csc\frac{3\pi}{2} = \frac{1}{-1} = -1 \qquad \cot\frac{3\pi}{2} = \frac{0}{-1} = 0.$$

The tangent and secant functions are undefined, since the meaningless expressions $\tan\theta = (-1)/0$ and $\sec\theta = 1/0$ occur when we substitute in the appropriate formulas.

Let us determine the signs associated with values of the trigonometric functions. If θ is in quadrant II and $P(x, y)$ is a point on the terminal side, then x is negative and y is positive. Hence, $\sin\theta = y/r$ and $\csc\theta = r/y$ are positive, and the other four trigonometric functions, which all involve x, are negative. Checking the remaining quadrants in a similar fashion, we obtain the following table.

▶ **Tutorial available at www.thomsonedu.com/login**

Signs of the Trigonometric Functions

Quadrant containing θ	Positive functions	Negative functions
I	all	none
II	sin, csc	cos, sec, tan, cot
III	tan, cot	sin, csc, cos, sec
IV	cos, sec	sin, csc, tan, cot

Figure 16
Positive trigonometric functions

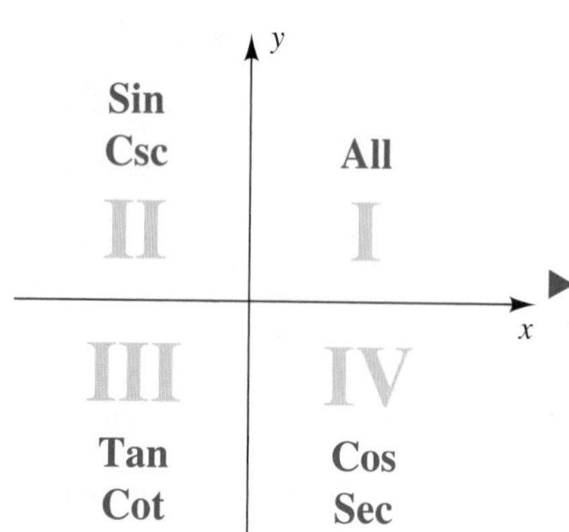

A mnemonic device for remembering the quadrants in which the trigonometric functions are positive is "A Smart Trig Class," which corresponds to All Sin Tan Cos.

The diagram in Figure 16 may be useful for remembering quadrants in which trigonometric functions are *positive*. If a function is not listed (such as cos in quadrant II), then that function is negative. We finish this section with three examples that require using the information in the preceding table.

EXAMPLE 10 **Finding the quadrant containing an angle**

Find the quadrant containing θ if both $\cos\theta > 0$ and $\sin\theta < 0$.

▶ **SOLUTION** Referring to the table of signs or Figure 16, we see that $\cos\theta > 0$ (cosine is positive) if θ is in quadrant I or IV and that $\sin\theta < 0$ (sine is negative) if θ is in quadrant III or IV. Hence, for both conditions to be satisfied, θ must be in quadrant IV.

EXAMPLE 11 **Finding values of trigonometric functions from prescribed conditions**

If $\sin\theta = \frac{3}{5}$ and $\tan\theta < 0$, use fundamental identities to find the values of the other five trigonometric functions.

▶ **SOLUTION** Since $\sin\theta = \frac{3}{5} > 0$ (positive) and $\tan\theta < 0$ (negative), θ is in quadrant II. Using the relationship $\sin^2\theta + \cos^2\theta = 1$ and the fact that $\cos\theta$ is negative in quadrant II, we have

$$\cos\theta = -\sqrt{1 - \sin^2\theta} = -\sqrt{1 - \left(\tfrac{3}{5}\right)^2} = -\sqrt{\tfrac{16}{25}} = -\tfrac{4}{5}.$$

Next we use the tangent identity to obtain

$$\tan\theta = \frac{\sin\theta}{\cos\theta} = \frac{3/5}{-4/5} = -\frac{3}{4}.$$

Finally, using the reciprocal identities gives us

$$\csc\theta = \frac{1}{\sin\theta} = \frac{1}{3/5} = \frac{5}{3}$$

$$\sec\theta = \frac{1}{\cos\theta} = \frac{1}{-4/5} = -\frac{5}{4}$$

$$\cot\theta = \frac{1}{\tan\theta} = \frac{1}{-3/4} = -\frac{4}{3}.$$

▶ Tutorial available at www.thomsonedu.com/login

EXAMPLE 12 **Using fundamental identities**

Rewrite $\sqrt{\cos^2\theta + \sin^2\theta + \cot^2\theta}$ in nonradical form without using absolute values for $\pi < \theta < 2\pi$.

▶ SOLUTION

$$\begin{aligned}\sqrt{\cos^2\theta + \sin^2\theta + \cot^2\theta} &= \sqrt{1+\cot^2\theta} && \cos^2\theta + \sin^2\theta = 1\\ &= \sqrt{\csc^2\theta} && 1 + \cot^2\theta = \csc^2\theta\\ &= |\csc\theta| && \sqrt{x^2} = |x|\end{aligned}$$

Since $\pi < \theta < 2\pi$, we know that θ is in quadrant III or IV. Thus, $\csc\theta$ is *negative,* and by the definition of absolute value, we have

$$|\csc\theta| = -\csc\theta.$$

5.2 *Exercises*

Exer. 1–2: Use common sense to match the variables and the values. (The triangles are drawn to scale, and the angles are measured in radians.)

1

z, β, x, α, y

(a) α (A) 7

(b) β (B) 0.28

(c) x (C) 24

(d) y (D) 1.29

(e) z (E) 25

2

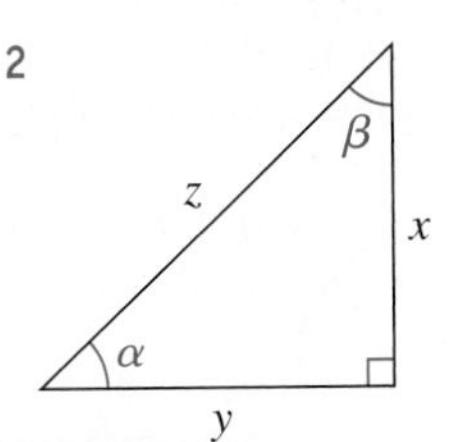

(a) α (A) 23.35

(b) β (B) 16

(c) x (C) 17

(d) y (D) 0.82

(e) z (E) 0.76

Exer. 3–10: Find the values of the six trigonometric functions for the angle θ.

3

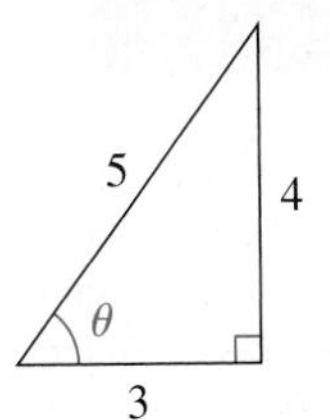

4

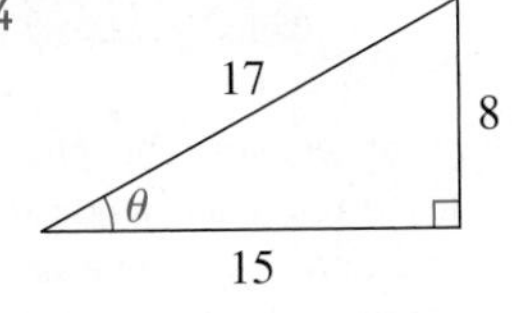

5

5, 2, θ

6

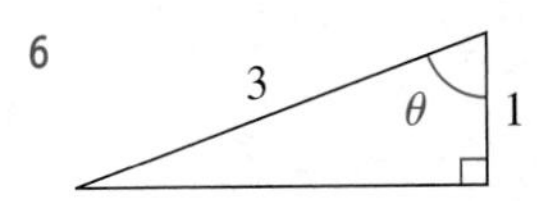

7

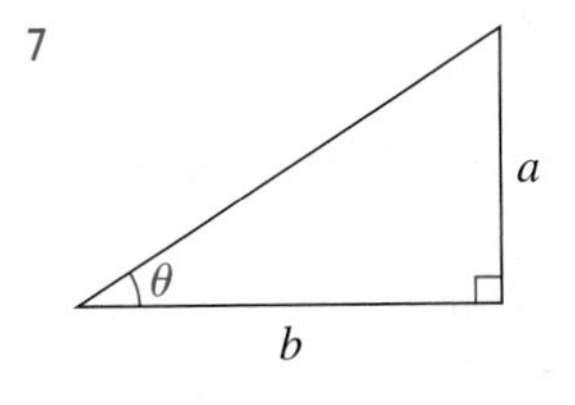

8

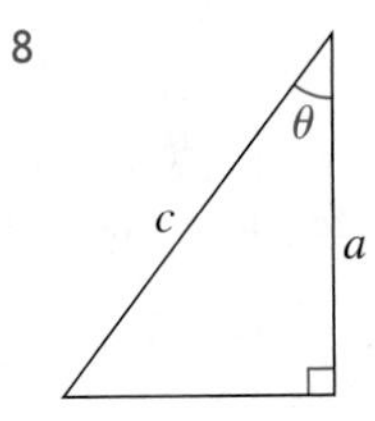

9

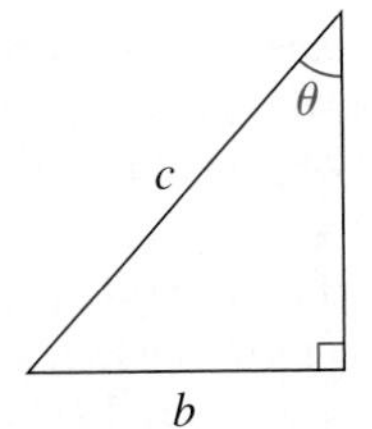

10

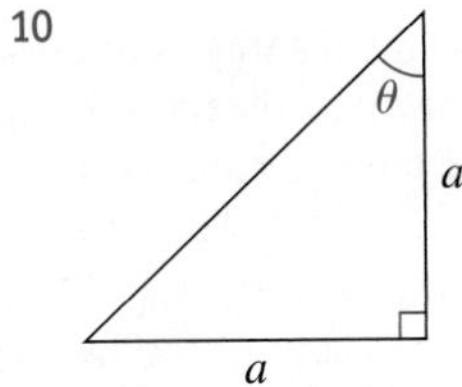

▶ Tutorial available at www.thomsonedu.com/login

Exer. 11–16: Find the exact values of x and y.

11
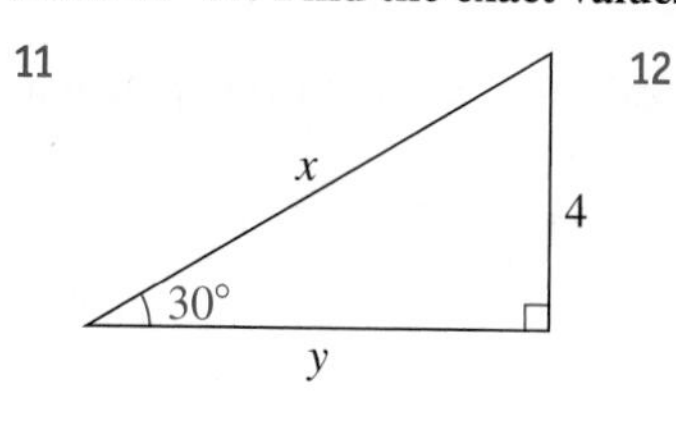

12
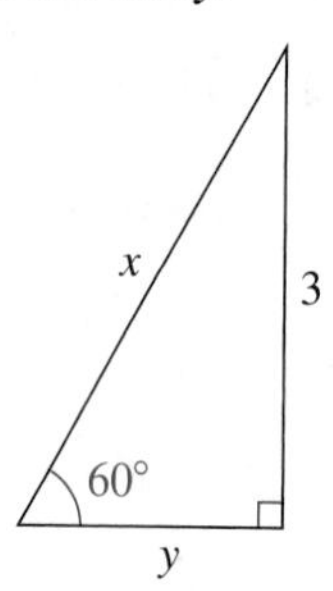

13
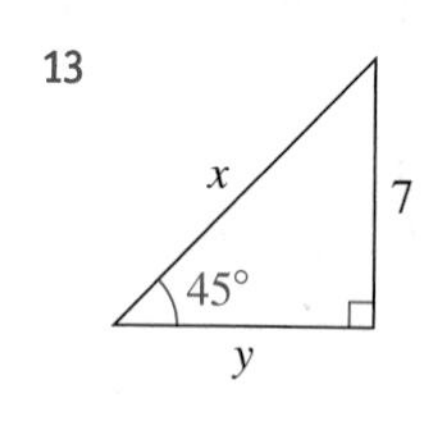

14
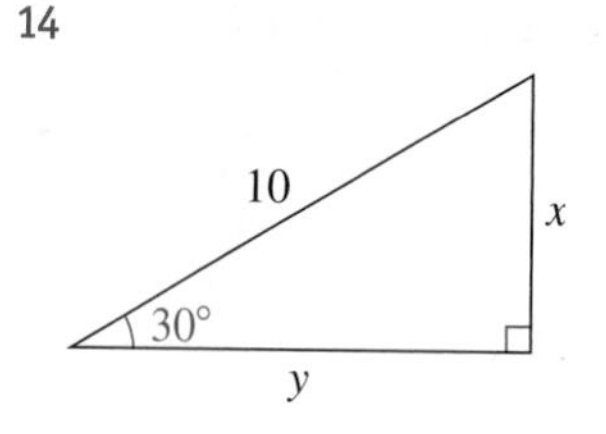

15
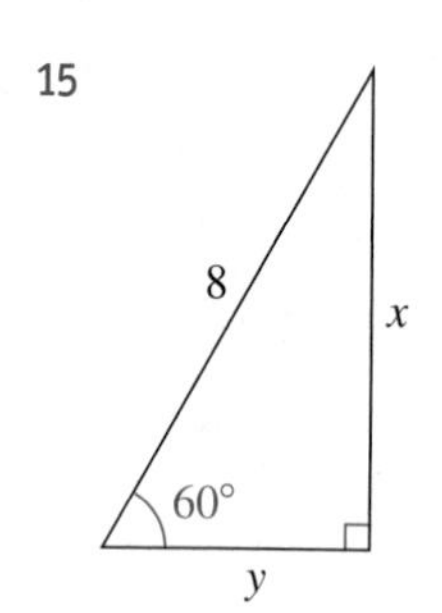

16
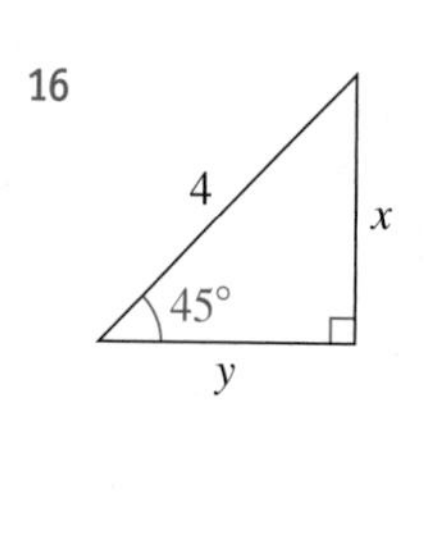

Exer. 17–22: Find the exact values of the trigonometric functions for the acute angle θ.

17 $\sin\theta = \frac{3}{5}$

18 $\cos\theta = \frac{8}{17}$

19 $\tan\theta = \frac{5}{12}$

20 $\cot\theta = \frac{7}{24}$

21 $\sec\theta = \frac{6}{5}$

22 $\csc\theta = 4$

23 **Height of a tree** A forester, 200 feet from the base of a redwood tree, observes that the angle between the ground and the top of the tree is 60°. Estimate the height of the tree.

24 **Distance to Mt. Fuji** The peak of Mt. Fuji in Japan is approximately 12,400 feet high. A trigonometry student, several miles away, notes that the angle between level ground and the peak is 30°. Estimate the distance from the student to the point on level ground directly beneath the peak.

25 **Stonehenge blocks** Stonehenge in Salisbury Plains, England, was constructed using solid stone blocks weighing over 99,000 pounds each. Lifting a single stone required 550 people, who pulled the stone up a ramp inclined at an angle of 9°. Approximate the distance that a stone was moved in order to raise it to a height of 30 feet.

26 **Advertising sign height** Added in 1990 and removed in 1997, the highest advertising sign in the world was a large letter I situated at the top of the 73-story First Interstate World Center building in Los Angeles. At a distance of 200 feet from a point directly below the sign, the angle between the ground and the top of the sign was 78.87°. Approximate the height of the top of the sign.

27 **Telescope resolution** Two stars that are very close may appear to be one. The ability of a telescope to separate their images is called its resolution. The smaller the resolution, the better a telescope's ability to separate images in the sky. In a refracting telescope, resolution θ (see the figure) can be improved by using a lens with a larger diameter D. The relationship between θ in degrees and D in meters is given by $\sin\theta = 1.22\lambda/D$, where λ is the wavelength of light in meters. The largest refracting telescope in the world is at the University of Chicago. At a wavelength of $\lambda = 550 \times 10^{-9}$ meter, its resolution is 0.000 037 69°. Approximate the diameter of the lens.

Exercise 27

28 **Moon phases** The phases of the moon can be described using the phase angle θ, determined by the sun, the moon, and Earth, as shown in the figure. Because the moon orbits Earth, θ changes during the course of a month. The area of the region A of the moon, which appears illuminated to

an observer on Earth, is given by $A = \frac{1}{2}\pi R^2(1 + \cos\theta)$, where $R = 1080$ mi is the radius of the moon. Approximate A for the following positions of the moon:

(a) $\theta = 0°$ (full moon) (b) $\theta = 180°$ (new moon)

(c) $\theta = 90°$ (first quarter) (d) $\theta = 103°$

Exercise 28

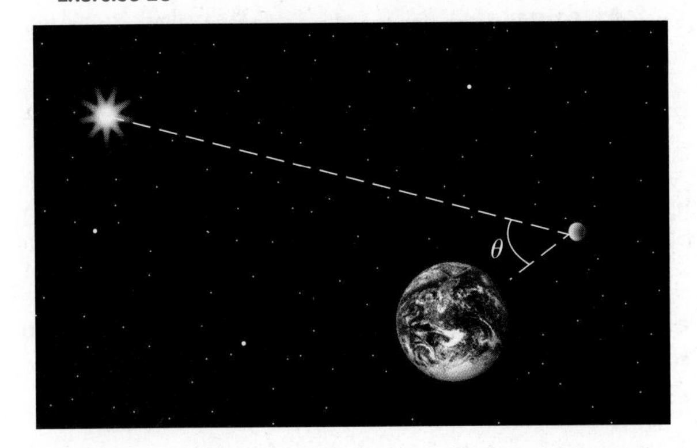

Exer. 29–34: Approximate to four decimal places, when appropriate.

29 (a) $\sin 42°$ (b) $\cos 77°$

(c) $\csc 123°$ (d) $\sec(-190°)$

30 (a) $\tan 282°$ (b) $\cot(-81°)$

(c) $\sec 202°$ (d) $\sin 97°$

31 (a) $\cot(\pi/13)$ (b) $\csc 1.32$

(c) $\cos(-8.54)$ (d) $\tan(3\pi/7)$

32 (a) $\sin(-0.11)$ (b) $\sec \frac{31}{27}$

(c) $\tan\left(-\frac{3}{13}\right)$ (d) $\cos 2.4\pi$

33 (a) $\sin 30°$ (b) $\sin 30$

(c) $\cos \pi°$ (d) $\cos \pi$

34 (a) $\sin 45°$ (b) $\sin 45$

(c) $\cos(3\pi/2)°$ (d) $\cos(3\pi/2)$

Exer. 35–38: Use the Pythagorean identities to write the expression as an integer.

35 (a) $\tan^2 4\beta - \sec^2 4\beta$ (b) $4\tan^2\beta - 4\sec^2\beta$

36 (a) $\csc^2 3\alpha - \cot^2 3\alpha$ (b) $3\csc^2\alpha - 3\cot^2\alpha$

37 (a) $5\sin^2\theta + 5\cos^2\theta$

(b) $5\sin^2(\theta/4) + 5\cos^2(\theta/4)$

38 (a) $7\sec^2\gamma - 7\tan^2\gamma$

(b) $7\sec^2(\gamma/3) - 7\tan^2(\gamma/3)$

Exer. 39–42: Simplify the expression.

39 $\dfrac{\sin^3\theta + \cos^3\theta}{\sin\theta + \cos\theta}$ 40 $\dfrac{\cot^2\alpha - 4}{\cot^2\alpha - \cot\alpha - 6}$

41 $\dfrac{2 - \tan\theta}{2\csc\theta - \sec\theta}$ 42 $\dfrac{\csc\theta + 1}{(1/\sin^2\theta) + \csc\theta}$

Exer. 43–48: Use fundamental identities to write the first expression in terms of the second, for any acute angle θ.

43 $\cot\theta$, $\sin\theta$ 44 $\tan\theta$, $\cos\theta$

45 $\sec\theta$, $\sin\theta$ 46 $\csc\theta$, $\cos\theta$

47 $\sin\theta$, $\sec\theta$ 48 $\cos\theta$, $\cot\theta$

Exer. 49–70: Verify the identity by transforming the left-hand side into the right-hand side.

49 $\cos\theta\sec\theta = 1$ 50 $\tan\theta\cot\theta = 1$

51 $\sin\theta\sec\theta = \tan\theta$ 52 $\sin\theta\cot\theta = \cos\theta$

53 $\dfrac{\csc\theta}{\sec\theta} = \cot\theta$ 54 $\cot\theta\sec\theta = \csc\theta$

55 $(1 + \cos 2\theta)(1 - \cos 2\theta) = \sin^2 2\theta$

56 $\cos^2 2\theta - \sin^2 2\theta = 2\cos^2 2\theta - 1$

57 $\cos^2\theta(\sec^2\theta - 1) = \sin^2\theta$

58 $(\tan\theta + \cot\theta)\tan\theta = \sec^2\theta$

59 $\dfrac{\sin(\theta/2)}{\csc(\theta/2)} + \dfrac{\cos(\theta/2)}{\sec(\theta/2)} = 1$

60 $1 - 2\sin^2(\theta/2) = 2\cos^2(\theta/2) - 1$

61 $(1 + \sin\theta)(1 - \sin\theta) = \dfrac{1}{\sec^2\theta}$

62 $(1 - \sin^2\theta)(1 + \tan^2\theta) = 1$

63 $\sec\theta - \cos\theta = \tan\theta\sin\theta$

64 $\dfrac{\sin\theta + \cos\theta}{\cos\theta} = 1 + \tan\theta$

65 $(\cot\theta + \csc\theta)(\tan\theta - \sin\theta) = \sec\theta - \cos\theta$

66 $\cot\theta + \tan\theta = \csc\theta\sec\theta$

67 $\sec^2 3\theta\csc^2 3\theta = \sec^2 3\theta + \csc^2 3\theta$

68 $\dfrac{1 + \cos^2 3\theta}{\sin^2 3\theta} = 2\csc^2 3\theta - 1$

69 $\log\csc\theta = -\log\sin\theta$

70 $\log\tan\theta = \log\sin\theta - \log\cos\theta$

Exer. 71–74: Find the exact values of the six trigonometric functions of θ if θ is in standard position and P is on the terminal side.

71 $P(4, -3)$

72 $P(-8, -15)$

73 $P(-2, -5)$

74 $P(-1, 2)$

Exer. 75–80: Find the exact values of the six trigonometric functions of θ if θ is in standard position and the terminal side of θ is in the specified quadrant and satisfies the given condition.

75 II; on the line $y = -4x$

76 IV; on the line $3y + 5x = 0$

77 I; on a line having slope $\frac{4}{3}$

78 III; bisects the quadrant

79 III; parallel to the line $2y - 7x + 2 = 0$

80 II; parallel to the line through $A(1, 4)$ and $B(3, -2)$

Exer. 81–82: Find the exact values of the six trigonometric functions of each angle, whenever possible.

81 (a) $90°$ (b) $0°$ (c) $7\pi/2$ (d) 3π

82 (a) $180°$ (b) $-90°$ (c) 2π (d) $5\pi/2$

Exer. 83–84: Find the quadrant containing θ if the given conditions are true.

83 (a) $\cos\theta > 0$ and $\sin\theta < 0$

(b) $\sin\theta < 0$ and $\cot\theta > 0$

(c) $\csc\theta > 0$ and $\sec\theta < 0$

(d) $\sec\theta < 0$ and $\tan\theta > 0$

84 (a) $\tan\theta < 0$ and $\cos\theta > 0$

(b) $\sec\theta > 0$ and $\tan\theta < 0$

(c) $\csc\theta > 0$ and $\cot\theta < 0$

(d) $\cos\theta < 0$ and $\csc\theta < 0$

Exer. 85–92: Use fundamental identities to find the values of the trigonometric functions for the given conditions.

85 $\tan\theta = -\frac{3}{4}$ and $\sin\theta > 0$

86 $\cot\theta = \frac{3}{4}$ and $\cos\theta < 0$

87 $\sin\theta = -\frac{5}{13}$ and $\sec\theta > 0$

88 $\cos\theta = \frac{1}{2}$ and $\sin\theta < 0$

89 $\cos\theta = -\frac{1}{3}$ and $\sin\theta < 0$

90 $\csc\theta = 5$ and $\cot\theta < 0$

91 $\sec\theta = -4$ and $\csc\theta > 0$

92 $\sin\theta = \frac{2}{5}$ and $\cos\theta < 0$

Exer. 93–98: Rewrite the expression in nonradical form without using absolute values for the indicated values of θ.

93 $\sqrt{\sec^2\theta - 1}$; $\pi/2 < \theta < \pi$

94 $\sqrt{1 + \cot^2\theta}$; $0 < \theta < \pi$

95 $\sqrt{1 + \tan^2\theta}$; $3\pi/2 < \theta < 2\pi$

96 $\sqrt{\csc^2\theta - 1}$; $3\pi/2 < \theta < 2\pi$

97 $\sqrt{\sin^2(\theta/2)}$; $2\pi < \theta < 4\pi$

98 $\sqrt{\cos^2(\theta/2)}$; $0 < \theta < \pi$

5.3 Trigonometric Functions of Real Numbers

The domain of each trigonometric function we have discussed is a set of angles. In calculus and in many applications, domains of functions consist of real numbers. To regard the domain of a trigonometric function as a subset of $\mathbb{R}$, we may use the following definition.

Definition of the Trigonometric Functions of Real Numbers	The **value of a trigonometric function at a real number t** is its value at an angle of t radians, provided that value exists.

Using this definition, we may interpret a notation such as sin 2 as *either* the sine of the real number 2 *or* the sine of an angle of 2 radians. As in Section 5.2, if degree measure is used, we shall write sin 2°. With this understanding,

$$\sin 2 \neq \sin 2°.$$

Figure 1

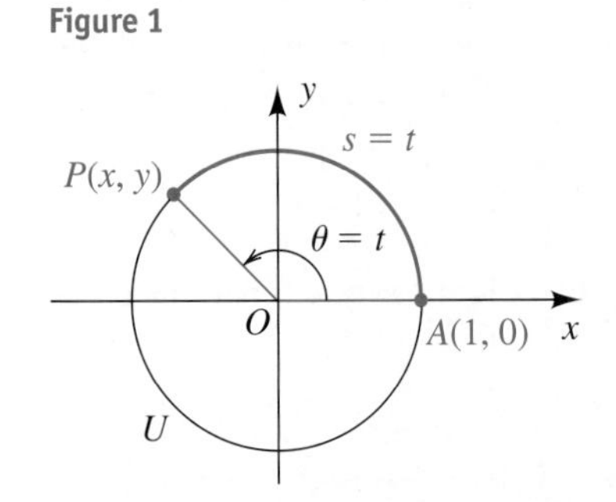

To find the values of trigonometric functions of real numbers with a calculator, we use the radian mode.

We may interpret trigonometric functions of real numbers geometrically by using a unit circle U—that is, a circle of radius 1, with center at the origin O of a rectangular coordinate plane. The circle U is the graph of the equation $x^2 + y^2 = 1$. Let t be a real number such that $0 < t < 2\pi$, and let θ denote the angle (in standard position) of radian measure t. One possibility is illustrated in Figure 1, where $P(x, y)$ is the point of intersection of the terminal side of θ and the unit circle U and where s is the length of the circular arc from $A(1, 0)$ to $P(x, y)$. Using the formula $s = r\theta$ for the length of a circular arc, with $r = 1$ and $\theta = t$, we see that

$$s = r\theta = 1(t) = t.$$

Thus, *t may be regarded either as the radian measure of the angle θ or as the length of the circular arc AP on U.*

Next consider *any* nonnegative real number t. If we regard the angle θ of radian measure t as having been generated by rotating the line segment OA

Figure 2
$\theta = t, t > 0$

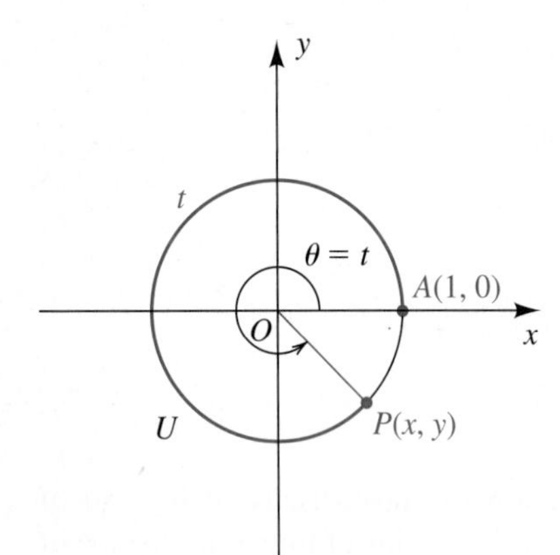

about O in the counterclockwise direction, then t is the distance along U that A travels before reaching its final position $P(x, y)$. In Figure 2 we have illustrated a case for $t < 2\pi$; however, if $t > 2\pi$, then A may travel around U several times in a counterclockwise direction before reaching $P(x, y)$.

If $t < 0$, then the rotation of OA is in the *clockwise* direction, and the distance A travels before reaching $P(x, y)$ is $|t|$, as illustrated in Figure 3.

Figure 3
$\theta = t, t < 0$

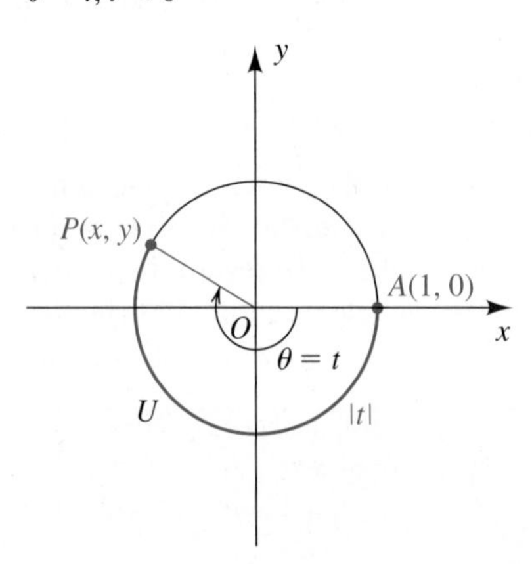

The preceding discussion indicates how *we may associate with each real number t a unique point* $P(x, y)$ *on* U. We shall call $P(x, y)$ the **point on the unit circle U that corresponds to t.** The coordinates (x, y) of P may be used to find the six trigonometric functions of t. Thus, by the definition of the trigonometric functions of real numbers together with the definition of the trigonometric functions of any angle (given in Section 5.2), we see that

$$\sin t = \sin \theta = \frac{y}{r} = \frac{y}{1} = y.$$

Using the same procedure for the remaining five trigonometric functions gives us the following formulas.

Definition of the Trigonometric Functions in Terms of a Unit Circle

If t is a real number and $P(x, y)$ is the point on the unit circle U that corresponds to t, then

$$\sin t = y \qquad \cos t = x \qquad \tan t = \frac{y}{x} \quad (\text{if } x \neq 0)$$

$$\csc t = \frac{1}{y} \quad (\text{if } y \neq 0) \qquad \sec t = \frac{1}{x} \quad (\text{if } x \neq 0) \qquad \cot t = \frac{x}{y} \quad (\text{if } y \neq 0).$$

The formulas in this definition express function values in terms of coordinates of a point P on a unit circle. For this reason, the trigonometric functions are sometimes referred to as the **circular functions.**

EXAMPLE 1 Finding values of the trigonometric functions

A point $P(x, y)$ on the unit circle U corresponding to a real number t is shown in Figure 4, for $\pi < t < 3\pi/2$. Find the values of the trigonometric functions at t.

Figure 4

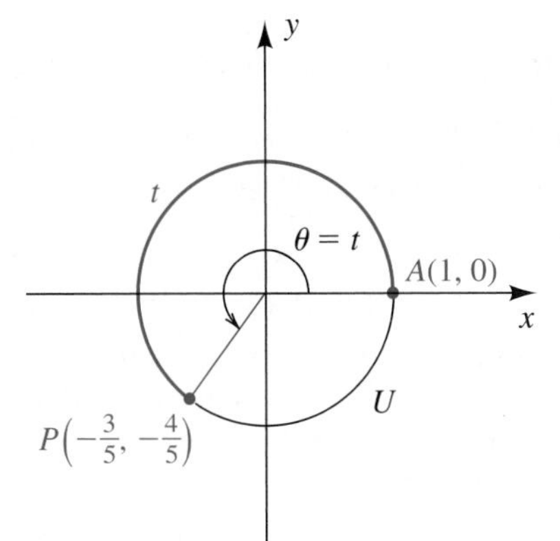

▶ SOLUTION Referring to Figure 4, we see that the coordinates of the point $P(x, y)$ are

$$x = -\tfrac{3}{5}, \qquad y = -\tfrac{4}{5}.$$

Using the definition of the trigonometric functions in terms of a unit circle gives us

$$\sin t = y = -\frac{4}{5} \qquad \cos t = x = -\frac{3}{5} \qquad \tan t = \frac{y}{x} = \frac{-\frac{4}{5}}{-\frac{3}{5}} = \frac{4}{3}$$

$$\csc t = \frac{1}{y} = \frac{1}{-\frac{4}{5}} = -\frac{5}{4} \quad \sec t = \frac{1}{x} = \frac{1}{-\frac{3}{5}} = -\frac{5}{3} \quad \cot t = \frac{x}{y} = \frac{-\frac{3}{5}}{-\frac{4}{5}} = \frac{3}{4}.$$

EXAMPLE 2 Finding a point on *U* relative to a given point

Let $P(t)$ denote the point on the unit circle U that corresponds to t for $0 \le t < 2\pi$. If $P(t) = \left(\frac{4}{5}, \frac{3}{5}\right)$, find

(a) $P(t + \pi)$ **(b)** $P(t - \pi)$ **(c)** $P(-t)$

SOLUTION

▶ **(a)** The point $P(t)$ on U is plotted in Figure 5(a), where we have also shown the arc AP of length t. To find $P(t + \pi)$, we travel a distance π in the *counterclockwise* direction along U from $P(t)$, as indicated by the blue arc in the figure. Since π is one-half the circumference of U, this gives us the point $P(t + \pi) = \left(-\frac{4}{5}, -\frac{3}{5}\right)$ diametrically opposite $P(t)$.

Figure 5

(a)

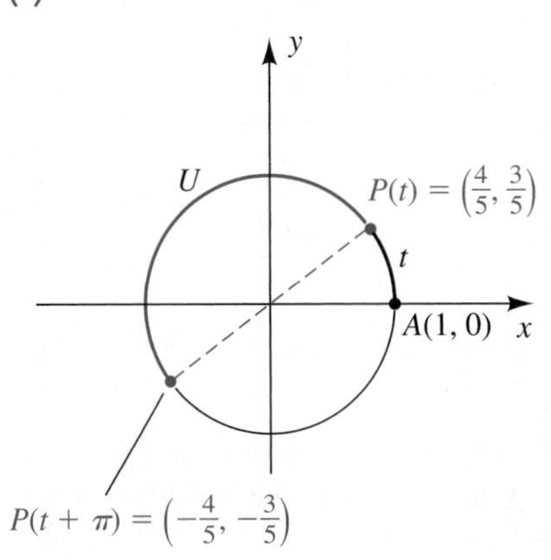

(b)

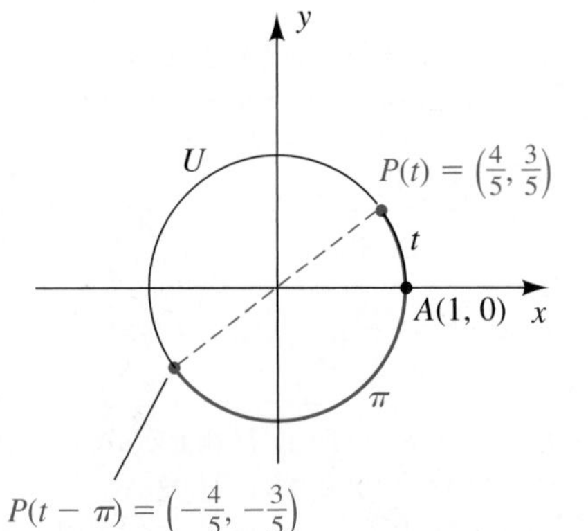

(c)

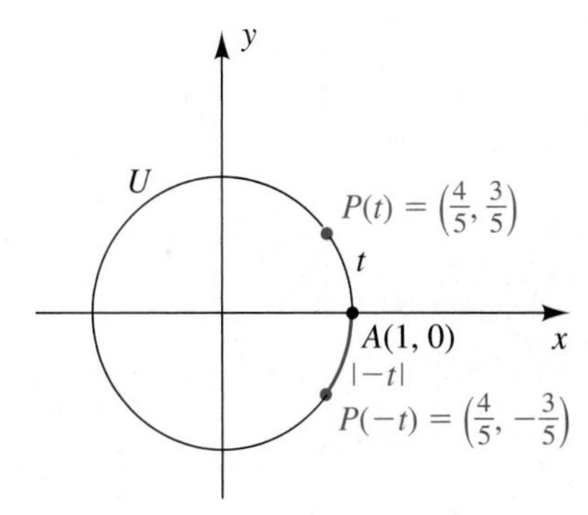

(continued)

▶ **Tutorial available at www.thomsonedu.com/login**

(b) To find $P(t - \pi)$, we travel a distance π in the *clockwise* direction along U from $P(t)$, as indicated in Figure 5(b). This gives us $P(t - \pi) = \left(-\frac{4}{5}, -\frac{3}{5}\right)$. Note that $P(t + \pi) = P(t - \pi)$.

▶ **(c)** To find $P(-t)$, we travel along U a distance $|-t|$ in the *clockwise* direction from $A(1, 0)$, as indicated in Figure 5(c). This is equivalent to reflecting $P(t)$ through the x-axis. Thus, we merely change the sign of the y-coordinate of $P(t) = \left(\frac{4}{5}, \frac{3}{5}\right)$ to obtain $P(-t) = \left(\frac{4}{5}, -\frac{3}{5}\right)$.

Figure 6

(a)

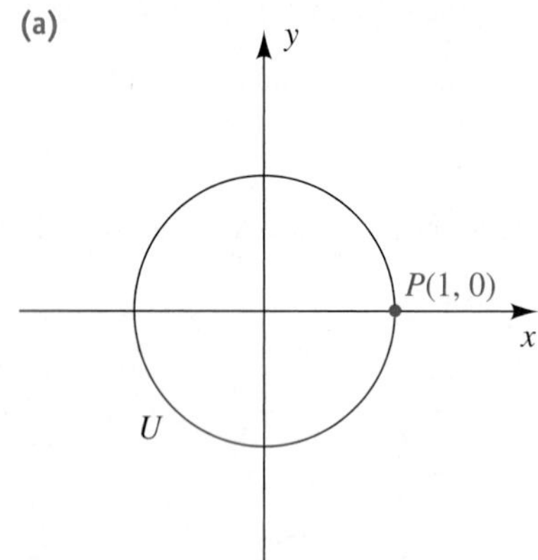

(b)

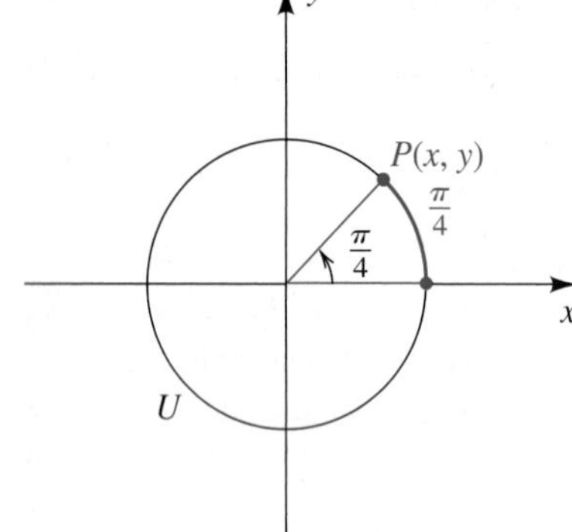

(c)

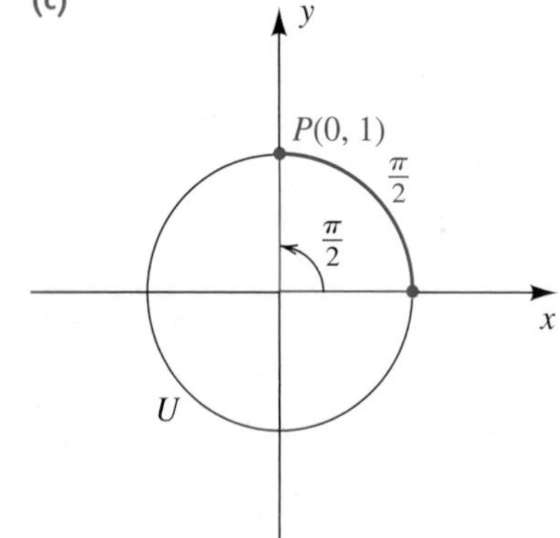

EXAMPLE 3 **Finding special values of the trigonometric functions**

Find the values of the trigonometric functions at t:

(a) $t = 0$ **(b)** $t = \dfrac{\pi}{4}$ **(c)** $t = \dfrac{\pi}{2}$

▶ **SOLUTION**

(a) The point P on the unit circle U that corresponds to $t = 0$ has coordinates $(1, 0)$, as shown in Figure 6(a). Thus, we let $x = 1$ and $y = 0$ in the definition of the trigonometric functions in terms of a unit circle, obtaining

$$\sin 0 = y = 0 \qquad \cos 0 = x = 1$$

$$\tan 0 = \frac{y}{x} = \frac{0}{1} = 0 \qquad \sec 0 = \frac{1}{x} = \frac{1}{1} = 1.$$

Note that csc 0 and cot 0 are undefined, since $y = 0$ is a denominator.

(b) If $t = \pi/4$, then the angle of radian measure $\pi/4$ shown in Figure 6(b) bisects the first quadrant and the point $P(x, y)$ lies on the line $y = x$. Since $P(x, y)$ is on the unit circle $x^2 + y^2 = 1$ and since $y = x$, we obtain

$$x^2 + x^2 = 1, \qquad \text{or} \qquad 2x^2 = 1.$$

Solving for x and noting that $x > 0$ gives us

$$x = \frac{1}{\sqrt{2}} = \frac{\sqrt{2}}{2}.$$

Thus, P is the point $\left(\sqrt{2}/2, \sqrt{2}/2\right)$. Letting $x = \sqrt{2}/2$ and $y = \sqrt{2}/2$ in the definition of the trigonometric functions in terms of a unit circle gives us

$$\sin\frac{\pi}{4} = \frac{\sqrt{2}}{2} \qquad \cos\frac{\pi}{4} = \frac{\sqrt{2}}{2} \qquad \tan\frac{\pi}{4} = \frac{\sqrt{2}/2}{\sqrt{2}/2} = 1$$

$$\csc\frac{\pi}{4} = \frac{2}{\sqrt{2}} = \sqrt{2} \qquad \sec\frac{\pi}{4} = \frac{2}{\sqrt{2}} = \sqrt{2} \qquad \cot\frac{\pi}{4} = \frac{\sqrt{2}/2}{\sqrt{2}/2} = 1.$$

(c) The point P on U that corresponds to $t = \pi/2$ has coordinates $(0, 1)$, as shown in Figure 6(c). Thus, we let $x = 0$ and $y = 1$ in the definition of the trigonometric functions in terms of a unit circle, obtaining

▶ **Tutorial available at www.thomsonedu.com/login**

$$\sin\frac{\pi}{2} = 1 \qquad \cos\frac{\pi}{2} = 0 \qquad \csc\frac{\pi}{2} = \frac{1}{1} = 1 \qquad \cot\frac{\pi}{2} = \frac{0}{1} = 0.$$

The tangent and secant functions are undefined, since $x = 0$ is a denominator in each case.

A summary of the trigonometric functions of special angles appears in Appendix IV.

We shall use the unit circle formulation of the trigonometric functions to help obtain their graphs. If t is a real number and $P(x, y)$ is the point on the unit circle U that corresponds to t, then by the definition of the trigonometric functions in terms of a unit circle,

$$x = \cos t \qquad \text{and} \qquad y = \sin t.$$

Figure 7

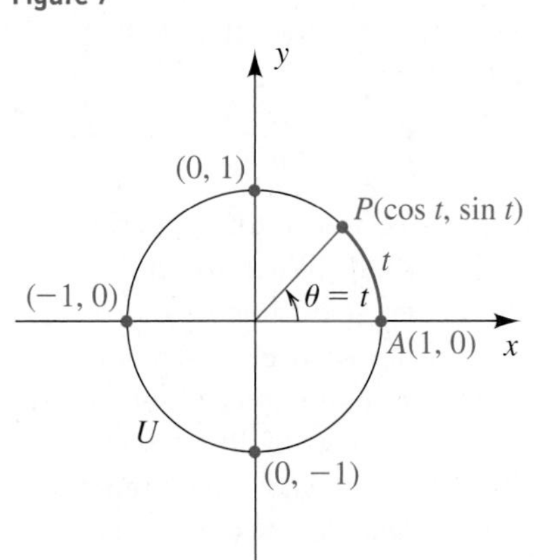

Thus, as shown in Figure 7, we may denote $P(x, y)$ by

$$P(\cos t, \sin t).$$

If $t > 0$, the real number t may be interpreted either as the radian measure of the angle θ or as the length of arc AP.

If we let t increase from 0 to 2π radians, the point $P(\cos t, \sin t)$ travels around the unit circle U one time in the counterclockwise direction. By observing the variation of the x- and y-coordinates of P, we obtain the next table. The notation $0 \to \pi/2$ in the first row of the table means that t increases from 0 to $\pi/2$, and the notation $(1, 0) \to (0, 1)$ denotes the corresponding variation of $P(\cos t, \sin t)$ as it travels along U from $(1, 0)$ to $(0, 1)$. If t increases from 0 to $\pi/2$, then $\sin t$ increases from 0 to 1, which we denote by $0 \to 1$. Moreover, $\sin t$ takes on every value between 0 and 1. If t increases from $\pi/2$ to π, then $\sin t$ decreases from 1 to 0, which is denoted by $1 \to 0$. Other entries in the table may be interpreted in similar fashion.

t	$P(\cos t, \sin t)$	$\cos t$	$\sin t$
$0 \to \frac{\pi}{2}$	$(1, 0) \to (0, 1)$	$1 \to 0$	$0 \to 1$
$\frac{\pi}{2} \to \pi$	$(0, 1) \to (-1, 0)$	$0 \to -1$	$1 \to 0$
$\pi \to \frac{3\pi}{2}$	$(-1, 0) \to (0, -1)$	$-1 \to 0$	$0 \to -1$
$\frac{3\pi}{2} \to 2\pi$	$(0, -1) \to (1, 0)$	$0 \to 1$	$-1 \to 0$

If t increases from 2π to 4π, the point $P(\cos t, \sin t)$ in Figure 7 traces the unit circle U again and the patterns for $\sin t$ and $\cos t$ are repeated—that is,

$$\sin(t + 2\pi) = \sin t \qquad \text{and} \qquad \cos(t + 2\pi) = \cos t$$

for every t in the interval $[0, 2\pi]$. The same is true if t increases from 4π to 6π, from 6π to 8π, and so on. In general, we have the following theorem.

Theorem on Repeated Function Values for sin and cos

If n is any integer, then

$$\sin(t + 2\pi n) = \sin t \qquad \text{and} \qquad \cos(t + 2\pi n) = \cos t.$$

The repetitive variation of the sine and cosine functions is *periodic* in the sense of the following definition.

Definition of Periodic Function

A function f is **periodic** if there exists a positive real number k such that

$$f(t + k) = f(t)$$

for every t in the domain of f. The least such positive real number k, if it exists, is the **period** of f.

You already have a common-sense grasp of the concept of the period of a function. For example, if you were asked on a Monday "What day of the week will it be in 15 days?" your response would be "Tuesday" due to your understanding that the days of the week repeat every 7 days and 15 is one day more than two complete periods of 7 days. From the discussion preceding the previous theorem, we see that the period of the sine and cosine functions is 2π.

We may now readily obtain the graphs of the sine and cosine functions. Since we wish to sketch these graphs on an xy-plane, let us replace the variable t by x and consider the equations

$$y = \sin x \qquad \text{and} \qquad y = \cos x.$$

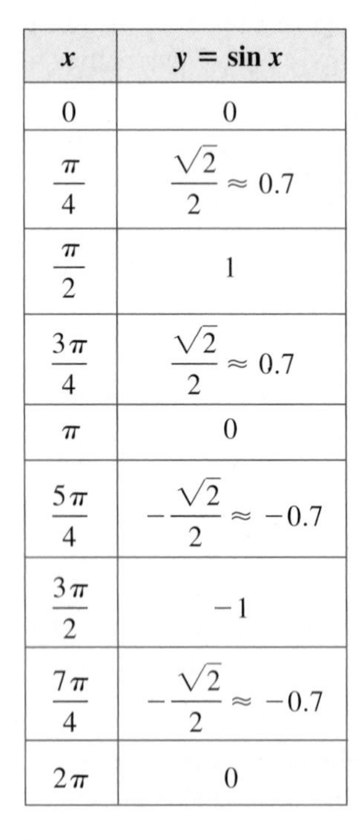

x	$y = \sin x$
0	0
$\frac{\pi}{4}$	$\frac{\sqrt{2}}{2} \approx 0.7$
$\frac{\pi}{2}$	1
$\frac{3\pi}{4}$	$\frac{\sqrt{2}}{2} \approx 0.7$
π	0
$\frac{5\pi}{4}$	$-\frac{\sqrt{2}}{2} \approx -0.7$
$\frac{3\pi}{2}$	-1
$\frac{7\pi}{4}$	$-\frac{\sqrt{2}}{2} \approx -0.7$
2π	0

We may think of x as the radian measure of any angle; however, in calculus, x is usually regarded as a real number. These are equivalent points of view, since the sine (or cosine) of an angle of x radians is the same as the sine (or cosine) of the real number x. The variable y denotes the function value that corresponds to x.

The table in the margin lists coordinates of several points on the graph of $y = \sin x$ for $0 \le x \le 2\pi$. Additional points can be determined using results on special angles, such as

$$\sin(\pi/6) = 1/2 \qquad \text{and} \qquad \sin(\pi/3) = \sqrt{3}/2 \approx 0.8660.$$

To sketch the graph for $0 \le x \le 2\pi$, we plot the points given by the table and remember that $\sin x$ increases on $[0, \pi/2]$, decreases on $[\pi/2, \pi]$ and $[\pi, 3\pi/2]$, and increases on $[3\pi/2, 2\pi]$. This gives us the sketch in Figure 8. Since the sine function is periodic, the pattern shown in Figure 8 is repeated to the right and to the left, in intervals of length 2π. This gives us the sketch in Figure 9.

Figure 8

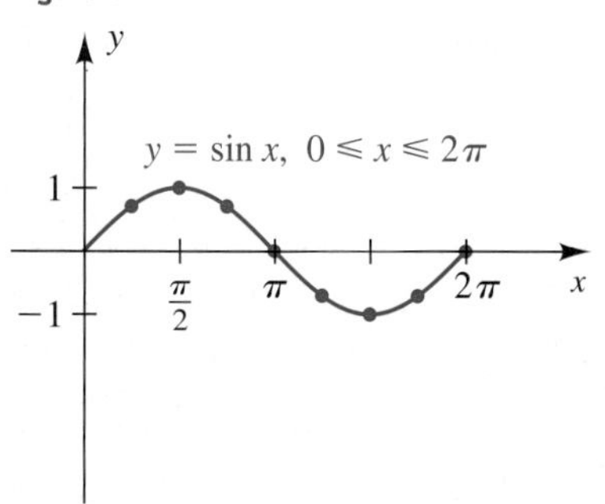

x	$y = \cos x$
0	1
$\frac{\pi}{4}$	$\frac{\sqrt{2}}{2} \approx 0.7$
$\frac{\pi}{2}$	0
$\frac{3\pi}{4}$	$-\frac{\sqrt{2}}{2} \approx -0.7$
π	-1
$\frac{5\pi}{4}$	$-\frac{\sqrt{2}}{2} \approx -0.7$
$\frac{3\pi}{2}$	0
$\frac{7\pi}{4}$	$\frac{\sqrt{2}}{2} \approx 0.7$
2π	1

Figure 9

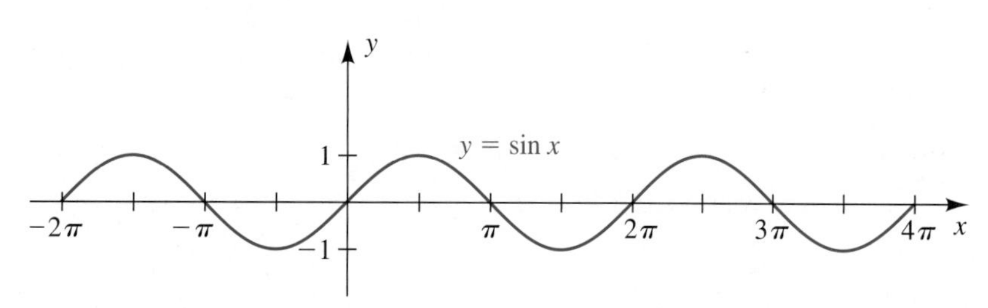

We can use the same procedure to sketch the graph of $y = \cos x$. The table in the margin lists coordinates of several points on the graph for $0 \leq x \leq 2\pi$. Plotting these points leads to the part of the graph shown in Figure 10. Repeating this pattern to the right and to the left, in intervals of length 2π, we obtain the sketch in Figure 11.

Figure 10

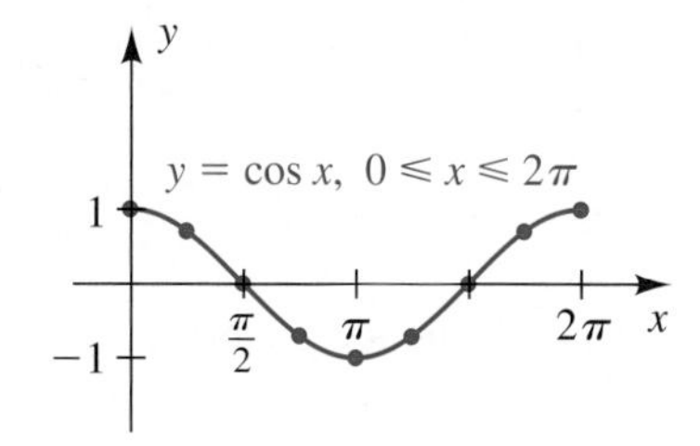

Figure 11

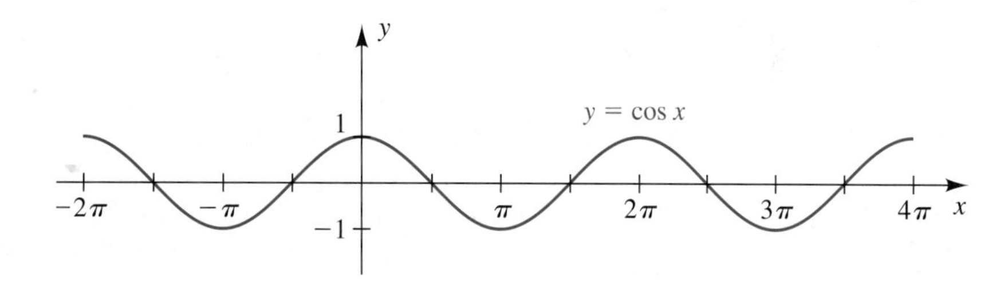

The part of the graph of the sine or cosine function corresponding to $0 \leq x \leq 2\pi$ is one **cycle.** We sometimes refer to a cycle as a **sine wave** or a **cosine wave.**

The range of the sine and cosine functions consists of all real numbers in the closed interval $[-1, 1]$. Since $\csc x = 1/\sin x$ and $\sec x = 1/\cos x$, it follows that the range of the cosecant and secant functions consists of all real numbers having absolute value greater than or equal to 1.

As we shall see, the range of the tangent and cotangent functions consists of all real numbers.

Before discussing graphs of the other trigonometric functions, let us establish formulas that involve functions of $-t$ for any t. Since a minus sign is involved, we call them *formulas for negatives.*

Formulas for Negatives

$$\sin(-t) = -\sin t \qquad \cos(-t) = \cos t \qquad \tan(-t) = -\tan t$$
$$\csc(-t) = -\csc t \qquad \sec(-t) = \sec t \qquad \cot(-t) = -\cot t$$

Figure 12

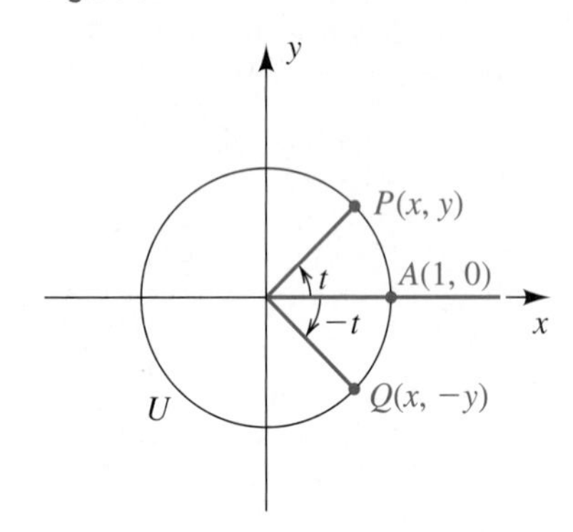

PROOFS Consider the unit circle U in Figure 12. As t increases from 0 to 2π, the point $P(x, y)$ traces the unit circle U once in the counterclockwise direction and the point $Q(x, -y)$, corresponding to $-t$, traces U once in the clockwise direction. Applying the definition of the trigonometric functions of any angle (with $r = 1$), we have

$$\sin(-t) = -y = -\sin t$$
$$\cos(-t) = x = \cos t$$
$$\tan(-t) = \frac{-y}{x} = -\frac{y}{x} = -\tan t.$$

The proofs of the remaining three formulas are similar.

In the following illustration, formulas for negatives are used to find an exact value for each trigonometric function.

ILLUSTRATION **Use of Formulas for Negatives**

- $\sin(-45°) = -\sin 45° = -\frac{\sqrt{2}}{2}$
- $\cos(-30°) = \cos 30° = \frac{\sqrt{3}}{2}$
- $\tan\left(-\frac{\pi}{3}\right) = -\tan\left(\frac{\pi}{3}\right) = -\sqrt{3}$
- $\csc(-30°) = -\csc 30° = -2$
- $\sec(-60°) = \sec 60° = 2$
- $\cot\left(-\frac{\pi}{4}\right) = -\cot\left(\frac{\pi}{4}\right) = -1$

We shall next use formulas for negatives to verify a trigonometric identity.

EXAMPLE 4 Using formulas for negatives to verify an identity

Verify the following identity by transforming the left-hand side into the right-hand side:

$$\sin(-x)\tan(-x) + \cos(-x) = \sec x$$

▶ **SOLUTION** We may proceed as follows:

$$\begin{aligned}
\sin(-x)\tan(-x) + \cos(-x) &= (-\sin x)(-\tan x) + \cos x && \text{formulas for negatives} \\
&= \sin x \frac{\sin x}{\cos x} + \cos x && \text{tangent identity} \\
&= \frac{\sin^2 x}{\cos x} + \cos x && \text{multiply} \\
&= \frac{\sin^2 x + \cos^2 x}{\cos x} && \text{add terms} \\
&= \frac{1}{\cos x} && \text{Pythagorean identity} \\
&= \sec x && \text{reciprocal identity}
\end{aligned}$$

We may use the formulas for negatives to prove the following theorem.

Theorem on Even and Odd Trigonometric Functions	**(1)** The cosine and secant functions are even. **(2)** The sine, tangent, cotangent, and cosecant functions are odd.

PROOFS We shall prove the theorem for the cosine and sine functions. If $f(x) = \cos x$, then

$$f(-x) = \cos(-x) = \cos x = f(x),$$

which means that the cosine function is even.

If $f(x) = \sin x$, then

$$f(-x) = \sin(-x) = -\sin x = -f(x).$$

Thus, the sine function is odd.

Since the sine function is odd, its graph is symmetric with respect to the origin (see Figure 13). Since the cosine function is even, its graph is symmetric with respect to the y-axis (see Figure 14).

Figure 13 sine is odd

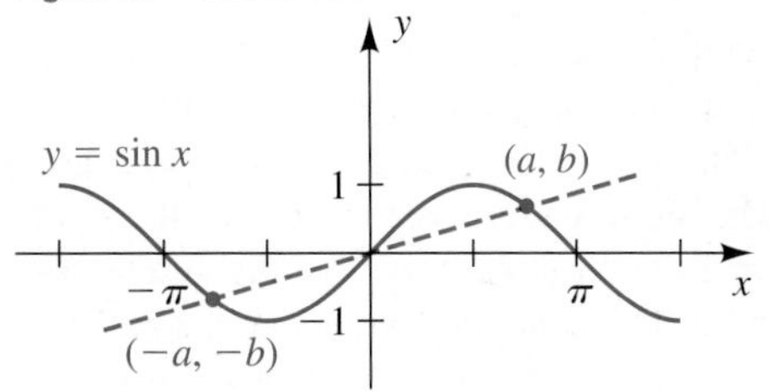

Figure 14 cosine is even

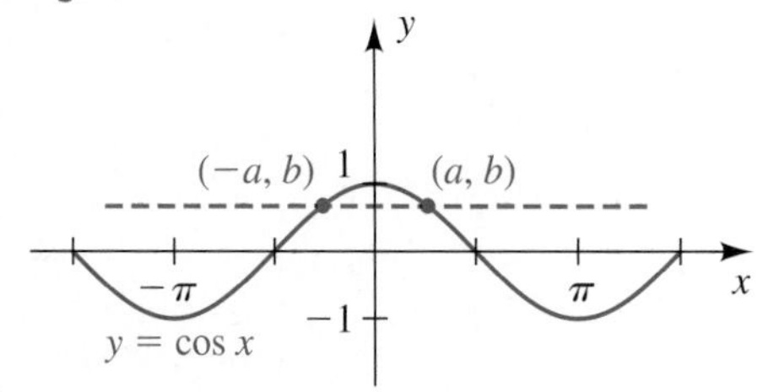

▶ **Tutorial available at www.thomsonedu.com/login**

x	$y = \tan x$
$-\dfrac{\pi}{3}$	$-\sqrt{3} \approx -1.7$
$-\dfrac{\pi}{4}$	-1
$-\dfrac{\pi}{6}$	$-\dfrac{\sqrt{3}}{3} \approx -0.6$
0	0
$\dfrac{\pi}{6}$	$\dfrac{\sqrt{3}}{3} \approx 0.6$
$\dfrac{\pi}{4}$	1
$\dfrac{\pi}{3}$	$\sqrt{3} \approx 1.7$

By the preceding theorem, the tangent function is odd, and hence the graph of $y = \tan x$ is symmetric with respect to the origin. The table in the margin lists some points on the graph if $-\pi/2 < x < \pi/2$. The corresponding points are plotted in Figure 15.

Figure 15

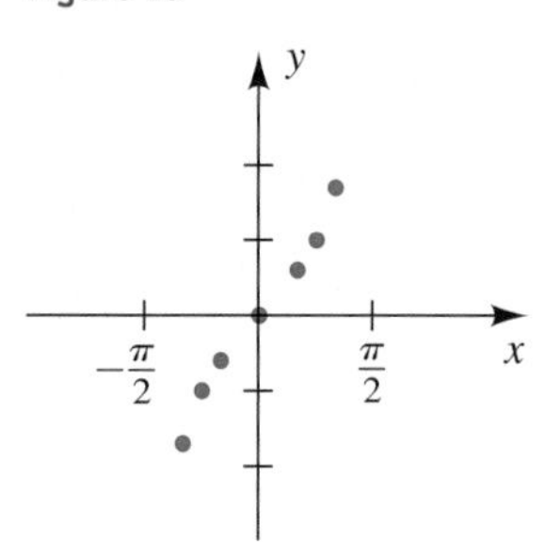

The values of $\tan x$ near $x = \pi/2$ require special attention. If we consider $\tan x = \sin x/\cos x$, then as x increases toward $\pi/2$, the numerator $\sin x$ approaches 1 and the denominator $\cos x$ approaches 0. Consequently, $\tan x$ takes on large positive values. Following are some approximations of $\tan x$ for x close to $\pi/2 \approx 1.5708$:

$$\begin{aligned}
\tan 1.57000 &\approx 1{,}255.8 \\
\tan 1.57030 &\approx 2{,}014.8 \\
\tan 1.57060 &\approx 5{,}093.5 \\
\tan 1.57070 &\approx 10{,}381.3 \\
\tan 1.57079 &\approx 158{,}057.9
\end{aligned}$$

Notice how rapidly $\tan x$ increases as x approaches $\pi/2$. We say that $\tan x$ *increases without bound as x approaches $\pi/2$ through values less than $\pi/2$.* Similarly, if x approaches $-\pi/2$ through values *greater* than $-\pi/2$, then $\tan x$ *decreases without bound.* We may denote this variation using the notation introduced for rational functions in Section 3.5:

$$\text{as} \quad x \to \frac{\pi}{2}^-, \quad \tan x \to \infty$$

$$\text{as} \quad x \to -\frac{\pi}{2}^+, \quad \tan x \to -\infty$$

This variation of $\tan x$ in the open interval $(-\pi/2, \pi/2)$ is illustrated in Figure 16 on the next page. This portion of the graph is called one **branch** of the tangent. The lines $x = -\pi/2$ and $x = \pi/2$ are vertical asymptotes for the graph. The same pattern is repeated in the open intervals $(-3\pi/2, -\pi/2)$, $(\pi/2, 3\pi/2)$, and $(3\pi/2, 5\pi/2)$ and in similar intervals of length π, as shown in the figure. Thus, *the tangent function is periodic with period π.*

Figure 16 $y = \tan x$

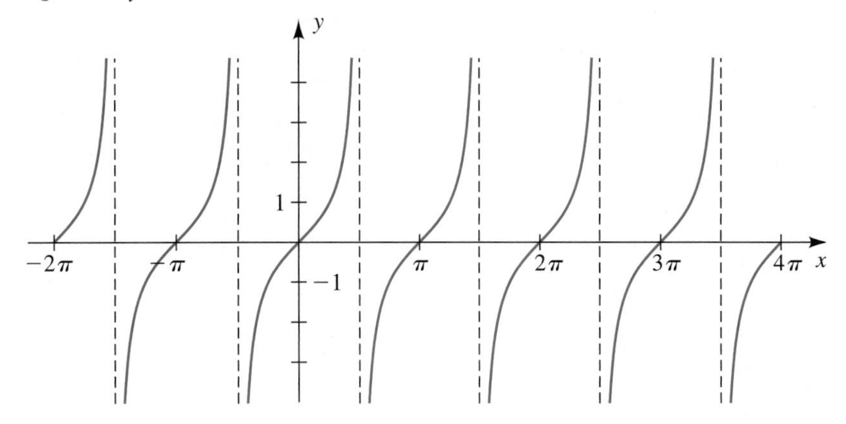

We may use the graphs of $y = \sin x$, $y = \cos x$, and $y = \tan x$ to help sketch the graphs of the remaining three trigonometric functions. For example, since $\csc x = 1/\sin x$, we may find the y-coordinate of a point on the graph of the cosecant function by taking the reciprocal of the corresponding y-coordinate on the sine graph for every value of x except $x = \pi n$ for any integer n. (If $x = \pi n$, $\sin x = 0$, and hence $1/\sin x$ is undefined.) As an aid to sketching the graph of the cosecant function, it is convenient to sketch the graph of the sine function (shown in red in Figure 17) and then take reciprocals to obtain points on the cosecant graph.

Figure 17 $y = \csc x$, $y = \sin x$

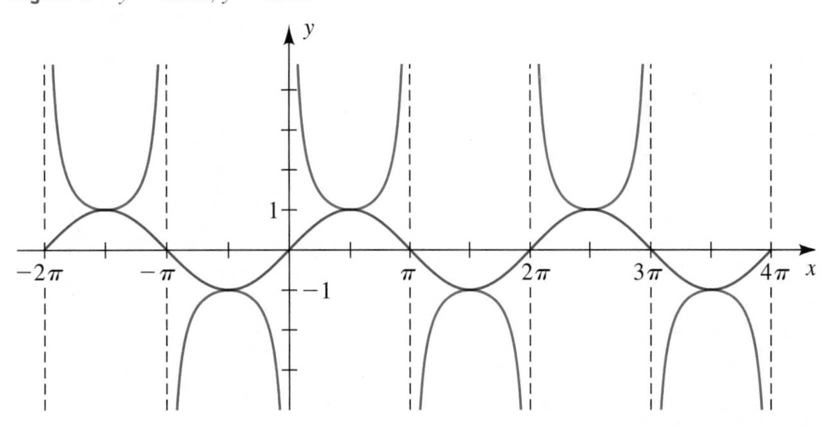

Notice the manner in which the cosecant function increases or decreases without bound as x approaches πn for any integer n. The graph has vertical asymptotes $x = \pi n$, as indicated in the figure. There is one **upper branch** of the cosecant on the interval $(0, \pi)$ and one **lower branch** on the interval $(\pi, 2\pi)$—together they compose one *cycle* of the cosecant.

Since $\sec x = 1/\cos x$ and $\cot x = 1/\tan x$, we may obtain the graphs of the secant and cotangent functions by taking reciprocals of y-coordinates of points on the graphs of the cosine and tangent functions, as illustrated in Figures 18 and 19.

A graphical summary of the six trigonometric functions and their inverses (discussed in Section 6.6) appears in Appendix III.

Figure 18 $y = \sec x$, $y = \cos x$

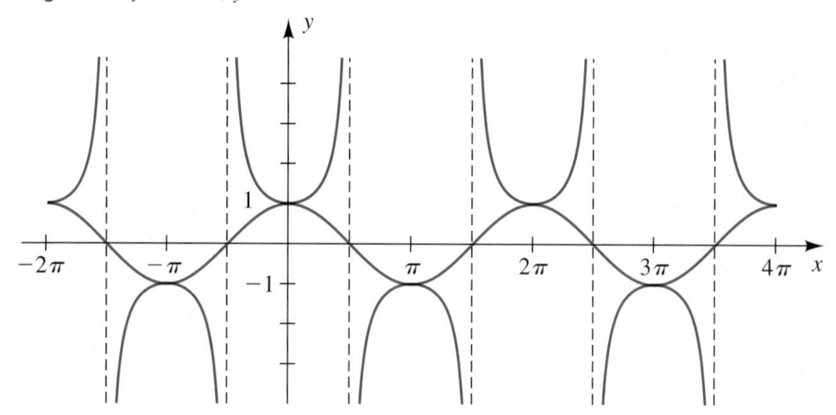

Figure 19 $y = \cot x$, $y = \tan x$

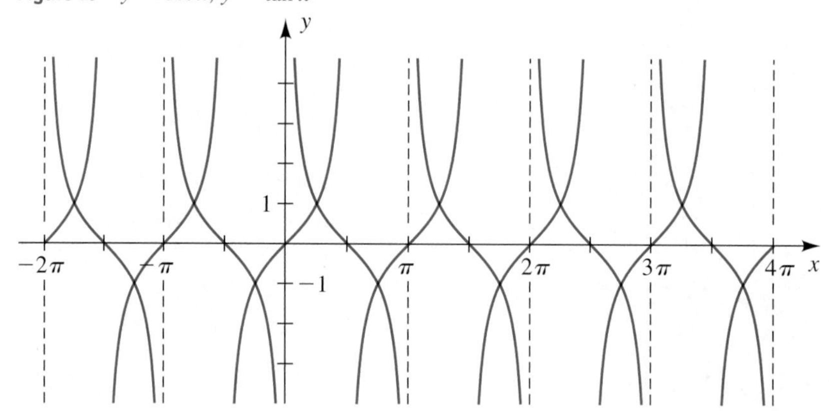

We have considered many properties of the six trigonometric functions of x, where x is a real number or the radian measure of an angle. The following chart contains a summary of important features of these functions (n denotes an arbitrary integer).

Summary of Features of the Trigonometric Functions and Their Graphs

Feature	$y = \sin x$	$y = \cos x$	$y = \tan x$	$y = \cot x$	$y = \sec x$	$y = \csc x$
Graph (one period)	$-\pi$, π, 1, -1	$-\frac{\pi}{2}$, $\frac{3\pi}{2}$, 1, -1	$x = -\frac{\pi}{2}$, $x = \frac{\pi}{2}$	$x = 0$, $x = \pi$	$x = -\frac{\pi}{2}$, $x = \frac{\pi}{2}$, $x = \frac{3\pi}{2}$, 1, -1	$x = -\pi$, $x = 0$, $x = \pi$, 1, -1
Domain	$\mathbb{R}$	$\mathbb{R}$	$x \neq \frac{\pi}{2} + \pi n$	$x \neq \pi n$	$x \neq \frac{\pi}{2} + \pi n$	$x \neq \pi n$
Vertical asymptotes	none	none	$x = \frac{\pi}{2} + \pi n$	$x = \pi n$	$x = \frac{\pi}{2} + \pi n$	$x = \pi n$
Range	$[-1, 1]$	$[-1, 1]$	$\mathbb{R}$	$\mathbb{R}$	$(-\infty, -1] \cup [1, \infty)$	$(-\infty, -1] \cup [1, \infty)$
x-intercepts	πn	$\frac{\pi}{2} + \pi n$	πn	$\frac{\pi}{2} + \pi n$	none	none
y-intercept	0	1	0	none	1	none
Period	2π	2π	π	π	2π	2π
Even or odd	odd	even	odd	odd	even	odd
Symmetry	origin	y-axis	origin	origin	y-axis	origin

EXAMPLE 5 Investigating the variation of csc x

Investigate the variation of $\csc x$ as

$$x \to \pi^-, \quad x \to \pi^+, \quad x \to \frac{\pi}{2}^-, \quad \text{and} \quad x \to \frac{\pi}{6}^+.$$

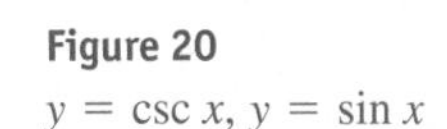

Figure 20
$y = \csc x$, $y = \sin x$

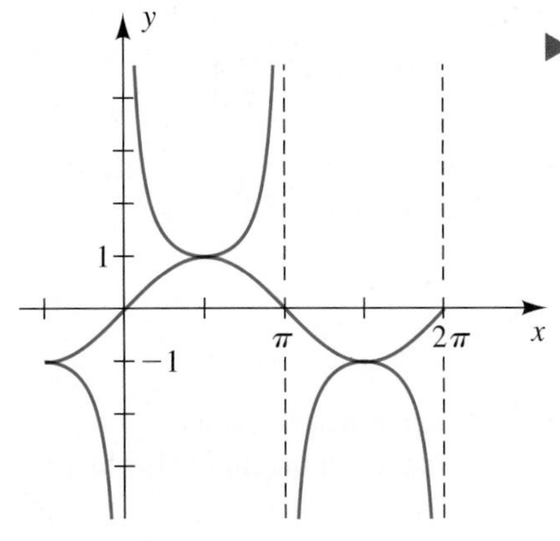

▶ **SOLUTION** Referring to the graph of $y = \csc x$ in Figure 20 and using our knowledge of the special values of the sine and cosecant functions, we obtain the following:

as $x \to \pi^-$,	$\sin x \to 0$ (through positive values)	and $\csc x \to \infty$
as $x \to \pi^+$,	$\sin x \to 0$ (through negative values)	and $\csc x \to -\infty$
as $x \to \frac{\pi}{2}^-$,	$\sin x \to 1$	and $\csc x \to 1$
as $x \to \frac{\pi}{6}^+$,	$\sin x \to \frac{1}{2}$	and $\csc x \to 2$

▶ **Tutorial available at www.thomsonedu.com/login**

EXAMPLE 6 **Solving equations and inequalities that involve a trigonometric function**

Find all values of x in the interval $[-2\pi, 2\pi]$ such that

(a) $\cos x = \frac{1}{2}$ **(b)** $\cos x > \frac{1}{2}$ **(c)** $\cos x < \frac{1}{2}$

▶ **SOLUTION** This problem can be easily solved by referring to the graphs of $y = \cos x$ and $y = \frac{1}{2}$, sketched on the same xy-plane in Figure 21 for $-2\pi \le x \le 2\pi$.

Figure 21

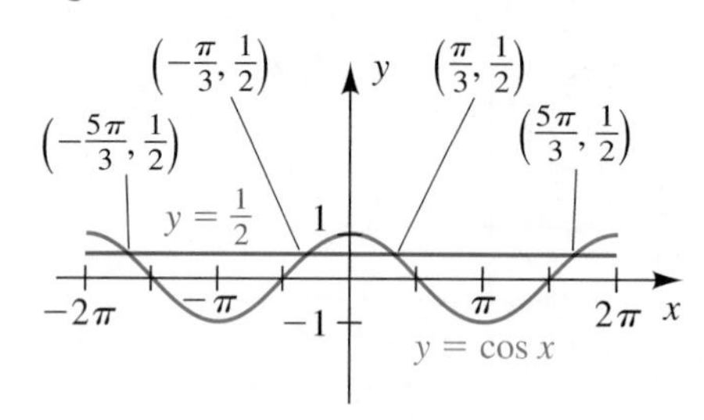

(a) The values of x such that $\cos x = \frac{1}{2}$ are the x-coordinates of the points at which the graphs intersect. Recall that $x = \pi/3$ satisfies the equation. By symmetry, $x = -\pi/3$ is another solution of $\cos x = \frac{1}{2}$. Since the cosine function has period 2π, the other values of x in $[-2\pi, 2\pi]$ such that $\cos x = \frac{1}{2}$ are

$$-\frac{\pi}{3} + 2\pi = \frac{5\pi}{3} \quad \text{and} \quad \frac{\pi}{3} - 2\pi = -\frac{5\pi}{3}.$$

(b) The values of x such that $\cos x > \frac{1}{2}$ can be found by determining where the graph of $y = \cos x$ in Figure 21 lies *above* the line $y = \frac{1}{2}$. This gives us the x-intervals

$$\left[-2\pi, -\frac{5\pi}{3}\right), \quad \left(-\frac{\pi}{3}, \frac{\pi}{3}\right), \quad \text{and} \quad \left(\frac{5\pi}{3}, 2\pi\right].$$

(c) To solve $\cos x < \frac{1}{2}$, we again refer to Figure 21 and note where the graph of $y = \cos x$ lies *below* the line $y = \frac{1}{2}$. This gives us the x-intervals

$$\left(-\frac{5\pi}{3}, -\frac{\pi}{3}\right) \quad \text{and} \quad \left(\frac{\pi}{3}, \frac{5\pi}{3}\right).$$

Another method of solving $\cos x < \frac{1}{2}$ is to note that the solutions are the open subintervals of $[-2\pi, 2\pi]$ that are *not* included in the intervals obtained in part (b).

▶ **Tutorial available at www.thomsonedu.com/login**

The result discussed in the next example plays an important role in advanced mathematics.

EXAMPLE 7 **Sketching the graph of $f(x) = (\sin x)/x$**

If $f(x) = (\sin x)/x$, sketch the graph of f on $[-\pi, \pi]$, and investigate the behavior of $f(x)$ as $x \to 0^-$ and as $x \to 0^+$.

SOLUTION Note that f is undefined at $x = 0$, because substitution yields the meaningless expression $0/0$.

We assign $(\sin x)/x$ to Y_1. Because our screen has a 3:2 (horizontal : vertical) proportion, we use the viewing rectangle $[-\pi, \pi]$ by $[-2.1, 2.1]$ (since $\frac{2}{3}\pi \approx 2.1$), obtaining a sketch similar to Figure 22. Using tracing and zoom features, we find it appears that

Figure 22
$[-\pi, \pi]$ by $[-2.1, 2.1]$

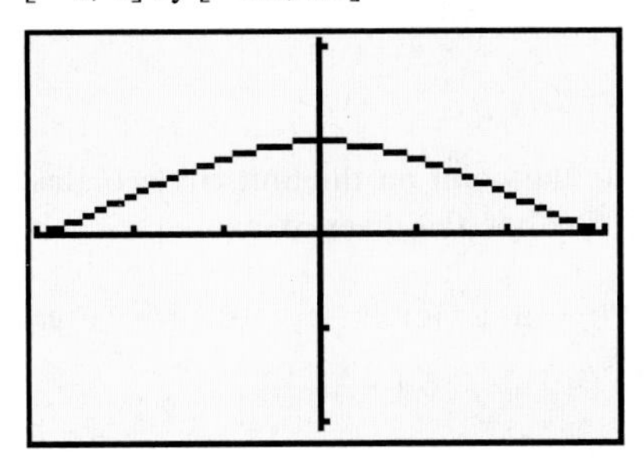

$$\text{as} \quad x \to 0^-, \quad f(x) \to 1 \quad \text{and as} \quad x \to 0^+, \quad f(x) \to 1.$$

There is a hole in the graph at the point (0, 1); however, most graphing utilities are not capable of showing this fact.

Our graphical technique does not *prove* that $f(x) \to 1$ as $x \to 0$, but it does make it appear highly probable. A rigorous proof, based on the definition of $\sin x$ and geometric considerations, can be found in calculus texts.

An interesting result obtained from Example 7 is that *if x is in radians* and

$$\text{if} \quad x \approx 0, \quad \text{then} \quad \frac{\sin x}{x} \approx 1, \quad \text{and so} \quad \sin x \approx x.$$

The last statement gives us an *approximation formula* for $\sin x$ if x is close to 0. To illustrate, using a calculator we find that

$$\sin(0.03) \approx 0.029\,995\,5 \approx 0.03$$

$$\sin(0.02) \approx 0.019\,998\,7 \approx 0.02$$

$$\sin(0.01) \approx 0.009\,999\,8 \approx 0.01.$$

We have now discussed two different approaches to the trigonometric functions. The development in terms of angles and ratios, introduced in Section 5.2, has many applications in the sciences and engineering. The definition in terms of a unit circle, considered in this section, emphasizes the fact that the trigonometric functions have domains consisting of real numbers. Such functions are the building blocks for calculus. In addition, the unit circle approach is useful for discussing graphs and deriving trigonometric identities. You should work to become proficient in the use of both formulations of the trigonometric functions, since each will reinforce the other and thus facilitate your mastery of more advanced aspects of trigonometry.

5.3 *Exercises*

Exer. 1–4: A point $P(x, y)$ is shown on the unit circle U corresponding to a real number t. Find the values of the trigonometric functions at t.

1

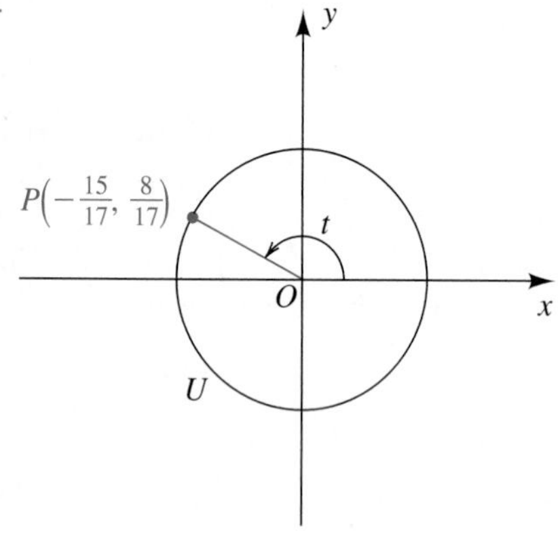

2

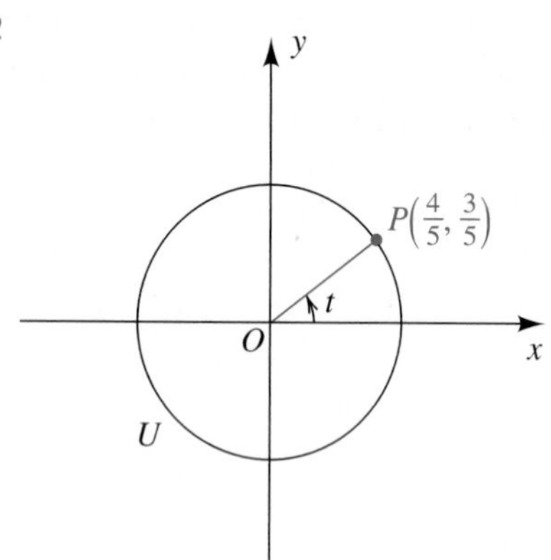

3

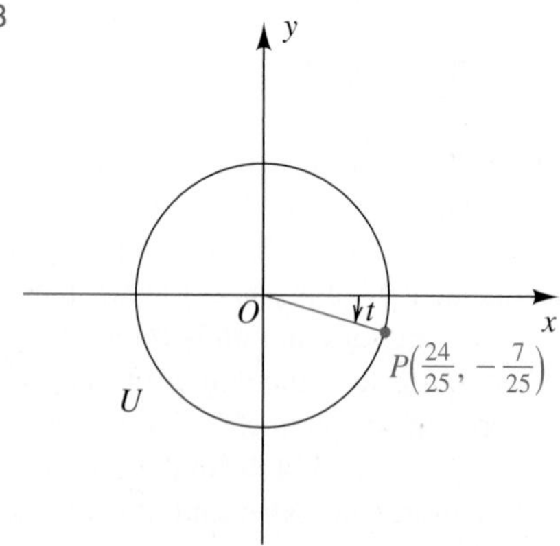

4

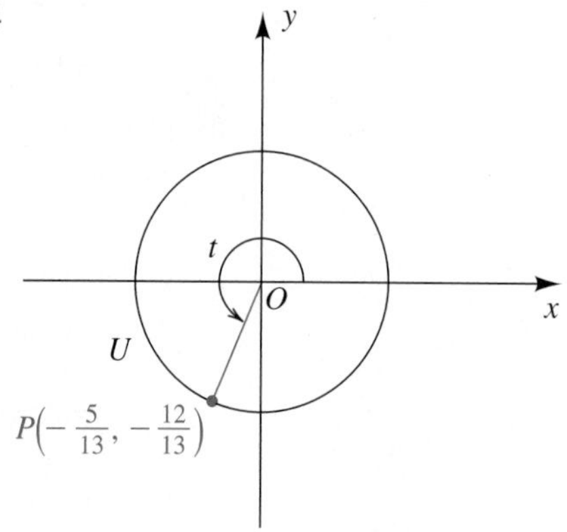

Exer. 5–8: Let $P(t)$ be the point on the unit circle U that corresponds to t. If $P(t)$ has the given rectangular coordinates, find

(a) $P(t + \pi)$ (b) $P(t - \pi)$ (c) $P(-t)$ (d) $P(-t - \pi)$

5 $\left(\frac{3}{5}, \frac{4}{5}\right)$

6 $\left(-\frac{8}{17}, \frac{15}{17}\right)$

7 $\left(-\frac{12}{13}, -\frac{5}{13}\right)$

8 $\left(\frac{7}{25}, -\frac{24}{25}\right)$

Exer. 9–16: Let P be the point on the unit circle U that corresponds to t. Find the coordinates of P and the exact values of the trigonometric functions of t, whenever possible.

9 (a) 2π (b) -3π

10 (a) $-\pi$ (b) 6π

11 (a) $3\pi/2$ (b) $-7\pi/2$

12 (a) $5\pi/2$ (b) $-\pi/2$

13 (a) $9\pi/4$ (b) $-5\pi/4$

14 (a) $3\pi/4$ (b) $-7\pi/4$

15 (a) $5\pi/4$ (b) $-\pi/4$

16 (a) $7\pi/4$ (b) $-3\pi/4$

Exer. 17–20: Use a formula for negatives to find the exact value.

17 (a) $\sin(-90°)$ (b) $\cos\left(-\frac{3\pi}{4}\right)$ (c) $\tan(-45°)$

18 (a) $\sin\left(-\frac{3\pi}{2}\right)$ (b) $\cos(-225°)$ (c) $\tan(-\pi)$

19 (a) $\cot\left(-\frac{3\pi}{4}\right)$ (b) $\sec(-180°)$ (c) $\csc\left(-\frac{3\pi}{2}\right)$

20 (a) $\cot(-225°)$ (b) $\sec\left(-\frac{\pi}{4}\right)$ (c) $\csc(-45°)$

Exer. 21–26: Verify the identity by transforming the left-hand side into the right-hand side.

21 $\sin(-x)\sec(-x) = -\tan x$

22 $\csc(-x)\cos(-x) = -\cot x$

23 $\dfrac{\cot(-x)}{\csc(-x)} = \cos x$

24 $\dfrac{\sec(-x)}{\tan(-x)} = -\csc x$

25 $\dfrac{1}{\cos(-x)} - \tan(-x)\sin(-x) = \cos x$

26 $\cot(-x)\cos(-x) + \sin(-x) = -\csc x$

Exer. 27–38: Complete the statement by referring to a graph of a trigonometric function.

27 (a) As $x \to 0^+$, $\sin x \to$ ___

(b) As $x \to (-\pi/2)^-$, $\sin x \to$ ___

28 (a) As $x \to \pi^+$, $\sin x \to$ ___

(b) As $x \to (\pi/6)^-$, $\sin x \to$ ___

29 (a) As $x \to (\pi/4)^+$, $\cos x \to$ ___

(b) As $x \to \pi^-$, $\cos x \to$ ___

30 (a) As $x \to 0^+$, $\cos x \to$ ___

(b) As $x \to (-\pi/3)^-$, $\cos x \to$ ___

31 (a) As $x \to (\pi/4)^+$, $\tan x \to$ ___

(b) As $x \to (\pi/2)^+$, $\tan x \to$ ___

32 (a) As $x \to 0^+$, $\tan x \to$ ___

(b) As $x \to (-\pi/2)^-$, $\tan x \to$ ___

33 (a) As $x \to (-\pi/4)^-$, $\cot x \to$ ___

(b) As $x \to 0^+$, $\cot x \to$ ___

34 (a) As $x \to (\pi/6)^+$, $\cot x \to$ ___

(b) As $x \to \pi^-$, $\cot x \to$ ___

35 (a) As $x \to (\pi/2)^-$, $\sec x \to$ ___

(b) As $x \to (\pi/4)^+$, $\sec x \to$ ___

36 (a) As $x \to (\pi/2)^+$, $\sec x \to$ ___

(b) As $x \to 0^-$, $\sec x \to$ ___

37 (a) As $x \to 0^-$, $\csc x \to$ ___

(b) As $x \to (\pi/2)^+$, $\csc x \to$ ___

38 (a) As $x \to \pi^+$, $\csc x \to$ ___

(b) As $x \to (\pi/4)^-$, $\csc x \to$ ___

Exer. 39–46: Refer to the graph of $y = \sin x$ or $y = \cos x$ to find the exact values of x in the interval $[0, 4\pi]$ that satisfy the equation.

39 $\sin x = -1$

40 $\sin x = 1$

41 $\sin x = \frac{1}{2}$

42 $\sin x = -\sqrt{2}/2$

43 $\cos x = 1$

44 $\cos x = -1$

45 $\cos x = \sqrt{2}/2$

46 $\cos x = -\frac{1}{2}$

Exer. 47–50: Refer to the graph of $y = \tan x$ to find the exact values of x in the interval $(-\pi/2, 3\pi/2)$ that satisfy the equation.

47 $\tan x = 1$

48 $\tan x = \sqrt{3}$

49 $\tan x = 0$

50 $\tan x = -1/\sqrt{3}$

Exer. 51–54: Refer to the graph of the equation on the specified interval. Find all values of x such that for the real number a, (a) $y = a$, (b) $y > a$, and (c) $y < a$.

51 $y = \sin x$; $[-2\pi, 2\pi]$; $a = \frac{1}{2}$

52 $y = \cos x$; $[0, 4\pi]$; $a = \sqrt{3}/2$

53 $y = \cos x$; $[-2\pi, 2\pi]$; $a = -\frac{1}{2}$

54 $y = \sin x$; $[0, 4\pi]$; $a = -\sqrt{2}/2$

Exer. 55–62: Use the graph of a trigonometric function to sketch the graph of the equation without plotting points.

55 $y = 2 + \sin x$

56 $y = 3 + \cos x$

57 $y = \cos x - 2$

58 $y = \sin x - 1$

59 $y = 1 + \tan x$

60 $y = \cot x - 1$

61 $y = \sec x - 2$

62 $y = 1 + \csc x$

Exer. 63–66: Find the intervals between -2π and 2π on which the given function is (a) increasing or (b) decreasing.

63 secant

64 cosecant

65 tangent

66 cotangent

67 Practice sketching the graph of the sine function, taking different units of length on the horizontal and vertical axes. Practice sketching graphs of the cosine and tangent functions in the same manner. Continue this practice until you reach the stage at which, if you were awakened from a sound sleep in the middle of the night and asked to sketch one of these graphs, you could do so in less than thirty seconds.

68 Work Exercise 67 for the cosecant, secant, and cotangent functions.

Exer. 69–72: Use the figure to approximate the following to one decimal place.

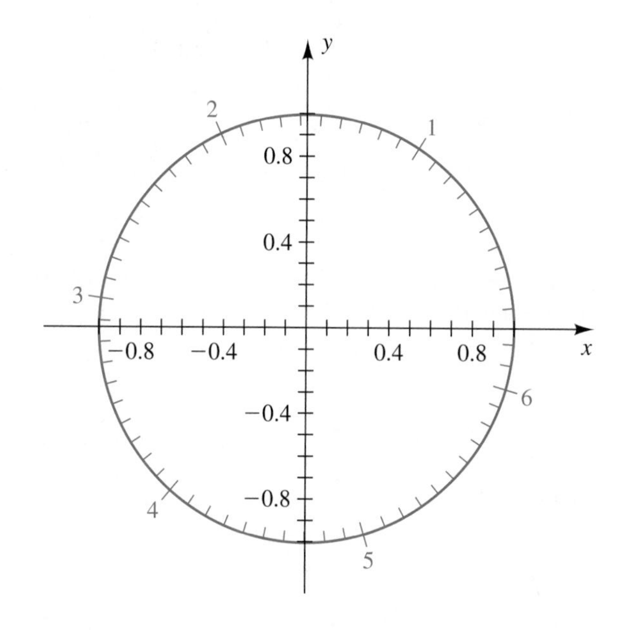

69 (a) $\sin 4$ (b) $\sin(-1.2)$

(c) All numbers t between 0 and 2π such that $\sin t = 0.5$

70 (a) $\sin 2$ (b) $\sin(-2.3)$

(c) All numbers t between 0 and 2π such that $\sin t = -0.2$

71 (a) $\cos 4$ (b) $\cos(-1.2)$

(c) All numbers t between 0 and 2π such that $\cos t = -0.6$

72 (a) $\cos 2$ (b) $\cos(-2.3)$

(c) All numbers t between 0 and 2π such that $\cos t = 0.2$

73 **Temperature-humidity relationship** On March 17, 1981, in Tucson, Arizona, the temperature in degrees Fahrenheit could be described by the equation

$$T(t) = -12 \cos\left(\frac{\pi}{12}t\right) + 60,$$

while the relative humidity in percent could be expressed by

$$H(t) = 20 \cos\left(\frac{\pi}{12}t\right) + 60,$$

where t is in hours and $t = 0$ corresponds to 6 A.M.

(a) Construct a table that lists the temperature and relative humidity every three hours, beginning at midnight.

(b) Determine the times when the maximums and minimums occurred for T and H.

(c) Discuss the relationship between the temperature and relative humidity on this day.

74 **Robotic arm movement** Trigonometric functions are used extensively in the design of industrial robots. Suppose that a robot's shoulder joint is motorized so that the angle θ increases at a constant rate of $\pi/12$ radian per second from an initial angle of $\theta = 0$. Assume that the elbow joint is always kept straight and that the arm has a constant length of 153 centimeters, as shown in the figure.

(a) Assume that $h = 50$ cm when $\theta = 0$. Construct a table that lists the angle θ and the height h of the robotic hand every second while $0 \le \theta \le \pi/2$.

(b) Determine whether or not a constant increase in the angle θ produces a constant increase in the height of the hand.

(c) Find the total distance that the hand moves.

Exercise 74

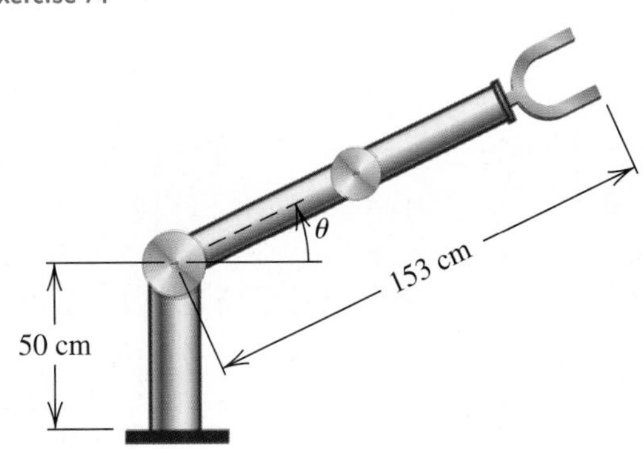

Exer. 75–76: Graph the equation, and estimate the values of x in the specified interval that correspond to the given value of y.

75 $y = \sin(x^2)$, $[-\pi, \pi]$; $y = 0.5$

76 $y = \tan(\sqrt{x})$, $[0, 25]$; $y = 5$

Exer. 77–78: Graph f on the interval $[-2\pi, 2\pi]$, and estimate the coordinates of the high and low points.

77 $f(x) = x \sin x$

78 $f(x) = \sin^2 x \cos x$

Exer. 79–84: As $x \to 0^+$, $f(x) \to L$ for some real number L. Use a graph to predict L.

79 $f(x) = \dfrac{1 - \cos x}{x}$

80 $f(x) = \dfrac{6x - 6 \sin x}{x^3}$

81 $f(x) = x \cot x$

82 $f(x) = \dfrac{x + \tan x}{\sin x}$

83 $f(x) = \dfrac{\tan x}{x}$

84 $f(x) = \dfrac{\cos\left(x + \frac{1}{2}\pi\right)}{x}$

5.4 Values of the Trigonometric Functions

In previous sections we calculated special values of the trigonometric functions by using the definition of the trigonometric functions in terms of either an angle or a unit circle. In practice we most often use a calculator to approximate function values.

We will next show how the value of any trigonometric function at an angle of θ degrees or at a real number t can be found from its value in the θ-interval $(0°, 90°)$ or the t-interval $(0, \pi/2)$, respectively. This technique is sometimes necessary when a calculator is used to find all angles or real numbers that correspond to a given function value.

We shall make use of the following concept.

Definition of Reference Angle

Let θ be a nonquadrantal angle in standard position. The **reference angle** for θ is the acute angle θ_R that the terminal side of θ makes with the x-axis.

Figure 1 illustrates the reference angle θ_R for a nonquadrantal angle θ, with $0° < \theta < 360°$ or $0 < \theta < 2\pi$, in each of the four quadrants.

Figure 1 Reference angles

(a) Quadrant I

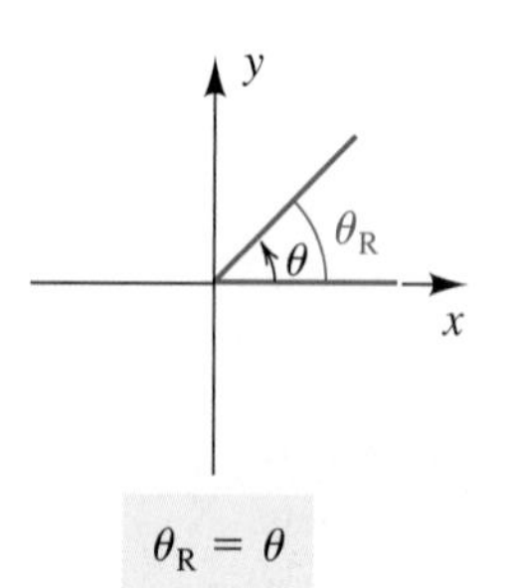

$\theta_R = \theta$

(b) Quadrant II

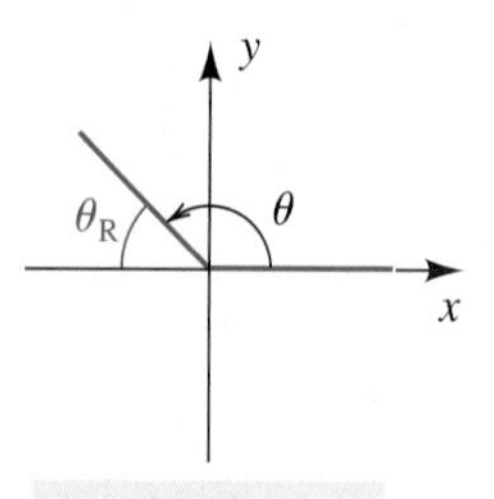

$\theta_R = 180° - \theta = \pi - \theta$

(c) Quadrant III

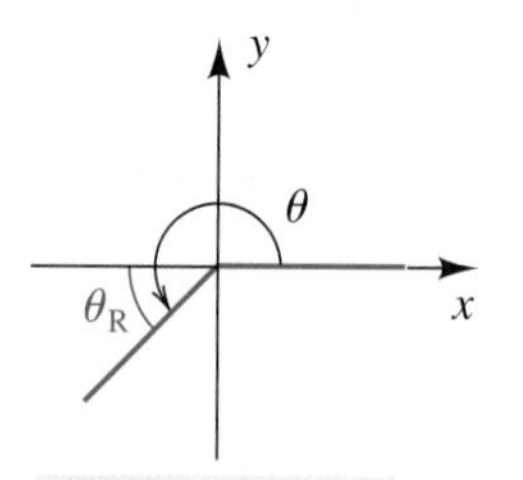

$\theta_R = \theta - 180° = \theta - \pi$

(d) Quadrant IV

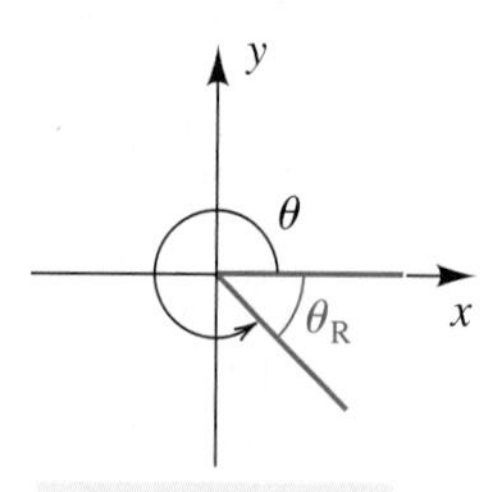

$\theta_R = 360° - \theta = 2\pi - \theta$

The formulas below the axes in Figure 1 may be used to find the degree or radian measure of θ_R when θ is in degrees or radians, respectively. *For a nonquadrantal angle greater than* $360°$ *or less than* $0°$, *first find the coterminal angle* θ *with* $0° < \theta < 360°$ *or* $0 < \theta < 2\pi$, and then use the formulas in Figure 1.

EXAMPLE 1 Finding reference angles

Find the reference angle θ_R for θ, and sketch θ and θ_R in standard position on the same coordinate plane.

(a) $\theta = 315°$ (b) $\theta = -240°$ (c) $\theta = \dfrac{5\pi}{6}$ (d) $\theta = 4$

Figure 2

(a)

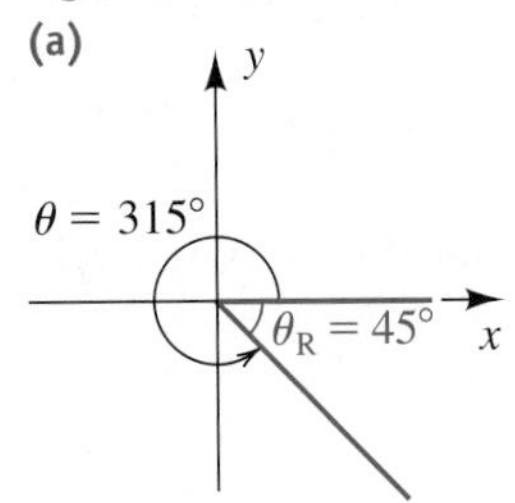

(b)

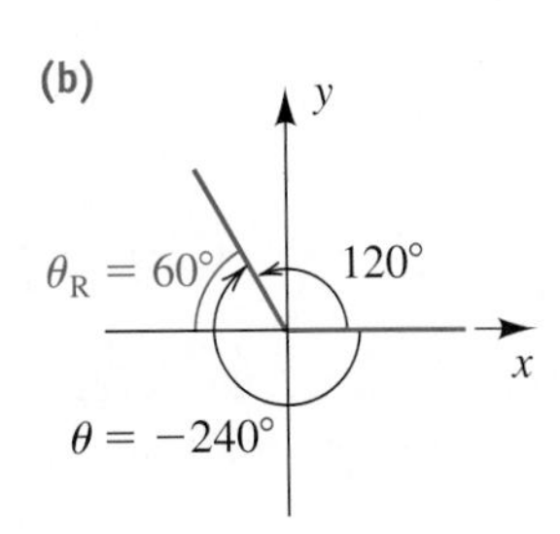

(c)

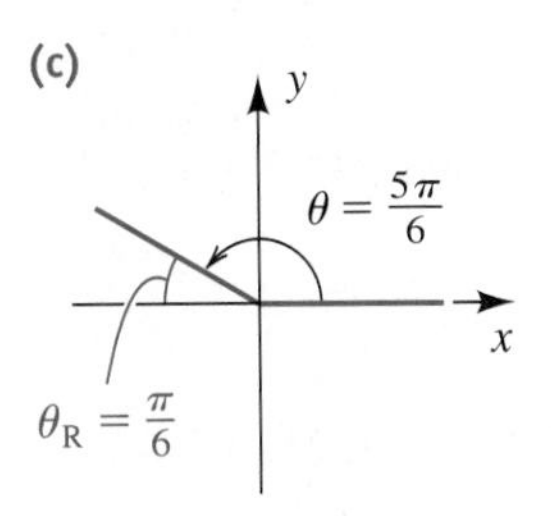

(d)

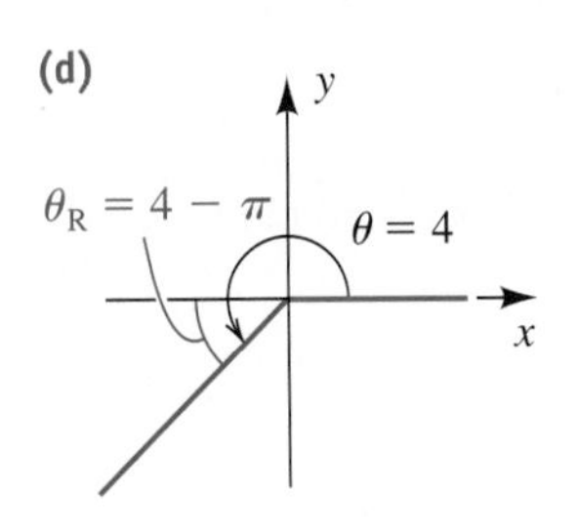

▶ **SOLUTION**

(a) The angle $\theta = 315°$ is in quadrant IV, and hence, as in Figure 1(d),

$$\theta_R = 360° - 315° = 45°.$$

The angles θ and θ_R are sketched in Figure 2(a).

(b) The angle between 0° and 360° that is coterminal with $\theta = -240°$ is

$$-240° + 360° = 120°,$$

which is in quadrant II. Using the formula in Figure 1(b) gives

$$\theta_R = 180° - 120° = 60°.$$

The angles θ and θ_R are sketched in Figure 2(b).

(c) Since the angle $\theta = 5\pi/6$ is in quadrant II, we have

$$\theta_R = \pi - \frac{5\pi}{6} = \frac{\pi}{6},$$

as shown in Figure 2(c).

(d) Since $\pi < 4 < 3\pi/2$, the angle $\theta = 4$ is in quadrant III. Using the formula in Figure 1(c), we obtain

$$\theta_R = 4 - \pi.$$

The angles are sketched in Figure 2(d).

We shall next show how reference angles can be used to find values of the trigonometric functions.

If θ is a nonquadrantal angle with reference angle θ_R, then we have $0° < \theta_R < 90°$ or $0 < \theta_R < \pi/2$. Let $P(x, y)$ be a point on the terminal side of θ, and consider the point $Q(x, 0)$ on the x-axis. Figure 3 illustrates a

Figure 3

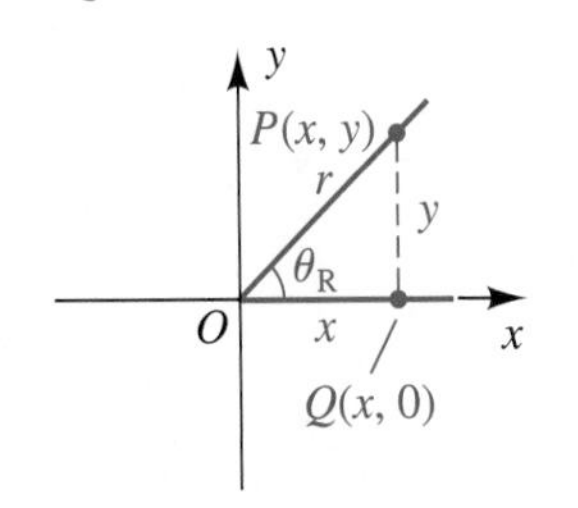

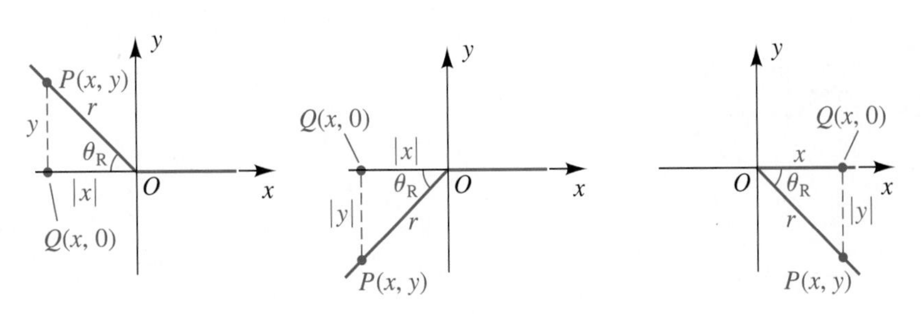

▶ **Tutorial available at www.thomsonedu.com/login**

typical situation for θ in each quadrant. In each case, the lengths of the sides of triangle OQP are

$$d(O, Q) = |x|, \quad d(Q, P) = |y|, \quad \text{and} \quad d(O, P) = \sqrt{x^2 + y^2} = r.$$

We may apply the definition of the trigonometric functions of any angle and also use triangle OQP to obtain the following formulas:

$$|\sin \theta| = \left|\frac{y}{r}\right| = \frac{|y|}{|r|} = \frac{|y|}{r} = \sin \theta_R$$

$$|\cos \theta| = \left|\frac{x}{r}\right| = \frac{|x|}{|r|} = \frac{|x|}{r} = \cos \theta_R$$

$$|\tan \theta| = \left|\frac{y}{x}\right| = \frac{|y|}{|x|} = \tan \theta_R$$

These formulas lead to the next theorem. If θ is a quadrantal angle, the definition of the trigonometric functions of any angle should be used to find values.

Theorem on Reference Angles

If θ is a nonquadrantal angle in standard position, then to find the value of a trigonometric function at θ, find its value for the reference angle θ_R and prefix the appropriate sign.

The "appropriate sign" referred to in the theorem can be determined from the table of signs of the trigonometric functions given on page 384.

EXAMPLE 2 Using reference angles

Use reference angles to find the exact values of $\sin \theta$, $\cos \theta$, and $\tan \theta$ if

(a) $\theta = \dfrac{5\pi}{6}$ (b) $\theta = 315°$

▶ **SOLUTION**

(a) The angle $\theta = 5\pi/6$ and its reference angle $\theta_R = \pi/6$ are sketched in Figure 4. Since θ is in quadrant II, $\sin \theta$ is positive and both $\cos \theta$ and $\tan \theta$ are negative. Hence, by the theorem on reference angles and known results about special angles, we obtain the following values:

Figure 4

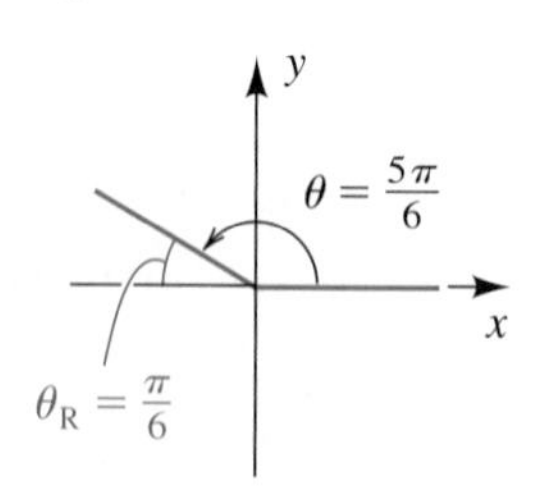

$$\sin \frac{5\pi}{6} = + \sin \frac{\pi}{6} = \frac{1}{2}$$

$$\cos \frac{5\pi}{6} = - \cos \frac{\pi}{6} = -\frac{\sqrt{3}}{2}$$

$$\tan \frac{5\pi}{6} = - \tan \frac{\pi}{6} = -\frac{\sqrt{3}}{3}$$

▶ **Tutorial available at www.thomsonedu.com/login**

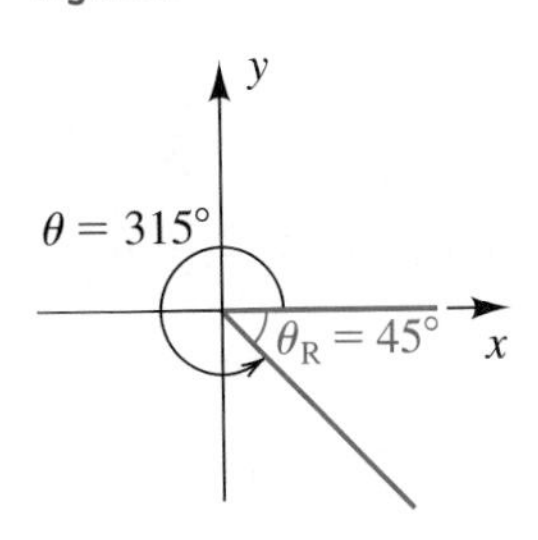

Figure 5

(b) The angle $\theta = 315°$ and its reference angle $\theta_R = 45°$ are sketched in Figure 5. Since θ is in quadrant IV, $\sin\theta < 0$, $\cos\theta > 0$, and $\tan\theta < 0$. Hence, by the theorem on reference angles, we obtain

$$\sin 315° = -\sin 45° = -\frac{\sqrt{2}}{2}$$

$$\cos 315° = +\cos 45° = \frac{\sqrt{2}}{2}$$

$$\tan 315° = -\tan 45° = -1.$$

If we use a calculator to approximate function values, reference angles are usually unnecessary (see Discussion Exercise 2 at the end of the chapter). As an illustration, to find sin 210°, we place the calculator in degree mode and obtain $\sin 210° = -0.5$, which is the exact value. Using the same procedure for 240°, we obtain a decimal representation:

$$\sin 240° \approx -0.8660$$

A calculator should not be used to find the *exact* value of sin 240°. In this case, we find the reference angle 60° of 240° and use the theorem on reference angles, together with known results about special angles, to obtain

$$\sin 240° = -\sin 60° = -\frac{\sqrt{3}}{2}.$$

Let us next consider the problem of solving an equation of the following type:

Problem: If θ is an acute angle and $\sin\theta = 0.6635$, approximate θ.

Most calculators have a key labeled [SIN⁻¹] that can be used to help solve the equation. With some calculators, it may be necessary to use another key or a keystroke sequence such as [INV] [SIN] (refer to the user manual for your calculator). We shall use the following notation when finding θ, where $0 \le k \le 1$:

$$\text{if} \quad \sin\theta = k, \quad \text{then} \quad \theta = \sin^{-1} k$$

This notation is similar to that used for the inverse function f^{-1} of a function f in Section 4.1, where we saw that under certain conditions,

$$\text{if} \quad f(x) = y, \quad \text{then} \quad x = f^{-1}(y).$$

For the problem $\sin\theta = 0.6635$, f is the sine function, $x = \theta$, and $y = 0.6635$. The notation $\sin^{-1}$ is based on the *inverse trigonometric functions* discussed in Section 6.6. At this stage of our work, *we shall regard* $\sin^{-1}$ *simply as an entry made on a calculator using a* [SIN⁻¹] *key.* Thus, for the stated problem, we obtain

$$\theta = \sin^{-1}(0.6635) \approx 41.57° \approx 0.7255.$$

As indicated, when finding an angle, we will usually round off degree measure to the nearest 0.01° and radian measure to four decimal places.

Similarly, given $\cos\theta = k$ or $\tan\theta = k$, where θ is acute, we write

$$\theta = \cos^{-1} k \qquad \text{or} \qquad \theta = \tan^{-1} k$$

to indicate the use of a COS⁻¹ or TAN⁻¹ key on a calculator.

Given $\csc\theta$, $\sec\theta$, or $\cot\theta$, we use a reciprocal relationship to find θ, as indicated in the following illustration.

ILLUSTRATION **Finding Acute Angle Solutions of Equations with a Calculator**

Equation	Calculator solution (degree and radian)
$\sin\theta = 0.5$	$\theta = \sin^{-1}(0.5) = 30^\circ \approx 0.5236$
$\cos\theta = 0.5$	$\theta = \cos^{-1}(0.5) = 60^\circ \approx 1.0472$
$\tan\theta = 0.5$	$\theta = \tan^{-1}(0.5) \approx 26.57^\circ \approx 0.4636$
$\csc\theta = 2$	$\theta = \sin^{-1}\left(\frac{1}{2}\right) = 30^\circ \approx 0.5236$
$\sec\theta = 2$	$\theta = \cos^{-1}\left(\frac{1}{2}\right) = 60^\circ \approx 1.0472$
$\cot\theta = 2$	$\theta = \tan^{-1}\left(\frac{1}{2}\right) \approx 26.57^\circ \approx 0.4636$

The same technique may be employed if θ is *any* angle or real number. Thus, using the SIN⁻¹ key, we obtain, in degree or radian mode,

$$\theta = \sin^{-1}(0.6635) \approx 41.57^\circ \approx 0.7255,$$

which is the reference angle for θ. If $\sin\theta$ is *negative,* then a calculator gives us the *negative* of the reference angle. For example,

$$\sin^{-1}(-0.6635) \approx -41.57^\circ \approx -0.7255.$$

Similarly, given $\cos\theta$ or $\tan\theta$, we find θ with a calculator by using COS⁻¹ or TAN⁻¹, respectively. The interval containing θ is listed in the next chart. It is important to note that if $\cos\theta$ is negative, then θ is *not* the negative of the reference angle, but instead is in the interval $\pi/2 < \theta \leq \pi$, or $90^\circ < \theta \leq 180^\circ$. The reasons for using these intervals are explained in Section 6.6. We may use reciprocal relationships to solve similar equations involving $\csc\theta$, $\sec\theta$, and $\cot\theta$.

Equation	Values of k	Calculator solution	Interval containing θ if a calculator is used
$\sin\theta = k$	$-1 \leq k \leq 1$	$\theta = \sin^{-1} k$	$-\frac{\pi}{2} \leq \theta \leq \frac{\pi}{2}$, or $-90^\circ \leq \theta \leq 90^\circ$
$\cos\theta = k$	$-1 \leq k \leq 1$	$\theta = \cos^{-1} k$	$0 \leq \theta \leq \pi$, or $0^\circ \leq \theta \leq 180^\circ$
$\tan\theta = k$	any k	$\theta = \tan^{-1} k$	$-\frac{\pi}{2} < \theta < \frac{\pi}{2}$, or $-90^\circ < \theta < 90^\circ$

The following illustration contains some specific examples for both degree and radian modes.

ILLUSTRATION **Finding Angles with a Calculator**

	Equation	Calculator solution (degree and radian)	
■	$\sin\theta = -0.5$	$\theta = \sin^{-1}(-0.5) = -30°$	≈ -0.5236
■	$\cos\theta = -0.5$	$\theta = \cos^{-1}(-0.5) = 120°$	≈ 2.0944
■	$\tan\theta = -0.5$	$\theta = \tan^{-1}(-0.5) \approx -26.57°$	≈ -0.4636

Figure 6

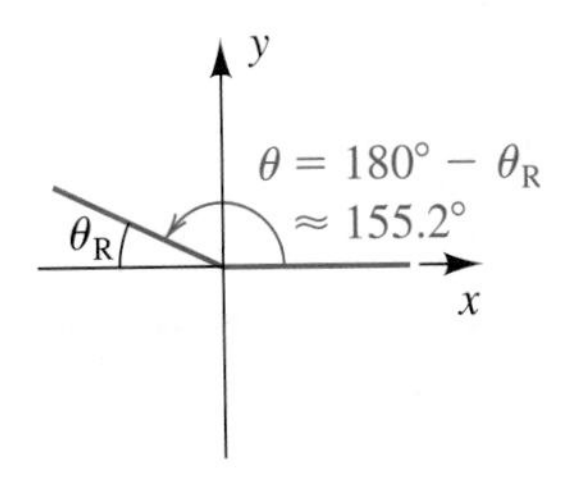

When using a calculator to find θ, be sure to keep the restrictions on θ in mind. If other values are desired, then reference angles or other methods may be employed, as illustrated in the next examples.

EXAMPLE 3 **Approximating an angle with a calculator**

If $\tan\theta = -0.4623$ and $0° \leq \theta < 360°$, find θ to the nearest $0.1°$.

▶ **SOLUTION** As pointed out in the preceding discussion, if we use a calculator (in degree mode) to find θ when $\tan\theta$ is negative, then the degree measure will be in the interval $(-90°, 0°)$. In particular, we obtain the following:

$$\theta = \tan^{-1}(-0.4623) \approx -24.8°$$

Figure 7

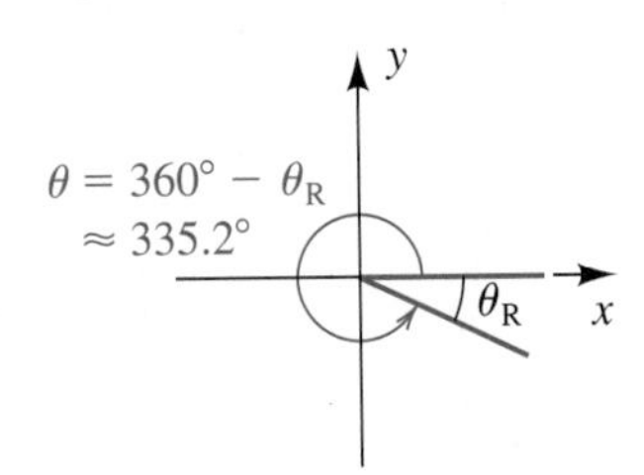

Since we wish to find values of θ between 0° and 360°, we use the (approximate) reference angle $\theta_R \approx 24.8°$. There are two possible values of θ such that $\tan\theta$ is negative—one in quadrant II, the other in quadrant IV. If θ is in quadrant II and $0° \leq \theta < 360°$, we have the situation shown in Figure 6, and

$$\theta = 180° - \theta_R \approx 180° - 24.8° = 155.2°.$$

If θ is in quadrant IV and $0° \leq \theta < 360°$, then, as in Figure 7,

$$\theta = 360° - \theta_R \approx 360° - 24.8° = 335.2°.$$

Figure 8

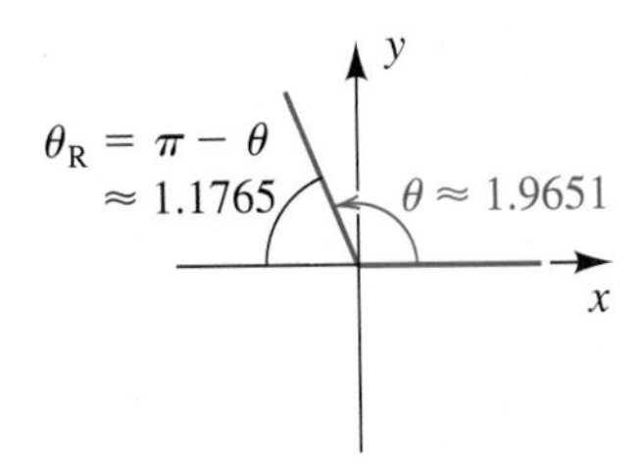

EXAMPLE 4 **Approximating an angle with a calculator**

If $\cos\theta = -0.3842$ and $0 \leq \theta < 2\pi$, find θ to the nearest 0.0001 radian.

▶ **SOLUTION** If we use a calculator (in radian mode) to find θ when $\cos\theta$ is negative, then the radian measure will be in the interval $[0, \pi]$. In particular, we obtain the following (shown in Figure 8):

$$\theta = \cos^{-1}(-0.3842) \approx 1.965\,137\,489$$

Figure 9

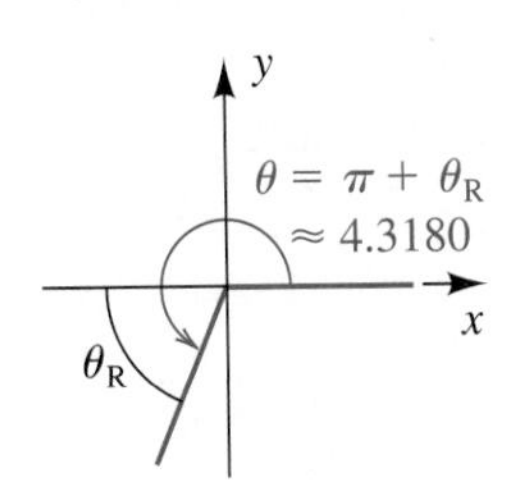

Since we wish to find values of θ between 0 and 2π, we use the (approximate) reference angle

$$\theta_R = \pi - \theta \approx 1.176\,455\,165.$$

There are two possible values of θ such that $\cos\theta$ is negative—the one we found in quadrant II and the other in quadrant III. If θ is in quadrant III, then

$$\theta = \pi + \theta_R \approx 4.318\,047\,819,$$

as shown in Figure 9.

(continued)

▶ **Tutorial available at www.thomsonedu.com/login**

Figure 10

```
cos⁻¹(-.3842)
          1.965137489
π-Ans
          1.176455165
π+Ans
          4.318047819
```

The calculator display in Figure 10 provides numerical support for the answers

$$\theta \approx 1.9651 \qquad \text{and} \qquad \theta \approx 4.3180.$$

We could also solve this problem graphically by finding the points of intersection of $Y_1 = \cos(X)$ and $Y_2 = -0.3842$ on the interval $[0, 2\pi)$. However, the purpose of this solution was to illustrate the use of reference angles.

5.4 Exercises

Exer. 1–6: Find the reference angle θ_R if θ has the given measure.

1 (a) $240°$ (b) $340°$ (c) $-202°$ (d) $-660°$

2 (a) $165°$ (b) $275°$ (c) $-110°$ (d) $400°$

3 (a) $3\pi/4$ (b) $4\pi/3$ (c) $-\pi/6$ (d) $9\pi/4$

4 (a) $7\pi/4$ (b) $2\pi/3$ (c) $-3\pi/4$ (d) $-23\pi/6$

5 (a) 3 (b) -2 (c) 5.5 (d) 100

6 (a) 6 (b) -4 ▶ (c) 4.5 (d) 80

Exer. 7–18: Find the exact value.

7 (a) $\sin(2\pi/3)$ (b) $\sin(-5\pi/4)$

8 (a) $\sin 210°$ (b) $\sin(-315°)$

9 (a) $\cos 150°$ (b) $\cos(-60°)$

10 (a) $\cos(5\pi/4)$ ▶ (b) $\cos(-11\pi/6)$

11 (a) $\tan(5\pi/6)$ (b) $\tan(-\pi/3)$

12 (a) $\tan 330°$ (b) $\tan(-225°)$

13 (a) $\cot 120°$ (b) $\cot(-150°)$

14 (a) $\cot(3\pi/4)$ (b) $\cot(-2\pi/3)$

15 (a) $\sec(2\pi/3)$ (b) $\sec(-\pi/6)$

16 (a) $\sec 135°$ (b) $\sec(-210°)$

17 (a) $\csc 240°$ (b) $\csc(-330°)$

18 (a) $\csc(3\pi/4)$ (b) $\csc(-2\pi/3)$

Exer. 19–24: Approximate to three decimal places.

19 (a) $\sin 73°20'$ (b) $\cos 0.68$

20 (a) $\cos 38°30'$ (b) $\sin 1.48$

21 (a) $\tan 21°10'$ (b) $\cot 1.13$

22 (a) $\cot 9°10'$ (b) $\tan 0.75$

23 (a) $\sec 67°50'$ (b) $\csc 0.32$

24 (a) $\csc 43°40'$ (b) $\sec 0.26$

Exer. 25–32: Approximate the acute angle θ to the nearest (a) 0.01° and (b) 1′.

25 $\cos\theta = 0.8620$

26 $\sin\theta = 0.6612$

27 $\tan\theta = 3.7$

28 $\cos\theta = 0.8$

29 $\sin\theta = 0.4217$

30 $\tan\theta = 4.91$

31 $\sec\theta = 4.246$

32 $\csc\theta = 11$

▶ **Tutorial available at www.thomsonedu.com/login**

Exer. 33–34: Approximate to four decimal places.

33 (a) $\sin 98°10'$ (b) $\cos 623.7°$ (c) $\tan 3$

(d) $\cot 231°40'$ (e) $\sec 1175.1°$ (f) $\csc 0.82$

34 (a) $\sin 496.4°$ (b) $\cos 0.65$ (c) $\tan 105°40'$

(d) $\cot 1030.2°$ (e) $\sec 1.46$ (f) $\csc 320°50'$

Exer. 35–36: Approximate, to the nearest 0.1°, all angles θ in the interval [0°, 360°) that satisfy the equation.

35 (a) $\sin \theta = -0.5640$ (b) $\cos \theta = 0.7490$

(c) $\tan \theta = 2.798$ (d) $\cot \theta = -0.9601$

(e) $\sec \theta = -1.116$ (f) $\csc \theta = 1.485$

▶ 36 (a) $\sin \theta = 0.8225$ (b) $\cos \theta = -0.6604$

(c) $\tan \theta = -1.5214$ (d) $\cot \theta = 1.3752$

(e) $\sec \theta = 1.4291$ (f) $\csc \theta = -2.3179$

Exer. 37–38: Approximate, to the nearest 0.01 radian, all angles θ in the interval $[0, 2\pi)$ that satisfy the equation.

37 (a) $\sin \theta = 0.4195$ (b) $\cos \theta = -0.1207$

(c) $\tan \theta = -3.2504$ (d) $\cot \theta = 2.6815$

(e) $\sec \theta = 1.7452$ (f) $\csc \theta = -4.8521$

38 (a) $\sin \theta = -0.0135$ (b) $\cos \theta = 0.9235$

(c) $\tan \theta = 0.42$ (d) $\cot \theta = -2.731$

(e) $\sec \theta = -3.51$ (f) $\csc \theta = 1.258$

39 Thickness of the ozone layer The thickness of the ozone layer can be estimated using the formula

$$\ln I_0 - \ln I = kx \sec \theta,$$

where I_0 is the intensity of a particular wavelength of light from the sun before it reaches the atmosphere, I is the intensity of the same wavelength after passing through a layer of ozone x centimeters thick, k is the absorption constant of ozone for that wavelength, and θ is the acute angle that the sunlight makes with the vertical. Suppose that for a wavelength of 3055×10^{-8} centimeter with $k \approx 1.88$, I_0/I is measured as 1.72 and $\theta = 12°$. Approximate the thickness of the ozone layer to the nearest 0.01 centimeter.

40 Ozone calculations Refer to Exercise 39. If the ozone layer is estimated to be 0.31 centimeter thick and, for a wavelength of 3055×10^{-8} centimeter, I_0/I is measured as 2.05, approximate the angle the sun made with the vertical at the time of the measurement.

41 Solar radiation The amount of sunshine illuminating a wall of a building can greatly affect the energy efficiency of the building. The solar radiation striking a vertical wall that faces east is given by the formula

$$R = R_0 \cos \theta \sin \phi,$$

where R_0 is the maximum solar radiation possible, θ is the angle that the sun makes with the horizontal, and ϕ is the direction of the sun in the sky, with $\phi = 90°$ when the sun is in the east and $\phi = 0°$ when the sun is in the south.

(a) When does the maximum solar radiation R_0 strike the wall?

(b) What percentage of R_0 is striking the wall when θ is equal to 60° and the sun is in the southeast?

42 Meteorological calculations In the mid-latitudes it is sometimes possible to estimate the distance between consecutive regions of low pressure. If ϕ is the latitude (in degrees), R is Earth's radius (in kilometers), and v is the horizontal wind velocity (in km/hr), then the distance d (in kilometers) from one low pressure area to the next can be estimated using the formula

$$d = 2\pi\left(\frac{vR}{0.52 \cos \phi}\right)^{1/3}.$$

(a) At a latitude of 48°, Earth's radius is approximately 6369 kilometers. Approximate d if the wind speed is 45 km/hr.

(b) If v and R are constant, how does d vary as the latitude increases?

43 Robot's arm Points on the terminal sides of angles play an important part in the design of arms for robots. Suppose a robot has a straight arm 18 inches long that can rotate about the origin in a coordinate plane. If the robot's hand is located at (18, 0) and then rotates through an angle of 60°, what is the new location of the hand?

44 Robot's arm Suppose the robot's arm in Exercise 43 can change its length in addition to rotating about the origin. If the hand is initially at (12, 12), approximately how many degrees should the arm be rotated and how much should its length be changed to move the hand to (−16, 10)?

▶ Tutorial available at www.thomsonedu.com/login

5.5

Trigonometric Graphs

In this section we consider graphs of the equations

$$y = a \sin (bx + c) \qquad \text{and} \qquad y = a \cos (bx + c)$$

for real numbers a, b, and c. Our goal is to sketch such graphs without plotting many points. To do so we shall use facts about the graphs of the sine and cosine functions discussed in Section 5.3.

Let us begin by considering the special case $c = 0$ and $b = 1$—that is,

$$y = a \sin x \qquad \text{and} \qquad y = a \cos x.$$

We can find y-coordinates of points on the graphs by multiplying y-coordinates of points on the graphs of $y = \sin x$ and $y = \cos x$ by a. To illustrate, if $y = 2 \sin x$, we multiply the y-coordinate of each point on the graph of $y = \sin x$ by 2. This gives us Figure 1, where for comparison we also show the graph of $y = \sin x$. The procedure is the same as that for vertically stretching the graph of a function, discussed in Section 2.5.

As another illustration, if $y = \frac{1}{2} \sin x$, we multiply y-coordinates of points on the graph of $y = \sin x$ by $\frac{1}{2}$. This multiplication vertically compresses the graph of $y = \sin x$ by a factor of 2, as illustrated in Figure 2.

Figure 1

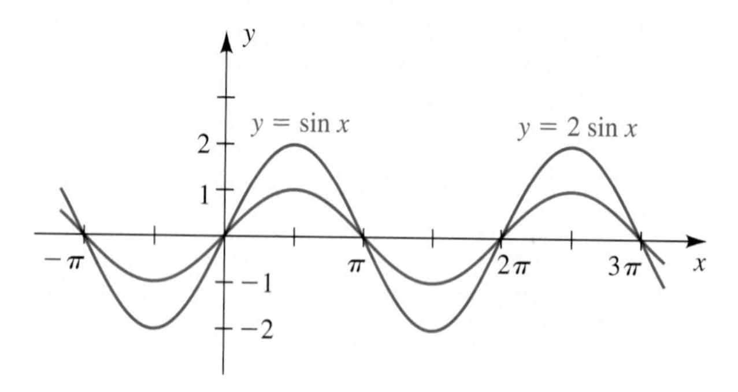

Figure 2

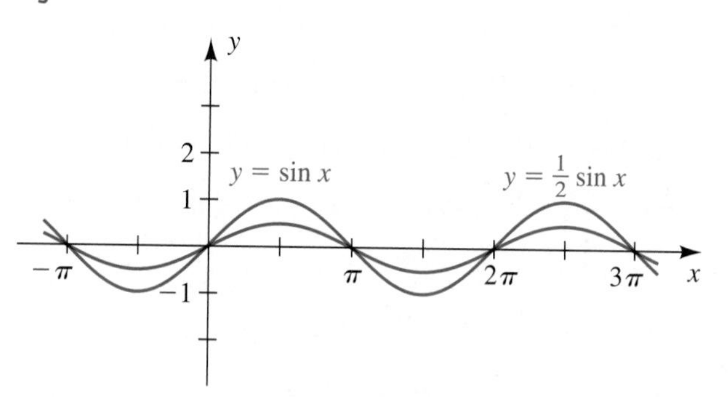

The following example illustrates a graph of $y = a \sin x$ with a negative.

EXAMPLE 1 Sketching the graph of an equation involving sin x

Sketch the graph of the equation $y = -2 \sin x$.

SOLUTION The graph of $y = -2 \sin x$ sketched in Figure 3 can be obtained by first sketching the graph of $y = \sin x$ (shown in the figure) and then multiplying y-coordinates by -2. An alternative method is to reflect the graph of $y = 2 \sin x$ (see Figure 1) through the x-axis.

Figure 3

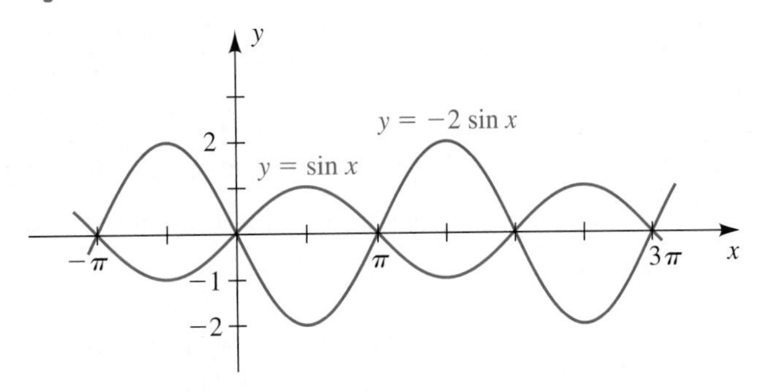

For any $a \neq 0$, the graph of $y = a \sin x$ has the general appearance of one of the graphs illustrated in Figures 1, 2, and 3. The amount of stretching of the graph of $y = \sin x$ and whether the graph is reflected are determined by the absolute value of a and the sign of a, respectively. The largest y-coordinate $|a|$ is the **amplitude of the graph** or, equivalently, the **amplitude of the function** ***f*** given by $f(x) = a \sin x$. In Figures 1 and 3 the amplitude is 2. In Figure 2 the amplitude is $\frac{1}{2}$. Similar remarks and techniques apply if $y = a \cos x$.

EXAMPLE 2 **Sketching the graph of an equation involving $\cos x$**

Find the amplitude and sketch the graph of $y = 3 \cos x$.

▶ **SOLUTION** By the preceding discussion, the amplitude is 3. As indicated in Figure 4, we first sketch the graph of $y = \cos x$ and then multiply y-coordinates by 3.

Figure 4

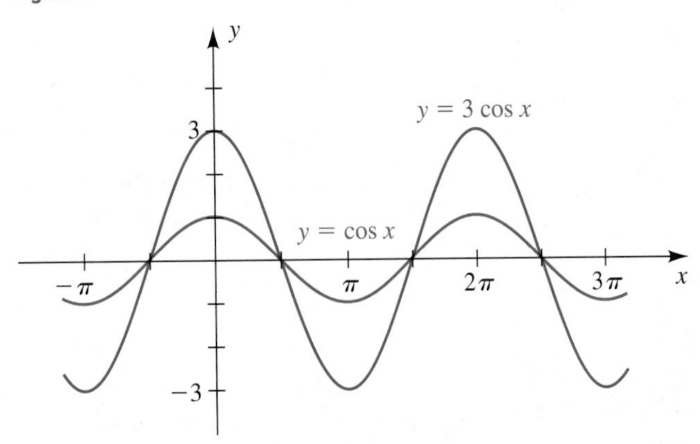

Let us next consider $y = a \sin bx$ and $y = a \cos bx$ for nonzero real numbers a and b. As before, the amplitude is $|a|$. If $b > 0$, then exactly one cycle occurs as bx increases from 0 to 2π or, equivalently, as x increases from 0 to $2\pi/b$. If $b < 0$, then $-b > 0$ and one cycle occurs as x increases from 0

▶ **Tutorial available at www.thomsonedu.com/login**

to $2\pi/(-b)$. Thus, the period of the function f given by $f(x) = a \sin bx$ or $f(x) = a \cos bx$ is $2\pi/|b|$. For convenience, we shall also refer to $2\pi/|b|$ as the period of the *graph of f*. The next theorem summarizes our discussion.

Theorem on Amplitudes and Periods	If $y = a \sin bx$ or $y = a \cos bx$ for nonzero real numbers a and b, then the graph has amplitude $|a|$ and period $\dfrac{2\pi}{|b|}$.

We can also relate the role of b to the discussion of horizontally compressing and stretching a graph in Section 2.5. If $|b| > 1$, the graph of $y = \sin bx$ or $y = \cos bx$ can be considered to be compressed horizontally by a factor b. If $0 < |b| < 1$, the graphs are stretched horizontally by a factor $1/b$. This concept is illustrated in the next two examples.

EXAMPLE 3 Finding an amplitude and a period

Find the amplitude and the period and sketch the graph of $y = 3 \sin 2x$.

Figure 5

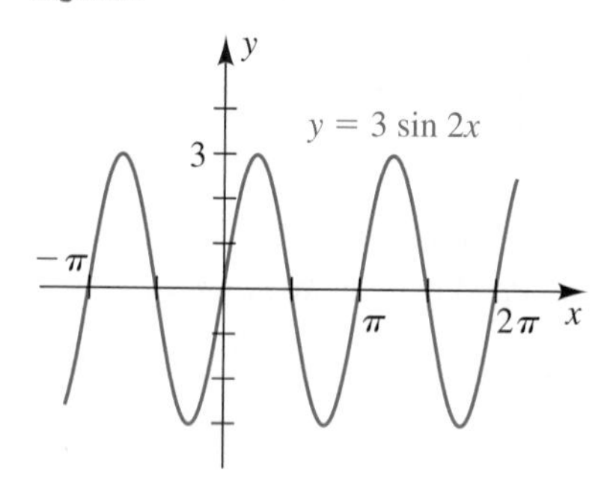

▶ **SOLUTION** Using the theorem on amplitudes and periods with $a = 3$ and $b = 2$, we obtain the following:

$$\text{amplitude: } |a| = |3| = 3$$

$$\text{period: } \frac{2\pi}{|b|} = \frac{2\pi}{|2|} = \frac{2\pi}{2} = \pi$$

Thus, there is exactly one sine wave of amplitude 3 on the x-interval $[0, \pi]$. Sketching this wave and then extending the graph to the right and left gives us Figure 5.

EXAMPLE 4 Finding an amplitude and a period

Find the amplitude and the period and sketch the graph of $y = 2 \sin \frac{1}{2}x$.

▶ **SOLUTION** Using the theorem on amplitudes and periods with $a = 2$ and $b = \frac{1}{2}$, we obtain the following:

Figure 6

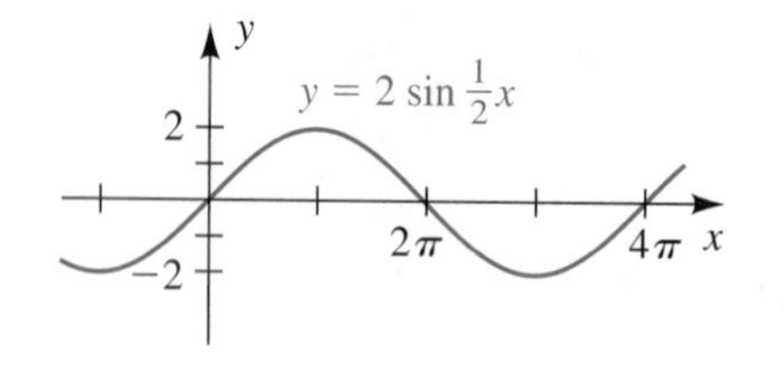

$$\text{amplitude: } |a| = |2| = 2$$

$$\text{period: } \frac{2\pi}{|b|} = \frac{2\pi}{|\frac{1}{2}|} = \frac{2\pi}{\frac{1}{2}} = 4\pi$$

Thus, there is one sine wave of amplitude 2 on the interval $[0, 4\pi]$. Sketching this wave and extending it left and right gives us the graph in Figure 6.

If $y = a \sin bx$ and if b is a large positive number, then the period $2\pi/b$ is small and the sine waves are close together, with b sine waves on the interval $[0, 2\pi]$. For example, in Figure 5, $b = 2$ and we have two sine waves on

▶ Tutorial available at www.thomsonedu.com/login

$[0, 2\pi]$. If b is a small positive number, then the period $2\pi/b$ is large and the waves are far apart. To illustrate, if $y = \sin \frac{1}{10}x$, then one-tenth of a sine wave occurs on $[0, 2\pi]$ and an interval 20π units long is required for one complete cycle. (See also Figure 6—for $y = 2 \sin \frac{1}{2}x$, one-half of a sine wave occurs on $[0, 2\pi]$.)

If $b < 0$, we can use the fact that $\sin(-x) = -\sin x$ to obtain the graph of $y = a \sin bx$. To illustrate, the graph of $y = \sin(-2x)$ is the same as the graph of $y = -\sin 2x$.

Figure 7

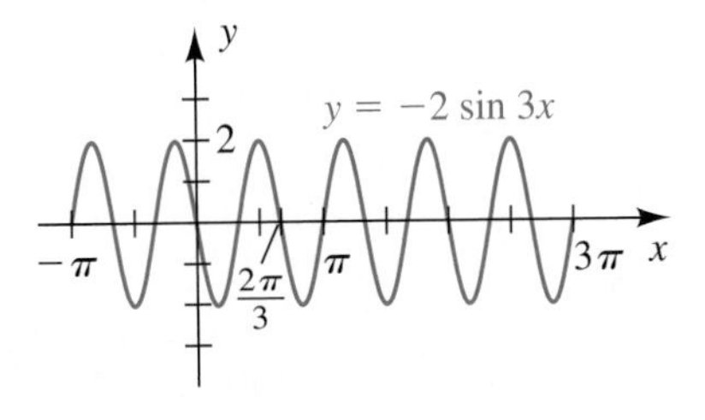

EXAMPLE 5 Finding an amplitude and a period

Find the amplitude and the period and sketch the graph of the equation $y = 2 \sin(-3x)$.

▶ SOLUTION Since the sine function is odd, $\sin(-3x) = -\sin 3x$, and we may write the equation as $y = -2 \sin 3x$. The amplitude is $|-2| = 2$, and the period is $2\pi/3$. Thus, there is one cycle on an interval of length $2\pi/3$. The negative sign indicates a reflection through the x-axis. If we consider the interval $[0, 2\pi/3]$ and sketch a sine wave of amplitude 2 (reflected through the x-axis), the shape of the graph is apparent. The part of the graph in the interval $[0, 2\pi/3]$ is repeated periodically, as illustrated in Figure 7.

Figure 8

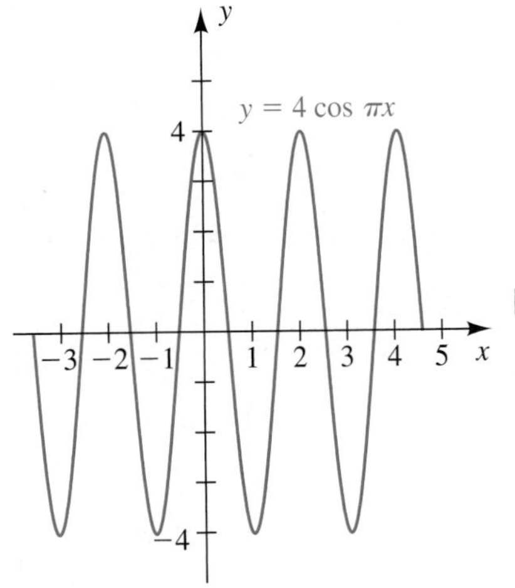

EXAMPLE 6 Finding an amplitude and a period

Find the amplitude and the period and sketch the graph of $y = 4 \cos \pi x$.

▶ SOLUTION The amplitude is $|4| = 4$, and the period is $2\pi/\pi = 2$. Thus, there is exactly one cosine wave of amplitude 4 on the interval $[0, 2]$. Since the period does not contain the number π, it makes sense to use integer ticks on the x-axis. Sketching this wave and extending it left and right gives us the graph in Figure 8.

As discussed in Section 2.5, if f is a function and c is a positive real number, then the graph of $y = f(x) + c$ can be obtained by shifting the graph of $y = f(x)$ vertically upward a distance c. For the graph of $y = f(x) - c$, we shift the graph of $y = f(x)$ vertically downward a distance of c. In the next example we use this technique for a trigonometric graph.

Figure 9

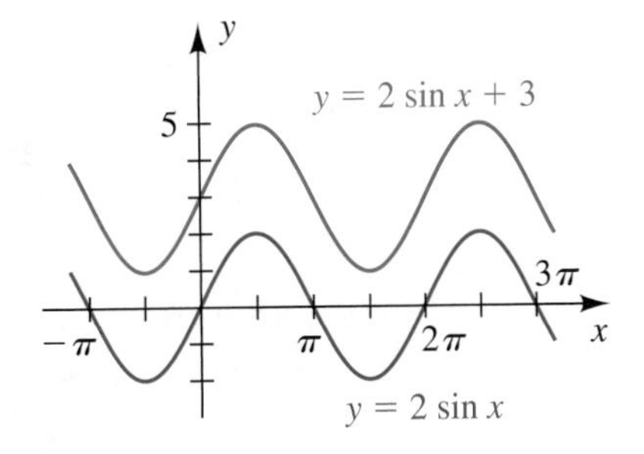

EXAMPLE 7 Vertically shifting a trigonometric graph

Sketch the graph of $y = 2 \sin x + 3$.

SOLUTION It is important to note that $y \neq 2 \sin(x + 3)$. The graph of $y = 2 \sin x$ is sketched in red in Figure 9. If we shift this graph vertically upward a distance 3, we obtain the graph of $y = 2 \sin x + 3$.

▶ Tutorial available at www.thomsonedu.com/login

Let us next consider the graph of

$$y = a \sin (bx + c).$$

As before, the amplitude is $|a|$, and the period is $2\pi/|b|$. One cycle occurs if $bx + c$ increases from 0 to 2π. Hence, we can find an interval containing exactly one sine wave by solving the following inequality for x:

$$\begin{aligned} 0 \le bx + c &\le 2\pi \\ -c \le bx \quad &\le 2\pi - c \qquad \text{subtract } c \\ -\frac{c}{b} \le x \quad &\le \frac{2\pi}{b} - \frac{c}{b} \qquad \text{divide by } b \end{aligned}$$

The number $-c/b$ is the **phase shift** associated with the graph. The graph of $y = a \sin (bx + c)$ may be obtained by shifting the graph of $y = a \sin bx$ to the left if the phase shift is negative or to the right if the phase shift is positive.

Analogous results are true for $y = a \cos (bx + c)$. The next theorem summarizes our discussion.

Theorem on Amplitudes, Periods, and Phase Shifts

If $y = a \sin (bx + c)$ or $y = a \cos (bx + c)$ for nonzero real numbers a and b, then

(1) the amplitude is $|a|$, the period is $\dfrac{2\pi}{|b|}$, and the phase shift is $-\dfrac{c}{b}$;

(2) an interval containing exactly one cycle can be found by solving the inequality

$$0 \le bx + c \le 2\pi.$$

We will sometimes write $y = a \sin (bx + c)$ in the equivalent form $y = a \sin \left[b\left(x + \dfrac{c}{b} \right) \right]$.

EXAMPLE 8 Finding an amplitude, a period, and a phase shift

Find the amplitude, the period, and the phase shift and sketch the graph of

$$y = 3 \sin \left(2x + \frac{\pi}{2} \right).$$

▶ **SOLUTION** The equation is of the form $y = a \sin (bx + c)$ with $a = 3$, $b = 2$, and $c = \pi/2$. Thus, the amplitude is $|a| = 3$, and the period is $2\pi/|b| = 2\pi/2 = \pi$.

By part (2) of the theorem on amplitudes, periods, and phase shifts, the phase shift and an interval containing one sine wave can be found by solving the following inequality:

Figure 10

$$0 \le 2x + \frac{\pi}{2} \le 2\pi$$

$$-\frac{\pi}{2} \le 2x \quad \le \frac{3\pi}{2} \qquad \text{subtract } \frac{\pi}{2}$$

$$-\frac{\pi}{4} \le \ x \quad \le \frac{3\pi}{4} \qquad \text{divide by 2}$$

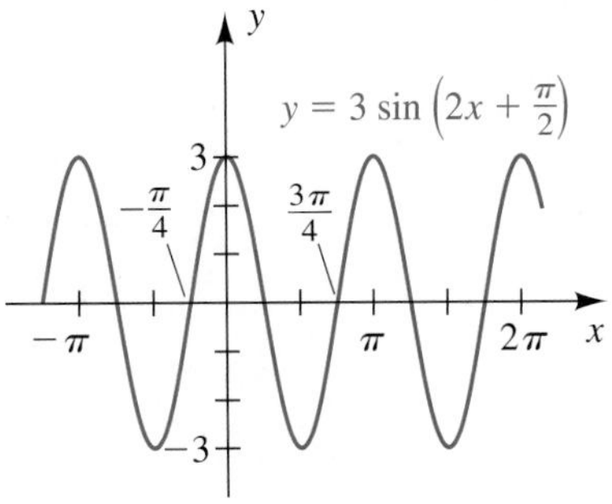

Thus, the phase shift is $-\pi/4$, and one sine wave of amplitude 3 occurs on the interval $[-\pi/4, 3\pi/4]$. Sketching that wave and then repeating it to the right and left gives us the graph in Figure 10.

EXAMPLE 9 **Finding an amplitude, a period, and a phase shift**

Find the amplitude, the period, and the phase shift and sketch the graph of $y = 2\cos(3x - \pi)$.

▶ **SOLUTION** The equation has the form $y = a\cos(bx + c)$ with $a = 2$, $b = 3$, and $c = -\pi$. Thus, the amplitude is $|a| = 2$, and the period is $2\pi/|b| = 2\pi/3$.

By part (2) of the theorem on amplitudes, periods, and phase shifts, the phase shift and an interval containing one cycle can be found by solving the following inequality:

$$0 \le 3x - \pi \le 2\pi$$

$$\pi \le 3x \quad \le 3\pi \qquad \text{add } \pi$$

$$\frac{\pi}{3} \le \ x \quad \le \pi \qquad \text{divide by 3}$$

Figure 11

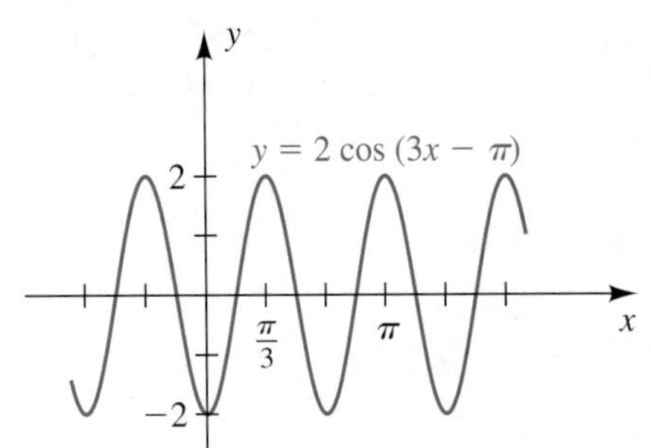

Hence, the phase shift is $\pi/3$, and one cosine-type cycle (from maximum to maximum) of amplitude 2 occurs on the interval $[\pi/3, \pi]$. Sketching that part of the graph and then repeating it to the right and left gives us the sketch in Figure 11.

If we solve the inequality

$$-\frac{\pi}{2} \le 3x - \pi \le \frac{3\pi}{2} \qquad \text{instead of} \qquad 0 \le 3x - \pi \le 2\pi,$$

we obtain the interval $\pi/6 \le x \le 5\pi/6$, which gives us a cycle between x-intercepts rather than a cycle between maximums.

▶ **Tutorial available at www.thomsonedu.com/login**

EXAMPLE 10 **Finding an equation for a sine wave**

Express the equation for the sine wave shown in Figure 12 in the form

$$y = a \sin (bx + c)$$

for $a > 0$, $b > 0$, and the least positive real number c.

Figure 12

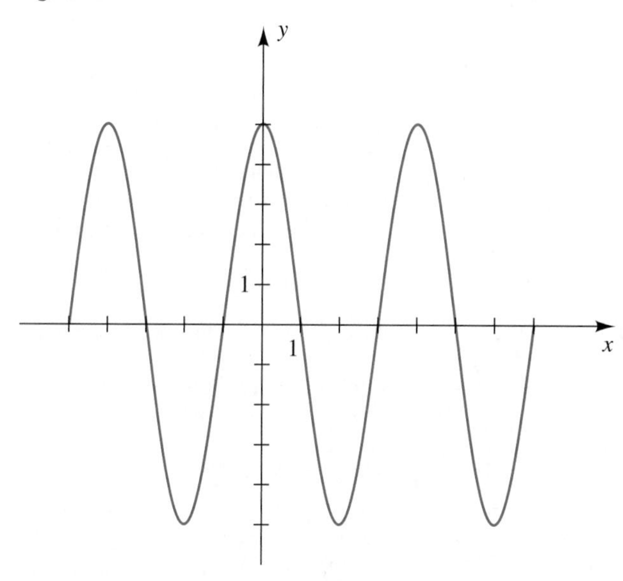

SOLUTION The largest and smallest y-coordinates of points on the graph are 5 and -5, respectively. Hence, the amplitude is $a = 5$.

Since one sine wave occurs on the interval $[-1, 3]$, the period has value $3 - (-1) = 4$. Hence, by the theorem on amplitudes, periods, and phase shifts (with $b > 0$),

$$\frac{2\pi}{b} = 4 \qquad \text{or, equivalently,} \qquad b = \frac{\pi}{2}.$$

The phase shift is $-c/b = -c/(\pi/2)$. Since c is to be positive, the phase shift must be *negative;* that is, the graph in Figure 12 must be obtained by shifting the graph of $y = 5 \sin [(\pi/2)x]$ to the *left.* Since we want c to be as small as possible, we choose the phase shift -1. Hence,

$$-\frac{c}{\pi/2} = -1 \qquad \text{or, equivalently,} \qquad c = \frac{\pi}{2}.$$

Thus, the desired equation is

$$y = 5 \sin \left(\frac{\pi}{2}x + \frac{\pi}{2}\right).$$

There are many other equations for the graph. For example, we could use the phase shifts $-5, -9, -13$, and so on, but these would not give us the *least* positive value for c. Two other equations for the graph are

$$y = 5 \sin\left(\frac{\pi}{2}x - \frac{3\pi}{2}\right) \quad \text{and} \quad y = -5 \sin\left(\frac{\pi}{2}x + \frac{3\pi}{2}\right).$$

However, neither of these equations satisfies the given criteria for a, b, and c, since in the first, $c < 0$, and in the second, $a < 0$ and c does not have its least positive value.

As an alternative solution, we could write

$$y = a \sin(bx + c) \quad \text{as} \quad y = a \sin\left[b\left(x + \frac{c}{b}\right)\right].$$

As before, we find $a = 5$ and $b = \pi/2$. Now since the graph has an x-intercept at $x = -1$, we can consider this graph to be a horizontal shift of the graph of $y = 5 \sin[(\pi/2)x]$ to the left by 1 unit—that is, replace x with $x + 1$. Thus, an equation is

$$y = 5 \sin\left[\frac{\pi}{2}(x + 1)\right], \quad \text{or} \quad y = 5 \sin\left(\frac{\pi}{2}x + \frac{\pi}{2}\right).$$

Many phenomena that occur in nature vary in a cyclic or rhythmic manner. It is sometimes possible to represent such behavior by means of trigonometric functions, as illustrated in the next two examples.

EXAMPLE 11 Analyzing the process of breathing

The rhythmic process of breathing consists of alternating periods of inhaling and exhaling. One complete cycle normally takes place every 5 seconds. If $F(t)$ denotes the air flow rate at time t (in liters per second) and if the maximum flow rate is 0.6 liter per second, find a formula of the form $F(t) = a \sin bt$ that fits this information.

SOLUTION If $F(t) = a \sin bt$ for some $b > 0$, then the period of F is $2\pi/b$. In this application the period is 5 seconds, and hence

$$\frac{2\pi}{b} = 5, \quad \text{or} \quad b = \frac{2\pi}{5}.$$

Since the maximum flow rate corresponds to the amplitude a of F, we let $a = 0.6$. This gives us the formula

$$F(t) = 0.6 \sin\left(\frac{2\pi}{5}t\right).$$

EXAMPLE 12 **Approximating the number of hours of daylight in a day**

The number of hours of daylight $D(t)$ at a particular time of the year can be approximated by

$$D(t) = \frac{K}{2} \sin\left[\frac{2\pi}{365}(t - 79)\right] + 12$$

for t in days and $t = 0$ corresponding to January 1. The constant K determines the total variation in day length and depends on the latitude of the locale.

(a) For Boston, $K \approx 6$. Sketch the graph of D for $0 \leq t \leq 365$.

(b) When is the day length the longest? the shortest?

SOLUTION

(a) If $K = 6$, then $K/2 = 3$, and we may write $D(t)$ in the form

$$D(t) = f(t) + 12,$$

where
$$f(t) = 3 \sin\left[\frac{2\pi}{365}(t - 79)\right].$$

We shall sketch the graph of f and then apply a vertical shift through a distance 12.

As in part (2) of the theorem on amplitudes, periods, and phase shifts, we can obtain a t-interval containing exactly one cycle by solving the following inequality:

$$0 \leq \frac{2\pi}{365}(t - 79) \leq 2\pi$$

$$0 \leq t - 79 \leq 365 \quad \text{multiply by } \frac{365}{2\pi}$$

$$79 \leq t \leq 444 \quad \text{add 79}$$

Figure 13

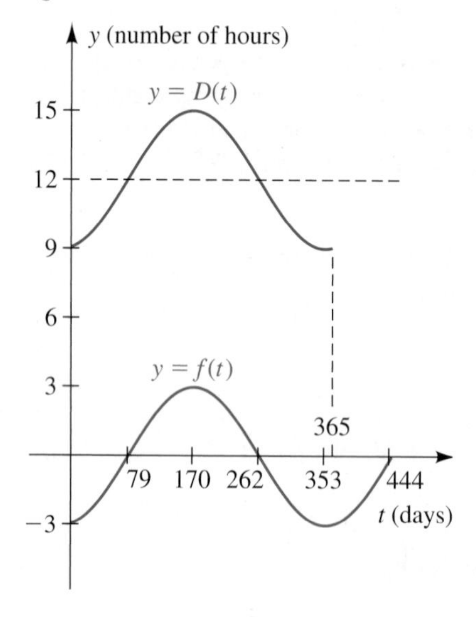

Hence, one sine wave occurs on the interval [79, 444]. Dividing this interval into four equal parts, we obtain the following table of values, which indicates the familiar sine wave pattern of amplitude 3.

t	79	170.25	261.5	352.75	444
$f(t)$	0	3	0	−3	0

If $t = 0$,

$$f(0) = 3 \sin\left[\frac{2\pi}{365}(-79)\right] \approx 3 \sin(-1.36) \approx -2.9.$$

Since the period of f is 365, this implies that $f(365) \approx -2.9$.

The graph of f for the interval [0, 444] is sketched in Figure 13, with different scales on the axes and t rounded off to the nearest day.

▶ **Tutorial available at www.thomsonedu.com/login**

Applying a vertical shift of 12 units gives us the graph of D for $0 \le t \le 365$ shown in Figure 13.

(b) The longest day—that is, the largest value of $D(t)$—occurs 170 days after January 1. Except for leap year, this corresponds to June 20. The shortest day occurs 353 days after January 1, or December 20.

In the next example we use a graphing utility to approximate the solution of an inequality that involves trigonometric expressions.

EXAMPLE 13 Approximating solutions of a trigonometric inequality

Approximate the solution of the inequality

$$\sin 3x < x + \sin x.$$

SOLUTION The given inequality is equivalent to

$$\sin 3x - x - \sin x < 0.$$

If we assign $\sin 3x - x - \sin x$ to Y_1, then the given problem is equivalent to finding where the graph of Y_1 is below the x-axis. Using the standard viewing rectangle gives us a sketch similar to Figure 14(a), where we see that the graph of Y_1 has an x-intercept c between -1 and 0. It appears that the graph is below the x-axis on the interval (c, ∞); however, this fact is not perfectly clear because of the small scale on the axes.

Figure 14

(a) $[-15, 15]$ by $[-10, 10]$

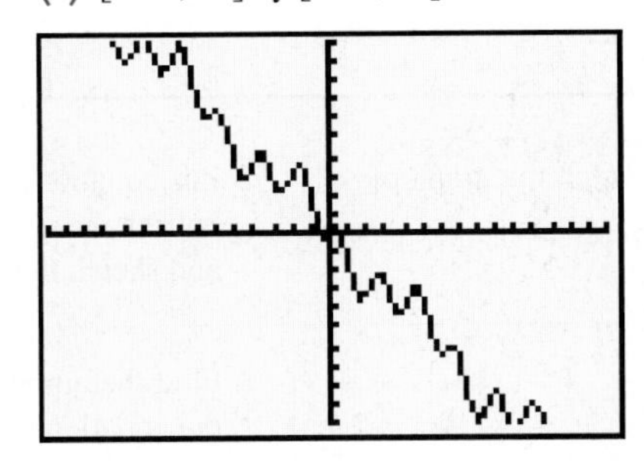

(b) $[-1.5, 1.5, 0.25]$ by $[-1, 1, 0.25]$

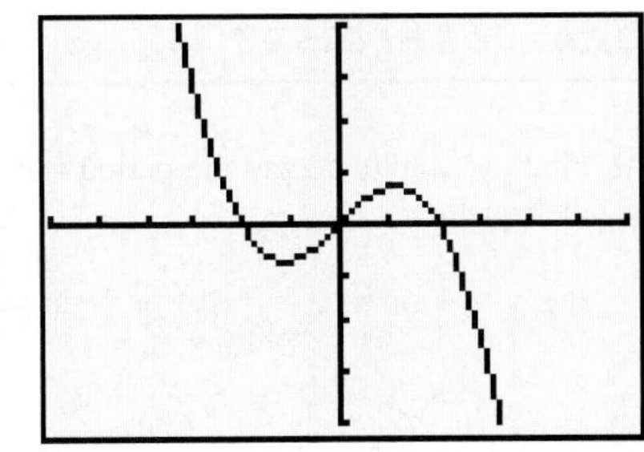

Using the viewing rectangle $[-1.5, 1.5, 0.25]$ by $[-1, 1, 0.25]$, we obtain Figure 14(b), where we see that the x-intercepts are approximately -0.5, 0, and 0.5. Using a root feature yields the more accurate positive value 0.51. Since the function involved is odd, the negative value is approximately -0.51. Hence, the solutions of the inequality are in the (approximate) intervals

$$(-0.51, 0) \cup (0.51, \infty).$$

EXAMPLE 14 **Investigating alternating current in an electrical circuit**

The current I (in amperes) in an alternating current circuit at time t (in seconds) is given by

$$I = 30 \sin\left(50\pi t - \frac{7\pi}{3}\right).$$

Approximate the smallest value of t for which $I = 15$.

SOLUTION Letting $I = 15$ in the given formula, we obtain

$$15 = 30 \sin\left(50\pi t - \frac{7\pi}{3}\right)$$

or, equivalently,

$$\sin\left(50\pi t - \frac{7\pi}{3}\right) - \frac{1}{2} = 0.$$

If we assign $\sin(50\pi x - 7\pi/3) - \frac{1}{2}$ to Y_1, then the given problem is equivalent to approximating the smallest x-intercept of the graph.

Since the period of Y_1 is

$$\frac{2\pi}{b} = \frac{2\pi}{50\pi} = \frac{1}{25} = 0.04$$

and since $-\frac{3}{2} \le Y_1 \le \frac{1}{2}$, we select the given viewing rectangle, obtaining a sketch similar to Figure 15. Using a root feature gives us $t \approx 0.01$ sec.

Figure 15
$[0, 0.04, 0.01]$ by $[-1.5, 0.5, 0.25]$

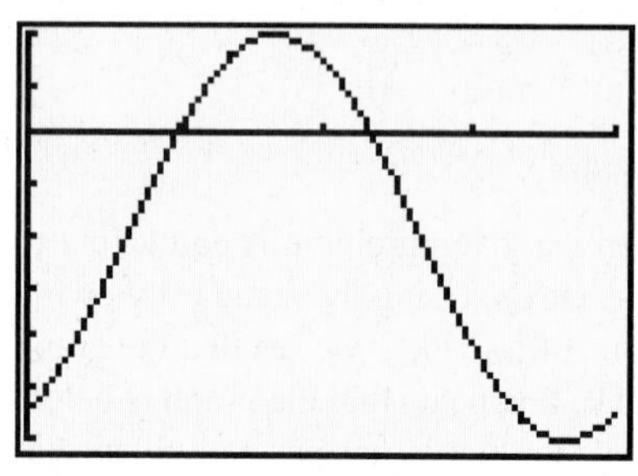

We will rework the preceding example in Section 6.2 and show how to find the exact value of t without the aid of a graphing utility.

5.5 *Exercises*

1 Find the amplitude and the period and sketch the graph of the equation:

(a) $y = 4 \sin x$ (b) $y = \sin 4x$

(c) $y = \frac{1}{4} \sin x$ (d) $y = \sin \frac{1}{4}x$

(e) $y = 2 \sin \frac{1}{4}x$ (f) $y = \frac{1}{2} \sin 4x$

(g) $y = -4 \sin x$ (h) $y = \sin(-4x)$

2 For equations analogous to those in (a)–(h) of Exercise 1 but involving the cosine, find the amplitude and the period and sketch the graph.

3 Find the amplitude and the period and sketch the graph of the equation:

(a) $y = 3 \cos x$ (b) $y = \cos 3x$

(c) $y = \frac{1}{3} \cos x$ (d) $y = \cos \frac{1}{3}x$

(e) $y = 2 \cos \frac{1}{3}x$ (f) $y = \frac{1}{2} \cos 3x$

(g) $y = -3 \cos x$ (h) $y = \cos(-3x)$

4 For equations analogous to those in (a)–(h) of Exercise 3 but involving the sine, find the amplitude and the period and sketch the graph.

Exer. 5–40: Find the amplitude, the period, and the phase shift and sketch the graph of the equation.

5 $y = \sin\left(x - \dfrac{\pi}{2}\right)$

6 $y = \sin\left(x + \dfrac{\pi}{4}\right)$

7 $y = 3\sin\left(x + \dfrac{\pi}{6}\right)$

8 $y = 2\sin\left(x - \dfrac{\pi}{3}\right)$

9 $y = \cos\left(x + \dfrac{\pi}{2}\right)$

▶ 10 $y = \cos\left(x - \dfrac{\pi}{3}\right)$

11 $y = 4\cos\left(x - \dfrac{\pi}{4}\right)$

12 $y = 3\cos\left(x + \dfrac{\pi}{6}\right)$

13 $y = \sin(2x - \pi) + 1$

14 $y = -\sin(3x + \pi) - 1$

15 $y = -\cos(3x + \pi) - 2$

▶ 16 $y = \cos(2x - \pi) + 2$

17 $y = -2\sin(3x - \pi)$

18 $y = 3\cos(3x - \pi)$

19 $y = \sin\left(\dfrac{1}{2}x - \dfrac{\pi}{3}\right)$

20 $y = \sin\left(\dfrac{1}{2}x + \dfrac{\pi}{4}\right)$

21 $y = 6\sin \pi x$

22 $y = 3\cos\dfrac{\pi}{2}x$

23 $y = 2\cos\dfrac{\pi}{2}x$

24 $y = 4\sin 3\pi x$

25 $y = \dfrac{1}{2}\sin 2\pi x$

26 $y = \dfrac{1}{2}\cos\dfrac{\pi}{2}x$

27 $y = 5\sin\left(3x - \dfrac{\pi}{2}\right)$

28 $y = -4\cos\left(2x + \dfrac{\pi}{3}\right)$

29 $y = 3\cos\left(\dfrac{1}{2}x - \dfrac{\pi}{4}\right)$

30 $y = -2\sin\left(\dfrac{1}{2}x + \dfrac{\pi}{2}\right)$

31 $y = -5\cos\left(\dfrac{1}{3}x + \dfrac{\pi}{6}\right)$

32 $y = 4\sin\left(\dfrac{1}{3}x - \dfrac{\pi}{3}\right)$

33 $y = 3\cos(\pi x + 4\pi)$

34 $y = -2\sin(2\pi x + \pi)$

35 $y = -\sqrt{2}\sin\left(\dfrac{\pi}{2}x - \dfrac{\pi}{4}\right)$

36 $y = \sqrt{3}\cos\left(\dfrac{\pi}{4}x - \dfrac{\pi}{2}\right)$

▶ 37 $y = -2\sin(2x - \pi) + 3$

38 $y = 3\cos(x + 3\pi) - 2$

39 $y = 5\cos(2x + 2\pi) + 2$

40 $y = -4\sin(3x - \pi) - 3$

Exer. 41–44: The graph of an equation is shown in the figure. (a) Find the amplitude, period, and phase shift. (b) Write the equation in the form $y = a\sin(bx + c)$ for $a > 0$, $b > 0$, and the least positive real number c.

41

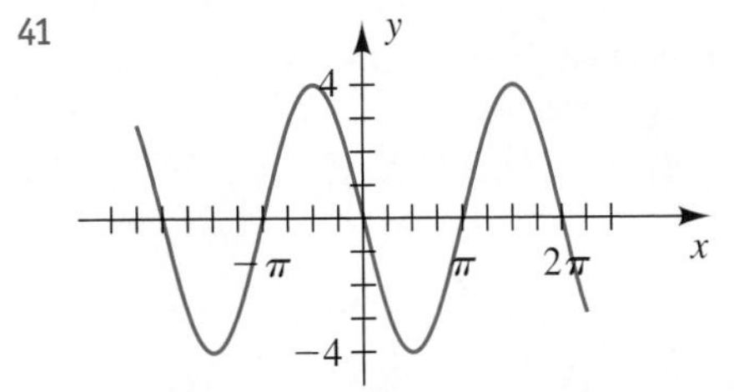

42

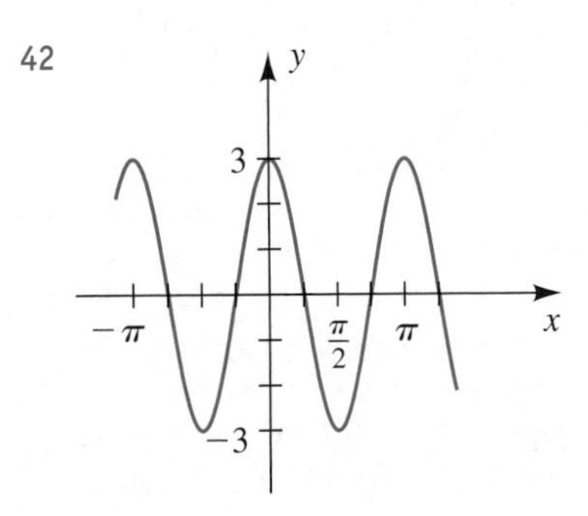

43

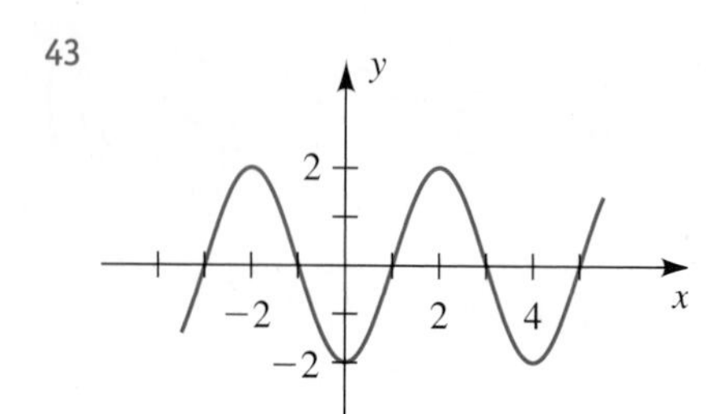

44

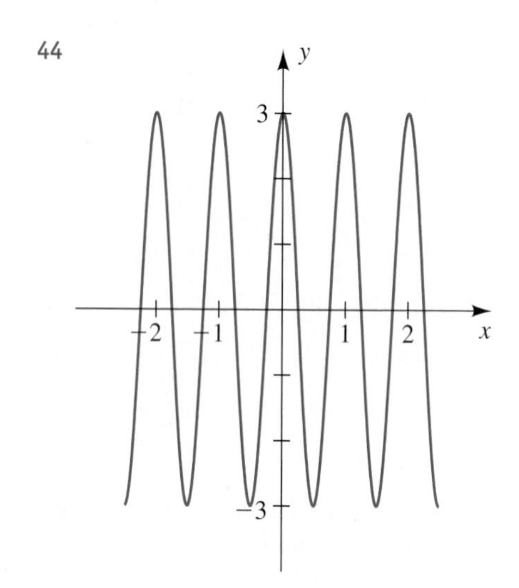

45 Electroencephalography Shown in the figure is an electroencephalogram of human brain waves during deep sleep. If we use $W = a \sin (bt + c)$ to represent these waves, what is the value of b?

Exercise 45

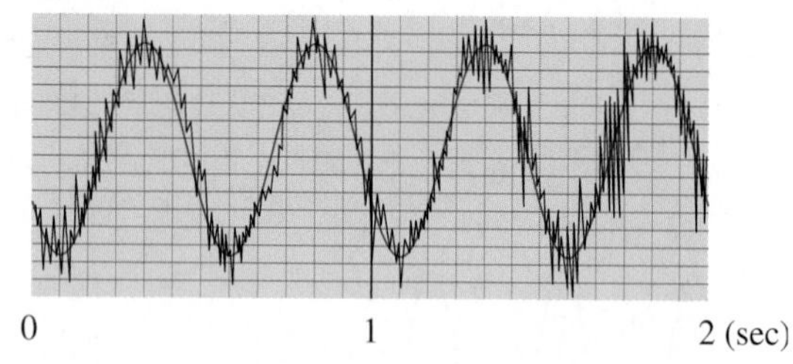

46 Intensity of daylight On a certain spring day with 12 hours of daylight, the light intensity I takes on its largest value of 510 calories/cm^2 at midday. If $t = 0$ corresponds to sunrise, find a formula $I = a \sin bt$ that fits this information.

47 Heart action The pumping action of the heart consists of the systolic phase, in which blood rushes from the left ventricle into the aorta, and the diastolic phase, during which the heart muscle relaxes. The function whose graph is shown in the figure is sometimes used to model one complete cycle of this process. For a particular individual, the systolic phase lasts $\frac{1}{4}$ second and has a maximum flow rate of 8 liters per minute. Find a and b.

Exercise 47

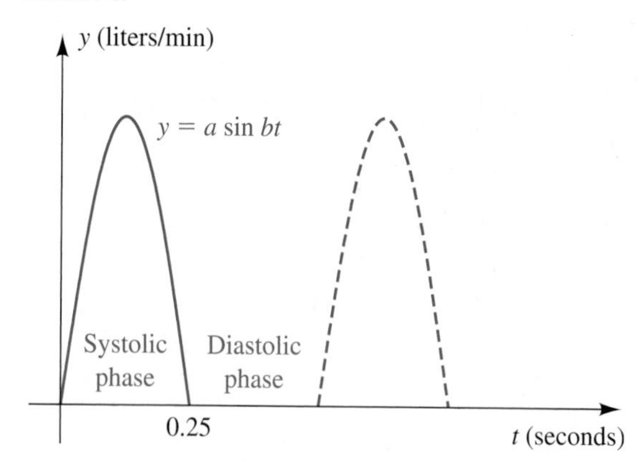

48 Biorhythms The popular biorhythm theory uses the graphs of three simple sine functions to make predictions about an individual's physical, emotional, and intellectual potential for a particular day. The graphs are given by $y = a \sin bt$ for t in days, with $t = 0$ corresponding to birth and $a = 1$ denoting 100% potential.

(a) Find the value of b for the physical cycle, which has a period of 23 days; for the emotional cycle (period 28 days); and for the intellectual cycle (period 33 days).

(b) Evaluate the biorhythm cycles for a person who has just become 21 years of age and is exactly 7670 days old.

49 Tidal components The height of the tide at a particular point on shore can be predicted by using seven trigonometric functions (called tidal components) of the form

$$f(t) = a \cos (bt + c).$$

The principal lunar component may be approximated by

$$f(t) = a \cos \left(\frac{\pi}{6} t - \frac{11\pi}{12} \right),$$

where t is in hours and $t = 0$ corresponds to midnight. Sketch the graph of f if $a = 0.5$ m.

50 Tidal components Refer to Exercise 49. The principal solar diurnal component may be approximated by

$$f(t) = a \cos\left(\frac{\pi}{12}t - \frac{7\pi}{12}\right).$$

Sketch the graph of f if $a = 0.2$ m.

51 Hours of daylight in Fairbanks If the formula for $D(t)$ in Example 12 is used for Fairbanks, Alaska, then $K \approx 12$. Sketch the graph of D in this case for $0 \leq t \leq 365$.

52 Low temperature in Fairbanks Based on years of weather data, the expected low temperature T (in °F) in Fairbanks, Alaska, can be approximated by

$$T = 36 \sin\left[\frac{2\pi}{365}(t - 101)\right] + 14,$$

where t is in days and $t = 0$ corresponds to January 1.

(a) Sketch the graph of T for $0 \leq t \leq 365$.

(b) Predict when the coldest day of the year will occur.

Exer. 53–54: Graph the equation $y = f(t)$ on the interval [0, 24]. Let y represent the outdoor temperature (in °F) at time t (in hours), where $t = 0$ corresponds to 9 A.M. Describe the temperature during the 24-hour interval.

53 $y = 20 + 15 \sin\frac{\pi}{12}t$

54 $y = 80 + 22 \cos\left[\frac{\pi}{12}(t - 3)\right]$

Exer. 55–58: Scientists sometimes use the formula

$$f(t) = a \sin(bt + c) + d$$

to simulate temperature variations during the day, with time t in hours, temperature $f(t)$ in °C, and $t = 0$ corresponding to midnight. Assume that $f(t)$ is decreasing at midnight.

(a) Determine values of a, b, c, and d that fit the information.

(b) Sketch the graph of f for $0 \leq t \leq 24$.

55 The high temperature is 10°C, and the low temperature of −10°C occurs at 4 A.M.

56 The temperature at midnight is 15°C, and the high and low temperatures are 20°C and 10°C.

57 The temperature varies between 10°C and 30°C, and the average temperature of 20°C first occurs at 9 A.M.

58 The high temperature of 28°C occurs at 2 P.M., and the average temperature of 20°C occurs 6 hours later.

59 Precipitation at South Lake Tahoe The average monthly precipitation P (in inches) at South Lake Tahoe, California, is listed in the table.

Month	P	Month	P	Month	P
Jan.	6.1	May	1.2	Sept.	0.5
Feb.	5.4	June	0.6	Oct.	2.8
March	3.9	July	0.3	Nov.	3.1
April	2.2	Aug.	0.2	Dec.	5.4

(a) Let t be time in months, with $t = 1$ corresponding to January, $t = 2$ to February, ... , $t = 12$ to December, $t = 13$ to January, and so on. Plot the data points for a two-year period.

(b) Find a function $P(t) = a \sin(bt + c) + d$ that approximates the average monthly precipitation. Plot the data and the function P on the same coordinate axes.

60 Thames River depth When a river flows into an ocean, the depth of the river varies near its mouth as a result of tides. Information about this change in depth is critical for safety. The following table gives the depth D (in feet) of the Thames River in London for a 24-hour period.

Time	D	Time	D	Time	D
12 A.M.	27.1	8 A.M.	20.0	4 P.M.	34.0
1 A.M.	30.1	9 A.M.	18.0	5 P.M.	32.4
2 A.M.	33.0	10 A.M.	18.3	6 P.M.	29.1
3 A.M.	34.3	11 A.M.	20.6	7 P.M.	25.2
4 A.M.	33.7	12 P.M.	24.2	8 P.M.	21.9
5 A.M.	31.1	1 P.M.	28.1	9 P.M.	19.6
6 A.M.	27.1	2 P.M.	31.7	10 P.M.	18.6
7 A.M.	23.2	3 P.M.	33.7	11 P.M.	19.6

(a) Plot the data, with time on the horizontal axis and depth on the vertical axis. Let $t = 0$ correspond to 12:00 A.M.

(b) Determine a function $D(t) = a \sin(bt + c) + d$, where $D(t)$ represents the depth of the water in the harbor at time t. Graph the function D with the data. (*Hint:* To determine b, find the time between maximum depths.)

(c) If a ship requires at least 24 feet of water to navigate the Thames safely, graphically determine the time interval(s) when navigation is *not* safe.

61 **Hours of daylight** The number of daylight hours D at a particular location varies with both the month and the latitude. The table lists the number of daylight hours on the first day of each month at 60°N latitude.

Month	D	Month	D	Month	D
Jan.	6.03	May	15.97	Sept.	14.18
Feb.	7.97	June	18.28	Oct.	11.50
March	10.43	July	18.72	Nov.	8.73
April	13.27	Aug.	16.88	Dec.	5.88

(a) Let t be time in months, with $t = 1$ corresponding to January, $t = 2$ to February, . . . , $t = 12$ to December, $t = 13$ to January, and so on. Plot the data for a two-year period.

(b) Find a function $D(t) = a \sin(bt + c) + d$ that approximates the number of daylight hours. Graph the function D with the data.

62 **Hours of daylight** Refer to Exercise 61. The maximum number of daylight hours at 40°N is 15.02 hours and occurs on June 21. The minimum number of daylight hours is 9.32 hours and occurs on December 22.

(a) Determine a function $D(t) = a \sin(bt + c) + d$ that models the number of daylight hours, where t is in months and $t = 1$ corresponds to January 1.

(b) Graph the function D using the viewing rectangle $[0.5, 24.5, 4]$ by $[0, 20, 4]$.

(c) Predict the number of daylight hours on February 1 and September 1. Compare your answers to the true values of 10.17 and 13.08 hours, respectively.

Exer. 63–66: Graph the equation on the interval $[-2, 2]$, and describe the behavior of y as $x \to 0^-$ and as $x \to 0^+$.

63 $y = \sin \dfrac{1}{x}$

64 $y = |x| \sin \dfrac{1}{x}$

65 $y = \dfrac{\sin 2x}{x}$

66 $y = \dfrac{1 - \cos 3x}{x}$

Exer. 67–68: Graph the equation on the interval $[-20, 20]$, and estimate the horizontal asymptote.

67 $y = x^2 \sin^2 \left(\dfrac{2}{x}\right)$

68 $y = \dfrac{1 - \cos^2 (2/x)}{\sin (1/x)}$

Exer. 69–70: Use a graph to solve the inequality on the interval $[-\pi, \pi]$.

69 $\cos 3x \geq \frac{1}{2}x - \sin x$

70 $\frac{1}{4} \tan\left(\frac{1}{3}x^2\right) < \frac{1}{2} \cos 2x + \frac{1}{5}x^2$

5.6 Additional Trigonometric Graphs

Methods we developed in Section 5.5 for the sine and cosine can be applied to the other four trigonometric functions; however, there are several differences. Since the tangent, cotangent, secant, and cosecant functions have no largest values, the notion of amplitude has no meaning. Moreover, we do not refer to cycles. For some tangent and cotangent graphs, we begin by sketching the portion between successive vertical asymptotes and then repeat that pattern to the right and to the left.

The graph of $y = a \tan x$ for $a > 0$ can be obtained by stretching or compressing the graph of $y = \tan x$. If $a < 0$, then we also use a reflection about the x-axis. Since the tangent function has period π, it is sufficient to sketch the branch between the two successive vertical asymptotes $x = -\pi/2$ and $x = \pi/2$. The same pattern occurs to the right and to the left, as in the next example.

EXAMPLE 1 Sketching the graph of an equation involving tan x

Sketch the graph of the equation:

(a) $y = 2 \tan x$ **(b)** $y = \frac{1}{2} \tan x$

SOLUTION We begin by sketching the graph of one branch of $y = \tan x$, as shown in red in Figures 1 and 2, between the vertical asymptotes $x = -\pi/2$ and $x = \pi/2$.

(a) For $y = 2 \tan x$, we multiply the y-coordinate of each point by 2 and then extend the resulting branch to the right and left, as shown in Figure 1.

Figure 1 $y = 2 \tan x$

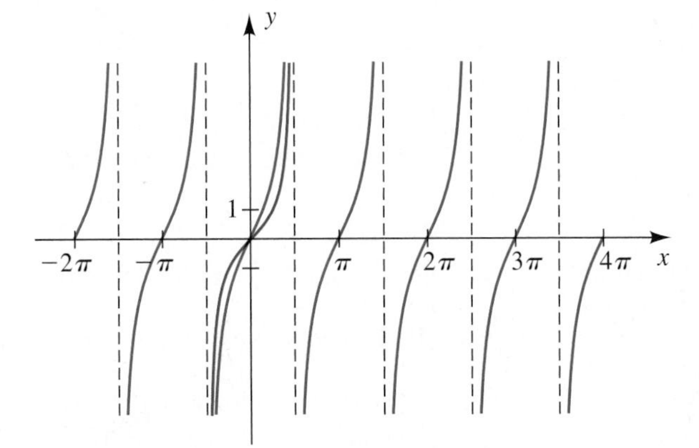

(b) For $y = \frac{1}{2} \tan x$, we multiply the y-coordinates by $\frac{1}{2}$, obtaining the sketch in Figure 2 on the next page.

Figure 2 $y = \frac{1}{2}\tan x$

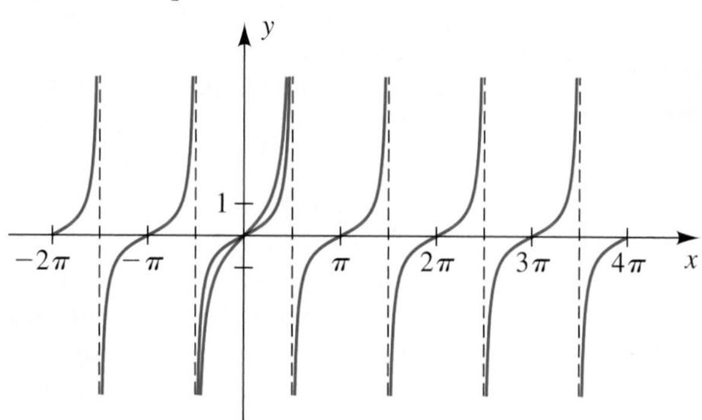

The method used in Example 1 can be applied to other functions. Thus, to sketch the graph of $y = 3 \sec x$, we could first sketch the graph of one branch of $y = \sec x$ and then multiply the y-coordinate of each point by 3.

The figure shown below is a typical graphing calculator graph of $y = \tan x$. It appears that the calculator has included the asymptotes, but the vertical lines actually result from the calculator's effort to connect consecutive pixels.

$[-\pi, \pi, \pi/4]$ by $[-2.1, 2.1]$

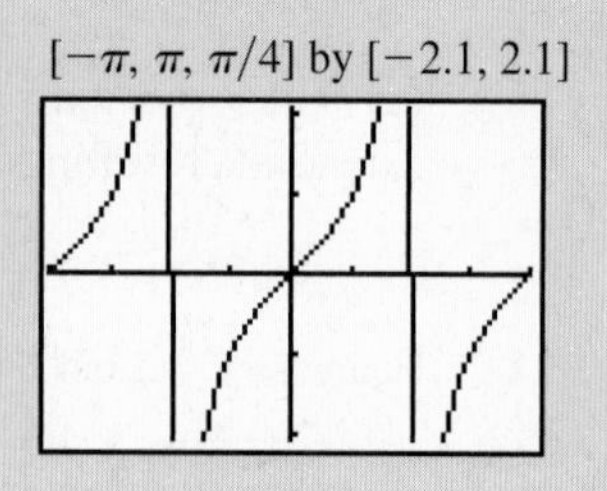

The next theorem is an analogue of the theorem on amplitudes, periods, and phase shifts stated in Section 5.5 for the sine and cosine functions.

Theorem on the Graph of $y = a \tan(bx + c)$

If $y = a \tan(bx + c)$ for nonzero real numbers a and b, then

(1) the period is $\dfrac{\pi}{|b|}$ and the phase shift is $-\dfrac{c}{b}$;

(2) successive vertical asymptotes for the graph of one branch may be found by solving the inequality

$$-\frac{\pi}{2} < bx + c < \frac{\pi}{2}.$$

EXAMPLE 2 **Sketching the graph of an equation of the form $y = a \tan(bx + c)$**

Find the period and sketch the graph of $y = \frac{1}{2}\tan\left(x + \frac{\pi}{4}\right)$.

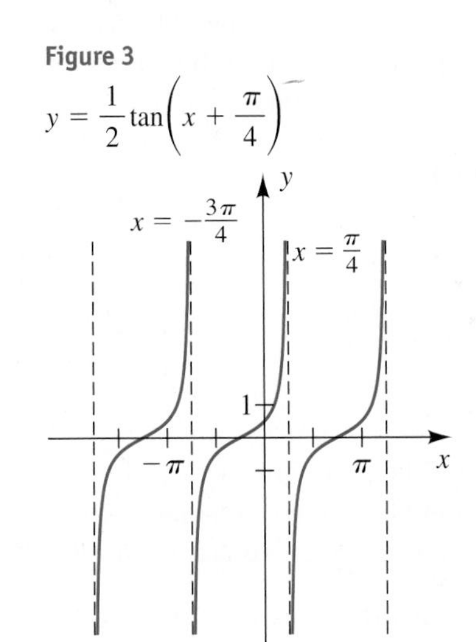

Figure 3
$y = \frac{1}{2}\tan\left(x + \frac{\pi}{4}\right)$

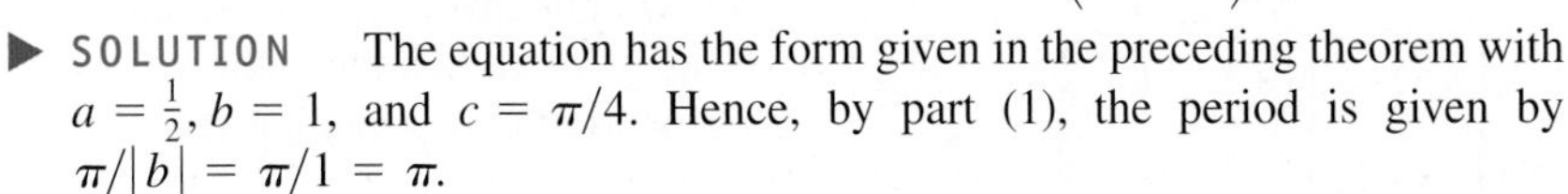

▶ **SOLUTION** The equation has the form given in the preceding theorem with $a = \frac{1}{2}$, $b = 1$, and $c = \pi/4$. Hence, by part (1), the period is given by $\pi/|b| = \pi/1 = \pi$.

As in part (2), to find successive vertical asymptotes we solve the following inequality:

$$-\frac{\pi}{2} \le x + \frac{\pi}{4} \le \frac{\pi}{2}$$

$$-\frac{3\pi}{4} \le x \qquad \le \frac{\pi}{4} \qquad \text{subtract } \frac{\pi}{4}$$

Because $a = \frac{1}{2}$, the graph of the equation on the interval $[-3\pi/4, \pi/4]$ has the shape of the graph of $y = \frac{1}{2}\tan x$ (see Figure 2). Sketching that branch and extending it to the right and left gives us Figure 3.

Note that since $c = \pi/4$ and $b = 1$, the phase shift is $-c/b = -\pi/4$. Hence, the graph can also be obtained by shifting the graph of $y = \frac{1}{2}\tan x$ in Figure 2 to the left a distance $\pi/4$.

If $y = a\cot(bx + c)$, we have a situation similar to that stated in the previous theorem. The only difference is part (2). Since successive vertical asymptotes for the graph of $y = \cot x$ are $x = 0$ and $x = \pi$ (see Figure 19 in Section 5.3), we obtain successive vertical asymptotes for the graph of one branch of $y = a\cot(bx + c)$ by solving the inequality

$$0 < bx + c < \pi.$$

EXAMPLE 3 **Sketching the graph of an equation of the form $y = a \cot(bx + c)$**

Find the period and sketch the graph of $y = \cot\left(2x - \frac{\pi}{2}\right)$.

▶ **SOLUTION** Using the usual notation, we see that $a = 1$, $b = 2$, and $c = -\pi/2$. The period is $\pi/|b| = \pi/2$. Hence, the graph repeats itself in intervals of length $\pi/2$.

As in the discussion preceding this example, to find two successive vertical asymptotes for the graph of one branch we solve the inequality:

$$0 \le 2x - \frac{\pi}{2} \le \pi$$

$$\frac{\pi}{2} \le 2x \qquad \le \frac{3\pi}{2} \qquad \text{add } \frac{\pi}{2}$$

$$\frac{\pi}{4} \le x \qquad \le \frac{3\pi}{4} \qquad \text{divide by 2}$$

(continued)

▶ **Tutorial available at www.thomsonedu.com/login**

Figure 4 $y = \cot\left(2x - \frac{\pi}{2}\right)$

Since a is positive, we sketch a cotangent-shaped branch on the interval $[\pi/4, 3\pi/4]$ and then repeat it to the right and left in intervals of length $\pi/2$, as shown in Figure 4.

Graphs involving the secant and cosecant functions can be obtained by using methods similar to those for the tangent and cotangent or by taking reciprocals of corresponding graphs of the cosine and sine functions.

EXAMPLE 4 Sketching the graph of an equation of the form $y = a \sec(bx + c)$

Sketch the graph of the equation:

(a) $y = \sec\left(x - \frac{\pi}{4}\right)$ **(b)** $y = 2\sec\left(x - \frac{\pi}{4}\right)$

SOLUTION

(a) The graph of $y = \sec x$ is sketched (without asymptotes) in red in Figure 5. The graph of $y = \cos x$ is sketched in black; notice that the asymptotes of $y = \sec x$ correspond to the zeros of $y = \cos x$. We can obtain the graph of $y = \sec\left(x - \frac{\pi}{4}\right)$ by shifting the graph of $y = \sec x$ to the right a distance $\pi/4$, as shown in blue in Figure 5.

(b) We can sketch this graph by multiplying the y-coordinates of the graph in part (a) by 2. This gives us Figure 6.

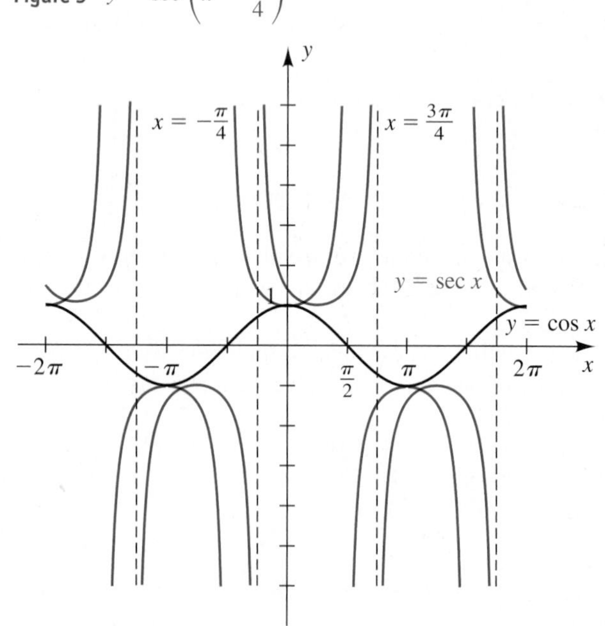

Figure 5 $y = \sec\left(x - \frac{\pi}{4}\right)$

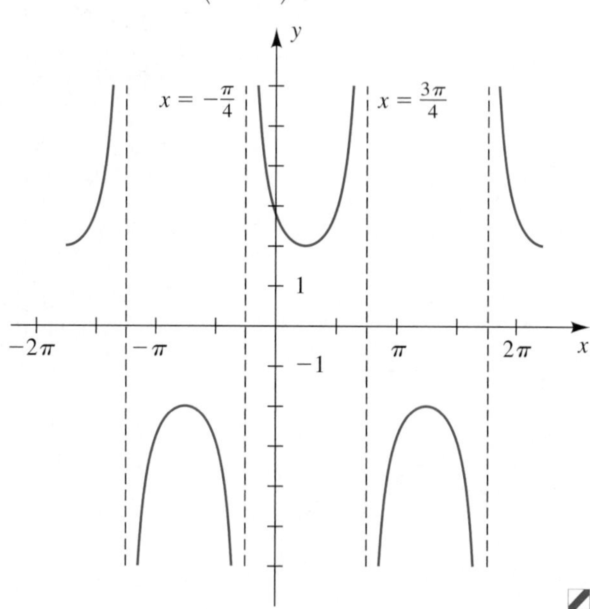

Figure 6 $y = 2\sec\left(x - \frac{\pi}{4}\right)$

Figure 7
$y = \csc(2x + \pi)$

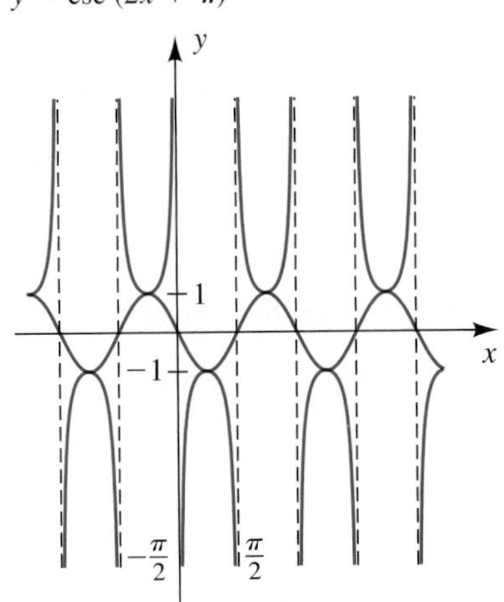

EXAMPLE 5 **Sketching the graph of an equation of the form $y = a \csc(bx + c)$**

Sketch the graph of $y = \csc(2x + \pi)$.

▶ **SOLUTION** Since $\csc\theta = 1/\sin\theta$, we may write the given equation as

$$y = \frac{1}{\sin(2x + \pi)}.$$

Thus, we may obtain the graph of $y = \csc(2x + \pi)$ by finding the graph of $y = \sin(2x + \pi)$ and then taking the reciprocal of the y-coordinate of each point. Using $a = 1$, $b = 2$, and $c = \pi$, we see that the amplitude of $y = \sin(2x + \pi)$ is 1 and the period is $2\pi/|b| = 2\pi/2 = \pi$. To find an interval containing one cycle, we solve the inequality

$$\begin{aligned} 0 \le 2x + \pi &\le 2\pi \\ -\pi \le 2x \quad &\le \pi \\ -\frac{\pi}{2} \le x \quad &\le \frac{\pi}{2}. \end{aligned}$$

This leads to the graph in red in Figure 7. Taking reciprocals gives us the graph of $y = \csc(2x + \pi)$ shown in blue in the figure. Note that the zeros of the sine curve correspond to the asymptotes of the cosecant graph.

Figure 8
(a)

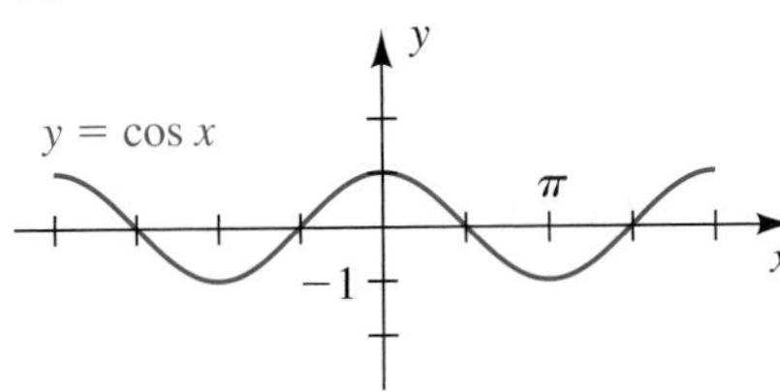

The next example involves the absolute value of a trigonometric function.

EXAMPLE 6 **Sketching the graph of an equation involving an absolute value**

Sketch the graph of $y = |\cos x| + 1$.

▶ **SOLUTION** We shall sketch the graph in three stages. First, we sketch the graph of $y = \cos x$, as in Figure 8(a).

Next, we obtain the graph of $y = |\cos x|$ by reflecting the negative y-coordinates in Figure 8(a) through the x-axis. This gives us Figure 8(b).

Finally, we vertically shift the graph in (b) upward 1 unit to obtain Figure 8(c).

We have used three separate graphs for clarity. In practice, we could sketch the graphs successively on one coordinate plane.

(b)

(c)

Mathematical applications often involve a function f that is a sum of two or more other functions. To illustrate, suppose

$$f(x) = g(x) + h(x),$$

where f, g, and h have the same domain D. Before graphing utilities were invented, a technique known as **addition of y-coordinates** was sometimes used

▶ **Tutorial available at www.thomsonedu.com/login**

Figure 9

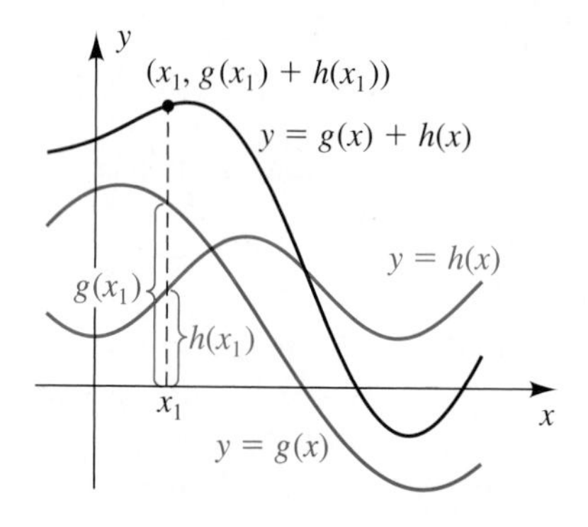

to sketch the graph of f. The method is illustrated in Figure 9, where for each x_1, the y-coordinate $f(x_1)$ of a point on the graph of f is the *sum* $g(x_1) + h(x_1)$ of the y-coordinates of points on the graphs of g and h. The graph of f is obtained by *graphically adding* a sufficient number of such y-coordinates, a task best left to a graphing utility.

It is sometimes useful to compare the graph of a sum of functions with the individual functions, as illustrated in the next example.

EXAMPLE 7 Sketching the graph of a sum of two trigonometric functions

Sketch the graph of $y = \cos x$, $y = \sin x$, and $y = \cos x + \sin x$ on the same coordinate plane for $0 \leq x \leq 3\pi$.

▶ **SOLUTION** We make the following assignments:

$$Y_1 = \cos x, \qquad Y_2 = \sin x, \qquad \text{and} \qquad Y_3 = Y_1 + Y_2$$

Figure 10
(a) $[0, 3\pi, \pi/4]$ by $[-\pi, \pi]$

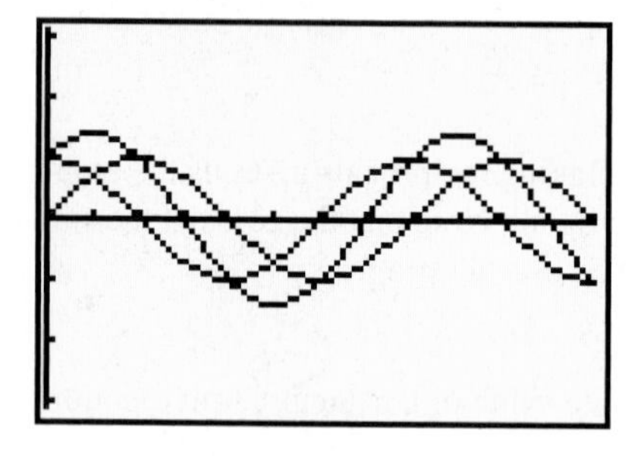

(b) $[0, 3\pi, \pi/4]$ by $[-1.5, 1.5]$

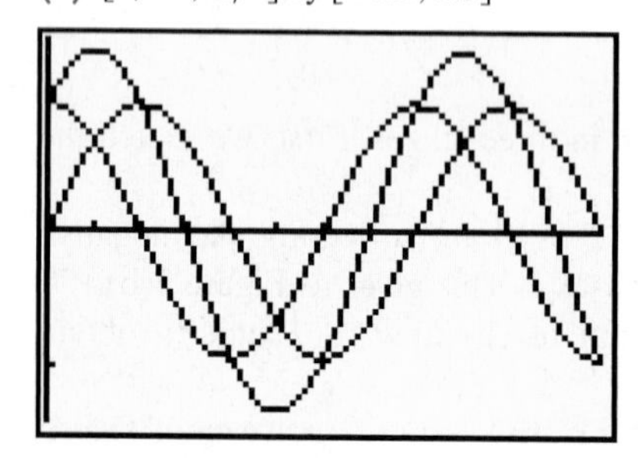

Since we desire a 3:2 (horizontal:vertical) screen proportion, we choose the viewing rectangle $[0, 3\pi, \pi/4]$ by $[-\pi, \pi]$, obtaining Figure 10(a). The clarity of the graph can be enhanced by changing the viewing rectangle to $[0, 3\pi, \pi/4]$ by $[-1.5, 1.5]$, as in Figure 10(b).

Note that the graph of Y_3 intersects the graph of Y_1 when $Y_2 = 0$, and the graph of Y_2 when $Y_1 = 0$. The x-intercepts for Y_3 correspond to the solutions of $Y_2 = -Y_1$. Finally, we see that the maximum and minimum values of Y_3 occur when $Y_1 = Y_2$ (that is, when $x = \pi/4, 5\pi/4$, and $9\pi/4$). These y-values are

$$\sqrt{2}/2 + \sqrt{2}/2 = \sqrt{2} \qquad \text{and} \qquad -\sqrt{2}/2 + \left(-\sqrt{2}/2\right) = -\sqrt{2}.$$

The graph of an equation of the form

$$y = f(x) \sin (ax + b) \qquad \text{or} \qquad y = f(x) \cos (ax + b),$$

where f is a function and a and b are real numbers, is called a **damped sine wave** or **damped cosine wave,** respectively, and $f(x)$ is called the **damping factor.** The next example illustrates a method for graphing such equations.

EXAMPLE 8 Sketching the graph of a damped sine wave

Sketch the graph of f if $f(x) = 2^{-x} \sin x$.

▶ **SOLUTION** We first examine the absolute value of f:

$$\begin{aligned}
|f(x)| &= |2^{-x} \sin x| && \text{absolute value of both sides} \\
&= |2^{-x}||\sin x| && |ab| = |a||b| \\
&\leq |2^{-x}| \cdot 1 && |\sin x| \leq 1 \\
|f(x)| &\leq 2^{-x} && |2^{-x}| = 2^{-x} \text{ since } 2^{-x} > 0 \\
-2^{-x} \leq f(x) &\leq 2^{-x} && |x| \leq a \Longleftrightarrow -a \leq x \leq a
\end{aligned}$$

▶ **Tutorial available at www.thomsonedu.com/login**

Figure 11

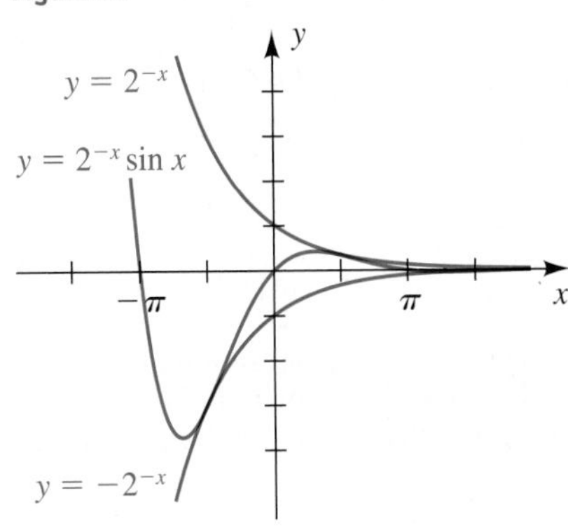

The last inequality implies that the graph of f lies between the graphs of the equations $y = -2^{-x}$ and $y = 2^{-x}$. The graph of f will coincide with one of these graphs if $|\sin x| = 1$—that is, if $x = (\pi/2) + \pi n$ for some integer n.

Since $2^{-x} > 0$, the x-intercepts on the graph of f occur at $\sin x = 0$—that is, at $x = \pi n$. Because there are an infinite number of x-intercepts, this is an example of a function that intersects its horizontal asymptote an infinite number of times. With this information, we obtain the sketch shown in Figure 11.

The damping factor in Example 8 is 2^{-x}. By using different damping factors, we can obtain other compressed or expanded variations of sine waves. The analysis of such graphs is important in physics and engineering.

5.6 *Exercises*

Exer. 1–52: Find the period and sketch the graph of the equation. Show the asymptotes.

1 $y = 4 \tan x$

2 $y = \frac{1}{4} \tan x$

3 $y = 3 \cot x$

4 $y = \frac{1}{3} \cot x$

5 $y = 2 \csc x$

6 $y = \frac{1}{2} \csc x$

7 $y = 3 \sec x$

8 $y = \frac{1}{4} \sec x$

9 $y = \tan\left(x - \frac{\pi}{4}\right)$

10 $y = \tan\left(x + \frac{\pi}{2}\right)$

▶ 11 $y = \tan 2x$

12 $y = \tan \frac{1}{2}x$

13 $y = \tan \frac{1}{4}x$

14 $y = \tan 4x$

15 $y = 2 \tan\left(2x + \frac{\pi}{2}\right)$

▶ 16 $y = \frac{1}{3} \tan\left(2x - \frac{\pi}{4}\right)$

17 $y = -\frac{1}{4} \tan\left(\frac{1}{2}x + \frac{\pi}{3}\right)$

18 $y = -3 \tan\left(\frac{1}{3}x - \frac{\pi}{3}\right)$

19 $y = \cot\left(x - \frac{\pi}{2}\right)$

20 $y = \cot\left(x + \frac{\pi}{4}\right)$

21 $y = \cot 2x$

22 $y = \cot \frac{1}{2}x$

23 $y = \cot \frac{1}{3}x$

24 $y = \cot 3x$

25 $y = 2 \cot\left(2x + \frac{\pi}{2}\right)$

26 $y = -\frac{1}{3} \cot (3x - \pi)$

27 $y = -\frac{1}{2} \cot\left(\frac{1}{2}x + \frac{\pi}{4}\right)$

▶ 28 $y = 4 \cot\left(\frac{1}{3}x - \frac{\pi}{6}\right)$

▶ Tutorial available at www.thomsonedu.com/login

29 $y = \sec\left(x - \frac{\pi}{2}\right)$

30 $y = \sec\left(x - \frac{3\pi}{4}\right)$

31 $y = \sec 2x$

32 $y = \sec \frac{1}{2}x$

33 $y = \sec \frac{1}{3}x$

34 $y = \sec 3x$

35 $y = 2 \sec\left(2x - \frac{\pi}{2}\right)$

36 $y = \frac{1}{2} \sec\left(2x - \frac{\pi}{2}\right)$

37 $y = -\frac{1}{3} \sec\left(\frac{1}{2}x + \frac{\pi}{4}\right)$

▶ 38 $y = -3 \sec\left(\frac{1}{3}x + \frac{\pi}{3}\right)$

39 $y = \csc\left(x - \frac{\pi}{2}\right)$

40 $y = \csc\left(x + \frac{3\pi}{4}\right)$

41 $y = \csc 2x$

42 $y = \csc \frac{1}{2}x$

43 $y = \csc \frac{1}{3}x$

44 $y = \csc 3x$

45 $y = 2 \csc\left(2x + \frac{\pi}{2}\right)$

46 $y = -\frac{1}{2} \csc (2x - \pi)$

47 $y = -\frac{1}{4} \csc\left(\frac{1}{2}x + \frac{\pi}{2}\right)$

48 $y = 4 \csc\left(\frac{1}{2}x - \frac{\pi}{4}\right)$

49 $y = \tan \frac{\pi}{2}x$

50 $y = \cot \pi x$

▶ 51 $y = \csc 2\pi x$

52 $y = \sec \frac{\pi}{8}x$

53 Find an equation using the cotangent function that has the same graph as $y = \tan x$.

54 Find an equation using the cosecant function that has the same graph as $y = \sec x$.

Exer. 55–60: Use the graph of a trigonometric function to aid in sketching the graph of the equation without plotting points.

55 $y = |\sin x|$

56 $y = |\cos x|$

57 $y = |\sin x| + 2$

58 $y = |\cos x| - 3$

59 $y = -|\cos x| + 1$

60 $y = -|\sin x| - 2$

Exer. 61–66: Sketch the graph of the equation.

61 $y = x + \cos x$

62 $y = x - \sin x$

63 $y = 2^{-x} \cos x$

64 $y = e^{x} \sin x$

65 $y = |x| \sin x$

66 $y = |x| \cos x$

Exer. 67–72: Graph the function f in the viewing rectangle $[-2\pi, 2\pi, \pi/2]$ by $[-4, 4]$. Use the graph of f to predict the graph of g. Verify your prediction by graphing g in the same viewing rectangle.

67 $f(x) = \tan 0.5x$; $\quad g(x) = \tan\left[0.5\left(x + \frac{\pi}{2}\right)\right]$

68 $f(x) = 0.5 \csc 0.5x$; $\quad g(x) = 0.5 \csc 0.5x - 2$

69 $f(x) = 0.5 \sec 0.5x$; $\quad g(x) = 0.5 \sec\left[0.5\left(x - \frac{\pi}{2}\right)\right] - 1$

70 $f(x) = \tan x - 1$; $\quad g(x) = -\tan x + 1$

71 $f(x) = 3 \cos 2x$; $\quad g(x) = |3 \cos 2x| - 1$

72 $f(x) = 1.2^{-x} \cos x$; $\quad g(x) = 1.2^{x} \cos x$

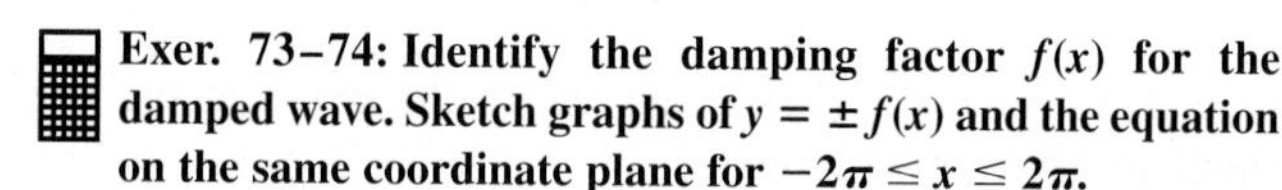

Exer. 73–74: Identify the damping factor $f(x)$ for the damped wave. Sketch graphs of $y = \pm f(x)$ and the equation on the same coordinate plane for $-2\pi \le x \le 2\pi$.

73 $y = e^{-x/4} \sin 4x$

74 $y = 3^{-x/5} \cos 2x$

Exer. 75–76: Graph the function f on $[-\pi, \pi]$, and estimate the high and low points.

75 $f(x) = \cos 2x + 2 \sin 4x - \sin x$

76 $f(x) = \tan \frac{1}{4}x - 2 \sin 2x$

Exer. 77–78: Use a graph to estimate the largest interval $[a, b]$, with $a < 0$ and $b > 0$, on which f is one-to-one.

77 $f(x) = \sin (2x + 2) \cos (1.5x - 1)$

78 $f(x) = 1.5 \cos\left(\frac{1}{2}x - 0.3\right) + \sin (1.5x + 0.5)$

Exer. 79–80: Use a graph to solve the inequality on the interval $[-\pi, \pi]$.

79 $\cos (2x - 1) + \sin 3x \ge \sin \frac{1}{3}x + \cos x$

80 $\frac{1}{2} \cos 2x + 2 \cos (x - 2) < 2 \cos (1.5x + 1) + \sin (x - 1)$

▶ **Tutorial available at www.thomsonedu.com/login**

81 **Radio signal intensity** Radio stations often have more than one broadcasting tower because federal guidelines do not usually permit a radio station to broadcast its signal in all directions with equal power. Since radio waves can travel over long distances, it is important to control their directional patterns so that radio stations do not interfere with one another. Suppose that a radio station has two broadcasting towers located along a north-south line, as shown in the figure. If the radio station is broadcasting at a wavelength λ and the distance between the two radio towers is equal to $\frac{1}{2}\lambda$, then the intensity I of the signal in the direction θ is given by

$$I = \tfrac{1}{2}I_0[1 + \cos(\pi \sin \theta)],$$

where I_0 is the maximum intensity. Approximate I in terms of I_0 for each θ.

(a) $\theta = 0$ (b) $\theta = \pi/3$ (c) $\theta = \pi/7$

Exercise 81

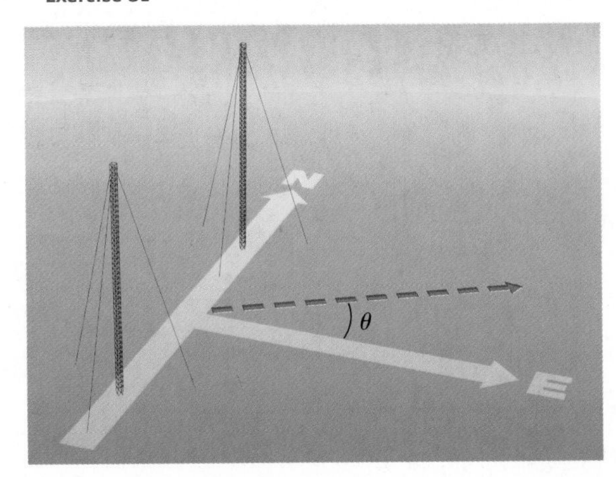

82 **Radio signal intensity** Refer to Exercise 81.

(a) Determine the directions in which I has maximum or minimum values.

(b) Graph I on the interval $[0, 2\pi)$. Graphically approximate θ to three decimal places, when I is equal to $\frac{1}{3}I_0$. (*Hint:* Let $I_0 = 1$.)

83 **Earth's magnetic field** The strength of Earth's magnetic field varies with the depth below the surface. The strength at depth z and time t can sometimes be approximated using the damped sine wave

$$S = A_0 e^{-\alpha z} \sin(kt - \alpha z),$$

where A_0, α, and k are constants.

(a) What is the damping factor?

(b) Find the phase shift at depth z_0.

(c) At what depth is the amplitude of the wave one-half the amplitude of the surface strength?

5.7 Applied Problems

Trigonometry was developed to help solve problems involving angles and lengths of sides of triangles. Problems of that type are no longer the most important applications; however, questions about triangles still arise in physical situations. When considering such questions in this section, we shall restrict our discussion to right triangles. Triangles that do not contain a right angle will be considered in Chapter 7.

We shall often use the following notation. The vertices of a triangle will be denoted by A, B, and C; the angles at A, B, and C will be denoted by α, β, and γ, respectively; and the lengths of the sides opposite these angles by a, b, and c, respectively. The triangle itself will be referred to as *triangle ABC* (or denoted $\triangle ABC$). If a triangle is a right triangle and if one of the acute angles and a side are known or if two sides are given, then we may find the remaining parts by using the formulas in Section 5.2 that express the trigonometric functions as ratios of sides of a triangle. We can refer to the process of finding the remaining parts as **solving the triangle.**

Figure 1

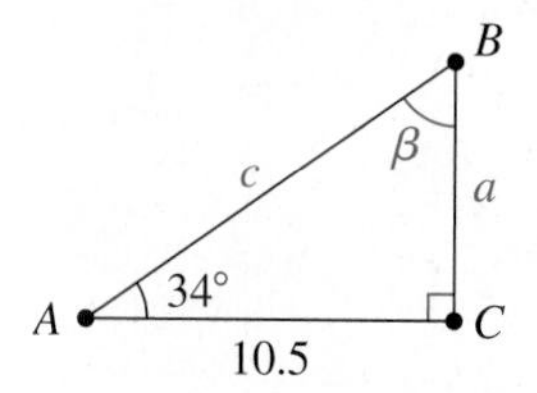

In all examples *it is assumed that you know how to find trigonometric function values and angles by using either a calculator or results about special angles.*

EXAMPLE 1 Solving a right triangle

Solve $\triangle ABC$, given $\gamma = 90°$, $\alpha = 34°$, and $b = 10.5$.

SOLUTION Since the sum of the three interior angles in a triangle is 180°, we have $\alpha + \beta + \gamma = 180°$. Solving for the unknown angle β gives us

$$\beta = 180° - \alpha - \gamma = 180° - 34° - 90° = 56°.$$

Referring to Figure 1, we obtain

$$\tan 34° = \frac{a}{10.5} \qquad \tan\alpha = \frac{\text{opp}}{\text{adj}}$$

$$a = (10.5)\tan 34° \approx 7.1. \qquad \text{solve for } a\text{; approximate}$$

To find side c, we can use either the cosine or the secant function, as follows in (1) or (2), respectively:

$$\textbf{(1)}\ \cos 34° = \frac{10.5}{c} \qquad \cos\alpha = \frac{\text{adj}}{\text{hyp}}$$

$$c = \frac{10.5}{\cos 34°} \approx 12.7 \qquad \text{solve for } c\text{; approximate}$$

$$\textbf{(2)}\ \sec 34° = \frac{c}{10.5} \qquad \sec\alpha = \frac{\text{hyp}}{\text{adj}}$$

$$c = (10.5)\sec 34° \approx 12.7 \qquad \text{solve for } c\text{; approximate}$$

Homework Helper

Organizing your work in a table makes it easy to see what parts remain to be found. Here are some snapshots of what a typical table might look like for Example 1.

After finding β:

Angles	*Opposite Sides*
$\alpha = 34°$	a
$\beta = 56°$	$b = 10.5$
$\gamma = 90°$	c

After finding a:

Angles	*Opposite Sides*
$\alpha = 34°$	$a \approx 7.1$
$\beta = 56°$	$b = 10.5$
$\gamma = 90°$	c

After finding c:

Angles	*Opposite Sides*
$\alpha = 34°$	$a \approx 7.1$
$\beta = 56°$	$b = 10.5$
$\gamma = 90°$	$c \approx 12.7$

As illustrated in Example 1, when working with triangles, we usually round off answers. One reason for doing so is that in most applications the lengths of sides of triangles and measures of angles are found by mechanical devices and hence are only approximations to the exact values. Consequently, a number such as 10.5 in Example 1 is assumed to have been rounded off to the nearest tenth. We cannot expect more accuracy in the calculated values for the remaining sides, and therefore they should also be rounded off to the nearest tenth.

In finding angles, answers should be rounded off as indicated in the following table.

Number of significant figures for sides	Round off degree measure of angles to the nearest
2	1°
3	0.1°, or 10′
4	0.01°, or 1′

Tutorial available at www.thomsonedu.com/login

Justification of this table requires a careful analysis of problems that involve approximate data.

EXAMPLE 2 Solving a right triangle

Solve $\triangle ABC$, given $\gamma = 90°$, $a = 12.3$, and $b = 31.6$.

Figure 2

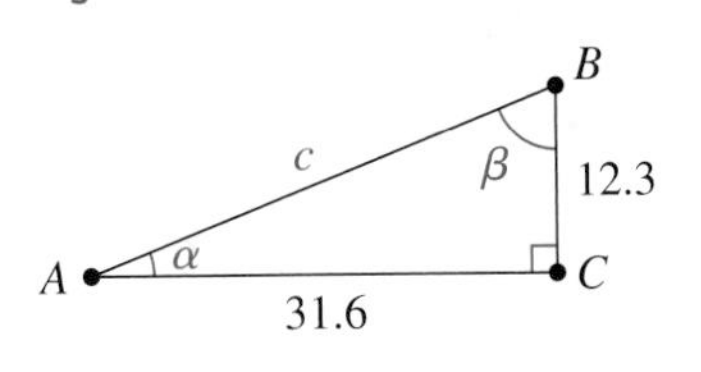

SOLUTION Referring to the triangle illustrated in Figure 2 gives us

$$\tan \alpha = \frac{12.3}{31.6}.$$

Since the sides are given with three significant figures, the rule stated in the preceding table tells us that α should be rounded off to the nearest 0.1°, or the nearest multiple of 10′. Using the degree mode on a calculator, we have

$$\alpha = \tan^{-1}\frac{12.3}{31.6} \approx 21.3° \qquad \text{or, equivalently,} \qquad \alpha \approx 21°20'.$$

Since α and β are complementary angles,

$$\beta = 90° - \alpha \approx 90° - 21.3° = 68.7°.$$

The only remaining part to find is c. We could use several relationships involving c to determine its value. Among these are

$$\cos \alpha = \frac{31.6}{c}, \quad \sec \beta = \frac{c}{12.3}, \quad \text{and} \quad a^2 + b^2 = c^2.$$

Whenever possible, it is best to use a relationship that involves only given information, since it doesn't depend on any previously calculated value. Hence, with $a = 12.3$ and $b = 31.6$, we have

$$c = \sqrt{a^2 + b^2} = \sqrt{(12.3)^2 + (31.6)^2} = \sqrt{1149.85} \approx 33.9.$$

Figure 3

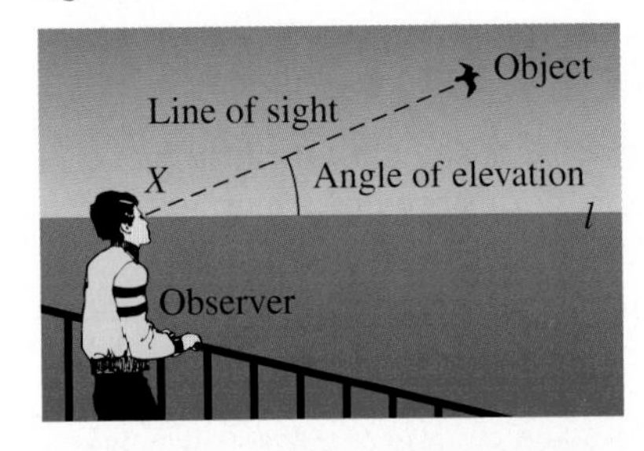

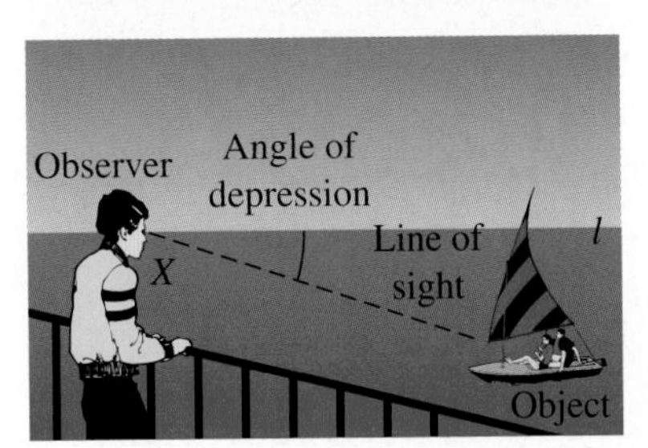

As illustrated in Figure 3, if an observer at point X sights an object, then the angle that the line of sight makes with the horizontal line l is the **angle of elevation** of the object, if the object is above the horizontal line, or the **angle of depression** of the object, if the object is below the horizontal line. We use this terminology in the next two examples.

EXAMPLE 3 Using an angle of elevation

From a point on level ground 135 feet from the base of a tower, the angle of elevation of the top of the tower is 57°20′. Approximate the height of the tower.

▶ SOLUTION If we let d denote the height of the tower, then the given facts are represented by the triangle in Figure 4. Referring to the figure, we obtain

$$\tan 57°20' = \frac{d}{135} \qquad \tan 57°20' = \frac{\text{opp}}{\text{adj}}$$

$$d = 135 \tan 57°20' \approx 211. \qquad \text{solve for } d\text{; approximate}$$

The tower is approximately 211 feet high.

Figure 4

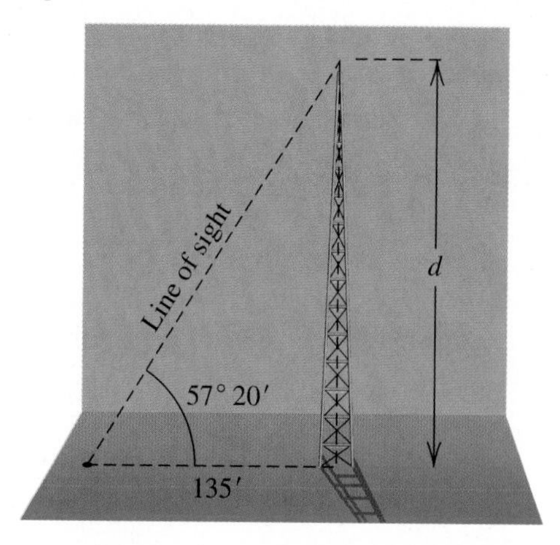

EXAMPLE 4 **Using angles of depression**

From the top of a building that overlooks an ocean, an observer watches a boat sailing directly toward the building. If the observer is 100 feet above sea level and if the angle of depression of the boat changes from 25° to 40° during the period of observation, approximate the distance that the boat travels.

▶ SOLUTION As in Figure 5, let A and B be the positions of the boat that correspond to the 25° and 40° angles, respectively. Suppose that the observer is at point D and that C is the point 100 feet directly below. Let d denote the distance the boat travels, and let k denote the distance from B to C. If α and β

Figure 5

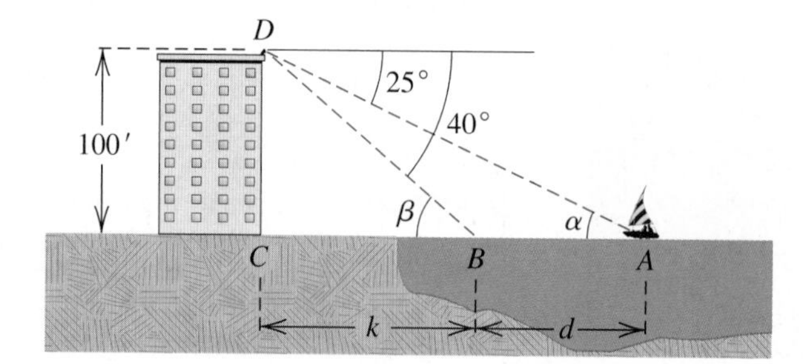

▶ Tutorial available at www.thomsonedu.com/login

denote angles DAC and DBC, respectively, then it follows from geometry (alternate interior angles) that $\alpha = 25°$ and $\beta = 40°$.

From triangle BCD:

$$\cot\beta = \cot 40° = \frac{k}{100} \qquad \cot\beta = \frac{\text{adj}}{\text{opp}}$$

$$k = 100\cot 40° \qquad \text{solve for } k$$

From triangle DAC:

$$\cot\alpha = \cot 25° = \frac{d+k}{100} \qquad \cot\alpha = \frac{\text{adj}}{\text{opp}}$$

$$d + k = 100\cot 25° \qquad \text{multiply by lcd}$$

$$d = 100\cot 25° - k \qquad \text{solve for } d$$

$$= 100\cot 25° - 100\cot 40° \qquad k = 100\cot 40°$$

$$= 100(\cot 25° - \cot 40°) \qquad \text{factor out 100}$$

$$\approx 100(2.145 - 1.192) \approx 95 \qquad \text{approximate}$$

Note that $d = \overline{AC} - \overline{BC}$, and if we use tan instead of cot, we get the equivalent equation

$$d = \frac{100}{\tan 25°} - \frac{100}{\tan 40°}.$$

Hence, the boat travels approximately 95 feet.

In certain navigation or surveying problems, the **direction,** or **bearing,** from a point P to a point Q is specified by stating the acute angle that segment PQ makes with the north-south line through P. We also state whether Q is north or south and east or west of P. Figure 6 illustrates four possibilities. The bearing from P to Q_1 is 25° east of north and is denoted by N25°E. We also refer to the **direction** N25°E, meaning the direction from P to Q_1. The bearings from P to Q_2, to Q_3, and to Q_4 are represented in a similar manner in the figure. Note that when this notation is used for bearings or directions, N or S always appears to the *left* of the angle and W or E to the *right.*

Figure 6

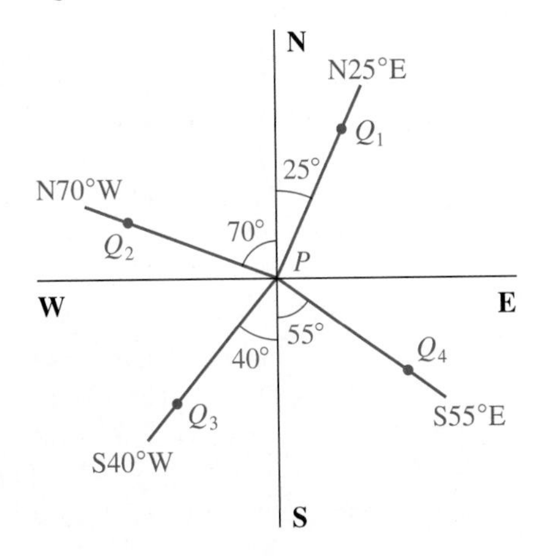

Figure 7

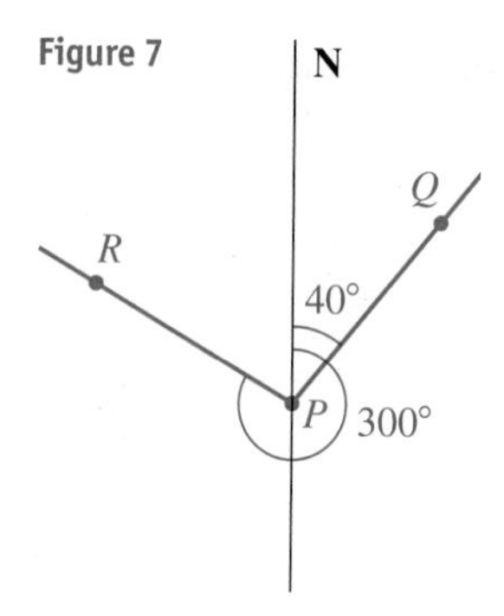

In air navigation, directions and bearings are specified by measuring from the north in a *clockwise* direction. In this case, a positive measure is assigned to the angle instead of the negative measure to which we are accustomed for clockwise rotations. Referring to Figure 7, we see that the direction of PQ is 40° and the direction of PR is 300°.

EXAMPLE 5 Using bearings

Two ships leave port at the same time, one ship sailing in the direction N23°E at a speed of 11 mi/hr and the second ship sailing in the direction S67°E at 15 mi/hr. Approximate the bearing from the second ship to the first, one hour later.

Figure 8

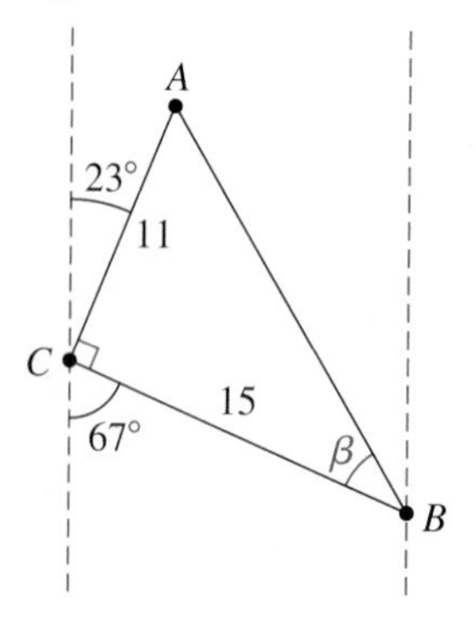

▶ SOLUTION The sketch in Figure 8 indicates the positions of the first and second ships at points A and B, respectively, after one hour. Point C represents the port. We wish to find the bearing from B to A. Note that

$$\angle ACB = 180° - 23° - 67° = 90°,$$

and hence triangle ACB is a right triangle. Thus,

$$\tan \beta = \frac{11}{15} \qquad \tan \beta = \frac{\text{opp}}{\text{adj}}$$

$$\beta = \tan^{-1}\tfrac{11}{15} \approx 36°. \qquad \text{solve for } \beta\text{; approximate}$$

We have rounded β to the nearest degree because the sides of the triangles are given with two significant figures.

Referring to Figure 9, we obtain the following:

Figure 9

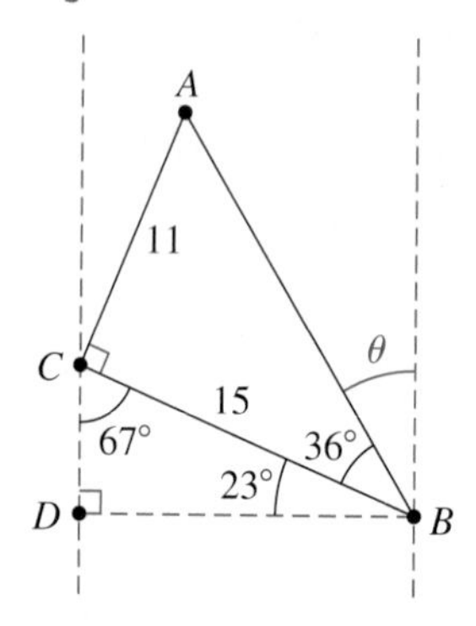

$$\angle CBD = 90° - \angle BCD = 90° - 67° = 23°$$
$$\angle ABD = \angle ABC + \angle CBD \approx 36° + 23° = 59°$$
$$\theta = 90° - \angle ABD \approx 90° - 59° = 31°$$

Thus, the bearing from B to A is approximately N31°W.

Trigonometric functions are useful in the investigation of vibratory or oscillatory motion, such as the motion of a particle in a vibrating guitar string or a spring that has been compressed or elongated and then released to oscillate back and forth. The fundamental type of particle displacement in these illustrations is *harmonic motion.*

Definition of Simple Harmonic Motion

A point moving on a coordinate line is in **simple harmonic motion** if its distance d from the origin at time t is given by either

$$d = a \cos \omega t \qquad \text{or} \qquad d = a \sin \omega t,$$

where a and ω are constants, with $\omega > 0$.

▶ Tutorial available at www.thomsonedu.com/login

In the preceding definition, the **amplitude** of the motion is the maximum displacement $|a|$ of the point from the origin. The **period** is the time $2\pi/\omega$ required for one complete oscillation. The reciprocal of the period, $\omega/(2\pi)$, is the number of oscillations per unit of time and is called the **frequency.**

A physical interpretation of simple harmonic motion can be obtained by considering a spring with an attached weight that is oscillating vertically relative to a coordinate line, as illustrated in Figure 10. The number d represents the coordinate of a fixed point Q in the weight, and we assume that the amplitude a of the motion is constant. In this case no frictional force is retarding the motion. If friction is present, then the amplitude decreases with time, and the motion is said to be *damped.*

Figure 10

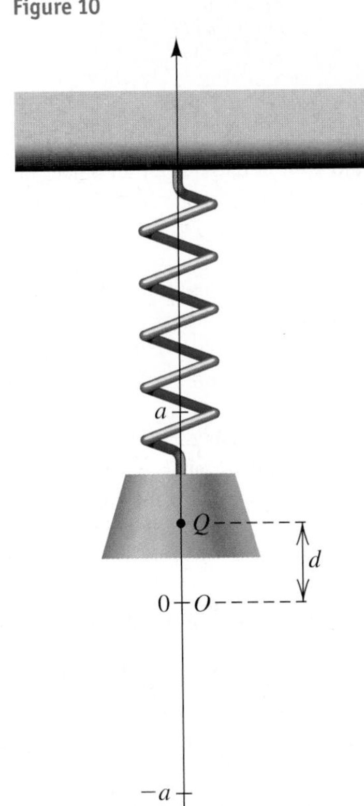

EXAMPLE 6 Describing harmonic motion

Suppose that the oscillation of the weight shown in Figure 10 is given by

$$d = 10\cos\left(\frac{\pi}{6}t\right),$$

with t measured in seconds and d in centimeters. Discuss the motion of the weight.

SOLUTION By definition, the motion is simple harmonic with amplitude $a = 10$ cm. Since $\omega = \pi/6$, we obtain the following:

$$\text{period} = \frac{2\pi}{\omega} = \frac{2\pi}{\pi/6} = 12$$

Thus, in 12 seconds the weight makes one complete oscillation. The frequency is $\frac{1}{12}$, which means that one-twelfth of an oscillation takes place each second. The following table indicates the position of Q at various times.

t	0	1	2	3	4	5	6
$\frac{\pi}{6}t$	0	$\frac{\pi}{6}$	$\frac{\pi}{3}$	$\frac{\pi}{2}$	$\frac{2\pi}{3}$	$\frac{5\pi}{6}$	π
$\cos\left(\frac{\pi}{6}t\right)$	1	$\frac{\sqrt{3}}{2}$	$\frac{1}{2}$	0	$-\frac{1}{2}$	$-\frac{\sqrt{3}}{2}$	-1
d	10	$5\sqrt{3} \approx 8.7$	5	0	-5	$-5\sqrt{3} \approx -8.7$	-10

The initial position of Q is 10 centimeters above the origin O. It moves downward, gaining speed until it reaches O. Note that Q travels approximately $10 - 8.7 = 1.3$ cm during the first second, $8.7 - 5 = 3.7$ cm during the next second, and $5 - 0 = 5$ cm during the third second. It then slows down until it reaches a point 10 centimeters below O at the end of 6 seconds. The direction of motion is then reversed, and the weight moves upward, gaining speed until it reaches O. Once it reaches O, it slows down until it returns to its original position at the end of 12 seconds. The direction of motion is then reversed again, and the same pattern is repeated indefinitely.

5.7 Exercises

Exer. 1–8: Given the indicated parts of triangle ABC with $\gamma = 90°$, find the exact values of the remaining parts.

1 $\alpha = 30°$, $b = 20$

2 $\beta = 45°$, $b = 35$

▶ 3 $\beta = 45°$, $c = 30$

4 $\alpha = 60°$, $c = 6$

5 $a = 5$, $b = 5$

▶ 6 $a = 4\sqrt{3}$, $c = 8$

7 $b = 5\sqrt{3}$, $c = 10\sqrt{3}$

8 $b = 7\sqrt{2}$, $c = 14$

Exer. 9–16: Given the indicated parts of triangle ABC with $\gamma = 90°$, approximate the remaining parts.

9 $\alpha = 37°$, $b = 24$

10 $\beta = 64°20'$, $a = 20.1$

11 $\beta = 71°51'$, $b = 240.0$

12 $\alpha = 31°10'$, $a = 510$

13 $a = 25$, $b = 45$

14 $a = 31$, $b = 9.0$

15 $c = 5.8$, $b = 2.1$

▶ 16 $a = 0.42$, $c = 0.68$

Exer. 17–24: Given the indicated parts of triangle ABC with $\gamma = 90°$, express the third part in terms of the first two.

17 $\alpha, c;\ b$

18 $\beta, c;\ b$

19 $\beta, b;\ a$

20 $\alpha, b;\ a$

21 $\alpha, a;\ c$

22 $\beta, a;\ c$

23 $a, c;\ b$

24 $a, b;\ c$

25 **Height of a kite** A person flying a kite holds the string 4 feet above ground level. The string of the kite is taut and makes an angle of 60° with the horizontal (see the figure). Approximate the height of the kite above level ground if 500 feet of string is payed out.

Exercise 25

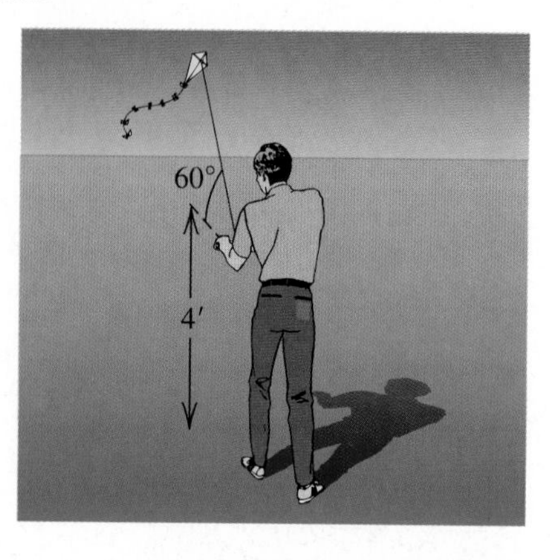

26 **Surveying** From a point 15 meters above level ground, a surveyor measures the angle of depression of an object on the ground at 68°. Approximate the distance from the object to the point on the ground directly beneath the surveyor.

27 **Airplane landing** A pilot, flying at an altitude of 5000 feet, wishes to approach the numbers on a runway at an angle of 10°. Approximate, to the nearest 100 feet, the distance from the airplane to the numbers at the beginning of the descent.

28 **Radio antenna** A guy wire is attached to the top of a radio antenna and to a point on horizontal ground that is 40.0 meters from the base of the antenna. If the wire makes an angle of 58°20′ with the ground, approximate the length of the wire.

▶ 29 **Surveying** To find the distance d between two points P and Q on opposite shores of a lake, a surveyor locates a point R that is 50.0 meters from P such that RP is perpendicular to PQ, as shown in the figure. Next, using a transit, the surveyor measures angle PRQ as 72°40′. Find d.

Exercise 29

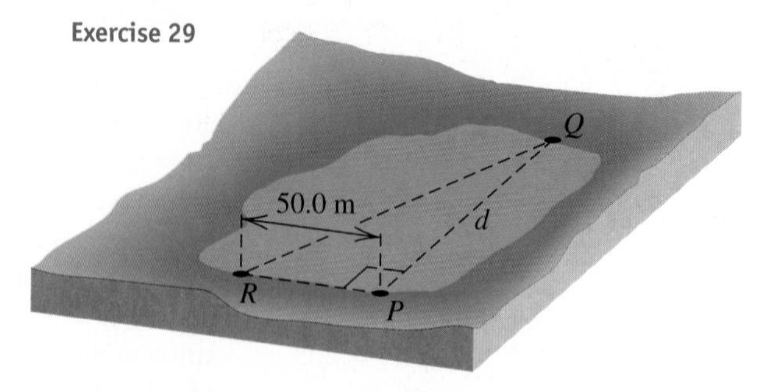

▶ Tutorial available at www.thomsonedu.com/login

30 **Meteorological calculations** To measure the height h of a cloud cover, a meteorology student directs a spotlight vertically upward from the ground. From a point P on level ground that is d meters from the spotlight, the angle of elevation θ of the light image on the clouds is then measured (see the figure).

(a) Express h in terms of d and θ.

(b) Approximate h if $d = 1000$ m and $\theta = 59°$.

Exercise 30

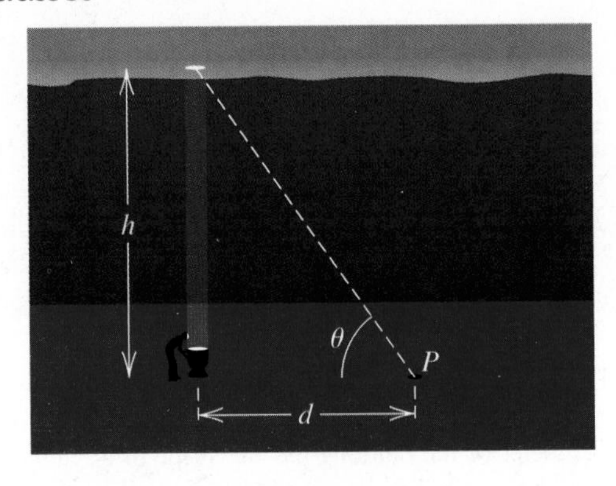

31 **Altitude of a rocket** A rocket is fired at sea level and climbs at a constant angle of 75° through a distance of 10,000 feet. Approximate its altitude to the nearest foot.

32 **Airplane takeoff** An airplane takes off at a 10° angle and travels at the rate of 250 ft/sec. Approximately how long does it take the airplane to reach an altitude of 15,000 feet?

33 **Designing a drawbridge** A drawbridge is 150 feet long when stretched across a river. As shown in the figure, the two sections of the bridge can be rotated upward through an angle of 35°.

(a) If the water level is 15 feet below the closed bridge, find the distance d between the end of a section and the water level when the bridge is fully open.

(b) Approximately how far apart are the ends of the two sections when the bridge is fully opened, as shown in the figure?

Exercise 33

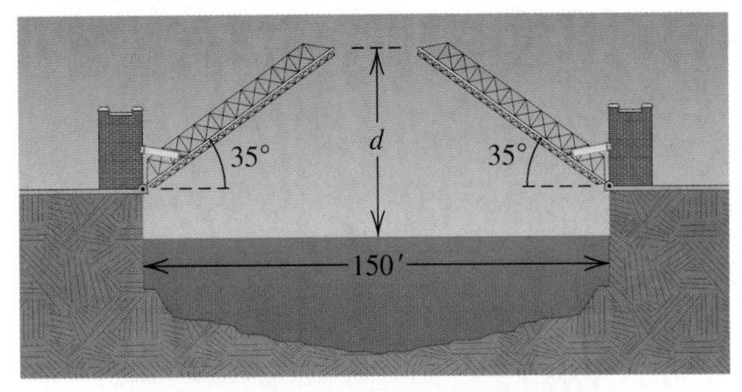

34 **Designing a water slide** Shown in the figure is part of a design for a water slide. Find the total length of the slide to the nearest foot.

Exercise 34

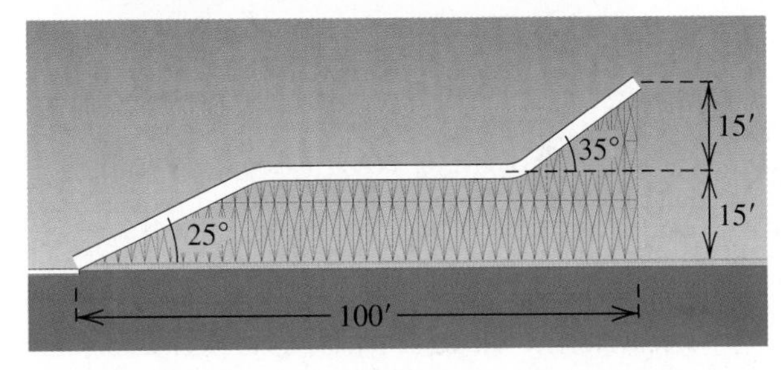

▶ 35 **Sun's elevation** Approximate the angle of elevation α of the sun if a person 5.0 feet tall casts a shadow 4.0 feet long on level ground (see the figure).

Exercise 35

36 **Constructing a ramp** A builder wishes to construct a ramp 24 feet long that rises to a height of 5.0 feet above level ground. Approximate the angle that the ramp should make with the horizontal.

37 **Video game** Shown in the figure is the screen for a simple video arcade game in which ducks move from A to B at

the rate of 7 cm/sec. Bullets fired from point O travel 25 cm/sec. If a player shoots as soon as a duck appears at A, at which angle φ should the gun be aimed in order to score a direct hit?

Exercise 37

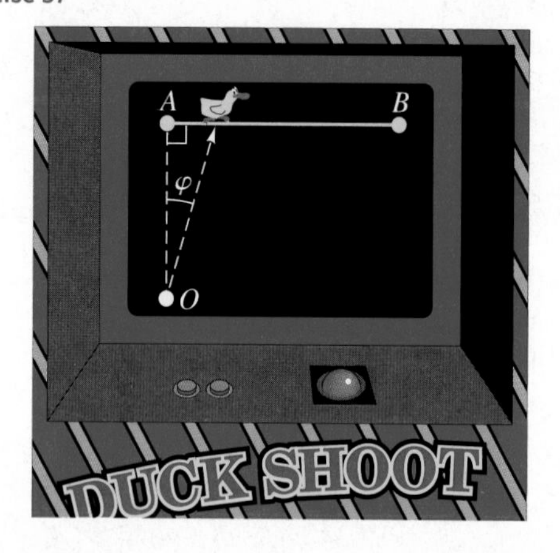

38 Conveyor belt A conveyor belt 9 meters long can be hydraulically rotated up to an angle of 40° to unload cargo from airplanes (see the figure).

(a) Find, to the nearest degree, the angle through which the conveyor belt should be rotated up to reach a door that is 4 meters above the platform supporting the belt.

(b) Approximate the maximum height above the platform that the belt can reach.

Exercise 38

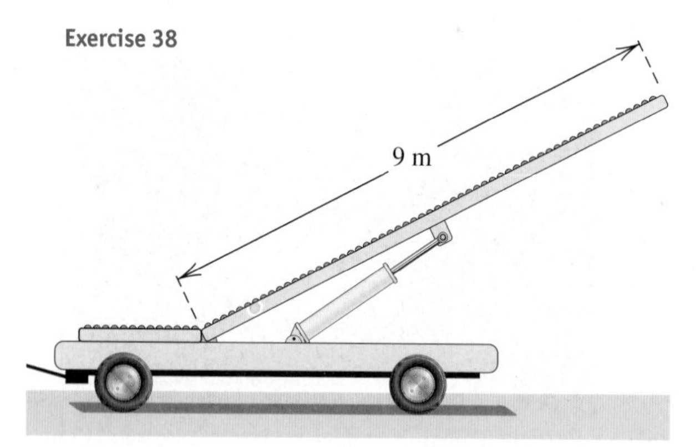

39 Tallest structure The tallest man-made structure in the world is a television transmitting tower located near Mayville, North Dakota. From a distance of 1 mile on level ground, its angle of elevation is 21°20′24″. Determine its height to the nearest foot.

40 Elongation of Venus The *elongation* of the planet Venus is defined to be the angle θ determined by the sun, Earth, and Venus, as shown in the figure. Maximum elongation of Venus occurs when Earth is at its minimum distance D_e from the sun and Venus is at its maximum distance D_v from the sun. If $D_e = 91{,}500{,}000$ mi and $D_v = 68{,}000{,}000$ mi, approximate the maximum elongation θ_{max} of Venus. Assume that the orbit of Venus is circular.

Exercise 40

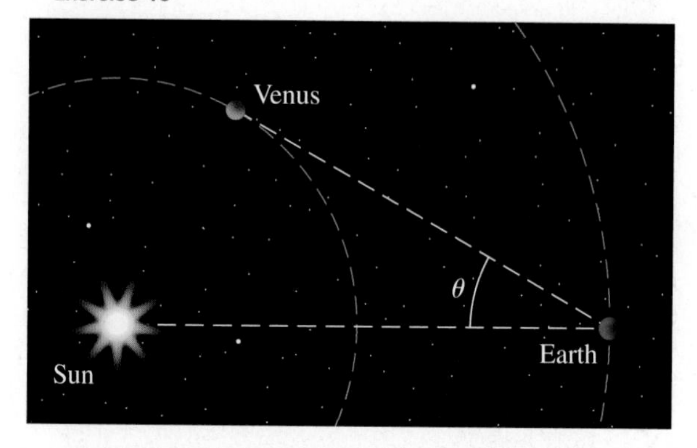

41 The Pentagon's ground area The Pentagon is the largest office building in the world in terms of ground area. The perimeter of the building has the shape of a regular pentagon with each side of length 921 feet. Find the area enclosed by the perimeter of the building.

42 A regular octagon is inscribed in a circle of radius 12.0 centimeters. Approximate the perimeter of the octagon.

43 A rectangular box has dimensions 8″ × 6″ × 4″. Approximate, to the nearest tenth of a degree, the angle θ formed by a diagonal of the base and the diagonal of the box, as shown in the figure.

Exercise 43

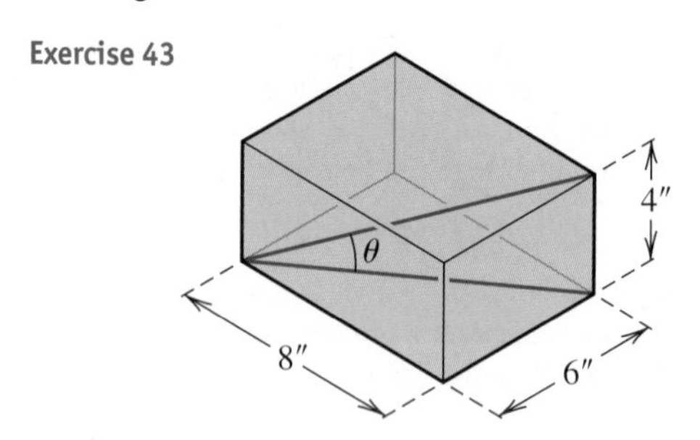

44 Volume of a conical cup A conical paper cup has a radius of 2 inches. Approximate, to the nearest degree, the angle β (see the figure) so that the cone will have a volume of 20 in^3.

Exercise 44

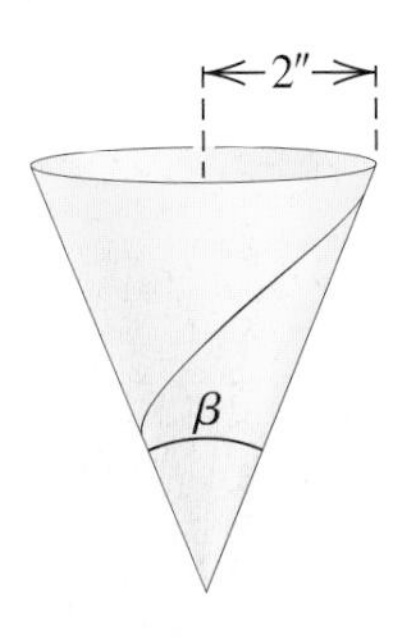

45 Height of a tower From a point P on level ground, the angle of elevation of the top of a tower is $26°50'$. From a point 25.0 meters closer to the tower and on the same line with P and the base of the tower, the angle of elevation of the top is $53°30'$. Approximate the height of the tower.

46 Ladder calculations A ladder 20 feet long leans against the side of a building, and the angle between the ladder and the building is 22°.

(a) Approximate the distance from the bottom of the ladder to the building.

(b) If the distance from the bottom of the ladder to the building is increased by 3.0 feet, approximately how far does the top of the ladder move down the building?

47 Ascent of a hot-air balloon As a hot-air balloon rises vertically, its angle of elevation from a point P on level ground 110 kilometers from the point Q directly underneath the balloon changes from $19°20'$ to $31°50'$ (see the figure). Approximately how far does the balloon rise during this period?

Exercise 47

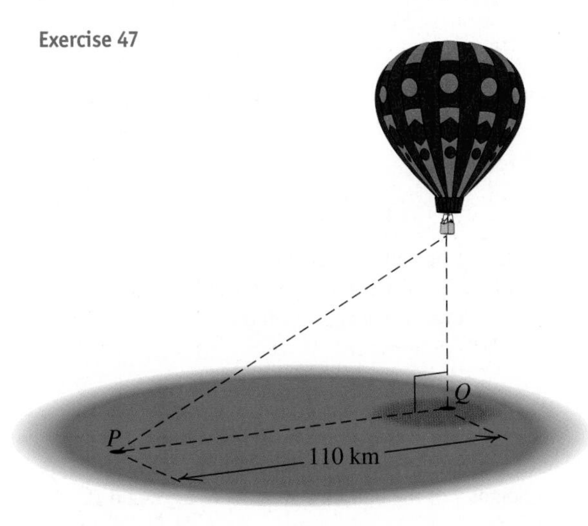

48 Height of a building From a point A that is 8.20 meters above level ground, the angle of elevation of the top of a building is $31°20'$ and the angle of depression of the base of the building is $12°50'$. Approximate the height of the building.

49 Radius of Earth A spacelab circles Earth at an altitude of 380 miles. When an astronaut views the horizon of Earth, the angle θ shown in the figure is 65.8°. Use this information to estimate the radius of Earth.

Exercise 49

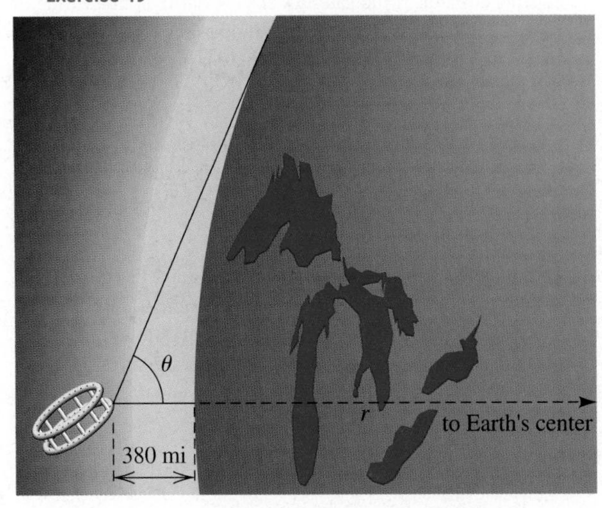

50 Length of an antenna A CB antenna is located on the top of a garage that is 16 feet tall. From a point on level ground that is 100 feet from a point directly below the antenna, the antenna subtends an angle of 12°, as shown in the figure. Approximate the length of the antenna.

Exercise 50

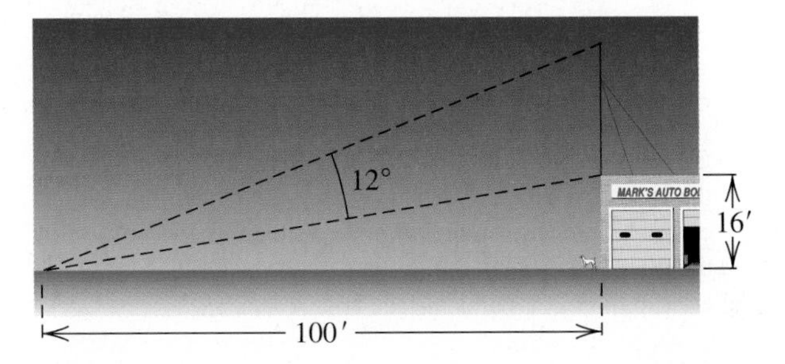

51 Speed of an airplane An airplane flying at an altitude of 10,000 feet passes directly over a fixed object on the ground. One minute later, the angle of depression of the object is 42°. Approximate the speed of the airplane to the nearest mile per hour.

52 Height of a mountain A motorist, traveling along a level highway at a speed of 60 km/hr directly toward a mountain, observes that between 1:00 P.M. and 1:10 P.M. the angle of elevation of the top of the mountain changes from 10° to 70°. Approximate the height of the mountain.

53 Communications satellite Shown in the left part of the figure is a communications satellite with an equatorial orbit—that is, a nearly circular orbit in the plane determined by Earth's equator. If the satellite circles Earth at an altitude of $a = 22{,}300$ mi, its speed is the same as the rotational speed of Earth; to an observer on the equator, the satellite appears to be stationary—that is, its orbit is synchronous.

(a) Using $R = 4000$ mi for the radius of Earth, determine the percentage of the equator that is within signal range of such a satellite.

(b) As shown in the right part of the figure, three satellites are equally spaced in equatorial synchronous orbits. Use the value of θ obtained in part (a) to explain why all points on the equator are within signal range of at least one of the three satellites.

Exercise 53

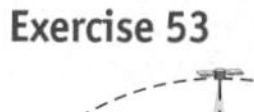

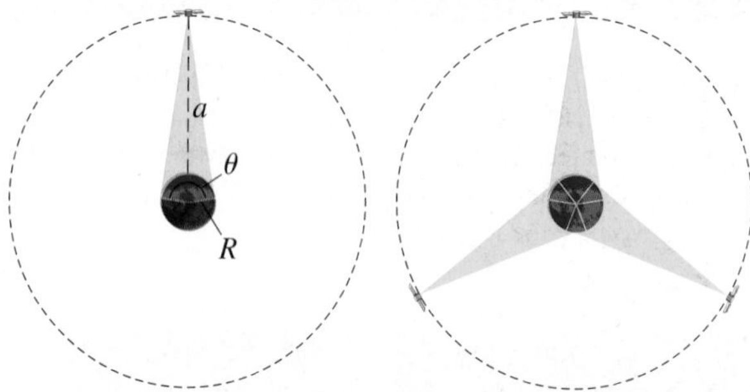

54 Communications satellite Refer to Exercise 53. Shown in the figure is the area served by a communications satellite circling a planet of radius R at an altitude a. The portion of the planet's surface within range of the satellite is a spherical cap of depth d and surface area $A = 2\pi Rd$.

(a) Express d in terms of R and θ.

(b) Estimate the percentage of the planet's surface that is within signal range of a single satellite in equatorial synchronous orbit.

Exercise 54

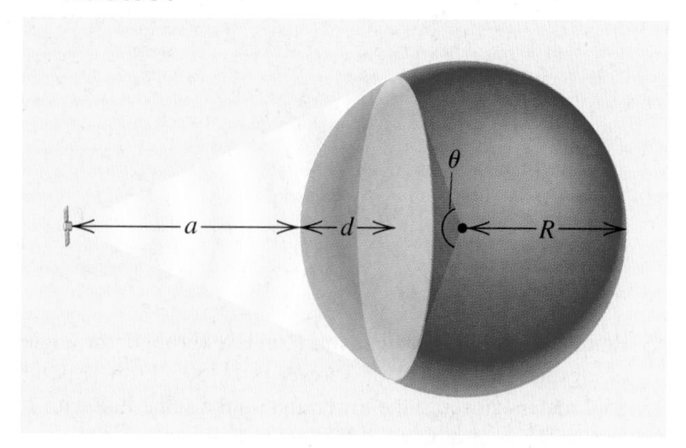

55 Height of a kite Generalize Exercise 25 to the case where the angle is α, the number of feet of string payed out is d, and the end of the string is held c feet above the ground. Express the height h of the kite in terms of α, d, and c.

56 Surveying Generalize Exercise 26 to the case where the point is d meters above level ground and the angle of depression is α. Express the distance x in terms of d and α.

57 Height of a tower Generalize Exercise 45 to the case where the first angle is α, the second angle is β, and the distance between the two points is d. Express the height h of the tower in terms of d, α, and β.

58 Generalize Exercise 42 to the case of an n-sided polygon inscribed in a circle of radius r. Express the perimeter P in terms of n and r.

59 Ascent of a hot-air balloon Generalize Exercise 47 to the case where the distance from P to Q is d kilometers and the angle of elevation changes from α to β.

60 Height of a building Generalize Exercise 48 to the case where point A is d meters above ground and the angles of elevation and depression are α and β, respectively. Express the height h of the building in terms of d, α, and β.

Exer. 61–62: Find the bearing from P to each of the points A, B, C, and D.

61

N
B
40°
A
20°
W
P
E
75°
25°
C
D
S

62

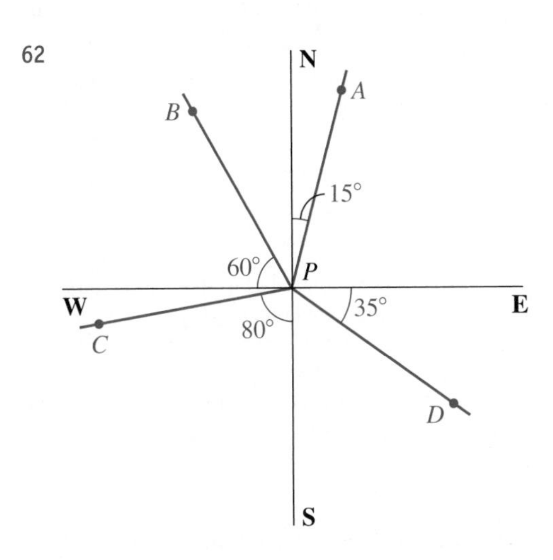

63 **Ship's bearings** A ship leaves port at 1:00 P.M. and sails in the direction N34°W at a rate of 24 mi/hr. Another ship leaves port at 1:30 P.M. and sails in the direction N56°E at a rate of 18 mi/hr.

(a) Approximately how far apart are the ships at 3:00 P.M.?

(b) What is the bearing, to the nearest degree, from the first ship to the second?

64 **Pinpointing a forest fire** From an observation point A, a forest ranger sights a fire in the direction S35°50′W (see the figure). From a point B, 5 miles due west of A, another ranger sights the same fire in the direction S54°10′E. Approximate, to the nearest tenth of a mile, the distance of the fire from A.

Exercise 64

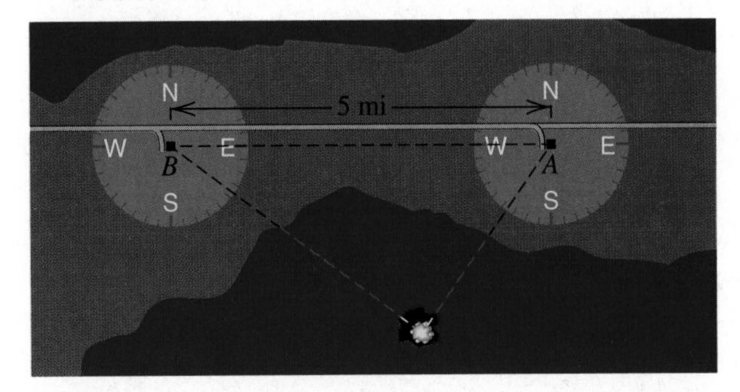

65 **Airplane flight** An airplane flying at a speed of 360 mi/hr flies from a point A in the direction 137° for 30 minutes and then flies in the direction 227° for 45 minutes. Approximate, to the nearest mile, the distance from the airplane to A.

66 **Airplane flight plan** An airplane flying at a speed of 400 mi/hr flies from a point A in the direction 153° for 1 hour and then flies in the direction 63° for 1 hour.

(a) In what direction does the plane need to fly in order to get back to point A?

(b) How long will it take to get back to point A?

Exer. 67–70: The formula specifies the position of a point P that is moving harmonically on a vertical axis, where t is in seconds and d is in centimeters. Determine the amplitude, period, and frequency, and describe the motion of the point during one complete oscillation (starting at $t = 0$).

67 $d = 10 \sin 6\pi t$

68 $d = \frac{1}{3} \cos \frac{\pi}{4} t$

69 $d = 4 \cos \frac{3\pi}{2} t$

70 $d = 6 \sin \frac{2\pi}{3} t$

71 A point P in simple harmonic motion has a period of 3 seconds and an amplitude of 5 centimeters. Express the motion of P by means of an equation of the form $d = a \cos \omega t$.

72 A point P in simple harmonic motion has a frequency of $\frac{1}{2}$ oscillation per minute and an amplitude of 4 feet. Express the motion of P by means of an equation of the form $d = a \sin \omega t$.

73 **Tsunamis** A tsunami is a tidal wave caused by an earthquake beneath the sea. These waves can be more than 100 feet in height and can travel at great speeds. Engineers sometimes represent such waves by trigonometric expressions of the form $y = a \cos bt$ and use these representations to estimate the effectiveness of sea walls. Suppose that a wave has height $h = 50$ ft and period 30 minutes and is traveling at the rate of 180 ft/sec.

Exercise 73

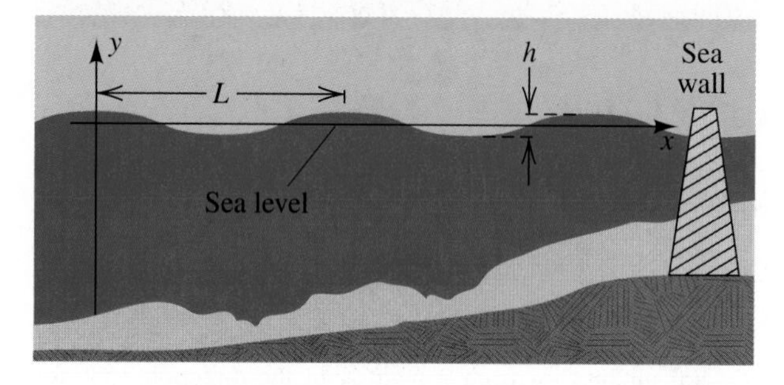

(a) Let (x, y) be a point on the wave represented in the figure. Express y as a function of t if $y = 25$ ft when $t = 0$.

(b) The wave length L is the distance between two successive crests of the wave. Approximate L in feet.

74 **Some Hawaiian tsunamis** For an interval of 45 minutes, the tsunamis near Hawaii caused by the Chilean earthquake of 1960 could be modeled by the equation $y = 8 \sin \frac{\pi}{6} t$, where y is in feet and t is in minutes.

(a) Find the amplitude and period of the waves.

(b) If the distance from one crest of the wave to the next was 21 kilometers, what was the velocity of the wave? (Tidal waves can have velocities of more than 700 km/hr in deep sea water.)

CHAPTER 5 REVIEW EXERCISES

▶ **Online support materials can be found at www.thomsonedu.com/login**

1 Find the radian measure that corresponds to each degree measure: 330°, 405°, −150°, 240°, 36°.

2 Find the degree measure that corresponds to each radian measure: $\frac{9\pi}{2}, -\frac{2\pi}{3}, \frac{7\pi}{4}, 5\pi, \frac{\pi}{5}$.

3 A central angle θ is subtended by an arc 20 centimeters long on a circle of radius 2 meters.

(a) Find the radian measure of θ.

(b) Find the area of the sector determined by θ.

4 (a) Find the length of the arc that subtends an angle of measure 70° on a circle of diameter 15 centimeters.

(b) Find the area of the sector in part (a).

5 **Angular speed of phonograph records** Two types of phonograph records, LP albums and singles, have diameters of 12 inches and 7 inches, respectively. The album rotates at a rate of $33\frac{1}{3}$ rpm, and the single rotates at 45 rpm. Find the angular speed (in radians per minute) of the album and of the single.

6 **Linear speed on phonograph records** Using the information in Exercise 5, find the linear speed (in ft/min) of a point on the circumference of the album and of the single.

Exer. 7–8: Find the exact values of x and y.

7

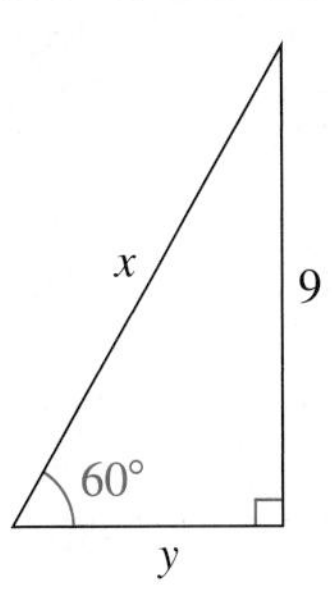

8

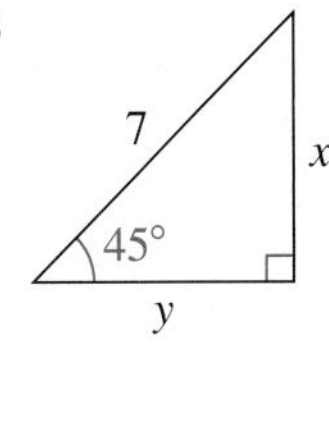

Exer. 9–10: Use fundamental identities to write the first expression in terms of the second, for any acute angle θ.

9 $\tan \theta, \quad \sec \theta$

10 $\cot \theta, \quad \csc \theta$

Exer. 11–20: Verify the identity by transforming the left-hand side into the right-hand side.

11 $\sin \theta (\csc \theta - \sin \theta) = \cos^2 \theta$

12 $\cos \theta (\tan \theta + \cot \theta) = \csc \theta$

13 $(\cos^2 \theta - 1)(\tan^2 \theta + 1) = 1 - \sec^2 \theta$

14 $\dfrac{\sec \theta - \cos \theta}{\tan \theta} = \dfrac{\tan \theta}{\sec \theta}$

15 $\dfrac{1 + \tan^2 \theta}{\tan^2 \theta} = \csc^2 \theta$

16 $\dfrac{\sec \theta + \csc \theta}{\sec \theta - \csc \theta} = \dfrac{\sin \theta + \cos \theta}{\sin \theta - \cos \theta}$

17 $\dfrac{\cot \theta - 1}{1 - \tan \theta} = \cot \theta$

18 $\dfrac{1 + \sec \theta}{\tan \theta + \sin \theta} = \csc \theta$

19 $\dfrac{\tan(-\theta) + \cot(-\theta)}{\tan \theta} = -\csc^2 \theta$

20 $\dfrac{1}{\csc(-\theta)} - \dfrac{\cot(-\theta)}{\sec(-\theta)} = \csc \theta$

21 If θ is an acute angle of a right triangle and if the adjacent side and hypotenuse have lengths 4 and 7, respectively, find the values of the trigonometric functions of θ.

22 Whenever possible, find the exact values of the trigonometric functions of θ if θ is in standard position and satisfies the stated condition.

(a) The point $(30, -40)$ is on the terminal side of θ.

(b) The terminal side of θ is in quadrant II and is parallel to the line $2x + 3y + 6 = 0$.

(c) The terminal side of θ is on the negative y-axis.

23 Find the quadrant containing θ if θ is in standard position.

(a) $\sec \theta < 0$ and $\sin \theta > 0$

(b) $\cot \theta > 0$ and $\csc \theta < 0$

(c) $\cos \theta > 0$ and $\tan \theta < 0$

24 Find the exact values of the remaining trigonometric functions if

(a) $\sin \theta = -\frac{4}{5}$ and $\cos \theta = \frac{3}{5}$

(b) $\csc \theta = \dfrac{\sqrt{13}}{2}$ and $\cot \theta = -\dfrac{3}{2}$

Exer. 25–26: $P(t)$ denotes the point on the unit circle U that corresponds to the real number t.

25 Find the rectangular coordinates of $P(7\pi)$, $P(-5\pi/2)$, $P(9\pi/2)$, $P(-3\pi/4)$, $P(18\pi)$, and $P(\pi/6)$.

26 If $P(t)$ has coordinates $\left(-\frac{3}{5}, -\frac{4}{5}\right)$, find the coordinates of $P(t + 3\pi)$, $P(t - \pi)$, $P(-t)$, and $P(2\pi - t)$.

27 (a) Find the reference angle for each radian measure: $\dfrac{5\pi}{4}, -\dfrac{5\pi}{6}, -\dfrac{9\pi}{8}$.

(b) Find the reference angle for each degree measure: 245°, 137°, 892°.

28 Without the use of a calculator, find the exact values of the trigonometric functions corresponding to each real number, whenever possible.

(a) $\frac{9\pi}{2}$ (b) $-\frac{5\pi}{4}$ (c) 0 (d) $\frac{11\pi}{6}$

29 Find the exact value.

(a) $\cos 225°$ (b) $\tan 150°$ (c) $\sin\left(-\frac{\pi}{6}\right)$

(d) $\sec\frac{4\pi}{3}$ (e) $\cot\frac{7\pi}{4}$ (f) $\csc 300°$

30 If $\sin\theta = -0.7604$ and $\sec\theta$ is positive, approximate θ to the nearest 0.1° for $0° \le \theta < 360°$.

31 If $\tan\theta = 2.7381$, approximate θ to the nearest 0.0001 radian for $0 \le \theta < 2\pi$.

32 If $\sec\theta = 1.6403$, approximate θ to the nearest 0.01° for $0° \le \theta < 360°$.

Exer. 33–40: Find the amplitude and period and sketch the graph of the equation.

33 $y = 5\cos x$

34 $y = \frac{2}{3}\sin x$

35 $y = \frac{1}{3}\sin 3x$

36 $y = -\frac{1}{2}\cos\frac{1}{3}x$

37 $y = -3\cos\frac{1}{2}x$

38 $y = 4\sin 2x$

39 $y = 2\sin\pi x$

40 $y = 4\cos\frac{\pi}{2}x - 2$

Exer. 41–44: The graph of an equation is shown in the figure. (a) Find the amplitude and period. (b) Express the equation in the form $y = a\sin bx$ or in the form $y = a\cos bx$.

41

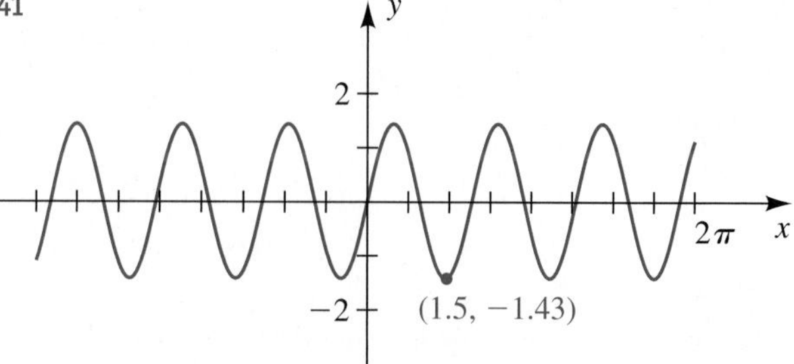

42

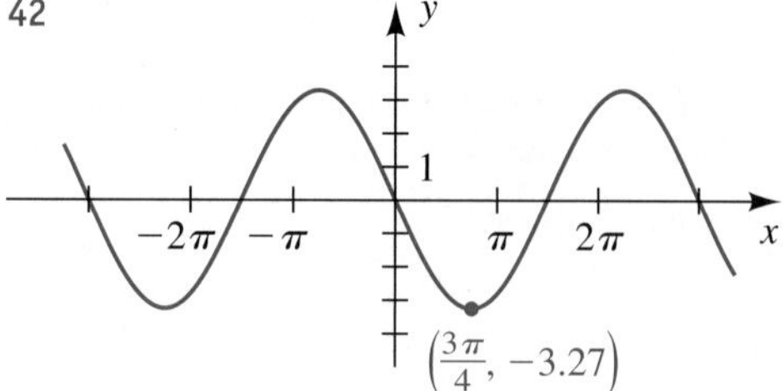

43

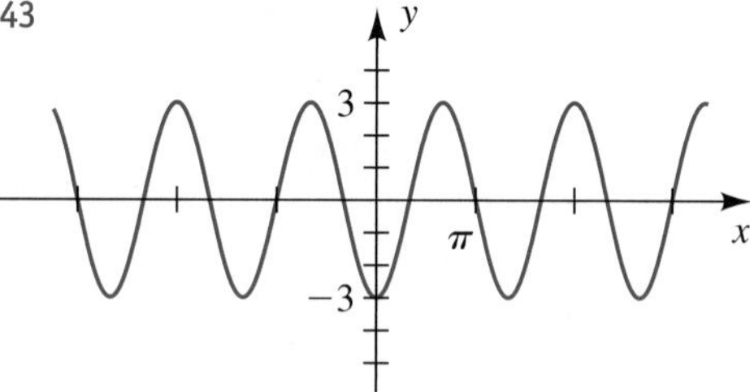

44

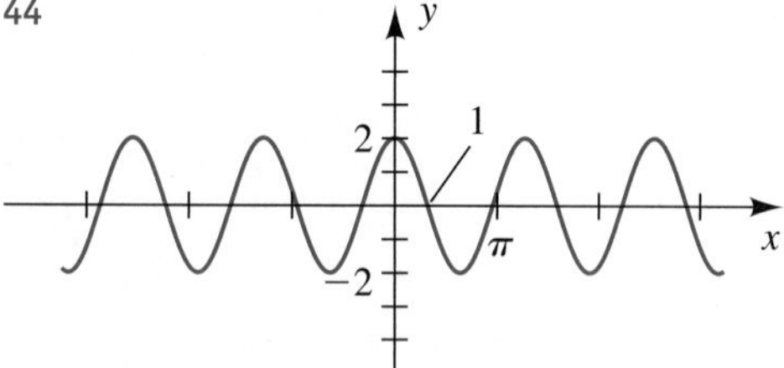

Exer. 45–56: Sketch the graph of the equation.

45 $y = 2\sin\left(x - \frac{2\pi}{3}\right)$

46 $y = -3\sin\left(\frac{1}{2}x - \frac{\pi}{4}\right)$

47 $y = -4\cos\left(x + \frac{\pi}{6}\right)$

48 $y = 5\cos\left(2x + \frac{\pi}{2}\right)$

49 $y = 2\tan\left(\frac{1}{2}x - \pi\right)$

50 $y = -3\tan\left(2x + \frac{\pi}{3}\right)$

51 $y = -4\cot\left(2x - \frac{\pi}{2}\right)$

52 $y = 2\cot\left(\frac{1}{2}x + \frac{\pi}{4}\right)$

53 $y = \sec\left(\frac{1}{2}x + \pi\right)$

54 $y = \sec\left(2x - \frac{\pi}{2}\right)$

55 $y = \csc\left(2x - \frac{\pi}{4}\right)$

56 $y = \csc\left(\frac{1}{2}x + \frac{\pi}{4}\right)$

Exer. 57–60: Given the indicated parts of triangle *ABC* with $\gamma = 90°$, approximate the remaining parts.

57 $\beta = 60°, \quad b = 40$

58 $\alpha = 54°40', \quad b = 220$

59 $a = 62, \quad b = 25$

60 $a = 9.0, \quad c = 41$

61 **Airplane propeller** The length of the largest airplane propeller ever used was 22 feet 7.5 inches. The plane was powered by four engines that turned the propeller at 545 revolutions per minute.

(a) What was the angular speed of the propeller in radians per second?

(b) Approximately how fast (in mi/hr) did the tip of the propeller travel along the circle it generated?

62 **The Eiffel Tower** When the top of the Eiffel Tower is viewed at a distance of 200 feet from the base, the angle of elevation is 79.2°. Estimate the height of the tower.

63 **Lasers and velocities** Lasers are used to accurately measure velocities of objects. Laser light produces an oscillating electromagnetic field E with a constant frequency f that can be described by

$$E = E_0 \cos(2\pi f t).$$

If a laser beam is pointed at an object moving toward the laser, light will be reflected toward the laser at a slightly higher frequency, in much the same way as a train whistle sounds higher when it is moving toward you. If Δf is this change in frequency and v is the object's velocity, then the equation

$$\Delta f = \frac{2fv}{c}$$

can be used to determine v, where $c = 186{,}000$ mi/sec is the velocity of the light. Approximate the velocity v of an object if $\Delta f = 10^8$ and $f = 10^{14}$.

64 **The Great Pyramid** The Great Pyramid of Egypt is 147 meters high, with a square base of side 230 meters (see the figure). Approximate, to the nearest degree, the angle φ formed when an observer stands at the midpoint of one of the sides and views the apex of the pyramid.

Exercise 64

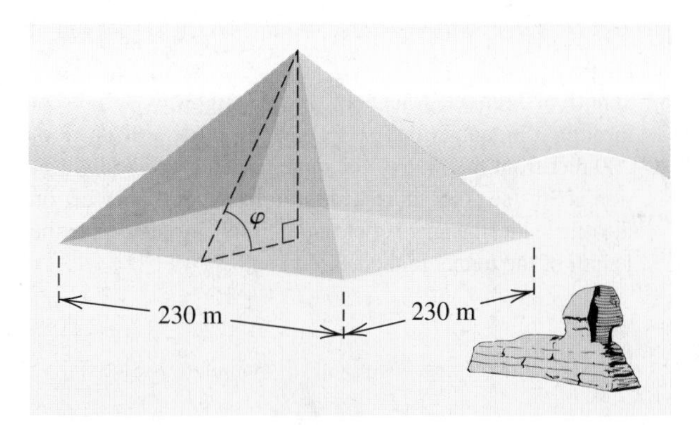

65 **Venus** When viewed from Earth over a period of time, the planet Venus appears to move back and forth along a line segment with the sun at its midpoint (see the figure). If *ES* is approximately 92,900,000 miles, then the maximum apparent distance of Venus from the sun occurs when angle *SEV* is approximately 47°. Assume that the orbit of Venus is circular and estimate the distance of Venus from the sun.

Exercise 65

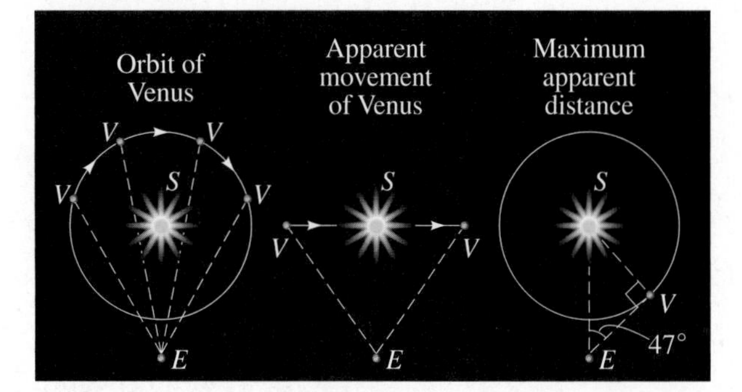

66 Constructing a conical cup A conical paper cup is constructed by removing a sector from a circle of radius 5 inches and attaching edge OA to OB (see the figure). Find angle AOB so that the cup has a depth of 4 inches.

Exercise 66

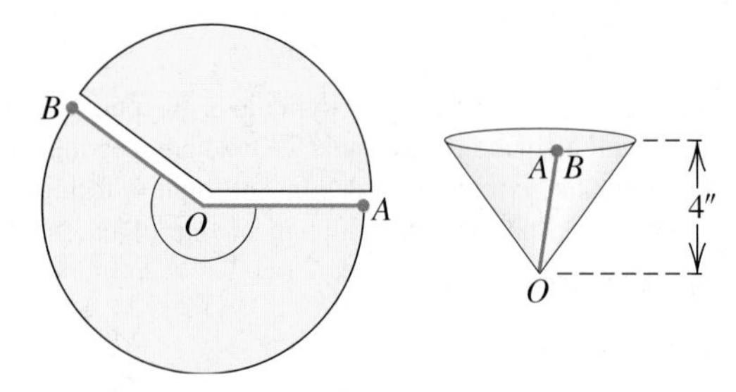

67 Length of a tunnel A tunnel for a new highway is to be cut through a mountain that is 260 feet high. At a distance of 200 feet from the base of the mountain, the angle of elevation is 36° (see the figure). From a distance of 150 feet on the other side, the angle of elevation is 47°. Approximate the length of the tunnel to the nearest foot.

Exercise 67

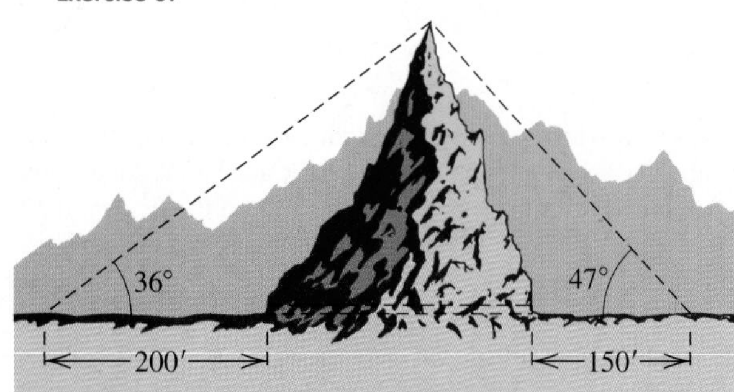

68 Height of a skyscraper When a certain skyscraper is viewed from the top of a building 50 feet tall, the angle of elevation is 59° (see the figure). When viewed from the street next to the shorter building, the angle of elevation is 62°.

(a) Approximately how far apart are the two structures?

(b) Approximate the height of the skyscraper to the nearest tenth of a foot.

Exercise 68

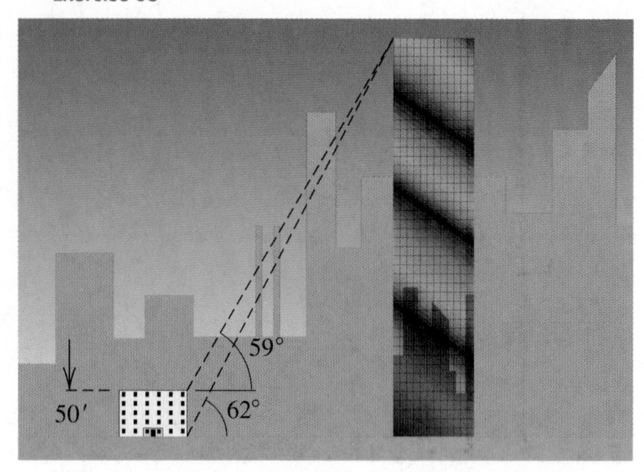

69 Height of a mountain When a mountaintop is viewed from the point P shown in the figure, the angle of elevation is α. From a point Q, which is d miles closer to the mountain, the angle of elevation increases to β.

(a) Show that the height h of the mountain is given by

$$h = \frac{d}{\cot \alpha - \cot \beta}.$$

(b) If $d = 2$ mi, $\alpha = 15°$, and $\beta = 20°$, approximate the height of the mountain.

Exercise 69

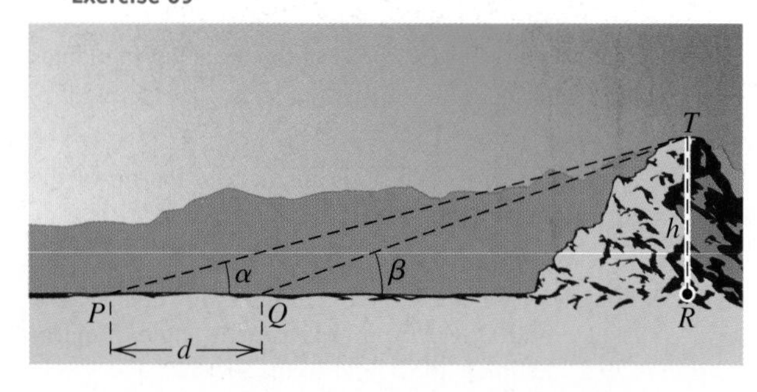

70 Height of a building An observer of height h stands on an incline at a distance d from the base of a building of height T, as shown in the figure. The angle of elevation from the observer to the top of the building is θ, and the incline makes an angle of α with the horizontal.

(a) Express T in terms of h, d, α, and θ.

(b) If $h = 6$ ft, $d = 50$ ft, $\alpha = 15°$, and $\theta = 31.4°$, estimate the height of the building.

Exercise 70

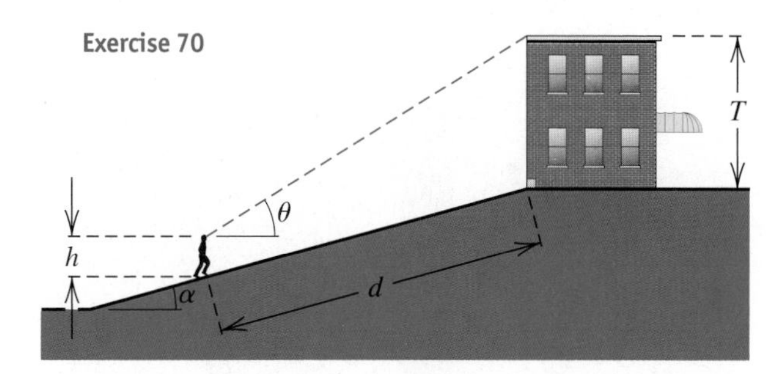

71 **Illuminance** A spotlight with intensity 5000 candles is located 15 feet above a stage. If the spotlight is rotated through an angle θ as shown in the figure, the illuminance E (in foot-candles) in the lighted area of the stage is given by

$$E = \frac{5000 \cos \theta}{s^2},$$

where s is the distance (in feet) that the light must travel.

(a) Find the illuminance if the spotlight is rotated through an angle of 30°.

(b) The maximum illuminance occurs when $\theta = 0°$. For what value of θ is the illuminance one-half the maximum value?

Exercise 71

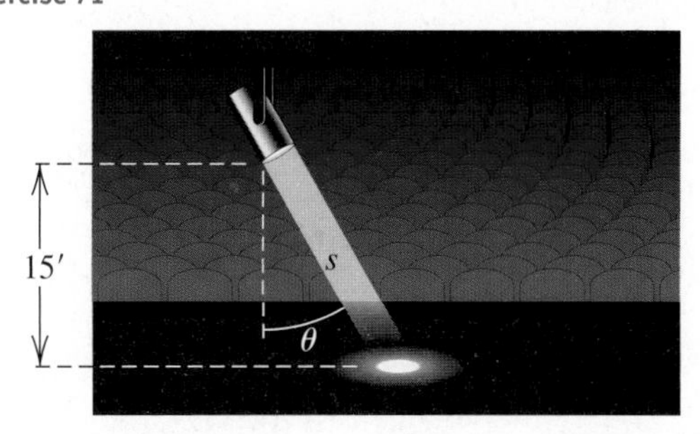

72 **Height of a mountain** If a mountaintop is viewed from a point P due south of the mountain, the angle of elevation is α (see the figure). If viewed from a point Q that is d miles east of P, the angle of elevation is β.

(a) Show that the height h of the mountain is given by

$$h = \frac{d \sin \alpha \sin \beta}{\sqrt{\sin^2 \alpha - \sin^2 \beta}}.$$

(b) If $\alpha = 30°$, $\beta = 20°$, and $d = 10$ mi, approximate h to the nearest hundredth of a mile.

Exercise 72

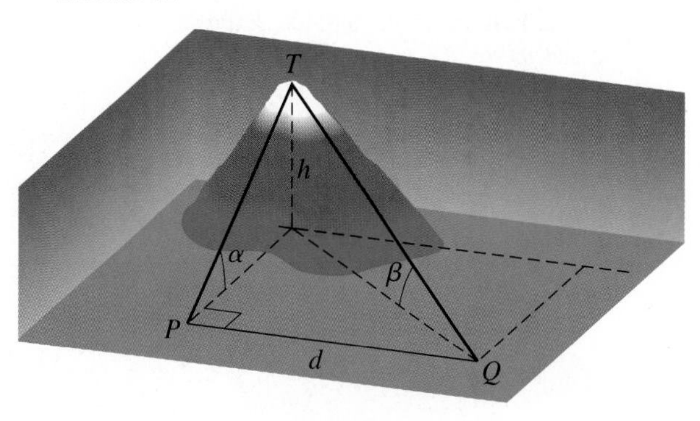

73 **Mounting a projection unit** The manufacturer of a computerized projection system recommends that a projection unit be mounted on the ceiling as shown in the figure. The distance from the end of the mounting bracket to the center of the screen is 85.5 inches, and the angle of depression is 30°.

(a) If the thickness of the screen is disregarded, how far from the wall should the bracket be mounted?

(b) If the bracket is 18 inches long and the screen is 6 feet high, determine the distance from the ceiling to the top edge of the screen.

Exercise 73

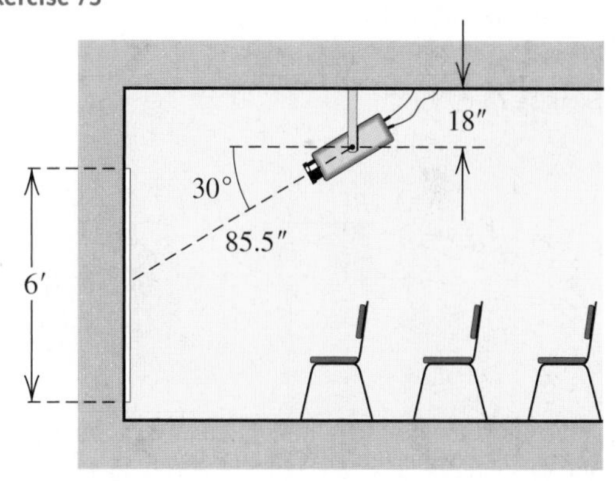

74 **Pyramid relationships** A pyramid has a square base and congruent triangular faces. Let θ be the angle that the altitude a of a triangular face makes with the altitude y of the pyramid, and let x be the length of a side (see the figure on the next page).

(a) Express the total surface area S of the four faces in terms of a and θ.

(b) The volume V of the pyramid equals one-third the area of the base times the altitude. Express V in terms of a and θ.

Exercise 74

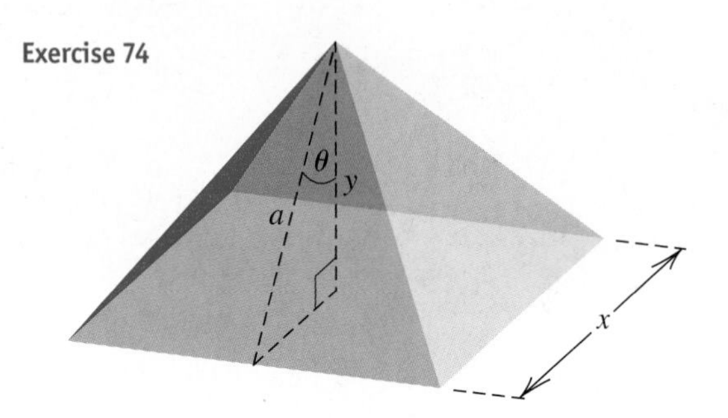

75 **Surveying a bluff** A surveyor, using a transit, sights the edge B of a bluff, as shown in the left part of the figure (not drawn to scale). Because of the curvature of Earth, the true elevation h of the bluff is larger than that measured by the surveyor. A cross-sectional schematic view of Earth is shown in the right part of the figure.

(a) If s is the length of arc PQ and R is the distance from P to the center C of Earth, express h in terms of R and s.

(b) If $R = 4000$ mi and $s = 50$ mi, estimate the elevation of the bluff in feet.

Exercise 75

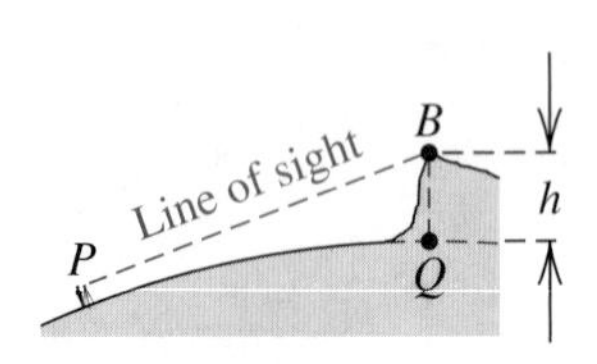

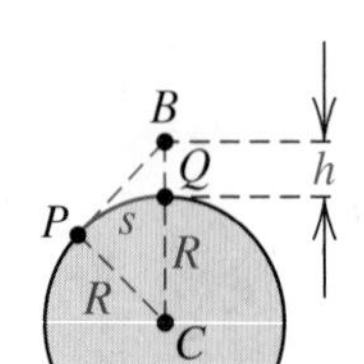

76 **Earthquake response** To simulate the response of a structure to an earthquake, an engineer must choose a shape for the initial displacement of the beams in the building. When the beam has length L feet and the maximum displacement is a feet, the equation

$$y = a - a \cos \frac{\pi}{2L} x$$

has been used by engineers to estimate the displacement y (see the figure). If $a = 1$ and $L = 10$, sketch the graph of the equation for $0 \leq x \leq 10$.

Exercise 76

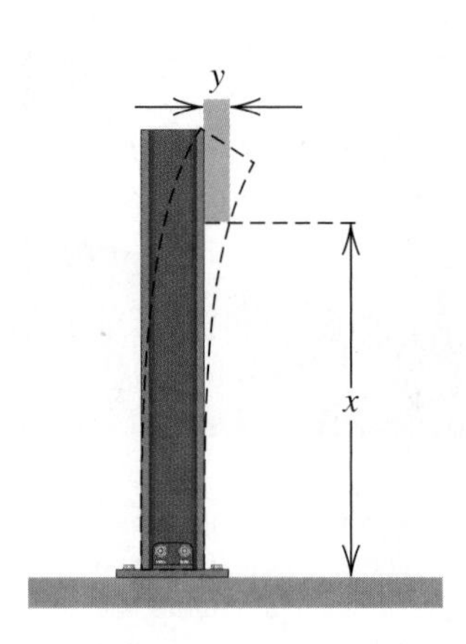

77 **Circadian rhythms** The variation in body temperature is an example of a circadian rhythm, a cycle of a biological process that repeats itself approximately every 24 hours. Body temperature is highest about 5 P.M. and lowest at 5 A.M. Let y denote the body temperature (in °F), and let $t = 0$ correspond to midnight. If the low and high body temperatures are 98.3° and 98.9°, respectively, find an equation having the form $y = 98.6 + a \sin (bt + c)$ that fits this information.

78 **Temperature variation in Ottawa** The annual variation in temperature T (in °C) in Ottawa, Canada, may be approximated by

$$T(t) = 15.8 \sin \left[\frac{\pi}{6}(t - 3) \right] + 5,$$

where t is the time in months and $t = 0$ corresponds to January 1.

(a) Sketch the graph of T for $0 \leq t \leq 12$.

(b) Find the highest temperature of the year and the date on which it occurs.

79 **Water demand** A reservoir supplies water to a community. During the summer months, the demand $D(t)$ for water (in ft^3/day) is given by

$$D(t) = 2000 \sin \frac{\pi}{90} t + 4000,$$

where t is time in days and $t = 0$ corresponds to the beginning of summer.

(a) Sketch the graph of D for $0 \leq t \leq 90$.

(b) When is the demand for water the greatest?

80 Bobbing cork A cork bobs up and down in a lake. The distance from the bottom of the lake to the center of the cork at time $t \geq 0$ is given by $s(t) = 12 + \cos \pi t$, where $s(t)$ is in feet and t is in seconds.

(a) Describe the motion of the cork for $0 \leq t \leq 2$.

(b) During what time intervals is the cork rising?

CHAPTER 5 DISCUSSION EXERCISES

1 Graph $y = \sin(ax)$ on $[-2\pi, 2\pi]$ by $[-1, 1]$ for $a = 15$, 30, and 45. Discuss the accuracy of the graphs and the graphing capabilities (in terms of precision) of your graphing calculator. (*Note:* If something strange doesn't occur for $a = 45$, keep increasing a until it does.)

2 Find the maximum integer k on your calculator such that $\sin(10^k)$ can be evaluated. Now discuss how you can evaluate $\sin(10^{k+1})$ on the same calculator, and then actually find that value.

3 Determine the number of solutions of the equation

$$\cos x + \cos 2x + \cos 3x = \pi.$$

4 Discuss the relationships among periodic functions, one-to-one functions, and inverse functions. With these relationships in mind, discuss what must happen for the trigonometric functions to have inverses.

5 Graph $y_1 = x$, $y_2 = \sin x$, and $y_3 = \tan x$ on $[-0.1, 0.1]$ by $[-0.1, 0.1]$. Create a table of values for these three functions, with small positive values (on the order of 10^{-10} or so). What conclusion can you draw from the graph and the table?

6 **Racetrack coordinates** Shown in the figure is a circular racetrack of diameter 2 kilometers. All races begin at S and proceed in a counterclockwise direction. Approximate, to four decimal places, the coordinates of the point at which the following races end relative to a rectangular coordinate system with origin at the center of the track and S on the positive x-axis.

Exercise 6

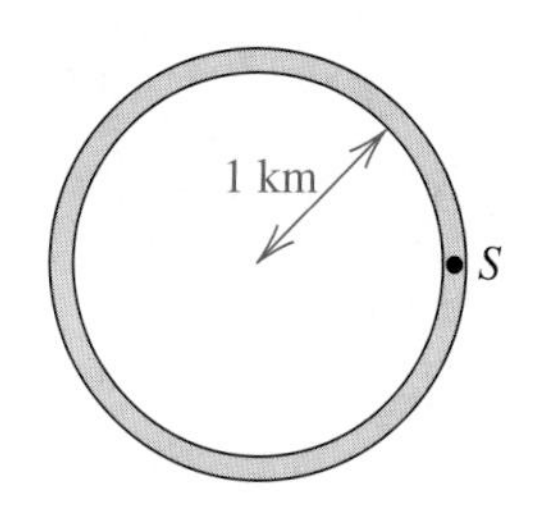

(a) A drag race of length 2 kilometers

(b) An endurance race of length 500 kilometers

7 **Racetrack coordinates** Work Exercise 6 for the track shown in the figure, if the origin of the rectangular coordinate system is at the center of the track and S is on the negative y-axis.

Exercise 7

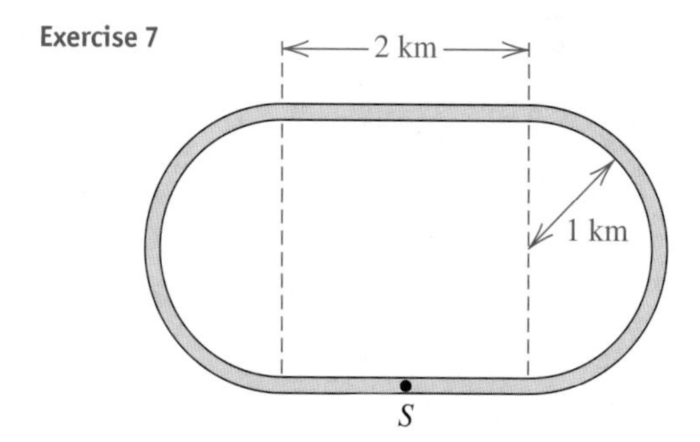

8 **Outboard motor propeller** A 90-horsepower outboard motor at full throttle will rotate its propeller at 5000 revolutions per minute.

(a) Find the angular speed ω of the propeller in radians per second.

(b) The center of a 10-inch-diameter propeller is located 18 inches below the surface of the water. Express the depth $D(t) = a\cos(\omega t + c) + d$ of a point on the edge of a propeller blade as a function of time t, where t is in seconds. Assume that the point is initially at a depth of 23 inches.

(c) Graphically determine the number of times the propeller rotates in 0.12 second.

6

Analytic Trigonometry

In advanced mathematics, the natural sciences, and engineering, it is sometimes necessary to simplify complicated trigonometric expressions and to solve equations that involve trigonometric functions. These topics are discussed in the first two sections of this chapter. We then derive many useful formulas with respect to sums, differences, and multiples; for reference they are listed on the inside back cover of the text. In addition to formal manipulations, we also consider numerous applications of these formulas. The last section contains the definitions and properties of the inverse trigonometric functions.

6.1 Verifying Trigonometric Identities

A **trigonometric expression** contains symbols involving trigonometric functions.

ILLUSTRATION Trigonometric Expressions

- $x + \sin x$
- $\dfrac{\sqrt{\theta} + 2^{\sin\theta}}{\cot\theta}$
- $\dfrac{\cos(3t + 1)}{t^2 + \tan^2(2 - t^2)}$

We assume that the domain of each variable in a trigonometric expression is the set of real numbers or angles for which the expression is meaningful. To provide manipulative practice in simplifying complicated trigonometric expressions, we shall use the fundamental identities (see page 377) and algebraic manipulations, as we did in Examples 5 and 6 of Section 5.2. In the first three examples our method consists of transforming the left-hand side of a given identity into the right-hand side, or vice versa.

EXAMPLE 1 Verifying an identity

Verify the identity $\sec\alpha - \cos\alpha = \sin\alpha\tan\alpha$.

▶ **SOLUTION** We transform the left-hand side into the right-hand side:

$$
\begin{aligned}
\sec\alpha - \cos\alpha &= \frac{1}{\cos\alpha} - \cos\alpha && \text{reciprocal identity} \\
&= \frac{1 - \cos^2\alpha}{\cos\alpha} && \text{add expressions} \\
&= \frac{\sin^2\alpha}{\cos\alpha} && \sin^2\alpha + \cos^2\alpha = 1 \\
&= \sin\alpha\left(\frac{\sin\alpha}{\cos\alpha}\right) && \text{equivalent expression} \\
&= \sin\alpha\tan\alpha && \text{tangent identity}
\end{aligned}
$$

In Section 5.2, we discussed providing numerical support for identities by examining a table of values. We can also provide graphical support for identities by examining the graphs of the left-hand side and the right-hand side of the proposed identity. If the graphs are equal (with the exception of holes in the graphs), we say that the graphs support the identity. If the graphs do not match, then the proposed identity is false.

▶ Tutorial available at www.thomsonedu.com/login

The graph in Figure 1 lends graphical support to our verification in Example 1. It is the graph (in radian and dot mode) of both

$$Y_1 = 1/\cos(X) - \cos(X) \qquad \text{and} \qquad Y_2 = \sin(X)\tan(X).$$

Figure 1
$[-2\pi, 2\pi, \pi/2]$ by $[-5, 5]$

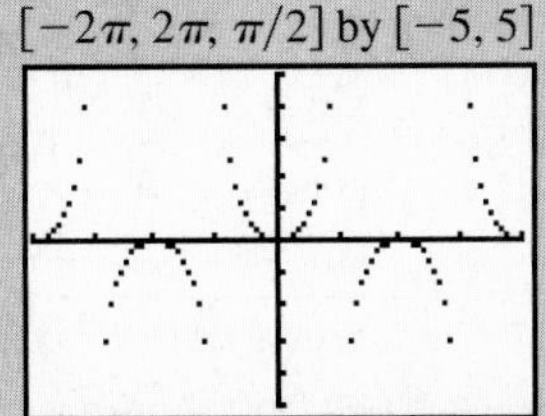

Figure 2

X	Y1	Y2
8.1158	-3.605	-3.605
8.3776	-1.5	-1.5
8.6394	-.7071	-.7071
8.9012	-.2887	-.2887
9.163	-.0694	-.0694
9.4248	0	-4E-26
9.6866	-.0694	-.0694

Y2=-4E-26

The values of Y_1 and Y_2 in Figure 2 also lend numerical support to our verification. There may be small discrepancies in the values, as the highlighted value illustrates.

Other Variations of Graphical Support for Example 1

(1) Graph Y_1 and $Y_3 = Y_2 + 1$, as shown in Figures 3 and 4. This allows us to see the graph of Y_2 shifted upward one unit, rather than on top of Y_1.

Figure 3

Figure 4
$[-2\pi, 2\pi, \pi/2]$ by $[-4, 4]$

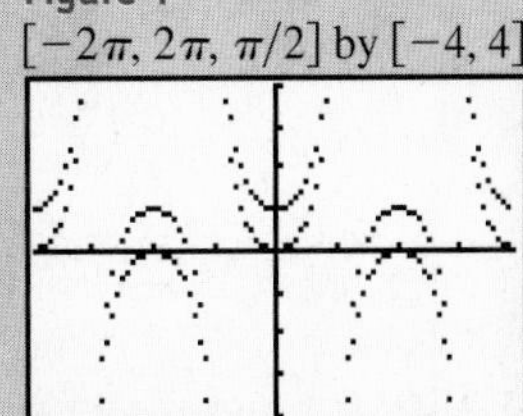

(2) Graph $Y_3 = Y_1 - Y_2 + 1$, as shown in Figures 5 and 6. If the proposed identity is true, then $Y_1 - Y_2$ will be zero, so the graph of Y_3 will be the graph of the line $y = 1$ with holes where Y_1 or Y_2 is undefined.

Figure 5

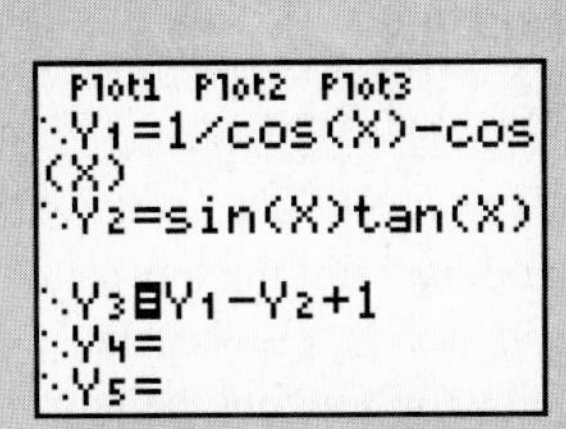

Figure 6
$[-2\pi, 2\pi, \pi/2]$ by $[-4, 4]$

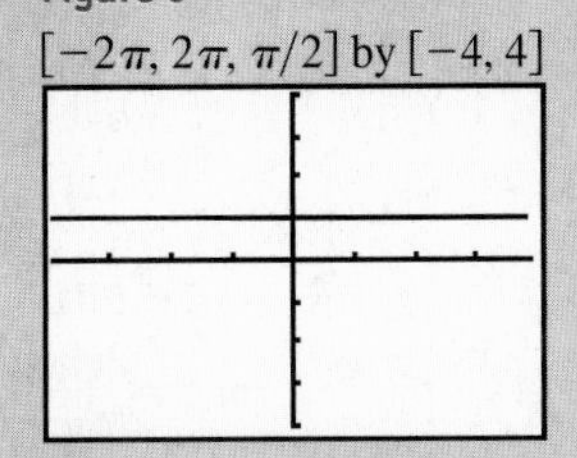

(3) Graph $Y_3 = (Y_1 = Y_2)$, as shown in Figures 7 and 8. When $Y_1 = Y_2$ is true, the value of Y_3 is 1. The graph of Y_3 will be the graph of the line $y = 1$ with holes where Y_1 or Y_2 is undefined.

Figure 7

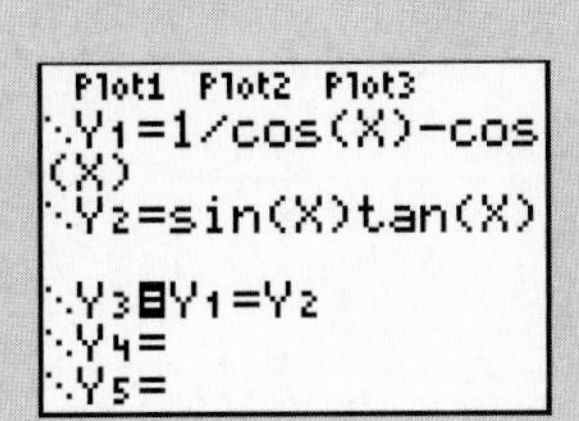

Figure 8
$[-2\pi, 2\pi, \pi/2]$ by $[-4, 4]$

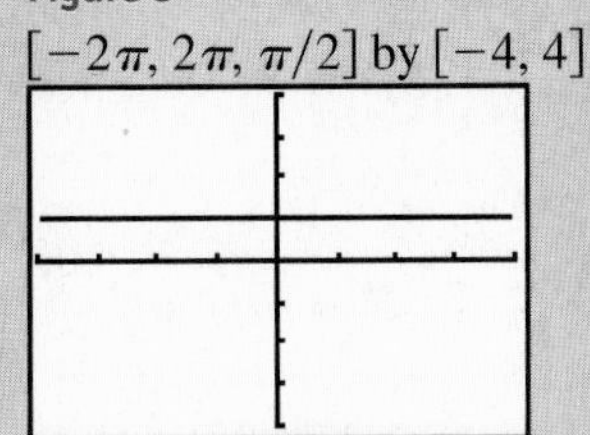

EXAMPLE 2 Verifying an identity

Verify the identity $\sec\theta = \sin\theta(\tan\theta + \cot\theta)$.

▶ **SOLUTION** Since the expression on the right-hand side is more complicated than that on the left-hand side, we transform the right-hand side into the left-hand side:

$$\begin{aligned}
\sin\theta(\tan\theta + \cot\theta) &= \sin\theta\left(\frac{\sin\theta}{\cos\theta} + \frac{\cos\theta}{\sin\theta}\right) && \text{tangent and cotangent identities}\\
&= \sin\theta\left(\frac{\sin^2\theta + \cos^2\theta}{\cos\theta\sin\theta}\right) && \text{add fractions}\\
&= \sin\theta\left(\frac{1}{\cos\theta\sin\theta}\right) && \text{Pythagorean identity}\\
&= \frac{1}{\cos\theta} && \text{cancel } \sin\theta\\
&= \sec\theta && \text{reciprocal identity}
\end{aligned}$$

Figure 9

X	Y_1	Y_2
-.5236	1.1547	1.1547
-.2618	1.0353	1.0353
0	1	ERROR
.2618	1.0353	1.0353
.5236	1.1547	1.1547
.7854	1.4142	1.4142
1.0472	2	2

Y_2=ERROR

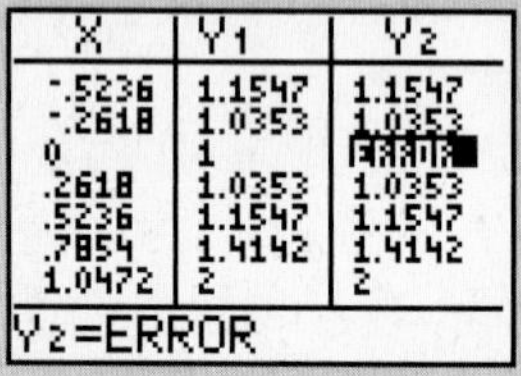

The table (with $\Delta\text{Tbl} = \pi/12$) in Figure 9 shows some values of

$$Y_1 = 1/\cos(X) \quad \text{and} \quad Y_2 = \sin(X)(\tan(X) + 1/\tan(X)),$$

the left-hand and right-hand sides of the identity in Example 2. Note that for $X = 0$, $Y_1 = 1$, but Y_2 has "ERROR." This results from the use of $1/\tan(X)$ for $\cot(X)$ in Y_2; for $X = 0$, we are trying to divide by zero.

EXAMPLE 3 Verifying an identity

Verify the identity $\dfrac{\cos x}{1 - \sin x} = \dfrac{1 + \sin x}{\cos x}$.

▶ **SOLUTION** Since the denominator of the left-hand side is a binomial and the denominator of the right-hand side is a monomial, we change the form of the fraction on the left-hand side by multiplying the numerator and denominator by the conjugate of the denominator and then use one of the Pythagorean identities:

$$\begin{aligned}
\frac{\cos x}{1 - \sin x} &= \frac{\cos x}{1 - \sin x}\cdot\frac{1 + \sin x}{1 + \sin x} && \text{multiply numerator and denominator by } 1 + \sin x\\
&= \frac{\cos x(1 + \sin x)}{1 - \sin^2 x} && \text{property of quotients}\\
&= \frac{\cos x(1 + \sin x)}{\cos^2 x} && \sin^2 x + \cos^2 x = 1\\
&= \frac{1 + \sin x}{\cos x} && \text{cancel } \cos x
\end{aligned}$$

▶ **Tutorial available at www.thomsonedu.com/login**

Another technique for showing that an equation $p = q$ is an identity is to begin by transforming the left-hand side p into another expression s, making sure that each step is *reversible*—that is, making sure it is possible to transform s back into p by reversing the procedure used in each step. In this case, the equation $p = s$ is an identity. Next, as a *separate* exercise, we show that the right-hand side q can also be transformed into the expression s by means of reversible steps and, therefore, that $q = s$ is an identity. It then follows that $p = q$ is an identity. This method is illustrated in the next example.

EXAMPLE 4 Verifying an identity

Verify the identity $(\tan\theta - \sec\theta)^2 = \dfrac{1 - \sin\theta}{1 + \sin\theta}$.

SOLUTION We shall verify the identity by showing that each side of the equation can be transformed into the same expression. First we work only with the left-hand side:

Work with the left-hand side.

$$(\tan\theta - \sec\theta)^2 = \tan^2\theta - 2\tan\theta\sec\theta + \sec^2\theta \quad \text{square expression}$$

$$= \left(\frac{\sin\theta}{\cos\theta}\right)^2 - 2\left(\frac{\sin\theta}{\cos\theta}\right)\left(\frac{1}{\cos\theta}\right) + \left(\frac{1}{\cos\theta}\right)^2 \quad \text{tangent and reciprocal identities}$$

$$= \frac{\sin^2\theta}{\cos^2\theta} - \frac{2\sin\theta}{\cos^2\theta} + \frac{1}{\cos^2\theta} \quad \text{equivalent expression}$$

$$= \frac{\sin^2\theta - 2\sin\theta + 1}{\cos^2\theta} \quad \text{add fractions}$$

At this point it may not be obvious how we can obtain the right-hand side of the given equation from the last expression. Thus, we next work with only the right-hand side and try to obtain the last expression. Multiplying numerator and denominator by the conjugate of the denominator gives us the following:

equivalent expressions

Work with the right-hand side.

$$\frac{1 - \sin\theta}{1 + \sin\theta} = \frac{1 - \sin\theta}{1 + \sin\theta} \cdot \frac{1 - \sin\theta}{1 - \sin\theta} \quad \text{multiply numerator and denominator by } 1 - \sin\theta$$

$$= \frac{1 - 2\sin\theta + \sin^2\theta}{1 - \sin^2\theta} \quad \text{property of quotients}$$

$$= \frac{1 - 2\sin\theta + \sin^2\theta}{\cos^2\theta} \quad \sin^2\theta + \cos^2\theta = 1$$

The last expression is the same as that obtained from $(\tan\theta - \sec\theta)^2$. Since all steps are reversible, the given equation is an identity.

In calculus it is sometimes convenient to change the form of certain algebraic expressions by making a **trigonometric substitution,** as illustrated in the following example.

Tutorial available at www.thomsonedu.com/login

EXAMPLE 5 Making a trigonometric substitution

Express $\sqrt{a^2 - x^2}$ in terms of a trigonometric function of θ, without radicals, by making the substitution $x = a \sin \theta$ for $-\pi/2 \leq \theta \leq \pi/2$ and $a > 0$.

▶ SOLUTION We proceed as follows:

$$\begin{aligned}
\sqrt{a^2 - x^2} &= \sqrt{a^2 - (a \sin \theta)^2} && \text{let } x = a \sin \theta \\
&= \sqrt{a^2 - a^2 \sin^2 \theta} && \text{law of exponents} \\
&= \sqrt{a^2(1 - \sin^2 \theta)} && \text{factor out } a^2 \\
&= \sqrt{a^2 \cos^2 \theta} && \sin^2 \theta + \cos^2 \theta = 1 \\
&= \sqrt{(a \cos \theta)^2} && c^2d^2 = (cd)^2 \\
&= |a \cos \theta| && \sqrt{c^2} = |c| \\
&= |a||\cos \theta| && |cd| = |c||d| \\
&= a \cos \theta && \text{see below}
\end{aligned}$$

The last equality is true because (1) if $a > 0$, then $|a| = a$, and (2) if $-\pi/2 \leq \theta \leq \pi/2$, then $\cos \theta \geq 0$ and hence $|\cos \theta| = \cos \theta$.

We may also use a geometric solution. If $x = a \sin \theta$, then $\sin \theta = x/a$, and the triangle in Figure 10 illustrates the problem for $0 < \theta < \pi/2$. The third side of the triangle, $\sqrt{a^2 - x^2}$, can be found by using the Pythagorean theorem. From the figure we can see that

$$\cos \theta = \frac{\sqrt{a^2 - x^2}}{a} \quad \text{or, equivalently,} \quad \sqrt{a^2 - x^2} = a \cos \theta.$$

Figure 10

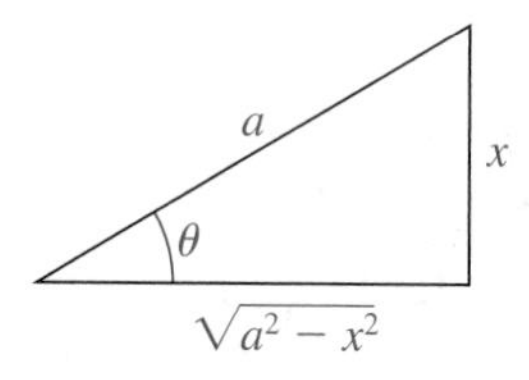

6.1 Exercises

Exer. 1–50: Verify the identity.

▶ 1 $\csc \theta - \sin \theta = \cot \theta \cos \theta$

2 $\sin x + \cos x \cot x = \csc x$

3 $\dfrac{\sec^2 2u - 1}{\sec^2 2u} = \sin^2 2u$

4 $\tan t + 2 \cos t \csc t = \sec t \csc t + \cot t$

5 $\dfrac{\csc^2 \theta}{1 + \tan^2 \theta} = \cot^2 \theta$

6 $(\tan u + \cot u)(\cos u + \sin u) = \csc u + \sec u$

7 $\dfrac{1 + \cos 3t}{\sin 3t} + \dfrac{\sin 3t}{1 + \cos 3t} = 2 \csc 3t$

8 $\tan^2 \alpha - \sin^2 \alpha = \tan^2 \alpha \sin^2 \alpha$

9 $\dfrac{1}{1 - \cos \gamma} + \dfrac{1}{1 + \cos \gamma} = 2 \csc^2 \gamma$

▶ 10 $\dfrac{1 + \csc 3\beta}{\sec 3\beta} - \cot 3\beta = \cos 3\beta$

11 $(\sec u - \tan u)(\csc u + 1) = \cot u$

12 $\dfrac{\cot \theta - \tan \theta}{\sin \theta + \cos \theta} = \csc \theta - \sec \theta$

13 $\csc^4 t - \cot^4 t = \csc^2 t + \cot^2 t$

14 $\cos^4 2\theta + \sin^2 2\theta = \cos^2 2\theta + \sin^4 2\theta$

▶ **Tutorial available at www.thomsonedu.com/login**

15 $\dfrac{\cos\beta}{1-\sin\beta} = \sec\beta + \tan\beta$

16 $\dfrac{1}{\csc y - \cot y} = \csc y + \cot y$

17 $\dfrac{\tan^2 x}{\sec x + 1} = \dfrac{1-\cos x}{\cos x}$

18 $\dfrac{\cot x}{\csc x + 1} = \dfrac{\csc x - 1}{\cot x}$

19 $\dfrac{\cot 4u - 1}{\cot 4u + 1} = \dfrac{1-\tan 4u}{1+\tan 4u}$

20 $\dfrac{1+\sec 4x}{\sin 4x + \tan 4x} = \csc 4x$

21 $\sin^4 r - \cos^4 r = \sin^2 r - \cos^2 r$

22 $\sin^4\theta + 2\sin^2\theta\cos^2\theta + \cos^4\theta = 1$

23 $\tan^4 k - \sec^4 k = 1 - 2\sec^2 k$

24 $\sec^4 u - \sec^2 u = \tan^2 u + \tan^4 u$

25 $(\sec t + \tan t)^2 = \dfrac{1+\sin t}{1-\sin t}$

26 $\sec^2\gamma + \tan^2\gamma = (1-\sin^4\gamma)\sec^4\gamma$

▶ 27 $(\sin^2\theta + \cos^2\theta)^3 = 1$

28 $\dfrac{\sin t}{1-\cos t} = \csc t + \cot t$

29 $\dfrac{1+\csc\beta}{\cot\beta+\cos\beta} = \sec\beta$

30 $\dfrac{\cos^3 x - \sin^3 x}{\cos x - \sin x} = 1 + \sin x\cos x$

31 $(\csc t - \cot t)^4(\csc t + \cot t)^4 = 1$

32 $(a\cos t - b\sin t)^2 + (a\sin t + b\cos t)^2 = a^2 + b^2$

33 $\dfrac{\sin\alpha\cos\beta + \cos\alpha\sin\beta}{\cos\alpha\cos\beta - \sin\alpha\sin\beta} = \dfrac{\tan\alpha+\tan\beta}{1-\tan\alpha\tan\beta}$

34 $\dfrac{\tan u - \tan v}{1+\tan u\tan v} = \dfrac{\cot v - \cot u}{\cot u\cot v + 1}$

35 $\dfrac{\tan\alpha}{1+\sec\alpha} + \dfrac{1+\sec\alpha}{\tan\alpha} = 2\csc\alpha$

36 $\dfrac{\csc x}{1+\csc x} - \dfrac{\csc x}{1-\csc x} = 2\sec^2 x$

37 $\dfrac{1}{\tan\beta + \cot\beta} = \sin\beta\cos\beta$

38 $\dfrac{\cot y - \tan y}{\sin y\cos y} = \csc^2 y - \sec^2 y$

39 $\sec\theta + \csc\theta - \cos\theta - \sin\theta = \sin\theta\tan\theta + \cos\theta\cot\theta$

40 $\sin^3 t + \cos^3 t = (1 - \sin t\cos t)(\sin t + \cos t)$

41 $(1-\tan^2\phi)^2 = \sec^4\phi - 4\tan^2\phi$

42 $\cos^4 w + 1 - \sin^4 w = 2\cos^2 w$

43 $\dfrac{\cot(-t) + \tan(-t)}{\cot t} = -\sec^2 t$

44 $\dfrac{\csc(-t) - \sin(-t)}{\sin(-t)} = \cot^2 t$

▶ 45 $\log 10^{\tan t} = \tan t$

46 $10^{\log|\sin t|} = |\sin t|$

47 $\ln\cot x = -\ln\tan x$

48 $\ln\sec\theta = -\ln\cos\theta$

49 $\ln|\sec\theta + \tan\theta| = -\ln|\sec\theta - \tan\theta|$

50 $\ln|\csc x - \cot x| = -\ln|\csc x + \cot x|$

Exer. 51–60: Show that the equation is *not* an identity. (*Hint:* Find one number for which the equation is false.)

51 $\cos t = \sqrt{1-\sin^2 t}$

52 $\sqrt{\sin^2 t + \cos^2 t} = \sin t + \cos t$

▶ 53 $\sqrt{\sin^2 t} = \sin t$

54 $\sec t = \sqrt{\tan^2 t + 1}$

55 $(\sin\theta + \cos\theta)^2 = \sin^2\theta + \cos^2\theta$

56 $\log\left(\dfrac{1}{\sin t}\right) = \dfrac{1}{\log\sin t}$

▶ **Tutorial available at www.thomsonedu.com/login**

57 $\cos(-t) = -\cos t$

58 $\sin(t + \pi) = \sin t$

59 $\cos(\sec t) = 1$

60 $\cot(\tan\theta) = 1$

Exer. 61–64: Either show that the equation *is* an identity or show that the equation *is not* an identity.

61 $(\sec x + \tan x)^2 = 2\tan x(\tan x + \sec x)$

62 $\dfrac{\tan^2 x}{\sec x - 1} = \sec x$

63 $\cos x(\tan x + \cot x) = \csc x$

64 $\csc^2 x + \sec^2 x = \csc^2 x \sec^2 x$

Exer. 65–68: Refer to Example 5. Make the trigonometric substitution $x = a\sin\theta$ for $-\pi/2 < \theta < \pi/2$ and $a > 0$. Use fundamental identities to simplify the resulting expression.

65 $(a^2 - x^2)^{3/2}$

▶ 66 $\dfrac{\sqrt{a^2 - x^2}}{x}$

67 $\dfrac{x^2}{\sqrt{a^2 - x^2}}$

68 $\dfrac{1}{x\sqrt{a^2 - x^2}}$

Exer. 69–72: Make the trigonometric substitution

$$x = a\tan\theta \quad \text{for } -\pi/2 < \theta < \pi/2 \text{ and } a > 0.$$

Simplify the resulting expression.

69 $\sqrt{a^2 + x^2}$

70 $\dfrac{1}{\sqrt{a^2 + x^2}}$

71 $\dfrac{1}{x^2 + a^2}$

72 $\dfrac{(x^2 + a^2)^{3/2}}{x}$

Exer. 73–76: Make the trigonometric substitution

$$x = a\sec\theta \quad \text{for } 0 < \theta < \pi/2 \text{ and } a > 0.$$

Simplify the resulting expression.

73 $\sqrt{x^2 - a^2}$

74 $\dfrac{1}{x^2\sqrt{x^2 - a^2}}$

75 $x^3\sqrt{x^2 - a^2}$

76 $\dfrac{\sqrt{x^2 - a^2}}{x^2}$

Exer. 77–80: Use the graph of f to find the simplest expression $g(x)$ such that the equation $f(x) = g(x)$ is an identity. Verify this identity.

77 $f(x) = \dfrac{\sin^2 x - \sin^4 x}{(1 - \sec^2 x)\cos^4 x}$

78 $f(x) = \dfrac{\sin x - \sin^3 x}{\cos^4 x + \cos^2 x \sin^2 x}$

79 $f(x) = \sec x(\sin x \cos x + \cos^2 x) - \sin x$

80 $f(x) = \dfrac{\sin^3 x + \sin x \cos^2 x}{\csc x} + \dfrac{\cos^3 x + \cos x \sin^2 x}{\sec x}$

6.2 Trigonometric Equations

A **trigonometric equation** is an equation that contains trigonometric expressions. Each identity considered in the preceding section is an example of a trigonometric equation with every number (or angle) in the domain of the variable a solution of the equation. If a trigonometric equation is not an identity, we often find solutions by using techniques similar to those used for algebraic equations. The main difference is that we first solve the trigonometric equation for $\sin x$, $\cos\theta$, and so on, and then find values of x or θ that satisfy the equation. Solutions may be expressed either as real numbers or as angles. Throughout our work we shall use the following rule: *If degree measure is not specified, then solutions of a trigonometric equation should be expressed in radian measure (or as real numbers).* If solutions in degree measure are desired, an appropriate statement will be included in the example or exercise.

▶ Tutorial available at www.thomsonedu.com/login

EXAMPLE 1 **Solving a trigonometric equation involving the sine function**

Find the solutions of the equation $\sin\theta = \frac{1}{2}$ if

(a) θ is in the interval $[0, 2\pi)$

(b) θ is any real number

▶ SOLUTION

(a) If $\sin\theta = \frac{1}{2}$, then the reference angle for θ is $\theta_R = \pi/6$. If we regard θ as an angle in standard position, then, since $\sin\theta > 0$, the terminal side is in either quadrant I or quadrant II, as illustrated in Figure 1. Thus, there are two solutions for $0 \le \theta < 2\pi$:

Figure 1

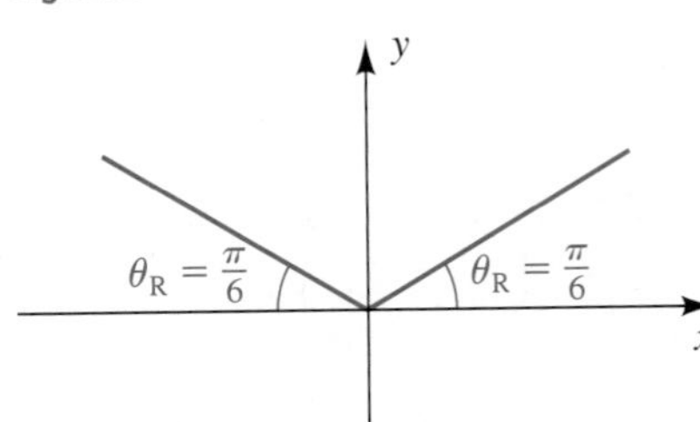

$$\theta = \frac{\pi}{6} \quad \text{and} \quad \theta = \pi - \frac{\pi}{6} = \frac{5\pi}{6}$$

(b) Since the sine function has period 2π, we may obtain all solutions by adding multiples of 2π to $\pi/6$ and $5\pi/6$. This gives us

$$\theta = \frac{\pi}{6} + 2\pi n \quad \text{and} \quad \theta = \frac{5\pi}{6} + 2\pi n \quad \text{for every integer } n.$$

Figure 2

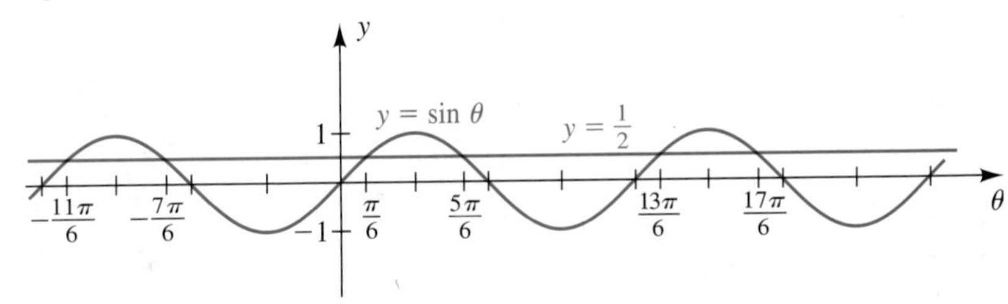

An alternative (graphical) solution involves determining where the graph of $y = \sin\theta$ intersects the horizontal line $y = \frac{1}{2}$, as illustrated in Figure 2.

EXAMPLE 2 **Solving a trigonometric equation involving the tangent function**

Find the solutions of the equation $\tan u = -1$.

▶ SOLUTION Since the tangent function has period π, it is sufficient to find one real number u such that $\tan u = -1$ and then add multiples of π.

A portion of the graph of $y = \tan u$ is sketched in Figure 3 on the next page. Since $\tan(3\pi/4) = -1$, one solution is $3\pi/4$; hence,

$$\text{if} \quad \tan u = -1, \quad \text{then} \quad u = \frac{3\pi}{4} + \pi n \quad \text{for every integer } n.$$

(continued)

▶ Tutorial available at www.thomsonedu.com/login

Figure 3 $y = \tan u$

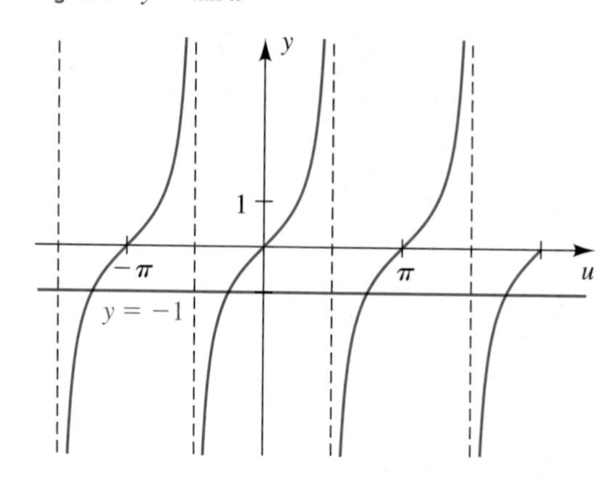

We could also have chosen $-\pi/4$ (or some other number u such that $\tan u = -1$) for the initial solution and written

$$u = -\frac{\pi}{4} + \pi n \quad \text{for every integer } n.$$

Figure 4

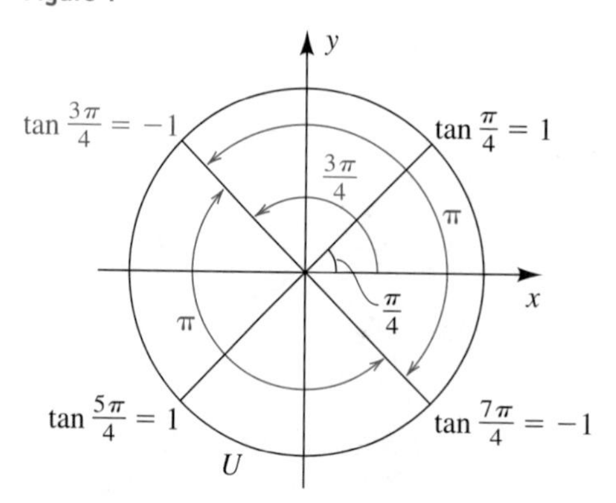

An alternative solution involves a unit circle. Using $\tan 3\pi/4 = -1$ and the fact that the period of the tangent is π, we can see from Figure 4 that the desired solutions are

$$u = \frac{3\pi}{4} + \pi n \quad \text{for every integer } n.$$

EXAMPLE 3 **Solving a trigonometric equation involving multiple angles**

(a) Solve the equation $\cos 2x = 0$, and express the solutions both in radians and in degrees.

(b) Find the solutions that are in the interval $[0, 2\pi)$ and, equivalently, $[0°, 360°)$.

▶ SOLUTION

(a) We proceed as follows, where n denotes any integer:

$\cos 2x = 0$	given
$\cos \theta = 0$	let $\theta = 2x$
$\theta = \dfrac{\pi}{2} + \pi n$	refer to Figure 5
$2x = \dfrac{\pi}{2} + \pi n$	$\theta = 2x$
$x = \dfrac{\pi}{4} + \dfrac{\pi}{2} n$	divide by 2

Figure 5

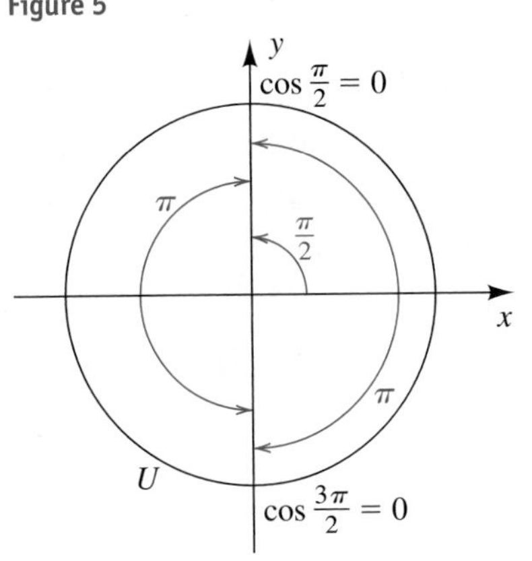

In degrees, we have $x = 45° + 90°n$.

(b) We may find particular solutions of the equation by substituting integers for n in either of the formulas for x obtained in part (a). Several such solutions are listed in the following table.

n	$\dfrac{\pi}{4} + \dfrac{\pi}{2} n$	$45° + 90°n$
-1	$\dfrac{\pi}{4} + \dfrac{\pi}{2}(-1) = -\dfrac{\pi}{4}$	$45° + 90°(-1) = -45°$
0	$\dfrac{\pi}{4} + \dfrac{\pi}{2}(0) = \dfrac{\pi}{4}$	$45° + 90°(0) = 45°$
1	$\dfrac{\pi}{4} + \dfrac{\pi}{2}(1) = \dfrac{3\pi}{4}$	$45° + 90°(1) = 135°$
2	$\dfrac{\pi}{4} + \dfrac{\pi}{2}(2) = \dfrac{5\pi}{4}$	$45° + 90°(2) = 225°$
3	$\dfrac{\pi}{4} + \dfrac{\pi}{2}(3) = \dfrac{7\pi}{4}$	$45° + 90°(3) = 315°$
4	$\dfrac{\pi}{4} + \dfrac{\pi}{2}(4) = \dfrac{9\pi}{4}$	$45° + 90°(4) = 405°$

Note that the solutions in the interval $[0, 2\pi)$ or, equivalently, $[0°, 360°)$ are given by $n = 0$, $n = 1$, $n = 2$, and $n = 3$. These solutions are

$$\frac{\pi}{4}, \frac{3\pi}{4}, \frac{5\pi}{4}, \frac{7\pi}{4} \qquad \text{or, equivalently,} \qquad 45°, 135°, 225°, 315°.$$

EXAMPLE 4 **Solving a trigonometric equation by factoring**

Solve the equation $\sin \theta \tan \theta = \sin \theta$.

▶ **Tutorial available at www.thomsonedu.com/login**

▶ SOLUTION

$$\begin{aligned} \sin\theta\tan\theta &= \sin\theta && \text{given} \\ \sin\theta\tan\theta - \sin\theta &= 0 && \text{make one side 0} \\ \sin\theta(\tan\theta - 1) &= 0 && \text{factor out } \sin\theta \\ \sin\theta = 0, \quad \tan\theta - 1 &= 0 && \text{zero factor theorem} \\ \sin\theta = 0, \quad \tan\theta &= 1 && \text{solve for } \sin\theta \text{ and } \tan\theta \end{aligned}$$

The solutions of the equation $\sin\theta = 0$ are $0, \pm\pi, \pm 2\pi, \ldots$. Thus,

$$\text{if} \quad \sin\theta = 0, \quad \text{then} \quad \theta = \pi n \quad \text{for every integer } n.$$

The tangent function has period π, and hence we find the solutions of the equation $\tan\theta = 1$ that are in the interval $(-\pi/2, \pi/2)$ and then add multiples of π. Since the only solution of $\tan\theta = 1$ in $(-\pi/2, \pi/2)$ is $\pi/4$, we see that

$$\text{if} \quad \tan\theta = 1, \quad \text{then} \quad \theta = \frac{\pi}{4} + \pi n \quad \text{for every integer } n.$$

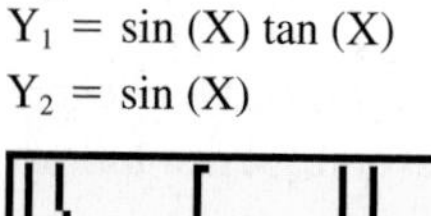

Figure 6
$Y_1 = \sin(X)\tan(X)$
$Y_2 = \sin(X)$

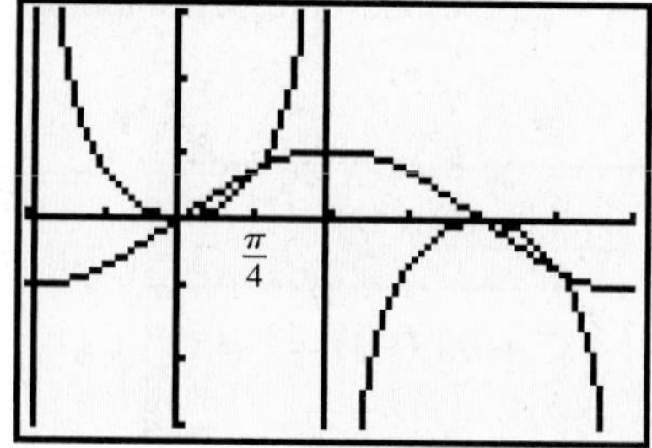

$[-\pi/2, 3\pi/2, \pi/4]$ by $[-3, 3]$

Thus, the solutions of the given equation are

$$\pi n \quad \text{and} \quad \frac{\pi}{4} + \pi n \quad \text{for every integer } n.$$

Some *particular* solutions, obtained by letting $n = 0$, $n = 1$, $n = 2$, and $n = -1$, are

$$0, \quad \frac{\pi}{4}, \quad \pi, \quad \frac{5\pi}{4}, \quad 2\pi, \quad \frac{9\pi}{4}, \quad -\pi, \quad \text{and} \quad -\frac{3\pi}{4}.$$

The graph in Figure 6 supports our conclusion.

In Example 4 it would have been incorrect to begin by dividing both sides by $\sin\theta$, since we would have lost the solutions of $\sin\theta = 0$.

EXAMPLE 5 Solving a trigonometric equation by factoring

Solve the equation $2\sin^2 t - \cos t - 1 = 0$, and express the solutions both in radians and in degrees.

▶ SOLUTION It appears that we have a quadratic equation in either $\sin t$ or $\cos t$. We do not have a simple substitution for $\cos t$ in terms of $\sin t$, but we do have one for $\sin^2 t$ in terms of $\cos^2 t$ ($\sin^2 t = 1 - \cos^2 t$), so we shall first express the equation in terms of $\cos t$ alone and then solve by factoring.

This is a quadratic equation in $\cos t$*, so you could use the quadratic formula at this point. If you do so, remember to solve for* cos t, not t.

$$\begin{aligned} 2\sin^2 t - \cos t - 1 &= 0 && \text{given} \\ 2(1 - \cos^2 t) - \cos t - 1 &= 0 && \sin^2 t + \cos^2 t = 1 \\ -2\cos^2 t - \cos t + 1 &= 0 && \text{simplify} \\ 2\cos^2 t + \cos t - 1 &= 0 && \text{multiply by } -1 \\ (2\cos t - 1)(\cos t + 1) &= 0 && \text{factor} \\ 2\cos t - 1 = 0, \quad \cos t + 1 &= 0 && \text{zero factor theorem} \\ \cos t = \tfrac{1}{2}, \quad \cos t &= -1 && \text{solve for } \cos t \end{aligned}$$

▶ **Tutorial available at www.thomsonedu.com/login**

Since the cosine function has period 2π, we may find all solutions of these equations by adding multiples of 2π to the solutions that are in the interval $[0, 2\pi)$.

If $\cos t = \frac{1}{2}$, the reference angle is $\pi/3$ (or 60°). Since $\cos t$ is positive, the angle of radian measure t is in either quadrant I or quadrant IV. Hence, in the interval $[0, 2\pi)$, we see that

$$\text{if} \quad \cos t = \frac{1}{2}, \quad \text{then} \quad t = \frac{\pi}{3} \quad \text{or} \quad t = 2\pi - \frac{\pi}{3} = \frac{5\pi}{3}.$$

Figure 7
$Y_1 = 2(\sin(X))^2 - \cos(X) - 1$

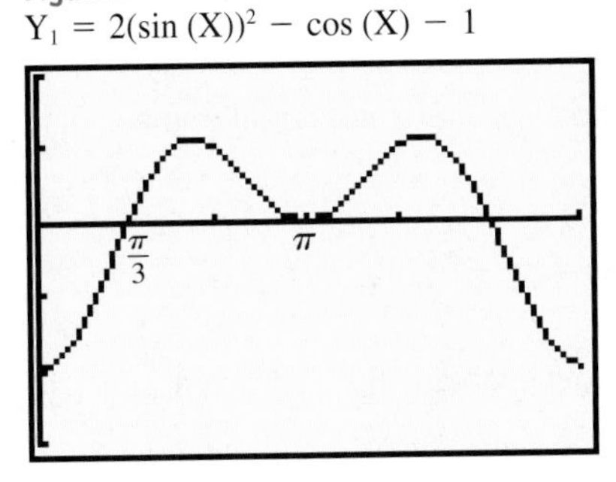

$[0, 2\pi, \pi/3]$ by $[-3, 2]$

Referring to the graph of the cosine function, we see that

$$\text{if} \quad \cos t = -1, \quad \text{then} \quad t = \pi.$$

Thus, the solutions of the given equation are the following, where n is any integer:

$$\frac{\pi}{3} + 2\pi n, \quad \frac{5\pi}{3} + 2\pi n, \quad \text{and} \quad \pi + 2\pi n$$

In degree measure, we have

$$60° + 360°n, \quad 300° + 360°n, \quad \text{and} \quad 180° + 360°n.$$

The graph in Figure 7 supports our conclusion.

EXAMPLE 6 Solving a trigonometric equation by factoring

Find the solutions of $4 \sin^2 x \tan x - \tan x = 0$ that are in the interval $[0, 2\pi)$.

SOLUTION

$4 \sin^2 x \tan x - \tan x = 0$	given
$\tan x (4 \sin^2 x - 1) = 0$	factor out $\tan x$
$\tan x = 0, \quad 4 \sin^2 x - 1 = 0$	zero factor theorem
$\tan x = 0, \quad \sin^2 x = \frac{1}{4}$	solve for $\tan x$, $\sin^2 x$
$\tan x = 0, \quad \sin x = \pm\frac{1}{2}$	solve for $\sin x$

Figure 8

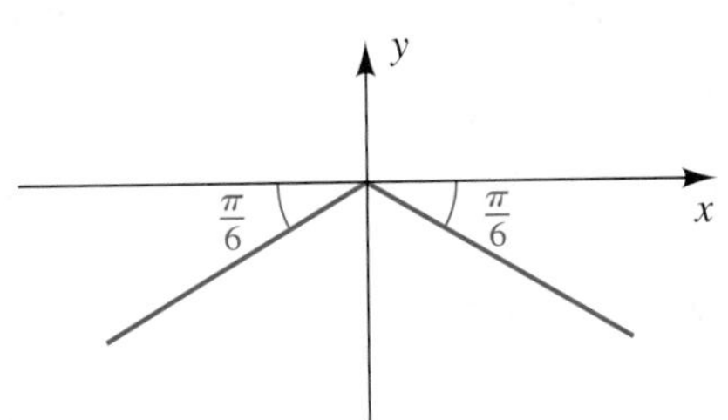

The reference angle $\pi/6$ for the third and fourth quadrants is shown in Figure 8. These angles, $7\pi/6$ and $11\pi/6$, are the solutions of the equation $\sin x = -\frac{1}{2}$ for $0 \le x < 2\pi$. The solutions of all three equations are listed in the following table.

Equation	Solutions in $[0, 2\pi)$	Refer to
$\tan x = 0$	$0, \pi$	Figure 3
$\sin x = \frac{1}{2}$	$\frac{\pi}{6}, \frac{5\pi}{6}$	Example 1
$\sin x = -\frac{1}{2}$	$\frac{7\pi}{6}, \frac{11\pi}{6}$	Figure 8 (use reference angle)

Thus, the given equation has the six solutions listed in the second column of the table.

▶ Tutorial available at www.thomsonedu.com/login

EXAMPLE 7 Solving a trigonometric equation involving multiple angles

Find the solutions of $\csc^4 2u - 4 = 0$.

▶ SOLUTION

$$\csc^4 2u - 4 = 0 \qquad \text{given}$$
$$(\csc^2 2u - 2)(\csc^2 2u + 2) = 0 \qquad \text{difference of two squares}$$
$$\csc^2 2u - 2 = 0, \quad \csc^2 2u + 2 = 0 \qquad \text{zero factor theorem}$$
$$\csc^2 2u = 2, \quad \csc^2 2u = -2 \qquad \text{solve for } \csc^2 2u$$
$$\csc 2u = \pm\sqrt{2}, \quad \csc 2u = \pm\sqrt{-2} \qquad \text{take square roots}$$

The second equation has no solution because $\sqrt{-2}$ is not a real number. The first equation is equivalent to

$$\sin 2u = \pm\frac{1}{\sqrt{2}} = \pm\frac{\sqrt{2}}{2}.$$

Since the reference angle for $2u$ is $\pi/4$, we obtain the following table, in which n denotes any integer.

Equation	Solution for $2u$	Solution for u
$\sin 2u = \frac{\sqrt{2}}{2}$	$2u = \frac{\pi}{4} + 2\pi n$	$u = \frac{\pi}{8} + \pi n$
	$2u = \frac{3\pi}{4} + 2\pi n$	$u = \frac{3\pi}{8} + \pi n$
$\sin 2u = -\frac{\sqrt{2}}{2}$	$2u = \frac{5\pi}{4} + 2\pi n$	$u = \frac{5\pi}{8} + \pi n$
	$2u = \frac{7\pi}{4} + 2\pi n$	$u = \frac{7\pi}{8} + \pi n$

The solutions of the given equation are listed in the last column. Note that *all* of these solutions can be written in the one form

$$u = \frac{\pi}{8} + \frac{\pi}{4}n.$$

The next example illustrates the use of a calculator in solving a trigonometric equation.

EXAMPLE 8 Approximating the solutions of a trigonometric equation

Approximate, to the nearest degree, the solutions of the following equation in the interval $[0°, 360°)$:

$$5 \sin\theta \tan\theta - 10\tan\theta + 3\sin\theta - 6 = 0$$

▶ **Tutorial available at www.thomsonedu.com/login**

▶ SOLUTION

$$
\begin{aligned}
5 \sin \theta \tan \theta - 10 \tan \theta + 3 \sin \theta - 6 &= 0 && \text{given}\\
(5 \sin \theta \tan \theta - 10 \tan \theta) + (3 \sin \theta - 6) &= 0 && \text{group terms}\\
5 \tan \theta (\sin \theta - 2) + 3(\sin \theta - 2) &= 0 && \text{factor each group}\\
(5 \tan \theta + 3)(\sin \theta - 2) &= 0 && \text{factor out } (\sin \theta - 2)\\
5 \tan \theta + 3 = 0, \qquad \sin \theta - 2 &= 0 && \text{zero factor theorem}\\
\tan \theta = -\tfrac{3}{5}, \qquad \sin \theta &= 2 && \text{solve for } \tan \theta \text{ and } \sin \theta
\end{aligned}
$$

The equation $\sin \theta = 2$ has no solution, since $-1 \leq \sin \theta \leq 1$ for every θ. For $\tan \theta = -\frac{3}{5}$, we use a calculator in degree mode, obtaining

$$\theta = \tan^{-1}\left(-\tfrac{3}{5}\right) \approx -31°.$$

Hence, the reference angle is $\theta_R \approx 31°$. Since θ is in either quadrant II or quadrant IV, we obtain the following solutions:

$$\theta = 180° - \theta_R \approx 180° - 31° = 149°$$

$$\theta = 360° - \theta_R \approx 360° - 31° = 329°$$

Let's take a look at how a graphing calculator could help us solve the equation in Example 8.

Approximating the Solutions of a Trigonometric Equation

TI-83/4 Plus | **TI-86**

Select radian mode and dot mode. Assign the left-hand side of the equation to Y_1.

Set the viewing rectangle to $[0, 2\pi]$ by $[-20, 20, 10]$. Graph Y_1.

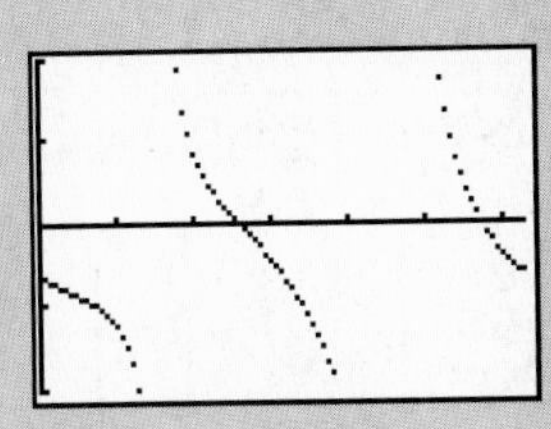

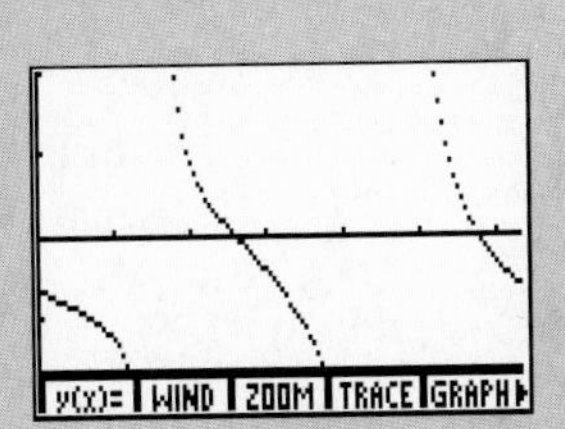

(continued)

▶ **Tutorial available at www.thomsonedu.com/login**

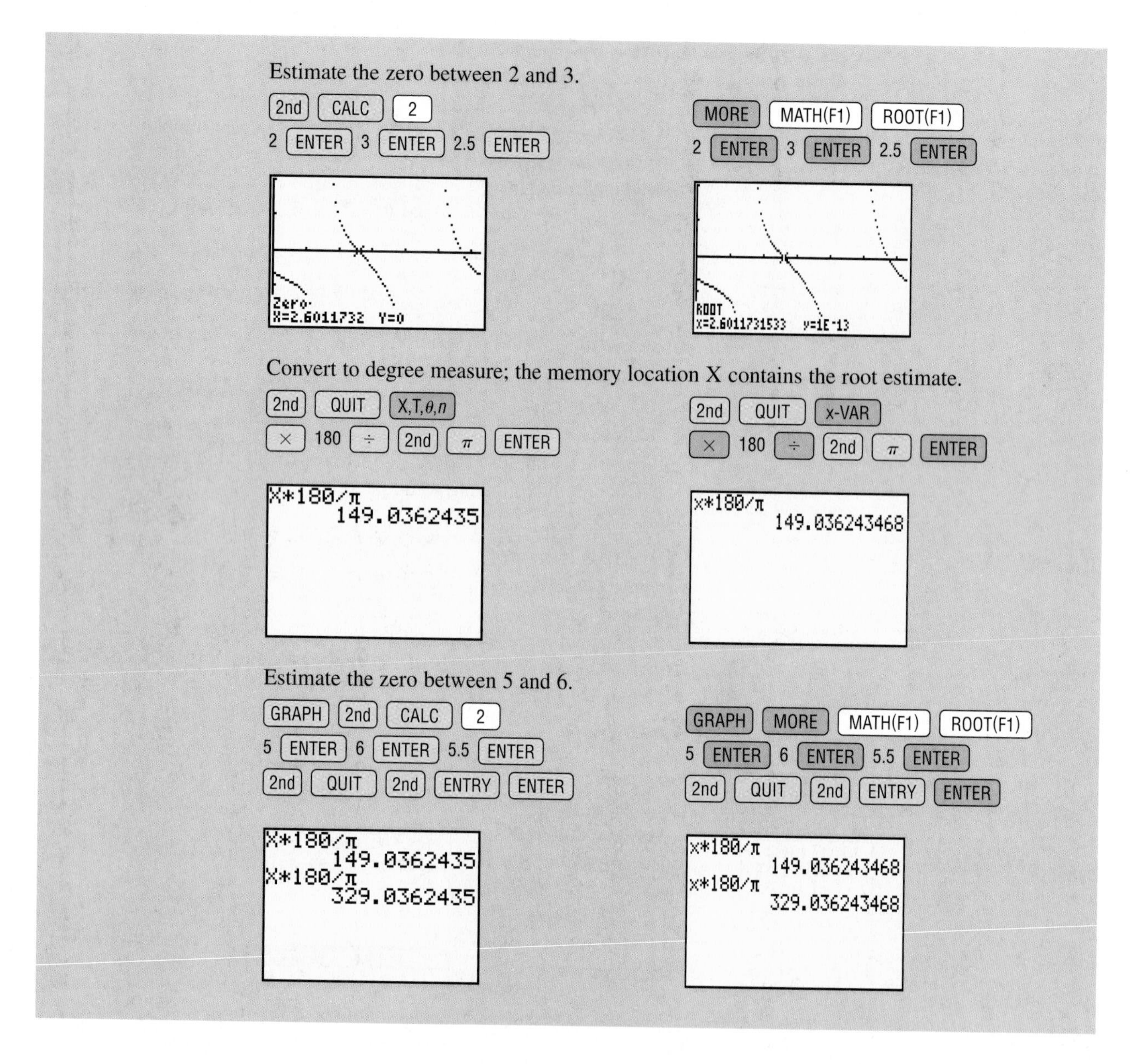

EXAMPLE 9 **Investigating the number of hours of daylight**

In Boston, the number of hours of daylight $D(t)$ at a particular time of the year may be approximated by

$$D(t) = 3\sin\left[\frac{2\pi}{365}(t - 79)\right] + 12,$$

with t in days and $t = 0$ corresponding to January 1. How many days of the year have more than 10.5 hours of daylight?

Figure 9

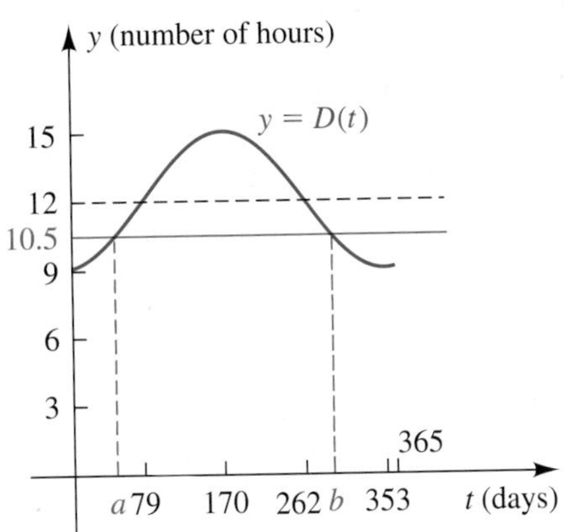

▶ **SOLUTION** The graph of D was discussed in Example 12 of Section 5.5 and is resketched in Figure 9. As illustrated in the figure, if we can find two numbers a and b with $D(a) = 10.5$, $D(b) = 10.5$, and $0 < a < b < 365$, then there will be more than 10.5 hours of daylight in the tth day of the year if $a < t < b$.

Let us solve the equation $D(t) = 10.5$ as follows:

$$3 \sin\left[\frac{2\pi}{365}(t - 79)\right] + 12 = 10.5 \qquad \text{let } D(t) = 10.5$$

$$3 \sin\left[\frac{2\pi}{365}(t - 79)\right] = -1.5 \qquad \text{subtract 12}$$

$$\sin\left[\frac{2\pi}{365}(t - 79)\right] = -0.5 = -\frac{1}{2} \qquad \text{divide by 3}$$

If $\sin \theta = -\frac{1}{2}$, then the reference angle is $\pi/6$ and the angle θ is in either quadrant III or quadrant IV. Thus, we can find the numbers a and b by solving the equations

$$\frac{2\pi}{365}(t - 79) = \frac{7\pi}{6} \quad \text{and} \quad \frac{2\pi}{365}(t - 79) = \frac{11\pi}{6}.$$

From the first of these equations we obtain

$$t - 79 = \frac{7\pi}{6} \cdot \frac{365}{2\pi} = \frac{2555}{12} \approx 213,$$

and hence $\qquad t \approx 213 + 79, \quad \text{or} \quad t \approx 292.$

Similarly, the second equation gives us $t \approx 414$. Since the period of the function D is 365 days (see Figure 9), we obtain

$$t \approx 414 - 365, \quad \text{or} \quad t \approx 49.$$

Thus, there will be at least 10.5 hours of daylight from $t = 49$ to $t = 292$—that is, for 243 days of the year.

A graphical solution of the next example was given in Example 14 of Section 5.5.

EXAMPLE 10 Finding the minimum current in an electrical circuit

The current I (in amperes) in an alternating current circuit at time t (in seconds) is given by

$$I = 30 \sin\left(50\pi t - \frac{7\pi}{3}\right).$$

Find the smallest exact value of t for which $I = 15$.

▶ **Tutorial available at www.thomsonedu.com/login**

▶ **SOLUTION** Letting $I = 15$ in the given formula, we obtain

$$15 = 30 \sin\left(50\pi t - \frac{7\pi}{3}\right) \quad \text{or, equivalently,} \quad \sin\left(50\pi t - \frac{7\pi}{3}\right) = \frac{1}{2}.$$

Thus, the reference angle is $\pi/6$, and consequently

$$50\pi t - \frac{7\pi}{3} = \frac{\pi}{6} + 2\pi n \qquad \text{or} \qquad 50\pi t - \frac{7\pi}{3} = \frac{5\pi}{6} + 2\pi n,$$

where n is any integer. Solving for t gives us

$$t = \frac{\frac{15}{6} + 2n}{50} \qquad \text{or} \qquad t = \frac{\frac{19}{6} + 2n}{50}.$$

The smallest positive value of t will occur when one of the numerators of these two fractions has its least positive value. Since $\frac{15}{6} = 2.5$, $\frac{19}{6} \approx 3.17$, and $2(-1) = -2$, we see that the smallest positive value of t occurs when $n = -1$ in the first fraction—that is, when

$$t = \frac{\frac{15}{6} + 2(-1)}{50} = \frac{1}{100}.$$

The next example illustrates how a graphing utility can aid in solving a complicated trigonometric equation.

EXAMPLE 11 Using a graph to determine solutions of a trigonometric equation

Find the solutions of the following equation that are in the interval $[0, 2\pi)$:

$$\sin x + \sin 2x + \sin 3x = 0$$

SOLUTION We assign $\sin x + \sin 2x + \sin 3x$ to Y_1. Since $|\sin\theta| \leq 1$ for $\theta = x$, $2x$, and $3x$, the left-hand side of the equation is between -3 and 3, and we choose the viewing rectangle $[0, 2\pi, \pi/4]$ by $[-3, 3]$ and obtain a sketch similar to Figure 10. Using a root feature, we obtain the following approximations for the x-intercepts—that is, the approximate solutions of the given equation in $[0, 2\pi)$:

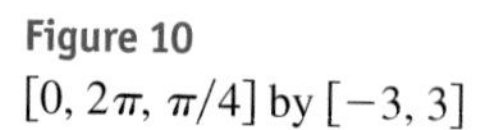

Figure 10
$[0, 2\pi, \pi/4]$ by $[-3, 3]$

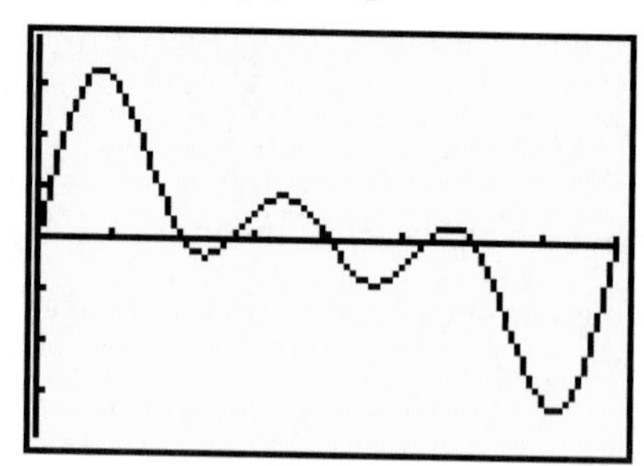

$$0, \quad 1.57, \quad 2.09, \quad 3.14, \quad 4.19, \quad 4.71$$

Changing to degree measure and rounding off to the nearest degree, we obtain

$$0°, \quad 90°, \quad 120°, \quad 180°, \quad 240°, \quad \text{and} \quad 270°.$$

Converting these degree measures to radian measures gives us

$$0, \quad \frac{\pi}{2}, \quad \frac{2\pi}{3}, \quad \pi, \quad \frac{4\pi}{3}, \quad \text{and} \quad \frac{3\pi}{2}.$$

Checking these values in the given equation, we see that all six are solutions. Figure 10 suggests that the graph has period 2π. After studying Section 6.4,

▶ Tutorial available at www.thomsonedu.com/login

you will be able to change the form of Y_1 and *prove* that the period is 2π and, therefore, that *all* solutions of the given equation can be obtained by adding integer multiples of 2π.

In the preceding example we were able to use a graphing utility to help us find the *exact* solutions of the equation. For many equations that occur in applications, however, it is only possible to approximate the solutions.

6.2 Exercises

Exer. 1–38: Find all solutions of the equation.

1 $\sin x = -\frac{\sqrt{2}}{2}$

2 $\cos t = -1$

3 $\tan \theta = \sqrt{3}$

4 $\cot \alpha = -\frac{1}{\sqrt{3}}$

5 $\sec \beta = 2$

6 $\csc \gamma = \sqrt{2}$

7 $\sin x = \frac{\pi}{2}$

8 $\cos x = -\frac{\pi}{3}$

9 $\cos \theta = \frac{1}{\sec \theta}$

10 $\csc \theta \sin \theta = 1$

11 $2 \cos 2\theta - \sqrt{3} = 0$

12 $2 \sin 3\theta + \sqrt{2} = 0$

13 $\sqrt{3} \tan \frac{1}{3} t = 1$

14 $\cos \frac{1}{4} x = -\frac{\sqrt{2}}{2}$

15 $\sin \left(\theta + \frac{\pi}{4}\right) = \frac{1}{2}$

16 $\cos \left(x - \frac{\pi}{3}\right) = -1$

17 $\sin \left(2x - \frac{\pi}{3}\right) = \frac{1}{2}$

18 $\cos \left(4x - \frac{\pi}{4}\right) = \frac{\sqrt{2}}{2}$

19 $2 \cos t + 1 = 0$

20 $\cot \theta + 1 = 0$

21 $\tan^2 x = 1$

22 $4 \cos \theta - 2 = 0$

23 $(\cos \theta - 1)(\sin \theta + 1) = 0$

24 $2 \cos x = \sqrt{3}$

25 $\sec^2 \alpha - 4 = 0$

26 $3 - \tan^2 \beta = 0$

27 $\sqrt{3} + 2 \sin \beta = 0$

28 $4 \sin^2 x - 3 = 0$

29 $\cot^2 x - 3 = 0$

30 $(\sin t - 1) \cos t = 0$

31 $(2 \sin \theta + 1)(2 \cos \theta + 3) = 0$

32 $(2 \sin u - 1)\left(\cos u - \sqrt{2}\right) = 0$

33 $\cos x + 1 = 2 \sin^2 x$

34 $2 \cos^2 x + \sin x = 1$

35 $\sin 2x (\csc 2x - 2) = 0$

36 $\tan \alpha + \tan^2 \alpha = 0$

37 $\cos (\ln x) = 0$

38 $\ln (\sin x) = 0$

Exer. 39–62: Find the solutions of the equation that are in the interval $[0, 2\pi)$.

39 $\cos \left(2x - \frac{\pi}{4}\right) = 0$

40 $\sin \left(3x - \frac{\pi}{4}\right) = 1$

41 $2 - 8\cos^2 t = 0$

42 $\cot^2 \theta - \cot \theta = 0$

43 $2\sin^2 u = 1 - \sin u$

44 $2\cos^2 t + 3\cos t + 1 = 0$

45 $\tan^2 x \sin x = \sin x$

46 $\sec \beta \csc \beta = 2\csc \beta$

47 $2\cos^2 \gamma + \cos \gamma = 0$

48 $\sin x - \cos x = 0$

49 $\sin^2 \theta + \sin \theta - 6 = 0$

50 $2\sin^2 u + \sin u - 6 = 0$

51 $1 - \sin t = \sqrt{3}\cos t$

52 $\cos \theta - \sin \theta = 1$

53 $\cos \alpha + \sin \alpha = 1$

54 $\sqrt{3}\sin t + \cos t = 1$

55 $2\tan t - \sec^2 t = 0$

56 $\tan \theta + \sec \theta = 1$

57 $\cot \alpha + \tan \alpha = \csc \alpha \sec \alpha$

58 $\sin x + \cos x \cot x = \csc x$

59 $2\sin^3 x + \sin^2 x - 2\sin x - 1 = 0$

60 $\sec^5 \theta = 4\sec \theta$

61 $2\tan t \csc t + 2\csc t + \tan t + 1 = 0$

62 $2\sin v \csc v - \csc v = 4\sin v - 2$

Exer. 63–68: Approximate, to the nearest 10′, the solutions of the equation in the interval [0°, 360°).

63 $\sin^2 t - 4\sin t + 1 = 0$

64 $\cos^2 t - 4\cos t + 2 = 0$

65 $\tan^2 \theta + 3\tan \theta + 2 = 0$

66 $2\tan^2 x - 3\tan x - 1 = 0$

67 $12\sin^2 u - 5\sin u - 2 = 0$

68 $5\cos^2 \alpha + 3\cos \alpha - 2 = 0$

69 **Tidal waves** A tidal wave of height 50 feet and period 30 minutes is approaching a sea wall that is 12.5 feet above sea level (see the figure). From a particular point on shore, the distance y from sea level to the top of the wave is given by

$$y = 25\cos \frac{\pi}{15}t,$$

with t in minutes. For approximately how many minutes of each 30-minute period is the top of the wave above the level of the top of the sea wall?

Exercise 69

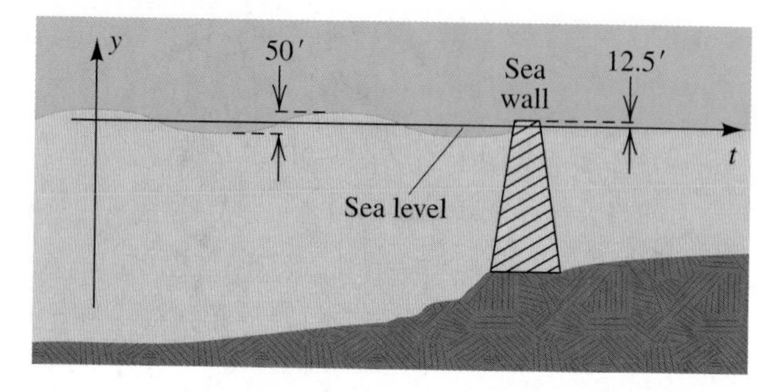

70 **Temperature in Fairbanks** The expected low temperature T (in °F) in Fairbanks, Alaska, may be approximated by

$$T = 36\sin\left[\frac{2\pi}{365}(t - 101)\right] + 14,$$

where t is in days, with $t = 0$ corresponding to January 1. For how many days during the year is the low temperature expected to be below −4°F?

71 **Temperature in Chicago** The average monthly high temperature T (in °F) in Chicago, Illinois, can be approximated using the function

$$T(t) = 26.5\sin\left(\frac{\pi}{6}t - \frac{2\pi}{3}\right) + 56.5,$$

where t is in months and $t = 1$ corresponds to January.

(a) Graph T over the two-year interval [1, 25].

(b) Calculate the average high temperature in July and in October.

(c) Graphically approximate the months when the average high temperature is 69°F or higher.

(d) Discuss why a sine function is an appropriate function to approximate these temperatures.

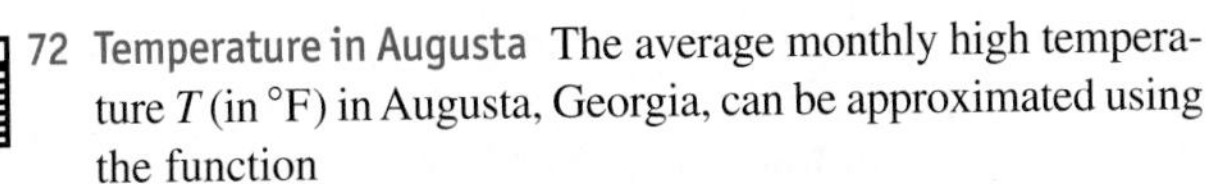

72 Temperature in Augusta The average monthly high temperature T (in °F) in Augusta, Georgia, can be approximated using the function

$$T(t) = 17\cos\left(\frac{\pi}{6}t - \frac{7\pi}{6}\right) + 75,$$

where t is in months and $t = 1$ corresponds to January.

(a) Graph T over the two-year interval $[1, 25]$.

(b) Calculate the average high temperature in April and in December.

(c) Graphically approximate the months when the average high temperature is 67°F or lower.

73 Intensity of sunlight On a clear day with D hours of daylight, the intensity of sunlight I (in calories/cm^2) may be approximated by

$$I = I_M \sin^3 \frac{\pi t}{D} \quad \text{for} \quad 0 \le t \le D,$$

where $t = 0$ corresponds to sunrise and I_M is the maximum intensity. If $D = 12$, approximately how many hours after sunrise is $I = \frac{1}{2}I_M$?

74 Intensity of sunlight Refer to Exercise 73. On cloudy days, a better approximation of the sun intensity I is given by

$$I = I_M \sin^2 \frac{\pi t}{D}.$$

If $D = 12$, how many hours after sunrise is $I = \frac{1}{2}I_M$?

75 Protection from sunlight Refer to Exercises 73 and 74. A dermatologist recommends protection from the sun when the intensity I exceeds 75% of the maximum intensity. If $D = 12$ hours, approximate the number of hours for which protection is required on

(a) a clear day (b) a cloudy day

76 Highway engineering In the study of frost penetration problems in highway engineering, the temperature T at time t hours and depth x feet is given by

$$T = T_0 e^{-\lambda x} \sin(\omega t - \lambda x),$$

where T_0, ω, and λ are constants and the period of T is 24 hours.

(a) Find a formula for the temperature at the surface.

(b) At what times is the surface temperature a minimum?

(c) If $\lambda = 2.5$, find the times when the temperature is a minimum at a depth of 1 foot.

77 Rabbit population Many animal populations, such as that of rabbits, fluctuate over ten-year cycles. Suppose that the number of rabbits at time t (in years) is given by

$$N(t) = 1000\cos\frac{\pi}{5}t + 4000.$$

(a) Sketch the graph of N for $0 \le t \le 10$.

(b) For what values of t in part (a) does the rabbit population exceed 4500?

78 River flow rate The flow rate (or water discharge rate) at the mouth of the Orinoco River in South America may be approximated by

$$F(t) = 26{,}000\sin\left[\frac{\pi}{6}(t - 5.5)\right] + 34{,}000,$$

where t is the time in months and $F(t)$ is the flow rate in m^3/sec. For approximately how many months each year does the flow rate exceed 55,000 m^3/sec?

79 Shown in the figure is a graph of $y = \frac{1}{2}x + \sin x$ for $-2\pi \le x \le 2\pi$. Using calculus, it can be shown that the x-coordinates of the turning points A, B, C, and D on the graph are solutions of the equation $\frac{1}{2} + \cos x = 0$. Determine the coordinates of these points.

Exercise 79

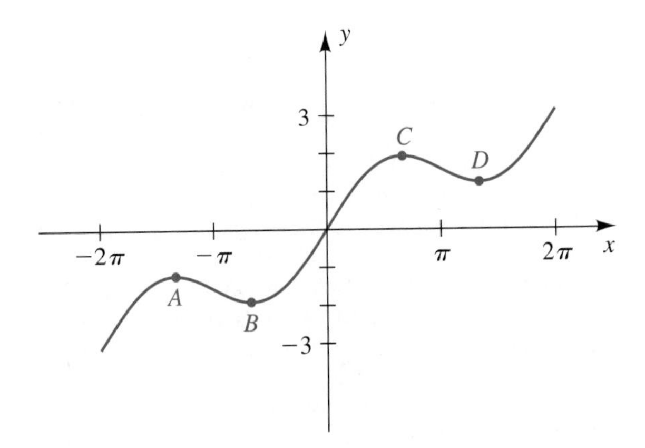

80 Shown in the figure is the graph of the equation

$$y = e^{-x/2} \sin 2x.$$

The x-coordinates of the turning points on the graph are solutions of $4 \cos 2x - \sin 2x = 0$. Approximate the x-coordinates of these points for $x > 0$.

Exercise 80

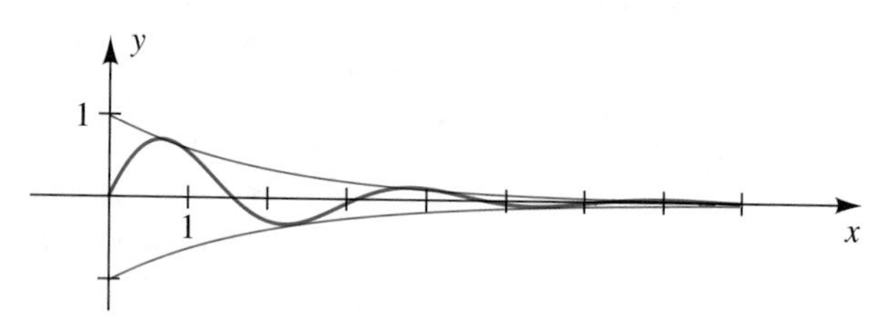

Exer. 81–82: If $I(t)$ is the current (in amperes) in an alternating current circuit at time t (in seconds), find the smallest exact value of t for which $I(t) = k$.

81 $I(t) = 20 \sin (60\pi t - 6\pi)$; $k = -10$

82 $I(t) = 40 \sin (100\pi t - 4\pi)$; $k = 20$

Exer. 83–86: Approximate the solution to each inequality on the interval $[0, 2\pi]$.

83 $\cos x \geq 0.3$

84 $\sin x < -0.6$

85 $\cos 3x < \sin x$

86 $\tan x \leq \sin 2x$

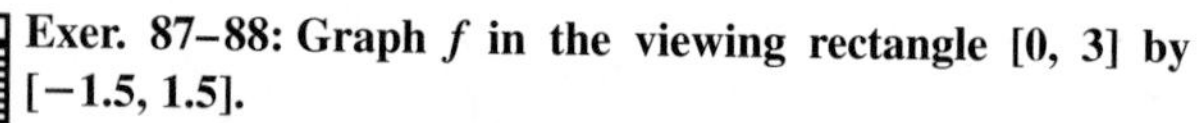

Exer. 87–88: Graph f in the viewing rectangle $[0, 3]$ by $[-1.5, 1.5]$.

(a) Approximate to within four decimal places the largest solution of $f(x) = 0$ on $[0, 3]$.

(b) Discuss what happens to the graph of f as x becomes large.

(c) Examine graphs of the function f on the interval $[0, c]$, where $c = 0.1, 0.01, 0.001$. How many zeros does f appear to have on the interval $[0, c]$, where $c > 0$?

87 $f(x) = \cos \dfrac{1}{x}$

88 $f(x) = \sin \dfrac{1}{x^2}$

Exer. 89–92: Because planets do not move in precisely circular orbits, the computation of the position of a planet requires the solution of Kepler's equation. Kepler's equation cannot be solved algebraically. It has the form $M = \theta + e \sin \theta$, where M is the mean anomaly, e is the eccentricity of the orbit, and θ is an angle called the eccentric anomaly. For the specified values of M and e, use graphical techniques to solve Kepler's equation for θ to three decimal places.

89 Position of Mercury $M = 5.241$, $e = 0.206$

90 Position of Mars $M = 4.028$, $e = 0.093$

91 Position of Earth $M = 3.611$, $e = 0.0167$

92 Position of Pluto $M = 0.09424$, $e = 0.255$

Exer. 93–98: Estimate the solutions of the equation in the interval $[-\pi, \pi]$.

93 $\sin 2x = 2 - x^2$

94 $\cos^3 x + \cos 3x - \sin^3 x = 0$

95 $\ln (1 + \sin^2 x) = \cos x$

96 $e^{\sin x} = \sec \left(\frac{1}{3}x - \frac{1}{2}\right)$

97 $3 \cos^4 x - 2 \cos^3 x + \cos x - 1 = 0$

98 $\cos 2x + \sin 3x - \tan \frac{1}{3}x = 0$

99 Weight at various latitudes The weight W of a person on the surface of Earth is directly proportional to the force of gravity g (in m/sec²). Because of rotation, Earth is flattened at the poles, and as a result weight will vary at different latitudes. If θ is the latitude, then g can be approximated by $g = 9.8066(1 - 0.00264 \cos 2\theta)$.

(a) At what latitude is $g = 9.8$?

(b) If a person weighs 150 pounds at the equator ($\theta = 0°$), at what latitude will the person weigh 150.5 pounds?

6.3 The Addition and Subtraction Formulas

In this section we derive formulas that involve trigonometric functions of $u + v$ or $u - v$ for any real numbers or angles u and v. These formulas are known as *addition* and *subtraction formulas*, respectively, or as *sum* and *difference identities*. The first formula that we will consider may be stated as follows.

Subtraction Formula for Cosine	$\cos(u - v) = \cos u \cos v + \sin u \sin v$

Figure 1

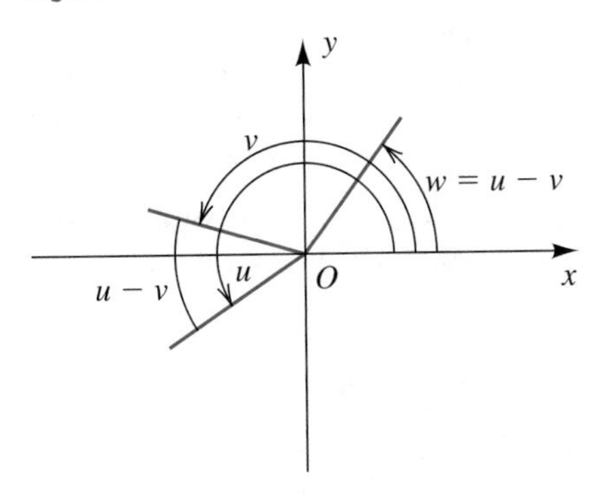

Figure 2

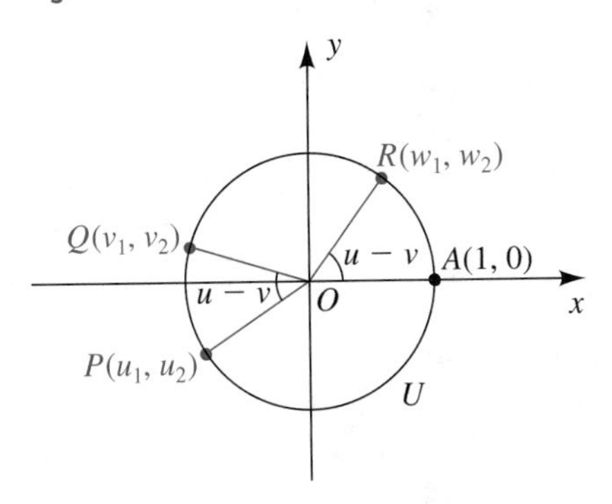

PROOF Let u and v be any real numbers, and consider angles of radian measure u and v. Let $w = u - v$. Figure 1 illustrates one possibility with the angles in standard position. For convenience we have assumed that both u and v are positive and that $0 \leq u - v < v$.

As in Figure 2, let $P(u_1, u_2)$, $Q(v_1, v_2)$, and $R(w_1, w_2)$ be the points on the terminal sides of the indicated angles that are each a distance 1 from the origin. In this case P, Q, and R are on the unit circle U with center at the origin. From the definition of trigonometric functions in terms of a unit circle,

$$\begin{array}{lll} \cos u = u_1 & \cos v = v_1 & \cos(u - v) = w_1 \\ \sin u = u_2 & \sin v = v_2 & \sin(u - v) = w_2. \end{array} \qquad (*)$$

We next observe that the distance between $A(1, 0)$ and R must equal the distance between Q and P, because angles AOR and QOP have the same measure, $u - v$. Using the distance formula yields

$$d(A, R) = d(Q, P)$$
$$\sqrt{(w_1 - 1)^2 + (w_2 - 0)^2} = \sqrt{(u_1 - v_1)^2 + (u_2 - v_2)^2}.$$

Squaring both sides and simplifying the expressions under the radicals gives us

$$w_1^2 - 2w_1 + 1 + w_2^2 = u_1^2 - 2u_1v_1 + v_1^2 + u_2^2 - 2u_2v_2 + v_2^2.$$

Since the points (u_1, u_2), (v_1, v_2), and (w_1, w_2) are on the unit circle U and since an equation for U is $x^2 + y^2 = 1$, we may substitute 1 for each of $u_1^2 + u_2^2$, $v_1^2 + v_2^2$, and $w_1^2 + w_2^2$. Doing this and simplifying, we obtain

$$2 - 2w_1 = 2 - 2u_1v_1 - 2u_2v_2,$$

which reduces to

$$w_1 = u_1v_1 + u_2v_2.$$

Substituting from the formulas stated in (*) gives us

$$\cos(u - v) = \cos u \cos v + \sin u \sin v,$$

which is what we wished to prove. It is possible to extend our discussion to all values of u and v.

The next example demonstrates the use of the subtraction formula in finding the *exact* value of $\cos 15°$. Of course, if only an approximation were desired, we could use a calculator.

EXAMPLE 1 Using a subtraction formula

Find the exact value of $\cos 15°$ by using the fact that $15° = 60° - 45°$.

▶ **SOLUTION** We use the subtraction formula for cosine with $u = 60°$ and $v = 45°$:

$$\begin{aligned}\cos 15° &= \cos(60° - 45°)\\ &= \cos 60° \cos 45° + \sin 60° \sin 45°\\ &= \frac{1}{2}\frac{\sqrt{2}}{2} + \frac{\sqrt{3}}{2}\frac{\sqrt{2}}{2}\\ &= \frac{\sqrt{2}+\sqrt{6}}{4}\end{aligned}$$

It is relatively easy to obtain a formula for $\cos(u + v)$. We begin by writing $u + v$ as $u - (-v)$ and then use the subtraction formula for cosine:

$$\begin{aligned}\cos(u + v) &= \cos[u - (-v)]\\ &= \cos u \cos(-v) + \sin u \sin(-v)\end{aligned}$$

Using the formulas for negatives, $\cos(-v) = \cos v$ and $\sin(-v) = -\sin v$, gives us the following addition formula for cosine.

Addition Formula for Cosine	$\cos(u + v) = \cos u \cos v - \sin u \sin v$

EXAMPLE 2 Using an addition formula

Find the exact value of $\cos \frac{7\pi}{12}$ by using the fact that $\frac{7\pi}{12} = \frac{\pi}{3} + \frac{\pi}{4}$.

▶ **SOLUTION** We apply the addition formula for cosine:

$$\begin{aligned}\cos\frac{7\pi}{12} &= \cos\left(\frac{\pi}{3} + \frac{\pi}{4}\right)\\ &= \cos\frac{\pi}{3}\cos\frac{\pi}{4} - \sin\frac{\pi}{3}\sin\frac{\pi}{4}\\ &= \frac{1}{2}\frac{\sqrt{2}}{2} - \frac{\sqrt{3}}{2}\frac{\sqrt{2}}{2}\\ &= \frac{\sqrt{2}-\sqrt{6}}{4}\end{aligned}$$

▶ **Tutorial available at www.thomsonedu.com/login**

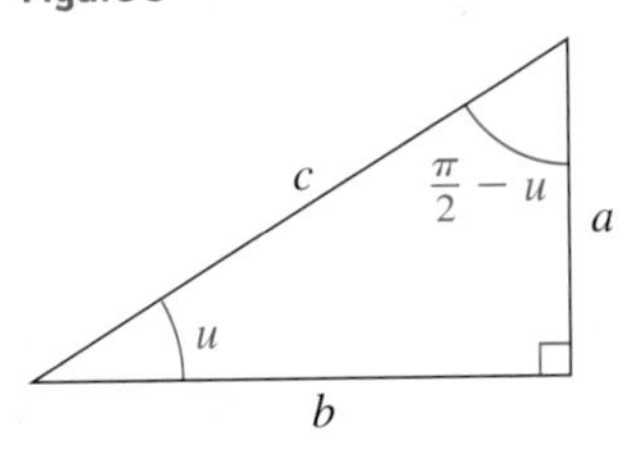

Figure 3

We refer to the sine and cosine functions as **cofunctions** of each other. Similarly, the tangent and cotangent functions are cofunctions, as are the secant and cosecant. If u is the radian measure of an acute angle, then the angle with radian measure $\pi/2 - u$ is complementary to u, and we may consider the right triangle shown in Figure 3. Using ratios, we see that

$$\sin u = \frac{a}{c} = \cos\left(\frac{\pi}{2} - u\right)$$

$$\cos u = \frac{b}{c} = \sin\left(\frac{\pi}{2} - u\right)$$

$$\tan u = \frac{a}{b} = \cot\left(\frac{\pi}{2} - u\right).$$

These three formulas and their analogues for $\sec u$, $\csc u$, and $\cot u$ state that *the function value of u equals the cofunction of the complementary angle $\pi/2 - u$.*

In the following formulas we use subtraction formulas to extend these relationships to any real number u, provided the function values are defined.

Cofunction Formulas

If u is a real number or the radian measure of an angle, then

(1) $\cos\left(\frac{\pi}{2} - u\right) = \sin u$ **(2)** $\sin\left(\frac{\pi}{2} - u\right) = \cos u$

(3) $\tan\left(\frac{\pi}{2} - u\right) = \cot u$ **(4)** $\cot\left(\frac{\pi}{2} - u\right) = \tan u$

(5) $\sec\left(\frac{\pi}{2} - u\right) = \csc u$ **(6)** $\csc\left(\frac{\pi}{2} - u\right) = \sec u$

PROOFS Using the subtraction formula for cosine, we have

$$\cos\left(\frac{\pi}{2} - u\right) = \cos\frac{\pi}{2}\cos u + \sin\frac{\pi}{2}\sin u$$
$$= (0)\cos u + (1)\sin u = \sin u.$$

This gives us formula 1.

If we substitute $\pi/2 - v$ for u in the first formula, we obtain

$$\cos\left[\frac{\pi}{2} - \left(\frac{\pi}{2} - v\right)\right] = \sin\left(\frac{\pi}{2} - v\right),$$

or

$$\cos v = \sin\left(\frac{\pi}{2} - v\right).$$

(continued)

Since the symbol v is arbitrary, this equation is equivalent to the second cofunction formula:

$$\sin\left(\frac{\pi}{2} - u\right) = \cos u$$

Using the tangent identity, cofunction formulas 1 and 2, and the cotangent identity, we obtain a proof for the third formula:

$$\tan\left(\frac{\pi}{2} - u\right) = \frac{\sin\left(\frac{\pi}{2} - u\right)}{\cos\left(\frac{\pi}{2} - u\right)} = \frac{\cos u}{\sin u} = \cot u$$

The proofs of the remaining three formulas are similar. ◢

An easy way to remember the cofunction formulas is to refer to the triangle in Figure 3.

We may now prove the following identities.

Addition and Subtraction Formulas for Sine and Tangent

(1) $\sin(u + v) = \sin u \cos v + \cos u \sin v$

(2) $\sin(u - v) = \sin u \cos v - \cos u \sin v$

(3) $\tan(u + v) = \dfrac{\tan u + \tan v}{1 - \tan u \tan v}$

(4) $\tan(u - v) = \dfrac{\tan u - \tan v}{1 + \tan u \tan v}$

PROOFS We shall prove formulas 1 and 3. Using the cofunction formulas and the subtraction formula for cosine, we can verify formula 1:

$$\begin{aligned}
\sin(u + v) &= \cos\left[\frac{\pi}{2} - (u + v)\right] \\
&= \cos\left[\left(\frac{\pi}{2} - u\right) - v\right] \\
&= \cos\left(\frac{\pi}{2} - u\right)\cos v + \sin\left(\frac{\pi}{2} - u\right)\sin v \\
&= \sin u \cos v + \cos u \sin v
\end{aligned}$$

To verify formula 3, we begin as follows:

$$\tan(u+v) = \frac{\sin(u+v)}{\cos(u+v)}$$
$$= \frac{\sin u \cos v + \cos u \sin v}{\cos u \cos v - \sin u \sin v}$$

Dividing by $\cos u \cos v$ *will give us an expression involving tangents; dividing by* $\sin u \sin v$ *would give us an expression involving cotangents.*

If $\cos u \cos v \neq 0$, then we may divide the numerator and the denominator by $\cos u \cos v$, obtaining

$$\tan(u+v) = \frac{\left(\dfrac{\sin u}{\cos u}\right)\left(\dfrac{\cos v}{\cos v}\right) + \left(\dfrac{\cos u}{\cos u}\right)\left(\dfrac{\sin v}{\cos v}\right)}{\left(\dfrac{\cos u}{\cos u}\right)\left(\dfrac{\cos v}{\cos v}\right) - \left(\dfrac{\sin u}{\cos u}\right)\left(\dfrac{\sin v}{\cos v}\right)}$$
$$= \frac{\tan u + \tan v}{1 - \tan u \tan v}.$$

If $\cos u \cos v = 0$, then either $\cos u = 0$ or $\cos v = 0$. In this case, either $\tan u$ or $\tan v$ is undefined and the formula is invalid. Proofs of formulas 2 and 4 are left as exercises.

EXAMPLE 3 **Using addition formulas to find the quadrant containing an angle**

Suppose $\sin\alpha = \frac{4}{5}$ and $\cos\beta = -\frac{12}{13}$, where α is in quadrant I and β is in quadrant II.

(a) Find the exact values of $\sin(\alpha+\beta)$ and $\tan(\alpha+\beta)$.

(b) Find the quadrant containing $\alpha + \beta$.

Figure 4

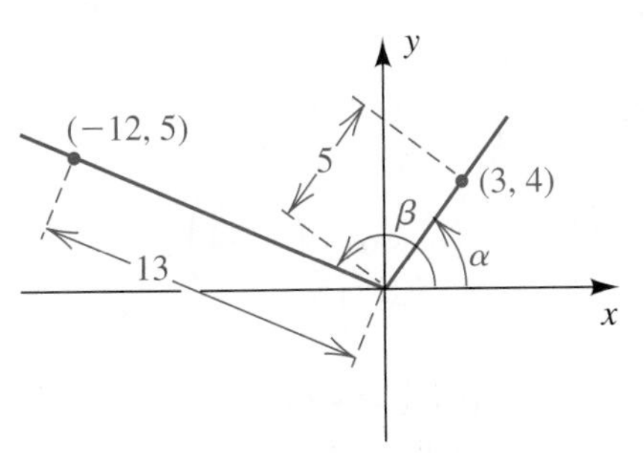

SOLUTION Angles α and β are illustrated in Figure 4. There is no loss of generality in regarding α and β as positive angles between 0 and 2π, as we have done in the figure. Since $\sin\alpha = \frac{4}{5}$, we may choose the point $(3, 4)$ on the terminal side of α. Similarly, since $\cos\beta = -\frac{12}{13}$, the point $(-12, 5)$ is on the terminal side of β. Referring to Figure 4 and using the definition of the trigonometric functions of any angle, we have

$$\cos\alpha = \tfrac{3}{5}, \quad \tan\alpha = \tfrac{4}{3}, \quad \sin\beta = \tfrac{5}{13}, \quad \tan\beta = -\tfrac{5}{12}.$$

(a) Addition formulas give us

$$\sin(\alpha+\beta) = \sin\alpha\cos\beta + \cos\alpha\sin\beta = \left(\tfrac{4}{5}\right)\left(-\tfrac{12}{13}\right) + \left(\tfrac{3}{5}\right)\left(\tfrac{5}{13}\right) = -\tfrac{33}{65}$$

$$\tan(\alpha+\beta) = \frac{\tan\alpha + \tan\beta}{1 - \tan\alpha\tan\beta} = \frac{\frac{4}{3} + \left(-\frac{5}{12}\right)}{1 - \left(\frac{4}{3}\right)\left(-\frac{5}{12}\right)} \cdot \frac{36}{36} = \frac{33}{56}.$$

(b) Since $\sin(\alpha+\beta)$ is negative and $\tan(\alpha+\beta)$ is positive, the angle $\alpha + \beta$ must be in quadrant III.

Tutorial available at www.thomsonedu.com/login

Figure 5

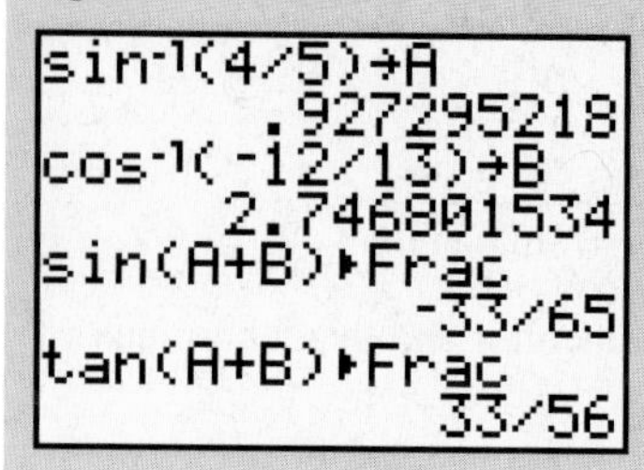

Here's how a graphing calculator can be applied to find the exact values in Example 3. Since α is in quadrant I, $\sin \alpha = \frac{4}{5}$ implies that $\alpha = \sin^{-1} \frac{4}{5}$; and since β is in quadrant II, $\cos \beta = -\frac{12}{13}$ implies that $\beta = \cos^{-1}\left(-\frac{12}{13}\right)$. (If the angles were in different quadrants, we could use reference angles as we did in Section 5.4.) In Figure 5, we stored the angles α and β in the locations A and B and then found the exact values of $\sin(\alpha + \beta)$ and $\tan(\alpha + \beta)$ as fractions. The values agree with those found in Example 3.

The next example illustrates a type of simplification of the difference quotient (introduced in Section 2.4) with the sine function. The resulting form is useful in calculus.

EXAMPLE 4 A formula used in calculus

If $f(x) = \sin x$ and $h \neq 0$, show that

$$\frac{f(x+h) - f(x)}{h} = \sin x \left(\frac{\cos h - 1}{h}\right) + \cos x \left(\frac{\sin h}{h}\right).$$

▶ **SOLUTION** We use the definition of f and the addition formula for sine:

$$\begin{aligned} \frac{f(x+h) - f(x)}{h} &= \frac{\sin(x+h) - \sin x}{h} \\ &= \frac{\sin x \cos h + \cos x \sin h - \sin x}{h} \\ &= \frac{\sin x (\cos h - 1) + \cos x \sin h}{h} \\ &= \sin x \left(\frac{\cos h - 1}{h}\right) + \cos x \left(\frac{\sin h}{h}\right) \end{aligned}$$

Addition formulas may also be used to derive **reduction formulas.** Reduction formulas may be used to change expressions such as

$$\sin\left(\theta + \frac{\pi}{2} n\right) \quad \text{and} \quad \cos\left(\theta + \frac{\pi}{2} n\right) \quad \text{for any integer } n$$

to expressions involving only $\sin \theta$ or $\cos \theta$. Similar formulas are true for the other trigonometric functions. Instead of deriving general reduction formulas, we shall illustrate two special cases in the next example.

▶ **Tutorial available at www.thomsonedu.com/login**

EXAMPLE 5 Obtaining reduction formulas

Express in terms of a trigonometric function of θ alone:

(a) $\sin\left(\theta - \frac{3\pi}{2}\right)$ **(b)** $\cos(\theta + \pi)$

▶ **SOLUTION** Using subtraction and addition formulas, we obtain the following:

(a) $$\sin\left(\theta - \frac{3\pi}{2}\right) = \sin\theta\cos\frac{3\pi}{2} - \cos\theta\sin\frac{3\pi}{2}$$

$$= \sin\theta \cdot (0) - \cos\theta \cdot (-1) = \cos\theta$$

(b) $$\cos(\theta + \pi) = \cos\theta\cos\pi - \sin\theta\sin\pi$$

$$= \cos\theta \cdot (-1) - \sin\theta \cdot (0) = -\cos\theta$$

EXAMPLE 6 Combining a sum involving the sine and cosine functions

Since $\cos u = \sin\left(\frac{\pi}{2} - u\right)$, we could also write the sum in terms of a sine function.

Let a and b be real numbers with $a > 0$. Show that for every x,

$$a\cos Bx + b\sin Bx = A\cos(Bx - C),$$

where $A = \sqrt{a^2 + b^2}$ and $\tan C = \frac{b}{a}$ with $-\frac{\pi}{2} < C < \frac{\pi}{2}$.

▶ **SOLUTION** Given $a\cos Bx + b\sin Bx$, let us consider $\tan C = b/a$ with $-\pi/2 < C < \pi/2$. Thus, $b = a\tan C$, and we may write

$$\begin{aligned} a\cos Bx + b\sin Bx &= a\cos Bx + (a\tan C)\sin Bx \\ &= a\cos Bx + a\frac{\sin C}{\cos C}\sin Bx \\ &= \frac{a}{\cos C}(\cos C\cos Bx + \sin C\sin Bx) \\ &= (a\sec C)\cos(Bx - C). \end{aligned}$$

We shall complete the proof by showing that $a\sec C = \sqrt{a^2 + b^2}$. Since $-\pi/2 < C < \pi/2$, it follows that $\sec C$ is positive, and hence

$$a\sec C = a\sqrt{1 + \tan^2 C}.$$

Using $\tan C = b/a$ and $a > 0$, we obtain

$$a\sec C = a\sqrt{1 + \frac{b^2}{a^2}} = \sqrt{a^2\left(1 + \frac{b^2}{a^2}\right)} = \sqrt{a^2 + b^2}.$$

EXAMPLE 7 An application of Example 6

If $f(x) = \cos x + \sin x$, use the formulas given in Example 6 to express $f(x)$ in the form $A\cos(Bx - C)$, and then sketch the graph of f.

▶ **Tutorial available at www.thomsonedu.com/login**

SOLUTION Letting $a = 1$, $b = 1$, and $B = 1$ in the formulas from Example 6, we have

$$A = \sqrt{a^2 + b^2} = \sqrt{1 + 1} = \sqrt{2} \quad \text{and} \quad \tan C = \frac{b}{a} = \frac{1}{1} = 1.$$

Since $\tan C = 1$ and $-\pi/2 < C < \pi/2$, we have $C = \pi/4$. Substituting for a, b, A, B, and C in the formula

$$a \cos Bx + b \sin Bx = A \cos (Bx - C)$$

gives us

$$f(x) = \cos x + \sin x = \sqrt{2} \cos \left(x - \frac{\pi}{4} \right).$$

Comparing the last formula with the equation $y = a \cos (bx + c)$, which we discussed in Section 5.5, we see that the amplitude of the graph is $\sqrt{2}$, the period is 2π, and the phase shift is $\pi/4$. The graph of f is sketched in Figure 6, where we have also shown the graphs of $y = \sin x$ and $y = \cos x$. Our sketch agrees with that obtained in Chapter 5 using a graphing utility. (See Figure 10 in Section 5.6.)

Figure 6

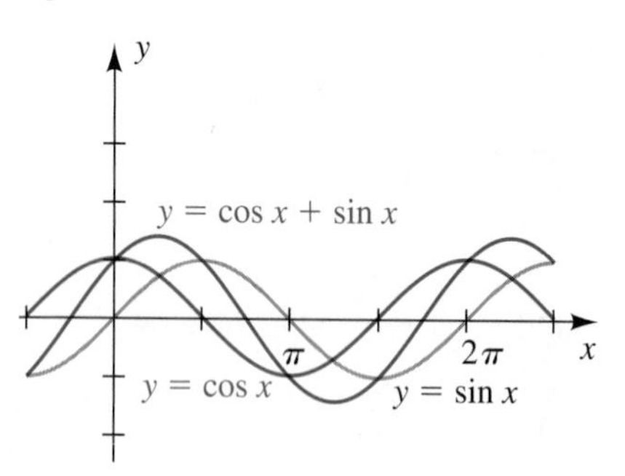

6.3 Exercises

Exer. 1–4: Express as a cofunction of a complementary angle.

1 (a) $\sin 46°37'$ (b) $\cos 73°12'$

(c) $\tan \dfrac{\pi}{6}$ (d) $\sec 17.28°$

2 (a) $\tan 24°12'$ (b) $\sin 89°41'$

(c) $\cos \dfrac{\pi}{3}$ (d) $\cot 61.87°$

3 (a) $\cos \dfrac{7\pi}{20}$ (b) $\sin \dfrac{1}{4}$

(c) $\tan 1$ (d) $\csc 0.53$

4 (a) $\sin \dfrac{\pi}{12}$ (b) $\cos 0.64$

(c) $\tan \sqrt{2}$ (d) $\sec 1.2$

Exer. 5–10: Find the exact values.

▶ 5 (a) $\cos \dfrac{\pi}{4} + \cos \dfrac{\pi}{6}$

(b) $\cos \dfrac{5\pi}{12}$ $\left(\text{use } \dfrac{5\pi}{12} = \dfrac{\pi}{4} + \dfrac{\pi}{6} \right)$

6 (a) $\sin \dfrac{2\pi}{3} + \sin \dfrac{\pi}{4}$

(b) $\sin \dfrac{11\pi}{12}$ $\left(\text{use } \dfrac{11\pi}{12} = \dfrac{2\pi}{3} + \dfrac{\pi}{4} \right)$

7 (a) $\tan 60° + \tan 225°$

(b) $\tan 285°$ (use $285° = 60° + 225°$)

8 (a) $\cos 135° - \cos 60°$

(b) $\cos 75°$ (use $75° = 135° - 60°$)

9 (a) $\sin \frac{3\pi}{4} - \sin \frac{\pi}{6}$

(b) $\sin \frac{7\pi}{12}$ $\left(\text{use } \frac{7\pi}{12} = \frac{3\pi}{4} - \frac{\pi}{6}\right)$

10 (a) $\tan \frac{3\pi}{4} - \tan \frac{\pi}{6}$

(b) $\tan \frac{7\pi}{12}$ $\left(\text{use } \frac{7\pi}{12} = \frac{3\pi}{4} - \frac{\pi}{6}\right)$

Exer. 11–16: Express as a trigonometric function of one angle.

▶ 11 $\cos 48° \cos 23° + \sin 48° \sin 23°$

12 $\cos 13° \cos 50° - \sin 13° \sin 50°$

13 $\cos 10° \sin 5° - \sin 10° \cos 5°$

14 $\sin 57° \cos 4° + \cos 57° \sin 4°$

15 $\cos 3 \sin (-2) - \cos 2 \sin 3$

16 $\sin (-5) \cos 2 + \cos 5 \sin (-2)$

17 If $\sin \alpha = -\frac{5}{13}$ and $\tan \alpha > 0$, find the exact value of $\sin\left(\alpha - \frac{\pi}{3}\right)$.

18 If $\cos \alpha = \frac{24}{25}$ and $\sin \alpha < 0$, find the exact value of $\cos\left(\alpha + \frac{\pi}{6}\right)$.

19 If α and β are acute angles such that $\cos \alpha = \frac{4}{5}$ and $\tan \beta = \frac{8}{15}$, find

(a) $\sin (\alpha + \beta)$ (b) $\cos (\alpha + \beta)$

(c) the quadrant containing $\alpha + \beta$

20 If α and β are acute angles such that $\csc \alpha = \frac{13}{12}$ and $\cot \beta = \frac{4}{3}$, find

(a) $\sin (\alpha + \beta)$ (b) $\tan (\alpha + \beta)$

(c) the quadrant containing $\alpha + \beta$

21 If $\sin \alpha = -\frac{4}{5}$ and $\sec \beta = \frac{5}{3}$ for a third-quadrant angle α and a first-quadrant angle β, find

(a) $\sin (\alpha + \beta)$ (b) $\tan (\alpha + \beta)$

(c) the quadrant containing $\alpha + \beta$

22 If $\tan \alpha = -\frac{7}{24}$ and $\cot \beta = \frac{3}{4}$ for a second-quadrant angle α and a third-quadrant angle β, find

(a) $\sin (\alpha + \beta)$ (b) $\cos (\alpha + \beta)$ (c) $\tan (\alpha + \beta)$

(d) $\sin (\alpha - \beta)$ (e) $\cos (\alpha - \beta)$ (f) $\tan (\alpha - \beta)$

23 If α and β are third-quadrant angles such that $\cos \alpha = -\frac{2}{5}$ and $\cos \beta = -\frac{3}{5}$, find

(a) $\sin (\alpha - \beta)$ (b) $\cos (\alpha - \beta)$

(c) the quadrant containing $\alpha - \beta$

24 If α and β are second-quadrant angles such that $\sin \alpha = \frac{2}{3}$ and $\cos \beta = -\frac{1}{3}$, find

(a) $\sin (\alpha + \beta)$ (b) $\tan (\alpha + \beta)$

(c) the quadrant containing $\alpha + \beta$

Exer. 25–36: Verify the reduction formula.

25 $\sin (\theta + \pi) = -\sin \theta$

26 $\sin\left(x + \frac{\pi}{2}\right) = \cos x$

27 $\sin\left(x - \frac{5\pi}{2}\right) = -\cos x$

▶ 28 $\sin\left(\theta - \frac{3\pi}{2}\right) = \cos \theta$

29 $\cos (\theta - \pi) = -\cos \theta$

30 $\cos\left(x + \frac{\pi}{2}\right) = -\sin x$

31 $\cos\left(x + \frac{3\pi}{2}\right) = \sin x$

32 $\cos\left(\theta - \frac{5\pi}{2}\right) = \sin \theta$

33 $\tan\left(x - \frac{\pi}{2}\right) = -\cot x$

34 $\tan (\pi - \theta) = -\tan \theta$

35 $\tan\left(\theta + \frac{\pi}{2}\right) = -\cot \theta$

36 $\tan (x + \pi) = \tan x$

Exer. 37–46: Verify the identity.

37 $\sin\left(\theta + \frac{\pi}{4}\right) = \frac{\sqrt{2}}{2}(\sin \theta + \cos \theta)$

▶ 38 $\cos\left(\theta + \frac{\pi}{4}\right) = \frac{\sqrt{2}}{2}(\cos \theta - \sin \theta)$

▶ **Tutorial available at www.thomsonedu.com/login**

39 $\tan\left(u + \frac{\pi}{4}\right) = \frac{1 + \tan u}{1 - \tan u}$

40 $\tan\left(x - \frac{\pi}{4}\right) = \frac{\tan x - 1}{\tan x + 1}$

41 $\cos(u + v) + \cos(u - v) = 2\cos u \cos v$

42 $\sin(u + v) + \sin(u - v) = 2\sin u \cos v$

43 $\sin(u + v) \cdot \sin(u - v) = \sin^2 u - \sin^2 v$

44 $\cos(u + v) \cdot \cos(u - v) = \cos^2 u - \sin^2 v$

45 $\frac{1}{\cot\alpha - \cot\beta} = \frac{\sin\alpha \sin\beta}{\sin(\beta - \alpha)}$

46 $\frac{1}{\tan\alpha + \tan\beta} = \frac{\cos\alpha \cos\beta}{\sin(\alpha + \beta)}$

47 Express $\sin(u + v + w)$ in terms of trigonometric functions of u, v, and w. (*Hint:* Write

$$\sin(u + v + w) \quad \text{as} \quad \sin[(u + v) + w]$$

and use addition formulas.)

48 Express $\tan(u + v + w)$ in terms of trigonometric functions of u, v, and w.

49 Derive the formula $\cot(u + v) = \frac{\cot u \cot v - 1}{\cot u + \cot v}$.

50 If α and β are complementary angles, show that

$$\sin^2\alpha + \sin^2\beta = 1.$$

51 Derive the subtraction formula for the sine function.

52 Derive the subtraction formula for the tangent function.

53 If $f(x) = \cos x$, show that

$$\frac{f(x + h) - f(x)}{h} = \cos x\left(\frac{\cos h - 1}{h}\right) - \sin x\left(\frac{\sin h}{h}\right).$$

54 If $f(x) = \tan x$, show that

$$\frac{f(x + h) - f(x)}{h} = \sec^2 x\left(\frac{\sin h}{h}\right)\frac{1}{\cos h - \sin h \tan x}.$$

Exer. 55–56: (a) Compare the decimal approximations of both sides of equation (1). (b) Find the acute angle x such that equation (2) is an identity. (c) How does equation (1) relate to equation (2)?

55 (1) $\sin 63° - \sin 57° = \sin 3°$

(2) $\sin(\alpha + \beta) - \sin(\alpha - \beta) = \sin\beta$

56 (1) $\sin 35° + \sin 25° = \cos 5°$

(2) $\sin(\alpha + \beta) + \sin(\alpha - \beta) = \cos\beta$

Exer. 57–62: Use an addition or subtraction formula to find the solutions of the equation that are in the interval $[0, \pi)$.

57 $\sin 4t \cos t = \sin t \cos 4t$

58 $\cos 5t \cos 3t = \frac{1}{2} + \sin(-5t)\sin 3t$

59 $\cos 5t \cos 2t = -\sin 5t \sin 2t$

60 $\sin 3t \cos t + \cos 3t \sin t = -\frac{1}{2}$

61 $\tan 2t + \tan t = 1 - \tan 2t \tan t$

62 $\tan t - \tan 4t = 1 + \tan 4t \tan t$

Exer. 63–66: (a) Use the formula from Example 6 to express f in terms of the cosine function. (b) Determine the amplitude, period, and phase shift of f. (c) Sketch the graph of f.

63 $f(x) = \sqrt{3}\cos 2x + \sin 2x$

64 $f(x) = \cos 4x + \sqrt{3}\sin 4x$

65 $f(x) = 2\cos 3x - 2\sin 3x$

66 $f(x) = 5\cos 10x - 5\sin 10x$

Exer. 67–68: For certain applications in electrical engineering, the sum of several voltage signals or radio waves

of the same frequency is expressed in the compact form $y = A \cos (Bt - C)$. Express the given signal in this form.

67 $y = 50 \sin 60\pi t + 40 \cos 60\pi t$

68 $y = 10 \sin \left(120\pi t - \dfrac{\pi}{2}\right) + 5 \sin 120\pi t$

69 **Motion of a mass** If a mass that is attached to a spring is raised y_0 feet and released with an initial vertical velocity of v_0 ft/sec, then the subsequent position y of the mass is given by

$$y = y_0 \cos \omega t + \frac{v_0}{\omega} \sin \omega t,$$

where t is time in seconds and ω is a positive constant.

(a) If $\omega = 1$, $y_0 = 2$ ft, and $v_0 = 3$ ft/sec, express y in the form $A \cos (Bt - C)$, and find the amplitude and period of the resulting motion.

(b) Determine the times when $y = 0$—that is, the times when the mass passes through the equilibrium position.

70 **Motion of a mass** Refer to Exercise 69. If $y_0 = 1$ and $\omega = 2$, find the initial velocities that result in an amplitude of 4 feet.

71 **Pressure on the eardrum** If a tuning fork is struck and then held a certain distance from the eardrum, the pressure $p_1(t)$ on the outside of the eardrum at time t may be represented by $p_1(t) = A \sin \omega t$, where A and ω are positive constants. If a second identical tuning fork is struck with a possibly different force and held a different distance from the eardrum (see the figure), its effect may be represented by the equation $p_2(t) = B \sin (\omega t + \tau)$, where B is a positive constant and $0 \le \tau \le 2\pi$. The total pressure $p(t)$ on the eardrum is given by

$$p(t) = A \sin \omega t + B \sin (\omega t + \tau).$$

(a) Show that $p(t) = a \cos \omega t + b \sin \omega t$, where

$$a = B \sin \tau \qquad \text{and} \qquad b = A + B \cos \tau.$$

(b) Show that the amplitude C of p is given by

$$C^2 = A^2 + B^2 + 2AB \cos \tau.$$

Exercise 71

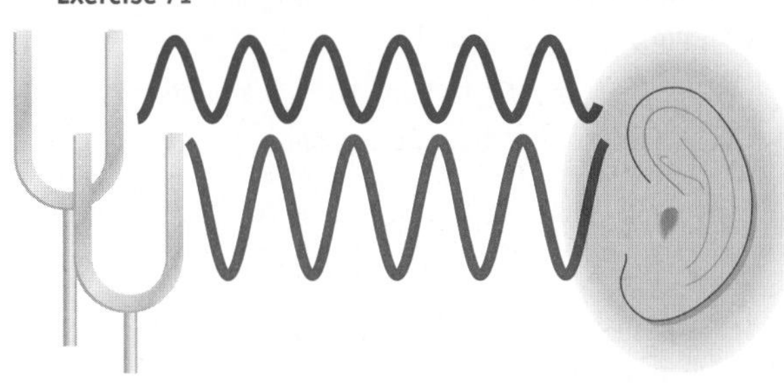

72 **Destructive interference** Refer to Exercise 71. Destructive interference occurs if the amplitude of the resulting sound wave is less than A. Suppose that the two tuning forks are struck with the same force—that is, $A = B$.

(a) When total destructive interference occurs, the amplitude of p is zero and no sound is heard. Find the least positive value of τ for which this occurs.

(b) Determine the τ-interval (a, b) for which destructive interference occurs and a has its least positive value.

73 **Constructive interference** Refer to Exercise 71. When two tuning forks are struck, constructive interference occurs if the amplitude C of the resulting sound wave is larger than either A or B (see the figure).

(a) Show that $C \le A + B$.

(b) Find the values of τ such that $C = A + B$.

(c) If $A \ge B$, determine a condition under which constructive interference will occur.

Exercise 73

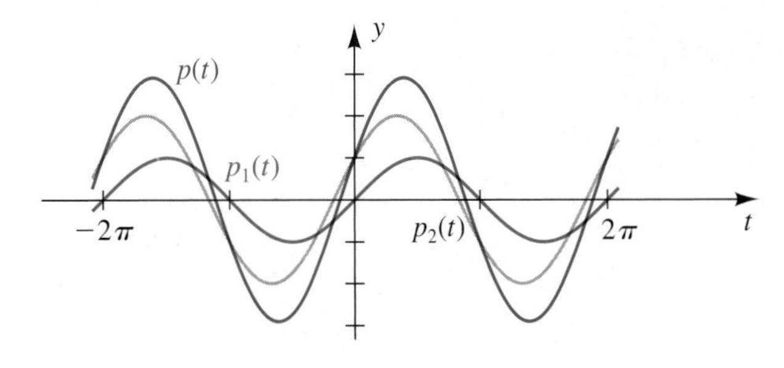

74 **Pressure on the eardrum** Refer to Exercise 71. If two tuning forks with different pitches are struck simultaneously

with different forces, then the total pressure $p(t)$ on the eardrum at time t is given by

$$p(t) = p_1(t) + p_2(t) = A \sin \omega_1 t + B \sin (\omega_2 t + \tau),$$

where A, B, ω_1, ω_2, and τ are constants.

(a) Graph p for $-2\pi \le t \le 2\pi$ if $A = B = 2$, $\omega_1 = 1$, $\omega_2 = 20$, and $\tau = 3$.

(b) Use the graph to describe the variation of the tone that is produced.

Exer. 75–76: Refer to Exercise 73. Graph the equation for $-\pi \le t \le \pi$, and estimate the intervals on which constructive interference occurs.

75 $y = 3 \sin 2t + 2 \sin (4t + 1)$

76 $y = 2 \sin t + 2 \sin (3t + 3)$

6.4 Multiple-Angle Formulas

We refer to the formulas considered in this section as **multiple-angle formulas.** In particular, the following identities are **double-angle formulas,** because they contain the expression $2u$.

Double-Angle Formulas

(1) $\sin 2u = 2 \sin u \cos u$

(2) **(a)** $\cos 2u = \cos^2 u - \sin^2 u$

(b) $\cos 2u = 1 - 2 \sin^2 u$

(c) $\cos 2u = 2 \cos^2 u - 1$

(3) $\tan 2u = \dfrac{2 \tan u}{1 - \tan^2 u}$

PROOFS Each of these formulas may be proved by letting $v = u$ in the appropriate addition formulas. If we use the formula for $\sin (u + v)$, then

$$\begin{aligned} \sin 2u &= \sin (u + u) \\ &= \sin u \cos u + \cos u \sin u \\ &= 2 \sin u \cos u. \end{aligned}$$

Using the formula for $\cos (u + v)$, we have

$$\begin{aligned} \cos 2u &= \cos (u + u) \\ &= \cos u \cos u - \sin u \sin u \\ &= \cos^2 u - \sin^2 u. \end{aligned}$$

To obtain the other two forms for $\cos 2u$ in 2(b) and 2(c), we use the fundamental identity $\sin^2 u + \cos^2 u = 1$. Thus,

$$\begin{aligned} \cos 2u &= \cos^2 u - \sin^2 u \\ &= (1 - \sin^2 u) - \sin^2 u \\ &= 1 - 2 \sin^2 u. \end{aligned}$$

Similarly, if we substitute for $\sin^2 u$ instead of $\cos^2 u$, we obtain

$$\begin{aligned}\cos 2u &= \cos^2 u - (1 - \cos^2 u)\\ &= 2\cos^2 u - 1.\end{aligned}$$

Formula 3 for $\tan 2u$ may be obtained by letting $v = u$ in the formula for $\tan(u + v)$.

EXAMPLE 1 Using double-angle formulas

If $\sin\alpha = \frac{4}{5}$ and α is an acute angle, find the exact values of $\sin 2\alpha$ and $\cos 2\alpha$.

Figure 1

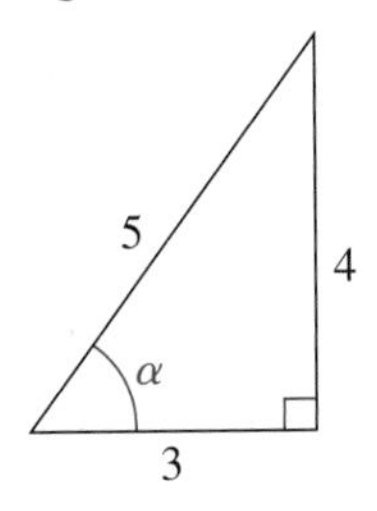

▶ **SOLUTION** If we regard α as an acute angle of a right triangle, as shown in Figure 1, we obtain $\cos\alpha = \frac{3}{5}$. We next substitute in double-angle formulas:

$$\sin 2\alpha = 2\sin\alpha\cos\alpha = 2\left(\tfrac{4}{5}\right)\left(\tfrac{3}{5}\right) = \tfrac{24}{25}$$

$$\cos 2\alpha = \cos^2\alpha - \sin^2\alpha = \left(\tfrac{3}{5}\right)^2 - \left(\tfrac{4}{5}\right)^2 = \tfrac{9}{25} - \tfrac{16}{25} = -\tfrac{7}{25}$$

Figure 2

```
sin⁻¹(4/5)→A
       .927295218
sin(2A)▸Frac
            24/25
cos(2A)▸Frac
            -7/25
```

Figure 2 shows one way to calculate the values in Example 1 on a calculator.

The next example demonstrates how to change a multiple-angle expression to a single-angle expression.

EXAMPLE 2 Changing the form of $\cos 3\theta$

Express $\cos 3\theta$ in terms of $\cos\theta$.

▶ **SOLUTION**

$\cos 3\theta = \cos(2\theta + \theta)$	$3\theta = 2\theta + \theta$
$= \cos 2\theta\cos\theta - \sin 2\theta\sin\theta$	addition formula
$= (2\cos^2\theta - 1)\cos\theta - (2\sin\theta\cos\theta)\sin\theta$	double-angle formulas
$= 2\cos^3\theta - \cos\theta - 2\cos\theta\sin^2\theta$	multiply
$= 2\cos^3\theta - \cos\theta - 2\cos\theta(1 - \cos^2\theta)$	$\sin^2\theta + \cos^2\theta = 1$
$= 4\cos^3\theta - 3\cos\theta$	simplify

We call each of the next three formulas a **half-angle identity,** because the number u is one-half the number $2u$.

Half-Angle Identities

(1) $\sin^2 u = \dfrac{1 - \cos 2u}{2}$ **(2)** $\cos^2 u = \dfrac{1 + \cos 2u}{2}$

(3) $\tan^2 u = \dfrac{1 - \cos 2u}{1 + \cos 2u}$

▶ **Tutorial available at www.thomsonedu.com/login**

PROOFS The first identity may be verified as follows:

$$\begin{aligned} \cos 2u &= 1 - 2\sin^2 u && \text{double-angle formula 2(b)} \\ 2\sin^2 u &= 1 - \cos 2u && \text{isolate } 2\sin^2 u \\ \sin^2 u &= \frac{1 - \cos 2u}{2} && \text{divide by 2} \end{aligned}$$

The second identity can be derived in similar fashion by starting with

$$\cos 2u = 2\cos^2 u - 1.$$

The third identity can be obtained from identities 1 and 2 by noting that

$$\tan^2 u = (\tan u)^2 = \left(\frac{\sin u}{\cos u}\right)^2 = \frac{\sin^2 u}{\cos^2 u}.$$

Half-angle identities may be used to express even powers of trigonometric functions in terms of functions with exponent 1, as illustrated in the next two examples.

EXAMPLE 3 Using half-angle identities to verify an identity

Verify the identity $\sin^2 x \cos^2 x = \frac{1}{8}(1 - \cos 4x)$.

SOLUTION

$$\begin{aligned} \sin^2 x \cos^2 x &= \left(\frac{1 - \cos 2x}{2}\right)\left(\frac{1 + \cos 2x}{2}\right) && \text{half-angle identities} \\ &= \tfrac{1}{4}(1 - \cos^2 2x) && \text{multiply} \\ &= \tfrac{1}{4}(\sin^2 2x) && \sin^2 2x + \cos^2 2x = 1 \\ &= \frac{1}{4}\left(\frac{1 - \cos 4x}{2}\right) && \text{half-angle identity with } u = 2x \\ &= \tfrac{1}{8}(1 - \cos 4x) && \text{multiply} \end{aligned}$$

EXAMPLE 4 Using half-angle identities to reduce a power of cos t

Express $\cos^4 t$ in terms of values of the cosine function with exponent 1.

▶ SOLUTION

$$\begin{aligned} \cos^4 t &= (\cos^2 t)^2 && \text{law of exponents} \\ &= \left(\frac{1 + \cos 2t}{2}\right)^2 && \text{half-angle identity} \\ &= \tfrac{1}{4}(1 + 2\cos 2t + \cos^2 2t) && \text{square} \\ &= \frac{1}{4}\left(1 + 2\cos 2t + \frac{1 + \cos 4t}{2}\right) && \text{half-angle identity with } u = 2t \\ &= \tfrac{3}{8} + \tfrac{1}{2}\cos 2t + \tfrac{1}{8}\cos 4t && \text{simplify} \end{aligned}$$

▶ **Tutorial available at www.thomsonedu.com/login**

Substituting $v/2$ for u in the three half-angle identities gives us

$$\sin^2 \frac{v}{2} = \frac{1 - \cos v}{2} \qquad \cos^2 \frac{v}{2} = \frac{1 + \cos v}{2} \qquad \tan^2 \frac{v}{2} = \frac{1 - \cos v}{1 + \cos v}.$$

Taking the square roots of both sides of each of these equations, we obtain the following, which we call the *half-angle formulas* in order to distinguish them from the half-angle identities.

Half-Angle Formulas

$$\textbf{(1)}\ \sin \frac{v}{2} = \pm\sqrt{\frac{1 - \cos v}{2}} \qquad \textbf{(2)}\ \cos \frac{v}{2} = \pm\sqrt{\frac{1 + \cos v}{2}}$$

$$\textbf{(3)}\ \tan \frac{v}{2} = \pm\sqrt{\frac{1 - \cos v}{1 + \cos v}}$$

When using a half-angle formula, we choose either the + or the −, depending on the quadrant containing the angle of radian measure $v/2$. Thus, for $\sin (v/2)$ we use + if $v/2$ is an angle in quadrant I or II or − if $v/2$ is in quadrant III or IV. For $\cos (v/2)$ we use + if $v/2$ is in quadrant I or IV, and so on.

EXAMPLE 5 Using half-angle formulas for the sine and cosine

Find exact values for

(a) $\sin 22.5°$ **(b)** $\cos 112.5°$

▶ **SOLUTION**

(a) We choose the positive sign because 22.5° is in quadrant I, and hence $\sin 22.5° > 0$.

$$\begin{aligned} \sin 22.5° &= +\sqrt{\frac{1 - \cos 45°}{2}} && \text{half-angle formula for sine with } v = 45° \\ &= \sqrt{\frac{1 - \sqrt{2}/2}{2}} && \cos 45° = \frac{\sqrt{2}}{2} \\ &= \frac{\sqrt{2 - \sqrt{2}}}{2} && \text{multiply radicand by } \frac{2}{2} \text{ and simplify} \end{aligned}$$

(b) Similarly, we choose the negative sign because 112.5° is in quadrant II, and so $\cos 112.5° < 0$.

$$\begin{aligned} \cos 112.5° &= -\sqrt{\frac{1 + \cos 225°}{2}} && \text{half-angle formula for cosine with } v = 225° \\ &= -\sqrt{\frac{1 - \sqrt{2}/2}{2}} && \cos 225° = -\frac{\sqrt{2}}{2} \\ &= -\frac{\sqrt{2 - \sqrt{2}}}{2} && \text{multiply radicand by } \frac{2}{2} \text{ and simplify} \end{aligned}$$

▶ **Tutorial available at www.thomsonedu.com/login**

We can obtain an alternative form for the half-angle formula for $\tan(v/2)$. Multiplying the numerator and denominator of the radicand in the third half-angle formula by $1 - \cos v$ gives us

$$\begin{aligned}\tan\frac{v}{2} &= \pm\sqrt{\frac{1-\cos v}{1+\cos v}\cdot\frac{1-\cos v}{1-\cos v}}\\ &= \pm\sqrt{\frac{(1-\cos v)^2}{1-\cos^2 v}}\\ &= \pm\sqrt{\frac{(1-\cos v)^2}{\sin^2 v}} = \pm\frac{1-\cos v}{\sin v}.\end{aligned}$$

We can eliminate the $\pm$ sign in the preceding formula. First note that the numerator $1 - \cos v$ is never negative. We can show that $\tan(v/2)$ and $\sin v$ always have the same sign. For example, if $0 < v < \pi$, then $0 < v/2 < \pi/2$, and consequently both $\sin v$ and $\tan(v/2)$ are positive. If $\pi < v < 2\pi$, then $\pi/2 < v/2 < \pi$, and hence both $\sin v$ and $\tan(v/2)$ are negative, which gives us the first of the next two identities. The second identity for $\tan(v/2)$ may be obtained by multiplying the numerator and denominator of the radicand in the third half-angle formula by $1 + \cos v$.

Half-Angle Formulas for the Tangent

(1) $\tan\dfrac{v}{2} = \dfrac{1-\cos v}{\sin v}$ **(2)** $\tan\dfrac{v}{2} = \dfrac{\sin v}{1+\cos v}$

EXAMPLE 6 Using a half-angle formula for the tangent

If $\tan\alpha = -\dfrac{4}{3}$ and α is in quadrant IV, find $\tan\dfrac{\alpha}{2}$.

Figure 3

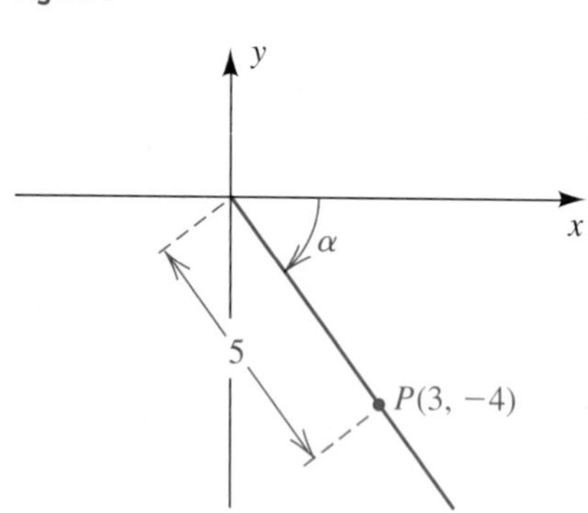

▶ SOLUTION If we choose the point $(3, -4)$ on the terminal side of α, as illustrated in Figure 3, then $\sin\alpha = -\frac{4}{5}$ and $\cos\alpha = \frac{3}{5}$. Applying the first half-angle formula for the tangent, we obtain

$$\tan\frac{\alpha}{2} = \frac{1-\cos\alpha}{\sin\alpha} = \frac{1-\frac{3}{5}}{-\frac{4}{5}} = -\frac{1}{2}.$$

EXAMPLE 7 Finding the x-intercepts of a graph

A graph of the equation $y = \cos 2x + \cos x$ for $0 \le x \le 2\pi$ is sketched in Figure 4. The x-intercepts appear to be approximately 1.1, 3.1, and 5.2. Find their exact values and three-decimal-place approximations.

▶ Tutorial available at www.thomsonedu.com/login

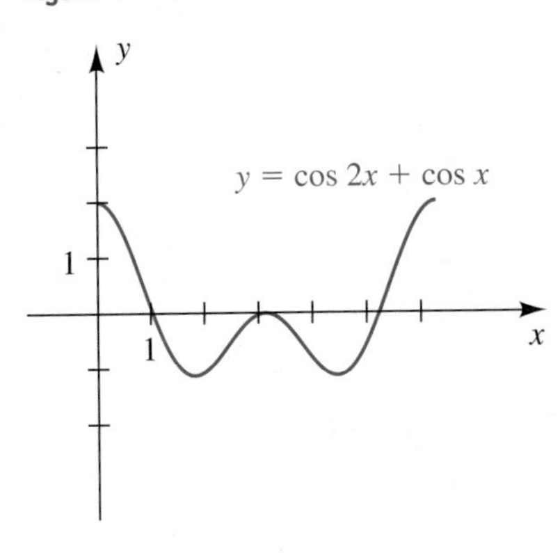

Figure 4

▶ SOLUTION To find the x-intercepts, we proceed as follows:

$$\cos 2x + \cos x = 0 \qquad \text{let } y = 0$$

$$(2\cos^2 x - 1) + \cos x = 0 \qquad \text{double-angle formula 2(c)}$$

$$2\cos^2 x + \cos x - 1 = 0 \qquad \text{equivalent equation}$$

$$(2\cos x - 1)(\cos x + 1) = 0 \qquad \text{factor}$$

$$2\cos x - 1 = 0, \quad \cos x + 1 = 0 \qquad \text{zero factor theorem}$$

$$\cos x = \tfrac{1}{2}, \quad \cos x = -1 \qquad \text{solve for } \cos x$$

The solutions of the last two equations in the interval $[0, 2\pi]$ give us the following exact and approximate x-intercepts:

$$\frac{\pi}{3} \approx 1.047, \quad \frac{5\pi}{3} \approx 5.236, \quad \pi \approx 3.142$$

EXAMPLE 8 Deriving a formula for the area of an isosceles triangle

An **isosceles triangle** has two equal sides of length a, and the angle between them is θ (see Figure 5). Express the area A of the triangle in terms of a and θ.

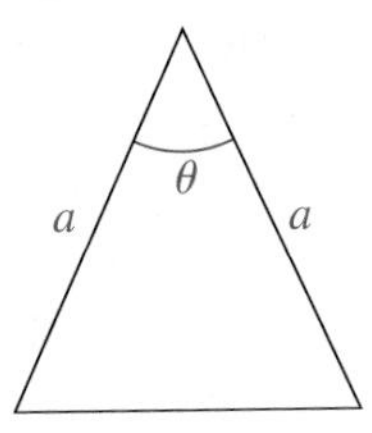

Figure 5

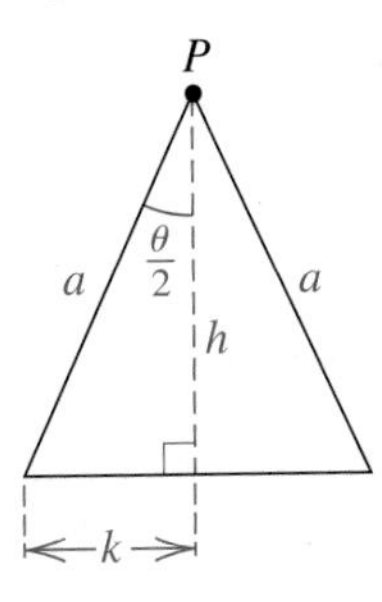

Figure 6

▶ SOLUTION From Figure 6 we see that the altitude from point P bisects θ and that $A = \frac{1}{2}(2k)h = kh$. Thus, we have the following, where $\theta/2$ is an acute angle:

$$\sin\frac{\theta}{2} = \frac{k}{a} \qquad \cos\frac{\theta}{2} = \frac{h}{a} \qquad \text{see Figure 6}$$

$$k = a\sin\frac{\theta}{2} \qquad h = a\cos\frac{\theta}{2} \qquad \text{solve for } k \text{ and } h$$

We next find the area:

$$A = a^2 \sin\frac{\theta}{2}\cos\frac{\theta}{2} \qquad \text{substitute in } A = kh \qquad (*)$$

$$= a^2\sqrt{\frac{1-\cos\theta}{2}}\sqrt{\frac{1+\cos\theta}{2}} \qquad \text{half-angle formulas with } \theta/2 \text{ in quadrant I}$$

$$= a^2\sqrt{\frac{1-\cos^2\theta}{4}} \qquad \text{law of radicals}$$

$$= a^2\sqrt{\frac{\sin^2\theta}{4}} \qquad \sin^2\theta + \cos^2\theta = 1$$

$$= \tfrac{1}{2}a^2|\sin\theta| \qquad \text{take the square root}$$

$$= \tfrac{1}{2}a^2\sin\theta \qquad \sin\theta > 0 \text{ for } 0^\circ < \theta < 180^\circ$$

(continued)

▶ **Tutorial available at www.thomsonedu.com/login**

Another method for simplifying (*) is to write the double-angle formula for the sine, $\sin 2u = 2 \sin u \cos u$, as

$$\sin u \cos u = \tfrac{1}{2} \sin 2u \qquad (**)$$

and proceed as follows:

$$\begin{aligned} A &= a^2 \sin \frac{\theta}{2} \cos \frac{\theta}{2} && \text{substitute in } A = kh \\ &= a^2 \cdot \frac{1}{2} \sin \left(2 \cdot \frac{\theta}{2}\right) && \text{let } u = \frac{\theta}{2} \text{ in } (**) \\ &= \tfrac{1}{2} a^2 \sin \theta && \text{simplify} \end{aligned}$$

6.4 Exercises

Exer. 1–4: Find the exact values of $\sin 2\theta$, $\cos 2\theta$, and $\tan 2\theta$ for the given values of θ.

1 $\cos \theta = \frac{3}{5}$; $0° < \theta < 90°$

▶ 2 $\cot \theta = \frac{4}{3}$; $180° < \theta < 270°$

3 $\sec \theta = -3$; $90° < \theta < 180°$

4 $\sin \theta = -\frac{4}{5}$; $270° < \theta < 360°$

Exer. 5–8: Find the exact values of $\sin (\theta/2)$, $\cos (\theta/2)$, and $\tan (\theta/2)$ for the given conditions.

5 $\sec \theta = \frac{5}{4}$; $0° < \theta < 90°$

6 $\csc \theta = -\frac{5}{3}$; $-90° < \theta < 0°$

7 $\tan \theta = 1$; $-180° < \theta < -90°$

8 $\sec \theta = -4$; $180° < \theta < 270°$

Exer. 9–10: Use half-angle formulas to find the exact values.

9 (a) $\cos 67°30'$ (b) $\sin 15°$ (c) $\tan \frac{3\pi}{8}$

10 (a) $\cos 165°$ (b) $\sin 157°30'$ (c) $\tan \frac{\pi}{8}$

Exer. 11–30: Verify the identity.

11 $\sin 10\theta = 2 \sin 5\theta \cos 5\theta$

12 $\cos^2 3x - \sin^2 3x = \cos 6x$

13 $4 \sin \frac{x}{2} \cos \frac{x}{2} = 2 \sin x$

14 $\frac{\sin^2 2\alpha}{\sin^2 \alpha} = 4 - 4 \sin^2 \alpha$

15 $(\sin t + \cos t)^2 = 1 + \sin 2t$

16 $\csc 2u = \frac{1}{2} \csc u \sec u$

17 $\sin 3u = \sin u\,(3 - 4 \sin^2 u)$

18 $\sin 4t = 4 \sin t \cos t\,(1 - 2 \sin^2 t)$

19 $\cos 4\theta = 8 \cos^4 \theta - 8 \cos^2 \theta + 1$

20 $\cos 6t = 32 \cos^6 t - 48 \cos^4 t + 18 \cos^2 t - 1$

21 $\sin^4 t = \frac{3}{8} - \frac{1}{2} \cos 2t + \frac{1}{8} \cos 4t$

22 $\cos^4 x - \sin^4 x = \cos 2x$

23 $\sec 2\theta = \frac{\sec^2 \theta}{2 - \sec^2 \theta}$

24 $\cot 2u = \frac{\cot^2 u - 1}{2 \cot u}$

25 $2 \sin^2 2t + \cos 4t = 1$

26 $\tan \theta + \cot \theta = 2 \csc 2\theta$

▶ Tutorial available at www.thomsonedu.com/login

27 $\tan 3u = \dfrac{\tan u\,(3 - \tan^2 u)}{1 - 3\tan^2 u}$

28 $\dfrac{1 + \sin 2v + \cos 2v}{1 + \sin 2v - \cos 2v} = \cot v$

29 $\tan \dfrac{\theta}{2} = \csc \theta - \cot \theta$

30 $\tan^2 \dfrac{\theta}{2} = 1 - 2\cot\theta\csc\theta + 2\cot^2\theta$

Exer. 31–34: Express in terms of the cosine function with exponent 1.

31 $\cos^4 \dfrac{\theta}{2}$

32 $\cos^4 2x$

33 $\sin^4 2x$

34 $\sin^4 \dfrac{\theta}{2}$

Exer. 35–42: Find the solutions of the equation that are in the interval $[0, 2\pi)$.

35 $\sin 2t + \sin t = 0$

36 $\cos t - \sin 2t = 0$

37 $\cos u + \cos 2u = 0$

38 $\cos 2\theta - \tan \theta = 1$

39 $\tan 2x = \tan x$

40 $\tan 2t - 2\cos t = 0$

41 $\sin \frac{1}{2}u + \cos u = 1$

42 $2 - \cos^2 x = 4\sin^2 \frac{1}{2}x$

43 If $a > 0$, $b > 0$, and $0 < u < \pi/2$, show that

$$a \sin u + b \cos u = \sqrt{a^2 + b^2}\,\sin(u + v)$$

for $0 < v < \pi/2$, with

$$\sin v = \frac{b}{\sqrt{a^2 + b^2}} \quad \text{and} \quad \cos v = \frac{a}{\sqrt{a^2 + b^2}}.$$

44 Use Exercise 43 to express $8 \sin u + 15 \cos u$ in the form $c \sin(u + v)$.

45 A graph of $y = \cos 2x + 2\cos x$ for $0 \le x \le 2\pi$ is shown in the figure.

(a) Approximate the x-intercepts to two decimal places.

(b) The x-coordinates of the turning points P, Q, and R on the graph are solutions of $\sin 2x + \sin x = 0$. Find the coordinates of these points.

Exercise 45

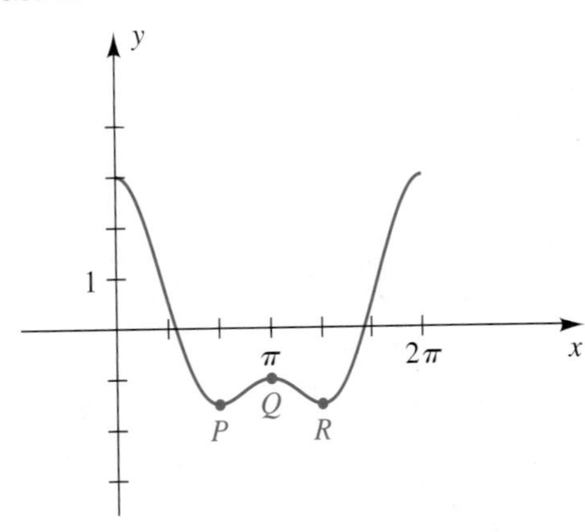

46 A graph of $y = \cos x - \sin 2x$ for $-2\pi \le x \le 2\pi$ is shown in the figure.

(a) Find the x-intercepts.

(b) The x-coordinates of the eight turning points on the graph are solutions of $\sin x + 2\cos 2x = 0$. Approximate these x-coordinates to two decimal places.

Exercise 46

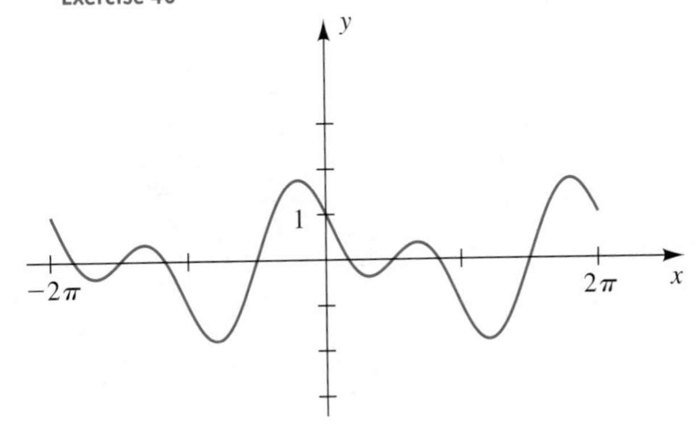

47 A graph of $y = \cos 3x - 3\cos x$ for $-2\pi \le x \le 2\pi$ is shown in the figure on the next page.

(a) Find the x-intercepts. (*Hint:* Use the formula for $\cos 3\theta$ given in Example 2.)

(b) The x-coordinates of the 13 turning points on the graph are solutions of $\sin 3x - \sin x = 0$. Find these x-coordinates. (*Hint:* Use the formula for $\sin 3u$ in Exercise 17.)

Exercise 47

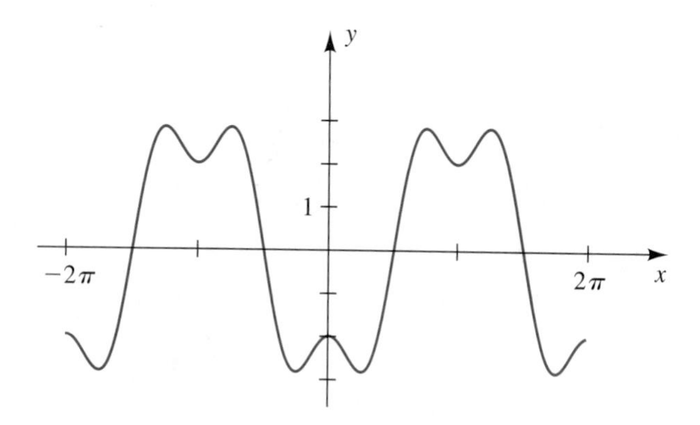

48 A graph of $y = \sin 4x - 4 \sin x$ for $-2\pi \leq x \leq 2\pi$ is shown in the figure. Find the x-intercepts. (*Hint:* Use the formula for $\sin 4t$ in Exercise 18.)

Exercise 48

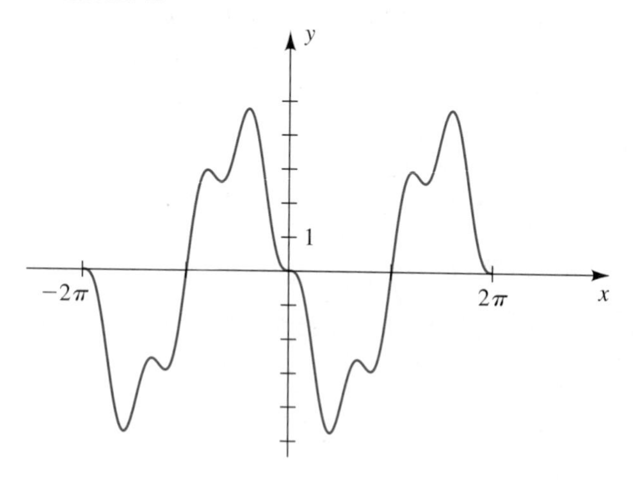

49 **Planning a railroad route** Shown in the figure is a proposed railroad route through three towns located at points A, B, and C. At B, the track will turn toward C at an angle θ.

(a) Show that the total distance d from A to C is given by $d = 20 \tan \frac{1}{2}\theta + 40$.

(b) Because of mountains between A and C, the turning point B must be at least 20 miles from A. Is there a route that avoids the mountains and measures exactly 50 miles?

Exercise 49

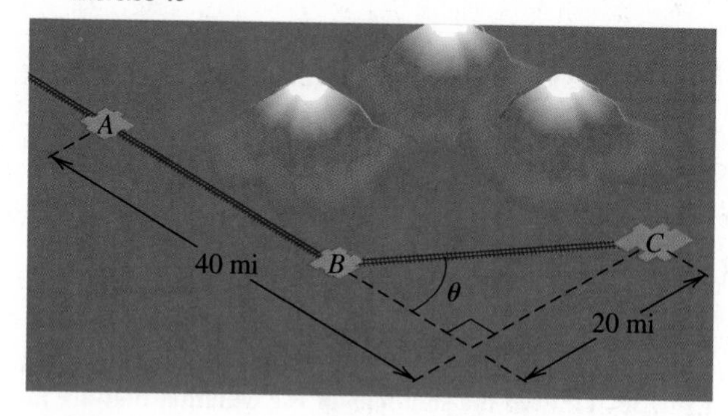

▶ 50 **Projectile's range** If a projectile is fired from ground level with an initial velocity of v ft/sec and at an angle of θ degrees with the horizontal, the range R of the projectile is given by

$$R = \frac{v^2}{16} \sin \theta \cos \theta.$$

If $v = 80$ ft/sec, approximate the angles that result in a range of 150 feet.

51 **Constructing a rain gutter** Shown in the figure is a design for a rain gutter.

(a) Express the volume V as a function of θ. (*Hint:* See Example 8.)

(b) Approximate the acute angle θ that results in a volume of 2 ft^3.

Exercise 51

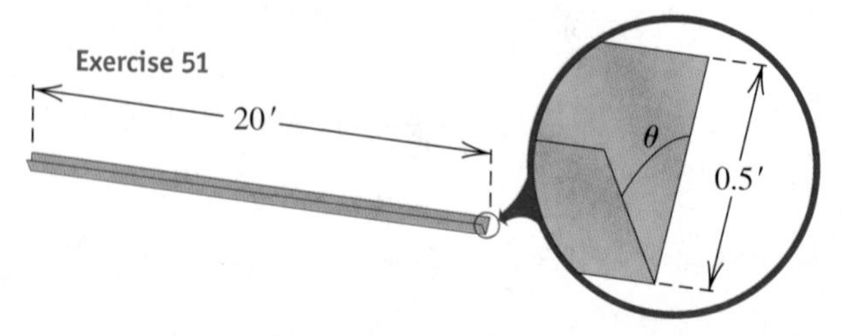

▶ **Tutorial available at www.thomsonedu.com/login**

52 Designing curbing A highway engineer is designing curbing for a street at an intersection where two highways meet at an angle ϕ, as shown in the figure. The curbing between points A and B is to be constructed using a circle that is tangent to the highway at these two points.

(a) Show that the relationship between the radius R of the circle and the distance d in the figure is given by the equation $d = R \tan(\phi/2)$.

(b) If $\phi = 45°$ and $d = 20$ ft, approximate R and the length of the curbing.

Exercise 52

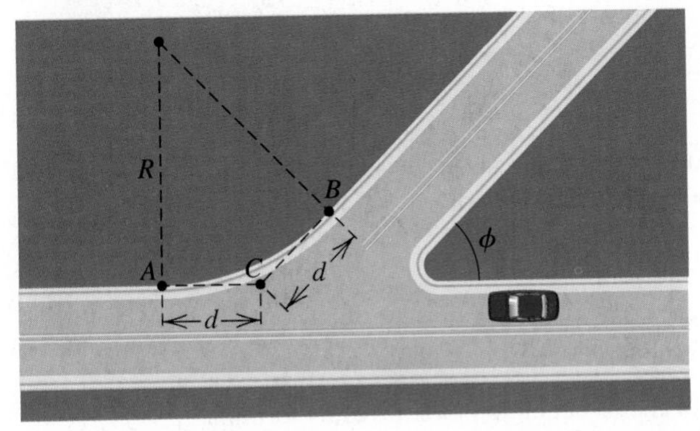

53 Arterial bifurcation A common form of cardiovascular branching is bifurcation, in which an artery splits into two smaller blood vessels. The bifurcation angle θ is the angle formed by the two smaller arteries. In the figure, the line through A and D bisects θ and is perpendicular to the line through B and C.

(a) Show that the length l of the artery from A to B is given by $l = a + \dfrac{b}{2} \tan \dfrac{\theta}{4}$.

(b) Estimate the length l from the three measurements $a = 10$ mm, $b = 6$ mm, and $\theta = 156°$.

Exercise 53

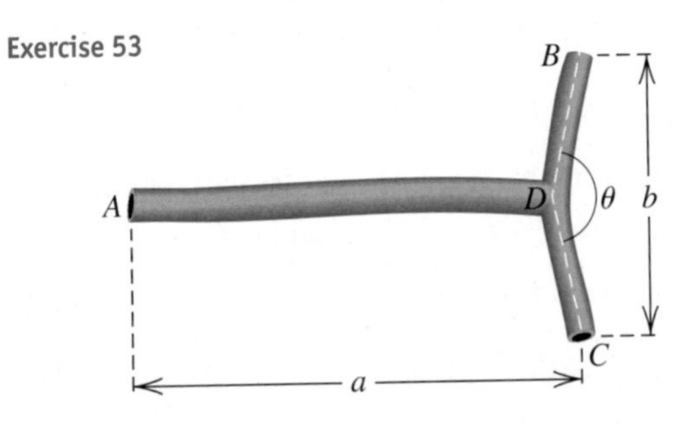

54 Heat production in an AC circuit By definition, the average value of $f(t) = c + a \cos bt$ for one or more complete cycles is c (see the figure).

(a) Use a double-angle formula to find the average value of $f(t) = \sin^2 \omega t$ for $0 \le t \le 2\pi/\omega$, with t in seconds.

(b) In an electrical circuit with an alternating current $I = I_0 \sin \omega t$, the rate r (in calories/sec) at which heat is produced in an R-ohm resistor is given by $r = RI^2$. Find the average rate at which heat is produced for one complete cycle.

Exercise 54

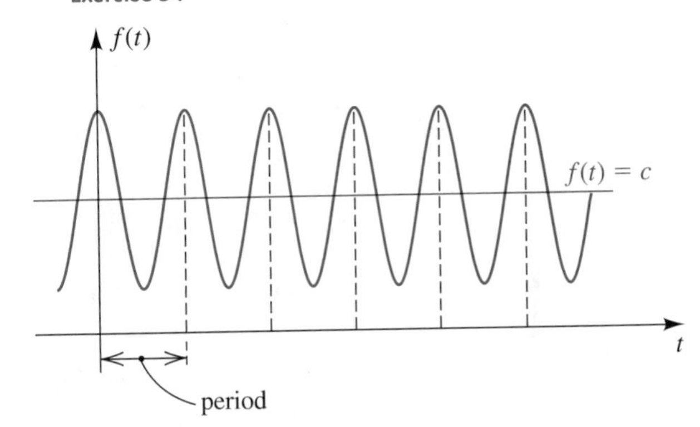

Exer. 55–56: Use the graph of f to find the simplest expression $g(x)$ such that the equation $f(x) = g(x)$ is an identity. Verify this identity.

55 $f(x) = \dfrac{\sin 2x + \sin x}{\cos 2x + \cos x + 1}$

56 $f(x) = \dfrac{\sin x\,(1 + \cos 2x)}{\sin 2x}$

Exer. 57–62: Graphically solve the trigonometric equation on the indicated interval to two decimal places.

57 $\tan\left(\frac{1}{2}x + 1\right) = \sin \frac{1}{2}x$; $[-2\pi, 2\pi]$

58 $\sec(2x + 1) = \cos \frac{1}{2}x + 1$; $[-\pi/2, \pi/2]$

59 $\csc\left(\frac{1}{4}x + 1\right) = 1.5 - \cos 2x$; $[-\pi, \pi]$

60 $3 \sin(2x) + 0.5 = 2 \sin\left(\frac{1}{2}x + 1\right)$; $[-\pi, \pi]$

61 $2 \cot \frac{1}{4}x = 1 - \sec \frac{1}{2}x$; $[-2\pi, 2\pi]$

62 $\tan\left(\frac{3}{2}x + \frac{1}{2}\right) = \frac{3}{2} \sin 2x$; $[-\pi, \pi]$

6.5 Product-to-Sum and Sum-to-Product Formulas

The following formulas may be used to change the form of certain trigonometric expressions from products to sums. We refer to these as **product-to-sum formulas** even though two of the formulas express a product as a difference, because any difference $x - y$ of two real numbers is also a sum $x + (-y)$. These formulas are frequently used in calculus as an aid in a process called *integration*.

Product-to-Sum Formulas

(1) $\sin u \cos v = \frac{1}{2}[\sin (u + v) + \sin (u - v)]$

(2) $\cos u \sin v = \frac{1}{2}[\sin (u + v) - \sin (u - v)]$

(3) $\cos u \cos v = \frac{1}{2}[\cos (u + v) + \cos (u - v)]$

(4) $\sin u \sin v = \frac{1}{2}[\cos (u - v) - \cos (u + v)]$

PROOFS Let us add the left-hand and right-hand sides of the addition and subtraction formulas for the sine function, as follows:

$$\begin{aligned} \sin (u + v) &= \sin u \cos v + \cos u \sin v \\ \sin (u - v) &= \sin u \cos v - \cos u \sin v \\ \hline \sin (u + v) + \sin (u - v) &= 2 \sin u \cos v \end{aligned}$$

Dividing both sides of the last equation by 2 gives us formula 1.

Formula 2 is obtained by *subtracting* the left- and right-hand sides of the addition and subtraction formulas for the sine function. Formulas 3 and 4 are developed in a similar fashion, using the addition and subtraction formulas for the cosine function.

EXAMPLE 1 Using product-to-sum formulas

Express as a sum:

(a) $\sin 4\theta \cos 3\theta$ (b) $\sin 3x \sin x$

▶ SOLUTION

(a) We use product-to-sum formula 1 with $u = 4\theta$ and $v = 3\theta$.

$$\begin{aligned} \sin 4\theta \cos 3\theta &= \tfrac{1}{2}[\sin (4\theta + 3\theta) + \sin (4\theta - 3\theta)] \\ &= \tfrac{1}{2}(\sin 7\theta + \sin \theta) \end{aligned}$$

We can also obtain this relationship by using product-to-sum formula 2.

(b) We use product-to-sum formula 4 with $u = 3x$ and $v = x$:

$$\begin{aligned} \sin 3x \sin x &= \tfrac{1}{2}[\cos (3x - x) - \cos (3x + x)] \\ &= \tfrac{1}{2}(\cos 2x - \cos 4x) \end{aligned}$$

▶ Tutorial available at www.thomsonedu.com/login

We may use the product-to-sum formulas to express a sum or difference as a product. To obtain forms that can be applied more easily, we shall change the notation as follows. If we let

$$u + v = a \qquad \text{and} \qquad u - v = b,$$

then $(u + v) + (u - v) = a + b$, which simplifies to

$$u = \frac{a + b}{2}.$$

Similarly, since $(u + v) - (u - v) = a - b$, we obtain

$$v = \frac{a - b}{2}.$$

We now substitute for $u + v$ and $u - v$ on the right-hand sides of the product-to-sum formulas and for u and v on the left-hand sides. If we then multiply by 2, we obtain the following sum-to-product formulas.

Sum-to-Product Formulas

(1) $\sin a + \sin b = 2 \sin \dfrac{a + b}{2} \cos \dfrac{a - b}{2}$

(2) $\sin a - \sin b = 2 \cos \dfrac{a + b}{2} \sin \dfrac{a - b}{2}$

(3) $\cos a + \cos b = 2 \cos \dfrac{a + b}{2} \cos \dfrac{a - b}{2}$

(4) $\cos a - \cos b = -2 \sin \dfrac{a + b}{2} \sin \dfrac{a - b}{2}$

EXAMPLE 2 **Using a sum-to-product formula**

Express $\sin 5x - \sin 3x$ as a product.

▶ **SOLUTION** We use sum-to-product formula 2 with $a = 5x$ and $b = 3x$:

$$\sin 5x - \sin 3x = 2 \cos \frac{5x + 3x}{2} \sin \frac{5x - 3x}{2}$$
$$= 2 \cos 4x \sin x$$

EXAMPLE 3 **Using sum-to-product formulas to verify an identity**

Verify the identity $\dfrac{\sin 3t + \sin 5t}{\cos 3t - \cos 5t} = \cot t.$

▶ **Tutorial available at www.thomsonedu.com/login**

SOLUTION We first use a sum-to-product formula for the numerator and one for the denominator:

$$\frac{\sin 3t + \sin 5t}{\cos 3t - \cos 5t} = \frac{2 \sin \dfrac{3t + 5t}{2} \cos \dfrac{3t - 5t}{2}}{-2 \sin \dfrac{3t + 5t}{2} \sin \dfrac{3t - 5t}{2}} \quad \text{sum-to-product formulas 1 and 4}$$

$$= \frac{2 \sin 4t \cos (-t)}{-2 \sin 4t \sin (-t)} \quad \text{simplify}$$

$$= \frac{\cos (-t)}{-\sin (-t)} \quad \text{cancel } 2 \sin 4t$$

$$= \frac{\cos t}{\sin t} \quad \text{formulas for negatives}$$

$$= \cot t \quad \text{cotangent identity}$$

EXAMPLE 4 **Using a sum-to-product formula to solve an equation**

Find the solutions of $\sin 5x + \sin x = 0$.

SOLUTION Changing a sum to a product allows us to use the zero factor theorem to solve the equation.

$$\sin 5x + \sin x = 0 \quad \text{given}$$

$$2 \sin \frac{5x + x}{2} \cos \frac{5x - x}{2} = 0 \quad \text{sum-to-product formula 1}$$

$$\sin 3x \cos 2x = 0 \quad \text{simplify and divide by 2}$$

$$\sin 3x = 0, \quad \cos 2x = 0 \quad \text{zero factor theorem}$$

The solutions of the last two equations are

$$3x = \pi n \quad \text{and} \quad 2x = \frac{\pi}{2} + \pi n \quad \text{for every integer } n.$$

Dividing by 3 and 2, respectively, we obtain

$$\frac{\pi}{3} n \quad \text{and} \quad \frac{\pi}{4} + \frac{\pi}{2} n \quad \text{for every integer } n.$$

EXAMPLE 5 **Finding the x-intercepts of a graph**

A graph of the equation $y = \cos x - \cos 3x - \sin 2x$ is shown in Figure 1. Find the 13 x-intercepts that are in the interval $[-2\pi, 2\pi]$.

Tutorial available at www.thomsonedu.com/login

Figure 1

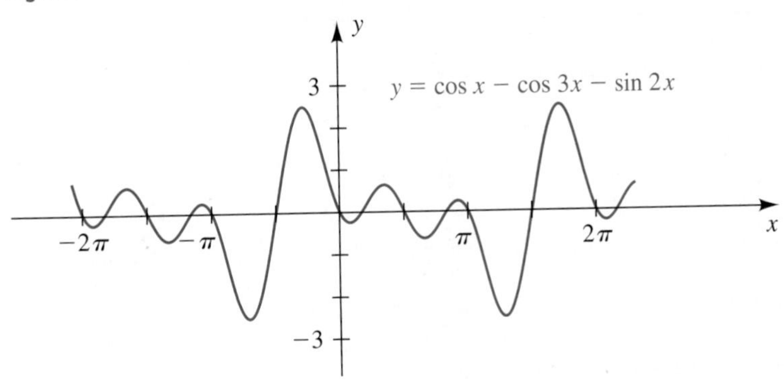

▶ SOLUTION To find the x-intercepts, we proceed as follows:

$$\cos x - \cos 3x - \sin 2x = 0 \qquad \text{let } y = 0$$

$$(\cos x - \cos 3x) - \sin 2x = 0 \qquad \text{group the first two terms}$$

$$-2 \sin \frac{x + 3x}{2} \sin \frac{x - 3x}{2} - \sin 2x = 0 \qquad \text{sum-to-product formula 4}$$

$$-2 \sin 2x \sin (-x) - \sin 2x = 0 \qquad \text{simplify}$$

$$2 \sin 2x \sin x - \sin 2x = 0 \qquad \text{formula for negatives}$$

$$\sin 2x\,(2 \sin x - 1) = 0 \qquad \text{factor out } \sin 2x$$

$$\sin 2x = 0, \quad 2 \sin x - 1 = 0 \qquad \text{zero factor theorem}$$

$$\sin 2x = 0, \quad \sin x = \tfrac{1}{2} \qquad \text{solve for } \sin x$$

The equation $\sin 2x = 0$ has solutions $2x = \pi n$, or, dividing by 2,

$$x = \frac{\pi}{2} n \quad \text{for every integer } n.$$

If we let $n = 0, \pm 1, \pm 2, \pm 3$, and ± 4, we obtain nine x-intercepts in $[-2\pi, 2\pi]$:

$$0, \quad \pm\frac{\pi}{2}, \quad \pm\pi, \quad \pm\frac{3\pi}{2}, \quad \pm 2\pi$$

The solutions of the equation $\sin x = \frac{1}{2}$ are

$$\frac{\pi}{6} + 2\pi n \quad \text{and} \quad \frac{5\pi}{6} + 2\pi n \quad \text{for every integer } n.$$

The four solutions in $[-2\pi, 2\pi]$ are obtained by letting $n = 0$ and $n = -1$:

$$\frac{\pi}{6}, \quad \frac{5\pi}{6}, \quad -\frac{11\pi}{6}, \quad -\frac{7\pi}{6}$$

▶ Tutorial available at www.thomsonedu.com/login

6.5 Exercises

Exer. 1–8: Express as a sum or difference.

1 $\sin 7t \sin 3t$

2 $\sin(-4x)\cos 8x$

3 $\cos 6u \cos(-4u)$

4 $\cos 4t \sin 6t$

5 $2\sin 9\theta \cos 3\theta$

6 $2\sin 7\theta \sin 5\theta$

▶ 7 $3\cos x \sin 2x$

8 $5\cos u \cos 5u$

Exer. 9–16: Express as a product.

9 $\sin 6\theta + \sin 2\theta$

10 $\sin 4\theta - \sin 8\theta$

11 $\cos 5x - \cos 3x$

▶ 12 $\cos 5t + \cos 6t$

13 $\sin 3t - \sin 7t$

14 $\cos\theta - \cos 5\theta$

15 $\cos x + \cos 2x$

16 $\sin 8t + \sin 2t$

Exer. 17–24: Verify the identity.

17 $\dfrac{\sin 4t + \sin 6t}{\cos 4t - \cos 6t} = \cot t$

18 $\dfrac{\sin\theta + \sin 3\theta}{\cos\theta + \cos 3\theta} = \tan 2\theta$

19 $\dfrac{\sin u + \sin v}{\cos u + \cos v} = \tan\dfrac{1}{2}(u + v)$

20 $\dfrac{\sin u - \sin v}{\cos u - \cos v} = -\cot\dfrac{1}{2}(u + v)$

21 $\dfrac{\sin u - \sin v}{\sin u + \sin v} = \dfrac{\tan\frac{1}{2}(u - v)}{\tan\frac{1}{2}(u + v)}$

22 $\dfrac{\cos u - \cos v}{\cos u + \cos v} = -\tan\dfrac{1}{2}(u + v)\tan\dfrac{1}{2}(u - v)$

23 $4\cos x \cos 2x \sin 3x = \sin 2x + \sin 4x + \sin 6x$

24 $\dfrac{\cos t + \cos 4t + \cos 7t}{\sin t + \sin 4t + \sin 7t} = \cot 4t$

Exer. 25–26: Express as a sum.

25 $(\sin ax)(\cos bx)$

26 $(\cos au)(\cos bu)$

Exer. 27–34: Use sum-to-product formulas to find the solutions of the equation.

27 $\sin 5t + \sin 3t = 0$

▶ 28 $\sin t + \sin 3t = \sin 2t$

29 $\cos x = \cos 3x$

30 $\cos 4x - \cos 3x = 0$

31 $\cos 3x + \cos 5x = \cos x$

32 $\cos 3x = -\cos 6x$

33 $\sin 2x - \sin 5x = 0$

34 $\sin 5x - \sin x = 2\cos 3x$

Exer. 35–36: Shown in the figure is a graph of the function f for $0 \le x \le 2\pi$. Use a sum-to-product formula to help find the x-intercepts.

▶ 35 $f(x) = \cos x + \cos 3x$

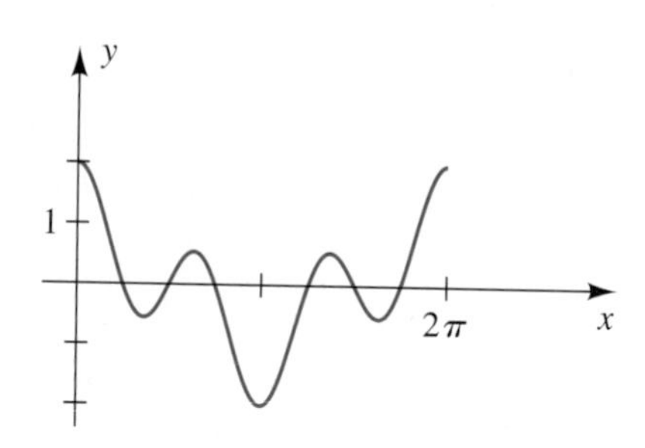

36 $f(x) = \sin 4x - \sin x$

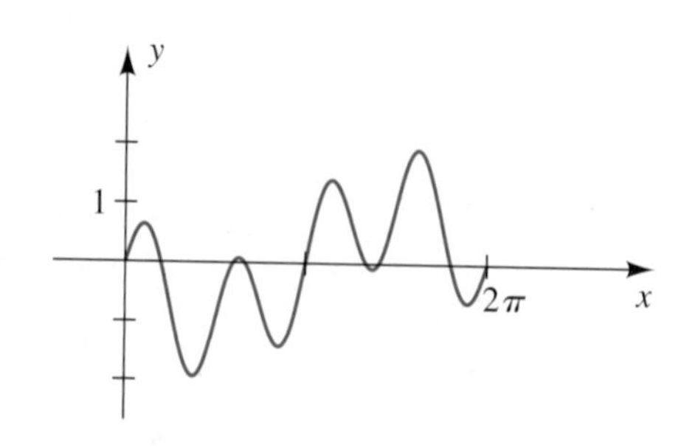

▶ Tutorial available at www.thomsonedu.com/login

37 Refer to Exercise 47 of Section 6.4. The graph of the equation $y = \cos 3x - 3\cos x$ has 13 turning points for $-2\pi \le x \le 2\pi$. The x-coordinates of these points are solutions of the equation $\sin 3x - \sin x = 0$. Use a sum-to-product formula to find these x-coordinates.

38 Refer to Exercise 48 of Section 6.4. The x-coordinates of the turning points on the graph of $y = \sin 4x - 4\sin x$ are solutions of $\cos 4x - \cos x = 0$. Use a sum-to-product formula to find these x-coordinates for $-2\pi \le x \le 2\pi$.

▶ 39 **Vibration of a violin string** Mathematical analysis of a vibrating violin string of length l involves functions such that

$$f(x) = \sin\left(\frac{\pi n}{l}x\right)\cos\left(\frac{k\pi n}{l}t\right),$$

where n is an integer, k is a constant, and t is time. Express f as a sum of two sine functions.

40 **Pressure on the eardrum** If two tuning forks are struck simultaneously with the same force and are then held at the same distance from the eardrum, the pressure on the outside of the eardrum at time t is given by

$$p(t) = a\cos\omega_1 t + a\cos\omega_2 t,$$

where a, ω_1, and ω_2 are constants. If ω_1 and ω_2 are almost equal, a tone is produced that alternates between loudness and virtual silence. This phenomenon is known as beats.

(a) Use a sum-to-product formula to express $p(t)$ as a product.

(b) Show that $p(t)$ may be considered as a cosine wave with approximate period $2\pi/\omega_1$ and variable amplitude $f(t) = 2a\cos\frac{1}{2}(\omega_1 - \omega_2)t$. Find the maximum amplitude.

(c) Shown in the figure is a graph of the equation

$$p(t) = \cos 4.5t + \cos 3.5t.$$

Near-silence occurs at points A and B, where the variable amplitude $f(t)$ in part (b) is zero. Find the coordinates of these points, and determine how frequently near-silence occurs.

(d) Use the graph to show that the function p in part (c) has period 4π. Conclude that the maximum amplitude of 2 occurs every 4π units of time.

Exercise 40

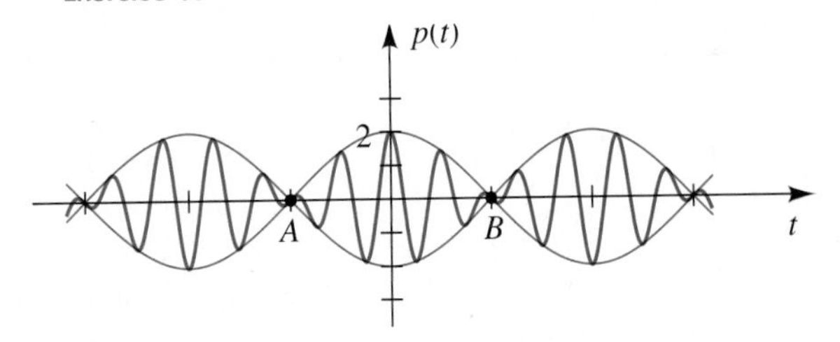

Exer. 41–42: Graph f on the interval $[-\pi, \pi]$. (a) Estimate the x-intercepts. (b) Use sum-to-product formulas to find the exact values of the x-intercepts.

41 $f(x) = \sin 4x + \sin 2x$

42 $f(x) = \cos 3x - \cos 2x$

Exer. 43–44: Use the graph of f to find the simplest expression $g(x)$ such that the equation $f(x) = g(x)$ is an identity. Verify this identity.

43 $f(x) = \dfrac{\sin x + \sin 2x + \sin 3x}{\cos x + \cos 2x + \cos 3x}$

44 $f(x) = \dfrac{\cos x - \cos 2x + \cos 3x}{\sin x - \sin 2x + \sin 3x}$

6.6 The Inverse Trigonometric Functions

Recall from Section 4.1 that to define the inverse function f^{-1} of a function f, it is essential that f be one-to-one; that is, if $a \ne b$ in the domain of f, then $f(a) \ne f(b)$. The inverse function f^{-1} *reverses* the correspondence given by f; that is,

$$u = f(v) \quad \text{if and only if} \quad v = f^{-1}(u).$$

The following general relationships involving f and f^{-1} were discussed in Section 4.1.

▶ Tutorial available at www.thomsonedu.com/login

Relationships Between f^{-1} and f

(1) $y = f^{-1}(x)$ if and only if $x = f(y)$, where x is in the domain of f^{-1} and y is in the domain of f

(2) domain of f^{-1} = range of f

(3) range of f^{-1} = domain of f

(4) $f(f^{-1}(x)) = x$ for every x in the domain of f^{-1}

(5) $f^{-1}(f(y)) = y$ for every y in the domain of f

(6) The point (a, b) is on the graph of f if and only if the point (b, a) is on the graph of f^{-1}.

(7) The graphs of f^{-1} and f are reflections of each other through the line $y = x$.

We shall use relationship 1 to define each of the inverse trigonometric functions.

The sine function is not one-to-one, since different numbers, such as $\pi/6$, $5\pi/6$, and $-7\pi/6$, yield the same function value $\left(\frac{1}{2}\right)$. If we restrict the domain to $[-\pi/2, \pi/2]$, then, as illustrated by the blue portion of the graph of $y = \sin x$ in Figure 1, we obtain a one-to-one (increasing) function that takes on every value of the sine function once and only once. We use this *new* function with domain $[-\pi/2, \pi/2]$ and range $[-1, 1]$ to define the *inverse sine function.*

Figure 1

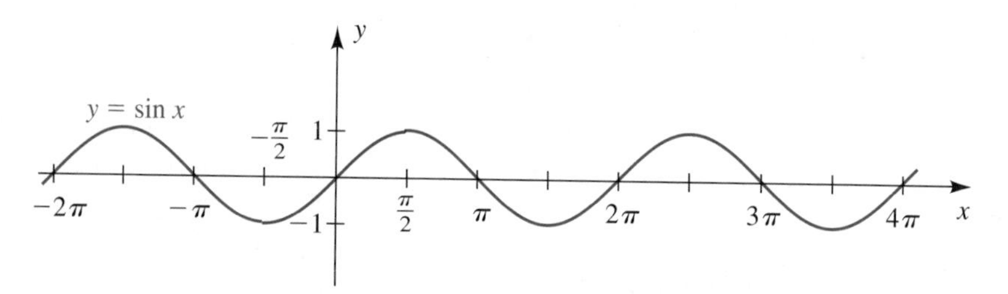

Definition of the Inverse Sine Function

The **inverse sine function,** denoted by $\sin^{-1}$, is defined by

$$y = \sin^{-1} x \quad \text{if and only if} \quad x = \sin y$$

for $-1 \leq x \leq 1$ and $-\dfrac{\pi}{2} \leq y \leq \dfrac{\pi}{2}$.

The domain of the inverse sine function is $[-1, 1]$, and the range is $[-\pi/2, \pi/2]$.

Note on notation:

While $(\sin x)^{-1} = \dfrac{1}{\sin x} = \csc x,$

none *of these equal* $\sin^{-1} x$.

The notation $y = \sin^{-1} x$ is sometimes read "y is the inverse sine of x." The equation $x = \sin y$ in the definition allows us to regard y as an angle, so $y = \sin^{-1} x$ may also be read "y is the angle whose sine is x" (with $-\pi/2 \le y \le \pi/2$).

The inverse sine function is also called the **arcsine function,** and arcsin x may be used in place of $\sin^{-1} x$. If $t = \arcsin x$, then $\sin t = x$, and t may be interpreted as an *arc length* on the unit circle U with center at the origin. We will use both notations—$\sin^{-1}$ and arcsin—throughout our work.

Several values of the inverse sine function are listed in the next chart.

Warning!

> It is *essential* to choose the value y in the range $[-\pi/2, \pi/2]$ of $\sin^{-1}$. Thus, even though $\sin(5\pi/6) = \frac{1}{2}$, the number $y = 5\pi/6$ is not the inverse function value $\sin^{-1}\frac{1}{2}$.

Equation	Equivalent statement	Solution
$y = \sin^{-1}\left(\dfrac{1}{2}\right)$	$\sin y = \dfrac{1}{2}$ and $-\dfrac{\pi}{2} \le y \le \dfrac{\pi}{2}$	$y = \dfrac{\pi}{6}$
$y = \sin^{-1}\left(-\dfrac{1}{2}\right)$	$\sin y = -\dfrac{1}{2}$ and $-\dfrac{\pi}{2} \le y \le \dfrac{\pi}{2}$	$y = -\dfrac{\pi}{6}$
$y = \sin^{-1}(1)$	$\sin y = 1$ and $-\dfrac{\pi}{2} \le y \le \dfrac{\pi}{2}$	$y = \dfrac{\pi}{2}$
$y = \arcsin(0)$	$\sin y = 0$ and $-\dfrac{\pi}{2} \le y \le \dfrac{\pi}{2}$	$y = 0$
$y = \arcsin\left(-\dfrac{\sqrt{3}}{2}\right)$	$\sin y = -\dfrac{\sqrt{3}}{2}$ and $-\dfrac{\pi}{2} \le y \le \dfrac{\pi}{2}$	$y = -\dfrac{\pi}{3}$

Figure 2

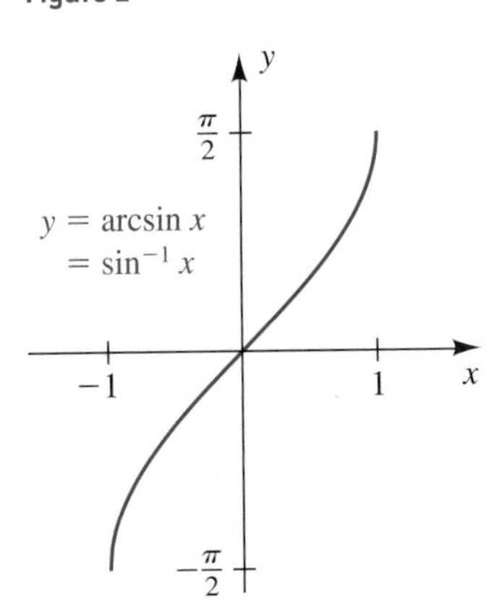

We have now justified the method of solving an equation of the form $\sin\theta = k$ as discussed in Chapter 5. We see that the calculator key SIN⁻¹ used to obtain $\theta = \sin^{-1} k$ gives us the value of the inverse sine function.

Relationship 7 for the graphs of f and f^{-1} tells us that we can sketch the graph of $y = \sin^{-1} x$ by reflecting the blue portion of Figure 1 through the line $y = x$. We can also use the equation $x = \sin y$ with the restriction $-\pi/2 \le y \le \pi/2$ to find points on the graph. This gives us Figure 2.

Relationship 4, $f(f^{-1}(x)) = x$, and relationship 5, $f^{-1}(f(y)) = y$, which hold for any inverse function f^{-1}, give us the following properties.

Properties of $\sin^{-1}$

(1) $\sin(\sin^{-1} x) = \sin(\arcsin x) = x$ if $-1 \le x \le 1$

(2) $\sin^{-1}(\sin y) = \arcsin(\sin y) = y$ if $-\dfrac{\pi}{2} \le y \le \dfrac{\pi}{2}$

EXAMPLE 1 Using properties of $\sin^{-1}$

Find the exact value:

(a) $\sin\left(\sin^{-1}\dfrac{1}{2}\right)$ (b) $\sin^{-1}\left(\sin\dfrac{\pi}{4}\right)$ (c) $\sin^{-1}\left(\sin\dfrac{2\pi}{3}\right)$

▶ SOLUTION

(a) The *difficult* way to find the value of this expression is to first find the angle $\sin^{-1}\frac{1}{2}$, namely $\pi/6$, and then evaluate $\sin(\pi/6)$, obtaining $\frac{1}{2}$. The *easy* way is to use property 1 of $\sin^{-1}$:

$$\text{since} \qquad -1 \le \tfrac{1}{2} \le 1, \qquad \sin\left(\sin^{-1}\tfrac{1}{2}\right) = \tfrac{1}{2}$$

(b) Since $-\pi/2 \le \pi/4 \le \pi/2$, we can use property 2 of $\sin^{-1}$ to obtain

$$\sin^{-1}\left(\sin\frac{\pi}{4}\right) = \frac{\pi}{4}.$$

(c) Be careful! Since $2\pi/3$ is *not* between $-\pi/2$ and $\pi/2$, we cannot use property 2 of $\sin^{-1}$. Instead, we first evaluate the inner expression, $\sin(2\pi/3)$, and then use the definition of $\sin^{-1}$, as follows:

$$\sin^{-1}\left(\sin\frac{2\pi}{3}\right) = \sin^{-1}\left(\frac{\sqrt{3}}{2}\right) = \frac{\pi}{3}$$

EXAMPLE 2 Finding a value of $\sin^{-1}$

Find the exact value of y if $y = \sin^{-1}\left(\tan\dfrac{3\pi}{4}\right)$.

▶ SOLUTION We first evaluate the inner expression—$\tan(3\pi/4)$—and then find the inverse sine of that number:

$$y = \sin^{-1}\left(\tan\frac{3\pi}{4}\right) = \sin^{-1}(-1)$$

In words, we have "y is the angle whose sine is -1." It may be helpful to recall the arcsine values by associating them with the angles corresponding to the blue portion of the unit circle shown in Figure 3. From the figure we see that $-\pi/2$ is the angle whose sine is -1. It follows that $y = -\pi/2$, and hence

$$y = \sin^{-1}\left(\tan\frac{3\pi}{4}\right) = -\frac{\pi}{2}.$$

Figure 3

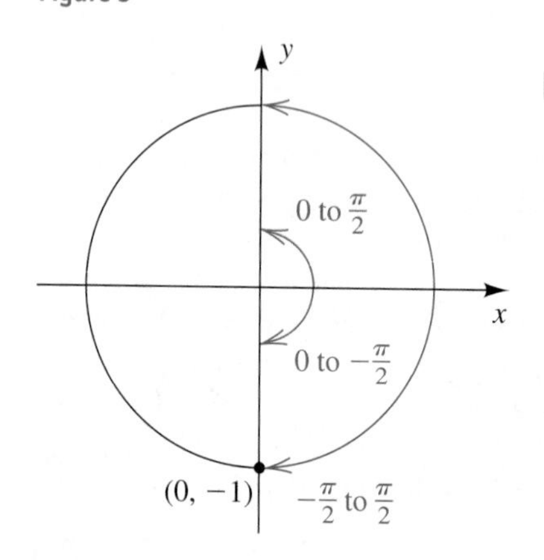

The other trigonometric functions may also be used to introduce inverse trigonometric functions. The procedure is first to determine a convenient subset of the domain in order to obtain a one-to-one function. If the domain of the cosine function is restricted to the interval $[0, \pi]$, as illustrated by the blue por-

▶ Tutorial available at www.thomsonedu.com/login

tion of the graph of $y = \cos x$ in Figure 4, we obtain a one-to-one (decreasing) function that takes on every value of the cosine function once and only once. Then, we use this *new* function with domain $[0, \pi]$ and range $[-1, 1]$ to define the *inverse cosine function.*

Figure 4

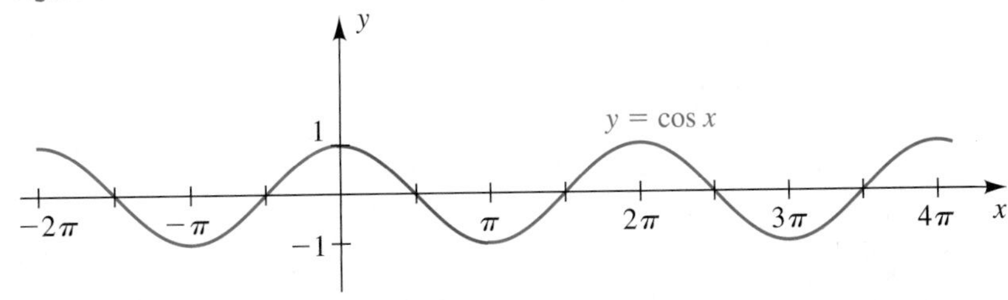

Definition of the Inverse Cosine Function

The **inverse cosine function,** denoted by $\cos^{-1}$, is defined by

$$y = \cos^{-1} x \quad \text{if and only if} \quad x = \cos y$$

for $-1 \leq x \leq 1$ and $0 \leq y \leq \pi$.

The domain of the inverse cosine function is $[-1, 1]$, and the range is $[0, \pi]$. Note that the range of $\cos^{-1}$ is *not* the same as the range of $\sin^{-1}$ but their domains are equal.

The notation $y = \cos^{-1} x$ may be read "y is the inverse cosine of x" or "y is the angle whose cosine is x" (with $0 \leq y \leq \pi$).

The inverse cosine function is also called the **arccosine function,** and the notation $\arccos x$ is used interchangeably with $\cos^{-1} x$.

Several values of the inverse cosine function are listed in the next chart.

It is *essential* to choose the value y in the range $[0, \pi]$ of $\cos^{-1}$.

Equation	Equivalent statement	Solution
$y = \cos^{-1}\left(\frac{1}{2}\right)$	$\cos y = \frac{1}{2}$ and $0 \leq y \leq \pi$	$y = \frac{\pi}{3}$
$y = \cos^{-1}\left(-\frac{1}{2}\right)$	$\cos y = -\frac{1}{2}$ and $0 \leq y \leq \pi$	$y = \frac{2\pi}{3}$
$y = \cos^{-1}(1)$	$\cos y = 1$ and $0 \leq y \leq \pi$	$y = 0$
$y = \arccos(0)$	$\cos y = 0$ and $0 \leq y \leq \pi$	$y = \frac{\pi}{2}$
$y = \arccos\left(-\frac{\sqrt{3}}{2}\right)$	$\cos y = -\frac{\sqrt{3}}{2}$ and $0 \leq y \leq \pi$	$y = \frac{5\pi}{6}$

We can sketch the graph of $y = \cos^{-1} x$ by reflecting the blue portion of Figure 4 through the line $y = x$. This gives us the sketch in Figure 5. We could also use the equation $x = \cos y$, with $0 \le y \le \pi$, to find points on the graph. As indicated by the graph, *the values of the inverse cosine function are never negative.*

As in Example 2 and Figure 3 for the arcsine, it may be helpful to associate the arccosine values with the angles corresponding to the blue arc in Figure 6.

Figure 5

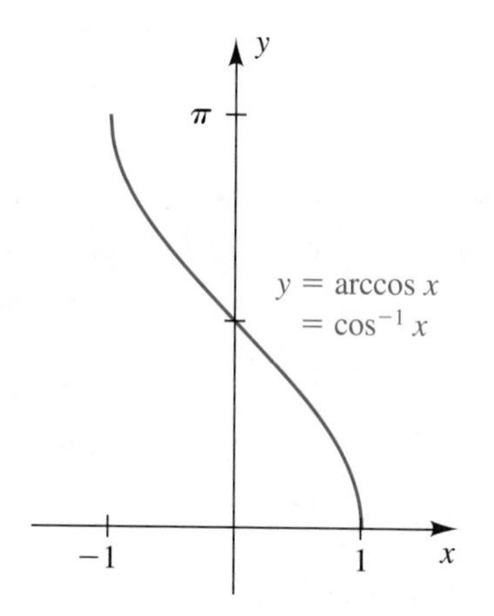

Figure 6

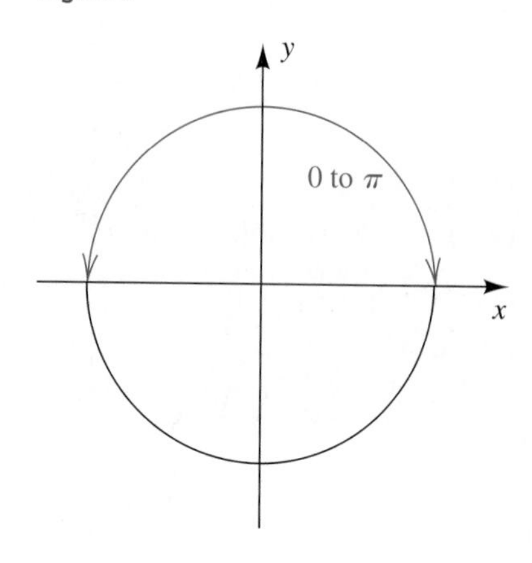

Using relationships 4 and 5 for general inverse functions f and f^{-1}, we obtain the following properties.

Properties of $\cos^{-1}$	(1) $\cos(\cos^{-1} x) = \cos(\arccos x) = x$ if $-1 \le x \le 1$ (2) $\cos^{-1}(\cos y) = \arccos(\cos y) = y$ if $0 \le y \le \pi$

EXAMPLE 3 Using properties of $\cos^{-1}$

Find the exact value:

(a) $\cos[\cos^{-1}(-0.5)]$ **(b)** $\cos^{-1}(\cos 3.14)$ **(c)** $\cos^{-1}\left[\sin\left(-\frac{\pi}{6}\right)\right]$

SOLUTION For parts (a) and (b), we may use properties 1 and 2 of $\cos^{-1}$, respectively.

▶ **(a)** Since $-1 \le -0.5 \le 1$, $\cos[\cos^{-1}(-0.5)] = -0.5$.

▶ **(b)** Since $0 \le 3.14 \le \pi$, $\cos^{-1}(\cos 3.14) = 3.14$.

▶ **Tutorial available at www.thomsonedu.com/login**

(c) We first find $\sin(-\pi/6)$ and then use the definition of $\cos^{-1}$, as follows:

$$\cos^{-1}\left[\sin\left(-\frac{\pi}{6}\right)\right] = \cos^{-1}\left(-\frac{1}{2}\right) = \frac{2\pi}{3}$$

EXAMPLE 4 **Finding a trigonometric function value**

Find the exact value of $\sin\left[\arccos\left(-\frac{2}{3}\right)\right]$.

SOLUTION If we let $\theta = \arccos\left(-\frac{2}{3}\right)$, then, using the definition of the inverse cosine function, we have

$$\cos\theta = -\tfrac{2}{3} \quad \text{and} \quad 0 \le \theta \le \pi.$$

Figure 7

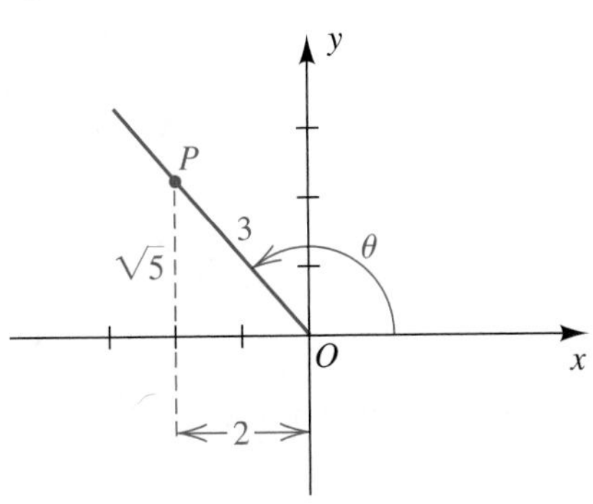

Hence, θ is in quadrant II, as illustrated in Figure 7. If we choose the point P on the terminal side with x-coordinate -2, the hypotenuse of the triangle in the figure must have length 3, since $\cos\theta = -\frac{2}{3}$. Thus, by the Pythagorean theorem, the y-coordinate of P is

$$\sqrt{3^2 - 2^2} = \sqrt{9 - 4} = \sqrt{5},$$

and therefore

$$\sin\left[\arccos\left(-\frac{2}{3}\right)\right] = \sin\theta = \frac{\sqrt{5}}{3}.$$

If we restrict the domain of the tangent function of the branch defined on the open interval $(-\pi/2, \pi/2)$, we obtain a one-to-one (increasing) function (see Figure 3 in Section 6.2). We use this *new* function to define the *inverse tangent function.*

Definition of the Inverse Tangent Function	The **inverse tangent function,** or **arctangent function,** denoted by $\tan^{-1}$ or arctan, is defined by $$y = \tan^{-1} x = \arctan x \quad \text{if and only if} \quad x = \tan y$$ for any real number x and for $-\dfrac{\pi}{2} < y < \dfrac{\pi}{2}$.

Figure 8

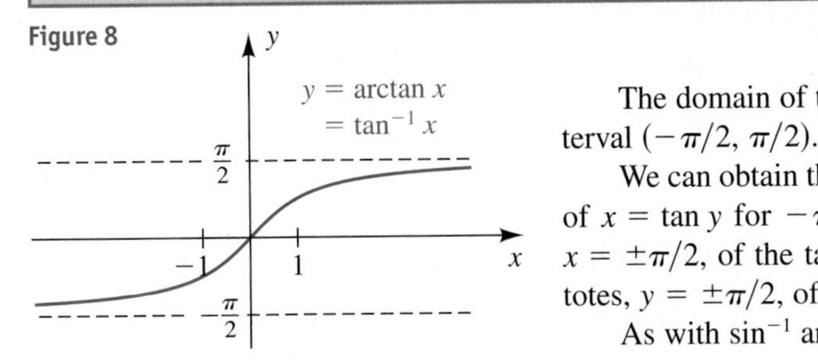

The domain of the arctangent function is $\mathbb{R}$, and the range is the open interval $(-\pi/2, \pi/2)$.

We can obtain the graph of $y = \tan^{-1} x$ in Figure 8 by sketching the graph of $x = \tan y$ for $-\pi/2 < y < \pi/2$. Note that the two *vertical* asymptotes, $x = \pm\pi/2$, of the tangent function correspond to the two *horizontal* asymptotes, $y = \pm\pi/2$, of the arctangent function.

As with $\sin^{-1}$ and $\cos^{-1}$, we have the following properties for $\tan^{-1}$.

Tutorial available at www.thomsonedu.com/login

Properties of $\tan^{-1}$	(1) $\tan(\tan^{-1} x) = \tan(\arctan x) = x$ for every x (2) $\tan^{-1}(\tan y) = \arctan(\tan y) = y$ if $-\dfrac{\pi}{2} < y < \dfrac{\pi}{2}$

EXAMPLE 5 **Using properties of $\tan^{-1}$**

Find the exact value:

(a) $\tan(\tan^{-1} 1000)$ (b) $\tan^{-1}\left(\tan \dfrac{\pi}{4}\right)$ (c) $\arctan(\tan \pi)$

▶ **SOLUTION**

(a) By property 1 of $\tan^{-1}$,

$$\tan(\tan^{-1} 1000) = 1000.$$

(b) Since $-\pi/2 < \pi/4 < \pi/2$, we have, by property 2 of $\tan^{-1}$,

$$\tan^{-1}\left(\tan \frac{\pi}{4}\right) = \frac{\pi}{4}.$$

(c) Since $\pi > \pi/2$, we cannot use the second property of $\tan^{-1}$. Thus, we first find $\tan \pi$ and then evaluate, as follows:

$$\arctan(\tan \pi) = \arctan 0 = 0$$

EXAMPLE 6 **Finding a trigonometric function value**

Find the exact value of $\sec\left(\arctan \frac{2}{3}\right)$.

▶ **SOLUTION** If we let $y = \arctan \frac{2}{3}$, then $\tan y = \frac{2}{3}$. We wish to find $\sec y$. Since $-\pi/2 < \arctan x < \pi/2$ for every x and $\tan y > 0$, it follows that $0 < y < \pi/2$. Thus, we may regard y as the radian measure of an angle of a right triangle such that $\tan y = \frac{2}{3}$, as illustrated in Figure 9. By the Pythagorean theorem, the hypotenuse is $\sqrt{3^2 + 2^2} = \sqrt{13}$. Referring to the triangle, we obtain

Figure 9

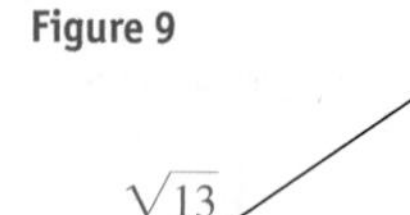

$$\sec\left(\arctan \frac{2}{3}\right) = \sec y = \frac{\sqrt{13}}{3}.$$

EXAMPLE 7 **Finding a trigonometric function value**

Find the exact value of $\sin\left(\arctan \frac{1}{2} - \arccos \frac{4}{5}\right)$.

▶ **Tutorial available at www.thomsonedu.com/login**

Figure 10

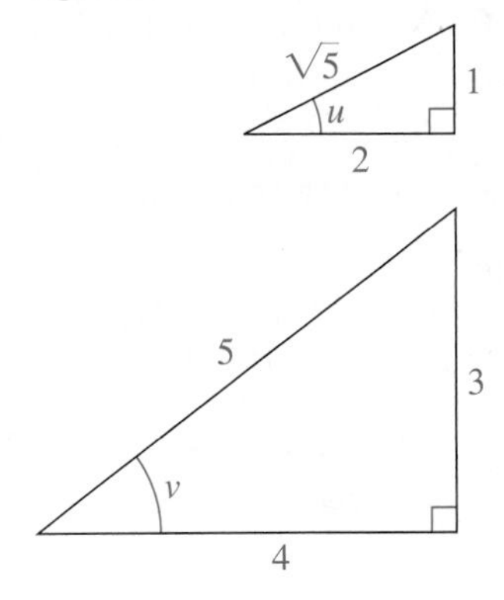

▶ SOLUTION If we let

$$u = \arctan \tfrac{1}{2} \qquad \text{and} \qquad v = \arccos \tfrac{4}{5},$$

then
$$\tan u = \tfrac{1}{2} \qquad \text{and} \qquad \cos v = \tfrac{4}{5}.$$

We wish to find $\sin(u - v)$. Since u and v are in the interval $(0, \pi/2)$, they can be considered as the radian measures of positive acute angles, and we may refer to the right triangles in Figure 10. This gives us

$$\sin u = \frac{1}{\sqrt{5}}, \quad \cos u = \frac{2}{\sqrt{5}}, \quad \sin v = \frac{3}{5}, \quad \text{and} \quad \cos v = \frac{4}{5}.$$

By the subtraction formula for sine,

$$\begin{aligned} \sin(u - v) &= \sin u \cos v - \cos u \sin v \\ &= \frac{1}{\sqrt{5}}\frac{4}{5} - \frac{2}{\sqrt{5}}\frac{3}{5} \\ &= \frac{-2}{5\sqrt{5}}, \quad \text{or} \quad \frac{-2\sqrt{5}}{25}. \end{aligned}$$

EXAMPLE 8 **Changing an expression involving $\sin^{-1} x$ to an algebraic expression**

If $-1 \le x \le 1$, rewrite $\cos(\sin^{-1} x)$ as an algebraic expression in x.

▶ SOLUTION Let

$$y = \sin^{-1} x \qquad \text{or, equivalently,} \qquad \sin y = x.$$

We wish to express $\cos y$ in terms of x. Since $-\pi/2 \le y \le \pi/2$, it follows that $\cos y \ge 0$, and hence (from $\sin^2 y + \cos^2 y = 1$)

$$\cos y = \sqrt{1 - \sin^2 y} = \sqrt{1 - x^2}.$$

Consequently,
$$\cos(\sin^{-1} x) = \sqrt{1 - x^2}.$$

Figure 11

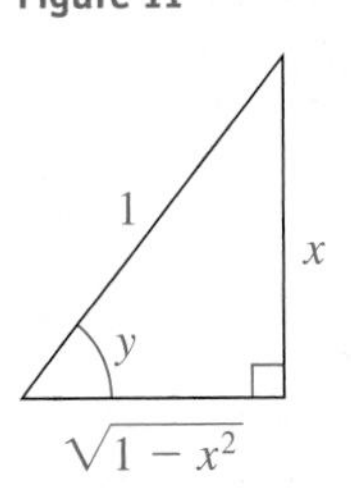

Note that $\sin y = \dfrac{x}{1} = x.$

The last identity is also evident geometrically if $0 < x < 1$. In this case $0 < y < \pi/2$, and we may regard y as the radian measure of an angle of a right triangle such that $\sin y = x$, as illustrated in Figure 11. (The side of length $\sqrt{1 - x^2}$ is found by the Pythagorean theorem.) Referring to the triangle, we have

$$\cos(\sin^{-1} x) = \cos y = \frac{\sqrt{1 - x^2}}{1} = \sqrt{1 - x^2}.$$

Most of the trigonometric equations we considered in Section 6.2 had solutions that were rational multiples of π, such as $\pi/3$, $3\pi/4$, π, and so on.

▶ **Tutorial available at www.thomsonedu.com/login**

Figure 12

(a)

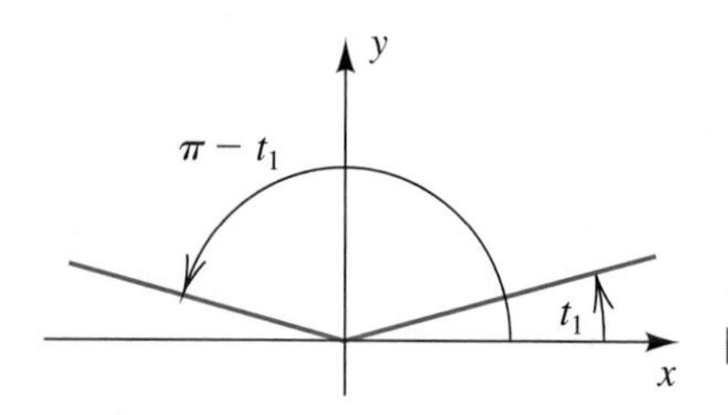

(b)

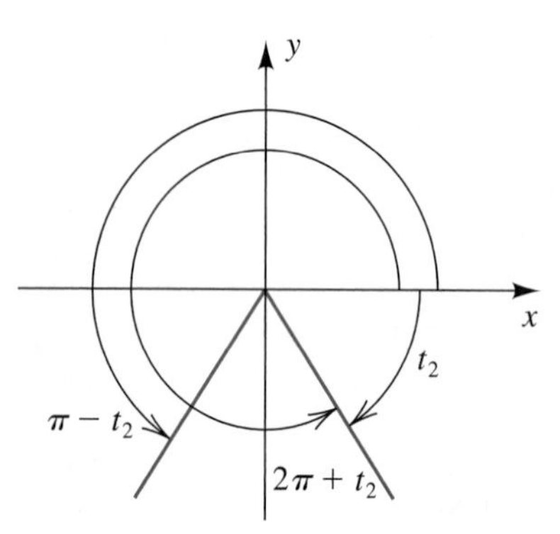

Figure 13

$[0, 2\pi]$ by $[-3, 8]$

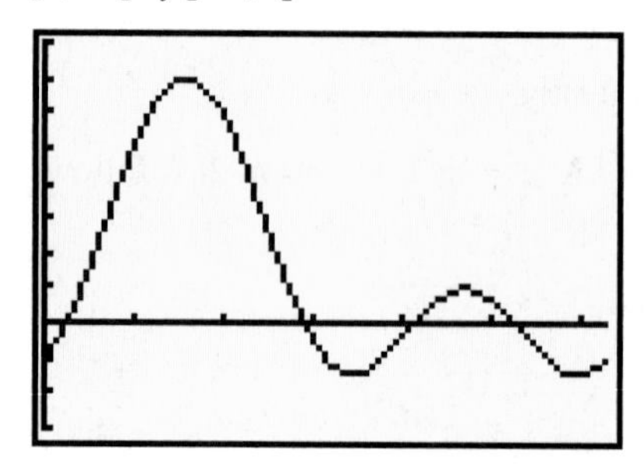

If solutions of trigonometric equations are not of that type, we can sometimes use inverse functions to express them in exact form, as illustrated in the next example.

EXAMPLE 9 Using inverse trigonometric functions to solve an equation

Find the solutions of $5 \sin^2 t + 3 \sin t - 1 = 0$ in $[0, 2\pi)$.

▶ SOLUTION The equation may be regarded as a quadratic equation in $\sin t$. Applying the quadratic formula gives us

$$\sin t = \frac{-3 \pm \sqrt{3^2 - 4(5)(-1)}}{2(5)} = \frac{-3 \pm \sqrt{29}}{10}.$$

Using the definition of the inverse sine function, we obtain the following solutions:

$$t_1 = \sin^{-1} \tfrac{1}{10}\left(-3 + \sqrt{29}\right) \approx 0.2408$$
$$t_2 = \sin^{-1} \tfrac{1}{10}\left(-3 - \sqrt{29}\right) \approx -0.9946$$

Since the range of arcsin is $[-\pi/2, \pi/2]$, we know that t_1 is in $[0, \pi/2]$ and t_2 is in $[-\pi/2, 0]$. Using t_1 as a reference angle, we also have $\pi - t_1$ as a solution in quadrant II, as shown in Figure 12(a). We can add 2π to t_2 to obtain a solution in quadrant IV, as shown in Figure 12(b). The solution in quadrant III is $\pi - t_2$, not $\pi + t_2$, because t_2 is negative.

Hence, with t_1 and t_2 as previously defined, the four exact solutions are

$$t_1, \quad \pi - t_1, \quad \pi - t_2, \quad \text{and} \quad 2\pi + t_2,$$

and the four approximate solutions are

$$0.2408, \quad 2.9008, \quad 4.1361, \quad \text{and} \quad 5.2886.$$

If only approximate solutions are required, we may use a graphing utility to find the x-intercepts of $Y_1 = 5 \sin^2 x + 3 \sin x - 1$. Graphing Y_1 as shown in Figure 13 and using a root feature, we obtain the same four approximate solutions as listed above.

The next example illustrates one of many identities that are true for inverse trigonometric functions.

EXAMPLE 10 Verifying an identity involving inverse trigonometric functions

Verify the identity $\sin^{-1} x + \cos^{-1} x = \dfrac{\pi}{2}$ for $-1 \le x \le 1$.

SOLUTION Let

$$\alpha = \sin^{-1} x \quad \text{and} \quad \beta = \cos^{-1} x.$$

▶ **Tutorial available at www.thomsonedu.com/login**

We wish to show that $\alpha + \beta = \pi/2$. From the definitions of $\sin^{-1}$ and $\cos^{-1}$,

$$\sin \alpha = x \quad \text{for} \quad -\frac{\pi}{2} \le \alpha \le \frac{\pi}{2}$$

and

$$\cos \beta = x \quad \text{for} \quad 0 \le \beta \le \pi.$$

Adding the two inequalities on the right, we see that

$$-\frac{\pi}{2} \le \alpha + \beta \le \frac{3\pi}{2}.$$

Note also that

$$\cos \alpha = \sqrt{1 - \sin^2 \alpha} = \sqrt{1 - x^2}$$

and

$$\sin \beta = \sqrt{1 - \cos^2 \beta} = \sqrt{1 - x^2}.$$

Using the addition formula for sine, we obtain

$$\begin{aligned} \sin(\alpha + \beta) &= \sin \alpha \cos \beta + \cos \alpha \sin \beta \\ &= x \cdot x + \sqrt{1 - x^2}\sqrt{1 - x^2} \\ &= x^2 + (1 - x^2) = 1. \end{aligned}$$

Since $\alpha + \beta$ is in the interval $[-\pi/2, 3\pi/2]$, the equation $\sin(\alpha + \beta) = 1$ has only one solution, $\alpha + \beta = \pi/2$, which is what we wished to show.

We may interpret the identity geometrically if $0 < x < 1$. If we construct a right triangle with one side of length x and hypotenuse of length 1, as illustrated in Figure 14, then angle β at B is an angle whose cosine is x; that is, $\beta = \cos^{-1} x$. Similarly, angle α at A is an angle whose sine is x; that is, $\alpha = \sin^{-1} x$. Since the acute angles of a right triangle are complementary, $\alpha + \beta = \pi/2$ or, equivalently,

Figure 14

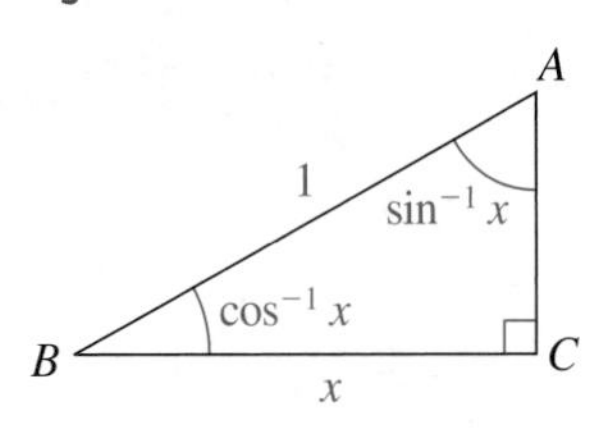

$$\sin^{-1} x + \cos^{-1} x = \frac{\pi}{2}.$$

Each of the remaining inverse trigonometric functions is defined in the same manner as the first three—by choosing a domain D in which the corresponding trigonometric function is one-to-one and then using the usual technique (where y is in D):

$$y = \cot^{-1} x \quad \text{if and only if} \quad x = \cot y$$

$$y = \sec^{-1} x \quad \text{if and only if} \quad x = \sec y$$

$$y = \csc^{-1} x \quad \text{if and only if} \quad x = \csc y$$

The function $\sec^{-1}$ is used in calculus; however, $\cot^{-1}$ and $\csc^{-1}$ are seldom used. Because of their limited use in applications, we will not consider examples or exercises pertaining to these functions. We will merely summarize typical domains, ranges, and graphs in the following chart. A similar summary for the six trigonometric functions and their inverses appears in Appendix III.

Summary of Features of $\cot^{-1}$, $\sec^{-1}$, and $\csc^{-1}$

Feature	$y = \cot^{-1} x$	$y = \sec^{-1} x$	$y = \csc^{-1} x$
Domain	$\mathbb{R}$	$\lvert x \rvert \geq 1$	$\lvert x \rvert \geq 1$
Range	$(0, \pi)$	$\left[0, \frac{\pi}{2}\right) \cup \left[\pi, \frac{3\pi}{2}\right)$	$\left(-\pi, -\frac{\pi}{2}\right] \cup \left(0, \frac{\pi}{2}\right]$
Graph			

It is often difficult to verify an identity involving inverse trigonometric functions, as we saw in Example 10. A graphing utility can be extremely helpful in determining whether an equation involving inverse trigonometric functions is an identity and, if it is not an identity, in finding any solutions of the equation. The next example illustrates this process.

EXAMPLE 11 **Investigating an equation**

We know that $\tan x = (\sin x)/\cos x$ is an identity. Determine whether the equation

$$\arctan x = \frac{\arcsin x}{\arccos x}$$

is an identity. If it is not an identity, approximate the values of x for which the equation is true—that is, solve the equation.

Figure 15
$[-1, 1, 0.1]$ by $[-\pi/2, \pi/2, 0.2]$

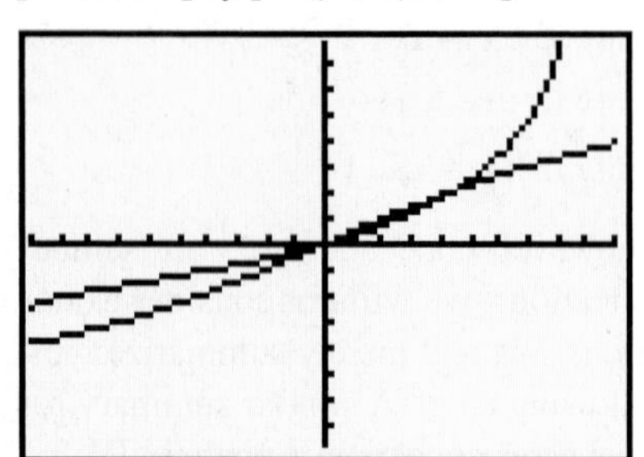

SOLUTION We begin by making the assignments

$$Y_1 = \tan^{-1} x \quad \text{and} \quad Y_2 = \sin^{-1} x/\cos^{-1} x.$$

Since the domain of $\sin^{-1}$ and $\cos^{-1}$ is $[-1, 1]$ and the range of $\tan^{-1}$ is $(-\pi/2, \pi/2)$, we choose the viewing rectangle shown in Figure 15.

Since the graphs representing Y_1 and Y_2 are not the same, we know that *the given equation is not an identity.* Because the graphs intersect twice, however, we know that the equation has two solutions. It appears that $x = 0$ is a solution, and a quick check in the given equation verifies that this is true. To estimate the point of intersection in the first quadrant, we use an intersect feature to determine that the point has the approximate coordinates (0.450, 0.423). Hence,

$$x = 0 \qquad \text{and} \qquad x \approx 0.450$$

are the values of x for which the given equation is true.

6.6 Exercises

Exer. 1–22: Find the exact value of the expression whenever it is defined.

1 (a) $\sin^{-1}\left(-\frac{\sqrt{2}}{2}\right)$ (b) $\cos^{-1}\left(-\frac{1}{2}\right)$

(c) $\tan^{-1}\left(-\sqrt{3}\right)$

2 (a) $\sin^{-1}\left(-\frac{1}{2}\right)$ (b) $\cos^{-1}\left(-\frac{\sqrt{2}}{2}\right)$

(c) $\tan^{-1}(-1)$

3 (a) $\arcsin\frac{\sqrt{3}}{2}$ (b) $\arccos\frac{\sqrt{2}}{2}$ (c) $\arctan\frac{1}{\sqrt{3}}$

4 (a) $\arcsin 0$ (b) $\arccos(-1)$ (c) $\arctan 0$

5 (a) $\sin^{-1}\frac{\pi}{3}$ (b) $\cos^{-1}\frac{\pi}{2}$ (c) $\tan^{-1} 1$

6 (a) $\arcsin\frac{\pi}{2}$ (b) $\arccos\frac{\pi}{3}$ (c) $\arctan\left(-\frac{\sqrt{3}}{3}\right)$

7 (a) $\sin\left[\arcsin\left(-\frac{3}{10}\right)\right]$ (b) $\cos\left(\arccos\frac{1}{2}\right)$

(c) $\tan(\arctan 14)$

8 (a) $\sin\left(\sin^{-1}\frac{2}{3}\right)$ (b) $\cos\left[\cos^{-1}\left(-\frac{1}{5}\right)\right]$

(c) $\tan\left[\tan^{-1}(-9)\right]$

9 (a) $\sin^{-1}\left(\sin\frac{\pi}{3}\right)$ (b) $\cos^{-1}\left[\cos\left(\frac{5\pi}{6}\right)\right]$

(c) $\tan^{-1}\left[\tan\left(-\frac{\pi}{6}\right)\right]$

10 (a) $\arcsin\left[\sin\left(-\frac{\pi}{2}\right)\right]$ (b) $\arccos(\cos 0)$

(c) $\arctan\left(\tan\frac{\pi}{4}\right)$

11 (a) $\arcsin\left(\sin\frac{5\pi}{4}\right)$ (b) $\arccos\left(\cos\frac{5\pi}{4}\right)$

(c) $\arctan\left(\tan\frac{7\pi}{4}\right)$

12 (a) $\sin^{-1}\left(\sin\frac{2\pi}{3}\right)$ (b) $\cos^{-1}\left(\cos\frac{4\pi}{3}\right)$

(c) $\tan^{-1}\left(\tan\frac{7\pi}{6}\right)$

13 (a) $\sin\left[\cos^{-1}\left(-\frac{1}{2}\right)\right]$ (b) $\cos\left(\tan^{-1} 1\right)$

(c) $\tan\left[\sin^{-1}(-1)\right]$

14 (a) $\sin\left(\tan^{-1}\sqrt{3}\right)$ (b) $\cos\left(\sin^{-1} 1\right)$

(c) $\tan\left(\cos^{-1} 0\right)$

15 (a) $\cot\left(\sin^{-1}\frac{2}{3}\right)$ (b) $\sec\left[\tan^{-1}\left(-\frac{3}{5}\right)\right]$

(c) $\csc\left[\cos^{-1}\left(-\frac{1}{4}\right)\right]$

16 (a) $\cot\left[\sin^{-1}\left(-\frac{2}{5}\right)\right]$ (b) $\sec\left(\tan^{-1}\frac{7}{4}\right)$

(c) $\csc\left(\cos^{-1}\frac{1}{5}\right)$

17 (a) $\sin\left(\arcsin\frac{1}{2} + \arccos 0\right)$

(b) $\cos\left[\arctan\left(-\frac{3}{4}\right) - \arcsin\frac{4}{5}\right]$

(c) $\tan\left(\arctan\frac{4}{3} + \arccos\frac{8}{17}\right)$

18 (a) $\sin\left[\sin^{-1}\frac{5}{13} - \cos^{-1}\left(-\frac{3}{5}\right)\right]$

(b) $\cos\left(\sin^{-1}\frac{4}{5} + \tan^{-1}\frac{3}{4}\right)$

(c) $\tan\left[\cos^{-1}\frac{1}{2} - \sin^{-1}\left(-\frac{1}{2}\right)\right]$

19 (a) $\sin\left[2\arccos\left(-\frac{3}{5}\right)\right]$ (b) $\cos\left(2\sin^{-1}\frac{15}{17}\right)$

(c) $\tan\left(2\tan^{-1}\frac{3}{4}\right)$

20 (a) $\sin\left(2\tan^{-1}\frac{5}{12}\right)$ (b) $\cos\left(2\arccos\frac{9}{41}\right)$

(c) $\tan\left[2\arcsin\left(-\frac{8}{17}\right)\right]$

21 (a) $\sin\left[\frac{1}{2}\sin^{-1}\left(-\frac{7}{25}\right)\right]$ (b) $\cos\left(\frac{1}{2}\tan^{-1}\frac{8}{15}\right)$

(c) $\tan\left(\frac{1}{2}\cos^{-1}\frac{3}{5}\right)$

22 (a) $\sin\left[\frac{1}{2}\cos^{-1}\left(-\frac{3}{5}\right)\right]$ (b) $\cos\left(\frac{1}{2}\sin^{-1}\frac{12}{13}\right)$

(c) $\tan\left(\frac{1}{2}\tan^{-1}\frac{40}{9}\right)$

Exer. 23–30: Write the expression as an algebraic expression in x for $x > 0$.

23 $\sin(\tan^{-1} x)$ 24 $\tan(\arccos x)$

25 $\sec\left(\sin^{-1}\dfrac{x}{\sqrt{x^2+4}}\right)$ 26 $\cot\left(\sin^{-1}\dfrac{\sqrt{x^2-9}}{x}\right)$

27 $\sin(2\sin^{-1} x)$ 28 $\cos(2\tan^{-1} x)$

29 $\cos\left(\frac{1}{2}\arccos x\right)$ 30 $\tan\left(\dfrac{1}{2}\cos^{-1}\dfrac{1}{x}\right)$

Exer. 31–32: Complete the statements.

31 (a) As $x \to -1^+$, $\sin^{-1} x \to$ ___

(b) As $x \to 1^-$, $\cos^{-1} x \to$ ___

(c) As $x \to \infty$, $\tan^{-1} x \to$ ___

32 (a) As $x \to 1^-$, $\sin^{-1} x \to$ ___

(b) As $x \to -1^+$, $\cos^{-1} x \to$ ___

(c) As $x \to -\infty$, $\tan^{-1} x \to$ ___

Exer. 33–42: Sketch the graph of the equation.

33 $y = \sin^{-1} 2x$ 34 $y = \frac{1}{2}\sin^{-1} x$

35 $y = \sin^{-1}(x + 1)$ 36 $y = \sin^{-1}(x - 2) + \dfrac{\pi}{2}$

37 $y = \cos^{-1}\frac{1}{2}x$ 38 $y = 2\cos^{-1} x$

39 $y = 2 + \tan^{-1} x$ 40 $y = \tan^{-1} 2x$

41 $y = \sin(\arccos x)$ 42 $y = \sin(\sin^{-1} x)$

Exer. 43–46: The given equation has the form $y = f(x)$. (a) Find the domain of f. (b) Find the range of f. (c) Solve for x in terms of y.

43 $y = \frac{1}{2}\sin^{-1}(x - 3)$ 44 $y = 3\tan^{-1}(2x + 1)$

45 $y = 4\cos^{-1}\frac{2}{3}x$ 46 $y = 2\sin^{-1}(3x - 4)$

Exer. 47–50: Solve the equation for x in terms of y if x is restricted to the given interval.

47 $y = -3 - \sin x$; $\left[-\frac{\pi}{2}, \frac{\pi}{2}\right]$

48 $y = 2 + 3 \sin x$; $\left[-\frac{\pi}{2}, \frac{\pi}{2}\right]$

49 $y = 15 - 2 \cos x$; $[0, \pi]$

50 $y = 6 - 3 \cos x$; $[0, \pi]$

Exer. 51–52: Solve the equation for x in terms of y if $0 < x < \pi$ and $0 < y < \pi$.

51 $\frac{\sin x}{3} = \frac{\sin y}{4}$

52 $\frac{4}{\sin x} = \frac{7}{\sin y}$

Exer. 53–64: Use inverse trigonometric functions to find the solutions of the equation that are in the given interval, and approximate the solutions to four decimal places.

53 $\cos^2 x + 2 \cos x - 1 = 0$; $[0, 2\pi)$

54 $\sin^2 x - \sin x - 1 = 0$; $[0, 2\pi)$

55 $2 \tan^2 t + 9 \tan t + 3 = 0$; $\left(-\frac{\pi}{2}, \frac{\pi}{2}\right)$

56 $3 \sin^2 t + 7 \sin t + 3 = 0$; $\left[-\frac{\pi}{2}, \frac{\pi}{2}\right]$

57 $15 \cos^4 x - 14 \cos^2 x + 3 = 0$; $[0, \pi]$

58 $3 \tan^4 \theta - 19 \tan^2 \theta + 2 = 0$; $\left(-\frac{\pi}{2}, \frac{\pi}{2}\right)$

59 $6 \sin^3 \theta + 18 \sin^2 \theta - 5 \sin \theta - 15 = 0$; $\left(-\frac{\pi}{2}, \frac{\pi}{2}\right)$

60 $6 \sin 2x - 8 \cos x + 9 \sin x - 6 = 0$; $\left(-\frac{\pi}{2}, \frac{\pi}{2}\right)$

61 $(\cos x)(15 \cos x + 4) = 3$; $[0, 2\pi)$

62 $6 \sin^2 x = \sin x + 2$; $[0, 2\pi)$

63 $3 \cos 2x - 7 \cos x + 5 = 0$; $[0, 2\pi)$

64 $\sin 2x = -1.5 \cos x$; $[0, 2\pi)$

Exer. 65–66: If an earthquake has a total horizontal displacement of S meters along its fault line, then the horizontal movement M of a point on the surface of Earth d kilometers from the fault line can be estimated using the formula

$$M = \frac{S}{2}\left(1 - \frac{2}{\pi} \tan^{-1} \frac{d}{D}\right),$$

where D is the depth (in kilometers) below the surface of the focal point of the earthquake.

65 Earthquake movement For the San Francisco earthquake of 1906, S was 4 meters and D was 3.5 kilometers. Approximate M for the stated values of d.

(a) 1 kilometer (b) 4 kilometers

(c) 10 kilometers

66 Earthquake movement Approximate the depth D of the focal point of an earthquake with $S = 3$ m if a point on the surface of Earth 5 kilometers from the fault line moved 0.6 meter horizontally.

67 A golfer's drive A golfer, centered in a 30-yard-wide straight fairway, hits a ball 280 yards. Approximate the largest angle the drive can have from the center of the fairway if the ball is to stay in the fairway (see the figure).

Exercise 67

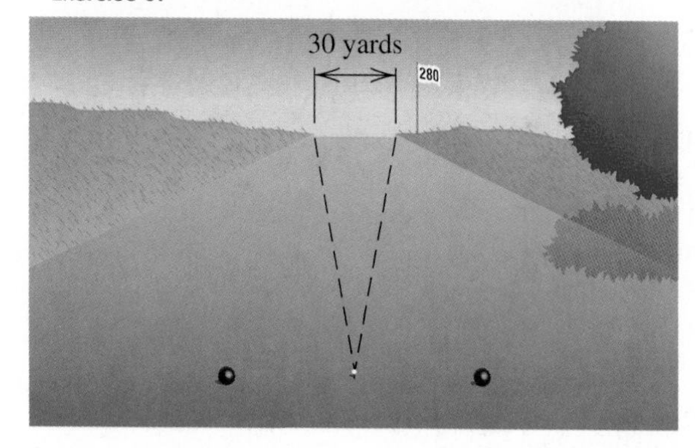

68 **Placing a wooden brace** A 14-foot piece of lumber is to be placed as a brace, as shown in the figure. Assuming all the lumber is 2 inches by 4 inches, find α and β.

Exercise 68

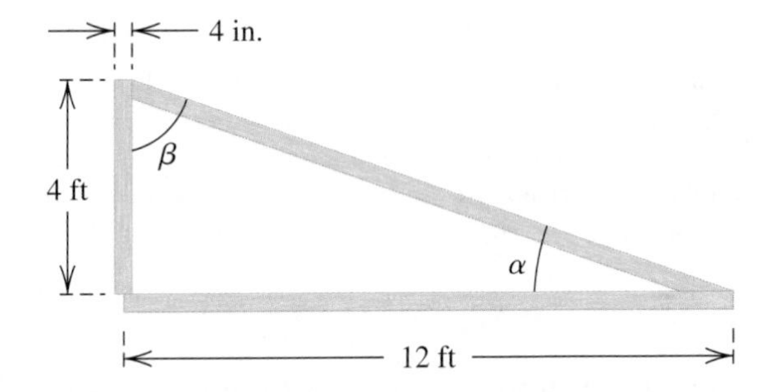

69 **Tracking a sailboat** As shown in the figure, a sailboat is following a straight-line course l. (Assume that the shoreline is parallel to the north-south line.) The shortest distance from a tracking station T to the course is d miles. As the boat sails, the tracking station records its distance k from T and its direction θ with respect to T. Angle α specifies the direction of the sailboat.

(a) Express α in terms of d, k, and θ.

(b) Estimate α to the nearest degree if $d = 50$ mi, $k = 210$ mi, and $\theta = 53.4°$.

Exercise 69

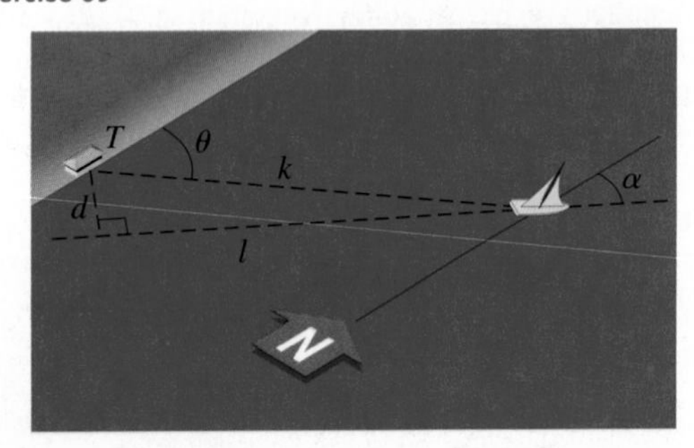

70 **Calculating viewing angles** An art critic whose eye level is 6 feet above the floor views a painting that is 10 feet in height and is mounted 4 feet above the floor, as shown in the figure.

(a) If the critic is standing x feet from the wall, express the viewing angle θ in terms of x.

(b) Use the addition formula for tangent to show that

$$\theta = \tan^{-1}\left(\frac{10x}{x^2 - 16}\right).$$

(c) For what value of x is $\theta = 45°$?

Exercise 70

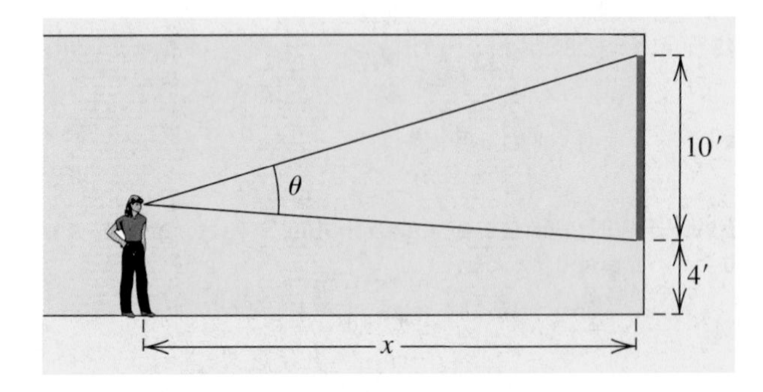

Exer. 71–76: Verify the identity.

71 $\sin^{-1} x = \tan^{-1} \dfrac{x}{\sqrt{1 - x^2}}$

72 $\arccos x + \arccos \sqrt{1 - x^2} = \dfrac{\pi}{2}, 0 \le x \le 1$

73 $\arcsin(-x) = -\arcsin x$

74 $\arccos(-x) = \pi - \arccos x$

75 $\arctan x + \arctan \dfrac{1}{x} = \dfrac{\pi}{2}, x > 0$

76 $2\cos^{-1} x = \cos^{-1}(2x^2 - 1), 0 \le x \le 1$

Exer. 77–78: Graph f, and determine its domain and range.

77 $f(x) = 2\sin^{-1}(x - 1) + \cos^{-1} \frac{1}{2}x$

78 $f(x) = \frac{1}{2}\tan^{-1}(1 - 2x) + 3\tan^{-1}\sqrt{x + 2}$

Exer. 79–80: Use a graph to estimate the solutions of the equation.

79 $\sin^{-1} 2x = \tan^{-1} (1 - x)$

80 $\cos^{-1} \left(x - \frac{1}{5}\right) = 2 \sin^{-1} \left(\frac{1}{2} - x\right)$

81 Designing a solar collector In designing a collector for solar power, an important consideration is the amount of sunlight that is transmitted through the glass into the water being heated. If the angle of incidence θ of the sun's rays is measured from a line perpendicular to the surface of the glass, then the fraction $f(\theta)$ of sunlight reflected off the glass can be approximated by

$$f(\theta) = \frac{1}{2}\left(\frac{\sin^2 \alpha}{\sin^2 \beta} + \frac{\tan^2 \alpha}{\tan^2 \beta}\right),$$

where

$$\alpha = \theta - \gamma, \quad \beta = \theta + \gamma, \quad \text{and} \quad \gamma = \sin^{-1}\left(\frac{\sin \theta}{1.52}\right).$$

Graph f for $0 < \theta < \pi/2$, and estimate θ when $f(\theta) = 0.2$.

82 Designing a solar collector The altitude of the sun is the angle ϕ that the sun's rays make with the horizon at a given time and place. Determining ϕ is important in tilting a solar collector to obtain maximum efficiency. On June 21 at a latitude of 51.7°N, the altitude of the sun can be approximated using the formula

$$\sin \phi = \sin 23.5° \sin 51.7° + \cos 23.5° \cos 51.7° \cos H,$$

where H is called the hour angle, with $H = -\pi/2$ at 6 A.M., $H = 0$ at noon, and $H = \pi/2$ at 6 P.M.

(a) Solve the formula for ϕ, and graph the resulting equation for $-\pi/2 \le H \le \pi/2$.

(b) Estimate the times when $\phi = 45°$.

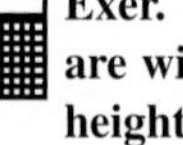

Exer. 83–86: Many calculators have viewing screens that are wider than they are high. The approximate ratio of the height to the width is often 2:3. Let the actual height of the calculator screen along the y-axis be 2 units, the actual width of the calculator screen along the x-axis be 3 units, and Xscl = Yscl = 1. Since the line $y = x$ must pass through the point (1, 1), the actual slope m_A of this line on the calculator screen is given by

$$m_A = \frac{\text{actual distance between tick marks on } y\text{-axis}}{\text{actual distance between tick marks on } x\text{-axis}}.$$

Using this information, graph $y = x$ in the given viewing rectangle and predict the actual angle θ that the graph makes with the x-axis on the viewing screen.

83 [0, 3] by [0, 2]

84 [0, 6] by [0, 2]

85 [0, 3] by [0, 4]

86 [0, 2] by [0, 2]

CHAPTER 6 REVIEW EXERCISES

▶ **Online support materials can be found at www.thomsonedu.com/login**

Exer. 1–22: Verify the identity.

1 $(\cot^2 x + 1)(1 - \cos^2 x) = 1$

2 $\cos \theta + \sin \theta \tan \theta = \sec \theta$

3 $\dfrac{(\sec^2 \theta - 1) \cot \theta}{\tan \theta \sin \theta + \cos \theta} = \sin \theta$

4 $(\tan x + \cot x)^2 = \sec^2 x \csc^2 x$

5 $\dfrac{1}{1 + \sin t} = (\sec t - \tan t) \sec t$

6 $\dfrac{\sin (\alpha - \beta)}{\cos (\alpha + \beta)} = \dfrac{\tan \alpha - \tan \beta}{1 - \tan \alpha \tan \beta}$

7 $\tan 2u = \dfrac{2 \cot u}{\csc^2 u - 2}$

8 $\cos^2 \dfrac{v}{2} = \dfrac{1 + \sec v}{2 \sec v}$

9 $\dfrac{\tan^3 \phi - \cot^3 \phi}{\tan^2 \phi + \csc^2 \phi} = \tan \phi - \cot \phi$

10 $\dfrac{\sin u + \sin v}{\csc u + \csc v} = \dfrac{1 - \sin u \sin v}{-1 + \csc u \csc v}$

11 $\left(\dfrac{\sin^2 x}{\tan^4 x}\right)^3 \left(\dfrac{\csc^3 x}{\cot^6 x}\right)^2 = 1$

12 $\dfrac{\cos \gamma}{1 - \tan \gamma} + \dfrac{\sin \gamma}{1 - \cot \gamma} = \cos \gamma + \sin \gamma$

13 $\dfrac{\cos(-t)}{\sec(-t) + \tan(-t)} = 1 + \sin t$

14 $\dfrac{\cot(-t) + \csc(-t)}{\sin(-t)} = \dfrac{1}{1 - \cos t}$

15 $\sqrt{\dfrac{1 - \cos t}{1 + \cos t}} = \dfrac{1 - \cos t}{|\sin t|}$

16 $\sqrt{\dfrac{1 - \sin\theta}{1 + \sin\theta}} = \dfrac{|\cos\theta|}{1 + \sin\theta}$

17 $\cos\left(x - \dfrac{5\pi}{2}\right) = \sin x$

18 $\tan\left(x + \dfrac{3\pi}{4}\right) = \dfrac{\tan x - 1}{1 + \tan x}$

19 $\frac{1}{4}\sin 4\beta = \sin\beta\cos^3\beta - \cos\beta\sin^3\beta$

20 $\tan\frac{1}{2}\theta = \csc\theta - \cot\theta$

21 $\sin 8\theta = 8\sin\theta\cos\theta(1 - 2\sin^2\theta)(1 - 8\sin^2\theta\cos^2\theta)$

22 $\arctan x = \dfrac{1}{2}\arctan\dfrac{2x}{1 - x^2}, \quad -1 < x < 1$

Exer. 23–40: Find the solutions of the equation that are in the interval $[0, 2\pi)$.

23 $2\cos^3\theta - \cos\theta = 0$

24 $2\cos\alpha + \tan\alpha = \sec\alpha$

25 $\sin\theta = \tan\theta$

26 $\csc^5\theta - 4\csc\theta = 0$

27 $2\cos^3 t + \cos^2 t - 2\cos t - 1 = 0$

28 $\cos x\cot^2 x = \cos x$

29 $\sin\beta + 2\cos^2\beta = 1$

30 $\cos 2x + 3\cos x + 2 = 0$

31 $2\sec u\sin u + 2 = 4\sin u + \sec u$

32 $\tan 2x\cos 2x = \sin 2x$

33 $2\cos 3x\cos 2x = 1 - 2\sin 3x\sin 2x$

34 $\sin x\cos 2x + \cos x\sin 2x = 0$

35 $\cos\pi x + \sin\pi x = 0$

36 $\sin 2u = \sin u$

37 $2\cos^2\frac{1}{2}\theta - 3\cos\theta = 0$

38 $\sec 2x\csc 2x = 2\csc 2x$

39 $\sin 5x = \sin 3x$

40 $\cos 3x = -\cos 2x$

Exer. 41–44: Find the exact value.

41 $\cos 75°$

42 $\tan 285°$

43 $\sin 195°$

44 $\csc\dfrac{\pi}{8}$

Exer. 45–56: If θ and ϕ are acute angles such that $\csc\theta = \frac{5}{3}$ and $\cos\phi = \frac{8}{17}$, find the exact value.

45 $\sin(\theta + \phi)$

46 $\cos(\theta + \phi)$

47 $\tan(\phi + \theta)$

48 $\tan(\theta - \phi)$

49 $\sin(\phi - \theta)$

50 $\sin(\theta - \phi)$

51 $\sin 2\phi$

52 $\cos 2\phi$

53 $\tan 2\theta$

54 $\sin\frac{1}{2}\theta$

55 $\tan\frac{1}{2}\theta$

56 $\cos\frac{1}{2}\phi$

57 Express as a sum or difference:

(a) $\sin 7t\sin 4t$

(b) $\cos\frac{1}{4}u\cos\left(-\frac{1}{6}u\right)$

(c) $6\cos 5x\sin 3x$

(d) $4\sin 3\theta\cos 7\theta$

58 Express as a product:

(a) $\sin 8u + \sin 2u$

(b) $\cos 3\theta - \cos 8\theta$

(c) $\sin \frac{1}{4}t - \sin \frac{1}{5}t$ (d) $3 \cos 2x + 3 \cos 6x$

Exer. 59–70: Find the exact value of the expression whenever it is defined.

59 $\cos^{-1}\left(\frac{\sqrt{3}}{2}\right)$

60 $\arcsin\left(\frac{\sqrt{2}}{2}\right)$

61 $\arctan \sqrt{3}$

62 $\arccos\left(\tan \frac{3\pi}{4}\right)$

63 $\arcsin\left(\sin \frac{5\pi}{4}\right)$

64 $\cos^{-1}\left(\cos \frac{5\pi}{4}\right)$

65 $\sin\left[\arccos\left(-\frac{\sqrt{3}}{2}\right)\right]$

66 $\tan(\tan^{-1} 2)$

67 $\sec\left(\sin^{-1} \frac{3}{2}\right)$

68 $\cos^{-1}(\sin 0)$

69 $\cos\left(\sin^{-1} \frac{15}{17} - \sin^{-1} \frac{8}{17}\right)$

70 $\cos\left(2 \sin^{-1} \frac{4}{5}\right)$

Exer. 71–74: Sketch the graph of the equation.

71 $y = \cos^{-1} 3x$

72 $y = 4 \sin^{-1} x$

73 $y = 1 - \sin^{-1} x$

74 $y = \sin\left(\frac{1}{2} \cos^{-1} x\right)$

75 Express $\cos(\alpha + \beta + \gamma)$ in terms of trigonometric functions of α, β, and γ.

76 **Force of a foot** When an individual is walking, the magnitude F of the vertical force of one foot on the ground (see the figure) can be described by

$$F = A(\cos bt - a \cos 3bt),$$

where t is time in seconds, $A > 0$, $b > 0$, and $0 < a < 1$.

Exercise 76

(a) Show that $F = 0$ when $t = -\pi/(2b)$ and $t = \pi/(2b)$. (The time $t = -\pi/(2b)$ corresponds to the moment when the foot first touches the ground and the weight of the body is being supported by the other foot.)

(b) The maximum force occurs when

$$3a \sin 3bt = \sin bt.$$

If $a = \frac{1}{3}$, find the solutions of this equation for the interval $-\pi/(2b) < t < \pi/(2b)$.

(c) If $a = \frac{1}{3}$, express the maximum force in terms of A.

77 Shown in the figure is a graph of the equation

$$y = \sin x - \tfrac{1}{2} \sin 2x + \tfrac{1}{3} \sin 3x.$$

The x-coordinates of the turning points are solutions of the equation $\cos x - \cos 2x + \cos 3x = 0$. Use a sum-to-product formula to find these x-coordinates.

Exercise 77

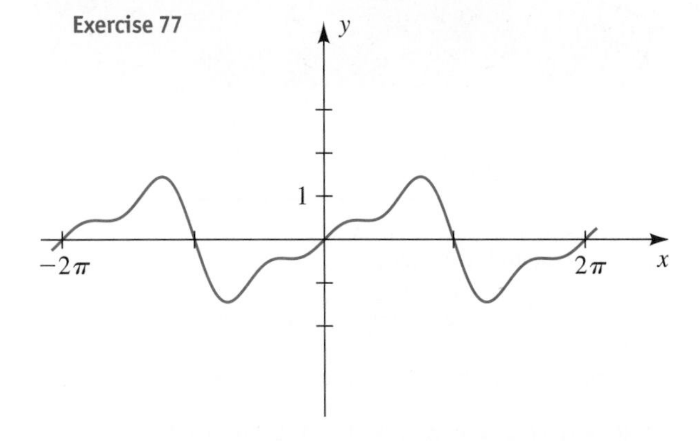

78 **Visual distinction** The human eye can distinguish between two distant points P and Q provided the angle of resolution θ is not too small. Suppose P and Q are x units apart and are d units from the eye, as illustrated in the figure.

(a) Express x in terms of d and θ.

(b) For a person with normal vision, the smallest distinguishable angle of resolution is about 0.0005 radian. If a pen 6 inches long is viewed by such an individual at a distance of d feet, for what values of d will the end points of the pen be distinguishable?

Exercise 78

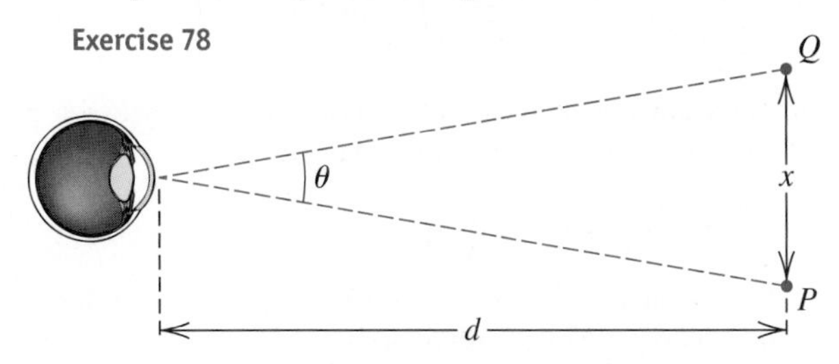

79 Satellites A satellite S circles a planet at a distance d miles from the planet's surface. The portion of the planet's surface that is visible from the satellite is determined by the angle θ indicated in the figure.

Exercise 79

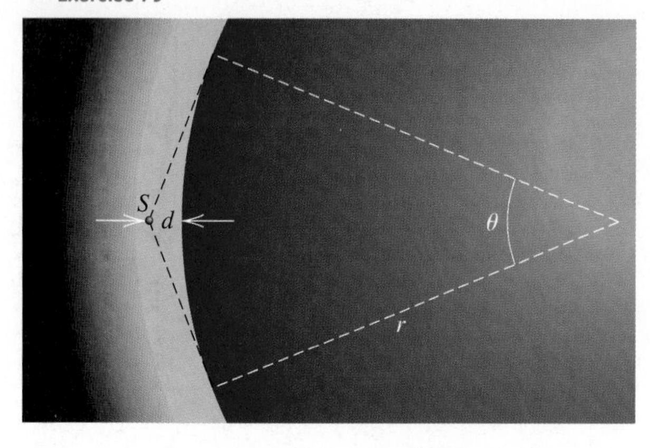

(a) Assuming that the planet is spherical in shape, express d in terms of θ and the radius r of the planet.

(b) Approximate θ for a satellite 300 miles from the surface of Earth, using $r = 4000$ mi.

80 Urban canyons Because of the tall buildings and relatively narrow streets in some inner cities, the amount of sunlight illuminating these "canyons" is greatly reduced. If h is the average height of the buildings and w is the width of the street, the narrowness N of the street is defined by $N = h/w$. The angle θ of the horizon is defined by $\tan\theta = N$. (The value $\theta = 63°$ may result in an 85% loss of illumination.) Approximate the angle of the horizon for the following values of h and w.

(a) $h = 400$ ft, $\quad w = 80$ ft

(b) $h = 55$ m, $\quad w = 30$ m

CHAPTER 6 DISCUSSION EXERCISES

1 Verify the following identity:

$$\frac{\tan x}{1 - \cot x} + \frac{\cot x}{1 - \tan x} = 1 + \sec x \csc x$$

(*Hint:* At some point, consider a special factoring.)

2 Refer to Example 5 of Section 6.1. Suppose $0 \le \theta < 2\pi$, and rewrite the conclusion using a piecewise-defined function.

3 How many solutions does the following equation have on $[0, 2\pi)$? Find the largest one.

$$3 \cos 45x + 4 \sin 45x = 5$$

4 Graph the difference quotient for $f(x) = \sin x$ and $h = 0.5$, 0.1, and 0.001 on the viewing rectangle $[0, 2\pi, \pi/2]$ by $[-1, 1]$. What generalization can you make from these graphs? Show that this quotient can be written as

$$\sin x \left(\frac{\cos h - 1}{h}\right) + \cos x \left(\frac{\sin h}{h}\right).$$

5 There are several interesting exact relationships between π and inverse trigonometric functions such as

$$\frac{\pi}{4} = 4 \tan^{-1}\left(\frac{1}{5}\right) - \tan^{-1}\left(\frac{1}{239}\right).$$

Use trigonometric identities to prove that this relationship is true. Two other relationships are

$$\frac{\pi}{4} = \tan^{-1}\left(\frac{1}{2}\right) + \tan^{-1}\left(\frac{1}{5}\right) + \tan^{-1}\left(\frac{1}{8}\right)$$

and $\quad \pi = \tan^{-1} 1 + \tan^{-1} 2 + \tan^{-1} 3.$

6 Shown in the figure is a function called a *sawtooth function*.

(a) Define an inverse sawtooth function **(arcsaw)**, including its domain and range.

(b) Find arcsaw (1.7) and arcsaw (−0.8).

(c) Formulate two properties of arcsaw (similar to the $\sin(\sin^{-1})$ property).

(d) Graph the arcsaw function.

Exercise 6

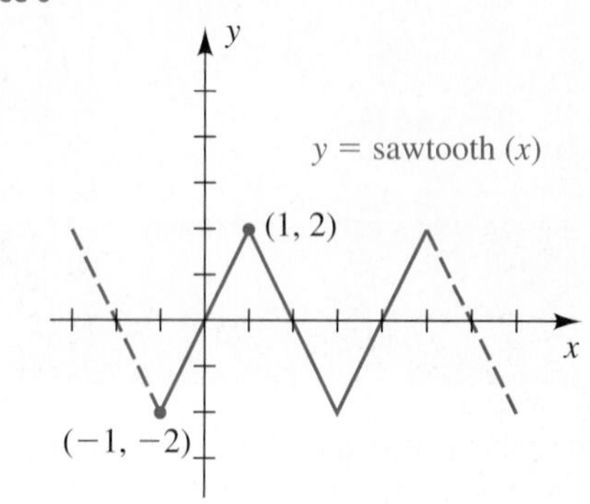

7

Applications of Trigonometry

In the first two sections of this chapter we consider methods of solving oblique triangles using the law of sines and the law of cosines. The next two sections contain an introduction to vectors—a topic that has many applications in engineering, the natural sciences, and advanced mathematics. We then introduce the trigonometric form for complex numbers and use it to find all n solutions of equations of the form $w^n = z$, where n is any positive integer and w and z are complex numbers.

7.1 The Law of Sines

An **oblique triangle** is a triangle that does not contain a right angle. We shall use the letters A, B, C, a, b, c, α, β, and γ for parts of triangles, as we did in Chapter 5. Given triangle ABC, let us place angle α in standard position so that B is on the positive x-axis. The case for α obtuse is illustrated in Figure 1; however, the following discussion is also valid if α is acute.

Figure 1

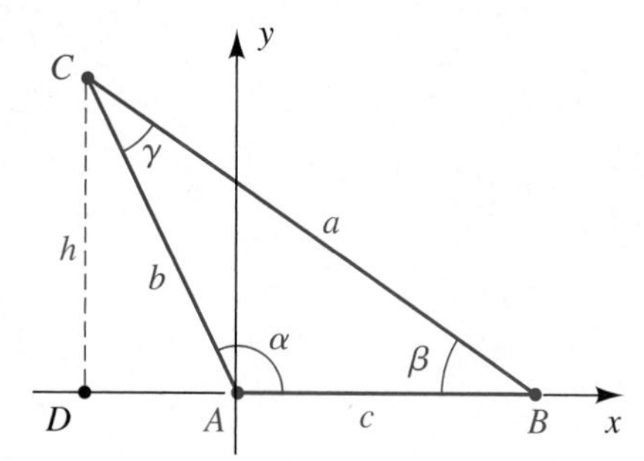

Consider the line through C parallel to the y-axis and intersecting the x-axis at point D. If we let $d(C, D) = h$, then the y-coordinate of C is h. From the definition of the trigonometric functions of any angle,

$$\sin\alpha = \frac{h}{b}, \quad \text{so} \quad h = b\sin\alpha.$$

Referring to right triangle BDC, we see that

$$\sin\beta = \frac{h}{a}, \quad \text{so} \quad h = a\sin\beta.$$

Equating the two expressions for h gives us

$$b\sin\alpha = a\sin\beta,$$

which we may write as

$$\frac{\sin\alpha}{a} = \frac{\sin\beta}{b}.$$

If we place α in standard position with C on the positive x-axis, then by the same reasoning,

$$\frac{\sin\alpha}{a} = \frac{\sin\gamma}{c}.$$

The last two equalities give us the following result.

The Law of Sines

If ABC is an oblique triangle labeled in the usual manner (as in Figure 1), then

$$\frac{\sin\alpha}{a} = \frac{\sin\beta}{b} = \frac{\sin\gamma}{c}.$$

Note that the law of sines consists of the following three formulas:

$$\textbf{(1)}\ \frac{\sin\alpha}{a} = \frac{\sin\beta}{b} \qquad \textbf{(2)}\ \frac{\sin\alpha}{a} = \frac{\sin\gamma}{c} \qquad \textbf{(3)}\ \frac{\sin\beta}{b} = \frac{\sin\gamma}{c}$$

To apply any one of these formulas to a specific triangle, we must know the values of three of the four variables. If we substitute these three values into the appropriate formula, we can then solve for the value of the fourth variable. It follows that the law of sines can be used to find the remaining parts of an oblique triangle whenever we know either of the following (the three letters in parentheses are used to denote the known parts, with S representing a side and A an angle):

(1) two sides and an angle *opposite* one of them (SSA)

(2) two angles and any side (AAS or ASA)

In the next section we will discuss the law of cosines and show how it can be used to find the remaining parts of an oblique triangle when given the following:

(1) two sides and the angle *between* them (SAS)

(2) three sides (SSS)

The law of sines cannot be applied directly to the last two cases.

The law of sines can also be written in the form

$$\frac{a}{\sin \alpha} = \frac{b}{\sin \beta} = \frac{c}{\sin \gamma}.$$

Instead of memorizing the three formulas associated with the law of sines, it may be more convenient to remember the following statement, which takes all of them into account.

The Law of Sines (General Form)	In any triangle, the ratio of the sine of an angle to the side opposite that angle is equal to the ratio of the sine of another angle to the side opposite that angle.

In examples and exercises involving triangles, we shall assume that known lengths of sides and angles have been obtained by measurement and hence are approximations to exact values. Unless directed otherwise, when finding parts of triangles we will round off answers according to the following rule: *If known sides or angles are stated to a certain accuracy, then unknown sides or angles should be calculated to the same accuracy.* To illustrate, if known sides are stated to the nearest 0.1, then unknown sides should be calculated to the nearest 0.1. If known angles are stated to the nearest $10'$, then unknown angles should be calculated to the nearest $10'$. Similar remarks hold for accuracy to the nearest 0.01, 0.1°, and so on.

Figure 2

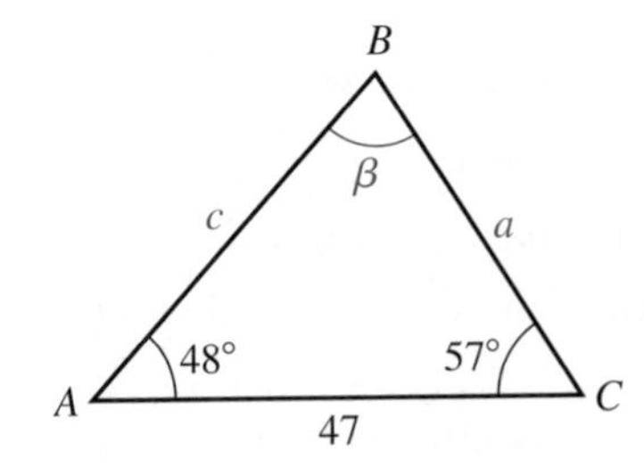

EXAMPLE 1 Using the law of sines (ASA)

Solve $\triangle ABC$, given $\alpha = 48°$, $\gamma = 57°$, and $b = 47$.

▶ **SOLUTION** The triangle is sketched in Figure 2. Since the sum of the angles of a triangle is 180°,

$$\beta = 180° - 57° - 48° = 75°.$$

(continued)

▶ Tutorial available at www.thomsonedu.com/login

Since side b and all three angles are known, we can find a by using a form of the law of sines that involves a, α, b, and β:

$$\begin{aligned} \frac{a}{\sin\alpha} &= \frac{b}{\sin\beta} && \text{law of sines} \\ a &= \frac{b\sin\alpha}{\sin\beta} && \text{solve for } a \\ &= \frac{47\sin 48^\circ}{\sin 75^\circ} && \text{substitute for } b,\ \alpha,\text{ and } \beta \\ &\approx 36 && \text{approximate to the nearest integer} \end{aligned}$$

To find c, we merely replace $\dfrac{a}{\sin\alpha}$ with $\dfrac{c}{\sin\gamma}$ in the preceding solution for a, obtaining

$$c = \frac{b\sin\gamma}{\sin\beta} = \frac{47\sin 57^\circ}{\sin 75^\circ} \approx 41.$$

Figure 3

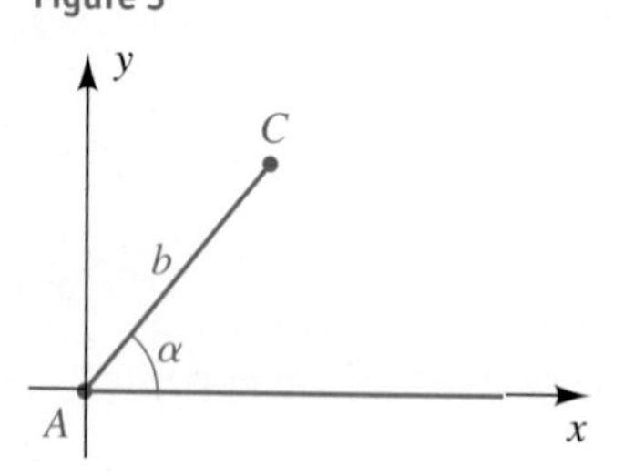

Data such as those in Example 1 lead to exactly one triangle ABC. However, if two sides and an angle *opposite* one of them are given, a unique triangle is not always determined. To illustrate, suppose that a and b are to be lengths of sides of triangle ABC and that a given angle α is to be opposite the side of length a. Let us examine the case for α acute. Place α in standard position and consider the line segment AC of length b on the terminal side of α, as shown in Figure 3. The third vertex, B, should be somewhere on the x-axis. Since the length a of the side opposite α is given, we may find B by striking off a circular arc of length a with center at C. The four possible outcomes are illustrated in Figure 4 (without the coordinate axes).

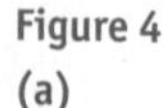
Figure 4

(a)

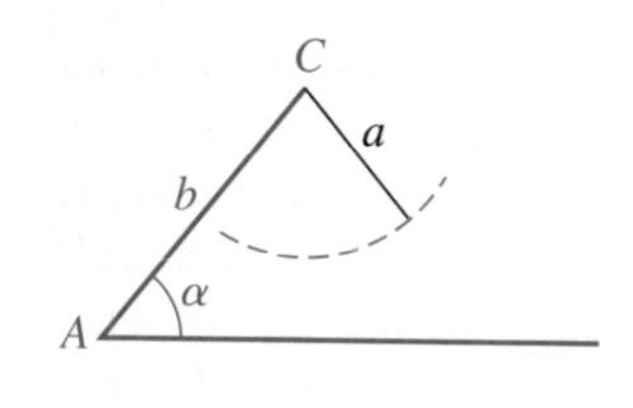

(b)

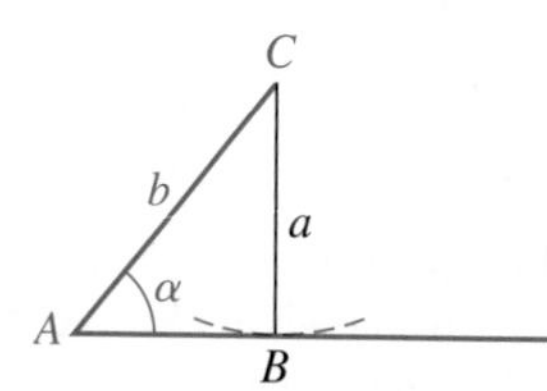

(c)

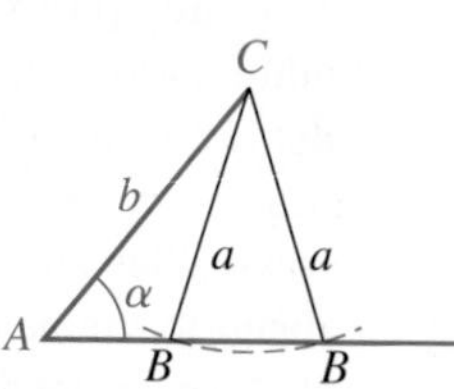

(d)

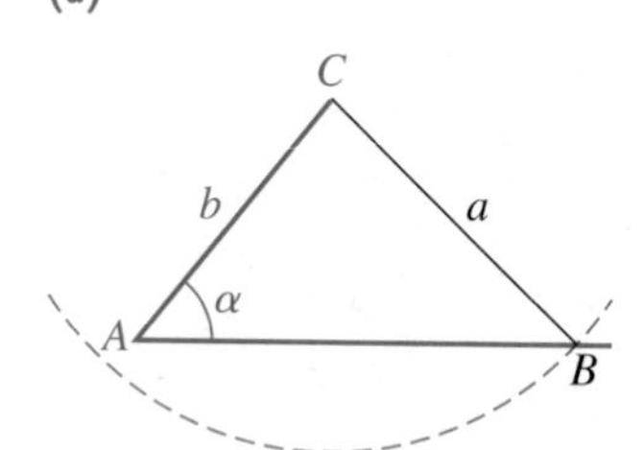

The four possibilities in the figure may be described as follows:

(a) The arc does not intersect the x-axis, and no triangle is formed.

(b) The arc is tangent to the x-axis, and a right triangle is formed.

(c) The arc intersects the positive x-axis in two distinct points, and two triangles are formed.

(d) The arc intersects both the positive and the nonpositive parts of the x-axis, and one triangle is formed.

Figure 5

(a) $a < b$

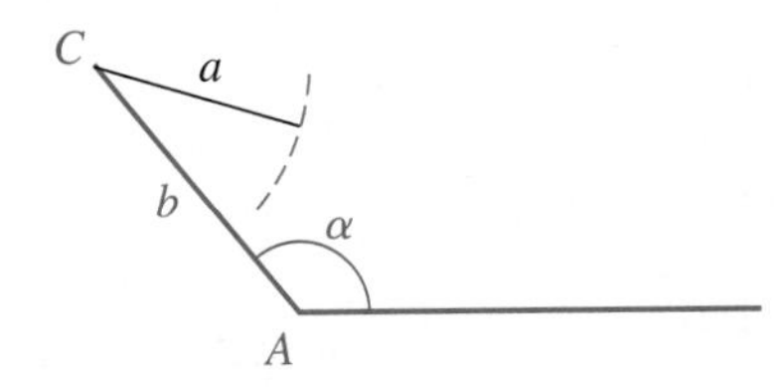

(b) $a > b$

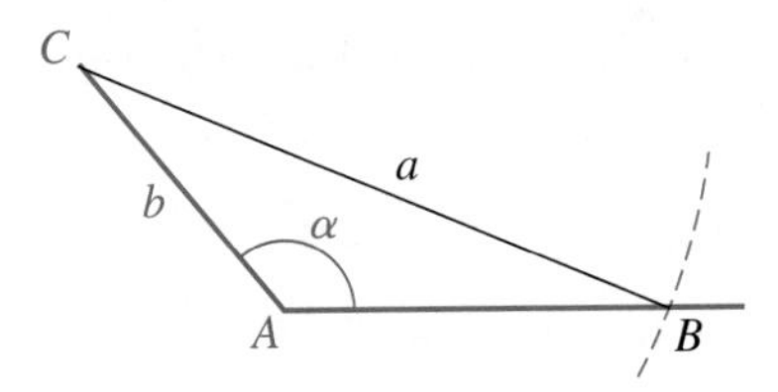

The particular case that occurs in a given problem will become evident when the solution is attempted. For example, if we solve the equation

$$\frac{\sin \alpha}{a} = \frac{\sin \beta}{b}$$

and obtain $\sin \beta > 1$, then no triangle exists and we have case (a). If we obtain $\sin \beta = 1$, then $\beta = 90°$ and hence (b) occurs. If $\sin \beta < 1$, then there are two possible choices for the angle β. By checking both possibilities, we may determine whether (c) or (d) occurs.

If the measure of α is greater than 90°, then a triangle exists if and only if $a > b$ (see Figure 5). Since we may have more than one possibility when two sides and an angle opposite one of them are given, this situation is sometimes called the **ambiguous case.**

EXAMPLE 2 **Using the law of sines (SSA)**

Solve $\triangle ABC$, given $\alpha = 67°$, $a = 100$, and $c = 125$.

▶ SOLUTION Since we know α, a, and c, we can find γ by using a form of the law of sines that involves a, α, c, and γ:

$\dfrac{\sin \gamma}{c} = \dfrac{\sin \alpha}{a}$	law of sines
$\sin \gamma = \dfrac{c \sin \alpha}{a}$	solve for $\sin \gamma$
$= \dfrac{125 \sin 67°}{100}$	substitute for c, α, and a
≈ 1.1506	approximate

Since $\sin \gamma$ *cannot* be greater than 1, no triangle can be constructed with the given parts.

EXAMPLE 3 **Using the law of sines (SSA)**

Solve $\triangle ABC$, given $a = 12.4$, $b = 8.7$, and $\beta = 36.7°$.

SOLUTION To find α, we proceed as follows:

$\dfrac{\sin \alpha}{a} = \dfrac{\sin \beta}{b}$	law of sines
$\sin \alpha = \dfrac{a \sin \beta}{b}$	solve for $\sin \alpha$
$= \dfrac{12.4 \sin 36.7°}{8.7}$	substitute for a, β, and b
≈ 0.8518	approximate

(continued)

▶ **Tutorial available at www.thomsonedu.com/login**

There are two possible angles α between 0° and 180° such that $\sin \alpha$ is approximately 0.8518. The reference angle α_R is

$$\alpha_R \approx \sin^{-1}(0.8518) \approx 58.4°.$$

Consequently, the two possibilities for α are

$$\alpha_1 \approx 58.4° \qquad \text{and} \qquad \alpha_2 = 180° - \alpha_1 \approx 121.6°.$$

Figure 6

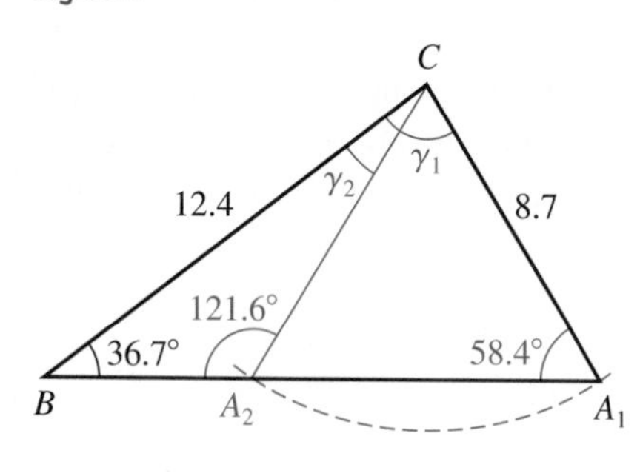

The angle $\alpha_1 \approx 58.4°$ gives us triangle A_1BC in Figure 6, and $\alpha_2 \approx 121.6°$ gives us triangle A_2BC.

If we let γ_1 and γ_2 denote the third angles of the triangles A_1BC and A_2BC corresponding to the angles α_1 and α_2, respectively, then

$$\gamma_1 = 180° - \alpha_1 - \beta \approx 180° - 58.4° - 36.7° \approx 84.9°$$

$$\gamma_2 = 180° - \alpha_2 - \beta \approx 180° - 121.6° - 36.7° \approx 21.7°.$$

If $c_1 = \overline{BA_1}$ is the side opposite γ_1 in triangle A_1BC, then

$$\frac{c_1}{\sin \gamma_1} = \frac{a}{\sin \alpha_1} \qquad \text{law of sines}$$

$$c_1 = \frac{a \sin \gamma_1}{\sin \alpha_1} \qquad \text{solve for } c_1$$

$$\approx \frac{12.4 \sin 84.9°}{\sin 58.4°} \approx 14.5. \qquad \text{substitute and approximate}$$

Thus, the remaining parts of triangle A_1BC are

$$\alpha_1 \approx 58.4°, \quad \gamma_1 \approx 84.9°, \quad \text{and} \quad c_1 \approx 14.5.$$

Similarly, if $c_2 = \overline{BA_2}$ is the side opposite γ_2 in $\triangle A_2BC$, then

$$c_2 = \frac{a \sin \gamma_2}{\sin \alpha_2} \approx \frac{12.4 \sin 21.7°}{\sin 121.6°} \approx 5.4,$$

and the remaining parts of triangle A_2BC are

$$\alpha_2 \approx 121.6°, \quad \gamma_2 \approx 21.7°, \quad \text{and} \quad c_2 \approx 5.4.$$

Figure 7

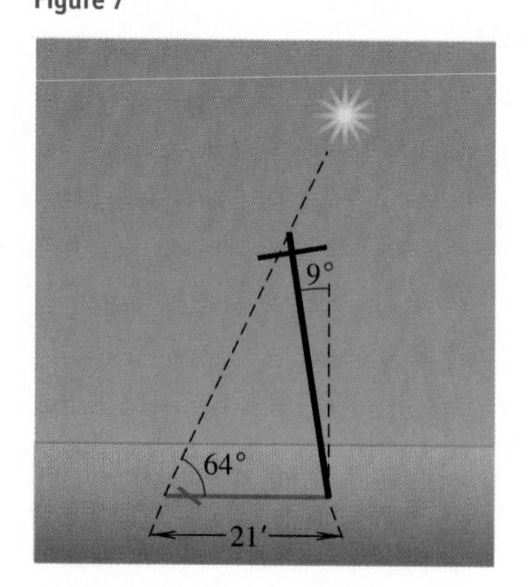

EXAMPLE 4 Using an angle of elevation

When the angle of elevation of the sun is 64°, a telephone pole that is tilted at an angle of 9° directly away from the sun casts a shadow 21 feet long on level ground. Approximate the length of the pole.

Figure 8

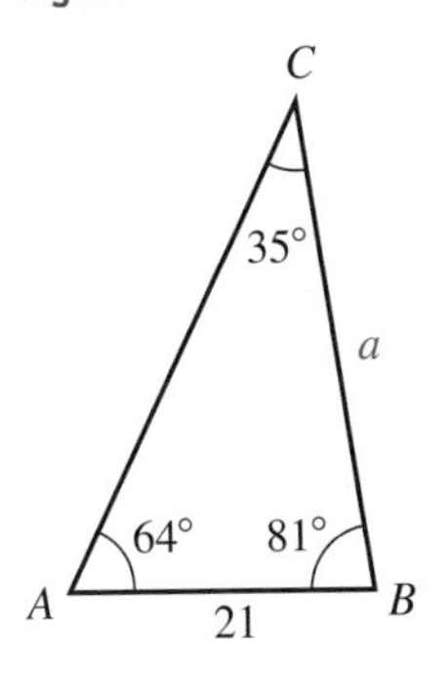

▶ **SOLUTION** The problem is illustrated in Figure 7. Triangle *ABC* in Figure 8 also displays the given facts. Note that in Figure 8 we have calculated the following angles:

$$\beta = 90° - 9° = 81°$$
$$\gamma = 180° - 64° - 81° = 35°$$

To find the length of the pole—that is, side *a* of triangle *ABC*—we proceed as follows:

$$\frac{a}{\sin 64°} = \frac{21}{\sin 35°} \qquad \text{law of sines}$$

$$a = \frac{21 \sin 64°}{\sin 35°} \approx 33 \qquad \text{solve for } a \text{ and approximate}$$

Thus, the telephone pole is approximately 33 feet in length.

EXAMPLE 5 Using bearings

Figure 9

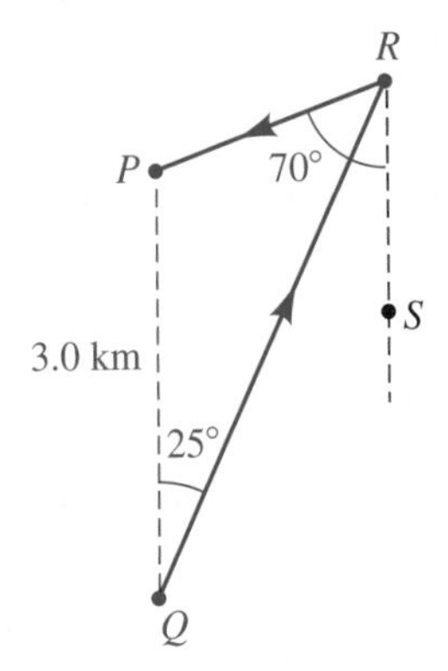

A point *P* on level ground is 3.0 kilometers due north of a point *Q*. A runner proceeds in the direction N25°E from *Q* to a point *R*, and then from *R* to *P* in the direction S70°W. Approximate the distance run.

▶ **SOLUTION** The notation used to specify directions was introduced in Section 5.7. The arrows in Figure 9 show the path of the runner, together with a north-south (dashed) line from *R* to another point *S*.

Since the lines through *PQ* and *RS* are parallel, it follows from geometry that the alternate interior angles *PQR* and *QRS* both have measure 25°. Hence,

$$\angle PRQ = \angle PRS - \angle QRS = 70° - 25° = 45°.$$

These observations give us triangle *PQR* in Figure 10 with

$$\angle QPR = 180° - 25° - 45° = 110°.$$

We apply the law of sines to find both *q* and *p*:

Figure 10

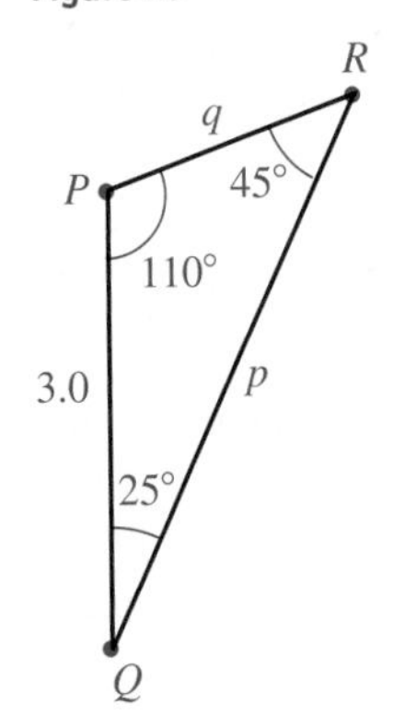

$$\frac{q}{\sin 25°} = \frac{3.0}{\sin 45°} \quad \text{and} \quad \frac{p}{\sin 110°} = \frac{3.0}{\sin 45°}$$

$$q = \frac{3.0 \sin 25°}{\sin 45°} \approx 1.8 \quad \text{and} \quad p = \frac{3.0 \sin 110°}{\sin 45°} \approx 4.0$$

The distance run, $p + q$, is approximately $4.0 + 1.8 = 5.8$ km.

EXAMPLE 6 Locating a school of fish

A commercial fishing boat uses sonar equipment to detect a school of fish 2 miles east of the boat and traveling in the direction of N51°W at a rate of 8 mi/hr (see Figure 11 on the next page).

▶ **Tutorial available at www.thomsonedu.com/login**

Figure 11

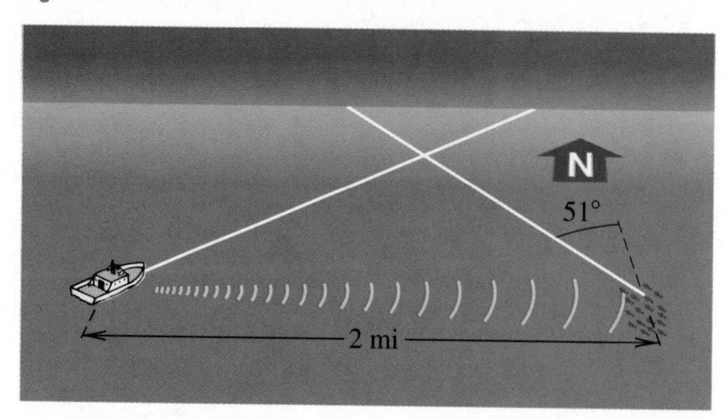

(a) If the boat travels at 20 mi/hr, approximate, to the nearest 0.1°, the direction it should head to intercept the school of fish.

(b) Find, to the nearest minute, the time it will take the boat to reach the fish.

▶ SOLUTION

Figure 12

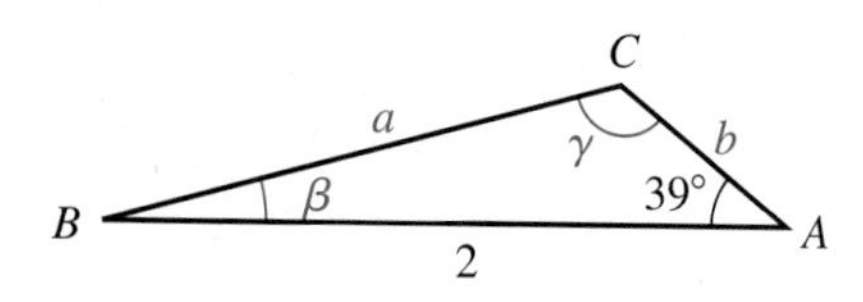

(a) The problem is illustrated by the triangle in Figure 12, with the school of fish at A, the boat at B, and the point of interception at C. Note that angle $\alpha = 90° - 51° = 39°$. To obtain β, we begin as follows:

$$\frac{\sin \beta}{b} = \frac{\sin 39°}{a} \qquad \text{law of sines}$$

$$\sin \beta = \frac{b}{a} \sin 39° \qquad \text{solve for } \sin \beta \qquad (*)$$

We next find b/a, letting t denote the amount of time required for the boat and fish to meet at C:

$$a = 20t, \qquad b = 8t \qquad \text{(distance) = (rate)(time)}$$

$$\frac{b}{a} = \frac{8t}{20t} = \frac{2}{5} \qquad \text{divide } b \text{ by } a$$

$$\sin \beta = \tfrac{2}{5} \sin 39° \qquad \text{substitute for } b/a \text{ in } (*)$$

$$\beta = \sin^{-1}\left(\tfrac{2}{5} \sin 39°\right) \approx 14.6° \qquad \text{approximate}$$

Since $90° - 14.6° = 75.4°$, the boat should travel in the (approximate) direction N75.4°E.

(b) We can find t using the relationship $a = 20t$. Let us first find the distance a from B to C. Since the only known side is 2, we need to find the angle γ opposite the side of length 2 in order to use the law of sines. We begin by noting that

$$\gamma \approx 180° - 39° - 14.6° = 126.4°.$$

▶ **Tutorial available at www.thomsonedu.com/login**

To find side a, we have

$$\frac{a}{\sin \alpha} = \frac{c}{\sin \gamma} \qquad \text{law of sines}$$

$$a = \frac{c \sin \alpha}{\sin \gamma} \qquad \text{solve for } a$$

$$\approx \frac{2 \sin 39^\circ}{\sin 126.4^\circ} \approx 1.56 \text{ mi.} \qquad \text{substitute and approximate}$$

Using $a = 20t$, we find the time t for the boat to reach C:

$$t = \frac{a}{20} \approx \frac{1.56}{20} \approx 0.08 \text{ hr} \approx 5 \text{ min}$$

7.1 Exercises

Exer. 1–16: Solve $\triangle ABC$.

▶ 1 $\alpha = 41^\circ$, $\gamma = 77^\circ$, $a = 10.5$

2 $\beta = 20^\circ$, $\gamma = 31^\circ$, $b = 210$

3 $\alpha = 27^\circ 40'$, $\beta = 52^\circ 10'$, $a = 32.4$

4 $\beta = 50^\circ 50'$, $\gamma = 70^\circ 30'$, $c = 537$

5 $\alpha = 42^\circ 10'$, $\gamma = 61^\circ 20'$, $b = 19.7$

▶ 6 $\alpha = 103.45^\circ$, $\gamma = 27.19^\circ$, $b = 38.84$

▶ 7 $\gamma = 81^\circ$, $c = 11$, $b = 12$

8 $\alpha = 32.32^\circ$, $c = 574.3$, $a = 263.6$

9 $\gamma = 53^\circ 20'$, $a = 140$, $c = 115$

10 $\alpha = 27^\circ 30'$, $c = 52.8$, $a = 28.1$

11 $\gamma = 47.74^\circ$, $a = 131.08$, $c = 97.84$

12 $\alpha = 42.17^\circ$, $a = 5.01$, $b = 6.12$

13 $\alpha = 65^\circ 10'$, $a = 21.3$, $b = 18.9$

14 $\beta = 113^\circ 10'$, $b = 248$, $c = 195$

15 $\beta = 121.624^\circ$, $b = 0.283$, $c = 0.178$

16 $\gamma = 73.01^\circ$, $a = 17.31$, $c = 20.24$

17 **Surveying** To find the distance between two points A and B that lie on opposite banks of a river, a surveyor lays off a line segment AC of length 240 yards along one bank and determines that the measures of $\angle BAC$ and $\angle ACB$ are $63^\circ 20'$ and $54^\circ 10'$, respectively (see the figure). Approximate the distance between A and B.

Exercise 17

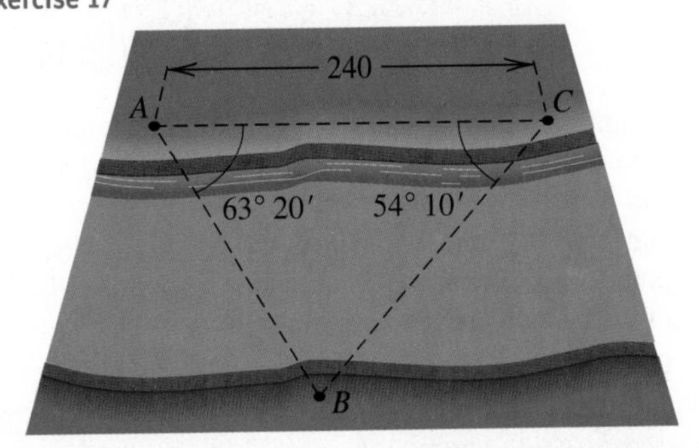

▶ 18 **Surveying** To determine the distance between two points A and B, a surveyor chooses a point C that is 375 yards from A and 530 yards from B. If $\angle BAC$ has measure $49^\circ 30'$, approximate the distance between A and B.

19 **Cable car route** As shown in the figure on the next page, a cable car carries passengers from a point A, which is 1.2 miles from a point B at the base of a mountain, to a point P at the top of the mountain. The angles of elevation of P from A and B are 21° and 65°, respectively.

(a) Approximate the distance between A and P.

(b) Approximate the height of the mountain.

▶ **Tutorial available at www.thomsonedu.com/login**

Exercise 19

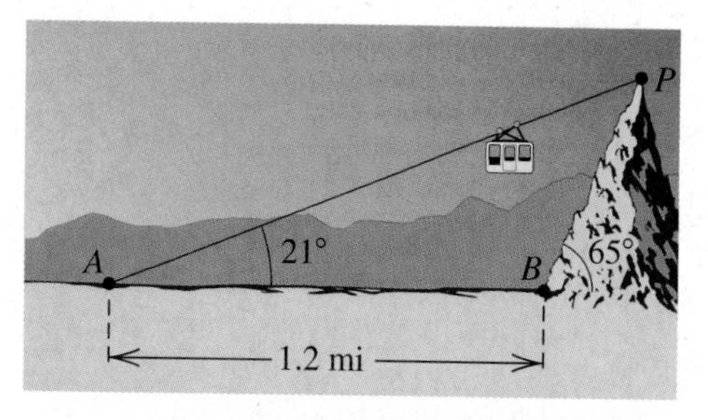

20 Length of a shadow A straight road makes an angle of 15° with the horizontal. When the angle of elevation of the sun is 57°, a vertical pole at the side of the road casts a shadow 75 feet long directly down the road, as shown in the figure. Approximate the length of the pole.

Exercise 20

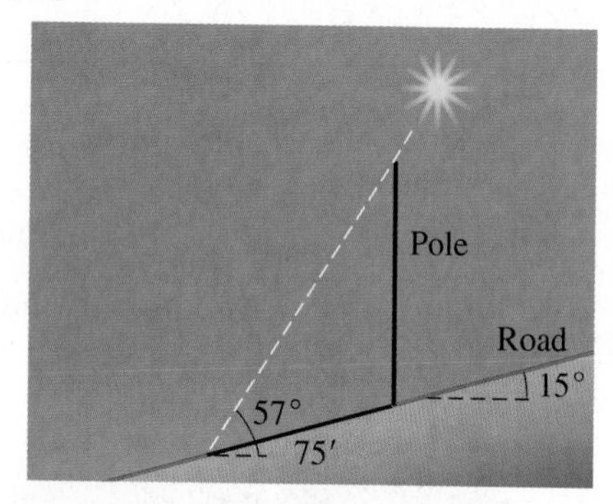

▶ **21 Height of a hot-air balloon** The angles of elevation of a balloon from two points A and B on level ground are 24°10′ and 47°40′, respectively. As shown in the figure, points A and B are 8.4 miles apart, and the balloon is between the points, in the same vertical plane. Approximate the height of the balloon above the ground.

Exercise 21

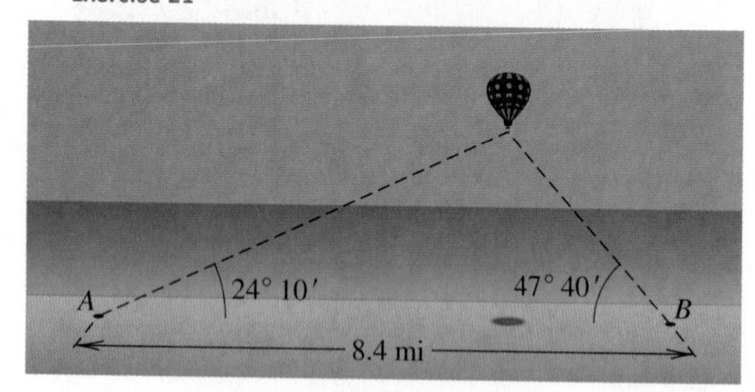

22 Installing a solar panel Shown in the figure is a solar panel 10 feet in width, which is to be attached to a roof that makes an angle of 25° with the horizontal. Approximate the length d of the brace that is needed for the panel to make an angle of 45° with the horizontal.

Exercise 22

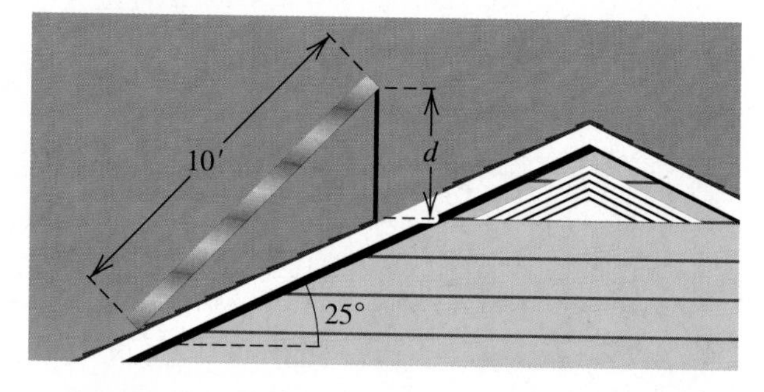

23 Distance to an airplane A straight road makes an angle of 22° with the horizontal. From a certain point P on the road, the angle of elevation of an airplane at point A is 57°. At the same instant, from another point Q, 100 meters farther up the road, the angle of elevation is 63°. As indicated in the figure, the points P, Q, and A lie in the same vertical plane. Approximate the distance from P to the airplane.

Exercise 23

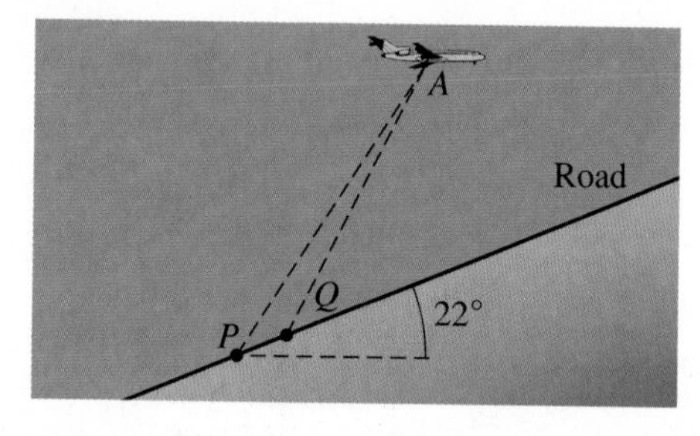

24 Surveying A surveyor notes that the direction from point A to point B is S63°W and the direction from A to point C is S38°W. The distance from A to B is 239 yards, and the distance from B to C is 374 yards. Approximate the distance from A to C.

25 Sighting a forest fire A forest ranger at an observation point A sights a fire in the direction N27°10′E. Another ranger at an observation point B, 6.0 miles due east of A, sights the same fire at N52°40′W. Approximate the distance from each of the observation points to the fire.

26 Leaning tower of Pisa The leaning tower of Pisa was originally perpendicular to the ground and 179 feet tall. Because of sinking into the earth, it now leans at a certain angle θ from the perpendicular, as shown in the figure. When the top of the tower is viewed from a point 150 feet from the center of its base, the angle of elevation is 53°.

▶ **Tutorial available at www.thomsonedu.com/login**

(a) Approximate the angle θ.

(b) Approximate the distance d that the center of the top of the tower has moved from the perpendicular.

Exercise 26

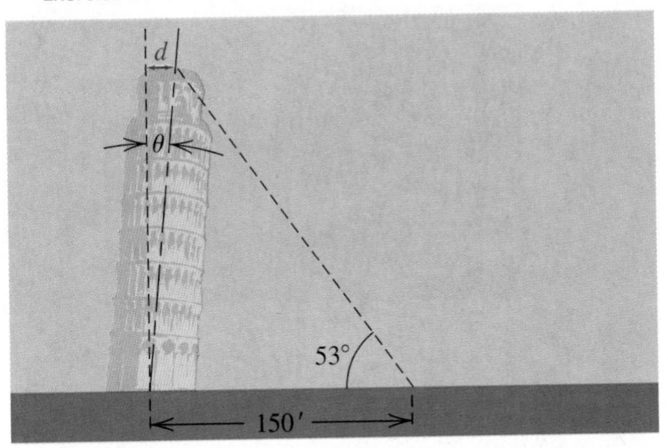

27 **Height of a cathedral** A cathedral is located on a hill, as shown in the figure. When the top of the spire is viewed from the base of the hill, the angle of elevation is 48°. When it is viewed at a distance of 200 feet from the base of the hill, the angle of elevation is 41°. The hill rises at an angle of 32°. Approximate the height of the cathedral.

Exercise 27

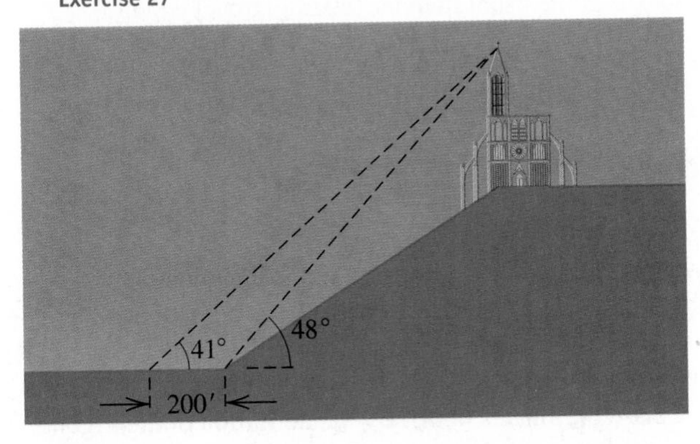

28 **Sighting from a helicopter** A helicopter hovers at an altitude that is 1000 feet above a mountain peak of altitude 5210 feet, as shown in the figure. A second, taller peak is viewed from both the mountaintop and the helicopter. From the helicopter, the angle of depression is 43°, and from the mountaintop, the angle of elevation is 18°.

(a) Approximate the distance from peak to peak.

(b) Approximate the altitude of the taller peak.

Exercise 28

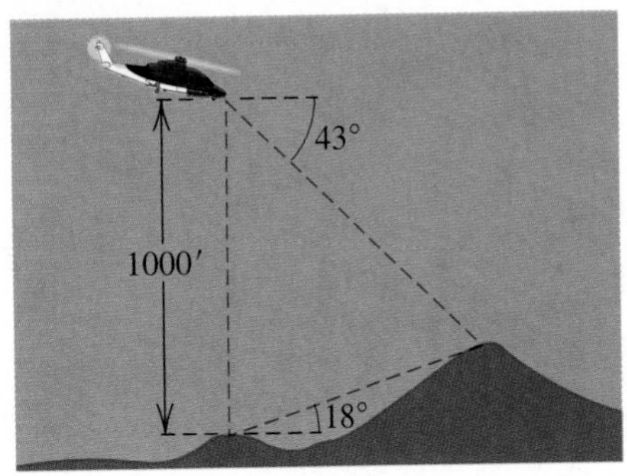

29 The volume V of the right triangular prism shown in the figure is $\frac{1}{3}Bh$, where B is the area of the base and h is the height of the prism.

(a) Approximate h.

(b) Approximate V.

Exercise 29

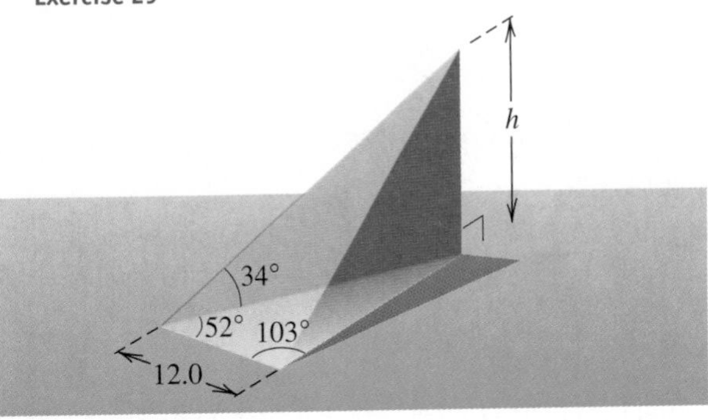

30 **Design for a jet fighter** Shown in the figure on the next page is a plan for the top of a wing of a jet fighter.

(a) Approximate angle ϕ.

(b) If the fuselage is 4.80 feet wide, approximate the wing span CC'.

(c) Approximate the area of triangle ABC.

Exercise 30

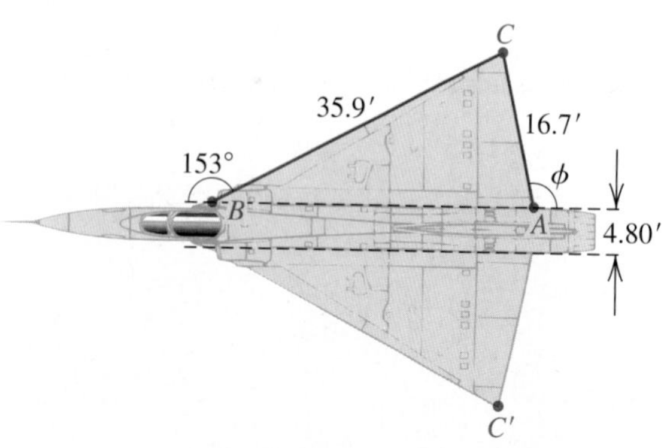

31 Software for surveyors Computer software for surveyors makes use of coordinate systems to locate geographic positions. An offshore oil well at point R in the figure is viewed from points P and Q, and $\angle QPR$ and $\angle RQP$ are found to be $55°50'$ and $65°22'$, respectively. If points P and Q have coordinates (1487.7, 3452.8) and (3145.8, 5127.5), respectively, approximate the coordinates of R.

Exercise 31

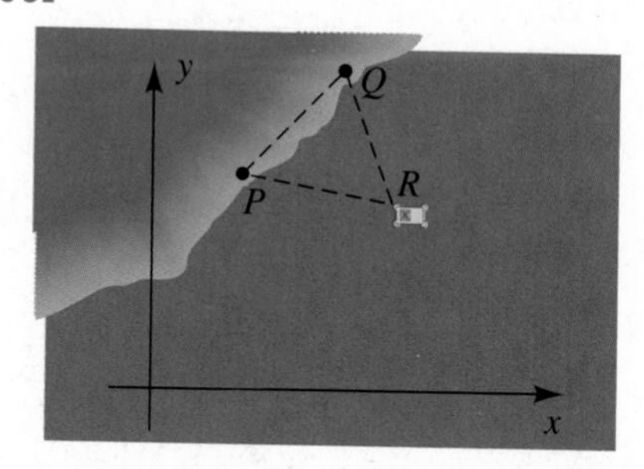

7.2 The Law of Cosines

In the preceding section we stated that the law of sines cannot be applied directly to find the remaining parts of an oblique triangle given either of the following:

(1) two sides and the angle *between* them (SAS)

(2) three sides (SSS)

For these cases we may apply the *law of cosines,* which follows.

The Law of Cosines

If ABC is a triangle labeled in the usual manner (as in Figure 1), then

(1) $a^2 = b^2 + c^2 - 2bc\cos\alpha$

(2) $b^2 = a^2 + c^2 - 2ac\cos\beta$

(3) $c^2 = a^2 + b^2 - 2ab\cos\gamma$

Figure 1

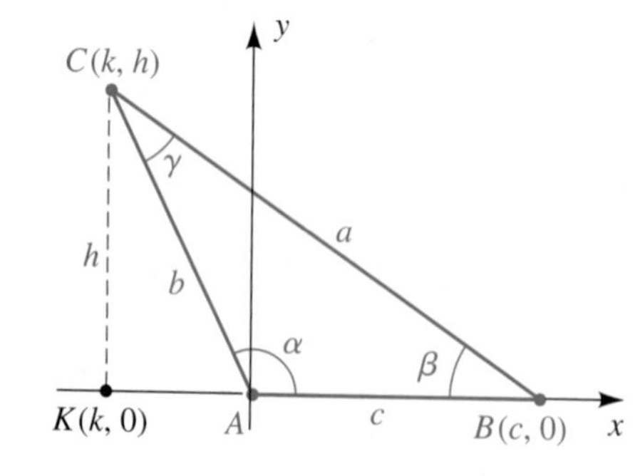

PROOF Let us prove the first formula. Given triangle ABC, place α in standard position, as illustrated in Figure 1. We have pictured α as obtuse; however, our discussion is also valid if α is acute. Consider the dashed line through C, parallel to the y-axis and intersecting the x-axis at the point $K(k, 0)$. If we let $d(C, K) = h$, then C has coordinates (k, h). By the definition of the trigonometric functions of any angle,

$$\cos\alpha = \frac{k}{b} \quad\text{and}\quad \sin\alpha = \frac{h}{b}.$$

Solving for k and h gives us

$$k = b\cos\alpha \quad\text{and}\quad h = b\sin\alpha.$$

Since the segment AB has length c, the coordinates of B are $(c, 0)$, and we obtain the following:

$$\begin{aligned} a^2 &= [d(B, C)]^2 = (k - c)^2 + (h - 0)^2 && \text{distance formula} \\ &= (b \cos \alpha - c)^2 + (b \sin \alpha)^2 && \text{substitute for } k \text{ and } h \\ &= b^2 \cos^2 \alpha - 2bc \cos \alpha + c^2 + b^2 \sin^2 \alpha && \text{square} \\ &= b^2(\cos^2 \alpha + \sin^2 \alpha) + c^2 - 2bc \cos \alpha && \text{factor the first and last terms} \\ &= b^2 + c^2 - 2bc \cos \alpha && \text{Pythagorean identity} \end{aligned}$$

Our result is the first formula stated in the law of cosines. The second and third formulas may be obtained by placing β and γ, respectively, in standard position on a coordinate system.

Note that if $\alpha = 90°$ in Figure 1, then $\cos \alpha = 0$ and the law of cosines reduces to $a^2 = b^2 + c^2$. This shows that the Pythagorean theorem is a special case of the law of cosines.

Instead of memorizing each of the three formulas of the law of cosines, it is more convenient to remember the following statement, which takes all of them into account.

The Law of Cosines (General Form)

The square of the length of any side of a triangle equals the sum of the squares of the lengths of the other two sides minus twice the product of the lengths of the other two sides and the cosine of the angle between them.

Given two sides and the included angle of a triangle, we can use the law of cosines to find the third side. We may then use the law of sines to find another angle of the triangle. Whenever this procedure is followed, it is best to find the angle opposite the shortest side, since that angle is always acute. In this way, we avoid the possibility of obtaining two solutions when solving a trigonometric equation involving that angle, as illustrated in the following example.

EXAMPLE 1 Using the law of cosines (SAS)

Solve $\triangle ABC$, given $a = 5.0$, $c = 8.0$, and $\beta = 77°$.

Figure 2

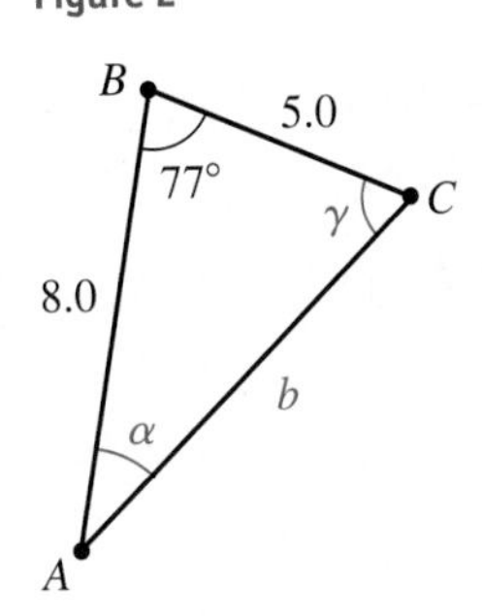

▶ **SOLUTION** The triangle is sketched in Figure 2. Since β is the angle *between* sides a and c, we begin by approximating b (the side opposite β) as follows:

$$\begin{aligned} b^2 &= a^2 + c^2 - 2ac \cos \beta && \text{law of cosines} \\ &= (5.0)^2 + (8.0)^2 - 2(5.0)(8.0) \cos 77° && \text{substitute for } a, c, \text{ and } \beta \\ &= 89 - 80 \cos 77° \approx 71.0 && \text{simplify and approximate} \\ b &\approx \sqrt{71.0} \approx 8.4 && \text{take the square root} \end{aligned}$$

(continued)

▶ **Tutorial available at www.thomsonedu.com/login**

Let us find another angle of the triangle using the law of sines. In accordance with the remarks preceding this example, we will apply the law of sines and find α, since it is the angle opposite the shortest side a.

$$\frac{\sin \alpha}{a} = \frac{\sin \beta}{b} \qquad \text{law of sines}$$

$$\sin \alpha = \frac{a \sin \beta}{b} \qquad \text{solve for } \sin \alpha$$

$$\approx \frac{5.0 \sin 77^\circ}{\sqrt{71.0}} \approx 0.5782 \qquad \text{substitute and approximate}$$

Since α is acute,

$$\alpha = \sin^{-1}(0.5782) \approx 35.3^\circ \approx 35^\circ.$$

Finally, since $\alpha + \beta + \gamma = 180^\circ$, we have

$$\gamma = 180^\circ - \alpha - \beta \approx 180^\circ - 35^\circ - 77^\circ = 68^\circ.$$

Given the three sides of a triangle, we can use the law of cosines to find *any* of the three angles. We shall always find the largest angle first—that is, *the angle opposite the longest side*—since this practice will guarantee that the remaining angles are acute. We may then find another angle of the triangle by using either the law of sines or the law of cosines. Note that when an angle is found by means of the law of cosines, there is no ambiguous case, since we always obtain a unique angle between 0° and 180°.

EXAMPLE 2 Using the law of cosines (SSS)

If triangle ABC has sides $a = 90$, $b = 70$, and $c = 40$, approximate angles α, β, and γ to the nearest degree.

▶ SOLUTION In accordance with the remarks preceding this example, we first find the angle opposite the longest side a. Thus, we choose the form of the law of cosines that involves α and proceed as follows:

$$a^2 = b^2 + c^2 - 2bc \cos \alpha \qquad \text{law of cosines}$$

$$\cos \alpha = \frac{b^2 + c^2 - a^2}{2bc} \qquad \text{solve for } \cos \alpha$$

$$= \frac{70^2 + 40^2 - 90^2}{2(70)(40)} = -\frac{2}{7} \qquad \text{substitute and simplify}$$

$$\alpha = \cos^{-1}\left(-\tfrac{2}{7}\right) \approx 106.6^\circ \approx 107^\circ \qquad \text{approximate } \alpha$$

We may now use either the law of sines or the law of cosines to find β. Let's use the law of cosines in this case:

$$b^2 = a^2 + c^2 - 2ac \cos \beta \qquad \text{law of cosines}$$

$$\cos \beta = \frac{a^2 + c^2 - b^2}{2ac} \qquad \text{solve for } \cos \beta$$

▶ **Tutorial available at www.thomsonedu.com/login**

$$= \frac{90^2 + 40^2 - 70^2}{2(90)(40)} = \frac{2}{3} \qquad \text{substitute and simplify}$$

$$\beta = \cos^{-1}\left(\tfrac{2}{3}\right) \approx 48.2° \approx 48° \qquad \text{approximate } \beta$$

At this point in the solution, we could find γ by using the relationship $\alpha + \beta + \gamma = 180°$. But if either α or β was incorrectly calculated, then γ would be incorrect. Alternatively, we can approximate γ and then check that the sum of the three angles is 180°. Thus,

$$\cos\gamma = \frac{a^2 + b^2 - c^2}{2ab}, \qquad \text{so} \qquad \gamma = \cos^{-1}\frac{90^2 + 70^2 - 40^2}{2(90)(70)} \approx 25°.$$

Note that $\alpha + \beta + \gamma = 107° + 48° + 25° = 180°$.

EXAMPLE 3 Approximating the diagonals of a parallelogram

A parallelogram has sides of lengths 30 centimeters and 70 centimeters and one angle of measure 65°. Approximate the length of each diagonal to the nearest centimeter.

Figure 3

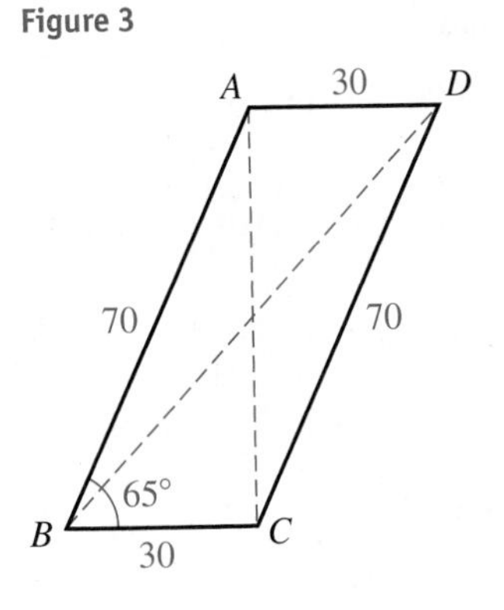

▶ SOLUTION The parallelogram $ABCD$ and its diagonals AC and BD are shown in Figure 3. Using triangle ABC with $\angle ABC = 65°$, we may approximate AC as follows:

$$(AC)^2 = 30^2 + 70^2 - 2(30)(70)\cos 65° \qquad \text{law of cosines}$$

$$\approx 900 + 4900 - 1775 = 4025 \qquad \text{approximate}$$

$$AC \approx \sqrt{4025} \approx 63 \text{ cm} \qquad \text{take the square root}$$

Similarly, using triangle BAD and $\angle BAD = 180° - 65° = 115°$, we may approximate BD as follows:

$$(BD)^2 = 30^2 + 70^2 - 2(30)(70)\cos 115° \approx 7575 \qquad \text{law of cosines}$$

$$BD \approx \sqrt{7575} \approx 87 \text{ cm} \qquad \text{take the square root}$$

EXAMPLE 4 Finding the length of a cable

A vertical pole 40 feet tall stands on a hillside that makes an angle of 17° with the horizontal. Approximate the minimal length of cable that will reach from the top of the pole to a point 72 feet downhill from the base of the pole.

Figure 4

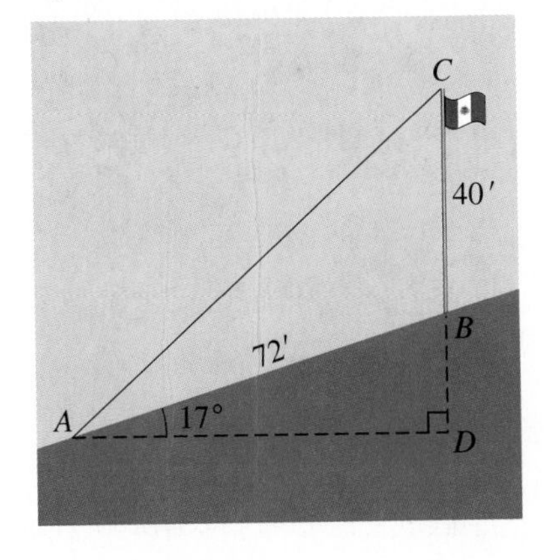

▶ SOLUTION The sketch in Figure 4 depicts the given data. We wish to find AC. Referring to the figure, we see that

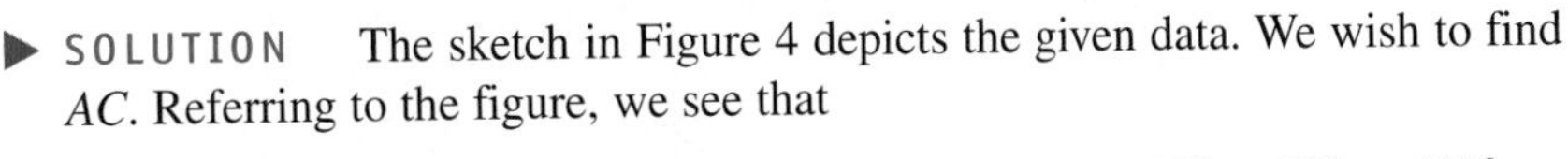

$$\angle ABD = 90° - 17° = 73° \qquad \text{and} \qquad \angle ABC = 180° - 73° = 107°.$$

Using triangle ABC, we may approximate AC as follows:

$$(AC)^2 = 72^2 + 40^2 - 2(72)(40)\cos 107° \approx 8468 \qquad \text{law of cosines}$$

$$AC \approx \sqrt{8468} \approx 92 \text{ ft} \qquad \text{take the square root}$$

The law of cosines can be used to derive a formula for the area of a triangle. Let us first prove a preliminary result.

▶ **Tutorial available at www.thomsonedu.com/login**

Figure 5

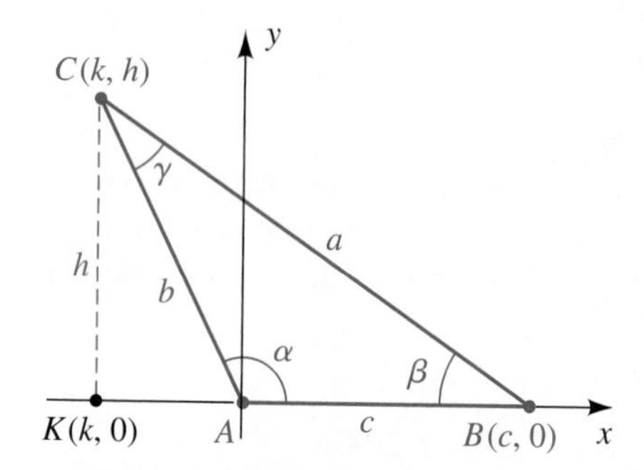

Given triangle ABC, place angle α in standard position (see Figure 5). As shown in the proof of the law of cosines, the altitude h from vertex C is $h = b \sin \alpha$. Since the area $\mathcal{A}$ of the triangle is given by $\mathcal{A} = \frac{1}{2}ch$, we see that

$$\mathcal{A} = \tfrac{1}{2}bc \sin \alpha.$$

Our argument is independent of the specific angle that is placed in standard position. By taking β and γ in standard position, we obtain the formulas

$$\mathcal{A} = \tfrac{1}{2}ac \sin \beta \qquad \text{and} \qquad \mathcal{A} = \tfrac{1}{2}ab \sin \gamma.$$

All three formulas are covered in the following statement.

Area of a Triangle	The area of a triangle equals one-half the product of the lengths of any two sides and the sine of the angle between them.

The next two examples illustrate uses of this result.

EXAMPLE 5 Approximating the area of a triangle

Approximate the area of triangle ABC if $a = 2.20$ cm, $b = 1.30$ cm, and $\gamma = 43.2°$.

Figure 6

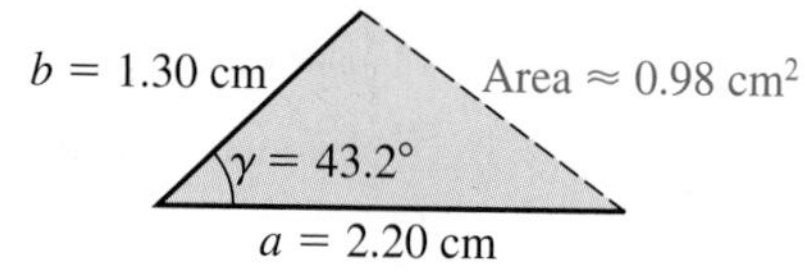

▶ **SOLUTION** Since γ is the angle between sides a and b as shown in Figure 6, we may use the preceding result directly, as follows:

$$\begin{aligned} \mathcal{A} &= \tfrac{1}{2}ab \sin \gamma && \text{area of a triangle formula} \\ &= \tfrac{1}{2}(2.20)(1.30) \sin 43.2° \approx 0.98 \text{ cm}^2 && \text{substitute and approximate} \end{aligned}$$

EXAMPLE 6 Approximating the area of a triangle

Approximate the area of triangle ABC if $a = 5.0$ cm, $b = 3.0$ cm, and $\alpha = 37°$.

▶ **SOLUTION** To apply the formula for the area of a triangle, we must find the angle γ between known sides a and b. Since we are given a, b, and α, let us first find β as follows:

$$\begin{aligned} \frac{\sin \beta}{b} &= \frac{\sin \alpha}{a} && \text{law of sines} \\ \sin \beta &= \frac{b \sin \alpha}{a} && \text{solve for } \sin \beta \\ &= \frac{3.0 \sin 37°}{5.0} && \text{substitute for } b, \alpha, \text{ and } a \\ \beta_R &= \sin^{-1}\left(\frac{3.0 \sin 37°}{5.0}\right) \approx 21° && \text{reference angle for } \beta \\ \beta \approx 21° \quad &\text{or} \quad \beta \approx 159° && \beta_R \text{ or } 180° - \beta_R \end{aligned}$$

▶ Tutorial available at www.thomsonedu.com/login

We reject $\beta \approx 159°$, because then $\alpha + \beta = 196° \geq 180°$. Hence, $\beta \approx 21°$ and

$$\gamma = 180° - \alpha - \beta \approx 180° - 37° - 21° = 122°.$$

Finally, we approximate the area of the triangle as follows:

$$\begin{aligned} \mathcal{A} &= \tfrac{1}{2}ab \sin \gamma && \text{area of a triangle formula} \\ &\approx \tfrac{1}{2}(5.0)(3.0) \sin 122° \approx 6.4 \text{ cm}^2 && \text{substitute and approximate} \end{aligned}$$

We will use the preceding result for the area of a triangle to derive *Heron's formula,* which expresses the area of a triangle in terms of the lengths of its sides.

Heron's Formula

The area $\mathcal{A}$ of a triangle with sides a, b, and c is given by

$$\mathcal{A} = \sqrt{s(s-a)(s-b)(s-c)},$$

where s is one-half the perimeter; that is, $s = \tfrac{1}{2}(a + b + c)$.

PROOF The following equations are equivalent:

$$\begin{aligned} \mathcal{A} &= \tfrac{1}{2}bc \sin \alpha \\ &= \sqrt{\tfrac{1}{4}b^2c^2 \sin^2 \alpha} \\ &= \sqrt{\tfrac{1}{4}b^2c^2(1 - \cos^2 \alpha)} \\ &= \sqrt{\tfrac{1}{2}bc(1 + \cos \alpha) \cdot \tfrac{1}{2}bc(1 - \cos \alpha)} \end{aligned}$$

We shall obtain Heron's formula by replacing the expressions under the final radical sign by expressions involving only a, b, and c. We solve formula 1 of the law of cosines for $\cos \alpha$ and then substitute, as follows:

$$\begin{aligned} \frac{1}{2}bc(1 + \cos \alpha) &= \frac{1}{2}bc\left(1 + \frac{b^2 + c^2 - a^2}{2bc}\right) \\ &= \frac{1}{2}bc\left(\frac{2bc + b^2 + c^2 - a^2}{2bc}\right) \\ &= \frac{2bc + b^2 + c^2 - a^2}{4} \\ &= \frac{(b + c)^2 - a^2}{4} \\ &= \frac{(b + c) + a}{2} \cdot \frac{(b + c) - a}{2} \end{aligned}$$

(continued)

We use the same type of manipulations on the second expression under the radical sign:

$$\frac{1}{2}bc(1 - \cos\alpha) = \frac{a - b + c}{2} \cdot \frac{a + b - c}{2}$$

If we now substitute for the expressions under the radical sign, we obtain

$$\mathscr{A} = \sqrt{\frac{b + c + a}{2} \cdot \frac{b + c - a}{2} \cdot \frac{a - b + c}{2} \cdot \frac{a + b - c}{2}}.$$

Letting $s = \frac{1}{2}(a + b + c)$, we see that

$$s - a = \frac{b + c - a}{2}, \quad s - b = \frac{a - b + c}{2}, \quad s - c = \frac{a + b - c}{2}.$$

Substitution in the above formula for $\mathscr{A}$ gives us Heron's formula.

EXAMPLE 7 **Using Heron's formula**

A triangular field has sides of lengths 125 yards, 160 yards, and 225 yards. Approximate the number of acres in the field. (One acre is equivalent to 4840 square yards.)

SOLUTION We first find one-half the perimeter of the field with $a = 125$, $b = 160$, and $c = 225$, as well as the values of $s - a$, $s - b$, and $s - c$:

$$\begin{aligned} s &= \tfrac{1}{2}(125 + 160 + 225) = \tfrac{1}{2}(510) = 255 \\ s - a &= 255 - 125 = 130 \\ s - b &= 255 - 160 = 95 \\ s - c &= 255 - 225 = 30 \end{aligned}$$

Substituting in Heron's formula gives us

$$\mathscr{A} = \sqrt{(255)(130)(95)(30)} \approx 9720 \text{ yd}^2.$$

Since there are 4840 square yards in one acre, the number of acres is $\frac{9720}{4840}$, or approximately 2.

7.2 Exercises

Exer. 1–2: Use common sense to match the variables and the values. (The triangles are drawn to scale, and the angles are measured in radians.)

1

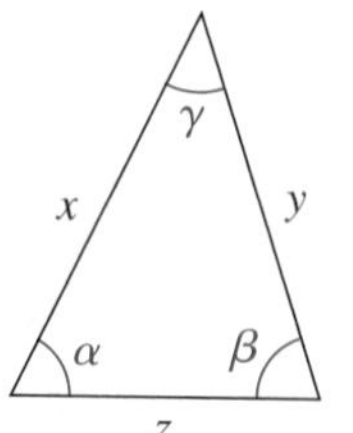

(a) α	(A) 12.60
(b) β	(B) 1.10
(c) γ	(C) 10
(d) x	(D) 0.79
(e) y	(E) 13.45
(f) z	(F) 1.26

2

(a) α	(A) 3
(b) β	(B) 0.87
(c) γ	(C) 8.24
(d) x	(D) 1.92
(e) y	(E) 6.72
(f) z	(F) 0.35

▶ **Tutorial available at www.thomsonedu.com/login**

Exer. 3–4: Given the indicated parts of $\triangle ABC$, what angle (α, β, or γ) or side (a, b, or c) would you find next, and what would you use to find it?

3 (a)

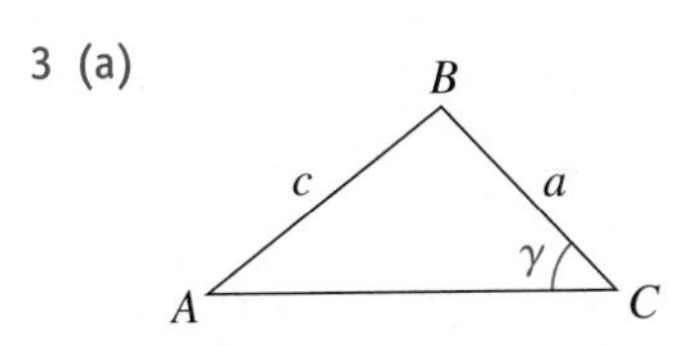

(b)

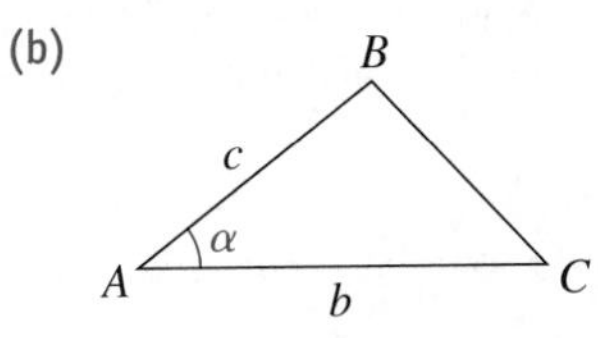

(c)

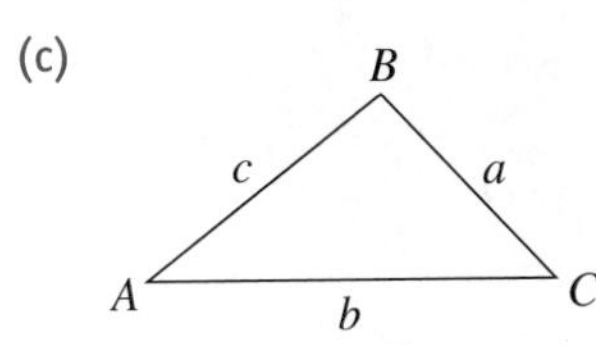

(d)

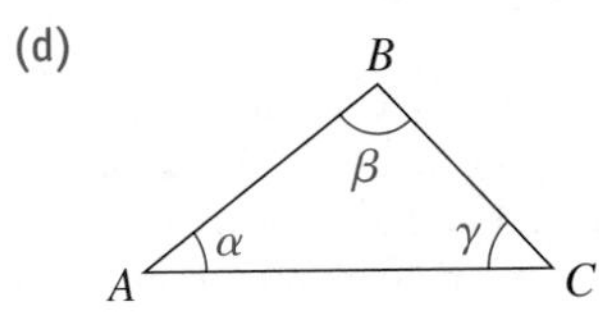

(e)

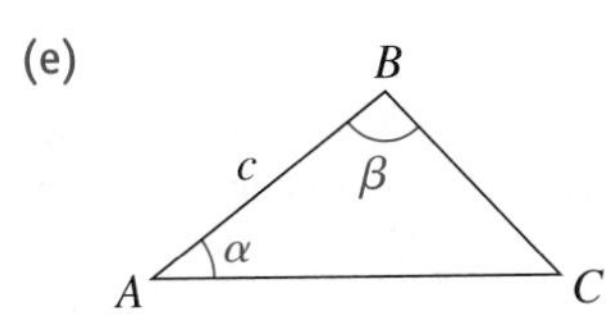

(f)

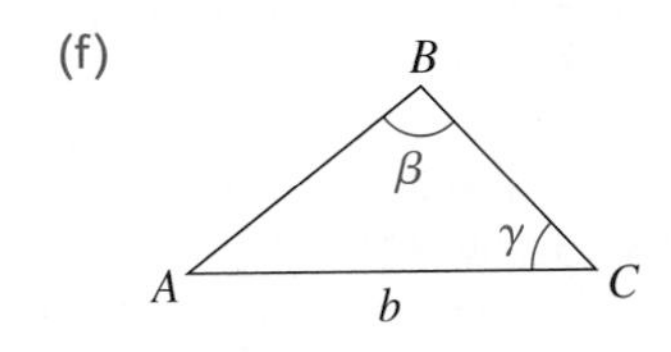

4 (a)

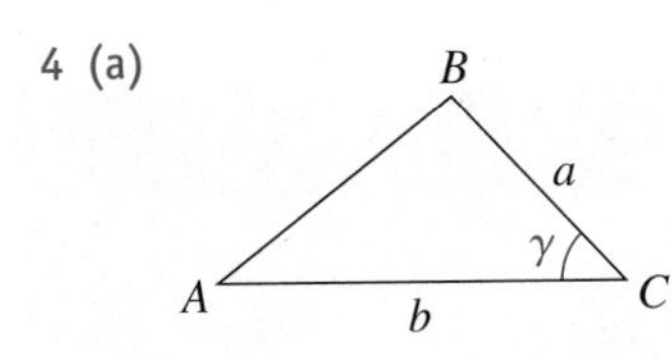

(b)

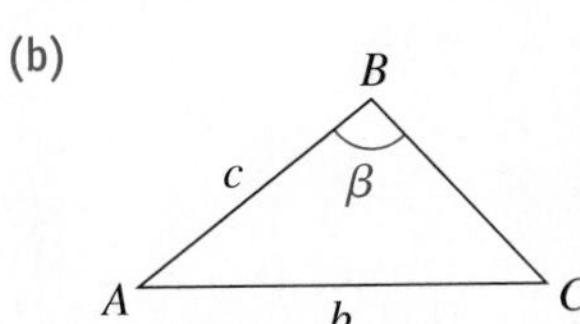

(c)

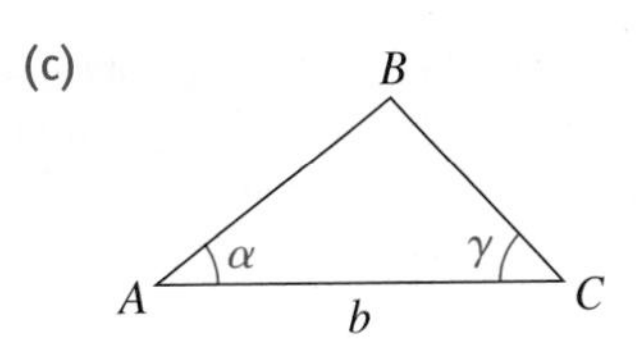

(d)

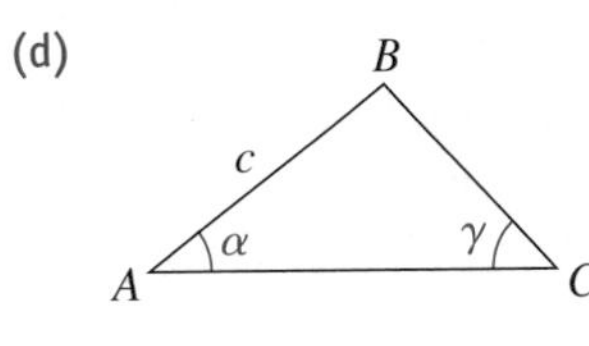

(e)

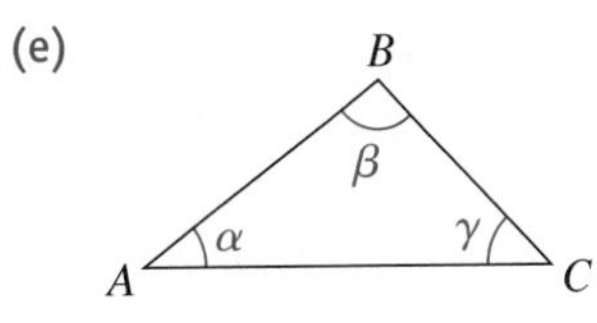

(f)

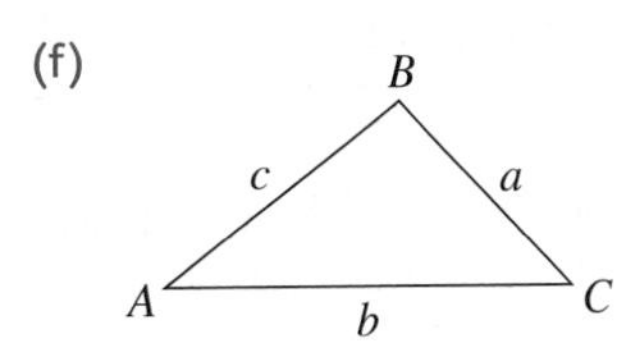

Exer. 5–14: Solve $\triangle ABC$.

5 $\alpha = 60°$, $b = 20$, $c = 30$

6 $\gamma = 45°$, $b = 10.0$, $a = 15.0$

7 $\beta = 150°$, $a = 150$, $c = 30$

8 $\beta = 73°50'$, $c = 14.0$, $a = 87.0$

9 $\gamma = 115°10'$, $a = 1.10$, $b = 2.10$

10 $\alpha = 23°40'$, $c = 4.30$, $b = 70.0$

11 $a = 2.0$, $b = 3.0$, $c = 4.0$

12 $a = 10$, $b = 15$, $c = 12$

13 $a = 25.0$, $b = 80.0$, $c = 60.0$

14 $a = 20.0$, $b = 20.0$, $c = 10.0$

▶ 15 **Dimensions of a triangular plot** The angle at one corner of a triangular plot of ground is 73°40′, and the sides that meet at this corner are 175 feet and 150 feet long. Approximate the length of the third side.

▶ **Tutorial available at www.thomsonedu.com/login**

16 Surveying To find the distance between two points A and B, a surveyor chooses a point C that is 420 yards from A and 540 yards from B. If angle ACB has measure $63°10'$, approximate the distance between A and B.

17 Distance between automobiles Two automobiles leave a city at the same time and travel along straight highways that differ in direction by 84°. If their speeds are 60 mi/hr and 45 mi/hr, respectively, approximately how far apart are the cars at the end of 20 minutes?

18 Angles of a triangular plot A triangular plot of land has sides of lengths 420 feet, 350 feet, and 180 feet. Approximate the smallest angle between the sides.

19 Distance between ships A ship leaves port at 1:00 P.M. and travels S35°E at the rate of 24 mi/hr. Another ship leaves the same port at 1:30 P.M. and travels S20°W at 18 mi/hr. Approximately how far apart are the ships at 3:00 P.M.?

20 Flight distance An airplane flies 165 miles from point A in the direction 130° and then travels in the direction 245° for 80 miles. Approximately how far is the airplane from A?

21 Jogger's course A jogger runs at a constant speed of one mile every 8 minutes in the direction S40°E for 20 minutes and then in the direction N20°E for the next 16 minutes. Approximate, to the nearest tenth of a mile, the straight-line distance from the endpoint to the starting point of the jogger's course.

22 Surveying Two points P and Q on level ground are on opposite sides of a building. To find the distance between the points, a surveyor chooses a point R that is 300 feet from P and 438 feet from Q and then determines that angle PRQ has measure $37°40'$ (see the figure). Approximate the distance between P and Q.

Exercise 22

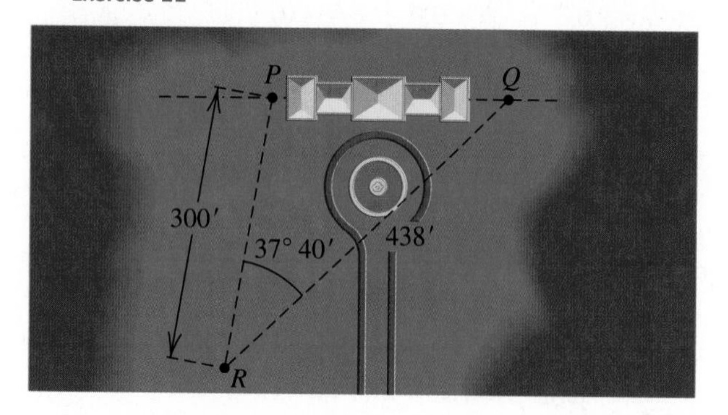

23 Motorboat's course A motorboat traveled along a triangular course having sides of lengths 2 kilometers, 4 kilometers, and 3 kilometers, respectively. The first side was traversed in the direction N20°W and the second in a direction Sθ°W, where $\theta°$ is the degree measure of an acute angle. Approximate, to the nearest minute, the direction in which the third side was traversed.

24 Angle of a box The rectangular box shown in the figure has dimensions $8'' \times 6'' \times 4''$. Approximate the angle θ formed by a diagonal of the base and a diagonal of the $6'' \times 4''$ side.

Exercise 24

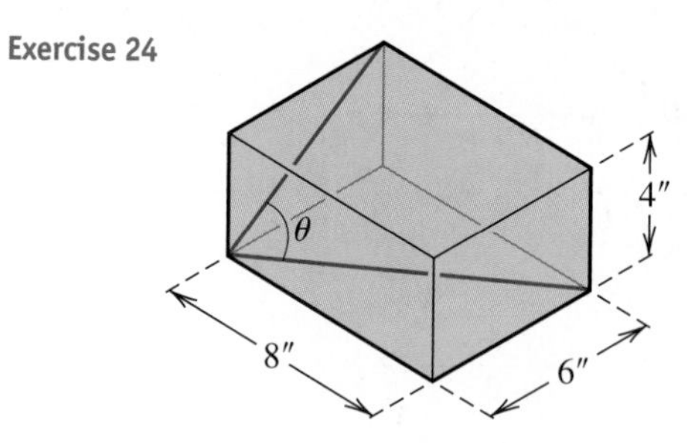

25 Distances in a baseball diamond A baseball diamond has four bases (forming a square) that are 90 feet apart; the pitcher's mound is 60.5 feet from home plate. Approximate the distance from the pitcher's mound to each of the other three bases.

26 A rhombus has sides of length 100 centimeters, and the angle at one of the vertices is 70°. Approximate the lengths of the diagonals to the nearest tenth of a centimeter.

27 Reconnaissance A reconnaissance airplane P, flying at 10,000 feet above a point R on the surface of the water, spots a submarine S at an angle of depression of 37° and a tanker T at an angle of depression of 21°, as shown in the figure. In addition, $\angle SPT$ is found to be 110°. Approximate the distance between the submarine and the tanker.

Exercise 27

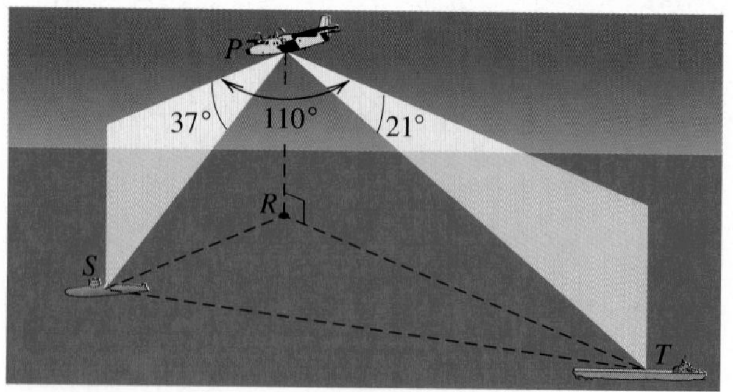

28 **Correcting a ship's course** A cruise ship sets a course N47°E from an island to a port on the mainland, which is 150 miles away. After moving through strong currents, the ship is off course at a position P that is N33°E and 80 miles from the island, as illustrated in the figure.

(a) Approximately how far is the ship from the port?

(b) In what direction should the ship head to correct its course?

Exercise 28

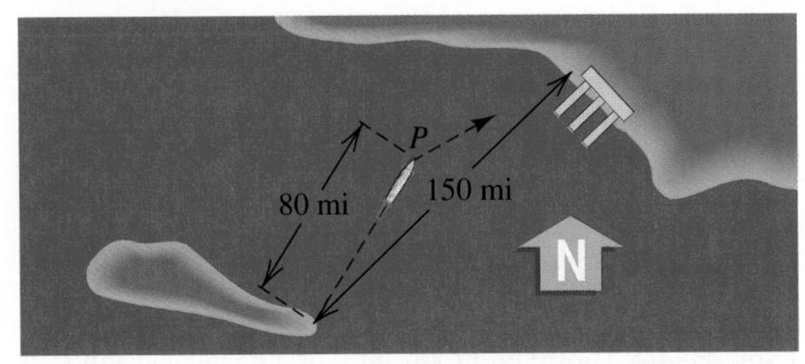

29 **Seismology** Seismologists investigate the structure of Earth's interior by analyzing seismic waves caused by earthquakes. If the interior of Earth is assumed to be homogeneous, then these waves will travel in straight lines at a constant velocity v. The figure shows a cross-sectional view of Earth, with the epicenter at E and an observation station at S. Use the law of cosines to show that the time t for a wave to travel through Earth's interior from E to S is given by

$$t = \frac{2R}{v}\sin\frac{\theta}{2},$$

where R is the radius of Earth and θ is the indicated angle with vertex at the center of Earth.

Exercise 29

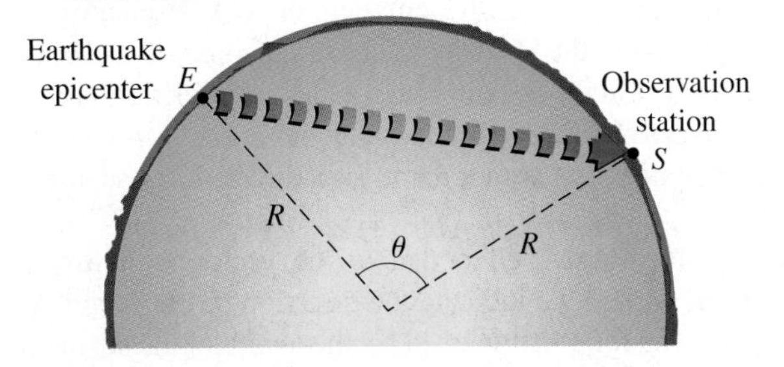

30 **Calculating distances** The distance across the river shown in the figure can be found without measuring angles. Two points B and C on the opposite shore are selected, and line segments AB and AC are extended as shown. Points D and E are chosen as indicated, and distances BC, BD, BE, CD, and CE are then measured. Suppose that $BC = 184$ ft, $BD = 102$ ft, $BE = 218$ ft, $CD = 236$ ft, and $CE = 80$ ft.

(a) Approximate the distances AB and AC.

(b) Approximate the shortest distance across the river from point A.

Exercise 30

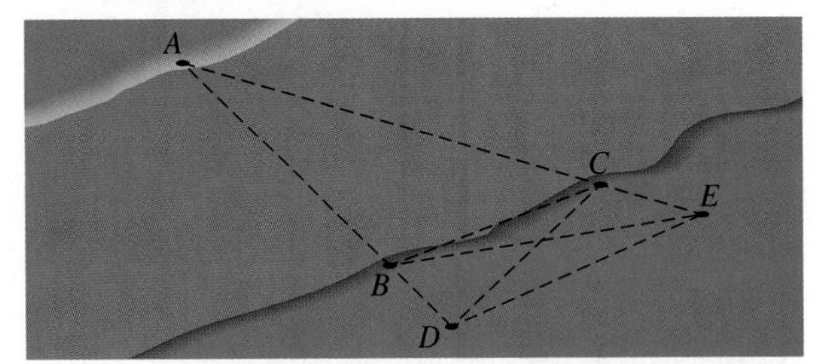

▶ 31 **Penrose tiles** Penrose tiles are formed from a rhombus $ABCD$ having sides of length 1 and an interior angle of 72°. First a point P is located that lies on the diagonal AC and is a distance 1 from vertex C, and then segments PB and PD are drawn to the other vertices of the diagonal, as shown in the figure. The two tiles formed are called a dart and a kite. Three-dimensional counterparts of these tiles have been applied in molecular chemistry.

(a) Find the degree measures of $\angle BPC$, $\angle APB$, and $\angle ABP$.

(b) Approximate, to the nearest 0.01, the length of segment BP.

(c) Approximate, to the nearest 0.01, the area of a kite and the area of a dart.

Exercise 31

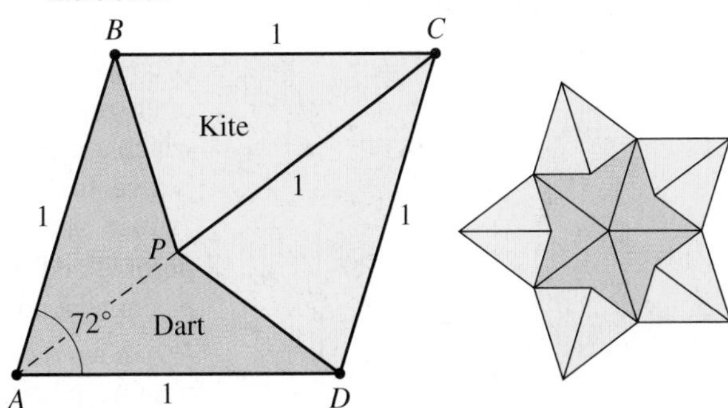

32 **Automotive design** The rear hatchback door of an automobile is 42 inches long. A strut with a fully extended length

of 24 inches is to be attached to the door and the body of the car so that when the door is opened completely, the strut is vertical and the rear clearance is 32 inches, as shown in the figure. Approximate the lengths of segments TQ and TP.

Exercise 32

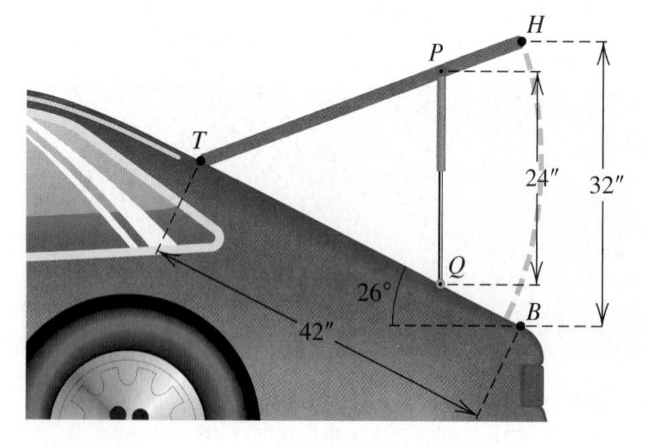

Exer. 33–40: Approximate the area of triangle ABC.

33 $\alpha = 60°$, $b = 20$, $c = 30$

34 $\gamma = 45°$, $b = 10.0$, $a = 15.0$

35 $\alpha = 40.3°$, $\beta = 62.9°$, $b = 5.63$

36 $\alpha = 35.7°$, $\gamma = 105.2°$, $b = 17.2$

37 $\alpha = 80.1°$, $a = 8.0$, $b = 3.4$

38 $\gamma = 32.1°$, $a = 14.6$, $c = 15.8$

39 $a = 25.0$, $b = 80.0$, $c = 60.0$

40 $a = 20.0$, $b = 20.0$, $c = 10.0$

Exer. 41–42: A triangular field has sides of lengths a, b, and c (in yards). Approximate the number of acres in the field (1 acre = 4840 yd^2).

41 $a = 115$, $b = 140$, $c = 200$

42 $a = 320$, $b = 350$, $c = 500$

Exer. 43–44: Approximate the area of a parallelogram that has sides of lengths a and b (in feet) if one angle at a vertex has measure θ.

▶ 43 $a = 12.0$, $b = 16.0$, $\theta = 40°$

44 $a = 40.3$, $b = 52.6$, $\theta = 100°$

7.3 Vectors

Quantities such as area, volume, length, temperature, and time have magnitude only and can be completely characterized by a single real number (with an appropriate unit of measurement such as in^2, ft^3, cm, deg, or sec). A quantity of this type is a **scalar quantity,** and the corresponding real number is a **scalar.** A concept such as velocity or force has both magnitude and direction and is often represented by a **directed line segment**—that is, a line segment to which a direction has been assigned. Another name for a directed line segment is a **vector.**

As shown in Figure 1, we use $\overrightarrow{PQ}$ to denote the vector with **initial point** P and **terminal point** Q, and we indicate the direction of the vector by placing the arrowhead at Q. The **magnitude** of $\overrightarrow{PQ}$ is the length of the segment PQ and is denoted by $\|\overrightarrow{PQ}\|$. As in the figure, we use boldface letters such as **u** and **v** to denote vectors whose endpoints are not specified. In handwritten work, a notation such as $\vec{u}$ or $\vec{v}$ is often used.

Vectors that have the same magnitude and direction are said to be **equivalent.** In mathematics, a vector is determined only by its magnitude and direc-

▶ Tutorial available at www.thomsonedu.com/login

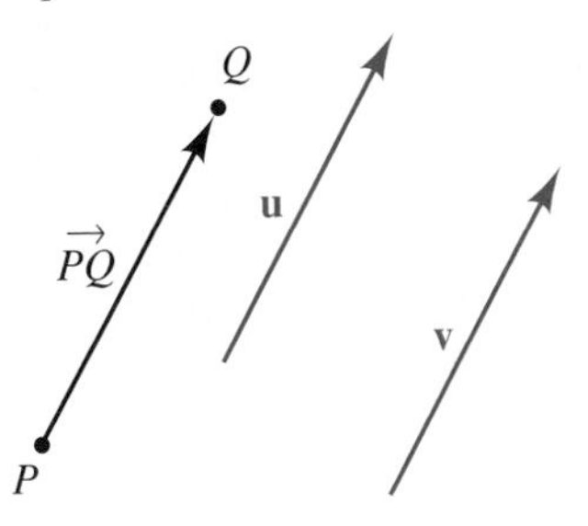

Figure 1
Equal vectors

tion, not by its location. Thus, we regard equivalent vectors, such as those in Figure 1, as **equal** and write

$$\mathbf{u} = \overrightarrow{PQ}, \quad \mathbf{v} = \overrightarrow{PQ}, \quad \text{and} \quad \mathbf{u} = \mathbf{v}.$$

Thus, *a vector may be translated from one location to another, provided neither the magnitude nor the direction is changed.*

We can represent many physical concepts by vectors. To illustrate, suppose an airplane is descending at a constant speed of 100 mi/hr and the line of flight makes an angle of 20° with the horizontal. Both of these facts are represented by the vector **v** of magnitude 100 in Figure 2. The vector **v** is a **velocity vector.**

Figure 2 Velocity vector

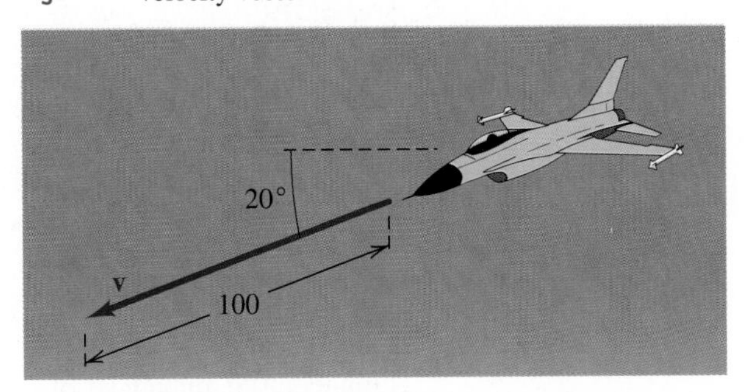

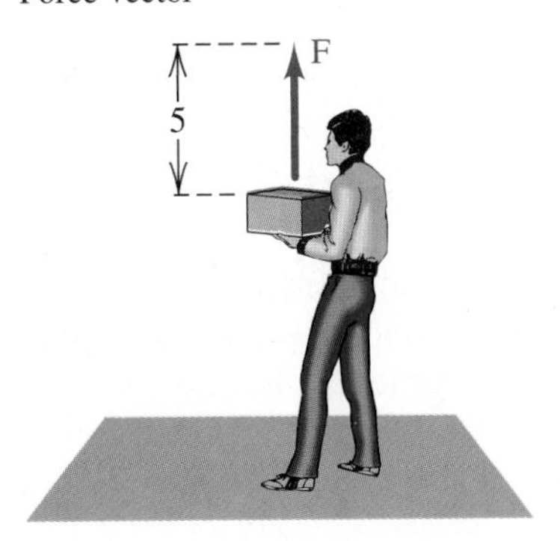

Figure 3
Force vector

A vector that represents a pull or push of some type is a **force vector.** The force exerted when a person holds a 5-pound weight is illustrated by the vector **F** of magnitude 5 in Figure 3. This force has the same magnitude as the force exerted on the weight by gravity, but it acts in the opposite direction. As a result, there is no movement upward or downward.

We sometimes use $\overrightarrow{AB}$ to represent the path of a point (or particle) as it moves along the line segment from A to B. We then refer to $\overrightarrow{AB}$ as a **displacement** of the point (or particle). As in Figure 4, a displacement $\overrightarrow{AB}$ followed by a displacement $\overrightarrow{BC}$ leads to the same point as the single displacement $\overrightarrow{AC}$. By definition, the vector AC is the **sum** of $\overrightarrow{AB}$ and $\overrightarrow{BC}$, and we write

$$\overrightarrow{AC} = \overrightarrow{AB} + \overrightarrow{BC}.$$

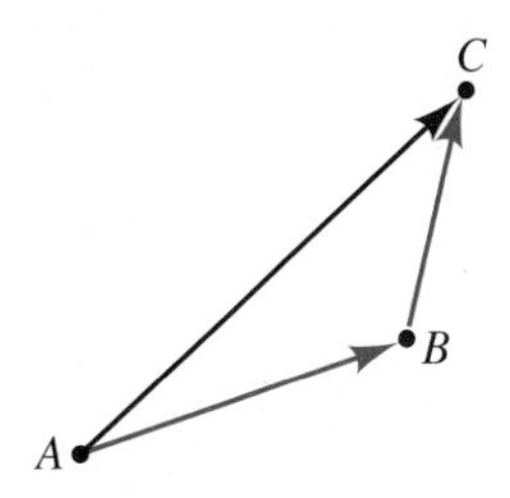

Figure 4
Sum of vectors

Since vectors may be translated from one location to another, *any* two vectors may be added by placing the initial point of the second vector on the terminal point of the first and then drawing the line segment from the initial point of the first to the terminal point of the second, as in Figure 4. We refer to this method of vector addition as using the **triangle law.**

Another way to find the sum is to choose vector PQ and vector PR that are equal to $\overrightarrow{AB}$ and $\overrightarrow{BC}$, respectively, and have the same initial point P, as shown in Figure 5. If we construct parallelogram $RPQS$, then, since $\overrightarrow{PR} = \overrightarrow{QS}$,

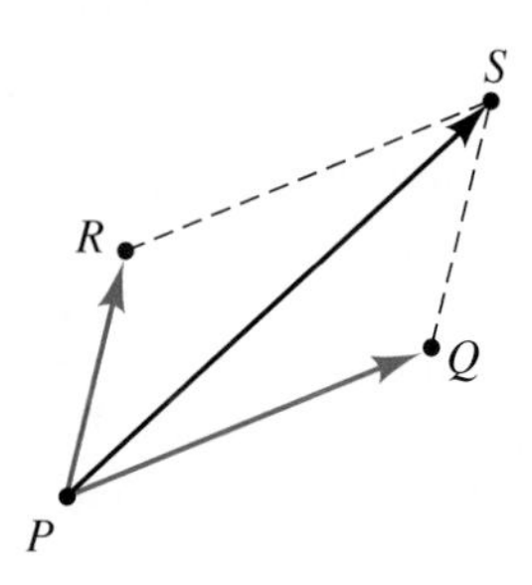

Figure 5 Resultant force

it follows that $\overrightarrow{PS} = \overrightarrow{PQ} + \overrightarrow{PR}$. If $\overrightarrow{PQ}$ and $\overrightarrow{PR}$ are two forces acting at P, then $\overrightarrow{PS}$ is the **resultant force**—that is, the single force that produces the same effect as the two combined forces. We refer to this method of vector addition as using the **parallelogram law.**

If m is a scalar and $\mathbf{v}$ is a vector, then $m\mathbf{v}$ is defined as a vector whose magnitude is $|m|$ times $\|\mathbf{v}\|$ (the magnitude of $\mathbf{v}$) and whose direction is either the same as that of $\mathbf{v}$ (if $m > 0$) or opposite that of $\mathbf{v}$ (if $m < 0$). Illustrations are given in Figure 6. We refer to $m\mathbf{v}$ as a **scalar multiple** of $\mathbf{v}$.

Figure 6 Scalar multiples

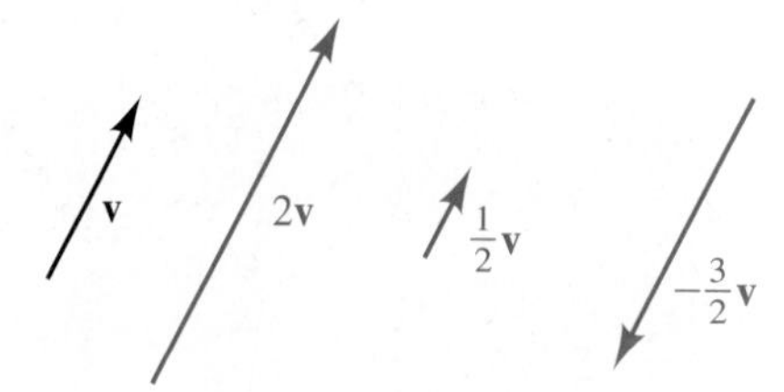

Figure 7

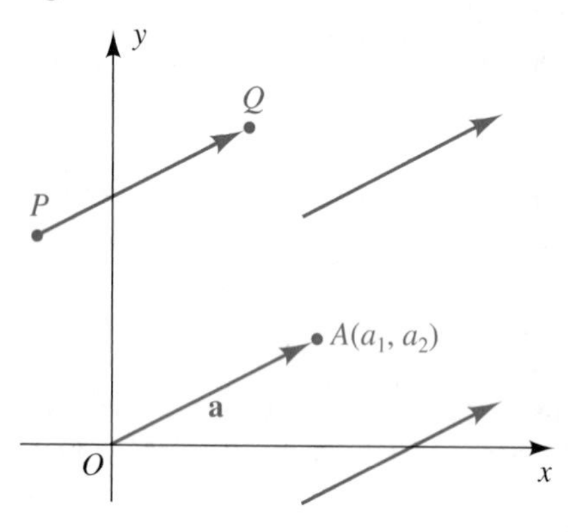

Throughout the remainder of this section we shall restrict our discussion to vectors that lie in an xy-plane. If $\overrightarrow{PQ}$ is such a vector, then, as indicated in Figure 7, there are many vectors that are equivalent to $\overrightarrow{PQ}$; however, there is exactly *one* equivalent vector $\mathbf{a} = \overrightarrow{OA}$ with initial point at the origin. In this sense, *each vector determines a unique ordered pair of real numbers,* the coordinates (a_1, a_2) of the terminal point A. Conversely, every ordered pair (a_1, a_2) determines the vector OA, where A has coordinates (a_1, a_2). Thus, *there is a one-to-one correspondence between vectors in an xy-plane and ordered pairs of real numbers.* This correspondence allows us to interpret a vector as both a directed line segment *and* an ordered pair of real numbers. To avoid confusion with the notation for open intervals or points, we use the symbol $\langle a_1, a_2 \rangle$ (referred to as *wedge notation*) for an ordered pair that represents a vector, and we denote it by a boldface letter—for example, $\mathbf{a} = \langle a_1, a_2 \rangle$. The numbers a_1 and a_2 are the **components** of the vector $\langle a_1, a_2 \rangle$. If A is the point (a_1, a_2), as in Figure 7, we call $\overrightarrow{OA}$ the **position vector** for $\langle a_1, a_2 \rangle$ or for the *point A*.

Figure 8 Magnitude $\|\mathbf{a}\|$

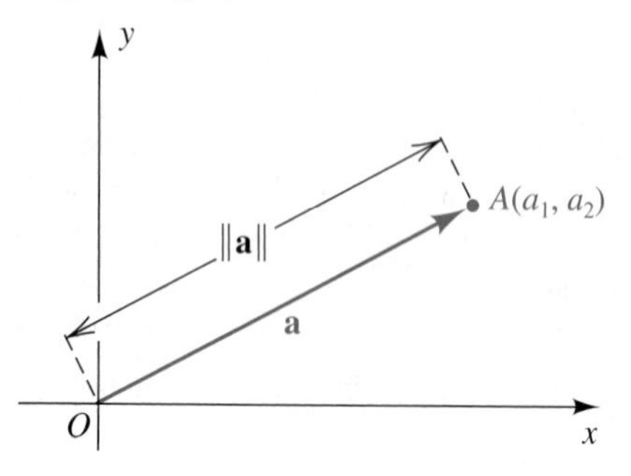

The preceding discussion shows that vectors have two different natures, one geometric and the other algebraic. Often we do not distinguish between the two. It should always be clear from our discussion whether we are referring to ordered pairs or directed line segments.

The *magnitude* of the vector $\mathbf{a} = \langle a_1, a_2 \rangle$ is, by definition, the length of its position vector OA, as illustrated in Figure 8.

Definition of the Magnitude of a Vector	The **magnitude** of the vector $\mathbf{a} = \langle a_1, a_2 \rangle$, denoted by $\|\mathbf{a}\|$, is given by $$\|\mathbf{a}\| = \|\langle a_1, a_2 \rangle\| = \sqrt{a_1^2 + a_2^2}.$$

EXAMPLE 1 Finding the magnitude of a vector

Sketch the vectors

$$\mathbf{a} = \langle -3, 2 \rangle, \quad \mathbf{b} = \langle 0, -2 \rangle, \quad \mathbf{c} = \left\langle \tfrac{4}{5}, \tfrac{3}{5} \right\rangle$$

on a coordinate plane, and find the magnitude of each vector.

Figure 9

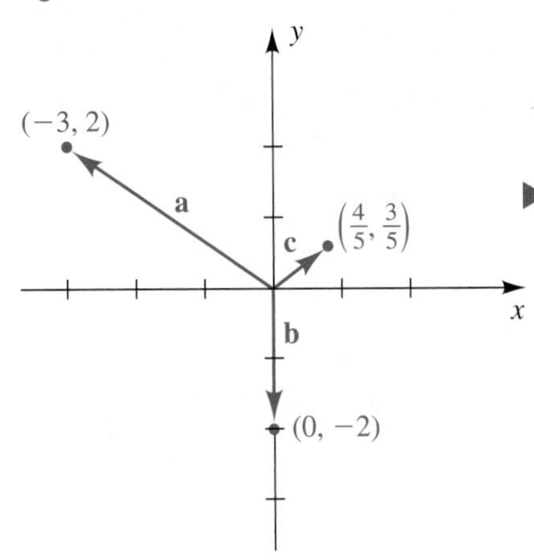

▶ SOLUTION The vectors are sketched in Figure 9. By the definition of the magnitude of a vector,

$$\|\mathbf{a}\| = \|\langle -3, 2 \rangle\| = \sqrt{(-3)^2 + 2^2} = \sqrt{13}$$
$$\|\mathbf{b}\| = \|\langle 0, -2 \rangle\| = \sqrt{0^2 + (-2)^2} = \sqrt{4} = 2$$
$$\|\mathbf{c}\| = \left\|\left\langle \tfrac{4}{5}, \tfrac{3}{5} \right\rangle\right\| = \sqrt{\left(\tfrac{4}{5}\right)^2 + \left(\tfrac{3}{5}\right)^2} = \sqrt{\tfrac{16}{25} + \tfrac{9}{25}} = \sqrt{\tfrac{25}{25}} = 1.$$

Consider the vector OA and the vector OB corresponding to $\mathbf{a} = \langle a_1, a_2 \rangle$ and $\mathbf{b} = \langle b_1, b_2 \rangle$, respectively, as illustrated in Figure 10. If $\overrightarrow{OC}$ corresponds to $\mathbf{c} = \langle a_1 + b_1, a_2 + b_2 \rangle$, we can show, using slopes, that the points O, A, C, and B are vertices of a parallelogram; that is,

$$\overrightarrow{OA} + \overrightarrow{OB} = \overrightarrow{OC}.$$

Figure 10

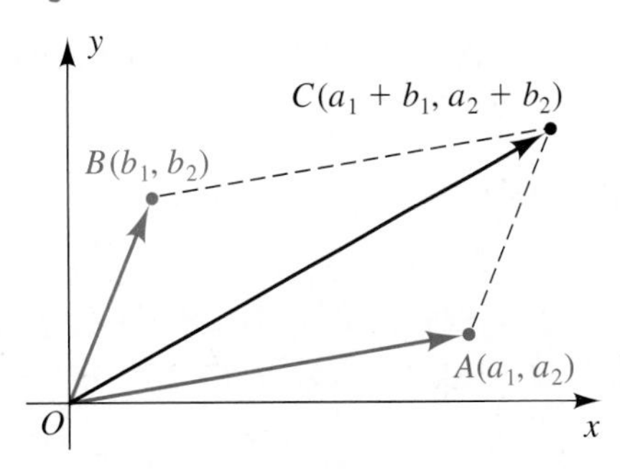

Expressing this equation in terms of ordered pairs leads to the following.

Definition of Addition of Vectors	$\langle a_1, a_2 \rangle + \langle b_1, b_2 \rangle = \langle a_1 + b_1, a_2 + b_2 \rangle$

Note that to add two vectors, we add corresponding components.

ILLUSTRATION **Addition of Vectors**

- $\langle 3, -4\rangle + \langle 2, 7\rangle = \langle 3 + 2, -4 + 7\rangle = \langle 5, 3\rangle$
- $\langle 5, 1\rangle + \langle -5, 1\rangle = \langle 5 + (-5), 1 + 1\rangle = \langle 0, 2\rangle$

It can also be shown that if m is a scalar and $\overrightarrow{OA}$ corresponds to $\mathbf{a} = \langle a_1, a_2\rangle$, then the ordered pair determined by $m\overrightarrow{OA}$ is (ma_1, ma_2), as illustrated in Figure 11 for $m > 1$. This leads to the next definition.

Figure 11

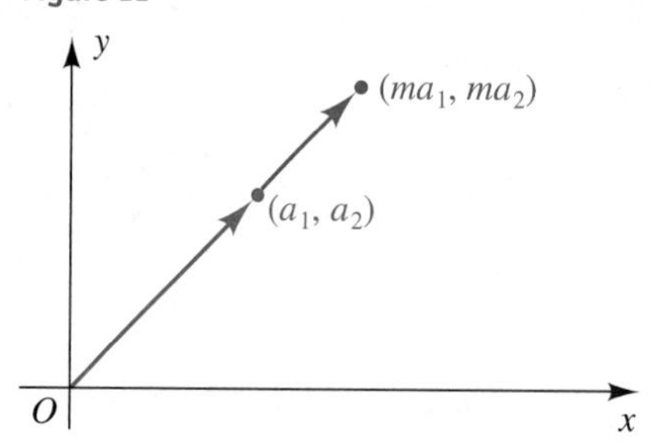

Definition of a Scalar Multiple of a Vector	$m\langle a_1, a_2\rangle = \langle ma_1, ma_2\rangle$

Thus, to find a scalar multiple of a vector, we multiply each component by the scalar.

ILLUSTRATION **Scalar Multiple of a Vector**

- $2\langle -3, 4\rangle = \langle 2(-3), 2(4)\rangle = \langle -6, 8\rangle$
- $-2\langle -3, 4\rangle = \langle (-2)(-3), (-2)(4)\rangle = \langle 6, -8\rangle$
- $1\langle 5, 2\rangle = \langle 1 \cdot 5, 1 \cdot 2\rangle = \langle 5, 2\rangle$

EXAMPLE 2 **Finding a scalar multiple of a vector**

If $\mathbf{a} = \langle 2, 1\rangle$, find $3\mathbf{a}$ and $-2\mathbf{a}$, and sketch each vector in a coordinate plane.

▶ SOLUTION Using the definition of scalar multiples of vectors, we find

$$3\mathbf{a} = 3\langle 2, 1\rangle = \langle 3 \cdot 2, 3 \cdot 1\rangle = \langle 6, 3\rangle$$
$$-2\mathbf{a} = -2\langle 2, 1\rangle = \langle (-2) \cdot 2, (-2) \cdot 1\rangle = \langle -4, -2\rangle.$$

The vectors are sketched in Figure 12 on the next page.

▶ **Tutorial available at www.thomsonedu.com/login**

Figure 12

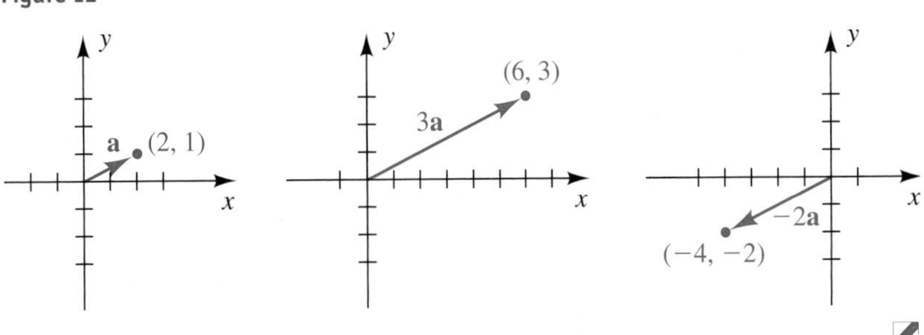

The **zero vector 0** and the **negative −a** of a vector $\mathbf{a} = \langle a_1, a_2 \rangle$ are defined as follows.

Definition of 0 and −a	$\mathbf{0} = \langle 0, 0 \rangle$ and $-\mathbf{a} = -\langle a_1, a_2 \rangle = \langle -a_1, -a_2 \rangle$

ILLUSTRATION **The Zero Vector and the Negative of a Vector**

- $\langle 3, 5 \rangle + \mathbf{0} = \langle 3, 5 \rangle + \langle 0, 0 \rangle = \langle 3 + 0, 5 + 0 \rangle = \langle 3, 5 \rangle$
- $-\langle 3, -5 \rangle = \langle -3, -(-5) \rangle = \langle -3, 5 \rangle$
- $\langle 3, -5 \rangle + \langle -3, 5 \rangle = \langle 3 + (-3), -5 + 5 \rangle = \langle 0, 0 \rangle = \mathbf{0}$
- $0\langle 2, 3 \rangle = \langle 0 \cdot 2, 0 \cdot 3 \rangle = \langle 0, 0 \rangle = \mathbf{0}$
- $5 \cdot \mathbf{0} = 5\langle 0, 0 \rangle = \langle 5 \cdot 0, 5 \cdot 0 \rangle = \langle 0, 0 \rangle = \mathbf{0}$

We next state properties of addition and scalar multiples of vectors for any vectors **a**, **b**, **c** and scalars m, n. You should have little difficulty in remembering these properties, since they resemble familiar properties of real numbers.

Properties of Addition and Scalar Multiples of Vectors		
	(1) $\mathbf{a} + \mathbf{b} = \mathbf{b} + \mathbf{a}$	(5) $m(\mathbf{a} + \mathbf{b}) = m\mathbf{a} + m\mathbf{b}$
	(2) $\mathbf{a} + (\mathbf{b} + \mathbf{c}) = (\mathbf{a} + \mathbf{b}) + \mathbf{c}$	(6) $(m + n)\mathbf{a} = m\mathbf{a} + n\mathbf{a}$
	(3) $\mathbf{a} + \mathbf{0} = \mathbf{a}$	(7) $(mn)\mathbf{a} = m(n\mathbf{a}) = n(m\mathbf{a})$
	(4) $\mathbf{a} + (-\mathbf{a}) = \mathbf{0}$	(8) $1\mathbf{a} = \mathbf{a}$
		(9) $0\mathbf{a} = \mathbf{0} = m\mathbf{0}$

PROOFS Let $\mathbf{a} = \langle a_1, a_2 \rangle$ and $\mathbf{b} = \langle b_1, b_2 \rangle$. To prove property 1, we note that

$$\mathbf{a} + \mathbf{b} = \langle a_1 + b_1, a_2 + b_2 \rangle = \langle b_1 + a_1, b_2 + a_2 \rangle = \mathbf{b} + \mathbf{a}.$$

(continued)

The proof of property 5 is as follows:

$$\begin{aligned} m(\mathbf{a} + \mathbf{b}) &= m\langle a_1 + b_1, a_2 + b_2\rangle && \text{definition of addition} \\ &= \langle m(a_1 + b_1), m(a_2 + b_2)\rangle && \text{definition of scalar multiple} \\ &= \langle ma_1 + mb_1, ma_2 + mb_2\rangle && \text{distributive property} \\ &= \langle ma_1, ma_2\rangle + \langle mb_1, mb_2\rangle && \text{definition of addition} \\ &= m\mathbf{a} + m\mathbf{b} && \text{definition of scalar multiple} \end{aligned}$$

Proofs of the remaining properties are similar and are left as exercises.

Vector subtraction (denoted by $-$) is defined by $\mathbf{a} - \mathbf{b} = \mathbf{a} + (-\mathbf{b})$. If we use the ordered pair notation for **a** and **b**, then $-\mathbf{b} = \langle -b_1, -b_2\rangle$, and we obtain the following.

Definition of Subtraction of Vectors	$\mathbf{a} - \mathbf{b} = \langle a_1, a_2\rangle - \langle b_1, b_2\rangle = \langle a_1 - b_1, a_2 - b_2\rangle$

Thus, to find $\mathbf{a} - \mathbf{b}$, we merely subtract the components of **b** from the corresponding components of **a**.

ILLUSTRATION **Subtraction of Vectors** If $\mathbf{a} = \langle 5, -4\rangle$ and $\mathbf{b} = \langle -3, 2\rangle$

- $\mathbf{a} - \mathbf{b} = \langle 5, -4\rangle - \langle -3, 2\rangle$
  $= \langle 5 - (-3), -4 - 2\rangle = \langle 8, -6\rangle$
- $2\mathbf{a} - 3\mathbf{b} = 2\langle 5, -4\rangle - 3\langle -3, 2\rangle$
  $= \langle 10, -8\rangle - \langle -9, 6\rangle = \langle 10 - (-9), -8 - 6\rangle = \langle 19, -14\rangle$

Figure 13

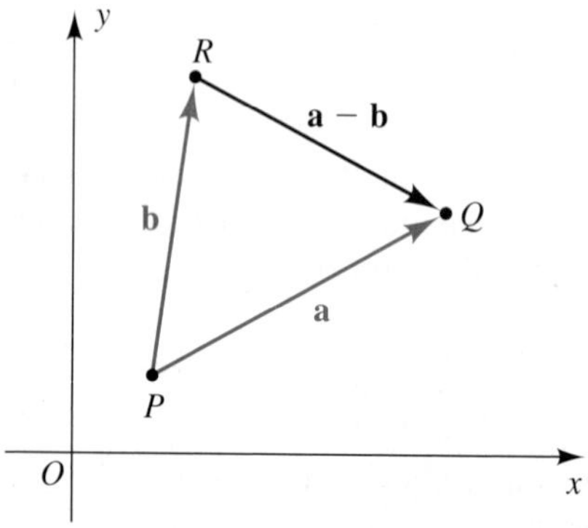

If **a** and **b** are arbitrary vectors, then

$$\mathbf{b} + (\mathbf{a} - \mathbf{b}) = \mathbf{a};$$

that is, $\mathbf{a} - \mathbf{b}$ *is the vector that, when added to* **b**, *gives us* **a**. If we represent **a** and **b** by vector PQ and vector PR *with the same initial point,* as in Figure 13, then $\overrightarrow{RQ}$ represents $\mathbf{a} - \mathbf{b}$.

Let's look at some of the operations with vectors on a graphing calculator.

TI-83/4 Plus

The TI-83/4 Plus does not have a specific vector mode, but *lists* will serve our purposes well. Visually, we merely replace the wedge notation used in this text by braces.

TI-86

The TI-86 *does* have a specific vector mode, with brackets used in place of wedges to denote a vector.

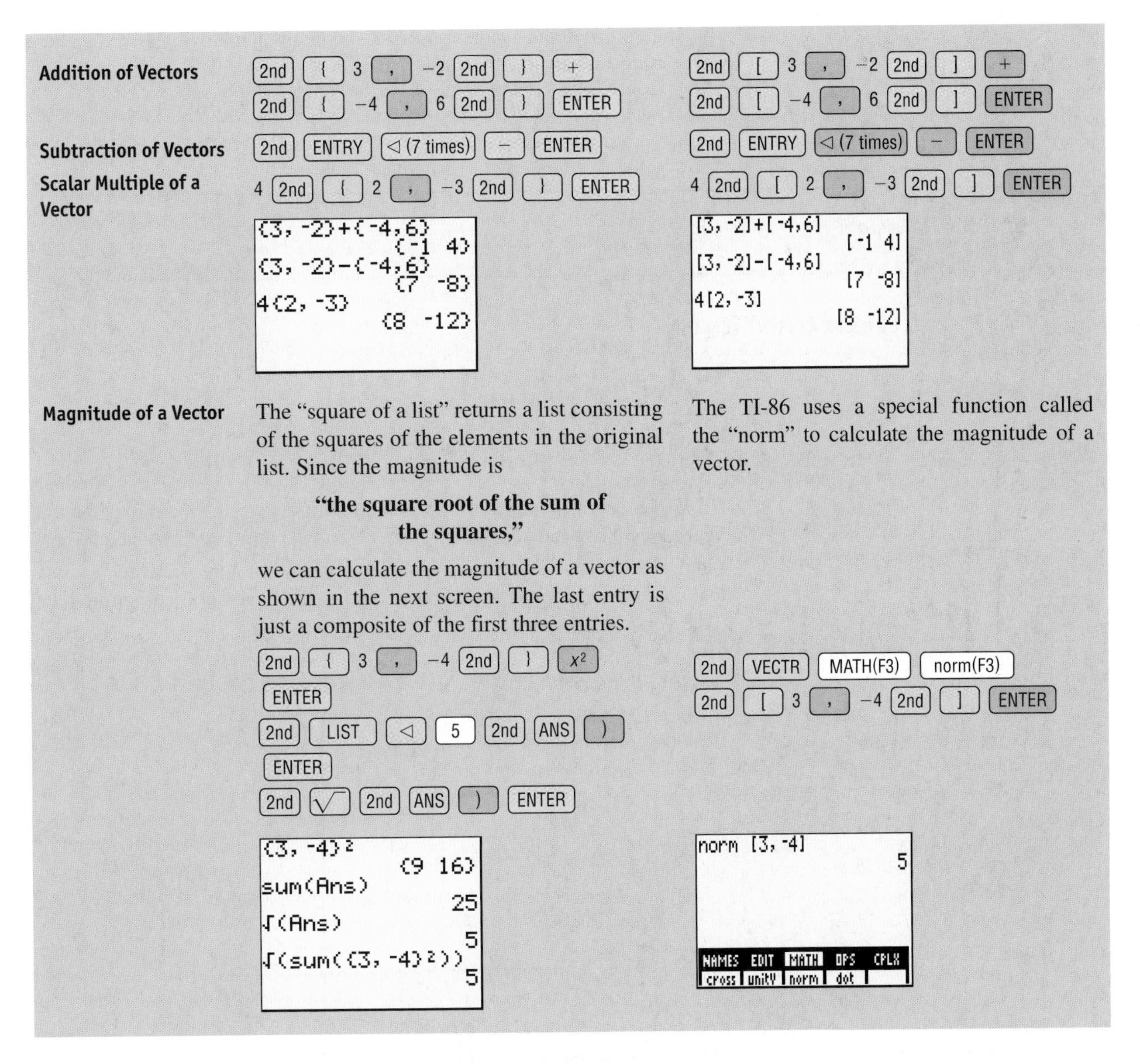

The special vectors **i** and **j** are defined as follows.

Definition of i and j	$\mathbf{i} = \langle 1, 0 \rangle, \qquad \mathbf{j} = \langle 0, 1 \rangle$

A **unit vector** is a vector of magnitude 1. The vectors **i** and **j** are unit vectors, as is the vector $\mathbf{c} = \left\langle \frac{4}{5}, \frac{3}{5} \right\rangle$ in Example 1.

The vectors **i** and **j** can be used to obtain an alternative way of denoting vectors. Specifically, if $\mathbf{a} = \langle a_1, a_2 \rangle$, then

$$\mathbf{a} = \langle a_1, 0 \rangle + \langle 0, a_2 \rangle = a_1\langle 1, 0 \rangle + a_2\langle 0, 1 \rangle.$$

This result gives us the following.

i, j Form for Vectors	$\mathbf{a} = \langle a_1, a_2 \rangle = a_1\mathbf{i} + a_2\mathbf{j}$

ILLUSTRATION **i, j Form**

- $\langle 5, 2 \rangle = 5\mathbf{i} + 2\mathbf{j}$
- $\langle -3, 4 \rangle = -3\mathbf{i} + 4\mathbf{j}$
- $\langle 0, -6 \rangle = 0\mathbf{i} + (-6)\mathbf{j} = -6\mathbf{j}$

Vectors corresponding to **i**, **j**, and an arbitrary vector **a** are illustrated in Figure 14. Since **i** and **j** are unit vectors, $a_1\mathbf{i}$ and $a_2\mathbf{j}$ may be represented by horizontal and vertical vectors of magnitudes $|a_1|$ and $|a_2|$, respectively, as illustrated in Figure 15. For this reason we call a_1 the **horizontal component** and a_2 the **vertical component** of the vector **a**.

Figure 14 $\mathbf{a} = \langle a_1, a_2 \rangle$

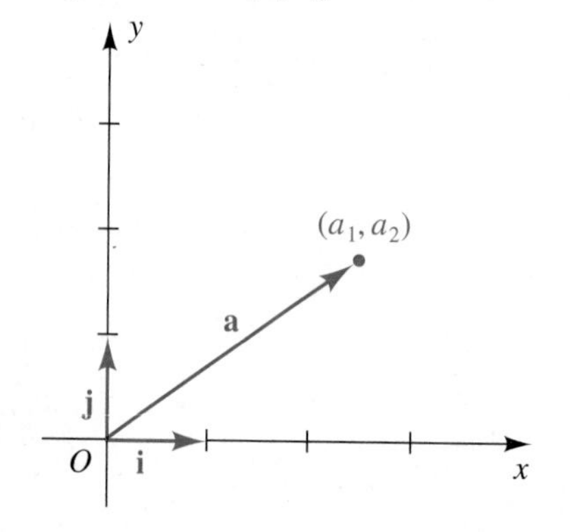

Figure 15 $\mathbf{a} = a_1\mathbf{i} + a_2\mathbf{j}$

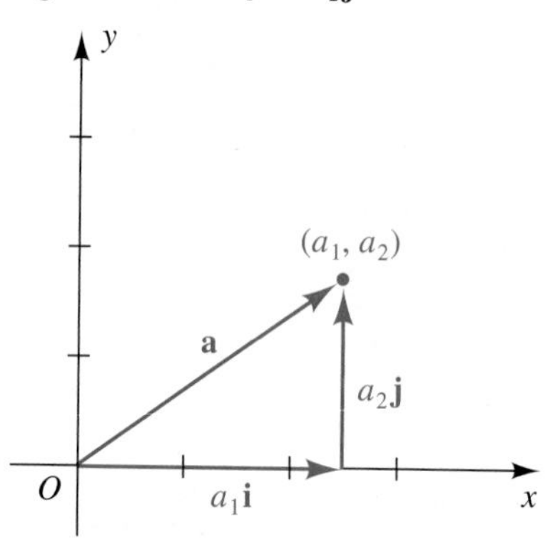

The vector sum $a_1\mathbf{i} + a_2\mathbf{j}$ is a **linear combination** of **i** and **j**. Rules for addition, subtraction, and multiplication by a scalar m may be written as follows, with $\mathbf{b} = \langle b_1, b_2 \rangle = b_1\mathbf{i} + b_2\mathbf{j}$:

$$(a_1\mathbf{i} + a_2\mathbf{j}) + (b_1\mathbf{i} + b_2\mathbf{j}) = (a_1 + b_1)\mathbf{i} + (a_2 + b_2)\mathbf{j}$$
$$(a_1\mathbf{i} + a_2\mathbf{j}) - (b_1\mathbf{i} + b_2\mathbf{j}) = (a_1 - b_1)\mathbf{i} + (a_2 - b_2)\mathbf{j}$$
$$m(a_1\mathbf{i} + a_2\mathbf{j}) = (ma_1)\mathbf{i} + (ma_2)\mathbf{j}$$

These formulas show that we may regard linear combinations of **i** and **j** as algebraic sums.

EXAMPLE 3 **Expressing a vector as a linear combination of i and j**

If $\mathbf{a} = 5\mathbf{i} + \mathbf{j}$ and $\mathbf{b} = 4\mathbf{i} - 7\mathbf{j}$, express $3\mathbf{a} - 2\mathbf{b}$ as a linear combination of $\mathbf{i}$ and $\mathbf{j}$.

▶ SOLUTION

$$\begin{aligned} 3\mathbf{a} - 2\mathbf{b} &= 3(5\mathbf{i} + \mathbf{j}) - 2(4\mathbf{i} - 7\mathbf{j}) \\ &= (15\mathbf{i} + 3\mathbf{j}) - (8\mathbf{i} - 14\mathbf{j}) \\ &= 7\mathbf{i} + 17\mathbf{j} \end{aligned}$$

Figure 16

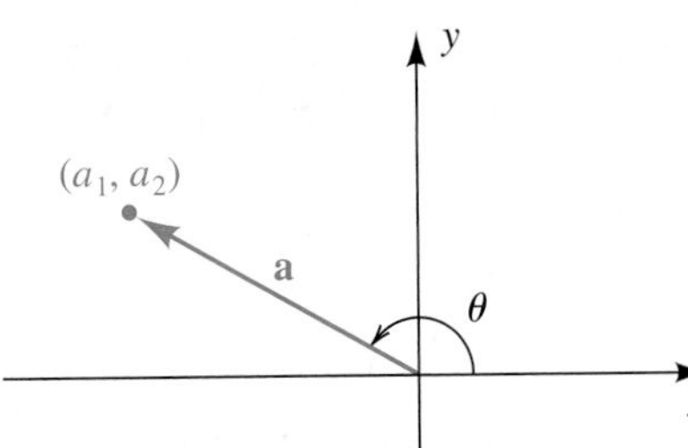

Let θ be an angle in standard position, measured from the positive x-axis to the vector $\mathbf{a} = \langle a_1, a_2 \rangle = a_1\mathbf{i} + a_2\mathbf{j}$, as illustrated in Figure 16. Since

$$\cos\theta = \frac{a_1}{\|\mathbf{a}\|} \quad \text{and} \quad \sin\theta = \frac{a_2}{\|\mathbf{a}\|},$$

we obtain the following formulas.

Formulas for Horizontal and Vertical Components of a = $\langle a_1, a_2 \rangle$	If the vector $\mathbf{a}$ and the angle θ are defined as above, then $$a_1 = \|\mathbf{a}\| \cos\theta \quad \text{and} \quad a_2 = \|\mathbf{a}\| \sin\theta.$$

Using these formulas, we have

$$\begin{aligned} \mathbf{a} = \langle a_1, a_2 \rangle &= \langle \|\mathbf{a}\| \cos\theta, \|\mathbf{a}\| \sin\theta \rangle \\ &= \|\mathbf{a}\| \cos\theta\,\mathbf{i} + \|\mathbf{a}\| \sin\theta\,\mathbf{j} \\ &= \|\mathbf{a}\|(\cos\theta\,\mathbf{i} + \sin\theta\,\mathbf{j}). \end{aligned}$$

EXAMPLE 4 **Expressing wind velocity as a vector**

If the wind is blowing at 12 mi/hr in the direction N40°W, express its velocity as a vector $\mathbf{v}$.

▶ SOLUTION The vector $\mathbf{v}$ and the angle $\theta = 90° + 40° = 130°$ are illustrated in Figure 17. Using the formulas for horizontal and vertical components with $\mathbf{v} = \langle v_1, v_2 \rangle$ gives us

$$v_1 = \|\mathbf{v}\| \cos\theta = 12 \cos 130°, \qquad v_2 = \|\mathbf{v}\| \sin\theta = 12 \sin 130°.$$

Figure 17

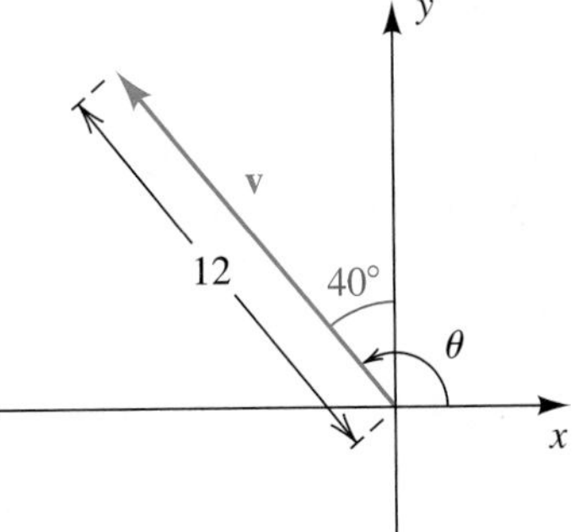

Hence,

$$\begin{aligned} \mathbf{v} &= v_1\mathbf{i} + v_2\mathbf{j} \\ &= (12 \cos 130°)\mathbf{i} + (12 \sin 130°)\mathbf{j} \\ &\approx (-7.7)\mathbf{i} + (9.2)\mathbf{j}. \end{aligned}$$

EXAMPLE 5 **Finding a vector of specified direction and magnitude**

Find a vector $\mathbf{b}$ in the opposite direction of $\mathbf{a} = \langle 5, -12 \rangle$ that has magnitude 6.

▶ **Tutorial available at www.thomsonedu.com/login**

Figure 18

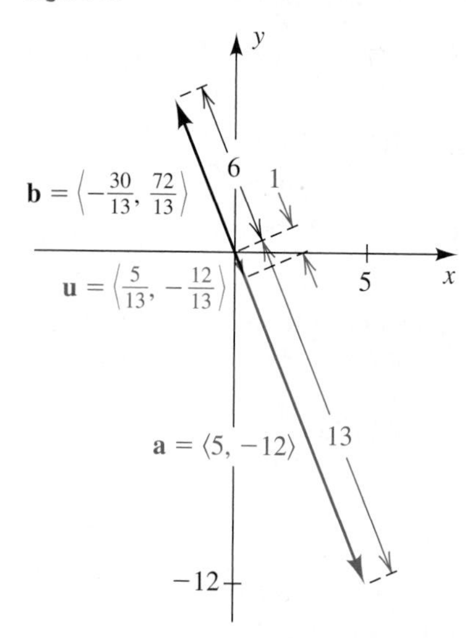

▶ SOLUTION The magnitude of **a** is given by

$$\|\mathbf{a}\| = \sqrt{5^2 + (-12)^2} = \sqrt{25 + 144} = \sqrt{169} = 13.$$

A unit vector **u** in the direction of **a** can be found by multiplying **a** by $1/\|\mathbf{a}\|$. Thus,

$$\mathbf{u} = \frac{1}{\|\mathbf{a}\|}\mathbf{a} = \frac{1}{13}\langle 5, -12 \rangle = \left\langle \frac{5}{13}, -\frac{12}{13} \right\rangle.$$

Multiplying **u** by 6 gives us a vector of magnitude 6 in the direction of **a**, so we'll multiply **u** by -6 to obtain the desired vector **b**, as shown in Figure 18:

$$\mathbf{b} = -6\mathbf{u} = -6\left\langle \frac{5}{13}, -\frac{12}{13} \right\rangle = \left\langle -\frac{30}{13}, \frac{72}{13} \right\rangle$$

EXAMPLE 6 Finding a resultant vector

Two forces $\overrightarrow{PQ}$ and $\overrightarrow{PR}$ of magnitudes 5.0 kilograms and 8.0 kilograms, respectively, act at a point P. The direction of $\overrightarrow{PQ}$ is N20°E, and the direction of $\overrightarrow{PR}$ is N65°E. Approximate the magnitude and direction of the resultant $\overrightarrow{PS}$.

Figure 19

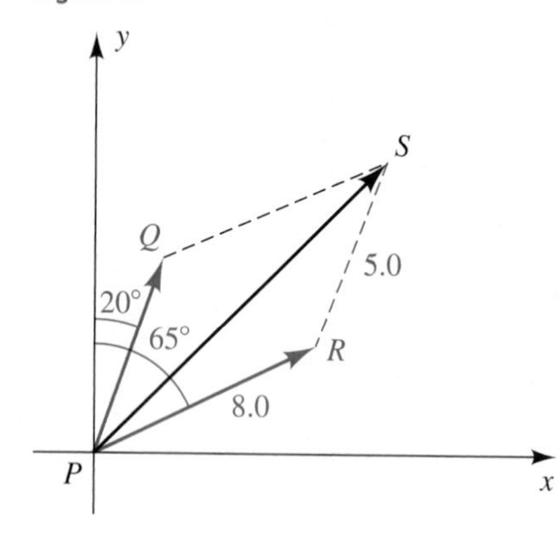

▶ SOLUTION The forces are represented geometrically in Figure 19. Note that the angles from the positive x-axis to $\overrightarrow{PQ}$ and $\overrightarrow{PR}$ have measures 70° and 25°, respectively. Using the formulas for horizontal and vertical components, we obtain the following:

$$\overrightarrow{PQ} = (5 \cos 70°)\mathbf{i} + (5 \sin 70°)\mathbf{j}$$
$$\overrightarrow{PR} = (8 \cos 25°)\mathbf{i} + (8 \sin 25°)\mathbf{j}$$

Since $\overrightarrow{PS} = \overrightarrow{PQ} + \overrightarrow{PR}$,

$$\begin{aligned}\overrightarrow{PS} &= (5 \cos 70° + 8 \cos 25°)\mathbf{i} + (5 \sin 70° + 8 \sin 25°)\mathbf{j} \\ &\approx 8.9606\mathbf{i} + 8.0794\mathbf{j} \approx (9.0)\mathbf{i} + (8.1)\mathbf{j}.\end{aligned}$$

Consequently,

$$\|\overrightarrow{PS}\| \approx \sqrt{(9.0)^2 + (8.1)^2} \approx 12.1.$$

We can also find $\|\overrightarrow{PS}\|$ by using the law of cosines (see Example 3 of Section 7.2). Since $\angle QPR = 45°$, it follows that $\angle PRS = 135°$, and hence

$$\|\overrightarrow{PS}\|^2 = (8.0)^2 + (5.0)^2 - 2(8.0)(5.0) \cos 135° \approx 145.6$$

and

$$\|\overrightarrow{PS}\| \approx \sqrt{145.6} \approx 12.1.$$

If θ is the angle from the positive x-axis to the resultant PS, then using the (approximate) coordinates (8.9606, 8.0794) of S, we obtain the following:

$$\tan \theta \approx \frac{8.0794}{8.9606} \approx 0.9017$$
$$\theta \approx \tan^{-1}(0.9017) \approx 42°$$

Hence, the direction of $\overrightarrow{PS}$ is approximately N(90° − 42°)E = N48°E.

▶ Tutorial available at www.thomsonedu.com/login

7.3 Exercises

Exer. 1–6: Find $\mathbf{a} + \mathbf{b}$, $\mathbf{a} - \mathbf{b}$, $4\mathbf{a} + 5\mathbf{b}$, $4\mathbf{a} - 5\mathbf{b}$, and $\|\mathbf{a}\|$.

1 $\mathbf{a} = \langle 2, -3\rangle$, $\mathbf{b} = \langle 1, 4\rangle$

2 $\mathbf{a} = \langle -2, 6\rangle$, $\mathbf{b} = \langle 2, 3\rangle$

3 $\mathbf{a} = -\langle 7, -2\rangle$, $\mathbf{b} = 4\langle -2, 1\rangle$

4 $\mathbf{a} = 2\langle 5, -4\rangle$, $\mathbf{b} = -\langle 6, 0\rangle$

5 $\mathbf{a} = \mathbf{i} + 2\mathbf{j}$, $\mathbf{b} = 3\mathbf{i} - 5\mathbf{j}$

6 $\mathbf{a} = -3\mathbf{i} + \mathbf{j}$, $\mathbf{b} = -3\mathbf{i} + \mathbf{j}$

Exer. 7–10: Sketch vectors corresponding to $\mathbf{a}$, $\mathbf{b}$, $\mathbf{a} + \mathbf{b}$, $2\mathbf{a}$, and $-3\mathbf{b}$.

7 $\mathbf{a} = 3\mathbf{i} + 2\mathbf{j}$, $\mathbf{b} = -\mathbf{i} + 5\mathbf{j}$

8 $\mathbf{a} = -5\mathbf{i} + 2\mathbf{j}$, $\mathbf{b} = \mathbf{i} - 3\mathbf{j}$

9 $\mathbf{a} = \langle -4, 6\rangle$, $\mathbf{b} = \langle -2, 3\rangle$

10 $\mathbf{a} = \langle 2, 0\rangle$, $\mathbf{b} = \langle -2, 0\rangle$

Exer. 11–16: Use components to express the sum or difference as a scalar multiple of one of the vectors a, b, c, d, e, or f shown in the figure.

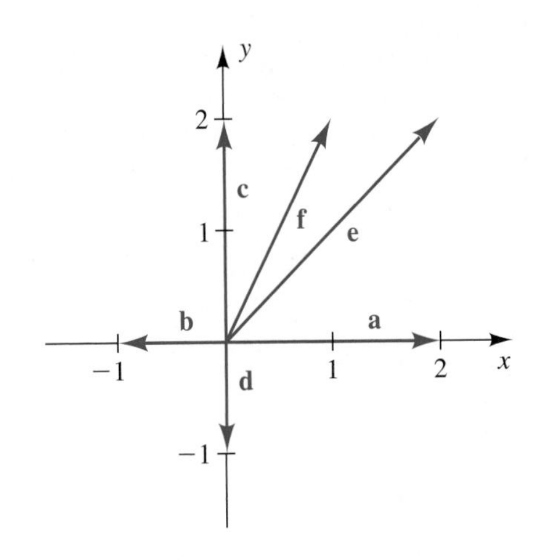

11 $\mathbf{a} + \mathbf{b}$

12 $\mathbf{c} - \mathbf{d}$

13 $\mathbf{b} + \mathbf{e}$

14 $\mathbf{f} - \mathbf{b}$

15 $\mathbf{b} + \mathbf{d}$

16 $\mathbf{e} + \mathbf{c}$

Exer. 17–26: If $\mathbf{a} = \langle a_1, a_2\rangle$, $\mathbf{b} = \langle b_1, b_2\rangle$, $\mathbf{c} = \langle c_1, c_2\rangle$, and m and n are real numbers, prove the stated property.

17 $\mathbf{a} + (\mathbf{b} + \mathbf{c}) = (\mathbf{a} + \mathbf{b}) + \mathbf{c}$

18 $\mathbf{a} + \mathbf{0} = \mathbf{a}$

19 $\mathbf{a} + (-\mathbf{a}) = \mathbf{0}$

20 $(m + n)\mathbf{a} = m\mathbf{a} + n\mathbf{a}$

21 $(mn)\mathbf{a} = m(n\mathbf{a}) = n(m\mathbf{a})$

22 $1\mathbf{a} = \mathbf{a}$

23 $0\mathbf{a} = \mathbf{0} = m\mathbf{0}$

24 $(-m)\mathbf{a} = -m\mathbf{a}$

25 $-(\mathbf{a} + \mathbf{b}) = -\mathbf{a} - \mathbf{b}$

26 $m(\mathbf{a} - \mathbf{b}) = m\mathbf{a} - m\mathbf{b}$

27 If $\mathbf{v} = \langle a, b\rangle$, prove that the magnitude of $2\mathbf{v}$ is twice the magnitude of $\mathbf{v}$.

28 If $\mathbf{v} = \langle a, b\rangle$ and k is any real number, prove that the magnitude of $k\mathbf{v}$ is $|k|$ times the magnitude of $\mathbf{v}$.

Exer. 29–36: Find the magnitude of the vector a and the smallest positive angle θ from the positive x-axis to the vector OP that corresponds to a.

29 $\mathbf{a} = \langle 3, -3\rangle$

30 $\mathbf{a} = \langle -2, -2\sqrt{3}\rangle$

31 $\mathbf{a} = \langle -5, 0\rangle$

32 $\mathbf{a} = \langle 0, 10\rangle$

▶ 33 $\mathbf{a} = -4\mathbf{i} + 5\mathbf{j}$

34 $\mathbf{a} = 10\mathbf{i} - 10\mathbf{j}$

35 $\mathbf{a} = -18\mathbf{j}$

36 $\mathbf{a} = 2\mathbf{i} - 3\mathbf{j}$

Exer. 37–40: The vectors a and b represent two forces acting at the same point, and θ is the smallest positive angle between a and b. Approximate the magnitude of the resultant force.

▶ 37 $\|\mathbf{a}\| = 40$ lb, $\|\mathbf{b}\| = 70$ lb, $\theta = 45°$

38 $\|\mathbf{a}\| = 5.5$ lb, $\|\mathbf{b}\| = 6.2$ lb, $\theta = 60°$

39 $\|\mathbf{a}\| = 2.0$ lb, $\|\mathbf{b}\| = 8.0$ lb, $\theta = 120°$

40 $\|\mathbf{a}\| = 30$ lb, $\|\mathbf{b}\| = 50$ lb, $\theta = 150°$

▶ Tutorial available at www.thomsonedu.com/login

Exer. 41–44: The magnitudes and directions of two forces acting at a point P are given in (a) and (b). Approximate the magnitude and direction of the resultant vector.

41 (a) 90 lb, N75°W (b) 60 lb, S5°E

42 (a) 20 lb, S17°W (b) 50 lb, N82°W

43 (a) 6.0 lb, 110° (b) 2.0 lb, 215°

44 (a) 70 lb, 320° (b) 40 lb, 30°

Exer. 45–48: Approximate the horizontal and vertical components of the vector that is described.

▶ 45 Releasing a football A quarterback releases a football with a speed of 50 ft/sec at an angle of 35° with the horizontal.

46 Pulling a sled A child pulls a sled through the snow by exerting a force of 20 pounds at an angle of 40° with the horizontal.

47 Biceps muscle The biceps muscle, in supporting the forearm and a weight held in the hand, exerts a force of 20 pounds. As shown in the figure, the muscle makes an angle of 108° with the forearm.

Exercise 47

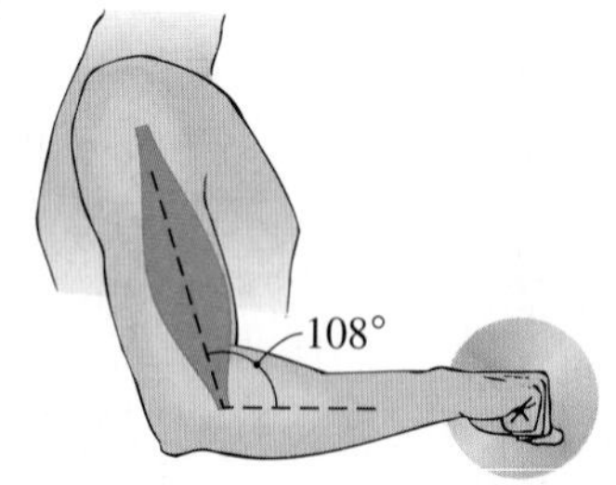

48 Jet's approach A jet airplane approaches a runway at an angle of 7.5° with the horizontal, traveling at a speed of 160 mi/hr.

Exer. 49–52: Find a unit vector that has (a) the same direction as the vector a and (b) the opposite direction of the vector a.

49 $\mathbf{a} = -8\mathbf{i} + 15\mathbf{j}$ ▶ 50 $\mathbf{a} = 5\mathbf{i} - 3\mathbf{j}$

51 $\mathbf{a} = \langle 2, -5 \rangle$ 52 $\mathbf{a} = \langle 0, 6 \rangle$

53 Find a vector that has the same direction as $\langle -6, 3 \rangle$ and

(a) twice the magnitude

(b) one-half the magnitude

54 Find a vector that has the opposite direction of $8\mathbf{i} - 5\mathbf{j}$ and

(a) three times the magnitude

(b) one-third the magnitude

55 Find a vector of magnitude 6 that has the opposite direction of $\mathbf{a} = 4\mathbf{i} - 7\mathbf{j}$.

56 Find a vector of magnitude 4 that has the opposite direction of $\mathbf{a} = \langle 2, -5 \rangle$.

Exer. 57–60: If forces $\mathbf{F}_1, \mathbf{F}_2, \ldots, \mathbf{F}_n$ act at a point P, the net (or resultant) force $\mathbf{F}$ is the sum $\mathbf{F}_1 + \mathbf{F}_2 + \cdots + \mathbf{F}_n$. If $\mathbf{F} = \mathbf{0}$, the forces are said to be in equilibrium. The given forces act at the origin O of an xy-plane.

(a) Find the net force F.

(b) Find an additional force G such that equilibrium occurs.

57 $\mathbf{F}_1 = \langle 4, 3 \rangle$, $\mathbf{F}_2 = \langle -2, -3 \rangle$, $\mathbf{F}_3 = \langle 5, 2 \rangle$

58 $\mathbf{F}_1 = \langle -3, -1 \rangle$, $\mathbf{F}_2 = \langle 0, -3 \rangle$, $\mathbf{F}_3 = \langle 3, 4 \rangle$

59

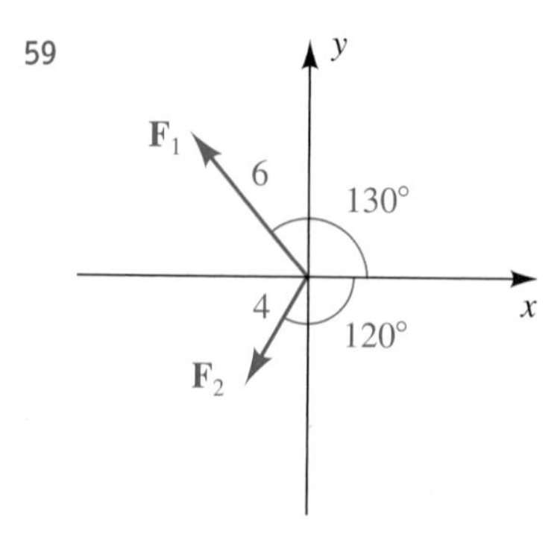

▶ Tutorial available at www.thomsonedu.com/login

60

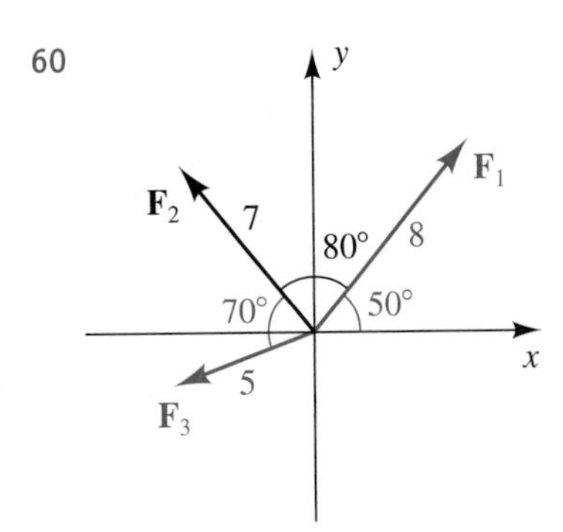

61 **Tugboat force** Two tugboats are towing a large ship into port, as shown in the figure. The larger tug exerts a force of 4000 pounds on its cable, and the smaller tug exerts a force of 3200 pounds on its cable. If the ship is to travel on a straight line l, approximate the angle θ that the larger tug must make with l.

Exercise 61

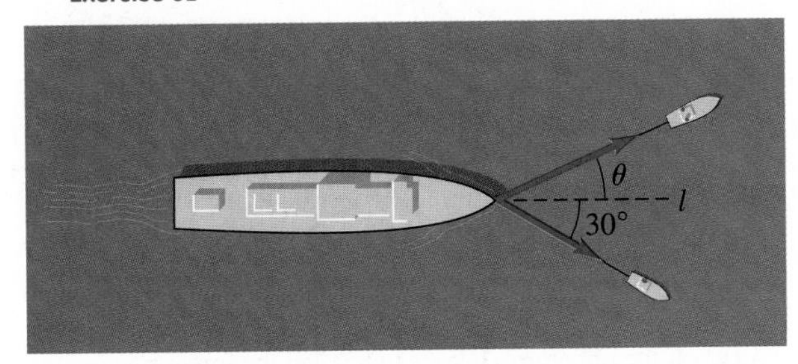

62 **Gravity simulation** Shown in the figure is a simple apparatus that may be used to simulate gravity conditions on other planets. A rope is attached to an astronaut who maneuvers on an inclined plane that makes an angle of θ degrees with the horizontal.

(a) If the astronaut weighs 160 pounds, find the x- and y-components of the downward force (see the figure for axes).

(b) The y-component in part (a) is the weight of the astronaut relative to the inclined plane. The astronaut would weigh 27 pounds on the moon and 60 pounds on Mars. Approximate the angles θ (to the nearest 0.01°) so that the inclined-plane apparatus will simulate walking on these surfaces.

Exercise 62

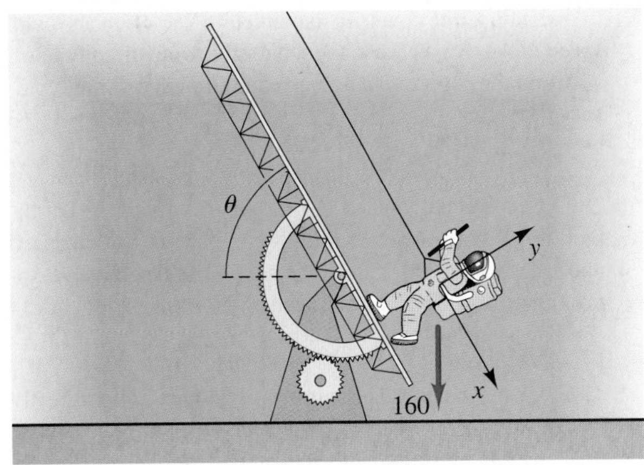

63 **Airplane course and ground speed** An airplane with an airspeed of 200 mi/hr is flying in the direction 50°, and a 40 mi/hr wind is blowing directly from the west. As shown in the figure, these facts may be represented by vectors **p** and **w** of magnitudes 200 and 40, respectively. The direction of the resultant **p** + **w** gives the true course of the airplane relative to the ground, and the magnitude $\|\mathbf{p} + \mathbf{w}\|$ is the ground speed of the airplane. Approximate the true course and ground speed.

Exercise 63

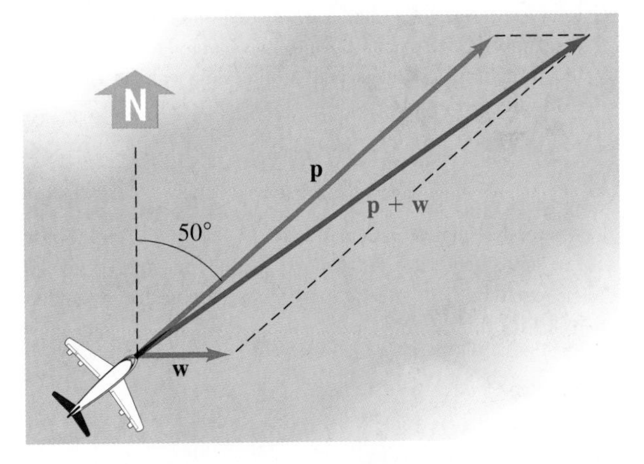

64 **Airplane course and ground speed** Refer to Exercise 63. An airplane is flying in the direction 140° with an airspeed of 500 mi/hr, and a 30 mi/hr wind is blowing in the direction 65°. Approximate the true course and ground speed of the airplane.

65 **Airplane course and ground speed** An airplane pilot wishes to maintain a true course in the direction 250° with a ground speed of 400 mi/hr when the wind is blowing directly north at 50 mi/hr. Approximate the required airspeed and compass heading.

66 **Wind direction and speed** An airplane is flying in the direction 20° with an airspeed of 300 mi/hr. Its ground speed and true course are 350 mi/hr and 30°, respectively. Approximate the direction and speed of the wind.

67 **Rowboat navigation** The current in a river flows directly from the west at a rate of 1.5 ft/sec. A person who rows a boat at a rate of 4 ft/sec in still water wishes to row directly north across the river. Approximate, to the nearest degree, the direction in which the person should row.

68 **Motorboat navigation** For a motorboat moving at a speed of 30 mi/hr to travel directly north across a river, it must aim at a point that has the bearing N15°E. If the current is flowing directly west, approximate the rate at which it flows.

69 **Flow of ground water** Ground-water contaminants can enter a community's drinking water by migrating through porous rock into the aquifer. If underground water flows with a velocity $\mathbf{v}_1$ through an interface between one type of rock and a second type of rock, its velocity changes to $\mathbf{v}_2$, and both the direction and the speed of the flow can be obtained using the formula

$$\frac{\|\mathbf{v}_1\|}{\|\mathbf{v}_2\|} = \frac{\tan \theta_1}{\tan \theta_2},$$

where the angles θ_1 and θ_2 are as shown in the figure. For sandstone, $\|\mathbf{v}_1\| = 8.2$ cm/day; for limestone, $\|\mathbf{v}_2\| = 3.8$ cm/day. If $\theta_1 = 30°$, approximate the vectors $\mathbf{v}_1$ and $\mathbf{v}_2$ in $\mathbf{i}, \mathbf{j}$ form.

Exercise 69

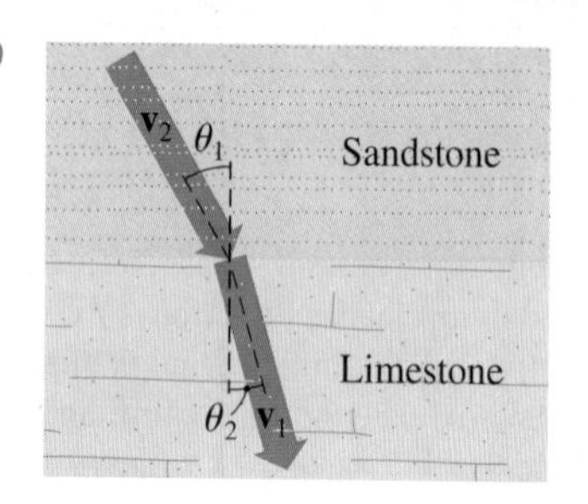

70 **Flow of ground water** Refer to Exercise 69. Contaminated ground water is flowing through silty sand with the direction of flow θ_1 and speed (in cm/day) given by the vector $\mathbf{v}_1 = 20\mathbf{i} - 82\mathbf{j}$. When the flow enters a region of clean sand, its rate increases to 725 cm/day. Find the new direction of flow by approximating θ_2.

71 **Robotic movement** Vectors are useful for describing movement of robots.

(a) The robot's arm illustrated in the first figure can rotate at the joint connections P and Q. The upper arm, represented by $\mathbf{a}$, is 15 inches long, and the forearm (including the hand), represented by $\mathbf{b}$, is 17 inches long. Approximate the coordinates of the point R in the hand by using $\mathbf{a} + \mathbf{b}$.

Exercise 71(a)

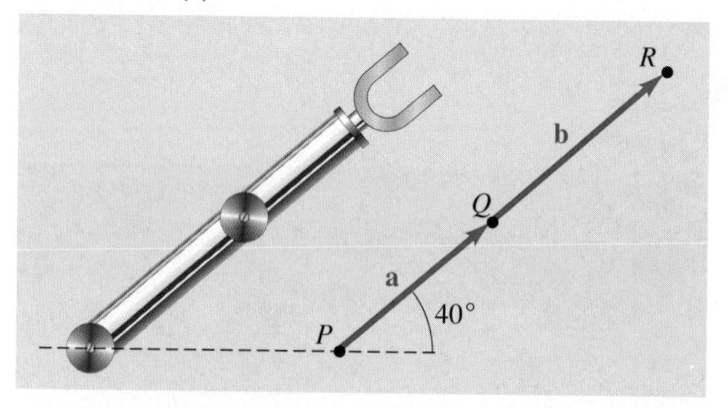

(b) If the upper arm is rotated 85° and the forearm is rotated an additional 35°, as illustrated in the second figure, approximate the new coordinates of R by using $\mathbf{c} + \mathbf{d}$.

Exercise 71(b)

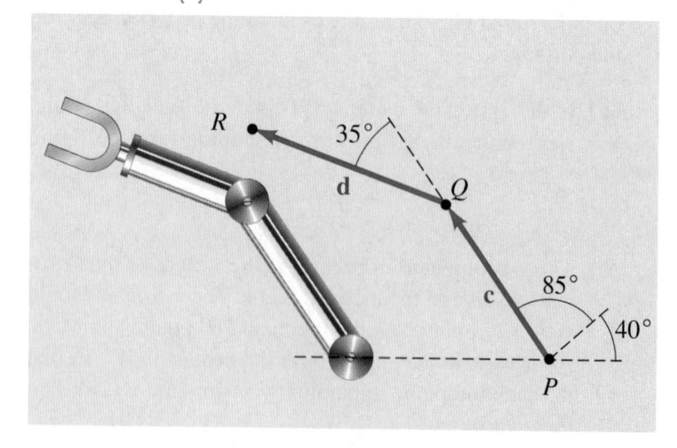

72 Robotic movement Refer to Exercise 71.

(a) Suppose the wrist joint of the robot's arm is allowed to rotate at the joint connection S and the arm is located as shown in the first figure. The upper arm has a length of 15 inches; the forearm, without the hand, has a length of 10 inches; and the hand has a length of 7 inches. Approximate the coordinates of R by using $\mathbf{a} + \mathbf{b} + \mathbf{c}$.

Exercise 72(a)

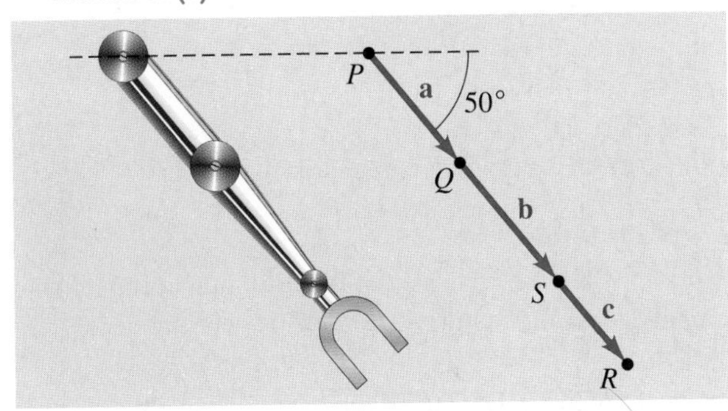

(b) Suppose the robot's upper arm is rotated 75°, and then the forearm is rotated −80°, and finally the hand is rotated an additional 40°, as shown in the second figure. Approximate the new coordinates of R by using $\mathbf{d} + \mathbf{e} + \mathbf{f}$.

Exercise 72(b)

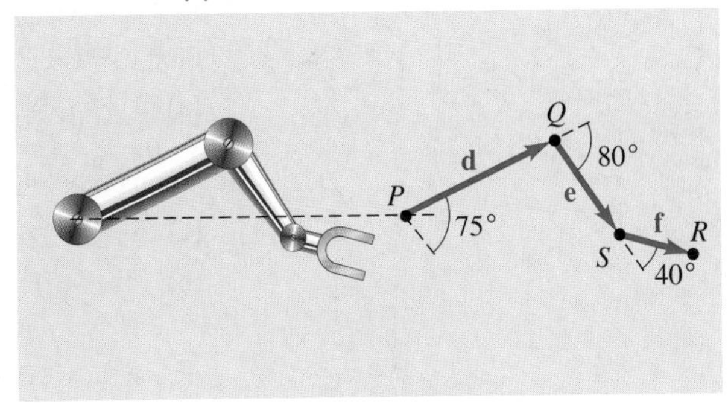

73 Stonehenge forces Refer to Exercise 25 in Section 5.2. In the construction of Stonehenge, groups of 550 people were used to pull 99,000-pound blocks of stone up ramps inclined at 9°. Ignoring friction, determine the force that each person had to contribute in order to move the stone up the ramp.

Exercise 73

550 people

9°

7.4 The Dot Product

The *dot product* of two vectors has many applications. We begin with an algebraic definition.

Definition of the Dot Product	Let $\mathbf{a} = \langle a_1, a_2 \rangle = a_i\mathbf{i} + a_2\mathbf{j}$ and $\mathbf{b} = \langle b_1, b_2 \rangle = b_i\mathbf{i} + b_2\mathbf{j}$. The **dot product** of **a** and **b**, denoted $\mathbf{a} \cdot \mathbf{b}$, is $$\mathbf{a} \cdot \mathbf{b} = \langle a_1, a_2 \rangle \cdot \langle b_1, b_2 \rangle = a_1b_1 + a_2b_2.$$

The symbol $\mathbf{a} \cdot \mathbf{b}$ is read "**a** dot **b**." We also refer to the dot product as the **scalar product** or the **inner product.** Note that $\mathbf{a} \cdot \mathbf{b}$ *is a real number and not a vector,* as illustrated in the following example.

EXAMPLE 1 **Finding the dot product of two vectors**

Find $\mathbf{a} \cdot \mathbf{b}$.

(a) $\mathbf{a} = \langle -5, 3 \rangle, \quad \mathbf{b} = \langle 2, 6 \rangle$ (b) $\mathbf{a} = 4\mathbf{i} + 6\mathbf{j}, \quad \mathbf{b} = 3\mathbf{i} - 7\mathbf{j}$

▶ SOLUTION

(a) $\langle -5, 3 \rangle \cdot \langle 2, 6 \rangle = (-5)(2) + (3)(6) = -10 + 18 = 8$

(b) $(4\mathbf{i} + 6\mathbf{j}) \cdot (3\mathbf{i} - 7\mathbf{j}) = (4)(3) + (6)(-7) = 12 - 42 = -30$

Finding a Dot Product

Let's find the dot product in Example 1(a) on a graphing calculator.

TI-83/4 Plus

The product of the lists $\{a_1, a_2\}$ and $\{b_1, b_2\}$ is the list $\{a_1b_1, a_2b_2\}$. Summing these elements gives us the dot product.

[2nd] [{] −5 [,] 3 [2nd] [}] [×]
[2nd] [{] 2 [,] 6 [2nd] [}] [ENTER]
[2nd] [LIST] [◁] [5] [2nd] [ANS] [)]
[ENTER]

```
{-5,3}*{2,6}
          {-10 18}
sum(Ans)
                 8
```

TI-86

The TI-86 has a dot function that accepts two vectors and returns their dot product.

[2nd] [VECTR] [MATH(F3)] [dot(F4)]
[2nd] [[] −5 [,] 3 [2nd] []] [,]
[2nd] [[] 2 [,] 6 [2nd] []] [)] [ENTER]

```
dot([-5,3],[2,6])
                 8
NAMES EDIT MATH OPS CPLX
cross unitV norm dot
```

Properties of the Dot Product

If **a**, **b**, and **c** are vectors and m is a real number, then

(1) $\mathbf{a} \cdot \mathbf{a} = \|\mathbf{a}\|^2$

(2) $\mathbf{a} \cdot \mathbf{b} = \mathbf{b} \cdot \mathbf{a}$

(3) $\mathbf{a} \cdot (\mathbf{b} + \mathbf{c}) = \mathbf{a} \cdot \mathbf{b} + \mathbf{a} \cdot \mathbf{c}$

(4) $(m\mathbf{a}) \cdot \mathbf{b} = m(\mathbf{a} \cdot \mathbf{b}) = \mathbf{a} \cdot (m\mathbf{b})$

(5) $\mathbf{0} \cdot \mathbf{a} = 0$

▶ **Tutorial available at www.thomsonedu.com/login**

PROOF The proof of each property follows from the definition of the dot product and the properties of real numbers. Thus, if $\mathbf{a} = \langle a_1, a_2 \rangle$, $\mathbf{b} = \langle b_1, b_2 \rangle$, and $\mathbf{c} = \langle c_1, c_2 \rangle$, then

$$\begin{aligned} \mathbf{a} \cdot (\mathbf{b} + \mathbf{c}) &= \langle a_1, a_2 \rangle \cdot \langle b_1 + c_1, b_2 + c_2 \rangle && \text{definition of addition} \\ &= a_1(b_1 + c_1) + a_2(b_2 + c_2) && \text{definition of dot product} \\ &= (a_1b_1 + a_2b_2) + (a_1c_1 + a_2c_2) && \text{real number properties} \\ &= \mathbf{a} \cdot \mathbf{b} + \mathbf{a} \cdot \mathbf{c}, && \text{definition of dot product} \end{aligned}$$

which proves property 3. The proofs of the remaining properties are left as exercises.

Any two nonzero vectors $\mathbf{a} = \langle a_1, a_2 \rangle$ and $\mathbf{b} = \langle b_1, b_2 \rangle$ may be represented in a coordinate plane by directed line segments from the origin O to the points $A(a_1, a_2)$ and $B(b_1, b_2)$, respectively. The **angle θ between a and b** is, by definition, $\angle AOB$ (see Figure 1). Note that $0 \leq \theta \leq \pi$ and that $\theta = 0$ if **a** and **b** have the same direction or $\theta = \pi$ if **a** and **b** have opposite directions.

Figure 1

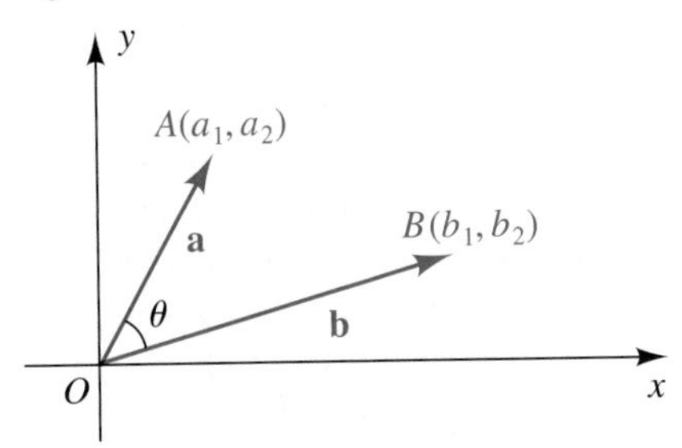

Definition of Parallel and Orthogonal Vectors

Let θ be the angle between two nonzero vectors **a** and **b**.

(1) **a** and **b** are **parallel** if $\theta = 0$ or $\theta = \pi$.

(2) **a** and **b** are **orthogonal** if $\theta = \dfrac{\pi}{2}$.

The vectors **a** and **b** in Figure 1 are parallel if and only if they lie on the same line that passes through the origin. In this case, $\mathbf{b} = m\mathbf{a}$ for some real number m. The vectors are orthogonal if and only if they lie on mutually perpendicular lines that pass through the origin. We assume that the zero vector **0** is parallel and orthogonal to *every* vector **a**.

The next theorem shows the close relationship between the angle between two vectors and their dot product.

Theorem on the Dot Product

If θ is the angle between two nonzero vectors $\mathbf{a}$ and $\mathbf{b}$, then

$$\mathbf{a} \cdot \mathbf{b} = \|\mathbf{a}\| \|\mathbf{b}\| \cos \theta.$$

PROOF If $\mathbf{a}$ and $\mathbf{b}$ are not parallel, we have a situation similar to that illustrated in Figure 1. We may then apply the law of cosines to triangle AOB. Since the lengths of the three sides of the triangle are $\|\mathbf{a}\|$, $\|\mathbf{b}\|$, and $d(A, B)$,

$$[d(A, B)]^2 = \|\mathbf{a}\|^2 + \|\mathbf{b}\|^2 - 2\|\mathbf{a}\| \|\mathbf{b}\| \cos \theta.$$

Using the distance formula and the definition of the magnitude of a vector, we obtain

$$(b_1 - a_1)^2 + (b_2 - a_2)^2 = (a_1^2 + a_2^2) + (b_1^2 + b_2^2) - 2\|\mathbf{a}\| \|\mathbf{b}\| \cos \theta,$$

which reduces to

$$-2a_1b_1 - 2a_2b_2 = -2\|\mathbf{a}\| \|\mathbf{b}\| \cos \theta.$$

Dividing both sides of the last equation by -2 gives us

$$a_1b_1 + a_2b_2 = \|\mathbf{a}\| \|\mathbf{b}\| \cos \theta,$$

which is equivalent to what we wished to prove, since the left-hand side is $\mathbf{a} \cdot \mathbf{b}$.

If $\mathbf{a}$ and $\mathbf{b}$ are parallel, then either $\theta = 0$ or $\theta = \pi$, and therefore $\mathbf{b} = m\mathbf{a}$ for some real number m with $m > 0$ if $\theta = 0$ and $m < 0$ if $\theta = \pi$. We can show, using properties of the dot product, that $\mathbf{a} \cdot (m\mathbf{a}) = \|\mathbf{a}\| \|m\mathbf{a}\| \cos \theta$, and hence the theorem is true for all nonzero vectors $\mathbf{a}$ and $\mathbf{b}$. ◪

Theorem on the Cosine of the Angle Between Vectors

If θ is the angle between two nonzero vectors $\mathbf{a}$ and $\mathbf{b}$, then

$$\cos \theta = \frac{\mathbf{a} \cdot \mathbf{b}}{\|\mathbf{a}\| \|\mathbf{b}\|}.$$

Figure 2

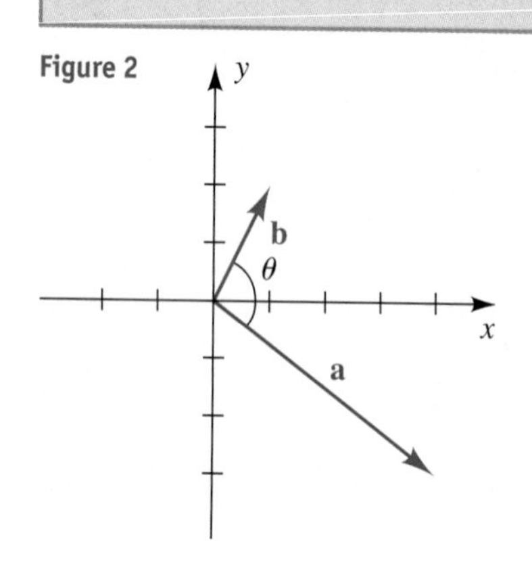

EXAMPLE 2 Finding the angle between two vectors

Find the angle between $\mathbf{a} = \langle 4, -3 \rangle$ and $\mathbf{b} = \langle 1, 2 \rangle$.

▶ SOLUTION The vectors are sketched in Figure 2. We apply the preceding theorem:

$$\cos \theta = \frac{\mathbf{a} \cdot \mathbf{b}}{\|\mathbf{a}\| \|\mathbf{b}\|} = \frac{(4)(1) + (-3)(2)}{\sqrt{16 + 9}\sqrt{1 + 4}} = \frac{-2}{5\sqrt{5}}, \quad \text{or} \quad \frac{-2\sqrt{5}}{25}$$

Hence,

$$\theta = \arccos\left(\frac{-2\sqrt{5}}{25}\right) \approx 100.3°.$$

◪

▶ Tutorial available at www.thomsonedu.com/login

EXAMPLE 3 **Showing that two vectors are parallel**

Let $\mathbf{a} = \frac{1}{2}\mathbf{i} - 3\mathbf{j}$ and $\mathbf{b} = -2\mathbf{i} + 12\mathbf{j}$.

(a) Show that **a** and **b** are parallel.

(b) Find the scalar m such that $\mathbf{b} = m\mathbf{a}$.

▶ SOLUTION

(a) By definition, the vectors **a** and **b** are parallel if and only if the angle θ between them is either 0 or π. Since

$$\cos\theta = \frac{\mathbf{a}\cdot\mathbf{b}}{\|\mathbf{a}\|\,\|\mathbf{b}\|} = \frac{\left(\frac{1}{2}\right)(-2) + (-3)(12)}{\sqrt{\frac{1}{4} + 9}\sqrt{4 + 144}} = \frac{-37}{37} = -1,$$

we conclude that

$$\theta = \arccos(-1) = \pi.$$

(b) Since **a** and **b** are parallel, there *is* a scalar m such that $\mathbf{b} = m\mathbf{a}$; that is,

$$-2\mathbf{i} + 12\mathbf{j} = m\left(\tfrac{1}{2}\mathbf{i} - 3\mathbf{j}\right) = \tfrac{1}{2}m\mathbf{i} - 3m\mathbf{j}.$$

Equating the coefficients of **i** and **j** gives us

$$-2 = \tfrac{1}{2}m \qquad \text{and} \qquad 12 = -3m.$$

Thus, $m = -4$; that is, $\mathbf{b} = -4\mathbf{a}$. Note that **a** and **b** have opposite directions, since $m < 0$. ✎

Using the formula $\mathbf{a}\cdot\mathbf{b} = \|\mathbf{a}\|\,\|\mathbf{b}\|\cos\theta$, together with the fact that two vectors are orthogonal if and only if the angle between them is $\pi/2$ (or one of the vectors is **0**), gives us the following result.

Theorem on Orthogonal Vectors	Two vectors **a** and **b** are orthogonal if and only if $\mathbf{a}\cdot\mathbf{b} = 0$.

EXAMPLE 4 **Showing that two vectors are orthogonal**

Show that the pair of vectors is orthogonal:

(a) **i**, **j** (b) $2\mathbf{i} + 3\mathbf{j}$, $6\mathbf{i} - 4\mathbf{j}$

▶ SOLUTION We may use the theorem on orthogonal vectors to prove orthogonality by showing that the dot product of each pair is zero:

(a) $\mathbf{i}\cdot\mathbf{j} = \langle 1, 0\rangle \cdot \langle 0, 1\rangle = (1)(0) + (0)(1) = 0 + 0 = 0$

(b) $(2\mathbf{i} + 3\mathbf{j})\cdot(6\mathbf{i} - 4\mathbf{j}) = (2)(6) + (3)(-4) = 12 - 12 = 0$ ✎

▶ **Tutorial available at www.thomsonedu.com/login**

Definition of $\text{comp}_{\mathbf{b}}\, \mathbf{a}$

Let θ be the angle between two nonzero vectors **a** and **b**. The **component of a along b**, denoted by $\text{comp}_{\mathbf{b}}\, \mathbf{a}$, is given by

$$\text{comp}_{\mathbf{b}}\, \mathbf{a} = \|\mathbf{a}\| \cos \theta.$$

The geometric significance of the preceding definition with θ acute or obtuse is illustrated in Figure 3, where the x- and y-axes are not shown.

Figure 3 $\text{comp}_{\mathbf{b}}\, \mathbf{a} = \|\mathbf{a}\| \cos \theta$

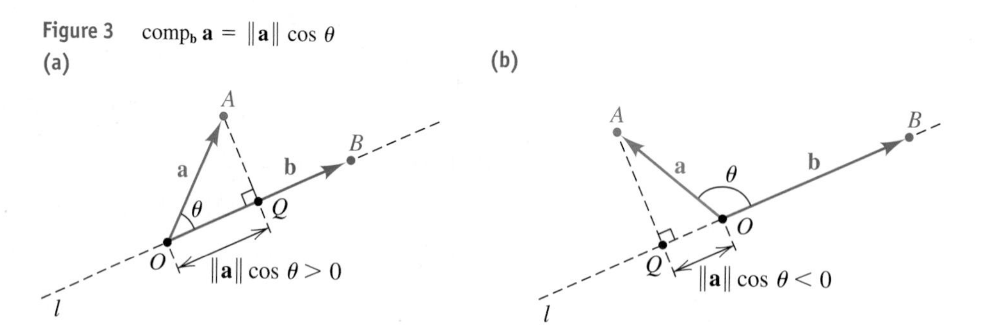

If angle θ is acute, then, as in Figure 3(a), we can form a right triangle by constructing a line segment AQ perpendicular to the line l through O and B. Note that $\overrightarrow{OQ}$ has the same direction as $\overrightarrow{OB}$. Referring to part (a) of the figure, we see that

$$\cos \theta = \frac{d(O, Q)}{\|\mathbf{a}\|} \quad \text{or, equivalently,} \quad \|\mathbf{a}\| \cos \theta = d(O, Q).$$

If θ is obtuse, then, as in Figure 3(b), we again construct AQ perpendicular to l. In this case, the direction of $\overrightarrow{OQ}$ is opposite that of $\overrightarrow{OB}$, and since $\cos \theta$ is negative,

$$\cos \theta = \frac{-d(O, Q)}{\|\mathbf{a}\|} \quad \text{or, equivalently,} \quad \|\mathbf{a}\| \cos \theta = -d(O, Q).$$

special cases for the component of **a** along **b**

(1) If $\theta = \pi/2$, then **a** is orthogonal to **b** and $\text{comp}_{\mathbf{b}}\, \mathbf{a} = 0$.

(2) If $\theta = 0$, then **a** has the same direction as **b** and $\text{comp}_{\mathbf{b}}\, \mathbf{a} = \|\mathbf{a}\|$.

(3) If $\theta = \pi$, then **a** and **b** have opposite directions and $\text{comp}_{\mathbf{b}}\, \mathbf{a} = -\|\mathbf{a}\|$.

The preceding discussion shows that the component of **a** along **b** may be found by *projecting* the endpoint of **a** onto the line l containing **b**. For this reason, $\|\mathbf{a}\| \cos \theta$ is sometimes called the **projection of a on b** and is denoted by $\text{proj}_{\mathbf{b}}\, \mathbf{a}$. The following formula shows how to compute this projection *without* knowing the angle θ.

Formula for $\text{comp}_{\mathbf{b}}\,\mathbf{a}$	If **a** and **b** are nonzero vectors, then $$\text{comp}_{\mathbf{b}}\,\mathbf{a} = \frac{\mathbf{a}\cdot\mathbf{b}}{\|\mathbf{b}\|}.$$

PROOF If θ is the angle between **a** and **b**, then, from the theorem on the dot product,

$$\mathbf{a}\cdot\mathbf{b} = \|\mathbf{a}\|\,\|\mathbf{b}\|\cos\theta.$$

Dividing both sides of this equation by $\|\mathbf{b}\|$ gives us

$$\frac{\mathbf{a}\cdot\mathbf{b}}{\|\mathbf{b}\|} = \|\mathbf{a}\|\cos\theta = \text{comp}_{\mathbf{b}}\,\mathbf{a}.$$

Figure 4

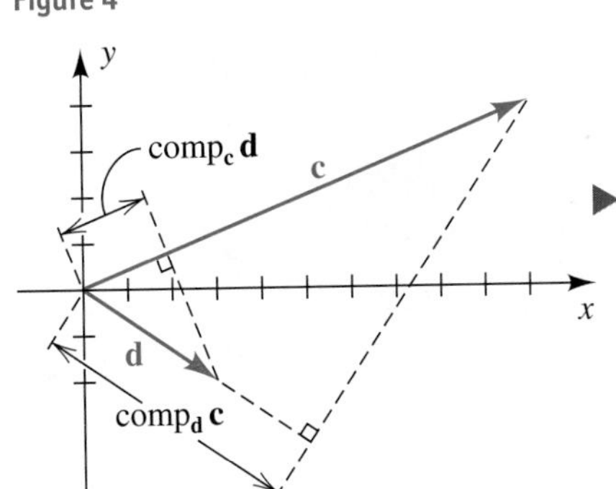

EXAMPLE 5 Finding the components of one vector along another

If $\mathbf{c} = 10\mathbf{i} + 4\mathbf{j}$ and $\mathbf{d} = 3\mathbf{i} - 2\mathbf{j}$, find $\text{comp}_{\mathbf{d}}\,\mathbf{c}$ and $\text{comp}_{\mathbf{c}}\,\mathbf{d}$, and illustrate these numbers graphically.

SOLUTION The vectors **c** and **d** and the desired components are illustrated in Figure 4. We use the formula for $\text{comp}_{\mathbf{b}}\,\mathbf{a}$, as follows:

$$\text{comp}_{\mathbf{d}}\,\mathbf{c} = \frac{\mathbf{c}\cdot\mathbf{d}}{\|\mathbf{d}\|} = \frac{(10)(3) + (4)(-2)}{\sqrt{3^2 + (-2)^2}} = \frac{22}{\sqrt{13}} \approx 6.10$$

$$\text{comp}_{\mathbf{c}}\,\mathbf{d} = \frac{\mathbf{d}\cdot\mathbf{c}}{\|\mathbf{c}\|} = \frac{(3)(10) + (-2)(4)}{\sqrt{10^2 + 4^2}} = \frac{22}{\sqrt{116}} \approx 2.04$$

We shall conclude this section with a physical application of the dot product. First let us briefly discuss the scientific concept of *work.*

A **force** may be thought of as the physical entity that is used to describe a push or pull on an object. For example, a force is needed to push or pull an object along a horizontal plane, to lift an object off the ground, or to move a charged particle through an electromagnetic field. Forces are often measured in pounds. If an object weighs 10 pounds, then, by definition, the force required to lift it (or hold it off the ground) is 10 pounds. A force of this type is a **constant force,** since its magnitude does not change while it is applied to the given object.

If a constant force F is applied to an object, moving it a distance d in the direction of the force, then, by definition, the **work** W done is

$$W = Fd.$$

If F is measured in pounds and d in feet, then the units for W are foot-pounds (ft-lb). In the cgs (centimeter-gram-second) system a **dyne** is used as the unit of force. If F is expressed in dynes and d in centimeters, then the unit for W is the dyne-centimeter, or **erg.** In the mks (meter-kilogram-second) system the **newton** is used as the unit of force. If F is in newtons and d is in meters, then the unit for W is the newton-meter, or **joule.**

EXAMPLE 6 Finding the work done by a constant force

Find the work done in pushing an automobile along a level road from a point A to another point B, 40 feet from A, while exerting a constant force of 90 pounds.

▶ SOLUTION The problem is illustrated in Figure 5, where we have pictured the road as part of a line l. Since the constant force is $F = 90$ lb and the distance the automobile moves is $d = 40$ feet, the work done is

$$W = (90)(40) = 3600 \text{ ft-lb}.$$

Figure 5

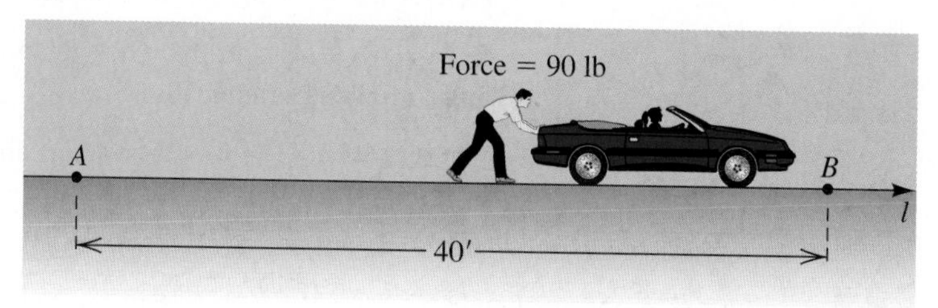

The formula $W = Fd$ is very restrictive, since it can be used only if the force is applied along the line of motion. More generally, suppose that a vector **a** represents a force and that its point of application moves along a vector **b**. This is illustrated in Figure 6, where the force **a** is used to pull an object along a level path from O to B, and $\mathbf{b} = \overrightarrow{OB}$.

Figure 6

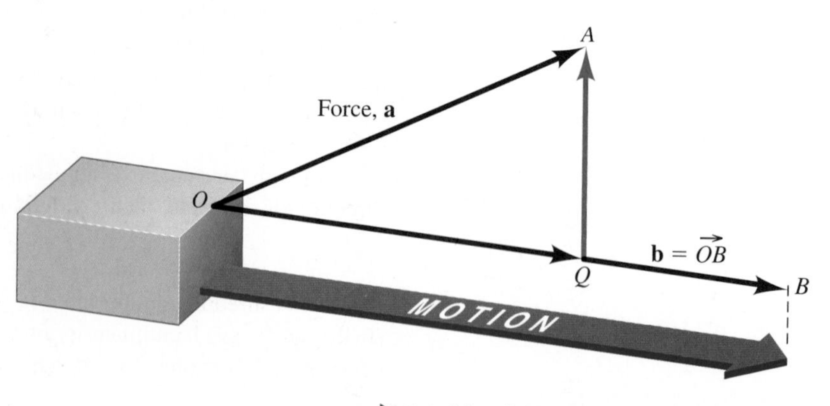

▶ Tutorial available at www.thomsonedu.com/login

The vector **a** is the sum of the vectors OQ and QA, where $\overrightarrow{QA}$ is orthogonal to **b**. Since $\overrightarrow{QA}$ does not contribute to the horizontal movement, we may assume that the motion from O to B is caused by $\overrightarrow{OQ}$ alone. Applying $W = Fd$, we know that the work is the product of $\|\overrightarrow{OQ}\|$ and $\|\mathbf{b}\|$. Since the magnitude $\|\overrightarrow{OQ}\| = \text{comp}_\mathbf{b}\,\mathbf{a}$, we obtain

$$W = (\text{comp}_\mathbf{b}\,\mathbf{a})\|\mathbf{b}\| = (\|\mathbf{a}\|\cos\theta)\|\mathbf{b}\| = \mathbf{a}\cdot\mathbf{b},$$

where θ represents $\angle AOQ$. This leads to the following definition.

Definition of Work	The **work** W done by a constant force **a** as its point of application moves along a vector **b** is $W = \mathbf{a}\cdot\mathbf{b}$.

Figure 7

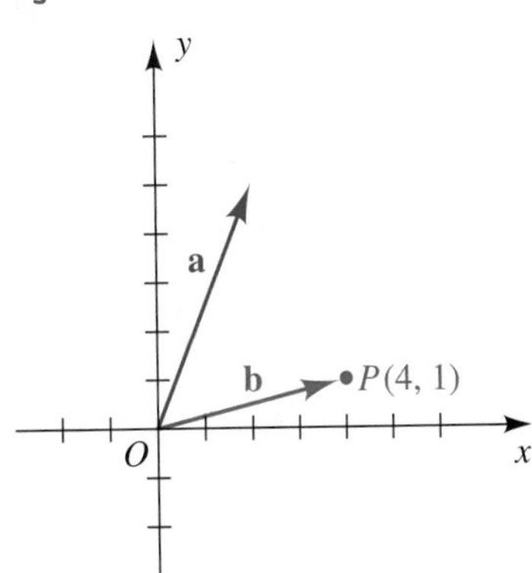

EXAMPLE 7 Finding the work done by a constant force

The magnitude and direction of a constant force are given by $\mathbf{a} = 2\mathbf{i} + 5\mathbf{j}$. Find the work done if the point of application of the force moves from the origin to the point $P(4, 1)$.

▶ **SOLUTION** The force **a** and the vector $\mathbf{b} = \overrightarrow{OP}$ are sketched in Figure 7. Since $\mathbf{b} = \langle 4, 1\rangle = 4\mathbf{i} + \mathbf{j}$, we have, from the preceding definition,

$$\begin{aligned} W = \mathbf{a}\cdot\mathbf{b} &= (2\mathbf{i} + 5\mathbf{j})\cdot(4\mathbf{i} + \mathbf{j}) \\ &= (2)(4) + (5)(1) = 13. \end{aligned}$$

If, for example, the unit of length is feet and the magnitude of the force is measured in pounds, then the work done is 13 ft-lb.

EXAMPLE 8 Finding the work done against gravity

A small cart weighing 100 pounds is pushed up an incline that makes an angle of 30° with the horizontal, as shown in Figure 8. Find the work done against gravity in pushing the cart a distance of 80 feet.

Figure 8

Figure 9

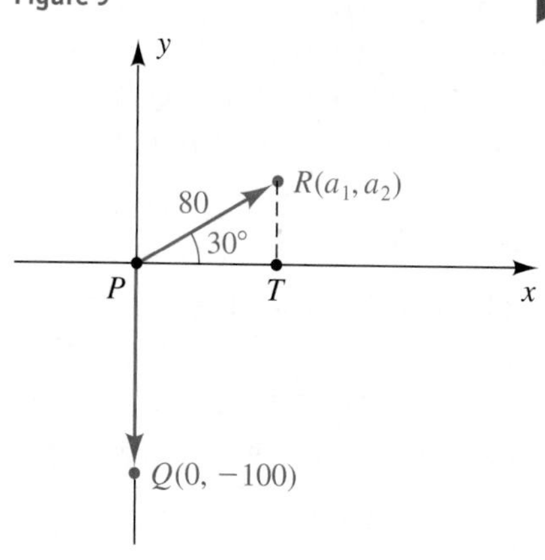

▶ SOLUTION Let us introduce an xy-coordinate system, as shown in Figure 9. The vector PQ represents the force of gravity acting vertically downward with a magnitude of 100 pounds. The corresponding vector $\mathbf{F}$ is $0\mathbf{i} - 100\mathbf{j}$. The point of application of this force moves along the vector PR of magnitude 80. If $\overrightarrow{PR}$ corresponds to $\mathbf{a} = a_1\mathbf{i} + a_2\mathbf{j}$, then, referring to triangle PTR, we see that

$$a_1 = 80 \cos 30° = 40\sqrt{3}$$
$$a_2 = 80 \sin 30° = 40,$$

and hence

$$\mathbf{a} = 40\sqrt{3}\mathbf{i} + 40\mathbf{j}.$$

Applying the definition, we find that the work done *by* gravity is

$$\mathbf{F} \cdot \mathbf{a} = (0\mathbf{i} - 100\mathbf{j}) \cdot \left(40\sqrt{3}\mathbf{i} + 40\mathbf{j}\right) = 0 - 4000 = -4000 \text{ ft-lb}.$$

The work done *against* gravity is

$$-\mathbf{F} \cdot \mathbf{a} = 4000 \text{ ft-lb}.$$

7.4 Exercises

Exer. 1–8: Find (a) the dot product of the two vectors and (b) the angle between the two vectors.

1 $\langle -2, 5\rangle$, $\langle 3, 6\rangle$

2 $\langle 4, -7\rangle$, $\langle -2, 3\rangle$

3 $4\mathbf{i} - \mathbf{j}$, $-3\mathbf{i} + 2\mathbf{j}$

4 $8\mathbf{i} - 3\mathbf{j}$, $2\mathbf{i} - 7\mathbf{j}$

5 $9\mathbf{i}$, $5\mathbf{i} + 4\mathbf{j}$

6 $6\mathbf{j}$, $-4\mathbf{i}$

7 $\langle 10, 7\rangle$, $\left\langle -2, -\frac{7}{5}\right\rangle$

8 $\langle -3, 6\rangle$, $\langle -1, 2\rangle$

Exer. 9–12: Show that the vectors are orthogonal.

9 $\langle 4, -1\rangle$, $\langle 2, 8\rangle$

10 $\langle 3, 6\rangle$, $\langle 4, -2\rangle$

▶ 11 $-4\mathbf{j}$, $-7\mathbf{i}$

12 $8\mathbf{i} - 4\mathbf{j}$, $-6\mathbf{i} - 12\mathbf{j}$

Exer. 13–16: Show that the vectors are parallel, and determine whether they have the same direction or opposite directions.

13 $\mathbf{a} = 3\mathbf{i} - 5\mathbf{j}$, $\mathbf{b} = -\frac{12}{7}\mathbf{i} + \frac{20}{7}\mathbf{j}$

14 $\mathbf{a} = -\frac{5}{2}\mathbf{i} + 6\mathbf{j}$, $\mathbf{b} = -10\mathbf{i} + 24\mathbf{j}$

15 $\mathbf{a} = \left\langle \frac{2}{3}, \frac{1}{2}\right\rangle$, $\mathbf{b} = \langle 8, 6\rangle$

16 $\mathbf{a} = \langle 6, 18\rangle$, $\mathbf{b} = \langle -4, -12\rangle$

Exer. 17–20: Determine m such that the two vectors are orthogonal.

17 $3\mathbf{i} - 2\mathbf{j}$, $4\mathbf{i} + 5m\mathbf{j}$

18 $4m\mathbf{i} + \mathbf{j}$, $9m\mathbf{i} - 25\mathbf{j}$

19 $9\mathbf{i} - 16m\mathbf{j}$, $\mathbf{i} + 4m\mathbf{j}$

20 $5m\mathbf{i} + 3\mathbf{j}$, $2\mathbf{i} + 7\mathbf{j}$

Exer. 21–28: Given that $\mathbf{a} = \langle 2, -3\rangle$, $\mathbf{b} = \langle 3, 4\rangle$, and $\mathbf{c} = \langle -1, 5\rangle$, find the number.

21 (a) $\mathbf{a} \cdot (\mathbf{b} + \mathbf{c})$ (b) $\mathbf{a} \cdot \mathbf{b} + \mathbf{a} \cdot \mathbf{c}$

22 (a) $\mathbf{b} \cdot (\mathbf{a} - \mathbf{c})$ (b) $\mathbf{b} \cdot \mathbf{a} - \mathbf{b} \cdot \mathbf{c}$

23 $(2\mathbf{a} + \mathbf{b}) \cdot (3\mathbf{c})$

24 $(\mathbf{a} - \mathbf{b}) \cdot (\mathbf{b} + \mathbf{c})$

25 $\text{comp}_{\mathbf{c}}\, \mathbf{b}$

26 $\text{comp}_{\mathbf{b}}\, \mathbf{c}$

27 $\text{comp}_{\mathbf{b}}\, (\mathbf{a} + \mathbf{c})$

28 $\text{comp}_{\mathbf{c}}\, \mathbf{c}$

Exer. 29–32: If c represents a constant force, find the work done if the point of application of c moves along the line segment from P to Q.

29 $\mathbf{c} = 3\mathbf{i} + 4\mathbf{j}$; $P(0, 0)$, $Q(5, -2)$

30 $\mathbf{c} = -10\mathbf{i} + 12\mathbf{j}$; $P(0, 0)$, $Q(4, 7)$

31 $\mathbf{c} = 6\mathbf{i} + 4\mathbf{j}$; $P(2, -1)$, $Q(4, 3)$
(*Hint:* Find a vector $\mathbf{b} = \langle b_1, b_2\rangle$ such that $\mathbf{b} = \overrightarrow{PQ}$.)

▶ Tutorial available at www.thomsonedu.com/login

32 $\mathbf{c} = -\mathbf{i} + 7\mathbf{j}$; $P(-2, 5)$, $Q(6, 1)$

33 A constant force of magnitude 4 has the same direction as $\mathbf{j}$. Find the work done if its point of application moves from $P(0, 0)$ to $Q(8, 3)$.

34 A constant force of magnitude 10 has the same direction as $-\mathbf{i}$. Find the work done if its point of application moves from $P(0, 1)$ to $Q(1, 0)$.

Exer. 35–40: Prove the property if a and b are vectors and *m* is a real number.

35 $\mathbf{a} \cdot \mathbf{a} = \|\mathbf{a}\|^2$

36 $\mathbf{a} \cdot \mathbf{b} = \mathbf{b} \cdot \mathbf{a}$

37 $(m\mathbf{a}) \cdot \mathbf{b} = m(\mathbf{a} \cdot \mathbf{b})$

38 $m(\mathbf{a} \cdot \mathbf{b}) = \mathbf{a} \cdot (m\mathbf{b})$

39 $\mathbf{0} \cdot \mathbf{a} = 0$

40 $(\mathbf{a} + \mathbf{b}) \cdot (\mathbf{a} - \mathbf{b}) = \mathbf{a} \cdot \mathbf{a} - \mathbf{b} \cdot \mathbf{b}$

41 Pulling a wagon A child pulls a wagon along level ground by exerting a force of 20 pounds on a handle that makes an angle of 30° with the horizontal, as shown in the figure. Find the work done in pulling the wagon 100 feet.

Exercise 41

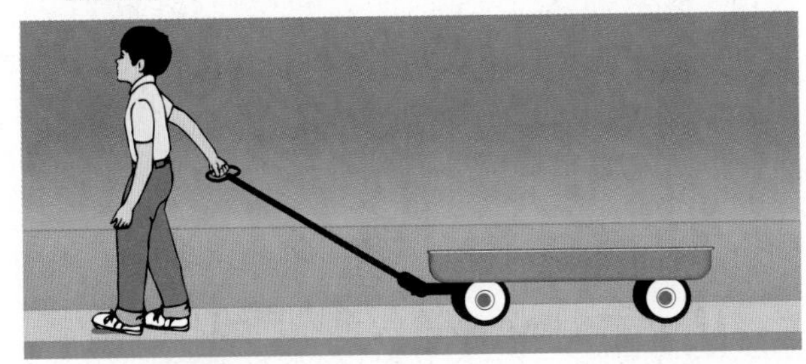

42 Pulling a wagon Refer to Exercise 41. Find the work done if the wagon is pulled, with the same force, 100 feet up an incline that makes an angle of 30° with the horizontal, as shown in the figure.

Exercise 42

43 The sun's rays The sun has a radius of 432,000 miles, and its center is 93,000,000 miles from the center of Earth. Let $\mathbf{v}$ and $\mathbf{w}$ be the vectors illustrated in the figure.

(a) Express $\mathbf{v}$ and $\mathbf{w}$ in $\mathbf{i}, \mathbf{j}$ form.

(b) Approximate the angle between $\mathbf{v}$ and $\mathbf{w}$.

Exercise 43

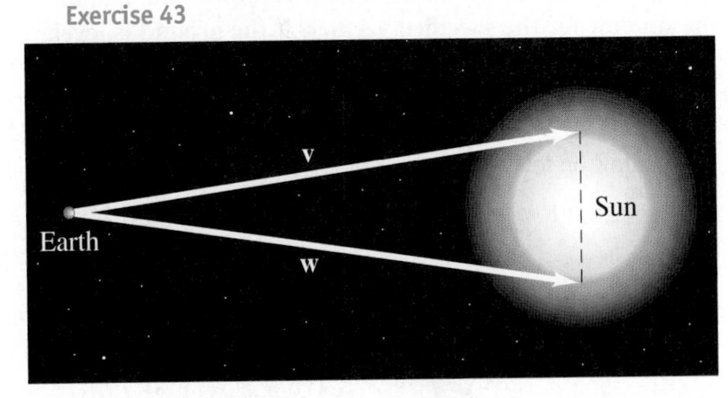

44 July sunlight The intensity I of sunlight (in watts/m^2) can be calculated using the formula $I = ke^{-c/\sin\phi}$, where k and c are positive constants and ϕ is the angle between the sun's rays and the horizon. The amount of sunlight striking a vertical wall facing the sun is equal to the component of the sun's rays along the horizontal. If, during July, $\phi = 30°$, $k = 978$, and $c = 0.136$, approximate the total amount of sunlight striking a vertical wall that has an area of 160 m^2.

Exer. 45–46: Vectors are used extensively in computer graphics to perform shading. When light strikes a flat surface, it is reflected, and that area should not be shaded. Suppose that an incoming ray of light is represented by a vector L and that N is a vector orthogonal to the flat surface, as shown in the figure. The ray of reflected light can be represented by the vector R and is calculated using the formula $\mathbf{R} = 2(\mathbf{N} \cdot \mathbf{L})\mathbf{N} - \mathbf{L}$. Compute R for the vectors L and N.

45 Reflected light $\mathbf{L} = \langle -\frac{4}{5}, \frac{3}{5} \rangle$, $\mathbf{N} = \langle 0, 1 \rangle$

46 Reflected light $\mathbf{L} = \langle \frac{12}{13}, -\frac{5}{13} \rangle$, $\mathbf{N} = \langle \frac{1}{2}\sqrt{2}, \frac{1}{2}\sqrt{2} \rangle$

Exercises 45–46

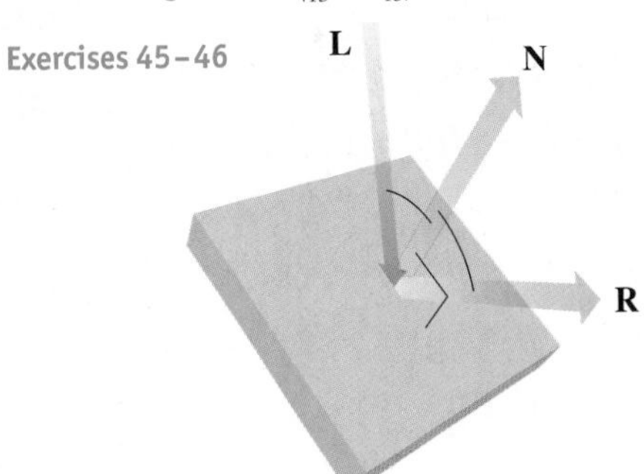

Exer. 47–48: Vectors are used in computer graphics to calculate the lengths of shadows over flat surfaces. The length of an object can sometimes be represented by a vector a. If a single light source is shining down on the object, then the length of its shadow on the ground will be equal to the absolute value of the component of the vector a along the direction of the ground, as shown in the figure. Compute the length of the shadow for the specified vector a if the ground is level.

47 Shadow on level ground $\mathbf{a} = \langle 2.6, 4.5 \rangle$

48 Shadow on level ground $\mathbf{a} = \langle -3.1, 7.9 \rangle$

Exercises 47–48

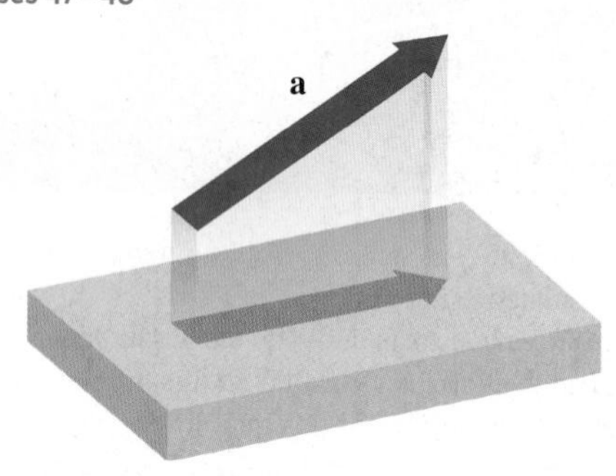

Exer. 49–50: Refer to Exercises 47 and 48. An object represented by a vector a is held over a flat surface inclined at an angle θ, as shown in the figure. If a light is shining directly downward, approximate the length of the shadow to two decimal places for the specified values of the vector a and θ.

49 Shadow on inclined plane $\mathbf{a} = \langle 25.7, -3.9 \rangle$, $\theta = 12°$

50 Shadow on inclined plane $\mathbf{a} = \langle -13.8, 19.4 \rangle$, $\theta = -17°$

Exercises 49–50

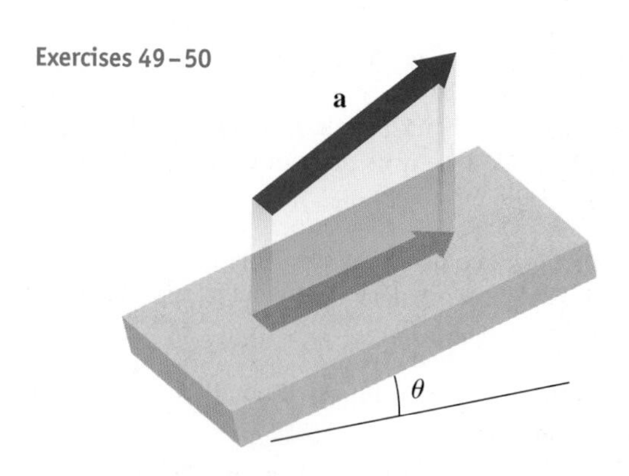

51 Determining horsepower The amount of horsepower P produced by an engine can be determined by using the formula $P = \frac{1}{550}(\mathbf{F} \cdot \mathbf{v})$, where $\mathbf{F}$ is the force (in pounds) exerted by the engine and $\mathbf{v}$ is the velocity (in ft/sec) of an object moved by the engine. An engine pulls with a force of 2200 pounds on a cable that makes an angle θ with the horizontal, moving a cart horizontally, as shown in the figure. Find the horsepower of the engine if the speed of the cart is 8 ft/sec when $\theta = 30°$.

Exercise 51

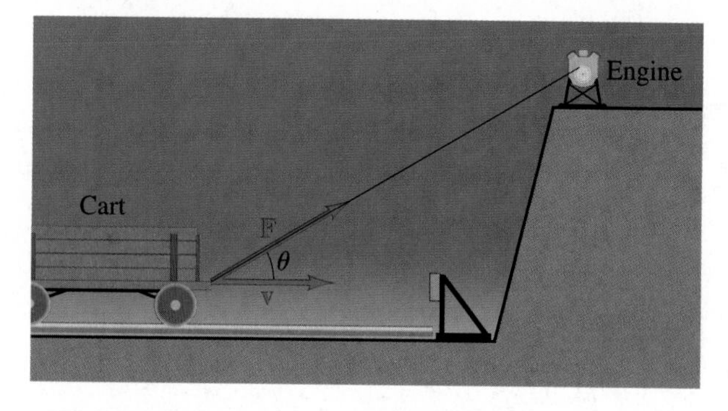

7.5 Trigonometric Form for Complex Numbers

In Section 1.1 we represented real numbers geometrically by using points on a coordinate line. We can obtain geometric representations for complex numbers by using points in a coordinate plane. Specifically, each complex number $a + bi$ determines a unique ordered pair (a, b). The corresponding point $P(a, b)$ in a coordinate plane is the **geometric representation** of $a + bi$. To emphasize that we are assigning complex numbers to points in a plane, we may label the point $P(a, b)$ as $a + bi$. A coordinate plane with a complex number assigned to each point is referred to as a **complex** (or Argand) **plane** instead of an xy-plane. The x-axis is the **real axis** and the y-axis is the **imaginary axis.** In Figure 1 (on the next page) we have represented several complex numbers geometrically. Note that to obtain the point corresponding to the conjugate $a - bi$ of any complex number $a + bi$, we simply reflect through the real axis.

Figure 1

Imaginary axis

$2 + 3i$

$-\frac{5}{2} + \sqrt{2}i$

$5 + i$

i

-3

$-i$

Real axis

$5 - i$

$-2 - 3i$

$2 - 3i$

$-5i$

The absolute value of a real number a (denoted $|a|$) is the distance between the origin and the point on the x-axis that corresponds to a. Thus, it is natural to interpret the absolute value of a complex number as the distance between the origin of a complex plane and the point (a, b) that corresponds to $a + bi$.

Definition of the Absolute Value of a Complex Number	If $z = a + bi$ is a complex number, then its **absolute value,** denoted by $|a + bi|$, is $$\sqrt{a^2 + b^2}.$$

EXAMPLE 1 **Finding the absolute value of a complex number**

Find

(a) $|2 - 6i|$ (b) $|3i|$

▶ SOLUTION We use the previous definition:

(a) $|2 - 6i| = \sqrt{2^2 + (-6)^2} = \sqrt{40} = 2\sqrt{10} \approx 6.3$

(b) $|3i| = |0 + 3i| = \sqrt{0^2 + 3^2} = \sqrt{9} = 3$

The points corresponding to all complex numbers that have a fixed absolute value k are on a circle of radius k with center at the origin in the complex plane. For example, the points corresponding to the complex numbers z with $|z| = 1$ are on a unit circle.

Figure 2
$z = a + bi = r(\cos\theta + i\sin\theta)$

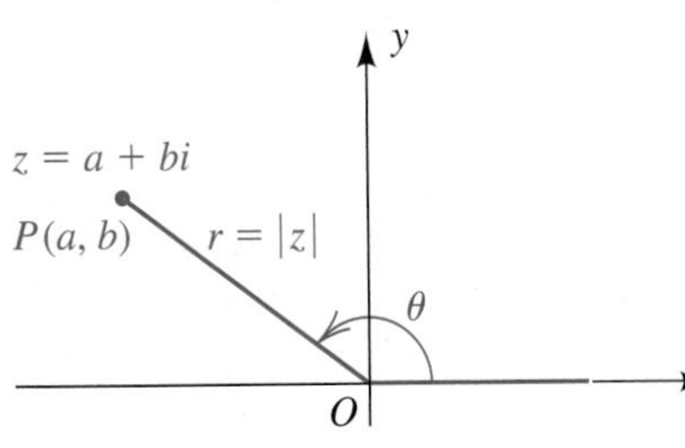

Let us consider a nonzero complex number $z = a + bi$ and its geometric representation $P(a, b)$, as illustrated in Figure 2. Let θ be any angle in standard position whose terminal side lies on the segment OP, and let $r = |z| = \sqrt{a^2 + b^2}$. Since $\cos\theta = a/r$ and $\sin\theta = b/r$, we see that $a = r\cos\theta$ and $b = r\sin\theta$. Substituting for a and b in $z = a + bi$, we obtain

$$z = a + bi = (r\cos\theta) + (r\sin\theta)i = r(\cos\theta + i\sin\theta).$$

▶ **Tutorial available at www.thomsonedu.com/login**

This expression is called the **trigonometric** (or **polar**) **form for the complex number $a + bi$.** A common abbreviation is

$$r(\underline{c}os\ \theta + \underline{i}\ \underline{s}in\ \theta) = r\ \underline{cis}\ \theta.$$

The trigonometric form for $z = a + bi$ is not unique, since there are an unlimited number of different choices for the angle θ. When the trigonometric form is used, the absolute value r of z is sometimes referred to as the **modulus** of z and an angle θ associated with z as an **argument** (or **amplitude**) of z.

We may summarize our discussion as follows.

Trigonometric (or Polar) Form for a Complex Number	Let $z = a + bi$. If $r = \lvert z \rvert = \sqrt{a^2 + b^2}$ and if θ is an argument of z, then $$z = r(\cos\theta + i\sin\theta) = r\ \text{cis}\ \theta.$$

Euler's formula,

$$\cos\theta + i\sin\theta = e^{i\theta},$$

gives us yet another form for the complex number $z = a + bi$, commonly called the **exponential form;** that is,

$$z = r(\cos\theta + i\sin\theta) = re^{i\theta}.$$

See Exercise 6 of the Discussion Exercises at the end of the chapter for some related problems.

EXAMPLE 2 Expressing a complex number in trigonometric form

Express the complex number in trigonometric form with $0 \le \theta < 2\pi$:

(a) $-4 + 4i$ (b) $2\sqrt{3} - 2i$ (c) $2 + 7i$ (d) $-2 + 7i$

▶ **SOLUTION** We begin by representing each complex number geometrically and labeling its modulus r and argument θ, as in Figure 3.

Figure 3

(a)

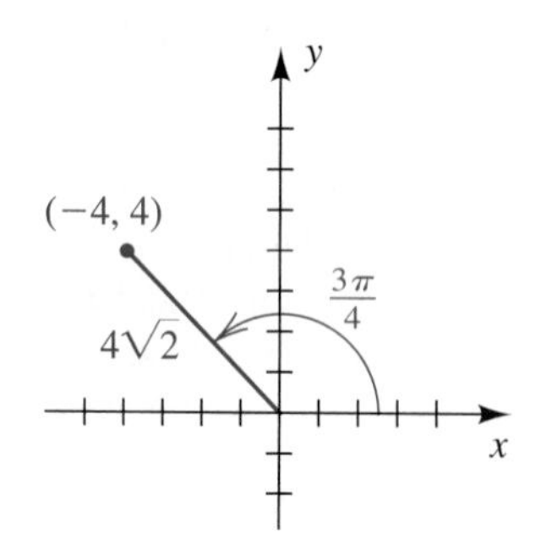

(b)

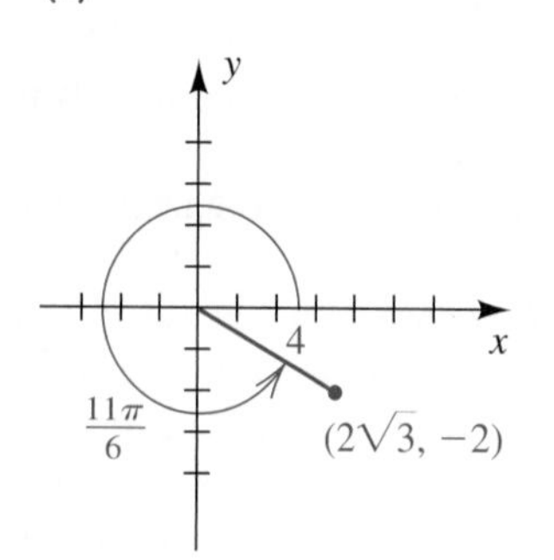

(c)

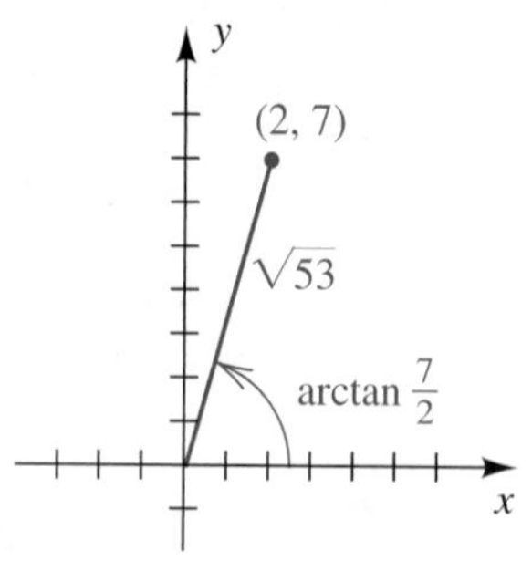

(d)

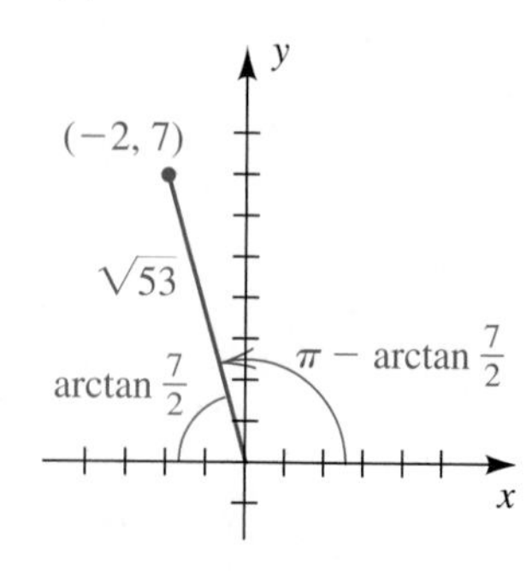

▶ Tutorial available at www.thomsonedu.com/login

We next substitute for r and θ in the trigonometric form:

(a) $-4 + 4i = 4\sqrt{2}\left(\cos\frac{3\pi}{4} + i\sin\frac{3\pi}{4}\right) = 4\sqrt{2}\operatorname{cis}\frac{3\pi}{4}$

(b) $2\sqrt{3} - 2i = 4\left(\cos\frac{11\pi}{6} + i\sin\frac{11\pi}{6}\right) = 4\operatorname{cis}\frac{11\pi}{6}$

(c) $2 + 7i = \sqrt{53}\left[\cos\left(\arctan\frac{7}{2}\right) + i\sin\left(\arctan\frac{7}{2}\right)\right] = \sqrt{53}\operatorname{cis}\left(\arctan\frac{7}{2}\right)$

(d) $-2 + 7i = \sqrt{53}\left[\cos\left(\pi - \arctan\frac{7}{2}\right) + i\sin\left(\pi - \arctan\frac{7}{2}\right)\right]$
$= \sqrt{53}\operatorname{cis}\left(\pi - \arctan\frac{7}{2}\right)$

Complex Number Operations

Let's take a look at how to find the absolute value and the argument of the complex number in Example 2(b) on a graphing calculator.

TI-83/4 Plus	**TI-86**
Assign $2\sqrt{3} - 2i$ to A.	
2 [2nd] [√] 3 [)] [−] 2 [2nd] [i]	[(] 2 [2nd] [√] 3 [,] −2 [)]
[STO ▷] [ALPHA] [A] [ENTER]	[STO ▷] [A] [ENTER]
Find the absolute value r.	
[MATH] [▷] [▷] [5]	[2nd] [CPLX] [abs(F4)]
[ALPHA] [A] [)] [ENTER]	[ALPHA] [A] [ENTER]
Find the argument θ (in degree mode).	
[MATH] [▷] [▷] [4]	[angle(F5)] [ALPHA] [A] [ENTER]
[ALPHA] [A] [)] [ENTER]	

```
2√(3)-2i→A
   3.464101615-2i
abs(A)
                4
angle(A)
              -30
```

```
(2√3,-2)→A
   (3.46410161514,-2)
abs A
                 4
angle A
               -30
conj real imag abs angle
```

Now we'll change the form of $2\sqrt{3} - 2i$ using the polar feature. The TI-83/4 Plus gives us the exponential form $re^{\theta i}$, and the TI-86 gives us the (magnitude∠angle) form. Note that $-30°$ is equivalent to $11\pi/6$ (the angle in Example 2(b) for $0 \le \theta < 2\pi$).

TI-83/4 Plus	**TI-86**
[ALPHA] [A] [MATH] [▷] [▷] [7] [ENTER]	[ALPHA] [A] [MORE] [▶Pol(F2)] [ENTER]

```
A▶Polar
        4e^(-30i)
```

```
A▶Pol
              (4∠-30)
▶Rec ▶Pol
```

If we allow arbitrary values for θ, there are many other trigonometric forms for the complex numbers in Example 2. Thus, for $-4 + 4i$ in part (a) we could use

$$\theta = \frac{3\pi}{4} + 2\pi n \quad \text{for any integer } n.$$

If, for example, we let $n = 1$ and $n = -1$, we obtain

$$4\sqrt{2} \operatorname{cis} \frac{11\pi}{4} \qquad \text{and} \qquad 4\sqrt{2} \operatorname{cis} \left(-\frac{5\pi}{4}\right),$$

respectively. In general, arguments for the same complex number always differ by a multiple of 2π.

If complex numbers are expressed in trigonometric form, then multiplication and division may be performed as indicated in the next theorem.

Theorem on Products and Quotients of Complex Numbers

If trigonometric forms for two complex numbers z_1 and z_2 are

$$z_1 = r_1(\cos \theta_1 + i \sin \theta_1) \qquad \text{and} \qquad z_2 = r_2(\cos \theta_2 + i \sin \theta_2),$$

then

(1) $z_1 z_2 = r_1 r_2[\cos (\theta_1 + \theta_2) + i \sin (\theta_1 + \theta_2)]$

(2) $\dfrac{z_1}{z_2} = \dfrac{r_1}{r_2}[\cos (\theta_1 - \theta_2) + i \sin (\theta_1 - \theta_2)],\ z_2 \neq 0$

PROOF We may prove (1) as follows:

$$\begin{aligned} z_1 z_2 &= r_1(\cos \theta_1 + i \sin \theta_1) \cdot r_2(\cos \theta_2 + i \sin \theta_2) \\ &= r_1 r_2[(\cos \theta_1 \cos \theta_2 - \sin \theta_1 \sin \theta_2) \\ &\qquad + i(\sin \theta_1 \cos \theta_2 + \cos \theta_1 \sin \theta_2)] \end{aligned}$$

Applying the addition formulas for $\cos (\theta_1 + \theta_2)$ and $\sin (\theta_1 + \theta_2)$ gives us (1). We leave the proof of (2) as an exercise.

Part (1) of the preceding theorem states that *the modulus of a product of two complex numbers is the product of their moduli, and an argument is the sum of their arguments.* An analogous statement can be made for (2).

EXAMPLE 3 Using trigonometric forms to find products and quotients

If $z_1 = 2\sqrt{3} - 2i$ and $z_2 = -1 + \sqrt{3}i$, use trigonometric forms to find **(a)** $z_1 z_2$ and **(b)** z_1/z_2. Check by using algebraic methods.

Figure 4

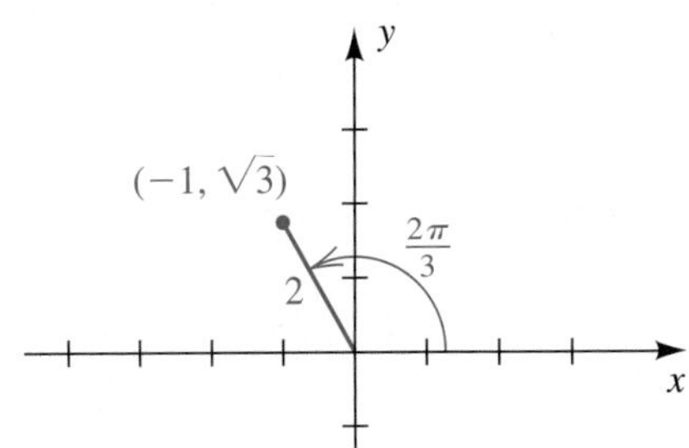

Figure 5

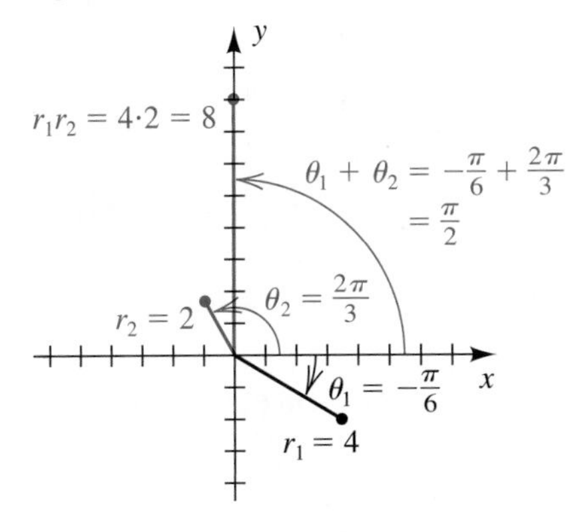

Figure 6

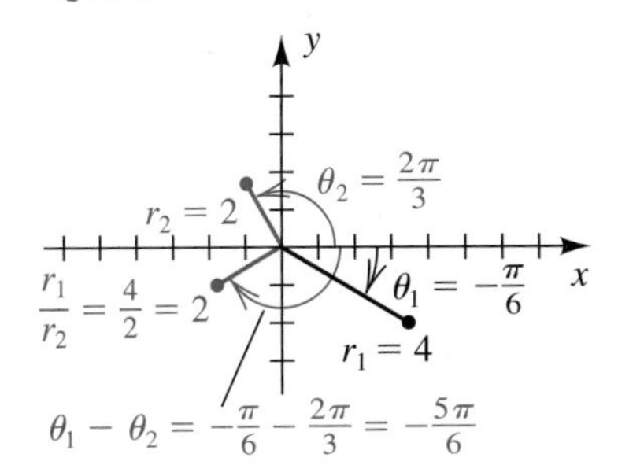

▶ SOLUTION The complex number $2\sqrt{3} - 2i$ is represented geometrically in Figure 3(b). If we use $\theta = -\pi/6$ in the trigonometric form, then

$$z_1 = 2\sqrt{3} - 2i = 4\left[\cos\left(-\frac{\pi}{6}\right) + i \sin\left(-\frac{\pi}{6}\right)\right].$$

The complex number $z_2 = -1 + \sqrt{3}i$ is represented geometrically in Figure 4. A trigonometric form is

$$z_2 = -1 + \sqrt{3}i = 2\left(\cos\frac{2\pi}{3} + i \sin\frac{2\pi}{3}\right).$$

(a) We apply part (1) of the theorem on products and quotients of complex numbers:

$$\begin{aligned} z_1z_2 &= 4 \cdot 2\left[\cos\left(-\frac{\pi}{6} + \frac{2\pi}{3}\right) + i \sin\left(-\frac{\pi}{6} + \frac{2\pi}{3}\right)\right] \\ &= 8\left(\cos\frac{\pi}{2} + i \sin\frac{\pi}{2}\right) = 8(0 + i) = 8i \end{aligned}$$

Figure 5 gives a geometric interpretation of the product z_1z_2.

Using algebraic methods to check our result, we have

$$\begin{aligned} z_1z_2 &= \left(2\sqrt{3} - 2i\right)\left(-1 + \sqrt{3}i\right) \\ &= \left(-2\sqrt{3} + 2\sqrt{3}\right) + (2 + 6)i = 0 + 8i = 8i. \end{aligned}$$

(b) We apply part (2) of the theorem:

$$\begin{aligned} \frac{z_1}{z_2} &= \frac{4}{2}\left[\cos\left(-\frac{\pi}{6} - \frac{2\pi}{3}\right) + i \sin\left(-\frac{\pi}{6} - \frac{2\pi}{3}\right)\right] \\ &= 2\left[\cos\left(-\frac{5\pi}{6}\right) + i \sin\left(-\frac{5\pi}{6}\right)\right] \\ &= 2\left[-\frac{\sqrt{3}}{2} + i\left(-\frac{1}{2}\right)\right] = -\sqrt{3} - i \end{aligned}$$

Figure 6 gives a geometric interpretation of the quotient z_1/z_2.

Using algebraic methods to check our result, we multiply the numerator and denominator by the conjugate of the denominator to obtain

$$\begin{aligned} \frac{z_1}{z_2} &= \frac{2\sqrt{3} - 2i}{-1 + \sqrt{3}i} \cdot \frac{-1 - \sqrt{3}i}{-1 - \sqrt{3}i} \\ &= \frac{\left(-2\sqrt{3} - 2\sqrt{3}\right) + (2 - 6)i}{(-1)^2 + \left(\sqrt{3}\right)^2} \\ &= \frac{-4\sqrt{3} - 4i}{4} = -\sqrt{3} - i. \end{aligned}$$

▶ Tutorial available at www.thomsonedu.com/login

7.5 Exercises

Exer. 1–10: Find the absolute value.

1 $|3 - 4i|$
2 $|5 + 8i|$
3 $|-6 - 7i|$
4 $|1 - i|$
5 $|8i|$
6 $|i^7|$
7 $|i^{500}|$
8 $|-15i|$
9 $|0|$
10 $|-15|$

Exer. 11–20: Represent the complex number geometrically.

11 $4 + 2i$
12 $-5 + 3i$
13 $3 - 5i$
14 $-2 - 6i$
15 $-(3 - 6i)$
16 $(1 + 2i)^2$
17 $2i(2 + 3i)$
18 $(-3i)(2 - i)$
19 $(1 + i)^2$
20 $4(-1 + 2i)$

Exer. 21–46: Express the complex number in trigonometric form with $0 \leq \theta < 2\pi$.

21 $1 - i$
22 $\sqrt{3} + i$
23 $-4\sqrt{3} + 4i$
24 $-2 - 2i$
25 $2\sqrt{3} + 2i$
26 $3 - 3\sqrt{3}i$
27 $-4 - 4i$
28 $-10 + 10i$
29 $-20i$
30 $-6i$
31 12
32 15
33 -7
34 -5
▶ 35 $6i$
36 $4i$
37 $-5 - 5\sqrt{3}i$
38 $\sqrt{3} - i$
39 $2 + i$
40 $3 + 2i$
41 $-3 + i$
42 $-4 + 2i$
43 $-5 - 3i$
44 $-2 - 7i$
45 $4 - 3i$
46 $1 - 3i$

Exer. 47–56: Express in the form $a + bi$, where a and b are real numbers.

47 $4\left(\cos \frac{\pi}{4} + i \sin \frac{\pi}{4}\right)$
48 $8\left(\cos \frac{7\pi}{4} + i \sin \frac{7\pi}{4}\right)$
49 $6\left(\cos \frac{2\pi}{3} + i \sin \frac{2\pi}{3}\right)$
50 $12\left(\cos \frac{4\pi}{3} + i \sin \frac{4\pi}{3}\right)$
51 $5(\cos \pi + i \sin \pi)$
52 $3\left(\cos \frac{3\pi}{2} + i \sin \frac{3\pi}{2}\right)$
53 $\sqrt{34} \operatorname{cis} \left(\tan^{-1} \frac{3}{5}\right)$
54 $\sqrt{53} \operatorname{cis} \left[\tan^{-1} \left(-\frac{2}{7}\right)\right]$
55 $\sqrt{5} \operatorname{cis} \left[\tan^{-1} \left(-\frac{1}{2}\right)\right]$
56 $\sqrt{10} \operatorname{cis} (\tan^{-1} 3)$

Exer. 57–66: Use trigonometric forms to find z_1z_2 and z_1/z_2.

57 $z_1 = -1 + i, \quad z_2 = 1 + i$
58 $z_1 = \sqrt{3} - i, \quad z_2 = -\sqrt{3} - i$
59 $z_1 = -2 - 2\sqrt{3}i, \quad z_2 = 5i$
▶ 60 $z_1 = -5 + 5i, \quad z_2 = -3i$
61 $z_1 = -10, \quad z_2 = -4$
62 $z_1 = 2i, \quad z_2 = -3i$
63 $z_1 = 4, \quad z_2 = 2 - i$
64 $z_1 = 7, \quad z_2 = 3 + 5i$
65 $z_1 = -5, \quad z_2 = 3 - 2i$
66 $z_1 = -3, \quad z_2 = 5 + 2i$

67 Prove (2) of the theorem on products and quotients of complex numbers.

68 (a) Extend (1) of the theorem on products and quotients of complex numbers to three complex numbers.

(b) Generalize (1) of the theorem to n complex numbers.

▶ **Tutorial available at www.thomsonedu.com/login**

Exer. 69–72: The trigonometric form of complex numbers is often used by electrical engineers to describe the current I, voltage V, and impedance Z in electrical circuits with alternating current. Impedance is the opposition to the flow of current in a circuit. Most common electrical devices operate on 115-volt, alternating current. The relationship among these three quantities is $I = V/Z$. Approximate the unknown quantity, and express the answer in rectangular form to two decimal places.

69 Finding voltage $I = 10 \text{ cis } 35°$, $Z = 3 \text{ cis } 20°$

70 Finding voltage $I = 12 \text{ cis } 5°$, $Z = 100 \text{ cis } 90°$

71 Finding impedance $I = 8 \text{ cis } 5°$, $V = 115 \text{ cis } 45°$

72 Finding current $Z = 78 \text{ cis } 61°$, $V = 163 \text{ cis } 17°$

73 Modulus of impedance The modulus of the impedance Z represents the total opposition to the flow of electricity in a circuit and is measured in ohms. If $Z = 14 - 13i$, compute $|Z|$.

74 Resistance and reactance The absolute value of the real part of Z represents the resistance in an electrical circuit; the absolute value of the complex part represents the reactance. Both quantities are measured in ohms. If $V = 220 \text{ cis } 34°$ and $I = 5 \text{ cis } 90°$, approximate the resistance and the reactance.

75 Actual voltage The real part of V represents the actual voltage delivered to an electrical appliance in volts. Approximate this voltage when $I = 4 \text{ cis } 90°$ and $Z = 18 \text{ cis } (-78°)$.

76 Actual current The real part of I represents the actual current delivered to an electrical appliance in amps. Approximate this current when $V = 163 \text{ cis } 43°$ and $Z = 100 \text{ cis } 17°$.

7.6 De Moivre's Theorem and nth Roots of Complex Numbers

If z is a complex number and n is a positive integer, then a complex number w is an ***n*th root** of z if $w^n = z$. We will show that every nonzero complex number has n different nth roots. Since $\mathbb{R}$ is contained in $\mathbb{C}$, it will also follow that every nonzero real number has n different nth (complex) roots. If a is a positive real number and $n = 2$, then we already know that the roots are $\sqrt{a}$ and $-\sqrt{a}$.

If, in the theorem on products and quotients of complex numbers, we let both z_1 and z_2 equal the complex number $z = r(\cos\theta + i\sin\theta)$, we obtain

$$\begin{aligned} z^2 &= r \cdot r[\cos(\theta + \theta) + i\sin(\theta + \theta)] \\ &= r^2(\cos 2\theta + i\sin 2\theta). \end{aligned}$$

Applying the same theorem to z^2 and z gives us

$$z^2 \cdot z = (r^2 \cdot r)[\cos(2\theta + \theta) + i\sin(2\theta + \theta)],$$

or

$$z^3 = r^3(\cos 3\theta + i\sin 3\theta).$$

Applying the theorem to z^3 and z, we obtain

$$z^4 = r^4(\cos 4\theta + i\sin 4\theta).$$

In general, we have the following result, named after the French mathematician Abraham De Moivre (1667–1754).

De Moivre's Theorem

For every integer n,

$$[r(\cos\theta + i\sin\theta)]^n = r^n(\cos n\theta + i\sin n\theta).$$

We will use only positive integers for n in examples and exercises involving De Moivre's theorem. However, for completeness, the theorem holds for $n = 0$ and n negative if we use the respective real number exponent definitions—that is, $z^0 = 1$ and $z^{-n} = 1/z^n$, where z is a nonzero complex number and n is a positive integer.

EXAMPLE 1 Using De Moivre's theorem

Use De Moivre's theorem to change $(1 + i)^{20}$ to the form $a + bi$, where a and b are real numbers.

Figure 1

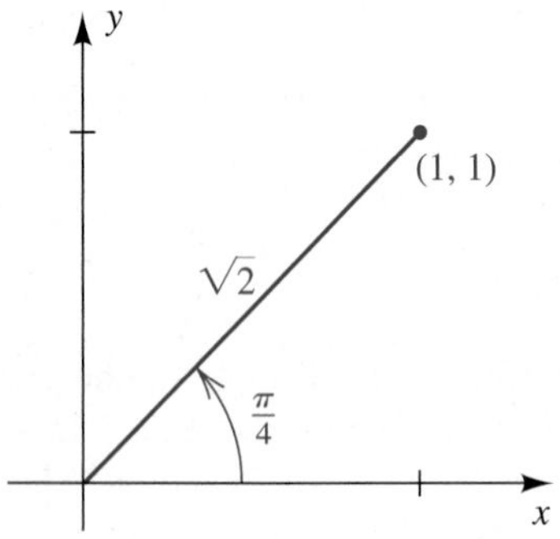

▶ **SOLUTION** It would be tedious to change $(1 + i)^{20}$ using algebraic methods. Let us therefore introduce a trigonometric form for $1 + i$. Referring to Figure 1, we see that

$$1 + i = \sqrt{2}\left(\cos\frac{\pi}{4} + i\sin\frac{\pi}{4}\right).$$

We now apply De Moivre's theorem:

$$\begin{aligned}(1 + i)^{20} &= (2^{1/2})^{20}\left[\cos\left(20 \cdot \frac{\pi}{4}\right) + i\sin\left(20 \cdot \frac{\pi}{4}\right)\right] \\ &= 2^{10}(\cos 5\pi + i\sin 5\pi) = 2^{10}(-1 + 0i) = -1024\end{aligned}$$

The number -1024 is of the form $a + bi$ with $a = -1024$ and $b = 0$.

If a nonzero complex number z has an nth root w, then $w^n = z$. If trigonometric forms for w and z are

$$w = s(\cos\alpha + i\sin\alpha) \qquad \text{and} \qquad z = r(\cos\theta + i\sin\theta), \qquad (*)$$

then applying De Moivre's theorem to $w^n = z$ yields

$$s^n(\cos n\alpha + i\sin n\alpha) = r(\cos\theta + i\sin\theta).$$

If two complex numbers are equal, then so are their absolute values. Consequently, $s^n = r$, and since s and r are nonnegative, $s = \sqrt[n]{r}$. Substituting s^n for r in the last displayed equation and dividing both sides by s^n, we obtain

$$\cos n\alpha + i\sin n\alpha = \cos\theta + i\sin\theta.$$

Since the arguments of equal complex numbers differ by a multiple of 2π, there is an integer k such that $n\alpha = \theta + 2\pi k$. Dividing both sides of the last equation by n, we see that

$$\alpha = \frac{\theta + 2\pi k}{n} \quad \text{for some integer } k.$$

Substituting in the trigonometric form for w (see $(*)$) gives us the formula

$$w = \sqrt[n]{r}\left[\cos\left(\frac{\theta + 2\pi k}{n}\right) + i\sin\left(\frac{\theta + 2\pi k}{n}\right)\right].$$

▶ **Tutorial available at www.thomsonedu.com/login**

If we substitute $k = 0, 1, \ldots, n - 1$ successively, we obtain n different nth roots of z. No other value of k will produce a new nth root. For example, if $k = n$, we obtain the angle $(\theta + 2\pi n)/n$, or $(\theta/n) + 2\pi$, which gives us the same nth root as $k = 0$. Similarly, $k = n + 1$ yields the same nth root as $k = 1$, and so on. The same is true for negative values of k. We have proved the following theorem.

Theorem on nth Roots

If $z = r(\cos\theta + i\sin\theta)$ is any nonzero complex number and if n is any positive integer, then z has exactly n different nth roots $w_0, w_1, w_2, \ldots, w_{n-1}$. These roots, for θ in radians, are

$$w_k = \sqrt[n]{r}\left[\cos\left(\frac{\theta + 2\pi k}{n}\right) + i\sin\left(\frac{\theta + 2\pi k}{n}\right)\right]$$

or, equivalently, for θ in degrees,

$$w_k = \sqrt[n]{r}\left[\cos\left(\frac{\theta + 360°k}{n}\right) + i\sin\left(\frac{\theta + 360°k}{n}\right)\right],$$

where $k = 0, 1, \ldots, n - 1$.

The nth roots of z in this theorem all have absolute value $\sqrt[n]{r}$, and hence their geometric representations lie on a circle of radius $\sqrt[n]{r}$ with center at O. Moreover, they are equispaced on this circle, since the difference in the arguments of successive nth roots is $2\pi/n$ (or $360°/n$).

EXAMPLE 2 **Finding the fourth roots of a complex number**

(a) Find the four fourth roots of $-8 - 8\sqrt{3}i$.

(b) Represent the roots geometrically.

Figure 2

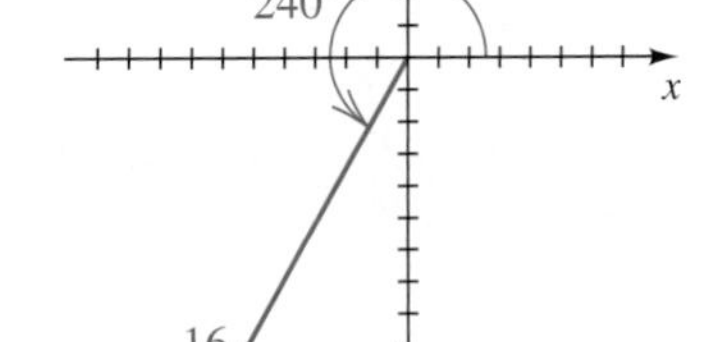

▶ **SOLUTION**

(a) The geometric representation of $-8 - 8\sqrt{3}i$ is shown in Figure 2. Introducing trigonometric form, we have

$$-8 - 8\sqrt{3}i = 16(\cos 240° + i\sin 240°).$$

Using the theorem on nth roots with $n = 4$ and noting that $\sqrt[4]{16} = 2$, we find that the fourth roots are

$$w_k = 2\left[\cos\left(\frac{240° + 360°k}{4}\right) + i\sin\left(\frac{240° + 360°k}{4}\right)\right]$$

(continued)

▶ **Tutorial available at www.thomsonedu.com/login**

Figure 3

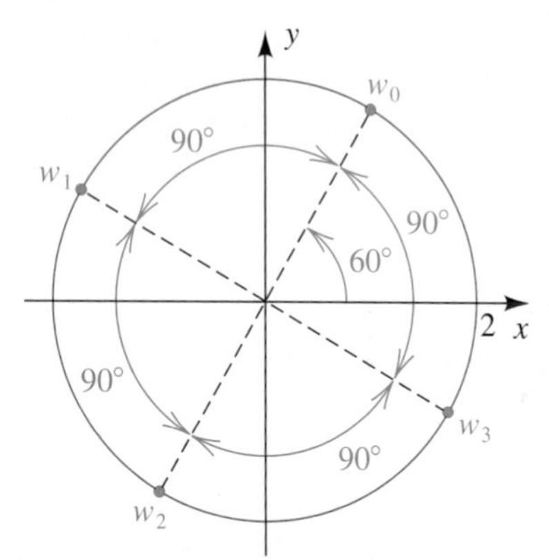

for $k = 0, 1, 2, 3$. This formula may be written

$$w_k = 2[\cos(60° + 90°k) + i \sin(60° + 90°k)].$$

Substituting 0, 1, 2, and 3 for k in $(60° + 90°k)$ gives us the four fourth roots:

$$w_0 = 2(\cos 60° + i \sin 60°) = 1 + \sqrt{3}i$$

$$w_1 = 2(\cos 150° + i \sin 150°) = -\sqrt{3} + i$$

$$w_2 = 2(\cos 240° + i \sin 240°) = -1 - \sqrt{3}i$$

$$w_3 = 2(\cos 330° + i \sin 330°) = \sqrt{3} - i$$

(b) By the comments preceding this example, all roots lie on a circle of radius $\sqrt[4]{16} = 2$ with center at O. The first root, w_0, has an argument of $60°$, and successive roots are spaced apart $360°/4 = 90°$, as shown in Figure 3.

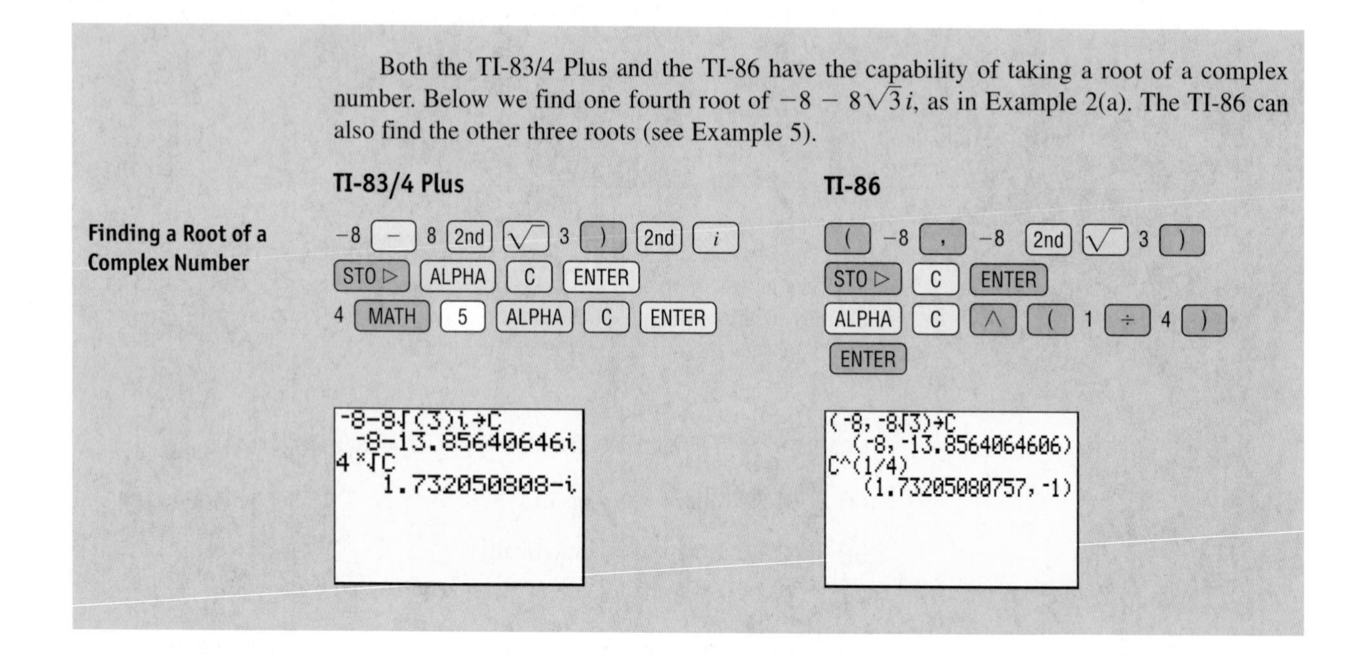

Finding a Root of a Complex Number

Both the TI-83/4 Plus and the TI-86 have the capability of taking a root of a complex number. Below we find one fourth root of $-8 - 8\sqrt{3}i$, as in Example 2(a). The TI-86 can also find the other three roots (see Example 5).

The special case in which $z = 1$ is of particular interest. The n distinct nth roots of 1 are called the ***n*th roots of unity.** In particular, if $n = 3$, we call these roots the **cube roots of unity.**

EXAMPLE 3 Finding the cube roots of unity

Find the three cube roots of unity.

▶ SOLUTION Writing $1 = 1(\cos 0 + i \sin 0)$ and using the theorem on nth roots with $n = 3$, we obtain

▶ **Tutorial available at www.thomsonedu.com/login**

$$w_k = 1\left[\cos\frac{2\pi k}{3} + i\sin\frac{2\pi k}{3}\right]$$

for $k = 0, 1, 2$. Substituting for k gives us the three roots:

$$w_0 = \cos 0 + i\sin 0 = 1$$

$$w_1 = \cos\frac{2\pi}{3} + i\sin\frac{2\pi}{3} = -\frac{1}{2} + \frac{\sqrt{3}}{2}i$$

$$w_2 = \cos\frac{4\pi}{3} + i\sin\frac{4\pi}{3} = -\frac{1}{2} - \frac{\sqrt{3}}{2}i$$

EXAMPLE 4 **Finding the sixth roots of a real number**

(a) Find the six sixth roots of -1.

(b) Represent the roots geometrically.

SOLUTION

(a) Writing $-1 = 1(\cos\pi + i\sin\pi)$ and using the theorem on nth roots with $n = 6$, we find that the sixth roots of -1 are given by

$$w_k = 1\left[\cos\left(\frac{\pi + 2\pi k}{6}\right) + i\sin\left(\frac{\pi + 2\pi k}{6}\right)\right]$$

for $k = 0, 1, 2, 3, 4, 5$. Substituting 0, 1, 2, 3, 4, 5 for k, we obtain the six sixth roots of -1:

$$w_0 = \cos\frac{\pi}{6} + i\sin\frac{\pi}{6} = \frac{\sqrt{3}}{2} + \frac{1}{2}i$$

$$w_1 = \cos\frac{\pi}{2} + i\sin\frac{\pi}{2} = i$$

$$w_2 = \cos\frac{5\pi}{6} + i\sin\frac{5\pi}{6} = -\frac{\sqrt{3}}{2} + \frac{1}{2}i$$

$$w_3 = \cos\frac{7\pi}{6} + i\sin\frac{7\pi}{6} = -\frac{\sqrt{3}}{2} - \frac{1}{2}i$$

$$w_4 = \cos\frac{3\pi}{2} + i\sin\frac{3\pi}{2} = -i$$

$$w_5 = \cos\frac{11\pi}{6} + i\sin\frac{11\pi}{6} = \frac{\sqrt{3}}{2} - \frac{1}{2}i$$

Figure 4

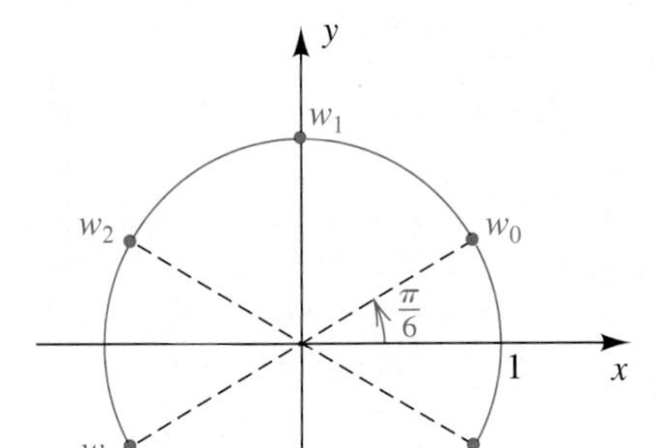

(b) Since $\sqrt[6]{1} = 1$, the points that represent the roots of -1 all lie on the unit circle shown in Figure 4. Moreover, they are equispaced on this circle by $\pi/3$ radians, or 60°.

Note that finding the nth roots of a complex number c, as we did in Examples 2–4, is equivalent to finding all the solutions of the equation

$$x^n = c, \qquad \text{or} \qquad x^n - c = 0.$$

We will use this concept in the next example as well as in Exercises 23–30.

Tutorial available at www.thomsonedu.com/login

EXAMPLE 5 **Finding roots by solving a polynomial equation**

Find the four fourth roots of $-8 - 8\sqrt{3}\,i$.

SOLUTION Let $c = -8 - 8\sqrt{3}\,i$. If x is any fourth root of c, then $x^4 = c$ and so $x^4 - c = 0$. The left-hand side of the last equation is a fourth-degree polynomial with coefficients 1, 0, 0, 0, $-c$. We'll use the polynomial-solving feature to find the fourth roots of c.

Using the TI-86 Poly Feature

Set the number of decimal places to 3.

[2nd] [MODE] [▽] [▷ (4 times)] [ENTER] [2nd] [QUIT]

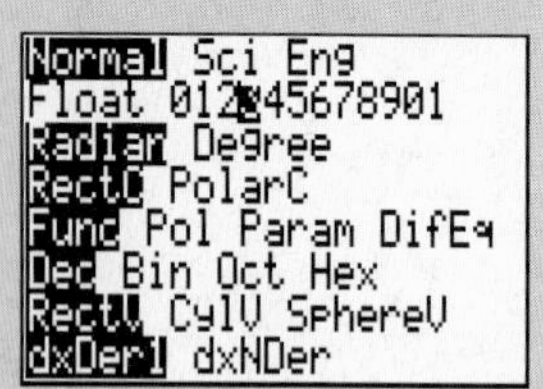

Store $-8 - 8\sqrt{3}\,i$ in C.

[(] −8 [,] −8 [2nd] [√] 3 [)] [STO▷] [C] [ENTER]

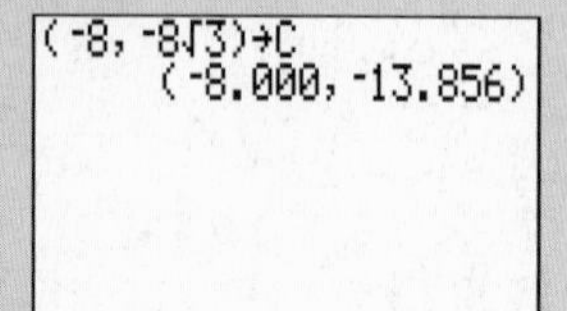

Declare the order of the polynomial.

[2nd] [POLY] 4 [ENTER]

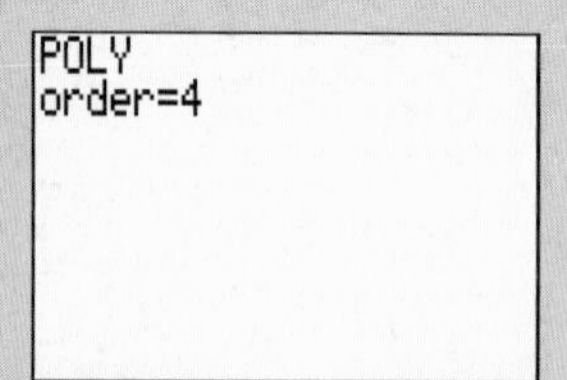

Enter the coefficients.

1 [▽] 0 [▽] 0 [▽] 0 [▽] [(−)] [ALPHA] [C]

Find the solutions.

[SOLVE(F5)]

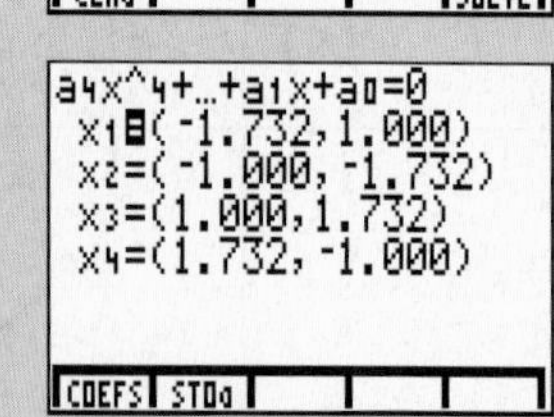

Comparing these solutions to those found in Example 2(a), we have

$$x_1 = w_1 = -\sqrt{3} + i$$
$$x_2 = w_2 = -1 - \sqrt{3}\,i$$
$$x_3 = w_0 = 1 + \sqrt{3}\,i$$
$$x_4 = w_3 = \sqrt{3} - i.$$

7.6 Exercises

Exer. 1–12: Use De Moivre's theorem to change the given complex number to the form $a + bi$, where a and b are real numbers.

1 $(3 + 3i)^5$

2 $(1 + i)^{12}$

3 $(1 - i)^{10}$

4 $(-1 + i)^8$

5 $(1 - \sqrt{3}i)^3$

6 $(1 - \sqrt{3}i)^5$

7 $\left(-\dfrac{\sqrt{2}}{2} + \dfrac{\sqrt{2}}{2}i\right)^{15}$

8 $\left(\dfrac{\sqrt{2}}{2} + \dfrac{\sqrt{2}}{2}i\right)^{25}$

9 $\left(-\dfrac{\sqrt{3}}{2} - \dfrac{1}{2}i\right)^{20}$

10 $\left(-\dfrac{\sqrt{3}}{2} - \dfrac{1}{2}i\right)^{50}$

11 $(\sqrt{3} + i)^7$

▶ 12 $(-2 - 2i)^{10}$

▶ 13 Find the two square roots of $1 + \sqrt{3}i$.

▶ 14 Find the two square roots of $-9i$.

15 Find the four fourth roots of $-1 - \sqrt{3}i$.

16 Find the four fourth roots of $-8 + 8\sqrt{3}i$.

17 Find the three cube roots of $-27i$.

▶ 18 Find the three cube roots of $64i$.

Exer. 19–22: Find the indicated roots, and represent them geometrically.

19 The six sixth roots of unity

20 The eight eighth roots of unity

21 The five fifth roots of $1 + i$

22 The five fifth roots of $-\sqrt{3} - i$

Exer. 23–30: Find the solutions of the equation.

▶ 23 $x^4 - 16 = 0$

24 $x^6 - 64 = 0$

25 $x^6 + 64 = 0$

26 $x^5 + 1 = 0$

▶ 27 $x^3 + 8i = 0$

28 $x^3 - 64i = 0$

29 $x^5 - 243 = 0$

30 $x^4 + 81 = 0$

31 Use Euler's formula to prove De Moivre's theorem.

CHAPTER 7 REVIEW EXERCISES

▶ **Online support materials can be found at www.thomsonedu.com/login**

Exer. 1–4: Find the exact values of the remaining parts of triangle ABC.

1 $\alpha = 60°$, $b = 6$, $c = 7$

2 $\gamma = 30°$, $a = 2\sqrt{3}$, $c = 2$

3 $\alpha = 60°$, $\beta = 45°$, $b = 100$

4 $a = 2$, $b = 3$, $c = 4$

Exer. 5–8: Approximate the remaining parts of triangle ABC.

5 $\beta = 67°$, $\gamma = 75°$, $b = 12$

6 $\alpha = 23°30'$, $c = 125$, $a = 152$

7 $\beta = 115°$, $a = 4.6$, $c = 7.3$

8 $a = 37$, $b = 55$, $c = 43$

▶ **Tutorial available at www.thomsonedu.com/login**

Exer. 9–10: Approximate the area of triangle ABC to the nearest 0.1 square unit.

9 $\alpha = 75°, \quad b = 20, \quad c = 30$

10 $a = 4, \quad b = 7, \quad c = 10$

11 If $\mathbf{a} = \langle -4, 5\rangle$ and $\mathbf{b} = \langle 2, -8\rangle$, sketch vectors corresponding to

(a) $\mathbf{a} + \mathbf{b}$ (b) $\mathbf{a} - \mathbf{b}$ (c) $2\mathbf{a}$ (d) $-\frac{1}{2}\mathbf{b}$

12 If $\mathbf{a} = 2\mathbf{i} + 5\mathbf{j}$ and $\mathbf{b} = 4\mathbf{i} - \mathbf{j}$, find the vector or number corresponding to

(a) $4\mathbf{a} + \mathbf{b}$ (b) $2\mathbf{a} - 3\mathbf{b}$

(c) $\|\mathbf{a} - \mathbf{b}\|$ (d) $\|\mathbf{a}\| - \|\mathbf{b}\|$

13 **A ship's course** A ship is sailing at a speed of 14 mi/hr in the direction S50°E. Express its velocity $\mathbf{v}$ as a vector.

14 The magnitudes and directions of two forces are 72 lb, S60°E and 46 lb, N74°E, respectively. Approximate the magnitude and direction of the resultant force.

15 Find a vector that has the opposite direction of $\mathbf{a} = 8\mathbf{i} - 6\mathbf{j}$ and twice the magnitude.

16 Find a vector of magnitude 4 that has the same direction as $\mathbf{a} = \langle -3, 7\rangle$.

17 If $\mathbf{a} = \langle a_1, a_2\rangle$, $\mathbf{r} = \langle x, y\rangle$, and $c > 0$, describe the set of all points $P(x, y)$ such that $\|\mathbf{r} - \mathbf{a}\| = c$.

18 If $\mathbf{a}$ and $\mathbf{b}$ are vectors with the same initial point and angle θ between them, prove that

$$\|\mathbf{a} - \mathbf{b}\|^2 = \|\mathbf{a}\|^2 + \|\mathbf{b}\|^2 - 2\|\mathbf{a}\|\|\mathbf{b}\| \cos \theta.$$

19 **Wind speed and direction** An airplane is flying in the direction 80° with an airspeed of 400 mi/hr. Its ground speed and true course are 390 mi/hr and 90°, respectively. Approximate the direction and speed of the wind.

20 If $\mathbf{a} = \langle 2, -3\rangle$ and $\mathbf{b} = \langle -1, -4\rangle$, find each of the following:

(a) $\mathbf{a} \cdot \mathbf{b}$ (b) the angle between $\mathbf{a}$ and $\mathbf{b}$

(c) $\text{comp}_{\mathbf{a}}\, \mathbf{b}$

21 If $\mathbf{a} = 6\mathbf{i} - 2\mathbf{j}$ and $\mathbf{b} = \mathbf{i} + 3\mathbf{j}$, find each of the following:

(a) $(2\mathbf{a} - 3\mathbf{b}) \cdot \mathbf{a}$

(b) the angle between $\mathbf{a}$ and $\mathbf{a} + \mathbf{b}$

(c) $\text{comp}_{\mathbf{a}}\, (\mathbf{a} + \mathbf{b})$

22 A constant force has the magnitude and direction of the vector $\mathbf{a} = 7\mathbf{i} + 4\mathbf{j}$. Find the work done when the point of application of $\mathbf{a}$ moves along the x-axis from $P(-5, 0)$ to $Q(3, 0)$.

Exer. 23–28: Express the complex number in trigonometric form with $0 \le \theta < 2\pi$.

23 $-10 + 10i$

24 $2 - 2\sqrt{3}\,i$

25 -17

26 $-12i$

27 $-5\sqrt{3} - 5i$

28 $4 + 5i$

Exer. 29–30: Express in the form $a + bi$, where a and b are real numbers.

29 $20\left(\cos \frac{11\pi}{6} + i \sin \frac{11\pi}{6}\right)$

30 $13 \text{ cis } \left(\tan^{-1} \frac{5}{12}\right)$

Exer. 31–32: Use trigonometric forms to find z_1z_2 and z_1/z_2.

31 $z_1 = -3\sqrt{3} - 3i, \quad z_2 = 2\sqrt{3} + 2i$

32 $z_1 = 2\sqrt{2} + 2\sqrt{2}\,i, \quad z_2 = -1 - i$

Exer. 33–36: Use De Moivre's theorem to change the given complex number to the form $a + bi$, where a and b are real numbers.

33 $\left(-\sqrt{3} + i\right)^9$

34 $\left(\frac{\sqrt{2}}{2} - \frac{\sqrt{2}}{2}i\right)^{30}$

35 $(3 - 3i)^5$

36 $\left(2 + 2\sqrt{3}\,i\right)^{10}$

37 Find the three cube roots of -27.

38 Let $z = 1 - \sqrt{3}\,i$.

(a) Find z^{24}. (b) Find the three cube roots of z.

39 Find the solutions of the equation $x^5 - 32 = 0$.

40 **Skateboard racecourse** A course for a skateboard race consists of a 200-meter downhill run and a 150-meter level portion. The angle of elevation of the starting point of the race from the finish line is 27.4°. What angle does the hill make with the horizontal?

41 Distances to planets The distances between Earth and nearby planets can be approximated using the phase angle α, as shown in the figure. Suppose that the distance between Earth and the sun is 93,000,000 miles and the distance between Venus and the sun is 67,000,000 miles. Approximate the distance between Earth and Venus to the nearest million miles when $\alpha = 34°$.

Exercise 41

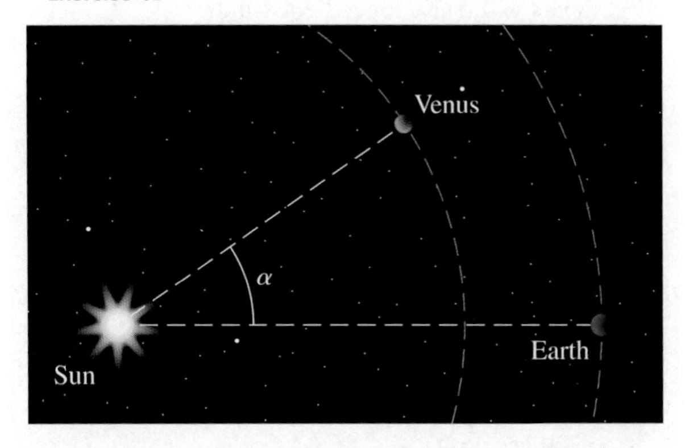

42 Height of a skyscraper If a skyscraper is viewed from the top of a 50-foot building, the angle of elevation is 59°. If it is viewed from street level, the angle of elevation is 62° (see the figure).

(a) Use the law of sines to approximate the shortest distance between the tops of the two buildings.

(b) Approximate the height of the skyscraper.

Exercise 42

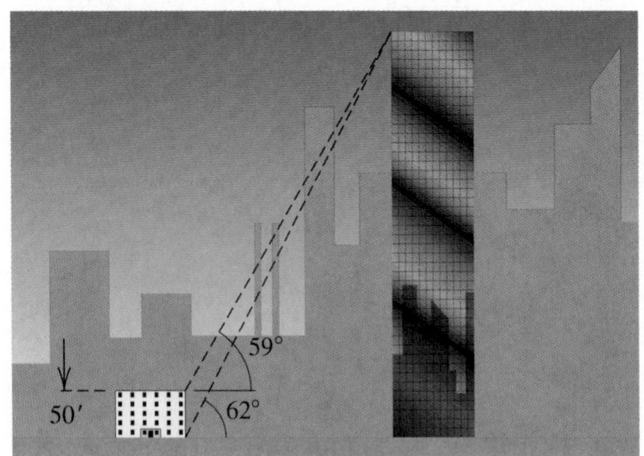

43 Distances between cities The beach communities of San Clemente and Long Beach are 41 miles apart, along a fairly straight stretch of coastline. Shown in the figure is the triangle formed by the two cities and the town of Avalon at the southeast corner of Santa Catalina Island. Angles ALS and ASL are found to be 66.4° and 47.2°, respectively.

(a) Approximate the distance from Avalon to each of the two cities.

(b) Approximate the shortest distance from Avalon to the coast.

Exercise 43

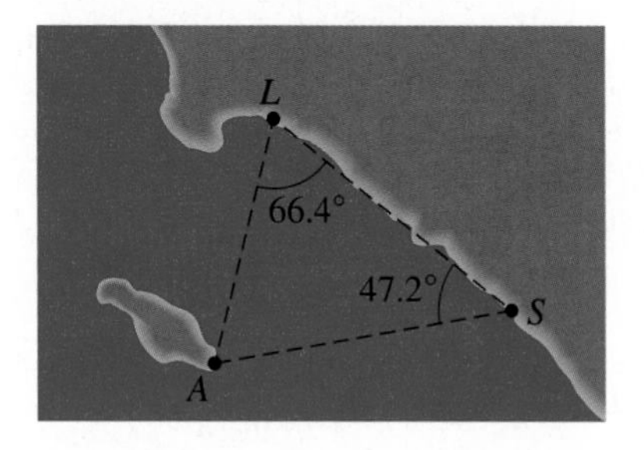

44 Surveying A surveyor wishes to find the distance between two inaccessible points A and B. As shown in the figure, two points C and D are selected from which it is possible to view both A and B. The distance CD and the angles ACD, ACB, BDC, and BDA are then measured. If $CD = 120$ ft, $\angle ACD = 115°$, $\angle ACB = 92°$, $\angle BDC = 125°$, and $\angle BDA = 100°$, approximate the distance AB.

Exercise 44

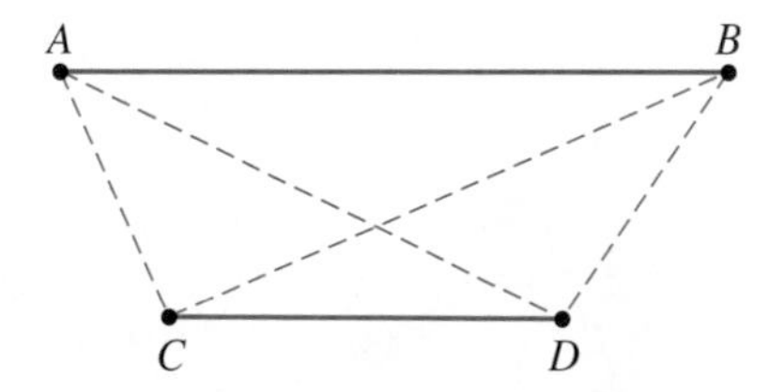

45 Radio contact Two girls with two-way radios are at the intersection of two country roads that meet at a 105° angle (see the figure on the next page). One begins walking in a northerly direction along one road at a rate of 5 mi/hr; at the same time the other walks east along the other road at the same rate. If each radio has a range of 10 miles, how long will the girls maintain contact?

Exercise 45

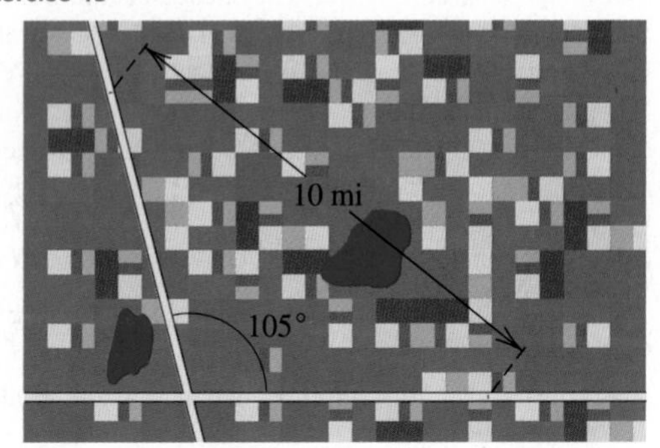

46 **Robotic design** Shown in the figure is a design for a robotic arm with two moving parts. The dimensions are chosen to emulate a human arm. The upper arm AC and lower arm CP rotate through angles θ_1 and θ_2, respectively, to hold an object at point $P(x, y)$.

(a) Show that $\angle ACP = 180° - (\theta_2 - \theta_1)$.

(b) Find $d(A, P)$, and then use part (a) and the law of cosines to show that

$$1 + \cos(\theta_2 - \theta_1) = \frac{x^2 + (y - 26)^2}{578}.$$

(c) If $x = 25$, $y = 4$, and $\theta_1 = 135°$, approximate θ_2.

Exercise 46

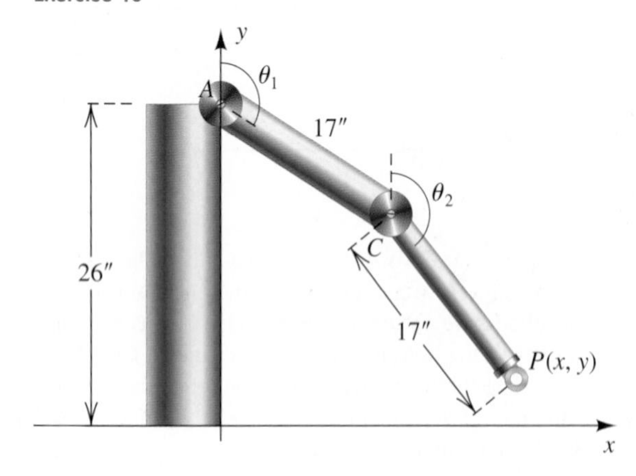

47 **Rescue efforts** A child is trapped 45 feet down an abandoned mine shaft that slants at an angle of 78° from the horizontal. A rescue tunnel is to be dug 50 feet from the shaft opening (see the figure).

(a) At what angle θ should the tunnel be dug?

(b) If the tunnel can be dug at a rate of 3 ft/hr, how many hours will it take to reach the child?

Exercise 47

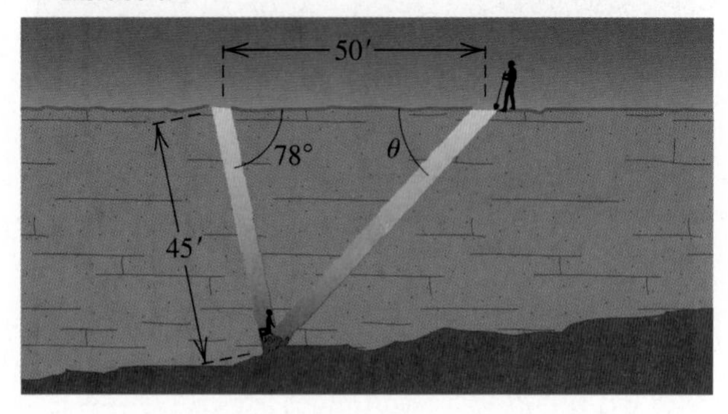

48 **Design for a jet fighter** Shown in the figure is a plan for the top of a wing of a jet fighter.

(a) Approximate angle ϕ.

(b) Approximate the area of quadrilateral $ABCD$.

(c) If the fuselage is 5.8 feet wide, approximate the wing span CC'.

Exercise 48

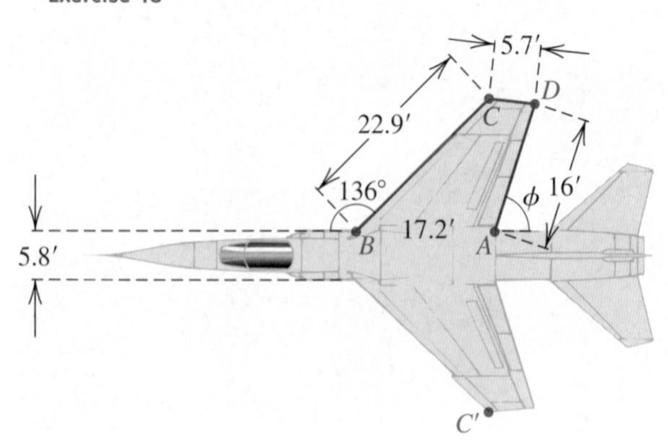

CHAPTER 7 DISCUSSION EXERCISES

1 **Mollweide's formula** The following equation, called *Mollweide's formula*, is sometimes used to check solutions to triangles because it involves all the angles and sides:

$$\frac{a+b}{c} = \frac{\cos \frac{1}{2}(\alpha - \beta)}{\sin \frac{1}{2}\gamma}$$

(a) Use the law of sines to show that

$$\frac{a+b}{c} = \frac{\sin \alpha + \sin \beta}{\sin \gamma}.$$

(b) Use a sum-to-product formula and a double-angle formula to verify Mollweide's formula.

2 Use the trigonometric form of a complex number to show that $z^{-n} = 1/z^n$, where n is a positive integer.

3 Discuss the algebraic and geometric similarities of the cube roots of any positive real number a.

4 Suppose that two vectors $\mathbf{v}$ and $\mathbf{w}$ have the same initial point, that the angle between them is θ, and that $\mathbf{v} \neq m\mathbf{w}$ (m is a real number).

(a) What is the geometric interpretation of $\mathbf{v} - \mathbf{w}$?

(b) How could you find $\|\mathbf{v} - \mathbf{w}\|$?

5 **A vector approach to the laws of sines and cosines**

(a) From the figure we see that $\mathbf{c} = \mathbf{b} + \mathbf{a}$. Use horizontal and vertical components to write $\mathbf{c}$ in terms of $\mathbf{i}$ and $\mathbf{j}$.

Exercise 5

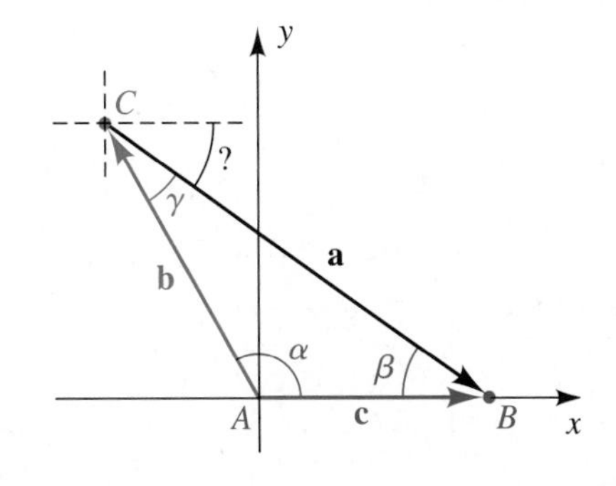

(b) Now find the magnitude of $\mathbf{c}$, using the answer to part (a), and simplify to the point where you have proved the law of cosines.

(c) If $\mathbf{c}$ lies on the x-axis, then its $\mathbf{j}$-component is zero. Use this fact to prove the law of sines.

6 **Euler's formula and other results** The following are some interesting and unexpected results involving complex numbers and topics that have been previously discussed.

(a) Leonhard Euler (1707–1783) gave us the following formula:

$$e^{i\theta} = \cos \theta + i \sin \theta$$

If we let $\theta = \pi$, we obtain $e^{i\pi} = -1$ or, equivalently,

$$e^{i\pi} + 1 = 0,$$

an equation relating five of the most important numbers in mathematics. Find $e^{2\pi i}$.

(b) We define the logarithm of a complex number $z \neq 0$ as follows:

$$\text{LN } z = \ln |z| + i(\theta + 2\pi n),$$

where ln is the natural logarithm function, θ is an argument of z, and n is an integer. The **principal value** of LN z is the value that corresponds to $n = 0$ and $-\pi < \theta \leq \pi$. Find the principal values of LN (-1) and LN i.

(c) We define the complex power w of a complex number $z \neq 0$ as follows:

$$z^w = e^{w \text{ LN } z}$$

We use principal values of LN z to find principal values of z^w. Find principal values of $\sqrt{i}$ and i^i.

7 **An interesting identity?** Suppose α, β, and γ are angles in an oblique triangle. Prove or disprove the following statement: The sum of the tangents of α, β, and γ is equal to the product of the tangents of α, β, and γ.

8 **Forces of hanging wires** A 5-pound ornament hangs from two wires as shown in the figure. Show that the magnitudes of the tensions (forces) in the wires are given by

$$\|\mathbf{T}_1\| = \frac{5 \cos \beta}{\sin (\alpha + \beta)} \quad \text{and} \quad \|\mathbf{T}_2\| = \frac{5 \cos \alpha}{\sin (\alpha + \beta)}.$$

Exercise 8

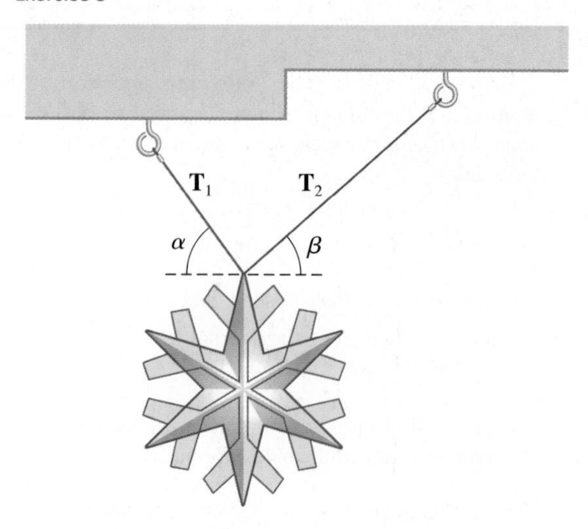

8

Systems of Equations and Inequalities

Applications of mathematics sometimes require working simultaneously with more than one equation in several variables—that is, with a system of equations. In this chapter we develop methods for finding solutions that are common to all the equations in a system. Of particular importance are the techniques involving matrices, because they are well suited for computer programs and can be readily applied to systems containing any number of linear equations in any number of variables. We shall also consider systems of inequalities and linear programming—topics that are of major importance in business applications and statistics. The last part of the chapter provides an introduction to the algebra of matrices and determinants.

8.1 Systems of Equations

Figure 1

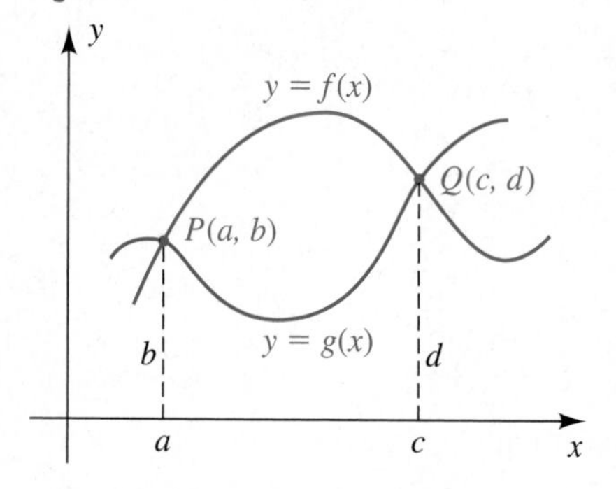

*In previous chapters we **estimated** solutions of systems by using the intersect feature on a graphing utility. We now concentrate on finding **exact** solutions.*

Consider the graphs of the two functions f and g, illustrated in Figure 1. In applications it is often necessary to find points such as $P(a, b)$ and $Q(c, d)$ at which the graphs intersect. Since $P(a, b)$ is on each graph, the pair (a, b) is a **solution** of *both* of the equations $y = f(x)$ and $y = g(x)$; that is,

$$b = f(a) \qquad \text{and} \qquad b = g(a).$$

We say that (a, b) is a solution of the **system of equations** (or simply **system**)

$$\begin{cases} y = f(x) \\ y = g(x) \end{cases}$$

where the brace is used to indicate that the equations are to be treated simultaneously. Similarly, the pair (c, d) is a solution of the system. To **solve** a system of equations means to find all the solutions.

As a special case, consider the system

$$\begin{cases} y = x^2 \\ y = 2x + 3 \end{cases}$$

The graphs of the equations are the parabola and line sketched in Figure 2. The following table shows that the points $(-1, 1)$ and $(3, 9)$ are on both graphs.

(x, y)	$y = x^2$	$y = 2x + 3$
$(-1, 1)$	$1 = (-1)^2$, or $1 = 1$	$1 = 2(-1) + 3$, or $1 = 1$
$(3, 9)$	$9 = 3^2$, or $9 = 9$	$9 = 2(3) + 3$, or $9 = 9$

Figure 2

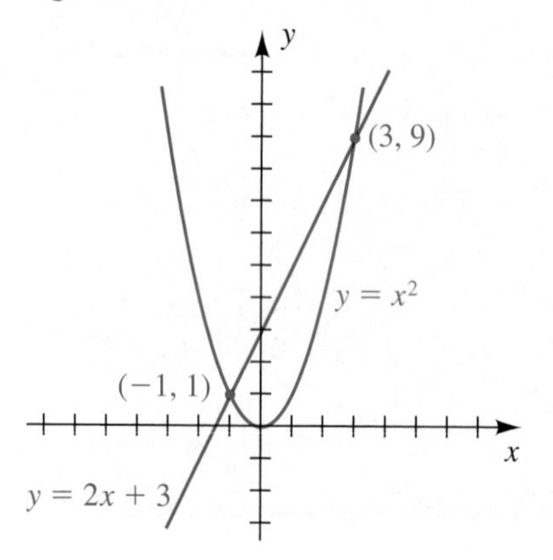

Hence, $(-1, 1)$ and $(3, 9)$ are solutions of the system.

The preceding discussion does not give us a strategy for actually finding the solutions. The next two examples illustrate how to find the solutions of the system using only algebraic methods.

EXAMPLE 1 Solving a system of two equations

Solve the system

$$\begin{cases} y = x^2 \\ y = 2x + 3 \end{cases}$$

▶ **SOLUTION** If (x, y) is a solution of the system, then the variable y in the equation $y = 2x + 3$ must satisfy the condition $y = x^2$. Hence, we *substitute* x^2 for y in $y = 2x + 3$:

$x^2 = 2x + 3$	substitute $y = x^2$ in $y = 2x + 3$
$x^2 - 2x - 3 = 0$	subtract $2x + 3$
$(x + 1)(x - 3) = 0$	factor
$x + 1 = 0,\quad x - 3 = 0$	zero factor theorem
$x = -1,\quad x = 3$	solve for x

▶ Tutorial available at www.thomsonedu.com/login

This gives us the x-values for the solutions (x, y) of the system. To find the corresponding y-values, we may use either $y = x^2$ or $y = 2x + 3$. Using $y = x^2$, we find that

$$\text{if} \quad x = -1, \quad \text{then} \quad y = (-1)^2 = 1$$

and

$$\text{if} \quad x = 3, \quad \text{then} \quad y = 3^2 = 9.$$

Hence, the solutions of the system are $(-1, 1)$ and $(3, 9)$.

We could also have found the solutions by substituting $y = 2x + 3$ in the *first* equation, $y = x^2$, obtaining

$$2x + 3 = x^2.$$

The remainder of the solution is the same.

Given the system in Example 1, we *could* have solved one of the equations for x in terms of y and then substituted in the other equation, obtaining an equation in y alone. Solving the latter equation would give us the y-values for the solutions of the system. The x-values could then be found using one of the given equations. In general, we may use the following guidelines, where u and v denote any two variables (*possibly* x and y). This technique is called the **method of substitution.**

Guidelines for the Method of Substitution for Two Equations in Two Variables

1. Solve one of the equations for one variable u in terms of the other variable v.
2. Substitute the expression for u found in guideline 1 in the other equation, obtaining an equation in v alone.
3. Find the solutions of the equation in v obtained in guideline 2.
4. Substitute the v-values found in guideline 3 in the equation of guideline 1 to find the corresponding u-values.
5. Check each pair (u, v) found in guideline 4 in the given system.

EXAMPLE 2 **Using the method of substitution**

Solve the following system and then sketch the graph of each equation, showing the points of intersection:

$$\begin{cases} x + y^2 = 6 \\ x + 2y = 3 \end{cases}$$

SOLUTION We must first decide which equation to solve and which variable to solve for. Let's examine the possibilities.

Solve the first equation for y: $y = \pm\sqrt{6 - x}$

Solve the first equation for x: $x = 6 - y^2$

Solve the second equation for y: $y = (3 - x)/2$

Solve the second equation for x: $x = 3 - 2y$

(continued)

Tutorial available at www.thomsonedu.com/login

Guideline 1 Looking ahead to guideline 2, we note that solving either equation for x will result in a simple substitution. Thus, we will use $x = 3 - 2y$ and follow the guidelines with $u = x$ and $v = y$.

Guideline 2 Substitute the expression for x found in guideline 1 in the first equation of the system:

$$(3 - 2y) + y^2 = 6 \quad \text{substitute } x = 3 - 2y \text{ in } x + y^2 = 6$$
$$y^2 - 2y - 3 = 0 \quad \text{simplify}$$

Guideline 3 Solve the equation in guideline 2 for y:

$$(y - 3)(y + 1) = 0 \quad \text{factor } y^2 - 2y - 3$$
$$y - 3 = 0, \quad y + 1 = 0 \quad \text{zero factor theorem}$$
$$y = 3, \quad y = -1 \quad \text{solve for } y$$

These are the only possible y-values for the solutions of the system.

Figure 3

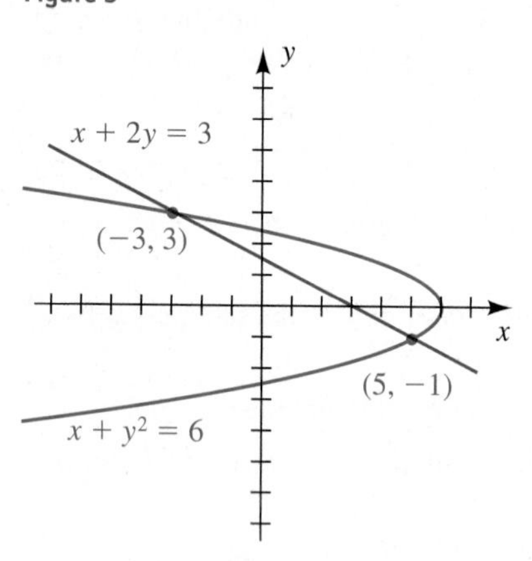

Guideline 4 Use the equation $x = 3 - 2y$ from guideline 1 to find the corresponding x-values:

$$\text{if } y = 3, \quad \text{then } x = 3 - 2(3) = 3 - 6 = -3$$
$$\text{if } y = -1, \quad \text{then } x = 3 - 2(-1) = 3 + 2 = 5$$

Thus, possible solutions are $(-3, 3)$ and $(5, -1)$.

Guideline 5 Substituting $x = -3$ and $y = 3$ in $x + y^2 = 6$, the first equation of the system, yields $-3 + 9 = 6$, a true statement. Substituting $x = -3$ and $y = 3$ in $x + 2y = 3$, the second equation of the system, yields $-3 + 6 = 3$, also a true statement. Hence, $(-3, 3)$ is a solution of the system. In a similar manner, we may check that $(5, -1)$ is also a solution.

The graphs of the two equations (a parabola and a line, respectively) are sketched in Figure 3, showing the two points of intersection.

In future examples we will not list the specific guidelines that are used in finding solutions of systems.

In solving certain systems using the method of substitution, it is convenient to let u or v in the guidelines denote an *expression* involving another variable. This technique is illustrated in the next example with $u = x^2$.

EXAMPLE 3 **Using the method of substitution**

Solve the following system and then sketch the graph of each equation, showing the points of intersection:

$$\begin{cases} x^2 + y^2 = 25 \\ x^2 + y = 19 \end{cases}$$

▶ **SOLUTION** We proceed as follows:

$$\begin{aligned} x^2 &= 19 - y && \text{solve } x^2 + y = 19 \text{ for } x^2 \\ (19 - y) + y^2 &= 25 && \text{substitute } x^2 = 19 - y \text{ in } x^2 + y^2 = 25 \\ y^2 - y - 6 &= 0 && \text{simplify} \\ (y - 3)(y + 2) &= 0 && \text{factor} \\ y - 3 = 0,\quad y + 2 &= 0 && \text{zero factor theorem} \\ y = 3,\quad y &= -2 && \text{solve for } y \end{aligned}$$

These are the only possible y-values for the solutions of the system. To find the corresponding x-values, we use $x^2 = 19 - y$:

$$\begin{aligned} &\text{If}\quad y = 3, &&\text{then}\quad x^2 = 19 - 3 = 16 &&\text{and}\quad x = \pm 4 \\ &\text{If}\quad y = -2, &&\text{then}\quad x^2 = 19 - (-2) = 21 &&\text{and}\quad x = \pm\sqrt{21} \end{aligned}$$

Figure 4

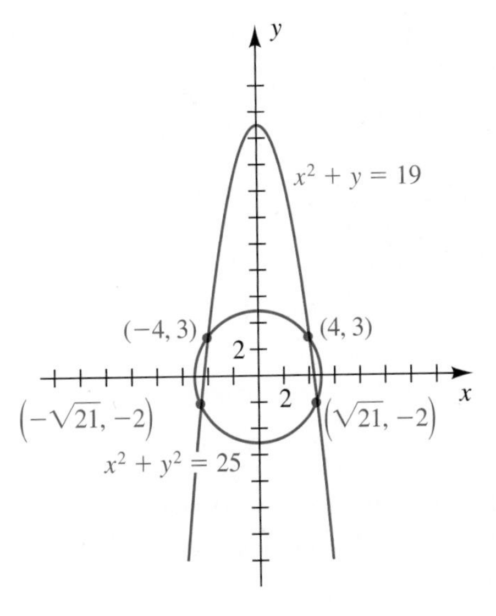

Thus, the only possible solutions of the system are

$$(4, 3),\quad (-4, 3),\quad \left(\sqrt{21}, -2\right),\quad \text{and}\quad \left(-\sqrt{21}, -2\right).$$

We can check by substitution in the given equations that all four pairs are solutions.

The graph of $x^2 + y^2 = 25$ is a circle of radius 5 with center at the origin, and the graph of $y = 19 - x^2$ is a parabola with a vertical axis. The graphs are sketched in Figure 4. The points of intersection correspond to the solutions of the system.

There are, of course, other ways to find the solutions. We could solve the first equation for x^2, $x^2 = 25 - y^2$, and then substitute in the second, obtaining $25 - y^2 + y = 19$. Another method is to solve the second equation for y, $y = 19 - x^2$, and substitute in the first.

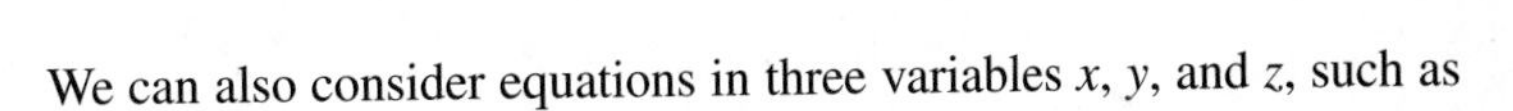

We can also consider equations in three variables x, y, and z, such as

$$x^2y + xz + 3^y = 4z^3.$$

Such an equation has a **solution** (a, b, c) if substitution of a, b, and c, for x, y, and z, respectively, yields a true statement. We refer to (a, b, c) as an **ordered triple** of real numbers. Systems of equations are **equivalent systems** provided they have the same solutions. A system of equations in three variables and the solutions of the system are defined as in the two-variable case. Similarly, we can consider systems of *any* number of equations in *any* number of variables.

The method of substitution can be extended to these more complicated systems. For example, given three equations in three variables, suppose that it is possible to solve one of the equations for one variable in terms of the remaining two variables. By substituting that expression in each of the other equations, we obtain a system of two equations in two variables. The solutions of the two-variable system can then be used to find the solutions of the original system.

▶ **Tutorial available at www.thomsonedu.com/login**

EXAMPLE 4 **Solving a system of three equations**

Solve the system

$$\begin{cases} x - y + z = 2 \\ xyz = 0 \\ 2y + z = 1 \end{cases}$$

▶ **SOLUTION** We proceed as follows:

$$z = 1 - 2y \qquad \text{solve } 2y + z = 1 \text{ for } z$$

$$\begin{cases} x - y + (1 - 2y) = 2 \\ xy(1 - 2y) = 0 \end{cases} \qquad \text{substitute } z = 1 - 2y \text{ in the first two equations}$$

$$\begin{cases} x - 3y - 1 = 0 \\ xy(1 - 2y) = 0 \end{cases} \qquad \text{equivalent system}$$

We now find the solutions of the last system:

$$x = 3y + 1 \qquad \text{solve } x - 3y - 1 = 0 \text{ for } x$$

$$(3y + 1)y(1 - 2y) = 0 \qquad \text{substitute } x = 3y + 1 \text{ in } xy(1 - 2y) = 0$$

$$3y + 1 = 0, \quad y = 0, \quad 1 - 2y = 0 \qquad \text{zero factor theorem}$$

$$y = -\tfrac{1}{3}, \quad y = 0, \quad y = \tfrac{1}{2} \qquad \text{solve for } y$$

These are the only possible y-values for the solutions of the system.

To obtain the corresponding x-values, we substitute for y in the equation $x = 3y + 1$, obtaining

$$x = 0, \qquad x = 1, \qquad \text{and} \qquad x = \tfrac{5}{2}.$$

Using $z = 1 - 2y$ gives us the corresponding z-values

$$z = \tfrac{5}{3}, \qquad z = 1, \qquad \text{and} \qquad z = 0.$$

Thus, the solutions (x, y, z) of the original system must be among the ordered triples

$$\left(0, -\tfrac{1}{3}, \tfrac{5}{3}\right), \qquad (1, 0, 1), \qquad \text{and} \qquad \left(\tfrac{5}{2}, \tfrac{1}{2}, 0\right).$$

Checking each shows that the three ordered triples are solutions of the system.

EXAMPLE 5 **An application of a system of equations**

Is it possible to construct an aquarium with a glass top and two square ends that holds 16 ft^3 of water and requires 40 ft^2 of glass? (Disregard the thickness of the glass.)

▶ **Tutorial available at www.thomsonedu.com/login**

SOLUTION We begin by sketching a typical aquarium and labeling it as in Figure 5, with x and y in feet. Referring to the figure and using formulas for volume and area, we see that

$$\text{volume of the aquarium} = x^2y \qquad \text{length} \times \text{width} \times \text{height}$$

$$\text{square feet of glass required} = 2x^2 + 4xy. \qquad \text{2 ends, 2 sides, top, and bottom}$$

Figure 5

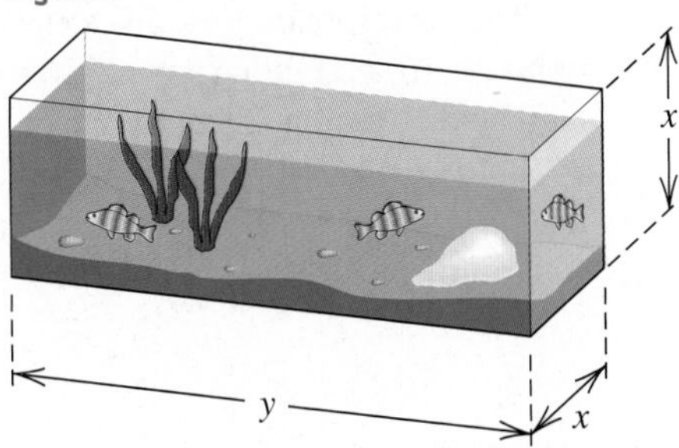

Since the volume is to be 16 ft^3 and the area of the glass required is 40 ft^2, we obtain the following system of equations:

$$\begin{cases} x^2y = 16 \\ 2x^2 + 4xy = 40 \end{cases}$$

We find the solutions as follows:

$$y = \frac{16}{x^2} \qquad \text{solve } x^2y = 16 \text{ for } y$$

$$2x^2 + 4x\left(\frac{16}{x^2}\right) = 40 \qquad \text{substitute } y = \frac{16}{x^2} \text{ in } 2x^2 + 4xy = 40$$

$$x^2 + \frac{32}{x} = 20 \qquad \text{cancel } x\text{, and divide by 2}$$

$$x^3 + 32 = 20x \qquad \text{multiply by } x \ (x \neq 0)$$

$$x^3 - 20x + 32 = 0 \qquad \text{subtract } 20x$$

Graphing $y = x^3 - 20x + 32$ shows two positive zeros. One appears to be 2 and the other slightly larger than 3.

We next look for rational solutions of the last equation. Dividing the polynomial $x^3 - 20x + 32$ synthetically by $x - 2$ gives us

$$\begin{array}{r|rrrr} 2 & 1 & 0 & -20 & 32 \\ & & 2 & 4 & -32 \\ \hline & 1 & 2 & -16 & 0 \end{array}$$

Thus, one solution of $x^3 - 20x + 32 = 0$ is 2, and the remaining two solutions are zeros of the quotient $x^2 + 2x - 16$—that is, roots of the depressed equation

$$x^2 + 2x - 16 = 0.$$

(continued)

By the quadratic formula,

$$x = \frac{-2 \pm \sqrt{2^2 - 4(1)(-16)}}{2(1)} = \frac{-2 \pm 2\sqrt{17}}{2} = -1 \pm \sqrt{17}.$$

Since x is positive, we may discard $x = -1 - \sqrt{17}$. Hence, the only possible values of x are

$$x = 2 \quad \text{and} \quad x = -1 + \sqrt{17} \approx 3.12.$$

The corresponding y-values can be found by substituting for x in the equation $y = 16/x^2$. Letting $x = 2$ gives us $y = \frac{16}{4} = 4$. Using these values, we obtain the dimensions 2 feet by 2 feet by 4 feet for the aquarium.

Letting $x = -1 + \sqrt{17}$, we obtain $y = 16/(-1 + \sqrt{17})^2$, which simplifies to $y = \frac{1}{8}(9 + \sqrt{17}) \approx 1.64$. Thus, approximate dimensions for another aquarium are 3.12 feet by 3.12 feet by 1.64 feet.

8.1 Exercises

Exer. 1–30: Use the method of substitution to solve the system.

1 $\begin{cases} y = x^2 - 4 \\ y = 2x - 1 \end{cases}$

▶ 2 $\begin{cases} y = x^2 + 1 \\ x + y = 3 \end{cases}$

3 $\begin{cases} y^2 = 1 - x \\ x + 2y = 1 \end{cases}$

4 $\begin{cases} y^2 = x \\ x + 2y + 3 = 0 \end{cases}$

5 $\begin{cases} 2y = x^2 \\ y = 4x^3 \end{cases}$

6 $\begin{cases} x - y^3 = 1 \\ 2x = 9y^2 + 2 \end{cases}$

7 $\begin{cases} x + 2y = -1 \\ 2x - 3y = 12 \end{cases}$

8 $\begin{cases} 3x - 4y + 20 = 0 \\ 3x + 2y + 8 = 0 \end{cases}$

9 $\begin{cases} 2x - 3y = 1 \\ -6x + 9y = 4 \end{cases}$

10 $\begin{cases} 4x - 5y = 2 \\ 8x - 10y = -5 \end{cases}$

11 $\begin{cases} x + 3y = 5 \\ x^2 + y^2 = 25 \end{cases}$

12 $\begin{cases} 3x - 4y = 25 \\ x^2 + y^2 = 25 \end{cases}$

13 $\begin{cases} x^2 + y^2 = 8 \\ y - x = 4 \end{cases}$

14 $\begin{cases} x^2 + y^2 = 25 \\ 3x + 4y = -25 \end{cases}$

15 $\begin{cases} x^2 + y^2 = 9 \\ y - 3x = 2 \end{cases}$

16 $\begin{cases} x^2 + y^2 = 16 \\ y + 2x = -1 \end{cases}$

▶ 17 $\begin{cases} x^2 + y^2 = 16 \\ 2y - x = 4 \end{cases}$

18 $\begin{cases} x^2 + y^2 = 1 \\ y + 2x = -3 \end{cases}$

19 $\begin{cases} (x - 1)^2 + (y + 2)^2 = 10 \\ x + y = 1 \end{cases}$

20 $\begin{cases} xy = 2 \\ 3x - y + 5 = 0 \end{cases}$

21 $\begin{cases} y = 20/x^2 \\ y = 9 - x^2 \end{cases}$

22 $\begin{cases} x = y^2 - 4y + 5 \\ x - y = 1 \end{cases}$

23 $\begin{cases} y^2 - 4x^2 = 4 \\ 9y^2 + 16x^2 = 140 \end{cases}$

24 $\begin{cases} 25y^2 - 16x^2 = 400 \\ 9y^2 - 4x^2 = 36 \end{cases}$

▶ 25 $\begin{cases} x^2 - y^2 = 4 \\ x^2 + y^2 = 12 \end{cases}$

26 $\begin{cases} 6x^3 - y^3 = 1 \\ 3x^3 + 4y^3 = 5 \end{cases}$

27 $\begin{cases} x + 2y - z = -1 \\ 2x - y + z = 9 \\ x + 3y + 3z = 6 \end{cases}$

28 $\begin{cases} 2x - 3y - z^2 = 0 \\ x - y - z^2 = -1 \\ x^2 - xy = 0 \end{cases}$

▶ 29 $\begin{cases} x^2 + z^2 = 5 \\ 2x + y = 1 \\ y + z = 1 \end{cases}$

30 $\begin{cases} x + 2z = 1 \\ 2y - z = 4 \\ xyz = 0 \end{cases}$

31 Find the values of b such that the system represented in the graph on the next page has
(a) one solution (b) two solutions (c) no solution

▶ Tutorial available at www.thomsonedu.com/login

Exercise 31

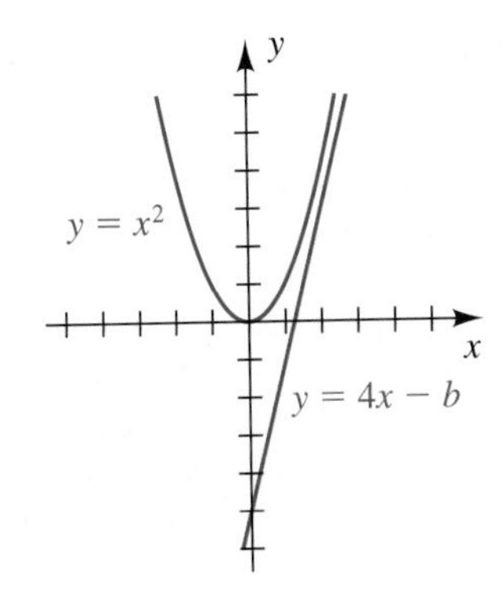

Interpret (a)–(c) graphically.

32 Find the values of b such that the system

$$\begin{cases} x^2 + y^2 = 4 \\ y = x + b \end{cases}$$

has

(a) one solution (b) two solutions

(c) no solution

Interpret (a)–(c) graphically.

33 Is there a real number x such that $x = 2^{-x}$? Decide by displaying graphically the system

$$\begin{cases} y = x \\ y = 2^{-x} \end{cases}$$

34 Is there a real number x such that $x = \log x$? Decide by displaying graphically the system

$$\begin{cases} y = x \\ y = \log x \end{cases}$$

35 Shown in the figure is the graph of $x = y^2$ and a line of slope m that passes through the point (4, 2). Find the value of m such that the line intersects the graph only at (4, 2) and interpret graphically.

Exercise 35

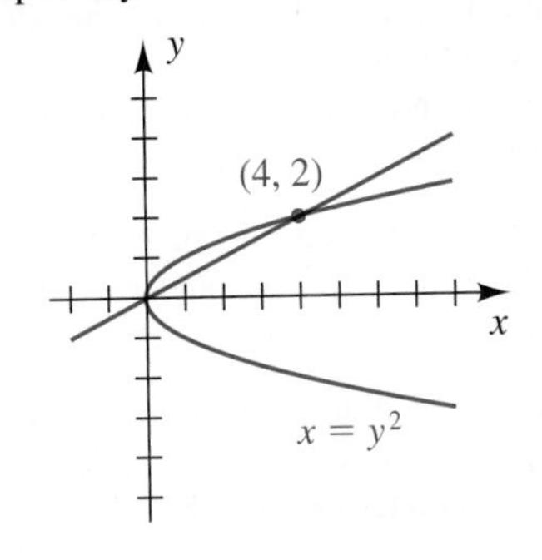

36 Shown in the figure is the graph of $y = x^2$ and a line of slope m that passes through the point (1, 1). Find the value of m such that the line intersects the graph only at (1, 1), and interpret graphically.

Exercise 36

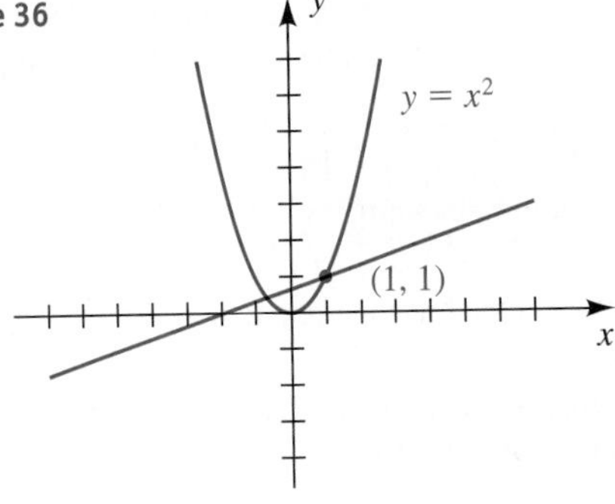

Exer. 37–38: Find an exponential function of the form $f(x) = ba^x + c$ for the graph.

37

38

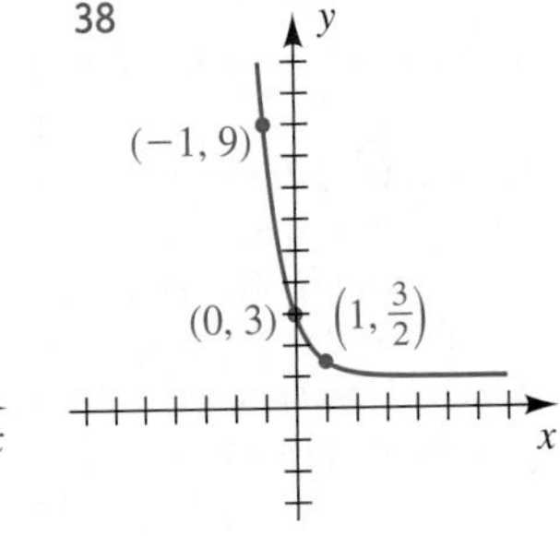

39 The perimeter of a rectangle is 40 inches, and its area is 96 in². Find its length and width.

40 Constructing tubing Sections of cylindrical tubing are to be made from thin rectangular sheets that have an area of 200 in² (see the figure). Is it possible to construct a tube that has a volume of 200 in³? If so, find r and h.

Exercise 40

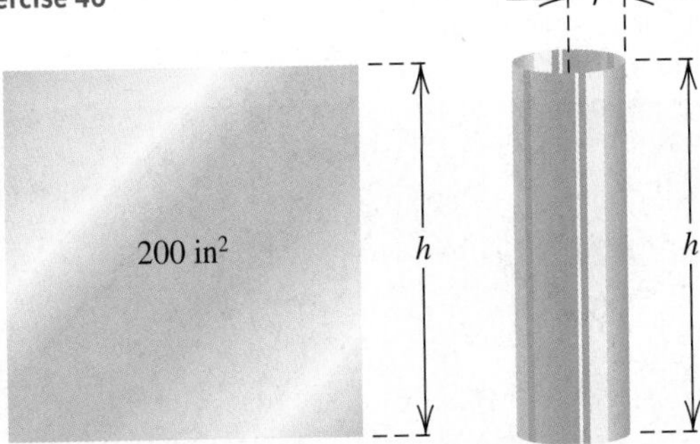

41 Fish population In fishery science, spawner-recruit functions are used to predict the number of adult fish R in next year's breeding population from an estimate S of the number of fish presently spawning.

(a) For a certain species of fish, $R = aS/(S + b)$. Estimate a and b from the data in the following table.

Year	2004	2005	2006
Number spawning	40,000	60,000	72,000

(b) Predict the breeding population for the year 2007.

42 Fish population Refer to Exercise 41. Ricker's spawner-recruit function is given by

$$R = aSe^{-bS}$$

for positive constants a and b. This relationship predicts low recruitment from very high stocks and has been found to be appropriate for many species, such as arctic cod. Rework Exercise 41 using Ricker's spawner-recruit function.

43 Competition for food A *competition model* is a collection of equations that specifies how two or more species interact in competition for the food resources of an ecosystem. Let x and y denote the numbers (in hundreds) of two competing species, and suppose that the respective rates of growth R_1 and R_2 are given by

$$R_1 = 0.01x(50 - x - y),$$
$$R_2 = 0.02y(100 - y - 0.5x).$$

Determine the population levels (x, y) at which both rates of growth are zero. (Such population levels are called *stationary points.*)

▶ **44 Fencing a region** A rancher has 2420 feet of fence to enclose a rectangular region that lies along a straight river. If no fence is used along the river (see the figure), is it possible to enclose 10 acres of land? Recall that 1 acre = 43,560 ft².

Exercise 44

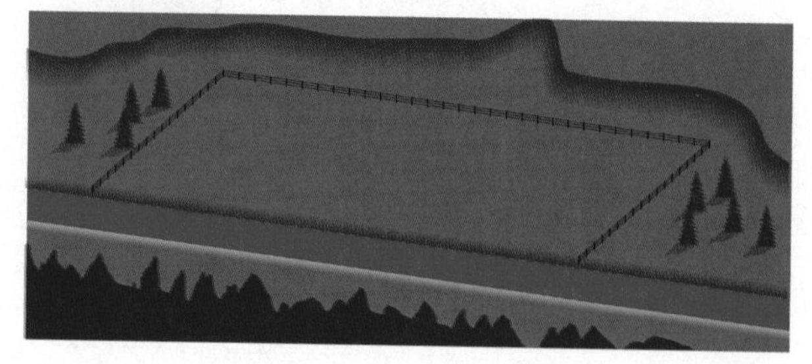

▶ **45 Constructing an aquarium** Refer to Example 5. Is it possible to construct a small aquarium with an *open* top and two square ends that holds 2 ft³ of water and requires 8 ft² of glass? If so, approximate the dimensions. (Disregard the thickness of the glass.)

46 Isoperimetric problem The isoperimetric problem is to prove that of all plane geometric figures with the same perimeter (isoperimetric figures), the circle has the greatest area. Show that no rectangle has both the same area and the same perimeter as any circle.

47 Moiré pattern A moiré pattern is formed when two geometrically regular patterns are superimposed. Shown in the figure is a pattern obtained from the family of circles $x^2 + y^2 = n^2$ and the family of horizontal lines $y = m$ for integers m and n.

(a) Show that the points of intersection of the circle $x^2 + y^2 = n^2$ and the line $y = n - 1$ lie on a parabola.

(b) Work part (a) using the line $y = n - 2$.

Exercise 47

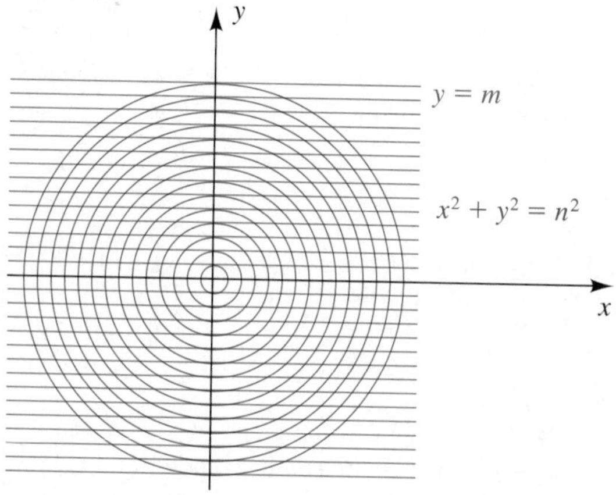

48 Dimensions of a pill A spherical pill has diameter 1 centimeter. A second pill in the shape of a right circular cylinder is to be manufactured with the same volume and twice the surface area of the spherical pill.

(a) If r is the radius and h is the height of the cylindrical pill, show that $6r^2h = 1$ and $r^2 + rh = 1$. Conclude that $6r^3 - 6r + 1 = 0$.

(b) The positive solutions of $6r^3 - 6r + 1 = 0$ are approximately 0.172 and 0.903. Find the corresponding heights, and interpret these results.

49 **Hammer throw** A hammer thrower is working on his form in a small practice area. The hammer spins, generating a circle with a radius of 5 feet, and when released, it hits a tall screen that is 50 feet from the center of the throwing area. Let coordinate axes be introduced as shown in the figure (not to scale).

(a) If the hammer is released at $(-4, -3)$ and travels in the tangent direction, where will it hit the screen?

(b) If the hammer is to hit at $(0, -50)$, where on the circle should it be released?

Exercise 49

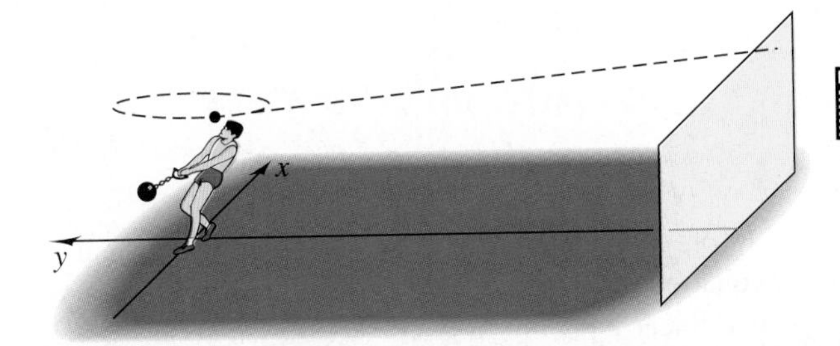

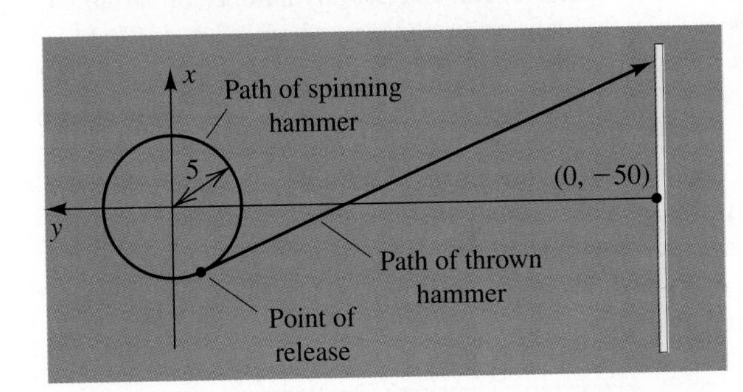

50 **Path of a tossed ball** A person throws a ball from the edge of a hill, at an angle of 45° with the horizontal, as illustrated in the figure. The ball lands 50 feet down the hill, which has slope $-\frac{3}{4}$. Using calculus, it can be shown that the path of the ball is given by $y = ax^2 + x + c$ for some constants a and c.

(a) Disregarding the height of the person, find an equation for the path.

(b) What is the maximum height of the ball *off the ground?*

Exercise 50

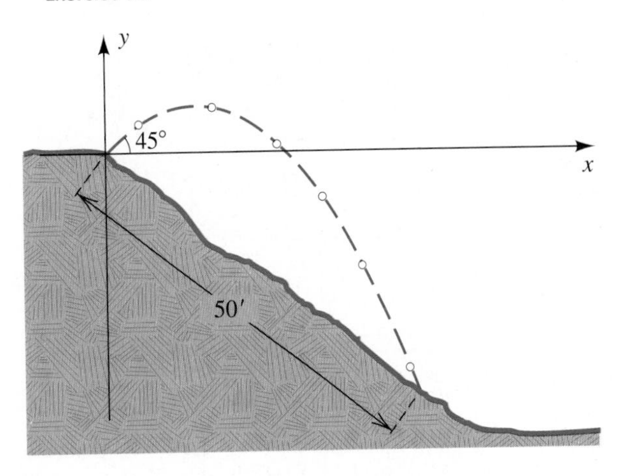

Exer. 51–52: Solve the system of equations graphically and algebraically. Compare your answers.

51 $x^2 + y^2 = 4; \quad x + y = 1$

52 $x^2y^2 = 9; \quad 2x + y = 0$

Exer. 53–56: Graph the two equations on the same coordinate plane, and estimate the coordinates of the points of intersection.

53 $y = 5x^3 - 5x; \quad x^2 + y^2 = 4$

54 $9x^2 + y^2 = 9; \quad y = e^x$

55 $|x + \ln|x|| - y^2 = 0; \quad \dfrac{x^2}{4} + \dfrac{y^2}{2.25} = 1$

56 $y^3 - e^{x/2} = x; \quad y + 0.85x^2 = 2.1$

Exer. 57–60: The data in the table are generated by the function *f*. Graphically approximate the unknown constants *a* and *b* to four decimal places.

57 $f(x) = ae^{-bx}$

x	$f(x)$
1	0.80487
2	0.53930
3	0.36136
4	0.24213

58 $f(x) = a \ln bx$

x	$f(x)$
1	−8.2080
2	−11.7400
3	−13.8061
4	−15.2720

59 $f(x) = ax^2 + e^{bx}$

x	$f(x)$
2	17.2597
3	40.1058
4	81.4579

60 $f(x) = \sqrt{ax + b}$

x	$f(x)$
2	3.8859
4	5.1284
6	6.1238

8.2 *Systems of Linear Equations in Two Variables*

An equation $ax + by = c$ (or, equivalently, $ax + by - c = 0$), with a and b not both zero, is a linear equation in two variables x and y. Similarly, the equation $ax + by + cz = d$ is a linear equation in three variables x, y, and z. We may also consider linear equations in four, five, or *any* number of variables. The most common systems of equations are those in which every equation is linear. In this section we shall consider only systems of two linear equations in two variables. Systems involving more than two variables are discussed in a later section.

Two systems of equations are equivalent if they have the same solutions. To find the solutions of a system, we may manipulate the equations until we obtain an equivalent system of simple equations for which the solutions can be found readily. Some manipulations (or *transformations*) that lead to equivalent systems are stated in the next theorem.

Theorem on Equivalent Systems

Given a system of equations, an equivalent system results if

(1) two equations are interchanged.

(2) an equation is multiplied or divided by a nonzero constant.

(3) a constant multiple of one equation is added to another equation.

A *constant multiple* of an equation is obtained by multiplying *each* term of the equation by the same nonzero constant k. When applying part (3) of the theorem, we often use the phrase *add to one equation k times any other equation*. To *add* two equations means to add corresponding sides of the equations.

The next example illustrates how the theorem on equivalent systems may be used to solve a system of linear equations.

EXAMPLE 1 **Using the theorem on equivalent systems**

Solve the system

$$\begin{cases} x + 3y = -1 \\ 2x - y = 5 \end{cases}$$

▶ **SOLUTION** We often multiply one of the equations by a constant that will give us the additive inverse of the coefficient of one of the variables in the other equation. Doing so enables us to add the two equations and obtain an equation in only one variable, as follows:

$$\begin{cases} x + 3y = -1 \\ 6x - 3y = 15 \end{cases} \quad \text{multiply the second equation by 3}$$

$$\begin{cases} x + 3y = -1 \\ 7x \qquad = 14 \end{cases} \quad \text{add the first equation to the second}$$

We see from the last system that $7x = 14$, and hence $x = \frac{14}{7} = 2$. To find the corresponding y-value, we substitute 2 for x in $x + 3y = -1$, obtaining $y = -1$. Thus, $(2, -1)$ is the only solution of the system.

There are many other ways to use the theorem on equivalent systems to find the solution. Another approach is to proceed as follows:

Figure 1

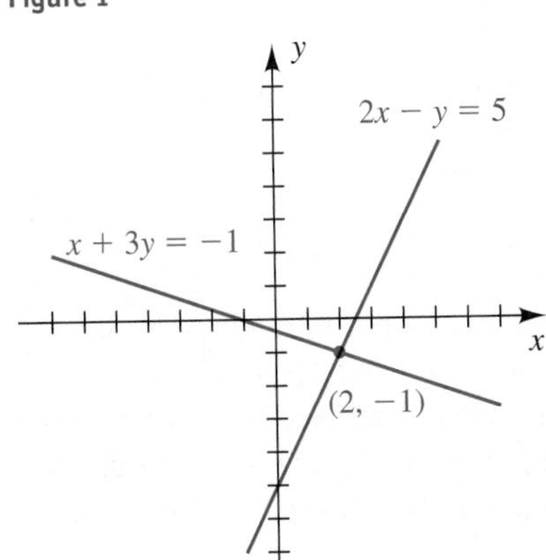

$$\begin{cases} x + 3y = -1 \\ 2x - y = 5 \end{cases} \quad \text{given}$$

$$\begin{cases} -2x - 6y = 2 \\ 2x - y = 5 \end{cases} \quad \text{multiply the first equation by } -2$$

$$\begin{cases} -2x - 6y = 2 \\ \qquad - 7y = 7 \end{cases} \quad \text{add the first equation to the second}$$

We see from the last system that $-7y = 7$, or $y = -1$. To find the corresponding x-value, we could substitute -1 for y in $x + 3y = -1$, obtaining $x = 2$. Hence, $(2, -1)$ is the solution.

The graphs of the two equations are lines that intersect at the point $(2, -1)$, as shown in Figure 1.

The technique used in Example 1 is called the **method of elimination,** since it involves the elimination of a variable from one of the equations. The method of elimination usually leads to solutions in fewer steps than does the method of substitution discussed in the preceding section.

▶ **Tutorial available at www.thomsonedu.com/login**

EXAMPLE 2 **A system of linear equations with an infinite number of solutions**

Solve the system

$$\begin{cases} 3x + \ \ y = \ 6 \\ 6x + 2y = 12 \end{cases}$$

▶ **SOLUTION** Multiplying the second equation by $\frac{1}{2}$ gives us

$$\begin{cases} 3x + y = 6 \\ 3x + y = 6 \end{cases}$$

Figure 2

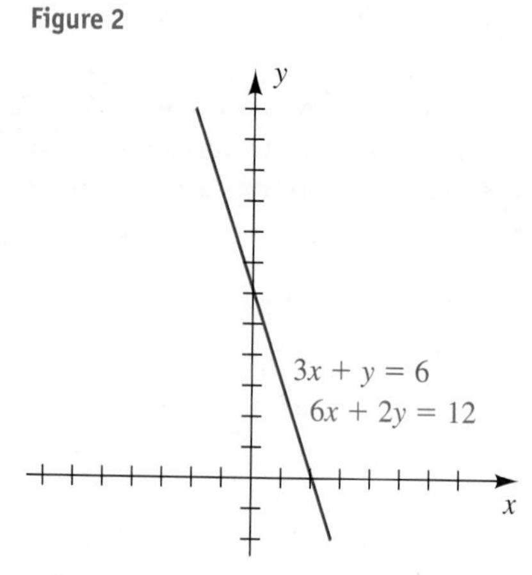

Thus, (a, b) is a solution if and only if $3a + b = 6$—that is, $b = 6 - 3a$. It follows that the solutions consist of ordered pairs of the form $(a, 6 - 3a)$, where a is any real number. If we wish to find particular solutions, we may substitute various values for a. A few solutions are $(0, 6)$, $(1, 3)$, $(3, -3)$, $(-2, 12)$, and $(\sqrt{2}, 6 - 3\sqrt{2})$.

It is incorrect to say that the solution is "all reals." It is correct to say that the solution is the set of all ordered pairs such that $3x + y = 6$, which can be written

$$\{(x, y) \mid 3x + y = 6\}.$$

The graph of each equation is the same line, as shown in Figure 2.

EXAMPLE 3 **A system of linear equations with no solutions**

Solve the system

$$\begin{cases} 3x + \ \ y = \ 6 \\ 6x + 2y = 20 \end{cases}$$

Figure 3

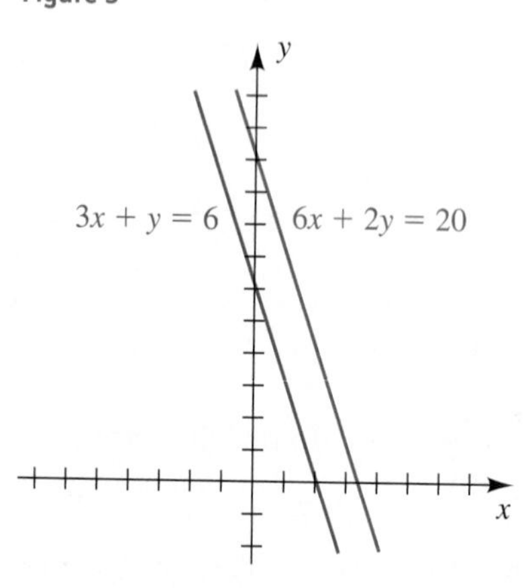

▶ **SOLUTION** If we add to the second equation -2 times the first equation, $-6x - 2y = -12$, we obtain the equivalent system

$$\begin{cases} 3x + y = 6 \\ \qquad 0 = 8 \end{cases}$$

The last equation can be written $0x + 0y = 8$, which is false for every ordered pair (x, y). Thus, the system has no solution.

The graphs of the two equations in the given system are lines that have the same slope and hence are parallel (see Figure 3). The conclusion that the system has no solution corresponds to the fact that these lines do not intersect.

The preceding three examples illustrate typical outcomes of solving a system of two linear equations in two variables: there is either exactly one solution, an infinite number of solutions, or no solution. A system is **consistent** if it has at least one solution. A system with an infinite number of solutions is **dependent and consistent.** A system is **inconsistent** if it has no solution.

▶ Tutorial available at www.thomsonedu.com/login

Since the graph of any linear equation $ax + by = c$ is a line, *exactly one* of the three cases listed in the following table holds for any system of two such equations.

Characteristics of a System of Two Linear Equations in Two Variables

Graphs	Number of solutions	Classification
Nonparallel lines	One solution	Consistent system
Identical lines	Infinite number of solutions	Dependent and consistent system
Parallel lines	No solution	Inconsistent system

In practice, there should be little difficulty determining which of the three cases occurs. The case of the unique solution will become apparent when suitable transformations are applied to the system, as illustrated in Example 1. The case of an infinite number of solutions is similar to that of Example 2, where one of the equations can be transformed into the other. The case of no solution is indicated by a contradiction, such as the statement $0 = 8$, which appeared in Example 3.

In the process of solving a system, suppose we obtain for x a rational number such as $-\frac{41}{29}$. Substituting $-\frac{41}{29}$ for x to find the value of y is cumbersome. It is easier to select a different multiplier for each of the original equations that will enable us to eliminate x and solve for y. This technique is illustrated in the next example.

Example 4 Solving a system

Solve the system

$$\begin{cases} 4x + 7y = 11 \\ 3x - 2y = -9 \end{cases}$$

▶ SOLUTION We select multipliers to eliminate y. (The least common multiple of 7 and 2 is 14.)

$$\begin{cases} 8x + 14y = 22 & \text{multiply the first equation by 2} \\ 21x - 14y = -63 & \text{multiply the second equation by 7} \end{cases}$$

Adding the first equation to the second gives us

$$29x = -41, \quad \text{so} \quad x = -\tfrac{41}{29}.$$

Next, we return to the original system and select multipliers to eliminate x. (The least common multiple of 4 and 3 is 12.)

$$\begin{cases} 4x + 7y = 11 \\ 3x - 2y = -9 \end{cases} \quad \text{original system}$$

$$\begin{cases} 12x + 21y = 33 & \text{multiply the first equation by 3} \\ -12x + 8y = 36 & \text{multiply the second equation by } -4 \end{cases}$$

(continued)

▶ Tutorial available at www.thomsonedu.com/login

Adding the equations gives us

$$29y = 69, \quad \text{so} \quad y = \tfrac{69}{29}.$$

Check $(x, y) = \left(-\tfrac{41}{29}, \tfrac{69}{29}\right)$

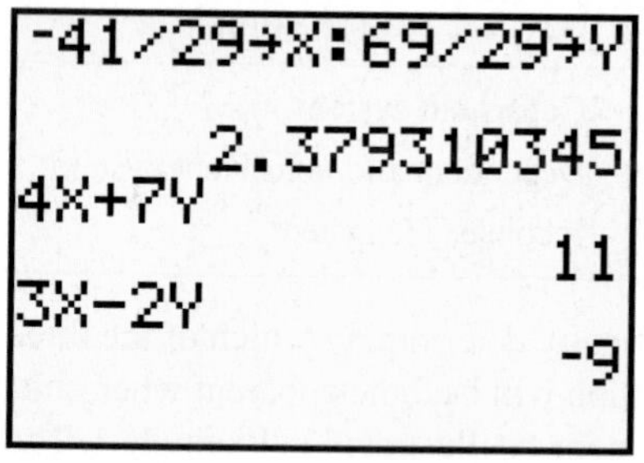

We substitute the values of x and y into the original equations.

$$4x + 7y = 4\left(-\tfrac{41}{29}\right) + 7\left(\tfrac{69}{29}\right) = -\tfrac{164}{29} + \tfrac{483}{29} = \tfrac{319}{29} = 11 \qquad \text{first equation checks}$$

$$3x - 2y = 3\left(-\tfrac{41}{29}\right) - 2\left(\tfrac{69}{29}\right) = -\tfrac{123}{29} - \tfrac{138}{29} = -\tfrac{261}{29} = -9 \qquad \text{so does the second}$$

Figure 4 shows a calculator check of the solution $\left(-\tfrac{41}{29}, \tfrac{69}{29}\right)$.

Figure 4

```
-41/29→X:69/29→Y
         2.379310345
4X+7Y
                  11
3X-2Y
                  -9
```

Certain applied problems can be solved by introducing systems of two linear equations, as illustrated in the next two examples.

EXAMPLE 5 An application of a system of linear equations

A produce company has a 100-acre farm on which it grows lettuce and cabbage. Each acre of cabbage requires 600 hours of labor, and each acre of lettuce needs 400 hours of labor. If 45,000 hours are available and if all land and labor resources are to be used, find the number of acres of each crop that should be planted.

SOLUTION Let us introduce variables to denote the unknown quantities as follows:

$$x = \text{number of acres of cabbage}$$
$$y = \text{number of acres of lettuce}$$

Thus, the number of hours of labor required for each crop can be expressed as follows:

$$600x = \text{number of hours required for cabbage}$$
$$400y = \text{number of hours required for lettuce}$$

Using the facts that the total number of acres is 100 and the total number of hours available is 45,000 leads to the following system:

$$\begin{cases} x + y = 100 \\ 600x + 400y = 45{,}000 \end{cases}$$

We next use the method of elimination:

$$\begin{cases} x + y = 100 \\ 6x + 4y = 450 \end{cases} \qquad \text{divide the second equation by 100}$$

$$\begin{cases} -6x - 6y = -600 \\ 6x + 4y = 450 \end{cases} \qquad \text{multiply the first equation by } -6$$

$$\begin{cases} -6x - 6y = -600 \\ -2y = -150 \end{cases} \qquad \text{add the first equation to the second}$$

Tutorial available at www.thomsonedu.com/login

We see from the last equation that $-2y = -150$, or $y = 75$. Substituting 75 for y in $x + y = 100$ gives us $x = 25$. Hence, the company should plant 25 acres of cabbage and 75 acres of lettuce.

Check Planting 25 acres of cabbage and 75 acres of lettuce requires $(25)(600) + (75)(400) = 45{,}000$ hours of labor. Thus, all 100 acres of land and 45,000 hours of labor are used.

EXAMPLE 6 **Finding the speed of the current in a river**

A motorboat, operating at full throttle, made a trip 4 miles upstream (against a constant current) in 15 minutes. The return trip (with the same current and at full throttle) took 12 minutes. Find the speed of the current and the equivalent speed of the boat in still water.

SOLUTION We begin by introducing variables to denote the unknown quantities. Thus, let

$$x = \text{speed of boat (in mi/hr)}$$
$$y = \text{speed of current (in mi/hr)}.$$

We plan to use the formula $d = rt$, where d denotes the distance traveled, r the rate, and t the time. Since the current slows the boat as it travels upstream but adds to its speed as it travels downstream, we obtain

$$\text{upstream rate} = x - y \quad \text{(in mi/hr)}$$
$$\text{downstream rate} = x + y \quad \text{(in mi/hr)}.$$

The time (in hours) traveled in each direction is

$$\text{upstream time} = \tfrac{15}{60} = \tfrac{1}{4}\text{ hr}$$
$$\text{downstream time} = \tfrac{12}{60} = \tfrac{1}{5}\text{ hr}.$$

The distance is 4 miles for each trip. Substituting in $d = rt$ gives us the system

$$\begin{cases} 4 = (x - y)\left(\tfrac{1}{4}\right) \\ 4 = (x + y)\left(\tfrac{1}{5}\right) \end{cases}$$

Applying the theorem on equivalent systems, we obtain

$$\begin{cases} x - y = 16 \\ x + y = 20 \end{cases} \quad \text{multiply the first equation by 4 and the second by 5}$$

$$\begin{cases} x - y = 16 \\ 2x \phantom{{}-y} = 36 \end{cases} \quad \text{add the first equation to the second}$$

We see from the last equation that $2x = 36$, or $x = 18$. Substituting 18 for x in $x + y = 20$ gives us $y = 2$. Hence, the speed of the boat in still water is 18 mi/hr, and the speed of the current is 2 mi/hr.

Check The upstream rate is $18 - 2 = 16$ mi/hr, and the downstream rate is $18 + 2 = 20$ mi/hr. An upstream 4-mile trip would take $\frac{4}{16} = \frac{1}{4}$ hr $= 15$ min, and a downstream 4-mile trip would take $\frac{4}{20} = \frac{1}{5}$ hr $= 12$ min.

Tutorial available at www.thomsonedu.com/login

8.2 Exercises

Exer. 1–20: Solve the system.

1 $\begin{cases} 2x + 3y = 2 \\ x - 2y = 8 \end{cases}$

2 $\begin{cases} 4x + 5y = 13 \\ 3x + y = -4 \end{cases}$

3 $\begin{cases} 2x + 5y = 16 \\ 3x - 7y = 24 \end{cases}$

4 $\begin{cases} 7x - 8y = 9 \\ 4x + 3y = -10 \end{cases}$

▶ 5 $\begin{cases} 3r + 4s = 3 \\ r - 2s = -4 \end{cases}$

6 $\begin{cases} 9u + 2v = 0 \\ 3u - 5v = 17 \end{cases}$

7 $\begin{cases} 5x - 6y = 4 \\ 3x + 7y = 8 \end{cases}$

8 $\begin{cases} 2x + 8y = 7 \\ 3x - 5y = 4 \end{cases}$

9 $\begin{cases} \frac{1}{3}c + \frac{1}{2}d = 5 \\ c - \frac{2}{3}d = -1 \end{cases}$

10 $\begin{cases} \frac{1}{2}t - \frac{1}{5}v = \frac{3}{2} \\ \frac{2}{3}t + \frac{1}{4}v = \frac{5}{12} \end{cases}$

11 $\begin{cases} \sqrt{3}x - \sqrt{2}y = 2\sqrt{3} \\ 2\sqrt{2}x + \sqrt{3}y = \sqrt{2} \end{cases}$

12 $\begin{cases} 0.11x - 0.03y = 0.25 \\ 0.12x + 0.05y = 0.70 \end{cases}$

13 $\begin{cases} 2x - 3y = 5 \\ -6x + 9y = 12 \end{cases}$

14 $\begin{cases} 3p - q = 7 \\ -12p + 4q = 3 \end{cases}$

▶ 15 $\begin{cases} 3m - 4n = 2 \\ -6m + 8n = -4 \end{cases}$

16 $\begin{cases} x - 5y = 2 \\ 3x - 15y = 6 \end{cases}$

17 $\begin{cases} 2y - 5x = 0 \\ 3y + 4x = 0 \end{cases}$

18 $\begin{cases} 3x + 7y = 9 \\ y = 5 \end{cases}$

▶ 19 $\begin{cases} \dfrac{2}{x} + \dfrac{3}{y} = -2 \\ \dfrac{4}{x} - \dfrac{5}{y} = 1 \end{cases}$ $\left(\textit{Hint}: \text{Let } u = \dfrac{1}{x} \text{ and } v = \dfrac{1}{y}.\right)$

20 $\begin{cases} \dfrac{3}{x-1} + \dfrac{4}{y+2} = 2 \\ \dfrac{6}{x-1} - \dfrac{7}{y+2} = -3 \end{cases}$

21 **Ticket sales** The price of admission to a high school play was \$3.00 for students and \$4.50 for nonstudents. If 450 tickets were sold for a total of \$1555.50, how many of each kind were purchased?

▶ 22 **Air travel** An airline that flies from Los Angeles to Albuquerque with a stopover in Phoenix charges a fare of \$90 to Phoenix and a fare of \$120 from Los Angeles to Albuquerque. A total of 185 passengers boarded the plane in Los Angeles, and fares totaled \$21,000. How many passengers got off the plane in Phoenix?

23 **Crayon dimensions** A crayon 8 centimeters in length and 1 centimeter in diameter will be made from 5 cm^3 of colored wax. The crayon is to have the shape of a cylinder surmounted by a small conical tip (see the figure). Find the length x of the cylinder and the height y of the cone.

Exercise 23

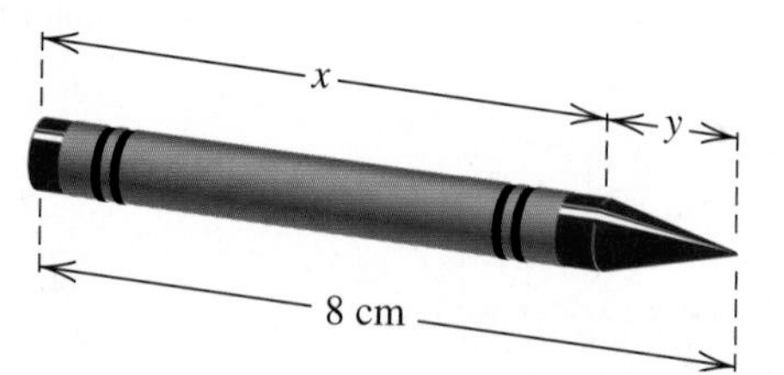

24 **Rowing a boat** A man rows a boat 500 feet upstream against a constant current in 10 minutes. He then rows 300 feet downstream (with the same current) in 5 minutes. Find the speed of the current and the equivalent rate at which he can row in still water.

25 **Table top dimensions** A large table for a conference room is to be constructed in the shape of a rectangle with two semicircles at the ends (see the figure). The table is to have a perimeter of 40 feet, and the area of the rectangular portion is to be twice the sum of the areas of the two ends. Find the length l and the width w of the rectangular portion.

Exercise 25

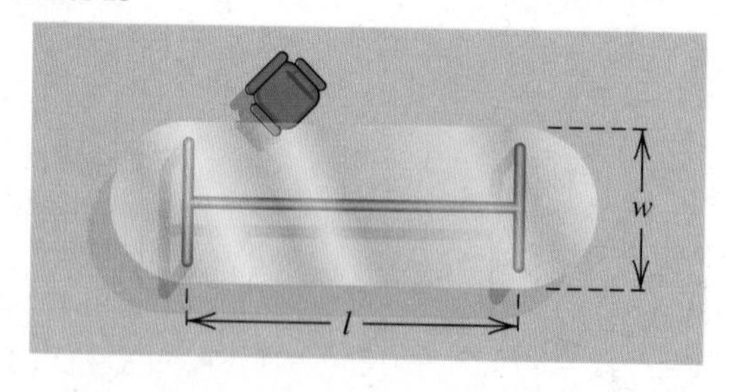

26 **Investment income** A woman has \$19,000 to invest in two funds that pay simple interest at the rates of 4% and 6% per

year. Interest on the 4% fund is tax-exempt; however, income tax must be paid on interest on the 6% fund. Being in a high tax bracket, the woman does not wish to invest the entire sum in the 6% account. Is there a way of investing the money so that she will receive \$1000 in interest at the end of one year?

27 **Bobcat population** A bobcat population is classified by age into kittens (less than 1 year old) and adults (at least 1 year old). All adult females, including those born the prior year, have a litter each June, with an average litter size of 3 kittens. The springtime population of bobcats in a certain area is estimated to be 6000, and the male-female ratio is one. Estimate the number of adults and kittens in the population.

28 **Flow rates** A 300-gallon water storage tank is filled by a single inlet pipe, and two identical outlet pipes can be used to supply water to the surrounding fields (see the figure). It takes 5 hours to fill an empty tank when both outlet pipes are open. When one outlet pipe is closed, it takes 3 hours to fill the tank. Find the flow rates (in gallons per hour) in and out of the pipes.

Exercise 28

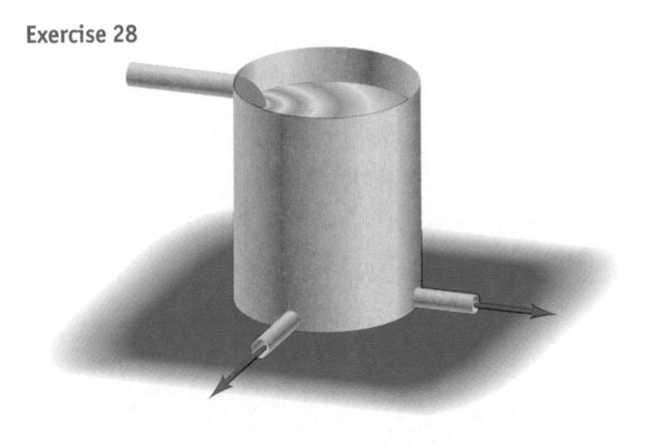

29 **Mixing a silver alloy** A silversmith has two alloys, one containing 35% silver and the other 60% silver. How much of each should be melted and combined to obtain 100 grams of an alloy containing 50% silver?

30 **Mixing nuts** A merchant wishes to mix peanuts costing \$3 per pound with cashews costing \$8 per pound to obtain 60 pounds of a mixture costing \$5 per pound. How many pounds of each variety should be mixed?

31 **Air travel** An airplane, flying with a tail wind, travels 1200 miles in 2 hours. The return trip, against the wind, takes $2\frac{1}{2}$ hours. Find the cruising speed of the plane and the speed of the wind (assume that both rates are constant).

32 **Filling orders** A stationery company sells two types of notepads to college bookstores, the first wholesaling for 50¢ and the second for 70¢. The company receives an order for 500 notepads, together with a check for \$286. If the order fails to specify the number of each type, how should the company fill the order?

33 **Acceleration** As a ball rolls down an inclined plane, its velocity $v(t)$ (in cm/sec) at time t (in seconds) is given by $v(t) = v_0 + at$ for initial velocity v_0 and acceleration a (in cm/sec^2). If $v(2) = 16$ and $v(5) = 25$, find v_0 and a.

34 **Vertical projection** If an object is projected vertically upward from an altitude of s_0 feet with an initial velocity of v_0 ft/sec, then its distance $s(t)$ above the ground after t seconds is

$$s(t) = -16t^2 + v_0t + s_0.$$

If $s(1) = 84$ and $s(2) = 116$, what are v_0 and s_0?

35 **Planning production** A small furniture company manufactures sofas and recliners. Each sofa requires 8 hours of labor and \$180 in materials, while a recliner can be built for \$105 in 6 hours. The company has 340 hours of labor available each week and can afford to buy \$6750 worth of materials. How many recliners and sofas can be produced if all labor hours and all materials must be used?

36 **Livestock diet** A rancher is preparing an oat-cornmeal mixture for livestock. Each ounce of oats provides 4 grams of protein and 18 grams of carbohydrates, and an ounce of cornmeal provides 3 grams of protein and 24 grams of carbohydrates. How many ounces of each can be used to meet the nutritional goals of 200 grams of protein and 1320 grams of carbohydrates per feeding?

37 **Services swap** A plumber and an electrician are each doing repairs on their offices and agree to swap services. The number of hours spent on each of the projects is shown in the following table.

	Plumber's office	Electrician's office
Plumber's hours	6	4
Electrician's hours	5	6

They would prefer to call the matter even, but because of tax laws, they must charge for all work performed. They agree to select hourly wage rates so that the bill on each project will match the income that each person would ordinarily receive for a comparable job.

(continued)

(a) If x and y denote the hourly wages of the plumber and electrician, respectively, show that

$$6x + 5y = 10x \quad \text{and} \quad 4x + 6y = 11y.$$

Describe the solutions to this system.

(b) If the plumber ordinarily makes \$35 per hour, what should the electrician charge?

38 Find equations for the altitudes of the triangle with vertices $A(-3, 2)$, $B(5, 4)$, and $C(3, -8)$, and find the point at which the altitudes intersect.

39 **Warming trend in Paris** As a result of urbanization, the temperatures in Paris have increased. In 1891 the average daily minimum and maximum temperatures were 5.8°C and 15.1°C, respectively. Between 1891 and 1968, these average temperatures rose 0.019°C/yr and 0.011°C/yr, respectively. Assuming the increases were linear, find the year when the difference between the minimum and maximum temperatures was 9°C, and determine the corresponding average maximum temperature.

40 **Long distance telephone rates** A telephone company charges customers a certain amount for the first minute of a long distance call and another amount for each additional minute. A customer makes two calls to the same city—a 36-minute call for \$2.93 and a 13-minute call for \$1.09.

(a) Determine the cost for the first minute and the cost for each additional minute.

(b) If there is a federal tax rate of 3.2% and a state tax rate of 7.2% on all long distance calls, find, to the nearest minute, the longest call to the same city whose cost will not exceed \$5.00.

41 **VCR taping** An avid tennis watcher wants to record 6 hours of a major tournament on a single tape. Her tape can hold 5 hours and 20 minutes at the LP speed and 8 hours at the slower SLP speed. The LP speed produces a better quality picture, so she wishes to maximize the time recorded at the LP speed. Find the amount of time to be recorded at each speed.

42 **Price and demand** Suppose consumers will buy 1,000,000 T-shirts if the selling price is \$15, but for each \$1 increase in price, they will buy 100,000 fewer T-shirts. Moreover, suppose vendors will order 2,000,000 T-shirts if the selling price is \$20, and for every \$1 increase in price, they will order an additional 150,000.

(a) Express the number Q of T-shirts consumers will buy if the selling price is p dollars.

(b) Express the number K of T-shirts vendors will order if the selling price is p dollars.

(c) Determine the market price—that is, the price when $Q = K$.

Exer. 43–46: Solve the system for a and b. (*Hint:* Treat terms such as e^{3x}, $\cos x$, and $\sin x$ as "constant coefficients.")

43 $\begin{cases} ae^{3x} + be^{-3x} = 0 \\ a(3e^{3x}) + b(-3e^{-3x}) = e^{3x} \end{cases}$

44 $\begin{cases} ae^{-x} + be^{4x} = 0 \\ -ae^{-x} + b(4e^{4x}) = 2 \end{cases}$

45 $\begin{cases} a\cos x + b\sin x = 0 \\ -a\sin x + b\cos x = \tan x \end{cases}$

46 $\begin{cases} a\cos x + b\sin x = 0 \\ -a\sin x + b\cos x = \sin x \end{cases}$

8.3 Systems of Inequalities

In Chapter 1 we restricted our discussion of inequalities to inequalities in one variable. We shall now consider inequalities in two variables x and y, such as those shown in the following illustration.

ILLUSTRATION Inequalities in x and y

- $y^2 < x + 4$
- $3x - 4y > 12$
- $x^2 + y^2 \le 16$

A **solution** of an inequality in x and y is an ordered pair (a, b) that yields a true statement if a and b are substituted for x and y, respectively. To **solve** an inequality in x and y means to find all the solutions. The **graph** of such an inequality is the set of all points (a, b) in an xy-plane that correspond to the solutions. Two inequalities are **equivalent** if they have the same solutions.

Given an inequality in x and y, if we replace the inequality symbol with an equal sign, we obtain an equation whose graph usually separates the xy-plane into two regions. We shall consider only equations having the property that if R is one such region and if a **test point** (p, q) in R yields a solution of the inequality, then *every* point in R yields a solution. The following guidelines may then be used to sketch the graph of the inequality.

Guidelines for Sketching the Graph of an Inequality in x and y

1. Replace the inequality symbol with an equal sign, and graph the resulting equation. Use dashes if the inequality symbol is $<$ or $>$ to indicate that no point on the graph yields a solution. Use a solid line or curve for $\leq$ or $\geq$ to indicate that solutions of the equation are also solutions of the inequality.
2. If R is a region of the xy-plane determined by the graph in guideline 1 and if a test point (p, q) in R yields a solution of the inequality, then every point in R yields a solution. Shade R to indicate this fact. If (p, q) is not a solution, then *no* point in R yields a solution and R is left unshaded.

The use of these guidelines is demonstrated in the next example.

Figure 1

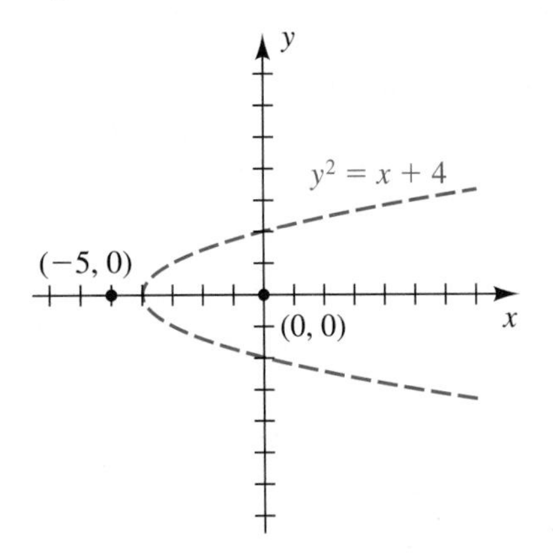

EXAMPLE 1 **Sketching the graph of an inequality**

Find the solutions and sketch the graph of the inequality $y^2 < x + 4$.

▶ **SOLUTION**

Guideline 1 We replace $<$ with $=$, obtaining $y^2 = x + 4$. The graph of this equation is a parabola, symmetric with respect to the x-axis and having x-intercept -4 and y-intercepts ± 2. Since the inequality symbol is $<$, we sketch the parabola using dashes, as in Figure 1.

Guideline 2 The graph in guideline 1 separates the xy-plane into two regions, one to the *left* of the parabola and the other to the *right.* Let us choose test points $(-5, 0)$ and $(0, 0)$ in the regions (see Figure 1) and substitute for x and y in $y^2 < x + 4$ as follows:

Test point $(-5, 0)$ LS: $0^2 = 0$
RS: $-5 + 4 = -1$

Since $0 < -1$ is a *false* statement, $(-5, 0)$ is *not* a solution of the inequality. Hence, *no* point to the left of the parabola is a solution, and we leave that region unshaded.

(continued)

▶ **Tutorial available at www.thomsonedu.com/login**

Figure 2

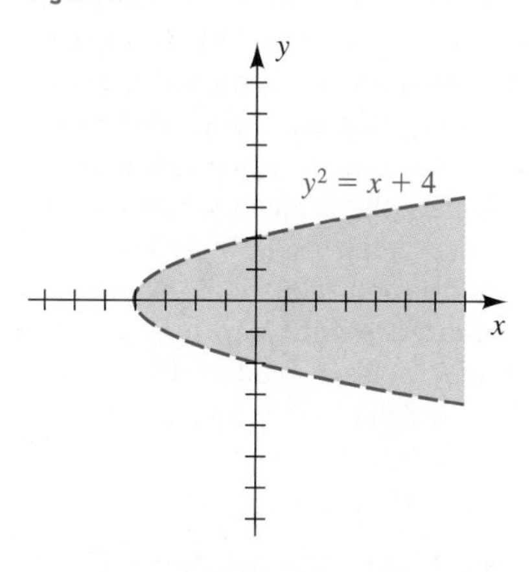

Test point $(0, 0)$ LS: $0^2 = 0$
RS: $0 + 4 = 4$

Since $0 < 4$ is a *true* statement, $(0, 0)$ *is* a solution of the inequality. Hence, *all* points to the right of the parabola are solutions, so we shade this region, as in Figure 2.

A **linear inequality** is an inequality that can be written in one of the following forms, where a, b, and c are real numbers:

$$ax + by < c, \quad ax + by > c, \quad ax + by \leq c, \quad ax + by \geq c$$

The line $ax + by = c$ separates the xy-plane into two **half-planes,** as illustrated in Figure 3. The solutions of the inequality consist of all points in *one* of these half-planes, where the line is included for $\leq$ or $\geq$ and is not included for $<$ or $>$. *For a linear inequality, only one test point* (p, q) *is required,* because if (p, q) is a solution, then the half-plane with (p, q) in it contains all the solutions, whereas if (p, q) is *not* a solution, then the *other* half-plane contains the solutions.

Figure 3

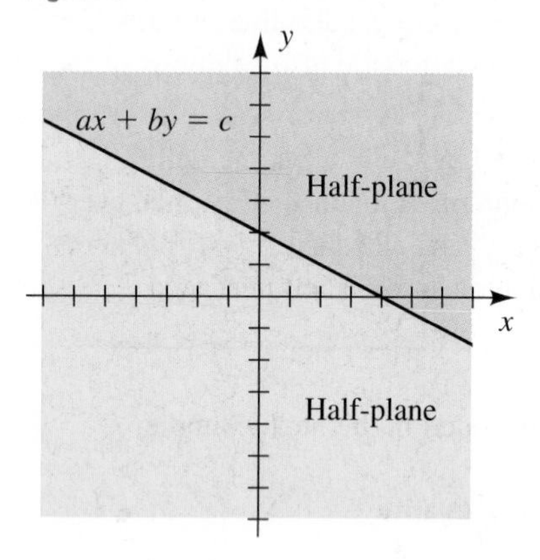

EXAMPLE 2 Sketching the graph of a linear inequality

Sketch the graph of the inequality $3x - 4y > 12$.

SOLUTION Replacing $>$ with $=$ gives us the line $3x - 4y = 12$, sketched with dashes in Figure 4. This line separates the xy-plane into two half-planes, one *above* the line and the other *below* the line. It is convenient to choose the test point $(0, 0)$ above the line and substitute in $3x - 4y > 12$, as follows:

Test point $(0, 0)$ LS: $3 \cdot 0 - 4 \cdot 0 = 0 - 0 = 0$
RS: 12

Since $0 > 12$ is a false statement, $(0, 0)$ is not a solution. Thus, no point above the line is a solution, and the solutions of $3x - 4y > 12$ are given by the points in the half-plane *below* the line. The graph is sketched in Figure 5.

Figure 4

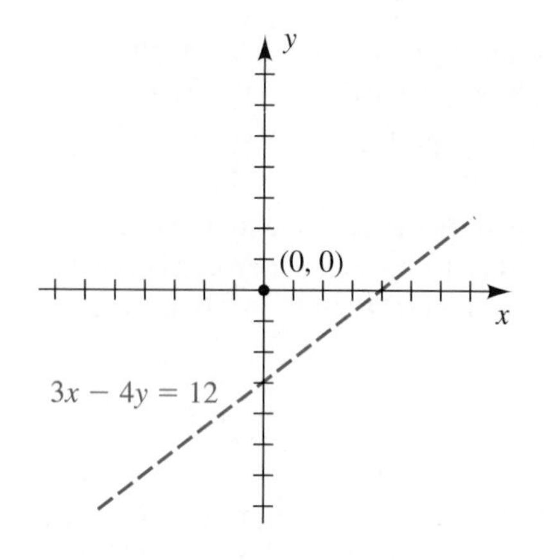

Figure 5

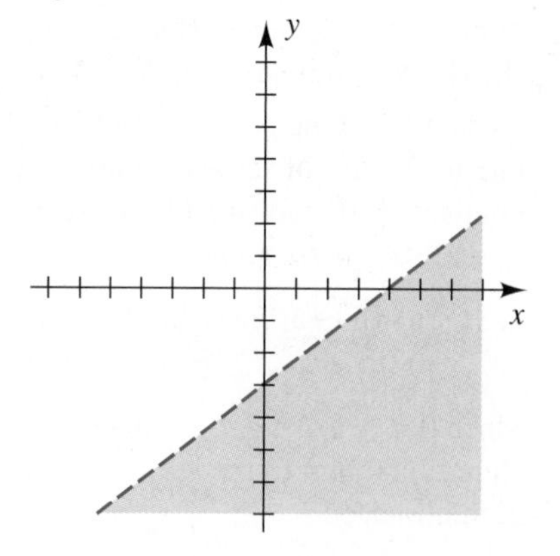

Tutorial available at www.thomsonedu.com/login

As we did with equations, we sometimes work simultaneously with several inequalities in two variables—that is, with a **system of inequalities.** The **solutions** of a system of inequalities are the solutions common to all inequalities in the system. The **graph** of a system of inequalities consists of the points corresponding to the solutions. The following examples illustrate a method for solving systems of inequalities.

Figure 6

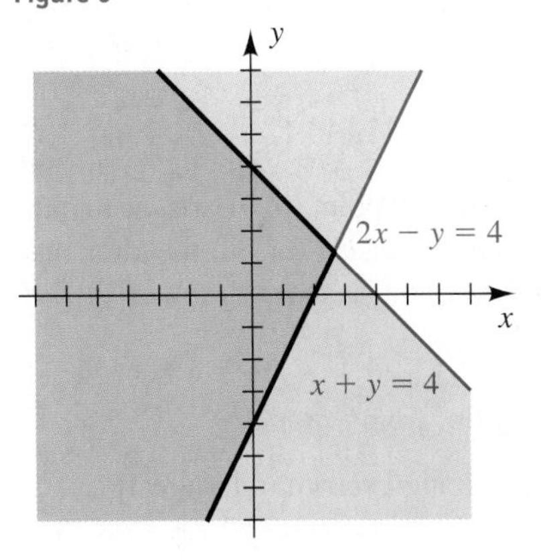

EXAMPLE 3 Solving a system of linear inequalities

Sketch the graph of the system

$$\begin{cases} x + y \leq 4 \\ 2x - y \leq 4 \end{cases}$$

▶ SOLUTION We replace each $\leq$ with $=$ and then sketch the resulting lines, as shown in Figure 6. Using the test point $(0, 0)$, we see that the solutions of the system correspond to the points *below* (and on) the line $x + y = 4$ and *above* (and on) the line $2x - y = 4$. Shading these half-planes with different colors, as in Figure 6, we have as the graph of the system the points that are in *both* regions, indicated by the purple portion of the figure.

Figure 7

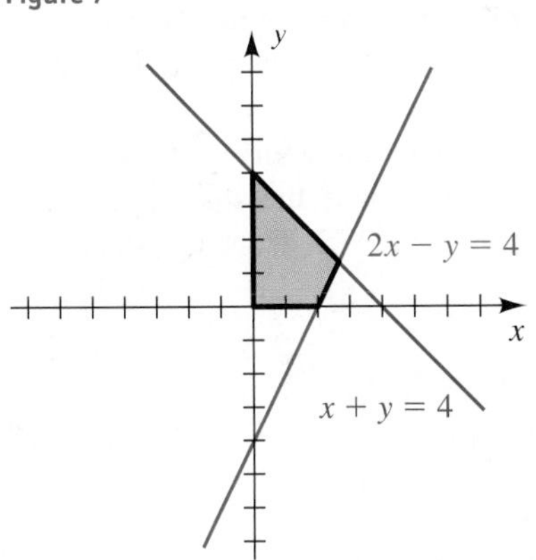

EXAMPLE 4 Solving a system of linear inequalities

Sketch the graph of the system

$$\begin{cases} x + y \leq 4 \\ 2x - y \leq 4 \\ x \geq 0 \\ y \geq 0 \end{cases}$$

▶ SOLUTION The first two inequalities are the same as those considered in Example 3, and hence the points on the graph of the system must lie within the purple region shown in Figure 6. In addition, the third and fourth inequalities in the system tell us that the points must lie in the first quadrant or on its boundaries. This gives us the region shown in Figure 7.

Figure 8

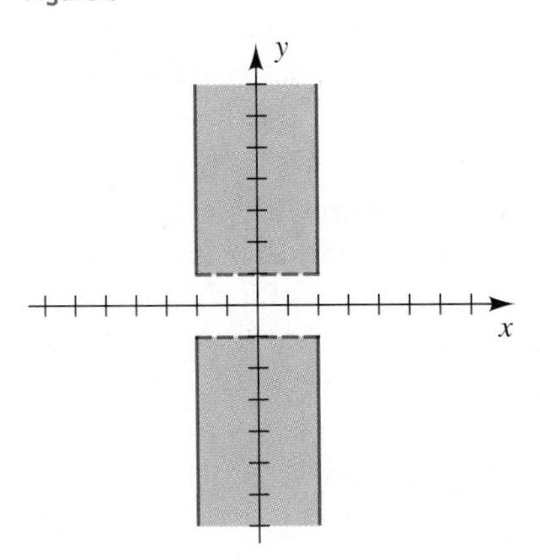

EXAMPLE 5 Solving a system of inequalities containing absolute values

Sketch the graph of the system

$$\begin{cases} |x| \leq 2 \\ |y| > 1 \end{cases}$$

SOLUTION Using properties of absolute values (listed on page 79), we see that (x, y) is a solution of the system if and only if *both* of the following conditions are true:

(1) $-2 \leq x \leq 2$

(2) $y < -1$ or $y > 1$

Thus, a point (x, y) on the graph of the system must lie between (or on) the vertical lines $x = \pm 2$ and also either below the horizontal line $y = -1$ or above the line $y = 1$. The graph is sketched in Figure 8.

▶ **Tutorial available at www.thomsonedu.com/login**

Figure 9

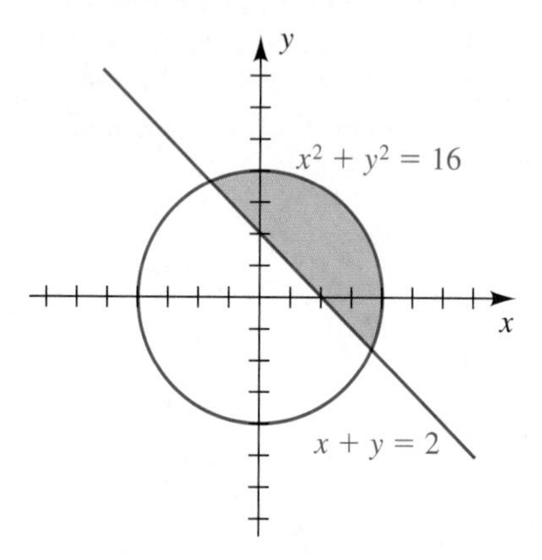

Figure 10

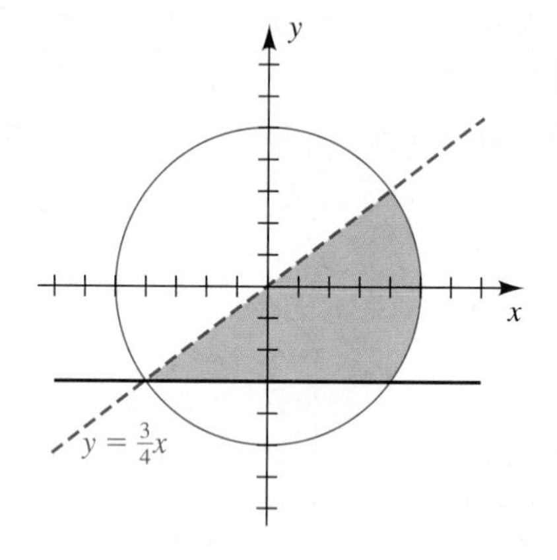

EXAMPLE 6 **Solving a system of inequalities**

Sketch the graph of the system

$$\begin{cases} x^2 + y^2 \leq 16 \\ x + y \geq 2 \end{cases}$$

▶ **SOLUTION** The graphs of $x^2 + y^2 = 16$ and $x + y = 2$ are the circle and line, respectively, shown in Figure 9. Using the test point (0, 0), we see that the points that yield solutions of the system must lie inside (or on) the circle and also above (or on) the line. This gives us the region sketched in Figure 9.

EXAMPLE 7 **Finding a system of inequalities from a graph**

Find a system of inequalities for the shaded region shown in Figure 10.

▶ **SOLUTION** An equation of the circle is $x^2 + y^2 = 5^2$. Since the *interior* of the solid circle is shaded, the shaded region (including the circle) can be described by $x^2 + y^2 \leq 25$. The *exterior* of the circle could be described by $x^2 + y^2 > 25$.

Because the shaded region is *below* the dashed line with equation $y = \frac{3}{4}x$, it is described by the inequality $y < \frac{3}{4}x$. Lastly, since the shaded region is *above* the solid horizontal line $y = -3$, we use $y \geq -3$. Thus, a system is

$$\begin{cases} x^2 + y^2 \leq 25 \\ y < \frac{3}{4}x \\ y \geq -3 \end{cases}$$

EXAMPLE 8 **An application of a system of inequalities**

The manager of a baseball team wishes to buy bats and balls costing \$20 and \$5 each, respectively. At least five bats and ten balls are required, and the total cost is not to exceed \$300. Find a system of inequalities that describes all possibilities, and sketch the graph.

▶ **SOLUTION** We begin by letting x denote the number of bats and y the number of balls. Since the cost of a bat is \$20 and the cost of a ball is \$5, we see that

$$20x = \text{cost of } x \text{ bats}$$
$$5y = \text{cost of } y \text{ balls.}$$

Since the total cost is not to exceed \$300, we must have

$$20x + 5y \leq 300$$

▶ Tutorial available at www.thomsonedu.com/login

Figure 11

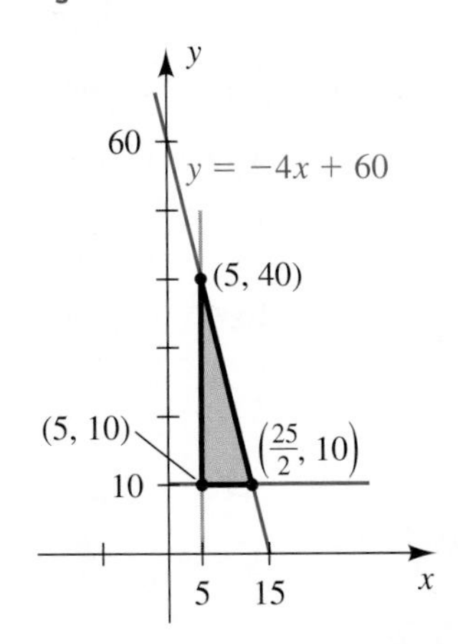

or, equivalently,

$$y \le -4x + 60.$$

Since at least five bats and ten balls are required, we also have

$$x \ge 5 \qquad \text{and} \qquad y \ge 10.$$

The graph of $y \le -4x + 60$ is the half-plane that lies *below* (or on) the line $y = -4x + 60$ shown in Figure 11.

The graph of $x \ge 5$ is the region to the right of (or on) the vertical line $x = 5$, and the graph of $y \ge 10$ is the region above (or on) the horizontal line $y = 10$.

The graph of the system—that is, the points common to the three half-planes—is the triangular region sketched in Figure 11.

EXAMPLE 9 Graphing an inequality

Graph the inequality $27y^3 \le 8 + x^3$.

SOLUTION We must first solve the associated *equality* for y:

$$27y^3 = 8 + x^3 \qquad \text{equality}$$

$$y^3 = \tfrac{1}{27}(8 + x^3) \qquad \text{divide by 27}$$

$$y = \tfrac{1}{3}\sqrt[3]{8 + x^3} \qquad \text{take the cube root of both sides}$$

We assign $\frac{1}{3}\sqrt[3]{8 + x^3}$ to Y_1 and graph Y_1 in the viewing rectangle $[-6, 6]$ by $[-4, 4]$, as shown in Figure 12. The test point $(0, 0)$ is in the solution region (since $0 \le 8$ is true), so we want to shade the region below the graph of Y_1. The commands for the TI-83/4 Plus and the TI-86 are shown.

Figure 12

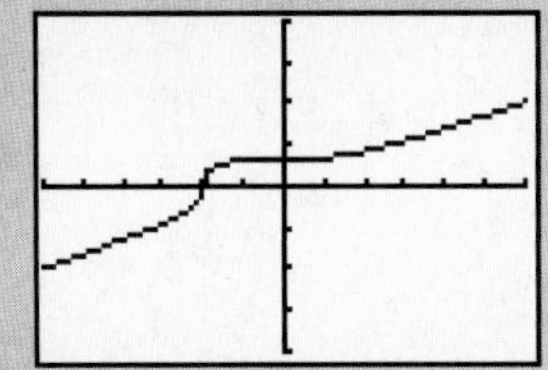

TI-83/4 Plus

2nd DRAW 7 −4 , VARS ▷ 1 1 , −6 , 6 , 1 , 3)

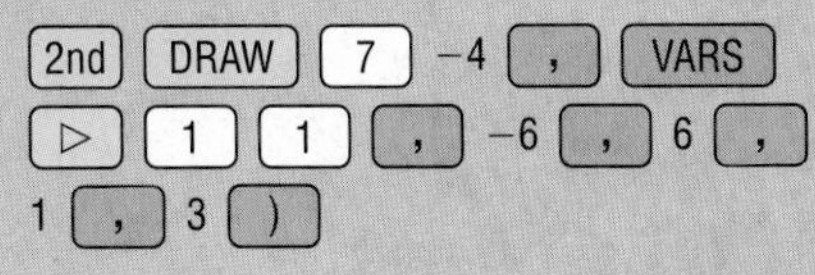

TI-86

GRAPH MORE DRAW(F2) Shade(F1) −4 , 2nd alpha Y 1 , −6 , 6 , 4 , 4)

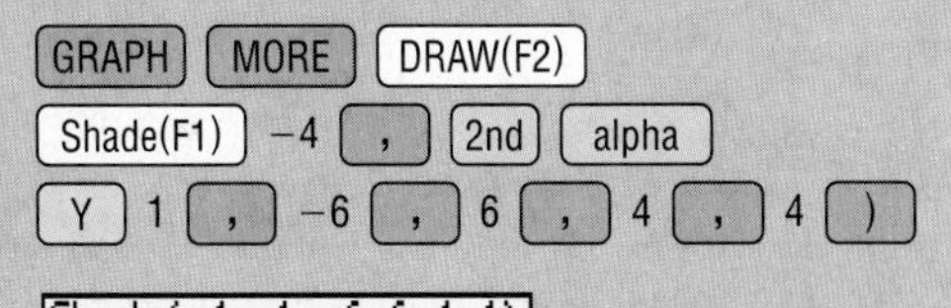

(continued)

The parameters for the Shade command are as follows:

-4 is the lower function for the shaded region—in this case, we simply use the value of Ymin.

Y_1 is the upper function for the shaded region.

-6 and 6 are Xmin and Xmax.

1 (or 4) is the shading pattern; there are four of them.

3 (or 4) shades every third (or fourth) pixel; you may specify an integer from 1 to 8.

Pressing ENTER gives the following graphs.

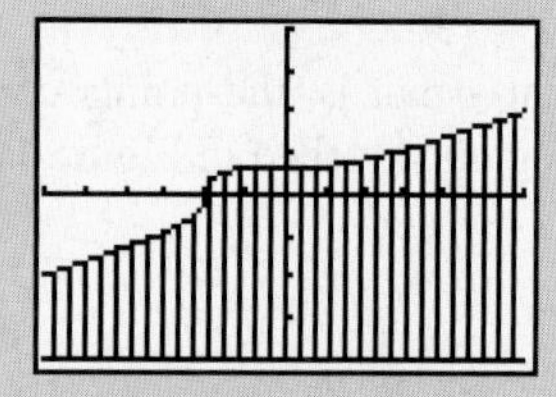

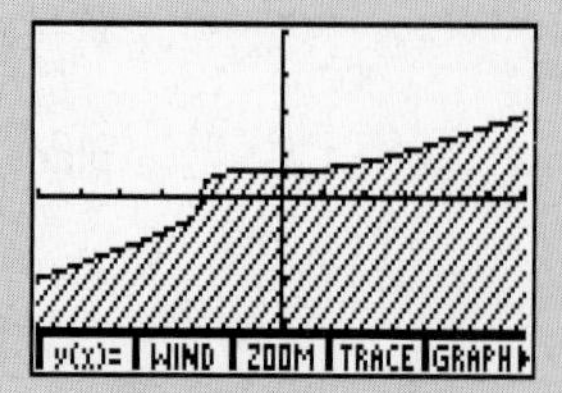

Alternative Method: There is an alternative method for shading available on each calculator. It can be executed by selecting a graphing style from the Y= or y(x)=(F1) screen.

Using the cursor keys, move the cursor to the left of "Y_1." Successively press ENTER to cycle through the seven graphing styles. Select the "shade below" style as shown in the figure. Pressing GRAPH produces a shaded figure as before.

With the cursor on the same line as "y1," press MORE. Successively press STYLE(F3) to cycle through the seven graphing styles. Select the "shade below" style as shown in the figure. Pressing 2nd GRAPH(M5) produces a shaded figure.

8.3 *Exercises*

Exer. 1–10: Sketch the graph of the inequality.

1 $3x - 2y < 6$

2 $4x + 3y < 12$

3 $2x + 3y \geq 2y + 1$

4 $2x - y > 3$

5 $y + 2 < x^2$

6 $y^2 - x \leq 0$

7 $x^2 + 1 \leq y$

8 $y - x^3 < 1$

9 $yx^2 \geq 1$

10 $x^2 + 4 \geq y$

Exer. 11–26: Sketch the graph of the system of inequalities.

11 $\begin{cases} 3x + y < 3 \\ 4 - y < 2x \end{cases}$

12 $\begin{cases} y + 2 < 2x \\ y - x > 4 \end{cases}$

13 $\begin{cases} y - x < 0 \\ 2x + 5y < 10 \end{cases}$

14 $\begin{cases} 2y - x \leq 4 \\ 3y + 2x < 6 \end{cases}$

15 $\begin{cases} 3x + y \le 6 \\ y - 2x \ge 1 \\ x \ge -2 \\ y \le 4 \end{cases}$

16 $\begin{cases} 3x - 4y \ge 12 \\ x - 2y \le 2 \\ x \ge 9 \\ y \le 5 \end{cases}$

17 $\begin{cases} x + 2y \le 8 \\ 0 \le x \le 4 \\ 0 \le y \le 3 \end{cases}$

18 $\begin{cases} 2x + 3y \ge 6 \\ 0 \le x \le 5 \\ 0 \le y \le 4 \end{cases}$

19 $\begin{cases} |x| \ge 2 \\ |y| < 3 \end{cases}$

20 $\begin{cases} |x| \ge 4 \\ |y| \ge 3 \end{cases}$

21 $\begin{cases} |x + 2| \le 1 \\ |y - 3| < 5 \end{cases}$

22 $\begin{cases} |x - 2| \le 5 \\ |y - 4| > 2 \end{cases}$

23 $\begin{cases} x^2 + y^2 \le 4 \\ x + y \ge 1 \end{cases}$

▶ 24 $\begin{cases} x^2 + y^2 > 1 \\ x^2 + y^2 < 4 \end{cases}$

25 $\begin{cases} x^2 \le 1 - y \\ x \ge 1 + y \end{cases}$

26 $\begin{cases} x - y^2 < 0 \\ x + y^2 > 0 \end{cases}$

Exer. 27–34: Find a system of inequalities whose graph is shown.

27

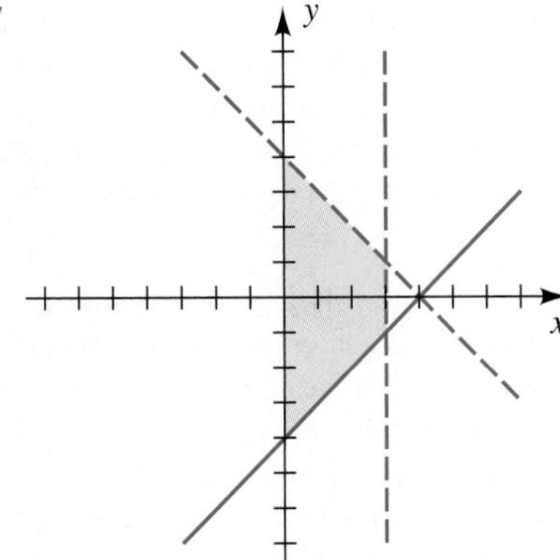

28

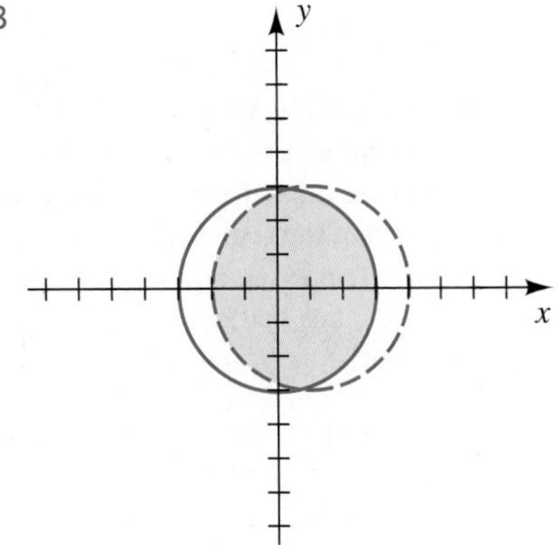

29

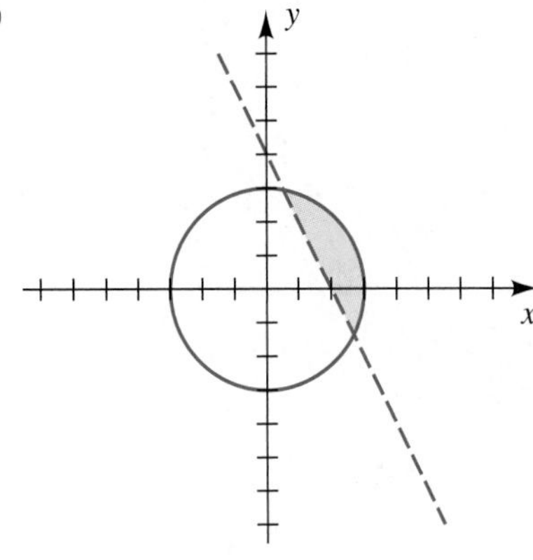

30

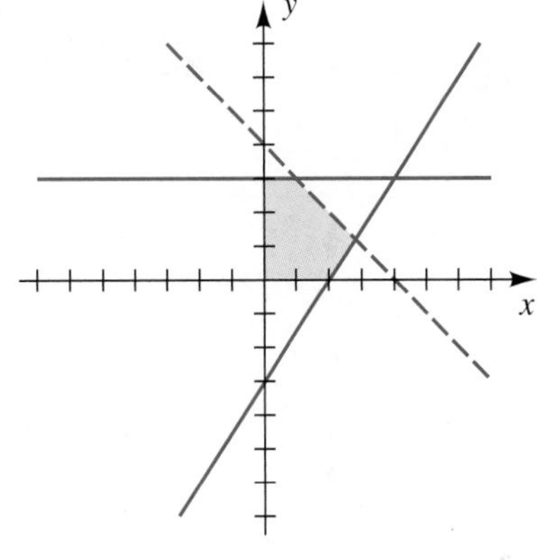

31

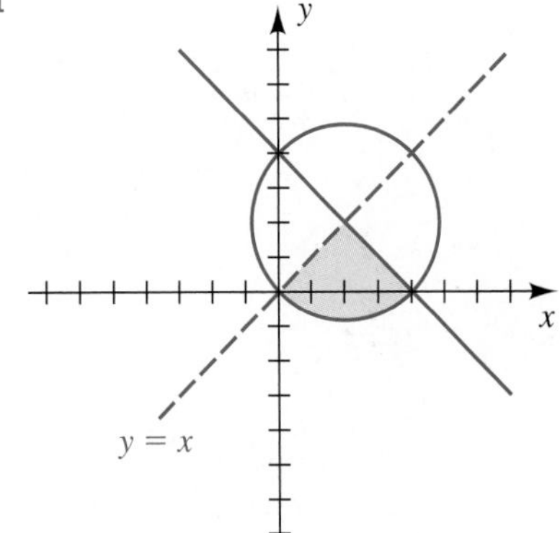

 32

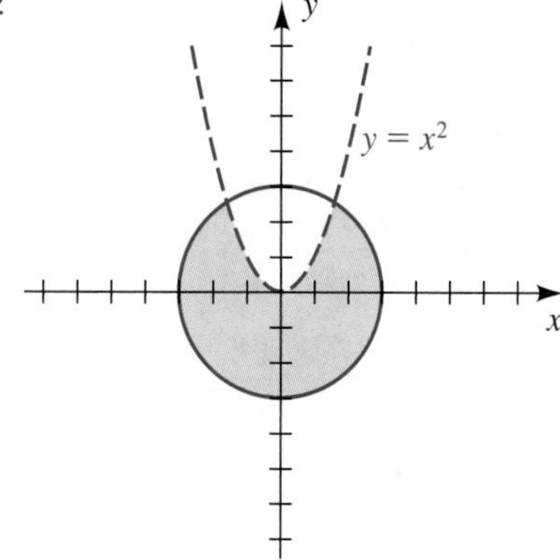

33

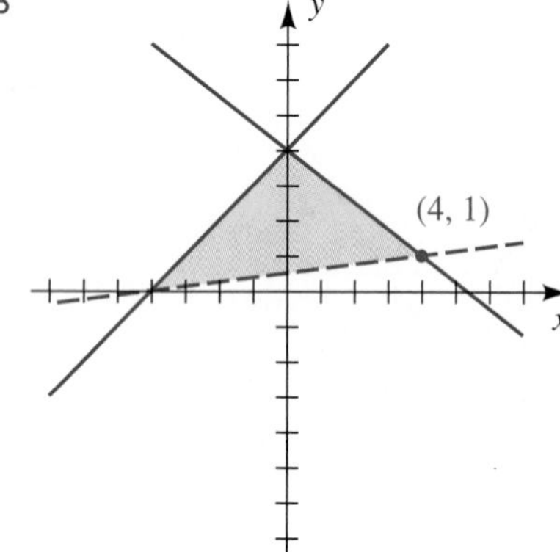

34

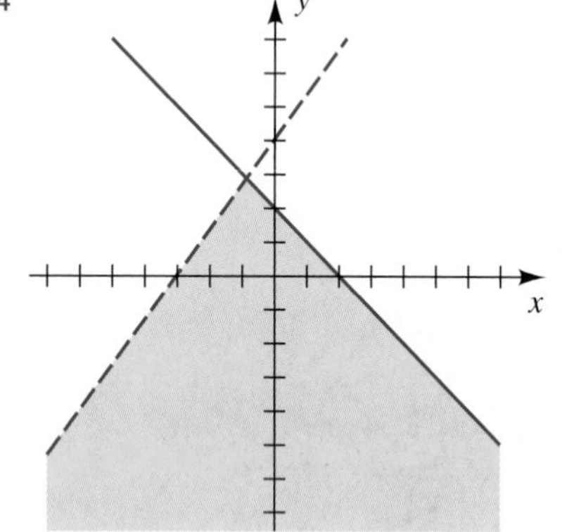

35 **Inventory levels** A store sells two brands of television sets. Customer demand indicates that it is necessary to stock at least twice as many sets of brand A as of brand B. It is also necessary to have on hand at least 10 sets of brand B. There is room for not more than 100 sets in the store. Find and graph a system of inequalities that describes all possibilities for stocking the two brands.

36 **Ticket prices** An auditorium contains 600 seats. For an upcoming event, tickets will be priced at \$8 for some seats and \$5 for others. At least 225 tickets are to be priced at \$5, and total sales of at least \$3000 are desired. Find and graph a system of inequalities that describes all possibilities for pricing the two types of tickets.

▶ 37 **Investment strategy** A woman with \$15,000 to invest decides to place at least \$2000 in a high-risk, high-yield investment and at least three times that amount in a low-risk, low-yield investment. Find and graph a system of inequalities that describes all possibilities for placing the money in the two investments.

38 **Inventory levels** The manager of a college bookstore stocks two types of notebooks, the first wholesaling for 55¢ and the second for 85¢. The maximum amount to be spent is \$600, and an inventory of at least 300 of the 85¢ variety and 400 of the 55¢ variety is desired. Find and graph a system of inequalities that describes all possibilities for stocking the two types of notebooks.

39 **Dimensions of a can** An aerosol can is to be constructed in the shape of a circular cylinder with a small cone on the top. The total height of the can including the conical top is to be no more than 9 inches, and the cylinder must contain at least 75% of the total volume. In addition, the height of the conical top must be at least 1 inch. Find and graph a system of inequalities that describes all possibilities for the relationship between the height y of the cylinder and the height x of the cone.

40 **Dimensions of a window** A stained-glass window is to be constructed in the form of a rectangle surmounted by a semicircle (see the figure). The total height h of the window can be no more than 6 feet, and the area of the rectangular part must be at least twice the area of the semicircle. In addition, the diameter d of the semicircle must be at least 2 feet. Find and graph a system of inequalities that describes all possibilities for the base and height of the rectangular part.

Exercise 40

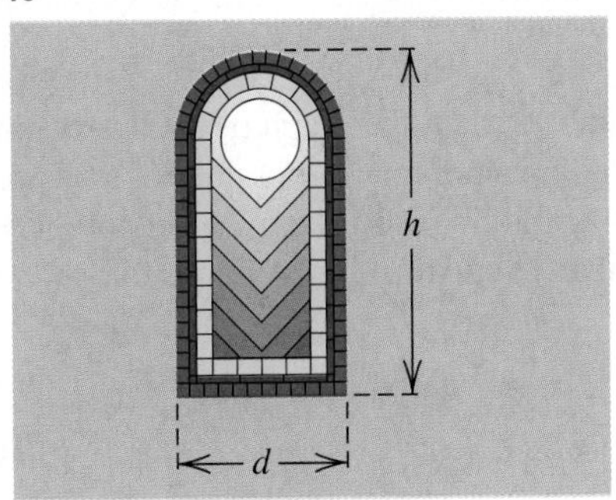

41 Locating a power plant A nuclear power plant will be constructed to serve the power needs of cities A and B. City B is 100 miles due east of A. The state has promised that the plant will be at least 60 miles from each city. It is not possible, however, to locate the plant south of either city because of rough terrain, and the plant must be within 100 miles of both A and B. Assuming A is at the origin, find and graph a system of inequalities that describes all possible locations for the plant.

42 Allocating space A man has a rectangular back yard that is 50 feet wide and 60 feet deep. He plans to construct a pool area and a patio area, as shown in the figure, where $y \geq 10$. He can spend at most \$67,500 on the project. The patio area must be at least as large as the pool area. The pool area will cost \$50 per square foot, and the patio will cost \$4 per square foot. Find and graph a system of inequalities that describes all possibilities for the width of the patio and pool areas.

Exercise 42

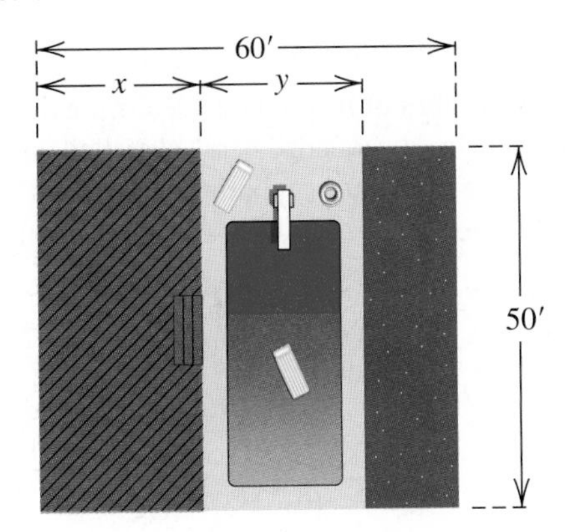

Exer. 43–44: Graph the inequality.

43 $64y^3 - x^3 \leq e^{1-2x}$

44 $e^{5y} - e^{-x} \geq x^4$

Exer. 45–48: Graph the system of inequalities.

45 $\begin{cases} 5^{1-y} \geq x^4 + x^2 + 1 \\ x + 3y \geq x^{5/3} \end{cases}$

46 $\begin{cases} x^4 + y^5 < 2^x \\ \ln(x^2 + 1) < y^3 \end{cases}$

47 $\begin{cases} x^4 - 2x < 3y \\ x + 2y < x^3 - 5 \end{cases}$

48 $\begin{cases} e^x + x^2 \leq 2^{x+2y} \\ 2^{x+2y} \leq x^3 2^y \\ x > 0 \end{cases}$

49 Forest growth Temperature and precipitation have a significant effect on plant life. If either the average annual temperature or the amount of precipitation is too low, trees and forests cannot grow. Instead, only grasslands and deserts will exist. The relationship between average annual temperature T (in °F) and average annual precipitation P (in inches) is a linear inequality. In order for forests to grow in a region, T and P must satisfy the inequality $29T - 39P < 450$, where $33 \leq T \leq 80$ and $13 \leq P \leq 45$.

(a) Determine whether forests can grow in Winnipeg, where $T = 37$°F and $P = 21.2$ in.

(b) Graph the inequality, with T on the horizontal axis and P on the vertical axis, in the viewing rectangle [33, 80, 5] by [0, 50, 5].

(c) Identify the region on the graph that represents where forests can grow.

50 Grassland growth Refer to Exercise 49. If the average annual precipitation P (in inches) is too low or the average annual temperature T (in °F) is too high, forests and grasslands become deserts. The conditions necessary for grasslands to grow are given by a linear inequality. T and P must satisfy $22P - 3T > 33$, where $33 \leq T \leq 80$ and $13 \leq P \leq 45$.

(a) Determine whether grasslands can grow in Phoenix, where $T = 70$°F and $P = 7.8$ in.

(b) Graph the inequality for forests and the inequality for grasslands on the same coordinate axes.

(c) Determine the region on the graph that represents where grasslands can exist but forests cannot.

8.4 Linear Programming

Figure 1

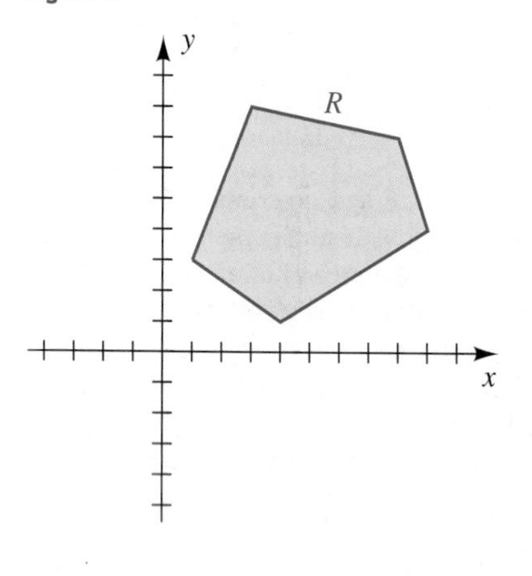

If a system of inequalities contains only linear inequalities of the form

$$ax + by \leq c \qquad \text{or} \qquad ax + by \geq c,$$

where a, b, and c are real numbers, then the graph of the system may be a region R in the xy-plane bounded by a polygon—possibly of the type illustrated in Figure 1 (for a specific illustration, see Example 4 and Figure 7 of Section 8.3). For problems in **linear programming,** we consider such systems together with an expression of the form

$$C = Ax + By + K,$$

where A, B, and K are real numbers and (x, y) is a point in R (that is, a solution of the system). Since for each (x, y) we obtain a specific value for C, we call C a *function of two variables x and y*. In linear programming, C is called an **objective function,** and the inequalities in the system are referred to as the **constraints** on C. The solutions of the system—that is, the pairs (x, y) corresponding to the points in R—are called the **feasible solutions** for the problem.

In typical business applications, the value of C may represent cost, profit, loss, or a physical resource, and the goal is to find a specific point (x, y) in R at which C takes on its maximum or minimum value. The methods of linear programming greatly simplify the task of finding this point. Specifically, it can be shown that *the maximum and minimum values of C occur at a vertex of R.* This fact is used in the next example.

Figure 2

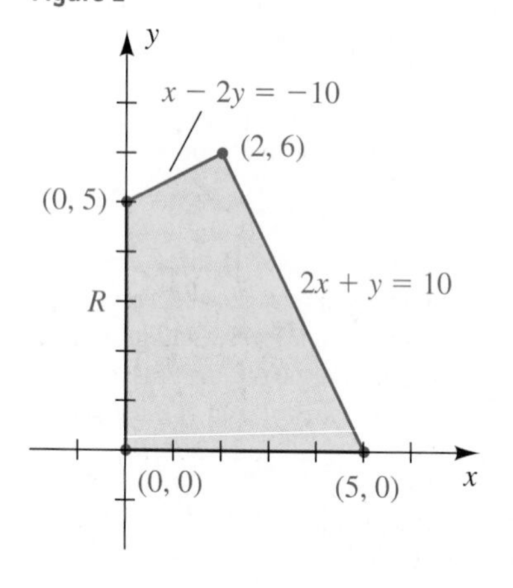

EXAMPLE 1 Finding the maximum and minimum values of an objective function

Find the maximum and minimum values of the objective function given by $C = 7x + 3y$ subject to the following constraints:

$$\begin{cases} x - 2y \geq -10 \\ 2x + y \leq 10 \\ x \geq 0 \\ y \geq 0 \end{cases}$$

▶ **SOLUTION** The graph of the system of inequalities determined by the constraints is the region R bounded by the quadrilateral sketched in Figure 2. From the preceding discussion, the maximum and minimum values of C must occur at a vertex of R. The values at the vertices are given in the following table.

Vertex	Value of $C = 7x + 3y$
(0, 0)	$7(0) + 3(0) = 0$
(0, 5)	$7(0) + 3(5) = 15$
(5, 0)	$7(5) + 3(0) = 35$
(2, 6)	$7(2) + 3(6) = 32$

Hence, the minimum value $C = 0$ occurs if $x = 0$ and $y = 0$. The maximum value $C = 35$ occurs if $x = 5$ and $y = 0$.

▶ Tutorial available at www.thomsonedu.com/login

Figure 3

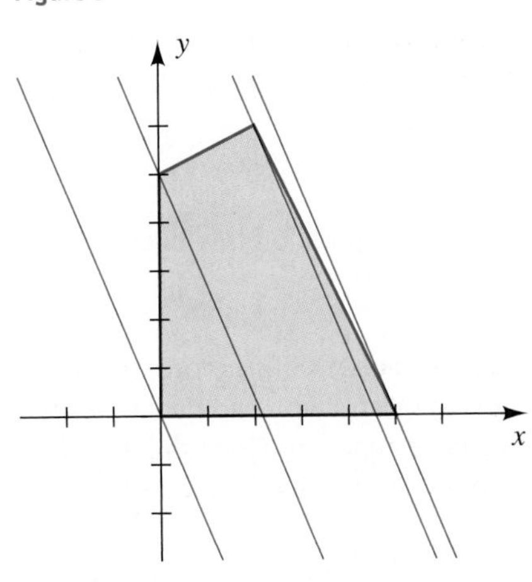

In the preceding example, we say that the maximum value of C *on R* occurs at the vertex (5, 0). To verify this fact, let us solve $C = 7x + 3y$ for y, obtaining

$$y = -\frac{7}{3}x + \frac{C}{3}.$$

For each C, the graph of this equation is a line of slope $-\frac{7}{3}$ and y-intercept $C/3$, as illustrated in Figure 3. To find the maximum value of C, we simply determine which of these lines that intersect the region has the largest y-intercept $C/3$. Referring to Figure 3, we see that the required line passes through (5, 0). Similarly, for the minimum value of C, we determine the line having equation $y = (-7/3)x + (C/3)$ that intersects the region and has the *smallest* y-intercept. This is the line through (0, 0).

We shall call a problem that can be expressed in the form of Example 1 a **linear programming problem.** To solve such problems, we may use the following guidelines.

Guidelines for Solving a Linear Programming Problem

1 Sketch the region R determined by the system of constraints.

2 Find the vertices of R.

3 Calculate the value of the objective function C at each vertex of R.

4 Select the maximum or minimum value(s) of C in guideline 3.

In the next example we encounter a linear programming problem in which the minimum value of the objective function occurs at more than one point.

EXAMPLE 2 Solving a linear programming problem

Find the minimum value of the objective function $C = 2x + 6y$ subject to the following constraints:

$$\begin{cases} 2x + 3y \geq 12 \\ x + 3y \geq 9 \\ x \geq 0 \\ y \geq 0 \end{cases}$$

Figure 4

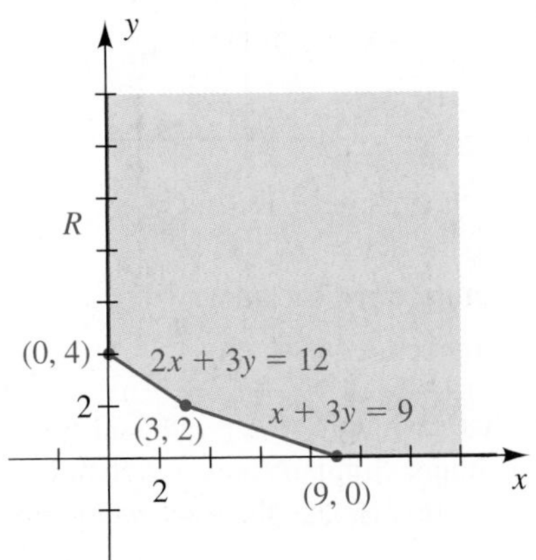

▶ **SOLUTION** We shall follow the guidelines.

Guideline 1 The graph of the system of inequalities determined by the constraints is the unbounded region R sketched in Figure 4.

Guideline 2 The vertices of R are (0, 4), (3, 2), and (9, 0), as shown in the figure.

Guideline 3 The value of C at each vertex of R is given in the following table.

Vertex	Value of $C = 2x + 6y$
(0, 4)	$2(0) + 6(4) = 24$
(3, 2)	$2(3) + 6(2) = 18$
(9, 0)	$2(9) + 6(0) = 18$

(continued)

▶ Tutorial available at www.thomsonedu.com/login

Guideline 4 The table in guideline 3 shows that the minimum value of C, 18, occurs at *two* vertices, (3, 2) and (9, 0). Moreover, if (x, y) is any point on the line segment joining these points, then (x, y) is a solution of the equation $x + 3y = 9$, and hence

$$C = 2x + 6y = 2(x + 3y) = 2(9) = 18.$$

Thus, the minimum value $C = 18$ occurs at *every* point on this line segment.

In the next two examples we consider applications of linear programming. For such problems it is necessary to use given information and data to formulate the system of constraints and the objective function. Once this has been accomplished, we may apply the guidelines as we did in the solution to Example 2.

EXAMPLE 3 Maximizing profit

A firm manufactures two products, X and Y. For each product, it is necessary to use three different machines, A, B, and C. To manufacture one unit of product X, machine A must be used for 3 hours, machine B for 1 hour, and machine C for 1 hour. To manufacture one unit of product Y requires 2 hours on A, 2 hours on B, and 1 hour on C. The profit on product X is \$500 per unit, and the profit on product Y is \$350 per unit. Machine A is available for a total of 24 hours per day; however, B can be used for only 16 hours and C for 9 hours. Assuming the machines are available when needed (subject to the noted total hour restrictions), determine the number of units of each product that should be manufactured each day in order to maximize the profit.

▶ **SOLUTION** The following table summarizes the data given in the statement of the problem.

Machine	Hours required for one unit of X	Hours required for one unit of Y	Hours available
A	3	2	24
B	1	2	16
C	1	1	9

Let us introduce the following variables:

x = number of units of X manufactured each day

y = number of units of Y manufactured each day

Using the first row of the table, we note that each unit of X requires 3 hours on machine A, and hence x units require $3x$ hours. Similarly, since each unit of Y requires 2 hours on A, y units require $2y$ hours. Hence, the total number of

Figure 5

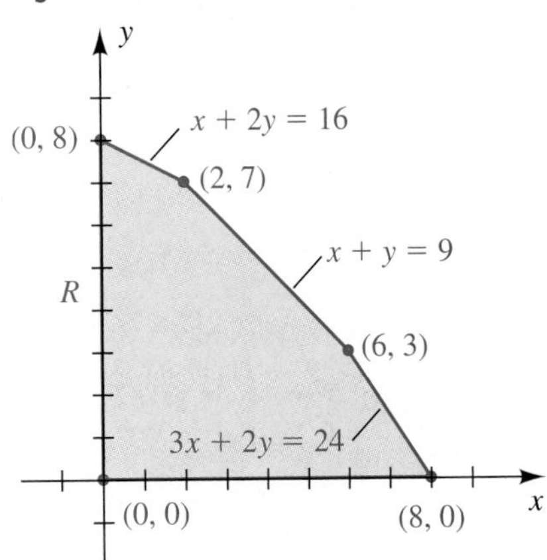

hours per day that machine A must be used is $3x + 2y$. This, together with the fact that A can be used for at most 24 hours per day, gives us the first constraint in the following system of inequalities—that is, $3x + 2y \leq 24$. The second and third constraints are obtained by using the same type of reasoning for rows 2 and 3 of the table. The last two constraints, $x \geq 0$ and $y \geq 0$, are true because x and y cannot be negative.

$$\begin{cases} 3x + 2y \leq 24 \\ x + 2y \leq 16 \\ x + y \leq 9 \\ x \geq 0 \\ y \geq 0 \end{cases}$$

The graph of this system is the region R in Figure 5.

Since the production of each unit of product X yields a profit of \$500 and each unit of product Y yields a profit of \$350, the profit P obtained by producing x units of X together with y units of Y is

$$P = 500x + 350y.$$

This is the objective function for the problem. The maximum value of P must occur at one of the vertices of R in Figure 5. The values of P at these vertices are given in the following table.

Vertex	Value of $P = 500x + 350y$
(0, 0)	$500(0) + 350(0) = 0$
(0, 8)	$500(0) + 350(8) = 2800$
(8, 0)	$500(8) + 350(0) = 4000$
(2, 7)	$500(2) + 350(7) = 3450$
(6, 3)	$500(6) + 350(3) = 4050$

We see from the table that a maximum profit of \$4050 occurs for a daily production of 6 units of product X and 3 units of product Y.

Example 3 illustrates maximization of profit. The next example demonstrates how linear programming can be used to minimize the cost in a certain situation.

EXAMPLE 4 Minimizing cost

A distributor of compact disc players has two warehouses, W_1 and W_2. There are 80 units stored at W_1 and 70 units at W_2. Two customers, A and B, order 35 units and 60 units, respectively. The shipping cost from each warehouse to A and B is determined according to the following table. How should the order be filled to minimize the total shipping cost?

(continued)

Warehouse	Customer	Shipping cost per unit
W_1	A	$ 8
W_1	B	12
W_2	A	10
W_2	B	13

▶ **SOLUTION** Let us begin by introducing the following variables:

$$x = \text{number of units sent to A from } W_1$$
$$y = \text{number of units sent to B from } W_1$$

Since A ordered 35 units and B ordered 60 units, we must have

$$35 - x = \text{number of units sent to A from } W_2$$
$$60 - y = \text{number of units sent to B from } W_2.$$

Our goal is to determine values for x and y that make the total shipping cost minimal.

The number of units shipped from W_1 cannot exceed 80, and the number shipped from W_2 cannot exceed 70. Expressing these facts in terms of inequalities gives us

$$\begin{cases} x + y \leq 80 \\ (35 - x) + (60 - y) \leq 70 \end{cases}$$

Simplifying, we obtain the first two constraints in the following system. The last two constraints are true because the largest values of x and y are 35 and 60, respectively.

$$\begin{cases} x + y \leq 80 \\ x + y \geq 25 \\ 0 \leq x \leq 35 \\ 0 \leq y \leq 60 \end{cases}$$

The graph of this system is the region R shown in Figure 6.

Let C denote the total cost (in dollars) of shipping the disc players to customers A and B. We see from the table of shipping costs that the following are true:

$$\text{cost of shipping 35 units to A} = 8x + 10(35 - x)$$
$$\text{cost of shipping 60 units to B} = 12y + 13(60 - y)$$

Hence, the total cost is

$$C = 8x + 10(35 - x) + 12y + 13(60 - y).$$

Simplifying gives us the following objective function:

$$C = 1130 - 2x - y$$

▶ **Tutorial available at www.thomsonedu.com/login**

Figure 6

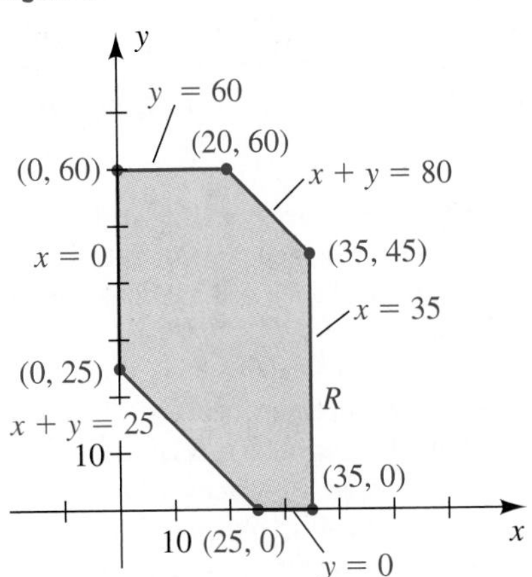

To determine the minimum value of C on R, we need check only the vertices shown in Figure 6, as in the following table.

Vertex	Value of $C = 1130 - 2x - y$
(0, 25)	$1130 - 2(0) - 25 = 1105$
(0, 60)	$1130 - 2(0) - 60 = 1070$
(20, 60)	$1130 - 2(20) - 60 = 1030$
(35, 45)	$1130 - 2(35) - 45 = 1015$
(35, 0)	$1130 - 2(35) - 0 = 1060$
(25, 0)	$1130 - 2(25) - 0 = 1080$

We see from the table that the minimal shipping cost, \$1015, occurs if $x = 35$ and $y = 45$. This means that the distributor should ship all of the disc players to A from W_1 and none from W_2. In addition, the distributor should ship 45 units to B from W_1 and 15 units to B from W_2. (Note that the *maximum* shipping cost will occur if $x = 0$ and $y = 25$—that is, if all 35 units are shipped to A from W_2 and if B receives 25 units from W_1 and 35 units from W_2.)

The examples in this section are elementary linear programming problems in two variables that can be solved by basic methods. The much more complicated problems in many variables that occur in practice may be solved by employing matrix techniques (discussed later) that are adapted for solution by computers.

8.4 *Exercises*

Exer. 1–2: Find the maximum and minimum values of the objective function C on the region in the figure.

▶ 1 $C = 3x + 2y + 5$

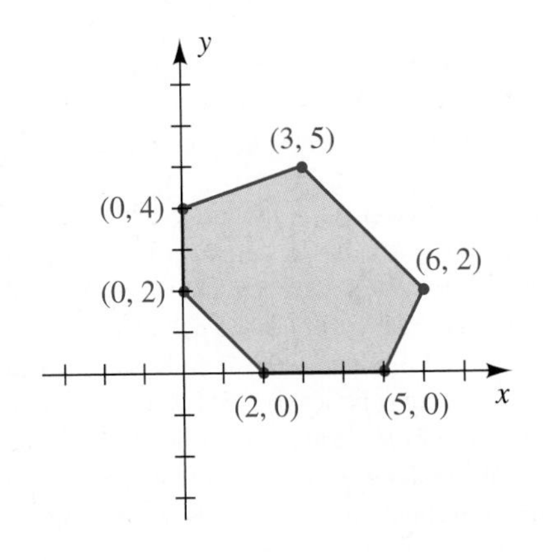

2 $C = 2x + 7y + 3$

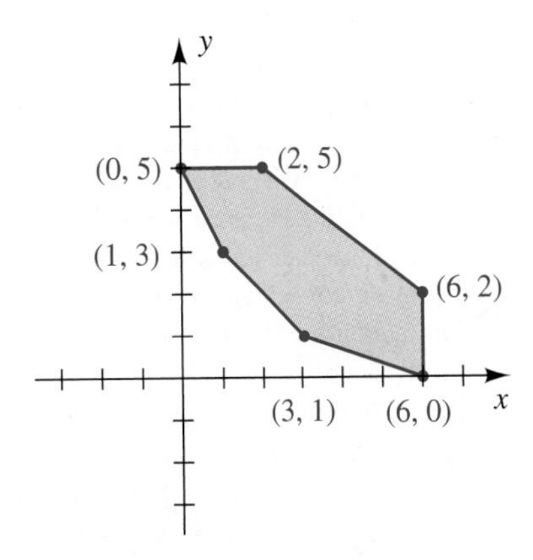

▶ **Tutorial available at www.thomsonedu.com/login**

Exer. 3–4: Sketch the region *R* determined by the given constraints, and label its vertices. Find the maximum value of *C* on *R*.

▶ 3 $C = 3x + y;$ $x \geq 0, y \geq 0,$
$3x - 4y \geq -12,$ $3x + 2y \leq 24,$ $3x - y \leq 15$

4 $C = 4x - 2y;$
$x - 2y \geq -8,$ $7x - 2y \leq 28,$ $x + y \geq 4$

Exer. 5–6: Sketch the region *R* determined by the given constraints, and label its vertices. Find the minimum value of *C* on *R*.

▶ 5 $C = 3x + 6y;$ $x \geq 0, y \geq 0,$
$2x + 3y \geq 12,$ $2x + 5y \geq 16$

6 $C = 6x + y;$ $y \geq 0,$
$3x + y \geq 3,$ $x + 5y \leq 15,$ $2x + y \leq 12$

Exer. 7–8: Sketch the region *R* determined by the given constraints, and label its vertices. Describe the set of points for which *C* is a maximum on *R*.

7 $C = 2x + 4y;$ $x \geq 0, y \geq 0,$
$x - 2y \geq -8,$ $\frac{1}{2}x + y \leq 6,$ $3x + 2y \leq 24$

▶ 8 $C = 6x + 3y;$ $x \geq 2, y \geq 1,$
$2x + 3y \leq 19,$ $x + 0.5y \leq 6.5$

9 **Production scheduling** A manufacturer of tennis rackets makes a profit of \$15 on each oversized racket and \$8 on each standard racket. To meet dealer demand, daily production of standard rackets should be between 30 and 80, and production of oversized rackets should be between 10 and 30. To maintain high quality, the total number of rackets produced should not exceed 80 per day. How many of each type should be manufactured daily to maximize the profit?

▶ 10 **Production scheduling** A manufacturer of cell phones makes a profit of \$25 on a deluxe model and \$30 on a standard model. The company wishes to produce at least 80 deluxe models and at least 100 standard models per day. To maintain high quality, the daily production should not exceed 200 radios. How many of each type should be produced daily in order to maximize the profit?

11 **Minimizing cost** Two substances, S and T, each contain two types of ingredients, I and G. One pound of S contains 2 ounces of I and 4 ounces of G. One pound of T contains 2 ounces of I and 6 ounces of G. A manufacturer plans to combine quantities of the two substances to obtain a mixture that contains at least 9 ounces of I and 20 ounces of G. If the cost of S is \$3 per pound and the cost of T is \$4 per pound, how much of each substance should be used to keep the cost to a minimum?

12 **Maximizing gross profit** A stationery company makes two types of notebooks: a deluxe notebook with subject dividers, which sells for \$4.00, and a regular notebook, which sells for \$3.00. The production cost is \$3.20 for each deluxe notebook and \$2.60 for each regular notebook. The company has the facilities to manufacture between 2000 and 3000 deluxe and between 3000 and 6000 regular notebooks, but not more than 7000 altogether. How many notebooks of each type should be manufactured to maximize the difference between the selling prices and the production costs?

13 **Minimizing shipping costs** Refer to Example 4 of this section. If the shipping costs are \$12 per unit from W_1 to A, \$10 per unit from W_2 to A, \$16 per unit from W_1 to B, and \$12 per unit from W_2 to B, determine how the order should be filled to minimize shipping cost.

14 **Minimizing cost** A coffee company purchases mixed lots of coffee beans and then grades them into premium, regular, and unusable beans. The company needs at least 280 tons of premium-grade and 200 tons of regular-grade coffee beans. The company can purchase ungraded coffee from two suppliers in any amount desired. Samples from the two suppliers contain the following percentages of premium, regular, and unusable beans:

Supplier	Premium	Regular	Unusable
A	20%	50%	30%
B	40%	20%	40%

If supplier A charges \$900 per ton and B charges \$1200 per ton, how much should the company purchase from each supplier to fulfill its needs at minimum cost?

15 **Planning crop acreage** A farmer, in the business of growing fodder for livestock, has 90 acres available for planting alfalfa and corn. The cost of seed per acre is \$32 for alfalfa and \$48 for corn. The total cost of labor will amount to \$60 per acre for alfalfa and \$30 per acre for corn. The expected revenue (before costs are subtracted) is \$500 per acre from alfalfa and \$700 per acre from corn. If the farmer does not wish to spend more than \$3840 for seed and \$4200 for labor, how many acres of each crop should be planted to obtain the maximum profit?

▶ **Tutorial available at www.thomsonedu.com/login**

16 Machinery scheduling A small firm manufactures bookshelves and desks for microcomputers. For each product it is necessary to use a table saw and a power router. To manufacture each bookshelf, the saw must be used for $\frac{1}{2}$ hour and the router for 1 hour. A desk requires the use of each machine for 2 hours. The profits are \$20 per bookshelf and \$50 per desk. If the saw can be used for 8 hours per day and the router for 12 hours per day, how many bookshelves and desks should be manufactured each day to maximize the profit?

17 Minimizing a mixture's cost Three substances, X, Y, and Z, each contain four ingredients, A, B, C, and D. The percentage of each ingredient and the cost in cents per ounce of each substance are given in the following table.

Substance	Ingredients A	B	C	D	Cost per ounce
X	20%	10%	25%	45%	25¢
Y	20%	40%	15%	25%	35¢
Z	10%	20%	25%	45%	50¢

If the cost is to be minimal, how many ounces of each substance should be combined to obtain a mixture of 20 ounces containing at least 14% A, 16% B, and 20% C? What combination would make the cost greatest?

18 Maximizing profit A man plans to operate a stand at a one-day fair at which he will sell bags of peanuts and bags of candy. He has \$2000 available to purchase his stock, which will cost \$2.00 per bag of peanuts and \$4.00 per bag of candy. He intends to sell the peanuts at \$3.00 and the candy at \$5.50 per bag. His stand can accommodate up to 500 bags of peanuts and 400 bags of candy. From past experience he knows that he will sell no more than a total of 700 bags. Find the number of bags of each that he should have available in order to maximize his profit. What is the maximum profit?

19 Maximizing passenger capacity A small community wishes to purchase used vans and small buses for its public transportation system. The community can spend no more than \$100,000 for the vehicles and no more than \$500 per month for maintenance. The vans sell for \$10,000 each and average \$100 per month in maintenance costs. The corresponding cost estimates for each bus are \$20,000 and \$75 per month. If each van can carry 15 passengers and each bus can accommodate 25 riders, determine the number of vans and buses that should be purchased to maximize the passenger capacity of the system.

20 Minimizing fuel cost Refer to Exercise 19. The monthly fuel cost (based on 5000 miles of service) is \$550 for each van and \$850 for each bus. Find the number of vans and buses that should be purchased to minimize the monthly fuel costs if the passenger capacity of the system must be at least 75.

21 Stocking a fish farm A fish farmer will purchase no more than 5000 young trout and bass from the hatchery and will feed them a special diet for the next year. The cost of food per fish will be \$0.50 for trout and \$0.75 for bass, and the total cost is not to exceed \$3000. At the end of the year, a typical trout will weigh 3 pounds, and a bass will weigh 4 pounds. How many fish of each type should be stocked in the pond in order to maximize the total number of pounds of fish at the end of the year?

22 Dietary planning A hospital dietician wishes to prepare a corn-squash vegetable dish that will provide at least 3 grams of protein and cost no more than 36¢ per serving. An ounce of creamed corn provides $\frac{1}{2}$ gram of protein and costs 4¢. An ounce of squash supplies $\frac{1}{4}$ gram of protein and costs 3¢. For taste, there must be at least 2 ounces of corn and at least as much squash as corn. It is important to keep the total number of ounces in a serving as small as possible. Find the combination of corn and squash that will minimize the amount of ingredients used per serving.

23 Planning storage units A contractor has a large building that she wishes to convert into a series of rental storage spaces. She will construct basic 8 ft × 10 ft units and deluxe 12 ft × 10 ft units that contain extra shelves and a clothes closet. Market considerations dictate that there be at least twice as many basic units as deluxe units and that the basic units rent for \$75 per month and the deluxe units for \$120 per month. At most 7200 ft^2 is available for the storage spaces, and no more than \$80,000 can be spent on construction. If each basic unit will cost \$800 to make and each deluxe unit will cost \$1600, how many units of each type should be constructed to maximize monthly revenue?

▶ **24 A moose's diet** A moose feeding primarily on tree leaves and aquatic plants is capable of digesting no more than 33 kilograms of these foods daily. Although the aquatic plants are lower in energy content, the animal must eat at least 17 kilograms to satisfy its sodium requirement. A kilogram of leaves provides four times as much energy as a kilogram of aquatic plants. Find the combination of foods that maximizes the daily energy intake.

▶ **Tutorial available at www.thomsonedu.com/login**

8.5

Systems of Linear Equations in More Than Two Variables

For systems of linear equations containing more than two variables, we can use either the method of substitution explained in Section 8.1 or the method of elimination developed in Section 8.2. The method of elimination is the shorter and more straightforward technique for finding solutions. In addition, it leads to the matrix technique, discussed in this section.

EXAMPLE 1 Using the method of elimination to solve a system of linear equations

Solve the system

$$\begin{cases} x - 2y + 3z = 4 \\ 2x + y - 4z = 3 \\ -3x + 4y - z = -2 \end{cases}$$

▶ **SOLUTION**

$$\begin{cases} x - 2y + 3z = 4 \\ 5y - 10z = -5 \\ -3x + 4y - z = -2 \end{cases}$$ add -2 times the first equation to the second equation

$$\begin{cases} x - 2y + 3z = 4 \\ 5y - 10z = -5 \\ -2y + 8z = 10 \end{cases}$$ add 3 times the first equation to the third equation

$$\begin{cases} x - 2y + 3z = 4 \\ y - 2z = -1 \\ y - 4z = -5 \end{cases}$$ multiply the second equation by $\frac{1}{5}$ and the third equation by $-\frac{1}{2}$

$$\begin{cases} x - 2y + 3z = 4 \\ y - 2z = -1 \\ -2z = -4 \end{cases}$$ add -1 times the second equation to the third equation

$$\begin{cases} x - 2y + 3z = 4 \\ y - 2z = -1 \\ z = 2 \end{cases}$$ multiply the third equation by $-\frac{1}{2}$

The solutions of the last system are easy to find by **back substitution.** From the third equation, we see that $z = 2$. Substituting 2 for z in the second equation, $y - 2z = -1$, we get $y = 3$. Finally, we find the x-value by substituting $y = 3$ and $z = 2$ in the first equation, $x - 2y + 3z = 4$, obtaining $x = 4$. Thus, there is one solution, (4, 3, 2).

Any system of three linear equations in three variables has either a *unique solution,* an *infinite number of solutions,* or *no solution.* As for two equations

▶ **Tutorial available at www.thomsonedu.com/login**

in two variables, the terminology used to describe these systems is *consistent, dependent and consistent,* or *inconsistent,* respectively.

If we analyze the method of solution in Example 1, we see that the symbols used for the variables are immaterial. The *coefficients* of the variables are what we must consider. Thus, if different symbols such as r, s, and t are used for the variables, we obtain the system

$$\begin{cases} r - 2s + 3t = 4 \\ 2r + s - 4t = 3 \\ -3r + 4s - t = -2 \end{cases}$$

The method of elimination can then proceed exactly as in the example. Since this is true, it is possible to simplify the process. Specifically, we introduce a scheme for keeping track of the coefficients in such a way that we do not have to write down the variables. Referring to the preceding system, we first check that variables appear in the same order in each equation and that terms not involving variables are to the right of the equal signs. We then list the numbers that are involved in the equations as follows:

$$\begin{bmatrix} 1 & -2 & 3 & 4 \\ 2 & 1 & -4 & 3 \\ -3 & 4 & -1 & -2 \end{bmatrix}$$

An array of numbers of this type is called a **matrix.** The **rows** of the matrix are the numbers that appear next to each other *horizontally:*

1	−2	3	4	first row, R_1
2	1	−4	3	second row, R_2
−3	4	−1	−2	third row, R_3

The **columns** of the matrix are the numbers that appear next to each other *vertically:*

first column, C_1	second column, C_2	third column, C_3	fourth column, C_4
1	−2	3	4
2	1	−4	3
−3	4	−1	−2

The matrix obtained from a system of linear equations in the preceding manner is the **matrix of the system.** If we delete the last column of this matrix, the remaining array of numbers is the **coefficient matrix.** Since the matrix of the system can be obtained from the coefficient matrix by adjoining one column, we call it the **augmented coefficient matrix** or simply the **augmented matrix.** Later, when we use matrices to find the solutions of a system of linear equations, we shall introduce a vertical line segment in the augmented matrix to indicate where the equal signs would appear in the corresponding system of equations, as in the next illustration.

ILLUSTRATION **Coefficient Matrix and Augmented Matrix**

system	coefficient matrix	augmented matrix
$\begin{cases} x - 2y + 3z = 4 \\ 2x + y - 4z = 3 \\ -3x + 4y - z = -2 \end{cases}$	$\begin{bmatrix} 1 & -2 & 3 \\ 2 & 1 & -4 \\ -3 & 4 & -1 \end{bmatrix}$	$\left[\begin{array}{ccc\|c} 1 & -2 & 3 & 4 \\ 2 & 1 & -4 & 3 \\ -3 & 4 & -1 & -2 \end{array}\right]$

Before discussing a matrix method of solving a system of linear equations, let us state a general definition of a matrix. We shall use a **double subscript notation,** denoting the number that appears in row i and column j by a_{ij}. The **row subscript** of a_{ij} is i, and the **column subscript** is j.

Definition of a Matrix

Let m and n be positive integers. An $\boldsymbol{m \times n}$ **matrix** is an array of the following form, where each a_{ij} is a real number:

$$\begin{bmatrix} a_{11} & a_{12} & a_{13} & \cdots & a_{1n} \\ a_{21} & a_{22} & a_{23} & \cdots & a_{2n} \\ a_{31} & a_{32} & a_{33} & \cdots & a_{3n} \\ \vdots & \vdots & \vdots & & \vdots \\ a_{m1} & a_{m2} & a_{m3} & \cdots & a_{mn} \end{bmatrix}$$

The notation $m \times n$ in the definition is read "m by n." We often say that the matrix *is* $m \times n$ and call $m \times n$ the **size** of the matrix. It is possible to consider matrices in which the symbols a_{ij} represent complex numbers, polynomials, or other mathematical objects. The rows and columns of a matrix are defined as before. Thus, the matrix in the definition has m rows and n columns. Note that a_{23} is in row 2 and column 3 and a_{32} is in row 3 and column 2. Each a_{ij} is an **element of the matrix.** If $m = n$, the matrix is a **square matrix of order n** and the elements $a_{11}, a_{22}, a_{33}, \ldots, a_{nn}$ are the **main diagonal elements.**

ILLUSTRATION $m \times n$ **Matrices**

2×3	2×2	1×3	3×2	3×1
$\begin{bmatrix} -5 & 3 & 1 \\ 7 & 0 & -2 \end{bmatrix}$	$\begin{bmatrix} 5 & -1 \\ 2 & 3 \end{bmatrix}$	$\begin{bmatrix} 3 & 1 & -2 \end{bmatrix}$	$\begin{bmatrix} 2 & -1 \\ 0 & 1 \\ 8 & 3 \end{bmatrix}$	$\begin{bmatrix} -4 \\ 0 \\ 5 \end{bmatrix}$

To find the solutions of a system of linear equations, we begin with the augmented matrix. If a variable does not appear in an equation, we assume that the coefficient is zero. We then work with the rows of the matrix *just as though they were equations.* The only items missing are the symbols for the variables, the addition or subtraction signs used between terms, and the equal signs. We simply keep in mind that the numbers in the first column are the coefficients of the first variable, the numbers in the second column are the coefficients of the second variable, and so on. The rules for transforming a matrix are formulated so that they always produce a matrix of an equivalent system of equations.

The next theorem is a restatement, in terms of matrices, of the theorem on equivalent systems in Section 8.2. In part (2) of the theorem, the terminology *a row is multiplied by a nonzero constant* means that each element in the row is multiplied by the constant. To *add* two rows of a matrix, as in part (3), we add corresponding elements in each row.

Theorem on Matrix Row Transformations

Given a matrix of a system of linear equations, a matrix of an equivalent system results if

(1) two rows are interchanged.

(2) a row is multiplied or divided by a nonzero constant.

(3) a constant multiple of one row is added to another row.

We refer to 1–3 as the **elementary row transformations** of a matrix. If a matrix is obtained from another matrix by one or more elementary row transformations, the two matrices are said to be **equivalent** or, more precisely, **row equivalent.** We shall use the symbols in the following chart to denote elementary row transformations of a matrix, where the arrow $\rightarrow$ may be read "replaces." Thus, for the transformation $kR_i \rightarrow R_i$, the constant multiple kR_i *replaces* R_i. Similarly, for $kR_i + R_j \rightarrow R_j$, the sum $kR_i + R_j$ *replaces* R_j. For convenience, we shall write $(-1)R_i$ as $-R_i$.

Elementary Row Transformations of a Matrix

Symbol	Meaning
$R_i \leftrightarrow R_j$	Interchange rows i and j
$kR_i \rightarrow R_i$	Multiply row i by k
$kR_i + R_j \rightarrow R_j$	Add k times row i to row j

We shall next rework Example 1 using matrices. You should compare the two solutions, since analogous steps are used in each case.

EXAMPLE 2 **Using matrices to solve a system of linear equations**

Solve the system

$$\begin{cases} x - 2y + 3z = 4 \\ 2x + y - 4z = 3 \\ -3x + 4y - z = -2 \end{cases}$$

▶ SOLUTION We begin with the matrix of the system—that is, with the augmented matrix:

$$\left[\begin{array}{rrr|r} 1 & -2 & 3 & 4 \\ 2 & 1 & -4 & 3 \\ -3 & 4 & -1 & -2 \end{array}\right]$$

We next apply elementary row transformations to obtain another (simpler) matrix of an equivalent system of equations. These transformations correspond to the manipulations used for equations in Example 1. We will place appropriate symbols between equivalent matrices.

$$\left[\begin{array}{rrr|r} 1 & -2 & 3 & 4 \\ 2 & 1 & -4 & 3 \\ -3 & 4 & -1 & -2 \end{array}\right] \begin{array}{r} \\ \mathbf{-2R_1 + R_2 \to R_2} \\ \mathbf{3R_1 + R_3 \to R_3} \end{array} \left[\begin{array}{rrr|r} 1 & -2 & 3 & 4 \\ 0 & 5 & -10 & -5 \\ 0 & -2 & 8 & 10 \end{array}\right] \quad \begin{array}{l} \\ \text{add } -2R_1 \text{ to } R_2 \\ \text{add } 3R_1 \text{ to } R_3 \end{array}$$

$$\begin{array}{r} \\ \mathbf{\tfrac{1}{5}R_2 \to R_2} \\ \mathbf{-\tfrac{1}{2}R_3 \to R_3} \end{array} \left[\begin{array}{rrr|r} 1 & -2 & 3 & 4 \\ 0 & 1 & -2 & -1 \\ 0 & 1 & -4 & -5 \end{array}\right] \quad \begin{array}{l} \\ \text{multiply } R_2 \text{ by } \tfrac{1}{5} \\ \text{multiply } R_3 \text{ by } -\tfrac{1}{2} \end{array}$$

$$\begin{array}{r} \\ \\ \mathbf{-R_2 + R_3 \to R_3} \end{array} \left[\begin{array}{rrr|r} 1 & -2 & 3 & 4 \\ 0 & 1 & -2 & -1 \\ 0 & 0 & -2 & -4 \end{array}\right] \quad \begin{array}{l} \\ \\ \text{add } -R_2 \text{ to } R_3 \end{array}$$

$$\begin{array}{r} \\ \\ \mathbf{-\tfrac{1}{2}R_3 \to R_3} \end{array} \left[\begin{array}{rrr|r} 1 & -2 & 3 & 4 \\ 0 & 1 & -2 & -1 \\ 0 & 0 & 1 & 2 \end{array}\right] \quad \begin{array}{l} \\ \\ \text{multiply } R_3 \text{ by } -\tfrac{1}{2} \end{array}$$

We use the last matrix to return to the system of equations

$$\left[\begin{array}{rrr|r} 1 & -2 & 3 & 4 \\ 0 & 1 & -2 & -1 \\ 0 & 0 & 1 & 2 \end{array}\right] \iff \begin{cases} x - 2y + 3z = 4 \\ \quad\;\; y - 2z = -1 \\ \qquad\quad\;\; z = 2 \end{cases}$$

which is equivalent to the original system. The solution $x = 4$, $y = 3$, $z = 2$ may now be found by back substitution, as in Example 1.

The final matrix in the solution of Example 2 is in **echelon form.** In general, a matrix is in echelon form if it satisfies the following conditions.

Echelon Form of a Matrix

(1) The first nonzero number in each row, reading from left to right, is 1.

(2) The column containing the first nonzero number in any row is to the left of the column containing the first nonzero number in the row below.

(3) Rows consisting entirely of zeros may appear at the bottom of the matrix.

▶ Tutorial available at www.thomsonedu.com/login

The following is an illustration of matrices in echelon form. The symbols a_{ij} represent real numbers.

ILLUSTRATION **Echelon Form**

■ $\begin{bmatrix} 1 & a_{12} & a_{13} & a_{14} \\ 0 & 1 & a_{23} & a_{24} \\ 0 & 0 & 1 & a_{34} \end{bmatrix}$

■ $\begin{bmatrix} 1 & a_{12} & a_{13} & a_{14} & a_{15} & a_{16} & a_{17} \\ 0 & 1 & a_{23} & a_{24} & a_{25} & a_{26} & a_{27} \\ 0 & 0 & 0 & 1 & a_{35} & a_{36} & a_{37} \\ 0 & 0 & 0 & 0 & 0 & 1 & a_{47} \\ 0 & 0 & 0 & 0 & 0 & 0 & 0 \\ 0 & 0 & 0 & 0 & 0 & 0 & 0 \end{bmatrix}$

The following guidelines may be used to find echelon forms.

Guidelines for Finding the Echelon Form of a Matrix

1 Locate the *first* column that contains nonzero elements, and apply elementary row transformations to get the number 1 into the first row of that column.

2 Apply elementary row transformations of the type $kR_1 + R_j \to R_j$ for $j > 1$ to get 0 underneath the number 1 obtained in guideline 1 in each of the remaining rows.

3 *Disregard the first row.* Locate the next column that contains nonzero elements, and apply elementary row transformations to get the number 1 into the *second* row of that column.

4 Apply elementary row transformations of the type $kR_2 + R_j \to R_j$ for $j > 2$ to get 0 underneath the number 1 obtained in guideline 3 in each of the remaining rows.

5 *Disregard the first and second rows.* Locate the next column that contains nonzero elements, and repeat the procedure.

6 Continue the process until the echelon form is reached.

Not all echelon forms contain rows consisting of only zeros (see Example 2).

We can use elementary row operations to transform the matrix of any system of linear equations to echelon form. The echelon form can then be used to produce a system of equations that is equivalent to the original system. The solutions of the given system may be found by back substitution. The next example illustrates this technique for a system of four linear equations.

EXAMPLE 3 **Using an echelon form to solve a system of linear equations**

Solve the system

$$\begin{cases} -2x + 3y + 4z \phantom{{}+2w} = -1 \\ x \phantom{{}+3y} - 2z + 2w = 1 \\ y + z - w = 0 \\ 3x + y - 2z - w = 3 \end{cases}$$

▶ SOLUTION We have arranged the equations so that the same variables appear in vertical columns. We begin with the augmented matrix and then obtain an echelon form as described in the guidelines.

$$\left[\begin{array}{rrrr|r} -2 & 3 & 4 & 0 & -1 \\ 1 & 0 & -2 & 2 & 1 \\ 0 & 1 & 1 & -1 & 0 \\ 3 & 1 & -2 & -1 & 3 \end{array}\right] \xrightarrow{\mathbf{R_1 \leftrightarrow R_2}} \left[\begin{array}{rrrr|r} 1 & 0 & -2 & 2 & 1 \\ -2 & 3 & 4 & 0 & -1 \\ 0 & 1 & 1 & -1 & 0 \\ 3 & 1 & -2 & -1 & 3 \end{array}\right]$$

$$\begin{array}{r} \\ \mathbf{2R_1 + R_2 \to R_2} \\ \\ \mathbf{-3R_1 + R_4 \to R_4} \end{array} \left[\begin{array}{rrrr|r} 1 & 0 & -2 & 2 & 1 \\ 0 & 3 & 0 & 4 & 1 \\ 0 & 1 & 1 & -1 & 0 \\ 0 & 1 & 4 & -7 & 0 \end{array}\right]$$

$$\xrightarrow{\mathbf{R_2 \leftrightarrow R_3}} \left[\begin{array}{rrrr|r} 1 & 0 & -2 & 2 & 1 \\ 0 & 1 & 1 & -1 & 0 \\ 0 & 3 & 0 & 4 & 1 \\ 0 & 1 & 4 & -7 & 0 \end{array}\right]$$

$$\begin{array}{r} \\ \\ \mathbf{-3R_2 + R_3 \to R_3} \\ \mathbf{-R_2 + R_4 \to R_4} \end{array} \left[\begin{array}{rrrr|r} 1 & 0 & -2 & 2 & 1 \\ 0 & 1 & 1 & -1 & 0 \\ 0 & 0 & -3 & 7 & 1 \\ 0 & 0 & 3 & -6 & 0 \end{array}\right]$$

$$\begin{array}{r} \\ \\ \\ \mathbf{R_3 + R_4 \to R_4} \end{array} \left[\begin{array}{rrrr|r} 1 & 0 & -2 & 2 & 1 \\ 0 & 1 & 1 & -1 & 0 \\ 0 & 0 & -3 & 7 & 1 \\ 0 & 0 & 0 & 1 & 1 \end{array}\right]$$

$$\begin{array}{r} \\ \\ \mathbf{-\tfrac{1}{3}R_3 \to R_3} \\ \\ \end{array} \left[\begin{array}{rrrr|r} 1 & 0 & -2 & 2 & 1 \\ 0 & 1 & 1 & -1 & 0 \\ 0 & 0 & 1 & -\frac{7}{3} & -\frac{1}{3} \\ 0 & 0 & 0 & 1 & 1 \end{array}\right]$$

The final matrix is in echelon form and corresponds to the following system of equations:

$$\begin{cases} x - 2z + 2w = 1 \\ y + z - w = 0 \\ z - \frac{7}{3}w = -\frac{1}{3} \\ w = 1 \end{cases}$$

We now use back substitution to find the solution. From the last equation we see that $w = 1$. Substituting in the third equation, $z - \frac{7}{3}w = -\frac{1}{3}$, we get

$$z - \tfrac{7}{3}(1) = -\tfrac{1}{3}, \qquad \text{or} \qquad z = \tfrac{6}{3} = 2.$$

▶ **Tutorial available at www.thomsonedu.com/login**

Substituting $w = 1$ and $z = 2$ in the second equation, $y + z - w = 0$, we obtain

$$y + 2 - 1 = 0, \qquad \text{or} \qquad y = -1.$$

Finally, from the first equation, $x - 2z + 2w = 1$, we have

$$x - 2(2) + 2(1) = 1, \qquad \text{or} \qquad x = 3.$$

Hence, the system has one solution, $x = 3$, $y = -1$, $z = 2$, and $w = 1$.

After obtaining an echelon form, it is often convenient to apply additional elementary row operations of the type $kR_i + R_j \to R_j$ so that 0 also appears *above* the first 1 in each row. We refer to the resulting matrix as being in **reduced echelon form.** The following is an illustration of matrices in reduced echelon form. (Compare them with the echelon forms on page 637.)

ILLUSTRATION **Reduced Echelon Form**

- $\begin{bmatrix} 1 & 0 & 0 & a_{14} \\ 0 & 1 & 0 & a_{24} \\ 0 & 0 & 1 & a_{34} \end{bmatrix}$
- $\begin{bmatrix} 1 & 0 & a_{13} & 0 & a_{15} & 0 & a_{17} \\ 0 & 1 & a_{23} & 0 & a_{25} & 0 & a_{27} \\ 0 & 0 & 0 & 1 & a_{35} & 0 & a_{37} \\ 0 & 0 & 0 & 0 & 0 & 1 & a_{47} \\ 0 & 0 & 0 & 0 & 0 & 0 & 0 \\ 0 & 0 & 0 & 0 & 0 & 0 & 0 \end{bmatrix}$

EXAMPLE 4 **Using a reduced echelon form to solve a system of linear equations**

Solve the system in Example 3 using reduced echelon form.

SOLUTION We begin with the echelon form obtained in Example 3 and apply additional row operations as follows:

$$\left[\begin{array}{cccc|c} 1 & 0 & -2 & 2 & 1 \\ 0 & 1 & 1 & -1 & 0 \\ 0 & 0 & 1 & -\frac{7}{3} & -\frac{1}{3} \\ 0 & 0 & 0 & 1 & 1 \end{array}\right] \begin{array}{r} -2R_4 + R_1 \to R_1 \\ R_4 + R_2 \to R_2 \\ \frac{7}{3}R_4 + R_3 \to R_3 \\ \end{array} \left[\begin{array}{cccc|c} 1 & 0 & -2 & 0 & -1 \\ 0 & 1 & 1 & 0 & 1 \\ 0 & 0 & 1 & 0 & 2 \\ 0 & 0 & 0 & 1 & 1 \end{array}\right]$$

$$\begin{array}{r} 2R_3 + R_1 \to R_1 \\ -R_3 + R_2 \to R_2 \\ \\ \end{array} \left[\begin{array}{cccc|c} 1 & 0 & 0 & 0 & 3 \\ 0 & 1 & 0 & 0 & -1 \\ 0 & 0 & 1 & 0 & 2 \\ 0 & 0 & 0 & 1 & 1 \end{array}\right]$$

The system of equations corresponding to the reduced echelon form gives us the solution *without* using back substitution:

$$x = 3, \quad y = -1, \quad z = 2, \quad w = 1$$

Tutorial available at www.thomsonedu.com/login

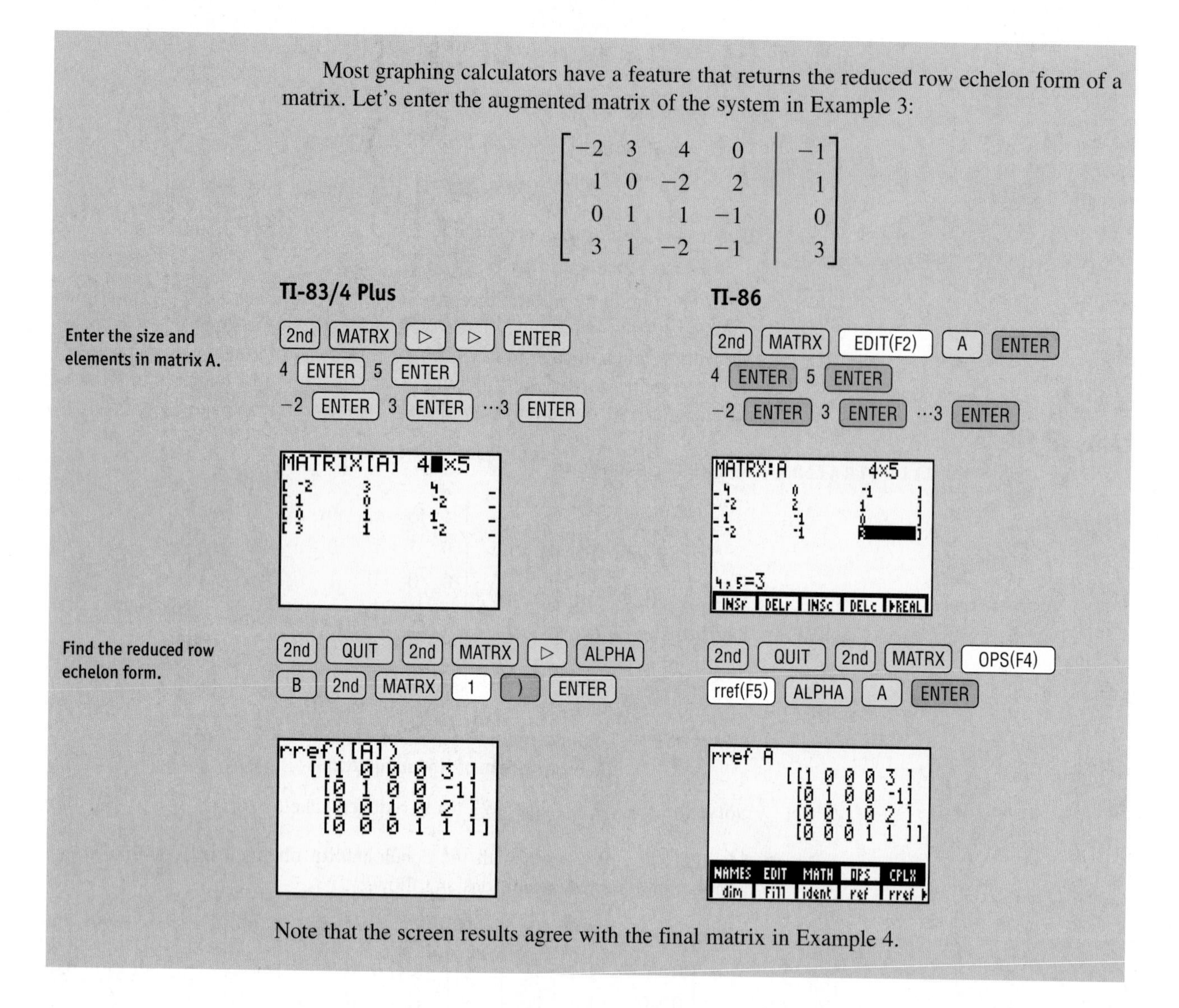

Sometimes it is necessary to consider systems in which the number of equations is not the same as the number of variables. The same matrix techniques are applicable, as illustrated in the next example.

EXAMPLE 5 **Solving a system of two linear equations in three variables**

Solve the system

$$\begin{cases} 2x + 3y + 4z = 1 \\ 3x + 4y + 5z = 3 \end{cases}$$

▶ **SOLUTION** We shall begin with the augmented matrix and then find a reduced echelon form. There are many different ways of getting the number 1

▶ **Tutorial available at www.thomsonedu.com/login**

into the first position of the first row. For example, the elementary row transformation $\frac{1}{2}R_1 \to R_1$ or $-\frac{1}{3}R_2 + R_1 \to R_1$ would accomplish this in one step. Another way, which does not involve fractions, is demonstrated in the following steps:

$$\left[\begin{array}{ccc|c} 2 & 3 & 4 & 1 \\ 3 & 4 & 5 & 3 \end{array}\right] \xrightarrow{\mathbf{R_1 \leftrightarrow R_2}} \left[\begin{array}{ccc|c} 3 & 4 & 5 & 3 \\ 2 & 3 & 4 & 1 \end{array}\right]$$

$$\mathbf{-R_2 + R_1 \to R_1} \left[\begin{array}{ccc|c} 1 & 1 & 1 & 2 \\ 2 & 3 & 4 & 1 \end{array}\right]$$

$$\mathbf{-2R_1 + R_2 \to R_2} \left[\begin{array}{ccc|c} 1 & 1 & 1 & 2 \\ 0 & 1 & 2 & -3 \end{array}\right]$$

$$\mathbf{-R_2 + R_1 \to R_1} \left[\begin{array}{ccc|c} 1 & 0 & -1 & 5 \\ 0 & 1 & 2 & -3 \end{array}\right]$$

The reduced echelon form is the matrix of the system

$$\begin{cases} x \quad\quad - z = 5 \\ \quad y + 2z = -3 \end{cases}$$

or, equivalently,

$$\begin{cases} x = z + 5 \\ y = -2z - 3 \end{cases}$$

There are an infinite number of solutions to this system; they can be found by assigning z any value c and then using the last two equations to express x and y in terms of c. This gives us

$$x = c + 5, \qquad y = -2c - 3, \qquad z = c.$$

Thus, the solutions of the system consist of all ordered triples of the form

$$(c + 5, -2c - 3, c)$$

for any real number c. The solutions may be checked by substituting $c + 5$ for x, $-2c - 3$ for y, and c for z in the two original equations.

We can obtain any number of solutions for the system by substituting specific real numbers for c. For example, if $c = 0$, we obtain $(5, -3, 0)$; if $c = 2$, we have $(7, -7, 2)$; and so on.

There are other ways to specify the general solution. For example, starting with $x = z + 5$ and $y = -2z - 3$, we could let $z = d - 5$ for any real number d. In this case,

$$x = z + 5 = (d - 5) + 5 = d$$

$$y = -2z - 3 = -2(d - 5) - 3 = -2d + 7,$$

and the solutions of the system have the form

$$(d, -2d + 7, d - 5).$$

These triples produce the same solutions as $(c + 5, -2c - 3, c)$. For example, if $d = 5$, we get $(5, -3, 0)$; if $d = 7$, we obtain $(7, -7, 2)$; and so on.

A system of linear equations is **homogeneous** if all the terms that do not contain variables—that is, the *constant terms*—are zero. A system of homogeneous equations always has the **trivial solution** obtained by substituting zero for each variable. Nontrivial solutions sometimes exist. The procedure for finding solutions is the same as that used for nonhomogeneous systems.

EXAMPLE 6 Solving a homogeneous system of linear equations

Solve the homogeneous system

$$\begin{cases} x - y + 4z = 0 \\ 2x + y - z = 0 \\ -x - y + 2z = 0 \end{cases}$$

▶ **SOLUTION** We begin with the augmented matrix and find a reduced echelon form:

$$\left[\begin{array}{rrr|r} 1 & -1 & 4 & 0 \\ 2 & 1 & -1 & 0 \\ -1 & -1 & 2 & 0 \end{array}\right] \begin{array}{r} \\ \mathbf{-2R_1 + R_2 \to R_2} \\ \mathbf{R_1 + R_3 \to R_3} \end{array} \left[\begin{array}{rrr|r} 1 & -1 & 4 & 0 \\ 0 & 3 & -9 & 0 \\ 0 & -2 & 6 & 0 \end{array}\right]$$

$$\begin{array}{r} \\ \mathbf{\tfrac{1}{3}R_2 \to R_2} \\ \mathbf{-\tfrac{1}{2}R_3 \to R_3} \end{array} \left[\begin{array}{rrr|r} 1 & -1 & 4 & 0 \\ 0 & 1 & -3 & 0 \\ 0 & 1 & -3 & 0 \end{array}\right]$$

$$\begin{array}{r} \mathbf{R_2 + R_1 \to R_1} \\ \\ \mathbf{-R_2 + R_3 \to R_3} \end{array} \left[\begin{array}{rrr|r} 1 & 0 & 1 & 0 \\ 0 & 1 & -3 & 0 \\ 0 & 0 & 0 & 0 \end{array}\right]$$

The reduced echelon form corresponds to the system

$$\begin{cases} x \qquad + \ z = 0 \\ \qquad y - 3z = 0 \end{cases}$$

or, equivalently,

$$\begin{cases} x = -z \\ y = \ \ 3z \end{cases}$$

Assigning any value c to z, we obtain $x = -c$ and $y = 3c$. The solutions consist of all ordered triples of the form $(-c, 3c, c)$ for any real number c.

EXAMPLE 7 A homogeneous system with only the trivial solution

Solve the system

$$\begin{cases} x + y + z = 0 \\ x - y + z = 0 \\ x - y - z = 0 \end{cases}$$

▶ **Tutorial available at www.thomsonedu.com/login**

SOLUTION We begin with the augmented matrix and find a reduced echelon form:

$$\left[\begin{array}{ccc|c} 1 & 1 & 1 & 0 \\ 1 & -1 & 1 & 0 \\ 1 & -1 & -1 & 0 \end{array}\right] \begin{array}{l} \\ -\mathbf{R}_1 + \mathbf{R}_2 \to \mathbf{R}_2 \\ -\mathbf{R}_1 + \mathbf{R}_3 \to \mathbf{R}_3 \end{array} \left[\begin{array}{ccc|c} 1 & 1 & 1 & 0 \\ 0 & -2 & 0 & 0 \\ 0 & -2 & -2 & 0 \end{array}\right]$$

$$\begin{array}{l} \\ -\frac{1}{2}\mathbf{R}_2 \to \mathbf{R}_2 \\ -\frac{1}{2}\mathbf{R}_3 \to \mathbf{R}_3 \end{array} \left[\begin{array}{ccc|c} 1 & 1 & 1 & 0 \\ 0 & 1 & 0 & 0 \\ 0 & 1 & 1 & 0 \end{array}\right]$$

$$\begin{array}{l} -\mathbf{R}_2 + \mathbf{R}_1 \to \mathbf{R}_1 \\ \\ -\mathbf{R}_2 + \mathbf{R}_3 \to \mathbf{R}_3 \end{array} \left[\begin{array}{ccc|c} 1 & 0 & 1 & 0 \\ 0 & 1 & 0 & 0 \\ 0 & 0 & 1 & 0 \end{array}\right]$$

$$\begin{array}{l} -\mathbf{R}_3 + \mathbf{R}_1 \to \mathbf{R}_1 \\ \\ \\ \end{array} \left[\begin{array}{ccc|c} 1 & 0 & 0 & 0 \\ 0 & 1 & 0 & 0 \\ 0 & 0 & 1 & 0 \end{array}\right]$$

The reduced echelon form is the matrix of the system

$$x = 0, \qquad y = 0, \qquad z = 0.$$

Thus, the only solution for the given system is the trivial one, (0, 0, 0).

The next two examples illustrate applied problems.

EXAMPLE 8 **Using a system of equations to determine maximum profit**

A manufacturer of electrical equipment has the following information about the weekly profit from the production and sale of an electric motor.

Production level x	25	50	100
Profit $P(x)$ (dollars)	5250	7500	4500

(a) Determine a, b, and c so that the graph of $P(x) = ax^2 + bx + c$ fits this information.

(b) According to the quadratic function P in part (a), how many motors should be produced each week for maximum profit? What is the maximum weekly profit?

▶ SOLUTION

(a) We see from the table that the graph of $P(x) = ax^2 + bx + c$ contains the points (25, 5250), (50, 7500), and (100, 4500). This gives us the system of equations

$$\begin{cases} 5250 = 625a + 25b + c \\ 7500 = 2500a + 50b + c \\ 4500 = 10{,}000a + 100b + c \end{cases}$$

It is easy to solve any of the equations for c, so we'll start solving the system by solving the first equation for c,

$$c = 5250 - 625a - 25b,$$

and then substituting that expression for c in the other two equations:

$$\begin{cases} 7500 = 2500a + 50b + (5250 - 625a - 25b) \\ 4500 = 10{,}000a + 100b + (5250 - 625a - 25b) \end{cases}$$

Note that we have reduced the system of three equations and three variables to two equations and two variables. Simplifying the system gives us

$$\begin{cases} 1875a + 25b = 2250 \\ 9375a + 75b = -750 \end{cases}$$

At this point we could divide the equations by 25, but we see that 75 is just 3 times 25, so we'll use the method of elimination to eliminate b:

*Note that we have used **both** the method of substitution and the method of elimination in solving this system of equations.*

$$\begin{cases} -5625a - 75b = -6750 \\ 9375a + 75b = -750 \end{cases} \quad \text{multiply the first equation by } -3$$

Adding the equations gives us $3750a = -7500$, so $a = -2$. We can verify that the solution is $a = -2$, $b = 240$, $c = 500$.

(b) From part (a),

$$P(x) = -2x^2 + 240x + 500.$$

Since $a = -2 < 0$, the graph of the quadratic function P is a parabola that opens downward. By the formula on page 177, the x-coordinate of the vertex (the highest point on the parabola) is

$$x = \frac{-b}{2a} = \frac{-240}{2(-2)} = \frac{-240}{-4} = 60.$$

Hence, for the maximum profit, the manufacturer should produce and sell 60 motors per week. The maximum weekly profit is

$$P(60) = -2(60)^2 + 240(60) + 500 = \$7700.$$

EXAMPLE 9 **Solving a mixture problem**

A merchant wishes to mix two grades of peanuts costing \$3 and \$4 per pound, respectively, with cashews costing \$8 per pound, to obtain 140 pounds of a

▶ Tutorial available at www.thomsonedu.com/login

mixture costing \$6 per pound. If the merchant also wants the amount of lower-grade peanuts to be twice that of the higher-grade peanuts, how many pounds of each variety should be mixed?

SOLUTION Let us introduce three variables, as follows:

x = number of pounds of peanuts at \$3 per pound

y = number of pounds of peanuts at \$4 per pound

z = number of pounds of cashews at \$8 per pound

We refer to the statement of the problem and obtain the following system:

$$\begin{cases} x + y + z = 140 & \text{weight equation} \\ 3x + 4y + 8z = 6(140) & \text{value equation} \\ x = 2y & \text{constraint} \end{cases}$$

You may verify that the solution of this system is $x = 40$, $y = 20$, $z = 80$. Thus, the merchant should use 40 pounds of the \$3/lb peanuts, 20 pounds of the \$4/lb peanuts, and 80 pounds of cashews.

Sometimes we can combine row transformations to simplify our work. For example, consider the augmented matrix

$$\begin{bmatrix} 11 & 3 & 8 & 9 \\ 7 & -2 & 2 & 1 \\ 0 & 87 & 80 & 94 \end{bmatrix}.$$

To obtain a 1 in the first column, it appears we have to multiply row 1 by $\frac{1}{11}$ or row 2 by $\frac{1}{7}$. However, we can multiply row 1 by 2 and row 2 by -3 and then add those two rows to obtain

$$2(11) + (-3)(7) = 22 + (-21) = 1$$

in column one, as shown in the next matrix:

$$\mathbf{2R_1 - 3R_2 \to R_1} \begin{bmatrix} 1 & 12 & 10 & 15 \\ 7 & -2 & 2 & 1 \\ 0 & 87 & 80 & 94 \end{bmatrix}$$

We can then proceed to find the reduced echelon form without the cumbersome use of fractions. This process is called using a **linear combination of rows.**

8.5 *Exercises*

Exer. 1–22: Use matrices to solve the system.

1 $\begin{cases} x - 2y - 3z = -1 \\ 2x + y + z = 6 \\ x + 3y - 2z = 13 \end{cases}$

2 $\begin{cases} x + 3y - z = -3 \\ 3x - y + 2z = 1 \\ 2x - y + z = -1 \end{cases}$

3 $\begin{cases} 5x + 2y - z = -7 \\ x - 2y + 2z = 0 \\ 3y + z = 17 \end{cases}$

4 $\begin{cases} 4x - y + 3z = 6 \\ -8x + 3y - 5z = -6 \\ 5x - 4y = -9 \end{cases}$

5 $\begin{cases} 2x + 6y - 4z = 1 \\ x + 3y - 2z = 4 \\ 2x + y - 3z = -7 \end{cases}$

6 $\begin{cases} x + 3y - 3z = -5 \\ 2x - y + z = -3 \\ -6x + 3y - 3z = 4 \end{cases}$

7 $\begin{cases} 2x - 3y + 2z = -3 \\ -3x + 2y + z = 1 \\ 4x + y - 3z = 4 \end{cases}$

8 $\begin{cases} 2x - 3y + z = 2 \\ 3x + 2y - z = -5 \\ 5x - 2y + z = 0 \end{cases}$

▶ 9 $\begin{cases} x + 3y + z = 0 \\ x + y - z = 0 \\ x - 2y - 4z = 0 \end{cases}$

10 $\begin{cases} 2x - y + z = 0 \\ x - y - 2z = 0 \\ 2x - 3y - z = 0 \end{cases}$

11 $\begin{cases} 2x + y + z = 0 \\ x - 2y - 2z = 0 \\ x + y + z = 0 \end{cases}$

12 $\begin{cases} x + y - 2z = 0 \\ x - y - 4z = 0 \\ y + z = 0 \end{cases}$

13 $\begin{cases} 3x - 2y + 5z = 7 \\ x + 4y - z = -2 \end{cases}$

14 $\begin{cases} 2x - y + 4z = 8 \\ -3x + y - 2z = 5 \end{cases}$

15 $\begin{cases} 4x - 2y + z = 5 \\ 3x + y - 4z = 0 \end{cases}$

▶ 16 $\begin{cases} 5x + 2y - z = 10 \\ y + z = -3 \end{cases}$

17 $\begin{cases} 5x + 2z = 1 \\ y - 3z = 2 \\ 2x + y = 3 \end{cases}$

18 $\begin{cases} 2x - 3y = 12 \\ 3y + z = -2 \\ 5x - 3z = 3 \end{cases}$

▶ 19 $\begin{cases} 4x - 3y = 1 \\ 2x + y = -7 \\ -x + y = -1 \end{cases}$

20 $\begin{cases} 2x + 3y = -2 \\ x + y = 1 \\ x - 2y = 13 \end{cases}$

21 $\begin{cases} 2x + 3y = 5 \\ x - 3y = 4 \\ x + y = -2 \end{cases}$

22 $\begin{cases} 4x - y = 2 \\ 2x + 2y = 1 \\ 4x - 5y = 3 \end{cases}$

23 **Mixing acid solutions** Three solutions contain a certain acid. The first contains 10% acid, the second 30%, and the third 50%. A chemist wishes to use all three solutions to obtain a 50-liter mixture containing 32% acid. If the chemist wants to use twice as much of the 50% solution as of the 30% solution, how many liters of each solution should be used?

24 **Filling a pool** A swimming pool can be filled by three pipes, A, B, and C. Pipe A alone can fill the pool in 8 hours. If pipes A and C are used together, the pool can be filled in 6 hours; if B and C are used together, it takes 10 hours. How long does it take to fill the pool if all three pipes are used?

25 **Production capability** A company has three machines, A, B, and C, that are each capable of producing a certain item. However, because of a lack of skilled operators, only two of the machines can be used simultaneously. The following table indicates production over a three-day period, using various combinations of the machines. How long would it take each machine, if used alone, to produce 1000 items?

Machines used	Hours used	Items produced
A and B	6	4500
A and C	8	3600
B and C	7	4900

26 **Electrical resistance** In electrical circuits, the formula $1/R = (1/R_1) + (1/R_2)$ is used to find the total resistance R if two resistors R_1 and R_2 are connected in parallel. Given three resistors, A, B, and C, suppose that the total resistance is 48 ohms if A and B are connected in parallel, 80 ohms if B and C are connected in parallel, and 60 ohms if A and C are connected in parallel. Find the resistances of A, B, and C.

27 **Mixing fertilizers** A supplier of lawn products has three types of grass fertilizer, G_1, G_2, and G_3, having nitrogen contents of 30%, 20%, and 15%, respectively. The supplier plans to mix them, obtaining 600 pounds of fertilizer with a 25% nitrogen content. The mixture is to contain 100 pounds more of type G_3 than of type G_2. How much of each type should be used?

28 **Particle acceleration** If a particle moves along a coordinate line with a constant acceleration a (in cm/sec^2), then at time t (in seconds) its distance $s(t)$ (in centimeters) from the origin is

$$s(t) = \tfrac{1}{2}at^2 + v_0t + s_0$$

for velocity v_0 and distance s_0 from the origin at $t = 0$. If the distances of the particle from the origin at $t = \frac{1}{2}$, $t = 1$, and $t = \frac{3}{2}$ are 7, 11, and 17, respectively, find a, v_0, and s_0.

▶ **Tutorial available at www.thomsonedu.com/login**

29 Electrical currents Shown in the figure is a schematic of an electrical circuit containing three resistors, a 6-volt battery, and a 12-volt battery. It can be shown, using Kirchhoff's laws, that the three currents I_1, I_2, and I_3 are solutions of the following system of equations:

$$\begin{cases} I_1 - I_2 + I_3 = 0 \\ R_1 I_1 + R_2 I_2 = 6 \\ R_2 I_2 + R_3 I_3 = 12 \end{cases}$$

Find the three currents if

(a) $R_1 = R_2 = R_3 = 3$ ohms

(b) $R_1 = 4$ ohms, $R_2 = 1$ ohm, and $R_3 = 4$ ohms

Exercise 29

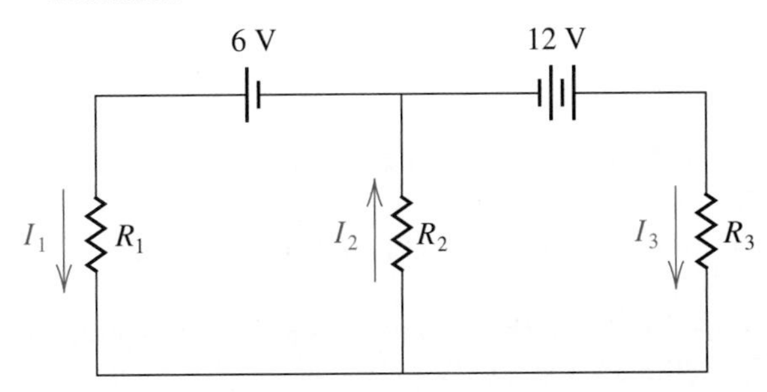

30 Bird population A stable population of 35,000 birds lives on three islands. Each year 10% of the population on island A migrates to island B, 20% of the population on island B migrates to island C, and 5% of the population on island C migrates to island A. Find the number of birds on each island if the population count on each island does not vary from year to year.

31 Blending coffees A shop specializes in preparing blends of gourmet coffees. From Colombian, Costa Rican, and Kenyan coffees, the owner wishes to prepare 1-pound bags that will sell for \$12.50. The cost per pound of these coffees is \$14, \$10, and \$12, respectively. The amount of Colombian is to be three times the amount of Costa Rican. Find the amount of each type of coffee in the blend.

32 Weights of chains There are three chains, weighing 450, 610, and 950 ounces, each consisting of links of three different sizes. Each chain has 10 small links. The chains also have 20, 30, and 40 medium links and 30, 40, and 70 large links, respectively. Find the weights of the small, medium, and large links.

33 Traffic flow Shown in the figure is a system of four one-way streets leading into the center of a city. The numbers in the figure denote the average number of vehicles per hour that travel in the directions shown. A total of 300 vehicles enter the area and 300 vehicles leave the area every hour. Signals at intersections A, B, C, and D are to be timed in order to avoid congestion, and this timing will determine traffic flow rates x_1, x_2, x_3, and x_4.

Exercise 33

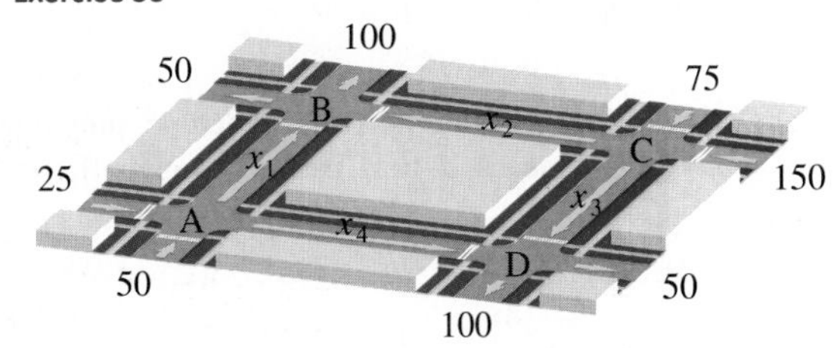

(a) If the number of vehicles entering an intersection per hour must equal the number leaving the intersection per hour, describe the traffic flow rates at each intersection with a system of equations.

(b) If the signal at intersection C is timed so that x_3 is equal to 100, find x_1, x_2, and x_4.

(c) Make use of the system in part (a) to explain why $75 \le x_3 \le 150$.

34 If $f(x) = ax^3 + bx + c$, determine a, b, and c such that the graph of f passes through the points $P(-3, -12)$, $Q(-1, 22)$, and $R(2, 13)$.

35 Air pollution Between 1850 and 1985 approximately 155 billion metric tons of carbon was added to the earth's atmosphere and the climate became about 0.5°C warmer, an indication of the *greenhouse effect.* It is estimated that doubling the carbon dioxide (CO_2) in the atmosphere would result in an average global temperature increase of 4–5°C. The future amount A of CO_2 in the atmosphere in parts per million is sometimes estimated using the formula $A = a + ct + ke^{rt}$, where a, c, and k are constants, r is the percentage increase in the emission of CO_2, and t is time in years, with $t = 0$ corresponding to 1990. Suppose it is estimated that in the year 2070, A will be 800 if $r = 2.5\%$ and A will be 560 if $r = 1.5\%$. If, in 1990, $A = 340$ and $r = 1\%$, find the year in which the amount of CO_2 in the atmosphere will have doubled.

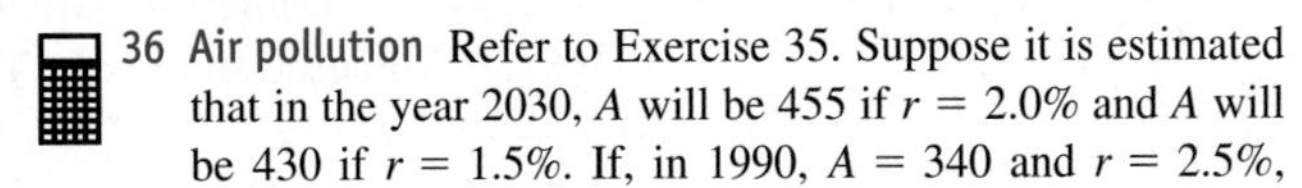

36 Air pollution Refer to Exercise 35. Suppose it is estimated that in the year 2030, A will be 455 if $r = 2.0\%$ and A will be 430 if $r = 1.5\%$. If, in 1990, $A = 340$ and $r = 2.5\%$, find the year in which the amount of CO_2 in the atmosphere will have doubled.

Exer. 37–38: Find an equation of the circle of the form $x^2 + y^2 + ax + by + c = 0$ that passes through the given points.

37 $P(2, 1)$, $Q(-1, -4)$, $R(3, 0)$

38 $P(-5, 5)$, $Q(-2, -4)$, $R(2, 4)$

Exer. 39–40: Find an equation of the cubic polynomial $f(x) = ax^3 + bx^2 + cx + d$ that passes through the given points.

39 $P(0, -6)$, $Q(1, -11)$, $R(-1, -5)$, $S(2, -14)$

40 $P(0, 4)$, $Q(1, 2)$, $R(-1, 10)$, $S(2, -2)$

41 If $f(x) = ax^3 + bx^2 + cx + d$, find a, b, c, and d if the graph of f is to pass through $(-1, 2)$, $(0.5, 2)$, $(1, 3)$, and $(2, 4.5)$.

42 If $f(x) = ax^4 + bx^3 + cx^2 + dx + e$, find a, b, c, d, and e if the graph of f is to pass through $(-2, 1.5)$, $(-1, -2)$, $(1, -3)$, $(2, -3.5)$, and $(3, -4.8)$.

8.6 The Algebra of Matrices

Matrices were introduced in Section 8.5 as an aid to finding solutions of systems of equations. In this section we discuss some of the properties of matrices. These properties are important in advanced fields of mathematics and in applications.

In the following definition, the symbol (a_{ij}) denotes an $m \times n$ matrix A of the type displayed in the definition on page 634. We use similar notations for the matrices B and C.

Definition of Equality and Addition of Matrices

Let $A = (a_{ij})$, $B = (b_{ij})$, and $C = (c_{ij})$ be $m \times n$ matrices.

(1) $A = B$ if and only if $a_{ij} = b_{ij}$ for every i and j.

(2) $C = A + B$ if and only if $c_{ij} = a_{ij} + b_{ij}$ for every i and j.

Note that two matrices are equal if and only if they have the same size and corresponding elements are equal.

ILLUSTRATION Equality of Matrices

$$\begin{bmatrix} 1 & 0 & 5 \\ \sqrt[3]{8} & 3^2 & -2 \end{bmatrix} = \begin{bmatrix} (-1)^2 & 0 & \sqrt{25} \\ 2 & 9 & -2 \end{bmatrix}$$

Using the parentheses notation for matrices, we may write the definition of addition of two $m \times n$ matrices as

$$(a_{ij}) + (b_{ij}) = (a_{ij} + b_{ij}).$$

Thus, to add two matrices, we add the elements in corresponding positions in each matrix. *Two matrices can be added only if they have the same size.*

ILLUSTRATION **Addition of Matrices**

- $$\begin{bmatrix} 4 & -5 \\ 0 & 4 \\ -6 & 1 \end{bmatrix} + \begin{bmatrix} 3 & 2 \\ 7 & -4 \\ -2 & 1 \end{bmatrix} = \begin{bmatrix} 4+3 & -5+2 \\ 0+7 & 4+(-4) \\ -6+(-2) & 1+1 \end{bmatrix} = \begin{bmatrix} 7 & -3 \\ 7 & 0 \\ -8 & 2 \end{bmatrix}$$
- $$\begin{bmatrix} 2 & 3 \\ -4 & 1 \end{bmatrix} + \begin{bmatrix} -2 & -3 \\ 4 & -1 \end{bmatrix} = \begin{bmatrix} 0 & 0 \\ 0 & 0 \end{bmatrix}$$
- $$\begin{bmatrix} 1 & 3 & -2 \\ 0 & -5 & 4 \end{bmatrix} + \begin{bmatrix} 0 & 0 & 0 \\ 0 & 0 & 0 \end{bmatrix} = \begin{bmatrix} 1 & 3 & -2 \\ 0 & -5 & 4 \end{bmatrix}$$

The $\boldsymbol{m \times n}$ **zero matrix,** denoted by O, is the matrix with m rows and n columns in which every element is 0.

ILLUSTRATION **Zero Matrices**

- $$\begin{bmatrix} 0 & 0 \\ 0 & 0 \end{bmatrix}$$
- $$\begin{bmatrix} 0 & 0 \\ 0 & 0 \\ 0 & 0 \end{bmatrix}$$
- $$\begin{bmatrix} 0 & 0 & 0 & 0 \\ 0 & 0 & 0 & 0 \end{bmatrix}$$

The **additive inverse** $-A$ of the matrix $A = (a_{ij})$ is the matrix $(-a_{ij})$ obtained by changing the sign of each nonzero element of A.

ILLUSTRATION **Additive Inverse**

- $$-\begin{bmatrix} 2 & -3 & 4 \\ -1 & 0 & 5 \end{bmatrix} = \begin{bmatrix} -2 & 3 & -4 \\ 1 & 0 & -5 \end{bmatrix}$$

The proof of the next theorem follows from the definition of addition of matrices.

Theorem on Matrix Properties

If A, B, and C are $m \times n$ matrices and if O is the $m \times n$ zero matrix, then

(1) $A + B = B + A$

(2) $A + (B + C) = (A + B) + C$

(3) $A + O = A$

(4) $A + (-A) = O$

Subtraction of two $m \times n$ matrices is defined by

$$A - B = A + (-B).$$

Using the parentheses notation, we have

$$\begin{aligned}(a_{ij}) - (b_{ij}) &= (a_{ij}) + (-b_{ij}) \\ &= (a_{ij} - b_{ij}).\end{aligned}$$

Thus, to subtract two matrices, we subtract the elements in corresponding positions.

ILLUSTRATION **Subtraction of Matrices**

■ $$\begin{bmatrix} 4 & -5 \\ 0 & 4 \\ -6 & 1 \end{bmatrix} - \begin{bmatrix} 3 & 2 \\ 7 & -4 \\ -2 & 1 \end{bmatrix} = \begin{bmatrix} 4-3 & -5-2 \\ 0-7 & 4-(-4) \\ -6-(-2) & 1-1 \end{bmatrix} = \begin{bmatrix} 1 & -7 \\ -7 & 8 \\ -4 & 0 \end{bmatrix}$$

Definition of the Product of a Real Number and a Matrix

The **product** of a real number c and an $m \times n$ matrix $A = (a_{ij})$ is

$$cA = (ca_{ij}).$$

Note that to find cA, we multiply each element of A by c.

ILLUSTRATION **Product of a Real Number and a Matrix**

■ $$3\begin{bmatrix} 4 & -1 \\ 2 & 3 \end{bmatrix} = \begin{bmatrix} 3\cdot 4 & 3\cdot(-1) \\ 3\cdot 2 & 3\cdot 3 \end{bmatrix} = \begin{bmatrix} 12 & -3 \\ 6 & 9 \end{bmatrix}$$

We can prove the following.

Theorem on Matrix Properties

If A and B are $m \times n$ matrices and if c and d are real numbers, then

(1) $c(A + B) = cA + cB$

(2) $(c + d)A = cA + dA$

(3) $(cd)A = c(dA)$

The next definition, of the product AB of two matrices, may seem unusual, but it has many uses in mathematics and applications. For multiplication, unlike addition, A and B may have different sizes; however, *the number of columns of A must be the same as the number of rows of B*. Thus, if A is $m \times n$, then B must be $n \times p$ for some p. As we shall see, the size of AB is then $m \times p$. If $C = AB$, then a method for finding the element c_{ij} in row i and column j of C is given in the following guidelines.

Guidelines for Finding c_{ij} in the Product $C = AB$ if A is $m \times n$ and B is $n \times p$

1 Single out the ith row, R_i, of A and the jth column, C_j, of B:

$$\begin{bmatrix} a_{11} & a_{12} & \cdots & a_{1n} \\ \vdots & \vdots & & \vdots \\ a_{i1} & a_{i2} & \cdots & a_{in} \\ \vdots & \vdots & & \vdots \\ a_{m1} & a_{m2} & \cdots & a_{mn} \end{bmatrix} \begin{bmatrix} b_{11} & \cdots & b_{1j} & \cdots & b_{1p} \\ b_{21} & \cdots & b_{2j} & \cdots & b_{2p} \\ \vdots & & \vdots & & \vdots \\ b_{n1} & \cdots & b_{nj} & \cdots & b_{np} \end{bmatrix}$$

2 *Simultaneously* move to the right along R_i and down C_j, multiplying pairs of elements, to obtain

$$a_{i1}b_{1j}, a_{i2}b_{2j}, a_{i3}b_{3j}, \ldots, a_{in}b_{nj}.$$

3 Add the products of the pairs in guideline 2 to obtain c_{ij}:

$$c_{ij} = a_{i1}b_{1j} + a_{i2}b_{2j} + a_{i3}b_{3j} + \cdots + a_{in}b_{nj}$$

Using the guidelines, we see that the element c_{11} in the first row and the first column of $C = AB$ is

$$c_{11} = a_{11}b_{11} + a_{12}b_{21} + a_{13}b_{31} + \cdots + a_{1n}b_{n1}.$$

The element c_{mp} in the last row and the last column of $C = AB$ is

$$c_{mp} = a_{m1}b_{1p} + a_{m2}b_{2p} + a_{m3}b_{3p} + \cdots + a_{mn}b_{np}.$$

The preceding discussion is summarized in the next definition.

Definition of the Product of Two Matrices

Let $A = (a_{ij})$ be an $m \times n$ matrix and let $B = (b_{ij})$ be an $n \times p$ matrix. The **product** AB is the $m \times p$ matrix $C = (c_{ij})$ such that

$$c_{ij} = a_{i1}b_{1j} + a_{i2}b_{2j} + a_{i3}b_{3j} + \cdots + a_{in}b_{nj}$$

for $i = 1, 2, 3, \ldots, m$ and $j = 1, 2, 3, \ldots, p$.

The following diagram may help you remember the relationship between sizes of matrices when working with a product AB.

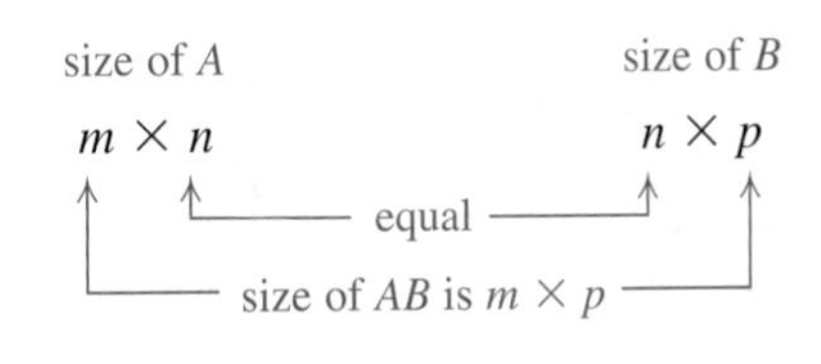

The next illustration contains some special cases.

ILLUSTRATION **Sizes of Matrices in Products**

	Size of A	Size of B	Size of AB
■	2×3	3×5	2×5
■	4×2	2×3	4×3
■	3×1	1×3	3×3
■	1×3	3×1	1×1
■	5×3	3×5	5×5
■	5×3	5×3	AB is not defined

In the following example we find the product of two specific matrices.

EXAMPLE 1 **Finding the product of two matrices**

Find the product AB if

$$A = \begin{bmatrix} 1 & 2 & -3 \\ 4 & 0 & -2 \end{bmatrix} \quad \text{and} \quad B = \begin{bmatrix} 5 & -4 & 2 & 0 \\ -1 & 6 & 3 & 1 \\ 7 & 0 & 5 & 8 \end{bmatrix}.$$

▶ **SOLUTION** The matrix A is 2×3, and the matrix B is 3×4. Hence, the product $C = AB$ is defined and is 2×4. We next use the guidelines to find the elements $c_{11}, c_{12}, \ldots, c_{24}$ of the product. For instance, to find the element c_{23} we single out the second row, R_2, of A and the third column, C_3, of B, as illustrated below, and then use guidelines 2 and 3 to obtain

$$c_{23} = 4 \cdot 2 + 0 \cdot 3 + (-2) \cdot 5 = -2.$$

$$\begin{bmatrix} 1 & 2 & -3 \\ 4 & 0 & -2 \end{bmatrix} \begin{bmatrix} 5 & -4 & 2 & 0 \\ -1 & 6 & 3 & 1 \\ 7 & 0 & 5 & 8 \end{bmatrix} = \begin{bmatrix} \blacksquare & \blacksquare & \blacksquare & \blacksquare \\ \blacksquare & \blacksquare & -2 & \blacksquare \end{bmatrix}$$

Similarly, to find the element c_{12} in row 1 and column 2 of the product, we proceed as follows:

$$c_{12} = 1 \cdot (-4) + 2 \cdot 6 + (-3) \cdot 0 = 8$$

$$\begin{bmatrix} 1 & 2 & -3 \\ 4 & 0 & -2 \end{bmatrix} \begin{bmatrix} 5 & -4 & 2 & 0 \\ -1 & 6 & 3 & 1 \\ 7 & 0 & 5 & 8 \end{bmatrix} = \begin{bmatrix} \blacksquare & 8 & \blacksquare & \blacksquare \\ \blacksquare & \blacksquare & -2 & \blacksquare \end{bmatrix}$$

▶ **Tutorial available at www.thomsonedu.com/login**

The remaining elements of the product are calculated as follows, where we have indicated the row of A and the column of B that are used when guideline 1 is applied.

Row of A	Column of B	Element of C
R_1	C_1	$c_{11} = 1 \cdot 5 + 2 \cdot (-1) + (-3) \cdot 7 = -18$
R_1	C_3	$c_{13} = 1 \cdot 2 + 2 \cdot 3 + (-3) \cdot 5 = -7$
R_1	C_4	$c_{14} = 1 \cdot 0 + 2 \cdot 1 + (-3) \cdot 8 = -22$
R_2	C_1	$c_{21} = 4 \cdot 5 + 0 \cdot (-1) + (-2) \cdot 7 = 6$
R_2	C_2	$c_{22} = 4 \cdot (-4) + 0 \cdot 6 + (-2) \cdot 0 = -16$
R_2	C_4	$c_{24} = 4 \cdot 0 + 0 \cdot 1 + (-2) \cdot 8 = -16$

Hence,

$$AB = \begin{bmatrix} 1 & 2 & -3 \\ 4 & 0 & -2 \end{bmatrix} \begin{bmatrix} 5 & -4 & 2 & 0 \\ -1 & 6 & 3 & 1 \\ 7 & 0 & 5 & 8 \end{bmatrix}$$
$$= \begin{bmatrix} -18 & 8 & -7 & -22 \\ 6 & -16 & -2 & -16 \end{bmatrix}.$$

Multiplying matrices on a graphing calculator is fairly straightforward. Let's check the result in Example 1. Enter the matrices A (2×3) and B (3×4):

$$A = \begin{bmatrix} 1 & 2 & -3 \\ 4 & 0 & -2 \end{bmatrix} \quad \text{and} \quad B = \begin{bmatrix} 5 & -4 & 2 & 0 \\ -1 & 6 & 3 & 1 \\ 7 & 0 & 5 & 8 \end{bmatrix}$$

Now enter the operation on the home screen.

TI-83/4 Plus

[2nd] [MATRX] [1] [×]
[2nd] [MATRX] [2] [ENTER]

```
[A]*[B]
[[-18 8    -7 -2...
 [6   -16 -2 -1...
```

TI-86

[ALPHA] [A] [×] [ALPHA] [B] [ENTER]

```
A*B
  [[-18 8   -7 -22]
   [6   -16 -2 -16]]
```

To see the elements in the fourth column, press the [▷] key.

A matrix is a **row matrix** if it has only one row. A **column matrix** has only one column. The following illustration contains some products involving row and column matrices. You should check each entry in the products.

ILLUSTRATION **Products Involving Row and Column Matrices**

- $\begin{bmatrix} -2 & 4 \\ 0 & -1 \\ 5 & 3 \end{bmatrix} \begin{bmatrix} -2 \\ 1 \end{bmatrix} = \begin{bmatrix} 8 \\ -1 \\ -7 \end{bmatrix}$
- $\begin{bmatrix} 3 & -1 & 2 \end{bmatrix} \begin{bmatrix} -2 & 4 \\ 0 & -1 \\ 5 & 3 \end{bmatrix} = \begin{bmatrix} 4 & 19 \end{bmatrix}$
- $\begin{bmatrix} -2 \\ 3 \end{bmatrix} \begin{bmatrix} 1 & 5 \end{bmatrix} = \begin{bmatrix} -2 & -10 \\ 3 & 15 \end{bmatrix}$
- $\begin{bmatrix} 1 & 5 \end{bmatrix} \begin{bmatrix} -2 \\ 3 \end{bmatrix} = \begin{bmatrix} 13 \end{bmatrix}$

The product operation for matrices is not commutative. For example, if A is 2×3 and B is 3×4, then AB may be found, since the number of columns of A is the same as the number of rows of B. However, BA is undefined, since the number of columns of B is different from the number of rows of A. Even if AB and BA are both defined, it is often true that these products are different. This is illustrated in the next example, along with the fact that the product of two nonzero matrices may equal a zero matrix.

EXAMPLE 2 **Matrix multiplication is not commutative**

If $A = \begin{bmatrix} 2 & 2 \\ -1 & -1 \end{bmatrix}$ and $B = \begin{bmatrix} 1 & 2 \\ 1 & 2 \end{bmatrix}$, show that $AB \neq BA$.

▶ **SOLUTION** Using the definition of the product of two matrices, we obtain the following:

$$AB = \begin{bmatrix} 2 & 2 \\ -1 & -1 \end{bmatrix} \begin{bmatrix} 1 & 2 \\ 1 & 2 \end{bmatrix} = \begin{bmatrix} 4 & 8 \\ -2 & -4 \end{bmatrix}$$

$$BA = \begin{bmatrix} 1 & 2 \\ 1 & 2 \end{bmatrix} \begin{bmatrix} 2 & 2 \\ -1 & -1 \end{bmatrix} = \begin{bmatrix} 0 & 0 \\ 0 & 0 \end{bmatrix}$$

Hence, $AB \neq BA$. Note that the last equality shows that *the product of two nonzero matrices can equal a zero matrix.*

Although matrix multiplication is not commutative, it is associative. Thus, if A is $m \times n$, B is $n \times p$, and C is $p \times q$, then

$$A(BC) = (AB)C.$$

The distributive properties also hold if the matrices involved have the proper number of rows and columns. If A_1 and A_2 are $m \times n$ matrices and if B_1 and B_2 are $n \times p$ matrices, then

$$A_1(B_1 + B_2) = A_1B_1 + A_1B_2$$
$$(A_1 + A_2)B_1 = A_1B_1 + A_2B_1.$$

▶ **Tutorial available at www.thomsonedu.com/login**

As a special case, if all matrices are square, of order n, then both the associative and the distributive property are true.

We conclude this section with an application of the product of two matrices.

EXAMPLE 3 **An application of a matrix product**

(a) Three investors, I_1, I_2, and I_3, each own a certain number of shares of four stocks, S_1, S_2, S_3, and S_4, according to matrix A. Matrix B contains the present value V of each share of each stock. Find AB, and interpret the meaning of its elements.

$$\text{investors}\begin{cases} I_1 \\ I_2 \\ I_3 \end{cases} \overbrace{\begin{array}{cccc} S_1 & S_2 & S_3 & S_4 \end{array}}^{\text{number of shares of stock}} \begin{bmatrix} 50 & 100 & 30 & 25 \\ 100 & 150 & 10 & 30 \\ 100 & 50 & 40 & 100 \end{bmatrix} = A, \qquad \text{stocks}\begin{cases} S_1 \\ S_2 \\ S_3 \\ S_4 \end{cases} \overbrace{V}^{\text{share value}} \begin{bmatrix} 20.37 \\ 16.21 \\ 90.80 \\ 42.75 \end{bmatrix} = B$$

(b) Matrix C contains the change in the value of each stock for the last week. Find AC, and interpret the meaning of its elements.

$$\text{stocks}\begin{cases} S_1 \\ S_2 \\ S_3 \\ S_4 \end{cases} \begin{bmatrix} +1.03 \\ -0.22 \\ -1.35 \\ +0.15 \end{bmatrix} = C$$

▶ **SOLUTION**

(a) Since A is a 3×4 matrix and B is a 4×1 matrix, the product AB is a 3×1 matrix:

$$AB = \begin{bmatrix} 50 & 100 & 30 & 25 \\ 100 & 150 & 10 & 30 \\ 100 & 50 & 40 & 100 \end{bmatrix} \begin{bmatrix} 20.37 \\ 16.21 \\ 90.80 \\ 42.75 \end{bmatrix} = \begin{bmatrix} 6432.25 \\ 6659.00 \\ 10{,}754.50 \end{bmatrix}$$

The first element in the product AB, 6432.25, was obtained from the computation

$$50(20.37) + 100(16.21) + 30(90.80) + 25(42.75)$$

and represents the total value that investor I_1 has in all four stocks. Similarly, the second and third elements represent the total value for investors I_2 and I_3, respectively.

(continued)

(b)

$$AC = \begin{bmatrix} 50 & 100 & 30 & 25 \\ 100 & 150 & 10 & 30 \\ 100 & 50 & 40 & 100 \end{bmatrix} \begin{bmatrix} +1.03 \\ -0.22 \\ -1.35 \\ +0.15 \end{bmatrix} = \begin{bmatrix} -7.25 \\ 61.00 \\ 53.00 \end{bmatrix}$$

The first element in the product AC, -7.25, indicates that the total value that investor I_1 has in all four stocks went down \$7.25 in the last week. The second and third elements indicate that the total value that investors I_2 and I_3 have in all four stocks went up \$61.00 and \$53.00, respectively.

8.6 Exercises

Exer. 1–8: Find, if possible, $A + B$, $A - B$, $2A$, and $-3B$.

▶ 1 $A = \begin{bmatrix} 5 & -2 \\ 1 & 3 \end{bmatrix}$, $B = \begin{bmatrix} 4 & 1 \\ -3 & 2 \end{bmatrix}$

2 $A = \begin{bmatrix} 3 & 0 \\ -1 & 2 \end{bmatrix}$, $B = \begin{bmatrix} 3 & -4 \\ 1 & 1 \end{bmatrix}$

▶ 3 $A = \begin{bmatrix} 6 & -1 \\ 2 & 0 \\ -3 & 4 \end{bmatrix}$, $B = \begin{bmatrix} 3 & 1 \\ -1 & 5 \\ 6 & 0 \end{bmatrix}$

▶ 4 $A = \begin{bmatrix} 0 & -2 & 7 \\ 5 & 4 & -3 \end{bmatrix}$, $B = \begin{bmatrix} 8 & 4 & 0 \\ 0 & 1 & 4 \end{bmatrix}$

5 $A = [4 \quad -3 \quad 2]$, $B = [7 \quad 0 \quad -5]$

6 $A = \begin{bmatrix} 7 \\ -16 \end{bmatrix}$, $B = \begin{bmatrix} -11 \\ 9 \end{bmatrix}$

7 $A = \begin{bmatrix} 3 & -2 & 2 \\ 0 & 1 & -4 \\ -3 & 2 & -1 \end{bmatrix}$, $B = \begin{bmatrix} 4 & 0 \\ 2 & -1 \\ -1 & 3 \end{bmatrix}$

8 $A = [2 \quad 1]$, $B = [3 \quad -1 \quad 5]$

Exer. 9–10: Find the given element of the matrix product $C = AB$ in the listed exercise.

9 c_{21}; Exercise 15

10 c_{23}; Exercise 16

Exer. 11–22: Find, if possible, AB and BA.

11 $A = \begin{bmatrix} 2 & 6 \\ 3 & -4 \end{bmatrix}$, $B = \begin{bmatrix} 5 & -2 \\ 1 & 7 \end{bmatrix}$

12 $A = \begin{bmatrix} 4 & -2 \\ -2 & 1 \end{bmatrix}$, $B = \begin{bmatrix} 2 & 1 \\ 4 & 2 \end{bmatrix}$

13 $A = \begin{bmatrix} 3 & 0 & -1 \\ 0 & 4 & 2 \\ 5 & -3 & 1 \end{bmatrix}$, $B = \begin{bmatrix} 1 & -5 & 0 \\ 4 & 1 & -2 \\ 0 & -1 & 3 \end{bmatrix}$

14 $A = \begin{bmatrix} 5 & 0 & 0 \\ 0 & -3 & 0 \\ 0 & 0 & 2 \end{bmatrix}$, $B = \begin{bmatrix} 3 & 0 & 0 \\ 0 & 4 & 0 \\ 0 & 0 & -2 \end{bmatrix}$

15 $A = \begin{bmatrix} 4 & -3 & 1 \\ -5 & 2 & 2 \end{bmatrix}$, $B = \begin{bmatrix} 2 & 1 \\ 0 & 1 \\ -4 & 7 \end{bmatrix}$

16 $A = \begin{bmatrix} 2 & 1 & -1 & 0 \\ 3 & -2 & 0 & 5 \\ -2 & 1 & 4 & 2 \end{bmatrix}$, $B = \begin{bmatrix} 5 & -3 & 1 \\ 1 & 2 & 0 \\ -1 & 0 & 4 \\ 0 & -2 & 3 \end{bmatrix}$

17 $A = \begin{bmatrix} 1 & 2 & 3 \\ 4 & 5 & 6 \\ 7 & 8 & 9 \end{bmatrix}$, $B = \begin{bmatrix} 1 & 0 & 0 \\ 0 & 1 & 0 \\ 0 & 0 & 1 \end{bmatrix}$

18 $A = \begin{bmatrix} 1 & 2 & 3 \\ 2 & 3 & 1 \\ 3 & 1 & 2 \end{bmatrix}$, $B = \begin{bmatrix} 2 & 0 & 0 \\ 0 & 2 & 0 \\ 0 & 0 & 2 \end{bmatrix}$

19 $A = [-3 \quad 7 \quad 2]$, $B = \begin{bmatrix} 1 \\ 4 \\ -5 \end{bmatrix}$

20 $A = [4 \quad 8]$, $B = \begin{bmatrix} -3 \\ 2 \end{bmatrix}$

▶ Tutorial available at www.thomsonedu.com/login

21 $A = \begin{bmatrix} 2 & 0 & 1 \\ -1 & 2 & 0 \end{bmatrix}$, $B = \begin{bmatrix} 1 & -1 & 2 \\ 3 & 1 & 0 \\ 0 & 2 & 1 \end{bmatrix}$

22 $A = [3 \quad -1 \quad 4]$, $B = \begin{bmatrix} -2 \\ 5 \end{bmatrix}$

Exer. 23–26: Find AB.

▶ 23 $A = \begin{bmatrix} 4 & -2 \\ 0 & 3 \\ -7 & 5 \end{bmatrix}$, $B = \begin{bmatrix} 3 \\ 4 \end{bmatrix}$

▶ 24 $A = \begin{bmatrix} 4 \\ -3 \\ 2 \end{bmatrix}$, $B = [5 \quad 1]$

25 $A = \begin{bmatrix} 2 & 1 & 0 & -3 \\ -7 & 0 & -2 & 4 \end{bmatrix}$, $B = \begin{bmatrix} 4 & -2 & 0 \\ 1 & 1 & -2 \\ 0 & 0 & 5 \\ -3 & -1 & 0 \end{bmatrix}$

26 $A = \begin{bmatrix} 1 & 2 & -3 \\ 4 & -5 & 6 \end{bmatrix}$, $B = \begin{bmatrix} 1 & -1 & 0 & 2 \\ -2 & 3 & 1 & 0 \\ 0 & 4 & 0 & -3 \end{bmatrix}$

Exer. 27–30: Let

$$A = \begin{bmatrix} 1 & 2 \\ 0 & -3 \end{bmatrix}, \quad B = \begin{bmatrix} 2 & -1 \\ 3 & 1 \end{bmatrix}, \quad C = \begin{bmatrix} 3 & 1 \\ -2 & 0 \end{bmatrix}.$$

Verify the statement.

27 $(A + B)(A - B) \neq A^2 - B^2$, where $A^2 = AA$ and $B^2 = BB$.

28 $(A + B)(A + B) \neq A^2 + 2AB + B^2$

29 $A(B + C) = AB + AC$

30 $A(BC) = (AB)C$

Exer. 31–34: Verify the identity for

$$A = \begin{bmatrix} a & b \\ c & d \end{bmatrix}, \quad B = \begin{bmatrix} p & q \\ r & s \end{bmatrix}, \quad C = \begin{bmatrix} w & x \\ y & z \end{bmatrix},$$

and real numbers m and n.

31 $m(A + B) = mA + mB$

32 $(m + n)A = mA + nA$

33 $A(B + C) = AB + AC$

34 $A(BC) = (AB)C$

Exer. 35–38: Let

$$A = \begin{bmatrix} 3 & -3 & 7 \\ 2 & 6 & -2 \\ 4 & 2 & 5 \end{bmatrix} \quad \text{and} \quad B = \begin{bmatrix} -9 & 5 & -8 \\ 3 & -7 & 1 \\ -1 & 2 & 6 \end{bmatrix}.$$

Evaluate the matrix expression.

35 $A^2 + B^2$

36 $3A - BA$

37 $A^2 - 5B$

38 $A + A^2 + B + B^2$

39 **Value of inventory** A store stocks these sizes of towels, each available in five colors: small, priced at $8.99 each; medium, priced at $10.99 each; and large, priced at $12.99 each. The store's current inventory is as follows:

	Colors				
Towel size	**White**	**Tan**	**Beige**	**Pink**	**Yellow**
Small	400	400	300	250	100
Medium	550	450	500	200	100
Large	500	500	600	300	200

(a) Organize these data into an inventory matrix A and a price matrix B so that the product $C = AB$ is defined.

(b) Find C.

(c) Interpret the meaning of element c_{51} in C.

▶ 40 **Building costs** A housing contractor has orders for 4 one-bedroom units, 10 two-bedroom units, and 6 three-bedroom units. The labor and material costs (in thousands of dollars) are given in the following table.

	1-Bedroom	2-Bedroom	3-Bedroom
Labor	70	95	117
Materials	90	105	223

(a) Organize these data into an order matrix A and a cost matrix B so that the product $C = AB$ is defined.

(b) Find C.

(c) Interpret the meaning of each element in C.

8.7

The Inverse of a Matrix

Throughout this section and the next two sections we shall restrict our discussion to *square* matrices. The symbol I_n will denote the square matrix of order n that has 1 in each position on the main diagonal and 0 elsewhere. We call I_n the **identity matrix of order n.**

ILLUSTRATION **Identity Matrices**

■ $I_2 = \begin{bmatrix} 1 & 0 \\ 0 & 1 \end{bmatrix}$ ■ $I_3 = \begin{bmatrix} 1 & 0 & 0 \\ 0 & 1 & 0 \\ 0 & 0 & 1 \end{bmatrix}$

We can show that if A is any square matrix of order n, then

$$AI_n = A = I_nA.$$

ILLUSTRATION $AI_2 = A = I_2A$

■ $$\begin{bmatrix} a_{11} & a_{12} \\ a_{21} & a_{22} \end{bmatrix}\begin{bmatrix} 1 & 0 \\ 0 & 1 \end{bmatrix} = \begin{bmatrix} a_{11} & a_{12} \\ a_{21} & a_{22} \end{bmatrix} = \begin{bmatrix} 1 & 0 \\ 0 & 1 \end{bmatrix}\begin{bmatrix} a_{11} & a_{12} \\ a_{21} & a_{22} \end{bmatrix}$$

Recall that when we are working with a nonzero real number b, the unique number b^{-1} (the multiplicative inverse of b) may be multiplied times b to obtain the multiplicative identity (the number 1)—that is,

$$b \cdot b^{-1} = 1.$$

We have a similar situation with matrices.

Definition of the Inverse of a Matrix

Let A be a square matrix of order n. If there exists a matrix B such that

$$AB = I_n = BA,$$

then B is called the **inverse** of A and is denoted A^{-1} (read "A inverse").

If a square matrix A has an inverse, then we say that A is **invertible.** If a matrix is not square, then it cannot have an inverse. For matrices (unlike real numbers), the symbol $1/A$ does not represent the inverse A^{-1}.

If A is invertible, we can calculate A^{-1} using elementary row operations. If $A = (a_{ij})$ is $n \times n$, we begin with the $n \times 2n$ matrix formed by *adjoining* I_n to A:

$$\left[\begin{array}{cccc|cccc} a_{11} & a_{12} & \cdots & a_{1n} & 1 & 0 & \cdots & 0 \\ a_{21} & a_{22} & \cdots & a_{2n} & 0 & 1 & \cdots & 0 \\ \vdots & \vdots & & \vdots & \vdots & \vdots & & \vdots \\ a_{n1} & a_{n2} & \cdots & a_{nn} & 0 & 0 & \cdots & 1 \end{array}\right]$$

We next apply a succession of elementary row transformations, as we did in Section 8.5 to find reduced echelon forms, until we arrive at a matrix of the form

$$\left[\begin{array}{cccc|cccc} 1 & 0 & \cdots & 0 & b_{11} & b_{12} & \cdots & b_{1n} \\ 0 & 1 & \cdots & 0 & b_{21} & b_{22} & \cdots & b_{2n} \\ \vdots & \vdots & & \vdots & \vdots & \vdots & & \vdots \\ 0 & 0 & \cdots & 1 & b_{n1} & b_{n2} & \cdots & b_{nn} \end{array}\right]$$

in which the identity matrix I_n appears to the left of the vertical rule. It can be shown that the $n \times n$ matrix (b_{ij}) is the inverse of A—that is, $B = A^{-1}$.

EXAMPLE 1 **Finding the inverse of a 2 × 2 matrix**

Find A^{-1} if $A = \begin{bmatrix} 3 & 5 \\ 1 & 4 \end{bmatrix}$.

▶ **SOLUTION** We begin with the matrix

$$\left[\begin{array}{cc|cc} 3 & 5 & 1 & 0 \\ 1 & 4 & 0 & 1 \end{array}\right].$$

Next we perform elementary row transformations until the identity matrix I_2 appears on the left of the vertical rule, as follows:

$$\left[\begin{array}{cc|cc} 3 & 5 & 1 & 0 \\ 1 & 4 & 0 & 1 \end{array}\right] \mathbf{R}_1 \leftrightarrow \mathbf{R}_2 \left[\begin{array}{cc|cc} 1 & 4 & 0 & 1 \\ 3 & 5 & 1 & 0 \end{array}\right]$$

$$-3\mathbf{R}_1 + \mathbf{R}_2 \to \mathbf{R}_2 \left[\begin{array}{cc|cc} 1 & 4 & 0 & 1 \\ 0 & -7 & 1 & -3 \end{array}\right]$$

$$-\tfrac{1}{7}\mathbf{R}_2 \to \mathbf{R}_2 \left[\begin{array}{cc|cc} 1 & 4 & 0 & 1 \\ 0 & 1 & -\frac{1}{7} & \frac{3}{7} \end{array}\right]$$

$$-4\mathbf{R}_2 + \mathbf{R}_1 \to \mathbf{R}_1 \left[\begin{array}{cc|cc} 1 & 0 & \frac{4}{7} & -\frac{5}{7} \\ 0 & 1 & -\frac{1}{7} & \frac{3}{7} \end{array}\right]$$

By the previous discussion,

$$A^{-1} = \begin{bmatrix} \frac{4}{7} & -\frac{5}{7} \\ -\frac{1}{7} & \frac{3}{7} \end{bmatrix} = \frac{1}{7}\begin{bmatrix} 4 & -5 \\ -1 & 3 \end{bmatrix}.$$

Let us verify that $AA^{-1} = I_2 = A^{-1}A$:

$$\begin{bmatrix} 3 & 5 \\ 1 & 4 \end{bmatrix}\begin{bmatrix} \frac{4}{7} & -\frac{5}{7} \\ -\frac{1}{7} & \frac{3}{7} \end{bmatrix} = \begin{bmatrix} 1 & 0 \\ 0 & 1 \end{bmatrix} = \begin{bmatrix} \frac{4}{7} & -\frac{5}{7} \\ -\frac{1}{7} & \frac{3}{7} \end{bmatrix}\begin{bmatrix} 3 & 5 \\ 1 & 4 \end{bmatrix}$$

EXAMPLE 2 **Finding the inverse of a 3 × 3 matrix**

Find A^{-1} if $A = \begin{bmatrix} -1 & 3 & 1 \\ 2 & 5 & 0 \\ 3 & 1 & -2 \end{bmatrix}$.

▶ **Tutorial available at www.thomsonedu.com/login**

▶ SOLUTION

$$\left[\begin{array}{ccc|ccc} -1 & 3 & 1 & 1 & 0 & 0 \\ 2 & 5 & 0 & 0 & 1 & 0 \\ 3 & 1 & -2 & 0 & 0 & 1 \end{array}\right] \begin{array}{l} -\mathbf{R_1 \to R_1} \\ \\ \\ \end{array} \left[\begin{array}{ccc|ccc} 1 & -3 & -1 & -1 & 0 & 0 \\ 2 & 5 & 0 & 0 & 1 & 0 \\ 3 & 1 & -2 & 0 & 0 & 1 \end{array}\right]$$

$$\begin{array}{r} \\ -\mathbf{2R_1 + R_2 \to R_2} \\ -\mathbf{3R_1 + R_3 \to R_3} \end{array} \left[\begin{array}{ccc|ccc} 1 & -3 & -1 & -1 & 0 & 0 \\ 0 & 11 & 2 & 2 & 1 & 0 \\ 0 & 10 & 1 & 3 & 0 & 1 \end{array}\right]$$

$$\begin{array}{r} \\ -\mathbf{R_3 + R_2 \to R_2} \\ \\ \end{array} \left[\begin{array}{ccc|ccc} 1 & -3 & -1 & -1 & 0 & 0 \\ 0 & 1 & 1 & -1 & 1 & -1 \\ 0 & 10 & 1 & 3 & 0 & 1 \end{array}\right]$$

$$\begin{array}{r} \mathbf{3R_2 + R_1 \to R_1} \\ \\ -\mathbf{10R_2 + R_3 \to R_3} \end{array} \left[\begin{array}{ccc|ccc} 1 & 0 & 2 & -4 & 3 & -3 \\ 0 & 1 & 1 & -1 & 1 & -1 \\ 0 & 0 & -9 & 13 & -10 & 11 \end{array}\right]$$

$$\begin{array}{r} \\ \\ -\frac{1}{9}\mathbf{R_3 \to R_3} \end{array} \left[\begin{array}{ccc|ccc} 1 & 0 & 2 & -4 & 3 & -3 \\ 0 & 1 & 1 & -1 & 1 & -1 \\ 0 & 0 & 1 & -\frac{13}{9} & \frac{10}{9} & -\frac{11}{9} \end{array}\right]$$

$$\begin{array}{r} -\mathbf{2R_3 + R_1 \to R_1} \\ -\mathbf{R_3 + R_2 \to R_2} \\ \\ \end{array} \left[\begin{array}{ccc|ccc} 1 & 0 & 0 & -\frac{10}{9} & \frac{7}{9} & -\frac{5}{9} \\ 0 & 1 & 0 & \frac{4}{9} & -\frac{1}{9} & \frac{2}{9} \\ 0 & 0 & 1 & -\frac{13}{9} & \frac{10}{9} & -\frac{11}{9} \end{array}\right]$$

Consequently,

$$A^{-1} = \begin{bmatrix} -\frac{10}{9} & \frac{7}{9} & -\frac{5}{9} \\ \frac{4}{9} & -\frac{1}{9} & \frac{2}{9} \\ -\frac{13}{9} & \frac{10}{9} & -\frac{11}{9} \end{bmatrix} = \frac{1}{9}\begin{bmatrix} -10 & 7 & -5 \\ 4 & -1 & 2 \\ -13 & 10 & -11 \end{bmatrix}.$$

You may verify that $AA^{-1} = I_3 = A^{-1}A$.

Not all square matrices are invertible. In fact, if the procedure used in Examples 1 and 2 does not lead to an identity matrix to the left of the vertical rule, then the matrix A has no inverse—that is, A is not invertible.

Finding the inverse of a square matrix on a graphing calculator is relatively easy. Enter the matrix A from Example 2:

$$A = \begin{bmatrix} -1 & 3 & 1 \\ 2 & 5 & 0 \\ 3 & 1 & -2 \end{bmatrix}$$

Now enter the inverse of A on the home screen.

▶ Tutorial available at www.thomsonedu.com/login

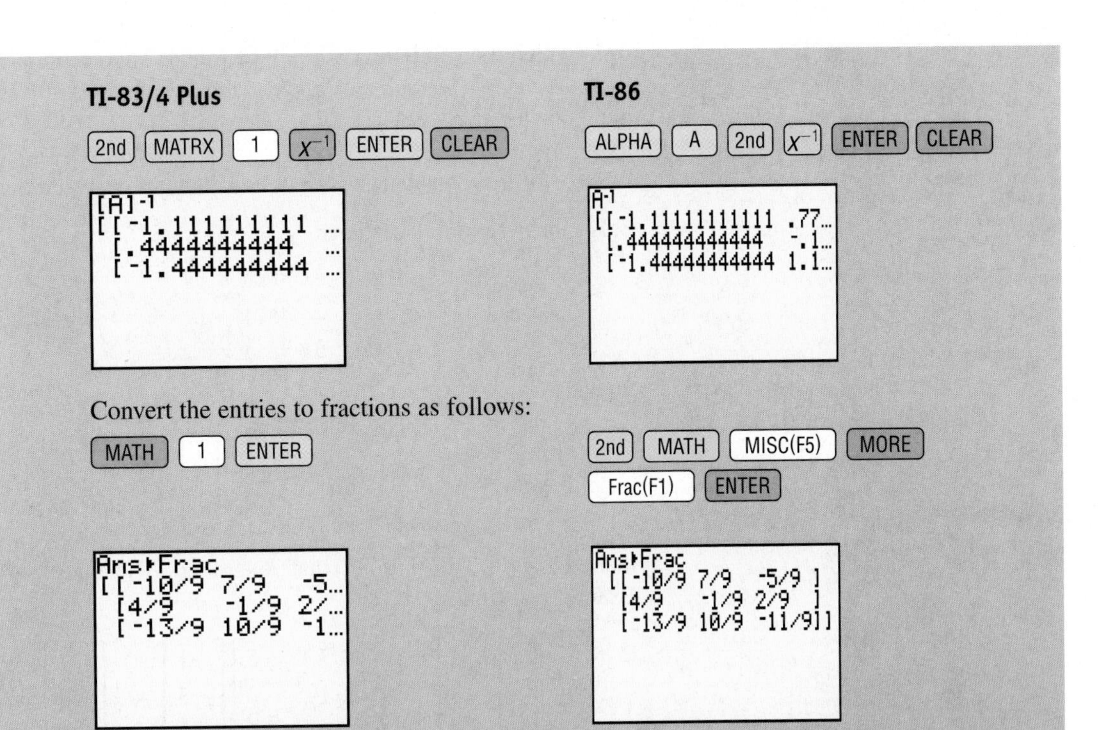

Note that you must use x^{-1} and not the notation A^(−1). If the matrix is not invertible, either calculator returns the error message SINGULAR MAT.

We may apply inverses of matrices to solutions of systems of linear equations. Consider the case of two linear equations in two unknowns:

$$\begin{cases} a_{11}x + a_{12}y = k_1 \\ a_{21}x + a_{22}y = k_2 \end{cases}$$

This system can be expressed in terms of matrices as

$$\begin{bmatrix} a_{11}x + a_{12}y \\ a_{21}x + a_{22}y \end{bmatrix} = \begin{bmatrix} k_1 \\ k_2 \end{bmatrix}.$$

If we let

$$A = \begin{bmatrix} a_{11} & a_{12} \\ a_{21} & a_{22} \end{bmatrix}, \quad X = \begin{bmatrix} x \\ y \end{bmatrix}, \quad \text{and} \quad B = \begin{bmatrix} k_1 \\ k_2 \end{bmatrix},$$

then a *matrix form* for the system is

$$AX = B.$$

If A^{-1} exists, then multiplying both sides of the last equation by A^{-1} gives us $A^{-1}AX = A^{-1}B$. Since $A^{-1}A = I_2$ and $I_2X = X$, this leads to

$$X = A^{-1}B,$$

from which the solution (x, y) may be found. This technique (which we refer to as the *inverse method*) may be extended to systems of n linear equations in n unknowns.

EXAMPLE 3 Solving a system of linear equations using the inverse method

Solve the system of equations:

$$\begin{cases} -x + 3y + z = 1 \\ 2x + 5y = 3 \\ 3x + y - 2z = -2 \end{cases}$$

▶ **SOLUTION** If we let

$$A = \begin{bmatrix} -1 & 3 & 1 \\ 2 & 5 & 0 \\ 3 & 1 & -2 \end{bmatrix}, \quad X = \begin{bmatrix} x \\ y \\ z \end{bmatrix}, \quad \text{and} \quad B = \begin{bmatrix} 1 \\ 3 \\ -2 \end{bmatrix},$$

then $AX = B$. This implies that $X = A^{-1}B$. The matrix A^{-1} was found in Example 2. Hence,

$$\begin{bmatrix} x \\ y \\ z \end{bmatrix} = \frac{1}{9}\begin{bmatrix} -10 & 7 & -5 \\ 4 & -1 & 2 \\ -13 & 10 & -11 \end{bmatrix}\begin{bmatrix} 1 \\ 3 \\ -2 \end{bmatrix} = \frac{1}{9}\begin{bmatrix} 21 \\ -3 \\ 39 \end{bmatrix} = \begin{bmatrix} \frac{7}{3} \\ -\frac{1}{3} \\ \frac{13}{3} \end{bmatrix}.$$

As expected, the calculator solution for Example 3 is quite simple—just enter $A^{-1} \times B$ to obtain the solution.

Thus, $x = \frac{7}{3}$, $y = -\frac{1}{3}$, $z = \frac{13}{3}$, and the ordered triple $\left(\frac{7}{3}, -\frac{1}{3}, \frac{13}{3}\right)$ is the solution of the given system.

If we are solving a system of linear equations without the aid of any computational device, then the inverse method of solution in Example 3 is beneficial only if A^{-1} is known (or can be easily computed) or if many systems with the same coefficient matrix are to be considered.

If we are using a computational device and if the coefficient matrix is not invertible, then the inverse method cannot be used, and the preferred method of solution is the matrix method discussed in Section 8.5. There are other important uses for the inverse of a matrix that arise in more advanced fields of mathematics and in applications of such fields.

8.7 Exercises

Exer. 1–2: Show that B is the inverse of A.

1 $A = \begin{bmatrix} 5 & 7 \\ 2 & 3 \end{bmatrix}$, $B = \begin{bmatrix} 3 & -7 \\ -2 & 5 \end{bmatrix}$

2 $A = \begin{bmatrix} 8 & -5 \\ -3 & 2 \end{bmatrix}$, $B = \begin{bmatrix} 2 & 5 \\ 3 & 8 \end{bmatrix}$

Exer. 3–12: Find the inverse of the matrix if it exists.

3 $\begin{bmatrix} 2 & -4 \\ 1 & 3 \end{bmatrix}$

4 $\begin{bmatrix} 3 & 2 \\ 4 & 5 \end{bmatrix}$

5 $\begin{bmatrix} 2 & 4 \\ 4 & 8 \end{bmatrix}$

▶ 6 $\begin{bmatrix} 3 & -1 \\ 6 & -2 \end{bmatrix}$

▶ **Tutorial available at www.thomsonedu.com/login**

▶ 7 $\begin{bmatrix} 3 & -1 & 0 \\ 2 & 2 & 0 \\ 0 & 0 & 4 \end{bmatrix}$

▶ 8 $\begin{bmatrix} 3 & 0 & 2 \\ 0 & 1 & 0 \\ -4 & 0 & 2 \end{bmatrix}$

9 $\begin{bmatrix} -2 & 2 & 3 \\ 1 & -1 & 0 \\ 0 & 1 & 4 \end{bmatrix}$

10 $\begin{bmatrix} 1 & 2 & 3 \\ -2 & 1 & 0 \\ 3 & -1 & 1 \end{bmatrix}$

▶ 11 $\begin{bmatrix} 2 & 0 & 0 \\ 0 & 4 & 0 \\ 0 & 0 & 6 \end{bmatrix}$

▶ 12 $\begin{bmatrix} 1 & 1 & 1 \\ 2 & 2 & 2 \\ 3 & 3 & 3 \end{bmatrix}$

13 State conditions on a and b that guarantee that the matrix $\begin{bmatrix} a & 0 \\ 0 & b \end{bmatrix}$ has an inverse, and find a formula for the inverse if it exists.

14 If $abc \neq 0$, find the inverse of $\begin{bmatrix} a & 0 & 0 \\ 0 & b & 0 \\ 0 & 0 & c \end{bmatrix}$.

15 If $A = \begin{bmatrix} a_{11} & a_{12} & a_{13} \\ a_{21} & a_{22} & a_{23} \\ a_{31} & a_{32} & a_{33} \end{bmatrix}$, show that $AI_3 = A = I_3A$.

16 Show that $AI_4 = A = I_4A$ for every square matrix A of order 4.

Exer. 17–20: Solve the system using the inverse method. Refer to Exercises 3–4 and 9–10.

▶ 17 $\begin{cases} 2x - 4y = c \\ x + 3y = d \end{cases}$

(a) $\begin{bmatrix} c \\ d \end{bmatrix} = \begin{bmatrix} 3 \\ 1 \end{bmatrix}$ (b) $\begin{bmatrix} c \\ d \end{bmatrix} = \begin{bmatrix} -2 \\ 5 \end{bmatrix}$

18 $\begin{cases} 3x + 2y = c \\ 4x + 5y = d \end{cases}$

(a) $\begin{bmatrix} c \\ d \end{bmatrix} = \begin{bmatrix} -1 \\ 1 \end{bmatrix}$ (b) $\begin{bmatrix} c \\ d \end{bmatrix} = \begin{bmatrix} 4 \\ 3 \end{bmatrix}$

19 $\begin{cases} -2x + 2y + 3z = c \\ x - y = d \\ y + 4z = e \end{cases}$

(a) $\begin{bmatrix} c \\ d \\ e \end{bmatrix} = \begin{bmatrix} 1 \\ 3 \\ -2 \end{bmatrix}$ (b) $\begin{bmatrix} c \\ d \\ e \end{bmatrix} = \begin{bmatrix} -1 \\ 0 \\ 4 \end{bmatrix}$

▶ 20 $\begin{cases} x + 2y + 3z = c \\ -2x + y = d \\ 3x - y + z = e \end{cases}$

(a) $\begin{bmatrix} c \\ d \\ e \end{bmatrix} = \begin{bmatrix} -1 \\ 4 \\ 2 \end{bmatrix}$ (b) $\begin{bmatrix} c \\ d \\ e \end{bmatrix} = \begin{bmatrix} -3 \\ -2 \\ 1 \end{bmatrix}$

Exer. 21–24: For each matrix A, approximate its inverse A^{-1} to five decimal places.

21 $A = \begin{bmatrix} 2 & -5 & 8 \\ 3 & 7 & -1 \\ 0 & 2 & 1 \end{bmatrix}$

22 $A = \begin{bmatrix} 0 & 1.2 & 4.1 \\ -1 & 0 & -1 \\ 5.9 & 2 & 0 \end{bmatrix}$

23 $A = \begin{bmatrix} 2 & -1 & 1 & 4 \\ 7 & 1.2 & -8 & 0 \\ 2.5 & 0 & 1.9 & 7.9 \\ 1 & -1 & 3 & 1 \end{bmatrix}$

24 $A = \begin{bmatrix} -3 & -7 & 4 & 0 \\ -7 & 0 & 5.5 & 9 \\ 3 & 1 & 0 & 0 \\ 9 & -11 & 4 & 1 \end{bmatrix}$

Exer. 25–28: (a) Express the system in the matrix form $AX = B$. (b) Approximate A^{-1}, using four-decimal-place accuracy for its elements. (c) Use $X = A^{-1}B$ to approximate the solution of the system to four-decimal-place accuracy.

25 $\begin{cases} 4.0x + 7.1y = 6.2 \\ 2.2x - 4.9y = 2.9 \end{cases}$

26 $\begin{cases} 5.1x + 8.7y + 2.5z = 1.1 \\ 9.9x + 15y + 12z = 3.8 \\ -4.3x - 2.2y - z = -7.1 \end{cases}$

▶ Tutorial available at www.thomsonedu.com/login

27 $\begin{cases} 3.1x + 6.7y - 8.7z = 1.5 \\ 4.1x - 5.1y + 0.2z = 2.1 \\ 0.6x + 1.1y - 7.4z = 3.9 \end{cases}$

28 $\begin{cases} 5.6x + 8.4y - 7.2z + 4.2w = 8.1 \\ 8.4x + 9.2y - 6.1z - 6.2w = 5.3 \\ -7.2x - 6.1y + 9.2z + 4.5w = 0.4 \\ 4.2x - 6.2y - 4.5z + 5.8w = 2.7 \end{cases}$

29 **Average low temperatures** Three average monthly low temperatures for Detroit are listed in the table.

Month	Temperature
Feb.	19°F
Aug.	59°F
Nov.	26°F

(a) Let $x = 1$ correspond to January, $x = 2$ to February, ..., and $x = 12$ to December. Determine a quadratic function $f(x) = ax^2 + bx + c$ that interpolates the data—that is, determine the constants a, b, and c such that $f(2) = 19$, $f(8) = 59$, and $f(11) = 26$.

(b) Graph f in the viewing rectangle $[1, 12]$ by $[-15, 70, 5]$.

(c) Use f to approximate the average monthly low temperatures in June and October. Compare your predictions to the actual temperatures of 58°F and 41°F, respectively.

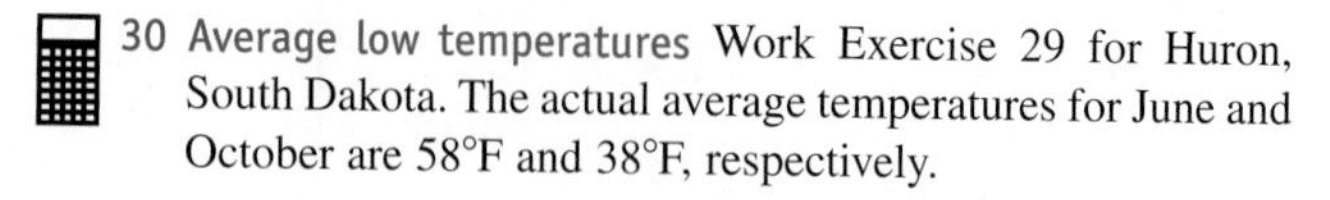

30 **Average low temperatures** Work Exercise 29 for Huron, South Dakota. The actual average temperatures for June and October are 58°F and 38°F, respectively.

Month	Temperature
Feb.	9°F
July	60°F
Nov.	21°F

8.8 Determinants

Associated with each square matrix A is a number called the **determinant of A,** denoted by $|A|$. This notation should not be confused with the symbol for the absolute value of a real number. To avoid any misunderstanding, the expression "det A" is sometimes used in place of $|A|$. We shall define $|A|$ by beginning with the case in which A has order 1 and then increasing the order one at a time. As we shall see in Section 8.9, these definitions arise in a natural way when systems of linear equations are solved.

If A is a square matrix of order 1, then A has only one element. Thus, $A = [a_{11}]$ and we define $|A| = a_{11}$. If A is a square matrix of order 2, then

$$A = \begin{bmatrix} a_{11} & a_{12} \\ a_{21} & a_{22} \end{bmatrix},$$

and the determinant of A is defined by

$$|A| = a_{11}a_{22} - a_{21}a_{12}.$$

Another notation for $|A|$ is obtained by replacing the brackets used for A with vertical bars, as follows.

Definition of the Determinant of a 2 × 2 Matrix A	$|A| = \begin{vmatrix} a_{11} & a_{12} \\ a_{21} & a_{22} \end{vmatrix} = a_{11}a_{22} - a_{21}a_{12}$

EXAMPLE 1 **Finding the determinant of a 2 × 2 matrix**

Find $|A|$ if $A = \begin{bmatrix} 2 & -1 \\ 4 & -3 \end{bmatrix}$.

▶ **SOLUTION** By definition,

$$|A| = \begin{vmatrix} 2 & -1 \\ 4 & -3 \end{vmatrix} = (2)(-3) - (4)(-1) = -6 + 4 = -2.$$

To assist in finding determinants for square matrices of order $n > 1$, we introduce the following terminology.

Definition of Minors and Cofactors

Let $A = (a_{ij})$ be a square matrix of order $n > 1$.

(1) The **minor** M_{ij} of the element a_{ij} is the determinant of the matrix of order $n - 1$ obtained by deleting row i and column j.

(2) The **cofactor** A_{ij} of the element a_{ij} is $A_{ij} = (-1)^{i+j}M_{ij}$.

To determine the minor of an element, we delete the row and column in which the element appears and then find the determinant of the resulting square matrix. This process is demonstrated in the following illustration, where deletions of rows and columns in a 3 × 3 matrix are indicated with horizontal and vertical line segments, respectively.

To obtain the cofactor of a_{ij} of a square matrix $A = (a_{ij})$, we find the minor and multiply it by 1 or -1, depending on whether the sum of i and j is even or odd, respectively, as demonstrated in the illustration.

ILLUSTRATION **Minors and Cofactors**

Matrix	Minor	Cofactor
$\begin{bmatrix} a_{11} & a_{12} & a_{13} \\ a_{21} & a_{22} & a_{23} \\ a_{31} & a_{32} & a_{33} \end{bmatrix}$ (row 1 and column 1 deleted)	$M_{11} = \begin{vmatrix} a_{22} & a_{23} \\ a_{32} & a_{33} \end{vmatrix} = a_{22}a_{33} - a_{32}a_{23}$	$A_{11} = (-1)^{1+1}M_{11} = M_{11}$
$\begin{bmatrix} a_{11} & a_{12} & a_{13} \\ a_{21} & a_{22} & a_{23} \\ a_{31} & a_{32} & a_{33} \end{bmatrix}$ (row 1 and column 2 deleted)	$M_{12} = \begin{vmatrix} a_{21} & a_{23} \\ a_{31} & a_{33} \end{vmatrix} = a_{21}a_{33} - a_{31}a_{23}$	$A_{12} = (-1)^{1+2}M_{12} = -M_{12}$
$\begin{bmatrix} a_{11} & a_{12} & a_{13} \\ a_{21} & a_{22} & a_{23} \\ a_{31} & a_{32} & a_{33} \end{bmatrix}$ (row 2 and column 3 deleted)	$M_{23} = \begin{vmatrix} a_{11} & a_{12} \\ a_{31} & a_{32} \end{vmatrix} = a_{11}a_{32} - a_{31}a_{12}$	$A_{23} = (-1)^{2+3}M_{23} = -M_{23}$

▶ Tutorial available at www.thomsonedu.com/login

For the matrix in the preceding illustration, there are six other minors—M_{13}, M_{21}, M_{22}, M_{31}, M_{32}, and M_{33}—that can be obtained in similar fashion.

Another way to remember the sign $(-1)^{i+j}$ associated with the cofactor A_{ij} is to consider the following checkerboard style of plus and minus signs:

$$\begin{bmatrix} + & - & + & - & \cdots \\ - & + & - & + & \cdots \\ + & - & + & - & \cdots \\ - & + & - & + & \cdots \\ \vdots & \vdots & \vdots & \vdots & \end{bmatrix}$$

EXAMPLE 2 Finding minors and cofactors

If $A = \begin{bmatrix} 1 & -3 & 3 \\ 4 & 2 & 0 \\ -2 & -7 & 5 \end{bmatrix}$, find M_{11}, M_{21}, M_{22}, A_{11}, A_{21}, and A_{22}.

▶ **SOLUTION** Deleting appropriate rows and columns of A, we obtain

$$M_{11} = \begin{vmatrix} 2 & 0 \\ -7 & 5 \end{vmatrix} = (2)(5) - (-7)(0) = 10$$

$$M_{21} = \begin{vmatrix} -3 & 3 \\ -7 & 5 \end{vmatrix} = (-3)(5) - (-7)(3) = 6$$

$$M_{22} = \begin{vmatrix} 1 & 3 \\ -2 & 5 \end{vmatrix} = (1)(5) - (-2)(3) = 11.$$

To obtain the cofactors, we prefix the corresponding minors with the proper signs. Thus, using the definition of cofactor, we have

$$A_{11} = (-1)^{1+1}M_{11} = (1)(10) = 10$$
$$A_{21} = (-1)^{2+1}M_{21} = (-1)(6) = -6$$
$$A_{22} = (-1)^{2+2}M_{22} = (1)(11) = 11.$$

We can also use the checkerboard style of plus and minus signs to determine the proper signs.

The determinant $|A|$ of a square matrix of order 3 is defined as follows.

Definition of the Determinant of a 3 × 3 Matrix A

$$|A| = \begin{vmatrix} a_{11} & a_{12} & a_{13} \\ a_{21} & a_{22} & a_{23} \\ a_{31} & a_{32} & a_{33} \end{vmatrix} = a_{11}A_{11} + a_{12}A_{12} + a_{13}A_{13}$$

▶ Tutorial available at www.thomsonedu.com/login

Since cofactors $A_{11} = (-1)^{1+1}M_{11} = M_{11}$, $A_{12} = (-1)^{1+2}M_{12} = -M_{12}$, and $A_{13} = (-1)^{1+3}M_{13} = M_{13}$, the preceding definition may also be written

$$|A| = a_{11}M_{11} - a_{12}M_{12} + a_{13}M_{13}.$$

If we express M_{11}, M_{12}, and M_{13} using elements of A and rearrange terms, we obtain the following formula for $|A|$:

$$|A| = a_{11}a_{22}a_{33} - a_{11}a_{23}a_{32} - a_{12}a_{21}a_{33} + a_{12}a_{23}a_{31} + a_{13}a_{21}a_{32} - a_{13}a_{22}a_{31}$$

The definition of $|A|$ for a square matrix A of order 3 displays a pattern of multiplying each element in row 1 by its cofactor and then adding to find $|A|$. This process is referred to as *expanding* $|A|$ *by the first row.* By actually carrying out the computations, we can show that $|A|$ *can be expanded in similar fashion by using any row or column.* As an illustration, the expansion by the second column is

$$\begin{aligned} |A| &= a_{12}A_{12} + a_{22}A_{22} + a_{32}A_{32} \\ &= a_{12}\left(-\begin{vmatrix} a_{21} & a_{23} \\ a_{31} & a_{33} \end{vmatrix}\right) + a_{22}\left(+\begin{vmatrix} a_{11} & a_{13} \\ a_{31} & a_{33} \end{vmatrix}\right) + a_{32}\left(-\begin{vmatrix} a_{11} & a_{13} \\ a_{21} & a_{23} \end{vmatrix}\right). \end{aligned}$$

Applying the definition to the determinants in parentheses, multiplying as indicated, and rearranging the terms in the sum, we could arrive at the formula for $|A|$ in terms of the elements of A. Similarly, the expansion by the third row is

$$|A| = a_{31}A_{31} + a_{32}A_{32} + a_{33}A_{33}.$$

Once again we can show that this result agrees with previous expansions.

EXAMPLE 3 **Finding the determinant of a 3 × 3 matrix**

Find $|A|$ if $A = \begin{bmatrix} -1 & 3 & 1 \\ 2 & 5 & 0 \\ 3 & 1 & -2 \end{bmatrix}$.

▶ **SOLUTION** Since the second row contains a zero, we shall expand $|A|$ by that row, because then we need to evaluate only two cofactors. Thus,

$$|A| = (2)A_{21} + (5)A_{22} + (0)A_{23}.$$

Using the definition of cofactors, we have

$$A_{21} = (-1)^{2+1}M_{21} = -\begin{vmatrix} 3 & 1 \\ 1 & -2 \end{vmatrix} = -[(3)(-2) - (1)(1)] = 7$$

$$A_{22} = (-1)^{2+2}M_{22} = \begin{vmatrix} -1 & 1 \\ 3 & -2 \end{vmatrix} = [(-1)(-2) - (3)(1)] = -1.$$

Consequently,

$$|A| = (2)(7) + (5)(-1) + (0)A_{23} = 14 - 5 + 0 = 9.$$

▶ **Tutorial available at www.thomsonedu.com/login**

Finding the determinant of a square matrix with real number entries is an easy task with a graphing calculator. First, enter the matrix A from Example 3:

$$A = \begin{bmatrix} -1 & 3 & 1 \\ 2 & 5 & 0 \\ 3 & 1 & -2 \end{bmatrix}$$

Now display A and find the determinant of A.

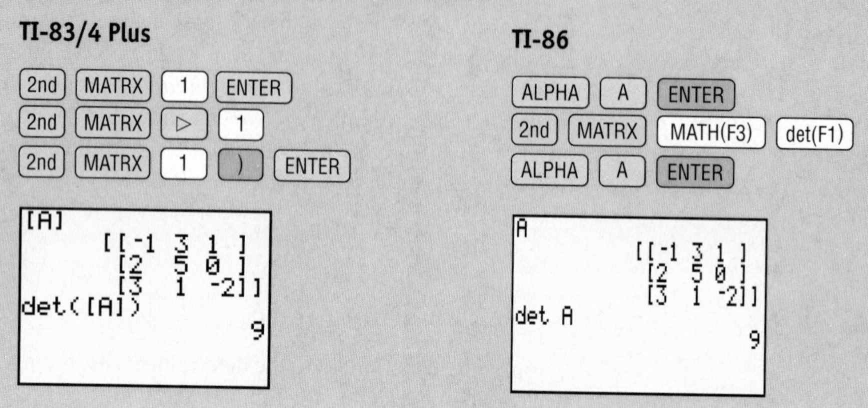

The following definition of the determinant of a matrix of arbitrary order n is patterned after that used for the determinant of a matrix of order 3.

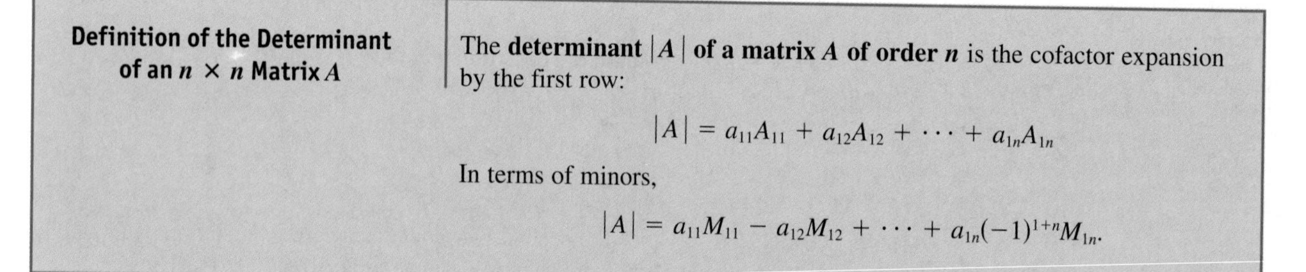

Definition of the Determinant of an $n \times n$ Matrix A

The **determinant $|A|$ of a matrix A of order n** is the cofactor expansion by the first row:

$$|A| = a_{11}A_{11} + a_{12}A_{12} + \cdots + a_{1n}A_{1n}$$

In terms of minors,

$$|A| = a_{11}M_{11} - a_{12}M_{12} + \cdots + a_{1n}(-1)^{1+n}M_{1n}.$$

The number $|A|$ may be found by using *any* row or column, as stated in the following theorem.

Theorem on Expansion of Determinants

If A is a square matrix of order $n > 1$, then the determinant $|A|$ may be found by multiplying the elements of any row (or column) by their respective cofactors and adding the resulting products.

This theorem is useful if many zeros appear in a row or column, as illustrated in the following example.

EXAMPLE 4 **Finding the determinant of a 4 × 4 matrix**

Find $|A|$ if $A = \begin{bmatrix} 1 & 0 & 2 & 5 \\ -2 & 1 & 5 & 0 \\ 0 & 0 & -3 & 0 \\ 0 & -1 & 0 & 3 \end{bmatrix}$.

▶ **SOLUTION** Note that all but one of the elements in the third row are zero. Hence, if we expand $|A|$ by the third row, there will be at most one nonzero term. Specifically,

$$|A| = (0)A_{31} + (0)A_{32} + (-3)A_{33} + (0)A_{34} = -3A_{33}$$

with $$A_{33} = (-1)^{3+3}M_{33} = M_{33} = \begin{vmatrix} 1 & 0 & 5 \\ -2 & 1 & 0 \\ 0 & -1 & 3 \end{vmatrix}.$$

We expand M_{33} by column 1:

$$M_{33} = (1)\begin{vmatrix} 1 & 0 \\ -1 & 3 \end{vmatrix} - (-2)\begin{vmatrix} 0 & 5 \\ -1 & 3 \end{vmatrix} + (0)\begin{vmatrix} 0 & 5 \\ 1 & 0 \end{vmatrix}$$
$$= (1)(3) - (-2)(5) + (0)(-5) = 3 + 10 + 0 = 13$$

Thus, $$|A| = -3A_{33} = (-3)(13) = -39.$$

In general, if all but one element in some row (or column) of A are zero and if the determinant $|A|$ is expanded by that row (or column), then all terms drop out except the product of that element with its cofactor. If *all* elements in a row (or column) are zero, we have the following.

Theorem on a Row of Zeros	If every element in a row (or column) of a square matrix A is zero, then $|A| = 0$.

PROOF If every element in a row (or column) of a square matrix A is zero, then the expansion by that row (or column) is a sum of terms that are zero (since each term is zero times its respective cofactor). Hence, this sum is equal to zero, and we conclude that $|A| = 0$.

In the previous section we found that if we could not obtain the identity matrix on the left side of the adjoined matrix, then the original matrix was not invertible. If we obtain a row of zeros in this process, we certainly cannot obtain the identity matrix. Combining this fact with the previous theorem leads to the following theorem.

Theorem on Matrix Invertibility	If A is a square matrix, then A is invertible if and only if $|A| \neq 0$.

▶ Tutorial available at www.thomsonedu.com/login

8.8 Exercises

Exer. 1–4: Find all the minors and cofactors of the elements in the matrix.

▶ 1 $\begin{bmatrix} 7 & -1 \\ 5 & 0 \end{bmatrix}$

2 $\begin{bmatrix} -6 & 4 \\ 3 & 2 \end{bmatrix}$

3 $\begin{bmatrix} 2 & 4 & -1 \\ 0 & 3 & 2 \\ -5 & 7 & 0 \end{bmatrix}$

4 $\begin{bmatrix} 5 & -2 & 1 \\ 4 & 7 & 0 \\ -3 & 4 & -1 \end{bmatrix}$

Exer. 5–8: Find the determinant of the matrix in the given exercise.

5 Exercise 1

6 Exercise 2

7 Exercise 3

8 Exercise 4

Exer. 9–20: Find the determinant of the matrix.

9 $\begin{bmatrix} -5 & 4 \\ -3 & 2 \end{bmatrix}$

▶ 10 $\begin{bmatrix} 6 & 4 \\ -3 & 2 \end{bmatrix}$

11 $\begin{bmatrix} a & -a \\ b & -b \end{bmatrix}$

12 $\begin{bmatrix} c & d \\ -d & c \end{bmatrix}$

▶ 13 $\begin{bmatrix} 3 & 1 & -2 \\ 4 & 2 & 5 \\ -6 & 3 & -1 \end{bmatrix}$

14 $\begin{bmatrix} 2 & -5 & 1 \\ -3 & 1 & 6 \\ 4 & -2 & 3 \end{bmatrix}$

15 $\begin{bmatrix} -5 & 4 & 1 \\ 3 & -2 & 7 \\ 2 & 0 & 6 \end{bmatrix}$

16 $\begin{bmatrix} 2 & 7 & -3 \\ 1 & 0 & 4 \\ 4 & -1 & -2 \end{bmatrix}$

17 $\begin{bmatrix} 3 & -1 & 2 & 0 \\ 4 & 0 & -3 & 5 \\ 0 & 6 & 0 & 0 \\ 1 & 3 & -4 & 2 \end{bmatrix}$

18 $\begin{bmatrix} 2 & 5 & 1 & 0 \\ -4 & 0 & -3 & 0 \\ 3 & -2 & 1 & 6 \\ -1 & 4 & 2 & 0 \end{bmatrix}$

▶ 19 $\begin{bmatrix} 0 & b & 0 & 0 \\ 0 & 0 & c & 0 \\ a & 0 & 0 & 0 \\ 0 & 0 & 0 & d \end{bmatrix}$

20 $\begin{bmatrix} a & u & v & w \\ 0 & b & x & y \\ 0 & 0 & c & z \\ 0 & 0 & 0 & d \end{bmatrix}$

Exer. 21–28: Verify the identity by expanding each determinant.

21 $\begin{vmatrix} a & b \\ c & d \end{vmatrix} = -\begin{vmatrix} c & d \\ a & b \end{vmatrix}$

22 $\begin{vmatrix} a & b \\ c & d \end{vmatrix} = -\begin{vmatrix} b & a \\ d & c \end{vmatrix}$

23 $\begin{vmatrix} a & kb \\ c & kd \end{vmatrix} = k\begin{vmatrix} a & b \\ c & d \end{vmatrix}$

24 $\begin{vmatrix} a & b \\ kc & kd \end{vmatrix} = k\begin{vmatrix} a & b \\ c & d \end{vmatrix}$

25 $\begin{vmatrix} a & b \\ c & d \end{vmatrix} = \begin{vmatrix} a & b \\ ka + c & kb + d \end{vmatrix}$

26 $\begin{vmatrix} a & b \\ c & d \end{vmatrix} = \begin{vmatrix} a & ka + b \\ c & kc + d \end{vmatrix}$

27 $\begin{vmatrix} a & b \\ c & d \end{vmatrix} + \begin{vmatrix} a & e \\ c & f \end{vmatrix} = \begin{vmatrix} a & b + e \\ c & d + f \end{vmatrix}$

28 $\begin{vmatrix} a & b \\ c & d \end{vmatrix} + \begin{vmatrix} a & b \\ e & f \end{vmatrix} = \begin{vmatrix} a & b \\ c + e & d + f \end{vmatrix}$

29 Let $A = (a_{ij})$ be a square matrix of order n such that $a_{ij} = 0$ if $i < j$. Show that $|A| = a_{11}a_{22} \cdots a_{nn}$.

30 If $A = (a_{ij})$ is any 2×2 matrix such that $|A| \neq 0$, show that A has an inverse, and find a general formula for A^{-1}.

Exer. 31–34: Let $I = I_2$ be the identity matrix of order 2, and let $f(x) = |A - xI|$. Find (a) the polynomial $f(x)$ and (b) the zeros of $f(x)$. (In the study of matrices, $f(x)$ is the *characteristic polynomial of* A, and the zeros of $f(x)$ are the *characteristic values (eigenvalues) of* A.)

31 $A = \begin{bmatrix} 1 & 2 \\ 3 & 2 \end{bmatrix}$

32 $A = \begin{bmatrix} 3 & 1 \\ 2 & 2 \end{bmatrix}$

33 $A = \begin{bmatrix} -3 & -2 \\ 2 & 2 \end{bmatrix}$

34 $A = \begin{bmatrix} 2 & -4 \\ -3 & 5 \end{bmatrix}$

Exer. 35–38: Let $I = I_3$ and let $f(x) = |A - xI|$. Find (a) the polynomial $f(x)$ and (b) the zeros of $f(x)$.

▶ 35 $A = \begin{bmatrix} 1 & 0 & 0 \\ 1 & 0 & -2 \\ -1 & 1 & -3 \end{bmatrix}$

36 $A = \begin{bmatrix} 2 & 1 & 0 \\ -1 & 0 & 0 \\ 1 & 3 & 2 \end{bmatrix}$

37 $A = \begin{bmatrix} 0 & 2 & -2 \\ -1 & 3 & 1 \\ -3 & 3 & 1 \end{bmatrix}$

38 $A = \begin{bmatrix} 3 & 2 & 2 \\ 1 & 0 & 2 \\ -1 & -1 & 0 \end{bmatrix}$

▶ **Tutorial available at www.thomsonedu.com/login**

Exer. 39–42: Express the determinant in the form $ai + bj + ck$ for real numbers a, b, and c.

39 $\begin{vmatrix} i & j & k \\ 2 & -1 & 6 \\ -3 & 5 & 1 \end{vmatrix}$

▶ 40 $\begin{vmatrix} i & j & k \\ 1 & -2 & 3 \\ 2 & 1 & -4 \end{vmatrix}$

41 $\begin{vmatrix} i & j & k \\ 5 & -6 & -1 \\ 3 & 0 & 1 \end{vmatrix}$

42 $\begin{vmatrix} i & j & k \\ 4 & -6 & 2 \\ -2 & 3 & -1 \end{vmatrix}$

Exer. 43–46: Find the determinant of the matrix.

43 $\begin{bmatrix} 29 & -17 & 90 \\ -34 & 91 & -34 \\ 48 & 7 & 10 \end{bmatrix}$

44 $\begin{bmatrix} -2 & 5.5 & 8 \\ -0.3 & 8.5 & 7 \\ 4.9 & 6.7 & 11 \end{bmatrix}$

45 $\begin{bmatrix} 4 & -7 & -3 & 13 \\ -17 & -0.8 & 5 & 0.9 \\ 1.1 & 0.2 & 10 & -4 \\ 3 & -6 & 2 & 1 \end{bmatrix}$

46 $\begin{bmatrix} 4.2 & 1.7 & -2 & -4 \\ -7 & 0.1 & 4.6 & 2.7 \\ 4.1 & -7 & 12 & 6.8 \\ 4.6 & 2 & 3.2 & 1.2 \end{bmatrix}$

Exer. 47–48: Let $I = I_3$ and let $f(x) = |A - xI|$. (a) Find the polynomial $f(x)$. (b) Graph f, and estimate the characteristic values of A.

47 $A = \begin{bmatrix} 1 & 0 & 1 \\ 0 & 2 & 1 \\ 1 & 1 & -2 \end{bmatrix}$

48 $A = \begin{bmatrix} 3 & -1 & -1 \\ -1 & 1 & 0 \\ -1 & 0 & -2 \end{bmatrix}$

8.9 Properties of Determinants

Evaluating a determinant by using the expansion theorem stated in Section 8.8 is inefficient for matrices of high order. For example, if a determinant of a matrix of order 10 is expanded by any row, a sum of 10 terms is obtained, and each term contains the determinant of a matrix of order 9, which is a cofactor of the original matrix. If any of the latter determinants is expanded by a row (or column), a sum of 9 terms is obtained, each containing the determinant of a matrix of order 8. Hence, at this stage there are 90 determinants of matrices of order 8 to evaluate. The process could be continued until only determinants of matrices of order 2 remain. You may verify that there are 1,814,400 such matrices of order 2! Unless many elements of the original matrix are zero, it is an enormous task to carry out all of the computations.

In this section we discuss rules that simplify the process of evaluating determinants. The main use for these rules is to introduce zeros into the determinant. They may also be used to change the determinant to **echelon form**—that is, to a form in which the elements below the main diagonal elements are all zero (see Section 8.5). The transformations on rows stated in the next theorem are the same as the elementary row transformations of a matrix introduced in Section 8.5. However, for determinants we may also use similar transformations on columns.

▶ Tutorial available at www.thomsonedu.com/login

Theorem on Row and Column Transformations of a Determinant

Let A be a square matrix of order n.

(1) If a matrix B is obtained from A by interchanging two rows (or columns), then $|B| = -|A|$.

(2) If B is obtained from A by multiplying every element of one row (or column) of A by a real number k, then $|B| = k|A|$.

(3) If B is obtained from A by adding k times any row (or column) of A to another row (or column) for a real number k, then $|B| = |A|$—that is, the determinants of B and A are equal.

When using the theorem, we refer to the rows (or columns) of the *determinant* in the obvious way. For example, property 3 may be phrased as follows: *Adding k times any row (or column) to another row (or column) of a determinant does not affect the value of the determinant.*

Row transformations of determinants will be specified by means of the symbols $R_i \leftrightarrow R_j$, $kR_i \rightarrow R_i$, and $kR_i + R_j \rightarrow R_j$, which were introduced in Section 8.5. Analogous symbols are used for column transformations. For example, $kC_i + C_j \rightarrow C_j$ means "add k times the ith column to the jth column."

Property 2 of the theorem on row and column transformations is useful for finding factors of determinants. To illustrate, for a determinant of a matrix of order 3, we have the following:

$$\begin{vmatrix} a_{11} & a_{12} & a_{13} \\ ka_{21} & ka_{22} & ka_{23} \\ a_{31} & a_{32} & a_{33} \end{vmatrix} = k\begin{vmatrix} a_{11} & a_{12} & a_{13} \\ a_{21} & a_{22} & a_{23} \\ a_{31} & a_{32} & a_{33} \end{vmatrix}$$

Similar formulas hold if k is a common factor of the elements of any other row or column. When referring to this manipulation, we often use the phrase *k is a common factor of the row (or column).*

The following are illustrations of the preceding theorem, with the reason for each equality stated at the right.

ILLUSTRATION Transformation of Determinants

- $$\begin{vmatrix} 2 & 0 & 1 \\ 6 & 4 & 3 \\ 0 & 3 & 5 \end{vmatrix} = -\begin{vmatrix} 6 & 4 & 3 \\ 2 & 0 & 1 \\ 0 & 3 & 5 \end{vmatrix}$$ $R_1 \leftrightarrow R_2$ (property 1)
- $$\begin{vmatrix} 2 & 0 & 1 \\ 6 & 4 & 3 \\ 0 & 3 & 5 \end{vmatrix} = 2\begin{vmatrix} 1 & 0 & 1 \\ 3 & 4 & 3 \\ 0 & 3 & 5 \end{vmatrix}$$ 2 is a common factor of column 1 (property 2)
- $$\begin{vmatrix} 1 & -3 & 4 \\ 2 & -1 & 0 \\ 3 & 1 & 6 \end{vmatrix} = \begin{vmatrix} -5 & -3 & 4 \\ 0 & -1 & 0 \\ 5 & 1 & 6 \end{vmatrix}$$ $2C_2 + C_1 \rightarrow C_1$ (property 3)

(continued)

$$\begin{vmatrix} 1 & -3 & 4 \\ 2 & -1 & 0 \\ 3 & 1 & 6 \end{vmatrix} = \begin{vmatrix} 1 & -3 & 4 \\ 0 & 5 & -8 \\ 0 & 10 & -6 \end{vmatrix} \quad \begin{array}{l} -2R_1 + R_2 \to R_2 \\ -3R_1 + R_3 \to R_3 \\ \text{(property 3 applied twice)} \end{array}$$ ■

Theorem on Identical Rows	If two rows (or columns) of a square matrix A are identical, then $|A| = 0$.

PROOF If B is the matrix obtained from A by interchanging the two identical rows (or columns), then B and A are the same and, consequently, $|B| = |A|$. However, by property 1 of the theorem on row and column transformations of a determinant, $|B| = -|A|$, and hence $-|A| = |A|$. Thus, $2|A| = 0$, and therefore $|A| = 0$.

EXAMPLE 1 Using row and column transformations

Find $|A|$ if $A = \begin{bmatrix} 2 & 3 & 0 & 4 \\ 0 & 5 & -1 & 6 \\ 1 & 0 & -2 & 3 \\ -3 & 2 & 0 & -5 \end{bmatrix}$.

▶ SOLUTION We plan to use property 3 of the theorem on row and column transformations of a determinant to introduce three zeros in some row or column. It is convenient to work with an element of the matrix that equals 1, since this enables us to avoid the use of fractions. If 1 is not an element of the original matrix, it is always possible to introduce the number 1 by using property 2 or 3 of the theorem. In this example, 1 appears in row 3, and we proceed as follows, with the reason for each equality stated at the right.

$$\begin{vmatrix} 2 & 3 & 0 & 4 \\ 0 & 5 & -1 & 6 \\ 1 & 0 & -2 & 3 \\ -3 & 2 & 0 & -5 \end{vmatrix} = \begin{vmatrix} 0 & 3 & 4 & -2 \\ 0 & 5 & -1 & 6 \\ 1 & 0 & -2 & 3 \\ 0 & 2 & -6 & 4 \end{vmatrix} \quad \begin{array}{l} -2R_3 + R_1 \to R_1 \\ 3R_3 + R_4 \to R_4 \end{array}$$

$$= (1) \cdot (-1)^{3+1} \begin{vmatrix} 3 & 4 & -2 \\ 5 & -1 & 6 \\ 2 & -6 & 4 \end{vmatrix} \quad \text{expand by the first column}$$

$$= \begin{vmatrix} 23 & 4 & 22 \\ 0 & -1 & 0 \\ -28 & -6 & -32 \end{vmatrix} \quad \begin{array}{l} 5C_2 + C_1 \to C_1 \\ 6C_2 + C_3 \to C_3 \end{array}$$

$$= (-1) \cdot (-1)^{2+2} \begin{vmatrix} 23 & 22 \\ -28 & -32 \end{vmatrix} \quad \text{expand by the second row}$$

$$= (-1)[(23)(-32) - (-28)(22)] \quad \text{definition of determinant}$$

$$= 120$$

▶ Tutorial available at www.thomsonedu.com/login

The next two examples illustrate the use of property 2 of the theorem on row and column transformations of a determinant.

EXAMPLE 2 Removing common factors from rows

Find $|A|$ if $A = \begin{bmatrix} 14 & -6 & 4 \\ 4 & -5 & 12 \\ -21 & 9 & -6 \end{bmatrix}$.

▶ **SOLUTION**

$$|A| = 2\begin{vmatrix} 7 & -3 & 2 \\ 4 & -5 & 12 \\ -21 & 9 & -6 \end{vmatrix} \qquad \text{2 is a common factor of row 1}$$

$$= (2)(-3)\begin{vmatrix} 7 & -3 & 2 \\ 4 & -5 & 12 \\ 7 & -3 & 2 \end{vmatrix} \qquad \text{−3 is a common factor of row 3}$$

$$= 0 \qquad \text{two rows are identical}$$

EXAMPLE 3 Removing a common factor from a column

Without expanding, show that $a - b$ is a factor of $|A|$ if

$$A = \begin{bmatrix} 1 & 1 & 1 \\ a & b & c \\ a^2 & b^2 & c^2 \end{bmatrix}.$$

▶ **SOLUTION**

$$\begin{vmatrix} 1 & 1 & 1 \\ a & b & c \\ a^2 & b^2 & c^2 \end{vmatrix} = \begin{vmatrix} 0 & 1 & 1 \\ a-b & b & c \\ a^2-b^2 & b^2 & c^2 \end{vmatrix} \qquad -C_2 + C_1 \rightarrow C_1$$

$$= (a-b)\begin{vmatrix} 0 & 1 & 1 \\ 1 & b & c \\ a+b & b^2 & c^2 \end{vmatrix} \qquad a-b \text{ is a common factor of column 1}$$

Hence, $|A|$ is equal to $a - b$ times the last determinant, and so $a - b$ is a factor of $|A|$.

Determinants arise in the study of solutions of systems of linear equations. To illustrate, let us consider two linear equations in two variables x and y:

$$\begin{cases} a_{11}x + a_{12}y = k_1 \\ a_{21}x + a_{22}y = k_2 \end{cases}$$

where at least one nonzero coefficient appears in each equation. We may assume that $a_{11} \neq 0$, for otherwise $a_{12} \neq 0$ and we could then regard y as the first

▶ **Tutorial available at www.thomsonedu.com/login**

variable instead of x. We shall use elementary row transformations to obtain the matrix of an equivalent system with $a_{21} = 0$, as follows:

$$\left[\begin{array}{cc|c} a_{11} & a_{12} & k_1 \\ a_{21} & a_{22} & k_2 \end{array}\right] \xrightarrow{-\frac{a_{21}}{a_{11}}\mathbf{R}_1 + \mathbf{R}_2 \to \mathbf{R}_2} \left[\begin{array}{cc|c} a_{11} & a_{12} & k_1 \\ 0 & a_{22} - \left(\dfrac{a_{21}a_{12}}{a_{11}}\right) & k_2 - \left(\dfrac{a_{21}k_1}{a_{11}}\right) \end{array}\right]$$

$$\xrightarrow{a_{11}\mathbf{R}_2 \to \mathbf{R}_2} \left[\begin{array}{cc|c} a_{11} & a_{12} & k_1 \\ 0 & (a_{11}a_{22} - a_{21}a_{12}) & (a_{11}k_2 - a_{21}k_1) \end{array}\right]$$

Thus, the given system is equivalent to

$$\begin{cases} a_{11}x + a_{12}y = k_1 \\ (a_{11}a_{22} - a_{21}a_{12})y = a_{11}k_2 - a_{21}k_1 \end{cases}$$

which may also be written

$$\begin{cases} a_{11}x + a_{12}y = k_1 \\ \begin{vmatrix} a_{11} & a_{12} \\ a_{21} & a_{22} \end{vmatrix} y = \begin{vmatrix} a_{11} & k_1 \\ a_{21} & k_2 \end{vmatrix} \end{cases}$$

If $\begin{vmatrix} a_{11} & a_{12} \\ a_{21} & a_{22} \end{vmatrix} \neq 0$, we can solve the second equation for y, obtaining

$$y = \frac{\begin{vmatrix} a_{11} & k_1 \\ a_{21} & k_2 \end{vmatrix}}{\begin{vmatrix} a_{11} & a_{12} \\ a_{21} & a_{22} \end{vmatrix}}.$$

The corresponding value for x may be found by substituting for y in the first equation, which leads to

$$x = \frac{\begin{vmatrix} k_1 & a_{12} \\ k_2 & a_{22} \end{vmatrix}}{\begin{vmatrix} a_{11} & a_{12} \\ a_{21} & a_{22} \end{vmatrix}}. \qquad (*)$$

The proof of this statement is left as Discussion Exercise 7 at the end of the chapter.

This proves that *if the determinant of the coefficient matrix of a system of two linear equations in two variables is not zero, then the system has a unique solution.* The last two formulas for x and y as quotients of determinants constitute **Cramer's rule** for two variables.

There is an easy way to remember Cramer's rule. Let

$$D = \begin{bmatrix} a_{11} & a_{12} \\ a_{21} & a_{22} \end{bmatrix}$$

be the coefficient matrix of the system, and let D_x denote the matrix obtained from D by replacing the coefficients a_{11}, a_{21} of x by the numbers k_1, k_2, respectively. Similarly, let D_y denote the matrix obtained from D by replacing the coefficients a_{12}, a_{22} of y by the numbers k_1, k_2, respectively. Thus,

$$D_x = \begin{bmatrix} k_1 & a_{12} \\ k_2 & a_{22} \end{bmatrix}, \quad D_y = \begin{bmatrix} a_{11} & k_1 \\ a_{21} & k_2 \end{bmatrix}.$$

If $|D| \neq 0$, the solution (x, y) is given by the following formulas.

Cramer's Rule for Two Variables

$$x = \frac{|D_x|}{|D|}, \quad y = \frac{|D_y|}{|D|}$$

EXAMPLE 4 **Using Cramer's rule to solve a system of two linear equations**

Use Cramer's rule to solve the system

$$\begin{cases} 2x - 3y = -4 \\ 5x + 7y = 1 \end{cases}$$

SOLUTION The determinant of the coefficient matrix is

$$|D| = \begin{vmatrix} 2 & -3 \\ 5 & 7 \end{vmatrix} = 29.$$

Using the notation introduced previously, we have

$$|D_x| = \begin{vmatrix} -4 & -3 \\ 1 & 7 \end{vmatrix} = -25, \qquad |D_y| = \begin{vmatrix} 2 & -4 \\ 5 & 1 \end{vmatrix} = 22.$$

Hence, $$x = \frac{|D_x|}{|D|} = \frac{-25}{29}, \qquad y = \frac{|D_y|}{|D|} = \frac{22}{29}.$$

Thus, the system has the unique solution $\left(-\frac{25}{29}, \frac{22}{29}\right)$.

Cramer's rule can be extended to systems of n linear equations in n variables $x_1, x_2, \ldots, x_n$, where the ith equation has the form

$$a_{i1}x_1 + a_{i2}x_2 + \cdots + a_{in}x_n = k_i.$$

To solve such a system, let D denote the coefficient matrix and let D_{x_j} denote the matrix obtained by replacing the coefficients of x_j in D by the numbers $k_1, \ldots, k_n$ that appear in the column to the right of the equal signs in the system. If $|D| \neq 0$, then the system has the following unique solution.

Cramer's Rule (General Form)

$$x_1 = \frac{|D_{x_1}|}{|D|}, \quad x_2 = \frac{|D_{x_2}|}{|D|}, \quad \ldots, \quad x_n = \frac{|D_{x_n}|}{|D|}$$

Tutorial available at www.thomsonedu.com/login

EXAMPLE 5 Using Cramer's rule to solve a system of three linear equations

Use Cramer's rule to solve the system

$$\begin{cases} x \qquad - 2z = 3 \\ \quad - y + 3z = 1 \\ 2x \qquad + 5z = 0 \end{cases}$$

▶ SOLUTION We shall merely list the various determinants. You should check the results.

$$|D| = \begin{vmatrix} 1 & 0 & -2 \\ 0 & -1 & 3 \\ 2 & 0 & 5 \end{vmatrix} = -9, \quad |D_x| = \begin{vmatrix} 3 & 0 & -2 \\ 1 & -1 & 3 \\ 0 & 0 & 5 \end{vmatrix} = -15$$

$$|D_y| = \begin{vmatrix} 1 & 3 & -2 \\ 0 & 1 & 3 \\ 2 & 0 & 5 \end{vmatrix} = 27, \quad |D_z| = \begin{vmatrix} 1 & 0 & 3 \\ 0 & -1 & 1 \\ 2 & 0 & 0 \end{vmatrix} = 6$$

By Cramer's rule, the solution is

$$x = \frac{|D_x|}{|D|} = \frac{-15}{-9} = \frac{5}{3}, \quad y = \frac{|D_y|}{|D|} = \frac{27}{-9} = -3, \quad z = \frac{|D_z|}{|D|} = \frac{6}{-9} = -\frac{2}{3}.$$

Cramer's rule is an inefficient method to apply if the system has a large number of equations, since many determinants of matrices of high order must be evaluated. Note also that Cramer's rule cannot be used directly if $|D| = 0$ or if the number of equations is not the same as the number of variables. For numerical calculations, the inverse method and the matrix method are superior to Cramer's rule; however, the Cramer's rule formulation is theoretically useful.

8.9 Exercises

Exer. 1–14: Without expanding, explain why the statement is true.

1 $\begin{vmatrix} 1 & 0 & 1 \\ 0 & 1 & 1 \\ 1 & 1 & 0 \end{vmatrix} = -\begin{vmatrix} 1 & 0 & 1 \\ 1 & 1 & 0 \\ 0 & 1 & 1 \end{vmatrix}$

2 $\begin{vmatrix} 1 & 0 & 1 \\ 0 & 1 & 1 \\ 1 & 1 & 0 \end{vmatrix} = -\begin{vmatrix} 1 & 1 & 0 \\ 0 & 1 & 1 \\ 1 & 0 & 1 \end{vmatrix}$

3 $\begin{vmatrix} 1 & 0 & 1 \\ 2 & 1 & 0 \\ 1 & 1 & 2 \end{vmatrix} = \begin{vmatrix} 1 & 0 & 1 \\ 2 & 1 & 0 \\ 0 & 1 & 1 \end{vmatrix}$

4 $\begin{vmatrix} 1 & 1 & 2 \\ 1 & 0 & 1 \\ 2 & 1 & 1 \end{vmatrix} = \begin{vmatrix} 0 & 1 & 1 \\ 1 & 0 & 1 \\ 2 & 1 & 1 \end{vmatrix}$

5 $\begin{vmatrix} 2 & 4 & 2 \\ 1 & 2 & 4 \\ 2 & 6 & 4 \end{vmatrix} = 4\begin{vmatrix} 1 & 2 & 1 \\ 1 & 2 & 4 \\ 1 & 3 & 2 \end{vmatrix}$

6 $\begin{vmatrix} 2 & 1 & 6 \\ 4 & 3 & 3 \\ 2 & 1 & 3 \end{vmatrix} = 6\begin{vmatrix} 1 & 1 & 2 \\ 2 & 3 & 1 \\ 1 & 1 & 1 \end{vmatrix}$

▶ Tutorial available at www.thomsonedu.com/login

7 $\begin{vmatrix} 1 & -1 & 2 \\ 1 & 2 & -1 \\ 1 & -1 & 2 \end{vmatrix} = 0$

8 $\begin{vmatrix} 1 & -1 & 1 \\ 0 & 1 & 0 \\ -1 & 1 & -1 \end{vmatrix} = 0$

9 $\begin{vmatrix} 1 & 5 \\ -3 & 2 \end{vmatrix} = -\begin{vmatrix} 1 & 5 \\ 3 & -2 \end{vmatrix}$

10 $\begin{vmatrix} 2 & -2 \\ 1 & 1 \end{vmatrix} = -\begin{vmatrix} -2 & 2 \\ 1 & 1 \end{vmatrix}$

11 $\begin{vmatrix} 0 & 0 & 1 \\ 1 & 0 & 0 \\ 0 & 0 & 2 \end{vmatrix} = 0$

12 $\begin{vmatrix} 1 & 0 & 1 \\ 0 & 0 & 0 \\ 1 & 1 & 0 \end{vmatrix} = 0$

13 $\begin{vmatrix} 1 & -1 & -2 \\ -1 & 2 & 1 \\ 0 & 1 & 1 \end{vmatrix} = \begin{vmatrix} 1 & -1 & 0 \\ -1 & 2 & -1 \\ 0 & 1 & 1 \end{vmatrix}$

14 $\begin{vmatrix} a & 0 & 0 \\ 0 & b & 0 \\ 0 & 0 & c \end{vmatrix} = -\begin{vmatrix} 0 & 0 & a \\ 0 & b & 0 \\ c & 0 & 0 \end{vmatrix}$

Exer. 15–24: Find the determinant of the matrix after introducing zeros, as in Example 1.

▶ 15 $\begin{bmatrix} 3 & 1 & 0 \\ -2 & 0 & 1 \\ 1 & 3 & -1 \end{bmatrix}$

16 $\begin{bmatrix} -3 & 0 & 4 \\ 1 & 2 & 0 \\ 4 & 1 & -1 \end{bmatrix}$

17 $\begin{bmatrix} 5 & 4 & 3 \\ -3 & 2 & 1 \\ 0 & 7 & -2 \end{bmatrix}$

18 $\begin{bmatrix} 0 & 2 & -6 \\ 5 & 1 & -3 \\ 6 & -2 & 5 \end{bmatrix}$

19 $\begin{bmatrix} 2 & 2 & -3 \\ 3 & 6 & 9 \\ -2 & 5 & 4 \end{bmatrix}$

▶ 20 $\begin{bmatrix} 3 & 8 & 5 \\ 5 & 3 & -6 \\ 2 & 4 & -2 \end{bmatrix}$

21 $\begin{bmatrix} 3 & 1 & -2 & 2 \\ 2 & 0 & 1 & 4 \\ 0 & 1 & 3 & 5 \\ -1 & 2 & 0 & -3 \end{bmatrix}$

22 $\begin{bmatrix} 3 & 2 & 0 & 4 \\ -2 & 0 & 5 & 0 \\ 4 & -3 & 1 & 6 \\ 2 & -1 & 2 & 0 \end{bmatrix}$

23 $\begin{bmatrix} 2 & -2 & 0 & 0 & -3 \\ 3 & 0 & 3 & 2 & -1 \\ 0 & 1 & -2 & 0 & 2 \\ -1 & 2 & 0 & 3 & 0 \\ 0 & 4 & 1 & 0 & 0 \end{bmatrix}$

24 $\begin{bmatrix} 2 & 0 & -1 & 0 & 2 \\ 1 & 3 & 0 & 0 & 1 \\ 0 & 4 & 3 & 0 & -1 \\ -1 & 2 & 0 & -2 & 0 \\ 0 & 1 & 5 & 0 & -4 \end{bmatrix}$

25 Show that

$$\begin{vmatrix} 1 & 1 & 1 \\ a & b & c \\ a^2 & b^2 & c^2 \end{vmatrix} = (a - b)(b - c)(c - a).$$

(*Hint:* See Example 3.)

26 Show that

$$\begin{vmatrix} 1 & 1 & 1 \\ a & b & c \\ a^3 & b^3 & c^3 \end{vmatrix} = (a - b)(b - c)(c - a)(a + b + c).$$

27 If

$$A = \begin{bmatrix} a_{11} & a_{12} & a_{13} & a_{14} \\ 0 & a_{22} & a_{23} & a_{24} \\ 0 & 0 & a_{33} & a_{34} \\ 0 & 0 & 0 & a_{44} \end{bmatrix},$$

show that $|A| = a_{11}a_{22}a_{33}a_{44}$.

28 If

$$A = \begin{bmatrix} a & b & 0 & 0 \\ c & d & 0 & 0 \\ 0 & 0 & e & f \\ 0 & 0 & g & h \end{bmatrix},$$

show that

$$|A| = \begin{vmatrix} a & b \\ c & d \end{vmatrix}\begin{vmatrix} e & f \\ g & h \end{vmatrix}.$$

29 If $A = (a_{ij})$ and $B = (b_{ij})$ are arbitrary square matrices of order 2, show that $|AB| = |A||B|$.

30 If $A = (a_{ij})$ is a square matrix of order n and k is any real number, show that $|kA| = k^n|A|$. (*Hint:* Use property 2 of the theorem on row and column transformations of a determinant.)

▶ **Tutorial available at www.thomsonedu.com/login**

31 Use properties of determinants to show that the following is an equation of a line through the points (x_1, y_1) and (x_2, y_2):

$$\begin{vmatrix} x & y & 1 \\ x_1 & y_1 & 1 \\ x_2 & y_2 & 1 \end{vmatrix} = 0$$

32 Use properties of determinants to show that the following is an equation of a circle through three noncollinear points (x_1, y_1), (x_2, y_2), and (x_3, y_3):

$$\begin{vmatrix} x^2 + y^2 & x & y & 1 \\ x_1^2 + y_1^2 & x_1 & y_1 & 1 \\ x_2^2 + y_2^2 & x_2 & y_2 & 1 \\ x_3^2 + y_3^2 & x_3 & y_3 & 1 \end{vmatrix} = 0$$

Exer. 33–42: Use Cramer's rule, whenever applicable, to solve the system.

▶ 33 $\begin{cases} 2x + 3y = 2 \\ x - 2y = 8 \end{cases}$

34 $\begin{cases} 4x + 5y = 13 \\ 3x + y = -4 \end{cases}$

35 $\begin{cases} 2x + 5y = 16 \\ 3x - 7y = 24 \end{cases}$

36 $\begin{cases} 7x - 8y = 9 \\ 4x + 3y = -10 \end{cases}$

37 $\begin{cases} 2x - 3y = 5 \\ -6x + 9y = 12 \end{cases}$

38 $\begin{cases} 3p - q = 7 \\ -12p + 4q = 3 \end{cases}$

▶ 39 $\begin{cases} x - 2y - 3z = -1 \\ 2x + y + z = 6 \\ x + 3y - 2z = 13 \end{cases}$

40 $\begin{cases} x + 3y - z = -3 \\ 3x - y + 2z = 1 \\ 2x - y + z = -1 \end{cases}$

▶ 41 $\begin{cases} 5x + 2y - z = -7 \\ x - 2y + 2z = 0 \\ 3y + z = 17 \end{cases}$

42 $\begin{cases} 4x - y + 3z = 6 \\ -8x + 3y - 5z = -6 \\ 5x - 4y = -9 \end{cases}$

43 Use Cramer's rule to solve the system for x.

$$\begin{cases} ax + by + cz = d \\ ex + fz = g \\ hx + iy = j \end{cases}$$

8.10 Partial Fractions

In this section we show how systems of equations can be used to help decompose rational expressions into sums of simpler expressions. This technique is useful in advanced mathematics courses.

We may verify that

$$\frac{2}{x^2 - 1} = \frac{1}{x - 1} + \frac{-1}{x + 1}$$

by adding the fractions $1/(x - 1)$ and $-1/(x + 1)$ to obtain $2/(x^2 - 1)$. The expression on the right-hand side of this equation is called the *partial fraction decomposition* of $2/(x^2 - 1)$.

It is theoretically possible to write *any* rational expression as a sum of rational expressions whose denominators involve powers of polynomials of degree not greater than two. Specifically, if $f(x)$ and $g(x)$ are polynomials *and the degree of $f(x)$ is less than the degree of $g(x)$*, it can be proved that

$$\frac{f(x)}{g(x)} = F_1 + F_2 + \cdots + F_r$$

such that each F_k has one of the forms

$$\frac{A}{(px + q)^m} \quad \text{or} \quad \frac{Ax + B}{(ax^2 + bx + c)^n},$$

where A and B are real numbers, m and n are nonnegative integers, and the quadratic polynomial $ax^2 + bx + c$ is irreducible over $\mathbb{R}$ (that is, has no real zero). The sum $F_1 + F_2 + \cdots + F_r$ is the **partial fraction decomposition** of $f(x)/g(x)$, and each F_k is a **partial fraction.**

▶ **Tutorial available at www.thomsonedu.com/login**

For the partial fraction decomposition of $f(x)/g(x)$ to be found, *it is essential that $f(x)$ have lower degree than $g(x)$*. If this is not the case, we can use long division to obtain such an expression. For example, given

$$\frac{x^3 - 6x^2 + 5x - 3}{x^2 - 1},$$

we obtain

$$\frac{x^3 - 6x^2 + 5x - 3}{x^2 - 1} = x - 6 + \frac{6x - 9}{x^2 - 1}.$$

We then find the partial fraction decomposition of $(6x - 9)/(x^2 - 1)$.

The following guidelines can be used to obtain decompositions.

Guidelines for Finding Partial Fraction Decompositions of $f(x)/g(x)$

1 If the degree of the numerator $f(x)$ is not lower than the degree of the denominator $g(x)$, use long division to obtain the proper form.

2 Factor the denominator $g(x)$ into a product of linear factors $px + q$ or irreducible quadratic factors $ax^2 + bx + c$, and collect repeated factors so that $g(x)$ is a product of *different* factors of the form $(px + q)^m$ or $(ax^2 + bx + c)^n$ for a nonnegative integer m or n.

3 Apply the following rules to the factors found in guideline 2.

Rule A: For each factor of the form $(px + q)^m$ with $m \geq 1$, the partial fraction decomposition contains a sum of m partial fractions of the form

$$\frac{A_1}{px + q} + \frac{A_2}{(px + q)^2} + \cdots + \frac{A_m}{(px + q)^m},$$

where each numerator A_k is a real number.

Rule B: For each factor of the form $(ax^2 + bx + c)^n$ with $n \geq 1$ and $ax^2 + bx + c$ irreducible, the partial fraction decomposition contains a sum of n partial fractions of the form

$$\frac{A_1x + B_1}{ax^2 + bx + c} + \frac{A_2x + B_2}{(ax^2 + bx + c)^2} + \cdots + \frac{A_nx + B_n}{(ax^2 + bx + c)^n},$$

where each A_k and each B_k is a real number.

4 Find the numbers A_k and B_k in guideline 3.

We shall apply the preceding guidelines in the following examples. For the sake of convenience, we will use the variables A, B, C, and so on, rather than the subscripted variables A_k and B_k given in the guidelines.

EXAMPLE 1 **A partial fraction decomposition in which each denominator is linear**

Find the partial fraction decomposition of

$$\frac{4x^2 + 13x - 9}{x^3 + 2x^2 - 3x}.$$

▶ **SOLUTION**

Guideline 1 The degree of the numerator, 2, is less than the degree of the denominator, 3, so long division is not required.

Guideline 2 We factor the denominator:

$$x^3 + 2x^2 - 3x = x(x^2 + 2x - 3) = x(x + 3)(x - 1)$$

Guideline 3 Each factor of the denominator has the form stated in Rule A with $m = 1$. Thus, to the factor x there corresponds a partial fraction of the form A/x. Similarly, to the factors $x + 3$ and $x - 1$ there correspond partial fractions of the form $B/(x + 3)$ and $C/(x - 1)$, respectively. The partial fraction decomposition has the form

$$\frac{4x^2 + 13x - 9}{x^3 + 2x^2 - 3x} = \frac{A}{x} + \frac{B}{x + 3} + \frac{C}{x - 1}.$$

Guideline 4 We find the values of A, B, and C in guideline 3. Multiplying both sides of the partial fraction decomposition by the least common denominator, $x(x + 3)(x - 1)$, gives us

$$\begin{aligned} 4x^2 + 13x - 9 &= A(x + 3)(x - 1) + Bx(x - 1) + Cx(x + 3) \\ &= A(x^2 + 2x - 3) + B(x^2 - x) + C(x^2 + 3x) \\ &= (A + B + C)x^2 + (2A - B + 3C)x - 3A. \end{aligned}$$

Equating the coefficients of like powers of x on each side of the last equation, we obtain the system of equations

$$\begin{cases} A + B + C = 4 \\ 2A - B + 3C = 13 \\ -3A = -9 \end{cases}$$

Using the methods of Section 8.5 yields the solution $A = 3$, $B = -1$, and $C = 2$. Hence, the partial fraction decomposition is

$$\frac{4x^2 + 13x - 9}{x(x + 3)(x - 1)} = \frac{3}{x} + \frac{-1}{x + 3} + \frac{2}{x - 1}.$$

There is an alternative way to find A, B, and C if all factors of the denominator are linear and nonrepeated, as in this example. Instead of equating coefficients and using a system of equations, we begin with the equation

$$4x^2 + 13x - 9 = A(x + 3)(x - 1) + Bx(x - 1) + Cx(x + 3).$$

(continued)

▶ **Tutorial available at www.thomsonedu.com/login**

We next substitute values for x that make the factors, x, $x - 1$, and $x + 3$, equal to zero. If we let $x = 0$ and simplify, we obtain

$$-9 = -3A, \qquad \text{or} \qquad A = 3.$$

Letting $x = 1$ in the equation leads to $8 = 4C$, or $C = 2$. Finally, if $x = -3$, then we have $-12 = 12B$, or $B = -1$.

EXAMPLE 2 **A partial fraction decomposition containing a repeated linear factor**

Find the partial fraction decomposition of

$$\frac{x^2 + 10x - 36}{x(x - 3)^2}.$$

SOLUTION

Guideline 1 The degree of the numerator, 2, is less than the degree of the denominator, 3, so long division is not required.

Guideline 2 The denominator, $x(x - 3)^2$, is already in factored form.

Guideline 3 By Rule A with $m = 1$, there is a partial fraction of the form A/x corresponding to the factor x. Next, applying Rule A with $m = 2$, we find that the factor $(x - 3)^2$ determines a sum of two partial fractions of the form $B/(x - 3)$ and $C/(x - 3)^2$. Thus, the partial fraction decomposition has the form

$$\frac{x^2 + 10x - 36}{x(x - 3)^2} = \frac{A}{x} + \frac{B}{x - 3} + \frac{C}{(x - 3)^2}.$$

Guideline 4 To find A, B, and C, we begin by multiplying both sides of the partial fraction decomposition in guideline 3 by the lcd, $x(x - 3)^2$:

$$\begin{aligned} x^2 + 10x - 36 &= A(x - 3)^2 + Bx(x - 3) + Cx \\ &= A(x^2 - 6x + 9) + B(x^2 - 3x) + Cx \\ &= (A + B)x^2 + (-6A - 3B + C)x + 9A \end{aligned}$$

We next equate the coefficients of like powers of x, obtaining the system

$$\begin{cases} A + B = 1 \\ -6A - 3B + C = 10 \\ 9A = -36 \end{cases}$$

This system of equations has the solution $A = -4$, $B = 5$, and $C = 1$. The partial fraction decomposition is therefore

$$\frac{x^2 + 10x - 36}{x(x - 3)^2} = \frac{-4}{x} + \frac{5}{x - 3} + \frac{1}{(x - 3)^2}.$$

As in Example 1, we could also obtain A and C by beginning with the equation

$$x^2 + 10x - 36 = A(x - 3)^2 + Bx(x - 3) + Cx$$

Tutorial available at www.thomsonedu.com/login

and then substituting values for x that make the factors, $x - 3$ and x, equal to zero. Thus, letting $x = 3$, we obtain $3 = 3C$, or $C = 1$. Letting $x = 0$ gives us $-36 = 9A$, or $A = -4$. The value of B may then be found by using one of the equations in the system.

EXAMPLE 3 A partial fraction decomposition containing an irreducible quadratic factor

Find the partial fraction decomposition of

$$\frac{4x^3 - x^2 + 15x - 29}{2x^3 - x^2 + 8x - 4}.$$

▶ SOLUTION

Guideline 1 The degree of the numerator, 3, is *equal* to the degree of the denominator. Thus, long division is required, and we obtain

$$\frac{4x^3 - x^2 + 15x - 29}{2x^3 - x^2 + 8x - 4} = 2 + \frac{x^2 - x - 21}{2x^3 - x^2 + 8x - 4}.$$

Guideline 2 The denominator may be factored by grouping, as follows:

$$2x^3 - x^2 + 8x - 4 = x^2(2x - 1) + 4(2x - 1) = (x^2 + 4)(2x - 1)$$

Guideline 3 Applying Rule B to the irreducible quadratic factor $x^2 + 4$ in guideline 2, we see that one partial fraction has the form $(Ax + B)/(x^2 + 4)$. By Rule A, there is also a partial fraction $C/(2x - 1)$ corresponding to $2x - 1$. Consequently,

$$\frac{x^2 - x - 21}{2x^3 - x^2 + 8x - 4} = \frac{Ax + B}{x^2 + 4} + \frac{C}{2x - 1}.$$

Guideline 4 Multiplying both sides of the partial fraction decomposition in guideline 3 by the lcd, $(x^2 + 4)(2x - 1)$, we obtain

$$\begin{aligned} x^2 - x - 21 &= (Ax + B)(2x - 1) + C(x^2 + 4) \\ &= 2Ax^2 - Ax + 2Bx - B + Cx^2 + 4C \\ &= (2A + C)x^2 + (-A + 2B)x - B + 4C. \end{aligned}$$

This leads to the system

$$\begin{cases} 2A \qquad\quad + C = 1 \\ -A + 2B \qquad\;\; = -1 \\ \qquad\; - B + 4C = -21 \end{cases}$$

This system has the solution $A = 3$, $B = 1$, and $C = -5$. Thus, the partial fraction decomposition in guideline 3 is

$$\frac{x^2 - x - 21}{2x^3 - x^2 + 8x - 4} = \frac{3x + 1}{x^2 + 4} + \frac{-5}{2x - 1},$$

(continued)

▶ **Tutorial available at www.thomsonedu.com/login**

and therefore the decomposition of the given expression (see guideline 1) is

$$\frac{4x^3 - x^2 + 15x - 29}{2x^3 - x^2 + 8x - 4} = 2 + \frac{3x + 1}{x^2 + 4} + \frac{-5}{2x - 1}.$$

EXAMPLE 4 **A partial fraction decomposition containing a repeated quadratic factor**

Find the partial fraction decomposition of

$$\frac{5x^3 - 3x^2 + 7x - 3}{(x^2 + 1)^2}.$$

▶ SOLUTION

Guideline 1 The degree of the numerator, 3, is less than the degree of the denominator, 4, so long division is not required.

Guideline 2 The denominator, $(x^2 + 1)^2$, is already in factored form.

Guideline 3 We apply Rule B with $n = 2$ to $(x^2 + 1)^2$, to obtain the partial fraction decomposition

$$\frac{5x^3 - 3x^2 + 7x - 3}{(x^2 + 1)^2} = \frac{Ax + B}{x^2 + 1} + \frac{Cx + D}{(x^2 + 1)^2}.$$

Guideline 4 Multiplying both sides of the decomposition in guideline 3 by $(x^2 + 1)^2$ gives us

$$\begin{aligned} 5x^3 - 3x^2 + 7x - 3 &= (Ax + B)(x^2 + 1) + Cx + D \\ &= Ax^3 + Bx^2 + (A + C)x + (B + D). \end{aligned}$$

Comparing the coefficients of x^3 and x^2, we obtain $A = 5$ and $B = -3$. From the coefficients of x, we see that $A + C = 7$. Thus, $C = 7 - A = 7 - 5 = 2$. Finally, comparing the constant terms gives us the equation $B + D = -3$, and so $D = -3 - B = -3 - (-3) = 0$. Therefore, the partial fraction decomposition is

$$\frac{5x^3 - 3x^2 + 7x - 3}{(x^2 + 1)^2} = \frac{5x - 3}{x^2 + 1} + \frac{2x}{(x^2 + 1)^2}.$$

8.10 Exercises

Exer. 1–28: Find the partial fraction decomposition.

▶ 1 $\dfrac{8x - 1}{(x - 2)(x + 3)}$

2 $\dfrac{x - 29}{(x - 4)(x + 1)}$

3 $\dfrac{x + 34}{x^2 - 4x - 12}$

4 $\dfrac{5x - 12}{x^2 - 4x}$

▶ 5 $\dfrac{4x^2 - 15x - 1}{(x - 1)(x + 2)(x - 3)}$

6 $\dfrac{x^2 + 19x + 20}{x(x + 2)(x - 5)}$

7 $\dfrac{4x^2 - 5x - 15}{x^3 - 4x^2 - 5x}$

8 $\dfrac{37 - 11x}{(x + 1)(x^2 - 5x + 6)}$

▶ Tutorial available at www.thomsonedu.com/login

▶ 9 $\dfrac{2x+3}{(x-1)^2}$

10 $\dfrac{5x^2-4}{x^2(x+2)}$

11 $\dfrac{19x^2+50x-25}{3x^3-5x^2}$

▶ 12 $\dfrac{10-x}{x^2+10x+25}$

13 $\dfrac{x^2-6}{(x+2)^2(2x-1)}$

14 $\dfrac{2x^2+x}{(x-1)^2(x+1)^2}$

15 $\dfrac{3x^3+11x^2+16x+5}{x(x+1)^3}$

16 $\dfrac{4x^3+3x^2+5x-2}{x^3(x+2)}$

17 $\dfrac{x^2+x-6}{(x^2+1)(x-1)}$

18 $\dfrac{x^2-x-21}{(x^2+4)(2x-1)}$

19 $\dfrac{9x^2-3x+8}{x^3+2x}$

20 $\dfrac{2x^3+2x^2+4x-3}{x^4+x^2}$

21 $\dfrac{4x^3-x^2+4x+2}{(x^2+1)^2}$

22 $\dfrac{3x^3+13x-1}{(x^2+4)^2}$

23 $\dfrac{2x^4-2x^3+6x^2-5x+1}{x^3-x^2+x-1}$

24 $\dfrac{x^3}{x^3-3x^2+9x-27}$

▶ 25 $\dfrac{3x^2-16}{x^2-4x}$

26 $\dfrac{2x^2+7x}{x^2+6x+9}$

▶ 27 $\dfrac{4x^3+4x^2-4x+2}{2x^2-x-1}$

28 $\dfrac{x^5-5x^4+7x^3-x^2-4x+12}{x^3-3x^2}$

CHAPTER 8 REVIEW EXERCISES

▶ **Online support materials can be found at www.thomsonedu.com/login**

Exer. 1–16: Solve the system.

1 $\begin{cases} 2x-3y=4 \\ 5x+4y=1 \end{cases}$

2 $\begin{cases} x-3y=4 \\ -2x+6y=2 \end{cases}$

3 $\begin{cases} y+4=x^2 \\ 2x+y=-1 \end{cases}$

4 $\begin{cases} x^2+y^2=25 \\ x-y=7 \end{cases}$

5 $\begin{cases} 9x^2+16y^2=140 \\ x^2-4y^2=4 \end{cases}$

6 $\begin{cases} 2x=y^2+3z \\ x=y^2+z-1 \\ x^2=xz \end{cases}$

7 $\begin{cases} \dfrac{1}{x}+\dfrac{3}{y}=7 \\ \dfrac{4}{x}-\dfrac{2}{y}=1 \end{cases}$

8 $\begin{cases} 2^x+3^{y+1}=10 \\ 2^{x+1}-3^y=5 \end{cases}$

9 $\begin{cases} 3x+y-2z=-1 \\ 2x-3y+z=4 \\ 4x+5y-z=-2 \end{cases}$

10 $\begin{cases} x+3y=0 \\ y-5z=3 \\ 2x+z=-1 \end{cases}$

11 $\begin{cases} 4x-3y-z=0 \\ x-y-z=0 \\ 3x-y+3z=0 \end{cases}$

12 $\begin{cases} 2x+y-z=0 \\ x-2y+z=0 \\ 3x+3y+2z=0 \end{cases}$

13 $\begin{cases} 4x+2y-z=1 \\ 3x+2y+4z=2 \end{cases}$

14 $\begin{cases} 2x+y=6 \\ x-3y=17 \\ 3x+2y=7 \end{cases}$

15 $\begin{cases} \dfrac{4}{x}+\dfrac{1}{y}+\dfrac{2}{z}=4 \\ \dfrac{2}{x}+\dfrac{3}{y}-\dfrac{1}{z}=1 \\ \dfrac{1}{x}+\dfrac{1}{y}+\dfrac{1}{z}=4 \end{cases}$

16 $\begin{cases} 2x-y+3z-w=-3 \\ 3x+2y-z+w=13 \\ x-3y+z-2w=-4 \\ -x+y+4z+3w=0 \end{cases}$

▶ **Tutorial available at www.thomsonedu.com/login**

Exer. 17–20: Sketch the graph of the system.

17 $\begin{cases} x^2 + y^2 < 16 \\ y - x^2 > 0 \end{cases}$

18 $\begin{cases} y - x \leq 0 \\ y + x \geq 2 \\ x \leq 5 \end{cases}$

19 $\begin{cases} x - 2y \leq 2 \\ y - 3x \leq 4 \\ 2x + y \leq 4 \end{cases}$

20 $\begin{cases} x^2 - y < 0 \\ y - 2x < 5 \\ xy < 0 \end{cases}$

Exer. 21–30: Express as a single matrix.

21 $\begin{bmatrix} 2 & -1 & 0 \\ 3 & 0 & -2 \end{bmatrix}\begin{bmatrix} 2 & -1 & 3 \\ 0 & 3 & 0 \\ 1 & 4 & 2 \end{bmatrix}$

22 $\begin{bmatrix} 4 & 2 \\ 5 & -3 \end{bmatrix}\begin{bmatrix} 3 \\ 7 \end{bmatrix}$

23 $\begin{bmatrix} 2 & 0 \\ 1 & 4 \\ -2 & 3 \end{bmatrix}\begin{bmatrix} 0 & 2 & -3 \\ 4 & 5 & 1 \end{bmatrix}$

24 $\begin{bmatrix} 0 & -2 & 3 \\ 4 & 1 & 2 \end{bmatrix}\begin{bmatrix} 2 & 0 \\ 3 & 8 \\ 2 & -7 \end{bmatrix}$

25 $2\begin{bmatrix} 0 & -1 & -4 \\ 3 & 2 & 1 \end{bmatrix} - 3\begin{bmatrix} 4 & -2 & 1 \\ 0 & 5 & -1 \end{bmatrix}$

26 $\begin{bmatrix} 1 & 3 \\ 2 & 4 \end{bmatrix}\begin{bmatrix} a & 0 \\ 0 & a \end{bmatrix}$

27 $\begin{bmatrix} a & 0 \\ 0 & b \end{bmatrix}\begin{bmatrix} 1 & 3 \\ 2 & 4 \end{bmatrix}$

28 $\begin{bmatrix} 3 & 2 \\ 0 & 0 \end{bmatrix}\begin{bmatrix} -2 & 0 \\ 3 & 0 \end{bmatrix}$

29 $\begin{bmatrix} 1 & 2 \\ 3 & 4 \end{bmatrix}\left\{\begin{bmatrix} 2 & -4 \\ 3 & 7 \end{bmatrix} + \begin{bmatrix} 1 & 5 \\ -2 & -3 \end{bmatrix}\right\}$

30 $\begin{bmatrix} 3 & 2 & 5 \\ -3 & 4 & 7 \\ 6 & 5 & 1 \end{bmatrix}\begin{bmatrix} 3 & 2 & 5 \\ -3 & 4 & 7 \\ 6 & 5 & 1 \end{bmatrix}^{-1}$

Exer. 31–34: Find the inverse of the matrix.

31 $\begin{bmatrix} 5 & -4 \\ -3 & 2 \end{bmatrix}$

32 $\begin{bmatrix} 2 & -1 & 0 \\ 1 & 4 & 2 \\ 3 & -2 & 1 \end{bmatrix}$

33 $\begin{bmatrix} 1 & 0 & 0 \\ 0 & 4 & 7 \\ 0 & 1 & 2 \end{bmatrix}$

34 $\begin{bmatrix} 2 & 0 & 5 \\ 0 & 3 & -1 \\ 3 & 4 & 0 \end{bmatrix}$

35 Use the result of Exercise 31 to solve the system

$$\begin{cases} 5x - 4y = 30 \\ -3x + 2y = -16 \end{cases}$$

36 Use the result of Exercise 32 to solve the system

$$\begin{cases} 2x - y = -5 \\ x + 4y + 2z = 15 \\ 3x - 2y + z = -7 \end{cases}$$

Exer. 37–46: Find the determinant of the matrix.

37 $[-6]$

38 $\begin{bmatrix} 3 & 4 \\ -6 & -5 \end{bmatrix}$

39 $\begin{bmatrix} 3 & -4 \\ 6 & 8 \end{bmatrix}$

40 $\begin{bmatrix} 0 & 4 & -3 \\ 2 & 0 & 4 \\ -5 & 1 & 0 \end{bmatrix}$

41 $\begin{bmatrix} 2 & -3 & 5 \\ -4 & 1 & 3 \\ 3 & 2 & -1 \end{bmatrix}$

42 $\begin{bmatrix} 3 & 1 & -2 \\ -5 & 2 & -4 \\ 7 & 3 & -6 \end{bmatrix}$

43 $\begin{bmatrix} 5 & 0 & 0 & 0 \\ 6 & -3 & 0 & 0 \\ 1 & 4 & -4 & 0 \\ 7 & 2 & 3 & 2 \end{bmatrix}$

44 $\begin{bmatrix} 1 & 2 & 0 & 3 & 1 \\ -2 & -1 & 4 & 1 & 2 \\ 3 & 0 & -1 & 0 & -1 \\ 2 & -3 & 2 & -4 & 2 \\ -1 & 1 & 0 & 1 & 3 \end{bmatrix}$

45 $\begin{bmatrix} 2 & 0 & 1 & 0 & -1 \\ 0 & 1 & 0 & 1 & 2 \\ 2 & -2 & 1 & -2 & 0 \\ 0 & 0 & -2 & 0 & 1 \\ 1 & -1 & 0 & -1 & 0 \end{bmatrix}$

46 $\begin{bmatrix} 1 & 2 & 0 & 0 & 0 \\ 3 & 4 & 0 & 0 & 0 \\ 0 & 0 & 1 & 2 & 3 \\ 0 & 0 & 2 & -1 & 1 \\ 0 & 0 & 1 & 3 & -1 \end{bmatrix}$

Exer. 47–48: Solve the equation $|A - xI| = 0$.

47 $A = \begin{bmatrix} 2 & 3 \\ 1 & -4 \end{bmatrix}, \quad I = I_2$

48 $A = \begin{bmatrix} 2 & -1 & 3 \\ 0 & 4 & 0 \\ 1 & 0 & -2 \end{bmatrix}, \quad I = I_3$

Exer. 49–50: Without expanding, explain why the statement is true.

49 $\begin{vmatrix} 2 & 4 & -6 \\ 1 & 4 & 3 \\ 2 & 2 & 0 \end{vmatrix} = 12 \begin{vmatrix} 1 & 1 & -1 \\ 1 & 2 & 1 \\ 2 & 1 & 0 \end{vmatrix}$

50 $\begin{vmatrix} a & b & c \\ d & e & f \\ g & h & k \end{vmatrix} = \begin{vmatrix} d & e & f \\ g & h & k \\ a & b & c \end{vmatrix}$

51 Find the determinant of the $n \times n$ matrix (a_{ij}) in which $a_{ij} = 0$ for $i \neq j$.

52 Without expanding, show that

$$\begin{vmatrix} 1 & a & b+c \\ 1 & b & a+c \\ 1 & c & a+b \end{vmatrix} = 0.$$

Exer. 53–54: Use Cramer's rule to solve the system.

53 $\begin{cases} 5x - 6y = 4 \\ 3x + 7y = 8 \end{cases}$

54 $\begin{cases} 2x - 3y + 2z = -3 \\ -3x + 2y + z = 1 \\ 4x + y - 3z = 4 \end{cases}$

Exer. 55–58: Find the partial fraction decomposition.

55 $\dfrac{4x^2 + 54x + 134}{(x + 3)(x^2 + 4x - 5)}$

56 $\dfrac{2x^2 + 7x + 9}{x^2 + 2x + 1}$

57 $\dfrac{x^2 + 14x - 13}{x^3 + 5x^2 + 4x + 20}$

58 $\dfrac{x^3 + 2x^2 + 2x + 16}{x^4 + 7x^2 + 10}$

59 **Watering a field** A rotating sprinkler head with a range of 50 feet is to be placed in the center of a rectangular field (see the figure). If the area of the field is 4000 ft^2 and the water is to just reach the corners, find the dimensions of the field.

Exercise 59

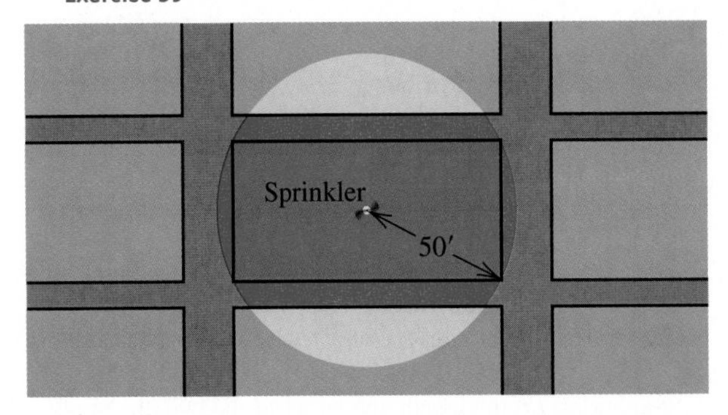

60 Find equations of the two lines that are tangent to the circle $x^2 + y^2 = 1$ and pass through the point $(0, 3)$. (*Hint:* Let $y = mx + 3$, and determine conditions on m that will ensure that the system has only one solution.)

61 **Payroll accounting** An accountant must pay taxes and payroll bonuses to employees from the company's profits of \$2,000,000. The total tax is 40% of the amount left after bonuses are paid, and the total paid in bonuses is 10% of the amount left after taxes. Find the total tax and the total bonus amount.

62 **Track dimensions** A circular track is to have a 10-foot-wide running lane around the outside (see the figure). The inside distance around the track is to be 90% of the outside distance. Find the dimensions of the track.

Exercise 62

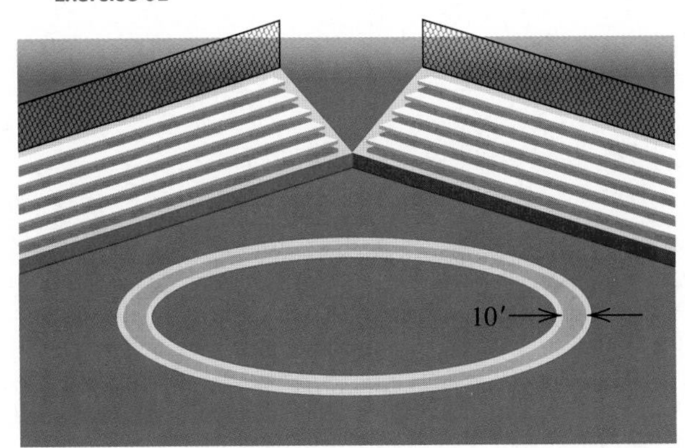

63 Flow rates Three inlet pipes, A, B, and C, can be used to fill a 1000-ft^3 water storage tank. When all three pipes are in operation, the tank can be filled in 10 hours. When only pipes A and B are used, the time increases to 20 hours. With pipes A and C, the tank can be filled in 12.5 hours. Find the individual flow rates (in ft^3/hr) for each of the three pipes.

64 Warehouse shipping charges To fill an order for 150 office desks, a furniture distributor must ship the desks from two warehouses. The shipping cost per desk is \$48 from the western warehouse and \$70 from the eastern warehouse. If the total shipping charge is \$8410, how many desks are shipped from each location?

65 Express-mail rates An express-mail company charges \$25 for overnight delivery of a letter, provided the dimensions of the standard envelope satisfy the following three conditions: (a) the length, the larger of the two dimensions, must be at most 12 inches; (b) the width must be at most 8 inches; (c) the width must be at least one-half the length. Find and graph a system of inequalities that describes all the possibilities for dimensions of a standard envelope.

66 Activities of a deer A deer spends the day in three basic activities: resting, searching for food, and grazing. At least 6 hours each day must be spent resting, and the number of hours spent searching for food will be at least two times the number of hours spent grazing. Using x as the number of hours spent searching for food and y as the number of hours spent grazing, find and graph the system of inequalities that describes the possible divisions of the day.

67 Production scheduling A company manufactures a power lawn mower and a power edger. These two products are of such high quality that the company can sell all the products it makes, but production capability is limited in the areas of machining, welding, and assembly. Each week the company has 600 hours available for machining, 300 hours for welding, and 550 hours for assembly. The number of hours required for the production of a single item is shown in the following table. The profits from the sale of a mower and an edger are \$100 and \$80, respectively. How many mowers and edgers should be made each week to maximize the profit?

Product	Machining	Welding	Assembly
Mower	6	2	5
Edger	4	3	5

68 Maximizing investment income A retired couple wishes to invest \$750,000, diversifying the investment in three areas: a high-risk stock that has an expected annual rate of return (or interest) of 12%, a low-risk stock that has an expected annual return of 8%, and government-issued bonds that pay annual interest of 4% and involve no risk. To protect the value of the investment, the couple wishes to place at least twice as much in the low-risk stock as in the high-risk stock and use the remainder to buy bonds. How should the money be invested to maximize the expected annual return?

CHAPTER 8 DISCUSSION EXERCISES

1 (a) It is easy to see that the system

$$\begin{cases} x + 2y = 4 \\ x + 2y = 5 \end{cases}$$

has no solution. Let $x + by = 5$ be the second equation, and solve the system for $b = 1.99$ and $b = 1.999$. Note that a small change in b produces a large change in x and y. Such a system is known as an *ill-conditioned system* (a precise definition is given in most numerical analysis texts).

(b) Solve this system for x and y in terms of b, and explain why a small change in b (for b near 2) produces a large change in x and y.

(c) If b gets very large, what happens to the solution of the system?

2 Bird migration trends Refer to Exercise 30 of Section 8.5. Suppose the initial bird populations on islands A, B, and C are 12,000, 9000, and 14,000, respectively.

(a) Represent the initial populations with a 1×3 matrix D. Represent the proportions of the populations that migrate to each island with a 3×3 matrix E. (*Hint:* The first row of E is 0.90, 0.10, and 0.00—indicating that 90% of the birds on A stay on A, 10% of the birds on A migrate to B, and no birds on A migrate to C.)

(b) Find the product $F = DE$, and interpret the meaning of the elements of F.

(c) Using a computational device, multiply F times E, and continue to multiply the result by E until a pattern becomes apparent. What is your conclusion?

(d) Suppose the initial population matrix D is equal to $[34{,}000 \quad 500 \quad 500]$. Multiply D times E, and continue to multiply the result by E until a pattern becomes apparent. What is your conclusion?

3 Explain why a nonsquare matrix A cannot have an inverse.

4 Distributing money A college president has received budgets from the athletic director (AD), dean of students (DS), and student senate president (SP), in which they propose to allocate department funds to the three basic areas of student scholarships, activities, and services, as shown in the table.

	Scholarships	Activities	Services
AD	50%	40%	10%
DS	30%	20%	50%
SP	20%	40%	40%

The Board of Regents has requested that the overall distribution of funding to these three areas be in the following proportions: scholarships, 34%; activities, 33%; and services, 33%. Determine what percentage of the total funds the president should allocate to each department so that the percentages spent in these three areas conform to the Board of Regents' requirements.

5 If $x^4 + ax^2 + bx + c = 0$ has roots $x = -1, 2$, and 3, find a, b, c, and the fourth root of the equation.

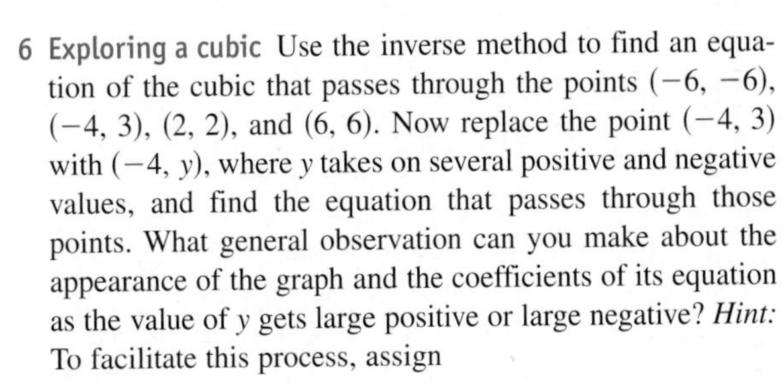

6 Exploring a cubic Use the inverse method to find an equation of the cubic that passes through the points $(-6, -6)$, $(-4, 3)$, $(2, 2)$, and $(6, 6)$. Now replace the point $(-4, 3)$ with $(-4, y)$, where y takes on several positive and negative values, and find the equation that passes through those points. What general observation can you make about the appearance of the graph and the coefficients of its equation as the value of y gets large positive or large negative? *Hint:* To facilitate this process, assign

$$[C](1, 1)x^3 + [C](2, 1)x^2 + [C](3, 1)x + [C](4, 1)$$

to Y_1, where $[C] = [A]^{-1} * [B]$.

7 Prove (*) on page 675.

8 Find, if possible, an equation of

(a) a line

(b) a circle

(c) a parabola with vertical axis

(d) a cubic

(e) an exponential

that passes through the three points $P(-1, 3)$, $Q(0, 4)$, and $R(3, 2)$.

9

Sequences, Series, and Probability

Sequences and summation notation, discussed in the first section, are very important in advanced mathematics and applications. Of special interest are arithmetic and geometric sequences, considered in Sections 9.2 and 9.3. We then discuss the method of mathematical induction, a process that is often used to prove that each statement in an infinite sequence of statements is true. As an application, we use it to prove the binomial theorem in Section 9.5. The last part of the chapter deals with counting processes that occur frequently in mathematics and everyday life. These include the concepts of permutations, combinations, and probability.

9.1

Infinite Sequences and Summation Notation

An arbitrary *infinite sequence* may be denoted as follows:

Infinite Sequence Notation	$a_1, a_2, a_3, \ldots, a_n, \ldots$

For convenience, we often refer to infinite sequences as *sequences.* We may regard an infinite sequence as a collection of real numbers that is in one-to-one correspondence with the positive integers. Each number a_k is a **term** of the sequence. The sequence is *ordered* in the sense that there is a **first term** a_1, a *second term* a_2, a *forty-fifth term* a_{45}, and, if n denotes an arbitrary positive integer, an ***n*th term** a_n. Infinite sequences are often defined by stating a formula for the nth term.

Infinite sequences occur frequently in mathematics. For example, the sequence

$$0.6, \quad 0.66, \quad 0.666, \quad 0.6666, \quad 0.66666, \quad \ldots$$

may be used to represent the rational number $\frac{2}{3}$. In this case the nth term gets closer and closer to $\frac{2}{3}$ as n increases.

We may regard an infinite sequence as a function. Recall from Section 2.4 that a function f is a correspondence that assigns to each number x in the domain D exactly one number $f(x)$ in the range R. If we restrict the domain to the positive integers $1, 2, 3, \ldots$, we obtain an infinite sequence, as in the following definition.

Definition of Infinite Sequence	An **infinite sequence** is a function whose domain is the set of positive integers.

In our work, the range of an infinite sequence will be a set of real numbers.

If a function f is an infinite sequence, then to each positive integer n there corresponds a real number $f(n)$. These numbers in the range of f may be represented by writing

$$f(1), f(2), f(3), \ldots, f(n), \ldots.$$

To obtain the subscript form of a sequence, as shown at the beginning of this section, we let $a_n = f(n)$ for every positive integer n.

If we regard a sequence as a function f, then we may consider its graph in an xy-plane. Since the domain of f is the set of positive integers, the only points on the graph are

$$(1, a_1), (2, a_2), (3, a_3), \ldots, (n, a_n), \ldots,$$

where a_n is the nth term of the sequence as shown in Figure 1. We sometimes use the graph of a sequence to illustrate the behavior of the nth term a_n as n increases without bound.

Figure 1 Graph of a sequence

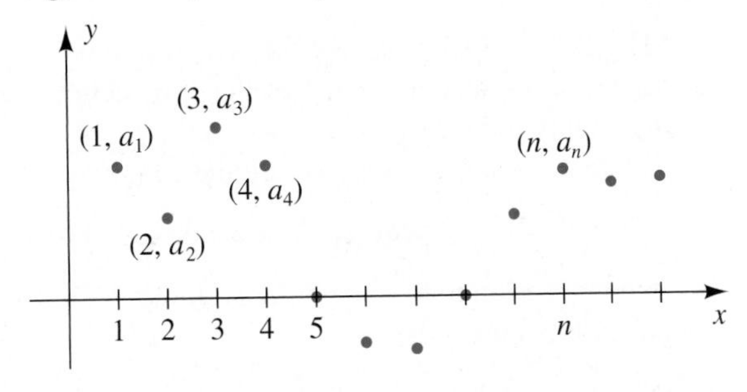

From the definition of equality of functions we see that a sequence

$$a_1, a_2, a_3, \ldots, a_n, \ldots$$

is **equal** to a sequence

$$b_1, b_2, b_3, \ldots, b_n, \ldots$$

if and only if $a_k = b_k$ for every positive integer k.

Another notation for a sequence with nth term a_n is $\{a_n\}$. For example, the sequence $\{2^n\}$ has nth term $a_n = 2^n$. Using sequence notation, we write this sequence as follows:

$$2^1, 2^2, 2^3, \ldots, 2^n, \ldots$$

By definition, the sequence $\{2^n\}$ is the function f with $f(n) = 2^n$ for every positive integer n.

EXAMPLE 1 Finding terms of a sequence

List the first four terms and the tenth term of each sequence:

(a) $\left\{\dfrac{n}{n+1}\right\}$ **(b)** $\{2 + (0.1)^n\}$ **(c)** $\left\{(-1)^{n+1}\dfrac{n^2}{3n-1}\right\}$ **(d)** $\{4\}$

▶ **SOLUTION** To find the first four terms, we substitute, successively, $n = 1$, 2, 3, and 4 in the formula for a_n. The tenth term is found by substituting 10 for n. Doing this and simplifying gives us the following:

(continued)

▶ **Tutorial available at www.thomsonedu.com/login**

	Sequence	nth term a_n	First four terms	Tenth term
(a)	$\left\{\frac{n}{n+1}\right\}$	$\frac{n}{n+1}$	$\frac{1}{2}, \frac{2}{3}, \frac{3}{4}, \frac{4}{5}$	$\frac{10}{11}$
(b)	$\{2 + (0.1)^n\}$	$2 + (0.1)^n$	2.1, 2.01, 2.001, 2.0001	2.000 000 000 1
(c)	$\left\{(-1)^{n+1}\frac{n^2}{3n-1}\right\}$	$(-1)^{n+1}\frac{n^2}{3n-1}$	$\frac{1}{2}, -\frac{4}{5}, \frac{9}{8}, -\frac{16}{11}$	$-\frac{100}{29}$
(d)	$\{4\}$	4	4, 4, 4, 4	4

The TI-83/4 Plus has a special sequence mode that is not available on the TI-86. The use of this mode is discussed at the end of this section. For now, we will consider generic methods that apply to both calculators.

To generate a sequence, we use the command

seq(expression, variable, begin, end, increment).

(If increment is omitted, the default value is 1.) Let's generate the first four terms of the sequence in Example 1(a).

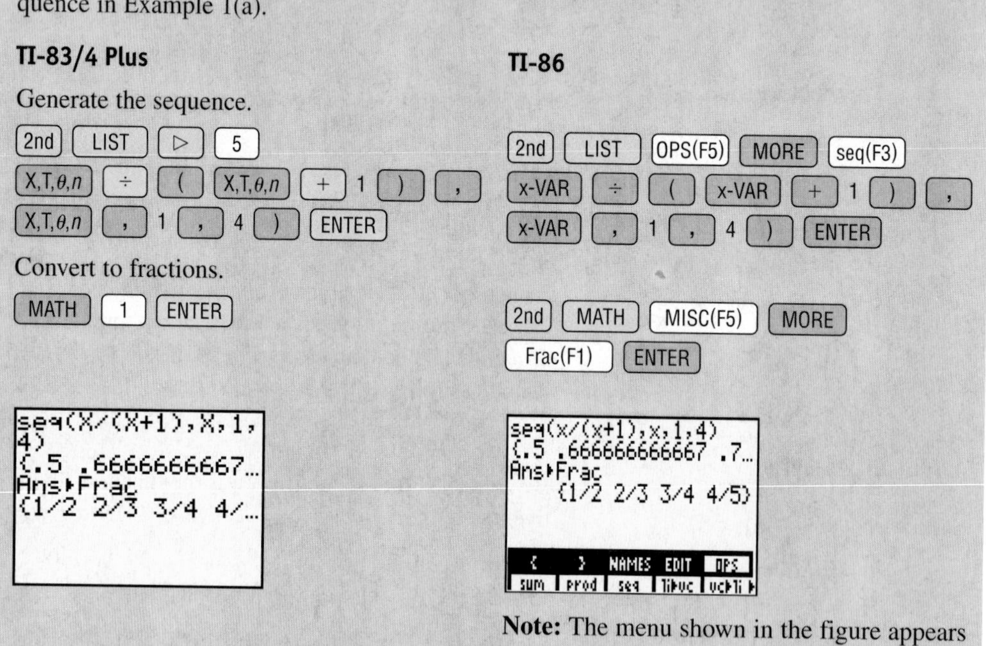

Note: The menu shown in the figure appears only after you enter the sequence command.

EXAMPLE 2 Graphing a sequence

Graph the sequence in Example 1(a)—that is,

$$\left\{\frac{n}{n+1}\right\}.$$

▶ SOLUTION The domain values are

$$1, 2, 3, \ldots, n, \ldots.$$

The range values are

$$\frac{1}{1+1}, \quad \frac{2}{2+1}, \quad \frac{3}{3+1}, \quad \ldots, \quad \frac{n}{n+1}, \quad \ldots$$

or, equivalently,

$$\frac{1}{2}, \quad \frac{2}{3}, \quad \frac{3}{4}, \quad \ldots, \quad \frac{n}{n+1}, \quad \ldots.$$

A plot of the ordered pairs $(n, n/(n+1))$ is shown in Figure 2.

Figure 2

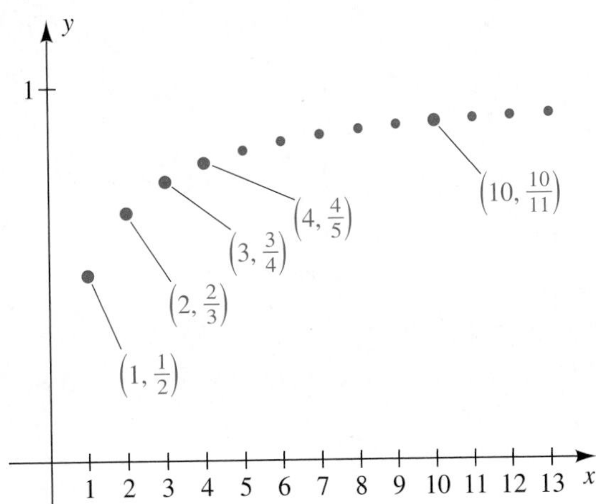

We'll make use of *lists* to graph the sequence in Example 2. (A demonstration of graphing in sequence mode on the TI-83/4 Plus appears at the end of the section.)

TI-83/4 Plus

Store the first n positive integers in a list (the domain values).

[2nd] [LIST] [▷] [5]
[X,T,θ,n] [,] [X,T,θ,n] [,] 1 [,] 4 [)]
[STO⇨] [2nd] [L1] [ENTER]

TI-86

[2nd] [LIST] [OPS(F5)] [MORE] [seq(F3)]
[x-VAR] [,] [x-VAR] [,] 1 [,] 4 [)]
[STO⇨] [2nd] [LIST] [NAMES(F3)]
[xStat(F2)] [ENTER]

Store the first n terms in a second list (the range values).

TI-83/4 Plus

[2nd] [LIST] [▷] [5]
[X,T,θ,n] [÷] [(] [X,T,θ,n] [+] 1 [)] [,]

TI-86

[2nd] [LIST] [OPS(F5)] [MORE] [seq(F3)]
[x-VAR] [÷] [(] [x-VAR] [+] 1 [)] [,]

(continued)

▶ **Tutorial available at www.thomsonedu.com/login**

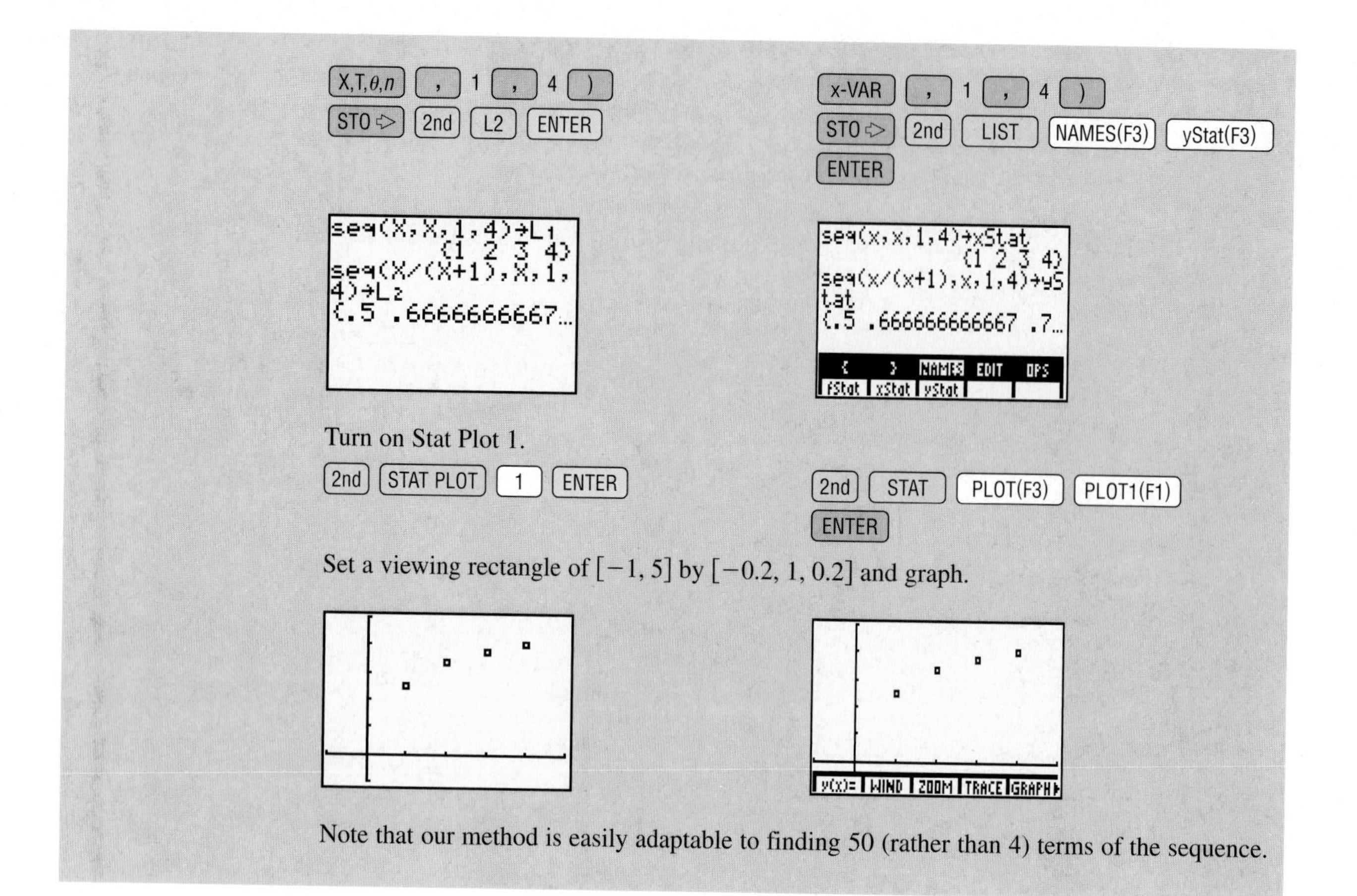

For some sequences we state the first term a_1, together with a rule for obtaining any term a_{k+1} from the preceding term a_k whenever $k \geq 1$. We call such a statement a **recursive definition,** and the sequence is said to be defined **recursively.**

EXAMPLE 3 Finding terms of a recursively defined sequence

Find the first four terms and the nth term of the infinite sequence defined recursively as follows:

$$a_1 = 3, \quad a_{k+1} = 2a_k \quad \text{for} \quad k \geq 1$$

▶ **SOLUTION** The first four terms are

$$\begin{aligned}
a_1 &= 3 && \text{given} \\
a_2 &= 2a_1 = 2 \cdot 3 = 6 && k = 1 \\
a_3 &= 2a_2 = 2 \cdot 2 \cdot 3 = 2^2 \cdot 3 = 12 && k = 2 \\
a_4 &= 2a_3 = 2 \cdot 2 \cdot 2 \cdot 3 = 2^3 \cdot 3 = 24. && k = 3
\end{aligned}$$

▶ **Tutorial available at www.thomsonedu.com/login**

We have written the terms as products to gain some insight into the nature of the nth term. Continuing, we obtain $a_5 = 2^4 \cdot 3$, $a_6 = 2^5 \cdot 3$, and, in general,

$$a_n = 2^{n-1} \cdot 3$$

for every positive integer n.

We can generate the terms of a recursively defined sequence by first storing a *seed* (or initial) value in a variable. Next, we write our recursive definition in terms of that variable and then store that result to the same variable. We can use any variable on the calculator, but the easiest one is the ANS location since the last calculated result is automatically stored there. Below are two examples for generating the terms in Example 3, one for the variable X and one for the ANS location. The keystrokes given are for the TI-83/4 Plus; just substitute [x-VAR] for [X,T,θ,n] for the TI-86. (The recursive capabilities of the TI-83/4 Plus are discussed at the end of this section.)

To generate a recursively defined sequence using the variable X, use

3 [STO⇨] [X,T,θ,n] [ENTER] 2 [×] [X,T,θ,n] [STO⇨] [X,T,θ,n] [ENTER] [ENTER] . . .

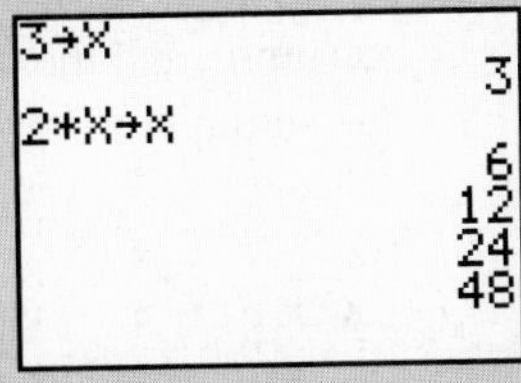

To generate a recursively defined sequence using ANS, use

3 [ENTER] 2 [×] [2nd] [ANS] [ENTER] [ENTER] . . .

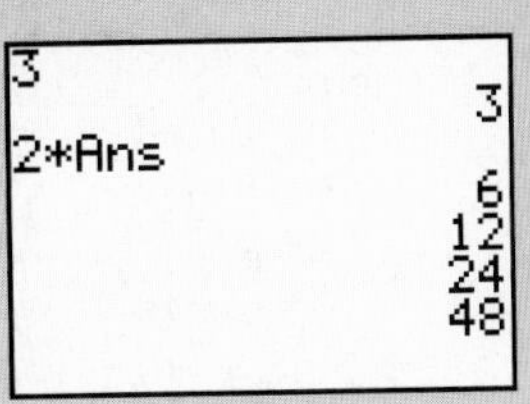

If only the first few terms of an infinite sequence are known, then it is impossible to predict additional terms. For example, if we were given 3, 6, 9, . . . and asked to find the fourth term, we could not proceed without further information. The infinite sequence with nth term

$$a_n = 3n + (1 - n)^3(2 - n)^2(3 - n)$$

has for its first four terms 3, 6, 9, and 120. It is possible to describe sequences in which the first three terms are 3, 6, and 9 and the fourth term is *any* given number. This shows that when we work with an infinite sequence it is essential to have either specific information about the nth term or a general scheme for obtaining each term from the preceding one. (See Exercise 1 of the Chapter 9 Discussion Exercises for a related problem.)

We sometimes need to find the sum of many terms of an infinite sequence. To express such sums easily, we use **summation notation.** Given an infinite sequence

$$a_1, a_2, a_3, \ldots, a_n, \ldots,$$

the symbol $\sum_{k=1}^{m} a_k$ represents the sum of the first m terms, as follows.

Summation Notation	$\sum_{k=1}^{m} a_k = a_1 + a_2 + a_3 + \cdots + a_m$

The Greek capital letter sigma, Σ, indicates a sum, and the symbol a_k represents the kth term. The letter k is the **index of summation,** or the **summation variable,** and the numbers 1 and m indicate the smallest and largest values of the summation variable, respectively.

EXAMPLE 4 Evaluating a sum

Find the sum $\sum_{k=1}^{4} k^2(k - 3)$.

▶ SOLUTION In this case, $a_k = k^2(k - 3)$. To find the sum, we merely substitute, in succession, the integers 1, 2, 3, and 4 for k and add the resulting terms:

$$\sum_{k=1}^{4} k^2(k - 3) = 1^2(1 - 3) + 2^2(2 - 3) + 3^2(3 - 3) + 4^2(4 - 3)$$

$$= (-2) + (-4) + 0 + 16 = 10$$

To find the sum in Example 4 on a graphing calculator, we simply sum a sequence.

TI-83/4 Plus

Generate the sequence.

[2nd] [LIST] [▷] [5]
[X,T,θ,n] [x²] [(] [X,T,θ,n] [−] 3 [)] [,]
[X,T,θ,n] [,] 1 [,] 4 [)] [ENTER]

TI-86

[2nd] [LIST] [OPS(F5)] [MORE] [seq(F3)]
[x-VAR] [x²] [(] [x-VAR] [−] 3 [)] [,]
[x-VAR] [,] 1 [,] 4 [)] [ENTER]

▶ Tutorial available at www.thomsonedu.com/login

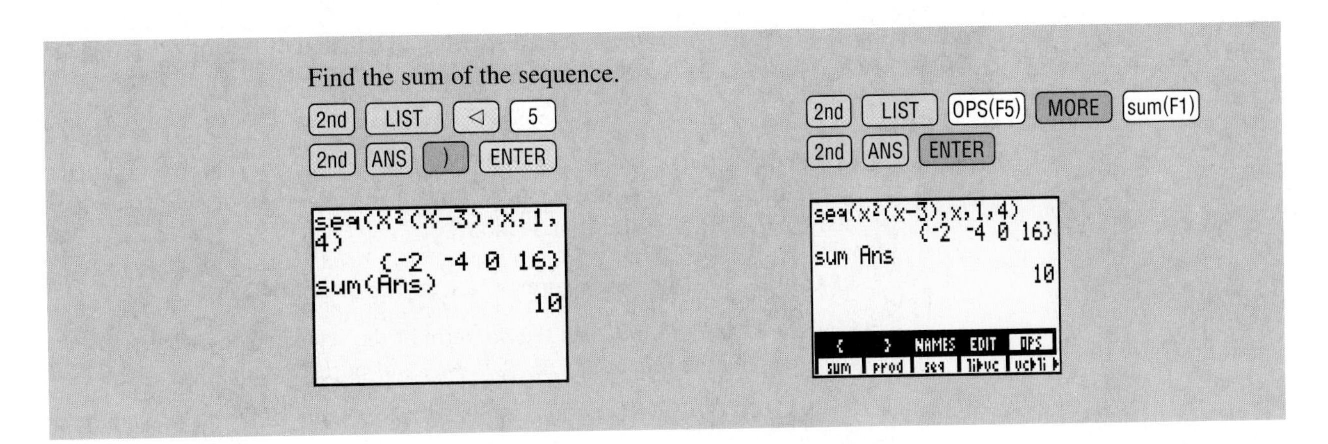

The letter we use for the summation variable is immaterial. To illustrate, if j is the summation variable, then

$$\sum_{j=1}^{m} a_j = a_1 + a_2 + a_3 + \cdots + a_m,$$

which is the same sum as $\Sigma_{k=1}^{m} a_k$. As a specific example, the sum in Example 4 can be written

$$\sum_{j=1}^{4} j^2(j - 3).$$

If n is a positive integer, then the sum of the first n terms of an infinite sequence will be denoted by S_n. For example, given the infinite sequence $a_1, a_2, a_3, \ldots, a_n, \ldots,$

$$\begin{aligned} S_1 &= a_1 \\ S_2 &= a_1 + a_2 \\ S_3 &= a_1 + a_2 + a_3 \\ S_4 &= a_1 + a_2 + a_3 + a_4 \end{aligned}$$

and, in general,

$$S_n = \sum_{k=1}^{n} a_k = a_1 + a_2 + \cdots + a_n.$$

Note that we can also write

$$\begin{aligned} S_1 &= a_1 \\ S_2 &= S_1 + a_2 \\ S_3 &= S_2 + a_3 \\ S_4 &= S_3 + a_4 \end{aligned}$$

and, for every $n > 1$,

$$S_n = S_{n-1} + a_n.$$

The real number S_n is called the **nth partial sum** of the infinite sequence $a_1, a_2, a_3, \ldots, a_n, \ldots$, and the sequence

$$S_1, S_2, S_3, \ldots, S_n, \ldots$$

is called a **sequence of partial sums.** Sequences of partial sums are important in the study of *infinite series,* a topic in calculus. We shall discuss some special types of infinite series in Section 9.3.

EXAMPLE 5 Finding the terms of a sequence of partial sums

Find the first four terms and the nth term of the sequence of partial sums associated with the sequence $1, 2, 3, \ldots, n, \ldots$ of positive integers.

▶ **SOLUTION** If we let $a_n = n$, then the first four terms of the sequence of partial sums are

$$\begin{aligned} S_1 &= a_1 = 1 \\ S_2 &= S_1 + a_2 = 1 + 2 = 3 \\ S_3 &= S_2 + a_3 = 3 + 3 = 6 \\ S_4 &= S_3 + a_4 = 6 + 4 = 10. \end{aligned}$$

The nth partial sum S_n (that is, the sum of $1, 2, 3, \ldots, n$) can be written in either of the following forms:

$$\begin{aligned} S_n &= 1 + 2 + 3 + \cdots + (n-2) + (n-1) + n \\ S_n &= n + (n-1) + (n-2) + \cdots + 3 + 2 + 1 \end{aligned}$$

Adding corresponding terms on each side of these equations gives us

$$2S_n = \underbrace{(n+1) + (n+1) + (n+1) + \cdots + (n+1) + (n+1) + (n+1)}_{n \text{ times}}.$$

Since the expression $(n + 1)$ appears n times on the right-hand side of the last equation, we see that

$$2S_n = n(n+1) \qquad \text{or, equivalently,} \qquad S_n = \frac{n(n+1)}{2}.$$

To find the terms of the sequence of partial sums in Example 5 on a graphing calculator, we use the cumulative sum feature.

TI-83/4 Plus

Generate the sequence.

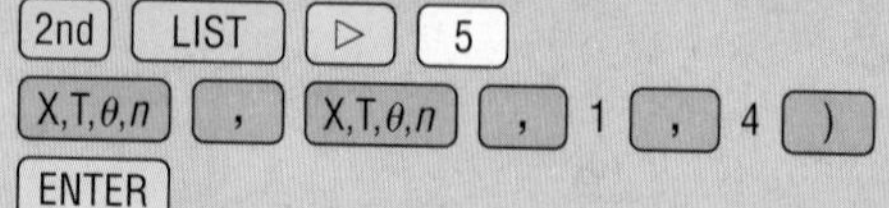

TI-86

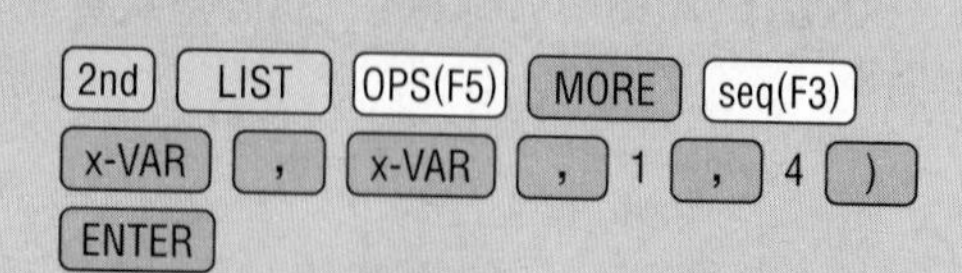

▶ Tutorial available at www.thomsonedu.com/login

Find the terms of the sequence of partial sums.

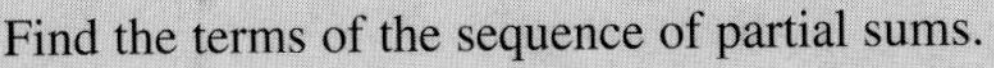

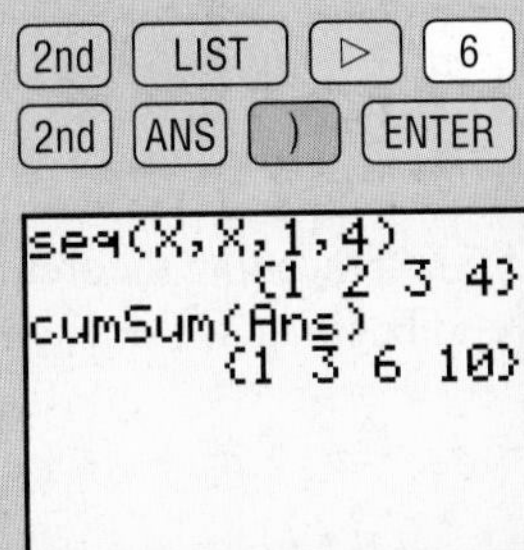

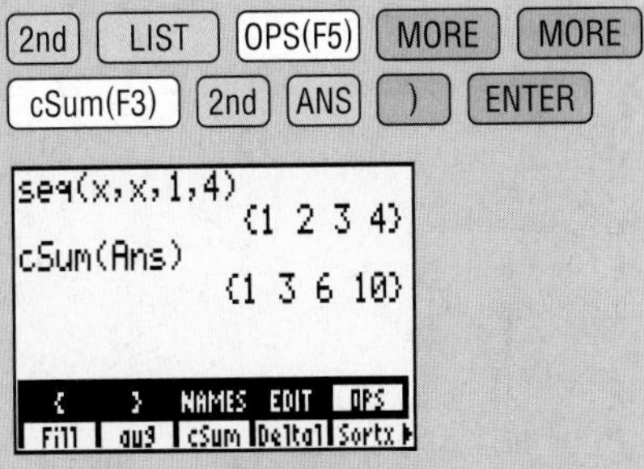

To graph the sequence of partial sums, we could store the first n positive integers and the cumulative sum in two lists and then graph them, as demonstrated in the calculator insert following Example 2.

If a_k is the same for every positive integer k—say $a_k = c$ for a real number c—then

$$\begin{aligned}\sum_{k=1}^{n} a_k &= a_1 + a_2 + a_3 + \cdots + a_n \\ &= c + c + c + \cdots + c = nc.\end{aligned}$$

We have proved property 1 of the following theorem.

Theorem on the Sum of a Constant

$$\text{(1)}\ \sum_{k=1}^{n} c = nc \qquad \text{(2)}\ \sum_{k=m}^{n} c = (n - m + 1)c$$

To prove property 2, we may write

$$\begin{aligned}\sum_{k=m}^{n} c &= \sum_{k=1}^{n} c - \sum_{k=1}^{m-1} c && \text{subtract the first } m-1 \text{ terms from the sum of } n \text{ terms} \\ &= nc - (m-1)c && \text{use property 1 for each sum} \\ &= [n - (m-1)]c && \text{factor out } c \\ &= (n - m + 1)c. && \text{simplify}\end{aligned}$$

ILLUSTRATION **Sum of a Constant**

- $\sum_{k=1}^{4} 7 = 4 \cdot 7 = 28$
- $\sum_{k=1}^{10} \pi = 10 \cdot \pi = 10\pi$

(continued)

- $\sum_{k=3}^{8} 9 = (8 - 3 + 1)(9) = 6(9) = 54$
- $\sum_{k=10}^{20} 5 = (20 - 10 + 1)(5) = 11(5) = 55$

As shown in property 2 of the preceding theorem, the domain of the summation variable does not have to begin at 1. For example,

$$\sum_{k=4}^{8} a_k = a_4 + a_5 + a_6 + a_7 + a_8.$$

As another variation, if the first term of an infinite sequence is a_0, as in

$$a_0, a_1, a_2, \ldots, a_n, \ldots,$$

then we may consider sums of the form

$$\sum_{k=0}^{n} a_k = a_0 + a_1 + a_2 + \cdots + a_n,$$

which is the sum of the first $n + 1$ terms of the sequence.

If the summation variable does not appear in the term a_k, then the *entire term* may be considered a constant. For example,

$$\sum_{j=1}^{n} a_k = n \cdot a_k,$$

since j does not appear in the term a_k.

Summation notation can be used to denote polynomials. Thus, if

$$f(x) = a_0 + a_1x + a_2x^2 + \cdots + a_nx^n,$$

then

$$f(x) = \sum_{k=0}^{n} a_kx^k.$$

The following theorem concerning sums has many uses.

Theorem on Sums

If $a_1, a_2, \ldots, a_n, \ldots$ and $b_1, b_2, \ldots, b_n, \ldots$ are infinite sequences, then for every positive integer n,

$$\textbf{(1)}\ \sum_{k=1}^{n} (a_k + b_k) = \sum_{k=1}^{n} a_k + \sum_{k=1}^{n} b_k$$

$$\textbf{(2)}\ \sum_{k=1}^{n} (a_k - b_k) = \sum_{k=1}^{n} a_k - \sum_{k=1}^{n} b_k$$

$$\textbf{(3)}\ \sum_{k=1}^{n} ca_k = c\left(\sum_{k=1}^{n} a_k\right) \text{ for every real number } c$$

PROOFS To prove formula 1, we first write

$$\sum_{k=1}^{n} (a_k + b_k) = (a_1 + b_1) + (a_2 + b_2) + (a_3 + b_3) + \cdots + (a_n + b_n).$$

Using commutative and associative properties of real numbers many times, we may rearrange the terms on the right-hand side to produce

$$\begin{aligned} \sum_{k=1}^{n}(a_k + b_k) &= (a_1 + a_2 + a_3 + \cdots + a_n) + (b_1 + b_2 + b_3 + \cdots + b_n) \\ &= \sum_{k=1}^{n} a_k + \sum_{k=1}^{n} b_k. \end{aligned}$$

For a proof of formula 3, we have

$$\begin{aligned} \sum_{k=1}^{n}(ca_k) &= ca_1 + ca_2 + ca_3 + \cdots + ca_n \\ &= c(a_1 + a_2 + a_3 + \cdots + a_n) \\ &= c\left(\sum_{k=1}^{n} a_k\right). \end{aligned}$$

The proof of formula 2 is left as an exercise.

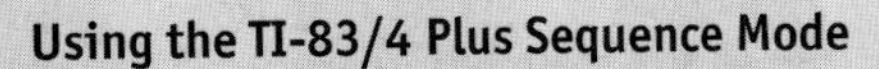

Using the TI-83/4 Plus Sequence Mode

Press MODE and use the cursor keys to highlight Seq and Dot. Turn off Stat Plot 1.

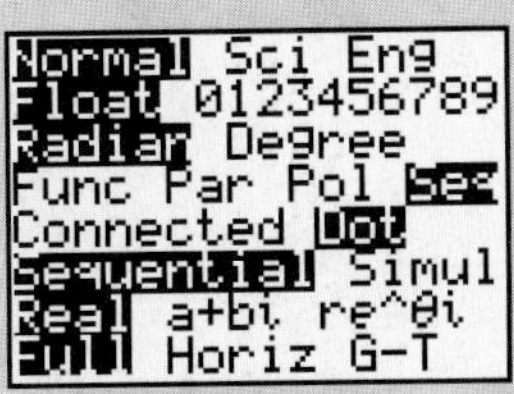

Listing and Graphing a Sequence

Enter the sequence in Example 1(a), $\left\{\dfrac{n}{n+1}\right\}$.

Y= X,T,θ,n ÷ (X,T,θ,n + 1)

Note: u(nMin) can be left blank.

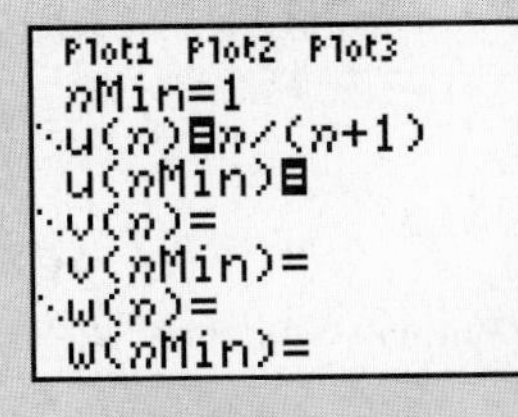

List the sequence.

2nd QUIT 2nd u (1 , 4) MATH 1 ENTER

List a specific term.

2nd u (3) MATH 1 ENTER

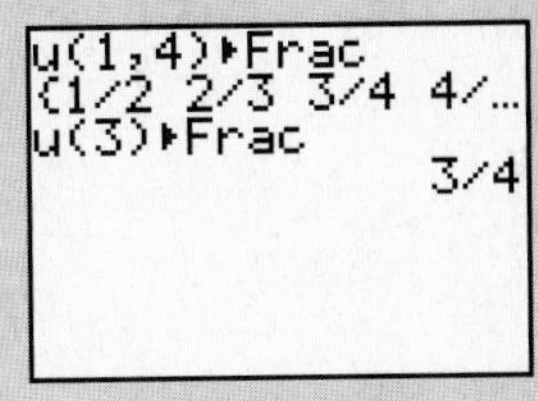

(continued)

Set the window variables to graph the first four terms of the sequence.

[WINDOW] 1 [▽] 4 [▽] 1 [▽] 1 [▽] −1 [▽] 5 [▽] 1 [▽] −.2 [▽] 1 [▽] .2

```
WINDOW
 nMin=1
 nMax=4
 PlotStart=1
 PlotStep=1
 Xmin=-1
 Xmax=5
 Xscl=1
 Ymin=-.2
 Ymax=1
 Yscl=.2
```

Graph the sequence by pressing [GRAPH]. Press [TRACE] and the left and right cursor keys to view the sequence values.

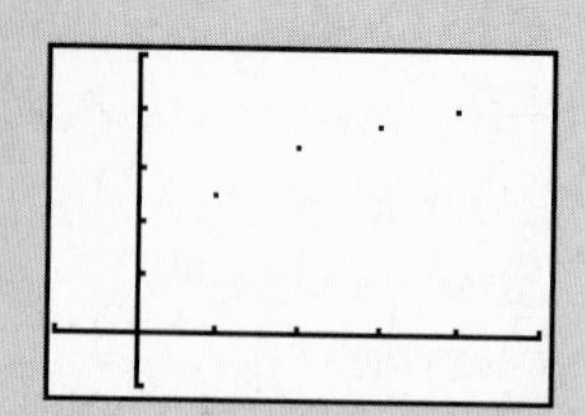

Generating a Recursively Defined Sequence

Recursively define the sequence in Example 3,

$$a_1 = 3, \quad a_{k+1} = 2a_k \quad \text{for} \quad k \geq 1.$$

[Y=] [CLEAR] 2 [×] [2nd] [u] [(] [X,T,θ,n] [−] 1 [)] [ENTER] 3 [ENTER]

```
Plot1 Plot2 Plot3
 nMin=1
 .u(n)■2*u(n-1)
  u(nMin)■{3}
 .v(n)=
  v(nMin)=
 .w(n)=
  w(nMin)=
```

List the first four terms of the sequence.

[2nd] [QUIT] [2nd] [u] [(] 1 [,] 4 [)] [ENTER]

```
u(1,4)
        {3 6 12 24}
```

Graphing a Sequence of Partial Sums

We can graph a sequence of partial sums by defining u to be the sequence of terms and v to be the sum of that sequence. We'll demonstrate with the sequence from Example 5—that is, $a_n = n$.

[Y=] [CLEAR] [X,T,θ,n] [▽] 1 [▽] [2nd] [LIST] [◁] [5] [2nd] [LIST] [▷] [5] [2nd] [u] [,] [X,T,θ,n] [,] 1 [,] [X,T,θ,n] [,] 1 [)] [)] [▽] 1 [▽]

```
Plot1 Plot2 Plot3
 nMin=1
 .u(n)■n
  u(nMin)■{1}
 .v(n)■sum(seq(u,
n,1,n,1))
  v(nMin)■{1}
 .w(n)=
```

Set the window variables to graph the first four terms of the sequences.

WINDOW 1 ▽ 4 ▽ 1 ▽ 1 ▽ −1 ▽ 5 ▽ 1 ▽ −1 ▽ 11 ▽ 1

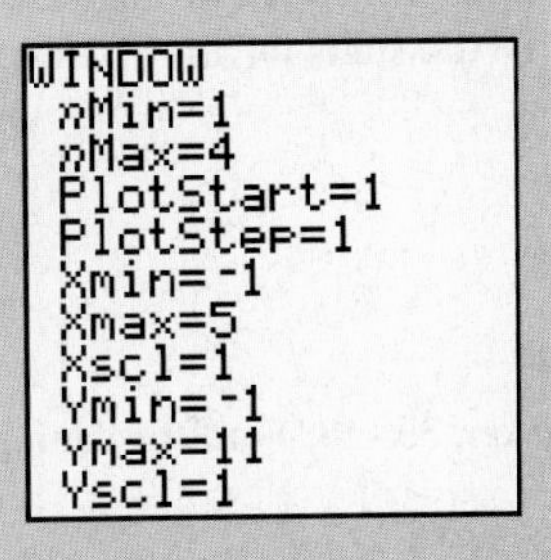

Graph the sequence and the sequence of partial sums by pressing GRAPH. Note that the first partial sum is equal to the first term of the sequence.

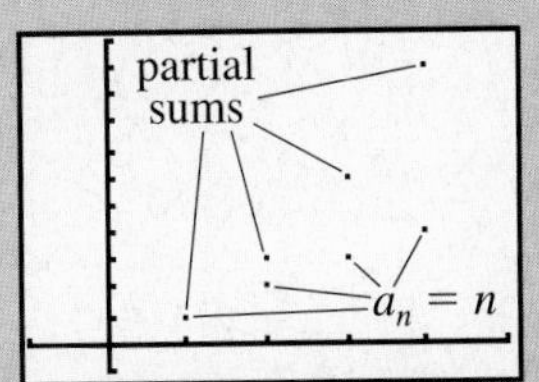

9.1 Exercises

Exer. 1–16: Find the first four terms and the eighth term of the sequence.

1 $\{12 - 3n\}$

2 $\left\{\dfrac{3}{5n - 2}\right\}$

3 $\left\{\dfrac{3n - 2}{n^2 + 1}\right\}$

4 $\left\{10 + \dfrac{1}{n}\right\}$

5 $\{9\}$

6 $\{\sqrt{2}\}$

7 $\{2 + (-0.1)^n\}$

8 $\{4 + (0.1)^n\}$

9 $\left\{(-1)^{n-1}\dfrac{n + 7}{2n}\right\}$

10 $\left\{(-1)^n\dfrac{6 - 2n}{\sqrt{n + 1}}\right\}$

▶ 11 $\{1 + (-1)^{n+1}\}$

12 $\{(-1)^{n+1} + (0.1)^{n-1}\}$

13 $\left\{\dfrac{2^n}{n^2 + 2}\right\}$

14 $\{(n - 1)(n - 2)(n - 3)\}$

15 a_n is the number of decimal places in $(0.1)^n$.

16 a_n is the number of positive integers less than n^3.

Exer. 17–20: Graph the sequence.

17 $\left\{\dfrac{1}{\sqrt{n}}\right\}$

18 $\left\{\dfrac{1}{n}\right\}$

19 $\{(-1)^{n+1}n^2\}$

20 $\{(-1)^n(2n + 1)\}$

Exer. 21–28: Find the first five terms of the recursively defined infinite sequence.

21 $a_1 = 2, \quad a_{k+1} = 3a_k - 5$

22 $a_1 = 5, \quad a_{k+1} = 7 - 2a_k$

23 $a_1 = -3, \quad a_{k+1} = a_k^2$

24 $a_1 = 128, \quad a_{k+1} = \frac{1}{4}a_k$

25 $a_1 = 5, \quad a_{k+1} = ka_k$

26 $a_1 = 3, \quad a_{k+1} = 1/a_k$

▶ 27 $a_1 = 2, \quad a_{k+1} = (a_k)^k$

28 $a_1 = 2, \quad a_{k+1} = (a_k)^{1/k}$

▶ Tutorial available at www.thomsonedu.com/login

Exer. 29–32: Find the first four terms of the sequence of partial sums for the given sequence.

29 $\{3 + \frac{1}{2}n\}$

30 $\{1/n^2\}$

31 $\{(-1)^n n^{-1/2}\}$

32 $\{(-1)^n(1/2)^n\}$

Exer. 33–48: Find the sum.

33 $\sum_{k=1}^{5} (2k - 7)$

34 $\sum_{k=1}^{6} (10 - 3k)$

▶ 35 $\sum_{k=1}^{4} (k^2 - 5)$

36 $\sum_{k=1}^{10} [1 + (-1)^k]$

37 $\sum_{k=0}^{5} k(k - 2)$

38 $\sum_{k=0}^{4} (k - 1)(k - 3)$

39 $\sum_{k=3}^{6} \frac{k - 5}{k - 1}$

40 $\sum_{k=1}^{6} \frac{3}{k + 1}$

41 $\sum_{k=1}^{5} (-3)^{k-1}$

42 $\sum_{k=0}^{4} 3(2^k)$

43 $\sum_{k=1}^{100} 100$

44 $\sum_{k=1}^{1000} 5$

45 $\sum_{k=253}^{571} \frac{1}{3}$

46 $\sum_{k=137}^{428} 2.1$

47 $\sum_{j=1}^{7} \frac{1}{2}k^2$

48 $\sum_{k=0}^{5} (3j + 2)$

49 Prove formula 2 of the theorem on sums.

50 Extend formula 1 of the theorem on sums to

$$\sum_{k=1}^{n} (a_k + b_k + c_k).$$

51 Consider the sequence defined recursively by $a_1 = 5$, $a_{k+1} = \sqrt{a_k}$ for $k \geq 1$. Describe what happens to the terms of the sequence as k increases.

52 Approximations to π may be obtained from the sequence

$$x_1 = 3, \quad x_{k+1} = x_k - \tan x_k.$$

Use the [TAN] key for tan.

(a) Find the first five terms of this sequence.

(b) What happens to the terms of the sequence when $x_1 = 6$?

53 **Bode's sequence** Bode's sequence, defined by

$$a_1 = 0.4, \quad a_k = 0.1(3 \cdot 2^{k-2} + 4) \quad \text{for} \quad k \geq 2,$$

can be used to approximate distances of planets from the sun. These distances are measured in astronomical units, with 1 AU = 93,000,000 mi. For example, the third term corresponds to Earth and the fifth term to the minor planet Ceres. Approximate the first five terms of the sequence.

54 **Growth of bacteria** The number of bacteria in a certain culture is initially 500, and the culture doubles in size every day.

(a) Find the number of bacteria present after one day, two days, and three days.

(b) Find a formula for the number of bacteria present after n days.

55 **The Fibonacci sequence** The Fibonacci sequence is defined recursively by

$$a_1 = 1, \quad a_2 = 1, \quad a_{k+1} = a_k + a_{k-1} \quad \text{for} \quad k \geq 2.$$

(a) Find the first ten terms of the sequence.

(b) The terms of the sequence $r_k = a_{k+1}/a_k$ give progressively better approximations to τ, the golden ratio. Approximate the first ten terms of this sequence.

56 **The Fibonacci sequence** The Fibonacci sequence can be defined by the formula

$$a_n = \frac{1}{\sqrt{5}}\left(\frac{1 + \sqrt{5}}{2}\right)^n - \frac{1}{\sqrt{5}}\left(\frac{1 - \sqrt{5}}{2}\right)^n.$$

Find the first eight terms, and show that they agree with those found using the definition in Exercise 55.

57 **Chlorine levels** Chlorine is often added to swimming pools to control microorganisms. If the level of chlorine rises above 3 ppm (parts per million), swimmers will experience burning eyes and skin discomfort. If the level drops below 1 ppm, there is a possibility that the water will turn green because of a large algae count. Chlorine must be added to pool water at regular intervals. If no chlorine is added to a pool during a 24-hour period, approximately 20% of the chlorine will dissipate into the atmosphere and 80% will remain in the water.

▶ **Tutorial available at www.thomsonedu.com/login**

(a) Determine a recursive sequence a_n that expresses the amount of chlorine present after n days if the pool has a_0 ppm of chlorine initially and no chlorine is added.

(b) If a pool has 7 ppm of chlorine initially, construct a table to determine the first day on which the chlorine level will drop below 3 ppm.

58 Chlorine levels Refer to Exercise 57. Suppose a pool has 2 ppm of chlorine initially, and 0.5 ppm of chlorine is added to the pool at the end of each day.

(a) Find a recursive sequence a_n that expresses the amount of chlorine present after n days.

(b) Determine the amount of chlorine in the pool after 15 days and after a long period of time.

(c) Estimate the amount of chlorine that needs to be added daily in order to stabilize the pool's chlorine level at 1.5 ppm.

59 Golf club costs A golf club company sells driver heads as follows:

Number of heads	1–4	5–9	10+
Cost per head	\$89.95	\$87.95	\$85.95

Find a piecewise-defined function C that specifies the total cost for n heads. Sketch a graph of C.

60 DVD player costs An electronics wholesaler sells DVD players at \$20 each for the first 4 units. All units after the first 4 sell for \$17 each. Find a piecewise-defined function C that specifies the total cost for n players. Sketch a graph of C.

Exer. 61–62: Some calculators use an algorithm similar to the following to approximate $\sqrt{N}$ for a positive real number N: Let $x_1 = N/2$ and find successive approximations $x_2, x_3, \ldots$ by using

$$x_2 = \frac{1}{2}\left(x_1 + \frac{N}{x_1}\right), \quad x_3 = \frac{1}{2}\left(x_2 + \frac{N}{x_2}\right), \quad \ldots$$

until the desired accuracy is obtained. Use this method to approximate the radical to six-decimal-place accuracy.

61 $\sqrt{5}$

62 $\sqrt{18}$

63 The equation $\frac{1}{3}\sqrt[3]{x} - x + 2 = 0$ has a root near 2. To approximate this root, rewrite the equation as $x = \frac{1}{3}\sqrt[3]{x} + 2$. Let $x_1 = 2$ and find successive approximations $x_2, x_3, \ldots$ by using the formulas

$$x_2 = \tfrac{1}{3}\sqrt[3]{x_1} + 2, \quad x_3 = \tfrac{1}{3}\sqrt[3]{x_2} + 2, \quad \ldots$$

until four-decimal-place accuracy is obtained.

64 The equation $2x + \dfrac{1}{x^4 + x + 2} = 0$ has a root near 0. Use a procedure similar to that in Exercise 63 to approximate this root to four-decimal-place accuracy.

Exer. 65–66: (a) Show that f takes on both positive and negative values on the interval [1, 2]. (b) Use the method of Exercise 63, with $x_1 = 1.5$, to approximate a zero of f to two-decimal-place accuracy.

65 $f(x) = x - 2 + \log x$

66 $f(x) = \log x - 10^{-x}$ (*Hint:* Solve for x in $\log x$.)

Exer. 67–70: For the given nth term $a_n = f(n)$ of a sequence, use the graph of $y = f(x)$ on the interval [1, 100] to verify that as n increases without bound, a_n approaches some real number c.

67 $a_n = \left(1 + \dfrac{1}{n} + \dfrac{1}{2n^2}\right)^n$

68 $a_n = n^{1/n}$

69 $a_n = \left(\dfrac{1}{n}\right)^{1/n}$

70 $a_n = (2.1^n + 1)^{1/n}$

Exer. 71–74: Graph the recursively defined sequence a_k in dot mode for $k = 1, 2, 3, \ldots, 20$ by plotting the value of k along the x-axis and the value of a_k along the y-axis. Trace the graph to determine the minimum k such that $a_k > 100$.

71 $a_1 = 0.25, \quad a_k = 1.7a_{k-1} + 0.5$

72 $a_1 = 6, \quad a_k = 1.05a_{k-1} + 4$

73 $a_1 = 7.25, \quad a_k = 0.1a_{k-1}^2 + 2$

74 $a_1 = 1, \quad a_k = 0.2a_{k-1}^2 + 1.5$

75 Insect population The sequence defined by

$$a_{k+1} = ca_k(1 - a_k)$$

is used in the study of insect population growth. The constant c is called the Malthusian factor. Suppose that $1000a_k$ equals the number of insects after k time intervals. If initially

$a_1 = 0.25$, describe the behavior of the insect population for each value of c.

(a) $c = 0.5$ (b) $c = 1.5$ (c) $c = 2.75$

76 **Insect population** Refer to Exercise 75. The Malthusian factor c affects the population a_k of insects dramatically, and c can be interpreted as the degree of fertility of the insects.

(a) Conjecture how c will affect the insect population if $0 < c < 1$.

(b) Test your conjecture using various values for c.

9.2 Arithmetic Sequences

In this section and the next we consider two special types of sequences: arithmetic and geometric. The first type may be defined as follows.

Definition of Arithmetic Sequence

A sequence $a_1, a_2, \ldots, a_n, \ldots$ is an **arithmetic sequence** if there is a real number d such that for every positive integer k,

$$a_{k+1} = a_k + d.$$

The number $d = a_{k+1} - a_k$ is called the **common difference** of the sequence.

Note that the common difference d is the difference of *any* two successive terms of an arithmetic sequence.

ILLUSTRATION **Arithmetic Sequence and Common Difference**

- $-3, 2, 7, 12, \ldots, 5n - 8, \ldots$ common difference $= 2 - (-3) = 5$
- $17, 10, 3, -4, \ldots, 24 - 7n, \ldots$ common difference $= 10 - 17 = -7$

EXAMPLE 1 **Showing that a sequence is arithmetic**

Show that the sequence

$$1, 4, 7, 10, \ldots, 3n - 2, \ldots$$

is arithmetic, and find the common difference.

▶ SOLUTION If $a_n = 3n - 2$, then for every positive integer k,

$$\begin{aligned} a_{k+1} - a_k &= [3(k + 1) - 2] - (3k - 2) \\ &= 3k + 3 - 2 - 3k + 2 = 3. \end{aligned}$$

Hence, the given sequence is arithmetic with common difference 3.

▶ Tutorial available at www.thomsonedu.com/login

Given an arithmetic sequence, we know that

$$a_{k+1} = a_k + d$$

for every positive integer k. This gives us a recursive formula for obtaining successive terms. Beginning with any real number a_1, we can obtain an arithmetic sequence with common difference d simply by adding d to a_1, then to $a_1 + d$, and so on, obtaining

$$a_1, \quad a_1 + d, \quad a_1 + 2d, \quad a_1 + 3d, \quad a_1 + 4d, \quad \ldots .$$

The nth term a_n of this sequence is given by the next formula.

The nth Term of an Arithmetic Sequence	$a_n = a_1 + (n - 1)d$

EXAMPLE 2 Finding a specific term of an arithmetic sequence

The first three terms of an arithmetic sequence are 20, 16.5, and 13. Find the fifteenth term.

▶ SOLUTION The common difference is

$$a_2 - a_1 = 16.5 - 20 = -3.5.$$

Substituting $n = 15$, $a_1 = 20$, and $d = -3.5$ in the formula for the nth term of an arithmetic sequence, $a_n = a_1 + (n - 1)d$, gives us

$$a_{15} = 20 + (15 - 1)(-3.5) = 20 - 49 = -29.$$

EXAMPLE 3 Finding a specific term of an arithmetic sequence

If the fourth term of an arithmetic sequence is 5 and the ninth term is 20, find the sixth term.

▶ SOLUTION We are given $a_4 = 5$ and $a_9 = 20$ and wish to find a_6. The following are equivalent systems of equations in the variables a_1 and d:

$$\begin{cases} a_4 = a_1 + (4 - 1)d & \text{let } n = 4 \text{ in } a_n = a_1 + (n - 1)d \\ a_9 = a_1 + (9 - 1)d & \text{let } n = 9 \text{ in } a_n = a_1 + (n - 1)d \end{cases}$$

$$\begin{cases} 5 = a_1 + 3d & a_4 = 5 \\ 20 = a_1 + 8d & a_9 = 20 \end{cases}$$

Alternatively, if we use the relationship

$$a_9 = a_4 + 5d,$$

we can obtain $d = 3$. Then using

$$a_6 = a_4 + 2d,$$

we get $a_6 = 11$ without ever finding a_1.

Subtracting the first equation of the system from the second equation gives us $15 = 5d$, or $d = 3$. Substituting 3 for d in the first equation, $5 = a_1 + 3d$, yields $a_1 = 5 - 3d = 5 - 3(3) = -4$. Hence, to find the sixth term we have

$$\begin{aligned} a_6 &= a_1 + (6 - 1)d & &\text{let } n = 6 \text{ in } a_n = a_1 + (n - 1)d \\ &= (-4) + (5)(3) = 11. & &a_1 = -4 \text{ and } d = 3 \end{aligned}$$

The next theorem contains a formula for the nth partial sum S_n of an arithmetic sequence.

▶ **Tutorial available at www.thomsonedu.com/login**

Theorem: Formulas for S_n

If $a_1, a_2, a_3, \ldots, a_n, \ldots$ is an arithmetic sequence with common difference d, then the nth partial sum S_n (that is, the sum of the first n terms) is given by either

$$S_n = \frac{n}{2}[2a_1 + (n-1)d] \quad \text{or} \quad S_n = \frac{n}{2}(a_1 + a_n).$$

PROOF We may write

$$\begin{aligned} S_n &= a_1 + a_2 + a_3 + \cdots + a_n \\ &= a_1 + (a_1 + d) + (a_1 + 2d) + \cdots + [a_1 + (n-1)d]. \end{aligned}$$

Employing the commutative and associative properties of real numbers many times, we obtain

$$S_n = (a_1 + a_1 + a_1 + \cdots + a_1) + [d + 2d + \cdots + (n-1)d],$$

with a_1 appearing n times within the first pair of parentheses. Thus,

$$S_n = na_1 + d[1 + 2 + \cdots + (n-1)].$$

The expression within brackets is the sum of the first $n - 1$ positive integers. Using the formula for the sum of the first n positive integers, $S_n = n(n+1)/2$, from Example 5 of Section 9.1, but with $n - 1$ in place of n and n in place of $n + 1$, we have

$$1 + 2 + \cdots + (n-1) = \frac{(n-1)n}{2}.$$

Substituting in the last equation for S_n and factoring out $n/2$ gives us

$$S_n = na_1 + d\frac{(n-1)n}{2} = \frac{n}{2}[2a_1 + (n-1)d].$$

Since $a_n = a_1 + (n-1)d$, the last equation is equivalent to

$$S_n = \frac{n}{2}(a_1 + a_n).$$

EXAMPLE 4 Finding a sum of even integers

Find the sum of all the even integers from 2 through 100.

SOLUTION This problem is equivalent to finding the sum of the first 50 terms of the arithmetic sequence

$$2, 4, 6, \ldots, 2n, \ldots.$$

Substituting $n = 50$, $a_1 = 2$, and $a_{50} = 100$ in $S_n = (n/2)(a_1 + a_n)$ produces

$$S_{50} = \tfrac{50}{2}(2 + 100) = 2550.$$

Tutorial available at www.thomsonedu.com/login

Figure 1

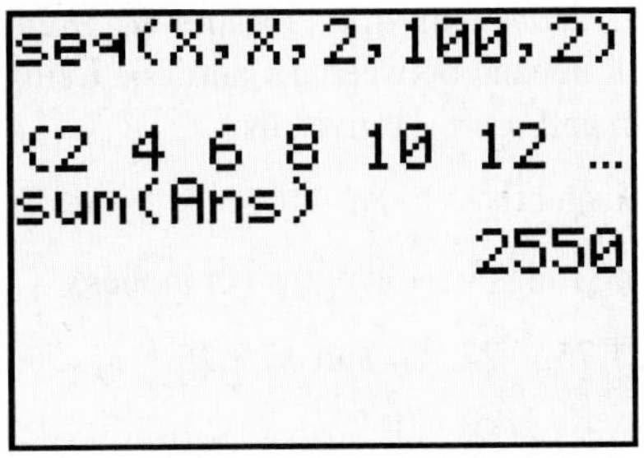

```
seq(X,X,2,100,2)
{2 4 6 8 10 12 ...
sum(Ans)
                2550
```

Alternatively, we may use $S_n = \frac{n}{2}[2a_1 + (n - 1)d]$ with $d = 2$:

$$S_{50} = \tfrac{50}{2}[2(2) + (50 - 1)2] = 25[4 + 98] = 2550$$

(See Figure 1 for calculator support of this result.)

The **arithmetic mean** of two numbers a and b is defined as $(a + b)/2$. This is the **average** of a and b. Note that the three numbers

$$a, \quad \frac{a+b}{2}, \quad \text{and} \quad b$$

constitute a (finite) arithmetic sequence with a common difference of $d = \frac{1}{2}(b - a)$. This concept may be generalized as follows: If $c_1, c_2, \ldots, c_k$ are real numbers such that

$$a, c_1, c_2, \ldots, c_k, b$$

is a (finite) arithmetic sequence, then $c_1, c_2, \ldots, c_k$ are ***k* arithmetic means** between the numbers a and b. The process of determining these numbers is referred to as *inserting k arithmetic means between a and b.*

EXAMPLE 5 Inserting arithmetic means

Insert three arithmetic means between 2 and 9.

SOLUTION We wish to find three real numbers c_1, c_2, and c_3 such that the following is a (finite) arithmetic sequence:

$$2, c_1, c_2, c_3, 9$$

We may regard this sequence as an arithmetic sequence with first term $a_1 = 2$ and fifth term $a_5 = 9$. To find the common difference d, we may proceed as follows:

$$\begin{aligned} a_5 &= a_1 + (5 - 1)d && \text{let } n = 5 \text{ in } a_n = a_1 + (n - 1)d \\ 9 &= 2 + 4d && a_5 = 9 \text{ and } a_1 = 2 \\ d &= \tfrac{7}{4} && \text{solve for } d \end{aligned}$$

Thus, the arithmetic means are

$$\begin{aligned} c_1 &= a_1 + d = 2 + \tfrac{7}{4} = \tfrac{15}{4} \\ c_2 &= c_1 + d = \tfrac{15}{4} + \tfrac{7}{4} = \tfrac{22}{4} = \tfrac{11}{2} \\ c_3 &= c_2 + d = \tfrac{22}{4} + \tfrac{7}{4} = \tfrac{29}{4}. \end{aligned}$$

EXAMPLE 6 An application of an arithmetic sequence

A carpenter wishes to construct a ladder with nine rungs whose lengths decrease uniformly from 24 inches at the base to 18 inches at the top. Determine the lengths of the seven intermediate rungs.

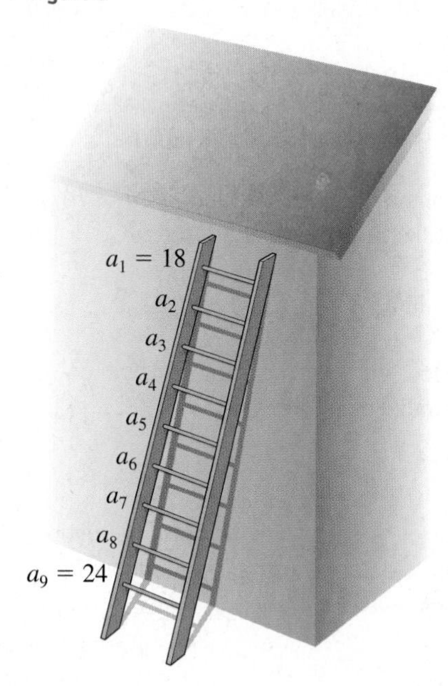

Figure 2

▶ **SOLUTION** The ladder is sketched in Figure 2. The lengths of the rungs are to form an arithmetic sequence $a_1, a_2, \ldots, a_9$ with $a_1 = 18$ and $a_9 = 24$. Hence, we need to insert seven arithmetic means between 18 and 24. Using $a_n = a_1 + (n - 1)d$ with $n = 9$, $a_1 = 18$, and $a_9 = 24$ gives us

$$24 = 18 + 8d \qquad \text{or, equivalently,} \qquad 8d = 6.$$

Hence, $d = \frac{6}{8} = 0.75$, and the intermediate rungs have lengths (in inches)

$$18.75, \quad 19.5, \quad 20.25, \quad 21, \quad 21.75, \quad 22.5, \quad \text{and} \quad 23.25.$$

It is sometimes desirable to express a sum in terms of summation notation, as illustrated in the next example.

EXAMPLE 7 Expressing a sum in summation notation

Express in terms of summation notation:

$$\frac{1}{4} + \frac{2}{9} + \frac{3}{14} + \frac{4}{19} + \frac{5}{24} + \frac{6}{29}$$

▶ **SOLUTION** The six terms of the sum do not form an arithmetic sequence; however, the numerators and denominators of the fractions, *considered separately,* are each an arithmetic sequence. Specifically, we have the following:

Numerators:	1, 2, 3, 4, 5, 6	common difference 1
Denominators:	4, 9, 14, 19, 24, 29	common difference 5

Using the formula for the nth term of an arithmetic sequence twice, we obtain the following nth term for each sequence:

$$a_n = a_1 + (n - 1)d = 1 + (n - 1)1 = n$$
$$a_n = a_1 + (n - 1)d = 4 + (n - 1)5 = 5n - 1$$

Hence, the nth term of the given sum is $n/(5n - 1)$, and we may write

$$\frac{1}{4} + \frac{2}{9} + \frac{3}{14} + \frac{4}{19} + \frac{5}{24} + \frac{6}{29} = \sum_{n=1}^{6} \frac{n}{5n - 1}.$$

9.2 *Exercises*

Exer. 1–2: Show that the given sequence is arithmetic, and find the common difference.

1 $-6, -2, 2, \ldots, 4n - 10, \ldots$

2 $53, 48, 43, \ldots, 58 - 5n, \ldots$

Exer. 3–10: Find the nth term, the fifth term, and the tenth term of the arithmetic sequence.

3 $2, 6, 10, 14, \ldots$

4 $16, 13, 10, 7, \ldots$

5 $3, 2.7, 2.4, 2.1, \ldots$

6 $-6, -4.5, -3, -1.5, \ldots$

7 $-7, -3.9, -0.8, 2.3, \ldots$

8 $x - 8, x - 3, x + 2, x + 7, \ldots$

9 $\ln 3, \ln 9, \ln 27, \ln 81, \ldots$

10 $\log 1000, \log 100, \log 10, \log 1, \ldots$

▶ **Tutorial available at www.thomsonedu.com/login**

Exer. 11–12: Find the common difference for the arithmetic sequence with the specified terms.

▶ 11 $a_2 = 21, a_6 = -11$

12 $a_4 = 14, a_{11} = 35$

Exer. 13–18: Find the specified term of the arithmetic sequence that has the two given terms.

13 a_{12}; $a_1 = 9.1$, $a_2 = 7.5$

14 a_{11}; $a_1 = 2 + \sqrt{2}$, $a_2 = 3$

15 a_1; $a_6 = 2.7$, $a_7 = 5.2$

16 a_1; $a_8 = 47$, $a_9 = 53$

▶ 17 a_{15}; $a_3 = 7$, $a_{20} = 43$

18 a_{10}; $a_2 = 1$, $a_{18} = 49$

Exer. 19–22: Find the sum S_n of the arithmetic sequence that satisfies the stated conditions.

▶ 19 $a_1 = 40$, $d = -3$, $n = 30$

20 $a_1 = 5$, $d = 0.1$, $n = 40$

21 $a_1 = -9$, $a_{10} = 15$, $n = 10$

22 $a_7 = \frac{7}{3}$, $d = -\frac{2}{3}$, $n = 15$

Exer. 23–28: Find the sum.

23 $\sum_{k=1}^{20} (3k - 5)$

24 $\sum_{k=1}^{12} (7 - 4k)$

25 $\sum_{k=1}^{18} \left(\frac{1}{2}k + 7\right)$

26 $\sum_{k=1}^{10} \left(\frac{1}{4}k + 3\right)$

27 $\sum_{k=126}^{592} (5k + 2j)$

28 $\sum_{k=88}^{371} (3j - 2k)$

Exer. 29–34: Express the sum in terms of summation notation. (Answers are not unique.)

29 $4 + 11 + 18 + 25 + 32$

30 $3 + 8 + 13 + 18 + 23$

31 $4 + 11 + 18 + \cdots + 466$

32 $3 + 8 + 13 + \cdots + 463$

33 $\frac{3}{7} + \frac{6}{11} + \frac{9}{15} + \frac{12}{19} + \frac{15}{23} + \frac{18}{27}$

34 $\frac{5}{13} + \frac{10}{11} + \frac{15}{9} + \frac{20}{7}$

Exer. 35–36: Express the sum in terms of summation notation and find the sum.

35 $8 + 19 + 30 + \cdots + 16{,}805$

36 $2 + 11 + 20 + \cdots + 16{,}058$

Exer. 37–40: Find the number of terms in the arithmetic sequence with the given conditions.

37 $a_1 = -2$, $d = \frac{1}{4}$, $S = 21$

38 $a_1 = -1$, $d = \frac{1}{5}$, $S = 21$

39 $a_1 = -\frac{29}{6}$, $d = \frac{1}{3}$, $S = -36$

40 $a_6 = -3$, $d = 0.2$, $S = -33$

41 Insert five arithmetic means between 2 and 10.

42 Insert three arithmetic means between 3 and -5.

43 (a) Find the number of integers between 32 and 395 that are divisible by 6.

(b) Find their sum.

44 (a) Find the number of negative integers greater than -500 that are divisible by 33.

(b) Find their sum.

45 **Log pile** A pile of logs has 24 logs in the bottom layer, 23 in the second layer, 22 in the third, and so on. The top layer contains 10 logs. Find the total number of logs in the pile.

46 **Stadium seating** The first ten rows of seating in a certain section of a stadium have 30 seats, 32 seats, 34 seats, and so on. The eleventh through the twentieth rows each contain 50 seats. Find the total number of seats in the section.

47 **Constructing a grain bin** A grain bin is to be constructed in the shape of a frustum of a cone (see the figure). The bin is to be 10 feet tall with 11 metal rings positioned uniformly around it, from the 4-foot opening at the bottom to the

24-foot opening at the top. Find the total length of metal needed to make the rings.

Exercise 47

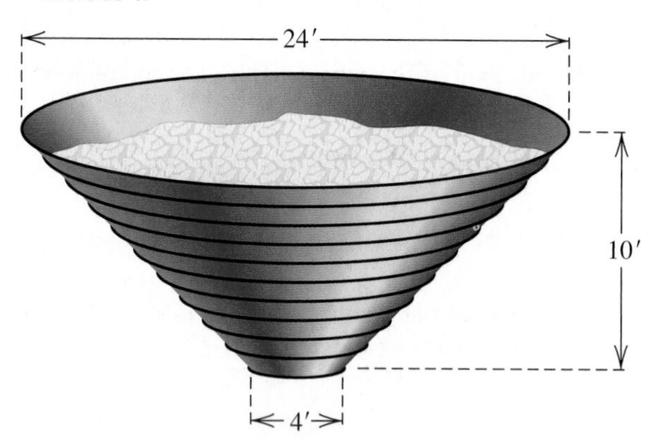

48 Coasting downhill A bicycle rider coasts downhill, traveling 4 feet the first second. In each succeeding second, the rider travels 5 feet farther than in the preceding second. If the rider reaches the bottom of the hill in 11 seconds, find the total distance traveled.

49 Prize money A contest will have five cash prizes totaling \$5000, and there will be a \$100 difference between successive prizes. Find the first prize.

50 Sales bonuses A company is to distribute \$46,000 in bonuses to its top ten salespeople. The tenth salesperson on the list will receive \$1000, and the difference in bonus money between successively ranked salespeople is to be constant. Find the bonus for each salesperson.

51 Distance an object falls Assuming air resistance is negligible, a small object that is dropped from a hot air balloon falls 16 feet during the first second, 48 feet during the second second, 80 feet during the third second, 112 feet during the fourth second, and so on. Find an expression for the distance the object falls in n seconds.

52 If f is a linear function, show that the sequence with nth term $a_n = f(n)$ is an arithmetic sequence.

53 Genetic sequence The sequence defined recursively by $x_{k+1} = x_k/(1 + x_k)$ occurs in genetics in the study of the elimination of a deficient gene from a population. Show that the sequence whose nth term is $1/x_n$ is arithmetic.

54 Dimensions of a maze Find the total length of the red-line curve in the figure if the width of the maze formed by the curve is 16 inches and all halls in the maze have width 1 inch. What is the length if the width of the maze is 32 inches?

Exercise 54

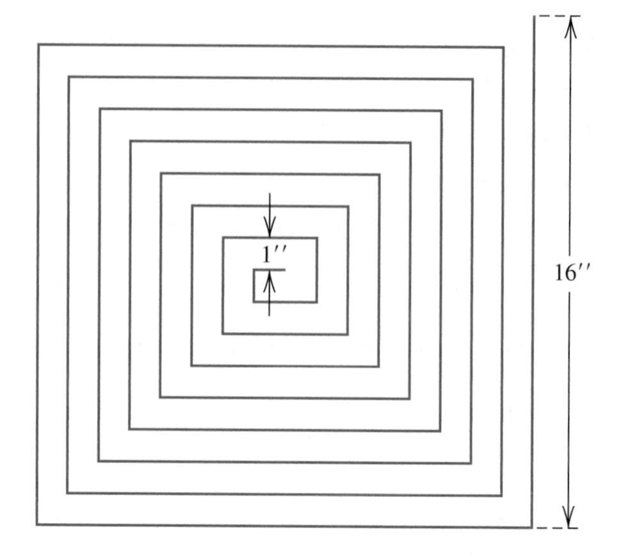

Exer. 55–56: Depreciation methods are sometimes used by businesses and individuals to estimate the value of an asset over a life span of n years. In the sum-of-year's-digits method, for each year $k = 1, 2, 3, \ldots, n$, the value of an asset is decreased by the fraction $A_k = \dfrac{n - k + 1}{T_n}$ of its initial cost, where $T_n = 1 + 2 + 3 + \cdots + n$.

55 (a) If $n = 8$, find $A_1, A_2, A_3, \ldots, A_8$.

(b) Show that the sequence in (a) is arithmetic, and find S_8.

(c) If the initial value of an asset is \$1000, how much has been depreciated after 4 years?

56 (a) If n is any positive integer, find $A_1, A_2, A_3, \ldots, A_n$.

(b) Show that the sequence in (a) is arithmetic, and find S_n.

9.3 Geometric Sequences

The second special type of sequence that we will discuss—the geometric sequence—occurs frequently in applications.

Definition of Geometric Sequence

A sequence $a_1, a_2, \ldots, a_n, \ldots$ is a **geometric sequence** if $a_1 \neq 0$ and if there is a real number $r \neq 0$ such that for every positive integer k,

$$a_{k+1} = a_k r.$$

The number $r = \dfrac{a_{k+1}}{a_k}$ is called the **common ratio** of the sequence.

Note that the common ratio $r = a_{k+1}/a_k$ is the ratio of *any* two successive terms of a geometric sequence.

ILLUSTRATION **Geometric Sequence and Common Ratio**

- $6, -12, 24, -48, \ldots, (-2)^{n-1}(6), \ldots$ common ratio $= \frac{-12}{6} = -2$
- $9, 3, 1, \frac{1}{3}, \ldots, (3)^{3-n}, \ldots$ common ratio $= \frac{3}{9} = \frac{1}{3}$

The formula $a_{k+1} = a_k r$ provides a recursive method for obtaining terms of a geometric sequence. Beginning with any nonzero real number a_1, we multiply by the number r successively, obtaining

$$a_1, \quad a_1 r, \quad a_1 r^2, \quad a_1 r^3, \quad \ldots.$$

The nth term a_n of this sequence is given by the following formula.

Formula for the nth Term of a Geometric Sequence

$$a_n = a_1 r^{n-1}$$

EXAMPLE 1 **Finding terms of a geometric sequence**

A geometric sequence has first term 3 and common ratio $-\frac{1}{2}$. Find the first five terms and the tenth term.

SOLUTION If we multiply $a_1 = 3$ successively by $r = -\frac{1}{2}$, then the first five terms are

$$3, \quad -\tfrac{3}{2}, \quad \tfrac{3}{4}, \quad -\tfrac{3}{8}, \quad \tfrac{3}{16}.$$

Using the formula $a_n = a_1 r^{n-1}$ with $n = 10$, we find that the tenth term is

$$a_{10} = a_1 r^9 = 3\left(-\tfrac{1}{2}\right)^9 = -\tfrac{3}{512}.$$

Tutorial available at www.thomsonedu.com/login

EXAMPLE 2 Finding a specific term of a geometric sequence

The third term of a geometric sequence is 5, and the sixth term is -40. Find the eighth term.

SOLUTION We are given $a_3 = 5$ and $a_6 = -40$ and wish to find a_8. The following are equivalent systems of equations in the variables a_1 and r:

$$\begin{cases} a_3 = a_1 r^{3-1} & \text{let } n = 3 \text{ in } a_n = a_1 r^{n-1} \\ a_6 = a_1 r^{6-1} & \text{let } n = 6 \text{ in } a_n = a_1 r^{n-1} \end{cases}$$

$$\begin{cases} 5 = a_1 r^2 & a_3 = 5 \\ -40 = a_1 r^5 & a_6 = -40 \end{cases}$$

Solving the first equation of the system for a_1 gives us $a_1 = 5/r^2$. Substituting this expression in the second equation yields

$$-40 = \frac{5}{r^2} \cdot r^5.$$

Alternatively, if we use the relationship

$$a_6 = a_3 r^3,$$

we can obtain $r = -2$. Then using

$$a_8 = a_6 r^2,$$

we get $a_8 = -160$ without ever finding a_1.

Simplifying, we get $r^3 = -8$, and hence $r = -2$. We next use $a_1 = 5/r^2$ to obtain

$$a_1 = \frac{5}{(-2)^2} = \frac{5}{4}.$$

Finally, using $a_n = a_1 r^{n-1}$ with $n = 8$ gives us

$$a_8 = a_1 r^7 = \left(\tfrac{5}{4}\right)(-2)^7 = -160.$$

The next theorem contains a formula for the nth partial sum S_n of a geometric sequence.

Theorem: Formula for S_n

The nth partial sum S_n of a geometric sequence with first term a_1 and common ratio $r \neq 1$ is

$$S_n = a_1 \frac{1 - r^n}{1 - r}.$$

PROOF By definition, the nth partial sum S_n of a geometric sequence is

$$S_n = a_1 + a_1 r + a_1 r^2 + \cdots + a_1 r^{n-2} + a_1 r^{n-1}. \quad \textbf{(1)}$$

Multiplying both sides of (1) by r, we obtain

$$rS_n = a_1 r + a_1 r^2 + a_1 r^3 + \cdots + a_1 r^{n-1} + a_1 r^n. \quad \textbf{(2)}$$

Tutorial available at www.thomsonedu.com/login

If we subtract equation (2) from equation (1), all but two terms on the right-hand side cancel and we obtain the following:

$$S_n - rS_n = a_1 - a_1r^n \quad \text{subtract (2) from (1)}$$
$$S_n(1 - r) = a_1(1 - r^n) \quad \text{factor both sides}$$
$$S_n = a_1\frac{1 - r^n}{1 - r} \quad \text{divide by } (1 - r)$$

Figure 1

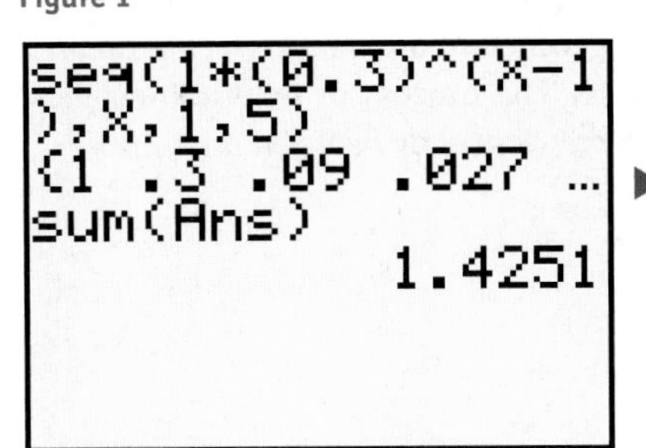

EXAMPLE 3 Finding a sum of terms of a geometric sequence

If the sequence 1, 0.3, 0.09, 0.027, ... is a geometric sequence, find the sum of the first five terms.

▶ SOLUTION If we let $a_1 = 1$, $r = 0.3$, and $n = 5$ in the formula for S_n stated in the preceding theorem, we obtain

$$S_5 = a_1\frac{1 - r^5}{1 - r} = (1)\frac{1 - (0.3)^5}{1 - 0.3} = 1.4251.$$

(See Figure 1 for calculator support of this result.)

EXAMPLE 4 The rapid growth of terms of a geometric sequence

A man wishes to save money by setting aside 1 cent the first day, 2 cents the second day, 4 cents the third day, and so on.

(a) If he continues to double the amount set aside each day, how much must he set aside on the fifteenth day?

(b) Assuming he does not run out of money, what is the total amount saved at the end of the 30 days?

▶ SOLUTION

(a) The amount (in cents) set aside on successive days forms a geometric sequence

$$1, 2, 4, 8, \ldots,$$

with first term 1 and common ratio 2. We find the amount to be set aside on the fifteenth day by using $a_n = a_1r^{n-1}$ with $a_1 = 1$ and $n = 15$:

$$a_{15} = a_1r^{14} = 1 \cdot 2^{14} = 16{,}384$$

Thus, \$163.84 should be set aside on the fifteenth day.

(b) To find the total amount saved after 30 days, we use the formula for S_n with $n = 30$, obtaining (in cents)

$$S_{30} = (1)\frac{1 - 2^{30}}{1 - 2} = 1{,}073{,}741{,}823.$$

Thus, the total amount saved is \$10,737,418.23.

▶ **Tutorial available at www.thomsonedu.com/login**

The terminology used with geometric sequences is analogous to that used with arithmetic sequences. If a and b are positive real numbers, then a positive number c is called the **geometric mean** of a and b if a, c, b is a geometric sequence. If the common ratio is r, then

$$r = \frac{c}{a} = \frac{b}{c}, \qquad \text{or} \qquad c^2 = ab.$$

Taking the square root of both sides of the last equation, we see that *the geometric mean of the positive numbers a and b is* $\sqrt{ab}$. As a generalization, k positive real numbers $c_1, c_2, \ldots, c_k$ are ***k* geometric means** between a and b if $a, c_1, c_2, \ldots, c_k, b$ is a geometric sequence. The process of determining these numbers is referred to as *inserting k geometric means between a and b.*

ILLUSTRATION Geometric Means

Numbers	Geometric mean
■ 20, 45	$\sqrt{20 \cdot 45} = \sqrt{900} = 30$
■ 3, 4	$\sqrt{3 \cdot 4} = \sqrt{12} \approx 3.46$

Given the geometric series with first term a_1 and common ratio $r \neq 1$, we may write the formula for S_n of the preceding theorem in the form

$$S_n = \frac{a_1}{1 - r} - \frac{a_1}{1 - r} r^n.$$

If $|r| < 1$, then r^n *approaches* 0 *as n increases* without bound. Thus, S_n approaches $a_1/(1 - r)$ as n increases without bound. Using the notation we developed for rational functions in Section 3.5, we have

$$S_n \to \frac{a_1}{1 - r} \quad \text{as} \quad n \to \infty.$$

The number $a_1/(1 - r)$ is called the *sum S* of the **infinite geometric series**

$$a_1 + a_1r + a_1r^2 + \cdots + a_1r^{n-1} + \cdots.$$

This gives us the following result.

Theorem on the Sum of an Infinite Geometric Series

If $|r| < 1$, then the infinite geometric series

$$a_1 + a_1r + a_1r^2 + \cdots + a_1r^{n-1} + \cdots$$

has the sum

$$S = \frac{a_1}{1 - r}.$$

The preceding theorem implies that as we add more terms of the indicated infinite geometric series, the sum gets closer to $a_1/(1 - r)$. The next example illustrates how the theorem can be used to show that every real number represented by a repeating decimal is rational.

EXAMPLE 5 **Expressing an infinite repeating decimal as a rational number**

Find a rational number that corresponds to $5.4\overline{27}$.

▶ **SOLUTION** We can write the decimal expression 5.4272727... as

$$5.4 + 0.027 + 0.00027 + 0.0000027 + \cdots.$$

Beginning with the second term, 0.027, the above sum has the form given in the theorem on the sum of an infinite geometric series, with $a_1 = 0.027$ and $r = 0.01$. Hence, the sum S of this infinite geometric series is

$$S = \frac{a_1}{1 - r} = \frac{0.027}{1 - 0.01} = \frac{0.027}{0.990} = \frac{27}{990} = \frac{3}{110}.$$

Thus, the desired number is

$$5.4 + \tfrac{3}{110} = \tfrac{594}{110} + \tfrac{3}{110} = \tfrac{597}{110}.$$

A check by division shows that $\frac{597}{110}$ corresponds to $5.4\overline{27}$.

In general, given any infinite sequence, $a_1, a_2, \ldots, a_n, \ldots$, the expression

$$a_1 + a_2 + \cdots + a_n + \cdots$$

is called an **infinite series** or simply a **series.** We denote this series by

$$\sum_{n=1}^{\infty} a_n.$$

Each number a_k is a **term** of the series, and a_n is the ***n*th term.** Since only *finite* sums may be added algebraically, it is necessary to define what is meant by an *infinite sum.* Consider the sequence of partial sums

$$S_1, S_2, \ldots, S_n, \ldots.$$

If there is a number S such that $S_n \to S$ as $n \to \infty$, then, as in our discussion of infinite geometric series, S is the **sum** of the infinite series and we write

$$S = a_1 + a_2 + \cdots + a_n + \cdots.$$

In the previous example we found that the infinite repeating decimal 5.4272727... corresponds to the rational number $\frac{597}{110}$. Since $\frac{597}{110}$ is the sum of an infinite series determined by the decimal, we may write

$$\tfrac{597}{110} = 5.4 + 0.027 + 0.00027 + 0.0000027 + \cdots.$$

▶ **Tutorial available at www.thomsonedu.com/login**

If the terms of an infinite sequence are alternately positive and negative, as in the expression

$$a_1 + (-a_2) + a_3 + (-a_4) + \cdots + [(-1)^{n+1}a_n] + \cdots$$

for positive real numbers a_k, then the expression is an **alternating infinite series** and we write it in the form

$$a_1 - a_2 + a_3 - a_4 + \cdots + (-1)^{n+1}a_n + \cdots.$$

The most common types of alternating infinite series are infinite geometric series in which the common ratio r is negative.

EXAMPLE 6 Finding the sum of an infinite geometric series

Find the sum S of the alternating infinite geometric series

$$\sum_{n=1}^{\infty} 3\left(-\tfrac{2}{3}\right)^{n-1} = 3 - 2 + \tfrac{4}{3} - \tfrac{8}{9} + \cdots + 3\left(-\tfrac{2}{3}\right)^{n-1} + \cdots.$$

▶ SOLUTION Using the formula for S in the theorem on the sum of an infinite geometric series, with $a_1 = 3$ and $r = -\frac{2}{3}$, we obtain

$$S = \frac{a_1}{1 - r} = \frac{3}{1 - \left(-\frac{2}{3}\right)} = \frac{3}{\frac{5}{3}} = \frac{9}{5}.$$

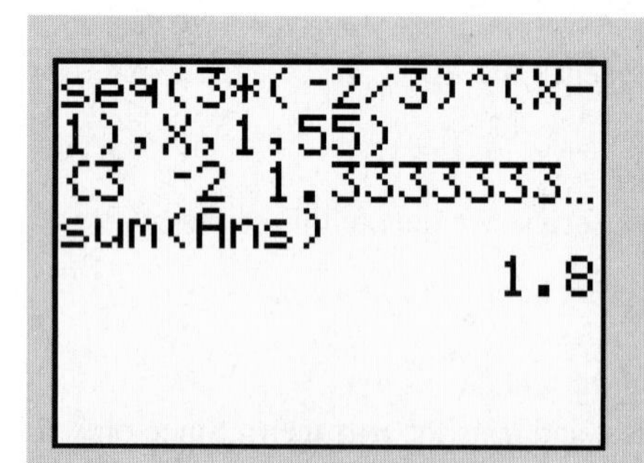

To check our result in Example 6, we can replace ∞ with a reasonably large number and find the sum of that geometric series. As shown in the figure, using 55 terms gives us 1.8, our previously obtained answer. **Note:** The calculator only lends support to our answer; the formula should be used to find sums of infinite geometric series.

EXAMPLE 7 An application of an infinite geometric series

A rubber ball is dropped from a height of 10 meters. Suppose it rebounds one-half the distance after each fall, as illustrated by the arrows in Figure 2. Find the total distance the ball travels.

Figure 2

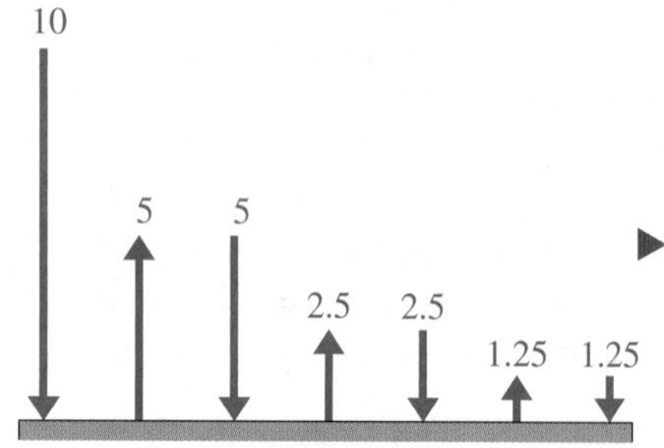

▶ SOLUTION The sum of the distances the ball travels downward and the sum of the distances it travels on the rebounds form two infinite geometric series:

Downward series: $10 + 5 + 2.5 + 1.25 + 0.625 + \cdots$

Upward series: $5 + 2.5 + 1.25 + 0.625 + \cdots$

▶ **Tutorial available at www.thomsonedu.com/login**

We assume that the total distance S the ball travels can be found by adding the sums of these infinite series. This gives us

$$S = 10 + 2[5 + 2.5 + 1.25 + 0.625 + \cdots]$$
$$= 10 + 2\left[5 + 5\left(\tfrac{1}{2}\right) + 5\left(\tfrac{1}{2}\right)^2 + 5\left(\tfrac{1}{2}\right)^3 + \cdots\right].$$

Using the formula $S = a_1/(1 - r)$ with $a_1 = 5$ and $r = \frac{1}{2}$, we obtain

$$S = 10 + 2\left(\frac{5}{1 - \frac{1}{2}}\right) = 10 + 2(10) = 30 \text{ m}.$$

A related problem: Does this ball ever come to rest? See Discussion Exercise 7 at the end of this chapter.

9.3 *Exercises*

Exer. 1–2: Show that the given sequence is geometric, and find the common ratio.

1 $5, -\frac{5}{4}, \frac{5}{16}, \ldots, 5\left(-\frac{1}{4}\right)^{n-1}, \ldots$

2 $\frac{1}{7}, \frac{3}{7}, \frac{9}{7}, \ldots, \frac{1}{7}(3)^{n-1}, \ldots$

Exer. 3–14: Find the nth term, the fifth term, and the eighth term of the geometric sequence.

3 $8, 4, 2, 1, \ldots$

4 $4, 1.2, 0.36, 0.108, \ldots$

5 $300, -30, 3, -0.3, \ldots$

6 $1, -\sqrt{3}, 3, -3\sqrt{3}, \ldots$

7 $5, 25, 125, 625, \ldots$

8 $2, 6, 18, 54, \ldots$

9 $4, -6, 9, -13.5, \ldots$

10 $162, -54, 18, -6, \ldots$

11 $1, -x^2, x^4, -x^6, \ldots$

12 $1, -\dfrac{x}{3}, \dfrac{x^2}{9}, -\dfrac{x^3}{27}, \ldots$

13 $2, 2^{x+1}, 2^{2x+1}, 2^{3x+1}, \ldots$

14 $10, 10^{2x-1}, 10^{4x-3}, 10^{6x-5}, \ldots$

Exer. 15–16: Find all possible values of r for a geometric sequence with the two given terms.

15 $a_4 = 3, a_6 = 9$

16 $a_3 = 4, a_7 = \frac{1}{4}$

17 Find the sixth term of the geometric sequence whose first two terms are 4 and 6.

18 Find the seventh term of the geometric sequence whose second and third terms are 2 and $-\sqrt{2}$.

19 Given a geometric sequence with $a_4 = 4$ and $a_7 = 12$, find r and a_{10}.

20 Given a geometric sequence with $a_2 = 3$ and $a_5 = -81$, find r and a_9.

Exer. 21–26: Find the sum.

21 $\sum_{k=1}^{10} 3^k$

22 $\sum_{k=1}^{9} \left(-\sqrt{5}\right)^k$

23 $\sum_{k=0}^{9} \left(-\frac{1}{2}\right)^{k+1}$

24 $\sum_{k=1}^{7} (3^{-k})$

25 $\sum_{k=16}^{26} (2^{k-14} + 5j)$

26 $\sum_{k=8}^{14} (3^{k-7} + 2j^2)$

Exer. 27–30: Express the sum in terms of summation notation. (Answers are not unique.)

27 $2 + 4 + 8 + 16 + 32 + 64 + 128$

28 $2 - 4 + 8 - 16 + 32 - 64$

29 $\frac{1}{4} - \frac{1}{12} + \frac{1}{36} - \frac{1}{108}$

30 $3 + \frac{3}{5} + \frac{3}{25} + \frac{3}{125} + \frac{3}{625}$

Exer. 31–38: Find the sum of the infinite geometric series if it exists.

31 $1 - \frac{1}{2} + \frac{1}{4} - \frac{1}{8} + \cdots$

32 $2 + \frac{2}{3} + \frac{2}{9} + \frac{2}{27} + \cdots$

33 $1.5 + 0.015 + 0.00015 + \cdots$

34 $1 - 0.1 + 0.01 - 0.001 + \cdots$

35 $\sqrt{2} - 2 + \sqrt{8} - 4 + \cdots$

36 $1 + \frac{3}{2} + \frac{9}{4} + \frac{27}{8} + \cdots$

▶ 37 $256 + 192 + 144 + 108 + \cdots$

38 $250 - 100 + 40 - 16 + \cdots$

Exer. 39–46: Find the rational number represented by the repeating decimal.

39 $0.\overline{23}$

40 $0.0\overline{71}$

41 $2.4\overline{17}$

42 $10.\overline{5}$

43 $5.\overline{146}$

44 $3.2\overline{394}$

45 $1.\overline{6124}$

46 $123.61\overline{83}$

47 Find the geometric mean of 12 and 48.

48 Find the geometric mean of 20 and 25.

49 Insert two geometric means between 4 and 500.

50 Insert three geometric means between 2 and 512.

51 **Using a vacuum pump** A vacuum pump removes one-half of the air in a container with each stroke. After 10 strokes, what percentage of the original amount of air remains in the container?

52 **Calculating depreciation** The yearly depreciation of a certain machine is 25% of its value at the beginning of the year. If the original cost of the machine is \$20,000, what is its value after 6 years?

▶ 53 **Growth of bacteria** A certain culture initially contains 10,000 bacteria and increases by 20% every hour.

(a) Find a formula for the number $N(t)$ of bacteria present after t hours.

(b) How many bacteria are in the culture at the end of 10 hours?

54 **Interest on savings** An amount of money P is deposited in a savings account that pays interest at a rate of r percent per year compounded quarterly; the principal and accumulated interest are left in the account. Find a formula for the total amount in the account after n years.

55 **Rebounding ball** A rubber ball is dropped from a height of 60 feet. If it rebounds approximately two-thirds the distance after each fall, use an infinite geometric series to approximate the total distance the ball travels.

56 **Motion of a pendulum** The bob of a pendulum swings through an arc 24 centimeters long on its first swing. If each successive swing is approximately five-sixths the length of the preceding swing, use an infinite geometric series to approximate the total distance the bob travels.

57 **Multiplier effect** A manufacturing company that has just located in a small community will pay two million dollars per year in salaries. It has been estimated that 60% of these salaries will be spent in the local area, and 60% of the money spent will again change hands within the community. This process, called the *multiplier effect,* will be repeated ad infinitum. Find the total amount of local spending that will be generated by company salaries.

58 **Pest eradication** In a pest eradication program, N sterilized male flies are released into the general population each day. It is estimated that 90% of these flies will survive a given day.

(a) Show that the number of sterilized flies in the population n days after the program has begun is

$$N + (0.9)N + (0.9)^2N + \cdots + (0.9)^{n-1}N.$$

(b) If the *long-range* goal of the program is to keep 20,000 sterilized males in the population, how many flies should be released each day?

59 **Drug dosage** A certain drug has a half-life of about 2 hours in the bloodstream. The drug is formulated to be administered in doses of D milligrams every 4 hours, but D is yet to be determined.

(a) Show that the number of milligrams of drug in the bloodstream after the nth dose has been administered is

$$D + \tfrac{1}{4}D + \cdots + \left(\tfrac{1}{4}\right)^{n-1}D$$

and that this sum is approximately $\frac{4}{3}D$ for large values of n.

(b) A level of more than 500 milligrams of the drug in the bloodstream is considered to be dangerous. Find the largest possible dose that can be given repeatedly over a long period of time.

60 **Genealogy** Shown in the figure is a family tree displaying the current generation (you) and 3 prior generations, with a total of 12 grandparents. If you were to trace your family history back 10 generations, how many grandparents would you find?

Exercise 60

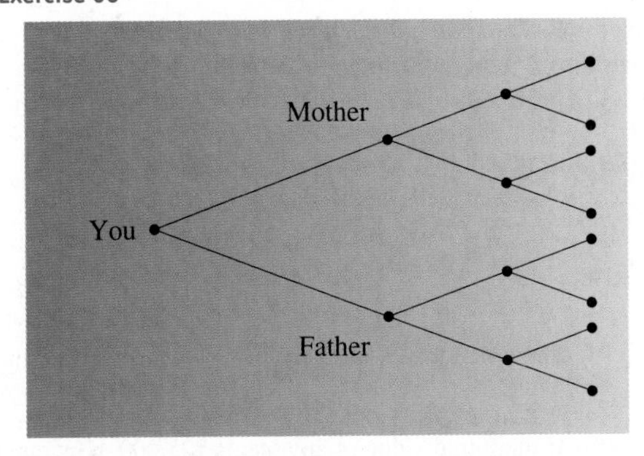

61 The first figure shows some terms of a sequence of squares $S_1, S_2, \ldots, S_k, \ldots$. Let a_k, A_k, and P_k denote the side, area, and perimeter, respectively, of the square S_k. The square S_{k+1} is constructed from S_k by connecting four points on S_k, with each point a distance of $\frac{1}{4}a_k$ from a vertex, as shown in the second figure.

(a) Find the relationship between a_{k+1} and a_k.

(b) Find a_n, A_n, and P_n.

(c) Calculate $\sum_{n=1}^{\infty} P_n$.

Exercise 61

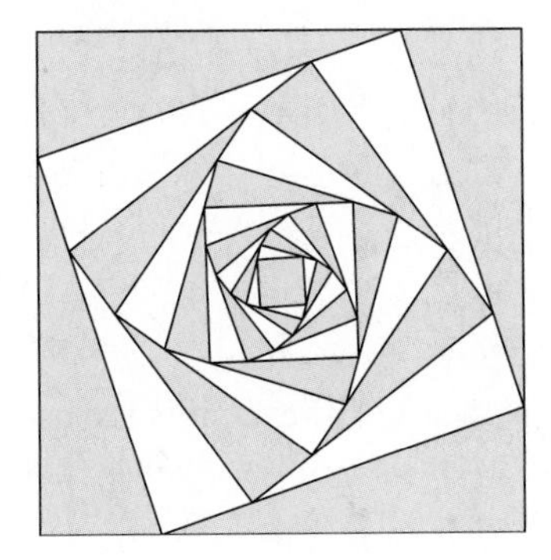

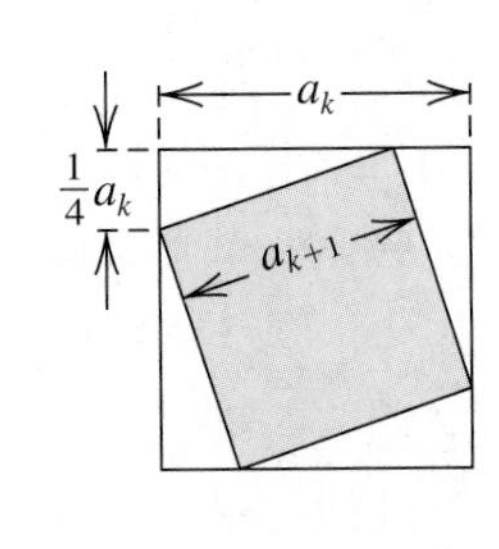

62 The figure shows several terms of a sequence consisting of alternating circles and squares. Each circle is inscribed in a square, and each square (excluding the largest) is inscribed in a circle. Let S_n denote the area of the nth square and C_n the area of the nth circle.

(a) Find the relationships between S_n and C_n and between C_n and S_{n+1}.

(b) What portion of the largest square is shaded in the figure?

Exercise 62

63 **Sierpinski sieve** The Sierpinski sieve, designed in 1915, is an example of a fractal. It can be constructed by starting with a solid black equilateral triangle. This triangle is divided into four congruent equilateral triangles, and the middle triangle is removed. On the next step, each of the three remaining equilateral triangles is divided into four congruent equilateral triangles, and the middle triangle in each of these triangles is removed, as shown in the first figure. On the third step, nine triangles are removed. If the process is continued indefinitely, the Sierpinski sieve results (see the second figure).

Exercise 63

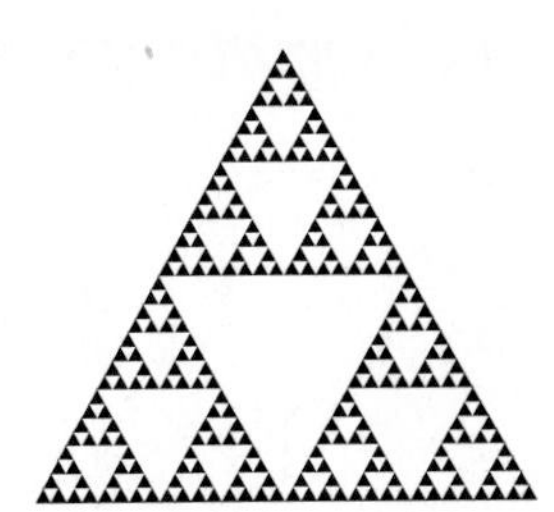

(a) Find a geometric sequence a_k that gives the number of triangles removed on the kth step.

(continued)

(b) Calculate the number of triangles removed on the fifteenth step.

(c) Suppose the initial triangle has an area of 1 unit. Find a geometric sequence b_k that gives the area removed on the kth step.

(d) Determine the area removed on the seventh step.

64 **Sierpinski sieve** Refer to Exercise 63.

(a) Write a geometric series that calculates the total number of triangles removed after n steps.

(b) Determine the total number of triangles removed after 12 steps.

(c) Write a geometric series that calculates the total area removed after n steps.

(d) Determine the total area removed after 12 steps.

65 **Annuity** If a deposit of \$100 is made on the first day of each month into an account that pays 6% interest per year compounded monthly, determine the amount in the account after 18 years.

66 **Annuity** Refer to Exercise 65. Show that if the monthly deposit is P dollars and the rate is r% per year compounded monthly, then the amount A in the account after n months is given by

$$A = P\left(\frac{12}{r} + 1\right)\left[\left(1 + \frac{r}{12}\right)^n - 1\right].$$

67 **Annuity** Use Exercise 66 to find A when $P = \$100$, $r = 8\%$, and $n = 60$.

68 **Annuity** Refer to Exercise 66. If $r = 10\%$, approximately how many years are required to accumulate \$100,000 if the monthly deposit P is

(a) \$100 (b) \$200

Exer. 69–70: The *double-declining balance method* is a method of depreciation in which, for each year $k = 1, 2, 3, \ldots, n$, the value of an asset is decreased by the fraction $A_k = \frac{2}{n}\left(1 - \frac{2}{n}\right)^{k-1}$ of its initial cost.

69 (a) If $n = 5$, find $A_1, A_2, \ldots, A_5$.

(b) Show that the sequence in (a) is geometric, and find S_5.

(c) If the initial value of an asset is \$25,000, how much of its value has been depreciated after 2 years?

70 (a) If n is any positive integer, find $A_1, A_2, \ldots, A_n$.

(b) Show that the sequence in (a) is geometric, and find S_n.

9.4 Mathematical Induction

If n is a positive integer and we let P_n denote the mathematical statement $(xy)^n = x^n y^n$, we obtain the following *infinite sequence of statements:*

$$\begin{aligned}
&\text{Statement } P_1\text{:} \quad (xy)^1 = x^1y^1\\
&\text{Statement } P_2\text{:} \quad (xy)^2 = x^2y^2\\
&\text{Statement } P_3\text{:} \quad (xy)^3 = x^3y^3\\
&\quad\vdots \qquad\qquad\qquad \vdots\\
&\text{Statement } P_n\text{:} \quad (xy)^n = x^ny^n\\
&\quad\vdots \qquad\qquad\qquad \vdots
\end{aligned}$$

It is easy to show that P_1, P_2, and P_3 are *true* statements. However, it is impossible to check the validity of P_n for *every* positive integer n. Showing that P_n is true for every n requires the following principle.

Principle of Mathematical Induction	If with each positive integer n there is associated a statement P_n, then all the statements P_n are true, provided the following two conditions are satisfied. **(1)** P_1 is true. **(2)** Whenever k is a positive integer such that P_k is true, then P_{k+1} is also true.

To help us understand this principle, we consider an infinite sequence of statements labeled

$$P_1, P_2, P_3, \ldots, P_n, \ldots$$

that satisfy conditions (1) and (2). By (1), statement P_1 is true. Since condition (2) holds, whenever a statement P_k is true the *next* statement P_{k+1} is also true. Hence, since P_1 is true, P_2 is also true, by (2). However, if P_2 is true, then, by (2), we see that the next statement P_3 is true. Once again, if P_3 is true, then, by (2), P_4 is also true. If we continue in this manner, we can argue that if n is any *particular* integer, then P_n is true, since we can use condition (2) one step at a time, eventually reaching P_n. Although this type of reasoning does not actually *prove* the principle of mathematical induction, it certainly makes it plausible. The principle is proved in advanced algebra using postulates for the positive integers.

When applying the principle of mathematical induction, we always follow two steps.

Steps in Applying the Principle of Mathematical Induction	*1* Show that P_1 is true. *2* *Assume* that P_k is true, and then prove that P_{k+1} is true.

Step 2 often causes confusion. Note that we do not *prove* that P_k is true (except for $k = 1$). Instead, we show that *if* P_k happens to be true, then the statement P_{k+1} is also true. We refer to the assumption that P_k is true as the **induction hypothesis.**

EXAMPLE 1 Using the principle of mathematical induction

Use mathematical induction to prove that for every positive integer n, the sum of the first n positive integers is

$$\frac{n(n+1)}{2}.$$

▶ **SOLUTION** If n is any positive integer, let P_n denote the statement

$$1 + 2 + 3 + \cdots + n = \frac{n(n+1)}{2}.$$

(continued)

▶ **Tutorial available at www.thomsonedu.com/login**

The following are some special cases of P_n.

If $n = 1$, then P_1 is

$$1 = \frac{1(1 + 1)}{2}; \quad \text{that is,} \quad 1 = 1.$$

If $n = 2$, then P_2 is

$$1 + 2 = \frac{2(2 + 1)}{2}; \quad \text{that is,} \quad 3 = 3.$$

If $n = 3$, then P_3 is

$$1 + 2 + 3 = \frac{3(3 + 1)}{2}; \quad \text{that is,} \quad 6 = 6.$$

Although it is instructive to check the validity of P_n for several values of n as we have done, it is unnecessary to do so. We need only apply the two-step process outlined prior to this example. Thus, we proceed as follows:

Step 1 If we substitute $n = 1$ in P_n, then the left-hand side contains only the number 1 and the right-hand side is $\frac{1(1 + 1)}{2}$, which also equals 1. Hence, P_1 is true.

Step 2 Assume that P_k is true. Thus, the induction hypothesis is

$$1 + 2 + 3 + \cdots + k = \frac{k(k + 1)}{2}.$$

Our goal is to show that P_{k+1} is true—that is, that

$$1 + 2 + 3 + \cdots + k + (k + 1) = \frac{(k + 1)[(k + 1) + 1]}{2}.$$

We may prove that the last formula is true by rewriting the left-hand side and using the induction hypothesis as follows:

$$\begin{aligned}
1 + 2 + 3 + \cdots + k + (k + 1) &= (1 + 2 + 3 + \cdots + k) + (k + 1) && \text{group the first } k \text{ terms} \\
&= \frac{k(k + 1)}{2} + (k + 1) && \text{induction hypothesis} \\
&= \frac{k(k + 1) + 2(k + 1)}{2} && \text{add terms} \\
&= \frac{(k + 1)(k + 2)}{2} && \text{factor out } k + 1 \\
&= \frac{(k + 1)[(k + 1) + 1]}{2} && \text{change form of } k + 2
\end{aligned}$$

This shows that P_{k+1} is true, and therefore the proof by mathematical induction is complete. ◪

EXAMPLE 2 Using the principle of mathematical induction

Prove that for each positive integer n,

$$1^2 + 3^2 + \cdots + (2n - 1)^2 = \frac{n(2n - 1)(2n + 1)}{3}.$$

SOLUTION For each positive integer n, let P_n denote the given statement. Note that this is a formula for the sum of the squares of the first n odd positive integers. We again follow the two-step procedure.

Step 1 Substituting 1 for n in P_n, we obtain

$$1^2 = \frac{(1)(2 - 1)(2 + 1)}{3} = \frac{3}{3} = 1.$$

This shows that P_1 is true.

Step 2 Assume that P_k is true. Thus, the induction hypothesis is

$$1^2 + 3^2 + \cdots + (2k - 1)^2 = \frac{k(2k - 1)(2k + 1)}{3}.$$

We wish to show that P_{k+1} is true—that is, that

$$1^2 + 3^2 + \cdots + [2(k + 1) - 1]^2 = \frac{(k + 1)[2(k + 1) - 1][2(k + 1) + 1]}{3}.$$

This equation simplifies to

$$1^2 + 3^2 + \cdots + (2k + 1)^2 = \frac{(k + 1)(2k + 1)(2k + 3)}{3}.$$

Remember that the next to last term on the left-hand side of the equation (the kth term) is $(2k - 1)^2$. In a manner similar to that used in the solution of Example 1, we may prove the formula for P_{k+1} by rewriting the left-hand side and using the induction hypothesis as follows:

$$\begin{aligned}
1^2 + 3^2 + \cdots + (2k + 1)^2 &= [1^2 + 3^2 + \cdots + (2k - 1)^2] + (2k + 1)^2 && \text{group the first } k \text{ terms} \\
&= \frac{k(2k - 1)(2k + 1)}{3} + (2k + 1)^2 && \text{induction hypothesis} \\
&= \frac{k(2k - 1)(2k + 1) + 3(2k + 1)^2}{3} && \text{add terms} \\
&= \frac{(2k + 1)[k(2k - 1) + 3(2k + 1)]}{3} && \text{factor out } 2k + 1 \\
&= \frac{(2k + 1)(2k^2 + 5k + 3)}{3} && \text{simplify} \\
&= \frac{(k + 1)(2k + 1)(2k + 3)}{3} && \text{factor and change order}
\end{aligned}$$

This shows that P_{k+1} is true, and hence P_n is true for every n.

Tutorial available at www.thomsonedu.com/login

EXAMPLE 3 Using the principle of mathematical induction

Prove that 2 is a factor of $n^2 + 5n$ for every positive integer n.

▶ **SOLUTION** For each positive integer n, let P_n denote the following statement:

$$2 \text{ is a factor of } n^2 + 5n$$

We shall follow the two-step procedure.

Step 1 If $n = 1$, then

$$n^2 + 5n = 1^2 + 5 \cdot 1 = 6 = 2 \cdot 3.$$

Thus, 2 is a factor of $n^2 + 5n$ for $n = 1$; that is, P_1 is true.

Step 2 Assume that P_k is true. Thus, the induction hypothesis is

$$2 \text{ is a factor of } k^2 + 5k$$

or, equivalently, $$k^2 + 5k = 2p$$

for some integer p.

We wish to show that P_{k+1} is true—that is, that

$$2 \text{ is a factor of } (k + 1)^2 + 5(k + 1).$$

We may do this as follows:

$$\begin{aligned}(k + 1)^2 + 5(k + 1) &= k^2 + 2k + 1 + 5k + 5 && \text{multiply}\\ &= (k^2 + 5k) + (2k + 6) && \text{rearrange terms}\\ &= 2p + 2(k + 3) && \text{induction hypothesis, factor } 2k + 6\\ &= 2(p + k + 3) && \text{factor out 2}\end{aligned}$$

Since 2 is a factor of the last expression, P_{k+1} is true, and hence P_n is true for every n.

Let j be a positive integer, and suppose that with each integer $n \geq j$ there is associated a statement P_n. For example, if $j = 6$, then the statements are numbered $P_6, P_7, P_8, \ldots$. The principle of mathematical induction may be extended to cover this situation. To prove that the statements P_n are true for $n \geq j$, we use the following two steps, in the same manner as we did for $n \geq 1$.

Steps in Applying the Extended Principle of Mathematical Induction for P_k, $k \geq j$	
	1 Show that P_j is true. *2* Assume that P_k is true with $k \geq j$, and then prove that P_{k+1} is true.

▶ **Tutorial available at www.thomsonedu.com/login**

EXAMPLE 4 Using the extended principle of mathematical induction

Let a be a nonzero real number such that $a > -1$. Prove that

$$(1 + a)^n > 1 + na$$

for every integer $n \geq 2$.

SOLUTION For each positive integer n, let P_n denote the inequality $(1 + a)^n > 1 + na$. Note that P_1 is *false,* since $(1 + a)^1 = 1 + (1)(a)$. However, we can show that P_n is true for $n \geq 2$ by using the extended principle with $j = 2$.

Step 1 We first note that $(1 + a)^2 = 1 + 2a + a^2$. Since $a \neq 0$, we have $a^2 > 0$, and so $1 + 2a + a^2 > 1 + 2a$ or, equivalently, $(1 + a)^2 > 1 + 2a$. Hence, P_2 is true.

Step 2 Assume that P_k is true. Thus, the induction hypothesis is

$$(1 + a)^k > 1 + ka.$$

We wish to show that P_{k+1} is true—that is, that

$$(1 + a)^{k+1} > 1 + (k + 1)a.$$

To prove the last inequality, we first observe the following:

$$\begin{aligned} (1 + a)^{k+1} &= (1 + a)^k(1 + a)^1 && \text{law of exponents} \\ &> (1 + ka)(1 + a) && \text{induction hypothesis and } 1 + a > 0 \end{aligned}$$

We next note that

$$\begin{aligned} (1 + ka)(1 + a) &= 1 + ka + a + ka^2 && \text{multiply} \\ &= 1 + (ka + a) + ka^2 && \text{group terms} \\ &= 1 + (k + 1)a + ka^2 && \text{factor out } a \\ &> 1 + (k + 1)a. && \text{since } ka^2 > 0 \end{aligned}$$

The last two inequalities give us

$$(1 + a)^{k+1} > 1 + (k + 1)a.$$

Thus, P_{k+1} is true, and the proof by mathematical induction is complete.

We have looked at several examples of proving statements by using the principle of mathematical induction. You may be wondering "Where do these statements come from?" These statements can often be "discovered" by observing patterns, combining results from several areas of mathematics, or recognizing certain types or categories of relationships. Two such statements are given in Exercises 37 and 38 in this section, and two additional (slightly more difficult) statements are given in Discussion Exercises 3 and 4 at the end of the chapter.

9.4 Exercises

Exer. 1–26: Prove that the statement is true for every positive integer n.

1 $2 + 4 + 6 + \cdots + 2n = n(n + 1)$

2 $1 + 4 + 7 + \cdots + (3n - 2) = \dfrac{n(3n - 1)}{2}$

3 $1 + 3 + 5 + \cdots + (2n - 1) = n^2$

▶ 4 $3 + 9 + 15 + \cdots + (6n - 3) = 3n^2$

5 $2 + 7 + 12 + \cdots + (5n - 3) = \frac{1}{2}n(5n - 1)$

▶ 6 $2 + 6 + 18 + \cdots + 2 \cdot 3^{n-1} = 3^n - 1$

7 $1 + 2 \cdot 2 + 3 \cdot 2^2 + \cdots + n \cdot 2^{n-1} = 1 + (n - 1) \cdot 2^n$

8 $(-1)^1 + (-1)^2 + (-1)^3 + \cdots + (-1)^n = \dfrac{(-1)^n - 1}{2}$

9 $1^2 + 2^2 + 3^2 + \cdots + n^2 = \dfrac{n(n + 1)(2n + 1)}{6}$

10 $1^3 + 2^3 + 3^3 + \cdots + n^3 = \left[\dfrac{n(n + 1)}{2}\right]^2$

▶ 11 $\dfrac{1}{1 \cdot 2} + \dfrac{1}{2 \cdot 3} + \dfrac{1}{3 \cdot 4} + \cdots + \dfrac{1}{n(n + 1)} = \dfrac{n}{n + 1}$

12 $\dfrac{1}{1 \cdot 2 \cdot 3} + \dfrac{1}{2 \cdot 3 \cdot 4} + \dfrac{1}{3 \cdot 4 \cdot 5} + \cdots + \dfrac{1}{n(n + 1)(n + 2)} = \dfrac{n(n + 3)}{4(n + 1)(n + 2)}$

▶ 13 $3 + 3^2 + 3^3 + \cdots + 3^n = \frac{3}{2}(3^n - 1)$

14 $1^3 + 3^3 + 5^3 + \cdots + (2n - 1)^3 = n^2(2n^2 - 1)$

▶ 15 $n < 2^n$

16 $1 + 2n \leq 3^n$

17 $1 + 2 + 3 + \cdots + n < \frac{1}{8}(2n + 1)^2$

18 If $0 < a < b$, then $\left(\dfrac{a}{b}\right)^{n+1} < \left(\dfrac{a}{b}\right)^n$.

▶ 19 3 is a factor of $n^3 - n + 3$.

20 2 is a factor of $n^2 + n$.

▶ 21 4 is a factor of $5^n - 1$.

22 9 is a factor of $10^{n+1} + 3 \cdot 10^n + 5$.

23 If a is greater than 1, then $a^n > 1$.

24 If $r \neq 1$, then

$$a + ar + ar^2 + \cdots + ar^{n-1} = \frac{a(1 - r^n)}{1 - r}.$$

25 $a - b$ is a factor of $a^n - b^n$.
(*Hint:* $a^{k+1} - b^{k+1} = a^k(a - b) + (a^k - b^k)b$.)

26 $a + b$ is a factor of $a^{2n-1} + b^{2n-1}$.

Exer. 27–32: Find the smallest positive integer j for which the statement is true. Use the extended principle of mathematical induction to prove that the formula is true for every integer greater than j.

27 $n + 12 \leq n^2$

28 $n^2 + 18 \leq n^3$

29 $5 + \log_2 n \leq n$

30 $n^2 \leq 2^n$

31 $2n + 2 \leq 2^n$

32 $n \log_2 n + 20 \leq n^2$

Exer. 33–36: Express the sum in terms of n.

33 $\sum_{k=1}^{n} (k^2 + 3k + 5)$

(*Hint:* Use the theorem on sums to write the sum as

$$\sum_{k=1}^{n} k^2 + 3\sum_{k=1}^{n} k + \sum_{k=1}^{n} 5.$$

Next use Exercise 9 above, Example 5 of Section 9.1, and the theorem on the sum of a constant.)

34 $\sum_{k=1}^{n} (3k^2 - 2k + 1)$

35 $\sum_{k=1}^{n} (2k - 3)^2$

36 $\sum_{k=1}^{n} (k^3 + 2k^2 - k + 4)$ (*Hint:* Use Exercise 10.)

Exer. 37–38: (a) Evaluate the given formula for the stated values of n, and solve the resulting system of equations for $a, b, c,$ and d. (This method can sometimes be used to obtain formulas for sums.) (b) Compare the result in part (a) with the indicated exercise, and explain why this method does not prove that the formula is true for every n.

37 $1^2 + 2^2 + 3^2 + \cdots + n^2 = an^3 + bn^2 + cn$;
$n = 1, 2, 3$ (Exercise 9)

38 $1^3 + 2^3 + 3^3 + \cdots + n^3 = an^4 + bn^3 + cn^2 + dn$;
$n = 1, 2, 3, 4$ (Exercise 10)

Exer. 39–42: Prove that the statement is true for every positive integer n.

39 $\sin(\theta + n\pi) = (-1)^n \sin\theta$

40 $\cos(\theta + n\pi) = (-1)^n \cos\theta$

41 Prove De Moivre's theorem:

$$[r(\cos\theta + i\sin\theta)]^n = r^n(\cos n\theta + i\sin n\theta)$$

for every positive integer n.

42 Prove that for every positive integer $n \geq 3$, the sum of the interior angles of an n-sided polygon is given by the expression $(n - 2) \cdot 180°$.

▶ **Tutorial available at www.thomsonedu.com/login**

9.5

The Binomial Theorem

A **binomial** is a sum $a + b$, where a and b represent numbers. If n is a positive integer, then a general formula for *expanding* $(a + b)^n$ (that is, for expressing it as a sum) is given by the **binomial theorem.** In this section we shall use mathematical induction to establish this general formula. The following special cases can be obtained by multiplication:

$$(a + b)^2 = a^2 + 2ab + b^2$$
$$(a + b)^3 = a^3 + 3a^2b + 3ab^2 + b^3$$
$$(a + b)^4 = a^4 + 4a^3b + 6a^2b^2 + 4ab^3 + b^4$$
$$(a + b)^5 = a^5 + 5a^4b + 10a^3b^2 + 10a^2b^3 + 5ab^4 + b^5$$

These expansions of $(a + b)^n$ for $n = 2, 3, 4,$ and 5 have the following properties.

(1) There are $n + 1$ terms, the first being a^n and the last b^n.

(2) As we proceed from any term to the next, the power of a decreases by 1 and the power of b increases by 1. For each term, the sum of the exponents of a and b is n.

(3) Each term has the form $(c)a^{n-k}b^k$, where the coefficient c is an integer and $k = 0, 1, 2, \ldots, n$.

(4) The following formula is true for each of the first n terms of the expansion:

$$\frac{(\text{coefficient of term}) \cdot (\text{exponent of } a)}{\text{number of term}} = \text{coefficient of next term}$$

The following table illustrates property 4 for the expansion of $(a + b)^5$.

Term	Number of term	Coefficient of term	Exponent of a	Coefficient of next term
a^5	1	1	5	$\frac{1 \cdot 5}{1} = 5$
$5a^4b$	2	5	4	$\frac{5 \cdot 4}{2} = 10$
$10a^3b^2$	3	10	3	$\frac{10 \cdot 3}{3} = 10$
$10a^2b^3$	4	10	2	$\frac{10 \cdot 2}{4} = 5$
$5ab^4$	5	5	1	$\frac{5 \cdot 1}{5} = 1$

Let us next consider $(a + b)^n$ for an arbitrary positive integer n. The first term is a^n, which has coefficient 1. If we assume that property 4 is true, we obtain the successive coefficients listed in the next table.

Term	Number of term	Coefficient of term	Exponent of a	Coefficient of next term
a^n	1	1	n	$\frac{1 \cdot n}{1} = n$
$\frac{n}{1}a^{n-1}b$	2	$\frac{n}{1}$	$n-1$	$\frac{n(n-1)}{2 \cdot 1}$
$\frac{n(n-1)}{2 \cdot 1}a^{n-2}b^2$	3	$\frac{n(n-1)}{2 \cdot 1}$	$n-2$	$\frac{n(n-1)(n-2)}{3 \cdot 2 \cdot 1}$
$\frac{n(n-1)(n-2)}{3 \cdot 2 \cdot 1}a^{n-3}b^3$	4	$\frac{n(n-1)(n-2)}{3 \cdot 2 \cdot 1}$	$n-3$	$\frac{n(n-1)(n-2)(n-3)}{4 \cdot 3 \cdot 2 \cdot 1}$

The pattern that appears in the fifth column leads to the following formula for the coefficient of the general term.

Coefficient of the $(k + 1)$st Term in the Expansion of $(a + b)^n$

$$\frac{n \cdot (n-1) \cdot (n-2) \cdot (n-3) \cdot \cdots \cdot (n-k+1)}{k \cdot (k-1) \cdot \cdots \cdot 3 \cdot 2 \cdot 1}, \quad k = 1, 2, \ldots, n$$

The $(k + 1)$st coefficient can be written in a compact form by using **factorial notation.** If n is any nonnegative integer, then the symbol $n!$ (*n factorial*) is defined as follows.

Definition of $n!$

(1) $n! = n(n-1)(n-2) \cdot \cdots \cdot 1 \quad \text{if} \quad n > 0$

(2) $0! = 1$

Thus, if $n > 0$, then $n!$ is the product of the first n positive integers. The definition $0! = 1$ is used so that certain formulas involving factorials are true for all *nonnegative* integers.

ILLUSTRATION ***n* Factorial**

- $1! = 1$
- $2! = 2 \cdot 1 = 2$
- $3! = 3 \cdot 2 \cdot 1 = 6$
- $4! = 4 \cdot 3 \cdot 2 \cdot 1 = 24$
- $5! = 5 \cdot 4 \cdot 3 \cdot 2 \cdot 1 = 120$
- $6! = 6 \cdot 5 \cdot 4 \cdot 3 \cdot 2 \cdot 1 = 720$
- $7! = 7 \cdot 6 \cdot 5 \cdot 4 \cdot 3 \cdot 2 \cdot 1 = 5040$
- $8! = 8 \cdot 7 \cdot 6 \cdot 5 \cdot 4 \cdot 3 \cdot 2 \cdot 1 = 40{,}320$

Notice the rapid growth of $n!$ as n increases.

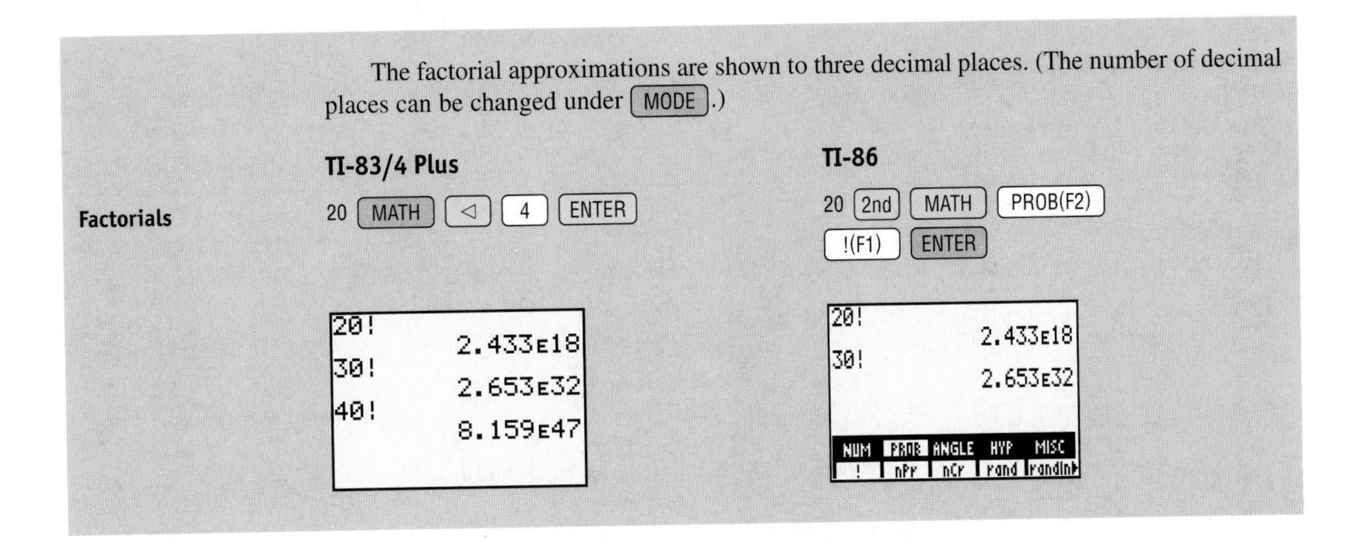

We sometimes wish to simplify quotients where both the numerator and the denominator contain factorials, as shown in the next illustration.

ILLUSTRATION **Simplifying Quotients of Factorials**

- $\dfrac{7!}{5!} = \dfrac{7 \cdot 6 \cdot 5!}{5!} = 7 \cdot 6 = 42$
- $\dfrac{10!}{6!} = \dfrac{10 \cdot 9 \cdot 8 \cdot 7 \cdot 6!}{6!} = 10 \cdot 9 \cdot 8 \cdot 7 = 5040$

As in the preceding illustration, if n and k are positive integers and $k < n$, then

$$\begin{aligned}\frac{n!}{(n-k)!} &= \frac{n \cdot (n-1) \cdot (n-2) \cdot \cdots \cdot (n-k+1) \cdot [(n-k)!]}{(n-k)!} \\ &= n \cdot (n-1) \cdot (n-2) \cdot \cdots \cdot (n-k+1),\end{aligned}$$

which is the numerator of the coefficient of the $(k+1)$st term of $(a+b)^n$. Dividing by the denominator $k!$ gives us the following alternative form for the $(k+1)$st coefficient:

$$\frac{n \cdot (n-1) \cdot (n-2) \cdot \cdots \cdot (n-k+1)}{k!} = \frac{n!}{k!\,(n-k)!}$$

These numbers are called **binomial coefficients** and are often denoted by either the symbol $\dbinom{n}{k}$ or the symbol $C(n, k)$. Thus, we have the following.

Coefficient of the $(k + 1)$st Term in the Expansion of $(a + b)^n$ (Alternative Form)	$\binom{n}{k} = C(n, k) = \dfrac{n!}{k!\,(n - k)!}, \quad k = 0, 1, 2, \ldots, n$

The symbols $\binom{n}{k}$ and $C(n, k)$ are sometimes read "n choose k."

EXAMPLE 1 **Evaluating $\binom{n}{k}$**

Find $\binom{5}{0}, \binom{5}{1}, \binom{5}{2}, \binom{5}{3}, \binom{5}{4}$, and $\binom{5}{5}$.

▶ SOLUTION These six numbers are the coefficients in the expansion of $(a + b)^5$, which we tabulated earlier in this section. By definition,

$$\binom{5}{0} = \frac{5!}{0!\,(5-0)!} = \frac{5!}{0!\,5!} = \frac{5!}{1 \cdot 5!} = 1$$

$$\binom{5}{1} = \frac{5!}{1!\,(5-1)!} = \frac{5!}{1!\,4!} = \frac{5!}{1 \cdot 4!} = \frac{5 \cdot 4!}{4!} = 5$$

$$\binom{5}{2} = \frac{5!}{2!\,(5-2)!} = \frac{5!}{2!\,3!} = \frac{5 \cdot 4 \cdot 3!}{2 \cdot 3!} = \frac{20}{2} = 10$$

$$\binom{5}{3} = \frac{5!}{3!\,(5-3)!} = \frac{5!}{3!\,2!} = \frac{5 \cdot 4 \cdot 3!}{3! \cdot 2} = \frac{20}{2} = 10$$

$$\binom{5}{4} = \frac{5!}{4!\,(5-4)!} = \frac{5!}{4!\,1!} = \frac{5!}{4! \cdot 1} = \frac{5 \cdot 4!}{4!} = 5$$

$$\binom{5}{5} = \frac{5!}{5!\,(5-5)!} = \frac{5!}{5!\,0!} = \frac{5!}{5! \cdot 1} = 1.$$

EXAMPLE 2 **Simplifying a quotient of factorials**

Rewrite $(3n + 3)!/(3n)!$ as an expression that does not contain factorials.

▶ SOLUTION By the definition of $n!$, we can write $(3n + 3)!$ as

$$(3n + 3)(3n + 2)(3n + 1)\underbrace{(3n)(3n - 1)(3n - 2) \cdots (3)(2)(1)}_{(3n)!}.$$

Thus,

$$\frac{(3n + 3)!}{(3n)!} = \frac{(3n + 3)(3n + 2)(3n + 1)\cancel{(3n)!}}{\cancel{(3n)!}} \qquad \text{definition of } n!$$

$$= (3n + 3)(3n + 2)(3n + 1). \qquad \text{cancel } (3n)! \neq 0$$

▶ Tutorial available at www.thomsonedu.com/login

The binomial theorem may be stated as follows.

The Binomial Theorem

$$(a + b)^n = a^n + \binom{n}{1}a^{n-1}b + \binom{n}{2}a^{n-2}b^2 + \cdots + \binom{n}{k}a^{n-k}b^k + \cdots + \binom{n}{n-1}ab^{n-1} + b^n$$

Using summation notation, we may write the binomial theorem

$$(a + b)^n = \sum_{k=0}^{n} \binom{n}{k} a^{n-k}b^k.$$

Note that there are $n + 1$ terms (not n terms) in the expansion of $(a + b)^n$, and so

$\binom{n}{k}a^{n-k}b^k$ is a formula for the $(k + 1)$st term of the expansion.

An alternative statement of the binomial theorem is as follows. (A proof is given at the end of this section.)

The Binomial Theorem (Alternative Form)

$$(a + b)^n = a^n + na^{n-1}b + \frac{n(n-1)}{2!}a^{n-2}b^2 + \cdots + \frac{n(n-1)(n-2)\cdots(n-k+1)}{k!}a^{n-k}b^k + \cdots + nab^{n-1} + b^n$$

The following examples may be solved either by using the general formulas for the binomial theorem or by repeated use of property 4, stated at the beginning of this section.

EXAMPLE 3 **Finding a binomial expansion**

Find the binomial expansion of $(2x + 3y^2)^4$.

▶ **SOLUTION** We use the binomial theorem with $a = 2x$, $b = 3y^2$, and $n = 4$:

$$\begin{aligned}(2x + 3y^2)^4 &= (2x)^4 + \binom{4}{1}(2x)^3(3y^2)^1 + \binom{4}{2}(2x)^2(3y^2)^2 + \binom{4}{3}(2x)^1(3y^2)^3 + (3y^2)^4 \\ &= 16x^4 + 4(8x^3)(3y^2) + 6(4x^2)(9y^4) + 4(2x)(27y^6) + 81y^8 \\ &= 16x^4 + 96x^3y^2 + 216x^2y^4 + 216xy^6 + 81y^8\end{aligned}$$

Examining the terms of the expansion from left to right, we see that the exponents on x decrease by 1 and that the exponents on y increase by 2. It is a good idea to check for exponent patterns after simplifying a binomial expansion.

▶ **Tutorial available at www.thomsonedu.com/login**

The next example illustrates that if either a or b is negative, then the terms of the expansion are alternately positive and negative.

EXAMPLE 4 Finding a binomial expansion

Expand $\left(\frac{1}{x} - 2\sqrt{x}\right)^5$.

SOLUTION The binomial coefficients for $(a + b)^5$ were calculated in Example 1. Thus, if we let $a = 1/x$, $b = -2\sqrt{x}$, and $n = 5$ in the binomial theorem, we obtain

$$\left(\frac{1}{x} - 2\sqrt{x}\right)^5 = \left(\frac{1}{x}\right)^5 + 5\left(\frac{1}{x}\right)^4(-2\sqrt{x})^1 + 10\left(\frac{1}{x}\right)^3(-2\sqrt{x})^2 + 10\left(\frac{1}{x}\right)^2(-2\sqrt{x})^3 + 5\left(\frac{1}{x}\right)^1(-2\sqrt{x})^4 + (-2\sqrt{x})^5,$$

which can be written as

$$\left(\frac{1}{x} - 2\sqrt{x}\right)^5 = \frac{1}{x^5} - \frac{10}{x^{7/2}} + \frac{40}{x^2} - \frac{80}{x^{1/2}} + 80x - 32x^{5/2}.$$

To find a specific term in the expansion of $(a + b)^n$, it is convenient to first find the exponent k that is to be assigned to b. Notice that, by the binomial theorem, *the exponent of b is always one less than the number of the term.* Once k is found, we know that the exponent of a is $n - k$ and the coefficient is $\binom{n}{k}$.

EXAMPLE 5 Finding a specific term of a binomial expansion

Find the fifth term in the expansion of $(x^3 + \sqrt{y})^{13}$.

SOLUTION Let $a = x^3$ and $b = \sqrt{y}$. The exponent of b in the fifth term is $k = 5 - 1 = 4$, and hence the exponent of a is $n - k = 13 - 4 = 9$. From the discussion of the preceding paragraph we obtain

$(k + 1)\text{st term} = \binom{n}{k}a^{n-k}b^k$

$$\binom{13}{4}(x^3)^9(\sqrt{y})^4 = \frac{13!}{4!\,(13-4)!}x^{27}y^2 = \frac{13 \cdot 12 \cdot 11 \cdot 10}{4!}x^{27}y^2 = 715x^{27}y^2.$$

EXAMPLE 6 Finding a specific term of a binomial expansion

Find the term involving q^{10} in the binomial expansion of $(\frac{1}{3}p + q^2)^{12}$.

SOLUTION From the statement of the binomial theorem with $a = \frac{1}{3}p$, $b = q^2$, and $n = 12$, each term in the expansion has the form

$$\binom{n}{k}a^{n-k}b^k = \binom{12}{k}\left(\frac{1}{3}p\right)^{12-k}(q^2)^k.$$

Tutorial available at www.thomsonedu.com/login

Since $(q^2)^k = q^{2k}$, we must let $k = 5$ to obtain the term involving q^{10}. Doing so gives us

$$\binom{12}{5}\left(\frac{1}{3}p\right)^{12-5}(q^2)^5 = \frac{12!}{5!\,(12-5)!}\left(\frac{1}{3}\right)^7 p^7q^{10} = \frac{88}{243}p^7q^{10}.$$

There is an interesting triangular array of numbers, called **Pascal's triangle,** that can be used to obtain binomial coefficients. The numbers are arranged as follows:

$$\begin{array}{ccccccccccccc}
 & & & & & & 1 & & & & & & \\
 & & & & & 1 & & 1 & & & & & \\
 & & & & 1 & & 2 & & 1 & & & & \\
 & & & 1 & & 3 & & 3 & & 1 & & & \\
 & & 1 & & 4 & & 6 & & 4 & & 1 & & \\
 & 1 & & 5 & & 10 & & 10 & & 5 & & 1 & \\
1 & & 6 & & 15 & & 20 & & 15 & & 6 & & 1 \\
\cdot & \cdot & \cdot & \cdot & \cdot & \cdot & \cdot & \cdot & \cdot & \cdot & \cdot & \cdot & \cdot \\
\cdot & \cdot & \cdot & \cdot & \cdot & \cdot & \cdot & \cdot & \cdot & \cdot & \cdot & \cdot & \cdot
\end{array}$$

The numbers in the second row are the coefficients in the expansion of $(a + b)^1$; those in the third row are the coefficients determined by $(a + b)^2$; those in the fourth row are obtained from $(a + b)^3$; and so on. Each number in the array that is different from 1 can be found by adding the two numbers in the previous row that appear above and immediately to the left and right of the number, as illustrated in the solution of the next example.

EXAMPLE 7 **Using Pascal's triangle**

Find the eighth row of Pascal's triangle, and use it to expand $(a + b)^7$.

SOLUTION Let us rewrite the seventh row and then use the process described above. In the following display the arrows indicate which two numbers in row seven are added to obtain the numbers in row eight.

$$\begin{array}{ccccccccccccccc}
 & 1 & & 6 & & 15 & & 20 & & 15 & & 6 & & 1 & \\
 & & \searrow\swarrow & & \searrow\swarrow & & \searrow\swarrow & & \searrow\swarrow & & \searrow\swarrow & & \searrow\swarrow & & \\
1 & & 7 & & 21 & & 35 & & 35 & & 21 & & 7 & & 1
\end{array}$$

The eighth row gives us the coefficients in the expansion of $(a + b)^7$:

$$(a + b)^7 = a^7 + 7a^6b + 21a^5b^2 + 35a^4b^3 + 35a^3b^4 + 21a^2b^5 + 7ab^6 + b^7$$

Pascal's triangle is useful for expanding small powers of $a + b$; however, for expanding large powers or finding a specific term, as in Examples 5 and 6, the general formula given by the binomial theorem is more useful.

We shall conclude this section by giving a proof of the binomial theorem using mathematical induction.

PROOF OF THE BINOMIAL THEOREM For each positive integer n, let P_n denote the statement given in the alternative form of the binomial theorem.

Step 1 If $n = 1$, the statement reduces to $(a + b)^1 = a^1 + b^1$. Consequently, P_1 is true.

Step 2 Assume that P_k is true. Thus, the induction hypothesis is

$$(a + b)^k = a^k + ka^{k-1}b + \frac{k(k-1)}{2!}a^{k-2}b^2 + \cdots + \frac{k(k-1)(k-2)\cdots(k-r+2)}{(r-1)!}a^{k-r+1}b^{r-1} + \frac{k(k-1)(k-2)\cdots(k-r+1)}{r!}a^{k-r}b^r + \cdots + kab^{k-1} + b^k.$$

We have shown both the rth term and the $(r + 1)$st term in the above expansion.

To prove that P_{k+1} is true, we first write

$$(a + b)^{k+1} = (a + b)^k(a + b).$$

Using the induction hypothesis to substitute for $(a + b)^k$ and then multiplying that expression by $a + b$, we obtain

$$(a + b)^{k+1} = \left[a^{k+1} + ka^kb + \frac{k(k-1)}{2!}a^{k-1}b^2 + \cdots + \frac{k(k-1)\cdots(k-r+1)}{r!}a^{k-r+1}b^r + \cdots + ab^k\right] + \left[a^kb + ka^{k-1}b^2 + \cdots + \frac{k(k-1)\cdots(k-r+2)}{(r-1)!}a^{k-r+1}b^r + \cdots + kab^k + b^{k+1}\right],$$

where the terms in the first pair of brackets result from multiplying the right-hand side of the induction hypothesis by a and the terms in the second pair of brackets result from multiplying by b. We next rearrange and combine terms:

$$(a + b)^{k+1} = a^{k+1} + (k + 1)a^kb + \left(\frac{k(k-1)}{2!} + k\right)a^{k-1}b^2 + \cdots + \left(\frac{k(k-1)\cdots(k-r+1)}{r!} + \frac{k(k-1)\cdots(k-r+2)}{(r-1)!}\right)a^{k-r+1}b^r + \cdots + (1 + k)ab^k + b^{k+1}$$

If the coefficients are simplified, we obtain statement P_n with $k + 1$ substituted for n. Thus, P_{k+1} is true, and therefore P_n holds for every positive integer n, which completes the proof.

9.5 Exercises

Exer. 1–12: Evaluate the expression.

▶ 1 $2!6!$

2 $3!4!$

3 $7!0!$

4 $5!0!$

▶ 5 $\dfrac{8!}{5!}$

6 $\dfrac{6!}{3!}$

7 $\dbinom{5}{5}$

▶ 8 $\dbinom{7}{0}$

9 $\dbinom{7}{5}$

10 $\dbinom{8}{4}$

11 $\dbinom{13}{4}$

12 $\dbinom{52}{2}$

Exer. 13–16: Rewrite as an expression that does not contain factorials.

13 $\dfrac{n!}{(n-2)!}$

14 $\dfrac{(n+1)!}{(n-1)!}$

15 $\dfrac{(2n+2)!}{(2n)!}$

▶ 16 $\dfrac{(3n+1)!}{(3n-1)!}$

Exer. 17–30: Use the binomial theorem to expand and simplify.

17 $(4x - y)^3$

18 $(x^2 + 2y)^3$

19 $(x + y)^6$

20 $(x + y)^4$

21 $(x - y)^7$

22 $(x - y)^5$

23 $(3t - 5s)^4$

24 $(2t - s)^5$

25 $\left(\frac{1}{3}x + y^2\right)^5$

26 $\left(\frac{1}{2}x + y^3\right)^4$

27 $\left(\dfrac{1}{x^2} + 3x\right)^6$

28 $\left(\dfrac{1}{x^3} - 2x\right)^5$

29 $\left(\sqrt{x} - \dfrac{1}{\sqrt{x}}\right)^5$

30 $\left(\sqrt{x} + \dfrac{1}{\sqrt{x}}\right)^5$

Exer. 31–46: Without expanding completely, find the indicated term(s) in the expansion of the expression.

31 $(3c^{2/5} + c^{4/5})^{25}$; first three terms

32 $(x^3 + 5x^{-2})^{20}$; first three terms

33 $(4z^{-1} - 3z)^{15}$; last three terms

34 $(s - 2t^3)^{12}$; last three terms

35 $\left(\dfrac{3}{c} + \dfrac{c^2}{4}\right)^7$; sixth term

36 $(3x^2 - \sqrt{y})^9$; fifth term

37 $\left(\frac{1}{3}u + 4v\right)^8$; seventh term

38 $(3x^2 - y^3)^{10}$; fourth term

39 $(x^{1/2} + y^{1/2})^8$; middle term

40 $(rs^2 + t)^7$; two middle terms

41 $(2y + x^2)^8$; term that contains x^{10}

42 $(x^2 - 2y^3)^5$; term that contains y^6

43 $(3y^3 - 2x^2)^4$; term that contains y^9

44 $(\sqrt{c} + \sqrt{d})^8$; term that contains c^2

▶ **Tutorial available at www.thomsonedu.com/login**

45 $\left(3x - \dfrac{1}{4x}\right)^6$; term that does not contain x

46 $(xy - 3y^{-3})^8$; term that does not contain y

47 Approximate $(1.2)^{10}$ by using the first three terms in the expansion of $(1 + 0.2)^{10}$, and compare your answer with that obtained using a calculator.

48 Approximate $(0.9)^4$ by using the first three terms in the expansion of $(1 - 0.1)^4$, and compare your answer with that obtained using a calculator.

Exer. 49–50: Simplify the expression using the binomial theorem.

49 $\dfrac{(x + h)^4 - x^4}{h}$

50 $\dfrac{(x + h)^5 - x^5}{h}$

51 Show that $\dbinom{n}{1} = \dbinom{n}{n - 1}$ for $n \geq 1$.

52 Show that $\dbinom{n}{0} = \dbinom{n}{n}$ for $n \geq 0$.

9.6 Permutations

Suppose that four teams are involved in a tournament in which first, second, third, and fourth places will be determined. For identification purposes, we label the teams A, B, C, and D. Let us find the number of different ways that first and second place can be decided. It is convenient to use a **tree diagram,** as in Figure 1. After the word START, the four possibilities for first place are listed. From each of these an arrow points to a possible second-place finisher. The final standings list the possible outcomes, from left to right. They are found by following the different paths (*branches* of the tree) that lead from the word START to the second-place team. The total number of outcomes is 12, which is the product of the number of choices (4) for first place and the number of choices (3) for second place (after first has been determined).

Figure 1

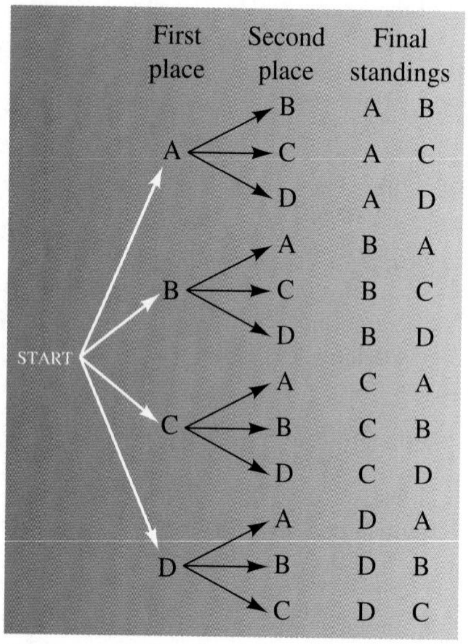

	First place	Second place	Final standings
START	A	B	A B
		C	A C
		D	A D
	B	A	B A
		C	B C
		D	B D
	C	A	C A
		B	C B
		D	C D
	D	A	D A
		B	D B
		C	D C

Let us now find the total number of ways that first, second, third, and fourth positions can be filled. To sketch a tree diagram, we may begin by drawing arrows from the word START to each possible first-place finisher A, B, C, or D. Next we draw arrows from those to possible second-place finishers, as was done in Figure 1. From each second-place position we then draw arrows indicating the possible third-place positions. Finally, we draw arrows to the fourth-place team. If we consider only the case in which team A finishes in first place, we have the diagram shown in Figure 2.

Figure 2

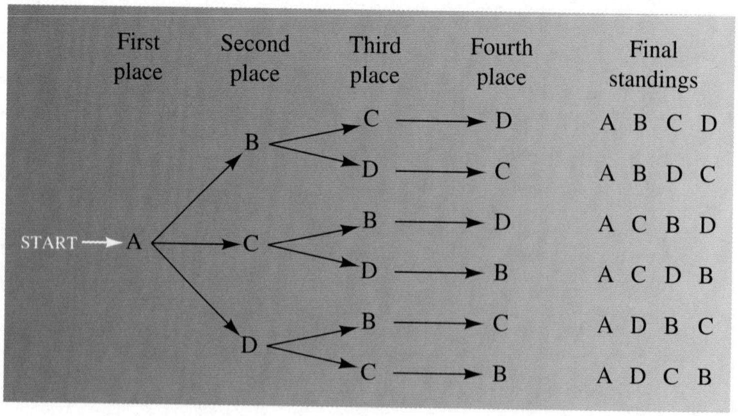

	First place	Second place	Third place	Fourth place	Final standings
START	A	B	C	D	A B C D
			D	C	A B D C
		C	B	D	A C B D
			D	B	A C D B
		D	B	C	A D B C
			C	B	A D C B

Note that there are six possible final standings in which team A occupies first place. In a complete tree diagram there would also be three other branches of this type corresponding to first-place finishes for B, C, and D. A complete diagram would display the following 24 possibilities for the final standings:

A first ABCD, ABDC, ACBD, ACDB, ADBC, ADCB,

B first BACD, BADC, BCAD, BCDA, BDAC, BDCA,

C first CABD, CADB, CBAD, CBDA, CDAB, CDBA,

D first DABC, DACB, DBAC, DBCA, DCAB, DCBA.

Note that the number of possibilities (24) is the product of the number of ways (4) that first place may occur, the number of ways (3) that second place may occur (after first place has been determined), the number of possible outcomes (2) for third place (after the first two places have been decided), and the number of ways (1) that fourth place can occur (after the first three places have been taken).

The preceding discussion illustrates the following general rule, which we accept as a basic axiom of counting.

Fundamental Counting Principle

Let $E_1, E_2, \ldots, E_k$ be a sequence of k events. If, for each i, the event E_i can occur in m_i ways, then the total number of ways all the events may take place is the product $m_1 m_2 \cdots m_k$.

Returning to our first illustration, we let E_1 represent the determination of the first-place team, so that $m_1 = 4$. If E_2 denotes the determination of the second-place team, then $m_2 = 3$. Hence, the number of outcomes for the sequence E_1, E_2 is $4 \cdot 3 = 12$, which is the same as that found by means of the tree diagram. If we proceed to E_3, the determination of the third-place team, then $m_3 = 2$, and hence $m_1 m_2 m_3 = 24$. Finally, if E_1, E_2, and E_3 have occurred, there is only one possible outcome for E_4. Thus, $m_4 = 1$, and $m_1 m_2 m_3 m_4 = 24$.

Instead of teams, let us now regard a, b, c, and d merely as symbols and consider the various *orderings,* or *arrangements,* that may be assigned to these symbols, taking them either two at a time, three at a time, or four at a time. By abstracting in this way we may apply our methods to other similar situations. The arrangements we have discussed are **arrangements without repetitions,** since a symbol may not be used more than once in an arrangement. In Example 1 we shall consider arrangements in which repetitions *are* allowed.

Previously we defined ordered pairs and ordered triples. Similarly, an *ordered 4-tuple* is a set containing four elements x_1, x_2, x_3, x_4 in which an ordering has been specified, so that one of the elements may be referred to as the *first element,* another as the *second element,* and so on. The symbol (x_1, x_2, x_3, x_4) is used for the ordered 4-tuple having first element x_1, second element x_2, third element x_3, and fourth element x_4. In general, for any positive integer r, we speak of the **ordered r-tuple**

$$(x_1, x_2, \ldots, x_r)$$

as a set of r elements in which x_1 is designated as the first element, x_2 as the second element, and so on.

EXAMPLE 1 Determining the number of r-tuples

Using only the letters a, b, c, and d, determine how many of the following can be obtained:

(a) ordered triples (b) ordered 4-tuples (c) ordered r-tuples

SOLUTION

▶ (a) We must determine the number of symbols of the form (x_1, x_2, x_3) that can be obtained using only the letters a, b, c, and d. This is not the same as listing first, second, and third place as in our previous illustration, since we have not ruled out the possibility of repetitions. For example, (a, b, a), (a, a, b), and (a, a, a) are different ordered triples. If, for $i = 1, 2, 3$, we let E_i represent the determination of x_i in the ordered triple (x_1, x_2, x_3), then, since repetitions are allowed, there are four possibilities—a, b, c, and d—for each of E_1, E_2, and E_3. Hence, by the fundamental counting principle, the total number of ordered triples is $4 \cdot 4 \cdot 4$, or 64.

(b) The number of possible ordered 4-tuples of the form (x_1, x_2, x_3, x_4) is $4 \cdot 4 \cdot 4 \cdot 4$, or 256.

(c) The number of ordered r-tuples is the product $4 \cdot 4 \cdot 4 \cdot \cdots \cdot 4$, with 4 appearing as a factor r times. That product equals 4^r.

EXAMPLE 2 Choosing class officers

A class consists of 60 girls and 40 boys. In how many ways can a president, vice-president, treasurer, and secretary be chosen if the treasurer must be a girl, the secretary must be a boy, and a student may not hold more than one office?

▶ SOLUTION If an event is specialized in some way (for example, the treasurer *must* be a girl), then that event should be considered before any nonspecialized events. Thus, we let E_1 represent the choice of treasurer and E_2 the choice of secretary. Next we let E_3 and E_4 denote the choices for president and vice-president, respectively. As in the fundamental counting principle, we let m_i denote the number of different ways E_i can occur for $i = 1, 2, 3$, and 4. It follows that $m_1 = 60$, $m_2 = 40$, $m_3 = 60 + 40 - 2 = 98$, and $m_4 = 97$. By the fundamental counting principle, the total number of possibilities is

$$m_1 m_2 m_3 m_4 = 60 \cdot 40 \cdot 98 \cdot 97 = 22{,}814{,}400.$$

When working with sets, we are usually not concerned about the order or arrangement of the elements. In the remainder of this section, however, the arrangement of the elements will be our main concern.

▶ Tutorial available at www.thomsonedu.com/login

Definition of Permutation	Let S be a set of n elements and let $1 \leq r \leq n$. A **permutation** of r elements of S is an arrangement, without repetitions, of r elements.

We also use the phrase **permutation of n elements taken r at a time.** The symbol $P(n, r)$ will denote the number of different permutations of r elements that can be obtained from a set of n elements. As a special case, $P(n, n)$ denotes the number of arrangements of n elements of S—that is, the number of ways of arranging *all* the elements of S.

In our first discussion involving the four teams A, B, C, and D, we had $P(4, 2) = 12$, since there are 12 different ways of arranging the four teams in groups of two. We also showed that the number of ways to arrange all the elements A, B, C, and D is 24. In permutation notation we would write this result as $P(4, 4) = 24$.

The next theorem gives us a general formula for $P(n, r)$.

Theorem on the Number of Different Permutations	Let S be a set of n elements and let $1 \leq r \leq n$. The number of different permutations of r elements of S is $$P(n, r) = n(n-1)(n-2)\cdots(n-r+1).$$

PROOF The problem of determining $P(n, r)$ is equivalent to determining the number of different r-tuples $(x_1, x_2, \ldots, x_r)$ such that each x_i is an element of S and no element of S appears twice in the same r-tuple. We may find this number by means of the fundamental counting principle. For each $i = 1, 2, \ldots, r$, let E_i represent the determination of the element x_i and let m_i be the number of different ways of choosing x_i. We wish to apply the sequence $E_1, E_2, \ldots, E_r$. We have n possible choices for x_1, and consequently $m_1 = n$. Since repetitions are not allowed, we have $n - 1$ choices for x_2, so $m_2 = n - 1$. Continuing in this manner, we successively obtain $m_3 = n - 2$, $m_4 = n - 3$, and ultimately $m_r = n - (r - 1)$ or, equivalently, $m_r = n - r + 1$. Hence, using the fundamental counting principle, we obtain the formula for $P(n, r)$. ■

Note that *the formula for $P(n, r)$ in the previous theorem contains exactly r factors on the right-hand side,* as shown in the following illustration.

ILLUSTRATION **Number of Different Permutations**

- $P(n, 1) = n$
- $P(n, 2) = n(n-1)$
- $P(n, 3) = n(n-1)(n-2)$
- $P(n, 4) = n(n-1)(n-2)(n-3)$

EXAMPLE 3 **Evaluating $P(n, r)$**

Find $P(5, 2)$, $P(6, 4)$, and $P(5, 5)$.

▶ **SOLUTION** We will use the formula for $P(n, r)$ in the preceding theorem. In each case, we first calculate the value of $(n - r + 1)$.

$$5 - 2 + 1 = \underline{4}, \quad \text{so} \quad P(5, 2) = 5 \cdot \underline{4} = 20$$
$$6 - 4 + 1 = \underline{3}, \quad \text{so} \quad P(6, 4) = 6 \cdot 5 \cdot 4 \cdot \underline{3} = 360$$
$$5 - 5 + 1 = \underline{1}, \quad \text{so} \quad P(5, 5) = 5 \cdot 4 \cdot 3 \cdot 2 \cdot \underline{1} = 120$$

EXAMPLE 4 **Arranging the batting order for a baseball team**

A baseball team consists of nine players. Find the number of ways of arranging the first four positions in the batting order if the pitcher is excluded.

SOLUTION We wish to find the number of permutations of 8 objects taken 4 at a time. Using the formula for $P(n, r)$ with $n = 8$ and $r = 4$, we have $n - r + 1 = 5$, and it follows that

$$P(8, 4) = 8 \cdot 7 \cdot 6 \cdot 5 = 1680.$$

The next result gives us a form for $P(n, r)$ that involves the factorial symbol.

Factorial Form for $P(n, r)$	If n is a positive integer and $1 \le r \le n$, then $$P(n, r) = \frac{n!}{(n - r)!}.$$

PROOF If we let $r = n$ in the formula for $P(n, r)$ in the theorem on permutations, we obtain the number of different arrangements of *all* the elements of a set consisting of n elements. In this case,

$$n - r + 1 = n - n + 1 = 1$$

and

$$P(n, n) = n(n - 1)(n - 2) \cdots 3 \cdot 2 \cdot 1 = n!.$$

Consequently, $P(n, n)$ is the product of the first n positive integers. This result is also given by the factorial form, for if $r = n$, then

$$P(n, n) = \frac{n!}{(n - n)!} = \frac{n!}{0!} = \frac{n!}{1} = n!.$$

If $1 \le r < n$, then

$$\frac{n!}{(n - r)!} = \frac{n(n - 1)(n - 2) \cdots (n - r + 1) \cdot [(n - r)!]}{(n - r)!}$$
$$= n(n - 1)(n - 2) \cdots (n - r + 1).$$

This agrees with the formula for $P(n, r)$ in the theorem on permutations.

▶ Tutorial available at www.thomsonedu.com/login

EXAMPLE 5 **Evaluating $P(n, r)$ using factorials**

Use the factorial form for $P(n, r)$ to find $P(5, 2)$, $P(6, 4)$, and $P(5, 5)$.

▶ SOLUTION

$$P(5, 2) = \frac{5!}{(5-2)!} = \frac{5!}{3!} = \frac{5 \cdot 4 \cdot 3!}{3!} = 5 \cdot 4 = 20$$

$$P(6, 4) = \frac{6!}{(6-4)!} = \frac{6!}{2!} = \frac{6 \cdot 5 \cdot 4 \cdot 3 \cdot 2!}{2!} = 6 \cdot 5 \cdot 4 \cdot 3 = 360$$

$$P(5, 5) = \frac{5!}{(5-5)!} = \frac{5!}{0!} = \frac{5!}{1} = 5 \cdot 4 \cdot 3 \cdot 2 \cdot 1 = 120$$

$P(n, r)$ is denoted nPr on many calculators. We can calculate the permutations in Example 5 as follows.

TI-83/4 Plus

5 [MATH] [◁] [2] 2 [ENTER]

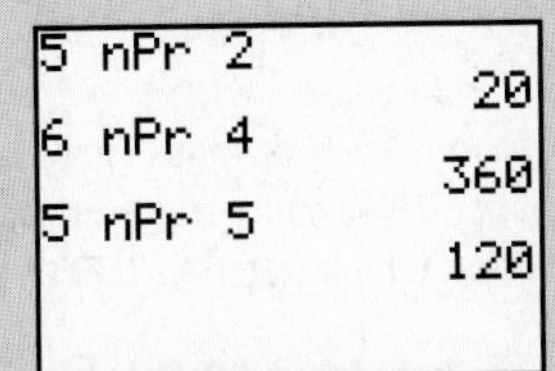

TI-86

5 [2nd] [MATH] [PROB(F2)] [nPr(F2)] 2 [ENTER]

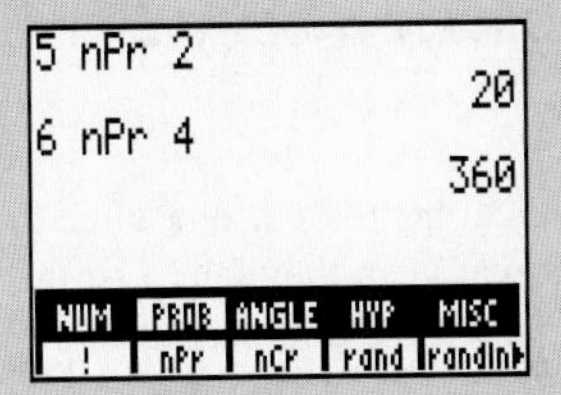

9.6 Exercises

Exer. 1–8: Find the number.

1 $P(7, 3)$

2 $P(8, 5)$

3 $P(9, 6)$

4 $P(5, 3)$

▶ 5 $P(5, 5)$

6 $P(4, 4)$

7 $P(6, 1)$

▶ 8 $P(5, 1)$

Exer. 9–12: Simplify the permutation.

9 $P(n, 0)$

10 $P(n, 1)$

11 $P(n, n - 1)$

12 $P(n, 2)$

13 How many three-digit numbers can be formed from the digits 1, 2, 3, 4, and 5 if repetitions

(a) are not allowed? (b) are allowed?

14 Work Exercise 13 for four-digit numbers.

15 How many numbers can be formed from the digits 1, 2, 3, and 4 if repetitions are not allowed? (*Note:* 42 and 231 are examples of such numbers.)

16 Determine the number of positive integers less than 10,000 that can be formed from the digits 1, 2, 3, and 4 if repetitions are allowed.

▶ **Tutorial available at www.thomsonedu.com/login**

▶ 17 **Basketball standings** If eight basketball teams are in a tournament, find the number of different ways that first, second, and third place can be decided, assuming ties are not allowed.

18 **Basketball standings** Work Exercise 17 for 12 teams.

19 **Wardrobe mix 'n' match** A girl has four skirts and six blouses. How many different skirt-blouse combinations can she wear?

20 **Wardrobe mix 'n' match** Refer to Exercise 19. If the girl also has three sweaters, how many different skirt-blouse-sweater combinations can she wear?

21 **License plate numbers** In a certain state, automobile license plates start with one letter of the alphabet, followed by five digits $(0, 1, 2, \ldots, 9)$. Find how many different license plates are possible if

(a) the first digit following the letter cannot be 0

(b) the first letter cannot be O or I and the first digit cannot be 0

22 **Tossing dice** Two dice are tossed, one after the other. In how many different ways can they fall? List the number of different ways the sum of the dots can equal

(a) 3 (b) 5 (c) 7 (d) 9 (e) 11

23 **Seating arrangement** A row of six seats in a classroom is to be filled by selecting individuals from a group of ten students.

(a) In how many different ways can the seats be occupied?

(b) If there are six boys and four girls in the group and if boys and girls are to be alternated, find the number of different seating arrangements.

24 **Scheduling courses** A student in a certain college may take mathematics at 8, 10, 11, or 2 o'clock; English at 9, 10, 1, or 2; and history at 8, 11, 2, or 3. Find the number of different ways in which the student can schedule the three courses.

▶ 25 **True-or-false test** In how many different ways can a test consisting of ten true-or-false questions be completed?

26 **Multiple-choice test** A test consists of six multiple-choice questions, and there are five choices for each question. In how many different ways can the test be completed?

27 **Seating arrangement** In how many different ways can eight people be seated in a row?

▶ 28 **Book arrangement** In how many different ways can ten books be arranged on a shelf?

29 **Semaphore** With six different flags, how many different signals can be sent by placing three flags, one above the other, on a flag pole?

30 **Selecting books** In how many different ways can five books be selected from a twelve-volume set of books?

31 **Radio call letters** How many four-letter radio station call letters can be formed if the first letter must be K or W and repetitions

(a) are not allowed? (b) are allowed?

32 **Fraternity designations** There are 24 letters in the Greek alphabet. How many fraternities may be specified by choosing three Greek letters if repetitions

(a) are not allowed? (b) are allowed?

33 **Phone numbers** How many ten-digit phone numbers can be formed from the digits $0, 1, 2, 3, \ldots, 9$ if the first digit may not be 0?

34 **Baseball batting order** After selecting nine players for a baseball game, the manager of the team arranges the batting order so that the pitcher bats last and the best hitter bats third. In how many different ways can the remainder of the batting order be arranged?

35 **ATM access code** A customer remembers that 2, 4, 7, and 9 are the digits of a four-digit access code for an automatic bank-teller machine. Unfortunately, the customer has forgotten the order of the digits. Find the largest possible number of trials necessary to obtain the correct code.

36 **ATM access code** Work Exercise 35 if the digits are 2, 4, and 7 and one of these digits is repeated in the four-digit code.

37 **Selecting theater seats** Three married couples have purchased tickets for a play. Spouses are to be seated next to each other, and the six seats are in a row. In how many ways can the six people be seated?

38 **Horserace results** Ten horses are entered in a race. If the possibility of a tie for any place is ignored, in how many ways can the first-, second-, and third-place winners be determined?

39 **Lunch possibilities** Owners of a restaurant advertise that they offer 1,114,095 different lunches based on the fact that they have 16 "free fixins" to go along with any of their 17 menu items (sandwiches, hot dogs, and salads). How did they arrive at that number?

40 **Shuffling cards**

(a) In how many ways can a standard deck of 52 cards be shuffled?

(b) In how many ways can the cards be shuffled so that the four aces appear on the top of the deck?

41 **Numerical palindromes** A palindrome is an integer, such as 45654, that reads the same backward and forward.

(a) How many five-digit palindromes are there?

(b) How many n-digit palindromes are there?

42 **Color arrangements** Each of the six squares shown in the figure is to be filled with any one of ten possible colors. How many ways are there of coloring the strip shown in the figure so that no two adjacent squares have the same color?

Exercise 42

43 This exercise requires a graphing utility that can graph $x!$.

(a) Graph $y = \dfrac{x!\, e^x}{x^x\sqrt{2\pi x}}$ on $(0, 20]$, and estimate the horizontal asymptote.

(b) Use the graph in part (a) to find an approximation for $n!$ if n is a large positive integer.

44 (a) What happens if a calculator is used to find $P(150, 50)$? Explain.

(b) Approximate r if $P(150, 50) = 10^r$ by using the following formula from advanced mathematics:

$$\log n! \approx \frac{n \ln n - n}{\ln 10}$$

9.7 Distinguishable Permutations and Combinations

Certain problems involve finding different arrangements of objects, some of which are indistinguishable. For example, suppose we are given five disks of the same size, of which three are black, one is white, and one is red. Let us find the number of ways they can be arranged in a row so that different color arrangements are obtained. If the disks were all different colors, then the number of arrangements would be 5!, or 120. However, since some of the disks have the same appearance, we cannot obtain 120 different arrangements. To clarify this point, let us write

B B B W R

for the arrangement having black disks in the first three positions in the row, the white disk in the fourth position, and the red disk in the fifth position. The first three disks can be arranged in 3!, or 6, different ways, but these arrangements cannot be distinguished from one another because the first three disks look alike. We say that those 3! permutations are **nondistinguishable.** Similarly, given any other arrangement, say

B R B W B,

there are 3! different ways of arranging the three black disks, but again each such arrangement is nondistinguishable from the others. Let us call two arrangements of objects **distinguishable permutations** if one arrangement cannot be obtained from the other by rearranging like objects. Thus, B B B W R and B R B W B are distinguishable permutations of the five disks. Let k denote the number of distinguishable permutations. Since to each such arrangement there correspond 3! *nondistinguishable* permutations, we must have $3!\,k = 5!$, the number of permutations of five *different* objects. Hence, $k = 5!/3! = 5 \cdot 4 = 20$. By the same type of reasoning we can obtain the following extension of this discussion.

First Theorem on Distinguishable Permutations

If r objects in a collection of n objects are alike and if the remaining objects are different from each other and from the r objects, then the number of distinguishable permutations of the n objects is

$$\frac{n!}{r!}.$$

We can generalize this theorem to the case in which there are several subcollections of nondistinguishable objects. For example, consider eight disks, of which four are black, three are white, and one is red. In this case, with each arrangement, such as

B W B W B W B R,

there are 4! arrangements of the black disks and 3! arrangements of the white disks that have no effect on the color arrangement. Hence, $4!\,3!$ possible arrangements of the disks will not produce distinguishable permutations. If we let k denote the number of *distinguishable* permutations, then $4!\,3!k = 8!$, since 8! is the number of permutations we would obtain if the disks were all different. Thus, the number of distinguishable permutations is

$$k = \frac{8!}{4!\,3!} = \frac{8 \cdot 7 \cdot 6 \cdot 5}{3!} \cdot \frac{4!}{4!} = 280.$$

The following general result can be proved.

Second Theorem on Distinguishable Permutations

If, in a collection of n objects, n_1 are alike of one kind, n_2 are alike of another kind, . . . , n_k are alike of a further kind, and

$$n = n_1 + n_2 + \cdots + n_k,$$

then the number of distinguishable permutations of the n objects is

$$\frac{n!}{n_1!\,n_2!\,\cdots\,n_k!}.$$

EXAMPLE 1 Finding a number of distinguishable permutations

Find the number of distinguishable permutations of the letters in the word *Mississippi.*

▶ SOLUTION In this example we are given a collection of eleven objects in which four are of one kind (the letter s), four are of another kind (i), two are of a third kind (p), and one is of a fourth kind (M). Hence, by the preceding theorem, we have $11 = 4 + 4 + 2 + 1$ and the number of distinguishable permutations is

$$\frac{11!}{4!\,4!\,2!\,1!} = 34{,}650.$$

When we work with permutations, our concern is with the orderings or arrangements of elements. Let us now ignore the order or arrangement of elements and consider the following question: Given a set containing n distinct elements, in how many ways can a subset of r elements be chosen with $r \leq n$? Before answering, let us state a definition.

Definition of Combination

Let S be a set of n elements and let $1 \leq r \leq n$. A **combination** of r elements of S is a subset of S that contains r distinct elements.

If S contains n elements, we also use the phrase **combination of n elements taken r at a time.** The symbol $C(n, r)$ will denote the number of combinations of r elements that can be obtained from a set of n elements.

Theorem on the Number of Combinations

The number of combinations of r elements that can be obtained from a set of n elements is

$$C(n, r) = \frac{n!}{(n - r)!\,r!}, \quad 1 \leq r \leq n.$$

▶ Tutorial available at www.thomsonedu.com/login

The formula for $C(n, r)$ is identical to the formula for the binomial coefficient $\binom{n}{r}$ in Section 9.5.

PROOF If S contains n elements, then, to find $C(n, r)$, we must find the total number of subsets of the form

$$\{x_1, x_2, \ldots, x_r\}$$

such that the x_i are *different* elements of S. Since the r elements $x_1, x_2, \ldots, x_r$ can be arranged in $r!$ different ways, each such subset produces $r!$ different r-tuples. Thus, the total number of different r-tuples is $r!\,C(n, r)$. However, in the previous section we found that the total number of r-tuples is

$$P(n, r) = \frac{n!}{(n-r)!}.$$

Hence,

$$r!\,C(n, r) = \frac{n!}{(n-r)!}.$$

Dividing both sides of the last equation by $r!$ gives us the formula for $C(n, r)$. ■

From the proof, note that

$$P(n, r) = r!\,C(n, r),$$

which means that there are *more permutations than combinations* when we choose a subset of r elements from a set of n elements. To remember this relationship, consider a presidency, say Bush-Quayle. There is only one group or combination of these two people, but when a president–vice-president ordering is associated with these two people, there are two permutations, and Bush-Quayle is clearly different from Quayle-Bush.

As you read the examples and work the exercises, keep the following in mind.

> If the order of selection *is* important, use a *permutation.*
>
> If the order of selection *is not* important, use a *combination.*

EXAMPLE 2 Choosing a baseball squad

A little league baseball squad has six outfielders, seven infielders, five pitchers, and two catchers. Each outfielder can play any of the three outfield positions, and each infielder can play any of the four infield positions. In how many ways can a team of nine players be chosen?

Remember—if the order of selection can be ignored, use a combination.

▶ SOLUTION The number of ways of choosing three outfielders from the six candidates is

$$C(6, 3) = \frac{6!}{(6-3)!\,3!} = \frac{6!}{3!\,3!} = \frac{6 \cdot 5 \cdot 4 \cdot 3!}{3 \cdot 2 \cdot 1 \cdot 3!} = \frac{6 \cdot 5 \cdot 4}{3 \cdot 2 \cdot 1} = 20.$$

▶ **Tutorial available at www.thomsonedu.com/login**

The number of ways of choosing the four infielders is

$$C(7, 4) = \frac{7!}{(7-4)!4!} = \frac{7!}{3!4!} = \frac{7 \cdot 6 \cdot 5 \cdot 4!}{3 \cdot 2 \cdot 1 \cdot 4!} = \frac{7 \cdot 6 \cdot 5}{3 \cdot 2 \cdot 1} = 35.$$

There are five ways of choosing a pitcher and two choices for the catcher. It follows from the fundamental counting principle that the total number of ways to choose a team is

$$20 \cdot 35 \cdot 5 \cdot 2 = 7000.$$

EXAMPLE 3 **Being dealt a full house**

In one type of poker, a five-card hand is dealt from a standard 52-card deck.

(a) How many hands are possible?

(b) A *full house* is a hand that consists of three cards of one denomination and two cards of another denomination. (The 13 denominations are 2's, 3's, 4's, 5's, 6's, 7's, 8's, 9's, 10's, J's, Q's, K's, and A's.) How many hands are full houses?

Figure 1

SOLUTION

(a) The order in which the five cards are dealt is not important, so we use a combination:

$$C(52, 5) = \frac{52!}{(52-5)!5!} = \frac{52 \cdot 51 \cdot 50 \cdot 49 \cdot 48 \cdot 47!}{47! \cdot 5 \cdot 4 \cdot 3 \cdot 2 \cdot 1} = 2{,}598{,}960$$

(b) We first determine how many ways we can be dealt a specific full house—say 3 aces and 2 kings (see Figure 1). There are four cards of each denomination and the order of selection can be ignored, so we use combinations:

The order of selection **is not** *important, so use* ***combinations.*** →

$$\text{number of ways to get 3 A's} = C(4, 3)$$
$$\text{number of ways to get 2 K's} = C(4, 2)$$

Now we must pick the two denominations. Since 3 A's and 2 K's is a different full house than 3 K's and 2 A's, the order of selecting the denominations is important, and so we use a permutation:

The order of selection **is** *important, so use a* ***permutation.*** →

$$\text{number of ways to select two denominations} = P(13, 2)$$

By the fundamental counting principle, the number of full houses is

$$C(4, 3) \cdot C(4, 2) \cdot P(13, 2) = 4 \cdot 6 \cdot 156 = 3744.$$

Tutorial available at www.thomsonedu.com/login

The keystrokes for calculating combinations are nearly identical to those for calculating permutations—just use nCr in place of nPr.

TI-83/4 Plus

5 [MATH] [◁] [3] 2 [ENTER]

```
5 nCr 2
                10
6 nCr 4
                15
5 nCr 5
                 1
```

TI-86

5 [2nd] [MATH] [PROB(F2)] [nCr(F3)] 2 [ENTER]

```
5 nCr 2
                10
6 nCr 4
                15
NUM  PROB  ANGLE  HYP  MISC
 !   nPr   nCr   rand  randIn
```

Note that if $r = n$, the formula for $C(n, r)$ becomes

$$C(n, n) = \frac{n!}{(n - n)!\,n!} = \frac{n!}{0!\,n!} = \frac{n!}{1 \cdot n!} = 1.$$

It is convenient to assign a meaning to $C(n, r)$ if $r = 0$. If the formula is to be true in this case, then we must have

$$C(n, 0) = \frac{n!}{(n - 0)!\,0!} = \frac{n!}{n!\,0!} = \frac{n!}{n! \cdot 1} = 1.$$

Hence, we *define* $C(n, 0) = 1$, which is the same as $C(n, n)$. Finally, for consistency, we also *define* $C(0, 0) = 1$. Thus, $C(n, r)$ has meaning for all nonnegative integers n and r with $r \leq n$.

EXAMPLE 4 Finding the number of subsets of a set

Let S be a set of n elements. Find the number of distinct subsets of S.

SOLUTION Let r be any nonnegative integer such that $r \leq n$. From our previous work, the number of subsets of S that consist of r elements is $C(n, r)$, or $\binom{n}{r}$. Hence, to find the total number of subsets, it suffices to find the sum

$$\binom{n}{0} + \binom{n}{1} + \binom{n}{2} + \binom{n}{3} + \cdots + \binom{n}{n}. \qquad (*)$$

Recalling the formula for the binomial theorem,

$$(a + b)^n = \sum_{k=0}^{n} \binom{n}{k} a^{n-k} b^k,$$

we can see that the indicated sum (∗) is precisely the binomial expansion of $(1 + 1)^n$. Thus, there are 2^n subsets of a set of n elements. In particular, a set of 3 elements has 2^3, or 8, different subsets. A set of 4 elements has 2^4, or 16, subsets. A set of 10 elements has 2^{10}, or 1024, subsets.

Pascal's triangle, introduced in Section 9.5, can easily be remembered by the following combination form:

$$\binom{0}{0}$$

$$\binom{1}{0} \quad \binom{1}{1}$$

$$\binom{2}{0} \quad \binom{2}{1} \quad \binom{2}{2}$$

$$\binom{3}{0} \quad \binom{3}{1} \quad \binom{3}{2} \quad \binom{3}{3}$$

$$\binom{4}{0} \quad \binom{4}{1} \quad \binom{4}{2} \quad \binom{4}{3} \quad \binom{4}{4}$$

$$\cdot \quad \cdot \quad \cdot \quad \cdot \quad \cdot \quad \cdot$$

$$\cdot \quad \cdot \quad \cdot \quad \cdot \quad \cdot \quad \cdot \quad \cdot$$

Figure 2

```
seq(3 nCr R,R,0,
3)
          {1 3 3 1}
seq(5 nCr R,R,0,
5)
  {1 5 10 10 5 1}
```

Combining this information with that in Example 4, we conclude that the third coefficient in the expansion of $(a + b)^4$, $\binom{4}{2}$, is exactly the same as the number of two-element subsets of a set that contains four elements. We leave it as an exercise to find a generalization of the last statement (see Discussion Exercise 6 at the end of the chapter). Note that we can use the sequence command to generate the rows of Pascal's triangle, as shown in Figure 2.

9.7 *Exercises*

Exer. 1–8: Find the number.

▶ 1 $C(7, 3)$

2 $C(8, 4)$

3 $C(9, 8)$

4 $C(6, 2)$

5 $C(n, n - 1)$

▶ 6 $C(n, 1)$

7 $C(7, 0)$

8 $C(5, 5)$

Exer. 9–10: Find the number of possible color arrangements for the 12 given disks, arranged in a row.

▶ 9 5 black, 3 red, 2 white, 2 green

10 3 black, 3 red, 3 white, 3 green

11 Find the number of distinguishable permutations of the letters in the word *bookkeeper*.

▶ 12 Find the number of distinguishable permutations of the letters in the word *moon*. List all the permutations.

13 **Choosing basketball teams** Ten people wish to play in a basketball game. In how many different ways can two teams of five players each be formed?

▶ **Tutorial available at www.thomsonedu.com/login**

14 Selecting test questions A student may answer any six of ten questions on an examination.

(a) In how many ways can six questions be selected?

(b) How many selections are possible if the first two questions must be answered?

Exer. 15–16: Consider any eight points such that no three are collinear.

15 How many lines are determined?

▶ 16 How many triangles are determined?

17 Book arrangement A student has five mathematics books, four history books, and eight fiction books. In how many different ways can they be arranged on a shelf if books in the same category are kept next to one another?

18 Selecting a basketball team A basketball squad consists of twelve players.

(a) Disregarding positions, in how many ways can a team of five be selected?

(b) If the center of a team must be selected from two specific individuals on the squad and the other four members of the team from the remaining ten players, find the number of different teams possible.

19 Selecting a football team A football squad consists of three centers, ten linemen who can play either guard or tackle, three quarterbacks, six halfbacks, four ends, and four fullbacks. A team must have one center, two guards, two tackles, two ends, two halfbacks, a quarterback, and a fullback. In how many different ways can a team be selected from the squad?

20 Arranging keys on a ring In how many different ways can seven keys be arranged on a key ring if the keys can slide completely around the ring?

▶ 21 Committee selection A committee of 3 men and 2 women is to be chosen from a group of 12 men and 8 women. Determine the number of different ways of selecting the committee.

22 Birth order Let the letters G and B denote a girl birth and a boy birth, respectively. For a family of three boys and three girls, one possible birth order is G G G B B B. How many birth orders are possible for these six children?

Exer. 23–24: Shown in each figure is a street map and a possible path from point *A* to point *B*. How many possible paths are there from *A* to *B* if moves are restricted to the right or up? (*Hint:* If R denotes a move one unit right and U denotes a move one unit up, then the path in Exercise 23 can be specified by R U U R R R U R.)

23

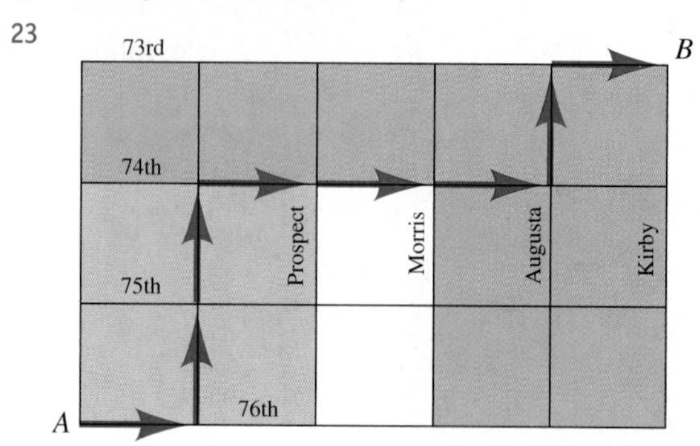

24

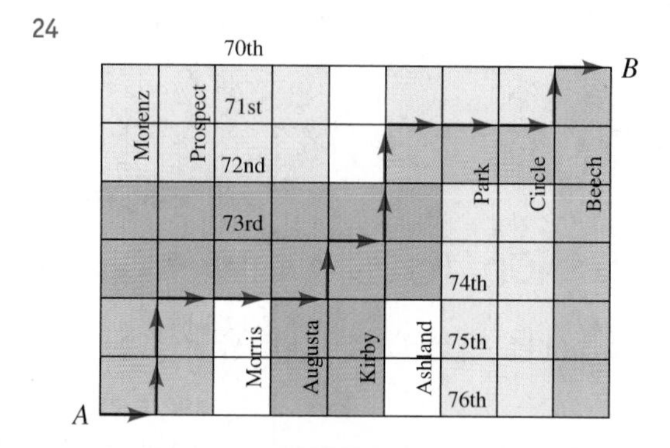

25 Lotto selections To win a state lottery game, a player must correctly select six numbers from the numbers 1 through 49.

(a) Find the total number of selections possible.

(b) Work part (a) if a player selects only even numbers.

26 Office assignments A mathematics department has ten faculty members but only nine offices, so one office must be shared by two individuals. In how many different ways can the offices be assigned?

▶ 27 Tennis tournament In a round-robin tennis tournament, every player meets every other player exactly once. How many players can participate in a tournament of 45 matches?

28 True-or-false test A true-or-false test has 20 questions.

(a) In how many different ways can the test be completed?

(b) In how many different ways can a student answer 10 questions correctly?

29 Basketball championship series The winner of the seven-game NBA championship series is the team that wins four games. In how many different ways can the series be extended to seven games?

30 A geometric design is determined by joining every pair of vertices of an octagon (see the figure).

(a) How many triangles in the design have their three vertices on the octagon?

(b) How many quadrilaterals in the design have their four vertices on the octagon?

Exercise 30

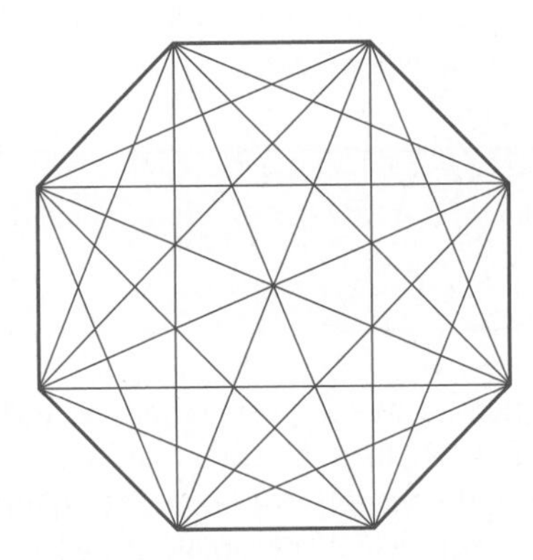

31 Ice cream selections An ice cream parlor stocks 31 different flavors and advertises that it serves almost 4500 different triple scoop cones, with each scoop being a different flavor. How was this number obtained?

32 Choices of hamburger condiments A fast food restaurant advertises that it offers any combination of 8 condiments on a hamburger, thus giving a customer 256 choices. How was this number obtained?

33 Scholarship selection A committee is going to select 30 students from a pool of 1000 to receive scholarships. How many ways could the students be selected if each scholarship is worth

(a) the same amount?

(b) a different amount?

34 Track rankings Twelve sprinters are running a heat; those with the best four times will advance to the finals.

(a) In how many ways can this group of four be selected?

(b) If the four best times will be seeded (ranked) in the finals, in how many ways can this group of four be selected and seeded?

▶ 35 Poker hands Refer to Example 3. How many hands will have exactly three kings?

36 Bridge hands How many 13-card hands dealt from a standard deck will have exactly seven spades?

Exer. 37–38: (a) Calculate the sum S_n for $n = 1, 2, 3, \ldots, 10$, where if $n < r$, then $\binom{n}{r} = 0$. (b) Predict a general formula for S_n.

37 $\binom{n}{1} + \binom{n}{3} + \binom{n}{5} + \binom{n}{7} + \cdots$

38 $(1)\binom{n}{1} - (2)\binom{n}{2} + (3)\binom{n}{3} - (4)\binom{n}{4} + (5)\binom{n}{5} - \cdots$

Exer. 39–42: (a) Graph $C(n, r)$ for the given value of n, where $r = 1, 2, 3, \ldots, n$. (b) Determine the maximum of $C(n, r)$ and the value(s) of r where this maximum occurs.

39 $n = 10$

40 $n = 13$

41 $n = 19$

42 $n = 20$

43 Show that $C(n, r - 1) + C(n, r) = C(n + 1, r)$. Interpret this formula in terms of Pascal's triangle.

9.8 Probability

If two dice are tossed, what are the chances of rolling a 7? If a person is dealt five cards from a standard deck of 52 playing cards, what is the likelihood of obtaining three aces? In the seventeenth century, similar questions about games of chance led to the study of *probability*. Since that time, the theory of probability has grown extensively. It is now used to predict outcomes of a large variety of situations that arise in the natural and social sciences.

Any chance process, such as flipping a coin, rolling a die, being dealt a card from a deck, determining if a manufactured item is defective, or finding the blood pressure of an individual, is an **experiment.** A result of an experiment is an **outcome.** We will restrict our discussion to experiments for which outcomes are **equally likely** unless stated otherwise. This means, for example, that if a coin is flipped, we assume that the possibility of obtaining a head is the same as that of obtaining a tail. Similarly, if a die is tossed, we assume that the die is *fair*—that is, there is an equal chance of obtaining either a 1, 2, 3, 4, 5, or 6. The set S of all possible outcomes of an experiment is the **sample space** of the experiment. Thus, if the experiment consists of flipping a coin and we let H or T denote the outcome of obtaining a head or tail, respectively, then the sample space S may be denoted by

$$S = \{H, T\}.$$

If a fair die is tossed as an experiment, then the set S of all possible outcomes (the sample space) is

$$S = \{1, 2, 3, 4, 5, 6\}.$$

The following definition expresses, in mathematical terms, the notion of obtaining *particular* outcomes of an experiment.

Definition of Event	Let S be the sample space of an experiment. An **event** associated with the experiment is any subset E of S.

Let us consider the experiment of tossing a single die, so that the sample space is $S = \{1, 2, 3, 4, 5, 6\}$. If $E = \{4\}$, then the event E associated with the experiment consists of the outcome of obtaining a 4 on the toss. Different events may be associated with the same experiment. For example, if we let $E = \{1, 3, 5\}$, then this event consists of obtaining an odd number on a toss of the die.

As another illustration, suppose the experiment consists of flipping two coins, one after the other. If we let HH denote the outcome in which two heads appear, HT that of a head on the first coin and a tail on the second, and so on, then the sample space S of the experiment may be denoted by

$$S = \{\text{HH}, \text{HT}, \text{TH}, \text{TT}\}.$$

If we let

$$E = \{\text{HT}, \text{TH}\},$$

then the event E consists of the appearance of a head on one of the coins and a tail on the other.

Next we shall define what is meant by the *probability* of an event. Throughout our discussion we will assume that the sample space S of an experiment contains only a finite number of elements. If E is an event, *the symbols $n(E)$ and $n(S)$ will denote the number of elements in E and S, respectively.* Keep in mind that E and S consist of outcomes that are equally likely.

Definition of the Probability of an Event

Let S be the sample space of an experiment and E an event. The **probability** $P(E)$ of E is given by

$$P(E) = \frac{n(E)}{n(S)}.$$

Since E is a subset of S, we see that

$$0 \leq n(E) \leq n(S).$$

Dividing by $n(S)$, we obtain

$$\frac{0}{n(S)} \leq \frac{n(E)}{n(S)} \leq \frac{n(S)}{n(S)} \quad \text{or, equivalently,} \quad 0 \leq P(E) \leq 1.$$

Note that $P(E) = 0$ if E contains no elements, and $P(E) = 1$ if $E = S$.

The next example provides three illustrations of the preceding definition if E contains exactly one element.

EXAMPLE 1 Finding the probability of an event

(a) If a coin is flipped, find the probability that a head will turn up.

(b) If a fair die is tossed, find the probability of obtaining a 4.

(c) If two coins are flipped, find the probability that both coins turn up heads.

▶ **SOLUTION** For each experiment we shall list sets S and E and then use the definition of probability of an event to find $P(E)$.

(a) $S = \{H, T\}$, $E = \{H\}$, $P(E) = \dfrac{n(E)}{n(S)} = \dfrac{1}{2}$

(b) $S = \{1, 2, 3, 4, 5, 6\}$, $E = \{4\}$, $P(E) = \dfrac{n(E)}{n(S)} = \dfrac{1}{6}$

(c) $S = \{HH, HT, TH, TT\}$, $E = \{HH\}$, $P(E) = \dfrac{n(E)}{n(S)} = \dfrac{1}{4}$

In part (a) of Example 1 we found that the probability of obtaining a head on a flip of a coin is $\frac{1}{2}$. We take this to mean that if a coin is flipped many times, the number of times that a head turns up should be approximately one-half the total number of flips. Thus, for 100 flips, a head should turn up approximately 50 times. It is unlikely that this number will be *exactly* 50. A probability of $\frac{1}{2}$ implies that if we let the number of flips increase, then the number of times a

▶ **Tutorial available at www.thomsonedu.com/login**

head turns up *approaches* one-half the total number of flips. Similar remarks can be made for parts (b) and (c) of Example 1.

In the next two examples we consider experiments in which an event contains more than one element.

EXAMPLE 2 **Finding probabilities when two dice are tossed**

If two dice are tossed, what is the probability of rolling a sum of

(a) 7? **(b)** 9?

▶ **SOLUTION** Let us refer to one die as *the first die* and the other as *the second die*. We shall use ordered pairs to represent outcomes as follows: (2, 4) denotes the outcome of obtaining a 2 on the first die and a 4 on the second; (5, 3) represents a 5 on the first die and a 3 on the second; and so on. Since there are six different possibilities for the first number of the ordered pair and, with each of these, six possibilities for the second number, the total number of ordered pairs is $6 \times 6 = 36$. Hence, if S is the sample space, then $n(S) = 36$.

(a) The event E corresponding to rolling a sum of 7 is given by

$$E = \{(1, 6), (2, 5), (3, 4), (4, 3), (5, 2), (6, 1)\},$$

and consequently $$P(E) = \frac{n(E)}{n(S)} = \frac{6}{36} = \frac{1}{6}.$$

(b) If E is the event corresponding to rolling a sum of 9, then

$$E = \{(3, 6), (4, 5), (5, 4), (6, 3)\}$$

and $$P(E) = \frac{n(E)}{n(S)} = \frac{4}{36} = \frac{1}{9}.$$

In the next example (and in the exercises), when it is stated that one or more cards are drawn from a deck, we mean that each card is removed from a standard 52-card deck and is *not* replaced before the next card is drawn.

EXAMPLE 3 **Finding the probability of drawing a certain hand of cards**

Suppose five cards are drawn from a deck of cards. Find the probability that all five cards are hearts.

▶ **SOLUTION** The sample space S of the experiment is the set of all possible five-card hands that can be formed from the 52 cards in the deck. It follows from our work in the preceding section that $n(S) = C(52, 5)$.

Since there are 13 cards in the heart suit, the number of different ways of obtaining a hand that contains five hearts is $C(13, 5)$. Hence, if E represents this event, then

$$P(E) = \frac{n(E)}{n(S)} = \frac{C(13, 5)}{C(52, 5)} = \frac{\frac{13!}{5!8!}}{\frac{52!}{5!47!}} = \frac{1287}{2{,}598{,}960} \approx 0.0005 = \frac{5}{10{,}000} = \frac{1}{2000}.$$

▶ **Tutorial available at www.thomsonedu.com/login**

This result implies that if the experiment is performed many times, a five-card heart hand should be drawn approximately once every 2000 times.

Suppose S is the sample space of an experiment and E_1 and E_2 are two events associated with the experiment. If E_1 and E_2 have no elements in common, they are called *disjoint sets* and we write $E_1 \cap E_2 = \varnothing$ (the *empty set*). In this case, if one event occurs, the other cannot occur; they are **mutually exclusive events.** Thus, if $E = E_1 \cup E_2$, then

$$n(E) = n(E_1 \cup E_2) = n(E_1) + n(E_2).$$

Hence,

$$P(E) = \frac{n(E_1) + n(E_2)}{n(S)} = \frac{n(E_1)}{n(S)} + \frac{n(E_2)}{n(S)},$$

or

$$P(E) = P(E_1) + P(E_2).$$

The probability of E is therefore the sum of the probabilities of E_1 and E_2. We have proved the following.

Theorem on Mutually Exclusive Events

If E_1 and E_2 are mutually exclusive events and $E = E_1 \cup E_2$, then

$$P(E) = P(E_1 \cup E_2) = P(E_1) + P(E_2).$$

The preceding theorem can be extended to any number of events E_1, $E_2, \ldots, E_k$ that are mutually exclusive in the sense that if $i \neq j$, then $E_i \cap E_j = \varnothing$. The conclusion of the theorem is then

$$P(E) = P(E_1 \cup E_2 \cup \cdots \cup E_k) = P(E_1) + P(E_2) + \cdots + P(E_k).$$

EXAMPLE 4 **Finding probabilities when two dice are tossed**

If two dice are tossed, find the probability of rolling a sum of either 7 or 9.

▶ **SOLUTION** Let E_1 denote the event of rolling 7 and E_2 that of rolling 9. Since E_1 and E_2 cannot occur simultaneously, they are mutually exclusive events. We wish to find the probability of the event $E = E_1 \cup E_2$. From Example 2 we know that $P(E_1) = \frac{6}{36}$ and $P(E_2) = \frac{4}{36}$. Hence, by the last theorem,

$$\begin{aligned} P(E) &= P(E_1) + P(E_2) \\ &= \tfrac{6}{36} + \tfrac{4}{36} = \tfrac{10}{36} = 0.2\overline{7}. \end{aligned}$$

If E_1 and E_2 are events that possibly have elements in common, then the following can be proved.

▶ **Tutorial available at www.thomsonedu.com/login**

Theorem on the Probability of the Occurrence of Either of Two Events

If E_1 and E_2 are any two events, then

$$P(E_1 \cup E_2) = P(E_1) + P(E_2) - P(E_1 \cap E_2).$$

Note that if E_1 and E_2 are mutually exclusive, then $E_1 \cap E_2 = \varnothing$ and $P(E_1 \cap E_2) = 0$. Hence, the last theorem includes, as a special case, the theorem on mutually exclusive events.

EXAMPLE 5 Finding the probability of selecting a certain card from a deck

If a single card is selected from a deck, find the probability that the card is either a jack or a spade.

▶ SOLUTION Let E_1 denote the event that the card is a jack and E_2 the event that it is a spade. The events E_1 and E_2 are *not* mutually exclusive, since there is one card—the jack of spades—in both events, and hence $P(E_1 \cap E_2) = \frac{1}{52}$. By the preceding theorem, the probability that the card is either a jack or a spade is

$$\begin{aligned} P(E_1 \cup E_2) &= P(E_1) + P(E_2) - P(E_1 \cap E_2) \\ &= \tfrac{4}{52} + \tfrac{13}{52} - \tfrac{1}{52} = \tfrac{16}{52} \approx 0.31. \end{aligned}$$

In solving probability problems, it is often helpful to categorize the outcomes of a sample space S into an event E and the set E' of elements of S that are not in E. We call E' the **complement** of E. Note that

$$E \cup E' = S \qquad \text{and} \qquad n(E) + n(E') = n(S).$$

Dividing both sides of the last equation by $n(S)$ gives us

$$\frac{n(E)}{n(S)} + \frac{n(E')}{n(S)} = 1.$$

Hence,

$$P(E) + P(E') = 1, \qquad \text{or} \qquad P(E) = 1 - P(E').$$

We shall use the last formula in the next example.

EXAMPLE 6 Finding the probability of drawing a certain hand of cards

If 13 cards are drawn from a deck, what is the probability that at least 2 of the cards are hearts?

▶ SOLUTION If $P(k)$ denotes the probability of getting k hearts, then the probability of getting *at least* two hearts is

$$P(2) + P(3) + P(4) + \cdots + P(13).$$

▶ **Tutorial available at www.thomsonedu.com/login**

Since the only remaining probabilities are $P(0)$ and $P(1)$, the desired probability is equal to

$$1 - [P(0) + P(1)].$$

To calculate $P(k)$ for any k, we may regard the deck as being split into two groups: hearts and non-hearts. For $P(0)$ we note that of the 13 hearts in the deck, we get none; and of the 39 non-hearts, we get 13. Since the number of ways to choose 13 cards from a 52-card deck is $C(52, 13)$, we see that

$$P(0) = \frac{n(0)}{n(S)} = \frac{C(13, 0) \cdot C(39, 13)}{C(52, 13)} \approx 0.0128.$$

The probability $P(1)$ corresponds to getting 1 of the hearts and 12 of the 39 non-hearts. Thus,

$$P(1) = \frac{n(1)}{n(S)} = \frac{C(13, 1) \cdot C(39, 12)}{C(52, 13)} \approx 0.0801.$$

Hence, the desired probability is

$$1 - [P(0) + P(1)] \approx 1 - [0.0128 + 0.0801] = 0.9071.$$

The words *probability* and *odds* are often used interchangeably. While knowing one allows us to calculate the other, they are quite different.

Definition of the Odds of an Event

Let S be the sample space of an experiment, E an event, and E' its complement. The **odds** $O(E)$ in favor of the event E occurring are given by

$$n(E) \quad \text{to} \quad n(E').$$

The odds $n(E)$ to $n(E')$ are sometimes denoted by $n(E):n(E')$.

We can think of the odds in favor of an event E as the number of ways E occurs compared to the number of ways E doesn't occur. Similarly, the odds *against* E occurring are given by $n(E')$ to $n(E)$.

EXAMPLE 7 Finding odds when two dice are tossed

If two dice are tossed and E is the event of rolling a sum of 7, what are the odds

(a) in favor of E? **(b)** against E?

SOLUTION From Example 2, we have $n(E) = 6$ and $n(S) = 36$, so

$$n(E') = n(S) - n(E) = 36 - 6 = 30.$$

▶ **(a)** The odds in favor of rolling a sum of 7 are $n(E)$ to $n(E')$ or

$$6 \quad \text{to} \quad 30 \quad \text{or, equivalently,} \quad 1 \quad \text{to} \quad 5.$$

(b) The odds against rolling a sum of 7 are $n(E')$ to $n(E)$ or

$$30 \quad \text{to} \quad 6 \quad \text{or, equivalently,} \quad 5 \quad \text{to} \quad 1.$$

▶ **Tutorial available at www.thomsonedu.com/login**

EXAMPLE 8 **Finding probabilities and odds**

(a) If $P(E) = 0.75$, find $O(E)$.

(b) If $O(E)$ are 6 to 5, find $P(E)$.

SOLUTION

(a) Since $P(E) = 0.75 = \frac{3}{4}$ and $P(E) = n(E)/n(S)$, we can let

$$n(E) = 3 \quad \text{and} \quad n(S) = 4.$$

Thus, $n(E') = n(S) - n(E) = 4 - 3 = 1$, and $O(E)$ are given by

$$n(E) \quad \text{to} \quad n(E'), \quad \text{or} \quad 3 \quad \text{to} \quad 1.$$

▶ **(b)** Since $O(E)$ are 6 to 5 and $O(E)$ are $n(E)$ to $n(E')$, we can let

$$n(E) = 6 \quad \text{and} \quad n(E') = 5.$$

Thus, $n(S) = n(E) + n(E') = 6 + 5 = 11$, and

$$P(E) = \frac{n(E)}{n(S)} = \frac{6}{11}.$$

Two events E_1 and E_2 are said to be **independent events** if the occurrence of one does not influence the occurrence of the other.

Theorem on Independent Events	If E_1 and E_2 are independent events, then $$P(E_1 \cap E_2) = P(E_1) \cdot P(E_2).$$

In words, the theorem states that if E_1 and E_2 are independent events, the probability that *both* E_1 and E_2 occur simultaneously is the product of their probabilities. Note that if two events E_1 and E_2 are mutually exclusive, then $P(E_1 \cap E_2) = 0$ and they cannot be independent. (We assume that both E_1 and E_2 are not empty.)

EXAMPLE 9 **An application of probability to an electrical system**

Figure 1

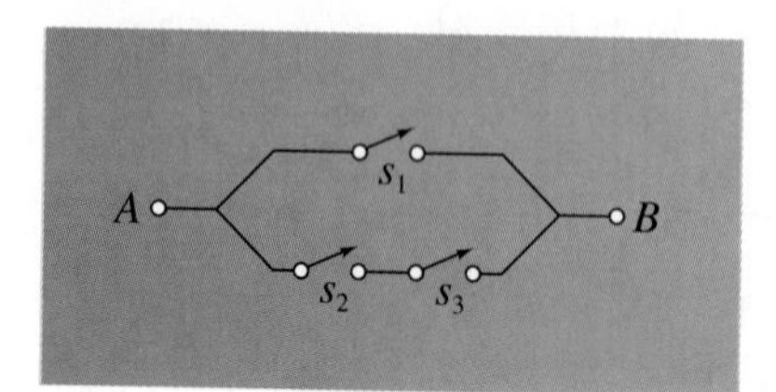

An electrical system has open-close switches s_1, s_2, and s_3, as shown in Figure 1. The switches operate independently of one another, and current will flow from A to B either if s_1 is closed or if *both* s_2 and s_3 are closed.

(a) If S_k denotes the event that s_k is closed, where $k = 1, 2, 3$, express, in terms of $P(S_1)$, $P(S_2)$, and $P(S_3)$, the probability p that current will flow from A to B.

(b) Find p if $P(S_k) = \frac{1}{2}$ for each k.

▶ **Tutorial available at www.thomsonedu.com/login**

▶ SOLUTION

(a) The probability p that either S_1 or *both* S_2 and S_3 occur is

$$p = P(S_1 \cup (S_2 \cap S_3)).$$

Using the theorem on the probability of the occurrence of either of two events S_1 or $S_2 \cap S_3$, we obtain

$$p = P(S_1) + P(S_2 \cap S_3) - P(S_1 \cap (S_2 \cap S_3)).$$

Applying the theorem on independent events twice gives us

$$p = P(S_1) + P(S_2) \cdot P(S_3) - P(S_1) \cdot P(S_2 \cap S_3).$$

Finally, using the theorem on independent events one more time, we see that

$$p = P(S_1) + P(S_2) \cdot P(S_3) - P(S_1) \cdot P(S_2) \cdot P(S_3).$$

(b) If $P(S_k) = \frac{1}{2}$ for each k, then from part (a) the probability that current will flow from A to B is

$$p = \tfrac{1}{2} + \tfrac{1}{2} \cdot \tfrac{1}{2} - \tfrac{1}{2} \cdot \tfrac{1}{2} \cdot \tfrac{1}{2} = \tfrac{5}{8} = 0.625.$$

EXAMPLE 10 **A continuation of Example 9**

Refer to Example 9. If the probability that s_k is closed is the same for each k, determine $P(S_k)$ such that $p = 0.99$.

SOLUTION Since the probability $P(S_k)$ is the same for each k, we let $P(S_k) = x$ for $k = 1, 2, 3$. Substituting in the formula for p obtained in part (a) of Example 9, we obtain

$$p = x + x \cdot x - x \cdot x \cdot x = -x^3 + x^2 + x.$$

Letting $p = 0.99$ gives us the equation

$$-x^3 + x^2 + x = 0.99.$$

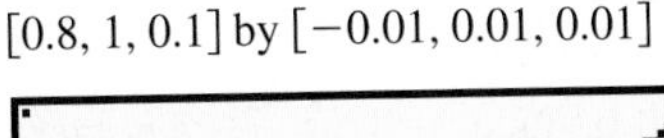

Figure 2
$[0.8, 1, 0.1]$ by $[-0.01, 0.01, 0.01]$

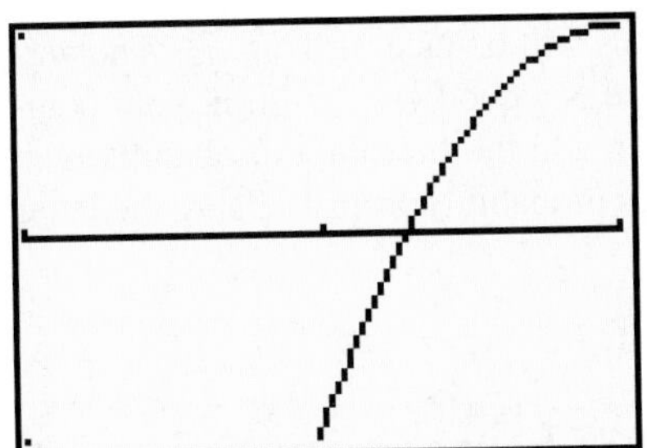

Graphing $y = -x^3 + x^2 + x - 0.99$ using a standard viewing rectangle, we see that there are three x-intercepts. The desired probability must lie between $x = 0$ and $x = 1$ and should be fairly close to 1. Using the viewing rectangle dimensions $[0.8, 1, 0.1]$ by $[-0.01, 0.01, 0.01]$, we obtain a sketch similar to Figure 2. Using a root or zero feature gives us $x \approx 0.93$. Hence, $P(S_k) \approx 0.93$.

Note that the probability that an individual switch is closed is *less* than the probability that current will flow through the system.

EXAMPLE 11 **Using a tree diagram to find a probability**

If two cards are drawn from a deck, what is the probability that at least one of the cards will be a face card?

▶ **Tutorial available at www.thomsonedu.com/login**

Figure 3

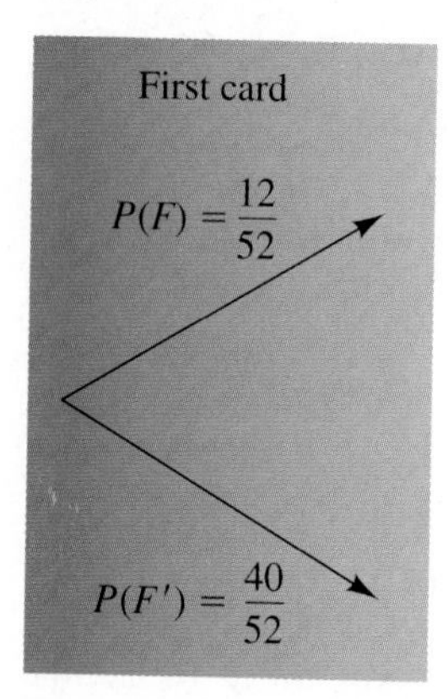

SOLUTION Let F denote the event of drawing a face card. There are 12 face cards in a 52-card deck, so $P(F) = \frac{12}{52}$. We can depict this probability, as well as the probability of its complement, with the *tree diagram* shown in Figure 3.

The probabilities for the second card depend on what the first card was. To cover all possibilities for the second card, we attach branches with similar probabilities to the end of each branch of the first tree diagram, as shown in Figure 4.

Figure 4

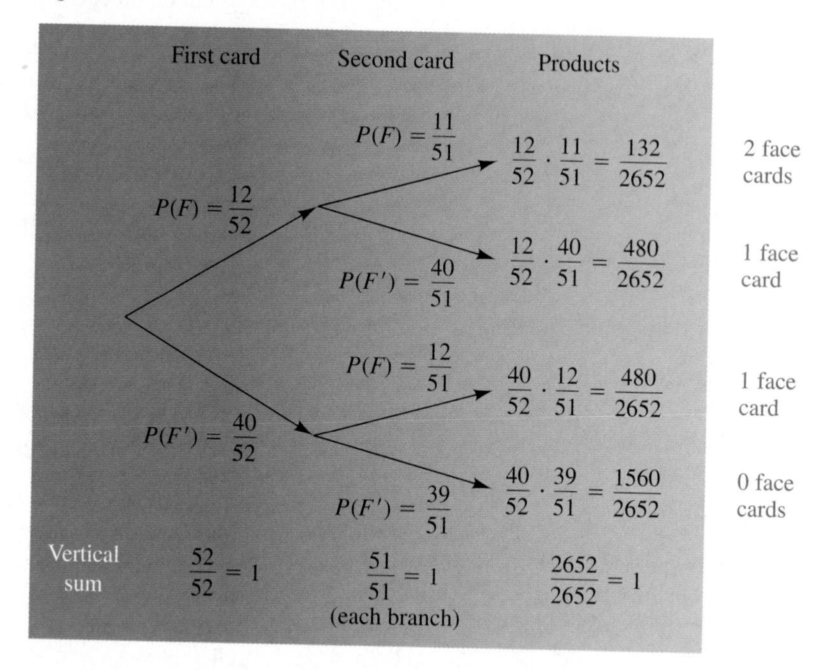

The Products column lists the probabilities for all two-card possibilities; for example, the probability that both cards will be face cards is $\frac{132}{2652}$. The vertical sums must equal 1—calculating these is a good way to check your computations. To answer the question, we can add the first three probabilities in the Products column or subtract the fourth probability from 1. Using the latter approach, we have

$$1 - \frac{1560}{2652} = \frac{1092}{2652} = \frac{7}{17} \approx 41\%.$$

It is often of interest to know what amount of return we can expect on an investment in a game of chance. The following definition will help us answer questions that fall in this category.

Definition of Expected Value

Suppose a variable can have payoff amounts $a_1, a_2, \ldots, a_n$ with corresponding probabilities $p_1, p_2, \ldots, p_n$. The **expected value** EV of the variable is given by

$$\text{EV} = a_1p_1 + a_2p_2 + \cdots + a_np_n = \sum_{k=1}^{n} a_kp_k.$$

EXAMPLE 12 Expected value of a single pull-tab

States that run lotteries often offer games in which a certain number of pull-tabs are printed, some being redeemable for money and the rest worthless. Suppose that in a particular game there are 4000 pull-tabs, 432 of which are redeemable according to the following table.

Number of pull-tabs	Value
4	\$100
8	50
20	20
400	2

Find the expected value of a pull-tab that sells for \$1.

▶ SOLUTION The payoff amounts \$100, \$50, \$20, and \$2 have probabilities $\frac{4}{4000}, \frac{8}{4000}, \frac{20}{4000}$, and $\frac{400}{4000}$, respectively. The remaining 3568 pull-tabs have a payoff amount of \$0. By the preceding definition, the expected value of a single pull-tab is

$$\begin{aligned} \text{EV} &= 100 \cdot \tfrac{4}{4000} + 50 \cdot \tfrac{8}{4000} + 20 \cdot \tfrac{20}{4000} + 2 \cdot \tfrac{400}{4000} + 0 \cdot \tfrac{3568}{4000} \\ &= \tfrac{2000}{4000} = \$0.50. \end{aligned}$$

Thus, after subtracting the \$1 cost of the pull-tab, we can expect to *lose* \$0.50 on each pull-tab we buy. Note that we cannot lose \$0.50 on any individual pull-tab, but we can expect to lose this amount on each pull-tab in the long run. This game yields a terribly poor return for the buyer and a healthy profit for the seller.

The expected value of \$0.50 obtained in Example 12 may be considered to be the amount we would expect to pay to play the game if the game were *fair*—that is, if we would not expect to win or lose any money after playing the game many times.

In this section we have merely introduced several basic concepts about probability. The interested person is referred to entire books and courses devoted to this branch of mathematics.

▶ Tutorial available at www.thomsonedu.com/login

9.8 *Exercises*

Exer. 1–2: A single card is drawn from a deck. Find the probability and the odds that the card is as specified.

1 (a) a king

(b) a king or a queen

(c) a king, a queen, or a jack

2 (a) a heart

(b) a heart or a diamond

(c) a heart, a diamond, or a club

Exer. 3–4: A single die is tossed. Find the probability and the odds that the die is as specified.

3 (a) a 4 (b) a 6 (c) a 4 or a 6

4 (a) an even number (b) a number divisible by 5

(c) an even number or a number divisible by 5

Exer. 5–6: An urn contains five red balls, six green balls, and four white balls. If a single ball is drawn, find the probability and the odds that the ball is as specified.

5 (a) red (b) green (c) red or white

6 (a) white (b) green or white (c) not green

Exer. 7–8: Two dice are tossed. Find the probability and the odds that the sum is as specified.

7 (a) 11 (b) 8 (c) 11 or 8

8 (a) greater than 9 (b) an odd number

Exer. 9–10: Three dice are tossed. Find the probability of the specified event.

9 A sum of 5

10 A 6 turns up on exactly one die

11 If three coins are flipped, find the probability that exactly two heads turn up.

12 If four coins are flipped, find the probability of obtaining two heads and two tails.

13 If $P(E) = \frac{5}{7}$, find $O(E)$ and $O(E')$.

14 If $P(E) = 0.4$, find $O(E)$ and $O(E')$.

15 If $O(E)$ are 9 to 5, find $O(E')$ and $P(E)$.

16 If $O(E')$ are 7 to 3, find $O(E)$ and $P(E)$.

Exer. 17–18: For the given value of $P(E)$, approximate $O(E)$ in terms of "X to 1."

17 $P(E) \approx 0.659$ 18 $P(E) \approx 0.822$

Exer. 19–24: Suppose five cards are drawn from a deck. Find the probability of obtaining the indicated cards.

19 Four of a kind (such as four aces or four kings)

20 Three aces and two kings

21 Four diamonds and one spade

22 Five face cards

23 A flush (five cards of the same suit)

24 A royal flush (an ace, king, queen, jack, and 10 of the same suit)

25 If a single die is tossed, find the probability of obtaining an odd number or a prime number.

26 A single card is drawn from a deck. Find the probability that the card is either red or a face card.

27 If the probability of a baseball player's getting a hit in one time at bat is 0.326, find the probability that the player gets no hits in 4 times at bat.

28 If the probability of a basketball player's making a free throw is 0.9, find the probability that the player makes at least 1 of 2 free throws.

Exer. 29–30: The outcomes 1, 2, . . . , 6 of an experiment and their probabilities are listed in the table.

Outcome	1	2	3	4	5	6
Probability	0.25	0.10	0.15	0.20	0.25	0.05

For the indicated events, find (a) $P(E_2)$, (b) $P(E_1 \cap E_2)$, (c) $P(E_1 \cup E_2)$, and (d) $P(E_2 \cup E_3')$.

29 $E_1 = \{1, 2\}$; $E_2 = \{2, 3, 4\}$; $E_3 = \{4, 6\}$

30 $E_1 = \{1, 2, 3, 6\}$; $E_2 = \{3, 4\}$; $E_3 = \{4, 5, 6\}$

Exer. 31–32: A box contains 10 red chips, 20 blue chips, and 30 green chips. If 5 chips are drawn from the box, find the probability of drawing the indicated chips.

31 (a) all blue

(b) at least 1 green

(c) at most 1 red

32 (a) exactly 4 green

(b) at least 2 red

(c) at most 2 blue

33 True-or-false test A true-or-false test consists of eight questions. If a student guesses the answer for each question, find the probability that

(a) eight answers are correct

(b) seven answers are correct and one is incorrect

(c) six answers are correct and two are incorrect

(d) at least six answers are correct

34 Committee selection A 6-member committee is to be chosen by drawing names of individuals from a hat. If the hat contains the names of 8 men and 14 women, find the probability that the committee will consist of 3 men and 3 women.

Exer. 35–36: Five cards are drawn from a deck. Find the probability of the specified event.

35 Obtaining at least one ace

36 Obtaining at least one heart

37 Card and die experiment Each suit in a deck is made up of an ace (A), nine numbered cards (2, 3, . . . , 10), and three face cards (J, Q, K). An experiment consists of drawing a single card from a deck followed by rolling a single die.

(a) Describe the sample space S of the experiment, and find $n(S)$.

(b) Let E_1 be the event consisting of the outcomes in which a numbered card is drawn and the number of dots on the die is the same as the number on the card. Find $n(E_1)$, $n(E_1')$, and $P(E_1)$.

(c) Let E_2 be the event in which the card drawn is a face card, and let E_3 be the event in which the number of dots on the die is even. Are E_2 and E_3 mutually exclusive? Are they independent? Find $P(E_2)$, $P(E_3)$, $P(E_2 \cap E_3)$, and $P(E_2 \cup E_3)$.

(d) Are E_1 and E_2 mutually exclusive? Are they independent? Find $P(E_1 \cap E_2)$ and $P(E_1 \cup E_2)$.

38 Letter and number experiment An experiment consists of selecting a letter from the alphabet and one of the digits 0, 1, . . . , 9.

(a) Describe the sample space S of the experiment, and find $n(S)$.

(b) Suppose the letters of the alphabet are assigned numbers as follows: $A = 1, B = 2, \ldots, Z = 26$. Let E_1 be the event in which the units digit of the number assigned to the letter of the alphabet is the same as the digit selected. Find $n(E_1)$, $n(E_1')$, and $P(E_1)$.

(c) Let E_2 be the event that the letter is one of the five vowels and E_3 the event that the digit is a prime number. Are E_2 and E_3 mutually exclusive? Are they independent? Find $P(E_2)$, $P(E_3)$, $P(E_2 \cap E_3)$, and $P(E_2 \cup E_3)$.

(d) Let E_4 be the event that the numerical value of the letter is even. Are E_2 and E_4 mutually exclusive? Are they independent? Find $P(E_2 \cap E_4)$ and $P(E_2 \cup E_4)$.

39 Tossing dice If two dice are tossed, find the probability that the sum is greater than 5.

40 Tossing dice If three dice are tossed, find the probability that the sum is less than 16.

41 Family makeup Assuming that girl-boy births are equally probable, find the probability that a family with five children has

(a) all boys (b) at least one girl

42 Slot machine A standard slot machine contains three reels, and each reel contains 20 symbols. If the first reel has five bells, the middle reel four bells, and the last reel two bells, find the probability of obtaining three bells in a row.

43 ESP experiment In a simple experiment designed to test ESP, four cards (jack, queen, king, and ace) are shuffled and then placed face down on a table. The subject then attempts to identify each of the four cards, giving a different name to each of the cards. If the individual is guessing, find the probability of correctly identifying

(a) all four cards (b) exactly two of the four cards

44 Tossing dice Three dice are tossed.

(a) Find the probability that all dice show the same number of dots.

(b) Find the probability that the numbers of dots on the dice are all different.

(c) Work parts (a) and (b) for n dice.

45 Trick dice For a normal die, the sum of the dots on opposite faces is 7. Shown in the figure is a pair of trick dice in which the *same* number of dots appears on opposite faces. Find the probability of rolling a sum of

(a) 7 (b) 8

Exercise 45

46 Carnival game In a common carnival game, three balls are rolled down an incline into slots numbered 1 through 9, as shown in the figure. Because the slots are so narrow, players have no control over where the balls collect. A prize is given if the sum of the three numbers is less than 7. Find the probability of winning a prize.

Exercise 46

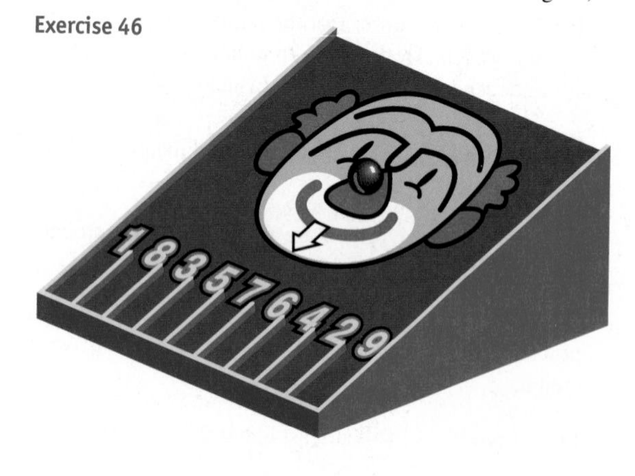

47 Smoking deaths In an average year during 1995–1999, smoking caused 442,398 deaths in the United States. Of these deaths, cardiovascular disease accounted for 148,605, cancer for 155,761, and respiratory diseases such as emphysema for 98,007.

(a) Find the probability that a smoking-related death was the result of either cardiovascular disease or cancer.

(b) Determine the probability that a smoking-related death was not the result of respiratory diseases.

48 Starting work times In a survey about what time people go to work, it was found that 8.2 million people go to work between midnight and 6 A.M., 60.4 million between 6 A.M. and 9 A.M., and 18.3 million between 9 A.M. and midnight.

(a) Find the probability that a person goes to work between 6 A.M. and midnight.

(b) Determine the probability that a person goes to work between midnight and 6 A.M.

49 Arsenic exposure and cancer In a certain county, 2% of the people have cancer. Of those with cancer, 70% have been exposed to high levels of arsenic. Of those without cancer, 10% have been exposed. What percentage of the people who have been exposed to high levels of arsenic have cancer? (*Hint*: Use a tree diagram.)

50 Computers and defective chips A computer manufacturer buys 30% of its chips from supplier A and the rest from supplier B. Two percent of the chips from supplier A are defective, as are 4% of the chips from supplier B. Approximately what percentage of the defective chips are from supplier B?

51 Probability demonstration Shown in the figure on the next page is a small version of a probability demonstration device. A small ball is dropped into the top of the maze and tumbles to the bottom. Each time the ball strikes an obstacle, there is a 50% chance that the ball will move to the left. Find the probability that the ball ends up in the slot

(a) on the far left (b) in the middle

Exercise 51

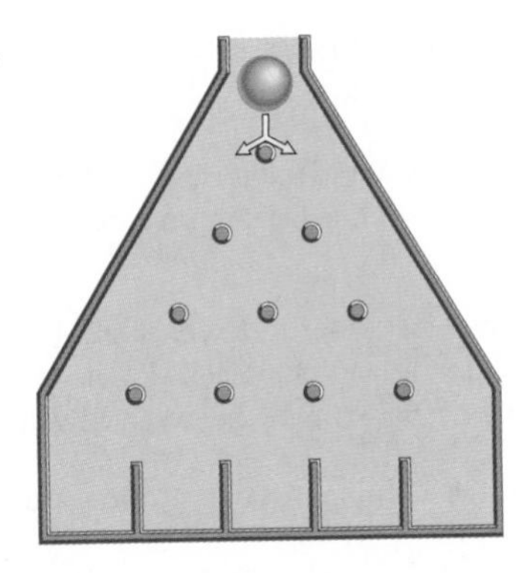

52 Roulette In the American version of roulette, a ball is spun around a wheel and has an equal chance of landing in any one of 38 slots numbered 0, 00, 1, 2, . . . , 36. Shown in the figure is a standard betting layout for roulette, where the color of the oval corresponds to the color of the slot on the wheel. Find the probability that the ball lands

(a) in a black slot

(b) in a black slot twice in succession

Exercise 52

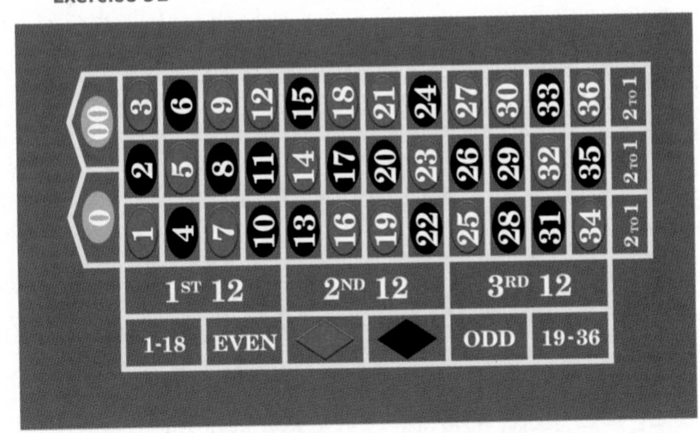

53 Selecting lottery numbers In one version of a popular lottery game, a player selects six of the numbers from 1 to 54. The agency in charge of the lottery also selects six numbers. What is the probability that the player will match the six numbers if two 50¢ tickets are purchased? (This jackpot is worth at least $2 million in prize money and grows according to the number of tickets sold.)

54 Lottery Refer to Exercise 53. The player can win about $1000 for matching five of the six numbers and about $40 for matching four of the six numbers. Find the probability that the player will win some amount of prize money on the purchase of one ticket.

55 Quality control In a quality control procedure to test for defective light bulbs, two light bulbs are randomly selected from a large sample without replacement. If either light bulb is defective, the entire lot is rejected. Suppose a sample of 200 light bulbs contains 5 defective light bulbs. Find the probability that the sample will be rejected. (*Hint:* First calculate the probability that neither bulb is defective.)

56 Life expectancy A man is 54 years old and a woman is 34 years old. The probability that the man will be alive in 10 years is 0.74, whereas the probability that the woman will be alive 10 years from now is 0.94. Assume that their life expectancies are unrelated.

(a) Find the probability that they will both be alive 10 years from now.

(b) Determine the probability that neither one will be alive 10 years from now.

(c) Determine the probability that at least one of the two will be alive 10 years from now.

57 Shooting craps In the game of *craps*, there are two ways a player can win a *pass line* bet. The player wins immediately if two dice are rolled and their sum is 7 or 11. If their sum is 4, 5, 6, 8, 9, or 10, the player can still win a pass line bet if this same number (called the *point*) is rolled again before a 7 is rolled. Find the probability that the player wins

(a) a pass line bet on the first roll

(b) a pass line bet with a 4 on the first roll

(c) on any pass line bet

58 Crapless craps Refer to Exercise 57. In the game of craps, a player loses a pass line bet if a sum of 2, 3, or 12 is obtained on the first roll (referred to as "craps"). In another version of the game, called *crapless craps,* the player does not lose by rolling craps and does not win by rolling an 11 on the first roll. Instead, the player wins if the first roll is a 7 or if the point (2–12, excluding 7) is repeated before a 7 is rolled. Find the probability that the player wins on a pass line bet in crapless craps.

Exer. 59–60: Refer to Examples 9 and 10. (a) Find p for the electrical system shown in the figure if $P(S_k) = 0.9$ for each k. (b) Use a graph to estimate $P(S_k)$ if $p = 0.99$.

59

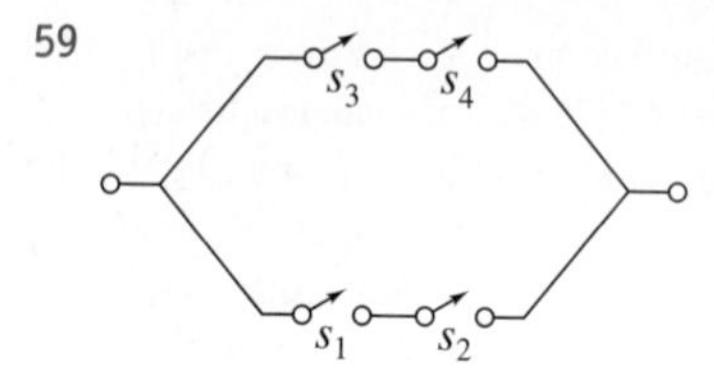

60

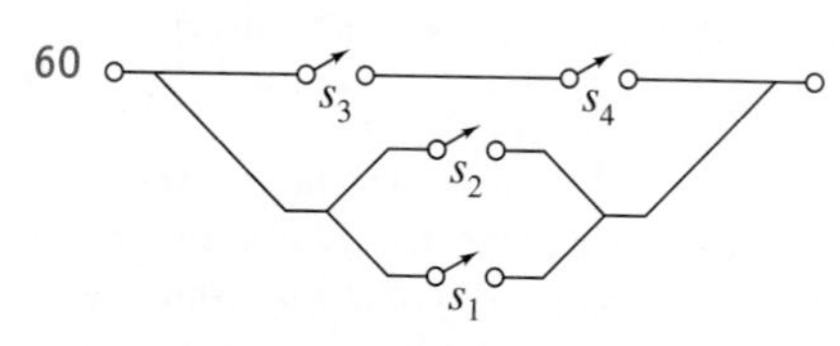

61 Birthday probability

(a) Show that the probability p that n people all have different birthdays is given by

$$p = \frac{365!}{365^n(365 - n)!}.$$

(b) If a room contains 32 people, approximate the probability that two or more people have the *same* birthday. (First approximate $\ln p$ by using the following formula from advanced mathematics:

$$\ln n! \approx n \ln n - n.)$$

62 Birthday probability Refer to Exercise 61. Find the smallest number of people in a room such that the probability that everyone has a different birthday is less than $\frac{1}{2}$. *Hint:* Rewrite the formula for p in part (a) of the previous exercise as

$$\frac{365}{365} \cdot \frac{364}{365} \cdot \frac{363}{365} \cdot \cdots \cdot \frac{365 - n + 1}{365}.$$

63 A bet in craps Refer to Exercise 57. A player receives \$2 for winning a \$1 pass line bet. Approximate the expected value of a \$1 bet.

64 A bet in roulette Refer to Exercise 52. If a player bets \$1 that the ball will land in a black slot, he or she will receive \$2 if it does. Approximate the expected value of a \$1 bet.

65 Contest prize winning A contest offers the following cash prizes:

Number of prizes	1	10	100	1000
Prize values	\$1,000,000	\$100,000	\$10,000	\$1000

If the sponsor expects 20 million contestants, find the expected value for a single contestant.

66 Tournament prize winnings A bowling tournament is handicapped so that all 80 bowlers are equally matched. The tournament prizes are listed in the table.

Place	1st	2nd	3rd	4th	5th–10th
Prize	\$1000	\$500	\$300	\$200	\$100

Find the expected winnings for one contestant.

CHAPTER 9 REVIEW EXERCISES

▶ **Online support materials can be found at www.thomsonedu.com/login**

Exer. 1–4: Find the first four terms and the seventh term of the sequence that has the given nth term.

1 $\left\{\dfrac{5n}{3 - 2n^2}\right\}$

2 $\{(-1)^{n+1} - (0.1)^n\}$

3 $\left\{1 + \left(-\frac{1}{2}\right)^{n-1}\right\}$

4 $\left\{\dfrac{2^n}{(n + 1)(n + 2)(n + 3)}\right\}$

Exer. 5–8: Find the first five terms of the recursively defined infinite sequence.

5 $a_1 = 10, \quad a_{k+1} = 1 + (1/a_k)$

6 $a_1 = 2, \quad a_{k+1} = a_k!$

7 $a_1 = 9, \quad a_{k+1} = \sqrt{a_k}$

8 $a_1 = 1, \quad a_{k+1} = (1 + a_k)^{-1}$

Exer. 9–12: Evaluate.

9 $\displaystyle\sum_{k=1}^{5} (k^2 + 4)$

10 $\displaystyle\sum_{k=2}^{6} \frac{2k - 8}{k - 1}$

11 $\displaystyle\sum_{k=7}^{100} 10$

12 $\displaystyle\sum_{k=1}^{4} (2^k - 10)$

Exer. 13–24: Express the sum in terms of summation notation. (Answers are not unique.)

13 $3 + 6 + 9 + 12 + 15$

14 $4 + 2 + 1 + \frac{1}{2} + \frac{1}{4} + \frac{1}{8}$

15 $\dfrac{1}{1 \cdot 2} + \dfrac{1}{2 \cdot 3} + \dfrac{1}{3 \cdot 4} + \cdots + \dfrac{1}{99 \cdot 100}$

16 $\dfrac{1}{1 \cdot 2 \cdot 3} + \dfrac{1}{2 \cdot 3 \cdot 4} + \dfrac{1}{3 \cdot 4 \cdot 5} + \cdots + \dfrac{1}{98 \cdot 99 \cdot 100}$

17 $\frac{1}{2} + \frac{2}{5} + \frac{3}{8} + \frac{4}{11}$

18 $\frac{1}{4} + \frac{2}{9} + \frac{3}{14} + \frac{4}{19}$

19 $100 - 95 + 90 - 85 + 80$

20 $1 - \frac{1}{2} + \frac{1}{3} - \frac{1}{4} + \frac{1}{5} - \frac{1}{6} + \frac{1}{7}$

21 $a_0 + a_1x^4 + a_2x^8 + \cdots + a_{25}x^{100}$

22 $a_0 + a_1x^3 + a_2x^6 + \cdots + a_{20}x^{60}$

23 $1 - \dfrac{x^2}{2} + \dfrac{x^4}{4} - \dfrac{x^6}{6} + \cdots + (-1)^n\dfrac{x^{2n}}{2n}$

24 $1 + x + \dfrac{x^2}{2} + \dfrac{x^3}{3} + \cdots + \dfrac{x^n}{n}$

25 Find the tenth term and the sum of the first ten terms of the arithmetic sequence whose first two terms are $4 + \sqrt{3}$ and 3.

26 Find the sum of the first eight terms of the arithmetic sequence in which the fourth term is 9 and the common difference is -5.

27 The fifth and thirteenth terms of an arithmetic sequence are 5 and 77, respectively. Find the first term and the tenth term.

28 Find the number of terms in the arithmetic sequence with $a_1 = 1$, $d = 5$, and $S = 342$.

29 Insert four arithmetic means between 20 and -10.

30 Find the tenth term of the geometric sequence whose first two terms are $\frac{1}{8}$ and $\frac{1}{4}$.

31 If a geometric sequence has 3 and -0.3 as its third and fourth terms, respectively, find the eighth term.

32 Given a geometric sequence with $a_3 = 16$ and $a_7 = 625$, find a_8.

33 Find the geometric mean of 4 and 8.

34 In a certain geometric sequence, the eighth term is 100 and the common ratio is $-\frac{3}{2}$. Find the first term.

35 Given an arithmetic sequence such that $S_{12} = 402$ and $a_{12} = 50$, find a_1 and d.

36 Given a geometric sequence such that $a_5 = \frac{1}{16}$ and $r = \frac{3}{2}$, find a_1 and S_5.

Exer. 37–40: Evaluate.

37 $\displaystyle\sum_{k=1}^{15} (5k - 2)$

38 $\displaystyle\sum_{k=1}^{10} \left(6 - \tfrac{1}{2}k\right)$

39 $\displaystyle\sum_{k=1}^{10} \left(2^k - \tfrac{1}{2}\right)$

40 $\displaystyle\sum_{k=1}^{8} \left(\tfrac{1}{2} - 2^k\right)$

41 Find the sum of the infinite geometric series

$$1 - \tfrac{2}{5} + \tfrac{4}{25} - \tfrac{8}{125} + \cdots.$$

42 Find the rational number whose decimal representation is $6.\overline{274}$.

Exer. 43–47: Prove that the statement is true for every positive integer *n*.

43 $2 + 5 + 8 + \cdots + (3n - 1) = \dfrac{n(3n + 1)}{2}$

44 $2^2 + 4^2 + 6^2 + \cdots + (2n)^2 = \dfrac{2n(2n + 1)(n + 1)}{3}$

45 $\dfrac{1}{1 \cdot 3} + \dfrac{1}{3 \cdot 5} + \dfrac{1}{5 \cdot 7} + \cdots + \dfrac{1}{(2n - 1)(2n + 1)} = \dfrac{n}{2n + 1}$

46 $1 \cdot 2 + 2 \cdot 3 + 3 \cdot 4 + \cdots + n(n + 1) = \dfrac{n(n + 1)(n + 2)}{3}$

47 3 is a factor of $n^3 + 2n$.

48 Prove that $n^2 + 3 < 2^n$ for every positive integer $n \geq 5$.

Exer. 49–50: Find the smallest positive integer j for which the statement is true. Use the extended principle of mathematical induction to prove that the formula is true for every integer greater than j.

49 $2^n \leq n!$

50 $10^n \leq n^n$

Exer. 51–52: Use the binomial theorem to expand and simplify the expression.

51 $(x^2 - 3y)^6$

52 $(2x + y^3)^4$

Exer. 53–56: Without expanding completely, find the indicated term(s) in the expansion of the expression.

53 $(x^{2/5} + 2x^{-3/5})^{20}$; first three terms

54 $\left(y^3 - \frac{1}{2}c^2\right)^9$; sixth term

55 $(4x^2 - y)^7$; term that contains x^{10}

56 $(2c^3 + 5c^{-2})^{10}$; term that does not contain c

57 **Building blocks** Ten-foot lengths of 2×2 lumber are to be cut into five pieces to form children's building blocks; the lengths of the five blocks are to form an arithmetic sequence.

(a) Show that the difference d in lengths must be less than 1 foot.

(b) If the smallest block is to have a length of 6 inches, find the lengths of the other four pieces.

58 **Constructing a ladder** A ladder is to be constructed with 16 rungs whose lengths decrease uniformly from 20 inches at the base to 16 inches at the top. Find the total length of material needed for the rungs.

59 Shown in the first figure is a broken-line curve obtained by taking two adjacent sides of a square, each of length s_n, decreasing the length of the side by a factor f with $0 < f < 1$, and forming two sides of a smaller square, each of length $s_{n+1} = f \cdot s_n$. The process is then repeated ad infinitum. If $s_1 = 1$ in the second figure, express the length of the resulting (infinite) broken-line curve in terms of f.

Exercise 59

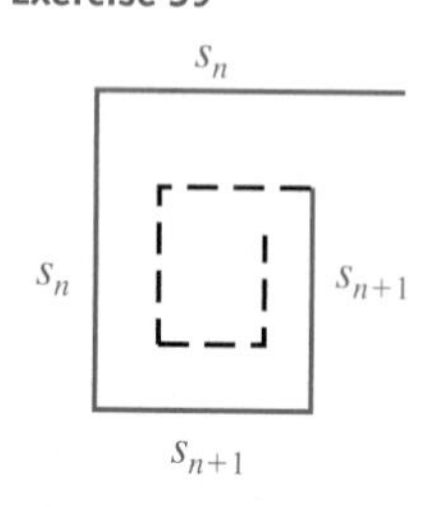

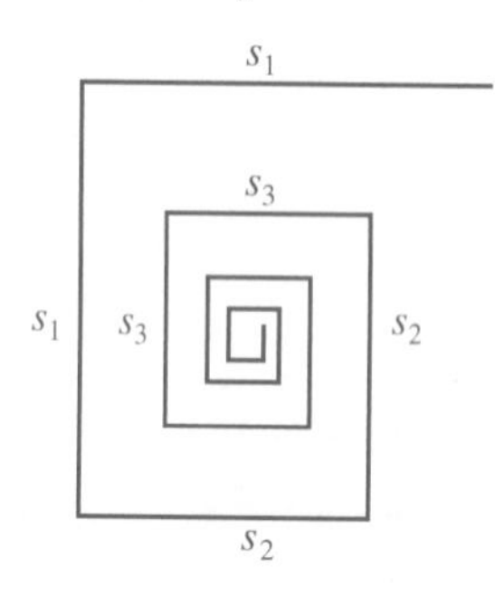

60 The commutative and associative laws of addition guarantee that the sum of integers 1 through 10 is independent of the order in which the numbers are added. In how many different ways can these integers be summed?

61 **Selecting cards**

(a) In how many ways can 13 cards be selected from a deck?

(b) In how many ways can 13 cards be selected to obtain five spades, three hearts, three clubs, and two diamonds?

62 How many four-digit numbers can be formed from the digits 1, 2, 3, 4, 5, and 6 if repetitions

(a) are not allowed?

(b) are allowed?

63 **Selecting test questions**

(a) If a student must answer 8 of 12 questions on an examination, how many different selections of questions are possible?

(b) How many selections are possible if the first three questions must be answered?

64 **Color arrangements** If six black, five red, four white, and two green disks are to be arranged in a row, what is the number of possible color arrangements?

65 If $O(E)$ are 8 to 5, find $O(E')$ and $P(E)$.

66 **Coin toss** Find the probability that the coins will match if

(a) two boys each toss a coin

(b) three boys each toss a coin

67 **Dealing cards** If four cards are dealt from a deck, find the probability that

(a) all four cards will be the same color

(b) the cards dealt will alternate red-black-red-black

68 **Raffle probabilities** If 1000 tickets are sold for a raffle, find the probability of winning if an individual purchases

(a) 1 ticket

(b) 10 tickets

(c) 50 tickets

69 **Coin toss** If four coins are flipped, find the probability and the odds of obtaining one head and three tails.

70 **True-or-false quiz** A quiz consists of six true-or-false questions; at least four correct answers are required for a passing grade. If a student guesses at each answer, what is the probability of

(a) passing? (b) failing?

71 **Die and card probabilities** If a single die is tossed and then a card is drawn from a deck, what is the probability of obtaining

(a) a 6 on the die and the king of hearts?

(b) a 6 on the die or the king of hearts?

72 **Population demographics** In a town of 5000 people, 1000 are over 60 years old and 2000 are female. It is known that 40% of the females are over 60. What is the probability that a randomly chosen individual from the town is either female or over 60?

73 **Backgammon moves** In the game of backgammon, players are allowed to move their counters the same number of spaces as the sum of the dots on two dice. However, if a double is rolled (that is, both dice show the same number of dots), then players may move their counters twice the sum of the dots. What is the probability that a player will be able to move his or her counters at least 10 spaces on a given roll?

74 **Games in a series** Two equally matched baseball teams are playing a series of games. The first team to win four games wins the series. Find the expected number of games in the series.

CHAPTER 9 DISCUSSION EXERCISES

1 A test question lists the first four terms of a sequence as 2, 4, 6, and 8 and asks for the fifth term. Show that the fifth term can be *any* real number a by finding the nth term of a sequence that has for its first five terms 2, 4, 6, 8, and a.

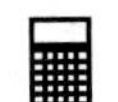

2 Decide whether $\square$ should be replaced by $\leq$ or $\geq$ in

$$n \; \square \; (\ln n)^3$$

for the statement to be true when $n \geq j$, where j is the smallest positive integer for which the statement is true. Find j.

Exer. 3–4: (a) Use the method of Exercises 37 and 38 in Section 9.4 to find a formula for the sum. (b) Verify that the formula found in part (a) is true for every n.

3 $1^4 + 2^4 + 3^4 + \cdots + n^4$

4 $2^3 + 4^3 + 6^3 + \cdots + (2n)^3$

5 Determine the largest factorial that your calculator can compute. Some typical values are 69! and 449!. Speculate as to why these numbers are the maximum values that your calculator can compute.

6 Find a relationship between the coefficients in the expansion of $(a + b)^n$ and the number of distinct subsets of an n-element set.

7 **Rebounding ball** When a ball is dropped from a height of h feet, it reaches the ground in $\sqrt{h}/4$ seconds. The ball rebounds to a height of d feet in $\sqrt{d}/4$ seconds. If a rubber ball is dropped from a height of 10 feet and rebounds to one-half of its height after each fall, for approximately how many seconds does the ball travel?

8 **Slot tournament** A slot tournament will be held over a 30-day month, eight hours each day, with 36 contestants each hour. The prize structure is as follows:

Place	1st	2nd	3rd	4th	5th
Prize $	4000	2000	1500	1000	800

Place	6th	7th	8th	9th	10th
Prize $	600	500	400	300	200

Place	11th–50th	51st–100th	101st–300th	301st–500th
Prize $	100	75	50	25

There is also a daily prize awarded as follows: \$250 for first, \$100 for second, and \$50 for third. How much would you expect to pay for an entry fee if the tournament is to be fair?

9 **Prize money** Suppose that the tenth prize of a \$1600 tournament will be \$100 and each place should be worth approximately 10% more than the next place. Discuss the realistic distribution of prize values if they are rounded to the nearest penny, dollar, five dollars, and ten dollars.

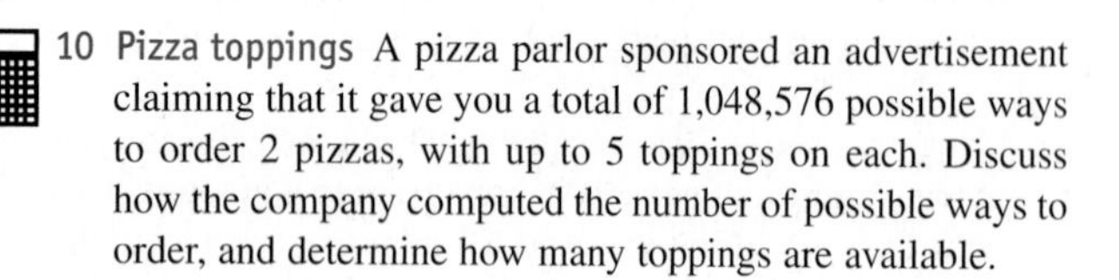

10 **Pizza toppings** A pizza parlor sponsored an advertisement claiming that it gave you a total of 1,048,576 possible ways to order 2 pizzas, with up to 5 toppings on each. Discuss how the company computed the number of possible ways to order, and determine how many toppings are available.

11 **Powerball** Powerball is a popular lottery game played in many states. The player selects five integers from 1 to 55 and one integer from 1 to 42. These numbers correspond to five white balls and one red Powerball drawn by the Multi-State Lottery Association. To win the jackpot, the player must match all six numbers. The prizes for all paying matches are listed in the table.

Match	Prize
5 white and red	jackpot
5 white	\$200,000
4 white and red	\$10,000
4 white	\$100
3 white and red	\$100
3 white	\$7
2 white and red	\$7
1 white and red	\$4
red only	\$3

(a) What is the probability of winning the jackpot?

(b) What is the probability of winning any prize?

(c) What is the expected value of the game without the jackpot?

(d) How much does the jackpot need to be worth for this lottery to be considered a fair game?

12 **Probability and odds confusion** Analyze the following statement: "There is a 20% chance that a male applicant will be admitted, but the odds are three times more favorable for a female applicant." What is the probability that a female applicant will be admitted?

13 Let $a = 0$ and $b = 1$ in

$$(a + b)^n = \sum_{k=0}^{n} \binom{n}{k} a^{n-k} b^k$$

and discuss the result.

14 Investigate the partial sums of

$$\sum_{n=0}^{\infty} (-1)^n \frac{3^{3/2}}{2^{3n+2}} \left(\frac{2}{3n + 1} + \frac{1}{3n + 2} \right)$$

and discuss them.

15 (a) Examine the following identities for $\tan nx$ in terms of $\tan x$:

$$\tan 2x = \frac{2 \tan x}{1 - \tan^2 x}$$

$$\tan 3x = \frac{3 \tan x - \tan^3 x}{1 - 3 \tan^2 x}$$

$$\tan 4x = \frac{4 \tan x - 4 \tan^3 x}{1 - 6 \tan^2 x + \tan^4 x}$$

By using a pattern formed by the three identities, predict an identity for $\tan 5x$ in terms of $\tan x$.

(b) Listed below are identities for $\cos 2x$ and $\sin 2x$:

$$\cos 2x = 1 \cos^2 x \qquad -1 \sin^2 x$$

$$\sin 2x = \qquad 2 \cos x \sin x$$

Write similar identities for $\cos 3x$ and $\sin 3x$ and then $\cos 4x$ and $\sin 4x$. Use a pattern to predict identities for $\cos 5x$ and $\sin 5x$.

10

Topics from Analytic Geometry

Plane geometry includes the study of figures—such as lines, circles, and triangles—that lie in a plane. Theorems are proved by reasoning deductively from certain postulates. In analytic geometry, plane geometric figures are investigated by introducing coordinate systems and then using equations and formulas. If the study of analytic geometry were to be summarized by means of one statement, perhaps the following would be appropriate: Given an equation, find its graph, and conversely, given a graph, find its equation. In this chapter we shall apply coordinate methods to several basic plane figures.

10.1
Parabolas

The *conic sections,* also called *conics,* can be obtained by intersecting a double-napped right circular cone with a plane. By varying the position of the plane, we obtain a *circle,* an *ellipse,* a *parabola,* or a *hyperbola,* as illustrated in Figure 1.

Figure 1

(a) Circle (b) Ellipse (c) Parabola (d) Hyperbola

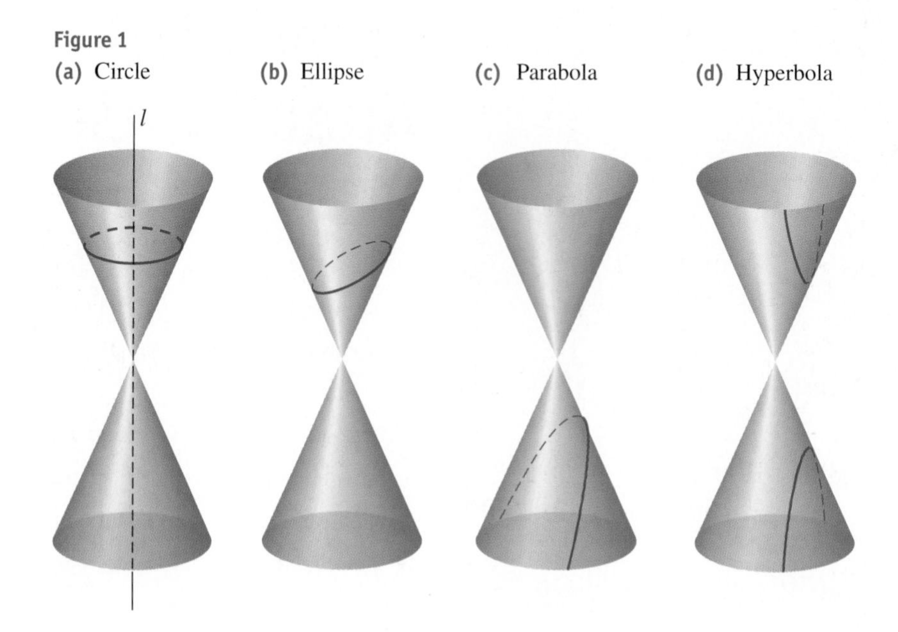

Degenerate conics are obtained if the plane intersects the cone in only one point or along either one or two lines that lie on the cone. Conic sections were studied extensively by the ancient Greeks, who discovered properties that enable us to state their definitions in terms of points and lines, as we do in our discussion.

From our work in Section 2.6, if $a \neq 0$, the graph of $y = ax^2 + bx + c$ is a *parabola* with a vertical axis. We shall next state a general definition of a parabola and derive equations for parabolas that have either a vertical axis or a horizontal axis.

Definition of a Parabola	A **parabola** is the set of all points in a plane equidistant from a fixed point F (the **focus**) and a fixed line l (the **directrix**) that lie in the plane.

We shall assume that F is not on l, for this would result in a line. If P is a point in the plane and P' is the point on l determined by a line through P that

Figure 2

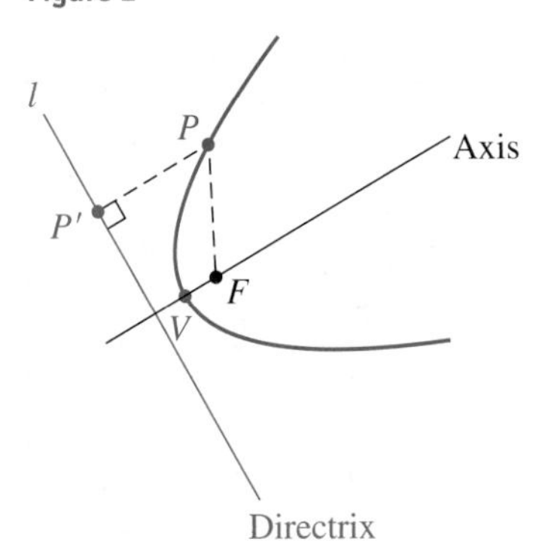

Figure 3

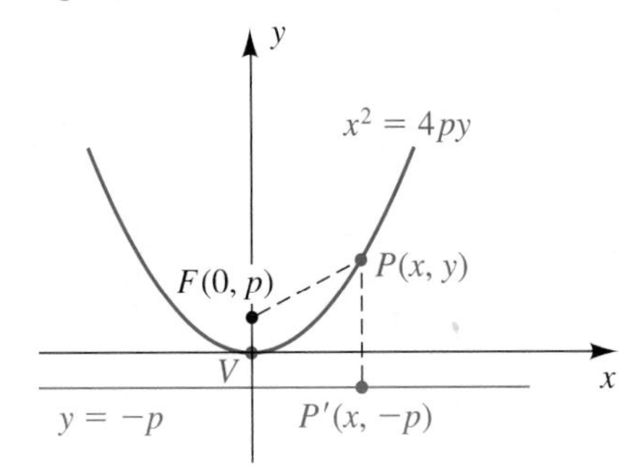

is perpendicular to l (see Figure 2), then, by the preceding definition, P is on the parabola if and only if the distances $d(P, F)$ and $d(P, P')$ are equal. The **axis** of the parabola is the line through F that is perpendicular to the directrix. The **vertex** of the parabola is the point V on the axis halfway from F to l. The vertex is the point on the parabola that is closest to the directrix.

To obtain a simple equation for a parabola, place the y-axis along the axis of the parabola, with the origin at the vertex V, as shown in Figure 3. In this case, the focus F has coordinates $(0, p)$ for some real number $p \neq 0$, and the equation of the directrix is $y = -p$. (The figure shows the case $p > 0$.) By the distance formula, a point $P(x, y)$ is on the graph of the parabola if and only if $d(P, F) = d(P, P')$—that is, if

$$\sqrt{(x - 0)^2 + (y - p)^2} = \sqrt{(x - x)^2 + (y + p)^2}.$$

We square both sides and simplify:

$$x^2 + (y - p)^2 = (y + p)^2$$

$$x^2 + y^2 - 2py + p^2 = y^2 + 2py + p^2$$

$$x^2 = 4py$$

An equivalent equation for the parabola is

$$y = \frac{1}{4p}x^2.$$

We have shown that the coordinates of every point (x, y) on the parabola satisfy $x^2 = 4py$. Conversely, if (x, y) is a solution of $x^2 = 4py$, then by reversing the previous steps we see that the point (x, y) is on the parabola.

If $p > 0$, the parabola opens upward, as in Figure 3. If $p < 0$, the parabola opens downward. The graph is symmetric with respect to the y-axis, since substitution of $-x$ for x does not change the equation $x^2 = 4py$.

If we interchange the roles of x and y, we obtain

$$y^2 = 4px \qquad \text{or, equivalently,} \qquad x = \frac{1}{4p}y^2.$$

This is an equation of a parabola with vertex at the origin, focus $F(p, 0)$, and opening right if $p > 0$ or left if $p < 0$. The equation of the directrix is $x = -p$.

For convenience we often refer to "the parabola $x^2 = 4py$" (or $y^2 = 4px$) instead of "the parabola with equation $x^2 = 4py$" (or $y^2 = 4px$).

The next chart summarizes our discussion.

Parabolas with Vertex $V(0, 0)$

Equation, focus, directrix	Graph for $p > 0$	Graph for $p < 0$
$x^2 = 4py$ or $y = \frac{1}{4p}x^2$ Focus: $F(0, p)$ Directrix: $y = -p$	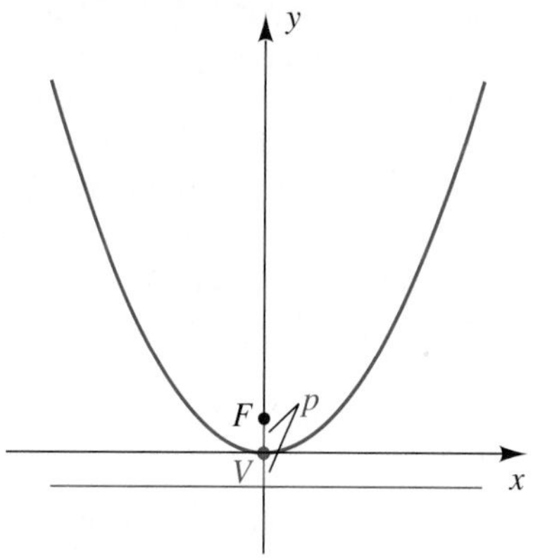 y, F, p, V, x	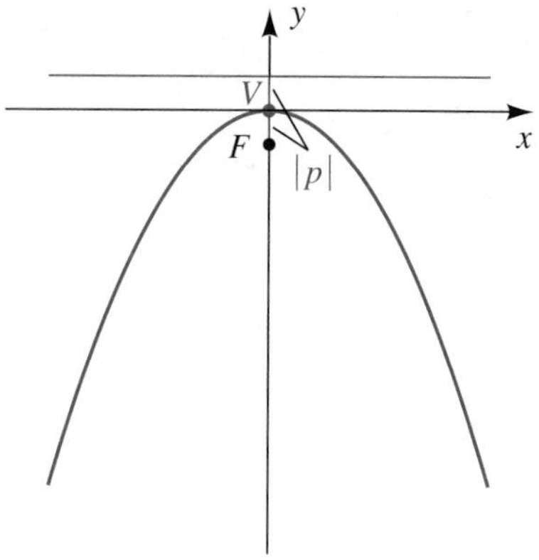 y, V, F, $\lvert p \rvert$, x
$y^2 = 4px$ or $x = \frac{1}{4p}y^2$ Focus: $F(p, 0)$ Directrix: $x = -p$	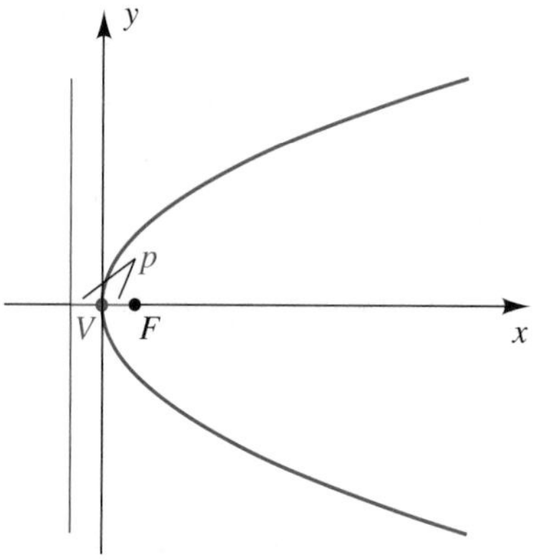 y, p, V, F, x	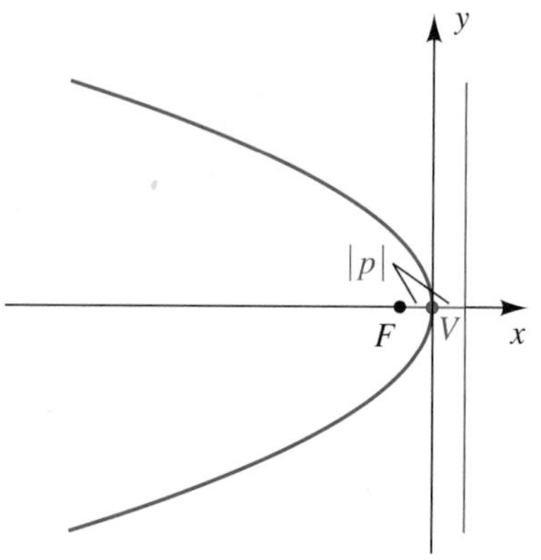 y, $\lvert p \rvert$, F, V, x

We see from the chart that for any nonzero real number a, the graph of the **standard equation** $y = ax^2$ or $x = ay^2$ is a parabola with vertex $V(0, 0)$. Moreover, $a = 1/(4p)$ or, equivalently, $p = 1/(4a)$, where $|p|$ is the distance between the focus F and vertex V. To find the directrix l, recall that l is also a distance $|p|$ from V.

EXAMPLE 1 Finding the focus and directrix of a parabola

Find the focus and directrix of the parabola $y = -\frac{1}{6}x^2$, and sketch its graph.

▶ SOLUTION The equation has the form $y = ax^2$, with $a = -\frac{1}{6}$. As in the preceding chart, $a = 1/(4p)$, and hence

$$p = \frac{1}{4a} = \frac{1}{4\left(-\frac{1}{6}\right)} = \frac{1}{-\frac{4}{6}} = -\frac{3}{2}.$$

▶ **Tutorial available at www.thomsonedu.com/login**

Figure 4

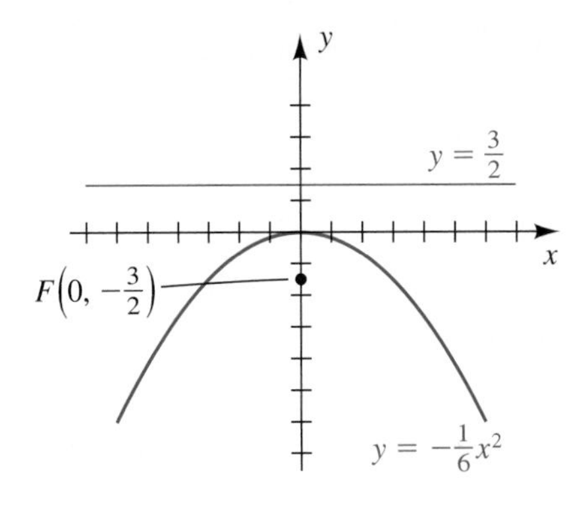

Thus, the parabola opens downward and has focus $F\left(0, -\frac{3}{2}\right)$, as illustrated in Figure 4. The directrix is the horizontal line $y = \frac{3}{2}$, which is a distance $\frac{3}{2}$ above V, as shown in the figure.

EXAMPLE 2 Finding an equation of a parabola satisfying prescribed conditions

(a) Find an equation of a parabola that has vertex at the origin, opens right, and passes through the point $P(7, -3)$.

(b) Find the focus of the parabola.

SOLUTION

Figure 5

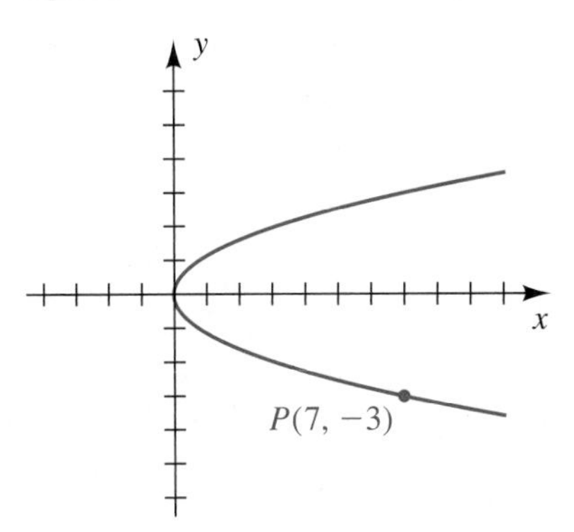

(a) The parabola is sketched in Figure 5. An equation of a parabola with vertex at the origin that opens right has the form $x = ay^2$ for some number a. If $P(7, -3)$ is on the graph, then we can substitute 7 for x and -3 for y to find a:

$$7 = a(-3)^2, \qquad \text{or} \qquad a = \tfrac{7}{9}$$

Hence, an equation for the parabola is $x = \frac{7}{9}y^2$.

(b) The focus is a distance p to the right of the vertex. Since $a = \frac{7}{9}$, we have

$$p = \frac{1}{4a} = \frac{1}{4\left(\frac{7}{9}\right)} = \frac{9}{28}.$$

Thus, the focus has coordinates $\left(\frac{9}{28}, 0\right)$.

If we take a standard equation of a parabola (of the form $x^2 = 4py$) and replace x with $x - h$ and y with $y - k$, then

$$x^2 = 4py \quad \text{becomes} \quad (x - h)^2 = 4p(y - k). \tag{$*$}$$

From our discussion of translations in Section 2.5, we recognize that the graph of the second equation can be obtained from the graph of the first equation by shifting it h units to the right and k units up—thereby moving the vertex from $(0, 0)$ to (h, k). Squaring the left-hand side of $(*)$ and simplifying leads to an equation of the form $y = ax^2 + bx + c$, where a, b, and c are real numbers.

Similarly, if we begin with $(y - k)^2 = 4p(x - h)$, it may be written in the form $x = ay^2 + by + c$. In the following chart $V(h, k)$ has been placed in the first quadrant, but the information given in the leftmost column holds true regardless of the position of V.

▶ Tutorial available at www.thomsonedu.com/login

Parabolas with Vertex $V(h, k)$

Equation, focus, directrix	Graph for $p > 0$	Graph for $p < 0$
$(x - h)^2 = 4p(y - k)$ or $y = ax^2 + bx + c$, where $p = \dfrac{1}{4a}$ Focus: $F(h, k + p)$ Directrix: $y = k - p$		
$(y - k)^2 = 4p(x - h)$ or $x = ay^2 + by + c$, where $p = \dfrac{1}{4a}$ Focus: $F(h + p, k)$ Directrix: $x = h - p$		

EXAMPLE 3 Sketching a parabola with a horizontal axis

Sketch the graph of $2x = y^2 + 8y + 22$.

▶ **SOLUTION** The equation can be written in the form shown in the second row of the preceding chart, $x = ay^2 + by + c$, so we see from the chart that the graph is a parabola with a horizontal axis. We first write the given equation as

$$y^2 + 8y + __ = 2x - 22 + __$$

and then complete the square by adding $\left[\frac{1}{2}(8)\right]^2 = 16$ to both sides:

$$y^2 + 8y + 16 = 2x - 6$$

$$(y + 4)^2 = 2(x - 3)$$

▶ Tutorial available at www.thomsonedu.com/login

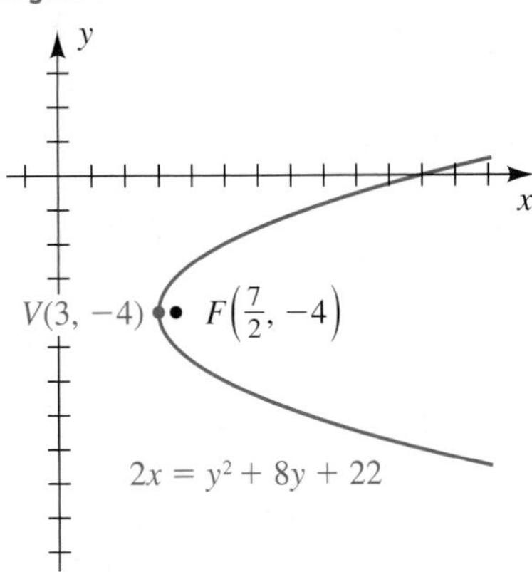

Figure 6

Referring to the last chart, we see that $h = 3$, $k = -4$, and $4p = 2$ or, equivalently, $p = \frac{1}{2}$. This gives us the following.

The vertex $V(h, k)$ is $V(3, -4)$.

The focus is $F(h + p, k) = F\left(3 + \frac{1}{2}, -4\right)$, or $F\left(\frac{7}{2}, -4\right)$.

The directrix is $x = h - p = 3 - \frac{1}{2}$, or $x = \frac{5}{2}$.

The parabola is sketched in Figure 6.

EXAMPLE 4 Finding an equation of a parabola given its vertex and directrix

A parabola has vertex $V(-4, 2)$ and directrix $y = 5$. Express the equation of the parabola in the form $y = ax^2 + bx + c$.

▶ SOLUTION The vertex and directrix are shown in Figure 7. The dashes indicate a possible position for the parabola. The last chart shows that an equation of the parabola is

$$(x - h)^2 = 4p(y - k),$$

with $h = -4$ and $k = 2$ and with p equal to *negative* 3, since V is 3 units *below* the directrix. This gives us

$$(x + 4)^2 = -12(y - 2).$$

The last equation can be expressed in the form $y = ax^2 + bx + c$, as follows:

$$\begin{aligned} x^2 + 8x + 16 &= -12y + 24 \\ 12y &= -x^2 - 8x + 8 \\ y &= -\tfrac{1}{12}x^2 - \tfrac{2}{3}x + \tfrac{2}{3} \end{aligned}$$

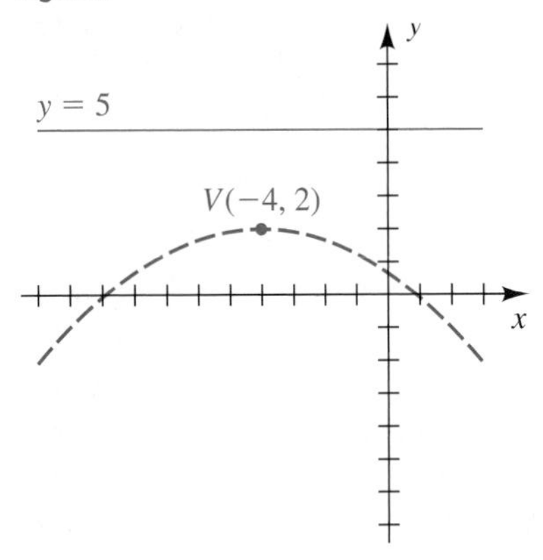

Figure 7

An important property is associated with a tangent line to a parabola. (A *tangent line* to a parabola is a line that has exactly one point in common with the parabola but does not cut through the parabola.) Suppose l is the tangent line at a point $P(x_1, y_1)$ on the graph of $y^2 = 4px$, and let F be the focus. As in Figure 8, let α denote the angle between l and the line segment FP, and let β denote the angle between l and the indicated horizontal half-line with endpoint P. It can be shown that $\alpha = \beta$. This *reflective property* has many applications. For example, the shape of the mirror in a searchlight is obtained by revolving a parabola about its axis. The resulting three-dimensional surface is said to be *generated* by the parabola and is called a **paraboloid.** The **focus** of the paraboloid is the same as the focus of the generating parabola. If a light source is placed at F, then, by a law of physics (the angle of reflection equals the angle of incidence), a beam of light will be reflected along a line parallel to the axis (see Figure 9(a)). The same principle is used in the construction of mirrors for telescopes or solar ovens—a beam of light coming toward the parabolic mirror and parallel to the axis will be reflected into the focus (see

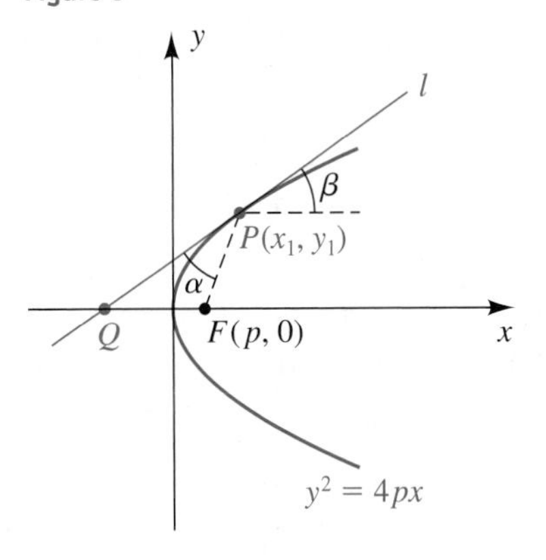

Figure 8

▶ Tutorial available at www.thomsonedu.com/login

Figure 9
(a) Searchlight mirror

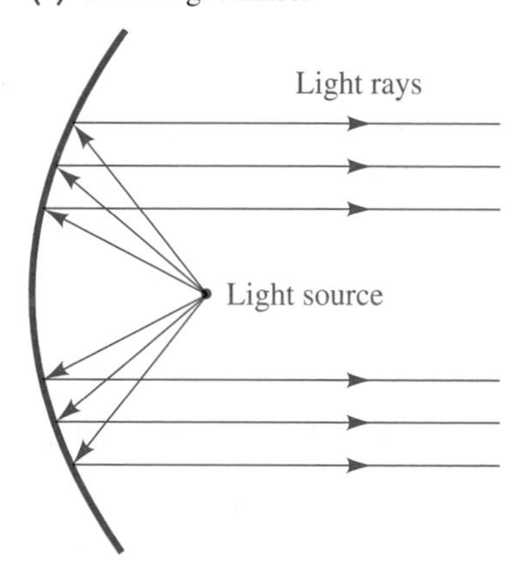

(b) Telescope mirror

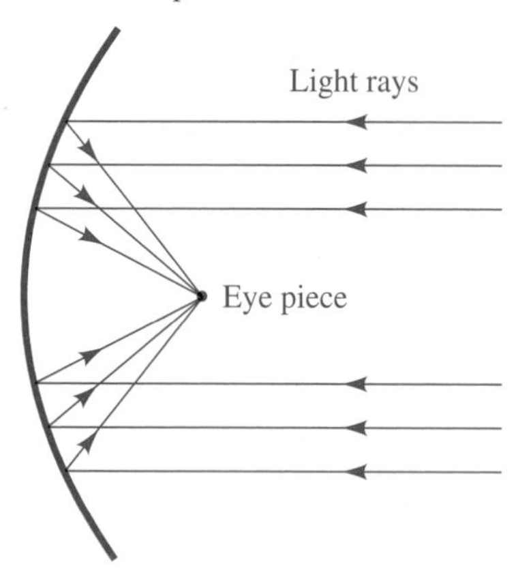

Figure 9(b)). Antennas for radar systems, radio telescopes, and field microphones used at football games also make use of this property.

EXAMPLE 5 Locating the focus of a satellite TV antenna

The interior of a satellite TV antenna is a dish having the shape of a (finite) paraboloid that has diameter 12 feet and is 2 feet deep, as shown in Figure 10. Find the distance from the center of the dish to the focus.

Figure 10

Figure 11

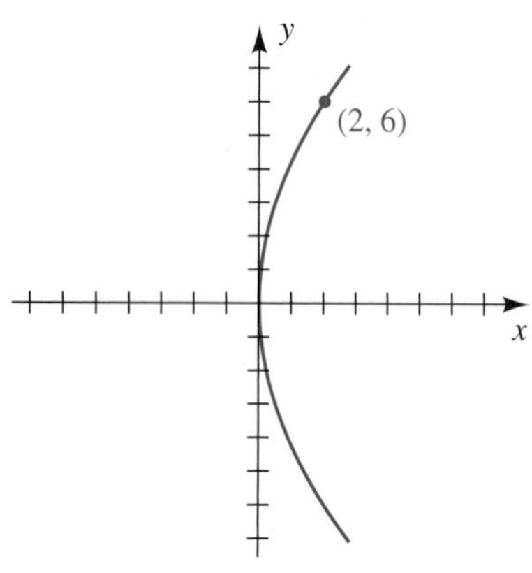

▶ SOLUTION The generating parabola is sketched on an xy-plane in Figure 11, where we have taken the vertex of the parabola at the origin and its axis along the x-axis. An equation of the parabola is $y^2 = 4px$, where p is the required distance from the center of the dish to the focus. Since the point (2, 6) is on the parabola, we obtain

$$6^2 = 4p \cdot 2, \qquad \text{or} \qquad p = \tfrac{36}{8} = 4.5 \text{ ft.}$$

In the next example we use a graphing utility to sketch the graph of a parabola with a horizontal axis.

EXAMPLE 6 Graphing half-parabolas

Graph $x = y^2 + 2y - 4$.

SOLUTION The graph is a parabola with a horizontal axis. Since y is not a function of x, we will solve the equation for y and obtain two equations (much as we did with circles in Example 11 of Section 2.2). We begin by solving the equivalent equation

$$y^2 + 2y - 4 - x = 0$$

▶ Tutorial available at www.thomsonedu.com/login

for y in terms of x by using the quadratic formula, with $a = 1$, $b = 2$, and $c = -4 - x$:

$$\begin{aligned} y &= \frac{-2 \pm \sqrt{2^2 - 4(1)(-4 - x)}}{2(1)} && \text{quadratic formula} \\ &= \frac{-2 \pm \sqrt{20 + 4x}}{2} && \text{simplify} \\ &= -1 \pm \sqrt{x + 5} && \text{factor out } \sqrt{4}\text{; simplify} \end{aligned}$$

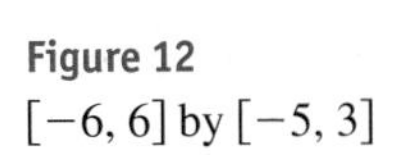

Figure 12
$[-6, 6]$ by $[-5, 3]$

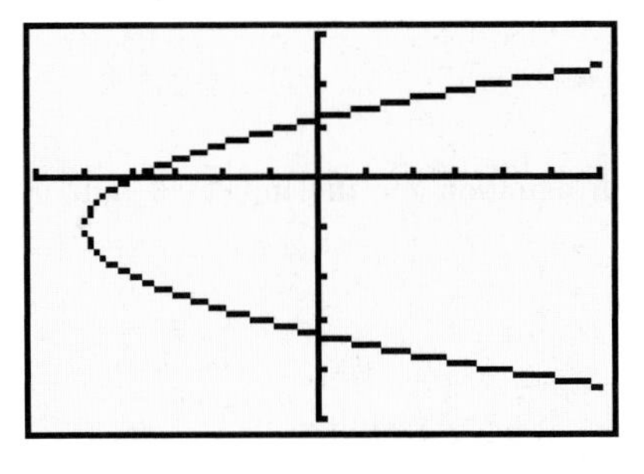

The last equation, $y = -1 \pm \sqrt{x + 5}$, represents the top half of the parabola $\left(y = -1 + \sqrt{x + 5}\right)$ and the bottom half $\left(y = -1 - \sqrt{x + 5}\right)$. Note that $y = -1$ is the axis of the parabola.

Next, we make the assignments

$$Y_1 = -1 + \sqrt{x + 5} \quad \text{and} \quad Y_2 = -1 - \sqrt{x + 5}.$$

Now graph Y_1 and Y_2 to obtain a display similar to Figure 12.

10.1 *Exercises*

Exer. 1–12: Find the vertex, focus, and directrix of the parabola. Sketch its graph, showing the focus and the directrix.

1 $8y = x^2$

▶ 2 $20x = y^2$

3 $2y^2 = -3x$

4 $x^2 = -3y$

5 $(x + 2)^2 = -8(y - 1)$

6 $(x - 3)^2 = \frac{1}{2}(y + 1)$

7 $(y - 2)^2 = \frac{1}{4}(x - 3)$

8 $(y + 1)^2 = -12(x + 2)$

▶ 9 $y = x^2 - 4x + 2$

10 $y^2 + 14y + 4x + 45 = 0$

11 $x^2 + 20y = 10$

12 $y^2 - 4y - 2x - 4 = 0$

Exer. 13–18: Find an equation for the parabola shown in the figure.

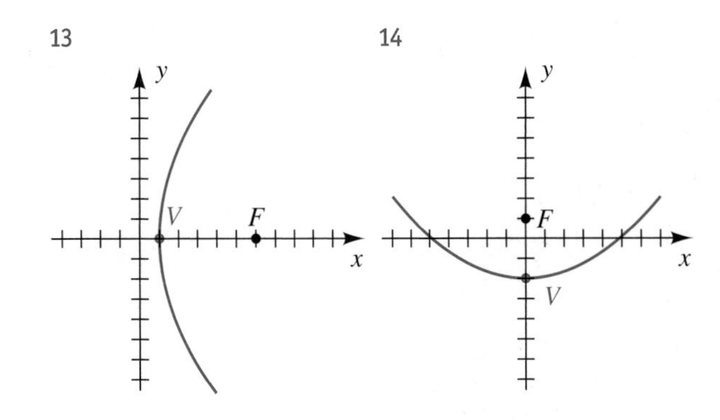

▶ **Tutorial available at www.thomsonedu.com/login**

15 16

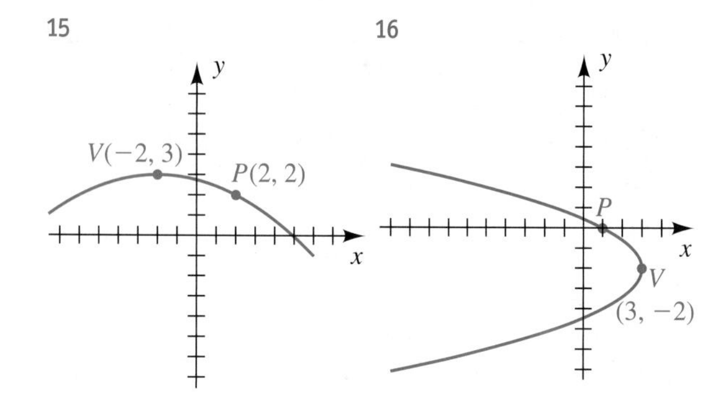

17 18

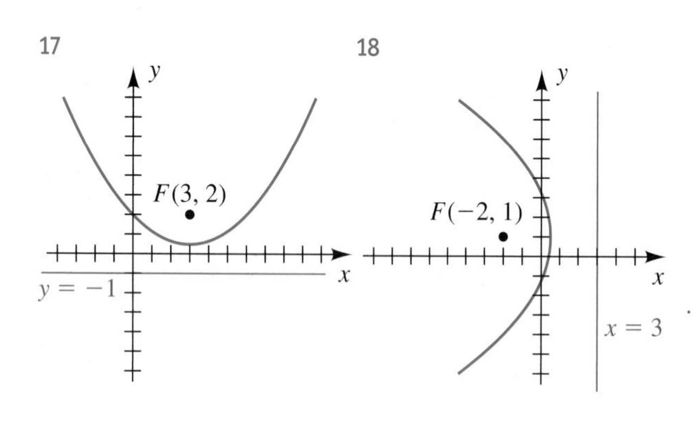

Exer. 19–30: Find an equation of the parabola that satisfies the given conditions.

19 Focus $F(2, 0)$, directrix $x = -2$

20 Focus $F(0, -4)$, directrix $y = 4$

21 Focus $F(6, 4)$, directrix $y = -2$

22 Focus $F(-3, -2)$, directrix $y = 1$

23 Vertex $V(3, -5)$, directrix $x = 2$

24 Vertex $V(-2, 3)$, directrix $y = 5$

25 Vertex $V(-1, 0)$ focus $F(-4, 0)$

26 Vertex $V(1, -2)$, focus $F(1, 0)$

▶ 27 Vertex at the origin, symmetric to the y-axis, and passing through the point $(2, -3)$

28 Vertex at the origin, symmetric to the y-axis, and passing through the point $(6, 3)$

29 Vertex $V(-3, 5)$, axis parallel to the x-axis, and passing through the point $(5, 9)$

30 Vertex $V(3, -2)$, axis parallel to the x-axis, and y-intercept 1

Exer. 31–34: Find an equation for the set of points in an xy-plane that are equidistant from the point P and the line l.

31 $P(0, 5)$; l: $y = -3$

32 $P(7, 0)$; l: $x = 1$

33 $P(-6, 3)$; l: $x = -2$

34 $P(5, -2)$; l: $y = 4$

Exer. 35–38: Find an equation for the indicated half of the parabola.

35 Lower half of $(y + 1)^2 = x + 3$

▶ 36 Upper half of $(y - 2)^2 = x - 4$

37 Right half of $(x + 1)^2 = y - 4$

38 Left half of $(x + 3)^2 = y + 2$

Exer. 39–40: Find an equation for the parabola that has a vertical axis and passes through the given points.

39 $P(2, 5)$, $Q(-2, -3)$, $R(1, 6)$

40 $P(3, -1)$, $Q(1, -7)$, $R(-2, 14)$

Exer. 41–42: Find an equation for the parabola that has a horizontal axis and passes through the given points.

41 $P(-1, 1)$, $Q(11, -2)$, $R(5, -1)$

42 $P(2, 1)$, $Q(6, 2)$, $R(12, -1)$

43 **Telescope mirror** A mirror for a reflecting telescope has the shape of a (finite) paraboloid of diameter 8 inches and depth 1 inch. How far from the center of the mirror will the incoming light collect?

Exercise 43

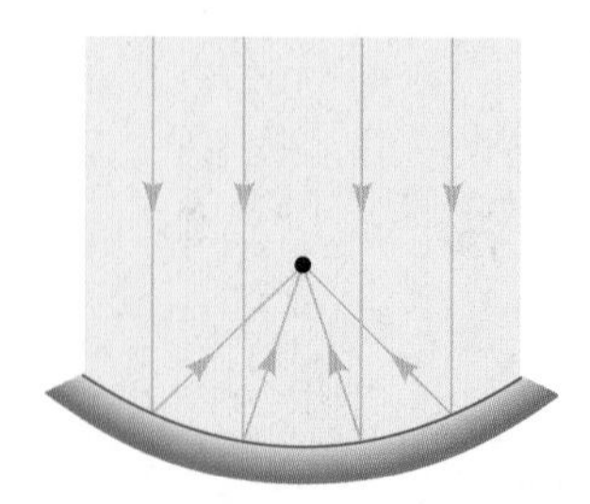

▶ **Tutorial available at www.thomsonedu.com/login**

44 **Antenna dish** A satellite antenna dish has the shape of a paraboloid that is 10 feet across at the open end and is 3 feet deep. At what distance from the center of the dish should the receiver be placed to receive the greatest intensity of sound waves?

45 **Searchlight reflector** A searchlight reflector has the shape of a paraboloid, with the light source at the focus. If the reflector is 3 feet across at the opening and 1 foot deep, where is the focus?

46 **Flashlight mirror** A flashlight mirror has the shape of a paraboloid of diameter 4 inches and depth $\frac{3}{4}$ inch, as shown in the figure. Where should the bulb be placed so that the emitted light rays are parallel to the axis of the paraboloid?

Exercise 46

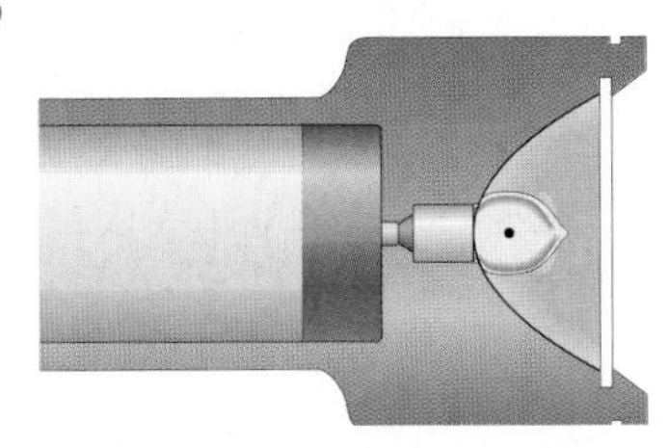

47 **Receiving dish** A sound receiving dish used at outdoor sporting events is constructed in the shape of a paraboloid, with its focus 5 inches from the vertex. Determine the width of the dish if the depth is to be 2 feet.

48 **Receiving dish** Work Exercise 47 if the receiver is 9 inches from the vertex.

▶ 49 **Parabolic reflector**

(a) The focal length of the (finite) paraboloid in the figure is the distance p between its vertex and focus. Express p in terms of r and h.

(b) A reflector is to be constructed with a focal length of 10 feet and a depth of 5 feet. Find the radius of the reflector.

Exercise 49

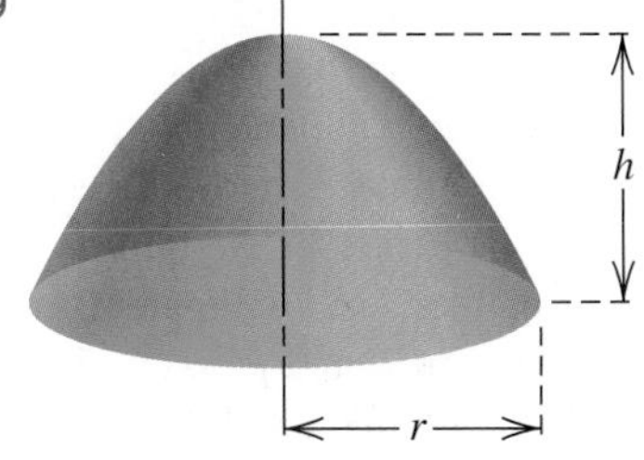

50 **Confocal parabolas** The parabola $y^2 = 4p(x + p)$ has its focus at the origin and axis along the x-axis. By assigning different values to p, we obtain a family of confocal parabolas, as shown in the figure. Such families occur in the study of electricity and magnetism. Show that there are exactly two parabolas in the family that pass through a given point $P(x_1, y_1)$ if $y_1 \neq 0$.

Exercise 50

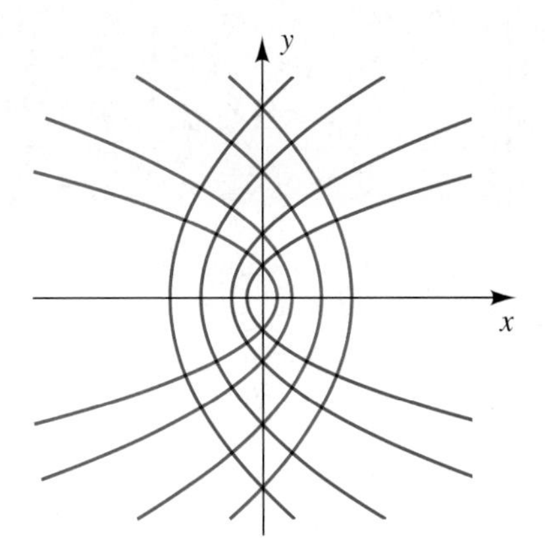

51 **Jodrell Bank radio telescope** A radio telescope has the shape of a paraboloid of revolution, with focal length p and diameter of base $2a$. From calculus, the surface area S available for collecting radio waves is

$$S = \frac{8\pi p^2}{3}\left[\left(1 + \frac{a^2}{4p^2}\right)^{3/2} - 1\right].$$

One of the largest radio telescopes, located in Jodrell Bank, Cheshire, England, has diameter 250 feet and focal length 75 feet. Approximate S to the nearest thousand square feet.

52 **Satellite path** A satellite will travel in a parabolic path near a planet if its velocity v in meters per second satisfies the equation $v = \sqrt{2k/r}$, where r is the distance in meters between the satellite and the center of the planet and k is a positive constant. The planet will be located at the focus of the parabola, and the satellite will pass by the planet once. Suppose a satellite is designed to follow a parabolic path and travel within 58,000 miles of Mars, as shown in the figure on the next page.

▶ **Tutorial available at www.thomsonedu.com/login**

Exercise 52

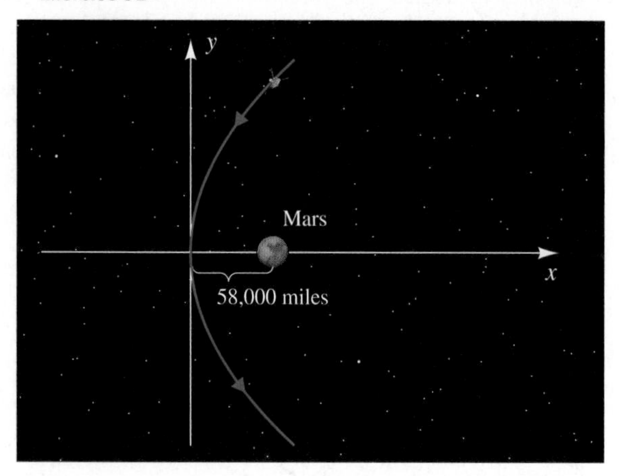

(a) Determine an equation of the form $x = ay^2$ that describes its flight path.

(b) For Mars, $k = 4.28 \times 10^{13}$. Approximate the maximum velocity of the satellite.

(c) Find the velocity of the satellite when its y-coordinate is 100,000 miles.

Exer. 53–54: Graph the equation.

53 $x = -y^2 + 2y + 5$

54 $x = 2y^2 + 3y - 7$

Exer. 55–56: Graph the parabolas on the same coordinate plane, and estimate the points of intersection.

55 $y = x^2 - 2.1x - 1$; $x = y^2 + 1$

56 $y = -2.1x^2 + 0.1x + 1.2$; $x = 0.6y^2 + 1.7y - 1.1$

10.2 Ellipses

An ellipse may be defined as follows. (*Foci* is the plural of *focus*.)

Definition of an Ellipse

An **ellipse** is the set of all points in a plane, the sum of whose distances from two fixed points (the **foci**) in the plane is a positive constant.

Figure 1

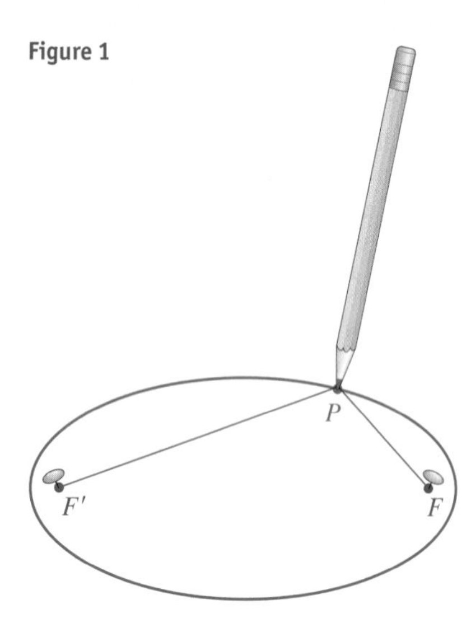

We can construct an ellipse on paper as follows: Insert two pushpins in the paper at any points F and F', and fasten the ends of a piece of string to the pins. After looping the string around a pencil and drawing it tight, as at point P in Figure 1, move the pencil, keeping the string tight. The sum of the distances $d(P, F)$ and $d(P, F')$ is the length of the string and hence is constant; thus, the pencil will trace out an ellipse with foci at F and F'. The midpoint of the segment $F'F$ is called the **center** of the ellipse. By changing the positions of F and F' while keeping the length of the string fixed, we can vary the shape of the ellipse considerably. If F and F' are far apart so that $d(F, F')$ is almost the same as the length of the string, the ellipse is flat. If $d(F, F')$ is close to zero, the ellipse is almost circular. If $F = F'$, we obtain a circle with center F.

To obtain a simple equation for an ellipse, choose the x-axis as the line through the two foci F and F', with the center of the ellipse at the origin. If F has coordinates $(c, 0)$ with $c > 0$, then, as in Figure 2, F' has coordinates $(-c, 0)$. Hence, the distance between F and F' is $2c$. The constant sum of the distances of P from F and F' will be denoted by $2a$. To obtain points

Figure 2

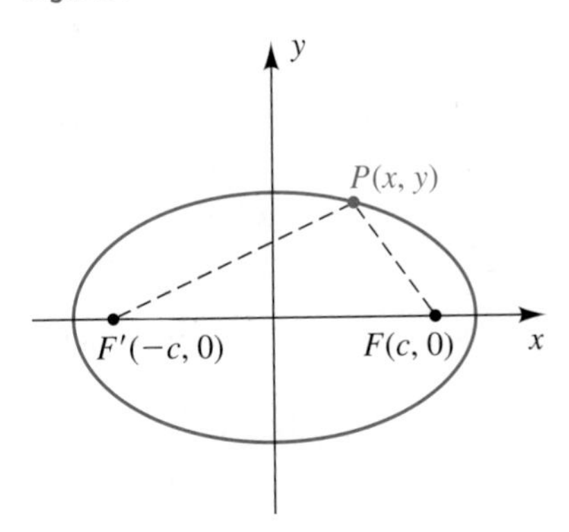

that are not on the x-axis, we must have $2a > 2c$—that is, $a > c$. By definition, $P(x, y)$ is on the ellipse if and only if the following equivalent equations are true:

$$d(P, F) + d(P, F') = 2a$$

$$\sqrt{(x-c)^2 + (y-0)^2} + \sqrt{(x+c)^2 + (y-0)^2} = 2a$$

$$\sqrt{(x-c)^2 + y^2} = 2a - \sqrt{(x+c)^2 + y^2}$$

Squaring both sides of the last equation gives us

$$x^2 - 2cx + c^2 + y^2 = 4a^2 - 4a\sqrt{(x+c)^2 + y^2} + x^2 + 2cx + c^2 + y^2,$$

or

$$a\sqrt{(x+c)^2 + y^2} = a^2 + cx.$$

Squaring both sides again yields

$$a^2(x^2 + 2cx + c^2 + y^2) = a^4 + 2a^2cx + c^2x^2,$$

or

$$x^2(a^2 - c^2) + a^2y^2 = a^2(a^2 - c^2).$$

Dividing both sides by $a^2(a^2 - c^2)$, we obtain

$$\frac{x^2}{a^2} + \frac{y^2}{a^2 - c^2} = 1.$$

Recalling that $a > c$ and therefore $a^2 - c^2 > 0$, we let

$$b = \sqrt{a^2 - c^2}, \qquad \text{or} \qquad b^2 = a^2 - c^2.$$

This substitution gives us the equation

$$\frac{x^2}{a^2} + \frac{y^2}{b^2} = 1.$$

Note that if $c = 0$, then $b^2 = a^2$, and we have a circle. Also note that if $c = a$, then $b = 0$, and we have a degenerate conic—*that is, a point.*

Since $c > 0$ and $b^2 = a^2 - c^2$, it follows that $a^2 > b^2$ and hence $a > b$.

We have shown that the coordinates of every point (x, y) on the ellipse in Figure 3 satisfy the equation $(x^2/a^2) + (y^2/b^2) = 1$. Conversely, if (x, y) is a solution of this equation, then by reversing the preceding steps we see that the point (x, y) is on the ellipse.

Figure 3

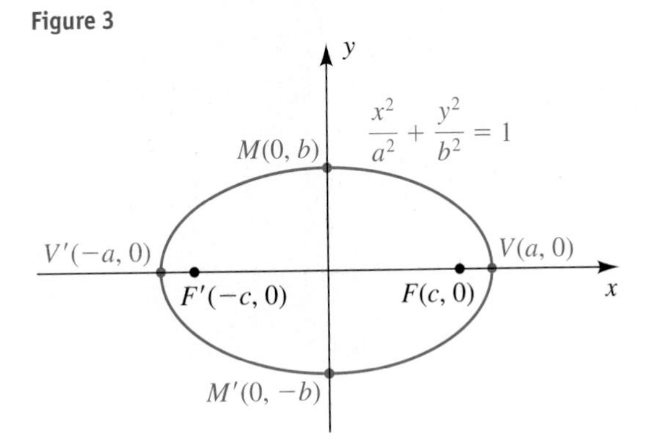

Figure 4

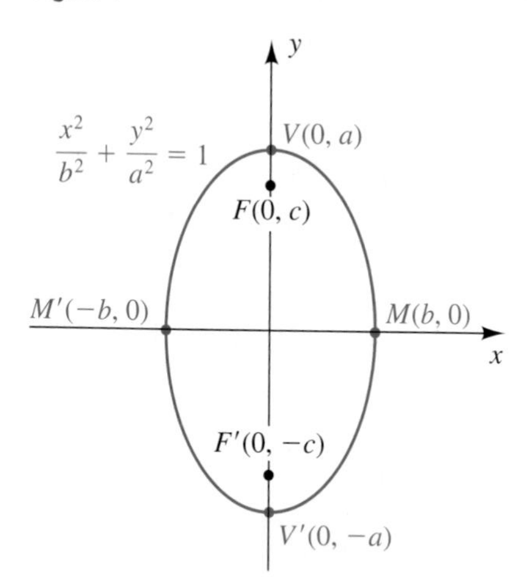

We may find the x-intercepts of the ellipse by letting $y = 0$ in the equation. Doing so gives us $x^2/a^2 = 1$, or $x^2 = a^2$. Consequently, the x-intercepts are a and $-a$. The corresponding points $V(a, 0)$ and $V'(-a, 0)$ on the graph are called the **vertices** of the ellipse (see Figure 3). The line segment $V'V$ is called the **major axis.** Similarly, letting $x = 0$ in the equation, we obtain $y^2/b^2 = 1$, or $y^2 = b^2$. Hence, the y-intercepts are b and $-b$. The segment between $M'(0, -b)$ and $M(0, b)$ is called the **minor axis** of the ellipse. The major axis is always longer than the minor axis, since $a > b$.

Applying tests for symmetry, we see that the ellipse is symmetric with respect to the x-axis, the y-axis, and the origin.

Similarly, if we take the foci on the y-axis, we obtain the equation

$$\frac{x^2}{b^2} + \frac{y^2}{a^2} = 1.$$

In this case, the vertices of the ellipse are $(0, \pm a)$ and the endpoints of the minor axis are $(\pm b, 0)$, as shown in Figure 4.

The preceding discussion may be summarized as follows.

Standard Equations of an Ellipse with Center at the Origin

The graph of

$$\frac{x^2}{a^2} + \frac{y^2}{b^2} = 1 \quad \text{or} \quad \frac{x^2}{b^2} + \frac{y^2}{a^2} = 1,$$

where $a > b > 0$, is an ellipse with center at the origin. The length of the major axis is $2a$, and the length of the minor axis is $2b$. The foci are a distance c from the origin, where $c^2 = a^2 - b^2$.

To help you remember the relationship for the foci, think of the right triangle formed by a ladder of length a leaning against a building, as shown in Figure 5. By the Pythagorean Theorem, $b^2 + c^2 = a^2$. In this position, the ends of the ladder are at a focus and an endpoint of the minor axis. If the ladder falls, the ends of the ladder will be at the center of the ellipse and an endpoint of the major axis.

Figure 5

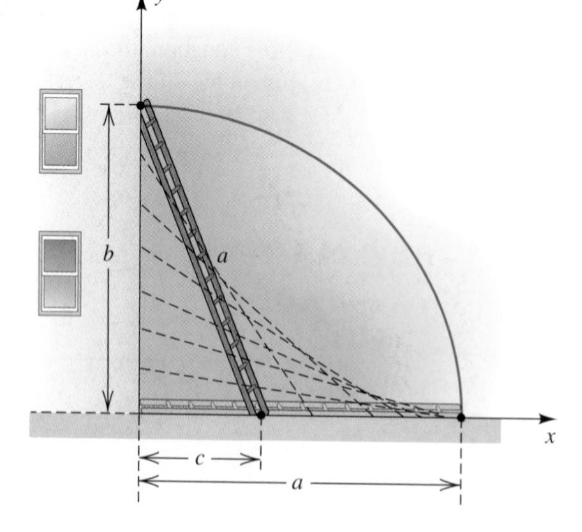

EXAMPLE 1 Sketching an ellipse with center at the origin

Sketch the graph of $2x^2 + 9y^2 = 18$, and find the foci.

▶ **SOLUTION** To write this equation in standard form, divide each term by 18 to obtain a constant of 1:

$$\frac{2x^2}{18} + \frac{9y^2}{18} = \frac{18}{18}, \qquad \text{or} \qquad \frac{x^2}{9} + \frac{y^2}{2} = 1$$

The graph is an ellipse with center at the origin and foci on a coordinate axis. From the last equation, since $9 > 2$, the major axis and the foci are on the x-axis. With $a^2 = 9$, we have $a = 3$, and the vertices are $V(3, 0)$ and $V'(-3, 0)$. Since $b^2 = 2$, $b = \sqrt{2}$, and endpoints of the minor axis are $M(0, \sqrt{2})$ and $M'(0, -\sqrt{2})$. Note that in this case, V and V' are also the x-intercepts, and M and M' are also the y-intercepts.

Figure 6

We now sketch the graph with major axis of length $2a = 2(3) = 6$ (shown in red in Figure 6) and minor axis of length $2b = 2\sqrt{2} \approx 2.8$ (shown in green).

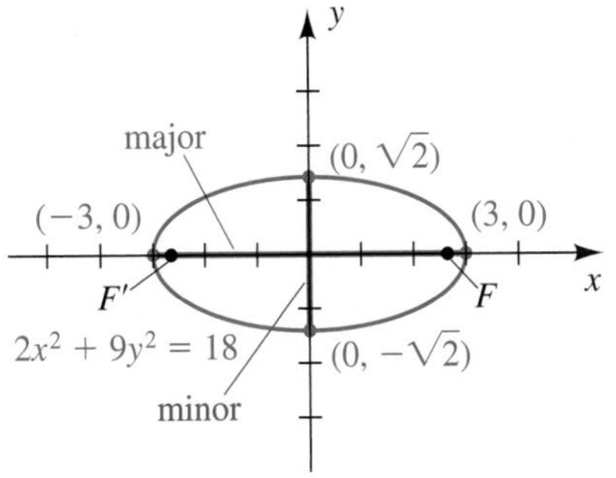

To find the foci, we let $a = 3$ and $b = \sqrt{2}$ and calculate

$$c^2 = a^2 - b^2 = 3^2 - (\sqrt{2})^2 = 7.$$

Thus, $c = \sqrt{7}$, and the foci are $F(\sqrt{7}, 0)$ and $F'(-\sqrt{7}, 0)$.

EXAMPLE 2 Sketching an ellipse with center at the origin

Sketch the graph of $9x^2 + 4y^2 = 25$, and find the foci.

▶ **SOLUTION** Divide each term by 25 to get the standard form:

$$\frac{9x^2}{25} + \frac{4y^2}{25} = \frac{25}{25}, \qquad \text{or} \qquad \frac{x^2}{\frac{25}{9}} + \frac{y^2}{\frac{25}{4}} = 1$$

The graph is an ellipse with center at the origin. Since $\frac{25}{4} > \frac{25}{9}$, the major axis and the foci are on the y-axis. With $a^2 = \frac{25}{4}$, $a = \frac{5}{2}$, and hence the vertices are $V(0, \frac{5}{2})$ and $V'(0, -\frac{5}{2})$ (also the y-intercepts). Since $b^2 = \frac{25}{9}$, $b = \frac{5}{3}$, and endpoints of the minor axis are $M(\frac{5}{3}, 0)$ and $M'(-\frac{5}{3}, 0)$ (also the x-intercepts).

Figure 7

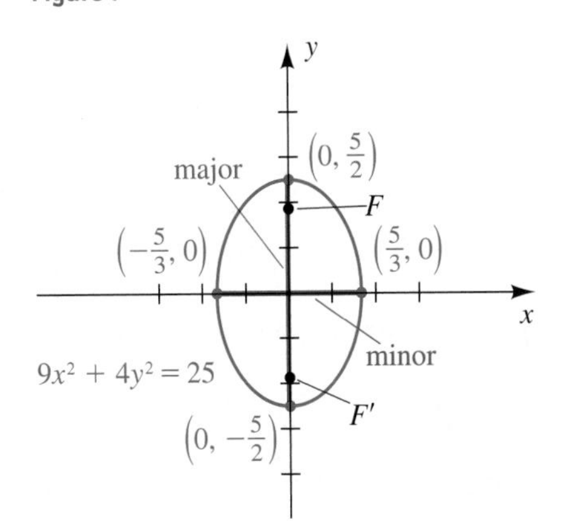

Sketch the graph with major axis of length $2a = 2(\frac{5}{2}) = 5$ (shown in red in Figure 7) and minor axis of length $2b = 2(\frac{5}{3}) = 3\frac{1}{3}$ (shown in green).

To find the foci, we let $a = \frac{5}{2}$ and $b = \frac{5}{3}$ and calculate

$$c^2 = a^2 - b^2 = (\tfrac{5}{2})^2 - (\tfrac{5}{3})^2 = \tfrac{125}{36}.$$

Thus, $c = \sqrt{125/36} = 5\sqrt{5}/6 \approx 1.86$, and the foci are approximately $F(0, 1.86)$ and $F'(0, -1.86)$.

▶ Tutorial available at www.thomsonedu.com/login

EXAMPLE 3 Finding an equation of an ellipse given its vertices and foci

Find an equation of the ellipse with vertices $(\pm 4, 0)$ and foci $(\pm 2, 0)$.

▶ SOLUTION Since the foci are on the x-axis and are equidistant from the origin, the major axis is on the x-axis and the ellipse has center $(0, 0)$. Thus, a general equation of an ellipse is

$$\frac{x^2}{a^2} + \frac{y^2}{b^2} = 1.$$

Since the vertices are $(\pm 4, 0)$, we conclude that $a = 4$. Since the foci are $(\pm 2, 0)$, we have $c = 2$. Hence,

$$b^2 = a^2 - c^2 = 4^2 - 2^2 = 12,$$

and an equation of the ellipse is

$$\frac{x^2}{16} + \frac{y^2}{12} = 1.$$

In certain applications it is necessary to work with only one-half of an ellipse. The next example indicates how to find equations in such cases.

EXAMPLE 4 Finding equations for half-ellipses

Find equations for the upper half, lower half, left half, and right half of the ellipse $9x^2 + 4y^2 = 25$.

▶ SOLUTION The graph of the entire ellipse was sketched in Figure 7. To find equations for the upper and lower halves, we solve for y in terms of x, as follows:

$$9x^2 + 4y^2 = 25 \qquad \text{given}$$

$$y^2 = \frac{25 - 9x^2}{4} \qquad \text{solve for } y^2$$

$$y = \pm\sqrt{\frac{25 - 9x^2}{4}} = \pm\frac{1}{2}\sqrt{25 - 9x^2} \qquad \text{take the square root}$$

Figure 8

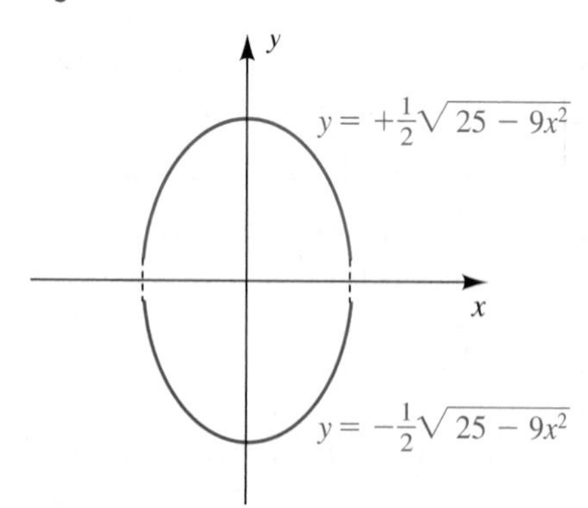

Since $\sqrt{25 - 9x^2} \geq 0$, it follows that equations for the upper and lower halves are $y = \frac{1}{2}\sqrt{25 - 9x^2}$ and $y = -\frac{1}{2}\sqrt{25 - 9x^2}$, respectively, as shown in Figure 8.

To find equations for the left and right halves, we use a procedure similar to that above and solve for x in terms of y, obtaining

$$x = \pm\sqrt{\frac{25 - 4y^2}{9}} = \pm\frac{1}{3}\sqrt{25 - 4y^2}.$$

▶ Tutorial available at www.thomsonedu.com/login

The left half of the ellipse has the equation $x = -\frac{1}{3}\sqrt{25 - 4y^2}$, and the right half is given by $x = \frac{1}{3}\sqrt{25 - 4y^2}$, as shown in Figure 9.

Figure 9

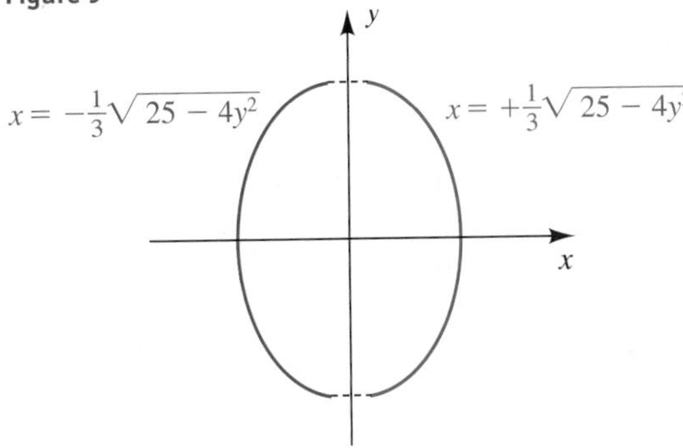

If we take a standard equation of an ellipse ($x^2/a^2 + y^2/b^2 = 1$) and replace x with $x - h$ and y with $y - k$, then

$$\frac{x^2}{a^2} + \frac{y^2}{b^2} = 1 \quad \text{becomes} \quad \frac{(x - h)^2}{a^2} + \frac{(y - k)^2}{b^2} = 1. \qquad (*)$$

The graph of $(*)$ is an ellipse with center (h, k). Squaring terms in $(*)$ and simplifying gives us an equation of the form

$$Ax^2 + Cy^2 + Dx + Ey + F = 0,$$

where the coefficients are real numbers and both A and C are positive. Conversely, if we start with such an equation, then by completing squares we can obtain a form that helps give us the center of the ellipse and the lengths of the major and minor axes. This technique is illustrated in the next example.

EXAMPLE 5 **Sketching an ellipse with center (h, k)**

Sketch the graph of the equation

$$16x^2 + 9y^2 + 64x - 18y - 71 = 0.$$

▶ **SOLUTION** We begin by grouping the terms containing x and those containing y:

$$(16x^2 + 64x) + (9y^2 - 18y) = 71$$

Next, we factor out the coefficients of x^2 and y^2 as follows:

$$16(x^2 + 4x + __) + 9(y^2 - 2y + __) = 71$$

We now complete the squares for the expressions within parentheses:

$$16(x^2 + 4x + 4) + 9(y^2 - 2y + 1) = 71 + \underline{16 \cdot 4} + \underline{9 \cdot 1}$$

By adding 4 to the expression within the first parentheses we have added 64 to the left-hand side of the equation, and hence we must compensate by adding 64 to the right-hand side. Similarly, by adding 1 to the expression within the

(continued)

▶ **Tutorial available at www.thomsonedu.com/login**

second parentheses we have added 9 to the left-hand side, and consequently we must also add 9 to the right-hand side. The last equation may be written

$$16(x + 2)^2 + 9(y - 1)^2 = 144.$$

Dividing by 144 to obtain 1 on the right-hand side gives us

$$\frac{(x + 2)^2}{9} + \frac{(y - 1)^2}{16} = 1.$$

The graph of the last equation is an ellipse with center $C(-2, 1)$ and major axis on the vertical line $x = -2$ (since $9 < 16$). Using $a = 4$ and $b = 3$ gives us the ellipse in Figure 10.

Figure 10

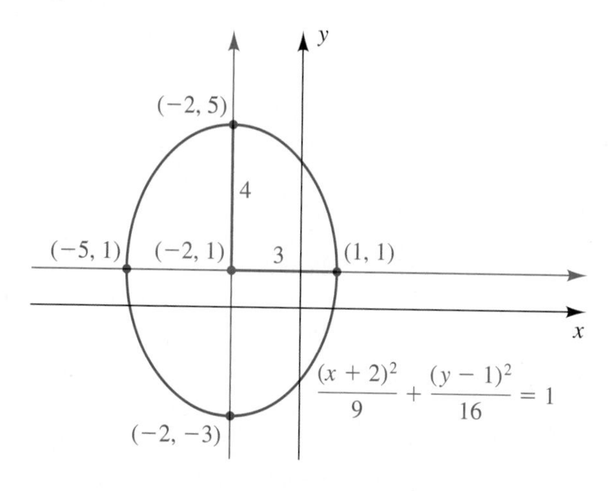

To find the foci, we first calculate

$$c^2 = a^2 - b^2 = 4^2 - 3^2 = 7.$$

The distance from the center of the ellipse to the foci is $c = \sqrt{7}$. Since the center is $(-2, 1)$, the foci are $(-2, 1 \pm \sqrt{7})$.

Graphing calculators and computer programs are sometimes unable to plot the graphs of an equation of the form

$$Ax^2 + Cy^2 + Dx + Ey + F = 0,$$

such as that considered in the last example. In these cases we must first solve the equation for y in terms of x and then plot the two resulting functions, as illustrated in the next example.

EXAMPLE 6 Graphing half-ellipses

Sketch the graph of $3x^2 + 4y^2 + 12x - 8y + 9 = 0$.

SOLUTION The equation may be regarded as a quadratic equation in y of the form $ay^2 + by + c = 0$ by rearranging terms as follows:

$$4y^2 - 8y + (3x^2 + 12x + 9) = 0$$

Applying the quadratic formula to the previous equation, with $a = 4$, $b = -8$, and $c = 3x^2 + 12x + 9$, gives us

$$\begin{aligned} y &= \frac{-(-8) \pm \sqrt{(-8)^2 - 4(4)(3x^2 + 12x + 9)}}{2(4)} \\ &= \frac{8 \pm \sqrt{64 - 16(3x^2 + 12x + 9)}}{8} \\ &= 1 \pm \tfrac{1}{8}\sqrt{64 - 16(3x^2 + 12x + 9)}. \end{aligned}$$

Note that we did not completely simplify the radicand, since we will be using a graphing calculator.

We now make the assignments

$$Y_1 = \tfrac{1}{8}\sqrt{64 - 16(3x^2 + 12x + 9)}, \quad Y_2 = 1 + Y_1, \quad \text{and} \quad Y_3 = 1 - Y_1.$$

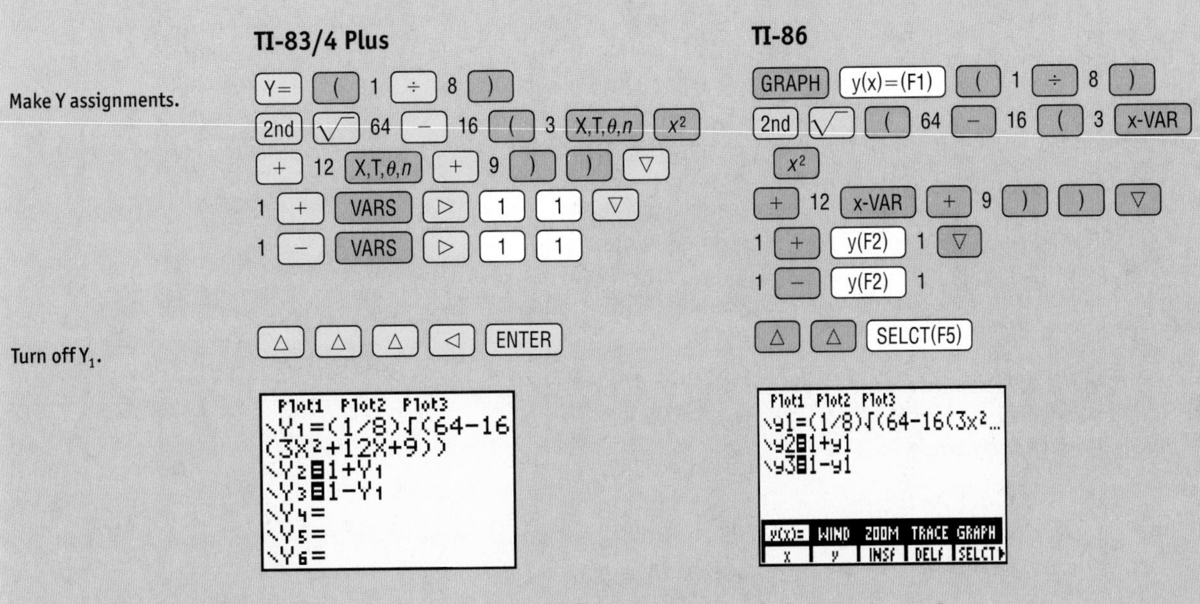

Now graph Y_2 and Y_3 in the viewing rectangle $[-5, 1]$ by $[-1, 3]$.

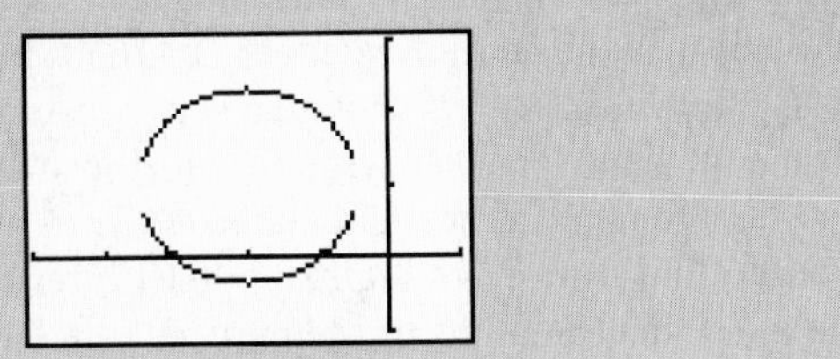

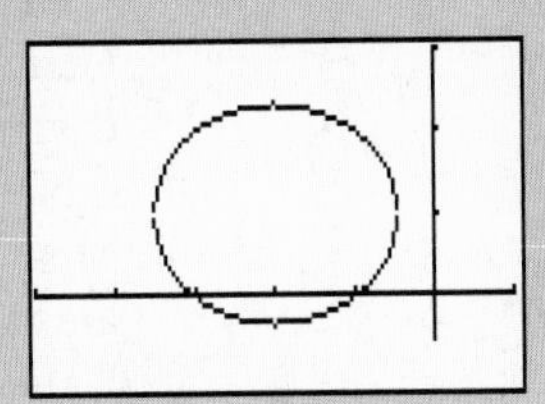

Ellipses can be very flat or almost circular. To obtain information about the *roundness* of an ellipse, we sometimes use the term *eccentricity,* which is defined as follows, with a, b, and c having the same meanings as before.

Definition of Eccentricity

The **eccentricity** e of an ellipse is

$$e = \frac{\text{distance from center to focus}}{\text{distance from center to vertex}} = \frac{c}{a} = \frac{\sqrt{a^2 - b^2}}{a}.$$

Consider the ellipse $(x^2/a^2) + (y^2/b^2) = 1$, and suppose that the length $2a$ of the major axis is fixed and the length $2b$ of the minor axis is variable (note that $0 < b < a$). Since b^2 is positive, $a^2 - b^2 < a^2$ and hence $\sqrt{a^2 - b^2} < a$. Dividing both sides of the last inequality by a gives us $\sqrt{a^2 - b^2}/a < 1$, or $0 < e < 1$. If b is close to 0 (c is close to a), then $\sqrt{a^2 - b^2} \approx a$, $e \approx 1$, and the ellipse is very flat. This case is illustrated in Figure 11(a), with $a = 2$, $b = 0.3$, and $e \approx 0.99$. If b is close to a (c is close to 0), then $\sqrt{a^2 - b^2} \approx 0$, $e \approx 0$, and the ellipse is almost circular. This case is illustrated in Figure 11(b), with $a = 2$, $b = 1.9999$, and $e \approx 0.01$.

In Figure 11(a), the foci are close to the vertices.

In Figure 11(b), the foci are close to the origin.

Note that the ellipse in Figure 5 on page 788 has eccentricity $\frac{5}{13} \approx 0.38$ and appears to be nearly circular.

Figure 11

(a) Eccentricity almost 1 **(b)** Eccentricity almost 0

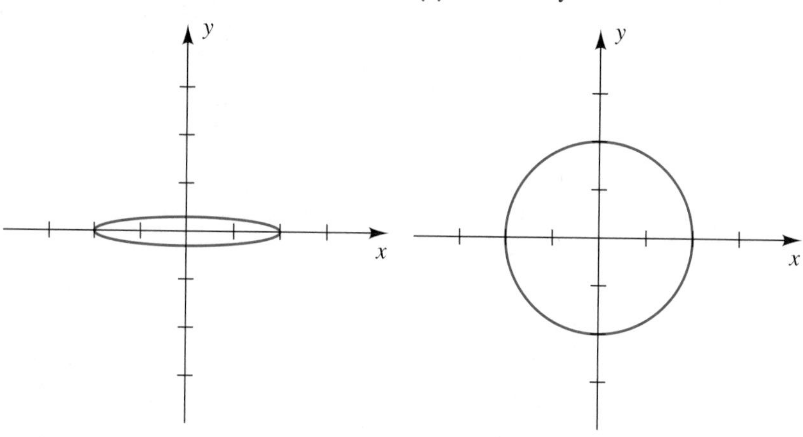

After many years of analyzing an enormous amount of empirical data, the German astronomer Johannes Kepler (1571–1630) formulated three laws that describe the motion of planets about the sun. Kepler's first law states that the orbit of each planet in the solar system is an ellipse with the sun at one focus. Most of these orbits are almost circular, so their corresponding eccentricities are close to 0. To illustrate, for Earth, $e \approx 0.017$; for Mars, $e \approx 0.093$; and for Uranus, $e \approx 0.046$. The orbits of Mercury and the dwarf planet Pluto are less circular, with eccentricities of 0.206 and 0.249, respectively.

Many comets have elliptical orbits with the sun at a focus. In this case the eccentricity e is close to 1, and the ellipse is very flat. In the next example we use the **astronomical unit** (AU)—that is, the average distance from Earth to the sun—to specify large distances (1 AU $\approx$ 93,000,000 mi).

EXAMPLE 7 Approximating a distance in an elliptical path

Halley's comet has an elliptical orbit with eccentricity $e = 0.967$. The closest that Halley's comet comes to the sun is 0.587 AU. Approximate the maximum distance of the comet from the sun, to the nearest 0.1 AU.

▶ **SOLUTION** Figure 12 illustrates the orbit of the comet, where c is the distance from the center of the ellipse to a focus (the sun) and $2a$ is the length of the major axis.

Figure 12

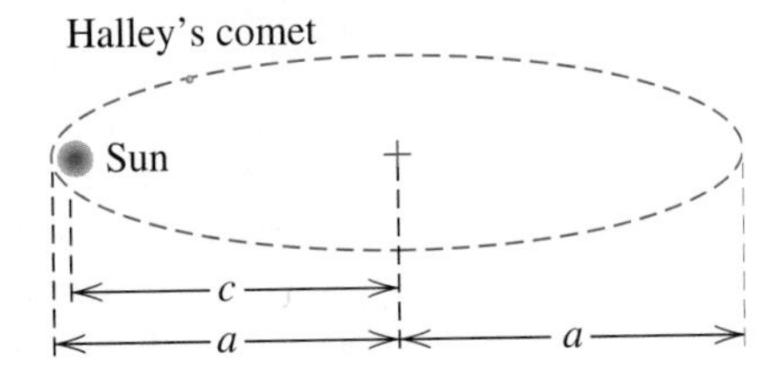

Since $a - c$ is the minimum distance between the sun and the comet, we have (in AU)

$$a - c = 0.587, \quad \text{or} \quad a = c + 0.587.$$

Since $e = c/a = 0.967$, we obtain the following:

$$\begin{aligned}
c &= 0.967a && \text{multiply by } a \\
&= 0.967(c + 0.587) && \text{substitute for } a \\
&\approx 0.967c + 0.568 && \text{multiply} \\
c - 0.967c &\approx 0.568 && \text{subtract } 0.967c \\
c(1 - 0.967) &\approx 0.568 && \text{factor out } c \\
c &\approx \frac{0.568}{0.033} \approx 17.2 && \text{solve for } c
\end{aligned}$$

Since $a = c + 0.587$, we obtain

$$a \approx 17.2 + 0.587 \approx 17.8,$$

and the maximum distance between the sun and the comet is

$$a + c \approx 17.8 + 17.2 = 35.0 \text{ AU}.$$

An ellipse has a *reflective property* analogous to that of the parabola discussed at the end of the previous section. To illustrate, let l denote the tangent

▶ Tutorial available at www.thomsonedu.com/login

Figure 13

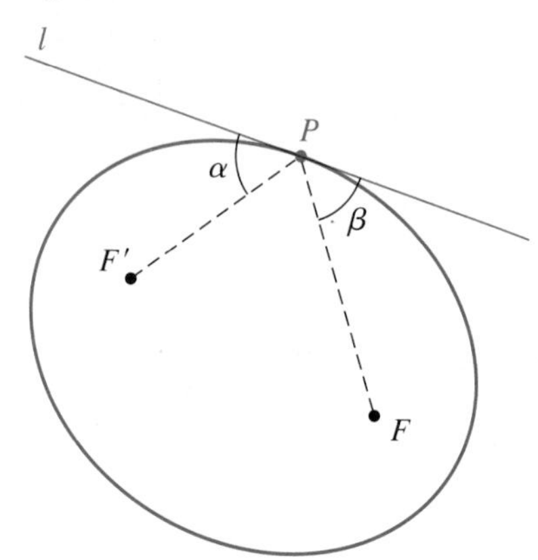

Figure 14

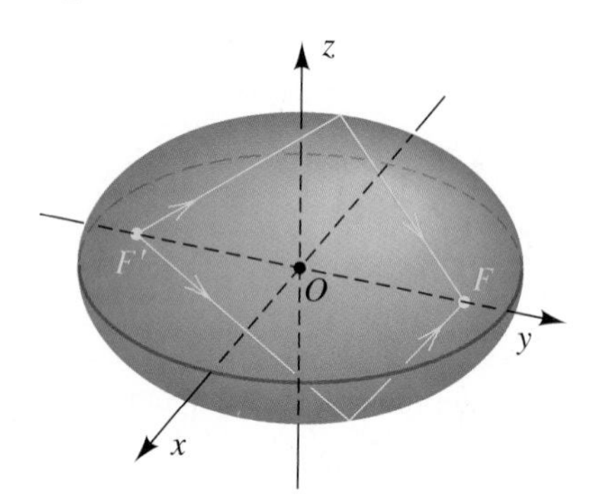

line at a point P on an ellipse with foci F and F', as shown in Figure 13. If α is the acute angle between $F'P$ and l and if β is the acute angle between FP and l, it can be shown that $\alpha = \beta$. Thus, if a ray of light or sound emanates from one focus, it is reflected to the other focus. This property is used in the design of certain types of optical equipment.

If the ellipse with center O and foci F' and F on the x-axis is revolved about the x-axis, as illustrated in Figure 14, we obtain a three-dimensional surface called an **ellipsoid**. The upper half or lower half is a **hemi-ellipsoid**, as is the right half or left half. Sound waves or other impulses that are emitted from the focus F' will be reflected off the ellipsoid into the focus F. This property is used in the design of *whispering galleries*—structures with ellipsoidal ceilings, in which a person who whispers at one focus can be heard at the other focus. Examples of whispering galleries may be found in the Rotunda of the Capitol Building in Washington, D.C., and in the Mormon Tabernacle in Salt Lake City.

The reflective property of ellipsoids (and hemi-ellipsoids) is used in modern medicine in a device called a *lithotripter,* which disintegrates kidney stones by means of high-energy underwater shock waves. After taking extremely accurate measurements, the operator positions the patient so that the stone is at a focus. Ultra–high frequency shock waves are then produced at the other focus, and reflected waves break up the kidney stone. Recovery time with this technique is usually 3–4 days, instead of the 2–3 weeks with conventional surgery. Moreover, the mortality rate is less than 0.01%, as compared to 2–3% for traditional surgery (see Exercises 51–52).

10.2 *Exercises*

Exer. 1–14: Find the vertices and foci of the ellipse. Sketch its graph, showing the foci.

1 $\dfrac{x^2}{9} + \dfrac{y^2}{4} = 1$

2 $\dfrac{x^2}{25} + \dfrac{y^2}{16} = 1$

3 $\dfrac{x^2}{15} + \dfrac{y^2}{16} = 1$

4 $\dfrac{x^2}{45} + \dfrac{y^2}{49} = 1$

5 $4x^2 + y^2 = 16$

6 $y^2 + 9x^2 = 9$

7 $4x^2 + 25y^2 = 1$

8 $10y^2 + x^2 = 5$

9 $\dfrac{(x-3)^2}{16} + \dfrac{(y+4)^2}{9} = 1$

10 $\dfrac{(x+2)^2}{25} + \dfrac{(y-3)^2}{4} = 1$

11 $4x^2 + 9y^2 - 32x - 36y + 64 = 0$

12 $x^2 + 2y^2 + 2x - 20y + 43 = 0$

13 $25x^2 + 4y^2 - 250x - 16y + 541 = 0$

14 $4x^2 + y^2 = 2y$

Exer. 15–18: Find an equation for the ellipse shown in the figure.

15

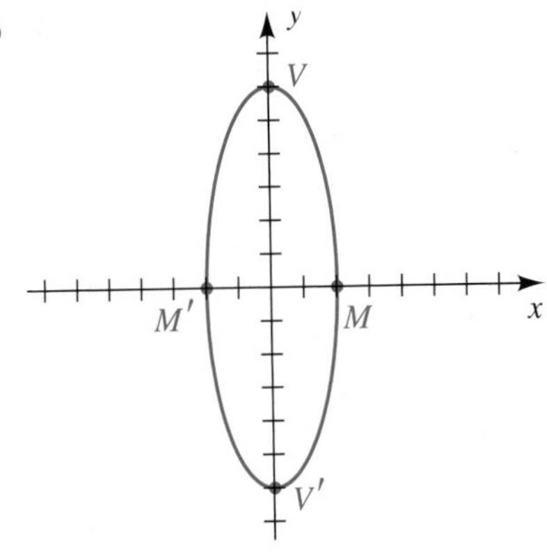

16

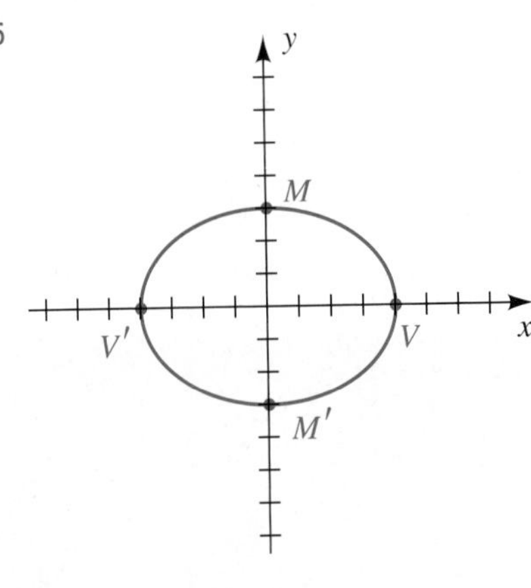

▶ 17

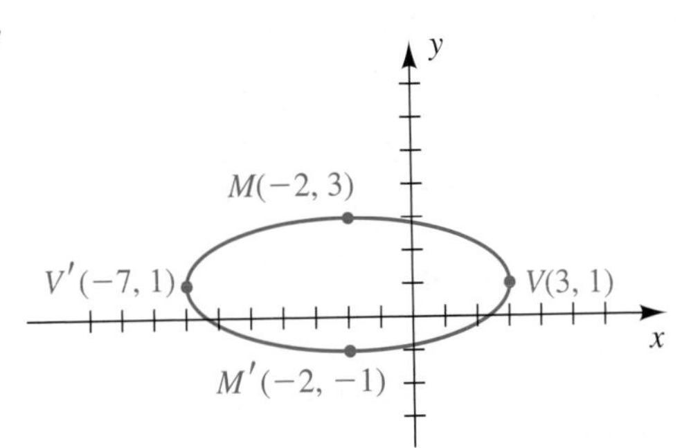

▶ 18

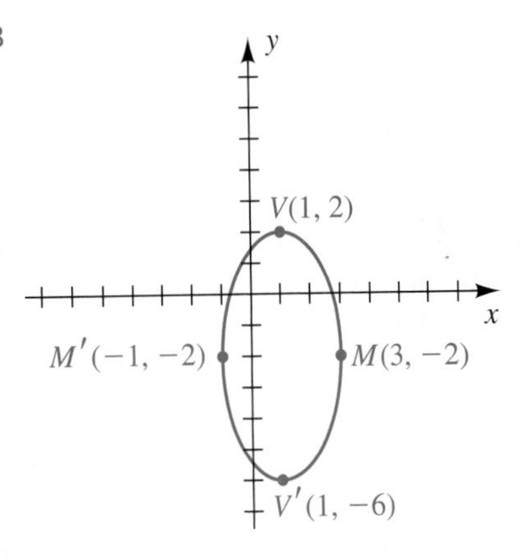

Exer. 19–30: Find an equation for the ellipse that has its center at the origin and satisfies the given conditions.

19 Vertices $V(\pm 8, 0)$, foci $F(\pm 5, 0)$

▶ 20 Vertices $V(0, \pm 7)$, foci $F(0, \pm 2)$

21 Vertices $V(0, \pm 5)$, minor axis of length 3

22 Foci $F(\pm 3, 0)$, minor axis of length 2

23 Vertices $V(0, \pm 6)$, passing through (3, 2)

24 Passing through (2, 3) and (6, 1)

25 Eccentricity $\frac{3}{4}$, vertices $V(0, \pm 4)$

26 Eccentricity $\frac{1}{2}$, vertices on the x-axis, passing through (1, 3)

27 x-intercepts ± 2, y-intercepts $\pm \frac{1}{3}$

28 x-intercepts $\pm \frac{1}{2}$, y-intercepts ± 4

29 Horizontal major axis of length 8, minor axis of length 5

30 Vertical major axis of length 7, minor axis of length 6

Exer. 31–32: Find the points of intersection of the graphs of the equations. Sketch both graphs on the same coordinate plane, and show the points of intersection.

31 $\begin{cases} x^2 + 4y^2 = 20 \\ x + 2y = 6 \end{cases}$

32 $\begin{cases} x^2 + 4y^2 = 36 \\ x^2 + y^2 = 12 \end{cases}$

Exer. 33–36: Find an equation for the set of points in an *xy*-plane such that the sum of the distances from *F* and *F′* is *k*.

33 $F(3, 0)$, $F'(-3, 0)$; $k = 10$

34 $F(12, 0)$, $F'(-12, 0)$; $k = 26$

35 $F(0, 15)$, $F'(0, -15)$; $k = 34$

36 $F(0, 8)$, $F'(0, -8)$; $k = 20$

Exer. 37–38: Find an equation for the ellipse with foci *F* and *F′* that passes through *P*. Sketch the ellipse.

37 38

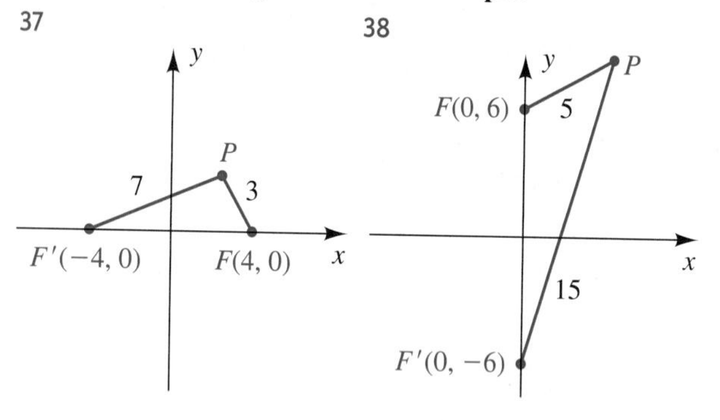

Exer. 39–46: Determine whether the graph of the equation is the upper, lower, left, or right half of an ellipse, and find an equation for the ellipse.

39 $y = 11\sqrt{1 - \dfrac{x^2}{49}}$

▶ 40 $y = -6\sqrt{1 - \dfrac{x^2}{25}}$

41 $x = -\frac{1}{3}\sqrt{9 - y^2}$

42 $x = \frac{4}{5}\sqrt{25 - y^2}$

▶ 43 $x = 1 + 2\sqrt{1 - \dfrac{(y+2)^2}{9}}$

44 $x = -2 - 5\sqrt{1 - \dfrac{(y-1)^2}{16}}$

45 $y = 2 - 7\sqrt{1 - \dfrac{(x+1)^2}{9}}$

46 $y = -1 + \sqrt{1 - \dfrac{(x-3)^2}{16}}$

47 **Dimensions of an arch** An arch of a bridge is semielliptical, with major axis horizontal. The base of the arch is 30 feet across, and the highest part of the arch is 10 feet above the horizontal roadway, as shown in the figure. Find the height of the arch 6 feet from the center of the base.

Exercise 47

48 **Designing a bridge** A bridge is to be constructed across a river that is 200 feet wide. The arch of the bridge is to be semielliptical and must be constructed so that a ship less than 50 feet wide and 30 feet high can pass safely through the arch, as shown in the figure on the next page.

(a) Find an equation for the arch.

(b) Approximate the height of the arch in the middle of the bridge.

Exercise 48

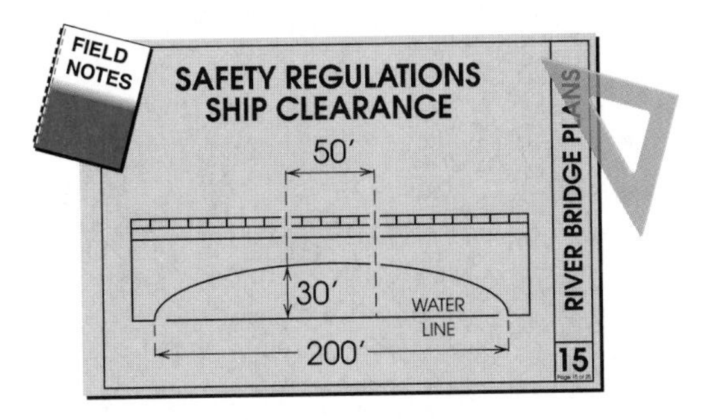

49 **Earth's orbit** Assume that the length of the major axis of Earth's orbit is 186,000,000 miles and that the eccentricity is 0.017. Approximate, to the nearest 1000 miles, the maximum and minimum distances between Earth and the sun.

50 **Mercury's orbit** The planet Mercury travels in an elliptical orbit that has eccentricity 0.206 and major axis of length 0.774 AU. Find the maximum and minimum distances between Mercury and the sun.

51 **Elliptical reflector** The basic shape of an elliptical reflector is a hemi-ellipsoid of height h and diameter k, as shown in the figure. Waves emitted from focus F will reflect off the surface into focus F'.

(a) Express the distances $d(V, F)$ and $d(V, F')$ in terms of h and k.

(b) An elliptical reflector of height 17 centimeters is to be constructed so that waves emitted from F are reflected to a point F' that is 32 centimeters from V. Find the diameter of the reflector and the location of F.

Exercise 51

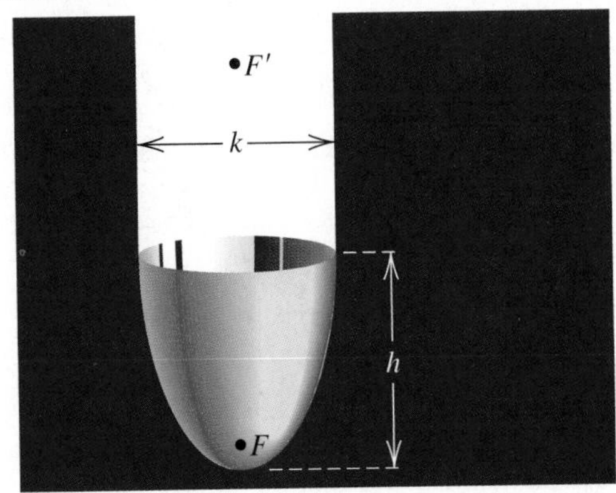

52 **Lithotripter operation** A lithotripter of height 15 centimeters and diameter 18 centimeters is to be constructed (see the figure). High-energy underwater shock waves will be emitted from the focus F that is closest to the vertex V.

(a) Find the distance from V to F.

(b) How far from V (in the vertical direction) should a kidney stone be located?

Exercise 52

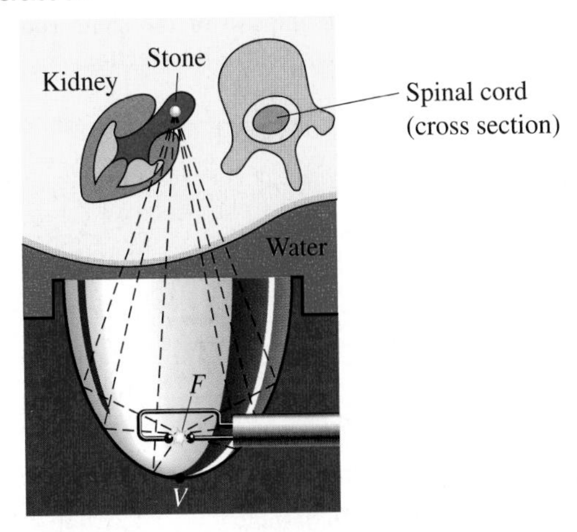

53 **Whispering gallery** The ceiling of a whispering gallery has the shape of the hemi-ellipsoid shown in Figure 14, with the highest point of the ceiling 15 feet above the elliptical floor and the vertices of the floor 50 feet apart. If two people are standing at the foci F' and F, how far from the vertices are their feet?

54 **Oval design** An artist plans to create an elliptical design with major axis 60″ and minor axis 24″, centered on a door that measures 80″ by 36″, using the method described by Figure 1. On a vertical line that bisects the door, approximately how far from each end of the door should the push-pins be inserted? How long should the string be?

Exercise 54

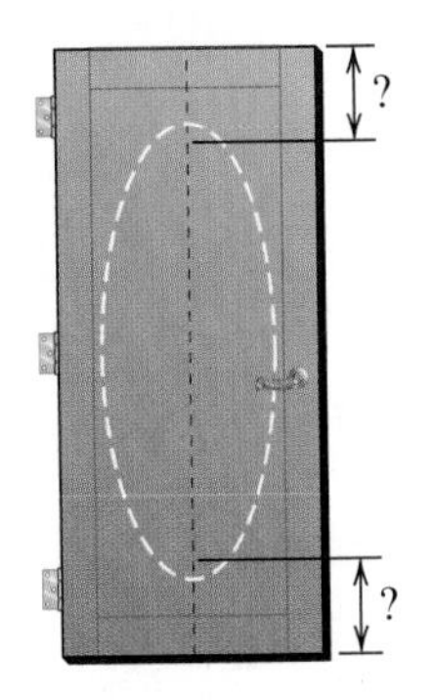

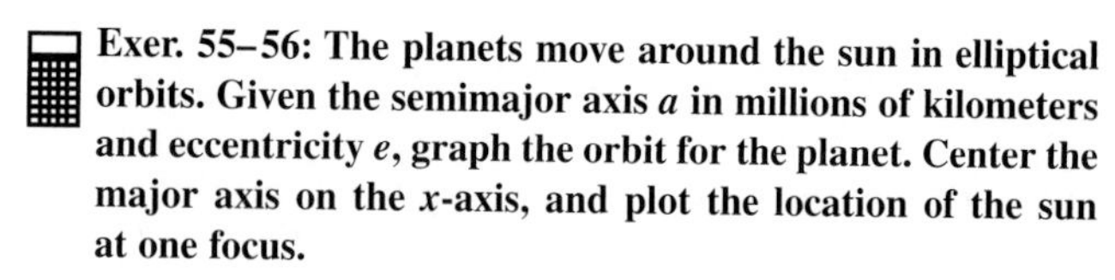

Exer. 55–56: The planets move around the sun in elliptical orbits. Given the semimajor axis a in millions of kilometers and eccentricity e, graph the orbit for the planet. Center the major axis on the x-axis, and plot the location of the sun at one focus.

55 Earth's path $a = 149.6, \quad e = 0.093$

56 Pluto's path $a = 5913, \quad e = 0.249$

Exer. 57–60: Graph the ellipses on the same coordinate plane, and estimate their points of intersection.

57 $\dfrac{x^2}{2.9} + \dfrac{y^2}{2.1} = 1; \quad \dfrac{x^2}{4.3} + \dfrac{(y - 2.1)^2}{4.9} = 1$

58 $\dfrac{x^2}{3.9} + \dfrac{y^2}{2.4} = 1; \quad \dfrac{(x + 1.9)^2}{4.1} + \dfrac{y^2}{2.5} = 1$

59 $\dfrac{(x + 0.1)^2}{1.7} + \dfrac{y^2}{0.9} = 1; \quad \dfrac{x^2}{0.9} + \dfrac{(y - 0.25)^2}{1.8} = 1$

60 $\dfrac{x^2}{3.1} + \dfrac{(y - 0.2)^2}{2.8} = 1; \quad \dfrac{(x + 0.23)^2}{1.8} + \dfrac{y^2}{4.2} = 1$

10.3 Hyperbolas

The definition of a hyperbola is similar to that of an ellipse. The only change is that instead of using the *sum* of distances from two fixed points, we use the *difference*.

Definition of a Hyperbola	A **hyperbola** is the set of all points in a plane, the difference of whose distances from two fixed points (the **foci**) in the plane is a positive constant.

Figure 1

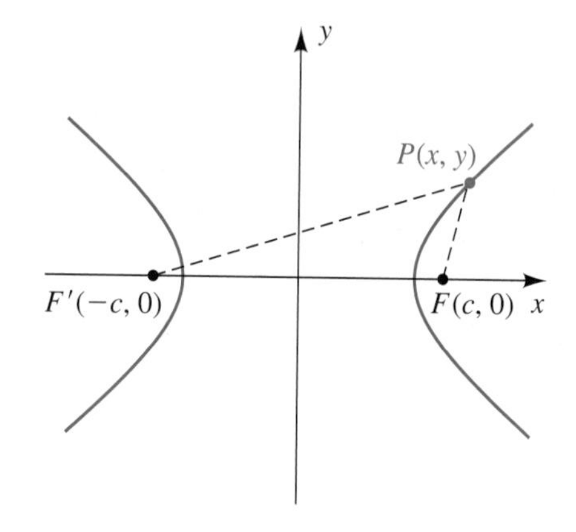

To find a simple equation for a hyperbola, we choose a coordinate system with foci at $F(c, 0)$ and $F'(-c, 0)$ and denote the (constant) distance by $2a$. The midpoint of the segment $F'F$ (the origin) is called the **center** of the hyperbola. Referring to Figure 1, we see that a point $P(x, y)$ is on the hyperbola if and only if either of the following is true:

(1) $d(P, F') - d(P, F) = 2a$ or **(2)** $d(P, F) - d(P, F') = 2a$

If P is not on the x-axis, then from Figure 1 we see that

$$d(P, F) < d(F', F) + d(P, F'),$$

because the length of one side of a triangle is always less than the sum of the lengths of the other two sides. Similarly,

$$d(P, F') < d(F', F) + d(P, F).$$

Equivalent forms for the previous two inequalities are

$$d(P, F) - d(P, F') < d(F', F) \quad \text{and} \quad d(P, F') - d(P, F) < d(F', F).$$

Since the differences on the left-hand sides of these inequalities both equal $2a$ and since $d(F', F) = 2c$, the last two inequalities imply that $2a < 2c$, or $a < c$. (Recall that for ellipses we had $a > c$.)

Next, equations (1) and (2) may be replaced by the single equation

$$|d(P, F) - d(P, F')| = 2a.$$

Using the distance formula to find $d(P, F)$ and $d(P, F')$, we obtain an equation of the hyperbola:

$$\left|\sqrt{(x-c)^2 + (y-0)^2} - \sqrt{(x+c)^2 + (y-0)^2}\right| = 2a$$

Employing the type of simplification procedure that we used to derive an equation for an ellipse, we can rewrite the preceding equation as

$$\frac{x^2}{a^2} - \frac{y^2}{c^2 - a^2} = 1.$$

Finally, if we let

$$b^2 = c^2 - a^2 \quad \text{with} \quad b > 0$$

in the preceding equation, we obtain

$$\frac{x^2}{a^2} - \frac{y^2}{b^2} = 1.$$

We have shown that the coordinates of every point (x, y) on the hyperbola in Figure 1 satisfy the equation $(x^2/a^2) - (y^2/b^2) = 1$. Conversely, if (x, y) is a solution of this equation, then by reversing steps we see that the point (x, y) is on the hyperbola.

Applying tests for symmetry, we see that the hyperbola is symmetric with respect to both axes and the origin. We may find the x-intercepts of the hyperbola by letting $y = 0$ in the equation. Doing so gives us $x^2/a^2 = 1$, or $x^2 = a^2$, and consequently the x-intercepts are a and $-a$. The corresponding points $V(a, 0)$ and $V'(-a, 0)$ on the graph are called the **vertices** of the hyperbola (see Figure 2). The line segment $V'V$ is called the **transverse axis.** The graph has no y-intercept, since the equation $-y^2/b^2 = 1$ has the *complex* solutions $y = \pm bi$. The points $W(0, b)$ and $W'(0, -b)$ are endpoints of the **conjugate axis** $W'W$. The points W and W' are not on the hyperbola; however, as we shall see, they are useful for describing the graph.

Figure 2

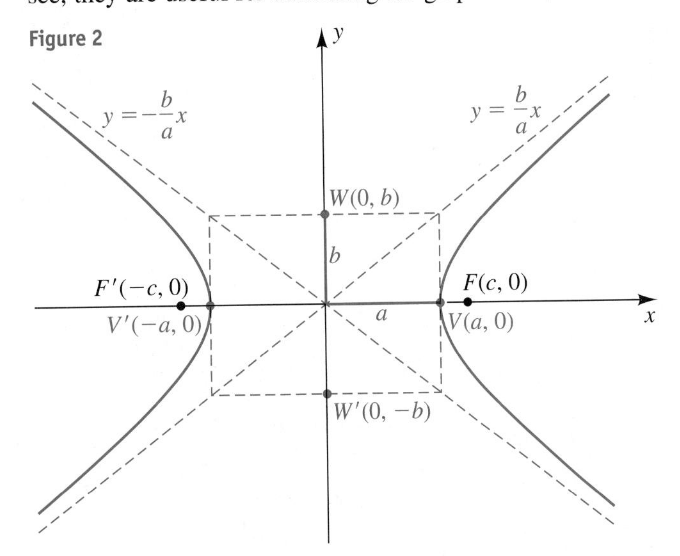

Solving the equation $(x^2/a^2) - (y^2/b^2) = 1$ for y gives us

$$y = \pm\frac{b}{a}\sqrt{x^2 - a^2}.$$

If $x^2 - a^2 < 0$ or, equivalently, $-a < x < a$, then there are no points (x, y) on the graph. There *are* points $P(x, y)$ on the graph if $x \geq a$ or $x \leq -a$.

It can be shown that *the lines* $y = \pm(b/a)x$ *are asymptotes for the hyperbola.* These asymptotes serve as excellent guides for sketching the graph. A convenient way to sketch the asymptotes is to first plot the vertices $V(a, 0)$, $V'(-a, 0)$ and the points $W(0, b)$, $W'(0, -b)$ (see Figure 2). If vertical and horizontal lines are drawn through these endpoints of the transverse and conjugate axes, respectively, then the diagonals of the resulting **auxiliary rectangle** have slopes b/a and $-b/a$. Hence, by extending these diagonals we obtain the asymptotes $y = \pm(b/a)x$. The hyperbola is then sketched as in Figure 2, using the asymptotes as guides. The two parts that make up the hyperbola are called the **right branch** and the **left branch** of the hyperbola.

Similarly, if we take the foci on the y-axis, we obtain the equation

$$\frac{y^2}{a^2} - \frac{x^2}{b^2} = 1.$$

In this case, the vertices of the hyperbola are $(0, \pm a)$ and the endpoints of the conjugate axis are $(\pm b, 0)$, as shown in Figure 3. The asymptotes are $y = \pm(a/b)x$ (*not* $y = \pm(b/a)x$, as in the previous case), and we now refer to the two parts that make up the hyperbola as the **upper branch** and the **lower branch** of the hyperbola.

Figure 3

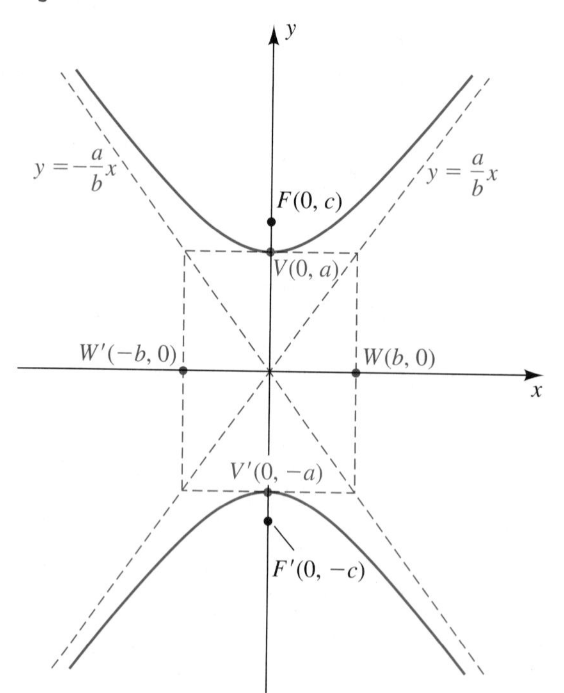

The preceding discussion may be summarized as follows.

Standard Equations of a Hyperbola with Center at the Origin

The graph of

$$\frac{x^2}{a^2} - \frac{y^2}{b^2} = 1 \quad \text{or} \quad \frac{y^2}{a^2} - \frac{x^2}{b^2} = 1$$

is a hyperbola with center at the origin. The length of the transverse axis is $2a$, and the length of the conjugate axis is $2b$. The foci are a distance c from the origin, where $c^2 = a^2 + b^2$.

Note that the vertices are on the x-axis if the x^2-term has a positive coefficient (the first equation in the above box) or on the y-axis if the y^2-term has a positive coefficient (the second equation).

EXAMPLE 1 Sketching a hyperbola with center at the origin

Sketch the graph of $9x^2 - 4y^2 = 36$. Find the foci and equations of the asymptotes.

▶ **SOLUTION** From the remarks preceding this example, the graph is a hyperbola with center at the origin. To express the given equation in a standard form, we divide both sides by 36 and simplify, obtaining

$$\frac{x^2}{4} - \frac{y^2}{9} = 1.$$

Figure 4

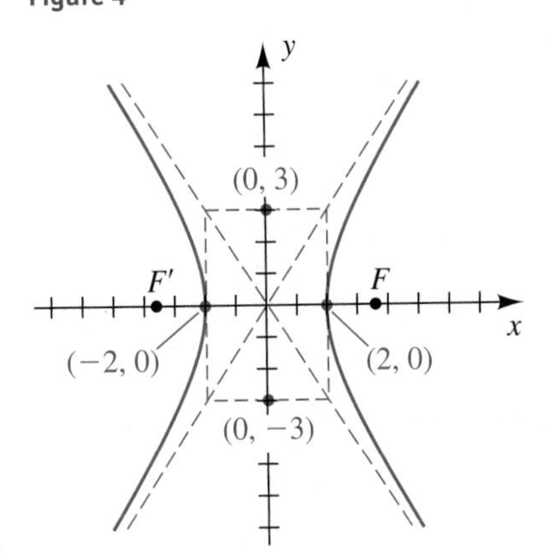

Comparing $(x^2/4) - (y^2/9) = 1$ to $(x^2/a^2) - (y^2/b^2) = 1$, we see that $a^2 = 4$ and $b^2 = 9$; that is, $a = 2$ and $b = 3$. The hyperbola has its vertices on the x-axis, since there are x-intercepts and no y-intercepts. The vertices $(\pm 2, 0)$ and the endpoints $(0, \pm 3)$ of the conjugate axis determine the auxiliary rectangle whose diagonals (extended) give us the asymptotes. The graph of the equation is sketched in Figure 4.

To find the foci, we calculate

$$c^2 = a^2 + b^2 = 4 + 9 = 13.$$

Thus, $c = \sqrt{13}$, and the foci are $F(\sqrt{13}, 0)$ and $F'(-\sqrt{13}, 0)$.

The equations of the asymptotes, $y = \pm\frac{3}{2}x$, can be found by referring to the graph or to the equations $y = \pm(b/a)x$.

The preceding example indicates that for hyperbolas it is not always true that $a > b$, as is the case for ellipses. In fact, we may have $a < b$, $a > b$, or $a = b$.

▶ **Tutorial available at www.thomsonedu.com/login**

EXAMPLE 2 **Sketching a hyperbola with center at the origin**

Sketch the graph of $4y^2 - 2x^2 = 1$. Find the foci and equations of the asymptotes.

▶ **SOLUTION** To express the given equation in a standard form, we write

$$\frac{y^2}{\frac{1}{4}} - \frac{x^2}{\frac{1}{2}} = 1.$$

Thus,

$$a^2 = \tfrac{1}{4}, \qquad b^2 = \tfrac{1}{2}, \qquad \text{and} \qquad c^2 = a^2 + b^2 = \tfrac{3}{4},$$

and consequently

$$a = \frac{1}{2}, \qquad b = \frac{1}{\sqrt{2}} = \frac{\sqrt{2}}{2}, \qquad \text{and} \qquad c = \frac{\sqrt{3}}{2}.$$

Figure 5

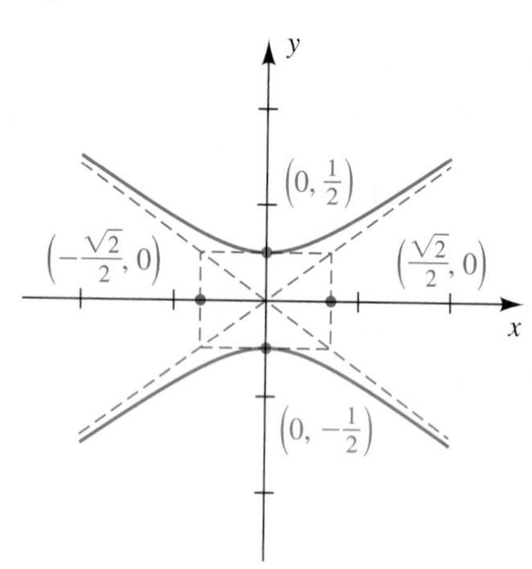

The hyperbola has its vertices on the y-axis, since there are y-intercepts and no x-intercepts. The vertices are $(0, \pm\frac{1}{2})$, the endpoints of the conjugate axes are $(\pm\sqrt{2}/2, 0)$, and the foci are $(0, \pm\sqrt{3}/2)$. The graph is sketched in Figure 5.

To find the equations of the asymptotes, we refer to the figure or use $y = \pm(a/b)x$, obtaining $y = \pm(\sqrt{2}/2)x$.

EXAMPLE 3 **Finding an equation of a hyperbola satisfying prescribed conditions**

A hyperbola has vertices $(\pm 3, 0)$ and passes through the point $P(5, 2)$. Find its equation, foci, and asymptotes.

▶ **SOLUTION** We begin by sketching a hyperbola with vertices $(\pm 3, 0)$ that passes through the point $P(5, 2)$, as in Figure 6.

Figure 6

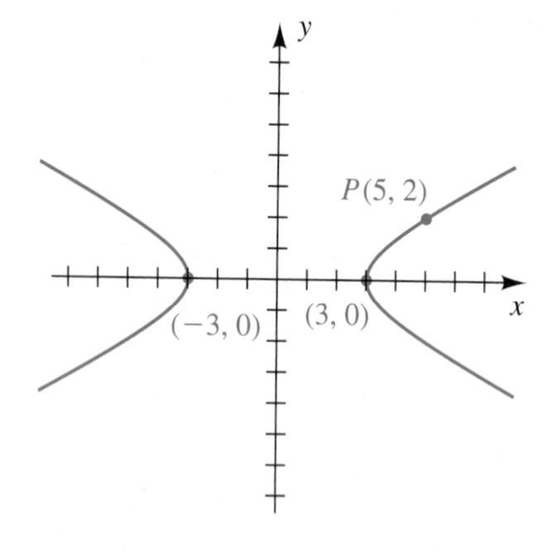

An equation of the hyperbola has the form

$$\frac{x^2}{3^2} - \frac{y^2}{b^2} = 1.$$

Since $P(5, 2)$ is on the hyperbola, the x- and y-coordinates satisfy this equation; that is,

$$\frac{5^2}{3^2} - \frac{2^2}{b^2} = 1.$$

Solving for b^2 gives us $b^2 = \frac{9}{4}$, and hence an equation for the hyperbola is

$$\frac{x^2}{9} - \frac{y^2}{\frac{9}{4}} = 1$$

or, equivalently,

$$x^2 - 4y^2 = 9.$$

▶ **Tutorial available at www.thomsonedu.com/login**

Figure 7

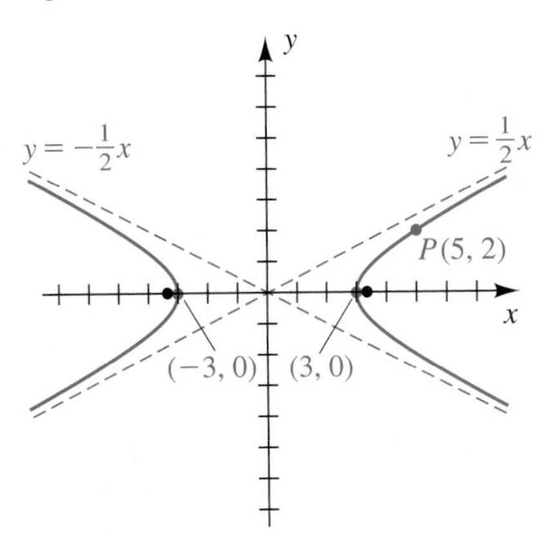

To find the foci, we first calculate

$$c^2 = a^2 + b^2 = 9 + \tfrac{9}{4} = \tfrac{45}{4}.$$

Hence, $c = \sqrt{\frac{45}{4}} = \frac{3}{2}\sqrt{5} \approx 3.35$, and the foci are $\left(\pm\frac{3}{2}\sqrt{5}, 0\right)$.

The general equations of the asymptotes are $y = \pm(b/a)x$. Substituting $a = 3$ and $b = \frac{3}{2}$ gives us $y = \pm\frac{1}{2}x$, as shown in Figure 7.

The next example indicates how to find equations for certain parts of a hyperbola.

EXAMPLE 4 Finding equations of portions of a hyperbola

The hyperbola $9x^2 - 4y^2 = 36$ was discussed in Example 1. Solve the equation as indicated, and describe the resulting graph.

(a) For x in terms of y **(b)** For y in terms of x

▶ **SOLUTION**

(a) We solve for x in terms of y as follows:

$$9x^2 - 4y^2 = 36 \quad \text{given}$$
$$x^2 = \frac{36 + 4y^2}{9} \quad \text{solve for } x^2$$
$$x = \pm\tfrac{2}{3}\sqrt{9 + y^2} \quad \text{factor out 4, and take the square root}$$

Figure 8

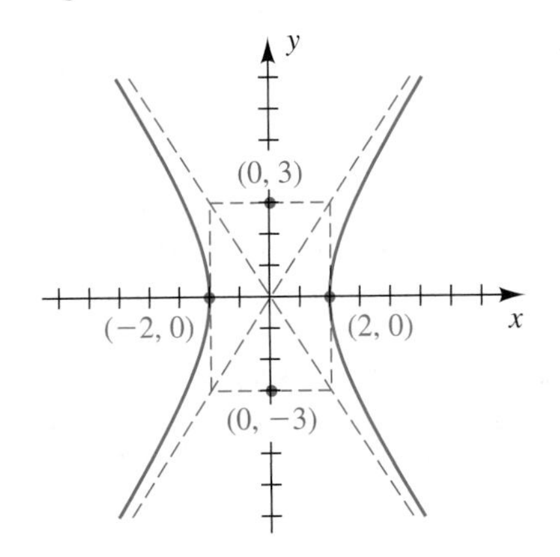

The graph of the equation $x = \frac{2}{3}\sqrt{9 + y^2}$ is the right branch of the hyperbola sketched in Figure 4 (and repeated in Figure 8), and the graph of $x = -\frac{2}{3}\sqrt{9 + y^2}$ is the left branch.

(b) We solve for y in terms of x as follows:

$$9x^2 - 4y^2 = 36 \quad \text{given}$$
$$y^2 = \frac{9x^2 - 36}{4} \quad \text{solve for } y^2$$
$$y = \pm\tfrac{3}{2}\sqrt{x^2 - 4} \quad \text{factor out 9, and take the square root}$$

The graph of $y = \frac{3}{2}\sqrt{x^2 - 4}$ is the upper half of the right and left branches, and the graph of $y = -\frac{3}{2}\sqrt{x^2 - 4}$ is the lower half of these branches.

As was the case for ellipses, we may use translations to help sketch hyperbolas that have centers at some point $(h, k) \neq (0, 0)$. The following example illustrates this technique.

EXAMPLE 5 Sketching a hyperbola with center (h, k)

Sketch the graph of the equation

$$9x^2 - 4y^2 - 54x - 16y + 29 = 0.$$

▶ **Tutorial available at www.thomsonedu.com/login**

SOLUTION We arrange our work using a procedure similar to that used for ellipses in Example 5 of the previous section:

$$(9x^2 - 54x) + (-4y^2 - 16y) = -29 \qquad \text{group terms}$$
$$9(x^2 - 6x + __) - 4(y^2 + 4y + __) = -29 \qquad \text{factor out 9 and } -4$$
$$9(x^2 - 6x + 9) - 4(y^2 + 4y + 4) = -29 + \underline{9 \cdot 9} - \underline{4 \cdot 4} \qquad \text{complete the squares}$$
$$9(x - 3)^2 - 4(y + 2)^2 = 36 \qquad \text{factor, and simplify}$$
$$\frac{(x - 3)^2}{4} - \frac{(y + 2)^2}{9} = 1 \qquad \text{divide by 36}$$

Figure 9

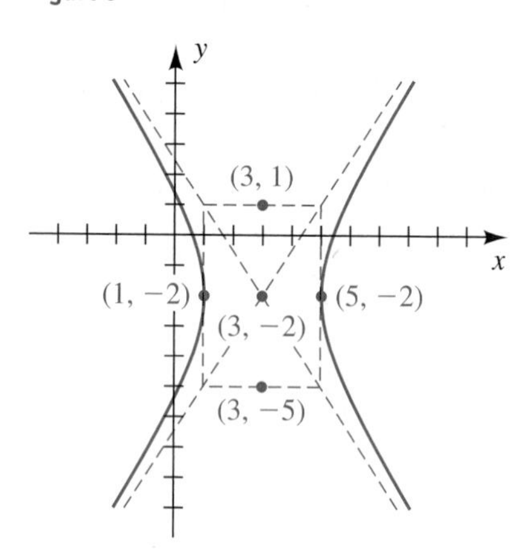

The last equation indicates that the hyperbola has center $C(3, -2)$ with vertices and foci on the horizontal line $y = -2$, because the term containing x is positive. We also know that

$$a^2 = 4, \qquad b^2 = 9, \qquad \text{and} \qquad c^2 = a^2 + b^2 = 13.$$

Hence,

$$a = 2, \qquad b = 3, \qquad \text{and} \qquad c = \sqrt{13}.$$

As illustrated in Figure 9, the vertices are $(3 \pm 2, -2)$—that is, $(5, -2)$ and $(1, -2)$. The endpoints of the conjugate axis are $(3, -2 \pm 3)$—that is, $(3, 1)$ and $(3, -5)$. The foci are $\left(3 \pm \sqrt{13}, -2\right)$, and equations of the asymptotes are

$$y + 2 = \pm\tfrac{3}{2}(x - 3).$$

The results of Sections 10.1 through 10.3 indicate that the graph of every equation of the form

$$Ax^2 + Cy^2 + Dx + Ey + F = 0$$

is a conic, except for certain degenerate cases in which a point, one or two lines, or no graph is obtained. Although we have considered only special examples, our methods can be applied to any such equation. If A and C are equal and not 0, then the graph, when it exists, is a circle or, in exceptional cases, a point. If A and C are unequal but have the same sign, an equation is obtained whose graph, when it exists, is an ellipse (or a point). If A and C have opposite signs, an equation of a hyperbola is obtained or possibly, in the degenerate case, two intersecting straight lines. If either A or C (but not both) is 0, the graph is a parabola or, in certain cases, a pair of parallel lines.

We shall conclude this section with an application involving hyperbolas.

EXAMPLE 6 Locating a ship

Coast Guard station A is 200 miles directly east of another station B. A ship is sailing on a line parallel to and 50 miles north of the line through A and B. Radio signals are sent out from A and B at the rate of 980 ft/μsec (microsecond). If, at 1:00 P.M., the signal from B reaches the ship 400 microseconds after the signal from A, locate the position of the ship at that time.

Figure 10
(a)

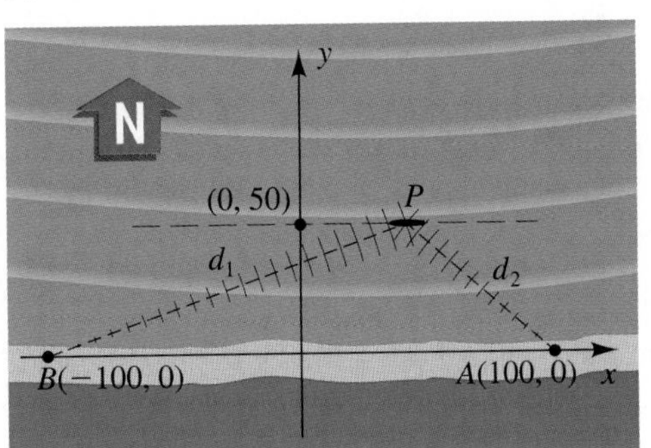

(b)

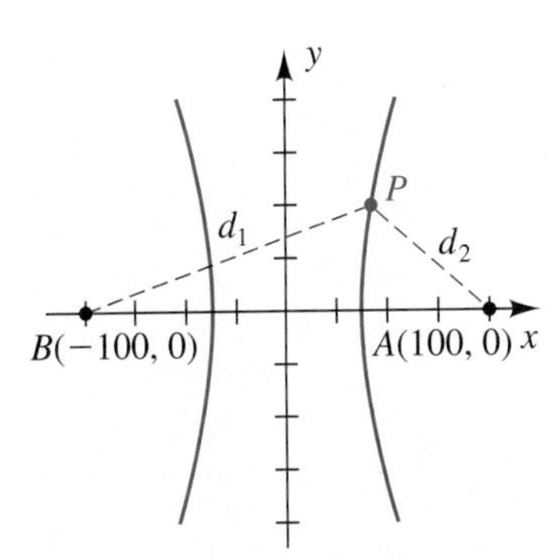

▶ SOLUTION Let us introduce a coordinate system, as shown in Figure 10(a), with the stations at points A and B on the x-axis and the ship at P on the line $y = 50$. Since at 1:00 P.M. it takes 400 microseconds longer for the signal to arrive from B than from A, the difference $d_1 - d_2$ in the indicated distances at that time is

$$d_1 - d_2 = (980)(400) = 392{,}000 \text{ ft.}$$

Dividing by 5280 (ft/mi) gives us

$$d_1 - d_2 = \frac{392{,}000}{5280} = 74.\overline{24} \text{ mi.}$$

At 1:00 P.M., point P is on the right branch of a hyperbola whose equation is $(x^2/a^2) - (y^2/b^2) = 1$ (see Figure 10(b)), consisting of all points whose difference in distances from the foci B and A is $d_1 - d_2$. In our derivation of the equation $(x^2/a^2) - (y^2/b^2) = 1$, we let $d_1 - d_2 = 2a$; it follows that in the present situation

$$a = \frac{74.\overline{24}}{2} = 37.\overline{12} \qquad \text{and} \qquad a^2 \approx 1378.$$

Since the distance c from the origin to either focus is 100,

$$b^2 = c^2 - a^2 \approx 10{,}000 - 1378, \qquad \text{or} \qquad b^2 \approx 8622.$$

Hence, an (approximate) equation for the hyperbola that has foci A and B and passes through P is

$$\frac{x^2}{1378} - \frac{y^2}{8622} = 1.$$

If we let $y = 50$ (the y-coordinate of P), we obtain

$$\frac{x^2}{1378} - \frac{2500}{8622} = 1.$$

(continued)

▶ **Tutorial available at www.thomsonedu.com/login**

Solving for x gives us $x \approx 42.16$. Rounding off to the nearest mile, we find that the coordinates of P are approximately (42, 50).

Figure 11

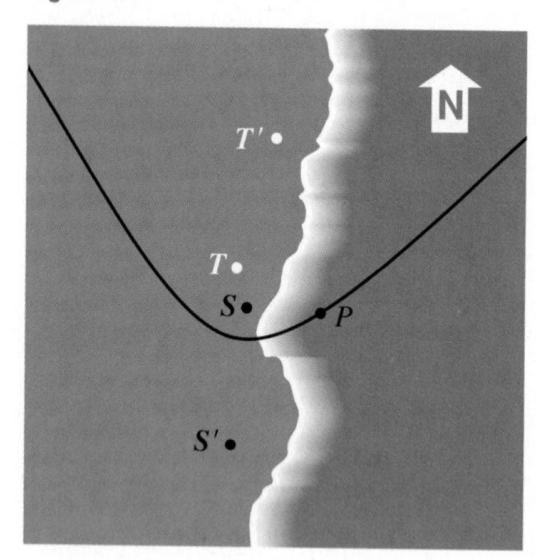

An extension of the method used in Example 6 is the basis for the navigational system LORAN (for Long Range Navigation). This system involves two pairs of radio transmitters, such as those located at T, T' and S, S' in Figure 11. Suppose that signals sent out by the transmitters at T and T' reach a radio receiver in a ship located at some point P. The difference in the times of arrival of the signals can be used to determine the difference in the distances of P from T and T'. Thus, P lies on one branch of a hyperbola with foci at T and T'. Repeating this process for the other pair of transmitters, we see that P also lies on one branch of a hyperbola with foci at S and S'. The intersection of these two branches determines the position of P.

A hyperbola has a *reflective property* analogous to that of the ellipse discussed in the previous section. To illustrate, let l denote the tangent line at a point P on a hyperbola with foci F and F', as shown in Figure 12. If α is the acute angle between $F'P$ and l and if β is the acute angle between FP and l, it can be shown that $\alpha = \beta$. If a ray of light is directed along the line l_1 toward F, it will be reflected back at P along the line l_2 toward F'. This property is used in the design of telescopes of the Cassegrain type (see Exercise 64).

Figure 12

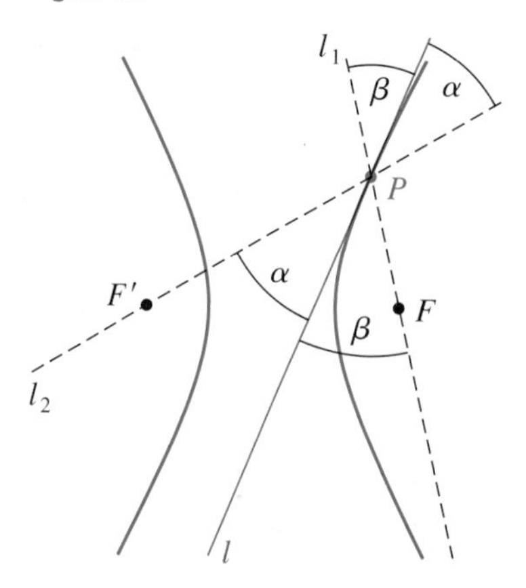

10.3 Exercises

Exer. 1–16: Find the vertices, the foci, and the equations of the asymptotes of the hyperbola. Sketch its graph, showing the asymptotes and the foci.

1 $\dfrac{x^2}{9} - \dfrac{y^2}{4} = 1$

2 $\dfrac{y^2}{49} - \dfrac{x^2}{16} = 1$

3 $\dfrac{y^2}{9} - \dfrac{x^2}{4} = 1$

4 $\dfrac{x^2}{49} - \dfrac{y^2}{16} = 1$

5 $x^2 - \dfrac{y^2}{24} = 1$

6 $y^2 - \dfrac{x^2}{15} = 1$

7 $y^2 - 4x^2 = 16$

8 $x^2 - 2y^2 = 8$

9 $16x^2 - 36y^2 = 1$

10 $y^2 - 16x^2 = 1$

11 $\dfrac{(y + 2)^2}{9} - \dfrac{(x + 2)^2}{4} = 1$

12 $\dfrac{(x - 3)^2}{25} - \dfrac{(y - 1)^2}{4} = 1$

13 $144x^2 - 25y^2 + 864x - 100y - 2404 = 0$

14 $y^2 - 4x^2 - 12y - 16x + 16 = 0$

15 $4y^2 - x^2 + 40y - 4x + 60 = 0$

16 $25x^2 - 9y^2 + 100x - 54y + 10 = 0$

Exer. 17–20: Find an equation for the hyperbola shown in the figure.

 17

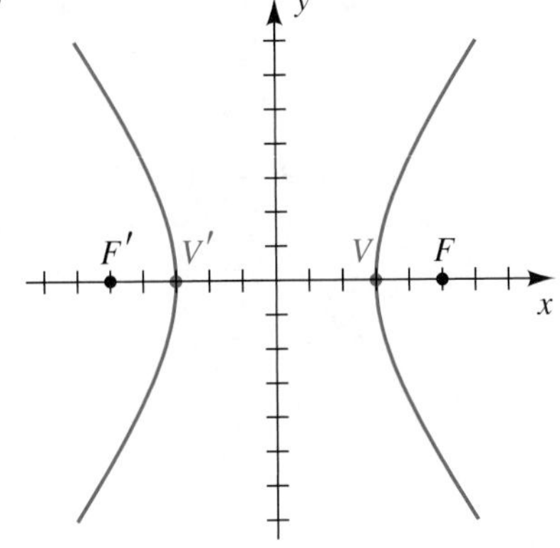

18

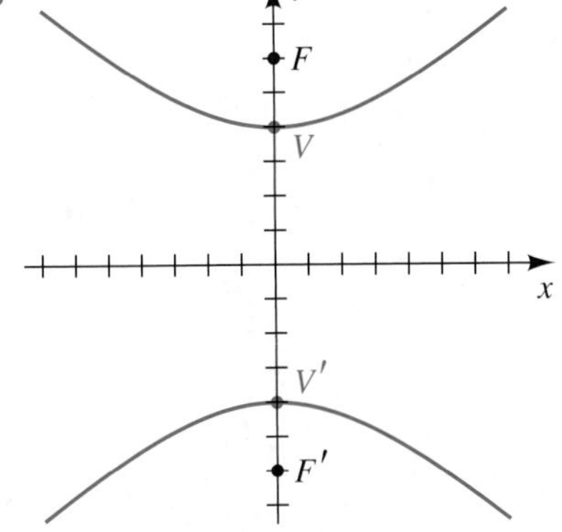

▶ 19

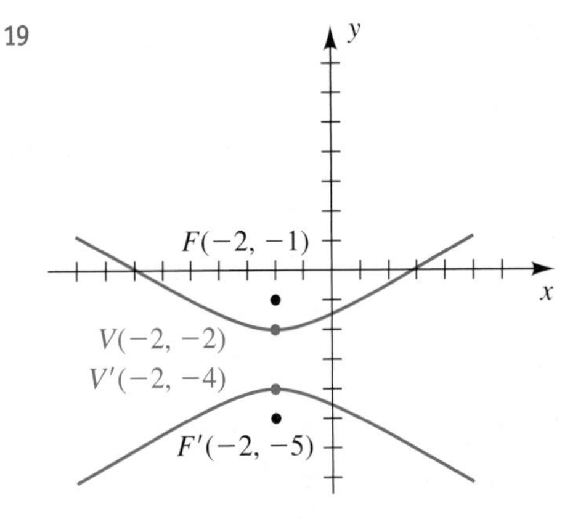

▶ 20

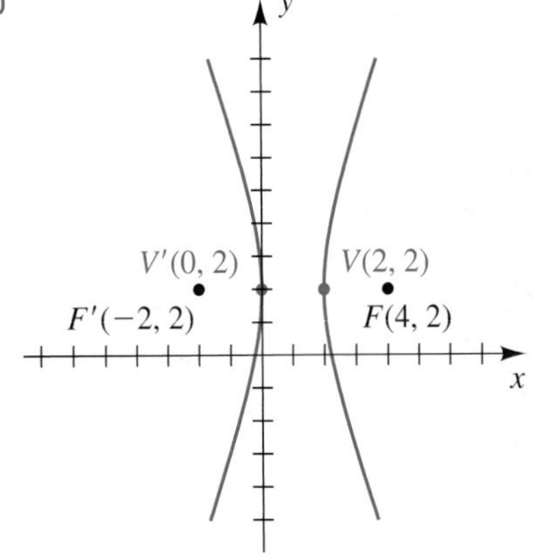

Exer. 21–32: Find an equation for the hyperbola that has its center at the origin and satisfies the given conditions.

21 Foci $F(0, \pm 4)$, vertices $V(0, \pm 1)$

22 Foci $F(\pm 8, 0)$, vertices $V(\pm 5, 0)$

23 Foci $F(\pm 5, 0)$, vertices $V(\pm 3, 0)$

24 Foci $F(0, \pm 3)$, vertices $V(0, \pm 2)$

25 Foci $F(0, \pm 5)$, conjugate axis of length 4

26 Vertices $V(\pm 4, 0)$, passing through $(8, 2)$

▶ **Tutorial available at www.thomsonedu.com/login**

27 Vertices $V(\pm 3, 0)$, asymptotes $y = \pm 2x$

28 Foci $F(0, \pm 10)$, asymptotes $y = \pm \frac{1}{3}x$

29 x-intercepts ± 5, asymptotes $y = \pm 2x$

30 y-intercepts ± 2, asymptotes $y = \pm \frac{1}{4}x$

31 Vertical transverse axis of length 10, conjugate axis of length 14

32 Horizontal transverse axis of length 6, conjugate axis of length 2

Exer. 33–42: Identify the graph of the equation as a parabola (with vertical or horizontal axis), circle, ellipse, or hyperbola.

33 $\frac{1}{3}(x + 2) = y^2$

34 $y^2 = \frac{14}{3} - x^2$

35 $x^2 + 6x - y^2 = 7$

36 $x^2 + 4x + 4y^2 - 24y = -36$

37 $-x^2 = y^2 - 25$

38 $x = 2x^2 - y + 4$

39 $4x^2 - 16x + 9y^2 + 36y = -16$

40 $x + 4 = y^2 + y$

41 $x^2 + 3x = 3y - 6$

42 $9x^2 - y^2 = 10 - 2y$

Exer. 43–44: Find the points of intersection of the graphs of the equations. Sketch both graphs on the same coordinate plane, and show the points of intersection.

43 $\begin{cases} y^2 - 4x^2 = 16 \\ y - x = 4 \end{cases}$

44 $\begin{cases} x^2 - y^2 = 4 \\ y^2 - 3x = 0 \end{cases}$

Exer. 45–48: Find an equation for the set of points in an xy-plane such that the difference of the distances from F and F' is k.

▶ 45 $F(13, 0)$, $F'(-13, 0)$; $k = 24$

46 $F(5, 0)$, $F'(-5, 0)$; $k = 8$

47 $F(0, 10)$, $F'(0, -10)$; $k = 16$

48 $F(0, 17)$, $F'(0, -17)$; $k = 30$

Exer. 49–50: Find an equation for the hyperbola with foci F and F' that passes through P. Sketch the hyperbola.

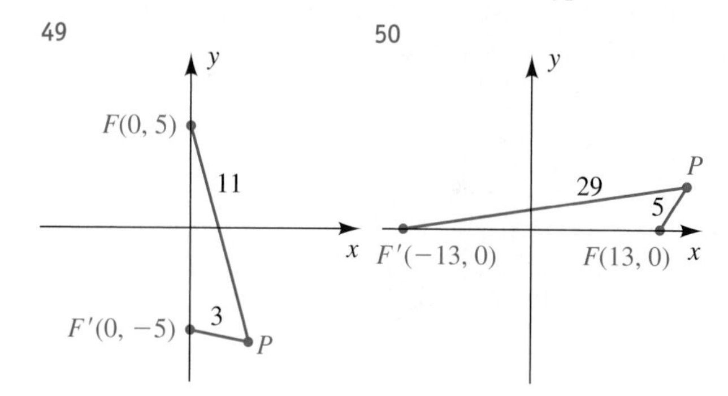

Exer. 51–58: Describe the part of a hyperbola given by the equation.

51 $x = \frac{5}{4}\sqrt{y^2 + 16}$

52 $x = -\frac{5}{4}\sqrt{y^2 + 16}$

53 $y = \frac{3}{7}\sqrt{x^2 + 49}$

54 $y = -\frac{3}{7}\sqrt{x^2 + 49}$

55 $y = -\frac{9}{4}\sqrt{x^2 - 16}$

56 $y = \frac{9}{4}\sqrt{x^2 - 16}$

57 $x = -\frac{2}{3}\sqrt{y^2 - 36}$

58 $x = \frac{2}{3}\sqrt{y^2 - 36}$

▶ **Tutorial available at www.thomsonedu.com/login**

59 The graphs of the equations

$$\frac{x^2}{a^2} - \frac{y^2}{b^2} = 1 \quad \text{and} \quad \frac{x^2}{a^2} - \frac{y^2}{b^2} = -1$$

are called *conjugate hyperbolas.* Sketch the graphs of both equations on the same coordinate plane, with $a = 5$ and $b = 3$, and describe the relationship between the two graphs.

60 Find an equation of the hyperbola with foci $(h \pm c, k)$ and vertices $(h \pm a, k)$, where

$$0 < a < c \quad \text{and} \quad c^2 = a^2 + b^2.$$

61 **Cooling tower** A cooling tower, such as the one shown in the figure, is a hyperbolic structure. Suppose its base diameter is 100 meters and its smallest diameter of 48 meters occurs 84 meters from the base. If the tower is 120 meters high, approximate its diameter at the top.

Exercise 61

62 **Airplane maneuver** An airplane is flying along the hyperbolic path illustrated in the figure. If an equation of the path is $2y^2 - x^2 = 8$, determine how close the airplane comes to a town located at $(3, 0)$. (*Hint:* Let S denote the square of the distance from a point (x, y) on the path to $(3, 0)$, and find the minimum value of S.)

Exercise 62

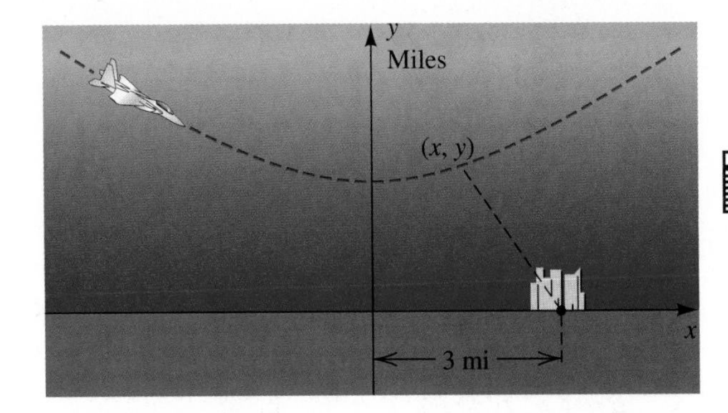

63 **Locating a ship** A ship is traveling a course that is 100 miles from, and parallel to, a straight shoreline. The ship sends out a distress signal that is received by two Coast Guard stations A and B, located 200 miles apart, as shown in the figure. By measuring the difference in signal reception times, it is determined that the ship is 160 miles closer to B than to A. Where is the ship?

Exercise 63

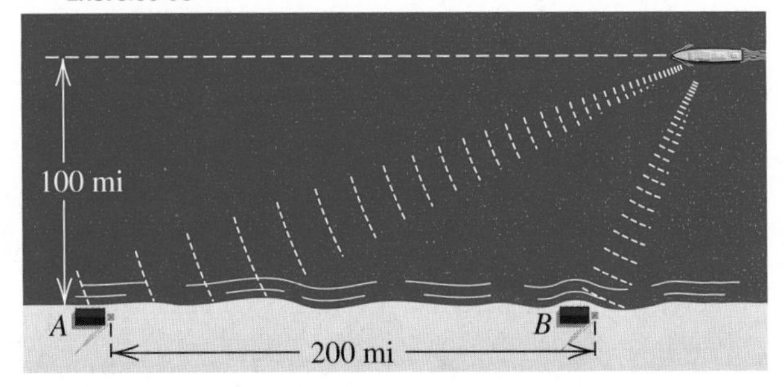

64 **Design of a telescope** The Cassegrain telescope design (dating back to 1672) makes use of the reflective properties of both the parabola and the hyperbola. Shown in the figure is a (split) parabolic mirror, with focus at F_1 and axis along the line l, and a hyperbolic mirror, with one focus also at F_1 and transverse axis along l. Where do incoming light waves parallel to the common axis finally collect?

Exercise 64

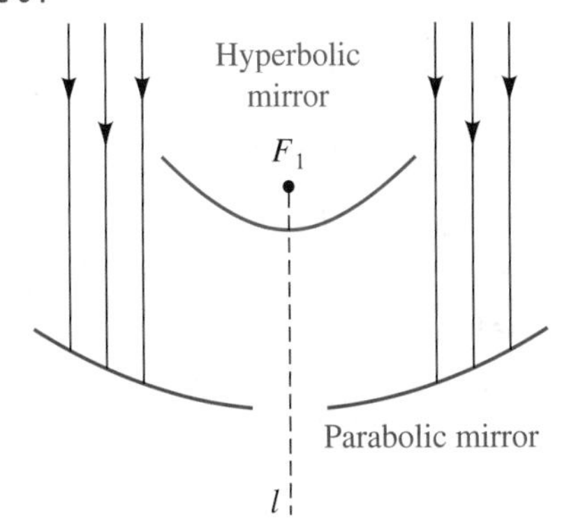

Exer. 65–66: Graph the hyperbolas on the same coordinate plane, and estimate their first-quadrant point of intersection.

65 $\dfrac{(y - 0.1)^2}{1.6} - \dfrac{(x + 0.2)^2}{0.5} = 1;$

$\dfrac{(y - 0.5)^2}{2.7} - \dfrac{(x - 0.1)^2}{5.3} = 1$

66 $\dfrac{(x-0.1)^2}{0.12} - \dfrac{y^2}{0.1} = 1;\ \dfrac{x^2}{0.9} - \dfrac{(y-0.3)^2}{2.1} = 1$

Exer. 67–68: Graph the hyperbolas on the same coordinate plane, and determine the number of points of intersection.

67 $\dfrac{(x-0.3)^2}{1.3} - \dfrac{y^2}{2.7} = 1;\ \dfrac{y^2}{2.8} - \dfrac{(x-0.2)^2}{1.2} = 1$

68 $\dfrac{(x+0.2)^2}{1.75} - \dfrac{(y-0.5)^2}{1.6} = 1;$

$\dfrac{(x-0.6)^2}{2.2} - \dfrac{(y+0.4)^2}{2.35} = 1$

69 **Comet's path** Comets can travel in elliptical, parabolic, or hyperbolic paths around the sun. If a comet travels in a parabolic or hyperbolic path, it will pass by the sun once and never return. Suppose that a comet's coordinates in miles can be described by the equation

$$\frac{x^2}{26 \times 10^{14}} - \frac{y^2}{18 \times 10^{14}} = 1 \quad \text{for} \quad x > 0,$$

where the sun is located at a focus, as shown in the figure.

(a) Approximate the coordinates of the sun.

(b) For the comet to maintain a hyperbolic trajectory, the minimum velocity v of the comet, in meters per second, must satisfy $v > \sqrt{2k/r}$, where r is the distance between the comet and the center of the sun in meters and $k = 1.325 \times 10^{20}$ is a constant. Determine v when r is minimum.

Exercise 69

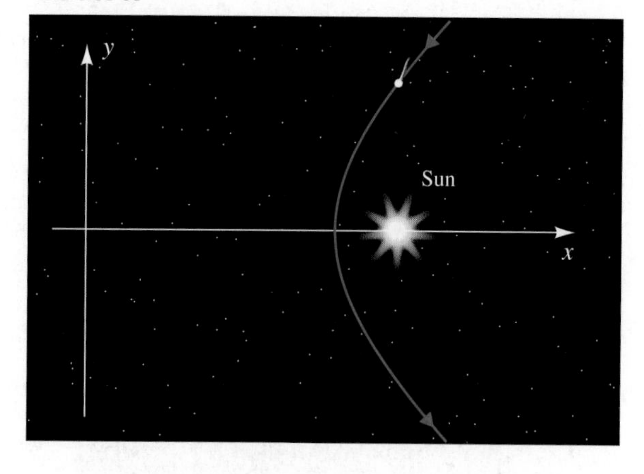

10.4 Plane Curves and Parametric Equations

If f is a function, the graph of the equation $y = f(x)$ is often called a *plane curve.* However, this definition is restrictive, because it excludes many useful graphs. The following definition is more general.

Definition of Plane Curve	A **plane curve** is a set C of ordered pairs $(f(t), g(t))$, where f and g are functions defined on an interval I.

For simplicity, we often refer to a plane curve as a **curve.** The **graph** of C in the preceding definition consists of all points $P(t) = (f(t), g(t))$ in an xy-plane, for t in I. We shall use the term *curve* interchangeably with *graph of a curve.* We sometimes regard the point $P(t)$ as tracing the curve C as t varies through the interval I.

The graphs of several curves are sketched in Figure 1, where I is a closed interval $[a, b]$—that is, $a \le t \le b$. In part (a) of the figure, $P(a) \ne P(b)$, and $P(a)$ and $P(b)$ are called the **endpoints** of C. The curve in (a) intersects

Figure 1

(a) Curve

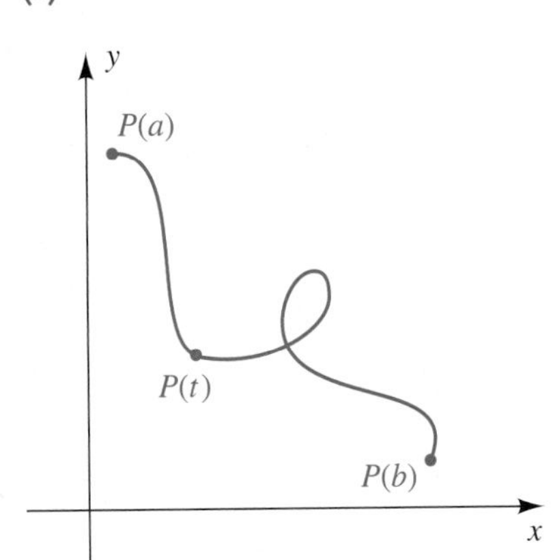

(b) Closed curve

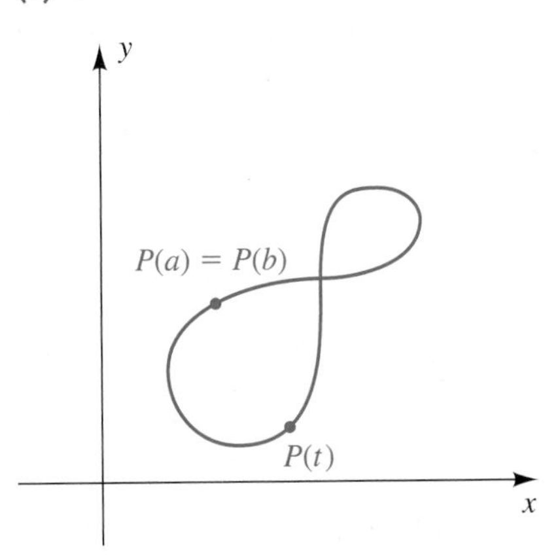

(c) Simple closed curve

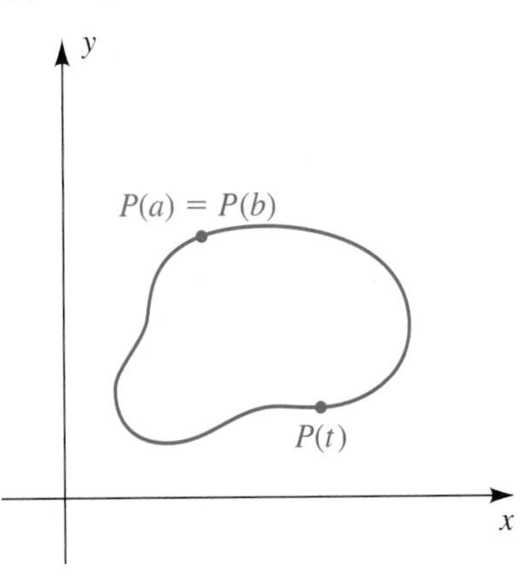

itself; that is, two different values of t produce the same point. If $P(a) = P(b)$, as in Figure 1(b), then C is a **closed curve.** If $P(a) = P(b)$ and C does not intersect itself at any other point, as in Figure 1(c), then C is a **simple closed curve.**

A convenient way to represent curves is given in the next definition.

Definition of Parametric Equations

Let C be the curve consisting of all ordered pairs $(f(t), g(t))$, where f and g are defined on an interval I. The equations

$$x = f(t), \quad y = g(t),$$

for t in I, are **parametric equations** for C with **parameter** t.

The curve C in this definition is referred to as a **parametrized curve,** and the parametric equations are a **parametrization** for C. We often use the notation

$$x = f(t), \quad y = g(t); \quad t \text{ in } I$$

to indicate the domain I of f and g. We can refer to these equations as the **x-equation** and the **y-equation.**

Sometimes it may be possible to eliminate the parameter and obtain a familiar equation in x and y for C. In simple cases we can sketch a graph of a parametrized curve by plotting points and connecting them in order of increasing t, as illustrated in the next example.

EXAMPLE 1 **Sketching the graph of a parametrized curve**

Sketch the graph of the curve C that has the parametrization

$$x = 2t, \quad y = t^2 - 1; \quad -1 \le t \le 2.$$

▶ SOLUTION We use the parametric equations to tabulate coordinates of points $P(x, y)$ on C, as follows.

t	-1	$-\frac{1}{2}$	0	$\frac{1}{2}$	1	$\frac{3}{2}$	2
x	-2	-1	0	1	2	3	4
y	0	$-\frac{3}{4}$	-1	$-\frac{3}{4}$	0	$\frac{5}{4}$	3

Figure 2
$x = 2t, y = t^2 - 1; -1 \le t \le 2$

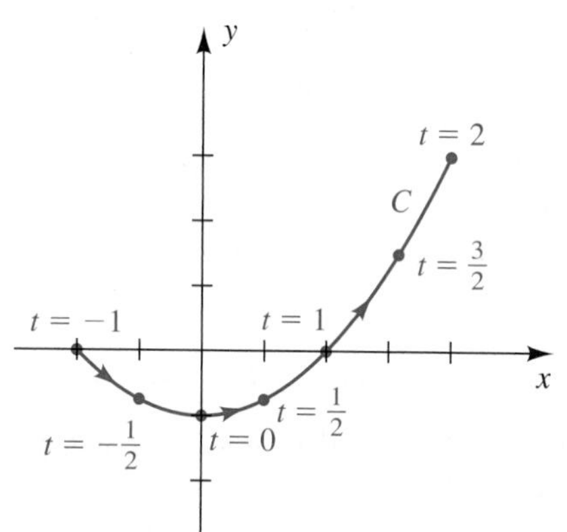

Plotting points leads to the sketch in Figure 2. The arrowheads on the graph indicate the direction in which $P(x, y)$ traces the curve as t *increases* from -1 to 2.

We may obtain a more familiar description of the graph by eliminating the parameter. Solving the x-equation for t, we obtain $t = \frac{1}{2}x$. Substituting this expression for t in the y-equation gives us

$$y = \left(\tfrac{1}{2}x\right)^2 - 1.$$

The graph of this equation in x and y is a parabola symmetric with respect to the y-axis with vertex $(0, -1)$. However, since $x = 2t$ and $-1 \le t \le 2$, we see that $-2 \le x \le 4$ for points (x, y) on C, and hence C is that part of the parabola between the points $(-2, 0)$ and $(4, 3)$ shown in Figure 2.

As indicated by the arrowheads in Figure 2, the point $P(x, y)$ traces the curve C from *left to right* as t increases. The parametric equations

$$x = -2t, \quad y = t^2 - 1; \quad -2 \le t \le 1$$

give us the same graph; however, as t increases, $P(x, y)$ traces the curve from *right to left*. For other parametrizations, the point $P(x, y)$ may oscillate back and forth as t increases.

The **orientation** of a parametrized curve C is the direction determined by *increasing* values of the parameter. We often indicate an orientation by placing arrowheads on C, as in Figure 2. If $P(x, y)$ moves back and forth as t increases, we may place arrows *alongside* of C.

As we have observed, a curve may have different orientations, depending on the parametrization. To illustrate, the curve C in Example 1 is given parametrically by any of the following:

$$x = 2t, \quad y = t^2 - 1; \quad -1 \le t \le 2$$
$$x = t, \quad y = \tfrac{1}{4}t^2 - 1; \quad -2 \le t \le 4$$
$$x = -t, \quad y = \tfrac{1}{4}t^2 - 1; \quad -4 \le t \le 2$$

▶ Tutorial available at www.thomsonedu.com/login

Example 2 Sketching graphs in parametric mode

Sketch the graph of the curve C that has the parametrization

$$x = t^2 - 3,\ y = 3t;\ -4 \le t \le 4.$$

SOLUTION

TI-83/4 Plus | **TI-86**

Set in parametric mode.

TI-83/4 Plus: MODE ▽ (3 times) ▷ ENTER

TI-86: 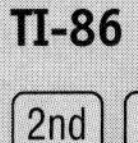 2nd MODE ▽ (4 times) ▷ ▷ ENTER

```
Normal Sci Eng
Float 0123456789
Radian Degree
Func Par Pol Seq
Connected Dot
Sequential Simul
Real a+bi re^θi
Full Horiz G-T
```

Assign the equations.

TI-83/4 Plus: Y= X,T,θ,n x^2 − 3 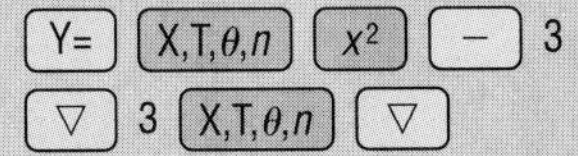 ▽ 3 X,T,θ,n ▽

TI-86: GRAPH E(t)=(F1) t(F1) x^2 − 3 ▽ 3 t(F1) ▽

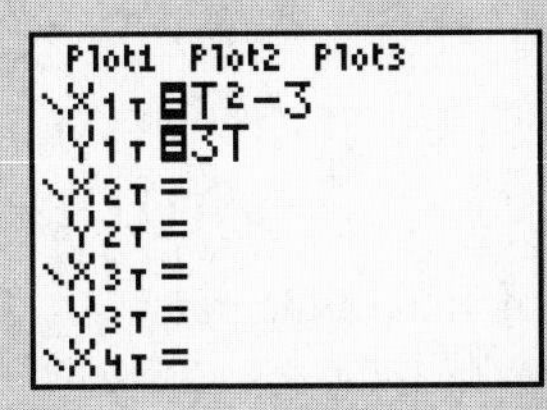

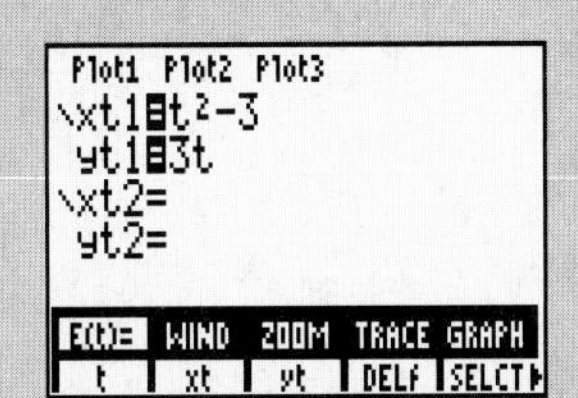

(The subscript 1T on X and Y indicates that X_{1T} and Y_{1T} represent the first *pair* of parametric equations.)

When graphing parametric equations, we need to assign minimum (Tmin) and maximum (Tmax) values to the parameter t, in addition to viewing rectangle dimensions. We also need to select an increment, or step value (Tstep), for t. A typical value for Tstep is 0.1. If a smaller value of Tstep is chosen, the accuracy of the sketch is increased, but so is the amount of time needed to sketch the graph.

Assign window values.

TI-83/4 Plus: WINDOW −4 ▽ 4 ▽ .1 ▽ −4 ▽ 15 ▽ 5 ▽ −15 ▽ 15 ▽ 5

TI-86: 2nd WIND(M2) −4 ▽ 4 ▽ .1 ▽ 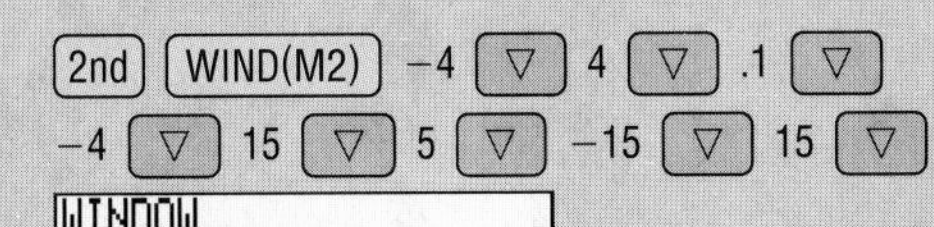 −4 ▽ 15 ▽ 5 ▽ −15 ▽ 15 ▽ 5

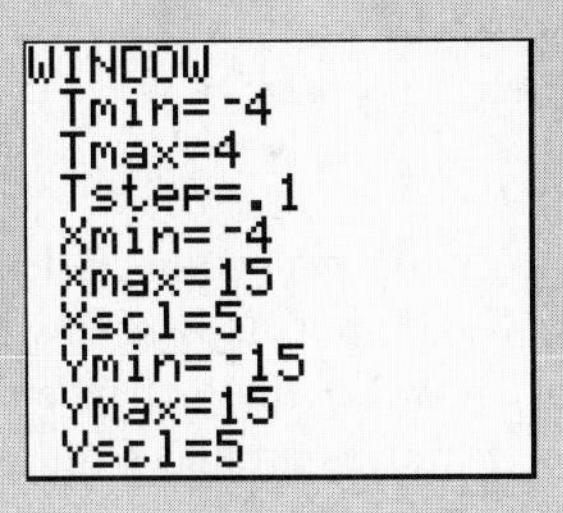


```
WINDOW
tMin=-4
tMax=4
tStep=.1
xMin=-4
xMax=15
xScl=5
yMin=-15
yMax=15
yScl=5
E(t)= WIND ZOOM TRACE GRAPH
```

(continued)

Graph the curve. GRAPH | GRAPH(F5)

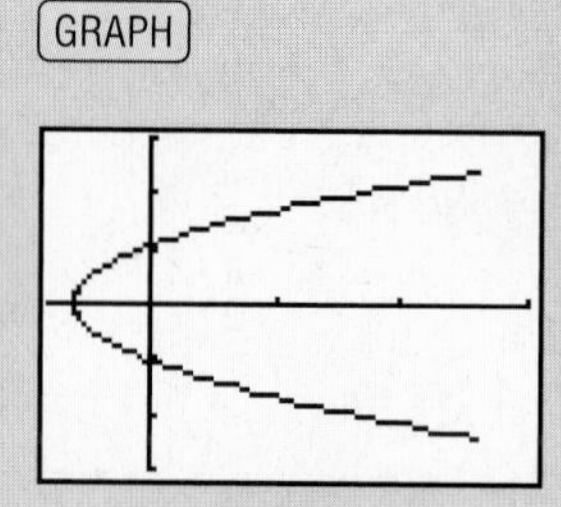

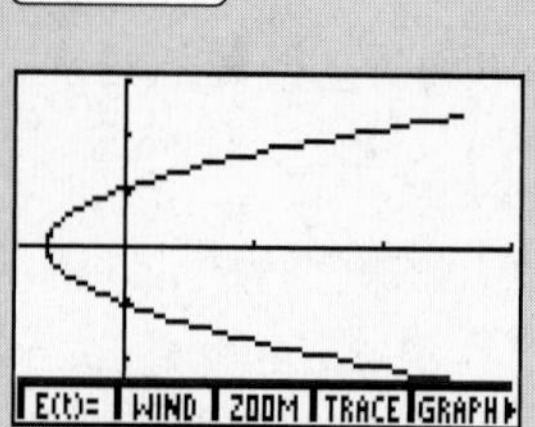

Observe the orientation of the curve.

Now press TRACE or TRACE(F4) and use the left and right cursor keys to trace C. Observe the values listed for T, X, and Y. Note how the values of T correspond to the choice of Tstep. Try graphing C with Tstep $= 1, 2, 4$, and 8.

The next example demonstrates that it is sometimes useful to eliminate the parameter *before* plotting points.

EXAMPLE 3 Describing the motion of a point

A point moves in a plane such that its position $P(x, y)$ at time t is given by

$$x = a\cos t, \quad y = a\sin t; \quad t \text{ in } \mathbb{R},$$

where $a > 0$. Describe the motion of the point.

SOLUTION When x and y contain trigonometric functions of t, we can often eliminate the parameter t by isolating the trigonometric functions, squaring both sides of the equations, and then using one of the Pythagorean identities, as follows:

$x = a\cos t, \quad y = a\sin t$	given
$\dfrac{x}{a} = \cos t, \quad \dfrac{y}{a} = \sin t$	isolate $\cos t$ and $\sin t$
$\dfrac{x^2}{a^2} = \cos^2 t, \quad \dfrac{y^2}{a^2} = \sin^2 t$	square both sides
$\dfrac{x^2}{a^2} + \dfrac{y^2}{a^2} = 1$	$\cos^2 t + \sin^2 t = 1$
$x^2 + y^2 = a^2$	multiply by a^2

Figure 3
$x = a\cos t,\ y = a\sin t;\ t$ in $\mathbb{R}$

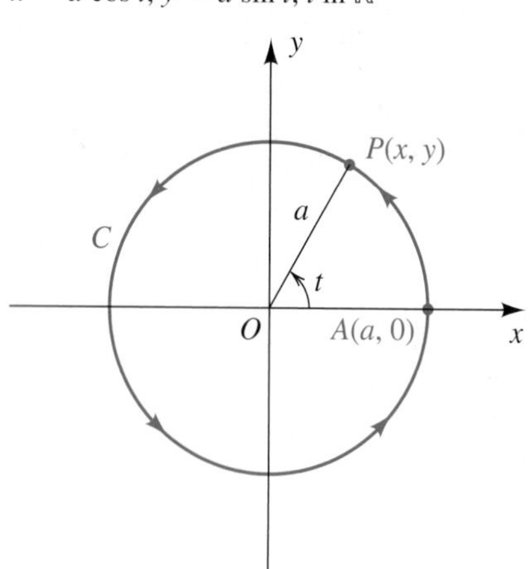

This shows that the point $P(x, y)$ moves on the circle C of radius a with center at the origin (see Figure 3). The point is at $A(a, 0)$ when $t = 0$, at $(0, a)$ when $t = \pi/2$, at $(-a, 0)$ when $t = \pi$, at $(0, -a)$ when $t = 3\pi/2$, and back at $A(a, 0)$ when $t = 2\pi$. Thus, P moves around C in a counterclockwise direction, making one revolution every 2π units of time. The orientation of C is indicated by the arrowheads in Figure 3.

Tutorial available at www.thomsonedu.com/login

Note that in this example we may interpret t geometrically as the radian measure of the angle generated by the line segment OP.

Sine and Cosine Values on the Unit Circle

Playing off the last example, you can use parametric equations as an aid to learning and remembering values of the sine and cosine functions. Set the calculator in the following modes: Degree, Par(ametric), and Dot. Make the function assignments cos(T) to X_{1T} and sin(T) to Y_{1T}. Next assign 0 to Tmin, 360 to Tmax, and 15 to Tstep. Graph in the window $[-3, 3]$ by $[-2, 2]$. Using the trace mode and cursor keys reveals many familiar values on the unit circle.

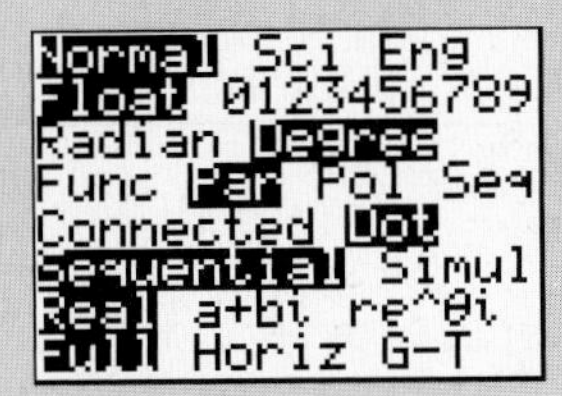

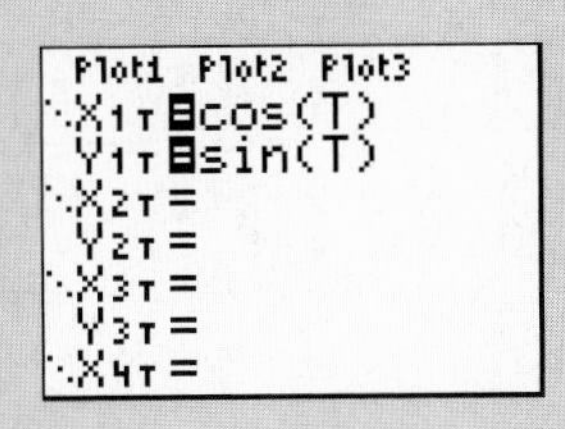

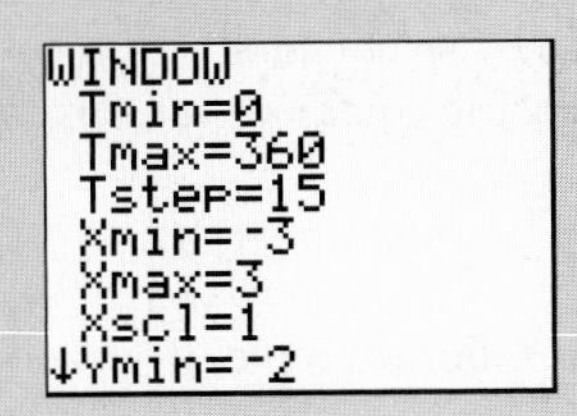

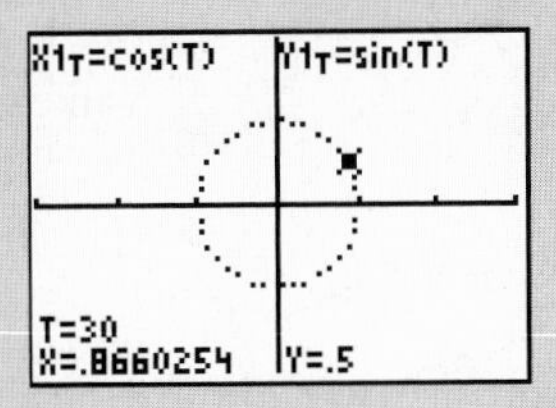

EXAMPLE 4 **Sketching the graph of a parametrized curve**

Sketch the graph of the curve C that has the parametrization

$$x = -2 + t^2, \quad y = 1 + 2t^2; \quad t \text{ in } \mathbb{R},$$

and indicate the orientation.

▶ **SOLUTION** To eliminate the parameter, we use the x-equation to obtain $t^2 = x + 2$ and then substitute for t^2 in the y-equation. Thus,

$$y = 1 + 2(x + 2).$$

The graph of the last equation is the line of slope 2 through the point $(-2, 1)$, as indicated by the dashes in Figure 4(a) on the next page. However, since $t^2 \geq 0$, we see from the parametric equations for C that

$$x = -2 + t^2 \geq -2 \quad \text{and} \quad y = 1 + 2t^2 \geq 1.$$

Thus, the graph of C is that part of the line to the right of $(-2, 1)$ (the point corresponding to $t = 0$), as shown in Figure 4(b). The orientation is indicated by the arrows alongside of C. As t increases in the interval $(-\infty, 0]$, $P(x, y)$ moves down the curve toward the point $(-2, 1)$. As t increases in $[0, \infty)$, $P(x, y)$ moves up the curve away from $(-2, 1)$.

(continued)

▶ **Tutorial available at www.thomsonedu.com/login**

Figure 4

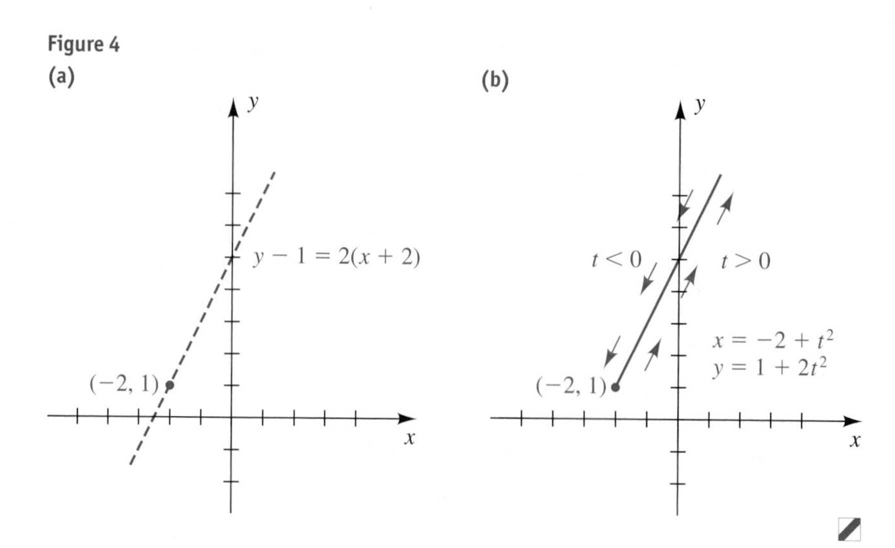

If a curve C is described by an equation $y = f(x)$ for some function f, then an easy way to obtain parametric equations for C is to let

$$x = t, \quad y = f(t),$$

where t is in the domain of f. For example, if $y = x^3$, then parametric equations are

$$x = t, \quad y = t^3; \quad t \text{ in } \mathbb{R}.$$

We can use many different substitutions for x, provided that as t varies through some interval, x takes on every value in the domain of f. Thus, the graph of $y = x^3$ is also given by

$$x = t^{1/3}, \quad y = t; \quad t \text{ in } \mathbb{R}.$$

Note, however, that the parametric equations

$$x = \sin t, \quad y = \sin^3 t; \quad t \text{ in } \mathbb{R}$$

give only that part of the graph of $y = x^3$ between the points $(-1, -1)$ and $(1, 1)$.

EXAMPLE 5 Finding parametric equations for a line

Find three parametrizations for the line of slope m through the point (x_1, y_1).

▶ SOLUTION By the point-slope form, an equation for the line is

$$y - y_1 = m(x - x_1). \tag{*}$$

▶ Tutorial available at www.thomsonedu.com/login

If we let $x = t$, then $y - y_1 = m(t - x_1)$ and we obtain the parametrization

$$x = t, \quad y = y_1 + m(t - x_1); \quad t \text{ in } \mathbb{R}.$$

We obtain another parametrization for the line if we let $x - x_1 = t$ in $(*)$. In this case $y - y_1 = mt$, and we have

$$x = x_1 + t, \quad y = y_1 + mt; \quad t \text{ in } \mathbb{R}.$$

As a third illustration, if we let $x - x_1 = \tan t$ in $(*)$, then

$$x = x_1 + \tan t, \quad y = y_1 + m \tan t; \quad -\frac{\pi}{2} < t < \frac{\pi}{2}.$$

There are many other parametrizations for the line.

In the next example, we use parametric equations to model the path of a projectile (object). These equations are developed by means of methods in physics and calculus. We assume that the object is moving near the surface of Earth under the influence of gravity alone; that is, air resistance and other forces that could affect acceleration are negligible. We also assume that the ground is level and the curvature of Earth is not a factor in determining the path of the object.

EXAMPLE 6 The path of a projectile

The path of a projectile at time t can be modeled using the parametric equations

$$x(t) = (s \cos \alpha)t, \quad y(t) = -\tfrac{1}{2}gt^2 + (s \sin \alpha)t + h; \quad t \geq 0, \qquad (1)$$

where, at $t = 0$, s is the speed of the projectile in ft/sec, α is the angle the path makes with the horizontal, and h is the height in feet. The acceleration due to gravity is $g = 32$ ft/sec^2. Suppose that the projectile is fired at a speed of 1024 ft/sec at an angle of 30° from the horizontal from a height of 2304 feet (see Figure 5 on the next page).

(a) Find parametric equations for the projectile.

(b) Find the range r of the projectile—that is, the horizontal distance it travels before hitting the ground.

(c) Find an equation in x and y for the projectile.

(d) Find the point and time at which the projectile reaches its maximum altitude.

Figure 5

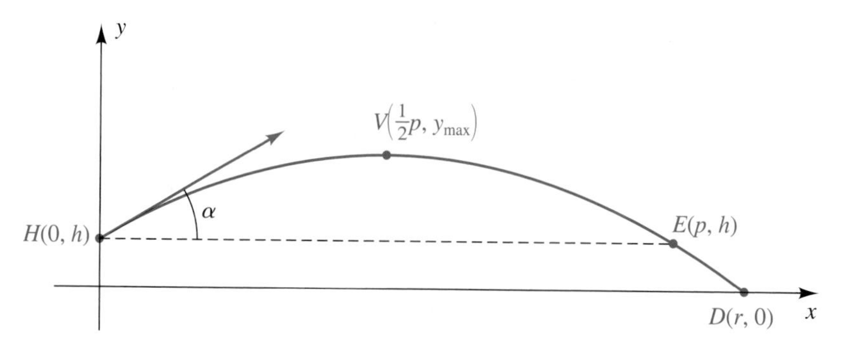

SOLUTION

(a) Substituting 1024 for s, 30° for α, 32 for g, and 2304 for h in the parametric equations in (1) gives

$$x = (1024 \cos 30°)t, \quad y = -\tfrac{1}{2}(32)t^2 + (1024 \sin 30°)t + 2304; \quad t \geq 0.$$

Simplifying yields

$$x = 512\sqrt{3}\,t, \quad y = -16t^2 + 512t + 2304; \quad t \geq 0. \tag{2}$$

▶ **(b)** To find the range r of the projectile, we must find the point D in Figure 5 at which the projectile hits the ground. Since the y-coordinate of D is 0, we let $y = 0$ in the y-equation of (2) and solve for t:

$$\begin{aligned} y &= -16t^2 + 512t + 2304 && \text{given in (2)} \\ 0 &= -16t^2 + 512t + 2304 && \text{let } y = 0 \\ 0 &= t^2 - 32t - 144 && \text{divide by } -16 \\ 0 &= (t - 36)(t + 4) && \text{factor} \end{aligned}$$

Since $t \geq 0$, we must have $t = 36$ sec. We can now use the x-equation of (2) to obtain the range:

$$x = 512\sqrt{3}\,t = 512\sqrt{3}(36) = 18{,}432\sqrt{3} \approx 31{,}925 \text{ ft}$$

▶ **(c)** To eliminate the parameter t, we solve the x-equation in (2) for t and substitute this expression for t in the y-equation in (2):

$$\begin{aligned} x = 512\sqrt{3}t \text{ implies } t &= \frac{x}{512\sqrt{3}} && \text{solve } x\text{-equation in (2) for } t \\ y &= -16t^2 + 512t + 2304 && y\text{-equation in (2)} \\ y &= -16\left(\frac{x}{512\sqrt{3}}\right)^2 + 512\left(\frac{x}{512\sqrt{3}}\right) + 2304 && \text{let } t = \frac{x}{512\sqrt{3}} \\ y &= -\frac{1}{49{,}152}x^2 + \frac{1}{\sqrt{3}}x + 2304 && \text{simplify} \end{aligned} \tag{3}$$

The last equation is of the form $y = ax^2 + bx + c$, showing that the path of the projectile is parabolic.

▶ **Tutorial available at www.thomsonedu.com/login**

▶ **(d)** The y-coordinate of point E in Figure 5 is 2304, so we can find the value of t at E by solving the equation $y = 2304$:

$$\begin{aligned} y &= -16t^2 + 512t + 2304 && \text{given in (2)} \\ 2304 &= -16t^2 + 512t + 2304 && \text{let } y = 2304 \\ 0 &= -16t^2 + 512t && \text{subtract 2304} \\ 0 &= -16t(t - 32) && \text{factor} \end{aligned}$$

So if $y = 2304$, $t = 0$ or $t = 32$. Since the path is parabolic, the x-coordinate of V is one-half of the x-coordinate p of E. Also, the value of t at V is one-half the value of t at E, so $t = \frac{1}{2}(32) = 16$ at V. We can find the x- and y-values at V by substituting 16 for t in (2):

$$x = 512\sqrt{3}\,t = 512\sqrt{3}(16) = 8192\sqrt{3} \approx 14{,}189 \text{ ft}$$

and

$$y = -16t^2 + 512t + 2304 = -16(16)^2 + 512(16) + 2304 = 6400 \text{ ft}$$

Thus, the projectile reaches its maximum altitude when $t = 16$ at approximately (14,189, 6400).

An alternative way of finding the maximum altitude is to use the theorem for locating the vertex of a parabola to find the x-value ($x = -b/(2a)$) of the highest point on the graph of equation (3) and then use the equations in (2) to find t and y.

See Discussion Exercises 7 and 8 at the end of the chapter for related problems concerning Example 6.

Parametric equations of the form

$$x = a \sin \omega_1 t, \quad y = b \cos \omega_2 t; \quad t \geq 0,$$

where a, b, ω_1, and ω_2 are constants, occur in electrical theory. The variables x and y usually represent voltages or currents at time t. The resulting curve is often difficult to sketch; however, using an oscilloscope and imposing voltages or currents on the input terminals, we can represent the graph, a **Lissajous figure,** on the screen of the oscilloscope. Graphing utilities are very helpful in obtaining these complicated graphs.

EXAMPLE 7 Graphing a Lissajous figure

Sketch the graph of the Lissajous figure that has the parametrization

$$x = \sin 2t, \quad y = \cos t; \quad 0 \leq t \leq 2\pi.$$

Determine the values of t that correspond to the curve in each quadrant.

SOLUTION We first need to set our graphing utility in a parametric mode. Next we make the assignments

$$X_{1T} = \sin 2t \qquad \text{and} \qquad Y_{1T} = \cos t.$$

(continued)

▶ **Tutorial available at www.thomsonedu.com/login**

Figure 6
$[-1.5, 1.5]$ by $[-1, 1]$

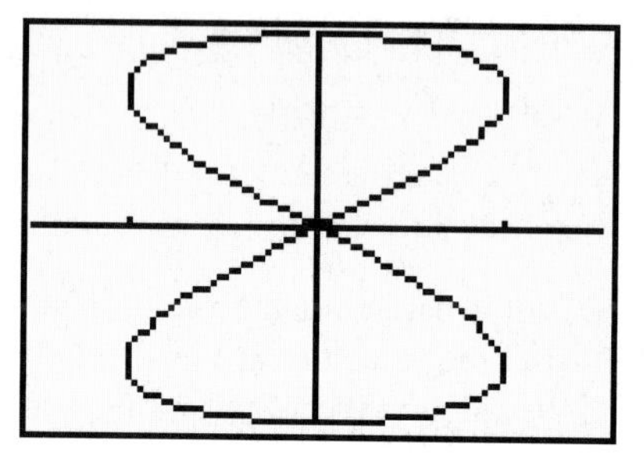

For this example, we use Tmin $= 0$, Tmax $= 2\pi$, and Tstep $= 0.1$. Since x and y are between -1 and 1, we will assign -1 to Ymin and 1 to Ymax. To maintain our 3:2 screen proportion, we select -1.5 for Xmin and 1.5 for Xmax, and then we graph X_{1T} and Y_{1T} to obtain the Lissajous figure in Figure 6.

Referring to the parametric equations, we see that as t increases from 0 to $\pi/2$, the point $P(x, y)$ starts at $(0, 1)$ and traces the part of the curve in quadrant I (in a generally clockwise direction). As t increases from $\pi/2$ to π, $P(x, y)$ traces the part in quadrant III (in a counterclockwise direction). For $\pi < t < 3\pi/2$, we obtain the part in quadrant IV; and $3\pi/2 < t < 2\pi$ gives us the part in quadrant II.

EXAMPLE 8 **Finding parametric equations for a cycloid**

The curve traced by a fixed point P on the circumference of a circle as the circle rolls along a line in a plane is called a **cycloid.** Find parametric equations for a cycloid.

▶ SOLUTION Suppose the circle has radius a and that it rolls along (and above) the x-axis in the positive direction. If one position of P is the origin, then Figure 7 depicts part of the curve and a possible position of the circle. The V-shaped part of the curve at $x = 2\pi a$ is called a **cusp.**

Figure 7

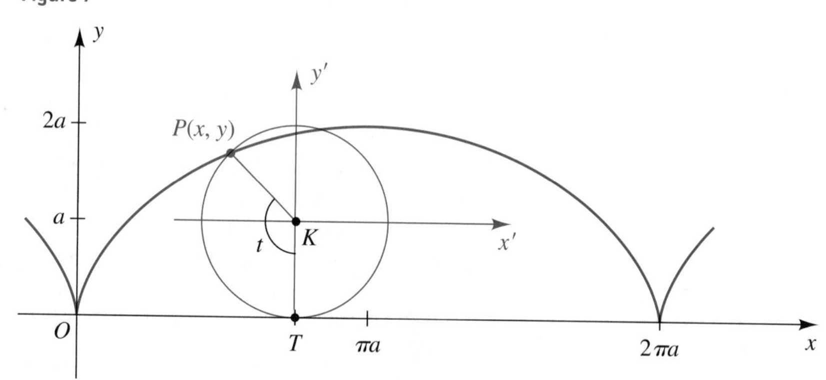

Let K denote the center of the circle and T the point of tangency with the x-axis. We introduce, as a parameter t, the radian measure of angle TKP. The distance the circle has rolled is $d(O, T) = at$ (formula for the length of a circular arc). Consequently, the coordinates of K are $(x, y) = (at, a)$. If we consider an $x'y'$-coordinate system with origin at $K(at, a)$ and if $P(x', y')$ denotes

▶ **Tutorial available at www.thomsonedu.com/login**

the point P relative to this system, then, by adding x' and y' to the x- and y-coordinates of K, we obtain

$$x = at + x', \qquad y = a + y'.$$

Figure 8

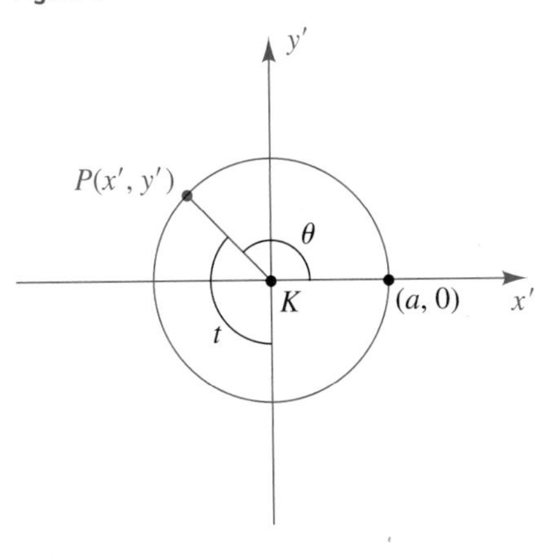

If, as in Figure 8, θ denotes an angle in standard position on the $x'y'$-plane, then $\theta + t = 3\pi/2$ or, equivalently, $\theta = (3\pi/2) - t$. Hence,

$$x' = a\cos\theta = a\cos\left(\frac{3\pi}{2} - t\right) = -a\sin t$$

$$y' = a\sin\theta = a\sin\left(\frac{3\pi}{2} - t\right) = -a\cos t,$$

and substitution in $x = at + x'$, $y = a + y'$ gives us parametric equations for the cycloid:

$$x = a(t - \sin t), \quad y = a(1 - \cos t); \quad t \text{ in } \mathbb{R}$$

If $a < 0$, then the graph of $x = a(t - \sin t)$, $y = a(1 - \cos t)$ is the inverted cycloid that results if the circle of Example 8 rolls *below* the x-axis. This curve has a number of important physical properties. To illustrate, suppose a thin wire passes through two fixed points A and B, as shown in Figure 9, and that the shape of the wire can be changed by bending it in any manner. Suppose further that a bead is allowed to slide along the wire and the only force acting on the bead is gravity. We now ask which of all the possible paths will allow the bead to slide from A to B in the least amount of time. It is natural to believe that the desired path is the straight line segment from A to B; however, this is not the correct answer. The path that requires the least time coincides with the graph of an inverted cycloid with A at the origin. Because the velocity of the bead increases more rapidly along the cycloid than along the line through A and B, the bead reaches B more rapidly, even though the distance is greater.

Figure 9

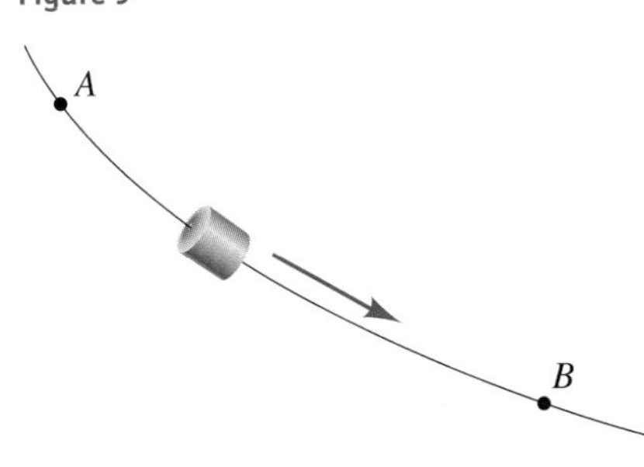

There is another interesting property of this **curve of least descent.** Suppose that A is the origin and B is the point with x-coordinate $\pi|a|$—that is, the lowest point on the cycloid in the first arc to the right of A. If the bead is released at *any* point between A and B, it can be shown that the *time* required for it to reach B is always the *same.*

Variations of the cycloid occur in applications. For example, if a motorcycle wheel rolls along a straight road, then the curve traced by a fixed point on one of the spokes is a cycloidlike curve. In this case the curve does not have cusps, nor does it intersect the road (the x-axis) as does the graph of a cycloid. If the wheel of a train rolls along a railroad track, then the curve traced by a fixed point on the circumference of the wheel (which extends below the track) contains loops at regular intervals. Other cycloids are defined in Exercises 45 and 46.

10.4 *Exercises*

Exer. 1–24: Find an equation in x and y whose graph contains the points on the curve C. Sketch the graph of C, and indicate the orientation.

1 $x = t - 2, \quad y = 2t + 3; \quad 0 \le t \le 5$

2 $x = 1 - 2t, \quad y = 1 + t; \quad -1 \le t \le 4$

3 $x = t^2 + 1, \quad y = t^2 - 1; \quad -2 \le t \le 2$

4 $x = t^3 + 1, \quad y = t^3 - 1; \quad -2 \le t \le 2$

5 $x = 4t^2 - 5, \quad y = 2t + 3; \quad t$ in $\mathbb{R}$

6 $x = \sqrt{t}, \quad y = 3t + 4; \quad t \ge 0$

7 $x = 4 \cos t + 1, \quad y = 3 \sin t; \quad 0 \le t \le 2\pi$

8 $x = 2 \sin t, \quad y = 3 \cos t; \quad 0 \le t \le 2\pi$

9 $x = 2 - 3 \sin t, \quad y = -1 - 3 \cos t; \quad 0 \le t \le 2\pi$

10 $x = \cos t - 2, \quad y = \sin t + 3; \quad 0 \le t \le 2\pi$

11 $x = \sec t, \quad y = \tan t; \quad -\pi/2 < t < \pi/2$

12 $x = \cos 2t, \quad y = \sin t; \quad -\pi \le t \le \pi$

13 $x = t^2, \quad y = 2 \ln t; \quad t > 0$

14 $x = \cos^3 t, \quad y = \sin^3 t; \quad 0 \le t \le 2\pi$

15 $x = \sin t, \quad y = \csc t; \quad 0 < t \le \pi/2$

16 $x = e^t, \quad y = e^{-t}; \quad t$ in $\mathbb{R}$

17 $x = t, \quad y = \sqrt{t^2 - 1}; \quad |t| \ge 1$

18 $x = -2\sqrt{1 - t^2}, \quad y = t; \quad |t| \le 1$

19 $x = t, \quad y = \sqrt{t^2 - 2t + 1}; \quad 0 \le t \le 4$

20 $x = 2t, \quad y = 8t^3; \quad -1 \le t \le 1$

21 $x = (t + 1)^3, \quad y = (t + 2)^2; \quad 0 \le t \le 2$

22 $x = t^3, \quad y = t^2; \quad t$ in $\mathbb{R}$

23 $x = e^t, \quad y = e^{-2t}; \quad t$ in $\mathbb{R}$

24 $x = \tan t, \quad y = 1; \quad -\pi/2 < t < \pi/2$

25 (a) Describe the graph of a curve C that has the parametrization

$$x = 3 + 2 \sin t, \quad y = -2 + 2 \cos t; \quad 0 \le t \le 2\pi.$$

(b) Change the parametrization to

$$x = 3 - 2 \sin t, \quad y = -2 + 2 \cos t; \quad 0 \le t \le 2\pi$$

and describe how this changes the graph from part (a).

(c) Change the parametrization to

$$x = 3 - 2 \sin t, \quad y = -2 - 2 \cos t; \quad 0 \le t \le 2\pi$$

and describe how this changes the graph from part (a).

26 (a) Describe the graph of a curve C that has the parametrization

$$x = -2 + 3 \sin t, \quad y = 3 - 3 \cos t; \quad 0 \le t \le 2\pi.$$

(b) Change the parametrization to

$$x = -2 - 3 \sin t, \quad y = 3 + 3 \cos t; \quad 0 \le t \le 2\pi$$

and describe how this changes the graph from part (a).

(c) Change the parametrization to

$$x = -2 + 3 \sin t, \quad y = 3 + 3 \cos t; \quad 0 \le t \le 2\pi$$

and describe how this changes the graph from part (a).

Exer. 27–28: Curves C_1, C_2, C_3, and C_4 are given parametrically, for t in $\mathbb{R}$. Sketch their graphs, and indicate orientations.

27 C_1: $x = t^2, \quad y = t$
C_2: $x = t^4, \quad y = t^2$
C_3: $x = \sin^2 t, \quad y = \sin t$
C_4: $x = e^{2t}, \quad y = -e^t$

28 C_1: $x = t, \quad y = 1 - t$
C_2: $x = 1 - t^2, \quad y = t^2$
C_3: $x = \cos^2 t, \quad y = \sin^2 t$
C_4: $x = \ln t - t, \quad y = 1 + t - \ln t; \ t > 0$

Exer. 29–30: The parametric equations specify the position of a moving point $P(x, y)$ at time t. Sketch the graph, and indicate the motion of P as t increases.

29 (a) $x = \cos t, \quad y = \sin t; \quad 0 \le t \le \pi$

(b) $x = \sin t, \quad y = \cos t; \quad 0 \le t \le \pi$

(c) $x = t, \quad y = \sqrt{1 - t^2}; \quad -1 \le t \le 1$

30 (a) $x = t^2, \quad y = 1 - t^2; \quad 0 \le t \le 1$

(b) $x = 1 - \ln t, \quad y = \ln t; \quad 1 \le t \le e$

(c) $x = \cos^2 t, \quad y = \sin^2 t; \quad 0 \le t \le 2\pi$

31 Show that

$$x = a \cos t + h, \quad y = b \sin t + k; \quad 0 \le t \le 2\pi$$

are parametric equations of an ellipse with center (h, k) and axes of lengths $2a$ and $2b$.

32 Show that

$$x = a \sec t + h, \quad y = b \tan t + k;$$
$$-\pi/2 < t < 3\pi/2 \text{ and } t \ne \pi/2$$

are parametric equations of a hyperbola with center (h, k), transverse axis of length $2a$, and conjugate axis of length $2b$. Determine the values of t for each branch.

Exer. 33–34: (a) Find three parametrizations that give the same graph as the given equation. (b) Find three parametrizations that give only a portion of the graph of the given equation.

▶ 33 $y = x^2$

34 $y = \ln x$

Exer. 35–38: Refer to the equations in (1) of Example 6. Find the range and maximum altitude for the given values.

35 $s = 256\sqrt{3}, \quad \alpha = 60°, \quad h = 400$

36 $s = 512\sqrt{2}, \quad \alpha = 45°, \quad h = 1088$

37 $s = 704, \quad \alpha = 45°, \quad h = 0$

38 $s = 2448, \quad \alpha = 30°, \quad h = 0$

▶ 39 Refer to Example 7.

(a) Describe the Lissajous figure given by $f(t) = a \sin \omega t$ and $g(t) = b \cos \omega t$ for $t \ge 0$ and $a \ne b$.

(b) Suppose $f(t) = a \sin \omega_1 t$ and $g(t) = b \sin \omega_2 t$, where ω_1 and ω_2 are positive rational numbers, and write ω_2/ω_1 as m/n for positive integers m and n. Show that if $p = 2\pi n/\omega_1$, then both $f(t + p) = f(t)$ and $g(t + p) = g(t)$. Conclude that the curve retraces itself every p units of time.

40 Shown in the figure is the Lissajous figure given by

$$x = 2 \sin 3t, \quad y = 3 \sin 1.5t; \quad t \ge 0.$$

Find the period of the figure—that is, the length of the smallest t-interval that traces the curve.

Exercise 40

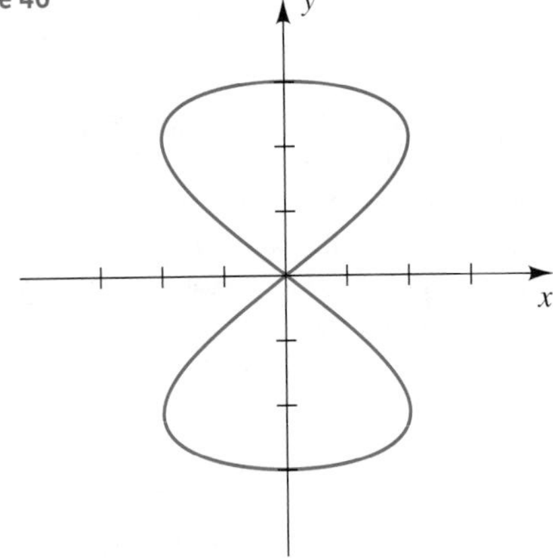

Exer. 41–44: Lissajous figures are used in the study of electrical circuits to determine the phase difference ϕ between a known voltage $V_1(t) = A \sin(\omega t)$ and an unknown voltage $V_2(t) = B \sin(\omega t + \phi)$ having the same frequency. The voltages are graphed parametrically as $x = V_1(t)$ and $y = V_2(t)$. If ϕ is acute, then

$$\phi = \sin^{-1} \frac{y_{int}}{y_{max}},$$

where y_{int} is the nonnegative y-intercept and y_{max} is the maximum y-value on the curve.

(a) Graph the parametric curve $x = V_1(t)$ and $y = V_2(t)$ for the specified range of t.

(b) Use the graph to approximate ϕ in degrees.

41 $V_1(t) = 3 \sin(240\pi t), \quad V_2(t) = 4 \sin(240\pi t);$
$0 \le t \le 0.01$

42 $V_1(t) = 6 \sin(120\pi t), \quad V_2(t) = 5 \cos(120\pi t);$
$0 \le t \le 0.02$

▶ Tutorial available at www.thomsonedu.com/login

43 $V_1(t) = 80 \sin(60\pi t)$, $V_2(t) = 70 \cos(60\pi t - \pi/3)$; $0 \le t \le 0.035$

44 $V_1(t) = 163 \sin(120\pi t)$, $V_2(t) = 163 \sin(120\pi t + \pi/4)$; $0 \le t \le 0.02$

Exer. 45–46: Graph the Lissajous figure in the viewing rectangle $[-1, 1]$ by $[-1, 1]$ for the specified range of t.

45 $x(t) = \sin(6\pi t)$, $y(t) = \cos(5\pi t)$; $0 \le t \le 2$

46 $x(t) = \sin(4t)$, $y(t) = \sin(3t + \pi/6)$; $0 \le t \le 6.5$

47 A circle C of radius b rolls on the outside of the circle $x^2 + y^2 = a^2$, and $b < a$. Let P be a fixed point on C, and let the initial position of P be $A(a, 0)$, as shown in the figure. If the parameter t is the angle from the positive x-axis to the line segment from O to the center of C, show that parametric equations for the curve traced by P (an *epicycloid*) are

$$x = (a + b)\cos t - b\cos\left(\frac{a+b}{b}t\right),$$

$$y = (a + b)\sin t - b\sin\left(\frac{a+b}{b}t\right); \quad 0 \le t \le 2\pi.$$

Exercise 47

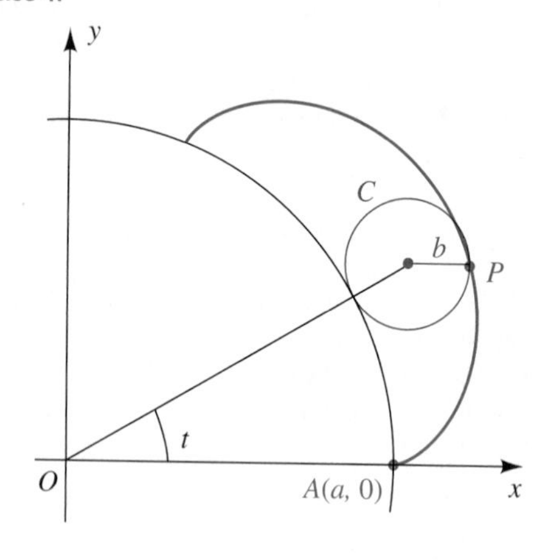

48 If the circle C of Exercise 47 rolls on the inside of the second circle (see the figure), then the curve traced by P is a *hypocycloid*.

(a) Show that parametric equations for this curve are

$$x = (a - b)\cos t + b\cos\left(\frac{a-b}{b}t\right),$$

$$y = (a - b)\sin t - b\sin\left(\frac{a-b}{b}t\right); \quad 0 \le t \le 2\pi.$$

(b) If $b = \frac{1}{4}a$, show that $x = a\cos^3 t$, $y = a\sin^3 t$, and sketch the graph.

Exercise 48

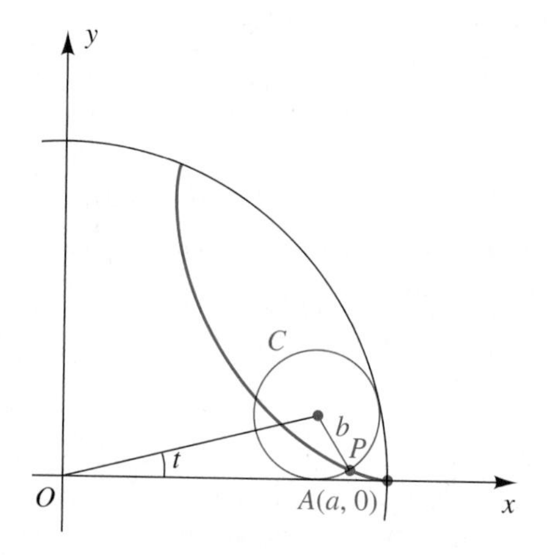

49 If $b = \frac{1}{3}a$ in Exercise 47, find parametric equations for the epicycloid and sketch the graph.

50 The radius of circle B is one-third that of circle A. How many revolutions will circle B make as it rolls around circle A until it reaches its starting point? (*Hint:* Use Exercise 49.)

Exer. 51–54: Graph the curve.

51 $x = 3\sin^5 t$, $y = 3\cos^5 t$; $0 \le t \le 2\pi$

52 $x = 8\cos t - 2\cos 4t$, $y = 8\sin t - 2\sin 4t$; $0 \le t \le 2\pi$

53 $x = 3t - 2\sin t$, $y = 3 - 2\cos t$; $-8 \le t \le 8$

54 $x = 2t - 3\sin t$, $y = 2 - 3\cos t$; $-8 \le t \le 8$

Exer. 55–58: Graph the given curves on the same coordinate plane, and describe the shape of the resulting figure.

55 C_1: $x = 2 \sin 3t$, $y = 3 \cos 2t$; $-\pi/2 \le t \le \pi/2$
C_2: $x = \frac{1}{4} \cos t + \frac{3}{4}$, $y = \frac{1}{4} \sin t + \frac{3}{2}$; $0 \le t \le 2\pi$
C_3: $x = \frac{1}{4} \cos t - \frac{3}{4}$, $y = \frac{1}{4} \sin t + \frac{3}{2}$; $0 \le t \le 2\pi$
C_4: $x = \frac{3}{4} \cos t$, $y = \frac{1}{4} \sin t$; $0 \le t \le 2\pi$
C_5: $x = \frac{1}{4} \cos t$, $y = \frac{1}{8} \sin t + \frac{3}{4}$; $\pi \le t \le 2\pi$

56 C_1: $x = \frac{3}{2} \cos t + 1$, $y = \sin t - 1$; $-\pi/2 \le t \le \pi/2$
C_2: $x = \frac{3}{2} \cos t + 1$, $y = \sin t + 1$; $-\pi/2 \le t \le \pi/2$
C_3: $x = 1$, $y = 2 \tan t$; $-\pi/4 \le t \le \pi/4$

57 C_1: $x = \tan t$, $y = 3 \tan t$; $0 \le t \le \pi/4$
C_2: $x = 1 + \tan t$, $y = 3 - 3 \tan t$; $0 \le t \le \pi/4$
C_3: $x = \frac{1}{2} + \tan t$, $y = \frac{3}{2}$; $0 \le t \le \pi/4$

58 C_1: $x = 1 + \cos t$, $y = 1 + \sin t$; $\pi/3 \le t \le 2\pi$
C_2: $x = 1 + \tan t$, $y = 1$; $0 \le t \le \pi/4$

10.5 Polar Coordinates

In a rectangular coordinate system, the ordered pair (a, b) denotes the point whose directed distances from the x- and y-axes are b and a, respectively. Another method for representing points is to use *polar coordinates.* We begin with a fixed point O (the **origin,** or **pole**) and a directed half-line (the **polar axis**) with endpoint O. Next we consider any point P in the plane different from O. If, as illustrated in Figure 1, $r = d(O, P)$ and θ denotes the measure of any angle determined by the polar axis and OP, then r and θ are **polar coordinates** of P and the symbols (r, θ) or $P(r, \theta)$ are used to denote P. As usual, θ is considered positive if the angle is generated by a counterclockwise rotation of the polar axis and negative if the rotation is clockwise. Either radian or degree measure may be used for θ.

Figure 1

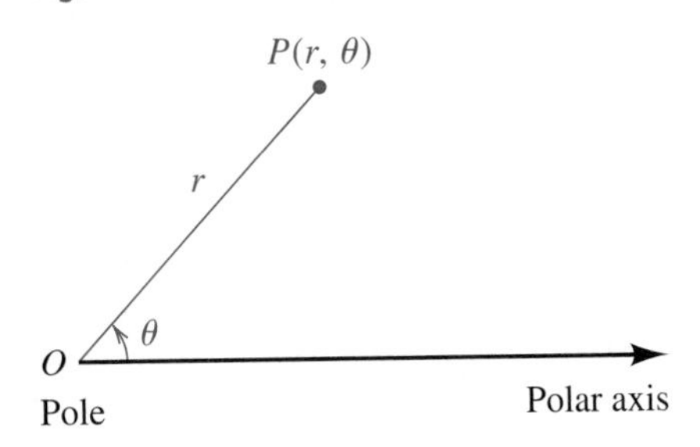

The polar coordinates of a point are not unique. For example, $(3, \pi/4)$, $(3, 9\pi/4)$, and $(3, -7\pi/4)$ all represent the same point (see Figure 2). We shall also allow r to be negative. In this case, instead of measuring $|r|$ units along the terminal side of the angle θ, we measure along the half-line with endpoint O that has direction *opposite* that of the terminal side. The points corresponding to the pairs $(-3, 5\pi/4)$ and $(-3, -3\pi/4)$ are also plotted in Figure 2 on the next page.

Figure 2

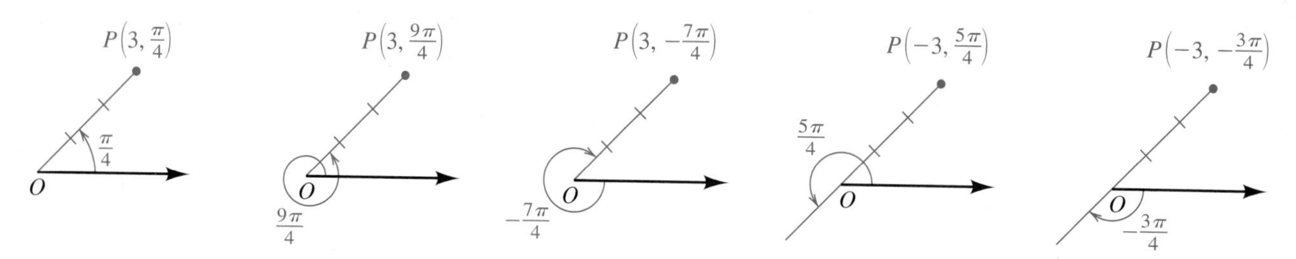

We agree that the pole O has polar coordinates $(0, \theta)$ for *any* θ. An assignment of ordered pairs of the form (r, θ) to points in a plane is a **polar coordinate system,** and the plane is an $\boldsymbol{r\theta}$**-plane.**

Let us next superimpose an xy-plane on an $r\theta$-plane so that the positive x-axis coincides with the polar axis. Any point P in the plane may then be assigned rectangular coordinates (x, y) or polar coordinates (r, θ). If $r > 0$, we have a situation similar to that illustrated in Figure 3(a); if $r < 0$, we have that shown in part (b) of the figure. In Figure 3(b), for later purposes, we have also plotted the point P', having polar coordinates $(|r|, \theta)$ and rectangular coordinates $(-x, -y)$.

Figure 3

(a) $r > 0$ (b) $r < 0$

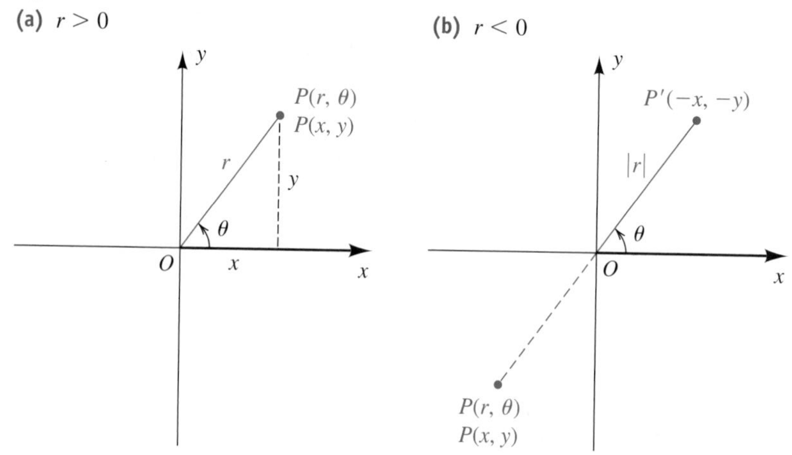

The following result specifies relationships between (x, y) and (r, θ), where it is assumed that the positive x-axis coincides with the polar axis.

Relationships Between Rectangular and Polar Coordinates

The rectangular coordinates (x, y) and polar coordinates (r, θ) of a point P are related as follows:

$$\textbf{(1)}\ x = r\cos\theta, \quad y = r\sin\theta$$

$$\textbf{(2)}\ r^2 = x^2 + y^2, \quad \tan\theta = \frac{y}{x} \quad \text{if } x \neq 0$$

PROOFS

(1) Although we have pictured θ as an acute angle in Figure 3, the discussion that follows is valid for all angles.

If $r > 0$, as in Figure 3(a), then $\cos\theta = x/r$ and $\sin\theta = y/r$, and hence

$$x = r\cos\theta, \qquad y = r\sin\theta.$$

If $r < 0$, then $|r| = -r$, and from Figure 3(b) we see that

$$\cos\theta = \frac{-x}{|r|} = \frac{-x}{-r} = \frac{x}{r}, \qquad \sin\theta = \frac{-y}{|r|} = \frac{-y}{-r} = \frac{y}{r}.$$

Multiplication by r gives us relationship 1, and therefore these formulas hold if r is either positive or negative.

If $r = 0$, then the point is the pole, and we again see that the formulas in (1) are true.

(2) The formulas in relationship 2 follow readily from Figure 3(a). By the Pythagorean theorem, $x^2 + y^2 = r^2$. From the definition of the trigonometric functions of any angle, $\tan\theta = y/x$ (if $x \neq 0$). If $x = 0$, then $\theta = (\pi/2) + \pi n$ from some integer n.

Figure 4

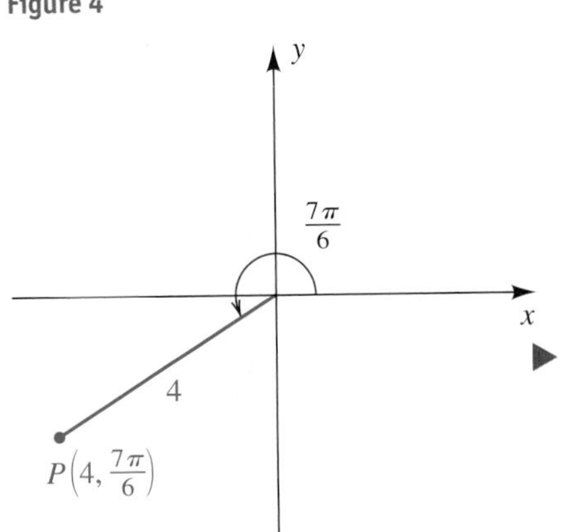

We may use the preceding result to change from one system of coordinates to the other.

EXAMPLE 1 **Changing polar coordinates to rectangular coordinates**

If $(r, \theta) = (4, 7\pi/6)$ are polar coordinates of a point P, find the rectangular coordinates of P.

▶ SOLUTION The point P is plotted in Figure 4. Substituting $r = 4$ and $\theta = 7\pi/6$ in relationship 1 of the preceding result, we obtain the following:

$$x = r\cos\theta = 4\cos(7\pi/6) = 4\left(-\sqrt{3}/2\right) = -2\sqrt{3}$$
$$y = r\sin\theta = 4\sin(7\pi/6) = 4(-1/2) = -2$$

Hence, the rectangular coordinates of P are $(x, y) = \left(-2\sqrt{3}, -2\right)$.

Let's confirm the results of Example 1 on a graphing calculator.

Polar to Rectangular Conversion

TI-83/4 Plus

We use the "given polar—return x" feature.

[2nd] [ANGLE] [7] 4 [,] 7
[2nd] [π] [÷] 6 [)] [ENTER]

TI-86

We use the magnitude ∠angle format to enter r and θ.

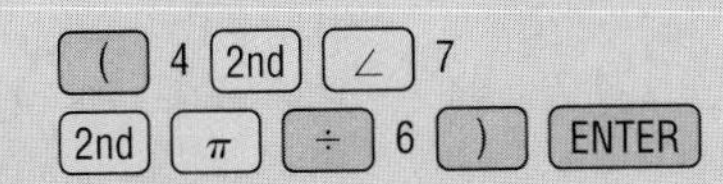

(continued)

▶ **Tutorial available at www.thomsonedu.com/login**

The second entry, $-2\sqrt{(3)}$, confirms the correctness of the x-value. Now we use the "given polar—return y" feature.

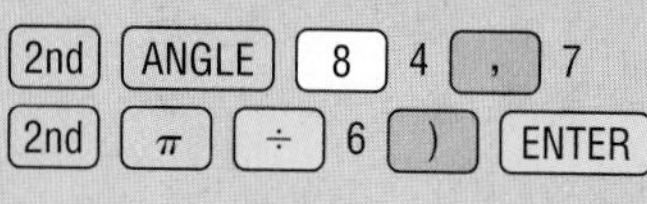

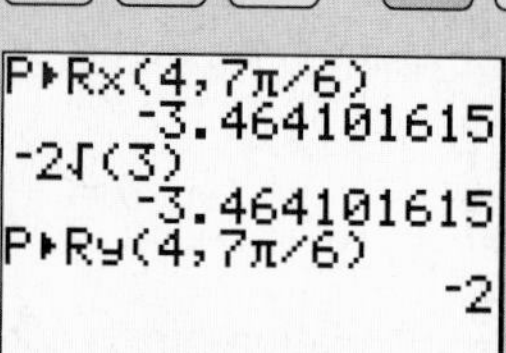

As an option, we can add the convert-to-rectangular command shown in the second line.

2nd ENTRY 2nd CPLX
MORE ▶Rec(F1) ENTER

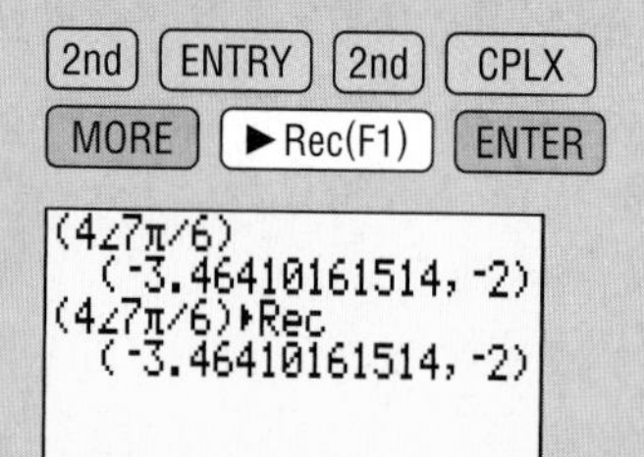

EXAMPLE 2 **Changing rectangular coordinates to polar coordinates**

If $(x, y) = \left(-1, \sqrt{3}\right)$ are rectangular coordinates of a point P, find three different pairs of polar coordinates (r, θ) for P.

Figure 5

(a)

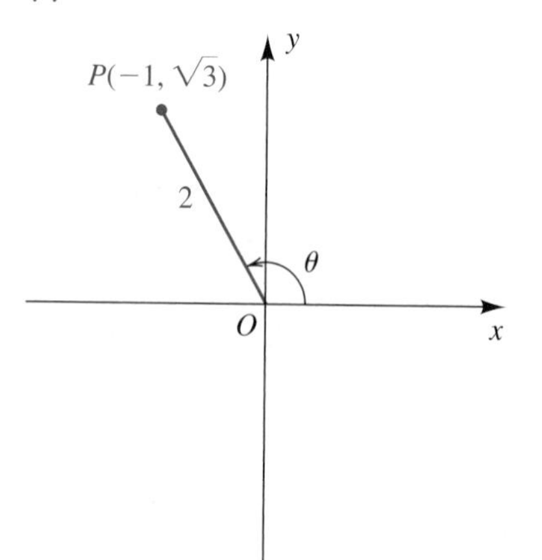

(b)

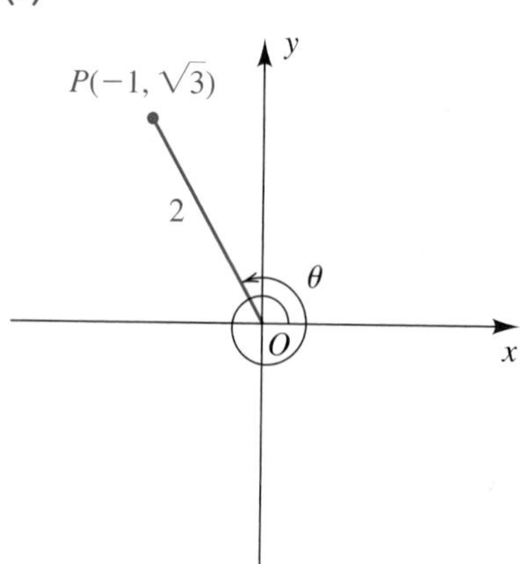

(c)

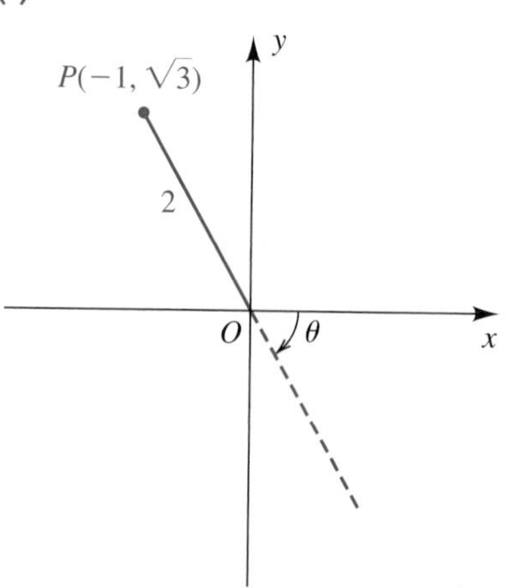

▶ **SOLUTION** Three possibilities for θ are illustrated in Figure 5(a)–(c). Using $x = -1$ and $y = \sqrt{3}$ in relationship 2 between rectangular and polar coordinates, we obtain

$$r^2 = x^2 + y^2 = (-1)^2 + \left(\sqrt{3}\right)^2 = 4,$$

and since r is positive in Figure 5(a), $r = 2$. Using

$$\tan\theta = \frac{y}{x} = \frac{\sqrt{3}}{-1} = -\sqrt{3},$$

▶ **Tutorial available at www.thomsonedu.com/login**

we see that the reference angle for θ is $\theta_R = \pi/3$, and hence

$$\theta = \pi - \frac{\pi}{3} = \frac{2\pi}{3}.$$

Thus, $(2, 2\pi/3)$ is one pair of polar coordinates for P.

Referring to Figure 5(b) and the values obtained for P in Figure 5(a), we get

$$r = 2 \qquad \text{and} \qquad \theta = \frac{2\pi}{3} + 2\pi = \frac{8\pi}{3}.$$

Hence, $(2, 8\pi/3)$ is another pair of polar coordinates for P.

In Figure 5(c), $\theta = -\pi/3$. In this case we use $r = -2$ to obtain $(-2, -\pi/3)$ as a third pair of polar coordinates for P.

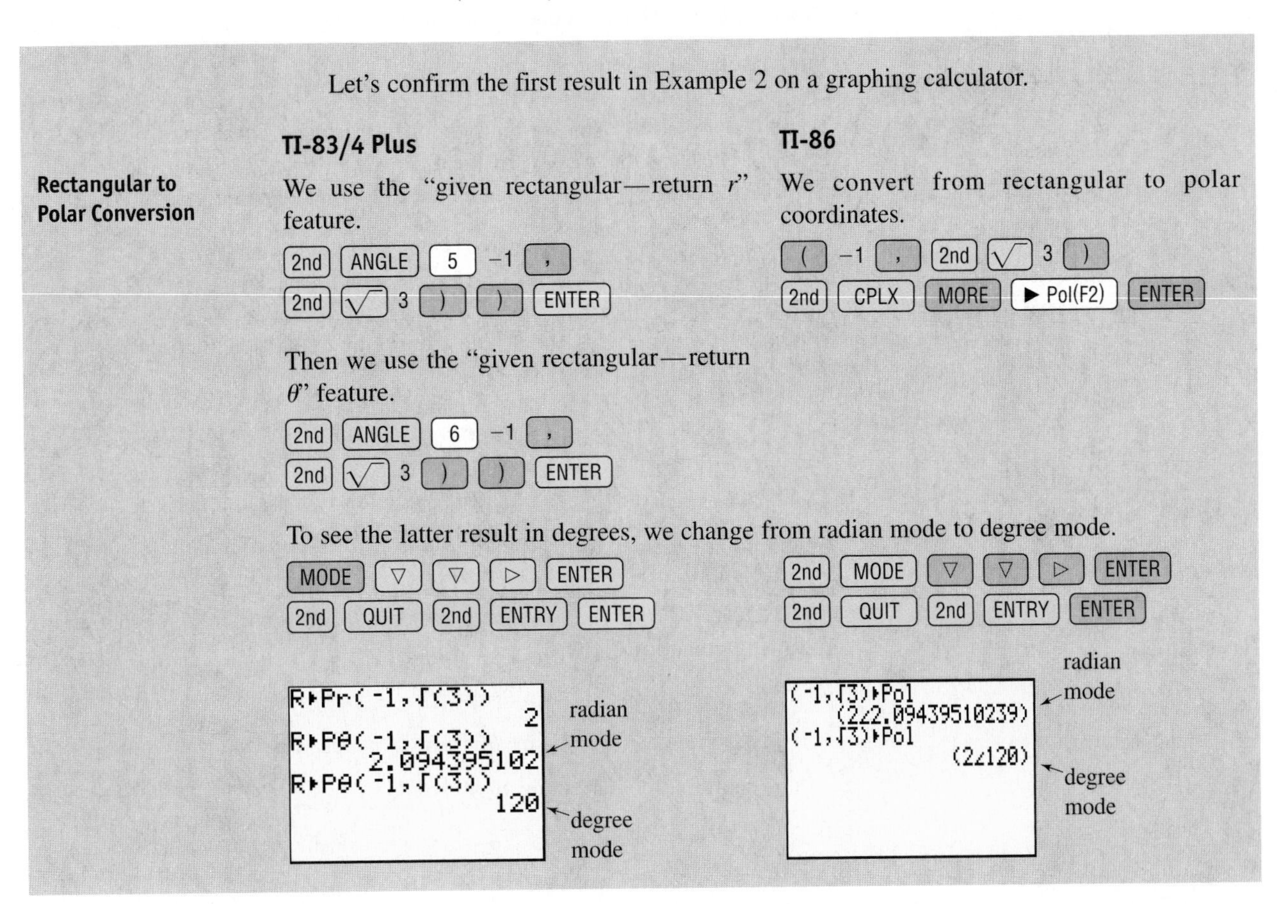

A **polar equation** is an equation in r and θ. A **solution** of a polar equation is an ordered pair (a, b) that leads to equality if a is substituted for r and b for θ. The **graph** of a polar equation is the set of all points (in an $r\theta$-plane) that correspond to the solutions.

Figure 6

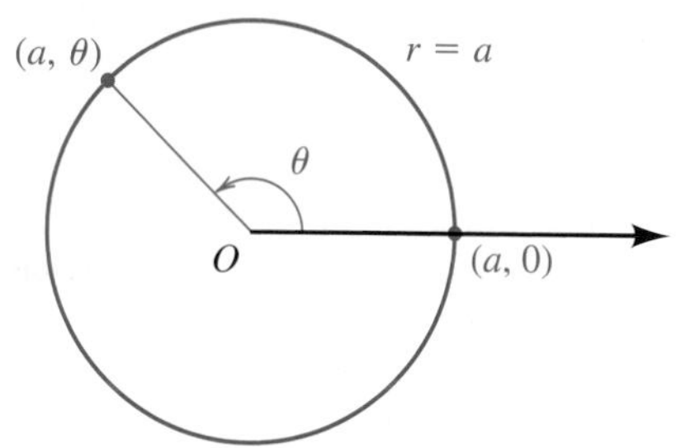

Figure 7

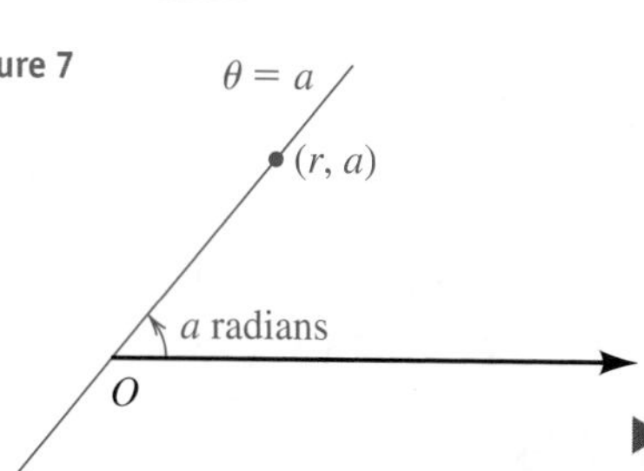

The simplest polar equations are $r = a$ and $\theta = a$, where a is a nonzero real number. Since the solutions of the polar equation $r = a$ are of the form (a, θ) for *any* angle θ, it follows that the graph is a circle of radius $|a|$ with center at the pole. A graph for $a > 0$ is sketched in Figure 6. The same graph is obtained for $r = -a$.

The solutions of the polar equation $\theta = a$ are of the form (r, a) for *any* real number r. Since the coordinate a (the angle) is constant, the graph of $\theta = a$ is a line through the origin, as illustrated in Figure 7 for an acute angle a.

We may use the relationships between rectangular and polar coordinates to transform a polar equation to an equation in x and y, and vice versa. This procedure is illustrated in the next three examples.

EXAMPLE 3 Finding a polar equation of a line

Find a polar equation of an arbitrary line.

SOLUTION Every line in an xy-coordinate plane is the graph of a linear equation that can be written in the form $ax + by = c$. Using the formulas $x = r\cos\theta$ and $y = r\sin\theta$ gives us the following equivalent polar equations:

$$ar\cos\theta + br\sin\theta = c \quad \text{substitute for } x \text{ and } y$$

$$r(a\cos\theta + b\sin\theta) = c \quad \text{factor out } r$$

If $a\cos\theta + b\sin\theta \neq 0$, the last equation may be written as follows:

$$r = \frac{c}{a\cos\theta + b\sin\theta}$$

EXAMPLE 4 Changing an equation in x and y to a polar equation

Find a polar equation for the hyperbola $x^2 - y^2 = 16$.

SOLUTION Using the formulas $x = r\cos\theta$ and $y = r\sin\theta$, we obtain the following polar equations:

$$(r\cos\theta)^2 - (r\sin\theta)^2 = 16 \quad \text{substitute for } x \text{ and } y$$

$$r^2\cos^2\theta - r^2\sin^2\theta = 16 \quad \text{square the terms}$$

$$r^2(\cos^2\theta - \sin^2\theta) = 16 \quad \text{factor out } r^2$$

$$r^2\cos 2\theta = 16 \quad \text{double-angle formula}$$

$$r^2 = \frac{16}{\cos 2\theta} \quad \text{divide by } \cos 2\theta$$

The division by $\cos 2\theta$ is allowable because $\cos 2\theta \neq 0$. (Note that if $\cos 2\theta = 0$, then $r^2\cos 2\theta \neq 16$.) We may also write the polar equation as $r^2 = 16\sec 2\theta$.

Tutorial available at www.thomsonedu.com/login

EXAMPLE 5 **Changing a polar equation to an equation in x and y**

Find an equation in x and y that has the same graph as the polar equation $r = a \sin \theta$, with $a \neq 0$. Sketch the graph.

▶ SOLUTION A formula that relates $\sin \theta$ and y is given by $y = r \sin \theta$. To introduce the expression $r \sin \theta$ into the equation $r = a \sin \theta$, we multiply both sides by r, obtaining

Figure 8

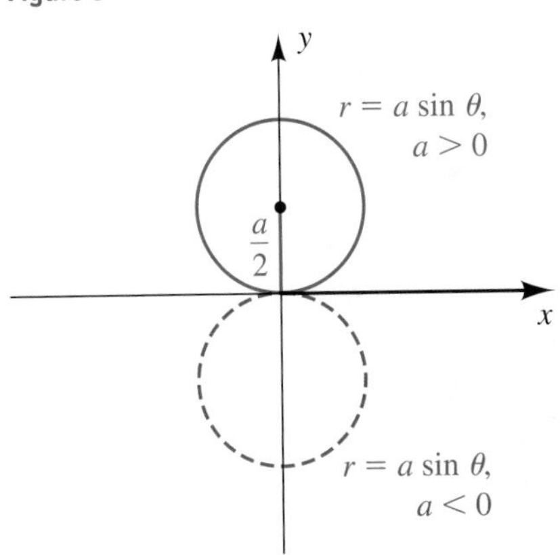

$$r^2 = ar \sin \theta.$$

Next, if we substitute $x^2 + y^2$ for r^2 and y for $r \sin \theta$, the last equation becomes

$$x^2 + y^2 = ay,$$

or

$$x^2 + y^2 - ay = 0.$$

Completing the square in y gives us

$$x^2 + y^2 - ay + \left(\frac{a}{2}\right)^2 = \left(\frac{a}{2}\right)^2,$$

or

$$x^2 + \left(y - \frac{a}{2}\right)^2 = \left(\frac{a}{2}\right)^2.$$

In the xy-plane, the graph of the last equation is a circle with center $(0, a/2)$ and radius $|a|/2$, as illustrated in Figure 8 for the case $a > 0$ (the solid circle) and $a < 0$ (the dashed circle).

Figure 9

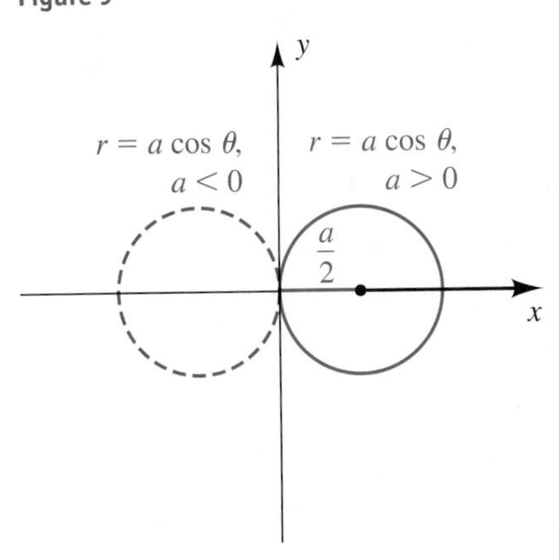

Using the same method as in the preceding example, we can show that the graph of $r = a \cos \theta$, with $a \neq 0$, is a circle of radius $a/2$ of the type illustrated in Figure 9.

In the following examples we obtain the graphs of polar equations by plotting points and examining the relationship between θ-intervals and r-intervals. As you proceed through this section, you should try to recognize forms of polar equations so that you will be able to sketch their graphs by plotting few, if any, points.

EXAMPLE 6 **Sketching the graph of a polar equation**

Sketch the graph of the polar equation $r = 4 \sin \theta$.

▶ SOLUTION The proof that the graph of $r = 4 \sin \theta$ is a circle was given in Example 5. The following table displays some solutions of the equation. We have included a third row in the table that contains one-decimal-place approximations to r.

(continued)

▶ **Tutorial available at www.thomsonedu.com/login**

θ	0	$\frac{\pi}{6}$	$\frac{\pi}{4}$	$\frac{\pi}{3}$	$\frac{\pi}{2}$	$\frac{2\pi}{3}$	$\frac{3\pi}{4}$	$\frac{5\pi}{6}$	π
r	0	2	$2\sqrt{2}$	$2\sqrt{3}$	4	$2\sqrt{3}$	$2\sqrt{2}$	2	0
r (approx.)	0	2	2.8	3.5	4	3.5	2.8	2	0

Figure 10

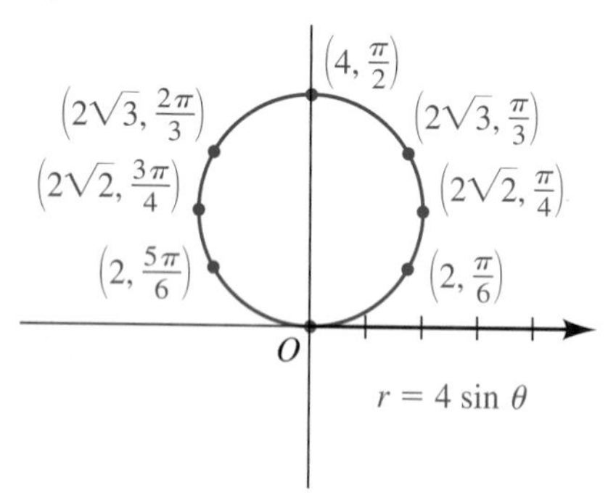

As an aid to plotting points in the $r\theta$-plane shown in Figure 10, we have extended the polar axis in the negative direction and introduced a vertical line through the pole (this line is the graph of the equation $\theta = \pi/2$). Additional points obtained by letting θ vary from π to 2π lie on the same circle. For example, the solution $(-2, 7\pi/6)$ gives us the same point as $(2, \pi/6)$; the point corresponding to $\left(-2\sqrt{2}, 5\pi/4\right)$ is the same as that obtained from $\left(2\sqrt{2}, \pi/4\right)$; and so on. If we let θ increase through all real numbers, we obtain the same points again and again because of the periodicity of the sine function.

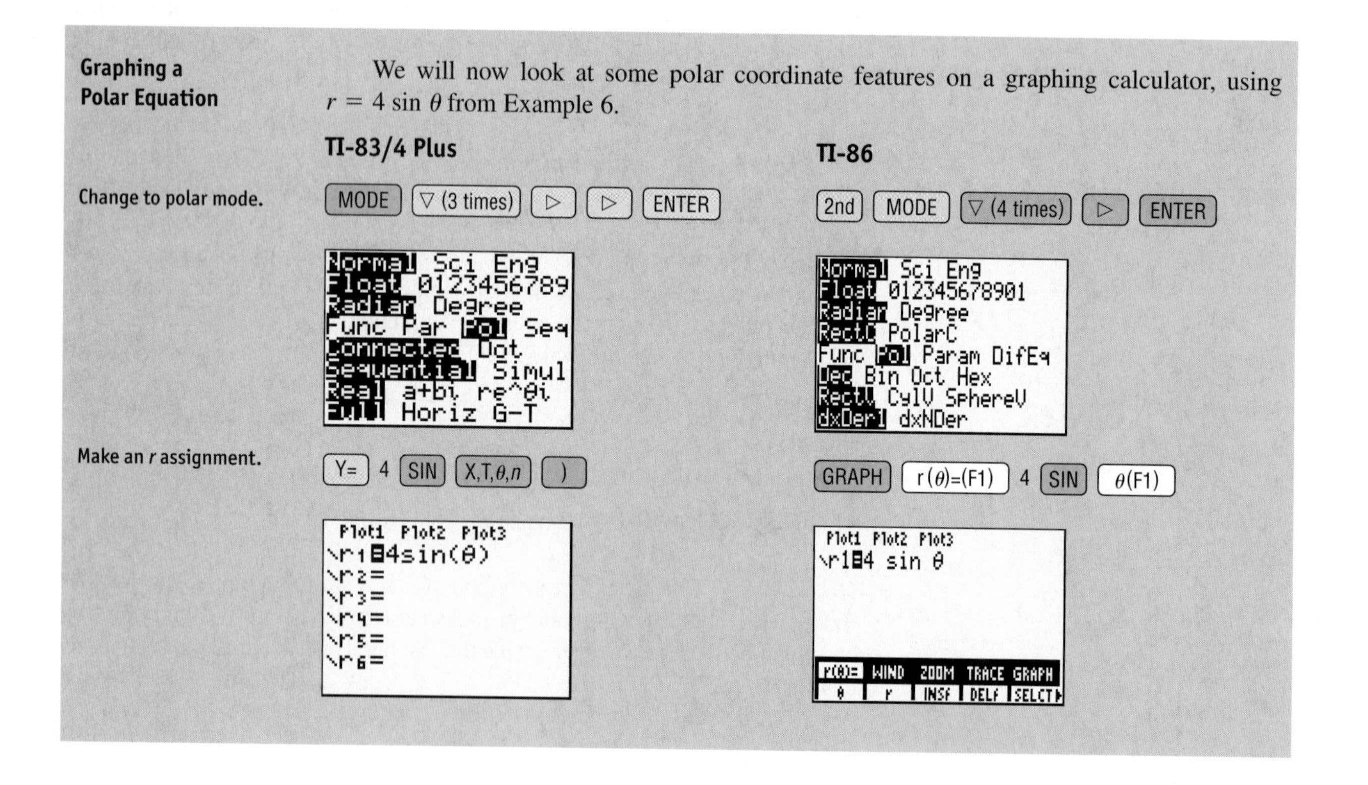

Set the window values.

We'll use θmin $= 0$ to θmax $= \pi$, since that gives us the circle. For θstep, we'll use 0.05. A smaller value such as 0.01 slows down the graphing process, and a larger value such as 0.5 yields a crude figure.

WINDOW 0 ▽ 2nd π ▽ .05 ▽ −4.5 ▽ 4.5 ▽ 1 ▽ −1 ▽ 5 ▽ 1

2nd WIND(M2) 0 ▽ 2nd π ▽ .05 ▽ −4.5 ▽ 4.5 ▽ 1 ▽ −1 ▽ 5 ▽ 1

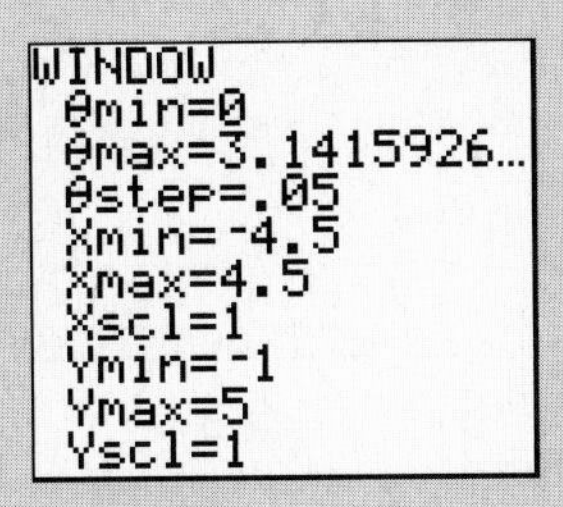

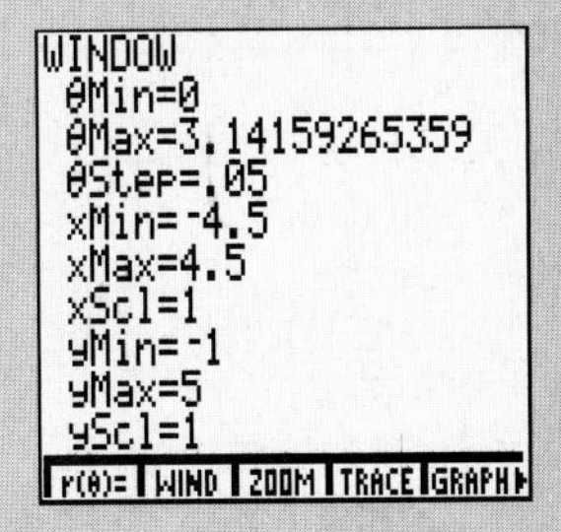

Graph the function.

GRAPH

GRAPH(F5)

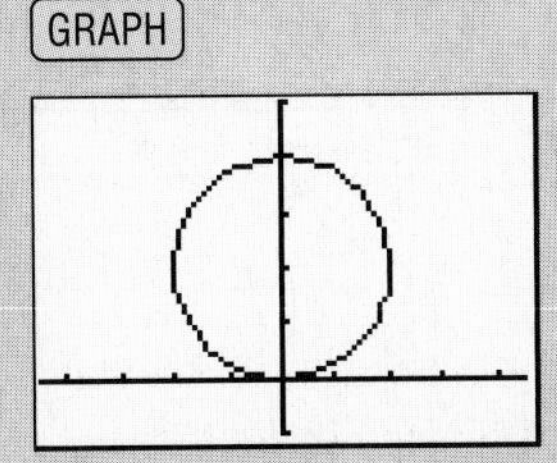

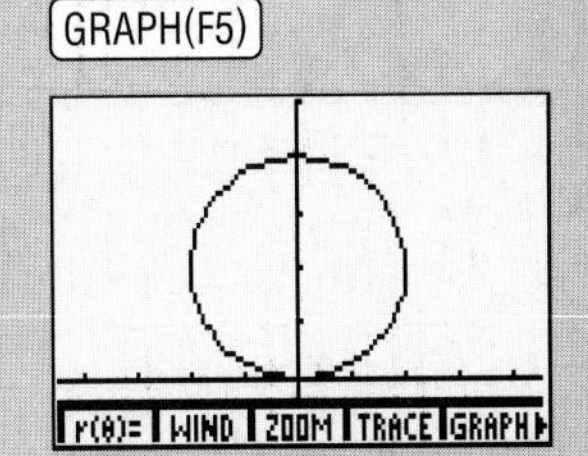

Trace the graph (rectangular mode).

Now we enter the trace mode and use the cursor keys to move around the circle. Note that the calculator displays the values of θ, X, and Y.

TRACE (▷ and ◁)

TRACE(F4) (▷ and ◁)

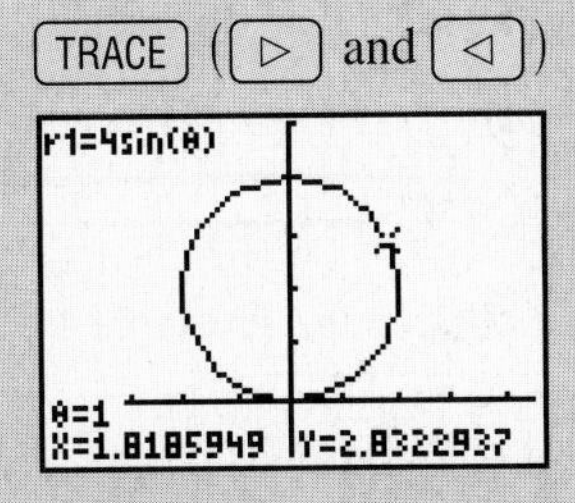

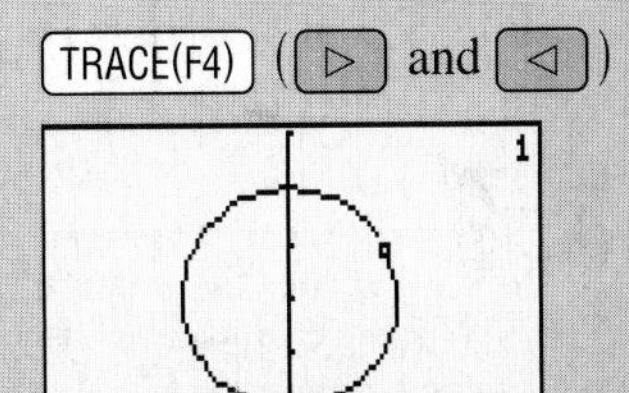

Evaluate the function for $\theta = 2$.

2nd CALC 1 2 ENTER

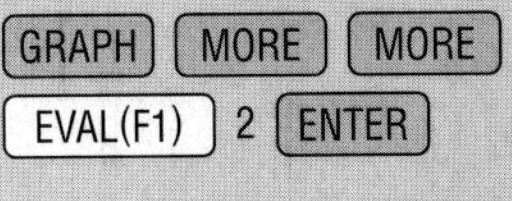

EVAL(F1) 2 ENTER

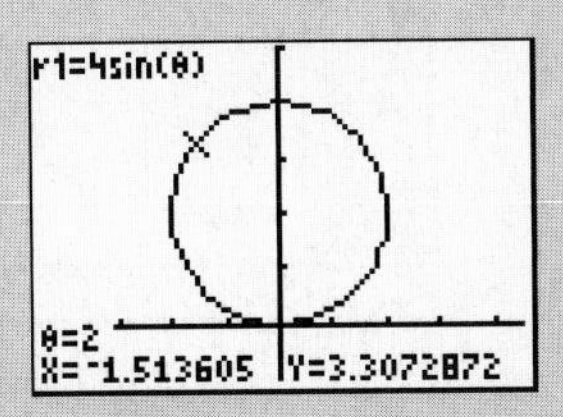

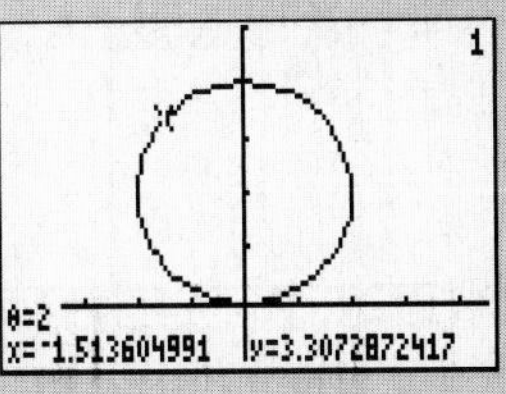

(continued)

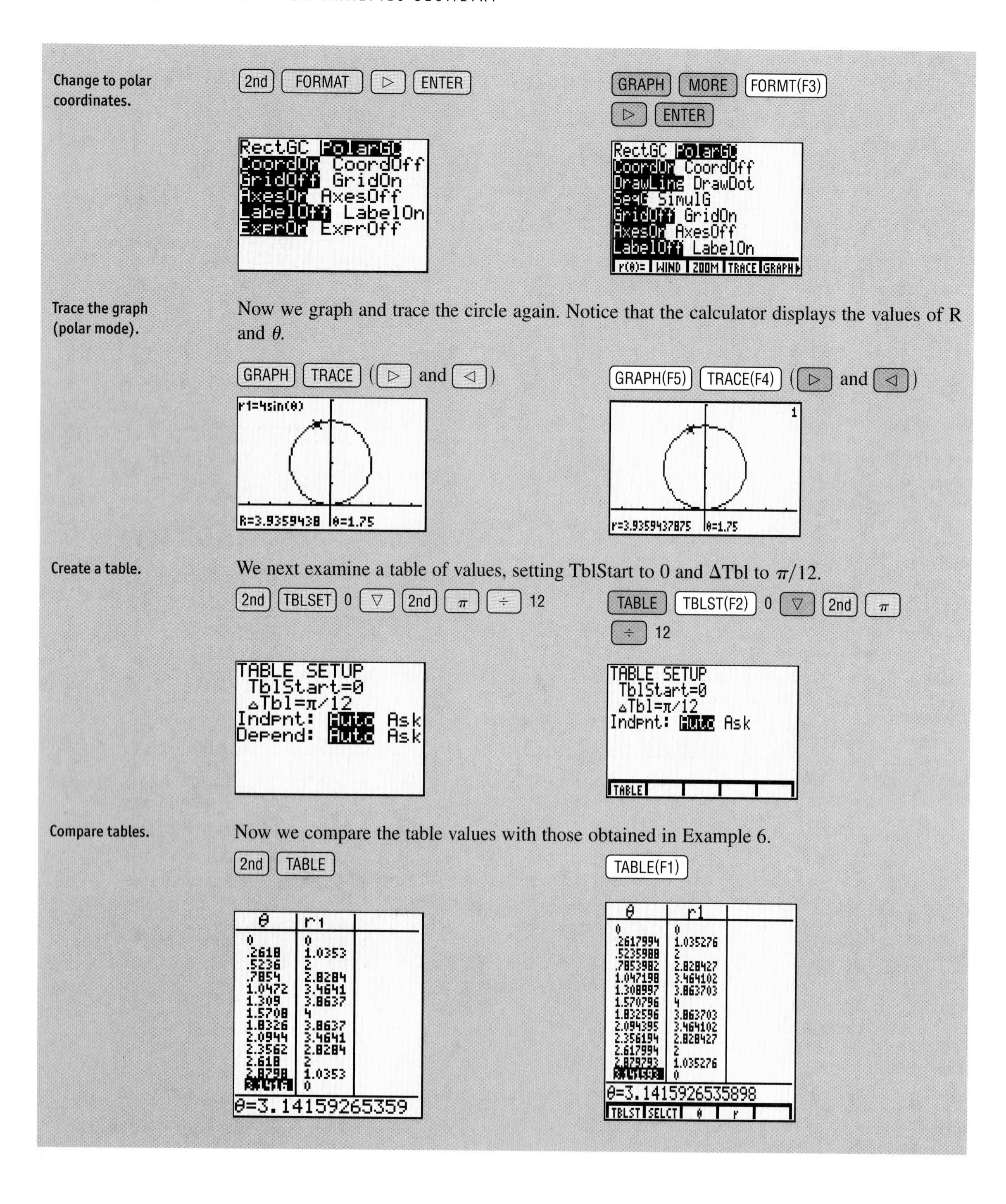
Change to polar coordinates.
2nd FORMAT ▷ ENTER
GRAPH MORE FORMT(F3) ▷ ENTER
RectGC PolarGC
CoordOn CoordOff
GridOff GridOn
AxesOn AxesOff
LabelOff LabelOn
ExprOn ExprOff
DrawLine DrawDot
SeqG SimulG
Trace the graph (polar mode).
Now we graph and trace the circle again. Notice that the calculator displays the values of R and θ.
GRAPH TRACE (▷ and ◁)
GRAPH(F5) TRACE(F4) (▷ and ◁)
r1=4sin(θ)
R=3.9359438 θ=1.75
Create a table.
We next examine a table of values, setting TblStart to 0 and ΔTbl to π/12.
2nd TBLSET 0 ▽ 2nd π ÷ 12
TABLE TBLST(F2) 0 ▽ 2nd π ÷ 12
TABLE SETUP
TblStart=0
ΔTbl=π/12
Indpnt: Auto Ask
Depend: Auto Ask
Compare tables.
Now we compare the table values with those obtained in Example 6.
2nd TABLE
TABLE(F1)
θ r1
0 0
.2618 1.0353
.5236 2
.7854 2.8284
1.0472 3.4641
1.309 3.8637
1.5708 4
1.8326 3.8637
2.0944 3.4641
2.3562 2.8284
2.618 2
2.8798 1.0353
θ=3.14159265359
θ=3.141592653589
TBLST SELCT θ r

EXAMPLE 7 Sketching the graph of a polar equation

Sketch the graph of the polar equation $r = 2 + 2\cos\theta$.

SOLUTION Since the cosine function decreases from 1 to -1 as θ varies from 0 to π, it follows that r decreases from 4 to 0 in this θ-interval. The following table exhibits some solutions of $r = 2 + 2\cos\theta$, together with one-decimal-place approximations to r.

θ	0	$\frac{\pi}{6}$	$\frac{\pi}{4}$	$\frac{\pi}{3}$	$\frac{\pi}{2}$	$\frac{2\pi}{3}$	$\frac{3\pi}{4}$	$\frac{5\pi}{6}$	π
r	4	$2+\sqrt{3}$	$2+\sqrt{2}$	3	2	1	$2-\sqrt{2}$	$2-\sqrt{3}$	0
r (approx.)	4	3.7	3.4	3	2	1	0.6	0.3	0

Plotting points in an $r\theta$-plane leads to the upper half of the graph sketched in Figure 11. (We have used polar coordinate graph paper, which displays lines through O at various angles and concentric circles with centers at the pole.)

Figure 11

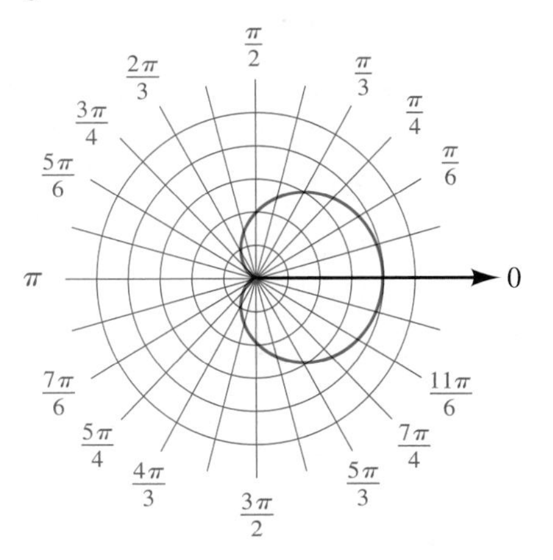

If θ increases from π to 2π, then $\cos\theta$ increases from -1 to 1, and consequently r increases from 0 to 4. Plotting points for $\pi \leq \theta \leq 2\pi$ gives us the lower half of the graph.

The same graph may be obtained by taking other intervals of length 2π for θ.

The heart-shaped graph in Example 7 is a **cardioid.** In general, the graph of any of the polar equations in Figure 12 on the next page, with $a \neq 0$, is a cardioid.

Figure 12

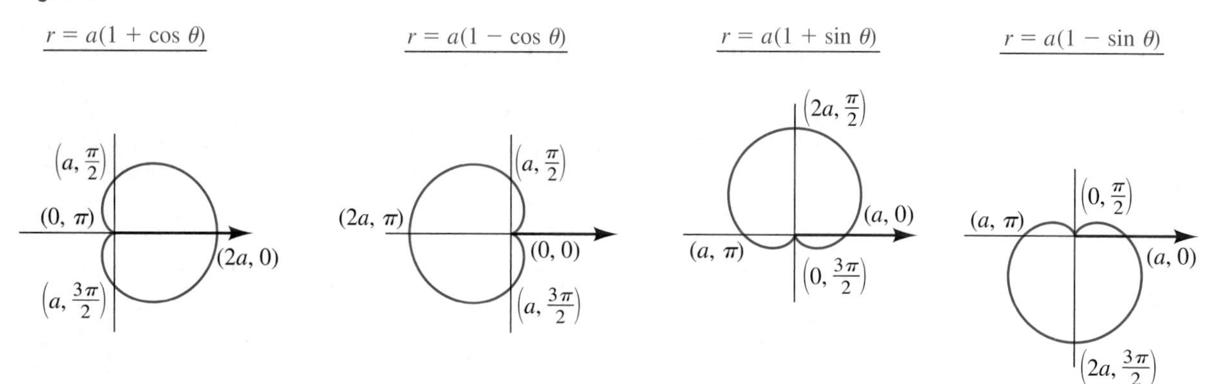

If a and b are not zero, then the graphs of the following polar equations are **limaçons:**

$$r = a + b \cos \theta \qquad r = a + b \sin \theta$$

Note that the special limaçons in which $|a| = |b|$ are cardioids.

Using the θ-interval $[0, 2\pi]$ (or $[-\pi, \pi]$) is usually sufficient to graph polar equations. For equations with more complex graphs, it is often helpful to graph by using subintervals of $[0, 2\pi]$ that are determined by the θ-values that make $r = 0$—that is, the **pole values.** We will demonstrate this technique in the next example.

EXAMPLE 8 Sketching the graph of a polar equation

Sketch the graph of the polar equation $r = 2 + 4 \cos \theta$.

Figure 13

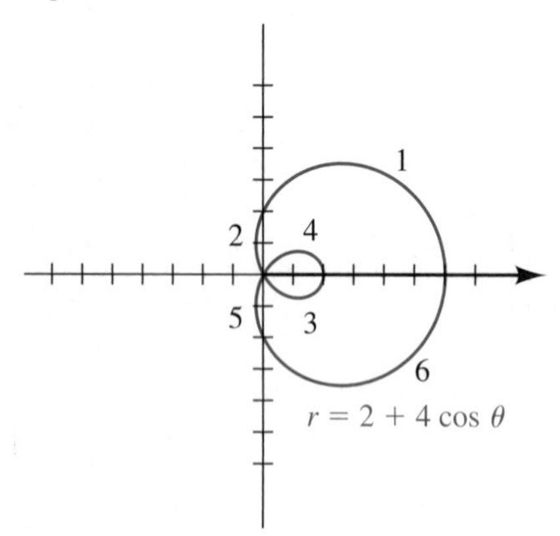

SOLUTION We first find the pole values by solving the equation $r = 0$:

$2 + 4 \cos \theta = 0$	let $r = 0$
$\cos \theta = -\frac{1}{2}$	solve for $\cos \theta$
$\theta = \frac{2\pi}{3}, \frac{4\pi}{3}$	solve for θ in $[0, 2\pi]$

We next construct a table of θ-values from 0 to 2π, using subintervals determined by the quadrantal angles and the pole values. The row numbers on the left-hand side correspond to the numbers in Figure 13.

	θ	$\cos\theta$	$4\cos\theta$	$r = 2 + 4\cos\theta$
(1)	$0 \rightarrow \pi/2$	$1 \rightarrow 0$	$4 \rightarrow 0$	$6 \rightarrow 2$
(2)	$\pi/2 \rightarrow 2\pi/3$	$0 \rightarrow -1/2$	$0 \rightarrow -2$	$2 \rightarrow 0$
(3)	$2\pi/3 \rightarrow \pi$	$-1/2 \rightarrow -1/2$	$-2 \rightarrow -4$	$0 \rightarrow -2$
(4)	$\pi \rightarrow 4\pi/3$	$-1 \rightarrow -1/2$	$-4 \rightarrow -2$	$-2 \rightarrow 0$
(5)	$4\pi/3 \rightarrow 3\pi/2$	$-1/2 \rightarrow 0$	$-2 \rightarrow 0$	$0 \rightarrow 2$
(6)	$3\pi/2 \rightarrow 2\pi$	$0 \rightarrow 1$	$0 \rightarrow 4$	$2 \rightarrow 6$

You should verify the table entries with the figure, especially for rows 3 and 4 (in which the value of r is negative). The graph is called a limaçon with an inner loop.

The following chart summarizes the four categories of limaçons according to the ratio of a and b in the listed general equations.

Limaçons $a \pm b\cos\theta$, $a \pm b\sin\theta$ ($a > 0$, $b > 0$)

Name	Limaçon with an inner loop	Cardioid	Limaçon with a dimple	Convex limaçon
Condition	$\frac{a}{b} < 1$	$\frac{a}{b} = 1$	$1 < \frac{a}{b} < 2$	$\frac{a}{b} \geq 2$
Specific graph				
Specific equation	$r = 2 + 4\cos\theta$	$r = 4 + 4\cos\theta$	$r = 6 + 4\cos\theta$	$r = 8 + 4\cos\theta$

EXAMPLE 9 Sketching the graph of a polar equation

Sketch the graph of the polar equation $r = a\sin 2\theta$ for $a > 0$.

SOLUTION The following table contains θ-intervals and the corresponding values of r. The row numbers on the left-hand side correspond to the numbers in Figure 14 on the next page.

(continued)

Figure 14

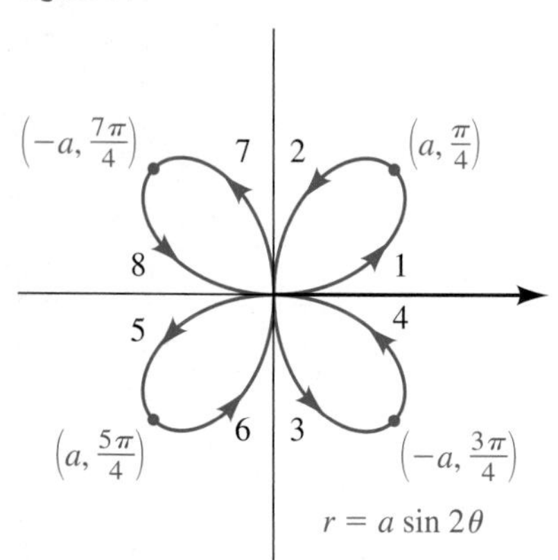

	θ	2θ	$\sin 2\theta$	$r = a \sin 2\theta$
(1)	$0 \to \pi/4$	$0 \to \pi/2$	$0 \to 1$	$0 \to a$
(2)	$\pi/4 \to \pi/2$	$\pi/2 \to \pi$	$1 \to 0$	$a \to 0$
(3)	$\pi/2 \to 3\pi/4$	$\pi \to 3\pi/2$	$0 \to -1$	$0 \to -a$
(4)	$3\pi/4 \to \pi$	$3\pi/2 \to 2\pi$	$-1 \to 0$	$-a \to 0$
(5)	$\pi \to 5\pi/4$	$2\pi \to 5\pi/2$	$0 \to 1$	$0 \to a$
(6)	$5\pi/4 \to 3\pi/2$	$5\pi/2 \to 3\pi$	$1 \to 0$	$a \to 0$
(7)	$3\pi/2 \to 7\pi/4$	$3\pi \to 7\pi/2$	$0 \to -1$	$0 \to -a$
(8)	$7\pi/4 \to 2\pi$	$7\pi/2 \to 4\pi$	$-1 \to 0$	$-a \to 0$

You should verify the table entries with the figure, especially for rows 3, 4, 7, and 8 (in which the value of r is negative).

The graph in Example 9 is a **four-leafed rose.** In general, a polar equation of the form

$$r = a \sin n\theta \qquad \text{or} \qquad r = a \cos n\theta$$

for any positive integer n greater than 1 and any nonzero real number a has a graph that consists of a number of loops through the origin. If n is even, there are $2n$ loops, and if n is odd, there are n loops.

The graph of the polar equation $r = a\theta$ for any nonzero real number a is a **spiral of Archimedes.** The case $a = 1$ is considered in the next example.

Figure 15

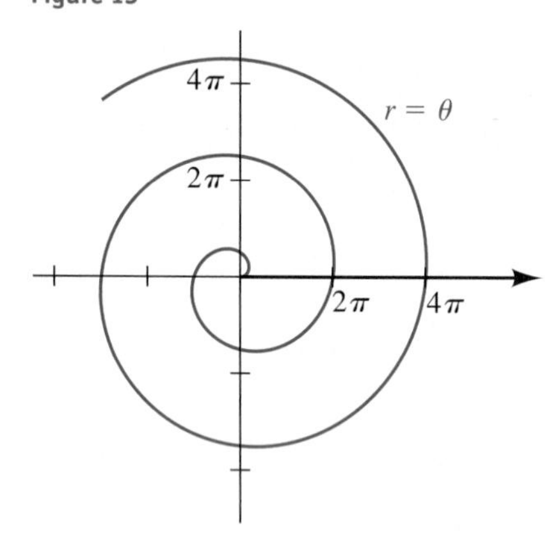

EXAMPLE 10 Sketching the graph of a spiral of Archimedes

Sketch the graph of the polar equation $r = \theta$ for $\theta \geq 0$.

SOLUTION The graph consists of all points that have polar coordinates of the form (c, c) for every real number $c \geq 0$. Thus, the graph contains the points $(0, 0)$, $(\pi/2, \pi/2)$, (π, π), and so on. As θ increases, r increases at the same rate, and the spiral winds around the origin in a counterclockwise direction, intersecting the polar axis at $0, 2\pi, 4\pi, \ldots,$ as illustrated in Figure 15.

If θ is allowed to be negative, then as θ decreases through negative values, the resulting spiral winds around the origin and is the symmetric image, with respect to the vertical axis, of the curve sketched in Figure 15.

If we superimpose an xy-plane on an $r\theta$-plane, then the graph of a polar equation may be symmetric with respect to the x-axis (the polar axis), the y-axis (the line $\theta = \pi/2$), or the origin (the pole). Some typical symmetries are illustrated in Figure 16. The next result summarizes these symmetries.

Figure 16 Symmetries of graphs of polar equations

(a) Polar axis

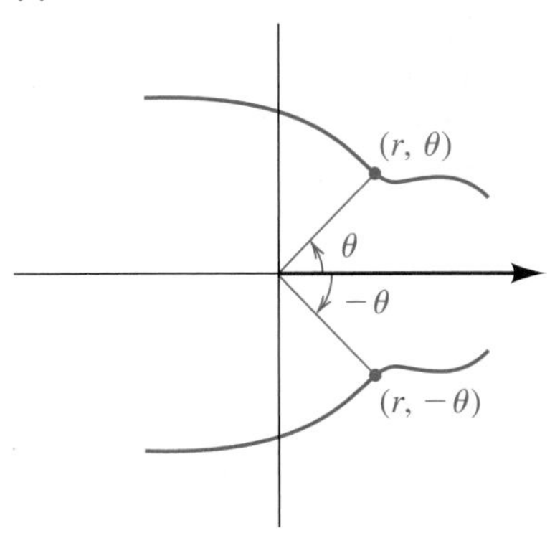

(b) Line $\theta = \pi/2$

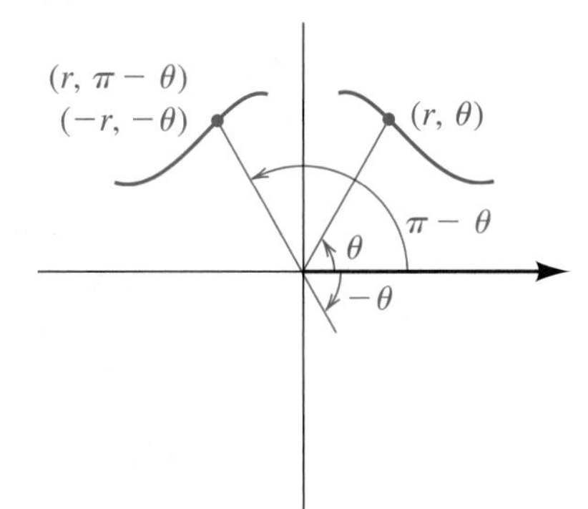

(c) Pole

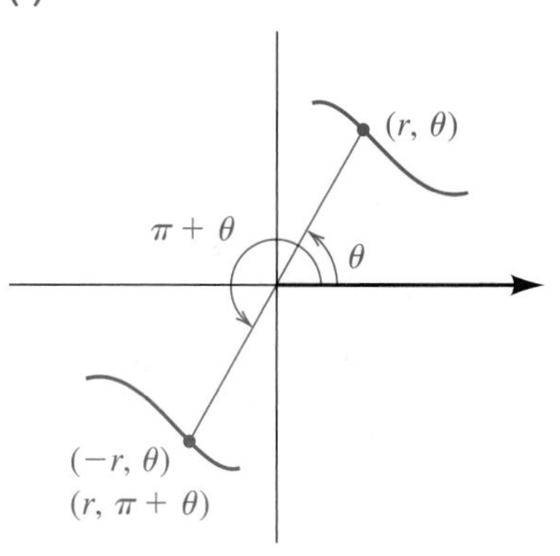

Tests for Symmetry

(1) The graph of $r = f(\theta)$ is symmetric with respect to the polar axis if substitution of $-\theta$ for θ leads to an equivalent equation.

(2) The graph of $r = f(\theta)$ is symmetric with respect to the vertical line $\theta = \pi/2$ if substitution of either (a) $\pi - \theta$ for θ or (b) $-r$ for r and $-\theta$ for θ leads to an equivalent equation.

(3) The graph of $r = f(\theta)$ is symmetric with respect to the pole if substitution of either (a) $\pi + \theta$ for θ or (b) $-r$ for r leads to an equivalent equation.

To illustrate, since $\cos(-\theta) = \cos\theta$, the graph of the polar equation $r = 2 + 4\cos\theta$ in Example 8 is symmetric with respect to the polar axis, by test 1. Since $\sin(\pi - \theta) = \sin\theta$, the graph in Example 6 is symmetric with respect to the line $\theta = \pi/2$, by test 2. The graph of the four-leafed rose in Example 9 is symmetric with respect to the polar axis, the line $\theta = \pi/2$, and the pole. Other tests for symmetry may be stated; however, those we have listed are among the easiest to apply.

Figure 17

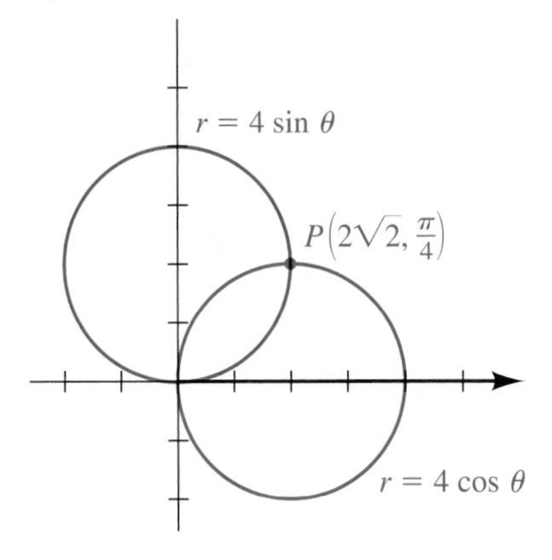

Unlike the graph of an equation in x and y, the graph of a polar equation $r = f(\theta)$ can be symmetric with respect to the polar axis, the line $\theta = \pi/2$, or the pole *without* satisfying one of the preceding tests for symmetry. This is true because of the many different ways of specifying a point in polar coordinates.

Another difference between rectangular and polar coordinate systems is that the points of intersection of two graphs cannot always be found by solving the polar equations simultaneously. To illustrate, from Example 6, the graph of $r = 4\sin\theta$ is a circle of diameter 4 with center at $(2, \pi/2)$ (see Figure 17). Similarly, the graph of $r = 4\cos\theta$ is a circle of diameter 4 with center at $(2, 0)$ on the polar axis. Referring to Figure 17, we see that the coordinates of the point of intersection $P(2\sqrt{2}, \pi/4)$ in quadrant I satisfy both equations; however, the origin O, which is on each circle, *cannot* be found by solving the equations simultaneously. Thus, in searching for points of intersection of polar graphs, it is sometimes necessary to refer to the graphs themselves, *in addition* to solving the two equations simultaneously.

An alternative method is to use different (equivalent) equations for the graphs. See Discussion Exercise 12 at the end of the chapter.

10.5 *Exercises*

1 Which polar coordinates represent the same point as $(3, \pi/3)$?

(a) $(3, 7\pi/3)$ (b) $(3, -\pi/3)$ (c) $(-3, 4\pi/3)$

(d) $(3, -2\pi/3)$ (e) $(-3, -2\pi/3)$ (f) $(-3, -\pi/3)$

2 Which polar coordinates represent the same point as $(4, -\pi/2)$?

(a) $(4, 5\pi/2)$ (b) $(4, 7\pi/2)$ (c) $(-4, -\pi/2)$

(d) $(4, -5\pi/2)$ (e) $(-4, -3\pi/2)$ (f) $(-4, \pi/2)$

Exer. 3–8: Change the polar coordinates to rectangular coordinates.

3 (a) $(3, \pi/4)$ (b) $(-1, 2\pi/3)$

4 (a) $(5, 5\pi/6)$ (b) $(-6, 7\pi/3)$

5 (a) $(8, -2\pi/3)$ (b) $(-3, 5\pi/3)$

6 (a) $(4, -\pi/4)$ (b) $(-2, 7\pi/6)$

▶ 7 $\left(6, \arctan \frac{3}{4}\right)$ 8 $\left(10, \arccos\left(-\frac{1}{3}\right)\right)$

Exer. 9–12: Change the rectangular coordinates to polar coordinates with $r > 0$ and $0 \le \theta \le 2\pi$.

9 (a) $(-1, 1)$ (b) $\left(-2\sqrt{3}, -2\right)$

10 (a) $\left(3\sqrt{3}, 3\right)$ (b) $(2, -2)$

11 (a) $\left(7, -7\sqrt{3}\right)$ ▶ (b) $(5, 5)$

12 (a) $\left(-2\sqrt{2}, -2\sqrt{2}\right)$ (b) $\left(-4, 4\sqrt{3}\right)$

Exer. 13–26: Find a polar equation that has the same graph as the equation in x and y.

13 $x = -3$ 14 $y = 2$

▶ 15 $x^2 + y^2 = 16$ 16 $x^2 + y^2 = 2$

17 $y^2 = 6x$ 18 $x^2 = 8y$

19 $x + y = 3$ 20 $2y = -x + 4$

21 $2y = -x$ 22 $y = 6x$

23 $y^2 - x^2 = 4$ 24 $xy = 8$

25 $(x - 1)^2 + y^2 = 1$

26 $(x + 2)^2 + (y - 3)^2 = 13$

Exer. 27–44: Find an equation in x and y that has the same graph as the polar equation. Use it to help sketch the graph in an $r\theta$-plane.

27 $r \cos \theta = 5$ 28 $r \sin \theta = -2$

29 $r - 6 \sin \theta = 0$ 30 $r = 2$

31 $\theta = \pi/4$ 32 $r = 4 \sec \theta$

▶ 33 $r^2(4 \sin^2 \theta - 9 \cos^2 \theta) = 36$

34 $r^2(\cos^2 \theta + 4 \sin^2 \theta) = 16$

▶ Tutorial available at www.thomsonedu.com/login

35 $r^2 \cos 2\theta = 1$

36 $r^2 \sin 2\theta = 4$

37 $r(\sin \theta - 2 \cos \theta) = 6$

38 $r(3 \cos \theta - 4 \sin \theta) = 12$

39 $r(\sin \theta + r \cos^2 \theta) = 1$

40 $r(r \sin^2 \theta - \cos \theta) = 3$

41 $r = 8 \sin \theta - 2 \cos \theta$

42 $r = 2 \cos \theta - 4 \sin \theta$

43 $r = \tan \theta$

44 $r = 6 \cot \theta$

Exer. 45–78: Sketch the graph of the polar equation.

45 $r = 5$

46 $r = -2$

47 $\theta = -\pi/6$

48 $\theta = \pi/4$

49 $r = 3 \cos \theta$

50 $r = -2 \sin \theta$

51 $r = 4 \cos \theta + 2 \sin \theta$

52 $r = 6 \cos \theta - 2 \sin \theta$

53 $r = 4(1 - \sin \theta)$

54 $r = 3(1 + \cos \theta)$

55 $r = -6(1 + \cos \theta)$

56 $r = 2(1 + \sin \theta)$

57 $r = 2 + 4 \sin \theta$

58 $r = 1 + 2 \cos \theta$

59 $r = \sqrt{3} - 2 \sin \theta$

60 $r = 2\sqrt{3} - 4 \cos \theta$

61 $r = 2 - \cos \theta$

62 $r = 5 + 3 \sin \theta$

63 $r = 4 \csc \theta$

64 $r = -3 \sec \theta$

65 $r = 8 \cos 3\theta$

66 $r = 2 \sin 4\theta$

67 $r = 3 \sin 2\theta$

68 $r = 8 \cos 5\theta$

69 $r^2 = 4 \cos 2\theta$ (lemniscate)

70 $r^2 = -16 \sin 2\theta$

71 $r = 2^\theta, \theta \geq 0$ (spiral)

72 $r = e^{2\theta}, \theta \geq 0$ (logarithmic spiral)

73 $r = 2\theta, \theta \geq 0$

74 $r\theta = 1, \theta > 0$ (spiral)

75 $r = 6 \sin^2 (\theta/2)$

76 $r = -4 \cos^2 (\theta/2)$

77 $r = 2 + 2 \sec \theta$ (conchoid)

78 $r = 1 - \csc \theta$

79 If $P_1(r_1, \theta_1)$ and $P_2(r_2, \theta_2)$ are points in an $r\theta$-plane, use the law of cosines to prove that

$$[d(P_1, P_2)]^2 = r_1^2 + r_2^2 - 2r_1r_2 \cos (\theta_2 - \theta_1).$$

80 Prove that the graph of each polar equation is a circle, and find its center and radius.

(a) $r = a \sin \theta, a \neq 0$ (b) $r = b \cos \theta, b \neq 0$

(c) $r = a \sin \theta + b \cos \theta, a \neq 0$ and $b \neq 0$

Exer. 81–82: Refer to Exercise 81 in Section 5.6. Suppose that a radio station has two broadcasting towers located along a north-south line and that the towers are separated by a distance of $\frac{1}{2}\lambda$, where λ is the wavelength of the station's broadcasting signal. Then the intensity I of the signal in the direction θ can be expressed by the given equation, where I_0 is the maximum intensity of the signal.

(a) Plot I using polar coordinates with $I_0 = 5$ for $\theta \in [0, 2\pi]$.

(b) Determine the directions in which the radio signal has maximum and minimum intensity.

81 $I = \frac{1}{2}I_0[1 + \cos (\pi \sin \theta)]$

82 $I = \frac{1}{2}I_0[1 + \cos (\pi \sin 2\theta)]$

Exer. 83–84: Graph the polar equation for the indicated values of θ, and use the graph to determine symmetries.

83 $r = 2 \sin^2 \theta \tan^2 \theta; \quad -\pi/3 \leq \theta \leq \pi/3$

84 $r = \dfrac{4}{1 + \sin^2 \theta}; \quad 0 \leq \theta \leq 2\pi$

Exer. 85–86: Graph the polar equations on the same coordinate plane, and estimate the points of intersection of the graphs.

85 $r = 8 \cos 3\theta, \quad r = 4 - 2.5 \cos \theta$

86 $r = 2 \sin^2 \theta, \quad r = \frac{3}{4}(\theta + \cos^2 \theta)$

10.6

Polar Equations of Conics

The following theorem combines the definitions of parabola, ellipse, and hyperbola into a unified description of the conic sections. The constant e in the statement of the theorem is the **eccentricity** of the conic. The point F is a **focus** of the conic, and the line l is a **directrix.** Possible positions of F and l are illustrated in Figure 1.

Figure 1

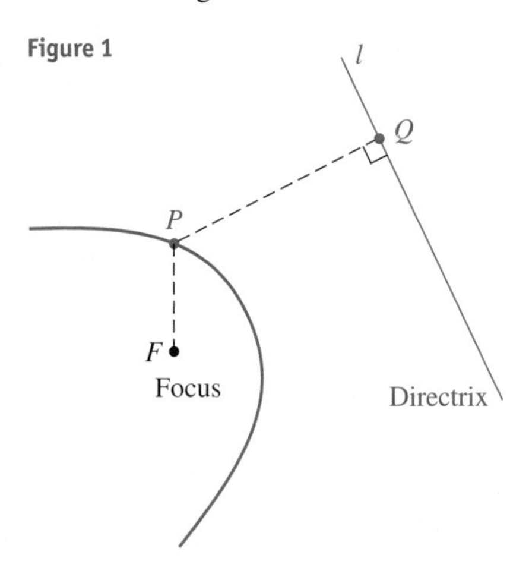

Theorem on Conics

Let F be a fixed point and l a fixed line in a plane. The set of all points P in the plane, such that the ratio $d(P, F)/d(P, Q)$ is a positive constant e with $d(P, Q)$ the distance from P to l, is a conic section. The conic is a parabola if $e = 1$, an ellipse if $0 < e < 1$, and a hyperbola if $e > 1$.

PROOF If $e = 1$, then $d(P, F) = d(P, Q)$, and, by definition, the resulting conic is a parabola with focus F and directrix l.

Suppose next that $0 < e < 1$. It is convenient to introduce a polar coordinate system in the plane with F as the pole and l perpendicular to the polar axis at the point $D(d, 0)$, with $d > 0$, as illustrated in Figure 2. If $P(r, \theta)$ is a point in the plane such that $d(P, F)/d(P, Q) = e < 1$, then P lies to the left of l. Let C be the projection of P on the polar axis. Since

$$d(P, F) = r \qquad \text{and} \qquad d(P, Q) = d - r\cos\theta,$$

it follows that P satisfies the condition in the theorem if and only if the following are true:

$$\begin{aligned} \frac{r}{d - r\cos\theta} &= e \\ r &= de - er\cos\theta \\ r(1 + e\cos\theta) &= de \\ r &= \frac{de}{1 + e\cos\theta} \end{aligned}$$

Figure 2

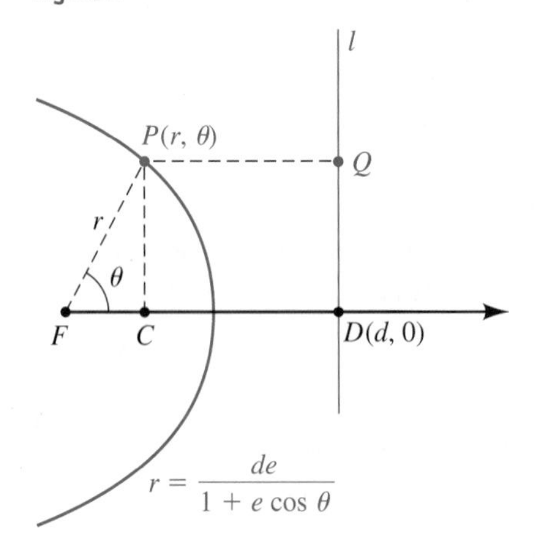

The same equations are obtained if $e = 1$; however, there is no point (r, θ) on the graph if $1 + \cos\theta = 0$.

An equation in x and y corresponding to $r = de - er\cos\theta$ is

$$\sqrt{x^2 + y^2} = de - ex.$$

Squaring both sides and rearranging terms leads to

$$(1 - e^2)x^2 + 2de^2x + y^2 = d^2e^2.$$

Completing the square and simplifying, we obtain

$$\left(x + \frac{de^2}{1 - e^2}\right)^2 + \frac{y^2}{1 - e^2} = \frac{d^2e^2}{(1 - e^2)^2}.$$

Finally, dividing both sides by $d^2e^2/(1 - e^2)^2$ gives us an equation of the form

$$\frac{(x - h)^2}{a^2} + \frac{y^2}{b^2} = 1,$$

with $h = -de^2/(1 - e^2)$. Consequently, the graph is an ellipse with center at the point $(h, 0)$ on the x-axis and with

$$a^2 = \frac{d^2e^2}{(1 - e^2)^2} \quad \text{and} \quad b^2 = \frac{d^2e^2}{1 - e^2}.$$

Since

$$c^2 = a^2 - b^2 = \frac{d^2e^4}{(1 - e^2)^2},$$

we obtain $c = de^2/(1 - e^2)$, and hence $|h| = c$. This proves that F is a focus of the ellipse. It also follows that $e = c/a$. A similar proof may be given for the case $e > 1$.

Figure 3

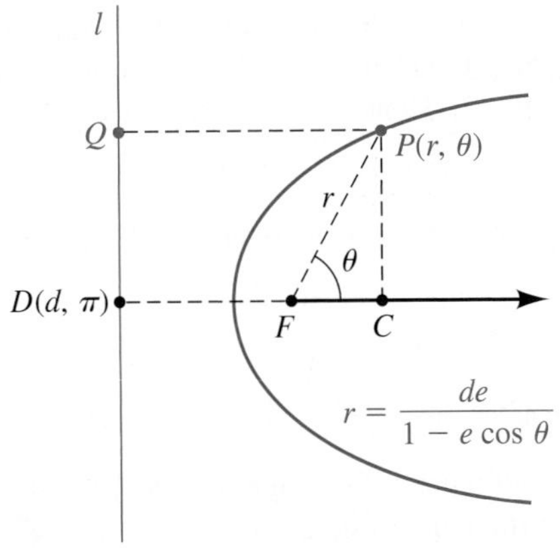

We also can show that every conic that is not degenerate may be described by means of the statement in the theorem on conics. This gives us a formulation of conic sections that is equivalent to the one used previously. Since the theorem includes all three types of conics, it is sometimes regarded as a definition for the conic sections.

If we had chosen the focus F to the *right* of the directrix, as illustrated in Figure 3 (with $d > 0$), then the equation $r = de/(1 - e\cos\theta)$ would have resulted. (Note the minus sign in place of the plus sign.) Other sign changes occur if d is allowed to be negative.

If we had taken l *parallel* to the polar axis through one of the points $(d, \pi/2)$ or $(d, 3\pi/2)$, as illustrated in Figure 4, then the corresponding equations would have contained $\sin\theta$ instead of $\cos\theta$.

Figure 4

(a)

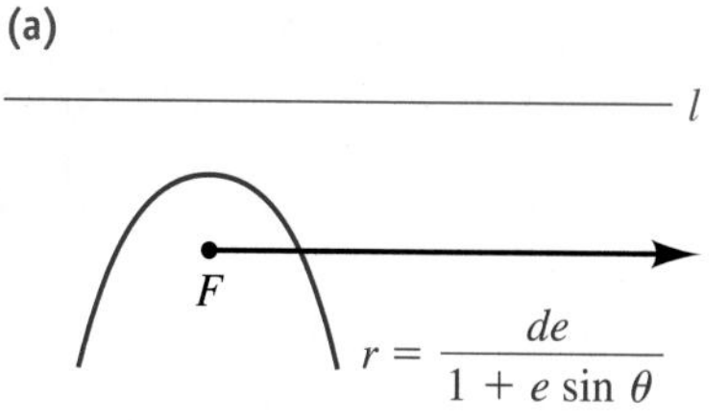

(b)

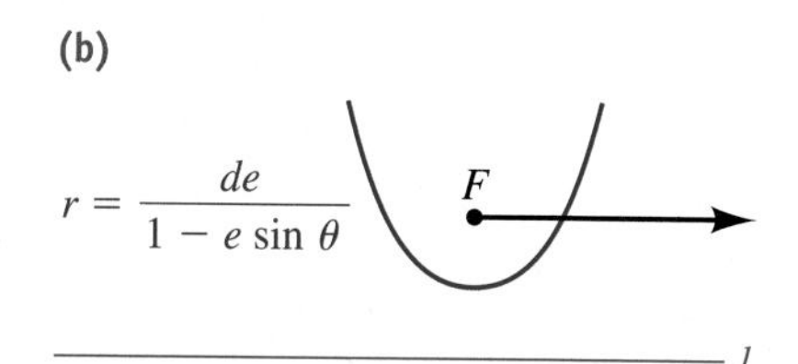

The following theorem summarizes our discussion.

Theorem on Polar Equations of Conics

A polar equation that has one of the four forms

$$r = \frac{de}{1 \pm e \cos \theta} \quad \text{or} \quad r = \frac{de}{1 \pm e \sin \theta}$$

is a conic section. The conic is a parabola if $e = 1$, an ellipse if $0 < e < 1$, or a hyperbola if $e > 1$.

EXAMPLE 1 Sketching the graph of a polar equation of an ellipse

Sketch the graph of the polar equation

$$r = \frac{10}{3 + 2 \cos \theta}.$$

SOLUTION We first divide the numerator and denominator of the fraction by 3 to obtain the constant term 1 in the denominator:

$$r = \frac{\frac{10}{3}}{1 + \frac{2}{3} \cos \theta}$$

This equation has one of the forms in the preceding theorem, with $e = \frac{2}{3}$. Thus, the graph is an ellipse with focus F at the pole and major axis along the polar axis. We find the endpoints of the major axis by letting $\theta = 0$ and $\theta = \pi$. This gives us the points $V(2, 0)$ and $V'(10, \pi)$. Hence,

$$2a = d(V', V) = 12, \quad \text{or} \quad a = 6.$$

Figure 5

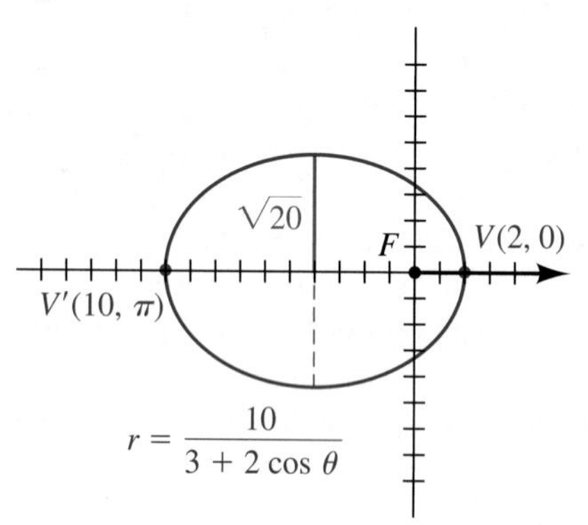

The center of the ellipse is the midpoint $(4, \pi)$ of the segment $V'V$. Using the fact that $e = c/a$, we obtain

$$c = ae = 6\left(\tfrac{2}{3}\right) = 4.$$

Hence, $$b^2 = a^2 - c^2 = 6^2 - 4^2 = 36 - 16 = 20.$$

Thus, $b = \sqrt{20}$. The graph is sketched in Figure 5. For reference, we have superimposed an xy-coordinate system on the polar system.

Tutorial available at www.thomsonedu.com/login

EXAMPLE 2 Sketching the graph of a polar equation of a hyperbola

Sketch the graph of the polar equation

$$r = \frac{10}{2 + 3\sin\theta}.$$

▶ SOLUTION To express the equation in the proper form, we divide the numerator and denominator of the fraction by 2:

$$r = \frac{5}{1 + \frac{3}{2}\sin\theta}$$

Thus, $e = \frac{3}{2}$, and, by the theorem on polar equations of conics, the graph is a hyperbola with a focus at the pole. The expression $\sin\theta$ tells us that the transverse axis of the hyperbola is perpendicular to the polar axis. To find the vertices, we let $\theta = \pi/2$ and $\theta = 3\pi/2$ in the given equation. This gives us the points $V(2, \pi/2)$ and $V'(-10, 3\pi/2)$. Hence,

$$2a = d(V, V') = 8, \text{ or } a = 4.$$

Figure 6

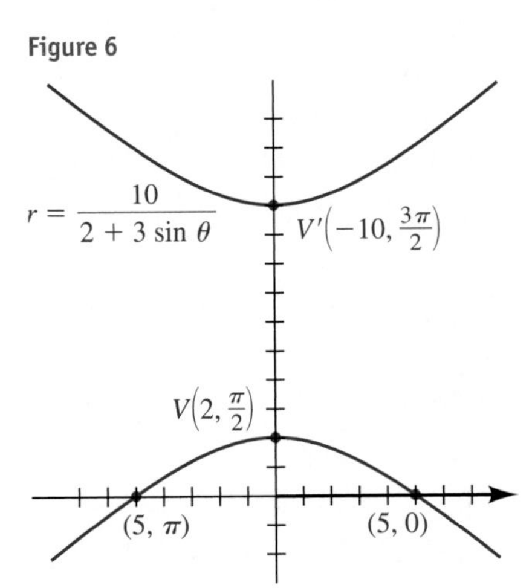

The points $(5, 0)$ and $(5, \pi)$ on the graph can be used to sketch the lower branch of the hyperbola. The upper branch is obtained by symmetry, as illustrated in Figure 6. If we desire more accuracy or additional information, we calculate

$$c = ae = 4\left(\tfrac{3}{2}\right) = 6$$

and $$b^2 = c^2 - a^2 = 6^2 - 4^2 = 36 - 16 = 20.$$

Asymptotes may then be constructed in the usual way.

EXAMPLE 3 Sketching the graph of a polar equation of a parabola

Sketch the graph of the polar equation

$$r = \frac{15}{4 - 4\cos\theta}.$$

▶ SOLUTION To obtain the proper form, we divide the numerator and denominator by 4:

$$r = \frac{\frac{15}{4}}{1 - \cos\theta}$$

Consequently, $e = 1$, and, by the theorem on polar equations of conics, the graph is a parabola with focus at the pole. We may obtain a sketch by plotting the points that correspond to the quadrantal angles indicated in the following table.

(continued)

▶ **Tutorial available at www.thomsonedu.com/login**

Figure 7

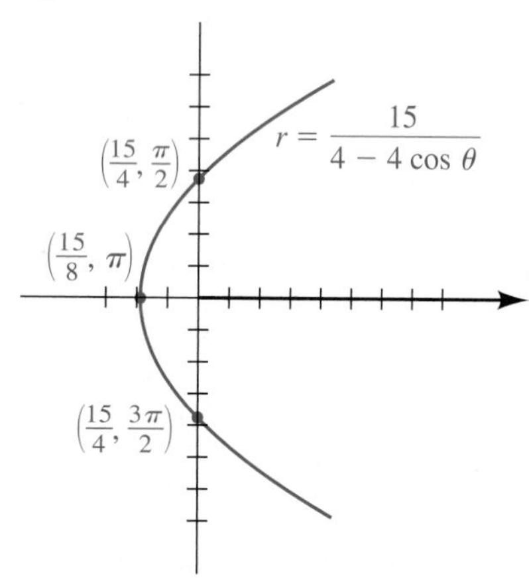

θ	0	$\frac{\pi}{2}$	π	$\frac{3\pi}{2}$
r	undefined	$\frac{15}{4}$	$\frac{15}{8}$	$\frac{15}{4}$

Note that there is no point on the graph corresponding to $\theta = 0$, since the denominator $1 - \cos\theta$ is 0 for that value. Plotting the three points and using the fact that the graph is a parabola with focus at the pole gives us the sketch in Figure 7.

If we desire only a rough sketch of a conic, then the technique employed in Example 3 is recommended. To use this method, we plot (if possible) points corresponding to $\theta = 0$, $\pi/2$, π, and $3\pi/2$. These points, together with the type of conic (obtained from the value of the eccentricity e), readily lead to the sketch.

EXAMPLE 4 **Expressing a polar equation of a conic in terms of x and y**

Find an equation in x and y that has the same graph as the polar equation

$$r = \frac{15}{4 - 4\cos\theta}.$$

SOLUTION

$$\begin{aligned} r(4 - 4\cos\theta) &= 15 && \text{multiply by the lcd} \\ 4r - 4r\cos\theta &= 15 && \text{distribute} \\ 4\left(\pm\sqrt{x^2 + y^2}\right) - 4x &= 15 && \text{substitute for } r \text{ and } r\cos\theta \\ 4\left(\pm\sqrt{x^2 + y^2}\right) &= 15 + 4x && \text{isolate the radical term} \\ 16(x^2 + y^2) &= 225 + 120x + 16x^2 && \text{square both sides} \\ 16y^2 &= 225 + 120x && \text{simplify} \end{aligned}$$

We may write the last equation as $x = \frac{16}{120}y^2 - \frac{225}{120}$ or, simplified, $x = \frac{2}{15}y^2 - \frac{15}{8}$. We recognize this equation as that of a parabola with vertex $V\left(-\frac{15}{8}, 0\right)$ and opening to the right. Its graph on an xy-coordinate system would be the same as the graph in Figure 7.

EXAMPLE 5 **Finding a polar equation of a conic satisfying prescribed conditions**

Find a polar equation of the conic with a focus at the pole, eccentricity $e = \frac{1}{2}$, and directrix $r = -3\sec\theta$.

SOLUTION The equation $r = -3\sec\theta$ of the directrix may be written $r\cos\theta = -3$, which is equivalent to $x = -3$ in a rectangular coordinate

Tutorial available at www.thomsonedu.com/login

system. This gives us the situation illustrated in Figure 3, with $d = 3$. Hence, a polar equation has the form

$$r = \frac{de}{1 - e\cos\theta}.$$

We now substitute $d = 3$ and $e = \frac{1}{2}$:

$$r = \frac{3(\frac{1}{2})}{1 - \frac{1}{2}\cos\theta} \quad \text{or, equivalently,} \quad r = \frac{3}{2 - \cos\theta}$$

10.6 *Exercises*

Exer. 1–12: Find the eccentricity, and classify the conic. Sketch the graph, and label the vertices.

▶ 1 $r = \dfrac{12}{6 + 2\sin\theta}$

2 $r = \dfrac{12}{6 - 2\sin\theta}$

3 $r = \dfrac{12}{2 - 6\cos\theta}$

4 $r = \dfrac{12}{2 + 6\cos\theta}$

5 $r = \dfrac{3}{2 + 2\cos\theta}$

6 $r = \dfrac{3}{2 - 2\sin\theta}$

7 $r = \dfrac{4}{\cos\theta - 2}$

8 $r = \dfrac{4\sec\theta}{2\sec\theta - 1}$

9 $r = \dfrac{6\csc\theta}{2\csc\theta + 3}$

10 $r = \dfrac{8\csc\theta}{2\csc\theta - 5}$

11 $r = \dfrac{4\csc\theta}{1 + \csc\theta}$

12 $r = \csc\theta\,(\csc\theta - \cot\theta)$

Exer. 13–24: Find equations in x and y for the polar equations in Exercises 1–12.

Exer. 25–32: Find a polar equation of the conic with focus at the pole that has the given eccentricity and equation of directrix.

25 $e = \frac{1}{3}$, $r = 2\sec\theta$

26 $e = 1$, $r\cos\theta = 5$

▶ 27 $e = \frac{4}{3}$, $r\cos\theta = -3$

28 $e = 3$, $r = -4\sec\theta$

29 $e = 1$, $r\sin\theta = -2$

30 $e = 4$, $r = -3\csc\theta$

31 $e = \frac{2}{5}$, $r = 4\csc\theta$

32 $e = \frac{3}{4}$, $r\sin\theta = 5$

Exer. 33–34: Find a polar equation of the parabola with focus at the pole and the given vertex.

▶ 33 $V\left(4, \dfrac{\pi}{2}\right)$

34 $V(5, 0)$

Exer. 35–36: An ellipse has a focus at the pole with the given center C and vertex V. Find (a) the eccentricity and (b) a polar equation for the ellipse.

▶ 35 $C\left(3, \dfrac{\pi}{2}\right)$, $V\left(1, \dfrac{3\pi}{2}\right)$

36 $C(2, \pi)$, $V(1, 0)$

37 **Kepler's first law** Kepler's first law asserts that planets travel in elliptical orbits with the sun at one focus. To find an equation of an orbit, place the pole O at the center of the sun and the polar axis along the major axis of the ellipse (see the figure).

(a) Show that an equation of the orbit is

$$r = \frac{(1 - e^2)a}{1 - e\cos\theta},$$

where e is the eccentricity and $2a$ is the length of the major axis.

▶ Tutorial available at www.thomsonedu.com/login

(b) The perihelion distance r_{per} and aphelion distance r_{aph} are defined as the minimum and maximum distances, respectively, of a planet from the sun. Show that

$$r_{per} = a(1 - e) \quad \text{and} \quad r_{aph} = a(1 + e).$$

Exercise 37

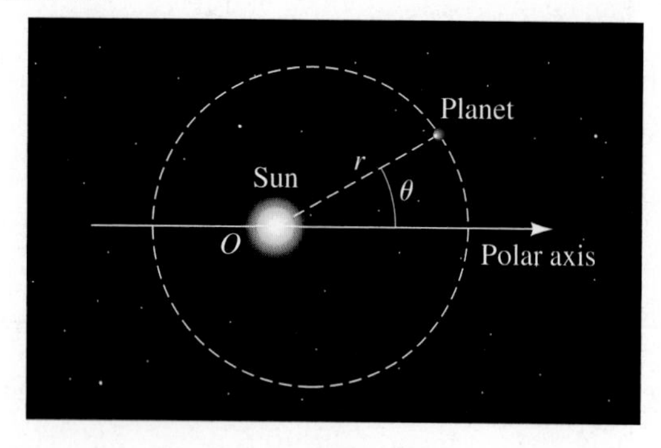

38 Kepler's first law Refer to Exercise 37. The planet Pluto travels in an elliptical orbit of eccentricity 0.249. If the perihelion distance is 29.62 AU, find a polar equation for the orbit and estimate the aphelion distance.

Exer. 39–42: Polar equations of conics can be used to describe the motion of comets. These paths can be graphed using the polar equation

$$r = \frac{r_{per}(1 + e)}{1 - e \cos \theta},$$

where e is the eccentricity of the conic and r_{per} is the perihelion distance measured in AU.

(a) For each comet, determine whether its trajectory is elliptical, parabolic, or hyperbolic.

(b) The orbit of Saturn has $r_{per} = 9.006$ and $e = 0.056$. Graph both the motion of the comet and the orbit of Saturn in the specified viewing rectangle.

39 Halley's Comet $r_{per} = 0.5871$, $e = 0.9673$, $[-36, 36, 3]$ by $[-24, 24, 3]$

40 Encke's Comet $r_{per} = 0.3317$, $e = 0.8499$, $[-18, 18, 3]$ by $[-12, 12, 3]$

41 Comet 1959 III $r_{per} = 1.251$, $e = 1.003$, $[-18, 18, 3]$ by $[-12, 12, 3]$

42 Comet 1973.99 $r_{per} = 0.142$, $e = 1.000$, $[-18, 18, 3]$ by $[-12, 12, 3]$

43 Earth's orbit The closest Earth gets to the sun is about 91,405,950 miles, and the farthest Earth gets from the sun is about 94,505,420 miles. Referring to the formulas in Exercise 37, find formulas for a and e in terms of r_{per} and r_{aph}.

CHAPTER 10 REVIEW EXERCISES

▶ Online support materials can be found at www.thomsonedu.com/login

Exer. 1–16: Find the vertices and foci of the conic, and sketch its graph.

1 $y^2 = 64x$

2 $y = 8x^2 + 32x + 33$

3 $9y^2 = 144 - 16x^2$

4 $9y^2 = 144 + 16x^2$

5 $x^2 - y^2 - 4 = 0$

6 $25x^2 + 36y^2 = 1$

7 $25y = 100 - x^2$

8 $3x^2 + 4y^2 - 18x + 8y + 19 = 0$

9 $x^2 - 9y^2 + 8x + 90y - 210 = 0$

10 $x = 2y^2 + 8y + 3$

11 $4x^2 + 9y^2 + 24x - 36y + 36 = 0$

12 $4x^2 - y^2 - 40x - 8y + 88 = 0$

13 $y^2 - 8x + 8y + 32 = 0$

14 $4x^2 + y^2 - 24x + 4y + 36 = 0$

15 $x^2 - 9y^2 + 8x + 7 = 0$

16 $y^2 - 2x^2 + 6y + 8x - 3 = 0$

Exer. 17–18: Find the standard equation of a parabola with a vertical axis that satisfies the given conditions.

17 x-intercepts -10 and -4, y-intercept 80

18 x-intercepts -11 and 3, passing through (2, 39)

Exer. 19–28: Find an equation for the conic that satisfies the given conditions.

19 Hyperbola, with vertices $V(0, \pm 7)$ and endpoints of conjugate axis $(\pm 3, 0)$

20 Parabola, with focus $F(-4, 0)$ and directrix $x = 4$

21 Parabola, with focus $F(0, -10)$ and directrix $y = 10$

22 Parabola, with vertex at the origin, symmetric to the x-axis, and passing through the point $(5, -1)$

23 Ellipse, with vertices $V(0, \pm 10)$ and foci $F(0, \pm 5)$

24 Hyperbola, with foci $F(\pm 10, 0)$ and vertices $V(\pm 5, 0)$

25 Hyperbola, with vertices $V(0, \pm 6)$ and asymptotes $y = \pm 9x$

26 Ellipse, with foci $F(\pm 2, 0)$ and passing through the point $(2, \sqrt{2})$

27 Ellipse, with eccentricity $\frac{2}{3}$ and endpoints of minor axis $(\pm 5, 0)$

28 Ellipse, with eccentricity $\frac{3}{4}$ and foci $F(\pm 12, 0)$

29 (a) Determine A so that the point $(2, -3)$ is on the conic $Ax^2 + 2y^2 = 4$.

(b) Is the conic an ellipse or a hyperbola?

30 If a square with sides parallel to the coordinate axes is inscribed in the ellipse $(x^2/a^2) + (y^2/b^2) = 1$, express the area A of the square in terms of a and b.

31 Find the standard equation of the circle that has center at the focus of the parabola $y = \frac{1}{8}x^2$ and passes through the origin.

32 Focal length and angular velocity A cylindrical container, partially filled with mercury, is rotated about its axis so that the angular speed of each cross section is ω radians/second. From physics, the function f, whose graph generates the inside surface of the mercury (see the figure), is given by

$$f(x) = \frac{1}{64}\omega^2 x^2 + k,$$

where k is a constant. Determine the angular speed ω that will result in a focal length of 2 feet.

Exercise 32

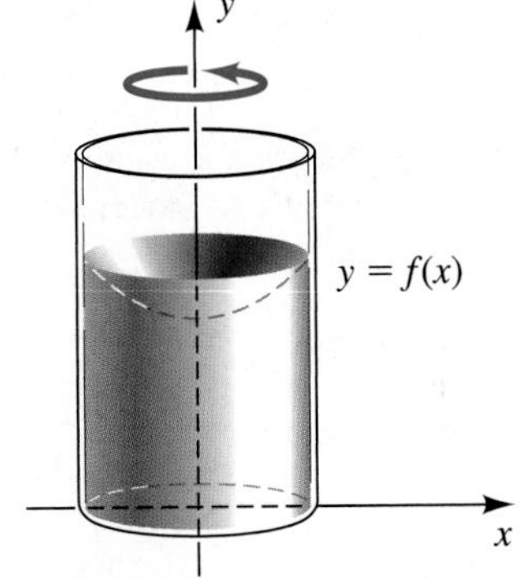

33 An ellipse has a vertex at the origin and foci $F_1(p, 0)$ and $F_2(p + 2c, 0)$, as shown in the figure. If the focus at F_1 is fixed and (x, y) is on the ellipse, show that y^2 approaches $4px$ as $c \to \infty$. (Thus, as $c \to \infty$, the ellipse takes on the shape of a parabola.)

Exercise 33

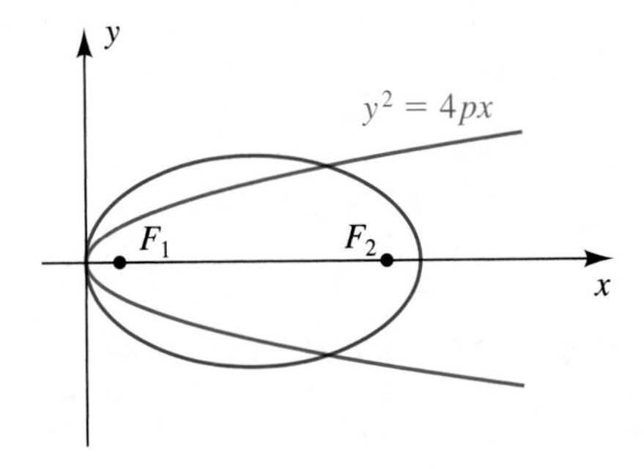

34 Alpha particles In 1911, the physicist Ernest Rutherford (1871–1937) discovered that if alpha particles are shot toward the nucleus of an atom, they are eventually repulsed away from the nucleus along hyperbolic paths. The figure illustrates the path of a particle that starts toward the origin

along the line $y = \frac{1}{2}x$ and comes within 3 units of the nucleus. Find an equation of the path.

Exercise 34

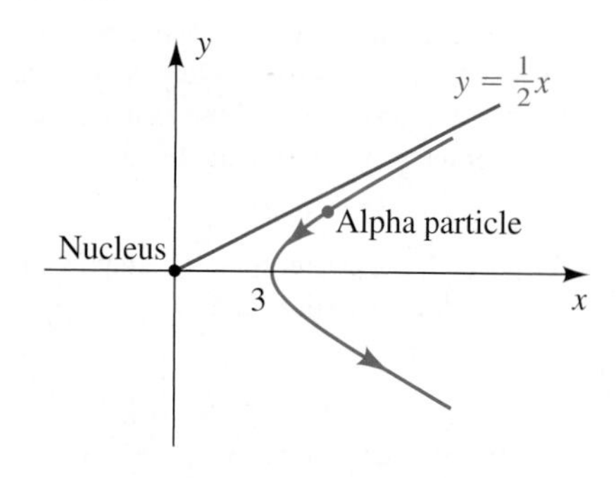

Exer. 35–39: Find an equation in x and y whose graph contains the points on the curve C. Sketch the graph of C, and indicate the orientation.

35 $x = 3 + 4t, \quad y = t - 1; \quad -2 \leq t \leq 2$

36 $x = \sqrt{-t}, \quad y = t^2 - 4; \quad t \leq 0$

37 $x = \cos^2 t - 2, \quad y = \sin t + 1; \quad 0 \leq t \leq 2\pi$

38 $x = \sqrt{t}, \quad y = 2^{-t}; \quad t \geq 0$

39 $x = \dfrac{1}{t} + 1, \quad y = \dfrac{2}{t} - t; \quad 0 < t \leq 4$

40 Curves C_1, C_2, C_3, and C_4 are given parametrically for t in $\mathbb{R}$. Sketch their graphs, and discuss their similarities and differences.

C_1: $x = t, \quad y = \sqrt{16 - t^2}$
C_2: $x = -\sqrt{16 - t}, \quad y = -\sqrt{t}$
C_3: $x = 4 \cos t, \quad y = 4 \sin t$
C_4: $x = e^t, \quad y = -\sqrt{16 - e^{2t}}$

41 Refer to the equations in (1) of Example 6 in Section 10.4. Find the range and maximum altitude for $s = 1024$, $\alpha = 30°$, and $h = 5120$.

42 List two polar coordinate points that represent the same point as $(2, \pi/4)$.

43 Change $(5, 7\pi/4)$ to rectangular coordinates.

44 Change $(2\sqrt{3}, -2)$ to polar coordinates with $r > 0$ and $0 \leq \theta < 2\pi$.

Exer. 45–48: Find a polar equation that has the same graph as the equation in x and y.

45 $y^2 = 4x$

46 $x^2 + y^2 - 3x + 4y = 0$

47 $2x - 3y = 8$

48 $x^2 + y^2 = 2xy$

Exer. 49–54: Find an equation in x and y that has the same graph as the polar equation.

49 $r^2 = \tan \theta$

50 $r = 2 \cos \theta + 3 \sin \theta$

51 $r^2 = 4 \sin 2\theta$

52 $\theta = \sqrt{3}$

53 $r = 5 \sec \theta + 3r \sec \theta$

54 $r^2 \sin \theta = 6 \csc \theta + r \cot \theta$

Exer. 55–66: Sketch the graph of the polar equation.

55 $r = -4 \sin \theta$

56 $r = 8 \sec \theta$

57 $r = 3 \sin 5\theta$

58 $r = 6 - 3 \cos \theta$

59 $r = 3 - 3 \sin \theta$

60 $r = 2 + 4 \cos \theta$

61 $r^2 = 9 \sin 2\theta$

62 $2r = \theta$

63 $r = \dfrac{8}{1 - 3 \sin \theta}$

64 $r = 6 - r \cos \theta$

65 $r = \dfrac{6}{3 + 2 \cos \theta}$

66 $r = \dfrac{-6 \csc \theta}{1 - 2 \csc \theta}$

CHAPTER 10 DISCUSSION EXERCISES

1 On a parabola, the line segment through the focus, perpendicular to the axis, and intercepted by the parabola is called the *focal chord* or *latus rectum.* The length of the focal chord is called the *focal width.* Find a formula for the focal width w in terms of the focal length $|p|$.

2 On the graph of a hyperbola with center at the origin O, draw a circle with center at the origin and radius $r = d(O, F)$, where F denotes a focus of the hyperbola. What relationship do you observe?

3 A point $P(x, y)$ is on an ellipse if and only if

$$d(P, F) + d(P, F') = 2a.$$

If $b^2 = a^2 - c^2$, derive the general equation of an ellipse—that is,

$$\frac{x^2}{a^2} + \frac{y^2}{b^2} = 1.$$

4 A point $P(x, y)$ is on a hyperbola if and only if

$$|d(P, F) - d(P, F')| = 2a.$$

If $c^2 = a^2 + b^2$, derive the general equation of a hyperbola—that is,

$$\frac{x^2}{a^2} - \frac{y^2}{b^2} = 1.$$

5 A point $P(x, y)$ is the same distance from $(4, 0)$ as it is from the circle $x^2 + y^2 = 4$, as illustrated in the figure. Show that the collection of all such points forms a branch of a hyperbola, and sketch its graph.

Exercise 5

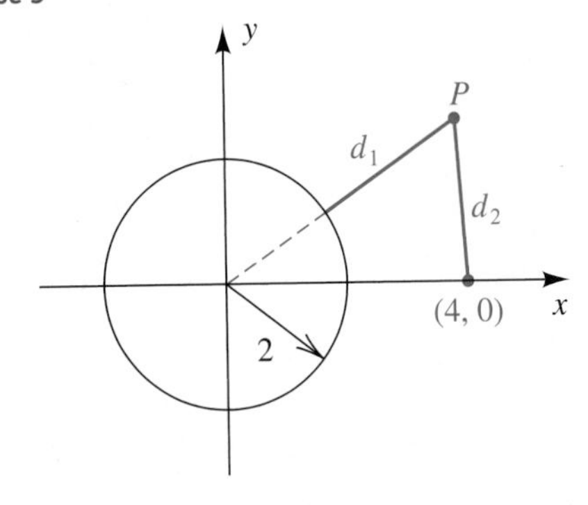

6 **Design of a telescope** Refer to Exercise 64 in Section 10.3. Suppose the upper branch of the hyperbola (shown) has equation $y = \frac{a}{b}\sqrt{x^2 + b^2}$ and an equation of the parabola is $y = dx^2$. Find d in terms of a and b.

Exercise 6

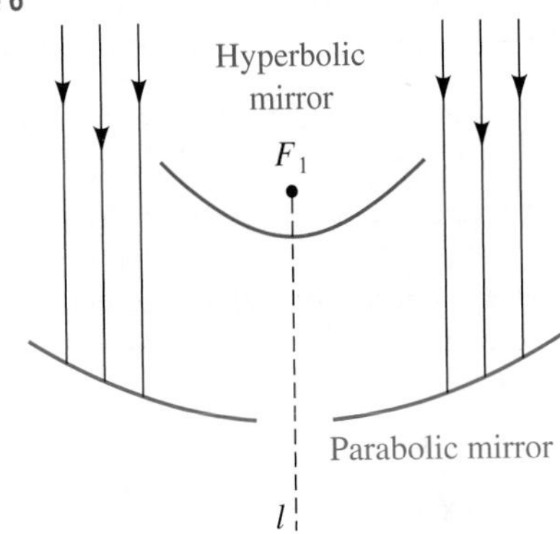

7 **Maximizing a projectile's range** As in Example 6 in Section 10.4, suppose a projectile is to be fired at a speed of 1024 ft/sec from a height of 2304 feet. Approximate the angle that maximizes the range.

8 **Generalizations for a projectile's path** If $h = 0$, the equations in (1) of Example 6 in Section 10.4 become

$$x(t) = (s \cos \alpha)t, \quad y(t) = -\tfrac{1}{2}gt^2 + (s \sin \alpha)t; \quad t \geq 0.$$

Show that each statement is true.

(a) The projectile strikes the ground when

$$t = \frac{2s \sin \alpha}{g}.$$

(b) The range r of the projectile is

$$r = \frac{s^2 \sin 2\alpha}{g}.$$

(c) The angle that maximizes the range r is 45°.

(d) The path of the projectile in rectangular coordinates is

$$y = -\frac{g}{2s^2 \cos^2 \alpha}x^2 + (\tan \alpha)x.$$

(continued)

(e) The time at which the maximum height is reached is

$$t = \frac{s \sin \alpha}{g}.$$

(f) The maximum height reached is

$$y = \frac{s^2 \sin^2 \alpha}{2g}.$$

9 **Investigating a Lissajous figure** Find an equation in x and y for the curve from Example 7 in Section 10.4 given by

$$x = \sin 2t, \quad y = \cos t; \quad 0 \leq t \leq 2\pi.$$

10 Sketch the graphs of the equations $r = f(\theta) = 2 + 4 \cos \theta$, $r = f(\theta - \alpha)$, and $r = f(\theta + \alpha)$ for $\alpha = \pi/4$. Try as many values of α as necessary to generalize results concerning the graphs of $r = f(\theta - \alpha)$ and $r = f(\theta + \alpha)$, where $\alpha > 0$.

11 **Generalized roses** Examine the graph of $r = \sin n\theta$ for odd values of n and even values of n. Derive an expression for the *leaf angle* (the number of degrees between consecutive pole values). What other generalizations do you observe? How do the graphs change if sin is replaced by cos?

12 Figure 17 of Section 10.5 shows the circles $r = 4 \sin \theta$ and $r = 4 \cos \theta$. Solve this system of equations for (r, θ) solutions. Now find equations in x and y that have the same graphs as the polar equations. Solve this system for (x, y) solutions, convert them to (r, θ) solutions, and explain why your answer to the first system did not reveal the solution at the pole.

Appendixes

APPENDIX I
Common Graphs and Their Equations

(Graphs of conics appear on the back endpaper of this text.)

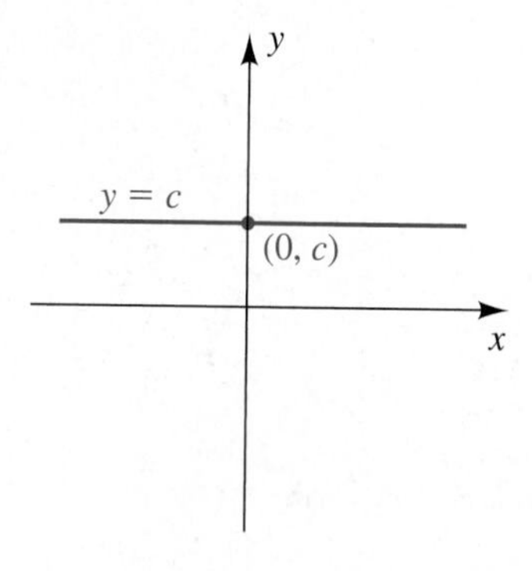

Horizontal line; constant function

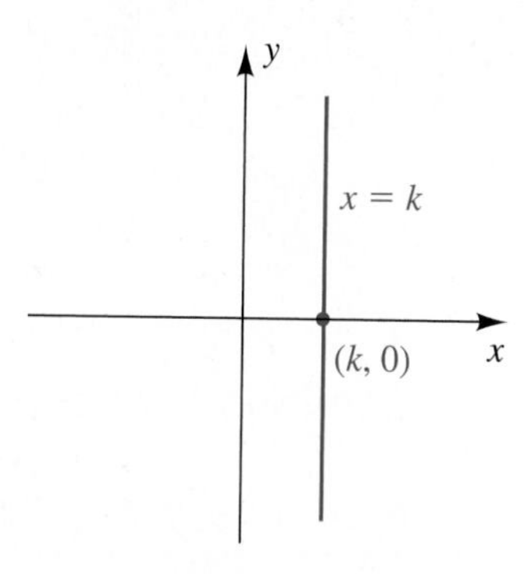

Vertical line

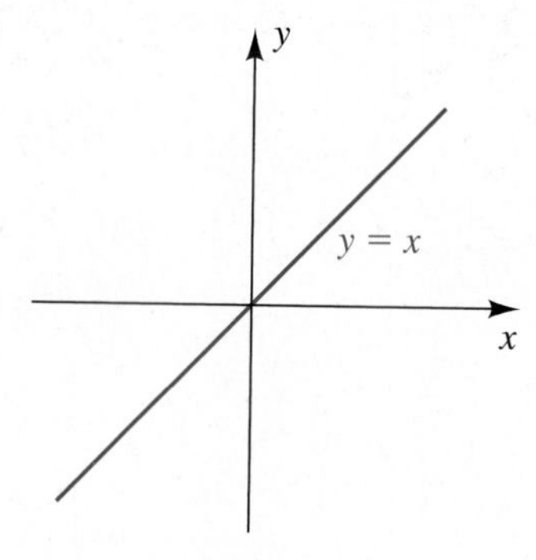

Identity function

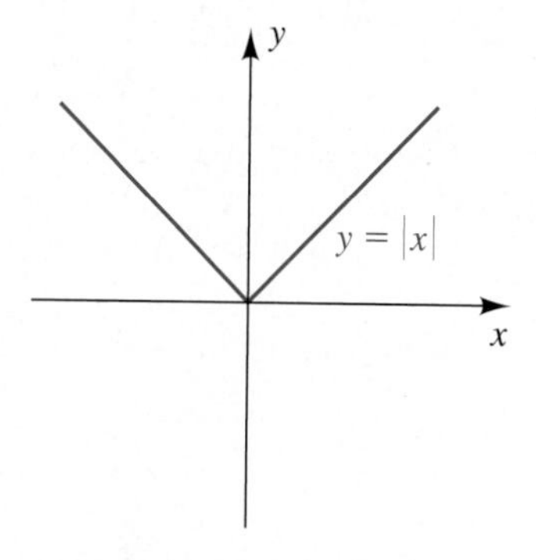

Absolute value function

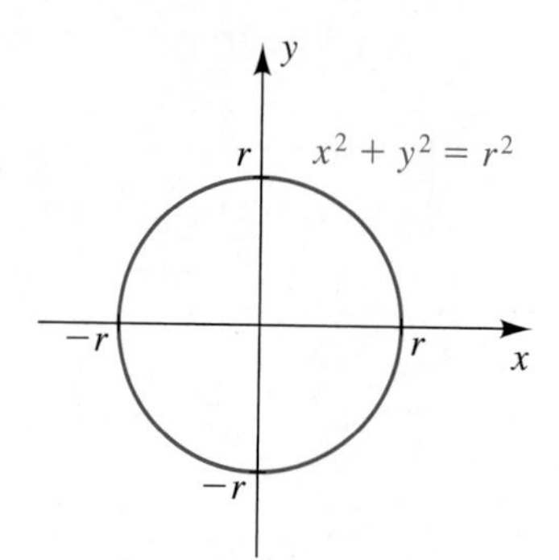

Circle with center (0, 0) and radius r

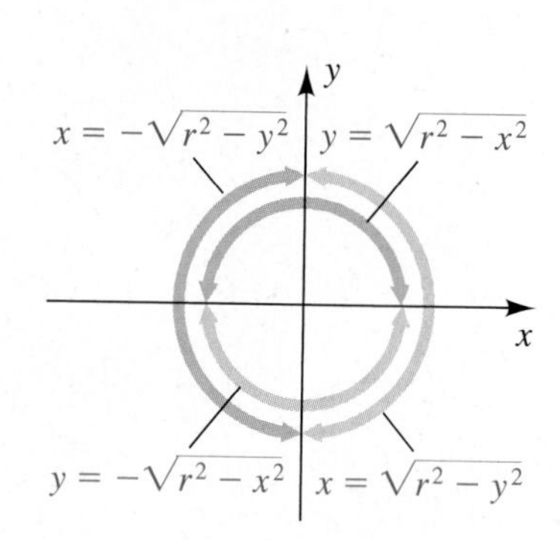

Semicircles

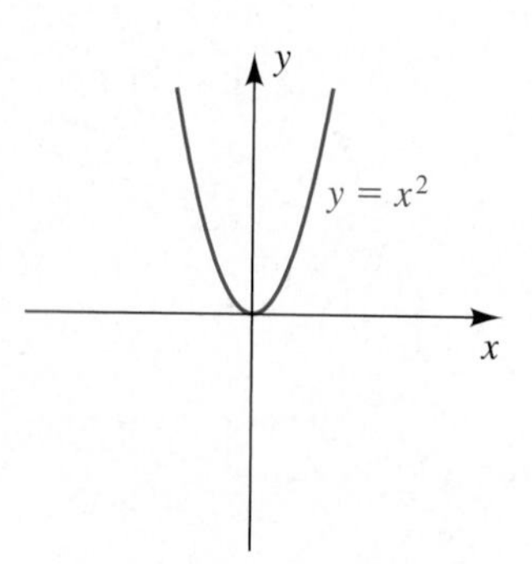

Parabola with vertical axis; squaring function

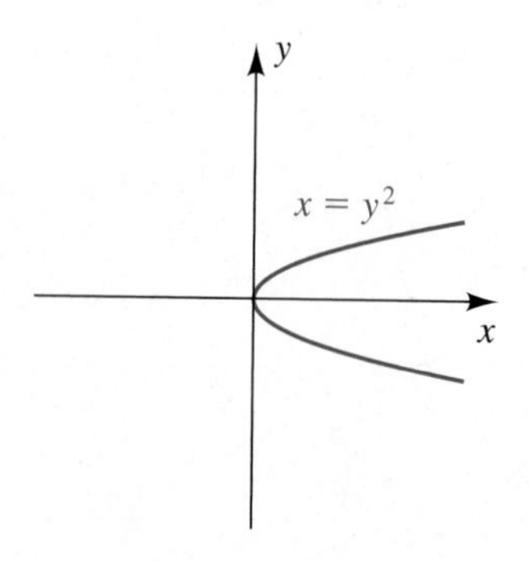

Parabola with horizontal axis

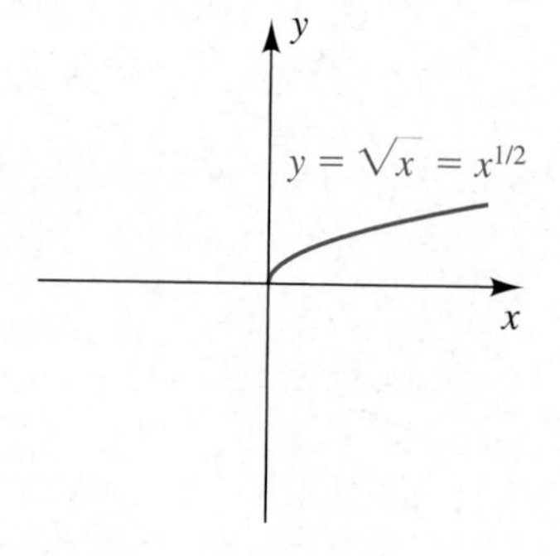

Square root function

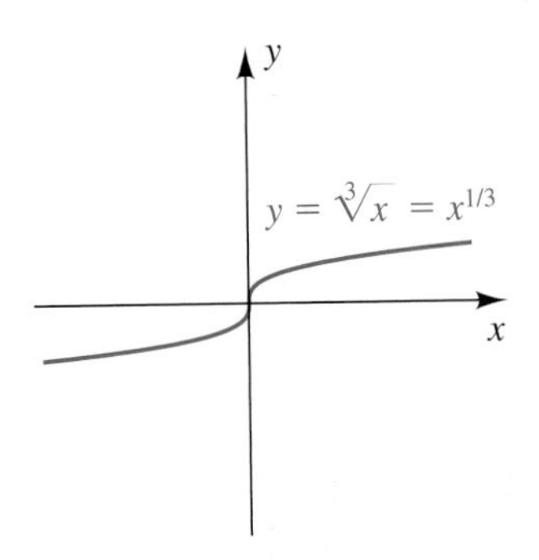

Cube root function

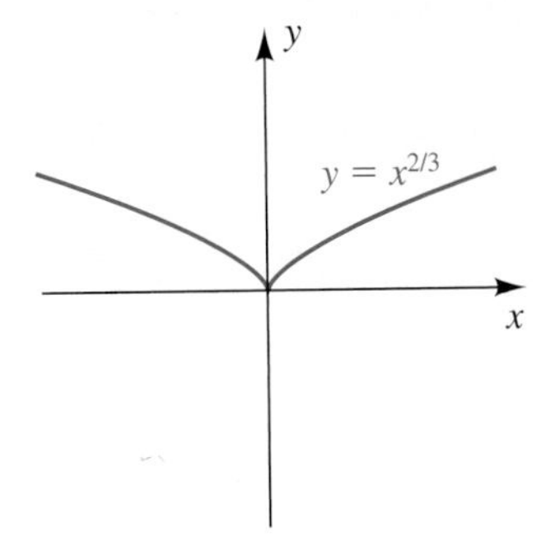

A graph with a cusp at the origin

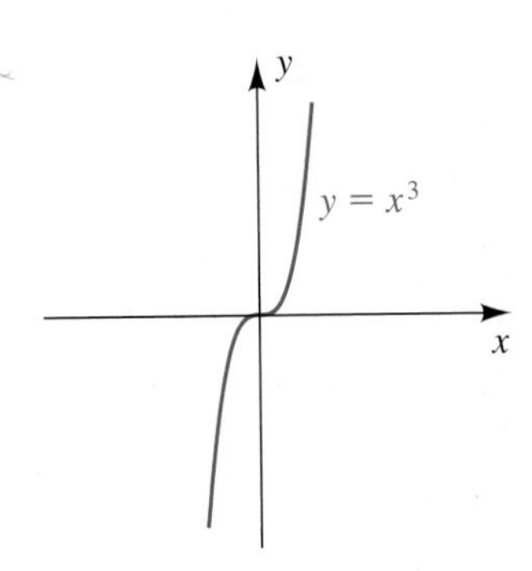

Cubing function

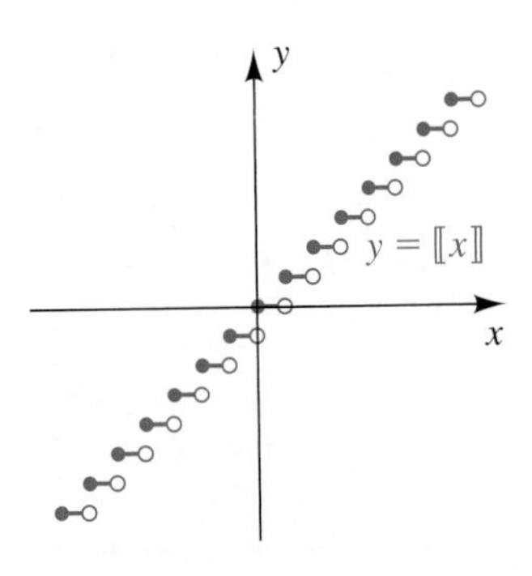

Greatest integer function

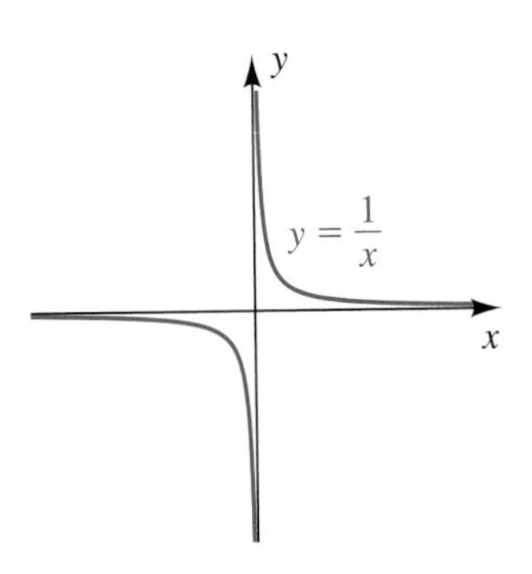

Reciprocal function

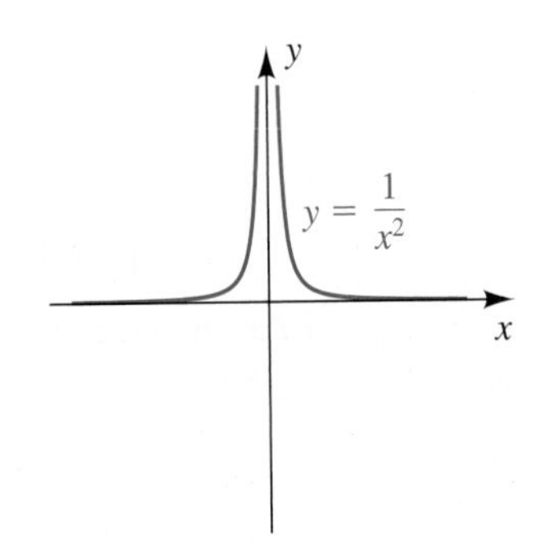

A rational function

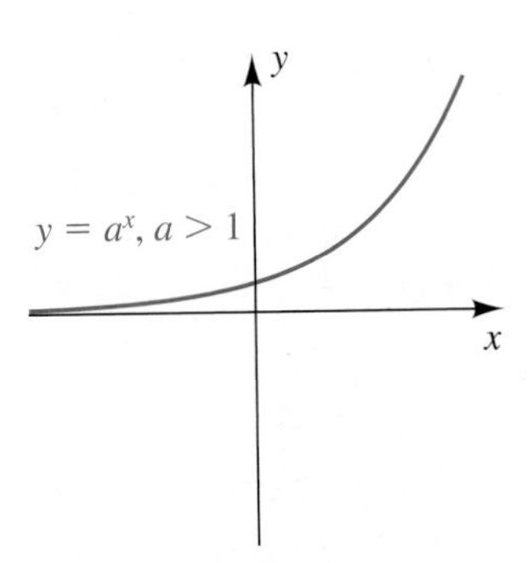

Exponential growth function (includes natural exponential function)

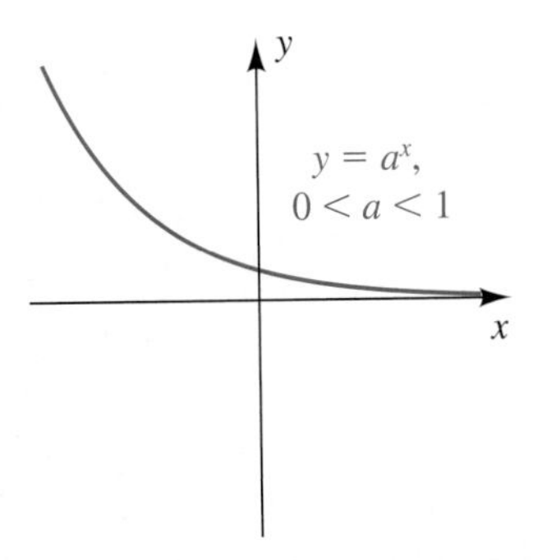

Exponential decay function

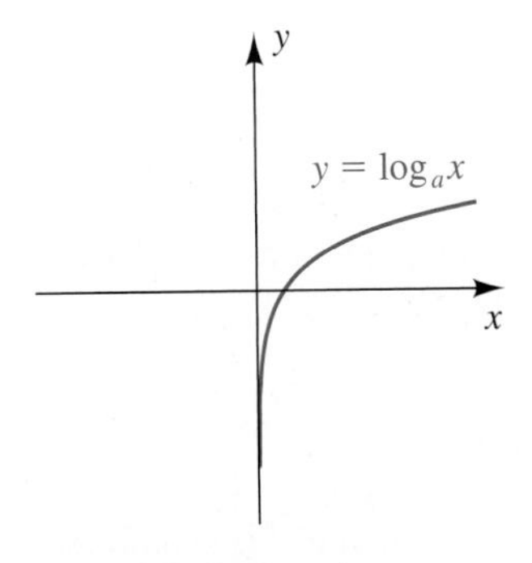

Logarithmic function (includes common and natural logarithmic functions)

APPENDIX II
A Summary of Graph Transformations

The graph of $y = f(x)$ is shown in black in each figure. The domain of f is $[-1, 3]$ and the range of f is $[-4, 3]$.

$$y = g(x) = f(x) + 3$$

The graph of f is shifted vertically upward 3 units.
Domain of g: $[-1, 3]$ Range of g: $[-1, 6]$

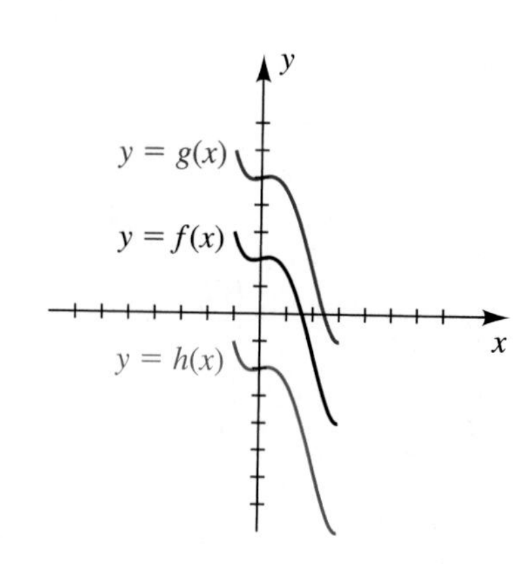

$$y = h(x) = f(x) - 4$$

The graph of f is shifted vertically downward 4 units.
Domain of h: $[-1, 3]$ Range of h: $[-8, -1]$

$$y = g(x) = f(x - 3)$$

The graph of f is shifted horizontally to the right 3 units.
Domain of g: $[2, 6]$ Range of g: $[-4, 3]$

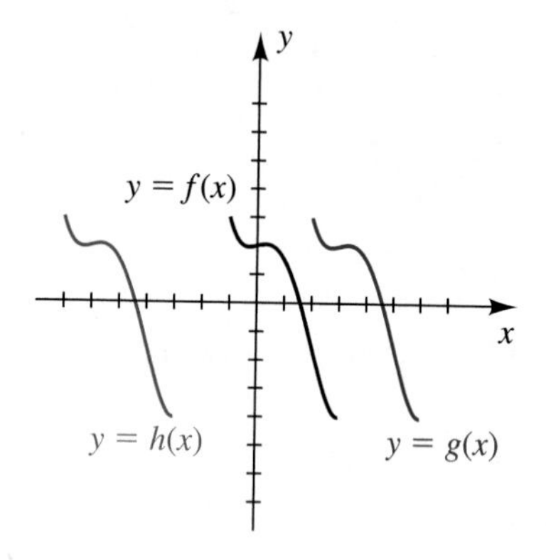

$$y = h(x) = f(x + 6)$$

The graph of f is shifted horizontally to the left 6 units.
Domain of h: $[-7, -3]$ Range of h: $[-4, 3]$

$$y = g(x) = 2f(x) \quad [2 > 1]$$

The graph of f is stretched vertically by a factor of 2.
Domain of g: $[-1, 3]$ Range of g: $[-8, 6]$

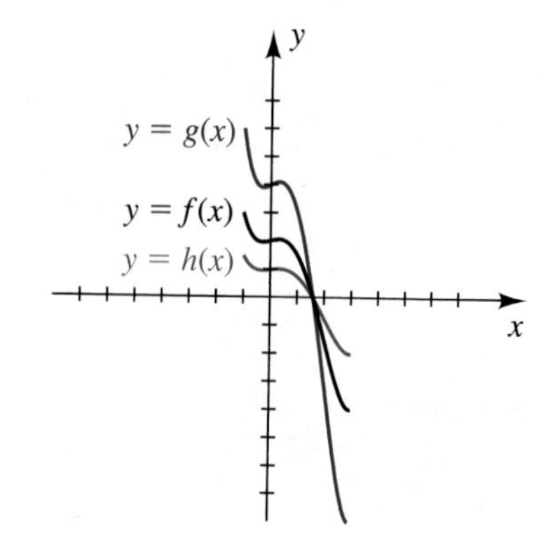

$$y = h(x) = \frac{1}{2}f(x) \quad \left[\frac{1}{2} < 1\right]$$

The graph of f is compressed vertically by a factor of 2.
Domain of h: $[-1, 3]$ Range of h: $\left[-2, \frac{3}{2}\right]$

$$y = g(x) = f(2x) \qquad [2 > 1]$$

The graph of f is compressed horizontally by a factor of 2.
Domain of g: $\left[-\frac{1}{2}, \frac{3}{2}\right]$ Range of g: $[-4, 3]$

$$y = h(x) = f\left(\tfrac{1}{2}x\right) \qquad \left[\tfrac{1}{2} < 1\right]$$

The graph of f is stretched horizontally by a factor of 2.
Domain of h: $[-2, 6]$ Range of h: $[-4, 3]$

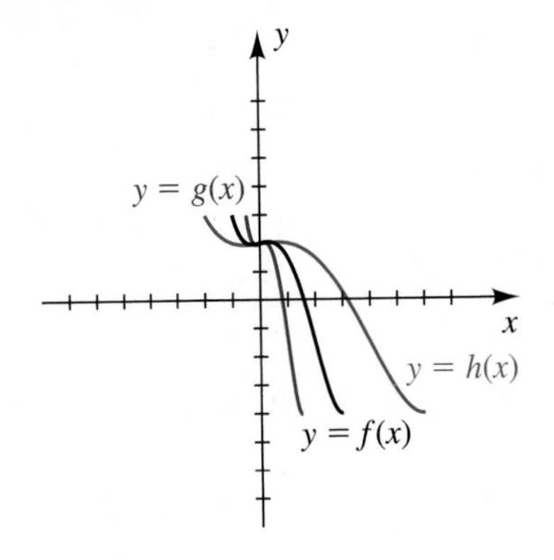

$$y = g(x) = -f(x)$$

The graph of f is reflected through the x-axis.
Domain of g: $[-1, 3]$ Range of g: $[-3, 4]$

$$y = h(x) = f(-x)$$

The graph of f is reflected through the y-axis.
Domain of h: $[-3, 1]$ Range of h: $[-4, 3]$

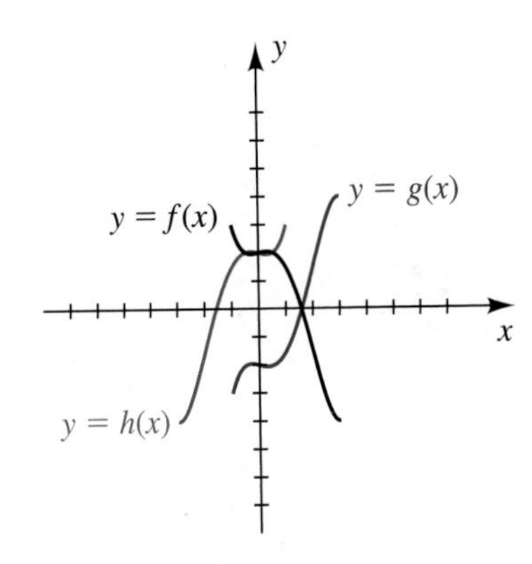

$$y = g(x) = |f(x)|$$

Reflect points on f with negative y-values through the x-axis.
Domain of g: $[-1, 3]$ Range of g: $[0, 4]$

$$y = h(x) = f(|x|)$$

Reflect points on f with positive x-values through the y-axis.
Domain of h: $[-3, 3]$ Range of h: $[-4, 3]$ at most.
In this case, the range is a subset of $[-4, 3]$.

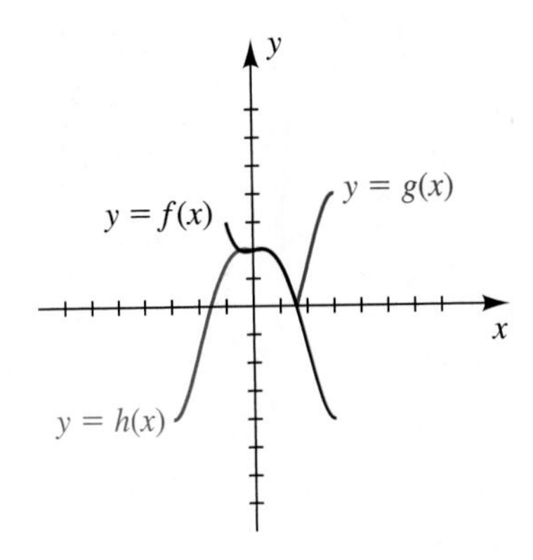

APPENDIX III

Graphs of Trigonometric Functions and Their Inverses

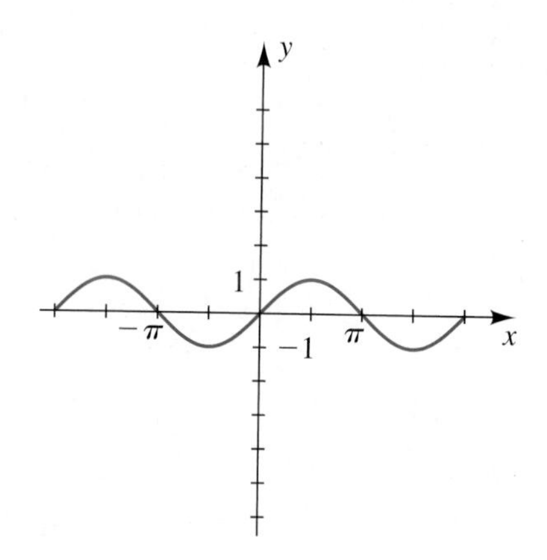

$y = \sin x$

Domain: $\mathbb{R}$

Range: $[-1, 1]$

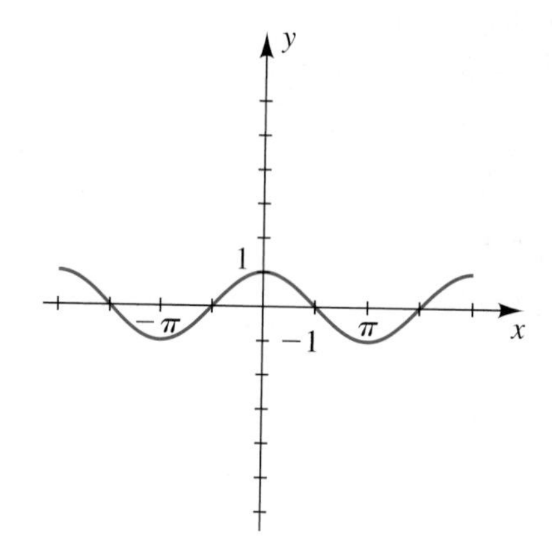

$y = \cos x$

Domain: $\mathbb{R}$

Range: $[-1, 1]$

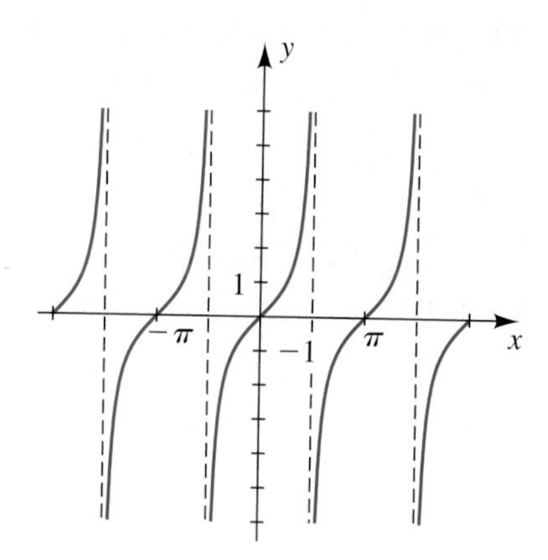

$y = \tan x$

Domain: $x \neq \dfrac{\pi}{2} + \pi n$

Range: $\mathbb{R}$

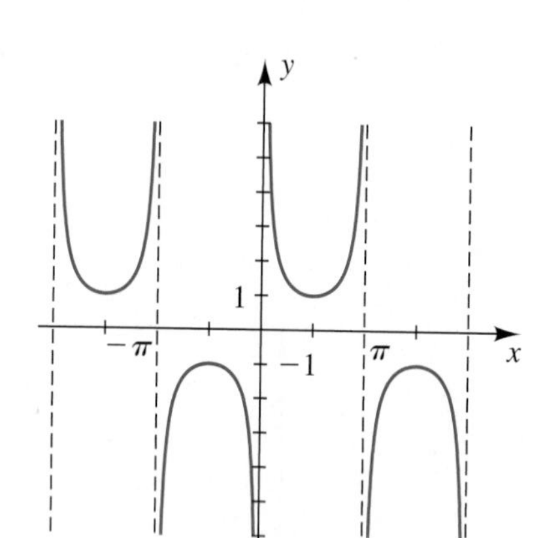

$y = \csc x$

Domain: $x \neq \pi n$

Range: $(-\infty, -1] \cup [1, \infty)$

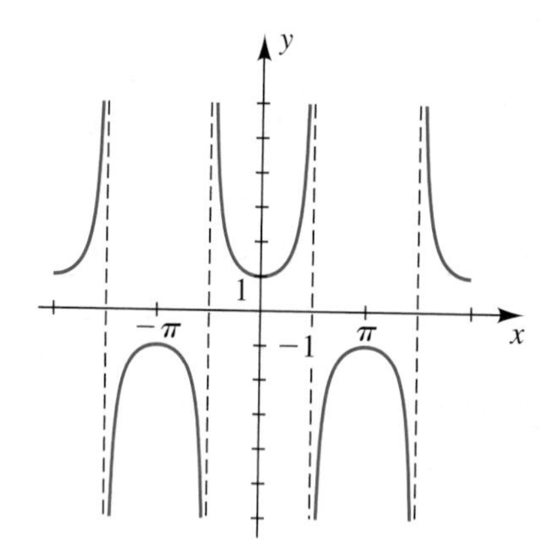

$y = \sec x$

Domain: $x \neq \dfrac{\pi}{2} + \pi n$

Range: $(-\infty, -1] \cup [1, \infty)$

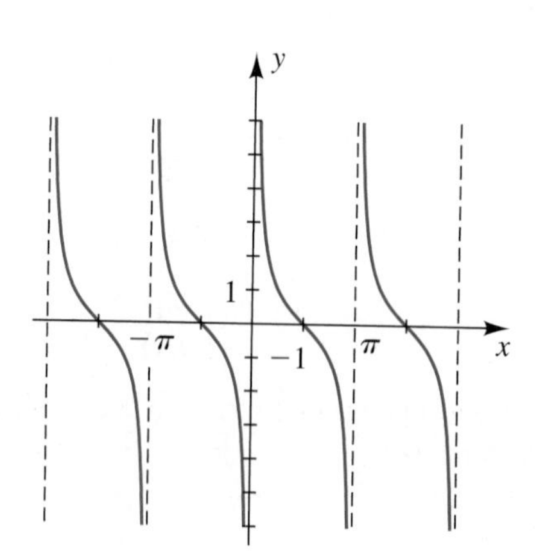

$y = \cot x$

Domain: $x \neq \pi n$

Range: $\mathbb{R}$

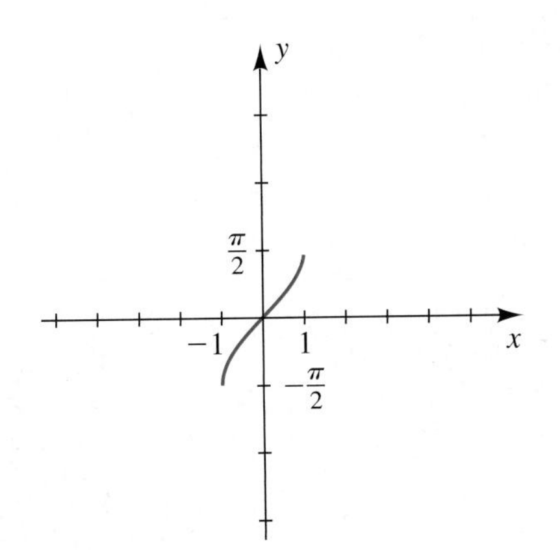

$y = \sin^{-1} x$

Domain: $[-1, 1]$

Range: $\left[-\frac{\pi}{2}, \frac{\pi}{2}\right]$

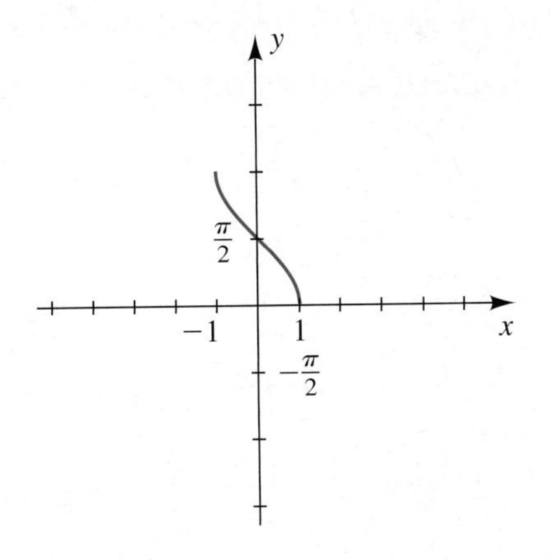

$y = \cos^{-1} x$

Domain: $[-1, 1]$

Range: $[0, \pi]$

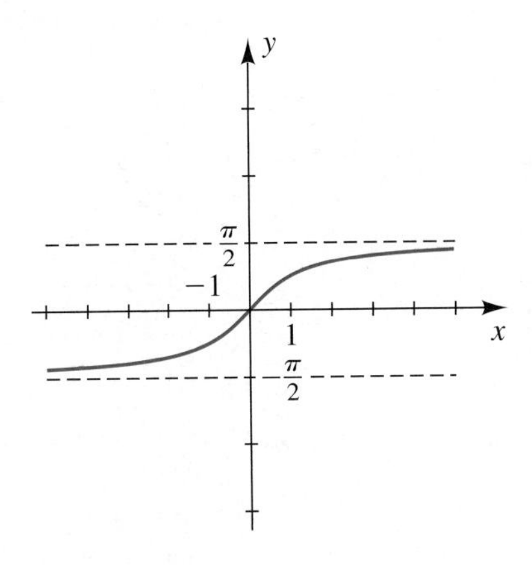

$y = \tan^{-1} x$

Domain: $\mathbb{R}$

Range: $\left(-\frac{\pi}{2}, \frac{\pi}{2}\right)$

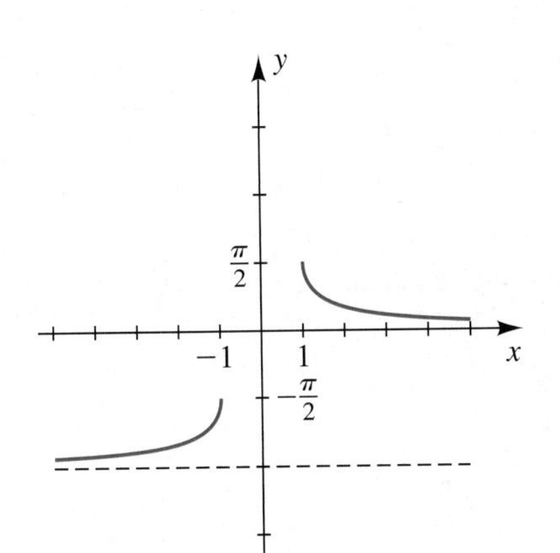

$y = \csc^{-1} x$

Domain: $(-\infty, -1] \cup [1, \infty)$

Range: $\left(-\pi, -\frac{\pi}{2}\right] \cup \left(0, \frac{\pi}{2}\right]$

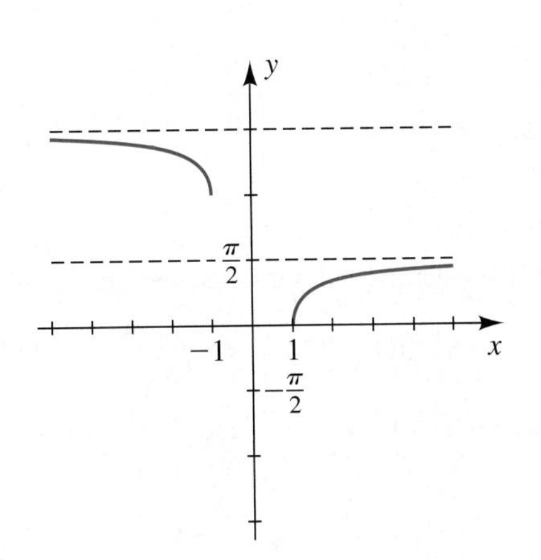

$y = \sec^{-1} x$

Domain: $(-\infty, -1] \cup [1, \infty)$

Range: $\left[0, \frac{\pi}{2}\right) \cup \left[\pi, \frac{3\pi}{2}\right)$

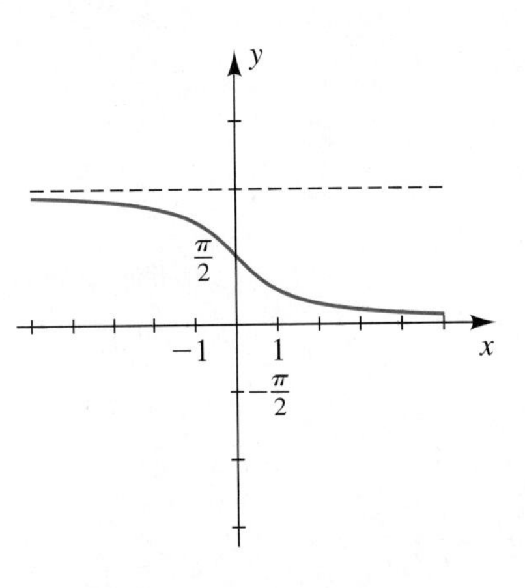

$y = \cot^{-1} x$

Domain: $\mathbb{R}$

Range: $(0, \pi)$

APPENDIX IV

Values of the Trigonometric Functions of Special Angles on a Unit Circle

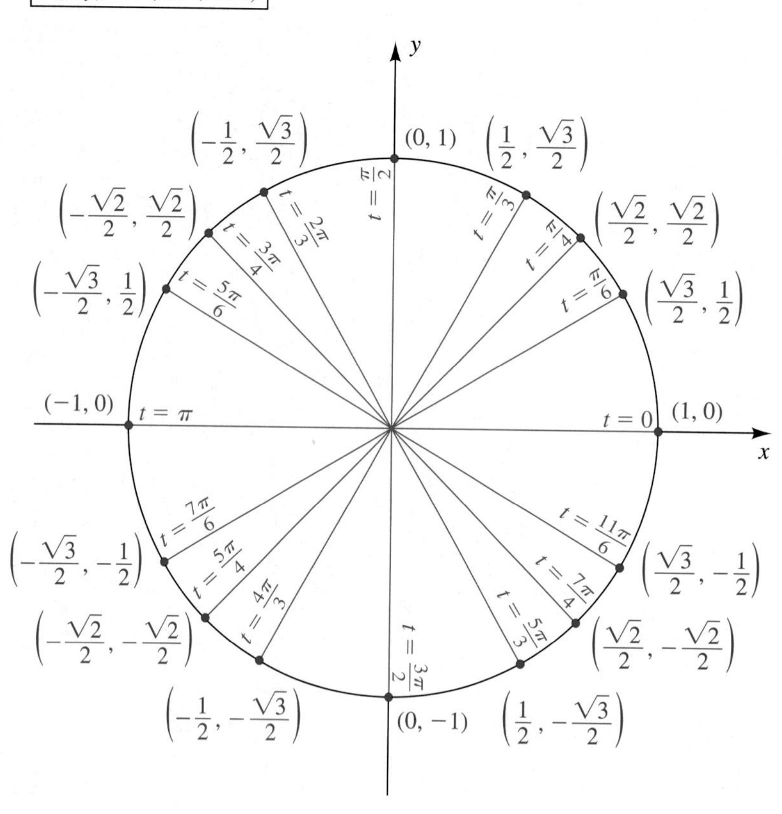

To find the values of the other trigonometric functions, use the following definitions:

$$\tan t = \frac{y}{x} \text{ (if } x \neq 0) \qquad \cot t = \frac{x}{y} \text{ (if } y \neq 0)$$

$$\sec t = \frac{1}{x} \text{ (if } x \neq 0) \qquad \csc t = \frac{1}{y} \text{ (if } y \neq 0)$$

Answers to Selected Exercises

A *Student's Solutions Manual* to accompany this textbook is available from your college bookstore. The guide contains detailed solutions to approximately one-half of the exercises, as well as strategies for solving other exercises in the text.

Chapter 1

EXERCISES 1.1

1 **(a)** Negative **(b)** Positive **(c)** Negative **(d)** Positive
3 **(a)** $<$ **(b)** $>$ **(c)** $=$
5 **(a)** $>$ **(b)** $>$ **(c)** $>$
7 **(a)** $x < 0$ **(b)** $y \geq 0$ **(c)** $q \leq \pi$ **(d)** $2 < d < 4$ **(e)** $t \geq 5$ **(f)** $-z \leq 3$ **(g)** $\dfrac{p}{q} \leq 7$ **(h)** $\dfrac{1}{w} \geq 9$ **(i)** $|x| > 7$
9 **(a)** 5 **(b)** 3 **(c)** 11
11 **(a)** -15 **(b)** -3 **(c)** 11
13 **(a)** $4 - \pi$ **(b)** $4 - \pi$ **(c)** $1.5 - \sqrt{2}$
15 **(a)** 4 **(b)** 12 **(c)** 12 **(d)** 8
17 **(a)** 10 **(b)** 9 **(c)** 9 **(d)** 19
19 $|7 - x| < 5$
21 $|-3 - x| \geq 8$
23 $|x - 4| \leq 3$
25 $-x - 3$
27 $2 - x$
29 $b - a$
31 $x^2 + 4$
33 $\neq$
35 $=$
37 $\neq$
39 $=$
41 **(a)** 8.4652 **(b)** 14.1428
43 **(a)** 6.557×10^{-1} **(b)** 6.708×10^{1}
45 Construct a right triangle with sides of lengths $\sqrt{2}$ and 1. The hypotenuse will have length $\sqrt{3}$. Next, construct a right triangle with sides of lengths $\sqrt{3}$ and $\sqrt{2}$. The hypotenuse will have length $\sqrt{5}$.
47 The large rectangle has area $a(b + c)$. The sum of the areas of the two small rectangles is $ab + ac$.
49 **(a)** 4.27×10^5 **(b)** 9.8×10^{-8} **(c)** 8.1×10^8
51 **(a)** 830,000 **(b)** 0.000 000 000 002 9 **(c)** 563,000,000
53 1.7×10^{-24}
55 5.87×10^{12}
57 1.678×10^{-24} g
59 4.1472×10^6 frames
61 **(a)** 201.6 lb **(b)** 32.256 tons

EXERCISES 1.2

1 $\dfrac{16}{81}$
3 $\dfrac{9}{8}$
5 $\dfrac{-47}{3}$
7 $\dfrac{1}{8}$
9 $\dfrac{1}{25}$
11 $8x^9$
13 $\dfrac{6}{x}$
15 $-2a^{14}$
17 $\dfrac{9}{2}$
19 $\dfrac{12u^{11}}{v^2}$
21 $\dfrac{4}{xy}$
23 $\dfrac{9y^6}{x^8}$
25 $\dfrac{81}{64}y^6$
27 $\dfrac{s^6}{4r^8}$
29 $\dfrac{20y}{x^3}$
31 $9x^{10}y^{14}$
33 $8a^2$
35 $24x^{3/2}$
37 $\dfrac{1}{9a^4}$
39 $\dfrac{8}{x^{1/2}}$
41 $4x^2y^4$
43 $\dfrac{3}{x^3y^2}$
45 1
47 $x^{3/4}$
49 $(a + b)^{2/3}$
51 $(x^2 + y^2)^{1/2}$
53 **(a)** $4x\sqrt{x}$ **(b)** $8x\sqrt{x}$
55 **(a)** $8 - \sqrt[3]{y}$ **(b)** $\sqrt[3]{8 - y}$
57 9
59 $-2\sqrt[5]{2}$
61 $\dfrac{1}{2}\sqrt[3]{4}$
63 $\dfrac{3y^3}{x^2}$
65 $\dfrac{2a^2}{b}$
67 $\dfrac{1}{2y^2}\sqrt{6xy}$
69 $\dfrac{xy}{3}\sqrt[3]{6y}$
71 $\dfrac{x}{3}\sqrt[4]{15x^2y^3}$
73 $\dfrac{1}{2}\sqrt[5]{20x^4y^2}$
75 $\dfrac{3x^5}{y^2}$
77 $\dfrac{2x}{y^2}\sqrt[5]{x^2y^4}$
79 $-3tv^2$
81 $|x^3|y^2$
83 $x^2|y - 1|^3$
85 $\neq$; $(a^r)^2 = a^{2r} \neq a^{(r^2)}$
87 $\neq$; $(ab)^{xy} = a^{xy}b^{xy} \neq a^xb^y$
89 $=$; $\sqrt[n]{\dfrac{1}{c}} = \left(\dfrac{1}{c}\right)^{1/n} = \dfrac{1^{1/n}}{c^{1/n}} = \dfrac{1}{\sqrt[n]{c}}$
91 **(a)** 1.5518 **(b)** 8.5499
93 **(a)** 2.0351 **(b)** 3.9670
95 \$232,825.78
97 2.82 m
99 The 120-kg lifter
101

Height	Weight	Height	Weight
64	137	72	168
65	141	73	172
66	145	74	176
67	148	75	180
68	152	76	184
69	156	77	188
70	160	78	192
71	164	79	196

EXERCISES 1.3

1 $6u^2 - 13u - 12$
3 $4y^2 - 5x$
5 $4x^2 - 9y^2$
7 $9x^2 + 12xy + 4y^2$
9 $x - y$
11 $x^3 - 6x^2y + 12xy^2 - 8y^3$
13 $(8x + 3)(x - 7)$
15 Irreducible
17 $(2x - 5)^2$
19 $x^2(x + 2)(x - 2)$
21 $(4x - y^2)(16x^2 + 4xy^2 + y^4)$
23 $(4x + 3)(16x^2 - 12x + 9)$
25 $3(x + 3)(x - 3)(x + 1)$
27 $(a + b)(a - b)(a^2 - ab + b^2)(a^2 + ab + b^2)$
29 $(x + 2 + 3y)(x + 2 - 3y)$
31 $\dfrac{y + 5}{y^2 + 5y + 25}$
33 $\dfrac{x}{x - 1}$
35 $\dfrac{6s - 7}{(3s + 1)^2}$

37 $\frac{5x^2+2}{x^3}$ **39** $\frac{4(2t+5)}{t+2}$ **41** $\frac{2(2x+3)}{3x-4}$

43 $\frac{2x-1}{x}$ **45** $\frac{11u^2+18u+5}{u(3u+1)}$ **47** $-\frac{x+5}{(x+2)^2}$

49 $a+b$ **51** $x+y$ **53** $\frac{rs}{r^2-s^2}$

55 $2x+h-3$ **57** $-\frac{3}{(x-1)(a-1)}$

59 $-\frac{3x^2+3xh+h^2}{x^3(x+h)^3}$ **61** $\frac{t+10\sqrt{t}+25}{t-25}$

63 $\frac{\sqrt[3]{a^2}+\sqrt[3]{ab}+\sqrt[3]{b^2}}{a-b}$ **65** $\frac{1}{(a+b)(\sqrt{a}+\sqrt{b})}$

67 $\frac{2}{\sqrt{2(x+h)+1}+\sqrt{2x+1}}$ **69** $4x^{4/3}-x^{1/3}+5x^{-2/3}$

71 $x^{-1}+4x^{-3}+4x^{-5}$ **73** $\frac{1+x^5}{x^3}$ **75** $\frac{1-x^2}{x^{1/2}}$

77 $(3x+2)^3(36x^2-37x+6)$

79 $\frac{(2x+1)^2(8x^2+x-24)}{(x^2-4)^{1/2}}$ **81** $\frac{(3x+1)^5(39x-89)}{(2x-5)^{1/2}}$

83 $\frac{27x^2-24x+2}{(6x+1)^4}$ **85** $\frac{4x(1-x^2)}{(x^2+2)^4}$ **87** $\frac{x^2+12}{(x^2+4)^{4/3}}$

89 $\frac{6(3-2x)}{(4x^2+9)^{3/2}}$

91

x	Y_1	Y_2
1	−0.6923	−0.6923
2	−26.12	−26.12
3	8.0392	8.0392
4	5.8794	5.8794
5	5.3268	5.3268

Might be equal

93 Area of I is $(x-y)x$, area of II is $(x-y)y$, and
$$A = x^2 - y^2 = (x-y)x + (x-y)y = (x-y)(x+y).$$

95 (a) 1525.7; 1454.7
(b) As people age, they require fewer calories. Coefficients of w and h are positive because large people require more calories.

EXERCISES 1.4

1 1 **3** $-\frac{24}{29}$ **5** $-\frac{3}{61}$

7 All real numbers except ± 2 **9** No solution

11 $-\frac{2}{3}, \frac{1}{5}$ **13** $-\frac{1}{2}$ **15** $\pm\frac{3}{5}$ **17** $3 \pm \sqrt{17}$

19 $-2 \pm \sqrt{2}$

21 $(x+6)(x-5)$ **23** $(2x-3)(6x+1)$

25 $-\frac{2}{3}, 2$ **27** No solution **29** $\pm\frac{2}{3}, 2$

31 0, 25 **33** $-\frac{57}{5}$ **35** 9

37 $\pm\frac{1}{10}\sqrt{70 \pm 10\sqrt{29}}$ **39** $\pm 2, \pm 3$ **41** $\frac{8}{27}, -8$

43 (a) 8 **(b)** ± 8 **(c)** No real solutions **(d)** 625
(e) No real solutions

45 (a) $x = \frac{y \pm \sqrt{2y^2-1}}{2}$ **(b)** $y = -2x \pm \sqrt{8x^2+1}$

47 (a) 0; −4,500,000 **(b)** 2.13×10^{-7}

49 $K = \frac{D-L}{E+T}$ **51** $Q = \frac{1}{M-1}$ **53** $r = \frac{A-P}{Pt}$

55 $q = \frac{p(1-S)}{S(1-p)}$ **57** $q = \frac{fp}{p-f}$

59 $v = \sqrt{\frac{2K}{m}}$ **61** $r = \frac{-\pi h + \sqrt{\pi^2 h^2 + 2\pi A}}{2\pi}$

63 $C = 2\sqrt{R^2-d^2}$ **65** 120 mo (or 10 yr)

67 (a) After 64 sec **(b)** 96 m and 128 m, respectively

69 1237.5 ft **71** 7 ft

73 (a) 40.96°F **(b)** 6909 ft **75** 37°F

77 (a) 206.25 ft **(b)** 40 mi/hr

79 (a) $d = 100\sqrt{20t^2+4t+1}$ **(b)** 3:30 P.M.

81 \$175 **83 (a)** 1 cm **(b)** $\frac{\pi}{8}$ cm³

85 $h \approx 97\%$ of L

87 There are two possible routes, corresponding to $x \approx 0.6743$ mi and $x \approx 2.2887$ mi. **89** (4)

91 (a) (2) **(b)** 860 min **93** $3.7 \times 3.7 \times 1.8$

EXERCISES 1.5

1 $2+4i$ **3** $18-3i$ **5** $41-11i$ **7** $17-i$

9 $21-20i$ **11** $-24-7i$ **13** 25 **15 (a)** $-i$

(b) 1 **17 (a)** i **(b)** -1 **19** $\frac{3}{10}-\frac{3}{5}i$

21 $\frac{1}{2}-i$ **23** $\frac{34}{53}+\frac{40}{53}i$ **25** $\frac{2}{5}+\frac{4}{5}i$

27 $-142-65i$ **29** $-2-14i$ **31** $-\frac{44}{113}+\frac{95}{113}i$

33 $\frac{21}{2}i$ **35** $x=4, y=-1$ **37** $x=3, y=-4$

39 $3 \pm 2i$ **41** $-2 \pm 3i$ **43** $\frac{5}{2} \pm \frac{1}{2}\sqrt{55}\,i$

45 $-\frac{1}{8} \pm \frac{1}{8}\sqrt{47}\,i$ **47** $-5, \frac{5}{2} \pm \frac{5}{2}\sqrt{3}\,i$

49 $\frac{5}{2}, -\frac{25}{26} \pm \frac{15}{26}\sqrt{3}i$ **51** $\pm 4, \pm 4i$ **53** $\pm 2i, \pm \frac{3}{2}i$

55 $0, -\frac{3}{2} \pm \frac{1}{2}\sqrt{7}i$

57
$$\begin{aligned}\overline{z+w} &= \overline{(a+bi)+(c+di)}\\ &= \overline{(a+c)+(b+d)i} = (a+c)-(b+d)i\\ &= (a-bi)+(c-di) = \overline{z}+\overline{w}\end{aligned}$$

59
$$\begin{aligned}\overline{z \cdot w} &= \overline{(a+bi)\cdot(c+di)}\\ &= \overline{(ac-bd)+(ad+bc)i}\\ &= (ac-bd)-(ad+bc)i\\ &= ac-adi-bd-bci\\ &= a(c-di)-bi(c-di)\\ &= (a-bi)\cdot(c-di) = \overline{z}\cdot\overline{w}\end{aligned}$$

61 If $\overline{z} = z$, then $a - bi = a + bi$ and hence $-bi = bi$, or $2bi = 0$. Thus, $b = 0$ and $z = a$ is real. Conversely, if z is real, then $b = 0$ and hence $\overline{z} = \overline{a + 0i} = a - 0i = a + 0i = z$.

EXERCISES 1.6

1 $(-\infty, -2)$ **3** $(-5, -3]$

−2 0 −5 −3 0

5 $-5 < x \le 8$ **7** $(12, \infty)$ **9** $[9, 19)$

11 $\left(-\frac{2}{3}, \infty\right)$ **13** $\left(\frac{4}{3}, \infty\right)$

15 All real numbers except 1 **17** $(-3.01, -2.99)$

19 $\left(-\infty, \frac{2}{3}\right] \cup [4, \infty)$ **21** $(-\infty, \infty)$

23 $(-\infty, 3) \cup (3, \infty)$ **25** $(-4, 4)$

27 $\left(-\frac{1}{3}, \frac{1}{2}\right)$ **29** $(-2, 3)$

31 $\left(-\infty, -\frac{5}{2}\right] \cup [1, \infty)$ **33** $\left(-\frac{3}{5}, \frac{3}{5}\right)$

35 $(-\infty, -2) \cup (-2, -1) \cup \{0\}$

37 $(-2, 0) \cup (0, 1]$ **39** $(-2, 2] \cup (5, \infty)$

41 $(-\infty, -3) \cup (0, 3)$ **43** $\left(\frac{3}{2}, \frac{7}{3}\right)$

45 $(-\infty, -1) \cup \left(2, \frac{7}{2}\right]$ **47** $\left(1, \frac{5}{3}\right) \cup [2, 5]$

49 $(-1, 0) \cup (1, \infty)$

51 **(a)** $-8, -2$ **(b)** $-8 < x < -2$ **(c)** $(-\infty, -8) \cup (-2, \infty)$

53 $|w - 148| \le 2$ **55** $4 \le p < 6$

57 $6\frac{2}{3}$ yr **59** $\frac{1}{2}$ sec **61** $0 \le v < 30$

63 **(a)** 5 ft 8 in. **(b)** $65.52 \le h \le 66.48$

65 $[-2, -1) \cup (1, 2) \cup (3, 3.5]$

CHAPTER 1 REVIEW EXERCISES

1 $-x - 3$ **2** $-(x-2)(x-3)$ **3** $\frac{b^3}{a^8}$ **4** $-\frac{p^8}{2q}$

5 $\frac{x^3z}{y^{10}}$ **6** $\frac{16x^2}{z^4y^6}$ **7** $\frac{b^6}{a^2}$ **8** $\frac{y - x^2}{x^2y}$ **9** $2xyz\sqrt[3]{x^2z}$

10 $2ab\sqrt{ac}$ **11** $\frac{1 - \sqrt{t}}{t}$ **12** c^2d^4

13 $\frac{2x}{y^2}$ **14** $\frac{x + 6\sqrt{x} + 9}{9 - x}$

15 $x^4 + x^3 - x^2 + x - 2$ **16** $-x^2 + 18x + 7$

17 $6a^2 + 11ab - 35b^2$ **18** $16r^4 - 24r^2s + 9s^2$

19 $169a^4 - 16b^2$ **20** $8a^3 + 12a^2b + 6ab^2 + b^3$

21 $81x^4 - 72x^2y^2 + 16y^4$

22 $a^2 + b^2 + c^2 + d^2 + 2(ab + ac + ad + bc + bd + cd)$

23 $10w(6x + 7)$ **24** $(4a^2 + 3b^2)^2$

25 $8(x + 2y)(x^2 - 2xy + 4y^2)$

26 $u^3v(v - u)(v^2 + uv + u^2)$

27 $(p^4 + q^4)(p^2 + q^2)(p + q)(p - q)$ **28** $x^2(x - 4)^2$

29 $(x - 7 + 7y)(x - 7 - 7y)$

30 $(x - 2)(x + 2)^2(x^2 - 2x + 4)$

31 $\frac{27}{(4x - 5)(10x + 1)}$ **32** $\frac{5x^2 - 6x - 20}{x(x + 2)^2}$ **33** $\frac{x^3 + 1}{x^2 + 1}$

34 $\frac{ab}{a + b}$ **35** $\frac{1}{x + 3}$ **36** $\frac{2(5x^2 + x + 4)}{(6x + 1)^{2/3}(4 - x^2)^2}$

37 $-\frac{5}{6}$ **38** $-4, \frac{3}{2}$ **39** $-\frac{2}{3} \pm \frac{1}{3}\sqrt{19}$

40 $\pm\frac{5}{2}, \pm\sqrt{2}$ **41** $\pm\frac{1}{2}\sqrt{7}, -\frac{2}{5}$

42 $-\frac{3}{2}, 2$ **43** $-5, 4$ **44** $\frac{1}{4}, \frac{1}{9}$ **45** 2

46 5 **47** $\left(\frac{2}{3}, \infty\right)$ **48** $\left(-\frac{11}{4}, \frac{9}{4}\right)$

49 $\left(-\infty, -\frac{3}{10}\right)$ **50** $\left(-7, \frac{7}{2}\right)$

51 $(-\infty, 1) \cup (5, \infty)$ **52** $\left(-\infty, \frac{11}{3}\right] \cup [7, \infty)$

53 $\left(-\infty, -\frac{3}{2}\right) \cup \left(\frac{2}{5}, \infty\right)$ **54** $[-2, 5]$

55 $(-\infty, -2) \cup \{0\} \cup [3, \infty)$ **56** $(-3, -1) \cup (-1, 2]$

57 $\left(-\infty, -\frac{3}{2}\right) \cup (2, 9)$ **58** $(-\infty, -5) \cup [-1, 5)$

59 $(1, \infty)$ **60** $(0, 1) \cup (2, 3)$ **61** $C = \frac{2}{P + N - 1}$

62 $D = \dfrac{CB^3}{(A+E)^3}$ **63** $R = \sqrt[4]{\dfrac{8FVL}{\pi P}}$

64 $r = \dfrac{-\pi hR + \sqrt{12\pi hV - 3\pi^2h^2R^2}}{2\pi h}$

65 $-55 + 48i$ **66** $\dfrac{9}{85} + \dfrac{2}{85}i$ **67** $-\dfrac{9}{53} - \dfrac{48}{53}i$

68 $-2 - 5i$ **69** 11.055% **70** 60.3 g

71 6 oz of vegetables and 4 oz of meat

72 80 gal of 20% solution and 40 gal of 50% solution

73 $\dfrac{640}{11} \approx 58.2$ mi/hr **74** 1 hr 40 min

75 **(a)** $d = \sqrt{2900t^2 - 200t + 4}$

(b) $t = \dfrac{5 + 2\sqrt{19{,}603}}{145} \approx 1.97$, or approximately 11:58 A.M.

76 There are two arrangements: 40 ft × 25 ft or 50 ft × 20 ft.

77 **(a)** $2\sqrt{2}$ ft **(b)** 2 ft

78 $4 \le p \le 8$

79 Over $100,000 **80** $T > 279.57$ K

81 36 to 38 trees/acre **82** $990 to $1040 **83** (3)

CHAPTER 1 DISCUSSION EXERCISES

1 1 gallon ≈ 0.13368 ft³; 586.85 ft²

2 Either $a = 0$ or $b = 0$

3 Add and subtract $10x$; $x + 5 \pm \sqrt{10x}$ are the factors.

4 The first expression can be evaluated at $x = 1$.

5 They get close to the ratio of leading coefficients as x gets larger.

7 If x is the age and y is the height, show that the final value is $100x + y$.

8 $V_{out} = \frac{1}{3}V_{in}$

9 No **10** $\dfrac{-b}{2a}$

11 **(a)** $\dfrac{ac+bd}{a^2+b^2} + \dfrac{ad-bc}{a^2+b^2}i$ **(b)** Yes

(c) a and b cannot both be 0

13 $a > 0, D \le 0$: $x \in \mathbb{R}$;
$a > 0, D > 0$: $(-\infty, x_1] \cup [x_2, \infty)$;
$a < 0, D < 0$: $\{\ \}$;
$a < 0, D = 0$: $x = \dfrac{-b}{2a}$;
$a < 0, D > 0$: $[x_1, x_2]$

14 **(a)** 11,006 ft **(b)** $h = \dfrac{1}{6}(2497D - 497G - 64{,}000)$

16 $1/10^{1000}$; $cx - 2/c$ must be nonnegative

17 **(a)** 109–45 **(b)** 1.88

Chapter 2

EXERCISES 2.1

1

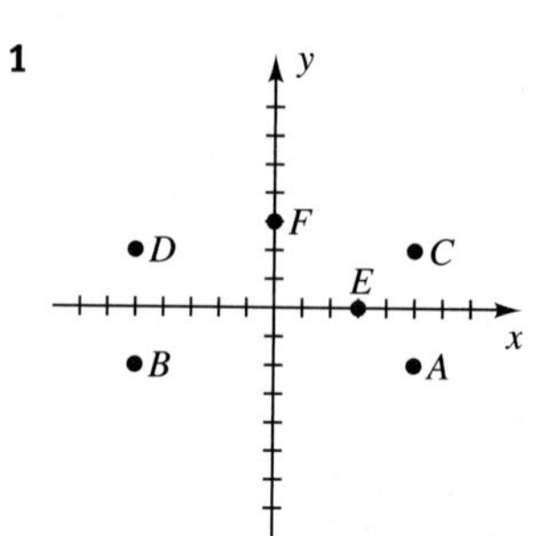

3 The line bisecting quadrants I and III

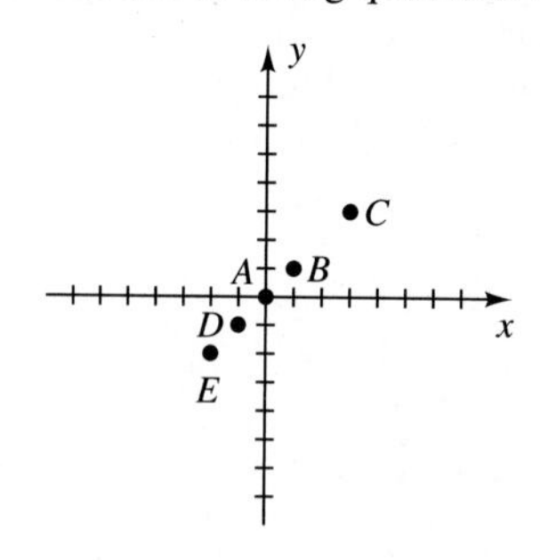

5 $A(3, 3), B(-3, 3), C(-3, -3), D(3, -3), E(3, 0), F(0, 3)$

7 **(a)** The line parallel to the y-axis that intersects the x-axis at $(-2, 0)$

(b) The line parallel to the x-axis that intersects the y-axis at $(0, 3)$

(c) All points to the right of and on the y-axis

(d) All points in quadrants I and III

(e) All points below the x-axis

(f) All points on the y-axis

9 **(a)** $\sqrt{29}$ **(b)** $\left(5, -\dfrac{1}{2}\right)$

11 **(a)** $\sqrt{13}$ **(b)** $\left(-\dfrac{7}{2}, -1\right)$

13 **(a)** 4 **(b)** $(5, -3)$

15 $d(A, C)^2 = d(A, B)^2 + d(B, C)^2$; area = 28

17 $d(A, B) = d(B, C) = d(C, D) = d(D, A)$ and $d(A, C)^2 = d(A, B)^2 + d(B, C)^2$

19 $(13, -28)$ **21** $d(A, C) = d(B, C) = \sqrt{145}$

23 $5x + 2y = 3$

25 $\sqrt{x^2 + y^2} = 5$; a circle of radius 5 with center at the origin

27 $(0, 3 + \sqrt{11}), (0, 3 - \sqrt{11})$ **29** $(-2, -1)$

31 $a < \dfrac{2}{5}$ or $a > 4$

33 Let M be the midpoint of the hypotenuse. Show that

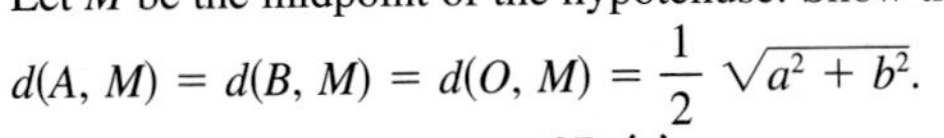

$$d(A, M) = d(B, M) = d(O, M) = \frac{1}{2}\sqrt{a^2 + b^2}.$$

35

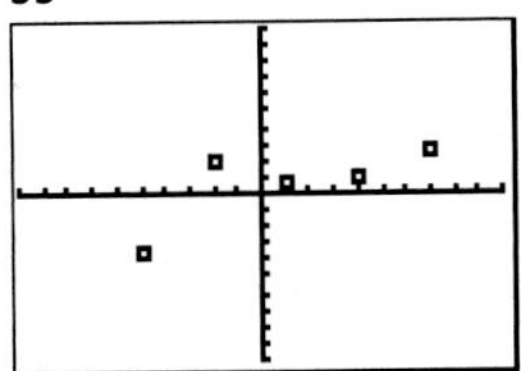

$[-10, 10]$ by $[-10, 10]$

37 (a)

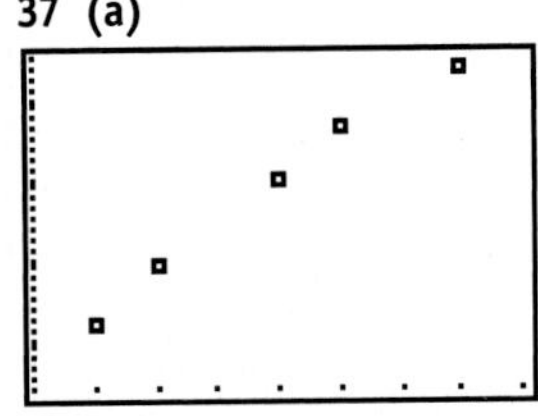

$[1996, 2004]$ by
$[35 \times 10^6, 63 \times 10^6, 10^6]$

(b) The number is increasing.

EXERCISES 2.2

Exer. 1–20: x-intercept(s) is listed, followed by y-intercept(s).

1 1.5; -3

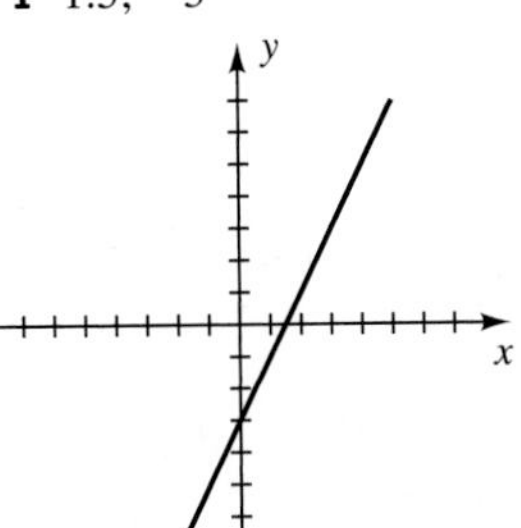

3 1; 1

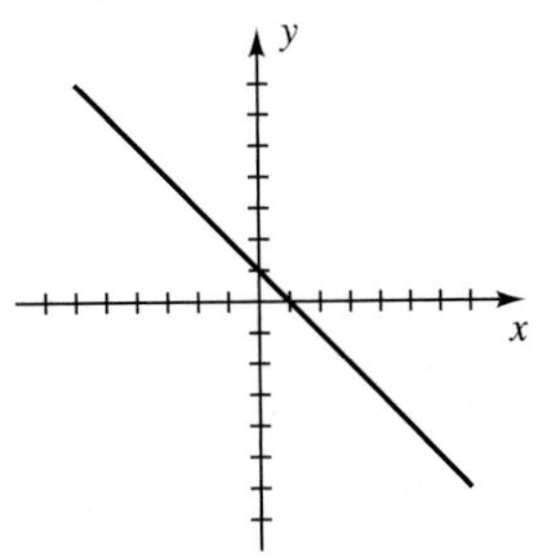

5 0; 0

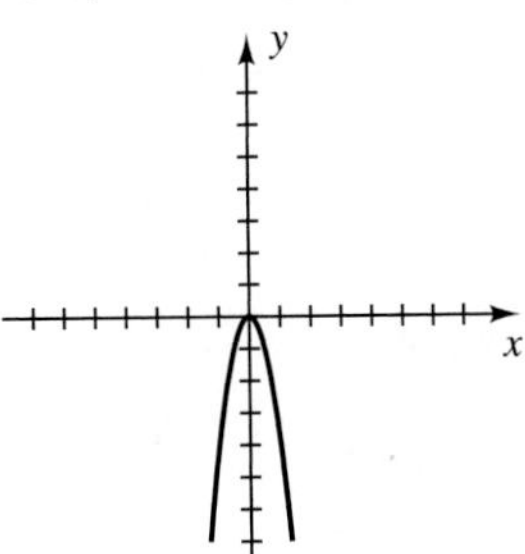

7 $\pm\frac{1}{2}\sqrt{2}$; -1

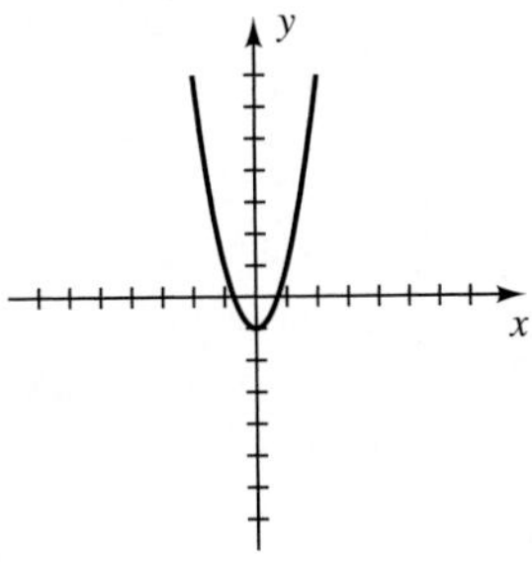

9 0; 0 **11** 3; $\pm\sqrt{3}$

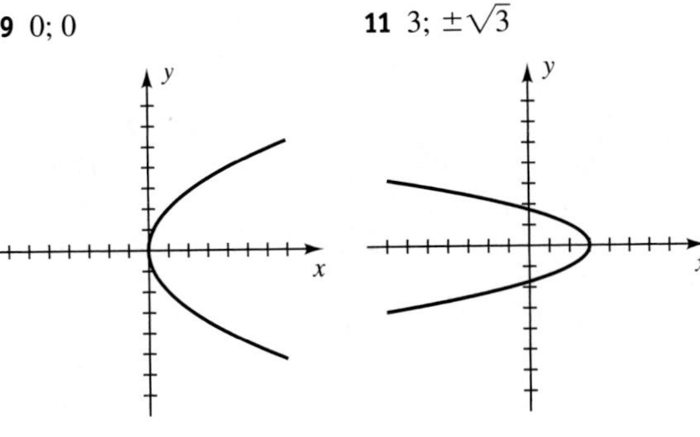

13 0; 0 **15** 2; -8

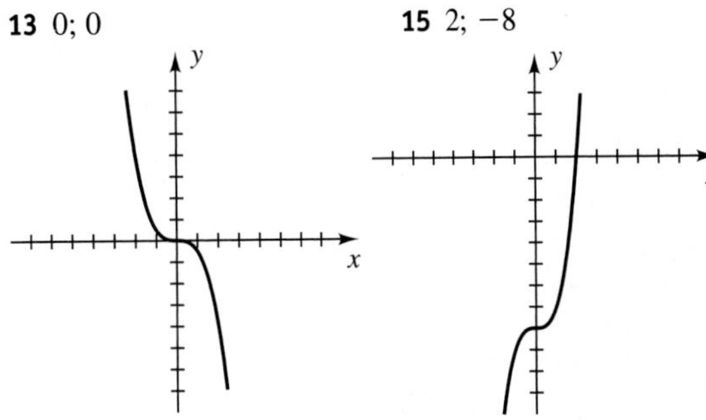

17 0; 0 **19** 16; -4

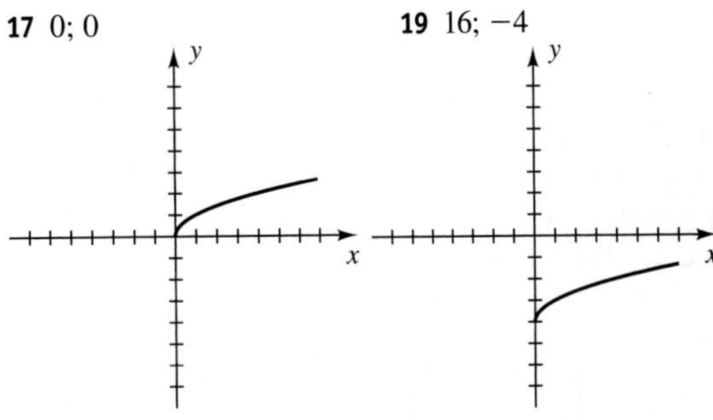

21 (a) 5, 7 **(b)** 9, 11 **(c)** 13

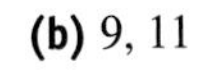

23 **25**

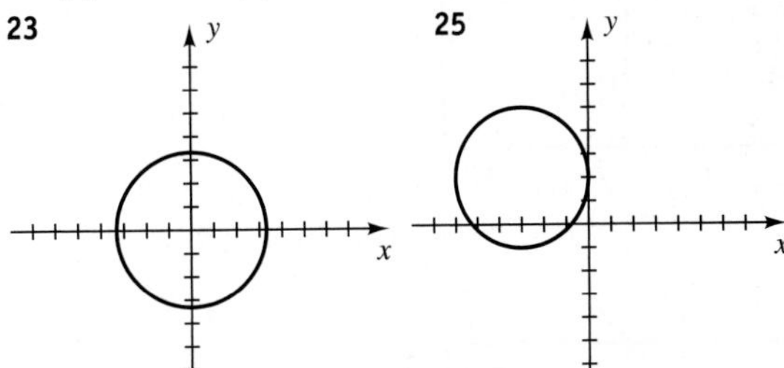

27 **29**

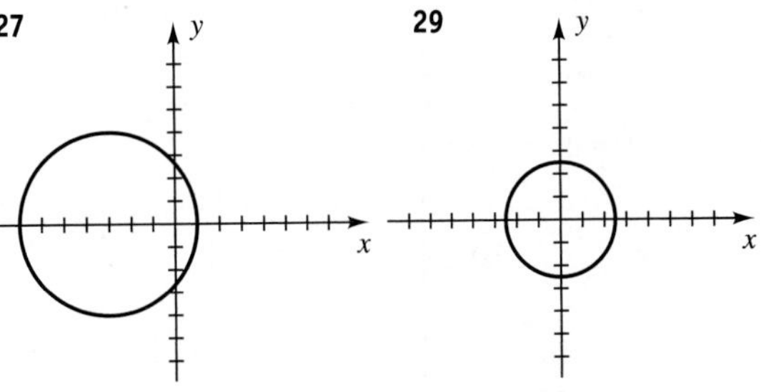

31 **33**

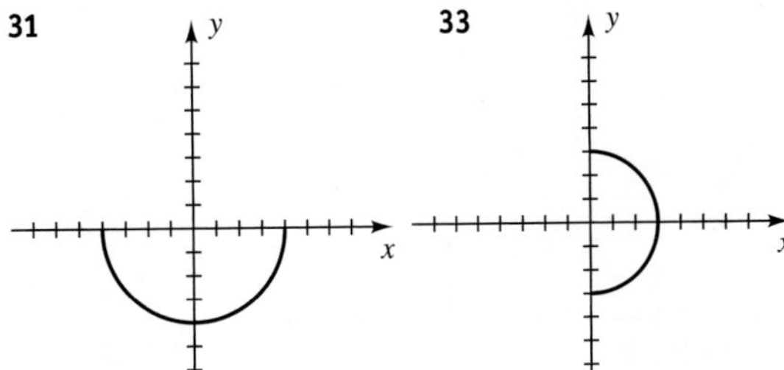

35 $(x - 2)^2 + (y + 3)^2 = 25$ **37** $\left(x - \frac{1}{4}\right)^2 + y^2 = 5$
39 $(x + 4)^2 + (y - 6)^2 = 41$
41 $(x + 3)^2 + (y - 6)^2 = 9$
43 $(x + 4)^2 + (y - 4)^2 = 16$
45 $(x - 1)^2 + (y - 2)^2 = 34$ **47** $C(2, -3)$; $r = 7$
49 $C(0, -2)$; $r = 11$ **51** $C(3, -1)$; $r = \frac{1}{2}\sqrt{70}$
53 $C(-2, 1)$; $r = 0$ (a point)
55 Not a circle, since r^2 cannot equal -2
57 $y = \sqrt{36 - x^2}$; $y = -\sqrt{36 - x^2}$; $x = \sqrt{36 - y^2}$; $x = -\sqrt{36 - y^2}$
59 $y = -1 + \sqrt{49 - (x - 2)^2}$; $y = -1 - \sqrt{49 - (x - 2)^2}$; $x = 2 + \sqrt{49 - (y + 1)^2}$; $x = 2 - \sqrt{49 - (y + 1)^2}$
61 $(x + 3)^2 + (y - 2)^2 = 4^2$ **63** $y = -\sqrt{4^2 - x^2}$
65 **(a)** Inside **(b)** On **(c)** Outside
67 **(a)** 2 **(b)** $3 \pm \sqrt{5}$
69 $(x + 2)^2 + (y - 3)^2 = 25$ **71** $\sqrt{5}$
73 $(-\infty, -3) \cup (2, \infty)$ **75** $(-1, 0) \cup (0, 1)$ **77** (2)
79 $-1.2, 0.5, 1.6$

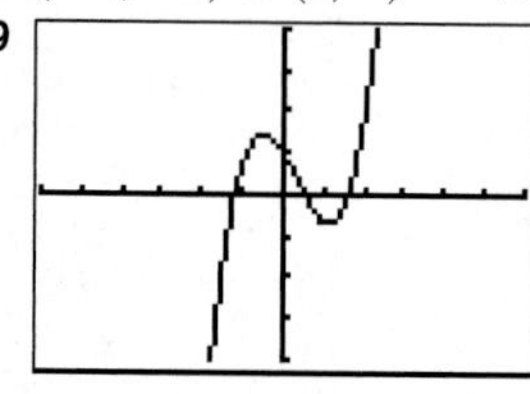

[−6, 6] by [−4, 4]

81 (0.6, 0.8), (−0.6, −0.8)

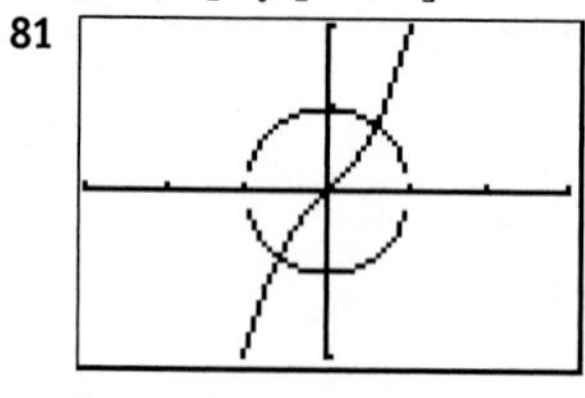

[−3, 3] by [−2, 2]

83 (0.999, 0.968), (0.251, 0.032)

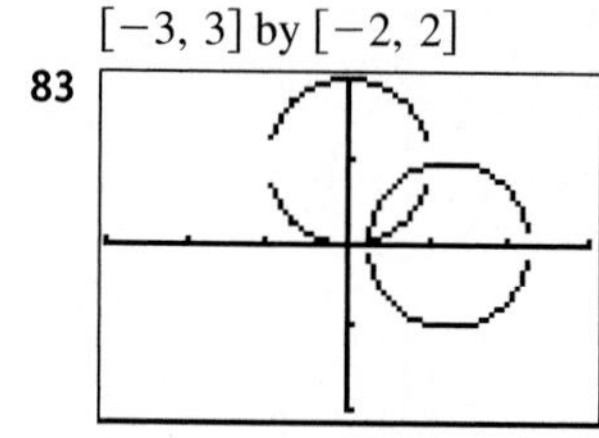

[−3, 3] by [−2, 2]

85

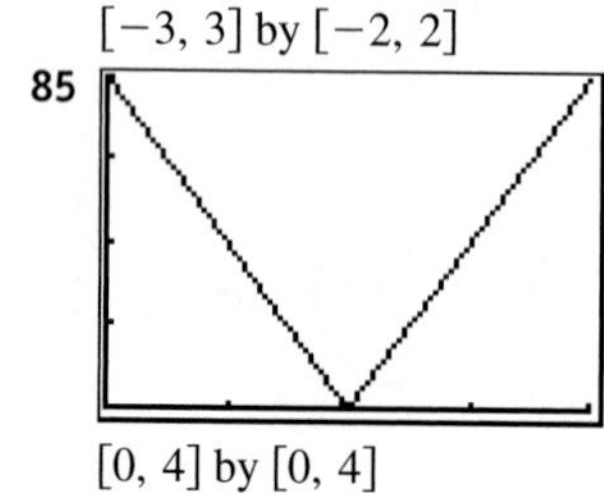

[0, 4] by [0, 4]

87 **(a)** 1126 ft/sec **(b)** −42°C

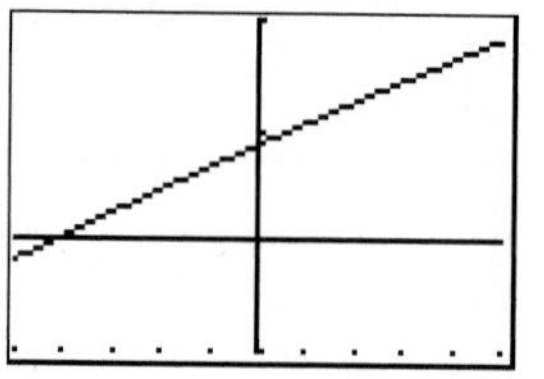

[−50, 50, 10] by [900, 1200, 100]

EXERCISES 2.3

1 $m = -\frac{3}{4}$ **3** $m = 0$

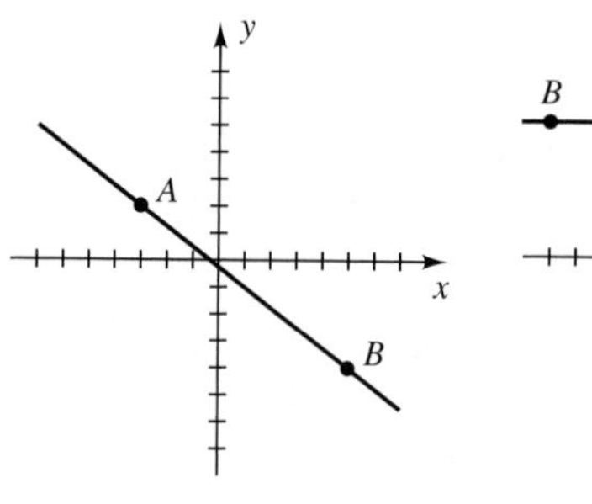

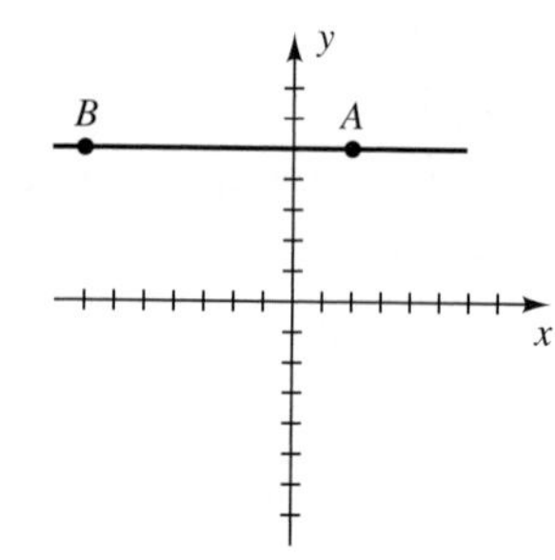

5 m is undefined

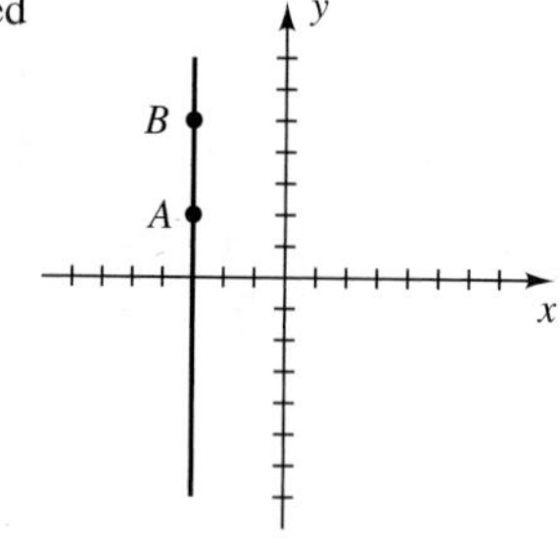

7 The slopes of opposite sides are equal.
9 The slopes of opposite sides are equal, and the slopes of two adjacent sides are negative reciprocals.
11 (−12, 0)
13

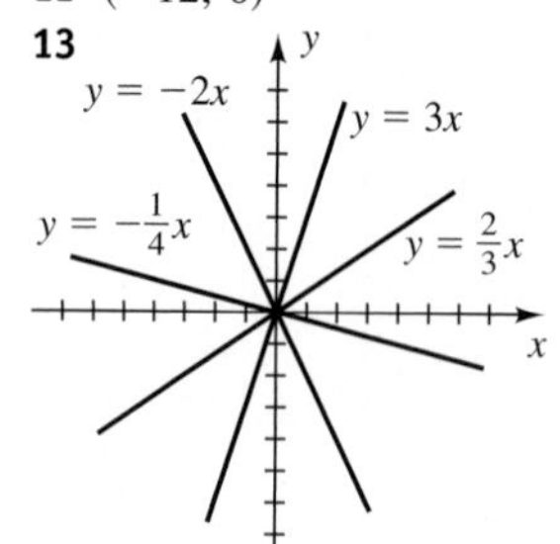

15

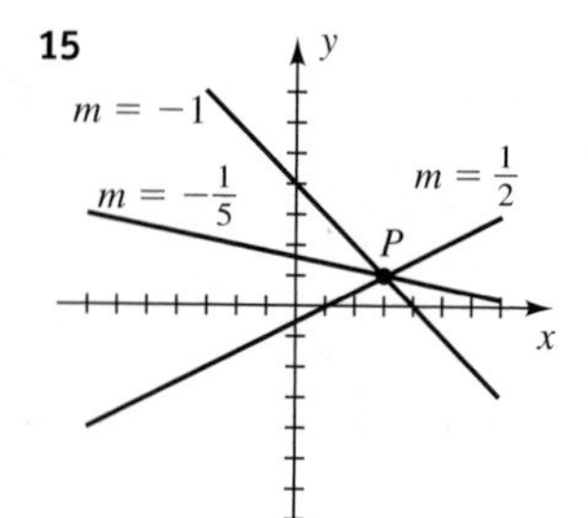

17 $y + 3 = \pm\frac{5}{4}(x - 2)$

19

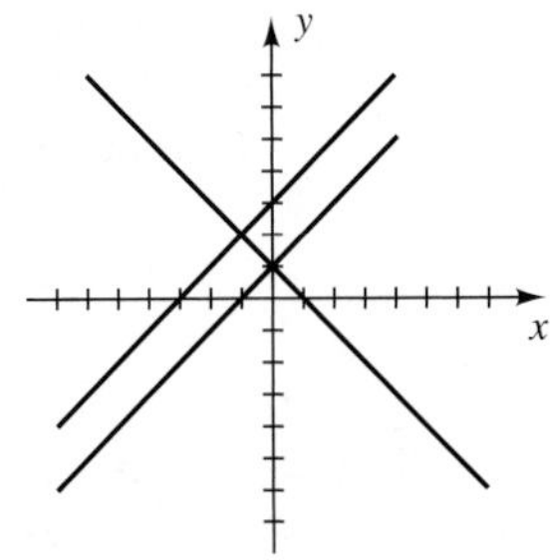

21 **(a)** $x = 5$ **(b)** $y = -2$ **23** $4x + y = 17$

25 $3x + y = 12$ **27** $11x + 7y = 9$

29 $5x - 2y = 18$ **31** $5x + 2y = 29$

33 $y = \frac{3}{4}x - 3$ **35** $y = -\frac{1}{3}x + \frac{11}{3}$

37 $5x - 7y = -15$ **39** $y = -x$

41 $m = -\frac{2}{3}, b = 5$ **43** $m = \frac{4}{3}, b = -3$

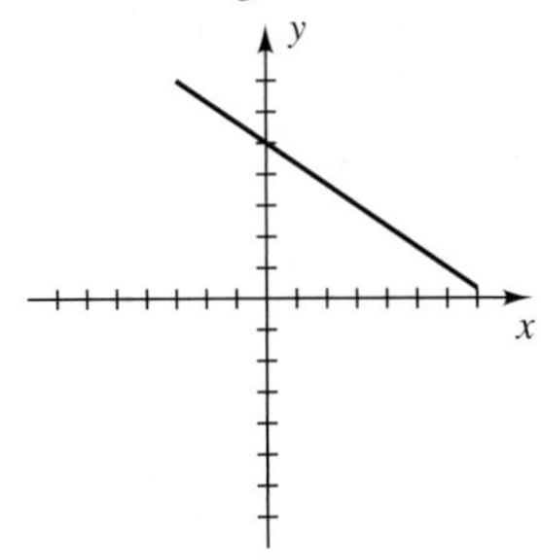

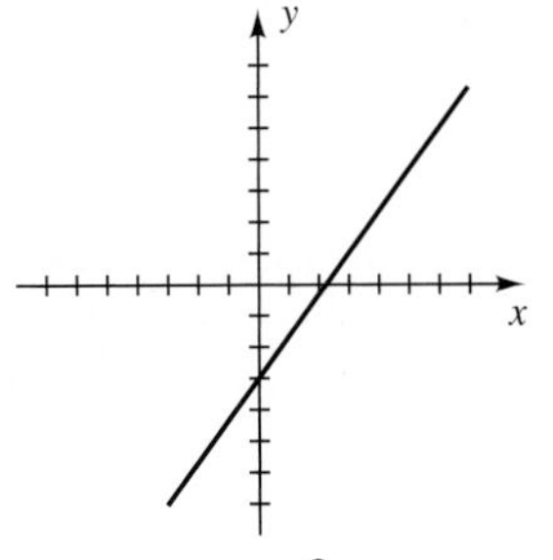

45 **(a)** $y = 3$ **(b)** $y = -\frac{1}{2}x$ **(c)** $y = -\frac{3}{2}x + 1$

(d) $y + 2 = -(x - 3)$

47 $\frac{x}{3/2} + \frac{y}{-3} = 1$ **49** $(x - 3)^2 + (y + 2)^2 = 49$

51 Approximately 23 weeks

53 **(a)** 25.2 tons **(b)** As large as 3.4 tons

55 **(a)** $y = \frac{5}{14}x$ **(b)** 58

57 **(a)** $W = \frac{20}{3}t + 10$ **(b)** 50 lb **(c)** 9 yr

(d)

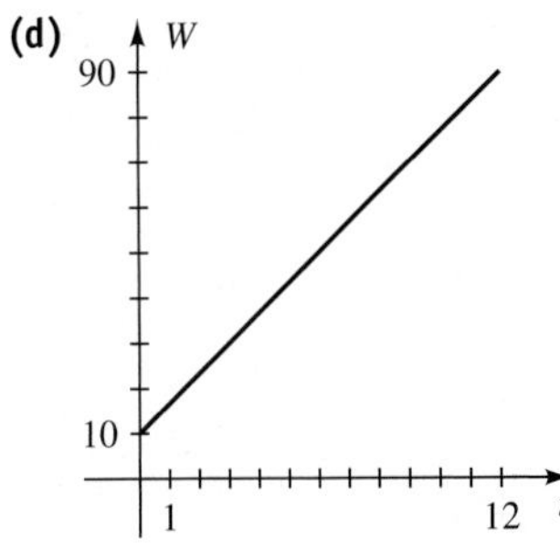

59 $H = -\frac{8}{3}T + \frac{7520}{3}$

61 **(a)** $T = 0.032t + 13.5$ **(b)** 16.54°C

63 **(a)** $E = 0.55R + 3600$ **(b)** $P = 0.45R - 3600$

(c) \$8000

65 **(a)** Yes: the creature at $x = 3$ **(b)** No

67 34.95 mi/hr **69** $a = 0.321$; $b = -0.9425$

71 $(-19, 13)$

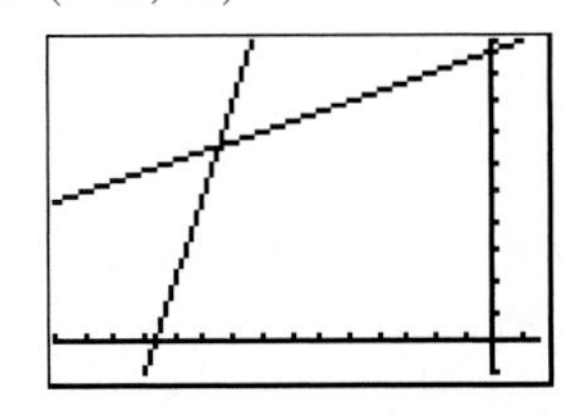

[−30, 3, 2] by [−2, 20, 2]

73 $(-0.8, -0.6)$, $(4.8, -3.4)$, $(2, 5)$; right isosceles triangle

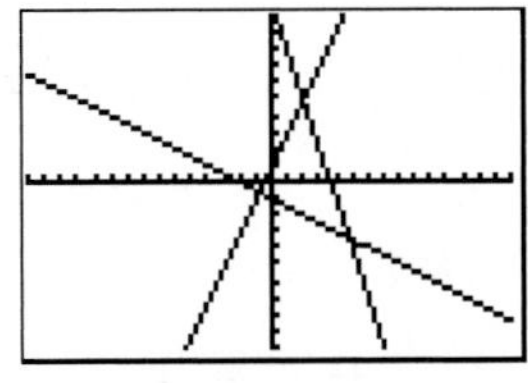

[−15, 15] by [−10, 10]

75 $y = 3.2x - 2.6$

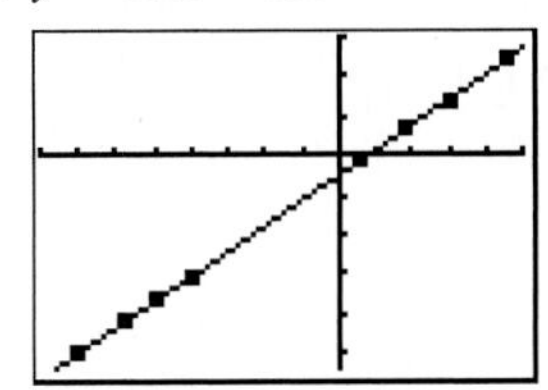

[−8, 5] by [−27, 15, 5]

77 **(b)** $y = 97.4x - 192{,}824$ (rounded)

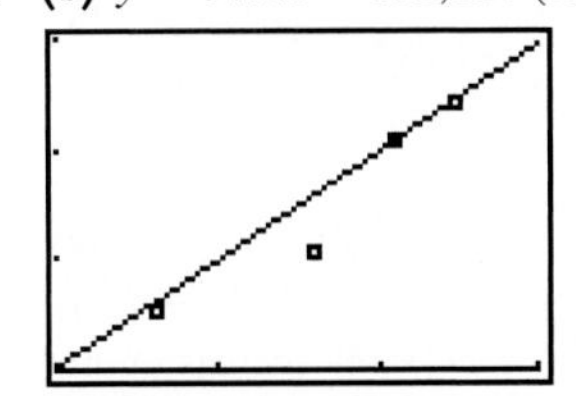

[1980, 2010, 10] by [0, 3000, 1000]

(c) \$2,107,895; \$2,205,263

EXERCISES 2.4

1 $-6, -4, -24$ **3** $-12, -22, -36$

5 **(a)** $5a - 2$ **(b)** $-5a - 2$ **(c)** $-5a + 2$

(d) $5a + 5h - 2$ **(e)** $5a + 5h - 4$ **(f)** 5

7 (a) $-a^2 + 4$ **(b)** $-a^2 + 4$ **(c)** $a^2 - 4$
(d) $-a^2 - 2ah - h^2 + 4$ **(e)** $-a^2 - h^2 + 8$
(f) $-2a - h$

9 (a) $a^2 - a + 3$ **(b)** $a^2 + a + 3$ **(c)** $-a^2 + a - 3$
(d) $a^2 + 2ah + h^2 - a - h + 3$
(e) $a^2 + h^2 - a - h + 6$ **(f)** $2a + h - 1$

11 (a) $\dfrac{4}{a^2}$ **(b)** $\dfrac{1}{4a^2}$ **(c)** $4a$ **(d)** $2a$

13 (a) $\dfrac{2a}{a^2 + 1}$ **(b)** $\dfrac{a^2 + 1}{2a}$ **(c)** $\dfrac{2\sqrt{a}}{a + 1}$
(d) $\dfrac{\sqrt{2a^3 + 2a}}{a^2 + 1}$

15 The graph is that of a function because it passes the vertical line test.

17 $D = [-4, 1] \cup [2, 4)$; $R = [-3, 3)$

19 (a) $[-3, 4]$ **(b)** $[-2, 2]$ **(c)** 0 **(d)** $-1, \frac{1}{2}, 2$
(e) $\left(-1, \frac{1}{2}\right) \cup (2, 4]$

21 $\left[-\frac{7}{2}, \infty\right)$ **23** $[-3, 3]$

25 All real numbers except -2, 0, and 2

27 $\left[\frac{3}{2}, 4\right) \cup (4, \infty)$ **29** $(2, \infty)$ **31** $[-2, 2]$

33 (a) $D = [-5, -3) \cup (-1, 1] \cup (2, 4]$;
$R = \{-3\} \cup [-1, 4]$
(b) Increasing on $[-4, -3) \cup [3, 4]$;
decreasing on $[-5, -4] \cup (2, 3]$;
constant on $(-1, 1]$

35

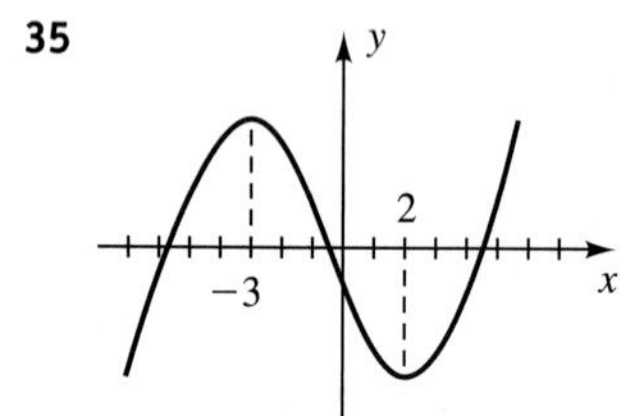

37 (a) [graph]

(b) $D = (-\infty, \infty)$,
$R = (-\infty, \infty)$
(c) Increasing on $(-\infty, \infty)$

39 (a) [graph]

(b) $D = (-\infty, \infty)$,
$R = (-\infty, 4]$
(c) Increasing on $(-\infty, 0]$,
decreasing on $[0, \infty)$

41 (a)

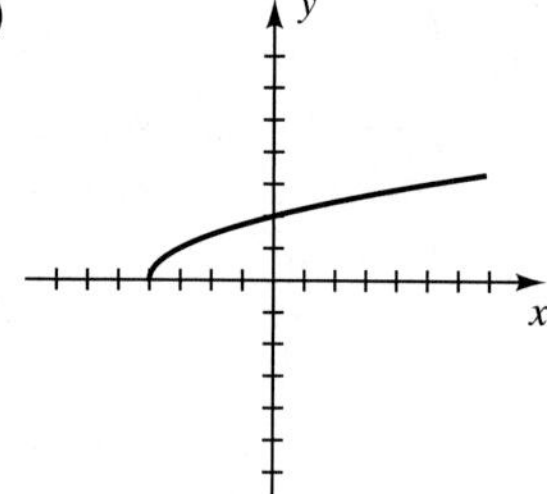

(b) $D = [-4, \infty)$,
$R = [0, \infty)$
(c) Increasing on $[-4, \infty)$

43 (a)

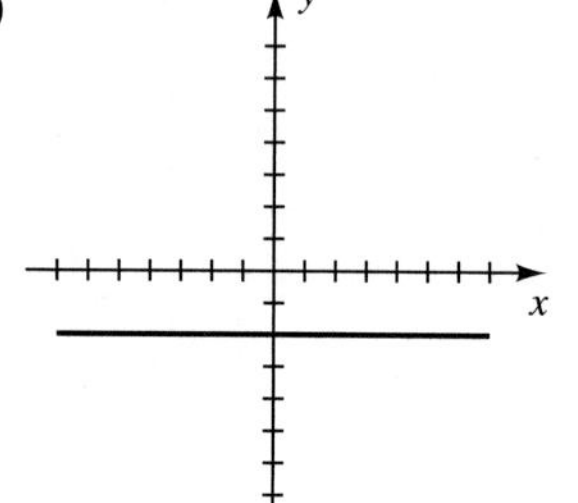

(b) $D = (-\infty, \infty)$,
$R = \{-2\}$
(c) Constant on $(-\infty, \infty)$

45 (a)

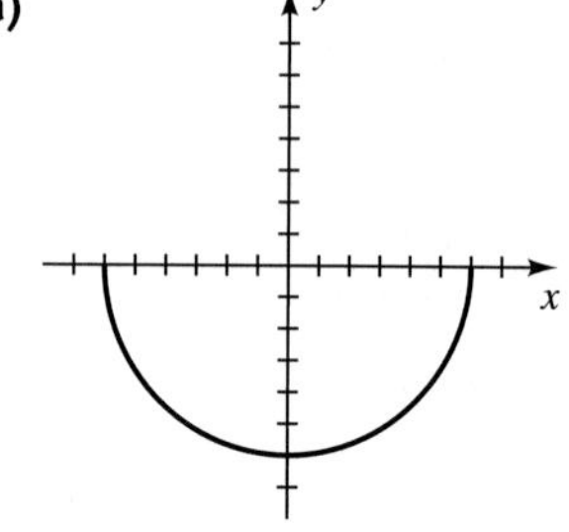

(b) $D = [-6, 6]$,
$R = [-6, 0]$
(c) Decreasing on $[-6, 0]$,
increasing on $[0, 6]$

47 $h + 1$

49 $2x + h$ **51** $\dfrac{1}{\sqrt{x - 3} + \sqrt{a - 3}}$

53 $f(x) = \dfrac{1}{6}x + \dfrac{3}{2}$ **55** Yes **57** No **59** Yes

61 No **63** No **65** $V(x) = 4x(15 - x)(10 - x)$

67 (a) $y(x) = \dfrac{500}{x}$ **(b)** $C(x) = 300x + \dfrac{100{,}000}{x} - 600$

69 $S(h) = 6h - 50$

71 **(a)** $y(t) = 2.5t + 33$

(b)

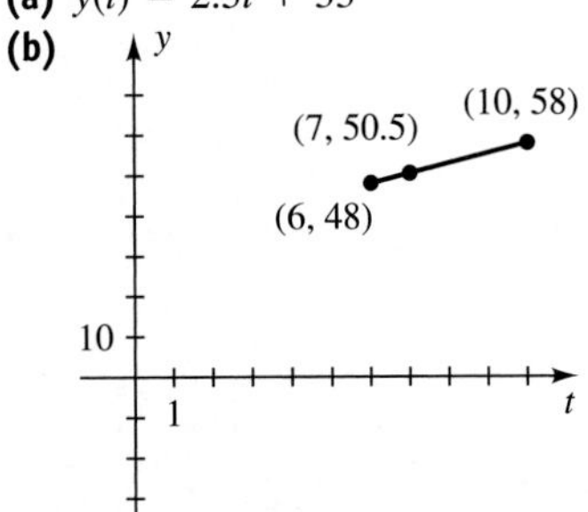

The yearly increase in height

(c) 58 in.

73 $d(t) = 2\sqrt{t^2 + 2500}$

75 **(a)** $y(h) = \sqrt{h^2 + 2hr}$ **(b)** 1280.6 mi

77 $d(x) = \sqrt{90{,}400 + x^2}$

79 **(a)**

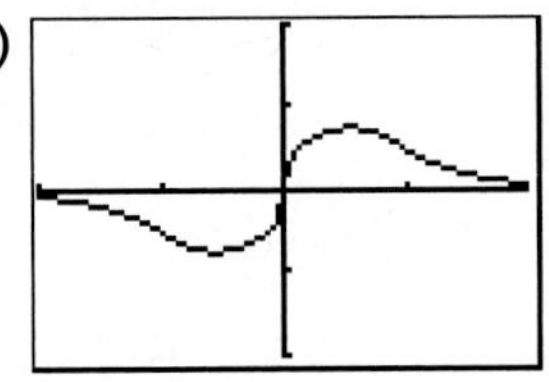

$[-2, 2]$ by $[-2, 2]$

(b) $[-0.75, 0.75]$
(c) Decreasing on $[-2, -0.55]$ and on $[0.55, 2]$, increasing on $[-0.55, 0.55]$

81 **(a)**

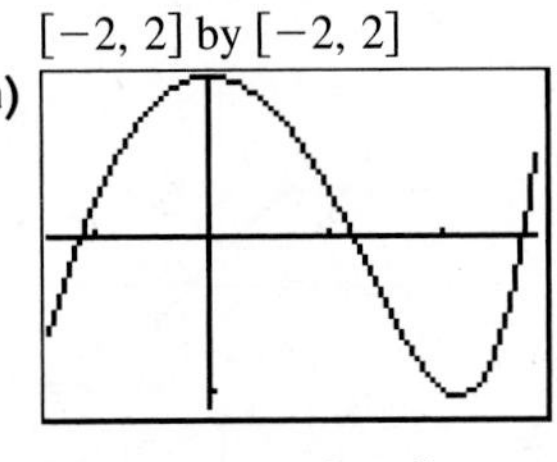

$[-0.7, 1.4, 0.5]$ by $[-1.1, 1]$

(b) $[-1.03, 1]$
(c) Increasing on $[-0.7, 0]$ and on $[1.06, 1.4]$, decreasing on $[0, 1.06]$

83 **(a)** 8 **(b)** ± 8 **(c)** No real solutions **(d)** 625
(e) No real solutions

85 **(a)** 5985 **(b)** At most 95

87 **(a)** $f(x) = \dfrac{3485}{7}x - \dfrac{6{,}827{,}508}{7}$

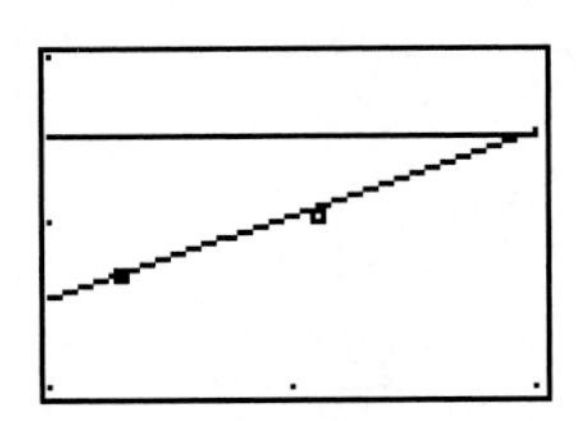

$[1990, 2010, 10]$ by $[10{,}000, 30{,}000, 10{,}000]$

(b) Average annual increase in the price paid for a new car
(c) 2009

EXERCISES 2.5

1 $f(-2) = 7$, $g(-2) = 6$

3 Odd **5** Even **7** Neither **9** Even **11** Odd

13

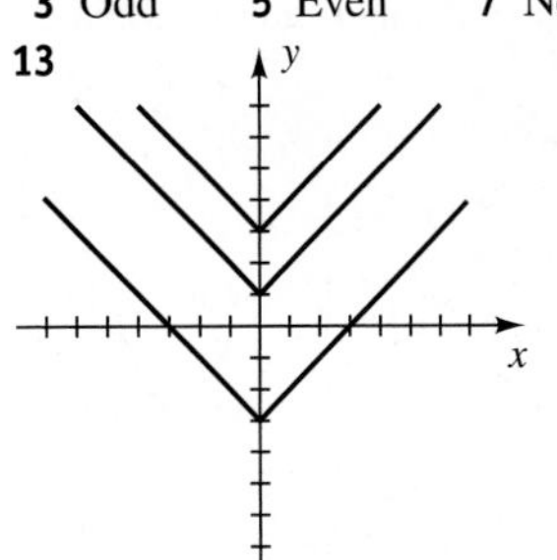

15

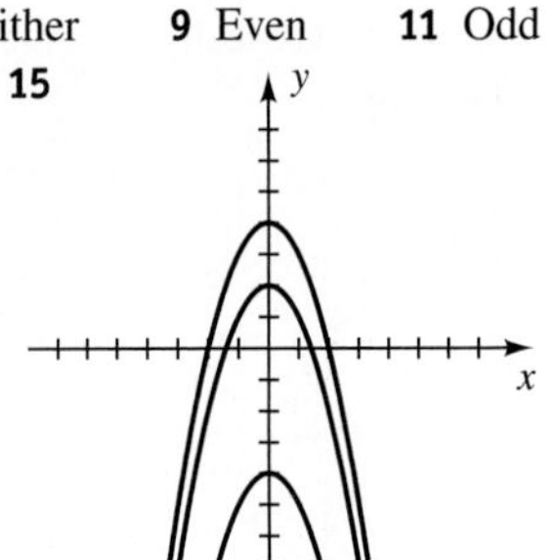

17

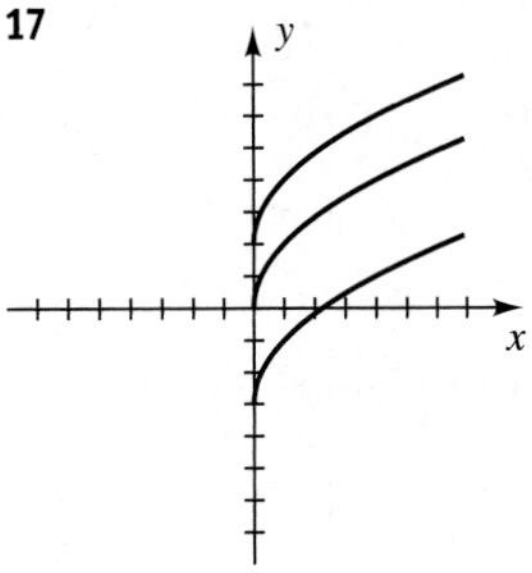

19

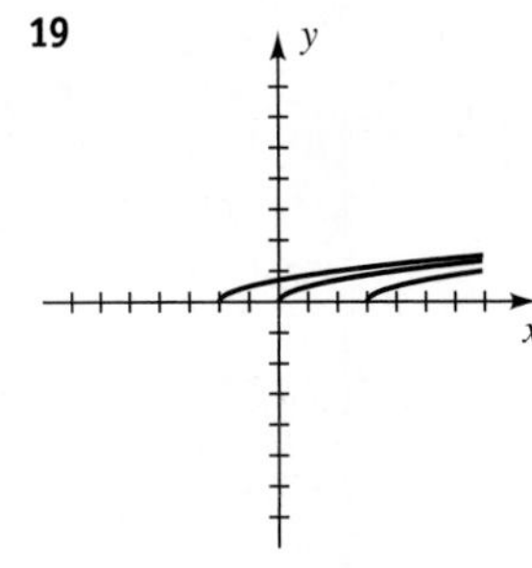

21

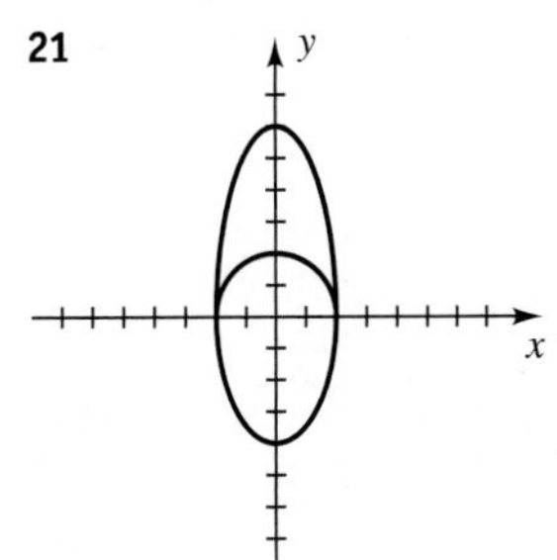

23

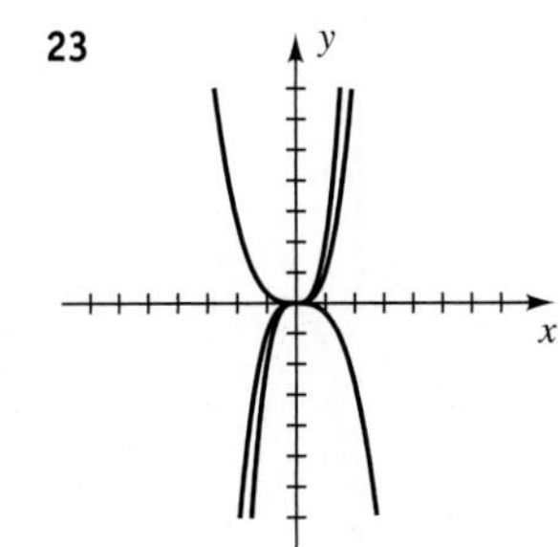

25

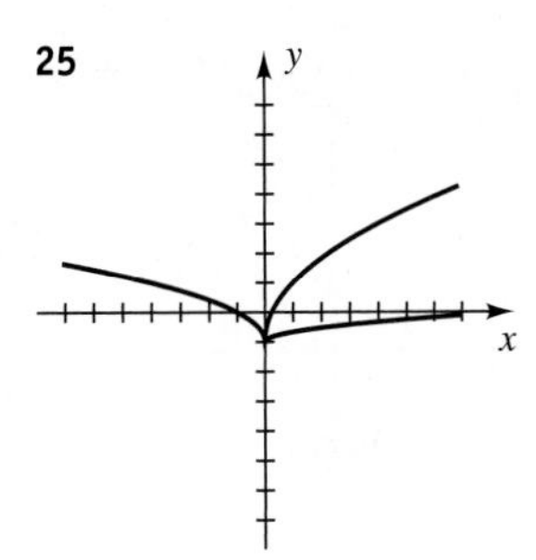

27 $(-2, 4)$ **29** $(7, -3)$ **31** $(6, 2)$

33 The graph of f is shifted 2 units to the right and 3 units up.

35 The graph of f is reflected about the y-axis and shifted 2 units down.

37 The graph of f is compressed vertically by a factor of 2 and reflected about the x-axis.

39 The graph of f is stretched horizontally by a factor of 3, stretched vertically by a factor of 2, and reflected about the x-axis.

41 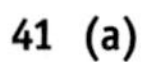

(a)

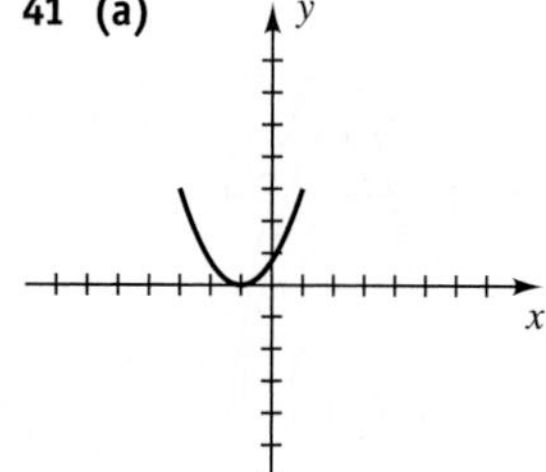

(b)

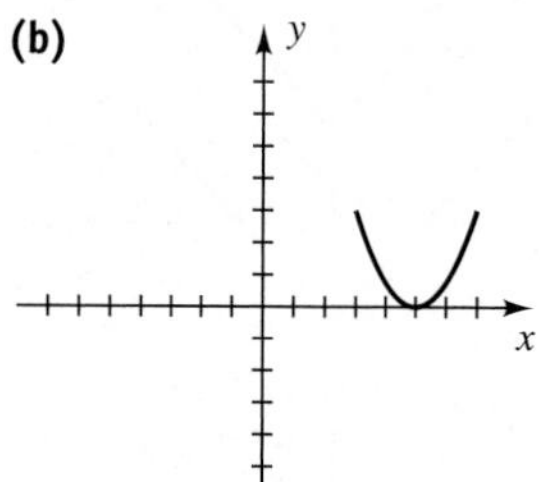

(c)

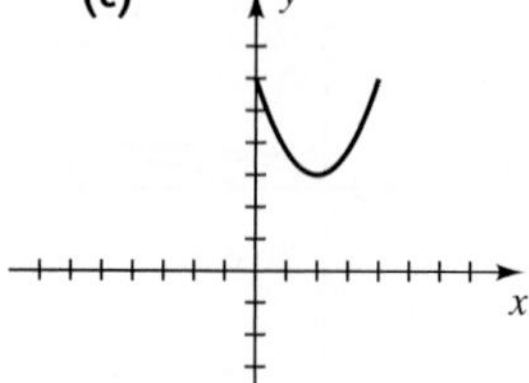

(d)

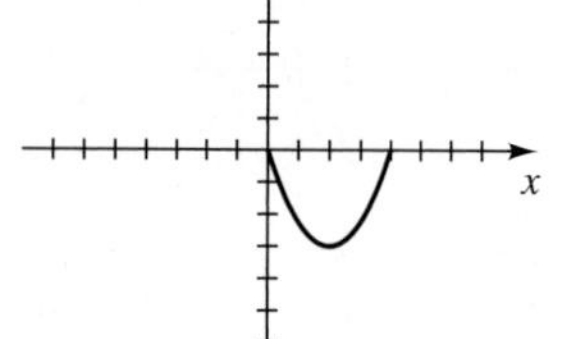

(e)

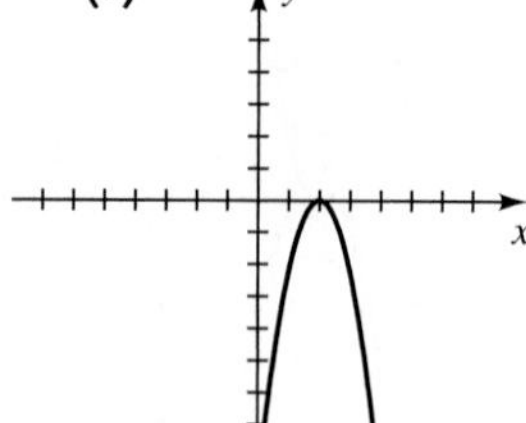

(f)

(g)

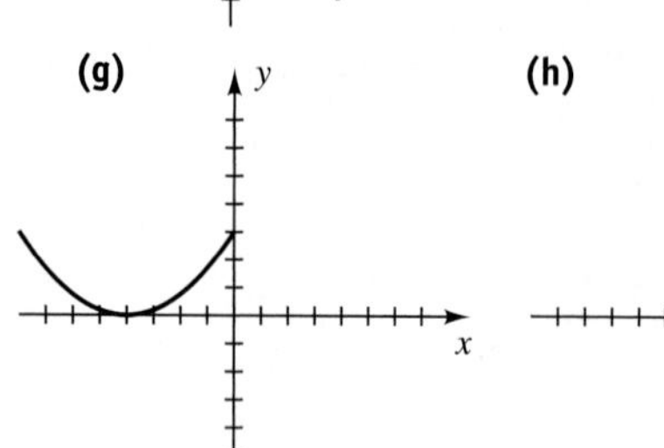

(h)

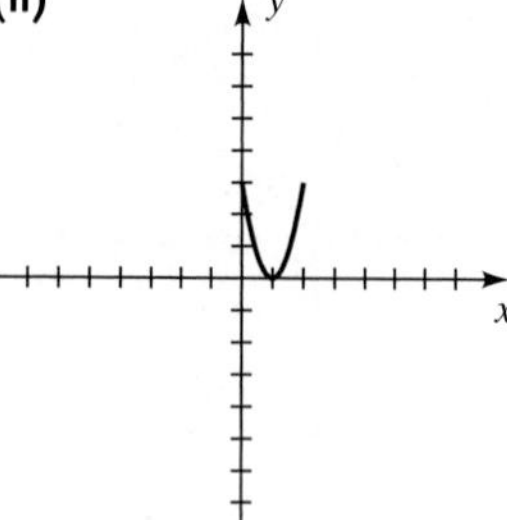

(i)

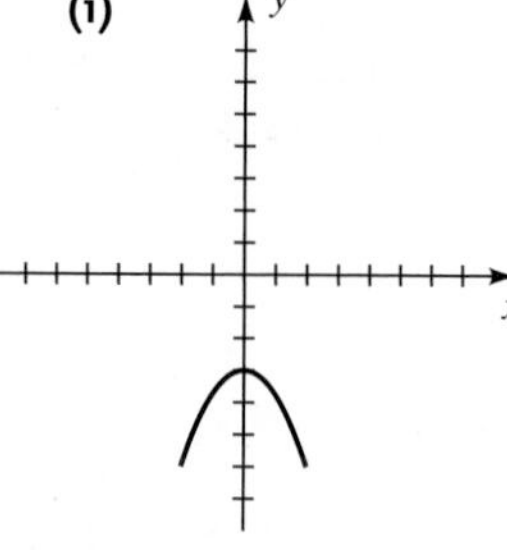

(j)

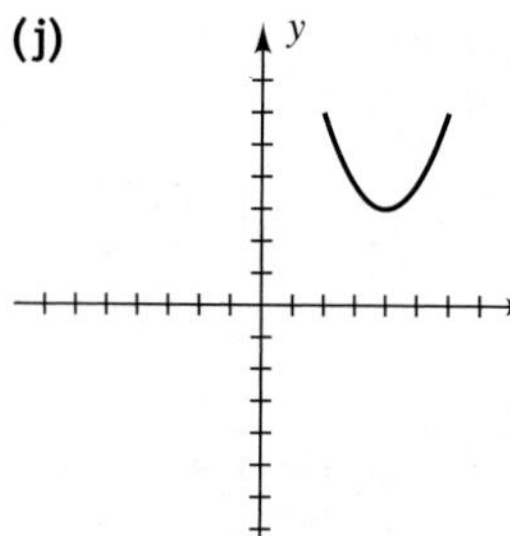

(k)

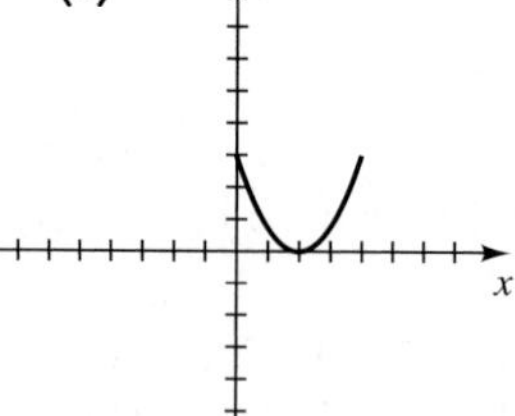

(l)

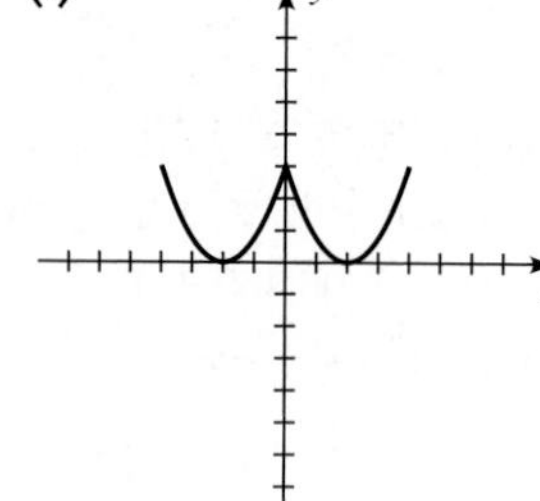

43 **(a)** $y = f(x + 9) + 1$ **(b)** $y = -f(x)$
(c) $y = -f(x + 7) - 1$

45 **(a)** $y = f(x + 4)$ **(b)** $y = f(x) + 1$ **(c)** $y = f(-x)$

47

49

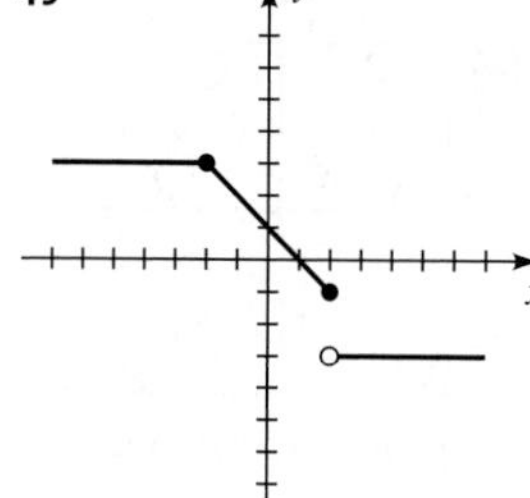

51

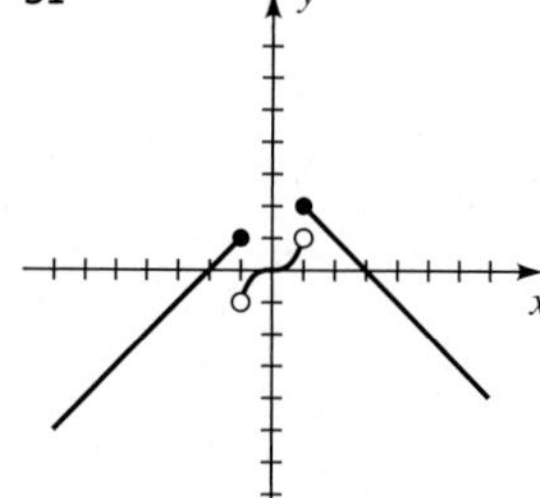

53 (a)

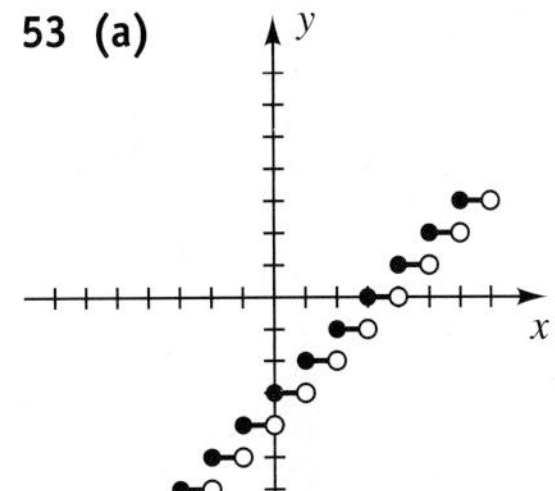

(b)

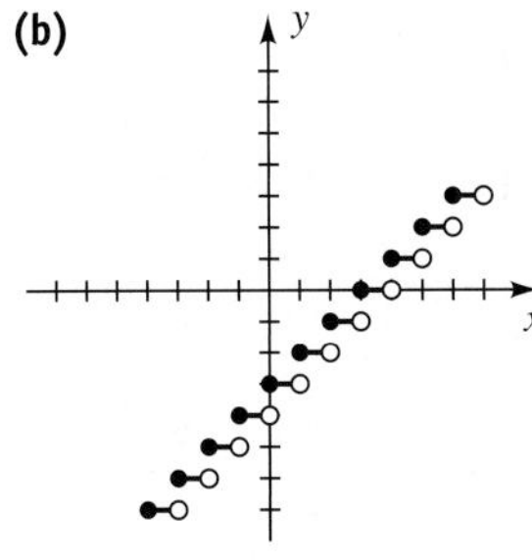

(c)

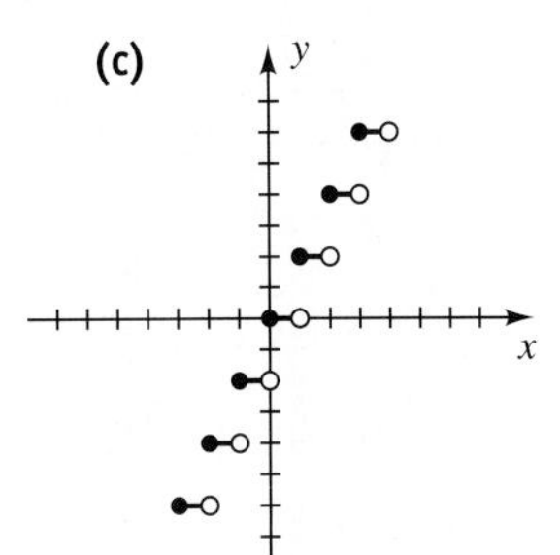

(d)

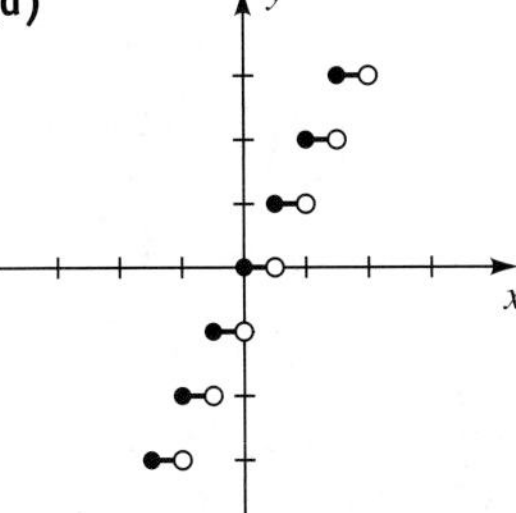

(e)

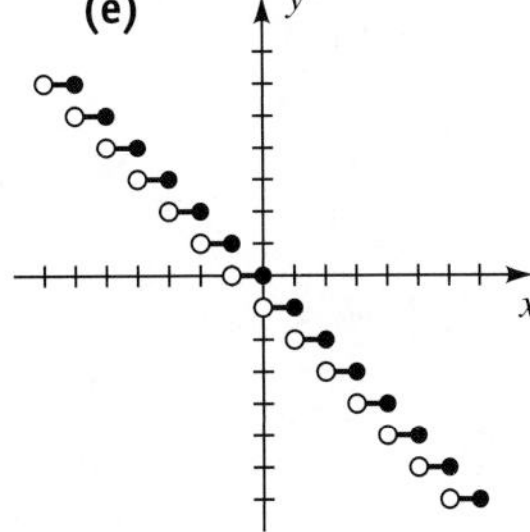

55 If $x > 0$, two different points on the graph have x-coordinate x.

57

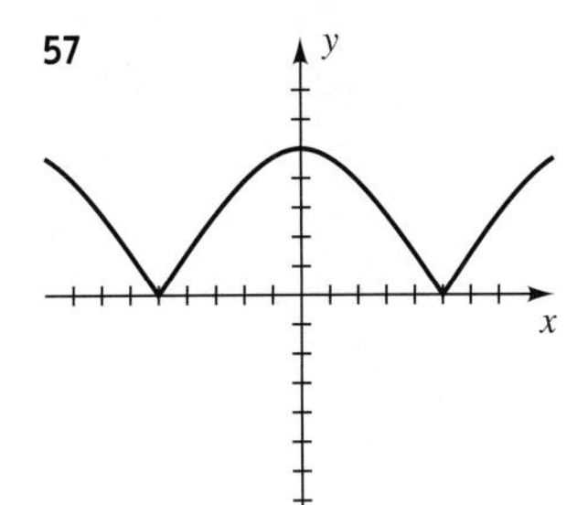

59

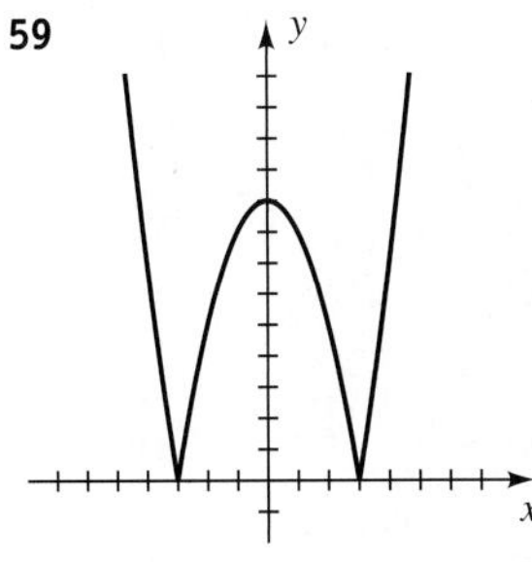

61

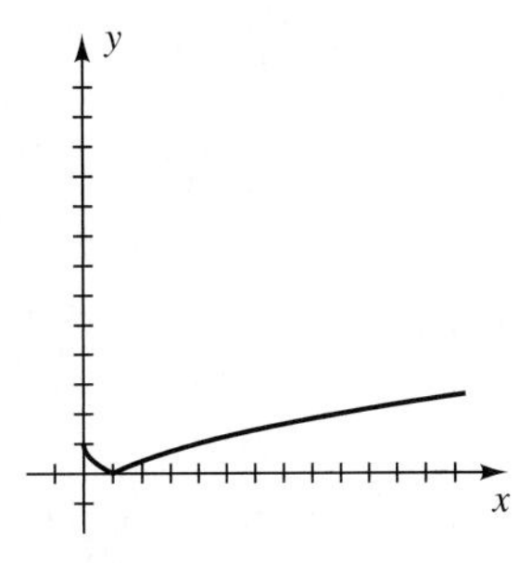

63 **(a)** $D = [-2, 6]$, $R = [-16, 8]$
(b) $D = [-4, 12]$, $R = [-4, 8]$
(c) $D = [1, 9]$, $R = [-3, 9]$
(d) $D = [-4, 4]$, $R = [-7, 5]$
(e) $D = [-6, 2]$, $R = [-4, 8]$
(f) $D = [-2, 6]$, $R = [-8, 4]$
(g) $D = [-6, 6]$, $R = [-4, 8]$
(h) $D = [-2, 6]$, $R = [0, 8]$

65 $T(x) = \begin{cases} 0.15x & \text{if } 0 \le x \le 20{,}000 \\ 0.20x - 1000 & \text{if } x > 20{,}000 \end{cases}$

67 $R(x) = \begin{cases} 1.20x & \text{if } 0 \le x \le 10{,}000 \\ 1.50x - 3000 & \text{if } 10{,}000 < x \le 15{,}000 \\ 1.80x - 7500 & \text{if } x > 15{,}000 \end{cases}$

69 $(-3.12, 22)$

71 $(-\infty, -3) \cup (-3, 1.87) \cup (4.13, \infty)$

73

$[-12, 12]$ by $[-8, 8]$

75

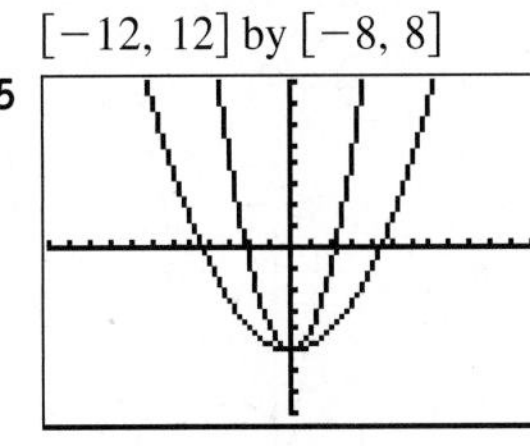

$[-12, 12]$ by $[-8, 8]$

77 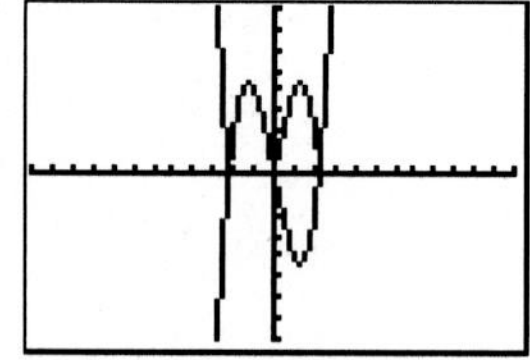

$[-12, 12]$ by $[-8, 8]$

79 (a) \$300, \$360

(b) $C_1(x) = \begin{cases} 180 & \text{if } 0 \le x \le 200 \\ 180 + 0.40(x - 200) & \text{if } x > 200 \end{cases}$

$C_2(x) = 235 + 0.25x$ for $x \ge 0$

(c)

x	Y_1	Y_2
100	180	260
200	180	285
300	220	310
400	260	335
500	300	360
600	340	385
700	380	410
800	420	435
900	460	460
1000	500	485
1100	540	510
1200	580	535

(d) I if $x \in [0, 900)$, II if $x > 900$

EXERCISES 2.6

1 $y = a(x + 3)^2 + 1$ **3** $y = ax^2 - 3$

5 $f(x) = -(x + 2)^2 - 4$ **7** $f(x) = 2(x - 3)^2 + 4$

9 $f(x) = -3(x + 1)^2 - 2$

11 $f(x) = -\frac{3}{4}(x - 6)^2 - 7$

13 (a) 0, 4
(b) Min: $f(2) = -4$
(c)

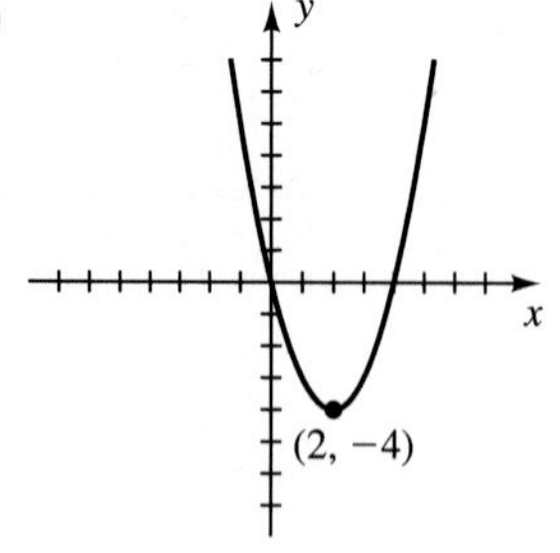

15 (a) $-\frac{3}{4}, \frac{5}{3}$
(b) Max: $f\left(\frac{11}{24}\right) = \frac{841}{48}$
(c)

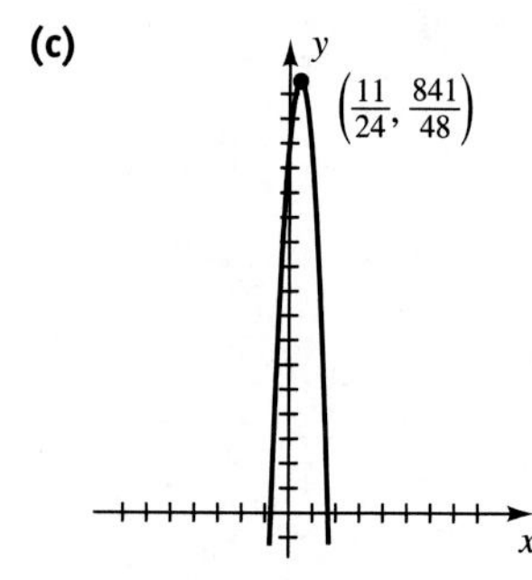

17 (a) $-\frac{4}{3}$

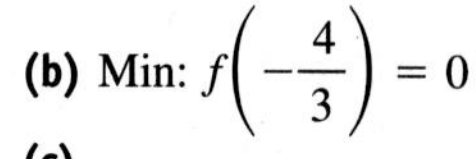

(b) Min: $f\left(-\frac{4}{3}\right) = 0$
(c)

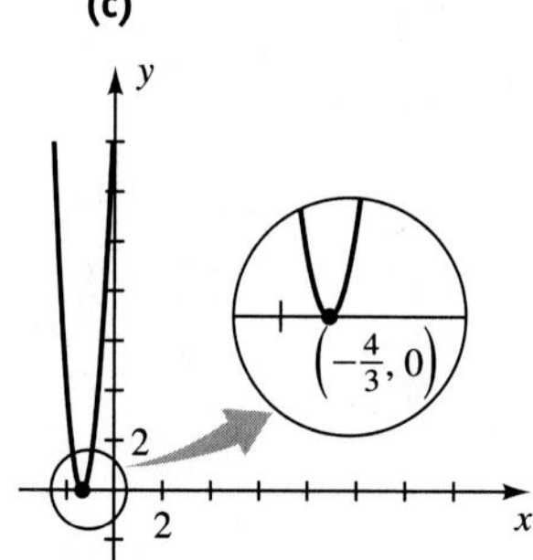

19 (a) None
(b) Min: $f(-2) = 5$
(c)

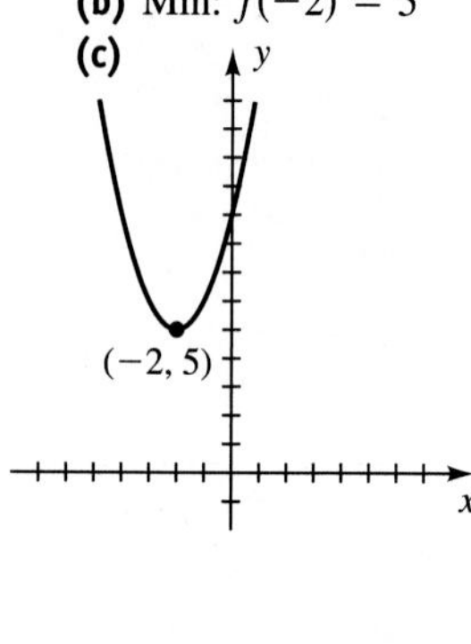

21 (a) $5 \pm \frac{1}{2}\sqrt{14} \approx 6.87, 3.13$ **(b)** Max: $f(5) = 7$
(c)

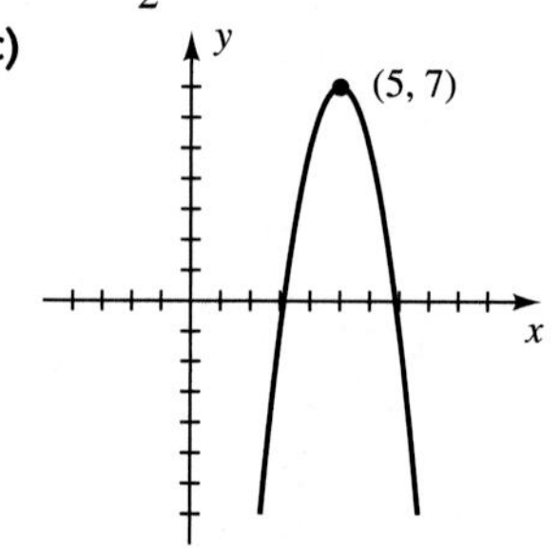

23 $y = \frac{1}{8}(x - 4)^2 - 1$ **25** $y = -\frac{4}{9}(x + 2)^2 + 4$

27 $y = -\frac{1}{2}(x + 2)(x - 4)$

29 $y = 3(x - 0)^2 - 2$ **31** $y = -\frac{5}{9}(x - 3)^2 + 5$

33 $y = -\frac{1}{4}(x - 1)^2 + 4$ **35** 6.125 **37** 24.72 km

39 10.5 lb **41 (a)** 424 ft **(b)** 100 ft **43** 20 and 20

45 **(a)** $y(x) = 250 - \frac{3}{4}x$ **(b)** $A(x) = x\left(250 - \frac{3}{4}x\right)$

(c) $166\frac{2}{3}$ ft by 125 ft

47 $y = -\frac{4}{27}\left(x - \frac{9}{2}\right)^2 + 3$

49 **(a)** $y = \frac{1}{500}x^2 + 10$ **(b)** 282 ft **51** 2 ft

53 500 pairs

55 **(a)** $R(x) = 200x(90 - x)$

(b) \$45

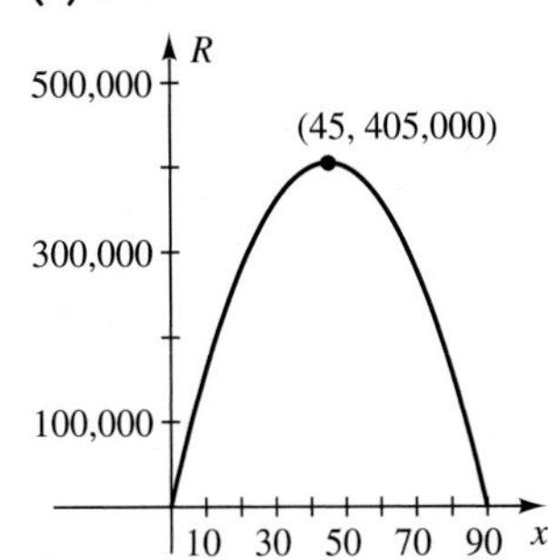

57 (−0.57, 0.64), (0.02, −0.27), (0.81, −0.41)

[−3, 3] by [−2, 2]

59 Smaller values of a result in a wider parabola; larger values of a result in a narrower parabola.

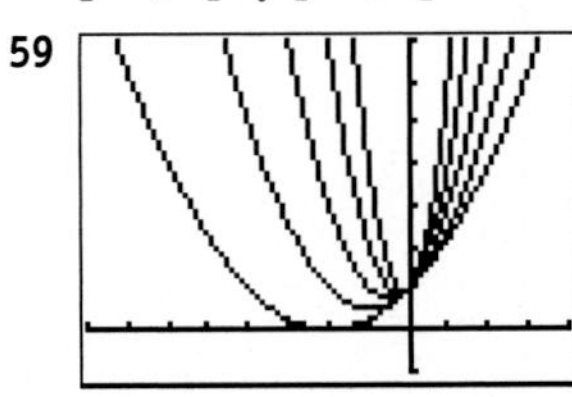

[−8, 4] by [−1, 7]

61 **(b)** $f(x) = 0.17(x - 7)^2 + 0.77$

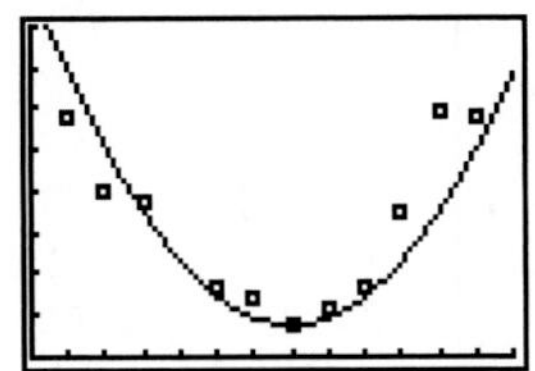

[0, 13] by [0, 8]

(c) 2.3 in.

63 **(a)** $f(x) = \begin{cases} \frac{4}{25}x + 80 & \text{if } -800 \le x < -500 \\ -\frac{1}{6250}x^2 + 40 & \text{if } -500 \le x \le 500 \\ -\frac{4}{25}x + 80 & \text{if } 500 < x \le 800 \end{cases}$

(b)

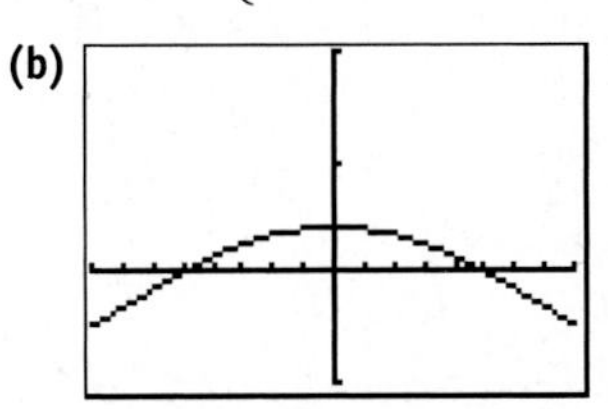

[−800, 800, 100] by [−100, 200, 100]

65 **(a)** $f(x) = -\frac{4}{225}x^2 + \frac{8}{3}x$

(b)

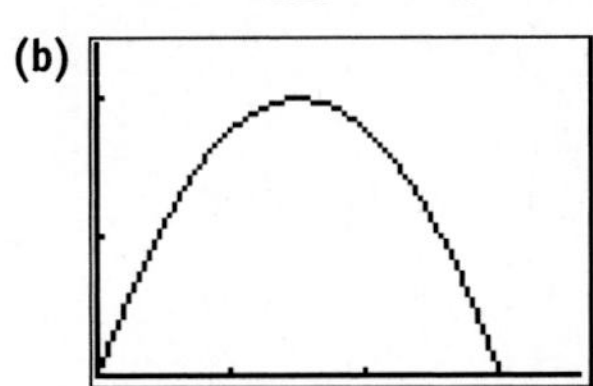

[0, 180, 50] by [0, 120, 50]

(c)

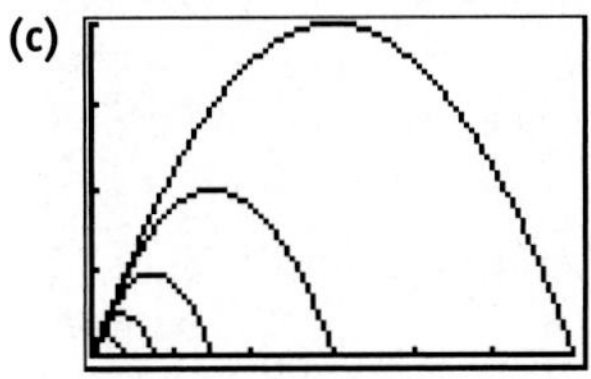

[0, 600, 50] by [0, 400, 50]
The value of k affects both the height and the distance traveled by a factor of $\frac{1}{k}$.

EXERCISES 2.7

1 **(a)** 15 **(b)** −3 **(c)** 54 **(d)** $\frac{2}{3}$

3 **(a)** $3x^2 + 1$; $3 - x^2$; $2x^4 + 3x^2 - 2$; $\frac{x^2 + 2}{2x^2 - 1}$

(b) $\mathbb{R}$ **(c)** All real numbers except $\pm\frac{1}{2}\sqrt{2}$

5 **(a)** $2\sqrt{x + 5}$; 0; $x + 5$; 1 **(b)** $[-5, \infty)$ **(c)** $(-5, \infty)$

7 **(a)** $\frac{3x^2 + 6x}{(x - 4)(x + 5)}$; $\frac{x^2 + 14x}{(x - 4)(x + 5)}$; $\frac{2x^2}{(x - 4)(x + 5)}$; $\frac{2(x + 5)}{x - 4}$

(b) All real numbers except -5 and 4
(c) All real numbers except -5, 0, and 4

9 (a) $-2x^2 - 1$ **(b)** $-4x^2 + 4x - 1$ **(c)** $4x - 3$ **(d)** $-x^4$

11 (a) $6x + 9$ **(b)** $6x - 8$ **(c)** -3 **(d)** 10

13 (a) $75x^2 + 4$ **(b)** $15x^2 + 20$ **(c)** 304 **(d)** 155

15 (a) $8x^2 - 2x - 5$ **(b)** $4x^2 + 6x - 9$ **(c)** 31 **(d)** 45

17 (a) $8x^3 - 20x$ **(b)** $128x^3 - 20x$ **(c)** -24 **(d)** 3396

19 (a) 7 **(b)** -7 **(c)** 7 **(d)** -7

21 (a) $x + 2 - 3\sqrt{x + 2}$; $[-2, \infty)$
(b) $\sqrt{x^2 - 3x + 2}$; $(-\infty, 1] \cup [2, \infty)$

23 (a) $3x - 4$; $[0, \infty)$
(b) $\sqrt{3x^2 - 12}$; $(-\infty, -2] \cup [2, \infty)$

25 (a) $\sqrt{\sqrt{x + 5} - 2}$; $[-1, \infty)$
(b) $\sqrt{\sqrt{x - 2} + 5}$; $[2, \infty)$

27 (a) $\sqrt{3 - \sqrt{x^2 - 16}}$; $[-5, -4] \cup [4, 5]$
(b) $\sqrt{-x - 13}$; $(-\infty, -13]$

29 (a) x; $\mathbb{R}$ **(b)** x; $\mathbb{R}$

31 (a) $\dfrac{1}{x^6}$; all nonzero real numbers
(b) $\dfrac{1}{x^6}$; all nonzero real numbers

33 (a) $\dfrac{1}{5 - x}$; all real numbers except 4 and 5
(b) $\dfrac{-2x + 5}{-3x + 7}$; all real numbers except 2 and $\dfrac{7}{3}$

35 $-3 \pm \sqrt{2}$

37 (a) 5 **(b)** 6 **(c)** 6 **(d)** 5 **(e)** Not possible

39 $20\sqrt{x^2 + 1}$ **41** Odd **43** 40.16

45 $A(t) = 36\pi t^2$ **47** $r(t) = 9\sqrt[3]{t}$

49 $h(t) = 5\sqrt{t^2 + 8t}$

51 $d(t) = \sqrt{90{,}400 + (500 + 150t)^2}$

Exer. 53–60: Answers are not unique.

53 $u = x^2 + 3x$, $y = u^{1/3}$ **55** $u = x - 3$, $y = u^{-4}$

57 $u = x^4 - 2x^2 + 5$, $y = u^5$

59 $u = \sqrt{x + 4}$, $y = \dfrac{u - 2}{u + 2}$ **61** 5×10^{-13}

63 (a) $Y_1 = x$, graph $Y_3 = -2Y_2$

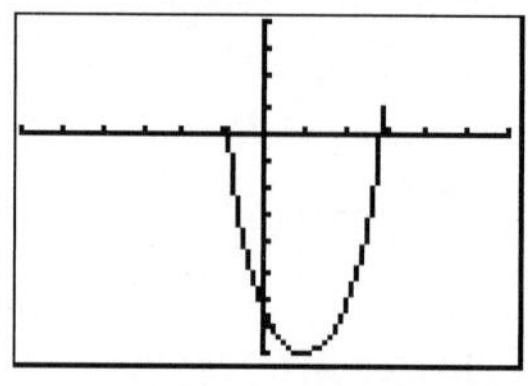

$[-12, 12, 2]$ by $[-16, 8, 2]$

(b) $Y_1 = 0.5x$, graph Y_2

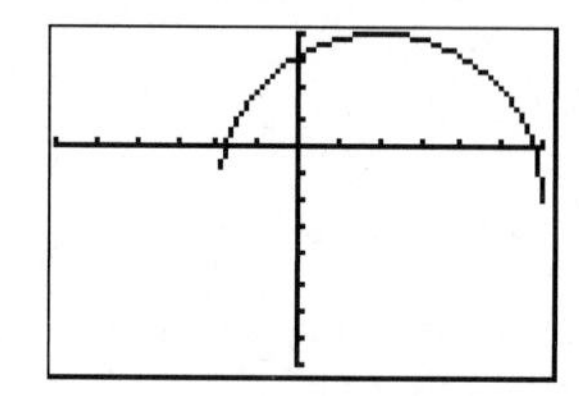

$[-12, 12, 2]$ by $[-16, 8, 2]$

(c) $Y_1 = x - 3$, graph $Y_3 = Y_2 + 1$

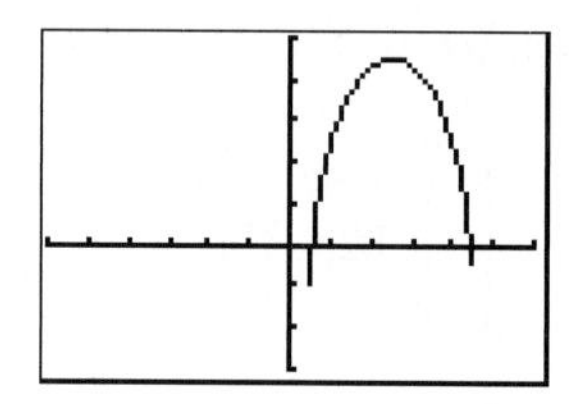

$[-12, 12, 2]$ by $[-6, 10, 2]$

(d) $Y_1 = x + 2$, graph $Y_3 = Y_2 - 3$

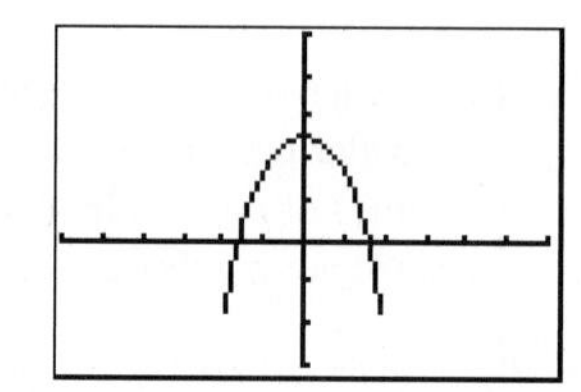

$[-12, 12, 2]$ by $[-6, 10, 2]$

(e) $Y_1 = -x$, graph Y_2

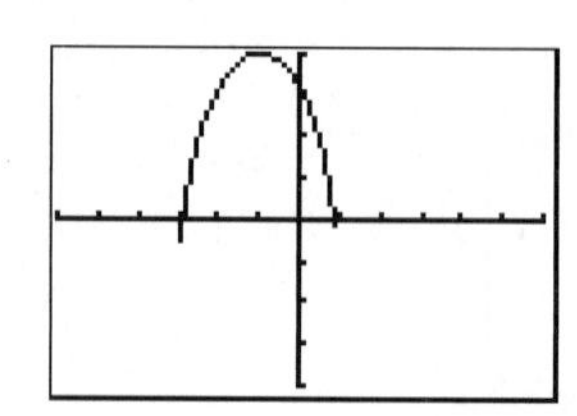

$[-12, 12, 2]$ by $[-8, 8, 2]$

(f) $Y_1 = x$, graph $Y_3 = -Y_2$

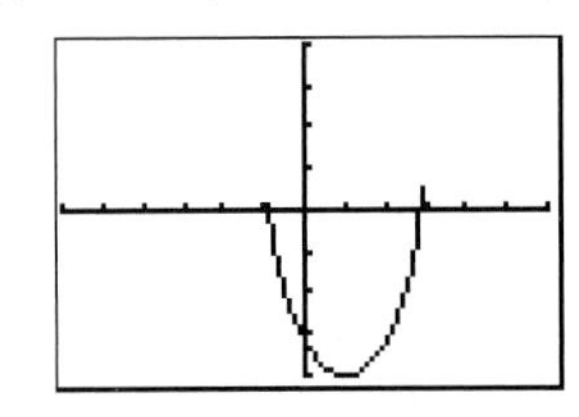

$[-12, 12, 2]$ by $[-8, 8, 2]$

(g) $Y_1 = \text{abs } x$, graph Y_2

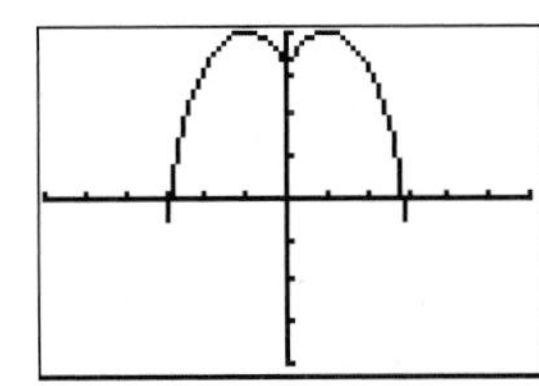

$[-12, 12, 2]$ by $[-8, 8, 2]$

(h) $Y_1 = x$, graph $Y_3 = \text{abs } Y_2$

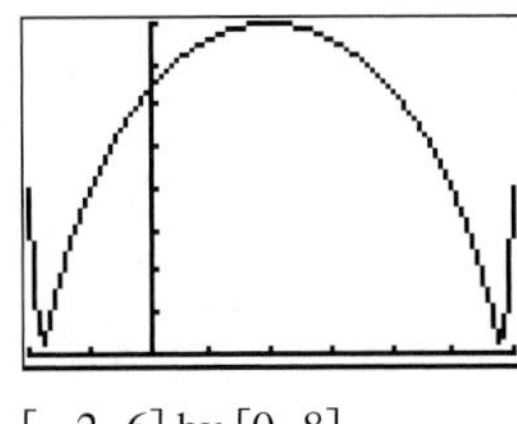

$[-2, 6]$ by $[0, 8]$

CHAPTER 2 REVIEW EXERCISES

1 The points in quadrants II and IV

2 $d(A, B)^2 + d(A, C)^2 = d(B, C)^2$; area $= 10$

3 **(a)** $\sqrt{265}$ **(b)** $\left(-\dfrac{13}{2}, 1\right)$ **(c)** $(-11, -23)$

4 $(0, 1), (0, 11)$ **5** $-2 < a < 1$

6 $(x - 7)^2 + (y + 4)^2 = 149$

7 $(x - 3)^2 + (y + 2)^2 = 169$

8 $x = -2 - \sqrt{9 - y^2}$ **9** $-\dfrac{11}{19}$

10 The slope of AD and BC is $\dfrac{2}{3}$.

11 **(a)** $18x + 6y = 7$ **(b)** $2x - 6y = 3$

12 $y = -\dfrac{8}{3}x + 8$ **13** $(x + 5)^2 + (y + 1)^2 = 81$

14 $x + y = -3$ **15** $5x - y = 23$

16 $2x - 3y = 5$ **17** $C(0, 6)$; $r = \sqrt{5}$

18 $C(-3, 2)$; $r = \dfrac{1}{2}\sqrt{13}$

19 **(a)** $\dfrac{1}{2}$ **(b)** $-\dfrac{1}{\sqrt{2}}$ **(c)** 0 **(d)** $-\dfrac{x}{\sqrt{3 - x}}$

(e) $-\dfrac{x}{\sqrt{x + 3}}$ **(f)** $\dfrac{x^2}{\sqrt{x^2 + 3}}$ **(g)** $\dfrac{x^2}{x + 3}$

20 Positive **21** Positive

22 **(a)** $\left[\dfrac{4}{3}, \infty\right)$; $[0, \infty)$

(b) All real numbers except -3; $(0, \infty)$

23 $-2a - h + 1$ **24** $-\dfrac{1}{(a + h + 2)(a + 2)}$

25 $f(x) = \dfrac{5}{2}x - \dfrac{1}{2}$

26 **(a)** Odd **(b)** Neither **(c)** Even

Exer. 27–40: x-intercept(s) is listed, followed by y-intercept(s).

27 -5; none **28** None; 3.5

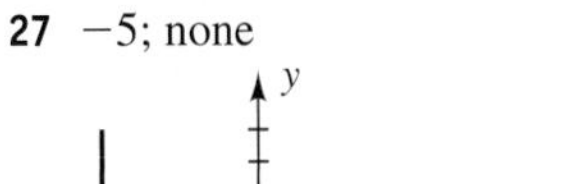

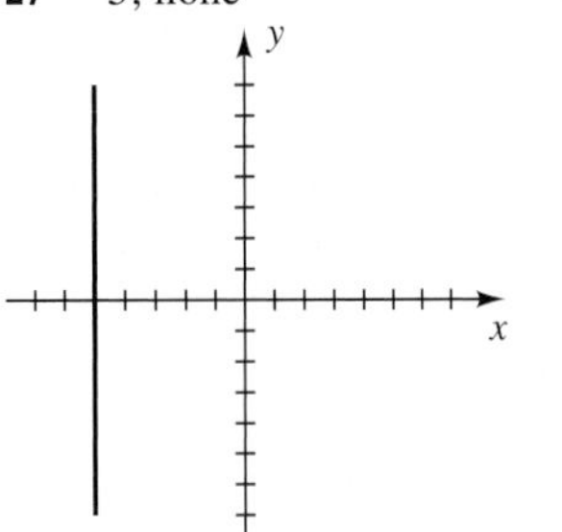

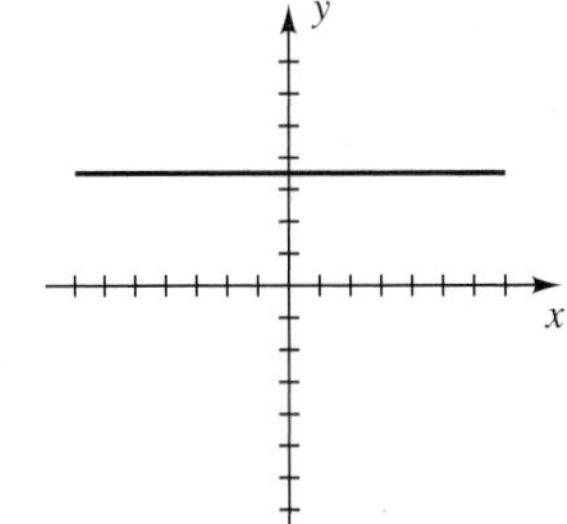

29 1.6; 4 **30** 4; $-\frac{4}{3}$

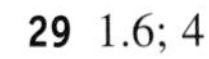

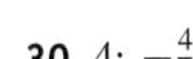

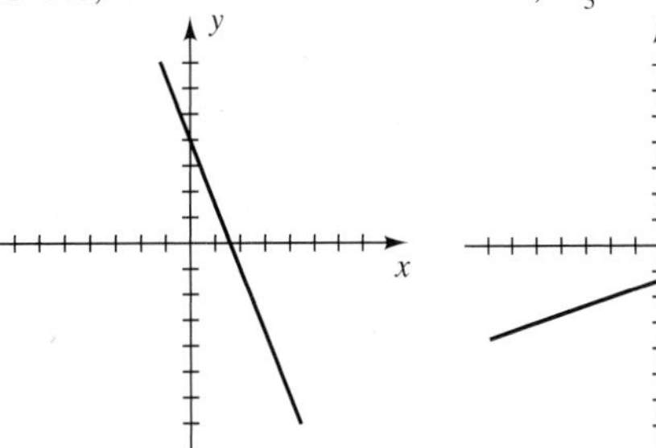

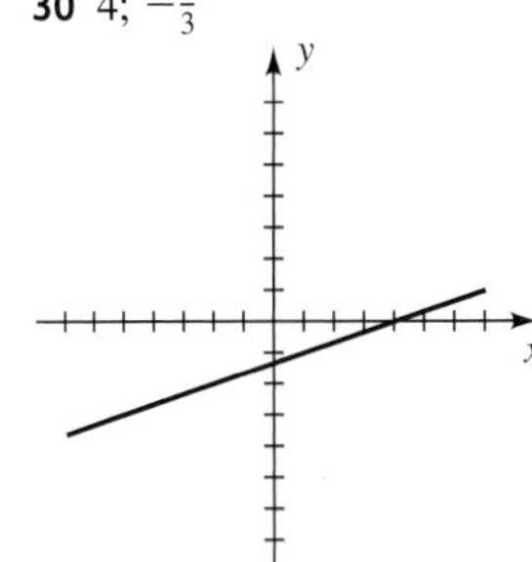

31 0; 0 **32** 0; 0

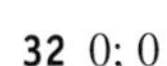

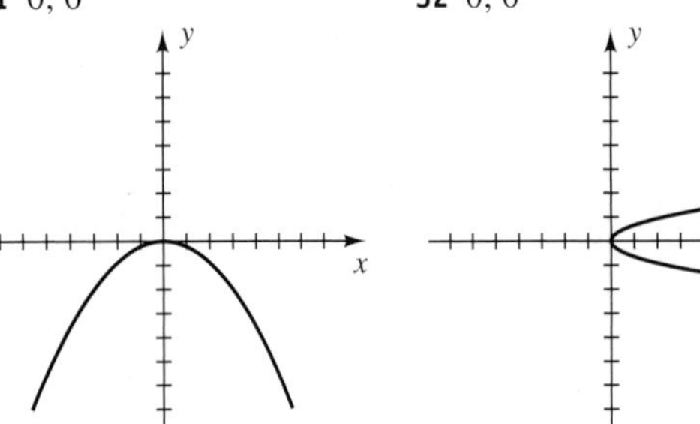

33 1; 1

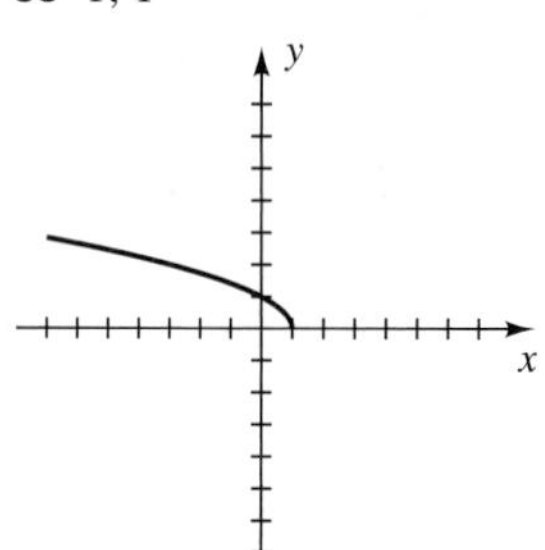

34 1; -1

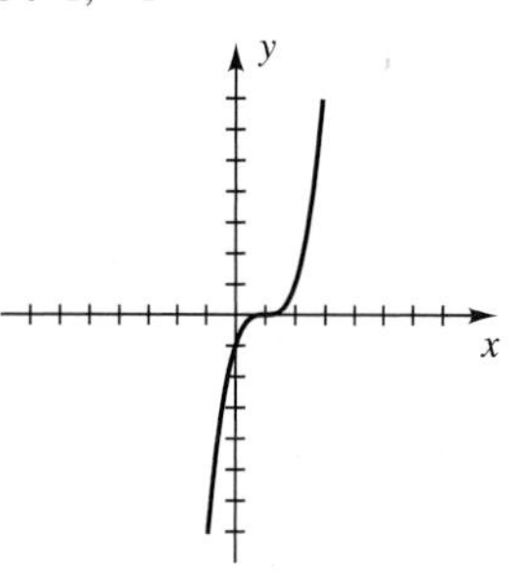

35 ± 4; ± 4

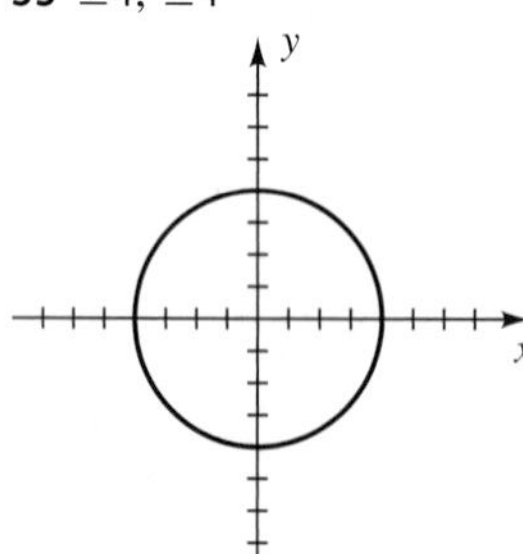

36 None; 8

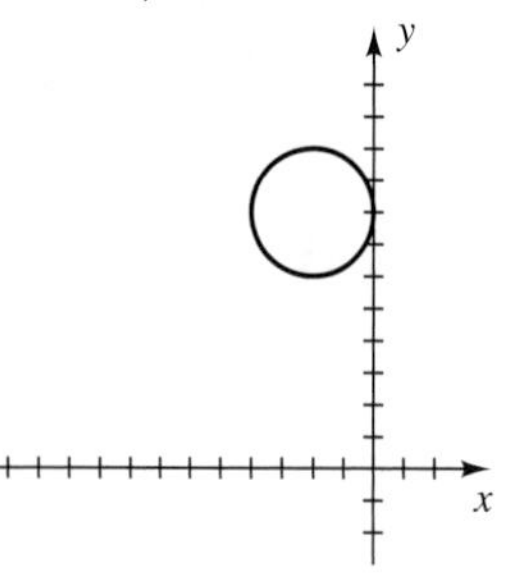

37 0, 8; 0

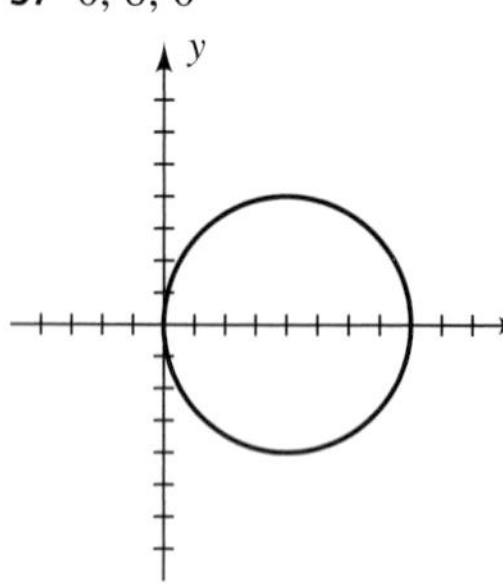

38 -3; ± 3

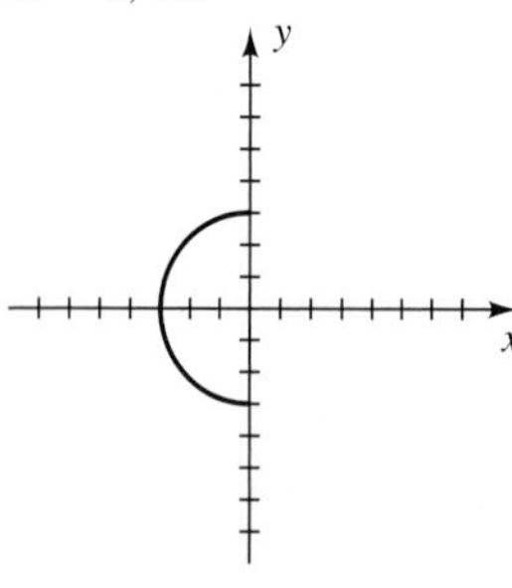

39 $3 \pm \sqrt{2}$; 7

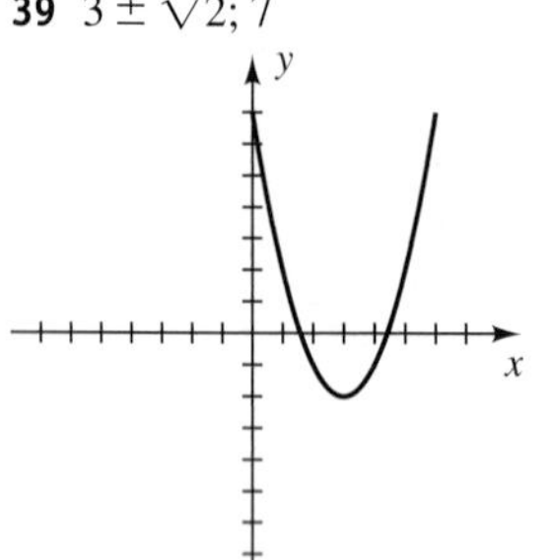

40 -3, 1; 3

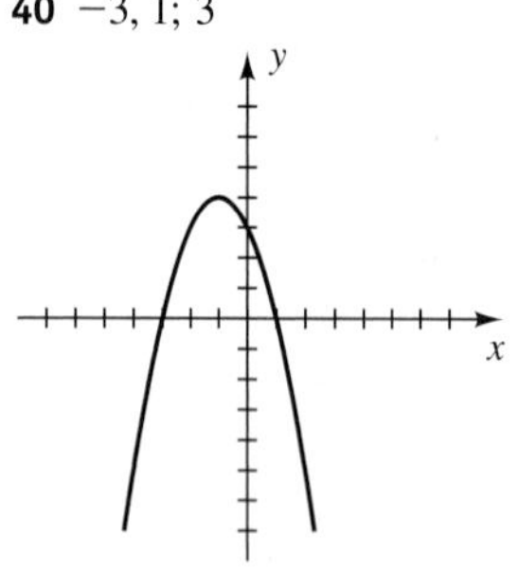

41 $\left(\sqrt{8}, \sqrt{8}\right)$

42 The graph of $y = -f(x - 2)$ is the graph of $y = f(x)$ shifted to the right 2 units and reflected about the x-axis.

43 (a)

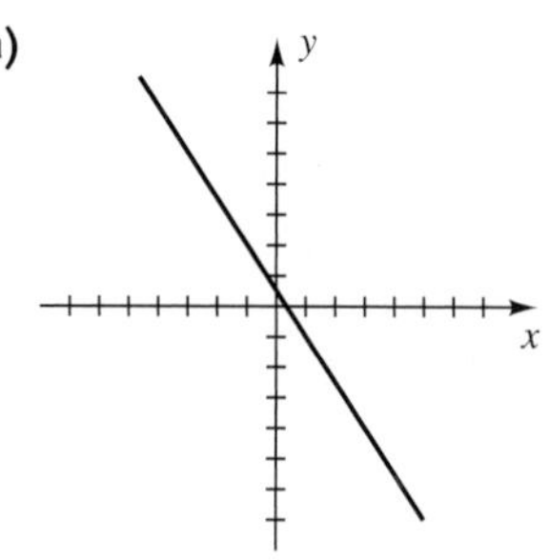

(b) $D = \mathbb{R}$; $R = \mathbb{R}$
(c) Decreasing on $(-\infty, \infty)$

44 (a)

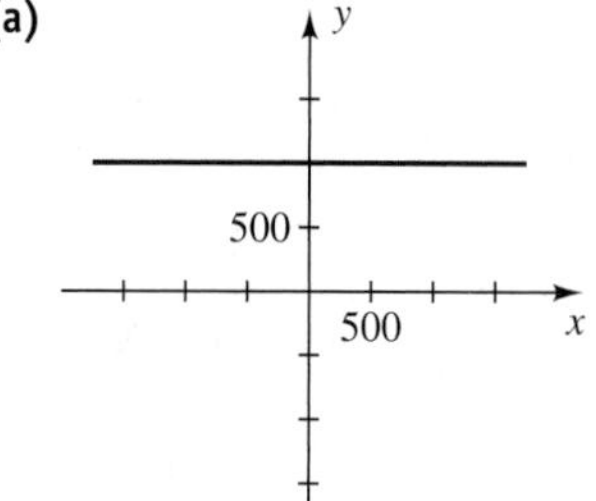

(b) $D = \mathbb{R}$; $R = \{1000\}$
(c) Constant on $(-\infty, \infty)$

45 (a)

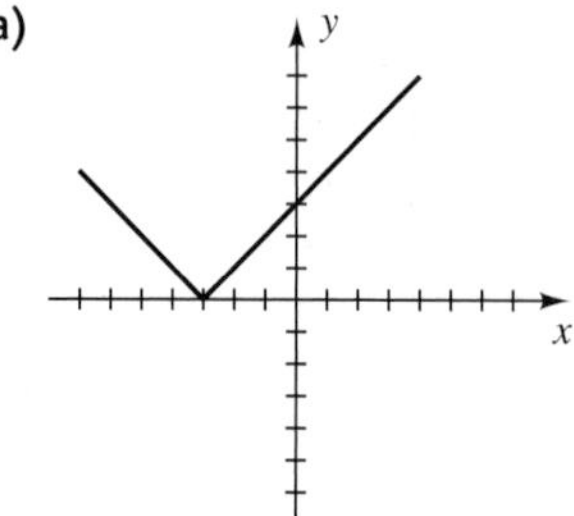

(b) $D = \mathbb{R}$; $R = [0, \infty)$
(c) Decreasing on $(-\infty, -3]$, increasing on $[-3, \infty)$

46 (a)

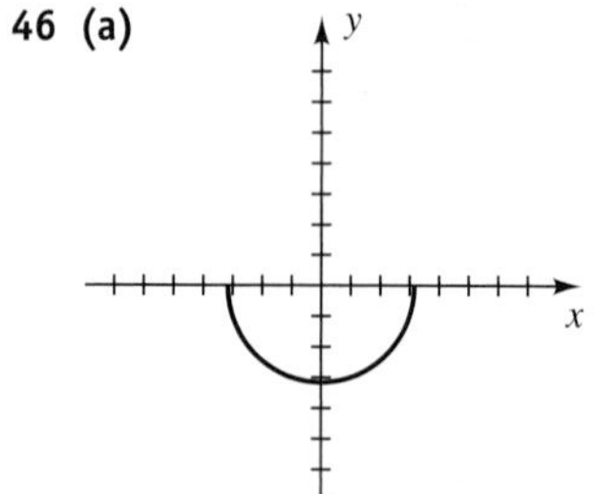

(b) $D = \left(-\sqrt{10}, \sqrt{10}\right)$; $R = \left(-\sqrt{10}, 0\right)$
(c) Decreasing on $\left[-\sqrt{10}, 0\right]$, increasing on $\left[0, \sqrt{10}\right]$

47 (a)

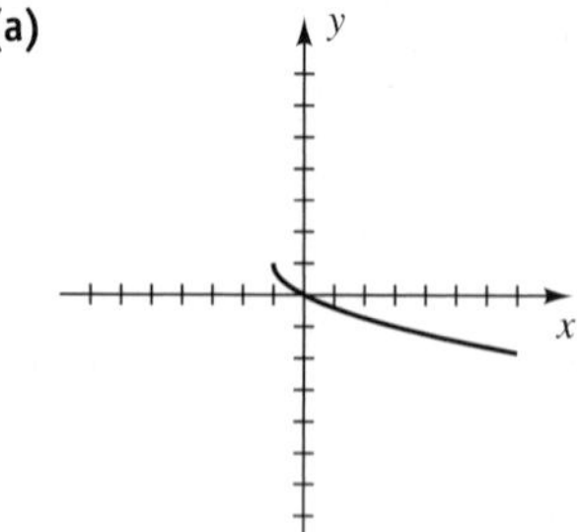

(b) $D = [-1, \infty)$; $R = (-\infty, 1]$
(c) Decreasing on $[-1, \infty)$

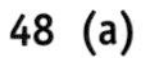
48 (a)

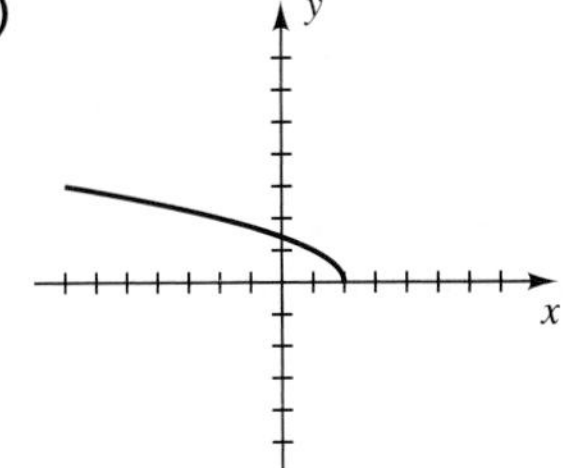

(b) $D = (-\infty, 2]$; $R = [0, \infty)$
(c) Decreasing on $(-\infty, 2]$

49 (a)

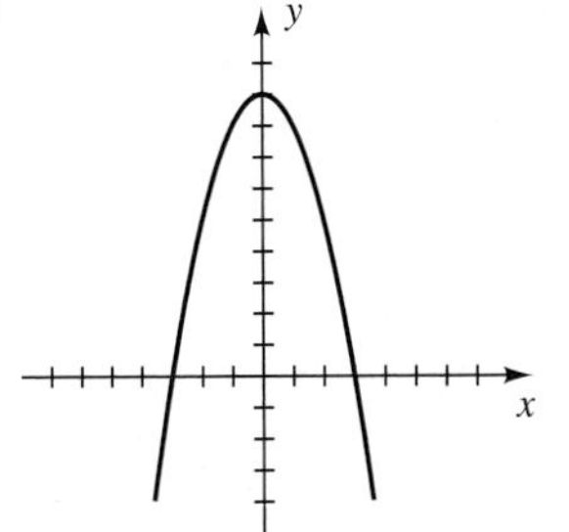

(b) $D = \mathbb{R}$; $R = (-\infty, 9]$
(c) Increasing on $(-\infty, 0]$, decreasing on $[0, \infty)$

50 (a)

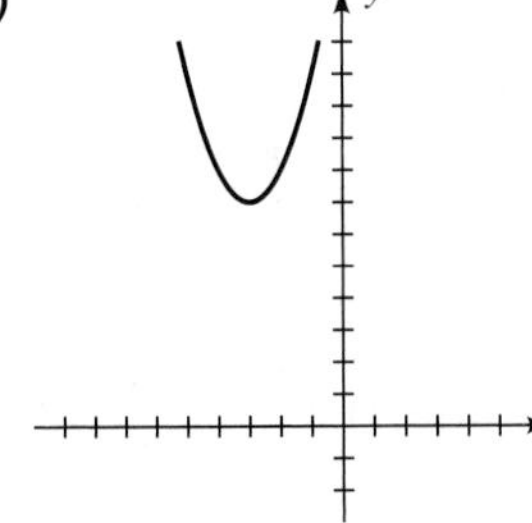

(b) $D = \mathbb{R}$; $R = [7, \infty)$
(c) Decreasing on $(-\infty, -3]$, increasing on $[-3, \infty)$

51 (a)

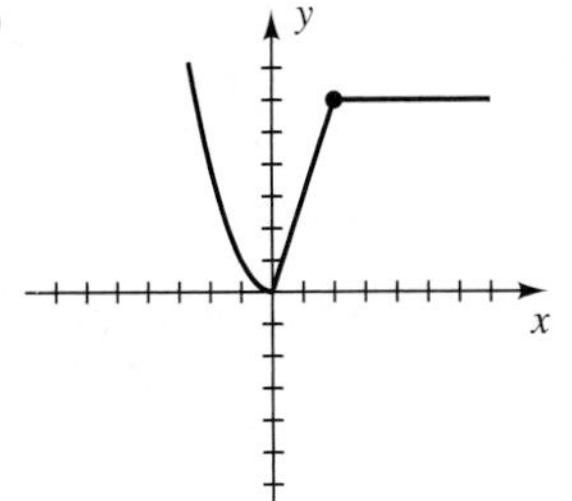

(b) $D = \mathbb{R}$; $R = [0, \infty)$
(c) Decreasing on $(-\infty, 0]$, increasing on $[0, 2]$, constant on $[2, \infty)$

52 (a)

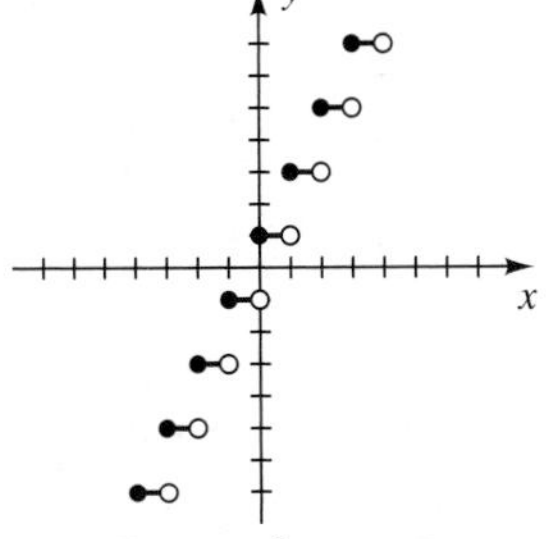

(b) $D = \mathbb{R}$; $R = \{\ldots, -3, -1, 1, 3, \ldots\}$
(c) Constant on $[n, n + 1)$, where n is any integer

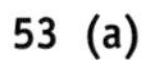
53 (a)

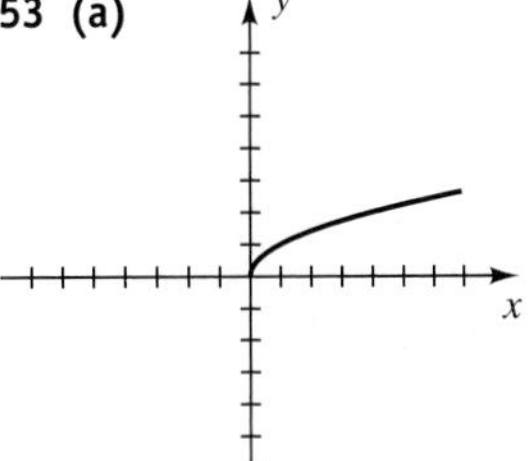

(b)

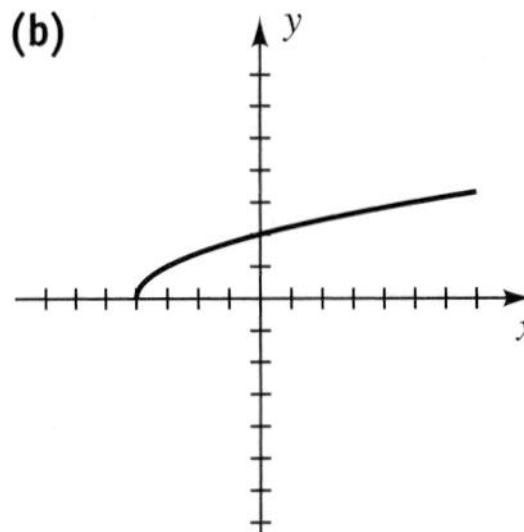

(c)

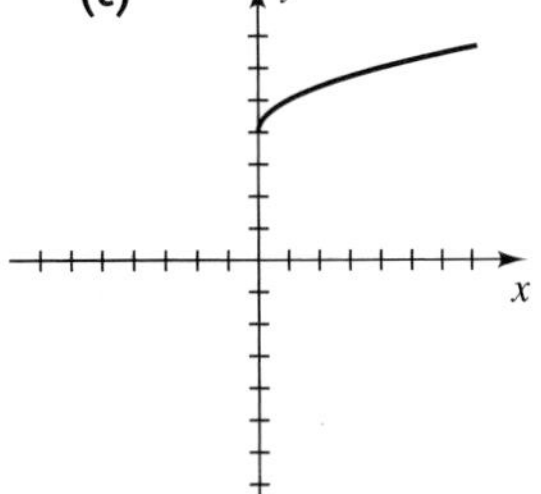

(d)

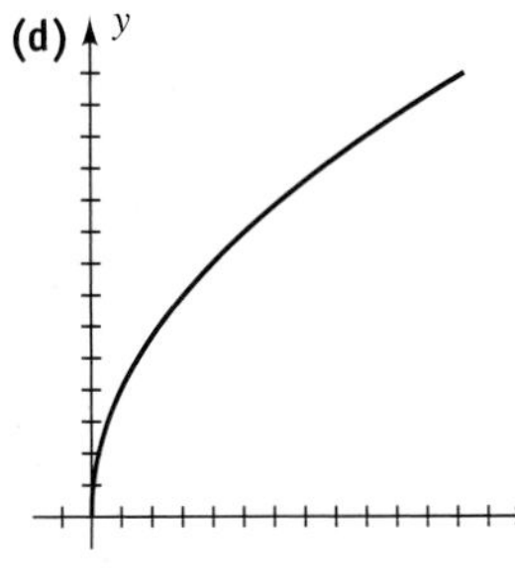

(e)

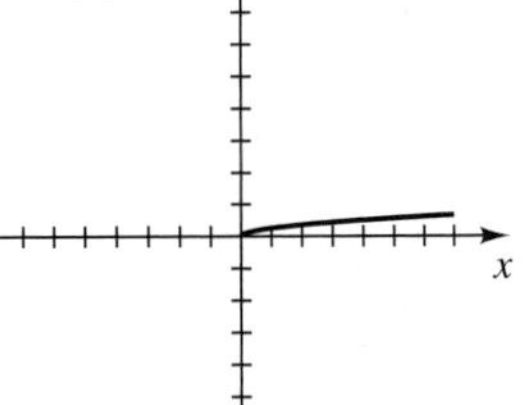

(f)

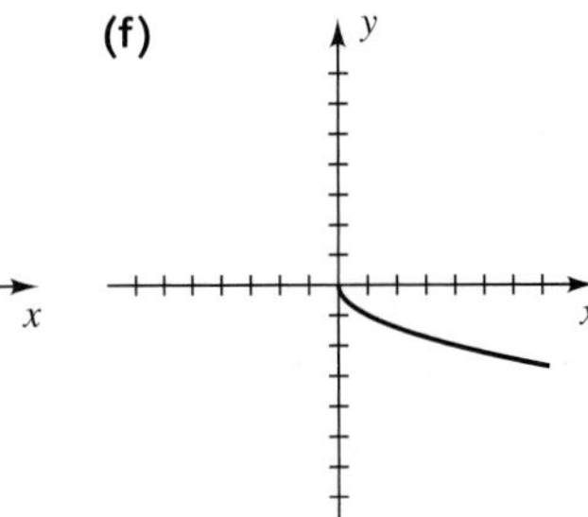

54 (a)

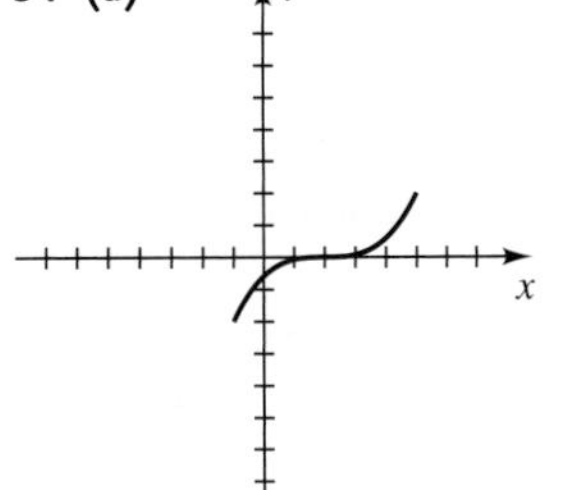

(b)

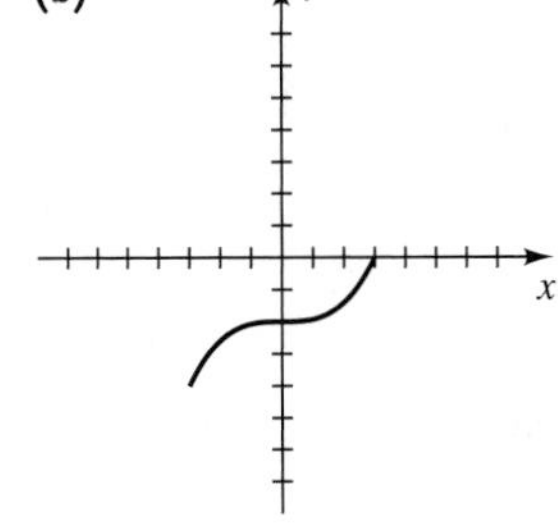

(c)

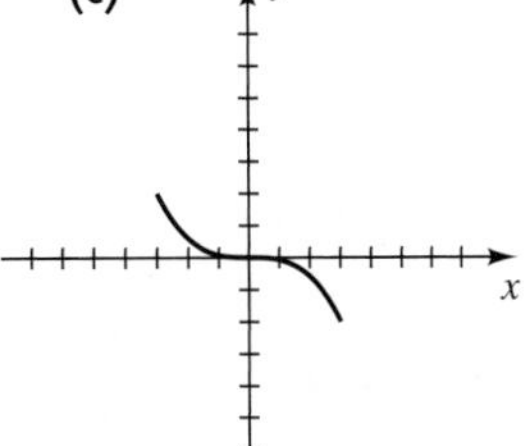

(d)

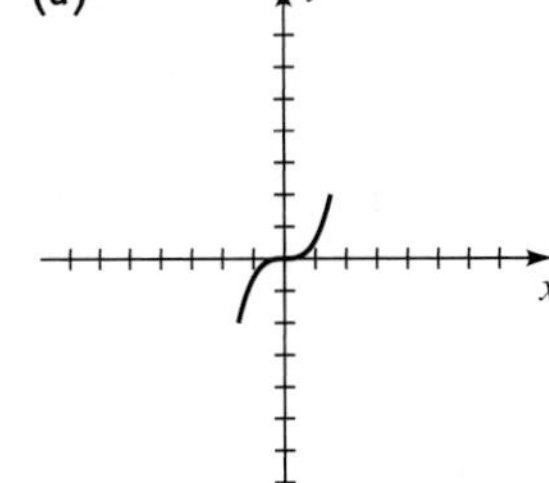

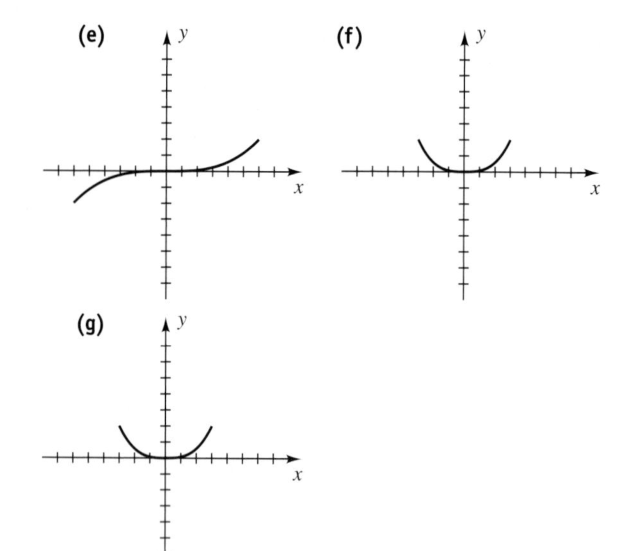

55 $2x - 5y = 10$ **56** $(x + 2)^2 + (y - 1)^2 = 25$

57 $y = \frac{1}{2}(x - 2)^2 - 4$ **58** $y = -|x - 2| - 1$

59 Min: $f(-3) = 4$ **60** Max: $f(5) = -7$

61 Max: $f(-1) = -37$ **62** Min: $f(4) = -108$

63 $f(x) = -2(x - 3)^2 + 4$ **64** $y = \frac{3}{2}(x - 3)^2 - 2$

65 (a) $[0, 2]$ **(b)** $(0, 2]$ **66 (a)** -1 **(b)** $\sqrt{13}$

67 (a) $18x^2 + 9x - 1$ **(b)** $6x^2 - 15x + 5$

68 (a) $\sqrt{\frac{3 + 2x^2}{x^2}}$ **(b)** $\frac{1}{3x + 2}$

69 (a) $\sqrt{28 - x}$; $[3, 28]$
(b) $\sqrt{\sqrt{25 - x^2} - 3}$; $[-4, 4]$

70 (a) $\frac{1}{x + 3}$; all real numbers except -3 and 0
(b) $\frac{6x + 4}{x}$; all real numbers except $-\frac{2}{3}$ and 0

71 $u = x^2 - 5x$, $y = \sqrt[3]{u}$ **72** Between 36.1 ft and 60.1 ft

73 (a) 245 ft **(b)** 2028

74 (a) $V = 6000t + 179{,}000$ **(b)** $2\frac{1}{3}$

75 (a) $F = \frac{9}{5}C + 32$ **(b)** 1.8°F

76 (a) $C_1(x) = \frac{3}{20}x$ **(b)** $C_2(x) = \frac{3}{22}x + 120$ **(c)** 8800

77 (a) $y(x) = -\frac{4}{5}x + 20$ **(b)** $V(x) = 4x\left(-\frac{4}{5}x + 20\right)$

78 $C(r) = \frac{3\pi(r^3 + 16)}{10r}$

79 (a) $V = 10t$
(b) $V = 200h^2$ for $0 \le h \le 6$;
$V = 7200 + 3200(h - 6)$ for $6 < h \le 9$
(c) $h = \sqrt{\frac{t}{20}}$ for $0 \le t \le 720$; $h = 6 + \frac{t - 720}{320}$ for $720 < t \le 1680$

80 (a) $r = \frac{1}{2}x$ **(b)** $y = \frac{5}{4\pi} - \frac{1}{48}x^3$

81 (a) $y(h) = \frac{bh}{a - b}$ **(b)** $V(h) = \frac{1}{3}\pi h(a^2 + ab + b^2)$
(c) $\frac{200}{7\pi} \approx 9.1$ ft

82 $\frac{18}{13}$ hr after 1:00 P.M., or about 2:23 P.M.

83 Radius of semicircle is $\frac{1}{8\pi}$ mi; length of rectangle is $\frac{1}{8}$ mi.

84 (a) 1 sec **(b)** 4 ft
(c) On the moon, 6 sec and 24 ft

85 (a) $(87.5, 17.5)$ **(b)** 30.625 units

CHAPTER 2 DISCUSSION EXERCISES

2 (a) $g(x) = -\frac{1}{2}x + 3$ **(b)** $g(x) = -\frac{1}{2}x - 3$
(c) $g(x) = -\frac{1}{2}x + 7$ **(d)** $g(x) = -\frac{1}{2}x$

4 $2ax + ah + b$ **5** m_{PQ}; the slope of the tangent line at P

6 $R(x_3, y_3) = \left(\left(1 - \frac{m}{n}\right)x_1 + \frac{m}{n}x_2, \left(1 - \frac{m}{n}\right)y_1 + \frac{m}{n}y_2\right)$

7 $h = -ad^2$ **8** $f(x) = 40 - 20[\![-x/15]\!]$

9 $x = \frac{0.4996 + \sqrt{(-0.4996)^2 - 4(0.0833)(3.5491 - D)}}{2(0.0833)}$

10 (b) $f(x) = \begin{cases} 0.132(x - 1)^2 + 0.7 & \text{if } 1 \le x \le 6 \\ -0.517x + 7.102 & \text{if } 6 < x \le 12 \end{cases}$

(c)

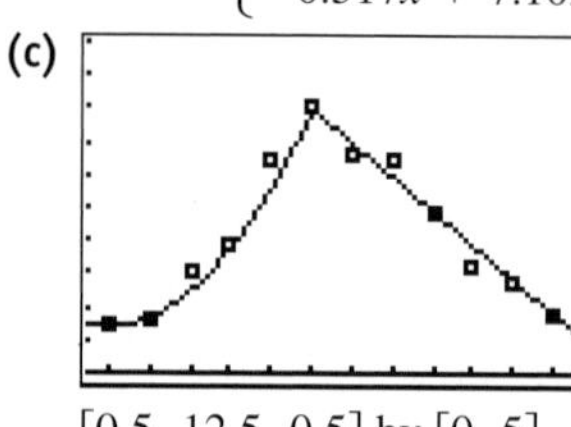

[0.5, 12.5, 0.5] by [0, 5]

Chapter 3

EXERCISES 3.1

1 **(a)**

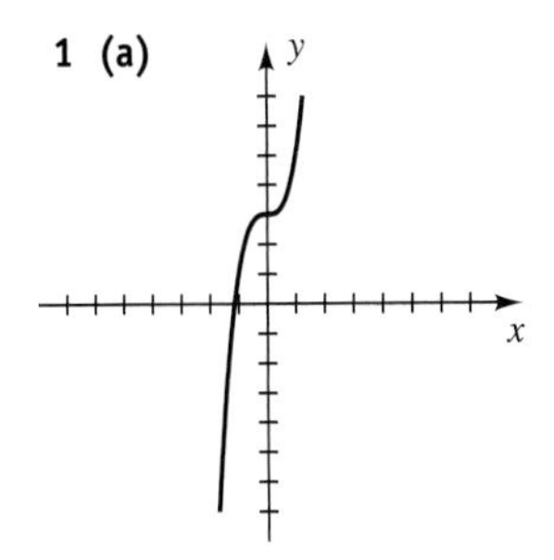

(b)

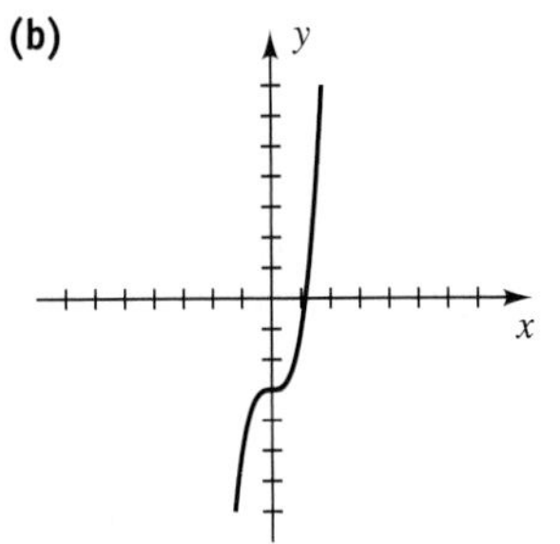

3 **(a)**

(b)

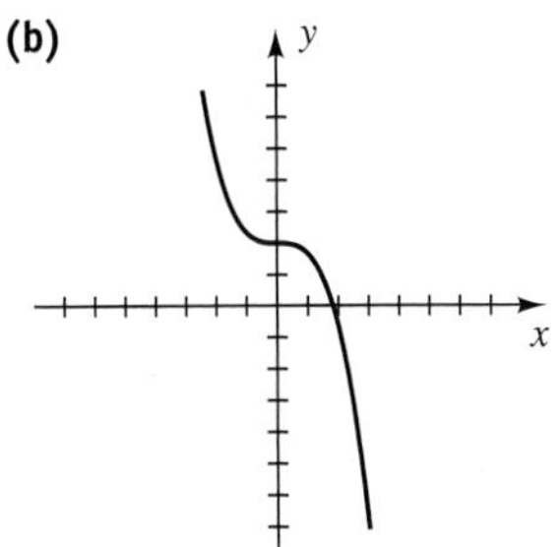

5 $f(3) = -2 < 0,\ f(4) = 10 > 0$

7 $f(2) = 5 > 0,\ f(3) = -5 < 0$

9 $f\left(-\frac{1}{2}\right) = \frac{19}{32} > 0,\ f(-1) = -1 < 0$

11 **(a)** C **(b)** D **(c)** B **(d)** A

13 $f(x) > 0$ if $x > 2$, $f(x) < 0$ if $x < 2$

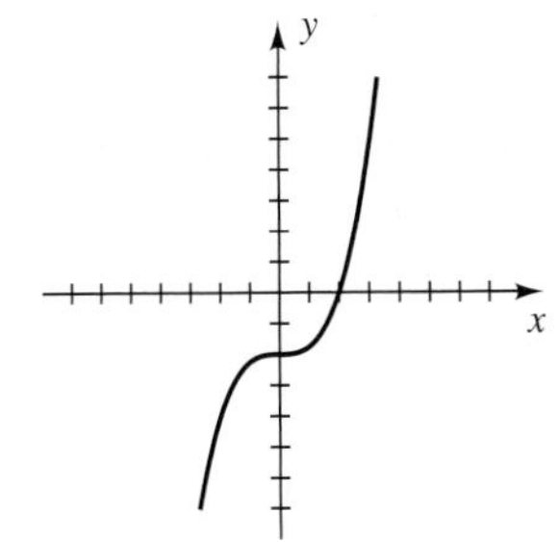

15 $f(x) > 0$ if $|x| < 2$, $f(x) < 0$ if $|x| > 2$

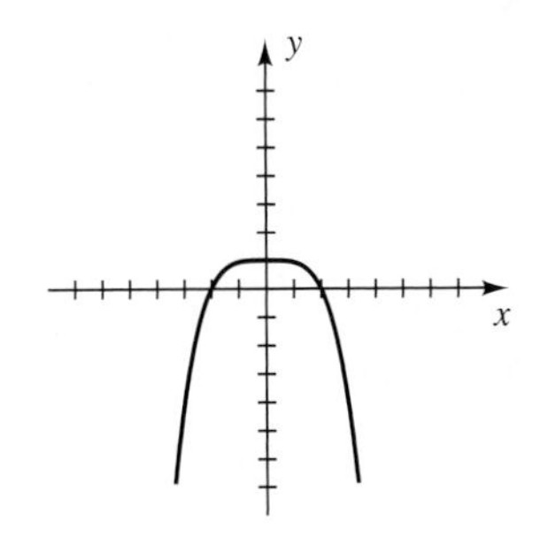

17 $f(x) > 0$ if $|x| > 2$, $f(x) < 0$ if $0 < |x| < 2$

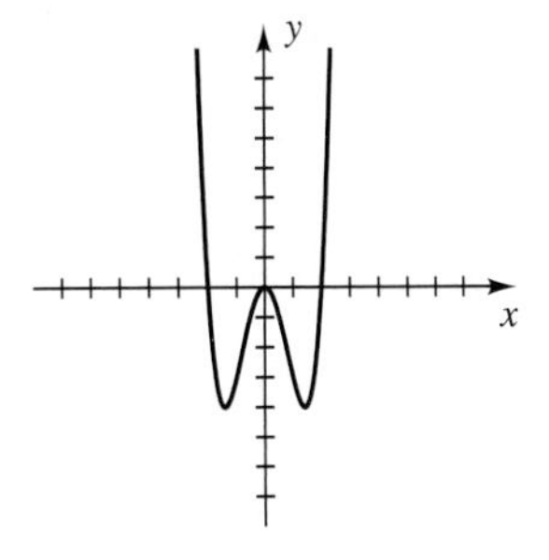

19 $f(x) > 0$ if $x < -2$ or $0 < x < 5$, $f(x) < 0$ if $-2 < x < 0$ or $x > 5$

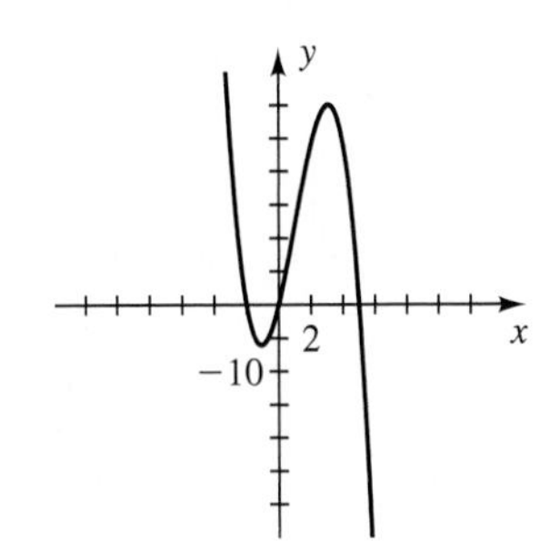

21 $f(x) > 0$ if $-2 < x < 3$ or $x > 4$, $f(x) < 0$ if $x < -2$ or $3 < x < 4$

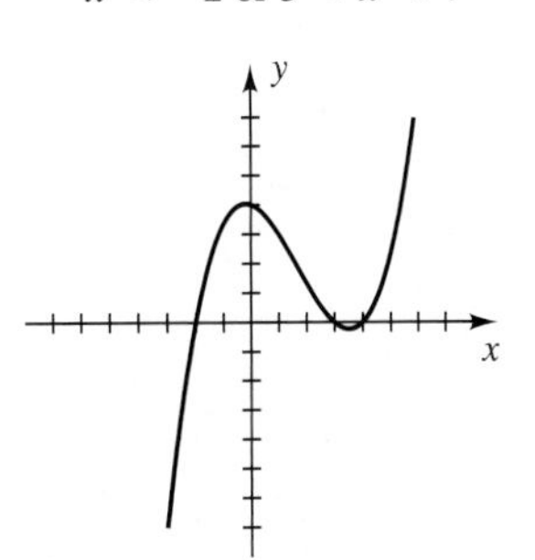

23 $f(x) > 0$ if $x > 2$, $f(x) < 0$ if $x < -2$ or $|x| < 2$

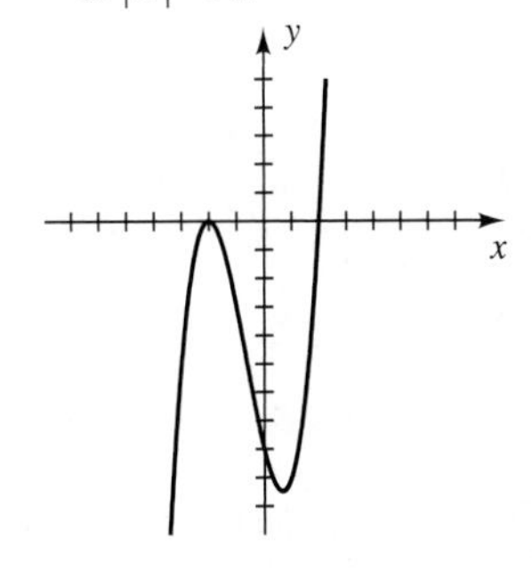

25 $f(x) > 0$ if $|x| > 2$ or $|x| < \sqrt{2}$, $f(x) < 0$ if $\sqrt{2} < |x| < 2$

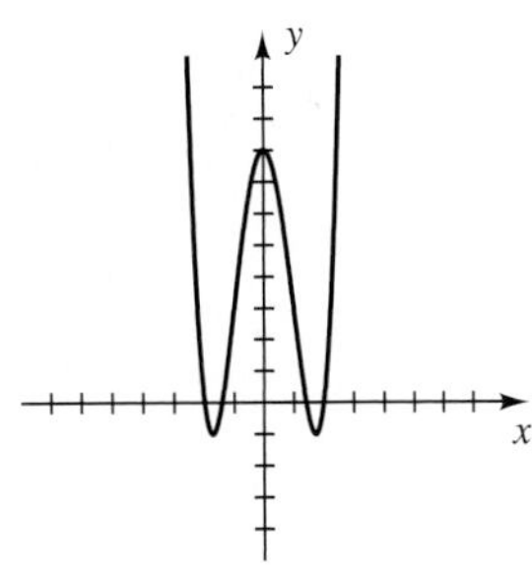

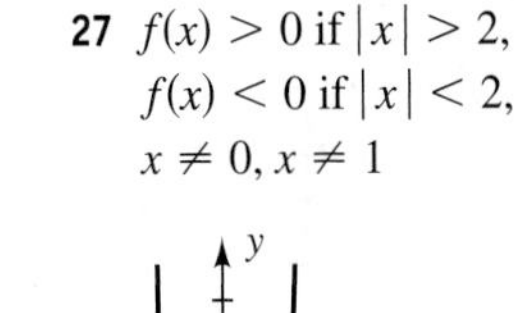

27 $f(x) > 0$ if $|x| > 2$, $f(x) < 0$ if $|x| < 2$, $x \neq 0, x \neq 1$

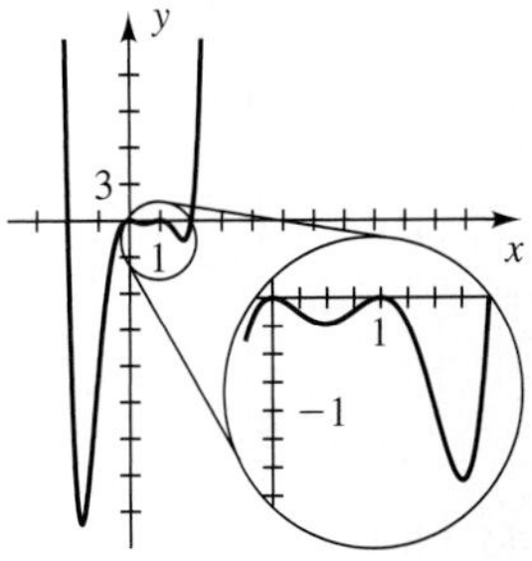

29

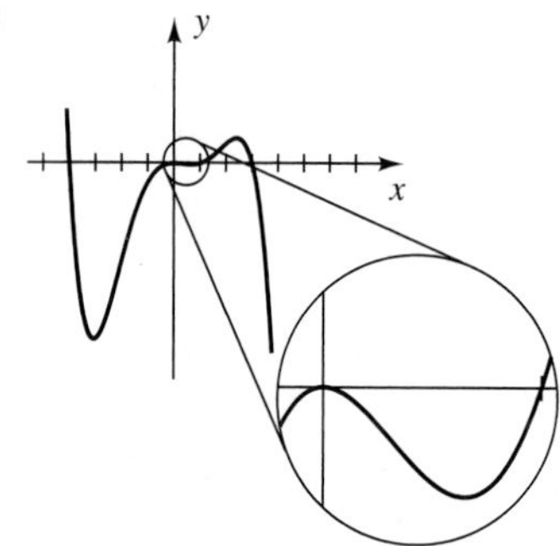

31 (a)

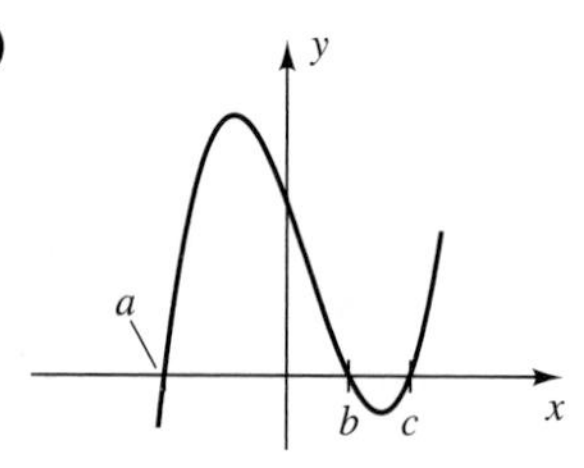

(b) $-abc$ **(c)** $(-\infty, a) \cup (b, c)$ **(d)** $[a, b] \cup [c, \infty)$

33 If n is even, then $(-x)^n = x^n$ and hence $f(-x) = f(x)$. Thus, f is an even function.

35 $-\dfrac{4}{3}$ **37** ± 4

39 $P(x) > 0$ on $\left(-\frac{1}{5}\sqrt{15}, 0\right)$ and $\left(\frac{1}{5}\sqrt{15}, \infty\right)$;

$P(x) < 0$ on $\left(-\infty, -\frac{1}{5}\sqrt{15}\right)$ and $\left(0, \frac{1}{5}\sqrt{15}\right)$

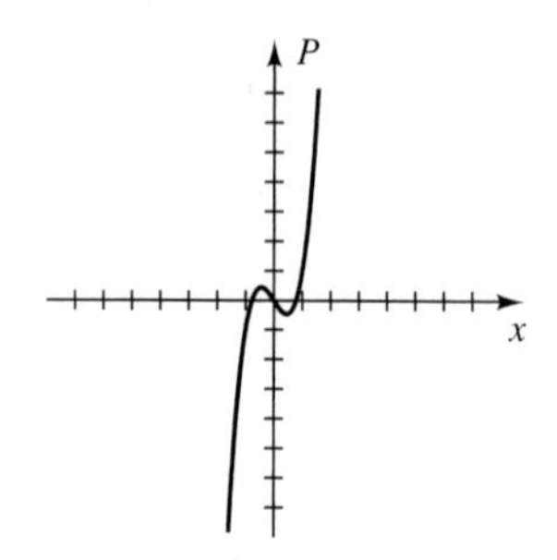

41 (b) $V(x) > 0$ on $(0, 10)$ and $(15, \infty)$; allowable values for x are in $(0, 10)$.

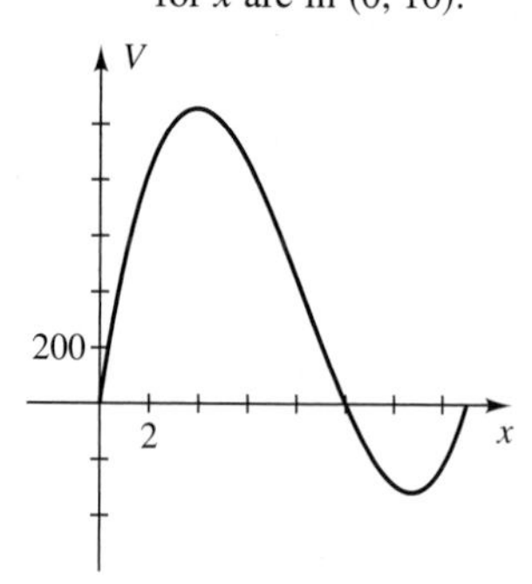

43 (a) $T > 0$ for $0 < t < 12$; $T < 0$ for $12 < t < 24$

(b)

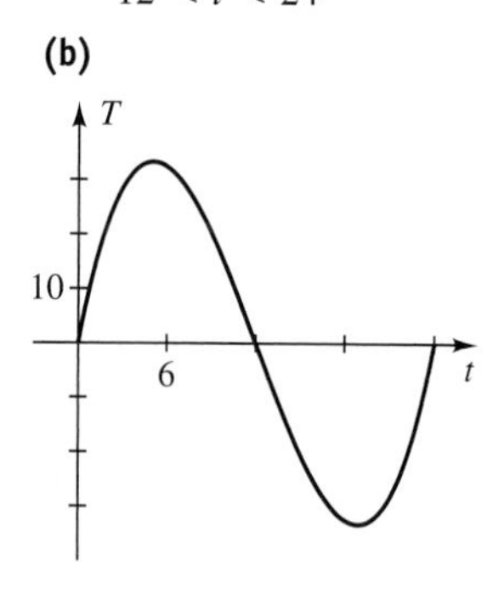

(c) $T(6) = 32.4 > 32$, $T(7) = 29.75 < 32$

45 (a) $N(t) > 0$ for $0 < t < 5$

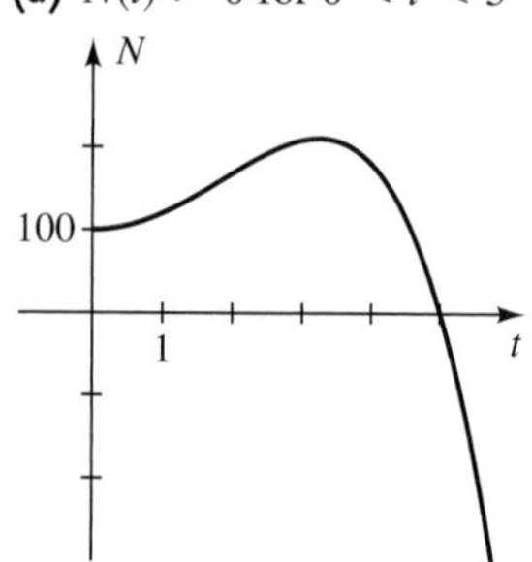

(b) The population becomes extinct after 5 years.

47 (a)

x	$f(x)$	$g(x)$	$h(x)$	$k(x)$
−60	25,920,000	25,902,001	25,937,999	26,135,880
−40	5,120,000	5,112,001	5,127,999	5,183,920
−20	320,000	318,001	321,999	327,960
20	320,000	318,001	321,999	312,040
40	5,120,000	5,112,001	5,127,999	5,056,080
60	25,920,000	25,902,001	25,937,999	25,704,120

(b) They become similar. **(c)** $2x^4$

49 (a) (1)

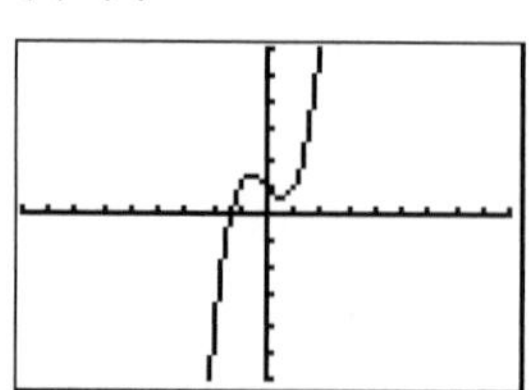

$[-9, 9]$ by $[-6, 6]$

(2)

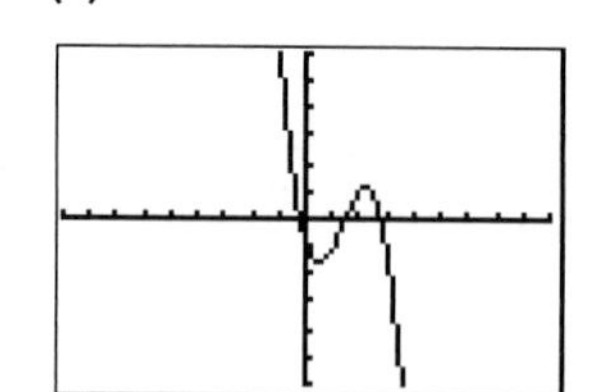

$[-9, 9]$ by $[-6, 6]$

(3)

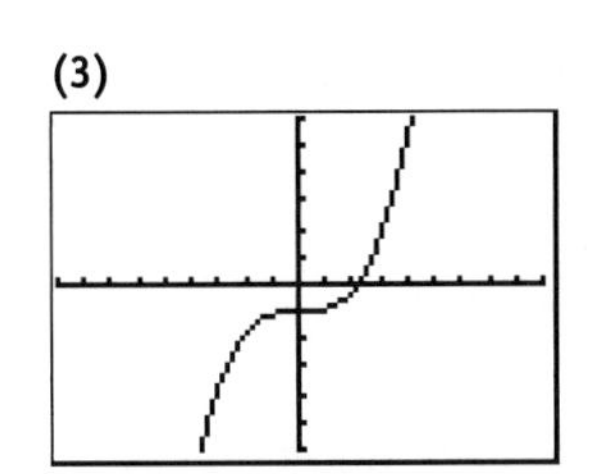

$[-9, 9]$ by $[-6, 6]$

(4)

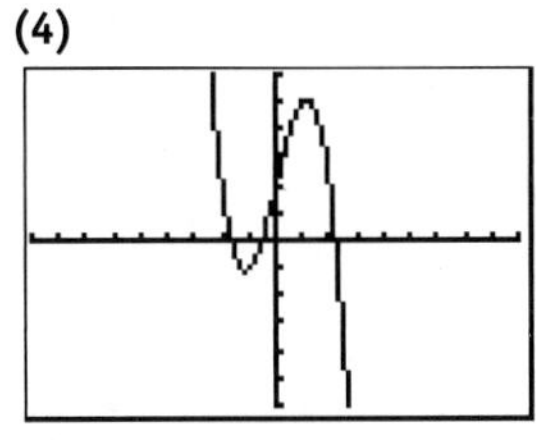

$[-9, 9]$ by $[-6, 6]$

(b) **(1)** As x approaches ∞, $f(x)$ approaches ∞; as x approaches $-\infty$, $f(x)$ approaches $-\infty$.
(2) As x approaches ∞, $f(x)$ approaches $-\infty$; as x approaches $-\infty$, $f(x)$ approaches ∞,
(3) As x approaches ∞, $f(x)$ approaches ∞; as x approaches $-\infty$, $f(x)$ approaches $-\infty$.
(4) As x approaches ∞, $f(x)$ approaches $-\infty$; as x approaches $-\infty$, $f(x)$ approaches ∞.

(c) For the cubic function $f(x) = ax^3 + bx^2 + cx + d$ with $a > 0$, $f(x)$ approaches ∞ as x approaches ∞ and $f(x)$ approaches $-\infty$ as x approaches $-\infty$. With $a < 0$, $f(x)$ approaches $-\infty$ as x approaches ∞ and $f(x)$ approaches ∞ as x approaches $-\infty$.

51

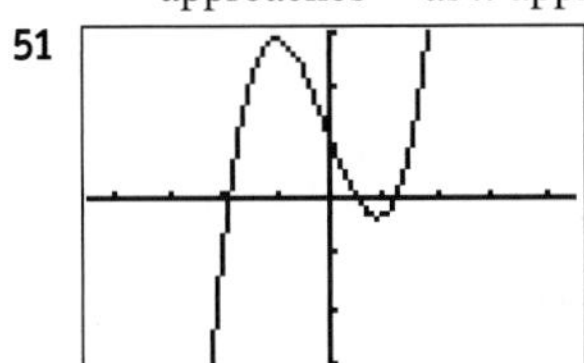

$[-4.5, 4.5]$ by $[-3, 3]$

$-1.89, 0.49, 1.20$

53

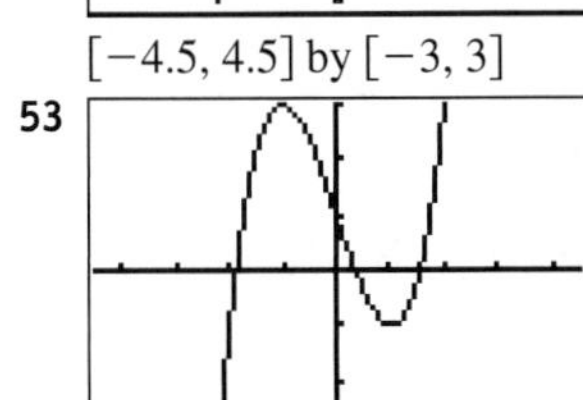

$[-4.5, 4.5]$ by $[-3, 3]$

$-1.88, 0.35, 1.53$

55

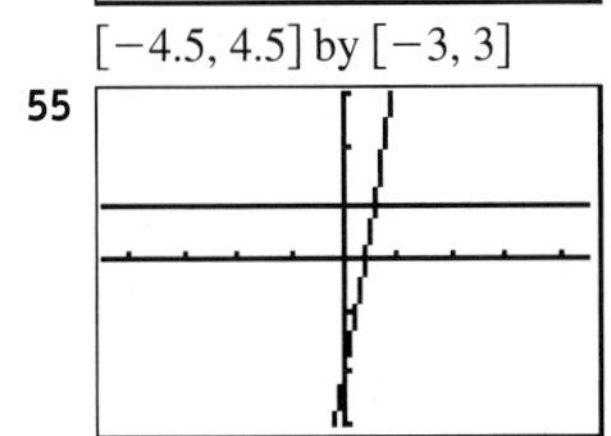

$[-4.5, 4.5]$ by $[-3, 3]$

$(0.56, \infty)$

57

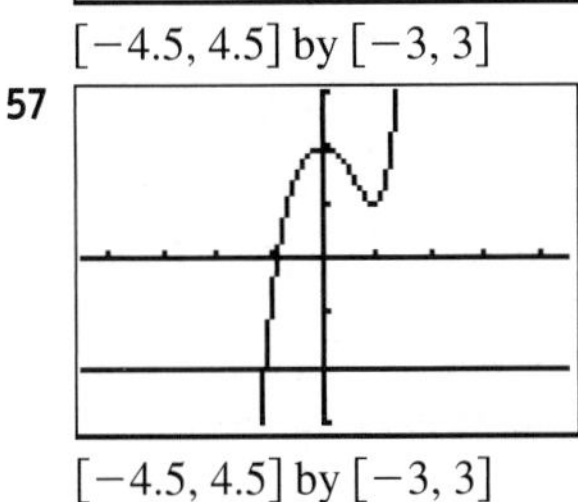

$[-4.5, 4.5]$ by $[-3, 3]$

$(-1.10, \infty)$

59

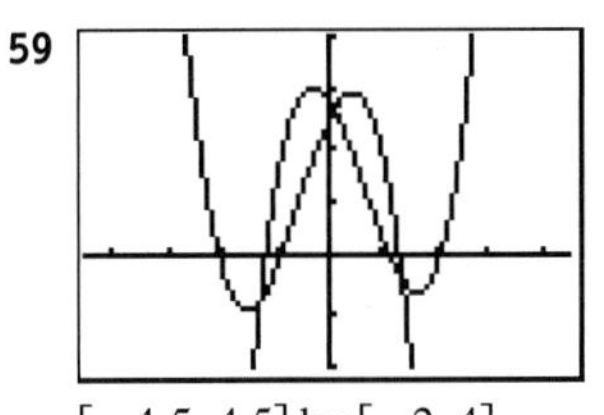

$[-4.5, 4.5]$ by $[-2, 4]$

$(-1.29, -0.77)$, $(0.085, 2.66)$, $(1.36, -0.42)$

61 **(a)** It has increased.

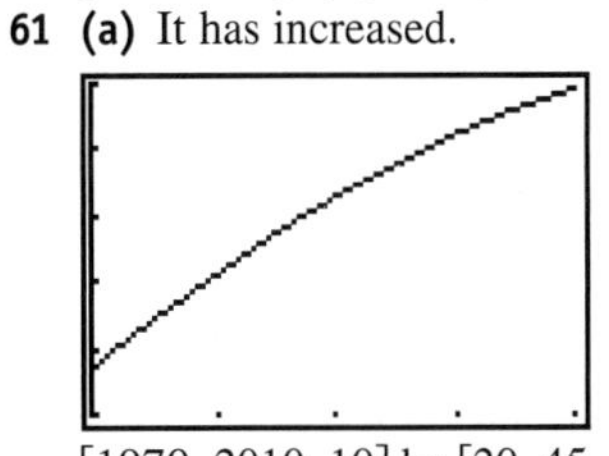

$[1970, 2010, 10]$ by $[20, 45, 5]$

(b) $y = 0.59x + 23.5$; linear

EXERCISES 3.2

1 $2x^2 - x + 3; 4x - 3$ **3** $\frac{3}{2}x; \frac{1}{2}x - 4$

5 $0; 7x + 2$ **7** $\frac{9}{2}; \frac{53}{2}$ **9** 26 **11** 7

13 $f(-3) = 0$ **15** $f(-2) = 0$ **17** $x^3 - 3x^2 - 10x$

19 $x^4 - 2x^3 - 9x^2 + 2x + 8$

21 $2x^2 + x + 6; 7$

23 $x^2 - 3x + 1; -8$

25 $3x^4 - 6x^3 + 12x^2 - 18x + 36; -65$

27 $4x^3 + 2x^2 - 4x - 2; 0$

29 73 **31** -0.0444

33 $8 + 7\sqrt{3}$

35 $f(-2) = 0$ **37** $f\left(\frac{1}{2}\right) = 0$

39 3, 5 **41** $f(c) > 0$ **43** -14

45 If $f(x) = x^n - y^n$ and n is even, then f(

47 **(a)** $V = \pi x^2(6 - x)$

(b) $\left(\frac{1}{2}(5 + \sqrt{45}), \frac{1}{\ }\right.$

49 **(a)** $A = 8$

51 -9.55

53 $-($

EX

7 $x^4 + 2x^3 - 23x^2 - 24x + 144$

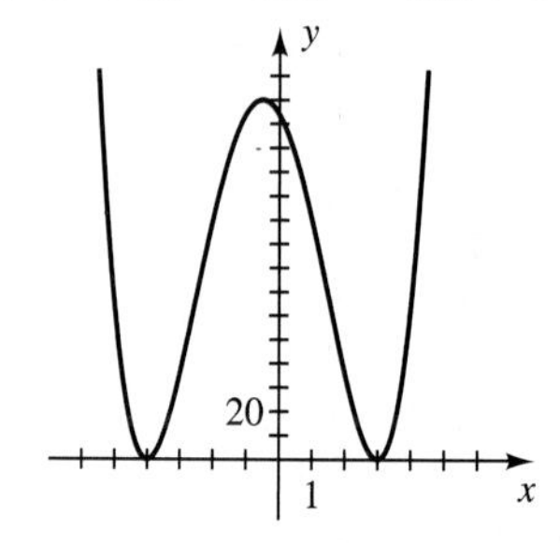

9 $3x^6 - 27x^5 + 81x^4 - 81x^3$

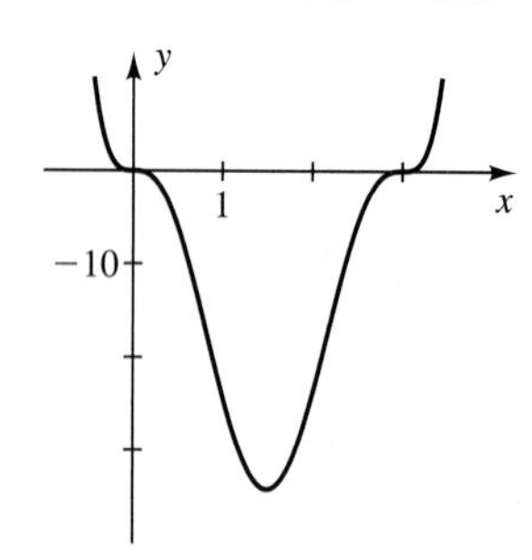

11 $f(x) = \frac{7}{9}(x + 1)\left(x - \frac{3}{2}\right)(x - 3)$

13 $f(x) = -1(x - 1)^2(x - 3)$

15 $-\frac{2}{3}$ (multiplicity 1); 0 (multiplicity 2); $\frac{5}{2}$ (multiplicity 3)

17 $-\frac{3}{2}$ (multiplicity 2); 0 (multiplicity 3)

19 -4 (multiplicity 3); -3 (multiplicity 2); 3 (multiplicity 5)

21 $\pm 4i$, ± 3 (each of multiplicity 1)

23 $f(x) = (x + 3)^2(x + 2)(x - 1)$

25 $f(x) = (x - 1)^5(x + 1)$

Exer. 27–34: The types of possible solutions are listed in the order positive, negative, nonreal complex.

27 3, 0, 0 or 1, 0, 2 **29** 0, 1, 2

31 2, 2, 0; 2, 0, 2; 0, 2, 2; 0, 0, 4

3, 0; 2, 1, 2; 0, 3, 2; 0, 1, 4

: lower, -2 **37** Upper, 2; lower, -2

er, -3

$- 1)(x - 2)^3$

43 **(a)** $f(x) = a(x + 3)^3(x + 1)(x - 2)^2$ **(b)** 108

45 $f(x) = (x + 4)(x + 2)(x - 1.5)^2(x - 3)$

47 No **49** Yes: $1.5(x - 2)(x - 5.2)(x - 10.1)$

51 $f(t) = \frac{5}{3528}t(t - 5)(t - 19)(t - 24)$

53

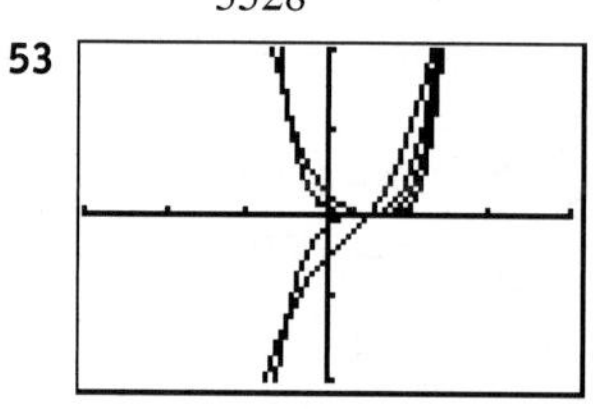

$[-3, 3]$ by $[-2, 2]$

As the multiplicity increases, the graph becomes more horizontal at (0.5, 0).

55

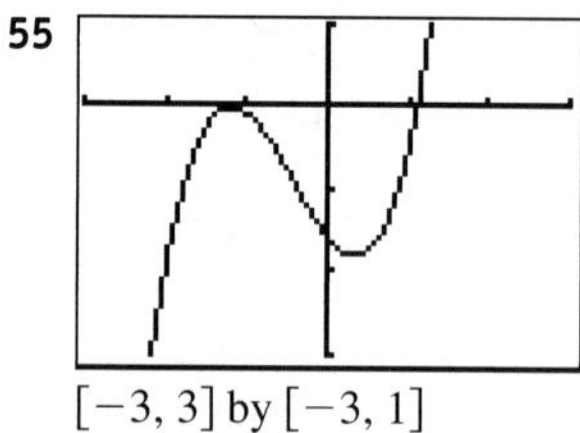

$[-3, 3]$ by $[-3, 1]$

-1.2 (multiplicity 2); 1.1 (multiplicity 1)

57 2007 (when $t \approx 27.1$)

59 **(a)** (3)

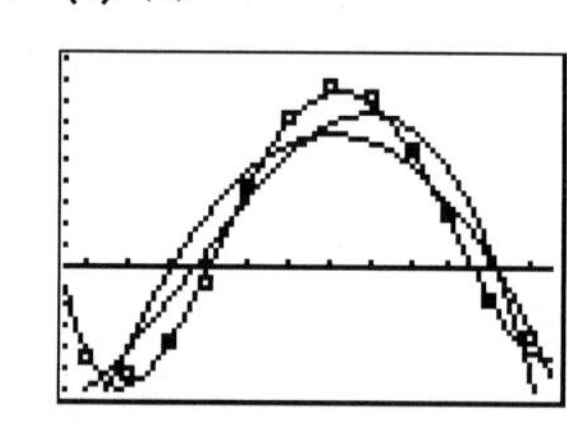

$[0.5, 12.5]$ by $[-30, 50, 5]$

(b) $4 \leq x \leq 5$ and $10 \leq x \leq 11$ **(c)** 4.02, 10.53

61 7.64 cm **63** 12 cm

EXERCISES 3.4

1 $x^2 - 6x + 13$ **3** $(x - 2)(x^2 + 4x + 29)$

5 $x(x + 1)(x^2 - 6x + 10)$

7 $(x^2 - 8x + 25)(x^2 + 4x + 5)$

9 $x(x^2 + 4)(x^2 - 2x + 2)$

Exer. 11–14: Show that none of the possible rational roots listed satisfy the equation.

11 $\pm 1, \pm 2, \pm 3, \pm 6$ **13** $\pm 1, \pm 2$ **15** $-2, -1, 4$

17 $-3, 2, \frac{5}{2}$ **19** $-7, \pm\sqrt{2}, 4$

21 $-3, -\frac{2}{3}$, 0 (multiplicity 2), $\frac{1}{2}$

23 $-\frac{3}{4}, -\frac{3}{4} \pm \frac{3}{4}\sqrt{7}i$

25 $f(x) = (3x + 2)(2x - 1)(x - 1)^2(x - 2)$

27 $f(x) = 2(x + 0.9)(x - 1.1)(x - 12.5)$

29 No. If i is a root, then $-i$ is also a root. Hence, the polynomial would have factors $x - 1, x + 1, x - i, x + i$ and therefore would be of degree greater than 3.

31 Since n is odd and nonreal complex zeros occur in conjugate pairs for polynomials with real coefficients, there must be at least one real zero.

33 **(a)** The two boxes correspond to $x = 5$ and $x = 5(2 - \sqrt{2})$.
(b) The box corresponding to $x = 5$

35 **(c)** In feet: 5, 12, and 13 **37** **(b)** 4 ft

39 None **41** $-1.2, 0.8, -\frac{1}{2} \pm \frac{\sqrt{3}}{2}i$ **43** 10,200 m

EXERCISES 3.5

1 **(a)**

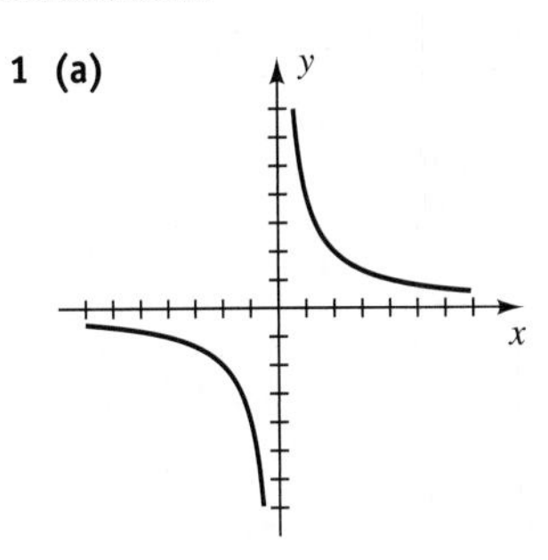

(b) D = all nonzero real numbers; $R = D$
(c) Decreasing on $(-\infty, 0)$ and on $(0, \infty)$

3 VA: $x = 3$;
HA: $y = -2$;
hole: $\left(6, -\frac{22}{3}\right)$

5

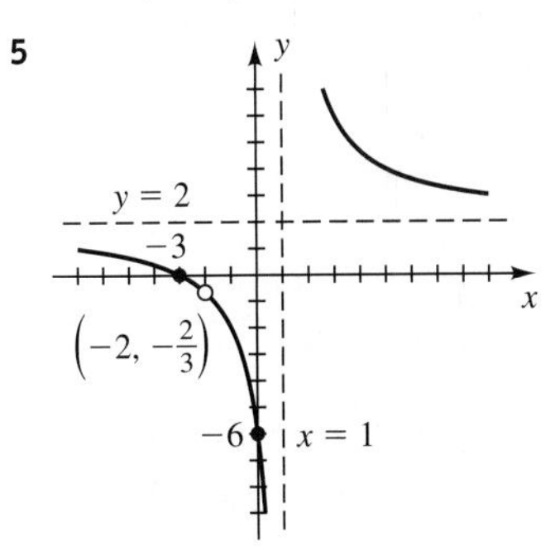

$$f(x) = \frac{2(x + 3)(x + 2)}{(x - 1)(x + 2)}$$

7

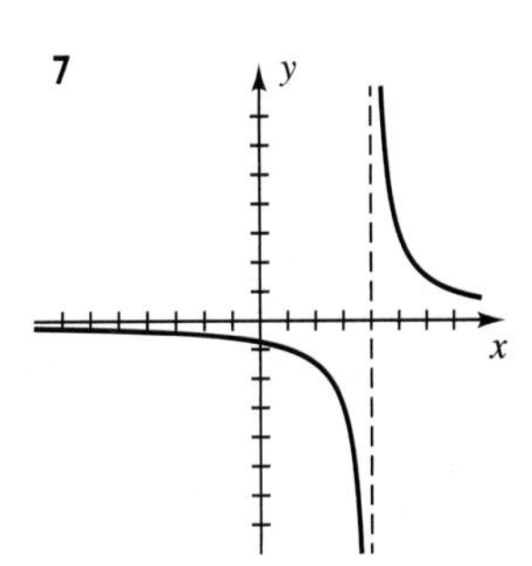

9

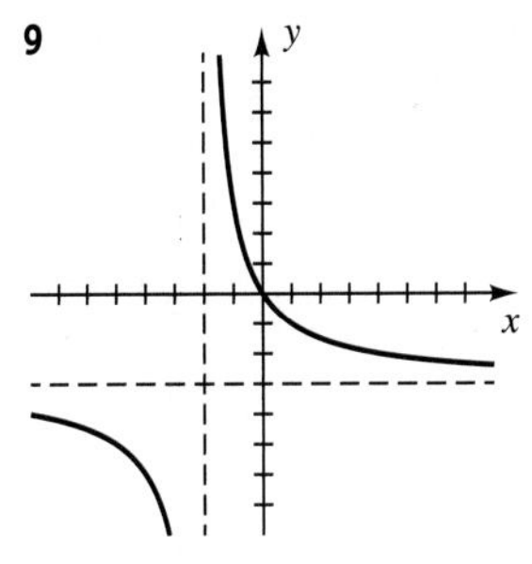

11

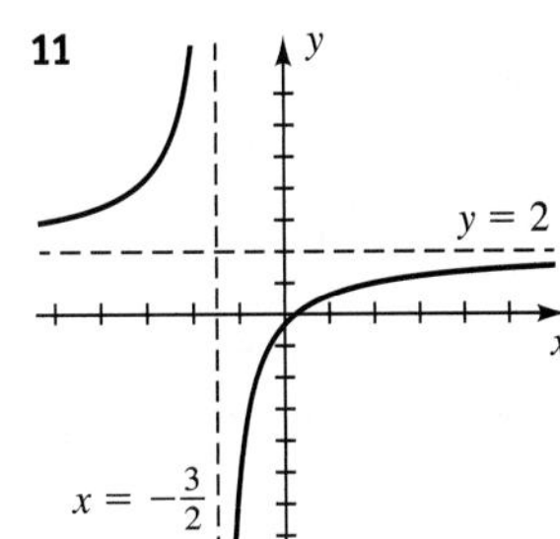

13

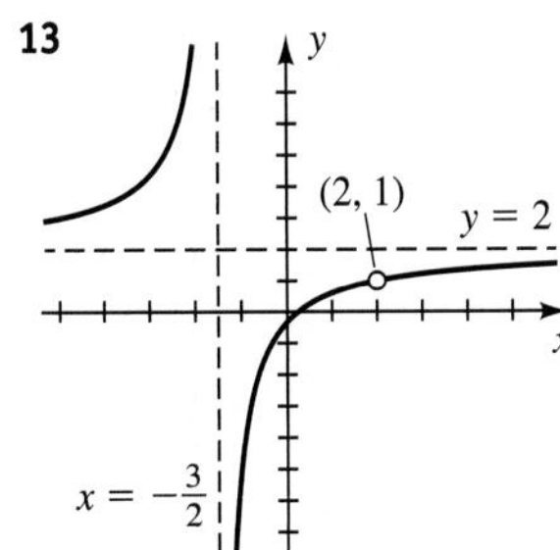

15

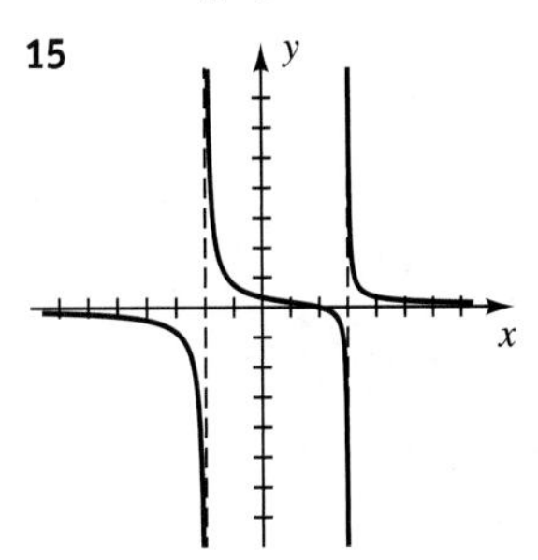

17

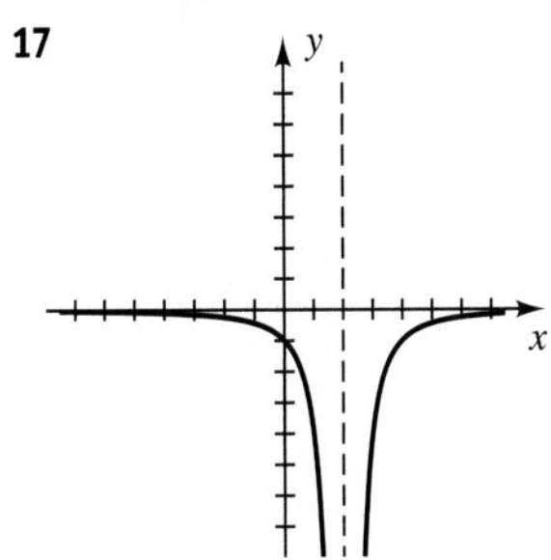

19

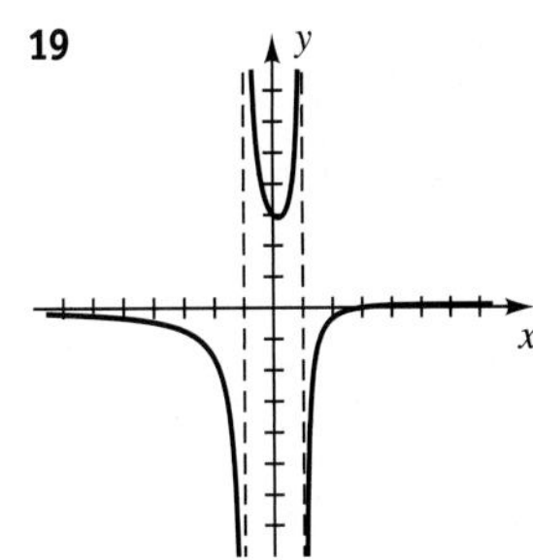

21

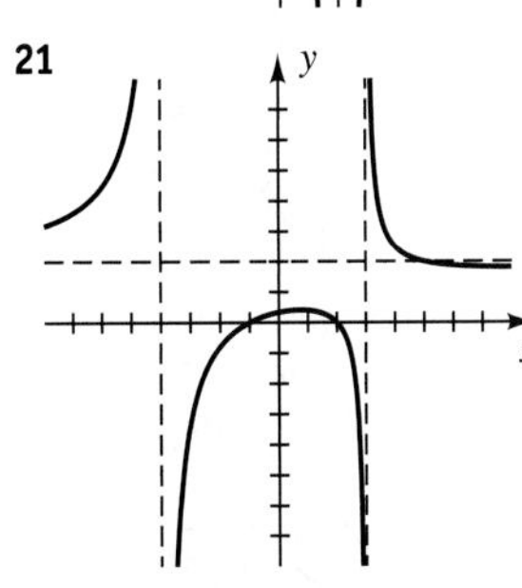

23

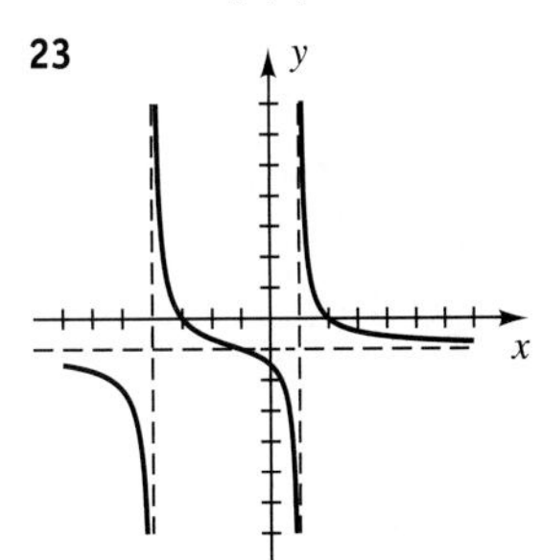

25

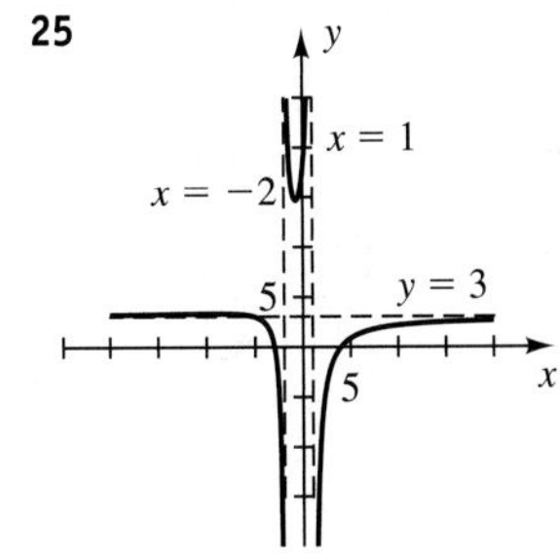

27

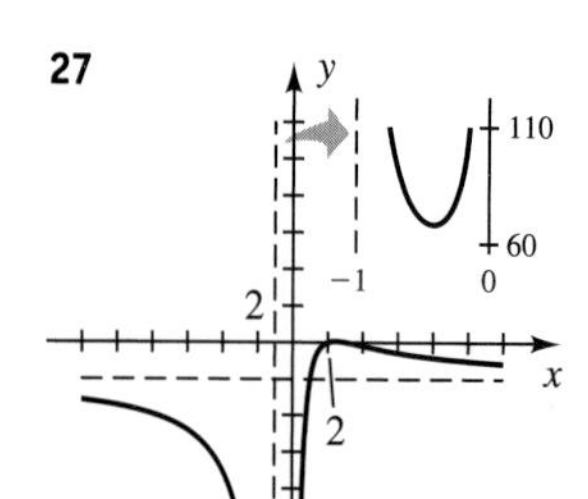

29

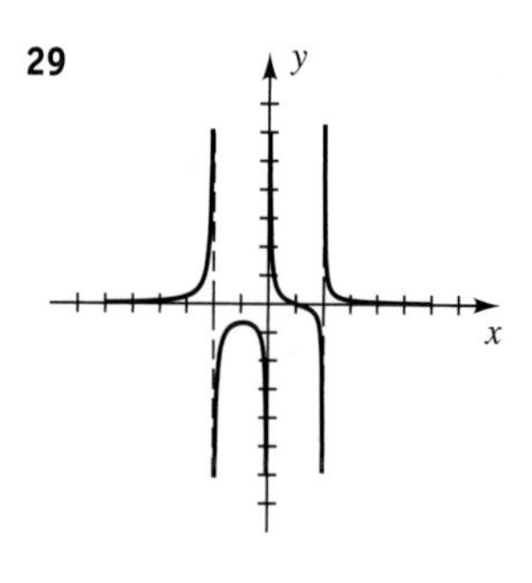

31

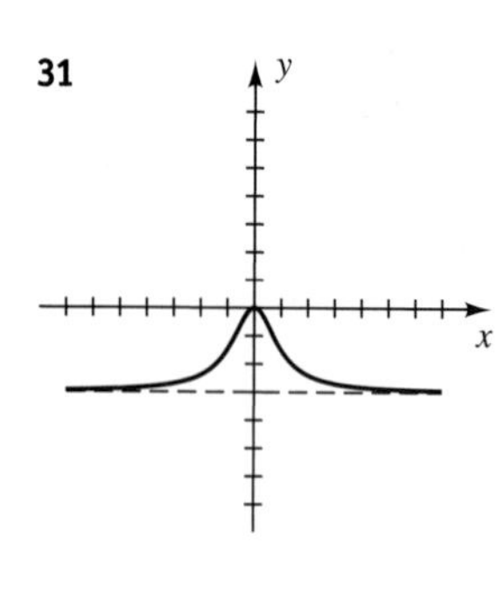

33 $y = x - 2$

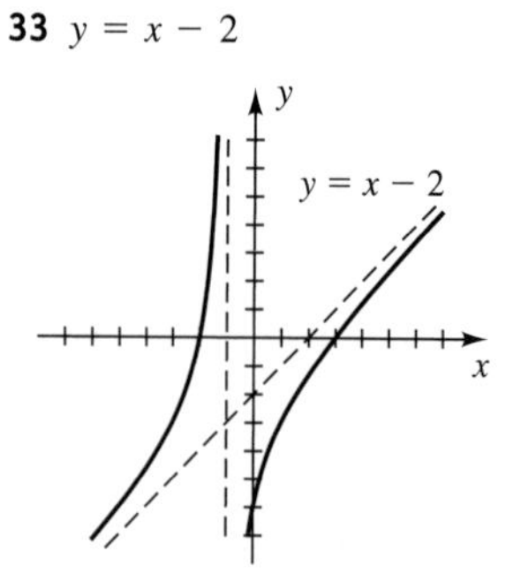

35 $y = -\dfrac{1}{2}x$

37 $f(x) = \dfrac{2x - 3}{x + 1}$ for $x \neq -2$

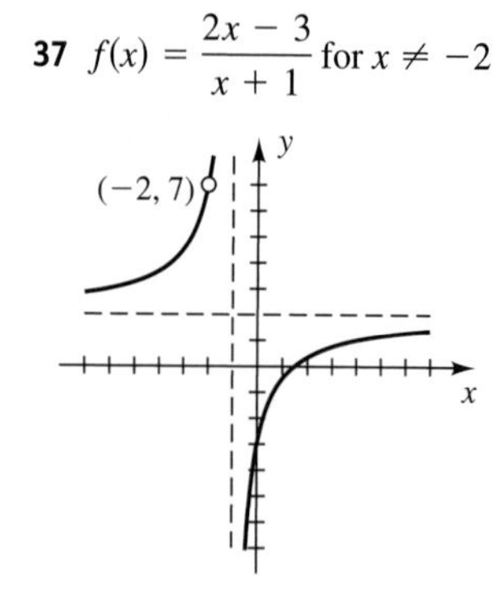

39 $f(x) = \dfrac{-1}{x + 1}$ for $x \neq 1$

41 $f(x) = x - 1$ for $x \neq -2$

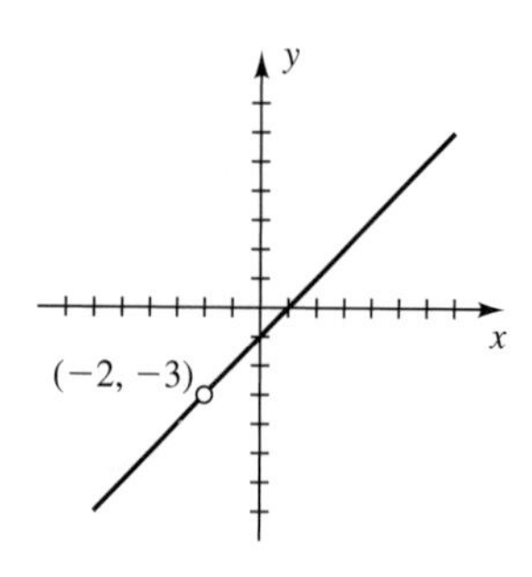

43 $f(x) = \dfrac{x + 2}{x + 1}$ for $x \neq -2$

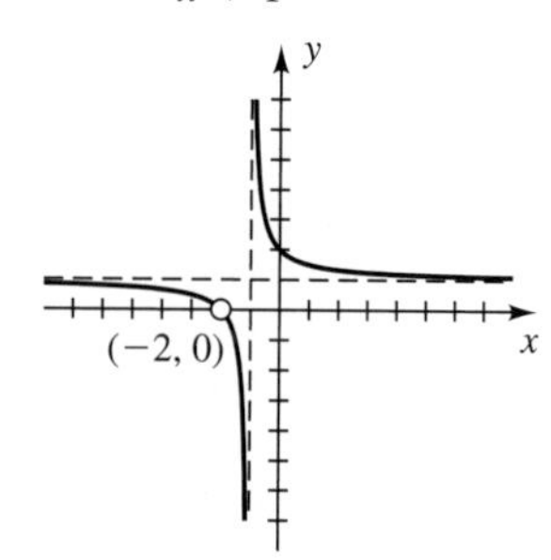

45 $f(x) = \dfrac{3 - x}{x - 4}$ **47** $f(x) = \dfrac{6x^2 - 6x - 12}{x^3 - 7x + 6}$

49 **(a)** $h = \dfrac{16}{(r + 0.5)^2} - 1$ **(b)** $V(r) = \pi r^2 h$
(c) Exclude $r \leq 0$ and $r \geq 3.5$.

51 **(a)** $V(t) = 50 + 5t, A(t) = 0.5t$ **(b)** $\dfrac{t}{10t + 100}$
(c) As $t \to \infty$, $c(t) \to 0.1$ lb of salt per gal.

53 **(a)** $0 < S < 4000$ **(b)** 4500 **(c)** 2000
(d) A 125% increase in S produces only a 12.5% increase in R.

55 None **57** $x = 0.999$

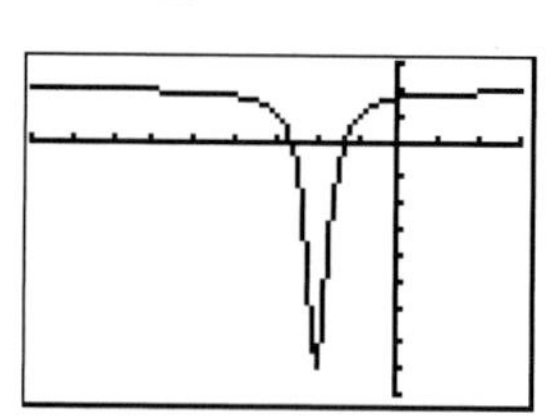

$[-9, 3]$ by $[-9, 3]$

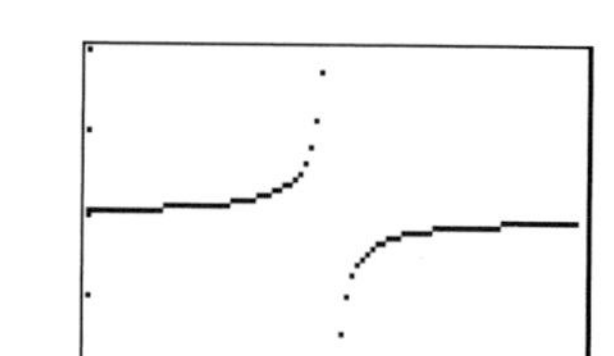

$[0.7, 1.3, 0.1]$ by $[0.8, 1.2, 0.1]$

59 **(a)** The graph of g is the horizontal line $y = 1$ with holes at $x = 0, \pm 1, \pm 2, \pm 3$.
(b) The graph of h is the graph of p with holes at $x = 0, \pm 1, \pm 2, \pm 3$.

61 **(a)** $y = \dfrac{132 - 48x}{x - 4}$ **(b)**

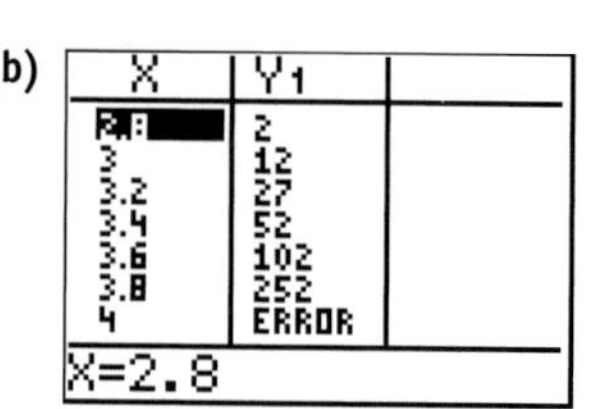

X	Y1	
2.8	2	
3	12	
3.2	27	
3.4	52	
3.6	102	
3.8	252	
4	ERROR	

X=2.8

(c)

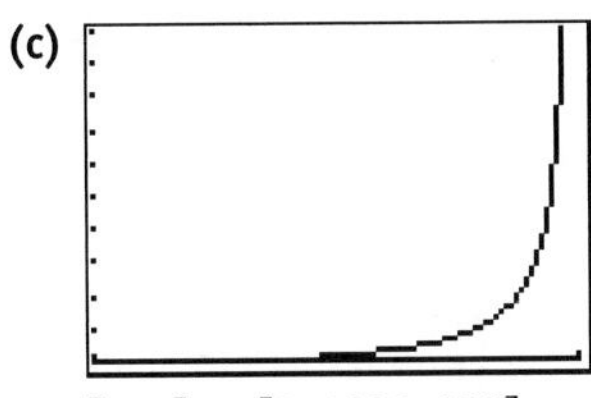

[2, 4] by [0, 1000, 100]

(d) $x = 4$

(e) Regardless of the number of additional credit hours obtained at 4.0, a cumulative GPA of 4.0 is not attainable.

EXERCISES 3.6

1 $u = kv;\ k = \dfrac{2}{5}$ **3** $r = k\dfrac{s}{t};\ k = -14$

5 $y = k\dfrac{x^2}{z^3};\ k = 27$ **7** $z = kx^2y^3;\ k = -\dfrac{2}{49}$

9 $y = k\dfrac{x}{z^2};\ k = 36$ **11** $y = k\dfrac{\sqrt{x}}{z^3};\ k = \dfrac{40}{3}$

13 (a) $P = kd$ **(b)** 59 **(c)** 295 lb/ft^2

(d)

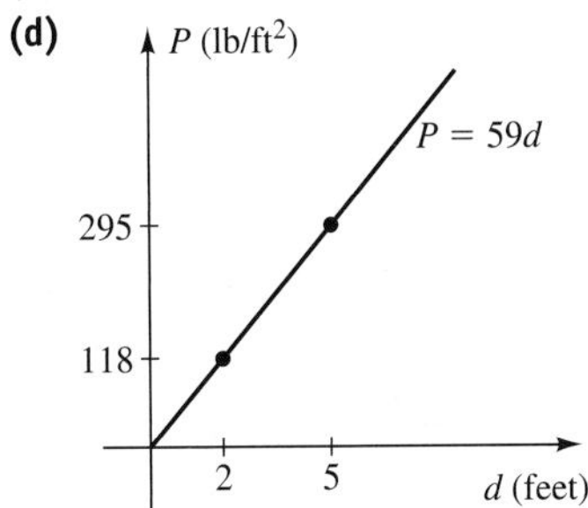

15 (a) $R = k\dfrac{l}{d^2}$ **(b)** $\dfrac{1}{40{,}000}$

(c)

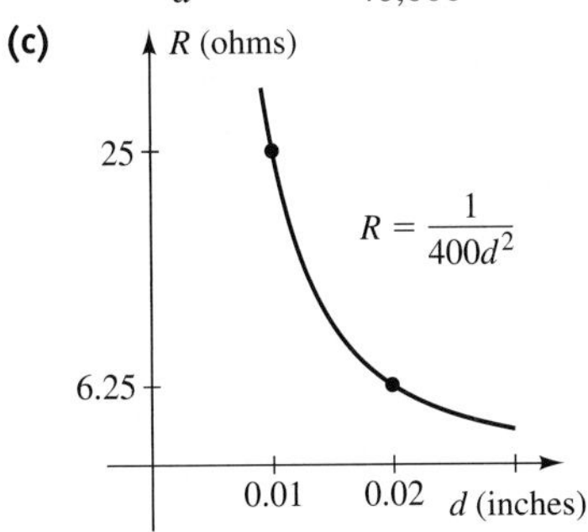

(d) $\dfrac{50}{9}$ ohms

17 (a) $P = k\sqrt{l}$ **(b)** $\dfrac{3}{4}\sqrt{2}$ **(c)** $\dfrac{3}{2}\sqrt{3}$ sec

19 (a) $T = kd^{3/2}$ **(b)** $\dfrac{365}{(93)^{3/2}}$ **(c)** 223.2 days

21 (a) $V = k\sqrt{L}$ **(b)** $\dfrac{7}{2}\sqrt{2}$ **(c)** 60.6 mi/hr

23 (a) $W = kh^3$ **(b)** $\dfrac{25}{27}$ **(c)** 154 lb

25 (a) $F = kPr^4$ **(b)** About 2.05 times as hard

27 Increases 250% **29** d is multiplied by 9.

31 $y = 1.2x$ **33** $y = -\dfrac{10.1}{x^2}$

35 (a) $k \approx 0.034$ **(b)**

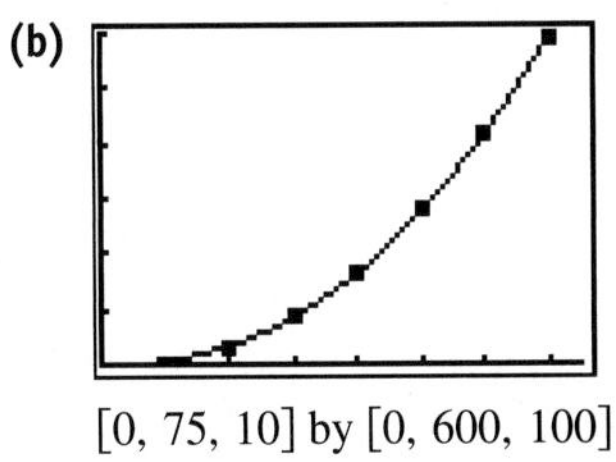

[0, 75, 10] by [0, 600, 100]

CHAPTER 3 REVIEW EXERCISES

1 $f(x) > 0$ if $x > -2$, $f(x) < 0$ if $x < -2$

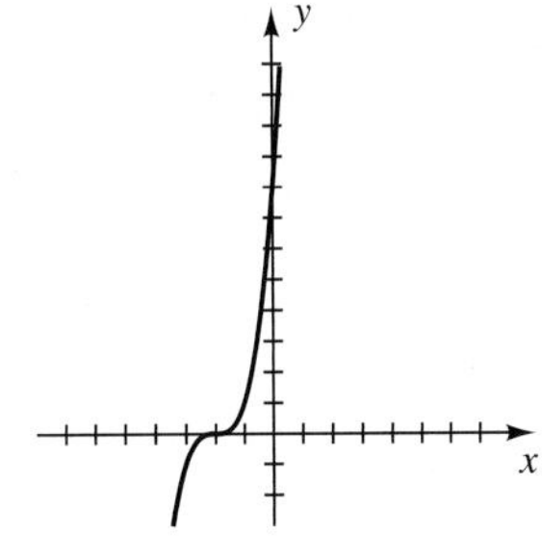

2

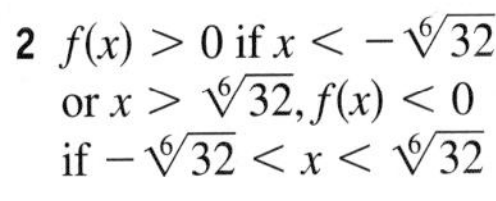

$f(x) > 0$ if $x < -\sqrt[6]{32}$ or $x > \sqrt[6]{32}$, $f(x) < 0$ if $-\sqrt[6]{32} < x < \sqrt[6]{32}$

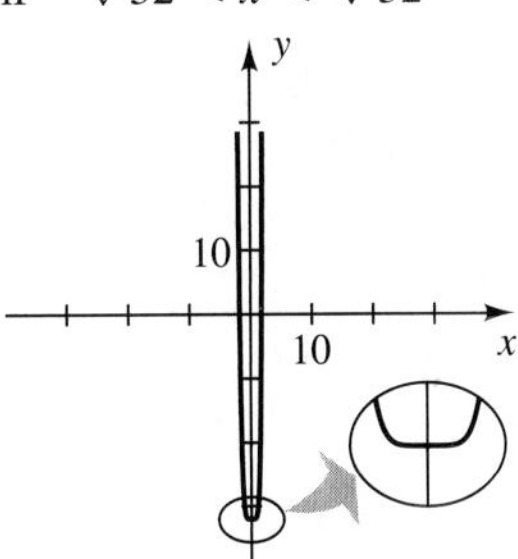

3

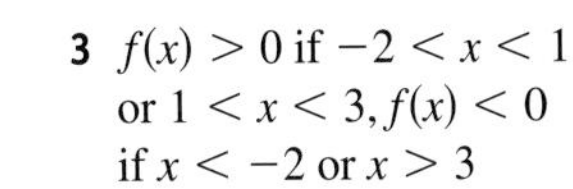

$f(x) > 0$ if $-2 < x < 1$ or $1 < x < 3$, $f(x) < 0$ if $x < -2$ or $x > 3$

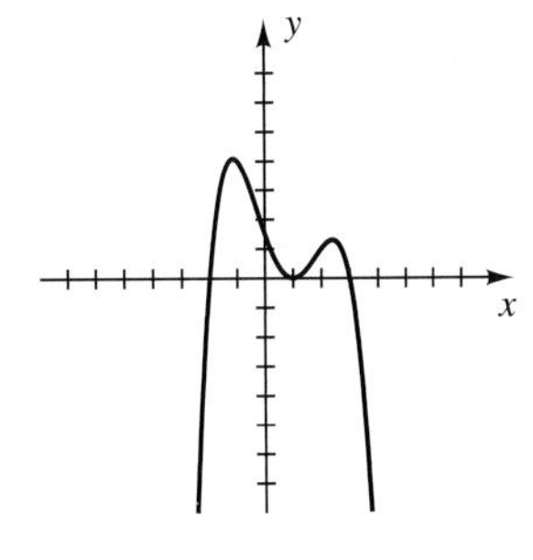

4

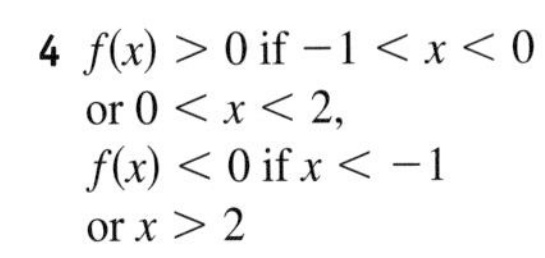

$f(x) > 0$ if $-1 < x < 0$ or $0 < x < 2$, $f(x) < 0$ if $x < -1$ or $x > 2$

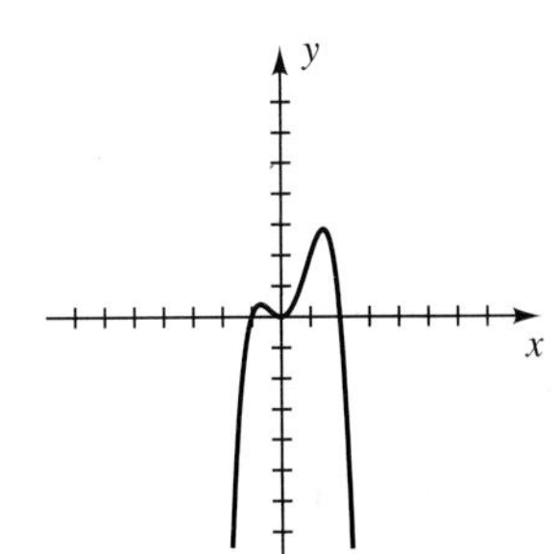

5 $f(x) > 0$ if $-4 < x < 0$ or $x > 2$, $f(x) < 0$ if $x < -4$ or $0 < x < 2$

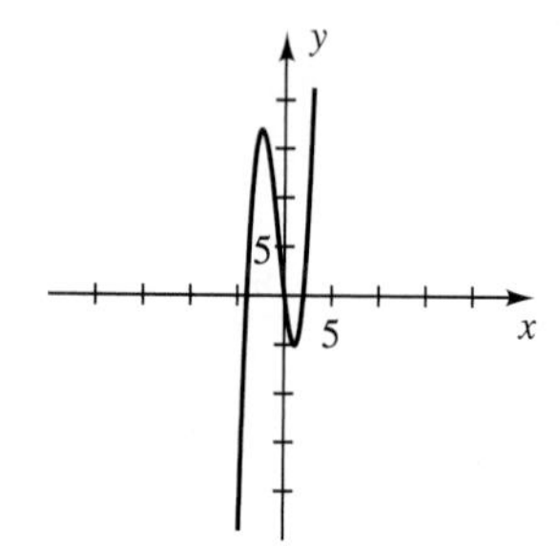

6 $f(x) > 0$ if $-4 < x < -2$, $0 < x < 2$, or $x > 4$, $f(x) < 0$ if $x < -4$, $-2 < x < 0$, or $2 < x < 4$

7 $f(0) = -9 < 100$ and $f(10) = 561 > 100$. By the intermediate value theorem for polynomial functions, f takes on every value between -9 and 561. Hence, there is at least one real number a in $[0, 10]$ such that $f(a) = 100$.

8 Let $f(x) = x^5 - 3x^4 - 2x^3 - x + 1$. $f(0) = 1 > 0$ and $f(1) = -4 < 0$. By the intermediate value theorem for polynomial functions, f takes on every value between -4 and 1. Hence, there is at least one real number a in $[0, 1]$ such that $f(a) = 0$.

9 $3x^2 + 2$; $-21x^2 + 5x - 9$ **10** $4x - 1$; $2x - 1$

11 -132 **12** $f(3) = 0$

13 $6x^4 - 12x^3 + 24x^2 - 52x + 104$; -200

14 $2x^2 + (5 + 2\sqrt{2})x + (2 + 5\sqrt{2})$; $11 + 2\sqrt{2}$

15 $\frac{2}{41}(x^2 + 6x + 34)(x + 1)$

16 $\frac{1}{4}x(x^2 - 2x + 2)(x - 3)$

17 $x^7 + 6x^6 + 9x^5$

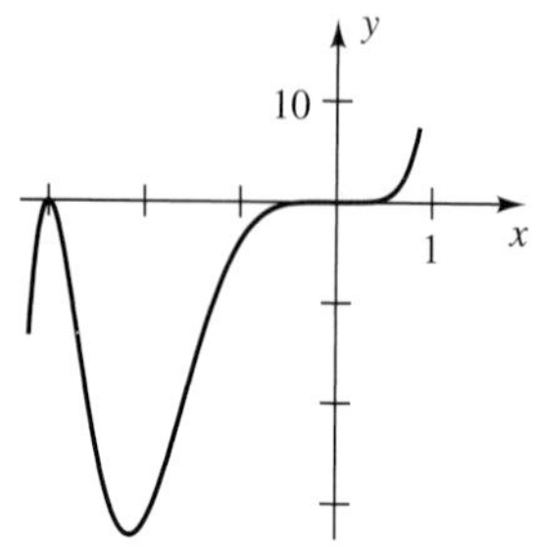

18 $(x - 2)^3(x + 3)(x - 1)$

19 1 (multiplicity 5); -3 (multiplicity 1)

20 0, $\pm i$ (all have multiplicity 2)

21 **(a)** Either 3 positive and 1 negative or 1 positive, 1 negative, and 2 nonreal complex
(b) Upper bound, 3; lower bound, -1

22 **(a)** Either 2 positive and 3 negative; 2 positive, 1 negative, and 2 nonreal complex; 3 negative and 2 nonreal complex; or 1 negative and 4 nonreal complex
(b) Upper bound, 2; lower bound, -3

23 Since there are only even powers, $7x^6 + 2x^4 + 3x^2 + 10 \geq 10$ for every real number x.

24 $-3, -2, -2 \pm i$ **25** $-\frac{1}{2}, \frac{1}{4}, \frac{3}{2}$ **26** $\pm\sqrt{6}, \pm 1$

27 $f(x) = -\frac{1}{6}(x + 2)^3(x - 1)^2(x - 3)$

28 $f(x) = \frac{1}{16}(x + 3)^2x^2(x - 3)^2$

29 VA: $x = 5$; HA: $y = \frac{4}{3}$; x-intercept: 1; y-intercept: $\frac{4}{15}$; hole: $\left(-2, \frac{4}{7}\right)$

30

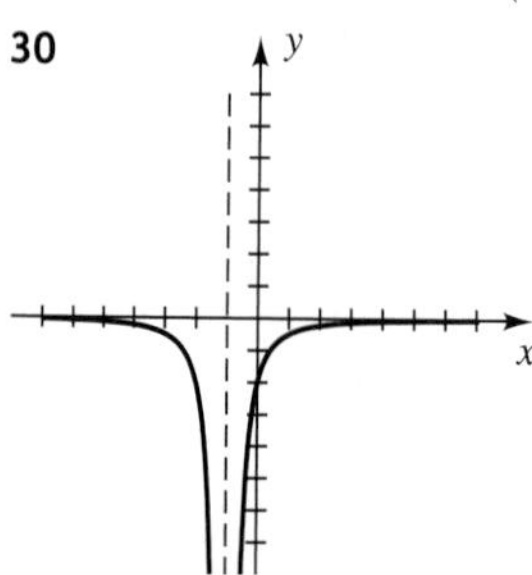

31

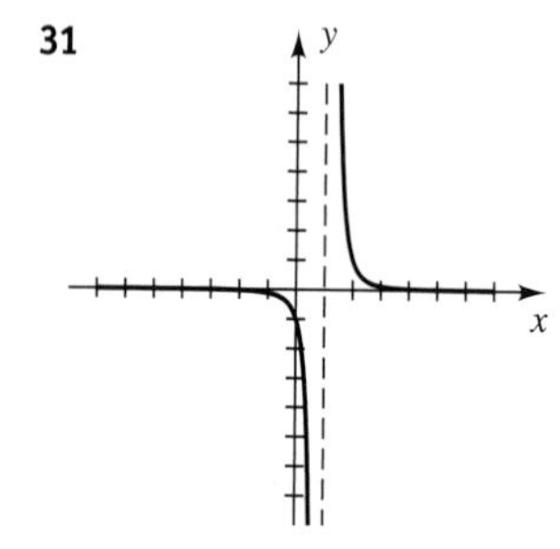

32

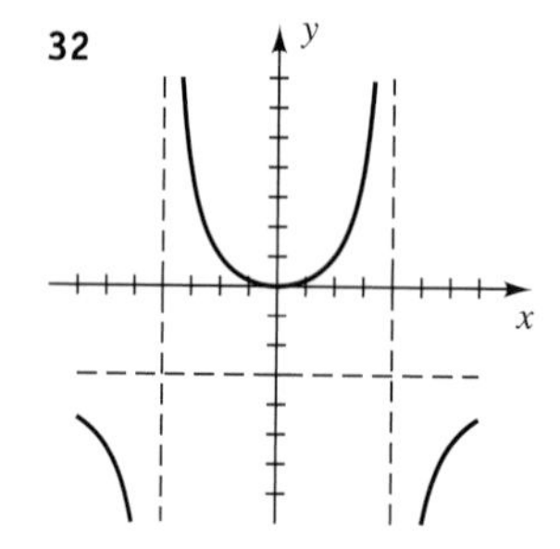

33

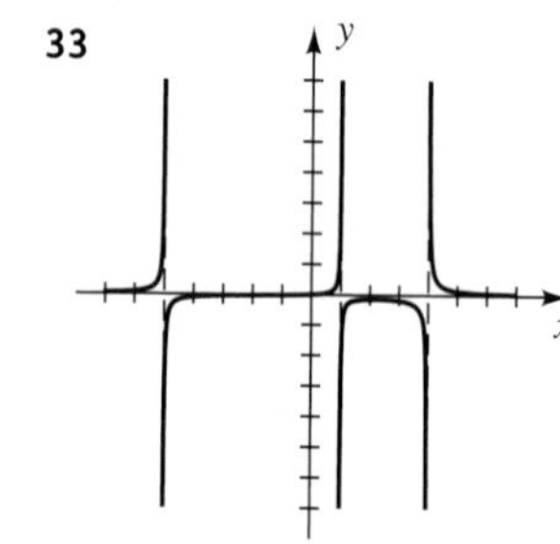

34

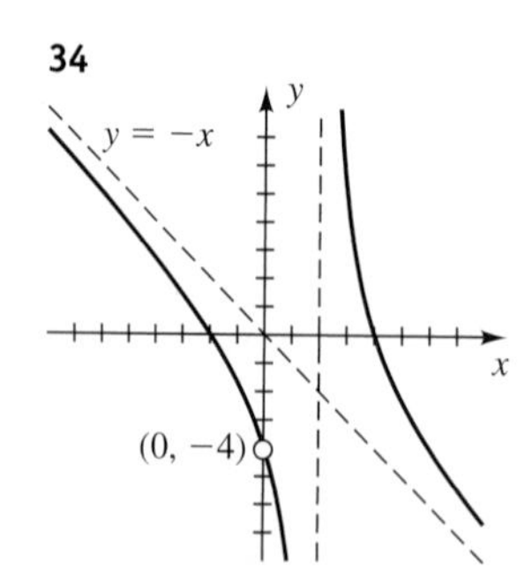

35

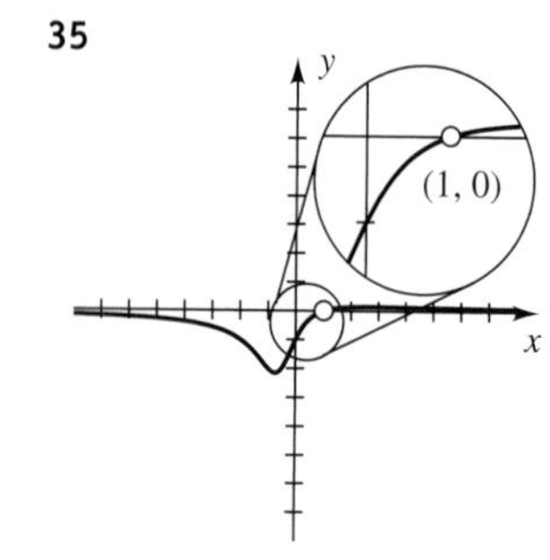

36

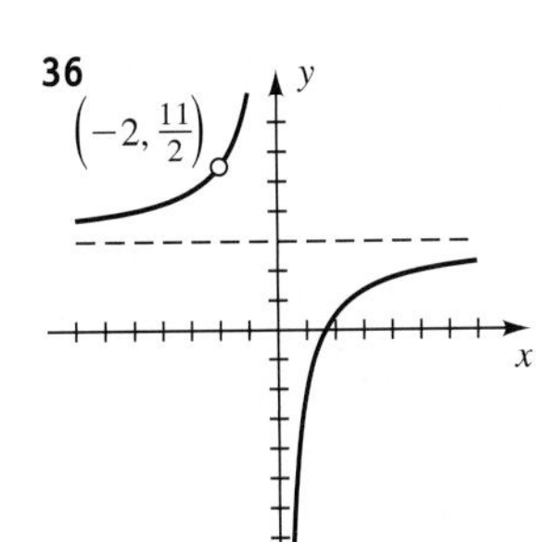

37

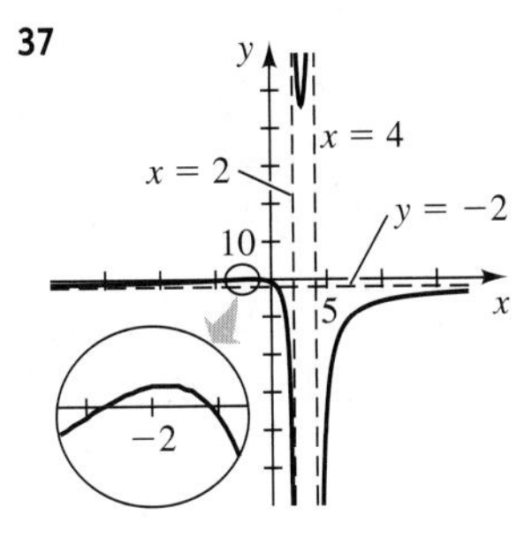

38

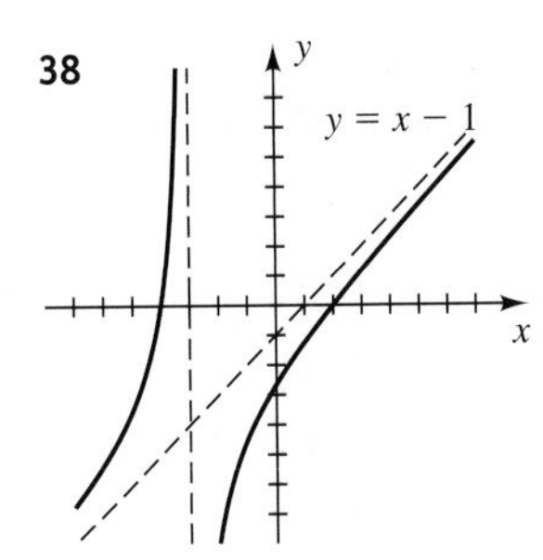

39

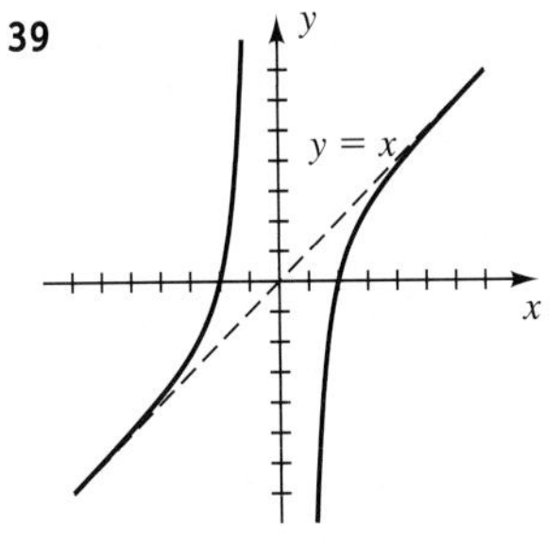

40 $f(x) = \dfrac{3(x-5)(x-2)}{2(x+3)(x-2)}$ or $f(x) = \dfrac{3x^2 - 21x + 30}{2x^2 + 2x - 12}$

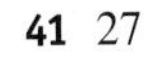

41 27 **42**

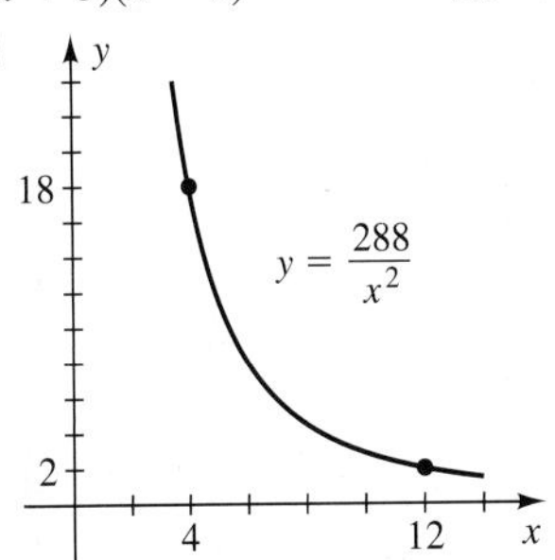

43 **(a)** $\dfrac{1}{15{,}000}$

(b) $y \approx 0.9754 < 1$ if $x = 6.1$, and $y \approx 1.0006 > 1$ if $x = 6.2$

44 **(a)** $V = \dfrac{1}{4\pi}x(l^2 - x^2)$

(b) If $x > 0$, $V > 0$ when $0 < x < l$.

45 $t = 4$ (10:00 A.M.) and $t = 16 - 4\sqrt{6} \approx 6.2020$ (12:12 P.M.)

46 $\sqrt{5} < t < 4$

47 **(a)** $R = k$

(b) k is the maximum rate at which the liver can remove alcohol from the bloodstream.

48 **(a)** $C(100) = \$30$ million and $C(90) \approx \$2.5$ million

(b)

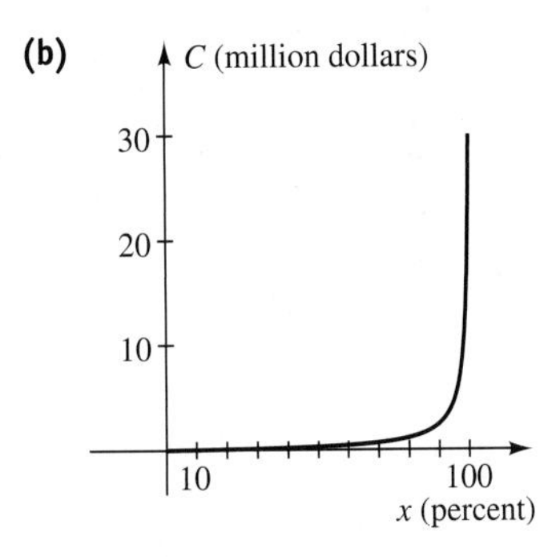

49 375 **50** 10,125 watts

CHAPTER 3 DISCUSSION EXERCISES

2 Yes **4** No **5** $n + 1$ **7** $f(x) = \dfrac{(x^2+1)(x-1)}{(x^2+1)(x-2)}$

8 **(a)** No

(b) Yes, when $x = \dfrac{cd - af}{ae - bd}$, provided the denominator is not zero

9 **(a)** \$1476

(b) Not valid for high confidence values

10 The second integer

11 **(a)** $R(I) = \dfrac{P + SI}{I}$ **(b)** R approaches S.

(c) As income gets larger, individuals pay more in taxes, but fixed tax amounts play a smaller role in determining their overall tax rate.

12 **(a)** 112.8 **(b)** 23 **(c)** 61 yards

Chapter 4

EXERCISES 4.1

1 **(a)** 4 **(b)** Not possible

3 **(a)** Yes **(b)** No **(c)** Not a function

5 Yes **7** No **9** Yes **11** No **13** No **15** Yes

Exer. 17–20: Show that $f(g(x)) = x = g(f(x))$.

17 **19**

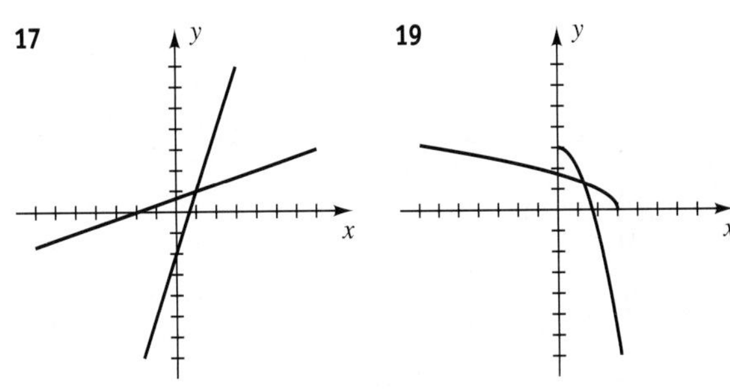

21 $(-\infty, 0) \cup (0, \infty)$; $(-\infty, 1) \cup (1, \infty)$

23 $\left(-\infty, \frac{4}{3}\right) \cup \left(\frac{4}{3}, \infty\right); \left(-\infty, \frac{8}{3}\right) \cup \left(\frac{8}{3}, \infty\right)$

25 $f^{-1}(x) = \dfrac{x-5}{3}$ **27** $f^{-1}(x) = \dfrac{2x+1}{3x}$

29 $f^{-1}(x) = \dfrac{5x+2}{2x-3}$ **31** $f^{-1}(x) = -\sqrt{\dfrac{2-x}{3}}$

33 $f^{-1}(x) = \sqrt[3]{\dfrac{x+5}{2}}$ **35** $f^{-1}(x) = 3 - x^2, x \geq 0$

37 $f^{-1}(x) = (x-1)^3$ **39** $f^{-1}(x) = x$

41 $f^{-1}(x) = 3 + \sqrt{x+9}$ **43** **(a)** 3 **(b)** -1 **(c)** 5

45 **(a)**

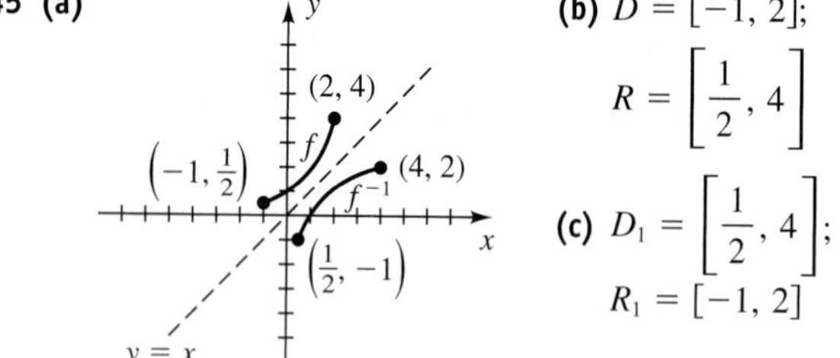

(b) $D = [-1, 2]$; $R = \left[\frac{1}{2}, 4\right]$

(c) $D_1 = \left[\frac{1}{2}, 4\right]$; $R_1 = [-1, 2]$

47 **(a)**

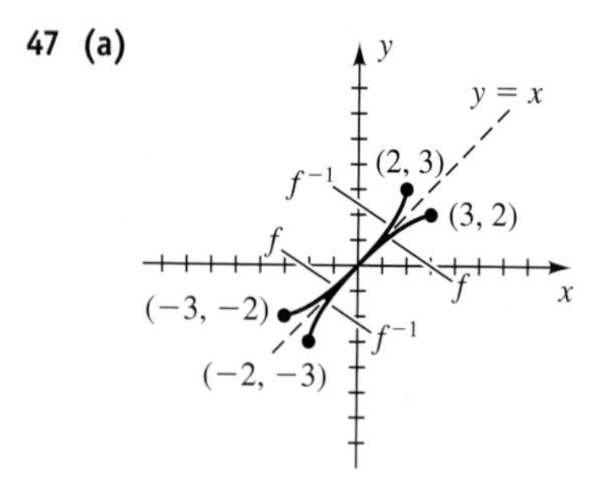

(b) $D = [-3, 3]$; $R = [-2, 2]$

(c) $D_1 = [-2, 2]$; $R_1 = [-3, 3]$

49 **(a)** Since f is one-to-one, an inverse exists; $f^{-1}(x) = \dfrac{x-b}{a}$

(b) No; not one-to-one

51 **(c)** The graph of f is symmetric about the line $y = x$. Thus, $f(x) = f^{-1}(x)$.

53 Yes

55

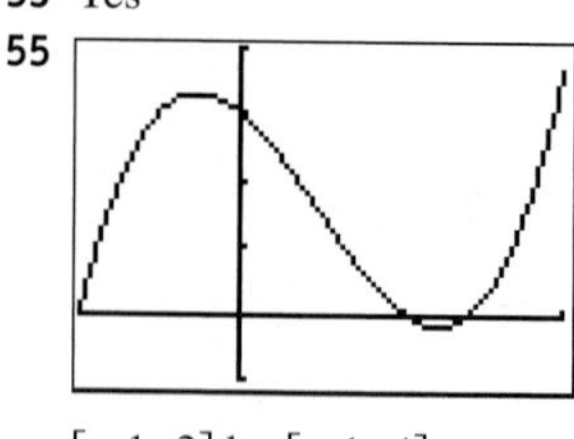

$[-1, 2]$ by $[-1, 4]$

(a) $[-0.27, 1.22]$

(b) $[-0.20, 3.31]$; $[-0.27, 1.22]$

57

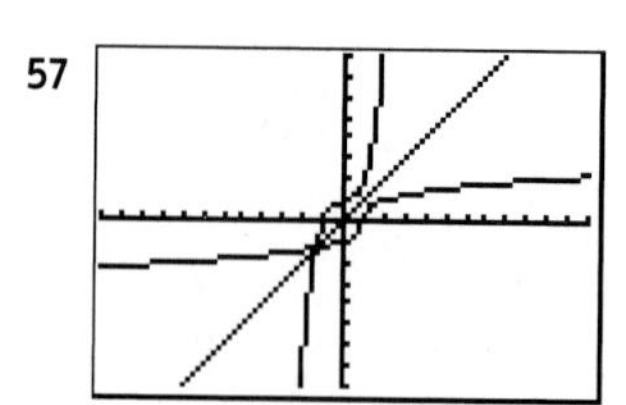

$[-12, 12]$ by $[-8, 8]$

$f^{-1}(x) = x^3 + 1$

59 **(a)** 805 ft³/min

(b) $V^{-1}(x) = \dfrac{1}{35}x$. Given an air circulation of x cubic feet per minute, $V^{-1}(x)$ computes the maximum number of people that should be in the restaurant at one time.

(c) 67

EXERCISES 4.2

1 5 **3** $-1, 3$ **5** $-\dfrac{4}{99}$ **7** $\dfrac{18}{5}$ **9** 3

11 **(a)**

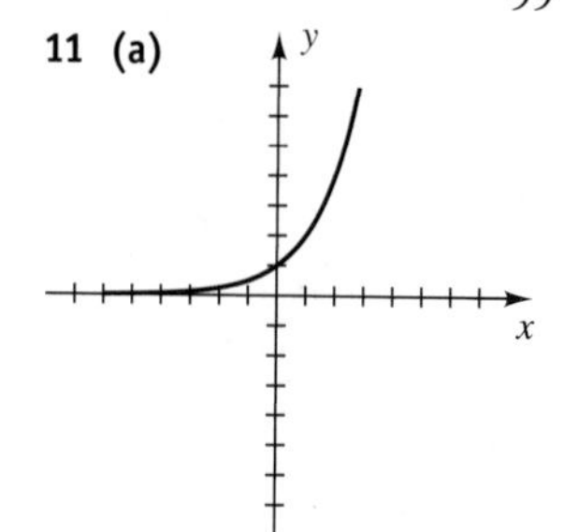

(b)

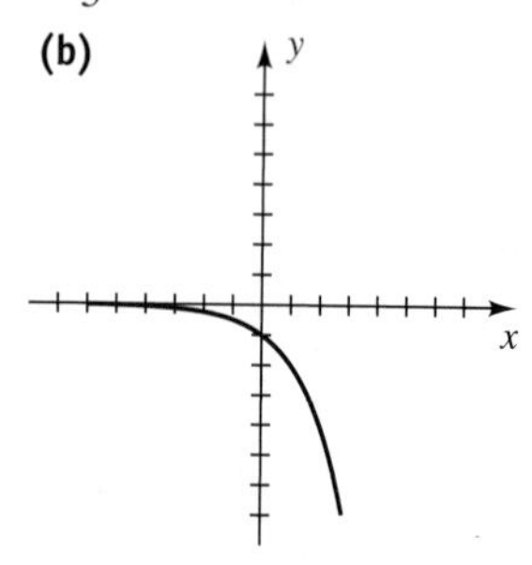

(c)

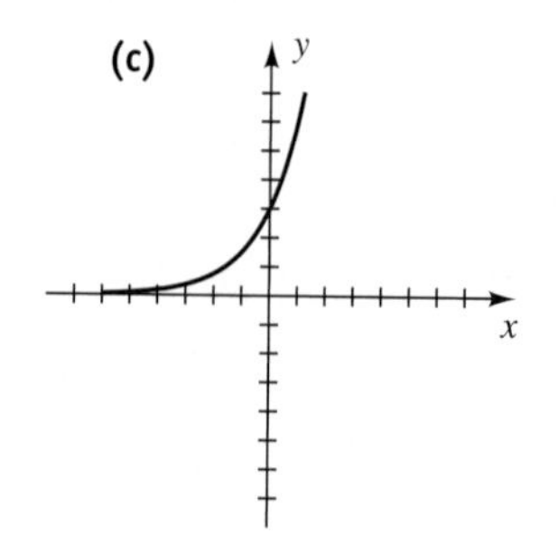

(d)

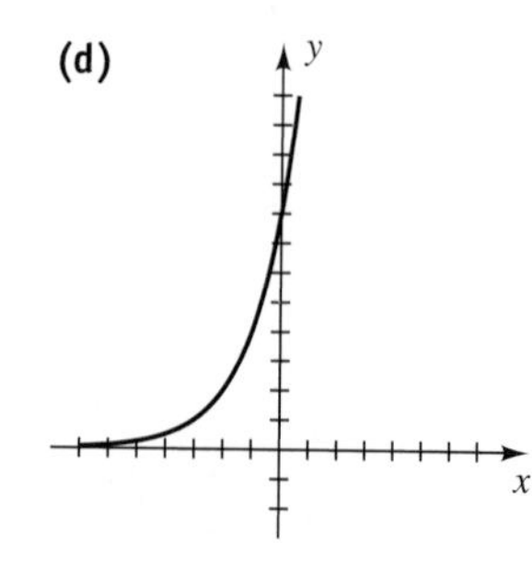

(e)

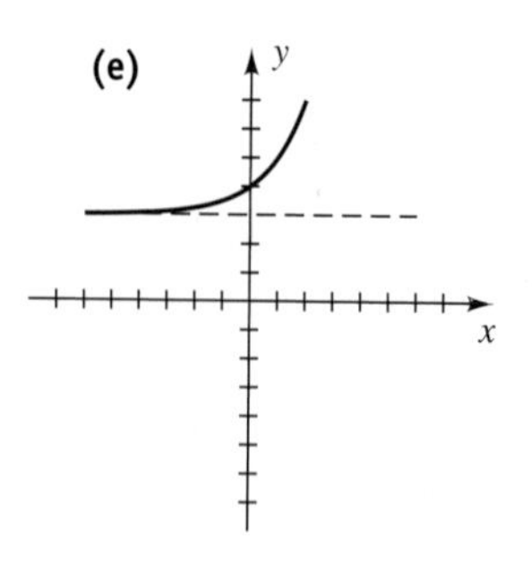

(f)

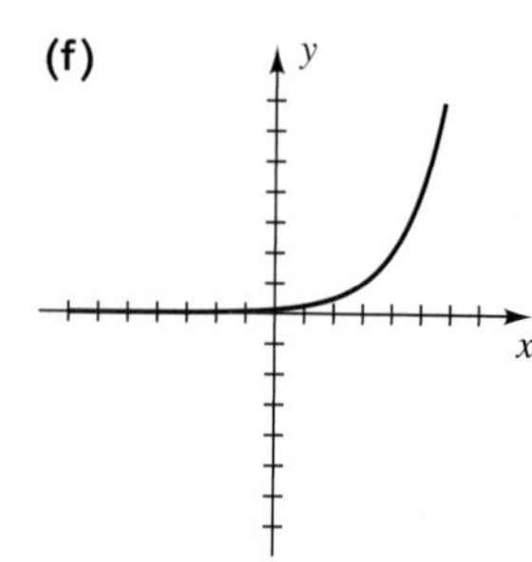

(g)

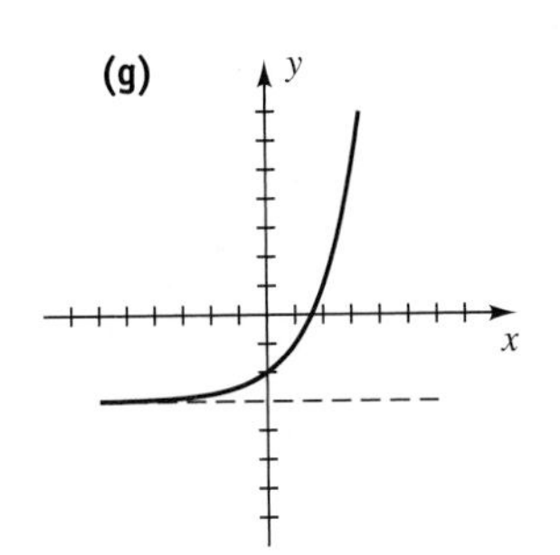

(h)

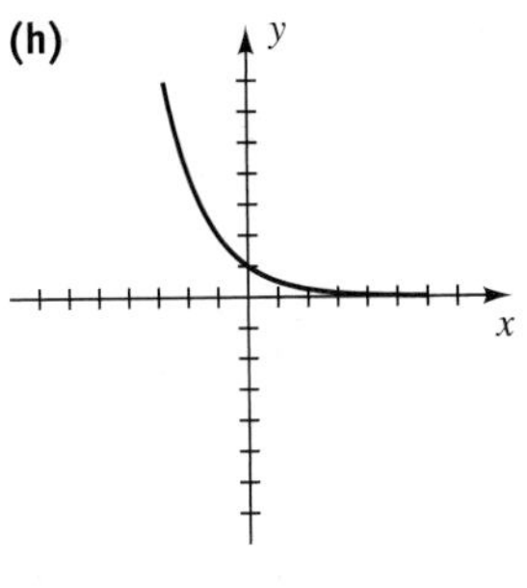

(i)

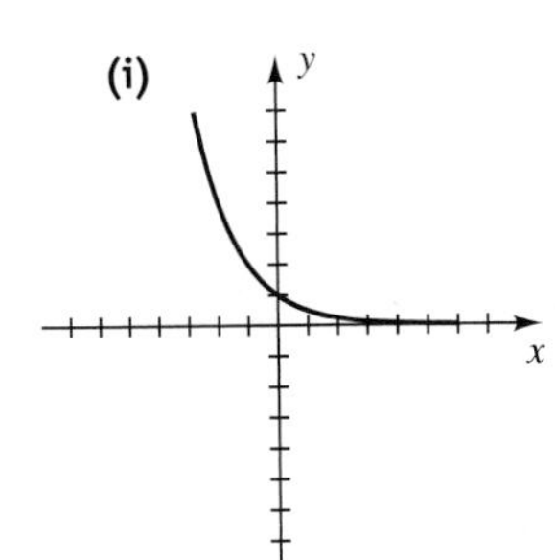

(j)

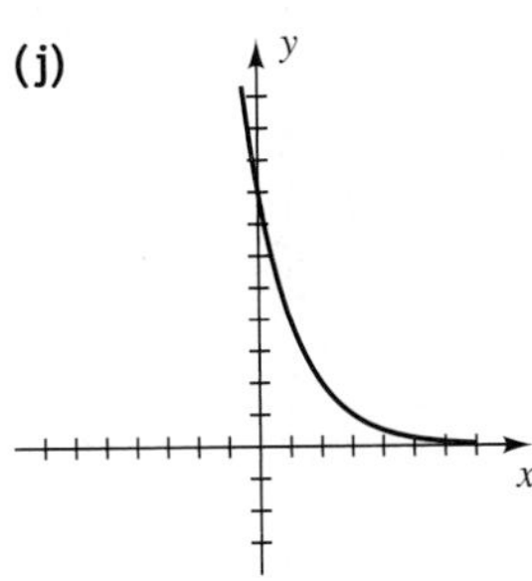

13

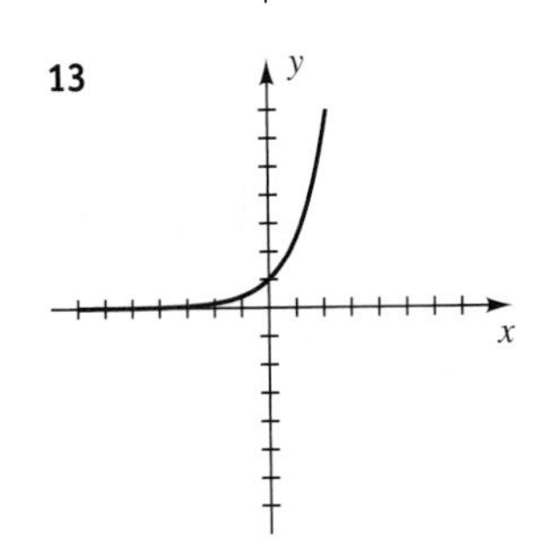

15

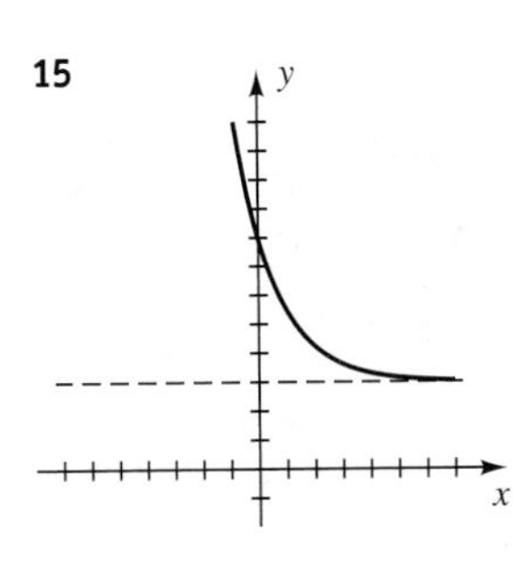

17

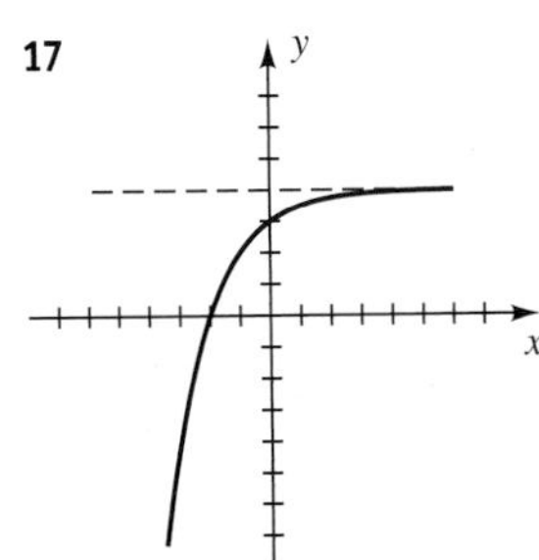

19

21

23

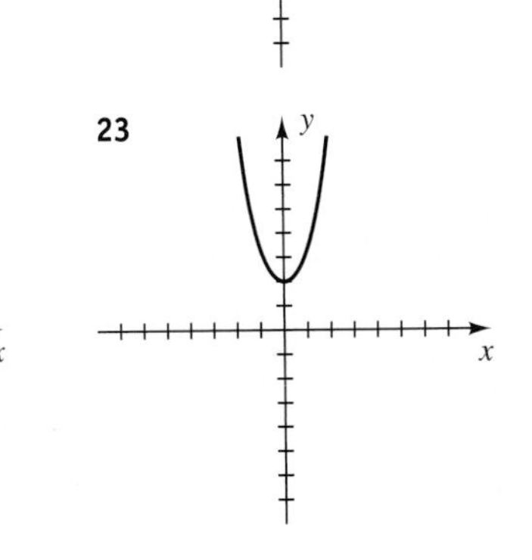

25 $f(x) = 2\left(\frac{5}{2}\right)^x$ **27** $f(x) = 2\left(\frac{2}{3}\right)^x - 3$

29 $f(x) = 8\left(\frac{1}{2}\right)^x$ **31** $f(x) = 180(1.5)^{-x} + 32$

33 **(a)** 90 **(b)** 59 **(c)** 35

35 **(a)** 1039; 3118; 5400

(b)

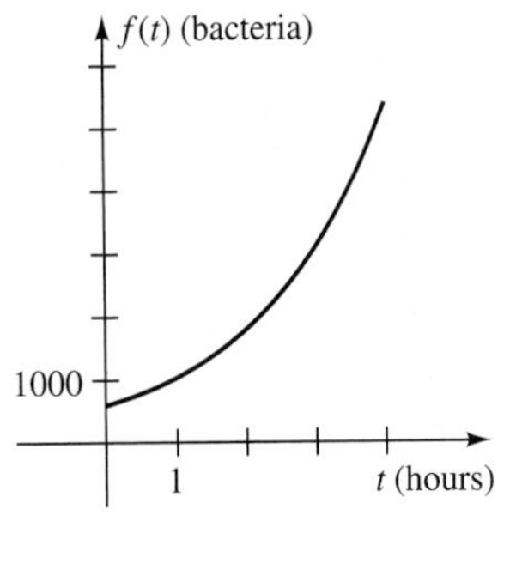

37 **(a)** 50 mg; 25 mg;

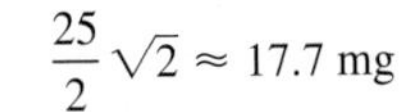
$\frac{25}{2}\sqrt{2} \approx 17.7$ mg

(b)

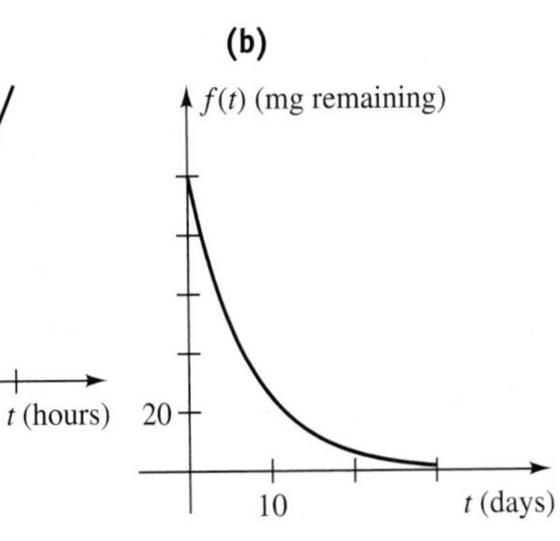

39 $-\frac{1}{1600}$

41 **(a)** \$1005.83 **(b)** \$1035.51 **(c)** \$1072.29 **(d)** \$4038.74

43 **(a)** \$19,500 **(b)** \$11,975 **(c)** \$7354

45 \$161,657,351,965.80

47 **(a)** Examine the pattern formed by the value y in the year n.

(b) Solve $s = (1 - a)^T y_0$ for a.

49 **(a)** \$1834.41 **(b)** \$410,387.60

51 \$15,495.62

53 **(a)** 180.1206 **(b)** 20.9758 **(c)** 7.3639

55

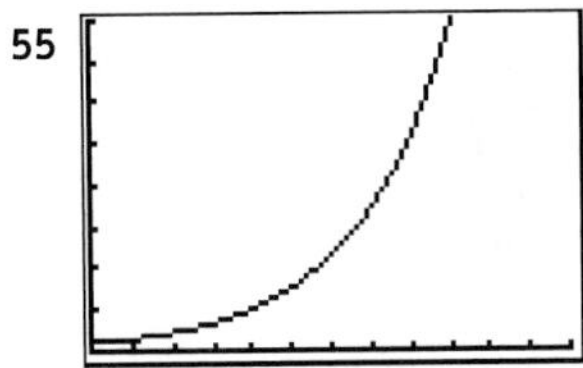

[0, 60, 5] by [0, 40, 5]

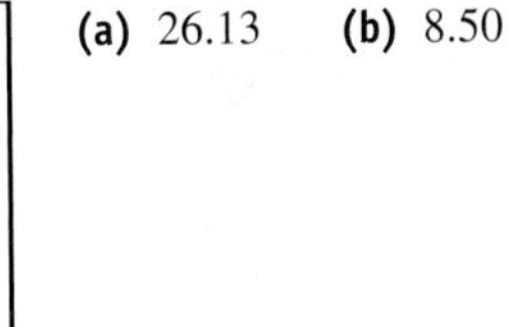
(a) 26.13 **(b)** 8.50

57 −1.02, 2.14, 3.62

59

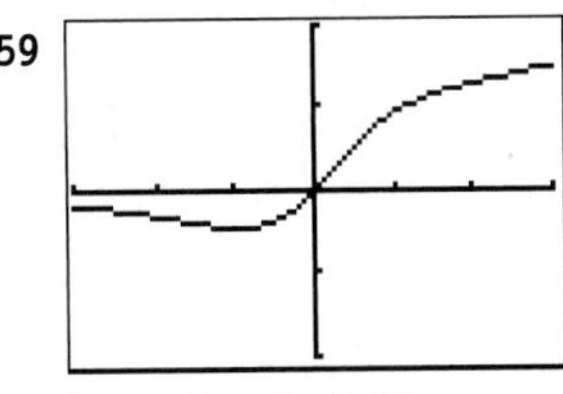
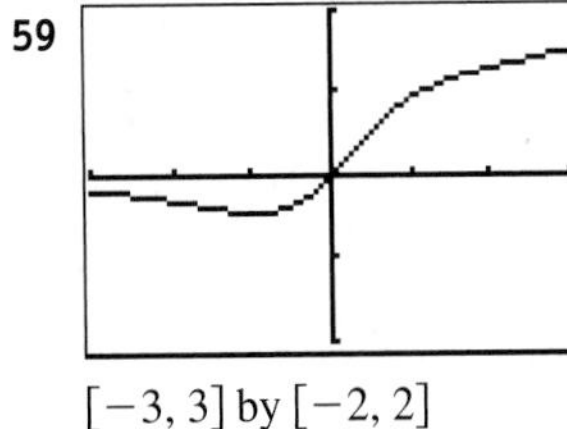

[−3, 3] by [−2, 2]

(a) Not one-to-one
(b) 0

61

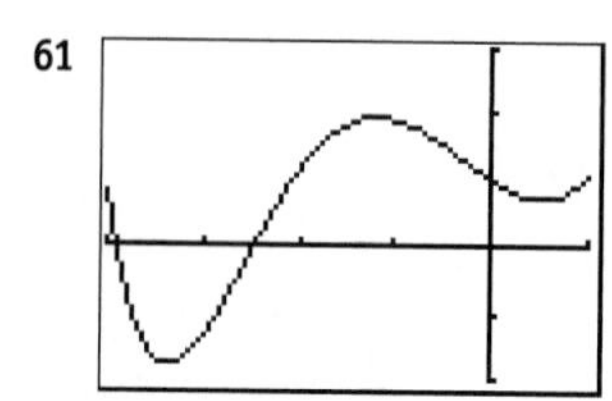

$[-4, 1]$ by $[-2, 3]$

(a) Increasing: $[-3.37, -1.19] \cup [0.52, 1]$; decreasing: $[-4, -3.37] \cup [-1.19, 0.52]$

(b) $[-1.79, 1.94]$

63 6.58 yr

65

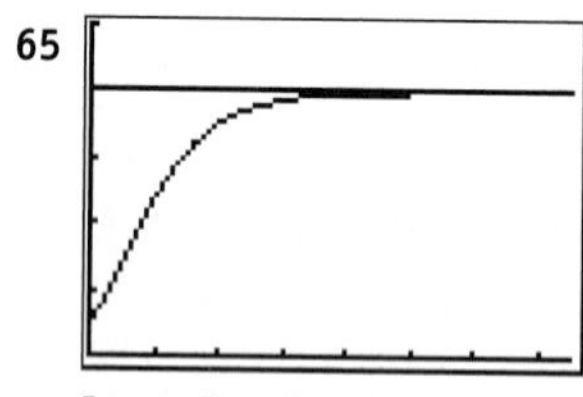

$[0, 7.5]$ by $[0, 5]$

The maximum number of sales approaches k.

67

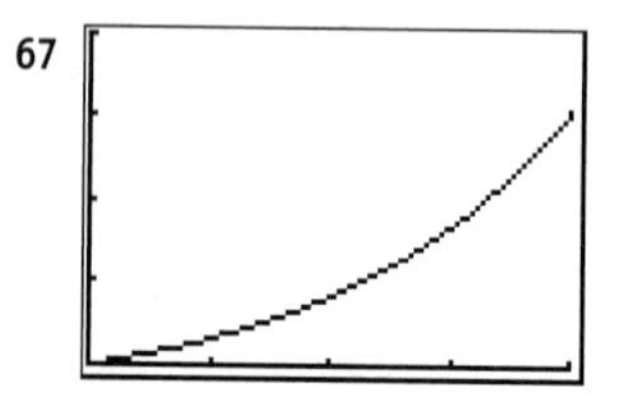

$[0, 40, 10]$ by $[0, 200{,}000, 50{,}000]$

After approximately 32.8 yr

69 **(a)**

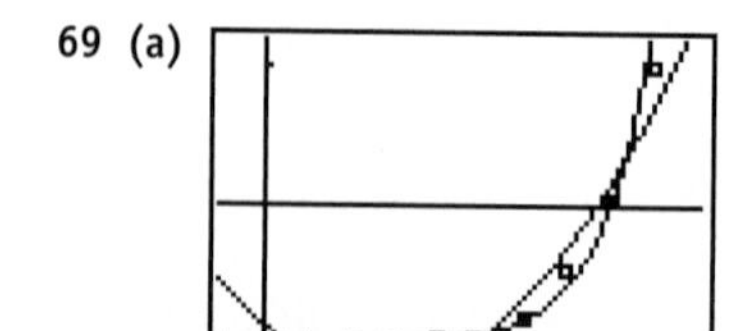

$[-10, 100, 10]$ by $[-200, 2200, 1000]$

(b) Exponential function f **(c)** 1989

71 $y = 0.03(1.0549)^t$; 48¢

73 **(a)** \$746,648.43; \$1,192,971 **(b)** 12.44%

(c) exponential; polynomial

EXERCISES 4.3

1 **(a)**

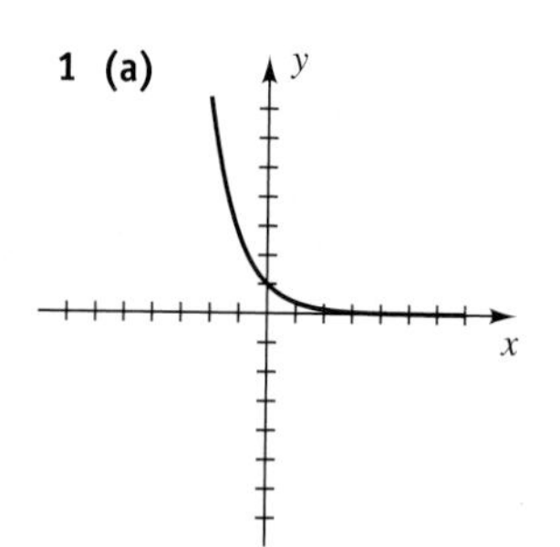

(b)

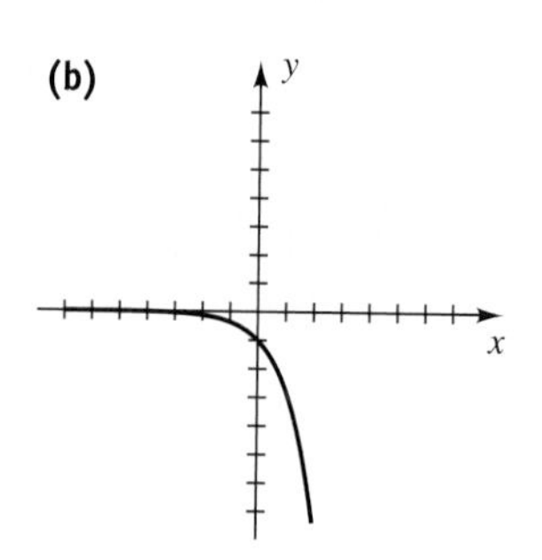

3 **(a)**

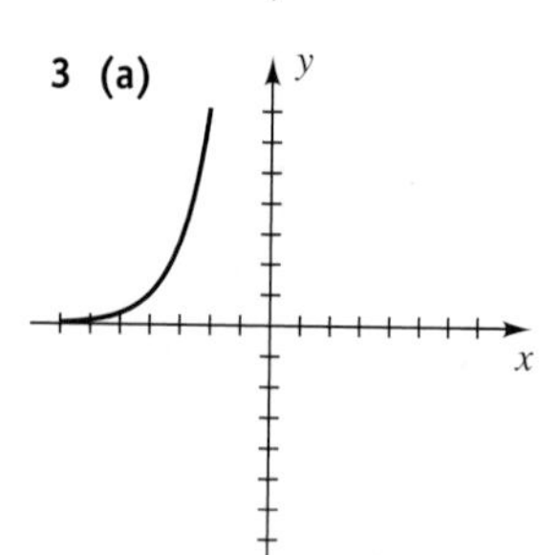

(b)

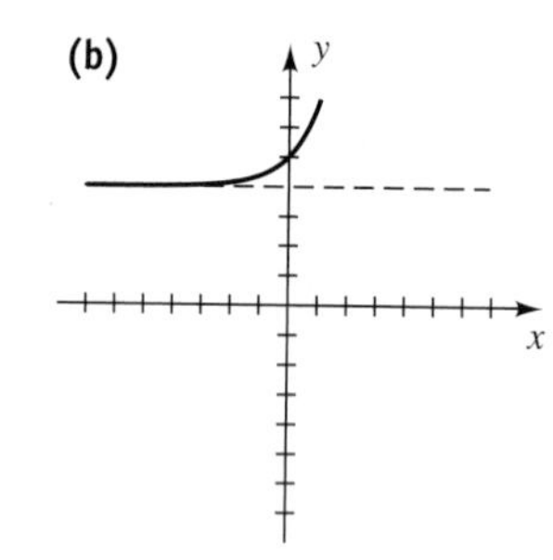

5 \$1510.59 **7** \$31,600.41 **9** 13% **11** 3, 4

13 -1 **15** $-\frac{3}{4}, 0$ **17** $\frac{4}{(e^x + e^{-x})^2}$ **19** 27.43 g

21 348.8 million **23** 13.5% **25** 41

27 7.44 in. **29** 75.77 cm; 15.98 cm/yr

31 \$11.25 per hr **33** **(a)** 7.19% **(b)** 7.25%

35

37

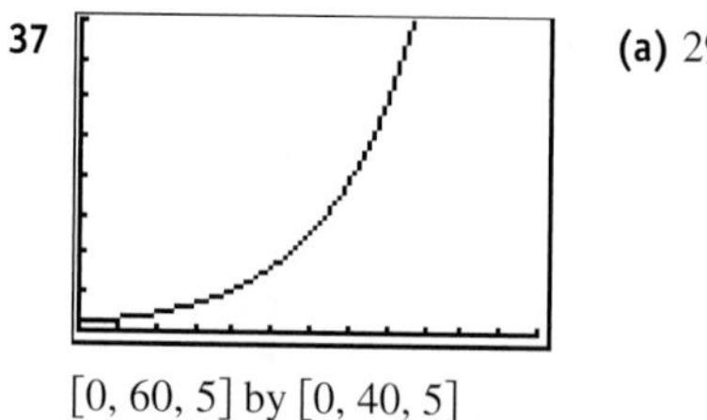

$[0, 60, 5]$ by $[0, 40, 5]$

(a) 29.96 **(b)** 8.15

39 (a)

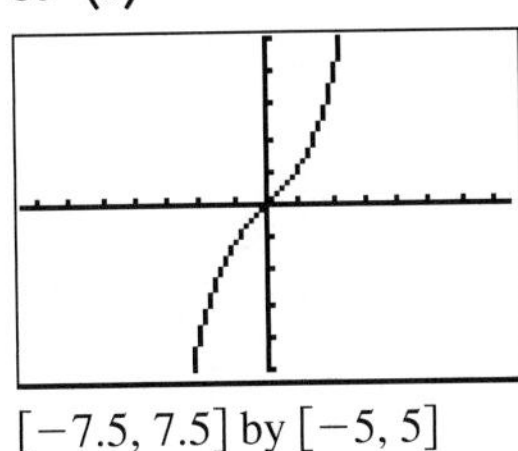

$[-7.5, 7.5]$ by $[-5, 5]$

(b)

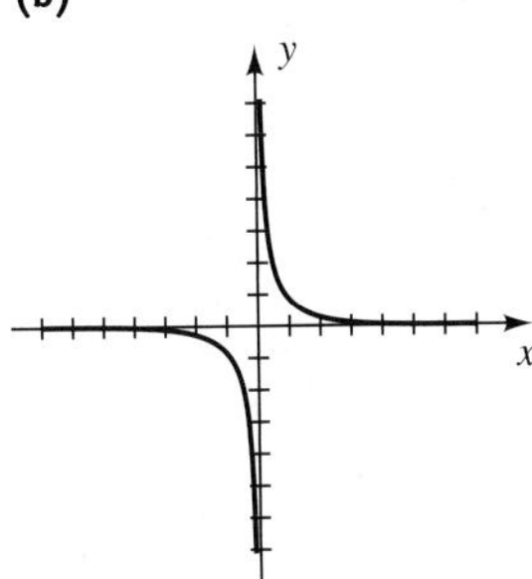

41 (a)

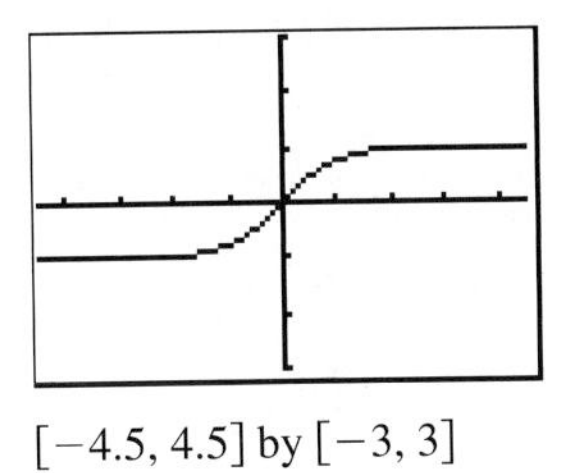

$[-4.5, 4.5]$ by $[-3, 3]$

(b)

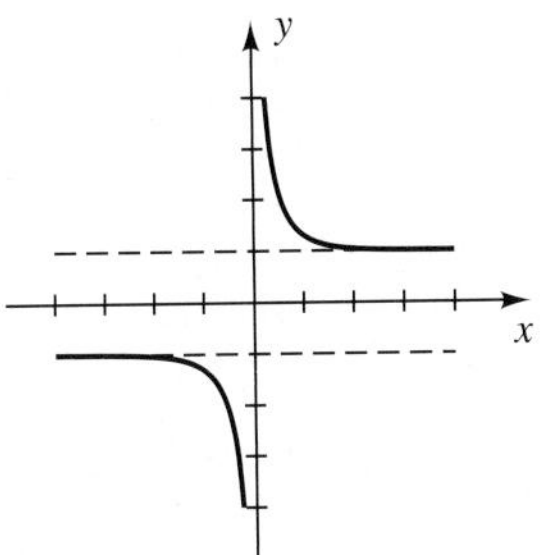

43

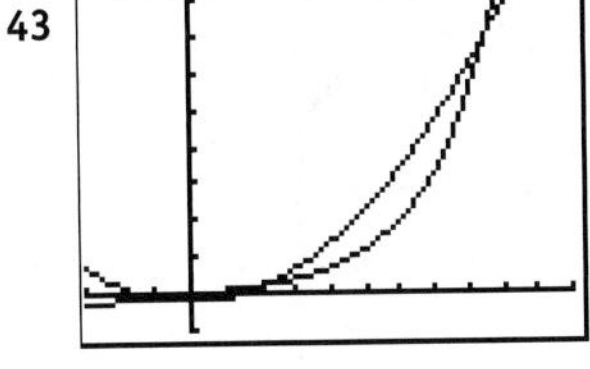

$[-3, 11]$ by $[-10, 80, 10]$

$-1.04, 2.11, 8.51$

45

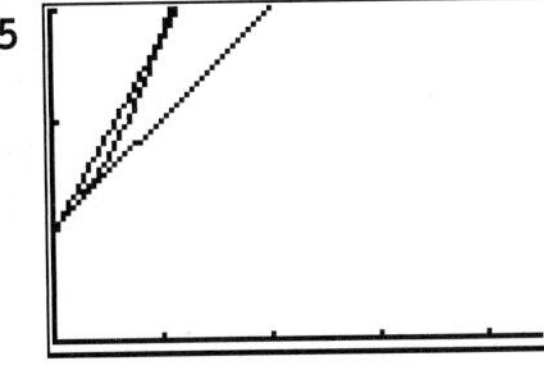

$[0, 4.5]$ by $[0, 3]$

$f(x)$ is closer to e^x if $x \approx 0$; $g(x)$ is closer to e^x if $x \approx 1$.

47

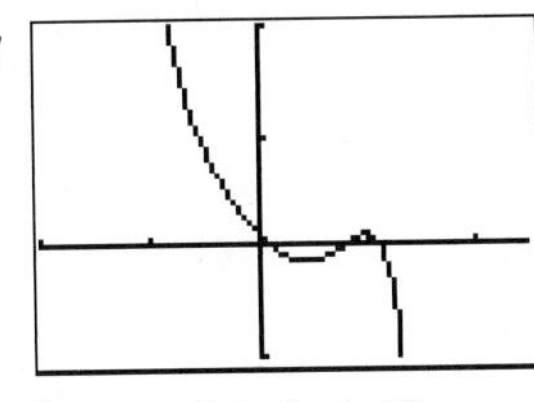

$[-2, 2.5]$ by $[-1, 2]$

$0.11, 0.79, 1.13$

49

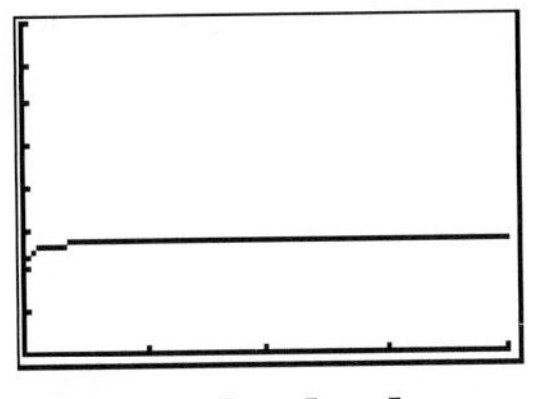

$[0, 200, 50]$ by $[0, 8]$

$y \approx 2.71 \approx e$

51 0.567

53

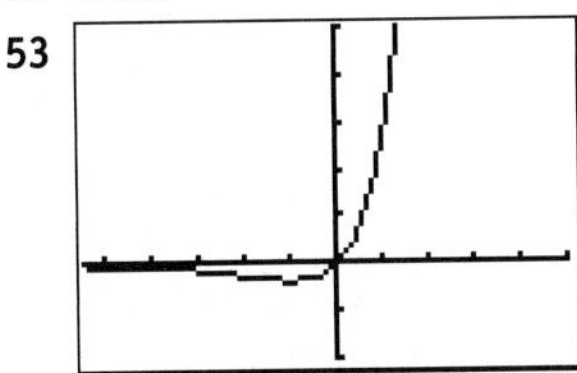

$[-5.5, 5]$ by $[-2, 5]$

Increasing on $[-1, \infty)$; decreasing on $(-\infty, -1]$

55 (a) As h increases, C decreases.
(b) As y increases, C decreases.

57 (a) $f(x) = 1.225e^{-0.0001085x}$

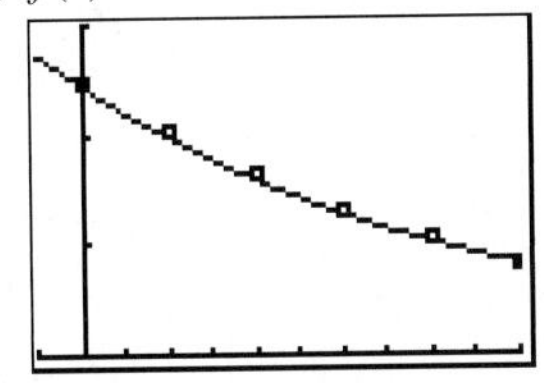

$[-1000, 10{,}000, 1000]$ by $[0, 1.5, 0.5]$

(b) $0.885, 0.461$

EXERCISES 4.4

1 (a) $\log_4 64 = 3$ **(b)** $\log_4 \frac{1}{64} = -3$
(c) $\log_t s = r$ **(d)** $\log_3 (4 - t) = x$
(e) $\log_5 \frac{a + b}{a} = 7t$ **(f)** $\log_{0.7} (5.3) = t$

3 (a) $2^5 = 32$ **(b)** $3^{-5} = \frac{1}{243}$ **(c)** $t^p = r$
(d) $3^5 = (x + 2)$ **(e)** $2^{3x+4} = m$ **(f)** $b^{3/2} = 512$

5 $t = 3 \log_a \frac{5}{2}$ **7** $t = \log_a\left(\frac{H - K}{C}\right)$

9 $t = \frac{1}{C} \log_a \left(\frac{A - D}{B}\right)$

11 (a) $\log 100{,}000 = 5$ **(b)** $\log 0.001 = -3$
(c) $\log (y + 1) = x$ **(d)** $\ln p = 7$
(e) $\ln (3 - x) = 2t$

13 (a) $10^{50} = x$ **(b)** $10^{20t} = x$ **(c)** $e^{0.1} = x$
(d) $e^{4+3x} = w$ **(e)** $e^{1/6} = z - 2$

15 **(a)** 0 **(b)** 1 **(c)** Not possible **(d)** 2 **(e)** 8 **(f)** 3 **(g)** -2

17 **(a)** 3 **(b)** 5 **(c)** 2 **(d)** -4 **(e)** 2 **(f)** -3 **(g)** $3e^2$

19 4 **21** No solution **23** $-1, -2$ **25** 13

27 27 **29** $\pm\dfrac{1}{e}$ **31** 3 **33** 3

35 **(a)**

(b)

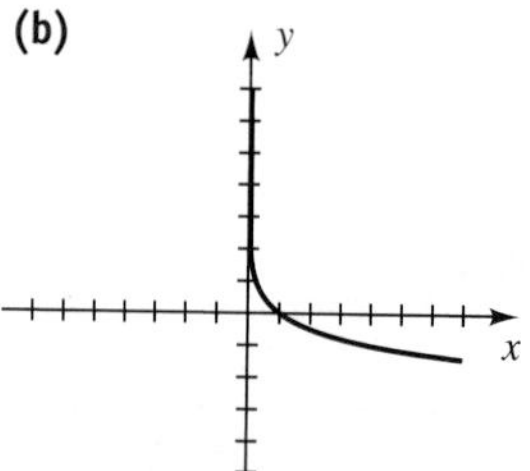

(c)

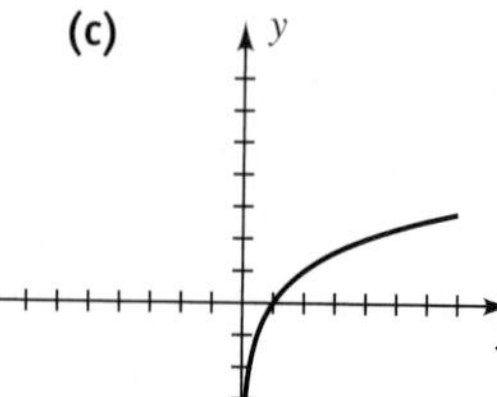

(d)

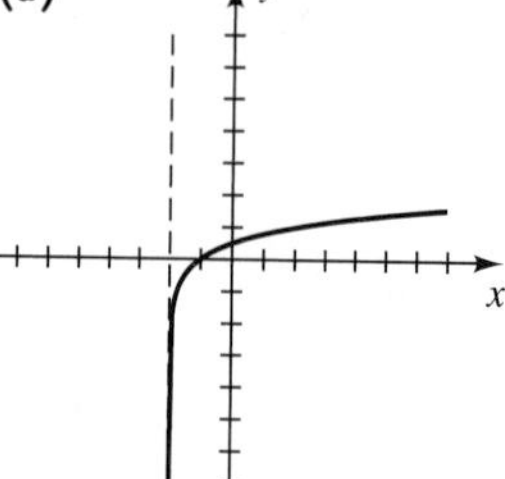

(e)

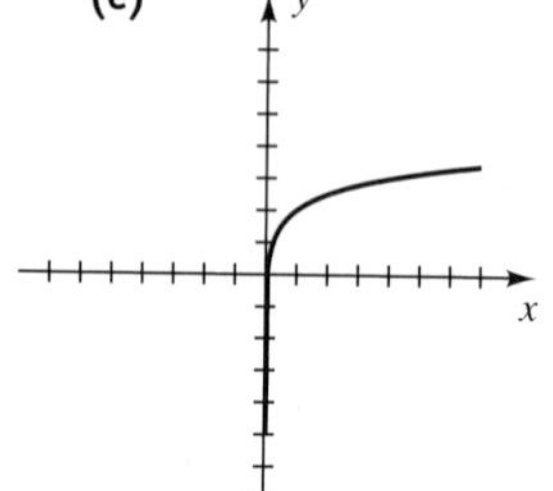

(f)

(g)

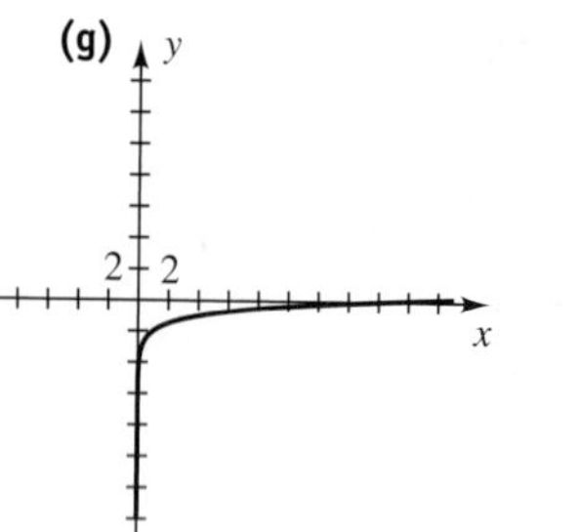

(h)

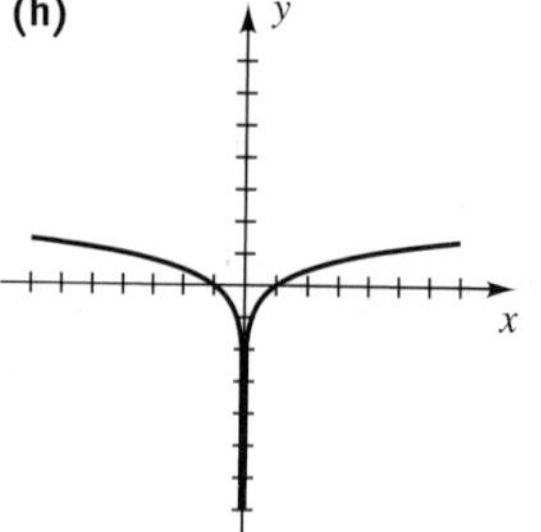

(i)

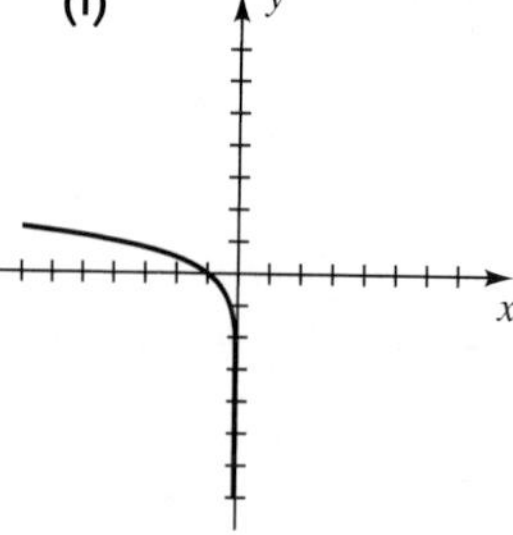

(j)

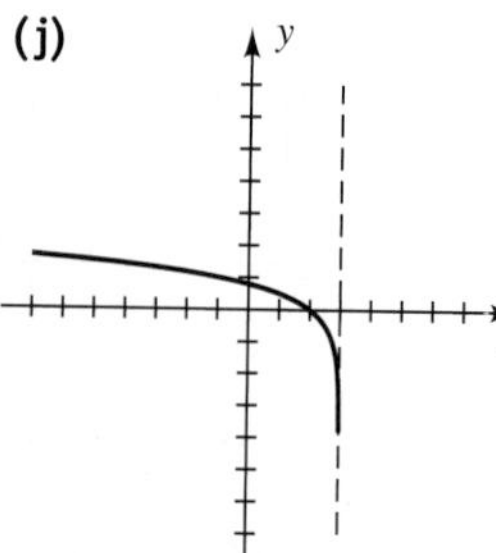

(k)

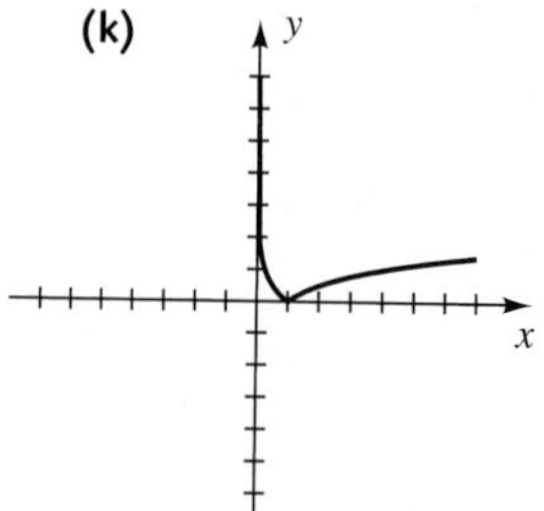

(l)

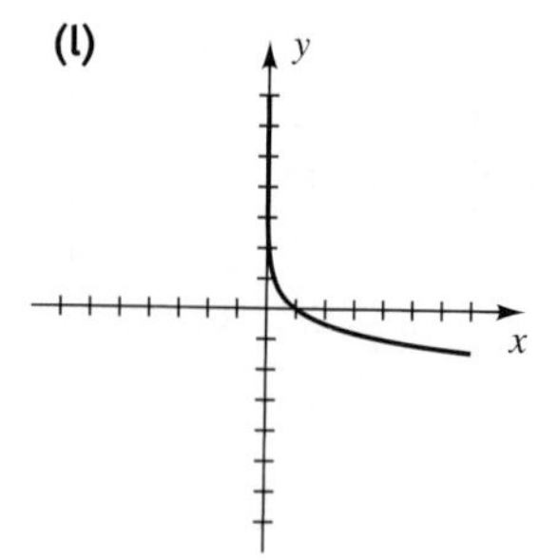

37

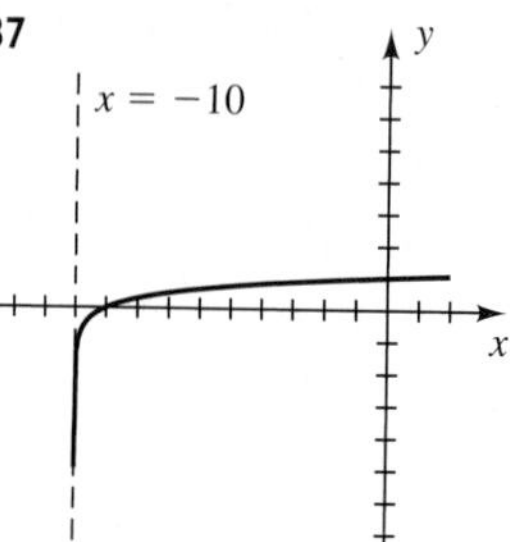

39

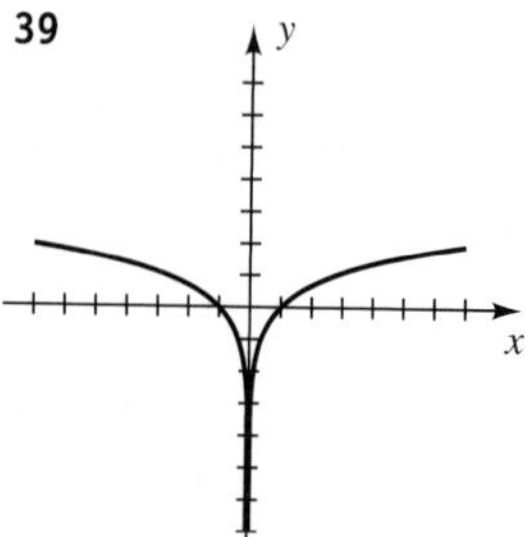

41

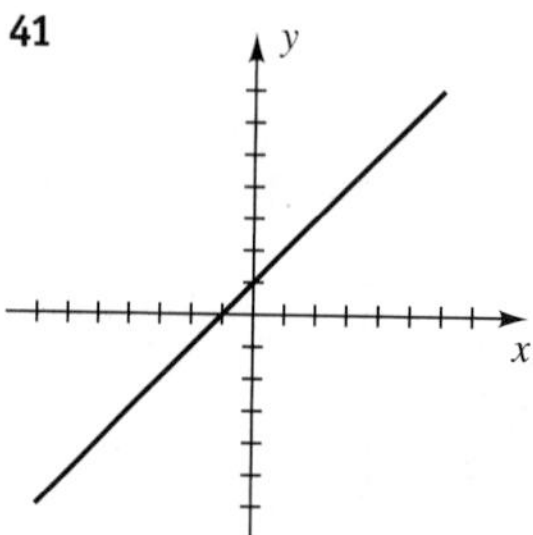

43 $f(x) = \log_3 x$

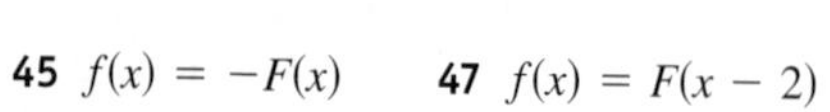

45 $f(x) = -F(x)$ **47** $f(x) = F(x-2)$

49 $f(x) = F(x) + 1$

51 **(a)** 4240 **(b)** 8.85 **(c)** 0.0237 **(d)** 9.97 **(e)** 1.05 **(f)** 0.202

53 $f(x) = 1000e^{x \ln 1.05}$; 4.88% **55** $t = -1600 \log_2\left(\dfrac{q}{q_0}\right)$

57 $t = -\dfrac{L}{R}\ln\left(\dfrac{I}{20}\right)$ **59** **(a)** 2 **(b)** 4 **(c)** 5

61 **(a)** 10 **(b)** 30 **(c)** 40 **63** In the year 2047

65 **(a)** $W = 2.4e^{1.84h}$ **(b)** 37.92 kg

67 **(a)** 10,007 ft **(b)** 18,004 ft

69 **(a)** 305.9 kg **(b)** (1) 20 yr (2) 19.8 yr

71 10.1 mi **73** $2^{1/8} \approx 1.09$

75 **(a)** Pedestrians have faster average walking speeds in large cities.

(b) 570,000

77 **(a)** 8.4877 **(b)** -0.0601

79 1.763 **81** (0, 14.90]

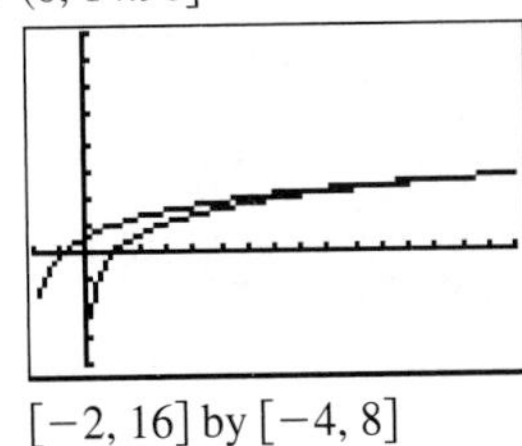

[−2, 16] by [−4, 8]

83 **(a)** 30% **(b)** 3.85

EXERCISES 4.5

1 **(a)** $\log_4 x + \log_4 z$ **(b)** $\log_4 y - \log_4 x$

(c) $\frac{1}{3}\log_4 z$

3 $3\log_a x + \log_a w - 2\log_a y - 4\log_a z$

5 $\frac{1}{3}\log z - \log x - \frac{1}{2}\log y$

7 $\frac{7}{4}\ln x - \frac{5}{4}\ln y - \frac{1}{4}\ln z$

9 **(a)** $\log_3 (5xy)$ **(b)** $\log_3 \frac{2z}{x}$ **(c)** $\log_3 y^5$

11 $\log_a \frac{x^2 \sqrt[3]{x-2}}{(2x+3)^5}$ **13** $\log \frac{y^{13/3}}{x^2}$ **15** $\ln x$ **17** $\frac{7}{2}$

19 $5\sqrt{5}$ **21** No solution **23** -7 **25** 1

27 -2 **29** $\frac{-1+\sqrt{65}}{2}$ **31** $-1 + \sqrt{1+e}$

33 $3 + \sqrt{10}$

35

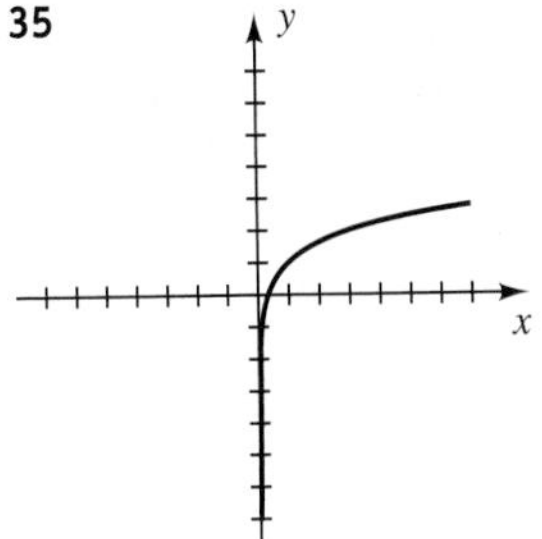

37

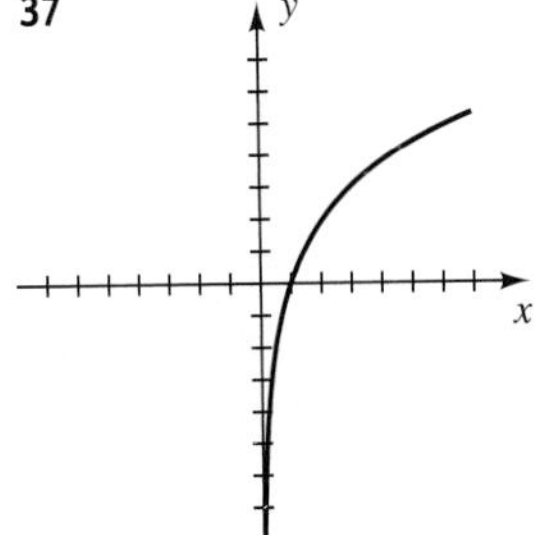

39

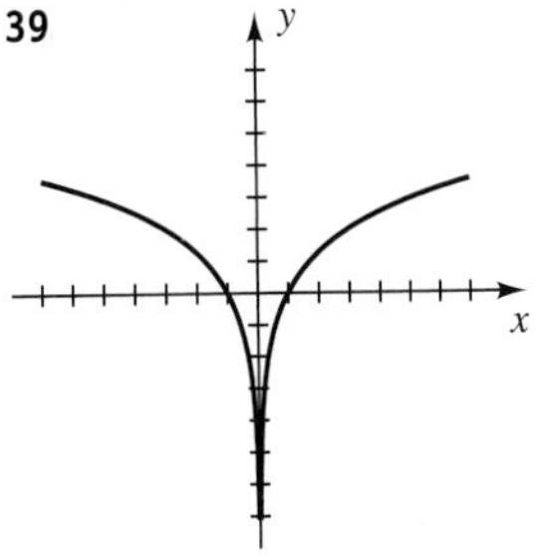

41

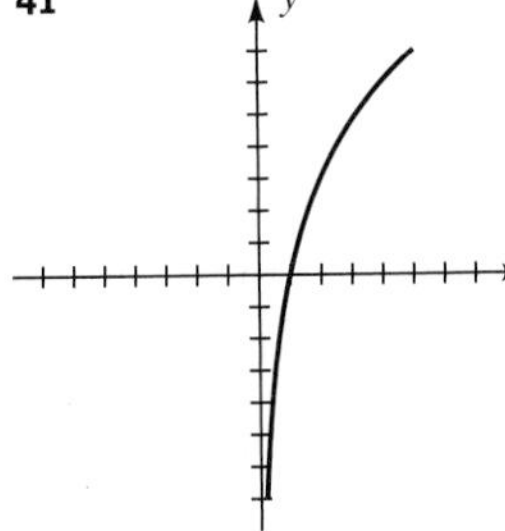

43

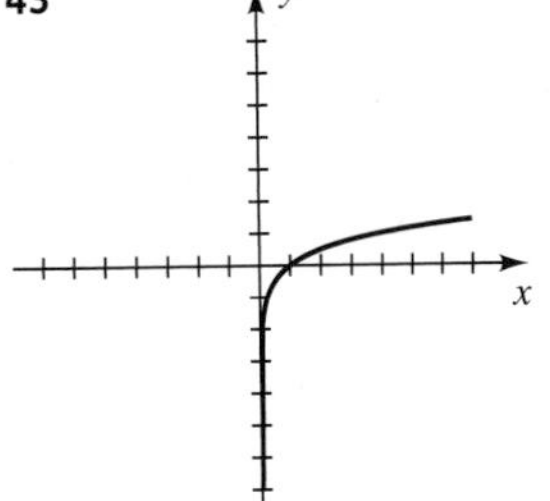

45

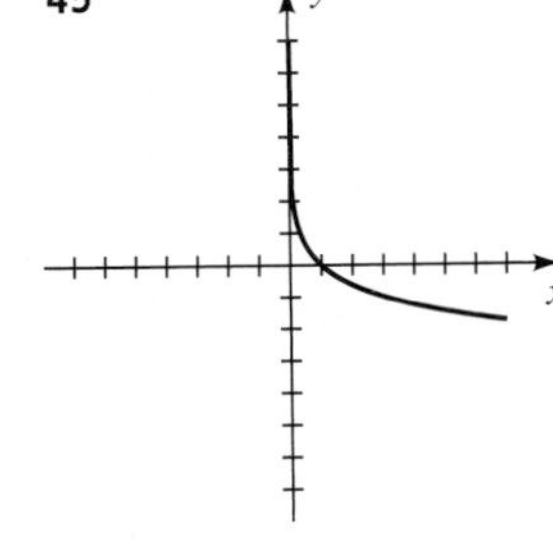

47 $f(x) = \log_2 x^2$ **49** $f(x) = \log_2 (8x)$ **51** $\approx +7$

53 $y = \frac{b}{x^k}$ **55**

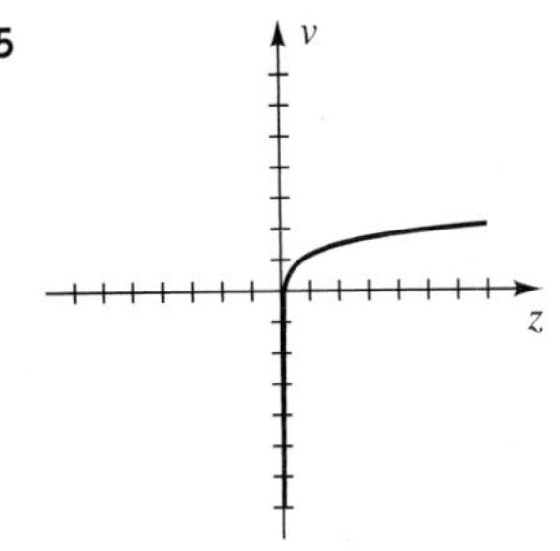

57 **(a)** 0 **(b)** $R(2x) = R(x) + a \log 2$ **59** 0.29 cm

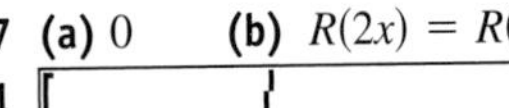

61 (0, 1.02] ∪ [2.40, ∞)

[0, 6] by [−1, 3]

63 1.41, 6.59

65

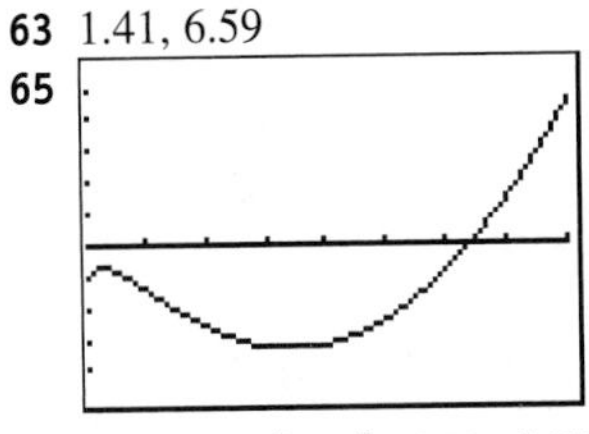

[0.2, 16, 2] by [−4.77, 5.77]

(a) Increasing on [0.2, 0.63] and [6.87, 16]; decreasing on [0.63, 6.87]

(b) 4.61; −3.31

67 6.94 **69** 115 m

EXERCISES 4.6

1 $\dfrac{\log 8}{\log 5} \approx 1.29$ **3** $4 - \dfrac{\log 5}{\log 3} \approx 2.54$ **5** 1.1133

7 -0.7325 **9** 2 **11** $\dfrac{\log (2/81)}{\log 24} \approx -1.16$

13 $\dfrac{\log (8/25)}{\log (4/5)} \approx 5.11$ **15** -3 **17** 5

19 $\dfrac{2}{3}\sqrt{\dfrac{101}{11}} \approx 2.02$ **21** 1, 2

23 $\dfrac{\log \left(4 + \sqrt{19}\right)}{\log 4} \approx 1.53$ **25** 1 or 100 **27** 10^{100}

29 10,000 **31** ln 3 **33** 7

35 $x = \log \left(y \pm \sqrt{y^2 - 1}\right)$

37 $x = \dfrac{1}{2} \log \left(\dfrac{1 + y}{1 - y}\right)$ **39** $x = \ln \left(y + \sqrt{y^2 + 1}\right)$

41 $x = \dfrac{1}{2} \ln \left(\dfrac{y + 1}{y - 1}\right)$

43 y-intercept $= \log_2 3 \approx 1.5850$ **45** x-intercept $= \log_4 3 \approx 0.7925$

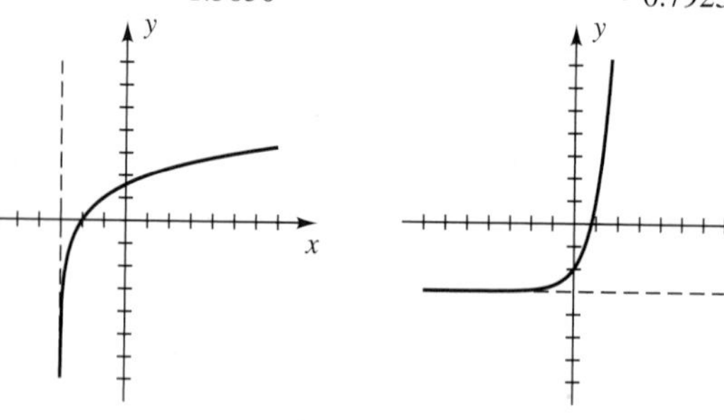

47 **(a)** 2.2 **(b)** 5 **(c)** 8.3

49 Basic if pH > 7, acidic if pH < 7

51 11.58 yr ≈ 11 yr 7 mo **53** 86.4 m

55 **(a)**

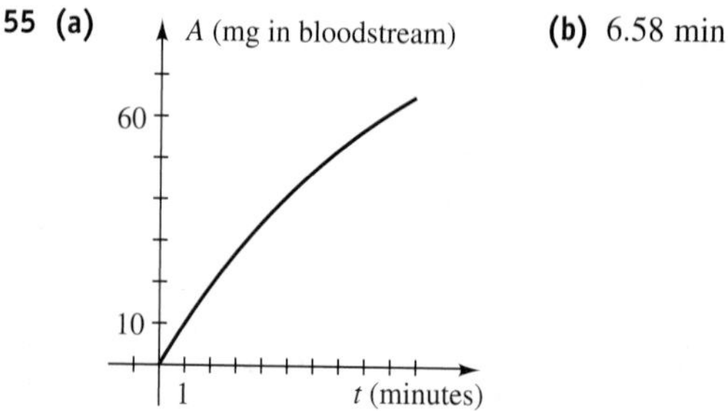

(b) 6.58 min

57 **(a)** $t = \dfrac{\log (F/F_0)}{\log (1 - m)}$ **(b)** After 13,863 generations

59 **(a)** 4.28 ft **(b)** 24.8 yr **61** $\dfrac{\ln (25/6)}{\ln (200/35)} \approx 0.82$

63 The suspicion is correct.

65 The suspicion is incorrect. **67** -0.5764 **69** None

71 1.37, 9.94

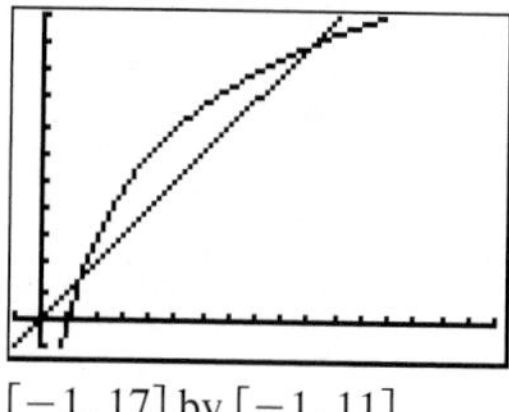

$[-1, 17]$ by $[-1, 11]$

73 $(-\infty, -0.32) \cup (1.52, 6.84)$

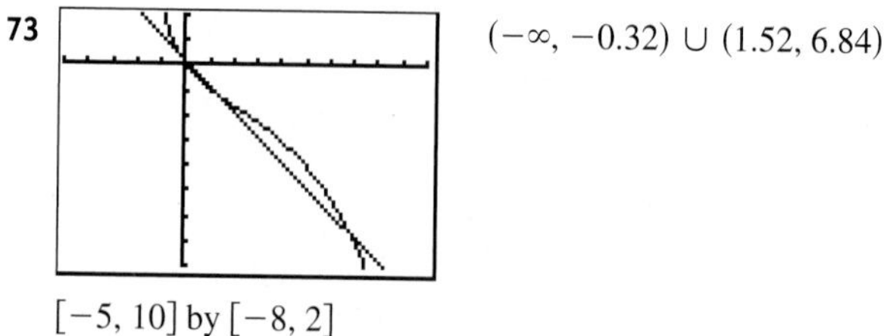

$[-5, 10]$ by $[-8, 2]$

75 (4)

CHAPTER 4 REVIEW EXERCISES

1 Yes

2

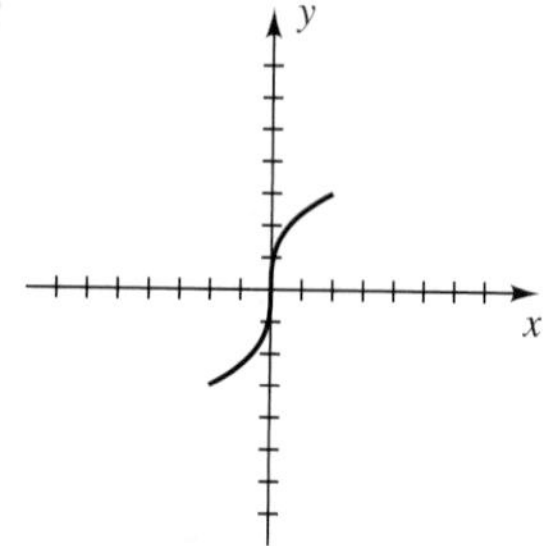

3 **(a)** $f^{-1}(x) = \dfrac{10 - x}{15}$

(b)

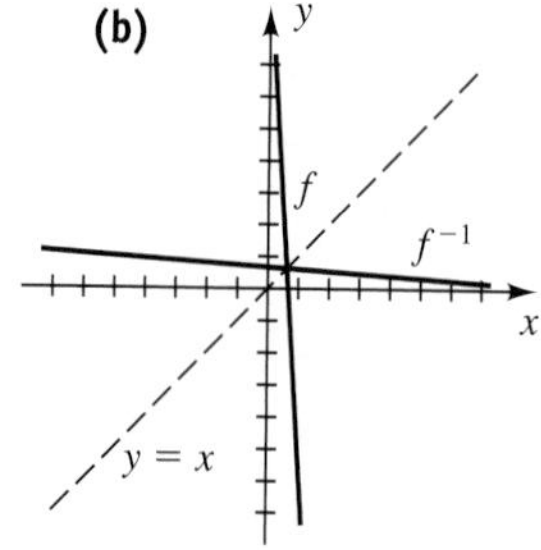

4 **(a)** $f^{-1}(x) = -\sqrt{\dfrac{9 - x}{2}}$

(b)

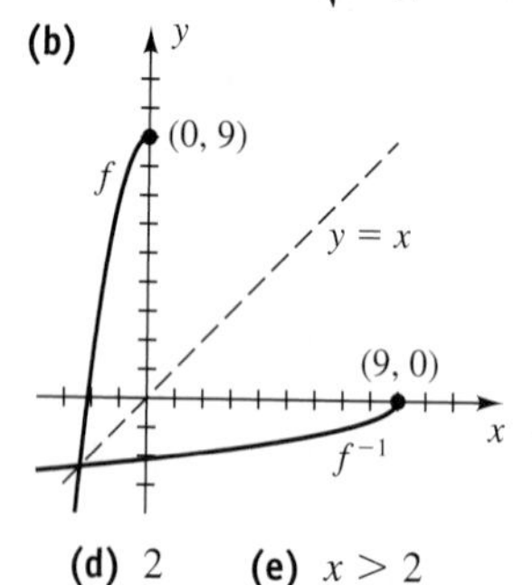

5 **(a)** 2 **(b)** 4 **(c)** 2 **(d)** 2 **(e)** $x > 2$

6 **(a)** 5 **(b)** 7 **(c)** 4

(d) Not enough information is given.

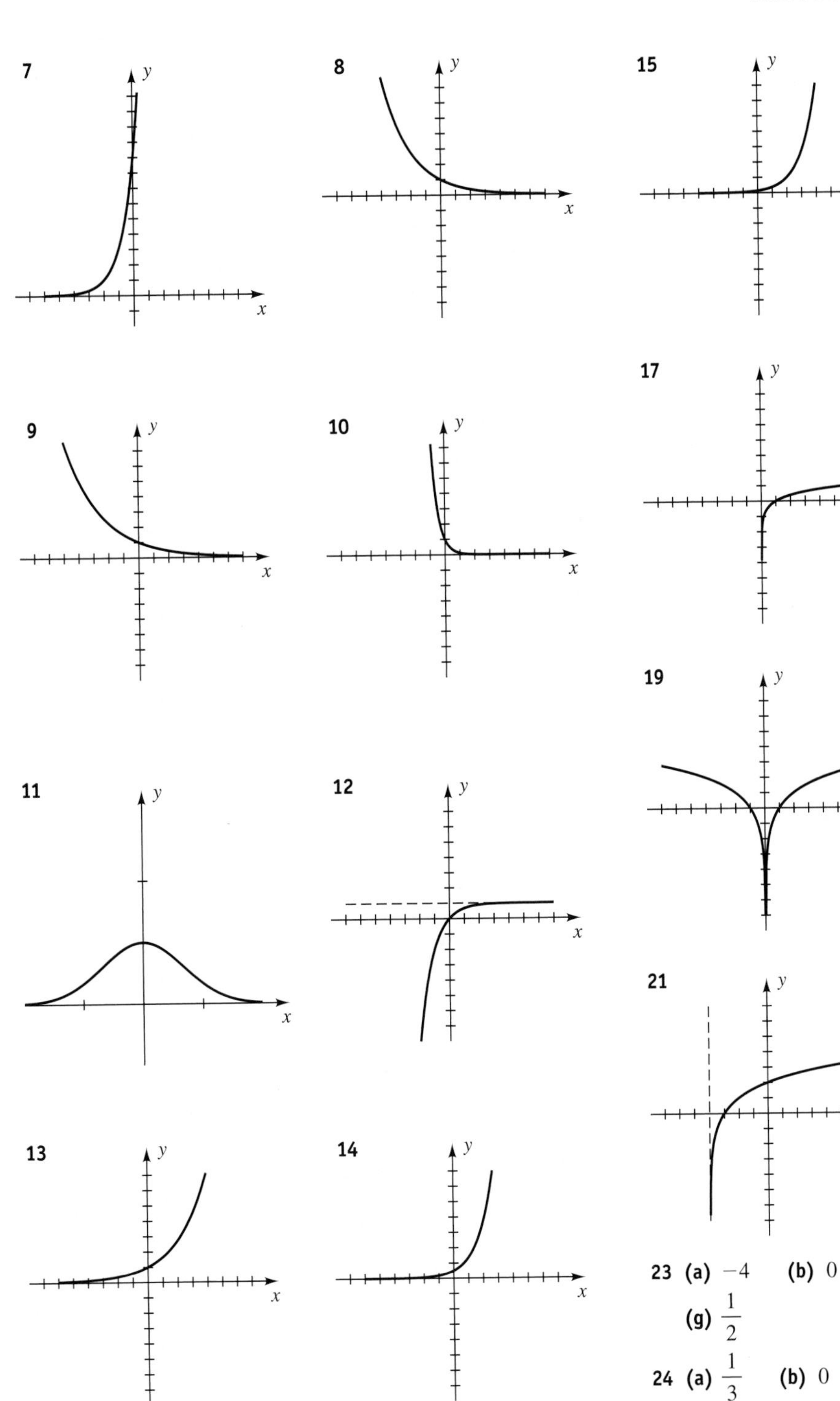

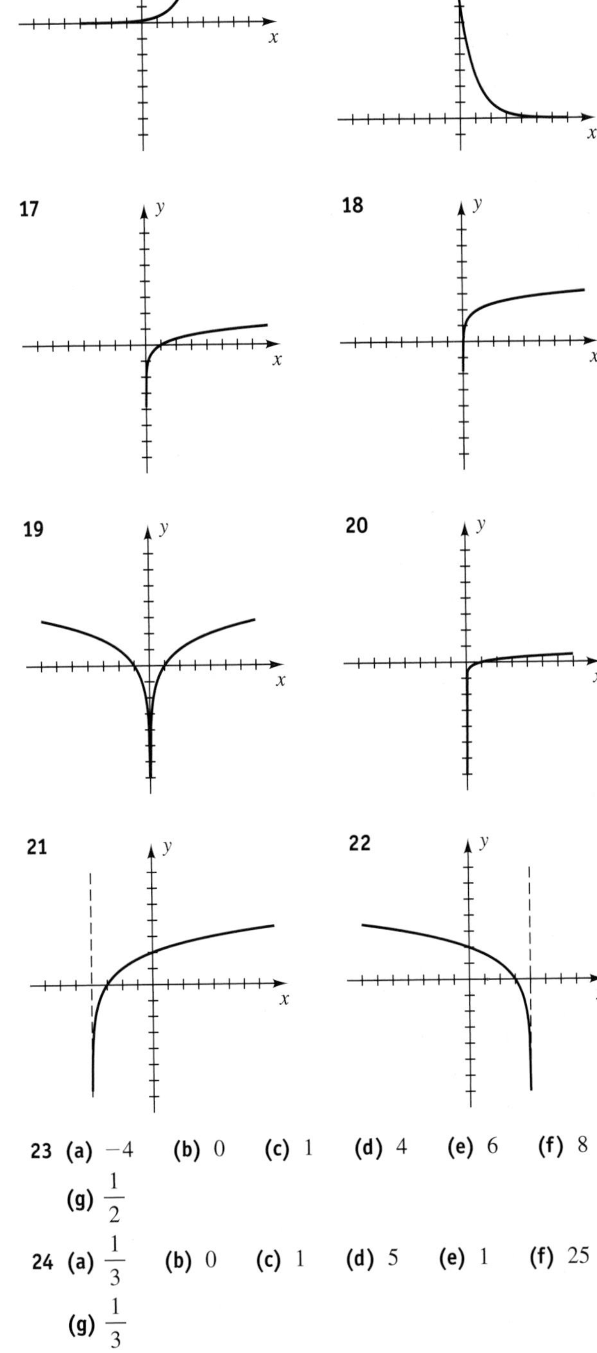

23 **(a)** -4 **(b)** 0 **(c)** 1 **(d)** 4 **(e)** 6 **(f)** 8

(g) $\frac{1}{2}$

24 **(a)** $\frac{1}{3}$ **(b)** 0 **(c)** 1 **(d)** 5 **(e)** 1 **(f)** 25

(g) $\frac{1}{3}$

25 0 **26** $-\frac{6}{5}$ **27** 9 **28** 9 **29** $\frac{33}{47}$ **30** 1

31 $-1 + \sqrt{3}$ **32** 99 **33** $5 - \frac{\log 6}{\log 2}$

34 $\pm\sqrt{\frac{\log 7}{\log 3}}$ **35** $\frac{\log (3/8)}{\log (32/9)}$ **36** 1 **37** $\frac{1}{4}$, 1, 4

38 No solution **39** $\sqrt{5}$ **40** 2 **41** 0, ± 1

42 ln 2 **43 (a)** $-3, 2$ **(b)** 2

44 (a) 8 **(b)** ± 4

45 $4 \log x + \frac{2}{3} \log y - \frac{1}{3} \log z$

46 $-\log (xy^2)$ **47** $f(x) = 6\left(\frac{4}{3}\right)^x$

48

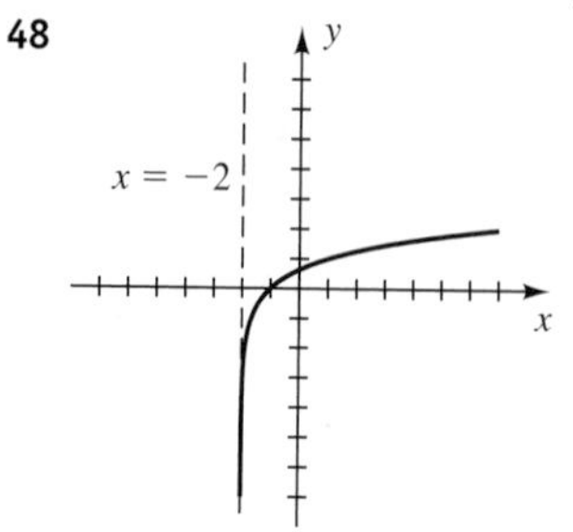

49 $x = \log \left(\frac{1 \pm \sqrt{1 - 4y^2}}{2y}\right)$

50 If $y < 0$, then $x = \log \left(\frac{1 - \sqrt{1 + 4y^2}}{2y}\right)$.

If $y > 0$, then $x = \log \left(\frac{1 + \sqrt{1 + 4y^2}}{2y}\right)$.

51 (a) 1.89 **(b)** 78.3 **(c)** 0.472

52 (a) 0.924 **(b)** 0.00375 **(c)** 6.05

53 (a) $D = (-1, \infty), R = \mathbb{R}$

(b) $y = 2^x - 1, D = \mathbb{R}, R = (-1, \infty)$

54 (a) $D = \mathbb{R}, R = (-2, \infty)$

(b) $y = 3 - \log_2 (x + 2), D = (-2, \infty), R = \mathbb{R}$

55 (a) 2000

(b) $2000(3^{1/6}) \approx 2401$; $2000(3^{1/2}) = 3464$; 6000

56 \$1082.43

57 (a) N (amount remaining); 60; 10; 1; 8 t (days) **(b)** 8 days

58 $N = 1000\left(\frac{3}{5}\right)^{t/3}$

59 (a) After 17.9 yr **(b)** 9.9 yr **60** 3.16%

61 $t = (\ln 100)\frac{L}{R} \approx 4.6\frac{L}{R}$

62 (a) $I = I_0 10^{\alpha/10}$

(b) Examine $I(\alpha + 1)$, where $I(\alpha)$ is the intensity corresponding to α decibels.

63 $t = -\frac{1}{k} \ln \left(\frac{a - L}{ab}\right)$ **64** $A = 10^{(R+5.1)/2.3} - 3000$

65 $\frac{A_1}{A_2} = \frac{10^{(R+5.1)/2.3} - 3000}{10^{(R+7.5)/2.3} - 34{,}000}$ **66** 26,615.9 mi^2

67 $h = \frac{\ln (29/p)}{0.000034}$ **68** $v = a \ln \left(\frac{m_1 + m_2}{m_1}\right)$

69 (a) $n = 10^{7.7 - 0.9R}$ **(b)** 12,589; 1585; 200

70 (a) $E = 10^{11.4 + 1.5R}$ **(b)** 7.9×10^{24} ergs **71** 110 days

72 86.8 cm; 9.715 cm/yr **73** $t = -\frac{L}{R} \ln \left(\frac{V - RI}{V}\right)$

74 (a) 26,749 yr **(b)** 30% **75** 31.5 yr

76 3196 yr

CHAPTER 4 DISCUSSION EXERCISES

1 (a)

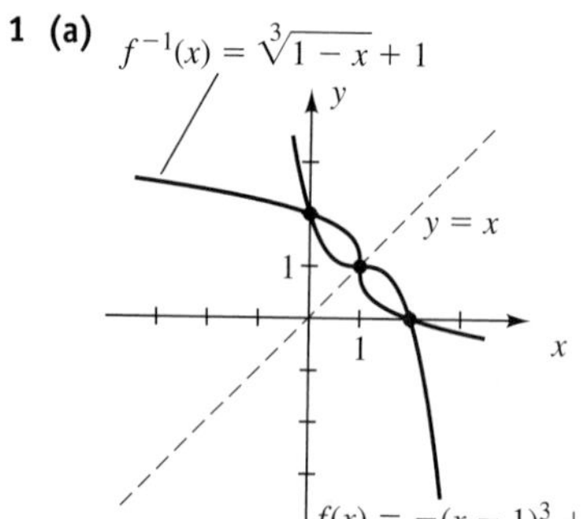

2

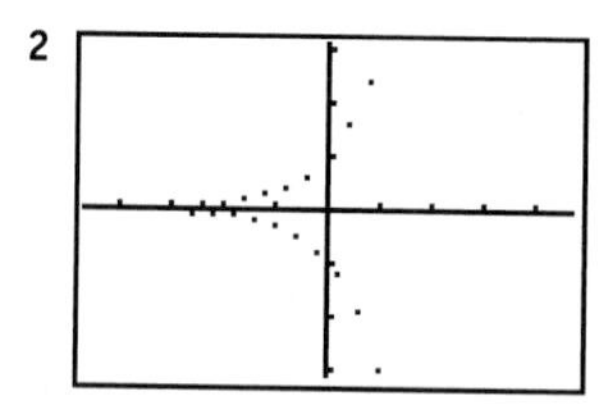

The base a must be positive so that the function $f(x) = a^x$ is defined for all values of x.

3 **(a)** Graph flattens
(b) $y = \frac{101}{2}(e^{x/101} + e^{-x/101}) - 71$

4 7.16 yr

5 **(a)** *Hint:* Take the natural logarithm of both sides first.
(b) 2.50 and 2.97
(c) Note that $f(e) = \frac{1}{e}$. Any horizontal line $y = k$, with $0 < k < \frac{1}{e}$, will intersect the graph at points $\left(x_1, \frac{\ln x_1}{x_1}\right)$ and $\left(x_2, \frac{\ln x_2}{x_2}\right)$, where $1 < x_1 < e$ and $x_2 > e$.

6 **(a)** The difference is in the compounding.
(b) Closer to the graph of the second function
(c) 29 and 8.2; 29.61 and 8.18

7 *Hint:* Check the restrictions for the logarithm laws.

8 **(a)** $U = P\left(1 + \frac{r}{12}\right)^{12t} - \frac{12M[(1 + r/12)^{12t} - 1]}{r}$
(b)
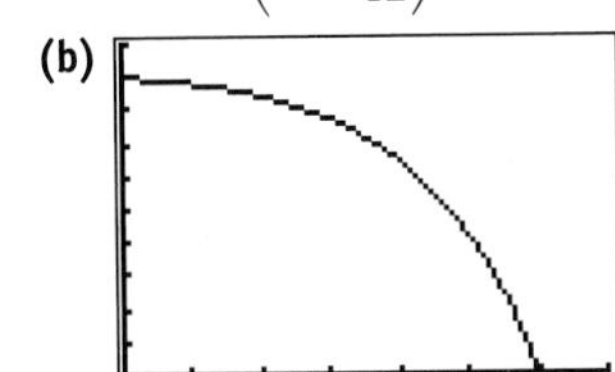
[0, 35, 5] by [0, 100,000, 10,000]
(c) \$84,076.50; 24.425 yr

9 $(-0.9999011, 0.00999001)$, $(-0.0001, 0.01)$, $(100, 0.01105111)$, and $(36{,}102.844, 4.6928 \times 10^{13})$. Exponential function values (with base > 1) are greater than polynomial function values (with leading term positive) for very large values of x.

10 (x, x) with $x \approx 0.44239443, 4.1770774$, and $5{,}503.6647$. The y-values for $y = x$ eventually will be larger than the y-values for $y = (\ln x)^n$.

11 8.447177%; \$1,025,156.25

12 **(a)** 3.5 earthquakes = 1 bomb, 425 bombs = 1 eruption
(b) 9.22; yes

13 April 15, 2010; about 7.31%

14 $y \approx 68.2(1.000353)^x$

15 **(a)**
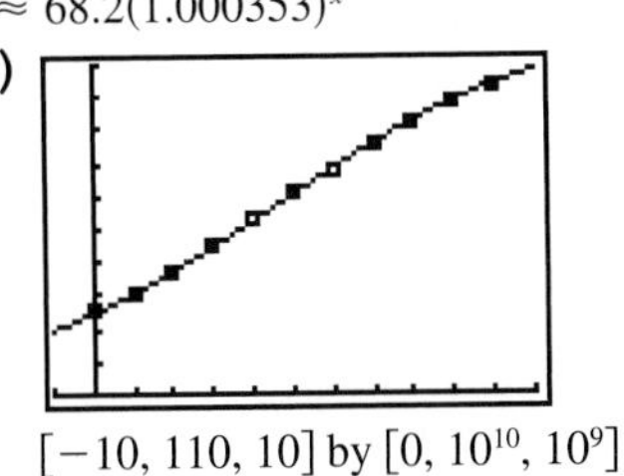
$[-10, 110, 10]$ by $[0, 10^{10}, 10^9]$
(b) Logistic
(c) $y \approx \frac{1.1542 \times 10^{10}}{1 + 3.6372e^{-0.0278x}}$; see the graph in part (a).
(d) 1.1542×10^{10}

16 e^b, with $b = \frac{11 \ln 5 \cdot \ln 7}{\ln 35}$

17 $f^{-1}(x) = \frac{x}{\sqrt{81 - x^2}}$. The vertical asymptotes are $x = \pm 9$. The horizontal asymptotes of f are $y = \pm 9$.

Chapter 5

EXERCISES 5.1

Exer. 1–4: The answers are not unique.

1 **(a)** 480°, 840°, −240°, −600°
(b) 495°, 855°, −225°, −585°
(c) 330°, 690°, −390°, −750°

3 **(a)** 260°, 980°, −100°, −460°
(b) $\frac{17\pi}{6}, \frac{29\pi}{6}, -\frac{7\pi}{6}, -\frac{19\pi}{6}$
(c) $\frac{7\pi}{4}, \frac{15\pi}{4}, -\frac{9\pi}{4}, -\frac{17\pi}{4}$

5 **(a)** 84°42′26″ **(b)** 57.5°

7 **(a)** 131°8′23″ **(b)** 43.58°

9 **(a)** $\frac{5\pi}{6}$ **(b)** $-\frac{\pi}{3}$ **(c)** $\frac{5\pi}{4}$

11 **(a)** $\frac{5\pi}{2}$ **(b)** $\frac{2\pi}{5}$ **(c)** $\frac{5\pi}{9}$

13 **(a)** 120° **(b)** 330° **(c)** 135°

15 **(a)** −630° **(b)** 1260° **(c)** 20°

17 114°35′30″ **19** 286°28′44″ **21** 37.6833°

23 115.4408° **25** 63°10′8″ **27** 310°37′17″

29 2.5 cm

31 **(a)** $2\pi \approx 6.28$ cm **(b)** $8\pi \approx 25.13$ cm^2

33 **(a)** 1.75; $\frac{315}{\pi} \approx 100.27°$ **(b)** 14 cm^2

35 **(a)** $\frac{20\pi}{9} \approx 6.98$ m **(b)** $\frac{80\pi}{9} \approx 27.93$ m^2

37 In miles: **(a)** 4189 **(b)** 3142 **(c)** 2094
(d) 698 **(e)** 70

39 $\frac{1}{8}$ radian $\approx 7°10'$ **41** 37.1%

43 7.29×10^{-5} rad/sec

45 **(a)** 80π rad/min **(b)** $\frac{100\pi}{3} \approx 104.72$ ft/min

47 **(a)** 400π rad/min **(b)** 38π cm/sec **(c)** 380 rpm
(d) $S(r) = \frac{1140}{r}$; inversely

49 (a) $\frac{21\pi}{8} \approx 8.25$ ft **(b)** $\frac{2}{3}d$

51 Large **53** 192.08 rev/min

EXERCISES 5.2

1 (a) B **(b)** D **(c)** A **(d)** C **(e)** E

Note: Answers are in the order *sin, cos, tan, cot, sec, csc* for any exercises that require the values of the six trigonometric functions.

3 $\frac{4}{5}, \frac{3}{5}, \frac{4}{3}, \frac{3}{4}, \frac{5}{3}, \frac{5}{4}$

5 $\frac{2}{5}, \frac{\sqrt{21}}{5}, \frac{2}{\sqrt{21}}, \frac{\sqrt{21}}{2}, \frac{5}{\sqrt{21}}, \frac{5}{2}$

7 $\frac{a}{\sqrt{a^2+b^2}}, \frac{b}{\sqrt{a^2+b^2}}, \frac{a}{b}, \frac{b}{a}, \frac{\sqrt{a^2+b^2}}{b}, \frac{\sqrt{a^2+b^2}}{a}$

9 $\frac{b}{c}, \frac{\sqrt{c^2-b^2}}{c}, \frac{b}{\sqrt{c^2-b^2}}, \frac{\sqrt{c^2-b^2}}{b}, \frac{c}{\sqrt{c^2-b^2}}, \frac{c}{b}$

11 $x = 8; y = 4\sqrt{3}$ **13** $x = 7\sqrt{2}; y = 7$

15 $x = 4\sqrt{3}; y = 4$

17 $\frac{3}{5}, \frac{4}{5}, \frac{3}{4}, \frac{4}{3}, \frac{5}{4}, \frac{5}{3}$ **19** $\frac{5}{13}, \frac{12}{13}, \frac{5}{12}, \frac{12}{5}, \frac{13}{12}, \frac{13}{5}$

21 $\frac{\sqrt{11}}{6}, \frac{5}{6}, \frac{\sqrt{11}}{5}, \frac{5}{\sqrt{11}}, \frac{6}{5}, \frac{6}{\sqrt{11}}$

23 $200\sqrt{3} \approx 346.4$ ft **25** 192 ft **27** 1.02 m

29 (a) 0.6691 **(b)** 0.2250 **(c)** 1.1924 **(d)** −1.0154

31 (a) 4.0572 **(b)** 1.0323 **(c)** −0.6335 **(d)** 4.3813

33 (a) 0.5 **(b)** −0.9880 **(c)** 0.9985 **(d)** −1

35 (a) −1 **(b)** −4

37 (a) 5 **(b)** 5

39 $1 - \sin\theta\cos\theta$ **41** $\sin\theta$

43 $\cot\theta = \frac{\sqrt{1-\sin^2\theta}}{\sin\theta}$ **45** $\sec\theta = \frac{1}{\sqrt{1-\sin^2\theta}}$

47 $\sin\theta = \frac{\sqrt{\sec^2\theta - 1}}{\sec\theta}$

Exer. 49–70: Typical verifications are given.

49 $\cos\theta\sec\theta = \cos\theta\,(1/\cos\theta) = 1$

51 $\sin\theta\sec\theta = \sin\theta\,(1/\cos\theta) = \sin\theta/\cos\theta = \tan\theta$

53 $\frac{\csc\theta}{\sec\theta} = \frac{1/\sin\theta}{1/\cos\theta} = \frac{\cos\theta}{\sin\theta} = \cot\theta$

55 $(1+\cos 2\theta)(1-\cos 2\theta) = 1-\cos^2 2\theta = \sin^2 2\theta$

57 $\cos^2\theta\,(\sec^2\theta - 1) = \cos^2\theta\,(\tan^2\theta)$
$= \cos^2\theta \cdot \frac{\sin^2\theta}{\cos^2\theta} = \sin^2\theta$

59 $\frac{\sin(\theta/2)}{\csc(\theta/2)} + \frac{\cos(\theta/2)}{\sec(\theta/2)} = \frac{\sin(\theta/2)}{1/\sin(\theta/2)} + \frac{\cos(\theta/2)}{1/\cos(\theta/2)}$
$= \sin^2(\theta/2) + \cos^2(\theta/2) = 1$

61 $(1+\sin\theta)(1-\sin\theta) = 1-\sin^2\theta = \cos^2\theta$
$= \frac{1}{\sec^2\theta}$

63 $\sec\theta - \cos\theta = \frac{1}{\cos\theta} - \cos\theta = \frac{1-\cos^2\theta}{\cos\theta} = \frac{\sin^2\theta}{\cos\theta}$
$= \frac{\sin\theta}{\cos\theta} \cdot \sin\theta = \tan\theta\sin\theta$

65 $(\cot\theta + \csc\theta)(\tan\theta - \sin\theta)$
$= \cot\theta\tan\theta - \cot\theta\sin\theta + \csc\theta\tan\theta$
$- \csc\theta\sin\theta$
$= \frac{1}{\tan\theta}\tan\theta - \frac{\cos\theta}{\sin\theta}\sin\theta + \frac{1}{\sin\theta}\frac{\sin\theta}{\cos\theta} - \frac{1}{\sin\theta}\sin\theta$
$= 1 - \cos\theta + \frac{1}{\cos\theta} - 1 = -\cos\theta + \sec\theta$
$= \sec\theta - \cos\theta$

67 $\sec^2 3\theta\csc^2 3\theta = (1+\tan^2 3\theta)(1+\cot^2 3\theta)$
$= 1 + \tan^2 3\theta + \cot^2 3\theta + 1$
$= \sec^2 3\theta + \csc^2 3\theta$

69 $\log\csc\theta = \log\left(\frac{1}{\sin\theta}\right) = \log 1 - \log\sin\theta$
$= 0 - \log\sin\theta = -\log\sin\theta$

71 $-\frac{3}{5}, \frac{4}{5}, -\frac{3}{4}, -\frac{4}{3}, \frac{5}{4}, -\frac{5}{3}$

73 $-\frac{5}{\sqrt{29}}, -\frac{2}{\sqrt{29}}, \frac{5}{2}, \frac{2}{5}, -\frac{\sqrt{29}}{2}, -\frac{\sqrt{29}}{5}$

75 $\frac{4}{\sqrt{17}}, -\frac{1}{\sqrt{17}}, -4, -\frac{1}{4}, -\sqrt{17}, \frac{\sqrt{17}}{4}$

77 $\frac{4}{5}, \frac{3}{5}, \frac{4}{3}, \frac{3}{4}, \frac{5}{3}, \frac{5}{4}$

79 $-\frac{7}{\sqrt{53}}, -\frac{2}{\sqrt{53}}, \frac{7}{2}, \frac{2}{7}, -\frac{\sqrt{53}}{2}, -\frac{\sqrt{53}}{7}$

Note: U denotes *undefined.*

81 (a) 1, 0, U, 0, U, 1 **(b)** 0, 1, 0, U, 1, U
(c) −1, 0, U, 0, U, −1 **(d)** 0, −1, 0, U, −1, U

83 (a) IV **(b)** III **(c)** II **(d)** III

85 $\frac{3}{5}, -\frac{4}{5}, -\frac{3}{4}, -\frac{4}{3}, -\frac{5}{4}, \frac{5}{3}$

87 $-\frac{5}{13}, \frac{12}{13}, -\frac{5}{12}, -\frac{12}{5}, \frac{13}{12}, -\frac{13}{5}$

89 $-\frac{\sqrt{8}}{3}, -\frac{1}{3}, \sqrt{8}, \frac{1}{\sqrt{8}}, -3, -\frac{3}{\sqrt{8}}$

91 $\frac{\sqrt{15}}{4}, -\frac{1}{4}, -\sqrt{15}, -\frac{1}{\sqrt{15}}, -4, \frac{4}{\sqrt{15}}$

93 $-\tan\theta$ **95** $\sec\theta$ **97** $-\sin\frac{\theta}{2}$

EXERCISES 5.3

1 $\frac{8}{17}, -\frac{15}{17}, -\frac{8}{15}, -\frac{15}{8}, -\frac{17}{15}, \frac{17}{8}$

3 $-\frac{7}{25}, \frac{24}{25}, -\frac{7}{24}, -\frac{24}{7}, \frac{25}{24}, -\frac{25}{7}$

5 (a) $\left(-\frac{3}{5}, -\frac{4}{5}\right)$ **(b)** $\left(-\frac{3}{5}, -\frac{4}{5}\right)$
(c) $\left(\frac{3}{5}, -\frac{4}{5}\right)$ **(d)** $\left(-\frac{3}{5}, \frac{4}{5}\right)$

7 (a) $\left(\frac{12}{13}, \frac{5}{13}\right)$ **(b)** $\left(\frac{12}{13}, \frac{5}{13}\right)$
(c) $\left(-\frac{12}{13}, \frac{5}{13}\right)$ **(d)** $\left(\frac{12}{13}, -\frac{5}{13}\right)$

Note: U denotes *undefined*.

9 (a) $(1, 0)$; $0, 1, 0$, U, 1, U
(b) $(-1, 0)$; $0, -1, 0$, U, -1, U

11 (a) $(0, -1)$; $-1, 0$, U, 0, U, -1
(b) $(0, 1)$; $1, 0$, U, 0, U, 1

13 (a) $\left(\frac{\sqrt{2}}{2}, \frac{\sqrt{2}}{2}\right); \frac{\sqrt{2}}{2}, \frac{\sqrt{2}}{2}, 1, 1, \sqrt{2}, \sqrt{2}$
(b) $\left(-\frac{\sqrt{2}}{2}, \frac{\sqrt{2}}{2}\right); \frac{\sqrt{2}}{2}, -\frac{\sqrt{2}}{2}, -1, -1, -\sqrt{2}, \sqrt{2}$

15 (a) $\left(-\frac{\sqrt{2}}{2}, -\frac{\sqrt{2}}{2}\right); -\frac{\sqrt{2}}{2}, -\frac{\sqrt{2}}{2}, 1, 1, -\sqrt{2}, -\sqrt{2}$
(b) $\left(\frac{\sqrt{2}}{2}, -\frac{\sqrt{2}}{2}\right); -\frac{\sqrt{2}}{2}, \frac{\sqrt{2}}{2}, -1, -1, \sqrt{2}, -\sqrt{2}$

17 (a) -1 **(b)** $-\frac{\sqrt{2}}{2}$ **(c)** -1

19 (a) 1 **(b)** -1 **(c)** 1

Exer. 21–26: Typical verifications are given.

21 $\sin(-x)\sec(-x) = (-\sin x)\sec x = (-\sin x)(1/\cos x) = -\tan x$

23 $\frac{\cot(-x)}{\csc(-x)} = \frac{-\cot x}{-\csc x} = \frac{\cos x/\sin x}{1/\sin x} = \cos x$

25 $\frac{1}{\cos(-x)} - \tan(-x)\sin(-x)$

$$= \frac{1}{\cos x} - (-\tan x)(-\sin x) = \frac{1}{\cos x} - \frac{\sin x}{\cos x}\sin x = \frac{1-\sin^2 x}{\cos x} = \frac{\cos^2 x}{\cos x} = \cos x$$

27 (a) 0 **(b)** -1 **29 (a)** $\frac{\sqrt{2}}{2}$ **(b)** -1

31 (a) 1 **(b)** $-\infty$ **33 (a)** -1 **(b)** ∞

35 (a) ∞ **(b)** $\sqrt{2}$ **37 (a)** $-\infty$ **(b)** 1

39 $\frac{3\pi}{2}, \frac{7\pi}{2}$ **41** $\frac{\pi}{6}, \frac{5\pi}{6}, \frac{13\pi}{6}, \frac{17\pi}{6}$ **43** $0, 2\pi, 4\pi$

45 $\frac{\pi}{4}, \frac{7\pi}{4}, \frac{9\pi}{4}, \frac{15\pi}{4}$ **47** $\frac{\pi}{4}, \frac{5\pi}{4}$ **49** $0, \pi$

51 (a) $-\frac{11\pi}{6}, -\frac{7\pi}{6}, \frac{\pi}{6}, \frac{5\pi}{6}$
(b) $-\frac{11\pi}{6} < x < -\frac{7\pi}{6}$ and $\frac{\pi}{6} < x < \frac{5\pi}{6}$
(c) $-2\pi \le x < -\frac{11\pi}{6}$, $-\frac{7\pi}{6} < x < \frac{\pi}{6}$, and $\frac{5\pi}{6} < x \le 2\pi$

53 (a) $-\frac{4\pi}{3}, -\frac{2\pi}{3}, \frac{2\pi}{3}, \frac{4\pi}{3}$
(b) $-2\pi \le x < -\frac{4\pi}{3}$, $-\frac{2\pi}{3} < x < \frac{2\pi}{3}$, and $\frac{4\pi}{3} < x \le 2\pi$
(c) $-\frac{4\pi}{3} < x < -\frac{2\pi}{3}$ and $\frac{2\pi}{3} < x < \frac{4\pi}{3}$

55 **57**

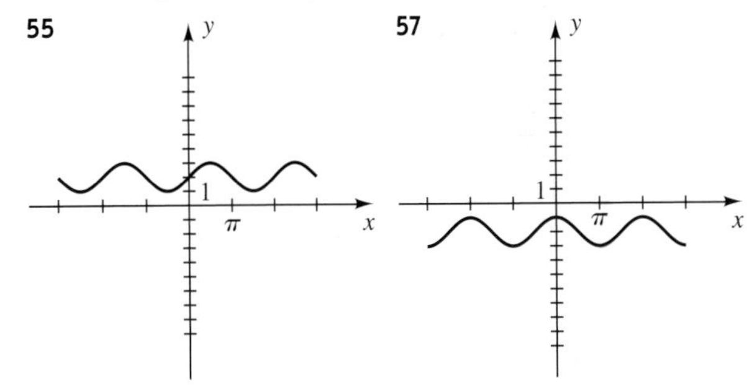

59 **61**

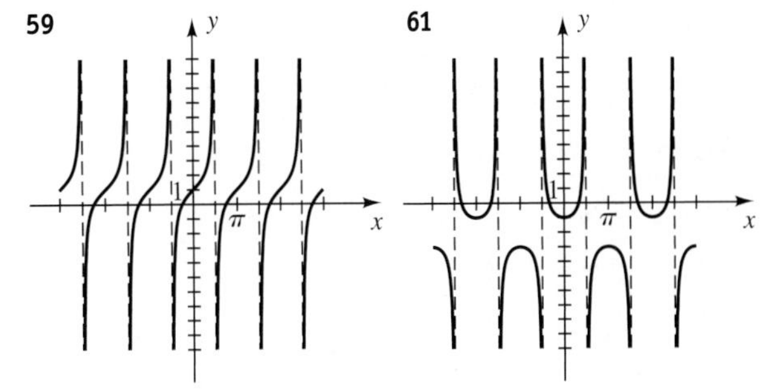

63 **(a)** $\left[-2\pi, -\frac{3\pi}{2}\right), \left(-\frac{3\pi}{2}, -\pi\right], \left[0, \frac{\pi}{2}\right), \left(\frac{\pi}{2}, \pi\right]$
(b) $\left[-\pi, -\frac{\pi}{2}\right), \left(-\frac{\pi}{2}, 0\right], \left[\pi, \frac{3\pi}{2}\right), \left(\frac{3\pi}{2}, 2\pi\right]$

65 **(a)** The tangent function increases on *all* intervals on which it is defined. Between -2π and 2π, these intervals are $\left[-2\pi, -\frac{3\pi}{2}\right), \left(-\frac{3\pi}{2}, -\frac{\pi}{2}\right), \left(-\frac{\pi}{2}, \frac{\pi}{2}\right), \left(\frac{\pi}{2}, \frac{3\pi}{2}\right)$, and $\left(\frac{3\pi}{2}, 2\pi\right]$.
(b) The tangent function is *never* decreasing on any interval for which it is defined.

69 **(a)** -0.8 **(b)** -0.9 **(c)** 0.5, 2.6

71 **(a)** -0.7 **(b)** 0.4 **(c)** 2.2, 4.1

73 **(a)**

Time	T	H	Time	T	H
12 A.M.	60	60	12 P.M.	60	60
3 A.M.	52	74	3 P.M.	68	46
6 A.M.	48	80	6 P.M.	72	40
9 A.M.	52	74	9 P.M.	68	46

(b) Max: 72°F at 6:00 P.M., 80% at 6:00 A.M.; min: 48°F at 6:00 A.M., 40% at 6:00 P.M.

75 $\pm 0.72, \pm 1.62, \pm 2.61, \pm 2.98$

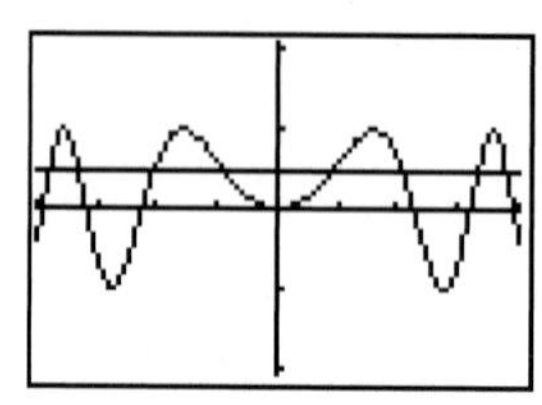

$[-\pi, \pi, \pi/4]$ by $[-2.09, 2.09]$

77 $(\pm 2.03, 1.82)$; $(\pm 4.91, -4.81)$

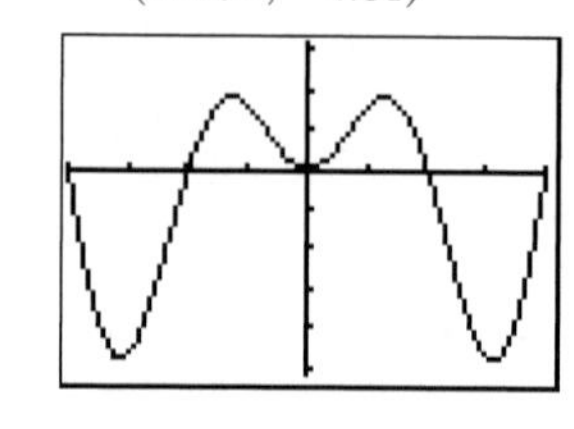

$[-2\pi, 2\pi, \pi/2]$ by $[-5.19, 3.19]$

79 0 **81** 1 **83** 1

EXERCISES 5.4

1 **(a)** 60° **(b)** 20° **(c)** 22° **(d)** 60°

3 **(a)** $\frac{\pi}{4}$ **(b)** $\frac{\pi}{3}$ **(c)** $\frac{\pi}{6}$ **(d)** $\frac{\pi}{4}$

5 **(a)** $\pi - 3 \approx 8.1°$ **(b)** $\pi - 2 \approx 65.4°$
(c) $2\pi - 5.5 \approx 44.9°$ **(d)** $32\pi - 100 \approx 30.4°$

7 **(a)** $\frac{\sqrt{3}}{2}$ **(b)** $\frac{\sqrt{2}}{2}$ **9** **(a)** $-\frac{\sqrt{3}}{2}$ **(b)** $\frac{1}{2}$

11 **(a)** $-\frac{\sqrt{3}}{3}$ **(b)** $-\sqrt{3}$ **13** **(a)** $-\frac{\sqrt{3}}{3}$ **(b)** $\sqrt{3}$

15 **(a)** -2 **(b)** $\frac{2}{\sqrt{3}}$ **17** **(a)** $-\frac{2}{\sqrt{3}}$ **(b)** 2

19 **(a)** 0.958 **(b)** 0.778 **21** **(a)** 0.387 **(b)** 0.472

23 **(a)** 2.650 **(b)** 3.179 **25** **(a)** 30.46° **(b)** 30°27′

27 **(a)** 74.88° **(b)** 74°53′

29 **(a)** 24.94° **(b)** 24°57′

31 **(a)** 76.38° **(b)** 76°23′

33 **(a)** 0.9899 **(b)** -0.1097 **(c)** -0.1425
(d) 0.7907 **(e)** -11.2493 **(f)** 1.3677

35 **(a)** 214.3°, 325.7° **(b)** 41.5°, 318.5°
(c) 70.3°, 250.3° **(d)** 133.8°, 313.8°
(e) 153.6°, 206.4° **(f)** 42.3°, 137.7°

37 **(a)** 0.43, 2.71 **(b)** 1.69, 4.59 **(c)** 1.87, 5.01
(d) 0.36, 3.50 **(e)** 0.96, 5.32 **(f)** 3.35, 6.07

39 0.28 cm

41 **(a)** The maximum occurs when the sun is rising in the east.
(b) $\frac{\sqrt{2}}{4} \approx 35\%$

43 $(9, 9\sqrt{3})$

EXERCISES 5.5

1 **(a)** $4, 2\pi$ **(b)** $1, \frac{\pi}{2}$

(c) $\frac{1}{4}, 2\pi$

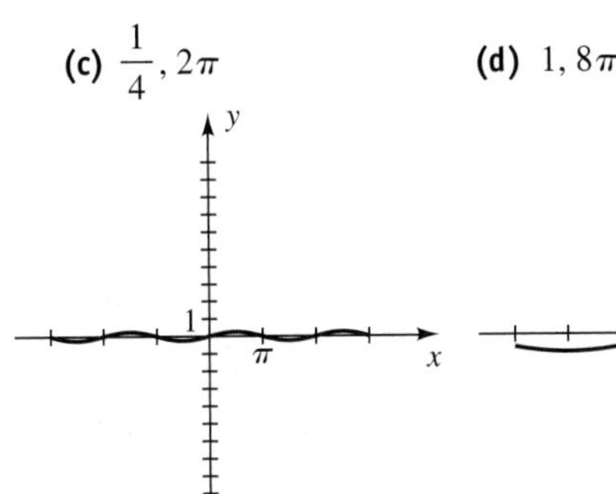

(d) $1, 8\pi$

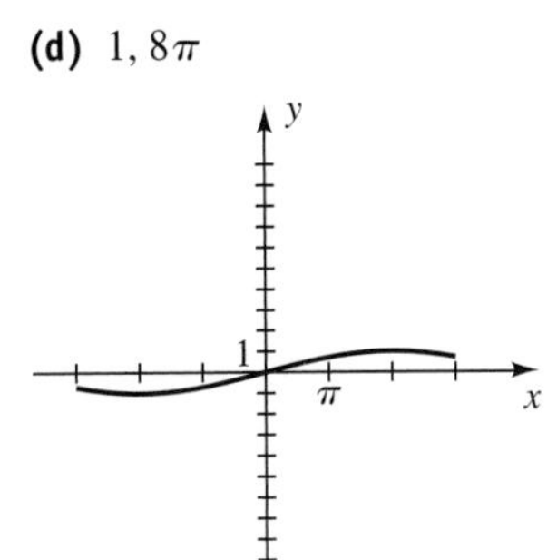

(e) $2, 8\pi$ **(f)** $\frac{1}{2}, \frac{\pi}{2}$

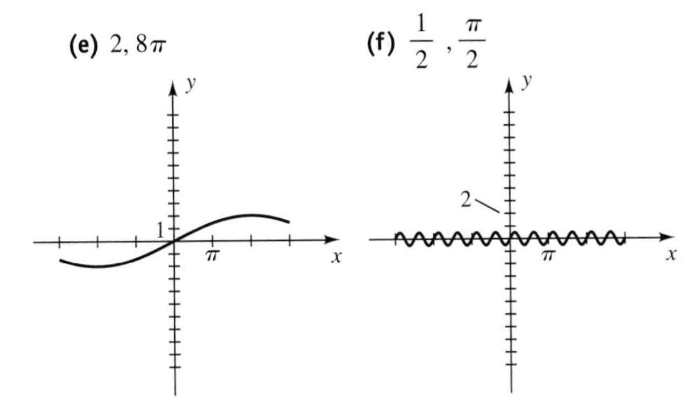

(g) $4, 2\pi$ **(h)** $1, \frac{\pi}{2}$

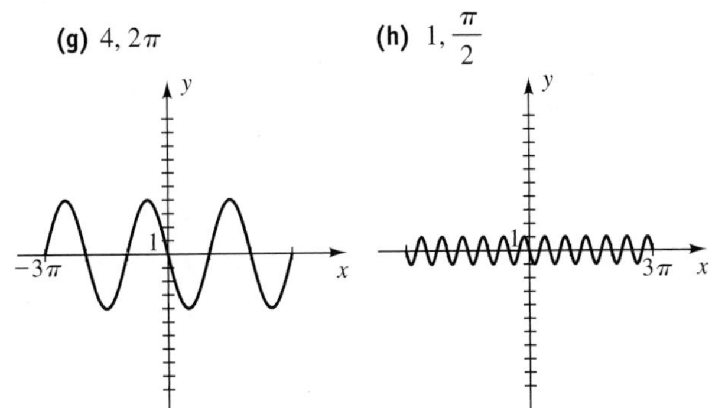

3 (a) $3, 2\pi$ **(b)** $1, \frac{2\pi}{3}$

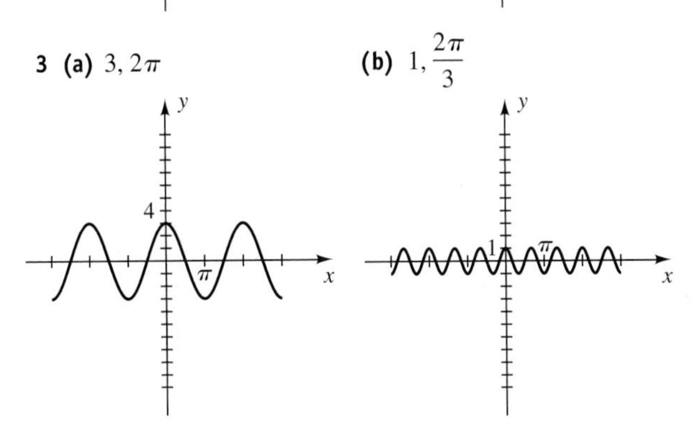

(c) $\frac{1}{3}, 2\pi$ **(d)** $1, 6\pi$

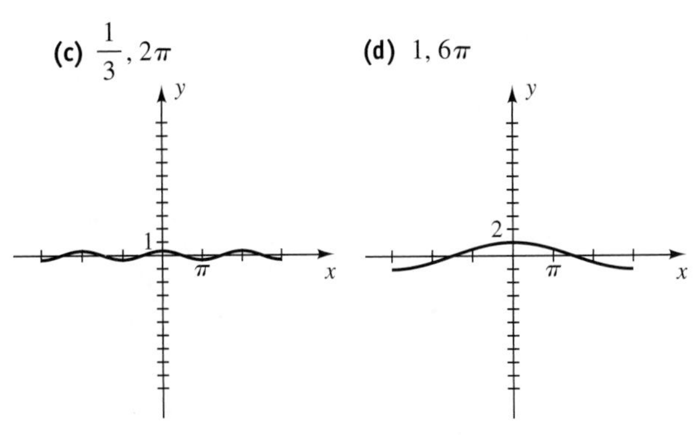

(e) $2, 6\pi$ **(f)** $\frac{1}{2}, \frac{2\pi}{3}$

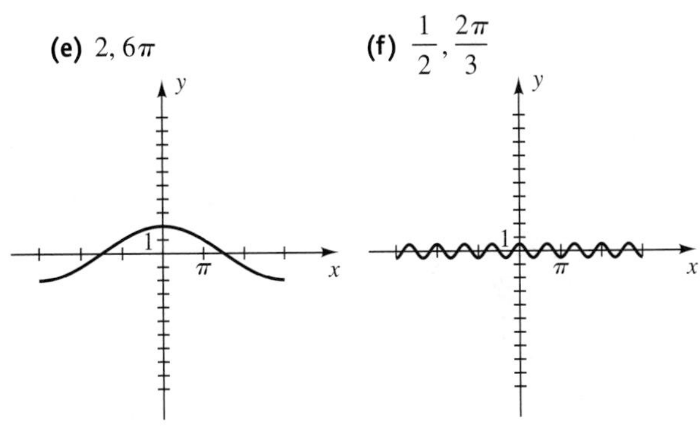

(g) $3, 2\pi$ **(h)** $1, \frac{2\pi}{3}$

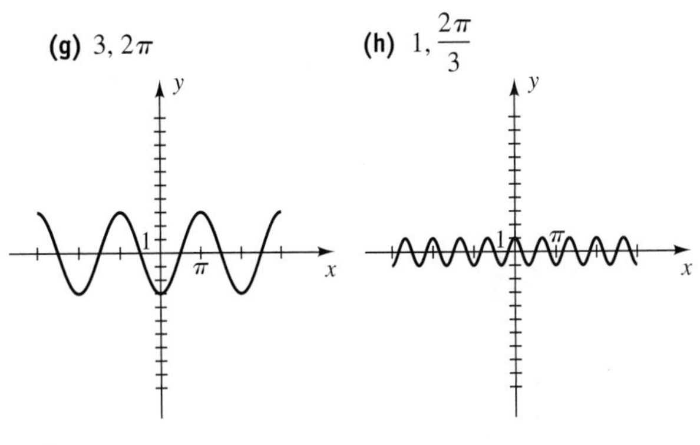

5 $1, 2\pi, \frac{\pi}{2}$ **7** $3, 2\pi, -\frac{\pi}{6}$

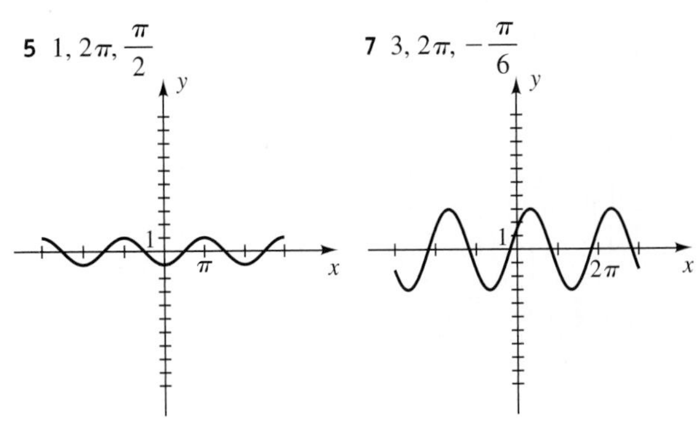

9 $1, 2\pi, -\frac{\pi}{2}$

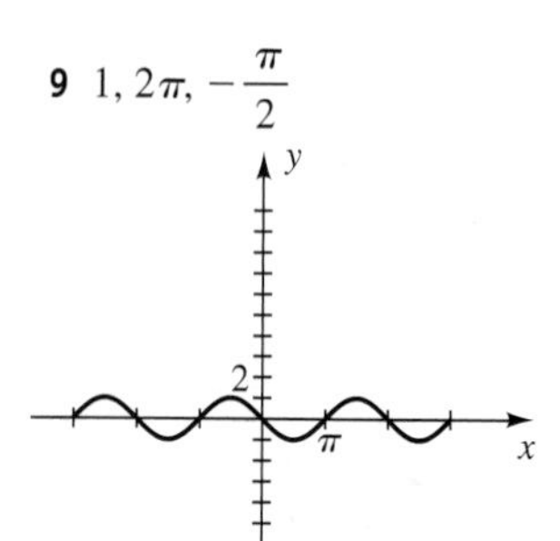

11 $4, 2\pi, \frac{\pi}{4}$

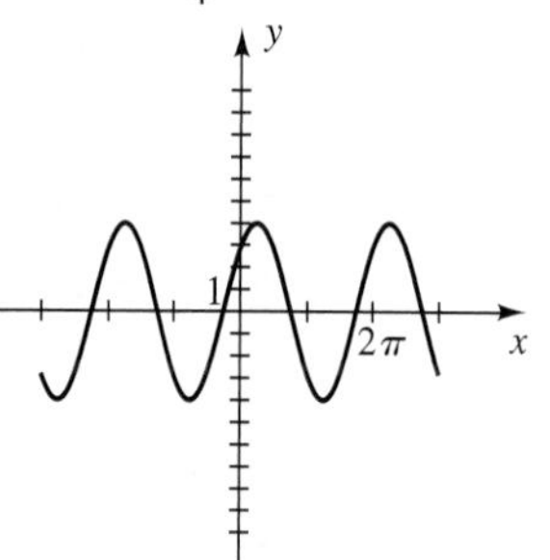

13 $1, \pi, \frac{\pi}{2}$

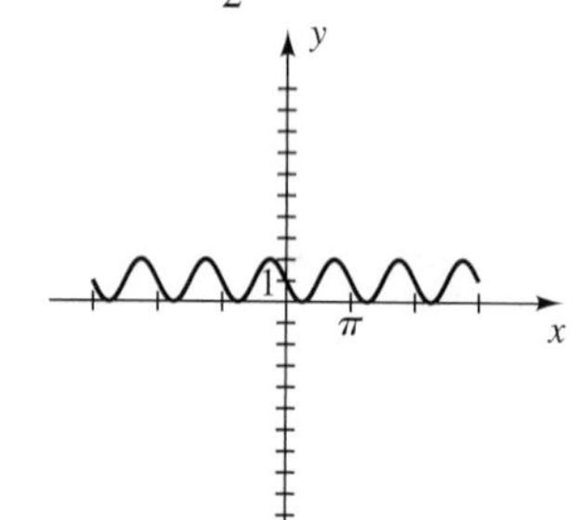

15 $1, \frac{2\pi}{3}, -\frac{\pi}{3}$

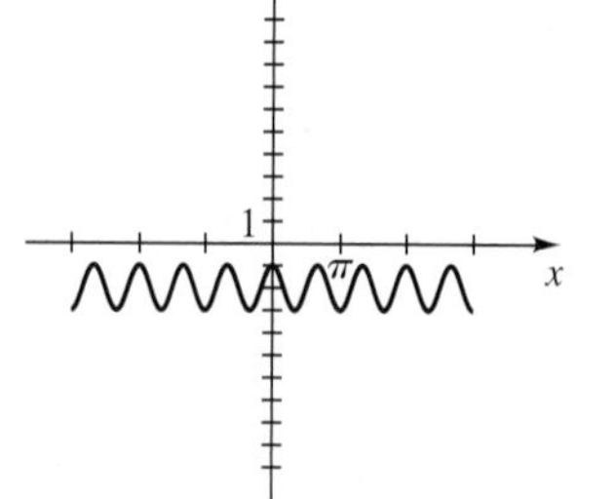

17 $2, \frac{2\pi}{3}, \frac{\pi}{3}$

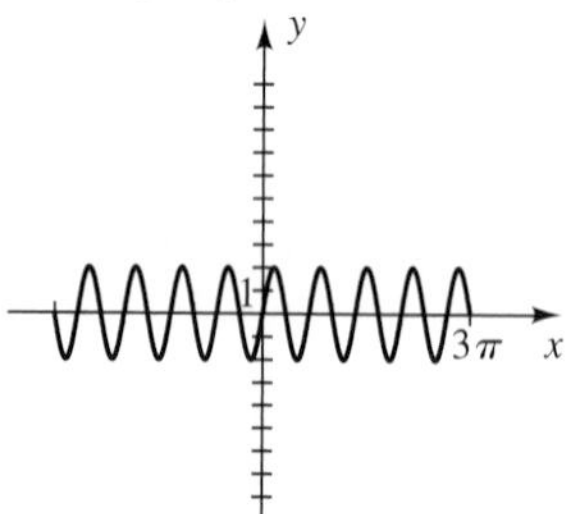

19 $1, 4\pi, \frac{2\pi}{3}$

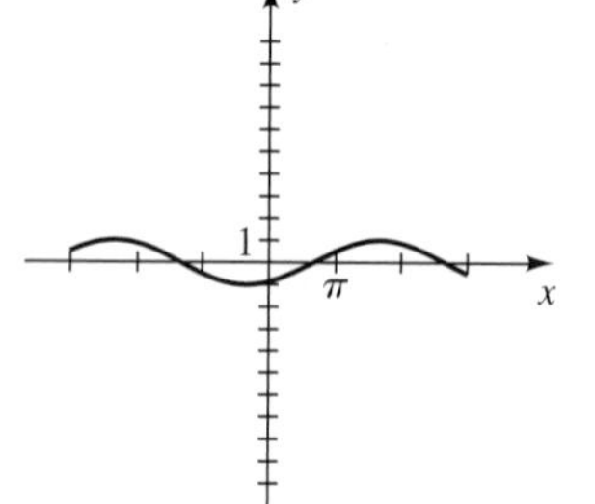

21 $6, 2, 0$

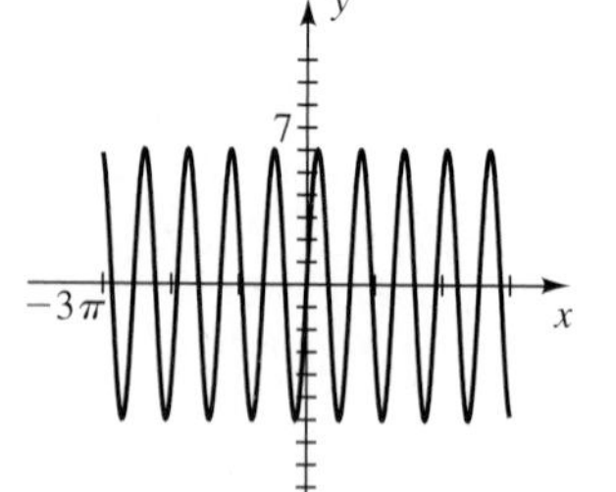

23 $2, 4, 0$

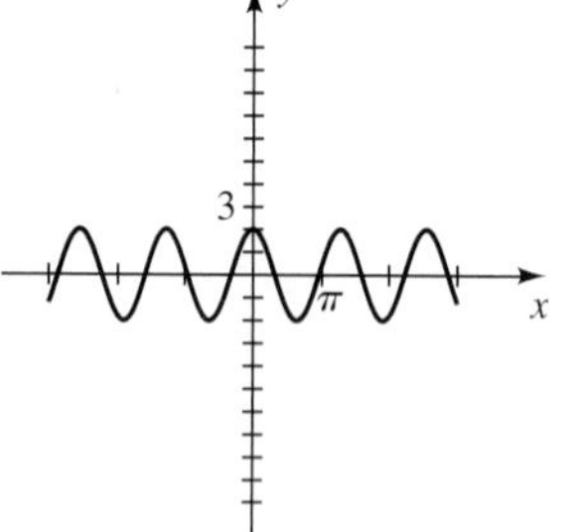

25 $\frac{1}{2}, 1, 0$

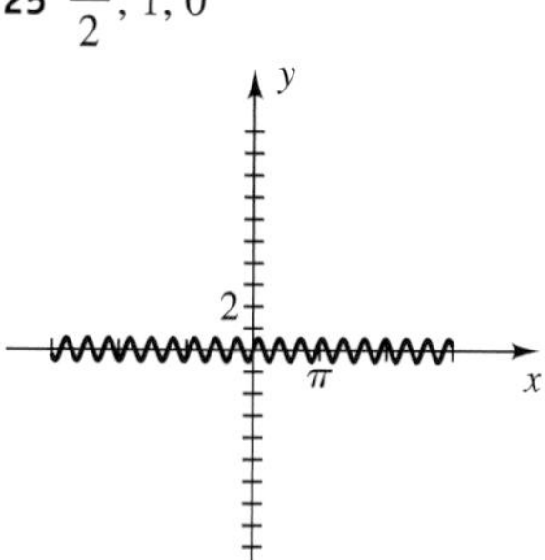

27 $5, \frac{2\pi}{3}, \frac{\pi}{6}$

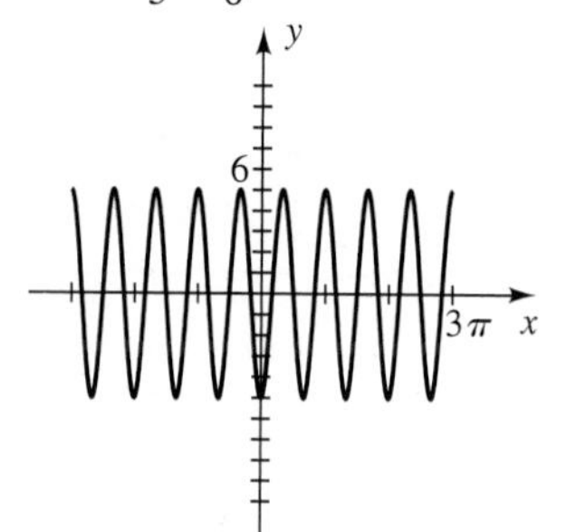

29 $3, 4\pi, \frac{\pi}{2}$

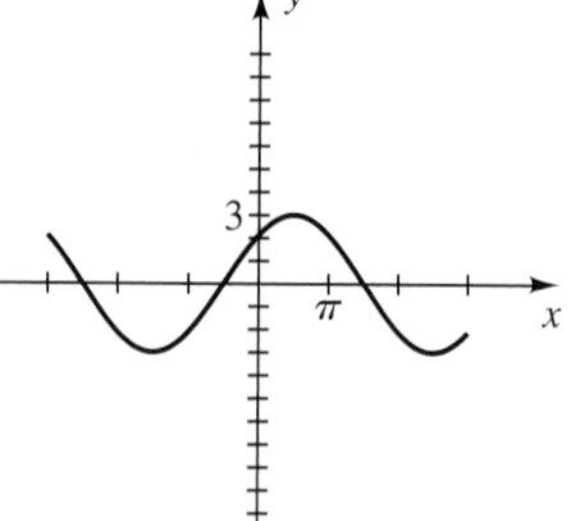

31 $5, 6\pi, -\frac{\pi}{2}$

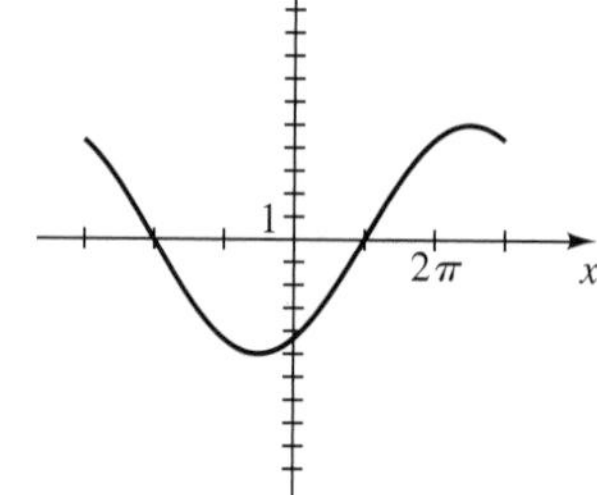

33 $3, 2, -4$

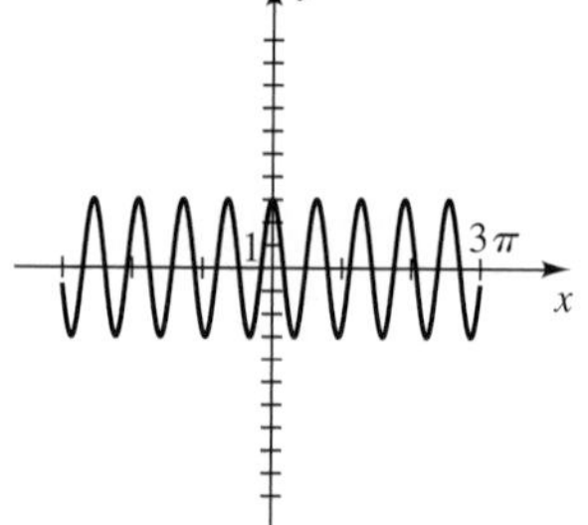

35 $\sqrt{2}, 4, \frac{1}{2}$

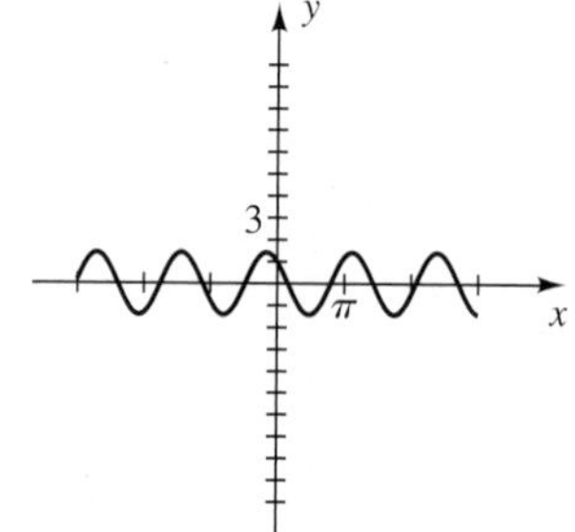

37 $2, \pi, \frac{\pi}{2}$

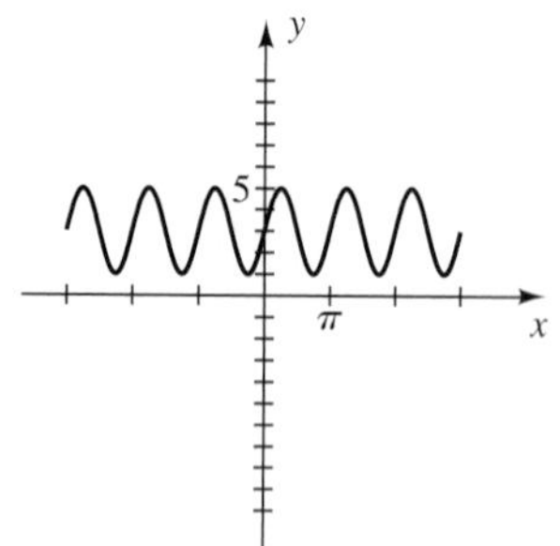

39 $5, \pi, -\pi$

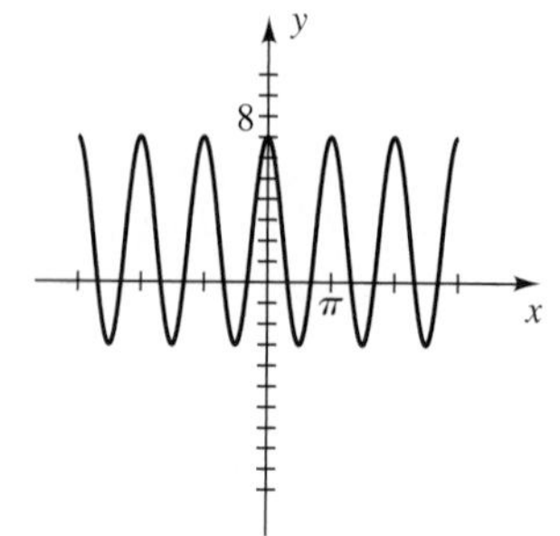

41 **(a)** $4, 2\pi, -\pi$ **(b)** $y = 4 \sin (x + \pi)$

43 **(a)** $2, 4, -3$ **(b)** $y = 2 \sin \left(\frac{\pi}{2}x + \frac{3\pi}{2}\right)$

45 4π **47** $a = 8, b = 4\pi$

49 **51**

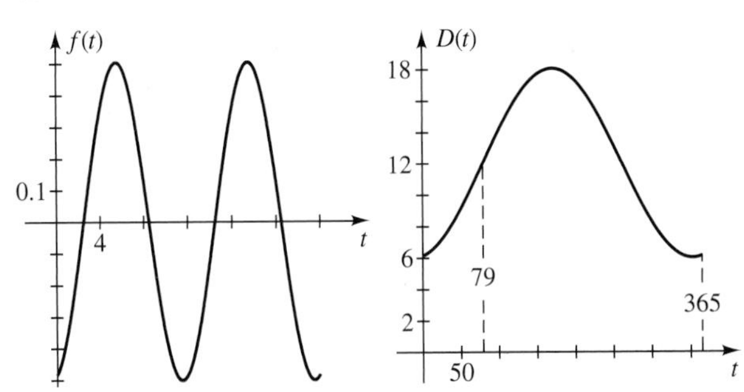

53 The temperature is 20°F at 9:00 A.M. ($t = 0$). It increases to a high of 35°F at 3:00 P.M. ($t = 6$) and then decreases to 20°F at 9:00 P.M. ($t = 12$). It continues to decrease to a low of 5°F at 3:00 A.M. ($t = 18$). It then rises to 20°F at 9:00 A.M. ($t = 24$).

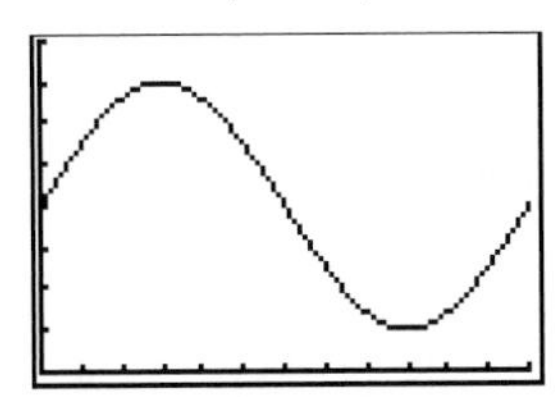

[0, 24, 2] by [0, 40, 5]

55 **(a)** $f(t) = 10 \sin \left[\frac{\pi}{12}(t - 10)\right] + 0$, with $a = 10$, $b = \frac{\pi}{12}$, $c = -\frac{5\pi}{6}$, $d = 0$

(b)

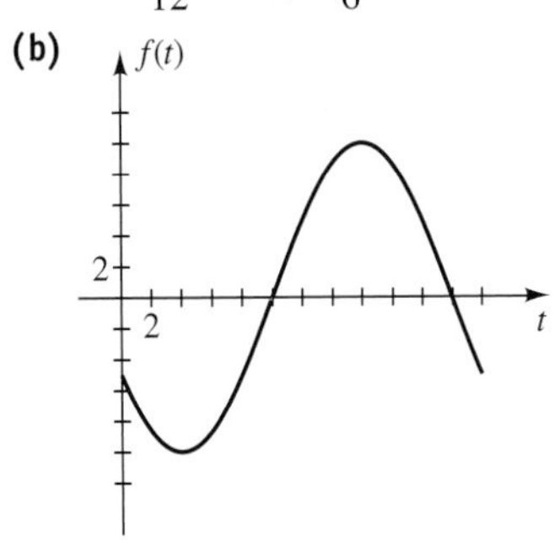

57 **(a)** $f(t) = 10 \sin \left[\frac{\pi}{12}(t - 9)\right] + 20$, with $a = 10$, $b = \frac{\pi}{12}$, $c = -\frac{3\pi}{4}$, $d = 20$

(b)

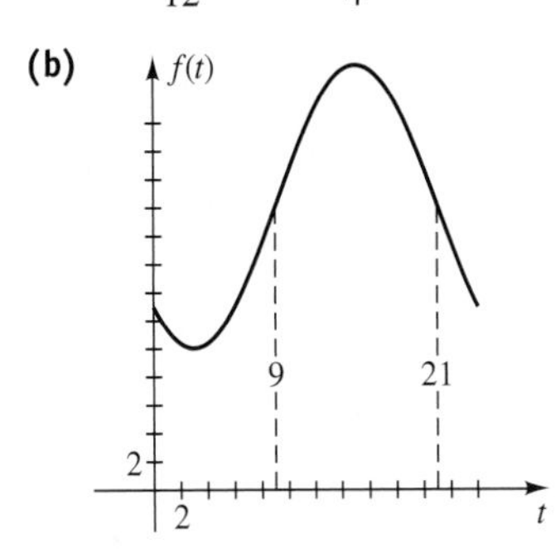

59 **(a)**

[0.5, 24.5, 5] by [−1, 8]

(b) $P(t) = 2.95 \sin \left(\frac{\pi}{6}t + \frac{\pi}{3}\right) + 3.15$

61 **(a)**

[0.5, 24.5, 5] by [0, 20, 2]

(b) $D(t) = 6.42 \sin \left(\frac{\pi}{6}t - \frac{2\pi}{3}\right) + 12.3$

63 As $x \to 0^-$ or as $x \to 0^+$, y oscillates between -1 and 1 and does not approach a unique value.

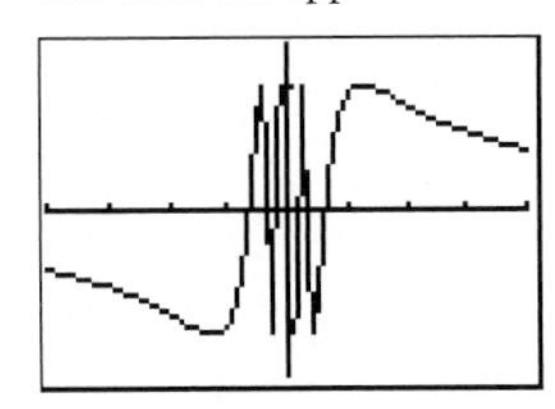

[−2, 2, 0.5] by [−1.33, 1.33, 0.5]

65 As $x \to 0^-$ or as $x \to 0^+$, y appears to approach 2.

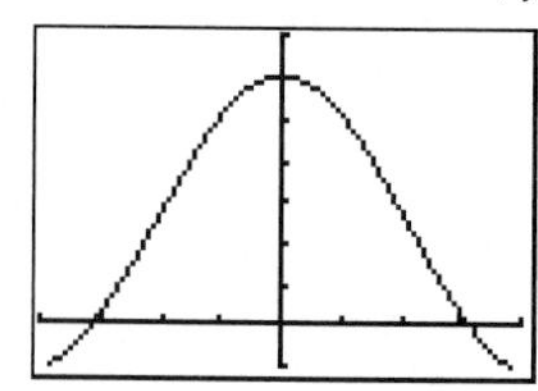

$[-2, 2, 0.5]$ by $[-0.33, 2.33, 0.33]$

67 $y = 4$

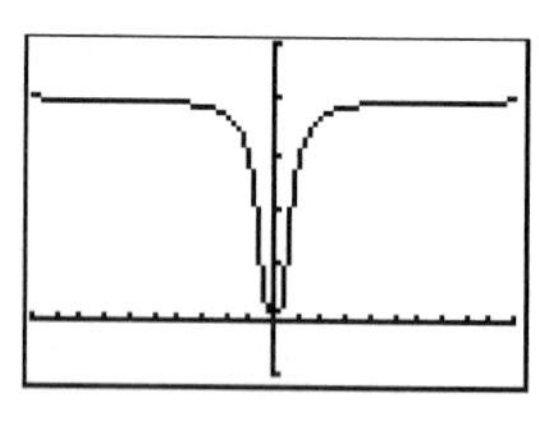

$[-20, 20, 2]$ by $[-1, 5]$

69 $[-\pi, -1.63] \cup [-0.45, 0.61] \cup [1.49, 2.42]$

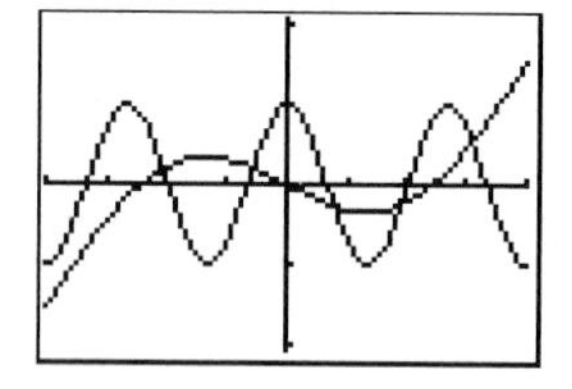

$[-\pi, \pi, \pi/4]$ by $[-2.09, 2.09]$

EXERCISES 5.6

1 π

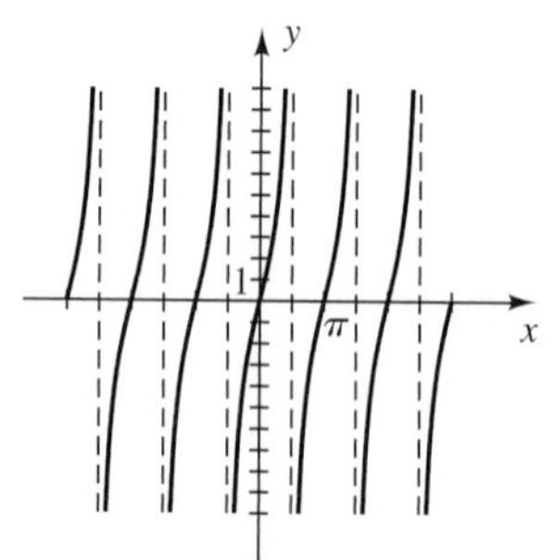

3 π

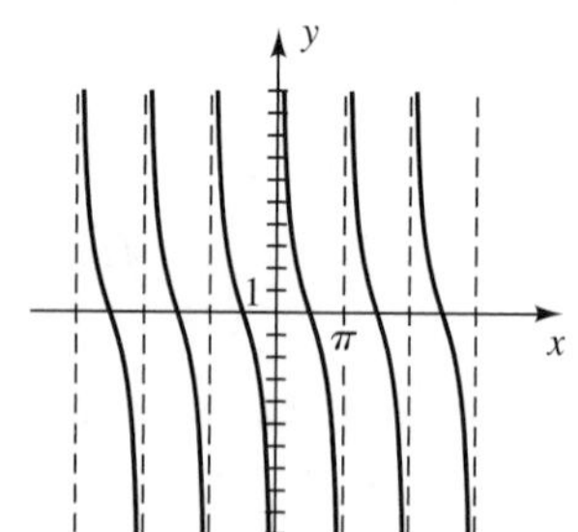

5 2π

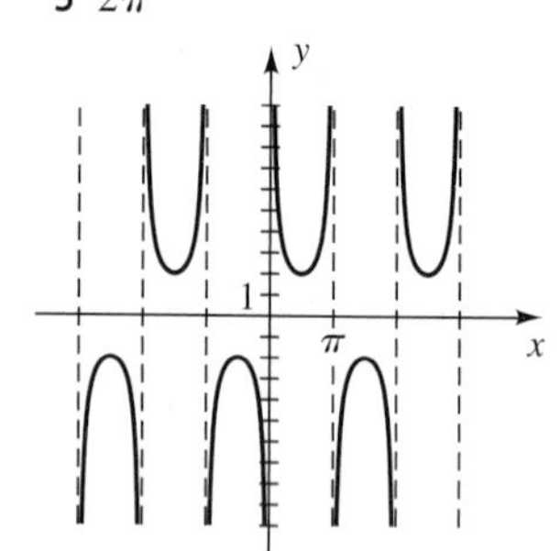

7 2π

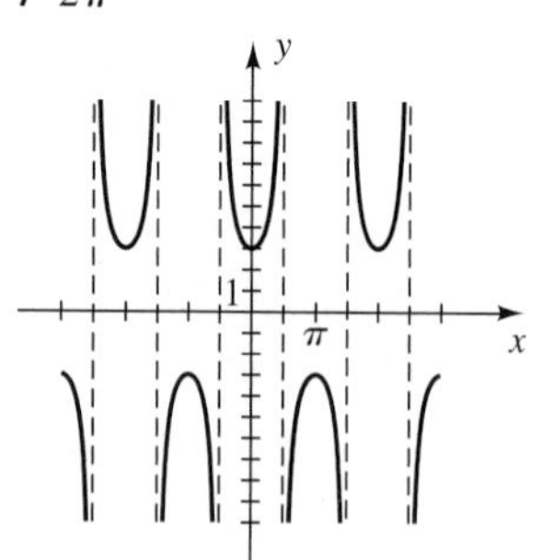

9 π

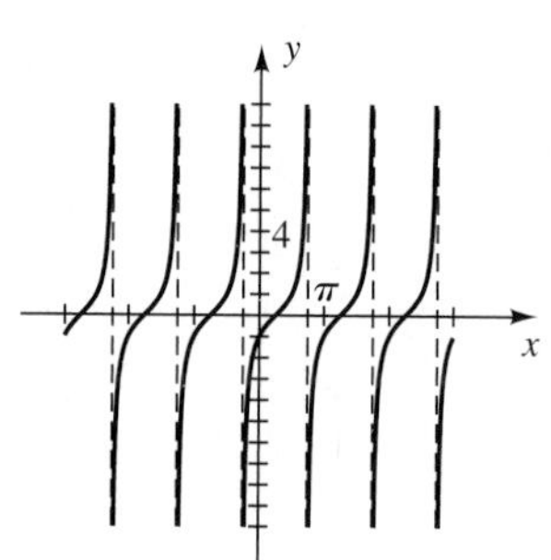

11 $\frac{\pi}{2}$

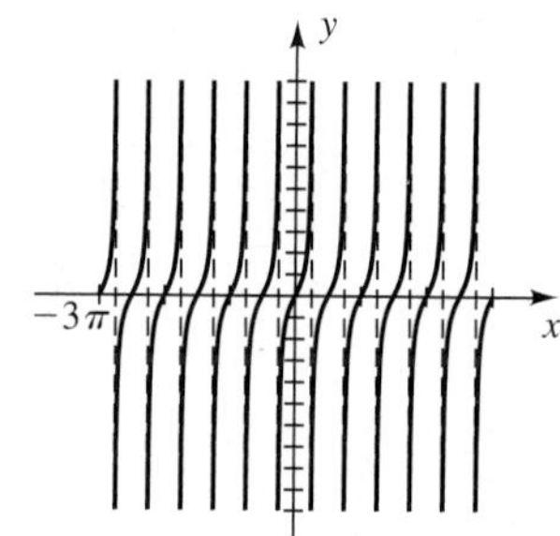

13 4π

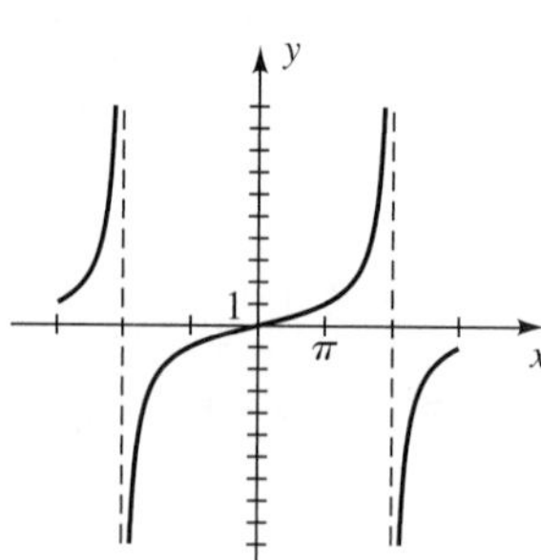

15 $\frac{\pi}{2}$

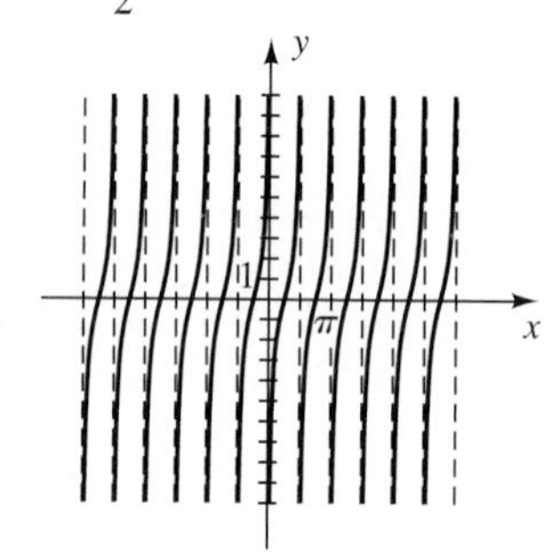

17 2π

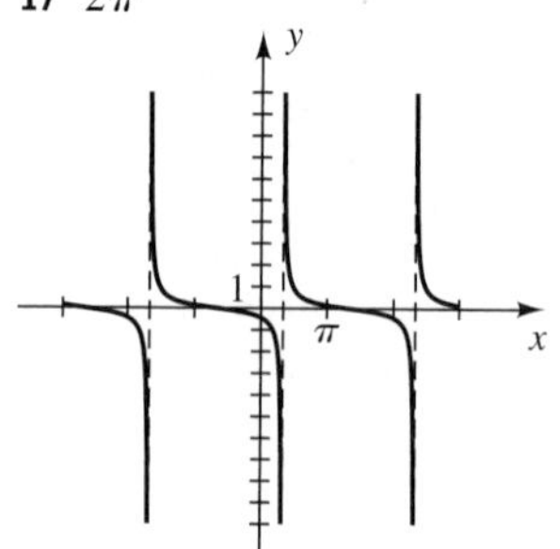

19 π

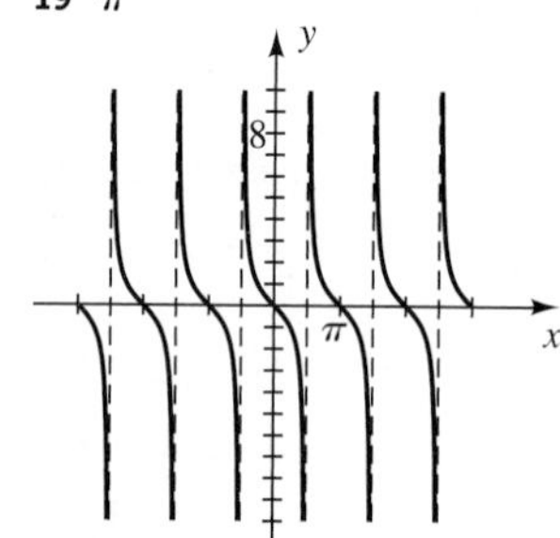

21 $\frac{\pi}{2}$

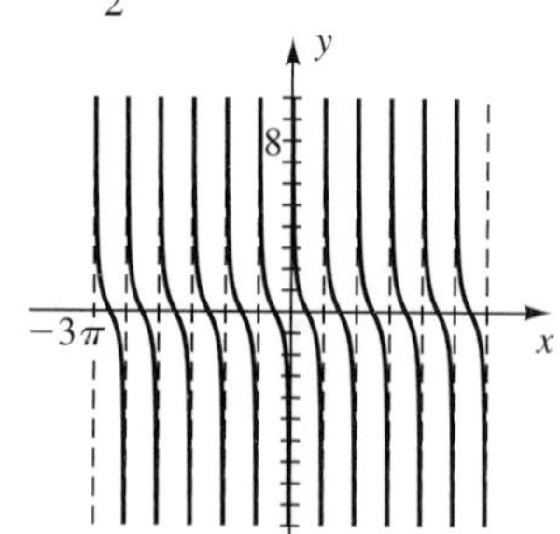

23 3π

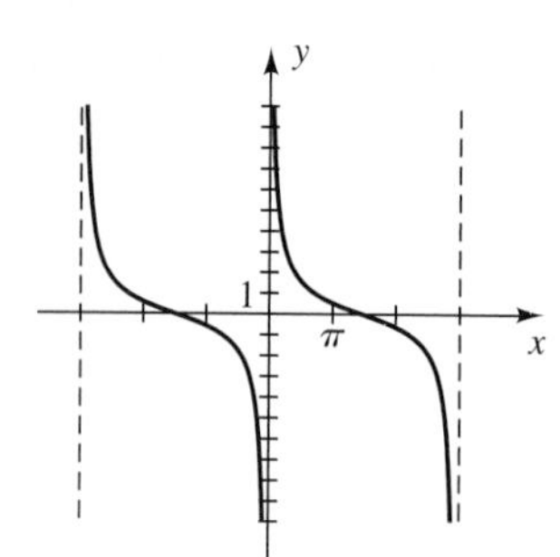

25 $\frac{\pi}{2}$

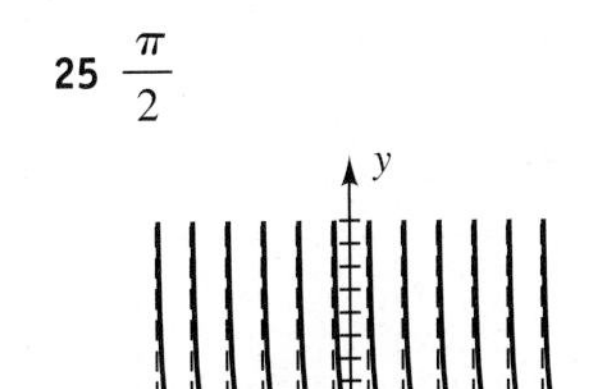

27 2π

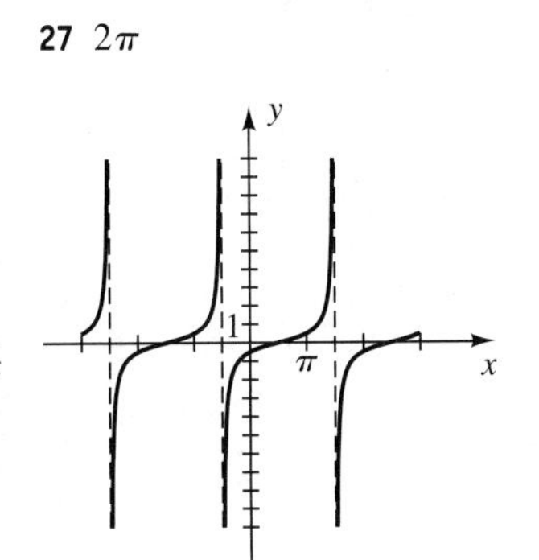

29 2π

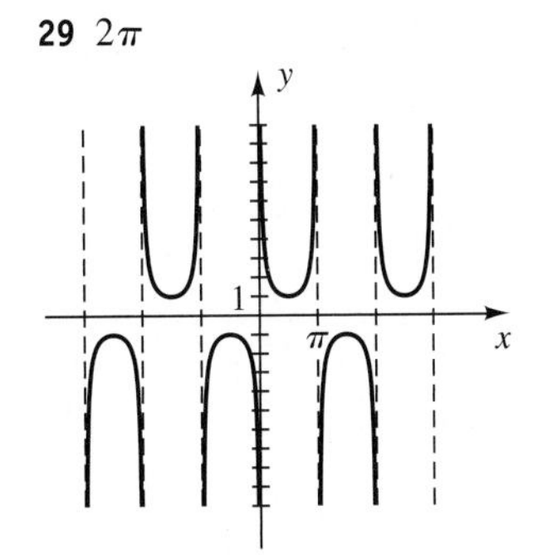

31 π

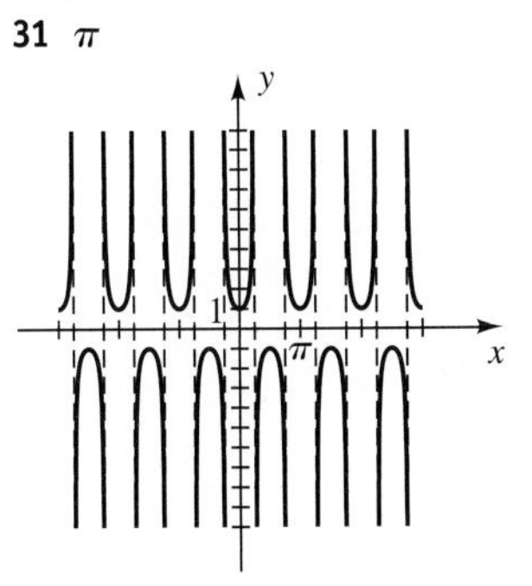

33 6π

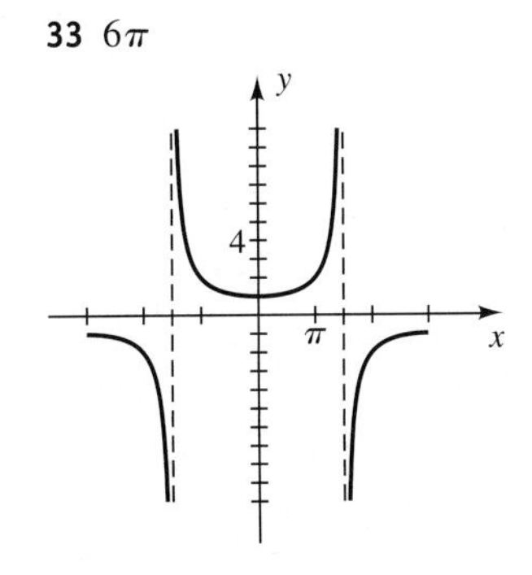

35 π

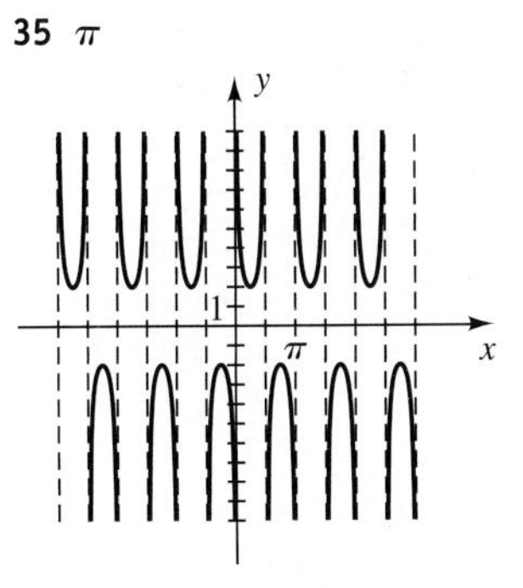

37 4π

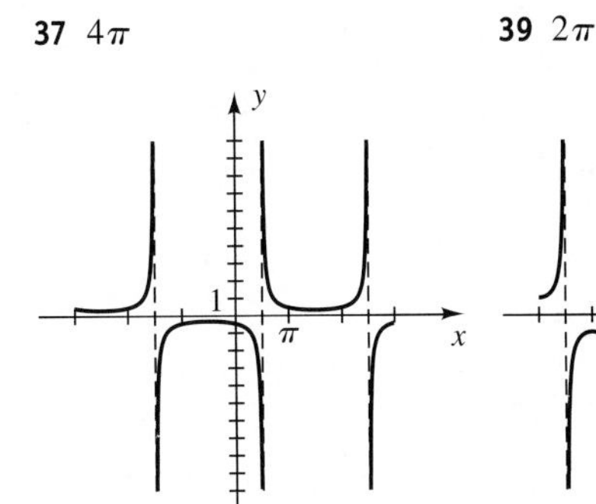

39 2π

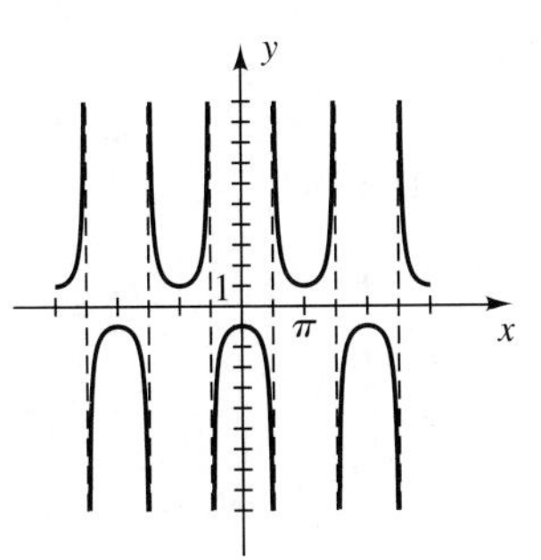

41 π

43 6π

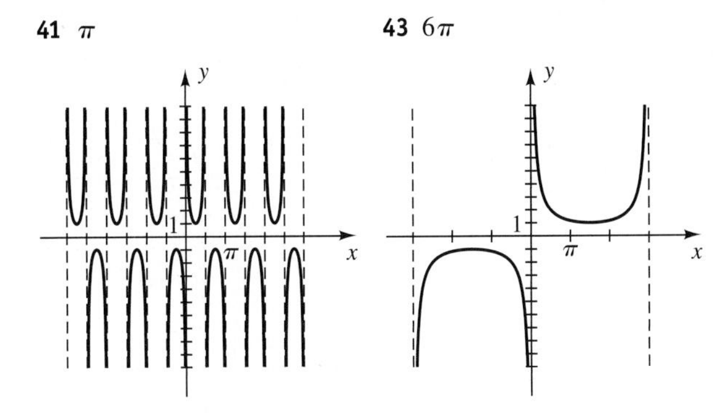

45 π

47 4π

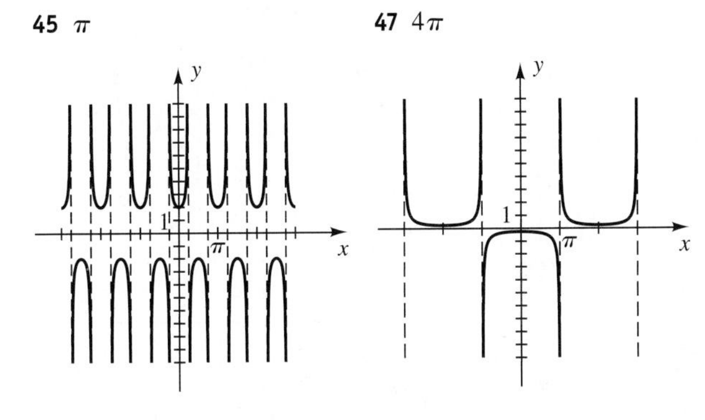

49 2

51 1

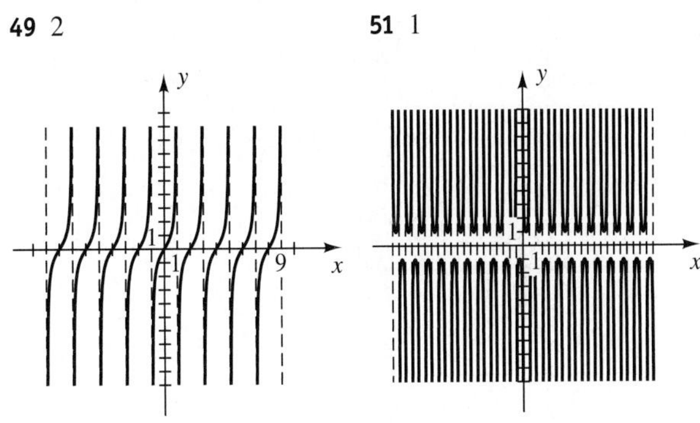

53 $y = -\cot\left(x + \frac{\pi}{2}\right)$

55

57

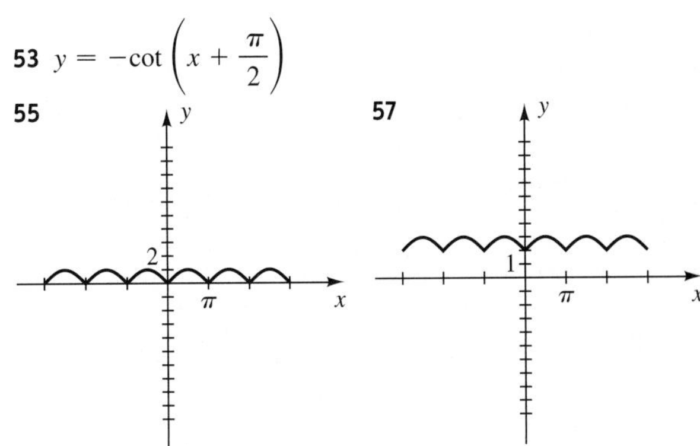

59

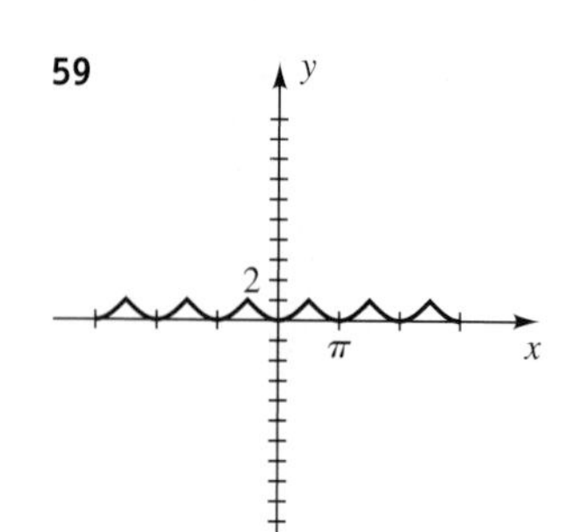

61

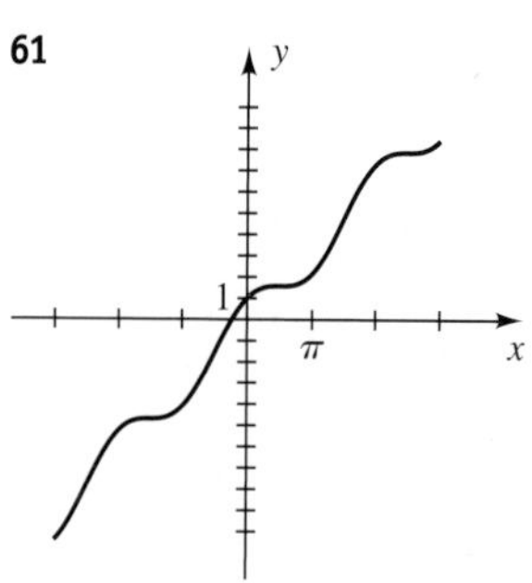

63

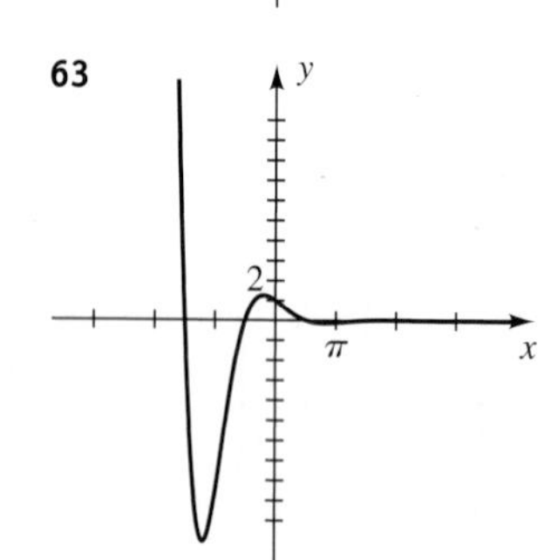

65

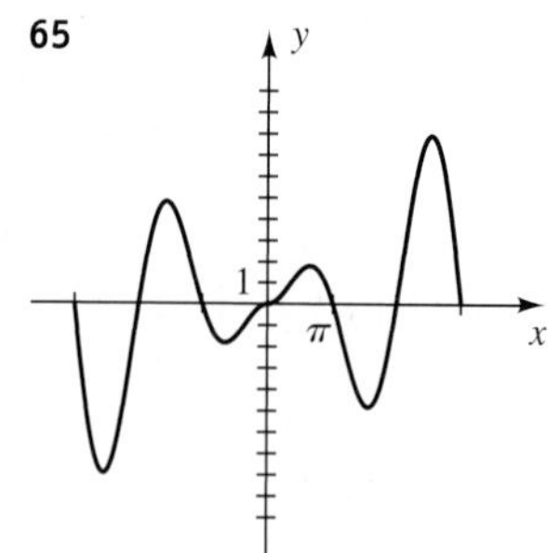

67

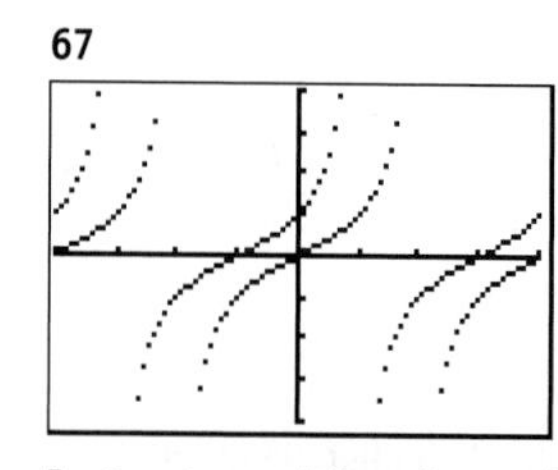

$[-2\pi, 2\pi, \pi/2]$ by $[-4, 4]$

69

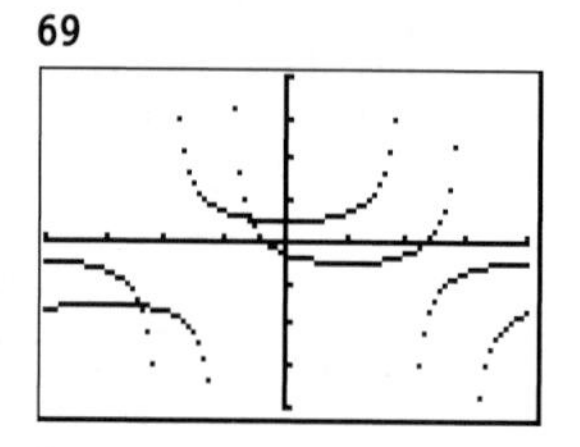

$[-2\pi, 2\pi, \pi/2]$ by $[-4, 4]$

71

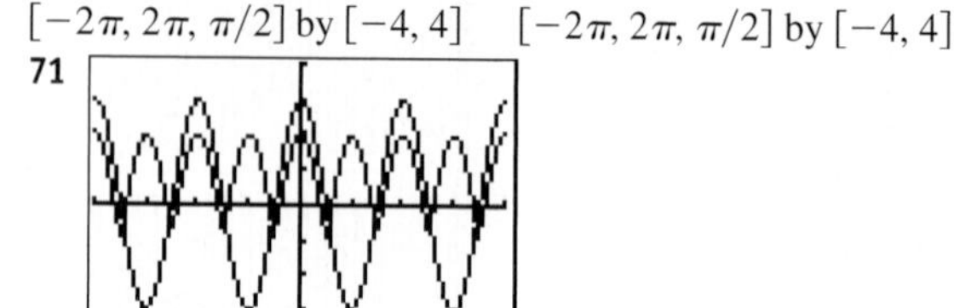

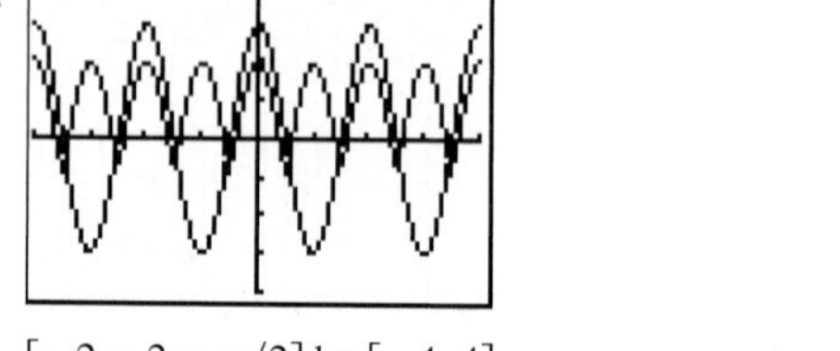

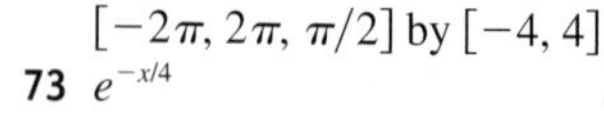

$[-2\pi, 2\pi, \pi/2]$ by $[-4, 4]$

73 $e^{-x/4}$

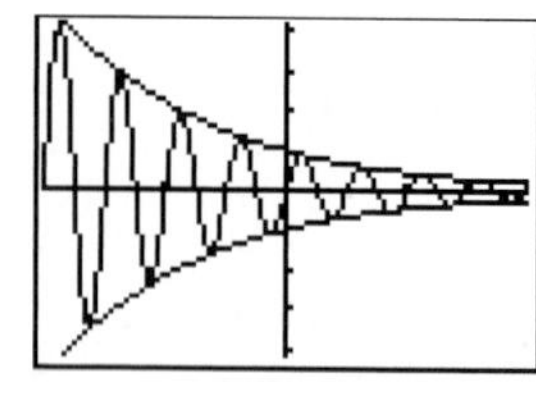

$[-2\pi, 2\pi, \pi/2]$ by $[-4.19, 4.19]$

75 $(-2.76, 3.09)$; $(1.23, -3.68)$

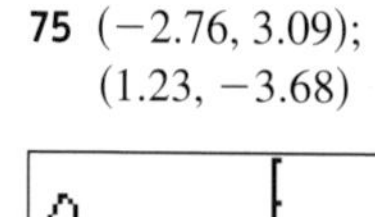

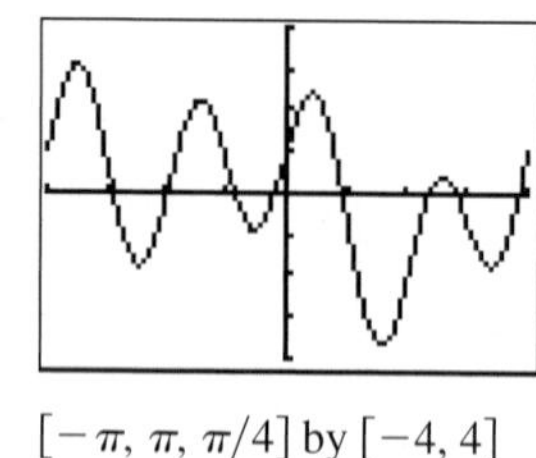

$[-\pi, \pi, \pi/4]$ by $[-4, 4]$

77 $[-0.70, 0.12]$

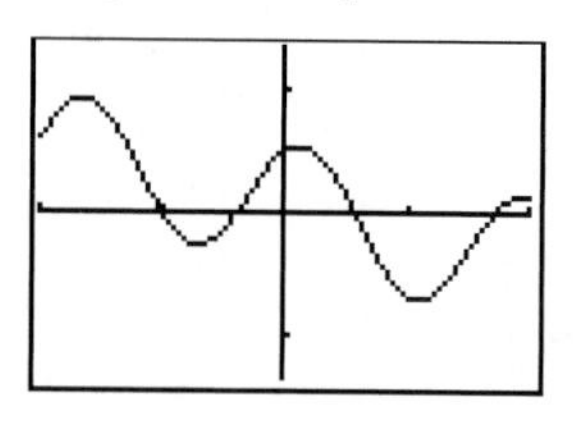

$[-2, 2]$ by $[-1.33, 1.33]$

79 $[-\pi, -1.31] \cup [0.11, 0.95] \cup [2.39, \pi]$

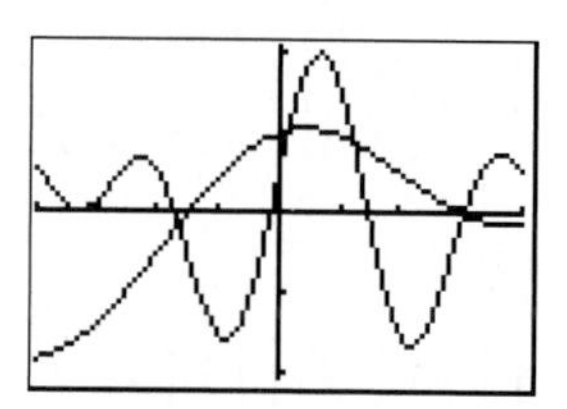

$[-\pi, \pi, \pi/4]$ by $[-2.09, 2.09]$

81 **(a)** I_0 **(b)** $0.044I_0$ **(c)** $0.603I_0$

83 **(a)** $A_0e^{-\alpha z}$ **(b)** $\dfrac{\alpha}{k}z_0$ **(c)** $\dfrac{\ln 2}{\alpha}$

EXERCISES 5.7

1 $\beta = 60°, a = \frac{20}{3}\sqrt{3}, c = \frac{40}{3}\sqrt{3}$

3 $\alpha = 45°, a = b = 15\sqrt{2}$

5 $\alpha = \beta = 45°, c = 5\sqrt{2}$

7 $\alpha = 60°, \beta = 30°, a = 15$

9 $\beta = 53°, a \approx 18, c \approx 30$

11 $\alpha = 18°9', a \approx 78.7, c \approx 252.6$

13 $\alpha \approx 29°, \beta \approx 61°, c \approx 51$

15 $\alpha \approx 69°, \beta \approx 21°, a \approx 5.4$ **17** $b = c \cos \alpha$

19 $a = b \cot \beta$ **21** $c = a \csc \alpha$

23 $b = \sqrt{c^2 - a^2}$

25 $250\sqrt{3} + 4 \approx 437$ ft **27** 28,800 ft **29** 160 m

31 9659 ft **33** **(a)** 58 ft **(b)** 27 ft **35** 51°20′

37 16.3° **39** 2063 ft **41** 1,459,379 ft^2

43 21.8° **45** 20.2 m **47** 29.7 km **49** 3944 mi

51 126 mi/hr

53 **(a)** 45%
(b) Each satellite has a signal range of more than 120°.

55 $h = d \sin \alpha + c$ **57** $h = \dfrac{d}{\cot \alpha - \cot \beta}$

59 $h = d(\tan \beta - \tan \alpha)$

61 N70°E; N40°W; S15°W; S25°E

63 **(a)** 55 mi **(b)** S63°E **65** 324 mi

67 Amplitude, 10 cm; period, $\frac{1}{3}$ sec; frequency, 3 osc/sec. The point is at the origin at $t = 0$. It moves upward with decreasing speed, reaching the point with coordinate 10 at $t = \frac{1}{12}$. It then reverses direction and moves downward, gaining speed until it reaches the origin at $t = \frac{1}{6}$. It continues downward with decreasing speed, reaching the point with coordinate -10 at $t = \frac{1}{4}$. It then reverses direction and moves upward with increasing speed, returning to the origin at $t = \frac{1}{3}$.

69 Amplitude, 4 cm; period, $\frac{4}{3}$ sec; frequency, $\frac{3}{4}$ osc/sec. The motion is similar to that in Exercise 67; however, the point starts 4 units above the origin and moves downward, reaching the origin at $t = \frac{1}{3}$ and the point with coordinate -4 at $t = \frac{2}{3}$. It then reverses direction and moves upward, reaching the origin at $t = 1$ and its initial point at $t = \frac{4}{3}$.

71 $d = 5\cos\frac{2\pi}{3}t$

73 **(a)** $y = 25\cos\frac{\pi}{15}t$

(b) 324,000 ft

CHAPTER 5 REVIEW EXERCISES

1 $\frac{11\pi}{6}, \frac{9\pi}{4}, -\frac{5\pi}{6}, \frac{4\pi}{3}, \frac{\pi}{5}$

2 810°, −120°, 315°, 900°, 36°

3 **(a)** 0.1 **(b)** 0.2 m^2

4 **(a)** $\frac{35\pi}{12}$ cm **(b)** $\frac{175\pi}{16}$ cm^2

5 $\frac{200\pi}{3}, 90\pi$ **6** $\frac{100\pi}{3}, \frac{105\pi}{4}$

7 $x = 6\sqrt{3}; y = 3\sqrt{3}$ **8** $x = \frac{7}{2}\sqrt{2}; y = \frac{7}{2}\sqrt{2}$

9 $\tan\theta = \sqrt{\sec^2\theta - 1}$ **10** $\cot\theta = \sqrt{\csc^2\theta - 1}$

Exer. 11–20: Typical verifications are given.

11 $\sin\theta(\csc\theta - \sin\theta) = \sin\theta\csc\theta - \sin^2\theta = 1 - \sin^2\theta = \cos^2\theta$

12 $\cos\theta(\tan\theta + \cot\theta) = \cos\theta\cdot\frac{\sin\theta}{\cos\theta} + \cos\theta\cdot\frac{\cos\theta}{\sin\theta} = \sin\theta + \frac{\cos^2\theta}{\sin\theta} = \frac{\sin^2\theta + \cos^2\theta}{\sin\theta} = \frac{1}{\sin\theta} = \csc\theta$

13 $(\cos^2\theta - 1)(\tan^2\theta + 1) = (\cos^2\theta - 1)(\sec^2\theta) = \cos^2\theta\sec^2\theta - \sec^2\theta = 1 - \sec^2\theta$

14 $\frac{\sec\theta - \cos\theta}{\tan\theta} = \frac{\frac{1}{\cos\theta} - \cos\theta}{\frac{\sin\theta}{\cos\theta}} = \frac{\frac{1 - \cos^2\theta}{\cos\theta}}{\frac{\sin\theta}{\cos\theta}} = \frac{\frac{\sin^2\theta}{\cos\theta}}{\frac{\sin\theta}{\cos\theta}} = \frac{\frac{\sin\theta}{\cos\theta}}{\frac{1}{\cos\theta}} = \frac{\tan\theta}{\sec\theta}$

15 $\frac{1 + \tan^2\theta}{\tan^2\theta} = \frac{1}{\tan^2\theta} + \frac{\tan^2\theta}{\tan^2\theta} = \cot^2\theta + 1 = \csc^2\theta$

16 $\frac{\sec\theta + \csc\theta}{\sec\theta - \csc\theta} = \frac{\frac{1}{\cos\theta} + \frac{1}{\sin\theta}}{\frac{1}{\cos\theta} - \frac{1}{\sin\theta}} = \frac{\frac{\sin\theta + \cos\theta}{\cos\theta\sin\theta}}{\frac{\sin\theta - \cos\theta}{\cos\theta\sin\theta}} = \frac{\sin\theta + \cos\theta}{\sin\theta - \cos\theta}$

17 $\frac{\cot\theta - 1}{1 - \tan\theta} = \frac{\frac{\cos\theta}{\sin\theta} - 1}{1 - \frac{\sin\theta}{\cos\theta}} = \frac{\frac{\cos\theta - \sin\theta}{\sin\theta}}{\frac{\cos\theta - \sin\theta}{\cos\theta}} = \frac{(\cos\theta - \sin\theta)\cos\theta}{(\cos\theta - \sin\theta)\sin\theta} = \frac{\cos\theta}{\sin\theta} = \cot\theta$

18 $\frac{1 + \sec\theta}{\tan\theta + \sin\theta} = \frac{1 + \frac{1}{\cos\theta}}{\frac{\sin\theta}{\cos\theta} + \frac{\sin\theta\cos\theta}{\cos\theta}} = \frac{\frac{\cos\theta + 1}{\cos\theta}}{\frac{\sin\theta(1 + \cos\theta)}{\cos\theta}} = \frac{1}{\sin\theta} = \csc\theta$

19 $\frac{\tan(-\theta) + \cot(-\theta)}{\tan\theta} = \frac{-\tan\theta - \cot\theta}{\tan\theta} = -\frac{\tan\theta}{\tan\theta} - \frac{\cot\theta}{\tan\theta} = -1 - \cot^2\theta = -(1 + \cot^2\theta) = -\csc^2\theta$

20 $-\dfrac{1}{\csc(-\theta)} - \dfrac{\cot(-\theta)}{\sec(-\theta)} = -\dfrac{1}{-\csc\theta} - \dfrac{-\cot\theta}{\sec\theta}$

$$= \sin\theta + \frac{\cos\theta/\sin\theta}{1/\cos\theta}$$

$$= \sin\theta + \frac{\cos^2\theta}{\sin\theta}$$

$$= \frac{\sin^2\theta + \cos^2\theta}{\sin\theta}$$

$$= \frac{1}{\sin\theta} = \csc\theta$$

21 $\dfrac{\sqrt{33}}{7}, \dfrac{4}{7}, \dfrac{\sqrt{33}}{4}, \dfrac{4}{\sqrt{33}}, \dfrac{7}{4}, \dfrac{7}{\sqrt{33}}$

22 **(a)** $-\dfrac{4}{5}, \dfrac{3}{5}, -\dfrac{4}{3}, -\dfrac{3}{4}, \dfrac{5}{3}, -\dfrac{5}{4}$

(b) $\dfrac{2}{\sqrt{13}}, -\dfrac{3}{\sqrt{13}}, -\dfrac{2}{3}, -\dfrac{3}{2}, -\dfrac{\sqrt{13}}{3}, \dfrac{\sqrt{13}}{2}$

(c) $-1, 0$, U, 0, U, -1

23 **(a)** II **(b)** III **(c)** IV

24 **(a)** $-\dfrac{4}{5}, \dfrac{3}{5}, -\dfrac{4}{3}, -\dfrac{3}{4}, \dfrac{5}{3}, -\dfrac{5}{4}$

(b) $\dfrac{2}{\sqrt{13}}, -\dfrac{3}{\sqrt{13}}, -\dfrac{2}{3}, -\dfrac{3}{2}, -\dfrac{\sqrt{13}}{3}, \dfrac{\sqrt{13}}{2}$

25 $(-1, 0)$; $(0, -1)$; $(0, 1)$; $\left(-\dfrac{\sqrt{2}}{2}, -\dfrac{\sqrt{2}}{2}\right)$; $(1, 0)$; $\left(\dfrac{\sqrt{3}}{2}, \dfrac{1}{2}\right)$

26 $\left(\dfrac{3}{5}, \dfrac{4}{5}\right)$; $\left(\dfrac{3}{5}, \dfrac{4}{5}\right)$; $\left(-\dfrac{3}{5}, \dfrac{4}{5}\right)$; $\left(-\dfrac{3}{5}, \dfrac{4}{5}\right)$

27 **(a)** $\dfrac{\pi}{4}, \dfrac{\pi}{6}, \dfrac{\pi}{8}$ **(b)** $65°, 43°, 8°$

28 **(a)** $1, 0$, U, 0, U, 1

(b) $\dfrac{\sqrt{2}}{2}, -\dfrac{\sqrt{2}}{2}, -1, -1, -\sqrt{2}, \sqrt{2}$

(c) $0, 1, 0$, U, 1, U

(d) $-\dfrac{1}{2}, \dfrac{\sqrt{3}}{2}, -\dfrac{\sqrt{3}}{3}, -\sqrt{3}, \dfrac{2}{\sqrt{3}}, -2$

29 **(a)** $-\dfrac{\sqrt{2}}{2}$ **(b)** $-\dfrac{\sqrt{3}}{3}$ **(c)** $-\dfrac{1}{2}$ **(d)** -2

(e) -1 **(f)** $-\dfrac{2}{\sqrt{3}}$

30 310.5° **31** 1.2206; 4.3622 **32** 52.44°; 307.56°

33 $5, 2\pi$

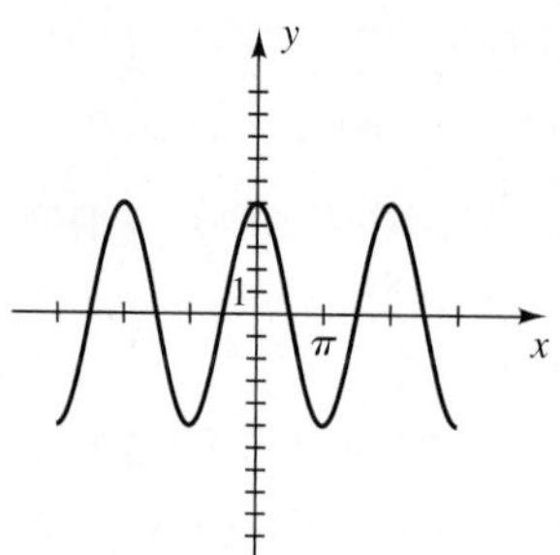

34 $\dfrac{2}{3}, 2\pi$

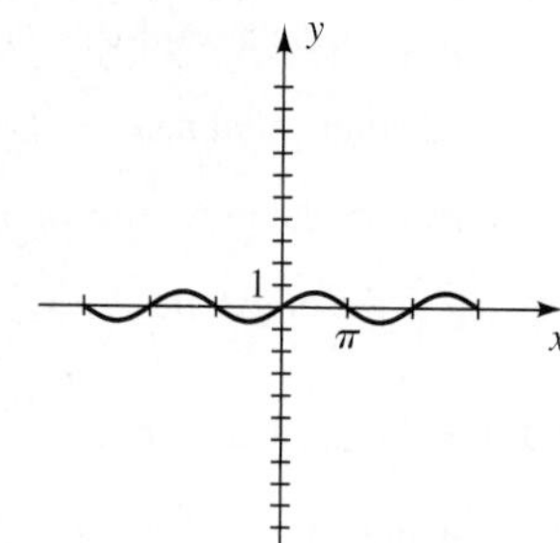

35 $\dfrac{1}{3}, \dfrac{2\pi}{3}$

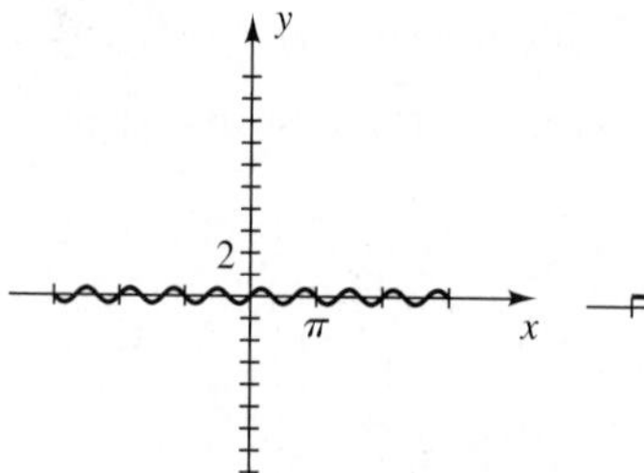

36 $\dfrac{1}{2}, 6\pi$

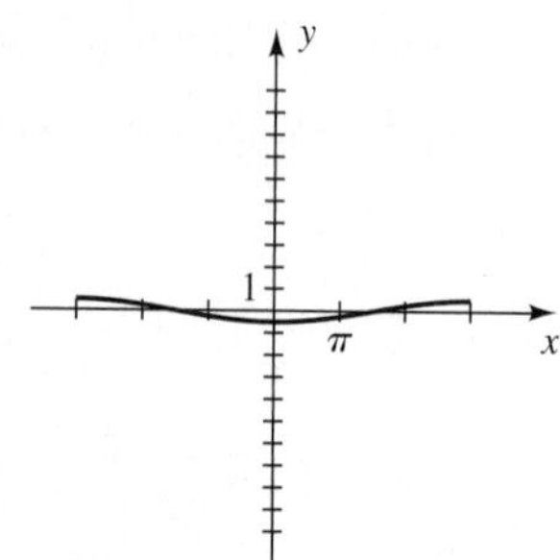

37 $3, 4\pi$

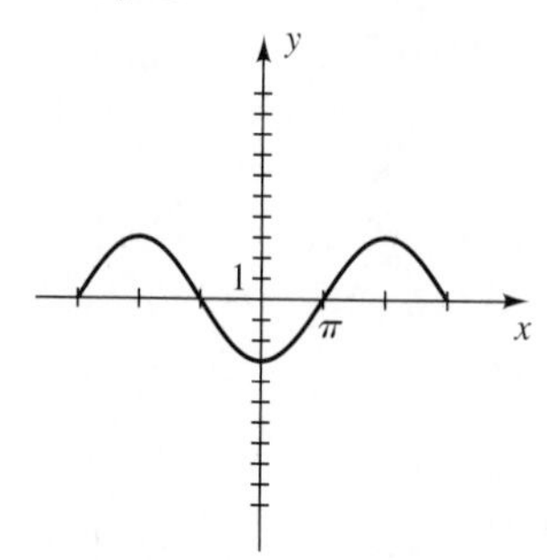

38 $4, \pi$

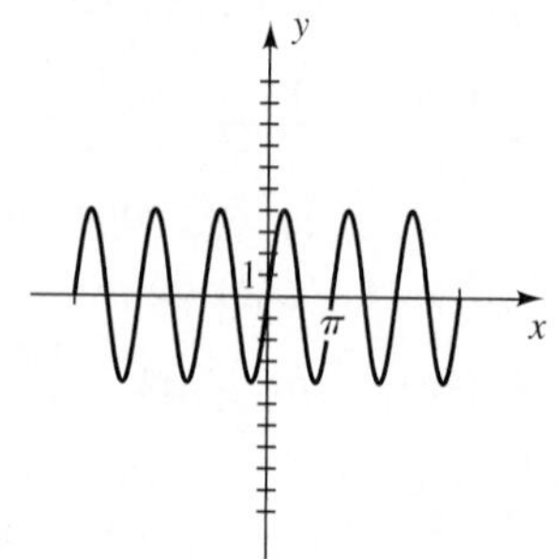

39 $2, 2$

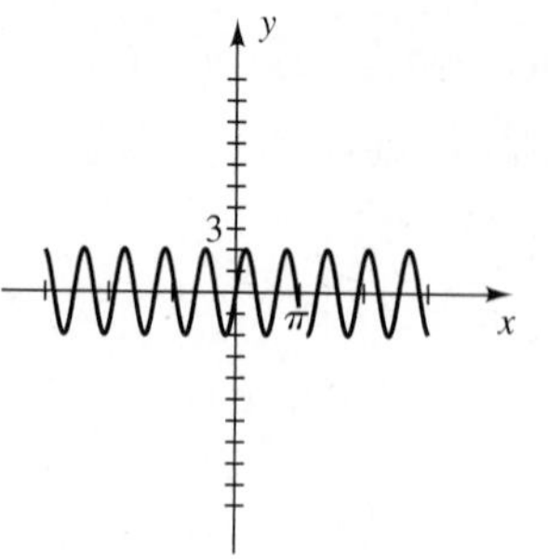

40 $4, 4$

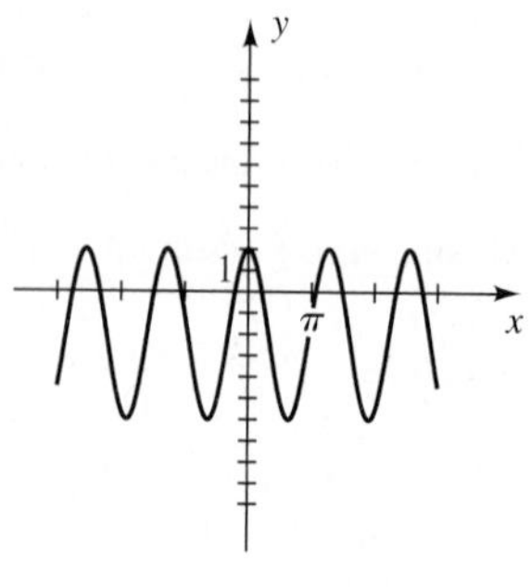

41 **(a)** 1.43, 2 **(b)** $y = 1.43 \sin \pi x$

42 **(a)** 3.27, 3π **(b)** $y = -3.27 \sin \frac{2}{3}x$

43 **(a)** 3, $\frac{4\pi}{3}$ **(b)** $y = -3 \cos \frac{3}{2}x$

44 **(a)** 2, 4 **(b)** $y = 2 \cos \frac{\pi}{2}x$

45

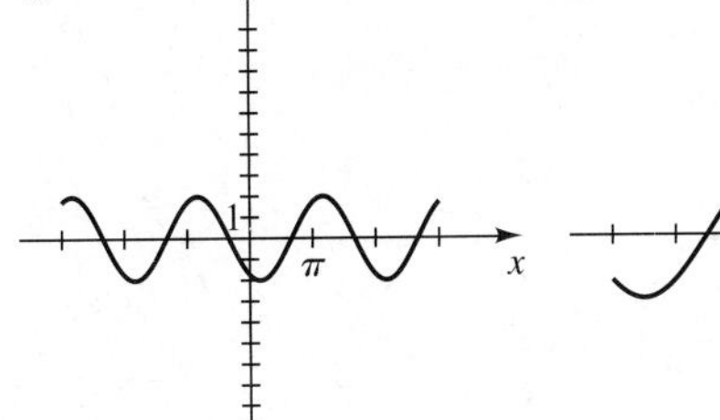

46

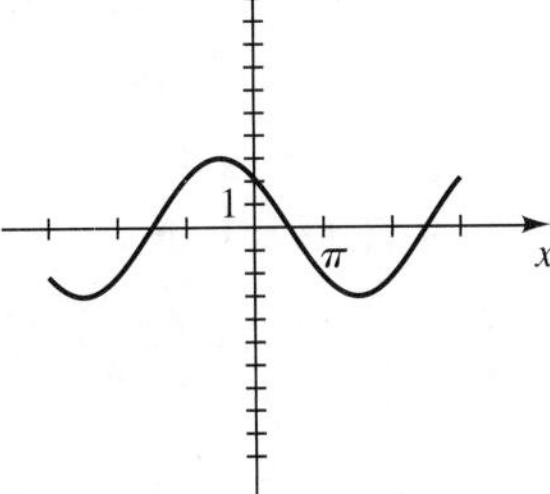

47

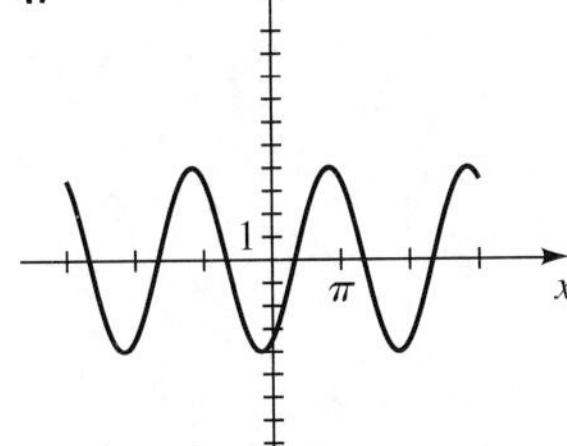

48

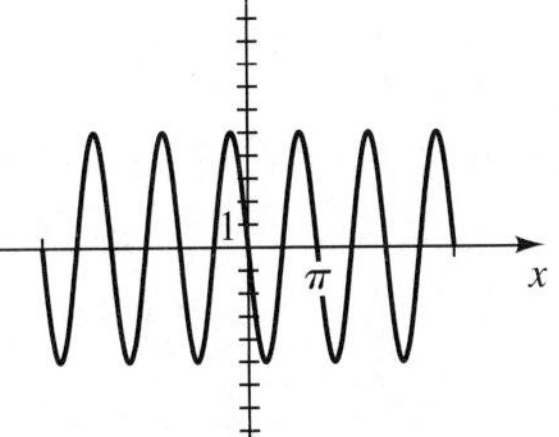

49

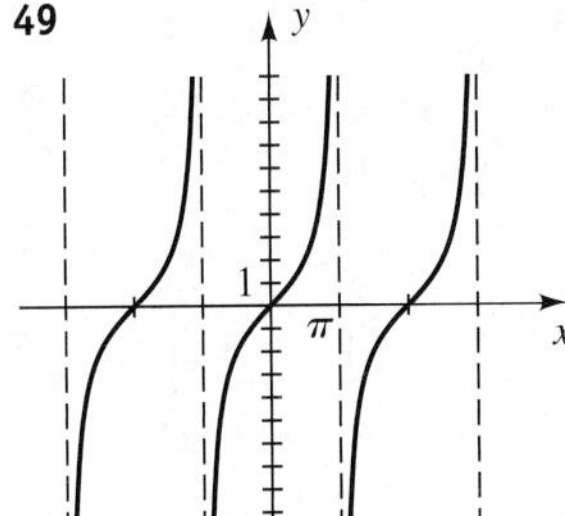

50

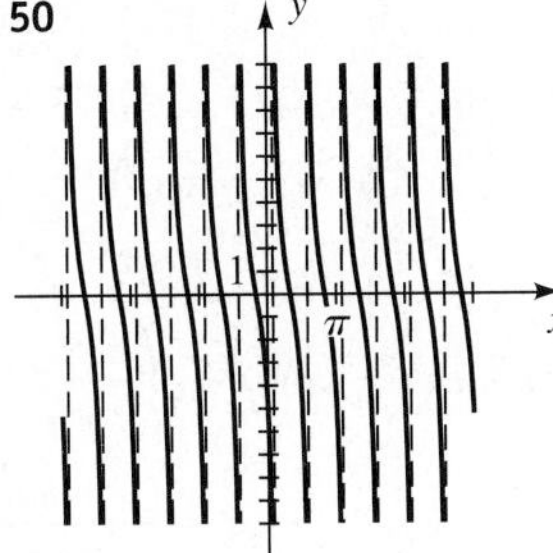

51

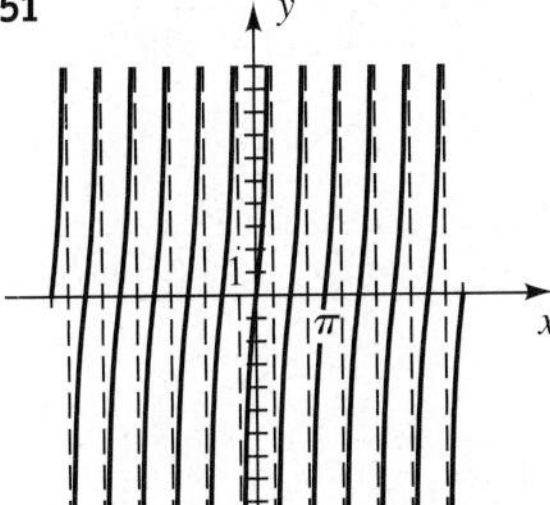

52

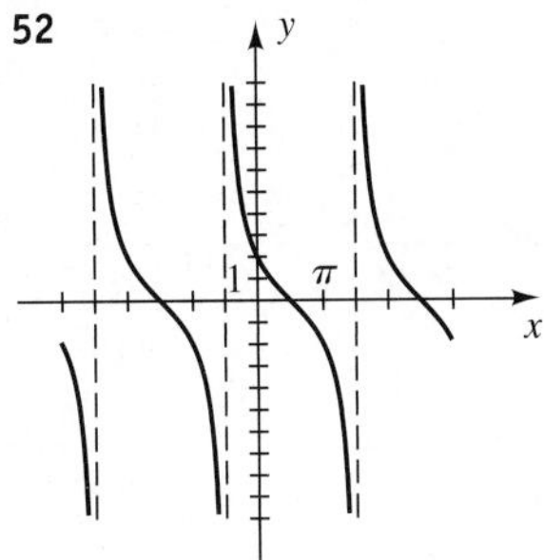

53

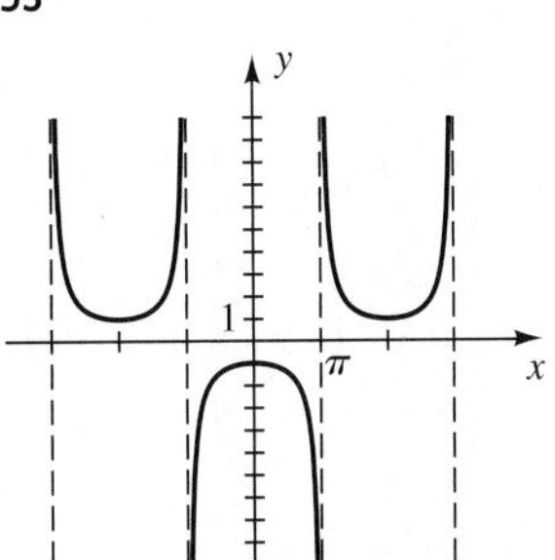

54

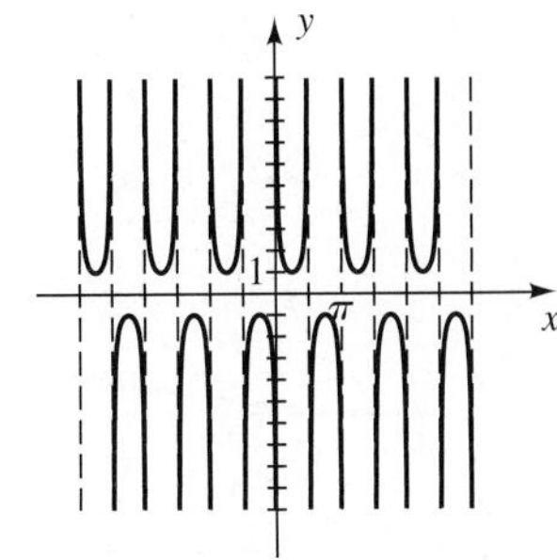

55

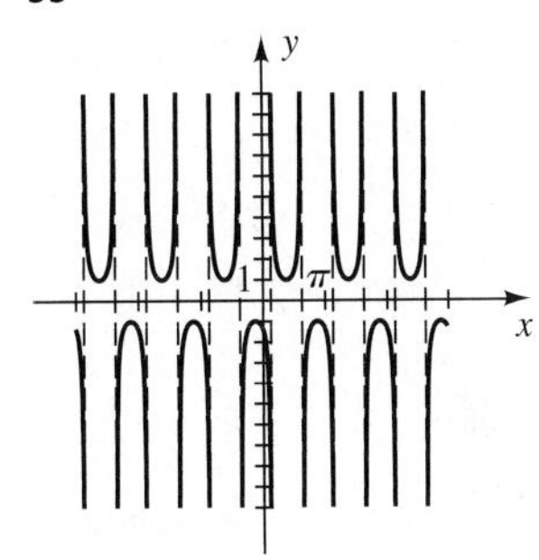

56

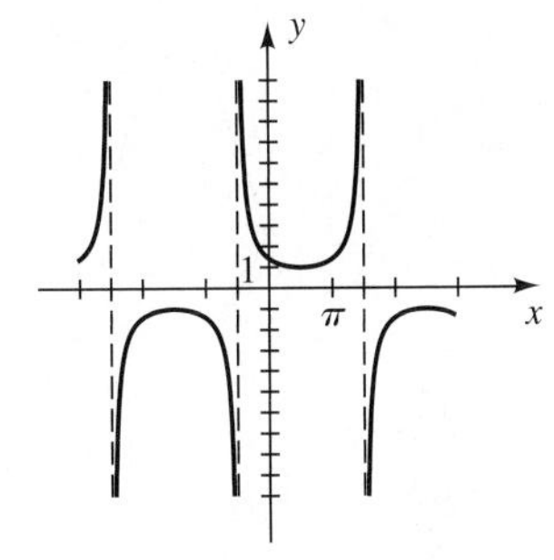

57 $\alpha = 30°, a \approx 23, c \approx 46$

58 $\beta = 35°20', a \approx 310, c \approx 380$

59 $\alpha \approx 68°, \beta \approx 22°, c \approx 67$

60 $\alpha \approx 13°, \beta \approx 77°, b = 40$

61 **(a)** $\frac{109\pi}{6}$ **(b)** 440.2 **62** 1048 ft

63 0.093 mi/sec **64** 52°

65 Approximately 67,900,000 mi

66 $\frac{6\pi}{5}$ radians = 216° **67** 250 ft

68 **(a)** 231.0 ft **(b)** 434.5 **69** **(b)** 2 mi

70 **(a)** $T = h + d(\cos \alpha \tan \theta - \sin \alpha)$ **(b)** 22.54 ft

71 **(a)** $\frac{25}{3}\sqrt{3} \approx 14.43$ ft-candles **(b)** 37.47°

72 **(b)** 4.69 **73** **(a)** 74.05 in. **(b)** 24.75 in.

74 **(a)** $S = 4a^2 \sin \theta$ **(b)** $V = \frac{4}{3}a^3 \sin^2 \theta \cos \theta$

75 **(a)** $h = R \sec \frac{s}{R} - R$ **(b)** $h \approx 1650$ ft

76

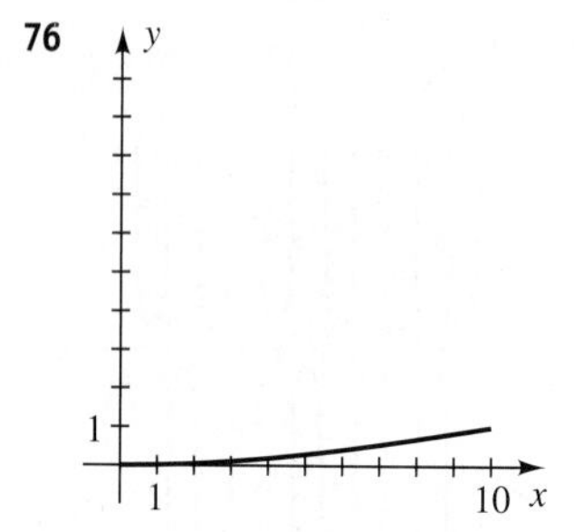

77 $y = 98.6 + (0.3)\sin\left(\frac{\pi}{12}t - \frac{11\pi}{12}\right)$

78 **(a)**

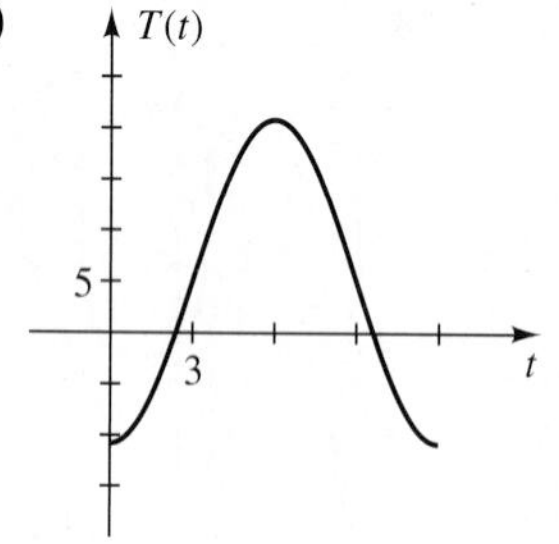

(b) 20.8°C on July 1

79 **(a)**

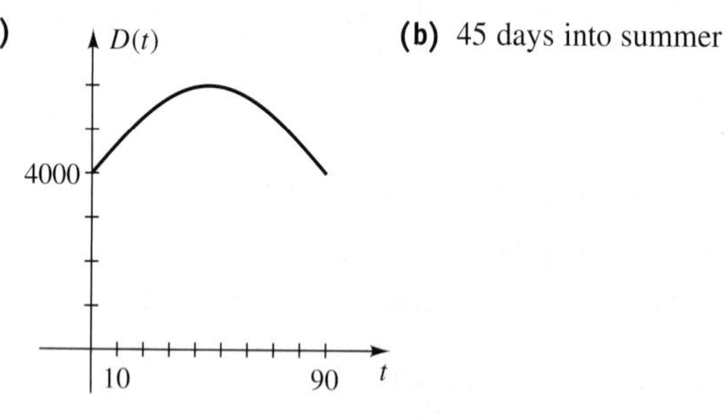

(b) 45 days into summer

80 **(a)** The cork is in simple harmonic motion.
(b) $1 \le t \le 2$

CHAPTER 5 DISCUSSION EXERCISES

3 None

5 The values of y_1, y_2, and y_3 are very close to each other near $x = 0$.

6 **(a)** $x \approx -0.4161,\ y \approx 0.9093$
(b) $x \approx -0.8838,\ y \approx -0.4678$

7 **(a)** $x \approx 1.8415,\ y \approx -0.5403$
(b) $x \approx -1.2624,\ y \approx 0.9650$

8 **(a)** $\frac{500\pi}{3}$ rad/sec **(b)** $D(t) = 5\cos\left(\frac{500\pi}{3}t\right) + 18$
(c) 10 revolutions

Chapter 6

EXERCISES 6.1

Exer. 1–50: Typical verifications are given for Exercises 1, 5, 9, . . . , 49.

1 $$\csc\theta - \sin\theta = \frac{1}{\sin\theta} - \sin\theta = \frac{1 - \sin^2\theta}{\sin\theta} = \frac{\cos^2\theta}{\sin\theta} = \frac{\cos\theta}{\sin\theta}\cos\theta = \cot\theta\cos\theta$$

5 $$\frac{\csc^2\theta}{1 + \tan^2\theta} = \frac{\csc^2\theta}{\sec^2\theta} = \frac{1/\sin^2\theta}{1/\cos^2\theta} = \frac{\cos^2\theta}{\sin^2\theta} = \left(\frac{\cos\theta}{\sin\theta}\right)^2 = \cot^2\theta$$

9 $$\frac{1}{1 - \cos\gamma} + \frac{1}{1 + \cos\gamma} = \frac{1 + \cos\gamma + 1 - \cos\gamma}{1 - \cos^2\gamma} = \frac{2}{\sin^2\gamma} = 2\csc^2\gamma$$

13 $$\csc^4 t - \cot^4 t = (\csc^2 t + \cot^2 t)(\csc^2 t - \cot^2 t) = (\csc^2 t + \cot^2 t)(1) = \csc^2 t + \cot^2 t$$

17 $$\frac{\tan^2 x}{\sec x + 1} = \frac{\sec^2 x - 1}{\sec x + 1} = \frac{(\sec x + 1)(\sec x - 1)}{\sec x + 1} = \sec x - 1 = \frac{1}{\cos x} - 1 = \frac{1 - \cos x}{\cos x}$$

21 $$\sin^4 r - \cos^4 r = (\sin^2 r - \cos^2 r)(\sin^2 r + \cos^2 r) = (\sin^2 r - \cos^2 r)(1) = \sin^2 r - \cos^2 r$$

25 $$(\sec t + \tan t)^2 = \left(\frac{1}{\cos t} + \frac{\sin t}{\cos t}\right)^2 = \left(\frac{1 + \sin t}{\cos t}\right)^2 = \frac{(1 + \sin t)^2}{\cos^2 t} = \frac{(1 + \sin t)^2}{1 - \sin^2 t} = \frac{(1 + \sin t)^2}{(1 + \sin t)(1 - \sin t)} = \frac{1 + \sin t}{1 - \sin t}$$

29 $$\frac{1 + \csc\beta}{\cot\beta + \cos\beta} = \frac{1 + \frac{1}{\sin\beta}}{\frac{\cos\beta}{\sin\beta} + \cos\beta} = \frac{\frac{\sin\beta + 1}{\sin\beta}}{\frac{\cos\beta + \cos\beta\sin\beta}{\sin\beta}} = \frac{\sin\beta + 1}{\cos\beta(1 + \sin\beta)} = \frac{1}{\cos\beta} = \sec\beta$$

33 $\text{RS} = \dfrac{\tan\alpha + \tan\beta}{1 - \tan\alpha\tan\beta} = \dfrac{\dfrac{\sin\alpha}{\cos\alpha} + \dfrac{\sin\beta}{\cos\beta}}{1 - \dfrac{\sin\alpha}{\cos\alpha}\cdot\dfrac{\sin\beta}{\cos\beta}}$

$= \dfrac{\dfrac{\sin\alpha\cos\beta + \cos\alpha\sin\beta}{\cos\alpha\cos\beta}}{\dfrac{\cos\alpha\cos\beta - \sin\alpha\sin\beta}{\cos\alpha\cos\beta}}$

$= \dfrac{\sin\alpha\cos\beta + \cos\alpha\sin\beta}{\cos\alpha\cos\beta - \sin\alpha\sin\beta}$

$= \text{LS}$

37 $\dfrac{1}{\tan\beta + \cot\beta} = \dfrac{1}{\dfrac{\sin\beta}{\cos\beta} + \dfrac{\cos\beta}{\sin\beta}} = \dfrac{1}{\dfrac{\sin^2\beta + \cos^2\beta}{\cos\beta\sin\beta}}$

$= \sin\beta\cos\beta$

41 $\text{RS} = \sec^4\phi - 4\tan^2\phi = (\sec^2\phi)^2 - 4\tan^2\phi$

$= (1 + \tan^2\phi)^2 - 4\tan^2\phi$

$= 1 + 2\tan^2\phi + \tan^4\phi - 4\tan^2\phi$

$= 1 - 2\tan^2\phi + \tan^4\phi$

$= (1 - \tan^2\phi)^2 = \text{LS}$

45 $\log 10^{\tan t} = \log_{10} 10^{\tan t} = \tan t$, since $\log_a a^x = x$.

49 $\ln|\sec\theta + \tan\theta| = \ln\left|\dfrac{(\sec\theta + \tan\theta)(\sec\theta - \tan\theta)}{\sec\theta - \tan\theta}\right|$

$= \ln\left|\dfrac{\sec^2\theta - \tan^2\theta}{\sec\theta - \tan\theta}\right|$

$= \ln\left|\dfrac{1}{\sec\theta - \tan\theta}\right|$

$= \ln|1| - \ln|\sec\theta - \tan\theta|$

$= -\ln|\sec\theta - \tan\theta|$

Exer. 51–62: A typical value of t or θ and the resulting nonequality are given.

51 $\pi, -1 \neq 1$ **53** $\dfrac{3\pi}{2}, 1 \neq -1$ **55** $\dfrac{\pi}{4}, 2 \neq 1$

57 $\pi, -1 \neq 1$ **59** $\dfrac{\pi}{4}, \cos\sqrt{2} \neq 1$

61 Not an identity **63** Identity

65 $a^3\cos^3\theta$ **67** $a\tan\theta\sin\theta$ **69** $a\sec\theta$

71 $\dfrac{1}{a^2}\cos^2\theta$ **73** $a\tan\theta$ **75** $a^4\sec^3\theta\tan\theta$

77 The graph of f appears to be that of $y = g(x) = -1$.

$\dfrac{\sin^2 x - \sin^4 x}{(1 - \sec^2 x)\cos^4 x} = \dfrac{\sin^2 x(1 - \sin^2 x)}{-\tan^2 x\cos^4 x}$

$= \dfrac{\sin^2 x\cos^2 x}{-(\sin^2 x/\cos^2 x)\cos^4 x}$

$= \dfrac{\sin^2 x\cos^2 x}{-\sin^2 x\cos^2 x} = -1$

79 The graph of f appears to be that of $y = g(x) = \cos x$.

$\sec x(\sin x\cos x + \cos^2 x) - \sin x$

$= \sec x\cos x(\sin x + \cos x) - \sin x$

$= (\sin x + \cos x) - \sin x = \cos x$

EXERCISES 6.2

Exer. 1–34: n denotes any integer.

1 $\dfrac{5\pi}{4} + 2\pi n, \dfrac{7\pi}{4} + 2\pi n$ **3** $\dfrac{\pi}{3} + \pi n$

5 $\dfrac{\pi}{3} + 2\pi n, \dfrac{5\pi}{3} + 2\pi n$

7 No solution, since $\dfrac{\pi}{2} > 1$.

9 All θ except $\theta = \dfrac{\pi}{2} + \pi n$

11 $\dfrac{\pi}{12} + \pi n, \dfrac{11\pi}{12} + \pi n$ **13** $\dfrac{\pi}{2} + 3\pi n$

15 $-\dfrac{\pi}{12} + 2\pi n, \dfrac{7\pi}{12} + 2\pi n$

17 $\dfrac{\pi}{4} + \pi n, \dfrac{7\pi}{12} + \pi n$ **19** $\dfrac{2\pi}{3} + 2\pi n, \dfrac{4\pi}{3} + 2\pi n$

21 $\dfrac{\pi}{4} + \dfrac{\pi}{2}n$ **23** $2\pi n, \dfrac{3\pi}{2} + 2\pi n$

25 $\dfrac{\pi}{3} + \pi n, \dfrac{2\pi}{3} + \pi n$ **27** $\dfrac{4\pi}{3} + 2\pi n, \dfrac{5\pi}{3} + 2\pi n$

29 $\dfrac{\pi}{6} + \pi n, \dfrac{5\pi}{6} + \pi n$ **31** $\dfrac{7\pi}{6} + 2\pi n, \dfrac{11\pi}{6} + 2\pi n$

33 $\dfrac{\pi}{3} + 2\pi n, \dfrac{5\pi}{3} + 2\pi n, \pi + 2\pi n$

35 $\dfrac{\pi}{12} + \pi n, \dfrac{5\pi}{12} + \pi n$ **37** $e^{(\pi/2)+\pi m}$

39 $\dfrac{3\pi}{8}, \dfrac{7\pi}{8}, \dfrac{11\pi}{8}, \dfrac{15\pi}{8}$ **41** $\dfrac{\pi}{3}, \dfrac{2\pi}{3}, \dfrac{4\pi}{3}, \dfrac{5\pi}{3}$

43 $\dfrac{\pi}{6}, \dfrac{5\pi}{6}, \dfrac{3\pi}{2}$ **45** $0, \pi, \dfrac{\pi}{4}, \dfrac{3\pi}{4}, \dfrac{5\pi}{4}, \dfrac{7\pi}{4}$

47 $\dfrac{\pi}{2}, \dfrac{3\pi}{2}, \dfrac{2\pi}{3}, \dfrac{4\pi}{3}$ **49** No solution **51** $\dfrac{11\pi}{6}, \dfrac{\pi}{2}$

53 $0, \dfrac{\pi}{2}$ **55** $\dfrac{\pi}{4}, \dfrac{5\pi}{4}$

57 All α in $[0, 2\pi)$ except $0, \dfrac{\pi}{2}, \pi$, and $\dfrac{3\pi}{2}$

59 $\dfrac{\pi}{2}, \dfrac{3\pi}{2}, \dfrac{7\pi}{6}, \dfrac{11\pi}{6}$ **61** $\dfrac{3\pi}{4}, \dfrac{7\pi}{4}$

63 15°30′, 164°30′ **65** 135°, 315°, 116°30′, 296°30′

67 41°50′, 138°10′, 194°30′, 345°30′ **69** 10

71 **(a)**

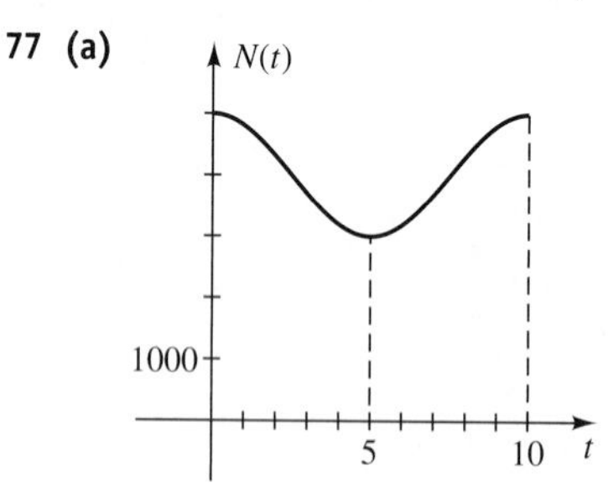

[1, 25, 5] by [0, 100, 10]

(b) July: 83°F; Oct.: 56.5°F **(c)** May through Sept.

73 $t \approx 3.50$ and $t \approx 8.50$ **75** **(a)** 3.29 **(b)** 4

77 **(a)** **(b)** $0 \le t < \frac{5}{3}$ and $\frac{25}{3} < t \le 10$

79 $A\left(-\frac{4\pi}{3}, -\frac{2\pi}{3} + \frac{1}{2}\sqrt{3}\right)$, $B\left(-\frac{2\pi}{3}, -\frac{\pi}{3} - \frac{1}{2}\sqrt{3}\right)$, $C\left(\frac{2\pi}{3}, \frac{\pi}{3} + \frac{1}{2}\sqrt{3}\right)$, $D\left(\frac{4\pi}{3}, \frac{2\pi}{3} - \frac{1}{2}\sqrt{3}\right)$

81 $\frac{7}{360}$ **83** $[0, 1.27] \cup [5.02, 2\pi]$

85 $(0.39, 1.96) \cup (2.36, 3.53) \cup (5.11, 5.50)$

87

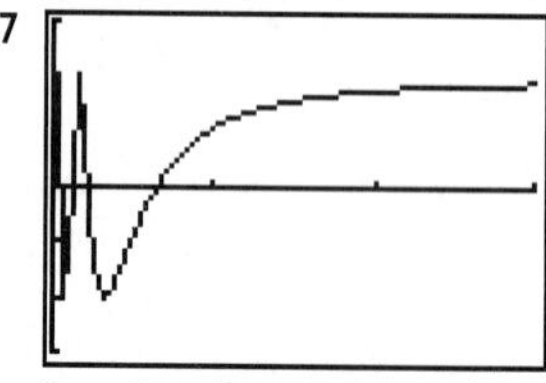

[0, 3] by [−1.5, 1.5, 0.5]

(a) 0.6366 **(b)** Approaches $y = 1$

(c) An infinite number of zeros

89 5.400 **91** 3.619 **93** −1.48, 1.08 **95** ±1.00

97 ±0.64, ±2.42 **99** **(a)** 37.6° **(b)** 52.5°

EXERCISES 6.3

1 **(a)** cos 43°23′ **(b)** sin 16°48′ **(c)** $\cot \frac{\pi}{3}$

(d) csc 72.72°

3 **(a)** $\sin \frac{3\pi}{20}$ **(b)** $\cos\left(\frac{2\pi - 1}{4}\right)$ **(c)** $\cot\left(\frac{\pi - 2}{2}\right)$

(d) $\sec\left(\frac{\pi}{2} - 0.53\right)$

5 **(a)** $\frac{\sqrt{2} + \sqrt{3}}{2}$ **(b)** $\frac{\sqrt{6} - \sqrt{2}}{4}$

7 **(a)** $\sqrt{3} + 1$ **(b)** $-2 - \sqrt{3}$

9 **(a)** $\frac{\sqrt{2} - 1}{2}$ **(b)** $\frac{\sqrt{6} + \sqrt{2}}{4}$

11 cos 25° **13** sin (−5°) **15** sin (−5)

17 $\frac{12\sqrt{3} - 5}{26}$ **19** **(a)** $\frac{77}{85}$ **(b)** $\frac{36}{85}$ **(c)** I

21 **(a)** $-\frac{24}{25}$ **(b)** $-\frac{24}{7}$ **(c)** IV

23 **(a)** $\frac{3\sqrt{21} - 8}{25} \approx 0.23$ **(b)** $\frac{4\sqrt{21} + 6}{25} \approx 0.97$ **(c)** I

25 $\sin(\theta + \pi) = \sin\theta\cos\pi + \cos\theta\sin\pi = \sin\theta(-1) + \cos\theta(0) = -\sin\theta$

27 $\sin\left(x - \frac{5\pi}{2}\right) = \sin x\cos\frac{5\pi}{2} - \cos x\sin\frac{5\pi}{2} = -\cos x$

29 $\cos(\theta - \pi) = \cos\theta\cos\pi + \sin\theta\sin\pi = -\cos\theta$

31 $\cos\left(x + \frac{3\pi}{2}\right) = \cos x\cos\frac{3\pi}{2} - \sin x\sin\frac{3\pi}{2} = \sin x$

33 $\tan\left(x - \frac{\pi}{2}\right) = \frac{\sin\left(x - \frac{\pi}{2}\right)}{\cos\left(x - \frac{\pi}{2}\right)} = \frac{\sin x\cos\frac{\pi}{2} - \cos x\sin\frac{\pi}{2}}{\cos x\cos\frac{\pi}{2} + \sin x\sin\frac{\pi}{2}} = \frac{-\cos x}{\sin x} = -\cot x$

35 $\tan\left(\theta + \frac{\pi}{2}\right) = \cot\left[\frac{\pi}{2} - \left(\theta + \frac{\pi}{2}\right)\right] = \cot(-\theta) = -\cot\theta$

37 $\sin\left(\theta + \frac{\pi}{4}\right) = \sin\theta\cos\frac{\pi}{4} + \cos\theta\sin\frac{\pi}{4} = \frac{\sqrt{2}}{2}\sin\theta + \frac{\sqrt{2}}{2}\cos\theta = \frac{\sqrt{2}}{2}(\sin\theta + \cos\theta)$

39 $\tan\left(u + \frac{\pi}{4}\right) = \frac{\tan u + \tan\frac{\pi}{4}}{1 - \tan u\tan\frac{\pi}{4}} = \frac{1 + \tan u}{1 - \tan u}$

41 $\cos(u+v) + \cos(u-v)$
$= (\cos u \cos v - \sin u \sin v) + (\cos u \cos v + \sin u \sin v)$
$= 2 \cos u \cos v$

43 $\sin(u+v) \cdot \sin(u-v)$
$= (\sin u \cos v + \cos u \sin v) \cdot (\sin u \cos v - \cos u \sin v)$
$= \sin^2 u \cos^2 v - \cos^2 u \sin^2 v$
$= \sin^2 u(1 - \sin^2 v) - (1 - \sin^2 u) \sin^2 v$
$= \sin^2 u - \sin^2 u \sin^2 v - \sin^2 v + \sin^2 u \sin^2 v$
$= \sin^2 u - \sin^2 v$

45 $\dfrac{1}{\cot\alpha - \cot\beta} = \dfrac{1}{\dfrac{\cos\alpha}{\sin\alpha} - \dfrac{\cos\beta}{\sin\beta}}$
$= \dfrac{1}{\dfrac{\cos\alpha\sin\beta - \cos\beta\sin\alpha}{\sin\alpha\sin\beta}}$
$= \dfrac{\sin\alpha\sin\beta}{\sin(\beta - \alpha)}$

47 $\sin u \cos v \cos w + \cos u \sin v \cos w + \cos u \cos v \sin w - \sin u \sin v \sin w$

49 $\cot(u+v) = \dfrac{\cos(u+v)}{\sin(u+v)}$
$= \dfrac{(\cos u \cos v - \sin u \sin v)(1/\sin u \sin v)}{(\sin u \cos v + \cos u \sin v)(1/\sin u \sin v)}$
$= \dfrac{\cot u \cot v - 1}{\cot v + \cot u}$

51 $\sin(u - v) = \sin[u + (-v)]$
$= \sin u \cos(-v) + \cos u \sin(-v)$
$= \sin u \cos v - \cos u \sin v$

53 $\dfrac{f(x+h) - f(x)}{h} = \dfrac{\cos(x+h) - \cos x}{h}$
$= \dfrac{\cos x \cos h - \sin x \sin h - \cos x}{h}$
$= \dfrac{\cos x \cos h - \cos x}{h} - \dfrac{\sin x \sin h}{h}$
$= \cos x\left(\dfrac{\cos h - 1}{h}\right) - \sin x\left(\dfrac{\sin h}{h}\right)$

55 **(a)** Each side ≈ 0.0523 **(b)** $\alpha = 60°$
(c) $\alpha = 60°, \beta = 3°$

57 $0, \dfrac{\pi}{3}, \dfrac{2\pi}{3}$ **59** $\dfrac{\pi}{6}, \dfrac{\pi}{2}, \dfrac{5\pi}{6}$

61 $\dfrac{\pi}{12}, \dfrac{5\pi}{12}; \dfrac{3\pi}{4}$ is extraneous

63 **(a)** $f(x) = 2\cos\left(2x - \dfrac{\pi}{6}\right)$ **(b)** $2, \pi, \dfrac{\pi}{12}$
(c)

65 **(a)** $f(x) = 2\sqrt{2}\cos\left(3x + \dfrac{\pi}{4}\right)$ **(b)** $2\sqrt{2}, \dfrac{2\pi}{3}, -\dfrac{\pi}{12}$
(c)

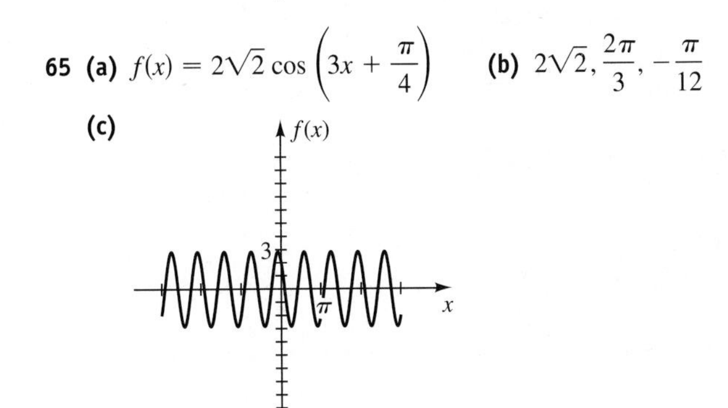

67 $y = 10\sqrt{41}\cos\left(60\pi t - \tan^{-1}\dfrac{5}{4}\right)$
$\approx 10\sqrt{41}\cos(60\pi t - 0.8961)$

69 **(a)** $y = \sqrt{13}\cos(t - C)$ with $\tan C = \dfrac{3}{2}$; $\sqrt{13}, 2\pi$
(b) $t = C + \dfrac{\pi}{2} + \pi n \approx 2.55 + \pi n$ for every nonnegative integer n

71 **(a)** $p(t) = A\sin\omega t + B\sin(\omega t + \tau)$
$= A\sin\omega t + B(\sin\omega t\cos\tau + \cos\omega t\sin\tau)$
$= (B\sin\tau)\cos\omega t + (A + B\cos\tau)\sin\omega t$
$= a\cos\omega t + b\sin\omega t$
with $a = B\sin\tau$ and $b = A + B\cos\tau$
(b) $C^2 = (B\sin\tau)^2 + (A + B\cos\tau)^2$
$= B^2\sin^2\tau + A^2 + 2AB\cos\tau + B^2\cos^2\tau$
$= A^2 + B^2(\sin^2\tau + \cos^2\tau) + 2AB\cos\tau$
$= A^2 + B^2 + 2AB\cos\tau$

73 **(a)** $C^2 = A^2 + B^2 + 2AB\cos\tau \le A^2 + B^2 + 2AB$, since $\cos\tau \le 1$ and $A > 0, B > 0$. Thus, $C^2 \le (A + B)^2$, and hence $C \le A + B$.
(b) $0, 2\pi$ **(c)** $\cos\tau > -B/(2A)$

75 $(-2.97, -2.69), (-1.00, -0.37), (0.17, 0.46), (2.14, 2.77)$

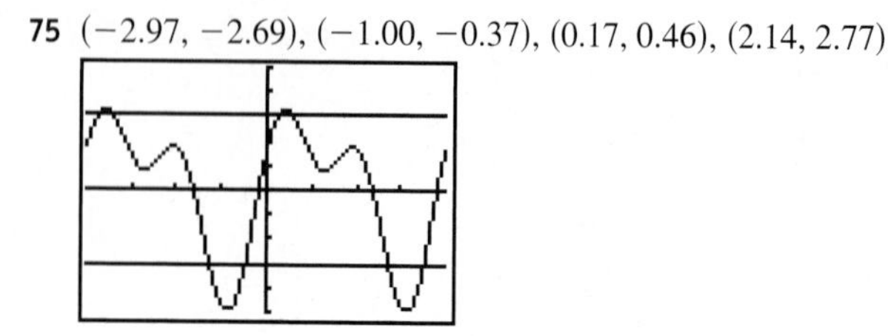

$[-3.14, 3.14, \pi/4]$ by $[-5, 5]$

EXERCISES 6.4

1 $\frac{24}{25}, -\frac{7}{25}, -\frac{24}{7}$ **3** $-\frac{4}{9}\sqrt{2}, -\frac{7}{9}, \frac{4}{7}\sqrt{2}$

5 $\frac{1}{10}\sqrt{10}, \frac{3}{10}\sqrt{10}, \frac{1}{3}$

7 $-\frac{1}{2}\sqrt{2+\sqrt{2}}, \frac{1}{2}\sqrt{2-\sqrt{2}}, -\sqrt{2}-1$

9 **(a)** $\frac{1}{2}\sqrt{2-\sqrt{2}}$ **(b)** $\frac{1}{2}\sqrt{2-\sqrt{3}}$ **(c)** $\sqrt{2}+1$

11 $\sin 10\theta = \sin(2 \cdot 5\theta) = 2\sin 5\theta \cos 5\theta$

13 $4\sin\frac{x}{2}\cos\frac{x}{2} = 2 \cdot 2\sin\frac{x}{2}\cos\frac{x}{2} = 2\sin\left(2 \cdot \frac{x}{2}\right) = 2\sin x$

15 $(\sin t + \cos t)^2 = \sin^2 t + 2\sin t\cos t + \cos^2 t = 1 + \sin 2t$

17 $\sin 3u = \sin(2u + u)$
$= \sin 2u\cos u + \cos 2u\sin u$
$= (2\sin u\cos u)\cos u + (1 - 2\sin^2 u)\sin u$
$= 2\sin u\cos^2 u + \sin u - 2\sin^3 u$
$= 2\sin u(1 - \sin^2 u) + \sin u - 2\sin^3 u$
$= 2\sin u - 2\sin^3 u + \sin u - 2\sin^3 u$
$= 3\sin u - 4\sin^3 u = \sin u(3 - 4\sin^2 u)$

19 $\cos 4\theta = \cos(2 \cdot 2\theta) = 2\cos^2 2\theta - 1$
$= 2(2\cos^2\theta - 1)^2 - 1$
$= 2(4\cos^4\theta - 4\cos^2\theta + 1) - 1$
$= 8\cos^4\theta - 8\cos^2\theta + 1$

21 $\sin^4 t = (\sin^2 t)^2 = \left(\frac{1-\cos 2t}{2}\right)^2$
$= \frac{1}{4}(1 - 2\cos 2t + \cos^2 2t)$
$= \frac{1}{4} - \frac{1}{2}\cos 2t + \frac{1}{4}\left(\frac{1+\cos 4t}{2}\right)$
$= \frac{1}{4} - \frac{1}{2}\cos 2t + \frac{1}{8} + \frac{1}{8}\cos 4t$
$= \frac{3}{8} - \frac{1}{2}\cos 2t + \frac{1}{8}\cos 4t$

23 $\sec 2\theta = \frac{1}{\cos 2\theta} = \frac{1}{2\cos^2\theta - 1} = \frac{1}{2\left(\frac{1}{\sec^2\theta}\right) - 1}$
$= \frac{1}{\frac{2-\sec^2\theta}{\sec^2\theta}} = \frac{\sec^2\theta}{2-\sec^2\theta}$

25 $2\sin^2 2t + \cos 4t = 2\sin^2 2t + \cos(2 \cdot 2t)$
$= 2\sin^2 2t + (1 - 2\sin^2 2t) = 1$

27 $\tan 3u = \tan(2u + u) = \frac{\tan 2u + \tan u}{1 - \tan 2u\tan u}$
$= \frac{\frac{2\tan u}{1-\tan^2 u} + \tan u}{1 - \frac{2\tan u}{1-\tan^2 u}\cdot\tan u}$
$= \frac{\frac{2\tan u + \tan u - \tan^3 u}{1-\tan^2 u}}{\frac{1 - \tan^2 u - 2\tan^2 u}{1-\tan^2 u}}$
$= \frac{3\tan u - \tan^3 u}{1 - 3\tan^2 u} = \frac{\tan u(3 - \tan^2 u)}{1 - 3\tan^2 u}$

29 $\tan\frac{\theta}{2} = \frac{1-\cos\theta}{\sin\theta} = \frac{1}{\sin\theta} - \frac{\cos\theta}{\sin\theta} = \csc\theta - \cot\theta$

31 $\frac{3}{8} + \frac{1}{2}\cos\theta + \frac{1}{8}\cos 2\theta$

33 $\frac{3}{8} - \frac{1}{2}\cos 4x + \frac{1}{8}\cos 8x$ **35** $0, \pi, \frac{2\pi}{3}, \frac{4\pi}{3}$

37 $\frac{\pi}{3}, \frac{5\pi}{3}, \pi$ **39** $0, \pi$ **41** $0, \frac{\pi}{3}, \frac{5\pi}{3}$

45 **(a)** $1.20, 5.09$
(b) $P\left(\frac{2\pi}{3}, -1.5\right), Q(\pi, -1), R\left(\frac{4\pi}{3}, -1.5\right)$

47 **(a)** $-\frac{3\pi}{2}, -\frac{\pi}{2}, \frac{\pi}{2}, \frac{3\pi}{2}$
(b) $0, \pm\pi, \pm 2\pi, \pm\frac{\pi}{4}, \pm\frac{3\pi}{4}, \pm\frac{5\pi}{4}, \pm\frac{7\pi}{4}$

49 **(b)** Yes, point B is 25 miles from A.

51 **(a)** $V = \frac{5}{2}\sin\theta$ **(b)** $53.13°$ **53** **(b)** 12.43 mm

55 The graph of f appears to be that of $y = g(x) = \tan x$.
$\frac{\sin 2x + \sin x}{\cos 2x + \cos x + 1} = \frac{2\sin x\cos x + \sin x}{(2\cos^2 x - 1) + \cos x + 1}$
$= \frac{\sin x(2\cos x + 1)}{\cos x(2\cos x + 1)} = \frac{\sin x}{\cos x} = \tan x$

57 $-3.55, 5.22$

59 $-2.03, -0.72, 0.58, 2.62$ **61** -2.59

EXERCISES 6.5

1 $\frac{1}{2}\cos 4t - \frac{1}{2}\cos 10t$ **3** $\frac{1}{2}\cos 2u + \frac{1}{2}\cos 10u$

5 $\sin 12\theta + \sin 6\theta$ **7** $\frac{3}{2}\sin 3x + \frac{3}{2}\sin x$

9 $2\sin 4\theta \cos 2\theta$ **11** $-2\sin 4x \sin x$

13 $-2\cos 5t \sin 2t$ **15** $2\cos\frac{3}{2}x\cos\frac{1}{2}x$

17 $\frac{\sin 4t + \sin 6t}{\cos 4t - \cos 6t} = \frac{2\sin 5t\cos t}{2\sin 5t\sin t} = \cot t$

19 $\frac{\sin u + \sin v}{\cos u + \cos v} = \frac{2\sin\frac{1}{2}(u+v)\cos\frac{1}{2}(u-v)}{2\cos\frac{1}{2}(u+v)\cos\frac{1}{2}(u-v)}$
$= \tan\frac{1}{2}(u+v)$

21 $\frac{\sin u - \sin v}{\sin u + \sin v} = \frac{2\cos\frac{1}{2}(u+v)\sin\frac{1}{2}(u-v)}{2\sin\frac{1}{2}(u+v)\cos\frac{1}{2}(u-v)}$
$= \cot\frac{1}{2}(u+v)\tan\frac{1}{2}(u-v)$
$= \frac{\tan\frac{1}{2}(u-v)}{\tan\frac{1}{2}(u+v)}$

23 $4\cos x\cos 2x\sin 3x = 2\cos 2x\,(2\sin 3x\cos x)$
$= 2\cos 2x\,(\sin 4x + \sin 2x)$
$= (2\cos 2x\sin 4x) + (2\cos 2x\sin 2x)$
$= [\sin 6x - \sin(-2x)] + (\sin 4x - \sin 0)$
$= \sin 2x + \sin 4x + \sin 6x$

25 $\frac{1}{2}\sin[(a+b)x] + \frac{1}{2}\sin[(a-b)x]$ **27** $\frac{\pi}{4}n$

29 $\frac{\pi}{2}n$ **31** $\frac{\pi}{2} + \pi n, \frac{\pi}{12} + \frac{\pi}{2}n, \frac{5\pi}{12} + \frac{\pi}{2}n$

33 $\frac{\pi}{7} + \frac{2\pi}{7}n, \frac{2\pi}{3}n$ **35** $\frac{\pi}{4}, \frac{3\pi}{4}, \frac{5\pi}{4}, \frac{7\pi}{4}, \frac{\pi}{2}, \frac{3\pi}{2}$

37 $0, \pm\pi, \pm 2\pi, \pm\frac{\pi}{4}, \pm\frac{3\pi}{4}, \pm\frac{5\pi}{4}, \pm\frac{7\pi}{4}$

39 $f(x) = \frac{1}{2}\sin\frac{\pi n}{l}(x+kt) + \frac{1}{2}\sin\frac{\pi n}{l}(x-kt)$

41 (a) $0, \pm 1.05, \pm 1.57, \pm 2.09, \pm 3.14$
(b) $0, \pm\frac{\pi}{3}, \pm\frac{\pi}{2}, \pm\frac{2\pi}{3}, \pm\pi$

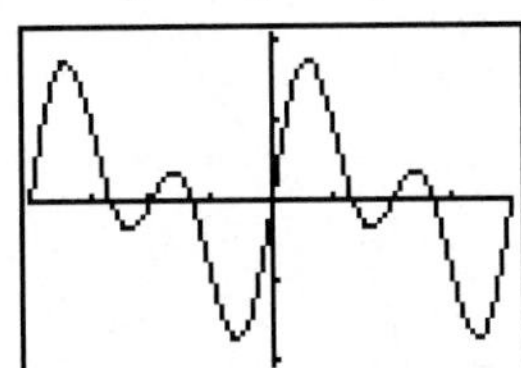

$[-3.14, 3.14, \pi/4]$ by $[-2.09, 2.09]$

43 The graph of f appears to be that of $y = g(x) = \tan 2x$.
$\frac{\sin x + \sin 2x + \sin 3x}{\cos x + \cos 2x + \cos 3x} = \frac{\sin 2x + (\sin 3x + \sin x)}{\cos 2x + (\cos 3x + \cos x)}$
$= \frac{\sin 2x + 2\sin 2x\cos x}{\cos 2x + 2\cos 2x\cos x}$
$= \frac{\sin 2x(1 + 2\cos x)}{\cos 2x(1 + 2\cos x)}$
$= \frac{\sin 2x}{\cos 2x} = \tan 2x$

EXERCISES 6.6

1 (a) $-\frac{\pi}{4}$ **(b)** $\frac{2\pi}{3}$ **(c)** $-\frac{\pi}{3}$

3 (a) $\frac{\pi}{3}$ **(b)** $\frac{\pi}{4}$ **(c)** $\frac{\pi}{6}$

5 (a) Not defined **(b)** Not defined **(c)** $\frac{\pi}{4}$

7 (a) $-\frac{3}{10}$ **(b)** $\frac{1}{2}$ **(c)** 14

9 (a) $\frac{\pi}{3}$ **(b)** $\frac{5\pi}{6}$ **(c)** $-\frac{\pi}{6}$

11 (a) $-\frac{\pi}{4}$ **(b)** $\frac{3\pi}{4}$ **(c)** $-\frac{\pi}{4}$

13 (a) $\frac{\sqrt{3}}{2}$ **(b)** $\frac{\sqrt{2}}{2}$ **(c)** Not defined

15 (a) $\frac{\sqrt{5}}{2}$ **(b)** $\frac{\sqrt{34}}{5}$ **(c)** $\frac{4}{\sqrt{15}}$

17 (a) $\frac{\sqrt{3}}{2}$ **(b)** 0 **(c)** $-\frac{77}{36}$

19 (a) $-\frac{24}{25}$ **(b)** $-\frac{161}{289}$ **(c)** $\frac{24}{7}$

21 (a) $-\frac{1}{10}\sqrt{2}$ **(b)** $\frac{4}{17}\sqrt{17}$ **(c)** $\frac{1}{2}$

23 $\frac{x}{\sqrt{x^2+1}}$ **25** $\frac{\sqrt{x^2+4}}{2}$ **27** $2x\sqrt{1-x^2}$

29 $\sqrt{\frac{1+x}{2}}$ **31 (a)** $-\frac{\pi}{2}$ **(b)** 0 **(c)** $\frac{\pi}{2}$

33

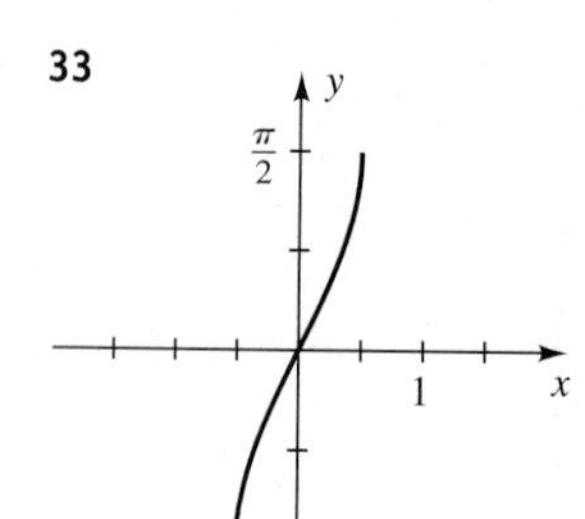

35

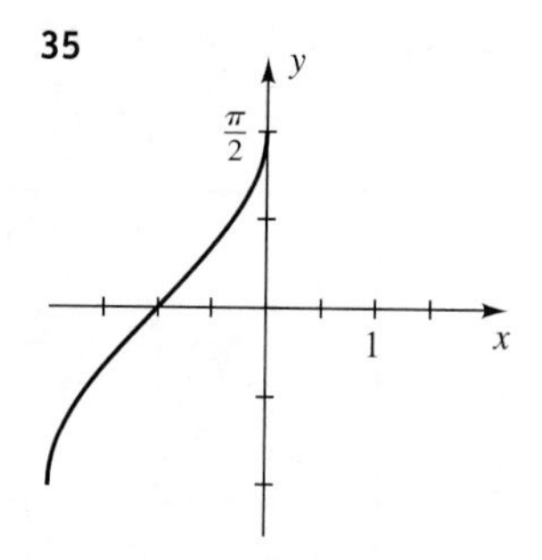

37

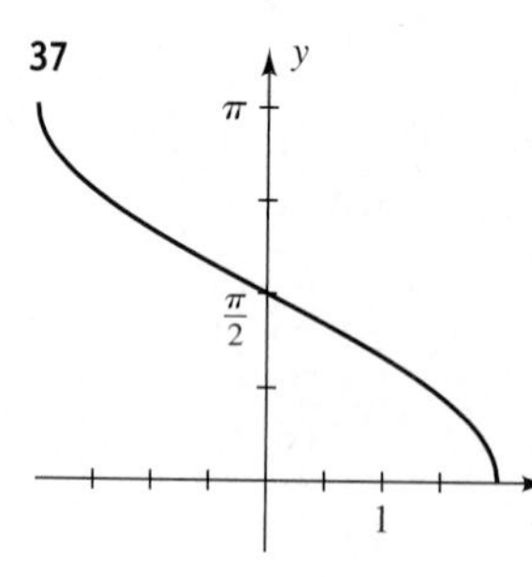

39

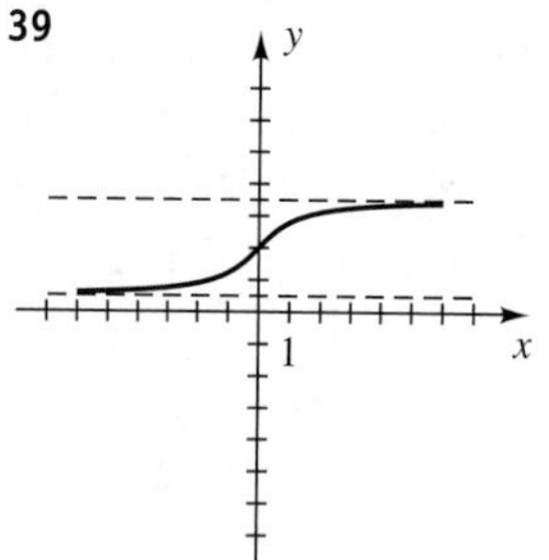

41

y 1 x

43 **(a)** $2 \le x \le 4$ **(b)** $-\frac{\pi}{4} \le y \le \frac{\pi}{4}$ **(c)** $x = \sin 2y + 3$

45 **(a)** $-\frac{3}{2} \le x \le \frac{3}{2}$ **(b)** $0 \le y \le 4\pi$ **(c)** $x = \frac{3}{2}\cos\frac{1}{4}y$

47 $x = \sin^{-1}(-y - 3)$ **49** $x = \cos^{-1}\left[\frac{1}{2}(15 - y)\right]$

51 $x = x_R$ or $x = \pi - x_R$, where $x_R = \sin^{-1}\left(\frac{3}{4}\sin y\right)$

53 $\cos^{-1}(-1 + \sqrt{2}) \approx 1.1437$, $2\pi - \cos^{-1}(-1 + \sqrt{2}) \approx 5.1395$

55 $\tan^{-1}\frac{1}{4}(-9 + \sqrt{57}) \approx -0.3478$, $\tan^{-1}\frac{1}{4}(-9 - \sqrt{57}) \approx -1.3337$

57 $\cos^{-1}\frac{1}{5}\sqrt{15} \approx 0.6847$, $\cos^{-1}\left(-\frac{1}{5}\sqrt{15}\right) \approx 2.4569$, $\cos^{-1}\frac{1}{3}\sqrt{3} \approx 0.9553$, $\cos^{-1}\left(-\frac{1}{3}\sqrt{3}\right) \approx 2.1863$

59 $\sin^{-1}\left(\pm\frac{1}{6}\sqrt{30}\right) \approx \pm 1.1503$

61 $\cos^{-1}\left(-\frac{3}{5}\right) \approx 2.2143$, $\cos^{-1}\frac{1}{3} \approx 1.2310$, $2\pi - \cos^{-1}\left(-\frac{3}{5}\right) \approx 4.0689$, $2\pi - \cos^{-1}\frac{1}{3} \approx 5.0522$

63 $\cos^{-1}\frac{2}{3} \approx 0.8411$, $2\pi - \cos^{-1}\frac{2}{3} \approx 5.4421$, $\frac{\pi}{3} \approx 1.0472$, $\frac{5\pi}{3} \approx 5.2360$

65 **(a)** 1.65 m **(b)** 0.92 m **(c)** 0.43 m **67** 3.07°

69 **(a)** $\alpha = \theta - \sin^{-1}\frac{d}{k}$ **(b)** 40°

71 Let $\alpha = \sin^{-1} x$ and $\beta = \tan^{-1}\frac{x}{\sqrt{1 + x^2}}$ with $-\frac{\pi}{2} < \alpha < \frac{\pi}{2}$ and $-\frac{\pi}{2} < \beta < \frac{\pi}{2}$. Thus, $\sin\alpha = x$ and $\sin\beta = x$. Since the sine function is one-to-one on $\left(-\frac{\pi}{2}, \frac{\pi}{2}\right)$, we have $\alpha = \beta$.

73 Let $\alpha = \arcsin(-x)$ and $\beta = \arcsin x$ with $-\frac{\pi}{2} \le \alpha \le \frac{\pi}{2}$ and $-\frac{\pi}{2} \le \beta \le \frac{\pi}{2}$. Thus, $\sin\alpha = -x$ and $\sin\beta = x$. Consequently, $\sin\alpha = -\sin\beta = \sin(-\beta)$. Since the sine function is one-to-one on $\left[-\frac{\pi}{2}, \frac{\pi}{2}\right]$, we have $\alpha = -\beta$.

75 Let $\alpha = \arctan x$ and $\beta = \arctan(1/x)$. Since $x > 0$, we have $0 < \alpha < \frac{\pi}{2}$ and $0 < \beta < \frac{\pi}{2}$, and hence $0 < \alpha + \beta < \pi$. Thus,

$$\tan(\alpha + \beta) = \frac{\tan\alpha + \tan\beta}{1 - \tan\alpha\tan\beta} = \frac{x + (1/x)}{1 - x \cdot (1/x)} = \frac{x + (1/x)}{0}.$$

Since the denominator is 0, $\tan(\alpha + \beta)$ is undefined and hence $\alpha + \beta = \frac{\pi}{2}$.

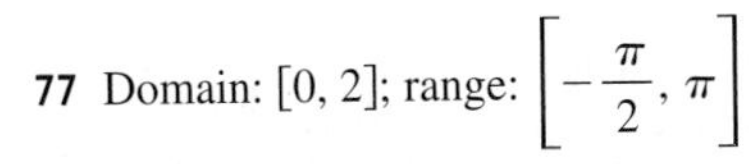

77 Domain: $[0, 2]$; range: $\left[-\frac{\pi}{2}, \pi\right]$

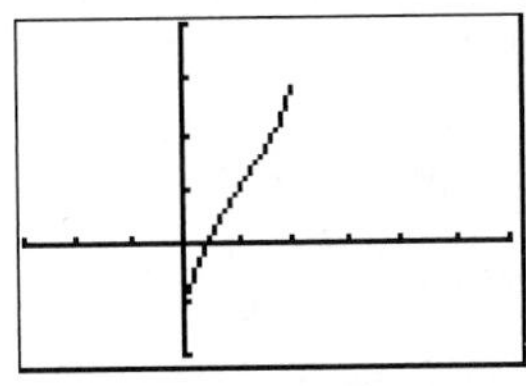

$[-3, 6]$ by $[-2, 4]$

79 0.29

81 $\theta \approx 1.25 \approx 72^\circ$

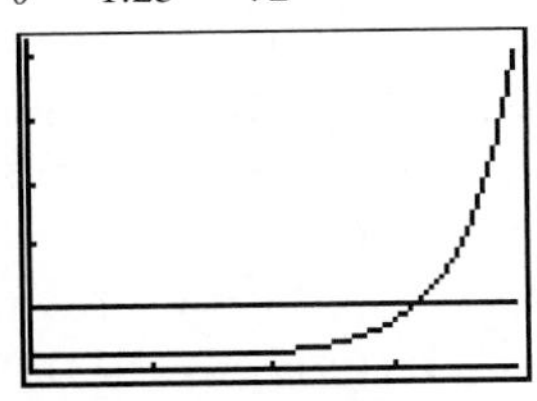

$[0, 1.57, \pi/8]$ by $[0, 1.05, 0.2]$

83 $\tan^{-1} 1 = 45^\circ$ **85** $\tan^{-1} \frac{1}{2} \approx 26.6^\circ$

CHAPTER 6 REVIEW EXERCISES

1 $(\cot^2 x + 1)(1 - \cos^2 x) = (\csc^2 x)(\sin^2 x) = 1$

2 $$\cos\theta + \sin\theta\tan\theta = \cos\theta + \sin\theta \cdot \frac{\sin\theta}{\cos\theta} = \frac{\cos^2\theta + \sin^2\theta}{\cos\theta} = \frac{1}{\cos\theta} = \sec\theta$$

3 $$\frac{(\sec^2\theta - 1)\cot\theta}{\tan\theta\sin\theta + \cos\theta} = \frac{(\tan^2\theta)\cot\theta}{\frac{\sin\theta}{\cos\theta}\cdot\sin\theta + \cos\theta} = \frac{\tan\theta}{\frac{\sin^2\theta + \cos^2\theta}{\cos\theta}} = \frac{\sin\theta/\cos\theta}{1/\cos\theta} = \sin\theta$$

4 $$(\tan x + \cot x)^2 = \left(\frac{\sin x}{\cos x} + \frac{\cos x}{\sin x}\right)^2 = \left(\frac{\sin^2 x + \cos^2 x}{\cos x \sin x}\right)^2 = \frac{1}{\cos^2 x \sin^2 x} = \sec^2 x \csc^2 x$$

5 $$\frac{1}{1 + \sin t}\cdot\frac{1 - \sin t}{1 - \sin t} = \frac{1 - \sin t}{1 - \sin^2 t} = \frac{1 - \sin t}{\cos^2 t} = \frac{1 - \sin t}{\cos t}\cdot\frac{1}{\cos t} = \left(\frac{1}{\cos t} - \frac{\sin t}{\cos t}\right)\cdot\sec t = (\sec t - \tan t)\sec t$$

6 $$\frac{\sin(\alpha - \beta)}{\cos(\alpha + \beta)} = \frac{(\sin\alpha\cos\beta - \cos\alpha\sin\beta)/\cos\alpha\cos\beta}{(\cos\alpha\cos\beta - \sin\alpha\sin\beta)/\cos\alpha\cos\beta} = \frac{\tan\alpha - \tan\beta}{1 - \tan\alpha\tan\beta}$$

7 $$\tan 2u = \frac{2\tan u}{1 - \tan^2 u} = \frac{2\cdot\frac{1}{\cot u}}{1 - \frac{1}{\cot^2 u}} = \frac{\frac{2}{\cot u}}{\frac{\cot^2 u - 1}{\cot^2 u}} = \frac{2\cot u}{\cot^2 u - 1} = \frac{2\cot u}{(\csc^2 u - 1) - 1} = \frac{2\cot u}{\csc^2 u - 2}$$

8 $$\cos^2\frac{v}{2} = \frac{1 + \cos v}{2} = \frac{1 + \frac{1}{\sec v}}{2} = \frac{\frac{\sec v + 1}{\sec v}}{2} = \frac{1 + \sec v}{2\sec v}$$

9 $$\frac{\tan^3\phi - \cot^3\phi}{\tan^2\phi + \csc^2\phi} = \frac{(\tan\phi - \cot\phi)[(\tan^2\phi + \tan\phi\cot\phi + \cot^2\phi)]}{[\tan^2\phi + (1 + \cot^2\phi)]} = \tan\phi - \cot\phi$$

10 $$\text{LS} = \frac{\sin u + \sin v}{\csc u + \csc v} = \frac{\sin u + \sin v}{\frac{1}{\sin u} + \frac{1}{\sin v}} = \frac{\sin u + \sin v}{\frac{\sin v + \sin u}{\sin u \sin v}} = \sin u \sin v$$

$$\text{RS} = \frac{1 - \sin u \sin v}{-1 + \csc u \csc v} = \frac{1 - \sin u \sin v}{-1 + \frac{1}{\sin u \sin v}} = \frac{1 - \sin u \sin v}{\frac{1 - \sin u \sin v}{\sin u \sin v}} = \sin u \sin v$$

Since the LS and RS equal the same expression and the steps are reversible, the identity is verified.

11 $$\left(\frac{\sin^2 x}{\tan^4 x}\right)^3\left(\frac{\csc^3 x}{\cot^6 x}\right)^2 = \left(\frac{\sin^6 x}{\tan^{12} x}\right)\left(\frac{\csc^6 x}{\cot^{12} x}\right) = \frac{(\sin x \csc x)^6}{(\tan x \cot x)^{12}} = \frac{(1)^6}{(1)^{12}} = 1$$

12 $\dfrac{\cos\gamma}{1-\tan\gamma}+\dfrac{\sin\gamma}{1-\cot\gamma}=\dfrac{\cos\gamma}{\dfrac{\cos\gamma-\sin\gamma}{\cos\gamma}}+\dfrac{\sin\gamma}{\dfrac{\sin\gamma-\cos\gamma}{\sin\gamma}}$

$=\dfrac{\cos^2\gamma}{\cos\gamma-\sin\gamma}+\dfrac{\sin^2\gamma}{\sin\gamma-\cos\gamma}$

$=\dfrac{\cos^2\gamma-\sin^2\gamma}{\cos\gamma-\sin\gamma}$

$=\dfrac{(\cos\gamma+\sin\gamma)(\cos\gamma-\sin\gamma)}{\cos\gamma-\sin\gamma}$

$=\cos\gamma+\sin\gamma$

13 $\dfrac{\cos(-t)}{\sec(-t)+\tan(-t)}=\dfrac{\cos t}{\sec t-\tan t}=\dfrac{\cos t}{\dfrac{1}{\cos t}-\dfrac{\sin t}{\cos t}}$

$=\dfrac{\cos t}{\dfrac{1-\sin t}{\cos t}}=\dfrac{\cos^2 t}{1-\sin t}=\dfrac{1-\sin^2 t}{1-\sin t}$

$=\dfrac{(1-\sin t)(1+\sin t)}{1-\sin t}=1+\sin t$

14 $\dfrac{\cot(-t)+\csc(-t)}{\sin(-t)}=\dfrac{-\cot t-\csc t}{-\sin t}=\dfrac{\dfrac{\cos t}{\sin t}+\dfrac{1}{\sin t}}{\sin t}$

$=\dfrac{\cos t+1}{\sin^2 t}=\dfrac{\cos t+1}{1-\cos^2 t}$

$=\dfrac{\cos t+1}{(1-\cos t)(1+\cos t)}=\dfrac{1}{1-\cos t}$

15 $\sqrt{\dfrac{1-\cos t}{1+\cos t}}=\sqrt{\dfrac{(1-\cos t)}{(1+\cos t)}\cdot\dfrac{(1-\cos t)}{(1-\cos t)}}$

$=\sqrt{\dfrac{(1-\cos t)^2}{1-\cos^2 t}}$

$=\sqrt{\dfrac{(1-\cos t)^2}{\sin^2 t}}=\dfrac{|1-\cos t|}{|\sin t|}=\dfrac{1-\cos t}{|\sin t|}$,

since $(1-\cos t)\geq 0$.

16 $\sqrt{\dfrac{1-\sin\theta}{1+\sin\theta}}=\sqrt{\dfrac{(1-\sin\theta)}{(1+\sin\theta)}\cdot\dfrac{(1+\sin\theta)}{(1+\sin\theta)}}$

$=\sqrt{\dfrac{1-\sin^2\theta}{(1+\sin\theta)^2}}$

$=\sqrt{\dfrac{\cos^2\theta}{(1+\sin\theta)^2}}$

$=\dfrac{|\cos\theta|}{|1+\sin\theta|}=\dfrac{|\cos\theta|}{1+\sin\theta}$,

since $(1+\sin\theta)\geq 0$.

17 $\cos\left(x-\dfrac{5\pi}{2}\right)=\cos x\cos\dfrac{5\pi}{2}+\sin x\sin\dfrac{5\pi}{2}=\sin x$

18 $\tan\left(x+\dfrac{3\pi}{4}\right)=\dfrac{\tan x+\tan\dfrac{3\pi}{4}}{1-\tan x\tan\dfrac{3\pi}{4}}=\dfrac{\tan x-1}{1+\tan x}$

19 $\dfrac{1}{4}\sin 4\beta=\dfrac{1}{4}\sin(2\cdot 2\beta)=\dfrac{1}{4}(2\sin 2\beta\cos 2\beta)$

$=\dfrac{1}{2}(2\sin\beta\cos\beta)(\cos^2\beta-\sin^2\beta)$

$=\sin\beta\cos^3\beta-\cos\beta\sin^3\beta$

20 $\tan\dfrac{1}{2}\theta=\dfrac{1-\cos\theta}{\sin\theta}=\dfrac{1}{\sin\theta}-\dfrac{\cos\theta}{\sin\theta}=\csc\theta-\cot\theta$

21 $\sin 8\theta=2\sin 4\theta\cos 4\theta$

$=2(2\sin 2\theta\cos 2\theta)(1-2\sin^2 2\theta)$

$=8\sin\theta\cos\theta(1-2\sin^2\theta)[1-2(2\sin\theta\cos\theta)^2]$

$=8\sin\theta\cos\theta(1-2\sin^2\theta)(1-8\sin^2\theta\cos^2\theta)$

22 Let $\alpha=\arctan x$ and $\beta=\arctan\dfrac{2x}{1-x^2}$. Because $-1<x<1$, $-\dfrac{\pi}{4}<\alpha<\dfrac{\pi}{4}$. Thus, $\tan\alpha=x$ and $\tan\beta=\dfrac{2x}{1-x^2}=\dfrac{2\tan\alpha}{1-\tan^2\alpha}=\tan 2\alpha$. Since the tangent function is one-to-one on $\left(-\dfrac{\pi}{2},\dfrac{\pi}{2}\right)$, we have $\beta=2\alpha$ or, equivalently, $\alpha=\dfrac{1}{2}\beta$.

23 $\dfrac{\pi}{2},\dfrac{3\pi}{2},\dfrac{\pi}{4},\dfrac{7\pi}{4},\dfrac{3\pi}{4},\dfrac{5\pi}{4}$ **24** $\dfrac{7\pi}{6},\dfrac{11\pi}{6}$ **25** $0,\pi$

26 $\dfrac{\pi}{4},\dfrac{3\pi}{4},\dfrac{5\pi}{4},\dfrac{7\pi}{4}$ **27** $0,\pi,\dfrac{2\pi}{3},\dfrac{4\pi}{3}$

28 $\dfrac{\pi}{2},\dfrac{3\pi}{2},\dfrac{\pi}{4},\dfrac{5\pi}{4},\dfrac{3\pi}{4},\dfrac{7\pi}{4}$ **29** $\dfrac{7\pi}{6},\dfrac{11\pi}{6},\dfrac{\pi}{2}$

30 $\dfrac{2\pi}{3},\dfrac{4\pi}{3},\pi$ **31** $\dfrac{\pi}{6},\dfrac{5\pi}{6},\dfrac{\pi}{3},\dfrac{5\pi}{3}$

32 All x in $[0,2\pi)$ except $\dfrac{\pi}{4},\dfrac{3\pi}{4},\dfrac{5\pi}{4},\dfrac{7\pi}{4}$

33 $\dfrac{\pi}{3},\dfrac{5\pi}{3}$ **34** $0,\dfrac{\pi}{3},\dfrac{2\pi}{3},\pi,\dfrac{4\pi}{3},\dfrac{5\pi}{3}$

35 $\dfrac{3}{4},\dfrac{7}{4},\dfrac{11}{4},\dfrac{15}{4},\dfrac{19}{4},\dfrac{23}{4}$ **36** $0,\pi,\dfrac{\pi}{3},\dfrac{5\pi}{3}$

37 $\dfrac{\pi}{3},\dfrac{5\pi}{3}$ **38** $\dfrac{\pi}{6},\dfrac{5\pi}{6},\dfrac{7\pi}{6},\dfrac{11\pi}{6}$

39 $0,\dfrac{\pi}{8},\dfrac{3\pi}{8},\dfrac{5\pi}{8},\dfrac{7\pi}{8},\pi,\dfrac{9\pi}{8},\dfrac{11\pi}{8},\dfrac{13\pi}{8},\dfrac{15\pi}{8}$

40 $\frac{\pi}{5}, \frac{3\pi}{5}, \pi, \frac{7\pi}{5}, \frac{9\pi}{5}$ **41** $\frac{\sqrt{6}-\sqrt{2}}{4}$

42 $-2-\sqrt{3}$ **43** $\frac{\sqrt{2}-\sqrt{6}}{4}$ **44** $\frac{2}{\sqrt{2-\sqrt{2}}}$

45 $\frac{84}{85}$ **46** $-\frac{13}{85}$ **47** $-\frac{84}{13}$ **48** $-\frac{36}{77}$

49 $\frac{36}{85}$ **50** $-\frac{36}{85}$ **51** $\frac{240}{289}$ **52** $-\frac{161}{289}$

53 $\frac{24}{7}$ **54** $\frac{1}{10}\sqrt{10}$ **55** $\frac{1}{3}$ **56** $\frac{5}{34}\sqrt{34}$

57 **(a)** $\frac{1}{2}\cos 3t - \frac{1}{2}\cos 11t$

(b) $\frac{1}{2}\cos\frac{1}{12}u + \frac{1}{2}\cos\frac{5}{12}u$

(c) $3\sin 8x - 3\sin 2x$ **(d)** $2\sin 10\theta - 2\sin 4\theta$

58 **(a)** $2\sin 5u \cos 3u$ **(b)** $2\sin\frac{11}{2}\theta\sin\frac{5}{2}\theta$

(c) $2\cos\frac{9}{40}t\sin\frac{1}{40}t$ **(d)** $6\cos 4x\cos 2x$

59 $\frac{\pi}{6}$ **60** $\frac{\pi}{4}$ **61** $\frac{\pi}{3}$ **62** π **63** $-\frac{\pi}{4}$

64 $\frac{3\pi}{4}$ **65** $\frac{1}{2}$ **66** 2 **67** Not defined **68** $\frac{\pi}{2}$

69 $\frac{240}{289}$ **70** $-\frac{7}{25}$

71

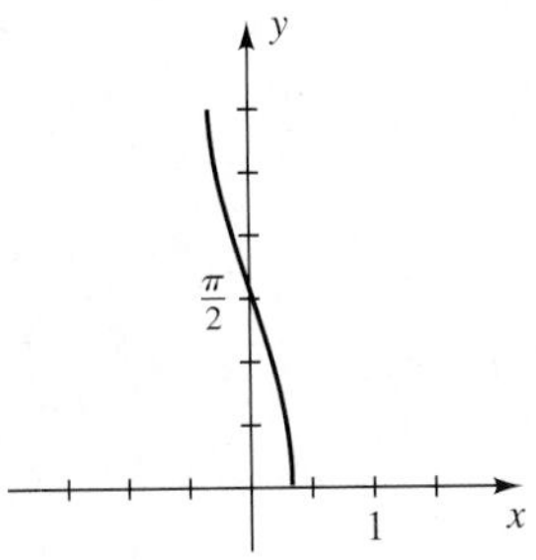

72

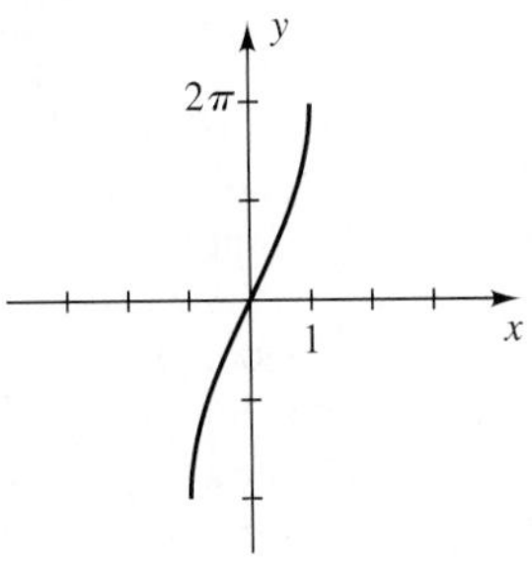

73

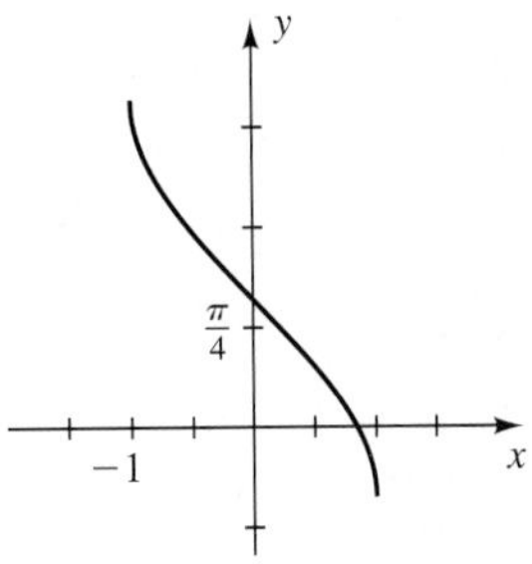

74

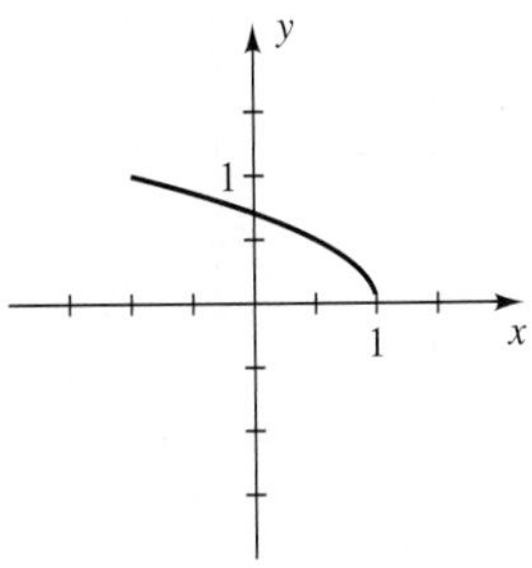

75 $\cos(\alpha+\beta+\gamma) = \cos[(\alpha+\beta)+\gamma]$
$= \cos(\alpha+\beta)\cos\gamma - \sin(\alpha+\beta)\sin\gamma$
$= (\cos\alpha\cos\beta - \sin\alpha\sin\beta)\cos\gamma -$
$(\sin\alpha\cos\beta + \cos\alpha\sin\beta)\sin\gamma$
$= \cos\alpha\cos\beta\cos\gamma - \sin\alpha\sin\beta\cos\gamma -$
$\sin\alpha\cos\beta\sin\gamma - \cos\alpha\sin\beta\sin\gamma$

76 **(b)** $t = 0, \pm\frac{\pi}{4b}$ **(c)** $\frac{2}{3}\sqrt{2}A$

77 $\pm\frac{\pi}{4}, \pm\frac{3\pi}{4}, \pm\frac{5\pi}{4}, \pm\frac{7\pi}{4}, \pm\frac{\pi}{3}, \pm\frac{5\pi}{3}$

78 **(a)** $x = 2d\tan\frac{1}{2}\theta$ **(b)** $d \le 1000$ ft

79 **(a)** $d = r\left(\sec\frac{1}{2}\theta - 1\right)$ **(b)** 43°

80 **(a)** 78.7° **(b)** 61.4°

CHAPTER 6 DISCUSSION EXERCISES

1 *Hint:* Factor $\sin^3 x - \cos^3 x$ as the difference of cubes.

2 $\sqrt{a^2-x^2}$
$$= \begin{cases} a\cos\theta & \text{if } 0 \le \theta \le \pi/2 \text{ or } 3\pi/2 \le \theta < 2\pi \\ -a\cos\theta & \text{if } \pi/2 < \theta < 3\pi/2 \end{cases}$$

3 45; approximately 6.164

4 The difference quotient for the sine function appears to be the cosine function.

5 *Hint:* Write the equation in the form $\frac{\pi}{4} + \alpha = 4\theta$, and take the tangent of both sides.

6 **(a)** The **inverse sawtooth function,** denoted by saw^{-1}, is defined by $y = \text{saw}^{-1} x$ iff $x = \text{saw}\, y$ for $-2 \le x \le 2$ and $-1 \le y \le 1$.
(b) 0.85; -0.4
(c) $\text{saw}(\text{saw}^{-1} x) = x$ if $-2 \le x \le 2$;
$\text{saw}^{-1}(\text{saw}\, y) = y$ if $-1 \le y \le 1$

(d)

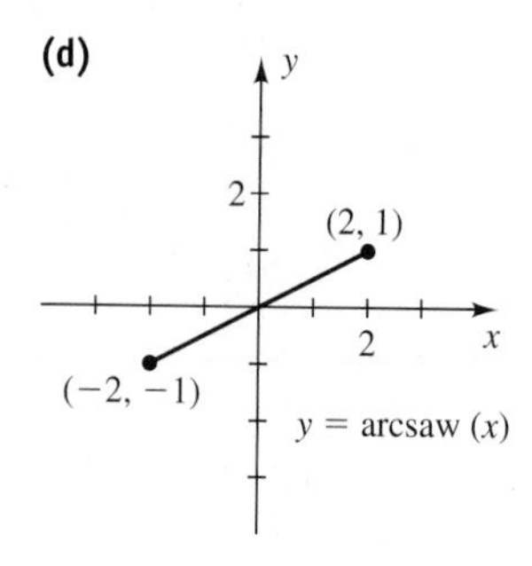

Chapter 7

EXERCISES 7.1

1 $\beta = 62°, b \approx 14.1, c \approx 15.6$
3 $\gamma = 100°10', b \approx 55.1, c \approx 68.7$
5 $\beta = 78°30', a \approx 13.6, c \approx 17.8$
7 No triangle exists.
9 $\alpha \approx 77°30', \beta \approx 49°10', b \approx 108;$
$\alpha \approx 102°30', \beta \approx 24°10', b \approx 59$
11 $\alpha \approx 82.54°, \beta \approx 49.72°, b \approx 100.85;$
$\alpha \approx 97.46°, \beta \approx 34.80°, b \approx 75.45$
13 $\beta \approx 53°40', \gamma \approx 61°10', c \approx 20.6$
15 $\alpha \approx 25.993°, \gamma \approx 32.383°, a \approx 0.146$ **17** 219 yd
19 (a) 1.6 mi **(b)** 0.6 mi **21** 2.7 mi **23** 628 m
25 3.7 mi from *A* and 5.4 mi from *B* **27** 350 ft
29 (a) 18.7 **(b)** 814 **31** (3949.9, 2994.2)

EXERCISES 7.2

1 (a) B **(b)** F **(c)** D **(d)** E
(e) A **(f)** C
3 (a) α, law of sines **(b)** a, law of cosines
(c) Any angle, law of cosines
(d) Not enough information given
(e) $\gamma, \alpha + \beta + \gamma = 180°$
(f) c, law of sines; or $\gamma, \alpha + \beta + \gamma = 180°$
5 $a \approx 26, \beta \approx 41°, \gamma \approx 79°$
7 $b \approx 180, \alpha \approx 25°, \gamma \approx 5°$
9 $c \approx 2.75, \alpha \approx 21°10', \beta \approx 43°40'$
11 $\alpha \approx 29°, \beta \approx 47°, \gamma \approx 104°$
13 $\alpha \approx 12°30', \beta \approx 136°30', \gamma \approx 31°00'$ **15** 196 ft
17 24 mi **19** 39 mi **21** 2.3 mi **23** N55°31′E
25 63.7 ft from first and third base; 66.8 ft from second base
27 37,039 ft ≈ 7 mi
29 *Hint:* Use the formula $\sin \frac{\theta}{2} = \sqrt{\frac{1 - \cos \theta}{2}}$.
31 (a) 72°, 108°, 36° **(b)** 0.62 **(c)** 0.59, 0.36

Exer. 33–40: The answer is in square units.

33 260 **35** 11.21 **37** 13.1 **39** 517.0
41 1.62 acres **43** 123.4 ft^2

EXERCISES 7.3

1 $\langle 3, 1\rangle, \langle 1, -7\rangle, \langle 13, 8\rangle, \langle 3, -32\rangle, \sqrt{13}$
3 $\langle -15, 6\rangle, \langle 1, -2\rangle, \langle -68, 28\rangle, \langle 12, -12\rangle, \sqrt{53}$
5 $4\mathbf{i} - 3\mathbf{j}, -2\mathbf{i} + 7\mathbf{j}, 19\mathbf{i} - 17\mathbf{j}, -11\mathbf{i} + 33\mathbf{j}, \sqrt{5}$
7 Terminal points are (3, 2), (−1, 5), (2, 7), (6, 4), (3, −15).
9 Terminal points are (−4, 6), (−2, 3), (−6, 9), (−8, 12), (6, −9).

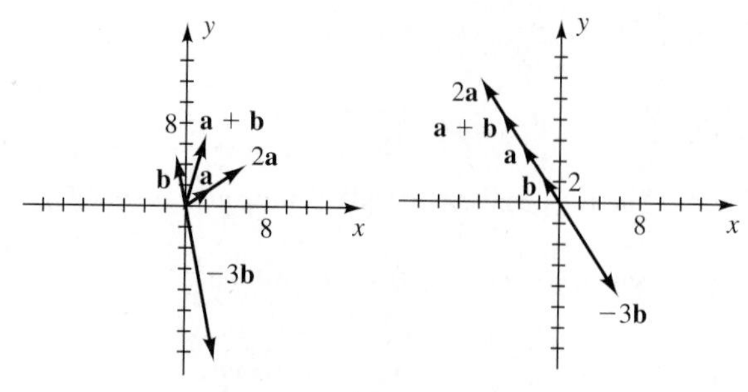

11 $-\mathbf{b}$ **13** $\mathbf{f}$ **15** $-\frac{1}{2}\mathbf{e}$

17
$$\begin{aligned} \mathbf{a} + (\mathbf{b} + \mathbf{c}) &= \langle a_1, a_2\rangle + (\langle b_1, b_2\rangle + \langle c_1, c_2\rangle) \\ &= \langle a_1, a_2\rangle + \langle b_1 + c_1, b_2 + c_2\rangle \\ &= \langle a_1 + b_1 + c_1, a_2 + b_2 + c_2\rangle \\ &= \langle a_1 + b_1, a_2 + b_2\rangle + \langle c_1, c_2\rangle \\ &= (\langle a_1, a_2\rangle + \langle b_1, b_2\rangle) + \langle c_1, c_2\rangle \\ &= (\mathbf{a} + \mathbf{b}) + \mathbf{c} \end{aligned}$$

19
$$\begin{aligned} \mathbf{a} + (-\mathbf{a}) &= \langle a_1, a_2\rangle + (-\langle a_1, a_2\rangle) \\ &= \langle a_1, a_2\rangle + \langle -a_1, -a_2\rangle \\ &= \langle a_1 - a_1, a_2 - a_2\rangle \\ &= \langle 0, 0\rangle = \mathbf{0} \end{aligned}$$

21
$$\begin{aligned} (mn)\mathbf{a} &= (mn)\langle a_1, a_2\rangle \\ &= \langle (mn)a_1, (mn)a_2\rangle \\ &= \langle mna_1, mna_2\rangle \\ &= m\langle na_1, na_2\rangle \quad \text{or} \quad n\langle ma_1, ma_2\rangle \\ &= m(n\langle a_1, a_2\rangle) \quad \text{or} \quad n(m\langle a_1, a_2\rangle) \\ &= m(n\mathbf{a}) \quad \text{or} \quad n(m\mathbf{a}) \end{aligned}$$

23 $0\mathbf{a} = 0\langle a_1, a_2\rangle = \langle 0a_1, 0a_2\rangle = \langle 0, 0\rangle = \mathbf{0}$.
Also, $m\mathbf{0} = m\langle 0, 0\rangle = \langle m0, m0\rangle = \langle 0, 0\rangle = \mathbf{0}$.

25
$$\begin{aligned} -(\mathbf{a} + \mathbf{b}) &= -(\langle a_1, a_2\rangle + \langle b_1, b_2\rangle) \\ &= -(\langle a_1 + b_1, a_2 + b_2\rangle) \\ &= \langle -(a_1 + b_1), -(a_2 + b_2)\rangle \\ &= \langle -a_1 - b_1, -a_2 - b_2\rangle \\ &= \langle -a_1, -a_2\rangle + \langle -b_1, -b_2\rangle \\ &= -\mathbf{a} + (-\mathbf{b}) = -\mathbf{a} - \mathbf{b} \end{aligned}$$

27 $\|2\mathbf{v}\| = \|2\langle a, b\rangle\| = \|\langle 2a, 2b\rangle\| = \sqrt{(2a)^2 + (2b)^2}$
$= \sqrt{4a^2 + 4b^2} = 2\sqrt{a^2 + b^2} = 2\|\langle a, b\rangle\|$
$= 2\|\mathbf{v}\|$

29 $3\sqrt{2}; \frac{7\pi}{4}$ **31** $5; \pi$ **33** $\sqrt{41}; \tan^{-1}\left(-\frac{5}{4}\right) + \pi$

35 $18; \frac{3\pi}{2}$ **37** 102 lb **39** 7.2 lb

41 89 lb; S66°W **43** 5.8 lb; 129°

45 40.96; 28.68 **47** −6.18; 19.02

49 (a) $-\frac{8}{17}\mathbf{i} + \frac{15}{17}\mathbf{j}$ **(b)** $\frac{8}{17}\mathbf{i} - \frac{15}{17}\mathbf{j}$

51 (a) $\left\langle \frac{2}{\sqrt{29}}, -\frac{5}{\sqrt{29}} \right\rangle$ **(b)** $\left\langle -\frac{2}{\sqrt{29}}, \frac{5}{\sqrt{29}} \right\rangle$

53 (a) $\langle -12, 6\rangle$ **(b)** $\left\langle -3, \frac{3}{2} \right\rangle$

55 $-\frac{24}{\sqrt{65}}\mathbf{i} + \frac{42}{\sqrt{65}}\mathbf{j}$

57 (a) $\mathbf{F} = \langle 7, 2\rangle$ **(b)** $\mathbf{G} = -\mathbf{F} = \langle -7, -2\rangle$

59 (a) $\mathbf{F} \approx \langle -5.86, 1.13\rangle$
(b) $\mathbf{G} = -\mathbf{F} \approx \langle 5.86, -1.13\rangle$

61 $\sin^{-1}(0.4) \approx 23.6°$ **63** 56°; 232 mi/hr

65 420 mi/hr; 244° **67** N22°W

69 $\mathbf{v}_1 \approx 4.1\mathbf{i} - 7.10\mathbf{j}$; $\mathbf{v}_2 \approx 0.98\mathbf{i} - 3.67\mathbf{j}$

71 (a) (24.51, 20.57) **(b)** (−24.57, 18.10)

73 28.2 lb/person

EXERCISES 7.4

1 (a) 24 **(b)** $\cos^{-1}\left(\frac{24}{\sqrt{29}\sqrt{45}}\right) \approx 48°22'$

3 (a) −14 **(b)** $\cos^{-1}\left(\frac{-14}{\sqrt{17}\sqrt{13}}\right) \approx 160°21'$

5 (a) 45 **(b)** $\cos^{-1}\left(\frac{45}{\sqrt{81}\sqrt{41}}\right) \approx 38°40'$

7 (a) $-\frac{149}{5}$ **(b)** $\cos^{-1}\left(\frac{-149/5}{\sqrt{149}\sqrt{149/25}}\right) = 180°$

9 $\langle 4, -1\rangle \cdot \langle 2, 8\rangle = 0$ **11** $(-4\mathbf{j}) \cdot (-7\mathbf{i}) = 0$

13 Opposite **15** Same **17** $\frac{6}{5}$ **19** $\pm\frac{3}{8}$

21 (a) −23 **(b)** −23 **23** −51

25 $17/\sqrt{26} \approx 3.33$ **27** 2.2 **29** 7

31 28 **33** 12

35 $\mathbf{a} \cdot \mathbf{a} = \langle a_1, a_2\rangle \cdot \langle a_1, a_2\rangle = a_1^2 + a_2^2$
$= \left(\sqrt{a_1^2 + a_2^2}\right)^2 = \|\mathbf{a}\|^2$

37 $(m\mathbf{a}) \cdot \mathbf{b} = (m\langle a_1, a_2\rangle) \cdot \langle b_1, b_2\rangle$
$= \langle ma_1, ma_2\rangle \cdot \langle b_1, b_2\rangle$
$= ma_1b_1 + ma_2b_2$
$= m(a_1b_1 + a_2b_2) = m(\mathbf{a} \cdot \mathbf{b})$

39 $\mathbf{0} \cdot \mathbf{a} = \langle 0, 0\rangle \cdot \langle a_1, a_2\rangle = 0(a_1) + 0(a_2)$
$= 0 + 0 = 0$

41 $1000\sqrt{3} \approx 1732$ ft-lb

43 (a) $\mathbf{v} = (93 \times 10^6)\mathbf{i} + (0.432 \times 10^6)\mathbf{j}$;
$\mathbf{w} = (93 \times 10^6)\mathbf{i} - (0.432 \times 10^6)\mathbf{j}$
(b) 0.53°

45 $\left\langle \frac{4}{5}, \frac{3}{5} \right\rangle$ **47** 2.6 **49** 24.33

51 $16\sqrt{3} \approx 27.7$ horsepower

EXERCISES 7.5

1 5 **3** $\sqrt{85}$ **5** 8 **7** 1 **9** 0

Note: Point P is the point corresponding to the geometric representation.

11 $P(4, 2)$ **13** $P(3, -5)$ **15** $P(-3, 6)$

17 $P(-6, 4)$ **19** $P(0, 2)$

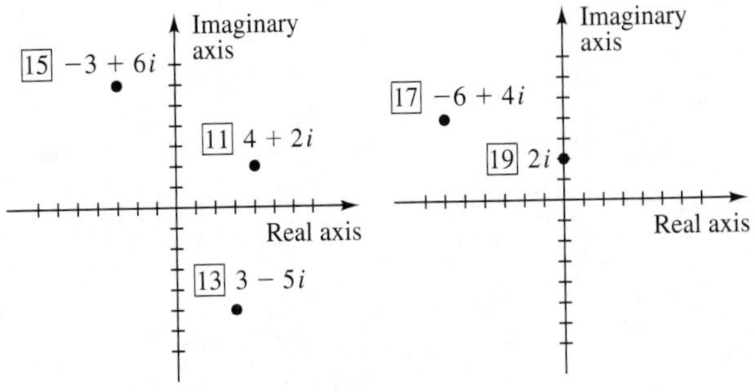

21 $\sqrt{2}\text{ cis }\frac{7\pi}{4}$ **23** $8\text{ cis }\frac{5\pi}{6}$ **25** $4\text{ cis }\frac{\pi}{6}$

27 $4\sqrt{2}\text{ cis }\frac{5\pi}{4}$ **29** $20\text{ cis }\frac{3\pi}{2}$ **31** 12 cis 0

33 $7\text{ cis }\pi$ **35** $6\text{ cis }\frac{\pi}{2}$ **37** $10\text{ cis }\frac{4\pi}{3}$

39 $\sqrt{5}\text{ cis }\left(\tan^{-1}\frac{1}{2}\right)$

41 $\sqrt{10}\text{ cis }\left[\tan^{-1}\left(-\frac{1}{3}\right) + \pi\right]$

43 $\sqrt{34}\text{ cis }\left(\tan^{-1}\frac{3}{5} + \pi\right)$

45 $5\text{ cis }\left[\tan^{-1}\left(-\frac{3}{4}\right) + 2\pi\right]$

47 $2\sqrt{2} + 2\sqrt{2}i$ **49** $-3 + 3\sqrt{3}i$ **51** −5

53 $5 + 3i$ **55** $2 - i$ **57** $-2, i$

59 $10\sqrt{3} - 10i, -\frac{2}{5}\sqrt{3} + \frac{2}{5}i$ **61** $40, \frac{5}{2}$

63 $8 - 4i, \frac{8}{5} + \frac{4}{5}i$ **65** $-15 + 10i, -\frac{15}{13} - \frac{10}{13}i$

69 $17.21 + 24.57i$ **71** $11.01 + 9.24i$
73 $\sqrt{365}$ ohms **75** 70.43 volts

EXERCISES 7.6

1 $-972 - 972i$ **3** $-32i$ **5** -8

7 $-\frac{1}{2}\sqrt{2} - \frac{1}{2}\sqrt{2}i$ **9** $-\frac{1}{2} - \frac{1}{2}\sqrt{3}i$

11 $-64\sqrt{3} - 64i$ **13** $\pm\left(\frac{1}{2}\sqrt{6} + \frac{1}{2}\sqrt{2}i\right)$

15 $\pm\left(\frac{\sqrt[4]{2}}{2} + \frac{\sqrt[4]{18}}{2}i\right), \pm\left(\frac{\sqrt[4]{18}}{2} - \frac{\sqrt[4]{2}}{2}i\right)$

17 $3i, \pm\frac{3}{2}\sqrt{3} - \frac{3}{2}i$

19 $\pm 1, \frac{1}{2} \pm \frac{1}{2}\sqrt{3}i,$
$-\frac{1}{2} \pm \frac{1}{2}\sqrt{3}i$

21 $\sqrt[10]{2}$ cis θ with $\theta = 9°$, 81°, 153°, 225°, 297°

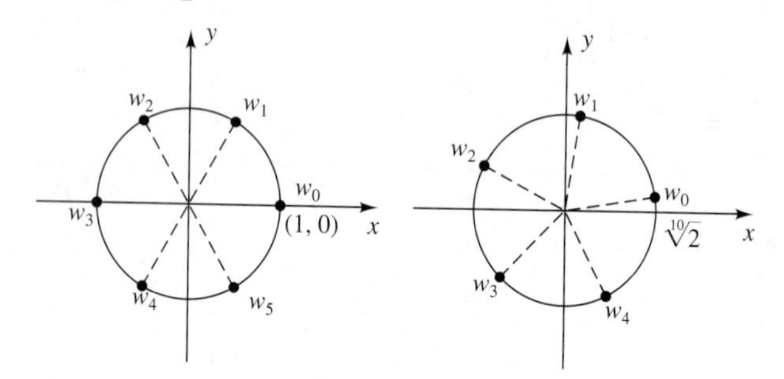

23 $\pm 2, \pm 2i$ **25** $\pm 2i, \sqrt{3} \pm i, -\sqrt{3} \pm i$

27 $2i, \pm\sqrt{3} - i$

29 3 cis θ with $\theta = 0°, 72°, 144°, 216°, 288°$

31
$$\begin{aligned}[r(\cos\theta + i\sin\theta)]^n &= [r(e^{i\theta})]^n \\ &= r^n(e^{i\theta})^n \\ &= r^n e^{i(n\theta)} \\ &= r^n(\cos n\theta + i\sin n\theta)\end{aligned}$$

CHAPTER 7 REVIEW EXERCISES

1 $a = \sqrt{43}, \beta = \cos^{-1}\left(\frac{4}{43}\sqrt{43}\right), \gamma = \cos^{-1}\left(\frac{5}{86}\sqrt{43}\right)$

2 $\alpha = 60°, \beta = 90°, b = 4; \alpha = 120°, \beta = 30°, b = 2$

3 $\gamma = 75°, a = 50\sqrt{6}, c = 50(1 + \sqrt{3})$

4 $\alpha = \cos^{-1}\left(\frac{7}{8}\right), \beta = \cos^{-1}\left(\frac{11}{16}\right), \gamma = \cos^{-1}\left(-\frac{1}{4}\right)$

5 $\alpha = 38°, a \approx 8.0, c \approx 13$

6 $\gamma \approx 19°10', \beta \approx 137°20', b \approx 258$

7 $\alpha \approx 24°, \gamma \approx 41°, b \approx 10.1$

8 $\alpha \approx 42°, \beta \approx 87°, \gamma \approx 51°$ **9** 290 **10** 10.9

11 Terminal points are $(-2, -3), (-6, 13), (-8, 10), (-1, 4)$.

12 **(a)** $12\mathbf{i} + 19\mathbf{j}$ **(b)** $-8\mathbf{i} + 13\mathbf{j}$ **(c)** $\sqrt{40} \approx 6.32$
(d) $\sqrt{29} - \sqrt{17} \approx 1.26$

13 $\langle 14\cos 40°, -14\sin 40°\rangle$ **14** 109 lb; S78°E

15 $-16\mathbf{i} + 12\mathbf{j}$

16 $\left\langle -\frac{12}{\sqrt{58}}, \frac{28}{\sqrt{58}}\right\rangle$

17 Circle with center (a_1, a_2) and radius c

18 The vectors **a**, **b**, and **a** − **b** form a triangle with the vector **a** − **b** opposite angle θ. The conclusion is a direct application of the law of cosines with sides $\|\mathbf{a}\|$, $\|\mathbf{b}\|$, and $\|\mathbf{a} - \mathbf{b}\|$.

19 183°; 70 mi/hr

20 **(a)** 10 **(b)** $\cos^{-1}\left(\frac{10}{\sqrt{13}\sqrt{17}}\right) \approx 47°44'$ **(c)** $\frac{10}{\sqrt{13}}$

21 **(a)** 80 **(b)** $\cos^{-1}\left(\frac{40}{\sqrt{40}\sqrt{50}}\right) \approx 26°34'$ **(c)** $\sqrt{40}$

22 56

23 $10\sqrt{2}$ cis $\frac{3\pi}{4}$ **24** 4 cis $\frac{5\pi}{3}$ **25** 17 cis π

26 12 cis $\frac{3\pi}{2}$ **27** 10 cis $\frac{7\pi}{6}$

28 $\sqrt{41}$ cis $\left(\tan^{-1}\frac{5}{4}\right)$ **29** $10\sqrt{3} - 10i$

30 $12 + 5i$ **31** $-12 - 12\sqrt{3}i, -\frac{3}{2}$

32 $-4\sqrt{2}i, -2\sqrt{2}$ **33** $-512i$ **34** i

35 $-972 + 972i$ **36** $-2^{19} - 2^{19}\sqrt{3}i$

37 $-3, \frac{3}{2} \pm \frac{3}{2}\sqrt{3}i$

38 **(a)** 2^{24} **(b)** $\sqrt[3]{2}$ cis θ with $\theta = 100°, 220°, 340°$

39 2 cis θ with $\theta = 0°, 72°, 144°, 216°, 288°$

40 47.6° **41** 53,000,000 mi

42 **(a)** 449 ft **(b)** 434 ft

43 **(a)** 33 mi, 41 mi **(b)** 30 mi **44** 204

45 1 hour and 16 minutes **46 (c)** 158°

47 **(a)** 47° **(b)** 20

48 **(a)** 72° **(b)** 181.6 ft^2 **(c)** 37.6 ft

CHAPTER 7 DISCUSSION EXERCISES

4 **(b)** *Hint:* Law of cosines

5 **(a)** $(\|\mathbf{b}\| \cos \alpha + \|\mathbf{a}\| \cos \beta)\mathbf{i} + (\|\mathbf{b}\| \sin \alpha - \|\mathbf{a}\| \sin \beta)\mathbf{j}$

6 **(a)** 1 **(b)** πi; $\frac{\pi}{2}i$ **(c)** $\frac{\sqrt{2}}{2} + \frac{\sqrt{2}}{2}i$; $e^{-\pi/2} \approx 0.2079$

7 The statement is true.

Chapter 8

EXERCISES 8.1

1 $(3, 5), (-1, -3)$ **3** $(1, 0), (-3, 2)$

5 $(0, 0), \left(\frac{1}{8}, \frac{1}{128}\right)$ **7** $(3, -2)$ **9** No solution

11 $(-4, 3), (5, 0)$ **13** $(-2, 2)$

15 $\left(-\frac{3}{5} + \frac{1}{10}\sqrt{86}, \frac{1}{5} + \frac{3}{10}\sqrt{86}\right)$, $\left(-\frac{3}{5} - \frac{1}{10}\sqrt{86}, \frac{1}{5} - \frac{3}{10}\sqrt{86}\right)$

17 $(-4, 0), \left(\frac{12}{5}, \frac{16}{5}\right)$ **19** $(0, 1), (4, -3)$

21 $(\pm 2, 5), (\pm\sqrt{5}, 4)$

23 $(\sqrt{2}, \pm 2\sqrt{3}), (-\sqrt{2}, \pm 2\sqrt{3})$

25 $(2\sqrt{2}, \pm 2), (-2\sqrt{2}, \pm 2)$ **27** $(3, -1, 2)$

29 $(1, -1, 2), (-1, 3, -2)$

31 **(a)** $b = 4$; tangent
(b) $b < 4$; intersect twice
(c) $b > 4$; no intersection

33 Yes; a solution occurs between 0 and 1.

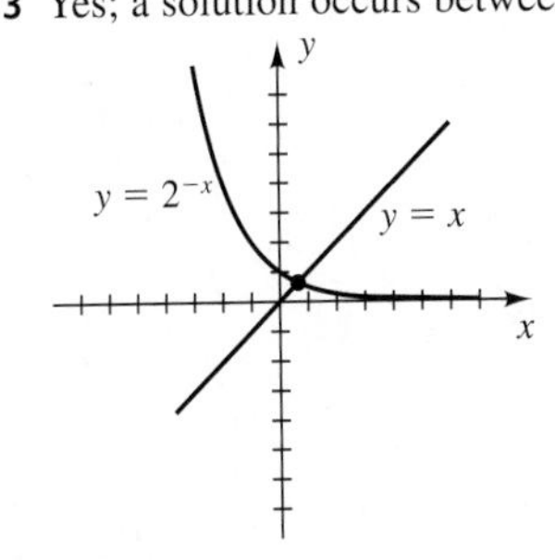

35 $\frac{1}{4}$; tangent **37** $f(x) = 2(3)^x + 1$

39 12 in. × 8 in.

41 **(a)** $a = 120{,}000$, $b = 40{,}000$ **(b)** 77,143

43 (0, 0), (0, 100), (50, 0); the fourth solution (−100, 150) is not meaningful.

45 Yes; 1 ft × 1 ft × 2 ft or $\frac{\sqrt{13} - 1}{2}$ ft × $\frac{\sqrt{13} - 1}{2}$ ft × $\frac{8}{(\sqrt{13} - 1)^2}$ ft ≈ 1.30 ft × 1.30 ft × 1.18 ft

47 The points are on the parabola **(a)** $y = \frac{1}{2}x^2 - \frac{1}{2}$ and **(b)** $y = \frac{1}{4}x^2 - 1$.

49 **(a)** $(31.25, -50)$
(b) $\left(-\frac{3}{2}\sqrt{11}, -\frac{1}{2}\right) \approx (-4.975, -0.5)$

51 $(\mp 0.82, \pm 1.82)$; $\left(\frac{1}{2} \pm \frac{\sqrt{7}}{2}, \frac{1}{2} \mp \frac{\sqrt{7}}{2}\right)$

53 $(\mp 0.56, \pm 1.92), (\mp 0.63, \pm 1.90), (\pm 1.14, \pm 1.65)$

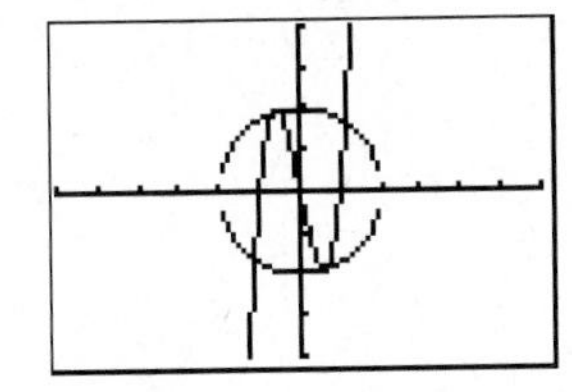

[−6, 6] by [−4, 4]

55 $(-1.44, \pm 1.04), (-0.12, \pm 1.50), (0{,}10, \pm 1.50),$ $(1.22, \pm 1.19)$

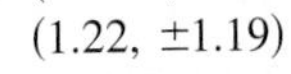

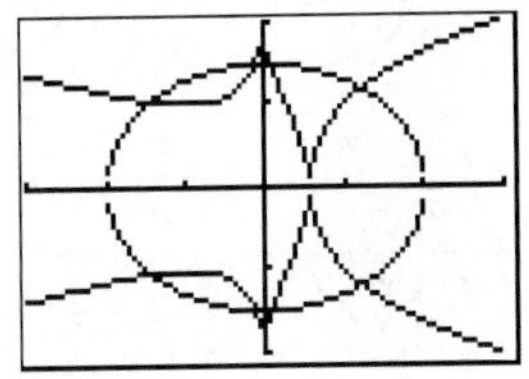

[−3, 3] by [−2, 2]

57 $a \approx 1.2012, b \approx 0.4004$ **59** $a \approx 2.8019, b \approx 0.9002$

EXERCISES 8.2

1 $(4, -2)$ **3** $(8, 0)$ **5** $\left(-1, \frac{3}{2}\right)$ **7** $\left(\frac{76}{53}, \frac{28}{53}\right)$

9 $\left(\frac{51}{13}, \frac{96}{13}\right)$ **11** $\left(\frac{8}{7}, -\frac{3}{7}\sqrt{6}\right)$ **13** No solution

15 All ordered pairs (m, n) such that $3m - 4n = 2$

17 $(0, 0)$ **19** $\left(-\frac{22}{7}, -\frac{11}{5}\right)$

21 313 students, 137 nonstudents

23 $x = \left(\frac{30}{\pi}\right) - 4 \approx 5.55$ cm, $y = 12 - \left(\frac{30}{\pi}\right) \approx 2.45$ cm

25 $l = 10$ ft, $w = \frac{20}{\pi}$ ft **27** 2400 adults, 3600 kittens

29 40 g of 35% alloy, 60 g of 60% alloy

31 540 mi/hr, 60 mi/hr **33** $v_0 = 10$, $a = 3$

35 20 sofas, 30 recliners

37 **(a)** $\left(c, \frac{4}{5}c\right)$ for an arbitrary $c > 0$ **(b)** $28 per hour

39 1928; 15.5°C **41** LP: 4 hr, SLP: 2 hr

43 $a = \frac{1}{6}$, $b = -\frac{1}{6}e^{6x}$ **45** $a = \cos x - \sec x$, $b = \sin x$

EXERCISES 8.3

1

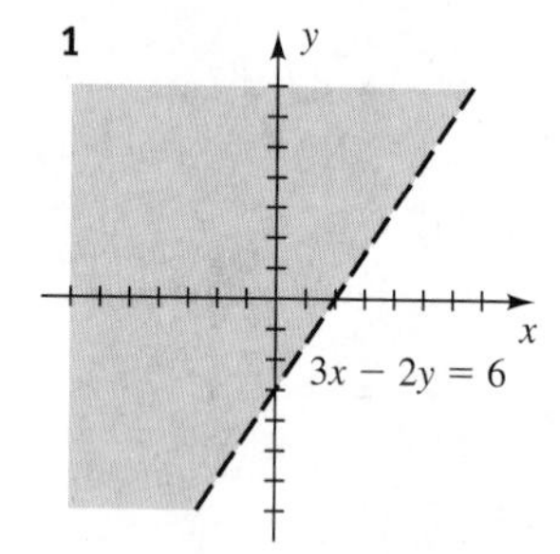

3

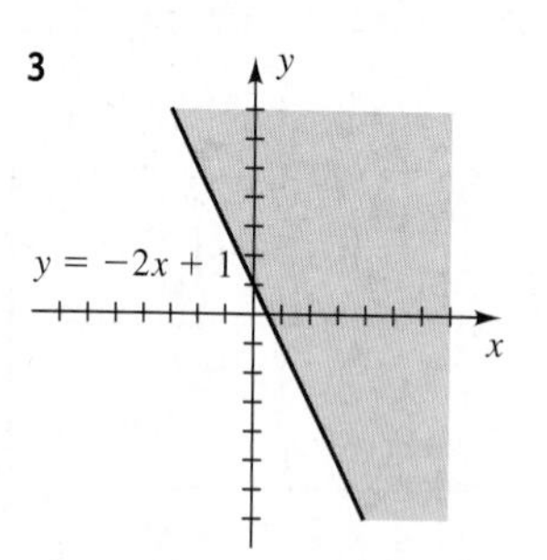

5

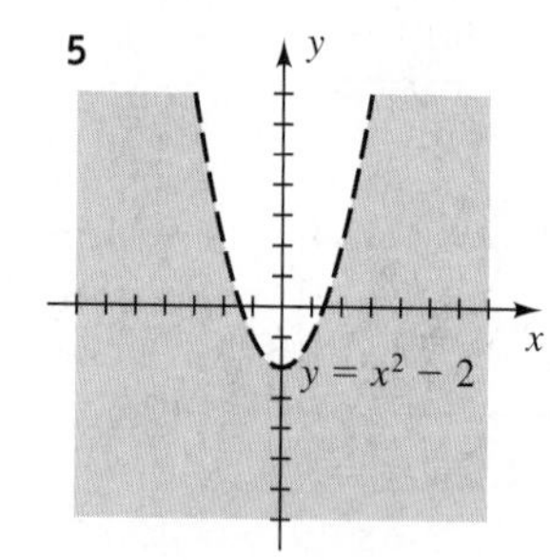

7

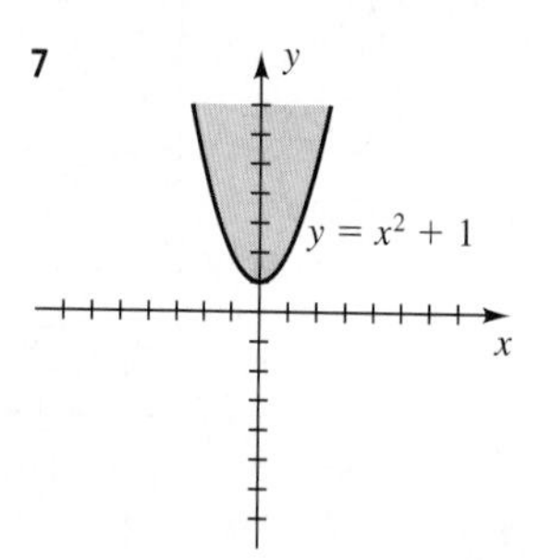

9

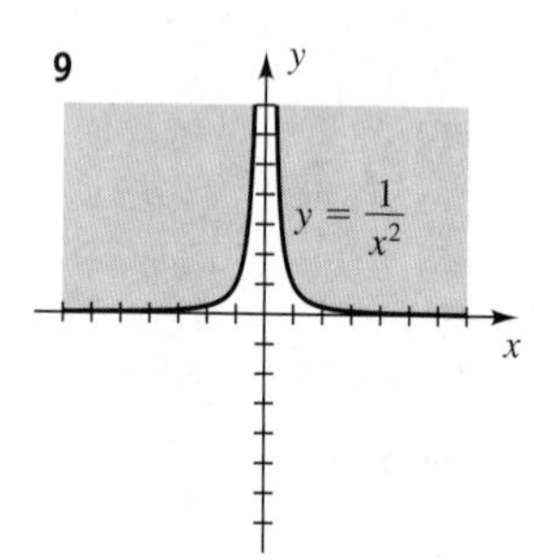

11

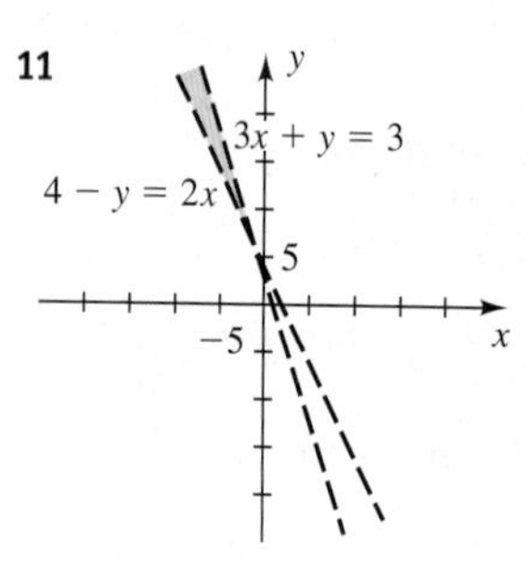

13

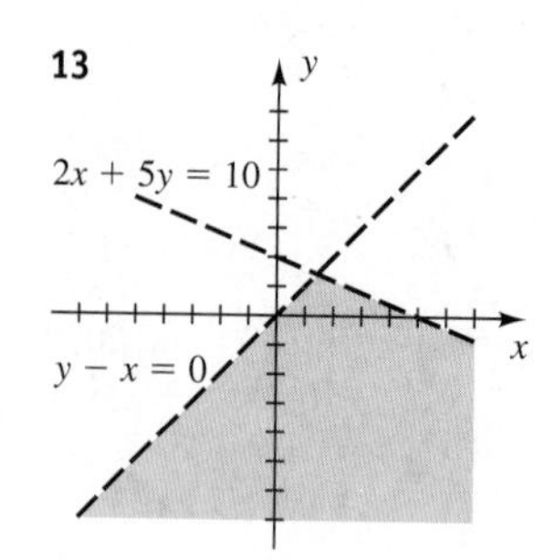

15

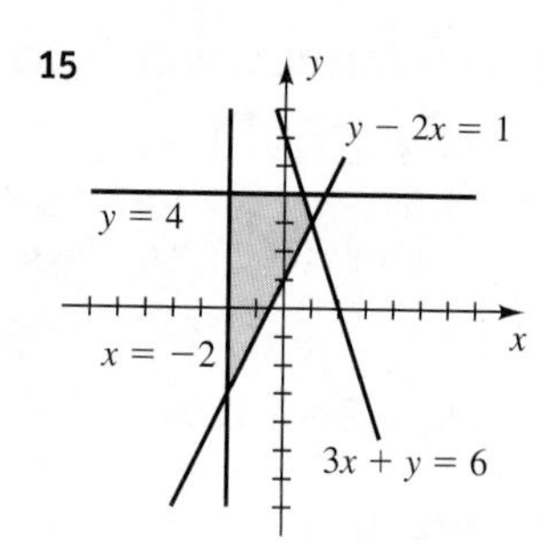

17

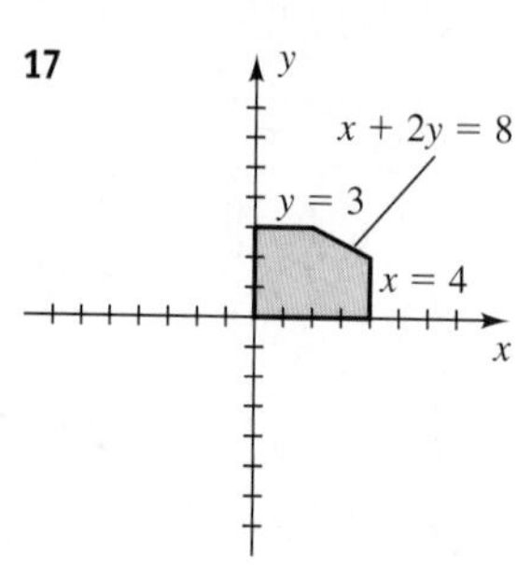

19

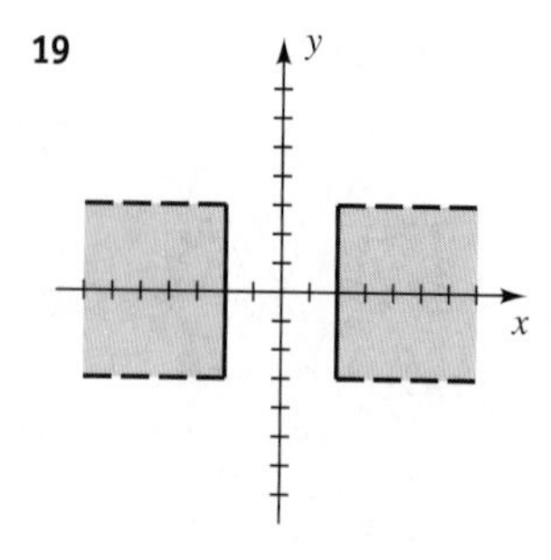

21

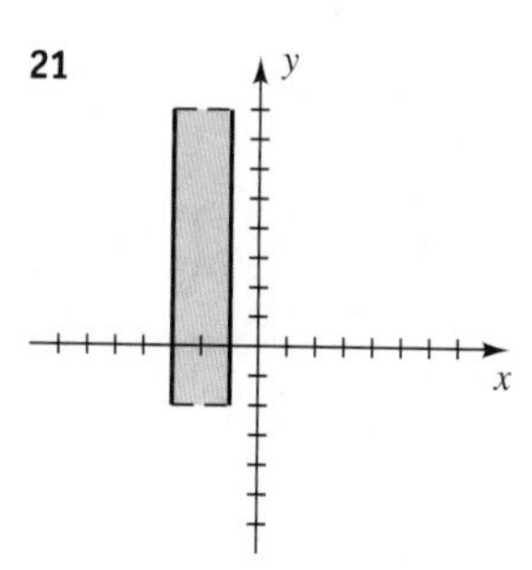

23

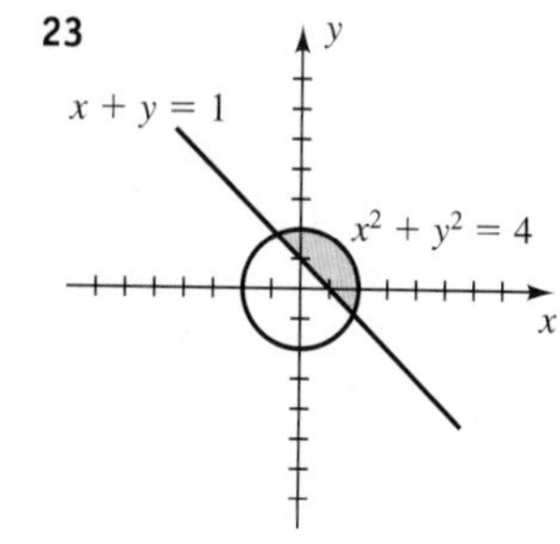

25

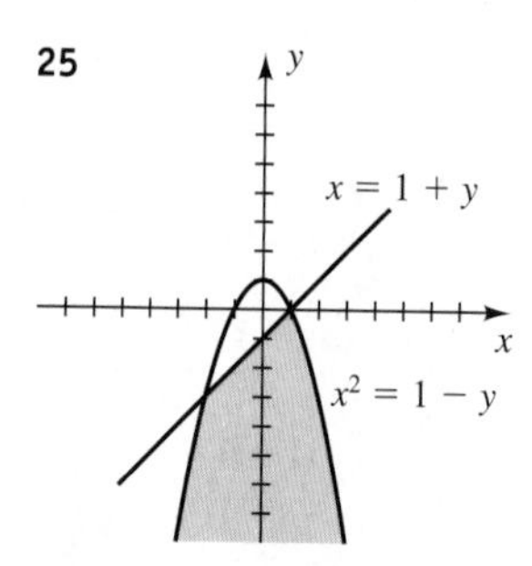

27 $0 \le x < 3$, $y < -x + 4$, $y \ge x - 4$

29 $x^2 + y^2 \le 9$, $y > -2x + 4$

31 $y < x$, $y \le -x + 4$, $(x - 2)^2 + (y - 2)^2 \le 8$

33 $y > \frac{1}{8}x + \frac{1}{2}$, $y \le x + 4$, $y \le -\frac{3}{4}x + 4$

35 If x and y denote the numbers of sets of brand A and brand B, respectively, then a system is $x \geq 20$, $y \geq 10$, $x \geq 2y$, $x + y \leq 100$.

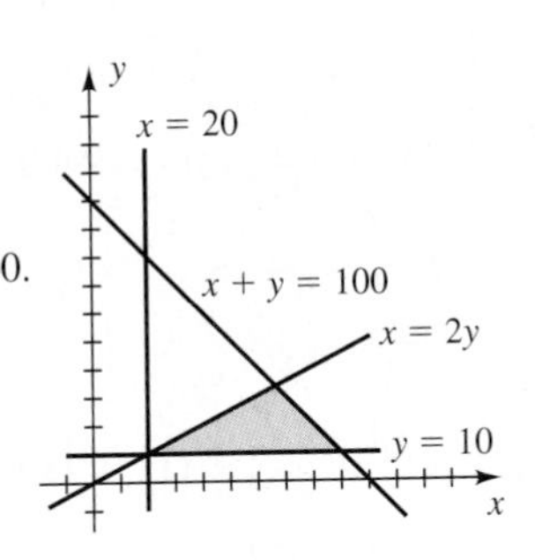

37 If x and y denote the amounts placed in the high-risk and low-risk investment, respectively, then a system is $x \geq 2000$, $y \geq 3x$, $x + y \leq 15{,}000$.

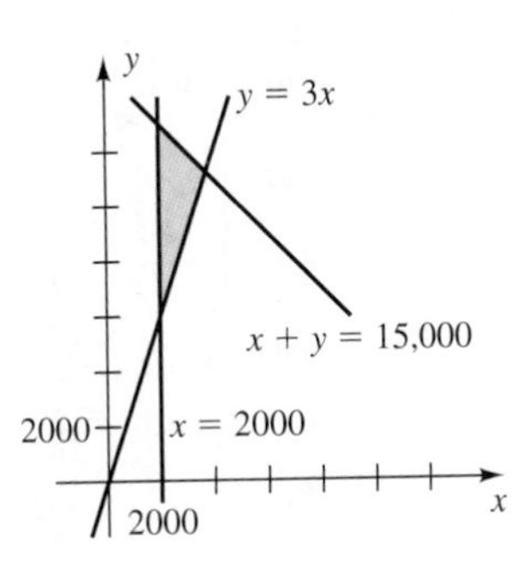

39 $x + y \leq 9$, $y \geq x$, $x \geq 1$

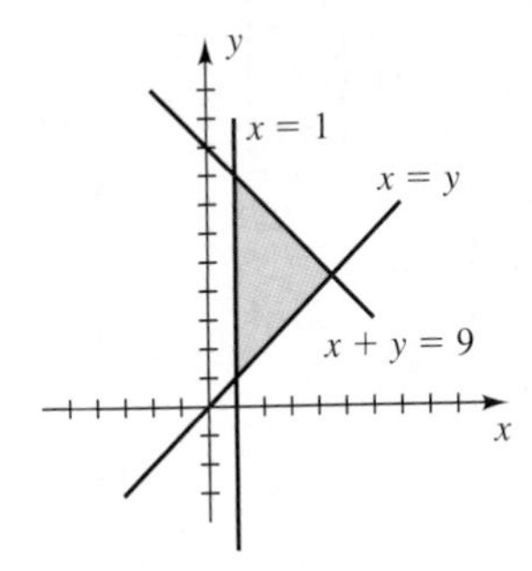

41 If the plant is located at (x, y), then a system is $60^2 \leq x^2 + y^2 \leq 100^2$, $60^2 \leq (x - 100)^2 + y^2 \leq 100^2$, $y \geq 0$.

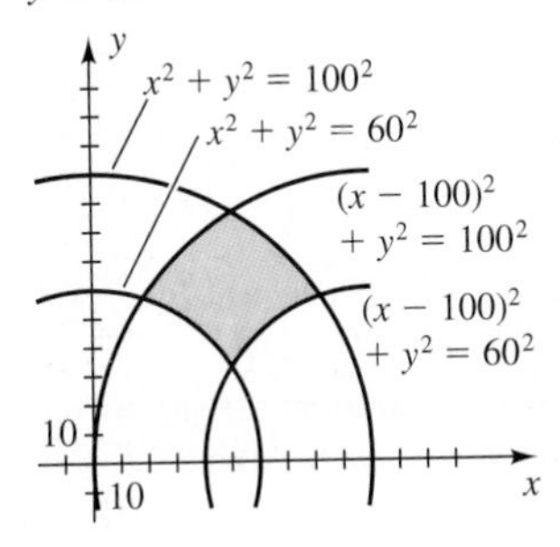

43

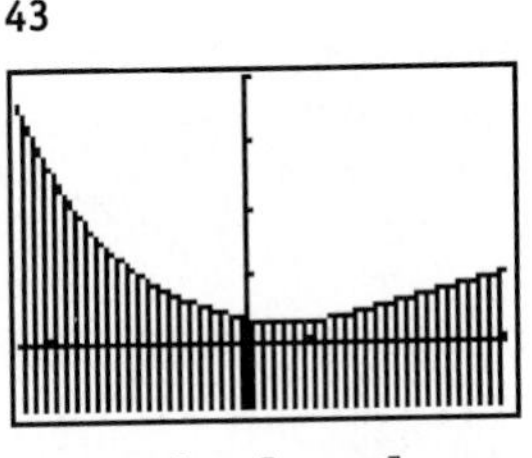

$[-3.5, 4]$ by $[-1, 4]$

45

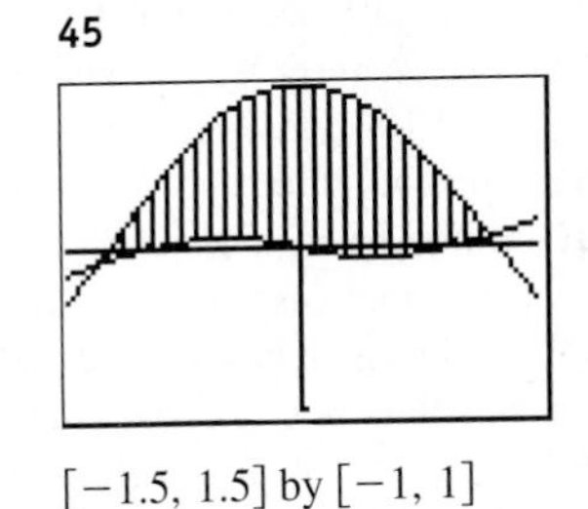

$[-1.5, 1.5]$ by $[-1, 1]$

47 There is no solution.

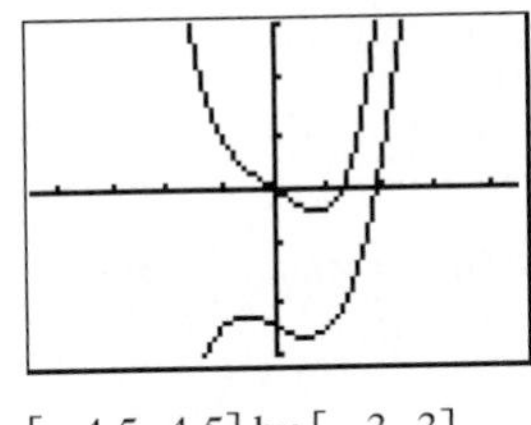

$[-4.5, 4.5]$ by $[-3, 3]$

49 **(a)** Yes

(b)

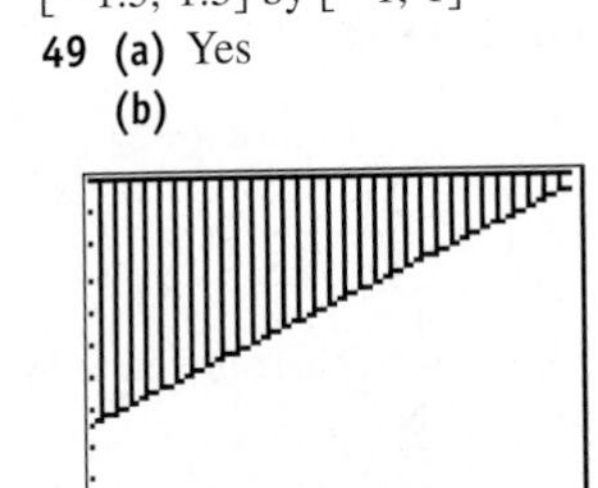

$[33, 80, 5]$ by $[0, 50, 5]$

(c) Region above the line

EXERCISES 8.4

1 Maximum of 27 at $(6, 2)$; minimum of 9 at $(0, 2)$

3 Maximum of 21 at $(6, 3)$

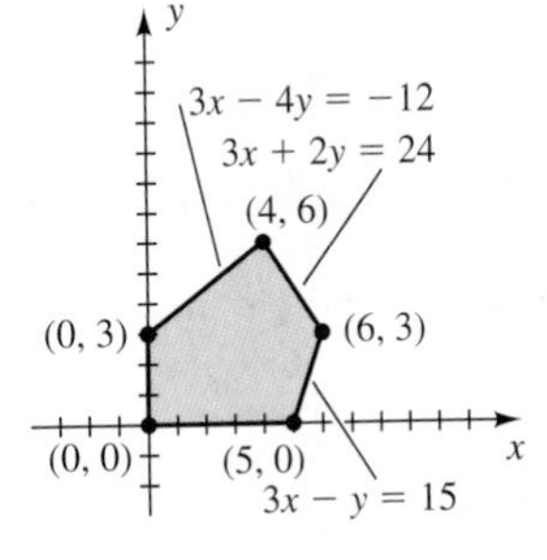

5 Minimum of 21 at $(3, 2)$

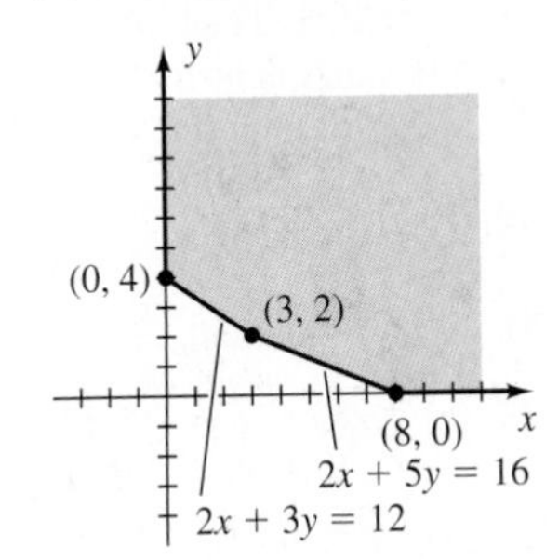

7 C has the maximum value 24 for any point on the line segment from $(2, 5)$ to $(6, 3)$.

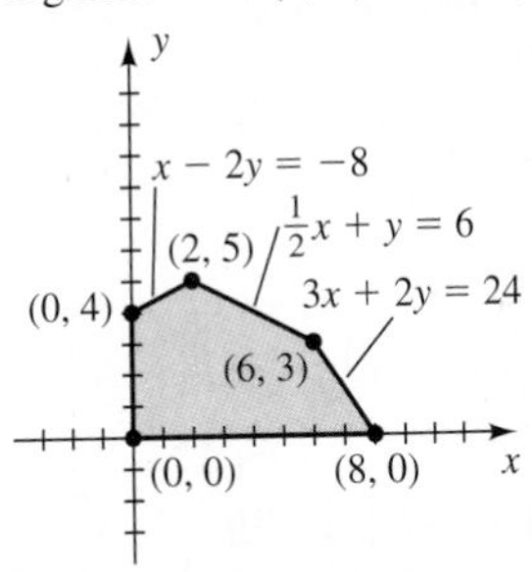

9 50 standard and 30 oversized

11 3.5 lb of S and 1 lb of T
13 Send 25 from W_1 to A and 0 from W_1 to B. Send 10 from W_2 to A and 60 from W_2 to B.
15 None of alfalfa and 80 acres of corn
17 Minimum cost: 16 oz X, 4 oz Y, 0 oz Z; maximum cost: 0 oz X, 8 oz Y, 12 oz Z
19 2 vans and 4 buses **21** 3000 trout and 2000 bass
23 60 small units and 20 deluxe units

EXERCISES 8.5

1 $(2, 3, -1)$ **3** $(-2, 4, 5)$ **5** No solution
7 $\left(\frac{2}{3}, \frac{31}{21}, \frac{1}{21}\right)$

Exer. 9–16: There are other forms for the answers; c is any real number.

9 $(2c, -c, c)$ **11** $(0, -c, c)$
13 $\left(\frac{12}{7} - \frac{9}{7}c, \frac{4}{7}c - \frac{13}{14}, c\right)$
15 $\left(\frac{7}{10}c + \frac{1}{2}, \frac{19}{10}c - \frac{3}{2}, c\right)$
17 $\left(\frac{1}{11}, \frac{31}{11}, \frac{3}{11}\right)$ **19** $(-2, -3)$ **21** No solution
23 17 of 10%, 11 of 30%, 22 of 50%
25 4 hr for A, 2 hr for B, 5 hr for C
27 380 lb of G_1, 60 lb of G_2, 160 lb of G_3
29 (a) $I_1 = 0, I_2 = 2, I_3 = 2$ **(b)** $I_1 = \frac{3}{4}, I_2 = 3, I_3 = \frac{9}{4}$
31 $\frac{3}{8}$ lb Colombian, $\frac{1}{8}$ lb Costa Rican, $\frac{1}{2}$ lb Kenyan
33 (a) A: $x_1 + x_4 = 75$, B: $x_1 + x_2 = 150$, C: $x_2 + x_3 = 225$, D: $x_3 + x_4 = 150$
(b) $x_1 = 25, x_2 = 125, x_4 = 50$
(c) $x_3 = 150 - x_4 \leq 150$; $x_3 = 225 - x_2 = 225 - (150 - x_1) = 75 + x_1 \geq 75$
35 2134 **37** $x^2 + y^2 - x + 3y - 6 = 0$
39 $f(x) = x^3 - 2x^2 - 4x - 6$
41 $a = -\frac{4}{9}, b = \frac{11}{9}, c = \frac{17}{18}, d = \frac{23}{18}$

EXERCISES 8.6

1 $\begin{bmatrix} 9 & -1 \\ -2 & 5 \end{bmatrix}, \begin{bmatrix} 1 & -3 \\ 4 & 1 \end{bmatrix}, \begin{bmatrix} 10 & -4 \\ 2 & 6 \end{bmatrix}, \begin{bmatrix} -12 & -3 \\ 9 & -6 \end{bmatrix}$

3 $\begin{bmatrix} 9 & 0 \\ 1 & 5 \\ 3 & 4 \end{bmatrix}, \begin{bmatrix} 3 & -2 \\ 3 & -5 \\ -9 & 4 \end{bmatrix}, \begin{bmatrix} 12 & -2 \\ 4 & 0 \\ -6 & 8 \end{bmatrix}, \begin{bmatrix} -9 & -3 \\ 3 & -15 \\ -18 & 0 \end{bmatrix}$

5 $[11 \;\; -3 \;\; -3], [-3 \;\; -3 \;\; 7], [8 \;\; -6 \;\; 4], [-21 \;\; 0 \;\; 15]$

7 Not possible, not possible, $\begin{bmatrix} 6 & -4 & 4 \\ 0 & 2 & -8 \\ -6 & 4 & -2 \end{bmatrix}, \begin{bmatrix} -12 & 0 \\ -6 & 3 \\ 3 & -9 \end{bmatrix}$

9 -18 **11** $\begin{bmatrix} 16 & 38 \\ 11 & -34 \end{bmatrix}, \begin{bmatrix} 4 & 38 \\ 23 & -22 \end{bmatrix}$

13 $\begin{bmatrix} 3 & -14 & -3 \\ 16 & 2 & -2 \\ -7 & -29 & 9 \end{bmatrix}, \begin{bmatrix} 3 & -20 & -11 \\ 2 & 10 & -4 \\ 15 & -13 & 1 \end{bmatrix}$

15 $\begin{bmatrix} 4 & 8 \\ -18 & 11 \end{bmatrix}, \begin{bmatrix} 3 & -4 & 4 \\ -5 & 2 & 2 \\ -51 & 26 & 10 \end{bmatrix}$

17 $\begin{bmatrix} 1 & 2 & 3 \\ 4 & 5 & 6 \\ 7 & 8 & 9 \end{bmatrix}, \begin{bmatrix} 1 & 2 & 3 \\ 4 & 5 & 6 \\ 7 & 8 & 9 \end{bmatrix}$

19 $[15], \begin{bmatrix} -3 & 7 & 2 \\ -12 & 28 & 8 \\ 15 & -35 & -10 \end{bmatrix}$

21 $\begin{bmatrix} 2 & 0 & 5 \\ 5 & 3 & -2 \end{bmatrix}$, not possible **23** $\begin{bmatrix} 4 \\ 12 \\ -1 \end{bmatrix}$

25 $\begin{bmatrix} 18 & 0 & -2 \\ -40 & 10 & -10 \end{bmatrix}$ **35** $\begin{bmatrix} 135 & -109 & 91 \\ -39 & 92 & -33 \\ 45 & 3 & 95 \end{bmatrix}$

37 $\begin{bmatrix} 76 & -38 & 102 \\ -5 & 61 & -13 \\ 41 & 0 & 19 \end{bmatrix}$

39 (a) $A = \begin{bmatrix} 400 & 550 & 500 \\ 400 & 450 & 500 \\ 300 & 500 & 600 \\ 250 & 200 & 300 \\ 100 & 100 & 200 \end{bmatrix}, B = \begin{bmatrix} \$8.99 \\ \$10.99 \\ \$12.99 \end{bmatrix}$

(b) $\begin{bmatrix} \$16{,}135.50 \\ \$15{,}036.50 \\ \$15{,}986.00 \\ \$\ 8342.50 \\ \$\ 4596.00 \end{bmatrix}$

(c) The $4596.00 represents the amount the store would receive if all the yellow towels were sold.

EXERCISES 8.7

1 Show that $AB = I_2$ and $BA = I_2$.

3 $\frac{1}{10}\begin{bmatrix} 3 & 4 \\ -1 & 2 \end{bmatrix}$ **5** Does not exist

7 $\frac{1}{8}\begin{bmatrix} 2 & 1 & 0 \\ -2 & 3 & 0 \\ 0 & 0 & 2 \end{bmatrix}$ **9** $\frac{1}{3}\begin{bmatrix} -4 & -5 & 3 \\ -4 & -8 & 3 \\ 1 & 2 & 0 \end{bmatrix}$

11 $\begin{bmatrix} \frac{1}{2} & 0 & 0 \\ 0 & \frac{1}{4} & 0 \\ 0 & 0 & \frac{1}{6} \end{bmatrix}$ **13** $ab \neq 0$; $\begin{bmatrix} \frac{1}{a} & 0 \\ 0 & \frac{1}{b} \end{bmatrix}$

17 (a) $\left(\frac{13}{10}, -\frac{1}{10}\right)$ **(b)** $\left(\frac{7}{5}, \frac{6}{5}\right)$

19 (a) $\left(-\frac{25}{3}, -\frac{34}{3}, \frac{7}{3}\right)$ **(b)** $\left(\frac{16}{3}, \frac{16}{3}, -\frac{1}{3}\right)$

21 $\begin{bmatrix} 0.11111 & 0.25926 & -0.62963 \\ -0.03704 & 0.02469 & 0.32099 \\ 0.07407 & -0.04938 & 0.35802 \end{bmatrix}$

23 $\begin{bmatrix} -0.22278 & 0.12932 & 0.06496 & 0.37796 \\ -1.17767 & 0.09503 & 0.55936 & 0.29171 \\ -0.37159 & 0.00241 & 0.14074 & 0.37447 \\ 0.15987 & -0.04150 & 0.07218 & -0.20967 \end{bmatrix}$

25 (a) $\begin{bmatrix} 4.0 & 7.1 \\ 2.2 & -4.9 \end{bmatrix}\begin{bmatrix} x \\ y \end{bmatrix} = \begin{bmatrix} 6.2 \\ 2.9 \end{bmatrix}$

(b) $\begin{bmatrix} 0.1391 & 0.2016 \\ 0.0625 & -0.1136 \end{bmatrix}$ **(c)** $x \approx 1.4472$, $y \approx 0.0579$

27 (a) $\begin{bmatrix} 3.1 & 6.7 & -8.7 \\ 4.1 & -5.1 & 0.2 \\ 0.6 & 1.1 & -7.4 \end{bmatrix}\begin{bmatrix} x \\ y \\ z \end{bmatrix} = \begin{bmatrix} 1.5 \\ 2.1 \\ 3.9 \end{bmatrix}$

(b) $\begin{bmatrix} 0.1474 & 0.1572 & -0.1691 \\ 0.1197 & -0.0696 & -0.1426 \\ 0.0297 & 0.0024 & -0.1700 \end{bmatrix}$

(c) $x \approx -0.1081$, $y \approx -0.5227$, $z \approx -0.6135$

29 (a) $a \approx -1.9630$, $b \approx 26.2963$, $c \approx -25.7407$

(b)

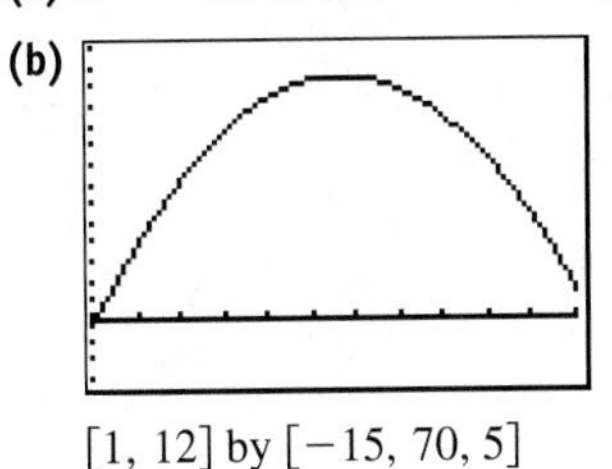

[1, 12] by [−15, 70, 5]

(c) June: 61°F; October: 41°F

EXERCISES 8.8

1 $M_{11} = 0 = A_{11}$; $M_{12} = 5$; $A_{12} = -5$; $M_{21} = -1$; $A_{21} = 1$; $M_{22} = 7 = A_{22}$

3 $M_{11} = -14 = A_{11}$; $M_{12} = 10$; $A_{12} = -10$; $M_{13} = 15 = A_{13}$; $M_{21} = 7$; $A_{21} = -7$; $M_{22} = -5 = A_{22}$; $M_{23} = 34$; $A_{23} = -34$; $M_{31} = 11 = A_{31}$; $M_{32} = 4$; $A_{32} = -4$; $M_{33} = 6 = A_{33}$

5 5 **7** −83 **9** 2 **11** 0 **13** −125 **15** 48

17 −216 **19** $abcd$ **31 (a)** $x^2 - 3x - 4$ **(b)** −1, 4

33 (a) $x^2 + x - 2$ **(b)** −2, 1

35 (a) $-x^3 - 2x^2 + x + 2$ **(b)** −2, −1, 1

37 (a) $-x^3 + 4x^2 + 4x - 16$ **(b)** −2, 2, 4

39 $-31i - 20j + 7k$ **41** $-6i - 8j + 18k$

43 −359,284 **45** 10,739.92

47 (a) $-x^3 + x^2 + 6x - 7$ **(b)** −2.51, 1.22, 2.29

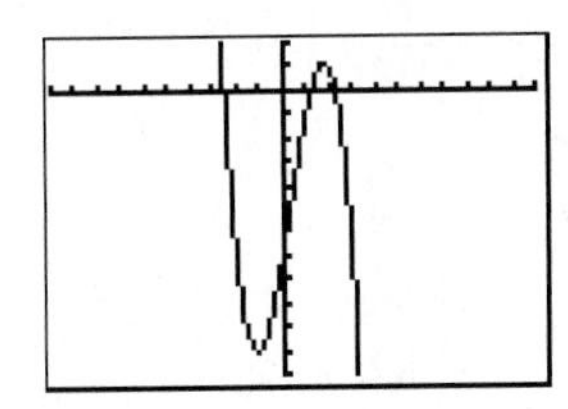

[−10, 11] by [−12, 2]

EXERCISES 8.9

1 $R_2 \leftrightarrow R_3$ **3** $-R_1 + R_3 \rightarrow R_3$

5 2 is a common factor of R_1 and R_3.

7 R_1 and R_3 are identical.

9 −1 is a common factor of R_2.

11 Every number in C_2 is 0. **13** $2C_1 + C_3 \rightarrow C_3$

15 −10 **17** −142 **19** −183 **21** 44 **23** 359

33 (4, −2) **35** (8, 0)

37 $|D| = 0$, so Cramer's rule cannot be used.

39 (2, 3, −1) **41** (−2, 4, 5)

43 $x = \dfrac{cgi - dfi + bfj}{cei - afi + bfh}$

EXERCISES 8.10

1 $\dfrac{3}{x-2}+\dfrac{5}{x+3}$ **3** $\dfrac{5}{x-6}-\dfrac{4}{x+2}$

5 $\dfrac{2}{x-1}+\dfrac{3}{x+2}-\dfrac{1}{x-3}$ **7** $\dfrac{3}{x}+\dfrac{2}{x-5}-\dfrac{1}{x+1}$

9 $\dfrac{2}{x-1}+\dfrac{5}{(x-1)^2}$ **11** $-\dfrac{7}{x}+\dfrac{5}{x^2}+\dfrac{40}{3x-5}$

13 $\dfrac{24/25}{x+2}+\dfrac{2/5}{(x+2)^2}-\dfrac{23/25}{2x-1}$

15 $\dfrac{5}{x}-\dfrac{2}{x+1}+\dfrac{3}{(x+1)^3}$ **17** $-\dfrac{2}{x-1}+\dfrac{3x+4}{x^2+1}$

19 $\dfrac{4}{x}+\dfrac{5x-3}{x^2+2}$ **21** $\dfrac{4x-1}{x^2+1}+\dfrac{3}{(x^2+1)^2}$

23 $2x+\dfrac{1}{x-1}+\dfrac{3x}{x^2+1}$ **25** $3+\dfrac{4}{x}+\dfrac{8}{x-4}$

27 $2x+3+\dfrac{2}{x-1}-\dfrac{3}{2x+1}$

CHAPTER 8 REVIEW EXERCISES

1 $\left(\dfrac{19}{23}, -\dfrac{18}{23}\right)$ **2** No solution **3** $(-3, 5), (1, -3)$

4 $(4, -3), (3, -4)$ **5** $(2\sqrt{3}, \pm\sqrt{2}), (-2\sqrt{3}, \pm\sqrt{2})$

6 $(-1, \pm 1, -1), \left(0, \pm\dfrac{1}{2}\sqrt{6}, -\dfrac{1}{2}\right)$ **7** $\left(\dfrac{14}{17}, \dfrac{14}{27}\right)$

8 $\left(\log_2\dfrac{25}{7}, \log_3\dfrac{15}{7}\right)$ **9** $\left(\dfrac{6}{11}, -\dfrac{7}{11}, 1\right)$

10 $\left(-\dfrac{6}{29}, \dfrac{2}{29}, -\dfrac{17}{29}\right)$

11 $(-2c, -3c, c)$ for any real number c **12** $(0, 0, 0)$

13 $\left(5c-1, -\dfrac{19}{2}c+\dfrac{5}{2}, c\right)$ for any real number c

14 $(5, -4)$ **15** $\left(-1, \dfrac{1}{2}, \dfrac{1}{3}\right)$ **16** $(3, -1, -2, 4)$

17

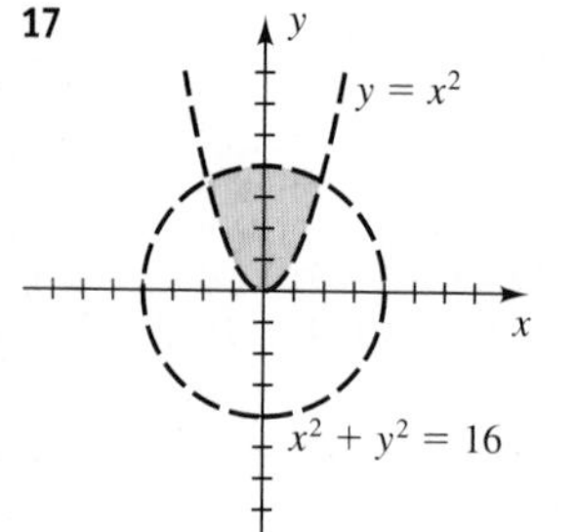

18

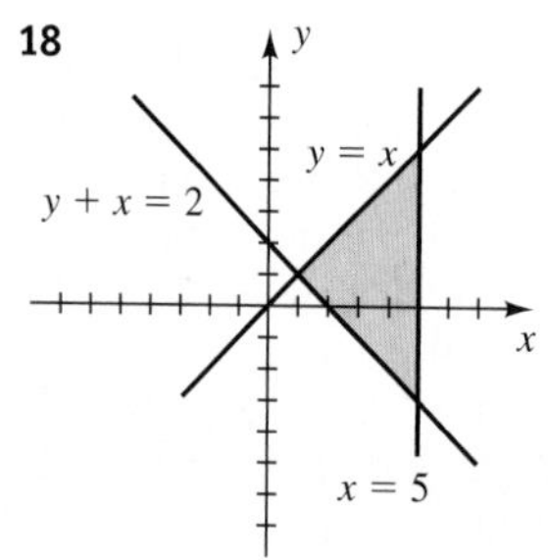

19 **20**

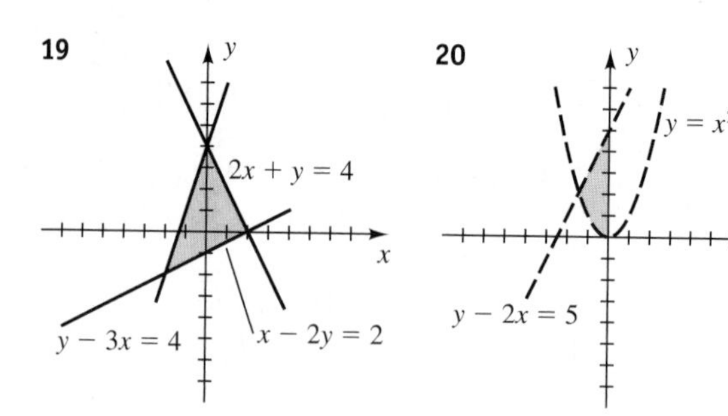

21 $\begin{bmatrix} 4 & -5 & 6 \\ 4 & -11 & 5 \end{bmatrix}$ **22** $\begin{bmatrix} 26 \\ -6 \end{bmatrix}$ **23** $\begin{bmatrix} 0 & 4 & -6 \\ 16 & 22 & 1 \\ 12 & 11 & 9 \end{bmatrix}$

24 $\begin{bmatrix} 0 & -37 \\ 15 & -6 \end{bmatrix}$ **25** $\begin{bmatrix} -12 & 4 & -11 \\ 6 & -11 & 5 \end{bmatrix}$

26 $\begin{bmatrix} a & 3a \\ 2a & 4a \end{bmatrix}$ **27** $\begin{bmatrix} a & 3a \\ 2b & 4b \end{bmatrix}$ **28** $\begin{bmatrix} 0 & 0 \\ 0 & 0 \end{bmatrix}$

29 $\begin{bmatrix} 5 & 9 \\ 13 & 19 \end{bmatrix}$ **30** $\begin{bmatrix} 1 & 0 & 0 \\ 0 & 1 & 0 \\ 0 & 0 & 1 \end{bmatrix}$ **31** $-\dfrac{1}{2}\begin{bmatrix} 2 & 4 \\ 3 & 5 \end{bmatrix}$

32 $\dfrac{1}{11}\begin{bmatrix} 8 & 1 & -2 \\ 5 & 2 & -4 \\ -14 & 1 & 9 \end{bmatrix}$ **33** $\begin{bmatrix} 1 & 0 & 0 \\ 0 & 2 & -7 \\ 0 & -1 & 4 \end{bmatrix}$

34 $\dfrac{1}{37}\begin{bmatrix} -4 & -20 & 15 \\ 3 & 15 & -2 \\ 9 & 8 & -6 \end{bmatrix}$ **35** $(2, -5)$ **36** $(-1, 3, 2)$

37 -6 **38** 9 **39** 48 **40** -86 **41** -84

42 0 **43** 120 **44** -76 **45** 0 **46** -50

47 $-1 \pm 2\sqrt{3}$ **48** $4, \pm\sqrt{7}$

49 2 is a common factor of R_1, 2 is a common factor of C_2, and 3 is a common factor of C_3.

50 Interchange R_1 with R_2 and then R_2 with R_3 to obtain the determinant on the right. The effect is to multiply by -1 twice.

51 $a_{11}a_{22}a_{33}\cdots a_{nn}$ **53** $\left(\dfrac{76}{53}, \dfrac{28}{53}\right)$ **54** $\left(\dfrac{2}{3}, \dfrac{31}{21}, \dfrac{1}{21}\right)$

55 $\dfrac{8}{x-1}-\dfrac{3}{x+5}-\dfrac{1}{x+3}$ **56** $2+\dfrac{3}{x+1}+\dfrac{4}{(x+1)^2}$

57 $-\dfrac{2}{x+5}+\dfrac{3x-1}{x^2+4}$ **58** $\dfrac{4}{x^2+2}+\dfrac{x-2}{x^2+5}$

59 $40\sqrt{5}$ ft $\times$ $20\sqrt{5}$ ft **60** $y = \pm 2\sqrt{2}x + 3$

61 Tax = \$750,000; bonus = \$125,000

62 Inside radius = 90 ft, outside radius = 100 ft

63 In ft^3/hr: A, 30; B, 20; C, 50

64 Western 95, eastern 55

65 If x and y denote the length and width, respectively, then a system is $x \le 12$, $y \le 8$, $y \ge \frac{1}{2}x$.

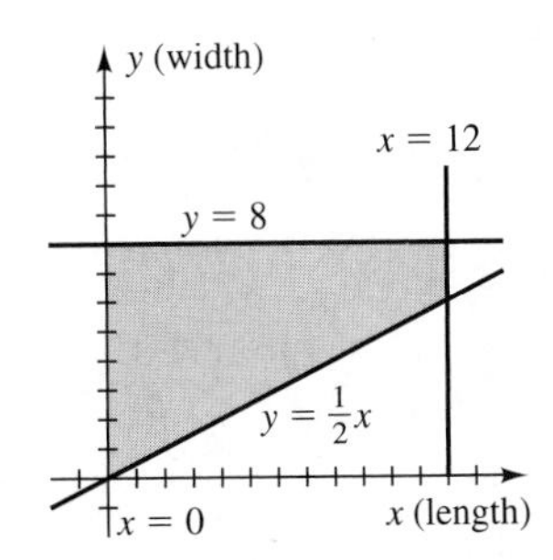

66 $x + y \le 18$, $x \ge 2y$, $x \ge 0$, $y \ge 0$

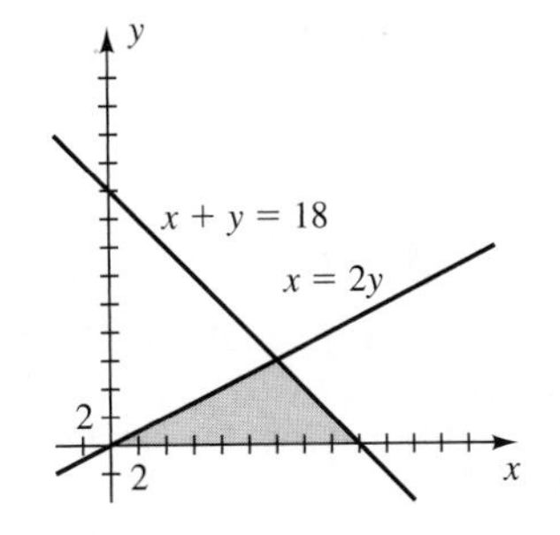

67 80 mowers and 30 edgers

68 High-risk \$250,000; low-risk \$500,000; bonds \$0

CHAPTER 8 DISCUSSION EXERCISES

1 **(a)** $b = 1.99$, $x = 204$, $y = -100$; $b = 1.999$, $x = 2004$, $y = -1000$
(b) $x = \dfrac{4b - 10}{b - 2}$, $y = \dfrac{1}{b - 2}$
(c) It gets close to (4, 0).

2 **(a)** $D = [12{,}000 \quad 9000 \quad 14{,}000]$;

$$E = \begin{bmatrix} 0.90 & 0.10 & 0.00 \\ 0.00 & 0.80 & 0.20 \\ 0.05 & 0.00 & 0.95 \end{bmatrix}$$

(b) The elements of $F = [11{,}500 \quad 8400 \quad 15{,}100]$ represent the populations on islands A, B, and C, respectively, after one year.
(c) The population stabilizes with 10,000 birds on A, 5000 birds on B, and 20,000 birds on C.
(d) Regardless of the initial population distribution of the 35,000 birds, the populations tend toward the distribution described in part (c).

3 *Hint:* Assign a size to A, and examine the definition of an inverse.

4 AD: 35%, DS: $33\frac{1}{3}$%, SP: $31\frac{2}{3}$%

5 $a = -15$, $b = 10$, $c = 24$; the fourth root is -4

6 $y = 0.058\overline{3}x^3 - 0.11\overline{6}x^2 - 1.1x + 4.2$

8 **(a)** Not possible **(b)** $x^2 + y^2 - 1.8x - 4.2y + 0.8 = 0$
(c) $f(x) = -\dfrac{5}{12}x^2 + \dfrac{7}{12}x + 4$
(d) $f(x) = ax^3 + \left(-2a - \dfrac{5}{12}\right)x^2 + \left(-3a + \dfrac{7}{12}\right)x + 4$, where a is any nonzero real number
(e) Not possible

Chapter 9

EXERCISES 9.1

1 9, 6, 3, 0; -12 **3** $\frac{1}{2}, \frac{4}{5}, \frac{7}{10}, \frac{10}{17}; \frac{22}{65}$

5 9, 9, 9, 9; 9

7 1.9, 2.01, 1.999, 2.0001; 2.000 000 01

9 $4, -\frac{9}{4}, \frac{5}{3}, -\frac{11}{8}; -\frac{15}{16}$ **11** 2, 0, 2, 0; 0

13 $\frac{2}{3}, \frac{2}{3}, \frac{8}{11}, \frac{8}{9}; \frac{128}{33}$ **15** 1, 2, 3, 4; 8

17 **19**

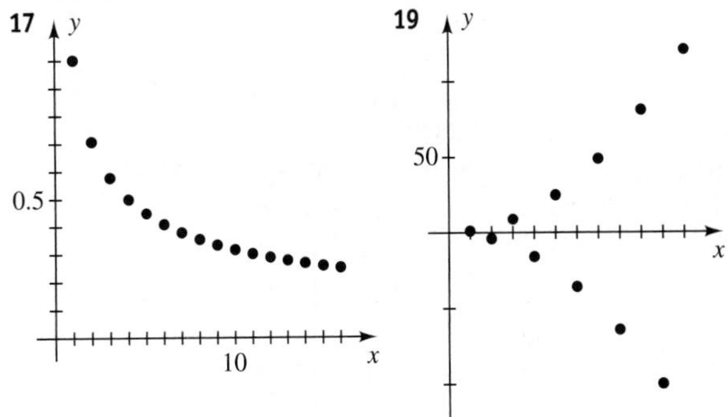

21 2, 1, -2, -11, -38 **23** $-3, 3^2, 3^4, 3^8, 3^{16}$

25 5, 5, 10, 30, 120 **27** $2, 2, 4, 4^3, 4^{12}$

29 $\frac{7}{2}, \frac{15}{2}, 12, 17$

31 $-1, -1 + \frac{1}{\sqrt{2}}, -1 + \frac{1}{\sqrt{2}} - \frac{1}{\sqrt{3}}, -\frac{1}{2} + \frac{1}{\sqrt{2}} - \frac{1}{\sqrt{3}}$

33 -5 **35** 10 **37** 25 **39** $-\frac{17}{15}$ **41** 61

43 10,000 **45** $\frac{319}{3}$ **47** $\frac{7}{2}k^2$

49 $\sum_{k=1}^{n} (a_k - b_k)$
$= (a_1 - b_1) + (a_2 - b_2) + \cdots + (a_n - b_n)$
$= (a_1 + a_2 + \cdots + a_n) + (-b_1 - b_2 - \cdots - b_n)$
$= (a_1 + a_2 + \cdots + a_n) - (b_1 + b_2 + \cdots + b_n)$
$= \sum_{k=1}^{n} a_k - \sum_{k=1}^{n} b_k$

51 As k increases, the terms approach 1.

53 0.4, 0.7, 1, 1.6, 2.8

55 (a) 1, 1, 2, 3, 5, 8, 13, 21, 34, 55
(b) 1, 2, 1.5, $1.\overline{6}$, 1.6, 1.625, 1.6153846, 1.6190476, 1.6176471, 1.6181818

57 (a) $a_n = 0.8a_{n-1}$ **(b)** The fourth day

59 $C(n) = \begin{cases} 89.95n & \text{if } 1 \le n \le 4 \\ 87.95n & \text{if } 5 \le n \le 9 \\ 85.95n & \text{if } n \ge 10 \end{cases}$

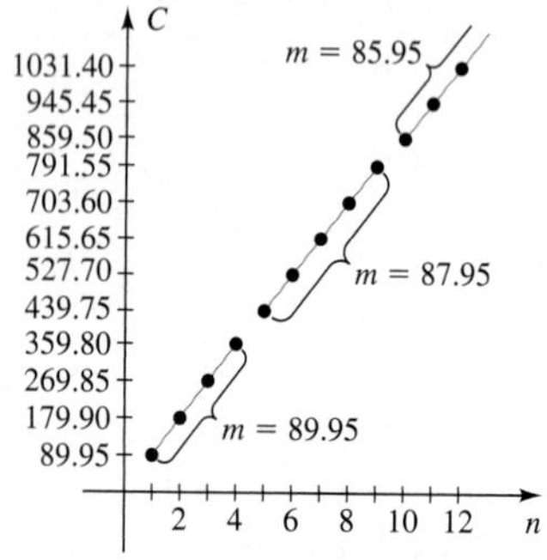

61 2.236068 **63** 2.4493

65 (a) $f(1) = -1 < 0$, $f(2) \approx 0.30 > 0$ **(b)** 1.76

67 a_n approaches e **69** a_n approaches 1

71 10

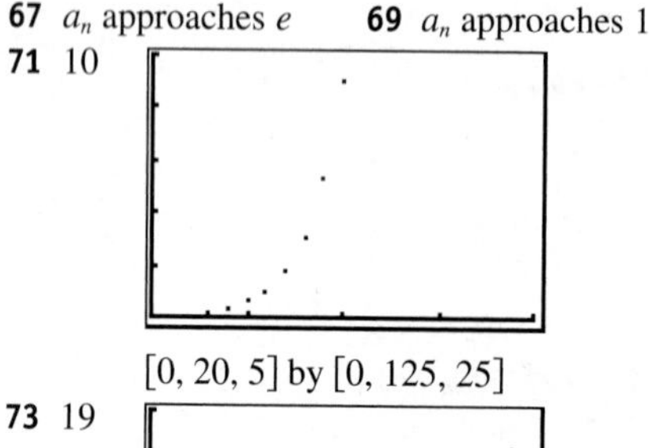

[0, 20, 5] by [0, 125, 25]

73 19

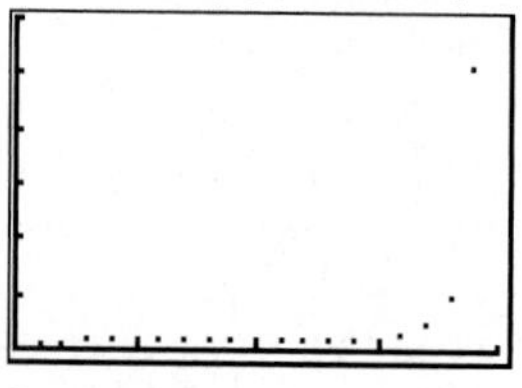

[0, 20, 5] by [0, 300, 50]

75 (a) Decreases from 250 insects to 0
(b) Stabilizes at 333 insects
(c) Stabilizes at 636 insects

EXERCISES 9.2

1 Show that $a_{k+1} - a_k = 4$. **3** $4n - 2$; 18; 38

5 $3.3 - 0.3n$; 1.8; 0.3 **7** $3.1n - 10.1$; 5.4; 20.9

9 $\ln 3^n$; $\ln 3^5$; $\ln 3^{10}$ **11** -8 **13** -8.5

15 -9.8 **17** $\frac{551}{17}$ **19** -105 **21** 30 **23** 530

25 $\frac{423}{2}$ **27** $934j + 838{,}265$

29 $\sum_{n=1}^{5} (7n - 3)$ or $\sum_{n=0}^{4} (4 + 7n)$

31 $\sum_{n=1}^{67} (7n - 3)$ or $\sum_{n=0}^{66} (4 + 7n)$

33 $\sum_{n=1}^{6} \frac{3n}{4n + 3}$ or $\sum_{n=0}^{5} \frac{3 + 3n}{7 + 4n}$

35 $\sum_{n=1}^{1528} (11n - 3) = 12{,}845{,}132$

37 24 **39** 12 or 18 **41** $\frac{10}{3}, \frac{14}{3}, 6, \frac{22}{3}, \frac{26}{3}$

43 (a) 60 **(b)** 12,780 **45** 255 **47** 154π ft

49 \$1200 **51** $16n^2$

53 Show that the $(n + 1)$st term is 1 greater than the nth term.

55 (a) $\frac{8}{36}, \frac{7}{36}, \frac{6}{36}, \ldots, \frac{1}{36}$ **(b)** $d = -\frac{1}{36}$; 1 **(c)** \$722.22

EXERCISES 9.3

1 Show that $\frac{a_{k+1}}{a_k} = -\frac{1}{4}$. **3** $8\left(\frac{1}{2}\right)^{n-1} = 2^{4-n}$; $\frac{1}{2}$; $\frac{1}{16}$

5 $300(-0.1)^{n-1}$; 0.03; -0.00003 **7** 5^n; 3125; 390,625

9 $4(-1.5)^{n-1}$; 20.25; -68.34375

11 $(-1)^{n-1}x^{2n-2}$; x^8; $-x^{14}$ **13** $2^{(n-1)x+1}$; 2^{4x+1}; 2^{7x+1}

15 $\pm\sqrt{3}$ **17** $\frac{243}{8}$ **19** $\sqrt[3]{3}$; 36 **21** 88,572

23 $-\frac{341}{1024}$ **25** $8188 + 55j$ **27** $\sum_{n=1}^{7} 2^n$

29 $\sum_{n=1}^{4} (-1)^{n+1} \frac{1}{4}\left(\frac{1}{3}\right)^{n-1}$ **31** $\frac{2}{3}$ **33** $\frac{50}{33}$

35 Since $|r| = \sqrt{2} > 1$, the sum does not exist.

37 1024 **39** $\frac{23}{99}$ **41** $\frac{2393}{990}$ **43** $\frac{5141}{999}$ **45** $\frac{16{,}123}{9999}$

47 24 **49** 4, 20, 100, 500 **51** $\frac{25}{256}\% \approx 0.1\%$

53 (a) $N(t) = 10{,}000(1.2)^t$ **(b)** 61,917 **55** 300 ft

57 \$3,000,000 **59 (b)** 375 mg

61 (a) $a_{k+1} = \dfrac{1}{4}\sqrt{10}a_k$

(b) $a_n = \left(\dfrac{1}{4}\sqrt{10}\right)^{n-1} a_1,\ A_n = \left(\dfrac{5}{8}\right)^{n-1} A_1,$

$P_n = \left(\dfrac{1}{4}\sqrt{10}\right)^{n-1} P_1$ **(c)** $\dfrac{16a_1}{4 - \sqrt{10}}$

63 (a) $a_k = 3^{k-1}$ **(b)** 4,782,969

(c) $b_k = \dfrac{3^{k-1}}{4^k} = \dfrac{1}{4}\left(\dfrac{3}{4}\right)^{k-1}$ **(d)** $\dfrac{729}{16{,}384} \approx 4.45\%$

65 \$38,929.00 **67** \$7396.67

69 (a) $\dfrac{2}{5}, \dfrac{6}{25}, \dfrac{18}{125}, \dfrac{54}{625}, \dfrac{162}{3125}$

(b) $r = \dfrac{3}{5}; \dfrac{2882}{3125} = 0.92224$ **(c)** \$16,000

EXERCISES 9.4

Exer. 1–32: A typical proof is given for Exercises 1, 5, 9, . . . , 29.

1 (1) P_1 is true, since $2(1) = 1(1 + 1) = 2$.
(2) Assume P_k is true:
$2 + 4 + 6 + \cdots + 2k = k(k + 1)$. Hence,

$$\begin{aligned} 2 + 4 + 6 + \cdots + 2k + 2(k + 1) &= k(k + 1) + 2(k + 1) \\ &= (k + 1)(k + 2) \\ &= (k + 1)(k + 1 + 1). \end{aligned}$$

Thus, P_{k+1} is true, and the proof is complete.

5 (1) P_1 is true, since $5(1) - 3 = \dfrac{1}{2}(1)[5(1) - 1] = 2$.
(2) Assume P_k is true:

$$2 + 7 + 12 + \cdots + (5k - 3) = \frac{1}{2}k(5k - 1).$$

Hence,

$$\begin{aligned} 2 + 7 + 12 + \cdots + (5k - 3) + 5(k + 1) - 3 &= \frac{1}{2}k(5k - 1) + 5(k + 1) - 3 \\ &= \frac{5}{2}k^2 + \frac{9}{2}k + 2 \\ &= \frac{1}{2}(5k^2 + 9k + 4) \\ &= \frac{1}{2}(k + 1)(5k + 4) \\ &= \frac{1}{2}(k + 1)[5(k + 1) - 1]. \end{aligned}$$

Thus, P_{k+1} is true, and the proof is complete.

9 (1) P_1 is true, since $(1)^1 = \dfrac{1(1 + 1)[2(1) + 1]}{6} = 1$.
(2) Assume P_k is true:

$$1^2 + 2^2 + 3^2 + \cdots + k^2 = \frac{k(k + 1)(2k + 1)}{6}.$$

Hence,

$$\begin{aligned} 1^2 + 2^2 + 3^2 + \cdots + k^2 + (k + 1)^2 &= \frac{k(k + 1)(2k + 1)}{6} + (k + 1)^2 \\ &= (k + 1)\left[\frac{k(2k + 1)}{6} + \frac{6(k + 1)}{6}\right] \\ &= \frac{(k + 1)(2k^2 + 7k + 6)}{6} \\ &= \frac{(k + 1)(k + 2)(2k + 3)}{6}. \end{aligned}$$

Thus, P_{k+1} is true, and the proof is complete.

13 (1) P_1 is true, since $3^1 = \dfrac{3}{2}(3^1 - 1) = 3$.
(2) Assume P_k is true:

$3 + 3^2 + 3^3 + \cdots + 3^k = \dfrac{3}{2}(3^k - 1)$. Hence,

$$\begin{aligned} 3 + 3^2 + 3^3 + \cdots + 3^k + 3^{k+1} &= \frac{3}{2}(3^k - 1) + 3^{k+1} \\ &= \frac{3}{2} \cdot 3^k - \frac{3}{2} + 3 \cdot 3^k \\ &= \frac{9}{2} \cdot 3^k - \frac{3}{2} \\ &= \frac{3}{2}(3 \cdot 3^k - 1) \\ &= \frac{3}{2}(3^{k+1} - 1). \end{aligned}$$

Thus, P_{k+1} is true, and the proof is complete.

17 (1) P_1 is true, since $1 < \frac{1}{8}[2(1) + 1]^2 = \frac{9}{8}$.

(2) Assume P_k is true:

$1 + 2 + 3 + \cdots + k < \frac{1}{8}(2k + 1)^2$. Hence,

$$\begin{aligned} 1 + 2 + 3 + \cdots + k + (k + 1) &< \frac{1}{8}(2k + 1)^2 + (k + 1) \\ &= \frac{1}{2}k^2 + \frac{3}{2}k + \frac{9}{8} \\ &= \frac{1}{8}(4k^2 + 12k + 9) \\ &= \frac{1}{8}(2k + 3)^2 \\ &= \frac{1}{8}[2(k + 1) + 1]^2. \end{aligned}$$

Thus, P_{k+1} is true, and the proof is complete.

21 (1) For $n = 1$, $5^n - 1 = 4$ and 4 is a factor of 4.

(2) Assume 4 is a factor of $5^k - 1$. The $(k + 1)$st term is

$$\begin{aligned} 5^{k+1} - 1 &= 5 \cdot 5^k - 1 \\ &= 5 \cdot 5^k - 5 + 4 \\ &= 5(5^k - 1) + 4. \end{aligned}$$

By the induction hypothesis, 4 is a factor of $5^k - 1$ and 4 is a factor of 4, so 4 is a factor of the $(k + 1)$st term. Thus, P_{k+1} is true, and the proof is complete.

25 (1) For $n = 1$, $a - b$ is a factor of $a^1 - b^1$.

(2) Assume $a - b$ is a factor of $a^k - b^k$. Following the hint for the $(k + 1)$st term,

$$\begin{aligned} a^{k+1} - b^{k+1} &= a^k \cdot a - b \cdot a^k + b \cdot a^k - b^k \cdot b \\ &= a^k(a - b) + (a^k - b^k)b. \end{aligned}$$

Since $(a - b)$ is a factor of $a^k(a - b)$ and since by the induction hypothesis $a - b$ is a factor of $(a^k - b^k)$, it follows that $a - b$ is a factor of the $(k + 1)$st term. Thus, P_{k+1} is true, and the proof is complete.

29 (1) P_8 is true, since $5 + \log_2 8 \le 8$.

(2) Assume P_k is true: $5 + \log_2 k \le k$. Hence,

$$\begin{aligned} 5 + \log_2 (k + 1) &< 5 + \log_2 (k + k) \\ &= 5 + \log_2 2k \\ &= 5 + \log_2 2 + \log_2 k \\ &= (5 + \log_2 k) + 1 \\ &\le k + 1. \end{aligned}$$

Thus, P_{k+1} is true, and the proof is complete.

33 $\dfrac{n^3 + 6n^2 + 20n}{3}$ **35** $\dfrac{4n^3 - 12n^2 + 11n}{3}$

37 **(a)** $a + b + c = 1$, $8a + 4b + 2c = 5$, $27a + 9b + 3c = 14$; $a = \frac{1}{3}$, $b = \frac{1}{2}$, $c = \frac{1}{6}$

(b) The method used in part (a) shows that the formula is true for only $n = 1, 2, 3$.

39 (1) For $n = 1$,

$$\begin{aligned} \sin(\theta + 1\pi) &= \sin\theta\cos\pi + \cos\theta\sin\pi \\ &= -\sin\theta = (-1)^1 \sin\theta. \end{aligned}$$

(2) Assume P_k is true: $\sin(\theta + k\pi) = (-1)^k \sin\theta$. Hence,

$$\begin{aligned} &\sin[\theta + (k + 1)\pi] \\ &= \sin[(\theta + k\pi) + \pi] \\ &= \sin(\theta + k\pi)\cos\pi + \cos(\theta + k\pi)\sin\pi \\ &= [(-1)^k \sin\theta]\cdot(-1) + \cos(\theta + k\pi)\cdot(0) \\ &= (-1)^{k+1}\sin\theta. \end{aligned}$$

Thus, P_{k+1} is true, and the proof is complete.

41 (1) For $n = 1$,

$[r(\cos\theta + i\sin\theta)]^1 = r^1[\cos(1\theta) + i\sin(1\theta)]$.

(2) Assume P_k is true:

$[r(\cos\theta + i\sin\theta)]^k = r^k(\cos k\theta + i\sin k\theta)$.

Hence,

$$\begin{aligned} &[r(\cos\theta + i\sin\theta)]^{k+1} \\ &= [r(\cos\theta + i\sin\theta)]^k[r(\cos\theta + i\sin\theta)] \\ &= r^k[\cos k\theta + i\sin k\theta][r(\cos\theta + i\sin\theta)] \\ &= r^{k+1}[(\cos k\theta\cos\theta - \sin k\theta\sin\theta) + i(\sin k\theta\cos\theta + \cos k\theta\sin\theta)] \\ &= r^{k+1}[\cos(k + 1)\theta + i\sin(k + 1)\theta]. \end{aligned}$$

Thus, P_{k+1} is true, and the proof is complete.

EXERCISES 9.5

1 1440 **3** 5040 **5** 336 **7** 1 **9** 21

11 715 **13** $n(n - 1)$ **15** $(2n + 2)(2n + 1)$

17 $64x^3 - 48x^2y + 12xy^2 - y^3$

19 $x^6 + 6x^5y + 15x^4y^2 + 20x^3y^3 + 15x^2y^4 + 6xy^5 + y^6$

21 $x^7 - 7x^6y + 21x^5y^2 - 35x^4y^3 + 35x^3y^4 - 21x^2y^5 + 7xy^6 - y^7$

23 $81t^4 - 540t^3s + 1350t^2s^2 - 1500ts^3 + 625s^4$

25 $\frac{1}{243}x^5 + \frac{5}{81}x^4y^2 + \frac{10}{27}x^3y^4 + \frac{10}{9}x^2y^6 + \frac{5}{3}xy^8 + y^{10}$

27 $x^{-12} + 18x^{-9} + 135x^{-6} + 540x^{-3} + 1215 + 1458x^3 + 729x^6$

29 $x^{5/2} - 5x^{3/2} + 10x^{1/2} - 10x^{-1/2} + 5x^{-3/2} - x^{-5/2}$

31 $3^{25}c^{10} + 25 \cdot 3^{24}c^{52/5} + 300 \cdot 3^{23}c^{54/5}$

33 $-1680 \cdot 3^{13}z^{11} + 60 \cdot 3^{14}z^{13} - 3^{15}z^{15}$ **35** $\frac{189}{1024}c^8$

37 $\frac{114,688}{9}u^2v^6$ **39** $70x^2y^2$ **41** $448y^3x^{10}$

43 $-216y^9x^2$ **45** $-\frac{135}{16}$ **47** 4.8, 6.19

49 $4x^3 + 6x^2h + 4xh^2 + h^3$

51 $\binom{n}{1} = \dfrac{n!}{(n-1)!\,1!} = n$ and

$$\binom{n}{n-1} = \frac{n!}{[n-(n-1)]!\,(n-1)!} = \frac{n!}{1!\,(n-1)!} = n$$

EXERCISES 9.6

1 210 **3** 60,480 **5** 120 **7** 6 **9** 1 **11** $n!$
13 (a) 60 **(b)** 125 **15** 64 **17** $P(8, 3) = 336$
19 24 **21 (a)** 2,340,000 **(b)** 2,160,000
23 (a) 151,200 **(b)** 5760 **25** 1024
27 $P(8, 8) = 40{,}320$ **29** $P(6, 3) = 120$
31 (a) 27,600 **(b)** 35,152 **33** 9,000,000,000
35 $P(4, 4) = 24$ **37** $3! \cdot 2^3 = 48$
39 $(2^{16} - 1) \cdot 17$
41 (a) 900
(b) If n is even, $9 \cdot 10^{(n/2)-1}$; if n is odd, $9 \cdot 10^{(n-1)/2}$.

43 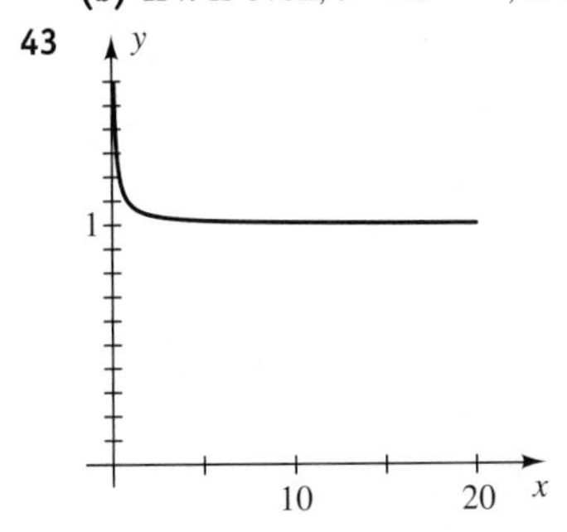

(a) $y = 1$ **(b)** $n! \approx \dfrac{n^n\sqrt{2\pi n}}{e^n}$

EXERCISES 9.7

1 35 **3** 9 **5** n **7** 1 **9** $\dfrac{12!}{5!\,3!\,2!\,2!} = 166{,}320$

11 $\dfrac{10!}{3!\,2!\,2!\,1!\,1!\,1!} = 151{,}200$ **13** $C(10, 5) = 252$

15 $C(8, 2) = 28$ **17** $(5! \cdot 4! \cdot 8!) \cdot 3! = 696{,}729{,}600$
19 $3 \cdot C(10, 2) \cdot C(8, 2) \cdot C(4, 2) \cdot C(6, 2) \cdot 3 \cdot 4 = 4{,}082{,}400$
21 $C(12, 3) \cdot C(8, 2) = 6160$ **23** $C(8, 3) = 56$
25 (a) $C(49, 6) = 13{,}983{,}816$ **(b)** $C(24, 6) = 134{,}596$
27 $C(n, 2) = 45$ and hence $n = 10$ **29** $C(6, 3) = 20$
31 By finding $C(31, 3) = 4495$
33 (a) $C(1000, 30) \approx 2.43 \times 10^{57}$
(b) $P(1000, 30) \approx 6.44 \times 10^{89}$
35 $C(4, 3) \cdot C(48, 2) = 4512$
37 (a) 1, 2, 4, 8, 16, 32, 64, 128, 256, 512
(b) $S_n = 2^{n-1}$

39 (a) 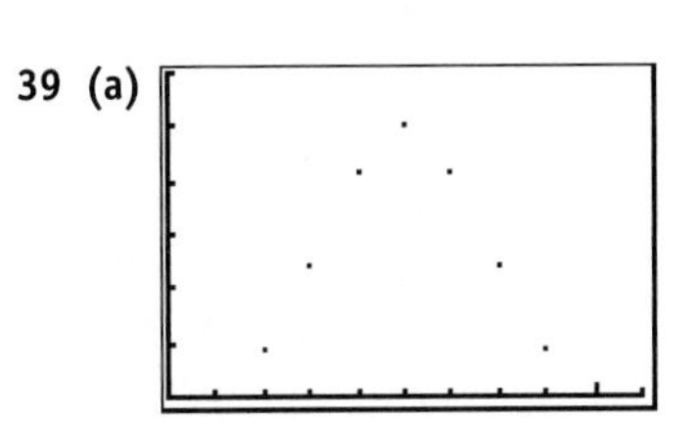

[0, 10] by [0, 300, 50]

(b) 252; 5

41 (a) 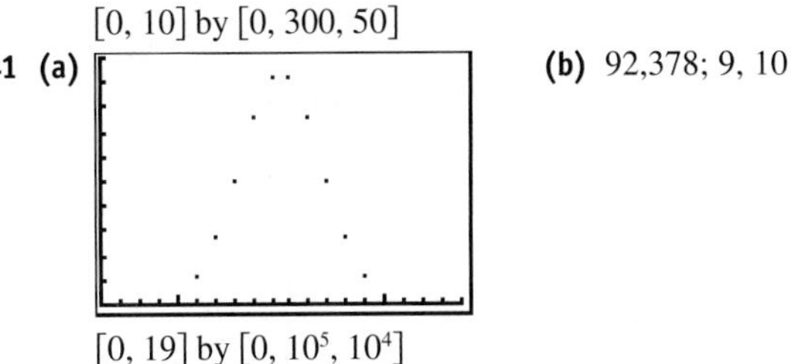

[0, 19] by $[0, 10^5, 10^4]$

(b) 92,378; 9, 10

43 The sum of two adjacent numbers is equal to the number below and between them.

EXERCISES 9.8

1 (a) $\dfrac{4}{52}$; 1 to 12 **(b)** $\dfrac{8}{52}$; 2 to 11 **(c)** $\dfrac{12}{52}$; 3 to 10

3 (a) $\dfrac{1}{6}$; 1 to 5 **(b)** $\dfrac{1}{6}$; 1 to 5 **(c)** $\dfrac{2}{6}$; 1 to 2

5 (a) $\dfrac{5}{15}$; 1 to 2 **(b)** $\dfrac{6}{15}$; 2 to 3 **(c)** $\dfrac{9}{15}$; 3 to 2

7 (a) $\dfrac{2}{36}$; 1 to 17 **(b)** $\dfrac{5}{36}$; 5 to 31 **(c)** $\dfrac{7}{36}$; 7 to 29

9 $\dfrac{6}{216}$ **11** $\dfrac{3}{8}$ **13** 5 to 2; 2 to 5 **15** 5 to 9; $\dfrac{9}{14}$

17 1.93 to 1 **19** $\dfrac{48 \cdot 13}{C(52, 5)} \approx 0.00024$

21 $\dfrac{C(13, 4) \cdot C(13, 1)}{C(52, 5)} \approx 0.00358$

23 $\dfrac{C(13, 5) \cdot 4}{C(52, 5)} \approx 0.00198$ **25** $\dfrac{4}{6}$

27 $(0.674)^4 \approx 0.2064$
29 (a) 0.45 **(b)** 0.10 **(c)** 0.70 **(d)** 0.95

31 (a) $\dfrac{C(20, 5) \cdot C(40, 0)}{C(60, 5)} \approx 0.0028$

(b) $1 - \dfrac{C(30, 0) \cdot C(30, 5)}{C(60, 5)} \approx 0.9739$

(c) $\dfrac{C(10, 0) \cdot C(50, 5)}{C(60, 5)} + \dfrac{C(10, 1) \cdot C(50, 4)}{C(60, 5)} \approx 0.8096$

33 (a) $\dfrac{C(8, 8)}{2^8} \approx 0.00391$ **(b)** $\dfrac{C(8, 7)}{2^8} = 0.03125$

(c) $\dfrac{C(8, 6)}{2^8} = 0.109375$

(d) $\dfrac{C(8, 6) + C(8, 7) + C(8, 8)}{2^8} \approx 0.14453$

35 $1 - \frac{C(48, 5)}{C(52, 5)} \approx 0.34116$

37 **(a)** A representative outcome is (nine of clubs, 3); 312
(b) 20; 292; $\frac{20}{312}$ **(c)** No; yes; $\frac{72}{312}; \frac{156}{312}; \frac{36}{312}; \frac{192}{312}$
(d) Yes; no; 0; $\frac{92}{312}$

39 $1 - \frac{10}{36} = \frac{26}{36}$ **41** **(a)** $\frac{1}{32}$ **(b)** $1 - \frac{1}{32} = \frac{31}{32}$

43 **(a)** $\frac{C(4, 4)}{4!} = \frac{1}{24}$ **(b)** $\frac{C(4, 2)}{4!} = \frac{1}{4}$

45 **(a)** 0 **(b)** $\frac{1}{9}$

47 **(a)** $\frac{304,366}{442,398} \approx 0.688$ **(b)** $\frac{344,391}{442,398} \approx 0.778$

49 12.5% **51** **(a)** $\frac{1}{16}$ **(b)** $\frac{C(4, 2)}{2^4} = \frac{6}{16}$

53 $\frac{2}{25,827,165}$ (about 1 chance in 13 million)

55 $\frac{1970}{39,800} \approx 0.0495$

57 **(a)** $\frac{8}{36}$ **(b)** $\frac{1}{36}$ **(c)** $\frac{244}{495} \approx 0.4929$

59 **(a)** 0.9639 **(b)** 0.95 **61** **(b)** 0.76 **63** \$0.99

65 \$0.20

CHAPTER 9 REVIEW EXERCISES

1 $5, -2, -1, -\frac{20}{29}; -\frac{7}{19}$

2 $0.9, -1.01, 0.999, -1.0001; 0.9999999$

3 $2, \frac{1}{2}, \frac{5}{4}, \frac{7}{8}; \frac{65}{64}$ **4** $\frac{1}{12}, \frac{1}{15}, \frac{1}{15}, \frac{8}{105}; \frac{8}{45}$

5 $10, \frac{11}{10}, \frac{21}{11}, \frac{32}{21}; \frac{53}{32}$ **6** 2, 2, 2, 2, 2

7 $9, 3, \sqrt{3}, \sqrt[4]{3}, \sqrt[8]{3}$ **8** $1, \frac{1}{2}, \frac{2}{3}, \frac{3}{5}, \frac{5}{8}$

9 75 **10** $-\frac{37}{10}$ **11** 940 **12** -10 **13** $\sum_{n=1}^{5} 3n$

14 $\sum_{n=1}^{6} 2^{3-n}$ **15** $\sum_{n=1}^{99} \frac{1}{n(n+1)}$ **16** $\sum_{n=1}^{98} \frac{1}{n(n+1)(n+2)}$

17 $\sum_{n=1}^{4} \frac{n}{3n-1}$ **18** $\sum_{n=1}^{4} \frac{n}{5n-1}$

19 $\sum_{n=1}^{5} (-1)^{n+1}(105 - 5n)$ **20** $\sum_{n=1}^{7} (-1)^{n-1} \frac{1}{n}$

21 $\sum_{n=0}^{25} a_n x^{4n}$ **22** $\sum_{n=0}^{20} a_n x^{3n}$ **23** $1 + \sum_{k=1}^{n} (-1)^k \frac{x^{2k}}{2k}$

24 $1 + \sum_{k=1}^{n} \frac{x^k}{k}$ **25** $-5 - 8\sqrt{3}; -5 - 35\sqrt{3}$

26 52 **27** -31; 50 **28** 12

29 $20, 14, 8, 2, -4, -10$ **30** 64 **31** -0.00003

32 1562.5 or -1562.5 **33** $4\sqrt{2}$ **34** $-\frac{12,800}{2187}$

35 17; 3 **36** $\frac{1}{81}; \frac{211}{1296}$ **37** 570 **38** 32.5

39 2041 **40** -506 **41** $\frac{5}{7}$ **42** $\frac{6268}{999}$

43 (1) P_1 is true, since $3(1) - 1 = \frac{1[3(1) + 1]}{2} = 2$.
(2) Assume P_k is true:

$$2 + 5 + 8 + \cdots + (3k - 1) = \frac{k(3k + 1)}{2}.$$

Hence,

$$\begin{aligned} 2 + 5 + 8 + \cdots + (3k - 1) + 3(k + 1) - 1 &= \frac{k(3k + 1)}{2} + 3(k + 1) - 1 \\ &= \frac{3k^2 + k + 6k + 4}{2} \\ &= \frac{3k^2 + 7k + 4}{2} \\ &= \frac{(k + 1)(3k + 4)}{2} \\ &= \frac{(k + 1)[3(k + 1) + 1]}{2}. \end{aligned}$$

Thus, P_{k+1} is true, and the proof is complete.

44 (1) P_1 is true, since $[2(1)]^2 = \frac{[2(1)][2(1) + 1][1 + 1]}{3} = 4$.
(2) Assume P_k is true:

$$2^2 + 4^2 + 6^2 + \cdots + (2k)^2 = \frac{(2k)(2k + 1)(k + 1)}{3}.$$

Hence,

$$\begin{aligned} 2^2 + 4^2 + 6^2 + \cdots + (2k)^2 + [2(k + 1)]^2 &= \frac{(2k)(2k + 1)(k + 1)}{3} + [2(k + 1)]^2 \\ &= (k + 1)\left(\frac{4k^2 + 2k}{3} + \frac{12(k + 1)}{3}\right) \\ &= \frac{(k + 1)(4k^2 + 14k + 12)}{3} \\ &= \frac{2(k + 1)(2k + 3)(k + 2)}{3}. \end{aligned}$$

Thus, P_{k+1} is true, and the proof is complete.

45 (1) P_1 is true, since $\dfrac{1}{[2(1)-1][2(1)+1]} = \dfrac{1}{2(1)+1} = \dfrac{1}{3}$.

(2) Assume P_k is true:

$$\frac{1}{1\cdot 3}+\frac{1}{3\cdot 5}+\frac{1}{5\cdot 7}+\cdots+\frac{1}{(2k-1)(2k+1)}=\frac{k}{2k+1}.$$

Hence,

$$\begin{aligned}\frac{1}{1\cdot 3}+\frac{1}{3\cdot 5}+\frac{1}{5\cdot 7}+\cdots+\frac{1}{(2k-1)(2k+1)}&\\ +\frac{1}{(2k+1)(2k+3)}&=\frac{k}{2k+1}+\frac{1}{(2k+1)(2k+3)}\\ &=\frac{k(2k+3)+1}{(2k+1)(2k+3)}\\ &=\frac{2k^2+3k+1}{(2k+1)(2k+3)}\\ &=\frac{(2k+1)(k+1)}{(2k+1)(2k+3)}\\ &=\frac{k+1}{2(k+1)+1}.\end{aligned}$$

Thus, P_{k+1} is true, and the proof is complete.

46 (1) P_1 is true, since $1(1+1) = \dfrac{(1)(1+1)(1+2)}{3} = 2$.

(2) Assume P_k is true:

$$1\cdot 2+2\cdot 3+3\cdot 4+\cdots+k(k+1)=\frac{k(k+1)(k+2)}{3}.$$

Hence,

$$\begin{aligned}1\cdot 2+2\cdot 3+3\cdot 4+\cdots+k(k+1)&\\ +(k+1)(k+2)&\\ &=\frac{k(k+1)(k+2)}{3}+(k+1)(k+2)\\ &=(k+1)(k+2)\left(\frac{k}{3}+1\right)\\ &=\frac{(k+1)(k+2)(k+3)}{3}.\end{aligned}$$

Thus, P_{k+1} is true, and the proof is complete.

47 (1) For $n = 1$, $n^3 + 2n = 3$ and 3 is a factor of 3.

(2) Assume 3 is a factor of $k^3 + 2k$. The $(k+1)$st term is

$$\begin{aligned}(k+1)^3+2(k+1)&=k^3+3k^2+5k+3\\ &=(k^3+2k)+(3k^2+3k+3)\\ &=(k^3+2k)+3(k^2+k+1).\end{aligned}$$

By the induction hypothesis, 3 is a factor of $k^3 + 2k$ and 3 is a factor of $3(k^2 + k + 1)$, so 3 is a factor of the $(k+1)$st term.

Thus, P_{k+1} is true, and the proof is complete.

48 (1) P_5 is true, since $5^2 + 3 < 2^5$.

(2) Assume P_k is true: $k^2 + 3 < 2^k$. Hence,

$$\begin{aligned}(k+1)^2+3&=k^2+2k+4\\ &=(k^2+3)+(k+1)\\ &<2^k+(k+1)\\ &<2^k+2^k\\ &=2\cdot 2^k=2^{k+1}.\end{aligned}$$

Thus, P_{k+1} is true, and the proof is complete.

49 (1) P_4 is true, since $2^4 \le 4!$.

(2) Assume P_k is true: $2^k \le k!$. Hence,

$2^{k+1} = 2\cdot 2^k \le 2\cdot k! < (k+1)\cdot k! = (k+1)!$.

Thus, P_{k+1} is true, and the proof is complete.

50 (1) P_{10} is true, since $10^{10} \le 10^{10}$.

(2) Assume P_k is true: $10^k \le k^k$. Hence,

$$\begin{aligned}10^{k+1}&=10\cdot 10^k\le 10\cdot k^k<(k+1)\cdot k^k\\ &<(k+1)\cdot(k+1)^k=(k+1)^{k+1}.\end{aligned}$$

Thus, P_{k+1} is true, and the proof is complete.

51 $x^{12} - 18x^{10}y + 135x^8y^2 - 540x^6y^3 + 1215x^4y^4 - 1458x^2y^5 + 729y^6$

52 $16x^4 + 32x^3y^3 + 24x^2y^6 + 8xy^9 + y^{12}$

53 $x^8 + 40x^7 + 760x^6$ **54** $-\dfrac{63}{16}y^{12}c^{10}$

55 $21{,}504x^{10}y^2$ **56** 52,500,000

57 **(a)** $d = 1 - \dfrac{1}{2}a_1$ **(b)** In ft: $1\frac{1}{4}, 2, 2\frac{3}{4}, 3\frac{1}{2}$

58 24 ft **59** $\dfrac{2}{1-f}$ **60** $P(10, 10) = 3{,}628{,}800$

61 **(a)** $P(52, 13) \approx 3.954 \times 10^{21}$
(b) $P(13, 5)\cdot P(13, 3)\cdot P(13, 3)\cdot P(13, 2) \approx 7.094 \times 10^{13}$

62 **(a)** $P(6, 4) = 360$ **(b)** $6^4 = 1296$

63 **(a)** $C(12, 8) = 495$ **(b)** $C(9, 5) = 126$

64 $\dfrac{17!}{6!\,5!\,4!\,2!} = 85{,}765{,}680$ **65** 5 to 8; $\dfrac{8}{13}$

66 **(a)** $\dfrac{2}{4}$ **(b)** $\dfrac{2}{8}$

67 **(a)** $\dfrac{P(26, 4)\cdot 2}{P(52, 4)} \approx 0.1104$ **(b)** $\dfrac{26^2\cdot 25^2}{P(52, 4)} \approx 0.0650$

68 **(a)** $\dfrac{1}{1000}$ **(b)** $\dfrac{10}{1000}$ **(c)** $\dfrac{50}{1000}$

69 $\dfrac{C(4, 1)}{2^4} = \dfrac{4}{16}$; 1 to 3

70 **(a)** $\dfrac{C(6, 4) + C(6, 5) + C(6, 6)}{2^6} = \dfrac{22}{64}$

(b) $1 - \dfrac{22}{64} = \dfrac{42}{64}$

71 **(a)** $\dfrac{1}{312}$ **(b)** $\dfrac{57}{312}$ **72** 0.44 **73** $\dfrac{8}{36}$

74 5.8125

CHAPTER 9 DISCUSSION EXERCISES

1 $a_n = 2n + \frac{1}{24}(n-1)(n-2)(n-3)(n-4)(a-10)$
(The answer is not unique.)

2 $\geq$; $j = 94$

3 **(a)** $\frac{1}{5}n^5 + \frac{1}{2}n^4 + \frac{1}{3}n^3 - \frac{1}{30}n$
(b) Use mathematical induction.

4 **(a)** $2n^4 + 4n^3 + 2n^2$
(b) Use mathematical induction.

5 Examine the number of digits in the exponent of the value in scientific notation.

6 The $(k+1)$st coefficient $(k = 0, 1, 2, \ldots, n)$ of the expansion of $(a+b)^n$, namely $\binom{n}{k}$, is the same as the number of k-element subsets of an n-element set.

7 4.61 **8** \$5.33

9 Penny amounts:

\$237.37 \$215.63 \$195.89 \$177.95 \$161.65
\$146.85 \$133.40 \$121.18 \$110.08 \$100.00

Realistic ten dollar amounts:

\$240.00 \$220.00 \$200.00 \$180.00 \$160.00
\$140.00 \$130.00 \$120.00 \$110.00 \$100.00

10 11 toppings are available.

11 **(a)** $\frac{1}{146{,}107{,}962}$ **(b)** $\frac{3{,}991{,}302}{146{,}107{,}962}$ (about 1 in 36.61)
(c) $\frac{28{,}800{,}030}{146{,}107{,}962} \approx 0.20$ **(d)** \$117,307,932

12 0.43 **13** $0^0 = 1$ **14** The sum equals π.

15 **(a)** $\tan 5x = \frac{5\tan x - 10\tan^3 x + \tan^5 x}{1 - 10\tan^2 x + 5\tan^4 x}$
(b) $\cos 5x = 1\cos^5 x - 10\cos^3 x \sin^2 x + 5\cos x \sin^4 x$;
$\sin 5x = 5\cos^4 x \sin x - 10\cos^2 x \sin^3 x + 1\sin^5 x$

Chapter 10

EXERCISES 10.1

1 $V(0, 0)$; $F(0, 2)$; $y = -2$

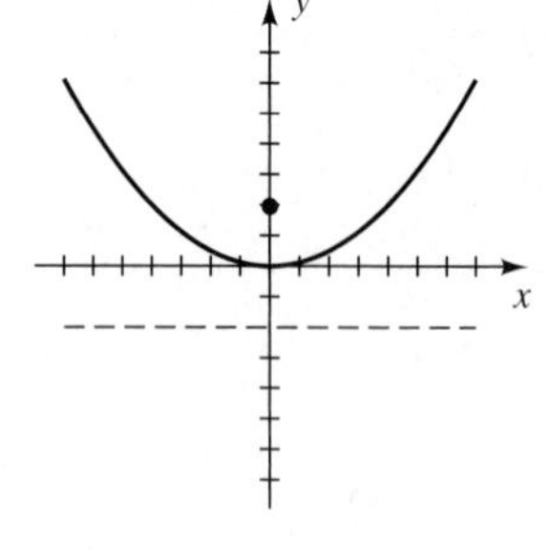

3 $V(0, 0)$; $F\left(-\frac{3}{8}, 0\right)$; $x = \frac{3}{8}$

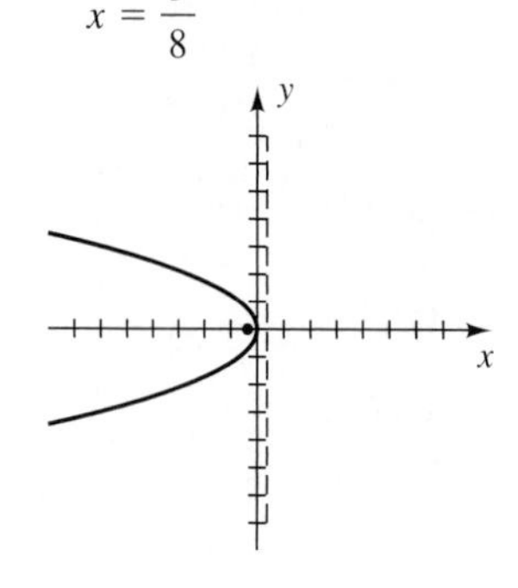

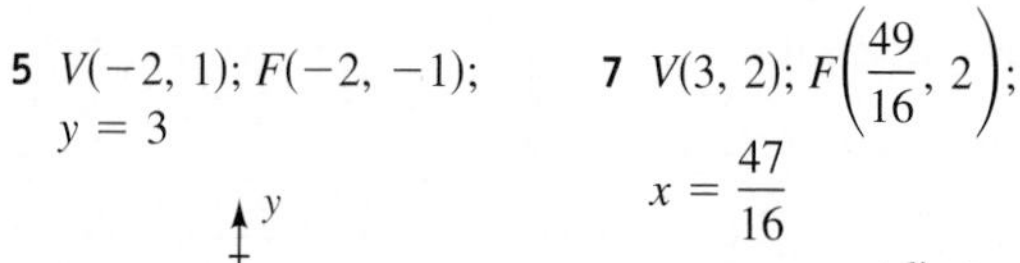

5 $V(-2, 1)$; $F(-2, -1)$; $y = 3$

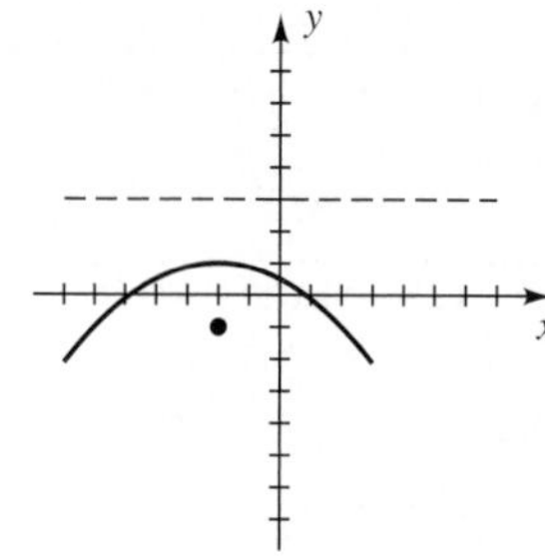

7 $V(3, 2)$; $F\left(\frac{49}{16}, 2\right)$; $x = \frac{47}{16}$

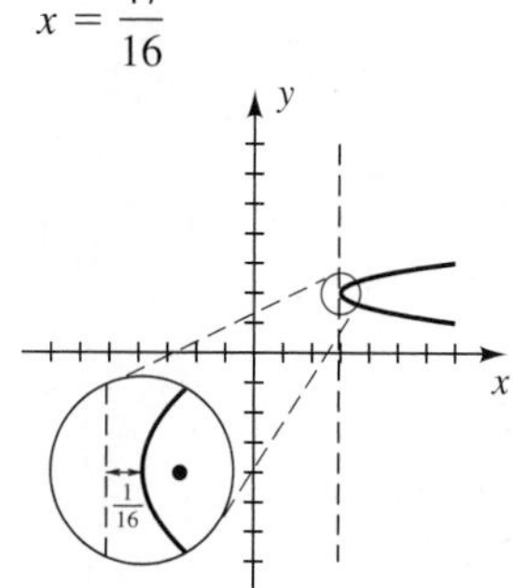

9 $V(2, -2)$; $F\left(2, -\frac{7}{4}\right)$; $y = -\frac{9}{4}$

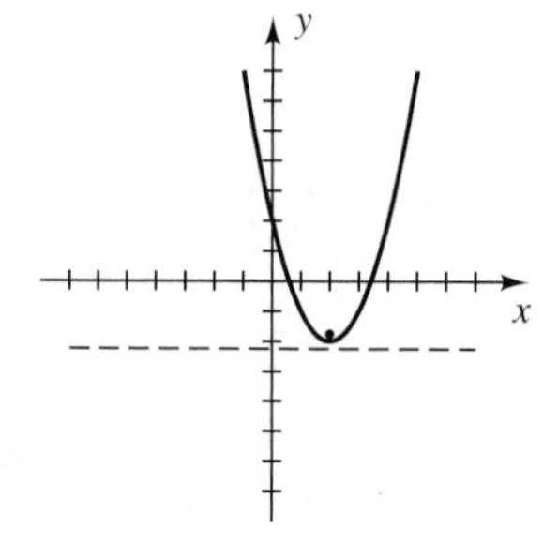

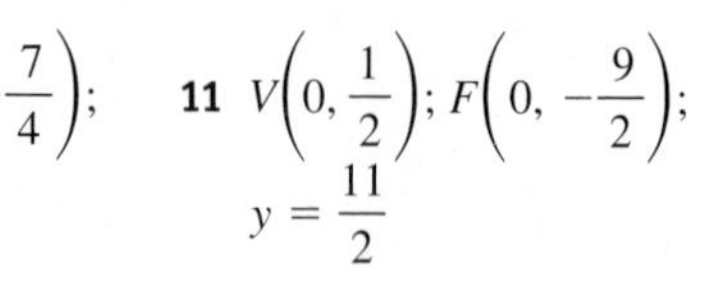

11 $V\left(0, \frac{1}{2}\right)$; $F\left(0, -\frac{9}{2}\right)$; $y = \frac{11}{2}$

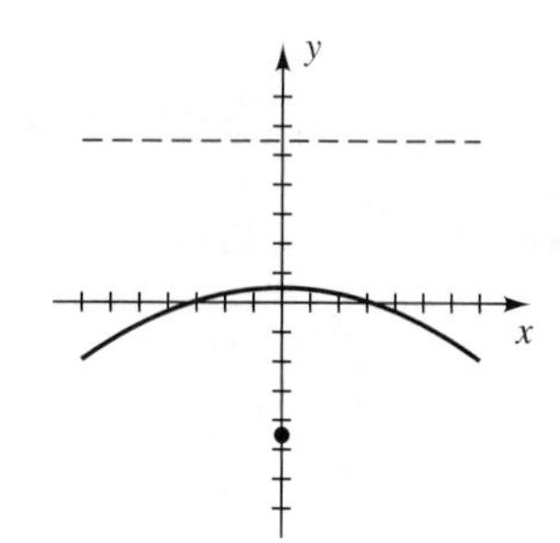

13 $y^2 = 20(x-1)$

15 $(x+2)^2 = -16(y-3)$

17 $(x-3)^2 = 6\left(y - \frac{1}{2}\right)$

19 $y^2 = 8x$

21 $(x-6)^2 = 12(y-1)$

23 $(y+5)^2 = 4(x-3)$

25 $y^2 = -12(x+1)$

27 $3x^2 = -4y$

29 $(y-5)^2 = 2(x+3)$

31 $x^2 = 16(y-1)$

33 $(y-3)^2 = -8(x+4)$

35 $y = -\sqrt{x+3} - 1$

37 $x = \sqrt{y-4} - 1$

39 $y = -x^2 + 2x + 5$

41 $x = y^2 - 3y + 1$

43 4 in.

45 $\frac{9}{16}$ ft from the center of the paraboloid

47 $2\sqrt{480} \approx 43.82$ in.

49 (a) $p = \frac{r^2}{4h}$ **(b)** $10\sqrt{2}$ ft **51** 57,000 ft^2

53

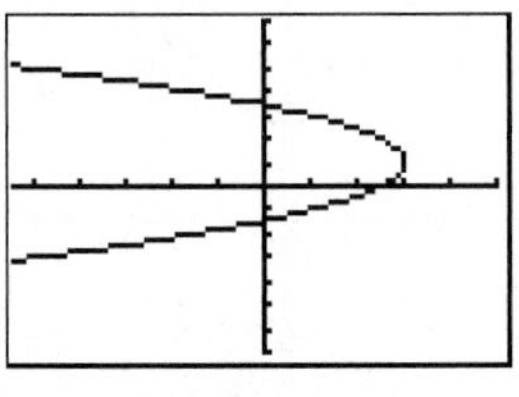

$[-11, 10, 2]$ by $[-7, 7]$

55 $(2.08, -1.04)$, $(2.92, 1.38)$

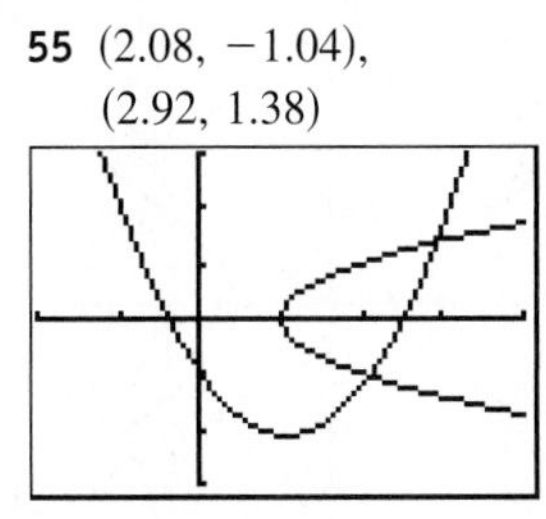

$[-2, 4]$ by $[-3, 3]$

EXERCISES 10.2

1 $V(\pm 3, 0)$; $F(\pm\sqrt{5}, 0)$

3 $V(0, \pm 4)$; $F(0, \pm 1)$

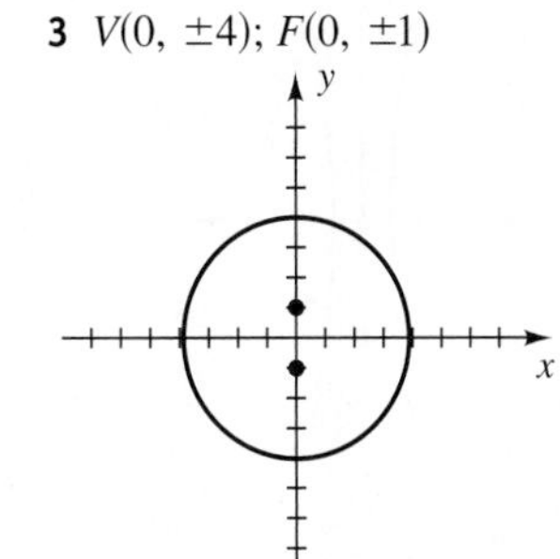

5 $V(0, \pm 4)$; $F(0, \pm 2\sqrt{3})$

7 $V\left(\pm\frac{1}{2}, 0\right)$; $F\left(\pm\frac{1}{10}\sqrt{21}, 0\right)$

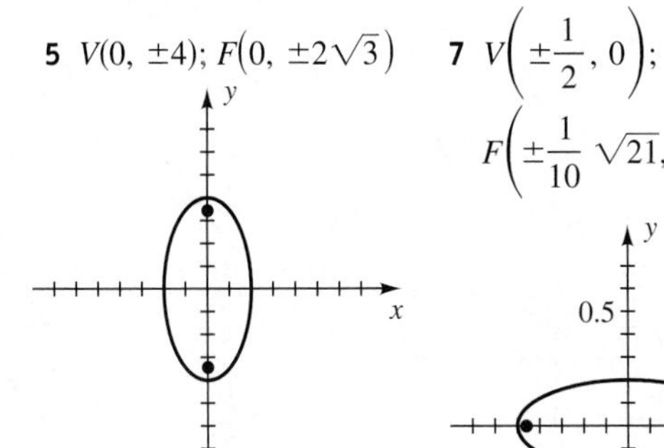

9 $V(3 \pm 4, -4)$; $F(3 \pm \sqrt{7}, -4)$

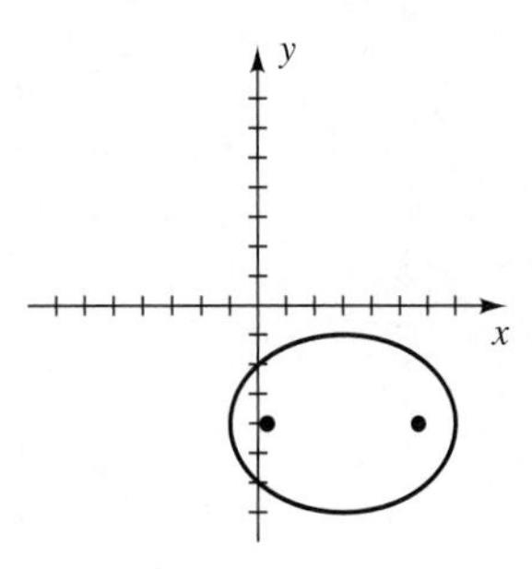

11 $V(4 \pm 3, 2)$; $F(4 \pm \sqrt{5}, 2)$

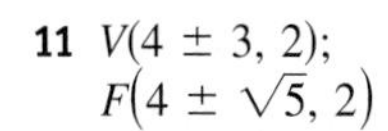

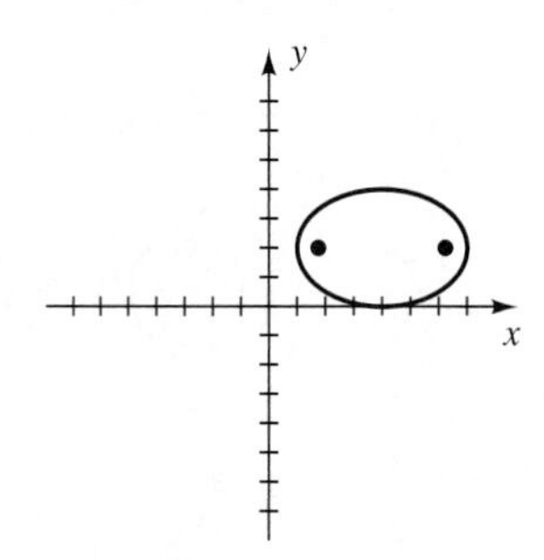

13 $V(5, 2 \pm 5)$; $F(5, 2 \pm \sqrt{21})$

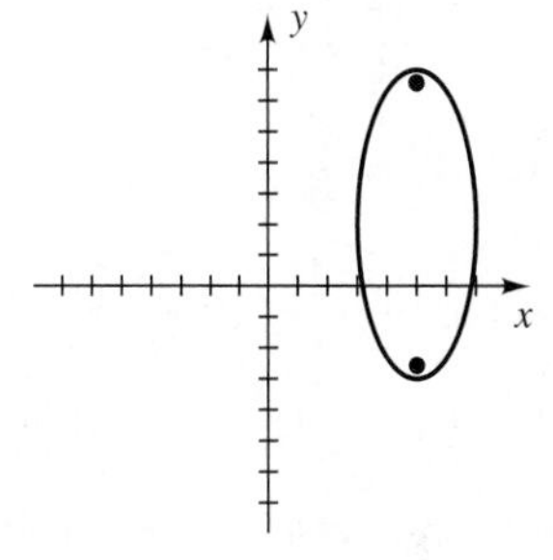

15 $\frac{x^2}{4} + \frac{y^2}{36} = 1$ **17** $\frac{(x + 2)^2}{25} + \frac{(y - 1)^2}{4} = 1$

19 $\frac{x^2}{64} + \frac{y^2}{39} = 1$ **21** $\frac{4x^2}{9} + \frac{y^2}{25} = 1$

23 $\frac{8x^2}{81} + \frac{y^2}{36} = 1$ **25** $\frac{x^2}{7} + \frac{y^2}{16} = 1$

27 $\frac{x^2}{4} + 9y^2 = 1$ **29** $\frac{x^2}{16} + \frac{4y^2}{25} = 1$

31 $(2, 2)$, $(4, 1)$

33 $\frac{x^2}{25} + \frac{y^2}{16} = 1$ **35** $\frac{x^2}{64} + \frac{y^2}{289} = 1$

37 $\dfrac{x^2}{25} + \dfrac{y^2}{9} = 1$

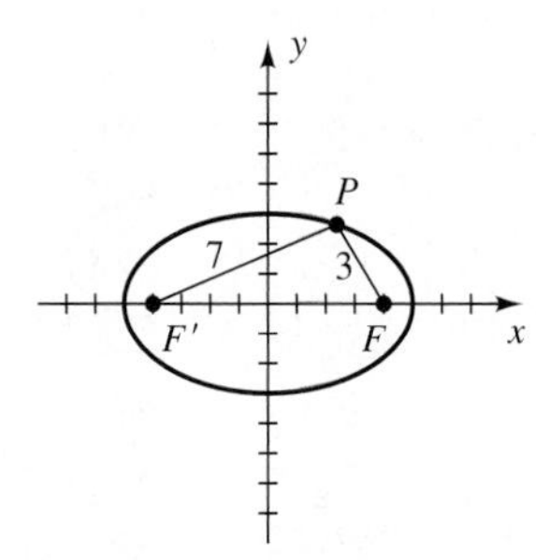

39 Upper half of $\dfrac{x^2}{49} + \dfrac{y^2}{121} = 1$

41 Left half of $x^2 + \dfrac{y^2}{9} = 1$

43 Right half of $\dfrac{(x-1)^2}{4} + \dfrac{(y+2)^2}{9} = 1$

45 Lower half of $\dfrac{(x+1)^2}{9} + \dfrac{(y-2)^2}{49} = 1$

47 $\sqrt{84} \approx 9.2$ ft **49** 94,581,000; 91,419,000

51 **(a)** $d = h - \sqrt{h^2 - \frac{1}{4}k^2}$; $d' = h + \sqrt{h^2 - \frac{1}{4}k^2}$

(b) 16 cm; 2 cm from V

53 5 ft

55

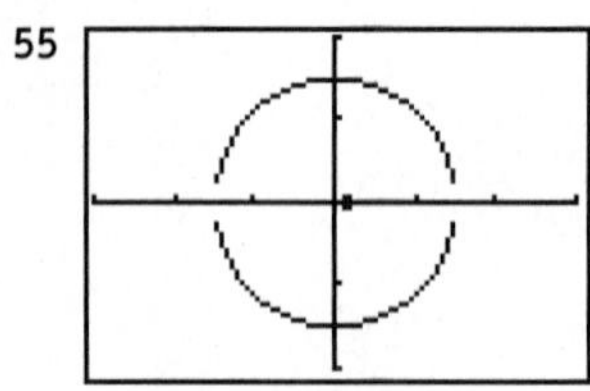

$[-300, 300, 100]$ by $[-200, 200, 100]$

57 $(\pm 1.540, 0.618)$

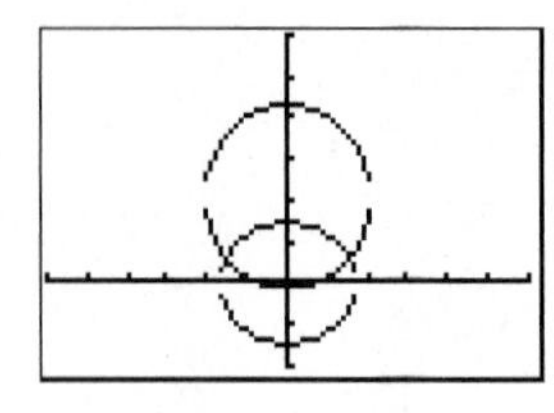

$[-6, 6]$ by $[-2, 6]$

59 $(-0.88, 0.76)$, $(-0.48, -0.91)$, $(0.58, -0.81)$, $(0.92, 0.59)$

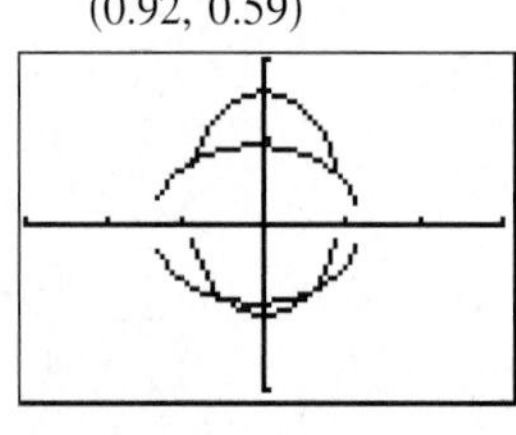

$[-3, 3]$ by $[-2, 2]$

EXERCISES 10.3

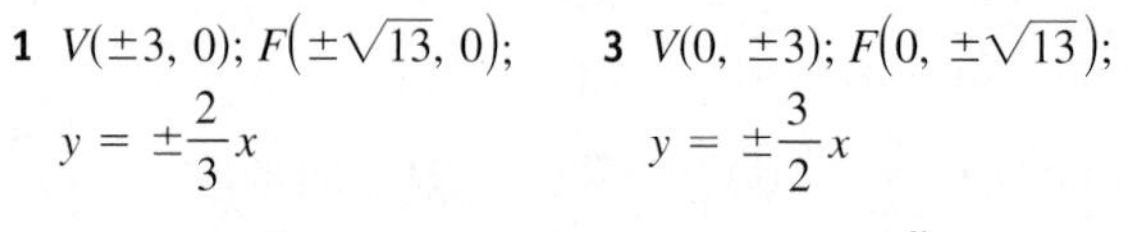

1 $V(\pm 3, 0)$; $F(\pm\sqrt{13}, 0)$; $y = \pm\frac{2}{3}x$

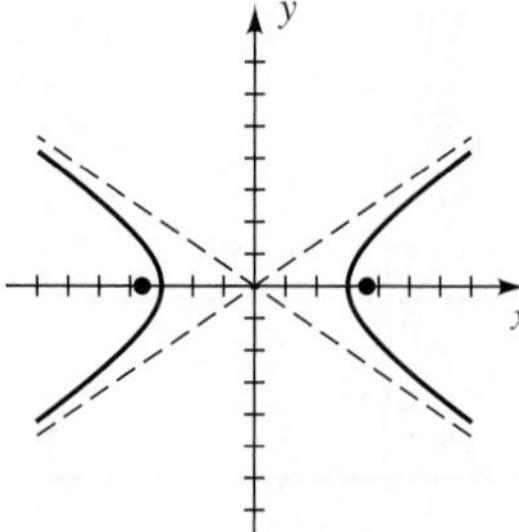

3 $V(0, \pm 3)$; $F(0, \pm\sqrt{13})$; $y = \pm\frac{3}{2}x$

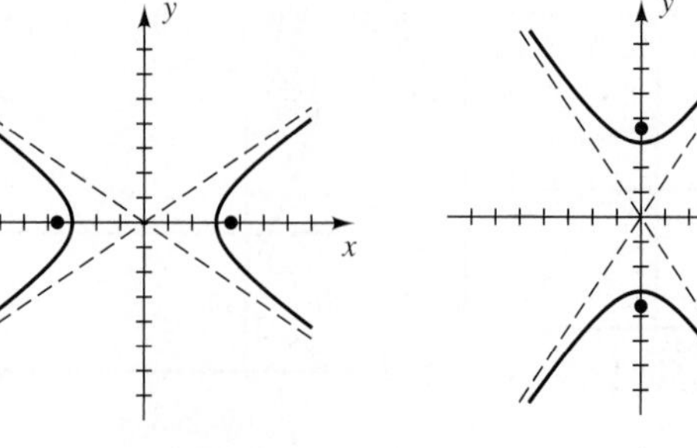

5 $V(\pm 1, 0)$; $F(\pm 5, 0)$; $y = \pm\sqrt{24}x$

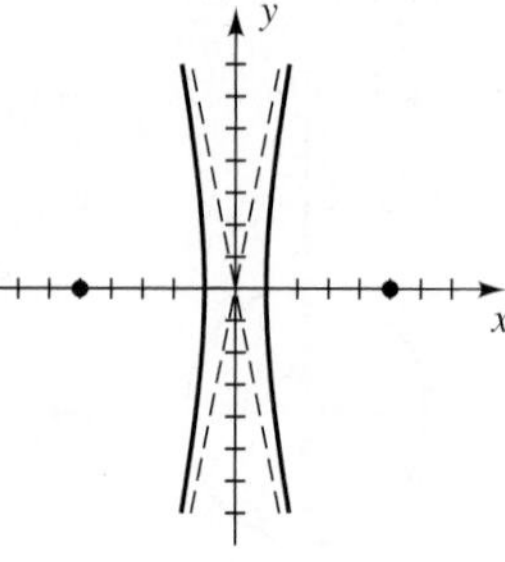

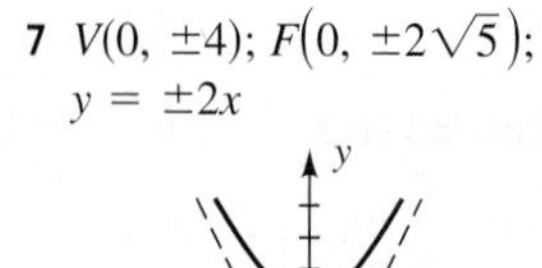

7 $V(0, \pm 4)$; $F(0, \pm 2\sqrt{5})$; $y = \pm 2x$

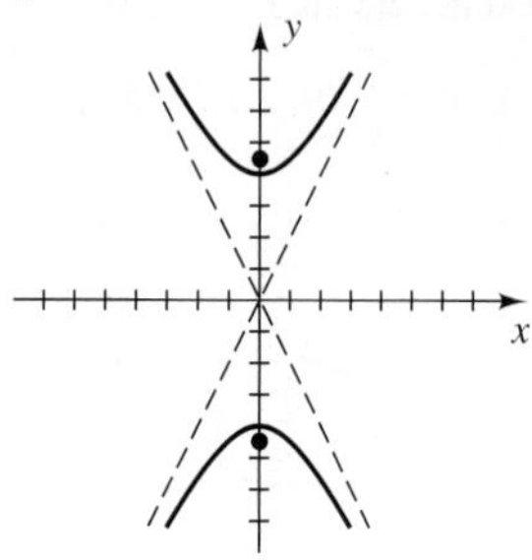

9 $V\left(\pm\frac{1}{4}, 0\right)$; $F\left(\pm\frac{1}{12}\sqrt{13}, 0\right)$; $y = \pm\frac{2}{3}x$

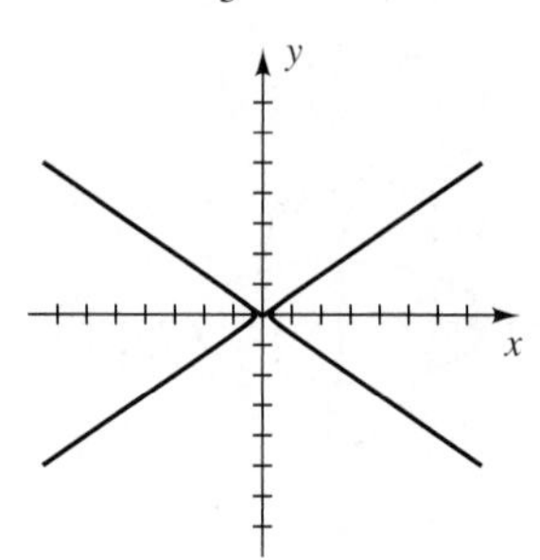

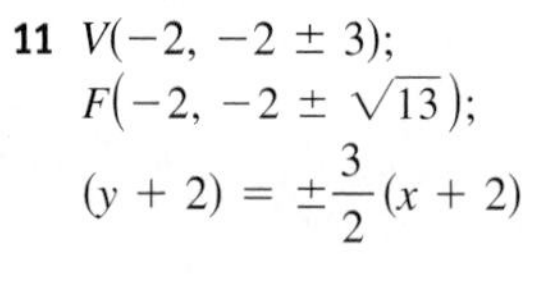

11 $V(-2, -2 \pm 3)$; $F(-2, -2 \pm \sqrt{13})$; $(y + 2) = \pm\frac{3}{2}(x + 2)$

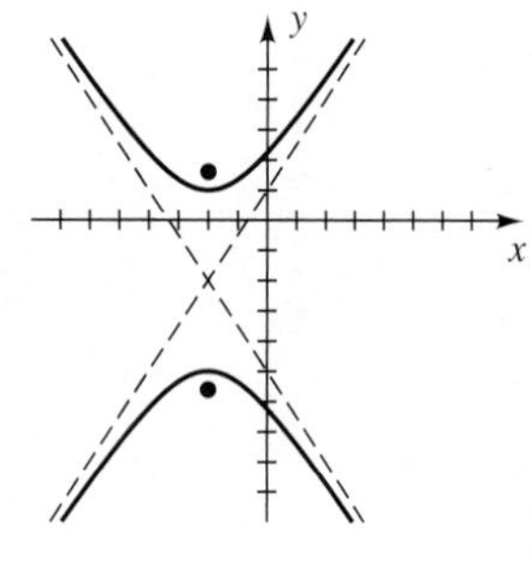

13 $V(-3 \pm 5, -2)$; $F(-3 \pm 13, -2)$; $(y + 2) = \pm\frac{12}{5}(x + 3)$

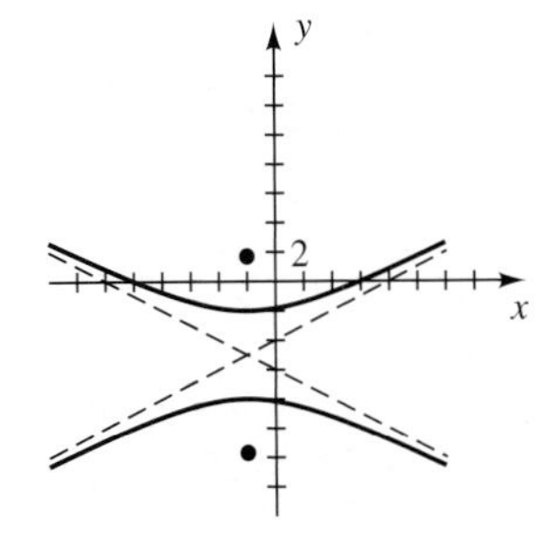

15 $V(-2, -5 \pm 3)$; $F(-2, -5 \pm 3\sqrt{5})$; $(y + 5) = \pm\frac{1}{2}(x + 2)$

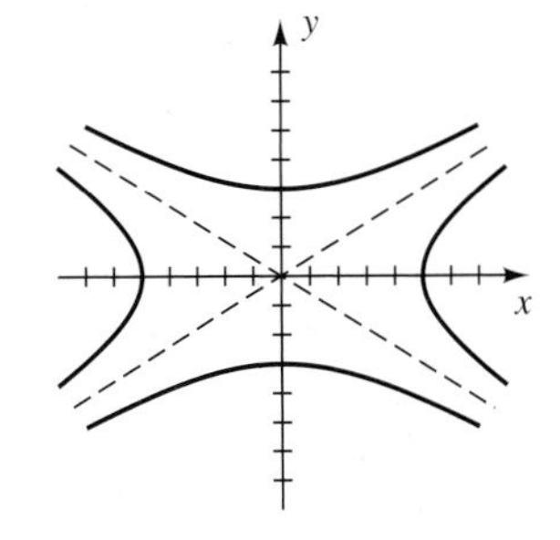

17 $\frac{x^2}{9} - \frac{y^2}{16} = 1$ **19** $(y + 3)^2 - \frac{(x + 2)^2}{3} = 1$

21 $y^2 - \frac{x^2}{15} = 1$ **23** $\frac{x^2}{9} - \frac{y^2}{16} = 1$ **25** $\frac{y^2}{21} - \frac{x^2}{4} = 1$

27 $\frac{x^2}{9} - \frac{y^2}{36} = 1$ **29** $\frac{x^2}{25} - \frac{y^2}{100} = 1$ **31** $\frac{y^2}{25} - \frac{x^2}{49} = 1$

33 Parabola with horizontal axis

35 Hyperbola **37** Circle **39** Ellipse

41 Parabola with vertical axis **43** $(0, 4), \left(\frac{8}{3}, \frac{20}{3}\right)$

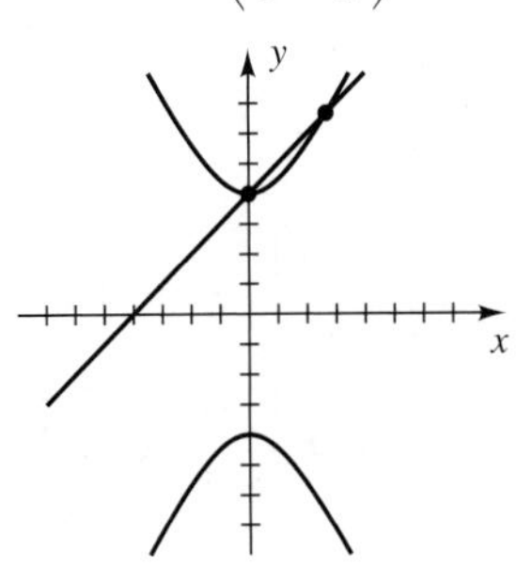

45 $\frac{x^2}{144} - \frac{y^2}{25} = 1$ **47** $\frac{y^2}{64} - \frac{x^2}{36} = 1$

49 $\frac{y^2}{16} - \frac{x^2}{9} = 1$

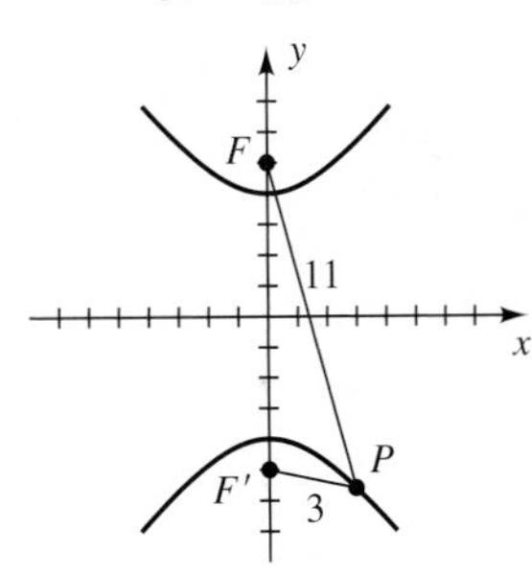

51 Right branch of $\frac{x^2}{25} - \frac{y^2}{16} = 1$

53 Upper branch of $\frac{y^2}{9} - \frac{x^2}{49} = 1$

55 Lower halves of the branches of $\frac{x^2}{16} - \frac{y^2}{81} = 1$

57 Left halves of the branches of $\frac{y^2}{36} - \frac{x^2}{16} = 1$

59 The graphs have the same asymptotes.

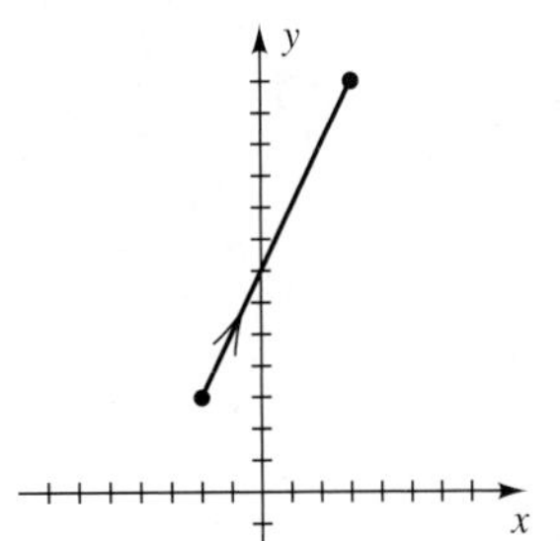

61 60.97 meters

63 If a coordinate system similar to that in Example 6 is introduced, then the ship's coordinates are $\left(\frac{80}{3}\sqrt{34}, 100\right) \approx (155.5, 100)$.

65 (0.741, 2.206)

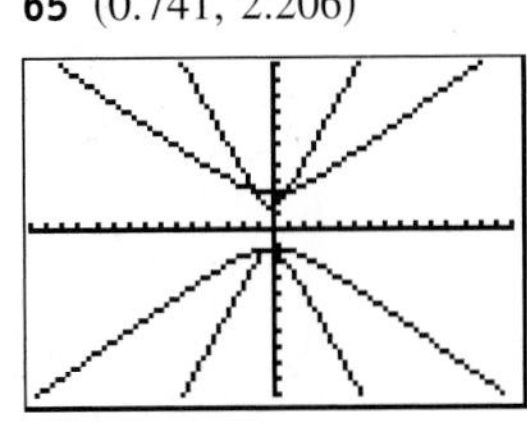

[−15, 15] by [−10, 10]

67 None

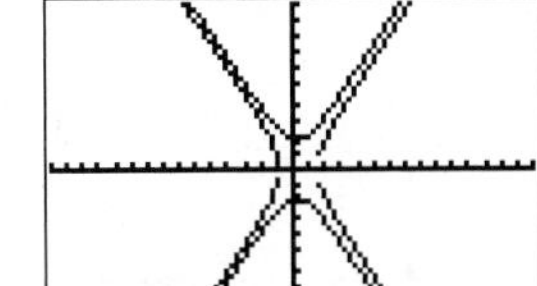

[−15, 15] by [−10, 10]

69 **(a)** $(6.63 \times 10^7, 0)$ **(b)** $v > 103{,}600$ m/sec

EXERCISES 10.4

1 $y = 2x + 7$

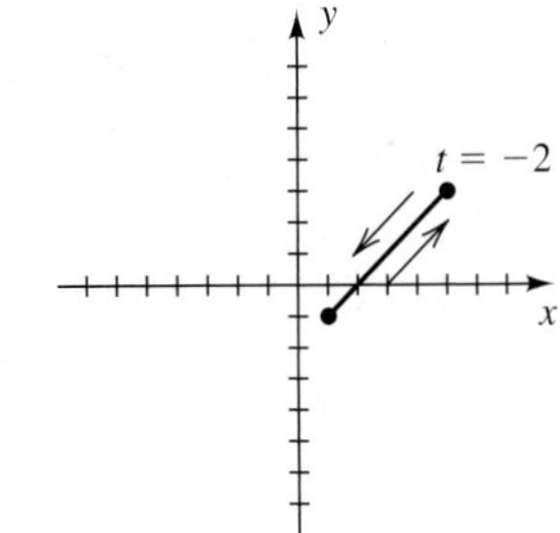

3 $y = x - 2$

5 $(y-3)^2 = x+5$

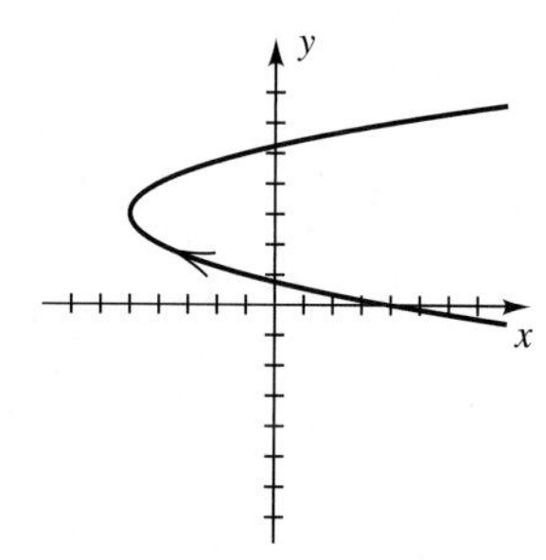

7 $\dfrac{(x-1)^2}{16} + \dfrac{y^2}{9} = 1$

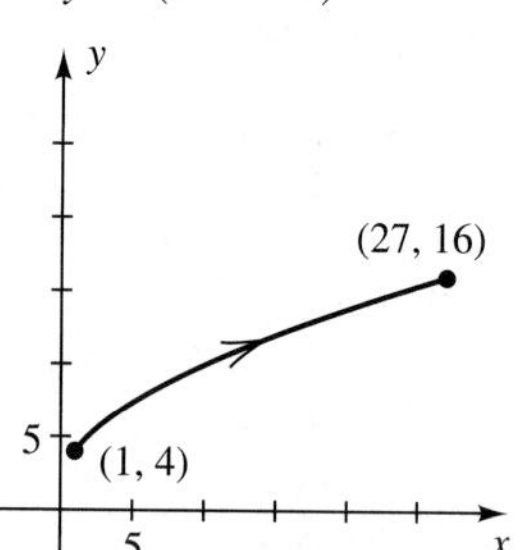

9 $(x-2)^2 + (y+1)^2 = 9$

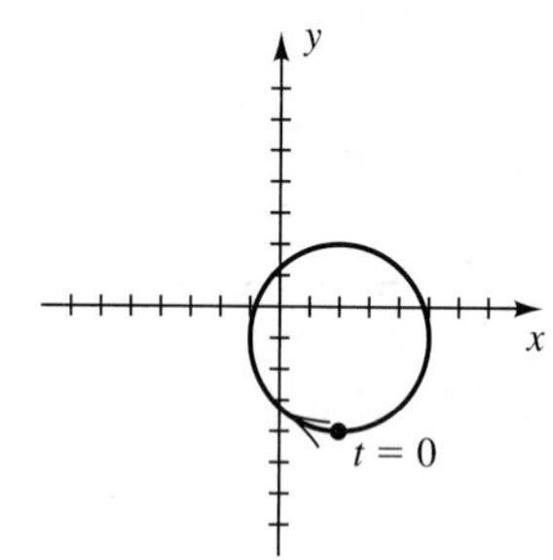

11 $x^2 - y^2 = 1$

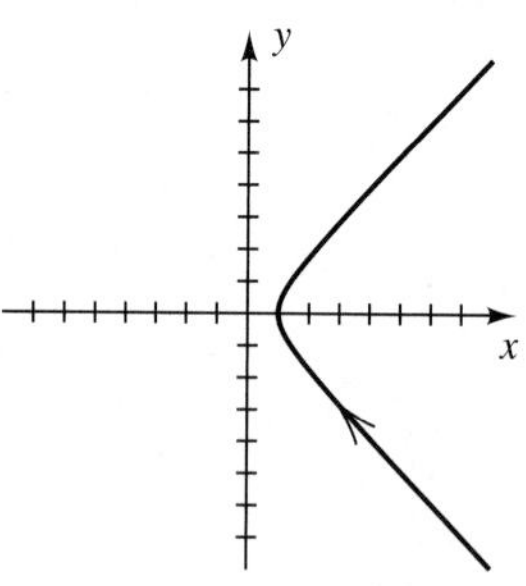

13 $y = \ln x$

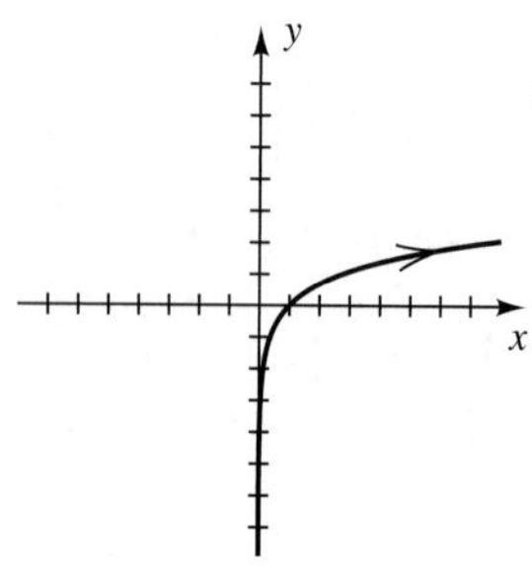

15 $y = 1/x$

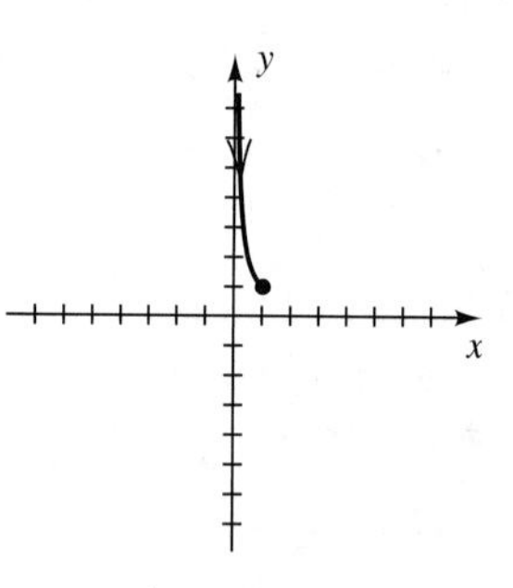

17 $y = \sqrt{x^2 - 1}$

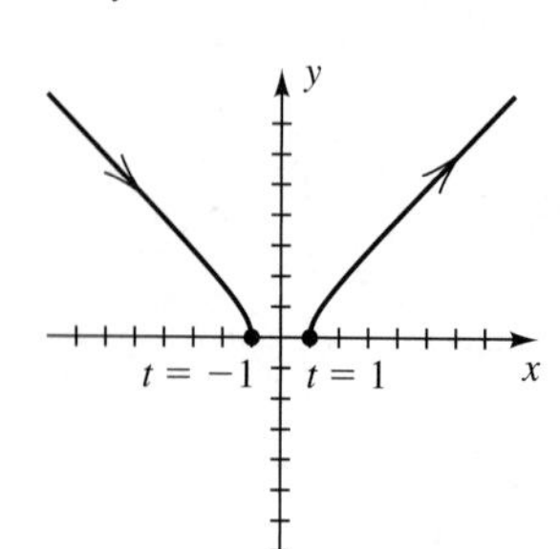

19 $y = |x - 1|$

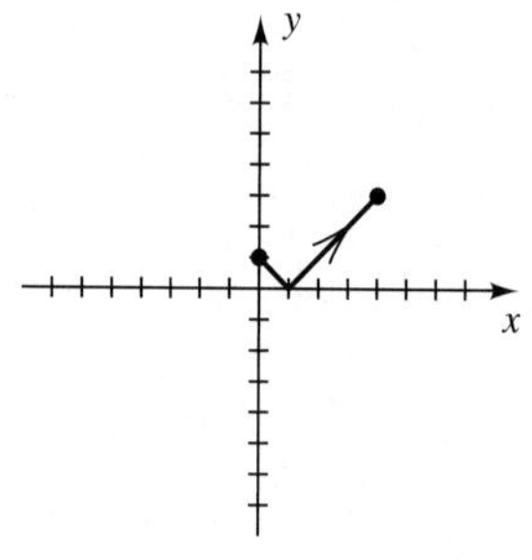

21 $y = (x^{1/3} + 1)^2$

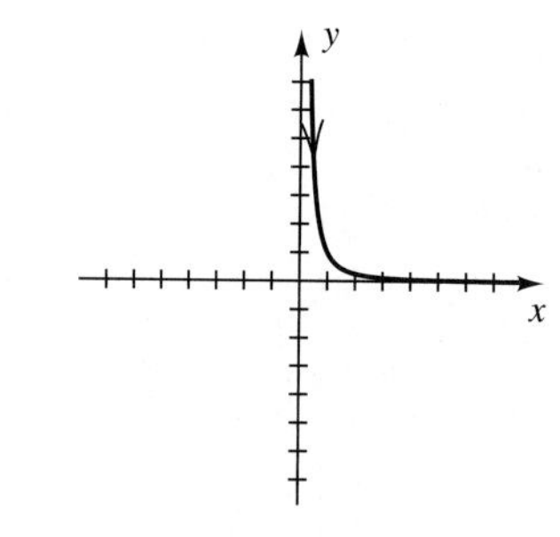

23 $y = 1/x^2$

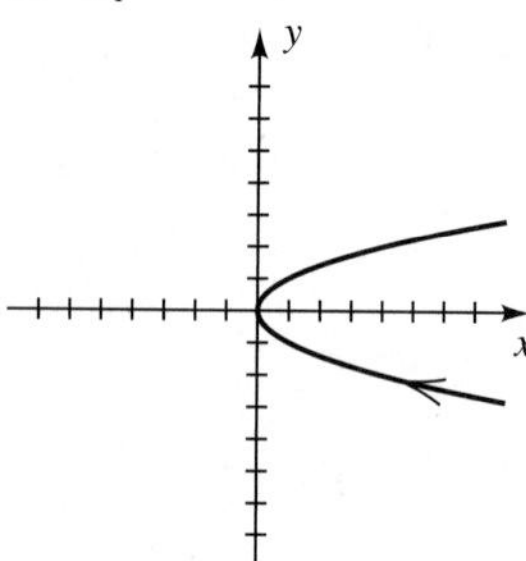

25 (a) The graph is a circle with center $(3, -2)$ and radius 2. Its orientation is clockwise, and it starts and ends at the point $(3, 0)$.

(b) The orientation changes to counterclockwise.

(c) The starting and ending point changes to $(3, -4)$.

27 C_1

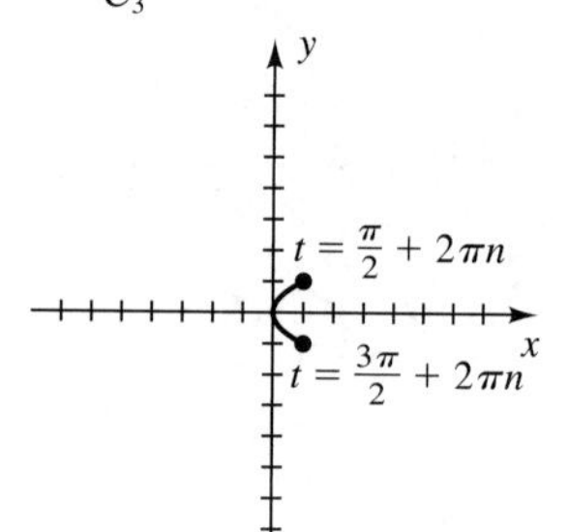

C_2

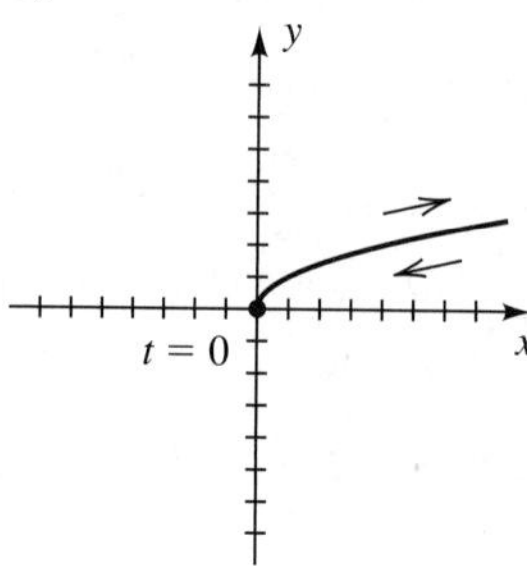

C_3

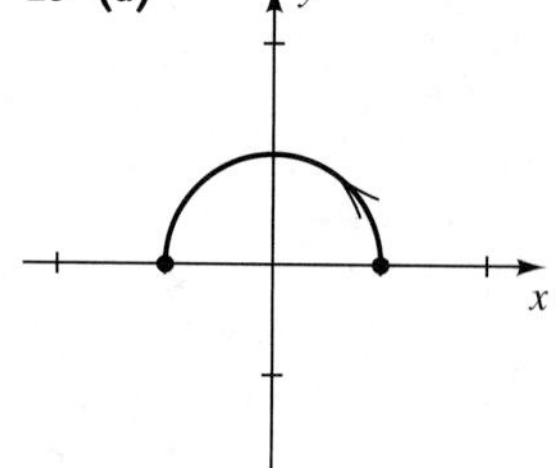

C_4

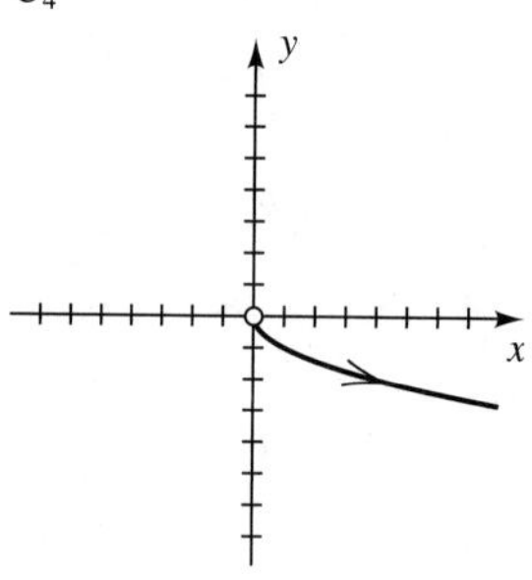

29 (a)

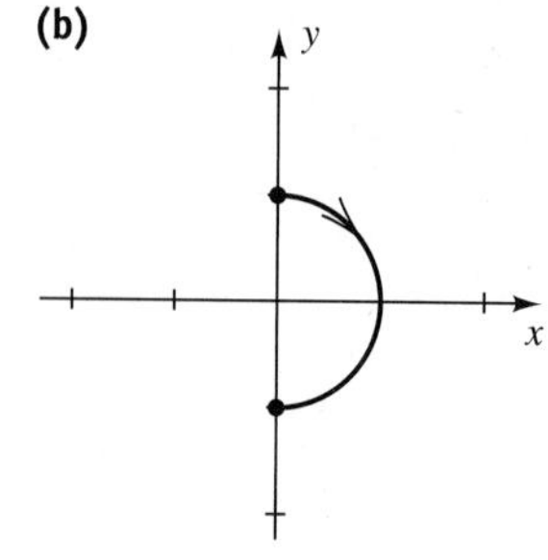

(b)

(c)

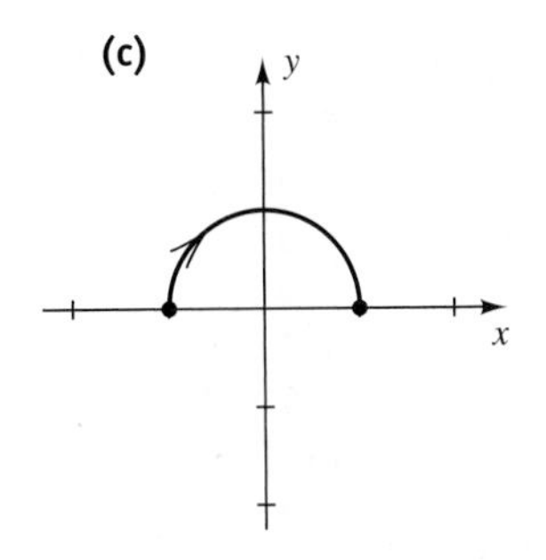

33 Answers are not unique.

(a) (1) $x = t,\quad y = t^2;\quad t \in \mathbb{R}$

(2) $x = \tan t,\quad y = \tan^2 t;\quad -\dfrac{\pi}{2} < t < \dfrac{\pi}{2}$

(3) $x = t^3,\quad y = t^6;\quad t \in \mathbb{R}$

(b) (1) $x = e^t,\quad y = e^{2t};\quad t \in \mathbb{R}$ (only gives $x > 0$)

(2) $x = \sin t,\quad y = \sin^2 t;\quad t \in \mathbb{R}$ (only gives $-1 \le x \le 1$)

(3) $x = \tan^{-1} t,\quad y = (\tan^{-1} t)^2;\quad t \in \mathbb{R}\left(\text{only gives } -\dfrac{\pi}{2} < x < \dfrac{\pi}{2}\right)$

35 $3200\sqrt{3}$; 2704 **37** 15,488; 3872

39 (a) The figure is an ellipse with center $(0, 0)$ and axes of length $2a$ and $2b$.

41 (a)

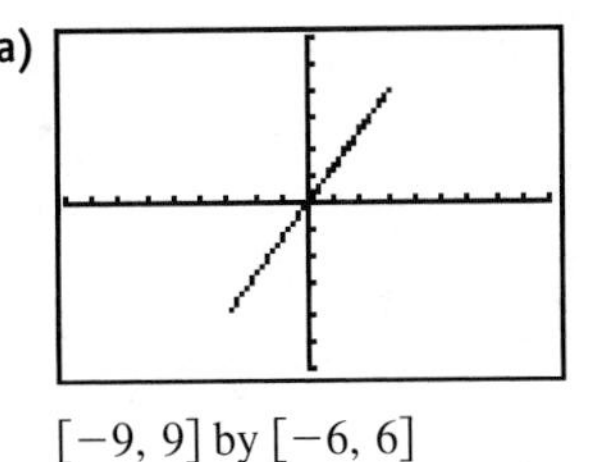

$[-9, 9]$ by $[-6, 6]$

(b) $0°$

43 (a)

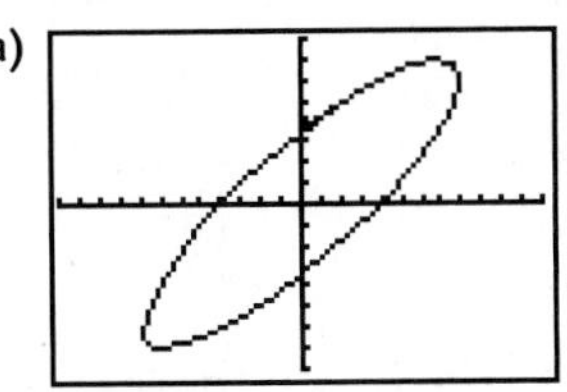

$[-120, 120, 10]$ by $[-80, 80, 10]$

(b) $30°$

45

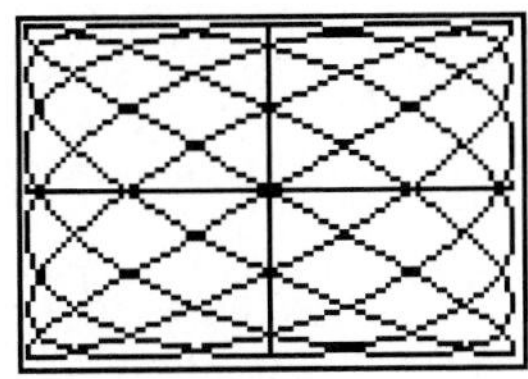

$[-1, 1]$ by $[-1, 1]$

49 $x = 4b \cos t - b \cos 4t$,
$y = 4b \sin t - b \sin 4t$

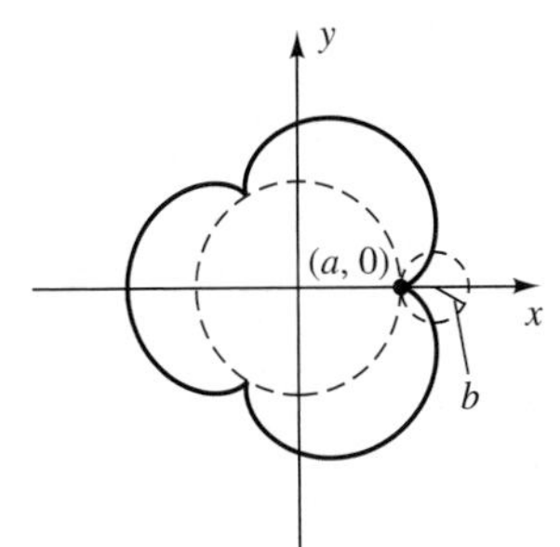

51

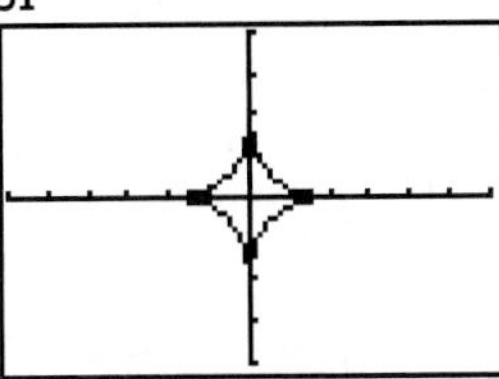

$[-6, 6]$ by $[-4, 4]$

53

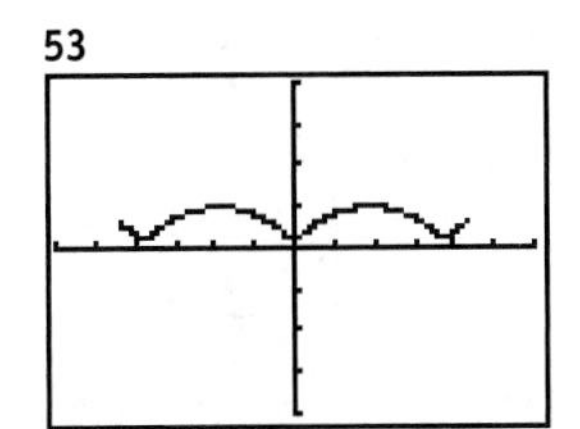

$[-30, 30, 5]$ by $[-20, 20, 5]$

55 A mask with a mouth, nose, and eyes

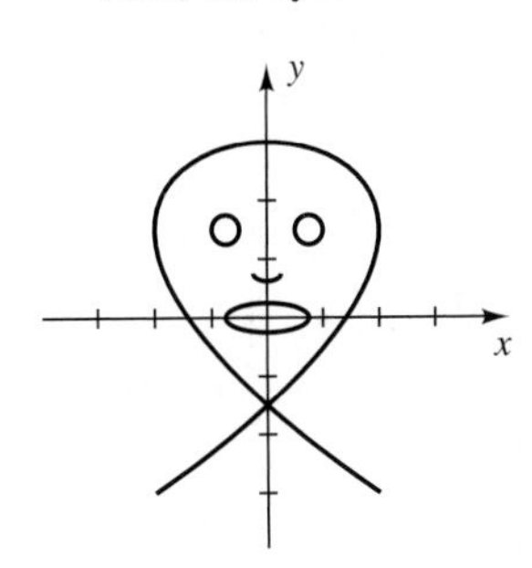

57 The letter A

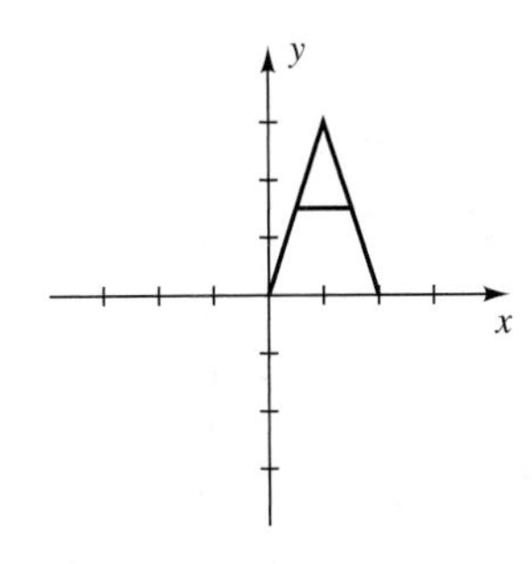

EXERCISES 10.5

1 (a), (c), (e)

3 (a) $\left(\frac{3}{2}\sqrt{2}, \frac{3}{2}\sqrt{2}\right)$ **(b)** $\left(\frac{1}{2}, -\frac{1}{2}\sqrt{3}\right)$

5 (a) $\left(-4, -4\sqrt{3}\right)$ **(b)** $\left(-\frac{3}{2}, \frac{3}{2}\sqrt{3}\right)$

7 $\left(\frac{24}{5}, \frac{18}{5}\right)$ **9 (a)** $\left(\sqrt{2}, \frac{3\pi}{4}\right)$ **(b)** $\left(4, \frac{7\pi}{6}\right)$

11 (a) $\left(14, \dfrac{5\pi}{3}\right)$ **(b)** $\left(5\sqrt{2}, \dfrac{\pi}{4}\right)$

13 $r = -3 \sec \theta$ **15** $r = 4$ **17** $r = 6 \cot \theta \csc \theta$

19 $r = \dfrac{3}{\cos \theta + \sin \theta}$ **21** $\theta = \tan^{-1}\left(-\dfrac{1}{2}\right)$

23 $r^2 = -4 \sec 2\theta$ **25** $r = 2 \cos \theta$

27 $x = 5$ **29** $x^2 + (y - 3)^2 = 9$

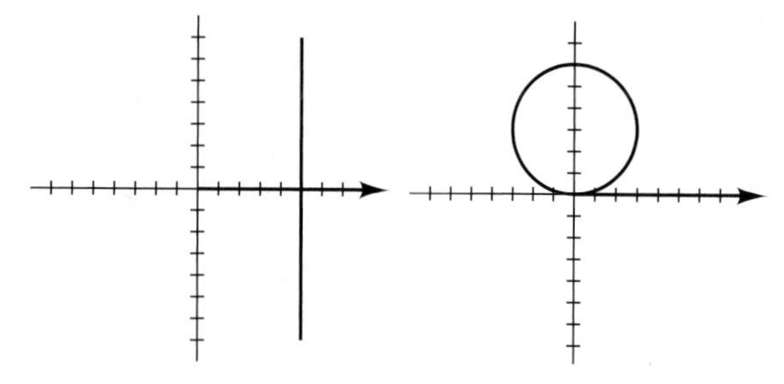

31 $y = x$ **33** $\dfrac{y^2}{9} - \dfrac{x^2}{4} = 1$

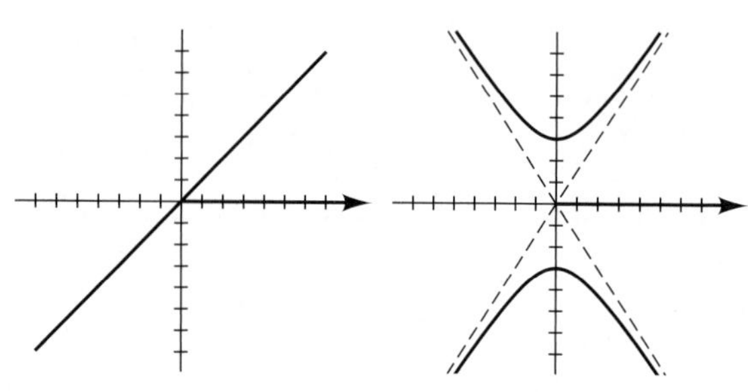

35 $x^2 - y^2 = 1$ **37** $y - 2x = 6$

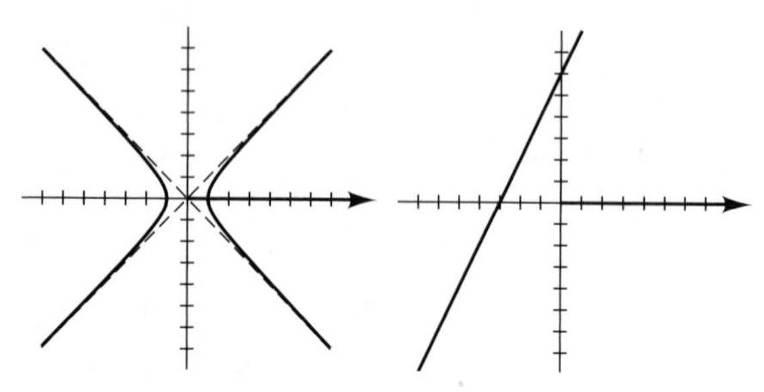

39 $y = -x^2 + 1$

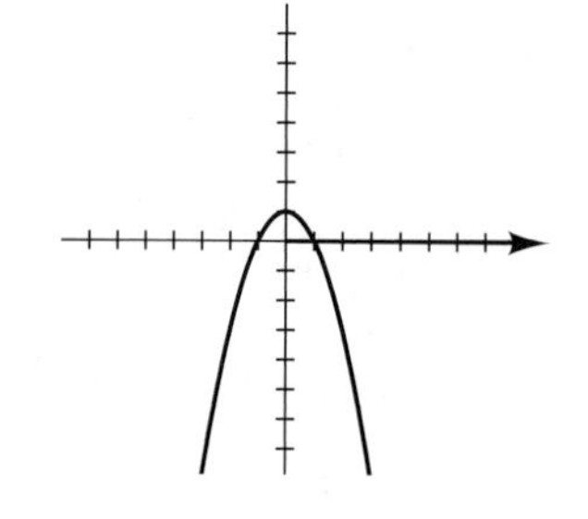

41 $(x + 1)^2 + (y - 4)^2 = 17$ **43** $y^2 = \dfrac{x^4}{1 - x^2}$

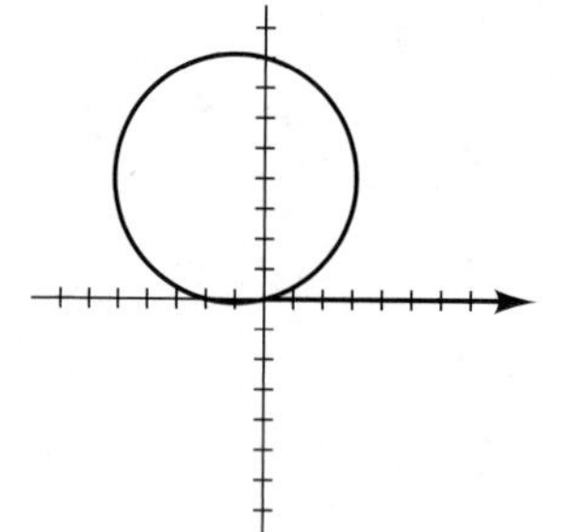

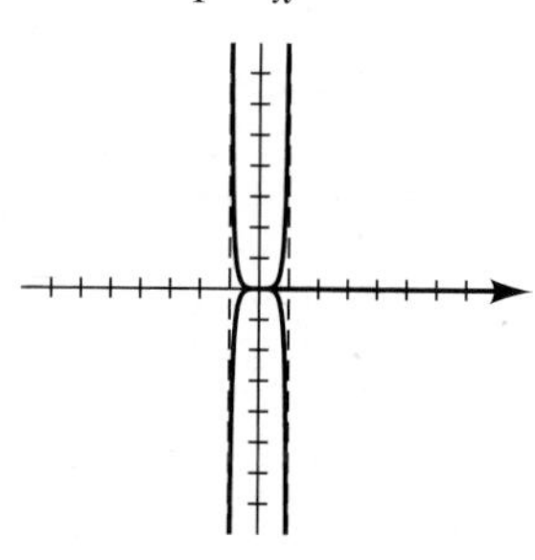

45

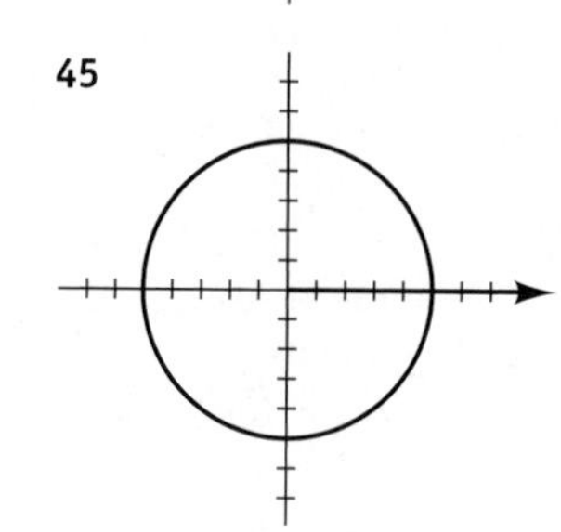

47

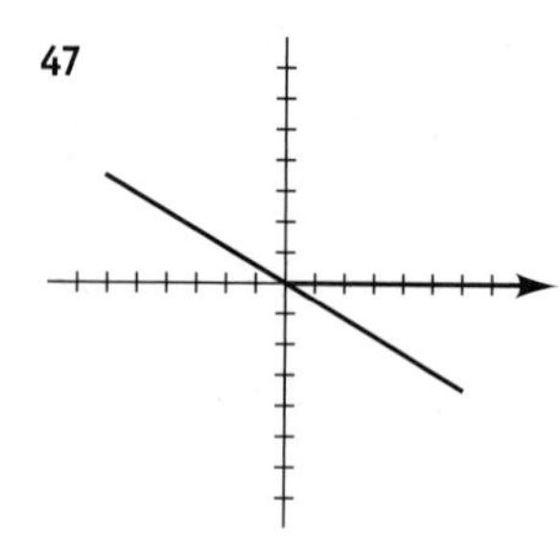

49

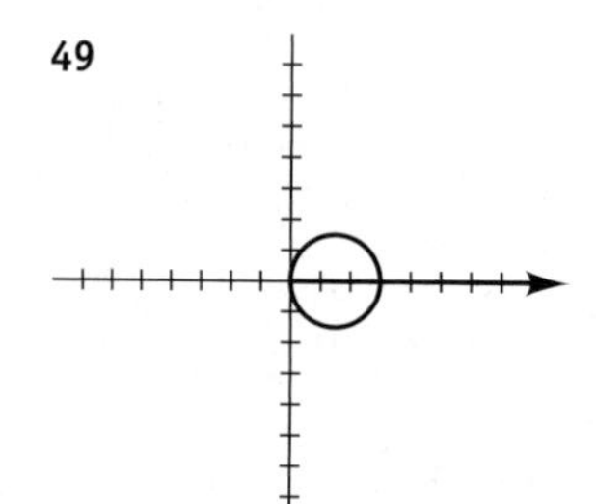

51

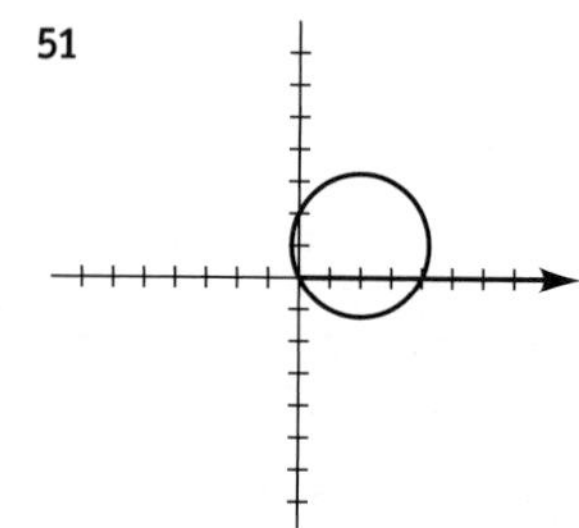

53

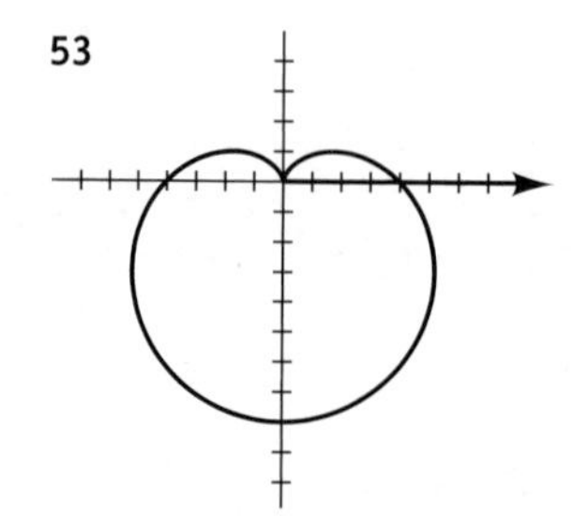

55

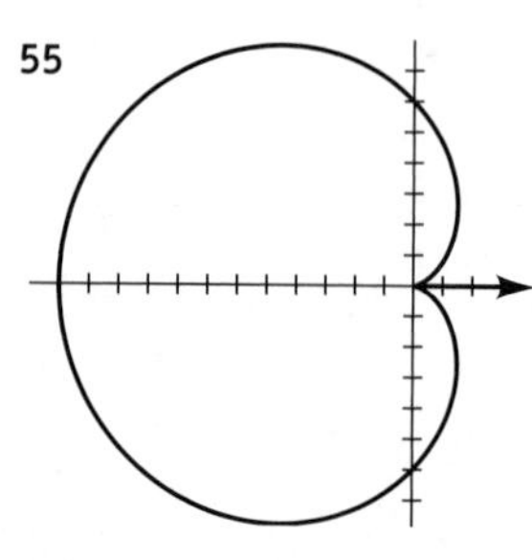

57

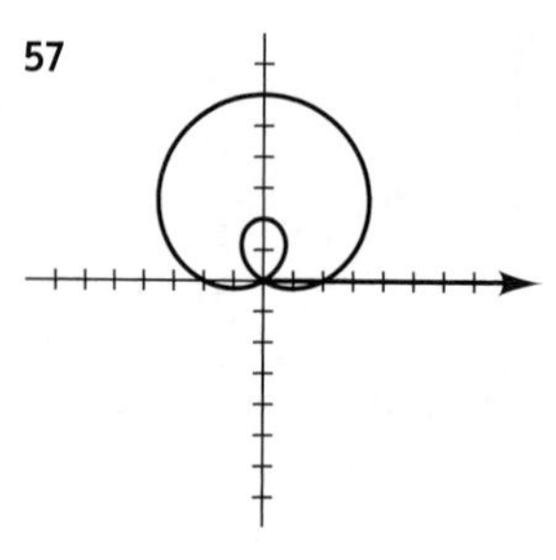

59

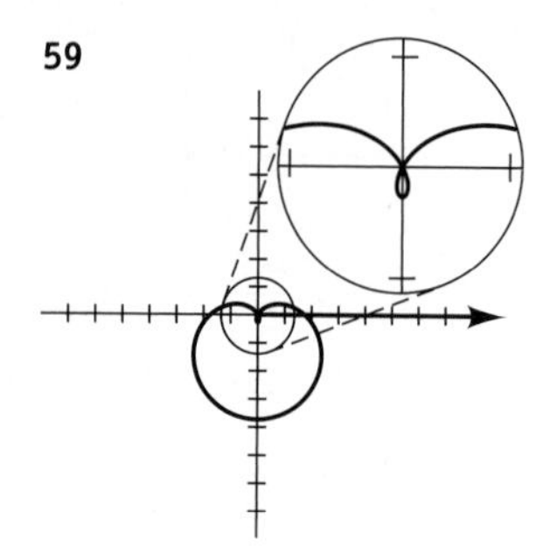

61

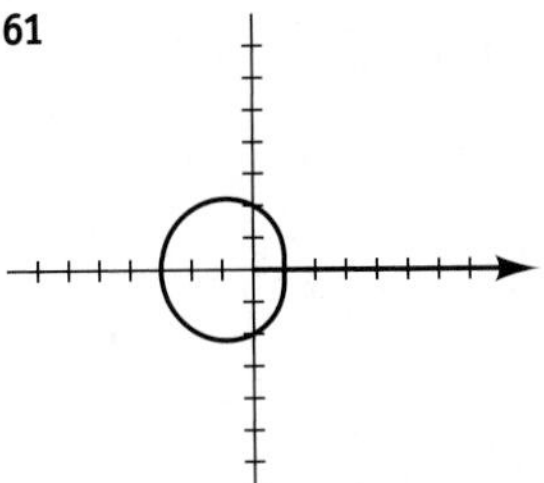

63

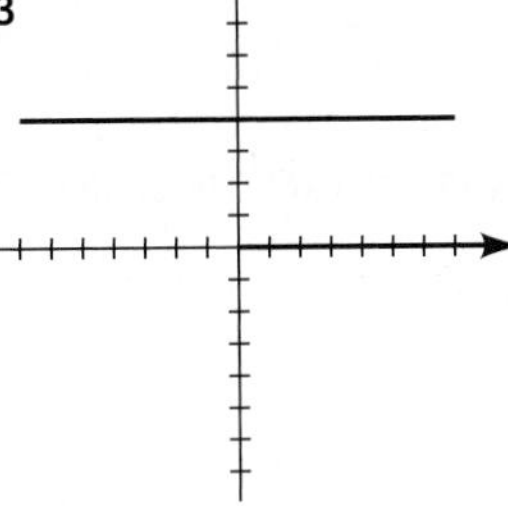

65

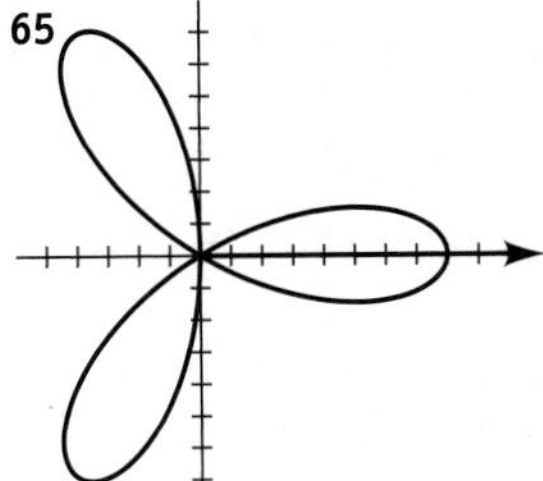

67

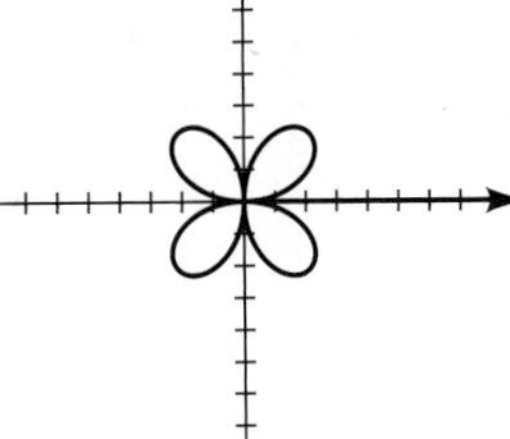

69

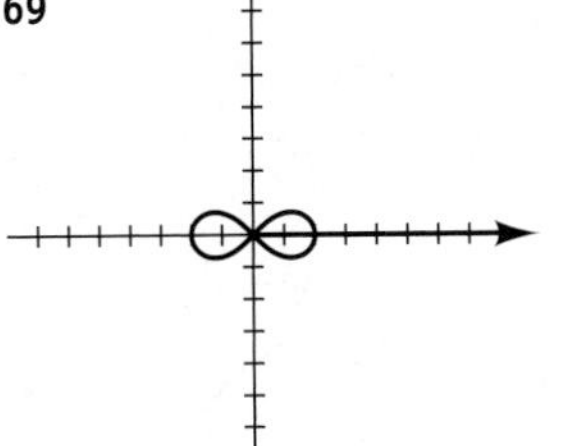

71

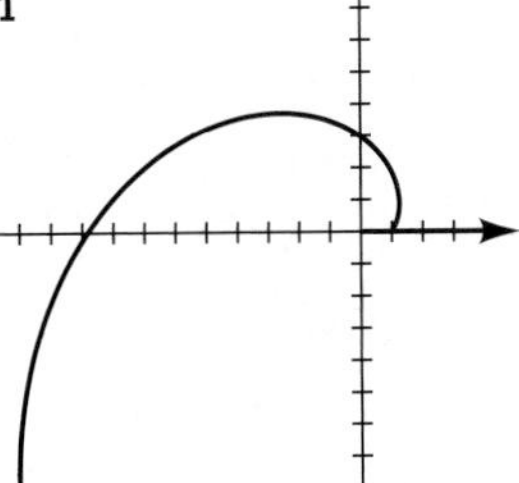

73

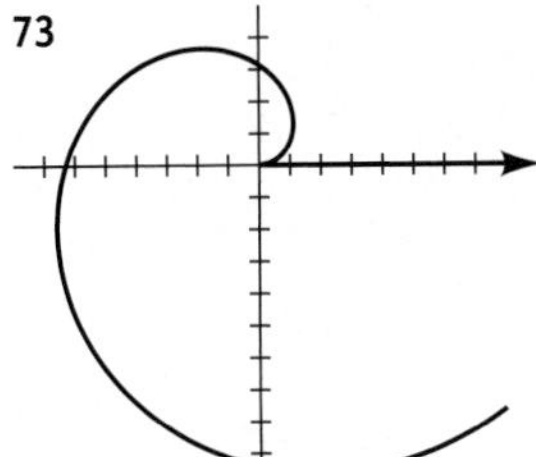

75

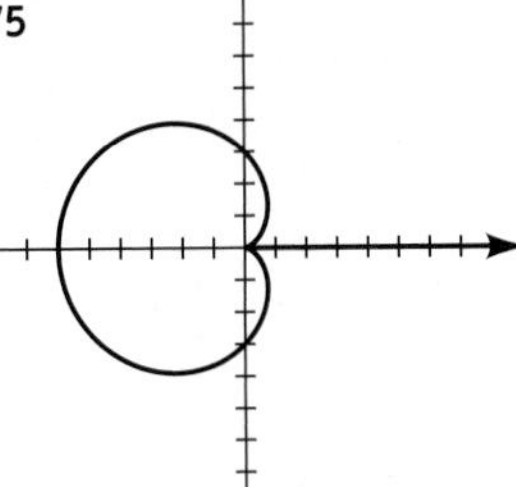

77

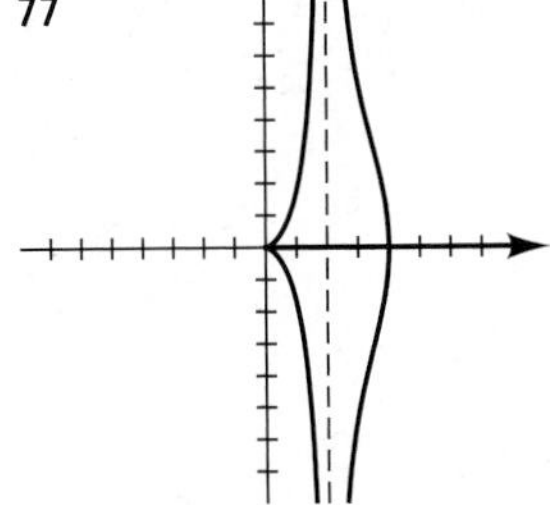

79 Let $P_1(r_1, \theta_1)$ and $P_2(r_2, \theta_2)$ be points in an $r\theta$-plane. Let $a = r_1$, $b = r_2$, $c = d(P_1, P_2)$, and $\gamma = \theta_2 - \theta_1$. Substituting into the law of cosines, $c^2 = a^2 + b^2 - 2ab \cos \gamma$, gives us the formula.

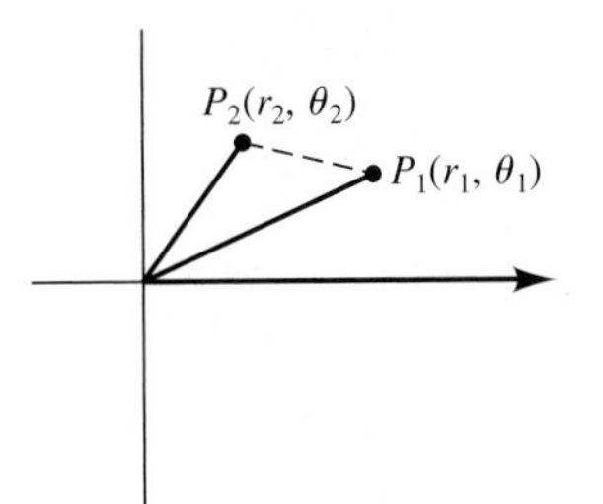

81 **(a)**

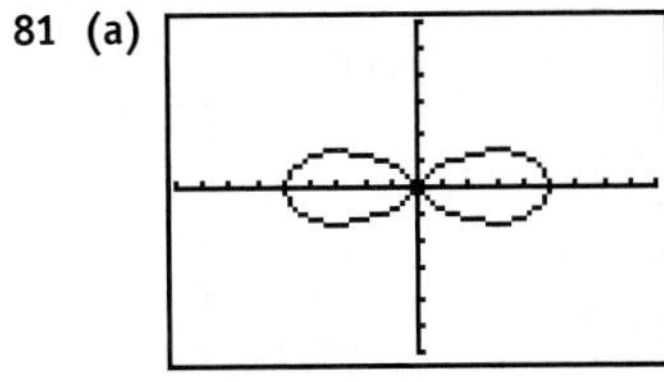

$[-9, 9]$ by $[-6, 6]$

(b) Max: east-west direction; min: north-south direction

83 Symmetric with respect to the polar axis

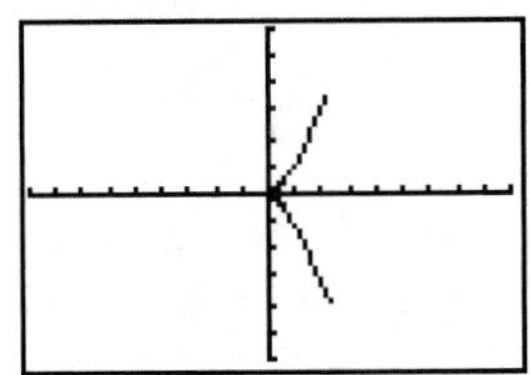

$[-9, 9]$ by $[-6, 6]$

85 The approximate polar coordinates are $(1.75, \pm 0.45)$, $(4.49, \pm 1.77)$, and $(5.76, \pm 2.35)$.

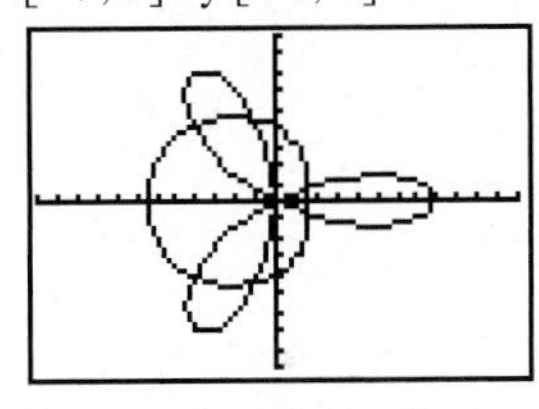

$[-12, 12]$ by $[-9, 9]$

EXERCISES 10.6

1 $\frac{1}{3}$, ellipse

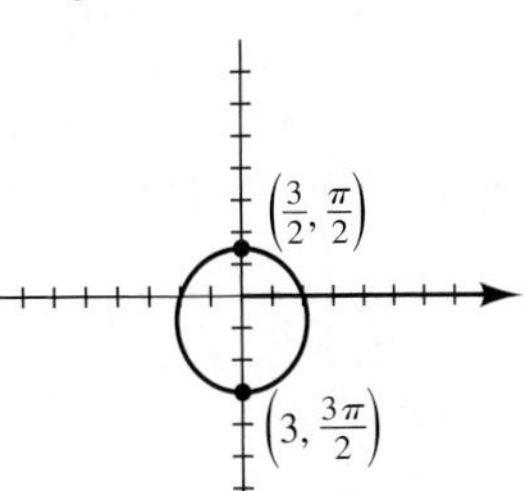

3 3, hyperbola

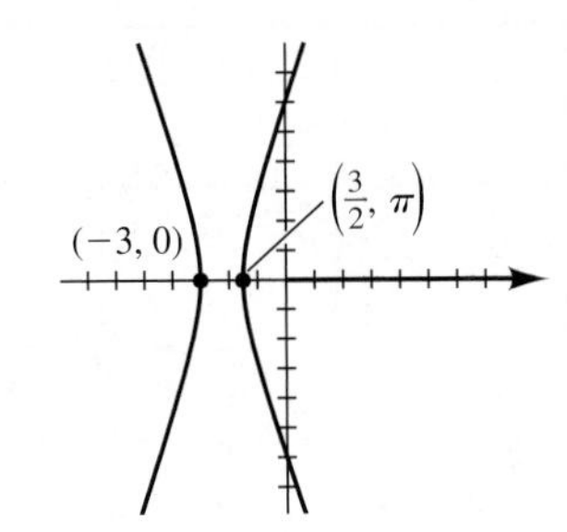

5 1, parabola

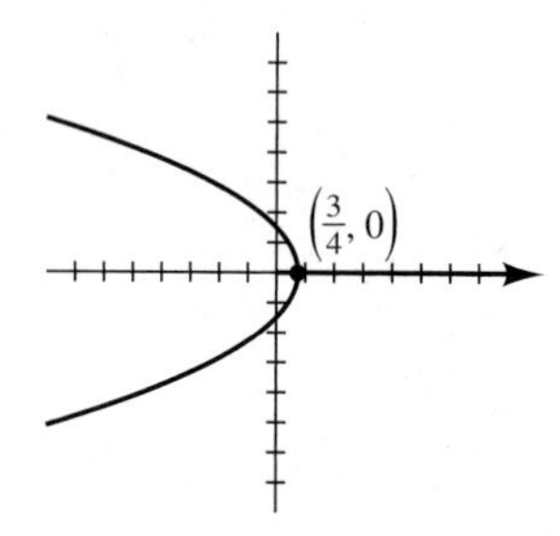

7 $\frac{1}{2}$, ellipse

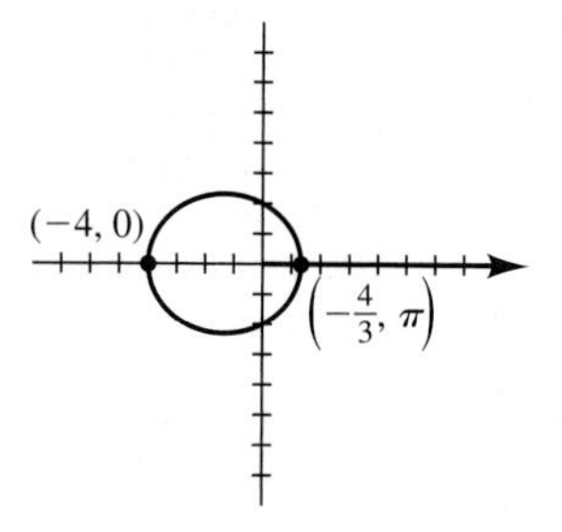

9 $\frac{3}{2}$, hyperbola

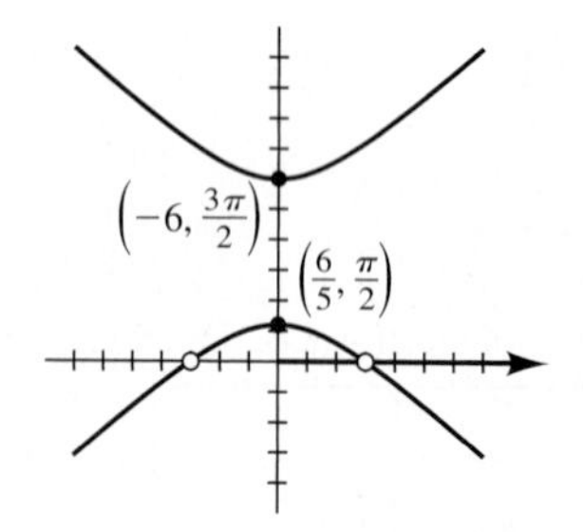

11 1, parabola

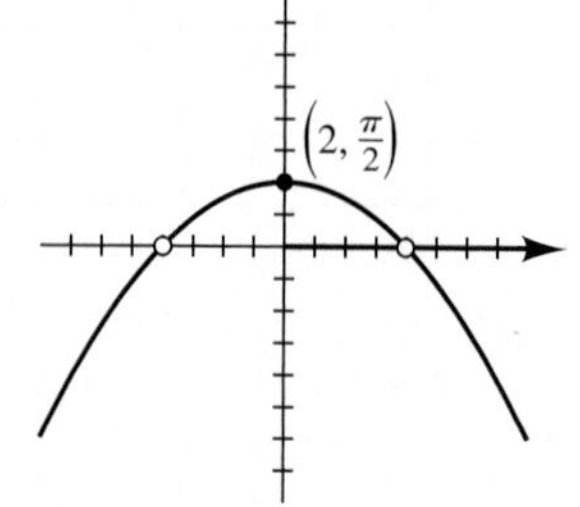

13 $9x^2 + 8y^2 + 12y - 36 = 0$

15 $8x^2 - y^2 + 36x + 36 = 0$

17 $4y^2 + 12x - 9 = 0$ **19** $3x^2 + 4y^2 + 8x - 16 = 0$

21 $4x^2 - 5y^2 + 36y - 36 = 0;\ x \neq \pm 3$

23 $x^2 + 8y - 16 = 0;\ x \neq \pm 4$ **25** $r = \dfrac{2}{3 + \cos\theta}$

27 $r = \dfrac{12}{3 - 4\cos\theta}$ **29** $r = \dfrac{2}{1 - \sin\theta}$

31 $r = \dfrac{8}{5 + 2\sin\theta}$ **33** $r = \dfrac{8}{1 + \sin\theta}$

35 **(a)** $\dfrac{3}{4}$ **(b)** $r = \dfrac{7}{4 - 3\sin\theta}$

39 **(a)** Elliptical
(b)

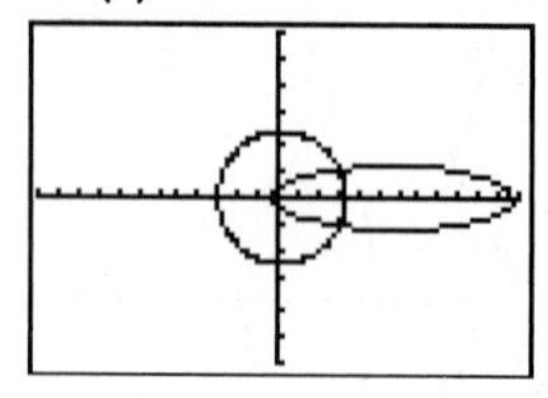

[−36, −36, 3] by
[−24, 24, 3]

41 **(a)** Hyperbolic
(b)

[−18, 18, 3] by
[−12, 12, 3]

43 $e = \dfrac{r_{aph} - r_{per}}{r_{aph} + r_{per}},\ a = \dfrac{r_{aph} + r_{per}}{2}$

CHAPTER 10 REVIEW EXERCISES

1 $V(0, 0);\ F(16, 0)$

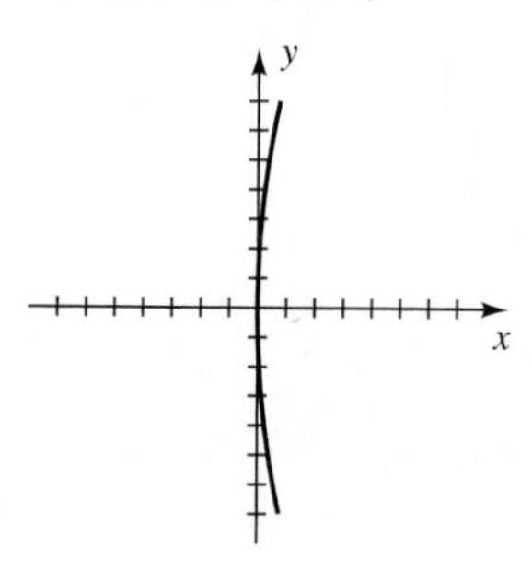

2 $V(-2, 1);\ F\left(-2, \dfrac{33}{32}\right)$

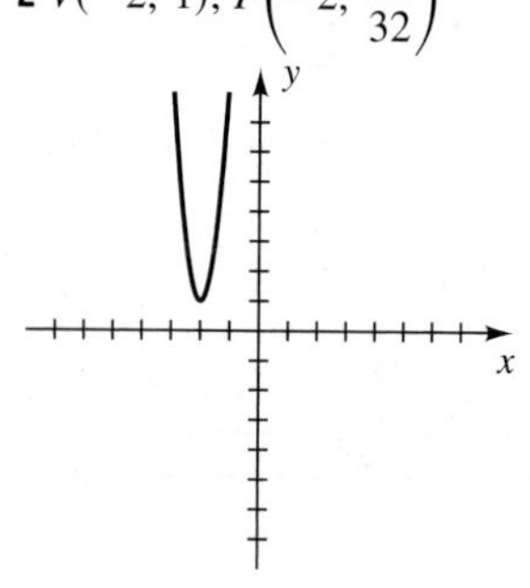

3 $V(0, \pm 4);\ F\left(0, \pm\sqrt{7}\right)$

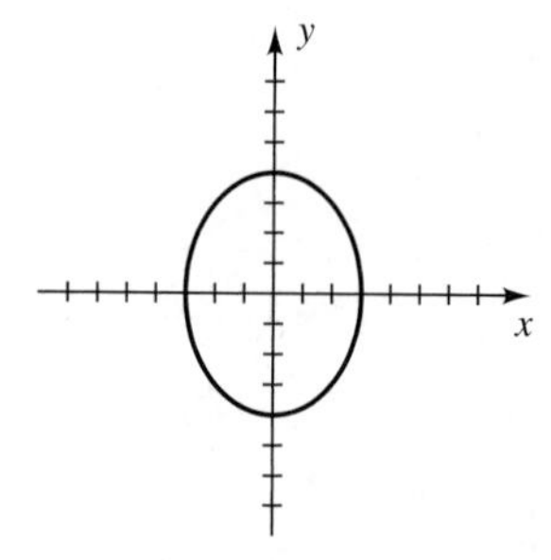

4 $V(0, \pm 4);\ F(0, \pm 5)$

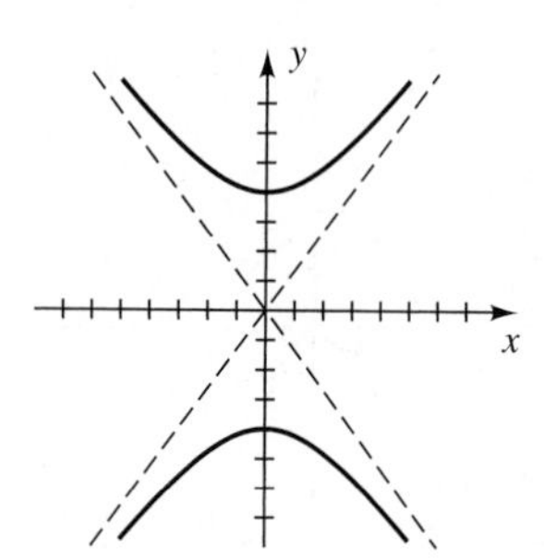

5 $V(\pm 2, 0);\ F\left(\pm 2\sqrt{2}, 0\right)$

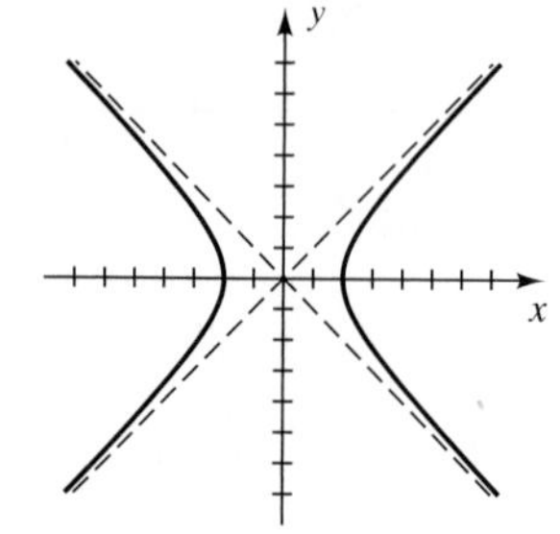

6 $V\left(\pm\dfrac{1}{5}, 0\right);$

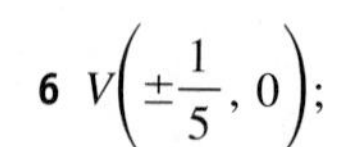
$F\left(\pm\dfrac{1}{30}\sqrt{11}, 0\right)$

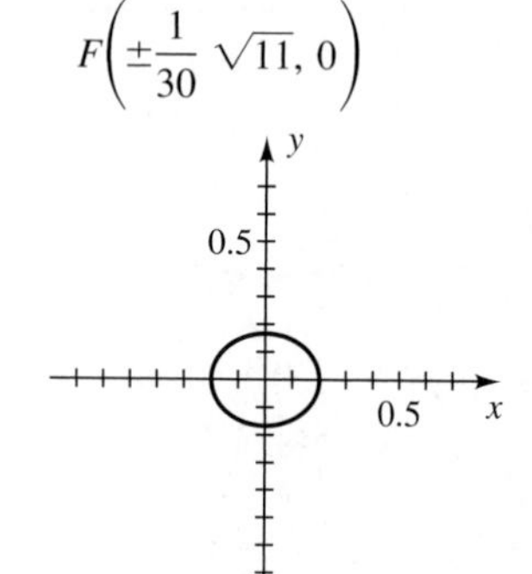

7 $V(0, 4);\ F\left(0, -\frac{9}{4}\right)$

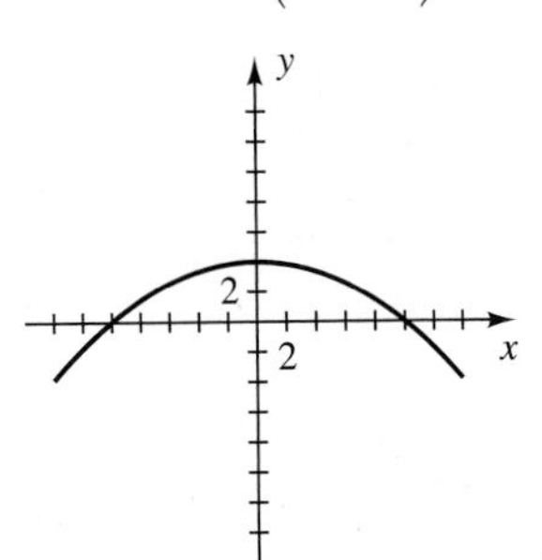

8 $V(3 \pm 2, -1);$
$F(3 \pm 1, -1)$

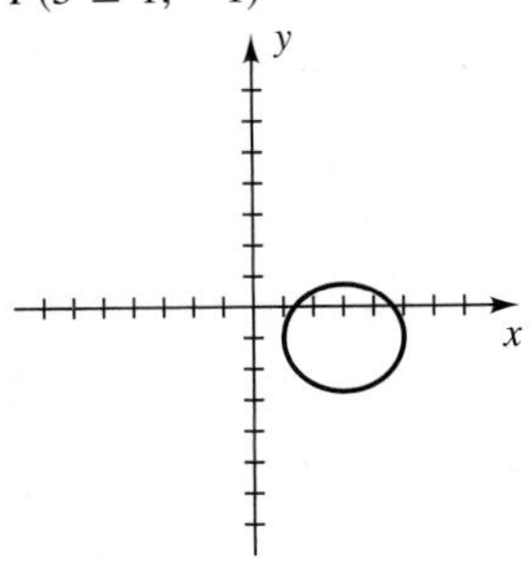

9 $V(-4 \pm 1, 5);$
$F\left(-4 \pm \frac{1}{3}\sqrt{10}, 5\right)$

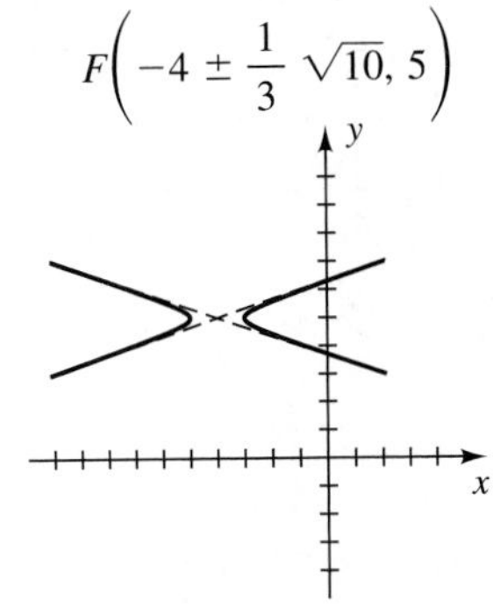

10 $V(-5, -2);$
$F\left(-\frac{39}{8}, -2\right)$

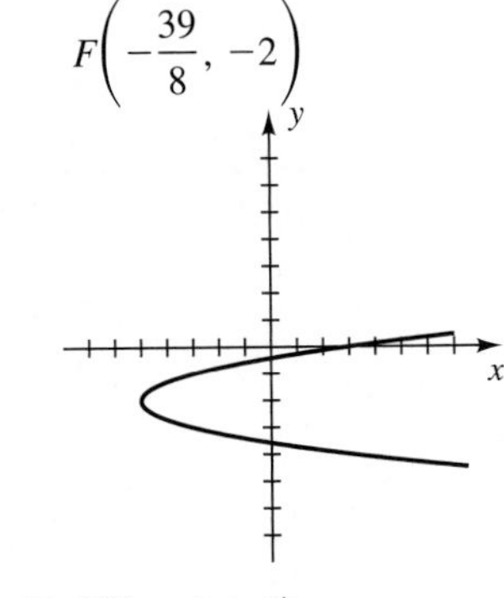

11 $V(-3 \pm 3, 2);$
$F\left(-3 \pm \sqrt{5}, 2\right)$

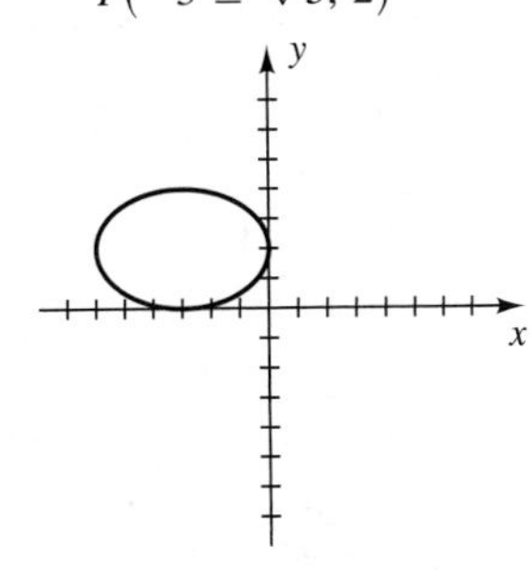

12 $V(5, -4 \pm 2);$
$F\left(5, -4 \pm \sqrt{5}\right)$

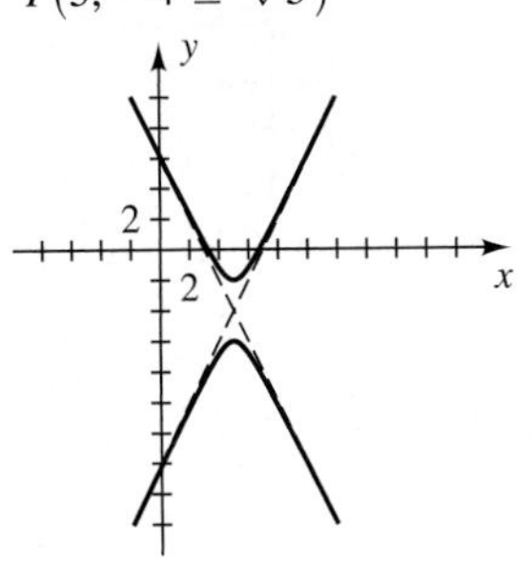

13 $V(2, -4);\ F(4, -4)$

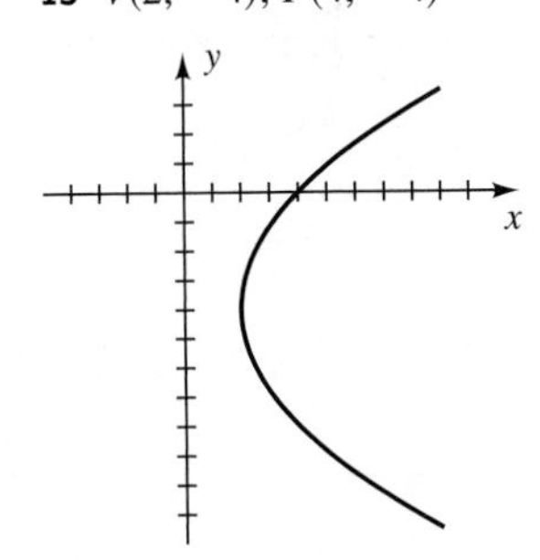

14 $V(3, -2 \pm 2);$
$F\left(3, -2 \pm \sqrt{3}\right)$

15 $V(-4 \pm 3, 0);$
$F\left(-4 \pm \sqrt{10}, 0\right)$

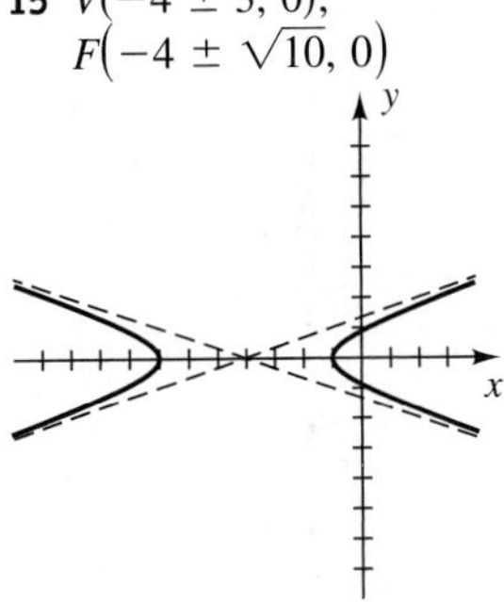

16 $V(2, -3 \pm 2);$
$F\left(2, -3 \pm \sqrt{6}\right)$

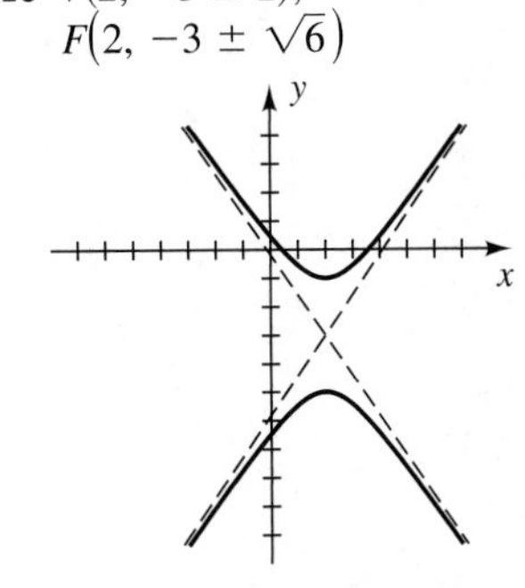

17 $y = 2(x + 7)^2 - 18$ **18** $y = -3(x + 4)^2 + 147$

19 $\frac{y^2}{49} - \frac{x^2}{9} = 1$ **20** $y^2 = -16x$ **21** $x^2 = -40y$

22 $x = 5y^2$ **23** $\frac{x^2}{75} + \frac{y^2}{100} = 1$ **24** $\frac{x^2}{25} - \frac{y^2}{75} = 1$

25 $\frac{y^2}{36} - \frac{x^2}{\frac{4}{9}} = 1$ **26** $\frac{x^2}{8} + \frac{y^2}{4} = 1$ **27** $\frac{x^2}{25} + \frac{y^2}{45} = 1$

28 $\frac{x^2}{256} + \frac{y^2}{112} = 1$ **29 (a)** $-\frac{7}{2}$ **(b)** Hyperbola

30 $A = \frac{4a^2b^2}{a^2 + b^2}$ **31** $x^2 + (y - 2)^2 = 4$

32 $2\sqrt{2}$ rad/sec ≈ 0.45 rev/sec

34 $x = \sqrt{9 + 4y^2}$

35 $x = 4y + 7$

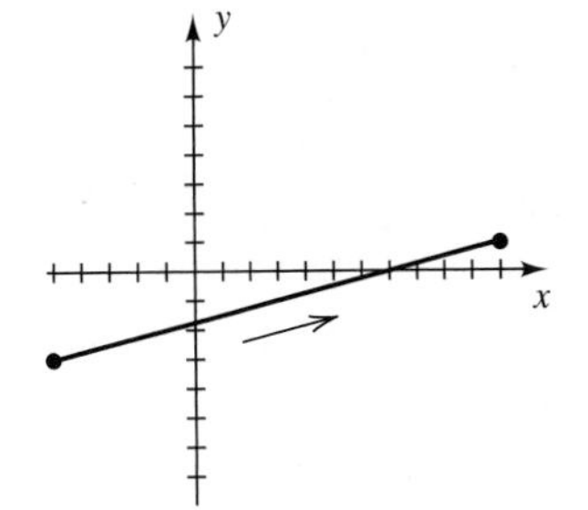

36 $y = x^4 - 4$

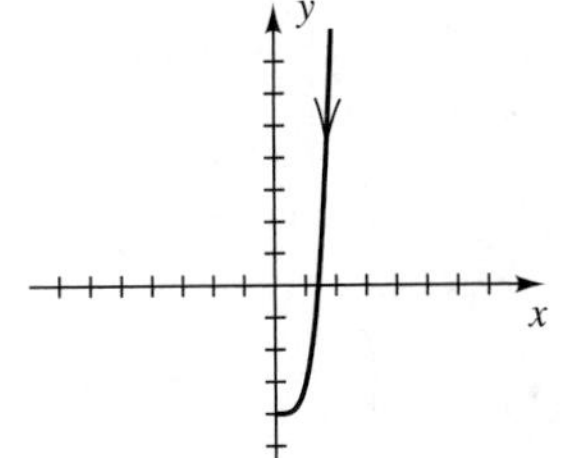

37 $(y-1)^2 = -(x+1)$

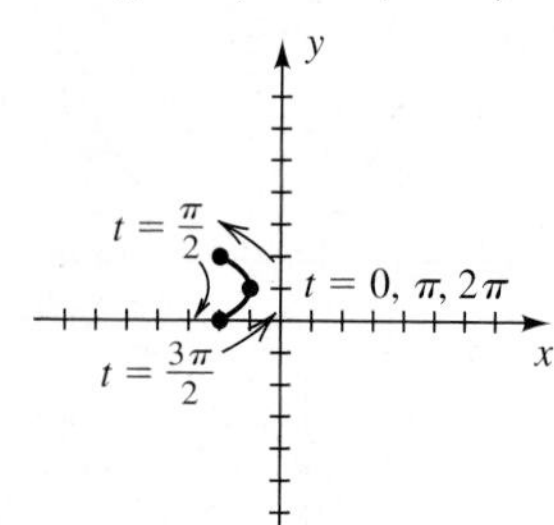

38 $y = 2^{-x^2}$

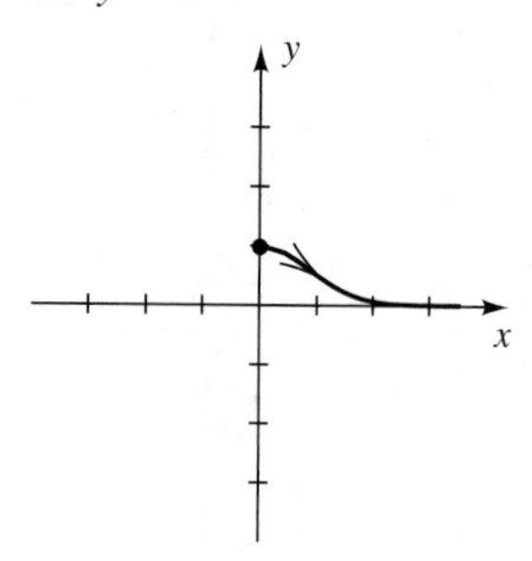

39 $y = \dfrac{2x^2 - 4x + 1}{x - 1}$

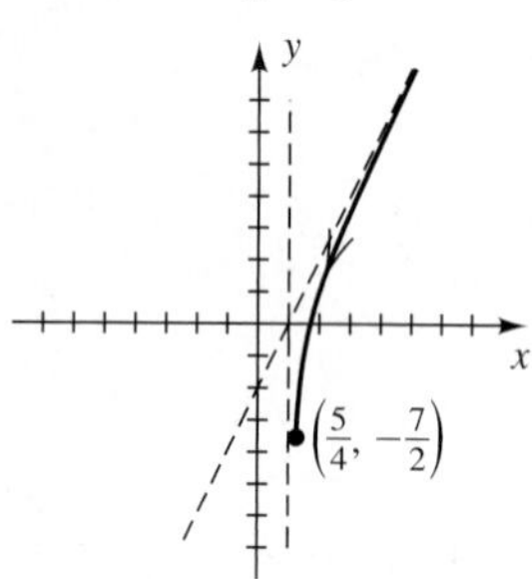

40 C_1

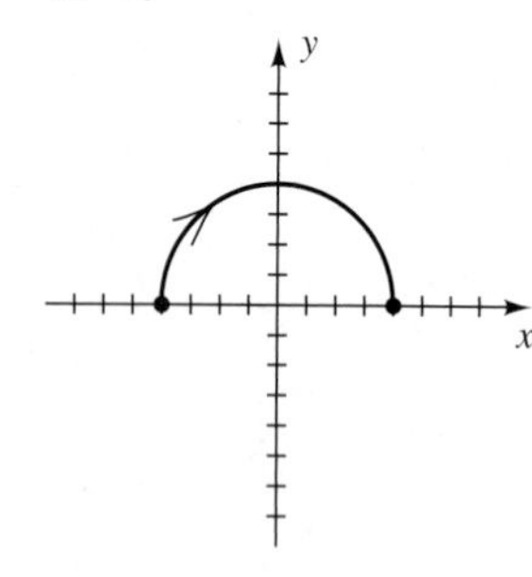

C_2

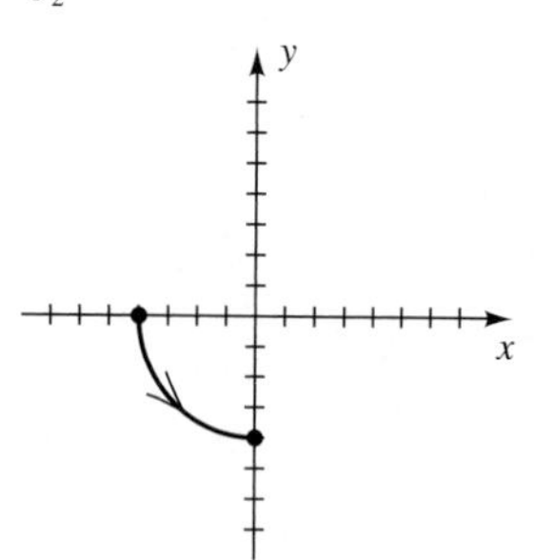

C_3

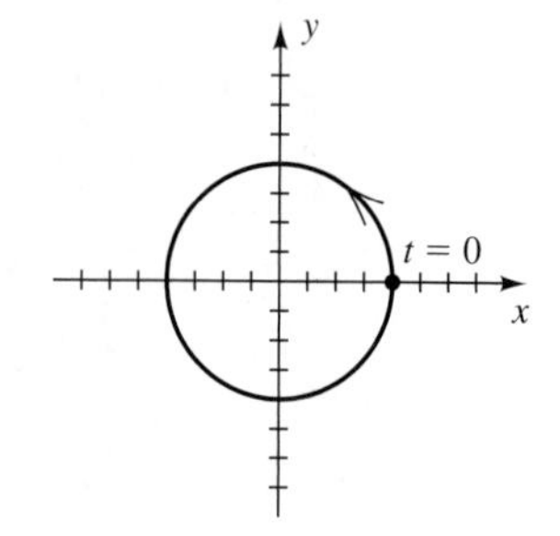

C_4

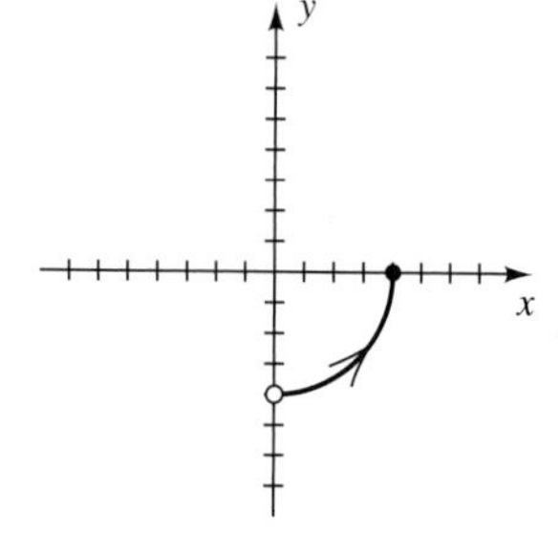

41 $20{,}480\sqrt{3}$; 9216 **42** $\left(-2, \dfrac{5\pi}{4}\right), \left(2, \dfrac{9\pi}{4}\right)$

43 $\left(\dfrac{5}{2}\sqrt{2}, -\dfrac{5}{2}\sqrt{2}\right)$ **44** $\left(4, \dfrac{11\pi}{6}\right)$

45 $r = 4 \cot\theta \csc\theta$ **46** $r = 3\cos\theta - 4\sin\theta$

47 $r(2\cos\theta - 3\sin\theta) = 8$ **48** $\theta = \dfrac{\pi}{4}$

49 $x^3 + xy^2 = y$ **50** $x^2 + y^2 = 2x + 3y$

51 $(x^2 + y^2)^2 = 8xy$ **52** $y = \left(\tan\sqrt{3}\right)x$

53 $8x^2 + 9y^2 + 10x - 25 = 0$ **54** $y^2 = 6 + x$

55

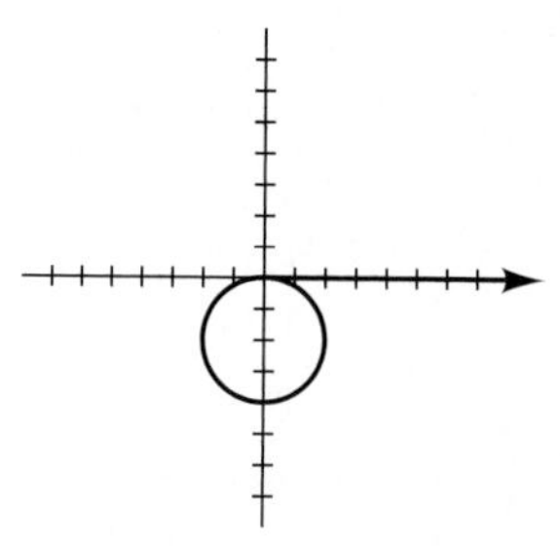

56

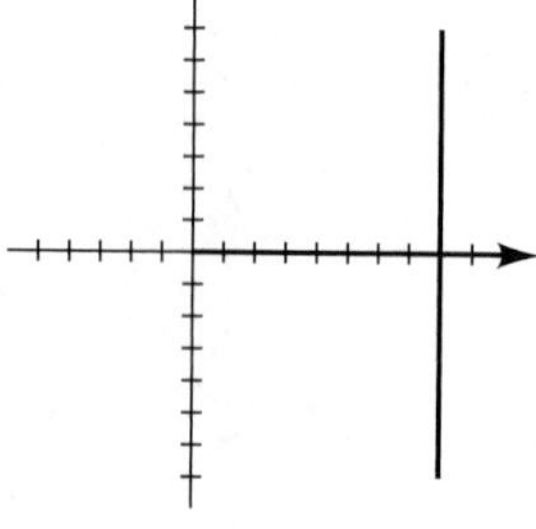

57

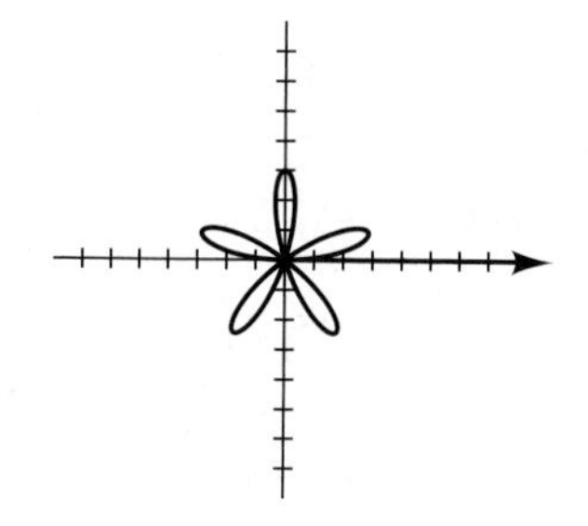

58

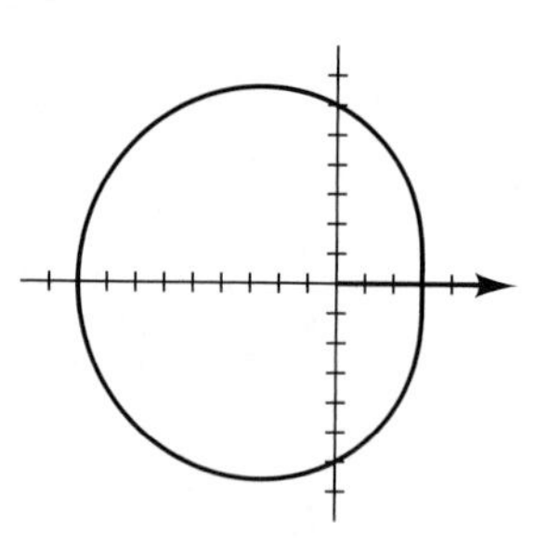

59

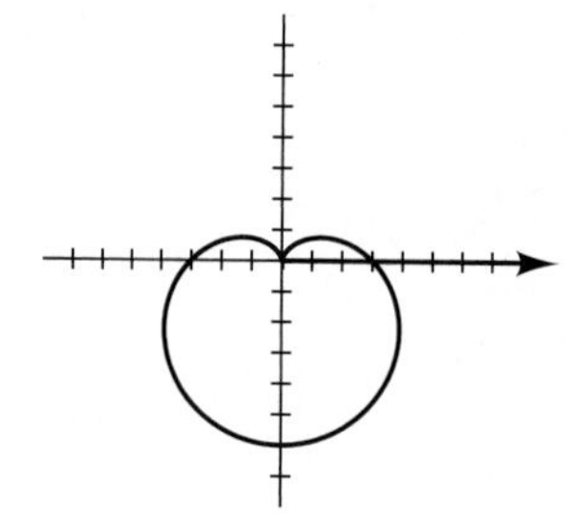

60

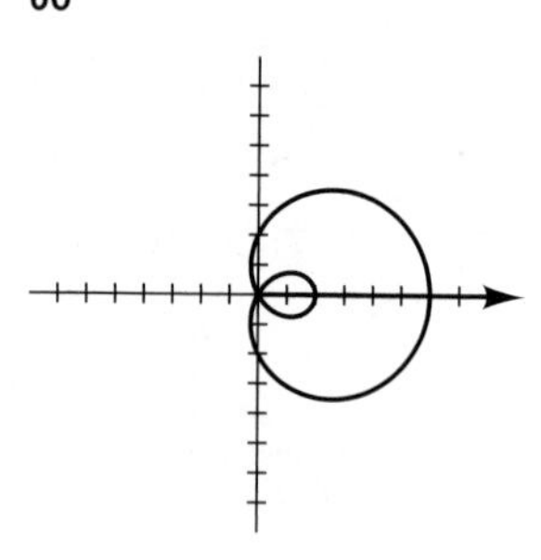

61

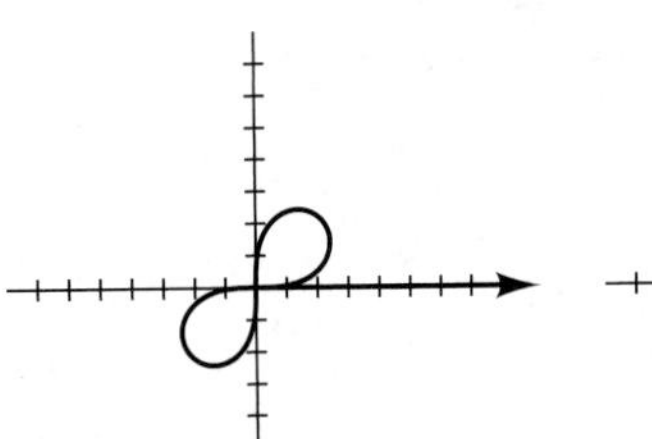

62

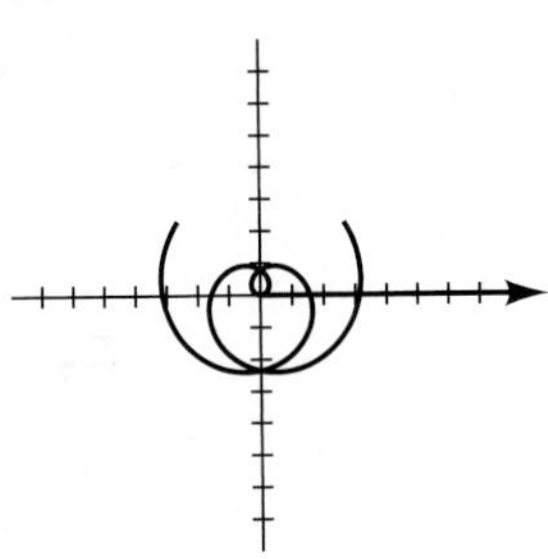

63

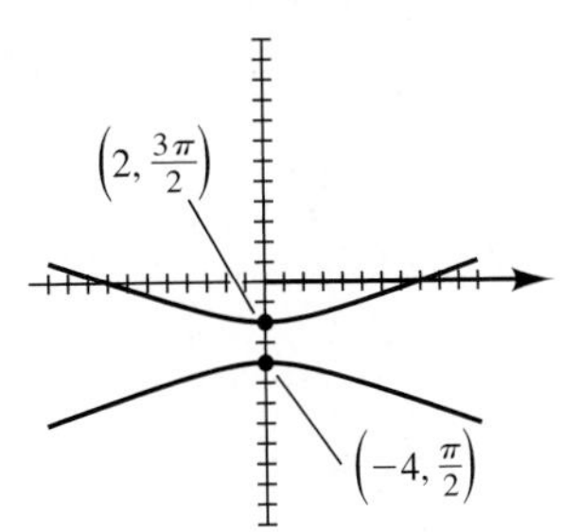

64

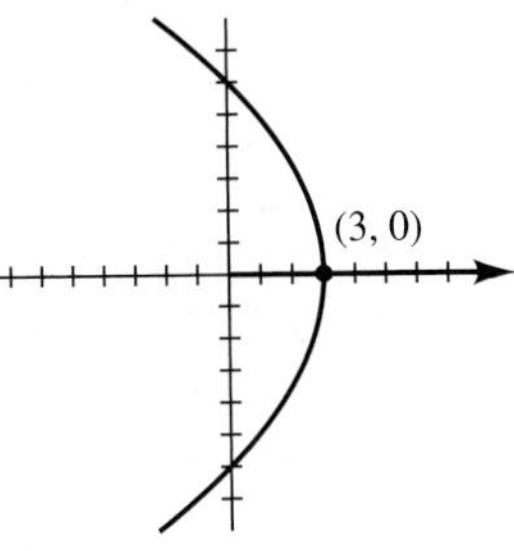

65 $\frac{2}{3}$, ellipse

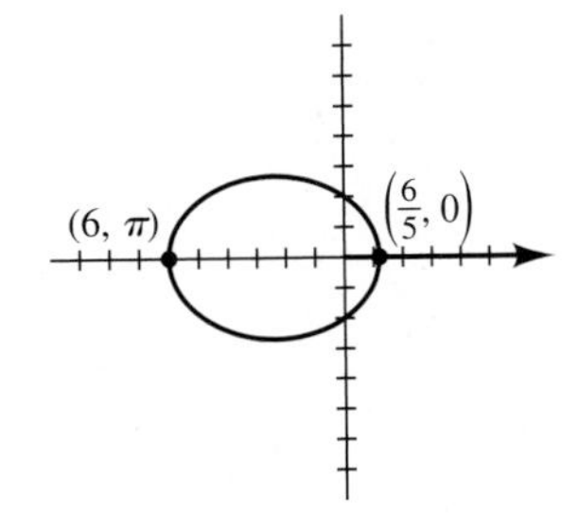

66 $\frac{1}{2}$, ellipse

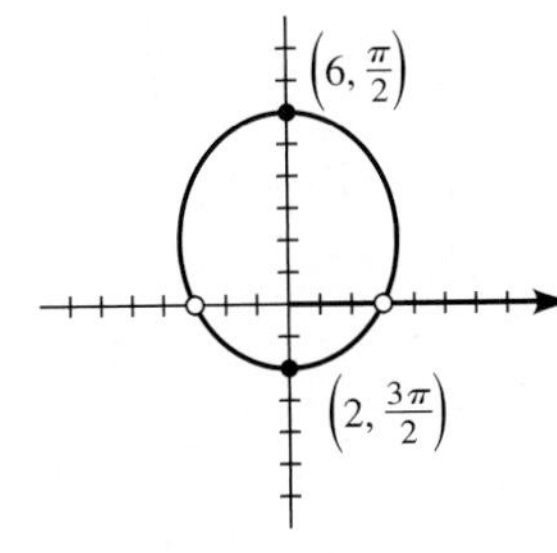

CHAPTER 10 DISCUSSION EXERCISES

1 $w = 4|p|$

2 The circle goes through both foci and all four vertices of the auxiliary rectangle.

5 $\frac{(x-2)^2}{1} - \frac{y^2}{3} = 1,\ x \geq 3,$

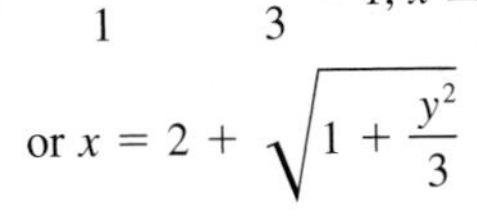

or $x = 2 + \sqrt{1 + \frac{y^2}{3}}$

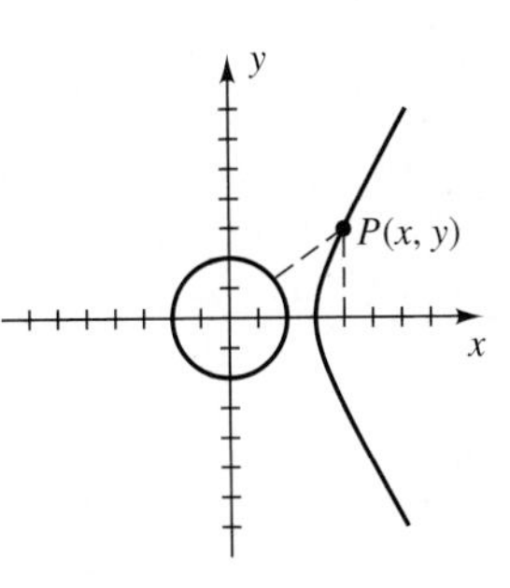

6 $d = \frac{1}{4\sqrt{a^2 + b^2}}$ **7** $43.12°$

9 $y = \pm\sqrt{\frac{1 \pm \sqrt{1 - x^2}}{2}}$

10 The graph of $r = f(\theta - \alpha)$ is the graph of $r = f(\theta)$ rotated counterclockwise through an angle α, whereas the graph of $r = f(\theta + \alpha)$ is rotated clockwise.

11 $(180/n)°$

12 $y = 2 \pm \sqrt{4 - x^2},\ y = \pm\sqrt{4 - (x-2)^2}$

INDEX OF APPLICATIONS

FOOD

GENERAL INTEREST

GEOMETRY

HEALTH/MEDICINE

OPERATIONS MANAGEMENT

PHYSICS/GENERAL SCIENCE

POLITICS/URBAN ISSUES

SPORTS

STATISTICS AND PROBABILITY

TRANSPORTATION

INDEX

D

E

K

L

M

N

Q

R

S

T

U

FORMULAS FROM TRIGONOMETRY

TRIGONOMETRIC FUNCTIONS

OF REAL NUMBERS

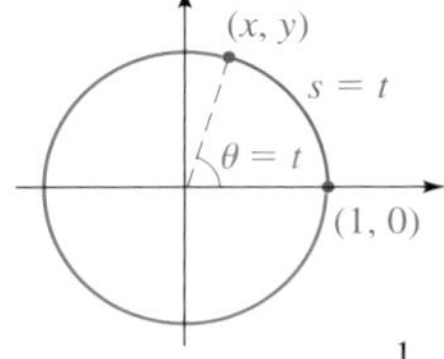

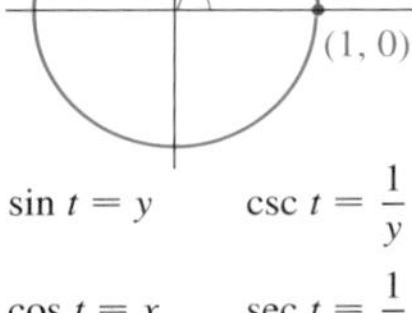

$\sin t = y \qquad \csc t = \dfrac{1}{y}$

$\cos t = x \qquad \sec t = \dfrac{1}{x}$

$\tan t = \dfrac{y}{x} \qquad \cot t = \dfrac{x}{y}$

OF ACUTE ANGLES

$\sin\theta = \dfrac{\text{opp}}{\text{hyp}} \qquad \csc\theta = \dfrac{\text{hyp}}{\text{opp}}$

$\cos\theta = \dfrac{\text{adj}}{\text{hyp}} \qquad \sec\theta = \dfrac{\text{hyp}}{\text{adj}}$

$\tan\theta = \dfrac{\text{opp}}{\text{adj}} \qquad \cot\theta = \dfrac{\text{adj}}{\text{opp}}$

LAW OF SINES

$$\frac{\sin\alpha}{a} = \frac{\sin\beta}{b} = \frac{\sin\gamma}{c}$$

OBLIQUE TRIANGLE

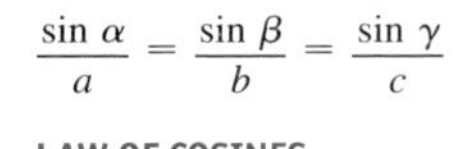

LAW OF COSINES

$a^2 = b^2 + c^2 - 2bc\cos\alpha$

$b^2 = a^2 + c^2 - 2ac\cos\beta$

$c^2 = a^2 + b^2 - 2ab\cos\gamma$

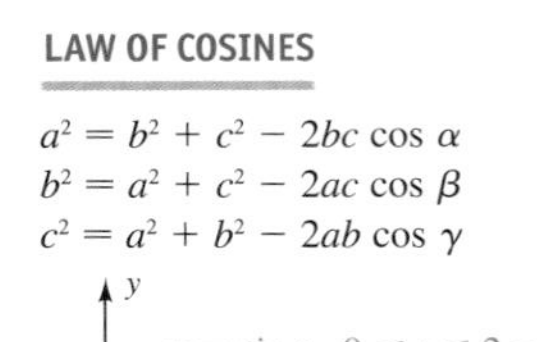

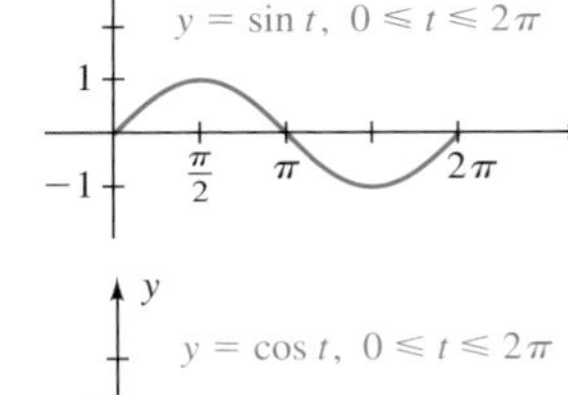

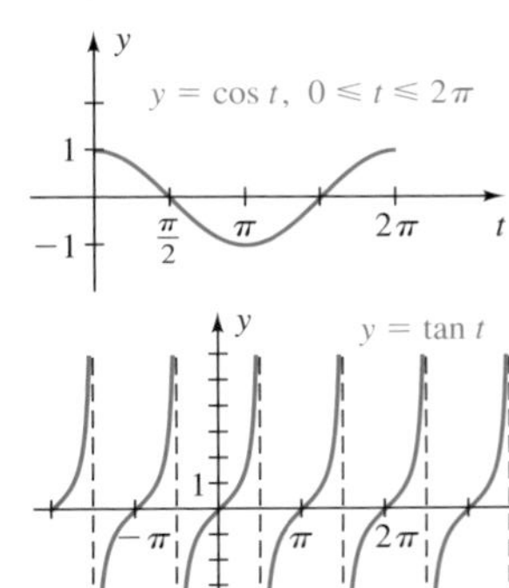

FORMULAS FROM TRIGONOMETRY

FUNDAMENTAL IDENTITIES

$\csc t = \dfrac{1}{\sin t}$

$\sec t = \dfrac{1}{\cos t}$

$\cot t = \dfrac{1}{\tan t}$

$\tan t = \dfrac{\sin t}{\cos t}$

$\cot t = \dfrac{\cos t}{\sin t}$

$\sin^2 t + \cos^2 t = 1$

$1 + \tan^2 t = \sec^2 t$

$1 + \cot^2 t = \csc^2 t$

FORMULAS FOR NEGATIVES

$\sin(-t) = -\sin t$

$\cos(-t) = \cos t$

$\tan(-t) = -\tan t$

$\cot(-t) = -\cot t$

$\sec(-t) = \sec t$

$\csc(-t) = -\csc t$

ADDITION FORMULAS

$\sin(u + v) = \sin u\cos v + \cos u\sin v$

$\cos(u + v) = \cos u\cos v - \sin u\sin v$

$\tan(u + v) = \dfrac{\tan u + \tan v}{1 - \tan u\tan v}$

SUBTRACTION FORMULAS

$\sin(u - v) = \sin u\cos v - \cos u\sin v$

$\cos(u - v) = \cos u\cos v + \sin u\sin v$

$\tan(u - v) = \dfrac{\tan u - \tan v}{1 + \tan u\tan v}$

HALF-ANGLE FORMULAS

$\sin\dfrac{u}{2} = \pm\sqrt{\dfrac{1 - \cos u}{2}}$

$\cos\dfrac{u}{2} = \pm\sqrt{\dfrac{1 + \cos u}{2}}$

$\tan\dfrac{u}{2} = \dfrac{1 - \cos u}{\sin u} = \dfrac{\sin u}{1 + \cos u}$

DOUBLE-ANGLE FORMULAS

$\sin 2u = 2\sin u\cos u$

$\cos 2u = \cos^2 u - \sin^2 u$

$\qquad = 1 - 2\sin^2 u$

$\qquad = 2\cos^2 u - 1$

$\tan 2u = \dfrac{2\tan u}{1 - \tan^2 u}$

QUICK REFERENCE CARD

THE SWOKOWSKI • COLE SERIES

FUNDAMENTALS OF COLLEGE ALGEBRA
Eleventh Edition

PRECALCULUS: FUNCTIONS AND GRAPHS
Eleventh Edition

ALGEBRA AND TRIGONOMETRY WITH ANALYTIC GEOMETRY
Twelfth Edition

ALGEBRA AND TRIGONOMETRY WITH ANALYTIC GEOMETRY
Classic Eleventh Edition

FORMULAS FROM GEOMETRY

area A perimeter P circumference C
volume V curved surface area S
altitude h radius r

RIGHT TRIANGLE

Pythagorean theorem:
$c^2 = a^2 + b^2$

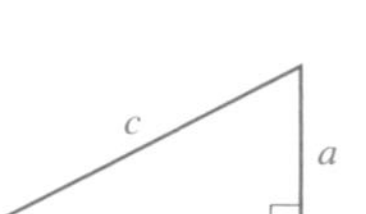

TRIANGLE

$A = \frac{1}{2}bh$ $\quad P = a + b + c$

CIRCLE

$A = \pi r^2$ $\quad C = 2\pi r$

SPHERE

$V = \frac{4}{3}\pi r^3$ $\quad S = 4\pi r^2$

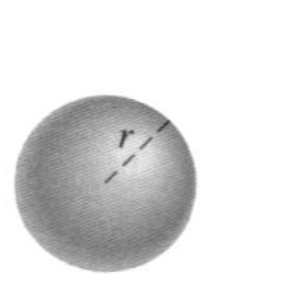

RIGHT CIRCULAR CYLINDER

$V = \pi r^2 h$ $\quad S = 2\pi r h$

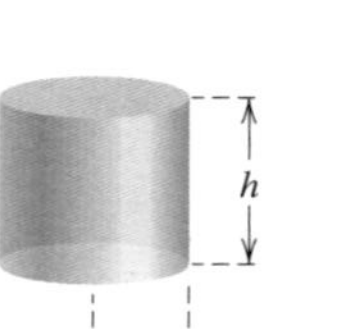

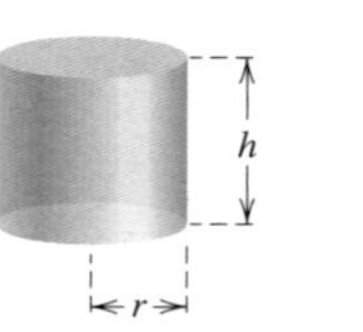

RIGHT CIRCULAR CONE

$V = \frac{1}{3}\pi r^2 h$ $\quad S = \pi r\sqrt{r^2 + h^2}$

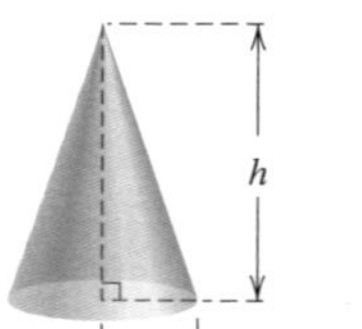

FORMULAS FROM ALGEBRA

QUADRATIC FORMULA

If $a \neq 0$, the roots of $ax^2 + bx + c = 0$ are

$$x = \frac{-b \pm \sqrt{b^2 - 4ac}}{2a}$$

SPECIAL FACTORING FORMULAS

$x^2 - y^2 = (x + y)(x - y)$
$x^2 + 2xy + y^2 = (x + y)^2$
$x^2 - 2xy + y^2 = (x - y)^2$
$x^3 - y^3 = (x - y)(x^2 + xy + y^2)$
$x^3 + y^3 = (x + y)(x^2 - xy + y^2)$

EXPONENTIALS AND LOGARITHMS

$y = \log_a x$ means $a^y = x$
$\log_a xy = \log_a x + \log_a y$
$\log_a \frac{x}{y} = \log_a x - \log_a y$
$\log_a x^r = r \log_a x$
$a^{\log_a x} = x$
$\log_a a^x = x$
$\log_a 1 = 0$
$\log_a a = 1$
$\log x = \log_{10} x$
$\ln x = \log_e x$
$\log_b u = \frac{\log_a u}{\log_a b}$

EXPONENTS AND RADICALS

$a^m a^n = a^{m+n}$ $\quad a^{1/n} = \sqrt[n]{a}$
$(a^m)^n = a^{mn}$ $\quad a^{m/n} = \sqrt[n]{a^m}$
$(ab)^n = a^n b^n$ $\quad a^{m/n} = (\sqrt[n]{a})^m$
$\left(\frac{a}{b}\right)^n = \frac{a^n}{b^n}$ $\quad \sqrt[n]{ab} = \sqrt[n]{a}\sqrt[n]{b}$
$\frac{a^m}{a^n} = a^{m-n}$ $\quad \sqrt[n]{\frac{a}{b}} = \frac{\sqrt[n]{a}}{\sqrt[n]{b}}$
$a^{-n} = \frac{1}{a^n}$ $\quad \sqrt[m]{\sqrt[n]{a}} = \sqrt[mn]{a}$

POINT-SLOPE FORM OF A LINE

$y - y_1 = m(x - x_1)$
m is the slope

SLOPE-INTERCEPT FORM OF A LINE

$y = mx + b$ $\quad m$ is the slope

CIRCLE

$(x - h)^2 + (y - k)^2 = r^2$

CONIC SECTIONS

PARABOLA

$x^2 = 4py$

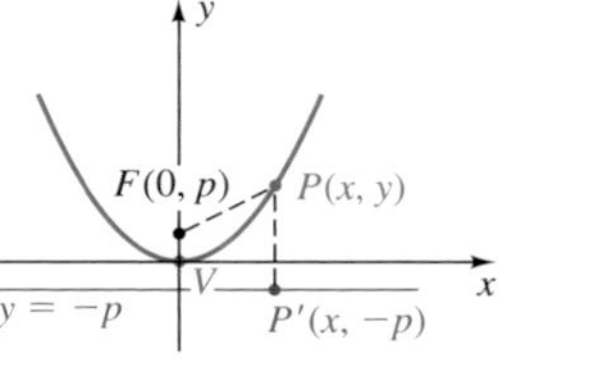

ELLIPSE

$\frac{x^2}{a^2} + \frac{y^2}{b^2} = 1$ with $a^2 = b^2 + c^2$

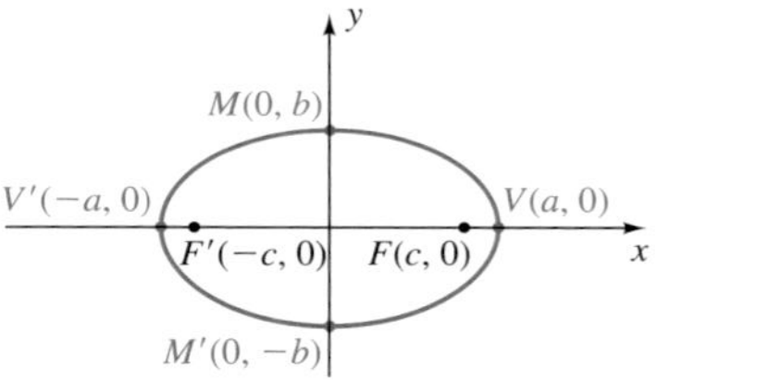

HYPERBOLA

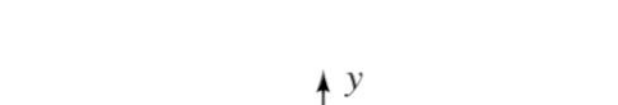

$\frac{x^2}{a^2} - \frac{y^2}{b^2} = 1$ with $c^2 = a^2 + b^2$

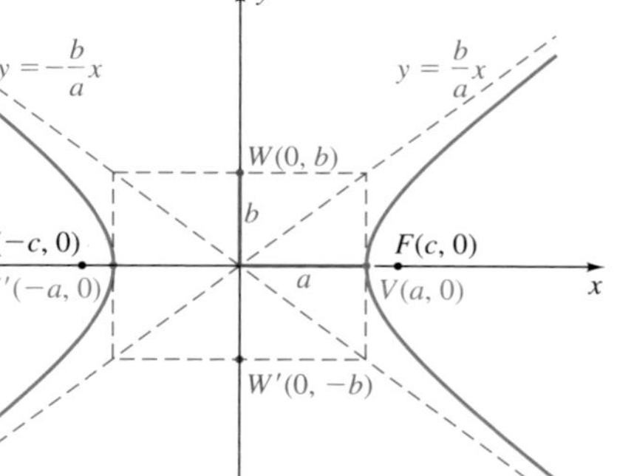

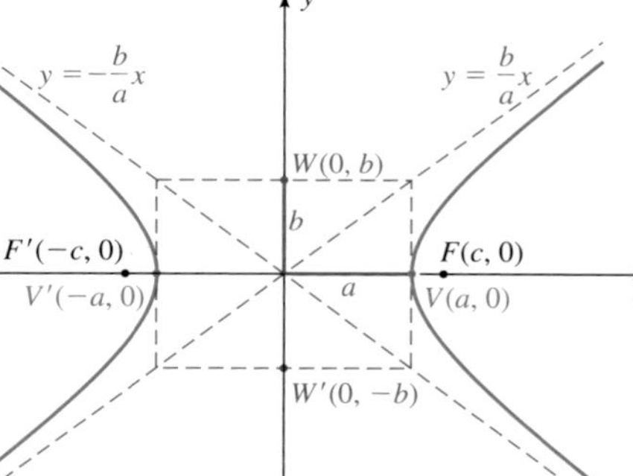

PARABOLA

$$x^2 = 4py$$

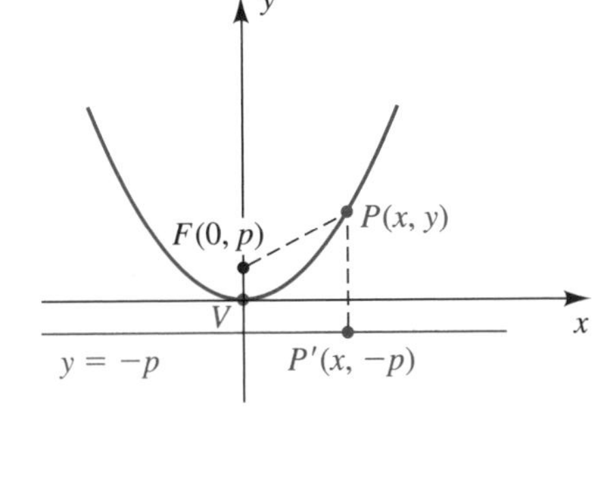

ELLIPSE

$$\frac{x^2}{a^2} + \frac{y^2}{b^2} = 1 \quad \text{with} \quad a^2 = b^2 + c^2$$

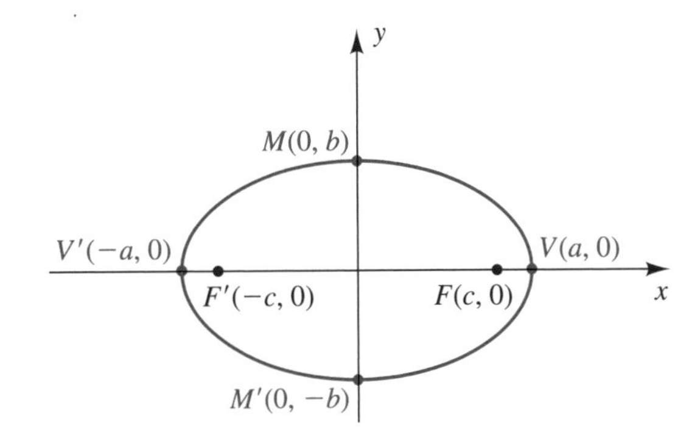

HYPERBOLA

$$\frac{x^2}{a^2} - \frac{y^2}{b^2} = 1 \quad \text{with} \quad c^2 = a^2 + b^2$$

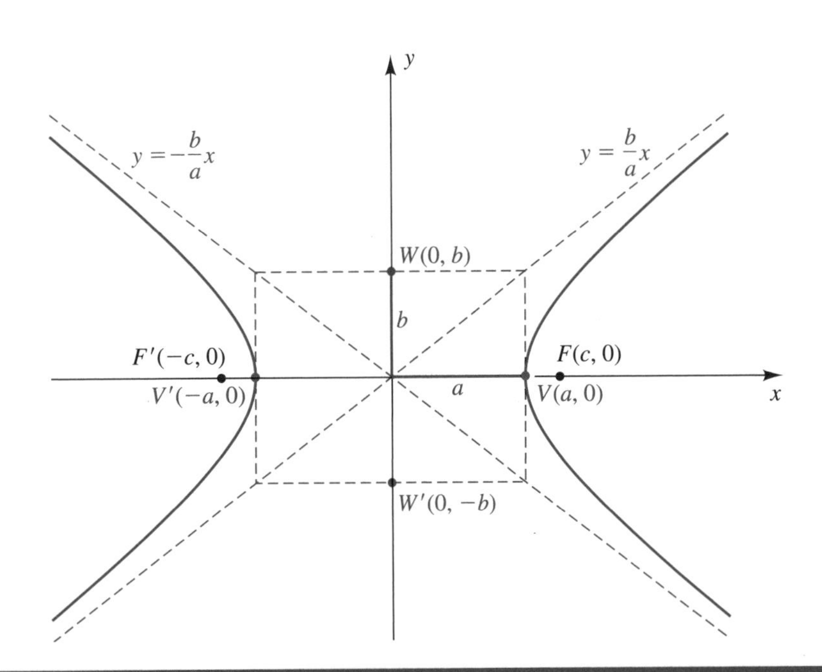

PLANE GEOMETRY

SIMILAR TRIANGLES

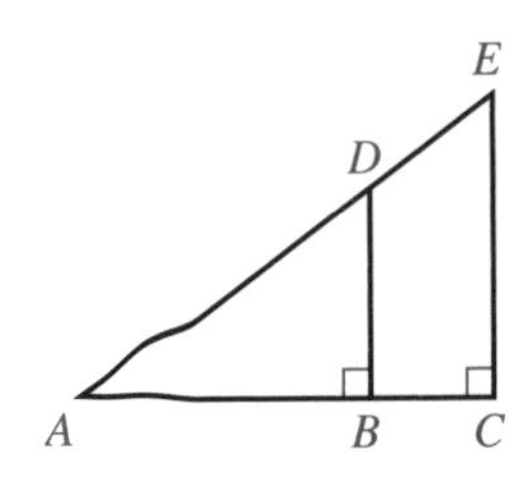

$$\frac{AB}{BD} = \frac{AC}{CE}$$

$$\frac{AB}{AD} = \frac{AC}{AE}$$

CONGRUENT ALTERNATE INTERIOR ANGLES

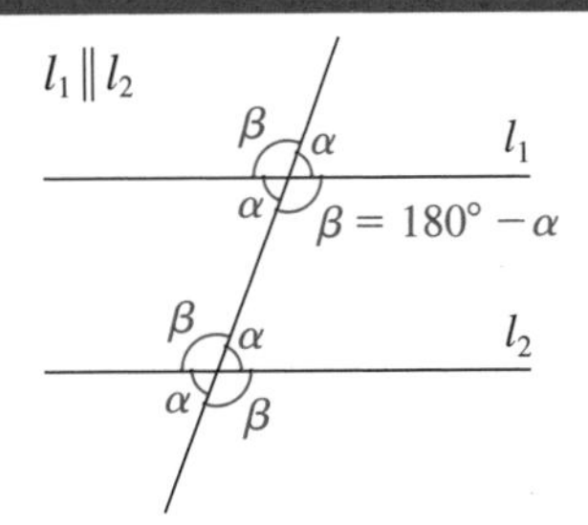

TRIGONOMETRIC FUNCTIONS

OF ACUTE ANGLES

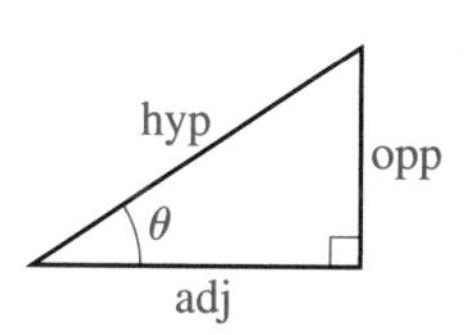

$$\sin\theta = \frac{\text{opp}}{\text{hyp}} \qquad \csc\theta = \frac{\text{hyp}}{\text{opp}}$$

$$\cos\theta = \frac{\text{adj}}{\text{hyp}} \qquad \sec\theta = \frac{\text{hyp}}{\text{adj}}$$

$$\tan\theta = \frac{\text{opp}}{\text{adj}} \qquad \cot\theta = \frac{\text{adj}}{\text{opp}}$$

OF ARBITRARY ANGLES

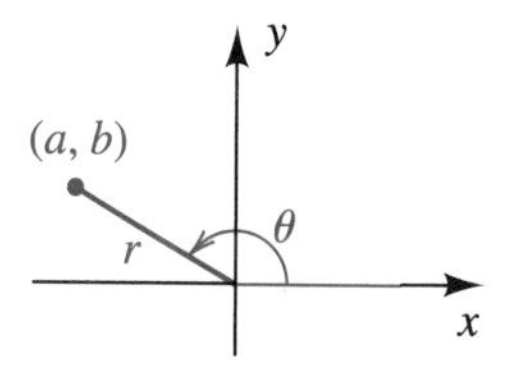

$$\sin\theta = \frac{b}{r} \qquad \csc\theta = \frac{r}{b}$$

$$\cos\theta = \frac{a}{r} \qquad \sec\theta = \frac{r}{a}$$

$$\tan\theta = \frac{b}{a} \qquad \cot\theta = \frac{a}{b}$$

OF REAL NUMBERS

$$\sin t = y \qquad \csc t = \frac{1}{y}$$

$$\cos t = x \qquad \sec t = \frac{1}{x}$$

$$\tan t = \frac{y}{x} \qquad \cot t = \frac{x}{y}$$

SPECIAL RIGHT TRIANGLES

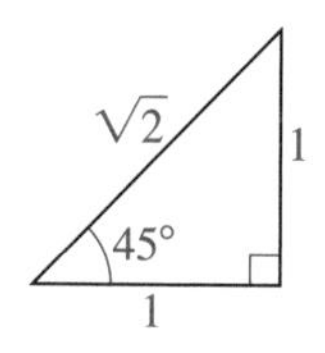

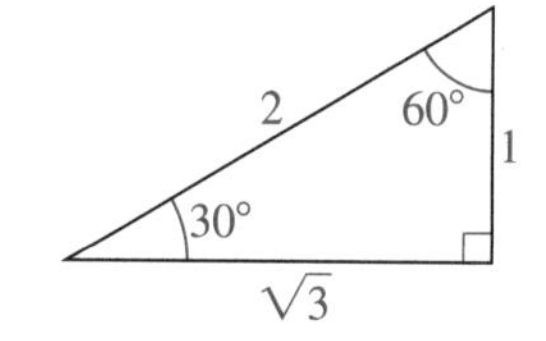

OBLIQUE TRIANGLE

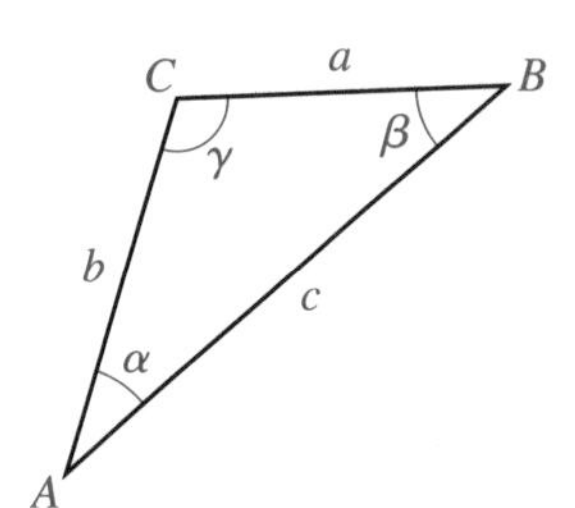

AREA

$$\mathscr{A} = \frac{1}{2}bc\sin\alpha$$

$$\mathscr{A} = \frac{1}{2}ac\sin\beta$$

$$\mathscr{A} = \frac{1}{2}ab\sin\gamma$$

$$\mathscr{A} = \sqrt{s(s-a)(s-b)(s-c)},$$

where $s = \frac{1}{2}(a + b + c)$ *(Heron's Formula)*

LAW OF COSINES

$$a^2 = b^2 + c^2 - 2bc\cos\alpha$$

$$b^2 = a^2 + c^2 - 2ac\cos\beta$$

$$c^2 = a^2 + b^2 - 2ab\cos\gamma$$

LAW OF SINES

$$\frac{\sin\alpha}{a} = \frac{\sin\beta}{b} = \frac{\sin\gamma}{c}$$

SPECIAL VALUES OF TRIGONOMETRIC FUNCTIONS

θ (degrees)	θ (radians)	$\sin\theta$	$\cos\theta$	$\tan\theta$	$\cot\theta$	$\sec\theta$	$\csc\theta$
0°	0	0	1	0	—	1	—
30°	$\frac{\pi}{6}$	$\frac{1}{2}$	$\frac{\sqrt{3}}{2}$	$\frac{\sqrt{3}}{3}$	$\sqrt{3}$	$\frac{2\sqrt{3}}{3}$	2
45°	$\frac{\pi}{4}$	$\frac{\sqrt{2}}{2}$	$\frac{\sqrt{2}}{2}$	1	1	$\sqrt{2}$	$\sqrt{2}$
60°	$\frac{\pi}{3}$	$\frac{\sqrt{3}}{2}$	$\frac{1}{2}$	$\sqrt{3}$	$\frac{\sqrt{3}}{3}$	2	$\frac{2\sqrt{3}}{3}$
90°	$\frac{\pi}{2}$	1	0	—	0	—	1

GREEK ALPHABET

Letter	Name	Letter	Name
Α α	alpha	Ν ν	nu
Β β	beta	Ξ ξ	xi
Γ γ	gamma	Ο ο	omicron
Δ δ	delta	Π π	pi
Ε ϵ	epsilon	Ρ ρ	rho
Ζ ζ	zeta	Σ σ	sigma
Η η	eta	Τ τ	tau
Θ θ	theta	Υ υ	upsilon
Ι ι	iota	Φ ϕ (φ)	phi
Κ κ	kappa	Χ χ	chi
Λ λ	lambda	Ψ ψ	psi
Μ μ	mu	Ω ω	omega